Peterson's Guide to Graduate Programs in the Physical Sciences and Mathematics 1992

Twenty-sixth Edition

Peterson's Annual Guides to Graduate Study: Book 4

Peterson's Guides
Princeton, New Jersey

Peterson's
GRADLINE

It's the perfect on-line companion to Peterson's Graduate Guides for recruiters, researchers, administrators, professors, and students!

GRADLINE contains complete information on the more than 1,440 accredited colleges and universities in the United States and Canada that offer over 31,000 degree-granting graduate and professional programs and provides the user with a wide variety of search and sort options.

GRADLINE is your instant access to the best graduate data on the market today.

Peterson's GRADLINE is available on DIALOG Information Retrieval Service as **File 273**. Call or write today.

DIALOG
3460 Hillview Avenue
Palo Alto, California 94304

800-3-DIALOG

GRADLINE is now also available on CD-ROM through SilverPlatter.
Call 800-EDU-DATA for information.

The colleges and universities represented in this book recognize that federal laws, where applicable, require compliance with Title IX (Education Amendments of 1972), Title VII (Civil Rights Act of 1964), and Section 504 of the Rehabilitation Act of 1973 as amended, prohibiting discrimination on the basis of sex, race, color, handicap, or national or ethnic origin in their educational programs and activities, including admissions and employment.

Editorial inquiries concerning this book should be addressed to:
Editor, Peterson's Guides, P.O. Box 2123, Princeton, New Jersey 08543-2123

Copyright © 1991 by Peterson's Guides, Inc.

Previous editions © 1966, 1967, 1968, 1969, 1970, 1971, 1972, 1973, 1974, 1975, 1976, 1977, 1978, 1979, 1980, 1981, 1982, 1983, 1984, 1985, 1986, 1987, 1988, 1989, 1990

All rights reserved. No part of this book may be reproduced, stored in a retrieval system, or transmitted, in any form or by any means—electronic, mechanical, photocopying, recording, or otherwise—except for citations of data for scholarly or reference purposes with full acknowledgment of title, edition, and publisher and written notification to Peterson's Guides prior to such use. All material and photographs provided by participating institutions are the property of Peterson's Guides.

ISSN 0894-9379
ISBN 1-56079-114-4

Composition and design by Peterson's Guides

Printed in the United States of America

10 9 8 7 6 5 4 3 2 1

Contents

Introduction	iv
The Graduate Adviser	**1**
Graduate Education in the Nineties	2
Applying to Graduate and Professional Schools	4
Financial Aid for Graduate and Professional Education	10
Tests Required of Applicants	16
Accreditation and Accrediting Agencies	17

Directory of Institutions with Programs in the Physical Sciences and Mathematics — 19

Academic and Professional Programs in the Physical Sciences and Mathematics

Section 1: Astronomy and Astrophysics	**29**
Program Directories	30
Astronomy	30
Astrophysics	35
Section 2: Chemistry	**39**
Program Directories	42
Analytical Chemistry	42
Chemistry	45
Inorganic Chemistry	71
Organic Chemistry	75
Physical Chemistry	79
Cross-Discipline Announcements	83
Full Descriptions	85
Section 3: Earth and Planetary Sciences	**183**
Program Directories	187
Earth Sciences	187
Geochemistry	191
Geodetic Sciences	193
Geology	194
Geophysics	210
Planetary Sciences	215
Cross-Discipline Announcement	216
Full Descriptions	217
Section 4: Marine Sciences/Oceanography	**301**
Program Directory	302
Cross-Discipline Announcements	305
Full Descriptions	307
Section 5: Mathematical Sciences	**335**
Program Directories	338
Applied Mathematics	338
Biometrics	345
Biostatistics	347
Mathematics	350
Statistics	375
Cross-Discipline Announcements	386
Full Descriptions	387
Section 6: Meteorology and Atmospheric Sciences	**519**
Program Directory	520
Cross-Discipline Announcement	523
Full Descriptions	525
Section 7: Physics	**537**
Program Directories	541
Acoustics	541
Applied Physics	541
Mathematical Physics	544
Optical Sciences	544
Physics	545
Plasma Physics	566
Cross-Discipline Announcements	567
Full Descriptions	569

Appendixes	**675**
Institutional Changes Since the 1991 Edition	676
Abbreviations Used in the Guides	677
Indexes	**687**
Index of Full Descriptions and Announcements	688
Index of Directories and Subject Areas in Books 2–6	693
Index of Directories and Subject Areas in This Book	698

Introduction

How to Use These Guides

OVERVIEW

The six volumes of Peterson's Annual Guides to Graduate Study, the only annually updated reference work of its kind, provide wide-ranging information on the graduate and professional programs offered by accredited colleges and universities in the United States and Canada. They are designed to be used by prospective graduate and professional students, placement counselors, faculty advisers, and all others interested in postbaccalaureate education.

Book 1, *Graduate and Professional Programs: An Overview,* contains information on institutions as a whole. Books 2 through 6 are devoted to specific academic and professional fields:

- Book 2—*Graduate Programs in the Humanities and Social Sciences*
- Book 3—*Graduate Programs in the Biological and Agricultural Sciences*
- Book 4—*Graduate Programs in the Physical Sciences and Mathematics*
- Book 5—*Graduate Programs in Engineering and Applied Sciences*
- Book 6—*Graduate Programs in Business, Education, Health, and Law*

The books may be used individually or as a set. For many readers—for example, those students who have chosen a field of study but do not know what institution they want to attend, as well as those who have a college or university in mind but have not decided what field to go into—the best place to begin is Book 1.

Book 1 presents several directories that help readers identify programs of study that might be of interest to them and that they can subsequently research further in Books 2 through 6. The directory Graduate and Professional Programs by Field lists the 309 fields for which there are program directories in Books 2 through 6 and gives the names of those institutions in the United States, Canada, and the U.S. territories that offer graduate degree programs in each. Degree levels are also indicated.

For geographical or financial reasons, some readers will be interested in attending a particular institution and will want to know what it has to offer. They should turn to the directory Institutions and Their Offerings, which lists the degree programs available at each institution, again, in the 309 academic and professional fields for which Books 2 through 6 have program directories. As in the Graduate and Professional Programs by Field directory, the level of degrees offered is also indicated.

Finally, the directory of Combined-Degree Programs lists the areas in which two graduate degrees may be earned concurrently and the schools that offer them.

CLASSIFICATION, FIELD DEFINITIONS, AND ACCREDITATION

Once you have identified the particular programs and institutions in which you are interested, you can use both Book 1 and the specialized volumes to obtain detailed information—Book 1 for information on the institutions overall and Books 2 through 6 for details regarding the smaller graduate units and their degree programs themselves.

Books 2 through 6 are divided into sections that contain one or more directories devoted to programs in a particular field. As indicated above, there are 309 program directories in all, and they list over 31,000 individual academic and professional units. Readers who do not find a directory devoted to the field they are interested in are urged to consult the Index of Directories and Subject Areas in Books 2–6 found in each book. Once you have identified the correct book, you should consult the Index of Directories and Subject Areas in This Book, which indicates (as does the more general directory) what directories cover subjects not specifically named in a directory or section title. This index in Book 2, for example, will tell you that if you are interested in sculpture, you should see the directory entitled Art/Fine Arts. The Art/Fine Arts entry will direct you to the proper page.

Books 2 through 6 have a number of comprehensive directories. These directories have entries for every institution granting graduate degrees in that field. For example, the comprehensive Education directory in Book 6 consists of profiles for schools and colleges of education and for departments or programs in institutions that do not have larger education units. General education programs offered by noneducation units, such as English departments, are also profiled. Cross-references identify specific education programs offered by noneducation units, such as English education programs in English departments, and note whether an overall education unit exists at that institution.

Comprehensive directories are followed by other directories, or sections in Books 3 and 5, that give more detailed information about programs in particular areas of the general field that has been covered. The comprehensive Education directory, in the example above, is followed by more than thirty directories in specific areas of education, such as English Education, Music Education, and Education of the Gifted. Where the parts of a book covered by a comprehensive directory exceed a single section, this is noted in the introductory information preceding it.

Because of the broad nature of many fields, any system of organization is bound to involve a certain amount of overlap. Environmental studies, for example, is a field whose various aspects are studied in several types of departments and schools. Readers interested in such studies will find information on relevant programs in Book 2 under City and Regional Planning, Environmental Policy and Resource Management, and Public Policy and Administration; in Book 3 under Ecology, Environmental Biology, and Natural Resources; in Book 5 under Energy Management and Policy and Environmental Engineering; and in Book 6 under Environmental and Occupational Health. In order to make it easy for readers to find all of the programs that may be of interest to them, the introduction to each section of Books 2 through 6 includes, if applicable, a paragraph suggesting other sections and directories with information on related areas of study to consult.

The introductory pages of the different sections of Books 2 through 6 also present information related to the field or fields covered therein. Brief statements describing all fields of study appear under the heading Field Definitions. Written by educators who are experts in their field, these statements describe areas of research and applied work as well as employment prospects for graduates. In Books 2, 3, 5, and 6, the introductions to some sections include short essays contributed by the agency that accredits programs in one or more of the fields the section is devoted to. These essays present information about how programs are accredited and discuss the importance that accreditation may have for students entering a particular field.

SCHOOL AND PROGRAM INFORMATION

In all of the books, information is presented in three forms: profiles—capsule summaries of basic information—and the announcements and full descriptions written by graduate school and program administrators. The format of the profiles is constant, making it easy to compare one institution with another and one program with another. A description of the information in the profiles in Books 2 through 6 may be found below; the Book 1 profile description is found immediately preceding the profiles in Book 1. A number of graduate school and program administrators have attached brief announcements to the end of their profile listings. In these, readers will find information that an institution or program wants to emphasize. The two-page full

INTRODUCTION

descriptions are by their very nature somewhat more expansive and flexible than the profiles, and the administrators who have written them may emphasize different aspects of their programs. All of these full descriptions are organized in the same way, however, and in each one the reader can count on finding information on the same basic topics, such as programs of study, research facilities, tuition and fees, financial aid, and application procedures. If an institution or program has submitted a full description, a boldface cross-reference appears below its profile. As is the case with the profile announcements, all of the full descriptions in the guides have been submitted by choice of administrators; the absence of an announcement or full description does not reflect any type of editorial judgment on the part of Peterson's Guides.

In addition to the regular directories that present profiles of programs in each field of study, many sections of Books 2 through 6 contain special notices under the heading Cross-Discipline Announcements. Appearing at the end of the profiles in many sections, these announcements call the reader's attention to programs described in a different section that he or she may find of interest. A biochemistry department, for example, may place a notice under Cross-Discipline Announcements in the Chemistry section (Book 4) to alert chemistry students to their current description in the Biochemistry section of Book 3. Cross-discipline announcements, also written by administrators wishing to highlight their programs, will be helpful to readers not only in finding out about programs in fields related to their own but also in locating departments that are actively recruiting students with a specific undergraduate major.

Profiles of Graduate Units (Books 2–6)

The profiles found in the 309 directories in Books 2 through 6 provide basic data about the graduate units in capsule form for quick reference. To make these directories as useful as possible, profiles are generally listed only for an institution's largest academic unit within a subject area. In other words, if an institution has a School of Physical Education that administers many related programs, the profile for the entire school, and not the individual programs, appears in the directory. If this listing is insufficient, additional profiles appear for that institution.

There are some programs that do not fit into any current directory and are not given individual profiles. The directory structure is reviewed annually in order to keep this number to a minimum and accommodate major trends in graduate education today.

The following outline describes the profile information found in the guides and explains how best to use that information. Any item that does not apply to or was not provided by a graduate unit is omitted from its listing.

Identifying Information. The institution's name, in boldface type, is followed by a complete listing of the administrative structure for that field of study. (For example, **University of Akron,** Buchtel College of Arts and Sciences, Department of Mathematical Sciences and Statistics, Program in Mathematics.) The last unit listed is the one to which all information in the profile pertains. The institution's address follows.

Offerings. Each field of study offered by the unit is listed with all postbaccalaureate degrees awarded. Degrees that are not preceded by a specific concentration are awarded in the general field listed in the unit name. Frequently, fields of study are broken down into subspecializations, and those appear following the degrees awarded, for example, "Offerings in secondary education (M Ed), including English education, mathematics education, science education." Students enrolled in the M.Ed. program would be able to specialize in any of the three fields mentioned.

Professional Accreditation. Profiles indicate whether a program is professionally accredited. Specific information on the accreditation status of a unit is obtained directly from the accreditation agency's most current listing at the time of publication. However, because it is possible for a program to receive or lose professional accreditation at any time, students entering fields in which accreditation is important to a career should verify the status of programs by contacting either the chairperson or the appropriate accrediting association (see Accreditation and Accrediting Agencies in each book).

Restricted and Suspended Admissions. Some programs admit only certain groups (such as students who were undergraduates at the institution), and other programs have suspended or permanently terminated admissions. Notes to this effect are printed after the list of degrees offered. Institutions that have restricted admission are often unable to handle queries, and some have requested that Peterson's Guides discourage applicants from writing directly to them; readers are advised, therefore, to write only if they qualify for admission under the stated restrictions. In many cases, programs with suspended admissions still have students completing degrees; enrollment figures as well as degree requirements are printed here in order to give an accurate reflection of each program's current status. Interested students should contact the unit head or dean to determine when and if admissions will be resumed if the profile states that the situation is temporary.

Jointly Offered Degrees. Explanatory statements concerning programs that are offered in cooperation with another institution also follow the list of degrees offered. This occurs most commonly on a regional basis (for example, two state universities offering a cooperative Ph.D. in special education) or where the specialized nature of the institutions encourages joint efforts (a J.D./M.B.A. offered by a law school at an institution with no formal business programs and an institution with a business school but lacking a law school). Only programs that are truly cooperative are listed; those involving only limited course work at another institution are not. Interested students should contact the unit head of such units for further information.

Part-Time and Evening/Weekend Programs. When information regarding the availability of part-time or evening/weekend study appears in the profile, it means that students are able to earn a degree exclusively through such study.

Faculty. Figures on the number of faculty members actively involved with graduate students through teaching or research are separated into full- and part-time as well as men and women whenever the information has been supplied.

Matriculated Students. Figures for the number of students enrolled in graduate degree programs pertain to the semester of highest enrollment from the 1990–91 academic year. These figures are broken down into full- and part-time and men and women whenever the data have been supplied. Information on the number of matriculated students enrolled in the unit who are members of a minority group or are foreign nationals appears here. The average age of the matriculated students is followed by the number of applicants and the percentage accepted for fall 1990.

Degrees Awarded in 1990. Under this section, all degrees are classified into one of four types: master's, doctoral, first professional, and other advanced degrees. A unit may award one or several degrees at a given level, however, the data are only collected by type and may therefore represent several different degree programs. In addition to the number of degrees awarded, this section contains information on the percentages of students who have gone on to continue full-time study, entered university research or teaching, or chosen other work related to their field. Many doctoral programs offer a terminal master's degree if students leave the program after completing only part of the requirements for a doctoral degree; this is indicated here.

Degree Requirements. The information in this section is also broken down by type of degree, and all information for a degree level pertains to all degrees of that type unless otherwise specified. Degree requirements are collected in a simplified form to provide some very basic information on the nature of the program. Information is provided on foreign language, computer language, and thesis or dissertation requirements. Many units also provide a short list of additional requirements, such as fieldwork or an internship. No information is listed on the number of courses or credits required for completion or whether a minimum or maximum number of years or semesters is needed. For complete information on graduation requirements, the graduate unit should be contacted.

Entrance Requirements. Entrance requirements are again broken down into the four degree levels of master's, doctoral, first professional, and other advanced degrees. Within each level, information may be provided in two basic categories, entrance exams and other requirements. The entrance exams use the standard acronyms used by the testing agencies unless they are not well known; a complete list of these acronyms appears in the appendix of Abbreviations Used in the Guides in each volume. Additional information on each of the common tests is provided in the section Tests Required of Applicants. The usual format in this part of the profile is a test name followed by a minimum score. When a minimum combined score is given for the GRE General Test, it is for the verbal and quantitative sections combined (without the analytical section) unless otherwise specified. More information on the scale and other aspects of the test may be obtained directly from the testing agency listed in the testing section. Other entrance requirements are quite varied, but they often contain an undergraduate or graduate grade point average (GPA). Unless otherwise stated, the GPA is calculated on a 4.0 scale and is listed as a minimum required for admission. The standard application deadline and any nonrefundable application fee may be listed here. Note that the deadline should be used for reference only; these dates are subject to change, and students interested in applying should contact the graduate unit directly about application procedures and deadlines.

Expenses. The cost of study for the 1991–92 academic year is given in two basic categories, tuition and fees. It is not possible to represent the complete tuition and fees schedule for each graduate unit, so a simplified version of the cost of studying in that unit is provided. In general, the costs of both full- and part-time study are listed if the unit allows for both types of programs and lists separate costs. For public institutions, the tuition and fees are listed for both state residents and nonresidents. Cost of study may be quite complex at a graduate institution. There are often sliding scales for part-time study, a different cost for first-year students, and other variables that make it impossible to completely cover the cost of study for each graduate program. To provide the most usable information, figures are given for full-time study for a full year where available and for part-time study in terms of a per-unit rate (per credit, per semester hour, etc.). If the tuition cannot be expressed in these terms, a minimum figure is provided for part-time study. This figure usually represents the minimum cost associated with one 3-credit course for one semester or quarter. Minimum figures are for comparison purposes only—your actual cost of study may differ. Although expenses usually rise from year to year, they may remain constant. In either case, they are usually subject to change; for exact costs at any given time, readers should contact schools and programs directly. It should also be kept in mind that the tuition of Canadian institutions is usually given in Canadian dollars.

Financial Aid. This section contains data on the number of awards given to graduate students during the 1990–91 academic year that are administered by the institution. The first figure given represents the total number of students receiving financial aid enrolled in that unit. That figure is followed by the total dollar amount of aid awarded by that unit. If the unit has provided information on graduate appointments, they are broken down into four major categories: *fellowships* give money to graduate students to cover the cost of study and living expenses and are not based on a work obligation or research commitment, *research assistantships* provide stipends to graduate students for assistance in a formal research project with a faculty member, *teaching assistantships* provide stipends to graduate students for teaching or for assisting faculty members in teaching undergraduate classes, and *other appointments* include a variety of awards that schools define in their own way. Within each category, figures are given for the total number of awards and the number given to first-year students. The average monthly stipend for a graduate assistantship, if the information is available or applicable, also follows this information.

In addition to graduate appointments, the availability of several other financial aid sources is covered in this section. *Tuition waivers* are routinely part of a graduate appointment, but units sometimes waive part or all of a student's tuition even if a graduate appointment is not available. *Federal work-study* is made available to students who demonstrate need and meet the federal guidelines; this form of aid normally includes 10 or more hours of work per week in an office of the institution. *Institutionally sponsored loans* are low-interest loans available to graduate students to cover both educational and living expenses. *Career-related internships or fieldwork* offer money to students who are participating in a formal off-campus research project or practicum. The availability of financial aid to part-time students is also indicated here.

Some programs list the financial aid application deadline and the forms that need to be completed for students to be eligible for financial aid. There are two common forms: the GAPSFAS form of the Graduate and Professional School Financial Aid Service and FAF, the Financial Aid Form of the College Scholarship Service.

Faculty Research. Each unit has the opportunity to list several keyword phrases describing the current research involving the faculty and graduate students. Space limitations prevent the unit from listing complete information on all research programs. The total budget for funded research from the previous academic year may also be included.

Unit Head and Application Contact. The head of the graduate program for each unit is listed with the academic title and telephone number if available. In addition to the unit head, many graduate programs list separate contacts for application and admission information, which follows the listing for the unit head. If no unit head or application contact is given, you should contact the overall institution for information on graduate admissions.

Data Collection and Editorial Procedures

DIRECTORIES AND PROFILES

The information published in the directories and profiles of all the books is collected through Peterson's Guides Annual Survey of Graduate Institutions, a series of questionnaires sent each spring and summer to the more than 1,480 accredited institutions in the United States and Canada offering postbaccalaureate degree programs. Deans and other administrators complete these surveys, providing information on programs in the 309 academic and professional fields covered in the guides as well as overall institutional information. Peterson's editorial staff then goes over each returned survey carefully and verifies or revises responses after further research and discussion with administrators at the institutions. Extensive files on past responses are kept from year to year.

While every effort has been made to ensure the accuracy and completeness of the data, information is sometimes unavailable or changes occur after publication deadlines. All usable information received in time for publication has been included. The omission of any particular item from a directory or profile signifies either that the item is not applicable to the institution or program or that information was not available. Profiles of programs scheduled to begin during the 1991–92 academic year cannot, obviously, include statistics on enrollment or, in many cases, the number of faculty members. If no usable data were submitted by an institution, its name, address, and program name where appropriate nonetheless appear in order to indicate the existence of graduate work.

ANNOUNCEMENTS AND FULL DESCRIPTIONS

While the profiles represent the result of Peterson's annual research project on U.S. and Canadian graduate education,

Peterson's Guide to Graduate Programs in the Physical Sciences and Mathematics 1992

INTRODUCTION

covering over 31,000 administrative and academic units at over 1,480 institutions and forming a database as complete as possible, the announcements and full descriptions are supplementary insertions submitted by deans, chairs, and other administrators who wish to make an additional, more individualized statement to readers. Those who have chosen to write these insertions are responsible for the accuracy of the content, but Peterson's editors have reserved the right to delete irrelevant material or questionable self-appraisals and to edit for style. Statements regarding a university's objectives and accomplishments are a reflection of its own beliefs and are not the opinions of the editors. Since inclusion of announcements and descriptions is by choice, their presence or absence in the guides should not be taken as an indication of status, quality, or approval.

Beverly vonVorys-Norton
Series Editor

Phil Williams
Data Editor

The Graduate Adviser

This section consists of three essays and information on admissions tests and accreditation. The first essay, Graduate Education in the Nineties, is by Mary G. Powers, Dean of the Graduate School of Arts and Sciences at Fordham University. It covers a number of topics of general interest, including the historical developments and trends in graduate education and the demand for graduate education and faculty in the nineties. The second essay, Applying to Graduate and Professional Schools, is by Jane E. Levy of Cornell University and Elinor R. Workman of the University of Chicago. It covers a number of points of interest to students considering postbaccalaureate work, including types of degrees, choosing a specialization and researching programs, applying, and some issues for returning, part-time, and foreign students. The third essay is Financial Aid for Graduate and Professional Education, by Patricia McWade of Georgetown University. It discusses how and when to apply for aid, determining financial need, and types of aid available. All three essays appear in each of the six Graduate Guides. Tests Required of Applicants lists all standardized admissions tests that are relevant to programs in the physical sciences and mathematics. Accreditation and Accrediting Agencies gives information on accreditation and its purpose and lists institutional accrediting agencies (other volumes of the Graduate Guides also list specialized accrediting agencies relevant to the programs in those books).

Graduate Education in the Nineties

Graduate schools prepare the experts needed in almost every field of endeavor. They prepare most administrators and instructors in schools, colleges, and universities and increasing numbers of professional employees in industry, government, and other settings. As disciplines become more specialized, graduate training is increasingly necessary and employers look for persons with advanced degrees. Recent college graduates who began careers as teachers, chemists, management trainees, social workers, and the like will find that, in order to stay abreast of developments in their fields, they must pursue graduate education, usually in the form of a degree-granting program, but also in the form of workshops, seminars, and mini courses.

Graduate schools in the United States have adapted well to these new needs and demands. In addition to providing traditional training for careers in scholarship and research, graduate schools in the United States also offer programs geared toward professional advancement and personal enrichment. Institutional adjustments, at all levels, have been made to respond to a much broader student body than in the past and to the knowledge explosion resulting in new programs and organizational structures.

Historical Developments and Trends

With more than a century of experience behind them, graduate schools in the United States are widely recognized as major centers of scholarship and research. Indeed, while one finds ample, and frequently undeserved, criticism of other segments of the educational system, it is widely acknowledged that graduate education in the United States is second to none.

The enrollment of full-time graduate students, especially in the humanities, declined during the seventies and early eighties, as the demand for Ph.D.'s declined, particularly in academia, with the enrollment of small cohorts of traditional-age full-time students. The surplus of qualified persons seeking positions in colleges and universities in the seventies forced many of them to move into careers in other areas, and the number of U.S. citizens seeking doctorates began a decline that continued throughout the eighties. This was a response to a period experiencing a recession, an energy crisis, and high unemployment levels in many professions. Everyone had a story of the cab driver with a Ph.D. College graduates turned to other pursuits and the number of Ph.D.'s produced annually declined from the early seventies through the mid-eighties.

During this period of slow growth, graduate schools, nonetheless, made innovative changes and expanded their programs in a variety of ways. Numerous centers and institutes for research were established as were interdisciplinary and interdepartmental programs. The number of part-time students increased as did the number of women and minority students. By 1989 over half the graduate students enrolled in schools that were members of the Council of Graduate Schools were enrolled part-time. This varied considerably by type of school, however. Students in major research institutions were still predominantly men and full-time in 1989, whereas women were 62 percent of the students in master's institutions where three quarters of the students attended part-time. During the last several years, full-time enrollment is again increasing more rapidly than part-time enrollment—perhaps because of the perception of increased employment opportunities in the nineties and on into the twenty-first century. Other major changes in the graduate school population of the late seventies and throughout the eighties included increasing numbers of minorities from all groups and increasing numbers of women and international students.

The Demand for Graduate Education in the Nineties

The next century will bring major changes in higher education and will provide opportunities and challenges that require increased numbers of persons with a graduate education. Several forces will come together to create a demand for personnel with graduate degrees. These include: (1) the growth of college-age populations after 1995 and the pattern of college enrollment of this population, (2) the aging of the population and the pending retirements of existing doctoral personnel both within academia and elsewhere, and (3) the numbers of advanced degrees awarded in the recent past and likely to be awarded in the near future, which suggest a shortage especially in the sciences. All of these forces will affect the demand for and supply of college faculty with the Ph.D. and persons with M.A. and Ph.D. degrees in other settings.

Enrollment in Colleges and Universities

Enrollment trends in colleges and universities in the nineties will differ from what has been experienced throughout the eighties. The most recent National Research Council report on doctorate recipients indicated that about 25 percent of all doctorates awarded in 1989 were earned by non-U.S. citizens. Most international students were concentrated in the sciences and engineering. There was considerable variation, however. Non-U.S. citizens earned over one third of all doctorates in the physical sciences, one quarter in the life sciences, but only 10 percent in education. While many still plan to return home for employment, the proportion planning to remain in the U.S. increased over the decade from 39 percent in 1979 to 58 percent in 1989.

In addition to the increase in international students, enrollment in higher education institutions in the nineties will include escalating proportions of women and minorities. In particular, the enrollment of women at all levels of higher education has increased dramatically from about 5 million in the mid-seventies to over 7 million in 1990. It is expected that the enrollment of women in graduate programs will continue to grow faster than the enrollment of men. Similarly, the 1980s saw an increase in the enrollment of minority students in virtually all levels of higher education, including graduate education. On a year-by-year basis, there were some fluctuations throughout the decade, but there has been a sizable net gain, particularly in African-American and Hispanic enrollment in graduate education, with increases expected to continue during the nineties.

In the same way that the characteristics of enrolled students at the undergraduate level have changed to incorporate a smaller proportion of traditional four-year, full-time students, so, too, has the graduate student population changed. Graduate programs are now targeted to personal enrichment and professional advancement as well as to traditional academic programs leading to the Ph.D. Because of this, larger proportions of college graduates are attracted to graduate programs but are attending part-time and for different reasons than in the past. Some individuals feel the need to keep up with the ever-increasing changes in disciplines such as computer science, engineering, and economics. Some prepare for second careers. Others simply explore interests outside their own fields that were not fully attended to while in a career-oriented program.

In any case, it appears that the decade of the nineties will see some growth in both undergraduate and graduate enrollment. According to William G. Bowen and Julie A. Sosa in *Prospects for Faculty in the Arts and Sciences* (Princeton: Princeton University Press, 1989), the level of growth in undergraduate enrollment is projected to be very low because of the composition of the population reaching age 18. It will include larger proportions of minority and immigrant youth who have not attended college at the same rate as the majority population. The National Center for Education Statistics, on the other hand, suggests a modest level of increase through the first decade of the next century. Growth at the graduate level has already begun and should help to prepare the faculty needed to teach the college students who will be enrolling in increasing numbers after 1995, just as professors are retiring in increasing numbers.

Demand for Faculty

Students interested in a particular discipline, such as physics, chemistry, mathematics, sociology, or English, will find increasing opportunities to work in that discipline in a university setting during the late nineties and into the twenty-first century. Several recent studies point to the potential shortage of faculty in the arts and sciences, which will be felt beginning in the second half of the 1990s. The underlying assumption is that the demand for faculty will be affected by the retirement of existing faculty, many of whom were hired in the sixties, and by enrollment trends noted previously, which will tend to rise slowly after 1995.

Several surveys indicate that, in the roughly 3,300 colleges and universities in the United States, the faculty was, on average, considerably older in 1990 than it was in the 1970s. Given the drop in the number of doctorates awarded in the seventies and eighties, the retirement of older faculty will lead to a shortage of faculty by the mid to late nineties, just as enrollments are projected to again increase.

Faculty shortages are already perceived in some fields, and most observers of the academic scene agree that major changes in the market are already under way. Whereas there were very few jobs in the humanities as recently as 1985, opportunities have increased dramatically in the past two years. Bowen and Sosa estimate that by 1997 there will be only about 7 qualified candidates for every 10 faculty openings in the humanities and social sciences and about 8 candidates for every 10 positions in the physical sciences and mathematics. Hence, the expectation of a faculty shortage at the end of the decade is real and should provide new opportunities for recipients of advanced degrees. Opportunities in sectors other than academia will also result from the retirement of persons in industry and government who earned their degrees in the fifties and sixties.

Nonacademic Demands

Research and development in American industry is heavily dependent on graduate education. It is expected that the demand for master's and doctoral personnel in industry will increase over the next decade. In 1987, for example, over half of all doctoral scientists and engineers were 45 years old or over. This suggests that during the nineties there also will be increasing numbers of retirements in nonacademic sectors. Even in the absence of growth in government and industry, there will be a demand for new doctorates as replacements for some of the retiring cohorts. This, coupled with the estimates of large numbers of faculty retirements and increased enrollments in colleges, indicates an increased demand for master's and Ph.D. recipients in most arts and sciences disciplines.

The Future in Graduate Education

While all of the evidence suggests that national needs require more persons with graduate-level degrees and that the value of a graduate degree will increase in the economy of the nineties, ultimately the decision to seek a graduate degree is an individual matter. For those people who have developed a compelling interest in some discipline or area of knowledge and must go on to obtain a greater understanding, the decision is relatively easy. This love of learning and search for greater knowledge for its own sake is what traditional graduate programs were and are about. They need only to seek the specific program that will provide them with the best experience in research and scholarship in their area of interest.

If, on the other hand, they have concrete concerns about career advancement and earnings, there are many programs that will enhance both. They may have to spend some time deciding whether a degree in accounting, engineering, business administration, or some other field best meets their needs and interests, but there are many such programs available. Similarly, if people have an interest in a career in education, social work, or other areas that emphasize social service, regardless of salary, graduate study provides the foundation and career training that will help them reach their career goals.

Graduate education in the United States is large, diverse, and competitive. It is an unplanned system of colleges and universities that provide the best research training in the world along with a vast array of programs oriented to professional development and personal enrichment. The program that best meets the individual's needs and interests is part of that larger enterprise.

Mary G. Powers, Dean
Graduate School of Arts and Sciences
Fordham University

Applying to Graduate and Professional Schools

The decision to attend graduate school and the choice of an institution and degree program require serious consideration. The time, money, and energy you will expend doing graduate work are significant, and you will want to analyze your options carefully. Before you begin filing applications, you should evaluate your interests and goals, know what programs are available, and be clear about your reasons for pursuing a particular degree.

There are two excellent reasons for attending graduate school, and if your decision is based on one of these, you probably have made the right choice. There are careers such as medicine, law, and college and university teaching that require specialized training and, therefore, necessitate advanced education. Another motivation for attending graduate school is to specialize in a particular discipline—to broaden your expertise in that area, to do research, and to specialize in a subject that you have decided is of great importance, either for career goals or for personal satisfaction.

Degrees

Traditionally, graduate education has involved acquiring and communicating knowledge gained through original research in a particular academic field. The highest earned academic degree, which requires the pursuit of original research, is the Doctor of Philosophy (Ph.D.). In contrast, professional training stresses the practical application of knowledge and skills; this is true, for example, in the fields of business, law, and medicine. At the doctoral level, degrees in these areas include the Doctor of Business Administration (D.B.A.), the Doctor of Jurisprudence (J.D.), and the Doctor of Medicine (M.D.).

Master's degrees are offered in most fields and may also be academic or professional in orientation. In many fields, the master's degree may be the only professional degree needed for employment. This is the case, for example, in fine arts (M.F.A.), library science (M.L.S.), and social work (M.S.W.). (For a list of the graduate and professional degrees currently being offered in the United States and Canada, readers may refer to the appendix of degree abbreviations at the back of each volume of these guides.)

Some people decide to earn a master's degree at one institution and then select a different university or a somewhat different program of study for doctoral work. This can be a way of acquiring a broad background: you can choose a master's program with one emphasis or orientation and a doctoral program with another. The total period of graduate study may be somewhat lengthened by proceeding this way, but probably not by much.

In recent years, the distinctions between traditional academic programs and professional programs have become blurred. The course of graduate education has changed direction in the last thirty years, and many programs have redefined their shape and focus. There are centers and institutes for research, many graduate programs are now interdepartmental and interdisciplinary, off-campus graduate programs have multiplied, part-time graduate programs have increased, and the demand for graduate education is clearly on the rise. Colleges and universities have also established combined-degree programs, in many cases in order to enable students to combine academic and professional studies; for example, they might earn an academic master's degree and an M.B.A. As a result of such changes, you now have considerable freedom in determining the program best suited to your current needs as well as your long-term goals.

Choosing a Specialization and Researching Programs

There are several sources of information you should make use of in choosing a specialization and a program. A good way to begin is to consult the appropriate directories of these Guides, which will tell you what programs exist in the field or fields you are interested in and, for each one, will give you information on degrees, research facilities, the faculty, financial aid resources, tuition and other costs, application requirements, and so on.

Talk with your college adviser and professors about your areas of interest and ask for their advice about the best programs to research. Besides being very well informed themselves, these faculty members may have colleagues at institutions you are investigating, and they can give you inside information about individual programs and the kind of background they seek in candidates for admission.

The valuable perspective of educators should not be overlooked. If the faculty members you know through your courses are not involved in your field of interest, do not hesitate to contact other appropriate professors at your institution or neighboring institutions to ask for advice on programs that might suit your goals. In addition, talk to graduate students studying in your field of interest; their advice can be valuable also.

Your decision about a field of study may be determined by your research interests or, if you choose to enter a professional school, by the appeal of a particular career. In either case, as you attempt to limit the number of institutions you will apply to, you will want to familiarize yourself with publications describing current research in your discipline. Find related professional journals and note who is publishing in the areas of specialization that interest you, as well as where they are teaching. Take note of the institutions represented on the publications' editorial boards also (they are usually listed on the inside cover); such representation usually reflects strength in the discipline.

Being aware of who the top people are and where they are will pay off in a number of ways. A graduate department's reputation rests heavily on the reputation of its faculty, and in some disciplines it is more important to study under someone of note than it is to study at a college or university with a prestigious name. In addition, in certain fields graduate funds are often tied to a particular research project and, as a result, to the faculty member directing that project. Finally, most Ph.D. candidates (and nonprofessional master's degree candidates) must pick an adviser and one or more other faculty members who form a committee that directs and approves their work. Many times this choice must be made during the first semester, so it is important to learn as much as you can about faculty members before you begin your studies. As you research the faculties of various departments, keep in mind the following questions: What is their academic training? What are their research activities? What kind of concern do they have for teaching and student development?

There are other important factors to consider in judging the educational quality of a program. First, what kind of students enroll in the program? What are their academic abilities, achievements, skills, geographical representation, and level of professional success upon completion of the program? Second, what are the program's resources? What kind of financial support does it have? How complete is the library? What laboratory equipment and computer facilities are available? And third, what does the program have to offer in terms of both curriculum and services? What are its purposes, its course offerings, and its job placement and student advisement services? What is the student-faculty ratio, and what kind of interaction is there between students and professors? What internships, assistantships, and other experiential education opportunities are available?

When evaluating a particular institution's reputation in a given field, you may also want to look at published graduate program ratings. There is no single rating that is universally accepted, so you would be well advised to read several and not place too much importance on any one. Most consist of what are known as "peer ratings"; that is, they are the results of polls of respected scholars who are asked to rate graduate departments in their field of expertise. Many academicians feel that these ratings are too heavily based upon traditional concepts of what constitutes quality—such as the publications of the faculty—and that they perpetuate the notion of a research-oriented department as the only model of excellence in graduate education. Depending on whether your own goals are research-oriented, you may want to attribute more or less importance to this type of rating.

If possible, visit the institutions that interest you and talk with faculty members and currently enrolled students. Be sure, however, to write or call the admissions office a week in advance to give the person in charge a chance to set up appointments for you with faculty members and students.

The Application Process

TIMETABLE

It is important to start gathering information early in order to be able to complete your applications on time. Most people should start the process a full year and a half before their anticipated date of matriculation. There are, however, some exceptions to this rule. The time frame will be different if you are applying for national scholarships or if your undergraduate institution has an evaluation committee through which you are applying to a health-care program or law school. In such a situation, you may have to begin the process two years before your date of matriculation in order to take your graduate admission test and arrange for letters of recommendation early enough to meet deadlines.

Application deadlines may range from August (before your senior year) for early decision programs of medical schools using the American Medical College Application Service (AMCAS) to late spring or summer (after your senior year) for a few programs with rolling admissions. Most deadlines for the entering class in the fall are between January and March. You should in all cases plan to meet formal deadlines; beyond this, you should be aware of the fact that many schools with rolling admissions encourage and act upon early applications. Applying early to a school with rolling admissions is usually advantageous, as it shows your enthusiasm for the program and gives admissions committees more time to evaluate the subjective components of your application, rather than just the "numbers." Applicants are not rejected early unless they are clearly below an institution's standards.

The timetable that appears below represents the ideal for most students.

Junior Year, Fall and Spring
- Research areas of interest, institutions, and programs.
- Talk to advisers about application requirements.
- Register and prepare for appropriate graduate admission tests.
- Investigate national scholarships.
- If appropriate, obtain letters of recommendation.

Junior Year, Summer
- Take required graduate admission tests.
- Write for application materials.
- Visit institutions of interest, if possible.
- Write your application essay.
- Check on application deadlines and rolling admissions policies.
- For medical, dental, osteopathy, podiatry, or law school, you may need to register for the national application or data assembly service most programs use.

Senior Year, Fall
- Obtain letters of recommendation.
- Take graduate admission tests if you haven't already.
- Send in completed applications.

Senior Year, Spring
- Register for Graduate and Professional School Financial Aid Service (GAPSFAS), if required.
- Check with all institutions before the deadline to make sure your file is complete.
- Visit institutions that accept you.
- Send a deposit to your institution of choice.
- Notify other colleges and universities that accepted you of your decision so that they may admit students on their waiting list.
- Send thank-you notes to people who wrote your recommendation letters, informing them of your success.

You may not be able to adhere to this timetable if your application deadlines are very early, as is the case with medical schools, or if you decide to attend graduate school at the last minute. In any case, keep in mind the various application requirements and be sure to meet all deadlines. If deadlines are impossible to meet, call the institution to see if a late application will be considered.

OBTAINING APPLICATION FORMS AND INFORMATION

To obtain the materials you need, a neatly typed or handwritten postcard requesting an application, a bulletin, and financial aid information is all that is necessary. However, you may want to request an application by writing a formal letter in which you briefly describe your training, experience, and specialized research interests. Either type of request should be sent to the admissions office directly. If you want to write to a particular faculty member about your background and interests in order to explore the possibility of an assistantship, you should also feel free to do so. However, do not ask a faculty member for an application, as this may cause a significant delay in your receipt of the forms.

NATIONAL APPLICATION SERVICES

In a few professional fields, there are national services that provide assistance with some part of the application process. These services are the Law School Data Assembly Service (LSDAS), American Medical College Application Service (AMCAS), American Association of Colleges of Osteopathic Medicine Application Service (AACOMAS), American Association of Colleges of Podiatric Medicine Application Service (AACPMAS), and American Association of Dental Schools Application Service (AADSAS). Many programs require applicants to use these services because they simplify the application process for both the professional programs' admissions committees and the applicant. The role these services play varies from one field to another. The LSDAS, for example, analyzes your transcript(s) and submits the analysis to the law schools to which you are applying, while the other services provide a more complete application service. More information and applications for these services can be obtained from your undergraduate institution.

Peterson's Guide to Graduate Programs in the Physical Sciences and Mathematics 1992

MEETING APPLICATION REQUIREMENTS

Requirements vary from one field to another and from one institution to another. Read each program's requirements carefully; the importance of this cannot be overemphasized!

Graduate Admission Tests

Colleges and universities usually require a specific graduate admission test, and departments sometimes have their own requirements as well. Scores are used in evaluating the likelihood of your success in a particular program (based upon the success rate of past students with similar scores). Most programs will not accept scores more than three to five years old. The various tests are described a little later in this book.

Transcripts

Admissions committees require official transcripts of your grades in order to evaluate your academic preparation for graduate study. Grade point averages are important but are not examined in isolation; the rigor of the courses you have taken, your course load, and the reputation of the undergraduate institution you have attended are also scrutinized. To have your college transcript sent to graduate institutions, contact your college registrar.

Letters of Recommendation

Choosing people to write recommendations can be difficult, and most graduate schools require two or three letters. While recommendations from faculty members are essential for academically oriented programs, professional programs may seriously consider nonacademic recommendations from professionals in the field. Indeed, often these nonacademic recommendations are as respected as those from faculty members.

To begin the process of choosing references, identify likely candidates from among those you know through your classes, extracurricular activities, and jobs. A good reference will meet several of the following criteria: he or she has a high opinion of you, knows you well in more than one area of your life, is familiar with the institutions to which you are applying as well as the kind of study you are pursuing, has taught or worked with a large number of students and can make a favorable comparison of you with your peers, is known by the admissions committee and is regarded as someone whose judgment should be given weight, and has good written communication skills. No one person is likely to satisfy all these criteria, so choose those people who come closest to the ideal.

If you are returning to school after working for several years, you may not be able to find professors at your undergraduate institution who remember you. If this is the case, contact the graduate schools you are applying to and see what their policies are regarding your situation. They may waive the requirement of recommendation letters, allow you to substitute letters from employment supervisors, or suggest you enroll in relevant courses at a neighboring institution and obtain letters from professors upon completion of the course work. Programs vary considerably in their policies, so it is best to check with each institution.

Once you have decided whom to ask for letters, you may wonder how to approach them. Ask them if they think they know you well enough to write a meaningful letter. Be aware that the later in the semester you ask, the more likely they are to hesitate because of time constraints; ask early in the fall semester of your senior year. Once those you ask to write letters agree in a suitably enthusiastic manner, make an appointment to talk with them. Go to the appointment with recommendation forms in hand, being sure to include addressed, stamped envelopes for their convenience. In addition, give them other supporting materials that will assist them in writing a good, detailed letter on your behalf. Such documents as transcripts, a résumé, a copy of your application essay, and a copy of a research paper can help them write a thorough recommendation.

On the recommendation form, you will be asked to indicate whether you wish to waive or retain the right to see the recommendation. Before you decide, discuss the confidentiality of the letter with each writer. Many faculty members will not write a letter unless it is confidential. This does not necessarily mean that they will write a negative letter but, rather, that they believe it will carry more weight as part of your application if it is confidential. Waiving the right to see a letter does, in fact, usually increase its validity.

Application Essays

Writing an essay, or personal statement, is often the most difficult part of the application process. Requirements vary widely in this regard. Some programs request only one or two paragraphs about why you want to pursue graduate study, while others require five or six separate essays in which you are expected to write at length about your motivation for graduate study, your strengths and weaknesses, your greatest achievements, and solutions to hypothetical problems. Business schools are notorious for requiring several time-consuming essays.

An essay or personal statement for an application should be essentially a statement of your ideas and goals. Usually it includes a certain amount of personal history, but, unless an institution specifically requests autobiographical information, you do not have to supply any. Even when the requirement is a "personal statement," the possibilities are almost unlimited. There is no set formula to follow, and, if you do write an autobiographical piece, it does not have to be arranged chronologically. Your aim should be a clear, succinct statement showing that you have a definite sense of what you want to do and enthusiasm for the field of study you have chosen. Your essay should reflect your writing abilities; more important, it should reveal the clarity, the focus, and the depth of your thinking.

Before writing anything, stop and consider what your reader might be looking for; the general directions or other parts of the application may give you an indication of this. Admissions committees may be trying to evaluate a number of things from your statement, including the following things about you:

- Motivation and commitment to a field of study
- Expectations with regard to the program and career opportunities
- Writing ability
- Major areas of interest
- Research and/or work experience
- Educational background
- Immediate and long-term goals
- Reasons for deciding to pursue graduate education in a particular field and at a particular institution
- Maturity
- Personal uniqueness—what you would add to the diversity of the entering class

There are two main approaches to organizing an essay. You can outline the points you want to cover and then expand on them, or you can put your ideas down on paper as they come to you, going over them, eliminating certain sentences, and moving others around until you achieve a logical sequence. Making an outline will probably lead to a well-organized essay, whereas writing spontaneously may yield a more inspired piece of writing. Use the approach you feel most comfortable with. Whichever approach you use, you will want someone to critique your essay. Your adviser and those who write your letters of recommendation may be very helpful to you in this regard. If they are in the field you plan to pursue, they will be able to tell you what things to stress and what things to keep brief. Do not be surprised, however, if you get differing opinions on the content of your essay. In the end, only you can decide on the best way of presenting yourself.

If there is information in your application that might reflect badly on you, such as poor grades or a low admission test score, it is better not to deal with it in your essay unless you are asked to. Keep your essay positive. You will need to explain anything that could be construed as negative in your application, however, as failure to do so may eliminate you from consideration. You can do this on a separate sheet entitled "Addendum," which you attach to the application, or in a cover letter that you enclose. In either form, your explanation should be short and to the point, avoiding long, tedious excuses. In addition to supplying your own explanation, you may find it appropriate to ask one or more of your recommenders to address the issue in their recommendation letter. Ask them to do this only if they are already familiar with your problem and could talk about it from a positive perspective.

In every case, essays should be typed. It is usually acceptable to attach pages to your application if the space provided is insufficient. Neatness, spelling, and grammar are important.

Interviews, Portfolios, and Auditions

Some graduate programs will require you to appear for an interview. In certain fields, you will have to submit a portfolio of your work or schedule an audition.

Interviews. Interviews are usually required by medical schools and are often required or suggested by business schools and other programs. An interview can be a very important opportunity for you to persuade an institution's admissions officer or committee that you would be an excellent doctor, dentist, manager, etc.

Interviewers will be interested in the way you think and approach problems and will probably concentrate on questions that enable them to assess your thinking skills, rather than questions that call upon your grasp of technical knowledge. Some interviewers will ask controversial questions, such as "What is your viewpoint on abortion?", or give you a hypothetical situation and ask how you would handle it. Bear in mind that the interviewer is more interested in *how* you think than in *what* you think. As in your essay, you may be asked to address such topics as your motivation for graduate study, personal philosophy, career goals, related research and work experience, and areas of interest.

You should prepare for a graduate school interview as you would for a job interview. Think about the questions you are likely to be asked and practice verbalizing your answers. Think too about what you want interviewers to know about you so that you can present this information when the opportunity is given. Dress as you would for an employment interview.

Portfolios. Many graduate programs in art, architecture, journalism, environmental design, and other fields involving visual creativity may require a portfolio as part of the application. The function of the portfolio is to show your skills and ability to do further work in a particular field, and it should reflect the scope of your cumulative training and experience. If you are applying to a program in graphic design, you may be required to submit a portfolio showing advertisements, posters, pamphlets, and illustrations you have prepared. In fine arts, applicants must submit a portfolio with pieces related to their proposed major.

Individual programs have very specific requirements regarding what your portfolio should contain and how it should be arranged and labeled. Many programs request an interview and ask you to present your portfolio at that time. They may not want you to send the portfolio in advance or leave it with them after the interview, as they are not insured against its loss. If you do send it, you usually do so at your own risk, and you should label all pieces with your name and address.

Auditions. Like a portfolio, the audition is a demonstration of your skills and talent, and it is often required by programs in music, theater, and dance. Although all programs require a reasonable level of proficiency, standards vary according to the field of study. In a nonperformance area like music education, you need only show that you have attained the level of proficiency normally acquired through an undergraduate program in that field. For a performance major, however, the audition is the most important element of the graduate application. Programs set specific requirements as to what material is appropriate, how long the performance should be, whether it should be memorized, and so on. The audition may be live or taped, but a live performance is usually preferred. In the case of performance students, a committee of professional musicians will view the audition and evaluate it according to prescribed standards.

MAILING COMPLETED APPLICATIONS

Graduate schools have established a wide variety of procedures for filing applications, so read each institution's instructions carefully. Some may request that you send all application materials in one package (including letters of recommendation). Others—medical schools, for example—may have a two-step application process. This system requires the applicant to file a preliminary application; if this is reviewed favorably, he or she submits a second set of documents and a second application fee. Pay close attention to each school's instructions.

Graduate schools generally require an application fee. Sometimes this fee may be waived if you meet certain financial criteria. Check with your undergraduate financial aid office and the graduate schools to which you are applying to see if you qualify.

ADMISSION DECISIONS

At most institutions, once the graduate school office has received all of your application materials, your file is sent directly to the academic department. A faculty committee (or the department chairperson) then makes a recommendation to the chief graduate school officer (usually a graduate dean or vice president), who is responsible for the final admission decision. Professional schools at most institutions act independently of the graduate school office; applications are submitted to them directly, and they make their own admission decisions.

Usually a student's grade point average, graduate admission test scores, and letters of recommendation are the primary factors considered by admissions committees. The appropriateness of the undergraduate degree, an interview, and evidence of creative talent may also be taken into account. Normally the student's total record is examined closely, and the weight assigned to specific factors fluctuates from program to program. Few, if any, institutions base their decisions purely on numbers, that is, admission test scores and grade point average. A study by the Graduate Record Examinations Board found that grades and recommendations by known faculty members were considered to be somewhat more important than GRE General Test scores and that GRE Subject Test scores were rated as relatively unimportant (Oltman and Hartnett, 1984). This indicates that some graduate admission test scores may be of less importance than is commonly believed, but this will of course differ from program to program.

Some of the common reasons applicants are rejected for admission to graduate schools are inappropriate undergraduate curriculum; poor grades or lack of academic prerequisites; low admission test scores; weak or ineffective recommendation letters; a poor interview, portfolio, or audition; and lack of extracurricular activities, volunteer experience, or research activities. To give yourself the best chances of being admitted where you apply, try to make a realistic assessment of an institution's admission standards and your own qualifications. Remember, too, that missing deadlines and filing an incomplete application can also be a cause for rejection; be sure that your transcripts and recommendation letters are received on time.

Returning Students

Many graduate programs not only accept the older, returning student but actually prefer these "seasoned" candidates. Programs in business administration, social work, law, and other professional fields value mature applicants with work experience, for they have found that these students often show a higher level of motivation and commitment and work harder than 21-year-olds. Many programs also seek the diversity older students bring to the student body, as differences in perspective and experience make for interesting—and often intense—class discussions. Nonprofessional programs also view older students favorably if their academic and experiential preparation is recent enough and sufficient for the proposed fields of study.

Many institutions have programs designed to make the transition to academic life easier for the returning student. Such programs include low-cost child-care centers, emotional support programs for both the returning student and his or her spouse, and review courses of various kinds.

Other than making the necessary changes in their life-style, older students report that the most difficult aspect of returning to school is recovering, or developing, appropriate study habits. Initially, older students often feel at a disadvantage compared to students fresh out of an undergraduate program and accustomed to preparing research papers and taking tests. This feeling can be overcome by taking advantage of noncredit courses in study skills and time management and review courses in math and writing, as well as by taking a tour of the library and becoming thoroughly familiar with it. By the end of the graduate program, most returning students feel that their life experience gave them an edge, because they could use concrete experiences to help them understand academic theory.

If you choose to go back to school, you are not alone. One out of 5 adults is currently enrolled in some kind of educational program in order to make his or her life or career more rewarding.

Part-Time Students

As graduate education has changed over the past thirty years, the number of part-time graduate programs has increased. Traditionally, graduate programs were completed by full-time students. Graduate schools instituted residence requirements, demanding that students take a full course load for a certain number of consecutive semesters, because it was felt that total immersion in the field of study and extensive interaction with the faculty were necessary in order to achieve mastery of an academic area.

In most academic Ph.D. programs as well as many health-care fields, this is still the only approach. However, many other programs now admit part-time students or allow a portion of the requirements to be completed on a part-time basis. Professional schools are more likely to allow part-time study because many students work full-time in the field and pursue their degree in order to enhance their career credentials. Other applicants choose part-time study because of financial considerations. By continuing to work full-time while attending school, they take fewer economic risks.

Part-time programs vary considerably in quality and admissions standards. When evaluating a part-time program, use the same criteria you would use in judging the reputation of any graduate program. Some schools use more adjunct faculty members with weaker academic training for their night and weekend courses, and this could lower the quality of the program; however, adjunct lecturers often have excellent experiential knowledge. Admissions standards may be lower for a part-time program than for an equivalent full-time program at the same school, but, again, your fellow students in the part-time program may be practicing in the field and have much to add to class discussions. Another concern is placement opportunities upon completion of the program. Some schools may not offer placement services to part-time students, and many employers do not value part-time training as highly as a full-time education. However, if a part-time program is the best option for you, do not hesitate to enroll after carefully researching available programs.

Foreign Students

If you are a foreign student, you will follow the same application procedures as other graduate school applicants. However, there are additional requirements you will have to meet in order to study in the United States.

Since your success as a graduate student will depend on your ability to understand, write, read, and speak English, you will be required to take the Test of English as a Foreign Language (TOEFL), or a similar test, if English is not your native language. Some schools will waive the language test requirement, however, if you have a degree from a college or university in a country where the native language is English or if you have studied two or more years in an undergraduate or graduate program in a country where the native language is English. As for all other tests, score requirements vary, but some schools admit students with lower scores on the condition that they enroll in an intensive English program before or during their graduate study. Some programs may also be willing to accept the TOEFL score as a substitute for that of the normally required graduate admission test. You should ask each school about its policies.

In addition to scores on your English test, or proof of competence in English, your formal application must be accompanied by a certified English translation of your academic transcripts. You may also be required to submit certain health certificates and documented evidence of financial support at the time of application. However, since you may apply for financial assistance from graduate schools as well as other sources, some institutions require evidence of financial support only as the last step in your formal admittance and may grant you conditional acceptance first.

Once you have been formally admitted into a graduate program, the school will send you Form I-20, Certificate of Eligibility for Non-Immigrant Student Status. You must present this document, along with a passport from your own government and certain health certificates, to a U.S. embassy or consulate in order to obtain a foreign student visa.

Your own government may have other requirements you must meet in order to study in the United States. Be sure to investigate those requirements as well.

Once all the paperwork has been completed and approved, you are ready to make your travel arrangements. If your port of entry into the United States will be New York's Kennedy Airport, you can arrange to be met and assisted by a representative of the YMCA International Student Service (ISS). This person, at no cost to you, will help you through customs and assist you in making travel connections. He or she can also help you find temporary overnight accommodations, if needed. If you are interested in this assistance, you should provide ISS with the following information: your name, age, sex, date and time of arrival, airline and flight number, college or university you will be attending, sponsoring agency (if any), and connecting flight information. Include a passport photo to help ISS find you, and note if you need overnight accommodations in New York. This information should be sent well in advance to International Student Service, Arrival Program, 356 West 34th Street, Third Floor, New York, New York 10001. You can also reach ISS by telex (ISS 620675), by telefax (212-563-3783), and by phone (212-563-0966).

When you arrive on your American college campus, you will want to contact the foreign student adviser. This person's job is to help international students in their academic and social adjustment. The foreign student adviser often coordinates special orientation programs for new foreign students, which may consist of lectures on American culture, intensive language instruction, campus tours, academic placement examinations, and visits to places of cultural interest in the community. The foreign student adviser will also help you with immigration and financial concerns.

A number of nonprofit educational organizations are available throughout the world to assist foreign students in planning graduate study in the United States. To learn how to contact these organizations for detailed information about international education, write to the U.S. embassy in your country.

Jane E. Levy
Associate Director for
Preprofessional Advising
University Career Center
Cornell University
and
Elinor R. Workman
Director of Alumni and
Career Management
Graduate School of Business
University of Chicago

Financial Aid for Graduate and Professional Education

If you are considering attending graduate school but do not have sufficient resources to finance such a costly endeavor, do not despair. Aid for graduate study does exist, and the prospects for future graduate education funding are brighter.

Basically, there are three ways to finance a graduate education—grants, loans, and work—and there are several sources of graduate support. Federal and state governments, private foundations, and, most significantly, universities themselves are all good sources of graduate aid. If you are seriously interested in graduate study, you should not be discouraged by a lack of personal funds; rather, you should energetically investigate *all* sources of graduate funding and apply for all the types of aid for which you may be eligible.

How and When to Apply

If, after estimating your expenses at the graduate schools that interest you, you conclude that you will need some help in meeting these costs, you should apply for financial aid. Do not write off any school as too expensive until you know what financial aid it offers.

Because every institution has its own application process, as well as its own system for awarding aid, you should communicate directly with each school and in some cases with each academic department that interests you. At the graduate and professional school level, financial aid sometimes is administered by the academic departments, unlike the way aid is handled at the undergraduate level. Also, you should read the school's application, brochures, and catalogs for information about financial aid.

Application deadlines vary. Some schools require you to apply for aid when applying for admission. Other schools require that you be admitted before applying for aid. Aid application instructions and deadlines should be clearly stated in each school's application material.

The proces of applying for aid can be confusing and time consuming, especially for the first-time applicant. You can alleviate some anxiety and even increase your chances of getting aid by doing the following:

- Apply to as many sources as you can find.
- Make sure you have completed all of the forms required by each school. Many schools use a document such as the Graduate and Professional School Financial Aid Service (GAPSFAS) form or the Financial Aid Form (FAF), provided by the College Scholarship Service (CSS), to assist them in determining your financial need. This process is called need analysis. GAPSFAS can be reached toll-free at 800-448-4631. Some schools require their own institutional forms in addition to the FAF or GAPSFAS form. There are other federally approved need analysis services used by graduate schools or by the school's central financial aid office. You may also be asked to submit copies of your federal income tax forms and financial aid transcripts from schools you attended previously. Be sure you read the schools' financial aid application instructions and complete the correct forms.
- Complete all forms legibly and accurately. Check your applications; errors and omissions often cause problems.
- Follow up on your aid application if you receive no response within a reasonable period of time.
- Keep copies of all forms.
- Apply for aid every year if you feel you need it. If you have special circumstances, communicate them directly to the financial aid officer both before and (if necessary) after the aid decision is made.
- In investigating financial aid opportunities, be aware that programs may change from year to year. It is up to you to find out if any changes have occurred.

Determining Financial Need

Most federal, most state, and some university aid is awarded on the basis of need. Need is normally defined as the difference between your basic educational budget—the cost of education including tuition and living expenses—and your resources. Resources may include such things as savings from summer earnings, earnings during the school year, spouses' earnings, and savings.

According to the federal government, you are independent if you are 24 years of age or over, or if you are under 24 and not claimed as a dependent for income tax purposes for the first calendar year of the year in which you are seeking financial aid. For example, those who indicate they will not be claimed as dependents in 1991 will be considered independent for the 1991-92 academic year.

Keep in mind, however, that while some schools do not view parents as a source of financial support, others require parental information and expect a contribution from those who are able to provide it, even if you are an independent student. Moreover, most schools expect the spouse of a married student to provide at least a contribution toward living expenses.

As noted above, many graduate schools use the GAPSFAS form or FAF. Both of these forms ask for information about your income, assets, and debts and, in some instances explained above, the income, assets, and debts of your parents. Once you submit the appropriate form to the processing center, a need analysis document is sent to the schools you have designated. Individual schools then make financial aid awards according to their own standards. (Neither GAPSFAS nor CSS awards financial aid.)

Types of Aid Available

The range of financial assistance available at the graduate level is very broad. There are three basic types of aid—grants and fellowships, work programs, and loans—and various sources—the federal government, state governments, educational institutions, foundations, corporations, and other private organizations such as professional associations.

GRANTS AND FELLOWSHIPS

Most grants and fellowships are outright awards that require no service in return. Often they provide the cost of tuition and fees plus a stipend to cover living expenses. Some are based exclusively on financial need, some exclusively on academic merit, and some on a combination of need and merit.

The meanings of these terms are often misunderstood. As a rule, grants are awarded to those with financial need, although they may require the recipient to have expertise in a certain field. The term "fellowship" (sometimes used interchangeably with "scholarship") connotes selectivity based on ability. Financial need is usually not a factor in awarding fellowships.

As a result of the Tax Reform Act of 1986, money provided by grants, scholarships, and fellowships is considered taxable income. That portion of the grant used for payment of tuition and course-required fees, books, supplies, and equipment is excludable from taxable income. Other student support, such as stipends and wages paid to research assistants and teaching assistants, is taxable income. Student loans are not taxable.

Federal Support

Several federal agencies fund fellowship and traineeship programs providing over $200-million in support for graduate and professional students. These agencies include the National Institutes of Health, National Science Foundation, National Aeronautics and Space Administration, U.S. Information Agency, and U.S. Departments of Agriculture, Defense, Education, and Energy. The amounts and types of assistance offered through federally funded fellowships vary considerably. For example, National Science Foundation Fellowships include tuition and fees plus an annual stipend (currently $14,000) for three years of graduate study in engineering, mathematics, the natural sciences, the social sciences, and the history and philosophy of science. The application deadline is in early November. For more information, contact the Fellowship Office, National Research Council, 2101 Constitution Avenue, Washington, D.C. 20418 (telephone: 202-334-2872).

The Jacob Javits Fellowship Program is a grant program for students in the arts, humanities, and social sciences to use at the school of their choice. Graduate students apply directly to the U.S. Department of Education. The school the Javits Fellow attends receives up to $6000 toward the cost of tuition. If the tuition exceeds $6000, the school is obliged to cover the additional tuition cost in the form of a grant. In addition, Javits Fellows receive as much as $10,000 in stipend, depending upon financial need and available funding. Application requests should be addressed to Dr. Allen Cissell, Director, Jacob Javits Fellowship Program, U.S. Department of Education, 400 Maryland Avenue, SW, ROB-3, Washington, D.C. 20202-5251 (telephone: 202-708-9415). The application deadline is in early February.

The National Institutes of Health offer training grants administered through schools' academic departments. Training grants provide tuition plus a monthly stipend of $708.

The G.I. bill has been replaced by a series of programs. Veterans may use their educational benefits for training at the graduate and professional level and should contact their regional office of the Veterans Administration for more details.

State Support

Some states offer support for graduate study. Those with the biggest programs are California, Michigan, New York, Oklahoma, and Texas. Over $40-million is awarded nationally in grants, loans, and work programs for graduate study each year; two thirds of the awards are need based, and one third are based upon academic merit. Due to fiscal constraints, some states have had to reduce or eliminate their financial aid programs for graduate study.

To qualify for a particular state's aid, you must be a resident of that state. Residency is established in most states after you have lived there for at least twelve consecutive months prior to enrolling in school. Many states provide funds for in-state students only; that is, funds are not transferable out of state. You should contact your state scholarship office directly to determine what aid it offers.

Institutional Aid

Educational institutions using their own funds provide between $1.5-billion and $2-billion in graduate assistance in the form of fellowships, tuition waivers, and assistantships. Consult each school's catalog for information about its aid programs. In graduate programs in the arts and sciences, much of the funding is awarded on a basis of merit. At the professional schools (law, business, medicine, etc.), much of the aid is need based.

Corporate Aid

Corporations provide several sources of support to students. One is in the form of the research they sponsor at universities. Another source is the tuition support they provide for their employees, usually so that they may attend school part-time while working. Most employees who receive this type of aid study at the master's level or take courses without enrolling in a particular degree program. *Corporate Tuition Aid Programs: A Directory of College Financial Aid for Employees at America's Largest Corporations,* by Joseph P. O'Neill (distributed for Conference University Press by Peterson's Guides, P.O. Box 2123, Princeton, New Jersey 08543-2123), contains information about the employee educational benefits provided by 735 of the largest banks, industrial firms, utilities, retailers, and transportation companies in the country.

Aid from Foundations

Foundations provide support in areas of interest to them. For example, the Howard Hughes Institute funds students in biomedical sciences and the Spencer Foundation funds dissertation research in the field of education.

The Foundation Center in New York City publishes several reference books on foundation support for graduate study. The center is located at 79 Fifth Avenue, 8th Floor, New York, New York 10003-3050 (telephone: 212-620-4230). They also have a computerized databank called Comsearch, which, for a fee, will produce a listing of grant possibilities in a variety of fields.

Researching Grants and Fellowships

Before undertaking graduate study, you should vigorously research all funding possibilities. The books listed below are good sources of information on grant and fellowship support for graduate education and should be consulted before you resort to borrowing.

Annual Register of Grant Support: A Directory of Funding Sources 1989-90. 23rd ed. Wilmette, Illinois: National Register Publishing, 1988. A comprehensive guide to grants and awards from government agencies, foundations, and business and professional organizations.

Corporate Foundation Profiles. 6th ed. New York: Foundation Center, 1990. An in-depth, analytical profile of 234 of the largest company-sponsored foundations in the United States. Brief descriptions of all 701 company-sponsored foundations are also included. There is an index of subjects, types of support, and geographical locations.

The Foundation Directory, 13th edition, edited by Stan Olson (New York: Foundation Center, 1990), with supplement, gives detailed information on U.S. foundations, with brief descriptions of their purpose and activities.

The Grants Register 1989-91. 11th ed. Edited by Roland Turner. New York: St. Martin's. Lists grant agencies alphabetically and gives information on awards available to graduate students, young professionals, and scholars for study and research.

Peterson's Grants for Graduate Study. 3rd ed. Princeton: Peterson's Guides, 1992. Nearly 700 grants and fellowships are described. Originally compiled by the Office of Research Affairs at the Graduate School of the University of Massachusetts at Amherst, this guide is updated periodically by Peterson's.

Scholarships, Fellowships and Loans. Volume 8. Edited by S. Norman and Marie Feingold. Bethesda: Bellman Publishing, 1986. Lists U.S. foundations and agencies offering financial support for undergraduate and graduate research and study. Sources are primarily for U.S. citizens or U.S. permanent residents, although some of the information may be helpful to foreign nationals.

Graduate schools sometimes publish listings of support sources in their catalogs, and some provide separate publications, such as the *Graduate Guide to Grants,* compiled by the Harvard Graduate School of Arts and Sciences (GSAS). Copies may be obtained for $18 (includes shipping and handling) by contacting GSAS, 8 Garden Street, Cambridge, Massachusetts 02138 (telephone: 617-495-1814).

WORK PROGRAMS

Certain types of support, such as teaching, research, and administrative assistantships, require recipients to provide service to the university in exchange for a salary or stipend; sometimes tuition is also provided or waived. For example, resident assistants live in an undergraduate dormitory in exchange for free or partially funded room and board. Assistantship responsibilities vary from school to school and from department to department.

Teaching Assistantships

If you pursue an advanced degree in a subject that is taught at the undergraduate level—for example, in the arts and sciences—you may have a good chance of securing a teaching assistantship. Such a position may involve delivering lectures, correcting classwork, grading papers, counseling students, and supervising laboratory groups; usually about 20 hours of work is required each week.

Teaching assistantships provide excellent educational experience as well as financial support. TAs generally receive a salary (now considered taxable income). Sometimes tuition is provided or waived as well. Appointments are based on academic qualifications and are subject to the availability of funds within a department. If you are interested in a teaching assistantship, you should contact the academic department that appoints them. Ordinarily you are not considered for such positions until you have been admitted to the graduate school.

Research Assistantships

Research assistantships usually require that you assist in the research activities of a faculty member. Appointments are ordinarily made for the academic year. They are rarely offered to first-year students. You should contact the academic department, describing your particular research interests. As is the case with teaching assistantships, research assistantships provide excellent academic training as well as practical experience and financial support.

Administrative Assistantships

These positions usually require 10 to 20 hours of work each week in an administrative office of the university. For example, those seeking a graduate degree in education may work in the admissions, financial aid, student affairs, or placement office of the school they are attending. Some administrative assistantships provide a tuition waiver, others a salary. Details concerning these positions can be found in the school catalog or by contacting the academic department directly.

College Work-Study Program

This federally funded program provides eligible students with employment opportunities, usually in public and private nonprofit organizations. Federal college work-study funds pay up to 70 percent of the wages, with the remainder paid by the employing agency. Work-study is available to graduate students who demonstrate financial need. Not all schools have work-study funds, and some limit funding to undergraduates. To qualify for work-study, you must be a U.S. citizen, national, or permanent resident; be enrolled in a degree or certificate program at least half-time (schools are allowed to use their work-study funds for students attending less than half-time, but most do not); and be making satisfactory academic progress, as determined by the graduate school.

Each school sets its own application deadline and work-study earnings limits. The dollar value of a work-study award depends upon your financial need, the amount of money the school has to offer, and the aid you receive from other sources. Wages vary and are related to the type of work done. Occasionally schools use work-study funds to pay teaching and research assistants. Work-study students may work part-time during the academic year and full-time in the summer.

Additional Employment Opportunities

In addition to the types of positions described above, many schools provide on-campus employment opportunities that do not require that you demonstrate financial need. The student employment office on most campuses assists students in securing jobs both on and off the campus.

LOANS

Loans, an important source of support for graduate students, should be approached carefully. Consult with the financial aid officer of the school you wish to attend before applying for a loan, and once the school has made its aid offer, borrow only what you really need.

You should submit your application several months before you need the loan. It can take from eight to twelve weeks for your application to be processed. If you have a loan from one lender, you may want to borrow again from the same lender to avoid repayments to multiple lenders. Loans have become a complicated, competitive business. If you have loan questions, contact your financial aid officer, the lender, or the loan guaranty agency that serves your state.

Various federal and private loan programs available for graduate study are described below.

Stafford Loans

The Stafford Loan Program is available to graduate and professional students who are citizens, nationals, or permanent residents of the United States and who are enrolled at least half-time in a degree or certificate program and making satisfactory academic progress. If you qualify, you may borrow as much as $7500 per year, to a total of $54,750 for both undergraduate and graduate or professional education. Interest payments are subsidized by the federal government, and you are not required to repay the loan until you leave school.

To qualify for a Stafford Loan you must demonstrate financial need, as determined by the school's financial aid office. Need is determined by a standard system of need analysis completed when you file the GAPSFAS form, FAF, or any other federally approved need analysis form. The maximum amount you borrow may not exceed your cost of attendance, minus your expected contribution and other estimated financial assistance from federal, state, or private sources.

Stafford Loans are available through participating banks, savings and loan associations, credit unions, pension funds, and insurance companies and, in some cases, directly through school or state guarantee agencies. When the loan is approved, you are required to pay an origination fee equal to 5.5 percent of the loan amount. For example, if you borrow $7500, a $412.50 origination fee is deducted from the loan. This fee, required by law, is used to offset a portion of the federal interest subsidy. In addition, you may be charged an insurance fee by the guarantee agency. This one-time fee varies by state, ranging from 0 to 3 percent of the loan amount.

Stafford Loans are disbursed in two installments for the academic year. The interest rate is 8 percent for new borrowers (7 or 9 percent for those who have Stafford Loans outstanding at 7 or 9 percent) until the beginning of the fifth year of repayment, when it increases to 10 percent.

Loan repayment is deferred, with the federal government paying the interest, while you are in school and for a grace period, for 8 and 9 percent loans, of up to six months after the completion of your studies (nine months for 7 percent loans). Loan repayments may be deferred for up to three years, for example, if you are a member of the U.S. Armed Forces, a Peace Corps volunteer, an active-duty member of the National Oceanic and Atmospheric Administration Corps, or a full-time teacher in an area the Department of Education has designated a teacher-shortage area or if you are temporarily disabled or unable to work because a dependent of yours is disabled. You may defer repayment for up to two years while serving in an eligible internship and for periods of unemployment if during those periods you are actively looking for a full-time job. You may also defer repayment for periods of full-time study and, for new borrowers, deferments are available if you return to school at least half-time. You should check with your lender to determine whether you are eligible for a deferment. The maximum repayment period is usually ten years, with repayment made in equal monthly installments. At 8 percent, the monthly amount to be repaid is $12.14 per $1000 borrowed. If you borrow a small amount, a shorter repayment period with minimum monthly installments of $50 is required.

Supplemental Loans for Students

Graduate and professional students may also apply for Supplemental Loans for Students (SLS). Like Stafford Loans, these loans are guaranteed by the federal government. But the federal government does not subsidize in-school interest payments on these loans. Moreover, SLS repayment may be required while you are in school, and the interest rate is set annually at the Treasury bill rate plus 3.75 percent, with a 12 percent cap. The 1991–92 interest rate is 9.34 percent.

Citizens, nationals, and permanent residents of the United States who are enrolled in eligible schools on at least a half-time basis and making satisfactory academic progress may qualify for Supplemental Loans. Graduate and professional students may borrow up to $4000 per academic year, to a maximum of $20,000, in addition to amounts borrowed under the Stafford Loan and other programs.

You may receive a Supplemental Loan regardless of your income. However, some lenders check applicants' creditworthiness (ability to repay the loan) before making Supplemental Loans.

Each year you may borrow up to the cost of attendance at the school in which you are enrolled or will attend, minus estimated financial assistance from other federal, state, and private sources, up to the $4000 maximum per year. You may use the Supplemental Loan in place of some or all of the amount of contribution you are expected to make according to the need analysis formula.

There is no origination fee for Supplemental Loans, but because they are insured by the federal government, you are likely to be charged an insurance fee by the guarantee agency. This one-time fee varies by state, ranging from 0 to 3 percent of the loan amount.

Repayment begins within sixty days of the time the loan is made, with certain possible exceptions. Lenders have, for example, the option of permitting graduate and professional students to (1) defer principal payments and pay only interest while in school or (2) defer both principal and interest payments while in school (with this option, interest accrues and may compound during the in-school period and is added to the principal, with payments beginning after you leave school). You may find either of these options preferable to making principal and interest payments while you are in school. You should keep in mind, however, the increased costs involved.

Lenders may require a $600 minimum annual repayment. Repayments are made in monthly installments. In general, lenders allow borrowers between five and ten years to repay these loans. Several deferment options are available. Check with your lender to determine if you are eligible for a deferment.

Supplemental Loans can be obtained through participating banks, savings and loan associations, credit unions, pension funds, and insurance companies. In some cases, they are available directly through state guarantee agencies.

Although Supplemental Loans may not be as desirable as Stafford Loans from the consumer's perspective, they are a useful source of support for those who may not qualify for the lower-interest loans or who need additional financial assistance.

If you are interested in applying for a Supplemental Loan, ask the banks in your community if they offer these loans. If you cannot find a lender, ask your state loan guarantee agency or your financial aid officer for the names of Supplemental Loan lenders in your state.

Carl D. Perkins Loans

The Carl D. Perkins Loan Program is a long-term-loan program with very low interest—5 percent—available for graduate students who demonstrate financial need. It is administered directly by the school, with 90 percent of the money coming from the federal government and 10 percent from the school. Not all schools have these funds, and some choose to award them to undergraduates only.

Depending upon your need, the availability of Perkins Loan funds at your school, and the amount of other aid you are receiving, you may borrow up to $18,000 as a graduate student. This includes any amount you borrowed through the program as an undergraduate or for another graduate program, even if your previous loan has been repaid. To qualify for a Perkins Loan you must be a U.S. citizen, national, or permanent resident; be enrolled at least half-time in a degree or certificate program; and be making satisfactory academic progress as determined by the graduate school. Each school sets its own application deadline. You should consult the school catalog and apply as early as possible. The school will notify you of your eligibility and arrange to have the loan credited to your account, usually in two installments.

Repayment begins nine months after you graduate, leave school, or drop below half-time status. Repayment can be arranged over a ten-year period, depending upon the size of your debt, but usually you must pay at least $30 per month. In special cases—for example, if you are unemployed or ill for a period of time—you may make arrangements with your school to pay a lesser monthly amount or you may extend your repayment period.

You may defer payment while you are attending an approved institution at least half-time. Students in the health professions may defer payment for two years of residency. A three-year deferment is available for borrowers with disabled dependents, those on active duty in the National Oceanic and Atmospheric Corps, members of the U.S. Armed Forces or the Commission Corps of the U.S. Public Health Service, Peace Corps volunteers, and participants in ACTION programs such as VISTA or in programs that the U.S. Department of Education has determined to be comparable to those of the Peace Corps or ACTION. Six-month deferments are available for parental leave, and twelve-month deferments may be arranged for mothers with preschool children.

You may also defer repayment because of extraordinary circumstances, such as unemployment or prolonged illness; however, interest on your loan will continue to accrue. This is not the case with other deferments.

Part of your loan can be canceled for each year you teach handicapped children, teach in a designated elementary or secondary school that serves low-income students, or work in specified Head Start programs. If you serve as an enlisted person in certain special areas of the U.S. Army, the Department of Defense will repay a portion of your loan.

For further details on this program, you should contact the financial aid office of the school you attend.

Health Professions Student Loans

The Health Professions Student Loan (HPSL) is available for full-time graduate students enrolled in programs required of those studying to be a physician, dentist, osteopath, optometrist, pharmacist, podiatrist, or veterinarian. It is administered by the school. Not all schools have HPSL, and some limit its use to undergraduates. You must be a U.S. citizen, national, or permanent resident; demonstrate financial need; and be making satisfactory academic progress in order to qualify. You may borrow the cost of tuition plus $2500 or the amount of your financial need, whichever amount is less. The interest rate is 9 percent. Loans are repayable over a ten-year period beginning one year after you complete or cease to pursue full-time study. Deferments are granted to those on active military duty, Peace Corps volunteers, and those pursuing advanced professional training. Further information about HPSL may be obtained from the financial aid office of the school you attend.

Health Education Assistance Loans

The Health Education Assistance Loan (HEAL) provides large insured loans to medical and other health professions students. The loans are made by participating banks. Application forms are available from your school's financial aid office, which will refer you to a participating lender.

Medical, dental, osteopathic, veterinary medicine, optometry, and podiatry students may borrow up to $20,000 per academic year, up to a total of $80,000. All other eligible students may obtain loans of $12,500 per year to an aggregate of $50,000. A previous HEAL borrower is permitted to continue borrowing through the HEAL program (within the aggregate limitation) during internship and residency in order to pay interest on the loans. An insurance premium of 8 percent per year is charged in advance.

HEAL quarterly interest is variable and tied to the Treasury bill rate. Interest accrues and may be compounded during the period before repayment begins. The lender must permit you to defer all payments toward interest and principal until you enter the repayment period. Instead of allowing the interest to capitalize, you can choose to pay it on an ongoing basis, which markedly decreases the overall cost of the loan because compounding of the interest does not occur.

Repayment begins nine months after the completion of training or withdrawal from school. You have up to twenty-five years to repay, excluding deferment periods. Repayment of principal may be deferred for up to four years for internship or residency training; service in the U.S. Armed Forces, Peace Corps, ACTION, or the National Health Service Corps; or further full-time study at an approved institution of higher education. Because of the high cost of these loans, it is strongly recommended that you seek loan counseling before borrowing through the HEAL program.

Alternative Loans

To supplement the various federal loan programs and to provide borrowing opportunities for students who do not qualify for federal loans, some lenders and schools and several professional school associations have designed loan programs for students.

For example, the Law School Admission Council/Law School Admission Services has developed a loan program to provide law students with access to a private loan. Known as the Law Access Loan, it is part of a program that allows law students to apply for a Stafford Loan, a Supplemental Loan, and a Law Access Loan, using one application. Law Access Loans, up to $14,500 each year, carry either a fixed or floating interest rate, based upon the Treasury bill rate plus 3.25 percent. Further information on Law Access Loans can be obtained by contacting the Law Plan, LSAC/LSAS, Box 2500, Newtown, Pennsylvania 18940-0990 (telephone: 800-282-1550). A similar program, Lawloans, is also available to law students. More information about Lawloans can be obtained by calling 800-366-5626. You can obtain advice from your law school financial aid officer about which alternative loan best suits your needs.

A similar program, Medloans, was developed by the Association of American Medical Colleges. The Medloans program offers access to Stafford Loans, SLS, HEAL, and the Alternative Loan Program (ALP), using one application and incorporating loan terms and conditions favorable to medical students. Further details on Medloans can be obtained by calling the Association of American Medical Colleges at 202-828-0600 or by contacting the financial aid office of the medical school you wish to attend.

M.B.A. students attending AACSB-accredited institutions at least half-time can now borrow through a program called MBA Loans. Designed specifically for M.B.A. students, this program offers a three-in-one application form to apply for Stafford Loans, Supplemental Loans for Students, and a new alternative loan through the Tuition Loan Program (TLP). M.B.A. students may borrow up to $13,500 each year. TLP has a variable interest rate indexed to the ninety-one-day T-bill plus 3.5 percent. Interest is capitalized only once, and a fixed-interest-rate option of T-bill plus 4.5 percent is available when repayment begins. There are no application, origination, or guarantee fees, but there is an insurance fee of 6.5–9 percent. No payments are due until six months after graduation. You have up to 12 years to repay. Loans of $5000 or more can be consolidated. Application forms are available by calling 800-366-6227.

In addition to the various federal and school loan programs, several banks and agencies are providing loans for graduate study.

The Education Resources Institute (TERI) sponsors a loan program called Professional Education Plan (PEP) for graduate students. Loans are available for study at any accredited institution of higher education in any state. To qualify, the applicant must demonstrate creditworthiness, and either the borrower or coborrower must be a U.S. citizen. There is no need test or upper income limit. The loan turnaround time is one week. Loans can be made for $2000–$20,000 per year. Interest is variable (bank's base rate plus 2 percent). There is no application, origination, or insurance fee, but there is a 6–8 percent guarantee fee. Repayment of principal and interest begins forty-five days after graduation or termination of study. While you are in school, only interest is due on the loan. You have up to twenty years to repay the loan. For more information, write to The Education Resources Institute (TERI), 330 Stuart Street, Boston, Massachusetts 02116 (telephone: 800-255-TERI).

The New England Education Loan Marketing Association (Nellie Mae) offers two privately funded graduate loan programs—GradEXCEL and GradSHARE. Graduate students can borrow from $2000 to $7500 a year (law students up to $12,000) or up to $20,000 annually with an eligible creditworthy coborrower. SHARE loans are available to students attending any of the thirty-two member schools of the Consortium on Financing Higher Education (COFHE). GradEXCEL loans are available to students attending any accredited, degree-granting college or university in the United States. Eligibility for these programs is based on creditworthiness, not need. Students borrowing through GradSHARE or GradEXCEL can choose either a monthly variable interest rate, not to exceed the prime rate plus 2 percent, or a one-year renewable interest rate, not to exceed prime plus 3 to 4 percent. Two repayment options are available, either monthly payments of principal and interest or interest-only payments with deferment of principal for up to four years while the student is continuously enrolled. Payments may be spread out for as many as twenty years, depending on the amount borrowed. There are no application, origination, or insurance fees, but there is a guarantee fee of 4 percent, which applicants have the option to borrow in addition to the annual loan amount. For more information, write to Nellie Mae, 50 Braintree Hill Park, Suite 300, Braintree, Massachusetts 02184 (telephone: 800-634-9308).

You should contact the graduate school you wish to attend to determine whether it participates in these programs.

Debt Management

You should explore all other sources of financial aid before borrowing, so as to keep your debt to a manageable level. The standard monthly installments for repaying borrowed principal at 5, 8, and 10 percent are indicated below.

Approximate Monthly Repayments for Student Loans (Ten-Year Repayment)

Principal	5% (Perkins)	8% (Stafford)	10% (*)
$ 2,000	30 †	50 †	‡
$ 6,000	64	73	79
$10,000	106	121	132
$14,000	148	170	185
$18,000	190	218	237
$25,000		303	330
$30,000		363	396
$35,000		424	462
$40,000		485	528
$45,000		545	594
$50,000		606	660

* Alternative loans

† The minimum repayment for Stafford Loans is $50; for Perkins Loans it is $30.

‡ Many have minimum repayments.

Consolidation

Students with more than $5000 in federal loans can consolidate their loans into one repayment package with a 9 percent interest rate or a weighted average of the rates on the loans consolidated. Repayment is extended up to twenty-five years depending on the aggregate amount borrowed. Lenders can offer a graduated or income-sensitive repayment option. Consult your lender for the types of consolidation provisions offered, but be aware that consolidating graduate loans can be a very expensive option.

AID FOR SPECIAL GROUPS

If you are a member of a minority group most severely underrepresented in the national Ph.D. population, there are several fellowships for which you may be eligible.

The Ford Foundation offers a Doctoral Fellowship for Minorities, which provides three-year doctoral fellowships and one-year dissertation fellowships to American Indians, blacks, Chicanos, Puerto Ricans, Alaskan Natives (Eskimo and Aleut), and Pacific Islanders (Polynesian/Micronesian). Designed to increase the presence of underrepresented minorities on college and university faculties, these fellowships provide higher education opportunities for members of minority groups. Fellowships are awarded in behavioral and social sciences, humanities, engineering, mathematics, physical sciences, and biological sciences and are tenable at any accredited nonprofit institution of higher education in the United States offering Ph.D.'s or Sc.D.'s in the fields eligible for support in this program. Each predoctoral fellowship includes an annual stipend of $11,500 to the Fellow and an annual institutional grant of $6000 to the fellowship institution in lieu of tuition and fees. Dissertation Fellows receive a stipend of $25,000 for a twelve-month period. Applications are due in early November. All inquiries concerning application materials and program administration should be addressed to the Fellowship Office, National Research Council, 2101 Constitution Avenue, Washington, D.C. 20418 (telephone: 202-334-2872).

Books such as the *Directory of Special Programs for Minority Group Members,* 4th edition (Garrett Park, Maryland: Garrett Park Press, 1986) describe programs for minorities offered by 2,100 national and local organizations, 450 federally funded programs, and hundreds sponsored by individual colleges and universities; the book costs $25.

If you register with the Minority Graduate Student Locater Service sponsored by Educational Testing Service in Princeton, New Jersey, you will be contacted by schools interested in increasing their minority student enrollment. Such schools may well have funds designated for members of minority groups.

The Bureau of Indian Affairs (BIA) offers aid to students who demonstrate financial need, who are at least one-fourth American Indian or Alaskan Native, and who are from a federally recognized tribe. This scholarship assistance may be used at any accredited postsecondary institution. To obtain more information, contact your tribal education officer at a BIA Area Office, or write to the Bureau of Indian Affairs, P.O. Box 8327, Albuquerque, New Mexico 87198.

The federal government provides support for minorities and women through the Patricia Roberts Harris Fellowship Program, which funds approximately 1,000 recipients nationally. Awards are made to institutions and are allocated to individual students by the graduate schools. These grants provide tuition and a stipend of $10,000 for living expenses.

Other federal programs include Patricia Roberts Harris Public Service Fellowships, offered on a competitive basis for those studying in fields of public policy and public administration, as well as programs for professional education in law. The Council on Legal Education Opportunity (CLEO) is an organization designed to help and encourage minority and disadvantaged students enter law school and become members of the legal profession. It sponsors seven summer institutes throughout the country to prepare economically and educationally disadvantaged students for law school. A stipend of approximately $2500 is provided. Participants who successfully complete the six-week institute and demonstrate a capability for success in law school are named as CLEO Fellows. As such, they are eligible for annual CLEO stipend awards for their three years of legal study, contingent upon CLEO's receipt of federal funding. For further information, contact CLEO, Suite 290, North Lobby, 1800 M Street, NW, Washington, D.C. 20036 (telephone: 202-785-4840).

Financial aid designated for women is available, especially for those interested in fields in which women are underrepresented—such as mathematics and engineering. *The Directory of Financial Aids for Women,* 4th edition, by Gail Ann Schlachter (Redwood City, California: Reference Service Press, 1987), lists sources of support and identifies foundations and other organizations interested in helping women to secure funding for graduate education. The Association for Women in Science publishes "Grants-at-a-Glance," a brochure highlighting fellowship opportunities for women in science. It can be obtained from the Association at 1522 K Street, NW, Suite 820, Washington, D.C. 20005 (telephone: 202-408-0742).

There are a number of organizations that provide financial aid to disabled students. You may contact the Vocational Rehabilitation Services in your home state for further details on such programs.

INTERNATIONAL EDUCATION AND STUDY ABROAD

There are several different sources of funding for those who wish to study abroad, as well as for foreign nationals who wish to study in the United States. The Institute of International Education, 809 United Nations Plaza, New York, New York 10017, assists students in locating such aid. It has recently published Funding *for U.S. Study—A Guide for Foreign Nationals* and authored *Financial Resources for International Study,* a guide to organizations offering awards for overseas study. The institute also publishes a free pamphlet, "Financial Resources for International Study: A Selected Bibliography," which is an excellent reference tool for locating directories that list sources of support for graduate study. Another such agency is the Council on International Educational Exchange, 205 East 42nd Street, New York, New York 10017, which publishes the *Student Travel Catalogue.* This publication lists fellowship sources and contains a detailed explanation of the council's services both for American students traveling abroad and for foreign students interested in coming to the United States.

The U.S. Department of Education administers programs that support approximately 700 fellowships related to international education and world area studies. Foreign Language and Area Studies Fellowships and Fulbright-Hays Doctoral Dissertation Awards are merit-based programs established to promote a wider knowledge and understanding of other countries and cultures. They offer support to graduate students interested in foreign languages and international relations.

Some schools have scholar exchange programs. You should discuss these and other possible opportunities for foreign study with the financial aid officer at the school you will attend.

A Final Note

While they vary from field to field as well as from year to year, many opportunities for financial assistance do exist. If you are interested in graduate or professional study, be sure to discuss your plans with faculty members and advisers. Explore all your options. Plan ahead, complete forms on time, and do not despair in your search for support for graduate or professional study. No matter what your financial situation, if you are academically qualified and are knowledgeable about the many different sources of aid, you should be able to attend the graduate school of your choice.

Patricia McWade
Dean of Financial Aid
Georgetown University

Tests Required of Applicants

Many graduate schools require that applicants submit scores on one or more standardized tests, often the Graduate Record Examinations (GRE) or the Miller Analogies Test (MAT). Virtually all graduate and professional schools ask students whose native language is not English to take the Test of English as a Foreign Language (TOEFL), and some also ask for the Test of Spoken English (TSE).

Brief descriptions of these tests and the addresses to write to for additional information are given below.

GRADUATE RECORD EXAMINATIONS

The GRE, given each year at many test centers in the United States and abroad, are administered by Educational Testing Service (ETS) of Princeton, New Jersey, under policies determined by the Graduate Record Examinations Board, an independent board affiliated with the Association of Graduate Schools and the Council of Graduate Schools.

The GRE General Test contains seven 30-minute sections designed to measure verbal, quantitative, and analytical abilities. The General Test is given in the morning on each international administration test date.

The Subject Tests are designed to measure knowledge and understanding of subject matter basic to graduate study in specific fields. Each Subject Test lasts 2 hours and 50 minutes, and only one may be taken on any given test date. Subject Tests are available in sixteen areas: biochemistry and cell and molecular biology, biology, chemistry, computer science, economics, education, engineering, geology, history, literature in English, mathematics, music, physics, political science, psychology, and sociology.

No Subject Tests are given in June. On all other dates, the Subject Tests are given in the afternoon of the same date as the General Test. The revised Subject Test in music is offered only in October and February.

The GRE schedule for 1991-92 is October 12, December 14, February 1, April 11, and (General Test only) June 6. Students who, for religious reasons, cannot take tests on Saturday may request a Monday administration immediately following a regular Saturday test date.

In the United States, U.S. territories, and Puerto Rico, the fee for the General Test is $44, and the fee for one Subject Test is $44. An additional fee of $25 is levied for each test date in India, Japan, Korea, and Taiwan; there is a fee of $20 for each test date in other locations. (The fees are subject to change.)

Additional information about the GRE is available from Graduate Record Examinations, Educational Testing Service, P.O. Box 6000, Princeton, New Jersey 08541-6000. In addition, the GRE phone lines in Princeton (609-771-7670) are open from 8:30 a.m. to 9 p.m. Eastern time for student inquiries, and the phone lines at the California Bay Area office of ETS (415-654-1200) are open from 8:30 a.m. to 4:30 p.m. Pacific time.

MILLER ANALOGIES TEST

The MAT is a high-level mental ability test that requires the solution of 100 problems stated in the form of analogies. The MAT is accepted by over 2,300 graduate school programs as part of their admission process. The test items use different types of analogies to sample a variety of fields, such as fine arts, literature, mathematics, natural science, and social science. Examinees are allowed 50 minutes to complete the test.

The MAT is offered at more than 600 test centers in the United States and Canada. For examinee convenience, the test is given on an as-needed basis at most test centers. Fees are also determined by each test center.

Additional information about the MAT, including preparatory materials and test center locations, is available from the Psychological Corporation, 555 Academic Court, San Antonio, Texas 78204. Telephone: 512-554-8183 (8:30 a.m. to 5 p.m., Central time).

TEST OF ENGLISH AS A FOREIGN LANGUAGE

The purpose of TOEFL is to evaluate the English proficiency of people whose native language is not English. Given in a single session of about 3 hours, the test consists of three sections: listening comprehension, structure and written expression, and vocabulary and reading comprehension. The test is given at over 1,200 centers in 170 countries and areas and is administered by Educational Testing Service (ETS) under the general direction of a policy council established by the College Board and the Graduate Record Examinations Board.

The 1991–92 test dates are July 12, August 3, September 27, October 26, November 16, December 13, January 11, February 14, March 14, April 10, May 9, and June 5. Examinees who take the test at the September, October, March, and May administrations are required to write a short essay (the Test of Written English). The fee for a test scheduled on a Saturday is $31, and the fee for a Friday test date is $39, which must be paid in U.S. dollars.

Registration material is available from TOEFL, P.O. Box 6151, Princeton, New Jersey 08541-6151, U.S.A. Telephone: 215-750-8050.

TEST OF SPOKEN ENGLISH

The major purpose of the TSE is to evaluate the spoken English proficiency of people whose native language is not English. The test, which takes about 30 minutes, requires examinees to demonstrate their ability to speak English by answering a variety of questions presented in printed and recorded form. All the answers to test questions are recorded on tape; no writing is required. TSE is given at selected TOEFL test centers worldwide. The test is administered by Educational Testing Service (ETS) under the general direction of a policy council established by the College Board and the Graduate Record Examinations Board.

The 1991–92 test dates are July 12, August 3, September 27, October 26, November 16, December 13, January 11, February 14, March 14, April 10, May 9, and June 5. There are two separate registration categories within the TSE program. TSE-A, which is for teaching and research assistant applicants at academic institutions or for other undergraduate or graduate school applicants, has a fee of $75. TSE-P, which is for all other individuals, such as those who are taking the TSE to obtain licensure or certification in a professional or occupational field, has a fee of $100. These fees must be paid in U.S. dollars.

Registration material is found in the *Bulletin of Information for TOEFL and TSE* available from TOEFL, P.O. Box 6151, Princeton, New Jersey 08541-6151, U.S.A. Telephone: 215-750-8050.

Accreditation and Accrediting Agencies

Colleges and universities in the United States, and their individual academic and professional programs, are accredited by nongovernmental agencies concerned with monitoring the quality of education in this country. Agencies with both regional and national jurisdictions grant accreditation to institutions as a whole, while specialized bodies acting on a nationwide basis—often national professional associations—grant accreditation to departments and programs in specific fields.

Institutional and specialized accrediting agencies share the same basic concerns: the purpose an academic unit—whether university or program—has set for itself and how well it fulfills that purpose, the adequacy of its financial and other resources, the quality of its academic offerings, and the level of services it provides. Agencies that grant institutional accreditation take a broader view, of course, and examine university- or college-wide services that a specialized agency may not concern itself with.

Both types of agencies follow the same general procedures when considering an application for accreditation. The academic unit prepares a self-evaluation, focusing on the concerns mentioned above and usually including an assessment of both its strengths and weaknesses; a team of representatives of the accrediting body reviews this evaluation, visits the campus, and makes its own report; and finally, the accrediting body makes a decision on the application. Often, even when accreditation is granted, the agency makes a recommendation regarding how the institution or program can improve. All institutions and programs are also reviewed every few years to determine whether they continue to meet established standards; if they do not, they may lose their accreditation.

Accrediting agencies themselves are reviewed and evaluated periodically by the U.S. Department of Education and a nongovernmental organization, the Council on Postsecondary Accreditation (COPA). Agencies recognized by COPA and the Department of Education adhere to certain standards and practices, and their authority in matters of accreditation is widely accepted in the educational community.

This does not mean, however, that accreditation is a simple matter, either for schools wishing to become accredited or for students deciding where to apply. Indeed, in certain fields the very meaning and methods of accreditation are the subject of a good deal of debate. For their part, those applying to graduate school should be aware of the safeguards provided by regional accreditation, especially in terms of degree acceptance and institutional longevity. Beyond this, applicants should understand the role that specialized accreditation plays in their field, as this varies considerably from one discipline to another. In certain professional fields, it is necessary to have graduated from a program that is accredited in order to be eligible for a license to practice, and in some fields the federal government also makes this a hiring requirement. In other disciplines, however, accreditation is not as essential, and there can be excellent programs that are not accredited. In fact, some programs choose not to seek accreditation, although most do.

Institutions and programs that present themselves for accreditation are sometimes granted the status of candidate for accreditation, or what is known as "preaccreditation." This may happen, for example, when an academic unit is too new to have met all the requirements for accreditation. Such status signifies initial recognition and indicates that the school or program in question is working to fulfill all requirements; it does not, however, guarantee that accreditation will be granted.

In order to help students understand the role of accreditation in their field, many of the specialized accrediting bodies have contributed short essays that are printed in the pages of these guides. They appear in Books 2, 3, 5, and 6, at the beginning of the appropriate section. Readers are also advised to contact agencies directly for answers to their questions about accreditation. The names and addresses of all agencies recognized by COPA and the U.S. Department of Education are listed below.

Institutional Accrediting Agencies—Regional

MIDDLE STATES ASSOCIATION OF COLLEGES AND SCHOOLS

Accredits institutions in Delaware, District of Columbia, Maryland, New Jersey, New York, Pennsylvania, Puerto Rico, and the Virgin Islands.

Howard L. Simmons, Executive Director
Commission on Higher Education
3624 Market Street
Philadelphia, Pennsylvania 19104
Telephone: 215-662-5606

NEW ENGLAND ASSOCIATION OF SCHOOLS AND COLLEGES

Accredits institutions in Connecticut, Maine, Massachusetts, New Hampshire, Rhode Island, and Vermont.

Charles M. Cook, Director of Evaluation
Commission on Institutions of Higher Education
Sanborn House
15 High Street
Winchester, Massachusetts 01890
Telephone: 617-729-6762

NORTH CENTRAL ASSOCIATION OF COLLEGES AND SCHOOLS

Accredits institutions in Arizona, Arkansas, Colorado, Illinois, Indiana, Iowa, Kansas, Michigan, Minnesota, Missouri, Nebraska, New Mexico, North Dakota, Ohio, Oklahoma, South Dakota, West Virginia, Wisconsin, and Wyoming.

Patricia Thrash, Executive Director
Commission on Institutions of Higher Education
159 North Dearborn Street
Chicago, Illinois 60601
Telephone: 312-263-0456

NORTHWEST ASSOCIATION OF SCHOOLS AND COLLEGES

Accredits institutions in Alaska, Idaho, Montana, Nevada, Oregon, Utah, and Washington.

Joseph A. Malik, Executive Director
Commission on Colleges
3700-B University Way, NE
Seattle, Washington 98105
Telephone: 206-543-0195

SOUTHERN ASSOCIATION OF COLLEGES AND SCHOOLS

Accredits institutions in Alabama, Florida, Georgia, Kentucky, Louisiana, Mississippi, North Carolina, South Carolina, Tennessee, Texas, and Virginia.

James T. Rogers, Executive Director
Commission on Colleges
1866 Southern Lane
Decatur, Georgia 30033-4097
Telephone: 404-329-6500

WESTERN ASSOCIATION OF SCHOOLS AND COLLEGES

Accredits institutions in American Samoa, California, Guam, Hawaii, and Trust Territory of the Pacific.

Stephen S. Weiner, Executive Director
Accrediting Commission for Senior Colleges and Universities
Mills College
P.O. Box 9990
Oakland, California 94613-0990
Telephone: 415-632-5000

ACCREDITATION AND ACCREDITING AGENCIES

Institutional Accrediting Agencies—Other

ASSOCIATION OF INDEPENDENT COLLEGES AND SCHOOLS
James M. Phillips, Executive Director
Accrediting Commission
One Dupont Circle, NW, Suite 350
Washington, D.C. 20036
Telephone: 202-659-2460

NATIONAL HOME STUDY COUNCIL
William A. Fowler, Executive Director
Accrediting Commission
1601 Eighteenth Street, NW
Washington, D.C. 20009
Telephone: 202-234-5100

Specialized Accrediting Agencies

[Only Book 1 of Peterson's Annual Guides to Graduate Study includes the complete list of specialized accrediting groups recognized by the U.S. Department of Education and COPA. The lists in Books 2, 3, 5, and 6 are abridged, and there are no such recognized specialized accrediting bodies for the programs in Book 4.]

Directory of Institutions with Programs in the Physical Sciences and Mathematics

This directory lists institutions in alphabetical order and includes beneath each name the academic fields in the physical sciences and mathematics in which each institution offers graduate programs. The degree level in each field is also indicated, provided that the institution has supplied that information in response to Peterson's Guides Annual Survey of Graduate Institutions. An *M* indicates that a master's degree program is offered; a *D* indicates that a doctoral degree program is offered; a *P* indicates that the first professional degree is offered; an *O* signifies that other advanced degrees (e.g., certificates or specialist degrees) are offered; and an * (asterisk) indicates that a two-page description and/or announcement is located in this volume. See the index for the page number of the two-page description and/or announcement.

DIRECTORY OF INSTITUTIONS

ABILENE CHRISTIAN UNIVERSITY
Chemistry — M

ACADIA UNIVERSITY
Chemistry — M
Geology — M
Mathematics — M

ADELPHI UNIVERSITY
Chemistry — M
Earth Sciences — M,O
Mathematics — M,D
Physics — M

AIR FORCE INSTITUTE OF TECHNOLOGY
Applied Mathematics — M,D

ALABAMA AGRICULTURAL AND MECHANICAL UNIVERSITY
Applied Physics — M,D
Optical Sciences — M,D
Physics — M,D

ALABAMA STATE UNIVERSITY
Mathematics — M,O

AMERICAN UNIVERSITY
Applied Mathematics — M
Chemistry — M,D
Mathematics — M*
Physics — M,D
Statistics — M,D*

ANDREWS UNIVERSITY
Physics — M

ANGELO STATE UNIVERSITY
Mathematics — M

APPALACHIAN STATE UNIVERSITY
Applied Physics — M
Chemistry — M
Mathematics — M
Physics — M

ARIZONA STATE UNIVERSITY
Applied Mathematics — M,D*
Applied Physics — D
Astronomy — D
Chemistry — M,D*
Geology — M,D
Geophysics — D
Mathematics — M,D*
Physics — M,D
Planetary Sciences — D
Statistics — M,D*

ARKANSAS STATE UNIVERSITY
Chemistry — M,O
Mathematics — M

AUBURN UNIVERSITY
Chemistry — M,D
Geology — M*
Mathematics — M,D
Physics — M,D*
Statistics — M,D

BALL STATE UNIVERSITY
Chemistry — M
Earth Sciences — M
Geology — M*
Mathematics — M
Physics — M
Statistics — M

BARUCH COLLEGE OF THE CITY UNIVERSITY OF NEW YORK
Statistics — M

BAYLOR UNIVERSITY
Chemistry — M,D*
Earth Sciences — M
Geology — M,D
Mathematics — M
Physics — M,D
Statistics — D

BISHOP'S UNIVERSITY
Chemistry — M
Physics — M

BOISE STATE UNIVERSITY
Geology — M
Geophysics — M

BOSTON COLLEGE
Analytical Chemistry — M,D*
Chemistry — M,D*
Geology — M
Geophysics — M
Inorganic Chemistry — M,D*
Marine Sciences/Oceanography — M
Mathematics — M
Organic Chemistry — M,D*
Physical Chemistry — M,D*
Physics — M,D

BOSTON UNIVERSITY
Applied Physics — D*
Astronomy — M,D
Biostatistics — M,D
Chemistry — M,D*
Geology — M,D
Inorganic Chemistry — M,D*
Mathematics — M,D
Organic Chemistry — M,D*
Physical Chemistry — M,D*
Physics — M,D*
Statistics — M,D

BOWLING GREEN STATE UNIVERSITY
Applied Mathematics — M
Astrophysics — M
Chemistry — M
Geology — M*
Mathematics — M,D
Statistics — M,D

BRADLEY UNIVERSITY
Chemistry — M

BRANDEIS UNIVERSITY
Chemistry — M,D*
Inorganic Chemistry — M,D*
Mathematics — M,D*
Organic Chemistry — M,D*
Physical Chemistry — M,D*
Physics — D*

BRIDGEWATER STATE COLLEGE
Chemistry — M,O
Earth Sciences — M,O
Physics — M,O

BRIGHAM YOUNG UNIVERSITY
Analytical Chemistry — M,D*
Astronomy — M
Chemistry — M,D*
Geology — M
Inorganic Chemistry — M,D*
Mathematics — M,D
Organic Chemistry — M,D*
Physical Chemistry — M,D*
Physics — M,D
Statistics — M

BROCK UNIVERSITY
Chemistry — M
Geology — M
Physics — M

BROOKLYN COLLEGE OF THE CITY UNIVERSITY OF NEW YORK
Applied Physics — M
Chemistry — M
Geology — M
Mathematics — M
Physics — M,D*

BROWN UNIVERSITY
Applied Mathematics — M,D*
Chemistry — M,D
Geology — M,D
Mathematics — M,D*
Physics — M,D*
Planetary Sciences — M,D*

BRYN MAWR COLLEGE
Chemistry — M,D
Geology — M,D
Mathematics — M,D
Physics — M,D

BUCKNELL UNIVERSITY
Chemistry — M
Mathematics — M

BUTLER UNIVERSITY
Chemistry — M

CALIFORNIA INSTITUTE OF TECHNOLOGY
Applied Mathematics — D*
Applied Physics — M,D
Astronomy — D
Astrophysics — D
Chemistry — M,D
Geochemistry — M,D*
Geology — M,D*
Geophysics — M,D*
Mathematics — D
Physics — D*
Planetary Sciences — M,D*
Plasma Physics — M,D

CALIFORNIA POLYTECHNIC STATE UNIVERSITY, SAN LUIS OBISPO
Chemistry — M
Mathematics — M

CALIFORNIA STATE POLYTECHNIC UNIVERSITY, POMONA
Applied Mathematics — M
Chemistry — M
Mathematics — M

CALIFORNIA STATE UNIVERSITY, BAKERSFIELD
Geology — M

CALIFORNIA STATE UNIVERSITY, CHICO
Chemistry — M
Earth Sciences — M
Geology — M
Mathematics — M
Physics — M

CALIFORNIA STATE UNIVERSITY, FRESNO
Chemistry — M
Geology — M*
Mathematics — M*
Physics — M*

CALIFORNIA STATE UNIVERSITY, FULLERTON
Analytical Chemistry — M
Chemistry — M*
Geochemistry — M
Inorganic Chemistry — M
Mathematics — M*
Organic Chemistry — M
Physical Chemistry — M
Statistics — M

CALIFORNIA STATE UNIVERSITY, HAYWARD
Applied Mathematics — M
Chemistry — M
Geology — M
Marine Sciences/Oceanography — M
Mathematics — M
Statistics — M

CALIFORNIA STATE UNIVERSITY, LONG BEACH
Applied Mathematics — M
Chemistry — M
Geology — M
Mathematics — M
Physics — M

CALIFORNIA STATE UNIVERSITY, LOS ANGELES
Analytical Chemistry — M
Applied Mathematics — M
Chemistry — M*
Geology — M
Inorganic Chemistry — M
Mathematics — M*
Organic Chemistry — M
Physical Chemistry — M
Physics — M*
Statistics — M

CALIFORNIA STATE UNIVERSITY, NORTHRIDGE
Chemistry — M
Geology — M
Mathematics — M
Physics — M

CALIFORNIA STATE UNIVERSITY, SACRAMENTO
Chemistry — M
Marine Sciences/Oceanography — M
Mathematics — M

CALIFORNIA UNIVERSITY OF PENNSYLVANIA
Earth Sciences — M
Mathematics — M

CARLETON UNIVERSITY
Chemistry — M,D
Earth Sciences — M,D*
Geology — M,D
Mathematics — M,D
Physics — M,D
Statistics — M,D

CARNEGIE MELLON UNIVERSITY
Applied Mathematics — M*
Applied Physics — D
Chemistry — M,D*
Mathematics — M,D*
Physics — M,D
Statistics — M,D*

CASE WESTERN RESERVE UNIVERSITY
Applied Mathematics — M
Applied Physics — M,D
Astronomy — M,D
Biometrics — M,D
Biostatistics — M,D*
Chemistry — M,D*
Geology — M,D
Mathematics — M,D
Physics — M,D

CATHOLIC UNIVERSITY OF AMERICA
Acoustics — M,D
Chemistry — M,D*
Mathematics — M,D
Physics — M,D

CATHOLIC UNIVERSITY OF PUERTO RICO
Chemistry — M

CENTENARY COLLEGE OF LOUISIANA
Geology — M

CENTRAL CONNECTICUT STATE UNIVERSITY
Chemistry — M
Earth Sciences — M
Mathematics — M
Physics — M
Statistics — M

CENTRAL MICHIGAN UNIVERSITY
Chemistry — M,D
Mathematics — M
Physics — M

CENTRAL MISSOURI STATE UNIVERSITY
Applied Mathematics — M
Mathematics — M

CENTRAL WASHINGTON UNIVERSITY
Chemistry — M
Mathematics — M

CHICAGO STATE UNIVERSITY
Mathematics — M

CITY COLLEGE OF THE CITY UNIVERSITY OF NEW YORK
Chemistry — M,D
Earth Sciences — M,D
Geology — M,D
Mathematics — M
Physics — M,D*

CLAREMONT GRADUATE SCHOOL
Applied Mathematics — M
Mathematics — M,D*
Statistics — M

CLARION UNIVERSITY OF PENNSYLVANIA
Mathematics — M

CLARK ATLANTA UNIVERSITY
Applied Mathematics — M
Inorganic Chemistry — M,D
Organic Chemistry — M,D
Physical Chemistry — M,D
Physics — M

CLARKSON UNIVERSITY
Analytical Chemistry — M,D
Chemistry — M,D
Inorganic Chemistry — M,D
Mathematics — M,D
Organic Chemistry — M,D
Physical Chemistry — M,D
Physics — M,D*

CLARK UNIVERSITY
Chemistry — M,D
Physics — M,D*

CLEMSON UNIVERSITY
Analytical Chemistry — M,D

Applied Mathematics	M,D
Astronomy	M,D*
Astrophysics	M,D*
Chemistry	M,D
Geology	M
Inorganic Chemistry	M,D
Mathematics	M,D*
Meteorology and Atmospheric Sciences	M,D*
Organic Chemistry	M,D
Physical Chemistry	M,D
Physics	M,D*
Statistics	M,D*

CLEVELAND STATE UNIVERSITY

Analytical Chemistry	M,D
Applied Mathematics	M
Chemistry	M,D
Inorganic Chemistry	M,D
Mathematics	M
Optical Sciences	M
Organic Chemistry	M
Physical Chemistry	M,D

COLGATE UNIVERSITY

| Geology | M |

COLLEGE OF STATEN ISLAND OF THE CITY UNIVERSITY OF NEW YORK

| Chemistry | D |

COLLEGE OF WILLIAM AND MARY

Chemistry	M
Marine Sciences/ Oceanography	M,D*
Mathematics	M*
Physics	M,D

COLORADO SCHOOL OF MINES

Applied Mathematics	M,D
Applied Physics	D
Chemistry	M,D
Geochemistry	M,D
Geology	M,D
Geophysics	M,D
Physics	M,D*

COLORADO STATE UNIVERSITY

Chemistry	M,D
Earth Sciences	M,D
Geology	M
Mathematics	M,D*
Meteorology and Atmospheric Sciences	M,D*
Physics	M,D
Statistics	M,D

COLUMBIA UNIVERSITY

Applied Mathematics	D
Applied Physics	M,D
Astronomy	M,D
Biostatistics	M,D
Chemistry	M,D*
Geochemistry	M,D
Geodetic Sciences	M,D
Geology	M,D
Geophysics	M,D
Inorganic Chemistry	M,D*
Marine Sciences/Oceanography	M,D
Mathematics	M,D*
Meteorology and Atmospheric Sciences	M,D*
Organic Chemistry	M,D*
Physical Chemistry	M,D*
Physics	M,D
Planetary Sciences	M,D*
Plasma Physics	M,D
Statistics	M,D

CONCORDIA UNIVERSITY (CANADA)

Applied Mathematics	M
Chemistry	M,D
Mathematics	M
Physics	M,D

CORNELL UNIVERSITY

Applied Mathematics	M,D*
Applied Physics	D
Astronomy	M,D*
Biometrics	M,D*
Chemistry	D
Earth Sciences	M,D
Geology	M,D*
Mathematics	M,D*
Meteorology and Atmospheric Sciences	M,D
Physics	M,D
Statistics	M,D

CREIGHTON UNIVERSITY

| Mathematics | M |

Meteorology and Atmospheric Sciences	M
Physics	M
Statistics	M

DALHOUSIE UNIVERSITY

Chemistry	M,D
Geology	M,D
Marine Sciences/ Oceanography	M,D*
Mathematics	M,D
Physics	M,D
Statistics	M,D

DARTMOUTH COLLEGE

Astronomy	M,D
Chemistry	M,D
Geology	M,D*
Mathematics	M,D*
Physics	M,D

DELAWARE STATE COLLEGE

| Chemistry | M |
| Physics | M |

DEPAUL UNIVERSITY

Applied Mathematics	M
Applied Physics	M
Chemistry	M*
Physics	M

DRAKE UNIVERSITY

| Physics | M |

DREXEL UNIVERSITY

Biostatistics	M,D*
Chemistry	M,D*
Mathematics	M,D*
Meteorology and Atmospheric Sciences	M,D*
Physics	M,D*

DUKE UNIVERSITY

Chemistry	D
Geology	M,D*
Mathematics	D
Physics	D
Statistics	D

DUQUESNE UNIVERSITY

| Chemistry | M,D |

EAST CAROLINA UNIVERSITY

Applied Mathematics	M
Applied Physics	M
Chemistry	M
Geology	M
Mathematics	M
Physics	M*

EASTERN ILLINOIS UNIVERSITY

| Chemistry | M |
| Mathematics | M |

EASTERN KENTUCKY UNIVERSITY

Chemistry	M
Geology	M
Mathematics	M
Physics	M

EASTERN MICHIGAN UNIVERSITY

Chemistry	M*
Mathematics	M
Physics	M

EASTERN NEW MEXICO UNIVERSITY

| Chemistry | M |
| Mathematics | M |

EASTERN WASHINGTON UNIVERSITY

| Geology | M |
| Mathematics | M |

EAST TENNESSEE STATE UNIVERSITY

| Chemistry | M |
| Mathematics | M |

EAST TEXAS STATE UNIVERSITY (COMMERCE)

Chemistry	M
Earth Sciences	M
Mathematics	M
Physics	M*

ÉCOLE POLYTECHNIQUE DE MONTRÉAL

| Applied Mathematics | M,D |
| Optical Sciences | M,D |

EMORY UNIVERSITY

Biostatistics	M,D*
Chemistry	M,D
Geophysics	M,D*
Mathematics	M,D*
Physics	M,D*
Plasma Physics	M,D*
Statistics	M

EMPORIA STATE UNIVERSITY

Chemistry	M
Earth Sciences	M
Mathematics	M
Physics	M

FAIRLEIGH DICKINSON UNIVERSITY, FLORHAM-MADISON CAMPUS

| Chemistry | M |
| Mathematics | M |

FAIRLEIGH DICKINSON UNIVERSITY, TEANECK-HACKENSACK CAMPUS

| Chemistry | M |
| Mathematics | M |

FISK UNIVERSITY

| Chemistry | M |
| Physics | M |

FLORIDA ATLANTIC UNIVERSITY

Chemistry	M,D
Geology	M
Mathematics	M,D
Physics	M,D

FLORIDA INSTITUTE OF TECHNOLOGY

Applied Mathematics	M,D
Chemistry	M,D
Marine Sciences/ Oceanography	M,D*
Physics	M,D
Planetary Sciences	M,D

FLORIDA INTERNATIONAL UNIVERSITY

Chemistry	M
Geology	M
Mathematics	M*
Physics	M*

FLORIDA STATE UNIVERSITY

Analytical Chemistry	M,D
Applied Mathematics	M,D
Chemistry	M,D*
Geology	M,D
Geophysics	D
Inorganic Chemistry	M,D*
Marine Sciences/ Oceanography	M,D*
Mathematics	M,D
Meteorology and Atmospheric Sciences	M,D
Organic Chemistry	M,D*
Physical Chemistry	M,D*
Physics	M,D*
Statistics	M,D

FORDHAM UNIVERSITY

| Mathematics | M |
| Physical Chemistry | M |

FORT HAYS STATE UNIVERSITY

Geology	M
Mathematics	M
Physics	M

FRAMINGHAM STATE COLLEGE

| Mathematics | M |

FURMAN UNIVERSITY

| Chemistry | M |

GEORGE MASON UNIVERSITY

Applied Physics	M
Chemistry	M
Mathematics	M
Statistics	M

GEORGETOWN UNIVERSITY

Analytical Chemistry	M,D
Biostatistics	M
Chemistry	M,D
Inorganic Chemistry	M,D
Organic Chemistry	M,D
Physical Chemistry	M,D

GEORGE WASHINGTON UNIVERSITY

Acoustics	M,D,O
Applied Mathematics	M
Biostatistics	M,D*
Chemistry	M,D
Geochemistry	M,D
Geology	M,D
Mathematics	M,D
Physics	M,D
Statistics	M,D*

GEORGIA INSTITUTE OF TECHNOLOGY

Applied Physics	M
Earth Sciences	M,D
Mathematics	M,D*
Physics	M,D*
Statistics	M

GEORGIAN COURT COLLEGE

| Mathematics | M |

GEORGIA SOUTHERN UNIVERSITY

| Mathematics | M |

GEORGIA STATE UNIVERSITY

Astronomy	D
Chemistry	M,D
Geology	M,O*
Statistics	M

GOVERNORS STATE UNIVERSITY

| Analytical Chemistry | M |

GRADUATE SCHOOL AND UNIVERSITY CENTER OF THE CITY UNIVERSITY OF NEW YORK

Chemistry	D
Earth Sciences	D
Mathematics	D
Physics	D

HAMPTON UNIVERSITY

Chemistry	M
Mathematics	M
Physics	M

HARVARD UNIVERSITY

Applied Mathematics	M,D
Applied Physics	M,D
Astronomy	M,D
Astrophysics	M,D
Biostatistics	M,D
Chemistry	M,D
Earth Sciences	M,D*
Geology	M,D*
Inorganic Chemistry	M,D
Mathematics	M,D
Organic Chemistry	M,D
Physical Chemistry	M,D
Physics	M,D*
Planetary Sciences	M,D*
Statistics	M,D*

HARVEY MUDD COLLEGE

| Mathematics | M |

HOFSTRA UNIVERSITY

| Applied Mathematics | M |

HOWARD UNIVERSITY

Analytical Chemistry	M,D
Applied Mathematics	M,D*
Chemistry	M,D
Inorganic Chemistry	M,D
Mathematics	M,D*
Organic Chemistry	M,D
Physical Chemistry	M,D
Physics	M,D

HUNTER COLLEGE OF THE CITY UNIVERSITY OF NEW YORK

Applied Mathematics	M
Mathematics	M
Physics	M

IDAHO STATE UNIVERSITY

Geology	M
Mathematics	M,D
Physics	M

ILLINOIS INSTITUTE OF TECHNOLOGY

Applied Mathematics	M
Chemistry	M,D
Mathematics	D
Physics	M,D
Statistics	M

*P—first professional degree; M—master's degree; D—doctorate; O—other advanced degree; * full description and/or announcement for program (see index)*

DIRECTORY OF INSTITUTIONS

ILLINOIS STATE UNIVERSITY
Chemistry — M
Mathematics — M,D

INDIANA STATE UNIVERSITY
Chemistry — M
Geology — M
Mathematics — M
Physics — M

INDIANA UNIVERSITY BLOOMINGTON
Applied Mathematics — M,D
Astronomy — M,D
Astrophysics — D*
Chemistry — M,D
Earth Sciences — M
Geochemistry — M,D
Geology — M,D*
Geophysics — M,D*
Mathematical Physics — M,D*
Physics — M,D*
Statistics — M,D

INDIANA UNIVERSITY OF PENNSYLVANIA
Chemistry — M
Mathematics — M
Physics — M

INDIANA UNIVERSITY–PURDUE UNIVERSITY AT FORT WAYNE
Applied Mathematics — M
Mathematics — M

INDIANA UNIVERSITY–PURDUE UNIVERSITY AT INDIANAPOLIS
Applied Mathematics — M,D
Chemistry — M,D*
Geology — M
Mathematics — M,D
Physics — M,D

INSTITUTE OF PAPER SCIENCE AND TECHNOLOGY
Chemistry — M,D*
Mathematics — M,D*
Physics — M,D*

INSTITUTO TECNOLÓGICO Y DE ESTUDIOS SUPERIORES DE MONTERREY
Chemistry — D
Organic Chemistry — M
Physical Chemistry — M

IOWA STATE UNIVERSITY OF SCIENCE AND TECHNOLOGY
Applied Mathematics — M,D
Astronomy — M,D
Chemistry — M,D
Earth Sciences — M,D*
Geodetic Sciences — M,D
Geology — M,D*
Mathematics — M,D
Statistics — M,D

JACKSON STATE UNIVERSITY
Analytical Chemistry — M
Chemistry — M
Inorganic Chemistry — M
Mathematics — M,O
Organic Chemistry — M
Physical Chemistry — M

JACKSONVILLE STATE UNIVERSITY
Mathematics — M

JAMES MADISON UNIVERSITY
Mathematics — M

JOHN CARROLL UNIVERSITY
Chemistry — M
Mathematics — M
Physics — M

JOHNS HOPKINS UNIVERSITY
Applied Mathematics — M,D*
Astronomy — M,D*
Biostatistics — M,D
Chemistry — M,D
Geochemistry — M,D
Geology — M,D
Geophysics — M,D
Marine Sciences/Oceanography — M,D*
Mathematics — M,D*
Physics — M,D*
Statistics — M,D*

KANSAS STATE UNIVERSITY
Analytical Chemistry — M,D

Chemistry — M,D
Geology — M,D
Inorganic Chemistry — M,D
Mathematics — M,D
Organic Chemistry — M,D
Physical Chemistry — M,D
Physics — M,D
Statistics — M,D

KENT STATE UNIVERSITY
Analytical Chemistry — M,D*
Chemistry — M,D*
Geology — M,D
Inorganic Chemistry — M,D*
Mathematics — M,D*
Organic Chemistry — M,D*
Physical Chemistry — M,D*
Physics — M,D*

KUTZTOWN UNIVERSITY OF PENNSYLVANIA
Mathematics — M

LAKEHEAD UNIVERSITY
Chemistry — M
Geology — M
Mathematics — M
Physics — M

LAMAR UNIVERSITY–BEAUMONT
Chemistry — M
Mathematics — M

LAURENTIAN UNIVERSITY
Applied Physics — M
Chemistry — M
Geology — M

LEHIGH UNIVERSITY
Analytical Chemistry — M,D*
Applied Mathematics — M,D*
Chemistry — M,D*
Geology — M,D
Inorganic Chemistry — M,D*
Mathematics — M,D*
Organic Chemistry — M,D*
Physical Chemistry — M,D*
Physics — M,D

LEHMAN COLLEGE OF THE CITY UNIVERSITY OF NEW YORK
Mathematics — M

LOMA LINDA UNIVERSITY
Biostatistics — M
Geology — M

LONG ISLAND UNIVERSITY, BROOKLYN CAMPUS
Chemistry — M

LONG ISLAND UNIVERSITY, C. W. POST CAMPUS
Applied Mathematics — M
Chemistry — M
Mathematics — M

LOUISIANA STATE UNIVERSITY AND AGRICULTURAL AND MECHANICAL COLLEGE
Astronomy — M,D*
Astrophysics — D*
Chemistry — M,D
Earth Sciences — M
Geology — M,D*
Geophysics — M,D*
Marine Sciences/Oceanography — M,D*
Mathematics — M,D
Physics — M,D*
Statistics — M

LOUISIANA STATE UNIVERSITY MEDICAL CENTER
Biometrics — M,D*
Chemistry — M,D

LOUISIANA TECH UNIVERSITY
Chemistry — M
Mathematics — M
Physics — M
Statistics — M

LOYOLA UNIVERSITY CHICAGO
Chemistry — M,D
Mathematics — M*

MAHARISHI INTERNATIONAL UNIVERSITY
Mathematics — M
Physics — M,D

MANKATO STATE UNIVERSITY
Chemistry — M
Mathematics — M
Physics — M
Statistics — M

MARQUETTE UNIVERSITY
Analytical Chemistry — M,D
Chemistry — M,D
Inorganic Chemistry — M,D
Mathematics — M,D
Organic Chemistry — M,D
Physical Chemistry — M,D
Physics — M
Statistics — M

MARSHALL UNIVERSITY
Chemistry — M
Mathematics — M
Physics — M

MASSACHUSETTS COLLEGE OF PHARMACY AND ALLIED HEALTH SCIENCES
Chemistry — M,D

MASSACHUSETTS INSTITUTE OF TECHNOLOGY
Applied Mathematics — D
Chemistry — M,D*
Earth Sciences — M,D
Geology — M,D
Geophysics — M,D
Inorganic Chemistry — M,D*
Marine Sciences/Oceanography — M,D
Mathematics — D
Meteorology and Atmospheric Sciences — M,D
Organic Chemistry — M,D*
Physical Chemistry — M,D*
Physics — M,D
Planetary Sciences — M,D
Plasma Physics — M

MASSACHUSETTS INSTITUTE OF TECHNOLOGY/WOODS HOLE OCEANOGRAPHIC INSTITUTION
Marine Sciences/Oceanography — M,D,O*

MCGILL UNIVERSITY
Biostatistics — M,D,O
Chemistry — M,D
Geology — M,D,O*
Geophysics — M,D,O*
Marine Sciences/Oceanography — M,D
Mathematics — M,D
Meteorology and Atmospheric Sciences — M,D*
Physics — M,D

MCMASTER UNIVERSITY
Chemistry — M,D
Earth Sciences — M,D*
Geochemistry — D
Geology — M,D
Mathematics — M,D
Physical Chemistry — M,D
Physics — M,D
Statistics — M,D

MCNEESE STATE UNIVERSITY
Chemistry — M
Mathematics — M
Statistics — M

MEDICAL COLLEGE OF WISCONSIN
Biostatistics — M

MEDICAL UNIVERSITY OF SOUTH CAROLINA
Biometrics — M,D*
Biostatistics — M,D*

MEMORIAL UNIVERSITY OF NEWFOUNDLAND
Chemistry — M,D
Earth Sciences — M,D
Geology — M,D
Geophysics — M,D
Marine Sciences/Oceanography — M,D
Mathematics — M,D
Physics — M,D
Statistics — M,D

MEMPHIS STATE UNIVERSITY
Applied Mathematics — M
Chemistry — M,D
Geology — M
Geophysics — M
Mathematics — M,D

Physics — M
Statistics — M,D

MIAMI UNIVERSITY
Chemistry — M,D
Geology — M,D*
Mathematics — M*
Physics — M
Statistics — M

MICHIGAN STATE UNIVERSITY
Analytical Chemistry — M,D
Applied Mathematics — M,D*
Astrophysics — M,D*
Chemistry — M,D
Geochemistry — M,D
Geology — M,D*
Geophysics — M,D*
Inorganic Chemistry — M,D
Mathematics — M,D
Organic Chemistry — M,D
Physical Chemistry — M,D
Physics — M,D*
Statistics — M,D

MICHIGAN TECHNOLOGICAL UNIVERSITY
Applied Physics — D
Chemistry — M,D
Geology — M,D
Geophysics — M
Mathematics — M*
Physics — M

MIDDLE TENNESSEE STATE UNIVERSITY
Chemistry — M,D
Mathematics — M

MISSISSIPPI STATE UNIVERSITY
Analytical Chemistry — M,D
Geology — M
Inorganic Chemistry — M,D
Mathematics — M,D
Organic Chemistry — M,D
Physical Chemistry — M,D
Physics — M
Statistics — M

MONTANA COLLEGE OF MINERAL SCIENCE AND TECHNOLOGY
Geochemistry — M*
Geology — M

MONTANA STATE UNIVERSITY
Chemistry — M,D
Earth Sciences — M
Mathematics — M,D
Physics — M,D
Statistics — M,D

MONTCLAIR STATE COLLEGE
Applied Mathematics — M
Chemistry — M
Geology — M
Mathematics — M
Statistics — M

MORGAN STATE UNIVERSITY
Mathematics — M

MOUNT ALLISON UNIVERSITY
Chemistry — M

MOUNT HOLYOKE COLLEGE
Chemistry — M

MOUNT SINAI SCHOOL OF MEDICINE OF THE CITY UNIVERSITY OF NEW YORK
Biometrics — D*

MURRAY STATE UNIVERSITY
Chemistry — M
Mathematics — M
Physics — M

NAVAL POSTGRADUATE SCHOOL
Marine Sciences/Oceanography — M,D
Mathematics — M,D
Meteorology and Atmospheric Sciences — M,D
Physics — M,D

NEW JERSEY INSTITUTE OF TECHNOLOGY
Applied Mathematics — M*
Applied Physics — M*
Chemistry — M

DIRECTORY OF INSTITUTIONS

NEW MEXICO HIGHLANDS UNIVERSITY
Chemistry M
Mathematics M

NEW MEXICO INSTITUTE OF MINING AND TECHNOLOGY
Astrophysics M,D
Chemistry M,D
Earth Sciences M
Geochemistry M,D*
Geology M,D*
Geophysics M,D*
Mathematics M
Meteorology and Atmospheric Sciences M,D
Physics M,D

NEW MEXICO STATE UNIVERSITY
Applied Mathematics M,D
Astronomy M,D*
Chemistry M,D
Earth Sciences M
Mathematics M,D
Physics M,D*
Statistics M,D

NEW YORK MEDICAL COLLEGE
Biostatistics M

NEW YORK UNIVERSITY
Chemistry M,D*
Inorganic Chemistry M,D*
Mathematics M,D*
Meteorology and Atmospheric Sciences M
Organic Chemistry M,D*
Physical Chemistry M,D*
Physics M,D
Statistics M,D,O

NICHOLLS STATE UNIVERSITY
Applied Mathematics M
Mathematics M

NORTH CAROLINA AGRICULTURAL AND TECHNICAL STATE UNIVERSITY
Applied Mathematics M
Chemistry M

NORTH CAROLINA CENTRAL UNIVERSITY
Chemistry M*
Mathematics M

NORTH CAROLINA STATE UNIVERSITY
Applied Mathematics M,D
Biometrics M,D*
Chemistry M,D
Geology M,D*
Geophysics M,D*
Marine Sciences/Oceanography M,D*
Mathematics M,D
Meteorology and Atmospheric Sciences M,D*
Physics M,D
Statistics M,D*

NORTH DAKOTA STATE UNIVERSITY
Chemistry M,D
Mathematics M,D
Physics M,D
Statistics M,D

NORTHEASTERN ILLINOIS UNIVERSITY
Chemistry M
Earth Sciences M
Mathematics M
Physics M

NORTHEASTERN UNIVERSITY
Analytical Chemistry M*
Chemistry M,D*
Inorganic Chemistry M*
Mathematics M,D
Organic Chemistry M*
Physical Chemistry M*
Physics M,D*

NORTHEAST LOUISIANA UNIVERSITY
Chemistry M
Earth Sciences M
Physics M

NORTHEAST MISSOURI STATE UNIVERSITY
Mathematics M

NORTHERN ARIZONA UNIVERSITY
Chemistry M
Earth Sciences M
Geology M
Mathematics M

NORTHERN ILLINOIS UNIVERSITY
Applied Physics M
Chemistry M,D
Geology M,D*
Geophysics M,D*
Mathematics M,D
Physics M*
Statistics M

NORTHERN MICHIGAN UNIVERSITY
Chemistry M

NORTHWESTERN UNIVERSITY
Applied Mathematics M,D*
Astronomy M,D*
Astrophysics M,D*
Chemistry M,D
Earth Sciences M,D
Geology M,D
Mathematics M,D*
Physics M,D*
Statistics M,D

NOVA UNIVERSITY
Marine Sciences/Oceanography M,D*

OAKLAND UNIVERSITY
Applied Mathematics M
Chemistry M,D
Mathematics M
Physics M,D
Statistics M,O

OHIO STATE UNIVERSITY (COLUMBUS)
Analytical Chemistry M,D*
Astronomy M,D
Biostatistics D
Chemistry M,D*
Geochemistry M,D
Geodetic Sciences M,D
Geology M,D
Inorganic Chemistry M,D*
Mathematics M,D*
Meteorology and Atmospheric Sciences M,D
Optical Sciences M,D
Organic Chemistry M,D*
Physical Chemistry M,D*
Physics M,D
Statistics M,D

OHIO UNIVERSITY (ATHENS)
Chemistry M,D
Geology M
Mathematics M,D
Physics M,D*

OKLAHOMA STATE UNIVERSITY
Chemistry M,D
Geology M
Mathematics M,D
Physics M,D
Statistics M,D

OLD DOMINION UNIVERSITY
Analytical Chemistry M
Applied Mathematics M,D*
Applied Physics D
Chemistry M
Geology M*
Marine Sciences/Oceanography M,D*
Mathematics M,D*
Organic Chemistry M
Physical Chemistry M
Physics M,D*
Statistics M,D*

OREGON GRADUATE INSTITUTE OF SCIENCE AND TECHNOLOGY
Applied Physics M,D
Chemistry M,D

OREGON STATE UNIVERSITY
Analytical Chemistry M,D*
Biometrics M,D*
Chemistry M,D*
Geochemistry M,D*
Geology M,D*
Geophysics M,D*
Inorganic Chemistry M,D*

Marine Sciences/Oceanography M,D
Mathematics M,D*
Meteorology and Atmospheric Sciences M,D
Organic Chemistry M,D*
Physical Chemistry M,D*
Physics M,D*
Statistics M,D*

PENNSYLVANIA STATE UNIVERSITY GREAT VALLEY GRADUATE CENTER
Mathematics M

PENNSYLVANIA STATE UNIVERSITY UNIVERSITY PARK CAMPUS
Acoustics M,D
Applied Mathematics M,D
Astronomy M,D
Astrophysics M,D
Chemistry M,D
Earth Sciences M,D
Geochemistry M,D*
Geology M,D*
Geophysics M,D
Mathematics M,D
Meteorology and Atmospheric Sciences M,D*
Physics M,D*
Statistics M,D*

PITTSBURG STATE UNIVERSITY
Applied Physics M
Chemistry M
Mathematics M
Physics M

PLYMOUTH STATE COLLEGE OF THE UNIVERSITY SYSTEM OF NEW HAMPSHIRE
Mathematics M

POLYTECHNIC UNIVERSITY, BROOKLYN CAMPUS
Applied Mathematics M,D
Chemistry M,D
Mathematics M,D
Physical Chemistry M,D
Physics M,D
Statistics M,D

POLYTECHNIC UNIVERSITY, FARMINGDALE CAMPUS
Applied Physics M,D

POLYTECHNIC UNIVERSITY, WESTCHESTER GRADUATE CENTER
Chemistry M

PORTLAND STATE UNIVERSITY
Chemistry M,D*
Earth Sciences M,D
Geology M,D*
Mathematics M,D
Physics M,D*

PRAIRIE VIEW A&M UNIVERSITY
Chemistry M
Mathematics M

PRINCETON UNIVERSITY
Applied Mathematics D
Applied Physics M,D
Astrophysics D
Chemistry M,D*
Geology D
Geophysics D
Marine Sciences/Oceanography D
Mathematical Physics D
Mathematics D
Meteorology and Atmospheric Sciences D
Physical Chemistry D
Physics D
Plasma Physics D
Statistics M,D*

PROVIDENCE COLLEGE
Chemistry M,D

PURDUE UNIVERSITY (WEST LAFAYETTE)
Analytical Chemistry M,D
Chemistry M,D
Earth Sciences M,D
Geochemistry M,D
Geology M,D
Geophysics M,D
Inorganic Chemistry M,D
Mathematics M,D*
Meteorology and Atmospheric Sciences M,D
Organic Chemistry M,D

Physical Chemistry M,D
Physics M,D
Statistics M,D*

PURDUE UNIVERSITY CALUMET
Applied Mathematics M

QUEENS COLLEGE OF THE CITY UNIVERSITY OF NEW YORK
Chemistry M
Earth Sciences M
Geology M
Mathematics M
Physics M

QUEEN'S UNIVERSITY AT KINGSTON
Biostatistics M
Chemistry M,D
Geology M,D
Mathematics M,D
Physics M,D
Statistics M,D

QUINNIPIAC COLLEGE
Chemistry M

RENSSELAER POLYTECHNIC INSTITUTE
Applied Mathematics M,D*
Applied Physics M,D
Chemistry M,D*
Geochemistry M,D*
Geology M,D*
Geophysics M,D*
Inorganic Chemistry M,D*
Mathematics M,D*
Organic Chemistry M,D*
Physical Chemistry M,D*
Physics M,D*
Planetary Sciences M,D*
Plasma Physics M,D
Statistics M,D*

RHODE ISLAND COLLEGE
Mathematics M,O

RICE UNIVERSITY
Applied Mathematics M,D
Astronomy M,D
Astrophysics M,D
Chemistry M,D
Geochemistry M,D
Geology M,D
Geophysics M,D
Mathematics M,D
Physics M,D*
Statistics M,D

RIVIER COLLEGE
Mathematics M

ROCHESTER INSTITUTE OF TECHNOLOGY
Chemistry M
Statistics M

ROOSEVELT UNIVERSITY
Applied Mathematics M
Chemistry M
Mathematics M

ROSE-HULMAN INSTITUTE OF TECHNOLOGY
Optical Sciences M

ROYAL ROADS MILITARY COLLEGE
Acoustics M
Marine Sciences/Oceanography M

RUTGERS, THE STATE UNIVERSITY OF NEW JERSEY, NEWARK
Analytical Chemistry M,D*
Chemistry M,D*
Geochemistry M
Geology M
Inorganic Chemistry M,D*
Organic Chemistry M,D*
Physical Chemistry M,D*

RUTGERS, THE STATE UNIVERSITY OF NEW JERSEY, NEW BRUNSWICK
Analytical Chemistry M,D*
Applied Mathematics M,D
Applied Physics M
Chemistry M,D*
Geology M,D
Inorganic Chemistry M,D*
Mathematics M,D*
Meteorology and Atmospheric Sciences M

*P—first professional degree; M—master's degree; D—doctorate; O—other advanced degree; * full description and/or announcement for program (see index)*

Peterson's Guide to Graduate Programs in the Physical Sciences and Mathematics 1992

DIRECTORY OF INSTITUTIONS

Rutgers, The State University of New Jersey, New Brunswick (continued)
Organic Chemistry	M,D*
Physical Chemistry	M,D*
Physics	M,D*
Statistics	M,D

ST. BONAVENTURE UNIVERSITY
Physics	M

ST. CLOUD STATE UNIVERSITY
Mathematics	M

ST. FRANCIS XAVIER UNIVERSITY
Chemistry	M
Physics	M

ST. JOHN'S UNIVERSITY (NY)
Applied Mathematics	M
Chemistry	M
Mathematics	M
Statistics	M

SAINT JOSEPH COLLEGE (CT)
Chemistry	M,O

SAINT JOSEPH'S UNIVERSITY
Chemistry	M

SAINT LOUIS UNIVERSITY
Chemistry	M
Geophysics	M,D
Mathematics	M,D
Meteorology and Atmospheric Sciences	M,D

SAINT MARY'S UNIVERSITY
Astronomy	M

SALEM STATE COLLEGE
Mathematics	M

SAM HOUSTON STATE UNIVERSITY
Chemistry	M
Mathematics	M
Physics	M

SAN DIEGO STATE UNIVERSITY
Applied Mathematics	M
Astronomy	M
Biostatistics	M
Chemistry	M,D
Geology	M
Mathematics	M
Physics	M
Statistics	M

SAN FRANCISCO STATE UNIVERSITY
Astrophysics	M
Chemistry	M
Mathematics	M
Physics	M*

SANGAMON STATE UNIVERSITY
Mathematics	M
Statistics	M

SAN JOSE STATE UNIVERSITY
Analytical Chemistry	M
Chemistry	M
Geology	M
Inorganic Chemistry	M
Marine Sciences/Oceanography	M
Mathematics	M
Meteorology and Atmospheric Sciences	M
Organic Chemistry	M
Physical Chemistry	M
Physics	M

SANTA CLARA UNIVERSITY
Applied Mathematics	M

SETON HALL UNIVERSITY
Analytical Chemistry	D
Chemistry	M,D*
Inorganic Chemistry	D
Mathematics	M
Organic Chemistry	D
Physical Chemistry	D

SHIPPENSBURG UNIVERSITY OF PENNSYLVANIA
Chemistry	M
Earth Sciences	M
Mathematics	M

SIMON FRASER UNIVERSITY
Applied Mathematics	M,D
Chemistry	M,D
Mathematics	M,D*
Physical Chemistry	M,D
Physics	M,D
Statistics	M,D*

SOUTH DAKOTA SCHOOL OF MINES AND TECHNOLOGY
Chemistry	M
Geology	M,D
Meteorology and Atmospheric Sciences	M,D
Physics	M

SOUTH DAKOTA STATE UNIVERSITY
Analytical Chemistry	M,D
Chemistry	M,D
Inorganic Chemistry	M,D
Mathematics	M*
Organic Chemistry	M,D
Physical Chemistry	M,D
Physics	M

SOUTHEAST MISSOURI STATE UNIVERSITY
Earth Sciences	M
Mathematics	M

SOUTHERN CONNECTICUT STATE UNIVERSITY
Chemistry	M

SOUTHERN ILLINOIS UNIVERSITY AT CARBONDALE
Chemistry	M,D
Geology	M,D*
Mathematics	M,D
Physics	M
Statistics	M

SOUTHERN ILLINOIS UNIVERSITY AT EDWARDSVILLE
Chemistry	M
Mathematics	M
Physics	M
Statistics	M

SOUTHERN METHODIST UNIVERSITY
Applied Mathematics	M
Chemistry	M
Geochemistry	M,D*
Geology	M,D*
Geophysics	M,D*
Mathematics	M,D
Planetary Sciences	M*
Statistics	M,D

SOUTHERN UNIVERSITY AND AGRICULTURAL AND MECHANICAL COLLEGE
Analytical Chemistry	M
Chemistry	M
Inorganic Chemistry	M
Mathematics	M
Organic Chemistry	M
Physical Chemistry	M

SOUTHWEST MISSOURI STATE UNIVERSITY
Mathematics	M

SOUTHWEST TEXAS STATE UNIVERSITY
Chemistry	M
Mathematics	M
Physics	M

STANFORD UNIVERSITY
Applied Physics	M,D
Chemistry	M,D*
Earth Sciences	M,D,O
Geology	M,D*
Geophysics	M,D
Mathematics	M,D
Physics	D
Statistics	M,D

STATE UNIVERSITY OF NEW YORK AT ALBANY
Chemistry	M,D
Geology	M,D
Mathematics	M,D
Meteorology and Atmospheric Sciences	M,D*
Physics	M,D
Statistics	M,D

STATE UNIVERSITY OF NEW YORK AT BINGHAMTON
Analytical Chemistry	D
Applied Physics	M
Chemistry	M,D
Geology	M,D
Geophysics	M,D
Inorganic Chemistry	D
Mathematics	M,D
Organic Chemistry	D
Physical Chemistry	D
Physics	M,D
Statistics	M,D

STATE UNIVERSITY OF NEW YORK AT BUFFALO
Applied Physics	M,D
Biometrics	M
Chemistry	M,D
Geology	M,D
Mathematics	M,D
Physics	M,D*
Statistics	M,D

STATE UNIVERSITY OF NEW YORK AT STONY BROOK
Applied Mathematics	M,D
Astronomy	M,D
Astrophysics	M,D
Chemistry	M,D*
Earth Sciences	M,D
Geochemistry	M,D
Geology	M,D
Geophysics	M,D
Marine Sciences/Oceanography	M,D*
Mathematics	M,D*
Meteorology and Atmospheric Sciences	D*
Physical Chemistry	M,D*
Physics	M,D*
Planetary Sciences	M,D
Statistics	M,D

STATE UNIVERSITY OF NEW YORK COLLEGE AT BROCKPORT
Mathematics	M*

STATE UNIVERSITY OF NEW YORK COLLEGE AT BUFFALO
Chemistry	M
Physics	M

STATE UNIVERSITY OF NEW YORK COLLEGE AT FREDONIA
Chemistry	M
Geology	M
Mathematics	M
Physics	M

STATE UNIVERSITY OF NEW YORK COLLEGE AT NEW PALTZ
Chemistry	M
Geology	M
Mathematics	M
Physics	M

STATE UNIVERSITY OF NEW YORK COLLEGE AT ONEONTA
Earth Sciences	M

STATE UNIVERSITY OF NEW YORK COLLEGE AT OSWEGO
Chemistry	M

STATE UNIVERSITY OF NEW YORK COLLEGE AT PLATTSBURGH
Chemistry	M

STATE UNIVERSITY OF NEW YORK COLLEGE AT POTSDAM
Mathematics	M

STATE UNIVERSITY OF NEW YORK COLLEGE OF ENVIRONMENTAL SCIENCE AND FORESTRY
Chemistry	M,D*

STEPHEN F. AUSTIN STATE UNIVERSITY
Chemistry	M
Geology	M
Mathematics	M
Physics	M
Statistics	M

STEVENS INSTITUTE OF TECHNOLOGY
Applied Mathematics	M,D
Chemistry	M,D*
Mathematics	M,D
Physics	M,D*

SUL ROSS STATE UNIVERSITY
Geology	M*

SYRACUSE UNIVERSITY
Analytical Chemistry	M,D
Chemistry	M,D
Geology	M,D*
Inorganic Chemistry	M,D
Mathematics	M,D
Organic Chemistry	M,D
Physical Chemistry	M,D
Physics	M,D*

TEACHERS COLLEGE, COLUMBIA UNIVERSITY
Statistics	M

TECHNICAL UNIVERSITY OF NOVA SCOTIA
Applied Mathematics	M,D

TEMPLE UNIVERSITY (PHILADELPHIA)
Chemistry	M,D
Geology	M
Mathematics	M,D*
Physics	M,D*
Statistics	M,D*

TEMPLE UNIVERSITY, AMBLER CAMPUS
Applied Mathematics	M
Statistics	M

TENNESSEE STATE UNIVERSITY
Chemistry	M
Mathematics	M

TENNESSEE TECHNOLOGICAL UNIVERSITY
Chemistry	M
Mathematics	M

TEXAS A&I UNIVERSITY
Chemistry	M
Geology	M
Mathematics	M

TEXAS A&M UNIVERSITY (COLLEGE STATION)
Analytical Chemistry	M,D*
Applied Mathematics	M,D
Chemistry	M,D*
Earth Sciences	M,D
Geology	M,D*
Geophysics	M,D*
Inorganic Chemistry	M,D*
Marine Sciences/Oceanography	M,D
Mathematics	M,D
Meteorology and Atmospheric Sciences	M
Organic Chemistry	M,D*
Physical Chemistry	M,D*
Physics	M,D*
Statistics	M,D

TEXAS CHRISTIAN UNIVERSITY
Chemistry	M,D
Geology	M
Mathematics	M
Physics	M,D

TEXAS SOUTHERN UNIVERSITY
Chemistry	M
Mathematics	M

TEXAS TECH UNIVERSITY
Applied Physics	M,D
Chemistry	M,D
Earth Sciences	M,D
Geology	M,D
Mathematics	M,D
Meteorology and Atmospheric Sciences	M,D
Physics	M,D
Statistics	M

TEXAS WOMAN'S UNIVERSITY
Chemistry	M
Mathematics	M

TRENTON STATE COLLEGE
Mathematics	M

TRENT UNIVERSITY
Chemistry	M
Physics	M

TRINITY COLLEGE (CT)
Mathematics — M

TUFTS UNIVERSITY
Analytical Chemistry — M,D
Astronomy — M,D*
Chemistry — M,D
Inorganic Chemistry — M,D
Mathematics — M,D
Organic Chemistry — M,D
Physical Chemistry — M,D
Physics — M,D*

TULANE UNIVERSITY
Applied Mathematics — M
Biostatistics — M,D
Chemistry — M,D
Geology — M,D*
Mathematics — M,D
Physics — M,D
Statistics — M

TUSKEGEE UNIVERSITY
Chemistry — M

UNIVERSITÉ DE MONCTON
Chemistry — M
Physics — M

UNIVERSITÉ DE MONTRÉAL
Chemistry — M,D
Geology — M,D
Mathematics — M,D
Physics — M,D

UNIVERSITÉ DE SHERBROOKE
Chemistry — M,D
Mathematics — M,D
Physics — M,D

UNIVERSITÉ DU QUÉBEC À CHICOUTIMI
Earth Sciences — M

UNIVERSITÉ DU QUÉBEC À MONTRÉAL
Chemistry — M,O
Earth Sciences — M
Mathematics — M,D
Meteorology and Atmospheric Sciences — M,O

UNIVERSITÉ DU QUÉBEC À RIMOUSKI
Marine Sciences/Oceanography — M,D,O

UNIVERSITÉ DU QUÉBEC À TROIS-RIVIÈRES
Chemistry — M
Physics — M

UNIVERSITÉ LAVAL
Chemistry — M,D
Geodetic Sciences — M,D
Geology — M,D
Mathematics — M,D
Physics — M,D

UNIVERSITY OF AKRON
Analytical Chemistry — M,D
Applied Mathematics — M
Chemistry — M,D
Geology — M
Inorganic Chemistry — M,D
Mathematics — M
Organic Chemistry — M,D
Physical Chemistry — M,D
Physics — M*
Statistics — M

UNIVERSITY OF ALABAMA (TUSCALOOSA)
Chemistry — M,D,O
Geology — M,D
Mathematics — M,D,O
Physics — M,D*
Statistics — D

UNIVERSITY OF ALABAMA AT BIRMINGHAM
Applied Mathematics — D
Biometrics — M,D*
Biostatistics — M,D*
Chemistry — M,D
Mathematics — M,D
Physics — M,D*

UNIVERSITY OF ALABAMA IN HUNTSVILLE
Applied Mathematics — M,D
Chemistry — M
Mathematics — M,D
Physics — M,D

UNIVERSITY OF ALASKA FAIRBANKS
Astronomy — M,D
Chemistry — M
Geology — M,D
Geophysics — M,D
Marine Sciences/Oceanography — M,D
Meteorology and Atmospheric Sciences — M,D
Physics — M,D*

UNIVERSITY OF ALBERTA
Chemistry — M,D
Geology — M,D,O
Geophysics — M,D
Mathematics — M,D
Physics — M,D
Statistics — M,D

UNIVERSITY OF ARIZONA
Applied Mathematics — M,D*
Astronomy — M,D
Chemistry — M,D*
Earth Sciences — M,D*
Geology — M,D
Mathematics — M,D*
Meteorology and Atmospheric Sciences — M,D
Optical Sciences — M,D*
Physics — M,D*
Planetary Sciences — M,D*
Statistics — M

UNIVERSITY OF ARKANSAS (FAYETTEVILLE)
Chemistry — M,D
Geology — M
Mathematics — M,D
Physics — M,D
Statistics — M

UNIVERSITY OF ARKANSAS AT LITTLE ROCK
Applied Mathematics — M
Chemistry — M

UNIVERSITY OF BRIDGEPORT
Chemistry — M
Mathematics — M
Physics — M

UNIVERSITY OF BRITISH COLUMBIA
Applied Mathematics — M,D
Astronomy — M,D
Chemistry — M,D
Geology — M,D*
Geophysics — M,D
Marine Sciences/Oceanography — M,D
Mathematics — M,D
Physics — M,D
Statistics — M,D

UNIVERSITY OF CALGARY
Analytical Chemistry — M,D
Astronomy — M,D
Chemistry — M,D
Geology — M,D*
Geophysics — M,D*
Inorganic Chemistry — M,D
Mathematics — M,D
Organic Chemistry — M,D
Physical Chemistry — M,D
Physics — M,D
Statistics — M,D

UNIVERSITY OF CALIFORNIA AT BERKELEY
Applied Mathematics — D
Astronomy — D
Biostatistics — M,D
Chemistry — M,D
Geology — M,D
Geophysics — M,D
Mathematics — M,D
Physics — M,D
Statistics — M,D*

UNIVERSITY OF CALIFORNIA, DAVIS
Applied Mathematics — M,D
Chemistry — M,D
Earth Sciences — M,D
Meteorology and Atmospheric Sciences — M,D*
Physics — M,D
Statistics — M,D*

UNIVERSITY OF CALIFORNIA, IRVINE
Chemistry — M,D
Mathematics — M,D
Physics — M,D

UNIVERSITY OF CALIFORNIA, LOS ANGELES
Applied Mathematics — M,D*
Astronomy — M,D
Astrophysics — M,D,O*
Biometrics — M,D*
Biostatistics — M,D*
Chemistry — M,D,O
Earth Sciences — M,D,O*
Geochemistry — M,D,O*
Geology — M,D,O*
Geophysics — M,D*
Mathematics — M,D,O
Meteorology and Atmospheric Sciences — M,D,O
Physics — M,D*
Planetary Sciences — M,D*
Statistics — M,D

UNIVERSITY OF CALIFORNIA, RIVERSIDE
Applied Mathematics — M,D*
Chemistry — M,D
Earth Sciences — M,D
Geology — M,D*
Mathematics — M,D*
Physics — M,D*
Statistics — M,D

UNIVERSITY OF CALIFORNIA, SAN DIEGO
Applied Mathematics — M
Chemistry — D
Earth Sciences — M,D*
Marine Sciences/Oceanography — M,D*
Mathematics — M,D
Physics — M,D
Statistics — M

UNIVERSITY OF CALIFORNIA, SANTA BARBARA
Applied Mathematics — M
Chemistry — M,D
Geochemistry — M,D*
Geology — M,D*
Geophysics — M,D*
Mathematics — M,D
Physics — D
Statistics — M,D*

UNIVERSITY OF CALIFORNIA, SANTA CRUZ
Astronomy — D
Astrophysics — D
Chemistry — M,D
Earth Sciences — M,D
Marine Sciences/Oceanography — M
Mathematics — M,D*
Physics — M,D

UNIVERSITY OF CENTRAL FLORIDA
Chemistry — M
Mathematics — M
Physics — M,D

UNIVERSITY OF CENTRAL OKLAHOMA (OK)
Applied Mathematics — M
Applied Physics — M

UNIVERSITY OF CHICAGO
Applied Mathematics — M,D
Astronomy — M,D
Astrophysics — M,D
Chemistry — M,D
Earth Sciences — M,D*
Geophysics — M,D*
Mathematics — M,D
Meteorology and Atmospheric Sciences — M,D*
Physics — M,D
Planetary Sciences — M,D*
Statistics — M,D

UNIVERSITY OF CINCINNATI
Analytical Chemistry — M,D
Applied Mathematics — M,D
Biostatistics — M
Chemistry — M,D
Geochemistry — M,D*
Geology — M,D*
Inorganic Chemistry — M,D
Mathematics — M,D*
Organic Chemistry — M,D
Physical Chemistry — M,D

UNIVERSITY OF COLORADO AT BOULDER
Applied Mathematics — M,D
Applied Physics — D
Astrophysics — D
Chemistry — M,D
Geology — M,D
Geophysics — D
Mathematical Physics — D
Mathematics — M,D
Meteorology and Atmospheric Sciences — M,D
Physics — M,D*
Plasma Physics — M,D

UNIVERSITY OF COLORADO AT COLORADO SPRINGS
Applied Mathematics — M
Physics — M

UNIVERSITY OF COLORADO AT DENVER
Applied Mathematics — M,D
Chemistry — M
Earth Sciences — M
Mathematics — M,D

UNIVERSITY OF COLORADO HEALTH SCIENCES CENTER
Biometrics — M,D

UNIVERSITY OF CONNECTICUT (STORRS)
Chemistry — M,D
Geology — M,D
Geophysics — M,D
Marine Sciences/Oceanography — M,D*
Mathematics — M,D
Physics — M,D*
Statistics — M,D

UNIVERSITY OF DAYTON
Applied Mathematics — M
Optical Sciences — M

UNIVERSITY OF DELAWARE
Applied Mathematics — M,D*
Astronomy — M,D
Chemistry — M,D
Geology — M,D
Marine Sciences/Oceanography — M,D
Mathematics — M,D*
Meteorology and Atmospheric Sciences — D
Physics — M,D
Statistics — M,D*

UNIVERSITY OF DENVER
Applied Mathematics — M
Chemistry — M
Mathematics — M,D*
Physics — M,D

UNIVERSITY OF DETROIT MERCY
Chemistry — M,D
Mathematics — M

UNIVERSITY OF FLORIDA
Analytical Chemistry — M,D*
Applied Mathematics — M,D*
Astronomy — M,D
Chemistry — M,D*
Geology — M,D*
Inorganic Chemistry — M,D*
Mathematics — M,D*
Organic Chemistry — M,D*
Physical Chemistry — M,D*
Physics — M,D*
Statistics — M,D

UNIVERSITY OF GEORGIA
Analytical Chemistry — M,D*
Applied Mathematics — M,D*
Geochemistry — M,D*
Geology — M,D*
Geophysics — M,D*
Inorganic Chemistry — M,D*
Mathematics — M,D*
Organic Chemistry — M,D*
Physical Chemistry — M,D*
Physics — M,D*
Statistics — M,D*

UNIVERSITY OF GUELPH
Applied Mathematics — D
Chemistry — M,D
Mathematics — M,D
Physics — M,D
Statistics — M,D

P—first professional degree; M—master's degree; D—doctorate; O—other advanced degree; * full description and/or announcement for program (see index)

DIRECTORY OF INSTITUTIONS

UNIVERSITY OF HARTFORD
Chemistry	M

UNIVERSITY OF HAWAII AT MANOA
Astronomy	M,D
Biostatistics	M,D
Chemistry	M,D
Geology	M,D*
Geophysics	M,D*
Marine Sciences/Oceanography	M,D*
Mathematics	M,D
Meteorology and Atmospheric Sciences	M,D
Physics	M,D

UNIVERSITY OF HOUSTON
Acoustics	M,D
Analytical Chemistry	M,D*
Applied Mathematics	M*
Chemistry	M,D*
Geology	M,D*
Geophysics	M,D*
Inorganic Chemistry	M,D*
Mathematics	M,D*
Organic Chemistry	M,D*
Physical Chemistry	M,D*
Physics	M,D*
Statistics	M,D

UNIVERSITY OF HOUSTON–CLEAR LAKE
Chemistry	M
Mathematics	M
Optical Sciences	M
Physics	M

UNIVERSITY OF IDAHO
Chemistry	M,D
Geology	M,D*
Geophysics	M
Mathematics	M,D
Physics	M,D
Statistics	M

UNIVERSITY OF ILLINOIS AT CHICAGO
Applied Mathematics	M,D*
Biostatistics	M,D
Chemistry	M,D*
Geochemistry	M,D*
Geology	M,D*
Geophysics	M,D*
Mathematics	M,D*
Physics	M,D
Statistics	M,D*

UNIVERSITY OF ILLINOIS AT URBANA-CHAMPAIGN
Applied Mathematics	M,D
Astronomy	M,D
Chemistry	M,D
Earth Sciences	M,D
Geochemistry	M,D
Geology	M,D
Geophysics	M,D
Mathematics	M,D
Physics	M,D
Statistics	M,D*

UNIVERSITY OF IOWA
Applied Mathematics	D
Astronomy	M
Biostatistics	M,D
Chemistry	M,D
Geology	M,D
Mathematics	M,D*
Physics	M,D
Statistics	M,D

UNIVERSITY OF KANSAS
Applied Mathematics	M,D
Astronomy	M,D*
Chemistry	M,D
Geology	M,D
Geophysics	M,D*
Mathematics	M,D
Meteorology and Atmospheric Sciences	M,D*
Organic Chemistry	M,D*
Physics	M,D
Statistics	M,D

UNIVERSITY OF KENTUCKY
Astronomy	M,D
Chemistry	M,D
Geology	M,D
Mathematics	M,D
Physics	M,D
Statistics	M,D*

UNIVERSITY OF LOUISVILLE
Analytical Chemistry	M,D
Applied Mathematics	M
Chemistry	M,D*
Inorganic Chemistry	M,D
Mathematics	M
Organic Chemistry	M,D
Physical Chemistry	M,D
Physics	M

UNIVERSITY OF MAINE (ORONO)
Astronomy	M,D
Chemistry	M,D*
Geology	M,D
Marine Sciences/Oceanography	M,D*
Mathematics	M
Physics	M,D

UNIVERSITY OF MANITOBA
Applied Mathematics	M
Chemistry	M,D
Geology	M,D
Geophysics	M,D
Mathematics	M,D
Physics	M,D
Statistics	M,D

UNIVERSITY OF MARYLAND COLLEGE PARK
Analytical Chemistry	M,D*
Applied Mathematics	M,D
Astronomy	M,D
Chemistry	M,D*
Geology	M,D*
Inorganic Chemistry	M,D*
Mathematics	M,D
Meteorology and Atmospheric Sciences	M,D*
Organic Chemistry	M,D*
Physical Chemistry	M,D*
Physics	M,D*
Plasma Physics	D
Statistics	M,D

UNIVERSITY OF MARYLAND EASTERN SHORE
Marine Sciences/Oceanography	M,D

UNIVERSITY OF MARYLAND GRADUATE SCHOOL, BALTIMORE
Applied Mathematics	M,D*
Applied Physics	M,D
Chemistry	M,D*
Marine Sciences/Oceanography	M,D
Physics	M,D*
Statistics	M,D*

UNIVERSITY OF MASSACHUSETTS AT AMHERST
Applied Mathematics	M
Astronomy	M,D
Chemistry	M,D
Geology	M,D
Physics	M,D
Statistics	M,D

UNIVERSITY OF MASSACHUSETTS AT BOSTON
Applied Physics	M
Chemistry	M

UNIVERSITY OF MASSACHUSETTS DARTMOUTH
Chemistry	M
Physics	M

UNIVERSITY OF MASSACHUSETTS, LOWELL
Applied Mathematics	M
Applied Physics	M,D*
Chemistry	M,D
Mathematics	M
Optical Sciences	M,D*
Physics	M,D*

UNIVERSITY OF MIAMI
Chemistry	M,D*
Geology	M,D
Geophysics	M,D
Inorganic Chemistry	D*
Marine Sciences/Oceanography	M,D
Mathematics	M,D
Meteorology and Atmospheric Sciences	M,D
Organic Chemistry	D*
Physical Chemistry	D*
Physics	M,D*

UNIVERSITY OF MICHIGAN (ANN ARBOR)
Analytical Chemistry	M,D*
Applied Physics	D
Astronomy	M,D
Biostatistics	M,D
Chemistry	M,D*
Geochemistry	M,D

UNIVERSITY OF MINNESOTA, DULUTH
Applied Mathematics	M*
Chemistry	M,D
Geology	M,D*
Physics	M*

UNIVERSITY OF MINNESOTA, TWIN CITIES CAMPUS
Astronomy	M,D*
Biometrics	M,D
Biostatistics	M,D*
Chemistry	M,D*
Geology	M,D*
Geophysics	M,D*
Mathematics	M,D
Physics	M,D
Plasma Physics	M,D
Statistics	M,D

UNIVERSITY OF MISSISSIPPI
Chemistry	M,D
Geology	M
Mathematics	M,D
Physics	M,D

UNIVERSITY OF MISSOURI–COLUMBIA
Analytical Chemistry	M,D
Applied Mathematics	M
Chemistry	M,D
Geology	M,D
Inorganic Chemistry	M,D
Mathematics	M,D
Meteorology and Atmospheric Sciences	M,D
Organic Chemistry	M,D
Physical Chemistry	M,D
Physics	M,D*
Statistics	M,D

UNIVERSITY OF MISSOURI–KANSAS CITY
Analytical Chemistry	M,D
Chemistry	M,D
Geology	M
Inorganic Chemistry	M,D
Mathematics	M,D
Organic Chemistry	M,D
Physical Chemistry	M,D
Physics	M

UNIVERSITY OF MISSOURI–ROLLA
Applied Mathematics	M
Chemistry	M,D
Geology	M,D*
Geophysics	M,D
Mathematics	M,D
Physics	M,D
Statistics	M,D

UNIVERSITY OF MISSOURI–ST. LOUIS
Chemistry	M,D*
Mathematics	M
Physics	M,D

UNIVERSITY OF MONTANA
Applied Mathematics	M,D
Chemistry	M,D
Geology	M,D
Inorganic Chemistry	D
Mathematics	M,D
Organic Chemistry	D
Physical Chemistry	D
Physics	M*
Statistics	M,D

UNIVERSITY OF NEBRASKA AT OMAHA
Mathematics	M

UNIVERSITY OF NEBRASKA–LINCOLN
Analytical Chemistry	D*
Applied Mathematics	M
Chemistry	M,D*
Geology	M,D
Inorganic Chemistry	D*
Mathematics	M,D
Organic Chemistry	D*
Physical Chemistry	D*

UNIVERSITY OF NEVADA
Physics	M,D*
Statistics	M,D

UNIVERSITY OF NEVADA, LAS VEGAS
Analytical Chemistry	M
Applied Mathematics	M
Chemistry	M
Geology	M
Mathematics	M
Physics	M,D*
Statistics	M

UNIVERSITY OF NEVADA, RENO
Chemistry	M,D
Geochemistry	M,D,O*
Geology	M,D,O*
Mathematics	M,D
Physics	M,D

UNIVERSITY OF NEW BRUNSWICK (FREDERICTON)
Chemistry	M,D
Geodetic Sciences	O
Geology	M,D
Mathematics	M,D
Physics	M,D
Statistics	M,D

UNIVERSITY OF NEW HAMPSHIRE (DURHAM)
Chemistry	M,D
Earth Sciences	M,D
Geology	M,D
Marine Sciences/Oceanography	M,D
Mathematics	M,D
Physics	M,D

UNIVERSITY OF NEW MEXICO
Astronomy	M,D*
Astrophysics	M,D
Chemistry	M,D
Geology	M,D*
Mathematics	M,D
Physics	M,D*
Statistics	M,D

UNIVERSITY OF NEW ORLEANS
Applied Physics	M
Chemistry	M,D
Geology	M
Geophysics	M
Mathematics	M
Physics	M

UNIVERSITY OF NORTH CAROLINA AT CHAPEL HILL
Astronomy	M,D
Astrophysics	M,D
Biometrics	M,D
Biostatistics	M,D*
Chemistry	M,D
Geology	M,D
Marine Sciences/Oceanography	M,D
Mathematics	M,D*
Physics	M,D*
Statistics	M,D

UNIVERSITY OF NORTH CAROLINA AT CHARLOTTE
Chemistry	M
Earth Sciences	M
Mathematics	M

UNIVERSITY OF NORTH CAROLINA AT GREENSBORO
Chemistry	M
Mathematics	M,O
Physics	M

UNIVERSITY OF NORTH CAROLINA AT WILMINGTON
Chemistry	M
Geology	M
Marine Sciences/Oceanography	M
Mathematics	M

UNIVERSITY OF NORTH DAKOTA
Chemistry	M,D
Geology	M,D
Mathematics	M
Physics	M,D

UNIVERSITY OF NORTHERN COLORADO
Chemistry	M,D
Earth Sciences	M
Mathematics	M,D
Statistics	M,D

DIRECTORY OF INSTITUTIONS

UNIVERSITY OF NORTHERN IOWA
Chemistry — M
Mathematics — M

UNIVERSITY OF NORTH FLORIDA
Mathematics — M*
Statistics — M

UNIVERSITY OF NORTH TEXAS
Chemistry — M,D
Mathematics — M,D
Physics — M,D*

UNIVERSITY OF NOTRE DAME
Chemistry — M,D
Inorganic Chemistry — M,D
Mathematics — M,D*
Organic Chemistry — M,D
Physical Chemistry — M,D
Physics — M,D*

UNIVERSITY OF OKLAHOMA
Astronomy — M,D
Chemistry — M,D*
Geology — M,D*
Geophysics — M*
Mathematics — M,D
Meteorology and Atmospheric Sciences — M,D
Physics — M,D

UNIVERSITY OF OKLAHOMA HEALTH SCIENCES CENTER
Biostatistics — M,D

UNIVERSITY OF OREGON
Chemistry — M,D
Geology — M,D*
Mathematics — M,D
Physics — M,D

UNIVERSITY OF OTTAWA
Chemistry — M,D
Geology — M,D
Mathematics — M,D
Physics — M,D

UNIVERSITY OF PENNSYLVANIA
Astronomy — M,D
Astrophysics — M,D
Chemistry — M,D
Geology — M,D*
Mathematics — M,D
Physics — M,D*
Statistics — M,D

UNIVERSITY OF PITTSBURGH
Analytical Chemistry — M,D*
Applied Mathematics — M
Astronomy — M,D*
Biostatistics — M,D
Chemistry — M,D*
Geochemistry — M,D*
Geology — M,D*
Geophysics — M,D*
Inorganic Chemistry — M,D*
Mathematics — M,D
Organic Chemistry — M,D*
Physical Chemistry — M,D*
Physics — M,D*
Planetary Sciences — M,D*
Statistics — M

UNIVERSITY OF PUERTO RICO, MAYAGÜEZ CAMPUS
Chemistry — M
Marine Sciences/Oceanography — M,D*
Mathematics — M
Physics — M

UNIVERSITY OF PUERTO RICO, RÍO PIEDRAS
Chemistry — M,D*
Mathematics — M
Physics — M,D*

UNIVERSITY OF REGINA
Analytical Chemistry — M,D
Chemistry — M,D
Geology — M
Inorganic Chemistry — M,D
Mathematics — M,D
Organic Chemistry — M
Physical Chemistry — M,D
Physics — M,D
Statistics — M

UNIVERSITY OF RHODE ISLAND
Applied Mathematics — D
Chemistry — M,D
Geology — M

Marine Sciences/Oceanography — M,D*
Mathematics — M,D
Physics — M,D
Statistics — M,D*

UNIVERSITY OF ROCHESTER
Astronomy — M,D*
Biostatistics — M
Chemistry — M,D
Geology — M,D*
Mathematics — M,D
Optical Sciences — M,D*
Physics — M,D
Statistics — M,D*

UNIVERSITY OF SAN DIEGO
Marine Sciences/Oceanography — M

UNIVERSITY OF SAN FRANCISCO
Analytical Chemistry — M*
Chemistry — M*
Inorganic Chemistry — M*
Organic Chemistry — M*
Physical Chemistry — M*

UNIVERSITY OF SASKATCHEWAN
Chemistry — M,D
Geology — M,D*
Geophysics — M,D*
Mathematics — M,D
Physics — M,D

UNIVERSITY OF SCRANTON
Chemistry — M

UNIVERSITY OF SOUTH ALABAMA
Mathematics — M

UNIVERSITY OF SOUTH CAROLINA (COLUMBIA)
Astronomy — M,D*
Biostatistics — M,D
Chemistry — M,D
Earth Sciences — M,D
Geology — M,D
Marine Sciences/Oceanography — M,D
Mathematics — M,D
Physics — M,D*
Statistics — M,D

UNIVERSITY OF SOUTH DAKOTA
Chemistry — M
Mathematics — M

UNIVERSITY OF SOUTHERN CALIFORNIA
Applied Mathematics — M,D*
Biometrics — M,D*
Chemistry — M,D
Geology — M,D*
Mathematics — M,D*
Physical Chemistry — D
Physics — M,D
Statistics — M*

UNIVERSITY OF SOUTHERN MAINE
Statistics — M

UNIVERSITY OF SOUTHERN MISSISSIPPI
Analytical Chemistry — M,D
Astronomy — M
Chemistry — M,D
Geology — M
Inorganic Chemistry — M,D
Marine Sciences/Oceanography — M
Mathematics — M
Organic Chemistry — M,D
Physical Chemistry — M,D
Physics — M

UNIVERSITY OF SOUTH FLORIDA
Analytical Chemistry — M,D*
Applied Mathematics — D
Chemistry — M,D*
Geology — M,O
Inorganic Chemistry — M,D*
Marine Sciences/Oceanography — M,D*
Mathematics — M,D*
Organic Chemistry — M,D*
Physical Chemistry — M,D*
Physics — M,D

UNIVERSITY OF SOUTHWESTERN LOUISIANA
Applied Physics — M
Chemistry — M
Geology — M
Mathematics — M,D*

Physics — M
Statistics — M,D*

UNIVERSITY OF TENNESSEE, KNOXVILLE
Analytical Chemistry — M,D
Chemistry — M,D
Geology — M,D
Mathematics — M,D
Organic Chemistry — M,D
Physical Chemistry — M,D
Physics — M,D*
Statistics — M

UNIVERSITY OF TENNESSEE SPACE INSTITUTE
Applied Mathematics — M
Physics — M,D

UNIVERSITY OF TEXAS AT ARLINGTON
Chemistry — M,D
Geology — M
Mathematics — M,D
Physics — M,D*

UNIVERSITY OF TEXAS AT AUSTIN
Analytical Chemistry — M,D*
Astronomy — M,D
Chemistry — M,D*
Geochemistry — M,D*
Geology — M,D*
Geophysics — M,D*
Inorganic Chemistry — M,D*
Marine Sciences/Oceanography — M,D*
Mathematics — M,D
Organic Chemistry — M,D*
Physical Chemistry — M,D*
Physics — M,D*
Statistics — M

UNIVERSITY OF TEXAS AT DALLAS
Applied Mathematics — M,D
Chemistry — M
Geochemistry — M,D*
Geology — M,D*
Geophysics — M,D*
Mathematics — M,D*
Physics — M,D
Statistics — M,D

UNIVERSITY OF TEXAS AT EL PASO
Chemistry — M
Geology — M,D
Geophysics — M
Mathematics — M
Physics — M

UNIVERSITY OF TEXAS AT SAN ANTONIO
Chemistry — M
Earth Sciences — M
Geology — M*
Mathematics — M
Statistics — M

UNIVERSITY OF TEXAS AT TYLER
Chemistry — M
Mathematics — M

UNIVERSITY OF TEXAS HEALTH SCIENCE CENTER AT HOUSTON
Biometrics — M,D

UNIVERSITY OF TEXAS OF THE PERMIAN BASIN
Geology — M

UNIVERSITY OF TEXAS–PAN AMERICAN
Mathematics — M

UNIVERSITY OF THE DISTRICT OF COLUMBIA
Mathematics — M

UNIVERSITY OF THE PACIFIC
Chemistry — M,D
Physics — M

UNIVERSITY OF TOLEDO
Analytical Chemistry — M,D
Applied Mathematics — M
Chemistry — M,D
Geochemistry — M
Geology — M*
Inorganic Chemistry — M,D
Mathematics — M,D
Organic Chemistry — M,D
Physical Chemistry — M,D

Physics — M,D
Statistics — M

UNIVERSITY OF TORONTO
Astronomy — M,D
Chemistry — M,D
Earth Sciences — M,D*
Geology — M,D*
Geophysics — M,D*
Mathematics — M,D
Physics — M,D
Statistics — M,D

UNIVERSITY OF TULSA
Applied Mathematics — M
Geochemistry — M,D
Geology — M,D
Geophysics — M,D
Mathematics — M

UNIVERSITY OF UTAH
Biostatistics — M
Chemistry — M,D
Geology — M,D
Geophysics — M,D
Mathematics — M,D*
Meteorology and Atmospheric Sciences — M,D
Physics — M,D
Statistics — M

UNIVERSITY OF VERMONT
Biostatistics — M
Chemistry — M,D
Geology — M
Mathematics — M
Physics — M
Statistics — M

UNIVERSITY OF VICTORIA
Applied Mathematics — M
Astronomy — M,D
Astrophysics — M,D
Chemistry — M,D
Geophysics — M,D
Mathematics — M
Physics — M,D
Statistics — M

UNIVERSITY OF VIRGINIA
Applied Mathematics — M,D
Astronomy — M,D*
Chemistry — M,D*
Mathematics — M,D*
Physics — M,D*

UNIVERSITY OF WASHINGTON
Applied Mathematics — M,D*
Applied Physics — M,D
Astronomy — M,D
Biostatistics — M,D*
Chemistry — M,D
Geology — M,D
Geophysics — M,D
Marine Sciences/Oceanography — M,D
Mathematics — M,D
Meteorology and Atmospheric Sciences — M,D
Physics — M,D*
Statistics — M,D

UNIVERSITY OF WATERLOO
Applied Mathematics — M,D
Chemistry — M,D
Earth Sciences — M,D*
Geology — M,D
Mathematics — M,D
Physics — M,D
Statistics — M,D

UNIVERSITY OF WESTERN ONTARIO
Applied Mathematics — M,D
Astronomy — M,D
Biostatistics — M,D
Chemistry — M,D
Geology — M,D
Geophysics — M,D
Mathematics — M,D
Physics — M,D
Statistics — M

UNIVERSITY OF WEST FLORIDA
Mathematics — M
Statistics — M

UNIVERSITY OF WINDSOR
Chemistry — M,D
Geology — M
Mathematics — M,D
Physics — M,D

*P—first professional degree; M—master's degree; D—doctorate; O—other advanced degree; * full description and/or announcement for program (see index)*

Peterson's Guide to Graduate Programs in the Physical Sciences and Mathematics 1992

UNIVERSITY OF WISCONSIN–MADISON

Analytical Chemistry	M
Astronomy	M,D
Biometrics	M
Chemistry	M,D
Geology	M,D
Geophysics	M,D
Marine Sciences/Oceanography	M,D
Mathematics	M,D
Meteorology and Atmospheric Sciences	M,D
Physics	M,D
Statistics	M,D

UNIVERSITY OF WISCONSIN–MILWAUKEE

Chemistry	M,D
Geology	M,D
Mathematics	M,D
Physics	M,D

UNIVERSITY OF WISCONSIN–OSHKOSH

Geophysics	M
Physics	M

UNIVERSITY OF WYOMING

Applied Mathematics	M,D*
Astronomy	M,D
Chemistry	M,D
Geology	M,D
Geophysics	M,D
Mathematics	M,D*
Meteorology and Atmospheric Sciences	M,D
Physics	M,D
Statistics	M,D

UTAH STATE UNIVERSITY

Astronomy	M
Astrophysics	M
Chemistry	M,D*
Earth Sciences	D
Geology	M
Mathematics	M,D*
Physics	M,D
Planetary Sciences	D
Statistics	M,D*

VANDERBILT UNIVERSITY

Astronomy	M
Chemistry	M,D
Geology	M
Mathematics	M,D*
Physics	M,D

VASSAR COLLEGE

Chemistry	M

VILLANOVA UNIVERSITY

Chemistry	M,D
Mathematics	M
Statistics	M

VIRGINIA COMMONWEALTH UNIVERSITY

Applied Mathematics	M
Applied Physics	M
Biostatistics	M,D
Chemistry	M,D
Mathematics	M
Physics	M
Statistics	M

VIRGINIA POLYTECHNIC INSTITUTE AND STATE UNIVERSITY

Applied Mathematics	M,D*
Chemistry	M,D*
Geochemistry	M,D
Geology	M,D
Geophysics	M,D
Mathematical Physics	M,D*
Mathematics	M,D*
Physics	M,D
Statistics	M,D*

VIRGINIA STATE UNIVERSITY

Geology	M
Mathematics	M
Physics	M

WAKE FOREST UNIVERSITY

Analytical Chemistry	M,D
Chemistry	M,D
Inorganic Chemistry	M,D
Mathematics	M
Organic Chemistry	M,D
Physical Chemistry	M,D
Physics	M,D

WASHINGTON STATE UNIVERSITY

Analytical Chemistry	M,D
Applied Mathematics	M,D
Chemistry	M,D
Geology	M,D
Inorganic Chemistry	M,D
Mathematics	M,D
Organic Chemistry	M,D
Physical Chemistry	M,D
Physics	M,D

WASHINGTON UNIVERSITY

Chemistry	M,D
Earth Sciences	M,D*
Geochemistry	D*
Geology	M,D*
Geophysics	D*
Mathematics	M,D
Physics	M,D
Planetary Sciences	M,D*
Statistics	M,D

WAYLAND BAPTIST UNIVERSITY

Chemistry	M

WAYNE STATE UNIVERSITY

Applied Mathematics	M,D
Chemistry	M,D
Geology	M
Mathematics	M,D
Physics	M,D
Statistics	M,D

WESLEYAN UNIVERSITY

Astronomy	M
Chemistry	M,D*
Earth Sciences	M
Inorganic Chemistry	M,D*
Mathematics	M,D*
Organic Chemistry	M,D*
Physical Chemistry	M,D*
Physics	M,D

WEST CHESTER UNIVERSITY OF PENNSYLVANIA

Astronomy	M
Chemistry	M
Geology	M
Mathematics	M

WESTERN CAROLINA UNIVERSITY

Chemistry	M
Mathematics	M
Physics	M

WESTERN CONNECTICUT STATE UNIVERSITY

Marine Sciences/Oceanography	M
Mathematics	M

WESTERN ILLINOIS UNIVERSITY

Chemistry	M
Mathematics	M
Physics	M

WESTERN KENTUCKY UNIVERSITY

Chemistry	M,D*
Mathematics	M
Organic Chemistry	D

WESTERN MICHIGAN UNIVERSITY

Applied Mathematics	M
Biostatistics	M
Chemistry	M
Earth Sciences	M
Geology	M
Mathematics	M,D
Physics	M,D
Statistics	M,D

WESTERN OREGON STATE COLLEGE

Mathematics	M

WESTERN WASHINGTON UNIVERSITY

Chemistry	M
Geology	M
Mathematics	M

WEST TEXAS STATE UNIVERSITY

Chemistry	M
Geology	M
Mathematics	M

WEST VIRGINIA UNIVERSITY

Chemistry	M,D
Geology	M,D*
Mathematics	M,D
Physics	M,D
Statistics	M

WICHITA STATE UNIVERSITY

Applied Mathematics	M,D
Chemistry	M,D
Geology	M
Mathematics	M,D
Physics	M

WILKES UNIVERSITY

Mathematics	M
Physics	M

WINTHROP COLLEGE

Mathematics	M

WORCESTER POLYTECHNIC INSTITUTE

Applied Mathematics	M
Chemistry	M,D
Physics	M,D

WRIGHT STATE UNIVERSITY

Applied Mathematics	M
Chemistry	M
Geology	M*
Mathematics	M
Physics	M*
Statistics	M

YALE UNIVERSITY

Applied Mathematics	D
Applied Physics	M,D*
Astronomy	M,D
Biostatistics	M,D
Chemistry	D*
Earth Sciences	D*
Geochemistry	D*
Geology	D*
Geophysics	D*
Inorganic Chemistry	D*
Marine Sciences/Oceanography	D*
Mathematics	M,D*
Meteorology and Atmospheric Sciences	D*
Organic Chemistry	D*
Physical Chemistry	D*
Physics	D
Plasma Physics	M,D
Statistics	M,D

YORK UNIVERSITY

Astronomy	M,D
Chemistry	M,D
Mathematics	M,D
Physics	M,D
Planetary Sciences	M,D
Statistics	M,D

YOUNGSTOWN STATE UNIVERSITY

Chemistry	M
Mathematics	M

Academic and Professional Programs in the Physical Sciences and Mathematics

This part of Book 4 consists of seven sections covering the physical sciences and mathematics. Each section has a table of contents (listing the program directories, announcements, and full descriptions); field definitions by graduate educators; program directories, which consist of brief profiles of programs in the relevant fields (and that include 50-word or 100-word announcements following the profiles if programs have chosen to include them); Cross-Discipline Announcements if any programs have chosen to submit such entries; and full two-page descriptions, which are more individualized statements included if programs have chosen to submit them.

Section 1
Astronomy and Astrophysics

This section contains directories of institutions offering graduate work in astronomy and astrophysics. Additional information about programs listed in the directories may be obtained by writing directly to the dean of a graduate school or chair of a department at the address given in the directory.

For programs offering related work, see also in this book: Earth and Planetary Sciences; Meteorology and Atmospheric Sciences; and Physics; in Book 3: Biology and Biomedical Sciences and Biophysics; and in Book 5: Engineering and Applied Sciences; Mechanical Engineering, Mechanics, and Aerospace/Aeronautical Engineering; and Nuclear Engineering.

CONTENTS

Program Directories

Astronomy	31
Astrophysics	35

Announcements

Cornell University	31
New Mexico State University	32
University of Minnesota, Twin Cities Campus	34
University of New Mexico	34
University of Virginia	34

Full Descriptions

See:

Boston University—Physics	571
Clemson University—Physics and Astronomy	581
Indiana University Bloomington—Physics	589
Johns Hopkins University—Physics and Astronomy	591
Louisiana State University—Physics and Astronomy	595
Michigan State University—Physics and Astronomy	597
Northwestern University—Physics and Astronomy	603
Tufts University—Physics and Astronomy	627
University of California, Los Angeles—Earth and Space Sciences	255
University of Kansas—Physics and Astronomy	649
University of Pittsburgh—Physics and Astronomy	659
University of Rochester—Physics and Astronomy	665
University of South Carolina—Physics and Astronomy	667

FIELD DEFINITIONS

In an effort to broaden prospective students' understanding of astronomy and astrophysics, educators in these fields have provided the following statements.

Astronomy

Astronomy is concerned with the study of the properties of all bodies external to the earth that can be examined by means of instruments located within the solar system and with the understanding of these properties in terms of physical processes that have their foundation in experimental physics. Current instrumentation allows the study of light emitted by external bodies over a considerable range in wavelength: less than 3,300 (ultraviolet, X-ray, and gamma-ray telescopes in space), 6,000–12,000 and 1.2–200 microns (infrared mountaintop telescopes), 3,000–6,000 (optical telescopes at all elevations on Earth and in space), and 1 millimeter to 20 meters (radio telescopes at all elevations). Objects of study range from galaxies, consisting of billions of stars and interstellar gas in bound systems, to stars and interstellar molecules within our own galaxy and in the local group of galaxies. Galaxies vary in type from quasars, which are thought to be bound systems possibly deriving their immense power from massive black holes that are swallowing stars, to ordinary galaxies such as our own, which contain light stars as old as the universe itself. Much of the matter in the universe, including most of the matter in galaxies, is thought to be in the form of "dark" matter, whose presence is known only by its gravitational effect on luminous objects. Stars vary in type from very young, bright stars that ultimately become supernovae to very condensed objects—such as white dwarfs, neutron stars, and black holes—which are thought to represent the final dying stages of stars that have exhausted their original store of nuclear fuel.

Careers for the Ph.D. astronomer may be found in the academic community (teaching and basic research), industry (instrument design and computer programming), and the space program (applied and basic research). In recent years, job opportunities for Ph.D.'s have been somewhat restricted, but there have consistently been positions available for those who have demonstrated a gift for innovative research. It is to be hoped that capable students with a real interest in and talent for astronomical research will not be dissuaded by past and future vagaries in the job market from entering a most fascinating and intellectually rewarding field.

Icko Iben Jr.
Departments of Astronomy and of Physics
University of Illinois at Urbana-Champaign

Astrophysics

Astrophysics is the study of the flow of matter and energy in astronomical objects. It is the study of the structure and evolution of interstellar gas clouds, of stars and clusters of stars, of galaxies and quasars, and of the universe. We are unable to produce, on Earth, the extremes of temperature, density, electromagnetic field, or gravitational field that exist in these objects, so the heavens provide the only laboratory to test much of modern physics. As an example consider a binary star system containing a pulsar. The periodic radio pulses emitted by the pulsar provide a way to read a "clock" in orbit. They can be used to test general relativity. The pulses themselves are created in the atmosphere of the pulsar (a neutron star) in a magnetic field millions of times stronger than the strongest field that can be created in earthbound plasma physics laboratories. The neutron star is believed to be composed almost entirely of nuclear matter. Both solid-state and nuclear physics are needed to explain its structure.

In the past few decades new technology has allowed us to examine astronomical objects in unprecedented detail. The development of radio astronomy, aperture synthesis, large optical telescopes and CCD imaging, orbiting observatories, and planetary probes has led to discoveries such as neutron stars, pulsars, quasars, radio-emitting jets, and the 2.78 K cosmic background radiation. We can expect this renaissance in astrophysics to continue for two reasons. First, technology continues to advance. Neutrinos have only recently been detected from an object outside the solar system. The first large orbiting optical observatory should be launched soon, designs for more powerful infrared and X-ray observatories are under development, and large area CCD detectors will soon be available. These advances will allow us to study in detail what we can only get a glimpse of today. Second, there are many unanswered questions in astrophysics. The list includes these: Do black holes exist? How did galaxies form? What is the nonluminous matter that we find in galaxies and clusters of galaxies? What is the energy source for the quasars? Why does the sun emit only one third of the predicted flux of neutrinos? Why is the cosmic background radiation so isotropic? What

SECTION 1: ASTRONOMY AND ASTROPHYSICS

Field Definition

phase transitions did the material of the early universe pass through as it cooled? What will the final state of the universe be? These questions and new ones will keep astrophysicists busy for many years.

Graduate programs in astrophysics exist as specialties in many physics and astronomy departments. Some universities have combined departments. A Ph.D. is required for a career in astrophysics. Most programs do not admit students working toward a master's degree only. Permanent academic positions in astrophysics at universities are scarce. Some research positions for Ph.D.'s are available at national laboratories, at national observatories, and in the space program. Astrophysicists also find that they can use their training in physics in related technical fields at industrial research laboratories.

Jeffrey B. Peterson
Department of Physics
Princeton University

Astronomy

Arizona State University, College of Liberal Arts and Sciences, Department of Physics, Tempe, AZ 85287. Offerings include astronomy (PhD). *Degree requirements:* Dissertation. *Entrance requirements:* GRE. Application fee: $25. *Tuition:* $1528 per year full-time, $80 per hour part-time for state residents; $6934 per year full-time, $289 per hour part-time for nonresidents. • Dr. John D. Dow, Chair, 602-965-3561.

Boston University, Graduate School, Department of Astronomy, Boston, MA 02215. Department awards MA, PhD. Faculty: 13 full-time (1 woman), 1 part-time (0 women). Matriculated students: 25 full-time (6 women), 0 part-time; includes 0 minority, 8 foreign. Average age 27. 56 applicants, 23% accepted. In 1990, 2 master's awarded (100% continued full-time study); 1 doctorate awarded (100% entered university research/teaching). Terminal master's awarded for partial completion of doctoral program. *Degree requirements:* For master's, 1 foreign language, comprehensive exam; for doctorate, 1 foreign language, dissertation. *Entrance requirements:* GRE General Test, GRE Subject Test, TOEFL (minimum score of 550 required). Application deadline: 6/1 (applications processed on a rolling basis). Application fee: $45. *Expenses:* Tuition of $15,950 per year full-time, $498 per credit part-time. Fees of $120 per year full-time, $35 per semester part-time. *Financial aid:* In 1990–91, 25 students received a total of $200,000 in aid awarded. 1 fellowship (to a first-year student), 10 research assistantships (1 to a first-year student), 6 teaching assistantships (3 to first-year students), 0 graduate assistantships were awarded; federal work-study also available. Average monthly stipend for a graduate assistantship: $1275. Financial aid application deadline: 1/15. *Faculty research:* Galactic and extragalactic astrophysics, dynamics of the Earth's magnetosphere and ionosphere, theoretical astrophysics and cosmology. Total annual research budget: $1.0-million. • Alan P. Marscher, Chairman, 617-353-5029. Application contact: Michael Mendillo, Graduate Admissions Chairman, 617-353-2629.

Boston University, Graduate School, Department of Physics, Boston, MA 02215. Offerings include astronomy and physics (MA, PhD). PhD in applied physics offered through the Division of Applied Sciences and Engineering. Terminal master's awarded for partial completion of doctoral program. *Degree requirements:* For master's, 1 foreign language, thesis, comprehensive exam; for doctorate, 1 foreign language, dissertation. *Entrance requirements:* GRE General Test, GRE Subject Test, TOEFL (minimum score of 550 required). Application deadline: 4/1. Application fee: $45. *Expenses:* Tuition of $15,950 per year full-time, $498 per credit part-time. Fees of $120 per year full-time, $35 per semester part-time. • Lawrence R. Sulak, Chairman, 617-353-2623. Application contact: Beverly Pacheco, Admissions Coordinator, 617-353-2623.

See full description on page 571.

Brigham Young University, College of Physical and Mathematical Sciences, Department of Physics and Astronomy, Provo, UT 84602. Department offers programs in physics (MS, PhD), physics and astronomy (PhD). Faculty: 30 full-time (0 women), 1 part-time (0 women). Matriculated students: 40 full-time (2 women), 5 part-time (0 women); includes 0 minority, 17 foreign. Average age 26. 70 applicants, 13% accepted. In 1990, 4 master's, 2 doctorates awarded. Terminal master's awarded for partial completion of doctoral program. *Degree requirements:* For master's, thesis; for doctorate, 1 foreign language, computer language, dissertation. *Entrance requirements:* For master's, GRE Subject Test (score in 40th percentile or higher required), minimum GPA of 3.0 during previous 60 credit hours; for doctorate, GRE Subject Test (score in 60th percentile or higher required). Application deadline: 2/15. Application fee: $30. *Tuition:* $1170 per semester (minimum) full-time, $130 per credit hour (minimum) part-time. *Financial aid:* In 1990–91, $388,000 in aid awarded. 5 fellowships (0 to first-year students), 15 research assistantships, 13 teaching assistantships (7 to first-year students) were awarded; institutionally sponsored loans also available. Aid available to part-time students. Financial aid application deadline: 2/15. *Faculty research:* Acoustics; astrophysics; atomic, nuclear, plasma, solid-state, and theoretical physics. • Dr. Daniel L. Decker, Chairman, 801-378-4361. Application contact: Dr. Dorian M. Hatch, Graduate Coordinator, 801-378-2427.

California Institute of Technology, Division of Physics, Mathematics and Astronomy, Department of Astronomy, Pasadena, CA 91125. Department awards PhD. Faculty: 17 full-time (1 woman), 0 part-time. Matriculated students: 24 full-time (4 women), 0 part-time; includes 0 minority, 11 foreign. Average age 25. 60 applicants, 8% accepted. In 1990, 4 doctorates awarded. Terminal master's awarded for partial completion of doctoral program. *Degree requirements:* For doctorate, 1 foreign language, dissertation, candidacy and final exams. *Entrance requirements:* For doctorate, GRE General Test, GRE Subject Test, TOEFL. Application deadline: 1/15. Application fee: $0. *Tuition:* $14,100 per year. *Financial aid:* In 1990–91, 24 students received a total of $600,000 in aid awarded. 4 fellowships (all to first-year students), 4 teaching assistantships (0 to first-year students), 3 outside awards (1 to a first-year student) were awarded; research assistantships, federal work-study, institutionally sponsored loans also available. Financial aid application deadline: 1/15. *Faculty research:* Observational and theoretical astrophysics, cosmology, radio astronomy, solar physics. Total annual research budget: $10.5-million. • A. C. R. Readhead, Executive Officer, 818-356-4972. Application contact: S. G. Djorgovski, Option Representative, 818-356-4415.

Case Western Reserve University, Department of Astronomy, Cleveland, OH 44106. Department awards MS, PhD. Part-time programs available. Faculty: 3 full-time (0 women), 0 part-time. Matriculated students: 0. 6 applicants, 33% accepted. In 1990, 1 master's awarded (100% continued full-time study); 0 doctorates awarded. *Degree requirements:* For doctorate, 2 foreign languages, dissertation. *Entrance requirements:* GRE General Test, GRE Subject Test, TOEFL (minimum score of 550 required). Application deadline: rolling. Application fee: $25. *Expenses:* Tuition of $13,600 per year full-time, $567 per credit part-time. Fees of $320 per year. *Financial aid:* Fellowships, research assistantships available. Financial aid applicants required to submit FAF. *Faculty research:* Ground-based optical astronomy, high and low-dispersion spectroscopy; theoretical astrophysics. Total annual research budget: $169,960. • Peter Pesch, Chairman, 216-368-3729.

Clemson University, College of Sciences, Department of Physics and Astronomy, Clemson, SC 29634. Department offers program in physics (MS, PhD). Faculty: 23 full-time (0 women), 0 part-time. Matriculated students: 45 full-time (4 women), 0 part-time; includes 2 minority (1 black American, 1 Hispanic American), 10 foreign. Average age 27. 92 applicants, 26% accepted. In 1990, 4 master's awarded (25% entered university research/teaching, 50% found other work related to degree, 25% continued full-time study); 3 doctorates awarded (67% entered university research/teaching, 33% found other work related to degree). Terminal master's awarded for partial completion of doctoral program. *Degree requirements:* For master's, thesis required (for some programs), foreign language not required; for doctorate, dissertation required, foreign language not required. *Entrance requirements:* GRE General Test, TOEFL. Application deadline: 6/1 (priority date, applications processed on a rolling basis). Application fee: $25. *Expenses:* Tuition of $102 per credit hour. Fees of $80 per semester full-time. *Financial aid:* In 1990–91, 40 students received a total of $421,400 in aid awarded. 2 fellowships (both to first-year students), 15 research assistantships (1 to a first-year student), 23 teaching assistantships (13 to first-year students) were awarded. Average monthly stipend for a graduate assistantship: $877. Financial aid application deadline: 6/1. *Faculty research:* Astrophysics, atmosphere physics, condensed matter, radiation physics, solid state physics. Total annual research budget: $457,126. • Dr. P. J. McNulty, Head, 803-656-3416. Application contact: J. A. Gilreath, Graduate Student Adviser, 803-656-3416.

See full description on page 581.

Columbia University, Graduate School of Arts and Sciences, Division of Natural Sciences, Department of Astronomy, New York, NY 10027. Department awards MA, M Phil, PhD. Part-time programs available. Faculty: 9 full-time, 2 part-time. Matriculated students: 12 full-time (3 women), 0 part-time; includes 6 foreign. Average age 25. 11 applicants, 36% accepted. In 1990, 0 master's, 5 doctorates awarded. *Degree requirements:* For master's, foreign language and thesis not required; for doctorate, dissertation, M Phil required, foreign language not required. *Entrance requirements:* GRE General Test, TOEFL, major in astronomy or physics. *Expenses:* Tuition of $7836 per semester for state residents; $426 per credit for nonresidents. Fees of $148 per semester for state residents. *Financial aid:* In 1990–91, $133,441 in aid awarded. 0 fellowships were awarded; teaching assistantships, federal work-study, institutionally sponsored loans also available. Financial aid application deadline: 1/5; applicants required to submit GAPSFAS. *Faculty research:* Theoretical astrophysics, x-ray astronomy, radio astronomy. • David Helfand, Chair, 212-854-2150.

Cornell University, Graduate Fields of Arts and Sciences, Field of Astronomy and Space Sciences, Ithaca, NY 14853. Field offers programs in astronomy (MS, PhD), astrophysics (MS, PhD), general space sciences (MS, PhD), infrared astronomy (MS, PhD), planetary studies (MS, PhD), radio astronomy (MS, PhD), radiophysics (MS, PhD). Faculty: 26 full-time, 0 part-time. Matriculated students: 26 full-time (4 women), 0 part-time; includes 0 minority, 6 foreign. 121 applicants, 10% accepted. In 1990, 7 master's, 3 doctorates awarded. Terminal master's awarded for partial completion of doctoral program. *Degree requirements:* Thesis/dissertation required, foreign language not required. *Entrance requirements:* GRE General Test, GRE Subject Test, TOEFL. Application deadline: 1/10. Application fee: $55. *Expenses:* Tuition of $16,170 per year. Fees of $28 per year. *Financial aid:* In 1990–91, 23 students received aid. 5 fellowships (2 to first-year students), 13 research assistantships (0 to first-year students), 5 teaching assistantships (0 to first-year students) awarded; full and partial tuition waivers, federal work-study, institutionally sponsored loans, and career-related internships or fieldwork also available. Financial aid application deadline: 1/10; applicants required to submit GAPSFAS. • Philip Nicholson, Graduate Faculty Representative, 607-255-5284. Application contact: Robert Brashear, Director of Admissions, 607-255-4884.

Announcement: Graduate studies in theoretical astrophysics, cosmology and relativistic astrophysics, radio and radar astronomy, optical and infrared astronomy, planetary sciences, exobiology, plasma astrophysics, celestial mechanics. Financial support for practically all graduate students. Research sponsored by Cornell's Center for Radiophysics and Space Research and the National Astronomy and Ionosphere Center (operated by Cornell for National Science Foundation).

Dartmouth College, School of Arts and Sciences, Department of Physics and Astronomy, Hanover, NH 03755. Department awards AM, PhD. Faculty: 14 full-time (1 woman), 0 part-time. Matriculated students: 30 full-time (5 women), 0 part-time; includes 1 minority (Asian American), 9 foreign. 191 applicants, 7% accepted. In 1990, 2 master's awarded (50% found work related to degree, 50% continued full-time study); 3 doctorates awarded. Terminal master's awarded for partial completion of doctoral program. *Degree requirements:* For master's, thesis required, foreign language not required; for doctorate, variable foreign language requirement, dissertation. *Entrance requirements:* GRE General Test, GRE Subject Test. Application deadline: 3/1. *Expenses:* Tuition of $16,230 per year. Fees of $650 per year. *Financial aid:* In 1990–91, $828,554 in aid awarded. 16 fellowships (6 to first-year students), 7 research assistantships were awarded; full tuition waivers, institutionally sponsored loans also available. Financial aid applicants required to submit GAPSFAS or FAF. *Total annual research budget:* $773,867. • Joseph Harris, Chairman, 603-646-2359.

Georgia State University, College of Arts and Sciences, Department of Physics and Astronomy, Atlanta, GA 30303. Offerings include astronomy (PhD). Terminal master's awarded for partial completion of doctoral program. Department faculty: 15 full-time (0 women), 0 part-time. *Degree requirements:* 2 foreign languages (computer language can substitute for one), dissertation. *Entrance requirements:* GRE General Test, GRE Subject Test, TOEFL (minimum score of 550 required), minimum GPA of 3.0. Application deadline: 7/15. Application fee: $10. *Expenses:* Tuition of $38 per quarter hour for state residents; $130 per quarter hour for nonresidents. Fees of $58 per quarter. • Dr. Martin Meder, Director of Graduate Studies, 404-651-2279.

Harvard University, Graduate School of Arts and Sciences, Department of Astronomy, Cambridge, MA 02138. Department offers programs in astronomy (AM, PhD), astrophysics (AM, PhD). Faculty: 21 full-time, 16 part-time. Matriculated students: 21 full-time (4 women), 0 part-time; includes 0 minority, 6 foreign. Average age 25. 73 applicants, 18% accepted. In 1990, 2 master's, 5 doctorates awarded. Terminal master's awarded for partial completion of doctoral program. *Degree requirements:* For doctorate, dissertation, research project, paper required, foreign language not required. *Entrance requirements:* For doctorate, GRE General Test, GRE Subject Test. Application deadline: 1/2. Application fee: $60. *Expenses:* Tuition of $14,860 per year. Fees of $550 per year. *Financial aid:* In 1990–91, $135,554 in aid awarded. 11 fellowships (7 to first-year students), 9 research assistantships, 7 teaching assistantships were awarded; federal work-study, institutionally sponsored loans, and career-related internships or fieldwork also available. Financial aid application deadline: 1/2; applicants required to submit GAPSFAS. *Faculty research:* Atomic and molecular physics, electromagnetism, solar physics, nuclear physics, fluid dynamics. • Dr. Mark Birkinshaw, Director of Graduate Study, 617-495-3752. Application contact: Office of Admissions and Financial Aid, 617-495-5315.

Indiana University Bloomington, College of Arts and Sciences, Department of Astronomy, Bloomington, IN 47405. Department offers programs in astronomy (MA, PhD), astrophysics (PhD). Part-time programs available. Faculty: 7 full-time (1 woman), 0 part-time. Matriculated students: 17 full-time (3 women), 0 part-time; includes 1 minority (black American), 2 foreign. 31 applicants, 35% accepted. In 1990, 1 master's awarded (100% continued full-time study); 2 doctorates awarded (100% entered university research/teaching). Terminal master's awarded for partial completion of doctoral program. *Degree requirements:* For master's, thesis, oral exam required, foreign language not required; for doctorate, dissertation, written qualifying exam required, foreign language not required. *Entrance requirements:* GRE General Test, GRE Subject Test (physics), BA or BS in a science. Application deadline:

Directory: Astronomy

2/1. Application fee: $25 ($35 for foreign students). *Tuition:* $99.85 per credit hour for state residents; $288 per credit hour for nonresidents. *Financial aid:* In 1990–91, $132,550 in aid awarded. 0 fellowships, 5 research assistantships (0 to first-year students), 7 teaching assistantships (1 to a first-year student) were awarded; full and partial tuition waivers, federal work-study also available. Aid available to part-time students. Financial aid application deadline: 2/1. *Faculty research:* Stellar astronomy, extragalactic astronomy, cataclysmic variables, high energy astrophysics, supercomputer simulations. Total annual research budget: $641,168. • Hollin Johnson, Chairman, 812-855-6195.

Iowa State University of Science and Technology, College of Liberal Arts and Sciences, Department of Physics and Astronomy, Ames, IA 50011. Department awards MS, PhD. Matriculated students: 95 full-time (13 women), 10 part-time (1 woman); includes 3 minority (2 Asian American, 1 black American), 59 foreign. In 1990, 5 master's, 17 doctorates awarded. *Degree requirements:* For master's, thesis optional; for doctorate, dissertation. *Entrance requirements:* For master's, GRE General Test. Application fee: $20 ($30 for foreign students). *Expenses:* Tuition of $1158 per semester full-time, $129 per credit part-time for state residents; $3340 per semester full-time, $372 per credit part-time for nonresidents. Fees of $10 per semester. *Financial aid:* In 1990–91, 1 fellowship (to a first-year student), 45 research assistantships (1 to a first-year student), 54 teaching assistantships (39 to first-year students), 5 scholarships (2 to first-year students) awarded. • Dr. Marshall Luban, Chair, 515-294-5442.

Johns Hopkins University, School of Arts and Sciences, Department of Physics and Astronomy, Baltimore, MD 21218. Department offers programs in astronomy (MA, PhD), physics (MA, PhD). Faculty: 27 full-time (3 women), 10 part-time (0 women). Matriculated students: 94 full-time (7 women), 0 part-time; includes 3 minority (1 Asian American, 2 Hispanic American), 30 foreign. Average age 25. 157 applicants, 38% accepted. In 1990, 15 master's awarded; 11 doctorates awarded (100% entered university research/teaching). Terminal master's awarded for partial completion of doctoral program. *Degree requirements:* For doctorate, 1 foreign language, dissertation, comprehensive exam. *Entrance requirements:* For master's, GRE General Test; for doctorate, GRE General Test, GRE Subject Test, TOEFL. Application deadline: 2/15. Application fee: $40. *Expenses:* Tuition of $15,500 per year full-time, $1550 per course part-time. Fees of $400 per year full-time. *Financial aid:* In 1990–91, $1.485-million in aid awarded. 93 fellowships (22 to first-year students), 56 research assistantships (3 to first-year students), 37 teaching assistantships (19 to first-year students) were awarded; full and partial tuition waivers, federal work-study, institutionally sponsored loans also available. Financial aid application deadline: 3/14; applicants required to submit FAF. *Faculty research:* Experimental, theoretical, atomic, plasma, particle, nuclear, and condensed matter physics. Total annual research budget: $4.5-million. • Dr. James C. Walker Jr., Chairman, 301-338-7347. Application contact: Barbara A. Staicer, Graduate Admissions Administrator, 301-338-7347.

See full description on page 591.

Louisiana State University and Agricultural and Mechanical College, College of Basic Sciences, Department of Physics and Astronomy, Baton Rouge, LA 70803. Department offers programs in astronomy (PhD), astrophysics (PhD), physics (MS, PhD). Faculty: 35 full-time (0 women), 0 part-time. Matriculated students: 65 full-time (15 women), 3 part-time (0 women); includes 3 minority (2 Asian American, 1 black American), 51 foreign. Average age 28. 100 applicants, 41% accepted. In 1990, 8 master's awarded (100% continued full-time study); 10 doctorates awarded (70% entered university research/teaching, 30% found other work related to degree). Terminal master's awarded for partial completion of doctoral program. *Degree requirements:* For master's, thesis or alternative required, foreign language not required; for doctorate, dissertation required, foreign language not required. *Entrance requirements:* For master's, GRE General Test, TOEFL (minimum score of 525 required for admission, 560 required for assistantships), minimum GPA of 3.0; for doctorate, GRE General Test, TOEFL (minimum score of 525 required), minimum GPA of 3.0. Application deadline: 7/1 (priority date, applications processed on a rolling basis). Application fee: $25. *Tuition:* $1020 per semester full-time for state residents; $2620 per semester full-time for nonresidents. *Financial aid:* In 1990–91, 64 students received a total of $757,068 in aid awarded. 3 fellowships (1 to a first-year student), 14 research assistantships (0 to first-year students), 46 teaching assistantships (11 to first-year students) were awarded; institutionally sponsored loans also available. Average monthly stipend for a graduate assistantship: $968. Financial aid application deadline: 3/15. *Faculty research:* Experimental atomic, nuclear, particle, cosmic-ray, low-temperature, and condensed-matter physics; theoretical atomic, nuclear, particle, and condensed-matter physics. • Dr. Jerry P. Draayer, Chair, 504-388-2261. Application contact: Dr. Michael Cherry, Chair, Assistantship Committee, 504-388-1194.

See full description on page 595.

New Mexico State University, College of Arts and Sciences, Department of Astronomy, Las Cruces, NM 88003. Department awards MS, PhD. Part-time programs available. Faculty: 9 full-time. Matriculated students: 23 full-time, 4 part-time; includes 1 foreign. In 1990, 4 master's, 3 doctorates awarded. *Degree requirements:* Thesis/dissertation or alternative. *Entrance requirements:* GRE, advanced physics recommended. Application deadline: 2/15. Application fee: $10. *Tuition:* $1608 per year full-time, $67 per credit hour part-time for state residents; $5304 per year full-time, $221 per credit hour part-time for nonresidents. *Financial aid:* Fellowships, research assistantships, teaching assistantships, minority fellowships available. • Dr. Jack O. Burns, Head, 505-646-4438.

Announcement: Graduate research in solar astrophysics, Jovian planetary atmospheres, solar-system space exploration, variable stars, interstellar medium, galactic structure, radio galaxies, and cosmology. Graduate student use of department's off-campus observatories is encouraged. New 3.5-meter telescope project with consortium of four other universities is under way; completion expected in early 1992.

Northwestern University, College of Arts and Sciences, Department of Physics and Astronomy, Evanston, IL 60208. Department offers programs in astronomy (MS, PhD), astrophysics (MS, PhD), physics (MS, PhD). Faculty: 34 full-time (2 women), 0 part-time. Matriculated students: 87 full-time (17 women), 0 part-time; includes 2 minority (both Asian American), 49 foreign. 160 applicants, 56% accepted. In 1990, 6 master's, 13 doctorates awarded. Terminal master's awarded for partial completion of doctoral program. *Degree requirements:* For master's, thesis or alternative, comprehensive exam required, foreign language not required; for doctorate, dissertation, qualifying exam, oral exam required, foreign language not required. *Entrance requirements:* For master's, GRE General Test; for doctorate, GRE General Test, GRE Subject Test (physics). Application deadline: 8/30. Application fee: $40 ($45 for foreign students). *Tuition:* $4665 per quarter full-time, $1704 per course part-time. *Financial aid:* In 1990–91, $491,669 in aid awarded. 13 fellowships, 46 research assistantships, 25 teaching assistantships were awarded; federal work-study, institutionally sponsored loans, and career-related internships or fieldwork also available. Financial aid application deadline: 1/15; applicants required to submit GAPSFAS. *Faculty research:* Condensed matter, high-energy physics, nuclear physics, elementary particle and fields, astrophysics. Total annual research budget: $5.2-million. • Dr. William P. Halpern, Chairperson, 708-491-5468. Application contact: David Taylor, Assistant Chairperson, 708-491-2053.

See full description on page 603.

Ohio State University, College of Mathematical and Physical Sciences, Department of Astronomy, Columbus, OH 43210. Department awards MS, PhD. Faculty: 16. Matriculated students: 14 full-time (2 women), 1 part-time (0 women); includes 1 minority (Asian American), 2 foreign. 71 applicants, 21% accepted. In 1990, 0 master's, 1 doctorate awarded. *Degree requirements:* For master's, thesis, comprehensive exam; for doctorate, dissertation, oral and written comprehensive exams. *Entrance requirements:* GRE General Test, TOEFL (minimum score of 550 required), minimum GPA of 2.7. Application deadline: 8/15 (applications processed on a rolling basis). Application fee: $0 ($25 for foreign students). *Tuition:* $1213 per quarter full-time, $364 per course part-time for state residents; $3143 per quarter full-time, $943 per course part-time for nonresidents. *Financial aid:* Fellowships, research assistantships, teaching assistantships, federal work-study, institutionally sponsored loans available. Aid available to part-time students. Financial aid applicants required to submit FAF. • Eugene R. Capriotti, Chairman, 614-292-1773.

Pennsylvania State University University Park Campus, College of Science, Department of Astronomy and Astrophysics, University Park, PA 16802. Department awards MS, PhD. Faculty: 12. Matriculated students: 22 full-time (4 women), 0 part-time. In 1990, 0 master's, 4 doctorates awarded. *Entrance requirements:* GRE General Test. Application fee: $35. *Tuition:* $203 per credit for state residents; $403 per credit for nonresidents. • France A. Cordova, Head, 814-865-0418.

Rice University, Wiess School of Natural Sciences, Department of Space Physics and Astronomy, Houston, TX 77251. Department offers program in space physics and astronomy (MS, PhD). Faculty: 14 full-time, 0 part-time. Matriculated students: 41 full-time (8 women), 0 part-time; includes 2 minority (both Asian American), 8 foreign. *Degree requirements:* Thesis/dissertation required, foreign language not required. *Entrance requirements:* For master's, GRE General Test, TOEFL (minimum score 0f 550 required), minimum GPA of 3.0; for doctorate, GRE General Test (score in 70th percentile or higher required), minimum GPA of 3.0. Application deadline: 3/1. Application fee: $0. *Expenses:* Tuition of $8300 per year full-time, $400 per credit hour part-time. Fees of $167 per year. *Financial aid:* Fellowships, research assistantships, full and partial tuition waivers available. Financial aid applicants required to submit FAF. *Faculty research:* Magnetospheric physics, planetary atmospheres, astrophysics. • Alexander J. Dessler, Chairman, 713-527-4939.

Saint Mary's University, Faculty of Science, Department of Astronomy, Halifax, NS B3H 3C3, Canada. Department awards M Sc. *Degree requirements:* Thesis required, foreign language not required. *Entrance requirements:* Honors B Sc. *Tuition:* $2150 per year full-time, $250 per course (minimum) part-time for Canadian residents; $3850 per year full-time, $340 per course (minimum) part-time for nonresidents.

San Diego State University, College of Sciences, Department of Astronomy, San Diego, CA 92182. Department awards MS. Faculty: 6 full-time, 2 part-time. Matriculated students: 0 full-time, 14 part-time (2 women); includes 1 minority (black American), 3 foreign. In 1990, 5 degrees awarded. *Degree requirements:* Foreign language and thesis not required. *Entrance requirements:* GRE General Test (minimum combined score of 950 required). Application deadline: 8/1 (applications processed on a rolling basis). Application fee: $55. *Expenses:* Tuition of $0 for state residents; $189 per unit for nonresidents. Fees of $1974 per year full-time, $692 per year part-time for state residents; $1074 per year full-time, $692 per year part-time for nonresidents. *Faculty research:* Binary stars, gaseous nebulae, galaxies. • Dr. Ronald Angione, Chair, 619-594-6182.

State University of New York at Stony Brook, College of Arts and Sciences, Division of Physical Sciences and Mathematics, Department of Earth and Space Sciences, Program in Astronomical Sciences, Stony Brook, NY 11794. Offers astronomical sciences (M Phil, MS, PhD), astrophysics (M Phil, MS, PhD). Faculty: 8 full-time. Matriculated students: 14 full-time (3 women), 0 part-time. *Degree requirements:* For master's, thesis or alternative required, foreign language not required; for doctorate, dissertation required, foreign language not required. *Entrance requirements:* GRE General Test, TOEFL, minimum GPA of 3.0. Application deadline: 2/1. Application fee: $35. *Expenses:* Tuition of $2450 per year full-time, $103 per credit part-time for state residents; $5766 per year full-time, $243 per credit part-time for nonresidents. Fees of $151 per year full-time, $10.45 per year (minimum) part-time. *Financial aid:* Fellowships available. • Dr. Gilbert Hansen, Chairman, Department of Earth and Space Sciences, 516-632-8200.

Tufts University, Graduate School of Arts and Sciences, Department of Physics and Astronomy, Medford, MA 02155. Department offers programs in astronomy (MS, PhD), physics (MS, PhD). Faculty: 21 full-time, 0 part-time. Matriculated students: 44 (5 women); includes 15 foreign. 52 applicants, 33% accepted. In 1990, 3 master's, 10 doctorates awarded. Terminal master's awarded for partial completion of doctoral program. *Degree requirements:* Thesis/dissertation required, foreign language not required. *Entrance requirements:* GRE General Test, GRE Subject Test, TOEFL (minimum score of 550 required). Application deadline: 2/15. Application fee: $50. *Expenses:* Tuition of $16,755 per year full-time, $2094 per course part-time. Fees of $885 per year. *Financial aid:* Research assistantships, teaching assistantships, federal work-study available. Financial aid application deadline: 2/15; applicants required to submit GAPSFAS. *Faculty research:* Search for nuclear decay, neutrino interactions, cosmology, superconductivity, protein structure. • David Weaver, Chair, 617-381-3515. Application contact: Anthony Mann, 617-381-3219.

See full description on page 627.

University of Arizona, College of Arts and Sciences, Faculty of Science, Department of Astronomy, Tucson, AZ 85721. Department awards MS, PhD. Faculty: 29 full-time (2 women), 3 part-time (0 women). Matriculated students: 33 full-time (9 women), 8 part-time (4 women); includes 4 minority (2 Asian American, 2 Hispanic American), 3 foreign. Average age 27. 122 applicants, 9% accepted. In 1990, 1 master's, 1 doctorate awarded. *Degree requirements:* For master's, foreign language and thesis not required; for doctorate, dissertation required, foreign language not required. *Entrance requirements:* For master's, GRE General Test, GRE Subject Test (minimum score of 600 required); for doctorate, GRE General Test, GRE Subject Test (minimum score of 700 required). Application deadline: 2/7 (applications processed on a rolling basis). Application fee: $25. *Expenses:* Tuition of $0 for state residents; $5406 per year full-time, $209 per credit hour part-time for nonresidents. Fees of $1528 per year full-time, $80 per credit hour part-time. *Financial aid:* In 1990–91, $182,934 in aid awarded. 4 fellowships, 19 research assistantships, 11 teaching assistantships, 2 scholarships were awarded. Financial aid applicants required to submit FAF. *Faculty research:* Astrophysics, submillimeter, infrared astronomy, SIRTE, NICMOS. Total annual research budget: $16.8-million. • Dr. Peter A. Strittmatter, Head Director, Astronomy/Steward Observatory, 602-621-6524. Application contact: Helen Bluestein, Administrative Assistant, 602-621-2289.

University of British Columbia, Faculty of Science, Department of Geophysics and Astronomy, Vancouver, BC V6T 1Z1, Canada. Offerings include astronomy (MA Sc, M Sc, PhD). *Degree requirements:* For master's, thesis; for doctorate, dissertation, comprehensive exam.

University of Calgary, Faculty of Science, Department of Physics and Astronomy, Calgary, AB T2N 1N4, Canada. Department awards M Sc, PhD. Part-time programs available. *Degree requirements:* Thesis/dissertation. *Entrance requirements:* GRE General Test, GRE Subject Test, TOEFL (minimum score of 550 required). Application deadline: 5/31. Application fee: $25. *Tuition:* $1705 per year full-time, $427 per course part-time for Canadian residents; $3410 per year full-time, $854 per course part-time for nonresidents. *Faculty research:* Astronomy and astrophysics, EPR, mass spectrometry, atmospheric physics, space physics.

University of California at Berkeley, College of Letters and Science, Department of Astronomy, Berkeley, CA 94720. Department awards PhD. Faculty: 13. Matriculated students: 32 full-time, 0 part-time; includes 4 minority (3 Asian American, 1 Hispanic American), 7 foreign. 79 applicants, 14% accepted. Terminal master's awarded for partial completion of doctoral program. *Degree requirements:* For doctorate, dissertation. *Entrance requirements:* For doctorate, GRE General Test, minimum GPA of 3.0. Application deadline: 2/11. Application fee: $40. *Expenses:* Tuition of $0. Fees of $1909 per year for state residents; $7825 per year for nonresidents. • John Arons, Acting Chair.

University of California, Los Angeles, College of Letters and Science, Department of Astronomy, Los Angeles, CA 90024. Department awards MA, MAT, PhD. Faculty: 10. Matriculated students: 22 full-time (6 women), 0 part-time; includes 1 minority, 4 foreign. 56 applicants, 34% accepted. In 1990, 2 master's, 0 doctorates awarded. *Degree requirements:* For master's, thesis, comprehensive exam required, foreign language not required; for doctorate, dissertation, written comprehensive exam, research project required, foreign language not required. *Entrance requirements:* GRE General Test, GRE Subject Test. Application fee: $40. *Expenses:* Tuition of $0 for state residents; $7699 per year for nonresidents. Fees of $2907 per year. *Financial aid:* In 1990–91, 20 students received a total of $317,696 in aid awarded. 19 fellowships, 16 research assistantships, 9 teaching assistantships were awarded; full and partial tuition waivers, federal work-study, institutionally sponsored loans also available. Financial aid application deadline: 3/1. • Dr. Michael Jura, Chair, 310-825-4434.

University of California, Santa Cruz, Division of Natural Sciences, Program in Astronomy and Astrophysics, Santa Cruz, CA 95064. Program awards PhD. Faculty: 20 full-time (2 women), 0 part-time. Matriculated students: 29 full-time (8 women), 0 part-time; includes 2 minority (both Asian American), 4 foreign. 85 applicants, 25% accepted. In 1990, 5 doctorates awarded. Terminal master's awarded for partial completion of doctoral program. *Degree requirements:* For doctorate, 1 foreign language (computer language can substitute), dissertation, qualifying exam. *Entrance requirements:* For doctorate, GRE General Test, GRE Subject Test. Application deadline: 2/1. Application fee: $40. *Expenses:* Tuition of $0 for state residents; $7699 per year for nonresidents. Fees of $3021 per year. *Financial aid:* In 1990–91, 3 fellowships, 16 research assistantships, 19 teaching assistantships awarded; federal work-study, institutionally sponsored loans also available. Average monthly stipend for a graduate assistantship: $1335. Financial aid applicants required to submit GAPSFAS. *Faculty research:* Stellar structure and evolution, stellar spectroscopy, the interstellar medium, galactic structure, external galaxies and quasars. • Dr. Stan Woosley, Chairperson, 408-459-2844.

University of Chicago, Division of the Physical Sciences, Department of Astronomy and Astrophysics, Chicago, IL 60637. Department awards SM, PhD. *Degree requirements:* For master's, foreign language and thesis not required; for doctorate, dissertation required, foreign language not required. *Entrance requirements:* GRE General Test, GRE Subject Test, TOEFL. Application deadline: 1/6. Application fee: $45. *Expenses:* Tuition of $16,275 per year full-time, $8140 per year part-time. Fees of $356 per year.

University of Delaware, College of Arts and Science, Department of Physics and Astronomy, Newark, DE 19716. Department awards MS, PhD. Part-time programs available. Faculty: 24 full-time (1 woman), 0 part-time. Matriculated students: 63 full-time (8 women), 1 part-time (0 women); includes 1 minority (Asian American), 37 foreign. 400 applicants, 6% accepted. In 1990, 3 master's, 2 doctorates awarded. Terminal master's awarded for partial completion of doctoral program. *Degree requirements:* Thesis/dissertation required, foreign language not required. *Entrance requirements:* GRE General Test. Application deadline: 7/1. Application fee: $40. *Tuition:* $179 per credit hour for state residents; $467 per credit hour for nonresidents. *Financial aid:* In 1990–91, $500,000 in aid awarded. 5 fellowships (1 to a first-year student), 28 research assistantships (2 to first-year students), 30 teaching assistantships (8 to first-year students) were awarded. *Total annual research budget:* $1-million. • Dr. James B. Mehl, Chair, 302-451-2661.

University of Florida, College of Liberal Arts and Sciences, Department of Astronomy, Gainesville, FL 32611. Department awards MS, MST, PhD. Faculty: 17 full-time (0 women), 0 part-time. Matriculated students: 22 full-time (7 women), 8 part-time (1 woman); includes 2 minority (1 black American, 1 Hispanic American), 11 foreign. 28 applicants, 39% accepted. In 1990, 4 master's, 0 doctorates awarded. Terminal master's awarded for partial completion of doctoral program. *Degree requirements:* For master's, thesis (for terminal MS) required, foreign language not required; for doctorate, 1 foreign language, dissertation. *Entrance requirements:* GRE General Test (minimum combined score of 1200 required), minimum GPA of 3.0. Application deadline: 6/1 (applications processed on a rolling basis). Application fee: $15. *Tuition:* $87 per credit hour for state residents; $289 per credit hour for nonresidents. *Financial aid:* In 1990–91, 6 research assistantships (0 to first-year students), 13 teaching assistantships (3 to first-year students) awarded; fellowships also available. Financial aid application deadline: 1/31. *Faculty research:* Radio astronomy, cosmology, photometry, variable and binary stars, dynamical and solar system astronomy. • Dr. Stephen Gottesman, Chairman, 904-392-2052. Application contact: Dr. George Lebo, Graduate Coordinator, 904-392-2052.

University of Hawaii at Manoa, College of Arts and Sciences, Department of Physics and Astronomy, Honolulu, HI 96822. Department awards MS, PhD. *Degree requirements:* For master's, qualifying exam or thesis; for doctorate, dissertation, master's degree, qualifying exam, oral comprehensive exam. *Entrance requirements:* GRE General Test, GRE Subject Test. *Tuition:* $800 per semester full-time for state residents; $2405 per semester full-time for nonresidents.

University of Illinois at Urbana-Champaign, College of Liberal Arts and Sciences, Department of Astronomy, Champaign, IL 61820. Department awards MS, PhD. Faculty: 18 full-time, 0 part-time. Matriculated students: 19 full-time (3 women), 0 part-time; includes 1 minority (Hispanic American), 7 foreign. 39 applicants, 8% accepted. In 1990, 0 master's, 3 doctorates awarded. *Degree requirements:* For master's, foreign language and thesis not required; for doctorate, dissertation required, foreign language not required. *Entrance requirements:* GRE General Test, GRE Subject Test. Application fee: $25. *Tuition:* $1838 per semester full-time, $708 per semester part-time for state residents; $4314 per semester full-time, $1673 per semester part-time for nonresidents. *Financial aid:* In 1990–91, 0 fellowships, 10 research assistantships, 8 teaching assistantships awarded. Financial aid application deadline: 2/15. • Ronald F. Webbink, Chair, 217-333-3090.

University of Iowa, College of Liberal Arts, Department of Physics and Astronomy, Program in Astronomy, Iowa City, IA 52242. Program awards MS. Matriculated students: 3 full-time (0 women), 3 part-time (1 woman); includes 0 minority, 1 foreign. 4 applicants, 50% accepted. In 1990, 1 degree awarded. *Application fee:* $20. *Expenses:* Tuition of $1158 per semester full-time, $387 per semester hour (minimum) part-time for state residents; $3372 per semester full-time, $387 per semester hour (minimum) part-time for nonresidents. Fees of $60 per semester (minimum). *Financial aid:* In 1990–91, 0 fellowships, 0 research assistantships, 4 teaching assistantships (0 to first-year students) awarded. • Dwight Nicholson, Chair, Department of Physics and Astronomy, 319-335-1686.

University of Kansas, College of Liberal Arts and Sciences, Department of Physics and Astronomy, Lawrence, KS 66045. Department offers programs in atmospheric science (MS), physics (MA, MS, PhD). Faculty: 25 full-time (1 woman), 2 part-time (0 women). Matriculated students: 38 full-time (6 women), 15 part-time (1 woman); includes 3 minority (1 Asian American, 2 Hispanic American), 20 foreign. In 1990, 4 master's, 3 doctorates awarded. *Degree requirements:* For master's, foreign language and thesis not required; for doctorate, computer language, dissertation. *Entrance requirements:* TOEFL (minimum score of 570 required). Application fee: $25. *Expenses:* Tuition of $1668 per year full-time, $56 per credit hour part-time for state residents; $5382 per year full-time, $179 per credit hour part-time for nonresidents. Fees of $338 per year full-time, $25 per credit hour part-time. *Financial aid:* Fellowships, research assistantships, teaching assistantships available. • Ray Ammar, Chairperson, 913-864-4626.

See full description on page 649.

University of Kentucky, Graduate School Programs from the College of Arts and Sciences, Program in Physics and Astronomy, Lexington, KY 40506-0032. Program awards MS, PhD. Faculty: 27 full-time (0 women), 0 part-time. Matriculated students: 27 full-time (6 women), 16 part-time (4 women); includes 2 minority (1 Asian American, 1 Hispanic American), 26 foreign. 180 applicants, 7% accepted. In 1990, 4 master's, 2 doctorates awarded. *Degree requirements:* For master's, 1 foreign language, comprehensive exam required, thesis optional; for doctorate, dissertation, comprehensive exam. *Entrance requirements:* For master's, GRE (verbal, quantitative, and analytical sections), minimum undergraduate GPA of 2.5; for doctorate, GRE (verbal, quantitative, and analytical sections), minimum graduate GPA of 3.0. Application deadline: 7/19 (applications processed on a rolling basis). Application fee: $20 ($25 for foreign students). *Tuition:* $1002 per semester full-time, $101 per credit hour part-time for state residents; $2782 per semester full-time, $299 per credit hour part-time for nonresidents. *Financial aid:* In 1990–91, 1 fellowship (2 to first-year students), 17 research assistantships, 26 teaching assistantships (4 to first-year students) awarded; federal work-study, institutionally sponsored loans also available. Aid available to part-time students. Financial aid applicants required to submit FAF. *Faculty research:* Interstellar cloud dynamics, interstellar magnetic fields, formation of interstellar molecules, circumstellar masers, maser excitation. • Jesse Weil, Director of Graduate Studies, 606-257-3997. Application contact: Dr. Constance L. Wood, Associate Dean for Academic Administration, 606-257-4905.

University of Maine, College of Sciences, Department of Physics and Astronomy, Orono, ME 04469. Department offers programs in engineering physics (M Eng), physics and astronomy (MS, PhD). Faculty: 17 full-time (1 woman), 1 part-time (0 women). Matriculated students: 26 full-time (3 women), 2 part-time (1 woman); includes 0 minority, 11 foreign. Average age 29. 42 applicants. In 1990, 2 master's awarded (100% found work related to degree); 4 doctorates awarded (100% entered university research/teaching). Terminal master's awarded for partial completion of doctoral program. *Degree requirements:* Thesis/dissertation required, foreign language not required. *Entrance requirements:* For master's, GRE General Test, GRE Subject Test, TOEFL (minimum score of 550 required). Application deadline: 12/15 (priority date, applications processed on a rolling basis). Application fee: $25. *Tuition:* $100 per credit hour for state residents; $275 per credit hour for nonresidents. *Financial aid:* In 1990–91, $230,000 in aid awarded. 2 fellowships (0 to first-year students), 9 research assistantships (3 to first-year students), 12 teaching assistantships (5 to first-year students) were awarded. *Faculty research:* Solid-state physics, fluids, biophysics, plasma physics, surface physics. Total annual research budget: $800,000. • Dr. Charles W. Smith, Chairperson, 207-581-1015. Application contact: Dr. Gerald S. Harmon, Graduate Coordinator, 207-581-1016.

University of Maryland College Park, College of Computer, Mathematical and Physical Sciences, Department of Physics and Astronomy, Program in Astronomy, College Park, MD 20742. Program awards MS, PhD. Faculty: 21 (1 woman). Matriculated students: 33 full-time (9 women), 3 part-time (1 woman); includes 0 minority, 11 foreign. 54 applicants, 37% accepted. In 1990, 3 master's, 1 doctorate awarded. *Degree requirements:* For master's, thesis or alternative required, foreign language not required; for doctorate, variable foreign language requirement, dissertation. *Entrance requirements:* For master's, GRE Subject Test, minimum GPA of 3.0; for doctorate, GRE Subject Test. Application deadline: rolling. Application fee: $25. *Expenses:* Tuition of $143 per credit hour for state residents; $256 per credit hour for nonresidents. Fees of $171.50 per semester. *Financial aid:* In 1990–91, 5 fellowships (3 to first-year students) awarded; research assistantships, teaching assistantships, and career-related internships or fieldwork also available. *Faculty research:* Solar radio astronomy, plasma and high-energy astrophysics, galactic and extragalactic astronomy. • Dr. Roger A. Bell, Director, 301-405-1508.

University of Massachusetts at Amherst, College of Arts and Sciences, Faculty of Natural Sciences and Mathematics, Department of Physics and Astronomy, Program in Astronomy, Amherst, MA 01003. Program awards MS, PhD. Part-time programs available. Faculty: 21 full-time (2 women), 0 part-time. Matriculated students: 7 full-time (4 women), 17 part-time (4 women); includes 2 minority (1 Asian American, 1 Hispanic American), 5 foreign. Average age 28. 66 applicants, 33% accepted. In 1990, 0 master's, 9 doctorates awarded. Terminal master's awarded for partial completion of doctoral program. *Degree requirements:* For master's, foreign language and thesis not required; for doctorate, dissertation required, foreign language not required. *Entrance requirements:* GRE General Test, GRE Subject Test. Application deadline: 3/1 (applications processed on a rolling basis). Application fee: $35. *Tuition:* $2568 per year full-time, $107 per credit part-time for state residents; $7920 per year full-time, $330 per credit part-time for nonresidents. *Financial aid:* In 1990–91, 1 fellowship, 18 research assistantships, 9 teaching assistantships awarded; federal work-study also available. Aid available to part-time students. Financial aid application deadline: 3/1; applicants required to submit FAF. • Dr. James F. Walker, Director, 413-545-1310. Application contact: Dr. Robert V. Krotkov, Chair, Admissions Committee, 413-545-2191.

Directory: Astronomy

University of Michigan, College of Literature, Science, and the Arts, Department of Astronomy, Ann Arbor, MI 48109. Department awards MS, PhD. Faculty: 9 full-time, 0 part-time. Matriculated students: 19 full-time (8 women), 0 part-time; includes 1 minority (Hispanic American), 1 foreign. 27 applicants, 22% accepted. In 1990, 3 master's, 3 doctorates awarded. *Degree requirements:* For master's, foreign language and thesis not required; for doctorate, dissertation, preliminary exam required, foreign language not required. *Entrance requirements:* GRE General Test, GRE Subject Test. Application deadline: 2/1 (applications processed on a rolling basis). Application fee: $30. *Tuition:* $3255 per semester full-time, $352 per credit (minimum) part-time for state residents; $6803 per semester full-time, $746 per credit (minimum) part-time for nonresidents. *Financial aid:* Fellowships, research assistantships, teaching assistantships available. Financial aid application deadline: 3/15. • Hugh D. Aller, Chair, 313-764-3440.

University of Minnesota, Twin Cities Campus, Institute of Technology, School of Physics and Astronomy, Department of Astronomy, Minneapolis, MN 55455. Department awards MS, PhD. *Degree requirements:* For master's, thesis optional; for doctorate, dissertation. *Entrance requirements:* GRE General Test, GRE Subject Test. *Expenses:* Tuition of $1084 per quarter full-time, $301 per credit part-time for state residents; $2168 per quarter full-time, $602 per credit part-time for nonresidents. Fees of $118 per quarter. • Thomas W. Jones, Chairman, 612-624-8546.

Announcement: Unusual research facilities in department include an ultrafast Automated Plate Scanner and associated image processing system; various networked computers and workstations, including MicroVAX IIs and IIIs, several Sun 3s and 4s with data analysis and image-processing software; and, at the Minnesota Supercomputer Institute, CRAY-2 and XMP supercomputers and a state-of-the-art image processing center. Department operates 1.5m telescope on Mt. Lemmon, Arizona, and 0.75m telescope 30 miles from Minneapolis, both equipped for optical and infrared photometry, spectroscopy, and polarimetry.

University of New Mexico, College of Arts and Sciences, Department of Physics and Astronomy, Albuquerque, NM 87131. Department offers programs in optical sciences (MS, PhD), physics (MS, PhD). Faculty: 28 full-time (1 woman), 10 part-time (0 women). *Entrance requirements:* GRE General Test. Application fee: $25. *Expenses:* Tuition of $467 per semester (minimum) full-time, $67.50 per credit hour part-time for state residents; $1549 per semester (minimum) full-time, $67.50 per credit hour part-time for nonresidents. Fees of $16 per semester. *Financial aid:* Fellowships, research assistantships, teaching assistantships available. • Daniel Finley, Chairman, 505-277-2616.

Announcement: Degrees in physics and optical sciences. Research concentrations in astrophysics, infrared astronomy, radio astronomy, space physics available. Research work is conducted on campus in newly remodeled labs of the department; research is also possible at neighboring national facilities, such as Sandia National Laboratories, Phillips Laboratory, Los Alamos National Laboratories, and the Very Large Array Radio Observatory. Research sponsored by DOE, NSF, NASA, NIH, AFOSR, ONR. Twenty-six tuition-exempt, 9-month teaching assistantships with stipends of $7500 to $8000 are available. A few stipends up to $10,000 are awarded to exceptional students. Research assistantships available for advanced students.

University of North Carolina at Chapel Hill, College of Arts and Sciences, Department of Physics and Astronomy, Chapel Hill, NC 27599. Department offers program in astronomy and astrophysics (MS, PhD). Faculty: 34 full-time, 2 part-time. *Degree requirements:* For master's, comprehensive exam required, foreign language not required; for doctorate, dissertation, comprehensive exam. *Entrance requirements:* GRE General Test (minimum combined score of 1000 required), minimum GPA of 3.0. Application deadline: 2/1. Application fee: $35. *Tuition:* $621 per semester full-time for state residents; $3555 per semester full-time for nonresidents. *Financial aid:* In 1990–91, 4 fellowships, 22 research assistantships, 33 teaching assistantships awarded. • Dr. Thomas B. Clegg, Chairman, 919-962-3016.

University of Oklahoma, College of Arts and Sciences, Department of Physics and Astronomy, Norman, OK 73019. Department offers programs in astronomy (M Nat Sci, MS, PhD), engineering physics (M Nat Sci, MS, PhD), physics (M Nat Sci, MS, PhD). Part-time programs available. Faculty: 25 full-time, 1 part-time. Matriculated students: 48 full-time (5 women), 3 part-time (1 woman); includes 1 minority (Hispanic American), 17 foreign. Average age 32. 164 applicants, 21% accepted. In 1990, 2 master's awarded (50% found work related to degree, 50% continued full-time study); 1 doctorate awarded (100% entered university research/teaching). Terminal master's awarded for partial completion of doctoral program. *Degree requirements:* For master's, comprehensive exam or thesis required, foreign language not required; for doctorate, dissertation, general exam required, foreign language not required. *Entrance requirements:* For master's, GRE General Test, GRE Subject Test, TOEFL (minimum score of 550 required), previous course work in physics; for doctorate, GRE General Test, GRE Subject Test, TOEFL (minimum score of 550 required). Application deadline: 4/1 (priority date, applications processed on a rolling basis). Application fee: $10. *Expenses:* Tuition of $63 per credit hour for state residents; $192 per credit hour for nonresidents. Fees of $67.50 per semester. *Financial aid:* In 1990–91, $288,000 in aid awarded. 17 research assistantships, 19 teaching assistantships (2 to first-year students) were awarded; federal work-study, institutionally sponsored loans also available. Aid available to part-time students. Financial aid application deadline: 3/1. *Faculty research:* Experimental and theoretical atomic and molecular physics, astronomy and astrophysics, high-energy physics, solid-state physics, atmospheric physics. Total annual research budget: $80,000. • Dr. Ryan Doezema, Chairperson, 405-325-3961. Application contact: Dr. Robert F. Petry, Chair, Graduate Studies Committee, 405-325-3961.

University of Pennsylvania, School of Arts and Sciences, Graduate Group in Astronomy and Astrophysics, Philadelphia, PA 19104. Group offers programs in astronomy (MS, PhD), astrophysics (MS, PhD). Faculty: 7 (0 women). Matriculated students: 6 full-time (0 women), 1 part-time (0 women). 20 applicants, 35% accepted. In 1990, 2 master's awarded (3% entered university research/teaching); 3 doctorates awarded. Terminal master's awarded for partial completion of doctoral program. *Degree requirements:* For master's, thesis or alternative required, foreign language not required; for doctorate, 1 foreign language, dissertation. *Entrance requirements:* GRE General Test, GRE Subject Test, TOEFL, TSE. Application fee: $40. *Expenses:* Tuition of $15,619 per year full-time, $1978 per course part-time. Fees of $965 per year full-time, $112 per course part-time. *Financial aid:* Application deadline 2/1. • Dr. Kenneth Lande, Chairman, 215-898-8177.

University of Pittsburgh, Faculty of Arts and Sciences, Department of Physics and Astronomy, Program in Astronomy, Pittsburgh, PA 15260. Program awards MS, PhD. *Degree requirements:* For master's, thesis or alternative required, foreign language not required; for doctorate, dissertation required, foreign language not required. *Entrance requirements:* For master's, GRE General Test, GRE Subject Test, TOEFL, minimum QPA of 3.0; for doctorate, GRE General Test, GRE General Test, TOEFL. Application deadline: 1/30 (priority date). Application fee: $15 ($25 for foreign students). *Expenses:* Tuition of $2920 per semester full-time, $241 per credit part-time for state residents; $5840 per semester full-time, $482 per credit part-time for nonresidents. Fees of $156 per year. *Financial aid:* Fellowships, research assistantships, teaching assistantships, federal work-study, institutionally sponsored loans available. Aid available to part-time students. Financial aid application deadline: 1/30; applicants required to submit FAF. • Application contact: Raymond S. Willey, Admissions Chairman, 412-624-9041.

See full description on page 659.

University of Rochester, College of Arts and Science, Department of Physics and Astronomy, Rochester, NY 14627-0001. Department awards MA, MS, PhD. Programs offered in astronomy (PhD), physics (MA, MS, PhD). Part-time programs available. Faculty: 37 full-time, 0 part-time. Matriculated students: 121 full-time (16 women), 1 part-time (0 women); includes 2 minority (both Native American), 63 foreign. 543 applicants, 11% accepted. In 1990, 21 master's, 18 doctorates awarded. Terminal master's awarded for partial completion of doctoral program. *Degree requirements:* For master's, thesis (for some programs), comprehensive exam required, foreign language not required; for doctorate, dissertation, comprehensive exam required, foreign language not required. *Entrance requirements:* For doctorate, GRE. Application deadline: 2/15 (priority date). Application fee: $25. *Expenses:* Tuition of $473 per credit hour. Fees of $243 per year full-time. *Financial aid:* Fellowships, research assistantships, teaching assistantships available. Financial aid application deadline: 2/15. *Faculty research:* Condensed matter, biophysics, quantum optics, astrophysics, observational astronomy. • Paul Slattery, Chair, 716-275-4344.

See full description on page 665.

University of South Carolina, Graduate School, College of Science and Mathematics, Department of Physics and Astronomy, Columbia, SC 29208. Department awards IMA, MAT, MS, PhD. IMA and MAT offered in cooperation with the College of Education. Part-time programs available. Faculty: 22 full-time (0 women), 2 part-time (0 women). Matriculated students: 33 full-time (6 women), 0 part-time; includes 0 minority, 11 foreign. Average age 29. 42 applicants, 10% accepted. In 1990, 1 master's awarded (100% continued full-time study); 4 doctorates awarded (100% entered university research/teaching). Terminal master's awarded for partial completion of doctoral program. *Degree requirements:* For master's, thesis required, foreign language not required; for doctorate, 1 foreign language, dissertation. *Entrance requirements:* GRE General Test. Application deadline: 8/1 (priority date, applications processed on a rolling basis). Application fee: $25. *Tuition:* $1404 per semester full-time. *Financial aid:* In 1990–91, 1 fellowship (0 to first-year students), 7 research assistantships (0 to first-year students), 24 teaching assistantships (5 to first-year students) awarded; federal work-study also available. Aid available to part-time students. *Faculty research:* Mechanics, electron spin resonance, magnetism, intermediate energy nuclear physics, high-energy physics. • Dr. Frank Avignone, Chairman, 803-777-4983. Application contact: H. A. Farach, Director of Graduate Studies, 803-777-6407.

See full description on page 667.

University of Southern Mississippi, College of Science and Technology, Department of Physics and Astronomy, Hattiesburg, MS 39406. Department awards MS. Faculty: 6 full-time, 2 part-time. *Entrance requirements:* GRE General Test (minimum combined score of 1000 required), minimum GPA of 2.75. Application deadline: 8/9 (priority date, applications processed on a rolling basis). Application fee: $0 ($25 for foreign students). *Expenses:* Tuition of $968 per semester full-time, $93 per semester hour part-time. Fees of $12 per semester part-time for state residents; $591 per year full-time, $12 per semester part-time for nonresidents. *Financial aid:* Application deadline 3/15. *Faculty research:* Polymers, atomic physics, fluid mechanics, liquid crystals, refractory materials. • Dr. Roger Hester, Chairman, 601-266-4934.

University of Texas at Austin, Graduate School, College of Natural Sciences, Department of Astronomy, Austin, TX 78712. Department awards MA, PhD. Matriculated students: 49 full-time (7 women), 0 part-time; includes 0 minority, 22 foreign. 66 applicants, 26% accepted. In 1990, 11 master's, 5 doctorates awarded. *Entrance requirements:* GRE. Application deadline: 2/1 (priority date, applications processed on a rolling basis). Application fee: $40 ($75 for foreign students). *Tuition:* $510.30 per semester for state residents; $1806 per semester for nonresidents. *Financial aid:* Fellowships, teaching assistantships available. Financial aid application deadline: 3/1. • Dr. Gregory A. Shields, Chairman, 512-471-3304. Application contact: Dr. James Douglas, Graduate Adviser, 512-471-8434.

University of Toronto, School of Graduate Studies, Physical Sciences Division, Department of Astronomy, Toronto, ON M5S 1A1, Canada. Department awards M Sc, PhD. Faculty: 20. Matriculated students: 26 full-time (5 women), 1 part-time (0 women); includes 7 foreign. 33 applicants, 45% accepted. In 1990, 3 master's, 3 doctorates awarded. *Degree requirements:* For master's, thesis optional; for doctorate, dissertation. *Entrance requirements:* For master's, GRE. Application deadline: 4/15. Application fee: $50. *Expenses:* Tuition of $2220 per year full-time, $666 per year part-time for Canadian residents; $10,198 per year full-time, $305.05 per year part-time for nonresidents. Fees of $277.56 per year full-time, $82.73 per year part-time. *Financial aid:* Application deadline 2/1. • E. R. Seaquist, Chair, 416-978-3150.

University of Victoria, Faculty of Arts and Science, Department of Physics, Victoria, BC V8W 2Y2, Canada. Offerings include astronomy and astrophysics (M Sc, PhD). Department faculty: 21 full-time (0 women), 0 part-time. *Degree requirements:* Thesis/dissertation required, foreign language not required. *Application deadline:* 5/31 (priority date, applications processed on a rolling basis). *Application fee:* $20. *Expenses:* Tuition of $754 per semester. Fees of $23 per year. • Dr. L. P. Robertson, Chair, 604-721-7698. Application contact: Dr. J. T. Weaver, Graduate Adviser, 604-721-7768.

University of Virginia, Graduate School of Arts and Sciences, Department of Astronomy, Charlottesville, VA 22906. Department awards MA, PhD. Faculty: 15 full-time (1 woman), 0 part-time. Matriculated students: 16 full-time (1 woman), 3 part-time (0 women); includes 0 minority, 1 foreign. Average age 26. 41 applicants, 29% accepted. In 1990, 2 master's, 2 doctorates awarded. *Degree requirements:* For master's, 1 foreign language, thesis; for doctorate, variable foreign language requirement, dissertation. *Entrance requirements:* GRE General Test, GRE Subject Test. Application deadline: 7/15. Application fee: $40. *Expenses:* Tuition of $2740 per year full-time, $904 per year (minimum) part-time for state residents; $8950 per year full-time, $2960 per year (minimum) part-time for nonresidents. Fees of $586 per year full-time, $342 per year part-time. *Financial aid:* Application deadline 2/1.

• Roger A. Chevalier, Chairman, 804-924-7494. Application contact: William A. Elwood, Associate Dean, 804-924-7184.

Announcement: Specializations include stellar evolution, supernovae, X-ray sources, stellar populations, active galaxy nuclei, cosmology, radio astronomy, space astronomy, and astrometry. Department operates 2 local observatories and 1 in Canberra, Australia, for astrometric research. Virginia Institute for Theoretical Astronomy is affiliated with the department. National Radio Astronomy Observatory staff scientists hold adjunct professorships and may supervise student research.

University of Washington, College of Arts and Sciences, Department of Astronomy, Seattle, WA 98195. Department awards MS, PhD. Part-time programs available. Faculty: 13 full-time (2 women), 1 part-time (0 women). Matriculated students: 22 full-time (3 women), 1 part-time (0 women); includes 4 minority (1 black American, 2 Hispanic American, 1 Native American), 1 foreign. Average age 29. 150 applicants, 7% accepted. In 1990, 2 master's awarded (100% entered university research/teaching); 2 doctorates awarded (100% entered university research/teaching). *Degree requirements:* For master's, foreign language and thesis not required; for doctorate, 1 foreign language, dissertation. *Entrance requirements:* GRE General Test, GRE Subject Test. Application deadline: 2/1. Application fee: $35. *Tuition:* $1129 per quarter full-time, $324 per credit (minimum) part-time for state residents; $2824 per quarter full-time, $809 per credit (minimum) part-time for nonresidents. *Financial aid:* In 1990–91, 7 research assistantships (1 to a first-year student), 10 teaching assistantships (3 to first-year students), 3 traineeships awarded; fellowships, federal work-study also available. *Faculty research:* Solar system dust, space astronomy, high-energy astrophysics, galactic and extragalactic astronomy, stellar astrophysics. Total annual research budget: $2-million. • Dr. Bruce Margon, Chairman, 206-543-2888.

University of Western Ontario, Physical Sciences Division, Department of Astronomy, London, ON N6A 3K7, Canada. Department awards M Sc, PhD. *Degree requirements:* For master's, thesis, written comprehensive exam required, foreign language not required; for doctorate, 1 foreign language, dissertation, written and oral comprehensive exams. *Entrance requirements:* For master's, minimum B average; for doctorate, minimum B average or M Sc. *Tuition:* $1015 per year full-time, $1050 per year part-time for Canadian residents; $4207 per year for nonresidents.

University of Wisconsin–Madison, College of Letters and Science, Department of Astronomy, Madison, WI 53706. Department awards MS, PhD. Faculty: 13 full-time, 0 part-time. Matriculated students: 17 full-time (2 women), 0 part-time; includes 0 minority, 5 foreign. 58 applicants, 14% accepted. In 1990, 0 master's, 2 doctorates awarded. *Application deadline:* rolling. *Application fee:* $20. *Financial aid:* In 1990–91, 17 students received aid. 0 fellowships, 12 research assistantships, 4 teaching assistantships, 1 project assistantship awarded; institutionally sponsored loans also available. Financial aid application deadline: 1/15; applicants required to submit FAF. • Edward B. Churchwell, Chairperson, 608-262-3071. Application contact: Joanne Nagy, Assistant Dean of the Graduate School, 608-262-2433.

University of Wyoming, College of Arts and Sciences, Department of Physics and Astronomy, Laramie, WY 82071. Department awards MS, MST, PhD. Faculty: 19 full-time (0 women), 1 part-time (0 women). Matriculated students: 24 full-time (5 women), 7 part-time (1 woman); includes 2 foreign. Average age 28. 150 applicants, 3% accepted. In 1990, 3 master's, 0 doctorates awarded. *Degree requirements:* Foreign language not required. *Entrance requirements:* GRE General Test, GRE Subject Test, minimum GPA of 3.0. Application deadline: 6/1 (priority date, applications processed on a rolling basis). Application fee: $30. *Tuition:* $1554 per year full-time, $74.25 per credit hour part-time for state residents; $4358 per year full-time, $74.25 per credit hour part-time for nonresidents. *Financial aid:* In 1990–91, 24 students received a total of $160,272 in aid awarded. 8 research assistantships (0 to first-year students), 15 teaching assistantships (4 to first-year students) were awarded; institutionally sponsored loans also available. Average monthly stipend for a graduate assistantship: $753. Financial aid application deadline: 4/15. *Total annual research budget:* $1.553-million. • Dr. Glen Rebka, Head, 307-766-6150.

Utah State University, College of Science, Department of Physics, Logan, UT 84322. Offerings include physics (MS, PhD), with options in astronomy and astrophysics (MS), earth and planetary sciences (PhD), engineering physics (PhD), physics (MS). Department faculty: 22 full-time (0 women), 3 part-time (0 women). *Degree requirements:* Thesis required, foreign language not required. *Entrance requirements:* GRE General Test (score in 40th percentile or higher required), minimum GPA of 3.0. Application deadline: 7/15 (priority date, applications processed on a rolling basis). Application fee: $25 ($30 for foreign students). *Tuition:* $426 per quarter (minimum) full-time, $184 per quarter (minimum) part-time for state residents; $1133 per quarter (minimum) full-time, $505 per quarter (minimum) part-time for nonresidents. • Dr. W. John Raitt, Head, 801-750-2848. Application contact: O. Harry Otteson, Assistant Head, 801-750-2850.

Vanderbilt University, Department of Physics, Program in Astronomy, Nashville, TN 37240. Program awards MA, MS. Faculty: 4 full-time (0 women), 0 part-time. Matriculated students: 2 full-time (0 women), 0 part-time; includes 0 minority, 0 foreign. Average age 23. 5 applicants, 40% accepted. In 1990, 0 degrees awarded. *Degree requirements:* 1 foreign language, thesis. *Entrance requirements:* GRE General Test. Application deadline: 1/15. Application fee: $25. *Expenses:* Tuition of $624 per semester hour. Fees of $196 per year. *Financial aid:* Fellowships, research assistantships, teaching assistantships, federal work-study, institutionally sponsored loans available. Financial aid application deadline: 1/15. • Douglas S. Hall, Director, 615-322-2455.

Wesleyan University, Department of Astronomy, Middletown, CT 06459. Department awards MA. Part-time programs available. Faculty: 2 full-time (0 women), 4 part-time (0 women). Matriculated students: 4 full-time (1 woman), 0 part-time; includes 1 minority, 1 foreign. Average age 25. 1 applicants, 100% accepted. In 1990, 0 degrees awarded. *Degree requirements:* Thesis required, foreign language not required. *Entrance requirements:* GRE General Test, GRE Subject Test. Application deadline: 3/1. Application fee: $0. *Expenses:* Tuition of $2035 per course. Fees of $627 per year full-time. *Financial aid:* Full and partial tuition waivers available. Financial aid application deadline: 4/15. *Faculty research:* Observational-theoretical astronomy and astrophysics. • Arthur R. Upqren, Chairman, 203-347-9411 Ext. 2829.

West Chester University of Pennsylvania, College of Arts and Sciences, Department of Geology and Astronomy, West Chester, PA 19383. Department offers program in physical science (MA). Faculty: 0 full-time, 4 part-time. Matriculated students: 5 full-time (3 women), 20 part-time (7 women); includes 3 minority (2 Asian American, 1 black American), 0 foreign. Average age 32. 8 applicants, 75% accepted. In 1990, 4 degrees awarded. *Degree requirements:* Comprehensive exam required, foreign language and thesis not required. *Application deadline:* rolling. *Application fee:* $20. *Tuition:* $127 per credit for state residents; $160 per credit for nonresidents. *Financial aid:* In 1990–91, 1 research assistantship awarded. Aid available to part-time students. Financial aid application deadline: 4/1. • Dr. Allen Johnson, Chair, 215-436-2727. Application contact: Dr. Sandra Pritchard, Graduate Coordinator, 215-436-2721.

Yale University, Graduate School of Arts and Sciences, Department of Astronomy, New Haven, CT 06520. Department awards MS, PhD. Faculty: 8 full-time (1 woman), 0 part-time. Matriculated students: 16 full-time (4 women), 0 part-time; includes 0 minority, 7 foreign. 18 applicants, 22% accepted. In 1990, 6 master's, 2 doctorates awarded. *Degree requirements:* For doctorate, dissertation required, foreign language not required. *Entrance requirements:* For doctorate, GRE General Test, GRE Subject Test. Application deadline: 1/2. Application fee: $45. *Tuition:* $15,160 per year full-time, $1895 per course part-time. • Dr. Augustus Oemler, Chairman, 203-432-3000. Application contact: Susan Webb, Coordinator of Admissions, 203-432-2770.

York University, Faculty of Science, Program in Physics and Astronomy, North York, ON M3J 1P3, Canada. Program awards M Sc, PhD. Part-time and evening/weekend programs available. Faculty: 36 full-time (2 women), 8 part-time (1 woman). Matriculated students: 37 full-time (5 women), 5 part-time (0 women); includes 3 foreign. 141 applicants, 9% accepted. In 1990, 2 master's, 3 doctorates awarded. *Degree requirements:* For master's, thesis optional, foreign language not required; for doctorate, dissertation required, foreign language not required. *Application fee:* $35. *Tuition:* $2436 per year for Canadian residents; $11,480 per year for nonresidents. *Financial aid:* In 1990–91, $531,339 in aid awarded. 2 fellowships, 46 research assistantships, 40 teaching assistantships were awarded. • Dr. R. P. McEachran, Director, 416-736-2100.

Astrophysics

Bowling Green State University, College of Arts and Sciences, Department of Physics and Astronomy, Bowling Green, OH 43403. Offerings include astrophysics (MS). Department faculty: 9 full-time (0 women), 0 part-time. *Application deadline:* 8/15. *Application fee:* $10. *Tuition:* $185 per credit hour for state residents; $357 per credit hour for nonresidents. • Dr. Robert I. Boughton, Chair, 419-372-2421. Application contact: Dr. Lewis Fulcher, Graduate Coordinator, 419-372-2635.

California Institute of Technology, Division of Physics, Mathematics and Astronomy, Department of Astronomy, Pasadena, CA 91125. Department awards PhD. Faculty: 17 full-time (1 woman), 0 part-time. Matriculated students: 24 full-time (4 women), 0 part-time; includes 0 minority, 11 foreign. Average age 25. 60 applicants, 8% accepted. In 1990, 4 doctorates awarded. Terminal master's awarded for partial completion of doctoral program. *Degree requirements:* For doctorate, 1 foreign language, dissertation, candidacy and final exams. *Entrance requirements:* For doctorate, GRE General Test, GRE Subject Test, TOEFL. Application deadline: 1/15. Application fee: $0. *Tuition:* $14,100 per year. *Financial aid:* In 1990–91, 24 students received a total of $600,000 in aid awarded. 4 fellowships (all to first-year students), 4 teaching assistantships (0 to first-year students), 3 outside awards (1 to a first-year student) were awarded; research assistantships, federal work-study, institutionally sponsored loans also available. Financial aid application deadline: 1/15. *Faculty research:* Observational and theoretical astrophysics, cosmology, radio astronomy, solar physics. Total annual research budget: $10.5-million. • A. C. R. Readhead, Executive Officer, 818-356-4972. Application contact: S. G. Djorgovski, Option Representative, 818-356-4415.

Clemson University, College of Sciences, Department of Physics and Astronomy, Clemson, SC 29634. Department offers program in physics (MS, PhD). Faculty: 23 full-time (0 women), 0 part-time. Matriculated students: 45 full-time (4 women), 0 part-time; includes 2 minority (1 black American, 1 Hispanic American), 10 foreign. Average age 27. 92 applicants, 26% accepted. In 1990, 4 master's awarded (25% entered university research/teaching, 50% found other work related to degree, 25% continued full-time study); 3 doctorates awarded (67% entered university research/teaching, 33% found other work related to degree). Terminal master's awarded for partial completion of doctoral program. *Degree requirements:* For master's, thesis required (for some programs), foreign language not required; for doctorate, dissertation required, foreign language not required. *Entrance requirements:* GRE General Test, TOEFL. Application deadline: 6/1 (priority date, applications processed on a rolling basis). Application fee: $25. *Expenses:* Tuition of $102 per credit hour. Fees of $80 per semester full-time. *Financial aid:* In 1990–91, 40 students received a total of $421,400 in aid awarded. 2 fellowships (both to first-year students), 15 research assistantships (1 to a first-year student), 23 teaching assistantships (13 to first-year students) were awarded. Average monthly stipend for a graduate assistantship: $877. Financial aid application deadline: 6/1. *Faculty research:* Astrophysics, atmosphere physics, condensed matter, radiation physics, solid state physics. Total annual research budget: $457,126. • Dr. P. J. McNulty, Head, 803-656-3416. Application contact: J. A. Gilreath, Graduate Student Adviser, 803-656-3416.

See full description on page 581.

Harvard University, Graduate School of Arts and Sciences, Department of Astronomy, Cambridge, MA 02138. Offerings include astrophysics (AM, PhD). Terminal master's awarded for partial completion of doctoral program. Department

Directory: Astrophysics

faculty: 21 full-time, 16 part-time. *Degree requirements:* For doctorate, dissertation, research project, paper required, foreign language not required. *Entrance requirements:* For doctorate, GRE General Test, GRE Subject Test. Application deadline: 1/2. Application fee: $60. *Expenses:* Tuition of $14,860 per year. Fees of $550 per year. • Dr. Mark Birkinshaw, Director of Graduate Study, 617-495-3752. Application contact: Office of Admissions and Financial Aid, 617-495-5315.

Indiana University Bloomington, College of Arts and Sciences, Department of Astronomy, Program in Astrophysics, Bloomington, IN 47405. Program awards PhD. Offered jointly with the Department of Physics. Faculty: 0 full-time, 9 part-time (1 woman). Matriculated students: 10 full-time (1 woman), 0 part-time; includes 0 minority, 1 foreign. In 1990, 1 degree awarded (100% found work related to degree). *Degree requirements:* Dissertation, written qualifying exams required, foreign language not required. *Entrance requirements:* GRE General Test, GRE Subject Test (physics), BA or BS in a science. Application deadline: 2/1. Application fee: $25 ($35 for foreign students). *Tuition:* $99.85 per credit hour for state residents; $288 per credit hour for nonresidents. *Financial aid:* In 1990–91, 4 research assistantships (0 to first-year students), 5 teaching assistantships (1 to a first-year student) awarded. *Faculty research:* Nuclear astrophysics, cosmic-ray physics, astrophysical fluid dynamics, active galactic nuclei, high-energy astrophysics. • Dr. Stuart Mufson, Chairperson, 812-855-6911.

See full description on page 589.

Louisiana State University and Agricultural and Mechanical College, College of Basic Sciences, Department of Physics and Astronomy, Baton Rouge, LA 70803. Offerings include astrophysics (PhD). Terminal master's awarded for partial completion of doctoral program. Department faculty: 35 full-time (0 women), 0 part-time. *Degree requirements:* Dissertation required, foreign language not required. *Entrance requirements:* GRE General Test, TOEFL (minimum score of 525 required), minimum GPA of 3.0. Application deadline: 7/1 (priority date, applications processed on a rolling basis). Application fee: $25. *Tuition:* $1020 per semester full-time for state residents; $2620 per semester full-time for nonresidents. • Dr. Jerry P. Draayer, Chair, 504-388-2261. Application contact: Dr. Michael Cherry, Chair, Assistantship Committee, 504-388-1194.

See full description on page 595.

Michigan State University, College of Natural Science, Department of Physics and Astronomy, East Lansing, MI 48824. Department offers program in physics (MAT, MS, PhD). Faculty: 41 full-time (1 woman), 0 part-time. Matriculated students: 170 full-time (20 women), 17 part-time (7 women); includes 4 minority (1 Asian American, 2 black American, 1 Hispanic American), 88 foreign. In 1990, 36 master's, 9 doctorates awarded. Terminal master's awarded for partial completion of doctoral program. *Degree requirements:* For master's, thesis or alternative required, foreign language not required; for doctorate, dissertation required, foreign language not required. *Application deadline:* 3/15 (applications processed on a rolling basis). *Application fee:* $25 ($40 for foreign students). *Tuition:* $104.75 per credit for state residents; $211.75 per credit for nonresidents. *Financial aid:* In 1990–91, 1 fellowship, 65 research assistantships, 41 teaching assistantships awarded. *Faculty research:* Nuclear and accelerator physics, high-energy physics, condensed-matter physics, astronomy and astrophysics. Total annual research budget: $23-million. • Dr. Gerard Crawley, Chairperson, 517-353-8662.

See full description on page 597.

New Mexico Institute of Mining and Technology, Department of Physics, Socorro, NM 87801. Offerings include physics (MS, PhD), with options in astrophysics (MS, PhD), atmospheric physics (MS, PhD), instrumentation (MS), mathematical physics (PhD). Department faculty: 15 full-time (1 woman), 1 part-time (0 women). *Degree requirements:* For master's, thesis required, foreign language not required; for doctorate, 1 foreign language, dissertation. *Entrance requirements:* For master's, GRE General Test, TOEFL (minimum score of 540 required); for doctorate, GRE General Test, GRE Subject Test, TOEFL (minimum score of 540 required). Application deadline: 6/1 (priority date, applications processed on a rolling basis). Application fee: $16. *Expenses:* Tuition of $617 per semester full-time, $51 per hour part-time for state residents; $2656 per semester full-time, $209 per hour part-time for nonresidents. Fees of $207 per semester. • Dr. David Raymond, Chairman, 505-835-5610. Application contact: Dr. J. A. Smoake, Dean, Graduate Studies, 505-835-5513.

Northwestern University, College of Arts and Sciences, Department of Physics and Astronomy, Evanston, IL 60208. Offerings include astrophysics (MS, PhD). Terminal master's awarded for partial completion of doctoral program. Department faculty: 34 full-time (2 women), 0 part-time. *Degree requirements:* For master's, thesis or alternative, comprehensive exam required, foreign language not required; for doctorate, dissertation, qualifying exam, oral exam required, foreign language not required. *Entrance requirements:* For master's, GRE General Test; for doctorate, GRE General Test, GRE Subject Test (physics). Application deadline: 8/30. Application fee: $40 ($45 for foreign students). *Tuition:* $4665 per quarter full-time, $1704 per course part-time. • Dr. William P. Halpern, Chairperson, 708-491-5468. Application contact: David Taylor, Assistant Chairperson, 708-491-2053.

See full description on page 603.

Pennsylvania State University University Park Campus, College of Science, Department of Astronomy and Astrophysics, University Park, PA 16802. Department awards MS, PhD. Faculty: 12. Matriculated students: 22 full-time (4 women), 0 part-time. In 1990, 0 master's, 4 doctorates awarded. *Entrance requirements:* GRE General Test. Application fee: $35. *Tuition:* $203 per credit for state residents; $403 per credit for nonresidents. • France A. Cordova, Head, 814-865-0418.

Princeton University, Department of Astrophysical Sciences, Princeton, NJ 08544. Department offers programs in astrophysical sciences (PhD), plasma physics (PhD). Faculty: 29 full-time (2 women), 4 part-time (0 women). Matriculated students: 50 full-time (6 women), 0 part-time; includes 2 minority (1 Asian American, 1 black American), 31 foreign. 67 applicants, 30% accepted. In 1990, 8 doctorates awarded (63% entered university research/teaching, 37% found other work related to degree). Terminal master's awarded for partial completion of doctoral program. *Degree requirements:* For doctorate, dissertation. *Entrance requirements:* For doctorate, GRE General Test, GRE Subject Test. Application deadline: 1/8. Application fee: $45 ($50 for foreign students). *Tuition:* $16,670 per year. *Financial aid:* Fellowships, research assistantships, teaching assistantships, federal work-study, institutionally sponsored loans available. Financial aid application deadline: 1/8. *Faculty research:* Theoretical astrophysics, cosmology, galaxy formation, galactic dynamics, interstellar and intergalactic matter. • Gillian Knapp, Director of Graduate Studies, 609-258-3824. Application contact: Michele Spreen, Director of Graduate Admissions, 609-258-3034.

Rice University, Wiess School of Natural Sciences, Department of Space Physics and Astronomy, Houston, TX 77251. Department offers program in space physics and astronomy (MS, PhD). Faculty: 14 full-time, 0 part-time. Matriculated students: 41 full-time (8 women), 0 part-time; includes 2 minority (both Asian American), 8 foreign. *Degree requirements:* Thesis/dissertation required, foreign language not required. *Entrance requirements:* For master's, GRE General Test, TOEFL (minimum score 0f 550 required), minimum GPA of 3.0; for doctorate, GRE General Test (score in 70th percentile or higher required), minimum GPA of 3.0. Application deadline: 3/1. Application fee: $0. *Expenses:* Tuition of $8300 per year full-time, $400 per credit hour part-time. Fees of $167 per year. *Financial aid:* Fellowships, research assistantships, full and partial tuition waivers available. Financial aid applicants required to submit FAF. *Faculty research:* Magnetospheric physics, planetary atmospheres, astrophysics. • Alexander J. Dessler, Chairman, 713-527-4939.

San Francisco State University, School of Science, Department of Physics and Astronomy, San Francisco, CA 94132. Department offers program in physics and astrophysics (MS). *Degree requirements:* Thesis or alternative. *Application deadline:* 11/30 (priority date, applications processed on a rolling basis). *Application fee:* $55. *Expenses:* Tuition of $0 for state residents; $250 per unit for nonresidents. Fees of $950 per year full-time, $350 per semester part-time for state residents; $1050 per year full-time, $350 per semester part-time for nonresidents. *Financial aid:* Research assistantships, teaching assistantships available. Financial aid application deadline: 3/1. *Faculty research:* Quark search, thin-films, dark matter detection, search for planetary systems, low temperature. • Dr. Gerald Fisher, Chair, 415-338-1659. Application contact: Dr. Oliver Johns, Graduate Coordinator, 415-338-1691.

State University of New York at Stony Brook, College of Arts and Sciences, Division of Physical Sciences and Mathematics, Department of Earth and Space Sciences, Program in Astronomical Sciences, Stony Brook, NY 11794. Offerings include astrophysics (M Phil, MS, PhD). Program faculty: 8 full-time. *Degree requirements:* For master's, thesis or alternative required, foreign language not required; for doctorate, dissertation required, foreign language not required. *Entrance requirements:* GRE General Test, TOEFL, minimum GPA of 3.0. Application deadline: 2/1. Application fee: $35. *Expenses:* Tuition of $2450 per year full-time, $103 per credit part-time for state residents; $5766 per year full-time, $243 per credit part-time for nonresidents. Fees of $151 per year full-time, $10.45 per year (minimum) part-time. • Dr. Gilbert Hansen, Chairman, Department of Earth and Space Sciences, 516-632-8200.

University of Alaska Fairbanks, College of Natural Sciences, Department of Physics, Fairbanks, AK 99775. Offerings include space physics (MS, PhD). Terminal master's awarded for partial completion of doctoral program. Department faculty: 25 full-time (1 woman), 0 part-time. *Degree requirements:* For master's, thesis; for doctorate, 1 foreign language (computer language can substitute), dissertation. *Entrance requirements:* For master's, GRE General Test, GRE Subject Test (minimum score of 550 required); for doctorate, GRE General Test, GRE Subject Test. Application deadline: 2/15. Application fee: $20. *Expenses:* Tuition of $1620 per year full-time, $90 per credit part-time for state residents; $3240 per year full-time, $180 per credit part-time for nonresidents. Fees of $464 per year full-time. • Dr. John Morack, Head, 907-474-7339.

University of California, Los Angeles, College of Letters and Science, Department of Earth and Space Sciences, Los Angeles, CA 90024. Department offers programs in geochemistry (MS, PhD, C Phil), geology (MS, PhD, C Phil), geophysics and space physics (MS, PhD). Faculty: 23. Matriculated students: 75 full-time (18 women), 0 part-time; includes 8 minority, 14 foreign. 120 applicants, 28% accepted. In 1990, 5 master's, 14 doctorates, 2 C Phils awarded. *Degree requirements:* For master's, thesis or comprehensive exams required, foreign language not required; for doctorate, dissertation, written and oral qualifying exams required, foreign language not required. *Entrance requirements:* For master's and doctorate, GRE General Test. Application fee: $40. *Expenses:* Tuition of $0 for state residents; $7699 per year for nonresidents. Fees of $2907 per year. *Financial aid:* In 1990–91, 76 students received a total of $1.04-million in aid awarded. 68 fellowships, 50 research assistantships, 37 teaching assistantships were awarded; full and partial tuition waivers, federal work-study, institutionally sponsored loans also available. Financial aid application deadline: 3/1. • Dr. Arthur Montana, Chair, 310-825-3880.

See full description on page 255.

University of California, Santa Cruz, Division of Natural Sciences, Program in Astronomy and Astrophysics, Santa Cruz, CA 95064. Program awards PhD. Faculty: 20 full-time (2 women), 0 part-time. Matriculated students: 29 full-time (8 women), 0 part-time; includes 2 minority (both Asian American), 4 foreign. 85 applicants, 25% accepted. In 1990, 5 doctorates awarded. Terminal master's awarded for partial completion of doctoral program. *Degree requirements:* For doctorate, 1 foreign language (computer language can substitute), dissertation, qualifying exam. *Entrance requirements:* For doctorate, GRE General Test, GRE Subject Test. Application deadline: 2/1. Application fee: $40. *Expenses:* Tuition of $0 for state residents; $7699 per year for nonresidents. Fees of $3021 per year. *Financial aid:* In 1990–91, 3 fellowships, 16 research assistantships, 19 teaching assistantships awarded; federal work-study, institutionally sponsored loans also available. Average monthly stipend for a graduate assistantship: $1335. Financial aid applicants required to submit GAPSFAS. *Faculty research:* Stellar structure and evolution, stellar spectroscopy, the interstellar medium, galactic structure, external galaxies and quasars. • Dr. Stan Woosley, Chairperson, 408-459-2844.

University of Chicago, Division of the Physical Sciences, Department of Astronomy and Astrophysics, Chicago, IL 60637. Department awards SM, PhD. *Degree requirements:* For master's, foreign language and thesis not required; for doctorate, dissertation required, foreign language not required. *Entrance requirements:* GRE General Test, GRE Subject Test, TOEFL. Application deadline: 1/6. Application fee: $45. *Expenses:* Tuition of $16,275 per year full-time, $8140 per year part-time. Fees of $356 per year.

University of Colorado at Boulder, College of Arts and Sciences, Department of Astrophysical, Planetary, and Atmospheric Sciences, Boulder, CO 80309. Offerings include astrophysics (MS, PhD). Terminal master's awarded for partial completion of doctoral program. Department faculty: 72 full-time (4 women). *Degree requirements:* For master's, thesis or alternative, comprehensive exam required, foreign language not required; for doctorate, 1 foreign language, dissertation. *Entrance requirements:* GRE General Test, GRE Subject Test. Application deadline: 3/1 (priority date, applications processed on a rolling basis). Application fee: $30 ($50 for foreign students). *Expenses:* Tuition of $2308 per year full-time, $387 per semester (minimum) part-time for state residents; $8730 per year full-time, $1455 per semester (minimum) part-time for nonresidents. Fees of $207 per semester full-time, $27.26 per semester (minimum) part-time. • Ellen Zweibel, Chairman, 303-492-8915.

University of New Mexico, College of Arts and Sciences, Center for Advanced Studies, Albuquerque, NM 87131. Center awards MS, PhD. *Application fee:* $25. *Expenses:* Tuition of $467 per semester (minimum) full-time, $67.50 per credit hour part-time for state residents; $1549 per semester (minimum) full-time, $67.50 per credit hour part-time for nonresidents. Fees of $16 per semester. • Dr. Marlan Scully, Director, 505-277-2616.

Directory: Astrophysics

University of North Carolina at Chapel Hill, College of Arts and Sciences, Department of Physics and Astronomy, Chapel Hill, NC 27599. Offerings include astronomy and astrophysics (MS, PhD). Department faculty: 34 full-time, 2 part-time. *Degree requirements:* For master's, comprehensive exam required, foreign language not required; for doctorate, dissertation, comprehensive exam. *Entrance requirements:* GRE General Test (minimum combined score of 1000 required), minimum GPA of 3.0. Application deadline: 2/1. Application fee: $35. *Tuition:* $621 per semester full-time for state residents; $3555 per semester full-time for nonresidents. • Dr. Thomas B. Clegg, Chairman, 919-962-3016.

University of Pennsylvania, School of Arts and Sciences, Graduate Group in Astronomy and Astrophysics, Philadelphia, PA 19104. Group offers programs in astronomy (MS, PhD), astrophysics (MS, PhD). Faculty: 7 (0 women). Matriculated students: 6 full-time (0 women), 1 part-time (0 women). 20 applicants, 35% accepted. In 1990, 2 master's awarded (3% entered university research/teaching); 3 doctorates awarded. Terminal master's awarded for partial completion of doctoral program. *Degree requirements:* For master's, thesis or alternative required, foreign language not required; for doctorate, 1 foreign language, dissertation. *Entrance requirements:* GRE General Test, GRE Subject Test, TOEFL, TSE. Application fee: $40. *Expenses:* Tuition of $15,619 per year full-time, $1978 per course part-time. Fees of $965 per year full-time, $112 per course part-time. *Financial aid:* Application deadline 2/1. • Dr. Kenneth Lande, Chairman, 215-898-8177.

University of Victoria, Faculty of Arts and Science, Department of Physics, Victoria, BC V8W 2Y2, Canada. Offerings include astronomy and astrophysics (M Sc, PhD). Department faculty: 21 full-time (0 women), 0 part-time. *Degree requirements:* Thesis/dissertation required, foreign language not required. *Application deadline:* 5/31 (priority date, applications processed on a rolling basis). *Application fee:* $20. *Expenses:* Tuition of $754 per semester. Fees of $23 per year. • Dr. L. P. Robertson, Chair, 604-721-7698. Application contact: Dr. J. T. Weaver, Graduate Adviser, 604-721-7768.

Utah State University, College of Science, Department of Physics, Logan, UT 84322. Offerings include physics (MS, PhD), with options in astronomy and astrophysics (MS), earth and planetary sciences (PhD), engineering physics (PhD), physics (MS). Department faculty: 22 full-time (0 women), 3 part-time (0 women). *Degree requirements:* Thesis required, foreign language not required. *Entrance requirements:* GRE General Test (score in 40th percentile or higher required), minimum GPA of 3.0. Application deadline: 7/15 (priority date, applications processed on a rolling basis). Application fee: $25 ($30 for foreign students). *Tuition:* $426 per quarter (minimum) full-time, $184 per quarter (minimum) part-time for state residents; $1133 per quarter (minimum) full-time, $505 per quarter (minimum) part-time for nonresidents. • Dr. W. John Raitt, Head, 801-750-2848. Application contact: O. Harry Otteson, Assistant Head, 801-750-2850.

Section 2
Chemistry

This section contains a directory of institutions offering graduate work in analytical chemistry, chemistry, inorganic chemistry, organic chemistry, and physical chemistry, followed by two-page entries submitted by institutions that chose to prepare detailed program descriptions. Additional information about programs listed in the directory but not augmented by a two-page entry may be obtained by writing directly to the dean of a graduate school or chair of a department at the address given in the directory.

For programs offering related work, see also in this book: Earth and Planetary Sciences and Physics; in Book 3: Biology and Biomedical Sciences; Biochemistry; Biophysics; Nutrition; and Pharmacology and Toxicology; in Book 5: Engineering and Applied Sciences; Agricultural Engineering; Chemical Engineering; Geological, Mineral/Mining, and Petroleum Engineering; and Materials Sciences and Engineering; and in Book 6: Pharmacy and Pharmaceutical Sciences.

CONTENTS

Program Directories

Analytical Chemistry	42
Chemistry	45
Inorganic Chemistry	71
Organic Chemistry	75
Physical Chemistry	79

Announcements

Boston University	45
California State University, Fullerton	46
California State University, Los Angeles	47
Catholic University of America	47
DePaul University	49
Institute of Paper Science and Technology	51
North Carolina Central University	54
Portland State University	56
Rutgers, The State University of New Jersey, Newark	57
Seton Hall University	57
University of Kansas	77
University of Louisville	64
University of Maine	64
University of Missouri-St. Louis	65
University of Oklahoma	66
University of Virginia	69
Virginia Polytechnic Institute and State University	70
Western Kentucky University	70

Cross-Discipline Announcements

Columbia University	83
Duke University	83
Harvard University	83
Johns Hopkins University	83
Louisiana State University	83
Massachusetts Institute of Technology	83
Northwestern University	83
Ohio State University	83
Purdue University	83
Rush University	83
University of California, San Francisco	83
University of Illinois at Urbana-Champaign	83
University of Kansas	83
University of Kentucky	84
University of Massachusetts at Amherst	84
University of Michigan	84
University of Notre Dame	84
University of Virginia	
Biochemistry	84
Biophysics	84
University of Washington	
Bioengineering	84
Materials Science and Engineering	84
University of Wisconsin-Madison	
Biophysics	84
Biotechnology	84
Pharmacy	84
Virginia Commonwealth University	
Biochemistry and Molecular Biophysics	84
Biomedical Engineering	84

Full Descriptions

Arizona State University	85
Baylor University	87
Boston College	89
Boston University	91
Brandeis University	93
Brigham Young University	95
Carnegie Mellon University	97
Case Western Reserve University	99
Columbia University	101
Eastern Michigan University	103
Florida State University	105
Indiana University-Purdue University at Indianapolis	107
Institute of Paper Science and Technology	109
Kent State University	111
Lehigh University	113
Massachusetts Institute of Technology	115
New York University	117
Northeastern University	119
Ohio State University	121
Oregon State University	123
Princeton University	125
Rensselaer Polytechnic Institute	127
Rutgers, The State University of New Jersey, Newark	129
Rutgers, The State University of New Jersey, New Brunswick	131
Stanford University	133
State University of New York at Stony Brook	135
State University of New York College of Environmental Science and Forestry	137
Stevens Institute of Technology	139
Texas A&M University	141
University of Arizona	143
University of Florida	145
University of Georgia	147
University of Houston	149
University of Illinois at Chicago	151
University of Maryland College Park	
Chemical Physics	153
Chemistry and Biochemistry	155
University of Maryland Graduate School, Baltimore	157
University of Miami	159
University of Michigan	161
University of Minnesota, Twin Cities Campus	163
University of Nebraska-Lincoln	165
University of Pittsburgh	167
University of Puerto Rico, Río Piedras	169
University of San Francisco	171
University of South Florida	173
University of Texas at Austin	175
Utah State University	177
Wesleyan University	179
Yale University	181

See also:

Drexel University—Chemistry, Mathematics and Computer Science, and Physics and Atmospheric Sciences	409

Peterson's Guide to Graduate Programs in the Physical Sciences and Mathematics 1992

39

SECTION 2: CHEMISTRY

Field Definitions

FIELD DEFINITIONS

In an effort to broaden prospective students' understanding of disciplines in chemistry, educators in the field have submitted the following statements on analytical chemistry, chemistry, inorganic chemistry, organic chemistry, and physical chemistry.

Analytical Chemistry

The primary purpose of analytical chemistry is to determine the constituents present in matter. If only the identification of constituents is desired, a qualitative analysis is performed; a quantitative analysis ascertains the amount of the constituent present in the sample. The field of analytical chemistry requires a fundamental knowledge of all areas of chemistry as well as the ability to use mathematics and think logically.

The work of the analytical chemist is broad, since analysis may be required of virtually any material. The amount of the sample or the constituent may be large or small, often very small. For example, it is not unusual to analyze for nanogram amounts of a substance using analytical instruments such as a high-performance liquid chromatograph or an atomic absorption spectrophotometer. The work can include analysis of metallurgical materials, plastics, hydrocarbons, cements, ceramics, fuels, foods, animal feed—practically anything known to mankind.

Analytical chemists are employed in a wide range of settings, including manufacturing facilities, industrial laboratories, hospital laboratories, food-testing facilities, agricultural laboratories, research centers, testing laboratories, pharmaceutical production facilities, forensic laboratories, government facilities, and academic institutions, and they are in very high demand. In an industrial setting, he or she is often responsible for keeping production going and maintaining quality control.

The individual analytical chemist may address questions of considerable import. Are accelerants present in the debris of a fire of suspicious origin? Is there asbestos present in a building (this can change the value of commercial property by millions of dollars)? Does a grain shipment meet export criteria? What is the level of dopants in a particular semiconductor? Does the air at a certain location meet ambient air-quality standards? What is the glucose level in a blood serum sample? Does a certain load of gasoline have the proper hydrocarbon blend?

Analytical chemists do work in laboratories, but they may also be out in the field supervising sample collection or checking other facilities. They develop new methods of analysis, supervise laboratories, work in management, serve as consultants, or own their own businesses. They may design new instruments or be employed in instrument or chemical sales. The work of the analytical chemist may be theoretical, practical, or a mixture of both.

Graduate education, particularly through the Ph.D., is highly desirable, as the complexity of the work mandates a strong background in chemistry. Occupational advancement depends not only on educational qualifications but also on the personality and ability of the individual. Graduate-level training can be very rewarding, both intellectually and financially. The field offers many challenges and a wide choice of opportunities for the practicing chemist.

Peter F. Lott
Chemistry Department
University of Missouri–Kansas City

Chemistry

Chemists are fond of describing their discipline as the central science. In research, it overlaps with physics and materials science at one end and with medicine and molecular biology at the other.

The traditional areas of chemistry—analytical, biochemistry, inorganic, organic, and physical—remain at the center of research and teaching. The 1985 National Academy of Sciences Publication "Opportunities in Chemistry" identified several areas as priority areas of the future. They include understanding chemical reactivity, chemical catalysis, chemistry of life processes, chemistry around us, and chemical behavior under extreme conditions.

Chemistry is maturing, and chemists are beginning to look at their area much more broadly. There is access to increasingly more sophisticated instruments, such as lasers, for the characterization of compounds and the measurement of processes. New insights have come from theory and dramatic applications from computers in calculations and analysis. Chemists have an increasingly sophisticated basis for making new compounds, including computer graphics and computer-aided design strategies. These developments have allowed chemists to explore interdisciplinary areas that cross the traditional boundaries with materials science, physics, and biology.

Chemistry plays an important role in the American economy. Business and industry employ more Ph.D.'s in chemistry than all the Ph.D.'s in the biological sciences, mathematics, physics, and astronomy combined. More than 90 percent of the funds for research and development are spent by industry. The problems that chemists work on in industry are also becoming increasingly more interdisciplinary in nature. Chemists often find themselves working with life scientists, materials scientists, or engineers in the development of new technologies.

Along with the growth in techniques and understanding, chemists are exploring new and fascinating areas of research. Exciting advances are occurring in the design and synthesis of pharmaceuticals, of model sites for enzymes, and of complex molecular assemblies. New laser techniques are being developed for observing chemical and physical processes on ultrashort time scales and separations techniques for the analysis of complex mixtures and of vanishingly small amounts of chemical materials. Chemists are leading the way in the development of novel polymers and new electronic materials and are busily engaged in preparing new superconductors.

Thomas J. Meyer
Kenan Professor of Chemistry
Department of Chemistry
University of North Carolina at Chapel Hill

Inorganic Chemistry

The scope of modern inorganic chemistry is so immense that it is very difficult to define succinctly. One of the best definitions has been offered by T. Moeller: "Inorganic chemistry is the experimental investigation and theoretical interpretation of the properties and reactions of all the elements and of all their compounds, except the hydrocarbons and most of their derivatives." (T. Moeller, *Inorganic Chemistry, A Modern Introduction*, p. 2. New York: Wiley, 1982.) Specific aspects of inorganic chemistry are integral to most scientific and science-based fields—among them geochemistry, pharmacology, medicine, metallurgy, and ceramic and glass technology, as well as apparently unrelated ones such as molecular biology, polymer production, energy development, waste disposal, and a host of others.

Nevertheless, it is possible to define this vast field into a few major areas of research pursued by inorganic chemists. Coordination chemistry is the study (synthesis and characterization) of small and large molecules that generally consist of a transition metal, or clusters of transition metals, coordinated to two or more ligands. An individual ligand may be as simple as a halide ion, a water molecule, or a nitrogen molecule or may be a very large and complex molecule such as a porphyrin or protein. Major advances have been made in the understanding of bonding, structure, reactivity, and reaction mechanisms in this class of compounds. This has led to a systematic investigation of coordination compounds in living systems and the rapid development of bioinorganic chemistry. Study of model complexes and natural metalloproteins has led to better understanding of vitamins, enzymes, and coenzymes, and models have been generated to duplicate the properties of oxygen-binding hemoproteins. Simple transition-metal complexes have been found to be active against certain cancerous growths.

Organometallic chemistry may be simply characterized as

the chemistry of organic groups bound to metals, especially transition metals. Research in this area has focused on bonding, structure, reactivity, and reaction mechanisms of the molecules, some of which are unique to organometallic chemistry. On the practical side, organometallic catalysts have become increasingly important in polymerization and other petrochemical processes, which accounts for the prominence of this field of research in inorganic chemistry.

Solid-state inorganic chemistry focuses on the synthesis and characterization of solids and is the most interdisciplinary area of inorganic chemistry. A background in crystallography, solid-state physics, and materials science is required to carry out research. Many technological breakthroughs—the development of transistors, phosphor materials used in color television, semiconductor materials, optical fibers, and magnetic recording, to name just a few—would not have been possible without the contribution of solid-state chemists. Most recently, solid-state chemists have been extremely active in the development of high-temperature superconducting materials.

There are many other areas that fit into one of these major areas at least loosely (inorganic polymers, surface science–solid state, etc.). Furthermore, theoretical research in inorganic chemistry has contributed to our understanding of bonding both on the molecular level and in the solid state. Modern tools used by inorganic chemists include X-ray diffraction; UV, visible, IR, Raman, ESR, NMR, and Mössbauer spectroscopies; magnetic susceptibility, electronic, and ionic conductivity measurements; and a host of the most recent techniques, such as X-ray photoelectron spectroscopy (XPS or ESCA, electron spectroscopy for chemical analysis).

Inorganic chemists gain employment in academic institutions and in industrial and government laboratories.

Martha Greenblatt
Professor of Chemistry
Chemistry Department
Rutgers University, New Brunswick

Organic Chemistry

The field of organic chemistry deals with the structure and reactivity of carbon-containing compounds. Since over 99 percent of all compounds known to man contain carbon, organic chemistry is fundamental to a wide variety of disciplines. Perhaps the best indication of its importance is that it is a required course for students seeking careers in chemistry, biology, chemical engineering, agriculture, nursing, medicine, dentistry, pharmacy, textiles, nutrition, and many others.

Organic chemistry is divided into many subdisciplines, many of them overlapping with other disciplines. Synthetic organic chemists deal with the conversion of one molecule into another. Their research efforts focus on the development of methods for the synthesis of molecules of practical and/or theoretical importance. Physical organic chemists study the intricate details of the structure and the reactions of organic molecules. Their research efforts often overlap with those of physical chemists. Bioorganic chemists and natural products chemists study the organic compounds associated with living systems. They interact heavily with biochemists, molecular biologists, and pharmacologists. Polymer chemists examine the synthesis and properties of high molecular weight materials. Their work often overlaps with that of materials scientists and chemical engineers. Organometallic chemists investigate compounds containing organic molecules bound to metal atoms. Their work overlaps heavily with that of inorganic chemists. Organic spectroscopists develop methods to determine the structure, purity, and identity of organic compounds. Their work overlaps heavily with that of analytical and physical chemists.

Organic chemistry is becoming increasingly dependent on advanced instrumentation and computers. In the last few years, a virtual revolution has taken place in the quality of instrumentation available to organic chemists. Many new types of experiments that were thought to be impossible ten years ago can now be done.

Careers for organic chemists who hold as their highest degree either the M.S. or the Ph.D. can be found in industry, academia, and the federal government. Recent demographic trends have created a shortage of people with advanced degrees, especially at the master's level. Most organic chemists seek careers in industry. Organic chemists typically work in the areas of chemical manufacture, pharmaceutical and agricultural products, plastics and synthetic fibers, and petroleum products. Rapidly growing areas such as biotechnology and materials science should provide organic chemists with exciting new opportunities. Careers in academic institutions typically focus on a combination of teaching and research. Careers with the federal government are chiefly in the areas of health, food production, and defense. Organic chemists have access to a wide variety of careers in diverse areas, and their knowledge is fundamental to research and development in many disciplines not directly related to organic chemistry.

James W. Herndon, Professor
Department of Chemistry and Biochemistry
University of Maryland College Park

Physical Chemistry

Physical chemistry, while traditionally classified as a subdivision of chemistry, is actually practiced by scientists in fields ranging from biochemistry to materials science. The unifying aspect of physical chemistry is the desire to understand chemical phenomena in terms of the basic laws of nature. The physical chemist provides a bridge between physics and chemistry, lending a great deal of excitement to the enterprise.

There are two broad areas in physical chemistry—theory and experiment. The first heading includes such diverse subjects as the ab initio calculation and prediction of molecular properties such as spectra, conformation, and potential energy surfaces, using today's powerful computers; the computer simulation of molecular motion in simple liquids and in polymers; the theory of rates of reactions; theories of the elastic and inelastic scattering of light from simple and complex liquids; theories of reactions in the intense electromagnetic fields generated by lasers; and the applications of statistical mechanics to problems such as phase transitions and the structures of polymers. Experimental physical chemistry includes the use of lasers to probe reaction dynamics and the motion of molecules in simple and in complex fluids, to probe the geometry of molecules (both small and large), and to probe the details of energy transfer in the excited states of molecules; the use of nuclear magnetic resonance to probe, on a time scale different from that which applies in laser research, the geometry and motion of molecules; the study of the properties of surfaces—particularly important in catalysis; and even the use of "old-fashioned" thermodynamic measurements to determine phase diagrams. These lists are not exhaustive. One important item not represented above is the development of new experimental and theoretical techniques in order to further these studies.

Many of the techniques mentioned have applications in other areas of chemistry. Organic chemists probe the mechanisms of organic reactions by using pulsed lasers. Biochemists use nuclear magnetic resonance to learn about the structure, and even the oxidation states of metals, in biologically important molecules. Some simple quantum mechanical ideas have been used to explain and predict the course of organic reactions. Because of the broad applicability of physical chemistry, employment opportunities are greater than commonly thought. Physical chemists work not only for academic institutions and chemical companies but also for oil industries and government laboratories, among others.

Preparation for graduate work in physical chemistry should include a strong background in physics and, especially for those of a theoretical bent, in mathematics, as well as in chemistry. For those interested in applying fundamental principles to other areas of chemistry, such as organic chemistry or biochemistry, a good acquaintance with these fields is also desirable.

Robert L. Fulton
Professor of Chemistry
Florida State University

SECTION 2: CHEMISTRY

Analytical Chemistry

Boston College, Graduate School of Arts and Sciences, Department of Chemistry, Chestnut Hill, MA 02167. Offerings include analytical chemistry (MS, PhD), chemistry (MS). *Degree requirements:* For master's, 1 foreign language, thesis; for doctorate, 1 foreign language, dissertation, foreign language exam, qualifying exam. *Entrance requirements:* GRE General Test, GRE Subject Test. *Tuition:* $412 per credit hour.
See full description on page 89.

Brigham Young University, College of Physical and Mathematical Sciences, Department of Chemistry, Provo, UT 84602. Offerings include analytical chemistry (MS, PhD). Department faculty: 34 full-time (1 woman), 0 part-time. *Degree requirements:* 2 foreign languages (computer language can substitute for one), thesis/dissertation. *Entrance requirements:* For master's, minimum GPA of 3.0 during previous 60 credit hours. Application deadline: 3/15 (priority date). Application fee: $30. *Tuition:* $1170 per semester (minimum) full-time, $130 per credit hour (minimum) part-time. • Dr. Earl M. Woolley, Chairman, 801-378-3667. Application contact: Francis R. Nordmeyer, Coordinator, Graduate Admissions, 801-378-4845.
See full description on page 95.

California State University, Fullerton, School of Natural Science and Mathematics, Department of Chemistry and Biochemistry, Fullerton, CA 92634. Offerings include analytical chemistry (MS). Department faculty: 20 full-time (5 women), 13 part-time (2 women). *Degree requirements:* Thesis, departmental qualifying exam required, foreign language not required. *Entrance requirements:* Minimum GPA of 2.5 in last 60 units, major in chemistry or related field. Application fee: $55. *Expenses:* Tuition of $0 for state residents; $246 per unit for nonresidents. Fees of $554 per semester full-time, $356 per unit part-time. • Dr. Glen Nagel, Chair, 714-773-3621. Application contact: Dr. Gene Hiegel, Adviser, 714-773-3624.

California State University, Los Angeles, School of Natural and Social Sciences, Department of Chemistry and Biochemistry, Los Angeles, CA 90032. Offerings include analytical chemistry (MS). Department faculty: 12 full-time, 25 part-time. *Degree requirements:* 1 foreign language (computer language can substitute), thesis or comprehensive exam. *Entrance requirements:* TOEFL (minimum score of 550 required). Application deadline: 8/7 (applications processed on a rolling basis). Application fee: $55. *Expenses:* Tuition of $0 for state residents; $164 per unit for nonresidents. Fees of $1046 per year full-time, $650 per semester (minimum) part-time. • Dr. Donald Paulson, Acting Chair, 213-343-2300.

Clarkson University, School of Science, Department of Chemistry, Potsdam, NY 13699. Offerings include analytical chemistry (MS, PhD). Department faculty: 12 full-time (0 women), 0 part-time. *Degree requirements:* For master's, foreign language and thesis not required; for doctorate, dissertation, departmental qualifying exam required, foreign language not required. *Application fee:* $10. *Expenses:* Tuition of $446 per credit hour. Fees of $75 per semester. • Dr. Phillip A. Christiansen, Chairman, 315-268-2389.

Clemson University, College of Sciences, Department of Chemistry, Clemson, SC 29634. Department awards MS, PhD. Faculty: 19 full-time (1 woman), 0 part-time. Matriculated students: 65 full-time (14 women), 0 part-time; includes 4 minority (1 Asian American, 3 black American), 30 foreign. Average age 25. 40 applicants, 38% accepted. In 1990, 4 master's awarded; 12 doctorates awarded (67% entered university research/teaching, 33% found other work related to degree). *Degree requirements:* 1 foreign language (computer language can substitute), thesis/dissertation. *Entrance requirements:* GRE General Test. Application deadline: 6/1. Application fee: $25. *Expenses:* Tuition of $102 per credit hour. Fees of $80 per semester full-time. *Financial aid:* In 1990–91, $500,000 in aid awarded. 1 fellowship (to a first-year student), 18 research assistantships (1 to a first-year student), 43 lab assistantships were awarded; teaching assistantships also available. *Faculty research:* Flourine chemistry, organic synthetic methods and natural products, metal and non-metal clusters, analytical spectroscopies, polymers. Total annual research budget: $2.45-million. • J. D. Petersen, Head, 803-656-5017.

Cleveland State University, College of Arts and Sciences, Department of Chemistry, Cleveland, OH 44115. Offerings include analytical chemistry (MS, PhD). Department faculty: 18 full-time (1 woman), 4 part-time (0 women). *Degree requirements:* For master's, thesis required (for some programs), foreign language not required; for doctorate, dissertation required, foreign language not required. *Entrance requirements:* For master's, GRE General Test, GRE Subject Test, TOEFL (minimum score of 525 required); for doctorate, GRE General Test (score in 50th percentile or higher required), GRE Subject Test, TOEFL (minimum score of 525 required). Application deadline: 9/1 (priority date, applications processed on a rolling basis). Application fee: $0. *Tuition:* $90 per credit for state residents; $180 per credit for nonresidents. • Dr. Robert Towns, Chair, 216-687-2004. Application contact: Dr. Thomas Flechtner, Chairman, Graduate Committee, 216-687-2458.

Florida State University, College of Arts and Sciences, Department of Chemistry, Tallahassee, FL 32306. Offerings include analytical chemistry (MS, PhD). Terminal master's awarded for partial completion of doctoral program. Department faculty: 39 full-time (2 women), 2 part-time (0 women). *Degree requirements:* For master's, diagnostic and cumulative exams required, foreign language not required; for doctorate, dissertation, diagnostic and cumulative exams required, foreign language not required. *Entrance requirements:* GRE General Test, minimum B average in undergraduate course work. Application deadline: 4/15. Application fee: $15. *Tuition:* $76.29 per credit hour for state residents; $238 per credit hour for nonresidents. • Dr. Edward K. Mellon, Chairman, 904-644-4074. Application contact: Dr. William T. Cooper, Graduate Adviser, 904-644-6875.
See full description on page 105.

Georgetown University, College of Arts and Sciences, Department of Chemistry, Washington, DC 20057. Offerings include analytical chemistry (MS, PhD). Terminal master's awarded for partial completion of doctoral program. Department faculty: 17 full-time, 4 part-time. *Degree requirements:* For master's, thesis required (for some programs), foreign language not required; for doctorate, 1 foreign language, dissertation. *Entrance requirements:* TOEFL (minimum score of 550 required, 600 for teaching assistants). Application deadline: 7/1 (applications processed on a rolling basis). *Expenses:* Tuition of $12,768 per year full-time, $532 per credit part-time. Fees of $142 per year full-time. • Dr. Michael T. Pope, Chairman, 202-687-6073.

Governors State University, College of Arts and Sciences, Division of Science, Program in Analytical Chemistry, University Park, IL 60466. Program awards MS. Part-time and evening/weekend programs available. Faculty: 6 full-time (2 women), 5 part-time (1 woman). Matriculated students: 2 full-time (1 woman), 24 part-time (5 women); includes 6 minority (all black American), 1 foreign. Average age 30. 17 applicants, 82% accepted. In 1990, 5 degrees awarded. *Degree requirements:* Computer language, thesis or alternative required, foreign language not required. *Application deadline:* 6/14 (priority date, applications processed on a rolling basis). Application fee: $0. *Expenses:* Tuition of $882 per semester full-time, $74 per hour part-time for state residents; $2646 per semester full-time, $221 per hour part-time for nonresidents. Fees of $25 per trimester. *Financial aid:* In 1990–91, 2 research assistantships (0 to first-year students), 1 scholarship (to a first-year student) awarded; full and partial tuition waivers, federal work-study, institutionally sponsored loans, and career-related internships or fieldwork also available. Aid available to part-time students. Average monthly stipend for a graduate assistantship: $400. Financial aid application deadline: 5/1; applicants required to submit FAF. *Faculty research:* Electrochemistry, photochemistry, spectrochemistry, biochemistry. • Dr. Edwin Cehelnik, Chairperson, Division of Science, 708-534-5000 Ext. 2402.

Howard University, Graduate School of Arts and Sciences, Department of Chemistry, 2400 Sixth Street, NW, Washington, DC 20059. Offerings include analytical chemistry (MS, PhD). *Degree requirements:* For master's, 1 foreign language (computer language can substitute), thesis, comprehensive exam; for doctorate, 2 foreign languages (computer language can substitute for one), dissertation, comprehensive exam. *Entrance requirements:* For master's, minimum GPA of 2.7. Application deadline: 4/1. Application fee: $25. *Expenses:* Tuition of $6100 per year full-time, $339 per credit hour part-time. Fees of $555 per year full-time, $245 per semester part-time.

Jackson State University, School of Science and Technology, Department of Chemistry, Jackson, MS 39217. Offerings include analytical chemistry (MS). Department faculty: 14 full-time (3 women), 0 part-time. *Degree requirements:* Thesis (for some programs). *Entrance requirements:* GRE General Test. Application deadline: rolling. Application fee: $0. *Expenses:* Tuition of $1114 per semester full-time, $113 per hour part-time for state residents; $1142 per semester full-time, $113 per hour part-time for nonresidents. Fees of $730 per semester for nonresidents. • Dr. Richard Sullivan, Chair, 601-968-2171.

Kansas State University, College of Arts and Sciences, Department of Chemistry, Manhattan, KS 66506. Offerings include analytical chemistry (MS, PhD). Department faculty: 19 full-time (0 women), 0 part-time. *Degree requirements:* For master's, thesis required, foreign language not required; for doctorate, dissertation. *Expenses:* Tuition of $1668 per year full-time, $51 per credit hour part-time for state residents; $5382 per year full-time, $142 per credit hour part-time for nonresidents. Fees of $305 per year full-time, $64.50 per semester part-time. • M. Dale Hawley, Head, 913-532-6668. Application contact: Robert Hammaker, Graduate Coordinator, 913-532-6671.

Kent State University, Department of Chemistry, Kent, OH 44242. Offerings include analytical chemistry (MS, PhD). Department faculty: 18 full-time, 0 part-time. *Degree requirements:* For doctorate, dissertation required, foreign language not required. *Application deadline:* 7/12 (applications processed on a rolling basis). *Application fee:* $25. *Tuition:* $1601 per semester full-time, $133.75 per hour part-time for state residents; $3101 per semester full-time, $258.75 per hour part-time for nonresidents. • Dr. Roger K. Gilpin, Chairman, 216-672-2032.
See full description on page 111.

Lehigh University, College of Arts and Sciences, Department of Chemistry, Bethlehem, PA 18015. Department awards MS, DA, PhD. Faculty: 22 full-time, 4 part-time. Matriculated students: 59 full-time (23 women), 24 part-time (6 women); includes 3 minority (1 Asian American, 2 black American), 7 foreign. 72 applicants, 72% accepted. In 1990, 12 master's, 4 doctorates awarded. *Degree requirements:* For master's, foreign language and thesis not required; for doctorate, 1 foreign language, dissertation. *Entrance requirements:* GRE, TOEFL. Application deadline: rolling. Application fee: $40. *Tuition:* $15,650 per year full-time, $655 per credit hour part-time. *Financial aid:* Fellowships, research assistantships, teaching assistantships, institutionally sponsored loans, and career-related internships or fieldwork available. *Faculty research:* Biochemistry; inorganic, organic, physical, and polymer chemistry. • Dr. John Larsen, Chairman, 215-758-3471. Application contact: Dr. Daniel Zeroka, Graduate Program Coordinator.
See full description on page 113.

Marquette University, College of Arts and Sciences, Department of Chemistry, Milwaukee, WI 53233. Offerings include analytical chemistry (MS, PhD). Terminal master's awarded for partial completion of doctoral program. Department faculty: 14 full-time (1 woman), 0 part-time. *Degree requirements:* For master's, thesis or alternative, comprehensive exam required, foreign language not required; for doctorate, dissertation, cumulative exams required, foreign language not required. *Entrance requirements:* TOEFL (minimum score of 550 required). Application fee: $25. *Tuition:* $300 per credit. • Dr. Michael McKinney, Chairman, 414-288-7065.

Michigan State University, College of Natural Science, Department of Chemistry, East Lansing, MI 48824. Offerings include analytical chemistry (MS, PhD). Terminal master's awarded for partial completion of doctoral program. Department faculty: 27 full-time (1 woman), 0 part-time. *Degree requirements:* For doctorate, 1 foreign language, dissertation. *Entrance requirements:* For doctorate, GRE General Test, TOEFL. Application deadline: rolling. Application fee: $25 ($40 for foreign students). *Tuition:* $104.75 per credit for state residents; $211.75 per credit for nonresidents. • Dr. Gerald Babcock, Chairperson, 517-353-9717. Application contact: Dr. William Reusch, Coordinator, Graduate Program, 517-355-9716.

Mississippi State University, College of Arts and Sciences, Department of Chemistry, Mississippi State, MS 39762. Offerings include analytical chemistry (MS, PhD). Terminal master's awarded for partial completion of doctoral program. Department faculty: 15 full-time (0 women), 0 part-time. *Degree requirements:* 1 foreign language (computer language can substitute), thesis/dissertation. *Entrance requirements:* GRE General Test, TOEFL. Application deadline: rolling. *Expenses:* Tuition of $891 per semester full-time, $99 per credit hour part-time for state residents; $1622 per semester full-time, $180 per credit hour part-time for nonresidents. Fees of $221 per semester full-time, $25 per semester (minimum) part-time. • Dr. Kenneth L. Brown, Head, 601-325-3584.

Northeastern University, Graduate School of Arts and Sciences, Department of Chemistry, Boston, MA 02115. Offerings include analytical chemistry (MS). Department faculty: 16 full-time (2 women), 0 part-time. *Degree requirements:* Thesis (for some programs). *Entrance requirements:* TOEFL (minimum score of 580 required). Application deadline: 4/15. Application fee: $40. *Expenses:* Tuition of $10,260 per year full-time, $285 per quarter hour part-time. Fees of $293 per year full-time. • Dr. John L. Roebber, Executive Officer, 617-437-2822. Application contact: Dr. Paul Vouros, Director of Graduate Affairs, 617-437-2822.
See full description on page 119.

Ohio State University, College of Mathematical and Physical Sciences, Department of Chemistry, Columbus, OH 43210. Department awards MS, PhD. Faculty: 49 full-time (3 women), 0 part-time. Matriculated students: 270 full-time (72 women), 14 part-time (4 women); includes 19 minority (13 Asian American, 4 black American, 1 Hispanic American, 1 Native American), 101 foreign. 203 applicants, 29% accepted. In 1990, 23 master's, 31 doctorates awarded. *Degree requirements:* For master's, thesis optional, foreign language not required; for doctorate, dissertation required, foreign language not required. *Entrance requirements:* For master's, GRE General Test, GRE Subject Test. Application deadline: 8/15 (applications processed on a rolling basis). Application fee: $0 ($25 for foreign students). *Tuition:* $1213 per quarter full-time, $364 per course part-time for state residents; $3143 per quarter full-time, $943 per course part-time for nonresidents. *Financial aid:* In 1990–91, $12,000 in aid awarded. 24 fellowships (10 to first-year students), 93 research assistantships (0 to first-year students), 150 teaching assistantships (50 to first-year students) were awarded; federal work-study, institutionally sponsored loans also available. Aid available to part-time students. Financial aid applicants required to submit FAF. *Total annual research budget:* $5.01-million. • Russell M. Pitzer, Chairman, 614-292-2251. Application contact: C. Weldon Mathews, Vice Chair for Graduate Study, 614-292-8917.

See full description on page 121.

Old Dominion University, College of Sciences, Department of Chemistry and Biochemistry, Norfolk, VA 23529. Offerings include analytical chemistry (MS). Department faculty: 13 full-time (2 women), 1 part-time (0 women). *Degree requirements:* Thesis required, foreign language not required. *Entrance requirements:* GRE General Test, GRE Subject Test, TOEFL, minimum GPA of 3.0 in major, 2.5 overall. Application deadline: 7/1 (applications processed on a rolling basis). Application fee: $20. *Expenses:* Tuition of $148 per credit hour for state residents; $375 per credit hour for nonresidents. Fees of $64 per year full-time. • Dr. Frank E. Scully Jr., Chairman, 804-683-4078. Application contact: Dr. Roy L. Williams, Graduate Program Director, 804-683-4078.

Oregon State University, Graduate School, College of Science, Department of Chemistry, Corvallis, OR 97331. Offerings include analytical chemistry (MS, PhD). Terminal master's awarded for partial completion of doctoral program. Department faculty: 26 full-time (0 women), 5 part-time (1 woman). *Degree requirements:* For doctorate, 1 foreign language, dissertation. *Entrance requirements:* For doctorate, TOEFL (minimum score of 550 required), minimum GPA of 3.0 in last 90 hours. Application deadline: 3/1. Application fee: $40. *Tuition:* $1140 per trimester full-time, $449 per year part-time for state residents; $1816 per trimester full-time, $674 per year part-time for nonresidents. • Dr. Carroll W. DeKock, Chair, 503-737-2081.

See full description on page 123.

Purdue University, School of Science, Department of Chemistry, West Lafayette, IN 47907. Offerings include analytical chemistry (MS, PhD). Terminal master's awarded for partial completion of doctoral program. Department faculty: 45 full-time, 1 part-time. *Degree requirements:* Thesis/dissertation required, foreign language not required. *Entrance requirements:* TOEFL (minimum score of 550 required). Application fee: $25. *Tuition:* $1162 per semester full-time, $83.50 per credit hour part-time for state residents; $3720 per semester full-time, $244.50 per credit hour part-time for nonresidents. • Dr. Harry Morrison, Head, 317-494-5200. Application contact: R. E. Wild, Chairman, Graduate Admissions, 317-494-5200.

Rutgers, The State University of New Jersey, Newark, Department of Chemistry, Newark, NJ 07102. Offerings include analytical chemistry (MS, PhD). Terminal master's awarded for partial completion of doctoral program. Department faculty: 13 full-time (1 woman), 5 part-time (1 woman). *Degree requirements:* For master's, cumulative exams required, thesis optional, foreign language not required; for doctorate, dissertation, cumulative exams required, foreign language not required. *Entrance requirements:* For master's, GRE General Test, GRE Subject Test, minimum undergraduate B average; for doctorate, GRE Subject Test, minimum undergraduate B average. Application deadline: 8/1. Application fee: $35. *Tuition:* $2216 per semester full-time for state residents; $3248 per semester full-time for nonresidents. • Dr. James Schlegel, Director, 201-648-5173.

See full description on page 129.

Rutgers, The State University of New Jersey, New Brunswick, Program in Chemistry, New Brunswick, NJ 08903. Offerings include analytical chemistry (MS, PhD). Terminal master's awarded for partial completion of doctoral program. Program faculty: 48 full-time (5 women), 0 part-time. *Degree requirements:* For doctorate, dissertation, cumulative exams required, foreign language not required. *Entrance requirements:* For doctorate, GRE General Test, GRE Subject Test, TOEFL (minimum score of 575 required). Application deadline: 7/1. Application fee: $35. *Expenses:* Tuition of $4432 per year full-time, $183 per credit part-time for state residents; $6496 per year full-time, $270 per credit part-time for nonresidents. Fees of $458 per year full-time, $117 per year part-time. • Dr. Martha Cotter, Director, 908-932-2259.

See full description on page 131.

San Jose State University, School of Science, Department of Chemistry, San Jose, CA 95192. Offerings include analytical chemistry (MS). Department faculty: 24 full-time (3 women), 5 part-time (2 women). *Degree requirements:* Thesis or alternative required, foreign language not required. *Entrance requirements:* GRE, minimum B average. Application deadline: 6/1 (applications processed on a rolling basis). Application fee: $55. *Expenses:* Tuition of $0 for state residents; $246 per unit for nonresidents. Fees of $592 per year. • Dr. Joseph Pesek, Chair, 408-924-5000. Application contact: Dr. Gerald Selter, Graduate Adviser, 408-924-4956.

Seton Hall University, College of Arts and Sciences, Department of Chemistry, South Orange, NJ 07079. Offerings include analytical chemistry (PhD). Terminal master's awarded for partial completion of doctoral program. Department faculty: 14 full-time (2 women). *Degree requirements:* 1 foreign language (computer language can substitute), dissertation, comprehensive exams. *Entrance requirements:* GRE General Test, GRE Subject Test, TOEFL, undergraduate major in chemistry or related field. Application fee: $30. *Expenses:* Tuition of $346 per credit. Fees of $105 per semester. • Dr. Robert L. Augustine, Chair, 201-761-9414.

South Dakota State University, Colleges of Arts and Science and Agriculture and Biological Sciences, Department of Chemistry, Brookings, SD 57007. Offerings include analytical chemistry (MS, PhD). *Degree requirements:* For doctorate, dissertation, research tool. *Entrance requirements:* For doctorate, TOEFL (minimum score of 500 required). *Tuition:* $553 per semester full-time, $61 per credit part-time for state residents; $1094 per semester full-time, $122 per credit part-time for nonresidents.

Southern University and Agricultural and Mechanical College, College of Sciences, Department of Chemistry, Baton Rouge, LA 70813. Offerings include analytical chemistry (MS). MS (environmental science) offered jointly with Louisiana State University. Department faculty: 5 full-time (0 women), 0 part-time. *Entrance requirements:* GRE General Test or GMAT, TOEFL. Application deadline: 7/1. Application fee: $5. *Tuition:* $1594 per year full-time, $299 per semester part-time for state residents; $3010 per year full-time, $299 per semester part-time for nonresidents. • Dr. Earl Doomes, Chairman, 504-771-3990.

State University of New York at Binghamton, School of Arts and Sciences, Department of Chemistry, Binghamton, NY 13902-6000. Offerings include analytical chemistry (PhD). Terminal master's awarded for partial completion of doctoral program. *Degree requirements:* Dissertation, cumulative exams required, foreign language not required. *Entrance requirements:* GRE General Test, GRE Subject Test, TOEFL. Application deadline: 4/15 (priority date). Application fee: $35. *Expenses:* Tuition of $2450 per year full-time, $102.50 per credit part-time for state residents; $5766 per year full-time, $242.50 per credit part-time for nonresidents. Fees of $77 per year full-time, $27.85 per semester (minimum) part-time. • Dr. Alistair J. Lees, Chairperson, 607-777-2362.

Syracuse University, College of Arts and Sciences, Department of Chemistry, Syracuse, NY 13244. Department awards MS, PhD. Faculty: 21 full-time. Matriculated students: 63 full-time (22 women), 1 part-time (0 women); includes 1 minority (Asian American), 36 foreign. Average age 28. 180 applicants, 21% accepted. In 1990, 5 master's, 8 doctorates awarded. *Degree requirements:* For master's, 1 foreign language; for doctorate, 1 foreign language, dissertation. *Entrance requirements:* GRE General Test, GRE Subject Test. Application deadline: rolling. Application fee: $35. *Expenses:* Tuition of $381 per credit. Fees of $289 per year full-time, $34.50 per semester part-time. *Financial aid:* In 1990–91, 4 fellowships, 35 research assistantships, 42 teaching assistantships awarded; partial tuition waivers, federal work-study also available. Financial aid application deadline: 3/1; applicants required to submit FAF. *Total annual research budget:* $1.5-million. • Laurence Nafie, Chair, 315-443-4109.

Texas A&M University, College of Science, Department of Chemistry, College Station, TX 77843. Offerings include analytical chemistry (MS, PhD). Department faculty: 60. *Entrance requirements:* GRE General Test, TOEFL. Application deadline: 7/15. Application fee: $25 ($50 for foreign students). *Expenses:* Tuition of $100 per semester (minimum) for state residents; $128 per credit hour for nonresidents. Fees of $459 per year full-time, $252 per semester part-time. • Michael B. Hall, Head, 409-845-2011.

See full description on page 141.

Tufts University, Graduate School of Arts and Sciences, Department of Chemistry, Medford, MA 02155. Offerings include analytical chemistry (MS, PhD). Terminal master's awarded for partial completion of doctoral program. Department faculty: 15 full-time, 2 part-time. *Degree requirements:* Thesis/dissertation required, foreign language not required. *Entrance requirements:* GRE General Test, GRE Subject Test, TOEFL (minimum score of 550 required). Application deadline: 2/15. Application fee: $50. *Expenses:* Tuition of $16,755 per year full-time, $2094 per course part-time. Fees of $885 per year. • David Walt, Head, 617-381-3470. Application contact: Edward Brush, 617-381-3475.

University of Akron, Buchtel College of Arts and Sciences, Department of Chemistry, Akron, OH 44325. Offerings include analytical chemistry (MS, PhD). Terminal master's awarded for partial completion of doctoral program. Department faculty: 20 full-time (1 woman), 3 part-time (0 women). *Degree requirements:* For master's, 1 foreign language (computer language can substitute), thesis, seminar presentation; for doctorate, 2 foreign languages (computer language can substitute for one), dissertation, cumulative exams. *Entrance requirements:* TOEFL. Application deadline: 3/1 (applications processed on a rolling basis). Application fee: $25. *Tuition:* $119.93 per credit hour for state residents; $210.93 per credit hour for nonresidents. • Dr. G. Edwin Wilson, Head, 216-972-7372.

University of Calgary, Faculty of Science, Department of Chemistry, Calgary, AB T2N 1N4, Canada. Offerings include analytical chemistry (M Sc, PhD). *Degree requirements:* For master's, thesis required, foreign language not required; for doctorate, dissertation, candidacy exam required, foreign language not required. *Application deadline:* 5/31. *Application fee:* $25. *Tuition:* $1705 per year full-time, $427 per course part-time for Canadian residents; $3410 per year full-time, $854 per course part-time for nonresidents.

University of Cincinnati, McMicken College of Arts and Sciences, Department of Chemistry, Cincinnati, OH 45221. Offerings include analytical chemistry (MS, PhD). Terminal master's awarded for partial completion of doctoral program. Department faculty: 26 full-time, 0 part-time. *Degree requirements:* Thesis/dissertation required, foreign language not required. *Entrance requirements:* GRE General Test, GRE Subject Test. Application deadline: 2/15. Application fee: $20. *Tuition:* $131 per credit hour for state residents; $261 per credit hour for nonresidents. • Dr. Bruce Ault, Head, 513-556-9200.

University of Florida, College of Liberal Arts and Sciences, Department of Chemistry, Gainesville, FL 32611. Department awards MS, MST, PhD. PhD offered jointly with Florida Atlantic University. Faculty: 51. Matriculated students: 149 full-time (38 women), 16 part-time (2 women); includes 13 minority (2 Asian American, 5 black American, 6 Hispanic American), 38 foreign. 101 applicants, 52% accepted. In 1990, 6 master's, 28 doctorates awarded. *Degree requirements:* For master's, thesis required, foreign language not required; for doctorate, 1 foreign language, dissertation. *Entrance requirements:* GRE General Test, minimum GPA of 3.0. Application deadline: 5/30 (priority date, applications processed on a rolling basis). Application fee: $15. *Tuition:* $87 per credit hour for state residents; $289 per credit hour for nonresidents. *Financial aid:* In 1990–91, 1 fellowship, 63 research assistantships, 71 teaching assistantships awarded; institutionally sponsored loans also available. Average monthly stipend for a graduate assistantship: $1000. *Faculty research:* Organic, analytical, physical, inorganic and biological chemistry. Total annual research budget: $4.9-million. • Dr. Michael Zerner, Chairman, 904-392-0541. Application contact: Dr. John F. Helling, Chair, Graduate Selections Committee, 904-392-0551.

See full description on page 145.

University of Georgia, College of Arts and Sciences, Department of Chemistry, Athens, GA 30602. Department offers programs in analytical chemistry (MS, PhD), inorganic chemistry (MS, PhD), organic chemistry (MS, PhD), physical chemistry (MS, PhD). Faculty: 27 full-time (0 women), 0 part-time. Matriculated students: 81 full-time (23 women), 7 part-time (1 woman); includes 2 minority (both black American), 44 foreign. 221 applicants, 20% accepted. In 1990, 4 master's, 7 doctorates awarded. *Degree requirements:* For master's, thesis required, foreign language not required; for doctorate, 1 foreign language (computer language can substitute), dissertation. *Entrance requirements:* GRE General Test. Application fee: $10. *Expenses:* Tuition of $598 per quarter full-time, $48 per quarter part-time for state residents; $1558 per quarter full-time, $144 per quarter part-time for nonresidents. Fees of $118 per quarter. *Financial aid:* Fellowships, research assistantships, teaching assistantships, assistantships available. • Dr. Henry F. Schaefer III, Graduate Coordinator, 404-542-2067.

See full description on page 147.

University of Houston, College of Natural Sciences and Mathematics, Department of Chemistry, 4800 Calhoun, Houston, TX 77004. Offerings include analytical chemistry (MS, PhD). Terminal master's awarded for partial completion of doctoral

SECTION 2: CHEMISTRY

Directory: Analytical Chemistry

program. Department faculty: 25 full-time (1 woman), 2 part-time (0 women). *Degree requirements:* Thesis/dissertation required, foreign language not required. *Entrance requirements:* GRE General Test (minimum combined score of 1100 required), TOEFL (minimum score of 575 required). Application deadline: 5/1 (applications processed on a rolling basis). Application fee: $25 ($75 for foreign students). *Expenses:* Tuition of $30 per hour for state residents; $134 per hour for nonresidents. Fees of $240 per year full-time, $125 per year part-time. • Dr. John L. Bear, Chairman, 713-749-2647. Application contact: Dr. Thomas Albright, Chair, Graduate Committee, 713-749-3721.

See full description on page 149.

University of Louisville, College of Arts and Sciences, Department of Chemistry, Louisville, KY 40292. Offerings include analytical chemistry (MS, PhD). Department faculty: 18 full-time (2 women), 2 part-time. *Degree requirements:* For master's, thesis; for doctorate, 1 foreign language, dissertation. *Entrance requirements:* GRE General Test, TOEFL (minimum score of 550 required). Application deadline: rolling. *Expenses:* Tuition of $1780 per year full-time, $99 per credit hour part-time for state residents; $5340 per year full-time, $297 per credit hour part-time for nonresidents. Fees of $60 per semester full-time, $12.50 per semester (minimum) part-time. • Dr. Richard P. Baldwin, Chair, 502-588-6798.

University of Maryland College Park, College of Life Sciences, Department of Chemistry and Biochemistry, College Park, MD 20742. Offerings include analytical chemistry (MS, PhD). Department faculty: 51 (5 women). *Degree requirements:* For master's, thesis or alternative required, foreign language not required; for doctorate, dissertation. *Entrance requirements:* For master's, GRE General Test, minimum GPA of 3.0; for doctorate, GRE General Test. Application deadline: rolling. Application fee: $25. *Expenses:* Tuition of $143 per credit hour for state residents; $256 per credit hour for nonresidents. Fees of $171.50 per semester. • Dr. Sandra Greer, Chairman, 301-405-1788.

See full description on page 155.

University of Michigan, College of Literature, Science, and the Arts, Department of Chemistry, Ann Arbor, MI 48109. Offerings include analytical chemistry (MS, PhD). Department faculty: 38. *Degree requirements:* For master's, foreign language and thesis not required; for doctorate, dissertation, preliminary exam required, foreign language not required. *Entrance requirements:* For master's, GRE General Test, GRE Subject Test; for doctorate, GRE General Test. Application deadline: 4/1 (applications processed on a rolling basis). Application fee: $30. *Tuition:* $3255 per semester full-time, $352 per credit (minimum) part-time for state residents; $6803 per semester full-time, $746 per credit (minimum) part-time for nonresidents. • M. David Curtis III, Chair, 313-764-7314.

See full description on page 161.

University of Missouri–Columbia, College of Arts and Sciences, Department of Chemistry, Columbia, MO 65211. Offerings include analytical chemistry (MS, PhD). Department faculty: 23 full-time, 1 part-time. *Degree requirements:* For master's, thesis required, foreign language not required; for doctorate, 1 foreign language, dissertation. *Entrance requirements:* GRE General Test, minimum GPA of 3.0. Application deadline: 8/1 (priority date, applications processed on a rolling basis). Application fee: $20 ($40 for foreign students). *Expenses:* Tuition of $89.90 per credit hour full-time, $98.35 per credit hour part-time for state residents; $244 per credit hour full-time, $252.45 per credit hour part-time for nonresidents. Fees of $123.55 per semester (minimum) full-time. • Dr. John Bauman, Director of Graduate Studies, 314-882-7720. Application contact: Gary L. Smith, Director of Admissions and Registrar, 314-882-7651.

University of Missouri–Kansas City, College of Arts and Sciences, Department of Chemistry, Kansas City, MO 64110. Offerings include analytical chemistry (MS, PhD). Department faculty: 15 full-time (0 women), 0 part-time. *Degree requirements:* For master's, thesis required, foreign language not required; for doctorate, 1 foreign language, dissertation, comprehensive exam. *Entrance requirements:* GRE. Application fee: $0. *Expenses:* Tuition of $2200 per year full-time, $92 per credit hour part-time for state residents; $5503 per year full-time, $229 per credit hour part-time for nonresidents. Fees of $122 per semester full-time, $9 per credit hour part-time. • Dr. Layton L. McCoy, Chairperson, 816-235-2280. Application contact: Dr. T. F. Thomas, Chairman, Graduate Admissions Committee, 816-235-2297.

University of Nebraska–Lincoln, College of Arts and Sciences, Department of Chemistry, Lincoln, NE 68588. Offerings include analytical chemistry (PhD). Terminal master's awarded for partial completion of doctoral program. Department faculty: 22 full-time (2 women), 0 part-time. *Degree requirements:* 1 foreign language, dissertation, comprehensive exams, departmental qualifying exam. *Entrance requirements:* GRE Subject Test, TOEFL (minimum score of 550 required), TSE (minimum score of 230 required). Application deadline: 5/1 (priority date, applications processed on a rolling basis). Application fee: $25. *Expenses:* Tuition of $75.75 per credit hour for state residents; $187.25 per credit hour for nonresidents. Fees of $161 per year full-time. • Dr. Pill-Soon Song Jr., Chairperson, 402-472-3501.

See full description on page 165.

University of Nevada, Las Vegas, College of Science and Mathematics, Department of Chemistry, Las Vegas, NV 89154. Offerings include environmental analytical chemistry (MS). Department faculty: 8 full-time (0 women), 0 part-time. *Degree requirements:* Foreign language not required. *Entrance requirements:* GRE General Test, minimum GPA of 2.75. Application deadline: 6/15. Application fee: $20. *Expenses:* Tuition of $66 per credit. Fees of $1800 per semester for nonresidents. • Dr. Boyd Earl, Chair, 702-739-3510. Application contact: Graduate Coordinator, 702-739-3753.

University of Pittsburgh, Faculty of Arts and Sciences, Department of Chemistry, Pittsburgh, PA 15260. Department awards MS, PhD. Part-time programs available. Faculty: 37 full-time (2 women), 2 part-time (0 women). Matriculated students: 174 full-time (48 women), 17 part-time (6 women); includes 3 minority (2 Asian American, 1 black American), 63 foreign. 351 applicants, 12% accepted. In 1990, 6 master's, 19 doctorates awarded. Terminal master's awarded for partial completion of doctoral program. *Degree requirements:* Thesis/dissertation required, foreign language not required. *Entrance requirements:* GRE General Test, GRE Subject Test, TOEFL. Application deadline: 3/31 (priority date, applications processed on a rolling basis). Application fee: $15 ($25 for foreign students). *Expenses:* Tuition of $2920 per semester full-time, $241 per credit part-time for state residents; $5840 per semester full-time, $482 per credit part-time for nonresidents. Fees of $156 per year. *Financial aid:* In 1990–91, 189 students received a total of $2.140-million in aid awarded. 16 fellowships (0 to first-year students), 89 research assistantships (1 to a first-year student), 84 teaching assistantships (40 to first-year students) were awarded; full tuition waivers, federal work-study, institutionally sponsored loans also available. Aid available to part-time students. Financial aid application deadline: 3/15; applicants required to submit FAF. *Faculty research:* Analytical chemistry, inorganic chemistry, organic chemistry, physical chemistry and surface science. Total annual research budget: $7.657-

million. • Dr. N. John Cooper, Chairman, 412-624-8500. Application contact: Rebecca Claycamp, Assistant Chairman, 412-624-8200.

See full description on page 167.

University of Regina, Faculty of Graduate Studies and Research, Faculty of Science, Department of Chemistry, Regina, SK S4S 0A2, Canada. Offerings include analytical chemistry (M Sc, PhD). Department faculty: 14 full-time, 0 part-time. *Degree requirements:* For master's, thesis, departmental qualifying exam required, foreign language not required; for doctorate, variable foreign language requirement, dissertation, departmental qualifying exam. *Application deadline:* 7/2 (applications processed on a rolling basis). *Tuition:* $1500 per year full-time, $242 per year (minimum) part-time. • Dr. W. D. Chandler, Head, 306-585-4146.

University of San Francisco, College of Arts and Sciences, Department of Chemistry, San Francisco, CA 94117-1080. Offerings include analytical chemistry (MS). Department faculty: 6 full-time (0 women), 0 part-time. *Degree requirements:* 1 foreign language, thesis. *Entrance requirements:* TOEFL (minimum score of 550 required). Application deadline: 5/15 (priority date). Application fee: $40 ($90 for foreign students). *Tuition:* $10,960 per year full-time, $432 per unit part-time. • Tom Gruhn, Chairman, 415-666-6486. Application contact: Jeff Curtis, Graduate Adviser, 415-666-6391.

See full description on page 171.

University of Southern Mississippi, College of Science and Technology, Department of Chemistry and Biochemistry, Hattiesburg, MS 39406. Offerings include analytical chemistry (MS, PhD). Department faculty: 18 full-time (2 women), 1 (woman) part-time. *Degree requirements:* For doctorate, 2 foreign languages (computer language can substitute for one), dissertation. *Entrance requirements:* For master's, GRE General Test, minimum GPA of 2.75; for doctorate, GRE General Test, minimum GPA of 3.5. Application deadline: 8/9 (priority date, applications processed on a rolling basis). Application fee: $0 ($25 for foreign students). *Expenses:* Tuition of $968 per semester full-time, $93 per semester hour part-time. Fees of $12 per semester part-time for state residents; $591 per year full-time, $12 per semester part-time for nonresidents. • Dr. David Creed, Chair, 601-266-4701. Application contact: Dr. David Wertz, Professor, 601-266-4702.

University of South Florida, College of Arts and Sciences, Department of Chemistry, Tampa, FL 33620. Offerings include analytical chemistry (MS, PhD). Department faculty: 26 full-time. *Degree requirements:* For master's, thesis; for doctorate, 1 foreign language, computer language, dissertation. *Entrance requirements:* For master's, GRE General Test (minimum combined score of 1000 required), minimum GPA of 3.0 (in last 60 credit hours). Application fee: $15. *Tuition:* $79.40 per credit hour for state residents; $241.33 per credit hour for nonresidents. • Stewart W. Schneller Jr., Chairperson, 813-974-2258. Application contact: Terence Oestereich, Graduate Program Director, 813-974-2534.

See full description on page 173.

University of Tennessee, Knoxville, College of Liberal Arts, Department of Chemistry, Knoxville, TN 37996. Offerings include analytical chemistry (MS, PhD). Terminal master's awarded for partial completion of doctoral program. Department faculty: 36 (1 woman). *Degree requirements:* For master's, thesis required, foreign language not required; for doctorate, 1 foreign language, dissertation. *Entrance requirements:* GRE General Test, TOEFL (minimum score of 525 required), minimum GPA of 2.5. Application deadline: 2/1 (priority date, applications processed on a rolling basis). Application fee: $15. *Tuition:* $1086 per semester full-time, $142 per credit hour part-time for state residents; $2768 per semester full-time, $308 per credit hour part-time for nonresidents. • Dr. Gleb Mamantov, Head, 615-974-3141.

University of Texas at Austin, Graduate School, College of Natural Sciences, Department of Chemistry and Biochemistry, Austin, TX 78712. Department offers programs in analytical chemistry (MA, PhD), biochemistry (MA, PhD), inorganic chemistry (MA, PhD), organic chemistry (MA, PhD), physical chemistry (MA, PhD). Matriculated students: 240 full-time (73 women), 0 part-time; includes 18 minority (10 Asian American, 2 black American, 4 Hispanic American, 2 Native American), 73 foreign. 197 applicants, 61% accepted. In 1990, 15 master's, 37 doctorates awarded. *Entrance requirements:* GRE. Application deadline: 2/1 (priority date, applications processed on a rolling basis). Application fee: $40 ($75 for foreign students). *Tuition:* $510.30 per semester for state residents; $1806 per semester for nonresidents. *Financial aid:* Fellowships, research assistantships, teaching assistantships available. Financial aid application deadline: 3/1. • Application contact: Dr. Peter J. Rossky, Graduate Adviser, 512-471-3890.

See full description on page 175.

University of Toledo, College of Arts and Sciences, Department of Chemistry, Toledo, OH 43606. Offerings include analytical chemistry (MS, PhD). Department faculty: 20 full-time (2 women), 0 part-time. *Degree requirements:* For master's, thesis required, foreign language not required; for doctorate, 1 foreign language, dissertation. *Entrance requirements:* GRE General Test, GRE Subject Test, TOEFL. Application deadline: 9/8 (priority date). Application fee: $30. *Tuition:* $122.59 per credit hour for state residents; $193.40 per credit hour for nonresidents. • Dr. James L. Fry, Chairman, 419-537-2109. Application contact: Dr. Max Funk, Chairman, Graduate Admissions Committee, 419-537-2254.

University of Wisconsin–Madison, Analytical Clinical Chemistry Program, Madison, WI 53706. Program awards MS. Faculty: 4. Matriculated students: 0 full-time, 1 (woman) part-time; includes 0 minority, 0 foreign. 1 applicants, 0% accepted. In 1990, 1 degree awarded. *Application deadline:* rolling. Application fee: $20. *Financial aid:* 0 students received aid. Institutionally sponsored loans available. Financial aid application deadline: 1/15; applicants required to submit FAF. • Merle Evenson, Chairperson, 608-263-7035. Application contact: Joanne Nagy, Assistant Dean of Graduate School, 608-262-2433.

Wake Forest University, Department of Chemistry, Winston-Salem, NC 27109. Offerings include analytical chemistry (MS, PhD). Department faculty: 10 full-time (0 women), 3 part-time (1 woman). *Degree requirements:* 1 foreign language (computer language can substitute), thesis/dissertation. *Entrance requirements:* GRE General Test, GRE Subject Test. Application deadline: 3/1. Application fee: $25. *Tuition:* $10,800 per year full-time, $280 per hour part-time. • Dr. Willie Hinze, Chairman, 919-759-5325.

Washington State University, College of Sciences and Arts, Division of Sciences, Department of Chemistry, Pullman, WA 99164. Offerings include analytical chemistry (MS, PhD). Terminal master's awarded for partial completion of doctoral program. Department faculty: 26 (2 women). *Degree requirements:* For master's, thesis, teaching experience required, foreign language not required; for doctorate, dissertation, teaching experience. *Entrance requirements:* GRE General Test, minimum GPA of 3.0. Application deadline: 3/1 (priority date, applications processed on a rolling basis). Application fee: $25. *Tuition:* $1694 per semester full-time, $169 per credit hour part-time for state residents; $4236 per semester full-time, $424 per credit hour part-time for nonresidents. • Dr. Roy Filby, Chair, 509-335-1516.

Chemistry

Abilene Christian University, College of Natural and Applied Sciences, Department of Chemistry, Abilene, TX 79699. Department awards MS. Part-time programs available. Faculty: 10 full-time, 0 part-time. Matriculated students: 1. In 1990, 4 degrees awarded. *Degree requirements:* 1 foreign language, thesis, comprehensive exam. *Entrance requirements:* GRE General Test. Application deadline: rolling. Application fee: $25. *Expenses:* Tuition of $172 per hour. Fees of $45 per hour. *Financial aid:* Research assistantships, teaching assistantships, federal work-study available. Aid available to part-time students. *Faculty research:* Surface science astrophysics, particle physics, inorganic synthesis, inorganic spectroscopy. • Dr. Thomas J. McCord, Chairman, 915-674-2176. Application contact: Dr. Clint Hurley, Graduate Dean, 915-674-2354.

Acadia University, Faculty of Pure and Applied Science, Department of Chemistry, Wolfville, NS B0P 1X0, Canada. Department awards M Sc. Faculty: 7 full-time (0 women), 0 part-time. Matriculated students: 3 full-time (1 woman), 0 part-time; includes 1 minority (Asian American). Average age 26. 6 applicants, 17% accepted. In 1990, 0 degrees awarded. *Degree requirements:* Thesis. *Application deadline:* 5/31. *Application fee:* $20. *Expenses:* Tuition of $2405 per year full-time, $495 per course part-time. Fees of $1700 per year full-time, $340 per year part-time for nonresidents. *Financial aid:* In 1990–91, $15,000 in aid awarded. 2 fellowships (1 to a first-year student) were awarded; teaching assistantships also available. *Faculty research:* Analytical studies of nonmetals, fungal metabolic studies, kinetics of gas-phase reactions, chemistry of thiols and thiolates, inhibitions of fruit ripening processes by flavoroids. Total annual research budget: $75,000. • Dr. Michael E. Peach, Head, 902-542-2201 Ext. 242.

Adelphi University, Graduate School of Arts and Sciences, Department of Chemistry, Garden City, NY 11530. Department offers programs in biochemistry (MS), chemistry (MS). Part-time and evening/weekend programs available. Matriculated students: 1 full-time (0 women), 23 part-time (5 women); includes 4 minority (3 Asian American, 1 black American), 4 foreign. Average age 30. 23 applicants, 74% accepted. In 1990, 3 degrees awarded. *Degree requirements:* 1 foreign language (computer language can substitute), thesis or alternative. *Application deadline:* 3/15 (priority date, applications processed on a rolling basis). *Application fee:* $50. *Expenses:* Tuition of $5300 per semester full-time, $315 per credit part-time. Fees of $159 per semester part-time. *Financial aid:* In 1990–91, 5 students received a total of $25,000 in aid awarded. 0 research assistantships, 5 teaching assistantships (4 to first-year students) were awarded; partial tuition waivers and career-related internships or fieldwork also available. Average monthly stipend for a graduate assistantship: $550. Financial aid applicants required to submit FAF. *Faculty research:* Molecular and supramolecular organization in eye tissues, organic pollutants, low-temperatures x-ray, novel techniques for toxicity testing. Total annual research budget: $300,000. • Dr. Frederick A. Bettelheim, Chairperson, 516-877-4135.

American University, College of Arts and Sciences, Department of Chemistry, Washington, DC 20016. Department awards MS, PhD. Part-time and evening/weekend programs available. Faculty: 7 full-time (2 women), 1 (woman) part-time. Matriculated students: 17 full-time (8 women), 32 part-time (10 women); includes 6 minority (5 black American, 1 Hispanic American), 20 foreign. Average age 32. 73 applicants, 60% accepted. In 1990, 7 master's, 3 doctorates awarded. *Degree requirements:* For master's, thesis (for some programs); for doctorate, 1 foreign language (computer language can substitute), dissertation. *Application deadline:* 2/15. *Application fee:* $50. *Expenses:* Tuition of $475 per semester hour. Fees of $20 per semester. *Financial aid:* In 1990–91, 9 students received a total of $91,188 in aid awarded. 6 fellowships, 1 teaching assistantship, 2 administrative assistants were awarded; research assistantships, federal work-study, institutionally sponsored loans, and career-related internships or fieldwork also available. Financial aid application deadline: 2/15; applicants required to submit FAF. • Dr. James E. Girard, Chair, 202-885-1750.

Appalachian State University, College of Arts and Sciences, Department of Chemistry, Boone, NC 28608. Department awards MA, MS. Part-time programs available. Faculty: 10 full-time (1 woman), 0 part-time. Matriculated students: 2 full-time (1 woman), 3 part-time (1 woman). 17 applicants, 47% accepted. *Degree requirements:* 1 foreign language, thesis, comprehensive exam. *Entrance requirements:* GRE General Test. Application deadline: 7/31 (priority date). Application fee: $15. *Tuition:* $598 per semester for state residents; $3125 per semester full-time, $6215 per semester part-time for nonresidents. *Financial aid:* Fellowships, research assistantships, teaching assistantships available. Financial aid application deadline: 7/31. • Dr. Lawrence Brown, Chairman, 704-262-3010.

Arizona State University, College of Liberal Arts and Sciences, Department of Chemistry and Biochemistry, Tempe, AZ 85287. Department awards MS, PhD. Matriculated students: 53 full-time (19 women), 57 part-time (21 women); includes 13 minority (5 Asian American, 1 black American, 6 Hispanic American, 1 Native American), 37 foreign. 129 applicants, 39% accepted. In 1990, 3 master's, 14 doctorates awarded. *Degree requirements:* For master's, thesis required, foreign language not required; for doctorate, 1 foreign language, dissertation. *Entrance requirements:* TOEFL (minimum score of 550 required), TSE (minimum score of 230 required). Application deadline: 2/1 (priority date). Application fee: $25. *Tuition:* $1528 per year full-time, $80 per hour part-time for state residents; $6934 per year full-time, $289 per hour part-time for nonresidents. *Financial aid:* Research assistantships, teaching assistantships, full tuition waivers available. *Faculty research:* Meteorite chemistry, structure of biopolymers, electron microprobe analysis of air pollutants, X-ray crystallography. • Dr. Morton E. Munk, Chair, 602-965-3461. Application contact: Director of Graduate Studies, 602-965-4664.

See full description on page 85.

Arkansas State University, College of Arts and Sciences, Department of Chemistry, State University, AR 72467. Department awards MS, MSE, SCCT. Faculty: 11 full-time (1 woman), 1 part-time (0 women). Matriculated students: 4 full-time (2 women), 3 part-time (2 women); includes 2 minority (both black American), 0 foreign. Average age 29. In 1990, 0 master's, 0 SCCTs awarded. *Degree requirements:* For master's, thesis, comprehensive exam; for SCCT, comprehensive exam required, foreign language and thesis not required. *Entrance requirements:* For master's, appropriate bachelor's degree; for SCCT, GRE General Test or MAT. Application deadline: 8/1 (priority date, applications processed on a rolling basis). Application fee: $0 ($25 for foreign students). *Tuition:* $790 per semester full-time, $67 per hour part-time for state residents; $1415 per semester full-time, $120 per hour part-time for nonresidents. *Financial aid:* Teaching assistantships available. Financial aid application deadline: 6/1. *Total annual research budget:* $85,000. • Dr. David Chittenden, Chairman, 501-972-3086.

Auburn University, College of Sciences and Mathematics, Department of Chemistry, Auburn University, AL 36849. Department awards MACT, MS, PhD. Part-time programs available. Faculty: 19 full-time, 0 part-time. Matriculated students: 59 (16 women); includes 1 minority (Hispanic American), 49 foreign. In 1990, 5 master's, 6 doctorates awarded. *Degree requirements:* For master's, thesis (MS) required, foreign language not required; for doctorate, 1 foreign language, computer language, dissertation, written and oral exams. *Entrance requirements:* For master's, GRE General Test, GRE Subject Test (MS); for doctorate, GRE General Test (minimum score of 400 on each section required), GRE Subject Test. Application deadline: 9/1 (priority date). Application fee: $15. *Expenses:* Tuition of $1596 per year full-time, $44 per credit hour part-time for state residents; $4788 per year full-time, $132 per credit hour part-time for nonresidents. Fees of $92 per quarter for state residents; $276 per quarter for nonresidents. *Financial aid:* Research assistantships, teaching assistantships available. Financial aid application deadline: 3/15. • Dr. J. Howard Hargis, Head, 205-844-4043. Application contact: Dr. Norman J. Doorenbos, Dean, Graduate School, 205-844-4700.

Ball State University, College of Sciences and Humanities, Department of Chemistry, 2000 University Avenue, Muncie, IN 47306. Department awards MA, MS. Faculty: 11. Matriculated students: 14. In 1990, 4 degrees awarded. *Degree requirements:* Foreign language not required. *Application fee:* $15. *Expenses:* Tuition of $1140 per semester (minimum) full-time for state residents; $2680 per semester (minimum) full-time for nonresidents. Fees of $6 per credit hour. *Financial aid:* In 1990–91, 9 research assistantships awarded. *Faculty research:* Synthetic and analytical chemistry, biochemistry, theoretical chemistry. • Dr. Scott Pattison, Chairman, 317-285-8060.

Baylor University, College of Arts and Sciences, Department of Chemistry, Waco, TX 76798. Department awards MS, PhD. Part-time programs available. Matriculated students: 3 full-time (1 woman), 23 part-time (8 women); includes 16 minority (11 Asian American, 4 black American, 1 Hispanic American). In 1990, 0 master's, 3 doctorates awarded. Terminal master's awarded for partial completion of doctoral program. *Degree requirements:* For master's, thesis required, foreign language not required; for doctorate, 1 foreign language, dissertation, comprehensive exam. *Entrance requirements:* GRE General Test, GRE Subject Test, TOEFL (minimum score of 550 required), minimum GPA of 3.0. Application deadline: 3/15 (applications processed on a rolling basis). Application fee: $0. *Expenses:* Tuition of $4440 per year full-time, $185 per credit hour part-time. Fees of $510 per year full-time. *Financial aid:* Fellowships, research assistantships, teaching assistantships, full tuition waivers, federal work-study, institutionally sponsored loans available. Financial aid applicants required to submit FAF. • Dr. Marianna A. Busch, Director of Graduate Studies, 817-755-3311.

See full description on page 87.

Bishop's University, Division of Natural Sciences, Department of Chemistry, Lennoxville, PQ J1M 1Z7, Canada. Department awards M Sc. *Application fee:* $25. *Tuition:* $883 per semester for Canadian residents; $3910 per semester for nonresidents. *Faculty research:* Synthetic approaches to maytansinoid synthons and to acylasides and derived carbamates. • Dr. R. B. Yeats, Chairperson, 819-822-9600 Ext. 365.

Boston College, Graduate School of Arts and Sciences, Department of Chemistry, Chestnut Hill, MA 02167. Department offers programs in analytical chemistry (MS, PhD), biochemistry (MS, PhD), chemistry (MS), inorganic chemistry (MS, PhD), organic chemistry (MS, PhD), physical chemistry (MS, PhD). *Degree requirements:* For master's, 1 foreign language, thesis; for doctorate, 1 foreign language, dissertation, foreign language exam, qualifying exam. *Entrance requirements:* GRE General Test, GRE Subject Test. *Tuition:* $412 per credit hour.

See full description on page 89.

Boston University, Graduate School, Department of Chemistry, Boston, MA 02215. Department offers programs in biochemistry (MA, PhD), inorganic chemistry (MA, PhD), organic chemistry (MA, PhD), physical chemistry (MA, PhD), theoretical chemistry (MA, PhD). Faculty: 20 full-time (0 women), 0 part-time. Matriculated students: 73 full-time (29 women), 3 part-time (1 woman); includes 4 minority (3 Asian American, 1 Hispanic American), 50 foreign. Average age 28. 300 applicants, 24% accepted. In 1990, 8 master's, 5 doctorates awarded. Terminal master's awarded for partial completion of doctoral program. *Degree requirements:* For master's, 1 foreign language required, thesis not required; for doctorate, 1 foreign language, dissertation. *Entrance requirements:* GRE General Test, GRE Subject Test, TOEFL (minimum score of 550 required). Application deadline: 4/1. Application fee: $45. *Expenses:* Tuition of $15,950 per year full-time, $498 per credit part-time. Fees of $120 per year full-time, $35 per semester part-time. *Financial aid:* Fellowships, research assistantships, teaching assistantships available. Financial aid application deadline: 1/15. • Guilford Jones II, Chairman, 617-353-2515.

Announcement: The chemistry department is housed in modern facilities, occupied in 1985, in the Science and Engineering Center. Departmental facilities include an extensive computer network, a spectroscopy laboratory (high-resolution NMR, MS, and FT-IR), and a biochemical preparation areas, as well as state-of-the-art instrumentation located in individual research laboratories.

See full description on page 91.

Boston University, Graduate School and College of Engineering, Division of Engineering and Applied Sciences, Boston, MA 02215. Offerings include applied chemistry (PhD). Division faculty: 70 full-time (6 women), 26 part-time (2 women). *Degree requirements:* 1 foreign language, dissertation. *Entrance requirements:* GRE General Test, GRE Subject Test, TOEFL. Application deadline: 7/1 (applications processed on a rolling basis). Application fee: $45. *Expenses:* Tuition of $15,950 per year full-time, $498 per credit part-time. Fees of $120 per year full-time, $35 per semester part-time. • John Baillieul, Director, 617-353-9848.

Bowling Green State University, College of Arts and Sciences, Department of Chemistry, Bowling Green, OH 43403. Department awards MAT, MS. Faculty: 14 full-time (1 woman), 0 part-time. Matriculated students: 28 full-time (10 women), 5 part-time (3 women); includes 2 minority (1 Asian American, 1 black American), 24 foreign. 159 applicants, 22% accepted. In 1990, 14 degrees awarded. *Degree requirements:* Thesis or alternative required, foreign language not required. *Entrance requirements:* GRE General Test, TOEFL (minimum score of 550 required). Application fee: $10. *Tuition:* $185 per credit hour for state residents; $357 per credit hour for nonresidents. *Financial aid:* In 1990–91, 1 fellowship, 22 teaching assistantships (9 to first-year students), 2 non-teaching assistantshi (1 to a first-year student) awarded. Financial aid application deadline: 2/15; applicants required to submit FAF. *Faculty research:* Interactions of nucleic acids with proteins, molten salt chemistry, organic synthesis, coordination chemistry, laser spectroscopy. Total annual research budget: $188,897. • Dr. Douglas C. Neckers, Chairman, 419-372-2031. Application contact: Dr. Thomas Kinstle, Graduate Coordinator, 419-372-2658.

Bradley University, College of Liberal Arts and Sciences, Department of Chemistry, Peoria, IL 61625. Department awards MS. Part-time and evening/weekend programs available. Faculty: 17 full-time, 0 part-time. Matriculated students: 0 full-time, 1 part-time; includes 0 minority. In 1990, 1 degree awarded. *Degree requirements:* Thesis, comprehensive exam required, foreign language not required.

Directory: Chemistry

Application deadline: 6/1. *Application fee:* $30. *Expenses:* Tuition of $8500 per year full-time, $231 per semester hour part-time. Fees of $4 per semester full-time. *Financial aid:* Application deadline 4/15. • Dr. A. D. Glover, Chairperson, 309-677-3028.

Brandeis University, Graduate School of Arts and Sciences, Program in Chemistry, Waltham, MA 02254. Offers inorganic chemistry (MA, PhD), organic chemistry (MA, PhD), physical chemistry (MA, PhD). Faculty: 16 full-time (2 women), 0 part-time. Matriculated students: 51 full-time (21 women), 0 part-time; includes 3 minority (2 Asian American, 1 Hispanic American), 30 foreign. Average age 25. 156 applicants, 8% accepted. In 1990, 6 master's awarded (33% found work related to degree, 67% continued full-time study); 8 doctorates awarded (50% entered university research/teaching, 50% found other work related to degree). Terminal master's awarded for partial completion of doctoral program. *Degree requirements:* For master's, 1 year residency required, foreign language and thesis not required; for doctorate, 1 foreign language (computer language can substitute), dissertation, 2 year residency, 2 seminars, qualifying exams, minimum GPA of 3.0. *Entrance requirements:* GRE, TOEFL. Application deadline: 2/28 (priority date, applications processed on a rolling basis). Application fee: $50. *Tuition:* $16,085 per year full-time, $2015 per course part-time. *Financial aid:* In 1990–91, 50 students received a total of $306,500 in aid awarded. 0 fellowships, 25 research assistantships, 24 teaching assistantships were awarded; institutionally sponsored loans also available. Average monthly stipend for a graduate assistantship: $970. Financial aid application deadline: 2/28; applicants required to submit GAPSFAS. *Faculty research:* Oscillating chemical reactions, molecular recognition systems, protein crystallography, synthesis of natural product spectroscopy and magnetic resonance. Total annual research budget: $1.676-million. • Dr. Peter Jordan, Chair, 617-736-2503. Application contact: Susan Isaacs, Secretary, 617-736-2593.

See full description on page 93.

Bridgewater State College, Division of Natural Sciences and Mathematics, Department of Chemical Sciences, Bridgewater, MA 02324. Department awards MA, MAT, CAGS. Evening/weekend programs available. Faculty: 4 full-time. Matriculated students: 0 full-time, 5 part-time. *Degree requirements:* For master's, 1 foreign language, thesis. *Entrance requirements:* For master's, GRE General Test, GRE Subject Test. Application deadline: 3/1. Application fee: $25. *Tuition:* $1446 per semester for state residents; $3303 per semester for nonresidents. • Dr. Vale Marganian, Chairperson, 508-697-1200.

Brigham Young University, College of Physical and Mathematical Sciences, Department of Chemistry, Provo, UT 84602. Department offers programs in analytical chemistry (MS, PhD), biochemistry (MS, PhD), inorganic chemistry (MS, PhD), organic chemistry (MS, PhD), physical chemistry (MS, PhD). Faculty: 34 full-time (1 woman), 0 part-time. Matriculated students: 56 full-time (13 women), 2 part-time (0 women); includes 0 minority, 22 foreign. Average age 29. 70 applicants, 26% accepted. In 1990, 7 master's awarded (14% entered university research/teaching, 14% found other work related to degree, 72% continued full-time study); 11 doctorates awarded (9% entered university research/teaching, 73% found other work related to degree, 18% continued full-time study). *Degree requirements:* 2 foreign languages (computer language can substitute for one), thesis/dissertation. *Entrance requirements:* For master's, minimum GPA of 3.0 during previous 60 credit hours. Application deadline: 3/15 (priority date). Application fee: $30. *Tuition:* $1170 per semester (minimum) full-time, $130 per credit hour (minimum) part-time. *Financial aid:* In 1990–91, 48 students received a total of $636,735 in aid awarded. 4 fellowships (0 to first-year students), 25 research assistantships (2 to first-year students), 23 teaching assistantships (9 to first-year students), 2 scholarships (1 to a first-year student) were awarded; full and partial tuition waivers, institutionally sponsored loans also available. Average monthly stipend for a graduate assistantship: $970. *Total annual research budget:* $3-million. • Dr. Earl M. Woolley, Chairman, 801-378-3667. Application contact: Francis R. Nordmeyer, Coordinator, Graduate Admissions, 801-378-4845.

See full description on page 95.

Brock University, Faculty of Mathematics and Science, Department of Chemistry, St. Catharines, ON L2S 3A1, Canada. Department awards M Sc. Part-time programs available. Faculty: 11 full-time, 1 part-time. Matriculated students: 14 full-time, 1 part-time; includes 11 foreign. Average age 26. 16 applicants, 31% accepted. In 1990, 8 degrees awarded. *Degree requirements:* Thesis optional, foreign language not required. *Entrance requirements:* Honor's B Sc. Application fee: $0. *Tuition:* $2667 per year full-time, $360 per trimester part-time for Canadian residents; $11,010 per year full-time for nonresidents. *Financial aid:* In 1990–91, 3 fellowships, 8 research assistantships (4 to first-year students), 12 teaching assistantships (4 to first-year students) awarded; career-related internships or fieldwork also available. Aid available to part-time students. *Faculty research:* Inorganic, organic, analytical, theoretical, and physical chemistry. Total annual research budget: $444,000. • Dr. S. Rothstein, Chair, 416-688-5550 Ext. 3405. Application contact: Dr. Gibson, Graduate Officer, 416-688-5550 Ext. 3848.

Brooklyn College of the City University of New York, Department of Chemistry, Bedford Avenue and Avenue H, Brooklyn, NY 11210. Department awards MA. Part-time programs available. Faculty: 19 full-time (1 woman), 0 part-time. Matriculated students: 6 full-time (2 women), 16 part-time (3 women); includes 2 minority (1 Asian American, 1 black American), 17 foreign. In 1990, 10 degrees awarded. *Degree requirements:* 1 foreign language (computer language can substitute), thesis or alternative. *Entrance requirements:* TOEFL. Application deadline: 4/1. Application fee: $30. *Expenses:* Tuition of $1202 per semester full-time for state residents; $2350 per semester full-time for nonresidents. Fees of $45 per semester full-time. • Dr. P. Gary Mennitt, Chairperson, 718-780-5458. Application contact: Darryl G. Howery, Graduate Deputy, 718-780-5457.

Brown University, Department of Chemistry, Providence, RI 02912. Department offers programs in biochemistry (PhD), chemistry (Sc M, PhD). Faculty: 20 full-time (3 women), 0 part-time. Matriculated students: 65 full-time (20 women), 1 (woman) part-time; includes 8 minority (5 Asian American, 1 black American, 2 Hispanic American), 30 foreign. 62 applicants, 34% accepted. In 1990, 4 master's, 16 doctorates awarded. *Degree requirements:* For master's, thesis required, foreign language not required; for doctorate, 1 foreign language, dissertation, cumulative exam. *Application deadline:* 1/2. *Application fee:* $40. *Expenses:* Tuition of $16,256 per year full-time, $2032 per course part-time. Fees of $372 per year. *Financial aid:* In 1990–91, $1.167-million in aid awarded. 3 fellowships (2 to first-year students), 35 research assistantships (1 to a first-year student), 36 teaching assistantships (13 to first-year students), 0 assistantships were awarded; full and partial tuition waivers, institutionally sponsored loans also available. Financial aid application deadline: 1/2; applicants required to submit GAPSFAS. • Peder Estrup, Chairman, 401-863-3588. Application contact: Ronald Lawler, Graduate Representative, 401-863-3385.

Bryn Mawr College, Graduate School of Arts and Sciences, Department of Chemistry, Bryn Mawr, PA 19010. Department offers programs in biochemistry (MA, PhD), chemistry (MA, PhD). Faculty: 10. Matriculated students: 7 full-time (5 women), 6 part-time (4 women); includes 3 foreign. 14 applicants, 64% accepted. In 1990, 0 master's, 1 doctorate awarded. *Degree requirements:* For master's, 1 foreign language, thesis; for doctorate, 2 foreign languages, dissertation. *Entrance requirements:* GRE General Test, GRE Subject Test. Application deadline: 8/20. Application fee: $35. *Tuition:* $13,900 per year full-time, $2350 per course part-time. *Financial aid:* In 1990–91, 0 fellowships, 0 research assistantships, 10 teaching assistantships awarded; federal work-study, institutionally sponsored loans also available. Aid available to part-time students. Financial aid application deadline: 2/1; applicants required to submit GAPSFAS. • Dr. Frank Mallory, Chairman, 215-526-5105. Application contact: Patricia Saukewitsch, Administrative Coordinator, 215-526-5072.

Bucknell University, College of Arts and Sciences, Department of Chemistry, Lewisburg, PA 17837. Department awards MA, MS. Faculty: 11 full-time. Matriculated students: 7. *Degree requirements:* Thesis required, foreign language not required. *Entrance requirements:* GRE General Test, GRE Subject Test, TOEFL (minimum score of 550 required), minimum GPA of 2.8. Application deadline: 6/1 (applications processed on a rolling basis). Application fee: $25. *Tuition:* $1800 per course. *Financial aid:* Assistantships available. Average monthly stipend for a graduate assistantship: $630. • Dr. Hans Veening, Head, 717-524-3258.

Butler University, College of Liberal Arts and Sciences, Department of Chemistry, Indianapolis, IN 46208. Department awards MS. Part-time and evening/weekend programs available. Faculty: 6 full-time (0 women), 4 part-time (1 woman). Matriculated students: 0 full-time, 13 part-time (3 women); includes 1 black American, 1 Hispanic American, 0 foreign. Average age 31. 5 applicants, 100% accepted. In 1990, 3 degrees awarded. *Degree requirements:* Thesis required (for some programs), foreign language not required. *Application fee:* $20. *Tuition:* $145 per credit (minimum). *Financial aid:* Career-related internships or fieldwork available. *Faculty research:* Organic synthesis, biophysical, theoretical, laser spectroscopy, environmental, analytical. • Dr. Joseph Kirsch, Head, 317-283-9405.

California Institute of Technology, Division of Chemistry and Chemical Engineering, Department of Chemistry, Pasadena, CA 91125. Department awards MS, PhD. Faculty: 26 full-time (3 women), 0 part-time. Matriculated students: 204 full-time (57 women), 0 part-time; includes 18 minority (16 Asian American, 2 Hispanic American), 58 foreign. Average age 24. 311 applicants, 40% accepted. In 1990, 2 master's awarded (100% found work related to degree); 21 doctorates awarded (57% entered university research/teaching, 43% found other work related to degree). Terminal master's awarded for partial completion of doctoral program. *Degree requirements:* Thesis/dissertation required, foreign language not required. *Application deadline:* 1/15. *Application fee:* $0. *Expenses:* Tuition of $14,100 per year. Fees of $8 per quarter. *Financial aid:* In 1990–91, 190 students received aid. 35 fellowships (9 to first-year students), 78 research assistantships (0 to first-year students), 67 teaching assistantships (28 to first-year students) awarded; federal work-study, institutionally sponsored loans also available. Average monthly stipend for a graduate assistantship: $1085. Financial aid application deadline: 1/15. *Faculty research:* Genetic structure and gene expression, organic synthesis, reagents and mechanisms, homogeneous and electrochemical catalysis. • Fred C. Anson, Chairman, Division of Chemistry and Chemical Engineering, 818-356-3646.

California Polytechnic State University, San Luis Obispo, School of Science and Mathematics, Department of Chemistry, San Luis Obispo, CA 93407. Department awards MS. Part-time programs available. Faculty: 28 full-time, 12 part-time. Matriculated students: 5 full-time (3 women), 0 part-time; includes 0 minority. In 1990, 3 degrees awarded. *Degree requirements:* Thesis optional, foreign language not required. *Entrance requirements:* Minimum GPA of 3.0. Application fee: $55. *Financial aid:* In 1990–91, 2 teaching assistantships (both to first-year students) awarded; federal work-study and career-related internships or fieldwork also available. Aid available to part-time students. *Faculty research:* Organic synthesis, electroanalytical methods, analysis for air particulates, antibody modification. Total annual research budget: $93,000. • Dr. Norman Eatough, Chair, 805-756-2693.

California State Polytechnic University, Pomona, College of Science, Department of Chemistry, Pomona, CA 91768. Department awards MS. Matriculated students: 1 full-time (0 women), 20 part-time (7 women); includes 5 minority (all Asian American). 5 applicants, 100% accepted. In 1990, 5 degrees awarded. *Degree requirements:* Thesis. *Application fee:* $55. *Expenses:* Tuition of $0 for state residents; $137 per unit for nonresidents. Fees of $880 per year full-time, $182 per quarter (minimum) part-time. *Financial aid:* Career-related internships or fieldwork available. Financial aid application deadline: 3/2. • Dr. Yu-Ping Hsia, Coordinator, 714-869-3670.

California State University, Chico, College of Natural Sciences, Department of Geology and Physical Sciences, Program in Physical Science, Chico, CA 95929. Offerings include chemistry (MS). Program faculty: 5 full-time (1 woman), 5 part-time (3 women). *Degree requirements:* Thesis required, foreign language not required. *Entrance requirements:* GRE General Test. Application deadline: 4/1 (applications processed on a rolling basis). Application fee: $55. *Expenses:* Tuition of $548 per semester full-time, $350 per semester part-time. Fees of $246 per unit for nonresidents. • Dr. K. R. Gina Rothe, Graduate Coordinator, 916-898-6269.

California State University, Fresno, Division of Graduate Studies and Research, School of Natural Sciences, Department of Chemistry, 5241 North Maple Avenue, Fresno, CA 93710. Department awards MS. Faculty: 21 full-time, 5 part-time. Matriculated students: 18. *Degree requirements:* Thesis or alternative required, foreign language not required. *Entrance requirements:* GRE General Test (minimum score of 450 on verbal or 430 on quantitative required). *Tuition:* $1098 per year full-time, $4485 per year part-time. • Kenneth Russell, Chairman, 209-278-2103. Application contact: Joe Gandler, Graduate Coordinator, 209-278-2103.

California State University, Fullerton, School of Natural Science and Mathematics, Department of Chemistry and Biochemistry, Fullerton, CA 92634. Department offers programs in analytical chemistry (MS), biochemistry (MS), geochemistry (MS), inorganic chemistry (MS), organic chemistry (MS), physical chemistry (MS). Part-time programs available. Faculty: 20 full-time (5 women), 13 part-time (2 women). Matriculated students: 2 full-time (1 woman), 23 part-time (4 women); includes 6 minority (all Asian American), 8 foreign. Average age 28. 33 applicants, 67% accepted. In 1990, 4 degrees awarded. *Degree requirements:* Thesis, departmental qualifying exam required, foreign language not required. *Entrance requirements:* Minimum GPA of 2.5 in last 60 units, major in chemistry or related field. Application fee: $55. *Expenses:* Tuition of $0 for state residents; $246 per unit for nonresidents. Fees of $554 per semester full-time, $356 per unit part-time. *Financial aid:* In 1990–91, 6 teaching assistantships, 0 state grants awarded; federal work-study, institutionally sponsored loans, and career-related internships or fieldwork also available. Aid available to part-time students. Financial aid application deadline: 3/1.

Total annual research budget: \$815,000. • Dr. Glen Nagel, Chair, 714-773-3621. Application contact: Dr. Gene Hiegel, Adviser, 714-773-3624.

Announcement: MS degree concentrations include analytical, inorganic, organic, and physical chemistry; biochemistry; and geochemistry. Highly productive faculty, modern facilities, and pleasant suburban location characterize the department. Assistantships are available. Evening classes and library thesis option offered. Graduates have excellent record of placement in PhD programs, teaching positions, and industry.

California State University, Hayward, School of Science, Department of Chemistry, Hayward, CA 94542. Department offers programs in biochemistry (MS), chemistry (MS). Faculty: 11. Matriculated students: 1 full-time (0 women), 16 part-time (6 women). Average age 30. In 1990, 4 degrees awarded. *Degree requirements:* Thesis or comprehensive exam required, foreign language not required. *Entrance requirements:* Minimum GPA of 2.5 in field in previous 2 years. Application deadline: 4/19 (priority date, applications processed on a rolling basis). Application fee: \$55. *Expenses:* Tuition of \$0 for state residents; \$137 per unit for nonresidents. Fees of \$895 per year full-time, \$188 per quarter part-time. *Financial aid:* Federal work-study, institutionally sponsored loans, and career-related internships or fieldwork available. Aid available to part-time students. Financial aid application deadline: 3/1. • Dr. Charles T. Perrino, Chair, 415-881-3452. Application contact: Dr. Robert Trinchero, Acting Associate Vice President, Admissions and Enrollment, 415-881-3828.

California State University, Long Beach, School of Natural Sciences, Department of Chemistry, 1250 Bellflower Boulevard, Long Beach, CA 90840. Department offers programs in biochemistry (MS), chemistry (MS). Matriculated students: 5 full-time (2 women), 7 part-time (2 women); includes 4 minority (all Asian American), 2 foreign. Average age 30. 6 applicants, 67% accepted. In 1990, 2 degrees awarded. *Degree requirements:* Thesis, departmental qualifying exam required, foreign language not required. *Application deadline:* 8/1 (applications processed on a rolling basis). *Application fee:* \$55. *Expenses:* Tuition of \$0 for state residents; \$246 per unit for nonresidents. Fees of \$1120 per year full-time, \$724 per year part-time. *Financial aid:* Application deadline 3/2; applicants required to submit FAF. • Dr. Kenneth Marsi, Chair, 310-985-4941.

California State University, Los Angeles, School of Natural and Social Sciences, Department of Chemistry and Biochemistry, Los Angeles, CA 90032. Department offers programs in analytical chemistry (MS), biochemistry (MS), chemistry (MS), inorganic chemistry (MS), organic chemistry (MS), physical chemistry (MS). Part-time and evening/weekend programs available. Faculty: 12 full-time, 25 part-time. Matriculated students: 5 full-time (2 women), 23 part-time (5 women); includes 12 minority (8 Asian American, 2 black American, 2 Hispanic American), 6 foreign. In 1990, 3 degrees awarded. *Degree requirements:* 1 foreign language (computer language can substitute), thesis or comprehensive exam. *Entrance requirements:* TOEFL (minimum score of 550 required). Application deadline: 8/7 (applications processed on a rolling basis). Application fee: \$55. *Expenses:* Tuition of \$0 for state residents; \$164 per unit for nonresidents. Fees of \$1046 per year full-time, \$650 per semester (minimum) part-time. *Financial aid:* Federal work-study available. Aid available to part-time students. Financial aid application deadline: 3/1; applicants required to submit FAF. *Faculty research:* Transition-metal chemistry, NMR studies on organoloaron compounds, computer modeling of reactions. • Dr. Donald Paulson, Acting Chair, 213-343-2300.

Announcement: MS degree concentrations include analytical, inorganic, organic, and physical chemistry and biochemistry. Evening classes and library thesis option available. Highly productive faculty, modern facilities, and pleasant suburban location characterize the department. Assistantships are available. Graduates have excellent record of placement in PhD programs, teaching positions, and industry.

California State University, Northridge, School of Science and Mathematics, Department of Chemistry, Northridge, CA 91330. Department awards MS. Faculty: 11. Matriculated students: 1 full-time (0 women), 25 part-time (10 women). Average age 30. 14 applicants, 71% accepted. In 1990, 7 degrees awarded. *Degree requirements:* Thesis required, foreign language not required. *Entrance requirements:* Minimum GPA of 2.5 or GRE General Test. Application deadline: 11/30. Application fee: \$55. *Expenses:* Tuition of \$0 for state residents; \$205 per unit for nonresidents. Fees of \$1128 per year full-time, \$366 per semester part-time. *Financial aid:* Teaching assistantships available. Aid available to part-time students. *Total annual research budget:* \$40,000. • Dr. Dean Skovlin, Chairman, 818-885-3381. Application contact: Dr. Francis L. Harris, Graduate Coordinator, 818-885-3371.

California State University, Sacramento, School of Arts and Sciences, Department of Chemistry, Sacramento, CA 95819. Department awards MS. Part-time programs available. *Degree requirements:* Thesis or alternative, writing proficiency exam, departmental qualifying exam required, foreign language not required. *Entrance requirements:* TOEFL (minimum score of 550 required), minimum GPA of 2.5 during previous 2 years and in math and science, BA in chemistry or equivalent. Application fee: \$45. *Expenses:* Tuition of \$0 for state residents; \$246 per unit for nonresidents. Fees of \$530 per semester full-time, \$332 per semester part-time.

Carleton University, Faculty of Science, Ottawa-Carleton Chemistry Institute, Ottawa, ON K1S 5B6, Canada. Institute awards M Sc, PhD. M Sc and PhD offered jointly with the University of Ottawa. Faculty: 36. Matriculated students: 92 full-time, 8 part-time; includes 9 foreign. In 1990, 5 master's, 16 doctorates awarded. *Degree requirements:* For master's, thesis; for doctorate, dissertation, comprehensive exam. *Entrance requirements:* For master's, TOEFL (minimum score of 550 required), honor's B Sc degree; for doctorate, TOEFL (minimum score of 550 required), M Sc. Application deadline: 7/1 (priority date, applications processed on a rolling basis). Application fee: \$15. *Tuition:* \$985 per semester full-time, \$284 per semester part-time for Canadian residents; \$3939 per semester full-time, \$1171 per semester part-time for nonresidents. *Financial aid:* Application deadline 3/1. *Faculty research:* Bioorganic chemistry, analytical toxicology, theoretical and physical chemistry, inorganic chemistry. • A. G. Fallis, Director, 613-788-3841.

Carnegie Mellon University, Mellon College of Science, Department of Chemistry, Pittsburgh, PA 15213. Department offers programs in chemical instrumentation (MS); chemistry (MS, PhD); colloids, polymers and surfaces (MS); polymer science (MS). Faculty: 22 full-time (1 woman), 0 part-time. Matriculated students: 54 full-time (12 women), 3 part-time (0 women); includes 0 minority, 16 foreign. Average age 25. 179 applicants, 31% accepted. In 1990, 10 master's, 12 doctorates awarded. Terminal master's awarded for partial completion of doctoral program. *Degree requirements:* For master's, thesis or oral exam, departmental qualifying exam required, foreign language not required; for doctorate, dissertation, cumulative exams, teaching experience, departmental qualifying exam required, foreign language not required. *Entrance requirements:* For master's, GRE General Test; for doctorate, GRE General Test, GRE Subject Test, TOEFL. Application deadline: 3/1 (priority date, applications processed on a rolling basis). Application fee: \$0. *Expenses:* Tuition of \$15,250 per year full-time, \$212 per unit part-time. Fees of \$80 per year. *Financial aid:* In 1990–91, 57 students received a total of \$1.311-million in aid awarded. 28 research assistantships, 26 teaching assistantships (10 to first-year students) were awarded; fellowships also available. Average monthly stipend for a graduate assistantship: \$950. *Faculty research:* Physical and theoretical chemistry, chemical synthesis, polymer science, biophysical/bioinorganic chemistry. Total annual research budget: \$3.6-million. • Dr. Guy C. Berry, Acting Head, 412-268-3125. Application contact: Dr. Charles L. Brooks, Chairman, Graduate Recruitment Committee, 412-268-3176.

See full description on page 97.

Case Western Reserve University, Department of Chemistry, Cleveland, OH 44106. Department awards MS, PhD. Part-time programs available. Faculty: 19 full-time (0 women), 0 part-time. Matriculated students: 16 full-time (3 women), 62 part-time (7 women); includes 2 minority (1 Asian American, 1 black American), 49 foreign. 361 applicants, 14% accepted. In 1990, 6 master's awarded (33% found work related to degree, 67% continued full-time study); 19 doctorates awarded (47% found work related to degree, 53% continued full-time study). Terminal master's awarded for partial completion of doctoral program. *Degree requirements:* For master's, foreign language and thesis not required; for doctorate, 1 foreign language, dissertation. *Entrance requirements:* GRE General Test, GRE Subject Test, TOEFL (minimum score of 550 required). *Expenses:* Tuition of \$13,600 per year full-time, \$567 per credit part-time. Fees of \$320 per year. *Financial aid:* In 1990–91, 0 students received aid. Fellowships, research assistantships, teaching assistantships available. Average monthly stipend for a graduate assistantship: \$900. Financial aid applicants required to submit FAF. *Faculty research:* Electrochemistry, synthetic chemistry, chemistry of living systems. • Gilles Klopman, Chairman, 216-368-6526. Application contact: Dr. Philip Paul Garner, Chairman, Graduate Admission, 216-368-3696.

See full description on page 99.

Catholic University of America, School of Arts and Sciences, Department of Chemistry, Washington, DC 20064. Department awards MS, PhD. Part-time programs available. Faculty: 9 full-time (2 women), 1 (woman) part-time. Matriculated students: 9 full-time (6 women), 15 part-time (7 women); includes 1 minority (black American), 19 foreign. Average age 34. 60 applicants, 10% accepted. In 1990, 3 master's, 1 doctorate awarded. Terminal master's awarded for partial completion of doctoral program. *Degree requirements:* For master's, 1 foreign language, thesis or alternative, comprehensive exam; for doctorate, 1 foreign language, dissertation, comprehensive exam. *Entrance requirements:* For master's, GRE General Test, GRE Subject Test, TOEFL; for doctorate, GRE General Test, GRE Subject Test. Application deadline: 8/1 (priority date, applications processed on a rolling basis). Application fee: \$30. *Expenses:* Tuition of \$11,626 per year full-time, \$445 per credit hour part-time. Fees of \$240 per year full-time, \$90 per year part-time. *Financial aid:* In 1990–91, 16 students received a total of \$109,200 in aid awarded. 3 research assistantships (0 to first-year students), 14 teaching assistantships (7 to first-year students), 2 scholarships (both to first-year students) were awarded; fellowships, institutionally sponsored loans, and career-related internships or fieldwork also available. Aid available to part-time students. Average monthly stipend for a graduate assistantship: \$911. Financial aid application deadline: 2/15; applicants required to submit GAPSFAS or FAF. *Faculty research:* Theoretical chemistry, bioinorganic chemistry, chemical kinetics, synthetic organic chemistry. Total annual research budget: \$459,200. • Dr. David Gutman, Chair, 202-319-5385. Application contact: Dr. Anthony Ponaras, Graduate Admissions Chairman, 202-319-6093.

Announcement: The department accepts a limited number of graduate students, maintaining small classes and emphasizing individualized instruction. All faculty members are research-active. Funding sources include NSF, DOE, Navy, NIH, and NASA. Distinguished adjunct faculty members offer additional research opportunities. Recently acquired instrumentation includes FT infrared and diode-array UV-visible spectrometers.

Catholic University of Puerto Rico, College of Sciences, Department of Chemistry, Ponce, PR 00732. Department awards MS. Part-time and evening/weekend programs available. Faculty: 0 full-time, 5 part-time (2 women). Matriculated students: 1 (woman) full-time, 22 part-time (14 women); includes 23 minority (all Hispanic American), 0 foreign. Average age 28. 7 applicants, 57% accepted. In 1990, 3 degrees awarded. *Degree requirements:* Thesis required, foreign language not required. *Entrance requirements:* GRE General Test, minimum GPA of 3.0, minimum of 37 credits in chemistry. Application deadline: 4/30 (priority date, applications processed on a rolling basis). Application fee: \$15. *Expenses:* Tuition of \$95 per credit hour. Fees of \$106 per year full-time, \$60 per year part-time. *Financial aid:* In 1990–91, 1 fellowship (0 to first-year students), 4 research assistantships (0 to first-year students) awarded; partial tuition waivers, federal work-study also available. Aid available to part-time students. Financial aid application deadline: 7/15. *Faculty research:* Chemical radiosensitization on cancerous tissues, biological action of nitrofurans, analysis of atmospheric particulates, radiation chemistry, synthesis and characterization of platinum complexes. • Carmen Velazquez, Director, 809-841-2000 Ext. 284. Application contact: Caridad C. Frau, Director of Admissions, 809-841-2000 Ext. 426.

Central Connecticut State University, School of Arts and Sciences, Department of Chemistry, New Britain, CT 06050. Department awards MS. Faculty: 1 (woman) full-time, 0 part-time. Matriculated students: 2 full-time (1 woman), 3 part-time (2 women); includes 0 minority, 0 foreign. 21 applicants, 19% accepted. In 1990, 0 degrees awarded. *Degree requirements:* Thesis or alternative, comprehensive exam. *Entrance requirements:* TOEFL (minimum score of 550 required for international students), minimum GPA of 2.7. Application deadline: 8/31. Application fee: \$20. *Expenses:* Tuition of \$1720 per year full-time, \$115 per credit part-time for state residents; \$4790 per year full-time, \$115 per credit part-time for nonresidents. Fees of \$1013 per year full-time, \$33 per semester part-time for state residents; \$1655 per year full-time, \$33 per semester part-time for nonresidents. *Financial aid:* In 1990–91, Research assistantships available. Financial aid application deadline: 3/15; applicants required to submit FAF. • Dr. John Mantzaris, Chair, 203-827-7293.

Central Michigan University, College of Arts and Sciences, Department of Chemistry, Mount Pleasant, MI 48859. Department awards MAT, MS, PhD. PhD offered jointly with Michigan Technological University. Faculty: 15 full-time (0 women). In 1990, 3 master's awarded. *Degree requirements:* For master's, thesis or alternative required, foreign language not required. *Application deadline:* 7/15 (priority date, applications processed on a rolling basis). *Application fee:* \$30. *Tuition:* \$96.50 per credit for state residents; \$210.50 per credit for nonresidents. *Financial aid:* In 1990–91, 9 teaching assistantships awarded; federal work-study also available. Financial aid applicants required to submit FAF. *Faculty research:* Bio, analytical, and organic-

Directory: Chemistry

inorganic chemistry; polymer chemistry. • Dr. Robert Kohrman, Chairperson, 517-774-3981.

Central Washington University, College of Letters, Arts, and Sciences, Department of Chemistry, Ellensburg, WA 98926. Department awards MS. Part-time programs available. Faculty: 7 full-time (0 women), 0 part-time. Matriculated students: 0. 0 applicants. In 1990, 0 degrees awarded. *Degree requirements:* Thesis or alternative required, foreign language not required. *Entrance requirements:* GRE General Test, GRE Subject Test, minimum GPA of 3.0. Application deadline: 7/1. Application fee: $25. *Expenses:* Tuition of $90 per credit for state residents; $273 per credit for nonresidents. Fees of $25 per year. *Financial aid:* 0 students received aid. Research assistantships, federal work-study available. Financial aid application deadline: 2/15; applicants required to submit FAF. • Dr. Robert Gaines, Chairman, 509-963-2811. Application contact: Britta Jo Hammond, Administrative Assistant, Graduate Studies and Research, 509-963-3103.

City College of the City University of New York, Graduate School, College of Liberal Arts and Science, Division of Science, Department of Chemistry, Convent Avenue at 138th Street, New York, NY 10031. Department offers programs in biochemistry (MA, PhD), chemistry (MA, PhD). Faculty: 23 full-time (2 women), 0 part-time. Matriculated students: 4 full-time (0 women), 30 part-time (9 women). 40 applicants, 88% accepted. Terminal master's awarded for partial completion of doctoral program. *Degree requirements:* For master's, foreign language and thesis not required; for doctorate, 1 foreign language, dissertation. *Entrance requirements:* For master's, TOEFL (minimum score of 300 required); for doctorate, GRE, TOEFL. Application deadline: 6/1 (priority date, applications processed on a rolling basis). Application fee: $30. *Expenses:* Tuition of $2204 per year full-time, $95 per credit part-time for state residents; $4700 per year full-time, $199 per credit part-time for nonresidents. Fees of $15 per semester. *Financial aid:* Federal work-study available. Financial aid application deadline: 6/1. *Total annual research budget:* $2.5-million. • Michael Green, Chairman, 212-650-8402. Application contact: Neil McKelvie, Graduate Adviser, 212-650-6063.

Clarkson University, School of Science, Department of Chemistry, Potsdam, NY 13699. Department offers programs in analytical chemistry (MS, PhD), inorganic chemistry (MS, PhD), organic chemistry (MS, PhD), physical chemistry (MS, PhD). Faculty: 12 full-time (0 women), 0 part-time. Matriculated students: 56 full-time (25 women), 0 part-time; includes 0 minority, 40 foreign. Average age 30. 82 applicants, 60% accepted. In 1990, 5 master's, 10 doctorates awarded. *Degree requirements:* For master's, foreign language and thesis not required; for doctorate, dissertation, departmental qualifying exam required, foreign language not required. *Application fee:* $10. *Expenses:* Tuition of $446 per credit hour. Fees of $75 per semester. *Financial aid:* In 1990–91, 1 fellowship, 39 research assistantships, 11 teaching assistantships awarded. *Faculty research:* Chemical kinetics, colloid and surface science, light scattering, environmental chemistry. Total annual research budget: $1.29-million. • Dr. Phillip A. Christiansen, Chairman, 315-268-2389.

Clark University, Department of Chemistry, Worcester, MA 01610. Department awards MA, PhD. Faculty: 10 full-time (1 woman), 0 part-time. Matriculated students: 17 (9 women); includes 13 foreign. In 1990, 0 master's, 0 doctorates awarded. Terminal master's awarded for partial completion of doctoral program. *Degree requirements:* For master's, thesis or alternative required, foreign language not required; for doctorate, 1 foreign language, dissertation. *Entrance requirements:* GRE General Test, TOEFL. Application deadline: 2/15 (priority date, applications processed on a rolling basis). Application fee: $40. *Tuition:* $15,000 per year full-time, $1875 per course part-time. *Financial aid:* Research assistantships, teaching assistantships available. *Faculty research:* Nuclear chemistry, molecular biology. Total annual research budget: $447,163. • Dr. Frederick Greenaway, Chair, 508-793-7116. Application contact: Dr. Donald J. Nelson, Coordinator, Graduate Admissions, 508-793-7121.

Clemson University, College of Sciences, Department of Chemistry, Clemson, SC 29634. Department awards MS, PhD. Faculty: 19 full-time (1 woman), 0 part-time. Matriculated students: 65 full-time (14 women), 0 part-time; includes 4 minority (1 Asian American, 3 black American), 30 foreign. Average age 25. 40 applicants, 38% accepted. In 1990, 4 master's awarded; 12 doctorates awarded (67% entered university research/teaching, 33% found other work related to degree). *Degree requirements:* 1 foreign language (computer language can substitute), thesis/dissertation. *Entrance requirements:* GRE General Test. Application deadline: 6/1. Application fee: $25. *Expenses:* Tuition of $102 per credit hour. Fees of $80 per semester full-time. *Financial aid:* In 1990–91, $500,000 in aid awarded. 1 fellowship (to a first-year student), 18 research assistantships (1 to a first-year student), 43 lab assistantships were awarded; teaching assistantships also available. *Faculty research:* Flourine chemistry, organic synthetic methods and natural products, metal and non-metal clusters, analytical spectroscopies, polymers. Total annual research budget: $2.45-million. • J. D. Petersen, Head, 803-656-5017.

Cleveland State University, College of Arts and Sciences, Department of Chemistry, Cleveland, OH 44115. Offerings include clinical chemistry (MS, PhD). Department faculty: 18 full-time (1 woman), 4 part-time (0 women). *Degree requirements:* For master's, thesis required (for some programs), foreign language not required; for doctorate, dissertation required, foreign language not required. *Entrance requirements:* For master's, GRE General Test, GRE Subject Test, TOEFL (minimum score of 525 required); for doctorate, GRE General Test (score in 50th percentile or higher required), GRE Subject Test, TOEFL (minimum score of 525 required). Application deadline: 9/1 (priority date, applications processed on a rolling basis). Application fee: $0. *Tuition:* $90 per credit for state residents; $180 per credit for nonresidents. • Dr. Robert Towns, Chair, 216-687-2004. Application contact: Dr. Thomas Flechtner, Chairman, Graduate Committee, 216-687-2458.

College of Staten Island of the City University of New York, Program in Polymer Chemistry, Staten Island, NY 10314. Program awards PhD. Offered jointly with City University Graduate Center and Brooklyn College of the City University of New York. Faculty: 5 full-time (1 woman), 0 part-time. Matriculated students: 13 full-time (11 women), 0 part-time; includes 8 foreign. Average age 23. 20 applicants, 15% accepted. In 1990, 3 doctorates awarded (100% found work related to degree). Terminal master's awarded for partial completion of doctoral program. *Degree requirements:* For doctorate, 1 foreign language, dissertation. *Entrance requirements:* For doctorate, GRE. Application deadline: rolling. Application fee: $30. *Expenses:* Tuition of $2204 per year full-time, $95 per credit part-time for state residents; $4700 per year full-time, $199 per credit part-time for nonresidents. Fees of $106 per year full-time, $27 per semester part-time. *Financial aid:* In 1990–91, $170,000 in aid awarded. 0 fellowships, 0 research assistantships, 13 teaching assistantships (3 to first-year students) were awarded; partial tuition waivers also available. Financial aid applicants required to submit FAF. *Faculty research:* Polymer synthesis, characterization and properties. Total annual research budget: $600,000. • Dr. George Odian, Coordinator, 718-390-7994. Application contact: Ramon Hulsey, Director of Admissions, 718-390-7557.

College of William and Mary, Faculty of Arts and Sciences, Department of Chemistry, Williamsburg, VA 23185. Department awards MA, MS. Faculty: 12 full-time, 1 part-time. Matriculated students: 7 full-time (2 women), 4 part-time (0 women); includes 1 minority (Asian American), 2 foreign. Average age 26. In 1990, 7 degrees awarded. *Degree requirements:* Thesis, comprehensive exam required, foreign language not required. *Entrance requirements:* Minimum GPA of 2.5. Application fee: $5. *Expenses:* Tuition of $2240 per year full-time, $120 per credit hour part-time for state residents; $8960 per year full-time, $320 per credit hour part-time for nonresidents. Fees of $1490 per year full-time. *Financial aid:* Research assistantships, teaching assistantships available. • Dr. David W. Thompson, Chairman, 804-253-4664. Application contact: Dr. Richard L. Kiefer, Graduate Director, 804-253-4667.

Colorado School of Mines, Department of Chemistry and Geochemistry, Golden, CO 80401. Department offers programs in applied chemistry (PhD), chemistry (MS), geochemistry (MS, PhD). Part-time programs available. Faculty: 15 full-time (0 women), 2 part-time (0 women). Matriculated students: 28 full-time (8 women), 24 part-time (12 women); includes 0 minority, 16 foreign. 27 applicants, 85% accepted. In 1990, 4 master's, 6 doctorates awarded. *Degree requirements:* For master's, thesis required, foreign language not required; for doctorate, dissertation. *Entrance requirements:* GRE General Test, minimum GPA of 3.0. Application deadline: 2/23. Application fee: $15 ($25 for foreign students). *Expenses:* Tuition of $3178 per year full-time, $124 per semester hour part-time for state residents; $10,304 per year full-time, $344 per semester hour part-time for nonresidents. Fees of $374 per year full-time. *Financial aid:* In 1990–91, 8 fellowships, 12 research assistantships, 10 teaching assistantships awarded. Financial aid application deadline: 2/23. *Faculty research:* Surface analysis, catalysis, aqueous geochemistry, analytical mass spectrometry, geomicrobiology. Total annual research budget: $789,350. • Dr. Stephen Daniel, Head, 303-273-3610.

Colorado State University, College of Natural Sciences, Department of Chemistry, Fort Collins, CO 80523. Department awards MS, PhD. Part-time programs available. Faculty: 21 full-time (2 women), 2 part-time (0 women). Matriculated students: 85 full-time (23 women), 0 part-time; includes 7 minority (4 Asian American, 2 black American, 1 Hispanic American), 11 foreign. Average age 25. 119 applicants, 32% accepted. In 1990, 1 master's, 24 doctorates awarded. Terminal master's awarded for partial completion of doctoral program. *Degree requirements:* For master's, thesis required, foreign language not required; for doctorate, dissertation. *Entrance requirements:* For master's, GRE General Test, GRE Subject Test, TOEFL, minimum GPA of 3.0; for doctorate, GRE General Test, TOEFL, minimum GPA of 3.0. Application deadline: 4/1 (priority date, applications processed on a rolling basis). Application fee: $30. *Tuition:* $1322 per semester full-time for state residents; $3673 per semester full-time for nonresidents. *Financial aid:* In 1990–91, 2 fellowships (both to first-year students), 39 research assistantships (11 to first-year students), 45 teaching assistantships (18 to first-year students), 3 traineeships (0 to first-year students) awarded. Average monthly stipend for a graduate assistantship: $698-750. *Faculty research:* Synthetic organic chemistry, orgamometallic electrochemistry, spectroscopy, NMR. Total annual research budget: $2.7-million. • Oren P. Anderson, Chairman, 303-491-5391.

Columbia University, Graduate School of Arts and Sciences, Division of Natural Sciences, Department of Chemistry, New York, NY 10027. Department awards M Phil, MS, PhD, MD/PhD. Programs offered in inorganic chemistry (M Phil, MS, PhD), organic chemistry (M Phil, MS, PhD), physical chemistry (M Phil, MS, PhD). Faculty: 17 full-time. Matriculated students: 111 full-time (30 women), 0 part-time; includes 45 foreign. 185 applicants, 41% accepted. In 1990, 14 master's, 17 doctorates awarded. *Degree requirements:* For master's, comprehensive exams (MS); 1 foreign language, teaching experience, oral and written exams (M Phil) required, thesis not required; for doctorate, 1 foreign language, dissertation, M Phil. *Entrance requirements:* GRE General Test, GRE Subject Test, TOEFL. Application deadline: 1/5. Application fee: $50. *Expenses:* Tuition of $7836 per semester for state residents; $426 per credit for nonresidents. Fees of $148 per semester for state residents. *Financial aid:* In 1990–91, $1.92-million in aid awarded. 7 fellowships (1 to a first-year student), 71 teaching assistantships (30 to first-year students) were awarded; federal work-study, institutionally sponsored loans also available. Financial aid application deadline: 1/5; applicants required to submit GAPSFAS. *Faculty research:* Chemical physics, biophysics. • Richard Bersohn, Chair, 212-854-2192.

See full description on page 101.

Concordia University, Faculty of Arts and Science, Department of Chemistry and Biochemistry, Montreal, PQ H3G 1M8, Canada. Offerings include chemistry (M Sc, PhD). *Degree requirements:* For master's, 1 foreign language, thesis; for doctorate, 2 foreign languages, dissertation. *Application fee:* $15. *Expenses:* Tuition of $10 per credit for Canadian residents; $195 per credit for nonresidents. Fees of $223 per year full-time, $38.85 per year part-time for Canadian residents; $118 per year full-time, $35.35 per year part-time for nonresidents. • Dr. P. Bird, Chair, 514-848-3366.

Cornell University, Graduate Fields of Arts and Sciences, Field of Chemistry, Ithaca, NY 14853. Field offers programs in analytical chemistry (PhD), bioorganic chemistry (PhD), biophysical chemistry (PhD), chemical physics (PhD), inorganic chemistry (PhD), organic chemistry (PhD), physical chemistry (PhD), polymer chemistry (PhD), theoretical chemistry (PhD). Faculty: 39 full-time, 0 part-time. Matriculated students: 184 full-time (64 women), 0 part-time; includes 15 minority (11 Asian American, 4 Hispanic American), 45 foreign. 473 applicants, 29% accepted. In 1990, 37 master's, 31 doctorates awarded. Terminal master's awarded for partial completion of doctoral program. *Degree requirements:* For doctorate, variable foreign language requirement, dissertation. *Entrance requirements:* For doctorate, GRE General Test, GRE Subject Test, TOEFL. Application deadline: 1/10. Application fee: $55. *Expenses:* Tuition of $16,170 per year. Fees of $28 per year. *Financial aid:* In 1990–91, 136 students received aid. 6 fellowships (0 to first-year students), 71 research assistantships (0 to first-year students), 58 teaching assistantships (32 to first-year students), 1 graduate assistantship (0 to first-year students) awarded; full and partial tuition waivers, federal work-study, institutionally sponsored loans also available. Financial aid application deadline: 1/10; applicants required to submit GAPSFAS. *Faculty research:* Organic, bioorganic, physical, biophysical, theoretical, analytical, inorganic, organometallic, solid state, polymer. • Peter Wolczanski, Graduate Faculty Representative, 607-255-4139. Application contact: Robert Brashear, Director of Admissions, 607-255-4884.

Dalhousie University, College of Arts and Science, Faculty of Science, Department of Chemistry, Halifax, NS B3H 4H6, Canada. Department awards M Sc, PhD. Part-time programs available. Faculty: 24 full-time, 3 part-time. Matriculated students: 42 full-time, 1 part-time. 101 applicants, 34% accepted. In 1990, 2 master's, 5 doctorates awarded. Terminal master's awarded for partial completion of doctoral program. *Degree requirements:* Thesis/dissertation. *Entrance requirements:* GRE Subject Test, TOEFL (minimum score of 550 required). Application deadline: 7/15 (applications processed on a rolling basis). Application fee: $20. *Tuition:* $2594 per year full-time for Canadian residents; $4294 per year full-time for nonresidents. *Financial aid:* In 1990–91, $575,464 in aid awarded. Fellowships available. *Faculty research:* Analytical, organic, physical, and theoretical chemistry. • Dr. J. Kwak, Chair, 902-494-3707.

Dartmouth College, School of Arts and Sciences, Department of Chemistry, Hanover, NH 03755. Department awards MS, PhD. Faculty: 16 full-time (2 women), 0 part-time. Matriculated students: 34 full-time (9 women), 0 part-time; includes 2 minority (1 Asian American, 1 Hispanic American), 3 foreign. 208 applicants, 12% accepted. In 1990, 2 master's awarded (50% found work related to degree, 50% continued full-time study); 6 doctorates awarded (83% entered university research/teaching, 17% found other work related to degree). Terminal master's awarded for partial completion of doctoral program. *Degree requirements:* For master's, thesis, departmental qualifying exam required, foreign language not required; for doctorate, variable foreign language requirement, dissertation, departmental qualifying exam. *Entrance requirements:* GRE. Application deadline: 3/1 (priority date). Application fee: $25 ($35 for foreign students). *Expenses:* Tuition of $16,230 per year. Fees of $650 per year. *Financial aid:* In 1990–91, 34 students received a total of $766,640 in aid awarded. 20 fellowships (7 to first-year students), 13 research assistantships were awarded; teaching assistantships, federal work-study, institutionally sponsored loans also available. Average monthly stipend for a graduate assistantship: $925. Financial aid applicants required to submit GAPSFAS or FAF. *Total annual research budget:* $939,388. • Gordon Gribble, Chairman, 603-646-2501.

Delaware State College, Department of Chemistry, Dover, DE 19901. Department offers programs in applied chemistry (MS), chemistry (MS). Part-time and evening/weekend programs available. Faculty: 0 full-time, 5 part-time (0 women). Matriculated students: 1 full-time (0 women), 7 part-time (6 women); includes 2 foreign. 7 applicants, 86% accepted. In 1990, 3 degrees awarded. *Degree requirements:* Foreign language not required. *Entrance requirements:* GRE, minimum GPA of 2.75 overall, 3.0 in major. Application deadline: 6/30 (priority date, applications processed on a rolling basis). Application fee: $10. *Tuition:* $95 per credit for state residents; $166 per credit for nonresidents. *Financial aid:* Full and partial tuition waivers, federal work-study, institutionally sponsored loans available. Aid available to part-time students. Financial aid applicants required to submit FAF. *Faculty research:* Chemiluminescence, environmental chemistry, forensic chemistry, heteropoly anions-anti-cancer and antiviral agents, low temperature infrared studies of lithium salts. • Dr. Ronald C. Machen, Chairperson, 302-739-5166.

DePaul University, College of Liberal Arts and Sciences, Department of Chemistry, Chicago, IL 60604. Department awards MS. Faculty: 9 full-time (1 woman), 0 part-time. Matriculated students: 35; includes 6 minority (5 Asian American, 1 black American), 3 foreign. 33 applicants, 52% accepted. In 1990, 8 degrees awarded. *Degree requirements:* Thesis optional, foreign language not required. *Entrance requirements:* Bachelor's degree in chemistry or equivalent. Application deadline: rolling. Application fee: $20. *Expenses:* Tuition of $215 per quarter hour. Fees of $10 per quarter hour (minimum). *Financial aid:* In 1990–91, 6 students received a total of $5000 in aid awarded. 6 research assistantships were awarded. • Dr. Avron Blumberg, Chairman, 312-362-8189.

Announcement: Program leading to MS in chemistry includes courses and research in analytical, biochemical, environmental, inorganic, organic, physical, and polymer chemistry, as well as in coatings technology. Courses for transition from other disciplines into chemistry are also offered. Nine full-time faculty, about 35 graduate students. Facilities for FT-IR, DSC, HPLC, NMR, etc.

Drexel University, College of Arts and Sciences, Department of Chemistry, 32nd and Chestnut Streets, Philadelphia, PA 19104. Department awards MS, PhD. Part-time programs available. Faculty: 16 full-time (1 woman), 1 part-time (0 women). Matriculated students: 30 full-time (12 women), 20 part-time (8 women); includes 4 minority (2 Asian American, 1 black American, 1 Hispanic American), 15 foreign. Average age 23. 173 applicants, 54% accepted. In 1990, 11 master's awarded; 5 doctorates awarded (20% entered university research/teaching, 60% found other work related to degree, 20% continued full-time study). Terminal master's awarded for partial completion of doctoral program. *Degree requirements:* For master's, thesis optional, foreign language not required; for doctorate, 1 foreign language, dissertation. *Entrance requirements:* For master's, GRE, TOEFL; TSE (for teaching assistants); for doctorate, GRE, TOEFL (minimum score of 550 required). Application deadline: 8/23 (applications processed on a rolling basis). Application fee: $25. *Expenses:* Tuition of $345 per credit hour. Fees of $81 per quarter full-time, $43 per quarter part-time. *Financial aid:* In 1990–91, $300,000 in aid awarded. 9 research assistantships (0 to first-year students), 22 teaching assistantships (5 to first-year students) were awarded; partial tuition waivers, federal work-study, institutionally sponsored loans, and career-related internships or fieldwork also available. Financial aid application deadline: 2/1. *Faculty research:* Inorganic, analytical, organic, physical, and atmospheric polymer chemistry. Total annual research budget: $1-million. • Dr. Robert O. Hutchins, Head, 215-895-2638. Application contact: Dr. Peter Wade, Chairman, Graduate Recruitment Committee, 215-895-2652.
See full description on page 409.

Duke University, Graduate School, Department of Chemistry, Durham, NC 27706. Department awards PhD. Faculty: 22 full-time, 17 part-time. Matriculated students: 91 full-time (29 women), 0 part-time; includes 7 minority (4 Asian American, 2 black American, 1 Hispanic American), 16 foreign. 138 applicants, 57% accepted. In 1990, 18 doctorates awarded. Terminal master's awarded for partial completion of doctoral program. *Degree requirements:* For doctorate, 1 foreign language, dissertation. *Entrance requirements:* For doctorate, GRE General Test. Application deadline: 1/31. Application fee: $50. *Expenses:* Tuition of $8640 per year full-time, $360 per unit part-time. Fees of $1356 per year full-time. *Financial aid:* In 1990–91, $648,376 in aid awarded. Financial aid application deadline: 1/31. • Linda McGown, Director of Graduate Studies, 919-684-2343.

Duquesne University, Graduate School of Liberal Arts and Sciences, Department of Chemistry, Pittsburgh, PA 15282. Department offers programs in biochemistry (MS, PhD), chemistry (MS, PhD). Part-time and evening/weekend programs available. Faculty: 12 full-time, 2 part-time. Matriculated students: 11 full-time (4 women), 32 part-time (8 women); includes 3 minority (all black American), 17 foreign. Average age 27. 50 applicants, 92% accepted. In 1990, 3 master's, 0 doctorates awarded. *Degree requirements:* For master's, thesis not required; for doctorate, 2 foreign languages, dissertation. *Entrance requirements:* For master's, GRE Subject Test, TOEFL; for doctorate, GRE Subject Test. Application fee: $20. *Expenses:* Tuition of $322 per credit. Fees of $20 per year. *Financial aid:* In 1990–91, 2 research assistantships (both to first-year students), 12 teaching assistantships (3 to first-year students) awarded; partial tuition waivers also available. Financial aid application deadline: 5/1. • Dr. Andrew J. Glaid, Chair, 412-434-6339.

East Carolina University, College of Arts and Sciences, Department of Chemistry, Greenville, NC 27858-4353. Department awards MS. Part-time programs available. Faculty: 15 full-time, 0 part-time. Matriculated students: 1 (woman) full-time, 13 part-time (3 women); includes 2 minority (1 black American, 1 Native American), 5 foreign. 18 applicants, 83% accepted. In 1990, 4 degrees awarded. *Degree requirements:* 1 foreign language (computer language can substitute), thesis, comprehensive exams. *Entrance requirements:* GRE General Test, TOEFL. Application deadline: 6/1 (priority date, applications processed on a rolling basis). Application fee: $25. *Tuition:* $627 per semester for state residents; $3154 per semester for nonresidents. *Financial aid:* In 1990–91, $74,400 in aid awarded. Teaching assistantships, federal work-study available. Financial aid application deadline: 6/1. *Faculty research:* Organometallic, natural-product syntheses; chemometrics; electroanalytical method development; microcomputer adaptations for handicapped students. • Dr. George Evans, Director of Graduate Studies, 919-757-6711. Application contact: Paul D. Tschetler, Assistant Dean, 919-757-6012.

Eastern Illinois University, College of Liberal Arts and Sciences, Department of Chemistry, 600 Lincoln Avenue, Charleston, IL 61920-3099. Department awards MS. Faculty: 13 full-time (2 women), 0 part-time. Matriculated students: 4 full-time (0 women), 5 part-time (2 women); includes 0 minority, 5 foreign. 32 applicants, 34% accepted. In 1990, 2 degrees awarded. *Degree requirements:* Thesis required, foreign language not required. *Entrance requirements:* GRE General Test. Application deadline: rolling. Application fee: $0. *Tuition:* $1220 per semester full-time for state residents; $2984 per semester full-time for nonresidents. *Financial aid:* In 1990–91, 7 research assistantships awarded. • Dr. David Buchanan, Chairperson, 217-581-3322. Application contact: Dr. Howard Black, Coordinator, 217-581-6225.

Eastern Kentucky University, College of Natural and Mathematical Sciences, Department of Chemistry, Richmond, KY 40475. Department awards MS. Part-time programs available. Matriculated students: 5 full-time (0 women), 4 part-time (3 women); includes 1 minority (black American), 3 foreign. Average age 25. In 1990, 2 degrees awarded. *Degree requirements:* Computer language, thesis. *Entrance requirements:* GRE General Test, minimum GPA of 2.5. Application fee: $0. *Tuition:* $1440 per year full-time, $88 per credit hour part-time for state residents; $4320 per year full-time, $248 per credit hour part-time for nonresidents. *Financial aid:* Research assistantships, teaching assistantships, federal work-study available. Aid available to part-time students. • Dr. Harry Smiley, Chair, 606-622-1457.

Eastern Michigan University, College of Arts and Sciences, Department of Chemistry, Ypsilanti, MI 48197. Department awards MS. Evening/weekend programs available. Faculty: 20 full-time (4 women). Matriculated students: 1 (woman) full-time, 37 part-time (14 women); includes 14 minority (8 Asian American, 2 black American, 1 Hispanic American, 3 Native American), 24 foreign. In 1990, 5 degrees awarded. *Degree requirements:* Thesis required (for some programs), foreign language not required. *Entrance requirements:* GRE General Test (minimum combined score of 1000 required). Application deadline: 6/15 (applications processed on a rolling basis). Application fee: $25. *Expenses:* Tuition of $89.50 per credit hour for state residents; $212 per credit hour for nonresidents. Fees of $90.25 per semester. *Financial aid:* Career-related internships or fieldwork available. Total annual research budget: $100,000. • Dr. Judith T. Levy, Head, 313-487-0106. Application contact: Edward L. Compere Jr., Coordinator, 313-487-2057.
See full description on page 103.

Eastern New Mexico University, College of Liberal Arts and Sciences, Department of Physical Sciences, Portales, NM 88130. Offerings include chemistry (MS). Department faculty: 3 full-time (0 women), 2 part-time (0 women). *Degree requirements:* Field exam required, thesis optional, foreign language not required. *Entrance requirements:* Minimum GPA of 2.5. Application deadline: rolling. Application fee: $10. *Tuition:* $711 per semester full-time, $59.25 per credit part-time for state residents; $2325 per semester full-time, $193.75 per credit part-time for nonresidents. • Dr. Andy Sae, Graduate Coordinator, 505-562-2463.

East Tennessee State University, College of Arts and Sciences and Biomedical Science Graduate Program, Department of Chemistry, Johnson City, TN 37614. Department awards MS. Part-time and evening/weekend programs available. Faculty: 5 full-time (0 women), 0 part-time. Matriculated students: 0 full-time, 10 part-time (5 women); includes 5 minority (all Asian American), 5 foreign. Average age 25. 36 applicants, 58% accepted. In 1990, 3 degrees awarded (100% found work related to degree). *Degree requirements:* Thesis, comprehensive exam required, foreign language not required. *Entrance requirements:* Bachelor's degree in ACS approved curriculum. Application deadline: 7/15. Application fee: $5. *Expenses:* Tuition of $1619 per year full-time, $82 per semester hour part-time for state residents; $4492 per year full-time, $207 per semester hour part-time for nonresidents. Fees of $14 per year full-time, $2 per year part-time. *Financial aid:* In 1990–91, $35,000 in aid awarded. 1 research assistantship (to a first-year student), 6 teaching assistantships (2 to first-year students) were awarded; federal work-study, institutionally sponsored loans also available. Average monthly stipend for a graduate assistantship: $555. *Faculty research:* Chemical kinetics, chemistry of coordination compounds, synthesis of novel molecules, reaction mechanisms. Total annual research budget: $80,000. • Dr. Thomas Huang, Chairman, 615-929-4367. Application contact: Dr. J. L. Miller, 615-929-4452.

East Texas State University, College of Arts and Sciences, Department of Chemistry, Commerce, TX 75429. Department awards MS. Faculty: 2 full-time (0 women), 1 part-time (0 women). Matriculated students: 4 full-time (0 women), 1 part-time (0 women); includes 0 minority, 4 foreign. In 1990, 1 degree awarded. *Degree requirements:* Thesis (for some programs), comprehensive exam. *Entrance requirements:* GRE General Test. Application deadline: rolling. Application fee: $0 ($25 for foreign students). *Tuition:* $430 per semester full-time for state residents; $1726 per semester full-time for nonresidents. *Financial aid:* Research assistantships, teaching assistantships, federal work-study, institutionally sponsored loans available. Financial aid applicants required to submit GAPSFAS or FAF. • Dr. Kenneth Ashley, Head, 903-886-5381.

Emory University, Graduate School of Arts and Sciences, Department of Chemistry, Atlanta, GA 30322. Department awards MS, PhD. Faculty: 20 full-time (1 woman), 0 part-time. Matriculated students: 107 full-time (46 women), 0 part-time; includes 9 minority (1 Asian American, 7 black American, 1 Hispanic American), 31 foreign. 260 applicants, 23% accepted. In 1990, 6 master's, 23 doctorates awarded. Terminal master's awarded for partial completion of doctoral program. *Degree requirements:* For doctorate, dissertation, comprehensive exam required, foreign language not required. *Entrance requirements:* For doctorate, GRE General Test, TOEFL. Application deadline: 1/20 (priority date, applications processed on a rolling basis). Application fee: $0 ($35 for foreign students). *Expenses:* Tuition of $7370 per semester full-time, $642 per semester hour part-time. Fees of $160 per year full-time. *Financial aid:* In 1990–91, 107 students received a total of $1.3-million in aid awarded. Fellowships, research assistantships, teaching assistantships, tuition scholarships, full and partial tuition waivers, federal work-study, institutionally sponsored loans available. Average monthly stipend for a graduate assistantship:

Directory: Chemistry

$1000. Financial aid application deadline: 2/1. *Faculty research:* New methods of chemical and neurochemical analysis; physical and chemical properties of mucleic acids, lipids and biological macromolecules; total synthesis and synthetic methodologies to organic compounds; design, synthesis, and investigation of new types of catalysts and high technology materials; theoretical chemistry and chemical dynamics. • Dr. Joel Bowman, Chairman, 404-727-6588. Application contact: Dr. Craig Hill, Director of Graduate Studies, 404-727-6639.

Emporia State University, School of Graduate Studies, College of Liberal Arts and Sciences, Division of Physical Sciences, Emporia, KS 66801-5087. Offerings include chemistry (MS). Division faculty: 14 full-time (0 women), 1 part-time (0 women). *Degree requirements:* Thesis or comprehensive exam required, foreign language not required. *Entrance requirements:* GRE General Test, TOEFL (minimum score of 550 required). Application deadline: 8/16 (priority date, applications processed on a rolling basis). Application fee: $0 ($50 for foreign students). *Tuition:* $858 per semester full-time, $62 per credit hour part-time for state residents; $2072 per semester full-time, $143 per credit hour part-time for nonresidents. • DeWayne Backhus, Chair, 316-343-5472.

Fairleigh Dickinson University, Florham-Madison Campus, Maxwell Becton College of Arts and Sciences, Department of Chemistry, 285 Madison Avenue, Madison, NJ 07940. Department awards MS. Faculty: 7 full-time (0 women), 3 part-time (0 women). Matriculated students: 0 full-time, 3 part-time (1 woman); includes 0 minority, 0 foreign. Average age 34. 4 applicants, 100% accepted. In 1990, 4 degrees awarded. *Degree requirements:* Thesis optional, foreign language not required. *Entrance requirements:* GRE General Test. Application deadline: rolling. Application fee: $35. *Expenses:* Tuition of $348 per credit. Fees of $205 per year full-time, $90 per year part-time. • Dr. Franc Lang, Chairperson, 201-593-8779.

Fairleigh Dickinson University, Teaneck-Hackensack Campus, College of Science and Engineering, Department of Chemistry, 1000 River Road, Teaneck, NJ 07666. Department offers programs in chemistry (MS), science (MA). Faculty: 8 full-time (1 woman), 4 part-time (1 woman). Matriculated students: 4 full-time (2 women), 29 part-time (12 women); includes 4 minority (1 Asian American, 2 black American, 1 Hispanic American), 5 foreign. Average age 31. In 1990, 12 degrees awarded. *Degree requirements:* Thesis optional, foreign language not required. *Entrance requirements:* GRE General Test. Application deadline: rolling. Application fee: $35. *Expenses:* Tuition of $348 per credit. Fees of $205 per year full-time, $90 per year part-time. • Dr. Edward Catanazaro, Chairperson, 201-692-2338.

Fisk University, Department of Chemistry, Nashville, TN 37208. Department awards MA. Faculty: 4 full-time (1 woman), 0 part-time. In 1990, 4 degrees awarded (25% found work related to degree, 75% continued full-time study). *Degree requirements:* Thesis, comprehensive exam required, foreign language not required. *Entrance requirements:* GRE, minimum GPA of 3.0. *Tuition:* $4950 per year full-time, $206 per credit hour part-time. *Financial aid:* Fellowships available. • Dr. Wesley Elliott, Chairman, 615-329-8628.

Florida Atlantic University, College of Science, Department of Chemistry, Boca Raton, FL 33431. Department awards MS, MST, PhD. Faculty: 16 full-time (4 women), 3 part-time (2 women). Matriculated students: 18 full-time (8 women), 7 part-time (4 women); includes 11 minority (9 Asian American, 2 black American), 5 foreign. Average age 23. In 1990, 4 master's awarded (50% found work related to degree, 50% continued full-time study). *Degree requirements:* 1 foreign language, thesis/dissertation. *Entrance requirements:* For master's, GRE General Test (minimum combined score of 1000 required), minimum GPA of 3.0; for doctorate, GRE Subject Test, minimum GPA of 3.0. Application fee: $15. *Tuition:* $89.28 per credit hour for state residents; $291 per credit hour for nonresidents. *Financial aid:* In 1990–91, $120,000 in aid awarded. 15 teaching assistantships (10 to first-year students) were awarded; fellowships, research assistantships, federal work-study also available. *Faculty research:* Polymer synthesis and characterization, spectroscopy, geochemistry. • Dr. Cyril Parkanyi, Chairman, 407-367-3390.

Florida Institute of Technology, College of Science and Liberal Arts, Department of Chemistry, Melbourne, FL 32901. Department awards MS, PhD. Part-time programs available. Faculty: 8 full-time (1 woman), 0 part-time. Matriculated students: 3 full-time (0 women), 2 part-time (0 women); includes 0 foreign. 30 applicants, 47% accepted. In 1990, 0 master's awarded. *Degree requirements:* For master's, thesis required, foreign language not required; for doctorate, 1 foreign language (computer language can substitute), dissertation, comprehensive exam. *Entrance requirements:* For master's, minimum GPA of 3.0; for doctorate, minimum GPA of 3.2. Application fee: $35. *Tuition:* $234 per credit hour. *Financial aid:* Research assistantships, teaching assistantships, federal work-study available. Financial aid application deadline: 3/1. *Faculty research:* Solar energy applications, aquatic and organic chemistry, natural products, stereo chemistry. • Dr. Michael W. Babich, Head, 407-768-8000 Ext. 8046. Application contact: Carolyn P. Farrior, Director of Graduate Admissions, 407-768-8000 Ext. 8027.

Florida International University, College of Arts and Sciences, Department of Chemistry, Miami, FL 33199. Department awards MS. Part-time and evening/weekend programs available. *Degree requirements:* 1 foreign language, thesis. *Entrance requirements:* GRE General Test, TOEFL. Application deadline: 4/1. Application fee: $15. *Tuition:* $462 per semester (minimum) full-time, $38.50 per credit part-time for state residents; $2217 per semester (minimum) full-time, $185 per credit part-time for nonresidents. *Faculty research:* Organic synthesis and reaction catalysis, environmental chemistry, molecular beam studies, organic geochemistry, bioinorganic and organometallic chemistry.

Florida State University, College of Arts and Sciences, Department of Chemistry, Tallahassee, FL 32306. Department offers programs in analytical chemistry (MS, PhD), biochemistry (MS, PhD), chemical physics (MS, PhD), inorganic chemistry (MS, PhD), organic chemistry (MS, PhD), physical chemistry (MS, PhD). Part-time programs available. Faculty: 39 full-time (2 women), 2 part-time (0 women). Matriculated students: 83 full-time (23 women), 3 part-time (1 woman); includes 2 minority (1 black American, 1 Hispanic American), 50 foreign. Average age 25. 424 applicants, 16% accepted. In 1990, 6 master's, 6 doctorates awarded. Terminal master's awarded for partial completion of doctoral program. *Degree requirements:* For master's, diagnostic and cumulative exams required, foreign language not required; for doctorate, dissertation, diagnostic and cumulative exams required, foreign language not required. *Entrance requirements:* GRE General Test, minimum B average in undergraduate course work. Application deadline: 4/15. Application fee: $15. *Tuition:* $76.29 per credit hour for state residents; $238 per credit hour for nonresidents. *Financial aid:* In 1990–91, $305,844 in aid awarded. 2 fellowships (0 to first-year students), 52 research assistantships (0 to first-year students), 29 teaching assistantships (21 to first-year students) were awarded; federal work-study, institutionally sponsored loans, and career-related internships or fieldwork also available. Financial aid application deadline: 2/15; applicants required to submit FAF. *Faculty research:* Spectroscopy, computational chemistry, nuclear chemistry. Total annual research budget: $1.6-million. • Dr. Edward K. Mellon, Chairman, 904-644-4074. Application contact: Dr. William T. Cooper, Graduate Adviser, 904-644-6875.
See full description on page 105.

Furman University, Department of Chemistry, Greenville, SC 29613. Department awards MS. Faculty: 7. *Degree requirements:* Thesis. *Entrance requirements:* GRE General Test, GRE Subject Test. Application deadline: rolling. Application fee: $10. *Tuition:* $117 per credit hour. • Lon Knight, Director of Graduate Program.

George Mason University, College of Arts and Sciences, Department of Chemistry, Fairfax, VA 22030. Department awards MS. Faculty: 11 (2 women). Matriculated students: 4 full-time (0 women), 19 part-time (8 women); includes 6 minority (3 Asian American, 2 black American, 1 Hispanic American), 1 foreign. 16 applicants, 75% accepted. In 1990, 1 degree awarded. *Degree requirements:* Computer language required, thesis optional. *Entrance requirements:* Minimum GPA of 3.0 in last 60 undergraduate hours. Application deadline: 5/1. Application fee: $25. *Expenses:* Tuition of $1872 per year full-time, $78 per semester hour part-time for state residents; $6264 per year full-time, $261 per semester hour part-time for nonresidents. Fees of $1080 per year full-time, $45 per semester hour part-time. *Financial aid:* Application deadline 3/1. • Dr. George Mushrush, Graduate Coordinator, 703-993-1080.

Georgetown University, College of Arts and Sciences, Department of Chemistry, Washington, DC 20057. Department offers programs in analytical chemistry (MS, PhD), biochemistry (MS, PhD), chemical physics (MS, PhD), inorganic chemistry (MS, PhD), organic chemistry (MS, PhD), physical chemistry (MS, PhD), theoretical chemistry (MS, PhD). Faculty: 17 full-time, 4 part-time. Matriculated students: 53 full-time (15 women), 15 part-time (5 women); includes 7 minority (5 Asian American, 2 Hispanic American). 345 applicants, 13% accepted. In 1990, 5 master's awarded (100% continued full-time study); 7 doctorates awarded (29% entered university research/teaching, 57% found other work related to degree, 14% continued full-time study). Terminal master's awarded for partial completion of doctoral program. *Degree requirements:* For master's, thesis required (for some programs), foreign language not required; for doctorate, 1 foreign language, dissertation. *Entrance requirements:* TOEFL (minimum score of 550 required, 600 for teaching assistants). Application deadline: 7/1 (applications processed on a rolling basis). *Expenses:* Tuition of $12,768 per year full-time, $532 per credit part-time. Fees of $142 per year full-time. *Financial aid:* In 1990–91, 59 students received a total of $966,000 in aid awarded. 30 fellowships (13 to first-year students), 14 research assistantships, 4 teaching assistantships (2 to first-year students) were awarded; full and partial tuition waivers, federal work-study, institutionally sponsored loans also available. Aid available to part-time students. Average monthly stipend for a graduate assistantship: $1030. Financial aid application deadline: 7/1. *Total annual research budget:* $2.6-million. • Dr. Michael T. Pope, Chairman, 202-687-6073.

George Washington University, Graduate School of Arts and Sciences, Department of Chemistry, Washington, DC 20052. Department awards MS, PhD. Part-time and evening/weekend programs available. Faculty: 4 full-time (0 women), 2 part-time (1 woman). Matriculated students: 8 full-time (3 women), 15 part-time (5 women); includes 0 minority, 14 foreign. Average age 30. 11 applicants, 55% accepted. In 1990, 0 master's, 3 doctorates awarded. Terminal master's awarded for partial completion of doctoral program. *Degree requirements:* For master's, computer language, thesis or alternative, comprehensive exam; for doctorate, computer language, dissertation, general exam. *Entrance requirements:* GRE General Test, minimum GPA of 3.0. Application deadline: 7/1. Application fee: $45. *Expenses:* Tuition of $490 per semester hour. Fees of $215 per year full-time, $125.60 per year (minimum) part-time. *Financial aid:* In 1990–91, $183,525 in aid awarded. 16 fellowships (4 to first-year students), 1 research assistantship, 11 teaching assistantships (4 to first-year students) were awarded; full and partial tuition waivers, federal work-study also available. Financial aid application deadline: 2/15. • Dr. David E. Ramaker, Chair, 202-994-6121.

Georgia State University, College of Arts and Sciences, Department of Chemistry, Atlanta, GA 30303. Department awards MAT, MS, PhD. PhD offered through the Laboratory for Microbial and Biochemical Sciences. Part-time and evening/weekend programs available. Faculty: 17 full-time (1 woman), 1 part-time (0 women). Matriculated students: 24 full-time (9 women), 17 part-time (8 women); includes 18 minority (9 Asian American, 6 black American, 3 Hispanic American), 4 foreign. Average age 29. 20 applicants, 80% accepted. In 1990, 1 master's awarded. Terminal master's awarded for partial completion of doctoral program. *Degree requirements:* For master's, 1 foreign language (computer language can substitute), thesis; for doctorate, 2 foreign languages (computer language can substitute for one), dissertation. *Entrance requirements:* For master's, GRE General Test, GRE Subject Test, TOEFL (minimum score of 550 required), minimum GPA of 3.0; for doctorate, GRE General Test, TOEFL (minimum score of 550 required), minimum GPA of 3.0. Application deadline: 7/15. Application fee: $10. *Expenses:* Tuition of $38 per quarter hour for state residents; $130 per quarter hour for nonresidents. Fees of $58 per quarter. *Financial aid:* In 1990–91, $12,600 in aid awarded. 8 research assistantships, 7 teaching assistantships, 15 assistantships were awarded; partial tuition waivers, federal work-study, institutionally sponsored loans, and career-related internships or fieldwork also available. Aid available to part-time students. Financial aid applicants required to submit FAF. *Faculty research:* DNA, AIDS, drug design, biothermodynamics, biological electron transfer and NMR applied to biochemical systems. Total annual research budget: $750,000. • Dr. David Boykin, Chair, 404-651-3120. Application contact: Dr. Harry Hopkins, Director of Graduate Studies, 404-651-3120.

Graduate School and University Center of the City University of New York, Program in Chemistry, New York, NY 10036. Program awards PhD. Faculty: 62 full-time (5 women), 0 part-time. Matriculated students: 137 full-time (49 women), 1 part-time (0 women); includes 14 minority (3 Asian American, 8 black American, 3 Hispanic American), 95 foreign. Average age 32. 109 applicants, 33% accepted. In 1990, 18 doctorates awarded (23% entered university research/teaching, 33% found other work related to degree, 0% continued full-time study). Terminal master's awarded for partial completion of doctoral program. *Degree requirements:* For doctorate, 1 foreign language, dissertation. *Entrance requirements:* For doctorate, GRE General Test, GRE Subject Test. Application deadline: 4/1. Application fee: $30. *Tuition:* $2204 per year full-time, $95 per credit part-time for state residents; $4700 per year full-time, $199 per credit part-time for nonresidents. *Financial aid:* In 1990–91, $1.01-million in aid awarded. 27 fellowships, 5 research assistantships, 71 teaching assistantships were awarded; full and partial tuition waivers, federal work-study, institutionally sponsored loans, and career-related internships or fieldwork also available. Financial aid application deadline: 2/1. • Dr. Richard Pizer, Executive Officer, 212-642-2451.

Hampton University, Department of Chemistry, Hampton, VA 23668. Department awards MS. Part-time and evening/weekend programs available. *Degree requirements:* Thesis required, foreign language not required. *Entrance requirements:* GRE General Test (minimum score of 450 on verbal section required).

Directory: Chemistry

Application deadline: 7/1. Application fee: $10. *Tuition:* $3050 per semester full-time, $155 per credit hour part-time. *Faculty research:* Element speciation.

Harvard University, Graduate School of Arts and Sciences, Committee on Chemical Physics, Cambridge, MA 02138. Offerings include chemistry (AM). Faculty: 21 full-time, 0 part-time. *Expenses:* Tuition of $14,860 per year. Fees of $550 per year. • Dr. Donald Ciappenelli, Director of Graduate Studies, 617-495-4076. Application contact: Office of Admissions and Financial Aid, 617-495-5315.

Harvard University, Graduate School of Arts and Sciences, Department of Chemistry, Cambridge, MA 02138. Department offers programs in biochemical chemistry (AM, PhD), inorganic chemistry (AM, PhD), organic chemistry (AM, PhD), physical chemistry (AM, PhD). Faculty: 20 full-time, 0 part-time. Matriculated students: 170 full-time (31 women), 0 part-time; includes 3 minority (1 black American, 2 Hispanic American). 284 applicants, 23% accepted. In 1990, 23 master's, 13 doctorates awarded. Terminal master's awarded for partial completion of doctoral program. *Degree requirements:* For doctorate, 2 foreign languages, dissertation, cumulative exams. *Entrance requirements:* For doctorate, GRE General Test, GRE Subject Test. Application deadline: 1/2. Application fee: $60. *Expenses:* Tuition of $14,860 per year. Fees of $550 per year. *Financial aid:* In 1990–91, $2.5-million in aid awarded. 58 fellowships (24 to first-year students), 117 research assistantships (28 to first-year students), 64 teaching assistantships (50 to first-year students) were awarded; federal work-study, institutionally sponsored loans, and career-related internships or fieldwork also available. Financial aid application deadline: 1/2; applicants required to submit GAPSFAS. • Dr. Donald J. Ciappenelli, Director of Graduate Studies, 617-495-4076.

Howard University, Graduate School of Arts and Sciences, Department of Chemistry, 2400 Sixth Street, NW, Washington, DC 20059. Department offers programs in analytical chemistry (MS, PhD), inorganic chemistry (MS, PhD), organic chemistry (MS, PhD), physical chemistry (MS, PhD). Part-time programs available. *Degree requirements:* For master's, 1 foreign language (computer language can substitute), thesis, comprehensive exam; for doctorate, 2 foreign languages (computer language can substitute for one), dissertation, comprehensive exam. *Entrance requirements:* For master's, minimum GPA of 2.7. Application deadline: 4/1. Application fee: $25. *Expenses:* Tuition of $6100 per year full-time, $339 per credit hour part-time. Fees of $555 per year full-time, $245 per semester part-time. *Faculty research:* Anti-cyanide drugs, DBH inhibition, agronomical and chemical studies.

Illinois Institute of Technology, Lewis College of Sciences and Letters, Department of Chemistry, Chicago, IL 60616. Offerings include chemistry (MS, PhD). Terminal master's awarded for partial completion of doctoral program. Department faculty: 5 full-time (0 women), 4 part-time (0 women). *Degree requirements:* For master's, thesis (for some programs), comprehensive exam required, foreign language not required; for doctorate, 1 foreign language, dissertation, comprehensive exam. *Entrance requirements:* GRE General Test, GRE Subject Test, TOEFL (minimum score of 500 required). Application deadline: 7/1 (applications processed on a rolling basis). Application fee: $30. *Expenses:* Tuition of $13,070 per year full-time, $435 per credit hour part-time. Fees of $20 per semester (minimum) full-time, $1 per credit hour part-time. • Dr. Robert Filler, Acting Chairman, 312-567-3425.

Illinois State University, College of Arts and Sciences, Department of Chemistry, Normal, IL 61761. Department awards MS. Faculty: 18 full-time (3 women), 0 part-time. Matriculated students: 32 full-time (11 women), 5 part-time (1 woman); includes 11 minority (9 Asian American, 1 black American, 1 Hispanic American). In 1990, 17 degrees awarded. *Degree requirements:* Thesis. *Entrance requirements:* GRE General Test, minimum GPA of 2.6 during last 60 hours. Application deadline: rolling. Application fee: $0. *Expenses:* Tuition of $1824 per year full-time, $76 per hour part-time for state residents; $5472 per year full-time, $228 per hour part-time for nonresidents. Fees of $630 per year full-time, $21 per hour (minimum) part-time. *Financial aid:* In 1990–91, $240,457 in aid awarded. 32 fellowships, 4 research assistantships, 28 teaching assistantships were awarded. Financial aid application deadline: 4/1. *Total annual research budget:* $54,996. • Michael Kurz, Chairperson, 309-438-7661.

Indiana State University, College of Arts and Sciences, Department of Chemistry, Terre Haute, IN 47809. Department awards MS. Part-time programs available. Faculty: 6 full-time (0 women), 3 part-time (0 women). Matriculated students: 0 full-time, 5 part-time (1 woman); includes 0 minority, 1 foreign. Average age 23. 16 applicants, 44% accepted. In 1990, 3 degrees awarded. *Degree requirements:* Thesis (for some programs), 2 research seminars required, foreign language not required. *Application deadline:* rolling. *Application fee:* $10 ($20 for foreign students). *Tuition:* $90 per hour for state residents; $199 per hour for nonresidents. *Financial aid:* In 1990–91, $2600 in aid awarded. 1 teaching assistantship (to a first-year student) was awarded. Financial aid application deadline: 3/1. *Faculty research:* Water pollution, enzymes, quantum chemistry, organometallics, forensics. • Dr. Arthur M. Halpern, Chairperson, 812-237-2240.

Indiana University Bloomington, College of Arts and Sciences, Department of Chemistry, Bloomington, IN 47405. Department awards MAT, MS, PhD. Faculty: 36 full-time, 0 part-time. Matriculated students: 160 full-time, 0 part-time; includes 9 minority (4 Asian American, 4 Hispanic American, 1 Native American), 21 foreign. Average age 24. 273 applicants, 48% accepted. In 1990, 6 master's, 26 doctorates awarded. Terminal master's awarded for partial completion of doctoral program. *Degree requirements:* Thesis/dissertation. *Application deadline:* 4/15. *Application fee:* $25 ($35 for foreign students). *Tuition:* $99.85 per credit hour for state residents; $288 per credit hour for nonresidents. *Financial aid:* Fellowships, research assistantships, teaching assistantships, institutionally sponsored loans available. *Total annual research budget:* $7-million. • Paul A. Grieco, Chairperson, 812-855-2268. Application contact: Dr. Adam Allerhand, Chair, Graduate Admissions, 812-855-2069.

Indiana University of Pennsylvania, College of Natural Sciences and Mathematics, Department of Chemistry, Indiana, PA 15705. Department awards MA, MS. Part-time programs available. Faculty: 8 full-time (1 woman), 0 part-time. Matriculated students: 8 full-time (0 women), 3 part-time (2 women); includes 1 minority (Hispanic American), 7 foreign. 37 applicants, 49% accepted. In 1990, 3 degrees awarded. *Degree requirements:* Thesis required (for some programs), foreign language not required. *Entrance requirements:* GRE General Test, TOEFL (minimum score of 500 required). Application deadline: 7/1 (priority date, applications processed on a rolling basis). Application fee: $20. *Expenses:* Tuition of $1139 per semester full-time, $127 per credit part-time for state residents; $1442 per semester full-time, $160 per credit part-time for nonresidents. Fees of $169 per semester full-time. *Financial aid:* In 1990–91, 7 research assistantships awarded; federal work-study also available. Aid available to part-time students. Financial aid application deadline: 3/15. • Dr. Neil Asting, Chairperson, 412-357-2361. Application contact: Dr. Augusta Syty, Director of Graduate Studies, 412-357-2361.

Indiana University–Purdue University at Indianapolis, School of Science, Department of Chemistry, Indianapolis, IN 46205. Department awards MS, PhD. Part-time and evening/weekend programs available. Faculty: 15 full-time (1 woman), 0 part-time. Matriculated students: 30 full-time (10 women), 24 part-time (12 women); includes 2 minority (1 Asian American, 1 Hispanic American), 6 foreign. Average age 25. 60 applicants, 33% accepted. In 1990, 5 master's awarded (100% found work related to degree). *Degree requirements:* For master's, thesis required (for some programs), foreign language not required; for doctorate, dissertation required, foreign language not required. *Entrance requirements:* GRE or minimum GPA of 3.0. Application deadline: 6/1 (priority date, applications processed on a rolling basis). Application fee: $20. *Tuition:* $100 per credit for state residents; $288 per credit for nonresidents. *Financial aid:* In 1990–91, 30 students received a total of $200,000 in aid awarded. 4 fellowships (2 to first-year students), 5 research assistantships (1 to a first-year student), 12 teaching assistantships (5 to first-year students), 6 co-op positions (3 to first-year students) were awarded; partial tuition waivers, institutionally sponsored loans, and career-related internships or fieldwork also available. Average monthly stipend for a graduate assistantship: $1000. Financial aid application deadline: 3/15. *Faculty research:* Analytical, biological, inorganic, organic, and physical chemistry. Total annual research budget: $900,000. • David Malik, Acting Chair, 317-274-6884.

See full description on page 107.

Institute of Paper Science and Technology, Program in Chemistry, Atlanta, GA 30318. Offers paper science and technology (MS, PhD). Part-time programs available. Terminal master's awarded for partial completion of doctoral program. *Degree requirements:* For master's, industrial experience required, foreign language and thesis not required; for doctorate, dissertation required, foreign language not required. *Entrance requirements:* For master's, GRE. Application deadline: 5/15 (priority date). Application fee: $0. *Tuition:* $0 full-time, $225 per credit hour part-time.

Announcement: Multidisciplinary programs in chemistry, engineering, physics, biology, mathematics, and pulp and paper science, plus electives. US and Canadian residents granted fellowships ($15,000 per year MS, $17,000 per year PhD) and tuition waivers. Admission geared to BS majors in the above disciplines. Summer employment provided. Good permanent positions with excellent starting salaries available.

See full description on page 109.

Instituto Tecnológico y de Estudios Superiores de Monterrey, Program in Chemistry, Monterrey, Nuevo León 64849, Mexico. Offerings include chemistry (PhD). Program faculty: 3 full-time (1 woman), 0 part-time. *Degree requirements:* Dissertation. *Entrance requirements:* Graduate admission exam, TOEFL. *Tuition:* $2350 per trimester full-time. • Dr. Xorge Dóminquez, Director, 83-58-3300 Ext. 4510. Application contact: Julieta Mier, Dean of Admissions, 83-58-4650.

Iowa State University of Science and Technology, College of Liberal Arts and Sciences, Department of Chemistry, Ames, IA 50011. Department awards MS, PhD. Matriculated students: 200 full-time (49 women), 13 part-time (6 women); includes 0 minority, 90 foreign. In 1990, 7 master's, 49 doctorates awarded. *Degree requirements:* For doctorate, dissertation. *Application fee:* $20 ($30 for foreign students). *Expenses:* Tuition of $1158 per semester full-time, $129 per credit part-time for state residents; $3340 per semester full-time, $372 per credit part-time for nonresidents. Fees of $10 per semester. *Financial aid:* In 1990–91, 1 fellowship (to a first-year student), 126 research assistantships (20 to first-year students), 85 teaching assistantships (59 to first-year students) awarded; scholarships also available. • Dr. James H. Espenson, Chair, 515-294-6342.

Jackson State University, School of Science and Technology, Department of Chemistry, Jackson, MS 39217. Department offers programs in analytical chemistry (MS), inorganic chemistry (MS), organic chemistry (MS), physical chemistry (MS). Evening/weekend programs available. Faculty: 14 full-time (3 women), 0 part-time. Matriculated students: 4 full-time (2 women), 3 part-time (1 woman); includes 5 minority (1 Asian American, 4 black American). In 1990, 1 degree awarded. *Degree requirements:* Thesis (for some programs). *Entrance requirements:* GRE General Test. Application deadline: rolling. Application fee: $0. *Expenses:* Tuition of $1114 per semester full-time, $113 per hour part-time for state residents; $1142 per semester full-time, $113 per hour part-time for nonresidents. Fees of $730 per semester for nonresidents. *Financial aid:* Federal work-study available. Financial aid application deadline: 5/1; applicants required to submit FAF. *Faculty research:* Electrochemical and spectroscopic studies on charge transfer and energy transfer processes, spectroscopy of trapped molecular ions, respirable mine dust. • Dr. Richard Sullivan, Chair, 601-968-2171.

John Carroll University, Department of Chemistry, University Heights, OH 44118. Department awards MS. Part-time and evening/weekend programs available. Faculty: 9 full-time (0 women), 0 part-time. Matriculated students: 6 full-time (3 women), 15 part-time (2 women); includes 2 minority (1 Asian American, 1 black American), 1 foreign. Average age 30. 8 applicants, 88% accepted. In 1990, 7 degrees awarded (100% found work related to degree). *Degree requirements:* 1 foreign language (computer language can substitute), thesis or research essay, comprehensive exam. *Application deadline:* 8/17 (priority date, applications processed on a rolling basis). *Application fee:* $20. *Tuition:* $300 per credit. *Financial aid:* In 1990–91, 7 students received a total of $54,000 in aid awarded. 6 teaching assistantships (2 to first-year students), 3 summer research supports were awarded; partial tuition waivers also available. Average monthly stipend for a graduate assistantship: $889. *Faculty research:* Prediction of rotational constants via Ab initio calculations, carbon clusters, organic synthesis via Pummerer reaction, Vanadium/sulfer complexes, FT Raman spectroscopy. • Dr. Nick Baumgartner, Chairperson, 216-397-4241.

Johns Hopkins University, School of Arts and Sciences, Department of Chemistry, Baltimore, MD 21218. Department awards MA, PhD. Faculty: 19 full-time (1 woman), 4 part-time (0 women). Matriculated students: 82 full-time (31 women), 0 part-time; includes 7 minority (5 Asian American, 1 black American, 1 Hispanic American), 27 foreign. Average age 25. 174 applicants, 35% accepted. In 1990, 10 master's awarded (100% continued full-time study); 12 doctorates awarded. Terminal master's awarded for partial completion of doctoral program. *Degree requirements:* For master's, 1 foreign language, oral exam; for doctorate, 1 foreign language, dissertation, oral exams. *Entrance requirements:* GRE General Test, GRE Subject Test. Application deadline: 2/1. Application fee: $40. *Expenses:* Tuition of $15,500 per year full-time, $1550 per course part-time. Fees of $400 per year full-time. *Financial aid:* In 1990–91, $1.239-million in aid awarded. 2 fellowships (0 to first-year students), 45 research assistantships (0 to first-year students), 42 teaching assistantships (18 to first-year students) were awarded; federal work-study, institutionally sponsored loans also available. Financial aid application deadline: 3/14; applicants required to submit FAF. *Total annual research budget:* $2.2-million. • Dr. Craig A. Townsend, Chairman, 301-338-7430.

Directory: Chemistry

Kansas State University, College of Arts and Sciences, Department of Chemistry, Manhattan, KS 66506. Department offers programs in analytical chemistry (MS, PhD), inorganic chemistry (MS, PhD), organic chemistry (MS, PhD), physical chemistry (MS, PhD). Faculty: 19 full-time (0 women), 0 part-time. Matriculated students: 55 full-time (7 women), 5 part-time (0 women); includes 1 minority (black American), 28 foreign. 95 applicants, 11% accepted. In 1990, 3 master's, 8 doctorates awarded. *Degree requirements:* For master's, thesis required, foreign language not required; for doctorate, dissertation. *Expenses:* Tuition of $1668 per year full-time, $51 per credit hour part-time for state residents; $5382 per year full-time, $142 per credit hour part-time for nonresidents. Fees of $305 per year full-time, $64.50 per semester part-time. *Financial aid:* In 1990–91, 19 research assistantships awarded; teaching assistantships also available. • M. Dale Hawley, Head, 913-532-6668. Application contact: Robert Hammaker, Graduate Coordinator, 913-532-6671.

Kent State University, Department of Chemistry, Kent, OH 44242. Department offers programs in analytical chemistry (MS, PhD), biochemistry (PhD), chemistry (MA, MS, PhD), inorganic chemistry (MS, PhD), organic chemistry (MS, PhD), physical chemistry (MS, PhD). Faculty: 18 full-time, 0 part-time. Matriculated students: 31 full-time (14 women), 4 part-time (2 women). 39 applicants, 74% accepted. In 1990, 4 master's, 3 doctorates awarded. *Degree requirements:* Thesis/dissertation required, foreign language not required. *Application deadline:* 7/12 (applications processed on a rolling basis). *Application fee:* $25. *Tuition:* $1601 per semester full-time, $133.75 per hour part-time for state residents; $3101 per semester full-time, $258.75 per hour part-time for nonresidents. *Financial aid:* Fellowships, research assistantships, teaching assistantships, full tuition waivers, federal work-study available. Financial aid application deadline: 2/1. • Dr. Roger K. Gilpin, Chairman, 216-672-2032.

See full description on page 111.

Lakehead University, Faculty of Arts and Science, Department of Chemistry, Thunder Bay, ON P7B 5E1, Canada. Department awards M Sc. Part-time and evening/weekend programs available. Faculty: 9 full-time (0 women), 0 part-time. Matriculated students: 4 full-time (2 women), 1 part-time (0 women); includes 1 foreign. 29 applicants, 14% accepted. In 1990, 1 degree awarded. *Degree requirements:* Thesis required, foreign language not required. *Entrance requirements:* TOEFL (minimum score of 550 required), minimum B+ average. Application fee: $0. *Expenses:* Tuition of $1969 per year full-time, $527 per year part-time for Canadian residents; $3712 per year full-time, $2618 per year part-time for nonresidents. Fees of $2735 per year full-time, $100 per year part-time for Canadian residents; $11,136 per year full-time, $100 per year part-time for nonresidents. *Financial aid:* In 1990–91, 4 students received a total of $430,488 in aid awarded. 4 teaching assistantships (3 to first-year students) were awarded; fellowships, entrance awards and research stipends also available. Average monthly stipend for a graduate assistantship: $680. Financial aid application deadline: 3/30. *Faculty research:* Environmental chemistry, polymer science, synthetic and reactivity studies in phosphorus heterocyclic chemistry, nuclear magnetic resonance, pulp and paper chemistry. Total annual research budget: $205,670. • Dr. Alan N. Hughes, Chairman, 807-343-8319. Application contact: Dr. Neil Weir, Graduate Coordinator, 807-343-8318.

Lamar University–Beaumont, College of Arts and Sciences, Department of Chemistry, Beaumont, TX 77705. Department awards MS. Part-time programs available. Faculty: 8 full-time (0 women), 1 part-time (0 women). Matriculated students: 10 full-time (1 woman), 5 part-time (1 woman); includes 2 minority (both Asian American), 12 foreign. In 1990, 12 degrees awarded (100% continued full-time study). *Degree requirements:* Computer language, thesis required, foreign language not required. *Entrance requirements:* GRE General Test (minimum combined score of 950 required), minimum GPA of 2.0 (in last 60 hours). Application deadline: rolling. Application fee: $0. *Expenses:* Tuition of $630 per year full-time, $378 per year part-time for state residents; $4350 per year full-time, $2610 per year part-time for nonresidents. Fees of $480 per year. *Financial aid:* In 1990–91, 15 teaching assistantships (6 to first-year students) awarded. • Dr. Keith Hansen, Chair, 409-880-8267.

Laurentian University, Programme in Chemistry, Sudbury, ON P3E 2C6, Canada. Program awards M Sc. Part-time programs available. Faculty: 9 full-time (0 women), 0 part-time. Matriculated students: 12 full-time (4 women), 1 part-time (0 women); includes 3 foreign. 14 applicants, 43% accepted. In 1990, 4 degrees awarded. *Degree requirements:* Thesis or alternative required, foreign language not required. *Entrance requirements:* Honors bachelor's with second class or better. *Tuition:* $1926 per year full-time, $364 per course part-time for Canadian residents; $11,138 per year full-time for nonresidents. *Financial aid:* In 1990–91, $80,470 in aid awarded. 3 fellowships (all to first-year students), 11 teaching assistantships were awarded; research assistantships, partial tuition waivers, institutionally sponsored loans also available. *Faculty research:* Analytical, inorganic, organic, biochemistry. Total annual research budget: $90,980. • Dr. F. Smith, Chairman, 705-675-1151 Ext. 2101. Application contact: Admissions Department, 705-675-1151 Ext. 3915.

Lehigh University, College of Arts and Sciences, Department of Chemistry, Bethlehem, PA 18015. Department awards MS, DA, PhD. Faculty: 22 full-time, 4 part-time. Matriculated students: 59 full-time (23 women), 24 part-time (6 women); includes 3 minority (1 Asian American, 2 black American), 7 foreign. 72 applicants, 72% accepted. In 1990, 12 master's, 4 doctorates awarded. *Degree requirements:* For master's, foreign language and thesis not required; for doctorate, 1 foreign language, dissertation. *Entrance requirements:* GRE, TOEFL. Application deadline: rolling. Application fee: $40. *Tuition:* $15,650 per year full-time, $655 per credit hour part-time. *Financial aid:* Fellowships, research assistantships, teaching assistantships, institutionally sponsored loans, and career-related internships or fieldwork available. *Faculty research:* Biochemistry; inorganic, organic, physical, and polymer chemistry. • Dr. John Larsen, Chairman, 215-758-3471. Application contact: Dr. Daniel Zeroka, Graduate Program Coordinator.

See full description on page 113.

Long Island University, Brooklyn Campus, Richard L. Conolly College of Liberal Arts and Sciences, Department of Chemistry, Brooklyn, NY 11201. Department awards MS. Part-time and evening/weekend programs available. Faculty: 9 full-time (1 woman), 8 part-time. Matriculated students: 11 full-time (6 women), 1 (woman) part-time; includes 5 minority (2 Asian American, 3 black American), 6 foreign. 20 applicants, 70% accepted. In 1990, 0 degrees awarded. *Degree requirements:* Thesis or alternative required, foreign language not required. *Application deadline:* rolling. *Application fee:* $30. *Tuition:* $310 per credit. *Financial aid:* In 1990–91, 10 students received a total of $53,300 in aid awarded. 3 research assistantships (all to first-year students), 5 teaching assistantships (all to first-year students) were awarded. Average monthly stipend for a graduate assistantship: $200-500. Financial aid application deadline: 8/1. *Faculty research:* Clinical chemistry, free radicals, heats of hydrogenation. • Dr. Donald Rogers, Chair, 718-403-1054.

Long Island University, C. W. Post Campus, School of Health Professions, Program in Medical Biology, Brookville, NY 11548. Offerings include clinical chemistry (MS). Program faculty: 6 full-time (3 women), 4 part-time (2 women). *Degree requirements:* Thesis required, foreign language not required. *Entrance requirements:* GRE General Test, GRE Subject Test. *Expenses:* Tuition of $310 per credit. Fees of $235 per semester full-time, $90 per semester part-time. • Dr. Francis Gizis, Chairman, 516-299-2485.

Louisiana State University and Agricultural and Mechanical College, College of Basic Sciences, Department of Chemistry, Baton Rouge, LA 70803. Department awards MS, PhD. Part-time programs available. Faculty: 31 full-time (2 women), 0 part-time. Matriculated students: 95 full-time (33 women), 7 part-time (2 women); includes 6 minority (3 Asian American, 2 black American, 1 Hispanic American), 55 foreign. Average age 29. 75 applicants, 57% accepted. In 1990, 3 master's awarded (100% continued full-time study); 20 doctorates awarded (50% entered university research/teaching, 50% found other work related to degree). Terminal master's awarded for partial completion of doctoral program. *Degree requirements:* For master's, thesis required (for some programs), foreign language not required; for doctorate, dissertation, general exam required, foreign language not required. *Entrance requirements:* GRE General Test, TOEFL, minimum GPA of 3.0. Application deadline: 7/1 (priority date, applications processed on a rolling basis). Application fee: $25. *Tuition:* $1020 per semester full-time for state residents; $2620 per semester full-time for nonresidents. *Financial aid:* In 1990–91, 94 students received a total of $1.28-million in aid awarded. 6 fellowships (0 to first-year students), 21 research assistantships (0 to first-year students), 65 teaching assistantships (14 to first-year students) were awarded. Average monthly stipend for a graduate assistantship: $1066. Financial aid application deadline: 7/1. *Faculty research:* Free radicals, bioinorganic chemistry, polymers, synthesis, spectroscopy. Total annual research budget: $3.6-million. • Dr. Frank Cartledge, Chair, 504-388-3361. Application contact: Dr. Steven Watkins, Director of Graduate Studies, 504-388-3467.

Louisiana State University Medical Center, School of Graduate Studies in New Orleans, Department of Pathology, 433 Bolivar Street, New Orleans, LA 70112. Offerings include pathology (MS, PhD), with option in clinical chemistry. Terminal master's awarded for partial completion of doctoral program. Department faculty: 29 full-time (4 women), 0 part-time. *Degree requirements:* Thesis/dissertation required, foreign language not required. *Entrance requirements:* GRE General Test (minimum combined score of 1000 required), TOEFL (minimum score of 550 required). Application deadline: 5/30. Application fee: $30. • Dr. Jack Strong, Head, 504-582-5487. Application contact: Dr. Peter Lehmann, Coordinator, Graduate Studies, 504-568-6057.

Louisiana Tech University, College of Arts and Sciences, Department of Chemistry, Ruston, LA 71272. Department awards MS. Part-time programs available. Faculty: 4 full-time (0 women), 0 part-time. Matriculated students: 4 full-time (1 woman), 1 part-time (0 women); includes 0 minority, 3 foreign. In 1990, 0 degrees awarded. *Degree requirements:* Computer language, thesis required, foreign language not required. *Entrance requirements:* GRE General Test. Application deadline: 8/13. Application fee: $5. *Tuition:* $613 per quarter full-time, $184 per semester (minimum) part-time for state residents; $999 per quarter full-time, $184 per semester (minimum) part-time for nonresidents. *Financial aid:* Fellowships, research assistantships available. Financial aid application deadline: 2/1; applicants required to submit GAPSFAS. • Dr. Harry Moseley, Head, 318-257-4911.

Loyola University Chicago, Graduate School, Department of Chemistry, 820 North Michigan Avenue, Chicago, IL 60611. Department awards MS, PhD. Faculty: 22 full-time (3 women), 0 part-time. Matriculated students: 32 full-time (17 women), 8 part-time (3 women); includes 0 minority, 11 foreign. Average age 24. 80 applicants, 44% accepted. In 1990, 9 doctorates awarded. *Degree requirements:* For master's, thesis required, foreign language not required; for doctorate, 1 foreign language (computer language can substitute), dissertation. *Entrance requirements:* GRE General Test, GRE Subject Test, TOEFL. Application deadline: 4/1. Application fee: $30. *Tuition:* $252 per credit hour. *Financial aid:* In 1990–91, 31 students received a total of $510,000 in aid awarded. 5 fellowships (0 to first-year students), 6 research assistantships (0 to first-year students), 20 teaching assistantships (5 to first-year students) were awarded; federal work-study also available. Aid available to part-time students. Average monthly stipend for a graduate assistantship: $850. Financial aid application deadline: 3/1; applicants required to submit FAF. *Faculty research:* Magnetic resonance of membrane/protein systems, organometallic catalysis, novel syntheses of natural products. Total annual research budget: $150,000. • Dr. Stephen F. Pavkovic, Chair, 312-508-3100. Application contact: Dr. Elliot Burrell, 312-508-3124.

Mankato State University, College of Natural Sciences, Mathematics and Home Economics, Department of Chemistry and Geology, South Road and Ellis Avenue, Mankato, MN 56002-8400. Department offers program in chemistry (MA, MS). Faculty: 11 full-time (2 women), 2 part-time (0 women). Matriculated students: 6 full-time (4 women), 6 part-time (1 woman); includes 0 minority, 7 foreign. Average age 29. 17 applicants, 100% accepted. In 1990, 1 degree awarded. *Degree requirements:* 1 foreign language, thesis or alternative, departmental qualifying exam, comprehensive exam. *Entrance requirements:* GRE General Test, GRE Subject Test, minimum GPA of 2.75 for last 2 years of undergraduate study. Application deadline: 2/3 (priority date, applications processed on a rolling basis). Application fee: $15. *Expenses:* Tuition of $52 per credit for state residents; $75 per credit for nonresidents. Fees of $6.50 per quarter. *Financial aid:* In 1990–91, 8 students received a total of $27,522 in aid awarded. 8 teaching assistantships (4 to first-year students) were awarded; federal work-study, institutionally sponsored loans, and career-related internships or fieldwork also available. Aid available to part-time students. Average monthly stipend for a graduate assistantship: $382. Financial aid application deadline: 7/1. *Faculty research:* General chemistry development. • Dr. Douglas E. Ralston, Chairman, 507-389-1963. Application contact: Dr. Teresa Salerno, Graduate Coordinator, 507-389-1963.

Marquette University, College of Arts and Sciences, Department of Chemistry, Milwaukee, WI 53233. Offerings include bioanalytical chemistry (MS, PhD), biophysical chemistry (MS, PhD). Terminal master's awarded for partial completion of doctoral program. Department faculty: 14 full-time (1 woman), 0 part-time. *Degree requirements:* For master's, thesis or alternative, comprehensive exam required, foreign language not required; for doctorate, dissertation, cumulative exams required, foreign language not required. *Entrance requirements:* TOEFL (minimum score of 550 required). Application fee: $25. *Tuition:* $300 per credit. • Dr. Michael McKinney, Chairman, 414-288-7065.

Marshall University, College of Science, Department of Chemistry, Huntington, WV 25755. Department awards MS. Faculty: 7 (0 women). Matriculated students: 7 full-time (2 women), 14 part-time (7 women); includes 0 minority, 3 foreign. In 1990, 4 degrees awarded. *Degree requirements:* Thesis required, foreign language not required. *Entrance requirements:* GRE General Test. Application fee: $0. *Tuition:*

$857 per semester full-time, $492 per semester part-time for state residents; $2207 per semester full-time, $1392 per semester part-time for nonresidents. *Financial aid:* Career-related internships or fieldwork available. • Dr. John W. Larson, Chairperson, 304-696-2430. Application contact: Dr. James Harless, Director of Admissions, 304-696-3160.

Massachusetts College of Pharmacy and Allied Health Sciences, Graduate Program in Chemistry, 179 Longwood Avenue, Boston, MA 02115. Program awards MS, PhD. Faculty: 8 full-time (1 woman), 0 part-time. Matriculated students: 15 full-time (7 women), 0 part-time; includes 0 minority, 13 foreign. Average age 24. 16 applicants, 50% accepted. In 1990, 0 master's, 0 doctorates awarded. Terminal master's awarded for partial completion of doctoral program. *Degree requirements:* For master's, oral defense of thesis required, foreign language not required; for doctorate, 1 foreign language (computer language can substitute), oral defense of dissertation, qualifying exam. *Entrance requirements:* GRE General Test (minimum combined score of 1600 on all three sections required), TOEFL (minimum score of 550 required), minimum QPA of 3.0. Application deadline: 2/1 (priority date). Application fee: $50. *Tuition:* $212 per credit. *Financial aid:* In 1990–91, $55,000 in aid awarded. 2 research assistantships, 5 teaching assistantships (2 to first-year students) were awarded. Financial aid application deadline: 2/1. *Faculty research:* Organic synthesis, analytical medicinal chemistry, biorganic and medicinal chemistry, phytochemistry. Total annual research budget: $10,000. • Dr. George Matelli, Chairman, 617-732-2933. Application contact: Dr. Vincent W. Bernardi, Coordinator, 617-732-2937.

Massachusetts Institute of Technology, School of Science, Department of Chemistry, Cambridge, MA 02139. Department offers programs in biochemistry (SM, PhD), biological chemistry (SM, PhD, Sc D), biophysical chemistry (SM, PhD), chemical physics (SM, PhD, Sc D), inorganic chemistry (SM, PhD, Sc D), organic chemistry (SM, PhD, Sc D), physical chemistry (SM, PhD, Sc D). Faculty: 35 full-time (2 women), 0 part-time. Matriculated students: 251 full-time (66 women), 0 part-time; includes 12 minority (7 Asian American, 2 black American, 3 Hispanic American), 48 foreign. Average age 26. 367 applicants, 35% accepted. In 1990, 5 master's, 47 doctorates awarded. Terminal master's awarded for partial completion of doctoral program. *Degree requirements:* For master's, thesis required, foreign language not required; for doctorate, dissertation, written exams, comprehensive oral exam, oral presentation required, foreign language not required. *Application deadline:* 1/16. *Application fee:* $45. *Tuition:* $16,900 per year. *Financial aid:* In 1990–91, $6.94-million in aid awarded. 1 fellowship (to a first-year student), 200 research assistantships (0 to first-year students), 57 teaching assistantships (53 to first-year students), 40 predoctoral fellowships were awarded. Financial aid application deadline: 1/14. *Faculty research:* Synthetic organic chemistry, enzymatic reaction mechanisms, inorganic and organometallic spectroscopy, high resolution NMR spectroscopy. Total annual research budget: $12.9-million. • Robert J. Silbey, Chairman, 617-253-1801.

See full description on page 115.

McGill University, Faculty of Graduate Studies and Research, Department of Chemistry, Montreal, PQ H3A 2T5, Canada. Department awards M Sc, PhD. *Degree requirements:* Thesis/dissertation. *Tuition:* $698.50 per semester full-time, $46.57 per credit part-time for Canadian residents; $3480 per semester full-time, $234 per semester part-time for nonresidents.

McMaster University, Faculty of Science, Department of Chemistry, Hamilton, ON L8S 4L8, Canada. Department offers programs in chemical physics (M Sc, PhD), chemistry (M Sc, PhD). *Degree requirements:* For master's, thesis required, foreign language not required; for doctorate, dissertation, comprehensive exam required, foreign language not required. *Expenses:* Tuition of $2250 per year full-time, $810 per semester part-time for Canadian residents; $10,340 per year full-time, $4050 per semester part-time for nonresidents. Fees of $76 per year full-time, $49 per semester part-time.

McNeese State University, College of Science, Department of Chemistry, Lake Charles, LA 70609. Department awards MS. Evening/weekend programs available. Faculty: 9 full-time (0 women), 1 (woman) part-time. Matriculated students: 4 full-time (1 woman), 10 part-time (4 women); includes 1 foreign. Average age 25. In 1990, 1 degree awarded. *Degree requirements:* 1 foreign language (computer language can substitute), thesis or alternative. *Entrance requirements:* GRE General Test. Application deadline: 7/15 (priority date, applications processed on a rolling basis). Application fee: $10 ($25 for foreign students). *Tuition:* $808 per semester full-time, $254 per course (minimum) part-time for state residents; $1583 per semester full-time, $254 per course (minimum) part-time for nonresidents. *Financial aid:* In 1990–91, $36,000 in aid awarded. 9 teaching assistantships (2 to first-year students) were awarded. Financial aid application deadline: 5/1. *Faculty research:* Environmental studies, carotenoids, polymers, chemical education. • Dr. Keith A. Stolzle, Head, 318-475-5776.

Memorial University of Newfoundland, School of Graduate Studies, Department of Chemistry, St. John's, NF A1C 5S7, Canada. Department awards M Sc, PhD. *Degree requirements:* Thesis/dissertation.

Memphis State University, College of Arts and Sciences, Department of Chemistry, Memphis, TN 38152. Department awards MS, PhD. Faculty: 12 full-time (0 women), 1 (woman) part-time. Matriculated students: 18 full-time (4 women), 13 part-time (2 women); includes 1 minority (black American), 17 foreign. In 1990, 2 master's, 3 doctorates awarded. *Degree requirements:* For master's, thesis, oral comprehensive exam; for doctorate, 1 foreign language, dissertation, oral comprehensive exam. *Entrance requirements:* For master's, GRE General Test, GRE Subject Test, 32 undergraduate hours in chemistry; for doctorate, GRE General Test, GRE Subject Test, MS in chemistry. Application deadline: 8/1 (applications processed on a rolling basis). Application fee: $5. *Expenses:* Tuition of $92 per credit hour for state residents; $239 per credit hour for nonresidents. Fees of $45 per year for state residents. *Financial aid:* In 1990–91, 3 research assistantships, 14 teaching assistantships awarded. *Faculty research:* Heterocyclic compounds, NMR of methylated decabosanes, Etoh oxidation research project, polymeric porphyrin synthesis, hollow cathode discharge source. • Dr. H. Graden Kirksey Jr., Chairman, 901-678-2621. Application contact: Dr. Peter K. Bridson, Coordinator, Graduate Studies, 901-678-4423.

Miami University, College of Arts and Sciences, Department of Chemistry, Oxford, OH 45056. Department offers programs in biochemistry (MA, MAT, MS, PhD), chemistry (MA, MAT, MS, PhD). Part-time programs available. Faculty: 25. Matriculated students: 35 full-time (11 women), 5 part-time (3 women); includes 1 minority (black American), 8 foreign. 72 applicants, 32% accepted. In 1990, 7 master's, 8 doctorates awarded. *Degree requirements:* For master's, thesis, final exam; for doctorate, 2 foreign languages, dissertation, comprehensive exam, final exam. *Entrance requirements:* For master's, minimum undergraduate GPA of 2.75 or 3.0 during previous 2 years; for doctorate, minimum undergraduate GPA of 2.75, 3.0 graduate. Application deadline: 3/1. Application fee: $30. *Expenses:* Tuition of $3196 per year full-time, $133 per semester (minimum) part-time for state residents; $7134 per year full-time, $312 per semester (minimum) part-time for nonresidents. Fees of $371 per year full-time, $165.50 per semester hour part-time. *Financial aid:* In 1990–91, $304,900 in aid awarded. Fellowships, research assistantships, teaching assistantships, full tuition waivers, federal work-study available. Financial aid application deadline: 3/1. • Dr. James A. Cox, Chair, 513-529-2813. Application contact: Dr. John Grunwell, Director of Graduate Studies.

Michigan State University, College of Natural Science, Department of Chemistry, East Lansing, MI 48824. Department offers programs in analytical chemistry (MS, PhD), chemistry (MAT), inorganic chemistry (MS, PhD), organic chemistry (MS, PhD), physical chemistry (PhD). Faculty: 27 full-time (1 woman), 0 part-time. Matriculated students: 195 full-time (48 women), 22 part-time (11 women); includes 8 minority (1 black American, 7 Hispanic American), 101 foreign. In 1990, 8 master's, 26 doctorates awarded. Terminal master's awarded for partial completion of doctoral program. *Degree requirements:* For master's, 1 foreign language, thesis (for some programs); for doctorate, 1 foreign language, dissertation. *Entrance requirements:* GRE General Test, TOEFL. Application deadline: rolling. Application fee: $25 ($40 for foreign students). *Tuition:* $104.75 per credit for state residents; $211.75 per credit for nonresidents. *Financial aid:* In 1990–91, $1.506-million in aid awarded. 19 fellowships, 44 research assistantships, 108 teaching assistantships were awarded; partial tuition waivers also available. *Faculty research:* Analytical instrumentation, laser spectroscopy, fundamental materials research, theoretical chemistry, biophysical and biorganic chemistry. Total annual research budget: $2.3-million. • Dr. Gerald Babcock, Chairperson, 517-353-9717. Application contact: Dr. William Reusch, Coordinator, Graduate Program, 517-355-9716.

Michigan Technological University, College of Engineering, Department of Chemistry, Houghton, MI 49931. Department awards MS, PhD. PhD program jointly offered with Central Michigan University. Part-time programs available. Faculty: 16 full-time (2 women), 1 (woman) part-time. Matriculated students: 31 full-time (20 women), 0 part-time; includes 0 minority, 20 foreign. Average age 29. 167 applicants, 17% accepted. In 1990, 3 master's, 3 doctorates awarded. *Degree requirements:* Thesis/dissertation required, foreign language not required. *Entrance requirements:* TOEFL (minimum score of 550 required), GRE. Application deadline: rolling. Application fee: $20. *Expenses:* Tuition of $852 per semester full-time, $71 per credit part-time for state residents; $2065 per semester full-time, $172 per credit part-time for nonresidents. Fees of $75 per semester. *Financial aid:* In 1990–91, 24 students received a total of $195,600 in aid awarded. 3 fellowships, 11 research assistantships (1 to a first-year student), 10 teaching assistantships (2 to first-year students) were awarded; federal work-study and career-related internships or fieldwork also available. Aid available to part-time students. Average monthly stipend for a graduate assistantship: $800-933. Financial aid application deadline: 3/1; applicants required to submit FAF. *Faculty research:* Synthetic and physical organic chemistry, polymer chemistry and processing of composites, inorganic and organic analytical chemistry, biochemistry of terpenoid and steroids. • Dr. John Adler, Head, 906-487-2047. Application contact: Dr. John Williams, Graduate Adviser, 906-487-2047.

Middle Tennessee State University, School of Basic and Applied Sciences, Department of Chemistry and Physics, Murfreesboro, TN 37132. Department offers programs in chemistry (MS, MST, DA), natural science (MS, MST). Faculty: 14 full-time (2 women), 0 part-time. Matriculated students: 3 full-time (0 women), 15 part-time (6 women); includes 1 minority (black American), 3 foreign. Average age 27. 10 applicants, 100% accepted. In 1990, 3 master's, 0 doctorates awarded. *Degree requirements:* For master's, 1 foreign language, thesis, comprehensive exams; for doctorate, dissertation, comprehensive exams required, foreign language not required. *Entrance requirements:* For master's, MAT, Cooperative English Test; for doctorate, GRE. Application deadline: 8/1 (priority date). Application fee: $5. *Expenses:* Tuition of $1760 per year full-time, $89 per semester hour part-time for state residents; $4964 per year full-time, $229 per semester hour part-time for nonresidents. Fees of $15 per year full-time, $5 per year part-time. *Financial aid:* In 1990–91, 12 teaching assistantships (7 to first-year students) awarded. • Dr. Dan Scott, Chairman, 615-898-2956.

Montana State University, College of Letters and Science, Department of Chemistry, 901 West Garfield Street, Bozeman, MT 59717. Department offers programs in biochemistry (MS, PhD), chemistry (MS, PhD). Faculty: 17 full-time (0 women), 0 part-time. Matriculated students: 38 full-time (8 women), 11 part-time (2 women); includes 2 minority (1 Asian American, 1 Hispanic American), 8 foreign. Average age 28. 17 applicants, 59% accepted. In 1990, 4 master's, 12 doctorates awarded. *Degree requirements:* For master's, thesis or alternative required, foreign language not required; for doctorate, dissertation required, foreign language not required. *Entrance requirements:* GRE General Test, TOEFL (minimum score of 570 required). Application deadline: 6/15 (applications processed on a rolling basis). Application fee: $20. *Tuition:* $1150 per year full-time, $74 per credit (minimum) part-time for state residents; $2824 per year full-time, $100 per credit (minimum) part-time for nonresidents. *Financial aid:* Research assistantships, teaching assistantships, and career-related internships or fieldwork available. Financial aid application deadline: 3/1. *Faculty research:* Analytical, physical, organic, inorganicand biochemical chemistry. • Dr. Edwin H. Abbott, Head, 406-994-4801.

Montclair State College, School of Mathematical and Natural Sciences, Department of Chemistry, Upper Montclair, NJ 07043. Department awards MA. Part-time and evening/weekend programs available. Faculty: 10 full-time, 0 part-time. In 1990, 2 degrees awarded. *Degree requirements:* Comprehensive exam required, foreign language and thesis not required. *Entrance requirements:* GRE General Test. Application deadline: 7/1 (priority date, applications processed on a rolling basis). Application fee: $25. *Expenses:* Tuition of $130 per credit for state residents; $165 per credit for nonresidents. Fees of $276 per year. *Financial aid:* In 1990–91, $3000 in aid awarded. 1 teaching assistantship was awarded. Financial aid application deadline: 3/15. • Roland Flynn, Chairperson, 201-893-5140.

Mount Allison University, Faculty of Science, Department of Chemistry, Sackville, NB E0A 3C0, Canada. Department awards M Sc. Faculty: 4 full-time (0 women), 0 part-time. Matriculated students: 1 full-time (0 women), 1 (woman) part-time; includes 0 foreign. Average age 24. 1 applicants. In 1990, 1 degree awarded (100% found work related to degree). *Degree requirements:* Thesis required, foreign language not required. *Entrance requirements:* Honors bachelor's degree in chemistry. *Tuition:* $300 per year (minimum) full-time, $425 per course part-time for Canadian residents; $2000 per year (minimum) full-time, $425 per course part-time for nonresidents. *Financial aid:* In 1990–91, 2 students received a total of $5000 in aid awarded. 1 fellowship (to a first-year student), 2 research assistantships (both to first-year students) were awarded. Average monthly stipend for a graduate assistantship: $250. *Faculty research:* Biophysical chemistry of model biomembranes, organic synthesis, fast-reaction kinetics, physical chemistry of micelles. • Dr. Vincent Conrad Reinsborough, Head, 506-364-2302. Application contact: Dr. R. Langler, Chairman, Graduate Studies Committee.

Directory: Chemistry

Mount Holyoke College, Graduate Studies, Department of Chemistry, South Hadley, MA 01075. Department awards MA. Faculty: 8 full-time (6 women), 1 part-time (0 women). Matriculated students: 2 full-time (both women), 0 part-time; includes 2 foreign. Average age 30. In 1990, 1 degree awarded (100% found work related to degree). *Degree requirements:* 1 foreign language, thesis. *Entrance requirements:* TOEFL. Application deadline: 2/15. Application fee: $0. *Expenses:* Tuition of $15,950 per year full-time, $500 per credit part-time. Fees of $100 per year. *Financial aid:* In 1990–91, 2 research assistantships (1 to a first-year student), 2 teaching assistantships awarded; full and partial tuition waivers also available. *Faculty research:* Logical, analytical, inorganic, physical, and organic chemistry. • Chair, 413-538-2214. Application contact: Hilary Shaw, Assistant to the Dean of Studies, 413-538-2599.

Murray State University, College of Sciences, Department of Chemistry, Murray, KY 42071. Department awards MAT, MS. Part-time programs available. Faculty: 11 full-time (0 women), 0 part-time. Matriculated students: 6 full-time (2 women), 4 part-time (1 woman); includes 5 minority (all Asian American), 5 foreign. 5 applicants, 100% accepted. In 1990, 5 degrees awarded. *Degree requirements:* Thesis required (for some programs), foreign language not required. *Entrance requirements:* GRE General Test, TOEFL (minimum score of 500 required). Application fee: $0 ($20 for foreign students). *Tuition:* $775 per semester full-time, $87 per credit hour part-time for state residents; $2215 per semester full-time, $246 per credit hour part-time for nonresidents. *Financial aid:* Research assistantships, teaching assistantships, federal work-study available. Average monthly stipend for a graduate assistantship: $335. Financial aid application deadline: 4/1; applicants required to submit GAPSFAS. • Dr. Oliver Muscio, Director, 502-762-2584.

New Jersey Institute of Technology, Department of Chemical Engineering, Chemistry, and Environmental Science, Newark, NJ 07102. Offerings include applied chemistry (MS). Department faculty: 32 full-time (4 women), 9 part-time (1 woman). *Application deadline:* 6/5 (priority date, applications processed on a rolling basis). *Application fee:* $30. *Tuition:* $2585 per semester full-time, $253 per credit part-time for state residents; $3864 per semester full-time, $349 per credit part-time for nonresidents. • Dr. Gordon Lewandowski, Chairman, 201-596-3573. Application contact: Petra Theodos, Director of Graduate Admissions, 201-596-3460.

New Mexico Highlands University, School of Science and Technology, Department of Physical Sciences, Las Vegas, NM 87701. Department offers program in applied chemistry (MS). Faculty: 5 full-time (0 women), 0 part-time. Matriculated students: 2 full-time (1 woman), 3 part-time (0 women); includes 5 minority (4 Asian American, 1 Hispanic American), 4 foreign. Average age 27. 34 applicants, 26% accepted. In 1990, 0 degrees awarded. *Degree requirements:* Foreign language not required. *Entrance requirements:* Minimum undergraduate GPA of 3.0. Application deadline: 8/1 (priority date, applications processed on a rolling basis). Application fee: $15 ($75 for foreign students). *Expenses:* Tuition of $1248 per year full-time, $52 per credit hour part-time for state residents; $4488 per year full-time, $187 per credit hour part-time for nonresidents. Fees of $20 per year. • Dr. Larry Sveum, Head, 505-454-3204. Application contact: Dr. Gilbert D. Rivera, Academic Vice President, 505-454-3311.

New Mexico Institute of Mining and Technology, Department of Chemistry, Socorro, NM 87801. Department offers programs in biochemistry (MS), chemistry (MS), explosives technology and atmospheric chemistry (PhD). Faculty: 8 full-time (1 woman), 0 part-time. Matriculated students: 15 full-time (3 women), 0 part-time; includes 1 minority (Hispanic American), 6 foreign. Average age 25. 21 applicants, 95% accepted. In 1990, 2 master's, 2 doctorates awarded. *Degree requirements:* 1 foreign language, thesis/dissertation. *Entrance requirements:* For master's, GRE General Test, TOEFL (minimum score of 540 required); for doctorate, GRE General Test, GRE Subject Test, TOEFL (minimum score of 540 required). Application deadline: 6/1 (priority date, applications processed on a rolling basis). Application fee: $16. *Expenses:* Tuition of $617 per semester full-time, $51 per hour part-time for state residents; $2656 per semester full-time, $209 per hour part-time for nonresidents. Fees of $207 per semester. *Financial aid:* In 1990–91, 14 students received a total of $49,708 in aid awarded. 0 fellowships, 6 research assistantships (2 to first-year students), 8 teaching assistantships (4 to first-year students) were awarded; federal work-study, institutionally sponsored loans also available. Average monthly stipend for a graduate assistantship: $600. Financial aid application deadline: 3/15; applicants required to submit GAPSFAS or FAF. *Faculty research:* Organic, analytical, environmental, atmospheric, and explosives chemistry. Total annual research budget: $33,300. • Dr. Donald K. Brandvold, Chairman, 505-835-5263. Application contact: Dr. J. A. Smoake, Dean, Graduate Studies, 505-835-5513.

New Mexico State University, College of Arts and Sciences, Department of Chemistry, Las Cruces, NM 88003. Department awards MS, PhD. Part-time programs available. Faculty: 24 full-time, 0 part-time. Matriculated students: 44 full-time (13 women), 4 part-time (1 woman). In 1990, 4 master's, 4 doctorates awarded. *Degree requirements:* Thesis/dissertation or alternative. *Entrance requirements:* GRE General Test. Application deadline: 7/1. Application fee: $10. *Tuition:* $1608 per year full-time, $67 per credit hour part-time for state residents; $5304 per year full-time, $221 per credit hour part-time for nonresidents. *Financial aid:* Fellowships, research assistantships, teaching assistantships, federal work-study available. Aid available to part-time students. Financial aid applicants required to submit GAPSFAS or FAF. • Dr. Dennis Darnell, Head, 505-646-2505.

New York University, Graduate School of Arts and Science, Department of Chemistry, New York, NY 10011. Department offers programs in bioorganic chemistry (MS, PhD), biophysical chemistry (MS, PhD), experimental physical chemistry (MS, PhD), inorganic chemistry (MS, PhD), organic chemistry (MS, PhD), theoretical physical chemistry (MS, PhD). Faculty: 26 full-time, 5 part-time. Matriculated students: 90 full-time, 18 part-time; includes 11 minority (8 Asian American, 2 black American, 1 Hispanic American), 66 foreign. Average age 24. 213 applicants, 64% accepted. In 1990, 26 master's awarded (15% found work related to degree, 85% continued full-time study); 12 doctorates awarded. *Degree requirements:* For master's, thesis required, foreign language not required; for doctorate, 1 foreign language, dissertation. *Entrance requirements:* GRE Subject Test, TSE. Application deadline: 1/15 (priority date, applications processed on a rolling basis). Application fee: $30. *Tuition:* $467 per credit. *Financial aid:* In 1990–91, $1.275-million in aid awarded. 5 fellowships (1 to a first-year student), 21 research assistantships, 44 teaching assistantships, 21 teaching fellowships (15 to first-year students) were awarded; full and partial tuition waivers, federal work-study also available. Financial aid application deadline: 1/15. *Faculty research:* Organic, bioorganic, and catalysis. Total annual research budget: $2-million. • Neville R. Kallenbach, Chairman, 212-998-8400. Application contact: David Schuster, Director of Graduate Studies, 212-998-8400.

See full description on page 117.

North Carolina Agricultural and Technical State University, Graduate School, College of Arts and Sciences, Department of Chemistry, Greensboro, NC 27411. Department awards MS. Part-time and evening/weekend programs available. Faculty: 11 full-time (3 women), 0 part-time. Matriculated students: 7 full-time (1 woman), 7 part-time (3 women); includes 13 minority (2 Asian American, 11 black American), 1 foreign. Average age 29. 6 applicants, 33% accepted. In 1990, 3 degrees awarded. *Degree requirements:* Thesis or alternative, qualifying exam, comprehensive exam required, foreign language not required. *Entrance requirements:* GRE General Test, minimum GPA of 3.0. Application deadline: 6/1 (priority date, applications processed on a rolling basis). Application fee: $15. *Tuition:* $614 per semester full-time for state residents; $3141 per semester full-time for nonresidents. *Financial aid:* In 1990–91, 2 fellowships (both to first-year students), 3 research assistantships (0 to first-year students), 6 teaching assistantships (0 to first-year students) awarded; career-related internships or fieldwork also available. Financial aid application deadline: 6/1; applicants required to submit GAPSFAS. *Faculty research:* Tobacco pesticide research. Total annual research budget: $55,541. • Alex Williamson, Acting Chairperson, 919-334-7601.

North Carolina Central University, Division of Academic Affairs, College of Arts and Sciences, Department of Chemistry, Durham, NC 27707. Department awards MS. Faculty: 8 full-time (0 women), 0 part-time. Matriculated students: 4 full-time (1 woman), 2 part-time (1 woman); includes 4 minority (all black American), 2 foreign. Average age 24. 3 applicants, 100% accepted. In 1990, 4 degrees awarded (100% found work related to degree). *Degree requirements:* 1 foreign language, computer language, thesis, comprehensive exam. *Entrance requirements:* Minimum cumulative GPA of 2.5, 3.0 in major. Application deadline: 8/1. Application fee: $15. *Financial aid:* In 1990–91, $11,000 in aid awarded. 2 research assistantships were awarded; fellowships, teaching assistantships, federal work-study, institutionally sponsored loans, and career-related internships or fieldwork also available. Aid available to part-time students. Financial aid application deadline: 5/1; applicants required to submit GAPSFAS or FAF. *Faculty research:* Synthesis of compounds of biomedical importance, synthesis and study of the chemical characteristics of derivatives of selected natural products. • Dr. John A. Myers, Chairperson, 919-560-6462. Application contact: Dr. Mary M. Townes, Dean, College of Arts and Sciences, 919-560-6368.

Announcement: Modern building and instrumentation. Research opportunities in biochemistry and inorganic, organic, analytical, or physical chemistry, with emphasis on organic synthesis and mechanisms and biomedically related projects. Graduates work in industry and government or continue with further study.

North Carolina State University, College of Physical and Mathematical Sciences, Department of Chemistry, Raleigh, NC 27695. Department awards MCH, MS, PhD. Part-time programs available. Faculty: 31 full-time (2 women), 0 part-time. Matriculated students: 73 full-time (20 women), 4 part-time (2 women); includes 5 minority (3 black American, 2 Hispanic American), 30 foreign. Average age 26. 89 applicants, 47% accepted. In 1990, 5 master's, 10 doctorates awarded. Terminal master's awarded for partial completion of doctoral program. *Degree requirements:* Thesis/dissertation required, foreign language not required. *Entrance requirements:* GRE General Test, minimum GPA of 3.0. Application deadline: 4/15 (priority date, applications processed on a rolling basis). Application fee: $35. *Tuition:* $1138 per year for state residents; $5805 per year for nonresidents. *Financial aid:* In 1990–91, $700,000 in aid awarded. 2 fellowships, 60 teaching assistantships were awarded; research assistantships also available. *Faculty research:* Solid-state chemistry, biotechnology, electrochemistry, excited-state systems, organometallic chemistry. Total annual research budget: $2.01-million. • Dr. Kenneth W. Hanck, Head, 919-737-2545. Application contact: Dr. Russ Linderman, Graduate Administrator, 919-737-2548.

North Dakota State University, College of Science and Mathematics, Department of Chemistry, Fargo, ND 58105. Department awards MS, PhD. Faculty: 14 full-time (0 women), 0 part-time. Matriculated students: 35 full-time (7 women), 0 part-time; includes 12 foreign. Average age 24. 20 applicants, 50% accepted. In 1990, 0 master's awarded; 5 doctorates awarded (100% found work related to degree). Terminal master's awarded for partial completion of doctoral program. *Degree requirements:* For master's, foreign language not required; for doctorate, dissertation required, foreign language not required. *Entrance requirements:* GRE General Test, GRE Subject Test, TOEFL (minimum score of 525 required). Application fee: $20. *Tuition:* $1411 per year full-time, $52 per credit hour part-time for state residents; $3571 per year full-time, $132 per credit hour part-time for nonresidents. *Financial aid:* In 1990–91, 33 students received aid. 0 fellowships, 17 research assistantships (2 to first-year students), 16 teaching assistantships (5 to first-year students) awarded; federal work-study, institutionally sponsored loans also available. Average monthly stipend for a graduate assistantship: $905. Financial aid application deadline: 4/15. *Faculty research:* Analytical, organic, inorganic, physical, and theoretical chemistry. Total annual research budget: $460,000. • Dr. Gregory D. Gillispie, Chair, 701-237-8244.

Northeastern Illinois University, College of Arts and Sciences, Department of Chemistry, Chicago, IL 60625. Department awards MS. Evening/weekend programs available. Faculty: 7 full-time (1 woman), 0 part-time. 8 applicants, 38% accepted. In 1990, 2 degrees awarded. *Degree requirements:* Thesis or alternative required, foreign language not required. *Application deadline:* rolling. *Application fee:* $0. *Expenses:* Tuition of $70 per credit hour for state residents; $210 per credit hour for nonresidents. Fees of $37 per credit hour. • Dr. Howard Murray, Chairperson, 312-583-4050.

Northeastern University, Graduate School of Arts and Sciences, Department of Chemistry, Boston, MA 02115. Department offers programs in analytical chemistry (MS), chemistry (PhD), clinical chemistry (MS), inorganic chemistry (MS), organic chemistry (MS), physical chemistry (MS). Part-time and evening/weekend programs available. Faculty: 16 full-time (2 women), 0 part-time. Matriculated students: 44 full-time (17 women), 32 part-time (13 women); includes 2 minority (1 Asian American, 1 black American), 30 foreign. Average age 23. 147 applicants, 24% accepted. In 1990, 2 master's, 10 doctorates awarded. Terminal master's awarded for partial completion of doctoral program. *Degree requirements:* For master's, thesis (for some programs); for doctorate, dissertation, qualifying exam required, foreign language not required. *Entrance requirements:* TOEFL (minimum score of 580 required). Application deadline: 4/15. Application fee: $40. *Expenses:* Tuition of $10,260 per year full-time, $285 per quarter hour part-time. Fees of $293 per year full-time. *Financial aid:* In 1990–91, $600,000 in aid awarded. 15 research assistantships (0 to first-year students), 21 teaching assistantships (7 to first-year students) were awarded; full tuition waivers and career-related internships or fieldwork also available. Financial aid application deadline: 4/15. *Total annual research budget:* $1.9-million. • Dr. John L. Roebber, Executive Officer, 617-437-2822. Application contact: Dr. Paul Vouros, Director of Graduate Affairs, 617-437-2822.

See full description on page 119.

Northeastern University, Graduate School of Pharmacy and Allied Health Professions, Programs in Biomedical Sciences, Boston, MA 02115. Offerings include clinical chemistry (MS). Faculty: 33 full-time (11 women), 15 part-time (5 women).

Directory: Chemistry

Degree requirements: Comprehensive exam required, foreign language not required. *Entrance requirements:* Minimum GPA of 3.0, bachelor's degree in a science. Application fee: $40. *Expenses:* Tuition of $10,260 per year full-time, $285 per quarter hour part-time. Fees of $293 per year full-time. • Application contact: Sandra Lally, Administrative Assistant, 617-437-3380.

Northeast Louisiana University, College of Pure and Applied Sciences, Department of Chemistry, Monroe, LA 71209. Department awards MS. Faculty: 10 full-time (2 women), 0 part-time. Matriculated students: 10 full-time (4 women), 4 part-time (2 women); includes 1 minority (black American), 9 foreign. 22 applicants, 68% accepted. In 1990, 6 degrees awarded. *Degree requirements:* 1 foreign language (computer language can substitute), thesis. *Entrance requirements:* GRE General Test. Application deadline: 7/1. Application fee: $5. *Tuition:* $816 per semester for state residents; $1608 per semester for nonresidents. *Financial aid:* In 1990–91, $61,200 in aid awarded. 8 teaching assistantships, 7 laboratory assistantships were awarded; research assistantships, federal work-study, and career-related internships or fieldwork also available. Financial aid application deadline: 7/13. *Faculty research:* Analysis of chemical processes, metabolism of steroids, nitrogen fixation in soybeans, metabolism of carnitine. Total annual research budget: $30,000. • Dr. Robert L. Holt, Head, 318-342-1825.

Northern Arizona University, College of Arts and Sciences, Department of Chemistry, Flagstaff, AZ 86011. Department awards MS. Part-time programs available. *Degree requirements:* Computer language, thesis, departmental qualifying exam required, foreign language not required. *Entrance requirements:* GRE Subject Test. Application deadline: 3/15 (priority date, applications processed on a rolling basis). Application fee: $0 ($25 for foreign students). *Expenses:* Tuition of $1528 per year full-time, $80 per credit part-time for state residents; $6180 per year full-time, $258 per credit part-time for nonresidents. Fees of $25 per semester. *Financial aid:* In 1990–91, 0 fellowships, 0 research assistantships, 2 teaching assistantships (1 to a first-year student) awarded; full and partial tuition waivers, federal work-study also available. Financial aid applicants required to submit FAF. *Faculty research:* Biochemistry of exercise, organic and inorganic mechanism studies, inhibition of ice mutation, polymer separation. • Dr. Michael Eastman, Chairman, 602-523-3008. Application contact: Dr. Gerald Caple, 602-523-2450.

Northern Illinois University, College of Liberal Arts and Sciences, Department of Chemistry, De Kalb, IL 60115. Department awards MS, PhD. Part-time programs available. Faculty: 21 full-time (1 woman), 0 part-time. Matriculated students: 46 full-time (17 women), 8 part-time (3 women); includes 5 minority (3 Asian American, 1 black American, 1 Hispanic American), 22 foreign. Average age 26. 259 applicants, 19% accepted. In 1990, 10 master's, 5 doctorates awarded. *Degree requirements:* For master's, comprehensive exam, thesis defense required, foreign language not required; for doctorate, 1 foreign language (computer language can substitute), comprehensive exam, thesis defense. *Entrance requirements:* For master's, GRE General Test, TOEFL, appropriate bachelor's degree, 2.75 GPA; for doctorate, GRE General Test, TOEFL, minimum GPA of 2.75, minimum graduate GPA of 3.2. Application deadline: 6/1. Application fee: $0. *Tuition:* $1339 per semester full-time for state residents; $3163 per semester full-time for nonresidents. *Financial aid:* In 1990–91, 9 research assistantships, 31 teaching assistantships awarded; fellowships, full tuition waivers, federal work-study, and career-related internships or fieldwork also available. Aid available to part-time students. Financial aid applicants required to submit FAF. *Total annual research budget:* $1.2-million. • Dr. Joseph W. Vaughn, Chair, 815-753-1181. Application contact: Dr. David M. Piatak, Director, Graduate Studies, 815-753-6895.

Northern Michigan University, School of Arts and Sciences, Department of Chemistry, Marquette, MI 49855. Department awards MA. Part-time programs available. Faculty: 8 full-time (1 woman), 0 part-time. Matriculated students: 0 full-time, 1 part-time (0 women); includes 0 minority, 0 foreign. In 1990, 0 degrees awarded. *Degree requirements:* Thesis required, foreign language not required. *Entrance requirements:* GRE General Test. Application deadline: 8/1 (priority date, applications processed on a rolling basis). Application fee: $0 ($25 for foreign students). *Tuition:* $92.65 per credit hour for state residents; $131.45 per credit hour for nonresidents. *Financial aid:* In 1990–91, 2 graduate assistantships awarded; federal work-study, institutionally sponsored loans, and career-related internships or fieldwork also available. Aid available to part-time students. Average monthly stipend for a graduate assistantship: $500. Financial aid application deadline: 3/1; applicants required to submit FAF. • Dr. Roger D. Barry, Head.

Northwestern University, College of Arts and Sciences, Department of Chemistry, Evanston, IL 60208. Department awards MS, PhD. Part-time programs available. Faculty: 29 full-time (0 women), 0 part-time. Matriculated students: 145 full-time (53 women), 0 part-time; includes 6 minority (4 Asian American, 1 black American, 1 Native American), 21 foreign. 198 applicants, 62% accepted. In 1990, 25 master's, 26 doctorates awarded. Terminal master's awarded for partial completion of doctoral program. *Degree requirements:* For master's, foreign language and thesis not required; for doctorate, dissertation required, foreign language not required. *Entrance requirements:* GRE General Test, GRE Subject Test. Application deadline: 8/30. Application fee: $40 ($45 for foreign students). *Tuition:* $4665 per quarter full-time, $1704 per course part-time. *Financial aid:* In 1990–91, $850,661 in aid awarded. 7 fellowships, 100 research assistantships, 35 teaching assistantships were awarded; federal work-study, institutionally sponsored loans, and career-related internships or fieldwork also available. Financial aid application deadline: 1/15; applicants required to submit GAPSFAS. *Faculty research:* Organic, inorganic, physical, analytical and biochemical chemistry. Total annual research budget: $6.51-million. • Dr. Mark A. Rathert, Chair, 708-491-5371.

Oakland University, College of Arts and Sciences, Department of Chemistry, Rochester, MI 48309. Department offers programs in chemistry (MS, PhD), health and environmental chemistry (PhD). *Degree requirements:* For master's, thesis required, foreign language not required; for doctorate, dissertation. *Entrance requirements:* For master's, minimum GPA of 3.0 for unconditional admission; for doctorate, GRE Subject Test. Application deadline: 7/15. Application fee: $25. *Expenses:* Tuition of $122 per credit hour for state residents; $270 per credit hour for nonresidents. Fees of $170 per year. • Dr. Paul Tomboulian, Chair, 313-370-2324. Application contact: Dr. Michael Sevilla, 313-370-2324.

Ohio State University, College of Mathematical and Physical Sciences, Department of Chemistry, Columbus, OH 43210. Department awards MS, PhD. Faculty: 49 full-time (3 women), 0 part-time. Matriculated students: 270 full-time (72 women), 14 part-time (4 women); includes 19 minority (13 Asian American, 4 black American, 1 Hispanic American, 1 Native American), 101 foreign. 203 applicants, 29% accepted. In 1990, 23 master's, 31 doctorates awarded. *Degree requirements:* For master's, thesis optional, foreign language not required; for doctorate, dissertation required, foreign language not required. *Entrance requirements:* For master's, GRE General Test, GRE Subject Test. Application deadline: 8/15 (applications processed on a rolling basis). Application fee: $0 ($25 for foreign students). *Tuition:* $1213 per quarter full-time, $364 per course part-time for state residents; $3143 per quarter full-time, $943 per course part-time for nonresidents. *Financial aid:* In 1990–91, $12,000 in aid awarded. 24 fellowships (10 to first-year students), 93 research assistantships (0 to first-year students), 150 teaching assistantships (50 to first-year students) were awarded; federal work-study, institutionally sponsored loans also available. Aid available to part-time students. Financial aid applicants required to submit FAF. *Total annual research budget:* $5.01-million. • Russell M. Pitzer, Chairman, 614-292-2251. Application contact: C. Weldon Mathews, Vice Chair for Graduate Study, 614-292-8917.

See full description on page 121.

Ohio University, Graduate Studies, College of Arts and Sciences, Department of Chemistry, Athens, OH 45701. Department awards MS, PhD. Faculty: 19 full-time (1 woman), 5 part-time (2 women). Matriculated students: 26 full-time, 2 part-time; includes 2 minority (both black American), 10 foreign. 72 applicants, 29% accepted. In 1990, 2 master's, 1 doctorate awarded. *Degree requirements:* For master's, thesis, exam required, foreign language not required; for doctorate, dissertation, exam. *Application fee:* $25. *Tuition:* $1112 per quarter hour full-time, $138 per credit hour part-time for state residents; $2227 per quarter hour full-time, $277 per credit hour part-time for nonresidents. *Financial aid:* In 1990–91, 31 research assistantships, 1 teaching assistantship awarded; fellowships, federal work-study, institutionally sponsored loans also available. Financial aid application deadline: 3/15; applicants required to submit FAF. *Total annual research budget:* $259,939. • Dr. Paul Sullivan, Chairman, 614-593-1737. Application contact: Dr. David Hendricker, Graduate Chair, 614-593-1733.

Oklahoma State University, College of Arts and Sciences, Department of Chemistry, Stillwater, OK 74078. Department awards MS, PhD. Faculty: 17 full-time, 0 part-time. Matriculated students: 55 (18 women); includes 0 minority, 26 foreign. In 1990, 6 master's, 11 doctorates awarded. *Degree requirements:* Thesis/dissertation required, foreign language not required. *Entrance requirements:* TOEFL (minimum score of 550 required). Application deadline: 7/1 (priority date). Application fee: $15. *Tuition:* $63.75 per credit for state residents; $138.25 per credit for nonresidents. *Financial aid:* In 1990–91, 25 research assistantships, 27 teaching assistantships awarded; fellowships, partial tuition waivers, federal work-study also available. Aid available to part-time students. Financial aid application deadline: 3/1; applicants required to submit GAPSFAS or FAF. *Faculty research:* Modern analytical methods, laser spectroscopy, theoretical chemistry, structural analysis (x-ray, NMR, MS), organic synthesis. • Dr. Horacio Mottola, Head, 405-744-5920.

Old Dominion University, College of Sciences, Department of Chemistry and Biochemistry, Norfolk, VA 23529. Offerings include clinical chemistry (MS). Department faculty: 13 full-time (2 women), 1 part-time (0 women). *Degree requirements:* Thesis required, foreign language not required. *Entrance requirements:* GRE General Test, GRE Subject Test, TOEFL, minimum GPA of 3.0 in major, 2.5 overall. Application deadline: 7/1 (applications processed on a rolling basis). Application fee: $20. *Expenses:* Tuition of $148 per credit hour for state residents; $375 per credit hour for nonresidents. Fees of $64 per year full-time. • Dr. Frank E. Scully Jr., Chairman, 804-683-4078. Application contact: Dr. Roy L. Williams, Graduate Program Director, 804-683-4078.

Oregon Graduate Institute of Science and Technology, Department of Chemical and Biological Sciences, Beaverton, OR 97006. Department offers programs in biochemistry and molecular biology (MS, PhD), chemistry (MS, PhD). Faculty: 14 full-time (2 women), 7 part-time (0 women). Matriculated students: 22 full-time (8 women), 2 part-time (both women); includes 0 minority, 14 foreign. Average age 28. 43 applicants, 37% accepted. In 1990, 0 master's awarded; 5 doctorates awarded (60% entered university research/teaching, 40% found other work related to degree). Terminal master's awarded for partial completion of doctoral program. *Degree requirements:* For master's, thesis required, foreign language not required; for doctorate, dissertation, comprehensive exam, oral defense required, foreign language not required. *Entrance requirements:* GRE General Test, GRE Subject Test, TOEFL (minimum score of 550 required). Application deadline: 3/1 (priority date, applications processed on a rolling basis). Application fee: $40. *Expenses:* Tuition of $9600 per year full-time, $240 per credit part-time. Fees of $60 per quarter. *Financial aid:* In 1990–91, 22 students received a total of $396,000 in aid awarded. 0 fellowships, 22 research assistantships (10 to first-year students) were awarded; institutionally sponsored loans also available. Average monthly stipend for a graduate assistantship: $833. Financial aid application deadline: 3/1. *Faculty research:* Fungal and plant biochemistry, membrane science, metallobiochemistry, molecular biology. Total annual research budget: $2.130-million. • Dr. Michael Gold, Chairman, 503-690-1076. Application contact: Admissions Adviser, 503-690-1070.

Oregon State University, Graduate School, College of Science, Department of Chemistry, Corvallis, OR 97331. Department awards MA, MAIS, MS, PhD. Programs offered in analytical chemistry (MS, PhD), chemistry (MA), inorganic chemistry (MS, PhD), nuclear and radiation chemistry (MS, PhD), organic chemistry (MS, PhD), physical chemistry (MS, PhD). Part-time programs available. Faculty: 26 full-time (0 women), 5 part-time (1 woman). Matriculated students: 73 full-time (19 women), 1 part-time (0 women); includes 5 minority (3 Asian American, 2 Hispanic American), 27 foreign. In 1990, 4 master's, 4 doctorates awarded. Terminal master's awarded for partial completion of doctoral program. *Degree requirements:* 1 foreign language, thesis/dissertation. *Entrance requirements:* TOEFL (minimum score of 550 required), minimum GPA of 3.0 in last 90 hours. Application deadline: 3/1. Application fee: $40. *Tuition:* $1140 per trimester full-time, $449 per year part-time for state residents; $1816 per trimester full-time, $674 per year part-time for nonresidents. *Financial aid:* In 1990–91, $323,505 in aid awarded. 2 fellowships (both to first-year students), 32 research assistantships (0 to first-year students), 38 teaching assistantships (22 to first-year students) were awarded; institutionally sponsored loans also available. Aid available to part-time students. Financial aid application deadline: 2/1; applicants required to submit FAF. *Faculty research:* Transition metal chemistry, enzyme reaction mechanisms, structure and dynamics of gas molecules, chemiluminescence, nonlinear optical spectroscopy. Total annual research budget: $1.73-million. • Dr. Carroll W. DeKock, Chair, 503-737-2081.

See full description on page 123.

Pennsylvania State University University Park Campus, College of Science, Department of Chemistry, University Park, PA 16802. Department awards MS, PhD. Faculty: 37. Matriculated students: 197 full-time (59 women), 17 part-time (11 women). In 1990, 3 master's, 23 doctorates awarded. *Entrance requirements:* GRE General Test. Application fee: $35. *Tuition:* $203 per credit for state residents; $403 per credit for nonresidents. • Dr. Barbara Garrison, Head, 814-865-6553.

Pittsburg State University, College of Arts and Sciences, Department of Chemistry, Pittsburg, KS 66762. Department awards MS. Faculty: 4 full-time (0 women), 0 part-time. Matriculated students: 7 full-time (0 women), 10 part-time (2 women); includes 0 minority, 3 foreign. In 1990, 3 degrees awarded. *Degree requirements:* Thesis or alternative required, foreign language not required. *Application fee:* $35. *Tuition:* $1670 per year full-time, $54 per credit hour part-time for state residents; $4108 per year full-time, $125 per credit hour part-time for nonresidents. *Financial aid:*

Directory: Chemistry

Teaching assistantships, federal work-study, and career-related internships or fieldwork available. • Dr. Clarence Pfluger, Chairman, 316-231-7000 Ext. 4747.

Polytechnic University, Brooklyn Campus, Division of Arts and Sciences, Department of Chemistry, 333 Jay Street, Brooklyn, NY 11201. Department offers programs in chemical physics (MS, PhD), chemistry (MS, PhD). Evening/weekend programs available. *Degree requirements:* For doctorate, 1 foreign language, dissertation. *Entrance requirements:* GRE General Test, GRE Subject Test. *Tuition:* $6000 per semester.

Polytechnic University, Westchester Graduate Center, Department of Chemistry, Hawthorne, NY 10532. Department awards MS. *Degree requirements:* Computer language. *Application fee:* $40.

Portland State University, College of Liberal Arts and Sciences, Department of Chemistry, Portland, OR 97207. Department awards MA, MS, PhD. PhD offered in conjunction with Interdisciplinary Program in Environmental Sciences and Resources. Part-time programs available. Faculty: 16 full-time (1 woman), 4 part-time (2 women). Matriculated students: 4 full-time (3 women), 2 part-time (1 woman); includes 0 minority, 3 foreign. Average age 29. 3 applicants, 33% accepted. In 1990, 1 master's awarded. *Degree requirements:* For master's, 1 foreign language, thesis; for doctorate, 1 foreign language, dissertation, cumulative exams, seminar presentations. *Entrance requirements:* For master's, TOEFL (minimum score of 550 required), minimum GPA of 3.0 in upper-division course work or 2.75 overall. Application deadline: 7/12 (priority date, applications processed on a rolling basis). Application fee: $40. *Tuition:* $1151 per trimester for state residents; $1827 per trimester for nonresidents. *Financial aid:* In 1990–91, 3 research assistantships (0 to first-year students), 12 teaching assistantships (6 to first-year students) awarded; federal work-study, institutionally sponsored loans, and career-related internships or fieldwork also available. Aid available to part-time students. Average monthly stipend for a graduate assistantship: $805. Financial aid application deadline: 3/1. *Faculty research:* Synthetic inorganic chemistry, atmospheric chemistry, electrochemistry, organic photochemistry, protein structure. Total annual research budget: $1-million. • Dr. Bruce Brown, Head, 503-725-3811.

Portland State University, College of Liberal Arts and Sciences, Interdisciplinary Program in Environmental Sciences and Resources, Portland, OR 97207. Offers environmental sciences/biology (PhD), environmental sciences/chemistry (PhD), environmental sciences/civil engineering (PhD), environmental sciences/geology (PhD), environmental sciences/physics (PhD). Part-time programs available. Faculty: 1 (woman) full-time, 0 part-time. Matriculated students: 24 full-time (7 women), 8 part-time (2 women); includes 0 minority, 11 foreign. Average age 35. 13 applicants, 100% accepted. In 1990, 3 degrees awarded. *Degree requirements:* Variable foreign language requirement, computer language, dissertation, qualifying exam, oral exam. *Entrance requirements:* TOEFL (minimum score of 550 required), minimum GPA of 3.0 in upper-division course work or 2.75 overall. Application deadline: 7/12. Application fee: $40. *Tuition:* $1151 per trimester for state residents; $1827 per trimester for nonresidents. *Financial aid:* In 1990–91, 1 research assistantship (to a first-year student) awarded; federal work-study, institutionally sponsored loans also available. Aid available to part-time students. Average monthly stipend for a graduate assistantship: $578. Financial aid application deadline: 1/31. *Faculty research:* Aquatic biology and chemistry, atmospheric pollution, natural resources, ecology, biophysics. Total annual research budget: $1.5-million. • Dr. Pavel Smejtek, Director, 503-725-4980.

Announcement: The PhD in environmental sciences and resources/chemistry is part of a multidisciplinary program sponsored by the Departments of Biology, Chemistry, Civil Engineering, Geology, and Physics. Chemistry thesis research is concerned with basic problems in chemistry that are environmentally relevant, including atmospheric chemistry, natural resources, energy resources, materials science, and analytical methods. Contact Dr. Pavel Smejtek, Program Director, 503-725-4980.

Prairie View A&M University, College of Arts and Sciences, Department of Chemistry, Prairie View, TX 77446. Department awards MS. Faculty: 6 full-time (0 women), 0 part-time. Matriculated students: 11 full-time (5 women), 0 part-time; includes 6 minority (all black American), 5 foreign. Average age 23. In 1990, 4 degrees awarded (50% found work related to degree, 50% continued full-time study). *Degree requirements:* 1 foreign language, thesis. *Entrance requirements:* GRE General Test. Application deadline: rolling. Application fee: $0. *Expenses:* Tuition of $162 per semester full-time, $144 per semester part-time for state residents; $1098 per semester full-time, $976 per semester part-time for nonresidents. Fees of $133 per year full-time, $125 per year part-time. *Financial aid:* In 1990–91, $88,200 in aid awarded. 4 research assistantships, 10 teaching assistantships were awarded; federal work-study, institutionally sponsored loans also available. Aid available to part-time students. Financial aid applicants required to submit FAF. *Total annual research budget:* $300,000. • Dr. John Williams, Head, 409-857-3910.

Princeton University, Department of Chemistry, Princeton, NJ 08544. Department offers programs in chemistry (PhD), physics and chemical physics (PhD), polymer sciences and materials (MSE, PhD). PhD (polymer sciences and materials) offered in conjunction with the Department of Chemical Engineering. Faculty: 23 full-time (1 woman), 0 part-time. Matriculated students: 131 full-time (48 women), 0 part-time; includes 13 minority (10 Asian American, 2 black American, 1 Hispanic American), 59 foreign. 178 applicants, 43% accepted. In 1990, 17 doctorates awarded (29% entered university research/teaching, 71% found other work related to degree). Terminal master's awarded for partial completion of doctoral program. *Degree requirements:* For doctorate, 1 foreign language, dissertation, cumulative and general exams. *Entrance requirements:* For doctorate, GRE General Test, GRE Subject Test. Application deadline: 1/8. Application fee: $45 ($50 for foreign students). *Tuition:* $16,670 per year. *Financial aid:* Fellowships, research assistantships, teaching assistantships, federal work-study, institutionally sponsored loans available. Financial aid application deadline: 1/8. *Faculty research:* Chemistry of interfaces, organic synthesis, organometallic chemistry, inorganic reactions, biostructural chemistry. • Director of Graduate Studies, 609-258-4116. Application contact: Michele Spreen, Director of Graduate Admissions, 609-258-3034.

See full description on page 125.

Providence College, Department of Chemistry, Providence, RI 02918. Department awards MS, PhD. Matriculated students: 0. In 1990, 0 master's, 0 doctorates awarded. *Degree requirements:* For master's, 1 foreign language (computer language can substitute), thesis; for doctorate, 2 foreign languages (computer language can substitute for one), dissertation. *Entrance requirements:* GRE. Application deadline: rolling. Application fee: $30. *Tuition:* $125 per credit. • Dr. James Belliveau, Chairman, 401-865-2379.

Purdue University, School of Science, Department of Chemistry, West Lafayette, IN 47907. Department offers programs in analytical chemistry (MS, PhD), biochemistry (MS, PhD), chemical education (MS), inorganic chemistry (MS, PhD), organic chemistry (MS, PhD), physical chemistry (MS, PhD). Faculty: 45 full-time, 1 part-time. Matriculated students: 285 full-time (89 women), 8 part-time (2 women); includes 20 minority (14 Asian American, 2 black American, 4 Hispanic American), 98 foreign. 454 applicants, 26% accepted. In 1990, 12 master's, 44 doctorates awarded. Terminal master's awarded for partial completion of doctoral program. *Degree requirements:* Thesis/dissertation required, foreign language not required. *Entrance requirements:* TOEFL (minimum score of 550 required). Application fee: $25. *Tuition:* $1162 per semester full-time, $83.50 per credit hour part-time for state residents; $3720 per semester full-time, $244.50 per credit hour part-time for nonresidents. *Financial aid:* In 1990–91, 285 students received aid. 36 fellowships (7 to first-year students), 67 research assistantships (1 to a first-year student), 171 teaching assistantships (42 to first-year students) awarded; partial tuition waivers also available. Average monthly stipend for a graduate assistantship: $1070. *Total annual research budget:* $8.1-million. • Dr. Harry Morrison, Head, 317-494-5200. Application contact: R. E. Wild, Chairman, Graduate Admissions, 317-494-5200.

Queens College of the City University of New York, Mathematics and Natural Sciences Division, Department of Chemistry, 65-30 Kissena Boulevard, Flushing, NY 11367. Department awards MA, MS Ed. Programs offered in biochemistry (MA), chemistry (MA). MS Ed awarded through the School of Education. Part-time and evening/weekend programs available. Matriculated students: 2 full-time (1 woman), 9 part-time (5 women). 38 applicants, 24% accepted. In 1990, 2 degrees awarded. *Degree requirements:* 1 foreign language required, thesis not required. *Entrance requirements:* GRE, TOEFL (minimum score of 500 required), previous course work in calculus and physics, minimum GPA of 3.0. Application deadline: 4/1 (applications processed on a rolling basis). Application fee: $35. *Tuition:* $2700 per year full-time, $45 per credit part-time for state residents; $4700 per year full-time, $199 per credit part-time for nonresidents. *Financial aid:* In 1990–91, 0 fellowships awarded; partial tuition waivers, federal work-study, institutionally sponsored loans, and career-related internships or fieldwork also available. Aid available to part-time students. Financial aid application deadline: 4/1. • Dr. George Axelrad, Chairperson, 718-520-7228. Application contact: Dr. Robert Engel, Graduate Adviser, 718-520-7228.

Queen's University at Kingston, Faculty of Arts and Sciences, Department of Chemistry, Kingston, ON K7L 3N6, Canada. Department awards MA, M Sc, PhD. MA offered jointly with Trent University. Part-time programs available. Matriculated students: 67 full-time, 8 part-time; includes 11 foreign. In 1990, 7 master's, 5 doctorates awarded. *Degree requirements:* For master's, thesis optional, foreign language not required; for doctorate, dissertation, comprehensive exam required, foreign language not required. *Entrance requirements:* TOEFL (minimum score of 550 required). Application deadline: 2/28 (priority date). Application fee: $35. *Tuition:* $2861 per year full-time, $426 per trimester part-time for Canadian residents; $10,613 per year full-time, $4998 per trimester part-time for nonresidents. *Financial aid:* Fellowships, research assistantships, teaching assistantships, institutionally sponsored loans available. Financial aid application deadline: 3/1. • Dr. J. A. Stone, Head, 613-545-2624.

Quinnipiac College, School of Allied Health and Natural Sciences, Programs in Medical Laboratory Sciences, Hamden, CT 06518. Offerings include medical laboratory sciences (MHS), with options in clinical chemistry, hematology, laboratory management, microbiology. Faculty: 20 full-time (2 women), 20 part-time (10 women). *Degree requirements:* Thesis or alternative, comprehensive exam. *Entrance requirements:* GRE or GMAT, minimum GPA of 2.5. Application deadline: rolling. Application fee: $30. *Expenses:* Tuition of $5250 per semester full-time, $260 per credit part-time. Fees of $10 per semester. • Dr. Kenneth Kaloustian, Director, 203-288-5251 Ext. 427.

Rensselaer Polytechnic Institute, School of Science, Department of Chemistry, Troy, NY 12180. Department offers programs in inorganic chemistry (MS, PhD), organic chemistry (MS, PhD), physical chemistry (MS, PhD). Part-time and evening/weekend programs available. Faculty: 26 full-time (3 women), 0 part-time. Matriculated students: 98 full-time (25 women), 11 part-time (6 women); includes 6 minority (4 Asian American, 1 black American, 1 Hispanic American), 37 foreign. 120 applicants, 38% accepted. In 1990, 13 master's, 9 doctorates awarded. *Degree requirements:* 1 foreign language, thesis/dissertation. *Entrance requirements:* GRE General Test, GRE Subject Test, TOEFL. Application deadline: 2/1. Application fee: $30. *Expenses:* Tuition of $455 per credit hour. Fees of $195.57 per semester. *Financial aid:* In 1990–91, 80 students received a total of $1-million in aid awarded. 4 fellowships, 30 research assistantships (8 to first-year students), 40 teaching assistantships (20 to first-year students) were awarded. *Faculty research:* Advanced materials, organic synthesis, medicinal chemistry, polymers, biophysics. • D. Aikens, Chairman, 518-276-8981. Application contact: Gerald Korenowski, Chairman, Graduate Committee, 518-276-8489.

See full description on page 127.

Rice University, Weiss School of Natural Sciences, Department of Chemistry, Houston, TX 77251. Department awards MA, PhD. Faculty: 19 full-time, 0 part-time. Matriculated students: 79 full-time (29 women), 0 part-time; includes 6 minority (all Hispanic American), 20 foreign. Terminal master's awarded for partial completion of doctoral program. *Degree requirements:* Thesis/dissertation required, foreign language not required. *Entrance requirements:* For master's, GRE General Test, TOEFL (minimum score 0f 550 required), minimum GPA of 3.0; for doctorate, GRE General Test (score in 70th percentile or higher required), minimum GPA of 3.0. Application deadline: 3/1. *Expenses:* Tuition of $8300 per year full-time, $400 per credit hour part-time. Fees of $167 per year. *Financial aid:* Fellowships, research assistantships available. Financial aid applicants required to submit FAF. *Faculty research:* NMR spectroscopy, organic synthesis, photochemistry, fluorine chemistry, transition-metal complexes. • W. E. Billups, Chairman, 713-527-3277.

Rochester Institute of Technology, College of Science, Department of Chemistry, Rochester, NY 14623. Department awards MS. Part-time and evening/weekend programs available. Matriculated students: 0 full-time, 18 part-time (6 women); includes 5 minority (all Asian American), 3 foreign. 20 applicants, 85% accepted. In 1990, 8 degrees awarded. *Entrance requirements:* American Chemical Society Graduate Placement Exam, GRE, TOEFL. Application deadline: 3/1 (priority date, applications processed on a rolling basis). Application fee: $35. *Tuition:* $12,657 per year full-time, $359 per hour part-time. *Financial aid:* Teaching assistantships, full and partial tuition waivers, federal work-study, institutionally sponsored loans, and career-related internships or fieldwork available. Aid available to part-time students. *Faculty research:* Organic polymer, magnetic resonance and imaging, inorganic coordination polymers, biophysical chemistry, physical polymer chemistry. Total annual research budget: $275,000. • Dr. Gerald Takacs, Head, 716-475-2497.

Roosevelt University, College of Arts and Sciences, Department of Chemistry, Chicago, IL 60605. Department offers programs in biochemistry (MS), chemistry (MS). Part-time and evening/weekend programs available. *Degree requirements:* Thesis optional. *Application deadline:* 8/1 (priority date, applications processed on a rolling basis). *Application fee:* $20 ($30 for foreign students). *Expenses:* Tuition of $302 per semester hour. Fees of $12 per year full-time, $6 per semester part-time. *Financial aid:* Application deadline 2/15. • Michael Prais, Chairman, 312-341-3685.

Application contact: Joan R. Ritter, Coordinator of Graduate Admissions, 312-341-3612.

Rutgers, The State University of New Jersey, Newark, Department of Chemistry, Newark, NJ 07102. Department offers programs in analytical chemistry (MS, PhD), inorganic chemistry (MS, PhD), organic chemistry (MS, PhD), physical chemistry (MS, PhD). Part-time and evening/weekend programs available. Faculty: 13 full-time (1 woman), 5 part-time (1 woman). Matriculated students: 30 full-time, 54 part-time; includes 28 minority (21 Asian American, 3 black American, 4 Hispanic American), 21 foreign. Average age 35. In 1990, 60 master's, 30 doctorates awarded. Terminal master's awarded for partial completion of doctoral program. *Degree requirements:* For master's, cumulative exams required, thesis optional, foreign language not required; for doctorate, dissertation, cumulative exams required, foreign language not required. *Entrance requirements:* For master's, GRE General Test, GRE Subject Test, minimum undergraduate B average; for doctorate, GRE Subject Test, minimum undergraduate B average. Application deadline: 8/1. Application fee: $35. *Tuition:* $2216 per semester full-time for state residents; $3248 per semester full-time for nonresidents. *Financial aid:* In 1990–91, 10 fellowships (1 to a first-year student), 23 teaching assistantships (5 to first-year students) awarded; research assistantships, federal work-study also available. Aid available to part-time students. Financial aid application deadline: 3/1; applicants required to submit GAPSFAS. *Faculty research:* Medicinal chemistry, natural products, isotope effects, enzyme mechanisms, biophysics and biorganic approaches to enzyme mechanisms. Total annual research budget: $600,000. • Dr. James Schlegel, Director, 201-648-5173.

Announcement: Opportunities are open to students pursuing interdisciplinary programs in the areas of biotechnology and neuroscience. There is close interaction with neighboring chemical and pharmaceutical industry. Special stipends and research support exist for minority graduate students.

See full description on page 129.

Rutgers, The State University of New Jersey, New Brunswick, Program in Chemistry, New Brunswick, NJ 08903. Offers analytical chemistry (MS, PhD), biological chemistry (PhD), chemistry education (MST), inorganic chemistry (MS, PhD), organic chemistry (MS, PhD), physical chemistry (MS, PhD). Part-time and evening/weekend programs available. Faculty: 48 full-time (5 women), 0 part-time. Matriculated students: 118 full-time (47 women), 40 part-time (9 women); includes 2 minority (both Hispanic American), 104 foreign. Average age 25. 293 applicants, 34% accepted. In 1990, 15 master's awarded (67% found work related to degree, 33% continued full-time study); 12 doctorates awarded (71% entered university research/teaching, 29% found other work related to degree). Terminal master's awarded for partial completion of doctoral program. *Degree requirements:* For master's, exam required, thesis optional, foreign language not required; for doctorate, dissertation, cumulative exams required, foreign language not required. *Entrance requirements:* For master's, GRE General Test, GRE Subject Test, TOEFL; for doctorate, GRE General Test, GRE Subject Test, TOEFL (minimum score of 575 required). Application deadline: 7/1. Application fee: $35. *Expenses:* Tuition of $4432 per year full-time, $183 per credit part-time for state residents; $6496 per year full-time, $270 per credit part-time for nonresidents. Fees of $458 per year full-time, $117 per year part-time. *Financial aid:* In 1990–91, $1.05-million in aid awarded. 16 fellowships (1 to a first-year student), 39 research assistantships (0 to first-year students), 71 teaching assistantships (24 to first-year students) were awarded; federal work-study also available. Financial aid application deadline: 3/1. *Faculty research:* Biophysical and bioorganic chemistry; chemical physics; theoretical chemistry; photochemistry; bioinorganic, solid state, and surface chemistry. Total annual research budget: $4.1-million. • Dr. Martha Cotter, Director, 908-932-2259.

See full description on page 131.

Rutgers, The State University of New Jersey, New Brunswick, Program in Environmental Science, New Brunswick, NJ 08903. Offerings include aquatic chemistry (MS, PhD). Terminal master's awarded for partial completion of doctoral program. Program faculty: 32 full-time (4 women), 0 part-time. *Degree requirements:* For master's, thesis or alternative required, foreign language not required; for doctorate, dissertation required, foreign language not required. *Entrance requirements:* GRE General Test. Application deadline: 3/1. Application fee: $35. *Expenses:* Tuition of $4432 per year full-time, $183 per credit part-time for state residents; $6496 per year full-time, $270 per credit part-time for nonresidents. Fees of $458 per year full-time, $117 per year part-time. • Christopher Uchrin, Director, 908-932-7991. Application contact: Peter F. Strom, Graduate Admissions Committee, 908-532-9185.

Rutgers, The State University of New Jersey, New Brunswick, Program in Radiation Science, New Brunswick, NJ 08903. Offerings include radiation chemistry (MS). Program faculty: 8 full-time (0 women), 0 part-time. *Degree requirements:* Essay or thesis, exam required, foreign language not required. *Entrance requirements:* GRE General Test, previous course work in chemistry, physics, biology, and calculus. Application deadline: 8/1. Application fee: $35. *Expenses:* Tuition of $4432 per year full-time, $183 per credit part-time for state residents; $6496 per year full-time, $270 per credit part-time for nonresidents. Fees of $458 per year full-time, $117 per year part-time. • Francis J. Haughey, Director, 908-932-2551.

St. Francis Xavier University, Program in Chemistry, Antigonish, NS B2G 1C0, Canada. Program awards M Sc. Faculty: 7 full-time. Matriculated students: 1 full-time (0 women), 0 part-time; includes 1 foreign. 1 applicants, 100% accepted. In 1990, 0 degrees awarded. *Degree requirements:* Thesis required, foreign language not required. *Application deadline:* 8/1 (priority date, applications processed on a rolling basis). *Application fee:* $25. *Tuition:* $400 per course. • Dr. D. L. Bunbury, Chair, 902-867-2141.

St. John's University, Graduate School of Arts and Sciences, Department of Chemistry, Jamaica, NY 11439. Department awards MS. Faculty: 9 full-time (0 women), 2 part-time (0 women). Matriculated students: 3 full-time (2 women), 6 part-time (2 women); includes 1 minority (Asian American), 4 foreign. Average age 26. In 1990, 1 degree awarded. *Degree requirements:* 1 foreign language. *Entrance requirements:* GRE General Test. Application deadline: rolling. Application fee: $20. *Expenses:* Tuition of $297 per credit. Fees of $130 per year full-time, $65 per semester part-time. • Dr. Istvan Lengyel, Chairman, 718-990-6161.

Saint Joseph College, Field of Natural Sciences, Department of Chemistry, West Hartford, CT 06117. Department offers program in chemistry and biological chemistry (MS, CAGS, Certificate). Faculty: 5 full-time (3 women), 2 part-time (1 woman). In 1990, 3 master's awarded. *Degree requirements:* For master's, thesis or alternative required, foreign language not required. *Application deadline:* 8/31. *Application fee:* $25. *Expenses:* Tuition of $225 per credit. Fees of $25 per year. *Financial aid:* Application deadline 8/31. • Dr. Harold T. Mckone, Chairman, Field of Natural Sciences, 230-252-9571 Ext. 241.

Saint Joseph's University, Department of Chemistry, Philadelphia, PA 19131. Department awards MS. Evening/weekend programs available. Matriculated students: 26 (6 women). In 1990, 4 degrees awarded. *Degree requirements:* Thesis optional. *Application deadline:* 7/15. *Application fee:* $30. *Tuition:* $300 per credit. *Financial aid:* Fellowships, graduate assistantships available. • Dr. E. Peter Zurbach, Chairman, 215-660-1780.

Saint Louis University, College of Arts and Sciences, Department of Chemistry, St. Louis, MO 63103. Department awards MS, MS(R). Faculty: 9 full-time, 0 part-time. Matriculated students: 9 full-time (4 women), 12 part-time (6 women); includes 4 minority (1 Asian American, 3 black American), 0 foreign. 10 applicants, 80% accepted. In 1990, 6 degrees awarded. *Degree requirements:* Comprehensive exam, thesis for MS(R) required, foreign language not required. *Entrance requirements:* GRE General Test. Application deadline: rolling. Application fee: $30. *Tuition:* $342 per credit hour. *Financial aid:* In 1990–91, 1 fellowship, 2 research assistantships, 6 teaching assistantships awarded. Financial aid application deadline: 4/1. • Dr. Vincent Spaziano, Chairman, 314-658-2850. Application contact: Dr. Robert J. Nikolai, Associate Dean of the Graduate School, 314-658-2240.

Sam Houston State University, College of Arts and Sciences, Chemistry Department, Huntsville, TX 77341. Department awards M Ed, MS. Part-time programs available. Faculty: 7 full-time (2 women), 0 part-time. Matriculated students: 16 full-time (5 women), 0 part-time; includes 1 minority (black American), 14 foreign. Average age 24. 24 applicants, 25% accepted. In 1990, 5 degrees awarded. *Degree requirements:* Thesis (for some programs). *Entrance requirements:* GRE General Test (minimum combined score of 800 required), TOEFL (minimum score of 550 required). Application deadline: 8/15 (priority date, applications processed on a rolling basis). Application fee: $0. *Expenses:* Tuition of $432 per year full-time, $216 per year part-time for state residents; $2880 per year full-time, $1440 per year part-time for nonresidents. Fees of $364 per year full-time, $220 per year part-time. *Financial aid:* In 1990–91, 15 students received a total of $60,000 in aid awarded. 6 research assistantships, 12 teaching assistantships (8 to first-year students) were awarded; partial tuition waivers, federal work-study, institutionally sponsored loans also available. Aid available to part-time students. Average monthly stipend for a graduate assistantship: $500. Financial aid application deadline: 8/15. *Faculty research:* Analytical, biochemical, inorganic, organic, and physical chemistry. • Dr. Paul A. Loeffler, Chairman, 409-294-1532. Application contact: Dr. Mary Plishker, Graduate Adviser, 409-294-1526.

San Diego State University, College of Sciences, Department of Chemistry, San Diego, CA 92182. Department offers programs in cell and molecular biology (PhD), chemistry (MA, MS, PhD). PhD offered jointly with University of California, San Diego. Faculty: 20 full-time, 1 part-time. Matriculated students: 5 full-time (4 women), 35 part-time (14 women); includes 4 minority (2 Asian American, 1 black American, 1 Native American), 5 foreign. In 1990, 6 master's, 2 doctorates awarded. *Degree requirements:* For master's, foreign language and thesis not required; for doctorate, dissertation required, foreign language not required. *Entrance requirements:* For master's, GRE General Test (minimum combined score of 950 required), GRE Subject Test, bachelor's degree in related field; for doctorate, GRE Subject Test. Application deadline: 8/1 (applications processed on a rolling basis). Application fee: $55. *Expenses:* Tuition of $0 for state residents; $189 per unit for nonresidents. Fees of $1974 per year full-time, $692 per year part-time for state residents; $1074 per year full-time, $692 per year part-time for nonresidents. *Financial aid:* In 1990–91, 5 fellowships awarded. *Faculty research:* Gas phase kinetics, synthesis and mechanism of organics, separation science, recombinant DNA and enzyme chemistry. Total annual research budget: $400,000. • Dr. Morey Ring, Adviser, 619-594-5595.

San Francisco State University, School of Science, Department of Chemistry and Biochemistry, San Francisco, CA 94132. Department awards MS. *Application deadline:* 11/30 (priority date, applications processed on a rolling basis). *Application fee:* $55. *Expenses:* Tuition of $0 for state residents; $250 per unit for nonresidents. Fees of $950 per year full-time, $350 per semester part-time for state residents; $1050 per year full-time, $350 per semester part-time for nonresidents. *Financial aid:* Application deadline 3/1. • Dr. Daniel Buttlaire, Chair, 415-338-1288.

San Jose State University, School of Science, Department of Chemistry, San Jose, CA 95192. Offerings include radiochemistry (MS). Department faculty: 24 full-time (3 women), 5 part-time (2 women). *Degree requirements:* Thesis or alternative required, foreign language not required. *Entrance requirements:* GRE, minimum B average. Application deadline: 6/1 (applications processed on a rolling basis). Application fee: $55. *Expenses:* Tuition of $0 for state residents; $246 per unit for nonresidents. Fees of $592 per year. • Dr. Joseph Pesek, Chair, 408-924-5000. Application contact: Dr. Gerald Selter, Graduate Adviser, 408-924-4956.

Seton Hall University, College of Arts and Sciences, Department of Chemistry, South Orange, NJ 07079. Department offers programs in analytical chemistry (PhD), biochemistry (PhD), chemistry (MS), inorganic chemistry (PhD), organic chemistry (PhD), physical chemistry (PhD). Faculty: 14 full-time (2 women). Matriculated students: 24 full-time (5 women), 84 part-time (33 women); includes 17 foreign. 64 applicants, 45% accepted. In 1990, 10 master's, 2 doctorates awarded. Terminal master's awarded for partial completion of doctoral program. *Degree requirements:* For master's, 1 foreign language (computer language can substitute), thesis (for some programs); for doctorate, 1 foreign language (computer language can substitute), dissertation, comprehensive exams. *Entrance requirements:* GRE General Test, GRE Subject Test, TOEFL, undergraduate major in chemistry or related field. Application fee: $30. *Expenses:* Tuition of $346 per credit. Fees of $105 per semester. *Financial aid:* In 1990–91, 5 research assistantships, 18 teaching assistantships (5 to first-year students) awarded; fellowships also available. *Faculty research:* Applied spectroscopy; inorganic, organic and biochemical reaction mechanisms; biophysical chemistry; polymers; enzymes. • Dr. Robert L. Augustine, Chair, 201-761-9414.

Announcement: Full-time and part-time programs with many courses offered evenings and Saturdays. Research and nonresearch paths for the MS degree (including one with a minor in business). Research (MS and PhD) in all areas plus collaborative research projects. Extensive interaction between students, faculty, and professional industrial chemists.

Shippensburg University of Pennsylvania, College of Arts and Sciences, Department of Chemistry, Shippensburg, PA 17257. Department awards MS. Faculty: 3 full-time (0 women), 0 part-time. Matriculated students: 0 full-time, 8 part-time (1 woman); includes 1 minority (Asian American), 0 foreign. In 1990, 0 degrees awarded. *Degree requirements:* Foreign language and thesis not required. *Entrance requirements:* Minimum GPA of 2.5 or GRE General Test. Application fee: $20. *Expenses:* Tuition of $1314 per semester full-time, $146 per credit part-time for state residents; $1677 per semester full-time, $186 per credit part-time for nonresidents. Fees of $240 per year full-time, $58 per semester (minimum) part-time. *Financial aid:* In

Directory: Chemistry

1990–91, 2 graduate assistantships (both to first-year students) awarded. Financial aid application deadline: 3/1. • Dr. Dan Mack, Chairperson, 717-532-1629.

Simon Fraser University, Faculty of Science, Department of Chemistry, Burnaby, BC V5A 1S6, Canada. Department offers programs in chemical physics (M Sc, PhD), chemistry (M Sc, PhD). Faculty: 26 full-time (2 women), 0 part-time. Matriculated students: 59 full-time (23 women), 4 part-time (1 woman). In 1990, 4 master's, 5 doctorates awarded. *Degree requirements:* Thesis/dissertation. *Application fee:* $0. *Expenses:* Tuition of $612 per trimester full-time, $306 per trimester part-time. Fees of $68 per trimester full-time, $34 per trimester part-time. *Financial aid:* In 1990–91, 17 fellowships awarded; research assistantships, teaching assistantships also available. *Faculty research:* Biochemistry, nuclear chemistry, physical chemistry, inorganic chemistry. • A. C. Oehlschlager, Chairman, 604-291-4884.

South Dakota School of Mines and Technology, Department of Chemistry and Chemical Engineering, Program in Chemistry, Rapid City, SD 57701. Program awards MS. Part-time programs available. Faculty: 6 full-time (1 woman), 2 part-time (1 woman). Matriculated students: 5 full-time (3 women), 2 part-time (1 woman); includes 0 minority, 5 foreign. Average age 28. In 1990, 4 degrees awarded. *Degree requirements:* Thesis or alternative required, foreign language not required. *Entrance requirements:* TOEFL (minimum score of 520 required). Application deadline: 7/15 (priority date). Application fee: $15 ($45 for foreign students). *Expenses:* Tuition of $1474 per year full-time, $61.40 per credit hour part-time for state residents; $2917 per year full-time, $122 per credit hour part-time for nonresidents. Fees of $871 per year full-time, $132 per semester (minimum) part-time. *Financial aid:* In 1990–91, 2 students received aid. 0 fellowships, 0 research assistantships, 2 teaching assistantships awarded; federal work-study, institutionally sponsored loans also available. Aid available to part-time students. Average monthly stipend for a graduate assistantship: $300. Financial aid application deadline: 5/15. *Faculty research:* Wood chemistry-coloration, synthesis-organic inorganic, de-icing chemicals, aluminum chemistry complexes. • Dr. J. M. Munro, Head, Department of Chemistry and Chemical Engineering, 605-394-2421. Application contact: Dr. Briant L. Davis, Dean, Graduate Division, 605-394-2493.

South Dakota State University, Colleges of Arts and Science and Agriculture and Biological Sciences, Department of Chemistry, Brookings, SD 57007. Department offers programs in analytical chemistry (MS, PhD), biochemistry (MS, PhD), chemistry (MS, MST, PhD), inorganic chemistry (MS, PhD), organic chemistry (MS, PhD), physical chemistry (MS, PhD). *Degree requirements:* For master's, thesis required, foreign language not required; for doctorate, dissertation, research tool. *Entrance requirements:* For master's, TOEFL (minimum score of 500 required), bachelor's degree in chemistry or equivalent; for doctorate, TOEFL (minimum score of 500 required). *Tuition:* $553 per semester full-time, $61 per credit part-time for state residents; $1094 per semester full-time, $122 per credit part-time for nonresidents. *Faculty research:* Trace metal metabolism, flow cytometry, rhenium chemistry, diffuse reflectance spectroscopy, analytical chemistry.

Southern Connecticut State University, School of Arts and Sciences, Department of Chemistry, New Haven, CT 06515. Department awards MS, MS Ed, MLS/MS. MLS/MS offered jointly with the Department of Library Science and Instructional Technology. Faculty: 4 full-time (1 woman), 2 part-time (0 women). Matriculated students: 19 (5 women). In 1990, 4 degrees awarded. *Degree requirements:* Thesis optional, foreign language not required. *Entrance requirements:* Interview. Application deadline: 7/15. Application fee: $20. *Expenses:* Tuition of $860 per semester full-time, $107 per credit part-time for state residents; $2395 per semester full-time, $107 per credit part-time for nonresidents. Fees of $409 per semester full-time, $20 per semester part-time for state residents; $730 per semester full-time, $20 per semester part-time for nonresidents. *Financial aid:* In 1990–91, 8 teaching assistantships awarded. • Chairman, 203-397-4608.

Southern Illinois University at Carbondale, College of Science, Department of Chemistry and Biochemistry, Carbondale, IL 62901. Offerings include chemistry (MS, PhD). Terminal master's awarded for partial completion of doctoral program. Department faculty: 24 full-time (0 women), 0 part-time. *Degree requirements:* For master's, 1 foreign language, thesis; for doctorate, variable foreign language requirement, dissertation. *Entrance requirements:* For master's, ACS Graduate Placement Exam (score in 50th percentile or higher required), TOEFL (minimum of 550 required), minimum GPA of 2.7; for doctorate, TOEFL (minimum score of 550 required), minimum GPA of 3.25. Application deadline: rolling. Application fee: $0. *Expenses:* Tuition of $1638 per year full-time, $204.75 per semester hour part-time for state residents; $4914 per year full-time, $614.25 per semester hour part-time for nonresidents. Fees of $700 per year full-time, $216 per year part-time. • Dr. David C. Schmulbach, Chairperson, 618-453-5721. Application contact: Dr. Steve Scheiner, Chairman, Graduate Admissions Committee, 618-453-5721.

Southern Illinois University at Edwardsville, School of Sciences, Department of Chemistry, Edwardsville, IL 62026. Department awards MS. Part-time programs available. Faculty: 13 full-time (2 women), 1 part-time (0 women). Matriculated students: 14 full-time (7 women), 6 part-time (2 women); includes 7 minority (all black American), 7 foreign. In 1990, 11 degrees awarded. *Degree requirements:* 1 foreign language, thesis or alternative, final exam. *Application fee:* $0. *Expenses:* Tuition of $1566 per year full-time, $43.40 per credit hour part-time for state residents; $4698 per year full-time, $130.20 per credit hour part-time for nonresidents. Fees of $291.75 per year full-time, $27.35 per year (minimum) part-time. *Financial aid:* In 1990–91, 4 research assistantships, 10 teaching assistantships, 4 assistantships awarded; fellowships, federal work-study, institutionally sponsored loans also available. Aid available to part-time students. • Dr. Emil Jason, Chairman, 618-692-2042.

Southern Methodist University, Dedman College, Department of Chemistry, Dallas, TX 75275. Department awards MS. Faculty: 10 full-time (1 woman), 0 part-time. Matriculated students: 1 full-time (0 women), 0 part-time; includes 0 minority, 1 foreign. Average age 30. In 1990, 1 degree awarded. *Degree requirements:* Thesis required, foreign language not required. *Entrance requirements:* GRE General Test, TOEFL (minimum score of 550 required). Application deadline: 7/1. Application fee: $25. *Expenses:* Tuition of $435 per credit. Fees of $664 per semester for state residents; $56 per year for nonresidents. *Financial aid:* In 1990–91, $26,000 in aid awarded. 2 fellowships were awarded; full and partial tuition waivers also available. *Faculty research:* Organic and inorganic synthesis, theoretical chemistry, organometallic chemistry, inorganic polymer chemistry. Total annual research budget: $700,000. • Dr. Edward R. Biehl, Chairman, 214-692-2480. Application contact: Dr. Mark Schell, Graduate Adviser, 214-692-2478.

Southern University and Agricultural and Mechanical College, College of Sciences, Department of Chemistry, Baton Rouge, LA 70813. Department offers programs in analytical chemistry (MS), biochemistry (MS), environmental sciences (MS), inorganic chemistry (MS), organic chemistry (MS), physical chemistry (MS). MS (environmental science) offered jointly with Louisiana State University. Faculty: 5 full-time (0 women), 0 part-time. Matriculated students: 3 full-time (1 woman), 10 part-time (3 women); includes 7 minority (all black American), 3 foreign. Average age 23. 4 applicants, 25% accepted. In 1990, 2 degrees awarded. *Entrance requirements:* GRE General Test or GMAT, TOEFL. Application deadline: 7/1. Application fee: $5. *Tuition:* $1594 per year full-time, $299 per semester part-time for state residents; $3010 per year full-time, $299 per semester part-time for nonresidents. *Faculty research:* Inorganic chemistry, organic chemistry, archives. • Dr. Earl Doomes, Chairman, 504-771-3990.

Southwest Texas State University, School of Science, Department of Chemistry, San Marcos, TX 78666. Department awards MA, M Ed, MS. Faculty: 12 full-time (0 women), 0 part-time. Matriculated students: 11 full-time (4 women), 3 part-time (2 women); includes 0 minority, 1 foreign. Average age 28. 4 applicants, 75% accepted. In 1990, 7 degrees awarded. *Degree requirements:* Thesis (for some programs), comprehensive exam required, foreign language not required. *Entrance requirements:* GRE General Test (minimum combined score of 950 required), TOEFL (minimum score of 550 required), minimum GPA of 2.75 in last 60 hours. Application deadline: 7/15 (priority date, applications processed on a rolling basis). Application fee: $0 ($50 for foreign students). *Expenses:* Tuition of $180 per semester (minimum) full-time, $100 per semester (minimum) part-time for state residents; $1152 per semester (minimum) full-time, $128 per semester (minimum) part-time for nonresidents. Fees of $217 per semester (minimum) full-time, $73 per semester (minimum) part-time. *Financial aid:* In 1990–91, 10 students received a total of $83,520 in aid awarded. 10 teaching assistantships (5 to first-year students) were awarded; research assistantships, assistantships, and career-related internships or fieldwork also available. Aid available to part-time students. Average monthly stipend for a graduate assistantship: $696. Financial aid application deadline: 3/1. *Faculty research:* Metal ions in biological systems cancer chemotherapy, adsorption of pesticides on solid surfaces, polymer chemistry, biochemistry of nucleic acids. Total annual research budget: $600,000. • Dr. Carl J. Carrano Jr., Chair, 512-245-2156. Application contact: Dr. James D. Irvin, Graduate Adviser, 512-245-2156.

Stanford University, School of Humanities and Sciences, Department of Chemistry, Stanford, CA 94305. Department awards MAT, PhD. Faculty: 20 (2 women). Matriculated students: 222 full-time (59 women), 0 part-time; includes 13 minority (9 Asian American, 1 black American, 3 Hispanic American), 40 foreign. Average age 26. 310 applicants, 41% accepted. In 1990, 2 master's, 18 doctorates awarded. Terminal master's awarded for partial completion of doctoral program. *Degree requirements:* For master's, thesis optional, foreign language not required; for doctorate, 2 foreign languages, dissertation. *Entrance requirements:* For doctorate, GRE General Test, GRE Subject Test, TOEFL. Application deadline: 1/1. Application fee: $55. *Expenses:* Tuition of $15,102 per year. Fees of $28 per quarter. *Financial aid:* Fellowships, research assistantships, teaching assistantships, institutionally sponsored loans available. Financial aid application deadline: 1/1; applicants required to submit GAPSFAS. • Application contact: Graduate Admissions Committee, 415-723-1525.

See full description on page 133.

State University of New York at Albany, College of Science and Mathematics, Department of Chemistry, Albany, NY 12222. Department awards MS, PhD. Evening/weekend programs available. Faculty: 17 full-time (0 women). Matriculated students: 20 full-time (5 women), 18 part-time (7 women); includes 2 minority (both Hispanic American), 22 foreign. Average age 26. In 1990, 4 master's awarded (75% found work related to degree, 25% continued full-time study); 9 doctorates awarded. *Degree requirements:* For master's, 1 foreign language (computer language can substitute), thesis, major field exam; for doctorate, 2 foreign languages (computer language can substitute for one), dissertation, cumulative exams and oral proposition. *Entrance requirements:* For doctorate, GRE. Application fee: $35. *Expenses:* Tuition of $2450 per year full-time, $103 per credit part-time for state residents; $5765 per year full-time, $243 per credit part-time for nonresidents. Fees of $25 per year full-time, $0.85 per credit part-time. *Financial aid:* In 1990–91, $267,800 in aid awarded. 8 research assistantships (0 to first-year students), 27 teaching assistantships (4 to first-year students) were awarded; minority assistantships also available. *Faculty research:* Synthetic, organic and inorganic chemistry; polymer chemistry; ESR and NMR spectroscopy; theoretical chemistry physical biochemistry. Total annual research budget: $1.545-million. • Eric Block, Chairman, 518-442-4400. Application contact: Shelton Bank, 518-442-4447.

State University of New York at Binghamton, School of Arts and Sciences, Department of Chemistry, Binghamton, NY 13902-6000. Department offers programs in analytical chemistry (PhD), chemistry (MA, MS), inorganic chemistry (PhD), organic chemistry (PhD), physical chemistry (PhD). Matriculated students: 42 full-time (11 women), 16 part-time (4 women); includes 2 minority (1 black American, 1 Hispanic American), 26 foreign. Average age 31. 50 applicants, 42% accepted. Terminal master's awarded for partial completion of doctoral program. *Degree requirements:* For master's, thesis or alternative, oral exam, seminar presentation required, foreign language not required; for doctorate, dissertation, cumulative exams required, foreign language not required. *Entrance requirements:* GRE General Test, GRE Subject Test, TOEFL. Application deadline: 4/15 (priority date). Application fee: $35. *Expenses:* Tuition of $2450 per year full-time, $102.50 per credit part-time for state residents; $5766 per year full-time, $242.50 per credit part-time for nonresidents. Fees of $77 per year full-time, $27.85 per semester (minimum) part-time. *Financial aid:* In 1990–91, 34 students received a total of $322,762 in aid awarded. 2 fellowships, 6 research assistantships, 14 teaching assistantships (1 to a first-year student), 12 graduate assistantships (6 to first-year students) were awarded; federal work-study, institutionally sponsored loans, and career-related internships or fieldwork also available. Aid available to part-time students. Average monthly stipend for a graduate assistantship: $949. Financial aid application deadline: 2/15. • Dr. Alistair J. Lees, Chairperson, 607-777-2362.

State University of New York at Buffalo, Graduate School, Faculty of Natural Sciences and Mathematics, Department of Chemistry, Buffalo, NY 14260. Department awards MA, PhD. Part-time programs available. Faculty: 27 full-time (2 women), 0 part-time. Matriculated students: 116 full-time (35 women), 9 part-time (1 woman); includes 4 minority (3 Asian American, 1 black American), 42 foreign. Average age 24. 92 applicants, 57% accepted. In 1990, 8 master's awarded (88% found work related to degree, 12% continued full-time study); 15 doctorates awarded (40% entered university research/teaching, 60% found other work related to degree). Terminal master's awarded for partial completion of doctoral program. *Degree requirements:* For master's, thesis or alternative, project required, foreign language not required; for doctorate, dissertation required, foreign language not required. *Entrance requirements:* GRE General Test, GRE Subject Test, TOEFL (minimum score of 550 required). Application deadline: 5/15 (priority date, applications processed on a rolling basis). Application fee: $35. *Expenses:* Tuition of $1600 per semester full-time, $134 per hour part-time for state residents; $3258 per semester full-time, $274 per hour part-time for nonresidents. Fees of $137 per semester full-time, $115 per semester (minimum) part-time. *Financial aid:* In 1990–91, $1.19-million in aid awarded. 2 fellowships (0 to first-year students), 41 research assistantships (0 to first-year students), 50 teaching assistantships (18 to first-year students), 11 graduate assistantships (4 to first-year students) were awarded;

federal work-study, institutionally sponsored loans also available. Financial aid application deadline: 6/15; applicants required to submit FAF. *Total annual research budget:* $3.60-million. • Dr. Joseph Tufariello, Chairman, 716-831-3015. Application contact: Dr. O. T. Beachley Jr., Director of Graduate Studies, 716-831-3020.

State University of New York at Buffalo, Graduate School, Graduate Programs in Biomedical Sciences at Roswell Park, Department of Chemistry at Roswell Park, Buffalo, NY 14263. Department awards MA, PhD. Faculty: 8 full-time (2 women), 0 part-time. Matriculated students: 1 (woman) full-time, 0 part-time; includes 0 minority, 1 foreign. Average age 27. 10 applicants, 0% accepted. In 1990, 0 master's, 1 doctorate awarded. Terminal master's awarded for partial completion of doctoral program. *Degree requirements:* Thesis/dissertation, comprehensive exam, departmental qualifying exam, project. *Entrance requirements:* GRE General Test, TOEFL (minimum score of 600 required). Application deadline: 6/1 (priority date, applications processed on a rolling basis). Application fee: $35. *Expenses:* Tuition of $1600 per semester full-time, $134 per hour part-time for state residents; $3258 per semester full-time, $274 per hour part-time for nonresidents. Fees of $137 per semester full-time, $115 per semester (minimum) part-time. *Financial aid:* In 1990–91, 2 students received a total of $21,000 in aid awarded. 2 fellowships (both to first-year students) were awarded; research assistantships, federal work-study, institutionally sponsored loans also available. Average monthly stipend for a graduate assistantship: $900. Financial aid application deadline: 2/1; applicants required to submit FAF. *Faculty research:* Chemistry of biological systems, medicinal chemistry, synthetic organic chemistry, mechanistic organic chemistry, steroid chemistry. • Dr. Jake Bello, Chairman, 716-845-5838.

State University of New York at Stony Brook, College of Arts and Sciences, Division of Physical Sciences and Mathematics, Department of Chemistry, Stony Brook, NY 11794. Department offers programs in chemical biology (M Phil, PhD); chemical physics (M Phil, PhD); chemistry (M Phil, MS, PhD); kinetics and structural inorganic chemistry (M Phil, PhD); nuclear and isotope chemistry (M Phil, PhD); quantum and statistical mechanics (M Phil, PhD); reaction dynamics (M Phil, PhD); spectroscopy and laser light scattering (M Phil, PhD); synthetic, mechanistic, biomimetic, and catalytic organic chemistry (M Phil, PhD). Faculty: 29 full-time, 1 part-time. Matriculated students: 133 full-time (45 women), 18 part-time (6 women); includes 16 minority (8 Asian American, 4 black American, 4 Hispanic American), 88 foreign. Average age 27. 269 applicants, 28% accepted. In 1990, 9 master's, 12 doctorates awarded. Terminal master's awarded for partial completion of doctoral program. *Degree requirements:* For master's, thesis required, foreign language not required; for doctorate, 1 foreign language, dissertation. *Entrance requirements:* GRE General Test, TOEFL. Application deadline: 2/1. Application fee: $35. *Expenses:* Tuition of $2450 per year full-time, $103 per credit part-time for state residents; $5766 per year full-time, $243 per credit part-time for nonresidents. Fees of $151 per year full-time, $10.45 per year (minimum) part-time. *Financial aid:* In 1990–91, 5 fellowships, 81 research assistantships, 46 teaching assistantships awarded. *Total annual research budget:* $4.23-million. • David Hanson, Chairman, 516-632-7884.

See full description on page 135.

State University of New York College at Buffalo, Faculty of Natural and Social Sciences, Department of Chemistry, Buffalo, NY 14222. Department offers programs in chemistry (MA); secondary education (MS Ed), including chemistry. Part-time and evening/weekend programs available. Faculty: 10 full-time (1 woman), 0 part-time. Matriculated students: 0 full-time, 8 part-time (0 women); includes 0 minority, 0 foreign. Average age 25. 10 applicants, 90% accepted. In 1990, 0 degrees awarded. *Degree requirements:* Thesis required (for some programs), foreign language not required. *Entrance requirements:* Minimum GPA of 2.6 in last 60 undergraduate hours, New York State provisional teaching certification (MS Ed). Application deadline: 5/1. Application fee: $35. *Tuition:* $2450 per year full-time, $103 per credit part-time for state residents; $5765 per year full-time, $243 per credit part-time for nonresidents. *Financial aid:* In 1990–91, $30,000 in aid awarded. 0 fellowships, 2 graduate assistantships were awarded; teaching assistantships, federal work-study, and career-related internships or fieldwork also available. Aid available to part-time students. Financial aid application deadline: 3/1; applicants required to submit FAF. *Total annual research budget:* $39,000. • Dr. Edward Schulman, Chairman, 716-878-5204.

State University of New York College at Fredonia, Chemistry Department, Fredonia, NY 14063. Department awards MS, MS Ed. Part-time and evening/weekend programs available. Faculty: 1 full-time (0 women), 0 part-time. Matriculated students: 0 full-time, 2 part-time (0 women); includes 0 minority, 0 foreign. 1 applicants, 0% accepted. In 1990, 2 degrees awarded. *Degree requirements:* Thesis required, foreign language not required. *Application deadline:* 7/5. *Application fee:* $35. *Tuition:* $1600 per semester full-time, $134 per credit part-time for state residents; $3288 per semester full-time, $274 per credit part-time for nonresidents. *Financial aid:* In 1990–91, 0 research assistantships, 1 teaching assistantship awarded; full and partial tuition waivers also available. Aid available to part-time students. Financial aid application deadline: 3/15; applicants required to submit FAF. *Faculty research:* Gas chromatography, organometallic synthesis, polymer chemistry. • Dr. Roy Keller, Chairman, 716-673-3281.

State University of New York College at New Paltz, Faculty of Liberal Arts and Sciences, Program in Chemistry, New Paltz, NY 12561. Program awards MA, MS Ed. Faculty: 7. Matriculated students: 0 full-time, 6 part-time (1 woman); includes 0 minority, 0 foreign. In 1990, 1 degree awarded. *Degree requirements:* Thesis. *Entrance requirements:* GRE General Test, minimum GPA of 3.0. Application deadline: 4/1 (priority date, applications processed on a rolling basis). Application fee: $35. *Tuition:* $1600 per semester full-time, $134 per credit part-time for state residents; $3258 per semester full-time, $274 per credit part-time for nonresidents. *Financial aid:* Research assistantships, teaching assistantships, federal work-study, institutionally sponsored loans available. Financial aid applicants required to submit FAF. • Angelos Patsis, Chairman, 914-257-3790.

State University of New York College at Oswego, Division of Arts and Sciences, Department of Chemistry, Oswego, NY 13126. Department awards MS. Part-time programs available. Faculty: 10 full-time (1 woman), 0 part-time. Matriculated students: 8 full-time (3 women), 15 part-time (1 woman); includes 0 minority, 3 foreign. Average age 25. 7 applicants, 71% accepted. In 1990, 2 degrees awarded (100% found work related to degree). *Degree requirements:* 1 foreign language (computer language can substitute), thesis, written comprehensive exams. *Entrance requirements:* GRE General Test, GRE Subject Test, BS or BA in chemistry. Application deadline: 7/15. Application fee: $35. *Tuition:* $1612 per semester full-time, $134 per credit hour part-time for state residents; $3270 per semester full-time, $274 per credit hour part-time for nonresidents. *Financial aid:* In 1990–91, 7 teaching assistantships (3 to first-year students) awarded; partial tuition waivers, federal work-study, institutionally sponsored loans, and career-related internships or fieldwork also available. Financial aid applicants required to submit FAF. *Total annual research budget:* $13,000. • Dr. Augustine Silveira, Chair, 315-341-3048. Application contact: Dr. Raymond O'Donnell, Graduate Coordinator, 315-341-3048.

State University of New York College at Plattsburgh, Faculty of Arts and Science, Department of Chemistry, Plattsburgh, NY 12901. Department offers programs in biochemistry (MA), chemistry (MA). Matriculated students: 1 full-time (0 women), 3 part-time (0 women). Average age 22. 8 applicants, 0% accepted. In 1990, 2 degrees awarded. *Degree requirements:* Thesis required, foreign language not required. *Entrance requirements:* GRE General Test. Application fee: $35. *Expenses:* Tuition of $1600 per semester full-time, $134 per credit hour part-time for state residents; $3258 per semester full-time, $274 per credit hour part-time for nonresidents. Fees of $200 per year full-time. *Financial aid:* In 1990–91, 1 student received a total of $6270 in aid awarded. 1 teaching assistantship (0 to first-year students) was awarded; federal work-study also available. Average monthly stipend for a graduate assistantship: $370. Financial aid application deadline: 5/1. • Dr. George F. Sheats, Chair, 518-564-2116.

State University of New York College of Environmental Science and Forestry, Faculty of Chemistry, Syracuse, NY 13210. Faculty awards MS, PhD. Faculty: 13 full-time (0 women), 0 part-time. Matriculated students: 17 full-time (4 women), 27 part-time (8 women); includes 1 minority (Native American), 19 foreign. 47 applicants, 53% accepted. In 1990, 2 master's, 4 doctorates awarded. Terminal master's awarded for partial completion of doctoral program. *Degree requirements:* For master's, thesis or alternative required, foreign language not required; for doctorate, variable foreign language requirement, dissertation. *Entrance requirements:* GRE General Test (minimum combined score of 1800 on all three sections required), GRE Subject Test (minimum score of 600 required), minimum GPA of 3.0. Application deadline: 7/15. Application fee: $35. *Tuition:* $1600 per semester full-time, $134 per credit hour part-time for state residents; $3258 per semester full-time, $274 per credit hour part-time for nonresidents. *Financial aid:* In 1990–91, 2 fellowships, 20 research assistantships, 5 teaching assistantships awarded; federal work-study and career-related internships or fieldwork also available. Average monthly stipend for a graduate assistantship: $853. *Faculty research:* Environmental and polymer chemistry, biochemistry. • Dr. Anatole Sarko, Chairperson, 315-470-6855. Application contact: Robert H. Frey, Dean, Instruction and Graduate Studies, 315-470-6599.

See full description on page 137.

Stephen F. Austin State University, School of Sciences and Mathematics, Department of Chemistry, Nacogdoches, TX 75962. Department awards MS. Part-time programs available. Faculty: 5 full-time (0 women), 0 part-time. Matriculated students: 4 full-time (1 woman), 2 part-time (both women); includes 0 minority, 0 foreign. Average age 25. 1 applicants, 100% accepted. In 1990, 5 degrees awarded (20% found work related to degree, 80% continued full-time study). *Degree requirements:* Comprehensive exam required, foreign language and thesis not required. *Entrance requirements:* GRE General Test (minimum combined score of 1000 required), minimum GPA of 2.5 overall, 2.8 in last half of major. Application deadline: 8/1 (applications processed on a rolling basis). Application fee: $0. *Expenses:* Tuition of $18 per semester hour for state residents; $122 per semester hour for nonresidents. Fees of $14 per semester hour. *Financial aid:* In 1990–91, 4 students received a total of $21,600 in aid awarded. 0 research assistantships, 3 teaching assistantships (1 to a first-year student) were awarded; fellowships, institutionally sponsored loans also available. Aid available to part-time students. Average monthly stipend for a graduate assistantship: $580. Financial aid application deadline: 8/1. *Faculty research:* Synthesis and chemistry of ferrate ion, properties of fluoroberyllates, polymer chemistry. Total annual research budget: $30,000. • Dr. Wayne Boring, Chairman, 409-568-3606.

Stevens Institute of Technology, Department of Chemistry and Chemical Engineering, Program in Chemistry and Chemical Biology, Hoboken, NJ 07030. Offers chemical biology (MS, PhD), chemistry (MS, PhD). Part-time and evening/weekend programs available. Terminal master's awarded for partial completion of doctoral program. *Degree requirements:* For master's, thesis or alternative required, foreign language not required; for doctorate, 1 foreign language, dissertation. *Application fee:* $25. *Tuition:* $4850 per semester full-time, $485 per credit part-time. *Faculty research:* Polymer chemistry, structural chemistry, medicinal chemistry, instrumentation methods and instrument design, molecular biology.

See full description on page 139.

Syracuse University, College of Arts and Sciences, Department of Chemistry, Syracuse, NY 13244. Department awards MS, PhD. Faculty: 21 full-time. Matriculated students: 63 full-time (22 women), 1 part-time (0 women); includes 1 minority (Asian American), 36 foreign. Average age 28. 180 applicants, 21% accepted. In 1990, 5 master's, 8 doctorates awarded. *Degree requirements:* For master's, 1 foreign language; for doctorate, 1 foreign language, dissertation. *Entrance requirements:* GRE General Test, GRE Subject Test. Application deadline: rolling. Application fee: $35. *Expenses:* Tuition of $381 per credit. Fees of $289 per year full-time, $34.50 per semester part-time. *Financial aid:* In 1990–91, 4 fellowships, 35 research assistantships, 42 teaching assistantships awarded; partial tuition waivers, federal work-study also available. Financial aid application deadline: 3/1; applicants required to submit FAF. *Total annual research budget:* $1.5-million. • Laurence Nafie, Chair, 315-443-4109.

Temple University, College of Arts and Sciences, Department of Chemistry, Philadelphia, PA 19122. Department awards MA, PhD. Faculty: 21 full-time (3 women), 0 part-time. Matriculated students: 67 (18 women); includes 10 minority (1 Asian American, 7 black American, 2 Hispanic American), 10 foreign. Average age 30. 89 applicants, 25% accepted. In 1990, 4 master's, 6 doctorates awarded. *Degree requirements:* For master's, 1 foreign language, thesis or alternative; for doctorate, 1 foreign language, dissertation, teaching experience or research. *Entrance requirements:* GRE General Test (minimum combined score of 1000 required), GRE Subject Test, minimum GPA of 2.8 overall, 3.0 during previous 2 years. Application deadline: 7/1. Application fee: $30. *Tuition:* $224 per credit for state residents; $283 per credit for nonresidents. *Financial aid:* In 1990–91, 3 fellowships, 9 research assistantships, 34 teaching assistantships awarded. • Dr. John Williams, Chair, 215-787-7968. Application contact: Dr. Kevin Mayo, Graduate Admissions Chair, 215-787-6665.

Tennessee State University, College of Arts and Sciences, Department of Chemistry, Nashville, TN 37209-1561. Department awards MS. Faculty: 5 full-time (0 women), 0 part-time. Matriculated students: 3 full-time (2 women), 6 part-time (2 women); includes 3 minority (all black American), 0 foreign. Average age 28. 12 applicants, 50% accepted. In 1990, 1 degree awarded. *Degree requirements:* Thesis required, foreign language not required. *Entrance requirements:* GRE General Test, GRE Subject Test, minimum GPA of 3.0, BS in engineering or science. Application deadline: 8/1 (applications processed on a rolling basis). Application fee: $5. *Tuition:* $1814 per year full-time, $92 per semester hour (minimum) part-time for state residents; $5018 per year full-time, $232 per semester hour (minimum) part-time for nonresidents. *Financial aid:* In 1990–91, $10,000 in aid awarded. 1 research assistantship (to a first-year student), 2 graduate assistantships (1 to a first-year student) were awarded; full tuition waivers, federal work-study, institutionally sponsored loans also available. Financial aid application deadline: 2/1; applicants

Directory: Chemistry

required to submit GAPSFAS or FAF. *Faculty research:* Binding benzol pyrenemetabolites to DNA, regulation of cyclic-3'-5'-nucleotide phosphod. • Dr. David Holder, Head, 615-320-3509.

Tennessee Technological University, College of Arts and Sciences, Department of Chemistry, Cookeville, TN 38505. Department awards MS. Part-time programs available. Faculty: 16 full-time (1 woman), 0 part-time. Matriculated students: 1 (woman) full-time, 10 part-time (5 women); includes 1 minority (black American), 2 foreign. Average age 28. 12 applicants, 67% accepted. In 1990, 3 degrees awarded. *Degree requirements:* Thesis required, foreign language not required. *Entrance requirements:* GRE General Test, TOEFL (minimum score of 525 required). Application deadline: 3/15 (priority date). Application fee: $5 ($30 for foreign students). *Tuition:* $2026 per year full-time, $102 per credit hour part-time for state residents; $5486 per year full-time, $253 per credit hour part-time for nonresidents. *Financial aid:* In 1990–91, 5 research assistantships (2 to first-year students), 5 teaching assistantships (1 to a first-year student) awarded; career-related internships or fieldwork also available. Average monthly stipend for a graduate assistantship: $473. Financial aid application deadline: 4/1. • Dr. Robert T. Swindell, Chairperson, 615-372-3421. Application contact: Dr. Rebecca Quattlebaum, Interim Dean, 615-372-3233.

Texas A&I University, College of Arts and Sciences, Department of Chemistry, Kingsville, TX 78363. Department awards MS. Part-time programs available. Faculty: 6 full-time, 0 part-time. Matriculated students: 1 (woman) full-time, 2 part-time (0 women); includes 2 minority (both Hispanic American). Average age 24. 2 applicants, 100% accepted. In 1990, 3 degrees awarded (33% entered university research/teaching, 33% found other work related to degree, 33% continued full-time study). *Degree requirements:* Thesis or alternative, comprehensive exam. *Entrance requirements:* GRE General Test (minimum combined score of 800 required), TOEFL (minimum score of 500 required), minimum GPA of 3.0. Application deadline: 6/1 (applications processed on a rolling basis). Application fee: $0 ($25 for foreign students). *Expenses:* Tuition of $180 per semester full-time, $120 per semester part-time for state residents; $1152 per semester full-time, $768 per semester part-time for nonresidents. Fees of $149 per semester full-time, $101 per semester part-time. *Financial aid:* In 1990–91, $6000 in aid awarded. 0 fellowships, 1 research assistantship, 1 teaching assistantship (to a first-year student) were awarded; institutionally sponsored loans also available. Financial aid application deadline: 5/15. *Faculty research:* Organic heterocycles, amino alcohol complexes, rare earth arsine complexes. Total annual research budget: $31,000. • Dr. Nicholas Beller, Graduate Coordinator, 512-595-2914.

Texas A&M University, College of Science, Department of Chemistry, College Station, TX 77843. Department offers programs in analytical chemistry (MS, PhD), inorganic chemistry (MS, PhD), organic chemistry (MS, PhD), physical chemistry (MS, PhD). Faculty: 60. Matriculated students: 328 full-time (93 women), 0 part-time; includes 57 minority (12 Asian American, 16 black American, 28 Hispanic American, 1 Native American), 105 foreign. 293 applicants, 65% accepted. In 1990, 3 master's, 36 doctorates awarded. *Entrance requirements:* GRE General Test, TOEFL. Application deadline: 7/15. Application fee: $25 ($50 for foreign students). *Expenses:* Tuition of $100 per semester (minimum) for state residents; $128 per credit hour for nonresidents. Fees of $459 per year full-time, $252 per semester part-time. *Financial aid:* Fellowships, research assistantships, teaching assistantships available. • Michael B. Hall, Head, 409-845-2011.

See full description on page 141.

Texas Christian University, Add Ran College of Arts and Sciences, Department of Chemistry, Fort Worth, TX 76129. Department awards MA, MS, PhD. Matriculated students: 21. In 1990, 1 master's, 3 doctorates awarded. *Degree requirements:* For master's, 1 foreign language required, thesis optional; for doctorate, dissertation. *Entrance requirements:* GRE. *Expenses:* Tuition of $244 per semester hour. Fees of $423 per semester full-time, $18.50 per semester hour part-time. *Financial aid:* Fellowships, teaching assistantships, graduate assistantships available. • Dr. Henry C. Kelly, Chairperson, 817-921-7195.

Texas Southern University, College of Arts and Sciences, Department of Chemistry, Houston, TX 77004. Department awards MS. Faculty: 8 full-time (1 woman), 0 part-time. Matriculated students: 17 full-time (7 women), 8 part-time (3 women); includes 17 foreign. Average age 25. 13 applicants, 31% accepted. In 1990, 3 degrees awarded (100% continued full-time study). *Degree requirements:* 1 foreign language (computer language can substitute), thesis, comprehensive exam. *Entrance requirements:* GRE General Test, TOEFL, minimum GPA of 2.5. Application deadline: 8/1 (applications processed on a rolling basis). Application fee: $25. *Financial aid:* In 1990–91, 15 students received a total of $67,500 in aid awarded. 8 research assistantships (4 to first-year students), 7 teaching assistantships (1 to a first-year student) were awarded; fellowships, institutionally sponsored loans also available. Average monthly stipend for a graduate assistantship: $500. Financial aid application deadline: 5/1. *Faculty research:* Analytical and physical chemistry, geochemistry, inorganic, biochemistry, organic. Total annual research budget: $469,000. • Dr. Curtis McDonald, Head, 713-527-7003.

Texas Tech University, Graduate School, College of Arts and Sciences, Department of Chemistry and Biochemistry, Lubbock, TX 79409. Offerings include chemistry (MS, PhD). Department faculty: 23. *Degree requirements:* Variable foreign language requirement, thesis/dissertation. *Entrance requirements:* GRE General Test. Application deadline: 4/15 (priority date, applications processed on a rolling basis). Application fee: $0 ($50 for foreign students). *Tuition:* $494 per semester full-time, $20 per credit hour part-time for state residents; $1790 per semester full-time, $455 per credit hour part-time for nonresidents. • Dr. David Knaff, Chairman, 806-742-3067.

Texas Woman's University, College of Arts and Sciences, Department of Chemistry and Physics, Denton, TX 76204-1925. Department offers programs in chemistry (MS), science education (MSSE). Part-time programs available. Faculty: 7 full-time (0 women), 1 part-time (0 women). Matriculated students: 2 full-time (both women), 18 part-time (15 women); includes 2 minority (1 Asian American, 1 black American), 2 foreign. Average age 28. 3 applicants, 33% accepted. In 1990, 2 degrees awarded (100% found work related to degree). *Degree requirements:* Thesis required (for some programs), foreign language not required. *Application fee:* $25. *Tuition:* $508 per semester full-time, $317 per semester part-time for state residents; $1804 per semester full-time, $955 per semester part-time for nonresidents. *Financial aid:* In 1990–91, $30,200 in aid awarded. 2 research assistantships (0 to first-year students), 8 teaching assistantships (4 to first-year students) were awarded; federal work-study, institutionally sponsored loans, and career-related internships or fieldwork also available. Aid available to part-time students. Financial aid application deadline: 4/1. *Faculty research:* Chromatographic theory, mechanisms of organic reactions, interhalogen fluorides, leucotrienes. Total annual research budget: $40,000. • Dr. Carlton Wendel, Chair, 817-898-2550.

Trent University, Program in Freshwater Science, Department of Chemistry, Peterborough, ON K9J 7B8, Canada. Department awards M Sc. Part-time programs available. *Degree requirements:* Thesis required, foreign language not required. *Application deadline:* 2/15 (priority date, applications processed on a rolling basis). *Application fee:* $0. *Expenses:* Tuition of $2326 per year full-time, $1163 per year part-time for Canadian residents; $9712 per year full-time, $1163 per year part-time for nonresidents. Fees of $225 per year full-time, $94 per year part-time. *Financial aid:* Fellowships, research assistantships, teaching assistantships available. *Faculty research:* Synthetic-organic chemistry, mass spectrometry and ion storage. • Dr. K. B. Oldham, Chairman, 705-748-1298. Application contact: Dr. W. F. J. Evans, Director, Freshwater Science Program, 705-748-1622.

Tufts University, Graduate School of Arts and Sciences, Department of Chemistry, Medford, MA 02155. Department offers programs in analytical chemistry (MS, PhD), bioorganic chemistry (MS, PhD), environmental chemistry (MS, PhD), inorganic chemistry (MS, PhD), organic chemistry (MS, PhD), physical chemistry (MS, PhD). Faculty: 15 full-time, 2 part-time. Matriculated students: 45 (24 women); includes 23 foreign. 83 applicants, 29% accepted. In 1990, 10 master's, 5 doctorates awarded. Terminal master's awarded for partial completion of doctoral program. *Degree requirements:* Thesis/dissertation required, foreign language not required. *Entrance requirements:* GRE General Test, GRE Subject Test, TOEFL (minimum score of 550 required). Application deadline: 2/15. Application fee: $50. *Expenses:* Tuition of $16,755 per year full-time, $2094 per course part-time. Fees of $885 per year. *Financial aid:* Research assistantships, teaching assistantships, federal work-study, and career-related internships or fieldwork available. Financial aid application deadline: 2/15; applicants required to submit GAPSFAS. • David Walt, Head, 617-381-3470. Application contact: Edward Brush, 617-381-3475.

Tulane University, Department of Chemistry, New Orleans, LA 70118. Department awards MAT, MS, PhD. Terminal master's awarded for partial completion of doctoral program. *Degree requirements:* Thesis/dissertation. *Entrance requirements:* For master's, GRE General Test, TOEFL (minimum score of 600 required) or TSE (minimum score of 220 required), minimum B average in undergraduate course work; for doctorate, GRE General Test, TOEFL (minimum score of 600 required) or TSE (minimum score of 220 required). Application deadline: 7/1. Application fee: $35. *Expenses:* Tuition of $16,750 per year full-time, $931 per hour part-time. Fees of $230 per year full-time, $40 per hour part-time. *Financial aid:* Fellowships, research assistantships, teaching assistantships, federal work-study, institutionally sponsored loans, and career-related internships or fieldwork available. Financial aid application deadline: 5/1; applicants required to submit GAPSFAS. • Dr. William Alworth, Chairman, 504-865-5573.

Tuskegee University, College of Arts and Sciences, Department of Chemistry, Tuskegee, AL 36088. Department awards MS. Faculty: 9 full-time (3 women), 0 part-time. Matriculated students: 0 full-time, 7 part-time (4 women). 9 applicants, 44% accepted. In 1990, 1 degree awarded (100% found work related to degree). *Degree requirements:* Thesis required, foreign language not required. *Entrance requirements:* GRE General Test. Application deadline: 7/15 (applications processed on a rolling basis). Application fee: $15. *Tuition:* $6250 per year full-time, $1380 per year (minimum) part-time. *Financial aid:* Fellowships, teaching assistantships, federal work-study, institutionally sponsored loans available. Aid available to part-time students. Financial aid application deadline: 4/15; applicants required to submit FAF. • Dr. Courtney Smith, Head, 205-727-8833.

Université de Moncton, Faculty of Graduate Studies and Research, Faculty of Science, Department of Chemistry and Biochemistry, Moncton, NB E1A 3E9, Canada. Department offers program in chemistry (M Sc). Faculty: 11 full-time (2 women), 1 (woman) part-time. Matriculated students: 6 full-time (0 women), 0 part-time; includes 1 foreign. Average age 25. 10 applicants, 30% accepted. In 1990, 0 degrees awarded. *Degree requirements:* 1 foreign language, thesis. *Entrance requirements:* Minimum GPA of 2.5. Application deadline: 1/6. *Expenses:* Tuition of $1915 per trimester full-time, $222 per course part-time. Fees of $108 per trimester full-time. *Financial aid:* In 1990–91, $70,000 in aid awarded. 5 teaching assistantships were awarded. *Faculty research:* Pesticides and herbicides studies, cytosquelet structure, mitochondrial DNA, protein and heromones cristacean, oxydation of alcaloides. Total annual research budget: $300,000. • Dr. Alan Fraser, Director, 506-858-4331.

Université de Montréal, Faculty of Arts and Sciences, Department of Chemistry, Montreal, PQ H3C 3J7, Canada. Department awards M Sc, PhD. Faculty: 29 full-time (0 women), 2 part-time (0 women). *Entrance requirements:* For doctorate, M Sc in chemistry or related field. *Faculty research:* Analytical, inorganic, physical, and organic chemistry. Total annual research budget: $3-million. • Joseph Hubert, Chairman, 514-343-6730. Application contact: Marcel Bourgon, 514-343-7058.

Université de Sherbrooke, Faculty of Sciences, Department of Chemistry, Sherbrooke, PQ J1K 2R1, Canada. Department awards M Sc, PhD. Faculty: 20 full-time (0 women), 0 part-time. Matriculated students: 49 full-time (13 women), 0 part-time. 15 applicants, 60% accepted. In 1990, 9 master's, 5 doctorates awarded. *Degree requirements:* Thesis/dissertation required, foreign language not required. *Entrance requirements:* For doctorate, master's degree. Application deadline: 6/30. Application fee: $15. *Expenses:* Tuition of $585 per trimester full-time, $43.34 per credit part-time for Canadian residents; $2900 per trimester full-time, $195 per credit part-time for nonresidents. Fees of $125 per trimester full-time, $7.75 per credit part-time for Canadian residents; $610 per year full-time, $7.50 per credit part-time for nonresidents. *Financial aid:* Fellowships, research assistantships, teaching assistantships available. *Faculty research:* Organic chemistry, electrochemistry, theoretical chemistry, physical chemistry. Total annual research budget: $2-million. • Hugues Ménard, Chairman, 819-821-7088.

Université du Québec à Montréal, Program in Applied Electrochemistry, Montreal, PQ H3C 3P8, Canada. Program awards Diploma. Offered jointly with Université de Montréal and École Polytechnique de Montréal. Admissions temporarily suspended. Part-time programs available. Matriculated students: 0 full-time, 1 (woman) part-time; includes 0 foreign. 0 applicants. In 1990, 0 degrees awarded. *Degree requirements:* Thesis not required. *Expenses:* Tuition of $555 per trimester full-time, $37 per credit part-time for Canadian residents; $3480 per trimester full-time, $234 per credit part-time for nonresidents. Fees of $57.50 per trimester full-time, $19.50 per trimester part-time. *Financial aid:* Fellowships, research assistantships, teaching assistantships available. • Francine Denizeau, Director, 514-987-8229.

Université du Québec à Montréal, Program in Chemistry, Montreal, PQ H3C 3P8, Canada. Program awards M Sc. Offered jointly with Université du Quebec à Trois-Rivières. Part-time programs available. Matriculated students: 13 full-time (4 women), 12 part-time (4 women); includes 1 foreign. 20 applicants, 40% accepted. In 1990, 8 degrees awarded. *Degree requirements:* Thesis. *Entrance requirements:* Appropriate bachelor's degree, proficiency in French. Application deadline: 5/1. Application fee: $15. *Expenses:* Tuition of $555 per trimester full-time, $37 per credit part-time for Canadian residents; $3480 per trimester full-time, $234 per credit part-time for nonresidents. Fees of $57.50 per trimester full-time, $19.50 per

Directory: Chemistry

trimester part-time. *Financial aid:* Fellowships, research assistantships, teaching assistantships available. • Francine Denizeau, Director, 514-987-8245. Application contact: Lucille Boisselle-Roy, Admissions Officer, 514-987-3128.

Université du Québec à Trois-Rivières, Program in Chemistry, Trois-Rivières, PQ G9A 5H7, Canada. Program awards M Sc. Offered jointly with Université du Québec à Montréal. Part-time programs available. Matriculated students: 1 (woman) full-time, 1 part-time (0 women); includes 1 foreign. 11 applicants, 27% accepted. In 1990, 0 degrees awarded. *Degree requirements:* Thesis. *Entrance requirements:* Appropriate bachelor's degree, proficiency in French. Application deadline: 4/1. Application fee: $15. *Expenses:* Tuition of $555 per trimester full-time, $37 per credit part-time for Canadian residents; $3480 per trimester full-time, $234 per credit part-time for nonresidents. Fees of $57.50 per trimester full-time, $19.50 per trimester part-time. *Financial aid:* Fellowships, research assistantships, teaching assistantships available. • Michel Ringuet, Director, 819-376-5053. Application contact: Michel Potvin, Office of Admissions, 819-376-5045.

Université Laval, Faculty of Sciences and Engineering, Department of Chemistry, Sainte-Foy, PQ G1K 7P4, Canada. Department awards M Sc, PhD. Matriculated students: 63 full-time (19 women), 8 part-time (1 woman). 51 applicants, 45% accepted. In 1990, 9 master's, 6 doctorates awarded. *Application deadline:* 3/1. *Application fee:* $15. *Expenses:* Tuition of $792 per year full-time for Canadian residents; $5914 per year full-time for nonresidents. Fees of $120 per year full-time. • Rodrigue Savoie, Director, 418-656-5135.

University of Akron, Buchtel College of Arts and Sciences, Department of Chemistry, Akron, OH 44325. Department offers programs in analytical chemistry (MS, PhD), biochemistry (MS, PhD), chemistry (MS, PhD), inorganic chemistry (MS, PhD), organic chemistry (MS, PhD), physical chemistry (MS, PhD). Part-time and evening/weekend programs available. Faculty: 20 full-time (1 woman), 3 part-time (0 women). Matriculated students: 74 full-time (20 women), 13 part-time (5 women); includes 3 minority (1 Asian American, 2 black American), 45 foreign. Average age 26. 360 applicants, 4% accepted. In 1990, 1 master's, 12 doctorates awarded. Terminal master's awarded for partial completion of doctoral program. *Degree requirements:* For master's, 1 foreign language (computer language can substitute), thesis, seminar presentation; for doctorate, 2 foreign languages (computer language can substitute for one), dissertation, cumulative exams. *Entrance requirements:* TOEFL. Application deadline: 3/1 (applications processed on a rolling basis). Application fee: $25. *Tuition:* $119.93 per credit hour for state residents; $210.93 per credit hour for nonresidents. *Financial aid:* In 1990–91, 1 fellowship, 27 research assistantships (8 to first-year students), 26 teaching assistantships (7 to first-year students) awarded; partial tuition waivers also available. Aid available to part-time students. *Faculty research:* NMR studies, catalyzing organic reactions, free radical chemistry, natural product synthesis, laser spectroscopy. • Dr. G. Edwin Wilson, Head, 216-972-7372.

University of Alabama, College of Arts and Sciences, Department of Chemistry, Tuscaloosa, AL 35487-0132. Department offers programs in biochemistry (MS), chemistry (MS, PhD, Ed S). Faculty: 16 full-time (0 women), 1 part-time (0 women). Matriculated students: 50 full-time (12 women), 0 part-time; includes 0 minority, 20 foreign. Average age 25. 40 applicants, 50% accepted. In 1990, 1 master's, 8 doctorates awarded. *Degree requirements:* For master's, thesis; for doctorate, 2 foreign languages (computer language can substitute for one), dissertation. *Entrance requirements:* For master's, GRE General Test (minimum combined score of 1000 required), MAT, ACS, minimum GPA of 3.0; for doctorate, GRE General Test or MAT, American Chemical Society exam. Application deadline: 7/6 (applications processed on a rolling basis). Application fee: $20. *Tuition:* $968 per year full-time, $82 per credit part-time for state residents; $2400 per year full-time, $218 per credit part-time for nonresidents. *Financial aid:* In 1990–91, $20,000 in aid awarded. 4 fellowships (2 to first-year students), 21 research assistantships (3 to first-year students), 24 teaching assistantships (14 to first-year students) were awarded; federal work-study also available. Financial aid application deadline: 7/14. *Faculty research:* Synthesis of anticancer agents, materials science, enzyme and protein structure, main group chemistry, x-ray crystallography. • Dr. Drury Caine, Chairman, 205-348-5954.

University of Alabama at Birmingham, Graduate School, School of Natural Sciences and Mathematics, Department of Chemistry, Birmingham, AL 35294. Department awards MS, PhD. Faculty: 15 full-time, 6 part-time. Matriculated students: 23 full-time (4 women), 3 part-time (1 woman); includes 3 minority (1 Asian American, 2 black American), 17 foreign. 95 applicants, 32% accepted. In 1990, 2 master's, 5 doctorates awarded. *Degree requirements:* For master's, foreign language and thesis not required; for doctorate, 1 foreign language, computer language, dissertation. *Entrance requirements:* For master's, GRE General Test (minimum combined score of 1000 required), minimum GPA of 3.0; for doctorate, GRE General Test (minimum combined score of 1100 required), minimum GPA of 3.0. Application deadline: rolling. Application fee: $25 ($50 for foreign students). *Tuition:* $66 per quarter for state residents; $132 per quarter for nonresidents. *Financial aid:* In 1990–91, 10 fellowships awarded. *Faculty research:* General and biochemical synthesis; spectroscopic studies of chemical systems; analysis using chromatography, GC/MS, and designed electrode system. • Dr. Larry K. Krannich, Chairman, 205-934-4747.

University of Alabama in Huntsville, College of Science, Department of Chemistry, Huntsville, AL 35899. Department awards MS. Part-time programs available. Faculty: 10 full-time (0 women), 1 part-time (0 women). Matriculated students: 9 full-time, 2 part-time; includes 1 minority (Asian American), 5 foreign. 12 applicants, 75% accepted. In 1990, 6 degrees awarded. *Degree requirements:* Thesis or alternative, oral and written exams. *Entrance requirements:* GRE General Test (minimum combined score of 1500 required), minimum GPA of 3.0. Application deadline: 5/16 (priority date, applications processed on a rolling basis). Application fee: $20. *Tuition:* $2500 per year full-time, $1250 per year part-time for state residents; $5000 per year full-time, $2500 per year part-time for nonresidents. *Financial aid:* In 1990–91, 29 students received a total of $77,148 in aid awarded. 2 fellowships, 19 research assistantships, 8 teaching assistantships were awarded; full and partial tuition waivers, institutionally sponsored loans, and career-related internships or fieldwork also available. Aid available to part-time students. Average monthly stipend for a graduate assistantship: $750. Financial aid application deadline: 3/1; applicants required to submit FAF. *Total annual research budget:* $931,102. • Dr. Clyde Riley, Chairman, 205-895-6153.

University of Alaska Fairbanks, College of Natural Sciences, Department of Chemistry, Fairbanks, AK 99775. Offerings include chemistry (MA, MAT, MS). Department faculty: 12 full-time (2 women), 0 part-time. *Degree requirements:* Thesis required, foreign language not required. *Entrance requirements:* GRE General Test. Application deadline: 3/1. Application fee: $20. *Expenses:* Tuition of $1620 per year full-time, $90 per credit part-time for state residents; $3240 per year full-time, $180 per credit part-time for nonresidents. Fees of $464 per year full-time. • Dr. L. Claron Hoskins, Head, 907-474-7525.

University of Alberta, Faculty of Graduate Studies and Research, Department of Chemistry, Edmonton, AB T6G 2J9, Canada. Department awards M Sc, PhD. Matriculated students: 127 full-time, 5 part-time. *Application fee:* $0. *Expenses:* Tuition of $1495 per year full-time, $748 per year part-time for Canadian residents; $2243 per year full-time, $1121 per year part-time for nonresidents. Fees of $301 per year full-time, $118 per year part-time. • Dr. B. G. Kratochvil, Chair, 403-492-3254.

University of Arizona, College of Arts and Sciences, Faculty of Science, Department of Chemistry, Tucson, AZ 85721. Department awards MA, MS, PhD. Faculty: 32 full-time (1 woman), 0 part-time. Matriculated students: 128 full-time (38 women), 14 part-time (2 women); includes 1 minority (Hispanic American), 41 foreign. Average age 27. 79 applicants, 67% accepted. In 1990, 8 master's, 26 doctorates awarded. Terminal master's awarded for partial completion of doctoral program. *Degree requirements:* For master's, thesis required, foreign language not required; for doctorate, 1 foreign language, dissertation. *Entrance requirements:* TOEFL (minimum score of 550 required), SPEAK or TSE (minimum score of 230 required). Application deadline: 2/1 (priority date, applications processed on a rolling basis). Application fee: $25. *Expenses:* Tuition of $0 for state residents; $5406 per year full-time, $209 per credit hour part-time for nonresidents. Fees of $1528 per year full-time, $80 per credit hour part-time. *Financial aid:* In 1990–91, 7 fellowships (all to first-year students), 50 research assistantships, 74 teaching assistantships, 10 scholarships awarded; partial tuition waivers, institutionally sponsored loans also available. Financial aid applicants required to submit FAF. *Faculty research:* Analytical, inorganic, organic, and physical chemistry. • Dr. Neal R. Armstrong, Head, 602-621-6354. Application contact: Edith K. Kleiss, Coordinator, Academic Affairs, 800-545-5814.

See full description on page 143.

University of Arkansas, J. William Fulbright College of Arts and Sciences, Department of Chemistry and Biochemistry, Fayetteville, AR 72701. Department awards MS, PhD. Faculty: 20 full-time (0 women), 0 part-time. Matriculated students: 47 full-time (14 women), 6 part-time (3 women); includes 4 minority (2 Asian American, 2 black American), 15 foreign. In 1990, 4 master's, 9 doctorates awarded. *Degree requirements:* 1 foreign language, thesis/dissertation. *Application fee:* $15. *Expenses:* Tuition of $2050 per year full-time, $103 per credit hour part-time for state residents; $4400 per year full-time, $220 per credit hour part-time for nonresidents. Fees of $50 per year full-time, $1.50 per credit hour part-time. *Financial aid:* In 1990–91, 32 teaching assistantships awarded; research assistantships also available. *Total annual research budget:* $750,000. • Dr. Collis P. Geren, Chairperson, 501-575-4648.

University of Arkansas at Little Rock, College of Sciences and Engineering Technology, Department of Chemistry, Little Rock, AR 72204. Department awards MA, MS. Part-time and evening/weekend programs available. Faculty: 2 full-time (0 women), 1 part-time (0 women). Matriculated students: 3 full-time (2 women), 11 part-time (2 women); includes 8 minority (5 Asian American, 2 black American, 1 Hispanic American), 4 foreign. Average age 30. 9 applicants, 56% accepted. In 1990, 2 degrees awarded. *Degree requirements:* Thesis optional, foreign language not required. *Entrance requirements:* Minimum GPA of 2.7. Application deadline: 8/1 (priority date, applications processed on a rolling basis). Application fee: $25 ($30 for foreign students). *Expenses:* Tuition of $2224 per year full-time, $104 per credit hour part-time for state residents; $4774 per year full-time, $223 per credit hour part-time for nonresidents. Fees of $40 per semester. *Financial aid:* In 1990–91, 8 students received a total of $36,600 in aid awarded. 6 teaching assistantships (5 to first-year students) were awarded; research assistantships, federal work-study, institutionally sponsored loans also available. Average monthly stipend for a graduate assistantship: $678. *Faculty research:* Organic, inorganic, and physical chemistry; analytical chemistry, polymer chemistry. Total annual research budget: $212,987. • Dr. Ali Shaikh, Chairperson, 501-569-3152. Application contact: Dr. Tito Viswanathan, Graduate Coordinator, 501-569-3152.

University of Bridgeport, College of Science and Engineering, Department of Chemistry, 380 University Avenue, Bridgeport, CT 06601. Department awards MS. Faculty: 5 full-time (0 women), 0 part-time. Matriculated students: 0 full-time, 5 part-time (3 women); includes 1 minority (black American), 2 foreign. 10 applicants, 60% accepted. In 1990, 3 degrees awarded. *Degree requirements:* Thesis optional, foreign language not required. *Entrance requirements:* GRE General Test. Application deadline: rolling. Application fee: $30 ($50 for foreign students). *Expenses:* Tuition of $6010 per semester full-time, $310 per credit part-time for state residents; $6010 per semester full-time, $610 per credit part-time for nonresidents. Fees of $50 per year full-time, $25 per year part-time. *Financial aid:* Federal work-study, institutionally sponsored loans, and career-related internships or fieldwork available. Aid available to part-time students. Financial aid application deadline: 6/1; applicants required to submit FAF. *Faculty research:* Spatially resolved measurements of concentrations and temperatures using laser-induced fluorescence spectroscopy. Total annual research budget: $50,000. • Dr. James V. Tucci, Chairman, 203-576-4271.

University of British Columbia, Faculty of Science, Department of Chemistry, Vancouver, BC V6T 1Z1, Canada. Department awards M Sc, PhD. *Degree requirements:* For master's, thesis; for doctorate, dissertation, comprehensive exam.

University of Calgary, Faculty of Science, Department of Chemistry, Calgary, AB T2N 1N4, Canada. Department offers programs in analytical chemistry (M Sc, PhD), applied chemistry (M Sc, PhD), inorganic chemistry (M Sc, PhD), organic chemistry (M Sc, PhD), physical chemistry (M Sc, PhD), polymer chemistry (M Sc, PhD), theoretical chemistry (M Sc, PhD). *Degree requirements:* For master's, thesis required, foreign language not required; for doctorate, dissertation, candidacy exam required, foreign language not required. *Application deadline:* 5/31. *Application fee:* $25. *Tuition:* $1705 per year full-time, $427 per course part-time for Canadian residents; $3410 per year full-time, $854 per course part-time for nonresidents. *Faculty research:* Alberta oil sands, biological chemistry, chemical analysis, chemical dynamics, energy.

University of California at Berkeley, College of Chemistry, Department of Chemistry, Berkeley, CA 94720. Department awards MS, PhD. Faculty: 56. Matriculated students: 363 full-time, 0 part-time; includes 37 minority (24 Asian American, 3 black American, 9 Hispanic American, 1 Native American), 54 foreign. 442 applicants, 49% accepted. Terminal master's awarded for partial completion of doctoral program. *Degree requirements:* For doctorate, dissertation. *Entrance requirements:* GRE General Test, minimum GPA of 3.0. Application deadline: 2/1. Application fee: $40. *Expenses:* Tuition of $0. Fees of $1909 per year for state residents; $7825 per year for nonresidents. • William H. Miller, Acting Chair.

University of California, Davis, Program in Agriculture and Environmental Chemistry, Davis, CA 95616. Program awards MS, PhD. Matriculated students: 43. 20 applicants, 50% accepted. In 1990, 1 master's, 8 doctorates awarded. *Degree requirements:* Thesis/dissertation. *Entrance requirements:* GRE General Test, minimum GPA of 3.0. Application fee: $40. *Expenses:* Tuition of $0 for state

SECTION 2: CHEMISTRY

Directory: Chemistry

residents; $7699 per year full-time, $3849 per year part-time for nonresidents. Fees of $2718 per year full-time, $1928 per year part-time. *Financial aid:* Fellowships, research assistantships available. • Graduate Adviser, 916-752-1415.

University of California, Davis, Program in Chemistry, Davis, CA 95616. Program awards MS, PhD. Faculty: 34 full-time, 2 part-time. Matriculated students: 127. 161 applicants. In 1990, 4 master's, 13 doctorates awarded. *Degree requirements:* Thesis/dissertation. *Entrance requirements:* GRE General Test, GRE Subject Test, minimum GPA of 3.0. Application fee: $40. *Expenses:* Tuition of $0 for state residents; $7699 per year full-time, $3849 per year part-time for nonresidents. Fees of $2718 per year full-time, $1928 per year part-time. *Financial aid:* Fellowships, research assistantships, teaching assistantships available. • Graduate Adviser, 916-752-0953.

University of California, Irvine, School of Physical Sciences, Department of Chemistry, Irvine, CA 92717. Department awards MS, PhD. Matriculated students: 131 (34 women); includes 16 minority (11 Asian American, 3 black American, 2 Hispanic American), 22 foreign. 181 applicants, 58% accepted. In 1990, 10 master's, 12 doctorates awarded. *Degree requirements:* For master's, thesis or alternative required, foreign language not required; for doctorate, computer language, dissertation required, foreign language not required. *Entrance requirements:* GRE General Test, GRE Subject Test. Application deadline: 2/1 (applications processed on a rolling basis). Application fee: $40. *Expenses:* Tuition of $0 for state residents; $7699 per year full-time, $3850 per year part-time for nonresidents. Fees of $2930 per year full-time, $2139 per year part-time. *Financial aid:* Fellowships, research assistantships, teaching assistantships available. • Richard Chamberlin, Graduate Admissions Committee Chair, 714-856-4261. Application contact: Robert Doedens, Associate Dean, 714-856-6507.

University of California, Los Angeles, College of Letters and Science, Department of Chemistry and Biochemistry, Los Angeles, CA 90024. Department offers programs in biochemistry (MS, PhD, C Phil), chemistry (MS, PhD, C Phil). Faculty: 44. Matriculated students: 251 full-time (97 women), 0 part-time; includes 53 minority, 49 foreign. 348 applicants, 41% accepted. In 1990, 22 master's, 37 doctorates, 37 C Phils awarded. *Degree requirements:* For master's, comprehensive exam or thesis; for doctorate, dissertation, written and oral qualifying exams, two foreign languages (chemistry). *Entrance requirements:* For master's and doctorate, GRE General Test, GRE Subject Test. Application fee: $40. *Expenses:* Tuition of $0 for state residents; $7699 per year for nonresidents. Fees of $2907 per year. *Financial aid:* In 1990–91, 244 students received a total of $3.9-million in aid awarded. 245 fellowships, 186 research assistantships, 103 teaching assistantships were awarded; full and partial tuition waivers, federal work-study, institutionally sponsored loans also available. Financial aid application deadline: 3/1. • Dr. Kenneth Trueblood, Chair, 310-825-4219.

University of California, Riverside, Graduate Division, College of Natural and Agricultural Sciences, Department of Chemistry, Riverside, CA 92521. Department awards MS, PhD. Part-time programs available. Terminal master's awarded for partial completion of doctoral program. *Degree requirements:* For master's, comprehensive exams or thesis required, foreign language not required; for doctorate, dissertation, qualifying exams, 3 quarters teaching experience, research proposition required, foreign language not required. *Entrance requirements:* GRE General Test, GRE Subject Test, TOEFL (minimum score of 550 required). Application deadline: 6/1. Application fee: $40. *Tuition:* $950 per quarter full-time, $264 per quarter part-time for state residents; $3517 per quarter full-time, $1758 per quarter part-time for nonresidents. *Faculty research:* Analytical, inorganic, organic, and physical chemistry.

University of California, San Diego, Department of Chemistry, 9500 Gilman Drive, La Jolla, CA 92093. Department offers programs in biochemistry (PhD), chemistry (PhD). Offered jointly with San Diego State University; PhD in biochemistry offered in cooperation with the Department of Biology. Faculty: 52. Matriculated students: 177 (60 women); includes 51 foreign. 322 applicants, 41% accepted. In 1990, 20 doctorates awarded. Terminal master's awarded for partial completion of doctoral program. *Degree requirements:* For doctorate, variable foreign language requirement, dissertation. *Entrance requirements:* For doctorate, GRE General Test, GRE Subject Test. Application fee: $40. *Expenses:* Tuition of $0 for state residents; $7699 per year full-time, $1283 per quarter part-time for nonresidents. Fees of $2798 per year full-time, $669 per quarter part-time. • Ernest Wenkert, Chair, 619-534-5489. Application contact: Applications Coordinator, 619-534-6870.

University of California, Santa Barbara, College of Letters and Science, Department of Chemistry, Santa Barbara, CA 93106. Department offers programs in biochemistry and molecular biology (PhD), chemistry (MA, MS, PhD). Matriculated students: 93 full-time (29 women), 0 part-time; includes 15 foreign. 89 applicants, 56% accepted. In 1990, 10 master's, 21 doctorates awarded. *Degree requirements:* For master's, thesis or alternative required, foreign language not required; for doctorate, dissertation required, foreign language not required. *Entrance requirements:* GRE, TOEFL (minimum score of 550 required). Application deadline: 5/1. Application fee: $40. *Expenses:* Tuition of $0 for state residents; $7699 per year for nonresidents. Fees of $2307 per year. *Financial aid:* Fellowships, research assistantships, teaching assistantships, full and partial tuition waivers, federal work-study, institutionally sponsored loans, and career-related internships or fieldwork available. Financial aid application deadline: 1/31. • Bernard Kirtman, Chair, 805-893-2056. Application contact: Chris Simms, Graduate Secretary, 805-893-2638.

University of California, Santa Cruz, Division of Natural Sciences, Department of Chemistry, Santa Cruz, CA 95064. Department awards MS, PhD. Faculty: 17 full-time (0 women), 2 part-time (0 women). Matriculated students: 64 full-time (23 women), 0 part-time; includes 12 minority (3 Asian American, 9 Hispanic American), 9 foreign. 54 applicants, 76% accepted. In 1990, 6 master's, 10 doctorates awarded. *Degree requirements:* For doctorate, 1 foreign language (computer language can substitute), dissertation, qualifying exam. *Entrance requirements:* GRE General Test, GRE Subject Test. Application deadline: 5/2. Application fee: $40. *Expenses:* Tuition of $0 for state residents; $7699 per year for nonresidents. Fees of $3021 per year. *Financial aid:* In 1990–91, 6 fellowships, 21 research assistantships, 100 teaching assistantships awarded; federal work-study, institutionally sponsored loans also available. Average monthly stipend for a graduate assistantship: $1335. Financial aid applicants required to submit GAPSFAS. *Faculty research:* Marine chemistry; biochemistry; inorganic, organic, and physical chemistry. • Dr. Anthony Fink, Chairperson, 408-459-4002.

University of Central Florida, College of Arts and Sciences, Program in Industrial Chemistry, Orlando, FL 32816. Program awards MS. Part-time and evening/weekend programs available. *Degree requirements:* Thesis or alternative required, foreign language not required. *Entrance requirements:* GRE General Test, minimum GPA of 3.0 (last 60 hours). Application fee: $15. *Expenses:* Tuition of $81 per credit hour for state residents; $364 per credit hour for nonresidents. Fees of $50 per semester.

University of Chicago, Division of the Physical Sciences, Department of Chemistry, Chicago, IL 60637. Department awards SM, PhD. Terminal master's awarded for partial completion of doctoral program. *Degree requirements:* For master's, 1 foreign language required, thesis not required; for doctorate, 1 foreign language, dissertation. *Entrance requirements:* GRE General Test, GRE Subject Test, TOEFL. Application deadline: 1/6. Application fee: $45. *Expenses:* Tuition of $16,275 per year full-time, $8140 per year part-time. Fees of $356 per year.

University of Cincinnati, McMicken College of Arts and Sciences, Department of Chemistry, Cincinnati, OH 45221. Department offers programs in analytical chemistry (MS, PhD), biochemistry (MS, PhD), inorganic chemistry (MS, PhD), organic chemistry (MS, PhD), physical chemistry (MS, PhD), polymer chemistry (MS, PhD). Part-time and evening/weekend programs available. Faculty: 26 full-time, 0 part-time. Matriculated students: 115 full-time (45 women), 20 part-time (2 women); includes 19 minority (13 Asian American, 3 black American, 3 Hispanic American), 51 foreign. 163 applicants, 19% accepted. In 1990, 18 master's, 20 doctorates awarded. Terminal master's awarded for partial completion of doctoral program. *Degree requirements:* Thesis/dissertation required, foreign language not required. *Entrance requirements:* GRE General Test, GRE Subject Test. Application deadline: 2/15. Application fee: $20. *Tuition:* $131 per credit hour for state residents; $261 per credit hour for nonresidents. *Financial aid:* In 1990–91, 61 teaching assistantships awarded; fellowships, research assistantships, full tuition waivers also available. Aid available to part-time students. Average monthly stipend for a graduate assistantship: $758. Financial aid application deadline: 5/1. *Faculty research:* Biomedical chemistry, laser chemistry, surface science, chemical sensors, synthesis. Total annual research budget: $3.8-million. • Dr. Bruce Ault, Head, 513-556-9200.

University of Colorado at Boulder, College of Arts and Sciences, Department of Chemistry and Biochemistry, Boulder, CO 80309. Department offers programs in biochemistry (PhD), chemistry (MS, PhD). Faculty: 49 full-time (2 women). Matriculated students: 162 full-time (61 women), 12 part-time (6 women); includes 13 minority (7 Asian American, 6 Hispanic American), 13 foreign. 183 applicants, 23% accepted. In 1990, 7 master's, 29 doctorates awarded. Terminal master's awarded for partial completion of doctoral program. *Degree requirements:* For master's, 1 foreign language, thesis or alternative, comprehensive exam; for doctorate, 1 foreign language, dissertation. *Entrance requirements:* GRE General Test. Application deadline: 3/1 (priority date, applications processed on a rolling basis). Application fee: $30 ($50 for foreign students). *Expenses:* Tuition of $2308 per year full-time, $387 per semester (minimum) part-time for state residents; $8730 per year full-time, $1455 per semester (minimum) part-time for nonresidents. Fees of $207 per semester full-time, $27.26 per semester (minimum) part-time. *Financial aid:* In 1990–91, $458,000 in aid awarded. 4 fellowships (0 to first-year students), 122 research assistantships (0 to first-year students), 56 teaching assistantships (0 to first-year students) were awarded; full tuition waivers also available. Aid available to part-time students. Financial aid application deadline: 3/1. Total annual research budget: $5.2-million. • Kevin Peters, Chairman, 303-492-6533.

University of Colorado at Denver, College of Liberal Arts and Sciences, Program in Chemistry, Denver, CO 80217. Program awards MS. Part-time programs available. Faculty: 7 full-time (2 women), 0 part-time. Matriculated students: 0 full-time, 11 part-time (2 women); includes 2 minority (1 black American, 1 Hispanic American), 1 foreign. In 1990, 2 degrees awarded. *Degree requirements:* Thesis or alternative. Application deadline: 6/1 (applications processed on a rolling basis). Application fee: $30 ($50 for foreign students). *Expenses:* Tuition of $1185 per semester full-time, $142 per semester hour part-time for state residents; $3969 per semester full-time, $476 per semester hour part-time for nonresidents. Fees of $103 per semester. *Financial aid:* Research assistantships, teaching assistantships, federal work-study available. Financial aid application deadline: 3/1. Total annual research budget: $80,000. • Douglas Dyckes, Chair, 303-556-4885. Application contact: Doris Kimbrough, Graduate Adviser, 303-556-4885.

University of Connecticut, College of Liberal Arts and Sciences, Field of Chemistry, Storrs, CT 06269. Field awards MS, PhD. Faculty: 25. Matriculated students: 98 full-time (41 women), 4 part-time (1 woman); includes 3 minority (1 Asian American, 1 black American, 1 Hispanic American), 65 foreign. Average age 29. 145 applicants, 24% accepted. In 1990, 7 master's, 10 doctorates awarded. Terminal master's awarded for partial completion of doctoral program. *Degree requirements:* For doctorate, dissertation. *Entrance requirements:* GRE General Test, GRE Subject Test. Application deadline: 6/1 (priority date, applications processed on a rolling basis). Application fee: $25. *Expenses:* Tuition of $3428 per year full-time, $571 per course part-time for state residents; $8914 per year full-time, $1486 per course part-time for nonresidents. Fees of $636 per year full-time, $87 per course part-time. *Financial aid:* In 1990–91, $896,669 in aid awarded. 67 fellowships (11 to first-year students), 29 research assistantships (2 to first-year students), 65 teaching assistantships (16 to first-year students) were awarded. Financial aid application deadline: 2/15; applicants required to submit GAPSFAS or FAF. • Edward S. Kostiner, Head, 203-486-3214.

University of Delaware, College of Arts and Science, Department of Chemistry and Biochemistry, Newark, DE 19716. Department offers programs in biochemistry (MA, MS, PhD), chemistry (MA, MS, PhD). Part-time programs available. Faculty: 33 full-time (3 women), 0 part-time. Matriculated students: 82 full-time (23 women), 15 part-time (3 women); includes 2 minority (both black American), 30 foreign. Average age 25. 198 applicants, 37% accepted. In 1990, 2 master's awarded (50% found work related to degree, 50% continued full-time study); 12 doctorates awarded (33% entered university research/teaching, 67% found other work related to degree). Terminal master's awarded for partial completion of doctoral program. *Degree requirements:* For master's, 1 foreign language, thesis (for some programs); for doctorate, 1 foreign language, dissertation, comprehensive exams. *Entrance requirements:* GRE General Test, GRE Subject Test. Application deadline: 7/1 (applications processed on a rolling basis). Application fee: $35. *Tuition:* $179 per credit hour for state residents; $467 per credit hour for nonresidents. *Financial aid:* In 1990–91, $12,000 in aid awarded. 3 fellowships (2 to first-year students), 26 research assistantships (1 to a first-year student), 48 teaching assistantships (21 to first-year students) were awarded. Financial aid application deadline: 3/31. *Faculty research:* Protein studies; mechanism of enzymes; synthesis, electronic structure, and bonding of inorganic and organometallic compounds; spectroscopy studies. • Dr. Jean Futrell, Chairman, 302-451-1247.

University of Denver, Graduate Studies, Faculty of Natural Sciences, Mathematics and Engineering, Department of Chemistry, Denver, CO 80208. Department awards MA, MS, PhD. Part-time programs available. Faculty: 10 full-time (2 women), 0 part-time. Matriculated students: 24 (9 women); includes 2 minority (1 Asian American, 1 black American), 6 foreign. 23 applicants, 91% accepted. In 1990, 2 master's, 2 doctorates awarded. Terminal master's awarded for partial completion of doctoral program. *Degree requirements:* For master's, thesis required, foreign language not required; for doctorate, 1 foreign language, dissertation. *Entrance requirements:* GRE

General Test, GRE Subject Test, TOEFL (minimum score of 550 required). Application deadline: rolling. Application fee: $30. *Tuition:* $12,852 per year full-time, $357 per credit hour part-time. *Financial aid:* In 1990–91, 12 students received a total of $161,810 in aid awarded. 10 teaching assistantships, 1 scholarship were awarded; research assistantships, federal work-study, institutionally sponsored loans, and career-related internships or fieldwork also available. Aid available to part-time students. Average monthly stipend for a graduate assistantship: $850. Financial aid application deadline: 3/1. • Dr. Robert D. Coombe, Chairperson, 303-871-2436. Application contact: Dr. T. Gregory Dewey, Chair, Graduate Committee, 303-871-3100.

University of Detroit Mercy, College of Engineering and Science, Department of Chemistry, Detroit, MI 48221. Department offers programs in economic aspects of chemistry (MSEC), macromolecular chemistry (MS, PhD). Evening/weekend programs available. Faculty: 6 full-time (2 women), 0 part-time. Matriculated students: 10 full-time (1 woman), 49 part-time (13 women); includes 10 minority (3 Asian American, 4 black American, 3 Hispanic American), 24 foreign. Average age 30. In 1990, 7 master's, 3 doctorates awarded. *Degree requirements:* Thesis/dissertation required, foreign language not required. *Entrance requirements:* For master's, GRE General Test, minimum GPA of 3.0; for doctorate, GRE Subject Test, minimum GPA of 3.0. Application deadline: 8/1 (priority date, applications processed on a rolling basis). Application fee: $25 ($35 for foreign students). *Tuition:* $360 per credit hour. *Financial aid:* In 1990–91, 11 teaching assistantships awarded; career-related internships or fieldwork also available. *Faculty research:* Polymer and physical chemistry, industrial aspects of chemistry. • Dr. John Mclean, Chairman, 313-927-1258.

University of Florida, College of Liberal Arts and Sciences, Department of Chemistry, Gainesville, FL 32611. Department awards MS, MST, PhD. PhD offered jointly with Florida Atlantic University. Faculty: 51. Matriculated students: 149 full-time (38 women), 16 part-time (2 women); includes 13 minority (2 Asian American, 5 black American, 6 Hispanic American), 38 foreign. 101 applicants, 52% accepted. In 1990, 6 master's, 28 doctorates awarded. *Degree requirements:* For master's, thesis required, foreign language not required; for doctorate, 1 foreign language, dissertation. *Entrance requirements:* GRE General Test, minimum GPA of 3.0. Application deadline: 5/30 (priority date, applications processed on a rolling basis). Application fee: $15. *Tuition:* $87 per credit hour for state residents; $289 per credit hour for nonresidents. *Financial aid:* In 1990–91, 1 fellowship, 63 research assistantships, 71 teaching assistantships awarded; institutionally sponsored loans also available. Average monthly stipend for a graduate assistantship: $1000. *Faculty research:* Organic, analytical, physical, inorganic and biological chemistry. Total annual research budget: $4.9-million. • Dr. Michael Zerner, Chairman, 904-392-0541. Application contact: Dr. John F. Helling, Chair, Graduate Selections Committee, 904-392-0551.

See full description on page 145.

University of Florida, College of Pharmacy and Graduate School, Graduate Programs in Pharmacy, Department of Medicinal Chemistry, Gainesville, FL 32611. Department awards MSP, PhD. Faculty: 5 full-time (1 woman), 0 part-time. Matriculated students: 17 full-time (2 women), 0 part-time; includes 2 minority (both Asian American), 9 foreign. Average age 29. 45 applicants, 7% accepted. In 1990, 0 master's, 1 doctorate awarded. Terminal master's awarded for partial completion of doctoral program. *Degree requirements:* Thesis/dissertation required, foreign language not required. *Entrance requirements:* GRE General Test, TOEFL, minimum GPA of 3.0. Application deadline: 6/1 (priority date, applications processed on a rolling basis). Application fee: $15. *Tuition:* $87 per credit hour for state residents; $289 per credit hour for nonresidents. *Financial aid:* In 1990–91, 17 students received a total of $172,176 in aid awarded. 17 research assistantships (3 to first-year students) were awarded. Average monthly stipend for a graduate assistantship: $844. *Faculty research:* Iron chelation, anticancer drug development, natural produce chemistry, drug metabolism and toxicity, dermal delivery of drug and prodrugs. • Dr. Raymond J. Bergeron, Graduate Coordinator, 904-392-5900.

University of Guelph, Guelph-Waterloo Centre for Graduate Work in Chemistry, Waterloo, ON N26 3G1, Canada. Center awards M Sc, PhD. Offered jointly with University of Waterloo. Faculty: 66. Matriculated students: 54; includes 12 foreign. In 1990, 4 master's awarded. *Degree requirements:* Thesis/dissertation. *Entrance requirements:* For master's, minimum B average during last two years; for doctorate, minimum B average. *Expenses:* Tuition of $898 per semester full-time, $450 per semester part-time for Canadian residents; $4053 per semester full-time, $2185 per semester part-time for nonresidents. Fees of $543 per semester full-time, $450 per semester part-time for Canadian residents; $2278 per semester full-time, $2185 per semester part-time for nonresidents. *Financial aid:* Fellowships, research assistantships, teaching assistantships available. *Faculty research:* Inorganic, analytical, biological, physical/theoretical, and polymer chemistry. • Dr. Henry, Chair, 519-824-4120 Ext. 6709. Application contact: Dr. Balahura, Graduate Coordinator, 519-824-4120 Ext. 2267.

University of Hartford, College of Arts and Sciences, Program in Chemistry, West Hartford, CT 06117. Program awards MS. Part-time and evening/weekend programs available. Faculty: 5 full-time (0 women), 0 part-time. Matriculated students: 0 full-time, 8 part-time (3 women); includes 1 minority (black American), 0 foreign. Average age 29. 12 applicants, 33% accepted. In 1990, 0 degrees awarded. *Degree requirements:* Thesis, comprehensive exam required, foreign language not required. *Entrance requirements:* GRE, TOEFL. Application deadline: 7/1 (priority date, applications processed on a rolling basis). Application fee: $35. *Expenses:* Tuition of $295 per credit hour. Fees of $30 per semester. *Financial aid:* In 1990–91, $5800 in aid awarded. 1 teaching assistantship (0 to first-year students) was awarded; partial tuition waivers and career-related internships or fieldwork also available. *Faculty research:* Fast reaction kinetics, polymer synthesis and modification, determination of trace amounts of chloramines in water, precise environmental measurement, computer interfacing. • Dr. E. T. Gray, Jr., Chairman, 203-243-4612. Application contact: Dr. Harry Workman, Professor, 230-243-4537.

University of Hawaii at Manoa, College of Arts and Sciences, Department of Chemistry, Honolulu, HI 96822. Department awards MS, PhD. *Degree requirements:* Thesis/dissertation. *Entrance requirements:* GRE General Test, GRE Subject Test. *Tuition:* $800 per semester full-time for state residents; $2405 per semester full-time for nonresidents.

University of Houston, College of Natural Sciences and Mathematics, Department of Chemistry, 4800 Calhoun, Houston, TX 77004. Department offers programs in analytical chemistry (MS, PhD), biological chemistry (MS, PhD), chemical physics (MS, PhD), inorganic chemistry (MS, PhD), organic chemistry (MS, PhD), physical chemistry (MS, PhD), theoretical chemistry (MS, PhD). Part-time and evening/weekend programs available. Faculty: 25 full-time (1 woman), 2 part-time (0 women). Matriculated students: 112 full-time (36 women), 3 part-time (1 woman); includes 2 minority (both Asian American), 86 foreign. Average age 25. 178 applicants, 21% accepted. In 1990, 9 master's, 8 doctorates awarded. Terminal master's awarded for partial completion of doctoral program. *Degree requirements:* Thesis/dissertation required, foreign language not required. *Entrance requirements:* GRE General Test (minimum combined score of 1100 required), TOEFL (minimum score of 575 required). Application deadline: 5/1 (applications processed on a rolling basis). Application fee: $25 ($75 for foreign students). *Expenses:* Tuition of $30 per hour for state residents; $134 per hour for nonresidents. Fees of $240 per year full-time, $125 per year part-time. *Financial aid:* In 1990–91, 110 students received aid. 25 fellowships (0 to first-year students), 41 research assistantships, 61 teaching assistantships (34 to first-year students) awarded; federal work-study, institutionally sponsored loans, and career-related internships or fieldwork also available. Average monthly stipend for a graduate assistantship: $950. Financial aid application deadline: 4/1. *Total annual research budget:* $3.5-million. • Dr. John L. Bear, Chairman, 713-749-2647. Application contact: Dr. Thomas Albright, Chair, Graduate Committee, 713-749-3721.

See full description on page 149.

University of Houston–Clear Lake, School of Natural and Applied Sciences, Program in Chemistry, Houston, TX 77058. Program awards MS. Matriculated students: 5 full-time (1 woman), 19 part-time (6 women); includes 6 minority (2 Asian American, 2 black American, 2 Hispanic American), 2 foreign. In 1990, 6 degrees awarded. *Degree requirements:* Foreign language not required. *Entrance requirements:* GRE General Test. Application fee: $0. *Tuition:* $40 per credit hour for state residents; $134 per credit hour for nonresidents. *Financial aid:* Application deadline 5/1. • Dr. Carroll B. Lassiter, Director. Application contact: Dr. Eldon Husband, Associate Dean and Director of Student Affairs, 713-283-3710.

University of Idaho, College of Graduate Studies, College of Letters and Science, Department of Chemistry, Moscow, ID 83843. Department offers programs in chemistry (M Nat Sci, M Nuc Sci, MS, PhD), chemistry education (MAT). Faculty: 18 full-time (4 women), 0 part-time. Matriculated students: 39 full-time (9 women), 18 part-time (9 women); includes 3 minority (1 black American, 2 Hispanic American), 18 foreign. In 1990, 6 master's, 3 doctorates awarded. *Degree requirements:* For master's, thesis or alternative required, foreign language not required; for doctorate, 1 foreign language, dissertation. *Entrance requirements:* For master's, minimum GPA of 2.8; for doctorate, minimum undergraduate GPA of 2.8, graduate GPA of 3.0. Application deadline: 8/1. Application fee: $20. *Expenses:* Tuition of $0 for state residents; $4146 per year for nonresidents. Fees of $818 per semester full-time, $82.75 per credit part-time. *Financial aid:* In 1990–91, 2 fellowships, 15 research assistantships, 22 teaching assistantships awarded. Financial aid application deadline: 3/1. • Dr. Peter R. Griffith, Head, 208-885-6552.

University of Illinois at Chicago, College of Pharmacy and Graduate College, Graduate Programs in Pharmacy, Chicago, IL 60680. Offerings include medicinal chemistry (MS, PhD). Terminal master's awarded for partial completion of doctoral program. Faculty: 42 full-time (4 women), 0 part-time. *Degree requirements:* Variable foreign language requirement, thesis/dissertation. *Entrance requirements:* GRE General Test, TOEFL. Application deadline: 7/5. Application fee: $20. *Expenses:* Tuition of $1369 per semester full-time, $521 per semester (minimum) part-time for state residents; $3840 per semester full-time, $1454 per semester (minimum) part-time for nonresidents. Fees of $458 per semester full-time, $398 per semester (minimum) part-time. • Dr. Michael E. Johnson, Associate Dean, Research and Graduate Education, 312-996-0796.

University of Illinois at Chicago, College of Liberal Arts and Sciences, Department of Chemistry, Chicago, IL 60680. Department awards MS, DA, PhD. Part-time programs available. Faculty: 29 full-time (2 women), 0 part-time. Matriculated students: 104 full-time (30 women), 35 part-time (12 women); includes 12 minority (8 Asian American, 2 black American, 2 Hispanic American), 79 foreign. 204 applicants, 28% accepted. In 1990, 12 master's, 16 doctorates awarded. Terminal master's awarded for partial completion of doctoral program. *Degree requirements:* For master's, thesis or cumulative exam required, foreign language not required; for doctorate, 1 foreign language, dissertation, cumulative exams. *Entrance requirements:* GRE Subject Test, TOEFL (minimum score of 580 required), minimum GPA of 4.0 on a 5.0 scale. Application deadline: 7/5. Application fee: $20. *Expenses:* Tuition of $1369 per semester full-time, $521 per semester (minimum) part-time for state residents; $3840 per semester full-time, $1454 per semester (minimum) part-time for nonresidents. Fees of $458 per semester full-time, $398 per semester (minimum) part-time. *Financial aid:* In 1990–91, 6 fellowships, 4 research assistantships, 72 teaching assistantships awarded; full tuition waivers, federal work-study, institutionally sponsored loans also available. • Jan Rocek, Head, 312-996-3179. Application contact: Ann Erskin, Graduate Secretary, 312-996-3161.

See full description on page 151.

University of Illinois at Urbana-Champaign, College of Liberal Arts and Sciences, School of Chemical Sciences, Department of Chemistry, Champaign, IL 61820. Department awards MS, PhD. Faculty: 61 full-time, 2 part-time. Matriculated students: 69 full-time (14 women), 1 (woman) part-time. 255 applicants, 27% accepted. In 1990, 11 master's, 11 doctorates awarded. *Degree requirements:* For master's, foreign language and thesis not required; for doctorate, 1 foreign language, dissertation. Application fee: $25. *Tuition:* $1838 per semester full-time, $708 per semester part-time for state residents; $4314 per semester full-time, $1673 per semester part-time for nonresidents. *Financial aid:* In 1990–91, 29 fellowships, 121 research assistantships, 146 teaching assistantships awarded. Financial aid application deadline: 2/15. • John Hummel, Acting Head, 217-333-0710.

University of Iowa, College of Liberal Arts, Department of Chemistry, Iowa City, IA 52242. Department awards MS, PhD. Faculty: 27 full-time, 0 part-time. Matriculated students: 51 full-time (18 women), 68 part-time (15 women); includes 5 minority (3 Asian American, 2 black American), 54 foreign. 103 applicants, 42% accepted. In 1990, 7 master's, 14 doctorates awarded. Application fee: $20. *Expenses:* Tuition of $1158 per semester full-time, $387 per semester hour (minimum) part-time for state residents; $3372 per semester full-time, $387 per semester hour (minimum) part-time for nonresidents. Fees of $60 per semester (minimum). *Financial aid:* In 1990–91, 3 fellowships, 61 research assistantships (0 to first-year students), 39 teaching assistantships (15 to first-year students) awarded. • Darrell P. Eyman, Acting Chair, 319-335-1359.

University of Kansas, College of Liberal Arts and Sciences, Department of Chemistry, Lawrence, KS 66045. Department awards MA, MS, PhD. Faculty: 21 full-time (2 women), 0 part-time. Matriculated students: 49 full-time (18 women), 34 part-time (8 women); includes 2 minority (1 Asian American, 1 Hispanic American), 47 foreign. In 1990, 3 master's, 6 doctorates awarded. *Degree requirements:* For master's, thesis required, foreign language not required; for doctorate, dissertation. *Entrance requirements:* GRE General Test, GRE Subject Test, TOEFL (minimum score of 570 required). Application fee: $25. *Expenses:* Tuition of $1668 per year full-time, $56 per credit hour part-time for state residents; $5382 per year full-time, $179 per credit hour part-time for nonresidents. Fees of $338 per year full-time, $25 per credit

Directory: Chemistry

hour part-time. *Financial aid:* Fellowships, research assistantships, teaching assistantships available. • Richard Givens, Chairperson, 913-864-4670.

University of Kentucky, Graduate School Programs from the College of Arts and Sciences, Program in Chemistry, Lexington, KY 40506-0032. Program awards MS, PhD. Faculty: 29 full-time (2 women), 0 part-time. Matriculated students: 40 full-time (11 women), 22 part-time (10 women); includes 3 minority (all Asian American), 28 foreign. 263 applicants, 10% accepted. In 1990, 5 master's, 6 doctorates awarded. *Degree requirements:* For master's, 1 foreign language, comprehensive exam required, thesis optional; for doctorate, 1 foreign language, dissertation, comprehensive exam. *Entrance requirements:* For master's, GRE (verbal, quantitative, and analytical sections), minimum undergraduate GPA of 2.5; for doctorate, GRE (verbal, quantitative, and analytical sections), minimum graduate GPA of 3.0. Application deadline: 7/19 (applications processed on a rolling basis). Application fee: $20 ($25 for foreign students). *Tuition:* $1002 per semester full-time, $101 per credit hour part-time for state residents; $2782 per semester full-time, $299 per credit hour part-time for nonresidents. *Financial aid:* In 1990–91, 9 fellowships (5 to first-year students), 19 research assistantships, 37 teaching assistantships (11 to first-year students) awarded; federal work-study, institutionally sponsored loans also available. Aid available to part-time students. Financial aid applicants required to submit FAF. *Faculty research:* Analytical, inorganic, organic, physical, biological, biophysical, and biomedical chemistry; nuclear and radiochemistry programs of research; electrochemistry; crystallography. • James O'Reilly, Director of Graduate Studies, 606-257-7078. Application contact: Dr. Constance L. Wood, Associate Dean for Academic Administration, 606-257-4905.

University of Louisville, College of Arts and Sciences, Department of Chemistry, Louisville, KY 40292. Department offers programs in analytical chemistry (MS, PhD), chemical physics (PhD), inorganic chemistry (MS, PhD), organic chemistry (MS, PhD), physical chemistry (MS, PhD). Faculty: 18 full-time (2 women), 2 part-time. Matriculated students: 40 full-time, 6 part-time. In 1990, 2 master's, 8 doctorates awarded. *Degree requirements:* For master's, thesis; for doctorate, 1 foreign language, dissertation. *Entrance requirements:* GRE General Test, TOEFL (minimum score of 550 required). Application deadline: rolling. *Expenses:* Tuition of $1780 per year full-time, $99 per credit hour part-time for state residents; $5340 per year full-time, $297 per credit hour part-time for nonresidents. Fees of $60 per semester full-time, $12.50 per semester (minimum) part-time. *Financial aid:* Fellowships, research assistantships, teaching assistantships available. *Faculty research:* Electrochemistry, spectroscopy, polymers, organometallics, bioinorganic chemistry. • Dr. Richard P. Baldwin, Chair, 502-588-6798.

Announcement: The department occupies a modern, well-equipped building and offers research opportunities in electrochemistry, spectroscopy, polymers, organometallics, bioinorganic chemistry, synthesis, magnetic resonance, and crystallography. Dr. Richard P. Baldwin, Chair, 502-588-6798.

University of Maine, College of Sciences, Department of Chemistry, Orono, ME 04469. Department awards MS, PhD. Faculty: 14 full-time (2 women), 0 part-time. Matriculated students: 25 full-time (10 women), 1 (woman) part-time; includes 0 minority, 18 foreign. In 1990, 2 master's awarded; 3 doctorates awarded (100% entered university research/teaching). Terminal master's awarded for partial completion of doctoral program. *Degree requirements:* For master's, thesis required, foreign language not required; for doctorate, dissertation, oral exam required, foreign language not required. *Entrance requirements:* For master's, GRE General Test, GRE Subject Test, TOEFL (minimum score of 520 required); for doctorate, GRE General Test, GRE Subject Test, TOEFL (minimum score of 550 required). Application deadline: 12/15 (priority date, applications processed on a rolling basis). Application fee: $25. *Tuition:* $100 per credit hour for state residents; $275 per credit hour for nonresidents. *Financial aid:* In 1990–91, $224,800 in aid awarded. 7 research assistantships (0 to first-year students), 17 teaching assistantships (6 to first-year students) were awarded; full tuition waivers also available. *Faculty research:* Quantum mechanics, natural products, insect chemistry, organic synthesis, electrochemistry. Total annual research budget: $645,000. • Dr. Raymond C. Fort Jr., Chairman, 207-581-1169. Application contact: Dr. Brian Green, Graduate Coordinator, 207-581-1176.

Announcement: Research programs in computer simulation and data reduction; biomedical mass spectrometry; chemical ecology; medicinal, natural products, wood, and paper chemistry; chemical oceanography; thermodynamics and statistical mechanics; molecular and inorganic spectroscopy; and surface chemistry. Interaction with physics or biochemistry department or the Laboratory for Surface Science and Technology possible. Financial aid available.

University of Manitoba, Faculty of Science, Department of Chemistry, Winnipeg, MB R3T 2N2, Canada. Department awards M Sc, PhD. *Degree requirements:* For master's, thesis required, foreign language not required; for doctorate, 1 foreign language, dissertation.

University of Maryland College Park, College of Life Sciences, Department of Chemistry and Biochemistry, College Park, MD 20742. Department offers programs in analytical chemistry (MS, PhD), biochemistry (MS, PhD), chemistry (MS, PhD), inorganic chemistry (MS, PhD), organic chemistry (MS, PhD), physical chemistry (MS, PhD). Faculty: 51 (5 women). Matriculated students: 127 full-time (53 women), 31 part-time (14 women); includes 23 minority (9 Asian American, 10 black American, 3 Hispanic American, 1 Native American), 58 foreign. 266 applicants, 35% accepted. In 1990, 10 master's, 25 doctorates awarded. *Degree requirements:* For master's, thesis or alternative required, foreign language not required; for doctorate, dissertation. *Entrance requirements:* For master's, GRE General Test, minimum GPA of 3.0; for doctorate, GRE General Test. Application deadline: rolling. Application fee: $25. *Expenses:* Tuition of $143 per credit hour for state residents; $256 per credit hour for nonresidents. Fees of $171.50 per semester. *Financial aid:* In 1990–91, 11 fellowships (6 to first-year students), 58 research assistantships, 109 teaching assistantships awarded. *Faculty research:* Environmental chemistry, geochemistry, nuclear chemistry. • Dr. Sandra Greer, Chairman, 301-405-1788.

See full description on page 155.

University of Maryland Graduate School, Baltimore, Graduate School, Department of Chemistry and Biochemistry, Baltimore, MD 21228. Department awards MS, PhD, MD/PhD. Programs offered in biochemistry (MS, PhD), including biochemistry (PhD), neuroscience (MS, PhD); chemistry (MS, PhD). Part-time and evening/weekend programs available. Faculty: 22. In 1990, 5 master's, 2 doctorates awarded. *Degree requirements:* For master's, thesis not required; for doctorate, dissertation required, foreign language not required. *Entrance requirements:* GRE General Test, GRE Subject Test, TOEFL, minimum GPA of 3.0. Application deadline: 7/1. Application fee: $25. *Tuition:* $134 per credit for state residents; $245 per credit for nonresidents. *Financial aid:* Fellowships, research assistantships, teaching assistantships available. • Dr. Catherine Fenselau, Chairperson, 301-455-2505.

See full description on page 157.

University of Maryland Graduate School, Baltimore, School of Pharmacy and Graduate School, Department of Biomedical Chemistry, Baltimore, MD 21228. Department awards MS, PhD. Faculty: 8 full-time (1 woman), 0 part-time. Matriculated students: 3 full-time (1 woman), 15 part-time (7 women); includes 1 minority (black American), 10 foreign. Average age 29. 14 applicants, 43% accepted. In 1990, 1 master's, 1 doctorate awarded. Terminal master's awarded for partial completion of doctoral program. *Degree requirements:* Thesis/dissertation required, foreign language not required. *Entrance requirements:* GRE General Test, TOEFL (minimum score of 550 required), minimum GPA of 3.0. Application deadline: 7/1. Application fee: $25. *Tuition:* $134 per credit for state residents; $245 per credit for nonresidents. *Financial aid:* In 1990–91, $30,000 in aid awarded. 0 fellowships, 7 research assistantships, 3 teaching assistantships were awarded. Aid available to part-time students. Financial aid application deadline: 2/15; applicants required to submit GAPSFAS. *Faculty research:* Biomedical mass spectrometry, pesticide biotransformation and biosynthesis, regulation and formation of secondary metabolites. Total annual research budget: $40,000. • Dr. Marilyn K. Speedie, Chairperson, 301-328-7541. Application contact: Dr. S. Edward Kri Korian, Program Director, 301-328-7442.

University of Massachusetts at Amherst, College of Arts and Sciences, Faculty of Natural Sciences and Mathematics, Department of Chemistry, Amherst, MA 01003. Department awards MS, PhD. Part-time programs available. Faculty: 33 full-time (1 woman), 0 part-time. Matriculated students: 38 full-time (15 women), 78 part-time (26 women); includes 7 minority (1 Asian American, 1 black American, 5 Hispanic American), 40 foreign. Average age 28. 142 applicants, 40% accepted. In 1990, 9 master's, 13 doctorates awarded. Terminal master's awarded for partial completion of doctoral program. *Degree requirements:* For master's, thesis required, foreign language not required; for doctorate, 1 foreign language, dissertation. *Entrance requirements:* GRE General Test, GRE Subject Test. Application deadline: 3/1 (applications processed on a rolling basis). Application fee: $35. *Tuition:* $2568 per year full-time, $107 per credit part-time for state residents; $7920 per year full-time, $330 per credit part-time for nonresidents. *Financial aid:* In 1990–91, 3 fellowships, 74 research assistantships, 58 teaching assistantships awarded; federal work-study also available. Aid available to part-time students. Financial aid application deadline: 3/1; applicants required to submit FAF. • Dr. Thomas R. Stengle, Director, 413-545-6079.

University of Massachusetts at Boston, College of Arts and Sciences, Program in Chemistry, Boston, MA 02125. Program awards MS. Matriculated students: 2 full-time (0 women), 13 part-time (7 women). *Degree requirements:* Thesis, comprehensive exams required, foreign language not required. *Entrance requirements:* Minimum GPA of 2.75. Application deadline: 6/1 (applications processed on a rolling basis). Application fee: $20. *Expenses:* Tuition of $2568 per year full-time, $107 per credit part-time for state residents; $7920 per year full-time, $330 per credit part-time for nonresidents. Fees of $767 per year full-time, $339 per year part-time. *Financial aid:* Research assistantships, teaching assistantships available. • Dr. Robert Carter, Director, 617-287-6130. Application contact: Lisa Lavely, Director of Graduate Admissions, 617-287-6400.

University of Massachusetts Dartmouth, Graduate School, College of Arts and Sciences, Department of Chemistry, North Dartmouth, MA 02747. Department awards MS. Part-time programs available. Faculty: 10 full-time (2 women), 0 part-time. Matriculated students: 8 full-time (2 women), 0 part-time; includes 0 minority, 8 foreign. 13 applicants, 92% accepted. In 1990, 2 degrees awarded. *Degree requirements:* Thesis (for some programs). *Entrance requirements:* GRE General Test, GRE Subject Test, TOEFL. Application deadline: 4/20 (priority date, applications processed on a rolling basis). *Expenses:* Tuition of $1368 per semester full-time, $7688 per credit part-time for state residents; $4388 per semester full-time, $244 per credit part-time for nonresidents. Fees of $771 per year full-time, $59 per year part-time. *Financial aid:* In 1990–91, $52,000 in aid awarded. 1 research assistantship, 8 teaching assistantships were awarded; full tuition waivers also available. Financial aid application deadline: 5/1. *Faculty research:* Raman spectroscopy, physical and theoretical organic chemistry, transition-metal chemistry, biochemistry of phospholipids, analytical radiochemistry. Total annual research budget: $350,000. • Dr. Donald Boerth, Director, 508-999-8244. Application contact: Carol A. Novo, Graduate Admissions Office, 508-999-8604.

University of Massachusetts, Lowell, College of Arts and Sciences, Department of Chemistry, 1 University Avenue, Lowell, MA 01854. Offerings include chemistry (MS, PhD). Terminal master's awarded for partial completion of doctoral program. *Degree requirements:* For master's, thesis required, foreign language not required; for doctorate, 2 foreign languages, computer language, dissertation. *Entrance requirements:* GRE General Test. Application deadline: 4/1. *Expenses:* Tuition of $87 per credit hour for state residents; $271 per credit hour for nonresidents. Fees of $114 per credit hour.

University of Massachusetts, Lowell, James B. Francis College of Engineering, Department of Plastics Engineering, 1 University Avenue, Lowell, MA 01854. Offerings include chemistry (PhD), with option in polymer science/plastics engineering. Terminal master's awarded for partial completion of doctoral program. *Application deadline:* 4/1. *Expenses:* Tuition of $87 per credit hour for state residents; $271 per credit hour for nonresidents. Fees of $114 per credit hour.

University of Miami, College of Arts and Sciences, Department of Chemistry, Coral Gables, FL 33124. Department offers programs in chemistry (MS), inorganic chemistry (PhD), organic chemistry (PhD), physical chemistry (PhD). Faculty: 10 full-time (1 woman), 0 part-time. Matriculated students: 47 full-time (17 women), 6 part-time (3 women); includes 10 minority (1 black American, 9 Hispanic American), 35 foreign. Average age 29. 300 applicants, 9% accepted. In 1990, 1 master's, 3 doctorates awarded. *Degree requirements:* For master's, foreign language and thesis not required; for doctorate, dissertation required, foreign language not required. *Entrance requirements:* GRE General Test (minimum combined score of 1000 required), TOEFL (minimum score of 550 required). Application deadline: 5/1 (priority date, applications processed on a rolling basis). Application fee: $0. *Expenses:* Tuition of $567 per credit hour. Fees of $87 per semester full-time. *Financial aid:* 45 students received aid. Fellowships, research assistantships, teaching assistantships, full tuition waivers available. Average monthly stipend for a graduate assistantship: $1000. Financial aid application deadline: 5/1. *Faculty research:* Molecular recognition, electron transfer processes, high-pressure chemistry, thermochemistry, electrochemistry. Total annual research budget: $996,245. • Dr. Cecil Criss, Chairman, 305-284-2174.

See full description on page 159.

University of Miami, Rosenstiel School of Marine and Atmospheric Science, Department of Marine and Atmospheric Chemistry, Coral Gables, FL 33124. Department awards MS, PhD. Faculty: 12 (10 women). Matriculated students: 15 full-time (8 women), 1 part-time (0 women); includes 0 minority, 7 foreign. 87 applicants, 15% accepted. In 1990, 0 master's, 0 doctorates awarded. *Degree requirements:* For doctorate, dissertation. *Entrance requirements:* GRE General Test, TOEFL (minimum score of 550 required). Application fee: $35. *Expenses:* Tuition of $567 per credit hour. Fees of $87 per semester full-time. • Dr. Rana Fine, Chairperson, 305-361-4722. Application contact: Dr. Frank Millero, Associate Dean, 305-361-4155.

University of Michigan, College of Pharmacy, Interdepartmental Program in Medicinal Chemistry, Ann Arbor, MI 48109. Program awards PhD. Faculty: 24. Matriculated students: 27 full-time (7 women), 0 part-time; includes 4 minority (1 Asian American, 2 black American, 1 Hispanic American), 3 foreign. 29 applicants, 21% accepted. In 1990, 1 degree awarded. *Degree requirements:* Dissertation, preliminary exam. *Entrance requirements:* GRE General Test. Application deadline: 3/15 (applications processed on a rolling basis). Application fee: $30. *Tuition:* $3607 per semester full-time, $391 per credit (minimum) part-time for state residents; $6996 per semester full-time, $768 per credit (minimum) part-time for nonresidents. *Financial aid:* Fellowships available. Financial aid application deadline: 3/15. • Leroy Townsend, Chair, 313-764-7547.

University of Michigan, College of Literature, Science, and the Arts, Department of Chemistry, Ann Arbor, MI 48109. Department offers programs in analytical chemistry (MS, PhD), inorganic chemistry (MS, PhD), organic chemistry (MS, PhD), physical chemistry (MS, PhD). Faculty: 38. Matriculated students: 164 full-time (45 women), 0 part-time; includes 10 minority (4 Asian American, 3 black American, 2 Hispanic American, 1 Native American), 63 foreign. 237 applicants, 54% accepted. In 1990, 18 master's, 25 doctorates awarded. *Degree requirements:* For master's, foreign language and thesis not required; for doctorate, dissertation, preliminary exam required, foreign language not required. *Entrance requirements:* For master's, GRE General Test, GRE Subject Test; for doctorate, GRE General Test. Application deadline: 4/1 (applications processed on a rolling basis). Application fee: $30. *Tuition:* $3255 per semester full-time, $352 per credit (minimum) part-time for state residents; $6803 per semester full-time, $746 per credit (minimum) part-time for nonresidents. *Financial aid:* Fellowships, research assistantships, teaching assistantships available. Financial aid application deadline: 3/15. • M. David Curtis III, Chair, 313-764-7314.

See full description on page 161.

University of Minnesota, Duluth, Graduate School, College of Science and Engineering, Department of Chemistry, Duluth, MN 55812. Department awards MS, PhD. PhD offered in cooperation with the University of Minnesota–Twin Cities. Part-time programs available. Faculty: 17 full-time (2 women), 0 part-time. Matriculated students: 14 full-time (7 women), 1 part-time (0 women); includes 5 foreign. Average age 23. 37 applicants, 70% accepted. In 1990, 3 master's awarded. *Degree requirements:* For master's, variable foreign language requirement, thesis. *Application deadline:* 7/15 (applications processed on a rolling basis). *Application fee:* $30. *Tuition:* $1184 per quarter full-time, $301 per credit (minimum) part-time for state residents; $2168 per quarter full-time, $602 per credit (minimum) part-time for nonresidents. *Financial aid:* In 1990–91, $120,000 in aid awarded. 4 fellowships, 11 teaching assistantships (5 to first-year students) were awarded; research assistantships, federal work-study, institutionally sponsored loans also available. *Total annual research budget:* $300,000. • J. C. Nichol, Director of Graduate Study, 218-726-7501.

University of Minnesota, Twin Cities Campus, Institute of Technology, Department of Chemistry, Minneapolis, MN 55455. Department awards MS, PhD. Faculty: 47. Matriculated students: 190. *Degree requirements:* For master's, thesis or alternative; for doctorate, dissertation, preliminary candidacy exams. *Entrance requirements:* GRE General Test, TOEFL. *Expenses:* Tuition of $1084 per quarter full-time, $301 per credit part-time for state residents; $2168 per quarter full-time, $602 per credit part-time for nonresidents. Fees of $118 per quarter. • W. Ronald Gentry, Chairman, 612-624-6000. Application contact: Graduate Admissions Coordinator, 612-624-8008.

See full description on page 163.

University of Mississippi, Graduate School, College of Liberal Arts, Department of Chemistry, University, MS 38677. Department awards MS, DA, PhD. Matriculated students: 28 full-time (14 women), 2 part-time (1 woman); includes 5 minority (1 Asian American, 3 black American, 1 Hispanic American), 19 foreign. In 1990, 8 master's, 9 doctorates awarded. *Degree requirements:* For master's, thesis required, foreign language not required; for doctorate, 1 foreign language, dissertation. *Entrance requirements:* For master's, GRE General Test, minimum GPA of 3.0; for doctorate, GRE General Test (minimum combined score of 900 required). Application deadline: 8/1. Application fee: $15 ($25 for foreign students). *Expenses:* Tuition of $1011 per semester full-time, $99 per semester part-time for state residents; $1842 per semester full-time, $180 per semester part-time for nonresidents. Fees of $219 per year full-time. *Financial aid:* Application deadline 3/1. • Dr. Andrew P. Stefani, Chairman, 601-232-7301.

University of Missouri–Columbia, College of Arts and Sciences, Department of Chemistry, Columbia, MO 65211. Department offers programs in analytical chemistry (MS, PhD), inorganic chemistry (MS, PhD), organic chemistry (MS, PhD), physical chemistry (MS, PhD). Faculty: 23 full-time, 1 part-time. Matriculated students: 57 full-time (20 women), 26 part-time (9 women); includes 1 minority (Asian American), 51 foreign. In 1990, 8 master's, 7 doctorates awarded. *Degree requirements:* For master's, thesis required, foreign language not required; for doctorate, 1 foreign language, dissertation. *Entrance requirements:* GRE General Test, minimum GPA of 3.0. Application deadline: 8/1 (priority date, applications processed on a rolling basis). Application fee: $20 ($40 for foreign students). *Expenses:* Tuition of $89.90 per credit hour full-time, $98.35 per credit hour part-time for state residents; $244 per credit hour full-time, $252.45 per credit hour part-time for nonresidents. Fees of $123.55 per semester (minimum) full-time. • Dr. John Bauman, Director of Graduate Studies, 314-882-7720. Application contact: Gary L. Smith, Director of Admissions and Registrar, 314-882-7651.

University of Missouri–Kansas City, College of Arts and Sciences, Department of Chemistry, Kansas City, MO 64110. Department offers programs in analytical chemistry (MS, PhD), inorganic chemistry (MS, PhD), organic chemistry (MS, PhD), physical chemistry (MS, PhD), polymer chemistry (MS, PhD). Part-time programs available. Faculty: 15 full-time (0 women), 0 part-time. Matriculated students: 15 full-time (5 women), 24 part-time (6 women); includes 3 minority (2 Asian American, 1 Hispanic American), 24 foreign. In 1990, 1 master's, 0 doctorates awarded. *Degree requirements:* For master's, thesis required, foreign language not required; for doctorate, 1 foreign language, dissertation, comprehensive exam. *Entrance requirements:* GRE. Application fee: $0. *Expenses:* Tuition of $2200 per year full-time, $92 per credit hour part-time for state residents; $5503 per year full-time, $229 per credit hour part-time for nonresidents. Fees of $122 per semester full-time, $9 per credit hour part-time. *Financial aid:* Fellowships, research assistantships, teaching assistantships, full and partial tuition waivers, federal work-study, institutionally sponsored loans available. Aid available to part-time students. *Faculty research:* Raman spectroscopy, polymer synthesis and properties, synthesis and reaction of nitrogen heterocytic compounds, ion selective electrodes, homogeneous catalysis by organometallic compounds. Total annual research budget: $91,121. • Dr. Layton L. McCoy, Chairperson, 816-235-2280. Application contact: Dr. T. F. Thomas, Chairman, Graduate Admissions Committee, 816-235-2297.

University of Missouri–Rolla, College of Arts and Sciences, Department of Chemistry, Rolla, MO 65401. Department awards MS, PhD. Faculty: 18 full-time (1 woman), 6 part-time (0 women). Matriculated students: 61 full-time (25 women), 0 part-time; includes 13 minority (11 Asian American, 1 black American, 1 Native American), 19 foreign. 37 applicants, 89% accepted. In 1990, 5 master's, 10 doctorates awarded. Terminal master's awarded for partial completion of doctoral program. *Degree requirements:* For master's, foreign language and thesis not required; for doctorate, 1 foreign language, dissertation. *Entrance requirements:* Minimum GPA of 3.0. Application deadline: 7/1 (applications processed on a rolling basis). Application fee: $20 ($40 for foreign students). *Expenses:* Tuition of $2090 per year full-time, $87.10 per credit hour part-time for state residents; $5582 per year full-time, $232.60 per credit hour part-time for nonresidents. Fees of $349 per year full-time, $61.63 per semester (minimum) part-time. *Financial aid:* In 1990–91, 57 students received a total of $476,038 in aid awarded. 7 fellowships (5 to first-year students), 19 research assistantships (2 to first-year students), 31 teaching assistantships (14 to first-year students) were awarded; institutionally sponsored loans also available. Average monthly stipend for a graduate assistantship: $1127. *Faculty research:* Structure and properties of inorganic materials, bioanalytical surface properties and chemical dynamics of organic systems. Total annual research budget: $992,391. • O. K. Manuel, Chairman, 314-341-4420.

University of Missouri–St. Louis, College of Arts and Sciences, Department of Chemistry, Normandy, MO 63121-4499. Department awards MS, PhD. Part-time and evening/weekend programs available. Faculty: 20 (2 women). Matriculated students: 35 full-time (11 women), 23 part-time (3 women); includes 5 minority (1 Asian American, 3 black American, 1 Hispanic American), 19 foreign. Average age 32. 37 applicants, 22% accepted. In 1990, 5 master's awarded (100% found work related to degree); 4 doctorates awarded (100% found work related to degree). Terminal master's awarded for partial completion of doctoral program. *Degree requirements:* For master's, thesis optional, foreign language not required; for doctorate, 1 foreign language, dissertation. *Entrance requirements:* GRE General Test, GRE Subject Test. Application deadline: 7/1 (applications processed on a rolling basis). Application fee: $0. *Expenses:* Tuition of $2157 per year full-time, $89.90 per credit hour part-time for state residents; $5856 per year full-time, $244 per credit hour part-time for nonresidents. Fees of $235 per year full-time, $9.80 per credit hour part-time. *Financial aid:* In 1990–91, $202,000 in aid awarded. 7 research assistantships, 22 teaching assistantships (8 to first-year students) were awarded. *Faculty research:* Organic, inorganic, and organometallic synthesis; physical organic chemistry; spectroscopy; polymer chemistry, solid-state chemistry and interfacial phenomena. Total annual research budget: $650,000. • Dr. Lawrence Barton, Chairman, 314-553-5311. Application contact: Dr. Gordon Anderson, Graduate Admissions Officer, 314-553-5311.

Announcement: Research programs in organic, physical, and inorganic chemistry, with emphases in organometallic, bioorganic, bioinorganic, computational, surfactant, and polymer chemistry, and laser spectroscopy. Modern laboratories contain computing facilities, molecular modeling, XRD, ORD-CD 300-MHz NMR, ESR, double focusing MS and GC/MS, FT-IR, and several UV-VIS and laser spectrometers. Nonthesis MS option available.

University of Montana, College of Arts and Sciences, Department of Chemistry, Missoula, MT 59812. Department offers programs in chemistry (MS, PhD), chemistry teaching (MST), environmental chemistry (PhD), inorganic chemistry (PhD), organic chemistry (PhD), physical chemistry (PhD). Part-time programs available. Terminal master's awarded for partial completion of doctoral program. *Degree requirements:* For master's, 1 foreign language, thesis (for some programs); for doctorate, 1 foreign language, dissertation. *Entrance requirements:* For master's, GRE General Test; for doctorate, GRE General Test, GRE Subject Test. Application deadline: 9/15. Application fee: $20. *Tuition:* $495 per quarter hour full-time for state residents; $1239 per quarter hour full-time for nonresidents. *Faculty research:* Reaction mechanisms and kinetics, excited electronic states, chemistry of natural products, control of steroid hormone biosynthesis, physical studies of biological macromolecules.

University of Nebraska–Lincoln, College of Arts and Sciences, Department of Chemistry, Lincoln, NE 68588. Department offers programs in analytical chemistry (PhD), chemistry (MS), inorganic chemistry (PhD), organic chemistry (PhD), physical chemistry (PhD). Faculty: 22 full-time (2 women), 0 part-time. Matriculated students: 96 full-time (21 women), 17 part-time (5 women); includes 10 minority (7 Asian American, 3 black American), 41 foreign. Average age 29. In 1990, 8 master's, 4 doctorates awarded. Terminal master's awarded for partial completion of doctoral program. *Degree requirements:* For master's, 1 foreign language, thesis (for some programs), departmental qualifying exam; for doctorate, 1 foreign language, dissertation, comprehensive exams, departmental qualifying exam. *Entrance requirements:* GRE Subject Test, TOEFL (minimum score of 550 required), TSE (minimum score of 230 required). Application deadline: 5/1 (priority date, applications processed on a rolling basis). Application fee: $25. *Expenses:* Tuition of $75.75 per credit hour for state residents; $187.25 per credit hour for nonresidents. Fees of $161 per year full-time. *Financial aid:* Fellowships, research assistantships, teaching assistantships, federal work-study available. Aid available to part-time students. Financial aid application deadline: 2/15. • Dr. Pill-Soon Song Jr., Chairperson, 402-472-3501.

See full description on page 165.

University of Nevada, Las Vegas, College of Science and Mathematics, Department of Chemistry, Las Vegas, NV 89154. Department offers program in environmental analytical chemistry (MS). Faculty: 8 full-time (0 women), 0 part-time. Matriculated students: 7 (2 women); includes 0 minority, 0 foreign. 2 applicants, 50% accepted. In 1990, 1 degree awarded. *Degree requirements:* Foreign language not required. *Entrance requirements:* GRE General Test, minimum GPA of 2.75. Application deadline: 6/15. Application fee: $20. *Expenses:* Tuition of $66 per credit. Fees of $1800 per semester for nonresidents. *Financial aid:* In 1990–91, 4 research assistantships awarded; teaching assistantships also available. Financial aid application deadline: 3/1. • Dr. Boyd Earl, Chair, 702-739-3510. Application contact: Graduate Coordinator, 702-739-3753.

University of Nevada, Reno, College of Arts and Science, Department of Chemistry, Reno, NV 89557. Department awards MS, PhD. Faculty: 15 (0 women). Matriculated students: 39 (7 women). Average age 26. In 1990, 2 master's, 3 doctorates awarded. Terminal master's awarded for partial completion of doctoral program. *Degree requirements:* For master's, thesis optional, foreign language not required; for doctorate, 1 foreign language, dissertation. *Entrance requirements:* For master's,

Directory: Chemistry

GRE General Test, GRE Subject Test, TOEFL, minimum GPA of 2.75; for doctorate, GRE General Test, GRE Subject Test, TOEFL, minimum GPA of 3.0. Application deadline: 8/1 (priority date, applications processed on a rolling basis). Application fee: $20. *Expenses:* Tuition of $0 for state residents; $3600 per year full-time, $66 per credit hour part-time for nonresidents. Fees of $66 per credit hour. *Financial aid:* In 1990–91, 13 research assistantships, 22 teaching assistantships awarded; federal work-study, institutionally sponsored loans also available. Average monthly stipend for a graduate assistantship: $1050. Financial aid applicants required to submit FAF. *Faculty research:* Synthetic and mechanistic organic, inorganic and organometallic, experimental and theoretical physical, bio-organic chemistry, chemical physics. Total annual research budget: $1.336-million. • Dr. Lawrence T. Scott, Chairman, 702-784-6041.

University of New Brunswick, Faculty of Science, Department of Chemistry, Fredericton, NB E3B 5A3, Canada. Department awards M Sc, PhD. *Degree requirements:* Thesis/dissertation. *Entrance requirements:* TOEFL, minimum GPA of 3.0. Application deadline: 3/1 (priority date). *Expenses:* Tuition of $2100 per year. Fees of $45 per year. *Financial aid:* Research assistantships, teaching assistantships available. • Dr. D. G. Brewer, Chairperson, 506-453-4774. Application contact: Zdenek Valenta, Director of Graduate Studies, 506-453-4781.

University of New Hampshire, College of Engineering and Physical Sciences, Department of Chemistry, Durham, NH 03824. Department awards MS, MST, PhD. Faculty: 16 full-time. Matriculated students: 26 full-time (10 women), 19 part-time (7 women); includes 13 foreign. 82 applicants, 34% accepted. In 1990, 8 master's, 4 doctorates awarded. Terminal master's awarded for partial completion of doctoral program. *Degree requirements:* For master's, thesis required, foreign language not required; for doctorate, 1 foreign language (computer language can substitute), dissertation. Application deadline: 7/1 (priority date, applications processed on a rolling basis). Application fee: $25. *Tuition:* $1645 per semester full-time, $183 per credit hour part-time for state residents; $4920 per semester full-time, $547 per credit hour part-time for nonresidents. *Financial aid:* In 1990–91, 0 fellowships, 12 research assistantships, 30 teaching assistantships awarded; full and partial tuition waivers, federal work-study also available. Aid available to part-time students. Financial aid application deadline: 2/15. *Faculty research:* Analytical, physical, organic, and inorganic chemistry. • Dr. Frank L. Pilar, Chairperson, 603-862-1550. Application contact: Dr. Howard Mayne, 603-862-1550.

University of New Mexico, College of Arts and Sciences, Department of Chemistry, Albuquerque, NM 87131. Department awards MS, PhD. Faculty: 18 full-time (10 women), 14 part-time (0 women). *Degree requirements:* For master's, thesis or alternative required, foreign language not required; for doctorate, dissertation required, foreign language not required. *Entrance requirements:* For master's, TOEFL (minimum score of 520 required). Application fee: $25. *Expenses:* Tuition of $467 per semester (minimum) full-time, $67.50 per credit hour part-time for state residents; $1549 per semester (minimum) full-time, $67.50 per credit hour part-time for nonresidents. Fees of $16 per semester. *Faculty research:* Analytical, inorganic, organic, and physical chemistry; biochemistry. • Richard Holder, Chairman, 505-277-6555.

University of New Orleans, College of Sciences, Department of Chemistry, New Orleans, LA 70148. Department awards MS, PhD. Faculty: 20 full-time (0 women), 0 part-time. Matriculated students: 34. In 1990, 2 master's awarded (50% entered university research/teaching, 50% found other work related to degree); 1 doctorate awarded. *Degree requirements:* Variable foreign language requirement, thesis/dissertation, departmental qualifying exam. *Entrance requirements:* GRE General Test. Application deadline: 7/1 (priority date, applications processed on a rolling basis). Application fee: $20. *Tuition:* $962 per quarter hour full-time for state residents; $2308 per quarter hour full-time for nonresidents. *Faculty research:* Synthesis and reactions of novel compounds, high-temperature kinetics, calculations of molecular electrostatic potentials, structures and reactions of metal complexes. Total annual research budget: $1.0-million. • Dr. Ralph Kern, Chairman, 504-286-6847.

University of North Carolina at Chapel Hill, College of Arts and Sciences, Department of Chemistry, Chapel Hill, NC 27599. Department awards MA, MS, PhD. Faculty: 38 full-time, 0 part-time. Matriculated students: 236 full-time (87 women), 0 part-time; includes 38 foreign. 322 applicants, 44% accepted. In 1990, 15 master's, 34 doctorates awarded. *Degree requirements:* For master's, thesis (for some programs), comprehensive exam required, foreign language not required; for doctorate, dissertation, comprehensive exam required, foreign language not required. *Entrance requirements:* GRE General Test (minimum combined score of 1000 required), GRE Subject Test, minimum GPA of 3.0. Application deadline: 2/15. Application fee: $35. *Tuition:* $621 per semester full-time for state residents; $3555 per semester full-time for nonresidents. *Financial aid:* In 1990–91, 10 fellowships, 130 research assistantships, 62 teaching assistantships awarded. • Dr. Joesph L. Templeton, Chairman, 919-962-4575. Application contact: Thomas Baer, Director of Graduate Studies, 919-967-4097.

University of North Carolina at Charlotte, College of Arts and Sciences, Department of Chemistry, Charlotte, NC 28223. Department awards MS. Part-time programs available. Faculty: 12 full-time (1 woman), 0 part-time. Matriculated students: 1 (woman) full-time, 11 part-time (6 women); includes 2 minority (1 Asian American, 1 black American), 6 foreign. Average age 27. 16 applicants, 88% accepted. In 1990, 3 degrees awarded. *Degree requirements:* 1 foreign language, thesis. *Entrance requirements:* GRE General Test or MAT, minimum GPA of 2.5 overall, 3.0 in undergraduate major. Application deadline: 7/1. Application fee: $15. *Tuition:* $574.50 per semester full-time for state residents; $3105 per semester full-time for nonresidents. *Financial aid:* In 1990–91, 0 research assistantships, 11 teaching assistantships (1 to a first-year student) awarded; federal work-study and career-related internships or fieldwork also available. Financial aid application deadline: 4/15; applicants required to submit FAF. *Faculty research:* Inorganic chemistry, analytical chemistry, biotechnology. • Dr. Paul Rillema, Chairman, 704-547-4765. Application contact: Kathi M. Baucom, Director of Admissions, 704-547-2213.

University of North Carolina at Greensboro, College of Arts and Sciences, Department of Chemistry, Greensboro, NC 27412. Department awards M Ed, MS. Faculty: 13 full-time (1 woman), 0 part-time. Matriculated students: 17 (6 women); includes 6 minority (5 Asian American, 1 black American). *Degree requirements:* 1 foreign language, thesis. *Entrance requirements:* For MS: GRE General Test; for M Ed: GRE General Test or NTE. Application fee: $35. *Tuition:* $751 per semester full-time for state residents; $3685 per semester full-time for nonresidents. *Faculty research:* Synthesis of novel cyclopentadienes, molybdenum hydroxylase-cata ladder polymers, vinyl silicones. • Dr. Michael Farona, Head, 919-334-5714.

University of North Carolina at Wilmington, College of Arts and Sciences, Department of Chemistry, Wilmington, NC 28403. Department awards MS. Part-time programs available. Faculty: 4 full-time (1 woman), 1 (woman) part-time. Matriculated students: 2 full-time (1 woman), 4 part-time (3 women); includes 1 minority (Native American), 0 foreign. Average age 31. 2 applicants, 100% accepted. In 1990, 0 degrees awarded. *Degree requirements:* Thesis, written and oral comprehensive exams. *Entrance requirements:* GRE General Test (minimum combined score of 1000 required), GRE Subject Test, minimum B average in undergraduate major. Application deadline: 7/1 (applications processed on a rolling basis). Application fee: $15. *Tuition:* $651 per semester full-time for state residents; $3178 per semester full-time for nonresidents. *Financial aid:* In 1990–91, $5000 in aid awarded. 5 teaching assistantships were awarded; federal work-study and career-related internships or fieldwork also available. Aid available to part-time students. Financial aid application deadline: 3/15; applicants required to submit FAF. • Dr. Jack Levy, Chairman, 919-395-3450. Application contact: Dr. Eric G. Bolen, Dean of Graduate School, 919-395-3135.

University of North Dakota, College of Arts and Sciences, Department of Chemistry, Grand Forks, ND 58202. Department awards MS, PhD. Faculty: 13 full-time (0 women), 0 part-time. Matriculated students: 20 full-time (3 women), 0 part-time. 19 applicants, 63% accepted. In 1990, 4 master's, 3 doctorates awarded. Terminal master's awarded for partial completion of doctoral program. *Degree requirements:* For master's, thesis; for doctorate, 1 foreign language, dissertation. *Entrance requirements:* GRE, TOEFL (minimum score of 550 required), minimum GPA of 3.0. Application deadline: 3/15 (priority date, applications processed on a rolling basis). Application fee: $20. *Tuition:* $2250 per year full-time, $94 per semester hour part-time for state residents; $5616 per year full-time, $234 per semester hour part-time for nonresidents. *Financial aid:* In 1990–91, 20 students received aid. 1 research assistantship, 17 teaching assistantships awarded; fellowships, full and partial tuition waivers, federal work-study, institutionally sponsored loans also available. Average monthly stipend for a graduate assistantship: $900. Financial aid application deadline: 3/15. • Dr. L. Radonovich, Chairperson, 701-777-2741.

University of Northern Colorado, College of Arts and Sciences, Department of Chemistry, Greeley, CO 80639. Department offers programs in chemical education (PhD), chemistry (MA). Faculty: 5 full-time (0 women), 3 part-time (0 women). Matriculated students: 7 full-time (1 woman), 1 part-time (0 women); includes 0 minority, 4 foreign. 11 applicants, 82% accepted. In 1990, 1 master's, 0 doctorates awarded. *Degree requirements:* For master's, thesis or alternative, comprehensive exams; for doctorate, dissertation, comprehensive exams. *Entrance requirements:* For doctorate, GRE General Test. Application deadline: rolling. Application fee: $30. *Expenses:* Tuition of $1900 per year full-time, $106 per credit hour part-time for state residents; $6078 per year full-time, $338 per credit hour part-time for nonresidents. Fees of $320 per year full-time, $18 per credit hour part-time. *Financial aid:* In 1990–91, 1 fellowship, 5 graduate assistantships (1 to a first-year student) awarded; research assistantships, teaching assistantships also available. Financial aid application deadline: 3/1. • Dr. David Pringle, Chairperson, 303-351-1292. Application contact: Richard Hyslop, Graduate Coordinator, 303-351-1288.

University of Northern Iowa, College of Natural Sciences, Department of Chemistry, Cedar Falls, IA 50614. Department awards MA. Part-time programs available. Faculty: 7 full-time (0 women), 0 part-time. Matriculated students: 2 full-time (0 women), 1 part-time (0 women); includes 0 minority, 0 foreign. Average age 33. 3 applicants, 67% accepted. In 1990, 1 degree awarded. *Degree requirements:* Thesis or alternative required, foreign language not required. *Entrance requirements:* GRE. Application deadline: 8/1 (priority date, applications processed on a rolling basis). Application fee: $20. *Tuition:* $2192 per year full-time, $240 per hour part-time for state residents; $5492 per year full-time, $240 per hour part-time for nonresidents. *Financial aid:* In 1990–91, 1 scholarship awarded; full and partial tuition waivers, federal work-study, and career-related internships or fieldwork also available. Aid available to part-time students. Financial aid application deadline: 3/1; applicants required to submit FAF. • Dr. LeRoy McGrew, Head, 319-273-2437.

University of North Texas, College of Arts and Sciences, Department of Chemistry, Denton, TX 76203. Department offers programs in biochemistry (MS, PhD), chemistry (MS, PhD). Faculty: 16 full-time, 0 part-time. Matriculated students: 57 full-time (14 women), 8 part-time (4 women); includes 3 minority (1 black American, 2 Hispanic American), 31 foreign. Average age 27. 150 applicants, 11% accepted. In 1990, 10 master's awarded (60% found work related to degree, 40% continued full-time study); 5 doctorates awarded (60% entered university research/teaching, 40% found other work related to degree). Terminal master's awarded for partial completion of doctoral program. *Degree requirements:* For master's, comprehensive exam required, foreign language not required; for doctorate, 1 foreign language, dissertation, comprehensive exams. *Entrance requirements:* For master's, GRE General Test, placement exams in 3 areas; for doctorate, GRE General Test, placement exams in 4 areas. Application deadline: 8/1. Application fee: $25. *Expenses:* Tuition of $40 per credit hour for state residents; $128 per credit hour for nonresidents. Fees of $298 per year full-time, $38 per year part-time. *Financial aid:* In 1990–91, $256,550 in aid awarded. 19 fellowships (1 to a first-year student), 18 research assistantships (1 to a first-year student), 26 teaching assistantships (10 to first-year students) were awarded; federal work-study, institutionally sponsored loans, and career-related internships or fieldwork also available. Financial aid application deadline: 4/1. *Faculty research:* Analytical, inorganic, physical, and organic chemistry. Total annual research budget: $3-million. • Dr. Jim Marshall, Acting Chair, 817-565-2713. Application contact: Dr. Martin Schwartz, Graduate Adviser, 817-565-2713.

University of Notre Dame, College of Science, Department of Chemistry and Biochemistry, Notre Dame, IN 46556. Department offers programs in biochemistry (MS, PhD), inorganic chemistry (MS, PhD), organic chemistry (MS, PhD), physical chemistry (MS, PhD). Faculty: 31 full-time (2 women), 0 part-time. Matriculated students: 99 full-time (31 women), 1 part-time (0 women); includes 3 minority (1 black American, 2 Hispanic American), 45 foreign. Average age 25. 194 applicants, 27% accepted. In 1990, 4 master's awarded (25% found work related to degree, 75% continued full-time study); 10 doctorates awarded (100% found work related to degree). Terminal master's awarded for partial completion of doctoral program. *Degree requirements:* For master's, thesis, comprehensive exam required, foreign language not required; for doctorate, dissertation, qualifying exam required, foreign language not required. *Entrance requirements:* GRE, TOEFL. Application deadline: 2/15 (priority date, applications processed on a rolling basis). Application fee: $25. *Tuition:* $13,385 per year full-time, $744 per credit hour part-time. *Financial aid:* In 1990–91, $1.049-million in aid awarded. 14 fellowships (4 to first-year students), 42 research assistantships (0 to first-year students), 37 teaching assistantships (21 to first-year students) were awarded; full and partial tuition waivers also available. Financial aid application deadline: 2/15. *Faculty research:* Protein, carbohydrate and lipid metabolism, structure and function; synthesis, structure, and reactivity of organometallic and cluster complexes; synthesis and structure determination of novel compounds. Total annual research budget: $1.6-million. • Dr. Paul M. Helquist, Chairman, 219-239-7487. Application contact: Dr. Michael Chetcuti, Coordinator, 219-239-5216.

University of Oklahoma, College of Arts and Sciences, Department of Chemistry and Biochemistry, Norman, OK 73019. Department awards MS, PhD. Part-time programs available. Faculty: 24 full-time, 0 part-time. Matriculated students: 88 full-time (29 women), 6 part-time (1 woman); includes 27 minority (25 Asian American, 1 black American, 1 Native American), 54 foreign. Average age 27. 154 applicants, 53% accepted. In 1990, 14 master's, 12 doctorates awarded. Terminal master's awarded for partial completion of doctoral program. *Degree requirements:* For master's, thesis optional; for doctorate, dissertation. *Entrance requirements:* For master's, GRE, TOEFL (minimum score of 550 required), BS in chemistry; for doctorate, GRE, TOEFL (minimum score of 550 required). Application deadline: 4/1 (priority date, applications processed on a rolling basis). Application fee: $10. *Expenses:* Tuition of $63 per credit hour for state residents; $192 per credit hour for nonresidents. Fees of $67.50 per semester. *Financial aid:* In 1990–91, $300,000 in aid awarded. 0 fellowships, 35 research assistantships (2 to first-year students), 40 teaching assistantships (15 to first-year students) were awarded; federal work-study, institutionally sponsored loans also available. Financial aid application deadline: 4/1. *Faculty research:* Analytic, organic, physical, inorganic chemistry, biochemistry. Total annual research budget: $2.2-million. • Dr. Glenn Dryhurst, Chair, 405-325-4811. Application contact: Dr. Stan Neely, Chair, Admissions Committee, 405-325-2946.

Announcement: The Department of Chemistry and Biochemistry offers programs leading to the MS and PhD degrees. Areas of specialization include analytical chemistry, biochemistry, inorganic chemistry, organic chemistry, physical chemistry, and chemical education. A moderate-size faculty (24) and graduate student body (about 100) ensure an individualized program of learning and research. A minimum of 21 hours of graduate course work, a general exam in the student's area of specialization, and an orginal research proposal that complements the dissertation research are required for the PhD.

University of Oklahoma, College of Arts and Sciences, Interdepartmental Program in Cellular and Molecular Biology, Norman, OK 73019. Offerings include chemistry (PhD), with option in biochemistry. *Degree requirements:* Dissertation. *Entrance requirements:* GRE General Test, GRE Subject Test, TOEFL (minimum score of 550 required). Application deadline: 2/15 (priority date, applications processed on a rolling basis). Application fee: $10. *Expenses:* Tuition of $63 per credit hour for state residents; $192 per credit hour for nonresidents. Fees of $67.50 per semester. • Application contact: Dr. Paul B. Bell Jr., Chair, Graduate Selections Committee, 405-325-4821.

University of Oregon, Graduate School, College of Arts and Sciences, Department of Chemistry, Eugene, OR 97403. Offerings include chemistry (MA, MS, PhD). Terminal master's awarded for partial completion of doctoral program. Department faculty: 26 full-time (3 women), 0 part-time. *Degree requirements:* For master's, foreign language and thesis not required; for doctorate, dissertation required, foreign language not required. *Entrance requirements:* GRE General Test, TOEFL (minimum score of 600 required). Application deadline: 4/15 (priority date, applications processed on a rolling basis). Application fee: $40. *Tuition:* $1171 per quarter full-time, $247 per credit part-time for state residents; $1980 per quarter full-time, $336 per credit part-time for nonresidents. • David Herrick, Head, 503-346-4601. Application contact: Marilyn Evans, Graduate Secretary, 800-782-4713.

University of Ottawa, Faculty of Science, Department of Chemistry, Ottawa, ON K1N 6N5, Canada. Department awards M Sc, PhD. Offered jointly with Carleton University. *Degree requirements:* Thesis/dissertation required, foreign language not required. *Entrance requirements:* For master's, honors bachelor's degree or equivalent, minimum B average; for doctorate, minimum B+ average. Application deadline: 3/1. Application fee: $10.

University of Pennsylvania, School of Arts and Sciences, Graduate Group in Chemistry, Philadelphia, PA 19104. Group awards MS, PhD. Faculty: 29 (2 women). Matriculated students: 159 full-time (61 women), 5 part-time (2 women). 385 applicants, 30% accepted. In 1990, 3 master's, 20 doctorates awarded. Terminal master's awarded for partial completion of doctoral program. *Degree requirements:* For master's, thesis or alternative required, foreign language not required; for doctorate, dissertation required, foreign language not required. *Entrance requirements:* TOEFL, previous course work in chemistry. Application fee: $40. *Expenses:* Tuition of $15,619 per year full-time, $1978 per course part-time. Fees of $965 per year full-time, $112 per course part-time. *Financial aid:* In 1990–91, 8 fellowships (4 to first-year students), 100 research assistantships (0 to first-year students), 45 teaching assistantships (35 to first-year students) awarded. Average monthly stipend for a graduate assistantship: $1050. Financial aid application deadline: 2/1. *Faculty research:* Biological chemistry, inorganic chemistry, materials chemistry, organic chemistry, experimental chemistry, theoretical physical chemistry. Total annual research budget: $8.0-million. • Amos Smith, Chairman, 215-898-8317.

University of Pittsburgh, Faculty of Arts and Sciences, Department of Chemistry, Pittsburgh, PA 15260. Department awards MS, PhD. Part-time programs available. Faculty: 37 full-time (2 women), 2 part-time (0 women). Matriculated students: 174 full-time (48 women), 17 part-time (6 women); includes 3 minority (2 Asian American, 1 black American), 63 foreign. 351 applicants, 12% accepted. In 1990, 6 master's, 19 doctorates awarded. Terminal master's awarded for partial completion of doctoral program. *Degree requirements:* Thesis/dissertation required, foreign language not required. *Entrance requirements:* GRE General Test, GRE Subject Test, TOEFL. Application deadline: 3/31 (priority date, applications processed on a rolling basis). Application fee: $15 ($25 for foreign students). *Expenses:* Tuition of $2920 per semester full-time, $241 per credit part-time for state residents; $5840 per semester full-time, $482 per credit part-time for nonresidents. Fees of $156 per year. *Financial aid:* In 1990–91, 189 students received a total of $2.140-million in aid awarded. 16 fellowships (0 to first-year students), 89 research assistantships (1 to a first-year student), 84 teaching assistantships (40 to first-year students) were awarded; full tuition waivers, federal work-study, institutionally sponsored loans also available. Aid available to part-time students. Financial aid application deadline: 3/15; applicants required to submit FAF. *Faculty research:* Analytical chemistry, inorganic chemistry, organic chemistry, physical chemistry and surface science. Total annual research budget: $7.657-million. • Dr. N. John Cooper, Chairman, 412-624-8500. Application contact: Rebecca Claycamp, Assistant Chairman, 412-624-8200.

See full description on page 167.

University of Puerto Rico, Mayagüez Campus, College of Arts and Sciences, Department of Chemistry, Mayagüez, PR 00709. Department awards MS. Faculty: 21 full-time (9 women), 0 part-time. Matriculated students: 20 full-time (7 women), 0 part-time; includes 13 minority (all Hispanic American), 7 foreign. 15 applicants, 67% accepted. In 1990, 5 degrees awarded. *Degree requirements:* 1 foreign language, thesis. Application deadline: 10/15. Application fee: $15. *Expenses:* Tuition of $45 per credit for commonwealth residents; $3000 per semester for nonresidents. Fees of $344 per semester. *Financial aid:* In 1990–91, 0 research assistantships, 11 teaching assistantships (4 to first-year students) awarded; federal work-study, institutionally sponsored loans also available. *Faculty research:* Biochemistry, spectroscopy, food chemistry, physical chemistry, electrochemistry. • Sylvia Pirazzi, Director, 809-832-4040 Ext. 3849.

University of Puerto Rico, Río Piedras, Faculty of Natural Sciences, Department of Chemistry, Río Piedras, PR 00931. Department awards MS, PhD. Part-time and evening/weekend programs available. Faculty: 34 (10 women). Matriculated students: 36 full-time (27 women), 36 part-time (23 women); includes 8 foreign. In 1990, 7 master's, 6 doctorates awarded. *Degree requirements:* Thesis/dissertation, comprehensive exam required, foreign language not required. *Entrance requirements:* GRE, minimum GPA of 2.75. Application deadline: 2/1. Application fee: $45. *Expenses:* Tuition of $55 per credit hour for commonwealth residents; $55 per credit hour (minimum) for nonresidents. Fees of $286 per year. *Financial aid:* Fellowships, research assistantships, teaching assistantships, partial tuition waivers, federal work-study, institutionally sponsored loans available. Financial aid application deadline: 5/31. *Faculty research:* Organometallic synthesis, transition metal chemistry, organic air pollutants, acylmetalloid. • Dr. Lillian Bird, Coordinator, 809-764-0000 Ext. 4782.

See full description on page 169.

University of Regina, Faculty of Graduate Studies and Research, Faculty of Science, Department of Chemistry, Regina, SK S4S 0A2, Canada. Offerings include X-ray crystallography (M Sc, PhD). Department faculty: 14 full-time, 0 part-time. *Degree requirements:* For master's, thesis, departmental qualifying exam required, foreign language not required; for doctorate, variable foreign language requirement, dissertation, departmental qualifying exam. *Application deadline:* 7/2 (applications processed on a rolling basis). *Tuition:* $1500 per year full-time, $242 per year (minimum) part-time. • Dr. W. D. Chandler, Head, 306-585-4146.

University of Rhode Island, College of Arts and Sciences, Department of Chemistry, Kingston, RI 02881. Department awards MS, PhD. Application deadline: 4/15. Application fee: $25. *Expenses:* Tuition of $2575 per year full-time, $120 per credit hour part-time for state residents; $5900 per year full-time, $274 per credit hour part-time for nonresidents. Fees of $696 per year full-time.

University of Rochester, College of Arts and Science, Department of Chemistry, Rochester, NY 14627-0001. Department awards MS, PhD. Part-time programs available. Faculty: 23 full-time, 3 part-time. Matriculated students: 122 full-time (43 women), 0 part-time; includes 9 minority (5 Asian American, 2 black American, 2 Hispanic American), 28 foreign. 251 applicants, 33% accepted. In 1990, 26 master's, 18 doctorates awarded. Terminal master's awarded for partial completion of doctoral program. *Degree requirements:* For master's, thesis not required; for doctorate, dissertation required, foreign language not required. *Entrance requirements:* For doctorate, GRE General Test, GRE Subject Test, TOEFL. Application deadline: 2/15 (priority date). Application fee: $25. *Expenses:* Tuition of $473 per credit hour. Fees of $243 per year full-time. *Financial aid:* Fellowships, research assistantships, teaching assistantships, federal work-study available. Financial aid application deadline: 2/15. *Faculty research:* Organic, inorganic, and physical chemistry. • Dr. David Whitten, Chair, 716-275-2525.

University of San Francisco, College of Arts and Sciences, Department of Chemistry, San Francisco, CA 94117-1080. Department offers programs in analytical chemistry (MS), biochemistry (MS), inorganic chemistry (MS), organic chemistry (MS), physical chemistry (MS). Part-time and evening/weekend programs available. Faculty: 6 full-time (0 women), 0 part-time. Matriculated students: 0 full-time, 8 part-time (5 women). In 1990, 2 degrees awarded (50% entered university research/teaching, 50% found other work related to degree). *Degree requirements:* 1 foreign language, thesis. *Entrance requirements:* TOEFL (minimum score of 550 required). Application deadline: 5/15 (priority date). Application fee: $40 ($90 for foreign students). *Tuition:* $10,960 per year full-time, $432 per unit part-time. *Financial aid:* In 1990–91, $30,000 in aid awarded. 12 fellowships (5 to first-year students), 3 research assistantships (2 to first-year students), 7 teaching assistantships (0 to first-year students) were awarded; federal work-study, institutionally sponsored loans, and career-related internships or fieldwork also available. Aid available to part-time students. Financial aid application deadline: 3/1. *Faculty research:* Microbial energetics, genetics of chromatic adaptation, electron-transfer processes in solution, metabolism of protein hormones. • Tom Gruhn, Chairman, 415-666-6486. Application contact: Jeff Curtis, Graduate Adviser, 415-666-6391.

See full description on page 171.

University of Saskatchewan, College of Arts and Sciences, Department of Chemistry, Saskatoon, SK S7N 0W0, Canada. Department awards M Sc, PhD. *Degree requirements:* Thesis/dissertation. *Entrance requirements:* TOEFL. Application fee: $0.

University of Scranton, Department of Chemistry, Scranton, PA 18510-4501. Department offers programs in biochemistry (MA, MS), chemistry (MA, MS), clinical chemistry (MA, MS). Part-time and evening/weekend programs available. Faculty: 10 full-time (3 women), 0 part-time. Matriculated students: 20 full-time (9 women), 23 part-time (12 women); includes 15 foreign. 44 applicants, 80% accepted. In 1990, 10 degrees awarded. *Degree requirements:* Thesis (for some programs), comprehensive exam required, foreign language not required. *Entrance requirements:* TOEFL (minimum score of 500 required), minimum GPA of 2.75. Application deadline: rolling. Application fee: $35. *Expenses:* Tuition of $283 per credit. Fees of $25 per semester. *Financial aid:* In 1990–91, 6 teaching assistantships (4 to first-year students), 13 teaching fellowships (6 to first-year students) awarded; federal work-study and career-related internships or fieldwork also available. Aid available to part-time students. Average monthly stipend for a graduate assistantship: $500. Financial aid application deadline: 3/15; applicants required to submit GAPSFAS. *Faculty research:* Heterogeneous equilibria, microbial extracellular proteases, use of proteins and peptides, analytical kinetic analysis, DNA-protein interactions. • Dr. Maurice I. Hart, Chair, 717-941-7511. Application contact: Dr. Joseph H. Dreisbach, Director, 717-941-7519.

University of South Carolina, Graduate School, College of Science and Mathematics, Department of Chemistry, Columbia, SC 29208. Department awards MS, PhD. Part-time programs available. Faculty: 34 full-time (4 women), 0 part-time. Matriculated students: 132 full-time (33 women), 0 part-time; includes 12 minority (6 Asian American, 3 black American, 3 Hispanic American), 34 foreign. 354 applicants, 20% accepted. In 1990, 3 master's, 24 doctorates awarded. Terminal master's awarded for partial completion of doctoral program. *Degree requirements:* Thesis/dissertation required, foreign language not required. *Entrance requirements:* GRE General Test. Application deadline: 4/15 (applications processed on a rolling basis). Application fee: $25. *Tuition:* $1404 per semester full-time. *Financial aid:* In 1990–91, 6 fellowships (all to first-year students), 54 teaching assistantships (26 to first-year students) awarded; full tuition waivers, federal work-study, institutionally sponsored loans also available. Average monthly stipend for a graduate assistantship: $917. Financial aid application deadline: 4/15; applicants required to submit FAF. *Faculty research:* Biochemistry, spectroscopy, crystallography, organic and organometallic synthesis. Total annual research budget: $1.70-million. • Dr. Jerome D. Odom,

SECTION 2: CHEMISTRY

Directory: Chemistry

Chairman, 803-777-5263. Application contact: Dr. John Dawson, Chairman, Graduate Admissions, 803-777-7234.

University of South Dakota, College of Arts and Sciences, Department of Chemistry, Vermillion, SD 57069-2390. Department awards MA, MNS. Faculty: 6 full-time (0 women), 1 part-time (0 women). Matriculated students: 6 full-time (1 woman), 1 part-time (0 women); includes 0 minority, 0 foreign. 7 applicants, 43% accepted. In 1990, 0 degrees awarded. *Degree requirements:* Thesis required, foreign language not required. *Application fee:* $15. *Expenses:* Tuition of $61 per hour for state residents; $122 per hour for nonresidents. Fees of $24.41 per hour. *Financial aid:* Teaching assistantships available. Aid available to part-time students. Financial aid applicants required to submit GAPSFAS or FAF. *Faculty research:* Electrochemistry, photochemistry, inorganic synthesis, environmental and solid-state chemistry. • Dr. Royce C. Engstrom, Chairman, 605-677-5487.

University of Southern California, Graduate School, College of Letters, Arts and Sciences, Division of Natural Sciences and Mathematics, Department of Chemistry, Program in Chemistry, Los Angeles, CA 90089. Program awards MA, MS, PhD. Matriculated students: 95 full-time (31 women), 2 part-time (0 women); includes 11 minority (9 Asian American, 1 black American, 1 Native American), 57 foreign. Average age 25. 96 applicants, 47% accepted. In 1990, 7 master's, 15 doctorates awarded. *Degree requirements:* For doctorate, dissertation. *Entrance requirements:* GRE General Test. Application deadline: 4/1 (priority date). Application fee: $50. *Expenses:* Tuition of $12,120 per year full-time, $505 per unit part-time. Fees of $280 per year. *Financial aid:* 95 students received aid. Fellowships, research assistantships, teaching assistantships, federal work-study, institutionally sponsored loans available. Average monthly stipend for a graduate assistantship: $878. Financial aid application deadline: 3/1. • Dr. Otto Schnepp, Chairman, Department of Chemistry, 213-740-7036.

University of Southern Mississippi, College of Science and Technology, Department of Chemistry and Biochemistry, Hattiesburg, MS 39406. Department offers programs in analytical chemistry (MS, PhD), biochemistry (MS, PhD), inorganic chemistry (MS, PhD), organic chemistry (MS, PhD), physical chemistry (MS, PhD). Faculty: 18 full-time (2 women), 1 (woman) part-time. Matriculated students: 30 full-time (11 women), 3 part-time (0 women); includes 2 minority (both black American), 13 foreign. In 1990, 1 master's awarded (100% continued full-time study); 5 doctorates awarded (100% found work related to degree). *Degree requirements:* For doctorate, 2 foreign languages (computer language can substitute for one), dissertation. *Entrance requirements:* For master's, GRE General Test, minimum GPA of 2.75; for doctorate, GRE General Test, minimum GPA of 3.5. Application deadline: 8/9 (priority date, applications processed on a rolling basis). Application fee: $0 ($25 for foreign students). *Expenses:* Tuition of $968 per semester full-time, $93 per semester hour part-time. Fees of $12 per semester part-time for state residents; $591 per year full-time, $12 per semester part-time for nonresidents. *Financial aid:* Application deadline 3/15. • Dr. David Creed, Chair, 601-266-4701. Application contact: Dr. David Wertz, Professor, 601-266-4702.

University of South Florida, College of Arts and Sciences, Department of Chemistry, Tampa, FL 33620. Department offers programs in analytical chemistry (MS, PhD), biochemistry (MS, PhD), inorganic chemistry (MS, PhD), organic chemistry (MS, PhD), physical chemistry (MS, PhD). Part-time and evening/weekend programs available. Faculty: 26 full-time. Matriculated students: 43 full-time (13 women), 15 part-time (4 women); includes 9 minority (2 Asian American, 5 black American, 2 Hispanic American), 9 foreign. Average age 30. 133 applicants, 43% accepted. In 1990, 6 master's, 9 doctorates awarded. *Degree requirements:* For master's, thesis; for doctorate, 1 foreign language, computer language, dissertation. *Entrance requirements:* For master's, GRE General Test (minimum combined score of 1000 required), minimum GPA of 3.0 (in last 60 credit hours). Application fee: $15. *Tuition:* $79.40 per credit hour for state residents; $241.33 per credit hour for nonresidents. *Financial aid:* In 1990–91, 43 students received a total of $67,021 in aid awarded. Fellowships, research assistantships, teaching assistantships, institutionally sponsored loans available. • Stewart W. Schneller Jr., Chairperson, 813-974-2258. Application contact: Terence Oestereich, Graduate Program Director, 813-974-2534.

See full description on page 173.

University of Southwestern Louisiana, College of Sciences, Department of Chemistry, Lafayette, LA 70504. Department awards MS. Part-time programs available. Faculty: 8 full-time (2 women), 0 part-time. Matriculated students: 2 full-time (1 woman), 3 part-time (1 woman); includes 2 minority (both black American), 2 foreign. Average age 23. 14 applicants, 29% accepted. In 1990, 0 degrees awarded. *Degree requirements:* Thesis required, foreign language not required. *Entrance requirements:* GRE General Test. Application deadline: 8/15. Application fee: $5. *Tuition:* $1560 per year full-time, $228 per credit (minimum) part-time for state residents; $3310 per year full-time, $228 per credit (minimum) part-time for nonresidents. *Financial aid:* In 1990–91, $21,000 in aid awarded. 0 research assistantships, 4 teaching assistantships (1 to a first-year student) were awarded; federal work-study also available. Financial aid application deadline: 5/1. *Faculty research:* Synthesis, electrochemistry, theoretical chemometrics, proteins. Total annual research budget: $15,000. • Dr. Karen Wiechelman, Head, 318-231-6734.

University of Tennessee, Knoxville, College of Liberal Arts, Department of Chemistry, Knoxville, TN 37996. Department offers programs in analytical chemistry (MS, PhD), chemical physics (PhD), environmental chemistry (MS, PhD), inorganic chemistry (MS, PhD), organic chemistry (MS, PhD), physical chemistry (MS, PhD), polymer chemistry (MS, PhD), theoretical chemistry (PhD). Part-time programs available. Faculty: 36 (1 woman). Matriculated students: 61 full-time (20 women), 45 part-time (17 women); includes 1 minority (black American), 19 foreign. 77 applicants, 53% accepted. In 1990, 5 master's, 11 doctorates awarded. Terminal master's awarded for partial completion of doctoral program. *Degree requirements:* For master's, thesis required, foreign language not required; for doctorate, 1 foreign language, dissertation. *Entrance requirements:* GRE General Test, TOEFL (minimum score of 525 required), minimum GPA of 2.5. Application deadline: 2/1 (priority date, applications processed on a rolling basis). Application fee: $15. *Tuition:* $1086 per semester full-time, $142 per credit hour part-time for state residents; $2768 per semester full-time, $308 per credit hour part-time for nonresidents. *Financial aid:* In 1990–91, 3 fellowships, 32 research assistantships, 64 teaching assistantships awarded; federal work-study, institutionally sponsored loans also available. Financial aid application deadline: 2/1; applicants required to submit FAF. • Dr. Gleb Mamantov, Head, 615-974-3141.

University of Texas at Arlington, College of Science, Department of Chemistry, Arlington, TX 76019. Department awards MS, D Sc, PhD. Matriculated students: 39 full-time (10 women), 5 part-time (1 woman); includes 1 minority (Hispanic American), 27 foreign. 105 applicants, 19% accepted. In 1990, 6 master's, 5 doctorates awarded. *Degree requirements:* For master's, comprehensive exam. *Entrance requirements:* For master's, GRE General Test. Application deadline: rolling. Application fee: $25. *Tuition:* $40 per hour for state residents; $148 per hour for nonresidents. *Financial aid:* In 1990–91, 4 research assistantships, 15 teaching assistantships awarded. *Total annual research budget:* $750,473. • Dr. Richard B. Timmons, Chairman, 817-273-3171. Application contact: Dr. John Reynolds, Graduate Adviser, 817-273-3171.

University of Texas at Austin, Graduate School, College of Natural Sciences, Department of Chemistry and Biochemistry, Austin, TX 78712. Department offers programs in analytical chemistry (MA, PhD), biochemistry (MA, PhD), inorganic chemistry (MA, PhD), organic chemistry (MA, PhD), physical chemistry (MA, PhD). Matriculated students: 240 full-time (73 women), 0 part-time; includes 18 minority (10 Asian American, 2 black American, 4 Hispanic American, 2 Native American), 73 foreign. 197 applicants, 61% accepted. In 1990, 15 master's, 37 doctorates awarded. *Entrance requirements:* GRE. Application deadline: 2/1 (priority date, applications processed on a rolling basis). Application fee: $40 ($75 for foreign students). *Tuition:* $510.30 per semester for state residents; $1806 per semester for nonresidents. *Financial aid:* Fellowships, research assistantships, teaching assistantships available. Financial aid application deadline: 3/1. • Application contact: Dr. Peter J. Rossky, Graduate Adviser, 512-471-3890.

See full description on page 175.

University of Texas at Dallas, School of Natural Sciences and Mathematics, Program in Chemistry, Richardson, TX 75083-0688. Offerings include chemistry (MS). Program faculty: 9 full-time (0 women), 1 part-time (0 women). *Degree requirements:* Thesis or internship required, foreign language not required. *Entrance requirements:* GRE General Test (minimum combined score of 1000 required), TOEFL (minimum score of 550 required), minimum GPA of 3.0 in upper level course work in field. Application deadline: 7/15 (applications processed on a rolling basis). Application fee: $0 ($75 for foreign students). *Expenses:* Tuition of $360 per semester full-time, $100 per semester (minimum) part-time for state residents; $2196 per semester full-time, $122 per semester hour (minimum) part-time for nonresidents. Fees of $338 per semester full-time, $22 per hour (minimum) part-time. • Dr. A. Dean Sherry, Head, 214-690-2901.

University of Texas at El Paso, College of Science, Department of Chemistry, 500 West University Avenue, El Paso, TX 79968. Department awards MS. Matriculated students: 16 full-time (6 women), 6 part-time (2 women); includes 4 minority (all Hispanic American), 14 foreign. In 1990, 2 degrees awarded. *Application deadline:* 7/1 (priority date, applications processed on a rolling basis). *Application fee:* $0 ($50 for foreign students). *Expenses:* Tuition of $360 per semester full-time, $100 per semester (minimum) part-time for state residents; $2304 per semester full-time, $128 per credit hour (minimum) part-time for nonresidents. Fees of $137 per semester full-time, $28.50 per semester (minimum) part-time. *Financial aid:* Application deadline 3/1. • William Herndon, Chair, 915-747-5701. Application contact: Diana Guerrero, Admissions Office, 915-747-5576.

University of Texas at San Antonio, College of Sciences and Engineering, Division of Earth and Physical Sciences, San Antonio, TX 78285. Offerings include chemistry (MS). *Entrance requirements:* GRE. Application deadline: 7/1 (applications processed on a rolling basis). Application fee: $20. *Expenses:* Tuition of $100 per semester hour (minimum) for state residents; $128 per semester hour (minimum) for nonresidents. Fees of $48 per semester hour (minimum). • Dr. Robert K. Smith, Director, 512-691-4455.

University of Texas at Tyler, School of Sciences and Mathematics, Department of Chemistry, Tyler, TX 75701. Department offers program in interdisciplinary studies (MS). Faculty: 1 full-time (0 women), 0 part-time. Matriculated students: 4. In 1990, 0 degrees awarded. *Degree requirements:* Thesis not required. *Entrance requirements:* GRE General Test (minimum combined score of 1000 required), BS in chemistry. Application fee: $0. *Expenses:* Tuition of $760 per year full-time, $150 per course part-time for state residents; $3300 per year full-time, $410 per course part-time for nonresidents. Fees of $340 per year full-time, $30 per course part-time. *Financial aid:* Application deadline 7/1. • Dr. Don McClaugherty, Chair, 903-566-7402. Application contact: Martha D. Wheat, Director of Admissions, 903-566-7201.

University of the Pacific, Department of Chemistry, Stockton, CA 95211. Department offers programs in biochemistry (MS, PhD), chemistry (MS, PhD). Faculty: 10 full-time (0 women), 0 part-time. Matriculated students: 8 full-time (all women), 4 part-time (2 women); includes 6 foreign. Average age 25. In 1990, 4 master's, 1 doctorate awarded. Terminal master's awarded for partial completion of doctoral program. *Degree requirements:* For master's, thesis required, foreign language not required; for doctorate, 1 foreign language (computer language can substitute), dissertation. *Entrance requirements:* GRE General Test, GRE Subject Test. Application deadline: 5/1 (applications processed on a rolling basis). Application fee: $30. *Expenses:* Tuition of $14,160 per year full-time, $485 per unit part-time. Fees of $616 per unit. *Financial aid:* In 1990–91, $172,400 in aid awarded. 4 research assistantships, 14 teaching assistantships (6 to first-year students) were awarded; institutionally sponsored loans also available. Aid available to part-time students. Financial aid application deadline: 3/1; applicants required to submit FAF. • Dr. Michael J. Minch, Chairman, 209-946-2271.

University of Toledo, College of Arts and Sciences, Department of Chemistry, Toledo, OH 43606. Department offers programs in analytical chemistry (MS, PhD), biological chemistry (MS, PhD), inorganic chemistry (MS, PhD), organic chemistry (MS, PhD), physical chemistry (MS, PhD). Faculty: 20 full-time (2 women), 0 part-time. Matriculated students: 43 full-time (17 women), 5 part-time (2 women); includes 2 minority (1 black American, 1 Hispanic American), 28 foreign. Average age 28. 33 applicants, 36% accepted. In 1990, 3 master's awarded; 5 doctorates awarded (100% found work related to degree). *Degree requirements:* For master's, thesis required, foreign language not required; for doctorate, 1 foreign language, dissertation. *Entrance requirements:* GRE General Test, GRE Subject Test, TOEFL. Application deadline: 9/8 (priority date). Application fee: $30. *Tuition:* $122.59 per credit hour for state residents; $193.40 per credit hour for nonresidents. *Financial aid:* In 1990–91, 6 fellowships, 9 research assistantships, 35 teaching assistantships awarded. Financial aid application deadline: 4/1; applicants required to submit GAPSFAS or FAF. *Faculty research:* Viruses lipoxygenase, enzymology, pharmaceuticals. Total annual research budget: $290,632. • Dr. James L. Fry, Chairman, 419-537-2109. Application contact: Dr. Max Funk, Chairman, Graduate Admissions Committee, 419-537-2254.

University of Toronto, School of Graduate Studies, Physical Sciences Division, Department of Chemistry, Toronto, ON M5S 1A1, Canada. Department awards M Sc, PhD. Faculty: 50. Matriculated students: 116 full-time (30 women), 16 part-time (3 women); includes 18 foreign. 218 applicants, 40% accepted. In 1990, 24 master's, 12 doctorates awarded. *Degree requirements:* Thesis/dissertation. *Application deadline:* 4/15. *Application fee:* $50. *Expenses:* Tuition of $2220 per year full-time, $666 per year part-time for Canadian residents; $10,198 per year full-time, $305.05 per year part-time for nonresidents. Fees of $277.56 per year full-time, $82.73 per year part-time. *Financial aid:* Application deadline 2/1. • M. J. Dignam, Chair, 416-978-3566.

University of Utah, College of Science, Department of Chemistry, Salt Lake City, UT 84112. Department offers programs in chemical physics (PhD), chemistry (MA, M Phil, MS, PhD). Faculty: 27 (0 women). Matriculated students: 144 full-time (27 women), 7 part-time (4 women); includes 4 minority (2 Asian American, 1 Hispanic American, 1 Native American), 51 foreign. Average age 28. In 1990, 6 master's, 23 doctorates awarded. Terminal master's awarded for partial completion of doctoral program. *Degree requirements:* For master's, variable foreign language requirement, thesis or alternative; for doctorate, dissertation, exams required, foreign language not required. *Application fee:* $25 ($50 for foreign students). *Tuition:* $195 per credit for state residents; $505 per credit for nonresidents. *Financial aid:* In 1990–91, 53 teaching assistantships awarded. *Faculty research:* Theoretical, inorganic, organic, and physical-analytical chemistry. • Peter J. Stang, Chairman, 801-581-6681. Application contact: William H. Breckenridge, Director of Graduate Studies, 801-581-8024.

University of Vermont, College of Arts and Sciences, Department of Chemistry, Burlington, VT 05405. Department awards MAT, MS, MST, PhD. Matriculated students: 44; includes 1 minority (black American), 8 foreign. 34 applicants, 50% accepted. In 1990, 0 master's, 5 doctorates awarded. *Degree requirements:* For master's, 1 foreign language (computer language can substitute), thesis; for doctorate, 2 foreign languages (computer language can substitute for one), dissertation. *Entrance requirements:* GRE General Test, TOEFL (minimum score of 550 required). Application deadline: 4/1 (priority date, applications processed on a rolling basis). Application fee: $25. *Expenses:* Tuition of $206 per credit for state residents; $564 per credit for nonresidents. Fees of $150 per semester full-time. *Financial aid:* In 1990–91, 8 fellowships, 0 research assistantships, 31 teaching assistantships (9 to first-year students) awarded. Financial aid application deadline: 3/1. • Dr. C. H. Bushweller, Chairperson, 802-656-2594. Application contact: Dr. W. Geiger, Coordinator, 802-656-2594.

University of Victoria, Faculty of Arts and Science, Department of Chemistry, Victoria, BC V8W 2Y2, Canada. Department offers programs in inorganic kinetics (M Sc, PhD), inorganic photochemistry (M Sc, PhD), molecular spectroscopy (M Sc, PhD), multinuclear NMR studies (M Sc, PhD), organometallic chemistry (M Sc, PhD), photoelectron spectroscopy (M Sc, PhD), synthetic organic chemistry (M Sc, PhD), transition-metal chemistry (M Sc, PhD). Faculty: 19 full-time (0 women), 0 part-time. Matriculated students: 27 full-time (8 women), 0 part-time; includes 16 foreign. Average age 28. 124 applicants, 14% accepted. In 1990, 2 master's, 3 doctorates awarded. *Degree requirements:* Thesis/dissertation required, foreign language not required. *Application deadline:* 5/31 (applications processed on a rolling basis). *Application fee:* $20. *Expenses:* Tuition of $754 per semester. Fees of $23 per year. *Financial aid:* Fellowships, research assistantships, teaching assistantships, institutionally sponsored loans, and career-related internships or fieldwork available. *Faculty research:* Photoelectron spectroscopy, natural products, pollution control, electrochemistry, theoretical chemistry. • Dr. T. A. Gough, Chair, 604-721-7150. Application contact: Dr. W. J. Balfour, Graduate Adviser, 604-721-7168.

University of Virginia, Graduate School of Arts and Sciences, Department of Chemistry, Charlottesville, VA 22906. Department awards MA, MAT, MS, PhD. Faculty: 29 full-time (1 woman), 2 part-time (0 women). Matriculated students: 90 full-time (31 women), 2 part-time (0 women); includes 4 minority (1 Asian American, 1 black American, 2 Hispanic American), 10 foreign. Average age 25. 197 applicants, 48% accepted. In 1990, 4 master's, 13 doctorates awarded. *Degree requirements:* Thesis/dissertation required, foreign language not required. *Entrance requirements:* GRE General Test, GRE Subject Test. Application deadline: 7/15. Application fee: $40. *Expenses:* Tuition of $2740 per year full-time, $904 per year (minimum) part-time for state residents; $8950 per year full-time, $2960 per year (minimum) part-time for nonresidents. Fees of $586 per year full-time, $342 per year part-time. *Financial aid:* Application deadline 2/1. • Robert E. Ireland, Chairman, 804-924-3344. Application contact: William A. Elwood, Associate Dean, 804-924-7184.

Announcement: Well-equipped department combines established research programs and vigorous young faculty with emphasis on interdisciplinary research. Group sizes of 4–10 students permit close student-faculty interaction. Areas of specialization include analytical, biological, biophysical, inorganic, organic, and physical chemistry. Major emphasis on spectroscopy, organometallic and biological applications of chemistry.

University of Washington, College of Arts and Sciences, Department of Chemistry, Seattle, WA 98195. Department awards MS, PhD. Part-time programs available. *Degree requirements:* For master's, thesis required (for some programs), foreign language not required; for doctorate, dissertation required, foreign language not required. *Entrance requirements:* GRE General Test, minimum GPA of 3.0. Application deadline: 7/31. Application fee: $35. *Tuition:* $1129 per quarter full-time, $324 per credit (minimum) part-time for state residents; $2824 per quarter full-time, $809 per credit (minimum) part-time for nonresidents. *Faculty research:* Biopolymers, spectroscopy, reaction mechanisms, catalysis, synthesis, instrumental analysis.

University of Waterloo, Guelph-Waterloo Centre for Graduate Work in Chemistry, Waterloo, ON N2L 3G1, Canada. Center awards M Sc, PhD. Faculty: 67 full-time, 0 part-time. Matriculated students: 106 full-time (29 women), 9 part-time (2 women); includes 27 foreign. 204 applicants, 24% accepted. In 1990, 17 master's, 10 doctorates awarded. *Degree requirements:* Thesis/dissertation. *Entrance requirements:* For master's, TOEFL (minimum score of 570 required), honor's degree, minimum B average; for doctorate, TOEFL (minimum score of 570 required), master's degree. Application fee: $25. *Expenses:* Tuition of $757 per year full-time, $530 per year part-time for Canadian residents; $3127 per year for nonresidents. Fees of $68 per year full-time, $17 per year part-time. *Financial aid:* Research assistantships, teaching assistantships available. *Faculty research:* Analytical, polymer, biochemistry, physical, theoretical. • Dr. F. J. Sharom, Director, 519-824-4120 Ext. 2247. Application contact: A. Wetmore, Administrative Assistant, 519-824-4120 Ext. 3447.

University of Western Ontario, Physical Sciences Division, Department of Chemistry, London, ON N6A 3K7, Canada. Department awards M Sc, PhD. *Degree requirements:* Thesis/dissertation. *Entrance requirements:* For master's, minimum B+ average, honors B Sc; for doctorate, M Sc or equivalent. *Tuition:* $1015 per year full-time, $1050 per year part-time for Canadian residents; $4207 per year for nonresidents.

University of Windsor, Faculty of Science, Department of Chemistry and Biochemistry, Windsor, ON N9B 3P4, Canada. Department offers programs in biochemistry (M Sc, PhD), chemistry (M Sc, PhD). Part-time programs available. Faculty: 18 full-time (2 women), 0 part-time. Matriculated students: 23 full-time, 0 part-time; includes 6 foreign. In 1990, 3 master's, 3 doctorates awarded. *Degree requirements:* Thesis/dissertation. *Entrance requirements:* For master's, GRE, TOEFL, minimum B average. Application deadline: 7/1 (priority date, applications processed on a rolling basis). Application fee: $0. *Tuition:* $819.15 per semester full-time for Canadian residents; $3646 per semester full-time for nonresidents. *Financial aid:* Research assistantships, teaching assistantships available. Average monthly stipend for a graduate assistantship: $700. • Dr. John E. Drake, Head. Application contact: Admissions Officer, 519-253-4232 Ext. 2108.

University of Wisconsin–Madison, Water Chemistry Program, Madison, WI 53706. Program awards MS, PhD. Matriculated students: 15 full-time (5 women), 4 part-time (2 women); includes 0 minority, 5 foreign. 14 applicants, 36% accepted. In 1990, 2 master's, 2 doctorates awarded. *Application deadline:* rolling. *Application fee:* $20. *Financial aid:* In 1990–91, 13 students received aid. 0 fellowships, 13 research assistantships awarded; teaching assistantships, institutionally sponsored loans also available. Financial aid application deadline: 1/15; applicants required to submit FAF. • Dr. D. E. Armstrong, Chairperson, 608-263-3264.

University of Wisconsin–Madison, College of Letters and Science, Department of Chemistry, Madison, WI 53706. Department awards MS, PhD. Faculty: 42 full-time, 0 part-time. Matriculated students: 256 full-time (56 women), 8 part-time (2 women); includes 21 minority (15 Asian American, 3 black American, 3 Hispanic American), 44 foreign. 317 applicants, 45% accepted. In 1990, 11 master's, 38 doctorates awarded. *Application deadline:* rolling. *Application fee:* $20. *Financial aid:* In 1990–91, 252 students received aid. 13 fellowships, 139 research assistantships, 93 teaching assistantships, 7 project assistantships, t awarded; institutionally sponsored loans also available. Financial aid application deadline: 1/15; applicants required to submit FAF. • Paul Treichel, Chairperson, 608-262-0806. Application contact: Joanne Nagy, Assistant Dean of the Graduate School, 608-262-2433.

University of Wisconsin–Milwaukee, College of Letters and Sciences, Department of Chemistry, Milwaukee, WI 53201. Department awards MS, PhD. Faculty: 18 full-time, 0 part-time. Matriculated students: 35 full-time (12 women), 36 part-time (11 women); includes 3 minority (all Asian American), 36 foreign. 99 applicants, 14% accepted. In 1990, 8 master's, 7 doctorates awarded. *Degree requirements:* For master's, thesis or alternative required, foreign language not required; for doctorate, dissertation required, foreign language not required. *Application deadline:* 3/1 (priority date, applications processed on a rolling basis). *Application fee:* $20. *Financial aid:* In 1990–91, 0 fellowships, 13 research assistantships, 41 teaching assistantships awarded. Financial aid application deadline: 4/15. • Walter England, Program Representative, 414-229-4411.

University of Wyoming, College of Arts and Sciences, Department of Chemistry, Laramie, WY 82071. Department awards MA, MAT, MS, PhD. Faculty: 17 full-time. Matriculated students: 27 full-time (9 women), 4 part-time (2 women); includes 6 foreign. In 1990, 7 master's, 1 doctorate awarded. *Entrance requirements:* GRE General Test, minimum GPA of 3.0. Application deadline: 6/1 (priority date, applications processed on a rolling basis). Application fee: $30. *Tuition:* $1554 per year full-time, $74.25 per credit hour part-time for state residents; $4358 per year full-time, $74.25 per credit hour part-time for nonresidents. *Financial aid:* Application deadline 4/15. • Dr. David A. Jaeger, Head, 307-766-4363.

Utah State University, College of Science, Department of Chemistry and Biochemistry, Logan, UT 84322. Department awards MS, PhD. Part-time programs available. Faculty: 18 full-time (5 women), 3 part-time (1 woman). Matriculated students: 9 full-time (2 women), 35 part-time (8 women); includes 24 foreign. Average age 25. 124 applicants, 36% accepted. In 1990, 6 master's awarded; 8 doctorates awarded (90% entered university research/teaching, 10% found other work related to degree). Terminal master's awarded for partial completion of doctoral program. *Degree requirements:* Thesis/dissertation, written and oral exams required, foreign language not required. *Entrance requirements:* GRE General Test (score in 40th percentile or higher required), TOEFL (minimum score of 550 required), minimum GPA of 3.0. Application deadline: 4/15 (priority date, applications processed on a rolling basis). Application fee: $25 ($30 for foreign students). *Tuition:* $426 per quarter (minimum) full-time, $184 per quarter (minimum) part-time for state residents; $1133 per quarter (minimum) full-time, $505 per quarter (minimum) part-time for nonresidents. *Financial aid:* In 1990–91, $10,320 in aid awarded. 13 teaching assistantships (8 to first-year students) were awarded; fellowships, research assistantships, partial tuition waivers, federal work-study, institutionally sponsored loans also available. Aid available to part-time students. Average monthly stipend for a graduate assistantship: $750. Financial aid application deadline: 4/1; applicants required to submit FAF. *Faculty research:* Analytical, inorganic, organic and physical chemistry; biochemistry. Total annual research budget: $813,922. • Dr. Vernon Parker, Head, 801-750-1619. Application contact: Dr. David B. Marshall, Admissions Chairman, 801-750-1628.

See full description on page 177.

Vanderbilt University, Department of Chemistry, Nashville, TN 37240. Department awards MA, MS, PhD. Faculty: 22 full-time (2 women), 4 part-time (1 woman). Matriculated students: 55 full-time (16 women), 0 part-time; includes 4 minority (2 black American, 2 Hispanic American), 23 foreign. Average age 26. 106 applicants, 24% accepted. In 1990, 7 master's, 7 doctorates awarded. *Degree requirements:* Thesis/dissertation required, foreign language not required. *Entrance requirements:* GRE General Test, GRE Subject Test. Application deadline: 1/15. Application fee: $25. *Expenses:* Tuition of $624 per semester hour. Fees of $196 per year. *Financial aid:* Fellowships, research assistantships, teaching assistantships, federal work-study, institutionally sponsored loans available. Financial aid application deadline: 1/15. • B. Andes Hess Jr., Chairman, 615-322-2861. Application contact: Charles M. Lukehurt, Director of Graduate Studies, 615-322-2935.

Vassar College, Department of Chemistry, Poughkeepsie, NY 12601. Department awards MA, MS. Faculty: 9 full-time, 0 part-time. Matriculated students: 0 full-time, 1 (woman) part-time; includes 0 minority, 0 foreign. 0 applicants. In 1990, 0 degrees awarded. *Degree requirements:* 1 foreign language, thesis. *Entrance requirements:* GRE, bachelor's degree in related field. *Expenses:* Tuition of $16,250 per year full-time, $1920 per unit part-time. Fees of $260 per year full-time. • Miriam Rossi, Chairman, 914-437-5746.

Villanova University, Graduate School of Arts and Sciences, Department of Chemistry, Villanova, PA 19085. Department offers programs in biochemistry (MS, PhD), chemistry (MA, MS, PhD). Part-time and evening/weekend programs available. Faculty: 18 full-time (0 women), 1 (woman) part-time. Matriculated students: 26 full-time (8 women), 57 part-time (19 women); includes 16 foreign. Average age 25. 54 applicants, 87% accepted. In 1990, 8 master's awarded (75% found work related to degree, 25% continued full-time study); 2 doctorates awarded (50% entered university research/teaching, 50% found other work related to degree). Terminal master's awarded for partial completion of doctoral program. *Degree requirements:* 1 foreign language, thesis/dissertation. *Entrance requirements:* GRE General Test, GRE Subject Test, minimum GPA of 3.0. Application deadline: 8/1. Application fee: $25. *Expenses:* Tuition of $300 per credit. Fees of $28 per semester. *Financial aid:* In 1990–91, $140,000 in aid awarded. 3 research assistantships (1 to a first-year student), 24 teaching assistantships (7 to first-year students) were awarded; federal

SECTION 2: CHEMISTRY

Directory: Chemistry

work-study also available. Financial aid application deadline: 4/1. *Total annual research budget:* $80,000. • Dr. John F. Wojcik, Chairperson, 215-645-4840.

Virginia Commonwealth University, College of Humanities and Sciences, Department of Chemistry, Richmond, VA 23284. Department awards MS, PhD. Part-time programs available. Faculty: 16 (2 women). Matriculated students: 38 full-time (16 women), 17 part-time (3 women); includes 11 minority (4 Asian American, 7 black American), 14 foreign. 115 applicants, 11% accepted. In 1990, 2 master's, 8 doctorates awarded. Terminal master's awarded for partial completion of doctoral program. *Degree requirements:* For master's, thesis required, foreign language not required; for doctorate, 1 foreign language, dissertation, comprehensive cumulative exams, research proposal. *Entrance requirements:* GRE General Test, GRE Subject Test. Application deadline: 3/15 (priority date). Application fee: $20. *Expenses:* Tuition of $2770 per year full-time, $154 per hour part-time for state residents; $7550 per year full-time, $419 per hour part-time for nonresidents. Fees of $717 per year full-time, $25.50 per hour part-time. *Financial aid:* Fellowships, research assistantships, teaching assistantships, institutionally sponsored loans, and career-related internships or fieldwork available. Aid available to part-time students. Financial aid application deadline: 7/1; applicants required to submit FAF. *Faculty research:* Physical, organic, inorganic, analytical, and polymer chemistry. Total annual research budget: $357,077. • Dr. Lawrence Winters, Chair, 804-367-1298. Application contact: Dr. Joseph A. Topich, Graduate Program Director, 804-367-1298.

Virginia Polytechnic Institute and State University, College of Arts and Sciences, Department of Chemistry, Blacksburg, VA 24061. Department awards MS, PhD. Part-time programs available. Faculty: 26 full-time (0 women). Matriculated students: 125 full-time (41 women), 21 part-time (8 women); includes 10 minority (7 Asian American, 3 black American), 42 foreign. Average age 25. 161 applicants, 42% accepted. In 1990, 10 master's, 20 doctorates awarded. Terminal master's awarded for partial completion of doctoral program. *Degree requirements:* For master's, thesis required (for some programs), foreign language not required; for doctorate, dissertation required, foreign language not required. *Entrance requirements:* GRE General Test, GRE Subject Test, TOEFL. Application deadline: 2/15 (priority date). Application fee: $10. *Tuition:* $1889 per semester full-time, $606 per credit hour part-time for state residents; $2627 per semester full-time, $853 per credit hour part-time for nonresidents. *Financial aid:* In 1990–91, $1.076-million in aid awarded. 5 fellowships (2 to first-year students), 44 research assistantships (2 to first-year students), 68 teaching assistantships (32 to first-year students), 8 assistantships were awarded; federal work-study, institutionally sponsored loans, and career-related internships or fieldwork also available. Aid available to part-time students. Financial aid application deadline: 7/1. *Faculty research:* Analytical, inorganic, organic, physical, and polymer chemistry. Total annual research budget: $3.8-million. • Dr. James F. Wolfe, Head, 703-231-5391.

Announcement: The National Science Foundation Science and Technology Center on High Performance Polymeric Adhesives Composite, located at Virginia Polytechnic Institute and State University, involves 21 faculty from the Departments of Chemistry; Chemical, Mechanical, and Materials Engineering; Engineering Science and Mechanics; and Forest Products. This integrated, interdisciplinary approach provides the means to understand, design, produce, and test high-performance polymeric adhesives and composites. MS and PhD education is available through each department, with the research centered on 1 of the 4 areas of synthesis and characterization, processing, interfacial phenomena, or mechanical behavior. The center has also developed many programs for educational outreach and technology transfer.

Wake Forest University, Department of Chemistry, Winston-Salem, NC 27109. Department offers programs in analytical chemistry (MS, PhD), inorganic chemistry (MS, PhD), organic chemistry (MS, PhD), physical chemistry (MS, PhD). Part-time programs available. Faculty: 10 full-time (0 women), 3 part-time (1 woman). Matriculated students: 19 full-time (5 women), 0 part-time; includes 1 black American, 8 foreign. Average age 28. 76 applicants, 14% accepted. In 1990, 2 master's awarded (100% found work related to degree); 1 doctorate awarded (100% found work related to degree). *Degree requirements:* 1 foreign language (computer language can substitute), thesis/dissertation. *Entrance requirements:* GRE General Test, GRE Subject Test. Application deadline: 3/1. Application fee: $25. *Tuition:* $10,800 per year full-time, $280 per hour part-time. *Financial aid:* In 1990–91, 19 students received a total of $305,300 in aid awarded. 0 fellowships, 0 research assistantships, 11 teaching assistantships (1 to a first-year student), 8 scholarships (1 to a first-year student) were awarded. Aid available to part-time students. Average monthly stipend for a graduate assistantship: $917. Financial aid application deadline: 3/1; applicants required to submit FAF. • Dr. Willie Hinze, Chairman, 919-759-5325.

Washington State University, College of Sciences and Arts, Division of Sciences, Department of Chemistry, Pullman, WA 99164. Department offers programs in analytical chemistry (MS, PhD), chemical physics (PhD), inorganic chemistry (MS, PhD), nuclear chemistry (MS, PhD), organic chemistry (MS, PhD), physical chemistry (MS, PhD). Faculty: 26 (2 women). Matriculated students: 45 full-time (12 women), 4 part-time (1 woman); includes 4 minority (3 Asian American, 1 black American), 10 foreign. Average age 25. In 1990, 8 master's, 7 doctorates awarded. Terminal master's awarded for partial completion of doctoral program. *Degree requirements:* For master's, thesis, teaching experience required, foreign language not required; for doctorate, dissertation, teaching experience. *Entrance requirements:* GRE General Test, minimum GPA of 3.0. Application deadline: 3/1 (priority date, applications processed on a rolling basis). Application fee: $25. *Tuition:* $1694 per semester full-time, $169 per credit hour part-time for state residents; $4236 per semester full-time, $424 per credit hour part-time for nonresidents. *Financial aid:* In 1990–91, 0 fellowships, 13 research assistantships, 25 teaching assistantships, 4 teaching associateships awarded; partial tuition waivers, federal work-study, institutionally sponsored loans also available. Average monthly stipend for a graduate assistantship: $975. Financial aid application deadline: 4/1. *Total annual research budget:* $1.04-million. • Dr. Roy Filby, Chair, 509-335-1516.

Washington University, Graduate School of Arts and Sciences, Department of Chemistry, St. Louis, MO 63130. Department awards MA, PhD. Matriculated students: 65 full-time (24 women), 0 part-time; includes 3 minority (1 Asian American, 1 black American, 1 Hispanic American), 21 foreign. In 1990, 12 master's, 9 doctorates awarded. Terminal master's awarded for partial completion of doctoral program. *Degree requirements:* For master's, thesis or alternative; for doctorate, dissertation. *Entrance requirements:* GRE General Test, GRE Subject Test. Application deadline: 1/15 (priority date, applications processed on a rolling basis). Application fee: $0. *Tuition:* $15,960 per year full-time, $665 per credit hour part-time. *Financial aid:* In 1990–91, 7 fellowships, 25 research assistantships, 32 teaching assistantships awarded; full and partial tuition waivers, federal work-study, institutionally sponsored loans also available. Financial aid application deadline: 1/15. • Dr. Joseph J. H. Ackerman, Chairman, 314-935-6593.

Wayland Baptist University, Program in Interdisciplinary Science, Plainview, TX 79072. Program awards MS. Part-time and evening/weekend programs available. Faculty: 6 full-time (1 woman), 2 part-time (1 woman). Matriculated students: 0 full-time, 76 part-time (55 women); includes 5 minority (1 black American, 4 Hispanic American). Average age 35. 30 applicants, 100% accepted. In 1990, 3 degrees awarded (100% found work related to degree). *Degree requirements:* Foreign language and thesis not required. *Entrance requirements:* GRE, GMAT, minimum GPA of 2.7. Application fee: $35. *Expenses:* Tuition of $145 per semester hour. Fees of $32 per semester. *Financial aid:* Fellowships, research assistantships, teaching assistantships, full and partial tuition waivers, federal work-study, institutionally sponsored loans, and career-related internships or fieldwork available. Aid available to part-time students. Financial aid applicants required to submit FAF. • Dr. J. Hoyt Bowers, Chairman, Division of Mathematics and Science, 806-296-5521 Ext. 382.

Wayne State University, College of Liberal Arts, Department of Chemistry, Detroit, MI 48202. Department awards MA, MS, PhD. Faculty: 36 (4 women). Matriculated students: 120 full-time (35 women), 37 part-time (12 women). 78 applicants, 74% accepted. In 1990, 5 master's, 15 doctorates awarded. *Degree requirements:* For doctorate, dissertation. *Application deadline:* 7/1. *Application fee:* $20 ($30 for foreign students). *Expenses:* Tuition of $119 per credit hour for state residents; $258 per credit hour for nonresidents. Fees of $50 per semester. *Faculty research:* Carcinogens, ACTH, synthesis of cyclopentanoids, melondialdehyde, DNA. Total annual research budget: $2.1-million. • Richard Lintvedt, Chairperson, 313-577-2591.

Wesleyan University, Department of Chemistry, Middletown, CT 06459. Department offers programs in biochemistry (MA, PhD), chemical physics (MA, PhD), inorganic chemistry (MA, PhD), organic chemistry (MA, PhD), physical chemistry (MA, PhD), theoretical chemistry (MA, PhD). Faculty: 14 full-time (1 woman), 2 part-time (0 women). Matriculated students: 31 full-time (12 women), 4 part-time (0 women); includes 2 minority (1 Asian American, 1 black American), 14 foreign. Average age 28. 46 applicants, 48% accepted. In 1990, 1 master's awarded (100% found work related to degree); 5 doctorates awarded (100% found work related to degree). Terminal master's awarded for partial completion of doctoral program. *Degree requirements:* 1 foreign language (computer language can substitute), thesis/dissertation. *Entrance requirements:* For master's, GRE General Test, GRE Subject Test; for doctorate, GRE Subject Test. Application deadline: 2/1. Application fee: $0. *Expenses:* Tuition of $2035 per course. Fees of $627 per year full-time. *Financial aid:* In 1990–91, $103,275 in aid awarded. 3 fellowships (0 to first-year students), 12 research assistantships (1 to a first-year student), 17 teaching assistantships (5 to first-year students) were awarded; institutionally sponsored loans also available. Average monthly stipend for a graduate assistantship: $875. *Total annual research budget:* $750,000. • Dr. Rex F. Pratt, Graduate Adviser, 203-347-9411 Ext. 2380.

See full description on page 179.

West Chester University of Pennsylvania, College of Arts and Sciences, Department of Chemistry, West Chester, PA 19383. Department offers programs in chemistry (MS), clinical chemistry (MS). Faculty: 0 full-time, 10 part-time. Matriculated students: 6 full-time (2 women), 15 part-time (7 women); includes 7 minority (all Asian American), 4 foreign. Average age 30. 24 applicants, 58% accepted. In 1990, 6 degrees awarded. *Degree requirements:* 1 foreign language (computer language can substitute), comprehensive exam required, thesis optional. *Application deadline:* rolling. *Application fee:* $20. *Tuition:* $127 per credit for state residents; $160 per credit for nonresidents. *Financial aid:* In 1990–91, 5 research assistantships awarded. Financial aid application deadline: 4/1. • Dr. Michael Moran, Chair, 215-436-2526. Application contact: Dr. Jamal Ghoroghchian, Graduate Coordinator, 215-436-2982.

Western Carolina University, School of Arts and Sciences, Department of Chemistry and Physics, Cullowhee, NC 28723. Department awards MA Ed, MS. Part-time and evening/weekend programs available. Faculty: 10 (0 women). Matriculated students: 5 full-time (2 women), 3 part-time (2 women); includes 1 minority (black American), 0 foreign. 8 applicants, 75% accepted. In 1990, 2 degrees awarded. *Degree requirements:* Variable foreign language requirement, thesis, comprehensive exams. *Entrance requirements:* GRE General Test, GRE Subject Test (for MS), NTE (for MA Ed). Application deadline: rolling. Application fee: $15. *Tuition:* $635 per semester full-time, $226 per course (minimum) part-time for state residents; $3162 per semester full-time, $1490 per course (minimum) part-time for nonresidents. *Financial aid:* In 1990–91, $24,750 in aid awarded. 0 fellowships, 1 research assistantship, 5 teaching assistantships were awarded; federal work-study, institutionally sponsored loans also available. Financial aid application deadline: 3/15; applicants required to submit FAF. • F. Glenn Liming, Head, 704-227-7260. Application contact: Kathleen Owen, Assistant to the Dean, 704-227-7398.

Western Illinois University, College of Arts and Sciences, Department of Chemistry, Macomb, IL 61455. Department awards MS. Part-time programs available. Faculty: 10 full-time (1 woman), 0 part-time. Matriculated students: 10 full-time (5 women), 3 part-time (0 women); includes 1 minority (Asian American), 8 foreign. Average age 29. 29 applicants, 24% accepted. In 1990, 4 degrees awarded. *Degree requirements:* Thesis or alternative required, foreign language not required. *Application fee:* $0. *Expenses:* Tuition of $870 per semester full-time, $520 per semester hour part-time for state residents; $2193 per semester full-time, $1402 per semester hour part-time for nonresidents. Fees of $537 per year full-time, $148 per year part-time. *Financial aid:* In 1990–91, $85,370 in aid awarded. 14 research assistantships were awarded. • Dr. W. Klopfenstein, Chairperson, 309-298-1538. Application contact: Barbara Baily, Assistant to the Dean, 309-298-4806.

Western Kentucky University, Ogden College of Science, Technology, and Health, Department of Chemistry, Bowling Green, KY 42101. Department offers programs in chemistry (MA Ed, MS), fossil fuel chemistry (PhD). PhD offered jointly with University of Louisville. Part-time programs available. Faculty: 13 full-time (0 women), 0 part-time. Matriculated students: 6 full-time (2 women), 9 part-time (4 women); includes 1 minority (Asian American), 5 foreign. 30 applicants, 50% accepted. In 1990, 6 master's awarded. *Degree requirements:* For master's, 1 foreign language (computer language can substitute), thesis. *Entrance requirements:* For master's, GRE General Test (minimum combined score of 1150 required), previous course work in chemistry (MS). Application deadline: 8/1 (priority date, applications processed on a rolling basis). *Tuition:* $1580 per year full-time, $87 per credit part-time for state residents; $4460 per year full-time, $247 per credit part-time for nonresidents. *Financial aid:* In 1990–91, 9 service awards (6 to first-year students) awarded; research assistantships, teaching assistantships, federal work-study, institutionally sponsored loans also available. Average monthly stipend for a graduate assistantship: $625. Financial aid application deadline: 4/1. • Dr. Donald Slocum, Head, 502-745-3457.

Announcement: The MS in Chemistry (Coal Chemistry Option) is a unique program that qualifies graduates for several options, including entrance into fuel science or chemistry PhD programs, managerial positions in industrial chemistry or coal testing laboratories, and research positions in chemistry or fossil fuel research laboratories.

Western Michigan University, College of Arts and Sciences, Department of Chemistry, Kalamazoo, MI 49008. Department awards MA. Matriculated students: 3 full-time (0 women), 18 part-time (5 women); includes 2 minority (1 Asian American, 1 black American), 14 foreign. 45 applicants, 31% accepted. In 1990, 1 degree awarded. *Degree requirements:* Thesis, oral exams, departmental qualifying exam required, foreign language not required. *Application deadline:* 2/15 (priority date, applications processed on a rolling basis). *Application fee:* $25. *Expenses:* Tuition of $100 per credit hour for state residents; $244 per credit hour for nonresidents. Fees of $178 per semester full-time, $76 per semester part-time. *Financial aid:* Fellowships, research assistantships, teaching assistantships, federal work-study available. Financial aid application deadline: 2/15. • Dr. Michael McCarville, Chairperson, 616-387-2846. Application contact: Paula J. Boodt, Director, Graduate Student Services, 616-387-3570.

Western Washington University, College of Arts and Sciences, Department of Chemistry, Bellingham, WA 98225. Department awards MS. Faculty: 14 (1 woman). Matriculated students: 5 full-time (1 woman), 2 part-time (0 women); includes 0 minority, 0 foreign. 4 applicants, 100% accepted. In 1990, 1 degree awarded. *Degree requirements:* Thesis required (for some programs), foreign language not required. *Entrance requirements:* GRE General Test, TOEFL (minimum score of 535 required), minimum GPA of 3.0 for last 60 semester hours or last 90 quarter hours. Application deadline: 6/1 (priority date, applications processed on a rolling basis). Application fee: $25. *Tuition:* $900 per quarter full-time, $90 per credit part-time for state residents; $2729 per quarter full-time, $273 per credit part-time for nonresidents. *Financial aid:* In 1990–91, 2 teaching assistantships awarded; partial tuition waivers, federal work-study, institutionally sponsored loans, and career-related internships or fieldwork also available. Aid available to part-time students. Financial aid application deadline: 3/31; applicants required to submit FAF. • Dr. Mark Wicholas, Chairperson, 206-676-3071. Application contact: Dr. Sal Russo, Graduate Program Adviser, 206-676-3134.

West Texas State University, College of Agriculture, Nursing, and Natural Sciences, Department of Mathematics and Physical Sciences, Program in Chemistry, Canyon, TX 79016. Program awards MS. Part-time programs available. Faculty: 0 part-time. Matriculated students: 11. In 1990, 3 degrees awarded. *Degree requirements:* Thesis or alternative required, foreign language not required. *Entrance requirements:* GRE General Test. Application deadline: rolling. *Expenses:* Tuition of $600 per year full-time, $20 per hour part-time for state residents; $3840 per year full-time, $128 per hour part-time for nonresidents. Fees of $448 per year full-time, $15 per hour part-time. *Financial aid:* Fellowships and career-related internships or fieldwork available. *Faculty research:* Biochemistry; inorganic, organic, and physical chemistry; vibrational spectroscopy; magnetic susceptibilities; carbene chemistry. • Dr. Gene Carlisle, Head, 806-656-2540.

West Virginia University, College of Arts and Sciences, Department of Chemistry, Morgantown, WV 26506. Department awards MS, PhD. Faculty: 30 full-time (2 women), 1 part-time (0 women). Matriculated students: 37 full-time (8 women), 4 part-time (3 women); includes 1 minority (Hispanic American), 23 foreign. Average age 26. 170 applicants, 19% accepted. In 1990, 3 master's, 3 doctorates awarded. Terminal master's awarded for partial completion of doctoral program. *Degree requirements:* For master's, thesis required, foreign language not required; for doctorate, 1 foreign language, dissertation, comprehensive exam. *Entrance requirements:* For master's, GRE General Test, TOEFL (minimum score of 550 required), minimum GPA of 2.5; for doctorate, GRE General Test, GRE Subject Test, TOEFL (minimum score of 550 required), minimum GPA of 2.5. Application deadline: 3/1 (priority date, applications processed on a rolling basis). Application fee: $25. *Expenses:* Tuition of $390 per year full-time for state residents; $1270 per year full-time for nonresidents. Fees of $1555 per year full-time for state residents; $3985 per year full-time for nonresidents. *Financial aid:* In 1990–91, $400,000 in aid awarded. 7 research assistantships, 38 teaching assistantships were awarded; full and partial tuition waivers, federal work-study, institutionally sponsored loans also available. Financial aid application deadline: 2/1; applicants required to submit FAF. *Faculty research:* Analytical, inorganic, organic, physical, coal, polymers. Total annual research budget: $600,000. • Dr. Paul Jagodzinski, Chairperson, 304-293-4742. Application contact: Dr. Jeffrey Petersen, Director of Graduate Studies, 304-293-3435.

Wichita State University, Fairmount College of Liberal Arts and Sciences, Department of Chemistry, Wichita, KS 67208. Department awards MS, PhD. Faculty: 12 full-time (1 woman), 0 part-time. Matriculated students: 15 full-time (6 women), 15 part-time (4 women); includes 1 minority (black American), 9 foreign. 15 applicants, 60% accepted. In 1990, 1 master's, 1 doctorate awarded. *Degree requirements:* For master's, variable foreign language requirement, thesis. *Entrance requirements:* For master's, TOEFL (minimum score of 550 required). Application deadline: 8/1 (priority date, applications processed on a rolling basis). Application fee: $0 ($25 for foreign students). *Expenses:* Tuition of $1590 per year full-time, $53 per credit part-time for state residents; $2574 per year full-time, $171.60 per credit part-time for nonresidents. Fees of $12.20 per credit. *Financial aid:* In 1990–91, $90,000 in aid awarded. 7 fellowships, 3 research assistantships, 1 graduate assistantship were awarded; teaching assistantships, federal work-study, institutionally sponsored loans also available. Financial aid application deadline: 4/1. • Dr. Jack McCormick, Chairperson, 316-689-3120.

Worcester Polytechnic Institute, Department of Chemistry, Worcester, MA 01609. Department awards MS, PhD. Faculty: 13 full-time (0 women), 0 part-time. Matriculated students: 12 full-time (6 women), 0 part-time. 53 applicants, 30% accepted. In 1990, 1 master's, 0 doctorates awarded. *Degree requirements:* Thesis/dissertation required, foreign language not required. *Entrance requirements:* TOEFL (minimum score of 550 required). Application deadline: 2/15. Application fee: $25. *Expenses:* Tuition of $460 per credit hour. Fees of $20 per year full-time. *Financial aid:* Teaching assistantships, institutionally sponsored loans available. Financial aid application deadline: 2/15. *Faculty research:* Photochemistry, enzyme structure, surface chemistry, molecular spectroscopy. Total annual research budget: $95,000. • Dr. J. W. Pavlik, Head, 508-831-5263.

Wright State University, College of Science and Mathematics, Department of Chemistry, Dayton, OH 45435. Department awards MS. Part-time and evening/weekend programs available. Faculty: 16 full-time (1 woman), 0 part-time. Matriculated students: 21 full-time (12 women), 13 part-time (3 women); includes 7 minority (3 Asian American, 2 black American, 2 Hispanic American), 10 foreign. Average age 28. 19 applicants, 84% accepted. In 1990, 11 degrees awarded (75% found work related to degree, 25% continued full-time study). *Degree requirements:* Thesis, oral defense of thesis required, foreign language not required. *Application fee:* $25. *Tuition:* $3342 per year full-time, $106 per credit hour part-time for state residents; $5991 per year full-time, $190 per credit hour part-time for nonresidents. *Financial aid:* In 1990–91, 0 fellowships, 0 research assistantships, 21 teaching assistantships, 1 graduate assistantships awarded. *Faculty research:* Polymer synthesis and characterization, laser-kinetics, organic and inorganic synthesis chromotography and mass spectrometry, gas solubilities. • Dr. Sue C. Cummings, Chair, 513-873-2855. Application contact: Dr. Paul Seybold, Graduate Adviser, 513-873-2407.

Yale University, Graduate School of Arts and Sciences, Department of Chemistry, New Haven, CT 06520. Department awards PhD, MD/PhD. Programs offered in biophysical chemistry (PhD), inorganic chemistry (PhD), organic chemistry (PhD), physical chemistry (PhD). Faculty: 22 full-time (1 woman), 0 part-time. Matriculated students: 135 full-time (51 women), 2 part-time (0 women); includes 13 minority (9 Asian American, 1 black American, 3 Hispanic American), 36 foreign. 210 applicants, 52% accepted. In 1990, 25 master's, 30 doctorates awarded. Terminal master's awarded for partial completion of doctoral program. *Degree requirements:* For doctorate, 1 foreign language, dissertation. *Entrance requirements:* For doctorate, GRE General Test, GRE Subject Test, TOEFL. Application deadline: 1/2. Application fee: $45. *Tuition:* $15,160 per year full-time, $1895 per course part-time. *Financial aid:* Fellowships, research assistantships, teaching assistantships available. • Dr. Frederick Ziegler, Chairman, 203-432-3916. Application contact: Susan Webb, Coordinator of Admissions, 203-432-2770.

See full description on page 181.

York University, Faculty of Science, Program in Chemistry, North York, ON M3J 1P3, Canada. Program awards M Sc, PhD. Part-time and evening/weekend programs available. Faculty: 26 full-time (1 woman), 3 part-time (0 women). Matriculated students: 31 full-time (7 women), 4 part-time (0 women); includes 2 foreign. 92 applicants, 22% accepted. In 1990, 6 master's, 1 doctorate awarded. *Degree requirements:* For master's, thesis optional, foreign language not required; for doctorate, dissertation required, foreign language not required. *Application deadline:* 2/15. *Application fee:* $35. *Tuition:* $2436 per year for Canadian residents; $11,480 per year for nonresidents. *Financial aid:* In 1990–91, $385,262 in aid awarded. 2 fellowships, 35 research assistantships, 28 teaching assistantships were awarded. • Dr. D. Stynes, Director, 416-736-5246.

Youngstown State University, College of Arts and Sciences, Department of Chemistry, Youngstown, OH 44555. Department awards MS. Part-time programs available. Faculty: 16 full-time (1 woman), 0 part-time. Matriculated students: 6 full-time (4 women), 7 part-time (4 women); includes 2 minority (1 black American, 1 Hispanic American), 3 foreign. 10 applicants, 70% accepted. In 1990, 1 degree awarded. *Degree requirements:* Thesis required, foreign language not required. *Entrance requirements:* Bachelor's degree in chemistry or equivalent, minimum GPA of 2.5. Application deadline: 8/15 (priority date, applications processed on a rolling basis). Application fee: $30. *Expenses:* Tuition of $1566 per year full-time, $58 per quarter hour part-time for state residents; $2808 per year full-time, $104 per quarter hour part-time for nonresidents. Fees of $432 per year full-time, $16 per quarter hour part-time. *Financial aid:* In 1990–91, 8 students received a total of $60,532 in aid awarded. 7 graduate assistantships (4 to first-year students) were awarded; federal work-study, institutionally sponsored loans also available. Aid available to part-time students. Average monthly stipend for a graduate assistantship: $833. Financial aid application deadline: 7/8. *Faculty research:* Metal and anion analysis, hydrogen bonding, protonation, solar energy, thermodynamics. • Dr. Thomas Dobbelstein, Chair, 216-742-3663. Application contact: Dr. Sally M. Hotchkiss, Dean of Graduate Studies, 216-742-3091.

Inorganic Chemistry

Boston College, Graduate School of Arts and Sciences, Department of Chemistry, Chestnut Hill, MA 02167. Offerings include inorganic chemistry (MS, PhD). *Degree requirements:* For master's, 1 foreign language, thesis; for doctorate, 1 foreign language, dissertation, foreign language exam, qualifying exam. *Entrance requirements:* GRE General Test, GRE Subject Test. *Tuition:* $412 per credit hour.

See full description on page 89.

Boston University, Graduate School, Department of Chemistry, Boston, MA 02215. Offerings include inorganic chemistry (MA, PhD). Terminal master's awarded for partial completion of doctoral program. Department faculty: 20 full-time (0 women), 0 part-time. *Degree requirements:* For master's, 1 foreign language required, thesis not required; for doctorate, 1 foreign language, dissertation. *Entrance requirements:* GRE General Test, GRE Subject Test, TOEFL (minimum score of 550 required). Application deadline: 4/1. Application fee: $45. *Expenses:* Tuition of $15,950 per year full-time, $498 per credit part-time. Fees of $120 per year full-time, $35 per semester part-time. • Guilford Jones II, Chairman, 617-353-2515.

See full description on page 91.

Brandeis University, Graduate School of Arts and Sciences, Program in Chemistry, Waltham, MA 02254. Offerings include inorganic chemistry (MA, PhD). Terminal master's awarded for partial completion of doctoral program. Program faculty: 16 full-time (2 women), 0 part-time. *Degree requirements:* For master's, 1 year residency required, foreign language and thesis not required; for doctorate, 1 foreign language (computer language can substitute), dissertation, 2 year residency, 2 seminars, qualifying exams, minimum GPA of 3.0. *Entrance requirements:* GRE, TOEFL. Application deadline: 2/28 (priority date, applications processed on a rolling

SECTION 2: CHEMISTRY

Directory: Inorganic Chemistry

basis). Application fee: $50. *Tuition:* $16,085 per year full-time, $2015 per course part-time. • Dr. Peter Jordan, Chair, 617-736-2503. Application contact: Susan Isaacs, Secretary, 617-736-2593.

See full description on page 93.

Brigham Young University, College of Physical and Mathematical Sciences, Department of Chemistry, Provo, UT 84602. Offerings include inorganic chemistry (MS, PhD). Department faculty: 34 full-time (1 woman), 0 part-time. *Degree requirements:* 2 foreign languages (computer language can substitute for one), thesis/dissertation. *Entrance requirements:* For master's, minimum GPA of 3.0 during previous 60 credit hours. Application deadline: 3/15 (priority date). Application fee: $30. *Tuition:* $1170 per semester (minimum) full-time, $130 per credit hour (minimum) part-time. • Dr. Earl M. Woolley, Chairman, 801-378-3667. Application contact: Francis R. Nordmeyer, Coordinator, Graduate Admissions, 801-378-4845.

See full description on page 95.

California State University, Fullerton, School of Natural Science and Mathematics, Department of Chemistry and Biochemistry, Fullerton, CA 92634. Offerings include inorganic chemistry (MS). Department faculty: 20 full-time (5 women), 13 part-time (2 women). *Degree requirements:* Thesis, departmental qualifying exam required, foreign language not required. *Entrance requirements:* Minimum GPA of 2.5 in last 60 units, major in chemistry or related field. Application fee: $55. *Expenses:* Tuition of $0 for state residents; $246 per unit for nonresidents. Fees of $554 per semester full-time, $356 per unit part-time. • Dr. Glen Nagel, Chair, 714-773-3621. Application contact: Dr. Gene Hiegel, Adviser, 714-773-3624.

California State University, Los Angeles, School of Natural and Social Sciences, Department of Chemistry and Biochemistry, Los Angeles, CA 90032. Offerings include inorganic chemistry (MS). Department faculty: 12 full-time, 25 part-time. *Degree requirements:* 1 foreign language (computer language can substitute), thesis or comprehensive exam. *Entrance requirements:* TOEFL (minimum score of 550 required). Application deadline: 8/7 (applications processed on a rolling basis). Application fee: $55. *Expenses:* Tuition of $0 for state residents; $164 per unit for nonresidents. Fees of $1046 per year full-time, $650 per semester (minimum) part-time. • Dr. Donald Paulson, Acting Chair, 213-343-2300.

Clark Atlanta University, School of Arts and Sciences, Department of Chemistry, Atlanta, GA 30314. Offerings include inorganic chemistry (MS, PhD). *Degree requirements:* For master's, 1 foreign language (computer language can substitute), thesis, comprehensive exam. *Entrance requirements:* For master's, minimum GPA of 2.5 in math and science. Application deadline: 4/1 (applications processed on a rolling basis). Application fee: $40. *Expenses:* Tuition of $4860 per year. Fees of $300 per year. • Dr. Alfred Spriggs, Chair, 404-880-8154. Application contact: Peggy Wade, Marketing and Recruitment, 404-880-8427.

Clarkson University, School of Science, Department of Chemistry, Potsdam, NY 13699. Offerings include inorganic chemistry (MS, PhD). Department faculty: 12 full-time (0 women), 0 part-time. *Degree requirements:* For master's, foreign language and thesis not required; for doctorate, dissertation, departmental qualifying exam required, foreign language not required. *Application fee:* $10. *Expenses:* Tuition of $446 per credit hour. Fees of $75 per semester. • Dr. Phillip A. Christiansen, Chairman, 315-268-2389.

Clemson University, College of Sciences, Department of Chemistry, Clemson, SC 29634. Department awards MS, PhD. Faculty: 19 full-time (1 woman), 0 part-time. Matriculated students: 65 full-time (14 women), 0 part-time; includes 4 minority (1 Asian American, 3 black American), 30 foreign. Average age 25. 40 applicants, 38% accepted. In 1990, 4 master's awarded; 12 doctorates awarded (67% entered university research/teaching, 33% found other work related to degree). *Degree requirements:* 1 foreign language (computer language can substitute), thesis/dissertation. *Entrance requirements:* GRE General Test. Application deadline: 6/1. Application fee: $25. *Expenses:* Tuition of $102 per credit hour. Fees of $80 per semester full-time. *Financial aid:* In 1990–91, $500,000 in aid awarded. 1 fellowship (to a first-year student), 18 research assistantships (1 to a first-year student), 43 lab assistantships were awarded; teaching assistantships also available. *Faculty research:* Flourine chemistry, organic synthetic methods and natural products, metal and non-metal clusters, analytical spectroscopies, polymers. Total annual research budget: $2.45-million. • J. D. Petersen, Head, 803-656-5017.

Cleveland State University, College of Arts and Sciences, Department of Chemistry, Cleveland, OH 44115. Offerings include inorganic chemistry (MS). Department faculty: 18 full-time (1 woman), 4 part-time (0 women). *Degree requirements:* Thesis required (for some programs), foreign language not required. *Entrance requirements:* GRE General Test, GRE Subject Test, TOEFL (minimum score of 525 required). Application deadline: 9/1 (priority date, applications processed on a rolling basis). Application fee: $0. *Tuition:* $90 per credit for state residents; $180 per credit for nonresidents. • Dr. Robert Towns, Chair, 216-687-2004. Application contact: Dr. Thomas Flechtner, Chairman, Graduate Committee, 216-687-2458.

Columbia University, Graduate School of Arts and Sciences, Division of Natural Sciences, Department of Chemistry, New York, NY 10027. Offerings include inorganic chemistry (M Phil, MS, PhD). Department faculty: 17 full-time. *Degree requirements:* For master's, comprehensive exams (MS); 1 foreign language, teaching experience, oral and written exams (M Phil) required, thesis not required; for doctorate, 1 foreign language, dissertation, M Phil. *Entrance requirements:* GRE General Test, GRE Subject Test, TOEFL. Application deadline: 1/5. Application fee: $50. *Expenses:* Tuition of $7836 per semester for state residents; $426 per credit for nonresidents. Fees of $148 per semester for state residents. • Richard Bersohn, Chair, 212-854-2192.

See full description on page 101.

Florida State University, College of Arts and Sciences, Department of Chemistry, Tallahassee, FL 32306. Offerings include inorganic chemistry (MS, PhD). Terminal master's awarded for partial completion of doctoral program. Department faculty: 39 full-time (2 women), 2 part-time (0 women). *Degree requirements:* For master's, diagnostic and cumulative exams required, foreign language not required; for doctorate, dissertation, diagnostic and cumulative exams required, foreign language not required. *Entrance requirements:* GRE General Test, minimum B average in undergraduate course work. Application deadline: 4/15. Application fee: $15. *Tuition:* $76.29 per credit hour for state residents; $238 per credit hour for nonresidents. • Dr. Edward K. Mellon, Chairman, 904-644-4074. Application contact: Dr. William T. Cooper, Graduate Adviser, 904-644-6875.

See full description on page 105.

Georgetown University, College of Arts and Sciences, Department of Chemistry, Washington, DC 20057. Offerings include inorganic chemistry (MS, PhD). Terminal master's awarded for partial completion of doctoral program. Department faculty: 17 full-time, 4 part-time. *Degree requirements:* For master's, thesis required (for some programs), foreign language not required; for doctorate, 1 foreign language, dissertation. *Entrance requirements:* TOEFL (minimum score of 550 required, 600 for teaching assistants). Application deadline: 7/1 (applications processed on a rolling basis). *Expenses:* Tuition of $12,768 per year full-time, $532 per credit part-time. Fees of $142 per year full-time. • Dr. Michael T. Pope, Chairman, 202-687-6073.

Harvard University, Graduate School of Arts and Sciences, Department of Chemistry, Cambridge, MA 02138. Offerings include inorganic chemistry (AM, PhD). Terminal master's awarded for partial completion of doctoral program. Department faculty: 20 full-time, 0 part-time. *Degree requirements:* For doctorate, 2 foreign languages, dissertation, cumulative exams. *Entrance requirements:* For doctorate, GRE General Test, GRE Subject Test. Application deadline: 1/2. Application fee: $60. *Expenses:* Tuition of $14,860 per year. Fees of $550 per year. • Dr. Donald J. Ciappenelli, Director of Graduate Studies, 617-495-4076.

Howard University, Graduate School of Arts and Sciences, Department of Chemistry, 2400 Sixth Street, NW, Washington, DC 20059. Offerings include inorganic chemistry (MS, PhD). *Degree requirements:* For master's, 1 foreign language (computer language can substitute), thesis, comprehensive exam; for doctorate, 2 foreign languages (computer language can substitute for one), dissertation, comprehensive exam. *Entrance requirements:* For master's, minimum GPA of 2.7. Application deadline: 4/1. Application fee: $25. *Expenses:* Tuition of $6100 per year full-time, $339 per credit hour part-time. Fees of $555 per year full-time, $245 per semester part-time.

Jackson State University, School of Science and Technology, Department of Chemistry, Jackson, MS 39217. Offerings include inorganic chemistry (MS). Department faculty: 14 full-time (3 women), 0 part-time. *Degree requirements:* Thesis (for some programs). *Entrance requirements:* GRE General Test. Application deadline: rolling. Application fee: $0. *Expenses:* Tuition of $1114 per semester full-time, $113 per hour part-time for state residents; $1142 per semester full-time, $113 per hour part-time for nonresidents. Fees of $730 per semester for nonresidents. • Dr. Richard Sullivan, Chair, 601-968-2171.

Kansas State University, College of Arts and Sciences, Department of Chemistry, Manhattan, KS 66506. Offerings include inorganic chemistry (MS, PhD). Department faculty: 19 full-time (0 women), 0 part-time. *Degree requirements:* For master's, thesis required, foreign language not required; for doctorate, dissertation. *Expenses:* Tuition of $1668 per year full-time, $51 per credit hour part-time for state residents; $5382 per year full-time, $142 per credit hour part-time for nonresidents. Fees of $305 per year full-time, $64.50 per semester part-time. • M. Dale Hawley, Head, 913-532-6668. Application contact: Robert Hammaker, Graduate Coordinator, 913-532-6671.

Kent State University, Department of Chemistry, Kent, OH 44242. Offerings include inorganic chemistry (MS, PhD). Department faculty: 18 full-time, 0 part-time. *Degree requirements:* For doctorate, dissertation required, foreign language not required. *Application deadline:* 7/12 (applications processed on a rolling basis). *Application fee:* $25. *Tuition:* $1601 per semester full-time, $133.75 per hour part-time for state residents; $3101 per semester full-time, $258.75 per hour part-time for nonresidents. • Dr. Roger K. Gilpin, Chairman, 216-672-2032.

See full description on page 111.

Lehigh University, College of Arts and Sciences, Department of Chemistry, Bethlehem, PA 18015. Department awards MS, DA, PhD. Faculty: 22 full-time, 4 part-time. Matriculated students: 59 full-time (23 women), 24 part-time (6 women); includes 3 minority (1 Asian American, 2 black American), 7 foreign. 72 applicants, 72% accepted. In 1990, 12 master's, 4 doctorates awarded. *Degree requirements:* For master's, foreign language and thesis not required; for doctorate, 1 foreign language, dissertation. *Entrance requirements:* GRE, TOEFL. Application deadline: rolling. Application fee: $40. *Tuition:* $15,650 per year full-time, $655 per credit hour part-time. *Financial aid:* Fellowships, research assistantships, teaching assistantships, institutionally sponsored loans, and career-related internships or fieldwork available. *Faculty research:* Biochemistry; inorganic, organic, physical, and polymer chemistry. • Dr. John Larsen, Chairman, 215-758-3471. Application contact: Dr. Daniel Zeroka, Graduate Program Coordinator.

See full description on page 113.

Marquette University, College of Arts and Sciences, Department of Chemistry, Milwaukee, WI 53233. Offerings include inorganic chemistry (MS, PhD). Terminal master's awarded for partial completion of doctoral program. Department faculty: 14 full-time (1 woman), 0 part-time. *Degree requirements:* For master's, thesis or alternative, comprehensive exam required, foreign language not required; for doctorate, dissertation, cumulative exams required, foreign language not required. *Entrance requirements:* TOEFL (minimum score of 550 required). Application fee: $25. *Tuition:* $300 per credit. • Dr. Michael McKinney, Chairman, 414-288-7065.

Massachusetts Institute of Technology, School of Science, Department of Chemistry, Cambridge, MA 02139. Offerings include inorganic chemistry (SM, PhD, Sc D). Terminal master's awarded for partial completion of doctoral program. Department faculty: 35 full-time (2 women), 0 part-time. *Degree requirements:* For master's, thesis required, foreign language not required; for doctorate, dissertation, written exams, comprehensive oral exam, oral presentation required, foreign language not required. *Application deadline:* 1/16. *Application fee:* $45. *Tuition:* $16,900 per year. • Robert J. Silbey, Chairman, 617-253-1801.

See full description on page 115.

Michigan State University, College of Natural Science, Department of Chemistry, East Lansing, MI 48824. Offerings include inorganic chemistry (MS, PhD). Terminal master's awarded for partial completion of doctoral program. Department faculty: 27 full-time (1 woman), 0 part-time. *Degree requirements:* For doctorate, 1 foreign language, dissertation. *Entrance requirements:* For doctorate, GRE General Test, TOEFL. Application deadline: rolling. Application fee: $25 ($40 for foreign students). *Tuition:* $104.75 per credit for state residents; $211.75 per credit for nonresidents. • Dr. Gerald Babcock, Chairperson, 517-353-9717. Application contact: Dr. William Reusch, Coordinator, Graduate Program, 517-355-9716.

Mississippi State University, College of Arts and Sciences, Department of Chemistry, Mississippi State, MS 39762. Offerings include inorganic chemistry (MS, PhD). Terminal master's awarded for partial completion of doctoral program. Department faculty: 15 full-time (0 women), 0 part-time. *Degree requirements:* 1 foreign language (computer language can substitute), thesis/dissertation. *Entrance requirements:* GRE General Test, TOEFL. Application deadline: rolling. *Expenses:* Tuition of $891 per semester full-time, $99 per credit hour part-time for state residents; $1622 per semester full-time, $180 per credit hour part-time for nonresidents. Fees of $221 per semester full-time, $25 per semester (minimum) part-time. • Dr. Kenneth L. Brown, Head, 601-325-3584.

New York University, Graduate School of Arts and Science, Department of Chemistry, New York, NY 10011. Offerings include inorganic chemistry (MS, PhD). Department faculty: 26 full-time, 5 part-time. *Degree requirements:* For master's, thesis required, foreign language not required; for doctorate, 1 foreign language, dissertation. *Entrance requirements:* GRE Subject Test, TSE. Application deadline: 1/15 (priority date, applications processed on a rolling basis). Application fee: $30. *Tuition:* $467 per credit. • Neville R. Kallenbach, Chairman, 212-998-8400. Application contact: David Schuster, Director of Graduate Studies, 212-998-8400.

See full description on page 117.

Northeastern University, Graduate School of Arts and Sciences, Department of Chemistry, Boston, MA 02115. Offerings include inorganic chemistry (MS). Department faculty: 16 full-time (2 women), 0 part-time. *Degree requirements:* Thesis (for some programs). *Entrance requirements:* TOEFL (minimum score of 580 required). Application deadline: 4/15. Application fee: $40. *Expenses:* Tuition of $10,260 per year full-time, $285 per quarter hour part-time. Fees of $293 per year full-time. • Dr. John L. Roebber, Executive Officer, 617-437-2822. Application contact: Dr. Paul Vouros, Director of Graduate Affairs, 617-437-2822.

See full description on page 119.

Ohio State University, College of Mathematical and Physical Sciences, Department of Chemistry, Columbus, OH 43210. Department awards MS, PhD. Faculty: 49 full-time (3 women), 0 part-time. Matriculated students: 270 full-time (72 women), 14 part-time (4 women); includes 19 minority (13 Asian American, 4 black American, 1 Hispanic American, 1 Native American), 101 foreign. 203 applicants, 29% accepted. In 1990, 23 master's, 31 doctorates awarded. *Degree requirements:* For master's, thesis optional, foreign language not required; for doctorate, dissertation required, foreign language not required. *Entrance requirements:* For master's, GRE General Test, GRE Subject Test. Application deadline: 8/15 (applications processed on a rolling basis). Application fee: $0 ($25 for foreign students). *Tuition:* $1213 per quarter full-time, $364 per course part-time for state residents; $3143 per quarter full-time, $943 per course part-time for nonresidents. *Financial aid:* In 1990–91, $12,000 in aid awarded. 24 fellowships (10 to first-year students), 93 research assistantships (0 to first-year students), 150 teaching assistantships (50 to first-year students) were awarded; federal work-study, institutionally sponsored loans also available. Aid available to part-time students. Financial aid applicants required to submit FAF. *Total annual research budget:* $5.01-million. • Russell M. Pitzer, Chairman, 614-292-2251. Application contact: C. Weldon Mathews, Vice Chair for Graduate Study, 614-292-8917.

See full description on page 121.

Oregon State University, Graduate School, College of Science, Department of Chemistry, Corvallis, OR 97331. Offerings include inorganic chemistry (MS, PhD). Terminal master's awarded for partial completion of doctoral program. Department faculty: 26 full-time (0 women), 5 part-time (1 woman). *Degree requirements:* For doctorate, 1 foreign language, dissertation. *Entrance requirements:* For doctorate, TOEFL (minimum score of 550 required), minimum GPA of 3.0 in last 90 hours. Application deadline: 3/1. Application fee: $40. *Tuition:* $1140 per trimester full-time, $449 per year part-time for state residents; $1816 per trimester full-time, $674 per year part-time for nonresidents. • Dr. Carroll W. DeKock, Chair, 503-737-2081.

See full description on page 123.

Purdue University, School of Science, Department of Chemistry, West Lafayette, IN 47907. Offerings include inorganic chemistry (MS, PhD). Terminal master's awarded for partial completion of doctoral program. Department faculty: 45 full-time, 1 part-time. *Degree requirements:* Thesis/dissertation required, foreign language not required. *Entrance requirements:* TOEFL (minimum score of 550 required). Application fee: $25. *Tuition:* $1162 per semester full-time, $83.50 per credit hour part-time for state residents; $3720 per semester full-time, $244.50 per credit hour part-time for nonresidents. • Dr. Harry Morrison, Head, 317-494-5200. Application contact: R. E. Wild, Chairman, Graduate Admissions, 317-494-5200.

Rensselaer Polytechnic Institute, School of Science, Department of Chemistry, Troy, NY 12180. Offerings include inorganic chemistry (MS, PhD). Department faculty: 26 full-time (3 women), 0 part-time. *Degree requirements:* 1 foreign language, thesis/dissertation. *Entrance requirements:* GRE General Test, GRE Subject Test, TOEFL. Application deadline: 2/1. Application fee: $30. *Expenses:* Tuition of $455 per credit hour. Fees of $195.57 per semester. • D. Aikens, Chairman, 518-276-8981. Application contact: Gerald Korenowski, Chairman, Graduate Committee, 518-276-8489.

See full description on page 127.

Rutgers, The State University of New Jersey, Newark, Department of Chemistry, Newark, NJ 07102. Offerings include inorganic chemistry (MS, PhD). Terminal master's awarded for partial completion of doctoral program. Department faculty: 13 full-time (1 woman), 5 part-time (1 woman). *Degree requirements:* For master's, cumulative exams required, thesis optional, foreign language not required; for doctorate, dissertation, cumulative exams required, foreign language not required. *Entrance requirements:* For master's, GRE General Test, GRE Subject Test, minimum undergraduate B average; for doctorate, GRE Subject Test, minimum undergraduate B average. Application deadline: 8/1. Application fee: $35. *Tuition:* $2216 per semester full-time for state residents; $3248 per semester full-time for nonresidents. • Dr. James Schlegel, Director, 201-648-5173.

See full description on page 129.

Rutgers, The State University of New Jersey, New Brunswick, Program in Chemistry, New Brunswick, NJ 08903. Offerings include inorganic chemistry (MS, PhD). Terminal master's awarded for partial completion of doctoral program. Program faculty: 48 full-time (5 women), 0 part-time. *Degree requirements:* For doctorate, dissertation, cumulative exams required, foreign language not required. *Entrance requirements:* For doctorate, GRE General Test, GRE Subject Test, TOEFL (minimum score of 575 required). Application deadline: 7/1. Application fee: $35. *Expenses:* Tuition of $4432 per year full-time, $183 per credit part-time for state residents; $6496 per year full-time, $270 per credit part-time for nonresidents. Fees of $458 per year full-time, $117 per year part-time. • Dr. Martha Cotter, Director, 908-932-2259.

See full description on page 131.

San Jose State University, School of Science, Department of Chemistry, San Jose, CA 95192. Offerings include inorganic chemistry (MS). Department faculty: 24 full-time (3 women), 5 part-time (2 women). *Degree requirements:* Thesis or alternative required, foreign language not required. *Entrance requirements:* GRE, minimum B average. Application deadline: 6/1 (applications processed on a rolling basis). Application fee: $55. *Expenses:* Tuition of $0 for state residents; $246 per unit for nonresidents. Fees of $592 per year. • Dr. Joseph Pesek, Chair, 408-924-5000. Application contact: Dr. Gerald Selter, Graduate Adviser, 408-924-4956.

Seton Hall University, College of Arts and Sciences, Department of Chemistry, South Orange, NJ 07079. Offerings include inorganic chemistry (PhD). Terminal master's awarded for partial completion of doctoral program. Department faculty: 14 full-time (2 women). *Degree requirements:* 1 foreign language (computer language can substitute), dissertation, comprehensive exams. *Entrance requirements:* GRE General Test, GRE Subject Test, TOEFL, undergraduate major in chemistry or related field. Application fee: $30. *Expenses:* Tuition of $346 per credit. Fees of $105 per semester. • Dr. Robert L. Augustine, Chair, 201-761-9414.

South Dakota State University, Colleges of Arts and Science and Agriculture and Biological Sciences, Department of Chemistry, Brookings, SD 57007. Offerings include inorganic chemistry (MS, PhD). *Degree requirements:* For doctorate, dissertation, research tool. *Entrance requirements:* For doctorate, TOEFL (minimum score of 500 required). *Tuition:* $553 per semester full-time, $61 per credit part-time for state residents; $1094 per semester full-time, $122 per credit part-time for nonresidents.

Southern University and Agricultural and Mechanical College, College of Sciences, Department of Chemistry, Baton Rouge, LA 70813. Offerings include inorganic chemistry (MS). MS (environmental science) offered jointly with Louisiana State University. Department faculty: 5 full-time (0 women), 0 part-time. *Entrance requirements:* GRE General Test or GMAT, TOEFL. Application deadline: 7/1. Application fee: $5. *Tuition:* $1594 per year full-time, $299 per semester part-time for state residents; $3010 per year full-time, $299 per semester part-time for nonresidents. • Dr. Earl Doomes, Chairman, 504-771-3990.

State University of New York at Binghamton, School of Arts and Sciences, Department of Chemistry, Binghamton, NY 13902-6000. Offerings include inorganic chemistry (PhD). Terminal master's awarded for partial completion of doctoral program. *Degree requirements:* Dissertation, cumulative exams required, foreign language not required. *Entrance requirements:* GRE General Test, GRE Subject Test, TOEFL. Application deadline: 4/15 (priority date). Application fee: $35. *Expenses:* Tuition of $2450 per year full-time, $102.50 per credit part-time for state residents; $5766 per year full-time, $242.50 per credit part-time for nonresidents. Fees of $77 per year full-time, $27.85 per semester (minimum) part-time. • Dr. Alistair J. Lees, Chairperson, 607-777-2362.

Syracuse University, College of Arts and Sciences, Department of Chemistry, Syracuse, NY 13244. Department awards MS, PhD. Faculty: 21 full-time. Matriculated students: 63 full-time (22 women), 1 part-time (0 women); includes 1 minority (Asian American), 36 foreign. Average age 28. 180 applicants, 21% accepted. In 1990, 5 master's, 8 doctorates awarded. *Degree requirements:* For master's, 1 foreign language; for doctorate, 1 foreign language, dissertation. *Entrance requirements:* GRE General Test, GRE Subject Test. Application deadline: rolling. Application fee: $35. *Expenses:* Tuition of $381 per credit. Fees of $289 per year full-time, $34.50 per semester part-time. *Financial aid:* In 1990–91, 4 fellowships, 35 research assistantships, 42 teaching assistantships awarded; partial tuition waivers, federal work-study also available. Financial aid application deadline: 3/1; applicants required to submit FAF. *Total annual research budget:* $1.5-million. • Laurence Nafie, Chair, 315-443-4109.

Texas A&M University, College of Science, Department of Chemistry, College Station, TX 77843. Offerings include inorganic chemistry (MS, PhD). Department faculty: 60. *Entrance requirements:* GRE General Test, TOEFL. Application deadline: 7/15. Application fee: $25 ($50 for foreign students). *Expenses:* Tuition of $100 per semester (minimum) for state residents; $128 per credit hour for nonresidents. Fees of $459 per year full-time, $252 per semester part-time. • Michael B. Hall, Head, 409-845-2011.

See full description on page 141.

Tufts University, Graduate School of Arts and Sciences, Department of Chemistry, Medford, MA 02155. Offerings include inorganic chemistry (MS, PhD). Terminal master's awarded for partial completion of doctoral program. Department faculty: 15 full-time, 2 part-time. *Degree requirements:* Thesis/dissertation required, foreign language not required. *Entrance requirements:* GRE General Test, GRE Subject Test, TOEFL (minimum score of 550 required). Application deadline: 2/15. Application fee: $50. *Expenses:* Tuition of $16,755 per year full-time, $2094 per course part-time. Fees of $885 per year. • David Walt, Head, 617-381-3470. Application contact: Edward Brush, 617-381-3475.

University of Akron, Buchtel College of Arts and Sciences, Department of Chemistry, Akron, OH 44325. Offerings include inorganic chemistry (MS, PhD). Terminal master's awarded for partial completion of doctoral program. Department faculty: 20 full-time (1 woman), 3 part-time (0 women). *Degree requirements:* For master's, 1 foreign language (computer language can substitute), thesis, seminar presentation; for doctorate, 2 foreign languages (computer language can substitute for one), dissertation, cumulative exams. *Entrance requirements:* TOEFL. Application deadline: 3/1 (applications processed on a rolling basis). Application fee: $25. *Tuition:* $119.93 per credit hour for state residents; $210.93 per credit hour for nonresidents. • Dr. G. Edwin Wilson, Head, 216-972-7372.

University of Calgary, Faculty of Science, Department of Chemistry, Calgary, AB T2N 1N4, Canada. Offerings include inorganic chemistry (M Sc, PhD). *Degree requirements:* For master's, thesis required, foreign language not required; for doctorate, dissertation, candidacy exam required, foreign language not required. *Application deadline:* 5/31. *Application fee:* $25. *Tuition:* $1705 per year full-time, $427 per course part-time for Canadian residents; $3410 per year full-time, $854 per course part-time for nonresidents.

University of Cincinnati, McMicken College of Arts and Sciences, Department of Chemistry, Cincinnati, OH 45221. Offerings include inorganic chemistry (MS, PhD). Terminal master's awarded for partial completion of doctoral program. Department faculty: 26 full-time, 0 part-time. *Degree requirements:* Thesis/dissertation required, foreign language not required. *Entrance requirements:* GRE General Test, GRE Subject Test. Application deadline: 2/15. Application fee: $20. *Tuition:* $131 per credit hour for state residents; $261 per credit hour for nonresidents. • Dr. Bruce Ault, Head, 513-556-9200.

University of Florida, College of Liberal Arts and Sciences, Department of Chemistry, Gainesville, FL 32611. Department awards MS, MST, PhD. PhD offered jointly with Florida Atlantic University. Faculty: 51. Matriculated students: 149 full-time (38 women), 16 part-time (2 women); includes 13 minority (2 Asian American, 5 black American, 6 Hispanic American), 38 foreign. 101 applicants, 52% accepted. In 1990, 6 master's, 28 doctorates awarded. *Degree requirements:* For master's, thesis required, foreign language not required; for doctorate, 1 foreign language, dissertation. *Entrance requirements:* GRE General Test, minimum GPA of 3.0. Application deadline: 5/30 (priority date, applications processed on a rolling basis). Application fee: $15. *Tuition:* $87 per credit hour for state residents; $289 per credit hour for nonresidents. *Financial aid:* In 1990–91, 1 fellowship, 63 research assistantships, 71 teaching assistantships awarded; institutionally sponsored loans also available. Average monthly stipend for a graduate assistantship: $1000. *Faculty research:* Organic, analytical, physical, inorganic and biological chemistry. Total annual research budget: $4.9-million. • Dr. Michael Zerner, Chairman, 904-392-

Directory: Inorganic Chemistry

0541. Application contact: Dr. John F. Helling, Chair, Graduate Selections Committee, 904-392-0551.

See full description on page 145.

University of Georgia, College of Arts and Sciences, Department of Chemistry, Athens, GA 30602. Offerings include inorganic chemistry (MS, PhD). Department faculty: 27 full-time (0 women), 0 part-time. *Degree requirements:* For master's, thesis required, foreign language not required; for doctorate, 1 foreign language (computer language can substitute), dissertation. *Entrance requirements:* GRE General Test. Application fee: $10. *Expenses:* Tuition of $598 per quarter full-time, $48 per quarter part-time for state residents; $1558 per quarter full-time, $144 per quarter part-time for nonresidents. Fees of $118 per quarter. • Dr. Henry F. Schaefer III, Graduate Coordinator, 404-542-2067.

See full description on page 147.

University of Houston, College of Natural Sciences and Mathematics, Department of Chemistry, 4800 Calhoun, Houston, TX 77004. Offerings include inorganic chemistry (MS, PhD). Terminal master's awarded for partial completion of doctoral program. Department faculty: 25 full-time (1 woman), 2 part-time (0 women). *Degree requirements:* Thesis/dissertation required, foreign language not required. *Entrance requirements:* GRE General Test (minimum combined score of 1100 required), TOEFL (minimum score of 575 required). Application deadline: 5/1 (applications processed on a rolling basis). Application fee: $25 ($75 for foreign students). *Expenses:* Tuition of $30 per hour for state residents; $134 per hour for nonresidents. Fees of $240 per year full-time, $125 per year part-time. • Dr. John L. Bear, Chairman, 713-749-2647. Application contact: Dr. Thomas Albright, Chair, Graduate Committee, 713-749-3721.

See full description on page 149.

University of Louisville, College of Arts and Sciences, Department of Chemistry, Louisville, KY 40292. Offerings include inorganic chemistry (MS, PhD). Department faculty: 18 full-time (2 women), 2 part-time. *Degree requirements:* For master's, thesis; for doctorate, 1 foreign language, dissertation. *Entrance requirements:* GRE General Test, TOEFL (minimum score of 550 required). Application deadline: rolling. *Expenses:* Tuition of $1780 per year full-time, $99 per credit hour part-time for state residents; $5340 per year full-time, $297 per credit hour part-time for nonresidents. Fees of $60 per semester full-time, $12.50 per semester (minimum) part-time. • Dr. Richard P. Baldwin, Chair, 502-588-6798.

University of Maryland College Park, College of Life Sciences, Department of Chemistry and Biochemistry, College Park, MD 20742. Offerings include inorganic chemistry (MS, PhD). Department faculty: 51 (5 women). *Degree requirements:* For master's, thesis or alternative required, foreign language not required; for doctorate, dissertation. *Entrance requirements:* For master's, GRE General Test, minimum GPA of 3.0; for doctorate, GRE General Test. Application deadline: rolling. Application fee: $25. *Expenses:* Tuition of $143 per credit hour for state residents; $256 per credit hour for nonresidents. Fees of $171.50 per semester. • Dr. Sandra Greer, Chairman, 301-405-1788.

See full description on page 155.

University of Miami, College of Arts and Sciences, Department of Chemistry, Coral Gables, FL 33124. Offerings include inorganic chemistry (PhD). Department faculty: 10 full-time (1 woman), 0 part-time. *Degree requirements:* Dissertation required, foreign language not required. *Entrance requirements:* GRE General Test (minimum combined score of 1000 required), TOEFL (minimum score of 550 required). Application deadline: 5/1 (priority date, applications processed on a rolling basis). Application fee: $0. *Expenses:* Tuition of $567 per credit hour. Fees of $87 per semester full-time. • Dr. Cecil Criss, Chairman, 305-284-2174.

See full description on page 159.

University of Michigan, College of Literature, Science, and the Arts, Department of Chemistry, Ann Arbor, MI 48109. Offerings include inorganic chemistry (MS, PhD). Department faculty: 38. *Degree requirements:* For master's, foreign language and thesis not required; for doctorate, dissertation, preliminary exam required, foreign language not required. *Entrance requirements:* For master's, GRE General Test, GRE Subject Test; for doctorate, GRE General Test. Application deadline: 4/1 (applications processed on a rolling basis). Application fee: $30. *Tuition:* $3255 per semester full-time, $352 per credit (minimum) part-time for state residents; $6803 per semester full-time, $746 per credit (minimum) part-time for nonresidents. • M. David Curtis III, Chair, 313-764-7314.

See full description on page 161.

University of Missouri–Columbia, College of Arts and Sciences, Department of Chemistry, Columbia, MO 65211. Offerings include inorganic chemistry (MS, PhD). Department faculty: 23 full-time, 1 part-time. *Degree requirements:* For master's, thesis required, foreign language not required; for doctorate, 1 foreign language, dissertation. *Entrance requirements:* GRE General Test, minimum GPA of 3.0. Application deadline: 8/1 (priority date, applications processed on a rolling basis). Application fee: $20 ($40 for foreign students). *Expenses:* Tuition of $89.90 per credit hour full-time, $98.35 per credit hour part-time for state residents; $244 per credit hour full-time, $252.45 per credit hour part-time for nonresidents. Fees of $123.55 per semester (minimum) full-time. • Dr. John Bauman, Director of Graduate Studies, 314-882-7720. Application contact: Gary L. Smith, Director of Admissions and Registrar, 314-882-7651.

University of Missouri–Kansas City, College of Arts and Sciences, Department of Chemistry, Kansas City, MO 64110. Offerings include inorganic chemistry (MS, PhD). Department faculty: 15 full-time (0 women), 0 part-time. *Degree requirements:* For master's, thesis required, foreign language not required; for doctorate, 1 foreign language, dissertation, comprehensive exam. *Entrance requirements:* GRE. Application fee: $0. *Expenses:* Tuition of $2200 per year full-time, $92 per credit hour part-time for state residents; $5503 per year full-time, $229 per credit hour part-time for nonresidents. Fees of $122 per semester full-time, $9 per credit hour part-time. • Dr. Layton L. McCoy, Chairperson, 816-235-2280. Application contact: Dr. T. F. Thomas, Chairman, Graduate Admissions Committee, 816-235-2297.

University of Montana, College of Arts and Sciences, Department of Chemistry, Missoula, MT 59812. Offerings include inorganic chemistry (PhD). Terminal master's awarded for partial completion of doctoral program. *Degree requirements:* 1 foreign language, dissertation. *Entrance requirements:* GRE General Test, GRE Subject Test. Application deadline: 9/15. Application fee: $20. *Tuition:* $495 per quarter hour full-time for state residents; $1239 per quarter hour full-time for nonresidents.

University of Nebraska–Lincoln, College of Arts and Sciences, Department of Chemistry, Lincoln, NE 68588. Offerings include inorganic chemistry (PhD). Terminal master's awarded for partial completion of doctoral program. Department faculty: 22 full-time (2 women), 0 part-time. *Degree requirements:* 1 foreign language, dissertation, comprehensive exams, departmental qualifying exam. *Entrance requirements:* GRE Subject Test, TOEFL (minimum score of 550 required), TSE (minimum score of 230 required). Application deadline: 5/1 (priority date, applications processed on a rolling basis). Application fee: $25. *Expenses:* Tuition of $75.75 per credit hour for state residents; $187.25 per credit hour for nonresidents. Fees of $161 per year full-time. • Dr. Pill-Soon Song Jr., Chairperson, 402-472-3501.

See full description on page 165.

University of Notre Dame, College of Science, Department of Chemistry and Biochemistry, Notre Dame, IN 46556. Offerings include inorganic chemistry (MS, PhD). Terminal master's awarded for partial completion of doctoral program. Department faculty: 31 full-time (2 women), 0 part-time. *Degree requirements:* For master's, thesis, comprehensive exam required, foreign language not required; for doctorate, dissertation, qualifying exam required, foreign language not required. *Entrance requirements:* GRE, TOEFL. Application deadline: 2/15 (priority date, applications processed on a rolling basis). Application fee: $25. *Tuition:* $13,385 per year full-time, $744 per credit hour part-time. • Dr. Paul M. Helquist, Chairman, 219-239-7487. Application contact: Dr. Michael Chetcuti, Coordinator, 219-239-5216.

University of Pittsburgh, Faculty of Arts and Sciences, Department of Chemistry, Pittsburgh, PA 15260. Department awards MS, PhD. Part-time programs available. Faculty: 37 full-time (2 women), 2 part-time (0 women). Matriculated students: 174 full-time (48 women), 17 part-time (6 women); includes 3 minority (2 Asian American, 1 black American), 63 foreign. 351 applicants, 12% accepted. In 1990, 6 master's, 19 doctorates awarded. Terminal master's awarded for partial completion of doctoral program. *Degree requirements:* Thesis/dissertation required, foreign language not required. *Entrance requirements:* GRE General Test, GRE Subject Test, TOEFL. Application deadline: 3/31 (priority date, applications processed on a rolling basis). Application fee: $15 ($25 for foreign students). *Expenses:* Tuition of $2920 per semester full-time, $241 per credit part-time for state residents; $5840 per semester full-time, $482 per credit part-time for nonresidents. Fees of $156 per year. *Financial aid:* In 1990–91, 189 students received a total of $2.140-million in aid awarded. 16 fellowships (0 to first-year students), 89 research assistantships (1 to a first-year student), 84 teaching assistantships (40 to first-year students) were awarded; full tuition waivers, federal work-study, institutionally sponsored loans also available. Aid available to part-time students. Financial aid application deadline: 3/15; applicants required to submit FAF. *Faculty research:* Analytical chemistry, inorganic chemistry, organic chemistry, physical chemistry and surface science. Total annual research budget: $7.657-million. • Dr. N. John Cooper, Chairman, 412-624-8500. Application contact: Rebecca Claycamp, Assistant Chairman, 412-624-8200.

See full description on page 167.

University of Regina, Faculty of Graduate Studies and Research, Faculty of Science, Department of Chemistry, Regina, SK S4S 0A2, Canada. Offerings include inorganic chemistry (M Sc, PhD). Department faculty: 14 full-time, 0 part-time. *Degree requirements:* For master's, thesis, departmental qualifying exam required, foreign language not required; for doctorate, variable foreign language requirement, dissertation, departmental qualifying exam. *Application deadline:* 7/2 (applications processed on a rolling basis). *Tuition:* $1500 per year full-time, $242 per year (minimum) part-time. • Dr. W. D. Chandler, Head, 306-585-4146.

University of San Francisco, College of Arts and Sciences, Department of Chemistry, San Francisco, CA 94117-1080. Offerings include inorganic chemistry (MS). Department faculty: 6 full-time (0 women), 0 part-time. *Degree requirements:* 1 foreign language, thesis. *Entrance requirements:* TOEFL (minimum score of 550 required). Application deadline: 5/15 (priority date). Application fee: $40 ($90 for foreign students). *Tuition:* $10,960 per year full-time, $432 per unit part-time. • Tom Gruhn, Chairman, 415-666-6486. Application contact: Jeff Curtis, Graduate Adviser, 415-666-6391.

See full description on page 171.

University of Southern Mississippi, College of Science and Technology, Department of Chemistry and Biochemistry, Hattiesburg, MS 39406. Offerings include inorganic chemistry (MS, PhD). Department faculty: 18 full-time (2 women), 1 (woman) part-time. *Degree requirements:* For doctorate, 2 foreign languages (computer language can substitute for one), dissertation. *Entrance requirements:* For master's, GRE General Test, minimum GPA of 2.75; for doctorate, GRE General Test, minimum GPA of 3.5. Application deadline: 8/9 (priority date, applications processed on a rolling basis). Application fee: $0 ($25 for foreign students). *Expenses:* Tuition of $968 per semester full-time, $93 per semester hour part-time. Fees of $12 per semester part-time for state residents; $591 per year full-time, $12 per semester part-time for nonresidents. • Dr. David Creed, Chair, 601-266-4701. Application contact: Dr. David Wertz, Professor, 601-266-4702.

University of South Florida, College of Arts and Sciences, Department of Chemistry, Tampa, FL 33620. Offerings include inorganic chemistry (MS, PhD). Department faculty: 26 full-time. *Degree requirements:* For master's, thesis; for doctorate, 1 foreign language, computer language, dissertation. *Entrance requirements:* For master's, GRE General Test (minimum combined score of 1000 required), minimum GPA of 3.0 (in last 60 credit hours). Application fee: $15. *Tuition:* $79.40 per credit hour for state residents; $241.33 per credit hour for nonresidents. • Stewart W. Schneller Jr., Chairperson, 813-974-2258. Application contact: Terence Oestereich, Graduate Program Director, 813-974-2534.

See full description on page 173.

University of Texas at Austin, Graduate School, College of Natural Sciences, Department of Chemistry and Biochemistry, Austin, TX 78712. Department offers programs in analytical chemistry (MA, PhD), biochemistry (MA, PhD), inorganic chemistry (MA, PhD), organic chemistry (MA, PhD), physical chemistry (MA, PhD). Matriculated students: 240 full-time (73 women), 0 part-time; includes 18 minority (10 Asian American, 2 black American, 4 Hispanic American, 2 Native American), 73 foreign. 197 applicants, 61% accepted. In 1990, 15 master's, 37 doctorates awarded. *Entrance requirements:* GRE. Application deadline: 2/1 (priority date, applications processed on a rolling basis). Application fee: $40 ($75 for foreign students). *Tuition:* $510.30 per semester for state residents; $1806 per semester for nonresidents. *Financial aid:* Fellowships, research assistantships, teaching assistantships available. Financial aid application deadline: 3/1. • Application contact: Dr. Peter J. Rossky, Graduate Adviser, 512-471-3890.

See full description on page 175.

University of Toledo, College of Arts and Sciences, Department of Chemistry, Toledo, OH 43606. Offerings include inorganic chemistry (MS, PhD). Department faculty: 20 full-time (2 women), 0 part-time. *Degree requirements:* For master's, thesis required, foreign language not required; for doctorate, 1 foreign language, dissertation. *Entrance requirements:* GRE General Test, GRE Subject Test, TOEFL. Application deadline: 9/8 (priority date). Application fee: $30. *Tuition:* $122.59 per credit hour for state residents; $193.40 per credit hour for nonresidents. • Dr. James L. Fry, Chairman, 419-537-2109. Application contact: Dr. Max Funk, Chairman, Graduate Admissions Committee, 419-537-2254.

Wake Forest University, Department of Chemistry, Winston-Salem, NC 27109. Offerings include inorganic chemistry (MS, PhD). Department faculty: 10 full-time (0 women), 3 part-time (1 woman). *Degree requirements:* 1 foreign language (computer language can substitute), thesis/dissertation. *Entrance requirements:* GRE General Test, GRE Subject Test. Application deadline: 3/1. Application fee: $25. *Tuition:* $10,800 per year full-time, $280 per hour part-time. • Dr. Willie Hinze, Chairman, 919-759-5325.

Washington State University, College of Sciences and Arts, Division of Sciences, Department of Chemistry, Pullman, WA 99164. Offerings include inorganic chemistry (MS, PhD). Terminal master's awarded for partial completion of doctoral program. Department faculty: 26 (2 women). *Degree requirements:* For master's, thesis, teaching experience required, foreign language not required; for doctorate, dissertation, teaching experience. *Entrance requirements:* GRE General Test, minimum GPA of 3.0. Application deadline: 3/1 (priority date, applications processed on a rolling basis). Application fee: $25. *Tuition:* $1694 per semester full-time, $169 per credit hour part-time for state residents; $4236 per semester full-time, $424 per credit hour part-time for nonresidents. • Dr. Roy Filby, Chair, 509-335-1516.

Wesleyan University, Department of Chemistry, Middletown, CT 06459. Offerings include inorganic chemistry (MA, PhD). Terminal master's awarded for partial completion of doctoral program. Department faculty: 14 full-time (1 woman), 2 part-time (0 women). *Degree requirements:* 1 foreign language (computer language can substitute), thesis/dissertation. *Entrance requirements:* For master's, GRE General Test, GRE Subject Test; for doctorate, GRE Subject Test. Application deadline: 2/1. Application fee: $0. *Expenses:* Tuition of $2035 per course. Fees of $627 per year full-time. • Dr. Rex F. Pratt, Graduate Adviser, 203-347-9411 Ext. 2380.

See full description on page 179.

Yale University, Graduate School of Arts and Sciences, Department of Chemistry, New Haven, CT 06520. Offerings include inorganic chemistry (PhD). Terminal master's awarded for partial completion of doctoral program. Department faculty: 22 full-time (1 woman), 0 part-time. *Degree requirements:* 1 foreign language, dissertation. *Entrance requirements:* GRE General Test, GRE Subject Test, TOEFL. Application deadline: 1/2. Application fee: $45. *Tuition:* $15,160 per year full-time, $1895 per course part-time. • Dr. Frederick Ziegler, Chairman, 203-432-3916. Application contact: Susan Webb, Coordinator of Admissions, 203-432-2770.

See full description on page 181.

Organic Chemistry

Boston College, Graduate School of Arts and Sciences, Department of Chemistry, Chestnut Hill, MA 02167. Offerings include organic chemistry (MS, PhD). *Degree requirements:* For master's, 1 foreign language, thesis; for doctorate, 1 foreign language, dissertation, foreign language exam, qualifying exam. *Entrance requirements:* GRE General Test, GRE Subject Test. *Tuition:* $412 per credit hour.

See full description on page 89.

Boston University, Graduate School, Department of Chemistry, Boston, MA 02215. Offerings include organic chemistry (MA, PhD). Terminal master's awarded for partial completion of doctoral program. Department faculty: 20 full-time (0 women), 0 part-time. *Degree requirements:* For master's, 1 foreign language required, thesis not required; for doctorate, 1 foreign language, dissertation. *Entrance requirements:* GRE General Test, GRE Subject Test, TOEFL (minimum score of 550 required). Application deadline: 4/1. Application fee: $45. *Expenses:* Tuition of $15,950 per year full-time, $498 per credit part-time. Fees of $120 per year full-time, $35 per semester part-time. • Guilford Jones II, Chairman, 617-353-2515.

See full description on page 91.

Brandeis University, Graduate School of Arts and Sciences, Program in Chemistry, Waltham, MA 02254. Offerings include organic chemistry (MA, PhD). Terminal master's awarded for partial completion of doctoral program. Program faculty: 16 full-time (2 women), 0 part-time. *Degree requirements:* For master's, 1 year residency required, foreign language and thesis not required; for doctorate, 1 foreign language (computer language can substitute), dissertation, 2 year residency, 2 seminars, qualifying exams, minimum GPA of 3.0. *Entrance requirements:* GRE, TOEFL. Application deadline: 2/28 (priority date, applications processed on a rolling basis). Application fee: $50. *Tuition:* $16,085 per year full-time, $2015 per course part-time. • Dr. Peter Jordan, Chair, 617-736-2503. Application contact: Susan Isaacs, Secretary, 617-736-2593.

See full description on page 93.

Brigham Young University, College of Physical and Mathematical Sciences, Department of Chemistry, Provo, UT 84602. Offerings include organic chemistry (MS, PhD). Department faculty: 34 full-time (1 woman), 0 part-time. *Degree requirements:* 2 foreign languages (computer language can substitute for one), thesis/dissertation. *Entrance requirements:* For master's, minimum GPA of 3.0 during previous 60 credit hours. Application deadline: 3/15 (priority date). Application fee: $30. *Tuition:* $1170 per semester (minimum) full-time, $130 per credit hour (minimum) part-time. • Dr. Earl M. Woolley, Chairman, 801-378-3667. Application contact: Francis R. Nordmeyer, Coordinator, Graduate Admissions, 801-378-4845.

See full description on page 95.

California State University, Fullerton, School of Natural Science and Mathematics, Department of Chemistry and Biochemistry, Fullerton, CA 92634. Offerings include organic chemistry (MS). Department faculty: 20 full-time (5 women), 13 part-time (2 women). *Degree requirements:* Thesis, departmental qualifying exam required, foreign language not required. *Entrance requirements:* Minimum GPA of 2.5 in last 60 units, major in chemistry or related field. Application fee: $55. *Expenses:* Tuition of $0 for state residents; $246 per unit for nonresidents. Fees of $554 per semester full-time, $356 per unit part-time. • Dr. Glen Nagel, Chair, 714-773-3621. Application contact: Dr. Gene Hiegel, Adviser, 714-773-3624.

California State University, Los Angeles, School of Natural and Social Sciences, Department of Chemistry and Biochemistry, Los Angeles, CA 90032. Offerings include organic chemistry (MS). Department faculty: 12 full-time, 25 part-time. *Degree requirements:* 1 foreign language (computer language can substitute), thesis or comprehensive exam. *Entrance requirements:* TOEFL (minimum score of 550 required). Application deadline: 8/7 (applications processed on a rolling basis). Application fee: $55. *Expenses:* Tuition of $0 for state residents; $164 per unit for nonresidents. Fees of $1046 per year full-time, $650 per semester (minimum) part-time. • Dr. Donald Paulson, Acting Chair, 213-343-2300.

Clark Atlanta University, School of Arts and Sciences, Department of Chemistry, Atlanta, GA 30314. Offerings include organic chemistry (MS, PhD). *Degree requirements:* For master's, 1 foreign language (computer language can substitute), thesis, comprehensive exam. *Entrance requirements:* For master's, minimum GPA of 2.5 in math and science. Application deadline: 4/1 (applications processed on a rolling basis). Application fee: $40. *Expenses:* Tuition of $4860 per year. Fees of $300 per year. • Dr. Alfred Spriggs, Chair, 404-880-8154. Application contact: Peggy Wade, Marketing and Recruitment, 404-880-8427.

Clarkson University, School of Science, Department of Chemistry, Potsdam, NY 13699. Offerings include organic chemistry (MS, PhD). Department faculty: 12 full-time (0 women), 0 part-time. *Degree requirements:* For master's, foreign language and thesis not required; for doctorate, dissertation, departmental qualifying exam required, foreign language not required. *Application fee:* $10. *Expenses:* Tuition of $446 per credit hour. Fees of $75 per semester. • Dr. Phillip A. Christiansen, Chairman, 315-268-2389.

Clemson University, College of Sciences, Department of Chemistry, Clemson, SC 29634. Department awards MS, PhD. Faculty: 19 full-time (1 woman), 0 part-time. Matriculated students: 65 full-time (14 women), 0 part-time; includes 4 minority (1 Asian American, 3 black American), 30 foreign. Average age 25. 40 applicants, 38% accepted. In 1990, 4 master's awarded; 12 doctorates awarded (67% entered university research/teaching, 33% found other work related to degree). *Degree requirements:* 1 foreign language (computer language can substitute), thesis/dissertation. *Entrance requirements:* GRE General Test. Application deadline: 6/1. Application fee: $25. *Expenses:* Tuition of $102 per credit hour. Fees of $80 per semester full-time. *Financial aid:* In 1990–91, $500,000 in aid awarded. 1 fellowship (to a first-year student), 18 research assistantships (1 to a first-year student), 43 lab assistantships were awarded; teaching assistantships also available. *Faculty research:* Flourine chemistry, organic synthetic methods and natural products, metal and non-metal clusters, analytical spectroscopies, polymers. Total annual research budget: $2.45-million. • J. D. Petersen, Head, 803-656-5017.

Cleveland State University, College of Arts and Sciences, Department of Chemistry, Cleveland, OH 44115. Offerings include organic chemistry (MS). Department faculty: 18 full-time (1 woman), 4 part-time (0 women). *Degree requirements:* Thesis required (for some programs), foreign language not required. *Entrance requirements:* GRE General Test, GRE Subject Test, TOEFL (minimum score of 525 required). Application deadline: 9/1 (priority date, applications processed on a rolling basis). Application fee: $0. *Tuition:* $90 per credit for state residents; $180 per credit for nonresidents. • Dr. Robert Towns, Chair, 216-687-2004. Application contact: Dr. Thomas Flechtner, Chairman, Graduate Committee, 216-687-2458.

Columbia University, Graduate School of Arts and Sciences, Division of Natural Sciences, Department of Chemistry, New York, NY 10027. Offerings include organic chemistry (M Phil, MS, PhD). Department faculty: 17 full-time. *Degree requirements:* For master's, comprehensive exams (MS); 1 foreign language, teaching experience, oral and written exams (M Phil) required, thesis not required; for doctorate, 1 foreign language, dissertation, M Phil. *Entrance requirements:* GRE General Test, GRE Subject Test, TOEFL. Application deadline: 1/5. Application fee: $50. *Expenses:* Tuition of $7836 per semester for state residents; $426 per credit for nonresidents. Fees of $148 per semester for state residents. • Richard Bersohn, Chair, 212-854-2192.

See full description on page 101.

Florida State University, College of Arts and Sciences, Department of Chemistry, Tallahassee, FL 32306. Offerings include organic chemistry (MS, PhD). Terminal master's awarded for partial completion of doctoral program. Department faculty: 39 full-time (2 women), 2 part-time (0 women). *Degree requirements:* For master's, diagnostic and cumulative exams required, foreign language not required; for doctorate, dissertation, diagnostic and cumulative exams required, foreign language not required. *Entrance requirements:* GRE General Test, minimum B average in undergraduate course work. Application deadline: 4/15. Application fee: $15. *Tuition:* $76.29 per credit hour for state residents; $238 per credit hour for nonresidents. • Dr. Edward K. Mellon, Chairman, 904-644-4074. Application contact: Dr. William T. Cooper, Graduate Adviser, 904-644-6875.

See full description on page 105.

Georgetown University, College of Arts and Sciences, Department of Chemistry, Washington, DC 20057. Offerings include organic chemistry (MS, PhD). Terminal master's awarded for partial completion of doctoral program. Department faculty: 17 full-time, 4 part-time. *Degree requirements:* For master's, thesis required (for some programs), foreign language not required; for doctorate, 1 foreign language, dissertation. *Entrance requirements:* TOEFL (minimum score of 550 required, 600 for teaching assistants). Application deadline: 7/1 (applications processed on a rolling basis). *Expenses:* Tuition of $12,768 per year full-time, $532 per credit part-time. Fees of $142 per year full-time. • Dr. Michael T. Pope, Chairman, 202-687-6073.

Harvard University, Graduate School of Arts and Sciences, Department of Chemistry, Cambridge, MA 02138. Offerings include organic chemistry (AM, PhD). Terminal master's awarded for partial completion of doctoral program. Department faculty: 20 full-time, 0 part-time. *Degree requirements:* For doctorate, 2 foreign languages, dissertation, cumulative exams. *Entrance requirements:* For doctorate, GRE General Test, GRE Subject Test. Application deadline: 1/2. Application fee: $60. *Expenses:* Tuition of $14,860 per year. Fees of $550 per year. • Dr. Donald J. Ciappenelli, Director of Graduate Studies, 617-495-4076.

Harvard University, Medical School and Graduate School of Arts and Sciences, Division of Medical Sciences, Tri-Department Program, Department of Biological Chemistry and Molecular Pharmacology, Boston, MA 02115. Department offers program in toxicology (PhD), including biological chemistry and molecular

Directory: Organic Chemistry

pharmacology. Matriculated students: 0 part-time. Terminal master's awarded for partial completion of doctoral program. *Degree requirements:* For doctorate, dissertation, qualifying exam required, foreign language not required. *Entrance requirements:* For doctorate, GRE General Test, GRE Subject Test. Application deadline: 1/2. Application fee: $60. *Tuition:* $15,410 per year. *Financial aid:* Fellowships, research assistantships, teaching assistantships, full tuition waivers, institutionally sponsored loans available. Financial aid application deadline: 1/2. *Faculty research:* Cellular and molecular mechanisms of drug action emphasis on basic approches to chemotherapy, neuropharmacology, membrane biology, endocrinology, and toxicology; molecular mechanisms of receptor and drug enzyme interactions; genetics and molecular biology of DNA application and transcription; molecular aspects of membrane protein function. • Dr. Christopher Walsh, Chair, 617-432-1715. Application contact: Dr. Donald M. Coen, Chair, Graduate Admissions Committee, 617-432-1691.

Howard University, Graduate School of Arts and Sciences, Department of Chemistry, 2400 Sixth Street, NW, Washington, DC 20059. Offerings include organic chemistry (MS, PhD). *Degree requirements:* For master's, 1 foreign language (computer language can substitute), thesis, comprehensive exam; for doctorate, 2 foreign languages (computer language can substitute for one), dissertation, comprehensive exam. *Entrance requirements:* For master's, minimum GPA of 2.7. Application deadline: 4/1. Application fee: $25. *Expenses:* Tuition of $6100 per year full-time, $339 per credit hour part-time. Fees of $555 per year full-time, $245 per semester part-time.

Instituto Tecnológico y de Estudios Superiores de Monterrey, Program in Chemistry, Monterrey, Nuevo León 64849, Mexico. Offerings include organic chemistry (MS). Program faculty: 3 full-time (1 woman), 0 part-time. *Degree requirements:* Thesis. *Entrance requirements:* Graduate admission exam (minimum score of 450 required). *Tuition:* $2350 per trimester full-time. • Dr. Xorge Domínquez, Director, 83-58-3300 Ext. 4510. Application contact: Julieta Mier, Dean of Admissions, 83-58-4650.

Jackson State University, School of Science and Technology, Department of Chemistry, Jackson, MS 39217. Offerings include organic chemistry (MS). Department faculty: 14 full-time (3 women), 0 part-time. *Degree requirements:* Thesis (for some programs). *Entrance requirements:* GRE General Test. Application deadline: rolling. Application fee: $0. *Expenses:* Tuition of $1114 per semester full-time, $113 per hour part-time for state residents; $1142 per semester full-time, $113 per hour part-time for nonresidents. Fees of $730 per semester for nonresidents. • Dr. Richard Sullivan, Chair, 601-968-2171.

Kansas State University, College of Arts and Sciences, Department of Chemistry, Manhattan, KS 66506. Offerings include organic chemistry (MS, PhD). Department faculty: 19 full-time (0 women), 0 part-time. *Degree requirements:* For master's, thesis required, foreign language not required; for doctorate, dissertation. *Expenses:* Tuition of $1668 per year full-time, $51 per credit hour part-time for state residents; $5382 per year full-time, $142 per credit hour part-time for nonresidents. Fees of $305 per year full-time, $64.50 per semester part-time. • M. Dale Hawley, Head, 913-532-6668. Application contact: Robert Hammaker, Graduate Coordinator, 913-532-6671.

Kent State University, Department of Chemistry, Kent, OH 44242. Offerings include organic chemistry (MS, PhD). Department faculty: 18 full-time, 0 part-time. *Degree requirements:* For doctorate, dissertation required, foreign language not required. *Application deadline:* 7/12 (applications processed on a rolling basis). *Application fee:* $25. *Tuition:* $1601 per semester full-time, $133.75 per hour part-time for state residents; $3101 per semester full-time, $258.75 per hour part-time for nonresidents. • Dr. Roger K. Gilpin, Chairman, 216-672-2032.

See full description on page 111.

Lehigh University, College of Arts and Sciences, Department of Chemistry, Bethlehem, PA 18015. Department awards MS, DA, PhD. Faculty: 22 full-time, 4 part-time. Matriculated students: 59 full-time (23 women), 24 part-time (6 women); includes 3 minority (1 Asian American, 2 black American), 7 foreign. 72 applicants, 72% accepted. In 1990, 12 master's, 4 doctorates awarded. *Degree requirements:* For master's, foreign language and thesis not required; for doctorate, 1 foreign language, dissertation. *Entrance requirements:* GRE, TOEFL. Application deadline: rolling. Application fee: $40. *Tuition:* $15,650 per year full-time, $655 per credit hour part-time. *Financial aid:* Fellowships, research assistantships, teaching assistantships, institutionally sponsored loans, and career-related internships or fieldwork available. *Faculty research:* Biochemistry; inorganic, organic, physical, and polymer chemistry. • Dr. John Larsen, Chairman, 215-758-3471. Application contact: Dr. Daniel Zeroka, Graduate Program Coordinator.

See full description on page 113.

Marquette University, College of Arts and Sciences, Department of Chemistry, Milwaukee, WI 53233. Offerings include organic chemistry (MS, PhD). Terminal master's awarded for partial completion of doctoral program. Department faculty: 14 full-time (1 woman), 0 part-time. *Degree requirements:* For master's, thesis or alternative, comprehensive exam required, foreign language not required; for doctorate, dissertation, cumulative exams required, foreign language not required. *Entrance requirements:* TOEFL (minimum score of 550 required). Application fee: $25. *Tuition:* $300 per credit. • Dr. Michael McKinney, Chairman, 414-288-7065.

Massachusetts Institute of Technology, School of Science, Department of Chemistry, Cambridge, MA 02139. Offerings include organic chemistry (SM, PhD, Sc D). Terminal master's awarded for partial completion of doctoral program. Department faculty: 35 full-time (2 women), 0 part-time. *Degree requirements:* For master's, thesis required, foreign language not required; for doctorate, dissertation, written exams, comprehensive oral exam, oral presentation required, foreign language not required. *Application deadline:* 1/16. *Application fee:* $45. *Tuition:* $16,900 per year. • Robert J. Silbey, Chairman, 617-253-1801.

See full description on page 115.

Michigan State University, College of Natural Science, Department of Chemistry, East Lansing, MI 48824. Offerings include organic chemistry (MS, PhD). Terminal master's awarded for partial completion of doctoral program. Department faculty: 27 full-time (1 woman), 0 part-time. *Degree requirements:* For doctorate, 1 foreign language, dissertation. *Entrance requirements:* For doctorate, GRE General Test, TOEFL. Application deadline: rolling. Application fee: $25 ($40 for foreign students). *Tuition:* $104.75 per credit for state residents; $211.75 per credit for nonresidents. • Dr. Gerald Babcock, Chairperson, 517-353-9717. Application contact: Dr. William Reusch, Coordinator, Graduate Program, 517-355-9716.

Mississippi State University, College of Arts and Sciences, Department of Chemistry, Mississippi State, MS 39762. Offerings include organic chemistry (MS, PhD). Terminal master's awarded for partial completion of doctoral program. Department faculty: 15 full-time (0 women), 0 part-time. *Degree requirements:* 1 foreign language (computer language can substitute), thesis/dissertation. *Entrance requirements:* GRE General Test, TOEFL. Application deadline: rolling. *Expenses:* Tuition of $891 per semester full-time, $99 per credit hour part-time for state residents; $1622 per semester full-time, $180 per credit hour part-time for nonresidents. Fees of $221 per semester full-time, $25 per semester (minimum) part-time. • Dr. Kenneth L. Brown, Head, 601-325-3584.

New York University, Graduate School of Arts and Science, Department of Chemistry, New York, NY 10011. Department offers programs in bioorganic chemistry (MS, PhD), biophysical chemistry (MS, PhD), experimental physical chemistry (MS, PhD), inorganic chemistry (MS, PhD), organic chemistry (MS, PhD), theoretical physical chemistry (MS, PhD). Faculty: 26 full-time, 5 part-time. Matriculated students: 90 full-time, 18 part-time; includes 11 minority (8 Asian American, 2 black American, 1 Hispanic American), 66 foreign. Average age 24. 213 applicants, 64% accepted. In 1990, 26 master's awarded (15% found work related to degree, 85% continued full-time study); 12 doctorates awarded. *Degree requirements:* For master's, thesis required, foreign language not required; for doctorate, 1 foreign language, dissertation. *Entrance requirements:* GRE Subject Test, TSE. Application deadline: 1/15 (priority date, applications processed on a rolling basis). Application fee: $30. *Tuition:* $467 per credit. *Financial aid:* In 1990–91, $1.275-million in aid awarded. 5 fellowships (1 to a first-year student), 21 research assistantships, 44 teaching assistantships, 21 teaching fellowships (15 to first-year students) were awarded; full and partial tuition waivers, federal work-study also available. Financial aid application deadline: 1/15. *Faculty research:* Organic, bioorganic, and catalysis. Total annual research budget: $2-million. • Neville R. Kallenbach, Chairman, 212-998-8400. Application contact: David Schuster, Director of Graduate Studies, 212-998-8400.

See full description on page 117.

Northeastern University, Graduate School of Arts and Sciences, Department of Chemistry, Boston, MA 02115. Offerings include organic chemistry (MS). Department faculty: 16 full-time (2 women), 0 part-time. *Degree requirements:* Thesis (for some programs). *Entrance requirements:* TOEFL (minimum score of 580 required). Application deadline: 4/15. Application fee: $40. *Expenses:* Tuition of $10,260 per year full-time, $285 per quarter hour part-time. Fees of $293 per year full-time. • Dr. John L. Roebber, Executive Officer, 617-437-2822. Application contact: Dr. Paul Vouros, Director of Graduate Affairs, 617-437-2822.

See full description on page 119.

Ohio State University, College of Mathematical and Physical Sciences, Department of Chemistry, Columbus, OH 43210. Department awards MS, PhD. Faculty: 49 full-time (3 women), 0 part-time. Matriculated students: 270 full-time (72 women), 14 part-time (4 women); includes 19 minority (13 Asian American, 4 black American, 1 Hispanic American, 1 Native American), 101 foreign. 203 applicants, 29% accepted. In 1990, 23 master's, 31 doctorates awarded. *Degree requirements:* For master's, thesis optional, foreign language not required; for doctorate, dissertation required, foreign language not required. *Entrance requirements:* For master's, GRE General Test, GRE Subject Test. Application deadline: 8/15 (applications processed on a rolling basis). Application fee: $0 ($25 for foreign students). *Tuition:* $1213 per quarter full-time, $364 per course part-time for state residents; $3143 per quarter full-time, $943 per course part-time for nonresidents. *Financial aid:* In 1990–91, $12,000 in aid awarded. 24 fellowships (10 to first-year students), 93 research assistantships (0 to first-year students), 150 teaching assistantships (50 to first-year students) were awarded; federal work-study, institutionally sponsored loans also available. Aid available to part-time students. Financial aid applicants required to submit FAF. *Total annual research budget:* $5.01-million. • Russell M. Pitzer, Chairman, 614-292-2251. Application contact: C. Weldon Mathews, Vice Chair for Graduate Study, 614-292-8917.

See full description on page 121.

Old Dominion University, College of Sciences, Department of Chemistry and Biochemistry, Norfolk, VA 23529. Offerings include organic chemistry (MS). Department faculty: 13 full-time (2 women), 1 part-time (0 women). *Degree requirements:* Thesis required, foreign language not required. *Entrance requirements:* GRE General Test, GRE Subject Test, TOEFL, minimum GPA of 3.0 in major, 2.5 overall. Application deadline: 7/1 (applications processed on a rolling basis). Application fee: $20. *Expenses:* Tuition of $148 per credit hour for state residents; $375 per credit hour for nonresidents. Fees of $64 per year full-time. • Dr. Frank E. Scully Jr., Chairman, 804-683-4078. Application contact: Dr. Roy L. Williams, Graduate Program Director, 804-683-4078.

Oregon State University, Graduate School, College of Science, Department of Chemistry, Corvallis, OR 97331. Offerings include organic chemistry (MS, PhD). Terminal master's awarded for partial completion of doctoral program. Department faculty: 26 full-time (0 women), 5 part-time (1 woman). *Degree requirements:* For doctorate, 1 foreign language, dissertation. *Entrance requirements:* For doctorate, TOEFL (minimum score of 550 required), minimum GPA of 3.0 in last 90 hours. Application deadline: 3/1. Application fee: $40. *Tuition:* $1140 per trimester full-time, $449 per year part-time for state residents; $1816 per trimester full-time, $674 per year part-time for nonresidents. • Dr. Carroll W. DeKock, Chair, 503-737-2081.

See full description on page 123.

Purdue University, School of Science, Department of Chemistry, West Lafayette, IN 47907. Offerings include organic chemistry (MS, PhD). Terminal master's awarded for partial completion of doctoral program. Department faculty: 45 full-time, 1 part-time. *Degree requirements:* Thesis/dissertation required, foreign language not required. *Entrance requirements:* TOEFL (minimum score of 550 required). Application fee: $25. *Tuition:* $1162 per semester full-time, $83.50 per credit hour part-time for state residents; $3720 per semester full-time, $244.50 per credit hour part-time for nonresidents. • Dr. Harry Morrison, Head, 317-494-5200. Application contact: R. E. Wild, Chairman, Graduate Admissions, 317-494-5200.

Rensselaer Polytechnic Institute, School of Science, Department of Chemistry, Troy, NY 12180. Offerings include organic chemistry (MS, PhD). Department faculty: 26 full-time (3 women), 0 part-time. *Degree requirements:* 1 foreign language, thesis/dissertation. *Entrance requirements:* GRE General Test, GRE Subject Test, TOEFL. Application deadline: 2/1. Application fee: $30. *Expenses:* Tuition of $455 per credit hour. Fees of $195.57 per semester. • D. Aikens, Chairman, 518-276-8981. Application contact: Gerald Korenowski, Chairman, Graduate Committee, 518-276-8489.

See full description on page 127.

Rutgers, The State University of New Jersey, Newark, Department of Chemistry, Newark, NJ 07102. Offerings include organic chemistry (MS, PhD). Terminal master's awarded for partial completion of doctoral program. Department faculty: 13 full-time (1 woman), 5 part-time (1 woman). *Degree requirements:* For master's, cumulative exams required, thesis optional, foreign language not required; for doctorate, dissertation, cumulative exams required, foreign language not required. *Entrance requirements:* For master's, GRE General Test, GRE Subject Test, minimum undergraduate B average; for doctorate, GRE Subject Test, minimum undergraduate B average. Application deadline: 8/1. Application fee: $35. *Tuition:* $2216 per semester full-time for state residents; $3248 per semester full-time for nonresidents. • Dr. James Schlegel, Director, 201-648-5173.

See full description on page 129.

Rutgers, The State University of New Jersey, New Brunswick, Program in Chemistry, New Brunswick, NJ 08903. Offerings include organic chemistry (MS, PhD). Terminal master's awarded for partial completion of doctoral program. Program faculty: 48 full-time (5 women), 0 part-time. *Degree requirements:* For doctorate, dissertation, cumulative exams required, foreign language not required. *Entrance requirements:* For doctorate, GRE General Test, GRE Subject Test, TOEFL (minimum score of 575 required). Application deadline: 7/1. Application fee: $35. *Expenses:* Tuition of $4432 per year full-time, $183 per credit part-time for state residents; $6496 per year full-time, $270 per credit part-time for nonresidents. Fees of $458 per year full-time, $117 per year part-time. • Dr. Martha Cotter, Director, 908-932-2259.

See full description on page 131.

San Jose State University, School of Science, Department of Chemistry, San Jose, CA 95192. Offerings include organic chemistry (MS). Department faculty: 24 full-time (3 women), 5 part-time (2 women). *Degree requirements:* Thesis or alternative required, foreign language not required. *Entrance requirements:* GRE, minimum B average. Application deadline: 6/1 (applications processed on a rolling basis). Application fee: $55. *Expenses:* Tuition of $0 for state residents; $246 per unit for nonresidents. Fees of $592 per year. • Dr. Joseph Pesek, Chair, 408-924-5000. Application contact: Dr. Gerald Selter, Graduate Adviser, 408-924-4956.

Seton Hall University, College of Arts and Sciences, Department of Chemistry, South Orange, NJ 07079. Offerings include organic chemistry (PhD). Terminal master's awarded for partial completion of doctoral program. Department faculty: 14 full-time (2 women). *Degree requirements:* 1 foreign language (computer language can substitute), dissertation, comprehensive exams. *Entrance requirements:* GRE General Test, GRE Subject Test, TOEFL, undergraduate major in chemistry or related field. Application fee: $30. *Expenses:* Tuition of $346 per credit. Fees of $105 per semester. • Dr. Robert L. Augustine, Chair, 201-761-9414.

South Dakota State University, Colleges of Arts and Science and Agriculture and Biological Sciences, Department of Chemistry, Brookings, SD 57007. Offerings include organic chemistry (MS, PhD). *Degree requirements:* For doctorate, dissertation, research tool. *Entrance requirements:* For doctorate, TOEFL (minimum score of 500 required). *Tuition:* $553 per semester full-time, $61 per credit part-time for state residents; $1094 per semester full-time, $122 per credit part-time for nonresidents.

Southern University and Agricultural and Mechanical College, College of Sciences, Department of Chemistry, Baton Rouge, LA 70813. Offerings include organic chemistry (MS). MS (environmental science) offered jointly with Louisiana State University. Department faculty: 5 full-time (0 women), 0 part-time. *Entrance requirements:* GRE General Test or GMAT, TOEFL. Application deadline: 7/1. Application fee: $5. *Tuition:* $1594 per year full-time, $299 per semester part-time for state residents; $3010 per year full-time, $299 per semester part-time for nonresidents. • Dr. Earl Doomes, Chairman, 504-771-3990.

State University of New York at Binghamton, School of Arts and Sciences, Department of Chemistry, Binghamton, NY 13902-6000. Offerings include organic chemistry (PhD). Terminal master's awarded for partial completion of doctoral program. *Degree requirements:* Dissertation, cumulative exams required, foreign language not required. *Entrance requirements:* GRE General Test, GRE Subject Test, TOEFL. Application deadline: 4/15 (priority date). Application fee: $35. *Expenses:* Tuition of $2450 per year full-time, $102.50 per credit part-time for state residents; $5766 per year full-time, $242.50 per credit part-time for nonresidents. Fees of $77 per year full-time, $27.85 per semester (minimum) part-time. • Dr. Alistair J. Lees, Chairperson, 607-777-2362.

Syracuse University, College of Arts and Sciences, Department of Chemistry, Syracuse, NY 13244. Department awards MS, PhD. Faculty: 21 full-time. Matriculated students: 63 full-time (22 women), 1 part-time (0 women); includes 1 minority (Asian American), 36 foreign. Average age 28. 180 applicants, 21% accepted. In 1990, 5 master's, 8 doctorates awarded. *Degree requirements:* For master's, 1 foreign language; for doctorate, 1 foreign language, dissertation. *Entrance requirements:* GRE General Test, GRE Subject Test. Application deadline: rolling. Application fee: $35. *Expenses:* Tuition of $381 per credit. Fees of $289 per year full-time, $34.50 per semester part-time. *Financial aid:* In 1990–91, 4 fellowships, 35 research assistantships, 42 teaching assistantships awarded; partial tuition waivers, federal work-study also available. Financial aid application deadline: 3/1; applicants required to submit FAF. *Total annual research budget:* $1.5-million. • Laurence Nafie, Chair, 315-443-4109.

Texas A&M University, College of Science, Department of Chemistry, College Station, TX 77843. Offerings include organic chemistry (MS, PhD). Department faculty: 60. *Entrance requirements:* GRE General Test, TOEFL. Application deadline: 7/15. Application fee: $25 ($50 for foreign students). *Expenses:* Tuition of $100 per semester (minimum) for state residents; $128 per credit hour for nonresidents. Fees of $459 per year full-time, $252 per semester part-time. • Michael B. Hall, Head, 409-845-2011.

See full description on page 141.

Tufts University, Graduate School of Arts and Sciences, Department of Chemistry, Medford, MA 02155. Offerings include organic chemistry (MS, PhD). Terminal master's awarded for partial completion of doctoral program. Department faculty: 15 full-time, 2 part-time. *Degree requirements:* Thesis/dissertation required, foreign language not required. *Entrance requirements:* GRE General Test, GRE Subject Test, TOEFL (minimum score of 550 required). Application deadline: 2/15. Application fee: $50. *Expenses:* Tuition of $16,755 per year full-time, $2094 per course part-time. Fees of $885 per year. • David Walt, Head, 617-381-3470. Application contact: Edward Brush, 617-381-3475.

University of Akron, Buchtel College of Arts and Sciences, Department of Chemistry, Akron, OH 44325. Offerings include organic chemistry (MS, PhD). Terminal master's awarded for partial completion of doctoral program. Department faculty: 20 full-time (1 woman), 3 part-time (0 women). *Degree requirements:* For master's, 1 foreign language (computer language can substitute), thesis, seminar presentation; for doctorate, 2 foreign languages (computer language can substitute for one), dissertation, cumulative exams. *Entrance requirements:* TOEFL. Application deadline: 3/1 (applications processed on a rolling basis). Application fee: $25. *Tuition:* $119.93 per credit hour for state residents; $210.93 per credit hour for nonresidents. • Dr. G. Edwin Wilson, Head, 216-972-7372.

University of Calgary, Faculty of Science, Department of Chemistry, Calgary, AB T2N 1N4, Canada. Offerings include organic chemistry (M Sc, PhD). *Degree requirements:* For master's, thesis required, foreign language not required; for doctorate, dissertation, candidacy exam required, foreign language not required. *Application deadline:* 5/31. *Application fee:* $25. *Tuition:* $1705 per year full-time, $427 per course part-time for Canadian residents; $3410 per year full-time, $854 per course part-time for nonresidents.

University of Cincinnati, McMicken College of Arts and Sciences, Department of Chemistry, Cincinnati, OH 45221. Offerings include organic chemistry (MS, PhD). Terminal master's awarded for partial completion of doctoral program. Department faculty: 26 full-time, 0 part-time. *Degree requirements:* Thesis/dissertation required, foreign language not required. *Entrance requirements:* GRE General Test, GRE Subject Test. Application deadline: 2/15. Application fee: $20. *Tuition:* $131 per credit hour for state residents; $261 per credit hour for nonresidents. • Dr. Bruce Ault, Head, 513-556-9200.

University of Florida, College of Liberal Arts and Sciences, Department of Chemistry, Gainesville, FL 32611. Department awards MS, MST, PhD. PhD offered jointly with Florida Atlantic University. Faculty: 51. Matriculated students: 149 full-time (38 women), 16 part-time (2 women); includes 13 minority (2 Asian American, 5 black American, 6 Hispanic American), 38 foreign. 101 applicants, 52% accepted. In 1990, 6 master's, 28 doctorates awarded. *Degree requirements:* For master's, thesis required, foreign language not required; for doctorate, 1 foreign language, dissertation. *Entrance requirements:* GRE General Test, minimum GPA of 3.0. Application deadline: 5/30 (priority date, applications processed on a rolling basis). Application fee: $15. *Tuition:* $87 per credit hour for state residents; $289 per credit hour for nonresidents. *Financial aid:* In 1990–91, 1 fellowship, 63 research assistantships, 71 teaching assistantships awarded; institutionally sponsored loans also available. Average monthly stipend for a graduate assistantship: $1000. *Faculty research:* Organic, analytical, physical, inorganic and biological chemistry. Total annual research budget: $4.9-million. • Dr. Michael Zerner, Chairman, 904-392-0541. Application contact: Dr. John F. Helling, Chair, Graduate Selections Committee, 904-392-0551.

See full description on page 145.

University of Georgia, College of Arts and Sciences, Department of Chemistry, Athens, GA 30602. Offerings include organic chemistry (MS, PhD). Department faculty: 27 full-time (0 women), 0 part-time. *Degree requirements:* For master's, thesis required, foreign language not required; for doctorate, 1 foreign language (computer language can substitute), dissertation. *Entrance requirements:* GRE General Test. Application fee: $10. *Expenses:* Tuition of $598 per quarter full-time, $48 per quarter part-time for state residents; $1558 per quarter full-time, $144 per quarter part-time for nonresidents. Fees of $118 per quarter. • Dr. Henry F. Schaefer III, Graduate Coordinator, 404-542-2067.

See full description on page 147.

University of Houston, College of Natural Sciences and Mathematics, Department of Chemistry, 4800 Calhoun, Houston, TX 77004. Offerings include organic chemistry (MS, PhD). Terminal master's awarded for partial completion of doctoral program. Department faculty: 25 full-time (1 woman), 2 part-time (0 women). *Degree requirements:* Thesis/dissertation required, foreign language not required. *Entrance requirements:* GRE General Test (minimum combined score of 1100 required), TOEFL (minimum score of 575 required). Application deadline: 5/1 (applications processed on a rolling basis). Application fee: $25 ($75 for foreign students). *Expenses:* Tuition of $30 per hour for state residents; $134 per hour for nonresidents. Fees of $240 per year full-time, $125 per year part-time. • Dr. John L. Bear, Chairman, 713-749-2647. Application contact: Dr. Thomas Albright, Chair, Graduate Committee, 713-749-3721.

See full description on page 149.

University of Kansas, School of Pharmacy, Department of Medicinal Chemistry, Lawrence, KS 66045. Department awards MS, PhD. Faculty: 6 full-time (1 woman), 0 part-time. Matriculated students: 27 full-time (6 women), 20 part-time (4 women); includes 0 minority, 32 foreign. In 1990, 7 master's, 1 doctorate awarded. *Degree requirements:* For doctorate, dissertation. *Entrance requirements:* TOEFL (minimum score of 570 required). Application fee: $50. *Expenses:* Tuition of $1668 per year full-time, $56 per credit hour part-time for state residents; $5382 per year full-time, $179 per credit hour part-time for nonresidents. Fees of $338 per year full-time, $25 per credit hour part-time. *Financial aid:* Fellowships, research assistantships, teaching assistantships available. *Faculty research:* Organic synthesis and drug design; antibiotics, polypeptides, and natural products; enzyme mechanism, inhibition, and model studies; drug metabolism and toxicology. • Dr. Lester Mitscher, Chairperson, 913-864-4495.

Announcement: The graduate program in medicinal chemistry emphasizes organic chemistry and its relationship to biological systems. Active research areas include organic synthesis and drug design; computational chemistry and graphics; enzyme mechanism and model studies; antibiotics, peptides, and natural products; and drug metabolism and toxicity. Excellent facilities and competitive stipends are available.

University of Louisville, College of Arts and Sciences, Department of Chemistry, Louisville, KY 40292. Offerings include organic chemistry (MS, PhD). Department faculty: 18 full-time (2 women), 2 part-time. *Degree requirements:* For master's, thesis; for doctorate, 1 foreign language, dissertation. *Entrance requirements:* GRE General Test, TOEFL (minimum score of 550 required). Application deadline: rolling. *Expenses:* Tuition of $1780 per year full-time, $99 per credit hour part-time for state residents; $5340 per year full-time, $297 per credit hour part-time for nonresidents. Fees of $60 per semester full-time, $12.50 per semester (minimum) part-time. • Dr. Richard P. Baldwin, Chair, 502-588-6798.

University of Maryland College Park, College of Life Sciences, Department of Chemistry and Biochemistry, College Park, MD 20742. Offerings include organic chemistry (MS, PhD). Department faculty: 51 (5 women). *Degree requirements:* For master's, thesis or alternative required, foreign language not required; for doctorate, dissertation. *Entrance requirements:* For master's, GRE General Test, minimum GPA of 3.0; for doctorate, GRE General Test. Application deadline: rolling. Application fee: $25. *Expenses:* Tuition of $143 per credit hour for state residents; $256 per credit hour for nonresidents. Fees of $171.50 per semester. • Dr. Sandra Greer, Chairman, 301-405-1788.

See full description on page 155.

University of Miami, College of Arts and Sciences, Department of Chemistry, Coral Gables, FL 33124. Offerings include organic chemistry (PhD). Department faculty: 10 full-time (1 woman), 0 part-time. *Degree requirements:* Dissertation required, foreign language not required. *Entrance requirements:* GRE General Test (minimum combined score of 1000 required), TOEFL (minimum score of 550 required). Application deadline: 5/1 (priority date, applications processed on a rolling basis). Application fee: $0. *Expenses:* Tuition of $567 per credit hour. Fees of $87 per semester full-time. • Dr. Cecil Criss, Chairman, 305-284-2174.

See full description on page 159.

SECTION 2: CHEMISTRY

Directory: Organic Chemistry

University of Michigan, College of Literature, Science, and the Arts, Department of Chemistry, Ann Arbor, MI 48109. Offerings include organic chemistry (MS, PhD). Department faculty: 38. *Degree requirements:* For master's, foreign language and thesis not required; for doctorate, dissertation, preliminary exam required, foreign language not required. *Entrance requirements:* For master's, GRE General Test, GRE Subject Test; for doctorate, GRE General Test. Application deadline: 4/1 (applications processed on a rolling basis). Application fee: $30. *Tuition:* $3255 per semester full-time, $352 per credit (minimum) part-time for state residents; $6803 per semester full-time, $746 per credit (minimum) part-time for nonresidents. • M. David Curtis III, Chair, 313-764-7314.

See full description on page 161.

University of Missouri–Columbia, College of Arts and Sciences, Department of Chemistry, Columbia, MO 65211. Offerings include organic chemistry (MS, PhD). Department faculty: 23 full-time, 1 part-time. *Degree requirements:* For master's, thesis required, foreign language not required; for doctorate, 1 foreign language, dissertation. *Entrance requirements:* GRE General Test, minimum GPA of 3.0. Application deadline: 8/1 (priority date, applications processed on a rolling basis). Application fee: $20 ($40 for foreign students). *Expenses:* Tuition of $89.90 per credit hour full-time, $98.35 per credit hour part-time for state residents; $244 per credit hour full-time, $252.45 per credit hour part-time for nonresidents. Fees of $123.55 per semester (minimum) full-time. • Dr. John Bauman, Director of Graduate Studies, 314-882-7720. Application contact: Gary L. Smith, Director of Admissions and Registrar, 314-882-7651.

University of Missouri–Kansas City, College of Arts and Sciences, Department of Chemistry, Kansas City, MO 64110. Offerings include organic chemistry (MS, PhD). Department faculty: 15 full-time (0 women), 0 part-time. *Degree requirements:* For master's, thesis required, foreign language not required; for doctorate, 1 foreign language, dissertation, comprehensive exam. *Entrance requirements:* GRE. Application fee: $0. *Expenses:* Tuition of $2200 per year full-time, $92 per credit hour part-time for state residents; $5503 per year full-time, $229 per credit hour part-time for nonresidents. Fees of $122 per semester full-time, $9 per credit hour part-time. • Dr. Layton L. McCoy, Chairperson, 816-235-2280. Application contact: Dr. T. F. Thomas, Chairman, Graduate Admissions Committee, 816-235-2297.

University of Montana, College of Arts and Sciences, Department of Chemistry, Missoula, MT 59812. Offerings include organic chemistry (PhD). Terminal master's awarded for partial completion of doctoral program. *Degree requirements:* 1 foreign language, dissertation. *Entrance requirements:* GRE General Test, GRE Subject Test. Application deadline: 9/15. Application fee: $20. *Tuition:* $495 per quarter hour full-time for state residents; $1239 per quarter hour full-time for nonresidents.

University of Nebraska–Lincoln, College of Arts and Sciences, Department of Chemistry, Lincoln, NE 68588. Offerings include organic chemistry (PhD). Terminal master's awarded for partial completion of doctoral program. Department faculty: 22 full-time (2 women), 0 part-time. *Degree requirements:* 1 foreign language, dissertation, comprehensive exams, departmental qualifying exam. *Entrance requirements:* GRE Subject Test, TOEFL (minimum score of 550 required), TSE (minimum score of 230 required). Application deadline: 5/1 (priority date, applications processed on a rolling basis). Application fee: $25. *Expenses:* Tuition of $75.75 per credit hour for state residents; $187.25 per credit hour for nonresidents. Fees of $161 per year full-time. • Dr. Pill-Soon Song Jr., Chairperson, 402-472-3501.

See full description on page 165.

University of Notre Dame, College of Science, Department of Chemistry and Biochemistry, Notre Dame, IN 46556. Offerings include organic chemistry (MS, PhD). Terminal master's awarded for partial completion of doctoral program. Department faculty: 31 full-time (2 women), 0 part-time. *Degree requirements:* For master's, thesis, comprehensive exam required, foreign language not required; for doctorate, dissertation, qualifying exam required, foreign language not required. *Entrance requirements:* GRE, TOEFL. Application deadline: 2/15 (priority date, applications processed on a rolling basis). Application fee: $25. *Tuition:* $13,385 per year full-time, $744 per credit hour part-time. • Dr. Paul M. Helquist, Chairman, 219-239-7487. Application contact: Dr. Michael Chetcuti, Coordinator, 219-239-5216.

University of Pittsburgh, Faculty of Arts and Sciences, Department of Chemistry, Pittsburgh, PA 15260. Department awards MS, PhD. Part-time programs available. Faculty: 37 full-time (2 women), 2 part-time (0 women). Matriculated students: 174 full-time (48 women), 17 part-time (6 women); includes 3 minority (2 Asian American, 1 black American), 63 foreign. 351 applicants, 12% accepted. In 1990, 6 master's, 19 doctorates awarded. Terminal master's awarded for partial completion of doctoral program. *Degree requirements:* Thesis/dissertation required, foreign language not required. *Entrance requirements:* GRE General Test, GRE Subject Test, TOEFL. Application deadline: 3/31 (priority date, applications processed on a rolling basis). Application fee: $15 ($25 for foreign students). *Expenses:* Tuition of $2920 per semester full-time, $241 per credit part-time for state residents; $5840 per semester full-time, $482 per credit part-time for nonresidents. Fees of $156 per year. *Financial aid:* In 1990–91, 189 students received a total of $2.140-million in aid awarded. 16 fellowships (0 to first-year students), 89 research assistantships (1 to a first-year student), 84 teaching assistantships (40 to first-year students) were awarded; full tuition waivers, federal work-study, institutionally sponsored loans also available. Aid available to part-time students. Financial aid application deadline: 3/15; applicants required to submit FAF. *Faculty research:* Analytical chemistry, inorganic chemistry, organic chemistry, physical chemistry and surface science. Total annual research budget: $7.657-million. • Dr. N. John Cooper, Chairman, 412-624-8500. Application contact: Rebecca Claycamp, Assistant Chairman, 412-624-8200.

See full description on page 167.

University of Regina, Faculty of Graduate Studies and Research, Faculty of Science, Department of Chemistry, Regina, SK S4S 0A2, Canada. Offerings include organic chemistry (M Sc). Department faculty: 14 full-time, 0 part-time. *Degree requirements:* Thesis, departmental qualifying exam required, foreign language not required. Application deadline: 7/2 (applications processed on a rolling basis). *Tuition:* $1500 per year full-time, $242 per year (minimum) part-time. • Dr. W. D. Chandler, Head, 306-585-4146.

University of San Francisco, College of Arts and Sciences, Department of Chemistry, San Francisco, CA 94117-1080. Offerings include organic chemistry (MS). Department faculty: 6 full-time (0 women), 0 part-time. *Degree requirements:* 1 foreign language, thesis. *Entrance requirements:* TOEFL (minimum score of 550 required). Application deadline: 5/15 (priority date). Application fee: $40 ($90 for foreign students). *Tuition:* $10,960 per year full-time, $432 per unit part-time. • Tom Gruhn, Chairman, 415-666-6486. Application contact: Jeff Curtis, Graduate Adviser, 415-666-6391.

See full description on page 171.

University of Southern Mississippi, College of Science and Technology, Department of Chemistry and Biochemistry, Hattiesburg, MS 39406. Offerings include organic chemistry (MS, PhD). Department faculty: 18 full-time (2 women), 1 (woman) part-time. *Degree requirements:* For doctorate, 2 foreign languages (computer language can substitute for one), dissertation. *Entrance requirements:* For master's, GRE General Test, minimum GPA of 2.75; for doctorate, GRE General Test, minimum GPA of 3.5. Application deadline: 8/9 (priority date, applications processed on a rolling basis). Application fee: $0 ($25 for foreign students). *Expenses:* Tuition of $968 per semester full-time, $93 per semester hour part-time. Fees of $12 per semester part-time for state residents; $591 per year full-time, $12 per semester part-time for nonresidents. • Dr. David Creed, Chair, 601-266-4701. Application contact: Dr. David Wertz, Professor, 601-266-4702.

University of South Florida, College of Arts and Sciences, Department of Chemistry, Tampa, FL 33620. Offerings include organic chemistry (MS, PhD). Department faculty: 26 full-time. *Degree requirements:* For master's, thesis; for doctorate, 1 foreign language, computer language, dissertation. *Entrance requirements:* For master's, GRE General Test (minimum combined score of 1000 required), minimum GPA of 3.0 (in last 60 credit hours). Application fee: $15. *Tuition:* $79.40 per credit hour for state residents; $241.33 per credit hour for nonresidents. • Stewart W. Schneller Jr., Chairperson, 813-974-2258. Application contact: Terence Oestereich, Graduate Program Director, 813-974-2534.

See full description on page 173.

University of Tennessee, Knoxville, College of Liberal Arts, Department of Chemistry, Knoxville, TN 37996. Offerings include environmental chemistry (MS, PhD), inorganic chemistry (MS, PhD), organic chemistry (MS, PhD). Terminal master's awarded for partial completion of doctoral program. Department faculty: 36 (1 woman). *Degree requirements:* For master's, thesis required, foreign language not required; for doctorate, 1 foreign language, dissertation. *Entrance requirements:* GRE General Test, TOEFL (minimum score of 525 required), minimum GPA of 2.5. Application deadline: 2/1 (priority date, applications processed on a rolling basis). Application fee: $15. *Tuition:* $1086 per semester full-time, $142 per credit hour part-time for state residents; $2768 per semester full-time, $308 per credit hour part-time for nonresidents. • Dr. Gleb Mamantov, Head, 615-974-3141.

University of Texas at Austin, Graduate School, College of Natural Sciences, Department of Chemistry and Biochemistry, Austin, TX 78712. Department offers programs in analytical chemistry (MA, PhD), biochemistry (MA, PhD), inorganic chemistry (MA, PhD), organic chemistry (MA, PhD), physical chemistry (MA, PhD). Matriculated students: 240 full-time (73 women), 0 part-time; includes 18 minority (10 Asian American, 2 black American, 4 Hispanic American, 2 Native American), 73 foreign. 197 applicants, 61% accepted. In 1990, 15 master's, 37 doctorates awarded. *Entrance requirements:* GRE. Application deadline: 2/1 (priority date, applications processed on a rolling basis). Application fee: $40 ($75 for foreign students). *Tuition:* $510.30 per semester for state residents; $1806 per semester for nonresidents. *Financial aid:* Fellowships, research assistantships, teaching assistantships available. Financial aid application deadline: 3/1. • Application contact: Dr. Peter J. Rossky, Graduate Adviser, 512-471-3890.

See full description on page 175.

University of Toledo, College of Arts and Sciences, Department of Chemistry, Toledo, OH 43606. Offerings include organic chemistry (MS, PhD). Department faculty: 20 full-time (2 women), 0 part-time. *Degree requirements:* For master's, thesis required, foreign language not required; for doctorate, 1 foreign language, dissertation. *Entrance requirements:* GRE General Test, GRE Subject Test, TOEFL. Application deadline: 9/8 (priority date). Application fee: $30. *Tuition:* $122.59 per credit hour for state residents; $193.40 per credit hour for nonresidents. • Dr. James L. Fry, Chairman, 419-537-2109. Application contact: Dr. Max Funk, Chairman, Graduate Admissions Committee, 419-537-2254.

Wake Forest University, Department of Chemistry, Winston-Salem, NC 27109. Offerings include organic chemistry (MS, PhD). Department faculty: 10 full-time (0 women), 3 part-time (1 woman). *Degree requirements:* 1 foreign language (computer language can substitute), thesis/dissertation. *Entrance requirements:* GRE General Test, GRE Subject Test. Application deadline: 3/1. Application fee: $25. *Tuition:* $10,800 per year full-time, $280 per hour part-time. • Dr. Willie Hinze, Chairman, 919-759-5325.

Washington State University, College of Sciences and Arts, Division of Sciences, Department of Chemistry, Pullman, WA 99164. Offerings include organic chemistry (MS, PhD). Terminal master's awarded for partial completion of doctoral program. Department faculty: 26 (2 women). *Degree requirements:* For master's, thesis, teaching experience required, foreign language not required; for doctorate, dissertation, teaching experience. *Entrance requirements:* GRE General Test, minimum GPA of 3.0. Application deadline: 3/1 (priority date, applications processed on a rolling basis). Application fee: $25. *Tuition:* $1694 per semester full-time, $169 per credit hour part-time for state residents; $4236 per semester full-time, $424 per credit hour part-time for nonresidents. • Dr. Roy Filby, Chair, 509-335-1516.

Wesleyan University, Department of Chemistry, Middletown, CT 06459. Offerings include organic chemistry (MA, PhD). Terminal master's awarded for partial completion of doctoral program. Department faculty: 14 full-time (1 woman), 2 part-time (0 women). *Degree requirements:* 1 foreign language (computer language can substitute), thesis/dissertation. *Entrance requirements:* For master's, GRE General Test, GRE Subject Test; for doctorate, GRE Subject Test. Application deadline: 2/1. Application fee: $0. *Expenses:* Tuition of $2035 per course. Fees of $627 per year full-time. • Dr. Rex F. Pratt, Graduate Adviser, 203-347-9411 Ext. 2380.

See full description on page 179.

Western Kentucky University, Ogden College of Science, Technology, and Health, Department of Chemistry, Bowling Green, KY 42101. Offerings include fossil fuel chemistry (PhD) PhD offered jointly with University of Louisville. Department faculty: 13 full-time (0 women), 0 part-time. Application deadline: 8/1 (priority date, applications processed on a rolling basis). *Tuition:* $1580 per year full-time, $87 per credit part-time for state residents; $4460 per year full-time, $247 per credit part-time for nonresidents. • Dr. Donald Slocum, Head, 502-745-3457.

Yale University, Graduate School of Arts and Sciences, Department of Chemistry, New Haven, CT 06520. Offerings include organic chemistry (PhD). Terminal master's awarded for partial completion of doctoral program. Department faculty: 22 full-time (1 woman), 0 part-time. *Degree requirements:* 1 foreign language, dissertation. *Entrance requirements:* GRE General Test, GRE Subject Test, TOEFL. Application deadline: 1/2. Application fee: $45. *Tuition:* $15,160 per year full-time, $1895 per course part-time. • Dr. Frederick Ziegler, Chairman, 203-432-3916. Application contact: Susan Webb, Coordinator of Admissions, 203-432-2770.

See full description on page 181.

Physical Chemistry

Boston College, Graduate School of Arts and Sciences, Department of Chemistry, Chestnut Hill, MA 02167. Offerings include physical chemistry (MS, PhD). *Degree requirements:* For master's, 1 foreign language, thesis; for doctorate, 1 foreign language, dissertation, foreign language exam, qualifying exam. *Entrance requirements:* GRE General Test, GRE Subject Test. *Tuition:* $412 per credit hour.
See full description on page 89.

Boston University, Graduate School, Department of Chemistry, Boston, MA 02215. Offerings include physical chemistry (MA, PhD). Terminal master's awarded for partial completion of doctoral program. Department faculty: 20 full-time (0 women), 0 part-time. *Degree requirements:* For master's, 1 foreign language required, thesis not required; for doctorate, 1 foreign language, dissertation. *Entrance requirements:* GRE General Test, GRE Subject Test, TOEFL (minimum score of 550 required). Application deadline: 4/1. Application fee: $45. *Expenses:* Tuition of $15,950 per year full-time, $498 per credit part-time. Fees of $120 per year full-time, $35 per semester part-time. • Guilford Jones II, Chairman, 617-353-2515.
See full description on page 91.

Brandeis University, Graduate School of Arts and Sciences, Program in Chemistry, Waltham, MA 02254. Offerings include physical chemistry (MA, PhD). Terminal master's awarded for partial completion of doctoral program. Program faculty: 16 full-time (2 women), 0 part-time. *Degree requirements:* For master's, 1 year residency required, foreign language and thesis not required; for doctorate, 1 foreign language (computer language can substitute), dissertation, 2 year residency, 2 seminars, qualifying exams, minimum GPA of 3.0. *Entrance requirements:* GRE, TOEFL. Application deadline: 2/28 (priority date, applications processed on a rolling basis). Application fee: $50. *Tuition:* $16,085 per year full-time, $2015 per course part-time. • Dr. Peter Jordan, Chair, 617-736-2503. Application contact: Susan Isaacs, Secretary, 617-736-2593.
See full description on page 93.

Brigham Young University, College of Physical and Mathematical Sciences, Department of Chemistry, Provo, UT 84602. Offerings include physical chemistry (MS, PhD). Department faculty: 34 full-time (1 woman), 0 part-time. *Degree requirements:* 2 foreign languages (computer language can substitute for one), thesis/dissertation. *Entrance requirements:* For master's, minimum GPA of 3.0 during previous 60 credit hours. Application deadline: 3/15 (priority date). Application fee: $30. *Tuition:* $1170 per semester (minimum) full-time, $130 per credit hour (minimum) part-time. • Dr. Earl M. Woolley, Chairman, 801-378-3667. Application contact: Francis R. Nordmeyer, Coordinator, Graduate Admissions, 801-378-4845.
See full description on page 95.

California State University, Fullerton, School of Natural Science and Mathematics, Department of Chemistry and Biochemistry, Fullerton, CA 92634. Offerings include physical chemistry (MS). Department faculty: 20 full-time (5 women), 13 part-time (2 women). *Degree requirements:* Thesis, departmental qualifying exam required, foreign language not required. *Entrance requirements:* Minimum GPA of 2.5 in last 60 units, major in chemistry or related field. Application fee: $55. *Expenses:* Tuition of $0 for state residents; $246 per unit for nonresidents. Fees of $554 per semester full-time, $356 per unit part-time. • Dr. Glen Nagel, Chair, 714-773-3621. Application contact: Dr. Gene Hiegel, Adviser, 714-773-3624.

California State University, Los Angeles, School of Natural and Social Sciences, Department of Chemistry and Biochemistry, Los Angeles, CA 90032. Offerings include physical chemistry (MS). Department faculty: 12 full-time, 25 part-time. *Degree requirements:* 1 foreign language (computer language can substitute), thesis or comprehensive exam. *Entrance requirements:* TOEFL (minimum score of 550 required). Application deadline: 8/7 (applications processed on a rolling basis). Application fee: $55. *Expenses:* Tuition of $0 for state residents; $164 per unit for nonresidents. Fees of $1046 per year full-time, $650 per semester (minimum) part-time. • Dr. Donald Paulson, Acting Chair, 213-343-2300.

Clark Atlanta University, School of Arts and Sciences, Department of Chemistry, Atlanta, GA 30314. Offerings include physical chemistry (MS, PhD). *Degree requirements:* For master's, 1 foreign language (computer language can substitute), thesis, comprehensive exam. *Entrance requirements:* For master's, minimum GPA of 2.5 in math and science. Application deadline: 4/1 (applications processed on a rolling basis). Application fee: $40. *Expenses:* Tuition of $4860 per year. Fees of $300 per year. • Dr. Alfred Spriggs, Chair, 404-880-8154. Application contact: Peggy Wade, Marketing and Recruitment, 404-880-8427.

Clarkson University, School of Science, Department of Chemistry, Potsdam, NY 13699. Offerings include physical chemistry (MS, PhD). Department faculty: 12 full-time (0 women), 0 part-time. *Degree requirements:* For master's, foreign language and thesis not required; for doctorate, dissertation, departmental qualifying exam required, foreign language not required. *Application fee:* $10. *Expenses:* Tuition of $446 per credit hour. Fees of $75 per semester. • Dr. Phillip A. Christiansen, Chairman, 315-268-2389.

Clemson University, College of Sciences, Department of Chemistry, Clemson, SC 29634. Department awards MS, PhD. Faculty: 19 full-time (1 woman), 0 part-time. Matriculated students: 65 full-time (14 women), 0 part-time; includes 4 minority (1 Asian American, 3 black American), 30 foreign. Average age 25. 40 applicants, 38% accepted. In 1990, 4 master's awarded; 12 doctorates awarded (67% entered university research/teaching, 33% found other work related to degree). *Degree requirements:* 1 foreign language (computer language can substitute), thesis/dissertation. *Entrance requirements:* GRE General Test. Application deadline: 6/1. Application fee: $25. *Expenses:* Tuition of $102 per credit hour. Fees of $80 per semester full-time. *Financial aid:* In 1990–91, $500,000 in aid awarded. 1 fellowship (to a first-year student), 18 research assistantships (1 to a first-year student), 43 lab assistantships were awarded; teaching assistantships also available. *Faculty research:* Flourine chemistry, organic synthetic methods and natural products, metal and non-metal clusters, analytical spectroscopies, polymers. Total annual research budget: $2.45-million. • J. D. Petersen, Head, 803-656-5017.

Cleveland State University, College of Arts and Sciences, Department of Chemistry, Cleveland, OH 44115. Offerings include physical chemistry (MS), structural analysis (MS, PhD). Department faculty: 18 full-time (1 woman), 4 part-time (0 women). *Degree requirements:* For master's, thesis required (for some programs), foreign language not required; for doctorate, dissertation required, foreign language not required. *Entrance requirements:* For master's, GRE General Test, GRE Subject Test, TOEFL (minimum score of 525 required); for doctorate, GRE General Test (score in 50th percentile or higher required), GRE Subject Test, TOEFL (minimum score of 525 required). Application deadline: 9/1 (priority date, applications processed on a rolling basis). Application fee: $0. *Tuition:* $90 per credit for state residents; $180 per credit for nonresidents. • Dr. Robert Towns, Chair, 216-687-2004. Application contact: Dr. Thomas Flechtner, Chairman, Graduate Committee, 216-687-2458.

Columbia University, Graduate School of Arts and Sciences, Division of Natural Sciences, Department of Chemistry, New York, NY 10027. Offerings include physical chemistry (M Phil, MS, PhD). Department faculty: 17 full-time. *Degree requirements:* For master's, comprehensive exams (MS); 1 foreign language, teaching experience, oral and written exams (M Phil) required, thesis not required; for doctorate, 1 foreign language, dissertation, M Phil. *Entrance requirements:* GRE General Test, GRE Subject Test, TOEFL. Application deadline: 1/5. Application fee: $50. *Expenses:* Tuition of $7836 per semester for state residents; $426 per credit for nonresidents. Fees of $148 per semester for state residents. • Richard Bersohn, Chair, 212-854-2192.
See full description on page 101.

Florida State University, College of Arts and Sciences, Department of Chemistry, Tallahassee, FL 32306. Offerings include chemical physics (MS, PhD), physical chemistry (MS, PhD). Terminal master's awarded for partial completion of doctoral program. Department faculty: 39 full-time (2 women), 2 part-time (0 women). *Degree requirements:* For master's, diagnostic and cumulative exams required, foreign language not required; for doctorate, dissertation, diagnostic and cumulative exams required, foreign language not required. *Entrance requirements:* GRE General Test, minimum B average in undergraduate course work. Application deadline: 4/15. Application fee: $15. *Tuition:* $76.29 per credit hour for state residents; $238 per credit hour for nonresidents. • Dr. Edward K. Mellon, Chairman, 904-644-4074. Application contact: Dr. William T. Cooper, Graduate Adviser, 904-644-6875.
See full description on page 105.

Florida State University, College of Arts and Sciences, Departments of Chemistry and Physics, Program in Chemical Physics, Tallahassee, FL 32306. Program awards MS, PhD. Faculty: 26 full-time (1 woman), 0 part-time. Matriculated students: 1 full-time (0 women), 0 part-time; includes 0 minority, 0 foreign. Average age 24. 0 applicants. In 1990, 0 master's, 0 doctorates awarded. *Degree requirements:* For master's, diagnostic and cumulative exams required, foreign language not required; for doctorate, dissertation, diagnostic and cumulative exams required, foreign language not required. *Entrance requirements:* GRE General Test (minimum combined score of 1100 required), minimum B average in undergraduate course work. Application deadline: 4/15 (applications processed on a rolling basis). Application fee: $15. *Tuition:* $76.29 per credit hour for state residents; $238 per credit hour for nonresidents. *Financial aid:* 0 students received aid. Teaching assistantships available. Financial aid application deadline: 2/15. *Faculty research:* Theoretical and experimental research in molecular and solid-state physics and chemistry, statistical mechanics. • S. A. Safron, Chairman, 904-644-5239.
See full description on page 105.

Fordham University, Graduate School of Arts and Sciences, Department of Chemistry, New York, NY 10458. Offerings include chemical physics (MS). *Degree requirements:* 1 foreign language required, thesis not required. *Entrance requirements:* GRE General Test, GRE Subject Test. Application fee: $35. *Tuition:* $352 per credit. • Rev. Robert Cloney SJ, Chairman, 212-579-2584.

Georgetown University, College of Arts and Sciences, Department of Chemistry, Washington, DC 20057. Offerings include chemical physics (MS, PhD), physical chemistry (MS, PhD). Terminal master's awarded for partial completion of doctoral program. Department faculty: 17 full-time, 4 part-time. *Degree requirements:* For master's, thesis required (for some programs), foreign language not required; for doctorate, 1 foreign language, dissertation. *Entrance requirements:* TOEFL (minimum score of 550 required, 600 for teaching assistants). Application deadline: 7/1 (applications processed on a rolling basis). *Expenses:* Tuition of $12,768 per year full-time, $532 per credit part-time. Fees of $142 per year full-time. • Dr. Michael T. Pope, Chairman, 202-687-6073.

Harvard University, Graduate School of Arts and Sciences, Committee on Chemical Physics, Cambridge, MA 02138. Committee offers programs in chemical physics (PhD), chemistry (AM), physics (AM). Faculty: 21 full-time, 0 part-time. Matriculated students: 4 full-time (0 women), 0 part-time; includes 0 minority. In 1990, 4 doctorates awarded. *Degree requirements:* For doctorate, 1 foreign language, dissertation, cumulative exams. *Entrance requirements:* For doctorate, GRE General Test, GRE Subject Test. *Expenses:* Tuition of $14,860 per year. Fees of $550 per year. *Financial aid:* In 1990–91, $104,331 in aid awarded. 2 fellowships (0 to first-year students), 14 research assistantships (3 to first-year students), 2 teaching assistantships (0 to first-year students) were awarded; federal work-study, institutionally sponsored loans, and career-related internships or fieldwork also available. Financial aid application deadline: 1/2; applicants required to submit GAPSFAS. • Dr. Donald Ciappenelli, Director of Graduate Studies, 617-495-4076. Application contact: Office of Admissions and Financial Aid, 617-495-5315.

Harvard University, Graduate School of Arts and Sciences, Department of Chemistry, Cambridge, MA 02138. Offerings include physical chemistry (AM, PhD). Terminal master's awarded for partial completion of doctoral program. Department faculty: 20 full-time, 0 part-time. *Degree requirements:* For doctorate, 2 foreign languages, dissertation, cumulative exams. *Entrance requirements:* For doctorate, GRE General Test, GRE Subject Test. Application deadline: 1/2. Application fee: $60. *Expenses:* Tuition of $14,860 per year. Fees of $550 per year. • Dr. Donald J. Ciappenelli, Director of Graduate Studies, 617-495-4076.

Howard University, Graduate School of Arts and Sciences, Department of Chemistry, 2400 Sixth Street, NW, Washington, DC 20059. Offerings include physical chemistry (MS, PhD). *Degree requirements:* For master's, 1 foreign language (computer language can substitute), thesis, comprehensive exam; for doctorate, 2 foreign languages (computer language can substitute for one), dissertation, comprehensive exam. *Entrance requirements:* For master's, minimum GPA of 2.7. Application deadline: 4/1. Application fee: $25. *Expenses:* Tuition of $6100 per year full-time, $339 per credit hour part-time. Fees of $555 per year full-time, $245 per semester part-time.

Instituto Tecnológico y de Estudios Superiores de Monterrey, Program in Chemistry, Monterrey, Nuevo León 64849, Mexico. Offerings include physical chemistry (MS). Program faculty: 3 full-time (1 woman), 0 part-time. *Degree requirements:* Thesis. *Entrance requirements:* Graduate admission exam (minimum score of 450 required). *Tuition:* $2350 per trimester full-time. • Dr. Xorge Domínquez, Director, 83-58-3300 Ext. 4510. Application contact: Julieta Mier, Dean of Admissions, 83-58-4650.

Jackson State University, School of Science and Technology, Department of Chemistry, Jackson, MS 39217. Offerings include physical chemistry (MS). Department faculty: 14 full-time (3 women), 0 part-time. *Degree requirements:* Thesis (for some programs). *Entrance requirements:* GRE General Test. Application deadline: rolling. Application fee: $0. *Expenses:* Tuition of $1114 per semester full-time, $113 per hour part-time for state residents; $1142 per semester full-time, $113 per hour part-time for nonresidents. Fees of $730 per semester for nonresidents. • Dr. Richard Sullivan, Chair, 601-968-2171.

SECTION 2: CHEMISTRY

Directory: Physical Chemistry

Kansas State University, College of Arts and Sciences, Department of Chemistry, Manhattan, KS 66506. Offerings include physical chemistry (MS, PhD). Department faculty: 19 full-time (0 women), 0 part-time. *Degree requirements:* For master's, thesis required, foreign language not required; for doctorate, dissertation. *Expenses:* Tuition of $1668 per year full-time, $51 per credit hour part-time for state residents; $5382 per year full-time, $142 per credit hour part-time for nonresidents. Fees of $305 per year full-time, $64.50 per semester part-time. • M. Dale Hawley, Head, 913-532-6668. Application contact: Robert Hammaker, Graduate Coordinator, 913-532-6671.

Kent State University, Department of Chemistry, Kent, OH 44242. Offerings include physical chemistry (MS, PhD). Department faculty: 18 full-time, 0 part-time. *Degree requirements:* For doctorate, dissertation required, foreign language not required. *Application deadline:* 7/12 (applications processed on a rolling basis). *Application fee:* $25. *Tuition:* $1601 per semester full-time, $133.75 per hour part-time for state residents; $3101 per semester full-time, $258.75 per hour part-time for nonresidents. • Dr. Roger K. Gilpin, Chairman, 216-672-2032.
See full description on page 111.

Lehigh University, College of Arts and Sciences, Department of Chemistry, Bethlehem, PA 18015. Department awards MS, DA, PhD. Faculty: 22 full-time, 4 part-time. Matriculated students: 59 full-time (23 women), 24 part-time (6 women); includes 3 minority (1 Asian American, 2 black American), 7 foreign. 72 applicants, 72% accepted. In 1990, 12 master's, 4 doctorates awarded. *Degree requirements:* For master's, foreign language and thesis not required; for doctorate, 1 foreign language, dissertation. *Entrance requirements:* GRE, TOEFL. Application deadline: rolling. Application fee: $40. *Tuition:* $15,650 per year full-time, $655 per credit hour part-time. *Financial aid:* Fellowships, research assistantships, teaching assistantships, institutionally sponsored loans, and career-related internships or fieldwork available. *Faculty research:* Biochemistry; inorganic, organic, physical, and polymer chemistry. • Dr. John Larsen, Chairman, 215-758-3471. Application contact: Dr. Daniel Zeroka, Graduate Program Coordinator.
See full description on page 113.

Marquette University, College of Arts and Sciences, Department of Chemistry, Milwaukee, WI 53233. Offerings include physical chemistry (MS, PhD). Terminal master's awarded for partial completion of doctoral program. Department faculty: 14 full-time (1 woman), 0 part-time. *Degree requirements:* For master's, thesis or alternative, comprehensive exam required, foreign language not required; for doctorate, dissertation, cumulative exams required, foreign language not required. *Entrance requirements:* TOEFL (minimum score of 550 required). Application fee: $25. *Tuition:* $300 per credit. • Dr. Michael McKinney, Chairman, 414-288-7065.

Massachusetts Institute of Technology, School of Science, Department of Chemistry, Cambridge, MA 02139. Offerings include chemical physics (SM, PhD, Sc D), physical chemistry (SM, PhD, Sc D). Terminal master's awarded for partial completion of doctoral program. Department faculty: 35 full-time (2 women), 0 part-time. *Degree requirements:* For master's, thesis required, foreign language not required; for doctorate, dissertation, written exams, comprehensive oral exam, oral presentation required, foreign language not required. *Application deadline:* 1/16. *Application fee:* $45. *Tuition:* $16,900 per year. • Robert J. Silbey, Chairman, 617-253-1801.
See full description on page 115.

McMaster University, Faculty of Science, Departments of Physics and Chemistry, Program in Chemical Physics, Hamilton, ON L8S 4L8, Canada. Program awards M Sc, PhD. *Expenses:* Tuition of $2250 per year full-time, $810 per semester part-time for Canadian residents; $10,340 per year full-time, $4050 per semester part-time for nonresidents. Fees of $76 per year full-time, $49 per semester part-time.

Michigan State University, College of Natural Science, Department of Chemistry, East Lansing, MI 48824. Offerings include physical chemistry (PhD). Terminal master's awarded for partial completion of doctoral program. Department faculty: 27 full-time (1 woman), 0 part-time. *Degree requirements:* 1 foreign language, dissertation. *Entrance requirements:* GRE General Test, TOEFL. Application deadline: rolling. Application fee: $25 ($40 for foreign students). *Tuition:* $104.75 per credit for state residents; $211.75 per credit for nonresidents. • Dr. Gerald Babcock, Chairperson, 517-353-9717. Application contact: Dr. William Reusch, Coordinator, Graduate Program, 517-355-9716.

Michigan State University, College of Natural Science, Program in Chemical Physics, East Lansing, MI 48824. Program awards MS, PhD. Matriculated students: 0. *Degree requirements:* For doctorate, dissertation. *Entrance requirements:* GRE General Test, TOEFL. Application deadline: rolling. Application fee: $25 ($40 for foreign students). *Tuition:* $104.75 per credit for state residents; $211.75 per credit for nonresidents. • Dr. James Harrison, Director, 517-355-9715 Ext. 342.

Mississippi State University, College of Arts and Sciences, Department of Chemistry, Mississippi State, MS 39762. Offerings include chemical physics (PhD), physical chemistry (MS, PhD). Terminal master's awarded for partial completion of doctoral program. Department faculty: 15 full-time (0 women), 0 part-time. *Degree requirements:* 1 foreign language (computer language can substitute), thesis/dissertation. *Entrance requirements:* GRE General Test, TOEFL. Application deadline: rolling. *Expenses:* Tuition of $891 per semester full-time, $99 per credit hour part-time for state residents; $1622 per semester full-time, $180 per credit hour part-time for nonresidents. Fees of $221 per semester full-time, $25 per semester (minimum) part-time. • Dr. Kenneth L. Brown, Head, 601-325-3584.

New York University, Graduate School of Arts and Science, Department of Chemistry, New York, NY 10011. Department offers programs in bioorganic chemistry (MS, PhD), biophysical chemistry (MS, PhD), experimental physical chemistry (MS, PhD), inorganic chemistry (MS, PhD), organic chemistry (MS, PhD), theoretical physical chemistry (MS, PhD). Faculty: 26 full-time, 5 part-time. Matriculated students: 90 full-time, 18 part-time; includes 11 minority (8 Asian American, 2 black American, 1 Hispanic American), 66 foreign. Average age 24. 213 applicants, 64% accepted. In 1990, 26 master's awarded (15% found work related to degree, 85% continued full-time study); 12 doctorates awarded. *Degree requirements:* For master's, thesis required, foreign language not required; for doctorate, 1 foreign language, dissertation. *Entrance requirements:* GRE Subject Test, TSE. Application deadline: 1/15 (priority date, applications processed on a rolling basis). Application fee: $30. *Tuition:* $467 per credit. *Financial aid:* In 1990–91, $1.275-million in aid awarded. 5 fellowships (1 to a first-year student), 21 research assistantships, 44 teaching assistantships, 21 teaching fellowships (15 to first-year students) were awarded; full and partial tuition waivers, federal work-study also available. Financial aid application deadline: 1/15. *Faculty research:* Organic, bioorganic, and catalysis. Total annual research budget: $2-million. • Neville R. Kallenbach, Chairman, 212-998-8400. Application contact: David Schuster, Director of Graduate Studies, 212-998-8400.
See full description on page 117.

Northeastern University, Graduate School of Arts and Sciences, Department of Chemistry, Boston, MA 02115. Offerings include physical chemistry (MS). Department faculty: 16 full-time (2 women), 0 part-time. *Degree requirements:* Thesis (for some programs). *Entrance requirements:* TOEFL (minimum score of 580 required). Application deadline: 4/15. Application fee: $40. *Expenses:* Tuition of $10,260 per year full-time, $285 per quarter hour part-time. Fees of $293 per year full-time. • Dr. John L. Roebber, Executive Officer, 617-437-2822. Application contact: Dr. Paul Vouros, Director of Graduate Affairs, 617-437-2822.
See full description on page 119.

Ohio State University, College of Mathematical and Physical Sciences, Department of Chemistry, Columbus, OH 43210. Department awards MS, PhD. Faculty: 49 full-time (3 women), 0 part-time. Matriculated students: 270 full-time (72 women), 14 part-time (4 women); includes 19 minority (13 Asian American, 4 black American, 1 Hispanic American, 1 Native American), 101 foreign. 203 applicants, 29% accepted. In 1990, 23 master's, 31 doctorates awarded. *Degree requirements:* For master's, thesis optional, foreign language not required; for doctorate, dissertation required, foreign language not required. *Entrance requirements:* For master's, GRE General Test, GRE Subject Test. Application deadline: 8/15 (applications processed on a rolling basis). Application fee: $0 ($25 for foreign students). *Tuition:* $1213 per quarter full-time, $364 per course part-time for state residents; $3143 per quarter full-time, $943 per course part-time for nonresidents. *Financial aid:* In 1990–91, $12,000 in aid awarded. 24 fellowships (10 to first-year students), 93 research assistantships (0 to first-year students), 150 teaching assistantships (50 to first-year students) were awarded; federal work-study, institutionally sponsored loans also available. Aid available to part-time students. Financial aid applicants required to submit FAF. *Total annual research budget:* $5.01-million. • Russell M. Pitzer, Chairman, 614-292-2251. Application contact: C. Weldon Mathews, Vice Chair for Graduate Study, 614-292-8917.
See full description on page 121.

Ohio State University, College of Mathematical and Physical Sciences, Program in Chemical Physics, Columbus, OH 43210. Program awards MS, PhD. Matriculated students: 18 full-time (4 women), 0 part-time; includes 1 minority (Native American), 13 foreign. 12 applicants, 33% accepted. In 1990, 0 master's, 1 doctorate awarded. *Degree requirements:* For master's, thesis optional; for doctorate, dissertation. *Entrance requirements:* GRE General Test, GRE Subject Test. Application deadline: 8/15 (applications processed on a rolling basis). Application fee: $0 ($25 for foreign students). *Tuition:* $1213 per quarter full-time, $364 per course part-time for state residents; $3143 per quarter full-time, $943 per course part-time for nonresidents. *Financial aid:* Fellowships, research assistantships, teaching assistantships, federal work-study, institutionally sponsored loans available. Aid available to part-time students. Financial aid applicants required to submit FAF. • Terry Miller, Chairman, 614-292-2569.

Old Dominion University, College of Sciences, Department of Chemistry and Biochemistry, Norfolk, VA 23529. Offerings include physical chemistry (MS). Department faculty: 13 full-time (2 women), 1 part-time (0 women). *Degree requirements:* Thesis required, foreign language not required. *Entrance requirements:* GRE General Test, GRE Subject Test, TOEFL, minimum GPA of 3.0 in major, 2.5 overall. Application deadline: 7/1 (applications processed on a rolling basis). Application fee: $20. *Expenses:* Tuition of $148 per credit hour for state residents; $375 per credit hour for nonresidents. Fees of $64 per year full-time. • Dr. Frank E. Scully Jr., Chairman, 804-683-4078. Application contact: Dr. Roy L. Williams, Graduate Program Director, 804-683-4078.

Oregon State University, Graduate School, College of Science, Department of Chemistry, Corvallis, OR 97331. Offerings include physical chemistry (MS, PhD). Terminal master's awarded for partial completion of doctoral program. Department faculty: 26 full-time (0 women), 5 part-time (1 woman). *Degree requirements:* For doctorate, 1 foreign language, dissertation. *Entrance requirements:* For doctorate, TOEFL (minimum score of 550 required), minimum GPA of 3.0 in last 90 hours. Application deadline: 3/1. Application fee: $40. *Tuition:* $1140 per trimester full-time, $449 per year part-time for state residents; $1816 per trimester full-time, $674 per year part-time for nonresidents. • Dr. Carroll W. DeKock, Chair, 503-737-2081.
See full description on page 123.

Polytechnic University, Brooklyn Campus, Division of Arts and Sciences, Department of Chemistry, 333 Jay Street, Brooklyn, NY 11201. Offerings include chemical physics (MS, PhD). *Degree requirements:* For doctorate, 1 foreign language, dissertation. *Entrance requirements:* GRE General Test, GRE Subject Test. *Tuition:* $6000 per semester.

Princeton University, Departments of Physics and Chemistry, Program in Physics and Chemical Physics, Princeton, NJ 08544. Program awards PhD. *Degree requirements:* 1 foreign language, dissertation. *Entrance requirements:* GRE General Test, GRE Subject Test. Application deadline: 1/8. Application fee: $45 ($50 for foreign students). *Tuition:* $16,670 per year. *Financial aid:* Federal work-study, institutionally sponsored loans available. Financial aid application deadline: 1/8. • Application contact: Michele Spreen, Director of Graduate Admissions, 609-258-3034.

Purdue University, School of Science, Department of Chemistry, West Lafayette, IN 47907. Offerings include physical chemistry (MS, PhD). Terminal master's awarded for partial completion of doctoral program. Department faculty: 45 full-time, 1 part-time. *Degree requirements:* Thesis/dissertation required, foreign language not required. *Entrance requirements:* TOEFL (minimum score of 550 required). Application fee: $25. *Tuition:* $1162 per semester full-time, $83.50 per credit hour part-time for state residents; $3720 per semester full-time, $244.50 per credit hour part-time for nonresidents. • Dr. Harry Morrison, Head, 317-494-5200. Application contact: R. E. Wild, Chairman, Graduate Admissions, 317-494-5200.

Rensselaer Polytechnic Institute, School of Science, Department of Chemistry, Troy, NY 12180. Offerings include physical chemistry (MS, PhD). Department faculty: 26 full-time (3 women), 0 part-time. *Degree requirements:* 1 foreign language, thesis/dissertation. *Entrance requirements:* GRE General Test, GRE Subject Test, TOEFL. Application deadline: 2/1. Application fee: $30. *Expenses:* Tuition of $455 per credit hour. Fees of $195.57 per semester. • D. Aikens, Chairman, 518-276-8981. Application contact: Gerald Korenowski, Chairman, Graduate Committee, 518-276-8489.
See full description on page 127.

Rutgers, The State University of New Jersey, Newark, Department of Chemistry, Newark, NJ 07102. Offerings include physical chemistry (MS, PhD). Terminal master's awarded for partial completion of doctoral program. Department faculty: 13 full-time (1 woman), 5 part-time (1 woman). *Degree requirements:* For master's, cumulative exams required, thesis optional, foreign language not required; for doctorate, dissertation, cumulative exams required, foreign language not required. *Entrance requirements:* For master's, GRE General Test, GRE Subject Test, minimum undergraduate B average; for doctorate, GRE Subject Test, minimum undergraduate B average. Application deadline: 8/1. Application fee: $35. *Tuition:* $2216 per semester full-time for state residents; $3248 per semester full-time for nonresidents. • Dr. James Schlegel, Director, 201-648-5173.

See full description on page 129.

Rutgers, The State University of New Jersey, New Brunswick, Program in Chemistry, New Brunswick, NJ 08903. Offerings include physical chemistry (MS, PhD). Terminal master's awarded for partial completion of doctoral program. Program faculty: 48 full-time (5 women), 0 part-time. *Degree requirements:* For doctorate, dissertation, cumulative exams required, foreign language not required. *Entrance requirements:* For doctorate, GRE General Test, GRE Subject Test, TOEFL (minimum score of 575 required). Application deadline: 7/1. Application fee: $35. *Expenses:* Tuition of $4432 per year full-time, $183 per credit part-time for state residents; $6496 per year full-time, $270 per credit part-time for nonresidents. Fees of $458 per year full-time, $117 per year part-time. • Dr. Martha Cotter, Director, 908-932-2259.

See full description on page 131.

San Jose State University, School of Science, Department of Chemistry, San Jose, CA 95192. Offerings include physical chemistry (MS). Department faculty: 24 full-time (3 women), 5 part-time (2 women). *Degree requirements:* Thesis or alternative required, foreign language not required. *Entrance requirements:* GRE, minimum B average. Application deadline: 6/1 (applications processed on a rolling basis). Application fee: $55. *Expenses:* Tuition of $0 for state residents; $246 per unit for nonresidents. Fees of $592 per year. • Dr. Joseph Pesek, Chair, 408-924-5000. Application contact: Dr. Gerald Selter, Graduate Adviser, 408-924-4956.

Seton Hall University, College of Arts and Sciences, Department of Chemistry, South Orange, NJ 07079. Offerings include physical chemistry (PhD). Terminal master's awarded for partial completion of doctoral program. Department faculty: 14 full-time (2 women). *Degree requirements:* 1 foreign language (computer language can substitute), dissertation, comprehensive exams. *Entrance requirements:* GRE General Test, GRE Subject Test, TOEFL, undergraduate major in chemistry or related field. Application fee: $30. *Expenses:* Tuition of $346 per credit. Fees of $105 per semester. • Dr. Robert L. Augustine, Chair, 201-761-9414.

Simon Fraser University, Faculty of Science, Department of Chemistry, Burnaby, BC V5A 1S6, Canada. Offerings include chemical physics (M Sc, PhD). Department faculty: 26 full-time (2 women), 0 part-time. *Degree requirements:* Thesis/dissertation. *Application fee:* $0. *Expenses:* Tuition of $612 per trimester full-time, $306 per trimester part-time. Fees of $68 per trimester full-time, $34 per trimester part-time. • A. C. Oehlschlager, Chairman, 604-291-4884.

Simon Fraser University, Faculty of Science, Department of Physics, Burnaby, BC V5A 1S6, Canada. Offerings include chemical physics (M Sc, PhD). Department faculty: 22 full-time (0 women), 0 part-time. *Degree requirements:* 1 foreign language, thesis/dissertation. *Application fee:* $0. *Expenses:* Tuition of $612 per trimester full-time, $306 per trimester part-time. Fees of $68 per trimester full-time, $34 per trimester part-time. • M. Plischke, Chairman, 604-291-3154.

South Dakota State University, Colleges of Arts and Science and Agriculture and Biological Sciences, Department of Chemistry, Brookings, SD 57007. Offerings include physical chemistry (MS, PhD). *Degree requirements:* For doctorate, dissertation, research tool. *Entrance requirements:* For doctorate, TOEFL (minimum score of 500 required). *Tuition:* $553 per semester full-time, $61 per credit part-time for state residents; $1094 per semester full-time, $122 per credit part-time for nonresidents.

Southern University and Agricultural and Mechanical College, College of Sciences, Department of Chemistry, Baton Rouge, LA 70813. Offerings include physical chemistry (MS). MS (environmental science) offered jointly with Louisiana State University. Department faculty: 5 full-time (0 women), 0 part-time. *Entrance requirements:* GRE General Test or GMAT, TOEFL. Application deadline: 7/1. Application fee: $5. *Tuition:* $1594 per year full-time, $299 per semester part-time for state residents; $3010 per year full-time, $299 per semester part-time for nonresidents. • Dr. Earl Doomes, Chairman, 504-771-3990.

State University of New York at Binghamton, School of Arts and Sciences, Department of Chemistry, Binghamton, NY 13902-6000. Offerings include physical chemistry (PhD). Terminal master's awarded for partial completion of doctoral program. *Degree requirements:* Dissertation, cumulative exams required, foreign language not required. *Entrance requirements:* GRE General Test, GRE Subject Test, TOEFL. Application deadline: 4/15 (priority date). Application fee: $35. *Expenses:* Tuition of $2450 per year full-time, $102.50 per credit part-time for state residents; $5766 per year full-time, $242.50 per credit part-time for nonresidents. Fees of $77 per year full-time, $27.85 per semester (minimum) part-time. • Dr. Alistair J. Lees, Chairperson, 607-777-2362.

State University of New York at Stony Brook, College of Arts and Sciences, Division of Physical Sciences and Mathematics, Department of Chemistry, Stony Brook, NY 11794. Offerings include chemical physics (M Phil, PhD). Terminal master's awarded for partial completion of doctoral program. Department faculty: 29 full-time, 1 part-time. *Degree requirements:* For doctorate, 1 foreign language, dissertation. *Entrance requirements:* For doctorate, GRE General Test, TOEFL. Application deadline: 2/1. Application fee: $35. *Expenses:* Tuition of $2450 per year full-time, $103 per credit part-time for state residents; $5766 per year full-time, $243 per credit part-time for nonresidents. Fees of $151 per year full-time, $10.45 per year (minimum) part-time. • David Hanson, Chairman, 516-632-7884.

See full description on page 135.

Syracuse University, College of Arts and Sciences, Department of Chemistry, Syracuse, NY 13244. Department awards MS, PhD. Faculty: 21 full-time. Matriculated students: 63 full-time (22 women), 1 part-time (0 women); includes 1 minority (Asian American), 36 foreign. Average age 28. 180 applicants, 21% accepted. In 1990, 5 master's, 8 doctorates awarded. *Degree requirements:* For master's, 1 foreign language; for doctorate, 1 foreign language, dissertation. *Entrance requirements:* GRE General Test, GRE Subject Test. Application deadline: rolling. Application fee: $35. *Expenses:* Tuition of $381 per credit. Fees of $289 per year full-time, $34.50 per semester part-time. *Financial aid:* In 1990–91, 4 fellowships, 35 research assistantships, 42 teaching assistantships awarded; partial tuition waivers, federal work-study also available. Financial aid application deadline: 3/1; applicants required to submit FAF. *Total annual research budget:* $1.5-million. • Laurence Nafie, Chair, 315-443-4109.

Texas A&M University, College of Science, Department of Chemistry, College Station, TX 77843. Offerings include physical chemistry (MS, PhD). Department faculty: 60. *Entrance requirements:* GRE General Test, TOEFL. Application deadline: 7/15. Application fee: $25 ($50 for foreign students). *Expenses:* Tuition of $100 per semester (minimum) for state residents; $128 per credit hour for nonresidents. Fees of $459 per year full-time, $252 per semester part-time. • Michael B. Hall, Head, 409-845-2011.

See full description on page 141.

Tufts University, Graduate School of Arts and Sciences, Department of Chemistry, Medford, MA 02155. Offerings include physical chemistry (MS, PhD). Terminal master's awarded for partial completion of doctoral program. Department faculty: 15 full-time, 2 part-time. *Degree requirements:* Thesis/dissertation required, foreign language not required. *Entrance requirements:* GRE General Test, GRE Subject Test, TOEFL (minimum score of 550 required). Application deadline: 2/15. Application fee: $50. *Expenses:* Tuition of $16,755 per year full-time, $2094 per course part-time. Fees of $885 per year. • David Walt, Head, 617-381-3470. Application contact: Edward Brush, 617-381-3475.

University of Akron, Buchtel College of Arts and Sciences, Department of Chemistry, Akron, OH 44325. Offerings include physical chemistry (MS, PhD). Terminal master's awarded for partial completion of doctoral program. Department faculty: 20 full-time (1 woman), 3 part-time (0 women). *Degree requirements:* For master's, 1 foreign language (computer language can substitute), thesis, seminar presentation; for doctorate, 2 foreign languages (computer language can substitute for one), dissertation, cumulative exams. *Entrance requirements:* TOEFL. Application deadline: 3/1 (applications processed on a rolling basis). Application fee: $25. *Tuition:* $119.93 per credit hour for state residents; $210.93 per credit hour for nonresidents. • Dr. G. Edwin Wilson, Head, 216-972-7372.

University of Calgary, Faculty of Science, Department of Chemistry, Calgary, AB T2N 1N4, Canada. Offerings include physical chemistry (M Sc, PhD). *Degree requirements:* For master's, thesis required, foreign language not required; for doctorate, dissertation, candidacy exam required, foreign language not required. *Application deadline:* 5/31. *Application fee:* $25. *Tuition:* $1705 per year full-time, $427 per course part-time for Canadian residents; $3410 per year full-time, $854 per course part-time for nonresidents.

University of Cincinnati, McMicken College of Arts and Sciences, Department of Chemistry, Cincinnati, OH 45221. Offerings include physical chemistry (MS, PhD). Terminal master's awarded for partial completion of doctoral program. Department faculty: 26 full-time, 0 part-time. *Degree requirements:* Thesis/dissertation required, foreign language not required. *Entrance requirements:* GRE General Test, GRE Subject Test. Application deadline: 2/15. Application fee: $20. *Tuition:* $131 per credit hour for state residents; $261 per credit hour for nonresidents. • Dr. Bruce Ault, Head, 513-556-9200.

University of Florida, College of Liberal Arts and Sciences, Department of Chemistry, Gainesville, FL 32611. Department awards MS, MST, PhD. PhD offered jointly with Florida Atlantic University. Faculty: 51. Matriculated students: 149 full-time (38 women), 16 part-time (2 women); includes 13 minority (2 Asian American, 5 black American, 6 Hispanic American), 38 foreign. 101 applicants, 52% accepted. In 1990, 6 master's, 28 doctorates awarded. *Degree requirements:* For master's, thesis required, foreign language not required; for doctorate, 1 foreign language, dissertation. *Entrance requirements:* GRE General Test, minimum GPA of 3.0. Application deadline: 5/30 (priority date, applications processed on a rolling basis). Application fee: $15. *Tuition:* $87 per credit hour for state residents; $289 per credit hour for nonresidents. *Financial aid:* In 1990–91, 1 fellowship, 63 research assistantships, 71 teaching assistantships awarded; institutionally sponsored loans also available. Average monthly stipend for a graduate assistantship: $1000. *Faculty research:* Organic, analytical, physical, inorganic and biological chemistry. Total annual research budget: $4.9-million. • Dr. Michael Zerner, Chairman, 904-392-0541. Application contact: Dr. John F. Helling, Chair, Graduate Selections Committee, 904-392-0551.

See full description on page 145.

University of Georgia, College of Arts and Sciences, Department of Chemistry, Athens, GA 30602. Offerings include physical chemistry (MS, PhD). Department faculty: 27 full-time (0 women), 0 part-time. *Degree requirements:* For master's, thesis required, foreign language not required; for doctorate, 1 foreign language (computer language can substitute), dissertation. *Entrance requirements:* GRE General Test. Application fee: $10. *Expenses:* Tuition of $598 per quarter full-time, $48 per quarter part-time for state residents; $1558 per quarter full-time, $144 per quarter part-time for nonresidents. Fees of $118 per quarter. • Dr. Henry F. Schaefer III, Graduate Coordinator, 404-542-2067.

See full description on page 147.

University of Houston, College of Natural Sciences and Mathematics, Department of Chemistry, 4800 Calhoun, Houston, TX 77004. Offerings include chemical physics (MS, PhD), physical chemistry (MS, PhD). Terminal master's awarded for partial completion of doctoral program. Department faculty: 25 full-time (1 woman), 2 part-time (0 women). *Degree requirements:* Thesis/dissertation required, foreign language not required. *Entrance requirements:* GRE General Test (minimum combined score of 1100 required), TOEFL (minimum score of 575 required). Application deadline: 5/1 (applications processed on a rolling basis). Application fee: $25 ($75 for foreign students). *Expenses:* Tuition of $30 per hour for state residents; $134 per hour for nonresidents. Fees of $240 per year full-time, $125 per year part-time. • Dr. John L. Bear, Chairman, 713-749-2647. Application contact: Dr. Thomas Albright, Chair, Graduate Committee, 713-749-3721.

See full description on page 149.

University of Louisville, College of Arts and Sciences, Department of Chemistry, Louisville, KY 40292. Offerings include physical chemistry (MS, PhD). Department faculty: 18 full-time (2 women), 2 part-time. *Degree requirements:* For master's, thesis; for doctorate, 1 foreign language, dissertation. *Entrance requirements:* GRE General Test, TOEFL (minimum score of 550 required). Application deadline: rolling. *Expenses:* Tuition of $1780 per year full-time, $99 per credit hour part-time for state residents; $5340 per year full-time, $297 per credit hour part-time for nonresidents. Fees of $60 per semester full-time, $12.50 per semester (minimum) part-time. • Dr. Richard P. Baldwin, Chair, 502-588-6798.

University of Maryland College Park, College of Life Sciences, Department of Chemistry and Biochemistry, College Park, MD 20742. Offerings include physical chemistry (MS, PhD). Department faculty: 51 (5 women). *Degree requirements:* For master's, thesis or alternative required, foreign language not required; for doctorate, dissertation. *Entrance requirements:* For master's, GRE General Test, minimum GPA of 3.0; for doctorate, GRE General Test. Application deadline: rolling. Application fee: $25. *Expenses:* Tuition of $143 per credit hour for state residents; $256 per credit hour for nonresidents. Fees of $171.50 per semester. • Dr. Sandra Greer, Chairman, 301-405-1788.

See full description on page 155.

University of Maryland College Park, College of Computer, Mathematical and Physical Sciences, Department of Physics and Astronomy, Program in Chemical Physics, College Park, MD 20742. Program awards MS, PhD. Matriculated students:

SECTION 2: CHEMISTRY

Directory: Physical Chemistry

22 full-time (6 women), 6 part-time (1 woman); includes 2 minority (both Asian American), 20 foreign. 14 applicants, 71% accepted. In 1990, 1 master's, 0 doctorates awarded. *Degree requirements:* For master's, thesis optional; for doctorate, dissertation. *Entrance requirements:* For master's, GRE Subject Test, minimum GPA of 3.3; for doctorate, GRE Subject Test. Application deadline: rolling. Application fee: $25. *Expenses:* Tuition of $143 per credit hour for state residents; $256 per credit hour for nonresidents. Fees of $171.50 per semester. *Financial aid:* In 1990–91, 5 fellowships (3 to first-year students), 0 research assistantships, 0 teaching assistantships awarded. • Dr. Thomas McIlrath, Director, 301-405-4781.
See full description on page 153.

University of Miami, College of Arts and Sciences, Department of Chemistry, Coral Gables, FL 33124. Offerings include physical chemistry (PhD). Department faculty: 10 full-time (1 woman), 0 part-time. *Degree requirements:* Dissertation required, foreign language not required. *Entrance requirements:* GRE General Test (minimum combined score of 1000 required), TOEFL (minimum score of 550 required). Application deadline: 5/1 (priority date, applications processed on a rolling basis). Application fee: $0. *Expenses:* Tuition of $567 per credit hour. Fees of $87 per semester full-time. • Dr. Cecil Criss, Chairman, 305-284-2174.
See full description on page 159.

University of Michigan, College of Literature, Science, and the Arts, Department of Chemistry, Ann Arbor, MI 48109. Offerings include physical chemistry (MS, PhD). Department faculty: 38. *Degree requirements:* For master's, foreign language and thesis not required; for doctorate, dissertation, preliminary exam required, foreign language not required. *Entrance requirements:* For master's, GRE General Test, GRE Subject Test; for doctorate, GRE General Test. Application deadline: 4/1 (applications processed on a rolling basis). Application fee: $30. *Tuition:* $3255 per semester full-time, $352 per credit (minimum) part-time for state residents; $6803 per semester full-time, $746 per credit (minimum) part-time for nonresidents. • M. David Curtis III, Chair, 313-764-7314.
See full description on page 161.

University of Missouri–Columbia, College of Arts and Sciences, Department of Chemistry, Columbia, MO 65211. Offerings include physical chemistry (MS, PhD). Department faculty: 23 full-time, 1 part-time. *Degree requirements:* For master's, thesis required, foreign language not required; for doctorate, 1 foreign language, dissertation. *Entrance requirements:* GRE General Test, minimum GPA of 3.0. Application deadline: 8/1 (priority date, applications processed on a rolling basis). Application fee: $20 ($40 for foreign students). *Expenses:* Tuition of $89.90 per credit hour full-time, $98.35 per credit hour part-time for state residents; $244 per credit hour full-time, $252.45 per credit hour part-time for nonresidents. Fees of $123.55 per semester (minimum) full-time. • Dr. John Bauman, Director of Graduate Studies, 314-882-7720. Application contact: Gary L. Smith, Director of Admissions and Registrar, 314-882-7651.

University of Missouri–Kansas City, College of Arts and Sciences, Department of Chemistry, Kansas City, MO 64110. Offerings include physical chemistry (MS, PhD). Department faculty: 15 full-time (0 women), 0 part-time. *Degree requirements:* For master's, thesis required, foreign language not required; for doctorate, 1 foreign language, dissertation, comprehensive exam. *Entrance requirements:* GRE. Application fee: $0. *Expenses:* Tuition of $2200 per year full-time, $92 per credit hour part-time for state residents; $5503 per year full-time, $229 per credit hour part-time for nonresidents. Fees of $122 per semester full-time, $9 per credit hour part-time. • Dr. Layton L. McCoy, Chairperson, 816-235-2280. Application contact: Dr. T. F. Thomas, Chairman, Graduate Admissions Committee, 816-235-2297.

University of Montana, College of Arts and Sciences, Department of Chemistry, Missoula, MT 59812. Offerings include physical chemistry (PhD). Terminal master's awarded for partial completion of doctoral program. *Degree requirements:* 1 foreign language, dissertation. *Entrance requirements:* GRE General Test, GRE Subject Test. Application deadline: 9/15. Application fee: $20. *Tuition:* $495 per quarter hour full-time for state residents; $1239 per quarter hour full-time for nonresidents.

University of Nebraska–Lincoln, College of Arts and Sciences, Department of Chemistry, Lincoln, NE 68588. Offerings include physical chemistry (PhD). Terminal master's awarded for partial completion of doctoral program. Department faculty: 22 full-time (2 women), 0 part-time. *Degree requirements:* 1 foreign language, dissertation, comprehensive exams, departmental qualifying exam. *Entrance requirements:* GRE Subject Test, TOEFL (minimum score of 550 required), TSE (minimum score of 230 required). Application deadline: 5/1 (priority date, applications processed on a rolling basis). Application fee: $25. *Expenses:* Tuition of $75.75 per credit hour for state residents; $187.25 per credit hour for nonresidents. Fees of $161 per year full-time. • Dr. Pill-Soon Song Jr., Chairperson, 402-472-3501.
See full description on page 165.

University of Notre Dame, College of Science, Department of Chemistry and Biochemistry, Notre Dame, IN 46556. Offerings include physical chemistry (MS, PhD). Terminal master's awarded for partial completion of doctoral program. Department faculty: 31 full-time (2 women), 0 part-time. *Degree requirements:* For master's, thesis, comprehensive exam required, foreign language not required; for doctorate, dissertation, qualifying exam required, foreign language not required. *Entrance requirements:* GRE, TOEFL. Application deadline: 2/15 (priority date, applications processed on a rolling basis). Application fee: $25. *Tuition:* $13,385 per year full-time, $744 per credit hour part-time. • Dr. Paul M. Helquist, Chairman, 219-239-7487. Application contact: Dr. Michael Chetcuti, Coordinator, 219-239-5216.

University of Pittsburgh, Faculty of Arts and Sciences, Department of Chemistry, Pittsburgh, PA 15260. Department awards MS, PhD. Part-time programs available. Faculty: 37 full-time (2 women), 2 part-time (0 women). Matriculated students: 174 full-time (48 women), 17 part-time (6 women); includes 3 minority (2 Asian American, 1 black American), 63 foreign. 351 applicants, 12% accepted. In 1990, 6 master's, 19 doctorates awarded. Terminal master's awarded for partial completion of doctoral program. *Degree requirements:* Thesis/dissertation required, foreign language not required. *Entrance requirements:* GRE General Test, GRE Subject Test, TOEFL. Application deadline: 3/31 (priority date, applications processed on a rolling basis). Application fee: $15 ($25 for foreign students). *Expenses:* Tuition of $2920 per semester full-time, $241 per credit part-time for state residents; $5840 per semester full-time, $482 per credit part-time for nonresidents. Fees of $156 per year. *Financial aid:* In 1990–91, 189 students received a total of $2.140-million in aid awarded. 16 fellowships (0 to first-year students), 89 research assistantships (1 to a first-year student), 84 teaching assistantships (40 to first-year students) were awarded; full tuition waivers, federal work-study, institutionally sponsored loans also available. Aid available to part-time students. Financial aid application deadline: 3/15; applicants required to submit FAF. *Faculty research:* Analytical chemistry, inorganic chemistry, organic chemistry, physical chemistry and surface science. Total annual research budget: $7.657-million. • Dr. N. John Cooper, Chairman, 412-624-8500. Application contact: Rebecca Claycamp, Assistant Chairman, 412-624-8200.
See full description on page 167.

University of Regina, Faculty of Graduate Studies and Research, Faculty of Science, Department of Chemistry, Regina, SK S4S 0A2, Canada. Offerings include physical chemistry (M Sc, PhD). Department faculty: 14 full-time, 0 part-time. *Degree requirements:* For master's, thesis, departmental qualifying exam required, foreign language not required; for doctorate, variable foreign language requirement, dissertation, departmental qualifying exam. *Application deadline:* 7/2 (applications processed on a rolling basis). *Tuition:* $1500 per year full-time, $242 per year (minimum) part-time. • Dr. W. D. Chandler, Head, 306-585-4146.

University of San Francisco, College of Arts and Sciences, Department of Chemistry, San Francisco, CA 94117-1080. Offerings include physical chemistry (MS). Department faculty: 6 full-time (0 women), 0 part-time. *Degree requirements:* 1 foreign language, thesis. *Entrance requirements:* TOEFL (minimum score of 550 required). Application deadline: 5/15 (priority date). Application fee: $40 ($90 for foreign students). *Tuition:* $10,960 per year full-time, $432 per unit part-time. • Tom Gruhn, Chairman, 415-666-6486. Application contact: Jeff Curtis, Graduate Adviser, 415-666-6391.
See full description on page 171.

University of Southern California, Graduate School, College of Letters, Arts and Sciences, Division of Natural Sciences and Mathematics, Department of Chemistry, Program in Chemical Physics, Los Angeles, CA 90089. Program awards PhD. Matriculated students: 3 full-time (0 women), 1 part-time (0 women); includes 0 minority, 3 foreign. Average age 28. 2 applicants, 0% accepted. In 1990, 0 degrees awarded. *Degree requirements:* Dissertation. *Entrance requirements:* GRE General Test. Application deadline: 4/1 (priority date). Application fee: $50. *Expenses:* Tuition of $12,120 per year full-time, $505 per unit part-time. Fees of $280 per year. *Financial aid:* 5 students received aid. Fellowships, research assistantships, teaching assistantships, federal work-study, institutionally sponsored loans available. Average monthly stipend for a graduate assistantship: $878. Financial aid application deadline: 3/1. • Dr. Otto Schnepp, Chairman, Department of Chemistry, 213-740-7036.

University of Southern Mississippi, College of Science and Technology, Department of Chemistry and Biochemistry, Hattiesburg, MS 39406. Offerings include physical chemistry (MS, PhD). Department faculty: 18 full-time (2 women), 1 (woman) part-time. *Degree requirements:* For doctorate, 2 foreign languages (computer language can substitute for one), dissertation. *Entrance requirements:* For master's, GRE General Test, minimum GPA of 2.75; for doctorate, GRE General Test, minimum GPA of 3.5. Application deadline: 8/9 (priority date, applications processed on a rolling basis). Application fee: $0 ($25 for foreign students). *Expenses:* Tuition of $968 per semester full-time, $93 per semester hour part-time. Fees of $12 per semester part-time for state residents; $591 per year full-time, $12 per semester part-time for nonresidents. • Dr. David Creed, Chair, 601-266-4701. Application contact: Dr. David Wertz, Professor, 601-266-4702.

University of South Florida, College of Arts and Sciences, Department of Chemistry, Tampa, FL 33620. Offerings include physical chemistry (MS, PhD). Department faculty: 26 full-time. *Degree requirements:* For master's, thesis; for doctorate, 1 foreign language, computer language, dissertation. *Entrance requirements:* For master's, GRE General Test (minimum combined score of 1000 required), minimum GPA of 3.0 (in last 60 credit hours). Application fee: $15. *Tuition:* $79.40 per credit hour for state residents; $241.33 per credit hour for nonresidents. • Stewart W. Schneller Jr., Chairperson, 813-974-2258. Application contact: Terence Oestereich, Graduate Program Director, 813-974-2534.
See full description on page 173.

University of Tennessee, Knoxville, College of Liberal Arts, Department of Chemistry, Knoxville, TN 37996. Offerings include chemical physics (PhD), physical chemistry (MS, PhD). Terminal master's awarded for partial completion of doctoral program. Department faculty: 36 (1 woman). *Degree requirements:* For master's, thesis required, foreign language not required; for doctorate, 1 foreign language, dissertation. *Entrance requirements:* GRE General Test, TOEFL (minimum score of 525 required), minimum GPA of 2.5. Application deadline: 2/1 (priority date, applications processed on a rolling basis). Application fee: $15. *Tuition:* $1086 per semester full-time, $142 per credit hour part-time for state residents; $2768 per semester full-time, $308 per credit hour part-time for nonresidents. • Dr. Gleb Mamantov, Head, 615-974-3141.

University of Texas at Austin, Graduate School, College of Natural Sciences, Department of Chemistry and Biochemistry, Austin, TX 78712. Department offers programs in analytical chemistry (MA, PhD), biochemistry (MA, PhD), inorganic chemistry (MA, PhD), organic chemistry (MA, PhD), physical chemistry (MA, PhD). Matriculated students: 240 full-time (73 women), 0 part-time; includes 18 minority (10 Asian American, 2 black American, 4 Hispanic American, 2 Native American), 73 foreign. 197 applicants, 61% accepted. In 1990, 15 master's, 37 doctorates awarded. *Entrance requirements:* GRE. Application deadline: 2/1 (priority date, applications processed on a rolling basis). Application fee: $40 ($75 for foreign students). *Tuition:* $510.30 per semester for state residents; $1806 per semester for nonresidents. *Financial aid:* Fellowships, research assistantships, teaching assistantships available. Financial aid application deadline: 3/1. • Application contact: Dr. Peter J. Rossky, Graduate Adviser, 512-471-3890.
See full description on page 175.

University of Toledo, College of Arts and Sciences, Department of Chemistry, Toledo, OH 43606. Offerings include physical chemistry (MS, PhD). Department faculty: 20 full-time (2 women), 0 part-time. *Degree requirements:* For master's, thesis required, foreign language not required; for doctorate, 1 foreign language, dissertation. *Entrance requirements:* GRE General Test, GRE Subject Test, TOEFL. Application deadline: 9/8 (priority date). Application fee: $30. *Tuition:* $122.59 per credit hour for state residents; $193.40 per credit hour for nonresidents. • Dr. James L. Fry, Chairman, 419-537-2109. Application contact: Dr. Max Funk, Chairman, Graduate Admissions Committee, 419-537-2254.

Wake Forest University, Department of Chemistry, Winston-Salem, NC 27109. Offerings include physical chemistry (MS, PhD). Department faculty: 10 full-time (0 women), 3 part-time (1 woman). *Degree requirements:* 1 foreign language (computer language can substitute), thesis/dissertation. *Entrance requirements:* GRE General Test, GRE Subject Test. Application deadline: 3/1. Application fee: $25. *Tuition:* $10,800 per year full-time, $280 per hour part-time. • Dr. Willie Hinze, Chairman, 919-759-5325.

Washington State University, College of Sciences and Arts, Division of Sciences, Department of Chemistry, Pullman, WA 99164. Offerings include chemical physics (PhD), physical chemistry (MS, PhD). Terminal master's awarded for partial completion of doctoral program. Department faculty: 26 (2 women). *Degree* *requirements:* For master's, thesis, teaching experience required, foreign language not required; for doctorate, dissertation, teaching experience. *Entrance requirements:* GRE General Test, minimum GPA of 3.0. Application deadline: 3/1 (priority date, applications processed on a rolling basis). Application fee: $25. *Tuition:* $1694 per semester full-time, $169 per credit hour part-time for state residents; $4236 per semester full-time, $424 per credit hour part-time for nonresidents. • Dr. Roy Filby, Chair, 509-335-1516.

Washington State University, College of Sciences and Arts, Division of Sciences, Department of Physics, Pullman, WA 99164. Offerings include chemical physics (PhD). Department faculty: 21 full-time (1 woman), 3 part-time (0 women). *Degree requirements:* Dissertation. *Entrance requirements:* GRE General Test, minimum GPA of 3.0. Application deadline: 3/1 (priority date, applications processed on a rolling basis). Application fee: $25. *Tuition:* $1694 per semester full-time, $169 per credit hour part-time for state residents; $4236 per semester full-time, $424 per credit hour part-time for nonresidents. • Dr. Michael Miller, Chair, 509-335-9531. Application contact: Dr. Y. Gupta, 509-335-3140.

Wesleyan University, Department of Chemistry, Middletown, CT 06459. Offerings include chemical physics (MA, PhD), physical chemistry (MA, PhD). Terminal master's awarded for partial completion of doctoral program. Department faculty: 14 full-time (1 woman), 2 part-time (0 women). *Degree requirements:* 1 foreign language (computer language can substitute), thesis/dissertation. *Entrance requirements:* For master's, GRE General Test, GRE Subject Test; for doctorate, GRE Subject Test. Application deadline: 2/1. Application fee: $0. *Expenses:* Tuition of $2035 per course. Fees of $627 per year full-time. • Dr. Rex F. Pratt, Graduate Adviser, 203-347-9411 Ext. 2380.

See full description on page 179.

Yale University, Graduate School of Arts and Sciences, Department of Chemistry, New Haven, CT 06520. Offerings include biophysical chemistry (PhD), physical chemistry (PhD). Terminal master's awarded for partial completion of doctoral program. Department faculty: 22 full-time (1 woman), 0 part-time. *Degree requirements:* 1 foreign language, dissertation. *Entrance requirements:* GRE General Test, GRE Subject Test, TOEFL. Application deadline: 1/2. Application fee: $45. *Tuition:* $15,160 per year full-time, $1895 per course part-time. • Dr. Frederick Ziegler, Chairman, 203-432-3916. Application contact: Susan Webb, Coordinator of Admissions, 203-432-2770.

See full description on page 181.

Cross-Discipline Announcements

Columbia University, Graduate School of Arts and Sciences, Division of Natural Sciences, Department of Biochemistry and Molecular Biophysics, New York, NY 10032.

The research activities of the Department of Biochemistry and Molecular Biophysics encompass a variety of topics of interest to chemistry majors, including X-ray diffraction and NMR studies of biological macromolecules, computer studies of proteins and nucleic acids and the structure and function of membrane receptors, recombinant DNA technology and neurochemistry. See full description in the Biochemistry Section in the Biological and Agricultural Sciences volume of this series.

Duke University, Graduate School, Department of Pharmacology, Durham, NC 27706.

Pharmacology is a science that uses biological and chemical concepts to determine how drugs affect cells, organ systems, and organisms. As such, it gives a unique perspective on understanding how these systems function. Because pharmacology is an interdisciplinary field, the department considers applicants with majors in biological, chemical, and neural or behavioral sciences.

Harvard University, Graduate School of Arts and Sciences, Committee on Biophysics, Cambridge, MA 02138.

The Committee on Higher Degrees in Biophysics at Harvard offers students with backgrounds in physics and chemistry graduate training in diverse areas of biophysical research with faculty from Biochemistry, Molecular, Cellular and Developmental Biology; Chemistry; Applied Physics; and the Medical Sciences Departments. Please see full description in the Biological and Agricultural Sciences volume of this series.

Johns Hopkins University, School of Medicine, Graduate Programs in Medicine, Program in Molecular Biophysics, Baltimore, MD 21218.

IPMB (The Inter-Campus Program in Molecular Biophysics) is staffed by about 35 faculty with interests in molecular biophysics. It offers special opportunities to applicants trained in the physical sciences or mathematics for graduate study in areas such as protein crystallography, NMR and ESR, thermodynamics, statistical mechanics, computer modeling, biophysical chemistry, and biochemistry. It emphasizes studies on macromolecules, or on interacting assemblies of macromolecules, for which a combination of approaches—molecular genetics and structural studies, for example—may be necessary for real progress. Collaborative projects between faculty are encouraged. For information, contact IPMB Office, 301-338-5197, fax 301-338-5199.

Louisiana State University and Agricultural and Mechanical College, College of Engineering, Department of Chemical Engineering, Baton Rouge, LA 70803.

An intensive summer transition program leading to regular admission to chemical engineering graduate study is offered to graduates in chemistry and other science and engineering disciplines. Graduation with an MS can be expected in 1½ to 2 years; PhD takes 3–5 years. Research assistantships and special distinguished fellowships of more than $15,000 per year are available to qualifying students. Contact Graduate Coordinator, 800-256-2084.

Massachusetts Institute of Technology, Whitaker College of Health Sciences and Technology, Division of Toxicology, Cambridge, MA 02139.

Program provides opportunities for specialization in toxicology, including studies in environmental carcinogenesis, metabolism of foreign compounds, molecular dosimetry, genetic toxicology, molecular aspects of interactions of mutagens and carcinogens with nucleic acids and proteins, molecular cloning, and characterization of transformation effector and suppressor genes. See program's full description in the Pharmacology and Toxicology Section in the Biological and Agricultural Sciences volume of this series.

Northwestern University, Robert R. McCormick School of Engineering and Applied Science, Department of Chemical Engineering, Evanston, IL 60208.

MS and PhD programs. Research areas include: kinetics and catalysis, polymer science and engineering, electronic materials, transport phenomena, computer-aided design and simulation, and biochemical and biomedical engineering. Typically, 5–10% of graduate students have a BS in chemistry. Financial aid is available. See full description in Book 5, Chemical Engineering Section.

Ohio State University, College of Pharmacy and Graduate School, Graduate Programs in Pharmacy, Division of Medicinal Chemistry and Pharmacognosy, Columbus, OH 43210.

Preparation for careers in design, synthesis, and isolation of drug molecules. Graduate program consists of studies in drug design, biochemistry, synthetic chemistry, spectroscopy, and analytical chemistry. Employment opportunities are available in academia, industry, and government laboratories.

Purdue University, Interdisciplinary Biochemistry Program, West Lafayette, IN 47907.

The Purdue University Biochemistry Program, an integrated PhD program administered by a diverse 60-member faculty from the Departments of Biochemistry, Biology, Chemistry, and Medicinal Chemistry, offers a core of basic biochemistry courses, highly individualized advanced courses, and faculty-guided research designed to develop the capacity for independent, creative research. See program's full description in the Biochemistry Section in the Biological and Agricultural Sciences volume of this series.

Rush University, Division of Biochemistry, Chicago, IL 60612.

The PhD program emphasizes basic research in human health and disease. Research areas include biochemistry of connective tissue, tumor invasion, endothelial cells, membrane lipids, metalloelements, regulation of gene expression, and development of clinical tests. Financial assistance available. Contact Dr. A. Bezkorovainy, 312-942-5429.

University of California, San Francisco, Department of Pharmaceutical Chemistry, San Francisco, CA 94143.

The PhD degree program in pharmaceutical chemistry is directed toward research at the interface between chemistry and biology. The program offers research in areas of bioorganic chemistry, macromolecular structure and function, medicinal chemistry, drug metabolism and biochemical toxicology, drug delivery, molecular parasitology, molecular pharmacology, and pharmacokinetics/pharmacodynamics.

University of Illinois at Urbana-Champaign, College of Agriculture, Department of Food Science, Champaign, IL 61820.

Degrees in food science (MS and PhD) with options in food chemistry. Areas of study include analytical chemistry; carbohydrate, lipid, and protein chemistry; biotechnology; enzymology; flavor chemistry; biochemistry and molecular biology; molecular basis of functionality. Fully equipped laboratories, including NMR, mass spectrometer, supercritical fluid chromatograph.

University of Kansas, School of Pharmacy, Department of Medicinal Chemistry, Lawrence, KS 66045.

The graduate program in medicinal chemistry emphasizes organic chemistry and its relationship to biological systems. Active research areas include organic synthesis and drug design and optimization; computational chemistry and graphics; enzyme mechanism and model studies; antibiotics, peptides, and natural products; drug metabolism and toxicity. Excellent facilities and instrumentation. Full details in Pharmaceutical Sciences Section in the Business, Education, Health, and Law volume of this series.

SECTION 2: CHEMISTRY

Cross-Discipline Announcements

University of Kentucky, Graduate School and College of Medicine, Graduate Programs in Medicine, Program in Biochemistry, Lexington, KY 40506-0032.

Doctoral program provides opportunities for specialization in biochemistry relating to cancer, aging, and parasite diseases, as well as in plant viruses and basic life processes. Students have diverse majors, with chemistry predominating. Excellent laboratory facilities and equipment available for students. Named fellowships and research assistantships awarded each year. See full description of Department of Biochemistry in the Biological and Agricultural Sciences volume of this series.

University of Massachusetts at Amherst, College of Arts and Sciences, Faculty of Natural Sciences and Mathematics, Department of Biochemistry, Program in Molecular and Cellular Biology, Amherst, MA 01003.

This program includes a strong chemistry component, with 5 faculty members in the Chemistry Department and members from other departments with strong chemistry backgrounds. Ongoing research includes DNA and protein chemistry and the study of metalloproteins and receptors, with techniques ranging from molecular biology to solid-state NMR. See full description in Section 7 of the Biological and Agricultural Sciences volume of this series.

University of Michigan, College of Engineering, Department of Materials Science and Engineering, Ann Arbor, MI 48109.

Interdisciplinary curriculum leads to MS and PhD degrees in materials science and engineering for chemistry students interested in solid-state chemistry, metals, ceramics, polymers, composites, and other engineering materials. Research assistantships and fellowships available on a competitive basis. See full description of the department in the Engineering and Applied Sciences volume of this series.

University of Notre Dame, College of Engineering, Department of Chemical Engineering, Notre Dame, IN 46556.

The department offers attractive stipends and excellent opportunities for well-qualified PhD candidates. Current research interests of the faculty include catalysis and surface science, supercritical fluids, thermodynamics, statistical mechanics, polymer physics, ceramics, fluid mechanics, process control, reaction engineering, and transport processes. For additional information, see profile in Chemical Engineering Section in the Engineering and Applied Sciences volume of this series.

University of Virginia, Graduate School of Arts and Sciences, Department of Biochemistry, Charlottesville, VA 22906.

Students with chemistry degrees are encouraged to apply to this program. Faculty research interests include studies of membranes, protein-membrane interactions, protein structure, protein–nucleic acid interactions, and nucleic acid structure. Students are encouraged to learn and apply a variety of chemical, biophysical, and molecular biological approaches.

University of Virginia, Graduate School of Arts and Sciences, Interdisciplinary Program in Biophysics, Charlottesville, VA 22906.

The Interdisciplinary Program in Biophysics at the University of Virginia offers training and research opportunities with 45 faculty in the Schools of Graduate Arts and Sciences, Engineering, and Medicine. Macromolecular structure and physical biochemistry, membrane biophysics, and radiological physics are areas of specific research strength. All students are financially supported.

University of Washington, College of Engineering and Graduate Programs in Medicine, Center for Bioengineering, Seattle, WA 98195.

The Center for Bioengineering is an interdisciplinary graduate program, integrating engineering and the physical sciences with biology and medicine. Research areas include bioinstrumentation, biomechanics, sensors, biosystems, simulation of bioprocesses, biomaterials, imaging, and molecular bioengineering. Students with degrees in engineering, computer science, or the physical sciences are encouraged to apply.

University of Washington, College of Engineering, Department of Materials Science and Engineering, Seattle, WA 98195.

Programs in the Department of Materials Science and Engineering apply solid-state chemistry to electronic, magnetic, and optical materials, ceramics, metals, and polymers. Current thrusts include research in superconducting ceramics, molecular-beam epitaxy in III–V compounds, colloidal ceramics, and inorganic polymer precursors for ceramic- and metal-matrix composites. See the department's description in the Materials Sciences and Engineering Section in the Engineering and Applied Sciences volume of this series.

University of Wisconsin–Madison, School of Pharmacy, Madison, WI 53706.

Graduate programs in medicinal chemistry, pharmaceutical biochemistry, pharmaceutics, and pharmacology focus on basic research and award only PhD. Students with degrees in pharmacy, chemistry, engineering, or biological sciences are encouraged to apply. Financial assistance available. Contact Ms. Linda Frei, Graduate Admissions.

University of Wisconsin–Madison, Biotechnology Training Program, Madison, WI 53706.

The University of Wisconsin–Madison offers a predoctoral training program in biotechnology. Trainees receive PhDs in their major field, for example, chemistry, while receiving extensive cross-disciplinary training through the minor degree. Trainees participate in industrial internships and a weekly student seminar series with other program participants. These experiences reinforce the cross-disciplinary nature of the program. Students choose a major and minor professor from a list of over 100 faculty in 30 different departments doing research related to biotechnology. See full description in the Biotechnolgy Section of the Biological and Agricultural Sciences volume of this series.

University of Wisconsin–Madison, Institute for Molecular Virology, Program in Biophysics, Madison, WI 53706.

The Biophysics Program offers research that revolves around the application of physical techniques to biological problems. It is an interdisciplinary program that includes 38 faculty members from 9 departments, laboratories, and institutes. See full description in the Biological and Agricultural Sciences volume of this series.

Virginia Commonwealth University, School of Graduate Studies and Medical College of Virginia-Professional Programs, School of Basic Health Sciences, Department of Biochemistry and Molecular Biophysics, Richmond, VA 23284.

Department of Biochemistry and Molecular Biophysics offers research opportunities in molecular biology, cancer, macromolecular structure determination, enzymology, membrane biochemistry, protein chemistry, neuroscience, and regulation of gene expression, leading to the PhD. Students working toward PhD receive tuition support and stipend (currently $16,000 per year).

Virginia Commonwealth University, School of Graduate Studies and Medical College of Virginia-Professional Programs, School of Basic Health Sciences, Program in Biomedical Engineering, Richmond, VA 23284.

The Biomedical Engineering Program offers tracks leading to the MS degree. Students may pursue course work in molecular modeling and graphics software development, medical imaging, signal processing, and bioinstrumentation. See full description in Book 5, Engineering and Applied Sciences.

SECTION 2: CHEMISTRY

ARIZONA STATE UNIVERSITY

College of Arts and Sciences
Department of Chemistry and Biochemistry

Programs of Study

Arizona State University offers both M.S. and Ph.D. degree programs in chemistry and biochemistry. These are viewed as research degrees, and graduate students are encouraged to identify their field of interest early in their career. A wide range of active research programs from synthetic and structural organic and inorganic chemistry and analytical techniques to studies of the function of biological molecules and high-level theoretical chemistry are under way in the department, including a number of interdisciplinary programs. The M.S. degree is not a prerequisite for the Ph.D. program.

Research Facilities

Chemistry is housed in several wings of the Bateman Center for Physical Sciences, along with the Departments of Mathematics, Physics, and Geology and the Center for Solid State Science. The proximity of these gives a special opportunity for interdisciplinary research and fosters creative exchange between the sciences. Chemistry is well equipped for modern research, with extensive instrumentation for electronic vibrational and magnetic resonance spectroscopies and laboratories for crystal growth and materials characterization, electron- and ion-beam microanalysis, and a variety of experiments in laser photochemistry and spectroscopy using pulsed and tunable lasers. The department also has a variety of microcomputer and minicomputer resources and access to the University IBM 3081 mainframe computing system. Finally, the physical sciences center is the home of the internationally recognized Center for High Resolution Electron Microscopy, with the world's largest collection of electron microscopes offering resolution down to the atomic scale.

Financial Aid

All domestic Ph.D. students are guaranteed at least four years of support at a minimum rate of $9260 per academic year in the form of teaching and research assistantships contingent upon satisfactory progress in the program. Summer support for the first year is provided. Guaranteed four-year research assistantships are also available on a competitive basis, and select applicants are guests of the department for a weekend visit in the spring.

Cost of Study

Graduate assistantships provide a waiver of out-of-state tuition. Registration fees are approximately $750 per semester.

Cost of Living

University housing in the Mariposa Graduate Residence Center and regular dormitories is available for single graduate students only. Requests for information should be addressed directly to the Housing Office of Arizona State University. A wide range of apartment housing is available within several blocks of the campus and throughout Tempe and the surrounding communities.

Student Group

The University attracts over 40,000 students each year, including a graduate enrollment of 12,000. Chemistry has approximately 110 students pursuing graduate degrees and 30 new students registering each year.

Location

Arizona State University is located in the city of Tempe, part of the metropolitan area of Phoenix. The recent economic development in the Valley of the Sun has transformed Phoenix into a major center for the electronics and aerospace industries. The atmosphere of intense change and dynamic growth is reflected at Arizona State, as the University realizes its potential as a major research institution and assumes its responsibility as the cultural and intellectual center of central Arizona.

The University

Arizona State University began as a training college for teachers in the Southwest just over a century ago and has developed into a vigorous research university with a variety of creative and challenging programs. The original college buildings still stand, now surrounded by the modern architecture of the present University departments. The 566-acre campus provides a pleasant shady oasis in the sunny desert environment.

Applying

Applicants requesting fall admission should apply by February 1, but later applications will also be given due consideration. International students are required to demonstrate English proficiency by submitting a minimum score of 550 on the TOEFL and 230 or better on the TSE.

Correspondence and Information

Director of Graduate Studies
Department of Chemistry
Arizona State University
Tempe, Arizona 85287-1604
Telephone: 602-965-4664
Fax: 602-965-2747

Peterson's Guide to Graduate Programs in the Physical Sciences and Mathematics 1992

SECTION 2: CHEMISTRY

Arizona State University

THE FACULTY AND THEIR RESEARCH

James Allen, Assistant Professor; Ph.D., Illinois, 1982. Biochemistry: biophysical studies of photosynthetic systems, crystallography of membrane proteins.

Austen Angell, Professor; Ph.D., London, 1961. Physical and solid-state chemistry: molecular and ionic dynamics in liquids and glasses, glass transition, liquids under tension and spinodal collapse phenomena, geochemical fluids.

Krishnan Balasubramanian, Professor; Ph.D., Johns Hopkins, 1980. Physical and solid-state chemistry: relativistic quantum chemistry of molecules containing heavy atoms, chemical applications of group theory and graph theory, methods in statistical mechanics, computers and chemistry.

Allan L. Bieber, Professor; Ph.D., Oregon State, 1961. Biochemistry: protein structure, dependence of function of toxins and enzymes on structure.

James P. Birk, Professor; Ph.D., Iowa State, 1967. Chemical education: kinetics and mechanisms of inorganic reactions, transition-metal and oxy-anion oxidation-reduction reactions, catalyzed substitution reactions, linkage isomerisms, chemical education.

Robert E. Blankenship, Professor; Ph.D., Berkeley, 1975. Biochemistry: biophysics of photosynthesis, biological electron transfer reactions, evolution of energy-conserving systems.

Theodore M. Brown, Professor; Ph.D., Iowa State, 1963. Inorganic chemistry: structure, kinetics, and properties of heavy transition-metal halide compounds (primarily groups IV, V, and VI).

Peter R. Buseck, Regents' Professor; Ph.D., Columbia, 1961. Geochemistry: high-resolution electron microscopy and electron diffraction of minerals, electron probe microanalysis, mineralogy and geochemistry of meteorites, ore genesis and process of deposition, air pollutant analysis.

John R. Cronin, Professor; Ph.D., Colorado, 1964. Biochemistry: analysis of organic constituents of carbonaceous meteorites; chemical evolution; analysis of amino acids and peptides, application of fluorescence methods to amino acid analysis.

Jacob Fuchs, Professor; Ph.D., Illinois, 1950. Analytical chemistry: instrumental analysis, spectrochemical and X-ray methods.

William S. Glaunsinger, Professor; Ph.D., Cornell, 1972. Physical and solid-state chemistry: chemical sensors for environmental analysis and process control, environmental applications of scanning tunneling microscopy, intercalation chemistry.

Douglas Grotjahn, Assistant Professor; Ph.D., Berkeley, 1988. Organic chemistry: metalloenzymes and organic synthesis.

J. Devens Gust, Professor; Ph.D., Princeton, 1974. Organic chemistry: design, synthesis, and spectroscopic studies of model systems for photosynthesis, synthesis, and NMR investigations of polyarylbenzenes.

John R. Holloway, Professor; Ph.D., Penn State, 1970. Geochemistry: experimental investigations of silicate systems and interactions of silicates with volatile species (H_2O, CO_2, HF) at high pressures (to 10,000 atmospheres) and high temperatures (to 1500°C); thermodynamics of multicomponent, multiphase systems at high pressures and temperatures.

Richard S. Juvet, Professor; Ph.D., UCLA, 1955. Analytical chemistry: separation methods emphasizing gas chromatography, computer interfacing with chemical instrumentation, analytical applications of photochemistry, fused-salt chemistry, chelate chemistry of carbohydrate derivatives.

Sheng H. Lin, Regents' Professor; Ph.D., Utah, 1964. Physical and solid-state chemistry: physical and theoretical chemistry; optical rotation, the Faraday effect and Kerr effect, molecular photochemistry, liquid theory, molecular spectra, atomic collision, react kinetics, solid-state chemistry.

C. H. Liu, Professor; Ph.D., Illinois, 1957. Analytical chemistry: electrochemistry, spectroscopy, and chemical reactions in molten salt and other nonaqueous media; computer techniques in solving overlapping chemical equilibria; chemical and electrochemical reactions of chelates and chelated ligands; synthesis and use of chelating agents.

Dennis E. Lohr, Professor; Ph.D., North Carolina, 1972. Biochemistry: coordinating relations between transcription and the structural organization of chromatin.

Paul F. McMillan, Associate Professor; Ph.D., Arizona State, 1981. Inorganic chemistry: micro-Raman and infrared spectroscopy in structural studies of amorphous and crystalline aluminosilicates; applications in solid state, materials science, and geochemistry.

Ana L. Moore, Associate Professor; Ph.D., Texas Tech, 1972. Organic chemistry: organic synthesis and photophysics of carotenoids.

Carleton B. Moore, Regents' Professor; Ph.D., Caltech, 1960. Geochemistry: chemical investigation of meteorites, lunar samples, and terrestrial rocks; geochemistry; cosmochemistry, environmental chemistry.

Thomas A. Moore, Professor; Ph.D., Texas Tech, 1975. Biochemistry: photobiology; biomimetic chemistry, artificial photosynthesis, photoacoustic spectroscopy.

Morton E. Munk, Professor; Ph.D., Wayne State, 1957. Organic chemistry: computer-assisted structure elucidation of biologically interesting natural products, cycloaddition, and cycloreversion reactions; conformational analysis; photochemical and electron-impact reactions.

Michael O'Keeffe, Professor; Ph.D., Bristol (England), 1958. Physical and solid-state chemistry: crystal chemistry, phase transitions, reactivity of solids.

Michael Pena, Assistant Professor; Ph.D., Colorado State, 1989. Organic chemistry: organic and organometallic chemistry with an emphasis on total synthesis of biologically active natural products.

George R. Pettit, Professor; Ph.D., Wayne State, 1956. Organic chemistry: chemistry of natural products (nucleotides, peptides, steroids, and triterpenes), cancer chemotherapy, general organic synthesis.

William T. Petuskey, Associate Professor; Sc.D., MIT, 1977. Physical and solid-state chemistry: synthesis chemistry and properties of oxides, carbides, and nitrides; ceramic chemistry; polycrystalline structures; atom transport; solid- and liquid-state electrochemistry.

Seth D. Rose, Professor; Ph.D., California, San Diego, 1974. Organic chemistry: developing suitable stains for electron microscopy of chromosome ultrastructure, photochemical damage to nucleic acids.

Edward B. Skibo, Associate Professor; Ph.D., California, San Francisco, 1980. Organic chemistry: reductive alkylation (quinones and purines), design of alkylating agents.

Timothy C. Steimle, Associate Professor; Ph.D., California, Santa Barbara, 1978. Physical and solid-state chemistry: high-resolution laser spectroscopy of nonequilibrium gas-phase chemical systems, nonlinear optical techniques and microwave techniques.

Bruce Wagner, Regents' Professor; Ph.D., Virginia, 1955. Physical and solid-state chemistry: solid-state ionics, diffusion and electrical properties of nonstoichiometric compounds and high-temperature oxidation.

Harry B. Whitehurst, Professor; Ph.D., Rice, 1950. Physical and solid-state chemistry: ionizing radiation effects in solids, photon excitation of electrons in ceramic oxides and semiconducting solids, kinetics of current carrier recombination after excitation.

Peter Williams, Professor; Ph.D., London, 1966. Analytical chemistry: mass spectrometry of involatile materials; laser ablation TOF mass spectrometry of DNA and proteins, ion microscopy; sputtering, ion, and electron-stimulated desorption.

George H. Wolf, Assistant Professor; Ph.D., Berkeley, 1981. Physical and solid-state chemistry: synthesis and study of materials at extreme conditions of pressure and temperature involving applications of diamond anvil cells, Raman and Brillouin scattering and X-ray diffraction.

Neal Woodbury, Assistant Professor; Ph.D., Washington (Seattle), 1987. Biochemistry: specific mutagenesis of bacterial reaction centers, photosynthesis, studies of ultrafast electron transfer reactions using subpicosecond resolution pulsed laser spectroscopy.

Lucy M. Ziurys, Assistant Professor; Ph.D., Berkeley, 1988. Physical and solid-state chemistry: interstellar chemistry, millimeter-wave high-resolution molecular spectroscopy, radio/millimeter-wave observational studies of interstellar molecules; millimeter and submillimeter devices; spectroscopy of transient species.

BAYLOR UNIVERSITY
Department of Chemistry

Programs of Study

The chemistry department offers a program of course work and research leading to the M.S. and Ph.D. degrees. Research interests of the faculty cover all major areas of chemistry, including analytical, inorganic, organic, and physical chemistry and biochemistry. A program leading to the M.S. in environmental chemistry is offered jointly with the Institute of Environmental Science. A favorable student-to-faculty ratio makes possible almost daily contact between graduate student and research professor, leading to a productive exchange of ideas. Since the traditional boundaries between the divisions of chemistry are rapidly disappearing, the program stresses breadth in understanding all aspects of chemistry and related disciplines. Cooperation between research groups within the department and between departments in allied fields is encouraged.

Research Facilities

The chemistry department is located in the three-story Marrs-McLean Science Building, which provides air-conditioned classrooms, research laboratories, offices, and such auxiliary services as the electronics and machine shops and the chemistry/physics library. The library subscribes to all major chemical publications, representing more than 300 current subscriptions. On-line computer-assisted literature searching is readily available. The presence of a full complement of instrumentation enables each graduate research student to acquire hands-on experience with the most modern research equipment available.

Financial Aid

A number of research and teaching assistantships are available. Assistantships include a full tuition scholarship, payment of all laboratory fees for required courses, and a stipend of up to $10,600 per academic year (1991–92). Various loan plans are also available. Assistantship applications should be directed to Dr. Marianna A. Busch, Director of Graduate Studies, Department of Chemistry. Loan information can be obtained from the Student Loan Officer, Financial Aid Office, BU Box 97028.

Cost of Study

Tuition is $185 per semester hour in 1991–92. Student fees are $18 per semester hour up to 8 hours, with a flat rate of $255 for 9 hours or more. Payment of student fees entitles the student to use University health-care services and recreational facilities.

Cost of Living

On-campus housing is not available for graduate students, except in the case of foreign students. A large number of apartments, many close to campus, offer a wide variety of living accommodations with prices typically starting at $275 per month.

Student Group

The total enrollment at Baylor University is about 11,500. There are approximately 30 graduate students in the chemistry department, representing several states and eight foreign countries. The scientific scope of the department is further enhanced by the presence of postdoctoral research associates from a number of universities around the world.

Location

Baylor University is located in Waco, a city of about 105,000 residents in the hill country of central Texas. Nearby Lake Waco and Lake Brazos provide year-round recreational opportunities. The city is only a few hours from such metropolitan centers as Dallas and Houston, the state capitol at Austin, and the historic city of San Antonio.

The University

Baylor University is a private institution affiliated with the Baptist General Convention of Texas. Established in 1845, it is the oldest university in continuous existence in Texas. The 350-acre Waco campus includes the College of Arts and Sciences, the Graduate School, and the Schools of Business, Education, Law, and Music. The Baylor College of Dentistry and School of Nursing are located in Dallas.

Applying

Applications are accepted at any time, but students requesting financial aid are advised to complete their applications at least 90 days before the first semester of study. Requirements for admission include a bachelor's degree, a GPA of at least 3.0 (B) in the student's major field, a GPA of at least 2.7 in all undergraduate work, and scores on the General Test and the Subject Test in chemistry of the Graduate Record Examinations. Applicants from most foreign countries are required to submit a score of at least 550 on the TOEFL. Students not meeting the minimum GPA for full admission can sometimes be admitted on probationary status.

Correspondence and Information

Dr. Marianna A. Busch
Director of Graduate Studies
Department of Chemistry
Box 97348
Baylor University
Waco, Texas 76798-7348
Telephone: 817-755-3311 Ext. 6875
Fax: 817-755-2403

SECTION 2: CHEMISTRY

Baylor University

THE FACULTY AND THEIR RESEARCH

Kenneth W. Busch, Professor; Ph.D., Florida State. Analytical spectroscopy and chemical instrumentation: design, evaluation, and theoretical modeling of new spectroscopic instrumentation for chemical analysis; development of infrared radiometers as detectors in gas and liquid chromatography; application of photon correlation spectroscopy to colloidal suspensions; effect of applied fields on colloidal suspensions.

Marianna A. Busch, Associate Professor and Director of Graduate Studies; Ph.D., Florida State. Physical inorganic chemistry and analytical spectroscopy: analytical applications of infrared emission spectroscopy; transform methods in spectroscopy; simultaneous multielement analysis; environmental analysis; chemometrics; colloid chemistry; effect of external fields on colloidal suspensions.

Mary Lynn Fink, Assistant Professor; Ph.D., Case Western Reserve. Biochemistry: mechanism of enzyme action; structure-activity studies and enzyme kinetics; design and synthesis of enzyme inhibitors as potential therapeutic agents; development of enzyme assays; metabolism of cross-linked proteins; peptide synthesis.

Thomas C. Franklin, Professor; Ph.D., Ohio State. Physical and analytical chemistry: effects of additives on processes occurring at the interface between phases; studies of catalytic and electrolytic processes, electrosynthesis, mechanism of electrode reactions, electroanalytical chemistry, phase transfer catalysis, and electrolytic destruction of toxic materials.

Charles M. Garner, Assistant Professor; Ph.D., Colorado. Organic chemistry: asymmetric synthesis, especially using new chiral organometallic reagents/catalysts; preparative and analytical separation of enantiomers; physical organic/organometallic studies (kinetics, thermodynamics, fluxionality) of molecules and reactions by multinuclear and variable-temperature NMR.

Stephen L. Gipson, Assistant Professor; Ph.D., Caltech. Inorganic and analytical chemistry: organometallic electrochemistry, electrocatalysis and electron transfer catalysis, studies of reaction rates and mechanisms through electroanalytical techniques.

Jesse W. Jones, Professor; Ph.D., Arizona State. Organic and medicinal chemistry: synthetic organic chemistry; reaction mechanisms; isolation and structural determination of naturally occurring nitrogen heterocycles.

Carlos E. Manzanares, Assistant Professor; Ph.D., Indiana. Physical chemistry and chemical physics: laser spectroscopy and photochemistry, laser photoacoustic spectroscopy, infrared laser-induced fluorescence in cryogenic liquids, vibrational energy transfer, analytical applications of lasers.

Donald F. Mullica, Research Crystallographer; Ph.D., Baylor. Physical chemistry: structural investigations on single crystals, polycrystalline and amorphous materials, ligand-substituted metal complexes, and organometallic compounds of the lanthanide and actinide series and the transition metals by X-ray, electron, and neutron diffraction methods.

John A. Olson, Assistant Professor; Ph.D., Florida. Chemical physics: gas-phase and gas-surface collision dynamics; emphasis on semiclassical descriptions of reactions involving more than one electronic potential energy surface, i.e., nonadiabatic reactions; treatment of potential energy surfaces, electron transfer, electronic transitions, nuclear rearrangements, and energy transfer between translational and vibrational modes.

David E. Pennington, Professor; Ph.D., Penn State. Inorganic chemistry: coordination chemistry, kinetics and mechanism of inorganic reactions, synthesis and characterization of inorganic complexes.

A. G. Pinkus, Professor; Ph.D., Ohio State. Organic chemistry: stereochemistry; reaction mechanisms; organometallic chemistry; polymer chemistry; restricted rotation studies; chemistry of glycolides, phenols, ketones, enols, esters, ethers, small-ring compounds, and isotope-substituted compounds; Grignard, Friedel-Crafts, and Fries reactions; organophosphorus, organosilicon, and organoaluminum compounds; organic peroxides; autoxidation.

William H. Scouten, Professor and Chairman; Ph.D., Pittsburgh. Biochemistry: immobilized enzymes, affinity chromatography, enzyme structure, multienzyme complexes, immunoassay development, fluorescence energy transfer/fluorescent probe development, biotechnology.

F. Gordon A. Stone, Robert A. Welch Distinguished Professor of Chemistry; Ph.D., Cambridge. Organometallic chemistry: synthetic and structural studies on organo-transition-metal complexes with metal-metal bonds, alkylidyne compounds, carbametallaboranes.

RECENT FACULTY PUBLICATIONS

Busch, K. W., E. Chibowski, S. Gopalakrishnan, and M. A. Busch. Residual variations in the zeta potential of TiO_2 (anatase) suspensions as a result of exposure to radiofrequency electric fields. *J. Colloid Interface Sci.* 139:43–54, 1990.

Busch, M. A., S. Ravishankar, D. C. Tilotta, and K. W. Busch. An element-specific, dual-channel, flame infrared emission, gas chromatography detector for chlorinated and fluorinated hydrocarbons. *Appl. Spectrosc.* 44:1247–58, 1990.

Fink, M. L., J. Shey, and Y. Y. Shao. Evaluation of substrates and inhibitors of transglutaminase by an HPLC assay. *J. Cell Biol.* 111:320, 1990.

Franklin, T. C., and S. A. Mathew. The effect of anionic additives on the volume of activation for the electrodeposition of nickel. *J. Electrochem. Soc.* 135:2725–28, 1988.

Garner, C. M., N. Quirós-Mendez, J. J. Kowalczyk, J. M. Fernández, K. Emerson, R. D. Larsen, and J. A. Gladysz. The selective binding and activation of one aldehyde enantioface by a chiral transition metal Lewis acid: synthesis, structure and reactivity of rhenium aldehyde complexes $[(\eta^5\text{-}C_5H_5)Re(NO)(PPh_3)(O=CHR)]^+X^-$. *J. Am. Chem. Soc.* 112:5146–60, 1990.

Gipson, S. L., Y. Y. Lau, and W. W. Huckabee. Ligand effects on the electrooxidation of molybdenum halide complexes of the type $CpMo(CO)_{3-n}(PR_3)_nX$ and $CpMo(CO)_2X$. *Inorg. Chim. Acta* 172:41–44, 1990.

Manzanares, C., N. L. S. Yamasaki, and E. Weitz. Spectroscopy of C-H stretching overtones in dimethylacetylene, dimethylcadmium, and dimethylmercury. *J. Phys. Chem.* 93:4733, 1989.

Mullica, D. F., E. L. Sappenfield, and T. A. Cunningham. Synthesis, spectroscopic, and structural investigation of ytterbium potassium hexacyanoferrate(II), $YbKFe(CN)_6 \cdot 3.5H_2O$. *J. Solid State Chem.* 91:98–104, 1991.

Olson, J. A., and S. Kristyan. Statistical mechanical treatment of the surface free enthalpy excess of solid chemical elements. Part 1. Temperature dependence. *J. Phys. Chem.* 95:921–32, 1991.

Pennington, D. E., J. E. Bradshaw, D. A. Grossie, and D. F. Mullica. Preparations and characterizations of μ_3-oxo-hexakis(μ_2-carboxylatopyridine-O,O) triaquachromium(III) perchlorates. *Inorg. Chim. Acta* 141:41–47, 1988.

Pinkus, A. G., R. Subramanyam, S. L. Clough, and T. C. Lairmore. Preparation of polymandelide by reaction of α-bromophenylacetic acid and triethylamine. *J. Polym. Sci. Part A., Polym. Chem.* 27:4291–96, 1989.

Scouten, W. H., S. E. Eckert-Tilotta, and J. Hines. A calcium-selective optrode based on the fluorescence of dansylated troponin. *Appl. Spectrosc.* 45:491–95, 1990.

Stone, F. G. A., N. Carr, and M. C. Gimeno. Chemistry of polynuclear metal complexes with bridging carbene or carbyne ligands. Part 99. Synthesis of the cluster compounds $[WMCoAu(\mu\text{-}CC_6H_4Me\text{-}4)(\mu_3\text{-}CR)(CO)_4(\eta\text{-}C_5H_5)(\eta\text{-}C_5Me_5)(\eta^5\text{-}C_2B_9H_9Me_2)]$ (M = Mo or W, R = C_6H_4Me-4; M = W, R = Me); crystal structure of the complex $[MoWCoAu(\mu\text{-}CC_6H_4Me\text{-}4)(\mu_3\text{-}CC_6H_4Me\text{-}4)(CO)_4(\eta\text{-}C_5H_5)(\eta\text{-}C_5Me_5)(\eta^5\text{-}C_2B_9H_9Me_2)]$. *J. Chem. Soc. Dalton Trans.* 1990:2247.

SECTION 2: CHEMISTRY

BOSTON COLLEGE
Department of Chemistry

Program of Study

Boston College offers a program of study leading to the Ph.D. and M.S. degrees with concentrations in physical, inorganic, organic, and analytical chemistry and in biochemistry. The Master of Science in Teaching (M.S.T.) is offered to students interested in secondary school teaching.

The Ph.D. dissertation, reporting on a well-organized, original research project, is the core of the Ph.D. program. An advanced chemistry curriculum, usually satisfied in two semesters, is offered to give the student breadth in the major branches of chemistry. Formal courses may be waived in whole or in part if the student demonstrates a mastery of the subject matter in qualifying examinations, given three times during the year. Under the guidance of a professor, the student pursues a program of seminars and advanced courses. The Ph.D. program normally takes four to five years to complete.

The M.S. degree is awarded on completion of 24 credit hours of satisfactory graduate course work and a research project presented as a formal thesis. Two to 2½ years are normally required to obtain a master's degree.

A graduate student–faculty ratio of 3:1 ensures a close working arrangement between the student and the research director throughout the study program. Frequent colloquia within the department and at neighboring institutions feature eminent scientists from university, industrial, and government laboratories.

Research Facilities

The department is located in a new building designed and built specifically for the study of chemistry. It contains research laboratories that are well equipped with modern instrumentation for NMR, ESR, IR, UV-VIS, atomic absorption, and mass spectrometry; laser spectroscopy; GC and HPLC; facilities for studies in magnetic susceptibility, electrochemistry, fermentation, and DNA sequencing; instrumentation for 300-MHz and 500-MHz multinuclear FT-NMRs, GC/MS, circular dichroism, and X-ray diffraction (area detector); and three VAX computers and an Evans & Sutherland molecular graphics system. Additional computing facilities in the department include a VAX-11/750, Apollo DN3000 and DN580 advanced workstations, and an IBM RISC/6000 computer. The University Computer Center, which has direct links to the Department of Chemistry, operates two IBM mainframe and four DEC VAX systems. The science section of the O'Neill Central Research Library contains an excellent collection of journals, monographs, and reference texts.

Financial Aid

Essentially all M.S. and Ph.D. students receive financial support as teaching or research assistants. Graduate teaching assistantships carry stipends beginning at $10,300 plus tuition remission for the 1991–92 academic year. Teaching or research assistantship support during the summer provides an additional $2060. Research assistantships carry stipends of $12,060 for the twelve months. Highly qualified students are awarded fellowships.

Cost of Study

The tuition costs for 1991–92 are $412 per semester hour ($7416 per academic year). Scholarships provide tuition remission for both teaching and research assistants and for fellowship holders.

Cost of Living

Students usually share apartments in the Boston area or rent private rooms at moderate cost in the beautiful residential area surrounding the campus. The College provides assistance to students in locating suitable accommodations.

Student Group

The enrollment at Boston College is about 14,000, including 3,800 graduate students. The student body is drawn from all parts of the United States and from many foreign countries. The chemistry department has approximately 50 graduate students and postdoctoral fellows.

Location

As a recreational, cultural, and scientific center, Boston is unsurpassed. The Boston Symphony Orchestra, live theater performances, opera, ballet, art and science museums, and a rich historic past coexist with a modern city now in the midst of a massive revitalization of its downtown area. The proximity of several universities and colleges and the establishment of many industrial research laboratories in the Boston area make this region the scientific center of the eastern United States. The mountains of New Hampshire, the forests and rocky coastline of Maine, the beaches of Cape Cod, and the old villages and lakes of inland Massachusetts provide a variety of easily accessible recreational areas. In addition, the renowned professional sports teams of Boston provide year-round spectator entertainment.

The College and The Department

Boston College, founded by the Jesuit order in 1863, is a university of ten colleges and schools dedicated to excellence in education. Faculty and students are recruited internationally, regardless of religion, race, or color. Situated in an attractive residential area on a hill overlooking Boston, the main campus is 6 miles west of the State House and downtown district.

In recent years, a stimulating climate of vigorous research activity in the chemistry department has resulted in a tremendous influx of external grant support, new equipment, and a substantial increase in departmental research publications. The low student-faculty ratio and favorable geographical location, combined with these exciting research developments, provide an excellent environment for graduate study.

Applying

A preliminary assessment of admission prospects may be obtained by submitting the application form, an unofficial transcript, and one letter of reference (no application fee required). Formal applications for admission in September are evaluated under a rolling admission policy beginning early in January. Thus, early submission is recommended, although late applications will be considered. Students are urged to take the Graduate Record Examinations (the General Test and the Subject Test in chemistry). Foreign students must submit TOEFL scores.

Correspondence and Information

Graduate Admissions Committee
Department of Chemistry
Box 100
Boston College
Chestnut Hill, Massachusetts 02167
Telephone: 617-552-3606

Boston College

THE FACULTY AND THEIR RESEARCH

James E. Anderson, Assistant Professor of Analytical Chemistry; Ph.D., Michigan, 1985. Electrochemistry and spectroelectrochemistry, electron transfer mechanisms of organometallic compounds in nonaqueous solvents.

O. Francis Bennett, Associate Professor of Organic Chemistry; Ph.D., Penn State, 1958. Nucleophilic displacement reactions, reactions and mechanisms of organosulfur compounds.

E. Joseph Billo Jr., Associate Professor of Inorganic and Analytical Chemistry; Ph.D., McMaster, 1967. Structural, thermodynamic, and kinetic studies of transition-metal complexes; macrocyclic complexes; solution equilibria.

Michael J. Clarke, Professor of Inorganic Chemistry; Ph.D., Stanford, 1974. Bioinorganic chemistry: interaction of transition-metal ions with nucleotides, nucleic acids, and coenzymes; radiopharmaceuticals and metal-containing anticancer drugs; electron transfer between metal ions and biologically important electron carriers; electrochemistry at solid electrodes.

Paul Davidovits, Professor of Physical Chemistry; Ph.D., Columbia, 1964. Study of gas molecules with liquid surfaces, using laser and mass spectrometry, with applications to atmospheric chemistry.

Amir H. Hoveyda, Assistant Professor of Organic Chemistry; Ph.D., Yale, 1986. Development of new reactions and methods for chemical synthesis, design and synthesis of substances of medicinal and biological significance, organometallic chemistry.

Evan R. Kantrowitz, Professor of Biochemistry; Ph.D., Harvard, 1976. Structure and function of proteins, the mechanism of cooperativity and use of recombinant DNA techniques to study proteins.

T. Ross Kelly, Professor of Organic Chemistry; Ph.D., Berkeley, 1968. Organic synthesis, natural products, cancer chemotherapy, medicinal chemistry, bioorganic chemistry, design of organic catalysts.

Lawrence B. Kool, Assistant Professor of Organic Chemistry; Ph.D., Massachusetts, 1986. Organometallic synthesis, new catalytic reactions, organometallic reagents in organic synthesis, polymer synthesis, nonlinear optical materials.

David L. McFadden, Professor of Physical Chemistry; Ph.D., MIT, 1972. Gas phase fast-reaction kinetics and dynamics, electron spin resonance and photoionization mass spectrometry applied to the study of elementary reactions of polyatomic free radicals and of electron capture reactions at high temperature.

Larry W. McLaughlin, Associate Professor of Bioorganic Chemistry; Ph.D., Alberta, 1979. Chemistry and biochemistry of nucleic acids, sequence-dependent nucleic-acid recognition by proteins, synthesis and modification of oligonucleotides and nucleic acids.

Udayan Mohanty, Assistant Professor of Physical Chemistry; Ph.D., Brown, 1981. Theoretical chemistry, equilibrium and nonequilibrium statistical mechanics of liquids and metastable systems.

Robert F. O'Malley, Emeritus Professor of Inorganic Chemistry; Ph.D., MIT, 1961. Partial and selective electrolytic fluorination of polycyclic aromatic hydrocarbons, fluoro derivatives of polycyclic aromatic hydrocarbons in the study of the mechanism of carcinogenesis.

Yuh-kang Pan, Professor of Theoretical Chemistry; Ph.D., Michigan State, 1966. Quantum theory of chemical reactivity, calculation of lifetimes of singlet and triplet states by perturbation theory, applications of Lie algebra to chemical dynamics and vibrational-rotational spectra.

Mary F. Roberts, Professor of Biochemistry and Biophysical Chemistry; Ph.D., Stanford, 1974. Biological NMR, model biomembranes, cell metabolic pathways, studies of archaebacteria.

Dennis J. Sardella, Professor of Physical Organic Chemistry; Ph.D., IIT, 1967. NMR spectroscopy, aromaticity; transmission of electronic effects in π-systems; chemical carcinogenesis; theoretical organic chemistry.

Martha M. Teeter, Associate Professor of Biochemistry; Ph.D., Penn State, 1973. Protein crystallography, biophysical chemistry, protein design and structure/function prediction.

George Vogel, Professor of Organic Chemistry; D.Sc., Prague Technical, 1950. Chemistry of 2-pyrones and 2-pyridones, hindered dienes; mass spectrometry.

BOSTON UNIVERSITY
Department of Chemistry

Programs of Study

The Department of Chemistry offers programs leading to the M.A. and Ph.D. degrees, with emphasis on the latter program. Areas of specialization include physical, theoretical, inorganic, organic, and bioorganic chemistry and biochemistry, although most research projects contribute to more than one of these areas. Two themes are receiving special attention: (1) aspects of electromagnetic interaction with matter, which include spectroscopy, photochemistry, radiation chemistry, laser surgery, and photopromoted solid catalysis and (2) the structure and chemistry of biological macromolecules. Following placement exams, students begin a program of core and advanced courses tailored to meet individual needs and interests. Course programs are usually completed by the end of the second year. Students choose a research adviser before the end of the first academic year. Ph.D. candidates take qualifying exams, which are completed by May of the second year. Ability to read scientific literature in an approved foreign language is also required. Ph.D. candidates complete a research program, write a dissertation, and defend the work in a final oral exam. Students entering with a bachelor's degree normally complete the Ph.D. requirements within four to five years. Boston University also offers a Ph.D. in applied chemistry; students accepted into this program must have completed a master's degree in chemistry or its equivalent. Postdoctoral positions are available; inquiries should be addressed to individual faculty members.

Research Facilities

The Department of Chemistry occupies a six-floor complex of the Science and Engineering Center and is equipped with modern instrumentation, including a 400-MHz NMR spectrometer, an FT-IR spectrometer, a Silicon Graphics system, a Finnigan MAT 90 high-performance mass spectrometer, and an ultracentrifugation facility. Three large laser laboratories house custom-designed equipment for research incorporating high-resolution spectroscopy, laser flash photolysis, and optically detected magnetic resonance. Several laboratories have implemented CAMAC (computer-assisted measurement and control) interfacing techniques for efficient operation of their data acquisition and analysis functions. Biochemical facilities include cold rooms, and instrumentation for recombinant DNA, tissue culture, radiochemical measurements, and electrophoresis. The X-ray crystallography laboratory houses a Syntex $P2_1$ automated diffractometer and associated equipment. Other equipment includes various nuclear magnetic resonance, electron spin resonance, infrared, UV-visible, and atomic absorption spectrometers; analytical and preparative chromatographs; and electrochemical instruments. The information-processing capabilities in the Department of Chemistry are exceptional. Three DEC VAX-11/750s, fifteen Sun-3 Workstations of different models, and six DEC PDP-11 minicomputers are interconnected using a DECnet and TCP/IP on Ethernet architecture. In addition, there are an Encore Multimax 310 system and Macintoshes and PCs in the department. These systems are used for a wide variety of activities, including instrument control, theoretical calculation, graphical display, and preparation of scientific documents. The Science and Engineering Library (SEL), which houses the chemistry collection of books, journals, videotapes, and reference materials, is open seven days a week. SEL also houses collections for biology, physics, computer science, mathematics, and engineering.

Financial Aid

Students accepted into the graduate program in chemistry are normally supported by teaching fellowships, research assistantships, special fellowships, or a combination of the three. This support includes both a living stipend and tuition scholarship. The majority of first-year students receive teaching fellowships. For a teaching fellow, the living stipend for the 1991–92 academic year is $9500; additional support is available for the summer in the form of teaching fellowships and research assistantships. During the 1991–92 academic year, graduate research assistants receive a stipend of $1187 per month; additional support is available for summer months. When financial aid is provided, it is usually sufficient to cover the costs of tuition, other fees, and living expenses.

Cost of Study

For 1991–92, full-time tuition is $15,950 for the academic year.

Cost of Living

A limited number of rooms and apartments in University residences are available for graduate students. Information may be obtained from the Boston University Office of Housing, 985 Commonwealth Avenue, Boston, Massachusetts 02215. The Boston University Office of Rental Property, 19 Deerfield Street, Boston, Massachusetts 02215, supplies information about off-campus housing.

Student Group

The Graduate School has approximately 2,000 students, of whom 45 percent are women and 26 percent come from abroad. The Department of Chemistry has about 80 graduate students, representing many different states and countries.

Location

Boston's outstanding qualities as an educational and cultural center are well known. Music, theater, and sporting events to suit all tastes are abundantly available. Boston University is itself a center for the arts. Mountain ski areas and some of the best ocean beaches in the United States are available within a driving distance of 1 to 2 hours. The student population of Boston University, in common with the total student population in the Boston area, is drawn from all fifty states of the United States and from all over the world. Boston provides a stimulating and congenial place to live for persons of every taste and life-style.

The University

Boston University is a private, nonsectarian, urban university. It was founded in 1839 and is located in the historic Back Bay section of Boston. The main campus encompasses approximately 63 acres along Commonwealth Avenue and the Charles River. The medical school campus is situated in the South End, about a 20-minute drive from the Charles River campus. Boston University has a full-time student population of about 19,500 and a faculty in excess of 2,500.

Applying

Applicants should have completed (by the time of entry) a bachelor's degree in chemistry or its equivalent. Applications for admission and financial aid for the fall semester are due in the Admissions Office by February 1. Although later applications are considered, early applicants are given preference for admission and awards.

Correspondence and Information

For program information:
Chairman, Graduate Admissions Committee
Department of Chemistry
Boston University
590 Commonwealth Avenue
Boston, Massachusetts 02215
Telephone: 617-353-2500
Fax: 617-353-6466
INTERNET: chemgrs@bu-chem.bu.edu

For admission application forms:
Admissions Office
Graduate School
Boston University
705 Commonwealth Avenue
Boston, Massachusetts 02215
Telephone: 617-353-2693

Boston University

THE FACULTY AND THEIR RESEARCH

Richard H. Clarke, Professor; Ph.D., Pennsylvania, 1969. Physical and biophysical chemistry: triplet states of photosynthetic systems; carcinogen-DNA complexes: structure and interactions; laser and microwave spectroscopy of organic molecules; medical applications of lasers.

David F. Coker, Assistant Professor; Ph.D., Australian National, 1986. Theoretical physical chemistry: classical and quantum statistical mechanics of condensed systems, atomic and molecular clusters and electrons in fluids, chemical reactions involving the transfer of electrons or protons.

Robert C. Davenport, Assistant Professor; Ph.D., MIT, 1986. X-ray crystallography of biological macromolecules, mechanisms of protein folding, enzymology of energy transduction, physical biochemistry, molecular biology of DNA recombination.

Dan Dill, Professor; Ph.D., Chicago, 1972. Theoretical molecular physics: analysis of spatial distributions of electrons scattered by molecules or ejected from them by light, theory of interconversions of electronic and rovibrational energy in molecules, motion of highly excited and unbound electrons in molecular fields.

Klaas Eriks, Professor; Ph.D., Amsterdam, 1948. Physical chemistry: structure determination, using single-crystal X-ray diffraction; calcium salts of amino acids and small polypeptides; amino acid complexes of transition-metal ions; metal olefin structures.

Warren P. Giering, Associate Professor; Ph.D., SUNY at Stony Brook, 1969. Organometallic chemistry: chemistry of odd electron transition-metal carbonyl complexes possessing acyl, alkyl, alkene, and alkyne ligands; structure and chemistry of complexes containing novel π-ligands.

Ronald L. Halterman, Assistant Professor; Ph.D., Berkeley, 1985. Organic chemistry: stereochemical control of organic and organometallic reactions, preparation of new chiral ligands for use in asymmetric organometallic-mediated transformations of interest in the context of organic synthesis.

Standish C. Hartman, Professor and Associate Chair for Graduate Studies; Ph.D., MIT, 1957. Biochemistry: recombinant DNA methodology for study of molecular mechanisms; genetic markers for cloning in animal cells; structures of proteins, using nucleotide sequences of their genes; modification of protein structure by mutagenesis of genetic material.

Morton Z. Hoffman, Professor and Associate Chair for Undergraduate Studies; Ph.D., Michigan, 1960. Physical/inorganic chemistry: intermolecular and intramolecular electron transfer in ground and excited states, photochemistry and photophysics of transition-metal coordination complexes, study of intermediates from excited-state electron-transfer reactions, free-radical reactions in lipid peroxidation processes.

Guilford Jones II, Professor and Chair; Ph.D., Wisconsin, 1970. Physical organic chemistry/photochemistry: photoinduced electron transfer and other reactions of organic molecules and ions, photoreactions for energy storage or chemical formation, luminescence and flash illumination techniques for study of aggregated dye systems.

Herman M. Kalckar, Distinguished Research Professor; M.D., 1933, Ph.D., 1938, Copenhagen. Biochemistry: regulation of the hexose transport system, protein synthesis, and oxidative energy metabolism.

Thomas F. Keyes, Professor; Ph.D., UCLA, 1971. Theoretical physical chemistry: motion of particles, fluids, sound, and electromagnetic radiation through dense disordered media; derivation of kinetic equations, use of the real-space renormalization group, and computer simulation for high-density scatterers.

Richard A. Laursen, Professor; Ph.D., Berkeley, 1961. Bioorganic chemistry: elucidation of the primary structure of proteins; chemical, physical, and molecular biological techniques for study of tertiary structure and evolution of proteins; methodology of protein sequencing; structure and function of human plasminogen.

Norman N. Lichtin, University Professor and Professor; Ph.D., Harvard, 1948. Organic and physical chemistry: photoassisted reactions, mechanisms involving electronically excited intermediates, photochemical conversion and storage of energy, chemical control of pollution of air and water.

Ronald M. Milburn, Professor; Ph.D., Duke, 1954. Inorganic and bioinorganic chemistry: structures and reactions of metal complexes in solution; reactions of coordinated ligands; cobalt (III) and other metal-ion-promoted phosphoryl transfer reactions of phosphate esters and polyphosphates, including ADP (adenosine-5'-diphosphate) and ATP (adenosine-5'-triphosphate).

Scott C. Mohr, Associate Professor; Ph.D., Harvard, 1968. Biophysical chemistry: application of physiochemical techniques to the study of biochemical systems, condensed (liquid crystalline) states of nucleic acids, polycyclic aromatic hydrocarbon interactions with DNA, computer-aided analysis of DNA conformation, gas-phase enzyme reactions.

James S. Panek, Assistant Professor; Ph.D., Kansas, 1984. Development of new organic methodology and application to the total synthesis of natural products, transition-metal-catalyzed cyclization reactions, stereo control in intramolecular cycloadditions.

Alfred Prock, Professor; Ph.D., Johns Hopkins, 1955. Physical chemistry: energy and charge transfer from excited molecules in solution (or molten state) to metal or semiconductor interfaces, mechanisms of dark conduction and photoconduction in fluids, diffusion lengths of triplet excitation in molten aromatic systems.

John K. Snyder, Associate Professor; Ph.D., Chicago, 1979. Organic chemistry/natural products: isolation and structure determination of biologically active natural products, particularly those from Chinese and other traditional medicines; techniques of isolation and structure determination, including new NMR procedures; strategies for the synthesis of natural products.

John E. Straub, Assistant Professor; Ph.D., Columbia, 1987. Theoretical physical and biophysical chemistry: chemical reaction dynamics in liquids and proteins; structure, dynamics, and thermal stability of proteins; energy transfer in liquids and biomolecules; ligand binding in hemeproteins.

The Boston University skyline, looking east along the Charles River.

SECTION 2: CHEMISTRY

BRANDEIS UNIVERSITY
Department of Chemistry

Programs of Study

The department offers the Doctor of Philosophy degree with specializations in chemical physics and inorganic, organic, physical, biophysical, and bioorganic chemistry.

There are no specific course requirements for the Ph.D. degree. New students, in consultation with a graduate advisory committee, design a program of study based upon their interests, experience, and background (as measured by a set of entrance examinations covering the basic chemistry curriculum). By the start of the second semester, the doctoral student begins a program of independent investigation supplemented by advanced courses, assigned readings, and seminars. Normally, the student will obtain the Ph.D. degree within five years.

The M.A. degree is not a prerequisite for the Ph.D., but it can be awarded after one year of full-time study, including both laboratory and course work.

Classes are small, and the ratio of graduate students to faculty is also favorable at about 3:1.

Research Facilities

Research facilities are housed in the Edison-Lecks Chemistry Building and the Kalman Science Building; they include well-equipped laboratories, a wide variety of modern instrumentation, and excellent machine and electronics shops. The computer center is adjacent, as is the Science Library, used both by the University and by nearby industries. Cooperative programs with the Departments of Physics and Biochemistry and with the Rosensteil Basic Medical Sciences Center are available.

Financial Aid

All students receive an annual stipend of $11,500–$12,500, according to merit. For first-year students the stipend is derived partly from a teaching assistantship involving two afternoons of instruction per week during the academic year. For advanced students, the stipend is provided from research assistantships (nonteaching) or University or industrial fellowships or teaching assistantships.

Cost of Study

For 1991–92, tuition for full-time graduate students is $16,085 per academic year. Additional fees include about $480 for health insurance and a final $275 doctoral fee. Tuition scholarships are awarded to students in good standing.

Cost of Living

Living accommodations are available off campus in Waltham and other nearby Greater Boston communities. In 1991–92, monthly rents range from $250–$400 per person for furnished or unfurnished apartments or houses. Students may eat in the University cafeterias on a single-meal or contract basis. A limited amount of unfurnished graduate housing is available at rents from $400 for efficiency units to $250 per person for two- or three-bedroom apartments, plus the cost of electric heat.

Student Group

About 2,900 undergraduates and 600 graduate students make up the student body at Brandeis. The graduate students form a cosmopolitan group. There are about 50 graduate students in chemistry, equally divided between men and women.

Location

Waltham is but one of a large group of independent small cities, like Cambridge, Wellesley, Lexington, and Concord, that are contiguous to Boston in a metropolitan area of about 2.5 million people. This Greater Boston community is justly famous for the richness and ferment of its intellectual and cultural life, arising from the many excellent universities in close contact, as well as from the wide offerings in concerts, museums, and theaters. This active urban center is geographically close-knit and internally accessible via excellent public transportation and high-speed highways. In addition to the many activities of the contemporary world, Greater Boston offers the charm and interest of its long history at the center of the United States' development. The surrounding countryside offers great scenic and recreational variety at remarkably close range, from the mountains, forests, and old villages inland to the fine beaches of Cape Cod to the south and the rocky coastline of Maine to the north.

The University

Brandeis, founded in 1948, is dedicated to the principles of Justice Louis Brandeis, who wrote of a university, "It must become truly a seat of learning, where research is pursued, books written and the creative instinct aroused, encouraged and developed." In the short time since its inception, Brandeis has achieved international recognition as a university of high quality. The University has deliberately maintained a small size in order to foster and encourage an active sense of intellectual community, but it is large enough, nevertheless, to thoroughly cover a broad range of scholarship.

Applying

Applicants who have completed a program leading to the bachelor's degree or its equivalent are considered. Students are urged to take the Graduate Record Examinations (General Test and the Subject Test in chemistry). When English is not the native language, the TOEFL is required. Early applicants are given preference for admission and awards. Applications completed after February 28 go on a waiting list. Prospective applicants are strongly encouraged to visit the University; a schedule of visits can be arranged with the department.

Correspondence and Information

Chairman
Department of Chemistry
Brandeis University
Waltham, Massachusetts 02254
Telephone: 617-736-2500

SECTION 2: CHEMISTRY

Brandeis University

THE FACULTY AND THEIR RESEARCH

In the listing that follows, each member of the staff has briefly indicated his or her principal interests. Detailed listings of current publications are contained in the American Chemical Society's *Directory of Graduate Research.* Any faculty member would be pleased to discuss specific questions either by correspondence or through a personal interview.

Peter C. Jordan, Professor and Chairman; Ph.D., Yale, 1960. Biophysical chemistry and chemical physics: membrane transport, electrostatic modeling of ion channels in biological membranes, molecular dynamics of biomolecules, ionic interactions with complex ionophores.

Iu-Yam Chan, Professor; Ph.D., Chicago, 1968. Chemical physics and physical chemistry: spectroscopy and magnetic resonance, spin-echo envelope modulation, charge transfer and energy transfer under high pressure.

James H. Davis, Assistant Professor; Ph.D., Vanderbilt, 1986. Inorganic and organometallic chemistry: activation of CO through coordination to main group centers, preparation and characterization of organometallic oxo- and alkoxide complexes.

Irving R. Epstein, Professor; Ph.D., Harvard, 1971. Physical chemistry and chemical physics: oscillating chemical reactions, chaos and pattern formation, aggregation, biochemical kinetics, neurobiology.

Bruce M. Foxman, Professor; Ph.D., MIT, 1968. Inorganic and solid-state chemistry: X-ray structure determination; coordination polymers; reactions in crystals; kinetics, mechanisms, and crystallography of rearrangement, polymerization, and decomposition reactions in the solid state.

Michael Henchman, Professor; Ph.D., Yale, 1961. Physical chemistry and chemical physics: kinetics and thermodynamics; the chemistry of ions in the gas phase and its application to chemistry in solution, plasmas, and interstellar space.

James B. Hendrickson, Professor; Ph.D., Harvard, 1955. Organic chemistry: synthesis of natural products, development of new synthetic reactions, computerized development of synthesis design.

Judith Herzfeld, Professor; Ph.D., MIT, 1972. Biophysics and chemical physics: solid-state NMR studies of membrane transport systems and related theory, statistical mechanical and optical studies of long-range order in self-assembling protein and surfactant systems.

Philip M. Keehn, Professor; Ph.D., Yale, 1969. Organic chemistry: organic synthesis of strained rings and theoretically interesting molecules, synthetic methods, application of NMR spectroscopy to organic systems, photooxidation, thermal chemistry, pure and applied laser chemistry of organic systems, host-guest chemistry, medicinals.

Kenneth Kustin, Professor; Ph.D., Minnesota, 1959. Inorganic biochemistry: vanadium and iron in tunicate blood cells and human tissues, fast reactions, oscillating reactions.

Gregory A. Petsko, Professor; D.Phil., Oxford, 1973. Biophysical chemistry: protein crystallography, structure-function relations of proteins, mechanism of enzyme action, low-temperature protein crystallography, site-directed mutagenesis, time-resolved protein crystallography, allergy.

Thomas C. Pochapsky, Assistant Professor; Ph.D., Illinois, 1986. Organic and bioorganic chemistry: design and synthesis of molecular recognition systems, NMR of biomacromolecules; protein structure and stability.

Dagmar Ringe, Associate Professor; Ph.D., Boston University, 1968. Biophysical chemistry: structure and function of proteins, enzyme mechanisms, protein-drug interactions, time-resolved protein crystallography, synthesis of enzyme inhibitors, transmembrane channel and receptor structure and mode of action.

Myron Rosenblum, Professor; Ph.D., Harvard, 1954. Organic and organometallic chemistry: new synthetic methods using organoiron complexes, natural product synthesis, electroactive organometallic polymers; chemistry of organometallic complexes.

Barry B. Snider, Professor; Ph.D., Harvard, 1973. Organic chemistry: synthesis of natural products; new synthetic methods; applications of the ene reaction and Lewis acid–induced reactions to synthesis, intramolecular ketene cycloadditions, oxidative free radical cyclizations.

Colin Steel, Professor; Ph.D., Edinburgh, 1958. Physical chemistry: chemistry of excited molecules and radicals, kinetics and mechanisms of photochemical and thermal reactions.

Robert Stevenson, Professor; Ph.D., Glasgow, 1952. Organic chemistry: isolation and structure of natural products, lignan synthesis, molecular rearrangements, heterocyclics, medicinal compounds (platelet-activating factor antagonists and lipoxygenase inhibitors).

Thomas R. Tuttle Jr., Professor; Ph.D., Washington (St. Louis), 1957. Physical chemistry and chemical physics: determination of species and their structures in liquid solutions by magnetic resonance spectrometry and optical absorption spectroscopy.

The Science Library.

The Science Quadrangle.

A pause from studies.

SECTION 2: CHEMISTRY

BRIGHAM YOUNG UNIVERSITY
Chemistry Department and Graduate Section of Biochemistry

Programs of Study

The Chemistry Department and the Graduate Section of Biochemistry offer courses of study leading to the Ph.D. and M.S. degrees in the areas of analytical, inorganic, organic, and physical chemistry and in biochemistry. Graduate students are involved in a broad range of research topics, including atomic and molecular spectroscopy, fundamental studies of supercritical fluids, analytical coal chemistry, organic photochemistry, high-resolution chromatography, cancer biochemistry, synthesis of biomedically important compounds, recombinant DNA techniques, stereoselective synthesis, metalloenzyme catalysis, high-sensitivity calorimetric measurements, atmospheric environmental chemistry, macrocycle synthesis and mediated transport, X-ray optics, and crystallography. As their research progresses, students have significant opportunities to interact with faculty members. All students must pass entrance exams in at least four of five subject areas before the end of their first year. A schedule of courses is established for each student based on his or her needs and interests. Most of this course work is taken in the first year, with the remainder generally completed in the second year. All students complete annual reviews, which for Ph.D. students includes a written comprehensive examination given when course work is complete. The research experience is the major element of all graduate programs in chemistry and biochemistry. Most students complete the M.S. degree in two years or the Ph.D. within four or five years.

M.S. and Ph.D. graduates enter careers in industry or accept academic positions.

Research Facilities

The department has available about 40,000 square feet of research space. Design and architectural work is being done for a new 100,000-square-foot science building that will provide modern and spacious facilities. Departmental research instrumentation includes a Varian Gemini 200-MHz FT-NMR spectrometer, a Varian VXR 500-MHz multinuclear RT-NMR spectrometer, a Finnigan-MAT Model 8430 double-focusing mass spectrometer, a Hewlett-Packard Quadrupole GC/MSD System, an Applied Biosystem 370A automated DNA sequencer, an Applied Biosystem 381A automated DNA synthesizer, a Nicolet Model R3 X-ray diffractometer with low-temperature accessory, a Mattson Sirius 100 FT-IR spectrometer, a Perkin-Elmer Plasma II ICP emission spectrometer, a Cary 118 UV-visible spectrophotometer, a Hewlett-Packard Tri-Carb 4000 liquid scintillation spectrometer, a Perkin-Elmer LS50 fluorimeter, a Lambda Physik EMG101ES excimer laser, a Lambda Physik FL2002E dye laser, a Nicomp 200D laser particle sizer, and an environmental analysis chamber.

In addition to the major instruments listed above, the department has numerous capabilities in individual research laboratories. A proton-induced X-ray emission/proton-induced gamma-ray emission (PIXE/PIGE) system utilizes a 2-MeV Van de Graaff accelerator and a Canberra Series 90 multichannel analyzer interfaced to a DEC MicroVAX II computer for high-sensitivity, multielement analysis. Chromatographic capabilities include numerous high-resolution gas, liquid, ion, supercritical-fluid and capillary electrophoresis instruments. Some of these were largely developed in the department. The department is well equipped to do thermodynamics research. Automated differential scanning and cryogenic adiabatic calorimeters are available for heat capacity, phase transition, and cell growth studies. Calorimeters for the measurement of thermodynamic properties of solutions include numerous titration and flow systems capable of operating over wide ranges of temperature, pressure, and sample size. Recent acquisitions for specialized spectroscopic research include a high-resolution monochromator, a boxcar signal averager, diode array detectors, an inductively coupled plasma source and spectrometer, and a stabilized diode laser system.

Equipment dedicated to biochemical and cancer research includes a DNA sequencer and a DNA synthesizer; a modern tissue culture facility; a recombinant DNA laboratory; an automated fluorimeter; a number of visible and UV absorbance spectrophotometers, including one with complete kinetics capability; several analytical and preparative ultracentrifuges; and high-performance liquid chromatographs.

The Harold B. Lee Library is located within a 2-minute walk of all chemistry and biochemistry areas.

Financial Aid

The department provides financial assistance to students through teaching and research assistantships, internships, and tuition scholarships. The stipend for beginning students for the 1991–92 year is $11,340. The amount of the stipend is adjusted annually. All qualified students who meet the application deadline receive financial assistance.

Cost of Study

Tuition scholarships, limited to the credit hours required for graduation, are provided by the department.

Cost of Living

The University Housing Office assists all students in locating satisfactory accommodations for both on- and off-campus housing. Monthly rent ranges from $140 to $350 a month, plus utilities.

Student Group

BYU has more than 27,000 full-time students, with 1,320 full-time graduate students. Students come from a variety of academic and ethnic backgrounds as well as from many different geographic areas. The Chemistry Department has a total of 56 graduate students, 12 in the M.S. program and 44 in the Ph.D. program.

Location

BYU's beautiful 536-acre campus is located in Provo, Utah, a semiurban area at the foot of Utah Valley's Wasatch Mountains. The University has an excellent lyceum and cultural program and is near numerous outdoor recreational areas for skiing (snow and water), hiking, and camping. The fine cultural resources in Salt Lake City, 45 miles to the north, include the Utah Symphony, Ballet West, and the Pioneer Theater.

The University and The School

Brigham Young University is the largest privately owned university in the United States. Established in 1875 as Brigham Young Academy and sponsored by the Church of Jesus Christ of Latter-Day Saints, BYU has a tradition of high standards in moral integrity and academic scholarship. Along with extensive undergraduate programs, BYU offers graduate degrees in a variety of disciplines through fifty-seven graduate departments. Chemistry, with its outstanding faculty and mature graduate program, is one of the leading research departments at BYU.

Applying

Applications and all supporting materials should be received by the Office of Graduate Studies no later than February 15 to be considered for admission the following fall. International students must pass the Test of English as a Foreign Language with a score of 550 or higher.

Correspondence and Information

Dr. Francis R. Nordmeyer
Coordinator, Graduate Admissions
225 Eyring Science Center
Brigham Young University
Provo, Utah 84602-1022
Telephone: 801-378-4845

SECTION 2: CHEMISTRY

Brigham Young University

THE FACULTY AND THEIR RESEARCH

James L. Bills, Professor; Ph.D., MIT, 1963. Electronic and molecular structure and bonding.

Juliana Boerio-Goates, Associate Professor; Ph.D., Michigan, 1979. Calorimetry of phase transitions in molecular crystals, thermodynamics of organic solid-state reactions, thermochemistry of bonding.

Jerald S. Bradshaw, Professor; Ph.D., UCLA, 1963. Synthesis of macrocyclic multidentate compounds; chemistry of polysiloxanes, particularly as applied to cross-linked stationary phases for chromatography.

Coran L. Cluff, Professor; Ph.D., Michigan, 1961. Laser chemistry and isotope separation.

N. Kent Dalley, Professor; Ph.D., Texas at Austin, 1968. Crystal structures of cyclic polyethers and their metal ion complexes, crystal structure of nucleosides.

Delbert J. Eatough, Professor; Ph.D., Brigham Young, 1967. Environmental atmospheric chemistry of SO, NO_x, and organics; analytical techniques, tracers, and source apportionment for pollution; indoor atmospheric chemistry; solution calorimetry applied to coal liquids and surface chemistry; development of calorimetric instruments.

Paul B. Farnsworth, Associate Professor; Ph.D., Wisconsin, 1981. Fundamental and applied measurements on inductively coupled plasmas, element-specific detectors for chromatography.

Steven A. Fleming, Assistant Professor; Ph.D., Wisconsin, 1984. Photochemistry of aromatic compounds, rearrangements of heterocycles, synthesis of natural products, mechanisms of thermal rearrangements and photorearrangements.

Steven R. Goates, Associate Professor; Ph.D., Michigan, 1981. Analysis of complex samples by laser-based methods, supersonic jet spectroscopy, spectroscopic study of the solid-state and chromatographic processes.

David M. Grant, Distinguished Professor; Ph.D., Utah, 1957. Carbon-13 magnetic resonance in fossil fuels and molecular biology, chemical shift theory and molecular structure, nuclear relaxation and molecular dynamics, NMR in solids.

Lee D. Hansen, Professor; Ph.D., Brigham Young, 1965. Calorimetry thermodynamics and kinetics, atmospheric chemistry of sulfur and nitrogen, characterization of coal ash and liquids, productivity and stress response of green plants.

Richard T. Hawkins, Professor; Ph.D., Illinois, 1959. Synthesis and properties of calixarenes.

Reed M. Izatt, Professor; Ph.D., Penn State, 1954. Thermodynamics of metal coordination with amino acids, macrocyclic compounds, carboxylic acids, etc.; microtitration calorimeters; macrocycle-mediated cation transport; heats of mixing in supercritical fluids; solute-solute interactions at high temperature.

John D. Lamb, Associate Professor; Ph.D., Brigham Young, 1978. Macrocyclic ligand chemistry, liquid membrane separations, ion chromatography, calorimetry.

Milton L. Lee, Professor; Ph.D., Indiana, 1975. Microcolumn chromatography, capillary column technology, gas chromatography, supercritical fluid chromatography, electrophoresis, mass spectrometry, study of coal-derived materials.

Nolan F. Mangelson, Professor; Ph.D., Berkeley, 1967. Element analysis by proton-induced X-ray and gamma-ray emission, EXAFS, and RBS; element analysis applied to environmental and biological problems.

John H. Mangum, Professor; Ph.D., Washington (Seattle), 1963. Characterization of enzymes in one-carbon metabolism, folate-dependent enzymes and antifolate-anticancer drugs, one-carbon metabolism in nerve and brain.

Francis R. Nordmeyer, Professor; Ph.D., Stanford, 1967. Ion chromatography.

J. Bevan Ott, Professor; Ph.D., Berkeley, 1959. Thermodynamics of nonelectrolyte solutions, phase equilibria, calorimetry.

Noel L. Owen, Professor; Ph.D., Cambridge, 1964. FT-IR, NMR, and Raman spectroscopy; structure and conformation of novel compounds; bonding between wood and various reagents; structure of wood polymers.

Edward G. Paul, Professor; Ph.D., Utah, 1962. Organic synthesis of oxygen heterocycles (coumarins and furanocoumarins).

Ronald J. Pugmire, Professor; Ph.D., Utah, 1966. NMR spectroscopy; carbon-13 resonance methods; study of fossil fuels, coal macerals, and coal petrography; organic geochemistry; nuclear relaxation and liquid dynamics; magnetic resonance imaging; in vivo spectroscopy.

Donald L. Robertson, Associate Professor; Ph.D., Washington (St. Louis), 1976. Biochemistry of RNA tumor virus gene expression, including RNA metabolism, protein synthesis, and recombinant DNA; vaccine development; oncogene biochemistry.

Morris J. Robins, J. Rex Goates Professor; Ph.D., Arizona State, 1965. Chemistry of nucleic acid components and nucleoside analogues, including natural-product nucleosides; design of enzyme inhibitors; anticancer and antiviral agents.

Bryant E. Rossiter, Assistant Professor; Ph.D., Stanford, 1981. Development of enantioselective systems, i.e., transition-metal–mediated organic reactions and stationary phases for chromatograhy; synthesis and analysis of enantiomerically enriched natural products and pharmaceuticals.

Randall B. Shirts, Associate Professor; Ph.D., Harvard, 1979. Theoretical chemistry, vibration-rotation and nonlinear dynamics, semiclassical quantization, laser-molecule interaction, quantum-classical correspondence.

Daniel L. Simmons, Assistant Professor; Ph.D., Wisconsin, 1986. Molecular mechanisms of neoplastic transformation by Rous sarcoma virus and the action of recessive oncogenes.

Marvin A. Smith, Professor; Ph.D., Wisconsin, 1964. Plant molecular biology: development, stress, and gene expression.

Richard L. Snow, Professor; Ph.D., Utah, 1957. Thermodynamics of nonelectrolyte solutions, quantum chemistry.

James M. Thorne, Professor; Ph.D., Berkeley, 1966. Cancer photoradiation therapy, X-ray diagnostics for laser fusion.

Gerald D. Watt, Professor; Ph.D., Brigham Young, 1966. Nitrogenase, ferritins, metalloproteins.

Byron J. Wilson, Professor; Ph.D., Washington (Seattle), 1961. Inorganic free radicals, reaction mechanisms and kinetics, deterioration of paper.

Earl M. Woolley, Professor; Ph.D., Brigham Young, 1969. Thermodynamics of reactions in mixed aqueous-organic solvents, of intermolecular hydrogen bonding, of electrolyte solutions, and of surfactants; calorimetric methods.

S. Scott Zimmerman, Associate Professor; Ph.D., Florida State, 1973. Structure and function of biological macromolecules; computer applications, simulations, and chemical education.

SECTION 2: CHEMISTRY

CARNEGIE MELLON UNIVERSITY

Mellon College of Science
Department of Chemistry

Programs of Study

The Department of Chemistry offers programs leading to the M.S. and Ph.D. degrees. Most students are admitted to the Ph.D. program, but terminal master's programs in polymer science and in colloids, polymers, and surfaces are offered. The graduate program is highly individualized to allow exploration of interdisciplinary interests. Research is carried out in biochemistry and biophysics; chemical physics; inorganic, nuclear, organic, physical, and theoretical chemistry; NMR spectroscopy; and polymer science.

To be admitted to the Ph.D. program, students must pass attainment examinations in inorganic, organic, and physical chemistry. These are given just before the beginning of the fall semester and in the spring semester. The student then chooses a research supervisor and commences thesis research. Students are expected to pass at least four advanced courses in science with an average grade of B, present a seminar during the second year, write and defend orally both a formal research proposal on a topic different from the thesis topic and on the thesis topic, pass two research literature examinations, and assist in undergraduate teaching for two semesters. The dissertation must embody the results of extended research and constitute an original contribution to knowledge that is worthy of publication.

The graduate program at CMU emphasizes close interaction with all the faculty. There are excellent opportunities for interdisciplinary programs with the Departments of Biological Sciences, Chemical Engineering, Materials Science, and Physics, along with the Biotechnology Program and the Center for Fluorescence Research in Biomedical Sciences.

Research Facilities

The Department of Chemistry is located with the Department of Biological Sciences in the Mellon Institute Building, a spacious and dramatic eight-story structure located near the main campus of CMU and directly adjacent to the University of Pittsburgh campus. The department houses state-of-the-art laser spectroscopy laboratories; a Langmuir-Blodgett film facility; the NIH-sponsored NMR Facility for Biomedical Studies, with its 620-MHz NMR spectrometer; two major computational facilities; modern light-scattering spectrometers for the study of polymers; and extensive analytical instrumentation, including two 300-MHz NMR spectrometers, one of which is equipped for solids. Nuclear chemistry research is carried out at major facilities in the United States and abroad. The Mellon Institute Building also houses the Center for Fluorescence Research in Biomedical Sciences and the NSF-sponsored Pittsburgh Supercomputing Center. The library of the Mellon Institute is especially strong.

Financial Aid

All U.S. doctoral students are guaranteed financial aid for the first academic year, usually as teaching assistants, with a stipend of at least $1000 per month and a tuition scholarship. In addition, competitive fellowships are available, which pay up to an additional $2000 per year. Research assistantships are also available for succeeding years. International students may be admitted without being granted financial aid.

Cost of Study

Tuition is $15,250 for the 1991–92 academic year.

Cost of Living

Pittsburgh provides an attractive and reasonably priced living environment. On-campus housing is limited, but the Off-Campus Housing Office assists students in finding suitable accommodations. Most graduate students prefer to live in nearby rooms and apartments, which are readily available.

Student Group

Graduate enrollment at CMU totals 2,708 and includes students from all parts of the United States and many foreign countries. All students in the Department of Chemistry are receiving financial aid. Upon completing the Ph.D., a few students (15 percent) go directly into academic positions, but most continue as postdoctoral fellows (40 percent) or take industrial jobs (45 percent).

Location

Pittsburgh is in a large metropolitan area of 2.3 million people. It was recently rated by the *Rand-McNally Places Rated Almanac* as the number one place to live in the United States. The city is the headquarters for twelve Fortune 500 corporations, and there is a large concentration of research laboratories in the area. Carnegie Mellon is located in the Oakland neighborhood, the cultural and civic center of Pittsburgh. The campus covers 90 acres and adjoins Schenley Park, the largest city park. The city's cultural and recreational opportunities are truly outstanding.

The University

Carnegie Mellon was first established in 1900 as the Carnegie Technical School through a gift from Andrew Carnegie. In 1912, the name was changed to Carnegie Institute of Technology. The Mellon Institute was founded by A. W. and R. B. Mellon; it carried out both pure research and applied research in cooperation with local industry. In 1967, the two entities merged to form Carnegie-Mellon University. Four colleges—the Carnegie Institute of Technology, the Mellon College of Science, the College of Fine Arts, and the College of Humanities and Social Sciences—offer both undergraduate and graduate programs. The Graduate School of Industrial Administration and the School of Urban and Public Affairs offer graduate programs only. The University has assets in excess of $932-million, a total enrollment of 7,216, and 1,004 faculty members.

Applying

Completed applications and credentials for graduate study in chemistry should be submitted by February 15 for decision by mid-April. However, admission decisions are made on a continuous basis, and applications are considered at any time. In addition to the application form, transcripts from all college-level institutions attended, three letters of recommendation, and an official report of the applicant's scores on the General Test and the Subject Test in chemistry of the Graduate Record Examinations are required. A full description of procedures and programs is given in the booklet *Graduate Studies in Chemistry*, which will be sent on request.

Correspondence and Information

Committee for Graduate Admissions
Department of Chemistry
Carnegie Mellon University
4400 Fifth Avenue
Pittsburgh, Pennsylvania 15213
Telephone: 412-268-3150

SECTION 2: CHEMISTRY

Carnegie Mellon University

THE FACULTY AND THEIR RESEARCH

Guy C. Berry, Professor; Ph.D., Michigan, 1960. Physical chemistry and polymer science, physical chemistry of macromolecules: photon correlation and integrated intensity light scattering, solution properties of flexible and rodlike polymers, rheology of polymers, properties of liquid crystalline polymers.

Charles L. Brooks III, Associate Professor; Ph.D., Purdue, 1982. Statistical mechanics: structure dynamics and thermodynamics of biomolecules, development of classical and quantum statistical mechanics and their application to problems associated with chemical reactivity in dense media, dynamics, thermodynamics and reactivity of biological systems in aqueous environment and small-molecule dynamics in liquids.

Albert A. Caretto Jr., Professor; Ph.D., Rochester, 1953. Chemical education, physical chemistry, nuclear chemistry, and radiochemistry: mechanism of high-energy nuclear reactions, meson-induced reactions, nuclear spectroscopy, nuclear properties, radio-electrochemical techniques for trace-element analysis.

Edward F. Casassa, Professor; Ph.D., MIT, 1952. Physical chemistry: physical chemistry of macromolecules, solution properties, thermodynamics, statistical mechanics, light scattering, conformations of confined polymer chains, theory of gel chromatography.

Terrence J. Collins, Associate Professor; Ph.D., Auckland (New Zealand), 1978. Inorganic chemistry, chemical synthesis, and oxidation chemistry; synthesis and reaction chemistry of highly oxidizing inorganic compounds; oxidation catalysis and mechanisms; bioinorganic chemistry of high-oxidation-state transition-metal species; structure-bonding-function relationships of strongly donating ligands.

Josef Dadok, Professor; Ph.D., Czechoslovak Academy of Sciences, 1963. NMR techniques and instrumentation: very high field, high-resolution NMR instrumentation and techniques, using superconducting magnets, rapid-scan correlation, and FT NMR techniques for biomedical applications.

Morton Kaplan, Professor; Ph.D., MIT, 1960. Nuclear chemistry and chemical physics: nuclear reactions of heavy ions and high-energy projectiles, recoil properties of radioactive products, Mössbauer resonance, perturbed angular correlations of gamma rays, statistical theory of nuclear reactions and light-particle emission.

Paul J. Karol, Associate Professor; Ph.D., Columbia, 1967. Nuclear chemistry and physical chemistry: high-energy nuclear reactions, rapid radiochemical separations, column chromatography.

Jonathan S. Lindsey, Associate Professor; Ph.D., Rockefeller, 1983. Bioorganic chemistry: synthesis of three-dimensional molecules for studies in artificial photosynthesis, synthesis of polymeric systems, molecular electronics, robotic chemistry.

Miguel Llinas, Professor; Ph.D., Berkeley, 1971. Molecular biophysics: structural dynamics of polypeptides in solution by NMR spectroscopy, plasminogen and blood coagulation proteins.

Krzysztof Matyjaszewski, Associate Professor; Ph.D., Polish Academy of Sciences, 1976. Polymer organic chemistry: kinetics and thermodynamics of ionic reactions, cationic polymerization, ring-opening polymerization, living polymers, inorganic and organometallic polymers, electronic materials.

Richard D. McCullough, Assistant Professor; Ph.D., Johns Hopkins, 1988. Organic and Materials Chemistry: Design, synthesis, and characterization of new organic conducting polymers and polymeric liquid crystals; development of organic and organometallic high spin molecules and materials; electrical, magnetic, and optical properties of novel solid-state materials; synthesis and study of organic molecules that can mimic biological catalysis.

Eckard Münck, Professor; Ph.D., Darmstadt Technical (Germany), 1967. Active sites of metalloproteins, in particular sites containing iron-sulfur clusters or oxo-bridged iron dimers; study of synthetic clusters which mimic structures in proteins; magnetochemistry: Heisenberg and double-exchange; Mössbauer spectroscopy and electron paramagnetic resonance.

Gary D. Patterson, Professor; Ph.D., Stanford, 1972. Chemical physics and polymer science: application of light-scattering spectroscopy to problems of structure and dynamics in amorphous materials, physics and chemistry of liquids and solutions, conformational statistics and molecular dynamics of polymers, nature and dynamics of the glass transition, nonlinear optical properties of organic materials, structure and dynamics of biopolymers.

Stuart W. Staley, Professor; Ph.D., Yale, 1964. Physical organic chemistry: synthetic and spectroscopic studies of carbanions; electron transfer reactions; theoretical, synthetic, and structural studies of small strained ring systems and molecules with Möbius topology; photoelectron and electron transmission spectroscopy; NMR spectroscopy; solid-state structure and nonlinear optical properties.

Robert F. Stewart, Professor; Ph.D., Caltech, 1962. Physical and theoretical chemistry: X-ray diffraction, high-energy electron-scattering calculations.

Charles H. Van Dyke, Associate Professor; Ph.D., Pennsylvania, 1964. Synthetic inorganic chemistry and chemical education.

James W. Whittaker, Assistant Professor; Ph.D., Minnesota, 1983. Bioinorganic chemistry and spectroscopy: structural and mechanistic studies of metal ion active sites, mechanisms of electron transfer in solids and proteins.

SECTION 2: CHEMISTRY

CASE WESTERN RESERVE UNIVERSITY
Department of Chemistry

Programs of Study

The Department of Chemistry offers programs leading to the M.S. and Ph.D. degrees. Students choose a curriculum of course work from a large array of offerings in chemistry and related areas of study. There is a 6-course core requirement that can be satisfied completely during the student's first year and includes one course in each of the three major chemistry disciplines (organic, inorganic, and physical). The Chemistry of Life Processes program offers students the opportunity to pursue a course of study that focuses on chemical aspects of biological systems. The Case Center for Electrochemical Sciences has its major focus in the Department of Chemistry and provides outstanding opportunities for interdisciplinary study and research in fundamental and applied aspects of electrochemistry.

The student's principal activity in working toward the Ph.D. degree is his or her original research under the supervision of a research adviser. Graduate students normally become affiliated with a research adviser at the end of their first semester. This adviser is the student's principal guide in course selection, preparation for examinations, conduct of research, composition of the thesis, and professional placement. During the second year, each student takes an oral qualifying examination before a review committee made up of three faculty members (excluding the research adviser). Upon passing this examination, the student is advanced to candidacy. Finally, the student presents and defends the thesis before a final examination committee made up of the review committee plus the research adviser and a faculty member from an outside department. A Ph.D. is typically earned after approximately four years of full-time graduate study.

Research Facilities

The Department of Chemistry is located in two adjacent buildings, the Morley Chemical Laboratories and the John Schoff Millis Science Center. Facilities for experimental and theoretical research are modern and extensive. They include diverse major instrumentation for use by faculty and students, as well as specialized equipment serving individual research groups. Among the former are nuclear magnetic resonance spectrometers for both high-resolution and broadline investigation at 200-, 300-, and 400-MHz proton frequencies, electron spin resonance spectrometers, X-ray diffractometers, a high-resolution mass spectrometer with FAB capability, and a complement of standard instrumentation that includes modern infrared and ultraviolet-visible spectrophotometers. Specialized research equipment includes ion cyclotron resonance spectrometers; stopped-flow and temperature-jump spectrometers; surface characterization techniques, including LEED, Auger, and ESCA spectroscopies; and an impressive array of electrochemical instrumentation. A facility based on a VAX-11/785 computer serves the computational needs of the department. This computer may be accessed from remote workstations and is used in on-line control of experiments and data acquisition. Direct access to a variety of supercomputer facilities is also available through a computerwide network.

Financial Aid

Graduate students studying for the Ph.D. degree are supported through teaching and research assistantships and fellowships. All graduate students who maintain good standing in the department receive full support for the duration of their studies. Students entering in 1991 receive taxable stipends of $11,925. Many supplemental awards of $1500 for the first year are made to highly qualified students. Prestigious four-year Michelson-Morley supplements of up to $5000 are awarded competitively. The stipend for fall 1992 is anticipated to be about $12,500.

Cost of Study

Tuition for all students receiving teaching or research assistantships is provided by the department. The health fee is $290 per academic year.

Cost of Living

Graduate and professional students wishing on-campus housing may live in a University dormitory or, if they are married, in one of a limited number of University apartments for married students. Most graduate students find privately owned apartments near the campus. Costs are about average for large urban areas, ranging from $300 to $450 a month plus utilities.

Student Group

The total University enrollment in the fall of 1990–91 was 8,557. Of this number, 3,067 were enrolled in the undergraduate colleges and 5,490 in the graduate and professional schools. The chemistry department has about 90 Ph.D. candidates and about 10 students enrolled in the M.S. program.

Location

Cleveland, an industrial and financial center on Lake Erie in northeastern Ohio with a metropolitan population of over 2 million, provides excellent opportunities for research, study, and employment. The University is located about 4 miles east of downtown Cleveland in an area known as University Circle, one of the nation's largest cultural and educational centers. This 500-acre Circle houses over thirty educational, scientific, medical, cultural, and religious institutions, including the world-famed Cleveland Orchestra, the Cleveland Museum of Art, and several excellent repertory theaters. The city is also the home of the Cleveland Browns, Indians, and Cavaliers. The camping, sailing, and skiing areas of Ohio, western Pennsylvania, and western New York are readily accessible.

The University

Case Western Reserve University, an independent coeducational university, was established in 1967 through the federation of Western Reserve University (founded in 1826) and Case Institute of Technology (founded in 1880). The strong science departments, with their close ties to the University's outstanding medical school and its affiliate, University Hospitals, contribute to the University's placement in the top rank of postsecondary institutions in this country.

Applying

Students are generally admitted to graduate study for the fall term only. Students are required to take the General Test of the Graduate Record Examinations, and taking the Subject Test in chemistry of the GRE is recommended. International students must demonstrate fluency in English by obtaining a score of at least 550 on the TOEFL. Offers are made beginning in January but applications will be considered until thirty days prior to the beginning of the semester.

Correspondence and Information

Graduate Admissions Committee
Department of Chemistry
Case Western Reserve University
10900 Euclid Avenue
Cleveland, Ohio 44106-7078
Telephone: 216-368-5914
Fax: 216-368-3006

SECTION 2: CHEMISTRY

Case Western Reserve University

THE FACULTY AND THEIR RESEARCH

Gilles Klopman, Charles F. Mabery Professor of Research in Chemistry, Professor of General Medical Sciences, Professor of Environmental Sciences, and Chairman of the Department; Ph.D., Brussels, 1960. Theoretical chemistry, interpretation and prediction of selectivity and reactivity in reactions, chemotherapy and carcinogenicity. Prediction of chemical carcinogenicity in rodents. *Mutagenesis* 5:425, 1990 (with H. S. Rosenkranz). Evaluation of an artificial intelligence system. *J. Math. Chem.* 5:389, 1990.

Alfred B. Anderson, Professor; Ph.D., Johns Hopkins, 1970. Quantum chemistry applied to surface science, catalysis and properties of materials, thermal generation of $CH_3\bullet$ from CH_3OH adsorbed on oxygen covered Mo(110). C-O bond strength considerations from molecular orbital theory. *J. Phys. Chem.* 112:7218, 1991 (with P. Shiller). Diffusion and sulfur segregation of carbon in α-Fe: molecular orbital theory. *Phys. Rev.* B40:7503, 1989 (with S. Y. Hong).

Douglas F. Dedolph, Instructor; Ph.D., Iowa State, 1983. Physical organic chemistry, radical anions, photochemistry, kinetics and mechanisms of free-radical reactions, advances in undergraduate education with computers. Addition, substitution, and deoxygenation reactions with anions of thiols and diethyl phosphite. *J. Org. Chem.* 56:663, 1991 (with Russell, et al.).

Robert C. Dunbar, Professor; Ph.D., Stanford, 1970. Photochemistry and spectroscopy of gas-phase ions, techniques of Fourier-transform mass spectrometry, new analytical approaches combining lasers with Fourier-transform mass spectrometers, kinetics of ion dissociation and association reactions of interest in interstellar chemistry. Slow dissociation of thiophenol molecular ion. *J. Am. Chem. Soc.* 112:7893, 1990 (with Faulk, et al.).

Philip Garner, Associate Professor; Ph.D., Pittsburgh, 1981. Synthetic organic chemistry; new strategies for the asymmetric synthesis of amino acids, carbohydrates, nucleoside antibiotics, and DNA-reactive alkaloids; development of novel antisense RNA surrogates. Stereo-controlled buildup of a penaldic acid equivalent. *J. Org. Chem.* 55:3772, 1990. An asymmetric approach to quinocarcin via 1,3-dipolar cycloaddition. *J. Org. Chem.* 55:3973, 1990.

Malcolm E. Kenney, Professor; Ph.D., Cornell, 1954. Phthalocyanines and naphthalocyanines for gas sensors, optical data storage, and phototherapy research; alkoxysiloxanes for the preparation of ceramics; organosilicon compounds. Synthesis of phthalocyanines for phototherapy research. *J. Am. Chem. Soc.* 112:8064, 1990 (with Rihter, et al.). Synthesis of alkoxysiloxanes for sol-gel ceramics. *Inorg. Chem.* 29:1216, 1990.

Stephen J. Klippenstein, Assistant Professor; Ph.D., Caltech, 1988. Theoretical chemical kinetics, unimolecular dissociations, statistical and dynamical treatments of rate constants and product state distributions, variational RRKM studies. Variational treatment of the dissociation of CH_2CO into $^1CH_2$ and CO. *J. Chem. Phys.* 93:2418, 1990 (with Marcus). Implementation of a bond length reaction coordinate in variational RRKM studies. *Chem. Phys. Lett.* 170:71, 1990.

Gheorghe D. Mateescu, Professor and Director of the Major Analytical Instruments Facility; Ph.D., Case Western Reserve, 1971. Oxygen-17 magnetic resonance imaging and spectroscopy; cell function by oxygen-17 detection of nascent mitochondrial water; isotopically labeled compounds; vibrational, photoelectron, and ion spectroscopies. Combined $^{17}O/^1H$ magnetic resonance microscopy in plants, animals, and materials. In *Synthesis and Applications of Isotopically Labeled Compounds*, eds. Baillie and Jones, pp. 499–508, 1989 (with Yvars, et al.).

Garnett R. McMillan, Professor; Ph.D., Rochester, 1958. Photochemistry, kinetics and mechanisms of free-radical reactions, photolysis at the gas-solid interface, photocatalysis, photochemical gas/polymer interactions.

Ignacio Ocasio, Assistant Professor; Ph.D., Puerto Rico, 1975. Physical chemistry.

Anthony J. Pearson, Rudolph and Susan Rense Professor; Ph.D., Aston (England), 1974. Total synthesis of natural products using organometallic complexes as synthetic building blocks. Development of new synthetic methods. *J. Chem. Soc. Perkin Trans. I*, 723, 1990. Stereocontrol during alkylation of enolates of π-allyl-molybdenum complexes. *J. Am. Chem. Soc.* 112:8034, 1990. Uses of organoiron systems in controlling stereochemistry. *J. Chem. Soc., Chem. Commun.* 392 and 394, 1990.

William M. Ritchey, Professor of Chemistry and Macromolecular Science; Ph.D., Ohio State, 1955. Nuclear magnetic resonance spectroscopy. Local chain dynamics of polyurethane elastomers. *J. Polym. Sci.: Part C: Polym. Lett.* 28:301, 1990. Correlation between macroscopic and microscopic polymer properties. *J. Appl. Polym. Sci.* 40:1717, 1990.

Robert G. Salomon, Professor; Ph.D., Wisconsin–Madison, 1971. Synthetic methods; synthesis and biological chemistry of physiologically active natural products, including spatane diterpenes, halichondrins, prostaglandin endoperoxides, and levuglandins. Levuglandin E_2 inhibits mitosis and microtubule assembly. *Prostaglandins* 39:611, 1990 (with Murthi, et al.). Total synthesis of spatol and other spatane diterpenes. *J. Am. Chem. Soc.* 113:3096, 1991 (with Basu, et al.).

Lawrence M. Sayre, Associate Professor; Ph.D., Berkeley, 1977. Physical organic chemistry, bioinorganic reaction mechanisms, mechanisms of chemical and biological oxidation of amines, metabolic activation of small-molecule neurotoxins, drug receptor interactions, metal ion catalysis of amide hydrolysis, copper-promoted oxidation and oxygenation reactions. Nonelectron transfer quinone mediated oxidated cleavage of cyclopropylamines. *J. Org. Chem.* 56:1353, 1991. Inhibition of mitochondrial respiration bis-pyridines related to MPP$^+$. *Arch. Biochem. Biophys.* 286:138, 1991 (with Singh, et al.).

Daniel A. Scherson, Associate Professor; Ph.D., California, Davis, 1979. Electrochemistry, electrode kinetics, electrocatalysis, in situ spectroscopic methods in electrochemistry. Oxidation state changes of molecules irreversibly adsorbed on an electrode surface. *Langmuir* 6:1234, 1990. In situ Mössbauer spectroscopy of a species irreversibly adsorbed on an electrode surface. *Langmuir* 6:1338, 1990.

John E. Stuehr, Professor of Chemistry and Biochemistry; Ph.D., Case Western Reserve, 1961. Rapid reactions in solution; metal ion interactions with nucleotides and amino acids, proton transfer kinetics, protein and enzyme dynamics. Kinetics of Ni(II) interactions with NAD and NADP. *Inorg. Chem.* 29:1143, 1990 (with J. Bidwell).

Chaim N. Sukenik, Professor; Ph.D., Caltech, 1976. Organic reactions in oriented media, structure/reactivity of self-assembled and Langmuir-Blodgett monolayer films, biomaterials. Transformation of self-assembled monolayers. *Langmuir* 6:1621, 1990 (with N. Balachander). Modulation of cell adhesion to titanium with covalently anchored monolayers. *J. Biomed. Mater. Res.* 24:1307, 1990 (with N. Balachander, et al.).

Terrence J. Swift, Professor of Chemistry and Biochemistry; Ph.D., Berkeley, 1962. Rapid-solution reactions, magnetic resonance.

Fred L. Urbach, Professor; Ph.D., Michigan State, 1964. Coordination chemistry of multidentate chelates of transition metal ions; biomimetic models for binuclear copper sites in proteins, such as hemocyanin and tyrosinase; electrochemical behavior of metal complexes. Redox activity in binucleating Schiff-base ligands. *Inorg. Chem. Acta* 164:123, 1989 (with Maloney, et al.). Copper catalyzed oxidations. *J. Am. Chem. Soc.* 112:2332, 1990 (with L. M. Sayre, et al.).

Donald R. Whitman, Professor and Associate Chairman; Ph.D., Yale, 1957. Theoretical chemistry.

Ernest B. Yeager, Emeritus Hovorka Professor and Director of the Case Center for Electrochemical Studies; Ph.D., Case Western Reserve, 1948. Electrochemistry: electrocatalytic aspects of the oxidation of small molecules; O_2 reduction and generation; electrochemistry of transition metal macrocycles; electrochemistry of single crystal electrodes, including STM, TEM, and FTIR studies; electrochemical sensors for CO_2. Electroreduction of carbon dioxide on platinum single crystal electrodes. *J. Electroanal. Chem.* 295:415, 1990 (with B. Nikolic, et al.). The electrochemistry of noble metal electrodes in aprotic organic solvents containing lithium salts. *J. Electroanal. Chem.* 297:225, 1991 (with D. Aurbach, et al.).

Associated Faculty

Marc S. Berridge, Associate Professor of Radiology and Chemistry; Ph.D., Washington (St. Louis), 1980.
Boris D. Cahan, Principal Research Scientist; Ph.D., Pennsylvania, 1968.
Gregory C. Hurst, Assistant Professor of Radiology and Chemistry; Ph.D., Rochester, 1983.
John J. Mieyal, Associate Professor of Pharmacology and Chemistry; Ph.D., Case Western Reserve, 1969.

SECTION 2: CHEMISTRY

COLUMBIA UNIVERSITY

Department of Chemistry

Program of Study

The Department of Chemistry at Columbia University runs one of the best graduate research and training programs in the country. The department is known for its lively and congenial intellectual atmosphere and for the intensity of the effort of its faculty and students. The reputation of the department, recognized by many honors to its faculty and graduates, is because of the high-quality of the students entering the graduate program each year. They come for the excitement of studying and performing scientific research in one of the best departments in one of the finest universities in the country.

The department offers graduate programs leading to the Ph.D. degree in organic, physical, inorganic, theoretical, and biochemical/biophysical chemistry. Columbia also has an interdepartmental program in chemical physics, administered by the Department of Chemistry.

The first year of graduate study is devoted mainly to course work, but to some extent during the first year, and increasingly thereafter, students are engaged in research for the doctoral dissertation. The selection of a research sponsor is most important. To help students with this decision, all members of the faculty discuss their research interests in colloquia held at the beginning of the first year; students choose a sponsor only after attending these colloquia and talking privately with at least three faculty members.

All Ph.D. candidates in chemistry are expected to complete their research and the other degree requirements—course work, some teaching in the undergraduate program, a set of cumulative exams on current topics in chemistry, and the design and defense of an original research proposal—within five years of residence at Columbia.

Research Facilities

The department has modern facilities for research in all areas of chemistry, maintained by a talented support staff with whom students often work closely. Departmental instruments include three FT NMR spectrometers; two mass spectrometers; an automated X-ray diffractometer; a Convex C210 minisupercomputer, a massively parallel FPS MCP 784/512EA computer, and a DEC VAXcluster; and FT-IR, IR, Raman, UV, CD, and ESR instrumentation. The department also employs a machinist, a glassblower, and an electronics engineer. The department is housed in three buildings: Havemeyer Hall, a magnificent limestone and brick structure built in 1897 and just renovated into modern laboratories in a manner respecting the dignity of the original structure; Chandler Laboratories, named for the famous Columbia chemist who designed Havemeyer Hall and then persuaded Mr. Havemeyer to pay for it; and Havemeyer Addition, a well-designed combination of teaching and research laboratories completed several years ago.

Financial Aid

The department awards every graduate student a fellowship that includes complete exemption from tuition and fees plus a twelve-month stipend. The stipend is adjusted each year for changes in the cost of living; for the twelve months beginning September 1, 1991, the stipend is $14,000.

Cost of Study

Students are exempted from tuition and fees, as described in the Financial Aid section, above.

Cost of Living

Every graduate student is guaranteed housing in a Columbia apartment at rents that are below the market price. New York prices for food, clothing, and entertainment range from absurdly low to very high. A single student can live modestly but comfortably on the stipend provided by the department and described in the Financial Aid section, above.

Student Group

Each year 25–30 students begin graduate study in chemistry at Columbia, and each year the department graduates about 25 Ph.D.'s. Since each student is at the University for four years or so, the students number about 110, a diverse and interesting group from all over the United States and from many foreign countries. The student body is about half male and half female.

Location

New York City is one of the world's great cities, a city that can be compared with Rome and Paris and London and Tokyo but with no place else in the United States, and contact with the life of New York—with its museums, its theater and music and dance, its restaurants and stores, its nervous energy—is in itself a liberal education. Columbia is located on the Upper West Side of Manhattan, 10 minutes from midtown, on Morningside Heights overlooking the Hudson River. The Heights is a pleasant and lively neighborhood, with various educational and religious institutions to the north; good shopping, college-style nightlife, a concentration of cheap and excellent restaurants to the south; and beautiful Riverside Park a block away to the west.

The University

Columbia University was founded in 1754, as King's College, with a charter from George II to provide instruction in "the Learned Languages and the Liberal Arts and Sciences." Over the years, the institution has grown into one of the leading research universities in the world by maintaining strong emphasis on the liberal arts and sciences. The University is now a remarkably stimulating place for graduate study; most of the academic departments at Columbia rank among the top ten in the nation in their respective disciplines, and the chemistry department, although small, is particularly strong, being generally regarded as one of the five strongest in the country.

Applying

Applications for the fall term must be received by January 15. The department rarely admits students for the spring term. All applicants must take the General Test of the Graduate Record Examinations and the Subject Test in chemistry. Applicants whose native language is not English must also take the Test of English as a Foreign Language (TOEFL).

Correspondence and Information

Professor Philip Pechukas
Chairman
Chemistry Admissions Committee
Box 368, Havemeyer Hall
Columbia University
New York, New York 10027
Telephone: 212-854-4231 or 2433

SECTION 2: CHEMISTRY

Columbia University

THE FACULTY AND THEIR RESEARCH

Brian E. Bent, Assistant Professor; Ph.D., Berkeley, 1986. Physical chemistry: surface reaction kinetics and mechanisms, including the role of transient species, studied by surface-sensitive spectroscopies combined with molecular-beam techniques.

Bruce J. Berne, Professor; Ph.D., Chicago, 1964. Theoretical chemistry: equilibrium and nonequilibrium statistical mechanics, computer simulation of condensed matter, structure and dynamics of aqueous solutions, theory of chemical reactions and vibrational relaxation in liquids, quantum Monte Carlo and molecular dynamic path integral techniques.

Richard Bersohn, Professor; Ph.D., Harvard, 1950. Physical chemistry: dynamics of laser photodissociation and elementary chemical reactions as reflected by energy partitioning into translation, rotation, and vibration of the products.

Ronald C. D. Breslow, Professor; Ph.D., Harvard, 1955. Organic chemistry; biomimetic synthesis using the template principle of enzyme action; enzyme model systems and artificial enzymes; aromaticity and antiaromaticity, organic electrochemistry, and the solid-state properties of stable triplets.

Kenneth B. Eisenthal, Professor; Ph.D., Harvard, 1959. Physical chemistry: ultrafast phenomena in bulk liquids using femtosecond and picosecond spectroscopy, equilibrium and dynamic phenomena at liquid interfaces studied by nonlinear optical techniques.

George W. Flynn, Professor; Ph.D., Harvard, 1965. Physical chemistry: laser studies of molecular energy transfer and chemical reaction in the gas phase, laser photochemistry on surfaces using scanning tunneling microscopy.

Richard A. Friesner, Professor; Ph.D., Berkeley, 1979. Theoretical chemistry: new algorithms for large-scale computation of chemical structure and dynamics, mechanism of photosynthetic charge separation, energy transfer in polymers, resonance Raman spectroscopy of biological chromophores.

David A. Horne, Assistant Professor; Ph.D., MIT, 1988. Bioorganic chemistry: synthetic methods to study the molecular basis of recognition in transfer RNA identity, protein biosynthesis, and peptidyl transferase activity of ribosomal RNA.

Thomas J. Katz, Professor; Ph.D., Harvard, 1959. Organic chemistry: metal-catalyzed reactions; organometallic reagents; synthesis of optically active helical-conjugated molecules and other new materials with novel electric, magnetic, and optical properties.

Robert M. Kennedy, Assistant Professor; Ph.D., Virginia Tech, 1986. Organic chemistry: total synthesis of polycyclic natural products, oxidation with high-valent transition metals.

Ann McDermott, Assistant Professor; Ph.D., Berkeley, 1987. Biophysical chemistry: structure and dynamics of enzymes in catalytic intermediates or inhibited complexes, characterized especially by solid-state NMR methods.

Koji Nakanishi, Professor; Ph.D., Nagoya (Japan), 1954. Organic chemistry: isolation, structural, and bioorganic studies on bioactive natural products; retinal proteins and the mechanism of visual processes; spectroscopic methods for microscale structure determination.

Gerard Parkin, Assistant Professor; D.Phil., Oxford, 1985. Inorganic chemistry: synthesis, structures, and reactivity of alkyl and hydride derivatives of s- and p-block elements; host-guest complexes of anion substrates; reactivity of transition metal complexes with "electron rich" metal centers.

Philip Pechukas, Professor; Ph.D., Chicago, 1966. Theoretical chemistry: phase space path integral methods for many-body quantum dynamics, quantum Brownian motion, energy transfer in nonlinear oscillators by frequency locking and dragging.

W. Clark Still, Professor; Ph.D., Emory, 1972. Organic chemistry: molecular recognition, organic synthesis, and molecular modeling directed to the development of methods for rational design and synthesis of binding agents and chemical catalysts.

Gilbert Stork, Professor; Ph.D., Wisconsin, 1945. Organic chemistry: synthesis of natural products, with particular emphasis on new methods and strategies for the solution of regiochemical and stereochemical problems.

Nicholas J. Turro, Professor; Ph.D., Caltech, 1963. Organic chemistry: photochemical techniques to study structure and dynamics of reactive intermediates and of molecules adsorbed on microheterogeneous systems (micelles, zeolites, DNA).

James J. Valentini, Professor; Ph.D., Berkeley, 1976. Physical chemistry: state-to-state dynamics of elementary chemical reactions and molecular photodissociation studied by time-resolved laser spectroscopy.

The Department of Chemistry is housed in three adjoining buildings; one of these is stately Havemeyer Hall, pictured above.

SECTION 2: CHEMISTRY

EASTERN MICHIGAN UNIVERSITY

Department of Chemistry

Programs of Study

The Department of Chemistry offers programs of study leading to a professional M.S. degree. The degree requires a minimum of 30 credits beyond the bachelor's degree. The professional M.S. degree provides training appropriate for industrial chemists and for students preparing for doctoral study. One advanced course from each of any three of the following areas of chemistry must be taken: physical, inorganic, analytical, organic, or biochemistry. The remaining credits are made up by seminar (1), chemical literature (1), cognate (0–6), research (6–10), and advanced courses elected to match the student's interest. A maximum of 4 credits to correct undergraduate deficiencies (if any) may be counted. A thesis based on original research by the student is required for all students except those whose current professional employment involves both laboratory and writing experience equivalent to the research and thesis requirement. A comprehensive examination is required for nonthesis students.

The academic year at Eastern Michigan University consists of two full 15-week semesters (fall and winter) plus two short 7½-week sessions (spring and summer).

Research Facilities

The department is housed in the modern, six-story Jefferson Science Complex. Among the major items of equipment available are research-quality FT-IR, UV, and visible spectrophotometers; an FT-NMR spectrometer; a laser-Raman spectrometer; a mass spectrometer; preparative and analytical gas chromatographs and HPLCs; X-ray diffraction apparatus, including both powder and single-crystal cameras; a Gouy magnetic susceptibility apparatus; an ultracentrifuge; a fluorometer; a DTA-TGA thermal analysis system; a high-speed membrane osmometer; and an autoviscometer. Specialized facilities and equipment for radiochemistry are also available. The chemistry library holdings, including both reference texts and current research journals, are housed in the adjacent University Library. An active weekly seminar program featuring a number of outside speakers also contributes to a better awareness of recent research developments. Computing facilities include a PDP-10 on campus with remote terminal time-sharing access in the chemistry building and ten minicomputers for student use.

Financial Aid

Teaching assistantships paying $6000–$6250 for ten months (1992–93) are available to qualified applicants. In addition, the University pays full tuition for 16 credit hours for each fiscal year of the assistantship as well as registration and health service fees. The total package amounts to $7500–$9500 and depends on whether the student is a resident of Michigan or not. The usual graduate teaching assistantship requires 20 hours per week of service (including preparation, student contact, and grading time) and allows the student to carry a full academic load of 7–9 hours of graduate credit. A limited number of teaching assistantships are also available for the summer session. These are awarded on a competitive basis.

Residence Hall Grants paying from $50 to $850 per year toward residence hall costs are available and are awarded on the basis of demonstrated financial need. Residence Hall Leadership Awards paying up to $850 per year toward residence hall costs are available to graduate students judged capable of providing leadership in the residence hall program. Both of the above residence hall programs require one term of prior residence in EMU residence halls. Loan programs available include Perkins Loans, Michigan State Direct Loans, and Michigan Guaranteed Student Loans. A College Work-Study Program provides hourly jobs and assistantships for students who demonstrate financial need.

Cost of Study

Tuition costs for 1991–92 are $93 per semester credit hour for in-state students and $223 per semester credit hour for out-of-state students. A registration fee of $40 for each of the fall and winter semesters and $30 for each of the spring and summer sessions is charged to each student. A health-service participation fee of $1.75 is charged for each credit hour taken. The graduation fee is $35.

Cost of Living

Housing on campus for married graduate students includes furnished one- and two-bedroom apartments renting for $359–$447 per month (single occupancy: $302–$807 per month) including utilities. Unmarried graduate students may live in the undergraduate residence halls ($1554–$2344 per semester for room and board). A wide variety of off-campus accommodations are also available.

Student Group

The total enrollment at Eastern Michigan University is approximately 25,000 students, including 5,500 graduate students. There are approximately 60 graduate students in chemistry, of whom 25 are enrolled full-time. The remainder are employed elsewhere and are pursuing their graduate studies on a part-time basis. About 60 percent of the chemistry graduate students are from foreign countries (recently China, Philippines, Korea, Iran, Somali, India, and Nigeria).

Location

Eastern Michigan University is located in Ypsilanti, a community of approximately 50,000 people. Although Ypsilanti is close to the cultural advantages of Detroit (30 miles) and Ann Arbor (7 miles), it is largely surrounded by farmland and open countryside. Nearby opportunities for outdoor recreation, such as boating, fishing, hunting, and skiing, are plentiful.

The University and The Department

Founded in 1849 as a normal school, Eastern Michigan University had teacher preparation as its primary purpose for its first 100 years, although the University's capabilities in the liberal arts and sciences developed steadily. In recent decades the growing recognition of Eastern as an excellent multipurpose university has enabled the chemistry department to attract an outstanding faculty committed to excellence in both teaching and research. The professional B.S. degree program is accredited by the American Chemical Society. Several department members are nationally and internationally recognized authorities in their disciplines. The modest size of the M.S. program facilitates extensive personal interactions among students and faculty.

Applying

Applications for admission and for teaching assistantships should be completed and returned to the graduate coordinator by March 15 for the fall semester and by November 1 for the winter semester. Submission of GRE scores (the General Test and the Subject Test in chemistry) is required and is especially helpful when other credentials are weak or when a student is applying from a foreign country. Placement examinations are not required. Special admission requirements for foreign students are listed in the application material.

Correspondence and Information

Graduate Coordinator
Department of Chemistry
Eastern Michigan University
Ypsilanti, Michigan 48197
Telephone: 313-487-2057

Peterson's Guide to Graduate Programs in the Physical Sciences and Mathematics 1992

SECTION 2: CHEMISTRY

Eastern Michigan University

THE FACULTY AND THEIR RESEARCH

Michael J. Brabec, Professor; Ph.D., Wyoming, 1970. Toxicology: mechanism of chemical reproductive toxicity, mitochondrial function in differentiating cells, effects of chemicals on mitochondrial function and biogenesis.

Stephen W. Brewer Jr., Professor; Ph.D., Wisconsin, 1969. Analytical chemistry: spectrochemistry, electron probe microanalysis.

Ronald W. Collins, Professor, University Provost, and Academic Affairs Vice President; Ph.D., Indiana, 1962. Inorganic chemistry: inorganic phosphates, structural chemistry of divalent tin, X-ray crystallography, educational use of computers.

Edward L. Compere Jr., Professor; Ph.D., Maryland, 1958. Inorganic-organic chemistry: synthesis and structure, metal complexes, acid-salts, organometallics; spectroscopic methods.

Ellene Tratras Contis, Associate Professor; M.S., Pittsburgh, 1971. Chemistry education: educational use of computers, analytical chemistry.

Arthur S. Howard, Associate Professor; Ph.D., Cambridge, 1964. Organic chemistry, natural products, and alkaloids; general synthetic methods and structural studies.

Judith T. Levy, Professor and Head; Ph.D., Johns Hopkins, 1971. Biochemistry: cardiac calcification inhibition, medical biochemical studies, enzymology.

Maria C. Miletti, Assistant Professor; Ph.D., Wisconsin, 1987. Inorganic and theoretical chemistry: structures and reactivity of organometallic compounds using MO calculations.

Elva Mae Nicholson, Professor; Ph.D., Harvard, 1965. Physical-organic chemistry, biochemistry.

Ross S. Nord, Assistant Professor; Ph.D., Iowa State, 1986. Physical chemistry: statistical mechanical theory of dynamical processes on lattices, development of formal theory via kinetic equations, computer programming of simulations of these processes.

Donald B. Phillips, Professor; Ed.D., Georgia, 1972. Science education.

Ralph R. Powell, Associate Professor; Ph.D., Purdue, 1965. Physical chemistry: quantum chemistry, educational use of computers.

O. Bertrand Ramsay, Professor; Ph.D., Pennsylvania, 1960. Organic chemistry: physical-organic chemistry, the history of chemistry, chemical education.

Krishnaswamy Rengan, Professor; Ph.D., Michigan, 1966. Nuclear chemistry: activation analysis, radiochemical separations, nuclear properties of radioactive isotopes.

Stephen E. Schullery, Professor; Ph.D., Cornell, 1970. Biophysical chemistry: characterization of lipid bilayer membranes: permeability, liquid-crystal phase behavior, ion binding, membrane fusion.

Ronald M. Scott, Professor; Ph.D., Illinois, 1959. Biochemistry: phenol-amine interactions, solvent-solute interactions, the structure of dioxane-water mixtures, measurement of gentamicin by an enzyme immunoelectrode.

John M. Sullivan, Professor; Ph.D., Michigan, 1959. Organic chemistry: conformational analysis of heterocyclic rings, synthesis of heterocyclic compounds.

Wade J. Tornquist, Assistant Professor; Ph.D., Minnesota, 1986. Analytical-physical chemistry: electrochemical and ultrahigh-vacuum surface studies of small-molecule absorbents, FT-infrared reflection–absorption spectroscopy.

Jose C. Vites, Assistant Professor; Ph.D., Notre Dame, 1984. Inorganic-organometallic chemistry: metal cluster and carbonyl complexes, computers as teaching and learning aids in chemistry.

Jerry R. Williamson, Associate Professor; Ph.D., Iowa, 1964. Organic-polymer chemistry: synthesis of heterocyclic, aromatic, thermally stable polymers; polymer characterization.

Stewart D. Work, Professor; Ph.D., Duke, 1963. Organic chemistry: condensation reactions of organic multiple anions, base-catalyzed cleavage of epoxides, organosilicon chemistry.

Masanobu Yamauchi, Professor; Ph.D., Michigan, 1961. Inorganic chemistry: synthesis, structure, and properties of inorganic triazoles.

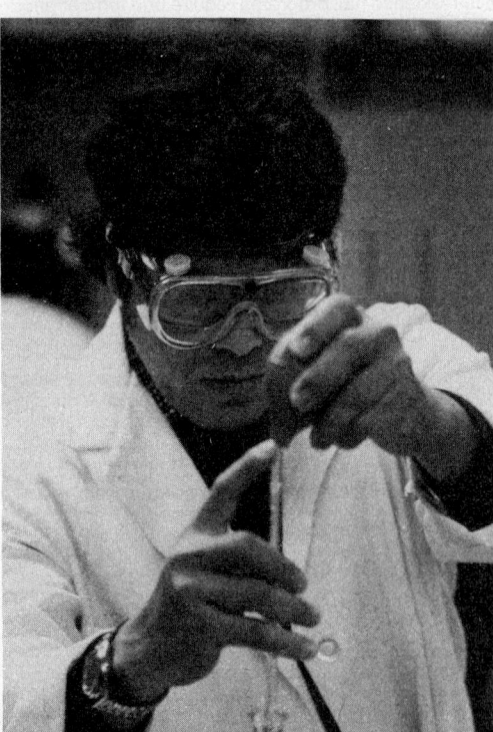

Research in chemistry.

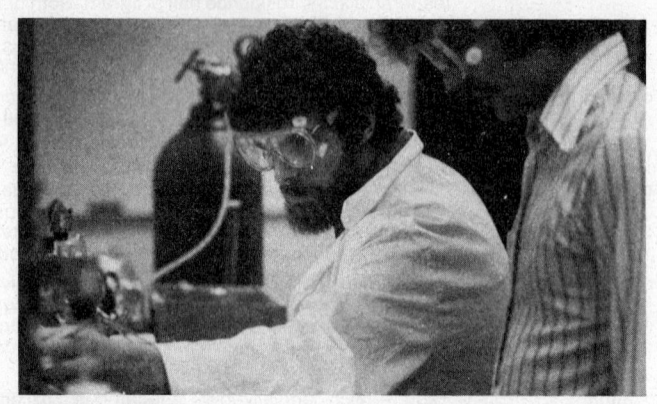

Learning by doing.

Mark Jefferson Science Complex.

SECTION 2: CHEMISTRY

FLORIDA STATE UNIVERSITY

Department of Chemistry

Programs of Study	The department offers programs leading to the M.S. and Ph.D. degrees in analytical, inorganic, nuclear, organic, and physical chemistry; biochemistry; or a combination of two areas. Interdisciplinary programs leading to advanced degrees in chemical physics and in molecular biophysics are offered in cooperation with the Departments of Physics and Biological Science. Ph.D. candidates who perform satisfactorily on the organic and physical chemistry entrance examinations may immediately begin advanced course work and research. Although studies are structured to meet individual needs and vary among the divisions, most Ph.D. programs incorporate nine to twelve 1-semester courses at the graduate level. Comprehensive examinations in the area of specialization are generally taken upon completion of the second year of residence, except in the organic division, which uses a series of cumulative examinations. A thesis or dissertation is required in all but the courses-only option of the master's degree program. The presence of about 30 postdoctoral and visiting faculty researchers, in addition to the low graduate student–faculty ratio, permits a high level of student-scientist interaction.
Research Facilities	Chemistry department research operations are housed mainly in the interconnected Dittmer Laboratory of Chemistry building and Institute of Molecular Biophysics building. Several adjacent structures serve other departmental teaching functions. Major items of research equipment include a variety of spectrometers and lasers for ultraviolet, visible, infrared, and Raman experiments; an ORD/CD spectrometer; computer-coupled EPR equipment; several high- and medium-resolution mass spectrometers with GC, EI, and CI capabilities; 60-, 150-, 200-, 270-, 300-, and 500-MHz nuclear magnetic resonance spectrometers; an automated four-circle X-ray diffractometer with complete in-house solution and refinement capability; and a bioanalytical facility containing automated DNA synthesis, peptide synthesis, and protein sequencing equipment. The department maintains excellently staffed electronics, glassworking, machine, photo, and woodworking shops in support of teaching and research programs. The departmental computer facility includes two DEC VAXstation computers with multiple input/output devices. Micro- and minicomputers are also available in a great number of research laboratories, and a centralized CYBER 760 and two CYBER 730s provide service for the entire campus. Several chemistry faculty members are members of the Supercomputer Computations Research Institute (SCRI), which maintains two supercomputers—a Cray YMP and Connection Machines CM-2. The department is a sponsoring academic unit for the new Florida National High Magnetic Field Laboratory (NHMFL), and it will have access to the facilities and instrumentation of the NHMFL when it opens in Tallahassee in 1992. The University's Strozier Library, which includes the Dirac Science Library, houses 1.2 million volumes and maintains 12,650 active journal subscriptions. The modern Dirac Science Library is located immediately adjacent to the Dittmer Laboratory of Chemistry.
Financial Aid	In addition to providing teaching and research assistantships, the department offers several special fellowships on a competitive basis. Nearly all graduate students are supported by one of these programs. Twelve-month teaching assistantships with stipends of $11,500 can be augmented by special fellowships ranging up to $3000 in 1991–92. Competitive fellowships on the university and college levels are also available. These have a $6000 stipend and require no teaching responsibilities. The department generally adds additional funds to this stipend.
Cost of Study	Tuition for in-state residents was $597 per semester and for out-of-state residents $2025 per semester in 1990–91. These tuition costs have normally been waived for teaching and research assistants, but the number of waivers available each year is determined in part by legislative appropriation. Nonwaivable fees, including the cost of health insurance ($94 per semester), are approximately $275 per semester in addition to the above tuition.
Cost of Living	Single students may share 2-person rooms in graduate apartment housing for $188 per month in 1991–92, including utilities and telephone service. The University also operates an apartment complex of 795 units and a trailer park for single and married students. Rents range from $175 to $284 plus utilities for furnished one- to three-bedroom apartments. Off-campus accommodations begin at about $300 per month.
Student Group	Graduate enrollment in chemistry is currently just under 100 students, almost all of whom are supported. The past high ratio of men to women is changing significantly. Although students come from all parts of the United States and numerous foreign countries, many are native to the eastern half of the United States.
Location	Metropolitan Tallahassee has a population of about 300,000 and is recognized for the scenic beauty of its rolling hills, abundant trees and flowers, and seasonal changes. As the capital of Florida, the city is host to many cultural affairs, including symphony, theater, and dance groups. Its location 30 minutes from the Gulf Coast, the area's many lakes, and an average annual temperature of 68°F make the region eminently suitable for a variety of year-round outdoor activities.
The University	Florida State University was founded in 1857 and is one of nine members of the State University System of Florida. Enrollment in 1991–92 was more than 28,000, including more than 5,000 graduate students. The University's rapid climb to prominence began with the adoption of an emphasis on graduate studies in 1947. The first doctoral degree was conferred in 1952. In addition to the College of Arts and Sciences, the University comprises Colleges of Business, Communication, Education, Home Economics, Law, and Social Sciences and Schools of Criminology, Library and Information Studies, Music, Social Work, Theatre, and Visual Arts. A School of Engineering is the latest addition to this list. A number of departments, including chemistry, enjoy national recognition.
Applying	Application for admission may be made at any time; however, initial inquiries concerning assistantships, especially fellowships, should be made nine to twelve months prior to the anticipated enrollment date. Later requests are considered as funds are available, up to the April 15 deadline for the following fall semester. Requests for forms, detailed requirements, and other information should be directed to the address below.
Correspondence and Information	Professor William T. Cooper Director, Graduate Recruiting and Admissions Department of Chemistry Florida State University Tallahassee, Florida 32306 Telephone: 904-644-6004

Peterson's Guide to Graduate Programs in the Physical Sciences and Mathematics 1992

Florida State University

THE FACULTY AND THEIR RESEARCH

Paul A. Bash, Assistant Professor; Ph.D., Berkeley, 1986. Computational molecular biophysics, enzyme-reaction mechanisms, protein folding, massively parallel computing.

Gregory R. Choppin, Professor; Ph.D., Texas, 1953. Chemical properties of organic and inorganic complexes of lanthanide and transuranic actinide elements, separation methods for actinides, environmental chemistry of actinides.

Jerzy Cioslowski, Assistant Professor; Ph.D., Georgetown, 1987. Computational quantum chemistry, ab initio electronic structure calculations, development of algorithms for supercomputers.

Ronald J. Clark, Professor; Ph.D., Kansas, 1958. Synthesis and characterization of superconducting materials, ceramics, metal carbonyl trifluorophosphine chemistry, precious metal thin films.

William T. Cooper, Associate Professor; Ph.D., Indiana, 1981. Chromatography, environmental geochemistry of organic compounds in natural waters, surface chemistry of minerals, organic geochemistry of recent sediments.

Timothy A. Cross, Associate Professor; Ph.D., Pennsylvania, 1981. Structure and dynamics of gramicidin A in a lipid bilayer by solid-state ^{15}N, ^{13}C, and ^{2}H nuclear magnetic resonance.

DeLos F. DeTar, Professor; Ph.D., Pennsylvania, 1944. Theoretical and experimental investigation of steric effects on rates and equilibria, including enzymatic catalysis.

Ralph Dougherty, Professor; Ph.D., Chicago, 1963. Immunoassay and mass spectrometry applied to problems in human and environmental health, absolute asymmetric synthesis.

Stephen C. Foster, Assistant Professor; Ph.D., Dalhousie, 1982. High-resolution gas-phase spectroscopy of radicals and molecular ions, IR- and UV-visible laser spectroscopy.

Earl Frieden, Professor; Ph.D., USC, 1949. Copper and iron metalloproteins and metalloenzymes, thyroid hormones and amphibian metamorphosis, comparative and developmental biochemistry, biochemistry of the essential elements.

Robert L. Fulton, Professor; Ph.D., Harvard, 1964. Theories of linear and nonlinear dielectric properties and their relation to molecular motion, theories of solvent effects on spectral properties, theories of relaxation.

Penny Gilmer, Associate Professor; Ph.D., Berkeley, 1972. Biochemistry, immunochemistry, biochemical nature of cell-cell recognition, lysosomal processing, metabolic diseases.

Richard E. Glick, Professor; Ph.D., UCLA, 1954. Biomass energy and energy by-product systems and utilization.

Virgil L. Goedken, Professor; Ph.D., Florida State, 1968. Transition-metal chemistry; synthesis, structural, and reactivity patterns of macrocyclic ligands; X-ray crystallography.

Kenneth A. Goldsby, Assistant Professor; Ph.D., North Carolina at Chapel Hill, 1983. Redox reactions of transition-metal complexes; directed electron transfer; mixed-valence complexes; electrochemistry.

Terry Gullion, Assistant Professor; Ph.D., William and Mary, 1986. Solid-state ^{13}C, ^{15}N, ^{19}F, and ^{31}P NMR of polymers and macromolecules; rotational echo double resonance.

Werner Herz, Professor; Ph.D., Colorado, 1947. Isolation and structure of natural products, chemotaxonomy of Compositae, structure and synthesis of sesquiterpenes, partial synthesis of diterpenes and related polycyclic substances, resin acid chemistry.

Edwin F. Hilinski, Associate Professor; Ph.D., Yale, 1982. Mechanistic studies of photochemical and thermal reactions of organic compounds in solution; picosecond laser spectroscopy.

Robert A. Holton, Professor; Ph.D., Florida State, 1971. Synthetic organic, organometallic, and bioorganic chemistry; total synthesis of natural products.

Russell H. Johnsen, Professor and Dean of Graduate Studies; Ph.D., Wisconsin, 1951. Chemical effects of ionizing radiation, energy partition and transfer in binary systems, ionization and excitation efficiencies, kinetics of radiation-induced reactions, EPR studies.

Michael Kasha, Professor; Ph.D., Berkeley, 1945. Molecular electronic spectroscopy, intermolecular electronic phenomena, theoretical photochemistry, molecular biophysics, singlet oxygen studies, laser studies of proton transfer.

Marie E. Krafft, Associate Professor; Ph.D., Virginia Tech, 1983. Synthetic organic and organometallic chemistry, natural products synthesis.

Keith S. Kyler, Associate Professor; Ph.D., Wyoming, 1983. Enzymatic organic synthesis; synthetic methodology involving carbanions bearing nitrogen-, sulfur-, and tin-stabilized heterogroups.

John E. Leffler, Professor Emeritus; Ph.D., Harvard, 1948. Theory and measurements of structural and solvent effects on reaction rates, free radicals and their reaction mechanisms, peroxides, surface chemistry.

Robley J. Light, Professor; Ph.D., Duke, 1960. Secondary metabolism in fungi and plants, lipid metabolism and function.

Bruno Linder, Professor; Ph.D., UCLA, 1955. Intermolecular forces, polarization phenomena, infrared and Raman intensities, physical adsorption, liquid crystals.

Leo Mandelkern, Professor Emeritus; Ph.D., Cornell, 1949. Chemical and physical properties of macromolecules: phase transition and crystallization mechanisms, morphology and physical properties; contractility and mechanochemistry of synthetic polymers and biopolymers; polypeptide conformation.

Charles K. Mann, Professor; Ph.D., Virginia, 1955. Quantitative analysis by Raman spectroscopy, numerical methods for resolution of spectroscopic interferences.

William F. Marzluff, Professor; Ph.D., Duke, 1971. RNA transcription in animal cells, small nuclear RNA genes, histone gene expression, gene expression in sea urchin embryos.

Edward Mellon, Professor and Chairman; Ph.D., Texas, 1963. Chemical education, inorganic chemistry.

William S. Rees Jr., Assistant Professor, joint with the Materials Research and Technology Center; Ph.D., UCLA, 1986. Preparation and characterization of main group inorganic and organometallic compounds suitable for use in chemical deposition of electronic materials; inorganic photochemistry; designed neutral hosts for anionic guests.

William C. Rhodes, Professor; Ph.D., Johns Hopkins, 1958. Quantum theory of molecular excitation and relaxation; dynamic and thermodynamic properties of coupled molecular systems, especially biological macromolecules.

Randolph L. Rill, Professor; Ph.D., Northwestern, 1971. Physical biochemistry, chromatin and DNA structure and function, DNA–small molecule interactions, biochemical applications of NMR spectroscopy.

Sanford A. Safron, Professor; Ph.D., Harvard, 1969. Dynamics of crystal surfaces, He atom-surface scattering experiments, models of chemical reactions.

Jack Saltiel, Professor; Ph.D., Caltech, 1964. Photochemistry of organic molecules, elucidation of the mechanisms of photochemical reactions by chemical and spectroscopic means.

Joseph B. Schlenoff, Assistant Professor; Ph.D., Massachusetts at Amherst, 1987. Synthesis and characterization of electrically conductive polymers and ceramics, electrochemical polymerization, properties of superconducting oxides.

Martin A. Schwartz, Professor; Ph.D., Stanford, 1965. Organic synthesis, biogenetic-type syntheses of natural products, synthesis of enzyme inhibitors.

Raymond K. Sheline, Professor; Ph.D., Berkeley, 1949. Nuclear spectroscopy and structure, decay schemes and nuclear reaction studies, experimental tests of nuclear models—recent emphasis on octupole shapes.

Harold E. Van Wart, Professor; Ph.D., Cornell, 1974. Physical biochemistry, protein structure, enzyme mechanisms, metalloenzymes.

Thomas J. Vickers, Professor and Associate Chairman; Ph.D., Florida, 1964. Spectrochemical analysis, Raman spectroscopy.

Harry M. Walborsky, Professor; Ph.D., Ohio State, 1949. Cyclopropyl carbanions, cyclopropyl radicals, asymmetric synthesis, organometallics.

SECTION 2: CHEMISTRY

INDIANA UNIVERSITY–PURDUE UNIVERSITY AT INDIANAPOLIS
Department of Chemistry

Programs of Study

The Department of Chemistry is part of the Purdue University School of Science at Indianapolis and offers Purdue M.S. and Ph.D. degrees. The graduate program is small with approximately 30 full-time and 40 part-time students. Most students are enrolled in the M.S. program, either in the thesis (full-time) option or nonthesis (part-time) option. The Ph.D. program is still relatively new but growing rapidly. Particular strengths include biological and computational chemistry, although the traditional areas of analytical, organic, inorganic, and physical chemistry are all offered.

All thesis students take five or six graduate courses during their studies. In their first semester of study, chemistry graduate students are given an opportunity to learn about possible thesis research projects with the faculty. Once a thesis adviser is chosen, students will prepare a written and oral presentation on their proposed thesis topic for approval by their individual graduate advisory committee. M.S. students will complete the program by carrying out their proposed research and writing and defending a master's thesis. Ph.D. students will, in addition, take a preliminary exam in their third year, consisting of both written and oral portions; upon passage of the preliminary exam, they will be advanced to candidacy. Completion and successful defense of a Ph.D. thesis completes the program. The master's program generally requires two years, while the Ph.D. usually requires four to six years of study.

Master's students may also participate in the unique Industrial Co-op Program. This program is the first M.S. co-op program in the country to provide for a parallel work-study experience. After the first semester, co-op students will spend half of each week at their participating industrial lab; the remainder of the week is devoted to course work and thesis research. The co-op program is limited to two years maximum. Past graduates of this program have either gone on to full-time positions in industry or continued with studies for the Ph.D. The co-op program is ideal for students who are undecided whether their future lies in the academic or the industrial setting.

Research Facilities

Both the department and its faculty are continually adding new instrumentation to support research activity. Major departmental equipment includes a 300-MHz NMR spectrometer and a gas chromatograph–mass spectrometer; FT-IR, UV, and fluorescence spectrophotometers; and a variety of chromatographic equipment. Access to a School of Science 500-MHz NMR spectrometer is also available. Computing facilities are particularly good, including state-of-the-art graphics equipment and a department-wide personal computer network. Chemistry shares a molecular modeling facility with the physics department that consists of two major Silicon Graphics workstations and associated software. Access to the statewide Indiana University Computing Network is also available; this network includes VAX and IBM mainframes as well as an IBM 3090 supercomputer. In the fall of 1992, the chemistry department will move into a new $40-million science and engineering complex located on the main IUPUI campus about three blocks from downtown Indianapolis.

Financial Aid

Most full-time students receive teaching assistantships ($12,000 per year), research assistantships ($12,000 per year), or University fellowships ($10,000–$12,000 per year) or are supported through the Industrial Co-op Program ($12,000 per year). Full-time students receive full fee remission; students with assistantships are also eligible for health insurance.

Cost of Study

All tuition and fees are waived for full-time students except for a nonrefundable fee of $216–$288 per semester, which cannot be remitted because of state law.

Cost of Living

Housing in Indianapolis is fairly inexpensive. Dormitory space is limited, and most graduate students prefer to live in University-subsidized apartments (approximately $300 per month for an unfurnished one-bedroom apartment) or off campus in private apartments (from $350 per month). Living expenses total about $7000 per year.

Student Group

Graduate students in the department come from all over the United States and countries around the world, including Germany, China, and India. Approximately 50 percent of the current group of graduate students are women. The research personnel in the department also include about 10 postdoctoral associates and several visiting scientists and professors every year.

Location

Indianapolis is the nation's fourteenth-largest city with a metro-area population of approximately 1.2 million people. In addition to being a major urban area, Indianapolis is also a collection of farmland, suburban communities, and wooded parks. Eagle Creek Park and Nature Preserve, located on the northwest side, consists of 3,500 acres of rolling terrain and a 1,300-acre reservoir and is one of the largest city parks in the country. The city also has twenty-two museums (including a world-renowned children's museum); international violin and piano competitions; world-class symphony, theater, and ballet companies; and the famous Indy 500 auto race.

The University

The IUPUI campus serves approximately 27,000 students, of which nearly one fourth are professional or graduate students. In 1969, the Indianapolis campuses of Indiana University and Purdue were merged to form IUPUI. Indiana University has fiscal responsibility for all functions of IUPUI, but some academic programs, including the chemistry graduate program, are administered by the Purdue graduate school. IUPUI is the most comprehensive public university in the state, with professional schools of medicine, dentistry, nursing, and law as well as undergraduate and graduate programs in the arts, sciences, engineering, and technology.

Applying

Applications may be submitted at any time, but it is recommended that applications for fall be received by mid-February. General requirements include at least 35 credit hours of chemistry and a minimum GPA of 3.0/4.0. GRE scores are strongly recommended for all students but are strictly required only for international applicants. In addition, a minimum TOEFL score of 550 is required of all international students.

Correspondence and Information

Director of Graduate Programs
Department of Chemistry
Indiana University–Purdue University at Indianapolis
1125 East 38th Street
Indianapolis, Indiana 46205-2810
Telephone: 317-274-6872
Fax: 317-274-4701

Indiana University–Purdue University at Indianapolis

THE FACULTY AND THEIR RESEARCH

Erwin Boschman, Professor and Associate Dean of Faculties; Ph.D., Colorado, 1968. Inorganic chemistry, chemical education.

Theodore W. Cutshall, Associate Professor; Ph.D., Norhwestern, 1964. Organic chemistry, chemical education.

Paul L. Dubin, Professor; Ph.D., Rutgers, 1970. Analytical chemistry, biopolymer separations, aqueous size exclusion chromatography, polyelectrolyte-colloid complexes.

Clifford E. Dykstra, Professor; Ph.D., Berkeley, 1976. Physical chemistry, theoretical and computational chemistry, chemical physics.

Wilmer K. Fife, Professor; Ph.D., Ohio State, 1960. Organic chemistry, synthesis and synthetic applications of pyridine derivatives, phase-managed synthesis, catalysis in multiphase systems.

Gordon H. Fricke, Associate Professor; Ph.D., Clarkson, 1970. Analytical chemistry, artificial intelligence, experimental design, optimization methods and data analysis.

Raima Larter, Associate Professor and Director of Graduate Programs; Ph.D., Indiana, 1980. Physical chemistry, nonlinear dynamics, oscillating reactions, membrane transport, chemical chaos.

Kenneth B. Lipkowitz, Professor; Ph.D., Montana State, 1975. Organic chemistry, computational chemistry, molecular modeling, molecular mechanics and computer-assisted molecular design.

Eric C. Long, Assistant Professor; Ph.D., Virginia, 1989. Bioinorganic chemistry, DNA interactions with peptides, metal complexes and antibiotics, nucleic acid strand scission, protein-DNA interactions, photoactivation and oxygen activation of metal complexes.

David J. Malik, Associate Professor and Chair; Ph.D., California, San Diego, 1976. Physical chemistry, quantum mechanics, intermolecular interactions, energy transfer, potential energy surfaces, effects of electric fields on molecules.

Barry B. Muhoberac, Associate Professor; Ph.D., Virginia, 1978. Physical chemistry and biochemistry, ligand binding and catalysis in metalloproteins, ethanol and drug interactions with biomembranes.

David Nurok, Associate Professor; Ph.D., Cape Town, 1986. Analytical chemistry, solvent effects in liquid chromatography, optimization of planar chromatography.

Martin J. O'Donnell, Professor; Ph.D., Yale, 1973. Organic chemistry, synthetic methods in amino acid and peptide chemistry, phase-transfer reactions, asymmetric synthesis.

Franklin A. Schultz, Professor; Ph.D., California, Riverside, 1967. Analytical chemistry, electrochemistry of transition-metal complexes, heterogeneous electron transfer kinetics, active-site model compounds of molybdenum-containing enzymes.

Stephanie E. Sen, Assistant Professor; Ph.D., SUNY at Stony Brook, 1989. Organic chemistry, design and synthesis of enzyme inhibitors, catalytic antibodies, protein structure-function relationships, insect biochemistry, pheromones.

Richard J. Wyma, Associate Professor; Ph.D., Michigan, 1964. Physical chemistry, chemical education, data acquisition for science laboratories.

Martel Zeldin, Professor; Ph.D., Penn State, 1966. Inorganic chemistry, synthesis of compounds of main groups 13 and 14 elements, inorganic and organometallic polymeric catalysts and reagents, degradation of organometallic polymers.

The new $40-million science and engineering complex on the main IUPUI campus. The chemistry department is scheduled to move into the wing under construction (at right) in the fall of 1992.

SECTION 2: CHEMISTRY

INSTITUTE OF PAPER SCIENCE AND TECHNOLOGY
Program in Chemistry

Program of Study

The Institute of Paper Science and Technology (IPST), an independent graduate school, offers graduate study in chemistry through a broad academic program in engineering and the sciences. The educational philosophy of the Institute is to develop scientific generalists who are well versed in several disciplines within the sciences and who can range across the boundaries of disciplines in their pursuit of knowledge and insight. The Institute awards the Master of Science and Doctor of Philosophy degrees in pulp and paper science with specializations in organic and physical chemistry as well as in chemical engineering, physics, biology, and mathematics.

In the master's program, the first year involves advanced work in engineering, chemistry, physics, biology, and mathematics, and students who have not previously studied chemical engineering complete elementary courses in this field. In the second year, emphasis is placed upon the interrelationships among fields and the integration of disciplines, and each student carries out an individual research program.

Students can be considered for admission to candidacy for the degree of Doctor of Philosophy after satisfactory completion of a program in Preparation for Research, in which the student searches the literature and plans a research program for each of a series of complex problems. The student then chooses a Ph.D. thesis topic in any area of faculty interest. The 75 students and 29 faculty members form a close-knit academic community where students are encouraged to develop their capabilities in their own way.

Research Facilities

Research studies at the Institute cover a wide spectrum from the fundamental to the applied or developmental project. Well-equipped laboratories, modern instrumentation, sophisticated specialized equipment, and state-of-the-art computing facilities are available for studies in the physical and biological sciences and in engineering and technology. Areas of research interest for which equipment is available include chemical reaction mechanisms; kinetics; polymer sciences; wood, carbohydrate, lignin, and extractive chemistry; surface and colloid science; thermodynamics; heat, mass, and momentum transfer, particularly in porous media and fibrous structures; fluid mechanics and fluid dynamics; process dynamics, simulation, and control; mathematical modeling; increased efficiency in energy utilization; mechanical, electrical, and optical properties of fibers and fibrous materials; pulping, bleaching, and chemical recovery technology; corrosion processes and control technology; chemical reactions of molten salts; system engineering; applied mechanics of containers; two-phase flow and laser anemometry; forest genetics; and tissue culture directed toward mass production of conifers.

Financial Aid

The Institute of Paper Science and Technology awards a fellowship and tuition scholarship, both of which are renewable, to each first-year student of U.S. or Canadian citizenship admitted to the graduate program. No services are required in return, and the student is expected to devote full time to graduate study. Fellowships are granted to M.S. students in the amount of $11,250 for the academic year or $15,000 for the calendar year. The fellowships provide $17,000 per calendar year for Ph.D. students. In addition, scholarships of $7500 per academic year are granted to cover the cost of tuition.

Cost of Study

Tuition costs are provided for by the arrangements explained above.

Cost of Living

Housing facilities are available through the Georgia Institute of Technology Housing Office.

Student Group

The fall term enrollment is approximately 75 students, including an entering class of about 25. Students have matriculated from more than 150 colleges and universities in the United States. Several foreign countries are represented in the student body as well.

Location

The Institute of Paper Science and Technology is located in Atlanta, Georgia. The Institute has established an alliance with Georgia Tech that involves a sharing of resources, thus enhancing the education opportunities available to students from both institutions who are interested in the pulp and paper industry and providing numerous opportunities for joint ventures in education, research, and academic services.

The Institute

The Institute of Paper Science and Technology, formerly known as the Institute of Paper Chemistry, is a small graduate research university related to the pulp and paper industry.

The Institute of Paper Science and Technology is the premier graduate education, research, and information services organization for an industry that employs about 1,000 technical graduates each year. The Institute supplies about one third of the technical manpower at the graduate level required by this dynamic industry. A very large percentage of IPST alumni fill upper and middle management positions in the pulp and paper industry.

In general, classes are small, and the atmosphere is one in which close student-faculty cooperation is maintained. Graduates are sought for responsible positions in research, development, technical service, sales, production, and management in many areas of the pulp and paper industry and in related industries. An increasing number have entered the teaching profession, and approximately half of the pulp and paper science programs in American colleges and universities are headed by IPST graduates.

Applying

Students are admitted only in September, and the deadline for applications is May 15. Applicants must submit scores on the Graduate Record Examinations. Admission requirements include organic and physical chemistry, physics, and calculus, with differential equations and advanced physics recommended.

Correspondence and Information

Dr. Terrance E. Conners
Director of Admissions
Institute of Paper Science and Technology
575 14th Street NW
Atlanta, Georgia 30318
Telephone: 404-853-9500

SECTION 2: CHEMISTRY

Institute of Paper Science and Technology

THE FACULTY

All 29 members of the faculty are listed below because of the interdisciplinary nature of the Institute's programs.

Chemical and Biological Sciences
Sujit Banerjee, Ph.D.
Barry W. Crouse, Ph.D.
Donald R. Dimmel, Ph.D.
Ronald J. Dinus, Ph.D.
Frank M. Etzler, Ph.D.
Earl W. Malcolm, Ph.D.
Arthur J. Ragauskas, Ph.D.
Nagmani Rangaswamy, Ph.D.
Alan W. Rudie, Ph.D.
Lucinda B. Sonnenberg, Ph.D.
Robert A. Stratton, Ph.D.
David Webb, Ph.D.

Engineering and Paper Materials
Per-Erik Ahlers, M.Sc.
Cyrus K. Aidun, Ph.D.
Pierre Brodeur, Ph.D.
Terrance E. Conners, Ph.D.
Richard Ellis, Ph.D.
Howard L. (Jeff) Empie, Ph.D.
Maclin S. Hall, Ph.D.
Robert Horton, Ph.D.
Gary L. Jones, Ph.D.
Jeffrey D. Lindsay, Ph.D.
Richard A. Matula, Ph.D.
Thomas J. McDonough, Ph.D.
Kenneth M. Nichols, Ph.D.
David I. Orloff, Ph.D.
James D. Rushton, Ph.D.
John F. Waterhouse, M.S.
Ronald A. Yeske, Ph.D.

SECTION 2: CHEMISTRY

KENT STATE UNIVERSITY
Department of Chemistry

Programs of Study

The Department of Chemistry offers programs leading to the Master of Science and Doctor of Philosophy degrees in analytical, inorganic, organic, and physical chemistry and biochemistry. Also available are interdisciplinary doctoral programs in chemical physics and molecular and cellular biology. In addition, there are opportunities to pursue research in the field of liquid crystals with members of the Liquid Crystal Institute.

Graduate students are required to complete a program of core courses in their area of specialization and at least one (for M.S. candidates) or two (for Ph.D. candidates) courses in other areas of chemistry. In addition to these courses, students may choose from a wide variety of electives. The program thus gives students considerable flexibility in curriculum design. At the end of the second year, doctoral candidates must pass a written examination in their field of specialization and present and subsequently defend an original research proposal on a topic different from their dissertation research. Students normally complete their doctoral program after four years.

Research Facilities

The Department of Chemistry currently occupies Williams Hall and part of the new Science Research Building, which also houses the Glenn H. Brown Liquid Crystal Institute. Williams Hall houses two large lecture halls, classrooms, undergraduate and research laboratories, the chemistry-physics library, a remote printing station and terminals connected with the University Computer Center, chemical stockrooms, and glass and electronics shops. A machine shop, which is jointly operated with the physics department, is located in nearby Smith Hall. Spectrometers include GE 300-MHz and Varian FT80 multinuclear NMR, Bomem and Nicolet FT-IR, photon-counting fluorometer, polarimeter, UV/VIS, and AA/AE. The department has HPLC, GC, and GC-MS facilities. Equipment available in specialty areas includes Durrum stopped-flow spectrophotometers, a variety of electrochemical instrumentation, an automated Siemens X-ray diffractometer, a positron annihilation lifetime spectrometer, and a liquid scintillation counter. The department houses a remote printing station and terminals for the University's IBM 3081 D mainframe computer and has its own PDP-11/44 computer. A variety of microcomputers and workstations are available throughout the department for data acquisition and analysis. The department has on-line access to the CRAY X-MP supercomputer and IMaG Center Molecular Graphics facilities at Ohio State University, which makes available a variety of chemical databases and molecular modeling and energy minimization facilities. There is also on-line access to the Chemical Abstracts Service, a chemical database facility operated by the American Chemical Society. The Chemistry-Physics Library contains up-to-date collections of all major chemistry and physics journals, all major abstracting and indexing series, and all books in chemistry and physics published by the major scientific publishing houses.

Financial Aid

Graduate students are supported through teaching and research assistantships and University fellowships. Students in good academic standing are guaranteed appointments for periods of 4½ years (Ph.D. candidates) or 2½ years (M.S. candidates). Stipends for 1991–92 range from $9600 to $10,267 for a twelve-month appointment.

Cost of Study

Graduate tuition and fees for the 1991–92 academic year are $3764, for which a tuition scholarship is provided to students in good academic standing.

Cost of Living

Rooms in the graduate hall of residence are $870 per semester; married students' apartments may be rented for $272–$288 per month. Information concerning off-campus housing may be obtained from the University Housing office. Costs vary widely, but a two-bedroom apartment typically rents for $350–$450 per month.

Student Group

Graduate students in chemistry currently number about 40. There are approximately 20,000 students enrolled at the main campus of Kent State University; 8,000 additional students attend the seven regional campuses.

Location

Kent, a city of about 28,000, is located 35 miles southeast of Cleveland and 12 miles east of Akron in a peaceful suburban setting. Kent offers the cultural advantages of a major metropolitan complex as well as the relaxed pace of semirural living. There are a number of theater and art groups at the University and in the community. Blossom Music Center, the summer home of the Cleveland Orchestra and the site of Kent State's cooperative programs in art, music, and theater, is only 15 miles from the main campus. The Akron and Cleveland art museums are also within easy reach of the campus. There are a wide variety of recreational facilities available on the campus and within the local area, including West Branch State Park and the Cuyahoga Valley National Recreation Area. Opportunities for outdoor activities such as summer sports, ice-skating, swimming, and downhill and cross-country skiing abound.

The University

Established in 1910, Kent State University is one of Ohio's largest state universities. The campus contains 820 acres of wooded hillsides plus an airport and an eighteen-hole golf course. There are approximately 100 buildings on the main campus. Bachelor's, master's, and doctoral degrees are offered in more than thirty subject areas. The faculty numbers approximately 800.

Applying

Forms for admission to the graduate programs are available on request from the address below. There is no formal deadline for admission, but graduate assistantships are normally awarded by May for the following fall. Applicants requesting assistantships should apply by March 1.

Correspondence and Information

Graduate Coordinator
Department of Chemistry
Kent State University
Kent, Ohio 44242
Telephone: 216-672-2032

SECTION 2: CHEMISTRY

Kent State University

THE FACULTY AND THEIR RESEARCH

Rathindra N. Bose, Associate Professor; Ph.D., Georgetown, 1982. Bioinorganic chemistry: mechanisms of interaction of metal drugs with nucleosides and nucleotides, metal ion catalysis of phosphate hydrolysis reactions, multinuclear magnetic resonance studies of metal complexes of biomolecules, mechanisms of chemical reactions coupled with electrochemical reactions.

Stephen E. Cabaniss, Assistant Professor; Ph.D., North Carolina, 1986. Analytical chemistry: environmental analysis and modeling, molecular fluorescence spectroscopy, metal binding to macromolecules and colloids, low-temperature aqueous geochemistry.

Norman V. Duffy, Professor; Ph.D., Georgetown, 1966. Inorganic chemistry: synthesis of dithiocarbamate complexes of transition metals, particularly those of iron, cobalt, indium, and gold; multinuclear NMR, Mossbäuer (Fe), infrared, magnetic susceptibility and EPR of these dithiocarbamate complexes and their selenium analogues.

Derry L. Fishel, Professor; Ph.D., Ohio State, 1959. Organic chemistry: syntheses of nitrogen heterocycles and mesomorphic compounds (liquid crystals), organo-nitrogen derivative rearrangements, ^{252}Cf fission-fragment and accelerated particle induced desorption (PID) mass spectrometry.

Julia E. Fulghum, Assistant Professor; Ph.D., North Carolina, 1987. Analytical chemistry: surface and microbeam analysis, X-ray photoelectron spectroscopy, secondary ion mass spectrometry, quantitative surface analysis, environmental analytical chemistry.

Roger K. Gilpin, Professor and Chairman; Ph.D., Arizona, 1973. Analytical chemistry: development of chemically modified surfaces for chromatography, chromatographic, NMR, and IR studies of the structure and dynamics of modified surfaces; separation of pharmaceutically and biomedically active compounds; chromatographic selectivity; micelles and the applications of micelles to chemical separations.

Edwin S. Gould, University Professor; Ph.D., UCLA, 1950. Inorganic chemistry: mechanisms of inorganic redox reactions; catalysis of redox reactions by organic species; electron-transfer reactions of flavin-related systems; reactions of cobalt, chromium, vanadium, titanium, europium, uranium, and ruthenium; reactions of water-soluble radical species.

Roger B. Gregory, Associate Professor; Ph.D., Sheffield (England), 1980. Biochemistry: role of conformational dynamics in protein function and stability, enzyme kinetics and mechanism, protein-solvent interactions, protein hydrogen isotope exchange, positron annihilation lifetime spectroscopy and its application in studies of proteins and liquid crystals, application of Laplace inversion techniques in kinetic data analysis.

Mieczyslaw Jaroniec, Professor; Ph.D., Lublin (Poland), 1976. Analytical/physical chemistry: thermodynamics of adsorption and chromatography at the gas/solid and liquid/solid interfaces; studies of the surface phenomena on heterogeneous, microporous, and fractal solids, such as microporous carbonaceous adsorbents, chemically modified silicas and carbon fibers, and composite solids; characterization of adsorbents, chromatographic packings and catalysts.

Khosrow Laali, Associate Professor; Ph.D., Manchester (England), 1977. Organic chemistry: mechanistic organic (organometallic) chemistry and synthetic applications, electrophilic aromatic substitution, chemistry in superacid media, fluorine chemistry, novel long-lived arenium ions, superacid catalysis, aromatic and aliphatic diazonium ions, vinyl cations, multinuclear NMR and modern mass spectrometry.

Vernon D. Neff, Professor; Ph.D., Syracuse, 1959. Physical chemistry: structure and phase transitions in liquid crystals, electrochemical oxidation and reduction of semiconducting intervalence compounds, molecular spectroscopy.

Thomas I. Pynadath, Professor; Ph.D., Georgetown, 1963. Biochemistry: effects of hormones and drugs on serum levels of lipoproteins and lipoprotein metabolism, induction of serum lipoproteins by drugs, effects of drugs on the synthesis of thromboxane and prostacyclin, effects and mechanism of action of drugs on platelet aggregation and thrombosis.

Gloria J. Pyrka, Assistant Professor; Ph.D., Arizona, 1988. Analytical chemistry: synthesis of quasi-one-dimensional conductors, electrochemistry, structure and electronic properties of 1-D metal and semiconductors, applications as electrode materials and chemical sensors.

John W. Reed, Associate Professor; Ph.D., Ohio State, 1956. Physical chemistry: crystal and molecular structure determination of organic and inorganic compounds by X-ray diffraction, structure of mesomorphic compounds in the crystalline state, relationships between structures of mesomorphic and crystalline states.

Richard J. Ruch, Professor; Ph.D., Iowa State, 1959. Physical chemistry: physical properties of colloidal dispersions and polymer solutions, stabilization mechanisms in colloidal dispersions, polymer coating formation, wetting, rheology and dielectrics in colloidal systems.

Paul Sampson, Assistant Professor; Ph.D., Birmingham (England), 1983. Organic chemistry: synthetic organic chemistry, total synthesis of natural and fluorine-containing "unnatural" products possessing important biological properties, development of new synthetic methods, new approaches for the construction of medium-sized and macrocyclic rings, synthetic (including stereoselective) organofluorine chemistry.

Chun-che Tsai, Professor; Ph.D., Indiana, 1968. Biochemistry: interaction of drugs with nucleic acids, structure and activity of anticancer drugs, antiviral agents and interferon inducers, structure and biological function relationships, X-ray diffraction and quantitative structure-activity relationships (QSAR).

Debbie F. Tuan, Professor; Ph.D., Yale, 1961. Physical chemistry: application of quantum mechanics to chemical problems, electronic structure calculations by self-consistent field Xα scattered wave and extended Hückel methods, electron correlation of atoms and molecules, calculation of physical properties by perturbation methods.

Frederick G. Walz, Professor; Ph.D., SUNY Downstate Medical Center, 1966. Biochemistry: cytochrome P-450 polymorphism, regulation, and genetics; site-directed mutagenesis studies on the folding and catalytic mechanism of simple ribonucleases; intracellular trafficking of proteins.

LEHIGH UNIVERSITY

Department of Chemistry

Programs of Study
The Department of Chemistry offers the M.S. and Ph.D. degrees in the areas of biochemistry and analytical, inorganic, organic, and physical chemistry; special programs are also available in clinical, polymer, and physiological chemistry. The M.S. degree requires 30 hours of course work plus demonstration of proficiency in two areas of chemistry. The Ph.D. program does not have formal course requirements. Proficiency in three areas of chemistry must be demonstrated. In addition, a doctoral examination, which consists of a written part and a proposal part, must be passed in the second year of residency. After successfully passing each component of the examination, students perform original research and write and defend a dissertation describing the research.

Research Facilities
Research facilities are situated in several locations. Most of the facilities are housed in the 90,000-square-foot, seven-story Seeley G. Mudd Building. The top three floors contain modern research laboratories. Most of the research laboratories in the Sinclair Laboratory are assigned to chemistry professors who specialize in research in surface and colloid chemistry. This laboratory houses the Scienta high-resolution electron spectrometer. Biochemistry research is located in Building A of the Mountaintop Campus.

The department has the instrumentation necessary for all of its areas of specialization. It is superbly equipped for surface analysis, with a very high resolution Scienta ESCA instrument and a variety of other surface spectrometers. Four NMR spectrometers are used by students: 90-MHz, 360-MHz, and 500-MHz solution spectrometers and a 300-MHz solids instrument. IR facilities include two FT instruments, well equipped for transmission, diffuse reflectance, internal reflectance, and photoacoustic spectroscopy. Numerous chromatographs, spectrometers, centrifuges, electrochemical instruments, and a pair of mass spectrometers are all in active use by graduate students. The department has an open policy in which students are trained and use the instruments necessary for their research. The chemistry department has its own electronics maintenance and instruction staff. Top-quality machine shop service is available.

The Fairchild-Martindale Science Library contains more than 21,000 monographs and 59,000 bound journals in chemistry and related areas, 222 subscriptions in chemistry, and 387 subscriptions in related areas.

The primary campus computers are a VAX 8530 and a CYBER 185. Access to each is possible over a University electronic network; computers off campus can be easily reached through the Telenet system. The chemistry department has a modern microcomputer lab housing eight IBM PS/2 microcomputers, a 4696 color plotter, and a 4105 Tektronix terminal for display of molecular structures.

Financial Aid
Most entering graduate students in chemistry are supported as half-time teaching assistants; the present stipend (1991–92) is $12,600 for a twelve-month period. In addition, teaching assistants receive tuition remission for 10 credits per semester and 3 credits during the summer. Students making satisfactory progress are normally supported until the completion of their degree program. Research assistantships are also held by graduate students and typically cover a twelve-month period. Fellowships are also available with stipends comparable with stipends for teaching assistants.

Cost of Study
Tuition for the 1991–92 academic year is $11,105. Tuition expenses are paid for teaching assistants and research assistants. There is also a graduate student activities fee of $24 per year.

Cost of Living
Total expenses range from $6500 to $7200. Students typically live in a wide variety of accommodations. Expenses can be reasonable, especially if accommodations are shared. The University operates a 148-unit five-building garden apartment complex for married and single graduate students, located in the nearby Saucon Valley Campus. Free bus service to the Asa Packer Campus is provided hourly.

Student Group
At the present time there are 96 graduate students—64 full-time and 32 part-time. Students come from many states and several foreign countries.

Location
Lehigh University is located in Bethlehem, Pennsylvania, 60 miles north of Philadelphia and 90 miles west of New York City. The location makes the cultural, entertainment, and transportation facilities of these cities readily accessible. Founded in 1741, Bethlehem has a rich cultural heritage. Historical buildings have been remarkably preserved and are in current use, giving the community a charming Colonial atmosphere. A weeklong music festival is held in late August. The Lehigh Valley is also an important commercial and industrial center with administrative, research, and manufacturing facilities for several major companies.

The University
Lehigh is an independent, nondenominational, coeducational university. Founded in 1865 as a predominantly technical four-year school also offering the liberal arts, the University now has approximately 4,500 undergraduate students within its three major units—the Colleges of Arts and Science, Engineering and Applied Science, and Business and Economics—and approximately 1,900 graduate students enrolled in the Graduate School and the graduate-level College of Education. Lehigh employs some 1,900 people, including 400 full-time faculty members. There are approximately 2,000 courses offered at Lehigh.

Lehigh's primary mission is to provide excellence in teaching, research, and the appreciation of knowledge for a useful life. While fulfilling its mission, Lehigh strives to earn national recognition as a small university of special distinction in teaching and research.

Most of the University's more than 100 buildings are located on the Asa Packer Campus. Research facilities are also available on the Mountaintop Campus, consisting of five buildings formerly constituting Bethlehem Steel's Homer Research Laboratories. A field house, playing fields, the Murray Goodman Football Stadium, and the Stabler Athletic and Convocation Center are also available.

The University has a wide range of athletic facilities and presents during the academic year a variety program of concerts, theater performances, films, lectures, art exhibits, and sporting events. Lehigh is committed to cultural diversity and human dignity as essential elements of a vibrant university.

Applying
A preliminary application is available for domestic students. For international students, the GRE General Test is required and a Subject Test recommended; a minimum TOEFL score of 550 is also required. There is a $40 fee for the formal application.

Correspondence and Information
Mary Ann Elgin, Secretary
Graduate Admissions Committee
Department of Chemistry #6
Lehigh University
Bethlehem, Pennsylvania 18015-3172
Telephone: 215-758-3470

SECTION 2: CHEMISTRY

Lehigh University

THE FACULTY AND THEIR RESEARCH

Jack A. Alhadeff, Professor; Ph.D., Oregon, 1972. Biochemistry: purification and characterization of glycosidases, biochemical basis of human diseases involving abnormal glycoprotein metabolism.

Michael J. Behe, Associate Professor; Ph.D., Pennsylvania, 1978. Biochemistry: structure of nucleic acids, specifically, oligopurine regions in eukaryotes and the B-Z transition in synthetic polydeoxynucleotides.

Gregory S. Ferguson, Assistant Professor; Ph.D., Cornell, 1984. Inorganic, surface, and polymer chemistry: chemistry of organic and inorganic surfaces.

Natalie Foster, Associate Professor; Ph.D., Lehigh, 1982. Analytical and organic chemistry: contrast enhancement agents for medical imaging NMR, structural dynamics and molecular associations of biologically significant molecules.

Ned D. Heindel, Howard S. Bunn Professor; Ph.D., Delaware, 1963. Organic and medicinal chemistry: synthesis of potential medicinal agents, diagnostic radioactive pharmaceuticals.

Leonard E. Klebanoff, Assistant Professor; Ph.D., Berkeley, 1985. Physical chemistry: interface magnetism and structure.

Kamil Klier, University Professor; Ph.D., Czechoslovak Academy of Sciences, 1961. Physical chemistry: catalysis, methanol synthesis.

Charles S. Kraihanzel, Professor; Ph.D., Wisconsin, 1962. Inorganic chemistry: reactions of coordinated phosphorus donor ligands, molecular modeling of inorganic and organic compounds.

John W. Larsen, Chair and Professor; Ph.D., Purdue, 1966. Organic chemistry: coal chemistry, thermodynamics of organic intermediates, organic chemistry in molten salts.

Roland W. Lovejoy, Professor; Ph.D., Washington State, 1960. Physical chemistry: very high resolution infrared spectroscopy.

Linda J. Lowe-Krentz, Assistant Professor; Ph.D., Northwestern, 1980. Biochemistry: heparan sulfate proteoglycans synthesized by endothelial cells.

Fortunato J. Micale, Professor; Ph.D., Lehigh, 1965. Physical chemistry: characterization of solid-gas and solid-liquid interfaces of fine particles by gas adsorption, solution adsorption, and electrokinetic techniques.

Steven L. Regen, Professor; Ph.D., MIT, 1972. Organic chemistry: organic and polymer chemistry, supramolecular assemblies, membrane structure and function, drug design.

James E. Roberts, Assistant Professor; Ph.D., Northwestern, 1982. Physical and analytical chemistry: applications of solid-state nuclear magnetic resonance.

Keith J. Schray, Assistant Chair and Professor; Ph.D., Penn State, 1970. Organic chemistry and biochemistry: protein-surface interactions, novel immunoassays.

Gary W. Simmons, Professor; Ph.D., Virginia, 1967. Physical chemistry: Auger spectroscopy, Mössbauer spectroscopy, low-energy electron diffraction, X-ray photoelectron spectroscopy.

Donald M. Smyth, Professor; Ph.D., MIT, 1954. Inorganic chemistry: defect chemistry and electrical properties of transition-metal oxides, superconducting ceramics.

James E. Sturm, Professor; Ph.D., Notre Dame, 1957. Physical chemistry: elementary processes in photochemistry, reactions of high-velocity atoms.

John W. Vanderhoff, Professor; Ph.D., Buffalo, 1951. Physical chemistry: mechanism and kinetics of emulsion polymerization, use of monodisperse latex particles as model colloids.

Daniel Zeroka, Professor; Ph.D., Pennsylvania, 1966. Physical chemistry: electronic structure of molecular species, solids and molecular species adsorbed onto solid surfaces, vibrational circular dichroism.

SECTION 2: CHEMISTRY

MASSACHUSETTS INSTITUTE OF TECHNOLOGY

Department of Chemistry

Programs of Study The Department of Chemistry offers programs leading to the S.M. and Ph.D. degrees, with emphasis on the latter. No qualifying examinations are given, and the S.M. degree is not a prerequisite for the Ph.D. There are no formal course requirements for the Ph.D. degree. Areas of concentration in the department include organic, inorganic, physical, biological, and biophysical chemistry and chemical physics.

Each student, with the advice of a research supervisor, pursues an individual program of study pertinent to his or her long-range research interests. Commencing with the second semester in residence, students are required to take monthly written examinations until six have been passed. A comprehensive oral examination in the candidate's major field of advanced study is given near the end of the third term of residence. Progress in research is examined at that time.

The main objective of the graduate program is to prepare a student for independent research; hence, performance in thesis research is of prime importance. A final oral presentation of the subject of the doctoral research is scheduled after the thesis has been submitted and tentatively evaluated by a committee of examiners.

Research Facilities The chemistry department has progressive facilities for research in all areas of chemistry. The departmental spectrometry laboratory has the latest instrumentation for NMR, EPR, IR, and mass spectrometry and other techniques. A departmental computer facility, which features a VAX-11/780, complements the department's research activities. Machine, electronics, and glass shops are also available within the department. Other on-campus facilities are accessible through various interdepartmental laboratories, including the MIT Spectroscopy Laboratory (especially its Tuneable Laser Laboratory), Information Processing Center, National Magnet Laboratory, Research Laboratory of Electronics, Center for Materials Science and Engineering, Center for Cancer Research, and Energy Laboratory.

Financial Aid Almost all first-year Ph.D. candidates are offered teaching assistantships, which carry taxable stipends of $1260 per month and a waiver of tuition in 1991–92. In later academic years, financial support is provided for students who maintain a satisfactory record, subject to the availability of funds. Most of these appointments are research assistantships, which carry stipends of $1140 per month as well as a grant that covers tuition. Financial aid is normally not offered to first-year S.M. candidates.

Cost of Study Tuition is $16,900 for the 1991–92 academic year.

Cost of Living Assistance in obtaining housing on or off campus is available from the MIT Housing Office. Housing costs range from $450 to $1000 per month for single students and from $600 to $1500 per month for married students.

Student Group MIT's total enrollment of approximately 9,600 is almost evenly divided between undergraduate and graduate students. During the academic year 1991–92, students came from fifty states and ninety-seven foreign countries. The proportion of non-U.S. citizens at the Institute, about 20 percent, is one of the highest in an American university. There are 250 graduate students in the chemistry department, including 72 women. Twenty-two percent of the enrolled students are married, 17 percent are foreign, and the average age is 26. Approximately 99 percent receive financial aid. Upon receiving their degree, about 7 percent of the department's Ph.D.'s enter positions in academia. The remaining number are divided about equally between those who enter industrial employment and those who spend one or more years in postdoctoral research before accepting permanent positions.

Location MIT is located in Cambridge, only minutes away from Boston. With more than fifty schools in the area, the concentration of academic, cultural, and intellectual activities is virtually unsurpassed. The mountains of Vermont and New Hampshire, the beaches of Cape Cod and Maine, the New England coast, and many historic places such as Salem, Sturbridge, Lexington, Concord, and Plymouth are a 1- or 2-hour drive by car. The four seasons of New England, combined with its varied landscape, offer almost unlimited possibilities for such recreational pursuits as skiing, mountain climbing, hiking, and sailing.

The Institute MIT was founded in 1865 for the purpose of creating a new kind of educational institution relevant to the times and the nation's needs, where students would be educated in the application as well as the acquisition of knowledge. Today, education and related research continue to be MIT's central purpose, with relevance to the practical world as a guiding principle. The Institute is an independent, coeducational, privately endowed university. The mix of ages, disciplines, and nationalities at MIT contributes to an academic environment that, while unusual for its singleness of method and purpose, is also notable for its diversity.

Applying Applications should be submitted by mid-February, but later applications are considered. Interviews are not required. Applicants are invited to visit the Institute but are advised to contact the department beforehand to arrange appointments.

Correspondence and Information
Chairman
Departmental Committee on Graduate Students
Chemistry Graduate Office 18-392
Massachusetts Institute of Technology
Cambridge, Massachusetts 02139
Telephone: 617-253-1845

Peterson's Guide to Graduate Programs in the Physical Sciences and Mathematics 1992

SECTION 2: CHEMISTRY

Massachusetts Institute of Technology

THE FACULTY AND THEIR RESEARCH

Robert A. Alberty, Professor; Ph.D., Wisconsin, 1947. Physical chemistry, thermodynamics and kinetics of organic reactions.

Moungi G. Bawendi, Assistant Professor; Ph.D., Chicago, 1988. Physical chemistry, structural characterization and spectroscopy of nanometer-size semiconductor clusters, exploration of materials between molecules and the bulk.

Glenn A. Berchtold, Professor; Ph.D., Indiana, 1959. Organic chemistry, synthetic organic chemistry.

Klaus Biemann, Professor; Ph.D., Innsbruck, 1951. Organic and analytical chemistry; application of mass spectrometry to problems in organic, analytical, and clinical chemistry and biochemistry; determination of protein structure by tandem mass spectrometry; computer acquisition and evaluation of mass spectra.

George H. Büchi, Professor; D.Sc., Zurich, 1947. Organic chemistry, natural products, photochemistry.

Stephen L. Buchwald, Associate Professor; Ph.D., Harvard, 1982. Organic and organometallic chemistry, organometallic methods and reagents for use in organic synthesis, organometallic synthesis, materials chemistry, main group organometallic chemistry.

Sylvia T. Ceyer, Professor; Ph.D., Berkeley, 1979. Physical chemistry, dynamics of molecule-surface interactions using molecular-beam techniques.

Rick L. Danheiser, Professor; Ph.D., Harvard, 1978. Organic chemistry, development of new synthetic methods and reagents, applications in the total synthesis of natural products and biologically significant molecules.

Alan Davison, Professor; Ph.D., Imperial College (London), 1962. Inorganic, technetium, and radiopharmaceutical chemistry.

John M. Deutch, Professor; Ph.D., MIT, 1966. Physical and theoretical chemistry, nonequilibrium statistical mechanics, structure of fluids, light scattering, polymer chemistry.

John M. Essigmann, Professor; Ph.D., MIT, 1976. Initiation mechanisms in chemical carcinogenesis and mutagenesis, mechanism of action of antitumor drugs.

Robert W. Field, Professor; Ph.D., Harvard, 1971. Physical chemistry; tunable laser spectroscopy, especially optical-optical double resonance electronic spectroscopy of diatomic molecules and semiempirical electronic structure models; structure and dynamics of vibrationally highly excited small polyatomic molecules; quantum chaos, isomerization, and large-amplitude vibrations.

Carl W. Garland, Professor; Ph.D., Berkeley, 1953. Physical chemistry, liquid crystal phase transitions, studies of solids near critical points, order-disorder phenomena, high-resolution AC calorimetry, ultrasonic investigation of critical dynamics.

Frederick D. Greene II, Professor; Ph.D., Harvard, 1952. Organic chemistry, mechanisms of organic reactions, small-ring compounds, free radical reactions.

Robert G. Griffin, Professor; Ph.D., Washington (St. Louis), 1969. Physical chemistry, magnetic resonance in condensed media, applications to biophysical systems.

Daniel S. Kemp, Professor; Ph.D., Harvard, 1964. Peptide chemistry, synthetic methods for peptide synthesis, synthesis of polypeptides, conformations of polypeptides and related molecules, mechanisms of peptide reactions.

H. Gobind Khorana, Professor; Ph.D., Liverpool, 1948. Biological membranes and nucleic acids, structure-function studies and reconstitution of integral membrane protein functions, mechanism of proton translocation by the light-dependent membrane protein bacteriorhodopsin, light transduction and biochemical processes in vertebrate and invertebrate vision.

Alexander M. Klibanov, Professor; Ph.D., Moscow, 1974. Enzyme technology, stability and stabilization of enzymes, enzymes as catalysts in organic chemistry.

Peter T. Lansbury Jr., Assistant Professor; Ph.D., Harvard, 1985. Organic and biological chemistry, peptide and protein conformation, synthetic and conformational studies of glycopeptides.

Stephen J. Lippard, Professor; Ph.D., MIT, 1965. Inorganic chemistry; bioinorganic chemistry; organometallic chemistry; transition metals as drugs, protein active sites, and catalysts.

Hans-Conrad zur Loye, Assistant Professor; Ph.D., Berkeley, 1988. Solid-state chemistry, inorganic chemistry, oxides, chalcogenides, catalysis, ionic conductors.

Satoru Masamune, Professor; Ph.D., Berkeley, 1957. Organic chemistry, development of synthetic methodology and syntheses of natural products, organometallic chemistry of main group elements, mechanisms of enzymatic carbon-carbon bond formation.

Mario J. Molina, Professor; Ph.D., Berkeley, 1972. Physical chemistry, chemical kinetics, atmospheric chemistry, global environmental effects of man-made pollutants.

Keith A. Nelson, Associate Professor; Ph.D., Stanford, 1981. Physical chemistry; picosecond spectroscopy of structural phase transitions and liquid-glass transitions; femtosecond spectroscopy of chemical reactions and transition states, liquid-state molecular dynamics, and vibrationally distorted molecules and crystals; ultrafast stimulated light scattering.

Irwin Oppenheim, Professor; Ph.D., Yale, 1956. Physical chemistry, statistical mechanics of equilibrium and transport properties, transport phenomena in fluids, light scattering, line broadening, nuclear and electron spin resonance, Brownian motion.

William H. Orme-Johnson III, Professor; Ph.D., Texas, 1964. Bioinorganic electron-transfer mechanisms; structure and function of cytochromes and ferredoxins and other iron-sulfur proteins, nickel, molybdenum, and copper proteins, particularly as studied by magnetic resonance; enzymology of nitrogen fixation, steroid metabolism, and methanogenesis.

Philip W. Phillips, Associate Professor; Ph.D., Washington (Seattle), 1982. Physical chemistry, quantum and statistical mechanics, electrons in fluids, electron transport in condensed phases, metal-insulator transitions, percolation.

Julius Rebek Jr., Professor; Ph.D., MIT, 1970. Organic chemistry, molecular recognition, enzyme models, asymmetric reagents and catalysis for organic synthesis.

Richard R. Schrock, Professor; Ph.D., Harvard, 1971. Inorganic and organometallic chemistry, synthetic organometallic chemistry, homogeneous catalysis, early transition-metal chemistry, alkyl and carbene complexes, controlled polymer synthesis.

Dietmar Seyferth, Professor; Ph.D., Harvard, 1955. Synthetic organometallic chemistry, main-group–transition-metal carbonyl clusters, acylmetal compounds (preparation and use in synthesis), preceramic polymer synthesis.

Robert J. Silbey, Professor; Ph.D., Chicago, 1965. Physical chemistry, quantum mechanics, statistical mechanics.

Jeffrey I. Steinfeld, Professor; Ph.D., Harvard, 1965. Physical chemistry; molecular spectroscopy; energy transfer and relaxation processes; chemical applications of lasers, including laser-induced surface photochemistry and atmospheric chemistry.

JoAnne Stubbe, Professor; Ph.D., Berkeley, 1971. Biochemistry; mechanisms of enzyme reactions: ribonucleotide reductases, purine biosynthetic enzymes, and α-ketoglutarate dioxygenases; mechanisms of DNA cleavers: bleomycin; design of mechanism-based inhibitors.

Steven R. Tannenbaum, Professor; Ph.D., MIT, 1962. Chemical carcinogenesis, analytical chemistry of carcinogens and their adducts.

John S. Waugh, Professor; Ph.D., Caltech, 1953. Physical chemistry, magnetic resonance, low temperatures.

James R. Williamson, Assistant Professor; Ph.D., Stanford, 1988. Bioorganic and biophysical chemistry, nucleic acid structure and function, telomeric DNA and catalytic RNA, NMR studies of nucleic acids.

Gerald N. Wogan, Professor; Ph.D., Illinois, 1957. Chemical carcinogenesis and naturally occurring carcinogens.

Mark S. Wrighton, Professor; Ph.D., Caltech, 1972. Inorganic chemistry, photoprocesses in metal-containing molecules, catalysis, photoelectrochemistry, surface chemistry, molecular electronics.

Douglas C. Youvan, Assistant Professor; Ph.D., Berkeley, 1981. Bioenergetics: genetic and biophysical analyses of the light reactions of photosynthesis; protein chemistry: folding and assembly of membrane proteins with prosthetic groups; spectroscopy/molecular biology interfaces: applications of imaging spectroscopy to chromophoric problems in biotechnology.

SECTION 2: CHEMISTRY

NEW YORK UNIVERSITY
Department of Chemistry

Program of Study

The Department of Chemistry offers the Ph.D. degree. The chemistry of the components of living systems is a major focus of the department; the faculty and students in all subdisciplines of chemistry at NYU work on problems relevant to the molecules of life. Spectroscopy, catalysis, organic mechanisms, and quantum calculations are additional areas of active research in which a student can pursue a degree.

The program is designed to train independent research scientists who can assume positions in teaching and research settings. These include universities; colleges; the chemical, pharmaceutical, and biotechnology industries; and public and private research institutions. A student participates in the seminars and journal clubs of one of three groups in the department: biomolecular chemistry (biophysical or bioorganic), organic chemistry, or physical chemistry (experimental, theoretical, or materials). The first year of study is typically spent in course work and in rotation through several research laboratories. Competence in the major field is established by a qualifying examination at the end of the first year, after which dissertation research is normally begun.

Research Facilities

The major equipment in the chemistry department includes a rotating anode X-ray diffractometer, a scanning tunneling microscope with a high-vacuum specimen-coating device, a state-of-the-art laser laboratory for fast time-resolved studies of chemical processes, an electrospray mass spectrometer, and a computational center. The department also has two 300-MHz NMR spectrometers, peptide and DNA synthesizers, a CD spectropolarimeter, three FT-IR spectrometers, photoacoustic and photothermal beam deflection spectrometers, microcalorimeters, and equipment for UV, ESR, fluorescence, and IR spectrometry. The W. M. Keck Laboratories for Biomolecular Imaging within the department have been completed recently.

The Academic Computing Center has, among other machines, high-end VAX's and a Convex vector machine. Within the department there are a large number of scientific workstations, a Stardent Graphics minisupercomputer, IBM RISC 6000s, MicroVAX's, and an Evans & Sutherland molecular modeling facility. Bobst Library is one of the largest open-stack research libraries in the nation.

Financial Aid

All students admitted to the Ph.D. program receive financial support in the form of teaching fellowships, research assistantships, or University fellowships. Students usually receive support for the duration of their studies. The basic stipend for the 1991–92 nine-month academic year is $12,000. Summer appointments as research or teaching assistants are available at comparable rates. Health insurance is included in the compensation package.

Cost of Study

Research and teaching appointments carry a waiver of tuition, which is $10,200, including fees, for 1991–92.

Cost of Living

It is estimated that the standard budget for a single student for the academic year 1991–92 (exclusive of tuition) is approximately $9500. This includes room and board, books, fees, and personal expenses. University, as well as privately owned, housing is available.

Student Group

The Graduate School of Arts and Science has an enrollment of 4,000 graduate students; about 85 are pursuing advanced degrees in chemistry. They represent a wide diversity of ethnic and national groups; over a third are women. Upon receiving their Ph.D.'s, about 10 percent of recent graduates entered positions in academia, and the others were approximately equally divided between those who accepted industrial employment and those who elected to gain postdoctoral research experience before accepting permanent positions.

Location

Greenwich Village, the home of the University, has long been famous for its contributions to the fine arts, literature, and drama and for its personalized, smaller-scale, European style of living. It is one of the most desirable places to live in the city. New York City is the business, cultural, artistic, and financial center of the nation, and its extraordinary resources enrich both the academic programs and the experience of living at NYU.

The University

New York University, a private university, awarded its first doctorate in chemistry in 1866. Ten years later, the American Chemical Society was founded in the original University building, and the head of the chemistry department, John W. Draper, assumed the presidency.

Applying

Application forms may be obtained by writing to the address below. Students beginning graduate study are accepted only for September admission. Applicants are expected to submit scores on the GRE General Test and the Subject Test in chemistry or a related discipline. Students whose native language is not English must submit a score on the Test of English as a Foreign Language (TOEFL). Applicants are invited to visit the University but are advised to contact the department beforehand to arrange an appointment.

Correspondence and Information

Director of Graduate Programs
Department of Chemistry
New York University
New York, New York 10003
Telephone: 212-998-8400
Fax: 212-260-7905

SECTION 2: CHEMISTRY

New York University

THE FACULTY AND THEIR RESEARCH

Zlatko Bacic, Assistant Professor; Ph.D., Utah, 1982. Theoretical chemistry: spectra and dynamics of highly vibrationally excited floppy molecules, dissociation dynamics of rare-gas clusters in collisions with solid surfaces.

Henry C. Brenner, Associate Professor; Ph.D., Chicago, 1972. Physical chemistry: effects of high pressure on electronic energy transfer in organic crystals, triplet energy transfer in molecular crystals, magnetic resonance of nucleic acids.

Edwin C. Campbell, Associate Professor; Ph.D., Berkeley, 1951. Theoretical chemistry: additive and nonadditive models for the energetics of hydrogen-bonded interactions in water.

James W. Canary, Assistant Professor; Ph.D., UCLA, 1988. Organic and bioorganic chemistry: synthetic receptors for complexation and catalysis, metalloprotein mimics, biomimetic molecular devices.

Michael L. Connolly, Research Assistant Professor; Ph.D., Berkeley, 1981. Theoretical biophysical chemistry: differential geometrical computation of structural properties of proteins of known three-dimensional structure.

Scott Courtney, Research Assistant Professor; Ph.D., Chicago, 1987. Time-resolved laser spectroscopic studies of condensed phase reactions.

Joel M. Friedman, Professor; Ph.D./M.D., Pennsylvania, 1975. Biophysical chemistry: use of time-resolved laser spectroscopic techniques to probe structure, function, and dynamics in biological macromolecules.

Paul J. Gans, Professor; Ph.D., Case Institute of Technology, 1959. Theoretical chemistry: determination of conformational and thermodynamic properties of macromolecules by Monte Carlo simulation.

Nicholas E. Geacintov, Professor; Ph.D., Syracuse, 1961. Physical and biophysical chemistry: interaction of polycyclic aromatic carcinogens with nucleic acids, laser studies of fluorescence mechanisms and photoinduced electron transfer.

Alvin D. Joran, Assistant Professor; Ph.D., Caltech, 1986. Organic chemistry: synthesis and physical characterization of peptides, porphyrins, and hybrid molecules that model electron transfer, enzyme catalysis, and energy transfer.

Neville R. Kallenbach, Professor; Ph.D., Yale, 1961. Biophysical chemistry of proteins and nucleic acids: structure, sequence, and site selectivity in DNA-drug interactions, protein folding, model helix and beta sheet structures.

Alvin I. Kosak, Professor; Ph.D., Ohio State, 1948. Organic chemistry: heterocyclic chemistry, polynuclear hydrocarbons.

Manfred J. D. Low, Professor; Ph.D., NYU, 1958. Physical chemistry: infrared studies of surface groups and reactions of carbons.

Luis A. Marky, Research Associate Professor; Ph.D., Rutgers, 1981. Physical and biophysical chemistry: thermodynamics of biomolecules and ligand interactions, hydration of nucleic acids as a function of sequence and conformation.

Edward J. McNelis, Professor; Ph.D., Columbia, 1960. Organic chemistry: oxidation as a route to synthetically useful substances, novel organometallic catalysts.

Jules W. Moskowitz, Professor; Ph.D., MIT, 1961. Theoretical chemistry: Monte Carlo methods applied to quantum chemistry.

Randall B. Murphy, Associate Professor; Ph.D., UCLA, 1975. Biophysical chemistry: chemistry of cell-surface receptors.

Louise Pape, Research Assistant Professor; Ph.D., Columbia, 1985. Transcriptional activation of the *S. pombe* ribosomal RNA genes.

Yorke E. Rhodes, Associate Professor; Ph.D., Illinois, 1964. Organic chemistry: alkyl group migrations and rearrangement mechanisms, interaction of cyclopropane with neighboring cationic centers, organic astrochemistry.

Tamar Schlick, Assistant Professor; Ph.D., NYU, 1987. Development of multivariate minimization and molecular dynamics algorithms and their applications to DNA structures.

David I. Schuster, Professor; Ph.D., Caltech, 1961. Organic and bioorganic chemistry: photochemistry of enones, effects of paramagnetic reagents on photocycloadditions, time-resolved photoacoustic calorimetry, photoaffinity labeling of neuroreceptors.

David C. Schwartz, Assistant Professor; Ph.D., Columbia, 1985. Physics and biology of large single DNA molecules, with applications to genomic analysis.

Steven D. Schwartz, Research Assistant Professor; Ph.D., Berkeley, 1984. Theoretical biophysical chemistry; quantum mechanical modeling of tertiary relaxation in hemoglobin and its effect on ligand reactivity.

Thomas W. Scott, Associate Professor; Ph.D., Cornell, 1981. Optical studies of the chemical and physical properties of condensed phase systems.

Nadrian C. Seeman, Professor; Ph.D., Pittsburgh, 1970. Biophysical chemistry: structural chemistry of recombination; design of nanometer-scale geometrical objects and devices from branched DNA; catenated and knotted DNA topologies; crystallography.

Robert Shapiro, Professor; Ph.D., Harvard, 1959. Organic and bioorganic chemistry: effects of mutagens on the structure and function of nucleic acids.

Benson R. Sundheim, Professor; Ph.D., Johns Hopkins, 1950. Physical and theoretical chemistry: computer simulations and statistical mechanics of transport processes, molecular dynamics of the interaction of photoemitted electrons with halocarbons.

Graham R. Underwood, Associate Professor; Ph.D., Melbourne, 1966. Organic and bioorganic chemistry: synthesis and study of carcinogen-DNA adducts, reactions at centers other than carbon.

Marc A. Walters, Assistant Professor; Ph.D., Princeton, 1981. Inorganic chemistry: kinetics and energetics of metal thiolate and selenolate formation, synthesis of complexes for the uptake of many-electron equivalents.

Stephen Wilson, Professor; Ph.D., Rice, 1972. Organic chemistry: total synthesis of natural products, new synthetic methodology, synthesis of enzyme mimics.

Donald J. Wink, Assistant Professor; Ph.D., Harvard, 1985. Inorganic chemistry: metal-assisted chiral synthesis, organic chemistry of chromium-olefin complexes, novel palladium hydrogenation catalysts.

John Z. H. Zhang, Assistant Professor; Ph.D., Houston, 1987. Theory study of molecular collision dynamics and chemical reactions in the gas phase and on surfaces.

SECTION 2: CHEMISTRY

NORTHEASTERN UNIVERSITY

Department of Chemistry

Programs of Study

The Department of Chemistry offers the M.S. and Ph.D. degrees with majors in physical, inorganic, analytical, and organic chemistry. In addition, various interdisciplinary doctoral programs can be arranged. The M.S. and Ph.D. thesis programs may be pursued only on a full-time basis while in residence. The principal requirement for the doctorate is a dissertation that presents publishable research results. A thesis director, who is chosen by the graduate student during his or her first year of residence, helps to design a program that will fulfill the student's academic needs and professional objectives.

Students entering with a bachelor's degree are normally required to take a minimum of twelve courses. These courses, each of which is one academic quarter in duration, are selected by the student with the aid of an academic adviser. In addition, doctoral candidates must pass a series of written qualifying examinations in the field of the student's interest. These are normally taken during the second year of residence, and students must pass four out of the eight examinations. There are no formal course requirements for students entering with a master's degree, and they take the qualifying examinations during their first year of residence.

A special option in the doctoral program is the Doctoral Internship Program, which allows selected second-year students to spend up to fifteen months in full-time employment at a government or industrial research facility in the Greater Boston area while maintaining graduate student status.

Research Facilities

An 80,000-square-foot, air-conditioned chemistry building houses most of the department's research laboratories, the departmental library, and most of its instructional facilities. The research instrumentation includes X-ray diffraction equipment, time-correlated single-photon spectrometers, excimer and Nd:yag lasers, dye lasers, a laser-Raman spectrometer, Mössbauer spectrometers, nuclear magnetic resonance equipment, mass spectrometers, gas chromatographs, high-pressure liquid chromatographs, high-performance capillary electrophoresis equipment, plasma emission spectrometers, and scanning electron microscopes. The University's computation facilities include a VAX 8650 computer. Additional research equipment is housed on campus in the Barnett Institute of Chemical Analysis and Materials Science. A high-speed data network links users and facilities on the central campus and on three satellite campuses. The campus network is also connected via the global Internet to computing resources around the world. At the University, students have access to Digital VAX systems, labs of microcomputers, a computer mail-and-conferencing system, and an array of specialized computing equipment.

Financial Aid

Northeastern University awards need-based financial aid to graduate students through the Perkins Loan, College Work-Study, and Stafford Loan programs. The University offers a limited number of minority fellowships and Martin Luther King Jr. Scholarships.

The graduate schools also offer financial assistance through teaching, research, and administrative assistantship awards, which include tuition remission and a stipend typically ranging between $7900 and $9000 (departmentally specific). These assistantships require a maximum of 20 hours of work per week. Also available are a limited number of tuition assistantships, which provide partial or full tuition remission and require a maximum of 10 hours of work per week.

The teaching assistantship stipend is $8900 for nine months in 1991–92. Students in good standing have the opportunity for full-year support at $11,870 per annum. A score of at least 580 in the Test of English as a Foreign Language is required for all teaching assistants from overseas.

Cost of Study

The cost of tuition for the 1991–92 academic year in the Graduate School of Arts and Sciences is $285 per quarter hour of credit. Where applicable, special tuition charges are made for teaching, practicums, and fieldwork. Other charges include the Student Center fee ($12.50 per quarter) and health and accident insurance fee ($450) required of all full-time students. There is an International Student fee of $100.

Cost of Living

On-campus living expenses are estimated at $620 per month, with on-campus housing available on a limited basis to newly accepted students. A referral service is available from the Office of Residential Life, which provides lists of apartments, rooms, and potential roommates. Off-campus living expenses are estimated at $1000 per month. A public transportation system services the Greater Boston area, and there are subway and bus services convenient to the University.

Student Group

Approximately 33,680 students are enrolled at Northeastern University; they represent a wide variety of academic, professional, geographic, and cultural backgrounds. The Graduate School of Arts and Sciences has approximately 850 students, 65 percent of whom attend on a full-time basis. In fall 1990, 48 full-time and 32 part-time students enrolled in the graduate chemistry program.

Location

Boston, the capital city of Massachusetts, offers students unusual academic, cultural, and recreational opportunities. In addition to the abundant resources available within Northeastern University, there are those of the other educational and cultural institutions of the Greater Boston area. Boston is a mixture of Colonial tradition and modern America, and is home to people of every intellectual, political, economic, racial, ethnic, and religious background. It is a place where the past is appreciated, the present enjoyed, and the future anticipated.

The University

Founded in 1898, Northeastern University is a privately endowed nonsectarian institution of higher learning that is one of the largest private universities in the country. Today, Northeastern University has eleven undergraduate schools and colleges, ten graduate and professional schools, several suburban campuses, and an extensive research division.

Applying

Applicants should direct inquiries to the address given below. Application forms, reference blanks, and a copy of the Graduate School bulletin will be mailed to the applicant. These forms, together with transcripts and the results of the Test of English as a Foreign Language (when appropriate), should be returned to the department as early as possible.

Correspondence and Information

Director, Graduate Affairs
Department of Chemistry-P9
Northeastern University
Boston, Massachusetts 02115
Telephone: 617-437-2383

Northeastern University

THE FACULTY AND THEIR RESEARCH

Geoffrey Davies, Distinguished Professor; Ph.D., D.Sc., Birmingham. Inorganic chemistry, catalytic activation of small abundant molecules, transmetalation chemistry.
David A. Forsyth, Professor; Ph.D., Berkeley. Organic chemistry, NMR spectroscopy.
Bill C. Giessen, Professor; Dr.Sci.Nat., Göttingen. Physical and inorganic chemistry, materials science, forensic chemistry.
Thomas R. Gilbert, Associate Professor; Ph.D., MIT. Environmental analysis, plasma spectroscopy.
David J. Jebaratnam, Assistant Professor; Ph.D., Stanford. Organic chemistry, bioorganic chemistry, DNA cleavage, synthetic methodology.
Barry L. Karger, James L. Waters Professor; Ph.D., Cornell. Analytical chemistry, analytical biotechnology, chromatography, capillary electrophoresis.
Lutfur R. Khundkar, Assistant Professor; Ph.D., Caltech. Physical chemistry, piosecond and femtosecond spectroscopy, electron transfer.
Rein U. Kirss, Assistant Professor; Ph.D., Wisconsin–Madison. Inorganic chemistry, organosilicon materials, organometallic complexes.
Ira S. Krull, Associate Professor; Ph.D., NYU. Analytical chemistry; trace organic and inorganic analysis; detectors for GC, LC, and flow injection analysis; liquid chromatography–electrochemistry.
Philip W. Le Quesne, Professor; Ph.D., D.Sc., Auckland. Organic chemistry, plant chemistry, organic synthesis.
Patricia A. Mabrouk, Assistant Professor; Ph.D., MIT. Analytical chemistry, electrochemistry of biomolecules, surface phenomena.
Kay D. Onan, Associate Professor; Ph.D., Duke. Structural chemistry, crystallography.
Mary J. Ondrechen, Professor; Ph.D., Northwestern. Physical chemistry, chemical physics.
William M. Reiff, Professor; Ph.D., Syracuse. Inorganic chemistry, magnetochemistry, Mössbauer spectroscopy.
John L. Roebber, Professor; Ph.D., Berkeley. Physical chemistry, excited states, spectroscopy.
Alfred Viola, Professor; Ph.D., Maryland. Organic chemistry, pericyclic reaction mechanisms.
Paul Vouros, Professor; Ph.D., MIT. Analytical chemistry, mass spectrometry (LC/MS, GC/MS).
Philip M. Warner, Professor and Chair; Ph.D., UCLA. Organic chemistry, organolithium species, carbenes and carbenoids, strained double bonds and ring systems.
Robert N. Wiener, Associate Professor; Ph.D., Pennsylvania. Physical chemistry.
Lawrence D. Ziegler, Associate Professor; Ph.D., Cornell. Laser spectroscopy, chemical physics.

Adjunct Professors

Robert N. Hanson, Professor of Medicinal Chemistry; Ph.D., Berkeley. Organic chemistry, synthesis of steroid derivatives, organotin derivatives, radiolabeling methodology.
John L. Neumeyer, Professor of Medicinal Chemistry; Ph.D., Wisconsin–Madison. Structure modification of natural products, anticancer chemotherapy, chemistry of CNS agents, receptor interactions.

Edward Hurtig Hall, the chemistry building.

OHIO STATE UNIVERSITY
Department of Chemistry

Programs of Study

Graduate courses and research programs leading to the M.S. and Ph.D. degrees in chemistry are offered in analytical, biological, inorganic, organic, physical, and theoretical chemistry. These programs include photochemistry; stereochemistry, electrochemistry; kinetics, including nanosecond and crossed-molecular-beam studies; theoretical structure and dynamics; statistical mechanics; organic synthesis; inorganic synthesis; carbohydrate chemistry; NMR, ESR, laser, and vacuum UV spectroscopy; pulse radiolysis; X-ray structures; multienzyme complexes; catalysis; mechanisms of action of enzymes and coenzymes; molecular biology; biomembrane studies; surface chemistry; and separations.

The first year is devoted mainly to advanced course work, with the opportunity to begin research in the latter part of the year. During the second and subsequent years, emphasis is on research for both M.S. and Ph.D. students. Ph.D. students begin their examinations for admission to Ph.D. candidacy in their second year. These examinations include both written and oral portions; they are designed to verify the student's competence as an independent scientist.

All M.S. and Ph.D. research is carried out under the direct supervision of a faculty adviser who serves as the student's preceptor. Many research groups are enriched by the presence of postdoctoral associates and visiting professors.

Research Facilities

The department is housed in four buildings that include space for research laboratories, offices, and classrooms. The department is well equipped for modern chemical research with multinuclear superconducting FT-NMR spectrometers and numerous other NMR spectrometers, several EPR spectrometers, an automated and computerized X-ray crystallography laboratory, a linear electron accelerator for pulse radiolysis studies, and a large complement of standard IR- and UV-visible spectrometers in individual research groups. The Laser Spectroscopy Facility has over twenty state-of-the-art lasers and spectrometer systems and a wide variety of support equipment. Modern separation equipment, including apparatus for GLC, HPLC, and electrophoretic separations, is available in various laboratories. The department employs a staff of technicians, engineers, and scientists who support the research programs by maintaining the department's equipment and operating the advanced NMR and mass spectroscopy laboratories. A major Chemical Instrumentation Center housing MS and NMR spectrometers also serves the department.

Financial Aid

Fellowships and teaching associateships are granted to applicants on a competitive basis. Teaching associateships require 8 contact hours of service per week for three academic quarters per year and provide a twelve-month stipend of $12,000 or more, depending upon qualifications, plus all tuition and fees. Many advanced students are supported as research associates, with stipends paid from their faculty adviser's research grants.

Cost of Study

Except in special circumstances, every student admitted is provided financial assistance that includes all tuition and fees with the exception of the original $50 acceptance fee.

Cost of Living

The University operates graduate residence halls and low-cost apartments for married students. Room and board contracts in residence halls range from $1185 to $1225 per quarter. The overall cost of living in the area is at the national average.

Student Group

The department enrolls about 275 graduate students with undergraduate degrees from colleges and universities all over the United States and in a number of other countries. All students in good academic standing receive financial aid. Since decisions on admission and financial aid are made on the basis of qualifications alone and without reference to race, sex, or national origin, all groups are well represented among the students. Graduates are employed by industrial and government laboratories and as research and teaching staff members at colleges and universities across the United States.

Location

The central campus of the University is located in Columbus, which offers the cultural opportunities normally found in major cities. Museums, active theaters, and an excellent orchestra provide some of the cultural flavor of Columbus. There are numerous sports events year-round that attract large followings, including professional minor-league baseball and soccer.

The University

The Ohio State University is a land-grant university and the major comprehensive university in Ohio. It enrolls over 50,000 students, of whom about 14,000 are in the graduate and professional schools. It encompasses twenty-seven undergraduate, graduate, and professional colleges and schools that provide a comprehensive range of educational opportunities. The Department of Chemistry is one of seven departments in the College of Mathematical and Physical Sciences.

Applying

Applications for admission in the autumn quarter and for financial aid in the form of teaching associateships should be submitted by February 1. Applications for admission are considered on the basis of scholastic merit. All applicants are required to submit GRE scores, and applicants whose native language is not English must also submit TOEFL scores.

Correspondence and Information

Barbara S. Cassity
Coordinator of Graduate Program
Department of Chemistry
The Ohio State University
120 West 18th Avenue
Columbus, Ohio 43210
Telephone: 614-292-8917

Ohio State University

THE FACULTY AND THEIR RESEARCH

Larry B. Anderson, Ph.D., Syracuse, 1964. Analytical chemistry: microscale electrochemistry, solar cells.

Charles Bender, Ph.D., Washington (Seattle), 1968. Theoretical chemistry: development and utilization of advanced computational methods in science.

Lawrence J. Berliner, Ph.D., Stanford, 1967. Magnetic resonance studies of structure and function in proteases, blood-clotting enzymes, and lactose snythetase complex; in vivo ESR imaging.

Bruce E. Bursten, Ph.D., Wisconsin, 1978. Inorganic and organometallic chemistry: correlation of theoretical and experimental data for transition-metal complexes.

Martin Caffrey, Ph.D., Cornell, 1982. Biochemistry: time-resolved X-ray diffraction of model membranes.

Matthew R. Callstrom, Ph.D., Minnesota, 1987. Organometallic catalysis, X-ray photoelectron and Auger spectroscopy.

Carolyn Carter, Ph.D., Purdue, 1987. Chemical education: how individuals develop understandings of chemical concepts and solve chemistry problems.

James V. Coe, Ph.D., Johns Hopkins, 1986. Generation and studies of molecular clusters with sub-Doppler spectroscopy.

James A. Cowan, Ph.D., Cambridge, 1986. Inorganic, bioinorganic, and bioorganic chemistry: use of synthesis, spectroscopy, and biology to explore problems at interface of chemistry and biology.

Anthony W. Czarnik, Ph.D., Illinois, 1981. Organic and bioorganic chemistry: artificial enzymes, catalysis.

Ross E. Dalbey, Ph.D., Washington State, 1983. Biochemistry, molecular biology: membrane protein structures, membrane targeting, membrane bioenergetics.

Prabir K. Dutta, Ph.D., Princeton, 1978. Heterogeneous catalysis, spectroscopic analysis of surfaces and zeolites.

Arthur Epstein, Ph.D., Pennsylvania, 1971. Chemical physics: characterization of conducting, superconducting, and magnetic properties of molecular, polymeric, and ceramic materials.

Richard F. Firestone, Ph.D., Wisconsin, 1954. Physical chemistry: kinetics and mechanism of fast reactions induced by pulsed beams of high-energy electrons.

Gideon Fraenkel, Ph.D., Harvard, 1957. Organic chemistry: use of NMR to study kinetics of inversion, carbon-metal bond exchange, and base-metal exchange in organometallic compounds.

Patrick Gallagher, Ph.D., Wisconsin, 1954. Inorganic chemistry and characterization, using thermal analysis, assorted spectroscopies, and X-ray diffraction.

Roger E. Gerkin, Ph.D., Berkeley, 1960. Physical chemistry: EPR spectroscopy of excited-state and ground-state paramagnetic species at low and high field strengths and cryogenic to ambient temperatures.

Terry Gustafson, Ph.D., Purdue, 1979. Analytical chemistry: structure and dynamics of photogenerated transient species in fluid solutions and in thin films.

David J. Hart, Ph.D., Berkeley, 1976. Organic chemistry: syntheses of terpenoids, alkaloids, and amino acid–based antibiotics; asymmetric syntheses.

Derek Horton, Ph.D., Birmingham (England), 1957. Organic and biological chemistry: synthesis, structure, and reactivities of carbohydrates and other natural products.

Marita M. King, Ph.D., South Florida, 1981. Biochemistry: enzymology of cyclic interconverting systems, regulation of protein kinases and phosphatases.

Michael H. Klapper, Ph.D., Berkeley, 1964. Biochemistry: theoretical and experimental aspects of structure-function relationships in enzymes, physiological studies with lower plants.

Daniel L. Leussing, Ph.D., Minnesota, 1953. Analytical and bioinorganic chemistry: kinetics and mechanisms of reactions of metal-ion coordination compounds, fast reactions, metal ions in biological systems.

Alan G. Marshall, Ph.D., Stanford, 1969. Analytical and biological spectroscopy, spectroscopic determination of molecular structure and flexibility of important biopolymers.

C. Weldon Mathews, Ph.D., Vanderbilt, 1965. High-resolution molecular spectroscopy of transients in the UV and IR.

Richard L. McCreery, Ph.D., Kansas, 1974. Analytical chemistry: electrochemistry, electronic and vibrational spectral probes of the electrode surface and adjacent solution layer.

C. William McCurdy Jr., Ph.D., Caltech, 1975. Physical chemistry: theoretical molecular dynamics.

Terry A. Miller, Ph.D., Cambridge, 1968. Physical chemistry: spectroscopic detection and characterization of short-lived molecular species, using pulsed-dye lasers to provide wide spectral coverage and sensitivity.

Susan V. Olesik, Ph.D., Wisconsin–Madison, 1982. Supercritical fluids and their use in separation techniques of complex mixtures of large molecules.

Robert J. Ouellette, Ph.D., Berkeley, 1962. Organic chemistry: mechanisms of oxidation of organic compounds by metal ions, conformational analysis of organometallic compounds by NMR.

Leo A. Paquette, Ph.D., MIT, 1959. Organic chemistry: synthesis and reactivity of unusual alicyclic and heterocyclic molecules, natural products chemistry, molecular rearrangements, stereoelectronic control, and physical-organic chemistry.

John M. Parson, Ph.D., Chicago, 1971. Characterization of reactive and inelastic collisions by molecular-beam techniques.

Russell M. Pitzer, Ph.D., Harvard, 1963. Physical chemistry: quantum chemistry, use of symmetry in the computation of molecular electronic structure.

Matthew S. Platz, Ph.D., Yale, 1977. Organic chemistry: kinetics and spectroscopy of radicals, biradicals, carbenes, and nitrenes; reactions of these species in intramolecular rearrangements and the synthesis of strained molecules.

Viresh H. Rawal, Ph.D., Pennsylvania, 1986. Organic chemistry: synthesis, new reaction for total synthesis of biologically important natural products, biomimetic chemistry.

Eugene P. Schram, Ph.D., Purdue, 1962. Inorganic chemistry: preparation and reactivity of inorganic heterocycles and clusters, catalytic processes involving functional group transfer and oxidative addition.

Isaiah Shavitt, Ph.D., Cambridge, 1957. Physical chemistry: quantum mechanical study of molecular structure, development of rigorous theoretical methods.

Harold Shechter, Ph.D., Purdue, 1946. Organic chemistry: intramolecular reactions of carbenes, nitrenes, and their respective cationic derivatives; syntheses of peri, single atom-bridged arenes.

Sheldon G. Shore, Ph.D., Michigan, 1957. Inorganic chemistry: structures and reactions of nonmetal and transition-metal cluster systems.

Sherwin J. Singer, Ph.D., Chicago, 1983. Theory of molecular liquids and solids, solvent effects on reaction rates and phase transitions.

Muttaiya Sundralingam, Ph.D., Pittsburgh, 1961. Biological macromolecular crystallography, molecular biophysics.

John S. Swenton, Ph.D., Wisconsin, 1965. Organic chemistry: synthetic and mechanistic organic photochemistry and electrochemistry, synthesis of natural products.

Ming-Daw Tsai, Ph.D., Purdue, 1978. Biochemistry: mechanisms of biochemical reactions and interactions involving phosphates, using stereochemistry and NMR.

Andrew Wojcicki, Ph.D., Northwestern, 1960. Synthetic and mechanistic organo-transition-metal chemistry.

Robin Ziebarth, Ph.D., Iowa State, 1987. Inorganic chemistry: synthesis and characterization of solid-state compounds.

SECTION 12: CHEMISTRY

OREGON STATE UNIVERSITY

Department of Chemistry

Programs of Study

The Department of Chemistry offers programs of study leading to the M.S. and Ph.D. degrees in the following areas of chemistry: analytical, inorganic, nuclear and radiation, organic, and physical.

Forty-five credit hours are required for the master's degree, including both course work and thesis research. No minimum hours are required for the Ph.D. degree; however, three years of full-time graduate work beyond the baccalaureate must be done, and each area of specialization has a core program that is required for students majoring in that area. Students select a course of study in conjunction with their major professor and with the approval of the doctoral committee. The program of study is designed to meet the needs and interests of the individual student. During the first year, students studying for either degree take two core courses in each of three areas chosen by them from the above-mentioned chemistry subdisciplines plus biochemistry and physics. During the first term, each student selects an adviser and begins research. In addition, during the first or second year, depending on the area of specialization, Ph.D. students begin taking a series of cumulative examinations on advanced topics. After successful completion of the cumulative exams and an oral exam, Ph.D. students complete research and present an oral defense of the written thesis.

The department offers an excellent training program for beginning teaching assistants.

Research Facilities

The Department of Chemistry is equipped with five nuclear magnetic resonance spectrometers, including a superconducting-magnet Bruker AM 400 Fourier-transform instrument with proton, ^{13}C, and multinuclear probe capability; facilities for far-infrared, infrared, visible, ultraviolet, vacuum ultraviolet, Raman (including CARS), electron, nuclear, plasma emission, and atomic absorption spectroscopy; X-ray and electron diffraction units; CW and pulsed tunable lasers for linear and nonlinear spectroscopy; ultrahigh vacuum surface spectroscopies; MS-50 and other mass spectrometers; gas chromatographs and high-performance liquid chromatographs; a well-equipped crystal-growing laboratory; a Triga III nuclear research reactor (at the Radiation Center); and a large variety of electronic instrumentation for both teaching and research. The University has several mainframe computers, and the department has a VAX and many microcomputers that are used extensively in both research and teaching. Well-equipped electronics, glass, and machine shops in the department provide for maintenance and construction of equipment.

Financial Aid

Graduate teaching assistantships are available with stipends of $11,506 for the calendar year 1991–92. Teaching assistants are exempt from tuition and fees, except for health service and incidental fees of $150 per term. Research assistantships are also available to students, normally after the first year. These carry approximately the same stipend as teaching assistantships and include a tuition waiver, except for the incidental fees of $150 per term.

Student loans are offered through the University student loan program, and opportunities are frequently available for employment of family members on campus.

Cost of Study

Tuition and fees for full-time graduate students are $1178 per quarter for Oregon residents and $1854 for nonresidents in 1991–92. Oregon residency can usually be obtained after the first year. Tuition for teaching and research assistants is waived except for the incidental fees of $150 per quarter. There is an application fee of $40, and a microfilming fee of $45 per thesis is also required. (Tuition and fees are subject to change.)

Cost of Living

Housing for unmarried students is available in the dormitories; the average cost is about $3050 for room and board for the 1991–92 academic year. Married student housing is available with rents ranging from $185 to $215 per month. Early application to the housing office is recommended. A wide variety of off-campus housing may also be found at rates of $300 to $550 per month.

Student Group

Current enrollment is 13,241 undergraduate students and 2,783 graduate students. The department has 68 graduate students, about 80 percent of whom are working toward the Ph.D.

Location

Corvallis, a city with a population of 43,713, is situated in the heart of the Willamette Valley. It has a temperate climate the year round—sunny, mild summers and snow-free winters. Corvallis offers an ideal location—Portland lies an hour and a half's drive to the north, the Cascade Mountains an hour's drive to the east, and the scenic Oregon coast an hour's drive to the west. Thus, the opportunities offered by a large metropolitan area and the recreational opportunities offered by the Cascades and the Pacific are readily available. The latter include hunting, fishing, mountain climbing, skiing, camping, hiking, picnicking, waterskiing, and all types of boating. The city of Corvallis itself has excellent parks and bicycle trails.

The University and The Department

Oregon State University was designated as Oregon's land-grant university in 1868. It is located on a beautiful 400-acre campus a few blocks from the center of Corvallis. The University provides many cultural attractions throughout the year.

In addition to maintaining its undergraduate teaching responsibilities, the chemistry department has developed a strong graduate program, with research activities under way in most aspects of chemistry. Research and teaching facilities are excellent.

Applying

Applications for admission and for research and teaching assistantships may be obtained from the department. There are no application deadlines, but early application is very helpful. Most assistantship offers are made by April 1, but offers may be made throughout the year if positions are available.

Oregon State University is an Equal Opportunity/Affirmative Action employer and complies with Section 504 of the Rehabilitation Act of 1973. Women and members of minority groups are especially encouraged to apply.

Correspondence and Information

Dr. Carroll W. DeKock, Chair
Department of Chemistry
Oregon State University
Gilbert Hall 153
Corvallis, Oregon 97331-4003
Telephone: 503-737-2081

Peterson's Guide to Graduate Programs in the Physical Sciences and Mathematics 1992

SECTION 12: CHEMISTRY

Oregon State University

THE FACULTY AND THEIR RESEARCH

Malcolm Daniels, Professor; Ph.D., Durham, 1955. Photochemistry, radiation chemistry, fast reactions, fluorescence, phosphorescence, time-resolved spectroscopy of excited states of DNA, low-level actinide speciation.

Carroll W. DeKock, Professor and Chairman, Department of Chemistry; Ph.D., Iowa State, 1965. Inorganic chemistry: transition-metal chemistry, metal oxide chemistry.

Glenn T. Evans, Professor; Ph.D., Brown, 1973. Theoretical physical chemistry, dynamics of simple fluids and polymers.

Peter K. Freeman, Professor; Ph.D., Colorado, 1958. Organic chemistry: reactive intermediates in organic chemistry, photochemistry, environmental chemistry.

Kevin P. Gable, Assistant Professor; Ph.D., Cornell, 1987. Organic chemistry: the mechanistic chemistry of organo–transition-metal compounds.

Gerald Jay Gleicher, Professor; Ph.D., Michigan, 1963. Physical-organic chemistry: free-radical reactions, structure-reactivity relationships.

Steven J. Gould, Professor; Ph.D., MIT, 1970. Organic chemistry: biosynthesis, synthesis and structure of natural products; enzyme reaction mechanisms; NMR spectroscopy; molecular genetics of antibiotic biosynthesis.

Stephen J. Hawkes, Professor; Ph.D., London, 1963. Analytical chemistry: chromatography.

Kenneth W. Hedberg, Professor Emeritus; Ph.D., Caltech, 1948. Physical chemistry: structures, dynamics, and vibrational force fields of gas molecules; internal motion in molecules.

Frederick H. Horne, Professor and Dean, College of Science; Ph.D., Kansas, 1962. Theoretical physical chemistry: nonequilibrium thermodynamics, statistical mechanics, thermal diffusion, membrane transport.

James D. Ingle, Professor; Ph.D., Michigan State, 1971. Analytical chemistry: kinetic methods; flow injection analysis, chemiluminescence and fluorescence, chemical speciation; fiber-optic sensors for environmental studies.

Douglas A. Keszler, Associate Professor; Ph.D., Northwestern, 1984. Inorganic chemistry: synthesis and characterization of solid-state materials.

James H. Krueger, Professor; Ph.D., Berkeley, 1961. Inorganic chemistry: kinetics and mechanisms of inorganic reactions, transition-metal–sulfur amino acid complexes.

Michael M. Lerner, Assistant Professor; Ph.D., Berkeley, 1988. Inorganic chemistry: synthesis and characterization of solid-state materials, polymer electrolytes; electrosynthesis.

John G. Loeser, Assistant Professor; Ph.D., Harvard, 1984. Theoretical physical chemistry: dimensional continuation methods applied to many-body effects in electronic structure and simple fluids.

Walter D. Loveland, Professor; Ph.D., Washington (Seattle), 1966. Nuclear chemistry: heavy-ion–induced nuclear reactions.

Joseph W. Nibler, Professor; Ph.D., Berkeley, 1966. Physical chemistry: study of molecular structures and spectra using high-resolution laser techniques (Raman, fluorescence, CARS, Raman gain, and two-photon electronic spectroscopies), low-temperature matrix isolation spectroscopy.

Edward H. Piepmeier, Professor; Ph.D., Illinois at Urbana-Champaign, 1966. Analytical chemistry: plasma spectroscopy and atomizers, laser-excited fluorescence and ionization spectroscopy, laser microprobes, plasma detectors for chromatography, plasma oscillation spectroscopy, high-resolution spectroscopy, instrumentation, chemometrics.

Roman A. Schmitt, Professor; Ph.D., Chicago, 1953. Nuclear chemistry: cosmochemistry and geochemistry, lunar sample research, activation analysis.

Michael W. Schuyler, Associate Professor; Ph.D., Indiana, 1970. Analytical chemistry: applications of microcomputers in research and teaching, chemical instrumentation.

Arthur W. Sleight, Professor; Ph.D., Connecticut, 1963. Inorganic chemistry: synthesis and characterization of solid-state materials.

Richard W. Thies, Associate Professor and Associate Dean, College of Science; Ph.D., Wisconsin–Madison, 1967. Organic chemistry: medium-sized-ring chemistry, thermal and solvolysis reactions, synthesis of large-ring hormone analogues, polymer synthesis.

T. Darrah Thomas, Distinguished Professor and Director, Center for Advanced Materials Research; Ph.D., Berkeley, 1957. Physical chemistry: electron spectroscopy of small molecules.

Philip R. Watson, Associate Professor; Ph.D., British Columbia, 1978. Physical chemistry: surface chemistry in catalysis and materials science.

Dwight D. Weller, Associate Professor; Ph.D., Berkeley, 1976. Organic chemistry: synthesis of alkaloids and oligonucleotides.

John C. Westall, Associate Professor; Ph.D., MIT, 1977. Analytical chemistry: ion-selective electrodes, surface chemistry, equilibria in multicomponent systems, aquatic chemistry.

James D. White, Professor; Ph.D., MIT, 1965. Organic chemistry: synthesis of natural products, new synthetic methods, heterocyclic chemistry, synthetic photochemistry, structural elucidation of natural products, simulation of enzymatic processes in vitro.

Douglas A. Barofsky, Adjunct Professor; Ph.D., Penn State, 1967. Analytical chemistry: mechanisms of particle-induced desorption ionization; mass-spectrometric analysis of biological molecules and natural products.

Max L. Deinzer, Adjunct Professor; Ph.D., Oregon, 1969. Organic mass spectrometry: electron attachment negative-ion mass spectrometry; fast-atom bombardment mass spectrometry of oligonucleotides and peptides.

Dr. Keszler and a graduate student align a crystal on the Rigaku AFC6R X-ray diffractometer.

A graduate student and Dr. Piepmeier discuss the operation of a laser.

A student discusses a microcomputer-interfacing exercise with Dr. Schuyler.

SECTION 2: CHEMISTRY

PRINCETON UNIVERSITY

Department of Chemistry

Programs of Study	The Department of Chemistry offers a program of study leading to the degree of Doctor of Philosophy. Upon entering, students take qualifying examinations in the fields of biochemistry, inorganic and organic chemistry, chemical physics, and physical chemistry and are expected to demonstrate a satisfactory level of proficiency in these areas by the end of the first year of study. The graduate program emphasizes research, and students enter a research group by the end of the first semester. There are no formal course requirements. However, students are expected to take six graduate courses in chemistry and allied areas early in their graduate program and to participate throughout their graduate career in the active lecture and seminar programs.
	To help evaluate the progress of graduate study, monthly cumulative examinations are given during the first two academic years. Each student must pass five of these exams. Early in the second year, the student takes a general examination that consists of an oral defense of a thesis-related subject. Upon satisfactory performance in the cumulative and general examinations, the student is readmitted for study toward the degree of Doctor of Philosophy in chemistry. The degree is awarded primarily on the basis of a thesis describing original research in one of the areas of chemistry. The normal length of the entire Ph.D. program is four years.
	In cooperation with the physics department, a program of graduate study in the field of chemical physics leading to a joint Ph.D. in physics and chemical physics is also offered. A program in biostructural chemistry is also offered.
Research Facilities	Research is conducted in Frick Chemical Laboratory and the adjoining Hoyt Laboratory. Extensive renovation and modernization of the laboratory has recently been completed. NMR, IR, ESR, FT-IR, atomic absorption, UV-visible, and vacuum UV spectrometers and departmental computers are available to students. In addition, high-resolution FT-NMR spectrometers and mass spectrometers are run by operators for any research group. There is a wide variety of equipment in individual research groups, including lasers of many kinds, high-resolution spectrographs, molecular beam instrumentation, a microwave spectrometer, computers, gas chromatographs, and ultrahigh-vacuum systems for surface studies. The department has an electronics shop, a machine shop, a student machine shop, and a glassblower for designing and building equipment. Extensive shop facilities are available on campus, and there is a large supercomputing center.
Financial Aid	Appointments as assistants in instruction (compensation taxable) are normally available to all entering students. The 1991–92 stipend is $10,700 for ten months (6 contact hours and 12–14 hours of preparation per week). Most entering students are also awarded a full tuition fellowship. Summer research stipends (also taxable) are also provided for the summer following the first year. Assistantships in research, fellowships, and student loans are available; they generally include support for the summer period.
Cost of Study	See Financial Aid section above.
Cost of Living	Rooms at the Graduate College cost from $1826 to $2889 for the 1991–92 academic year of thirty-five weeks. Several meal plans are available, priced from $1635 to $2427. University apartments for married students currently rent for $295 to $792 a month. Accommodations are also available in the surrounding community.
Student Group	The total number of graduate students in chemistry is currently about 130. Postdoctoral students number about 70. A wide variety of academic, ethnic, and national backgrounds are represented among these students.
Location	Princeton University and the surrounding community together provide an ideal environment for learning and research. From the point of view of a chemist in the University, the engineering, physics, mathematics, and molecular biology departments, as well as the Forrestal campus, provide valuable associates, supplementary facilities, and sources of special knowledge. Many corporations have located their research laboratories near Princeton, and interactions with them have been fruitful.
	Because of the nature of the institutions located here, the small community of Princeton has a very high proportion of professional people. To satisfy the needs of this unusual community, the intellectual and cultural activities approach the number and variety ordinarily found only in large cities, but with the advantage that everything is within walking distance. There are many film series, a resident repertory theater, orchestras, ballet, and chamber music and choral groups. Scientific seminars and other symposia bring prominent visitors from every field of endeavor. Princeton's picturesque and rural countryside provides a pleasant area for work and recreation, yet New York City and Philadelphia are each only about an hour away.
The University	Princeton University was founded in 1746 as the College of New Jersey. At its 150th anniversary in 1896, the trustees changed the name to Princeton University. The Graduate School was organized in 1901 and has since won international recognition in mathematics, the natural sciences, philosophy, and the humanities.
Applying	Applications, specifying the chemistry department, must be made on a form available on request from the Admissions Office, Princeton Graduate School, Box 270. A nonrefundable fee of $45 for U.S. residents and $50 for international students must accompany each application. Consideration is given to applications received before January 10. All assistantships and fellowships for entering students are awarded to applicants at this time. Applications received after January 10 are normally not considered. Admission consideration is open to all qualified candidates without regard to race, color, national origin, religion, sex, or handicap.
	Because the financial aid offered to chemistry students is in the form of assistantships, it is not necessary for students to submit a GAPSFAS form unless they are requesting a loan.
Correspondence and Information	Director of Graduate Studies Department of Chemistry Princeton University Princeton, New Jersey 08544

Peterson's Guide to Graduate Programs in the Physical Sciences and Mathematics 1992

SECTION 2: CHEMISTRY

Princeton University

THE FACULTY AND THEIR RESEARCH

Although all major areas of chemistry are represented by the faculty members, the department is small and housed in Frick Chemical Laboratory and the adjoining Hoyt Laboratory, so that fruitful collaborations develop. The following list briefly indicates the areas of interest of each faculty member.

Professor L. C. Allen. Theoretical chemistry: electronic structure theory; applications to physical, inorganic, organic, and biochemical problems.

Professor S. L. Bernasek. Chemical physics of surfaces: basic studies of chemisorption and reaction on well-characterized transition-metal surfaces, using electron diffraction and electron spectroscopy; surface reaction dynamics; heterogeneous catalysis; electronic materials.

Professor A. B. Bocarsly. Inorganic photochemistry, photoelectrochemistry, chemically modified electrodes, electrocatalysis, sensors, applications to solar energy conversion.

Professor J. Carey. Biochemistry: protein and nucleic acid structure, function, and interactions; protein folding and stability.

Professor G. C. Dismukes. Molecular spectroscopy of biological molecules and mechanisms of their activity, photosynthetic electron transfer reactions, metalloenzyme electronic structure, synthetic models of metalloenzymes, magnetic resonance spectroscopy.

Professor J. T. Groves. Organic and inorganic chemistry: synthetic and mechanistic studies of reactions of biological interest, metalloenzymes, transition-metal redox catalysis.

Professor M. Hecht. Biochemistry: sequence determinants of protein structure and design of novel proteins.

Professor M. Jones Jr. Reactions and spin states of carbenes, arynes, twisted π systems, and other reactive intermediates; carborane chemistry.

Professor D. Kahne. Organic chemistry: synthetic methods, with an emphasis on the construction of oligosaccharides; studies on carbohydrate structure and function in biological molecules.

Professor K. K. Lehmann. High-resolution molecular spectroscopy, intramolecular energy redistribution, determination of potential energy surfaces, ultrasensitive methods of spectroscopic detection.

Professor Emeritus D. S. McClure. Electronic spectroscopy of molecules and crystals, one- and two-photon electronic spectroscopy of inorganic complexes; photoionization and electron transfer in crystals, radiationless processes.

Professor R. A. Naumann. Chemical physics, nuclear physics: nuclear structure from nuclear spectroscopy, electromagnetic separation of short-lived radioisotopes, ultrafast radiochemical separation methods, meson capture by atoms and molecules, nuclear properties associated with neutrino detectors, new radiation detection methods.

Professor R. A. Pascal Jr. Catalytic mechanisms of redox enzymes, enzyme inhibitor design and synthesis, bioorganic chemistry, biochemistry of parasitic organisms, chemistry of distorted aromatic hydrocarbons.

Professor H. Rabitz. Physical chemistry: atomic and molecular collisions, theory of chemical reactions and chemical kinetics, time- and space-dependent relaxation processes, heterogeneous phenomena, control of molecular motion.

Professor C. E. Schutt. Structural biology: structure and function of proteins and cellular organelles, X-ray crystallography.

Professor J. Schwartz. Applications of organometallic chemistry to organic synthesis, organozirconium chemistry, organometallic reaction mechanisms, oxide-supported organometallic complexes.

Professor G. Scoles. Chemical physics: laser spectroscopy, chemical dynamics and cluster studies with molecular beams; structure, dynamics, and spectroscopic properties of overlayers adsorbed on crystal surfaces.

Professor M. F. Semmelhack. Organometallic and electrogenerated intermediates in organic synthesis, synthesis of unusual ring systems in natural and unnatural molecules.

Professor Z. G. Soos. Chemical physics: electronic states of π-molecular crystals and conjugated polymers, paramagnetic and charge-transfer excitons, linear chain crystals, energy transfer.

Professor T. G. Spiro. Biological structure and dynamics from spectroscopic probes; role of metals in biology; metalloporphyrins in photocatalysis; chemically modified electrodes.

Professor E. C. Taylor. Organic synthesis; medicinal chemistry, heterocyclic chemistry.

Professor M. E. Thompson. Solid-state organometallic chemistry, materials for nonlinear optics, intercalation chemistry of layered solids.

Professor W. S. Warren. Laser spectroscopy in gaseous and solid phases, coherence effects, multiphoton processes, nuclear magnetic resonance.

Associated Faculty

Professor A. Navrotsky, Department of Geological and Geophysical Sciences. High-temperature solid-state chemistry, crystal chemistry and chemical bonding in solids, calorimetry, mineral thermodynamics, silicate glasses and melts, structure, bonding, phase transitions, amorphous films and powders, oxide superconductors.

Professor J. Stock, Department of Biology. Structure and function of membrane receptors and signal transduction proteins.

Frick Chemical Laboratory.

SECTION 2: CHEMISTRY

RENSSELAER POLYTECHNIC INSTITUTE
Department of Chemistry

Programs of Study

The Department of Chemistry faculty work closely with each student to develop his or her interest in one of Rensselaer's very strong traditional research programs or in one of numerous innovative cross-disciplinary areas. The Master of Science and Doctor of Philosophy degrees are offered with courses and research in analytical chemistry, biochemistry and biophysics, electrochemistry, inorganic chemistry, coordination and organometallic chemistry, medicinal chemistry, natural products synthesis, material chemistry, nuclear chemistry and radiochemistry, organic and bioorganic chemistry, photochemistry (including laser techniques), physical chemistry, physical organic chemistry, polymer chemistry (synthesis and physical properties), solid state chemistry and crystal growth, spectroscopy (laser, microwave, NMR, vibrational, fluorescence, in situ environmental probes) and surface science. Courses and research are available for interdisciplinary programs.

For the degree of Doctor of Philosophy, a candidate must complete a minimum of 90 credit hours beyond the bachelor's degree, satisfy preliminary examination requirements, pass the candidacy examination, and complete the research thesis and final thesis defense. The degree of Master of Science is based upon 30 credits of courses and research, and a research thesis; the residence requirement is at least two terms.

Research Facilities

Shared departmental research equipment is based in the Major Instrumentation Center operated by professional staff and available to all research students. Instrumentation includes state-of-the-art solution and solid-state NMR spectrometer equipment and FT-IR, GC/MS, and single-crystal X-ray diffraction equipment. Individual laboratories contain a wide variety of GC, HPLC, IR, Vis-UV, electrochemical, laser, ESR, and thermal analysis equipment and other instruments. The Molecular Modeling Center has up-to-date computer hardware and software for molecular modeling and quantum mechanical calculations.

Research is also supported by such state-of-the-art facilities as the Folsom Library, whose automated retrieval systems link users to its own holdings as well as 5,600 libraries and 250 databases and the Voorhees Computing Center, which offers students exceptionally wide access to a diverse array of advanced-function workstations, microcomputers, and mainframes that are interconnected to each other and to national networks via the campus network.

Financial Aid

Financial aid is available in the form of fellowships, research or teaching assistantships, and scholarships. The stipend for assistantships ranges up to $11,000 for a nine-month academic year. In addition, full tuition remission is granted. Stipends for the summer months are $1800 or more. Outstanding students may qualify for industry- and university-supported Rensselaer Scholar Fellowships, which carry a stipend of $13,500 and a full tuition scholarship. Low-interest, deferred-repayment graduate loans are also available to U.S. citizens with demonstrated need.

Cost of Study

Tuition for 1991–92 is $455 per credit hour. Other fees amount to approximately $230 per semester. The application fee for admission is $30. Books and supplies cost about $930 per year.

Cost of Living

The cost of rooms for single students in residence halls or apartments varies from $2400 to $3800 for the 1991–92 academic year. A twenty-meal-per-week board plan costs $2500 for the academic year. Apartments for married students, with monthly rents of $360 to $590, are available on campus.

Student Group

There are about 4,500 undergraduates and 2,000 graduate students representing all fifty states and over sixty-five countries at Rensselaer.

Location

Rensselaer is situated on a scenic 260-acre hillside campus in Troy, New York, across the Hudson River from the state capital of Albany. Troy's central Northeast location provides students with a supportive, active, medium-sized community in which to live and an easy commute to Boston, New York, and Montreal. The Capital Region has one of the largest concentrations of academic institutions in the United States. Sixty thousand students attend fourteen area colleges and benefit from shared activities and courses, as well as access to the Saratoga Performing Arts Center, the Empire Plaza, the Rensselaer Technology Park, and some of the country's finest outdoor recreation, including Lake George, Lake Placid, and the Adirondack, Catskill, Berkshire, and Vermont mountains.

The Institute

Founded in 1824 and the first American college to award degrees in engineering and science, Rensselaer Polytechnic Institute today is accredited by the Middle States Association of Colleges and Schools and is a private, nonsectarian, coeducational university. Rensselaer's five schools—Architecture, Engineering, Management, Science, and Humanities and Social Sciences—offer a total of eighty-seven graduate degrees in forty fields.

Applying

Admissions applications and all supporting credentials should be submitted well in advance of the preferred semester of entry to allow sufficient time for departmental review and processing. Since the first departmental awards are made in February and March for the next full academic year, applicants requesting financial aid are encouraged to submit all required credentials by February 1 to ensure that they will receive consideration. GRE General Test scores are required.

Correspondence and Information

For further information about graduate work:
Department of Chemistry
Graduate Admissions Committee
Rensselaer Polytechnic Institute
Troy, New York 12180-3590
Telephone: 518-276-6356

For applications and admissions information:
Director of Graduate Admissions
Graduate Center
Rensselaer Polytechnic Institute
Troy, New York 12180-3590
Telephone: 518-276-6789

Peterson's Guide to Graduate Programs in the Physical Sciences and Mathematics 1992

SECTION 2: CHEMISTRY

Rensselaer Polytechnic Institute

THE FACULTY AND THEIR RESEARCH

D. A. Aikens, Professor and Chairman; Ph.D., MIT. Analytical chemistry: electrode reaction mechanisms.

S. Archer, Research Professor of Pharmaceutical Chemistry; Ph.D., Penn State. Chemistry and pharmacology of analgesics and chemotherapeutic agents.

R. A. Bailey, Professor; Ph.D., McGill. Inorganic chemistry: inorganic coordination complexes, electrochemistry and spectroscopy of metal ions in fused salt solutions, chemical reactions in molten salts.

J. Bell, Professor; Ph.D., Cornell. Biochemistry and biophysics: X-ray crystallography, enzyme kinetics, protein physical chemistry, genetic engineering.

C. M. Breneman, Assistant Professor; Ph.D., California, Santa Barbara. Physical organic chemistry: structure reactivity relationships, molecular modeling, organic synthesis.

J. V. Crivello, Professor; Ph.D., Notre Dame. Organic chemistry: organic nitrations, oxidations and arylations, polyimides, silicones, and new photo and thermal initiators for cationic polymerizations.

A. R. Cutler, Associate Professor; Ph.D., Brandeis. Organometallic chemistry: synthetic and mechanistic inorganic and transition organometallic chemistry, homogeneous catalysis, metallobiochemical processes, metal-containing oligomers and polymeric materials.

G. D. Daves Jr., Professor; Ph.D., MIT. Organic chemistry: organopalladium chemistry, C-glycoside synthesis, applications of mass and nuclear magnetic resonance spectrometries.

J. P. Ferris, Professor; Ph.D., Indiana. Organic chemistry and biochemistry: chemistry of the origins of life, cyanocarbon chemistry, oxidase enzymes, photochemistry, synthesis of nucleoside antitumor drugs.

L. L. Frye, Assistant Professor; Ph.D., Johns Hopkins. Bioorganic chemistry: design and synthesis of enzyme inhibitors, mechanism of new methods for the selective fluorination of organic molecules.

C. W. Gillies, Associate Professor; Ph.D., Michigan. Physical chemistry: microwave spectroscopy, organic reaction mechanisms, ozone chemistry.

N. F. Hepfinger, Associate Professor and Associate Chairman; Ph.D., Pittsburgh. Organic chemistry: applications of NMR spectroscopy, reaction mechanisms, unstable intermediates in heterocyclic systems, cationic polymerization of cyclic ethers.

H. B. Hollinger, Professor; Ph.D., Wisconsin–Madison. Physical chemistry: kinetic theory of dense gases, nonequilibrium statistical mechanics.

C. P. Horwitz, Assistant Professor; Ph.D., Northwestern. Inorganic chemistry: inorganic and organometallic electrochemistry and photochemistry.

L. V. Interrante, Professor; Ph.D., Illinois at Urbana-Champaign. Inorganic chemistry: solid-state chemistry, metal-organic precursors to ceramics and electronic materials, coordination compounds with extended solid-state interactions.

G. M. Korenowski, Associate Professor; Ph.D., Cornell. Physical chemistry: surface science, spectroscopy and its applications to molecular and chemical dynamics.

S. Krause, Professor; Ph.D., Berkeley. Physical chemistry: dilute solution properties of polymers, block copolymers, polymer compatibility, transient electric birefringence, muscle proteins.

K. J. Miller, Professor; Ph.D., Iowa State. Theoretical chemistry: molecular quantum mechanics, studies of chemical reaction paths, theoretical investigations of binding in DNA and the correlation with antitumor and carcinogenic activity, computer graphics.

J. A. Moore, Associate Professor; Ph.D., Polytechnic of Brooklyn. Organic polymer chemistry: preparation and chemical (photochemical) transformations of synthetic polymers and biopolymers, structure-property relationships in polymers, synthetic organic chemistry.

K. T. Potts, Professor; D.Phil., D.Sc., Oxford. Organic chemistry: synthesis and investigation of the chemistry of bicyclic heterocycles, nonclassical and classical heteroaromatics, photochemistry, application of spectral methods and molecular orbital theory to structural and mechanistic problems, organic electrochemistry.

I. L. Preiss, Professor; Ph.D., Arkansas. Nuclear chemistry: nuclear reaction mechanisms including reactions between complex nuclei, photo- and neutron-induced reactions, decay schemes of new radioisotopes, problems of nuclear structure as related to direct interactions, activation analysis, X-ray fluorescence studies.

R. R. Reeves, Professor; Ph.D., RPI. Physical-analytical chemistry: reaction kinetics, atmospheric chemistry and photochemistry, atom chemistry.

H. H. Richtol, Professor and Dean of the Undergraduate College; Ph.D., NYU. Analytical chemistry: fluorescence and phosphorescence, photochemistry, analytical applications of liquid crystals.

A. G. Schultz, Professor; Ph.D., Rochester. Organic chemistry: natural-products synthesis, synthetic and mechanistic organic photochemistry, development of new organic synthetic methods.

R. L. Strong, Professor; Ph.D., Wisconsin–Madison. Physical chemistry: photochemistry and flash photolysis; atom recombination in gas and solution systems; halogen atom charge-transfer complexes; kinetic, spectral, and optical rotatory dispersion behavior of excited states.

S. C. Wait, Professor and Associate Dean of the School of Science; Ph.D., RPI. Physical chemistry: spectroscopy, vibrational analysis, fine-structural analyses, theoretical methods, simple and polyatomic systems.

B. A. Wallace, Professor; Ph.D., Yale. Biochemical and biophysical chemistry: crystallography of membrane proteins, circular dichroism, protein NMR, theoretical studies on absorption and light scattering, chemical modification and synthesis of model polypeptides, low angle X-ray and neutron scattering, prediction algorithms for membrane protein folding, molecular graphics modeling.

J. T. Warden, Associate Professor; Ph.D., Minnesota. Biophysical chemistry: physicochemical studies in photosynthesis, applications of ESR to biological electron transport, structure and function of hemoproteins, flash photolysis–ESR spectroscopy, photophysics, solar energy.

H. Wiedemeier, Professor; D.Sc., Münster. Solid-state and high-temperature chemistry: single-crystal growth of transition-metal Group VI compounds, thermodynamics and kinetics of vaporization processes.

G. E. Wnek, Associate Professor; Ph.D., Massachusetts at Amherst. Organic polymer chemistry: synthesis and characterization of polymers, optical and electrical conductivity properties.

SECTION 2: CHEMISTRY

RUTGERS, THE STATE UNIVERSITY OF NEW JERSEY, NEWARK
Department of Chemistry

Programs of Study

The Department of Chemistry at Rutgers in Newark offers programs leading to the degrees of Master of Science and Doctor of Philosophy. (Separate M.S. and Ph.D. programs in chemistry are also offered at the New Brunswick campus. The description on this page refers only to the programs at Newark.)

All entering students must pass with a grade of at least B a core of six courses (12 credits). All students must pass written examinations that are cumulative in nature. In addition, students enrolled in the Ph.D. program must satisfactorily present a departmental seminar on a current research topic and must present and defend in an oral examination a research proposal. Thirty hours of course and research credit are required for the master's degree; 72 hours of course and research credit are required for the Ph.D.

Within the four traditional areas of chemistry, many fields of specialization are represented—biophysical and bioorganic approaches to enzyme mechanisms, molecular modeling and drug design, medicinal chemistry, neurochemistry, biological membranes, and X-ray crystallography, among others.

Research Facilities

Research in chemistry is conducted in the Carl A. Olson Memorial Laboratories, a modern facility housing state-of-the-art instrumentation. Major items of equipment include Fourier-transform infrared and NMR spectrometers (200-MHz and 400-MHz multinuclear NMR); mass spectrometers; a circular dichroism spectrophotometer; a wide selection of modern UV, IR, fluorescence, and phosphorescence spectrometers; an automatic X-ray diffractometer; a fast-reaction laboratory with t-jump and stopped-flow capabilities; high-pressure liquid chromatographs and gas chromatographs; electrochemistry units; and an Evans and Sutherland molecular modeling system including software for small-molecule and macromolecule modeling. Other equipment used to support a biotechnology laboratory includes an automatic peptide synthesizer, a GC-mass spectrometer, ultracentrifuges, a pilot-scale fermenter, a scintillation counter, and a transmission electron microscope.

There are a number of PCs and a MicroVAX II in the department and two VAX-11/780s (VMS) and a Pyramid 90x (UNIX) on campus. There is a DECnet connection by fiber-optic cable to an AS/9000 (IBM-type mainframe) and to VAX 8650s on the New Brunswick campus, and there is access to the supercomputer at Princeton University. Also, there are more than 100 microcomputers on campus.

Financial Aid

In 1991–92, graduate teaching assistantships provide $9400 (plus remission of tuition fees) for the academic year and $10,810 for twelve months; merit increases are given for subsequent years. Fellowships providing stipends of up to $12,000 per year for four years (plus tuition remission) are also available. More advanced students often receive research assistantships under the supervision of their research directors. Most students receive a summer stipend as either a graduate teaching assistant or graduate research assistant. Thirty full-time students are supported by teaching or research assistantships.

Cost of Study

In 1991–92, tuition for full-time study is $2033 per term for New Jersey state residents and $2980 per term for nonresidents. Part-time study costs are $168 per credit for state residents and $247 per credit for nonresidents. Tuition is remitted for assistantship and fellowship recipients. The activity fee is $151.25 per semester for full-time students and $38.50 per semester for part-time students. The nonrefundable application fee is $35.

Cost of Living

Graduate student housing, consisting of apartments, each with a kitchen, is available to full-time students at $2406 for the academic year and $2891 for twelve months. Rooms and apartments are also available in the immediate urban area and in the surrounding suburban area.

Student Group

During the past academic year, more than 10,000 students were enrolled in all programs on the Newark campus. In 1991–92, there are 1,100 graduate students enrolled at Newark. Each program carefully selects its graduate students from among applicants from the New York metropolitan area, various parts of the United States, and abroad. Most students have strong liberal arts backgrounds, and a few hold master's degrees. Many of the students in the graduate program in chemistry are already working in the vast chemical industry of northern New Jersey.

The Department of Chemistry has, in addition to 35 full-time graduate students, about 90 part-time students who work during the day and attend classes in the evening. There is also 1 postdoctoral fellow in the department.

Location

Newark is New Jersey's largest city and, although located 30 minutes by road and rail from midtown Manhattan, is a major commercial center in its own right. Surrounded by communities that become increasingly suburban and rural with distance, Newark also lies at the center of the nation's largest concentration of chemical industries. Many Rutgers chemistry graduates find employment in the area.

The University

Rutgers University has several academic units on the Newark campus, including the Graduate School, the School of Law, the Graduate School of Management, the undergraduate Newark College of Arts and Sciences, the College of Nursing, and University College. The Newark units are part of a complex of higher education that includes Essex County College, New Jersey Institute of Technology, and the University of Medicine and Dentistry of New Jersey.

Applying

Applications for the fall semester should normally be submitted before July 1. All forms, as well as a booklet describing the Department of Chemistry, may be obtained from the address given below.

Correspondence and Information

Director of Graduate Program in Chemistry
Department of Chemistry
Rutgers University
Newark, New Jersey 07102
Telephone: 201-648-5282 or 5173

SECTION 2: CHEMISTRY

Rutgers, The State University of New Jersey, Newark

THE FACULTY AND THEIR RESEARCH

R. Ian Fryer, Professor; Ph.D., Manchester (England). Heterocyclic/medicinal chemistry, design and synthesis of novel benzodiazepines and other heterocycles, molecular modeling and drug design.

Stan S. Hall, Professor; Ph.D., MIT. Synthetic methods, total synthesis, tandem reactions, (η^3-allyl)palladium chemistry.

William Phillip Huskey, Assistant Professor; Ph.D., Kansas. Mechanisms of enzymic reactions, isotope effects.

Frank Jordan, Professor; Ph.D., Pennsylvania. Experimental and theoretical bioorganic chemistry; enzyme mechanisms; protein NMR.

Rudolph W. Kluiber, Professor; Ph.D., Wisconsin–Madison. Reactivity of coordinated organic ligands.

Roger A. Lalancette, Professor; Ph.D., Fordham. X-ray diffraction and the structure of solids, synthesis and characterization of nitrogen and sulfur complexes, hydrogen bonding in keto-carboxylic acids.

Darrell McCaslin, Assistant Professor; Ph.D., Duke. Structure-function relationships in soluble and membrane-bound proteins.

Richard Mendelsohn, Professor and Chairman; Ph.D., MIT. Biophysical chemistry, lipid-protein interactions in biological membranes, phospholipid phase transitions, development of FT-IT spectroscopy for biomedical applications.

Ernst U. Monse, Professor; Ph.D., Max Planck Institute (Mainz). Isotope effects and their applications to theoretical chemistry.

Gilbert S. Panson, Professor Emeritus; Ph.D., Columbia. Molecular interactions in liquids, hydrogen bonding, mechanisms of aryl substitution in nonpolar media.

Susanne Raynor, Associate Professor; Ph.D., Georgetown. Quantum mechanics of molecular solids and clusters, collision dynamics.

Irvin Rothberg, Professor; Ph.D., Pennsylvania. Carbonium ion reactions, synthetic techniques, natural products.

James M. Schlegel, Professor; Ph.D., Iowa State. Electroanalytical chemistry, kinetics and mechanism of electrode reactions.

John Sheridan, Assistant Professor; Ph.D., Bristol (England). Transition-metal organometallic chemistry; synthesis, structure, and mechanism; applications to organic synthesis.

Merry Rubin Sherman, Professor; Ph.D., Berkeley. Endocrine and physical biochemistry, characterization of hormone receptors and proteolytic enzymes.

Hugh W. Thompson, Professor and Graduate Program Coordinator; Ph.D., MIT. Mechanisms and stereochemical courses of organic reactions, compounds of unusual symmetry and stereochemistry, development of new synthetic methods.

C. Edwin Weill, Professor Emeritus; Ph.D., Columbia. Carbohydrate chemistry, carbohydrate enzyme systems.

The Carl A. Olson Memorial Chemistry Laboratories.

Entrance to the campus.

The campus plaza.

RUTGERS, THE STATE UNIVERSITY OF NEW JERSEY, NEW BRUNSWICK
Graduate Program in Chemistry

Programs of Study

The Graduate Program in Chemistry at Rutgers in New Brunswick offers programs leading to the degrees of Master of Science, Master of Science for Teachers, and Doctor of Philosophy.

The principal requirement for the Ph.D. degree is the completion and successful oral defense of a thesis based on original research. A wide variety of research specializations are available in the traditional areas of chemistry—analytical, inorganic, organic, and physical—as well as in related areas and subdisciplines, including bioanalytical, bioinorganic, bioorganic, and biophysical chemistry; chemical physics; theoretical chemistry; and solid-state and surface chemistry.

The M.S. degree may be taken with or without a research thesis. The principal requirements are completion of 30 credits of graduate courses, a passing grade on the master's examination, and a master's essay or thesis. When the thesis option is chosen, 6 of the 30 credits may be in research.

Research Facilities

The research facilities of the program, located in the Wright and Rieman chemistry laboratories on the Busch campus, include a comprehensive chemistry library; glassworking, electronics, and machine shops; and extensive state-of-the-art instrumentation. Instruments of particular note include 500-, 400-, and 200-MHz NMR spectrometers with 2-D and solid-state capabilities, computer-interfaced single-crystal and powder X-ray diffractometers, a nanosecond laser flash photolysis system, a temperature-programmable ORD-CD spectropolarimeter, an automated DNA synthesizer, a SQUID magnetometer, several ultrahigh-vacuum surface analysis systems, scanning tunneling microscopes, molecular beam and supersonic jet apparatuses, Mössbauer spectrometers, and extensive laser instrumentation, crystal-growing facilities, and microcalorimetric equipment. The chemistry department's computer facilities include VAX-11/780 minicomputers, a Convex C-210 minisupercomputer, and MicroVAX II workstations linked via Ethernet networks to national supercomputer facilities. Molecular graphics facilities include an Evans & Sutherland PS-390 color vector graphics system, a Silicon Graphics IRIS workstation, color raster graphics terminals, and laser printers. University computing facilities include a pair of VAX 8650s, and an NAS 9000-2 mainframe computer.

Financial Aid

Full-time Ph.D. students receive financial assistance in the form of fellowships, research assistantships, teaching assistantships, or a combination of these. Financial assistance for entering students ranges from approximately $11,000 to $16,000 plus tuition remission for a calendar-year appointment. This includes the J. R. L. Morgan fellowships awarded annually to outstanding applicants.

Cost of Study

In 1991–92, the full-time tuition (remitted for assistantship and fellowship recipients) is $2220 per semester for New Jersey residents and $3250 per semester for out-of-state residents. There is a fee for full-time students of $190 per semester. All of these fees are subject to change for the next academic year.

Cost of Living

A furnished double room in the University residence halls or apartments rents for approximately $2400 to $3000 per person for the 1991–92 academic year. Married student apartments rent for approximately $315 to $475 per month. Current information may be obtained from the Department of Housing (908-932-2215), which also has information on private housing in the New Brunswick area.

Student Group

Total University enrollment is more than 47,000. Graduate and professional enrollment is approximately 13,500, of whom about 8,500 are in the Graduate School. Enrollment of graduate students in chemistry totals 160. Of these, about three fourths are full-time students. Students come from all parts of the United States as well as from other countries. In addition, there are approximately 35 postdoctoral research associates in residence.

Location

New Brunswick, with a population of about 42,000, is located in central New Jersey, roughly midway between New York City and Philadelphia. The cultural facilities of these two cities are easily accessible by automobile or regularly scheduled bus and train service. Within a 1½ hours' drive of New Brunswick are the recreational areas of the Pocono Mountains of Pennsylvania and the beaches of the New Jersey shore. The University also offers a rich program of cultural, recreational, and social activities.

The University

Graduate instruction and research in chemistry are conducted on the University's Busch campus, which has a predominantly rural environment and is a few minutes' drive from downtown New Brunswick. On the same campus, within walking distance, are the Hill Center for the Mathematical Sciences (home of the University's computer center), the Library of Science and Medicine, the physics and biology laboratories, the Waksman Institute of Microbiology, the Robert Wood Johnson Medical School, the Center for Advanced Biotechnology and Medicine, and the College of Engineering. The University provides a free shuttle-bus service between the Busch and New Brunswick campuses.

Applying

Applications for assistantships and fellowships should be made at the same time as applications to the Graduate School. All forms, as well as a booklet describing the current research programs, may be obtained from the address given below. Admission consideration is open to all qualified candidates without regard to race, color, national origin, religion, sex, or handicap.

Correspondence and Information

Executive Officer
Graduate Program in Chemistry at New Brunswick
Wright-Rieman Laboratories
Rutgers, The State University of New Jersey
P.O. Box 939
Piscataway, New Jersey 08855
Telephone: 908-932-3223

SECTION 2: CHEMISTRY

Rutgers, The State University of New Jersey, New Brunswick

THE FACULTY AND THEIR RESEARCH

William H. Adams, Professor. Fundamental problems in the quantum theory of intermolecular interactions.

Stephen Anderson, Associate Professor. Protein folding and molecular recognition.

Edward Arnold, Assistant Professor. Structure and function of molecules in living systems; X-ray crystallographic studies of macromolecules and aggregates, including proteins and viruses; role of three-dimensional structure in biological molecules.

Jean Baum, Assistant Professor. Structural studies of proteins by nuclear magnetic resonance techniques.

Helen Berman, Professor. X-ray crystallographic and molecular modeling studies of biological molecules.

George Bird, Professor. Mechanisms of dye sensitization of solar photovoltaic cells and silver halide imaging systems, photoactive materials for laser systems, chemical image-forming systems, high-resolution spectroscopy.

Robert S. Boikess, Professor and Director, Graduate Program. Chemical education.

John G. Brennan, Assistant Professor. Molecular and solid-state inorganic chemistry, lanthanide chalcogenides and pnictides, semiconductor thin films and nanometer-sized clusters.

Kenneth J. Breslauer, Professor. Characterization of the molecular forces that dictate the affinity and specificity of drug binding to DNA and RNA duplexes as well as the stability and conformational transitions of nucleic acids and proteins.

Kuang-Yu Chen, Professor. Molecular mechanism of cytodifferentiation, tumor reversion, and cellular aging; metabolism and function of polyamines; cell-surface proteins.

Martha A. Cotter, Professor and Executive Officer, Graduate Program. Theoretical investigations of liquid crystals and micellar solutions, phase transitions in simple model systems, theory of liquids.

Donald B. Denney, Professor. Organic reaction mechanisms, organophosphorus chemistry, high-resolution NMR spectroscopy, preparation of hypervalent molecules, preparation and studies of the properties of polymers.

Richard H. Ebright, Assistant Professor. Relationship of protein structure to protein function; protein-ligand recognition, especially protein–nucleic acid recognition; allostery; site-directed and random mutagenesis.

Eric L. Garfunkel, Associate Professor. Surface science and catalysis; ultrahigh-vacuum studies of basic chemical processes on metal surfaces; LEED, photoelectron, thermal desorption, and electron loss spectroscopies; scanning tunneling microscopy.

Alan S. Goldman, Assistant Professor. Inorganic and organometallic reaction mechanisms, photochemistry; transition-metal-mediated activation and functionalization of carbon-hydrogen bonds.

Lionel Goodman, Professor. Laser spectroscopy, particularly involving multiphoton excitation in supersonic jets; coupling of vibrational and electronic motions in molecules; molecular potential surfaces.

Martha Greenblatt, Professor. Solid-state inorganic chemistry, transition-metal oxides and chalcogenides, superionic conductors, intercalation compounds, glasses, high-T_c superconductors.

Gene S. Hall, Associate Professor. Radioanalytical chemistry, interactions of positrons and positronium with matter, trace-element analysis of biological and environmental samples by proton-induced X-ray emission (PIXE) and by scanning PIXE.

Rolfe H. Herber, Professor. Chemical physics of the solid state, structure and bonding in organometallics and covalent solids, variable-temperature FT infrared spectroscopy, phase transitions and ordering phenomena in solids, radio and nuclear chemistry.

Gregory F. Herzog, Professor. Origin and evolution of meteorites, cosmogenic radioisotopes by accelerator mass spectrometry.

Stephan S. Isied, Professor. Peptide synthesis using inorganic protecting groups, intramolecular electron transfer in proteins and binuclear metal complexes, oxygen evolution catalyzed by transition-metal complexes.

Leslie S. Jimenez, Assistant Professor. Molecular recognition, synthesis and characterization of analogues of antitumor antibiotics, synthesis and kinetic studies of coenzyme analogues.

Roger A. Jones, Professor. New methods of oligonucleotide synthesis, synthetic transformations of nucleosides and nucleoside antibiotics, synthesis and characterization of oligonucleotides containing modified or isotopically labeled nucleosides.

Spencer Knapp, Professor. Total synthesis of natural products; design and synthesis of enzyme models and inhibitors and of complex ligands, new synthetic methods.

Joachim Kohn, Assistant Professor. Development of structurally new polymers as biomaterials for medical applications, drug delivery, and immunology; new synthetic approaches for the preparation of peptide hormone analogues.

John Krenos, Associate Professor. Chemical physics, energy transfer in hyperthermal collisions and collisions involving electronically excited reactants, molecular beam chemiluminescence, model calculations of chemical reactions.

Karsten Krogh-Jespersen, Associate Professor. Computational chemistry, electronic structure of organic and metalloorganic molecules, properties of ground and excited electronic states, solvent effects.

Ronald M. Levy, Professor. Biophysical/physical chemistry, computer simulations of macromolecules, theory of NMR relaxation and crystallographic refinement, semiclassical methods in spectroscopy, protein-ligand binding thermodynamics, liquid-state theory.

Theodore E. Madey, Professor. Structure and reactivity of surfaces, electron- and photon-induced surface processes.

Gerald S. Manning, Professor. Theory of polyelectrolyte solutions, ionic and elastic effects on biopolymer structure and configuration.

Gaetano Montelione, Assistant Professor. Nuclear magnetic resonance studies of proteins, protein molecular design.

Robert A. Moss, Professor. Bioorganic chemistry: membranes, vesicles, and micelles; physical organic chemistry: reactive intermediates, nanosecond kinetics.

Wilma K. Olson, Professor. The relationship of structure, conformation, and function in biological macromolecules, with emphasis at present on nucleic acids and proteins.

Joseph A. Potenza, Professor. Molecular structure, X-ray diffraction, magnetic resonance.

Laurence S. Romsted, Associate Professor. Theory of micellar effects on reaction rates and equilibria, ion binding at aqueous interfaces, organic reaction mechanisms, oscillating reactions.

Heinz Roth, Professor. Chemistry of reactive intermediates, nuclear spin polarization, electron spin resonance.

Ronald R. Sauers, Professor. Synthesis of highly strained molecules, organic photochemistry, synthesis of new dyes for lasers and photovoltaic cells, intramolecular energy transfer, force-field computations.

Harvey J. Schugar, Professor. Inorganic and bioinorganic chemistry, modeling of copper chromophores in biological systems, charge-transfer spectra of metal ions in biological systems, long-range electron transfer.

Stanley Stein, Adjunct Professor. Methods development research on protein analysis, synthesis of biologically active peptides and oligonucleotides.

George Strauss, Professor. Biophysical chemistry; binding of solutes, fusion, and energy transfer in lipid bilayer membranes; fluorescence and light-scattering techniques.

John W. Taylor, Associate Professor. Bioactive peptide design and synthesis, multicyclic peptides, peptide conformation, protein engineering, peptide ligand-receptor interactions.

Irwin Tobias, Professor. Theoretical chemical physics, particularly the theory of the phenomena associated with the interaction of light with molecules; topology of biological molecules.

Sidney Toby, Professor. Chemical kinetics, photochemistry, and chemiluminescence, particularly of ozone reactions.

Alexander M. Yacynych, Associate Professor. Analytical chemistry, electrochemistry, chemically modified electrodes as analytical sensors, enzyme electrodes, biosensors, clinical analysis, and flow injection analysis.

Martin L. Yarmush, Professor. Applied immunology, protein chemistry and engineering, bioseparations, artificial organs.

SECTION 2: CHEMISTRY

STANFORD UNIVERSITY
Department of Chemistry

Program of Study

The Department of Chemistry strives to promote excellence in education and research. Only candidates for the Ph.D. degree are accepted. The department has a relatively small faculty, and this promotes outstanding interactions betweeen faculty, students, and staff. The faculty has achieved broad national and international recognition for its outstanding research contributions, and over a third of its members belong to the National Academy of Sciences. The graduate program is based strongly on research, and students enter a research group by the end of the winter quarter of their first year. Students are also expected to complete a rigorous set of core courses in various areas of chemistry in their first year and to complement these courses later by studying upper-level subjects of their choice. Qualifying examinations are administered early in the fall quarter of the first year in inorganic, organic, physical, and biophysical chemistry. Students with deficiencies in undergraduate training in these areas are identified and work with the faculty to make them up. There are no other departmental examinations or orals required of students progressing toward the Ph.D. degree. Much of the department's instruction is informal and includes diverse and active seminar programs, group meetings, and discussions with visiting scholars and with colleagues in other departments of the University. Stanford Ph.D. recipients are particularly well prepared for advanced scientific and technological study, and chemistry graduates typically accept positions on highly regarded university faculties or enter a wide variety of positions in industry.

Research Facilities

The department occupies five buildings with about 165,000 square feet. The department has a strong commitment to obtaining and maintaining state-of-the-art instrumentation for analysis and spectroscopy. Equipment available includes 200-, 300-, 400-, and 500-MHz NMR spectrometers; two FT-IR spectrometers; several mass spectrometers; X-ray crystallography facilities; picosecond absorption and fluorescence spectroscopy facilities; ultrahigh-resolution laser spectroscopy facilities; ion cyclotron resonance facilities; dynamic light-scattering spectroscopy facilities; tissue culture facilities; electrochemical systems; ultrahigh-vacuum facilities for surface analysis, including ESCA, Auger, EELS, LEED, and UPS; laser-Raman facilities; a superconducting magnetometer; and a Varian E-11 EPR spectrometer. Extensive computing capabilities are available in all research groups, and these are supplemented by a department computer network and the campus's IBM mainframe computers. Additional major instrumentation and expertise are available in the Center for Materials Research, the Stanford Synchrotron Radiation Laboratory, and the Stanford Magnetic Resonance Laboratory.

Financial Aid

Financial support of graduate students is provided in the form of teaching assistantships, research assistantships, and fellowships. All graduate students in good standing receive full financial support (tuition and stipend) for the duration of their graduate studies. The stipend for the 1991–92 year is $13,650; the amount of the stipend is adjusted annually to allow for inflation. Typical appointments involve teaching assistantships in the first year and research assistantships in subsequent years. Supplements are provided to holders of outside fellowships.

Cost of Study

Holders of teaching assistantships or research assistantships pay no academic tuition. In 1991–92, full tuition is $20,136 for the year (four quarters).

Cost of Living

Both University-owned and privately owned housing accommodations are available. Due to the residential nature of the surrounding area, it is not uncommon for several graduate students to share in a house rental. Escondido Village, an apartment development on campus, provides one- to three-bedroom apartments for married students and single parents. A recent survey of all graduate students in the department indicated that the median monthly expenditure for rent and utilities was $450.

Student Group

The total enrollment at Stanford University is 13,441, and there are 6,886 graduate students. The Department of Chemistry has 220 graduate students in its Ph.D. program.

Location

Stanford University is located in Palo Alto, a community of 60,000 about 35 miles south of San Francisco. Extensive cultural and recreational opportunities are available at the University and in surrounding areas, as well as in San Francisco. To the west lie the Santa Cruz Mountains and the Pacific Ocean, and to the east, the Sierra Nevada range with its many national parks, hiking and skiing trails, and redwood forests.

The University

Stanford is a private university founded in 1885 and ranked in the top few for academic excellence in liberal arts, humanities, physical and natural sciences, and engineering. The campus occupies 8,800 acres of land, of which 5,200 acres are in general academic use. In all disciplines, the University has a primary commitment to excellence in research and education.

Applying

Admission to the chemistry department is by competitive application. Application forms are available from the Graduate Admissions Office and should be filed before January 1 for admission in the fall quarter. All applicants are required to submit GRE scores from the verbal, quantitative, and analytical tests and the Subject Test in chemistry, as well as transcripts and three letters of recommendation. Applicants are notified of admission decisions before March 15. In unusual circumstances, late applications or a deferred enrollment will be considered.

Correspondence and Information

Professor Michael D. Fayer, Chairman
Graduate Admissions Committee
Department of Chemistry
Stanford University
Stanford, California 94305-5080
Telephone: 415-723-4867

Stanford University

THE FACULTY AND THEIR RESEARCH

Wesley D. Allen, Assistant Professor; Ph.D., Berkeley, 1987. Ab initio quantum chemistry, excited states of polyatomic molecules, strained organic molecules, vibration-rotation spectra, development of theoretical methods.

Hans C. Andersen, Professor; Ph.D., MIT, 1966. Physical chemistry: statistical mechanics, theories of structure and properties of liquids, computer simulation methods, glass transition, water, exciton migration in random solids and polymers, phase transitions in lipid bilayer membranes, nonlinear spectroscopy.

Michel Boudart, Professor; Ph.D., Princeton, 1950. Adsorption and catalysis: chemical dynamics in homogeneous and heterogeneous systems, adsorption at solid surfaces and catalysis.

Steven G. Boxer, Professor; Ph.D., Chicago, 1976. Physical and biophysical chemistry: primary photochemistry of photosynthesis in real and model systems, magnetic field effects to probe mechanisms, CIDNP studies of protein and nucleic acid structure, spectroscopy of synthetic chromophore-protein complexes.

John I. Brauman, Professor; Ph.D., Berkeley, 1963. Organic physical chemistry: structure and reactivity of ions in the gas phase, photochemistry and spectroscopy of gas phase ions, electron photodetachment spectroscopy, photodetachment of electrons, electron affinities, reaction mechanisms.

James P. Collman, Professor; Ph.D., Illinois, 1958. Organometallic chemistry: synthetic analogues of the active sites in hemoproteins, homogeneous catalysis, multielectron redox catalysts, organo–transition-metal chemistry, multiple metal-metal bonds, organic metals.

Carl Djerassi, Professor; Ph.D., Wisconsin–Madison, 1945. Organic chemistry: chemistry of steroids, terpenes, and alkaloids with major emphasis on marine sources, application of chiroptical methods—especially circular dichroism and magnetic circular dichroism—to organic and biochemical problems, organic chemical applications of mass spectrometry, use of computer artificial-intelligence techniques in structure elucidation of organic compounds.

Dale G. Drueckhammer, Assistant Professor; Ph.D., Texas A&M, 1987. Bioorganic chemistry: mechanism of enzymatic reactions, applications of enzymes in organic synthesis.

Michael D. Fayer, Professor; Ph.D., Berkeley, 1974. Physical chemistry and chemical physics: dynamics in molecular condensed phases; laser spectroscopy; picosecond nonlinear and holographic grating techniques; narrow-band methods; crystals, polymers, glasses, solutions, and biologically related materials; radiationless and radiative processes.

John H. Griffin, Assistant Professor; Ph.D., Caltech, 1989. Bioorganic chemistry: studies in molecular recognition, new catalytic antibodies.

Keith O. Hodgson, Professor; Ph.D., Berkeley, 1972. Inorganic and structural chemistry: chemistry and structure of metal sites in biomolecules, molecular and crystal structure analysis, protein crystallography, extended X-ray absorption, fine-structure spectroscopy.

Wray H. Huestis, Professor; Ph.D., Caltech, 1972. Biophysical chemistry: chemistry of cell-surface receptors, membrane-mediated control mechanisms, biochemical studies of membrane protein complexes in situ, ^{19}F-magnetic resonance studies of conformation and function in soluble proteins.

William S. Johnson, Professor Emeritus; Ph.D., Harvard, 1940. Organic chemistry: new synthetic methods; synthesis of natural products, e.g., steroids, insect hormones, terpenoids, biomimetic olefinic cyclizations.

Robert J. Madix, Professor; Ph.D., Berkeley, 1964. Surface and interface science: relationships between surface composition structure and heterogeneous reactivity of metal and metalloid surfaces, catalysis, organometallic surface chemistry, electrochemistry and corrosion, reaction dynamics.

Harden M. McConnell, Professor; Ph.D., Caltech, 1951. Physical chemistry, biophysics, immunology: membrane biophysics and immunology, cell-surface recognition, spin labels, surface diffraction of X rays, membrane phase transitions.

Lisa McElwee-White, Assistant Professor; Ph.D., Caltech, 1983. Physical organic and organometallic chemistry, mechanisms of organometallic photoreactions, electronic structure theory applied to metal-carbon bonding.

Harry S. Mosher, Professor Emeritus; Ph.D., Penn State, 1942. Organic chemistry: stereochemistry, asymmetric organic reactions, Grignard reaction mechanisms, animal toxins.

Robert Pecora, Professor; Ph.D., Columbia, 1962. Physical chemistry: statistical mechanics of fluids and macromolecules, molecular motions in fluids, light-scattering spectroscopy of liquids, macromolecules and biological systems.

John Ross, Professor; Ph.D., MIT, 1951. Physical chemistry: experimental and theoretical studies of chemical kinetics, chemical instabilities, oscillatory reactions, efficiency of thermal engines and reaction rate processes, glycolysis.

Edward I. Solomon, Professor; Ph.D., Princeton, 1972. Physical inorganic and bioinorganic chemistry: inorganic spectroscopy and ligand field theory, active sites, spectral and magnetic studies on bioinorganic systems directed toward understanding the structural origins of their activity, photoelectron spectroscopic studies on surfaces and heterogeneous catalysts.

T. Daniel P. Stack, Assistant Professor; Ph.D., Harvard, 1988. Inorganic/bioinorganic chemistry, coordination chemistry of transition-metal complexes relevant to active sites in biological systems.

Henry Taube, Professor Emeritus; Ph.D., Berkeley, 1940. Inorganic chemistry: mechanisms of inorganic reactions and reactivity of inorganic substances, new aquo ions, dinitrogen as ligand, back-bonding as affecting properties including reactivity of ligands, mixed-valence molecules.

Barry M. Trost, Professor; Ph.D., MIT, 1965. Organometallic chemistry, new synthetic methods, natural product synthesis and structure determinations, insect chemistry, potentially antiaromatic unsaturated hydrocarbons, chemistry of ylides.

Robert M. Waymouth, Assistant Professor; Ph.D., Caltech, 1987. Mechanistic and synthetic chemistry of the early transition elements, inorganic polymers, mechanisms of olefin polymerization.

Paul A. Wender, Professor; Ph.D., Yale, 1973. Organic chemistry: synthetic organic chemistry, formalisms for synthesis design, new reactions and reagents, mechanism of action of tumor promoters.

Richard N. Zare, Professor; Ph.D., Harvard, 1964. Physical chemistry, chemical physics: application of lasers to chemical problems, molecular structure and molecular reaction dynamics.

STATE UNIVERSITY OF NEW YORK AT STONY BROOK
Department of Chemistry

Programs of Study

The Department of Chemistry at the State University of New York at Stony Brook offers programs of study leading to the M.S., M.A.T., and Ph.D. degrees in chemistry.

Students pursuing the Ph.D. may choose to conduct their dissertation research in any one of the diverse areas of chemistry represented by the interests of the departmental faculty. Interdisciplinary study under the guidance of a faculty member in another department is also possible. There are numerous coordinated activities with the Departments of Biochemistry, Biology, Computer Sciences, Earth and Space Sciences, Materials Science, Pharmacological Sciences, and Physics and with the College of Engineering and Applied Science, including formal degree options in chemical physics and chemical biology.

Research Facilities

The Graduate Chemistry Building is a modern seven-story structure containing spacious, well-equipped laboratories and offices overlooking Long Island Sound. General facilities available to support research include a well-staffed, comprehensive chemical library located within the chemistry building; glassblowing, machine, and electronics shops; a variety of minicomputers for on-line data gathering and instrumentation control; and departmental computers and array processors with high-resolution color graphics workstations.

Departmental instruments include a wide range of ultraviolet, visible, electron spin resonance, nuclear magnetic resonance (including one operating at 600 MHz), infrared, and mass spectrometers; graphics workstations; a multiprocessor computer network; low-temperature magnetic susceptibility apparatus; an electrochemical system; preparative and analytical ultracentrifuges; temperature-jump and stop-flow kinetics spectrometers; X-ray diffractometers; and a full range of vapor and liquid chromatographs.

Specialized instrumentation and facilities employed in ongoing research programs include high-pressure (10,000 atm) apparatus for the study of organic reactions; six Fourier-transform nuclear magnetic resonance spectrometers, including superconducting high-field instruments; supersonic beam multiphoton ionization spectrometers; pulsed nitrogen, excimer, Nd:yag, and CW dye laser systems; Stark spectroscopy systems; a high-resolution spectrometer-spectrograph; laser-Raman systems; a Fourier-transform computer system; intensity-fluctuation and Brillouin spectrometers; and a "hot lab" equipped for the preparation of radioactive compounds. Much research is done utilizing the synchrotron radiation facilities at nearby Brookhaven National Laboratory.

Financial Aid

The research program of the department is supported both by University funds and by grants and fellowships from governmental and industrial sponsors. Students are normally awarded teaching and research assistantships and fellowships that carry stipends of $12,500–$16,500 in 1991–92 for twelve months (including summer research appointments).

Cost of Study

Current University policy is to grant tuition scholarships for full-time teaching and research assistants. Other required fees amount to approximately $500 per academic year.

Cost of Living

Graduate students, including married couples, can be accommodated in the residence halls. The room charge for single students living on campus was approximately $1300 per semester in 1990–91. On-campus garden apartments also provide housing for nearly 1,000 graduate students and their families. Off-campus apartments and houses can be found in the vicinity of the University.

Student Group

Total University enrollment is more than 16,000, including both graduate and undergraduate students. The Department of Chemistry has more than 140 full-time graduate students and 45 postdoctoral fellows from all parts of the United States and the world.

Location

Stony Brook is located about 60 miles east of Manhattan on the wooded North Shore of Long Island, convenient to New York City's cultural life and Suffolk County's tranquil recreational countryside and seashores. Brookhaven National Laboratory and Cold Spring Harbor Laboratory are nearby. Long Island's hundreds of miles of magnificent coastline attract many swimming, boating, and fishing enthusiasts from around the world.

The University

Established thirty years ago as New York's comprehensive State University Center for the downstate-metropolitan area, the State University of New York at Stony Brook is recognized as one of the nation's finest research universities. Stony Brook offers excellent programs in a broad range of academic subjects and conducts major research and public service projects. An internationally renowned faculty offers courses from the undergraduate to the doctoral level through seventy-one undergraduate and graduate departmental and interdisciplinary majors. The Health Sciences Center includes nationally recognized medical and dental schools and a teaching hospital.

Stony Brook's bustling academic community is situated on 1,000 acres of fields and woodland. Bicycle paths, an apple orchard, a duck pond, and spacious plazas complement modern laboratories, classroom buildings, and the Fine Arts Center, giving Stony Brook spirit and cultural vitality.

Applying

Applications for September admission to graduate study in chemistry should be completed by March 15 in order to ensure consideration. Late applications will be considered if positions are available. Applications for admission, assistantships, and fellowships should be made at the same time. All application forms and a sixty-page booklet describing in detail the current research programs can be obtained from the address below.

Correspondence and Information

Chairperson, Graduate Admissions Committee
Department of Chemistry
State University of New York at Stony Brook
Stony Brook, New York 11794-3400
Telephone: 516-632-7886

SECTION 2: CHEMISTRY

State University of New York at Stony Brook

THE FACULTY AND THEIR RESEARCH

John M. Alexander, Professor; Ph.D., MIT, 1956. Reactions between heavy nuclei: scattering, transfer, and complete fusion; role of energy and angular momentum; fission and fragmentation; light particle emission.

Scott L. Anderson, Associate Professor; Ph.D., Berkeley, 1981. Experimental studies of state-to-state reaction dynamics and metal cluster reactions and spectroscopy.

Rodney A. Bednar, Joint Assistant Professor; Ph.D., Delaware, 1981. Mechanisms of enzyme action, affinity labeling and suicide enzyme inactivators, rational design of drugs.

Thomas W. Bell, Professor; Ph.D., University College, London, 1980. Synthesis and study of new cation-complexing molecules, synthetic methods, isolation and synthesis of insect pheromones.

Francis T. Bonner, Professor; Ph.D., Yale, 1945. Inorganic nitrogen chemistry, kinetics and mechanism studies in aqueous-solution and gas-solution systems, isotopic tracer and isotope exchange application, N-15 NMR.

Cynthia J. Burrows, Associate Professor; Ph.D., Cornell, 1982. Design and synthesis of compounds for complexation of organic anions and cations, design and synthesis of transition-metal ligands, biomimetic reactions and catalysis, small molecule–DNA interactions.

Benjamin Chu, Professor; Ph.D., Cornell, 1959. Laser and small-angle X-ray scattering, using synchrotron radiation; configuration and dynamics of macromolecules; conducting polymers; gels; pulsed field gel electrophoresis of DNA; spinodal decomposition and nucleation in polymer blends.

Frank W. Fowler, Professor; Ph.D., Colorado, 1967. Synthetic chemistry: development of new synthetic methods and application to total synthesis of important compounds, design and preparation of new solid-state materials, application of thermal methods.

Harold L. Friedman, Professor; Ph.D., Chicago, 1949. Molecular interpretation of equilibrium and dynamic properties of solutions, solvation, excess functions, transport and relaxation coefficients, spectral line shapes, scattering phenomena.

Theodore D. Goldfarb, Associate Professor; Ph.D., Berkeley, 1959. Environmental effects of the use of agricultural chemicals and of waste disposal technologies, science and public policy.

Albert Haim, Professor; Ph.D., USC, 1960. Mechanistic studies of thermal and photochemical reactions of transition-metal complexes.

David M. Hanson, Professor and Chairperson; Ph.D., Caltech, 1968. Studies of the physical and chemical processes induced by laser and synchrotron radiation in molecules, on surfaces, and in molecular solids.

Gerard S. Harbison, Associate Professor; Ph.D., Harvard, 1984. Solid-state NMR studies of biological systems.

Patrick J. Herley, Professor of Materials Science and Chemistry; Ph.D., 1964, D.Sc., 1983, Imperial College (London). Processes occurring in the thermal and photolytic decomposition of inorganic solids, electron microscopy of nanometer metallic particles.

Takanobu Ishida, Professor; Ph.D., MIT, 1964. Experimental and theoretical studies of equilibrium and kinetic isotope effects, stable isotope fractionation processes, electrochemical catalysis.

Franco P. Jona, Professor of Material Sciences and Chemistry; Ph.D., Zurich, 1949. Surface crystallography, low-energy electron diffraction, multilayer relaxation, gross characterization of ultrathin epitaxial films, electronic structure photoemission.

Francis Johnson, Professor of Pharmacology and Chemistry; Ph.D., Glasgow, 1954. Chemistry of mutagenic and carcinogenic events at the level of modified DNA, development of potential new drugs for AIDS, new synthetic methods, heterocyclic chemistry.

Philip M. Johnson, Professor; Ph.D., Cornell, 1966. Laser spectroscopy and the development of techniques for measurement of molecular structure and excited-state molecular dynamics, multiple optical resonance and multiphoton ionization spectroscopy.

Robert C. Kerber, Professor; Ph.D., Purdue, 1965. Organo–transition-metal complexes: synthesis, structure, mechanisms.

Stephen A. Koch, Associate Professor; Ph.D., MIT, 1975. Metal sulfide and oxide cluster compounds in bioinorganic chemistry, relationships between solid-state and molecular compounds.

Joseph W. Lauher, Professor; Ph.D., Northwestern, 1974. Synthetic, theoretical, and structural studies of new inorganic compounds: transition-metal carbonyl clusters, gold compounds, and complexes of macrocyclic ligands.

William J. le Noble, Professor; Ph.D., Chicago, 1957. Chemistry of highly compressed solutions, with applications such as the concertedness of multiple bond-reorganization reactions, encumbrance of transient intermediates, and biomimetic processes and synthesis; new approach to stereochemistry based on adamantanes.

Andreas Mayr, Associate Professor; Ph.D., Munich, 1978. Inorganic and organometallic chemistry; synthesis, reactivity, and properties of new transition-metal complexes, metal-ligand multiple bonds.

Michelle Millar, Research Associate Professor; Ph.D., MIT, 1975. Transition-metal complexes, organometallic/bioinorganic chemistry.

Marshall Newton, Adjunct Professor; Ph.D., Harvard, 1966. Theoretical chemistry, prediction and analysis of structure and energetics.

Iwao Ojima, Professor; Ph.D., Tokyo, 1973. Synthetic, bioorganic, and organometallic chemistry; asymmetric synthesis; homogeneous catalysis of transition-metal complexes.

Richard N. Porter, Professor; Ph.D., Illinois, 1960. Theoretical chemistry, quantum theory of reaction complexes, theory of small polyatomic molecules, field-theoretical treatment of electron correlation, dynamical mechanisms of gas-phase reactions.

Glenn D. Prestwich, Professor; Ph.D., Stanford, 1974. Biological and organic chemistry; hormone and pheromone receptors in insects, cholesterol biosynthesis in human liver, inositol phosphate and neuropeptide photoaffinity labels, modification of polysaccharides.

Steven E. Rokita, Associate Professor; Ph.D., MIT, 1983. Structural dependence of DNA modification, sequence-specific derivatization of nucleic acids, mechanisms of aromatic substitution reactions catalyzed by enzymes.

Stanley Seltzer, Adjunct Professor; Ph.D., Harvard, 1958. Elucidation of enzyme and organic reaction mechanisms.

Scott McN. Sieburth, Assistant Professor; Ph.D., Harvard, 1983. New synthetic methods and strategies, novel chemistry for enzyme inhibition, probes for small molecule–receptor interactions.

Charles S. Springer, Professor; Ph.D., Ohio State, 1967. In vivo nuclear magnetic resonance spectroscopy and imaging.

George Stell, Professor of Chemistry and Mechanical Engineering; Ph.D., NYU, 1961. Molecular theory of the fluid state, ionic fluid structural properties, transport in multiphase systems.

Daniel Strongin, Assistant Professor; Ph.D., Berkeley, 1988. Experimental surface science, laser-induced desorption, catalysis and electronic materials.

Sei Sujishi, Professor; Ph.D., Purdue, 1949. Organosilicon–transition-metal compounds; synthesis, new reactions, and bonding.

Hans Thomann, Adjunct Assistant Professor; Ph.D., SUNY at Stony Brook, 1982. Magnetic resonance in disordered heterogeneous and amorphous condensed matter.

Frank Webster, Assistant Professor; Ph.D., Chicago, 1987. Theory and computer modeling and simulation of condensed phase quantum molecular dynamics: problems of solvation, vibrational relaxation, and dissociation.

Arnold Wishnia, Associate Professor; Ph.D., NYU, 1957. Physicochemical studies of macromolecules, mechanism of initiation and role of cations in protein synthesis, polyelectrolyte interactions, applications of pressure-jump kinetics and NMR.

SECTION 2: CHEMISTRY

STATE UNIVERSITY OF NEW YORK COLLEGE OF ENVIRONMENTAL SCIENCE AND FORESTRY
Faculty of Chemistry

Programs of Study

The Faculty of Chemistry offers programs in chemistry leading to the degrees of Master of Science and Doctor of Philosophy. The student makes an appropriate selection of courses and chooses a research problem in consultation with his or her major professor within general guidelines established by the Faculty. The course work requirements are a minimum of 30 credits for the Ph.D. and 18 credits for the M.S. In addition, a significant research effort, resulting in a written thesis, is required in both programs. Courses are chosen from those offered by the chemistry and other faculties at the College as well as from offerings of the physical and biological science departments of Syracuse University, whose campus adjoins that of the College. Optional interdisciplinary programs in chemical ecology, biotechnology, and environmental systems science are available and recommended.

Major research areas of the Faculty are in synthetic and natural polymers, membrane science, organic materials science, environmental chemistry, natural products, ecological chemistry, and biochemistry. The fields of specialization are biochemistry and analytical, organic, and physical chemistry. The usual fundamental courses in physics, mathematics, biology, and advanced chemistry appropriate for chemists and biochemists are required of all students and are available at the College and at Syracuse University. Specialized courses in polymer chemistry, membrane science, wood and natural products chemistry, environmental chemistry, spectrometric identification, and biochemistry are also available.

Research Facilities

The Faculty of Chemistry is located in the Hugh P. Baker Laboratory, a 130,000-square-foot building well equipped for chemical, biochemical, wood, and polymer research. Research equipment and facilities available include recording ultraviolet, infrared, FT-IR, nuclear magnetic resonance, and mass spectrometers; X-ray equipment; small-angle and standard light-scattering instrumentation; differential scanning calorimeters; analytical and preparative ultracentrifuges; rapid membrane and vapor phase osmometers; an optical rotatory dispersion polarimeter; radiochemical laboratories with counters for solids, liquids, and gases; and extensive facilities for both light and electron microscopy. A large number of modern analytical instruments, including liquid and gas chromatographs, are also available. Extensive mainframe, microcomputer, and graphics computer resources are available and easily accessible. The Cellulose Research Institute and the Polymer Research Institute are associated with the Faculty's program.

Financial Aid

Essentially all M.S. and Ph.D. candidates in the Faculty are financially supported by fellowships, teaching assistantships, or government, industrial, or other research grants. Stipends vary from $9500 to $12,000 per year, in some cases with additional allowance for dependents. Tuition for fellows and assistants is waived. Teaching requirements are light.

Cost of Study

The tuition for 1990–91 was $1075 per semester for state residents and $2732.50 for out-of-state residents.

Cost of Living

Syracuse University dormitories are available for single students. The University also operates rental units for married students with or without children. Rooms and apartments can be found in the neighborhood of the College. Meal tickets are available from the University.

Student Group

The student body at Syracuse University currently numbers 21,000. The number of graduate students in the department is approximately 60. In addition, there are more than 10 postdoctoral fellows and visiting scientists.

Location

The city of Syracuse is the center of an urban region with a population of about 500,000. In addition to the State University College of Environmental Science and Forestry and Syracuse University, the State University Health Sciences Center at Syracuse, LeMoyne College, and Onondaga Community College are located in Syracuse. Both the city of Syracuse and Syracuse University offer a great variety of cultural and recreational opportunities. Situated on hills, near several lakes, the city offers good facilities for both summer and winter sports. The attractive Finger Lakes region is less than 20 miles away, the Thousand Islands are 90 miles, and the Adirondack and Catskill mountains and Niagara Falls are all within 150 miles. Travel time to New York City by car is 5 hours.

The College

The College of Environmental Science and Forestry is part of the State University of New York, which is, with its 230,000 students and seventy units, the largest university system in the United States. The College, founded in 1911, at present has 1,000 undergraduates, about 500 graduate students, and 100 faculty members. The campus has six major buildings. The Faculty of Chemistry was founded in 1919 and has awarded more than 80 M.S. degrees and 100 Ph.D.'s in chemistry during the past twenty years.

Applying

The usual requirement for entrance is a bachelor's degree in chemistry, biochemistry, or chemical engineering at the level of the American Chemical Society Accredited Curriculum, but deficiencies can be made up later. All applicants must take the Graduate Record Examinations. Typically, a student should have taken general chemistry, qualitative and quantitative analysis, instrumental analysis, organic chemistry, physical chemistry, at least a year of physics, and preferably a year and a half of calculus. There are no specific deadlines for applications. It is desirable, however, if admission is to be in September, that applications be submitted during the first few months of the year.

Correspondence and Information

Graduate Admissions
SUNY College of Environmental Science and Forestry
Syracuse, New York 13210
Telephone: 315-470-6599

SECTION 2: CHEMISTRY

State University of New York College of Environmental Science and Forestry

THE FACULTY AND THEIR RESEARCH

Gregory L. Boyer, Assistant Professor of Biochemistry; Ph.D., Wisconsin–Madison, 1980. Characterization and biochemistry of toxins, hormones, siderophores, and other biologically active natural products from plants and algae; biochemistry of iron.

Israel Cabasso, Professor of Polymer Chemistry and Director of the Polymer Research Institute; Ph.D., Weizmann (Israel), 1973. Polymer membranes, polymer blends, diffusion and transport in polymer matrices, biodegradable polymers, ion-exchange polymers, inorganic polymers, conductive polymers, electrochemical processes.

Paul M. Caluwe, Associate Professor of Polymer Chemistry; Ph.D., Louvain (Belgium), 1967. Synthesis of new monomers, reactions on polymers, heterocyclic chemistry, polycyclic aromatic hydrocarbons.

John P. Hassett, Associate Professor of Environmental Chemistry; Ph.D., Wisconsin–Madison, 1978. Behavior and analysis of natural and anthropogenic organic compounds in aquatic and soil environments.

David L. Johnson, Associate Professor of Environmental Chemistry; Ph.D., Rhode Island, 1973. Analytical methods development, automated SEM/Image analysis, computerized individual particle characterizations, heavy-metal speciation.

David J. Kieber, Assistant Professor of Environmental Chemistry; Ph.D., Miami (Florida), 1988. Environmental organic chemistry, aquatic photochemistry, analytical methods development, chemical oceanography.

Robert T. LaLonde, Professor of Chemistry; Ph.D., Colorado, 1957. Isolation, structure determination, and synthesis of natural products, emphasizing molecules significant to aquatic ecology; chlorinated products from lignin and humics.

Anatole Sarko, Professor of Polymer Chemistry and Chairman of the Faculty; Ph.D., SUNY at Syracuse, 1966. Structure determination of natural polymers by X-ray crystallography and theoretical model analysis, molecular dynamics of polymers, solution conformation of polymers, polymer morphology by solid-state light scattering, computer techniques in chemistry.

Conrad Schuerch, Distinguished Professor of Chemistry Emeritus; Ph.D., MIT, 1947. Stereospecific synthesis of glycosides and polysaccharides; synthetic carbohydrate antigens; mechanisms of stereospecific polymerization; wood chemistry.

Robert Milton Silverstein, Professor of Ecological Chemistry Emeritus; Ph.D., NYU, 1949. Spectrometric identification of organic compounds; isolation, characterization, and synthesis of insect and mammalian pheromones and other natural products of ecological significance.

Johannes Smid, Professor of Polymer Chemistry; Ph.D., SUNY at Syracuse, 1957. Specific solvent-solute interactions, structure of ion pairs, mechanism of anionic polymerization, ion binding on polymeric molecules, polymer catalysis, solid ion-containing and ion-conducting polymers, polysiloxanes, X-ray contrast polymers.

Kenneth J. Smith Jr., Professor of Polymer Chemistry; Ph.D., Duke, 1962. Physical chemistry of polymers, thermodynamics and conformational behavior of macromolecules, elasticity and optical properties of cross-linked polymer networks, semicrystalline polymers and polyelectrolyte gels, structure-property relationships of fibrous polymers, biomedical polymers.

Stuart W. Tanenbaum, Professor of Microbial Chemistry and Biotechnology; Ph.D., Columbia, 1951. Isolation and characterization of fungal secondary metabolites, molecular biology of the action of cytotoxic natural products, biomass conversions, structure-function biochemistry of microbial enzymes, bioremediation of toxic pollutants.

Tore E. Timell, Professor of Forest Chemistry; Ph.D., Royal Institute of Technology (Stockholm), 1950. Isolation, structure, and physical properties of hemicelluloses and pectin present in wood and bark; distribution of polysaccharides and lignin in wood cells; formation, ultrastructure, and chemistry of reaction wood.

Francis Xavier Webster, Assistant Professor of Ecological Chemistry; Ph.D., SUNY-ESF, 1986. Isolation, identification, and synthesis of insect, mammalian, and plant semiochemicals; synthesis of radiolabeled semiochemicals and analogues for pheromone biochemical studies.

William T. Winter, Associate Professor of Polymer Chemistry; Ph.D., SUNY at Syracuse, 1974. Polymer morphology, X-ray and solid-state NMR studies on biopolymers, microbial polysaccharides, molecular modeling, plant gums.

Adjunct Faculty

David A. Driscoll, Ph.D. SUNY-ESF, Analytical and Technical Services.
Harry L. Frisch, Ph.D. SUNY at Albany.
D. Graiver, Ph.D. Dow Corning Company.
Donald E. Nettleton, Ph.D. Bristol-Myers Company.
S. Alexander Stern, Ph.D. Syracuse University.

SECTION 2: CHEMISTRY

STEVENS INSTITUTE OF TECHNOLOGY

Chemistry and Chemical Biology Programs

Programs of Study

Graduate programs in chemistry and chemical biology, offered by the combined Department of Chemistry and Chemical Engineering, lead to the M.S. and Ph.D. degrees.

Students in chemistry programs are required to complete five core courses in the following areas: chemical thermodynamics, advanced physical chemistry, advanced organic chemistry, instrumental analysis, and advanced inorganic chemistry. Chemical biology students substitute physiology for the advanced inorganic chemistry course. Additional elective courses are chosen from the student's major concentration area.

Thirty credits are required for the master's degree, which must include either a thesis (5 to 10 credits) or a special research problem (3 credits). An additional 60 credits are required for the doctoral degree, of which 15 are normally taken in formal courses. There are no special examinations for the master's degree. Doctoral examinations include the qualifying examination, which must be passed within ten months of matriculating in the doctoral program, the preliminary examination, and a foreign language examination.

Many areas of specialization are available, including polymer synthesis and characterization, natural-products and medicinal chemistry, instrumental analysis and instrument design, X-ray crystallography, computational methods, theoretical chemistry, neuromuscular physiology, physiological chemistry, and molecular biology.

Most graduate courses are scheduled in the evening for the convenience of both full-time students, who can concentrate on research during the daytime hours, and part-time students, most of whom are employed at nearby laboratories during the day.

Research Facilities

Research in this combined department provides access to sophisticated equipment and instrumentation not usually available within a single department. Major facilities include 200-MHz multinuclear FT-NMR, FT-IR, automated X-ray diffractometers; low-angle X-ray scattering, electron impact, and chemical ionization mass spectrometers; a scanning tunneling microscope (STM); GC/MS equipment; analytical and preparative ultracentrifuges; a tissue-culture laboratory; automated scintillation counters; small-animal facilities; a wide selection of spectrophotometers (including atomic absorption) and gas and liquid chromatographs; and extensive facilities for polymer characterization.

Financial Aid

Appointments as teaching or research assistants are awarded competitively, based on qualifications. A nine-month assistantship in 1991–92, beginning at $14,910 for first-year students, includes a stipend of $7515 and tuition and fees for 15 credits. Supplementary stipends are available from various sources to augment assistantships. Stipends for doctoral students and doctoral candidates are appropriately higher.

Cost of Study

In 1991–92, tuition is $485 per credit. Most graduate courses are 2½ credits. A normal full-time load is 7½–10 credits per semester; part-time students may enroll in one or two courses per semester. The general fee is $60 per semester. The application fee is $35.

Cost of Living

A wide variety of accommodations are available; costs vary depending on the students' circumstances and life-style. The Married Students' Apartments are located on campus. Single graduate students are accommodated in Stevens-owned apartments located adjacent to the campus. In addition, privately owned off-campus rooms and apartments are available in the immediate area.

Student Group

Total graduate enrollment in science, engineering, and management is 1,837, and the undergraduate enrollment is 1,274. Forty-four students are enrolled in graduate programs in chemistry and chemical biology. There are also 11 postdoctoral research associates. Approximately 65 percent of the graduate students are U.S. citizens. Approximately 21 percent of the graduate students are women.

Location

Hoboken today is a center of urban renaissance. Once a major Eastern Seaboard port, Hoboken has been rediscovered by a new generation of professionals associated with Stevens Institute or employed across the Hudson River in New York City who find it an attractive place to live. Art galleries, restaurants, and a chamber orchestra add to its cultural attractions, yet century-old bakeries, specialty shops, and ornate residences provide stability in the midst of rapid change. Hoboken is also a hub for surface transportation, lying at the terminus of several commuter railroads and major highways and only minutes away from Newark International Airport to the west and New York City to the east.

The Institute

Founded in 1870, Stevens Institute has been a major force in technological education. The fields of engineering economics and industrial psychology were created at Stevens, and its undergraduate science and engineering programs have been noted for their breadth and intensity. Graduate education is a major component of the Institute's mission. The department ranks eighth in the nation in the number of master's degrees granted in chemical engineering and frequently ranks among the top fifteen in the number of master's degrees granted in chemistry. Close ties are maintained with the vast chemical and pharmaceutical industries, medical schools, and research hospitals centered in the New York–New Jersey area. Research in biotechnology and in polymer science plays an important role in New Jersey's outstanding progress in technology.

Applying

Applications for admission can be considered at any time, but at least four weeks should be allowed for processing after all required documents have been received. Applications for financial aid should be made at the same time as applications for admission and should be received, along with all documentation, by February 1. GRE scores are strongly recommended. International applicants whose native language is not English must include TOEFL scores. Admission consideration is open to all qualified applicants without regard to race, color, national origin, religion, sex, or handicap.

Correspondence and Information

Professor Maghar S. Manhas
Department of Chemistry and Chemical Engineering
Stevens Institute of Technology
Castle Point on the Hudson
Hoboken, New Jersey 07030
Telephone: 201-216-5526

SECTION 2: CHEMISTRY

Stevens Institute of Technology

THE FACULTY AND THEIR RESEARCH

Ajay K. Bose, George Meade Bond Professor; Ph.D., MIT, 1950. Organic and bioorganic chemistry; synthesis of penicillins and other antibiotics; steroids and heterocycles; natural products; carbon-13, deuterium, and nitrogen-15 NMR spectroscopy; chemical ionization mass spectrometry; stable-isotope labeling for biosynthetic and metabolic studies.

Donald F. DeWitt, Assistant Professor; Ph.D., Michigan, 1980. Physiology, cardiovascular physiology, protection of the ischemic myocardium, control of coronary blood flow.

Walter C. Ermler, Professor; Ph.D., Ohio State, 1972. Theoretical chemistry; electronic structure of atoms and molecules, using first-principles methods; relativistic effects in molecules composed of heavy atoms; pseudopotentials; vibration-rotation theory.

Francis T. Jones, Professor and Head of the Department; Ph.D. Polytechnic of Brooklyn, 1960. Physical chemistry, mass spectroscopy, reaction kinetics, radiation chemistry and photochemistry, decomposition of halogenated compounds.

Nuran M. Kumbaraci, Associate Professor; Ph.D., Columbia, 1977. Physiology, neuromuscular physiology, biochemistry and biophysics of muscle contraction, nerve conduction, synaptic transmission.

Edmund R. Malinowski, Professor; Ph.D., Stevens, 1961. Physical and analytical chemistry, nuclear magnetic resonance, applications of factor analysis to chemistry, spectroscopic solvent effects.

Maghar S. Manhas, Professor; Ph.D., Allahabad (India), 1950. Organic chemistry, synthesis of heterocyclic compounds, penicillins and cephalosporins, compounds of medicinal interest, NMR and mass spectroscopy in structure determination, optical rotatory dispersion.

Ernest W. Robb, Professor; Ph.D., Harvard, 1957. Organic chemistry, reaction mechanisms, NMR spectroscopy, computational chemistry.

Salvatore S. Stivala, Rene Wasserman Professor; Ph.D., Pennsylvania, 1960. Physical chemistry of macromolecules, light scattering, X-ray scattering, ultracentrifugation, solution properties of synthetic and biological polymers, kinetics of polymer degradation.

SECTION 2: CHEMISTRY

TEXAS A&M UNIVERSITY
Department of Chemistry

Program of Study

The Department of Chemistry offers the M.S. and Ph.D. degrees in the traditional areas of chemistry and in chemical physics and biochemical, catalytic, electrochemical, environmental, nuclear, polymer, solid-state, spectroscopic, structural, and theoretical fields. This work is supervised by a faculty that has achieved broad national and international recognition for its outstanding research contributions and includes a Nobel laureate, a National Medal of Science awardee, holders of international medals in a variety of chemistry subdisciplines, and members of both the National Academy of Sciences and the Royal Society. The department has a faculty of 51, whose efforts are supported by approximately 90 postdoctoral and visiting faculty researchers. This student-to-scientist ratio of 2:1 provides an intensive, personalized learning environment.

The M.S. degree requires 18 and the Ph.D. approximately 25-30 semester hours of lecture course work. Both are research degrees. Requirements for the Ph.D. usually include one basic course in each of the four major areas of chemistry and a comprehensive examination in the major field of study.

The graduate student selects a research supervisor and, in consultation with the supervisor, a committee. The results of the research investigation must be summarized in a dissertation suitable for later publication. There is no language requirement at the M.S. level; the Ph.D. language requirement is determined by the student's committee to meet individual needs related to research. The average period required to complete the degrees is two years for the M.S. and approximately four years for the Ph.D. Students are encouraged to work directly toward the Ph.D. degree.

Research Facilities

The chemistry complex has 224,000 square feet for teaching and research in four buildings, with major institutes housed in three other buildings. It maintains professionally staffed laboratories for high-resolution mass spectrometry, solution NMR, solid-state NMR, single-crystal and powder X-ray diffraction, and departmental computing. Departmental instrumentation includes ESCA, SIMS, Auger, and other surface science instruments; a GC/MS mass spectrometer; a Kratos MS-50 triple analyzer including fast-atom bombardment ionization, extended mass range, and tandem mass spectrometry configuration; and a variety of EPR, infrared, far-infrared, Raman, UV-visible, fluorescence, atomic absorption, gamma-ray, and photoelectron spectrometers. Other campus facilities include the Nuclear Science Center (1-MW reactor) and the Cyclotron Institute with its new superconducting cyclotron.

Financial Aid

All graduate students in good standing receive full financial support for the duration of their studies. Teaching assistantships are available to all qualified entering students. The 1991–92 stipend for a nine-month teaching assistantship is $10,800. Teaching or research assistantship support during the summer provides an additional $2700–$3300. Research assistantship stipends average $10,800 per year. The Robert A. Welch Foundation and the Industry-University Cooperative Chemistry Program sponsor prestigious twelve-month fellowships for outstanding applicants. These stipends are $16,000 for twelve months. University Minority Merit Scholarships are available for black and Hispanic students.

Cost of Study

In 1991–92, tuition and fees for graduate students are $474 per semester (12 credit hours) for Texas residents. Students on assistantships and fellowships are considered in-state students for tuition purposes.

Cost of Living

University apartments are available, although applications for University housing should be made early. Private rooms, apartments, and houses are available close to the campus. In 1991–92, University apartments for students rent for as low as $170 per month, while private apartments range from $200 to $400 per month.

Student Group

The student body consists of more than 40,000 resident students enrolled in over a dozen colleges and schools. Approximately 15 percent of the students are enrolled in the many programs of the Graduate College. Both the enrollment and the percentage of students in graduate work have been increasing rapidly. All of the states and numerous foreign countries are normally represented in the student body. The Department of Chemistry has about 300 graduate students.

Location

As a university town, College Station has a high proportion of professional people and enjoys many of the advantages of a cosmopolitan center without the disadvantages of a congested urban environment. There are many film series, a symphony, chamber music, and choral groups. Situated in the middle of a triangle formed by Dallas, Houston, and Austin, the symphonies, ballets, sporting events, museums, and concerts of these cities are within easy driving distance.

Mild, sunny winters make the region eminently suitable for year-round outdoor activities from fishing and hiking in the beautiful pine woods of eastern Texas to boating and camping in the magnificent Texas hill country. There are over 100 state parks within a day's drive of College Station.

The University and The Department

Founded in 1876, Texas A&M University is the state's oldest public institution of higher education. Vigorous research programs in engineering, physics, mathematics, and medicine provide the chemist with supplementary facilities and sources of specialized information. The Evans Library houses 1.6 million volumes and maintains subscriptions to approximately 8,000 scientific and technical journals.

An active faculty of 51 generated approximately 312 publications and over $10-million in external grant funding last year. The department is considered one of the top twenty in the country.

Applying

Inquiries regarding admission to the University, as well as information about facilities for advanced studies, research, and requirements for graduate work in chemistry, should be addressed to the Department of Chemistry. Application for admission should be filed no later than four weeks prior to the opening of the semester. Applications for assistantships and fellowships are accepted for both regular semesters and the summer session. Awards are made as long as funds are available, but early application is advisable.

Correspondence and Information

Department of Chemistry
Texas A&M University
College Station, Texas 77843
Telephone: 409-845-5345
              800-334-1082 (toll-free)

SECTION 2: CHEMISTRY

Texas A&M University

THE FACULTY AND THEIR RESEARCH

Emory T. Adams, Professor; Ph.D., Wisconsin–Madison, 1962. Biophysical chemistry.
Thomas O. Baldwin, Professor; Ph.D., Texas, 1971. Protein chemistry.
Derek H. R. Barton, Distinguished Professor; Ph.D., Imperial College (London), 1940. Organic chemistry.
David E. Bergbreiter, Professor; Ph.D., MIT, 1974. Organic chemistry.
John W. Bevan, Professor; Ph.D., London, 1975. Physical chemistry.
John O'M. Bockris, Distinguished Professor; Ph.D., Imperial College (London), 1945. Physical chemistry.
Lawrence S. Brown, Assistant Professor; Ph.D., Princeton, 1986. Physical chemistry.
Abraham Clearfield, Professor; Ph.D., Rutgers, 1954. Inorganic chemistry.
Dwight C. Conway, Professor; Ph.D., Chicago, 1956. Physical chemistry.
F. Albert Cotton, Doherty-Welch Distinguished Professor; Ph.D., Harvard, 1955. Inorganic chemistry.
Donald J. Darensbourg, Professor; Ph.D., Illinois at Urbana-Champaign, 1968. Organometallic chemistry.
Marcetta Y. Darensbourg, Professor; Ph.D., Illinois at Urbana-Champaign, 1967. Organometallic chemistry.
John P. Fackler Jr., Distinguished Professor and Dean; Ph.D., MIT, 1960. Inorganic chemistry.
Karl A. Gingerich, Professor; Ph.D., Freiburg (Germany), 1957. Physical chemistry.
D. Wayne Goodman, Professor; Ph.D.; Texas at Austin, 1974. Physical chemistry.
Michael B. Hall, Professor and Head; Ph.D., Wisconsin–Madison, 1971. Inorganic chemistry.
Kenn E. Harding, Professor; Ph.D., Stanford, 1968. Organic chemistry.
James W. Haw, Associate Professor; Ph.D., Virginia Tech, 1982. Analytical chemistry.
Cornelis A. J. Hoeve, Professor; D.Sc., Pretoria (South Africa), 1955. Polymer chemistry.
John L. Hogg, Professor; Ph.D., Kansas, 1974. Bioorganic chemistry.
Timothy Hughbanks, Assistant Professor; Ph.D., Cornell, 1983. Inorganic chemistry.
Jeffery W. Kelly, Assistant Professor; Ph.D., North Carolina, 1986. Organic and bioorganic chemistry.
Jaan Laane, Professor; Ph.D., MIT, 1967. Physical chemistry.
Paul A. Lindahl, Assistant Professor; Ph.D., MIT, 1985. Inorganic chemistry.
Robert R. Lucchese, Associate Professor; Ph.D., Caltech, 1982. Theoretical chemistry.
Jack H. Lunsford, Professor; Ph.D., Rice, 1962. Physical chemistry.
Ronald D. Macfarlane, Professor; Ph.D., Carnegie Tech, 1959. Nuclear and physical chemistry.
Kenneth N. Marsh, Professor; Ph.D., New England (Australia), 1968. Physical chemistry.
Arthur E. Martell, Distinguished Professor; Ph.D., NYU, 1941. Inorganic chemistry.
William E. McMullen, Assistant Professor; Ph.D., UCLA, 1985. Physical chemistry.
Edward A. Meyers, Professor; Ph.D., Minnesota, 1955. Physical chemistry.
Joseph B. Natowitz, Professor; Ph.D., Pittsburgh, 1965. Nuclear chemistry.
Daniel H. O'Brien, Associate Professor; Ph.D., Virginia, 1961. Organic chemistry.
Frank M. Raushel, Professor; Ph.D., Wisconsin–Madison, 1976. Biochemistry.
Alan S. Rodgers, Associate Professor; Ph.D., Colorado, 1960. Physical chemistry.
Michael P. Rosynek, Professor; Ph.D., Rice, 1972. Physical chemistry.
Marvin W. Rowe, Professor; Ph.D., Arkansas, 1966. Analytical cosmochemistry.
David H. Russell, Professor; Ph.D., Nebraska–Lincoln, 1978. Analytical chemistry.
Donald T. Sawyer, Distinguished Professor; Ph.D., UCLA, 1956. Analytical and inorganic chemistry.
Richard P. Schmitt, Professor; Ph.D., Berkeley, 1978. Nuclear chemistry.
Emile A. Schweikert, Professor; Ph.D., Paris IV (Sorbonne), 1964. Activation analysis and analytical chemistry.
A. Ian Scott, Davidson Distinguished Professor; Ph.D., Glasgow, 1952. Organic chemistry, biochemistry.
Daniel A. Singleton, Assistant Professor; Ph.D., Minnesota, 1986. Organic chemistry.
Manuel P. Soriaga, Associate Professor; Ph.D., Hawaii, 1978. Analytical chemistry.
Gary Sulikowski, Assistant Professor; Ph.D., Pennsylvania, 1989. Organic chemistry.
Gyula Vigh, Associate Professor; Ph.D., Veszprém (Hungary), 1975. Analytical chemistry.
Rand L. Watson, Professor; Ph.D., Berkeley, 1966. Nuclear chemistry.
Kevin L. Wolf, Professor; Ph.D., Washington (Seattle), 1969. Nuclear chemistry.
Danny L. Yeager, Professor; Ph.D., Caltech, 1975. Theoretical chemistry.
Ralph A. Zingaro, Professor; Ph.D., Kansas, 1950. Inorganic chemistry.

SECTION 2: CHEMISTRY

UNIVERSITY OF ARIZONA

College of Arts and Sciences
Department of Chemistry

Programs of Study	The University of Arizona offers programs of study leading to the degrees of Master of Arts, Master of Science, and Doctor of Philosophy in chemistry. Although advanced course work is an integral part of the M.S. and Ph.D. degrees, strong emphasis is placed on developing a student's ability to initiate and carry out an original research project. Each student combines courses, seminars, and research in a program that is personally appropriate. Teaching and research programs encompass the areas of analytical, inorganic, organic, and physical chemistry. In addition to these usual divisions, active programs are under way in polymer chemistry, biochemistry, chemical physics, quantum chemistry, molecular structure, and other areas that extend over several divisions or disciplines. Faculty members participate in interdisciplinary programs with other units on campus, such as the Optical Sciences Center, Department of Biochemistry, and College of Medicine.
Research Facilities	The department is housed in four buildings, with a fifth under construction. One, the Carl S. Marvel Laboratory, is devoted exclusively to research. The University Science Library is adjacent. The department has its own glass, machine, and electronics shops. Modern instrumentation includes one 500-, two 250-, and one 90-MHz FT heteronuclear NMR spectrometers; an X-ray diffractometer; two high-resolution photoelectron spectrometers; an Auger spectrometer; a minisupercomputer; a multitechnique surface analysis system (SPS, UPS, AES); inductively coupled plasmas; two picosecond laser systems; ring dye lasers; several argon ion and pulsed solid-state lasers; supersonic jet spectrometers; Raman spectrometers; molecular beam spectrometers; and a variety of HPLCs, mass spectrometers, etc. The technical staff includes professional specialists in X-ray crystallography, nuclear magnetic resonance, surface science, lasers and optics, and computer science. The facilities for the synthesis of air-sensitive compounds, thermally stable polymers, and biologically active compounds are outstanding.
Financial Aid	Graduate teaching assistantships with a minimum stipend of $10,130 for the academic year 1991–92 are available to qualified students. Assistantships may be extended through the summer session with additional remuneration. Research assistantships are available for some students who have completed their first year of graduate study. Stipends are comparable to those for teaching assistantships. Out-of-state tuition is waived for graduate students with assistantships. The department holds a scholarship competition every spring. Invited participants are the guests of the chemistry department. A variety of scholarships are awarded, including the Dupont/Marvel Fellowship, Chemistry Department Fellowships, Gregson Fellowships, and the Graduate College Fellowships. Many of these fellowships are awarded to continuing graduate students throughout their period of study.
Cost of Study	Graduate assistantships include a waiver of out-of-state tuition. For 1991–92, full-time students are assessed a registration fee of $795 per semester.
Cost of Living	Dormitory housing for single students in 1991–92 ranges from $1355 to $1800 for the academic year. Married students may apply for apartments in Family Housing, located in northeast Tucson about 12–15 minutes from campus, with rents that vary from $240 to $410 per month. Privately owned off-campus housing is plentiful and available at similar rates. Early application for all campus housing is advised.
Student Group	The University of Arizona, with a total student body of over 34,000, draws students from every state and over fifty foreign countries. More than 125 students are currently pursuing graduate studies in the chemistry department. Admission to the graduate program in chemistry is highly competitive, with about 35 new students registering each fall. The department has approximately 50 postdoctoral fellows.
Location	The city of Tucson has a population of over 700,000. Located at the northern end of the great Sonoran plateau, Tucson is surrounded by four mountain ranges providing spectacular outdoor recreation sites. Hiking, mountain climbing, boating, snow skiing, and horseback riding are nearby diversions, while the Gulf of California, Grand Canyon, and numerous other national parks and monuments are within a day's drive. Since tourism is one of the principal sources of its income, Tucson has unusually fine restaurants, shopping centers, and cultural activities. The University is one of the largest single contributors to the economy, and many of the city's activities revolve around it.
The University	The University of Arizona, a land-grant university established in 1885, consists of 319 acres with over 130 buildings. The University has continually encouraged vigorous development of the sciences; the Departments of Chemistry, Physics, Mathematics, Astronomy, and Biochemistry are recognized as outstanding centers of graduate study. The University ranks ninth among all American public colleges and universities in total research funding for the physical sciences.
Applying	Applications for fall admission should be received by February 1, but later applications will also be given due consideration. International students are required to demonstrate their proficiency in English with a score of 230 or better on either the TSE or the University's SPEAK exam and a minimum score of 550 on the TOEFL. Application materials may be obtained from the Office of Academic Affairs in the Department of Chemistry.
Correspondence and Information	Department of Chemistry Academic Affairs Office University of Arizona Tucson, Arizona 85721 Telephone: 800-545-5814 (toll-free)

SECTION 2: CHEMISTRY

University of Arizona

THE FACULTY AND THEIR RESEARCH

Analytical Chemistry

Neal R. Armstrong, Ph.D., New Mexico, 1974. Photoelectrochemistry/surface electrochemistry, photoconduction of thin-film organic materials, microsensors; surface chemistry of active-metal surfaces; development of analytical methodology in surface-analysis spectroscopies: quantitation of composition, molecular lineshape analysis.

Steven W. Buckner, Ph.D., Purdue, 1988. Chemical aspects of mass spectrometry, ion-molecule reactions, photochemistry.

Michael F. Burke, Ph.D., Virginia Tech, 1965. Chromatographic separations, surface modification and characterization of adsorbents, detector systems for chromatography, use of digital computers for chemical research, field flow fractionation.

M. Bonner Denton, Ph.D., Illinois, 1972. Trace-level spectroscopic techniques; advanced detectors in spectroscopic analysis; instrumentation design, automation, and computer control; new design principles for quadrupole mass spectrometers; human genome mapping.

Quintus Fernando, Ph.D., Louisville, 1953. Molecular and X-ray emission spectroscopy, structural and kinetic aspects of metal-ligand interactions, analysis of trace components in environmental and forensic samples.

Henry Freiser, Ph.D., Duke, 1944. Chemistry of metal chelates; analytical separation processes; analysis and recovery of metals; ion-selective electrodes, trace analysis, environmental analytical methodology; liquid-liquid interfacial chemistry.

Jeanne E. Pemberton, Ph.D., North Carolina, 1981. Electrochemistry, surface electrochemistry, surface Raman spectroscopy, surface and interface analysis, electron spectroscopy, materials characterization.

S. Scott Saavedra, Ph.D., Duke, 1986. Bioanalytical chemistry, biosensors, bioadhesion, new optical materials and technologies.

Inorganic Chemistry

John H. Enemark, Ph.D., Harvard, 1966. Molybdenum chemistry; bioinorganic chemistry; metal nitrosyls; X-ray crystallography, heteronuclear NMR.

Robert D. Feltham, Ph.D., Berkeley, 1957. Synthesis and spectroscopic properties of transition-metal complexes of NO_x, dinitrogen, tertiary arsines, and sulfur ligands; high-temperature superconductors.

Philip C. Keller, Ph.D., Indiana, 1966. Boron hydride chemistry, aluminum chemistry, reactions of heterocycles and polydentate bases with boranes.

Dennis L. Lichtenberger, Ph.D., Wisconsin, 1974. Organometallic chemistry, catalysis, gas phase and surface photoelectron spectroscopy (UPS, XPS, synchrotron) of metal species with small organic molecules and fragments, metal-metal bonding, metal complex surfaces.

John V. Rund, Ph.D., Cornell, 1962. Reaction mechanisms of coordination compounds and organometallics, luminescence of transition-metal compounds, hydroboration of aromatic heterocycles.

F. Ann Walker, Ph.D., Brown, 1966. Metalloporphyrins; heme proteins; bioinorganic chemistry; NMR and EPR spectroscopy; electrochemistry.

David E. Wigley, Ph.D., Purdue, 1983. Inorganic and organometallic synthesis; metal-ligand multiple bonds; metallacycles and cyclization reactions; new synthetic methods.

Organic Chemistry

Robert B. Bates, Ph.D., Wisconsin, 1957. Mode of action of antitumor agents and antibiotics; resonance-stabilized carbanions; structure proof, synthesis, and biogenesis of natural products, especially antitumor agents; X-ray crystallography; NMR.

Daniel P. Dolata, Ph.D., California, Santa Cruz, 1983. Use of artificial intelligence, symbolic logic, and axiomatic theories in conformational analysis of transition-metal complexes, oligosaccharides, small peptides, and zeolites.

Richard S. Glass, Ph.D., Harvard, 1966. Synthetic methods, natural product synthesis, organosulfur chemistry, bioorganic mechanisms, organometallic chemistry.

Henry K. Hall Jr., Ph.D., Illinois, 1949. Synthesis and polymerization of novel monomers and piezoelectric and NLO polymers, mechanism of spontaneous polymerization and cycloadditions of electron-rich olefins with electron-poor olefins.

Victor J. Hruby (joint appointment with Biochemistry), Ph.D., Cornell, 1965. Design and synthesis of new amino acids, peptides, and proteins with specific conformational and topographical properties; conformation–biological activity relationships; brain chemistry; NMR for conformation and dynamic studies of peptides and peptide-macromolecular interactions; intercellular communication.

Eugene Mash Jr., Ph.D., Utah, 1980. Organic and bioorganic chemistry, asymmetric synthesis of natural products and macromolecules, new synthetic methods, mechanisms of enzyme-catalyzed reactions, chemical toxicology.

James E. Mulvaney, Ph.D., Polytechnic of Brooklyn, 1959. Polymer synthesis and stereochemistry; organolithium compounds; donor-acceptor polymeric interactions, conductive polymers.

David F. O'Brien, (joint appointment with Biochemistry), Ph.D., Illinois, 1962. Chemistry of membranes; synthesis of amphiphiles; formation, polymerization, and characterization of supramolecular assemblies; reconstitution of biological function.

Robin L. Polt, Ph.D., Columbia, 1986. New methods in organic synthesis (enolates and vinyl anions), the total synthesis of natural products (amino sugars and gangliosides).

Physical Chemistry

Ludwik Adamowicz, Ph.D., Polish Academy of Sciences (Warsaw), 1977. Ab initio quantum chemistry, many-body perturbation theory and coupled cluster methods for electronic structure calculation.

George H. Atkinson (joint appointment with Optical Sciences Center), Ph.D., Indiana, 1971. Picosecond spectroscopy and photochemistry of gas phase polyatomic molecules, time-resolved Raman spectroscopy, intracavity laser detection, chemical vapor deposition of new materials.

Michael Barfield, Ph.D., Utah, 1962. Experimental and theoretical aspects of magnetic resonance spectroscopy.

Peter F. Bernath, Ph.D., MIT, 1980. High-resolution spectroscopy of small molecules, laser spectroscopy and nonlinear optics, Fourier-transform spectroscopy, atmospheric and extraterrestrial molecules.

Michael F. Brown (joint appointment with Biochemistry), Ph.D., California, Santa Cruz, 1975. NMR spectroscopy; molecular physics of liquid crystals, biophysical chemistry of membranes, structure and dynamics of biological macromolecules, molecular basis of vision.

Stephen G. Kukolich, Ph.D., MIT, 1966. Molecular-beam microwave measurements on structure and interactions in small molecules and complexes, measurements of chemical shift tensors, microwave-infrared double resonance experiments, EPR studies on biologically important systems.

Walter B. Miller, Ph.D., Harvard, 1970. Chemical kinetics in crossed molecular beams.

W. R. Salzman, Ph.D., UCLA, 1967. Semiclassical theory of the interaction of radiation with matter, methods in time-dependent quantum mechanics, perturbation theory and approximation methods in quantum chemistry, theory of optical activity.

Mark A. Smith, Ph.D., Colorado, 1982. Gas phase reaction dynamics, ion- and electron-molecule reaction mechanisms, ultralow-energy collisons, molecular energy transfer, free jet flow theory.

G. Krishna Vemulapalli, Ph.D., Penn State, 1961. Electronic spectroscopy, supercritical phenomena.

Joint Appointments

Michael A. Cusanovich, Professor of Biochemistry; Ph.D., California, San Diego, 1967.

John A. Rupley, Professor of Biochemistry; Ph.D., Washington (Seattle), 1959.

Richard L. Shoemaker, Professor of Optical Sciences; Ph.D., Illinois, 1971.

Gordon Tollin, Professor of Biochemistry; Ph.D., Iowa State, 1956.

SECTION 2: CHEMISTRY

UNIVERSITY OF FLORIDA

Department of Chemistry

Programs of Study The Department of Chemistry offers programs of study leading to the M.S. and Ph.D. degrees in the general areas of analytical, inorganic, organic, and physical chemistry. More specialized research programs, such as chemical physics and theoretical, biological, polymer, organometallic, and nuclear chemistry, are available within the major areas.

The M.S. and Ph.D. degree requirements include a course of study, attendance at and presentation of a series of seminars, and completion and defense of a research topic worthy of publication. Candidates for the Ph.D. degree must also demonstrate a reading ability of one foreign language and show satisfactory performance on a qualifying examination. The M.S. degree is not a prerequisite for the Ph.D. degree. A nonthesis degree program leading to the M.S.T. degree is offered for teachers.

Students are encouraged to begin their research shortly after selecting a research director, who is the chairman of the supervisory committee that guides the student through his or her graduate career.

Research Facilities The chemistry department occupies 166,000 square feet of space in six buildings: Leigh Hall, the Chemical Research Building, Bryant Hall, Williamson Hall, the Nuclear Science Building, and a 55,000-square-foot unit completed in May 1990. The Marston Science Library is located near the chemistry facilities. The University library system holds more than 2.7 million volumes.

The major instrumentation includes ultraviolet-visible, infrared, fluorescence, Raman, nuclear magnetic resonance, electron spin resonance, X-ray, ESCA, and mass spectrometers. Many are equipped with temperature-control and Fourier-transform attachments, and some have laser sources. Data-storage and data-acquiring minicomputers are interfaced to some of the instruments, such as the recently constructed quadrupole resonance mass spectrometer. The chemistry department has VAX-11/780, VAX-11/750, and Sun-3/280 computers as well as multiple terminals connected to IBM machines in the main computer center on campus.

The departmental technical services include two well-equipped stockrooms and glassblowing, electronics, and machine shops to assist in equipment design, fabrication, and maintenance.

Financial Aid Most graduate students are given financial support in the form of teaching and research assistantships. Twelve-month stipends range from $12,500 to $15,800 for 1991–92. Assistantship holders pay fees of about $700 per calendar year. A limited number of full or supplemental fellowships are available to superior candidates.

Cost of Study In 1990–91, in-state students paid a registration fee of $76.78 per credit hour for each semester; out-of-state students paid an additional $161.93 ($238.71 per credit hour each semester).

Cost of Living The Gainesville area has a cost of living somewhat below the national average for the United States and housing is abundant. Although some graduate students reside in University-owned dormitories or apartments, the majority live in nearby apartments or homes. The cost varies with the quality of the accommodations. In 1991–92, a student should expect to spend about $185 per month for a shared apartment.

Student Group The graduate student body of the department totals about 170, and the graduate faculty currently numbers 42.

Location Gainesville, a heavily wooded city of about 100,000, is situated in north-central Florida about halfway between the cities of Jacksonville and Tampa, each about 100 miles away. The city is 75 miles from the nearest point on the Atlantic Ocean and 55 miles from the nearest point on the Gulf of Mexico. Gainesville has excellent shopping and recreational facilities as well as four major hospitals. It is served by four airlines. The mild climate and natural beauty of the area make it a particularly appealing place to live and study.

The University The University of Florida is the main institution of higher education in Florida, with 28,000 undergraduates and 7,000 graduate students. It consists of eighteen colleges, including the Colleges of Medicine, Dentistry, Veterinary Medicine, and Law. Many cultural and sporting events are offered on campus. There are several swimming pools, a golf course, and many tennis, handball, and racquetball courts in numerous locations around the campus. The Florida Museum of Natural History, the Harn Art Museum, and an 1,800-seat theater for the performing arts are also found on campus.

Applying Application forms for admission and for financial aid may be obtained by writing to the address below. All applicants must submit scores on the GRE General Test. International applicants whose native language is not English are required to score at least 550 on the TOEFL, and, if financial aid is desired, they must also achieve a score of at least 220 on the Test of Spoken English.

Correspondence and Information
Chairman, Graduate Selection Committee
Department of Chemistry
University of Florida
Gainesville, Florida 32611
Telephone: 904-392-0551

Peterson's Guide to Graduate Programs in the Physical Sciences and Mathematics 1992

University of Florida

THE FACULTY AND THEIR RESEARCH

Rodney J. Bartlett, Graduate Research Professor; Ph.D., Florida, 1971. Physical and quantum chemistry: ab initio predictions of molecular potential-energy surfaces, surface chemistry.

Merle A. Battiste, Professor; Ph.D., Columbia, 1959. Organic chemistry: nonclassical carbonium ions, nonbenzenoid aromatics, pheromones.

James M. Boncella, Assistant Professor; Ph.D., Berkeley, 1984. Inorganic chemistry: organometallic chemistry.

Anna Brajter-Toth, Assistant Professor; Ph.D., Southern Illinois, 1979. Analytical chemistry: chemically modified electrodes, electroanalytical determination of biologically active compounds.

Wallace S. Brey Jr., Professor; Ph.D., Pennsylvania, 1948. Physical chemistry: magnetic resonance and surface chemistry.

Philip J. Brucat, Assistant Professor; Ph.D., Stanford, 1984. Physical chemistry: experimental study of clusters and surfaces, band-gap structure in semiconductors.

Samuel O. Colgate, Associate Professor; Ph.D., MIT, 1959. Physical chemistry: thin film metastables, chemical vapor deposition.

Michael J. S. Dewar, Graduate Research Professor; D.Phil., Oxford, 1942. Theoretical chemistry: development of new theoretical procedures, applications of theory to chemical problems.

James A. Deyrup, Professor; Ph.D., Illinois, 1961. Organic chemistry: small-ring heterocycles.

William R. Dolbier Jr., Professor; Ph.D., Cornell, 1965. Organic chemistry: cycloadditions, thermal reorganizations, mechanistic photochemistry.

Russell S. Drago, Graduate Research Professor; Ph.D., Ohio State, 1954. Inorganic chemistry: transition-metal ion chemistry, catalysis, theoretical inorganic chemistry.

Randolph S. Duran, Assistant Professor; Ph.D., Strasbourg (France), 1987. Physical chemistry: polymer liquid crystals, polymer properties and structures.

Eric Enholm, Assistant Professor; Ph.D., Utah, 1985. Organic chemistry: synthetic methodology, total syntheses.

John R. Eyler, Professor; Ph.D., Stanford, 1972. Physical chemistry: laser-induced gaseous chemistry.

Robert J. Hanrahan, Professor; Ph.D., Wisconsin, 1957. Physical chemistry: radiation chemistry, photochemistry and mass spectroscopy.

Willard W. Harrison, Professor and Dean; Ph.D., Illinois, 1964. Analytical chemistry: glow discharge mass spectrometry, trace-element analysis, sputter atomization.

John F. Helling, Professor; Ph.D., Ohio State, 1960. Organic chemistry: organometallic chemistry.

William M. Jones, Distinguished Service Professor; Ph.D., USC, 1955. Organic chemistry: carbene chemistry, metal-carbene complexes.

Alan R. Katritzky, Kenan Professor; D.Phil., Oxford, 1954. Organic chemistry: heterocyclic chemistry.

Robert T. Kennedy, Assistant Professor; Ph.D., North Carolina, 1988. Analytical chemistry: separations, application of trace analysis to biological systems.

Per-Olov Löwdin, Graduate Research Professor; Ph.D., Uppsala (Sweden), 1948. Physical and quantum chemistry: quantum theory of atoms, molecules, and solid state.

David A. Micha, Professor; Ph.D., Uppsala (Sweden), 1966. Physical chemistry: electronic structure and dynamics of molecular systems.

Luis M. Muga, Professor; Ph.D., Texas, 1957. Physical chemistry: investigations of ternary fission.

Gardiner H. Myers, Associate Professor; Ph.D., Berkeley, 1965. Physical chemistry: gas kinetics, atmospheric chemistry, chemical education.

N. Yngve Öhrn, Professor; Ph.D., Uppsala (Sweden), 1966. Physical and quantum chemistry: many-electron properties of molecules.

Gus J. Palenik, Professor; Ph.D., USC, 1960. Inorganic chemistry: structure determination by X-ray diffraction.

Willis B. Person, Professor; Ph.D., Berkeley, 1953. Physical chemistry: molecular spectroscopy.

John R. Reynolds, Associate Professor; Ph.D., Massachusetts, 1985. Organic chemistry: polymer chemistry.

Nigel G. Richards, Assistant Professor; Ph.D., Cambridge, 1985. Biological chemistry: proteins, chiral syntheses, molecular modeling.

David Richardson, Associate Professor; Ph.D., Stanford, 1981. Inorganic chemistry: model compounds for oxygen binding to nonheme iron proteins, reactions of oxygen with metal complexes.

Kirk S. Schanze, Assistant Professor; Ph.D., North Carolina, 1983. Organic chemistry: photochemistry, micelles and vesicles as model membranes.

Gerhard M. Schmid, Associate Professor; Ph.D., Innsbruck (Austria), 1958. Analytical chemistry: interfacial electrochemistry, absorption, electrode kinetics.

R. Carl Stoufer, Associate Professor; Ph.D., Ohio State, 1959. Inorganic chemistry: coordination chemistry—synthesis, structure, and properties.

Daniel R. Talham, Assistant Professor; Ph.D., Johns Hopkins, 1985. Inorganic chemistry: charge-transfer complexes, synthesis and properties of electrically conducting compounds.

Martin T. Vala, Professor; Ph.D., Chicago, 1964. Physical chemistry: spectroscopy and photochemistry.

Kenneth Wagener, Associate Professor; Ph.D., Florida, 1973. Organic chemistry: polymer chemistry.

William Weltner Jr., Professor; Ph.D., Berkeley, 1950. Physical chemistry: molecular spectroscopy and EPR spectroscopy.

James D. Winefordner, Graduate Research Professor; Ph.D., Illinois, 1958. Analytical chemistry: analytical atomic and molecular spectroscopy.

Richard A. Yost, Professor; Ph.D., Michigan State, 1979. Analytical chemistry: computer-controlled triple quadrupole mass spectroscopy, trace analysis, structure elucidation.

Vaneica Y. Young, Assistant Professor; Ph.D., Missouri–Kansas City, 1976. Analytical chemistry: surface chemistry, photoelectron spectroscopy.

Michael C. Zerner, Professor and Chairman; Ph.D., Harvard, 1966. Physical and quantum chemistry: calculations of molecular properties, structures of polynuclear transition-metal complexes and heterocyclic systems.

John A. Zoltewicz, Professor; Ph.D., Princeton, 1960. Organic chemistry: heterocyclic chemistry, nucleophilic substitution.

UNIVERSITY OF GEORGIA
Department of Chemistry

Programs of Study

The Department of Chemistry offers programs of study leading to the Ph.D. or M.S. degree in four major areas: analytical, inorganic, organic, and physical chemistry, including projects within a number of multidisciplinary centers of excellence: the Center for Complex Carbohydrate Research, the Center for Computational Quantum Chemistry, and the Center for Metalloenzyme Studies.

Research advisers are chosen by incoming students during their first year, after they have interviewed all faculty members in their chosen area. An advisory committee is then selected to advise the student on the preparation of a program of study. At least 45 quarter hours of credit are required. Students must demonstrate proficiency in each of the four areas of chemistry by passing either a standardized American Chemical Society test or an appropriate graduate course. A preliminary examination, including defense of an original research proposal, is administered by the student's advisory committee before the student's seventh quarter in residence. After the student passes the preliminary examination and is admitted to candidacy, the emphasis is on research, culminating in the writing of a dissertation, which is defended by the student in front of the advisory committee and other interested faculty members. Ph.D. graduates generally complete the program in four to five years. The Northeast Regional Section of the American Chemical Society, based in Athens, provides monthly programs featuring well-known speakers, social gatherings, and an annual award banquet. The department maintains an active seminar program, featuring mainly outside speakers. All students participate in this program.

Research Facilities

The Department of Chemistry occupies nearly 200,000 square feet of space, housing combined teaching, laboratory, and research areas. Much of the department's major instrumentation is located in a new central instrumentation facility. Graduate students are normally expected to have hands-on training on these instruments. Among the facilities generally available in the department are a Bruker AC250 250-MHz NMR spectrometer, a Bruker AM300 300-MHz NMR spectrometer, a Bruker ACF400 400-MHz NMR spectrometer, a Finnigan 4000 Series gas chromatograph–mass spectrometer system, an ENRAF-NONIUS X-ray diffraction system, and a Jasco recording spectropolarimeter. Various faculty members have equipment of their own, including several major laser systems; numerous monochromators and laser-excited fluorescence analysis instruments; several UV-visible spectrometers, including a Hewlett-Packard 8451 diode array spectrometer; time-of-flight, FT-ICR, and quadrupole mass spectrometers; two pulsed supersonic molecular beam machines; numerous minicomputers and microcomputers for data acquisition, etc.; ultrahigh-vacuum surface analysis instrumentation for LEED, Auger electron spectroscopy, XPS, and TDS; and five Fourier-transform infrared spectrometers. Finally, the department's computing facilities consist of a VAX-11/750 minicomputer connected through a local area network to several MicroVAX computers, some of which are in individual research laboratories. The local area network accesses a campuswide broadband network for use of the mainframe and supercomputer. Also available is the Biological Sequence/Structure Computation Facility, consisting of a VAX 6210 minicomputer and two real-time molecular graphics workstations (Silicon Graphics 4D70GT and 4D20GT) for molecular modeling and sequence studies. The University System's resources in computing include a CDC CYBER 205 supercomputer, a CDC 850 computer, an IBM 3090-400 E/VF computer, an IBM 4381 computer, and several others.

Financial Aid

All graduate students are given stipends by the department in the form of teaching and research assistantships. These range from about $9430 to $12,575 per year. Fellowship awards of up to $15,000 are available for superior students.

Cost of Study

All graduate students are charged tuition on an in-state resident basis. Fees for the 1990–91 academic year were $140 per quarter. This is all-inclusive, covering registration, tuition, health, etc. There are no fees for graduate teaching assistants.

Cost of Living

Athens is an average-cost-of-living town in almost every respect. Since considerable apartment and condominium construction has occurred in the past year, housing is relatively easy to find in all price ranges. The University has a housing office, and local real estate agents are eager to assist graduate students.

Student Group

The total enrollment of nearly 28,000 includes about 7,000 graduate students. The department currently has about 90 full-time graduate students, with a graduate faculty of 28. The department expects to add several new faculty members over the next few years.

Location

Athens is located in the hills of northeast Georgia at an elevation of approximately 800 feet. Only about 65 miles from downtown Atlanta, the area enjoys the conveniences of travel, the arts, shopping, and professional sports provided by a large city, while maintaining a comfortable, small-town atmosphere. Numerous outdoor activities are available. Approximately 3 hours to the north are the Blue Ridge and Great Smoky mountains with winter sports, while the beautiful beaches of the Georgia-Carolina coast are only 4 hours or so to the east. Local activities also include the full range of intercollegiate sports in the highly competitive Southeastern Conference, as well as a well-organized intramural program for students, faculty, and staff.

The University

The University of Georgia is the flagship unit of the University System of Georgia. It is the oldest state university in the United States, having been chartered in 1785. There are thirteen schools and colleges on campus, and the Franklin College of Arts and Sciences, of which the chemistry department is a member, is the oldest. The dominant spirit of the department is the spirit of inquiry, a spirit that holds that teaching and learning are two aspects of the same thing.

Applying

The University and the department do not discriminate with regard to race, nationality, religion, or gender; all qualified students are urged to apply. A completed application will contain an application form, Graduate Record Examinations (GRE) scores, transcripts from every institution of higher learning attended, and three letters of recommendation. Foreign students whose mother tongue is not English must furnish evidence of English proficiency in the form of Test of Spoken English (TSE) and Test of English as a Foreign Language (TOEFL) scores.

Correspondence and Information

Professor Michael A. Duncan, Graduate Coordinator
Department of Chemistry
University of Georgia
Athens, Georgia, 30602
Telephone: 404-542-1936

SECTION 2: CHEMISTRY

University of Georgia

THE FACULTY AND THEIR RESEARCH

Nigel G. Adams, Ph.D., Birmingham (England), 1966. Physical chemistry: interstellar chemistry as modeled by gas-phase ion-molecule reaction in a flowing afterglow apparatus.

Peter Albersheim, Ph.D., Caltech, 1959. Biochemistry: structure and function of complex carbohydrates.

Norman L. Allinger, Ph.D., UCLA, 1954. Organic chemistry: theory of structure of organic compounds.

I. Jonathan Amster, Ph.D., Cornell, 1986. Analytical chemistry: Fourier-transform mass spectrometry.

James L. Anderson, Ph.D., Wisconsin, 1974. Analytical chemistry: environment and biochemical trace-element analysis.

J. Phillip Bowen, Ph.D., Emory, 1984. Organic chemistry: rational molecular design, mechanistic studies, conformational analysis, structure-function relationships using computational chemistry and computer graphics.

Lionel A. Carreira, Ph.D., MIT, 1969. Physical chemistry: chemical applications of coherent anti-Stokes Raman spectroscopy (CARS), detection of trace species by ultrasensitive photothermal techniques.

James A. de Haseth, Ph.D., North Carolina at Chapel Hill, 1977. Analytical chemistry: various aspects of FT-IR spectroscopy coupled with gas and liquid chromatographic techniques.

Richard A. Dluhy, Ph.D., Rutgers, 1982. Analytical chemistry: biological applications of FT-IR spectrometry.

Michael A. Duncan, Ph.D., Rice, 1981. Physical chemistry: laser spectroscopy and mass spectroscopy of metal clusters.

John F. Garst, Ph.D., Iowa State, 1957. Physical organic chemistry: formation, structure, and reactivities of Grignard reagents; reactions of allenes with metal hydrides.

Richard K. Hill, Ph.D., Harvard, 1954. Organic chemistry: stereochemistry, synthesis and mechanisms, cycloadditions and sigmatropic rearrangements, stereochemistry of enzymatic reactions at prochiral centers.

Michael K. Johnson, Ph.D., East Anglia (England), 1977. Inorganic chemistry: elucidation of structural, electronic, and magnetic properties of transition-metal clusters in metalloproteins.

Allen D. King Jr., Ph.D., Texas, 1963. Physical chemistry: surface and colloidal density, gas adsorption and other properties of polar-nonpolar interfaces.

R. Bruce King, Ph.D., Harvard, 1961. Inorganic chemistry: synthesis, spectra, and reactivity of organic derivatives of various metals and metalloids; chemical applications of graph theory.

John J. Kozak, Dean, College of Arts and Sciences; Ph.D., Princeton, 1965. . Theoretical chemistry: theory of condensed media, phase transitions, energy conversion and storage.

Donald M. Kurtz Jr., Ph.D., Northwestern, 1977. Inorganic chemistry: mechanism of assembly of Fe, S, and Mo clusters in metalloproteins; properties of the non-heme iron oxygen-carrying protein, hemerythrin.

Charles Kutal, Head; Ph.D., Illinois, 1970. Inorganic chemistry: organic and organometallic photochemistry, photocatalysis of organic reactions, gas phase photochemistry of transition-metal compounds.

George F. Majetich, Ph.D., Pittsburgh, 1979. Organic chemistry: total synthesis of natural products and their analogues, development of organosilicon synthetic methods, application of NMR techniques to structural analysis.

Charles E. Melton, Ph.D., Notre Dame, 1954. Physical chemistry: geochemistry and geophysics, origin of life, evolution of the oceans and the atmosphere.

M. Gary Newton, Ph.D., Georgia Tech, 1966. Organic chemistry: structure and reactivity of organic phosphorus compounds, computer modeling of structure, X-ray structure determination.

S. William Pelletier, Ph.D., Cornell, 1950. Organic chemistry: isolation and elucidation of structure and stereochemistry of alkaloids; synthesis of alkaloids, terpenes, photoalexines, and antitumor agents.

Robert S. Phillips, Ph.D., Georgia Tech, 1979. Organic chemistry: enzymatic reaction mechanisms, use of enzymes in organic synthesis.

Henry F. Schaefer III, Director of the Center for Computational Quantum Chemistry; Ph.D., Stanford, 1969. Physical chemistry: structure of various molecules, such as SiC_2, silaethylene, and HO_2 dimer; prediction of potential energy surfaces and electronic spectra.

Paul von R. Schleyer, Ph.D., Harvard, 1957. Organic chemistry: experimental and theoretical studies of organic compounds, carbocations, lithium compounds, identification of molecular structures from NMR.

Robert A. Scott, Presidential Young Investigator; Ph.D., Caltech, 1980; . Physical and inorganic chemistry: structure of active sites of metalloproteins and metalloenzymes using X-ray absorption spectroscopy, dynamics of long-range electron transfer in biological systems.

John L. Stickney, Ph.D., California, Santa Barbara, 1984. Analytical chemistry: surface chemistry of electrodes; surface studies of catalysts using LEED, Auger spectrometry, ESCA, and other techniques.

Herman van Halbeek, Ph.D., Utrecht (Netherlands), 1982. Biochemistry: structural analysis of complex carbohydrates using NMR spectroscopy.

SECTION 2: CHEMISTRY

UNIVERSITY OF HOUSTON

Department of Chemistry

Programs of Study

The Department of Chemistry offers programs leading to Master of Science and Doctor of Philosophy degrees in the areas of analytical, biological, inorganic, organic, physical, and theoretical chemistry, as well as chemical physics. A minimum of 54 total credit hours of graduate-level work is necessary for the Doctor of Philosophy degree, and a minimum of 30 total credit hours is required for the Master of Science degree. Of these totals, a minimum of 24 credit hours of the Doctor of Philosophy program and 18 credit hours of the Master of Science program must be in graduate-level lecture course work. Entering students are required to take diagnostic examinations, which are administered immediately preceding the first semester of enrollment in the graduate program. Each graduate student must take the appropriate divisional candidacy examination within two calendar years after admission to the program. The candidate must pass an oral final examination after the research problem has been completed and the dissertation or thesis has been written. In addition, each candidate for a Ph.D. in chemistry must successfully present one oral proposition to the dissertation committee. All degree requirements must be met within five calendar years after admission to the graduate program. A minimum of twelve months of continuous full-time residence at the University of Houston is required.

Research Facilities

The department occupies approximately 60,000 square feet in the Lamar Fleming Building and 18,000 square feet in the Science and Research I Building. Current external research funding is $4.4-million. Instrumentation includes mass spectrometers (including LC and GC); UV visible and IR spectrophotometers; Raman spectrometers: NMR, ESR, FT-IR, and AA spectrometers; X-ray and electron-beam equipment; ultracentrifuges; spectropolarimeters; gas and liquid chromatographs; ESCA, SIMS, Auger, and other surface science instruments; and a VAX 8650 computer with an FPS 264 vector processor and extensive graphics facilities, as well as access to the campus mainframe, an AS 9000. The department employs full-time technical support personnel: glassblower, machinist, electronics engineer and technician, crystallographer, instrument operators, and computer specialists. The University's central campus library contains over 2 million volumes and 600,000 microfilms, plus large quantities of manuscripts, maps, and similar materials. The University has developed an extensive network of computer systems to satisfy the academic, research, and administrative requirements for computing services.

Financial Aid

Teaching fellowships, research assistantships, and loans are available. Teaching fellowships range from $8000 to $9000 in 1991–92 for a nine-month period. Stipends for research assistantships vary. Scholarships from Bayer-Mobay, Dow, Exxon, Pennzoil, and Shell are awarded yearly to outstanding graduate students in the department. Those who qualify for the Crockett Fellowship receive up to $15,000. Lynn Murray awards are given to one or more graduate students in the amount of $1000 per student. Teaching assistants' awards are given each spring in the amounts of $200 and $500 per year. Other forms of financial assistance for graduate students are also available.

Cost of Study

Tuition per semester for 1991–92 is about $800 for Texas residents and about $2300 for out-of-state residents for 15 semester hours and includes all fees except laboratory fees, books, and parking. Teaching fellows are exempt from out-of-state tuition fees.

Cost of Living

Dormitory rates for single students, including room and board, average less than $4000 for the 1991–92 academic year. Apartment costs in the Houston area begin at approximately $250–$300 per month.

Student Group

Enrollment at the University is approximately 33,000, including about 5,000 graduate students. The 120-member graduate student body in the Department of Chemistry includes an increasing number of out-of-state students and a number of foreign students.

Location

Houston is the nation's fourth-largest city and the largest in the South and Southwest. Houston's weather is generally mild, which encourages outdoor activities. The cost of living is the seventh lowest out of the forty major metropolitan areas of the United States. The city has 260 municipal parks, encompassing 5,747 acres, and 70 miles of Gulf Coast beaches are only an hour's drive from downtown Houston. Houston has many outstanding cultural attractions, which include the nationally acclaimed Houston Symphony Orchestra, the Houston Grand Opera, the Houston Ballet, and numerous museums.

The University

The University of Houston was established in 1927. It became a state-supported university in 1963 and is one of the largest state-supported universities in Texas. Accessible from all parts of the city by public transportation and freeway systems, the 390-acre University Park campus of the University of Houston is located approximately 3 miles from downtown Houston. The natural attractions of this convenient setting are being steadily augmented through long-range planning, which emphasizes architectural harmony and appropriate landscaping.

Applying

All applicants are required to submit GRE General Test scores, transcripts, and three letters of reference. Applicants are strongly encouraged, but not required, to take the GRE Subject Test in chemistry. Foreign applicants must submit TOEFL and TSE scores. All applicants are considered for fellowships. Early application is advisable to ensure consideration for financial support. Applicants are notified after the completed application is reviewed. Some financial assistance is available for students who wish to visit the department; arrangements should be made through the contact below. Application and other information may be obtained from the address below.

Correspondence and Information

Director of Graduate Admissions
Department of Chemistry
University of Houston
Houston, Texas 77204-5641
Telephone: 713-749-4695

Peterson's Guide to Graduate Programs in the Physical Sciences and Mathematics 1992

SECTION 2: CHEMISTRY

University of Houston

THE FACULTY AND THEIR RESEARCH

Thomas A. Albright, Associate Professor of Organic and Inorganic Chemistry; Ph.D., Delaware, 1975. Theoretical organic and inorganic chemistry, structure and reactivity of organometallic molecules and solid-state materials.

John L. Bear, Professor of Inorganic Chemistry and Chairman; Ph.D., Texas Tech, 1960. Chemistry of dinuclear transition-metal complexes, electrochemistry, photochemistry and chemical reactivity.

Ralph S. Becker, Professor of Physical Chemistry; Ph.D., Florida State, 1955. Spectroscopy, photophysics, photochemistry, laser-flash photolysis, photoelectrochemistry, liquid crystals and photochromism.

Ivan Bernal, Professor of Inorganic Chemistry; Ph.D., Columbia, 1963. Conformational and configurational problems of organometallics containing metals as chiral centers, optical induction by distal ligand centers upon the metal of an organometallic or coordination compound, stereochemistry of the disulfide linkage in organic and organometallic compounds, mechanism(s) of self-resolution in chemistry.

Edwin Carrasquillo-Molina, Assistant Professor of Physical Chemistry; Ph.D., Chicago, 1984. Spectroscopy and molecular dynamics from single quantum states; photophysics, photodissociation, and collisional energy transfer of vibrationally energized molecules; triplet state spectroscopy and photochemistry, intersystem crossing.

James R. Cox Jr., Associate Professor of Organic Chemistry; Ph.D., Harvard, 1958. Mechanisms of organic and biological reactions.

Roman S. Czernuszewicz, Assistant Professor of Analytical Chemistry; Ph.D., Marquette, 1981. Vibrational spectroscopy; resonance Raman effect; structure, functions, and role of metal sites in biological molecules; coordination, compounds, and intermediates of biological interest; developing spectroscopic techniques for probing metalloproteins; metalloporphyrins and oxidized metalloporphyrins; surface adsorbed species; Raman spectroelectrochemistry and surface enhanced Raman scattering.

Stanley N. Deming, Professor of Analytical Chemistry; Ph.D., Purdue, 1971. Application of computers to analytical chemistry, automated development of analytical chemical methods, optimization techniques applied to the development of methods, high-performance liquid chromatography, surface and interfacial phenomena.

Donald Elthon, Associate Professor of Analytical Chemistry; Ph.D., Columbia, 1980. Microanalysis and materials characterization by electron beam, X-ray, and ion beam (SIMS) techniques; phase equilibria and thermodynamics at high temperatures and pressures; crystal-liquid equilibria of oxides and silicates.

Royal B. Freas III, Assistant Professor of Physical Chemistry; Ph.D., Delaware, 1984. Gas-phase ion/molecule reactions, structure and reactivity of metal clusters, mass spectrometry, collision spectroscopy, photodissociation spectroscopy, development of instrumentation and desorption ionization methods, analytical and biomedical applications of liquid chromatography/mass spectrometry.

Russell A. Geanangel, Professor of Inorganic Chemistry; Ph.D., Ohio State, 1968. Synthesis of inorganic and organometallic compounds with structure and bonding interest, use of multinuclear NMR and Mössbauer spectroscopy in the elucidation of new structures, acceptor-donor interactions, boron-nitrogen and divalent tin compounds.

David M. Hoffman, Associate Professor of Inorganic Chemistry; Ph.D., Cornell, 1982. Synthesis of transition-metal inorganic and organometallic compounds, preparation of thin films from organometallic and inorganic precursors by chemical vapor deposition.

Allan J. Jacobson, Robert A. Welch Professor of Chemistry; Ph.D., Oxford, 1969. Synthesis and structural characterization of mixed-metal oxides, ionic and electronic conductivity in solids, synthesis of micropores and inorganic-organic composite structures, heterogeneous catalysis.

Marvin T. Jones, Professor of Physical Chemistry; Ph.D., Washington (St. Louis), 1961. Chemical and physical problems related to molecular and electronic structure; magnetic resonance techniques for free radical solids, especially low-dimensional electrical conducting materials.

Karl M. Kadish, Professor of Analytical Chemistry; Ph.D., Penn State, 1970. Electroanalytical and bioelectroanalytical chemistry, rates and mechanisms of electron transfer in biologically important compounds, chemistry and electrochemistry of metalloporphyrins, redox reactions of transition-metal complexes, spectroelectrochemistry.

Larry Kevan, Cullen Professor of Chemistry; Ph.D., UCLA, 1963. Development of electron magnetic resonance methods and their application to disordered media, especially electron spin echo spectrometry and double resonance; paramagnetic probes of zeolite surfaces; radical orientation and adsorbate interactions on oxide surfaces; photoionization processes in micelles and vesicles; electron solvation and localization; ion and atom solvation geometry; peroxy and nitroxide radical probes of molecular motion in polymers and on surfaces; X-ray photoelectron spectroscopy studies of metal ions in oxides; electron magnetic resonance studies of copper oxide superconductors.

Jay K. Kochi, Robert A. Welch Professor of Chemistry; Ph.D., Iowa State, 1952. Mechanisms of organic reactions catalyzed by metal complexes, oxidation and reduction mechanisms, electrochemistry and photochemistry of organometallic compounds, application of ESR spectroscopy to organic and organometallic free radicals and the mechanism of homolytic and catalytic reactions.

Harold L. Kohn, Professor of Organic Chemistry; Ph.D., Penn State, 1971. Mechanisms of biochemical and medicinal processes, synthesis and investigation of heterocyclic compounds and new drug candidates.

Donald J. Kouri, Professor of Theoretical Chemistry; Ph.D., Wisconsin–Madison, 1965. Quantum theory of atomic and molecular collisions, few-body problem, approximate methods for calculating cross sections, theory of reactive scattering, molecule-surface collisions, applications of supercomputers for solving quantum mechanical molecular scattering.

Robert L. Matcha, Professor of Theoretical Chemistry; Ph.D., Wisconsin–Madison, 1965. Relativistic interactions in small molecular systems, development of numerical techniques for computing molecular wave function, application of perturbation theory to the study of molecular properties and models.

James Andrew McCammon, M. D. Anderson Professor of Theoretical, Physical, and Biological Chemistry; Ph.D., Harvard, 1976. Statistical mechanics of macromolecules and liquids; theory of protein structure, dynamics, and function; development and application of computer models and simulation methods for molecular systems.

Mamie W. Moy, Professor of Chemical Education; M.S., Houston, 1952. Chemical education, teaching techniques and methodology.

B. Montgomery Pettitt, Associate Professor of Physical and Theoretical Chemistry; Ph.D., Houston, 1980. Structure and thermodynamics of fluids and aqueous solutions, theory of molecular fluids and biomolecules in solution, structure of synthetic metal catalysts, design of antivirals and other pharmaceuticals.

John Wayne Rabalais, Professor of Physical Chemistry; Ph.D., LSU, 1970. UV and X-ray (ESCA) photoelectron spectroscopy, with applications of ESCA to surface reactions and catalysis; applied quantum chemistry; ion-scattering spectrometry; secondary ion mass spectrometry; reactions of active ions with surfaces; dynamics of ion-surface collisions.

Joseph Paul Street, Assistant Professor of Organic Chemistry; Ph.D., California, Davis, 1983. Physical organic chemistry, with main interests in biological reaction mechanisms and enzyme catalysis.

Randolph P. Thummel, Professor of Organic Chemistry; Ph.D., California, Santa Barbara, 1971. Organic synthesis and mechanism; annelated aromatic compounds; small-ring and heterocyclic chemistry, synthesis and transition-metal chemistry of novel chelating systems.

Wayne E. Wentworth, Professor of Physical-Analytical Chemistry; Ph.D., Florida State, 1957. Electron attachment phenomena, electron-capture/detector flame inhibition.

Albert Zlatkis, Professor of Analytical and Organic Chemistry; Ph.D., Wayne State, 1952. Gas chromatographic methods of analysis, metabolites of biological fluids, environmental problems.

SECTION 2: CHEMISTRY

UNIVERSITY OF ILLINOIS AT CHICAGO

Department of Chemistry

Programs of Study

The department offers a Ph.D. program in chemistry with specializations in analytical chemistry, biochemistry, inorganic chemistry, organic chemistry, physical chemistry, and theoretical chemistry. There is considerable overlap between these areas, as well as opportunity for interdisciplinary work involving other disciplines, such as biology and physics. For those whose principal goal is teaching, the department offers a Doctor of Arts (D.A.) program.

The doctoral programs require 96 semester hours beyond the B.S. degree. A thesis is the center of the program, and the course work is determined according to the student's background and interests. Candidates must successfully complete a program of cumulative examinations. The Ph.D. program is usually completed in four to five years.

The school year (1991–92) is divided into two equivalent semesters plus a shorter semester session in which no graduate courses are offered. Most students begin their graduate studies in the fall semester, which starts in late August. The department does not have evening graduate programs.

Research Facilities

The research activities are housed in modern, well-equipped laboratories and are supported by department-staffed electronics, machine, and glassblowing shops. Major special facilities are in the areas of surface chemistry, laser photochemical kinetics, computer graphics, vibrational circular dichroism, computational chemistry, light scattering, time-resolved fluorescence spectroscopy, very low temperature magnetic measurements, optical spectroscopy, X-ray diffraction, fast kinetics, photoelectron spectroscopy, and molecular beam studies. The science library is in the same building as the chemistry department. The department has shared facilities, including a well-equipped NMR laboratory and a mass spectroscopy lab, for carrying out experimental and theoretical research in wide areas of chemistry and biochemistry.

The Health Sciences Center, which has additional shared research facilities, is just 5 minutes away by shuttle bus, and Argonne National Laboratory and Fermilab are less than an hour's drive away, as are several major research institutions and industrial centers.

Financial Aid

Financial support is available to all qualified graduate students. A limited number of fellowships for outstanding students, as well as dedicated fellowships for minority students, are available through the department and the University. Support for newly admitted teaching assistants starts at $12,200 for the 1991–92 calendar year. Most advanced graduate students are supported by research assistantships.

Cost of Study

The University waives tuition and service fees for all graduate assistants and fellows on a quarter-time to two-thirds-time appointment (graduate students have a half-time appointment); these students pay $292 per semester for health insurance, health services, and a general fee.

Cost of Living

Housing is available in residence halls of the University; the cost for twelve months in 1991–92 is $3800–$4400. Many graduate students secure housing in the immediate neighborhood, around town, or in the suburbs at a wide range of rents. Chicago's rental prices are moderate for a major urban area.

Student Group

The campus has over 25,000 students, approximately 7,000 of whom are enrolled in graduate programs. The Department of Chemistry has approximately 130 graduate students. The department has an excellent record of success in placing its students, particularly in industrial positions in the Chicago area.

Location

Chicago, with a metropolitan population of about 7 million, is the cultural center of the Midwest, famous for its symphony, opera, theater companies, and museums. All forms of sports and indoor and outdoor recreation are readily accessible. The campus is located just a few blocks from the Loop, the business district of Chicago, and less than 2 miles from Lake Michigan. Transportation to the campus from the city and the suburbs is excellent.

The University

The University of Illinois at Chicago is one of two research campuses of the University of Illinois and is the major public university in Chicago. UIC is classified as a Research I university and ranks fifty-eighth in the country in terms of total funds for research and development. It has fifteen colleges, including the nation's largest college of medicine, and a number of special institutes, centers, and laboratories. The university occupies seventy-eight buildings on approximately 187 acres.

Applying

Applications for admission may be submitted at any time, but should be received at least three months before the beginning of the desired term. Graduate students may start in fall or (in special cases) spring, and credit may be received for previous graduate work. Applicants from foreign institutions must provide scores on the TOEFL, and all students must submit scores on the chemistry Subject Test of the GRE.

Correspondence and Information

Further information and application forms may be obtained from:
Director of Graduate Studies
Department of Chemistry (M/C 111)
University of Illinois at Chicago
P.O. Box 4348
Chicago, Illinois 60680

Peterson's Guide to Graduate Programs in the Physical Sciences and Mathematics 1992

SECTION 2: CHEMISTRY

University of Illinois at Chicago

THE FACULTY AND THEIR RESEARCH

Professors

R. L. Carlin, Ph.D., Illinois. Inorganic chemistry—magnetism at low temperatures; susceptibilities, specific heats, and EPR; antiferromagnetism and magnetothermodynamics.

E. A. Gislason, Ph.D., Harvard. Physical chemistry—theoretical studies of ion-molecule reactions, neutral reactions, vibrational excitation, collisional dissociation, intermolecular potentials and vibrational effects in electronic transitions, negative ion resonances.

R. J. Gordon, Ph.D., Harvard. Physical chemistry—experimental studies of molecular reaction dynamics, gas phase kinetics, and photochemistry.

C. J. Jameson, Ph.D., Illinois. Physical and theoretical chemistry—nuclear magnetic resonance; theoretical calculations of NMR chemical shifts and coupling constants, polarizabilities, and other molecular electronic properties.

J. Kagan, Ph.D., Rice. Organic and biological chemistry—photochemistry and photobiochemistry, reaction mechanisms, heterocyclic chemistry.

R. Kassner, Ph.D., Yale. Biochemistry—structure and properties of metalloproteins; model and protein studies of redox, ligand binding, spin-state, and catalytic properties of heme proteins; metalloporphyrin formation in heme and chlorophyll biosynthesis.

T. Keiderling, Ph.D., Princeton. Physical chemistry—vibrational optical activity applications to theory and molecular structural analyses: protein, peptide, and nucleic acid conformational studies; new spectroscopic instrumentation, electronic and vibrational magnetic circular dichroism, electronic spectroscopy of transition metal complexes, and nonlinear spectroscopy.

P. LeBreton, Ph.D., Harvard. Physical and biophysical chemistry—UV photoelectron studies of biological molecules: fluorescence probes of interactions of hydrocarbon metabolites with nucleotides.

W. L. Mock, Ph.D., Harvard. Organic chemistry—structure-reactivity relationships, synthetic methods, concerted reactions, organosulfur chemistry, biorganic chemistry.

R. M. Moriarty, Ph.D., Princeton. Organic chemistry—mechanism of oxygen fixation, singlet dioxygen oxidation, photolyses in lipid bilayer membranes, organic syntheses with lasers, polycyclic aromatic hydrocarbons, steroid synthesis, hypervalent iodine in synthesis, mechanism cholinesterase activity, alkyl azides.

J. Rocek, Head of the Department; Ph.D., Prague Technical. Physical organic chemistry—mechanism of oxidation reactions.

M. Sinnott, Ph.D., Bristol. Organic and bioorganic chemistry—mechanisms of enzymic and nonenzyme glycosyl transfer; suicide enzyme inactivation, especially by active site generation of diazonium ions; stereoelectronic effects.

B. K. Teo, Ph.D., Wisconsin. Inorganic chemistry—synthesis, structure, bonding, and properties of large metal clusters; magic numbers; X-ray crystallography and metal tetrathiolenes; EXAFS and synchrotron research.

P. R. Young, Ph.D., South Florida. Physical organic chemistry and biochemistry—the mechanism and driving force for general acid-base catalysis; nucleophilic substitution at tricoordinate sulfur; evidence for sulfurane intermediates; biochemistry of methionine redox enzymes.

Associate Professors

A. S. Benight, Ph.D., Georgia Tech. Biophysical chemistry—dynamic light-scattering investigations of DNA, proteins, and DNA/protein complexes; experimental and theoretical investigations of DNA thermal denaturation.

R. P. Burns, Ph.D., Chicago. Physical chemistry—kinetics of evaporation, condensation, and chemical reaction between adsorbed species; high-temperature chemistry; mass spectroscopy.

D. Crich, Ph.D., London. Organic chemistry—diastereoselective free-radical reactions, acyl radical cyclizations, asymmetric synthesis of amino acids.

R. Elber, Ph.D., Hebrew (Jerusalem). Physical and biophysical chemistry—theoretical studies of protein dynamics, kinetics of ligand diffusion, kinetics of protein folding.

W. A. Freeman, Ph.D., Michigan. Inorganic chemistry—X-ray crystallographic investigation of ligand conformation in chelates and the correlation of such structural results with circular dichroism and other spectroscopic information.

M. Kahn, Ph.D., Yale. Organic chemistry—design and synthesis of molecules that mimic the activity of biologically important peptides and proteins.

J. A. Morrison, Ph.D., Maryland. Inorganic chemistry—synthesis of inorganic and organometallic compounds and the study of their structure and reactivity; preparative reactions using discharge and high-temperature techniques; chemistry of the boron subhalides and fluorinated organometallic compounds.

M. Trenary, Ph.D., MIT. Physical chemistry—structure, dynamics, and reactions of molecules adsorbed on well-defined crystal surfaces; adsorbate phase transitions; kinetics of surface diffusion; infrared spectroscopy of molecular adsorbates.

Assistant Professors

V. Buch, Ph.D., Hebrew (Jerusalem). Physical chemistry—gas-surface dynamics (sticking, spectra of chemisorbed molecules), interstellar surface chemistry.

W. Cho, Ph.D., Chicago. Biochemistry—protein engineering of phospholipase A_2 and its inhibitory proteins, development of anti-inflammatory agents, biophysical studies of surface-active proteins and peptides, design of novel protein catalysts.

G. Gould, Ph.D., Yale. Inorganic chemistry—organometallic chemistry, molecular semiconductors, fluoroacetylenes and fluorocarbynes, metal dimers, metal alkylidenes.

L. Hanley, Ph.D., SUNY at Stony Brook. Analytical chemistry—experimental studies of atomic and molecular ion-surface interactions at low collision energies, employing mass, Auger electron, and infrared reflection-absorption spectroscopies.

SECTION 2: CHEMISTRY

UNIVERSITY OF MARYLAND COLLEGE PARK

Institute for Physical Science and Technology
Chemical Physics Program

Program of Study

M.S. and Ph.D. degrees are offered in the field of chemical physics. The Chemical Physics Program provides an academic path for those candidates wishing to establish a professional career for which in-depth knowledge of both physics and chemistry is desirable. The Ph.D. program of study takes approximately six years to complete. Students may opt to study chemical physics with a concentration in physics, chemistry, chemical engineering, electrical engineering, mechanical engineering, or meteorology. Specialization is pursued in a wide variety of research areas including the following: atomic and molecular structure, atmospheric chemistry, atmospheric remote sensing, biophysics, fluctuation phenomena, intermolecular energy transfer and low energy scattering, laser spectroscopy, molecular dynamics, phase transitions, properties of fluids, statistical mechanics, surface science, ultrafast laser applications, working fluid mixtures and thermodynamic cycles, and X-ray physics. This program has been designed with maximum flexibility so that a student can achieve a strong background in his or her chosen field of specialization. A cooperative program with the National Institutes of Health (NIH) allows chemical physics students to pursue their thesis work with NIH scientists specializing in biophysical research.

Research Facilities

The Chemical Physics Program provides the facilities of a major research university, including modern equipment for Fourier-transform spectroscopy, 500-MHz NMR, laser light scattering, ultrafast and very high intensity laser systems, SQUIDs, VUV high-resolution spectroscopy, electron spectroscopy, surface analysis, lidar, and tunneling electron microscopy. Opportunities for scientific interaction and exchange are enhanced by access to the National Aeronautics and Space Administration, Goddard Space Flight Center, Naval Research Laboratory, National Institutes of Health, National Institute of Standards and Technology, U.S. Department of Agriculture, and Applied Physics Laboratory.

Financial Aid

Support is available for full-time graduate students in the form of teaching assistantships, fellowships, and research assistantships. The annual stipend for entering graduate teaching assistants in the fall of 1991 is $10,560. Graduate fellowships are available for qualified students, ranging from $2000 to $12,000 in 1991–92. There is currently a Cooperative Graduate Program in Biophysics at the National Institutes of Health and the University of Maryland.

Cost of Study

Tuition charges are waived for up to 10 credits of study for students with teaching assistantships and fellowships.

Cost of Living

Off-campus housing ranges from $250 up for single rooms and from $450 up for three-bedroom apartments. University apartments, though very limited, are available to graduate students.

Student Group

There are 25 students pursuing advanced degrees in chemical physics. Students come from all over the United States and many foreign countries. Ten Ph.D. degrees and twelve M.S. degrees have been awarded to students from 1983 to 1991.

Location

The campus is at the northeastern edge of the Washington, D.C., metropolitan area. It is 8 miles from the Capitol Mall, 40 miles from Chesapeake Bay, 80 miles from the Shenandoah Valley, and 120 miles from the Atlantic Ocean beaches.

The Program

The Chemical Physics Program is under the joint sponsorship of the Institute for Physical Science and Technology, the Chemistry Department, the Meteorology Department, the Department of Physics and Astronomy, and the College of Engineering. There are currently 38 faculty members from the aforementioned departments who are active in the research areas listed above.

Applying

Applicants seeking financial support should submit applications for admission by February 1 for the fall semester and by October 1 for the spring semester. Deadlines for admission to the Graduate School are July 1 for the fall semester and November 1 for the spring semester. The program requires that students submit GRE scores, both General and Subject. The application fee is $25. International students should apply at least seven months prior to the semester in which they plan to enter and must submit official TOEFL scores.

Correspondence and Information

Director, Chemical Physics Program
Room 1115
I.P.S.T. Building
University of Maryland
College Park, Maryland 20742
Telephone: 301-405-4780

Peterson's Guide to Graduate Programs in the Physical Sciences and Mathematics 1992

SECTION 2: CHEMISTRY

University of Maryland College Park

THE FACULTY AND THEIR RESEARCH

Professors
Millard Alexander, Ph.D., Paris, 1967. Molecular collisions and energy transfer.
William M. Benesch, Ph.D., Johns Hopkins, 1952. Molecular spectroscopy, atmospheric and astrophysical spectroscopy.
Michael A. Coplan, Ph.D., Yale, 1963. Electron and ion impact spectroscopy, space physics.
Sankar DasSarma, Ph.D., Brown, 1979. Equilibrium and nonequilibrium statistical mechanics, surface science, semiconductor science, computer simulation.
Christopher C. Davis, Ph.D., Manchester, 1970. Quantum electronics, molecular energy transfer, atmospheric trace monitoring.
J. Robert Dorfman, Ph.D., Johns Hopkins, 1961. Nonequilibrium statistical mechanics, molecular hydrodynamics.
Theodore L. Einstein, Ph.D., Pennsylvania, 1973. Surface science: phase transitions of adsorbates, surface spectroscopy.
Richard A. Ferrell, Ph.D., Princeton, 1952. Phase transitions and condensed-matter science.
Michael E. Fisher, Ph.D., King's College (London), 1957. Statistical mechanics, condensed-matter theory, physical chemistry and associated mathematics.
James W. Gentry, Ph.D., Texas, 1969. Aerosol physics.
Marshall L. Ginter, Ph.D., Vanderbilt, 1961. Molecular electronic spectroscopy.
Sandra C. Greer, Ph.D., Chicago, 1969. Thermodynamics and phase transitions of fluids and fluid mixtures.
Ashwani K. Gupta, Ph.D., 1973, D.Sc. (Tech.), 1986, Sheffield. Experimental research on combustion, pollution, alternative fuels, laser probes and diagnostics.
Urs E. Hochuli, Ph.D., Catholic University, 1962. Quantum electronics and gas lasers.
Raj K. Khanna, Ph.D., Indian Institute of Science, 1962. Infrared and Raman spectroscopy.
Theodore R. Kirkpatrick, Ph.D., Rockefeller, 1981. Nonequilibrium statistical mechanics, quantum fluids, molecular hydrodynamics.
Chi H. Lee, Ph.D., Harvard, 1968. Nonlinear optics, picosecond and femtosecond laser spectroscopy.
Jeffrey W. Lynn, Ph.D., Georgia Tech, 1974. Neutron scattering and condensed-matter science.
Thomas J. McIlrath, Ph.D., Princeton, 1966. Nonlinear optical processes in gases, high-intensity-laser physics, X-ray and vacuum ultraviolet physics.
Alice C. Mignerey, Ph.D., Rochester, 1975. Heavy-ion-induced nuclear reactions.
Gerald R. Miller, Ph.D., Illinois, 1962. Nuclear magnetic resonance.
John H. Moore, Ph.D., Johns Hopkins, 1967. Electron and ion scattering, molecular spectroscopy, space physics.
Jan V. Sengers, Ph.D., Amsterdam, 1962. Phase transitions, fluctuation phenomena in fluids, thermophysical properties of fluids.
John A. Tossell, Ph.D., Harvard, 1972. Molecular quantum mechanics, photoelectron spectroscopy.
John D. Weeks, Ph.D., Chicago, 1969. Static and dynamic properties of interfaces, pattern formation during crystal growth, theories for the structure and dynamics of liquids.
John Weiner, Ph.D., Chicago, 1970. Laser-induced atomic and molecular collisions.
Thomas D. Wilkerson, Ph.D., Michigan, 1962. Spectroscopy of atmospheric molecules, laser sensing of atmospheric constituents.

Associate Professors
Richard V. Calabrese, Ph.D., Massachusetts, 1976. Fluid mechanics of single-phase and multiphase systems.
Mario Dagenais, Ph.D., Rochester, 1978. Nonlinear optical interactions in condensed matter and optical switching.
Russell R. Dickerson, Ph.D., Michigan, 1980. Development and use of analytical techniques in atmospheric chemistry.
Robert Ellingson, Ph.D., Florida State, 1972. Atmospheric radiation.
Robert W. Gammon, Ph.D., Johns Hopkins, 1967. Laser light scattering: Brillouin, Rayleigh, and Raman scattering, critical phenomena.
Wendell T. Hill III, Ph.D., Stanford, 1980. Atomic and molecular structure, laser spectroscopy.
Reinhard K. Radermacher, Ph.D., Munich Technical, 1981. Experimental and theoretical analysis of advanced energy conversion cycles.
Devarajan Thirumalai, Ph.D., Minnesota, 1981. Applications of quantum statistical mechanics.
Ellen D. Williams, Ph.D., Caltech, 1982. Experimental research on the properties of solid surfaces.

Assistant Professors
Keith E. Herold, Ph.D., Ohio State, 1985. Thermodynamic properties of mixtures, mixture equations of state.
Howard M. Milchberg, Ph.D., Princeton, 1985. High-intensity laser-matter interactions.
Janice E. Reutt-Robey, Ph.D., Berkeley, 1986. Experimental research on transient surface chemistry processes—energy transfer, diffusion, reaction.

Adjunct Professor
Ralph J. Nossal, Ph.D., Michigan, 1963. Biophysics.

SECTION 2: CHEMISTRY

UNIVERSITY OF MARYLAND COLLEGE PARK

Department of Chemistry and Biochemistry

Programs of Study

M.S. and Ph.D. degrees are offered with specialization in the fields of analytical chemistry, biochemistry, bioorganic chemistry, chemical physics (in cooperation with the Institute for Physical Sciences and Technology and the Department of Physics and Astronomy), environmental chemistry, inorganic chemistry, nuclear chemistry, organic chemistry, and physical chemistry. The graduate programs have been designed with maximum flexibility so that a student can achieve a strong background in his or her chosen field of specialization.

Research Facilities

The Department of Chemistry and Biochemistry operates a broad spectrum of modern research equipment, including NMR facilities (60, 90, 200, 400, and 500 MHz), high-resolution mass spectrometers, an electron paramagnetic resonance spectrometer, computer-controlled X-ray spectrometers, X-ray fluorescence systems, an ESCA system, a well-equipped gamma-ray detection and data reduction facility, an ES molecular modeling system, numerous analytical spectroscopy systems, and numerous computing facilities, including a VAX-11/780. Support facilities include excellent departmental glassblowing, machine, and electronics shops; a microcomputer laboratory; and a self-service chemical storeroom. Additional campus facilities include the large mainframe computer, high-precision machine shops, and the electron microscope laboratory. The Charles White Memorial Chemistry Library is an integral part of the chemistry complex and provides on-line search capability.

Financial Aid

Support is available for full-time graduate students in the form of teaching assistantships, fellowships, research assistantships, and federally or industrially supported traineeships. The annual stipend for entering graduate teaching assistants in the fall of 1991 is $11,040. Fellowships, which range from $2500 to $9000 in 1991–92, are available for entering students.

Cost of Study

Tuition charges are waived for up to 10 credits of study for students with teaching assistantships and fellowships.

Cost of Living

Off-campus housing ranges from $200 up for single rooms and from $500 up for one-bedroom apartments. University apartments are available to graduate students.

Student Group

There are about 190 graduate students pursuing advanced degrees in chemistry and biochemistry. Students come from all over the United States and many foreign countries. The department's Graduate Student Association organizes both social and professional activities. Approximately twenty-five Ph.D. degrees and ten M.S. degrees are awarded annually.

Location

The campus is at the northeastern edge of the Washington, D.C., metropolitan area. It is 8 miles from the Capitol Mall, 40 miles from Chesapeake Bay, 80 miles from the Shenandoah Valley, and 120 miles from the Atlantic Ocean beaches. With the area's large concentration of high-quality government research laboratories, and hence its large scientific community, the opportunities for scientific interaction and exchange are surpassed by few areas in the country. There are a large number of local scientific societies that hold biweekly and monthly evening meetings that are in part social as well as scientific.

The Department

The Department of Chemistry and Biochemistry has a large, diverse faculty whose approximately 50 members are active in the research areas listed above, involved in wide-ranging collaborations with faculty in other departments and in many area research laboratories, and involved in many national and international professional activities. The department is housed in a five-wing complex that was completed in 1975. Graduate study is organized into five divisions of up to 60 students each so that individual attention can be given to each student's course of study.

For additional general information about the University, readers should see the page on the Graduate School in Book 1 of this series.

Applying

Applicants seeking financial support should submit applications for admission by March 15 (preferably earlier) for the fall semester and by November 1 for the spring semester. Deadlines for admission to the Graduate School through the department are May 1 for the summer session and fall semester and November 1 for the spring semester. Applications received after the deadlines are processed on a "best efforts" and space-available basis. The application fee is $26.

Correspondence and Information

Associate Chairman for Graduate Studies
Department of Chemistry and Biochemistry
University of Maryland
College Park, Maryland 20742
Telephone: 301-405-7022

Peterson's Guide to Graduate Programs in the Physical Sciences and Mathematics 1992

SECTION 2: CHEMISTRY

University of Maryland College Park

THE FACULTY AND THEIR RESEARCH

Millard Alexander, Ph.D., Paris, 1967. Physical chemistry.
Herman L. Ammon, Ph.D., Washington (Seattle), 1963. Organic chemistry.
Richard N. Armstrong, Ph.D., Marquette, 1975. Organic chemistry.
Jon M. Bellama, Ph.D., Pennsylvania, 1966. Inorganic chemistry.
*William M. Benesch, Ph.D., Johns Hopkins, 1952. Chemical physics.
Alfred C. Boyd, Ph.D., Purdue, 1957. Inorganic chemistry.
Gilbert Castellan, Ph.D., Catholic University, 1949. Physical chemistry.
*Michael A. Coplan, Ph.D., Yale, 1963. Chemical physics.
Philip R. DeShong, Sc.D., MIT, 1976. Organic chemistry.
Howard DeVoe, Ph.D., Harvard, 1960. Physical chemistry.
Debra Dunaway-Mariano, Ph.D., Texas A&M, 1975. Biochemistry.
Bryan Eichhorn, Ph.D., Indiana, 1987. Inorganic chemistry.
Daniel E. Falvey, Ph.D., Illinois, 1988. Organic chemistry.
David H. Freeman, Ph.D., MIT, 1957. Analytical chemistry.
*Robert W. Gammon, Ph.D., Johns Hopkins, 1967. Chemical physics.
John Gerlt, Ph.D., Harvard, 1974. Biochemistry.
*Marshall L. Ginter, Ph.D., Vanderbilt, 1961. Chemical physics.
Glen Gordon, Ph.D., Berkeley, 1960. Nuclear chemistry.
Sandra C. T. Greer, Ph.D., Chicago, 1969. Chemical physics.
Samuel C. Grim, Ph.D., MIT, 1960. Inorganic chemistry.
J. Norman Hansen, Ph.D., UCLA, 1968. Biochemistry.
George Helz, Ph.D., Penn State, 1970. Geochemistry.
James W. Herndon, Ph.D., Princeton, 1983. Organic chemistry.
James E. Huheey, Ph.D., Illinois, 1961. Inorganic chemistry.
Bruce B. Jarvis, Ph.D., Colorado, 1966. Organic chemistry.
Douglas A. Julin, Ph.D., Berkeley, 1984. Biochemistry.
Franz J. Kasler, Ph.D., Vienna, 1959. Analytical chemistry.
Raj K. Khanna, Ph.D., Indian Institute of Science, 1962. Physical chemistry.
John Kozarich, Ph.D., MIT, 1975. Organic chemistry.
Patrick S. Mariano, Ph.D., Wisconsin–Madison, 1969. Organic chemistry.
Paul H. Mazzocchi, Ph.D., Fordham, 1966. Organic chemistry.
Alice Mignerey, Ph.D., Rochester, 1975. Nuclear chemistry.
Cary J. Miller, Ph.D., Berkeley, 1987. Analytical chemistry.
Gerald R. Miller, Ph.D., Illinois, 1962. Physical chemistry.
John H. Moore, Ph.D., Johns Hopkins, 1967. Physical chemistry.
Thomas J. Murphy, Ph.D., Rockefeller, 1968. Physical chemistry.
Thomas C. O'Haver, Ph.D., Florida, 1968. Analytical chemistry.
John M. Ondov, Ph.D., Maryland, 1974. Environmental chemistry.
Rinaldo Poli, Ph.D., Scuola Normale Superiore di Pisa, 1985. Inorganic chemistry.
Cyril Ponnamperuma, Ph.D., Berkeley, 1962. Chemical evolution.
Janice Reutt-Robey, Ph.D., Berkeley, 1986. Physical chemistry.
Joseph Sampugna, Ph.D., Connecticut, 1968. Biochemistry.
*Jan V. Sengers, Ph.D., Amsterdam, 1962. Chemical physics.
James M. Stewart, Ph.D., Washington (Seattle), 1958. Physical chemistry.
Devarajan Thirumalai, Ph.D., Minnesota, 1981. Physical chemistry.
John A. Tossell, Ph.D., Harvard, 1972. Geochemistry.
William B. Walters, Ph.D., Illinois, 1964. Nuclear chemistry.
*John D. Weeks, Ph.D., Chicago, 1969. Chemical physics.
John Weiner, Ph.D., Chicago, 1970. Chemical physics.
Sarah A. Woodson, Ph.D., Yale, 1987. Biochemistry.

---

*Members of the Institute for Physical Sciences and Technology.

SECTION 2: CHEMISTRY

UNIVERSITY OF MARYLAND GRADUATE SCHOOL, BALTIMORE
Department of Chemistry and Biochemistry

Program of Study

The Department of Chemistry and Biochemistry offers a program of graduate study in chemistry and biochemistry leading to the M.S. and the Ph.D. degrees. The chemistry department also participates in a biotechnology M.S. program (Applied Molecular Biology) in collaboration with the Department of Biological Sciences and an interdisciplinary Ph.D. program in molecular and cellular biology. The department's graduate programs in biochemistry and in molecular and cellular biology benefit from being part of larger intercampus joint programs in those fields sponsored in conjunction with departments at the Medical, Dental, and Pharmacy Schools of the downtown Baltimore campus.

Upon entering the M.S. program, students are required to take a set of placement examinations designed to test their proficiency at the senior undergraduate level and to indicate any areas of deficiency. Under the guidance of an advisory committee, they next complete a group of courses, constituting a core curriculum, which has been designed to bring them to a minimum level of proficiency in each of the major areas of chemistry. In addition, they are expected to take specialized courses in their field of interest. To fulfill the course requirements normally requires one to two years, depending upon the student's initial level of proficiency as demonstrated by the placement examinations. To qualify for the degree, students must pass a set of qualifying examinations. These are given at regular intervals and each student is allowed several opportunities to pass them. Thesis research must be approved by a thesis committee, which also administers an oral examination based on the thesis research. Completing the course requirements, passing the qualifying and oral examinations, and gaining approval of the thesis constitute fulfillment of the M.S. degree requirements. Candidates for the M.S. also have the option of substituting additional course work for the thesis.

Graduate students in the Ph.D. program are expected to pass a comprehensive examination, prepare an acceptable research proposal, present a thesis based upon original research, and pass an oral thesis defense.

The principal areas of thesis research for both the M.S. and Ph.D. degrees include bioinorganic chemistry, enzymology, mass spectrometry, models for enzymic reactions, organic mechanisms, organic synthesis, protein and nucleic acid chemistry, and theoretical chemistry.

Research Facilities

Extensive facilities are available for modern research. The specialized research instrumentation available includes ultracentrifugation, calorimetry, gas chromatography, high-performance liquid chromatography, stopped-flow and temperature-jump kinetics, nanosecond fluorometry, nuclear magnetic resonance spectroscopy (80-MHz and 500-MHz instruments), electron spin resonance spectroscopy, circular dichroism, X-ray diffraction, infrared spectroscopy, laser fluorescence spectroscopy, atomic absorption, gas chromatography–mass spectrometry, and scintillation and gamma counting apparatus as well as extensive computer facilities. The National Science Foundation supports a national Center for Structural Biochemistry in the UMBC chemistry department; this center specializes in the structural analysis of biological molecules (biopolymers, peptides, glycoproteins, etc.). The center houses one of only four tandem mass spectrometers located in academic institutions worldwide. The main University library contains more than 2,500 volumes of chemistry texts, as well as 150 periodicals. In addition, the department reference room provides access to the principal journals.

Financial Aid

A number of teaching assistantships are available for graduate students. These involve laboratory instruction for an average of 8 hours a week. In addition, extensive grant support is available for research fellowships.

Cost of Study

Tuition in 1991–92 is $141 per credit for Maryland residents and $252 per credit for nonresidents. Students holding teaching assistantships receive a waiver of tuition.

Cost of Living

There are a limited number of on-campus dormitory rooms available for graduate students. Most graduate students are housed in apartments or rooming houses in the nearby communities of Arbutus and Catonsville. A single graduate student can expect living and educational expenses of approximately $11,000 to $13,000 a year.

Student Group

In 1991–92, the department's graduate students include 24 men and 19 women. All of the students plan careers in chemistry and biochemistry in either teaching or research. Approximately 80 percent of the students are receiving some form of financial aid.

Location

The University has a scenic location on the periphery of the Baltimore metropolitan area. Downtown Baltimore can be reached in 20 minutes by car, while Washington is an hour away. Both Baltimore and Washington have very extensive cultural facilities, including eight major universities, a number of museums and art galleries of international reputation, two major symphony orchestras, and numerous theaters. All the usual recreational facilities are available in both cities. Several excellent beaches are accessible by car.

The University

The University of Maryland Baltimore County (UMBC) was established in 1966 on a 500-acre campus. The Department of Chemistry and Biochemistry of the University of Maryland Graduate School, Baltimore, is located at the UMBC campus. The University has about 9,000 students drawn primarily from Maryland, although an increasing number have enrolled from other states and foreign countries. The undergraduates are predominantly interested in professional or business careers. Because a high percentage of the students are the first members of their families to attend college, UMBC has a very different atmosphere from that encountered in older institutions and has made a particular effort to attract minority students, who now account for about 18 percent of the undergraduate student body.

Applying

Applications should include an academic transcript, three references, and the results of the General Test of the Graduate Record Examinations.

Correspondence and Information

Interested persons should contact:
Graduate Program Director
Department of Chemistry and Biochemistry
University of Maryland Baltimore County
5401 Wilkens Avenue
Baltimore, Maryland 21228
Telephone: 301-455-2491

SECTION 2: CHEMISTRY

University of Maryland Graduate School, Baltimore

THE FACULTY AND THEIR RESEARCH

Dorothy Beckett, Assistant Professor; Ph.D., Illinois at Urbana-Champaign, 1986; postdoctoral studies at MIT and Johns Hopkins. Regulation of genes, regulation of biotin biosynthesis in *E. coli*, molecular biology and biophysical chemistry.

C. Allen Bush, Professor; Ph.D., Berkeley, 1965; postdoctoral studies at Cornell. Biophysical chemistry: conformation and dynamics of carbohydrates, glycoproteins, glycopeptides, and polysaccharides by NMR spectroscopy, computer modeling, and circular dichroism.

Donald Creighton, Professor; Ph.D., UCLA, 1972; postdoctoral studies at the Institute for Cancer Research. Enzyme mechanisms and protein structure, studies on sulfhydryl proteases and glyoxalase enzymes.

Paul E. Dietze, Assistant Professor; Ph.D., NYU, 1983; postdoctoral studies at Brandeis. The mechanism of solvolysis and group transfer reactions, use of structure-reactivity relationships to characterize transition states.

Catherine C. Fenselau, Professor; Ph.D., Stanford, 1965; postdoctoral studies at Berkeley. Mass spectrometry and its applications in biochemistry and biotechnology, immobilized enzyme systems for biosynthesis and metabolism.

Fred Gornick, Professor; Ph.D., Pennsylvania, 1959; postdoctoral studies at the National Bureau of Standards. Conformational transitions in synthetic polypeptides, physical chemistry of biological mineralization processes.

Ramachandra S. Hosmane, Associate Professor; Ph.D., South Florida, 1978; postdoctoral studies at Illinois. Reagents for organic synthesis; synthesis of heterocycles, natural products, and nucleosides.

Martin Hulce, Assistant Professor; Ph.D., Johns Hopkins, 1983. New synthetic methods, order effects in organic reactions, catalysis and stereochemistry.

Arthur S. Hyman, Associate Professor; Ph.D., Rutgers, 1964; postdoctoral studies at the National Bureau of Standards. Physical chemistry: X-ray scattering, theoretical chemistry.

Richard L. Karpel, Associate Professor; Ph.D., Brandeis, 1970; postdoctoral studies at Princeton. Interactions of helix destabilizing proteins with nucleic acids and the involvement of such proteins in various aspects of RNA function, metal ion–nucleic acid interactions.

Joel F. Liebman, Professor; Ph.D., Princeton, 1970; postdoctoral studies at Cambridge and the National Bureau of Standards. Strained organic compounds and their energetics, gaseous ions, noble gas and fluorine chemistry, mathematical and quantum chemistry.

Ralph M. Pollack, Professor; Ph.D., Berkeley, 1968; postdoctoral studies at Northwestern. Enzyme reactions, model systems for enzyme mechanisms, organic reaction mechanisms.

Robert F. Steiner, Professor; Ph.D., Harvard, 1950; postdoctoral studies at the Naval Medical Institute. Physical chemistry of protein and nucleic acids, fluorescence, and control mechanisms of allosteric enzymes.

Michael F. Summers, Assistant Professor; Ph.D., Emory, 1984; postdoctoral studies at Center for Drugs and Biologics, FDA, Bethesda, Maryland. Elucidation of structural, dynamic, and thermodynamic features of metallobiomolecules utilizing advanced two-dimensional and multinuclear NMR methods.

James S. Vincent, Associate Professor; Ph.D., Harvard, 1963; postdoctoral studies at Caltech. Infrared and Raman spectroscopy of phospholipid membrane systems, magnetic spectroscopy of transition-metal complexes.

Dale L. Whalen, Professor; Ph.D., Berkeley, 1965; postdoctoral studies at UCLA. Reactions of carcinogenic polycyclic aromatic hydrocarbon epoxides, organic reaction mechanisms.

Marek Wojciechowski, Assistant Professor; Ph.D., Warsaw, 1977; postdoctoral studies at SUNY at Buffalo. Electroanalytical chemistry: pulse voltammetry, square-wave stripping voltammetry, microelectrodes, flow detectors.

UNIVERSITY OF MIAMI
Department of Chemistry

Program of Study

The Department of Chemistry offers the Master of Science (M.S.) and Doctor of Philosophy (Ph.D.) degrees, with emphasis on the Ph.D. Specialization is possible in inorganic, organic, or physical chemistry. There are several interdisciplinary research programs in such areas as biophysical chemistry, bioorganic chemistry, physical inorganic chemistry, and marine and atmospheric chemistry. Although there is no formal division in analytical chemistry, students interested in such training may enroll in the physical chemistry division.

All students take standardized background examinations to determine if they need remedial work before beginning the regular program. Entering students enroll in the two of the three core courses offered in the first semester (inorganic, organic, and physical chemistry) that are most suited to their interests. They also take a seminar course in which the faculty members describe their research interests. On the basis of this course and personal interviews, students choose a research director by the start of the second semester. During the second semester, students take the core course in analytical chemistry and a second course in their chosen subdiscipline. The final core course and, if necessary, additional study in the subdiscipline are undertaken during the second year. Students are expected to attend throughout their graduate careers a weekly series of seminars given by visiting scientists and other students.

To earn the Ph.D., students must present a seminar in the second year and a final (usually fourth-year) research seminar, pass four cumulative examinations, pass an oral qualifying examination at the end of the second year, submit (usually in the fourth year) an original proposal for research unrelated to their dissertation, and prepare and successfully defend a dissertation. There is no language requirement.

To complete the M.S. degree, students may choose to take a comprehensive examination or prepare and defend a research thesis. Those who select the thesis option must also present their research in a seminar.

Research Facilities

Major instrumentation includes a complement of UV-visible and infrared spectrometers; a mass spectrometer; two FT-IR spectrometers; and carbon and hydrogen (60- and 80-MHz), multinuclear (90-MHz), and multinuclear high-field (400-MHz) nuclear magnetic resonance spectrometers. Atomic absorption, scanning tunneling microscopy, ORD-CD, high-precision calorimetry, HPLC, and automated cyclic voltammetry and polarography can all be carried out in house. An electron microscopy facility is available jointly with the Rosenstiel Marine School campus nearby.

Financial Aid

Beginning full-time students may be awarded a teaching assistantship that carries a twelve-month stipend of $12,300 in 1991–92. Students who have completed the first academic year may be awarded a research assistantship. The University offers full tuition scholarships for teaching and research assistants. Maytag and University of Miami fellowships are available to outstanding applicants and are similarly valued. Application for financial aid is made on the application form for admission.

Cost of Study

In 1991–92, the tuition for one semester of full-time graduate study (9 credits) is $5190.

Cost of Living

The cost of a double room in the dormitories is $1630 per person per semester. Typical rents for off-campus apartments are $475–$625 per month for one bedroom, $550–$700 for two bedrooms, and $675–$800 for three bedrooms.

Student Group

Currently about 55 graduate students and several postdoctoral fellows from throughout the United States and several foreign countries are enrolled. There are about an equal number of students in organic and physical chemistry; slightly fewer are pursuing studies in inorganic chemistry.

Location

Metropolitan Miami, the largest urban area in Florida, is a center of Latin culture and the commercial gateway to Central and South America. It offers all the amenities of a large and prosperous city, as well as an excellent oceanside climate of warm winters and moderate summers. There are many excellent restaurants and a full schedule of theater, museum exhibits, and concerts. Professional sports events and unrivaled water recreation are also among Miami's attractions.

The University

The University of Miami was founded in 1925 and held its first classes in the fall of 1926. The University is accredited by the Southern Association of Colleges and Schools, and individual programs by a total of twelve professional accrediting agencies. The largest private university in the Southeast, it has a full-time enrollment of more than 13,500, including 3,000 graduate students and 2,000 law and medical students. Two colleges and ten schools are located on the main campus in Coral Gables. The University's medical school, the fourth-largest in the United States, is situated in Miami's civic center. The Rosenstiel School of Marine and Atmospheric Science is located on Virginia Key in Biscayne Bay. Funded research activities totaled $133.7-million in 1989.

The Coral Gables campus occupies 260 beautifully landscaped acres in a predominantly residential area. Downtown Miami is readily accessible by Metrorail train, which has a stop next to the main campus.

Applying

Consideration is given to applicants who have successfully completed general chemistry (two semesters), organic chemistry (two semesters), physical chemistry (two semesters), and the related laboratories. A course in advanced inorganic chemistry is strongly recommended, and remedial work in this area may be required of students who have not taken such a course. The mathematics and physics courses that are normally included in a B.S. program in chemistry are also required. Applicants must obtain a minimum combined score of 1000 on the verbal and quantitative portions of the General Test of the Graduate Record Examinations. Those from non-English-speaking countries must, in addition, obtain a minimum score of 550 on the Test of English as a Foreign Language (TOEFL). Three letters of recommendation are required for all applicants.

Application for financial aid is made at the time of application for admission. Complete application forms, a brochure describing the Department of Chemistry, and other pertinent information can be obtained from the address below.

Correspondence and Information

Director of Graduate Studies
Department of Chemistry
University of Miami
Coral Gables, Florida 33124
Telephone: 305-284-2174

SECTION 2: CHEMISTRY

University of Miami

THE FACULTY AND THEIR RESEARCH

Inorganic Chemistry
Curtis Hare, Ph.D.; Carl Hoff, Ph.D.; Nita Lewis, Ph.D.; William Purcell, Ph.D.
Organometallic chemistry, especially thermochemistry of reactive intermediates; electron transfer in novel metal complexes; high-energy heterocycles; selectivity in the formation of isomeric transition-metal complexes; bioinorganic chemistry—models for redox proteins; high-pressure chemistry in solution.

Organic Chemistry
Robert Gawley, Ph.D.; George Gokel, Ph.D.; Eugene Man, Ph.D.; Carl Snyder, Ph.D.; Keith Wellman, Ph.D.
Stereochemistry and asymmetric syntheses, asymmetric alkylation, and total synthesis of natural products; development of novel synthetic methods; applications of organometallic species in organic synthesis; synthesis and studies of macrocyclic crown polyethers, lariat ethers, bibracchial lariat ethers, molecular boxes, and self-assembling cyclophanes and membranes; isolation and characterization of natural products; amino acid chemistry; analysis of brain proteins for chemical signs to correlate with clinical degradation.

Physical Chemistry
Cecil Criss, Ph.D.; Walter Drost-Hansen, D.Sc.; Luis Echegoyen, Ph.D.; Ariel Fernandez, Ph.D.; Angel Kaifer, Ph.D.; Frank Millero, Ph.D.; Alfred Mills, Ph.D.
High-precision, high-temperature calorimetry; high-resolution multinuclear NMR of complexes; cation complexation and transport; electrochemistry in micelles, in vesicles, and in membranes; electrochemical switching of lariat ethers and related species; fundamental studies in electron transfer; physical chemistry of water at phase boundaries; electron paramagnetic resonance of radical ions and radical complexes; statistical mechanics of self-assembling systems; marine physical chemistry.

The most recent addition to the department's instruments is a Digital Instruments Nanoscope II scanning tunneling microscope.

A student examining spectra taken from a varian VXR 400 NMR, equipped with a 5200 Sun Workstation and complete multinuclear capabilities.

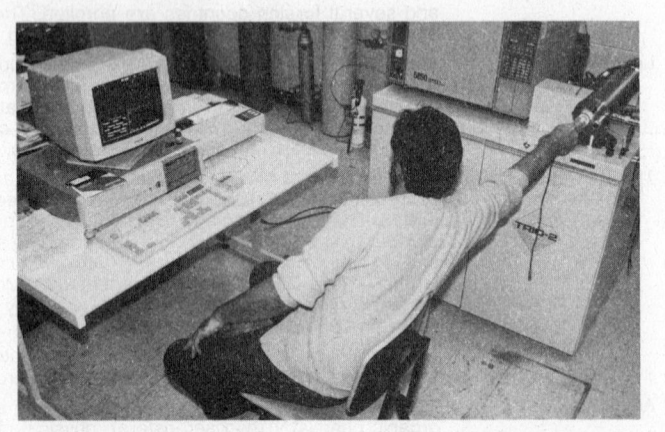

A graduate student operating a VG Trio 2 gas chromatograph–mass spectrometer.

SECTION 2: CHEMISTRY

UNIVERSITY OF MICHIGAN

Department of Chemistry

Programs of Study Programs of study are offered that lead to the M.S. and Ph.D. degrees, with emphasis on the Ph.D. Areas of study are analytical, bioinorganic, bioorganic, inorganic, organic, and physical chemistry; additional opportunities are available through joint programs in macromolecular science, medicinal chemistry, and biophysics. Other interdisciplinary programs, e.g., with the Department of Chemical Engineering (catalysis and surface science), can be arranged on an individual basis.

Qualifying examinations in analytical, inorganic, organic, and physical chemistry are given to incoming students during their first semester. One purpose of these examinations is to identify weaknesses in a student's background so that appropriate courses can be selected to guarantee a strong academic foundation for future research. Students are required to take courses to meet the requirements in their major area, including two courses outside the Department of Chemistry. Formal class work is enriched through lectures, seminars, and colloquia given throughout the year by internationally renowned scholars.

The heart of the graduate program is the research group. Group meetings, faculty advisory committees, and interaction with fellow graduate and postdoctoral students combine to form a support network for the graduate student working toward a Ph.D. Students are encouraged to select a research director and begin research as soon as possible. This typically takes place after the first or second semester of the first year. Research results must be summarized in a formal dissertation suitable for publication, and the thesis must be defended in an oral examination before the Ph.D. degree is granted.

Graduates take positions in academia or in industry where they can apply their skills to new research problems at the frontiers of chemistry and related disciplines. The department maintains an excellent job placement office, which interacts strongly with representatives of many industrial firms seeking to hire chemists.

Research Facilities The chemistry department has recently added a new 270,000-square-foot $45-million building. In addition, renovations are underway to modernize the existing 115,000-square-foot facility. Advanced research hardware complements the department's new center for the chemical sciences. Below is a sampling of major departmental research instrumentation: two 500-MHz NMR; one 360-MHz NMR; one 300-MHz NMR; three 200-MHz NMR; three data stations for off-line NMR processing; one SQUID magnetometer; one ESR; two X-ray diffractometers with four dedicated computers for data analysis; one high-resolution mass spectrometer capable of electron, chemical, or fast-atom bombardment ionization; one low-resolution gas chromatograph–mass spectrometer; a C, H, N microanalysis and atomic absorption facility; several Fourier-transform infrared spectrometers, including one instrument coupled to a gas chromatograph; one circular dichroism spectopolarimeter; one ultraviolet-visible spectrophotometer; one frequency-doubled picosecond Nd:yag laser with dye laser; one Nd:yag laser with dye laser and wavelength extension; one excimer laser and dye laser; and one Raman instrument consisting of an argon ion laser with a photodiode array detector. The department is particularly well equipped with the latest in computer technology. The heart of the research computer system consists of two machines: an IBM 550 and a Silicon Graphics 4D/340 GTX. Located in the departmental computer room are several other systems, including five DECstations, one VAXstation, one Hewlett-Packard workstation, and one Evans & Sutherland PS390 graphics computer for molecular modeling. The entire chemistry complex is wired with both Ethernet and Appletalk networks to allow efficient communication among different computers in the department and the outside world. The Chemistry Library is conveniently located within the complex. The chemistry collection consists of over 50,000 bound volumes and subscriptions to nearly 500 scientific journals, serials, and monographs. Computer-aided literature searches and current-awareness services are available to the entire chemistry community at Michigan.

Financial Aid Financial support for graduate students is provided in the form of teaching assistantships and fellowships. All graduate students receive full support for the duration of their studies. The basic chemistry stipend for the 1991–92 year is slightly more than $12,500. This consists of support for twelve months and a fringe benefit package that includes full medical insurance coverage. A typical appointment is as a teaching assistant for the first year or two and subsequently as a research assistant. Fellowships providing from $15,000 to $17,000 are also available on a competitive basis to meritorious applicants. Supplements are provided for holders of outside fellowships.

Cost of Study In 1990–91, teaching assistants or research assistants had all tuition, worth $12,704 per year out of state or $6074 in state, paid by the University.

Cost of Living The cost of living in Ann Arbor is comparable with other midwestern cities.

Student Group In 1990, there were 167 graduate students enrolled in the Department of Chemistry.

Location The University of Michigan is located in Ann Arbor, a town of about 120,000 people. It is 40 miles west of Detroit and within half an hour of a major international airport. Ann Arbor was recently rated in the top ten communities in the country for its excellent quality of life. There are a rich cultural life, a wide variety of sports activities, and unusually good parks and recreation services. Within an easy drive are many lakes and parks. Although it is near a city, Ann Arbor is still separated from the metropolis by farms and small towns. High-tech industrial parks are being developed to augment the government and industrial research laboratories that make up the industrial base of Ann Arbor, but the University is the town's largest single employer.

The University The University was founded in 1817 and moved to Ann Arbor in 1837. It is a state institution, much of its funding coming from the Michigan legislature. There are four campuses in Ann Arbor: Central Campus, where the chemistry department is housed; North Campus, a 900-acre area that is being developed (the latest addition being a new Engineering Campus); the Athletic Campus, southeast of town, with the famous Michigan stadium; and the Medical School Campus, where a multimillion-dollar hospital was recently completed.

Applying The Department of Chemistry is part of the Horace H. Rackham School of Graduate Studies. Application information can be obtained by writing directly to the chairman of the admissions committee of the chemistry department. All applicants are required to submit GRE scores as well as transcripts and three letters of recommendation. Students for whom English is a second language must submit TOEFL or MELAB test results. All applicants are considered for fellowships. Early application is advisable. Applicants are notified of admission decisions after each completed application is reviewed by the Admissions Committee.

Correspondence and Information
Graduate Admissions Committee
Department of Chemistry
University of Michigan
Ann Arbor, Michigan 48109-1055
Telephone: 313-764-7278

SECTION 2: CHEMISTRY

University of Michigan

THE FACULTY AND THEIR RESEARCH

Arthur J. Ashe III, Professor; Ph.D., Yale, 1966. Organic chemistry: aromaticity, organometallic chemistry of the main group elements.

John R. Barker, Professor; Ph.D., Carnegie-Mellon, 1969. Physical chemistry: chemical kinetics, energy transfer, atmospheric chemistry, laser-induced chemistry.

Lawrence S. Bartell, Professor; Ph.D., Michigan, 1952. Physical chemistry: molecular structure and intramolecular forces, clusters, supersonic jets, electron diffraction, holography, Monte Carlo computations on condensed matter.

S. M. Blinder, Professor; Ph.D., Harvard, 1959. Theoretical chemistry: application of quantum mechanics to atomic and molecular problems, Coulomb Green's functions.

Dimitri Coucouvanis, Professor; Ph.D., Case Western Reserve, 1967. Inorganic chemistry: transition-metal complexes, heterogeneous and homogeneous catalysis.

James K. Coward, Professor; Ph.D., SUNY at Buffalo, 1967. Medicinal and bioorganic chemistry: rational design, synthesis, and biochemical evaluation of enzyme inhibitors; applications in cancer research.

M. David Curtis, Professor; Ph.D., Northwestern, 1965. Inorganic chemistry: organometallic chemistry of transition metals, metal-metal bonding, heterogeneous and homogeneous catalysis.

Thomas M. Dunn, Professor; Ph.D., University College, London, 1957. Physical chemistry: molecular spectroscopy, electronic transitions and molecular excited states.

Seyhan N. Ege, Professor; Ph.D., Michigan, 1956. Organic chemistry: photochemical reactions of heterocyclic compounds.

Billy Joe Evans, Professor; Ph.D., Chicago, 1968. Inorganic chemistry: solid-state chemistry, crystal structures, and magnetic materials, structure/property relations in electronic and magnetic materials.

Anthony H. Francis, Professor; Ph.D., Michigan, 1969. Physical chemistry: solid-state optical spectroscopy and ESR; semiconductor photoluminescence, surface states, impurity centers in crystals.

John Gland, Professor; Ph.D., Berkeley, 1973. Solid-state and surface chemistry.

Gary Glick, Assistant Professor; Ph.D., Columbia, 1988. Bioorganic chemistry and molecular recognition.

Adon A. Gordus, Professor; Ph.D., Wisconsin, 1956. Analytical chemistry: neutron activation and other physical methods applied to archaeology and criminology and to the study of the history of pollution.

Henry C. Griffin, Professor; Ph.D., MIT, 1962. Nuclear chemistry: gamma-ray spectroscopy, in-beam spectroscopy of heavy-ion reactions, radioisotopes for medicine.

Paul Knochel, Professor; Ph.D., E. T. H. Zurich (Switzerland), 1982; Ph.D., University of Pierre and Marie Curie (France), 1984. Organic-organometallic chemistry, dimetallics and functionalized organometallics in organic synthesis.

Raoul Kopelman, Professor; Ph.D., Columbia, 1960. Physical chemistry: energy transfer in molecular and biomolecular aggregates, heterogeneous materials and kinetics, supercomputer simulations of low-dimensional reactions, laser ultramicroscopy.

Masato Koreeda, Professor; Ph.D., Tohoku (Japan), 1970. Organic and bioorganic chemistry: structural and synthetic natural-product chemistry, synthetic methods, mechanism of chemical carcinogenesis.

Carol Korzeniewski, Assistant Professor; Ph.D., Utah, 1987. Analytical chemistry: infrared spectroscopy as a probe of processes in electroanalytical chemistry.

Robert L. Kuczkowski, Professor; Ph.D., Harvard, 1964. Physical inorganic chemistry: application of microwave spectroscopy to chemical problems, molecular structures, mechanism of ozonolysis and catalytic reactions.

Richard G. Lawton, Professor; Ph.D., Wisconsin, 1962. Organic chemistry: chemistry of peptide and protein chains, interactions of close and remote functional groups.

Stephen Lee, Assistant Professor; Ph.D., Chicago, 1985. Inorganic and solid-state chemistry: synthesis of new solid-state phases; correlation of atomic and magnetic structure to electronic structure.

Lawrence L. Lohr, Professor; Ph.D., Harvard, 1964. Physical chemistry: relativistic quantum chemistry, theoretical studies of molecular structure, spectra, and reactivity.

David Lubman, Associate Professor; Ph.D., Stanford, 1979. Physical chemistry: analytical uses of lasers in mass spectrometry.

Joseph P. Marino, Professor; Ph.D., Harvard, 1967. Organic chemistry: development of synthetic methods for various natural products of potential medicinal value.

M. E. Meyerhoff, Professor; Ph.D., SUNY at Buffalo, 1979. Analytical chemistry: chemical sensors, biosensors, immobilized enzymes and antibodies, competitive bonding assays.

Jeffrey Moore, Assistant Professor; Ph.D., Illinois, 1989. Molecular engineering of polymeric materials, synthesis and characterizations of new macromolecules.

Michael D. Morris, Professor; Ph.D., Harvard, 1964. Analytical chemistry: laser spectroscopic techniques.

Christer E. Nordman, Professor; Ph.D., Minnesota, 1953. Physical chemistry: X-ray diffraction studies of biologically interesting compounds, development of improved methods for crystallographic structure determination.

William H. Pearson, Associate Professor; Ph.D., Wisconsin, 1982. Organic chemistry: development of new synthetic methods, heterocyclic chemistry, asymmetric synthesis.

Vincent L. Pecoraro, Associate Professor; Ph.D., Berkeley, 1981. Inorganic chemistry: role of metal ions in biological electron transfer and oxygen metabolism.

James E. Penner-Hahn, Associate Professor; Ph.D., Stanford, 1984. Biophysical chemistry: X-ray absorption spectroscopy, metalloprotein metal–site structure elucidation.

Paul G. Rasmussen, Professor; Ph.D., Michigan State, 1964. Inorganic chemistry/polymer chemistry: mixed valence and conducting complexes, polymers with acceptor groups, cyanoazacarbons.

Richard D. Sacks, Professor; Ph.D., Wisconsin, 1969. Analytical chemistry: development of new high-temperature radiation sources for spectrochemical analysis of trace elements.

Robert R. Sharp, Professor; Ph.D., Case Western Reserve, 1967. Physical chemistry: magnetic resonance spectroscopy of the photosynthetic membrane and of enzymes of photosynthetic relevance, NMR spectroscopy of biological systems.

Leroy B. Townsend, Professor; Ph.D., Arizona State, 1965. Organic chemistry: heterocyclic and nucleoside synthesis of biologically active compounds.

John R. Wiseman, Professor; Ph.D., Stanford, 1965. Organic chemistry: synthesis of compounds with medicinal properties, mechanisms of organic reactions.

Charles F. Yocum, Professor; Ph.D., Indiana, 1971. Biological chemistry: structure/function of the photosynthetic oxygen evolving reaction.

SECTION 2: CHEMISTRY

UNIVERSITY OF MINNESOTA
Department of Chemistry

Programs of Study

The graduate program in the Department of Chemistry centers on the research of students and leads to the Ph.D. degree. The research interests of the faculty cover all major areas, including analytical, biological, inorganic, materials, organic, and physical chemistry. There is also a program in chemical physics.

Upon entering the program, each student takes proficiency exams in analytical, inorganic, organic, and physical chemistry. Most students complete their course requirements (36 quarter credits—usually nine courses) during the first year, and they are encouraged to pick a research adviser early during this period and to begin research. The preliminary candidacy exam for the Ph.D. consists of a written and an oral section, each of which is completed during the second academic year in residence. After the first year, the majority of students' effort is directed toward their thesis research.

The excellent educational and research background prepares students for positions in academia or industry. During the fall of each year, the Department of Chemistry hosts a large number of interviewers from the major chemical companies in the United States. Many students accept job offers as a result of these preliminary on-campus interviews.

An M.S. degree is also offered, which requires 28 quarter credits with a thesis or 44 quarter credits without a thesis.

Research Facilities

The Department of Chemistry is housed in two large buildings, Smith and Kolthoff halls. Smith Hall was completely remodeled in 1987. The University library is adjacent and connected to Smith Hall. Many of the design and manufacturing needs of the chemistry program are carried out directly in the department in sophisticated machine, electronics, and glassblowing shops. The extensive instrumentation for research housed in the department includes multinuclear NMR spectrometers (one 100-, one 200-, two 300-, and one 500-MHz, one fitted with autosampler); a new VG 7070 high-resolution mass spectrometer with fast-atom bombardment (FAB) capability, a Finnigan GC/MS with chemical ionization and electron impact ionization capability, and an AEI MS30 high-resolution spectrometer; an Enraf-Nonius CAD4 single-crystal X-ray diffractometer; and an ESCA/Auger/SIMS spectrometer for surface analysis. Computing capabilities include a VAX-11/780 computer and ready access to the extensive University mainframes and the Cray supercomputer. Many other instruments and facilities, such as FT-IR spectrophotometers, high-energy pulsed lasers, instrumentation for studying reactions in molecular beams, chromatographs, and cold rooms, are used and maintained by individual research groups.

Financial Aid

Research assistantships and teaching assistantships are both available, carrying nine-month stipends of about $10,100 and a waiver of tuition in 1991–92. A number of fellowships are awarded each year to highly qualified students by the Department of Chemistry, the Graduate School, and the Institute of Technology. Summer support is independent of academic-year funding and may come in any of the three forms mentioned above.

Cost of Study

Tuition is approximately $1084 per quarter, but is waived for students holding assistantships and/or fellowships. Students also are assessed a quarterly student services fee of approximately $115.

Cost of Living

Dormitory room and board rates vary from about $980 per quarter for small double rooms to about $1100 per quarter for large single rooms and suites. A variety of rooming houses and apartments are available close to campus. Excellent bus service in the metropolitan and suburban areas also makes the University readily accessible from other areas of the Twin Cities. Depending on the area, rents can range from approximately $250 per month for a room in a private home to $400 or more per month for a one-bedroom apartment.

Student Group

The Department of Chemistry has about 200 undergraduates and 230 graduate students; the graduate students come from many sections of the United States and several foreign countries. The vast majority of the graduate students in the Department of Chemistry receive financial assistance. Recent graduates of the department have been employed by universities, industrial firms, and government laboratories.

Location

The Twin Cities area, with a population of about 2 million people, has all the advantages of a major metropolitan area but few of the usual urban drawbacks. It offers two flourishing downtowns, sophisticated educational and cultural institutions, countless recreational opportunities, and citizens known for their friendliness. Within the city limits of both Minneapolis and St. Paul are numerous large lakes, parks, bicycle and walking trails, and parkways. The Twin Cities are home to professional baseball, basketball, football, and hockey teams, in addition to the numerous University teams. The Guthrie Theatre's reputation as a top regional theater is firmly established throughout the country. The Twin Cities are also the home of two internationally respected musical ensembles: the St. Paul Chamber Orchestra, which performs in the recently completed Ordway Center, and the Minnesota Orchestra, which performs in Orchestra Hall in downtown Minneapolis.

The University

The University of Minnesota was chartered in 1851, seven years before the Minnesota Territory became a state. Today, the University has 4,500 full-time faculty members and 58,000 students enrolled in day classes, with tens of thousands more students in evening, continuing education, and noncredit courses. As one of the largest public institutions of higher learning in the United States, the University offers a rich variety of highly respected programs leading to baccalaureate, graduate, and professional degrees.

Applying

Interested students should complete both an application for admission to the Graduate School and an application for financial aid. These forms may be obtained from the department at the address listed below and should normally be returned, accompanied by the necessary supporting documents, by January 1 for fall quarter admission. Transcripts from all colleges or universities attended and three letters of recommendation are required. Scores on the TOEFL and GRE General and Subject test are required of all foreign applicants.

A strong background in the fundamental areas of chemistry, along with supporting course work in mathematics and physics, is highly recommended for admission to graduate work.

Correspondence and Information

Graduate Admissions Coordinator
Department of Chemistry
University of Minnesota
207 Pleasant Street, SE
Minneapolis, Minnesota 55455
Telephone: 612-626-7444
           800-777-2431 (toll-free)

SECTION 2: CHEMISTRY

University of Minnesota

THE FACULTY AND THEIR RESEARCH

Jan Almlöf, Professor; Ph.D., Uppsala (Sweden), 1974. Theoretical and computational chemistry.
George Barany, Associate Professor; Ph.D., Rockefeller, 1977. Bioorganic and peptide synthesis, organosulfur chemistry.
Paul F. Barbara, Associate Professor; Ph.D., Brown, 1978. Photochemistry, spectroscopy.
Victor A. Bloomfield, Professor; Ph.D., Wisconsin–Madison, 1962. Biophysical chemistry.
Larry D. Bowers, Associate Professor; Ph.D., Georgia, 1975. Bioanalytical and clinical chemistry.
Doyle Britton, Professor; Ph.D., Caltech, 1955. Crystal structure, intermolecular interactions.
Peter W. Carr, Professor; Ph.D., Penn State, 1969. Analytical chemistry, chromatography, biochemical analysis.
John S. Dahler, Professor; Ph.D., Wisconsin–Madison, 1955. Theoretical chemistry.
H. Ted Davis, Professor; Ph.D., Chicago, 1962. Physical chemistry.
John E. Ellis, Professor; Ph.D., MIT, 1971. Synthetic inorganic and organometallic chemistry.
Margaret C. Etter, Associate Professor; Ph.D., Minnesota, 1974. Organic solid-state chemistry, solid-state NMR, hydrogen bonding.
John F. Evans, Associate Professor; Ph.D., Delaware, 1977. Surface chemistry, plasma chemistry, electrochemistry.
Paul G. Gassman, Regents' Professor; Ph.D., Cornell, 1960. Physical organic and synthetic chemistry, organometallic chemistry.
W. Ronald Gentry, Professor; Ph.D., Berkeley, 1967. Chemical physics, molecular beams.
Wayne L. Gladfelter, Professor; Ph.D., Penn State, 1978. Organometallic and inorganic chemistry, metal clusters, catalysis.
Gary R. Gray, Professor; Ph.D., Iowa, 1969. Chemical immunology, polysaccharide structure determination.
Thomas R. Hoye, Professor; Ph.D., Harvard, 1976. Organic synthesis, natural products, symmetry, organometallics.
Essie Kariv-Miller, Professor; Ph.D., Weizmann (Israel), 1967. Organic electrochemistry, catalytic electroreduction, radical cyclization.
Steven R. Kass, Assistant Professor; Ph.D., Yale, 1984. Organic chemistry, gas phase reactions.
Maurice M. Kreevoy, Professor; Ph.D., MIT, 1954. Chemical kinetics, isotope effects, synthetic membranes.
Edward Leete, Professor; Ph.D., Leeds (England), 1950. Biosynthesis and metabolism of natural products.
Doreen Geller Leopold, Assistant Professor; Ph.D., Harvard, 1983. Physical chemistry, organometallic ion-molecule reactions.
Kenneth R. Leopold, Assistant Professor; Ph.D., Harvard, 1983. Physical chemistry, Van der Waals molecules.
Sanford Lipsky, Professor; Ph.D., Chicago, 1954. Physical chemistry, photophysics, photochemistry, electronic spectroscopy.
Hung-Wen Liu, Assistant Professor; Ph.D., Columbia, 1981. Bioorganic chemistry, immunochemistry, biosynthesis, enzyme mechanisms.
Timothy P. Lodge, Associate Professor; Ph.D., Wisconsin–Madison, 1980. Polymer and analytical chemistry, conformational dynamics, diffusion.
Kent R. Mann, Associate Professor; Ph.D., Caltech, 1976. Inorganic and organometallic chemistry, photochemistry.
C. Alden Mead, Professor; Ph.D., Washington (St. Louis), 1957. Molecular quantum mechanics, symmetry in chemistry.
Larry L. Miller, Professor; Ph.D., Illinois, 1964. Organic chemistry, electrochemistry.
Wilmer G. Miller, Professor; Ph.D., Wisconsin–Madison, 1958. Polymer, surfactant, and physical chemistry.
Albert Moscowitz, Professor; Ph.D., Harvard, 1957. Vibrational circular dichroism, electronic structure, stereochemistry.
Eckard Münck, Professor; Ph.D., Darmstadt (Germany), 1967. Biophysical and bioinorganic chemistry, Mössbauer spectroscopy.
Wayland E. Noland, Professor; Ph.D., Harvard, 1952. Organic chemistry, nitrogen heterocycles, nitro rearrangements.
Louis H. Pignolet, Professor; Ph.D., Princeton, 1969. Inorganic chemistry, organometallic chemistry, homogeneous catalysis.
Stephen Prager, Professor; Ph.D., Cornell, 1951. Theoretical chemistry, statistical mechanics, polymer chemistry.
Lawrence Que Jr., Professor; Ph.D., Minnesota, 1973. Bioinorganic chemistry, iron proteins, oxygen activation.
Michael A. Raftery, Professor; Ph.D., 1960, D.Sci., 1971, Ireland. Isolation and characterization of receptors and ion channels of the nervous system.
Jeffrey T. Roberts, Assistant Professor; Ph.D., Harvard, 1988. Surface chemistry of transition metals and semiconductors, physical chemistry.
Scott D. Rychnovsky, Assistant Professor; Ph.D., Columbia, 1986. Organic chemistry and organic synthesis.
Marian T. Stankovich, Associate Professor; Ph.D., Texas at Austin, 1975. Bioanalytical chemistry, spectroelectrochemistry, electron transfer, flavoproteins.
Harold S. Swofford Jr., Professor; Ph.D., Illinois, 1962. Electrochemistry, electroanalytical chemistry, mechanisms and rates of electrooxidation and electroreduction.
William B. Tolman, Assistant Professor; Ph.D., Berkeley, 1987. Synthetic inorganic and organometallic chemistry, biomimetic transition-metal complexes, chiral ligands.
Donald G. Truhlar, Professor; Ph.D., Caltech, 1970. Theoretical chemical dynamics, quantum chemistry, kinetics.

SECTION 2: CHEMISTRY

UNIVERSITY OF NEBRASKA–LINCOLN

Department of Chemistry

Programs of Study

The University of Nebraska–Lincoln has had an outstanding history in the field of chemistry, having granted advanced degrees in chemistry since 1899. The Department of Chemistry currently offers graduate programs leading to the Ph.D. degree in the areas of biochemistry and analytical, inorganic, organic, and physical chemistry. A solid mastery of fundamentals is required of every student, but this does not prevent students from undertaking a program designed for their own needs and interests. Although many students elect to work in one of the traditional areas, joint programs encompassing more than one of these areas are not uncommon. All entering students are required to take a series of proficiency examinations to aid in the planning of their program. In addition to completing required course work, Ph.D. candidates are required to pass six cumulative exams, a research oral exam, and an original research proposal examination.

Although all qualified students are strongly urged to proceed to the Ph.D. degree, interested students may work toward the M.S. degree. Students wishing more details concerning the master's program should write to the address given below.

Research Facilities

The Department of Chemistry occupies a modern nine-story building, Hamilton Hall, which is equipped with all routine instrumentation associated with current chemical research, including scintillation counters; FTIR, visible UV spectrophotometers; ultracentrifuges; a photoelectron spectrometer (PES); an ESR spectrometer; 60-MHz and 90-MHz NMR spectrometers; and 200-MHz, 300-MHz, 360-MHz, and 500-MHz Fourier-transform NMR spectrometers. Specialized research instrumentation is also available, such as rapid kinetic instrumentation, a light-scattering spectrometer, a LEED, Auger spectrometers, electrochemical instrumentation, high-temperature and high-pressure reaction equipment, and resonance Raman, SRGS (simulated Raman gain spectroscopy), and a solid-state NMR spectrometer. A nuclear reactor facility is located in nearby Omaha. The department houses the Midwest Center for Mass Spectrometry, which is one of the nation's most completely equipped laboratories for mass spectrometry. It has a triple-focusing mass spectrometer, as well as several other mass spectrometers and ion cyclotron resonance spectrometers. The first four-sector mass spectrometer is currently being built for this UNL facility. Special instruments not commercially available are constructed in the department's electronics, machine, and glassblowing shops. The UNL library contains an extensive selection of journals and monographs on the fourth floor of Hamilton Hall. The University computer has remote terminals in Hamilton Hall and has a large remote facility in an adjacent building.

Financial Aid

The Department of Chemistry provides support for qualified advanced-degree candidates through assistantships and fellowships. Entering graduate students are typically supported as teaching assistants, although each year several outstanding candidates are appointed to research assistant positions. Teaching assistants can expect to assist in undergraduate laboratories and discussion hours for 7 contact hours per week during the academic semesters. Teaching assistants receive a ten-month stipend totaling $10,500. Research assistant stipends are comparable, and both kinds of assistants are supported at comparable levels during the summer months. For outstanding candidates, departmental fellowships are derived from the Avery, Broderson, and Parker Fellowship Endowments. The Parker Fellowship Endowment has currently accumulated a principal of $1-million. Certain candidates are also eligible for special UNL minority graduate fellowships and Department of Education GAANN fellowships, which provide a $15,000 assistantship/fellowship.

In addition to providing direct financial support in the form of assistantships and fellowships, the Department of Chemistry also pays out-of-state tuition totaling approximately $4000 per year for both teaching and research assistants.

Cost of Study

Graduate teaching and research assistants have tuition remitted for up to 12 hours of credit per semester during the academic year and total remission for the summer session. Fees are approximately $280 per year in 1991–92.

Cost of Living

There is a reasonably ample supply of pleasant apartments and other accommodations off campus, with monthly rents ranging from $200 to $350. The University provides some housing for married students, and single students may live in graduate dormitory areas.

Student Group

About 100 graduate students are enrolled in chemistry. The students are drawn from all areas of the United States and from many foreign countries.

Location

Lincoln, the capital of Nebraska, is consistently rated as one of the most desirable American cities of its size in which to live. With a population of around 200,000, it has many cultural and recreational opportunities; the surrounding countryside provides fine opportunities for fishing and hunting and an excellent series of water-conservation lakes for water activities. The city has maintained a vibrant core and has a public transportation system and a superior air terminal. The larger urban center of Omaha is only 50 miles away by interstate highway.

The University

The University of Nebraska was established in 1869. Over 22,000 students are enrolled on the Lincoln campus, with over 3,700 in the Graduate College. It was the first public university in the United States to have an organized graduate school. The University is composed of colleges that encompass all the major areas of scholarship and the professions. The standard of quality of the University is recognized by its membership in the Association of American Universities. Cultural activity is extensive. The newly opened Lied Center for the Performing Arts offers an outstanding schedule of musical and theatrical performances. The Sheldon Gallery has one of the largest collections of twentieth-century American art in the country. Big Eight sporting events, as well as other University-sponsored events, are available to graduate students at reduced prices.

Applying

For application forms or further information, correspondence should be directed to the address below. Scores on the Graduate Record Examinations are not generally required but are recommended. Graduates from foreign institutions must provide both GRE scores and TOEFL scores.

Correspondence and Information

Graduate Admissions Committee
Department of Chemistry
University of Nebraska–Lincoln
Lincoln, Nebraska 68588-0304
Telephone: 800-332-0261 (toll-free in Nebraska)
 800-228-4514 (toll-free outside Nebraska)

SECTION 2: CHEMISTRY

University of Nebraska–Lincoln

THE FACULTY AND THEIR RESEARCH

Henry E. Baumgarten, Foundation Professor Emeritus; Ph.D., Rice. Electrochemical methods in organic nitrogen chemistry, computer-assisted control and analysis, chemistry of optically active α-lactams, synthesis and reactions of new types of transition-metal complexes.

David B. Berkowitz, Assistant Professor; Ph.D., Harvard. Design and synthesis of transition state analog inhibitors for phosphatases, glucosidases, and amino acid decarboxylases; the role of stereoelectronic control in enzymatic behavior.

William H. Braunlin, Assistant Professor; Ph.D., Wisconsin–Madison. Cation-induced DNA structural transitions, multinuclear NMR studies of cation binding to oligomeric and polymeric nucleic acids.

James D. Carr, Professor; Ph.D., Purdue. Measurement and interpretation of reaction kinetics of ions in aqueous solution, oxidation kinetics of aqueous solutes by Fe(VI).

Norman H. Cromwell, Regents Professor Emeritus; Ph.D., Minnesota. Chemistry of novel three- and four-membered ring nitrogen heterocyclics, chemical carcinogenesis.

Victor Day, Professor; Ph.D., Cornell. Single-crystal X-ray structural studies, catalysis, structure-function relationships in biological systems.

John R. Demuth, Professor Emeritus; Ph.D., Illinois at Urbana-Champaign. Chemical education.

Patrick H. Dussault, Assistant Professor; Ph.D., Caltech. Synthetic and bioorganic hydroperoxide chemistry, bioorganic chemistry of surfactant aggregates.

Craig J. Eckhardt, Professor; Ph.D., Yale. Solid-state chemistry; energy transfer in solids and liquids; liquid crystals; elastic and inelastic light scattering; reflection, piezomodulation, photoacoustic, and laser spectroscopy; electronic properties of organic metals; phase transitions and lattice dynamics.

Gordon A. Gallup, Professor; Ph.D., Kansas. Theoretical calculations of intermolecular forces between small molecular systems, molecular scattering, application of valence bond theory, theory of chemical shifts.

T. Adrian George, Professor; Ph.D., Sussex. Fixation and conversion of atmospheric nitrogen, preparation of organometallic complexes with coordinated molecular nitrogen and inorganic synthesis, synthesis of novel microporous inorganic solids.

Mark A. Griep, Assistant Professor; Ph.D., Minnesota. Initiation of replication, DNA-binding proteins and the structure and function of primase.

Michael L. Gross, C. P. Peterson Professor; Ph.D., Minnesota. Mass spectrometry, trace analysis, structure determination of biomolecules, Fourier-transform mass spectrometry, rates and mechanisms of reactions of gas-phase organic ions.

Naba K. Gupta, Professor; Ph.D., Michigan. Mechanism of protein synthesis initiation in animal cells, isolation of protein factors.

David S. Hage, Assistant Professor; Ph.D., Iowa State. Theory and development of high-performance liquid chromatography (HPLC), especially high performance affinity chromatography (HPAC); chemical separations; separation of biomolecules.

Robert H. Harris, Professor; Ph.D., Caltech. Chemical education and inorganic synthesis.

Robert T. Hembre, Assistant Professor; Ph.D., Colorado State. Synthesis and reactivity of transition-metal complexes relevant to the modeling of metalloenzyme centers and/or organometallic catalysts, activation of small molecules of interest in organic synthesis and industrial catalysis.

Henry F. Holtzclaw Jr., Foundation Professor Emeritus; Ph.D., Illinois at Urbana-Champaign. Synthesis and stereochemistry of metal complexes, metal chelate polymers.

Robert B. Johnston, Professor Emeritus; Ph.D., Chicago. Purification, structure, regulation, and mechanism of action enzymes; active peptides found in brain and their derivatives.

Charles A. Kingsbury, Professor; Ph.D., UCLA. Stereochemistry of organic molecules, correlation of NMR data with molecular mechanics; polymeric systems.

Marjorie A. Langell, Associate Professor; Ph.D., Princeton. Gas adsorption and electronic properties of transition-metal compounds; catalysis; photochemical conversion with Auger, thermal desorption mass spectrometry; LEED and HREELS.

James H. Looker, Professor; Ph.D., Ohio State. Preparation: chemistry of oxygen heterocycles, especially flavonoids; partial synthesis of γ-pyrones and γ-pyridones.

Marion H. O'Leary, Professor; Ph.D., MIT. Mechanisms of biochemical processes, especially enzymes that utilize carbon dioxide as a substrate or product (carboxylases and decarboxylases).

Lawrence J. Parkhurst, Professor; Ph.D., Yale. Ligand-binding kinetics of diverse hemoglobins; kinetics of ribosomes and their interactions with tRNA molecules; DNA-cell-surface interactions; light-scattering, stopped-flow, temperature-jump, and laser photolysis.

Carolyn M. Price, Assistant Professor; Ph.D., Colorado. Structure and function of telomeres, the molecular ends of chromosomes, DNA-protein interactions; unusual DNA structures, de novo synthesis of telomeres.

Edward P. Rack, Professor; Ph.D., Michigan. Stereochemical consequences of energetic recoil atom reactions, isomeric-transition chemistry, molecular neutron activation analysis in medicine and in analysis of environmental samples.

Jody G. Redepenning, Assistant Professor; Ph.D., Colorado State. Electrochemistry: chemically modified electrode surfaces, electrocatalysis; electrocrystallization of calcium phosphate coatings for orthopedic implants; intramolecular charge transfer reactions, mixed-valence metal complexes.

Reuben D. Rieke, Wilson Professor; Ph.D., Wisconsin–Madison. Preparation and chemistry of highly reactive metal powders, synthesis of organic metals, electrochemical studies of metal carbonyl complexes and carbene metal carbonyl complexes.

John J. Scholz, Professor Emeritus; Ph.D., Illinois at Urbana-Champaign. Adsorption of gases on crystal surfaces, film-solid interactions.

Pill-Soon Song, Professor and Chair; Ph.D., California, Davis. Interactions between light and organisms; photobiology, photochemistry, and biophysical chemistry; action of phytochrome, photomovement in Stentor; psoralen intercalation and DNA conformation.

John J. Stezowski, Professor; Ph.D., Michigan State. Interrelationships between chemical, physical, and biological properties of organic systems; X-ray crystallography; molecular modeling using experimental and theoretical methods; temperature-dependent studies of molecular crystals.

George D. Sturgeon, Associate Professor; Ph.D., Michigan State. High-temperature chemistry of nonstoichiometric solids, particularly borides; synthesis and properties of metal fluorides.

James M. Takacs, Associate Professor; Ph.D., Caltech. Synthetic organic chemistry, the application of organotransition-metal catalysis to asymmetric organic synthesis, design of practical chiral auxiliaries for heterocyclization reactions, the design and synthesis of novel organic materials for nonlinear optics.

Chin Hsien (Jim) Wang, M. Clark Professor; Ph.D., MIT. Polymer chemistry and polymer physics, dynamic light scattering and nonlinear optics and laser spectroscopy of amorphous polymers.

Desmond M. S. Wheeler, Professor; Ph.D., National University of Ireland. Natural products: synthesis of anthracyclines and diterpenoid lactones with anticancer activity, isolation and structural studies of anticarcinogenic compounds.

SECTION 2: CHEMISTRY

UNIVERSITY OF PITTSBURGH
Department of Chemistry

Programs of Study The department offers programs of study leading to the M.S. and Ph.D. degrees in analytical, inorganic, organic, and physical chemistry. Interdisciplinary research is currently conducted in the areas of surface science, natural product synthesis, laser spectroscopy, drug design, biosensors, and theoretical chemistry. Both of the advanced degree programs involve original research and course work. Other requirements include a comprehensive examination, a thesis, a seminar, and, for the Ph.D. candidate, a proposal. For the typical Ph.D. candidate, this process takes approximately four years.

Representative of current research activities in the department in analytical chemistry are techniques in electroanalytical chemistry, photoelectrochemistry, surfactants and microemulsions, liquid chromatography, analysis of surfaces, UV resonance Raman spectroscopy, polymer analysis, electron spectroscopy (ESCA), laser microprobe mass analysis (LAMMA), secondary-ion spectrometry (SIMS), and time-of-flight secondary-ion mass spectrometry (TOF-SIMS). In inorganic chemistry, studies are being conducted on unusual coordination compounds, organometallics, redox reactions, complexes of biological interest, and circular dichroism; in organic chemistry, on reaction mechanisms, ion transport, total synthesis, molecular recognition, natural products synthesis, bioorganic chemistry, synthetic methodology, organometallics, enzyme mechanisms, and epoxidation reagents. Research areas in physical chemistry include Raman, photoelectron, Auger, NMR, EPR, infrared, and mass spectroscopy; electron-stimulated desorption ion angular distribution (ESDIAD); laser spectroscopy on ultracold molecules in a supersonic jet; high-resolution laser spectroscopy; molecular spectroscopy; electron and molecular beam scattering; electronic emission spectroscopy; atmospheric chemistry; and catalysis. Theoretical fields of research include electronic structure, reaction mechanisms, and scattering dynamics. Research on computer applications to chemistry is under way in a variety of areas.

Research Facilities The Department of Chemistry is housed in a modern $20-million chemistry complex. The main fifteen-story laboratory tower contains a vast array of modern research instruments and also offers in-house machine, electronics, and glassblowing shops. The Chemistry Library is a spacious 6,000-square-foot facility that contains more than 25,000 monographs and bound periodicals and more than 190 maintained journal subscriptions. Three other chemistry libraries are nearby. A laboratory instrumentation manager supervises and staffs the instrument shops, regulates the use of shared instrumentation (Cambridge 90B scanning electron microscope, VG 7070 and Varian CH-5 mass spectrometers, NICOLET R3M/E single-crystal X-ray diffraction systems, and one 500-MHz Bruker AM-500 and three 300-MHz Bruker AM-300 NMR spectrometers), and directs the training of research personnel in the operation of major instrumentation.

Financial Aid Seventy-five teaching assistantships and teaching fellowships are available. The former provide $12,735 in 1991–92 for the three trimesters of the year; the fellowships (awarded to superior students after their first year) carry an annual stipend of $13,380. A large number of advanced students are supported by research assistantships and fellowships, which pay up to $1150 per month. All teaching assistantships, fellowships, and research assistantships include a full scholarship that covers all tuition, fees, and medical insurance. Special Kaufman Fellowships provide up to an additional $4000 award to truly outstanding incoming Ph.D. candidates. In addition, Mobay Center of Excellence awards and Ashe Fellowships provide supplements to the three-term teaching assistantships that range from $1000 to $5000. In some cases, these supplements may be used to begin research in the summer prior to the formal initiation of graduate study. Chemistry graduate students are also eligible for Andrew Mellon Fellowships, which are awarded by the Graduate School on a competitive basis, and for NSF fellowships. The department also has a current grant from the Department of Education in support of Graduate Assistants in an Area of National Need, which provides prestigious national fellowships for 8 students a year within the program.

Cost of Study All graduate assistants and fellows receive full tuition scholarships. Estimated tuition and fees for full-time study in 1991–92 are $5310 per term for out-of-state students and $2655 for state residents.

Cost of Living Most graduate students prefer private housing, which is available in a wide range of apartments and rooms in areas of Pittsburgh near the campus. The University maintains a housing office to assist students seeking off-campus housing. Living costs compare favorably with other urban areas.

Student Group The University enrolls about 25,000 students, including about 9,500 graduate and professional school students. Most parts of the United States and many foreign countries are represented. About 200 full-time graduate chemistry students are supported by the various sources listed under Financial Aid. The University is coeducational in all schools and divisions; about 25 percent of the graduate chemistry students are women. An honorary chemistry society promotes a social program for all faculty and graduate students in the department.

Location Deservedly, Pittsburgh is currently ranked "among the most livable cities in the United States" by Rand McNally. It is recognized for its outstanding blend of cultural, educational, and technological resources. Pittsburgh's famous Golden Triangle is enclosed by the Allegheny and Monongahela rivers, which meet at the Point in downtown Pittsburgh to form the Ohio River. Pittsburgh has enjoyed a dynamic renaissance in the last few years. The city's cultural resources include the Pittsburgh Ballet, Opera Company, Symphony Orchestra, Civic Light Opera, and Public Theatre and the Three Rivers Shakespeare Festival. Many outdoor activities, such as rock climbing, rafting, sailing, skiing, and hunting, are also available within a 50-mile radius.

The University The University of Pittsburgh, founded in 1787, is one of the oldest schools west of the Allegheny Mountains. Although privately endowed and controlled, the University is state-related to permit lower tuition rates for Pennsylvania residents and to provide a steady source of funds for all of its programs. Attracting more than $143-million in sponsored research annually, the University has continued to increase in stature.

Applying Applications for September admission and assistantships should be made prior to March 1. However, special cases may be considered throughout the year. A background that includes a B.S. degree in chemistry, with courses in mathematics through integral calculus, is preferred. GRE scores, including the chemistry Subject Test, are required. Foreign applicants must submit TOEFL results and GRE scores.

Correspondence and Information
Graduate Admissions
Department of Chemistry
University of Pittsburgh
Pittsburgh, Pennsylvania 15260
Telephone: 412-624-8500

SECTION 2: CHEMISTRY

University of Pittsburgh

THE FACULTY AND THEIR RESEARCH

S. A. Asher, Professor; Ph.D., Berkeley, 1977. Analytical chemistry: resonance Raman spectroscopy; biophysical chemistry development of techniques for probing macromolecules and surface-adsorbed species, porphyrins, heme proteins, colloids.

H. A. Bent, Professor; Ph.D., Berkeley, 1952. Physical chemistry: molecular modeling of organic, inorganic, and metallic compounds and reaction mechanisms; education in chemistry.

R. A. Butera, Professor; Ph.D., Berkeley, 1963. Physical chemistry: thermodynamics of magnetic materials in high magnetic fields; low-temperature heat capacity measurements; chemical processes occurring at metal-metal, metal-semiconductor, and metal-compound interfaces; heat capacity studies of critical magnetic ordering.

T. M. Chapman, Associate Professor; Ph.D., Polytechnic of Brooklyn, 1965. Organic chemistry: peptide synthesis, peptide conformation, oligonucleotide synthesis, new supports for solid-phase synthesis, polyurethane chemistry.

R. D. Coalson, Assistant Professor; Ph.D., Harvard, 1984. Physical chemistry: quantum theory of rate processes, optical spectroscopy, molecular scattering, multidimensional tunneling, computational techniques for quantum dynamics.

T. Cohen, Professor; Ph.D., USC, 1955. Organic chemistry: new synthetic methods; organosulfur chemistry; natural product synthesis; synthetic uses of 3- and 4-member rings, carbenes, organolithium compounds, and sulfur-stabilized carbocations.

N. J. Cooper, Professor and Chair; D.Phil., Oxford, 1976. Inorganic chemistry: synthetic and mechanistic inorganic and organometallic chemistry, heteroallene activation; anionic complexes containing metals in negative oxidation states, synthesis and reactivity of cationic alkylidene complexes of transition metals, organometallic photochemistry.

D. P. Curran, Professor; Ph.D., Rochester, 1979. Organic chemistry: natural products total synthesis and new synthetic methodology, heterocycles in organic synthesis, synthesis via free-radical reactions.

P. Dowd, Professor; Ph.D., Columbia, 1962. Organic chemistry: trimethylene-methane, diradicals, bioorganic chemistry, mechanism of action of vitamin B_{12}.

M. F. Golde, Associate Professor; Ph.D., Cambridge, 1972. Physical chemistry: kinetic and spectroscopic studies of mechanisms of formation and removal of electronically excited atoms and small molecules, noble-gas halide excimers and similar species.

J. J. Grabowski, Associate Professor; Ph.D., Colorado, 1983. Physical organic chemistry: reaction mechanism, reactive intermediates, gas-phase techniques, flash photolitic methods.

W. K. Hall, Senior Visiting Professor; Ph.D., Pittsburgh, 1956. Physical chemistry: surface chemistry, chemisorption, catalysis.

A. D. Hamilton, Associate Professor; Ph.D., Cambridge, 1979. Organic chemistry: organic synthesis applied to biological problems, molecular models of enzymes, host-guest chemistry, novel macrocyclic ligands, porphyrin chemistry.

D. M. Hercules, Professor; Ph.D., MIT, 1957. Analytical chemistry: analytical chemistry of surfaces; electron spectroscopy (ESCA), Auger electron spectroscopy, secondary-ion mass spectrometry (SIMS), ion-scattering spectrometry (ISS), heterogeneous catalysis, surface characterization and reactions, solid-state mass spectrometry.

M. D. Hopkins, Assistant Professor; Ph.D., Caltech, 1986. Inorganic chemistry: electronically excited states of transition-metal complexes, linear-chain metal clusters, photochemistry and photophysics of multiple metal-metal and metal-ligand bonds.

K. C. Janda, Professor; Ph.D., Harvard, 1977. Physical chemistry: structure and dynamics of weakly interacting atoms, molecules, and surfaces; laser-excited fluorescence of rare gas dihalide molecules; single-photon IR desorption from surfaces.

K. D. Jordan, Professor; Ph.D., MIT, 1974. Physical chemistry: theoretical studies of the electronic structure of molecules, electron transmission and electron energy loss spectroscopy, electronic structure of metal clusters, computer simulations.

A. C. Michael, Assistant Professor; Ph.D., Emory, 1987. Analytical chemistry: in vivo voltametry, voltametry in supercritical fluids, voltametry using microelectrodes, enzyme voltametry.

D. W. Pratt, Professor; Ph.D., Berkeley, 1967. Physical chemistry: structural and dynamic properties of molecular excited states, magnetic resonance in the gas phase and in the solid state, supersonic jet spectroscopy, photochemistry and photophysics.

R. E. Shepherd, Associate Professor; Ph.D., Stanford, 1971. Inorganic chemistry: reaction mechanisms and properties of biochemically related transition-metal complexes, electron transfer reactions, catalysis mechanisms.

P. E. Siska, Professor; Ph.D., Harvard, 1970. Physical chemistry: crossed molecular beam studies of intermolecular forces and chemical reaction dynamics.

D. K. Straub, Associate Professor; Ph.D., Illinois, 1961. Inorganic chemistry: iron-sulfur complexes, Mössbauer spectroscopy.

D. H. Waldeck, Associate Professor; Ph.D., Chicago, 1983. Physical chemistry: ultrafast spectroscopy as used to investigate the condensed phase, liquid-state dynamics, charge transfer events at the semiconductor-electrolyte interface.

S. G. Weber, Associate Professor; Ph.D., McGill, 1979. Bioanalytical chemistry: photoelectrochemistry as an analytical technique, electrochemistry effects of electrode surfaces, nonhydrophobic effects in reversed-phase liquid chromatography, zwitterionic surfactant chemistry.

C. S. Wilcox, Associate Professor; Ph.D., Caltech, 1979. Organic chemistry: biomimesis and synthetic organic chemistry, synthetic receptors and functional group arrays, enzyme regulatory carbohydrate and nucleoside analogs, chemistry of O-silylketene acetals.

P. Wipf, Assistant Professor; Ph.D., Zurich, 1987. Organic chemistry: synthesis and reactivity of bioactive molecules, peptidomimetics, synthesis of reactive functionalities.

J. T. Yates Jr., Mellon Chair Professor and Director, Surface Science Laboratory; Ph.D., MIT, 1960. Surface science: kinetics of surface processes; vibrational spectroscopy of surface species; electronic spectroscopy of surfaces; catalytic and surface chemistry on model clusters, oxides, and single crystals; semiconductor surfaces.

Adjunct Faculty

A. A. Bothner-By, Professor; Ph.D., Harvard, 1949. Organic chemistry: nuclear magnetic resonance spectroscopy.

T. R. Chay, Associate Professor; Ph.D., Utah, 1961. Biophysics: statistical mechanisms.

D. Farcasiu, Professor; Ph.D., Polytechnic Institute of Timisoara (Romania), 1969. Organic chemistry, carbonium ion chemistry.

J. H. Magill, Professor; Ph.D., Queen's (Belfast), 1956. Physical chemistry: physical properties of polymers.

UNIVERSITY OF PUERTO RICO, RÍO PIEDRAS
Department of Chemistry

Programs of Study

The Department of Chemistry of the University of Puerto Rico, Rio Piedras campus, offers programs of study leading to the M.S. and Ph.D. degrees. Both degree programs require the presentation and oral defense of a dissertation. Students should have a knowledge of Spanish and English. All students select research advisers and begin research by the end of the first year. Areas of concentration are analytical, inorganic, organic, physical, and biological chemistry. An interdepartmental doctoral program in chemical physics is also offered.

M.S. students normally complete the dissertation and the required 24 semester credits in courses during the second year. Students pursuing the Ph.D. degree must pass qualifying examinations in three major areas by the end of the first year, after which they must pass four cumulative examinations and present an oral research proposal. Forty-five semester credits in courses and 24 in research are required for the Ph.D.

Research Facilities

The building housing the graduate program contains all the laboratories, classrooms, and offices used by the program, as well as the Science Library. Most rooms in the building are air conditioned. The major equipment and instruments necessary for research are available in the department. These include an X-ray diffractometer and FT-NMR, FT-IR, and MS/GC spectrometers. Undergraduate chemistry classes and laboratories are held in another building. Computing facilities are in a nearby building, and special facilities of the Puerto Rico Center for Energy and Environmental Research and of the University's School of Tropical Medicine are available for use by the department.

Financial Aid

The research program of the department is supported by funds from the University and from government and industrial grants. Both teaching and research assistantships are available from these sources. Support for students from Latin America may be obtained from foundations and from the OAS. Support of $500 or $550 per month plus remission of tuition and fees is available, depending on the qualifications of the student.

Cost of Study

For 1991–92, tuition and fees for students who are not teaching or research assistants is $55 per credit hour for residents. There is a nonresident fee, which is $1750 per semester for foreign students or equal to the nonresident fee charged by the state university in the state where the student resides.

Cost of Living

University housing is very limited, but private apartments and rooms are available in the University area. Housing costs vary from $150 to $300 per month per student.

Student Group

The student body of the Río Piedras campus consists of 20,000 full-time students in eight colleges.

About 80 students are enrolled in the graduate programs in chemistry, a number that permits careful supervision of each student's progress and needs.

Location

The University is located in a residential suburb of San Juan, the capital and cultural center of Puerto Rico. The numerous historical sites and the carefully restored buildings of Old San Juan give the city a highly individual character. With its perennially pleasant climate, excellent beaches, and convenient transportation to North and South America and all the Caribbean, San Juan is the center of tourism in the Caribbean, and its residents enjoy a wide variety of entertainment and recreational facilities.

The University

The University of Puerto Rico, founded in 1903, is supported by the Commonwealth of Puerto Rico. Río Piedras, the oldest and largest campus, includes the Colleges of Natural Sciences, Social Sciences, Humanities, General Studies, Law, Education, and Business Administration, and a School of Architecture, a graduate School of Planning, and a School of Library Science. The Medical School is located near the Río Piedras campus and the Engineering Schools at the Mayagüez campus of the University. Facilities at the 288-acre Río Piedras campus are rapidly being expanded and remodeled.

Applying

The application for admission, including results of the General Test and chemistry Subject Test of the Graduate Record Examinations, should be addressed to the Department of Chemistry and returned no later than February 15. The application for financial assistance also should be sent to the Department of Chemistry.

Correspondence and Information

Graduate Program Coordinator
Department of Chemistry
University of Puerto Rico
Río Piedras, Puerto Rico 00931
Telephone: 809-764-0000 Ext. 2445

SECTION 2: CHEMISTRY

University of Puerto Rico, Río Piedras

THE FACULTY AND THEIR RESEARCH

Rafael Arce Quintero, Professor and Dean of Natural Sciences; Ph.D., Wisconsin, 1971. Physical chemistry (photochemistry and radiation): photochemistry and photolysis of purine bases and their derivatives, heterogeneous photochemistry of polyhalogenated pollutants, ESR, flash photolysis.

Josefina Arce de Sanabia, Professor; Ph.D., Puerto Rico, 1975. Organic chemistry: organic reaction mechanisms; radicals, diradicals, and radical anion chemistry.

Lillian Bird de Pulgar, Associate Professor and Chairman; Ph.D., Puerto Rico, 1981. Inorganic chemistry.

Carlos R. Cabrera, Assistant Professor; Ph.D., Cornell, 1987. Analytical chemistry: photoelectrochemistry, electrochemiluminescence, electrochemistry in supercritical fluids.

Nestor Carballeira, Assistant Professor; Ph.D., Würzburg, 1983. Organic chemistry: organic reaction mechanisms, photochemistry, isolation and biosynthesis of marine natural products.

Osvaldo Cox, Professor; Ph.D., Ohio State, 1968. Organic chemistry: synthesis of biologically active molecules, aromatic heterocyclic compounds, photochemistry mesocyclic enones, isolation of natural products from Puerto Rican flora.

Ana R. Guadalupe, Assistant Professor; Ph.D., Cornell, 1987. Analytical chemistry: electroanalysis of metal ions in solution with polymer-modified electrodes, stripping voltammetry, HPLC with electrochemical detection, electrocatalysis, electron transfer reactions in proteins.

Yasuyuki Ishikawa, Associate Professor; Ph.D., Iowa, 1976. Theoretical chemistry: relativistic effects in atoms and molecules, investigated by relativistic SCF theory; Monte Carlo simulations of quantum many-body systems.

Yong Ji Li, Associate Professor; Ph.D., SUNY at Buffalo, 1986. Inorganic chemistry: organometallic synthesis, structure elucidation, and reactivity, X-ray crystallography, experimental determination of electron distribution of transition-metal complexes.

Reginald Morales, Professor; Ph.D., Rutgers, 1976. Biochemistry: phospholipid organization on cell surfaces, phospholipases as probes of membrane structure, structure-function relationships of lipid analogues, phospholipid synthesis and analysis.

Mariel M. Muir, Professor; Ph.D., Northwestern, 1965. Inorganic chemistry: photochemical and rate studies of transition-metal complexes; structure, reactivity, and stability of complexes of biological importance.

Jose A. Prieto, Associate Professor; Ph.D., Puerto Rico, 1982. Organic chemistry: organic synthesis, synthesis of biologically active compounds, Lewis acid catalysis.

Edwin Quiñones, Assistant Professor; Ph.D., Puerto Rico, 1986. Physical chemistry: dynamics of unimolecular and bimolecular elementary reactions, laser-induced reactions, energy-transfer collisions, photochemistry and chemiluminescence, spectroscopy of small molecules, photochemistry and photophysics of microheterogeneous solutions.

Pío R. Rechani, Associate Professor; Ph.D., Puerto Rico, 1972. Inorganic chemistry: analysis of environmental pollutants in air particulates and of human and animal tissue by chromatography, electrochemistry, and atomic absorption.

Abimael Rodriguez, Assistant Professor; Ph.D., Johns Hopkins, 1983. Organic chemistry: isolation, characterization, and synthesis of marine natural products.

Osvaldo Rosario, Associate Professor; Ph.D., Puerto Rico, 1978. Analytical chemistry: development of methods for the analysis of environmental pollutants, analysis of air pollutants, gas chromatography–mass spectrometry, artifacts during sampling.

John A. Soderquist, Professor; Ph.D., Colorado, 1977. Organic chemistry: organometallic reagents in organic synthesis and natural product chemistry, stereochemically defined functional derivatives of main group organometallics, silicon-containing analogues of natural products, metal-metalloid combinations for organic synthesis.

Manuel Torrens, Professor and Graduate Program Coordinator; Ph.D., Pittsburgh, 1970. Analytical chemistry.

Basil Vassos, Professor; Ph.D., Michigan, 1965. Analytical chemistry: instrumentation, biomedical engineering, surface studies by electrochemistry, educational instrumentation.

Luis A. Veguilla-Berdecía, Professor; Ph.D., Howard, 1964. Quantum chemistry.

Brad R. Weiner, Assistant Professor; Ph.D., California, Davis, 1986. Physical chemistry: laser studies of molecular reaction dynamics, photochemistry and photophysics with lasers, radical kinetics in gaseous phase, nonlinear processes, molecular energy transfer.

SECTION 2: CHEMISTRY

UNIVERSITY OF SAN FRANCISCO
Department of Chemistry

Program of Study

The University of San Francisco offers a program leading to the Master of Science degree in chemistry. Graduate research programs are available in selected areas of organic, physical, analytical, inorganic, and organometallic chemistry as well as biochemistry.

The M.S. degree is research oriented and is awarded on the basis of completion of 24 semester units and the presentation of a thesis resulting from a research investigation. The individual research groups tend to be modest in size, and the student-professor interaction is highly valued.

No entrance or qualifying examinations are required. Each student selects his or her research supervisor early in the first semester, and together they work out a specific program in consultation with the graduate adviser. Completion of the program generally requires eighteen months to two years. Since some courses are offered in the evening, it is possible to undertake a program on a part-time basis, in which case the completion time is extended.

Research Facilities

The Department of Chemistry is located in the Harney Science Center. The proximity of the various sciences has been particularly suitable for interdisciplinary research programs. Major research equipment includes FT-IR, GC, HPLC, electrochemical, and UV-Vis-NIR instruments as well as a 200-MHz IBM FT-NMR spectrometer.

Financial Aid

Teaching assistantships for the 1991-92 academic year carry stipends up to $5000 for two laboratory sessions. Teaching assistants are generally responsible for recitation and lab sections in the lower-division chemistry courses. Research fellowships or assistantships may be provided through grants to faculty members by such agencies as NIH and NSF. Scholarships are also available each year; many students receive $3000, and some as much as $5000.

Cost of Study

Tuition in 1991-92 is $432 per unit.

Cost of Living

Housing is available in the immediate vicinity of the University at moderate rates. The central location of the University makes it readily accessible to students who live off campus.

Student Group

There are approximately 8-12 graduate students in the department. In addition, a few graduate students are employed in local industries and attend on a part-time basis. The majority of graduate students hold teaching assistantships. Students come to USF from virtually every other part of the world as well as from the United States, making the campus truly international.

Location

The University of San Francisco is situated in the heart of the city of San Francisco—the location itself is one of the great educational assets of the institution. San Francisco, with its world-famous natural setting, is the commercial center of the Pacific Coast. This industrial metropolis, with its tradition of promoting the arts and its cosmopolitan population, provides an environment for culture and for business contacts that is unobtainable in a small community. The climate is world renowned, and every conceivable type of recreation can be found in or near the Bay Area.

The University

The University of San Francisco, known for more than three quarters of a century as Saint Ignatius College, began its existence almost simultaneously with the city of San Francisco. The school was opened in 1855 during the glamorous Gold Rush days by a group of bold Jesuits. In 1906, fire and earthquake totally destroyed the majestic institution, its laboratories, its libraries, and its art treasures. Rebuilding began immediately with characteristic pioneer courage, and today the University consists of the College of Liberal Arts, the College of Science, the College of Business Administration, the School of Law, and the School of Nursing. Enrollment at present is about 6,300 students.

Applying

Applications should be received by April 1 for the fall semester and November 1 for the spring semester. Later applications will be given due consideration. Applicants should specify their areas of research interest to the department.

Correspondence and Information

For more information on teaching assistantships, research fellowships, and the department's programs, students should write to the following address.

Dr. Jeff C. Curtis
Graduate Advisor
Department of Chemistry
University of San Francisco
San Francisco, California 94117-1080
Telephone: 415-666-6391

University of San Francisco

THE FACULTY AND THEIR RESEARCH

John G. Cobley, Professor; Ph.D., Bristol (England), 1972; Fellow of the European Molecular Biology Organization, 1973. Molecular biology of chromatic adaptation in blue-green bacteria.

Jeff C. Curtis, Associate Professor; Ph.D., North Carolina, 1980. Mechanistic/synthetic/physical inorganic chemistry: optically and thermally induced electron-transfer processes, solvent-solute interactions.

Thomas A. Gruhn, Professor; Ph.D., Berkeley, 1967. Physical chemistry of solid-state nuclear track detectors.

Theodore H. D. Jones, Professor; Ph.D., MIT, 1966. Biochemistry: biosynthetic pathways, enzymology, hormone metabolism.

Robert J. Seiwald, Emeritus Professor; Ph.D., Saint Louis, 1954. Organic chemistry: derivatives of uracil, protonation of weak bases in concentrated sulfuric acid.

Tami Spector, Assistant Professor; Ph.D., Dartmouth, 1987. Organic chemistry: synthesis and NMR investigations of perfluorocyclooctatetraene derivatives.

Kim D. Summerhays, Professor; Ph.D., California, Davis, 1971. Analytical and physical chemistry, computer applications in chemistry.

An aerial view of the USF campus.

SECTION 2: CHEMISTRY

UNIVERSITY OF SOUTH FLORIDA

Department of Chemistry

Programs of Study

Master of Science and Doctor of Philosophy degrees are offered in biochemistry and analytical, inorganic, organic, and physical chemistry. Within these general areas, students are free to choose from interdisciplinary and more specialized studies in bioinorganic, bioorganic, or clinical chemistry; computer modeling; electrochemistry; environmental or marine chemistry; enzymes; fluorescence; heterocyclic or macrocyclic chemistry; lasers; mass spectrometry; medicinal chemistry; 2-D NMR; nucleic acids and nucleosides; photochemistry; polymer chemistry; proteins; surface science; thermodynamics; and transition-metal chemistry.

All incoming graduate students take diagnostic examinations prior to enrollment to assist in planning a course of study that will complement the individual student's academic abilities, stated goals, and academic preferences. Each student receives close personal attention and guidance from a faculty of 30.

Completion of the M.S. degree requirements is expected within two years and requires (1) a minimum of 30 credit hours (18 in formal courses and the remaining 12 in seminar, research, thesis, and instructional methods), (2) a written comprehensive examination, and (3) an orally defended thesis.

Completion of the Ph.D. degree requirements is expected within five years and requires (1) a minimum of 90 credit hours (60 beyond the M.S.), of which 18 must be in formal courses and the remaining 72 in seminar, research, dissertation, instructional methods, and special topics; (2) proficiency in two languages (one foreign and one computer); (3) a comprehensive examination, which differs from the M.S. examination; (4) a departmental colloquium on the dissertation research; and (5) a dissertation that is defended orally. The M.S. degree is not required for pursuing the Ph.D.

Research Facilities

Modern well-equipped laboratories are located in the Science Center and the Physics Building. A new Bio-Sciences Facility, currently under construction, will provide ultramodern research facilities for organic chemistry and biochemistry. Major instrumentation includes two 60-MHz, two multinuclear FT, and a multinuclear 360-MHz superconducting NMR spectrometers; an X-ray photoelectron spectrometer (ESCA) and LEED/Auger system; and two research mass spectrometers. Other instrumentation includes UV-visible, FT-IR, flame, tunable organic-dye laser, and circular dichroism spectrophotometers; a stopped-flow kinetic apparatus with temperature-jump accessory; a scanning polarimeter; a single-photon counting apparatus; a flow microcalorimeter; X-ray diffractometers; radiochemical counting equipment; and various GC and HPLC systems. Instrument use is supported by faculty and technical staff.

The University maintains a data processing center with two IBM 3081 mainframes that are available to faculty and students. The College of Arts and Sciences has VAX and Sun systems. Macintosh and IBM personal computers are available in the department. Standard reference works and journals are available in the University Library and the Medical Center Library. The department has on-line access to *Chemical Abstracts*.

Financial Aid

Teaching and research assistantships ranging from $7500 to $9000 for the academic year are available to qualified applicants. Additional summer support is assured. Tuition and fee waivers are available to graduate assistants and other qualified students. University fellowships are available for minority students and other qualified students.

Cost of Study

In 1990–91, tuition was $79 per credit hour for Florida residents and $241 per credit hour for nonresidents; a tuition waiver reduces this cost to $20 and $28, respectively, per credit hour.

Cost of Living

Living expenses in Tampa are among the more moderate in the country. A wide selection of off-campus housing is readily available within easy walking or bicycling distance.

Student Group

The University has an enrollment of 32,000, the second largest in Florida; there are 60 graduate students in the department. Graduates enjoy successful careers in academia, government, and industry.

Location

The 1,748-acre campus is located in Tampa, 8 miles northeast of downtown. Numerous lakes and bays are interspersed throughout the Tampa Bay area, which has about 2 million residents. The year-round climate is excellent, and recreational and cultural activities abound.

The University

The University of South Florida was founded in 1956 and was the first state university to be built entirely in the twentieth century. Today the University consists of eight colleges, including Engineering, Medicine, and Arts and Sciences. Graduate degrees are offered in more than 100 areas.

Applying

Applicants must (1) hold a baccalaureate degree in chemistry or its equivalent from an accredited college or university, (2) have at least a B average for the last two years of undergraduate work, and (3) score a total of at least 1000 on the GRE General Test. Admission decisions are based on the applicant's grade point average, GRE scores, and letters of recommendation. Applicants whose native language is not English must submit a TOEFL score of at least 550, and those applying for assistantships must submit a satisfactory TSE score.

Correspondence and Information

Chemistry Graduate Office
Department of Chemistry
University of South Florida
4202 East Fowler Avenue
Tampa, Florida 33620-5250

Telephone: 813-974-2534
Fax: 813-974-3203

SECTION 2: CHEMISTRY

University of South Florida

THE FACULTY AND THEIR RESEARCH

Jesse S. Binford Jr., Professor; Ph.D., Utah, 1955. Physical chemistry: thermodynamics of phospholipid vesicles, enthalpy titrations with amphiphilic molecules, solubilization of vesicles and biopolymers with surfactants, fusogens, fusion of vesicles, micelles.

Robert S. Braman, Professor; Ph.D., Northwestern, 1956. Analytical chemistry: analytical chemical instrumentation, environmental chemical analysis.

Raymond N. Castle, Graduate Research Professor; Ph.D., Colorado, 1944. Organic chemistry: medicinal chemistry, organic synthesis, heterocyclic molecules, 2-D NMR spectroscopy, polycyclic S, N-heterocycles, mutagenic activity.

Alfred T. D'Agostino, Assistant Professor; Ph.D., Utah State, 1984; Postdoctoral Fellow, Lawrence Berkeley Laboratory, 1984–86. Analytical chemistry: surface chemistry, interfacial electrochemistry, corrosion, microelectronics and sensors, X-ray photoelectron and Auger electron spectroscopies, low-energy electron diffraction.

Jefferson C. Davis Jr., Professor and Associate Chairman; Ph.D., Berkeley, 1959. Physical chemistry: nuclear magnetic resonance, molecular spectroscopy, chemical education.

Jack E. Fernandez, Professor; Ph.D., Florida, 1954. Organic chemistry: polymer synthesis and properties, electrically conductive polymers, reaction mechanisms and kinetics.

Steven H. Grossman, Associate Professor; Ph.D., Purdue, 1972; NIH Postdoctoral Fellow, Wisconsin, 1972–74. Biochemistry: studies on creatine kinase, subunit interactions in multimeric proteins, fluorescence analysis.

Milton D. Johnston Jr., Associate Professor; Ph.D., Princeton, 1970; Postdoctoral Research Associate, Arizona, 1970–71; Robert A. Welch Postdoctoral Fellow, Texas A&M, 1971–73. Physical chemistry: nuclear magnetic resonance spectroscopy, molecular orbital theory of NMR spectral parameters, intermolecular forces, lanthanide-induced shifts, conformational analysis, computational chemistry.

George R. Jurch Jr., Professor; Ph.D., California, San Diego, 1965; Postdoctoral Fellow, Yale, 1966. Organic chemistry: physical-organic chemistry, organic reaction mechanisms, radical cations, free radicals, model systems in citrus chemistry, sulfur chemistry, nonaqueous solvent interactions with biological systems, NMR spectroscopy.

Leon Mandell, Professor; Ph.D., Harvard, 1951. Synthetic organic chemistry: natural products, electrochemical methods for forming carbon-carbon bonds.

Dean F. Martin, Professor, Associate Chairman, and Director, Institute for Environmental Studies; Ph.D., Penn State, 1958; NSF Fellow, University College, London, 1958–59. Inorganic and environmental chemistry: coordination compounds, coordination chemistry of the environment (red-tide outbreaks, aquatic plants), land factors and disease.

P. Calvin Maybury, Professor; Ph.D., Johns Hopkins, 1952; Research Associate, Johns Hopkins, 1952–54. Physical chemistry: chemistry of boron hydrides and metal borohydrides, heterogeneous catalysts involving transition-metal borides, controlled drug delivery systems.

Gerhard G. Meisels, Professor and Provost; Ph.D., Notre Dame, 1956; AEC Postdoctoral Fellow, Notre Dame, 1955–56. First- and second-order reactions of energy-selected gaseous ions; analytical mass spectrometry; chemical ionization; collision-induced mass spectrometry; ion-molecule reactions; radiation chemistry; structure of heavy-ion tracks; range of low-energy electrons.

Li-June Ming, Assistant Professor; Ph.D., UCLA, 1988; Postdoctoral Fellow, Minnesota, 1988–91. Bioinorganic chemistry: metalloproteins, 2-D NMR spectroscopy, protein active sites, metal-protein and protein-ligand interactions.

George R. Newkome, Professor and Vice President for Research; Ph.D., Kent State, 1966. Organic chemistry: heterocyclic chemistry, synthesis and structure determination of natural products, medicinal chemistry, inorganic complexes of biological interest, organometallics, biosensors and microelectronics.

Eugene D. Olsen, Professor; Ph.D., Wisconsin–Madison, 1960; NIH Postdoctoral Fellow, Yale, 1971–72. Analytical-clinical chemistry: atomic absorption, automation in analysis, radiochemistry, microsampling methods in clinical chemistry, chelation and ion exchange reactions.

Rebecca O'Malley, Associate Professor; Ph.D., Sheffield (England), 1970; Postdoctoral Fellow, Sheffield, 1970–72. Physical chemistry: laser desorption mass spectrometry, multiphoton ionization, unimolecular gas phase reactions, ion cyclotron resonance.

Terence C. Owen, Professor; Ph.D., Manchester (England), 1954. Organic chemistry: heterocyclic synthesis, bioorganic mechanisms and enzyme models.

Robert L. Potter, Assistant Professor; Ph.D., California, San Diego, 1979. Biochemistry: regulation of cell division, protein phosphorylation-dephosphorylation, photoaffinity labeling, high-performance liquid chromatography of peptides and proteins, microinjection, photocross-linking, food chemistry and immunoassay development.

Towner B. Scheffler, Assistant Professor; Ph.D., Mississippi, 1984; NRC Postdoctoral Fellow, Frank J. Seiler Research Laboratory, USAF Academy, 1984–85. Analytical chemistry: electroanalytical chemistry, chemistry of molten salts, electrochemistry in nonaqueous solvents, analytical spectroscopy and separations, environmental analysis.

Stewart W. Schneller, Professor and Chairman; Ph.D., Indiana, 1968; NIH Postdoctoral Fellow, Stanford, 1968–70. Medicinal chemistry: synthetic heterocyclic chemistry, nucleoside and nucleotide synthesis, antiviral agents, inhibition of enzymes involved in nucleic acid metabolism.

Joseph A. Stanko, Associate Professor; Ph.D., Illinois at Urbana-Champaign, 1966. Inorganic chemistry: X-ray crystal structure determination of transition-metal complexes, platinum complexes in cancer chemotherapy, one-dimensional conductors based on partially oxidized platinum compounds, X-ray photoelectron spectroscopy of surfaces, copper-oxide superconductors.

Brian Stevens, Graduate Research Professor; D.Phil., 1953, D.Sc., 1977, Oxford; Postdoctoral Fellow, National Research Council, Canada, 1953. Physical chemistry: solute reencounter probabilities, electron–donor-acceptor orbital correlations, singlet oxygen formation and relaxation, intermolecular electron transfer dynamics in liquids.

Jon E. Weinzierl, Associate Professor; Ph.D., Caltech, 1968; Postdoctoral Fellow, MIT, 1971–73. Biochemistry: linear programming, computer modeling of biological systems.

George R. Wenzinger, Associate Professor; Ph.D., Rochester, 1960; Research Fellow, Washington (Seattle), 1960; NSF Fellow, Yale, 1962. Organic chemistry: tricyclic rearrangements, conjugate eliminations, dinucleotide methyl phosphonate chiral synthesis.

Robert D. Whitaker, Professor; Ph.D., Florida, 1959. Inorganic chemistry: molecular adducts, chemistry of dinitrogen tetroxide.

Eric Wickstrom, Professor; Ph.D., Berkeley, 1972; Postdoctoral Research Associate, Colorado, Boulder, 1973–74. Biochemistry: antisense DNA therapeutics, ribonucleic acid structure and function, ribonucleic acid–protein interactions.

Jay H. Worrell, Professor; Ph.D., Ohio State, 1966; Postdoctoral Fellow, SUNY at Stony Brook, 1967–68. Inorganic chemistry: preparations, properties, and theory of transition-metal coordination compounds; asymmetric interactions between optically active coordination compounds; kinetics of oxidation-reduction reactions; industrial chemistry.

SECTION 2: CHEMISTRY

UNIVERSITY OF TEXAS AT AUSTIN

Department of Chemistry and Biochemistry

Programs of Study

The Department of Chemistry and Biochemistry offers programs leading to both the M.A. and the Ph.D. degrees in five major divisions of chemistry—biochemistry and analytical, inorganic, organic, and physical chemistry. In each division, there are numerous programs of research directed by renowned chemists, many of whom are leaders in their respective fields. The faculty includes several members of the National Academy of Sciences and a Nobel Prize winner. Some specific areas of interest are organometallics, homogeneous and heterogeneous catalysis, natural product synthesis, electrochemistry and photoelectrochemistry, protein crystallography, enzyme and protein chemistry, DNA replication and repair, surface science, kinetics and reaction mechanisms, marine sciences, statistical and quantum mechanics, and chemical physics.

Requirements for the M.A. degree include a total of 30 hours of course work and research culminating in a written thesis. M.A. degrees are usually completed in 1½ to 2 years.

Although advanced course work is also an integral part of the doctoral candidate's program, no set minimum number of semester hours is required for the degree. The Ph.D. is awarded on the basis of the candidate's demonstration of mastery of the selected field and the ability to pursue independent scholarship and research in it. In addition to completing the qualifying course work in the major area representing the most intensive field of study, candidates must take supporting work in two other areas of chemistry to supplement their field of specialization. A major examination, which is devised and administered separately in each area of specialization, must be passed. The average time needed to earn the Ph.D. degree is four to five years, the final three of which are devoted almost exclusively to research work.

Research Facilities

The classrooms and research laboratories comprise approximately 500,000 square feet of modern, air-conditioned, fully equipped space. The research facilities include completely equipped professional machine, glassblowing, and electronics shops with trained staffs and technicians. Graduate students may also use the equipment of the student machine shop. The department's laboratories house a variety of specialized research instruments, which include a full range (80–500 MHz) of nuclear magnetic resonance spectrometers, a high/low-resolution mass spectrometer as well as a gas chromatograph–mass spectrometer, a Fourier-transform infrared spectrometer, electron spectrometers for chemical analysis (ESCAs), molecular graphics workstations, and X-ray diffractometers. A CDC Dual CYBER 170/750 computer system and numerous minicomputers and PCs are available within the department and the University. In addition, the department has ongoing and ready access to the UT System Center for High Performance Computing, which has two CRAY X-MP Model 24 supercomputers.

Financial Aid

The research programs are supported by funds provided by the University through the state of Texas in addition to the grants that are awarded to individual faculty members by the federal government and other outside sources. Teaching assistantships and fellowships are also provided by the department. After the first year, graduate students are typically supported by either teaching or research assistantships. In addition, University fellowships, which allow time for dedicated research, are available for outstanding entering and advanced students.

Cost of Study

For 1991–92, in-state tuition and fees of approximately $350 per semester are charged to Texas residents as well as all students receiving teaching or research assistantships or fellowships. Out-of-state tuition adds approximately $1000 to the per-semester cost.

Cost of Living

The University provides housing for married students near the campus within easy walking distance of convenient University shuttle bus routes, at rents that range from $250 to $450 per month in 1991–92. In addition to these facilities, there are mobile-home lots renting for $59, and dormitory space is available for single students at $3288 per academic year. Rental properties such as apartments, duplexes, and houses are abundant around the campus and throughout the city at locations within easy walking distance of the University shuttle routes.

Student Group

The University of Texas has approximately 50,000 students, 12,000 of whom are graduate students registered in advanced programs in all areas of the sciences and humanities. The graduate student population comes from all parts of the nation and around the world. The Department of Chemistry and Biochemistry has approximately 260 graduate students and approximately 100 postdoctoral fellows.

Location

Austin, the capital of Texas and a thriving city of 500,000, is surrounded by lakes and the natural beauty of the hill country of central Texas. Despite the urban sophistication, a pleasant relaxed atmosphere prevails. The cultural and recreational attractions of the campus—and the city—extend from dramatic and musical performances to intercollegiate athletics and informal outdoor sports.

The University

The University of Texas, which is widely regarded as one of the top institutions of higher education in the United States, is the largest university in the Southwest and one of the oldest in Texas. It celebrated its centennial year in 1983. In addition to its main campus, which covers a vast area in the heart of Austin, the University also has several off-campus sites with research facilities.

Applying

Inquiries concerning admission to the Graduate School, as well as questions concerning specific degree requirements and financial support, should be directed to the Graduate Advisor's Office at the address below. Applications for assistantships and fellowships are accepted for both regular semesters and the summer session. Because of the large number of applications received each year, prospective graduate students should apply as early as possible.

Correspondence and Information

Chairman, Graduate Admissions
Graduate Advisor's Office
Department of Chemistry and Biochemistry
University of Texas at Austin
Austin, Texas 78712
Telephone: 512-471-4538

SECTION 2: CHEMISTRY

University of Texas at Austin

THE FACULTY AND THEIR RESEARCH

Creed W. Abell, Professor; Ph.D., Wisconsin, 1962. Molecular biology of proteins in the nervous system.
Eric Anslyn, Assistant Professor; Ph.D., Caltech, 1987. Bioorganic chemistry.
Dean R. Appling, Assistant Professor; Ph.D., Vanderbilt, 1981. Enzymology, regulation of one-carbon metabolism.
Allen J. Bard, Professor; Ph.D., Harvard, 1958. Electrochemical methods, photoelectrochemistry.
Nathan L. Bauld, Professor; Ph.D., Illinois, 1959. Cation radical pericyclic reactions: theory, mechanism, and synthetic uses.
Mark Berg, Assistant Professor; Ph.D., Berkeley, 1985. Femtosecond spectroscopy, dynamics of molecules in solution, dynamic nonlinear spectroscopies.
James E. Boggs, Professor; Ph.D., Michigan, 1953. Molecular structure and spectroscopy, quantum chemistry.
Jennifer Brodbelt, Assistant Professor; Ph.D., Purdue, 1988. Quadrapole ion trap mass spectrometry, ion molecule reactions.
Alan Campion, Professor; Ph.D., UCLA, 1977. Surface physics and chemistry.
Alan H. Cowley, Professor; Ph.D., Manchester, 1958. Synthetic structural and theoretical main-group and organometallic chemistry.
Patrick Davis, Associate Professor; Ph.D., Iowa, 1976. Microbial transformations, biotechnology and biomedicinal chemistry.
Raymond E. Davis, Professor; Ph.D., Yale, 1965. Packing in molecular crystals, molecular structure studies, X-ray crystallography.
Marye Anne Fox, Professor; Ph.D., Dartmouth, 1974. Physical-organic chemistry.
William C. Gardiner Jr., Professor; Ph.D., Harvard, 1960. Gas kinetics of high-temperature chemistry, kinetics of RNA replication.
John C. Gilbert, Professor; Ph.D., Yale, 1965. Reaction mechanisms, reactive intermediates.
Marvin L. Hackert, Professor; Ph.D., Iowa State, 1970. Protein crystallography, structure-function relationships of biomacromolecules.
Boyd A. Hardesty, Professor; Ph.D., Caltech, 1961. Regulation of protein phosphatases and kinases, function of ribosomes.
Adam Heller, Professor; Ph.D., Hebrew (Jerusalem), 1961. Transparent metal films, electrical microengineering of enzymes.
James A. Holcombe, Professor; Ph.D., Michigan, 1974. Graphite furnace atomic analysis, trace-metal speciation, high-temperature secondary-ion mass spectrometry.
Laurence Hurley, Professor; Ph.D., Purdue, 1970. Bioorganic chemistry, mechanism of action of DNA reactive drugs.
Brent L. Iverson, Assistant Professor; Ph.D., Caltech, 1987. Bioorganic chemistry, production of catalytic monoclonal antibodies.
Richard A. Jones, Professor; Ph.D., Imperial College (London), 1978. Organometallic chemistry, homogeneous catalysis.
G. Barrie Kitto, Professor; Ph.D., Brandeis, 1966. Evolution of protein structure, applied biochemistry.
Thomas Kodadek, Assistant Professor; Ph.D., Stanford, 1985. Organometallic catalysis, protein-DNA interactions.
Denis A. Kohl, Associate Professor; Ph.D., Indiana, 1967. Vibration/rotation in polyatomics, surface studies, photochemistry.
Richard J. Lagow, Professor; Ph.D., Rice, 1969. Polylithium organic compounds; fluorine, polymer, and organometallic chemistry.
J. J. Lagowski, Professor; Ph.D., Michigan State, 1957; Ph.D., Cambridge, 1959. Liquid ammonia chemistry, organometallic π-complexes.
David A. Laude Jr., Assistant Professor; Ph.D., California, Riverside, 1984. Analytical development of Fourier-transform mass spectrometry and NMR.
Philip D. Magnus, Professor; Ph.D., 1968, D.Sc., 1982, Imperial College (London). Organic synthesis of antitumor antibiotics.
Thomas E. Mallouk, Associate Professor; Ph.D., Berkeley, 1983. Solid-state chemistry, electrochemistry, organized molecular structures at interfaces.
Stephen F. Martin, Professor; Ph.D., Princeton, 1972. Synthesis of biologically active natural and nonnatural products.
John T. McDevitt, Assistant Professor; Ph.D., Stanford, 1985. High-temperature superconductors, macromolecular electronics and solid-state materials.
Leon O. Morgan, Professor; Ph.D., Berkeley, 1948. Nuclear magnetic resonance of biologically active metal ion complex species.
Petr Munk, Professor; Ph.D., Czechoslovak Academy of Sciences, 1960. Thermodynamics of polymers, polymer characterization, inverse gas chromatography.
Patrick L. Parker, Professor; Ph.D., Arkansas, 1960. Stable isotope geochemistry, organic geochemistry.
Ilya Prigogine, Professor; Ph.D., Brussels, 1941. Nonlinear thermodynamics, statistical mechanics.
Joanne M. Ravel, Professor; Ph.D., Texas, 1954. Protein biosynthesis and its regulation.
Scott A. Raybuck, Assistant Professor; Ph.D., Caltech, 1987. Molecular recognition, bioorganic and organometallic catalysis, molecular and polymeric materials.
Lester J. Reed, Professor; Ph.D., Illinois, 1946. Biochemistry and molecular biology of alpha-keto acid dehydrogenase multienzyme complexes.
Jon D. Robertus, Professor; Ph.D., California, San Diego, 1972. Protein crystallography, genetic engineering.
Peter J. Rossky, Professor; Ph.D., Harvard, 1977. Statistical mechanics of liquids.
Jonathan L. Sessler, Associate Professor; Ph.D., Stanford, 1982. Macrocyclic ligands, electron transfer, DNA analogues.
Hugo Steinfink, Professor; Ph.D., Polytechnic of Brooklyn, 1954. Solid-state chemistry, crystal structures, physical properties.
William W. Wade, Professor; Ph.D., Texas, 1955. Surfactant interfacial phenomena.
Kuan Wang, Professor; Ph.D., Yale, 1974. Structure and function of contractile proteins.
Stephen E. Webber, Professor; Ph.D., Chicago, 1965. Photophysics of polymers (optical spectroscopy and energy transfer).
John M. White, Professor; Ph.D., Illinois, 1966. Spectroscopy and kinetics of surface species, photochemistry, materials chemistry.
James K. Whitesell, Professor; Ph.D., Harvard, 1971. Synthesis of substances with biological activity, preparation of ordered materials for optical switches.
Robert E. Wyatt, Professor; Ph.D., Johns Hopkins, 1965. Theoretical chemistry, reactive scattering, laser molecule energy transfer, neural networks.
Daniel M. Ziegler, Professor; Ph.D., Loyola of Chicago, 1955. Molecular toxicology, mammalian drug oxidases.

SECTION 2: CHEMISTRY

UTAH STATE UNIVERSITY

College of Science
Department of Chemistry and Biochemistry

Programs of Study

Doctoral and master's degree programs in chemistry and biochemistry are administered through the Department of Chemistry and Biochemistry. Degrees in molecular biology are administered through the molecular biology graduate program, an interdepartmental program with participation by the Department of Chemistry and Biochemistry. Students in the chemistry programs may choose to specialize in analytical, inorganic, organic, or physical chemistry. All degrees are research degrees, awarded only upon completion of a thesis or dissertation describing an original contribution to science. The M.S. is not a prerequisite to the Ph.D.; most students elect to pursue the Ph.D. degree directly following the bachelor's degree.

To qualify as a degree candidate, the student must first complete a core curriculum which provides a broad base at the graduate level within a discipline. This includes 18 quarter hours in the four major branches of the science. A total of 24 quarter hours of nonresearch graduate credit, which include seminar credits, are required for the M.S. and 45 for the Ph.D. There is no language requirement. For Ph.D. candidacy, students must also pass written and oral comprehensive examinations to demonstrate both mastery of the fundamentals of a chosen specialty and an ability to formulate scientifically significant questions and a plan for answering them.

Research Facilities

The Department of Chemistry and Biochemistry is housed in two adjoining buildings containing research and teaching laboratories, lecture halls, and instrumentation for chemical analysis. In addition to the specialized research equipment, the department has separation and spectroscopic equipment used in routine chemical analysis. The latter includes high-field NMR, X-ray crystallography, cryogenic electron paramagnetic resonance, dispersive and Fourier-transform infrared (FT-IR) spectrophotometers, atomic absorption spectrometers, VIS-UV spectrophotometers and fluorimeters, high-performance liquid chomatographs, a mass spectrometer, and several gas chromatographs. Excellent pulsed laser spectroscopic facilities are also available in the department.

The chemistry stockroom is well equipped in both chemicals and glassware. Stockroom inventory is computer-based for easy location of specific items. Additional stockroom services include a full-time glassblower and an electronics technician. A machine shop is available in the department for student use; excellent professional machine shop support is available on campus.

There are numerous personal computers reserved for student access. Many of these are linked to the University's central VAX computer, which in turn is connected to the departmental Sun-4/280 computer and the Ardent Titan graphics supercomputer.

Financial Aid

Several fellowships, assistantships, and competitive awards are available through the department, the College, and the University. First-year support is usually through teaching assistantships. Financial aid through research assistantships is normally obtained beginning in the second year. Research fellowships are competitive and are usually awarded to students in their third or fourth year of study, after demonstration of research competency. In addition to the assistantships and fellowships, the department gives several competitive awards. These awards are given to recognize good teaching or research creativity.

Cost of Study

The 1991–92 full-time (9 quarter hours) graduate student tuition is $426 for Utah residents, $1133 for nonresidents. The nonresident tuition portion is waived for graduate students on assistantships. Several waivers of resident tuition are available on a competitive basis. In addition to these costs, the students normally spend $100 per quarter for books and supplies.

Cost of Living

The cost of living in the state of Utah is generally low compared to that in the rest of the country. Married student housing ranges from $259 to $355 per month. University apartments rent for $365 per quarter. Dormitory room and board is $930 per quarter. There are also many off-campus rooms, apartments, and houses for rent.

Student Group

The student group currently consists of 44 full time matriculated students, 9 of whom are women. Approximately 50 percent of the students are international. The department received 100 graduate applications in the 1990–91 year. Of these, about 50 percent were accepted for fall 1991. Most matriculated students receive financial aid. Recent graduates have found employment in industry, research laboratories, and academic institutions.

Location

The University is in a rural setting. It is in a valley situated between two mountain ranges of wilderness-designated land-use status. Three hours to the north are the Tetons and Yellowstone. One hour to the east is Bear Lake. Southeastern Utah's spectacular canyonlands desert contains five national parks, all within an easy day's drive. Ample locations for Nordic and Alpine skiing in the winter and wilderness hiking and mountain climbing in the summer are immediately adjacent to campus.

The University

Founded in 1888 as a land-grant agricultural college, Utah State has expanded to become a major university with forty-five academic departments, featuring modern buildings, a spacious library, excellent laboratories, an open, attractive campus, extensive recreation facilities, and outstanding cultural attractions.

With nearly 13,000 full-time students, the University has the advantages of both large and small schools. The student body is cosmopolitan; more than fifty foreign countries are represented. Western friendliness and informality is the life-style.

Utah State University is an accredited institution listed by the Association of American Universities.

Applying

Students may apply directly to the department. Transcripts, three letters of reference, and GRE General Test scores are required. International applications must include scores on the TOEFL. There is a $25 U.S./$30 international application fee. All applications are automatically considered for financial aid. The deadline for financial aid consideration is April 15.

Correspondence and Information

Chairman, Graduate Admissions
Department of Chemistry
Utah State University
Logan, Utah 84322-0300
Telephone: 801-750-1792

SECTION 2: CHEMISTRY

Utah State University

THE FACULTY AND THEIR RESEARCH

Ann E. Aust, Research Assistant Professor (biochemistry); Ph.D., Michigan State, 1975. Mechanisms of mutagenic carcinogens, asbestos DNA damage, mechanism for oxygen radical activation of oncogenes.

Steven D. Aust, Professor (biochemistry) and Director of the Biotechnology Center; Ph.D., Illinois, 1965. Biodegradation of environmental pollutants, role of iron in lipid peroxidase.

Stephen E. Bialkowski, Associate Professor (analytical); Ph.D., Utah, 1978. Infrared spectroscopy, laser-based molecular spectroscopy, optical and digital signal processing.

Danny J. Blubaugh, Assistant Professor (biochemistry); Ph.D., Illinois, 1987. Assembly, regulation, and structure-function relationships in photosynthesis.

Mitchell S. Chinn, Assistant Professor (inorganic); Ph.D., Yale, 1989. Main-group and transition-metal complexes as models for surface chemistry.

Eric D. Edstrom, Assistant Professor (organic); Ph.D., Minnesota, 1987. Synthetic methodologies for biological and medically important molecules.

Thomas Emery, Professor (biochemistry); Ph.D., Berkeley, 1960. Discovery, isolation, purification, and structural determination of iron binding "siderophores."

David Farrelly, Assistant Professor (physical); Ph.D., Manchester, 1980. Semiclassical quantization of highly nonseparable systems, theory of surface-structure elucidation techniques (STM, AFM).

John L. Hubbard, Assistant Professor (inorganic); Ph.D., Arizona, 1982. Novel a-halo(alkyl) transition metal complexes.

Gayle Knapp, Assistant Professor (biochemistry); Ph.D., Illinois, 1977. RNA biosynthesis, RNA protein involved in tRNA, molecular structure–based RNA biosynthesis.

Jack R. Lancaster Jr., Associate Professor (biochemistry); Ph.D., Tennessee, 1974. Biochemical pathways in methanogenic bacteria, investigations of tumor necrosis factor (TNF) protein pathways.

David B. Marshall, Associate Professor (physical/analytical); Ph.D., Utah, 1980. Surface chemistry, liquid/solid interface chemistry, theories of chromatography from the perspective of liquid/solid interactions.

Edward A. McCullough Jr., Professor (physical); Ph.D., Texas at Austin, 1971. Exact solutions to Schrodinger equation, numerical techniques for enzyme ESR spectrum prediction.

William M. Moore, Professor (physical); Ph.D., Iowa State, 1959. Chemical kinetics, photochemical reduction of riboflavin by EDTA, photochemical reduction of polycyclic aromatic compounds on surfaces.

Joseph G. Morse, Associate Professor (inorganic) and Director, University Honors Program; Ph.D., Michigan, 1967. Phosphorus-based ligands/lewis acid interactions.

Karen W. Morse, Professor (inorganic) and Provost; Ph.D., Michigan, 1967. Hydroborate metal complexes, boran derivatives of α-aminoalkylphosphonates, boron-containing agents for neutron capture therapy.

Richard K. Olsen, Professor (organic); Ph.D., Illinois, 1964. Natural product synthesis, cyclic peptide structure and synthesis.

Vernon D. Parker, Professor (physical organic); Ph.D., Stanford, 1964. Electrochemistry, electron transfer reactions, reactions of charge transfer intermediates, radical cation chemistry.

Lawrence H. Piette, Professor (biochemistry) and Dean of the School of Graduate Studies; Ph.D., Stanford, 1957. Free radical mechanisms of chemical carcinogenesis, electron spin resonance spin-trapping and spin-labeling for biomolecular structure determination.

Linda S. Powers, Professor (physical/biochemistry); Ph.D., Harvard, 1976. Spectroscopic characterization of enzyme active sites, synthetic enzyme analogs, molecular electronics, biosensors.

Cindra Widrig, Assistant Professor (analytical); Ph.D., Berkeley, 1988. Surface modified electrodes, biomimetic electrode coatings, electron transfer studies.

Michael E. Wright, Associate Professor (organic); Ph.D., Arizona, 1983. Transition-metal catalysis, design, synthesis, and evaluation of chiral ligands; metal-catalyzed polymer synthesis.

SECTION 2: CHEMISTRY

WESLEYAN UNIVERSITY
Department of Chemistry

Program of Study

The Department of Chemistry offers a program of study leading to the Ph.D. degree. Students are awarded this degree upon demonstration of creativity and scholarly achievement. This demands intensive specialization in one field of chemistry as well as broad knowledge of related areas. The department provides coverage of physical, organic, inorganic, bioorganic, and biophysical chemistry.

The first year of graduate study contains much of the required course work, although most students also choose a research adviser and begin a research program at the beginning of the second semester. Students are expected to demonstrate knowledge of five core areas of chemistry, either by taking the appropriate course or by passing a placement examination. In addition, students take advanced courses in their area of specialization. Classes are small (5–10 students) and emphasize interaction and discussion. Student seminar presentations are also emphasized. Election of interdisciplinary programs in chemical physics and molecular biophysics in conjunction with the Departments of Physics and Molecular Biology & Biochemistry, respectively, is also possible. Students are admitted to Ph.D. candidacy, generally in the second year, by demonstrating proficiency in the core course curriculum, passing a specified number of regularly scheduled progress exams, demonstrating an aptitude for original research, and defending a research proposal. The progress and development of a student is monitored throughout by a 3-member faculty advisory committee. The Ph.D. program, culminating in the completion of a Ph.D. thesis, is normally completed within four to five years. Two semesters of teaching in undergraduate courses is required, where the load is, on average, about 5 hours per week during the academic year. This requirement is normally met in the first year.

An M.A. degree may be awarded on completion of two years of study, as described above, and with a written research thesis. Students who wish only a master's degree, however, are generally not admitted.

Research Facilities

The Hall-Atwater Laboratory is equipped with a wide variety of modern instrumentation appropriate to the research interests of the faculty. There are excellent machine and electronics shops and glassblowing facilities. A departmental computer network consisting of three MicroVAX computers, a shared disk system, an array processor, and a color vector graphics unit and connected to University and national computer networks is available to students. The Science Library, containing an excellent collection of journals, monographs, and reference materials, is located in a building directly adjacent to the Hall-Atwater Laboratory.

Financial Aid

All students receive a 12-month stipend, which, for 1991–92, is $11,160. In the first year, this stipend derives from a teaching assistantship. In later academic years, students may be supported by research assistantships where funds are available, or by further teaching assistantships.

Cost of Study

Tuition for 1991–92 is $2035 per graduate course, but remission of this is granted to all holders of teaching and research assistantships.

Cost of Living

Most graduate students, both single and married, live in houses administered and maintained by the University, with rents in the $300–$400 per month range.

Student Group

The student body at Wesleyan is composed of some 2,600 undergraduates and 150 graduate students. Of the latter, most are in the sciences, with about 35, equally divided between men and women, in the chemistry department. Most graduates obtain industrial positions, although some choose academic careers, normally after postdoctoral experience in each case.

Location

Middletown is a small city on the west bank of the Connecticut River, 15 miles south of Hartford, the state capital. New Haven is 24 miles to the southwest; New York City and Boston are 2 hours away by automobile. Middletown's population of 50,000 is spread over an area of 43 square miles, much of which is rural. Although Wesleyan is the primary source of cultural activity in Middletown, the city is not a "college town" but serves as a busy commercial center for the region between Hartford and the coast. Water sports, skiing, hiking, and other outdoor activities can be enjoyed in the hills, lakes, and river nearby.

The University

For more than 150 years, Wesleyan University has been identified with the highest aspirations and achievements of private liberal arts higher education. Wesleyan's commitment to the sciences dates from the founding of the University, when natural sciences and modern languages were placed on an equal footing with traditional classical studies. In order to maintain and strengthen this commitment, graduate programs leading to the Ph.D. degree in the sciences were established in the late 1960s. The program in chemistry was designed to be small, distinctive, and personal, emphasizing research, acquisition of a broad knowledge of advanced chemistry, and creative thinking.

Applying

By and large, a rolling admissions policy is in place, although applicants seeking admission in September are advised to submit applications (no application fee) as early as possible in the calendar year. Three letters of recommendation are required, and applicants are strongly advised to take the Graduate Record Examinations. Students whose native language is not English should take the TOEFL. Applicants are strongly encouraged to visit the University after arrangements are made with the department.

Correspondence and Information

Chairman
Graduate Admissions Committee
Department of Chemistry
Wesleyan University
Middletown, Connecticut 06459

Peterson's Guide to Graduate Programs in the Physical Sciences and Mathematics 1992

Wesleyan University

THE FACULTY AND THEIR RESEARCH

David L. Beveridge, Professor; Ph.D., Cincinnati, 1965. Theoretical physical chemistry and molecular biophysics: statistical thermodynamics and computer simulation studies of hydrated biological molecules, nucleic acid and protein structure, environmental effects on conformational stability, organization of water in crystal hydrates.

Philip H. Bolton, Professor; Ph.D., California, San Diego, 1976. Biochemistry and physical chemistry: NMR and modeling studies of duplex DNA, the structure of DNA containing abasic sites, the structure and activity of wild-type and mutant forms of the enzyme staphylococcal nuclease; development of NMR methodology.

Joseph W. Bruno, Associate Professor; Ph.D., Northwestern, 1983. Inorganic and organometallic chemistry: synthetic and mechanistic studies of transition-metal compounds; organometallic photochemistry, metal-mediated reactions of unsaturated organics.

Albert J. Fry, Professor; Ph.D., Wisconsin, 1963. Organic chemistry: mechanisms of organic electrode processes, development of synthetically useful organic electrochemical reactions.

Peter A. Jacobi, Professor; Ph.D., Princeton, 1973. Organic chemistry: synthetic organic chemistry, natural products and materials of biological or theoretical importance.

Joseph L. Knee, Assistant Professor; Ph.D., SUNY at Stony Brook, 1983. Chemical physics: investigation of ultrafast energy redistribution in molecules using picosecond laser techniques, emphasis on isolated molecule processes including unimolecular photodissociation reaction rates.

Stewart E. Novick, Associate Professor; Ph.D., Harvard, 1973. Physical chemistry: molecular-beam spectroscopy, structure and dynamics of weakly bound complexes, conformations of floppy molecules.

George A. Petersson, Professor; Ph.D., Caltech, 1970. Theoretical chemistry: development of improved methods for electronic structure calculations, with applications to small molecular systems and chemical reactions.

Rex F. Pratt, Professor; Ph.D., Melbourne (Australia), 1969. Bioorganic chemistry: enzyme mechanisms and inhibitor design, beta-lactam antibiotics and beta-lactamases, protein chemistry.

Wallace C. Pringle, Professor; Ph.D., MIT, 1966. Physical chemistry: Spectroscopic studies of internal interactions in small molecules, collision-induced spectra, environmental chemistry.

Ganesan Ravishanker, Adjunct Lecturer of Chemistry and Manager of Computer Graphics Systems; Ph.D., CUNY, 1984. Molecular dynamics and Monte Carlo simulation studies on biomolecular systems, application of computer graphics techniques for the analysis of molecular dynamics simulations, complete correlation analysis of conformational parameters of DNA.

John W. Sease, Emeritus Professor; Ph.D., Caltech, 1946. Organic and analytical chemistry: mechanism of electrochemical reduction of the carbon-halogen bond, product analysis and voltammetry, chemical instrumentation.

Susan B. Sobolov, Assistant Professor; Ph.D., Yale, 1989. Bioorganic chemistry: enzymes in organic synthesis; relationships between enzyme structure, function, and substrate specificity.

H. David Todd, Adjunct Associate Professor of Chemistry and Director of the Computer Center; Ph.D., Johns Hopkins, 1971. Physical chemistry: computational problems in chemistry and quantum chemistry.

T. David Westmoreland, Assistant Professor; Ph.D., North Carolina, 1985. Inorganic and bioinorganic chemistry: electronic structure and mechanism in molybdenum-containing enzymes, EPR spectroscopy of transition-metal complexes, fundamental aspects of atom transfer reactions in solution.

The Hall-Atwater Laboratory, which houses the chemistry department.

SECTION 2: CHEMISTRY

YALE UNIVERSITY
Department of Chemistry

Programs of Study

The Department of Chemistry admits students for graduate study leading to the Ph.D. degree. Students may earn an M.S. while working toward the doctorate, but no students are admitted who wish to terminate their studies at the master's level.

The department is divided into four divisions: inorganic, organic, physical, and biophysical. Entering students select the division in which they wish to get their degrees at the beginning of their first semester of study. There are two stages to the Ph.D. program: qualification and the thesis. In order to qualify, the student must complete a certain number of courses with satisfactory grades, demonstrate proficiency in reading the chemical literature in either German, French, or Russian, and pass a qualifying examination. The courses to be taken vary from one division of the department to another and are arranged for each student, taking his or her previous experience into account. The format of the qualifying examination also depends on the division in which the student is enrolled. Qualification is normally completed by the end of the second year. All students are expected to participate in teaching by serving as a teaching assistant in departmental courses for two semesters. This requirement is met by most students in their first year.

Thesis research is unquestionably the most important part of a student's training in the Ph.D. program. By the beginning of the second term of study, students select the faculty member in whose laboratory they will do their research. Full-time research is carried out in the summer between the first and second years. Most students do research part-time while engaged in the final stages of qualifying for the degree in their second year of study and full-time thereafter.

Research Facilities

The chemistry department is located in adjoining buildings, the Sterling Chemistry Laboratory and the Kline Chemistry Laboratory. A substantial amount of this laboratory space has been renovated recently. Most of the department's major research instrumentation is contained in the Yale Chemical Instrumentation Center, which is part of Sterling. Included in the center are two 500-MHz NMR spectrometers, one of them a Bruker WM500 and the second built by the personnel of the center. In addition, there are two Bruker 250-MHz NMR spectrometers, a Kratos mass spectrometer, and equipment for X-ray crystallography. The center also maintains UV, visible, IR, and CD spectrometers along with a laser spectroscopy laboratory. The laboratories of individual faculty members contain the more specialized instruments specific to their research.

The department operates a variety of computers, including a new superminicomputer for the use of its faculty and students. These machines are connected to University and national computing networks. A small library devoted to chemical journals and monographs is housed in Sterling Chemistry, and the main science library covering all disciplines is a few steps away.

Financial Aid

Graduate students in the department, including the entering class, receive financial aid. Entering students who do not have outside support receive a nine-month University Fellowship that pays tuition plus a stipend. The details of this support are clearly spelled out in the letter of admission. After the spring of the first year, most students are supported by the research grants of their research director as assistants in research. Those students who are not supported in their later years by their directors' grants are awarded University Fellowships in an amount comparable to the aid they received as first-year students. The 1991–92 stipend for those supported by teaching or as assistants in research is $12,800 for twelve months, plus tuition. Onsager, Pfizer, and Kent fellowships for outstanding applicants increase stipends by modest amounts.

Cost of Study

Tuition is normally covered by a tuition fellowship or tuition remission from a research grant.

Cost of Living

It is estimated that the standard budget for a single student for the academic year 1991–92 (exclusive of tuition) is $10,530. This includes room and board, transportation, and academic and personal expenses. The estimated standard budget for a married couple is $15,360; this figure increases by approximately $2000 for each dependent.

Student Group

The total number of graduate students in the chemistry department is 140. Postdoctoral students number about 50. A wide variety of ethnic and national backgrounds are represented by these students.

Location

Yale enjoys many of the advantages of a cosmopolitan center without the disadvantages of a congested urban environment. The Greater New Haven area has a population of about 360,000. One of the most extensive urban redevelopment programs ever undertaken in a city of comparable size has made significant improvements to the center of New Haven and its very character. New Haven is 90 minutes from New York by train or car, and Boston is only 3 hours away. The University is an integral part of the worldwide community of scholars, and its central location makes it possible for many distinguished visitors to come to Yale, both from the United States and from abroad.

The University

Yale began to offer graduate education in 1847 and in 1861 conferred the first Ph.D. degrees awarded in North America. In 1876, Yale became the first American university to award the Ph.D. to a black American. With the appointment of a dean in 1892, the Graduate School was formally established. In the same year, women were first admitted as candidates for the doctorate. Today, the Graduate School community includes a growing number of postdoctoral students, fellows, and research associates in addition to its faculty and the many students enrolled in regular degree programs. Besides the Graduate School of Arts and Sciences, Yale's Schools of Architecture, Art, Music, Drama, Forestry and Environmental Studies, Organization and Management, Divinity, Nursing, Medicine, and Law offer graduate-level and professional programs.

Applying

Applications for admission may be obtained by writing to the address below. A nonrefundable fee of $45 must accompany each application. The application deadline is January 2; however, applications received after this date will be considered when possible. All applicants must take the GRE General Test and the Subject Test in chemistry. It is recommended that these tests be taken in September, if possible, and not later than December. Students whose native language is not English are also required to take the Test of English as a Foreign Language (TOEFL).

Correspondence and Information

Department of Chemistry
Yale University
P.O. Box 6666
New Haven, Connecticut 06511
Telephone: 203-432-3915

Yale University

THE FACULTY AND THEIR RESEARCH

Jerome A. Berson, Professor; Ph.D., Columbia, 1949. Organic chemistry: reaction mechanisms, synthesis of molecules of theoretical interest, biradicals and other reactive intermediates.

Gary Brudvig, Associate Professor; Ph.D., Caltech, 1981. Biophysical chemistry: structural and mechanistic studies of photosynthetic water oxidation, application of electron paramagnetic resonance spectroscopy to the study of metalloproteins, biological electron transfer reactions, manganese–enzyme active site model compounds.

John P. Caradonna, Assistant Professor; Ph.D., Columbia, 1985. Inorganic/bioinorganic chemistry: structural and mechanistic studies of transition-metal-based atom transfer reactions in biological systems; determination of the structure and function of metal-mediated DNA-binding proteins.

William A. Chupka, Professor; Ph.D., Chicago, 1951. Physical chemistry: photoionization mass spectrometry, laser multiphoton and photoelectron spectroscopy, vacuum-ultraviolet spectroscopy, ion-molecule reactions, chemi-ionization phenomena.

Robert H. Crabtree, Professor; Ph.D., Sussex, 1973. Inorganic chemistry: exploratory synthetic studies in transition-metal chemistry.

R. James Cross Jr., Professor; Ph.D., Harvard, 1965. Physical chemistry: detailed dynamics of organic reactions, using crossed molecular beams; theory of inelastic scattering.

Donald M. Crothers, Professor; Ph.D., California, San Diego, 1963. Biophysical chemistry: nucleic acid structure and protein interactions, mechanism of control of gene expression.

Samuel Danishefsky, Professor; Ph.D., Harvard, 1962. Organic chemistry: development of new synthetic strategies and their application to the synthesis and modification of biologically interesting natural products.

John W. Faller, Professor; Ph.D., MIT, 1967. Inorganic chemistry: synthesis and elucidation of structure and bonding of inorganic and organometallic compounds; mechanism of catalysis; stereoselective and asymmetric synthesis, utilizing transition-metal complexes.

Gary L. Haller, Professor; Ph.D., Northwestern, 1966. Chemical engineering and chemistry: heterogeneous catalysis and surface structure and mechanism of catalyzed reactions; catalyst characterization utilizes classical chemisorption and reaction kinetics of sample reactions with several physical methods: XPS, IR, EELS, EXAFS, NMR, and ESR spectroscopies.

D. Michael Heinekey, Associate Professor; Ph.D., Alberta, 1982. Inorganic chemistry: transition-metal organometallic chemistry, reaction mechanisms, stereochemistry and reactivity of metal-metal bonds.

Mark A. Johnson, Associate Professor; Ph.D., Stanford, 1983. Physical chemistry: spectroscopy and reaction dynamics of molecular clusters, photoelectron spectroscopy of negative ion clusters, photodissociation dynamics of trapped species in ionic clusters.

William L. Jorgensen, Professor; Ph.D., Harvard, 1975. Organic chemistry: applications of molecular dynamics, quantum mechanics, and statistical mechanics in organic chemistry and biochemistry; computer simulations of reactions in solution; protein dynamics; computer-assisted synthetic analysis; molecular recognition.

J. Michael McBride, Professor; Ph.D., Harvard, 1966. Organic chemistry: physical and chemical properties of organic solids, free-radical reactions, crystal growth.

Peter B. Moore, Professor; Ph.D., Harvard, 1966. Biophysical chemistry: determination of the structure and function of ribonucleoproteins, especially ribosomes, using neutron scattering, NMR, and X-ray crystallography.

James H. Prestegard, Professor; Ph.D., Caltech, 1971. Biophysical chemistry: structure-function studies of biological membranes and related systems, using NMR.

Martin Saunders, Professor; Ph.D., Harvard, 1956. Organic chemistry: structures and rearrangement processes in carbocations in stable solution, isotopic perturbation, new method for molecular mechanics.

Alanna Schepartz, Assistant Professor; Ph.D., Columbia, 1987. Organic/bioorganic chemistry: development and application of protein affinity cleavage reagents, molecular recognition of RNA, design of self-assembling structures.

Robert G. Shulman, Professor; Ph.D., Columbia, 1949. Molecular biophysics and biochemistry and chemistry: study of metabolism in vivo by high-resolution NMR; living systems studied range from *Escherichia coli* to humans and the nuclei followed are ^{31}P, ^{13}C, ^{1}H, ^{23}Na, and ^{39}K; generally, metabolic fluxes are determined and the control of these pathways ascertained.

Oktay Sinanoglu, Professor; Ph.D., Berkeley, 1959. Theoretical chemistry: quantum theory of the stability of atomic and molecular species, theory of electronic spectra and electron correlation in atoms and molecules, intermolecular forces in liquids and theory of solvent effects, atomic and molecular processes in scattering, potential energy surfaces, group theory.

Thomas A. Steitz, Professor; Ph.D., Harvard, 1966. Molecular biophysics and biochemistry and chemistry: macromolecular structure and function; protein–nucleic acid interaction, including structures of proteins involved in control of gene expression, DNA replication, and recombination; enzyme mechanism, specificity, and conformational changes; membrane protein folding; protein crystallography.

Patrick H. Vaccaro, Assistant Professor; Ph.D., MIT, 1986. Physical chemistry: state-selective preparation and characterization of energetic molecular species; state-to-state studies of reaction dynamics and relaxation phenomena; development and application of multiple-resonance laser techniques.

H. H. Wasserman, Professor; Ph.D., Harvard, 1949. Organic chemistry: development of new reactions useful in synthesis of products having biological interest.

Kenneth B. Wiberg, Professor; Ph.D., Columbia, 1950. Organic chemistry: synthesis and reactions of small-ring hydrocarbons, quantum chemistry, thermochemistry, IR and UV spectroscopy.

Frederick E. Ziegler, Professor; Ph.D., Columbia, 1964. Organic chemistry: synthetic methods, natural products synthesis.

Kurt W. Zilm, Associate Professor; Ph.D., Utah, 1981. Physical chemistry: solid-state NMR, zero-field NMR, applications to matrix isolation chemistry of reactive species, surface chemistry, transition-metal polyhydrides, main group multiple bonding.

Sterling Chemistry Laboratory is part of the Science Hill quadrangle.

Section 3
Earth and Planetary Sciences

This section contains directories of institutions offering graduate work in earth sciences, geochemistry, geodetic sciences, geology, geophysics, and planetary sciences, followed by two-page entries submitted by institutions that chose to prepare detailed program descriptions. Additional information about programs listed in the directories but not augmented by a two-page entry may be obtained by writing directly to the dean of a graduate school or chair of a department at the address given in the directory.

For programs offering related work, see all other areas in this book; in Book 2, see: Geography; in Book 3: Agricultural Sciences; Biology and Biomedical Sciences; Biophysics; Botany and Plant Sciences; and Natural Resource Sciences; in Book 5: Agricultural Engineering; Civil and Environmental Engineering; Engineering and Applied Sciences; Geological, Mineral/Mining, and Petroleum Engineering; Mechanical Engineering, Mechanics, and Aerospace/Aeronautical Engineering; and Nuclear Engineering.

CONTENTS

Program Directories

Earth Sciences	187
Geochemistry	191
Geodetic Sciences	193
Geology	194
Geophysics	210
Planetary Sciences	215

Announcements

Auburn University	194
Ball State University	194
Brown University	215
California State University, Fresno	195
Carleton University	187
Cornell University	196
Dartmouth College	196
Duke University	196
Georgia State University	197
McMaster University	188
Miami University	198
Michigan State University	198
Montana College of Mineral Science and Technology	191
New Mexico Institute of Mining and Technology	
Geochemistry	192
Geology	199
Geophysics	211
Old Dominion University	200
Pennsylvania State University	192, 200, 211
Portland State University	200
Southern Illinois University at Carbondale	201
Sul Ross State University	202
Syracuse University	202
Texas A&M University	
Geology	202
Geophysics	212
University of British Columbia	203
University of California, Riverside	203
University of Delaware	204
University of Florida	204
University of Hawaii at Manoa	204
University of Idaho	204
University of Minnesota, Duluth	205
University of Missouri-Rolla	205
University of Oregon	206
University of Pennsylvania	207
University of Rochester	207
University of Southern California	207
University of Texas at San Antonio	208
University of Waterloo	190
West Virginia University	209
Yale University	210

Cross-Discipline Announcement

University of Virginia	216

Full Descriptions

Bowling Green State University	217
California Institute of Technology	219
Harvard University	221
Indiana University Bloomington	223
Iowa State University	225
Louisiana State University	227
McGill University	229
Michigan State University	231
Northern Illinois University	233
Oregon State University	235
Pennsylvania State University	237
Rensselaer Polytechnic Institute	239
Southern Methodist University	241
Stanford University	243
Texas A&M University	245
Tulane University	247
University of Arizona	
Geosciences	249
Planetary Sciences	251
University of Calgary	253
University of California, Los Angeles	255
University of California, Santa Barbara	257
University of Chicago	259
University of Cincinnati	261
University of Georgia	263
University of Hawaii at Manoa	265
University of Houston	267
University of Illinois at Chicago	269
University of Maryland College Park	271
University of Minnesota, Twin Cities Campus	273
University of Nevada, Reno	275
University of New Mexico	277
University of Oklahoma	279
University of Pittsburgh	281
University of Saskatchewan	283
University of Texas at Austin	285
University of Texas at Dallas	287
University of Toledo	289
University of Toronto	291
University of Washington	293
Washington University	295
Wright State University	297
Yale University	299

    See also:

Columbia University—Atmospheric and Planetary Science	527
Emory University—Physics	583
North Carolina State University—Marine, Earth, and Atmospheric Sciences	317
University of California, San Diego—Oceanography	321
University of Kansas—Physics and Astronomy	649

FIELD DEFINITIONS

In an effort to broaden prospective students' understanding of disciplines in earth and planetary sciences, educators in the field have submitted the following statements on earth sciences, geochemistry, geodetic science, geology, geophysics, and planetary sciences.

SECTION 3: EARTH AND PLANETARY SCIENCES

Field Definitions

Earth Sciences

The earth sciences today connote a very diverse area of study, including the various pursuits of traditional geology (mineralogy, petrology, geomorphology, glacial geology, volcanology, sedimentation, stratigraphy, tectonics, etc.) as well as geochemistry, geophysics, glaciology, hydrology, oceanography (chemical, geological, and physical), and paleontology. These disciplines cover the study of the solid, liquid, and gaseous parts of the earth and other planets of the solar system, including studies of their magnetic and gravitational fields. Virtually all of the earth sciences are interdisciplinary in that they are the application of chemistry, biology, mathematics, and physics to their special area of study.

All of the disciplines of the earth sciences have an orientation toward the study of processes. These studies vary widely from erosion, sedimentation, and volcanism on earth, other planets, and their satellites to organic evolution and to the processes that move the earth's crustal plates, produce ocean tides and currents, and modify the weather (long and short term). Most of the disciplines have a historical aspect leading to the history and evolution of the earth, its continents, mountains, oceans, etc., as well as the other planets and their major features.

Much research in the earth sciences has been directed to exploration for nonrenewable essential minerals and fossil fuels (oil, gas, and coal), as well as to alternative energy sources. Most recently, because there has been a great concern for decreasing supplies of clean surface and underground water as well as for pollution of this necessary resource, demand for specialists in hydrology and hydrogeology has been very high.

Herbert Tischler, Chairman
Department of Earth Sciences
University of New Hampshire

Geochemistry

Geochemistry, broadly construed, is the application of the principles of chemistry to earth, environmental, and planetary sciences. We may broadly distinguish two main categories, organic and inorganic, although many terrestrial subjects clearly involve both types.

Geochemistry has enjoyed a remarkable flowering in the past thirty years, chiefly because of the increased capabilities brought about by advances in instrumentation and technology. We are now able to sustain static pressures in the laboratory exceeding half a million bars, attain plasma temperatures of several thousand degrees C, and detect atomic and isotopic abundances in favorable circumstances to less than a part per billion.

Individual fields have grown up around analytical capabilities, such as mass spectrometry; gas chromatography; X-ray diffraction and fluorescence; instrumental and radiochemical neutron activation; atomic absorption; Raman, Mössbauer, optical, plasma, and infrared spectroscopy; and synchrotron beam, electron, and ion microanalysis, among others. Materials investigated include living systems; fossil flora and fauna; petroleum; natural gas; coal; igneous, sedimentary, and metamorphic rocks; volcanic emanations; hydrothermal and connate fluids; ore deposits; lunar rocks; tectites; and meteorites. Particular disciplines involve the study of major and trace-element distribution, radioactive and stable isotope fractionations, experimental phase equilibria, and thermochemical computation and numerical modeling.

Employment opportunities, as might be anticipated, are as diverse as the subject covered by the term geochemistry. Geochemists are utilized in the mineral resource industries (working on fissionable and fossil fuels and ore deposits), earth and space sciences, environmental sciences (including hydrology, groundwater geology, aqueous geochemistry, and fluid dynamics), and several branches of metallurgy, ceramics, and solid-state chemical science and technology.

W. G. Ernst, Dean
School of Earth Sciences
Stanford University

Geodetic Science

The task of geodetic science is to determine the size and shape of the Earth and carry out measurements and computations necessary for accurate mapping of the Earth's surface. With the conquest of space, this task extends itself to celestial bodies other than the Earth.

Geodetic science techniques find many applications in engineering, architecture, geology, geodynamics and other areas of geophysics, agriculture, archaeology, and geography, to mention but a few.

The study of geodetic science can be built only on a solid foundation, which includes mathematics, physics, astronomy, and geophysics.

The field of geodetic science usually includes geodesy, photogrammetry, and cartography.

Some consider geodesy to be the oldest earth science. It has the general goals of precisely determining positions on the Earth's surface, determining the Earth's size and shape, determining the Earth's gravity field, and measuring and representing geodynamic phenomena.

Photogrammetry obtains reliable information about the properties of surfaces and objects without physical contact. The fundamental task is to reconstruct, measure, and analyze three-dimensional objects from two-dimensional photographs or digital imagery. In the case of digital imagery, the techniques of digital image processing, pattern recognition, and computer vision are used. A conventional application of photogrammetry has been the compilation of topographic maps represented graphically— the classical paper map—or digitally. Today digital maps are the most important source for geographic information systems. Close-range applications of photogrammetry include medicine, industry, and architecture.

Cartography describes an environment by graphical and/or digital means. The irreversible trend in cartography today is toward computer-assisted methods of mapping and data manipulation. The cartographer's goal is to organize the data of an environment into an efficient land (or geographic) information system. For example, a graphical or analog land (or geographic) information system would be a topographic or thematic map. A digital (or geographic) information system would exist in a computer and would be accessible via appropriate terminals.

Geodetic science forms the basis of studies in the mapping sciences. Until a few years ago, the only maps available were those printed on material such as paper. Today the mapping sciences stand on the threshold of an exciting and ever-expanding future. The computer now allows the mapping scientist to produce maps on computer graphics terminals. Some maps reside in a computer, in complete numerical (or digital) form, from which a user can extract needed information. The computer and its peripherals, such as digitizers, plotters, file management systems, and graphics terminals, are used today in all stages of map production.

Ivan I. Mueller, Professor and Chairman
Department of Geodetic Science and Surveying
Ohio State University

Geology

Geology is the study of the Earth, especially the solid part of the Earth and its interactions with the fluids and gases within and above it. The concerns of geology, now frequently called geological sciences or included in earth and planetary sciences, have traditionally been with the rocks and structures of the crust of the Earth as exposed at the surface or studied in bore holes or mines up to a few kilometers beneath the surface. In the last generations, the subject has come to include the geochemistry and geophysics of the Earth and its relation to the other planets as integral parts of the geological sciences. Modern geologists not only study their rocks directly in the field but study them by remote sensing from satellites and by using geophysical and geochemical exploration techniques. They carry the rocks to the laboratory and there analyze them physically and chemically. Experiments on rock synthesis, deformation, and destruction constitute an important part of the field.

There are two major aspects to the study of geology: how

the Earth operates as a dynamic system today and its historical evolution as a planet. The first aspect concerns the current forces shaping the surface and interior of the Earth, from mountains to the floors of the oceans, from the sedimentary processes by which the products of erosion are deposited as sedimentary layers to the dynamics of volcanism and earthquakes, and all of the other processes that continue to move and modify the structure and form of the Earth. The second aspect, usually lumped under the subject historical geology, includes the study of the Earth through its 4½-billion-year history. From historical geological studies, we are able to learn the dynamics of very slow processes. In almost all of the subfields of geology—stratigraphy, paleontology, sedimentology, mineralogy and petrology, structural geology, and geomorphology—there are new and exciting developments, many of them linked to the major new theory that has swept the geological sciences in the past twenty years—plate tectonics. Geophysics and geochemistry have been broadly joined to geology, particularly in the study of the development of the ocean basins, the origin of mountain belts, and the evolution of continental landmasses.

Researches in geology have extreme practical value: the energy and mineral resources of the Earth are largely contained in its crustal layers. Geologists, in particular those engaged in the hunt for fossil fuels (oil and gas, coal, shale oil, and tar sands) and those hunting for the mineral resources that we use for much of our industry (ores of various metals, nonmetallic mineral resources, such as phosphate, and others), are in great demand by the energy and mineral resource industry. There is a great deal of research going on into both fundamental and applied aspects of the origin of various economically important deposits. Mining has been extended to the seafloor, and so, as with many other areas of geology, the study of the oceans has become an important division of the subject: marine geology, geophysics, and geochemistry are integral parts of the field.

Because of the great breadth of the geological sciences and their relation to all of the other sciences, the advanced student needs to know not only a great deal about the Earth but also physics, chemistry, and mathematics in order to be able to carry on the newer researches in the field. With a good background in related sciences and in geological sciences, geologists are having enormous successes in understanding the ways in which the Earth is formed and the ways in which its economically useful products may be recovered.

*Raymond Siever
Professor of Geology
Department of Earth and Planetary Sciences
Harvard University*

Geophysics

Geophysics means physics of the Earth. Because the Earth is a vast and complicated object, geophysics is a broad subject and includes, for example, meteorology (the science of the weather) and oceanography (the science of the oceans, their currents, waves, and tides). Solid-earth geophysics is concerned with the problems of the internal constitution of the Earth and its evolution. What causes mountains to form, volcanoes to erupt? What are earthquakes? What accounts for the heat that seeps out of the Earth, or for its magnetic field? What causes continents to move, or the seafloor to be constantly renewed as it spreads from places where new crust is formed to places where it sinks back down into the depths? These are some of the questions that geophysics tries to answer in studies of the structure and internal composition of the Earth; seismology (the science of earthquakes and of the propagation through the Earth of the elastic waves they generate); studies of the gravitational, electrical, and magnetic fields of the Earth; studies of heat sources, heat transfer, and temperature distribution within the Earth; and laboratory studies of the physical and chemical properties of minerals under the conditions of high pressure and temperature that are present within the Earth.

Geophysics is clearly an interdisciplinary field. On the one hand, it is very closely related to geology, the study of rocks and the record of the Earth's history that can be read from them. On the other hand, it is closely related to physics as it attempts to explain geological observations in terms of experimentally verifiable physical laws. It is also related to chemistry and to the branch of that science (geochemistry) that deals with the distribution of chemical elements and chemical compounds (minerals) that form the Earth. Mathematics is an indispensable tool of geophysics, as it is of physics.

Geophysics also has some highly practical applications, closely related to engineering. Much of the search for economically valuable deposits of metals and petroleum, for example, is conducted by geophysical means, such as by detecting the effects these bodies have on gravity or on the electrical or magnetic fields on the Earth's surface. Geophysicists are called upon to study the mechanical properties of rocks on which the foundations of dams or other large structures may be laid; they evaluate earthquake hazards and design instruments for the exploration of the moon and other planets. They find employment in academic institutions in teaching and research, in government and industrial research laboratories, and in companies involved in exploration for petroleum and mineral deposits. They work in the laboratory and in the field and spend much time exploring the more distant corners of the Earth, from the rift valleys of equatorial Africa or the volcanoes of the Pacific to the glaciers of Antarctica. Geophysics, perhaps more than any other science, is essentially an international effort, inasmuch as all geophysicists, regardless of nationality, are concerned with one and the same Earth.

*Lane R. Johnson, Professor
Department of Geology and Geophysics
University of California at Berkeley*

Planetary Sciences

Planetary sciences refer to the study of the nature, origin, and history of the solar system, applying the principles of all the other branches of the physical sciences, especially those that play important roles in the geosciences and astronomy. Our understanding of the solar system is benefiting greatly from the wealth of data returned from space probes, which have now visited all but Pluto and have opened up whole new "planetary systems" (the satellite and ring systems of Jupiter, Saturn, Uranus, and Neptune) for detailed study. Space-probe data are combined with observations from the ground and from Earth orbit, laboratory measurements, and theoretical analyses in an attempt to understand the scientific issues posed by our nearby cosmic environment. Planetary sciences also involve studies of the chemistry, mineralogy, and history of extraterrestrial materials (meteorites, comets, and lunar samples) and an analysis of extraterrestrial processes that have modified Earth history (impacts by meteorites or comets, tidal influences, changes in the Earth's orbit caused by other planets, solar variability, and other possible causes of climatic variation).

Much of planetary sciences is interdisciplinary and defies categorization into subdisciplines. However, specializations within the field are often one of four areas that define regions of a body: the external region of plasmas and fields, the atmosphere, the surface (if there is one), and the interior. Scientists concerned with the external region study the behavior of charged plasmas in space and how they interact with the planetary magnetic field and the outermost regions of the planet. Their tools include radio astronomy and the direct measurement of magnetic fields and plasmas by spacecraft. Atmospheric scientists include dynamicists and chemists; they wish to understand the thermal structure, composition, energy balance, spatial and temporal variability, global circulation, winds, atmospheric waves, clouds, and interactions of the atmosphere with the regions above and below. Their database includes IR and UV spectra, imaging at visible wavelengths, and (in the cases of Mars and Venus) direct sampling of the atmosphere. Planetary geologists interpret the surfaces of solid bodies using techniques developed in terrestrial geomorphology

SECTION 3: EARTH AND PLANETARY SCIENCES

Field Definition

as well as other techniques developed within planetary sciences, such as interpretation of the number density and morphology of impact craters. Their data include high-resolution imaging and infrared mapping by spacecraft. Scientists who work on the properties of planetary interiors must synthesize the information from other areas together with gravity field, heat flow data, and laboratory and theoretical information on the properties of materials at high pressures. In each subdiscipline, "planet" can also mean "satellite," because there are many satellites in the solar system with histories and structures as complex as a conventionally defined planet. Asteroids and comets also have a sufficiently rich phenomenology to demand specialized attention. All of these areas of interest require a comparative approach in which similarities and differences between planets not only provide the basis for a cosmic synthesis but also help us to understand the Earth better.

Exciting possibilities exist in planetary science for people who are talented in physics, mathematics, and chemistry. In addition to ongoing and planned planetary missions (Galileo, Magellan Radar Mapper, Mars Observer, CRAF, and Cassini), new data continue to be gathered from ground-based and Earth-orbit observations. People with training in planetary science are also well suited to a variety of geoscience or astronomy careers. The long-term stability of funding for planetary science and NASA missions is uncertain, but there is the profound hope that planetary exploration is much too important a human endeavor to be permanently affected by these uncertainties.

David J. Stevenson
Professor of Planetary Science
Division of Geological and Planetary Sciences
California Institute of Technology

Earth Sciences

Adelphi University, Graduate School of Arts and Sciences, Department of Earth Sciences, Garden City, NY 11530. Department offers programs in earth sciences (MS), environmental management (Certificate). Part-time and evening/weekend programs available. Matriculated students: 0 full-time, 21 part-time (4 women); includes 0 minority, 0 foreign. Average age 34. 2 applicants, 0% accepted. In 1990, 9 master's awarded. *Degree requirements:* For master's, computer language required, thesis optional, foreign language not required. *Application deadline:* rolling. *Application fee:* $50. *Expenses:* Tuition of $5300 per semester full-time, $315 per credit part-time. Fees of $159 per semester part-time. *Financial aid:* In 1990–91, 1 teaching assistantship awarded; career-related internships or fieldwork also available. *Faculty research:* Environmental and marine sciences, hydrogeology, stratigraphy, palynology, paleontology. • Dr. Les Sirkin, Chairperson, 516-877-4170.

Ball State University, College of Sciences and Humanities, Department of Geography, Program in Earth Sciences, 2000 University Avenue, Muncie, IN 47306. Program awards MA. Matriculated students: 2. In 1990, 3 degrees awarded. *Degree requirements:* Foreign language not required. *Application fee:* $15. *Expenses:* Tuition of $1140 per semester (minimum) full-time for state residents; $2680 per semester (minimum) full-time for nonresidents. Fees of $6 per credit hour. • Dr. Michael Sullivan, Chairman, Department of Geography, 317-285-1776.

Baylor University, College of Arts and Sciences, Department of Geology, Waco, TX 76798. Offerings include earth science (MA). *Application deadline:* rolling. *Application fee:* $0. *Expenses:* Tuition of $4440 per year full-time, $185 per credit hour part-time. Fees of $510 per year full-time. • Dr. H. H. Beaver, Chairman, 817-755-2361.

Bridgewater State College, Division of Natural Sciences and Mathematics, Department of Earth Sciences and Geography, Bridgewater, MA 02324. Department offers programs in earth sciences (MAT, CAGS), geography (MAT, CAGS). *Entrance requirements:* For master's, GRE General Test. Application deadline: 3/1. Application fee: $25. *Tuition:* $1446 per semester for state residents; $3303 per semester for nonresidents. • Marilyn W. Barry, Graduate Dean, Graduate School, 508-697-1300.

California State University, Chico, College of Natural Sciences, Department of Geology and Physical Sciences, Program in Physical Science, Option in Earth Sciences, Chico, CA 95929. Option awards MS. Faculty: 5 full-time (1 woman), 5 part-time (3 women). Matriculated students: 0 full-time, 1 part-time (0 women); includes 0 minority, 0 foreign. Average age 31. In 1990, 0 degrees awarded. *Degree requirements:* Thesis required, foreign language not required. *Entrance requirements:* GRE General Test. Application deadline: 4/1 (applications processed on a rolling basis). Application fee: $55. *Expenses:* Tuition of $548 per semester full-time, $350 per semester part-time. Fees of $246 per unit for nonresidents. • Dr. K. R. Gina Rothe, Graduate Coordinator, Program in Physical Science, 916-898-6269.

California University of Pennsylvania, School of Liberal Arts, Program in Geography and Earth Sciences, 250 University Avenue, California, PA 15419. Offers earth science (MS), geography (MA, M Ed). Part-time and evening/weekend programs available. Faculty: 0 full-time, 5 part-time (0 women). Matriculated students: 17 full-time (3 women), 7 part-time (4 women); includes 3 minority (all black American), 1 foreign. 8 applicants, 100% accepted. In 1990, 24 degrees awarded. *Degree requirements:* Comprehensive exam required, foreign language not required. *Entrance requirements:* MAT (minimum score of 33 required), TOEFL (minimum score of 550 required), minimum GPA of 2.5, teaching certificate (M Ed). Application deadline: rolling. Application fee: $20. *Tuition:* $2550 per year full-time, $127 per credit part-time for state residents; $3156 per year full-time, $160 per credit part-time for nonresidents. *Financial aid:* In 1990–91, 22 research assistantships awarded. • Dr. Larry Moses, Head, 412-938-4130.

Carleton University, Faculty of Science, Ottawa-Carleton Geoscience Centre, Ottawa, ON K1S 5B6, Canada. Center awards M Sc, PhD. M Sc and PhD offered jointly with the University of Ottawa. Evening/weekend programs available. Faculty: 56 full-time. Matriculated students: 78 full-time, 24 part-time; includes 10 foreign. Average age 23. 53 applicants, 26% accepted. In 1990, 18 master's, 9 doctorates awarded. *Degree requirements:* For master's, thesis; for doctorate, dissertation, comprehensive exam. *Entrance requirements:* For master's, honor's degree in science; for doctorate, M Sc. Application deadline: 7/1. Application fee: $10. *Tuition:* $985 per semester full-time, $284 per semester part-time for Canadian residents; $3939 per semester full-time, $1171 per semester part-time for nonresidents. *Financial aid:* In 1990–91, $446,653 in aid awarded. 72 fellowships (14 to first-year students), 64 research assistantships (18 to first-year students), 78 teaching assistantships (18 to first-year students) were awarded. Financial aid application deadline: 3/1. *Faculty research:* Precambrian geology, mineral resources, structural geology, geodynamics, Northern studies. Total annual research budget: $1.4-million. • Keiko Hattori, Director, 613-564-8169. Application contact: Helene De Gouffe, Secretary, 613-564-6834.

Announcement: Ottawa-Carleton Geoscience Centre is one of the largest graduate teaching and research institutions of earth sciences in Canada. The Centre offers M Sc and PhD degrees in various areas of earth science. Members of the Centre include 56 full-time faculty in geology, geography, physics and civil engineering departments at two universities and 20 adjunct center members at the Geological Survey of Canada, National Museum of Nature, and Canada Centre for Mineral and Energy Technology. Special strength of the Centre lies in northern studies, Precambrian studies, resource geology, tectonics, and environmental geoscience and geochemistry. Excellent analytical facilities include 1 solid-source and 3 gas-source mass spectrometers, scanning electron microscopy, electron microprobe, X-ray fluorescence spectrometer, induced coupled plasma spectrometer (ICP-ES), and geotechnical and equipment-testing laboratories.

Central Connecticut State University, School of Arts and Sciences, Department of Physics and Earth Science, New Britain, CT 06050. Department offers programs in earth science (MS), general science (MS), physics (MS). Faculty: 3 full-time (1 woman), 1 (woman) part-time. Matriculated students: 5 full-time (2 women), 18 part-time (10 women); includes 0 minority, 0 foreign. 10 applicants, 50% accepted. In 1990, 7 degrees awarded. *Degree requirements:* Thesis or alternative, comprehensive exam. *Entrance requirements:* TOEFL (minimum score of 550 required for international students), minimum GPA of 2.7. Application deadline: 8/31. Application fee: $20. *Expenses:* Tuition of $1720 per year full-time, $115 per credit part-time for state residents; $4790 per year full-time, $115 per credit part-time for nonresidents. Fees of $1013 per year full-time, $33 per semester part-time for state residents; $1655 per year full-time, $33 per semester part-time for nonresidents. *Financial aid:* Application deadline 3/15; applicants required to submit FAF. • Dr. Ali Antar, Chair, 203-827-7228.

City College of the City University of New York, Graduate School, College of Liberal Arts and Science, Division of Science, Department of Earth and Planetary Sciences, Convent Avenue at 138th Street, New York, NY 10031. Department offers programs in earth and environmental science (PhD), geology (MA). PhD offered through the Graduate School and University Center of the City University of New York. Matriculated students: 0. 4 applicants, 25% accepted. In 1990, 2 master's awarded. *Degree requirements:* For master's, thesis, comprehensive exam required, foreign language not required. *Entrance requirements:* For master's, TOEFL (minimum score of 300 required), appropriate bachelor's degree. Application deadline: 6/1 (priority date, applications processed on a rolling basis). Application fee: $30. *Expenses:* Tuition of $2204 per year full-time, $95 per credit part-time for state residents; $4700 per year full-time, $199 per credit part-time for nonresidents. Fees of $15 per semester. *Financial aid:* Fellowships and career-related internships or fieldwork available. *Faculty research:* Water resources, high-temperature geochemistry, sedimentary basin analysis, tectonics. • Dr. Dennis Weiss, Chairman, 212-650-6984. Application contact: O. Lehn Franke, Adviser, 212-650-6984.

Colorado State University, College of Forestry and Natural Resources, Department of Earth Resources, Fort Collins, CO 80523. Department awards MF, MS, PhD. Programs offered in earth resources (PhD), geology (MS), watershed sciences (MS). MF offered through the Department of Forest and Wood Sciences. Part-time programs available. Faculty: 16 full-time (1 woman), 1 part-time (0 women). Matriculated students: 23 full-time (9 women), 17 part-time (6 women); includes 0 minority, 14 foreign. Average age 27. 50 applicants, 68% accepted. In 1990, 15 master's, 6 doctorates awarded. *Degree requirements:* For master's, computer language, thesis required, foreign language not required; for doctorate, 1 foreign language (computer language can substitute), dissertation. *Entrance requirements:* GRE General Test, TOEFL (minimum score of 550 required), minimum GPA of 3.0. Application deadline: 4/1 (priority date, applications processed on a rolling basis). Application fee: $30. *Tuition:* $1322 per semester full-time for state residents; $3673 per semester full-time for nonresidents. *Financial aid:* In 1990–91, 2 fellowships (both to first-year students), 2 research assistantships (1 to a first-year student), 3 teaching assistantships (1 to a first-year student), 3 traineeships (0 to first-year students) awarded; federal work-study also available. Average monthly stipend for a graduate assistantship: $698-750. *Faculty research:* Fluvial geomorphology, economic geology, structural geology, watershed sciences. Total annual research budget: $1.5-million. • Dr. Donald O. Doehring, Head, 303-491-5661. Application contact: Barbara Holtz, 301-491-5662.

Cornell University, Graduate Fields of Arts and Sciences, Field of Astronomy and Space Sciences, Ithaca, NY 14853. Field offers programs in astronomy (MS, PhD), astrophysics (MS, PhD), general space sciences (MS, PhD), infrared astronomy (MS, PhD), planetary studies (MS, PhD), radio astronomy (MS, PhD), radiophysics (MS, PhD). Faculty: 26 full-time, 0 part-time. Matriculated students: 26 full-time (4 women), 0 part-time; includes 0 minority, 6 foreign. 121 applicants, 10% accepted. In 1990, 7 master's, 3 doctorates awarded. Terminal master's awarded for partial completion of doctoral program. *Degree requirements:* Thesis/dissertation required, foreign language not required. *Entrance requirements:* GRE General Test, GRE Subject Test, TOEFL. Application deadline: 1/10. Application fee: $55. *Expenses:* Tuition of $16,170 per year. Fees of $28 per year. *Financial aid:* In 1990–91, 23 students received aid. 5 fellowships (2 to first-year students), 13 research assistantships (0 to first-year students), 5 teaching assistantships (0 to first-year students) awarded; full and partial tuition waivers, federal work-study, institutionally sponsored loans, and career-related internships or fieldwork also available. Financial aid application deadline: 1/10; applicants required to submit GAPSFAS. • Philip Nicholson, Graduate Faculty Representative, 607-255-5284. Application contact: Robert Brashear, Director of Admissions, 607-255-4884.

East Texas State University, College of Arts and Sciences, Department of Earth Science, Commerce, TX 75429. Department awards M Ed, MS. Faculty: 6 full-time (1 woman), 0 part-time. Matriculated students: 5 full-time (1 woman), 10 part-time (5 women); includes 0 minority, 0 foreign. In 1990, 5 degrees awarded. *Degree requirements:* Thesis (for some programs), comprehensive exam. *Entrance requirements:* GRE General Test. Application deadline: rolling. Application fee: $0 ($25 for foreign students). *Tuition:* $430 per semester full-time for state residents; $1726 per semester full-time for nonresidents. *Financial aid:* Research assistantships, teaching assistantships, federal work-study, institutionally sponsored loans available. Financial aid applicants required to submit GAPSFAS or FAF. • Dr. James Humphries, Head, 903-886-5433.

Emporia State University, School of Graduate Studies, College of Liberal Arts and Sciences, Division of Physical Sciences, Emporia, KS 66801-5087. Offerings include earth science (MS). Division faculty: 14 full-time (0 women), 1 part-time (0 women). *Degree requirements:* Thesis or comprehensive exam required, foreign language not required. *Entrance requirements:* GRE General Test, TOEFL (minimum score of 550 required). Application deadline: 8/16 (priority date, applications processed on a rolling basis). Application fee: $0 ($50 for foreign students). *Tuition:* $858 per semester full-time, $62 per credit hour part-time for state residents; $2072 per semester full-time, $143 per credit hour part-time for nonresidents. • DeWayne Backhus, Chair, 316-343-5472.

Georgia Institute of Technology, College of Sciences, School of Earth and Atmospheric Sciences, Atlanta, GA 30332. School awards MS, PhD. Part-time programs available. Faculty: 11 full-time (0 women), 0 part-time. Matriculated students: 52 full-time (13 women), 11 part-time (3 women); includes 9 minority (1 Asian American, 5 black American, 3 Hispanic American), 23 foreign. Average age 30. 93 applicants, 23% accepted. In 1990, 12 master's, 7 doctorates awarded. Terminal master's awarded for partial completion of doctoral program. *Degree requirements:* For master's, thesis or alternative required, foreign language not required; for doctorate, dissertation required, foreign language not required. *Entrance requirements:* GRE General Test, TOEFL (minimum score of 550 required), minimum GPA of 2.7. Application deadline: 8/1. Application fee: $15. *Expenses:* Tuition of $574 per semester full-time, $48 per credit part-time for state residents; $1380 per semester full-time, $115 per credit part-time for nonresidents. Fees of $132 per semester full-time. *Financial aid:* In 1990–91, 3 fellowships, 42 research assistantships, 0 teaching assistantships awarded; partial tuition waivers, federal work-study, institutionally sponsored loans, and career-related internships or fieldwork also available. Financial aid application deadline: 2/15. *Faculty research:* Geophysics, geochemistry, atmospheric sciences, physical meteorology, sedimentology. Total annual research budget: $2-million. • Dr. William L. Chameides, Director, 404-894-3893. Application contact: Rita Bryan, Administrative Coordinator, 404-894-3893.

Graduate School and University Center of the City University of New York, Program in Earth and Environmental Sciences, New York, NY 10036. Program awards PhD. Faculty: 34 full-time (4 women), 0 part-time. Matriculated students: 30 full-time (8 women), 0 part-time; includes 0 minority, 8 foreign. Average age 34. 31 applicants, 52% accepted. In 1990, 3 doctorates awarded (33% entered university research/teaching, 0% found other work related to degree, 33% continued full-time study). Terminal master's awarded for partial completion of doctoral program. *Degree requirements:* For doctorate, 1 foreign language (computer language can substitute), dissertation, comprehensive exam. *Entrance requirements:* For doctorate, GRE

Directory: Earth Sciences

General Test. Application deadline: 4/1. Application fee: $30. *Tuition:* $2204 per year full-time, $95 per credit part-time for state residents; $4700 per year full-time, $199 per credit part-time for nonresidents. *Financial aid:* In 1990–91, $172,000 in aid awarded. 10 fellowships, 3 research assistantships, 6 teaching assistantships were awarded; full and partial tuition waivers, federal work-study, institutionally sponsored loans, and career-related internships or fieldwork also available. Financial aid application deadline: 2/1. • Dr. Daniel Habib, Executive Officer, 212-642-2202.

Harvard University, Graduate School of Arts and Sciences, Department of Earth and Planetary Sciences, Cambridge, MA 02138. Department awards AM, PhD. Faculty: 12 full-time (0 women), 2 part-time (1 woman). Matriculated students: 31 full-time (7 women), 0 part-time; includes 0 minority, 10 foreign. In 1990, 2 master's awarded (100% continued full-time study); 6 doctorates awarded (33% entered university research/teaching, 67% continued full-time study). Terminal master's awarded for partial completion of doctoral program. *Degree requirements:* For master's, 1 foreign language, thesis or alternative; for doctorate, 1 foreign language, dissertation. *Entrance requirements:* GRE General Test, GRE Subject Test. Application deadline: 1/2. Application fee: $60. *Expenses:* Tuition of $14,860 per year. Fees of $550 per year. *Financial aid:* In 1990–91, $180,233 in aid awarded. 17 fellowships (9 to first-year students), 31 research assistantships (5 to first-year students), 24 teaching assistantships (9 to first-year students) were awarded; federal work-study, institutionally sponsored loans, and career-related internships or fieldwork also available. Financial aid application deadline: 1/2; applicants required to submit GAPSFAS. *Faculty research:* Economic geography, geochemistry, geophysics, mineralogy, crystallography. • Dr. Rick O'Connell, Director of Graduate Studies, 617-495-2351. Application contact: Office of Admissions and Financial Aid, 617-495-5315.

See full description on page 221.

Indiana University Bloomington, School of Public and Environmental Affairs, Environmental Programs, Bloomington, IN 47405. Offerings include applied earth science (MSES). MSES/MA offered jointly with the Departments of Biology and Geology. Faculty: 8 full-time (0 women), 2 part-time (1 woman). *Degree requirements:* Computer language required, thesis optional, foreign language not required. *Entrance requirements:* GRE General Test. Application deadline: 6/1. Application fee: $20. *Tuition:* $99.85 per credit hour for state residents; $288 per credit hour for nonresidents. • Dr. Daniel E. Willard, Director, 812-855-9485. Application contact: Jeanne Heeb, Coordinator of Student Recruitment, 812-855-2840.

Iowa State University of Science and Technology, College of Liberal Arts and Sciences, Department of Geological and Atmospheric Sciences, Ames, IA 50011. Department offers programs in earth science (MS, PhD), geology (MS, PhD), meteorology (MS, PhD). Matriculated students: 17 full-time (2 women), 11 part-time (1 woman); includes 1 minority (Asian American), 15 foreign. In 1990, 3 master's, 4 doctorates awarded. *Degree requirements:* For master's, thesis optional; for doctorate, dissertation. *Application fee:* $20 ($30 for foreign students). *Expenses:* Tuition of $1158 per semester full-time, $129 per credit part-time for state residents; $3340 per semester full-time, $372 per credit part-time for nonresidents. Fees of $10 per semester. *Financial aid:* In 1990–91, 1 fellowship (0 to first-year students), 19 research assistantships (6 to first-year students), 19 teaching assistantships (10 to first-year students), 9 scholarships (3 to first-year students) awarded. • Dr. Karl E. Seifert, Chair, 515-294-4478.

See full description on page 225.

Louisiana State University and Agricultural and Mechanical College, College of Basic Sciences, Master of Natural Sciences Program, Baton Rouge, LA 70803. Program awards MNS. Part-time programs available. Matriculated students: 2 full-time (both women), 2 part-time (0 women); includes 0 minority, 0 foreign. 4 applicants, 50% accepted. In 1990, 2 degrees awarded. *Degree requirements:* Comprehensive exam required, foreign language and thesis not required. *Entrance requirements:* GRE General Test. Application deadline: 7/1 (priority date, applications processed on a rolling basis). Application fee: $25. *Tuition:* $1020 per semester full-time for state residents; $2620 per semester full-time for nonresidents. *Financial aid:* In 1990–91, 2 students received a total of $8688 in aid awarded. • Dr. R. G. Hussey, Coordinator, 504-388-4200.

Massachusetts Institute of Technology, School of Science, Department of Earth, Atmospheric, and Planetary Sciences, Cambridge, MA 02139. Department offers programs in geology (SM, PhD, Sc D), geophysics (SM, PhD, Sc D), meteorology and physical oceanography (SM, PhD, Sc D), planetary science (SM, PhD). Faculty: 37 full-time (3 women), 0 part-time. Matriculated students: 163 full-time (45 women), 0 part-time; includes 3 minority (1 Asian American, 1 black American, 1 Hispanic American), 54 foreign. Average age 24. 189 applicants, 15% accepted. In 1990, 12 master's awarded; 30 doctorates awarded (70% entered university research/teaching, 30% found other work related to degree). Terminal master's awarded for partial completion of doctoral program. *Degree requirements:* For master's, thesis required, foreign language not required; for doctorate, dissertation, general exam required, foreign language not required. *Entrance requirements:* GRE General Test, GRE Subject Test. Application deadline: 1/15. Application fee: $45. *Tuition:* $16,900 per year. *Financial aid:* In 1990–91, 20 fellowships (2 to first-year students), 93 research assistantships (17 to first-year students), 14 teaching assistantships (1 to a first-year student) awarded; institutionally sponsored loans also available. Financial aid application deadline: 1/15. *Faculty research:* Evolution of main features of the planetary system; origin, composition, structure, and state of the atmospheres, oceans surfaces, and interiors of planets; dynamics of planets and satellite motions. Total annual research budget: $12-million. • Dr. Thomas H. Jordan, Chairman, 617-253-3382. Application contact: Anita Killian, Graduate Administrator, 617-253-3381.

McMaster University, Faculty of Science, Department of Geography, Hamilton, ON L8S 4L8, Canada. Department offers programs in human geography (M Sc, PhD), physical geography (M Sc, PhD). *Degree requirements:* For master's, thesis or alternative required, foreign language not required; for doctorate, dissertation, comprehensive exam required, foreign language not required. *Expenses:* Tuition of $2250 per year full-time, $810 per semester part-time for Canadian residents; $10,340 per year full-time, $4050 per semester part-time for nonresidents. Fees of $76 per year full-time, $49 per semester part-time.

Announcement: Human geography areas: behavioral geography, environment and regional development, historical geography, industrial geography, mathematical modelling and theory, medical geography, migration, social geography, urban geography. Physical geography areas: biogeography, climatology (radiation balance, subarctic microclimate), geomorphology (coastal, karst, alpine), hydrology (snow, permafrost, wetlands, water chemistry). Substantial fieldwork and computer modelling. Scholarships and teaching assistantships of $11,665–$13,165 (1991–92). Write to Chairman, Department of Geography.

Memorial University of Newfoundland, School of Graduate Studies, Department of Earth Sciences, St. John's, NF A1C 5S7, Canada. Department offers program in earth sciences (M Sc, PhD), including geology (M Sc, PhD), geophysics (M Sc, PhD). *Degree requirements:* For master's, thesis; for doctorate, dissertation, comprehensive exam.

Montana State University, College of Letters and Science, Department of Earth Sciences, 901 West Garfield Street, Bozeman, MT 59717. Department awards MS. Faculty: 9 full-time (1 woman), 0 part-time. Matriculated students: 10 full-time (4 women), 12 part-time (4 women); includes 2 minority (1 Asian American, 1 Native American), 1 foreign. Average age 30. 23 applicants, 61% accepted. In 1990, 6 degrees awarded. *Degree requirements:* Thesis required, foreign language not required. *Entrance requirements:* GRE General Test (minimum combined score of 1100 required), TOEFL (minimum score of 550 required), minimum GPA of 2.8. Application deadline: 6/15 (applications processed on a rolling basis). Application fee: $20. *Tuition:* $1150 per year full-time, $74 per credit (minimum) part-time for state residents; $2824 per year full-time, $100 per credit (minimum) part-time for nonresidents. *Financial aid:* Research assistantships, teaching assistantships available. Financial aid application deadline: 3/1. *Faculty research:* Earth surface processes, biogeography, landscape analysis, sedimentary basin analysis, applied hydrology. • Dr. David R. Lageson, Head, 406-994-3331.

New Mexico Institute of Mining and Technology, Department of Geoscience, Socorro, NM 87801. Department offers programs in geochemistry (MS, PhD), geology (MS, PhD), geophysics (MS, PhD), hydrology (MS, PhD). Faculty: 11 full-time (0 women), 2 part-time (0 women). Matriculated students: 92 full-time (18 women), 0 part-time; includes 0 minority, 23 foreign. Average age 25. 108 applicants, 73% accepted. In 1990, 24 master's, 5 doctorates awarded. *Degree requirements:* For master's, thesis optional, foreign language not required; for doctorate, 1 foreign language, dissertation. *Entrance requirements:* For master's, GRE General Test, TOEFL (minimum score of 540 required); for doctorate, GRE General Test, GRE Subject Test, TOEFL (minimum score of 540 required). Application deadline: 6/1 (priority date, applications processed on a rolling basis). Application fee: $16. *Expenses:* Tuition of $617 per semester full-time, $51 per hour part-time for state residents; $2656 per semester full-time, $209 per hour part-time for nonresidents. Fees of $207 per semester. *Financial aid:* In 1990–91, 55 students received a total of $163,094 in aid awarded. 39 research assistantships (16 to first-year students), 16 teaching assistantships (6 to first-year students) were awarded; fellowships, federal work-study, institutionally sponsored loans also available. Average monthly stipend for a graduate assistantship: $600. Financial aid application deadline: 3/15; applicants required to submit GAPSFAS or FAF. *Faculty research:* Crust-mantle geochemistry, volcanology (Southwest and Antarctic); ore genesis, stratigraphy and sedimentation; coal petrology; fluid inclusions; stable isotope geochemistry. • Dr. John Schlue Jr., Chairman, 505-835-5426. Application contact: Dr. J. A. Smoake, Dean, Graduate Studies, 505-835-5513.

New Mexico State University, College of Arts and Sciences, Department of Earth Science, Las Cruces, NM 88003. Department awards MAG, MS. Part-time programs available. Faculty: 10 full-time, 0 part-time. Matriculated students: 27 full-time (10 women), 20 part-time (7 women); includes 0 foreign. In 1990, 1 degree awarded. *Degree requirements:* Thesis (for some programs). *Entrance requirements:* GRE General Test, GRE Subject Test (biology). Application deadline: 7/1. Application fee: $10. *Tuition:* $1608 per year full-time, $67 per credit hour part-time for state residents; $5304 per year full-time, $221 per credit hour part-time for nonresidents. *Financial aid:* Fellowships, research assistantships, teaching assistantships, federal work-study available. Aid available to part-time students. Financial aid applicants required to submit GAPSFAS or FAF. • Dr. Robert Czerniak, Head, 505-646-2708.

Northeastern Illinois University, College of Arts and Sciences, Department of Earth Science, Chicago, IL 60625. Department awards MS. Evening/weekend programs available. Faculty: 7 full-time (1 woman), 2 part-time (1 woman). 5 applicants, 80% accepted. In 1990, 5 degrees awarded. *Degree requirements:* Thesis or alternative, comprehensive exam required, foreign language not required. *Application deadline:* rolling. *Application fee:* $0. *Expenses:* Tuition of $70 per credit hour for state residents; $210 per credit hour for nonresidents. Fees of $37 per credit hour. • Dr. Charles Sabica, Chairperson, 312-583-4050.

Northeast Louisiana University, College of Pure and Applied Sciences, Department of Geosciences, Monroe, LA 71209. Department awards MS. Faculty: 11 full-time (0 women), 1 part-time (0 women). Matriculated students: 13 full-time (5 women), 12 part-time (3 women); includes 2 minority (1 black American, 1 Native American), 4 foreign. 5 applicants, 100% accepted. In 1990, 2 degrees awarded. *Degree requirements:* Thesis required, foreign language not required. *Entrance requirements:* GRE General Test (minimum combined score of 900 required), minimum GPA of 2.8 during previous two years or 3.0 in 21 hours of geosciences. Application deadline: 7/1 (priority date, applications processed on a rolling basis). Application fee: $5. *Tuition:* $816 per semester for state residents; $1608 per semester for nonresidents. *Financial aid:* In 1990–91, 16 students received a total of $72,000 in aid awarded. 4 research assistantships, 8 teaching assistantships, 4 laboratory assistantships were awarded; federal work-study also available. Average monthly stipend for a graduate assistantship: $500. *Faculty research:* Sedimetology, planetary, micropaleontology, environmental petroleum. Total annual research budget: $20,000. • Dr. Mervin Kontrovitz, Head, 318-342-1878.

Northern Arizona University, College of Arts and Sciences, Program in Quaternary Studies, Flagstaff, AZ 86011. Program awards MS. In 1990, 1 degree awarded. *Degree requirements:* Foreign language not required. *Application deadline:* 3/15 (priority date, applications processed on a rolling basis). *Application fee:* $0 ($25 for foreign students). *Expenses:* Tuition of $1528 per year full-time, $80 per credit part-time for state residents; $6180 per year full-time, $258 per credit part-time for nonresidents. Fees of $25 per semester. *Financial aid:* Research assistantships, full and partial tuition waivers, federal work-study, and career-related internships or fieldwork available. • Dr. Larry Agenbroad, Director, 602-523-4561. Application contact: Dr. Jim Mead, Adviser, 602-523-7184.

Northern Arizona University, College of Arts and Sciences, Department of Geology, Program in Earth Science, Flagstaff, AZ 86011. Program awards MAT, MS. *Degree requirements:* Thesis required, foreign language not required. *Entrance requirements:* GRE General Test, GRE Subject Test. Application deadline: 3/15 (priority date, applications processed on a rolling basis). Application fee: $0 ($25 for foreign students). *Expenses:* Tuition of $1528 per year full-time, $80 per credit part-time for state residents; $6180 per year full-time, $258 per credit part-time for nonresidents. Fees of $25 per semester. *Financial aid:* Fellowships, research assistantships, teaching assistantships, scholarships, full and partial tuition waivers, federal work-study, and career-related internships or fieldwork available. Financial aid applicants required to submit FAF. • Dr. Stanley S. Bens, Graduate Adviser, 602-523-7179.

Northwestern University, College of Arts and Sciences, Department of Geological Sciences, Evanston, IL 60208. Department awards MS, PhD. Part-time programs available. Faculty: 12 full-time (1 woman), 0 part-time. Matriculated students: 18 full-time (4 women), 0 part-time; includes 1 minority (black American), 6 foreign. 15 applicants, 73% accepted. In 1990, 3 master's, 9 doctorates awarded. Terminal master's awarded for partial completion of doctoral program. *Degree requirements:* For doctorate, dissertation. *Entrance requirements:* For doctorate, GRE General Test, TOEFL. Application deadline: 8/30. Application fee: $40 ($45 for foreign students). *Tuition:* $4665 per quarter full-time, $1704 per course part-time. *Financial aid:* In 1990–91, $86,804 in aid awarded. 4 fellowships, 10 research assistantships, 7 teaching assistantships were awarded; federal work-study, institutionally sponsored loans, and career-related internships or fieldwork also available. Financial aid application deadline: 1/15; applicants required to submit GAPSFAS. *Faculty research:* Geophysics, geochemistry, petrology/mineralogy, paleoclimate, stratigraphy/structure. Total annual research budget: $1-million. • Seth Stein, Chairman, 708-491-3238. Application contact: Richard Gordon, Graduate Adviser, 708-491-1170.

Pennsylvania State University University Park Campus, College of Earth and Mineral Sciences, University Park, PA 16802. College awards M Eng, MS, PhD. Matriculated students: 409 full-time (90 women), 78 part-time (14 women). In 1990, 72 master's, 48 doctorates awarded. *Application fee:* $35. *Tuition:* $203 per credit for state residents; $403 per credit for nonresidents. *Financial aid:* Fellowships available. • Dr. John A. Dutton, Dean, 814-865-6546.

Portland State University, College of Liberal Arts and Sciences, Department of Geology, Portland, OR 97207. Department awards MA, MS, PhD. PhD offered in conjunction with Interdisciplinary Program in Environmental Sciences and Resources. Part-time programs available. Faculty: 7 full-time (0 women), 6 part-time (1 woman). Matriculated students: 12 full-time (3 women), 6 part-time (2 women); includes 1 minority (Native American), 1 foreign. Average age 29. 11 applicants, 82% accepted. In 1990, 5 master's awarded. *Degree requirements:* For master's, computer language, thesis (for some programs), field comprehensive required, foreign language not required; for doctorate, computer language, dissertation, two years residence. *Entrance requirements:* For master's, TOEFL (minimum score of 550 required), minimum GPA of 3.0 in upper-division course work or 2.75 overall, undergraduate degree in geology or equivalant; for doctorate, GRE General Test, GRE Subject Test. Application deadline: 4/15 (priority date, applications processed on a rolling basis). Application fee: $40. *Tuition:* $1151 per trimester for state residents; $1827 per trimester for nonresidents. *Financial aid:* In 1990–91, 2 research assistantships (both to first-year students), 8 teaching assistantships (2 to first-year students) awarded; federal work-study, institutionally sponsored loans, and career-related internships or fieldwork also available. Aid available to part-time students. Average monthly stipend for a graduate assistantship: $435. Financial aid application deadline: 3/1. *Faculty research:* Volcanic stratigraphy, stratigraphy of Columbia River basalt, Neogene tectonics, economic geology, low and medium temperature alteration. Total annual research budget: $175,000. • Dr. Ansel Johnson, Head, 503-725-3022.

Purdue University, School of Science, Department of Earth and Atmospheric Sciences, West Lafayette, IN 47907. Department awards MS, PhD. Faculty: 24 full-time, 1 part-time. Matriculated students: 74 full-time (10 women), 6 part-time (3 women); includes 4 minority (1 Asian American, 1 black American, 1 Hispanic American, 1 Native American), 27 foreign. Average age 28. 87 applicants, 23% accepted. In 1990, 19 master's, 6 doctorates awarded. *Degree requirements:* For master's, thesis required, foreign language not required; for doctorate, 1 foreign language, dissertation. *Entrance requirements:* GRE General Test. Application deadline: 6/1. Application fee: $25. *Tuition:* $1162 per semester full-time, $83.50 per credit hour part-time for state residents; $3720 per semester full-time, $244.50 per credit hour part-time for nonresidents. *Financial aid:* In 1990–91, 3 fellowships (2 to first-year students), 22 research assistantships (1 to a first-year student), 26 teaching assistantships (3 to first-year students) awarded. • Dr. E. M. Agee, Head, 317-494-0251.

Queens College of the City University of New York, Mathematics and Natural Sciences Division, Department of Geology, 65-30 Kissena Boulevard, Flushing, NY 11367. Department awards MA, MS Ed. MS Ed awarded through the School of Education. Part-time and evening/weekend programs available. Matriculated students: 0 full-time, 11 part-time (3 women). 10 applicants, 70% accepted. In 1990, 0 degrees awarded. *Degree requirements:* Thesis, comprehensive exam required, foreign language not required. *Entrance requirements:* GRE, TOEFL (minimum score of 550 required), previous course work in calculus, physics, and chemistry, minimum GPA of 3.0. Application deadline: 4/1 (applications processed on a rolling basis). Application fee: $35. *Tuition:* $2700 per year full-time, $45 per credit part-time for state residents; $4700 per year full-time, $199 per credit part-time for nonresidents. *Financial aid:* In 1990–91, 4 fellowships (0 to first-year students) awarded; partial tuition waivers, federal work-study, institutionally sponsored loans, and career-related internships or fieldwork also available. Aid available to part-time students. Financial aid application deadline: 4/1. *Faculty research:* Sedimentology/stratigraphy, paleontology, field petrology. • Dr. Allan Ludman, Chairperson, 718-997-3300. Application contact: Dr. Robert Finks, Graduate Adviser, 718-997-3300.

Shippensburg University of Pennsylvania, College of Arts and Sciences, Department of Geography and Earth Science, Shippensburg, PA 17257. Department offers program in geoenvironmental studies (MS). Faculty: 3 full-time (0 women), 0 part-time. Matriculated students: 19 full-time (8 women), 15 part-time (7 women); includes 0 minority, 0 foreign. 100% of applicants accepted. In 1990, 14 degrees awarded. *Degree requirements:* Thesis optional, foreign language not required. *Entrance requirements:* Minimum GPA of 2.5 or GRE General Test. Application fee: $20. *Expenses:* Tuition of $1314 per semester full-time, $146 per credit part-time for state residents; $1677 per semester full-time, $186 per credit part-time for nonresidents. Fees of $240 per year full-time, $58 per semester (minimum) part-time. *Financial aid:* In 1990–91, 10 graduate assistantships (6 to first-year students) awarded. Financial aid application deadline: 3/1. • Dr. John E. Benhart, Chairman, 717-532-1685.

Southeast Missouri State University, Department of Earth Sciences, Cape Girardeau, MO 63701. Department awards MNS. Faculty: 7 full-time (1 woman). Matriculated students: 7; includes 0 minority, 0 foreign. Average age 24. In 1990, 1 degree awarded. *Degree requirements:* Thesis or alternative. *Application deadline:* 3/1. *Application fee:* $10. *Expenses:* Tuition of $1860 per year full-time, $75 per credit part-time for state residents; $3420 per year full-time, $140 per credit part-time for nonresidents. Fees of $25 per year. *Financial aid:* In 1990–91, 3 research assistantships (2 to first-year students), 2 teaching assistantships (both to first-year students) awarded. *Faculty research:* Earthquake studies, remote sensing. • Nicholas Tibbs, Chairperson, 314-651-2168.

Stanford University, School of Earth Sciences, Stanford, CA 94305. School awards MS, PhD, Eng. Faculty: 34 (2 women). Matriculated students: 190 full-time (40 women), 0 part-time; includes 9 minority (2 Asian American, 7 Hispanic American), 83 foreign. Average age 28. 145 applicants, 39% accepted. In 1990, 25 master's, 19 doctorates, 0 Engs awarded. Terminal master's awarded for partial completion of doctoral program. *Degree requirements:* For master's and doctorate, thesis/dissertation; for Eng, computer language, thesis required, foreign language not required. *Entrance requirements:* GRE General Test, TOEFL. Application fee: $55. *Expenses:* Tuition of $15,102 per year. Fees of $28 per quarter. *Financial aid:* Fellowships, research assistantships, teaching assistantships, federal work-study, institutionally sponsored loans available. Financial aid applicants required to submit GAPSFAS or FAF.

State University of New York at Stony Brook, College of Arts and Sciences, Division of Physical Sciences and Mathematics, Department of Earth and Space Sciences, Stony Brook, NY 11794. Department offers programs in astronomical sciences (M Phil, MS, PhD), including astronomical sciences, astrophysics; geological sciences (M Phil, MS, PhD), including geochemistry, geology, geophysics. Faculty: 23 full-time, 0 part-time. Matriculated students: 48 full-time (14 women), 4 part-time (2 women); includes 1 minority (Asian American), 26 foreign. 116 applicants, 28% accepted. In 1990, 4 master's, 1 doctorate awarded. Terminal master's awarded for partial completion of doctoral program. *Degree requirements:* For master's, thesis or alternative required, foreign language not required; for doctorate, dissertation required, foreign language not required. *Entrance requirements:* GRE General Test, TOEFL, minimum GPA of 3.0. Application deadline: 2/1. Application fee: $35. *Expenses:* Tuition of $2450 per year full-time, $103 per credit part-time for state residents; $5766 per year full-time, $243 per credit part-time for nonresidents. Fees of $151 per year full-time, $10.45 per year (minimum) part-time. *Financial aid:* In 1990–91, 0 fellowships, 27 research assistantships, 21 teaching assistantships awarded. *Faculty research:* Astronomy, theoretical and observational astrophysics, paleontology, petrology, crystallography. Total annual research budget: $3.2-million. • Dr. Gilbert Hansen, Chairman, 516-632-8200.

State University of New York College at Oneonta, Department of Earth Science, Oneonta, NY 13820. Department awards MA. Faculty: 9 full-time (0 women), 0 part-time. Matriculated students: 3 full-time (0 women), 0 part-time. In 1990, 1 degree awarded. *Degree requirements:* Thesis. *Entrance requirements:* GRE General Test. Application deadline: 4/15. Application fee: $35. *Expenses:* Tuition of $2450 per year full-time, $103 per credit hour part-time for state residents; $5765 per year full-time, $248 per credit hour part-time for nonresidents. Fees of $25 per year full-time, $0.85 per credit hour part-time. *Financial aid:* Fellowships available. • Dr. P. Jay Fleisher, Chair, 607-431-3707.

Texas A&M University, College of Geosciences, College Station, TX 77843. College awards MS, PhD. Faculty: 109. Matriculated students: 374 full-time (98 women), 0 part-time; includes 17 minority (3 Asian American, 3 black American, 10 Hispanic American, 1 Native American), 113 foreign. 178 applicants, 87% accepted. In 1990, 42 master's, 20 doctorates awarded. *Entrance requirements:* GRE General Test, TOEFL. Application deadline: 7/15. Application fee: $25 ($50 for foreign students). *Expenses:* Tuition of $100 per semester (minimum) for state residents; $128 per credit hour for nonresidents. Fees of $459 per year full-time, $252 per semester part-time. *Financial aid:* Fellowships, research assistantships, teaching assistantships available. • Melvin Friedman, Dean, 409-845-3651.

Texas Tech University, Graduate School, College of Arts and Sciences, Department of Geosciences, Lubbock, TX 79409. Department offers programs in atmospheric sciences (MS, PhD), geology (MS, PhD). Faculty: 15 full-time. Matriculated students: 62. In 1990, 10 master's, 3 doctorates awarded. *Degree requirements:* Thesis/dissertation. *Application deadline:* 4/15 (priority date, applications processed on a rolling basis). *Application fee:* $0 ($50 for foreign students). *Tuition:* $494 per semester full-time, $20 per credit hour part-time for state residents; $1790 per semester full-time, $455 per credit hour part-time for nonresidents. *Faculty research:* Structural geology, geophysics, sedimentary geology, paleontology, petroleum geology. • Dr. Alonzo Jacka, Chairman, 806-742-3102.

Université du Québec à Chicoutimi, Program in Earth Sciences, Chicoutimi, PQ G7H 2B1, Canada. Program awards M Sc A. Part-time programs available. Matriculated students: 9 full-time (0 women), 7 part-time (3 women); includes 3 foreign. 9 applicants, 78% accepted. In 1990, 10 degrees awarded. *Degree requirements:* Thesis. *Entrance requirements:* Appropriate bachelor's degree, proficiency in French. Application deadline: 5/1. Application fee: $15. *Expenses:* Tuition of $555 per trimester full-time, $37 per credit part-time for Canadian residents; $3480 per trimester full-time, $234 per credit part-time for nonresidents. Fees of $57.50 per trimester full-time, $19.50 per trimester part-time. *Financial aid:* Fellowships, research assistantships, teaching assistantships available. • Pierre-A. Cousineau, Director, 418-545-5289. Application contact: Laurent Massé, Director of Admissions, 418-545-5005.

Université du Québec à Montréal, Program in Earth Sciences, Montreal, PQ H3C 3P8, Canada. Program awards M Sc. Part-time programs available. Matriculated students: 11 full-time (3 women), 13 part-time (4 women); includes 6 foreign. 24 applicants, 38% accepted. In 1990, 13 degrees awarded. *Degree requirements:* Thesis. *Entrance requirements:* Appropriate bachelor's degree, proficiency in French. Application deadline: 5/1 (priority date). Application fee: $15. *Expenses:* Tuition of $555 per trimester full-time, $37 per credit part-time for Canadian residents; $3480 per trimester full-time, $234 per credit part-time for nonresidents. Fees of $57.50 per trimester full-time, $19.50 per trimester part-time. *Financial aid:* Fellowships, research assistantships, teaching assistantships available. • Norman Goulet, Director, 514-987-3370. Application contact: Lucille Boisselle-Roy, Admissions Officer, 514-987-3128.

University of Arizona, College of Arts and Sciences, Faculty of Science, Department of Geosciences, Tucson, AZ 85721. Department awards MS, PhD. Part-time programs available. Faculty: 31 full-time (3 women), 0 part-time. Matriculated students: 82 full-time (21 women), 40 part-time (12 women); includes 3 minority (all Hispanic American), 19 foreign. Average age 31. 191 applicants, 25% accepted. In 1990, 23 master's, 17 doctorates awarded. *Degree requirements:* For master's, thesis or alternative required, foreign language not required; for doctorate, 1 foreign language, dissertation. *Entrance requirements:* GRE General Test. Application deadline: 2/15 (applications processed on a rolling basis). Application fee: $25. *Expenses:* Tuition of $0 for state residents; $5406 per year full-time, $209 per credit hour part-time for nonresidents. Fees of $1528 per year full-time, $80 per credit hour part-time. *Financial aid:* In 1990–91, $503,412 in aid awarded. 1 fellowship, 79 research assistantships, 76 teaching assistantships, 12 scholarships were awarded; full tuition waivers, institutionally sponsored loans also available. Financial aid application deadline: 2/15; applicants required to submit FAF. *Faculty research:* Geology, geochemistry, geophysics. Total annual research budget: $1.881-million. • Dr. Clement C. Chase, Head, 602-621-4051. Application contact: Graduate Program Office, 602-621-6004.

See full description on page 249.

University of California, Davis, Program in Hydrologic Sciences, Davis, CA 95616. Program awards MS, PhD. Matriculated students: 7. *Degree requirements:* Thesis/dissertation. *Entrance requirements:* For master's, GRE General Test, minimum GPA of 3.0. Application fee: $40. *Expenses:* Tuition of $0 for state residents; $7699 per year full-time, $3849 per year part-time for nonresidents. Fees of $2718 per year full-time, $1928 per year part-time. *Financial aid:* Fellowships, research assistantships available. • Graduate Adviser, 916-752-1669.

University of California, Los Angeles, College of Letters and Science, Department of Earth and Space Sciences, Los Angeles, CA 90024. Department offers programs in geochemistry (MS, PhD, C Phil), geology (MS, PhD, C Phil), geophysics and space physics (MS, PhD). Faculty: 23. Matriculated students: 75 full-time (18 women), 0 part-time; includes 8 minority, 14 foreign. 120 applicants, 28% accepted. In 1990, 5 master's, 14 doctorates, 2 C Phils awarded. *Degree requirements:* For master's, thesis or comprehensive exams required, foreign language not required; for doctorate, dissertation, written and oral qualifying exams required, foreign language not required. *Entrance requirements:* For master's and doctorate, GRE General Test. Application fee: $40. *Expenses:* Tuition of $0 for state residents; $7699 per year for nonresidents. Fees of $2907 per year. *Financial aid:* In 1990–91, 76 students received a total of $1.04-million in aid awarded. 68 fellowships, 50 research assistantships, 37 teaching assistantships were awarded; full and partial tuition waivers, federal work-study, institutionally sponsored loans also available. Financial aid application deadline: 3/1. • Dr. Arthur Montana, Chair, 310-825-3880.

See full description on page 255.

University of California, Riverside, Graduate Division, College of Natural and Agricultural Sciences, Department of Earth Sciences, Riverside, CA 92521. Department offers programs in geography (MA, MS, PhD), geological sciences (MS, PhD). Part-time programs available. Terminal master's awarded for partial completion of doctoral program. *Degree requirements:* For master's, thesis or comprehensive exams; for doctorate, dissertation, qualifying exams. *Entrance requirements:* GRE General Test, TOEFL (minimum score of 550 required). Application deadline: 6/1. Application fee: $40. *Tuition:* $950 per quarter full-time, $264 per quarter part-time for state residents; $3517 per quarter full-time, $1758 per quarter part-time for nonresidents. *Faculty research:* Cultural and physical geography, applied geophysics, geochemistry, paleontology, sedimentology.

University of California, San Diego, Scripps Institution of Oceanography, 9500 Gilman Drive, La Jolla, CA 92093. Institution offers programs in applied ocean science (MS, PhD), biological oceanography (MS, PhD), geochemistry and marine chemistry (MS, PhD), marine biology (MS, PhD), physical oceanography and geological sciences (MS, PhD). Faculty: 73. Matriculated students: 183 (57 women); includes 59 foreign. 273 applicants, 26% accepted. In 1990, 14 master's, 29 doctorates awarded. Terminal master's awarded for partial completion of doctoral program. *Entrance requirements:* GRE General Test, GRE Subject Test (marine biology). Application fee: $40. *Expenses:* Tuition of $0 for state residents; $7699 per year full-time, $1283 per quarter part-time for nonresidents. Fees of $2798 per year full-time, $669 per quarter part-time. *Financial aid:* Fellowships, research assistantships available. • J. Freeman Gilbert, Chair, 619-534-3208. Application contact: Betty Stover, Graduate Coordinator, 619-534-3206.

See full description on page 321.

University of California, Santa Cruz, Division of Natural Sciences, Program in Earth Sciences, Santa Cruz, CA 95064. Program awards MS, PhD. Faculty: 21 full-time (4 women), 0 part-time. Matriculated students: 72 full-time (24 women), 0 part-time; includes 6 minority (1 Asian American, 5 Hispanic American), 15 foreign. 52 applicants, 46% accepted. In 1990, 5 master's, 9 doctorates awarded. *Degree requirements:* For master's, thesis; for doctorate, 1 foreign language (computer language can substitute), dissertation, qualifying exam. *Entrance requirements:* GRE General Test, GRE Subject Test. Application deadline: 1/15. Application fee: $40. *Expenses:* Tuition of $0 for state residents; $7699 per year for nonresidents. Fees of $3021 per year. *Financial aid:* In 1990–91, 6 fellowships, 16 research assistantships, 63 teaching assistantships awarded; federal work-study, institutionally sponsored loans, and career-related internships or fieldwork also available. Average monthly stipend for a graduate assistantship: $1335. Financial aid applicants required to submit GAPSFAS. *Faculty research:* Evolution of continental margins and orogenic belts, geologic processes occurring at plate boundaries, deep-sea sediment diagenesis, paleoecology, hydrogeology. • Dr. Kenneth Cameron, Chairperson, 408-459-2504.

University of Chicago, Division of the Physical Sciences, Department of the Geophysical Sciences, Chicago, IL 60637. Offerings include earth sciences (SM, PhD). *Degree requirements:* For master's, thesis, seminar; for doctorate, 1 foreign language, dissertation, seminar. *Entrance requirements:* For master's, TOEFL; for doctorate, GRE, TOEFL (minimum score of 550 required). Application deadline: 1/6. Application fee: $45. *Expenses:* Tuition of $16,275 per year full-time, $8140 per year part-time. Fees of $356 per year.

See full description on page 259.

University of Colorado at Denver, College of Liberal Arts and Sciences, Program in Basic Science, Denver, CO 80217. Program awards MBS. Part-time and evening/weekend programs available. Matriculated students: 2 full-time, 4 part-time; includes 0 minority, 0 foreign. In 1990, 1 degree awarded. *Degree requirements:* Thesis or alternative. *Application deadline:* 6/1 (applications processed on a rolling basis). *Application fee:* $30 ($50 for foreign students). *Expenses:* Tuition of $1185 per semester full-time, $142 per semester hour part-time for state residents; $3969 per semester full-time, $476 per semester hour part-time for nonresidents. Fees of $103 per semester. *Financial aid:* Research assistantships, teaching assistantships, federal work-study available. Financial aid application deadline: 3/1. • Zenis Hartvigson, Director, 303-556-8464. Application contact: Suzie Perez, Senior Administrative Clerk, 303-556-2704.

University of Illinois at Urbana-Champaign, College of Liberal Arts and Sciences, Department of Geology, Champaign, IL 61820. Offerings include earth sciences (MS, PhD). Department faculty: 19 full-time (1 woman), 5 part-time (0 women). *Degree requirements:* Thesis/dissertation required, foreign language not required. *Entrance requirements:* GRE General Test, TOEFL. Application fee: $25. *Tuition:* $1838 per semester full-time, $708 per semester part-time for state residents; $4314 per semester full-time, $1673 per semester part-time for nonresidents. • Dr. R. James Kirkpatrick, Head, 217-333-3542. Application contact: Dr. Wang-Ping Chen, Graduate Admissions, 217-244-4065.

University of New Hampshire, College of Engineering and Physical Sciences, Department of Earth Sciences, Durham, NH 03824. Department offers programs in earth sciences geology (MS, PhD), earth sciences oceanography (MS, PhD), hydrology (MS). Faculty: 26 full-time. Matriculated students: 40 full-time (12 women), 7 part-time (1 woman); includes 9 foreign. In 1990, 2 master's, 1 doctorate awarded. *Degree requirements:* For master's, thesis required, foreign language not required; for doctorate, 1 foreign language, dissertation. *Entrance requirements:* GRE General Test. Application deadline: 7/1 (priority date, applications processed on a rolling basis). Application fee: $25. *Tuition:* $1645 per semester full-time, $183 per credit hour part-time for state residents; $4920 per semester full-time, $547 per credit hour part-time for nonresidents. *Financial aid:* In 1990–91, 1 fellowship, 12 research assistantships, 9 teaching assistantships, 4 scholarships awarded; full and partial tuition waivers, federal work-study, and career-related internships or fieldwork also available. Aid available to part-time students. Financial aid application deadline: 2/15. *Faculty research:* Geology, oceanography, geochemical systems. • Dr. S. Lawrence Dingman, Chairperson, 603-862-1718. Application contact: Dr. Francis Birch, 603-862-1718.

University of North Carolina at Charlotte, College of Arts and Sciences, Department of Geography and Earth Sciences, Charlotte, NC 28223. Department awards MA. Part-time and evening/weekend programs available. Faculty: 15 full-time (1 woman), 0 part-time. Matriculated students: 13 full-time (4 women), 12 part-time (4 women); includes 1 minority (black American), 1 foreign. Average age 27. 23 applicants, 83% accepted. In 1990, 4 degrees awarded. *Degree requirements:* Foreign language not required. *Entrance requirements:* GRE General Test or MAT, Doppelt Math Reasoning Test, minimum GPA of 2.5 overall, 3.0 in undergraduate major. Application deadline: 7/1. Application fee: $15. *Tuition:* $574.50 per semester full-time for state residents; $3105 per semester full-time for nonresidents. *Financial aid:* In 1990–91, 2 research assistantships (0 to first-year students), 14 teaching assistantships (4 to first-year students) awarded; federal work-study and career-related internships or fieldwork also available. Financial aid application deadline: 4/15; applicants required to submit FAF. *Faculty research:* Petrology, hydrology, transportation. • Dr. Wayne A. Walcott, Chairman, 704-547-2293. Application contact: Kathi M. Baucom, Director of Admissions, 704-547-2213.

University of Northern Colorado, College of Arts and Sciences, Department of Earth Sciences, Greeley, CO 80639. Department awards MA. Faculty: 5 full-time (0 women), 0 part-time. Matriculated students: 7 full-time (2 women), 3 part-time (1 woman); includes 0 minority, 0 foreign. 3 applicants, 100% accepted. In 1990, 1 degree awarded. *Degree requirements:* Comprehensive exams required, thesis not required. *Application deadline:* rolling. *Application fee:* $30. *Expenses:* Tuition of $1900 per year full-time, $106 per credit hour part-time for state residents; $6078 per year full-time, $338 per credit hour part-time for nonresidents. Fees of $320 per year full-time, $18 per credit hour part-time. *Financial aid:* In 1990–91, 2 fellowships (1 to a first-year student), 1 teaching assistantship (to a first-year student), 3 graduate assistantships awarded; research assistantships also available. Financial aid application deadline: 3/1. • Dr. William D. Nesse, Chairperson, 303-351-2830. Application contact: Dr. Lee Shropshire, Graduate Coordinator, 303-351-2285.

University of South Carolina, Graduate School, College of Science and Mathematics, Department of Geological Sciences, Columbia, SC 29208. Department awards MS, PhD. Faculty: 24 full-time (1 woman), 0 part-time. Matriculated students: 80 full-time (30 women), 10 part-time; includes 3 minority (2 black American, 1 Hispanic American), 15 foreign. 90 applicants, 22% accepted. In 1990, 15 master's awarded (60% found work related to degree, 40% continued full-time study); 9 doctorates awarded. Terminal master's awarded for partial completion of doctoral program. *Degree requirements:* For master's, thesis required, foreign language not required; for doctorate, dissertation, published paper required, foreign language not required. *Entrance requirements:* GRE General Test. Application deadline: 3/1 (priority date, applications processed on a rolling basis). Application fee: $25. *Tuition:* $1404 per semester full-time. *Financial aid:* In 1990–91, 40 research assistantships (10 to first-year students), 28 teaching assistantships (10 to first-year students) awarded; federal work-study and career-related internships or fieldwork also available. Financial aid application deadline: 3/1. *Faculty research:* Sedimentary geology, geochemistry, tectonics, marine geology, petrology. • Dr. Robert Thunell Jr., Chairman, 803-777-7593. Application contact: Dr. Pradeep Talwani, Director of Graduate Studies, 803-777-4519.

University of Texas at San Antonio, College of Sciences and Engineering, Division of Earth and Physical Sciences, San Antonio, TX 78285. Division offers programs in chemistry (MS), geology (MS), natural resources (MS). Matriculated students: 66 (24 women); includes 19 minority (1 Asian American, 3 black American, 15 Hispanic American). In 1990, 17 degrees awarded. *Entrance requirements:* GRE. Application deadline: 7/1 (applications processed on a rolling basis). Application fee: $20. *Expenses:* Tuition of $100 per semester hour (minimum) for state residents; $128 per semester hour (minimum) for nonresidents. Fees of $48 per semester hour (minimum). *Financial aid:* Research assistantships, teaching assistantships available. • Dr. Robert K. Smith, Director, 512-691-4455.

University of Toronto, School of Graduate Studies, Physical Sciences Division, Department of Geology, Program in the Earth Sciences, Toronto, ON M5S 1A1, Canada. Program awards MA Sc, M Sc, PhD. *Application deadline:* 4/15. *Application fee:* $50. *Expenses:* Tuition of $2220 per year full-time, $666 per year part-time for Canadian residents; $10,198 per year full-time, $305.05 per year part-time for nonresidents. Fees of $277.56 per year full-time, $82.73 per year part-time. *Financial aid:* Application deadline 2/1. • J. Westgate, Chair, Department of Geology, 416-978-3021.

See full description on page 291.

University of Waterloo, Faculty of Science, Department of Earth Sciences, Waterloo, ON N2L 3G1, Canada. Department awards M Sc, PhD. Faculty: 28 full-time, 3 part-time. Matriculated students: 119 full-time (28 women), 19 part-time (2 women); includes 37 foreign. In 1990, 20 master's, 4 doctorates awarded. *Degree requirements:* Thesis/dissertation. *Entrance requirements:* For master's, TOEFL (minimum score of 530 required), honor's degree, minimum B average; for doctorate, TOEFL (minimum score of 530 required), master's degree. Application deadline: 8/1. Application fee: $25. *Expenses:* Tuition of $757 per year full-time, $530 per year part-time for Canadian residents; $3127 per year for nonresidents. Fees of $68 per year full-time, $17 per year part-time. *Financial aid:* In 1990–91, 261 students received aid. 180 research assistantships, 81 teaching assistantships awarded; institutionally sponsored loans and career-related internships or fieldwork also available. Average monthly stipend for a graduate assistantship: $1000. *Faculty research:* Hydrogeology, engineering geology, quaternary geology, geophysics, low temperature and isotope geochemistry. Total annual research budget: $5.3-million. • Dr. J. P. Greenhouse, Chairman, 519-885-1211 Ext. 3791. Application contact: S. Fischer, Graduate Secretary, 519-885-1211 Ext. 6870.

Announcement: Department offers M Sc and PhD in most areas of earth science. Special strength in engineering geology, engineering geophysics and hydrogeology, hydrogeochemistry, isotope hydrology, isotope geochemistry, waste disposal, water pollution, mathematical modelling, rock and soil mechanics, sedimentation, Quaternary and Paleozoic geology, paleontology and palynology, Precambrian geology, mineral deposits. Excellent facilities; financial support available. Department houses Ontario Centre for Groundwater Research and the Quaternary Sciences Institute.

Utah State University, College of Science, Department of Physics, Logan, UT 84322. Offerings include physics (MS, PhD), with options in astronomy and astrophysics (MS), earth and planetary sciences (PhD), engineering physics (PhD), physics (MS).

Terminal master's awarded for partial completion of doctoral program. Department faculty: 22 full-time (0 women), 3 part-time (0 women). *Degree requirements:* Dissertation required, foreign language not required. *Entrance requirements:* GRE General Test (score in 40th percentile or higher required), minimum GPA of 3.0. Application deadline: 7/15 (priority date, applications processed on a rolling basis). Application fee: $25 ($30 for foreign students). *Tuition:* $426 per quarter (minimum) full-time, $184 per quarter (minimum) part-time for state residents; $1133 per quarter (minimum) full-time, $505 per quarter (minimum) part-time for nonresidents. • Dr. W. John Raitt, Head, 801-750-2848. Application contact: O. Harry Otteson, Assistant Head, 801-750-2850.

Washington University, Graduate School of Arts and Sciences, Department of Earth and Planetary Sciences, St. Louis, MO 63130. Department offers programs in earth and planetary sciences (MA), geochemistry (PhD), geology (MA, PhD), geophysics (PhD), planetary sciences (PhD). Matriculated students: 32 full-time (7 women), 0 part-time; includes 0 minority, 7 foreign. In 1990, 3 master's, 3 doctorates awarded. Terminal master's awarded for partial completion of doctoral program. *Degree requirements:* Thesis/dissertation. *Entrance requirements:* GRE General Test. Application deadline: 1/15 (priority date, applications processed on a rolling basis). Application fee: $0. *Tuition:* $15,960 per year full-time, $665 per credit hour part-time. *Financial aid:* In 1990–91, 8 fellowships, 15 research assistantships, 9 teaching assistantships awarded; full and partial tuition waivers, federal work-study, institutionally sponsored loans also available. Financial aid application deadline: 1/15. • Dr. Larry A. Haskin, Chairman, 314-935-5610.

See full description on page 295.

Wesleyan University, Department of Earth Sciences, Middletown, CT 06459. Department awards MA. Faculty: 7 full-time (2 women), 0 part-time. Matriculated students: 2 full-time (1 woman), 0 part-time; includes 0 foreign. Average age 28. 0 applicants. In 1990, 1 degree awarded (100% continued full-time study). *Degree requirements:* Thesis. *Entrance requirements:* GRE General Test, GRE Subject Test. Application deadline: 3/1. Application fee: $0. *Expenses:* Tuition of $2035 per course. Fees of $627 per year full-time. *Financial aid:* In 1990–91, $17,212 in aid awarded. 4 teaching assistantships (3 to first-year students) were awarded; partial tuition waivers also available. Average monthly stipend for a graduate assistantship: $875. *Faculty research:* Tectonics, volcanology, stratigraphy, coastal processes, geochemistry. Total annual research budget: $5308. • Jelle Z. deBoer, Chair, 203-347-9411 Ext. 2833.

Western Michigan University, College of Arts and Sciences, Department of Geology, Program in Earth Science, Kalamazoo, MI 49008. Program awards MS. Matriculated students: 1 full-time (0 women), 7 part-time (2 women); includes 0 minority, 0 foreign. 4 applicants, 100% accepted. In 1990, 1 degree awarded. *Degree requirements:* Oral exam required, foreign language not required. *Application deadline:* 2/15 (priority date, applications processed on a rolling basis). *Application fee:* $25. *Expenses:* Tuition of $100 per credit hour for state residents; $244 per credit hour for nonresidents. Fees of $178 per semester full-time, $76 per semester part-time. *Financial aid:* Fellowships, research assistantships, teaching assistantships, federal work-study available. Financial aid application deadline: 2/15. • Application contact: Paula J. Boodt, Director, Graduate Student Services, 616-387-3570.

Yale University, Graduate School of Arts and Sciences, Department of Geology and Geophysics, New Haven, CT 06520. Department offers programs in geochemistry (PhD), geophysics (PhD), meteorology (PhD), mineralogy and crystallography (PhD), oceanography (PhD), paleoecology (PhD), paleontology and stratigraphy (PhD), petrology (PhD), structural geology (PhD). Faculty: 24 full-time (2 women), 1 part-time (0 women). Matriculated students: 44 full-time (5 women), 0 part-time; includes 0 minority, 25 foreign. 54 applicants, 30% accepted. In 1990, 5 master's, 10 doctorates awarded. Terminal master's awarded for partial completion of doctoral program. *Degree requirements:* For doctorate, dissertation. *Entrance requirements:* For doctorate, GRE General Test, GRE Subject Test. Application deadline: 1/2. Application fee: $45. *Tuition:* $15,160 per year full-time, $1895 per course part-time. • Dr. B. Saltzman, Chairman, 203-432-3114. Application contact: Susan Webb, Coordinator of Admissions, 203-432-2770.

See full description on page 299.

Geochemistry

California Institute of Technology, Division of Geological and Planetary Sciences, Pasadena, CA 91125. Offerings include cosmochemistry (PhD), geochemistry (MS, PhD). Division faculty: 28 full-time (0 women), 0 part-time. *Entrance requirements:* GRE General Test, GRE Subject Test. Application deadline: 1/15. Application fee: $0. *Expenses:* Tuition of $14,100 per year. Fees of $8 per quarter. • Dr. David J. Stevenson, Chairman, 818-356-6108.

See full description on page 219.

California State University, Fullerton, School of Natural Science and Mathematics, Department of Chemistry and Biochemistry, Fullerton, CA 92634. Offerings include geochemistry (MS). Department faculty: 20 full-time (5 women), 13 part-time (2 women). *Degree requirements:* Thesis, departmental qualifying exam required, foreign language not required. *Entrance requirements:* Minimum GPA of 2.5 in last 60 units, major in chemistry or related field. Application fee: $55. *Expenses:* Tuition of $0 for state residents; $246 per unit for nonresidents. Fees of $554 per semester full-time, $356 per unit part-time. • Dr. Glen Nagel, Chair, 714-773-3621. Application contact: Dr. Gene Hiegel, Adviser, 714-773-3624.

Colorado School of Mines, Department of Chemistry and Geochemistry, Golden, CO 80401. Department offers programs in applied chemistry (PhD), chemistry (MS), geochemistry (MS, PhD). Part-time programs available. Faculty: 15 full-time (0 women), 2 part-time (0 women). Matriculated students: 28 full-time (8 women), 24 part-time (12 women); includes 0 minority, 16 foreign. 27 applicants, 85% accepted. In 1990, 4 master's, 6 doctorates awarded. *Degree requirements:* For master's, thesis required, foreign language not required; for doctorate, dissertation. *Entrance requirements:* GRE General Test, minimum GPA of 3.0. Application deadline: 2/23. Application fee: $15 ($25 for foreign students). *Expenses:* Tuition of $3178 per year full-time, $124 per semester hour part-time for state residents; $10,304 per year full-time, $344 per semester hour part-time for nonresidents. Fees of $374 per year full-time. *Financial aid:* In 1990–91, 8 fellowships, 12 research assistantships, 10 teaching assistantships awarded. Financial aid application deadline: 2/23. *Faculty research:* Surface analysis, catalysis, aqueous geochemistry, analytical mass spectrometry, geomicrobiology. Total annual research budget: $789,350. • Dr. Stephen Daniel, Head, 303-273-3610.

Columbia University, Graduate School of Arts and Sciences, Division of Natural Sciences, Department of Geological Sciences, New York, NY 10027. Offerings include geochemistry (MA, M Phil, PhD). Department faculty: 21 full-time, 19 part-time. *Degree requirements:* For master's, thesis or alternative, fieldwork, written exam required, foreign language not required; for doctorate, 1 foreign language, dissertation. *Entrance requirements:* GRE General Test, GRE Subject Test, TOEFL, major in natural or physical science. Application deadline: 1/5. Application fee: $50. *Expenses:* Tuition of $7836 per semester for state residents; $426 per credit for nonresidents. Fees of $148 per semester for state residents. • Dennis Hayes, Chair, 212-854-4525.

George Washington University, Graduate School of Arts and Sciences, Department of Geology, Program in Geochemistry, Washington, DC 20052. Program awards MS, PhD. Part-time and evening/weekend programs available. Matriculated students: 0 full-time, 2 part-time (1 woman); includes 0 minority, 0 foreign. Average age 42. 1 applicants, 100% accepted. In 1990, 1 master's, 0 doctorates awarded. Terminal master's awarded for partial completion of doctoral program. *Degree requirements:* For master's, computer language, thesis, comprehensive exam; for doctorate, computer language, dissertation, general exam. *Entrance requirements:* For master's, GRE General Test, GRE Subject Test, bachelor's degree in field, minimum GPA of 3.0; for doctorate, GRE General Test, GRE Subject Test, minimum GPA of 3.0. Application deadline: 7/1. Application fee: $45. *Expenses:* Tuition of $490 per semester hour. Fees of $215 per year full-time, $125.60 per year (minimum) part-time. *Financial aid:* Application deadline 2/15. • Dr. Frederick R. Siegel, Director, 202-994-6194.

Indiana University Bloomington, College of Arts and Sciences, Department of Geological Sciences, Program in Geochemistry, Bloomington, IN 47405. Program awards MS, PhD. Faculty: 1 full-time. *Degree requirements:* For master's, 1 foreign language, computer language; for doctorate, 2 foreign languages (computer language can substitute for one), dissertation. *Entrance requirements:* GRE General Test. Application deadline: 4/15. Application fee: $25 ($35 for foreign students). *Tuition:* $99.85 per credit hour for state residents; $288 per credit hour for nonresidents. *Financial aid:* Application deadline 2/15. • Lee J. Suttner, Chairman, Department of Geological Sciences, 812-855-5581. Application contact: David G. Towell, Chairman, 812-855-7214.

Johns Hopkins University, School of Arts and Sciences, Department of Earth and Planetary Sciences, Program in Geochemistry, Baltimore, MD 21218. Program awards MA, PhD. Faculty: 2 full-time (0 women). Matriculated students: 2 full-time (1 woman), 0 part-time. Average age 24. Terminal master's awarded for partial completion of doctoral program. *Degree requirements:* For doctorate, 1 foreign language, dissertation. *Entrance requirements:* GRE General Test. Application deadline: 2/1. Application fee: $40. *Expenses:* Tuition of $15,500 per year full-time, $1550 per course part-time. Fees of $400 per year full-time. *Financial aid:* Federal work-study, institutionally sponsored loans available. Financial aid application deadline: 3/14; applicants required to submit FAF. • Bruce Marsh, Chairman, Department of Earth and Planetary Sciences, 301-338-7135.

McMaster University, Faculty of Science, Department of Geology, Hamilton, ON L8S 4L8, Canada. Offerings include geochemistry (PhD). *Degree requirements:* 1 foreign language, dissertation, comprehensive exam. *Expenses:* Tuition of $2250 per year full-time, $810 per semester part-time for Canadian residents; $10,340 per year full-time, $4050 per semester part-time for nonresidents. Fees of $76 per year full-time, $49 per semester part-time.

Michigan State University, College of Natural Science, Department of Geological Sciences, East Lansing, MI 48824-1115. Department awards MAT, MS, PhD. Faculty: 16 full-time, 0 part-time. Matriculated students: 28 full-time (5 women), 13 part-time (5 women); includes 1 minority, 4 foreign. 23 applicants, 40% accepted. In 1990, 6 master's, 2 doctorates awarded. *Degree requirements:* Thesis/dissertation. *Application deadline:* rolling. *Application fee:* $25 ($40 for foreign students). *Tuition:* $104.75 per credit for state residents; $211.75 per credit for nonresidents. *Financial aid:* In 1990–91, $180,782 in aid awarded. 9 fellowships (2 to first-year students), 3 research assistantships (0 to first-year students), 17 teaching assistantships (11 to first-year students) were awarded. • Thomas Vogel, Acting Chair, 517-355-4626.

See full description on page 231.

Montana College of Mineral Science and Technology, Graduate School, Chemistry and Geochemistry Program, Butte, MT 59701. Offers geoscience (MS), including geochemistry. Part-time programs available. Faculty: 5 full-time (0 women), 3 part-time (1 woman). Matriculated students: 2 full-time (0 women), 0 part-time; includes 0 minority, 0 foreign. Average age 25. 2 applicants, 100% accepted. In 1990, 4 degrees awarded (50% found work related to degree, 50% continued full-time study). *Degree requirements:* Thesis optional, foreign language not required. *Entrance requirements:* GRE General Test, GRE Subject Test, TOEFL (minimum score of 525 required), minimum B average. Application deadline: 3/31 (priority date, applications processed on a rolling basis). Application fee: $20. *Tuition:* $524.25 per credit hour full-time, $310.50 per credit hour part-time for state residents; $1362 per credit hour full-time, $310.50 per credit hour part-time for nonresidents. *Financial aid:* In 1990–91, 3 students received a total of $16,509 in aid awarded. 0 fellowships, 1 research assistantship (0 to first-year students), 1 teaching assistantship (to a first-year student), 2 fee waivers (1 to a first-year student) were awarded; federal work-study, institutionally sponsored loans, and career-related internships or fieldwork also available. Average monthly stipend for a graduate assistantship: $500. Financial aid application deadline: 3/31. *Faculty research:* Coal characterization, mine water effluent studies, quartz crystal growth, environmental geochemistry, molecular spectroscopy. • Dr. Doug Drew, Department Head, 406-

496-4202. Application contact: Mary J. Seccombe, Graduate School Administrator, 406-496-4128.

Announcement: Thesis and nonthesis options offered. Research areas include neutron activation analysis, coal geochemistry, geochemical exploration, geothermal water systems, and extraction of metals from aqueous solutions. Graduate course program established by student's committee with objective of creating a balance between geology and chemistry and of supporting student's research interest. For further information, contact Dr. Douglas Coe.

New Mexico Institute of Mining and Technology, Department of Geoscience, Program in Geochemistry, Socorro, NM 87801. Program awards MS, PhD. Faculty: 2 full-time (0 women), 0 part-time. Matriculated students: 7 full-time (2 women), 0 part-time; includes 0 minority, 1 foreign. Average age 25. 9 applicants, 33% accepted. In 1990, 1 master's, 1 doctorate awarded. *Degree requirements:* For master's, thesis optional, foreign language not required; for doctorate, 1 foreign language, dissertation. *Entrance requirements:* For master's, GRE General Test, TOEFL (minimum score of 540 required); for doctorate, GRE General Test, GRE Subject Test, TOEFL (minimum score of 540 required). Application deadline: 6/1 (priority date, applications processed on a rolling basis). Application fee: $16. *Expenses:* Tuition of $617 per semester full-time, $51 per hour part-time for state residents; $2656 per semester full-time, $209 per hour part-time for nonresidents. Fees of $207 per semester. *Financial aid:* In 1990–91, 4 students received a total of $13,577 in aid awarded. 3 research assistantships (0 to first-year students), 1 teaching assistantship (0 to first-year students) were awarded; federal work-study, institutionally sponsored loans also available. Average monthly stipend for a graduate assistantship: $600. Financial aid application deadline: 3/15; applicants required to submit GAPSFAS or FAF. *Faculty research:* Crust-mantle evolution, ore deposits, isotope geochemistry, volatiles in magmas. Total annual research budget: $377,886. • Dr. Phillip Kyle, Coordinator, 505-835-5995. Application contact: Dr. J. A. Smoake, Dean, Graduate Studies, 505-835-5513.

Announcement: MS and PhD programs also available in geology (structural geology, depositional environments, ore deposits, fluid inclusions), geophysics (exploration geophysics, earthquake seismology, geothermics), and hydrology (groundwater transport and contamination, modeling, isotope hydrology). Courses, research, and support augmented by 14 adjuncts, most of whom are professionals with the on-campus NM Bureau of Mines and Mineral Resources. See full description of NMIMT in the Graduate and Professional Programs (An Overview) volume of this series.

Ohio State University, College of Mathematical and Physical Sciences, Department of Geology, Columbus, OH 43210. Offerings include geochemistry (MS, PhD). Department faculty: 25. *Degree requirements:* For master's, thesis required, foreign language not required; for doctorate, 1 foreign language, dissertation. *Entrance requirements:* GRE General Test, GRE Subject Test. Application deadline: 8/15 (applications processed on a rolling basis). Application fee: $0 ($25 for foreign students). *Tuition:* $1213 per quarter full-time, $364 per course part-time for state residents; $3143 per quarter full-time, $943 per course part-time for nonresidents. • Peter Webb, Chairman, 614-292-2721.

Oregon State University, Graduate School, College of Science, Department of Geosciences, Corvallis, OR 97331. Department awards MA, MS, PhD. Part-time programs available. Faculty: 24 full-time (2 women), 12 part-time (0 women). In 1990, 29 master's, 2 doctorates awarded. Terminal master's awarded for partial completion of doctoral program. *Degree requirements:* For master's, variable foreign language requirement, thesis optional; for doctorate, 1 foreign language, dissertation. *Entrance requirements:* GRE General Test, GRE Subject Test, TOEFL (minimum score of 550 required), minimum GPA of 3.0 in last 90 hours. Application fee: $40. *Tuition:* $1140 per trimester full-time, $449 per year part-time for state residents; $1816 per trimester full-time, $674 per year part-time for nonresidents. *Financial aid:* Fellowships, research assistantships, teaching assistantships, federal work-study, institutionally sponsored loans available. Aid available to part-time students. Financial aid application deadline: 2/14; applicants required to submit GAPSFAS or FAF. • Dr. Cyrus W. Field, Chair. Application contact: Graduate Admissions Coordinator, 503-737-1201.

See full description on page 235.

Pennsylvania State University University Park Campus, College of Earth and Mineral Sciences, Department of Geosciences, University Park, PA 16802. Department offers programs in geochemistry (MS, PhD), geology (MS, PhD), geophysics (MS, PhD). Faculty: 39. Matriculated students: 85 full-time (21 women), 16 part-time (2 women). In 1990, 18 master's, 4 doctorates awarded. *Entrance requirements:* GRE General Test, TOEFL. Application fee: $35. *Tuition:* $203 per credit for state residents; $403 per credit for nonresidents. • Dr. David Eggler, Chair, 814-865-6393.

Announcement: Penn State offers MS and PhD research opportunities in aqueous geochemistry, igneous and metamorphic petrology, isotope studies, mineralogy, ore deposits, and paleoclimate chemistry.

See full description on page 237.

Purdue University, School of Science, Department of Earth and Atmospheric Sciences, West Lafayette, IN 47907. Department awards MS, PhD. Faculty: 24 full-time, 1 part-time. Matriculated students: 74 full-time (10 women), 6 part-time (3 women); includes 4 minority (1 Asian American, 1 black American, 1 Hispanic American, 1 Native American), 27 foreign. Average age 28. 87 applicants, 23% accepted. In 1990, 19 master's, 6 doctorates awarded. *Degree requirements:* For master's, thesis required, foreign language not required; for doctorate, 1 foreign language, dissertation. *Entrance requirements:* GRE General Test. Application deadline: 6/1. Application fee: $25. *Tuition:* $1162 per semester full-time, $83.50 per credit hour part-time for state residents; $3720 per semester full-time, $244.50 per credit hour part-time for nonresidents. *Financial aid:* In 1990–91, 3 fellowships (2 to first-year students), 22 research assistantships (1 to a first-year student), 26 teaching assistantships (3 to first-year students) awarded. • Dr. E. M. Agee, Head, 317-494-0251.

Rensselaer Polytechnic Institute, School of Science, Department of Earth and Environmental Sciences, Troy, NY 12180. Offerings include geochemistry (MS, PhD). Department faculty: 8 full-time (0 women), 2 part-time (both women). *Degree requirements:* For master's, thesis optional; for doctorate, dissertation. *Entrance requirements:* GRE General Test, TOEFL. Application deadline: 2/1. Application fee: $30. *Expenses:* Tuition of $455 per credit hour. Fees of $195.57 per semester. • E. B. Watson, Chairman, 518-276-6474. Application contact: M. Brian Bayly, Graduate Coordinator, 518-276-6494.

See full description on page 239.

Rice University, Wiess School of Natural Sciences, Department of Geology and Geophysics, Houston, TX 77251. Offerings include geochemistry (MA, PhD). Department faculty: 15 full-time (0 women), 0 part-time. *Degree requirements:* Thesis/dissertation required, foreign language not required. *Entrance requirements:* GRE General Test, GRE Subject Test, TOEFL (minimum score of 550 required), minimum GPA of 3.0. Application deadline: 3/1. Application fee: $0. *Expenses:* Tuition of $8300 per year full-time, $400 per credit hour part-time. Fees of $167 per year. • J. C. Stormer, Chairman, 713-527-4880.

Rutgers, The State University of New Jersey, Newark, Department of Geology, Newark, NJ 07102. Offerings include geochemistry (MS). Department faculty: 5 full-time (0 women), 6 part-time (2 women). *Degree requirements:* Comprehensive exam required, thesis optional. *Entrance requirements:* GRE, minimum undergraduate B average. Application deadline: 6/1. Application fee: $35. *Tuition:* $2216 per semester full-time for state residents; $3248 per semester full-time for nonresidents. • Dr. George Theokritoff, Director, 201-648-5100.

Southern Methodist University, Dedman College, Department of Geological Sciences, Program in Geochemistry, Dallas, TX 75275. Program awards MS, PhD. *Degree requirements:* For master's, thesis, qualifying exam required, foreign language not required. *Entrance requirements:* For master's, GRE General Test (minimum combined score of 1200 required), minimum GPA of 3.0. Application deadline: 2/15. Application fee: $25. *Expenses:* Tuition of $435 per credit. Fees of $664 per semester for state residents; $56 per year for nonresidents. *Financial aid:* Application deadline 2/15. • Dr. Michael J. Holdaway, Chairman, Department of Geological Sciences, 214-692-2770. Application contact: Dr. Peter Scholle, Graduate Adviser, 214-692-4011.

See full description on page 241.

State University of New York at Stony Brook, College of Arts and Sciences, Division of Physical Sciences and Mathematics, Department of Earth and Space Sciences, Stony Brook, NY 11794. Department offers programs in astronomical sciences (M Phil, MS, PhD), including astronomical sciences, astrophysics; geological sciences (M Phil, MS, PhD), including geochemistry, geology, geophysics. Faculty: 23 full-time, 0 part-time. Matriculated students: 48 full-time (14 women), 4 part-time (2 women); includes 1 minority (Asian American), 26 foreign. 116 applicants, 28% accepted. In 1990, 4 master's, 1 doctorate awarded. Terminal master's awarded for partial completion of doctoral program. *Degree requirements:* For master's, thesis or alternative required, foreign language not required; for doctorate, dissertation required, foreign language not required. *Entrance requirements:* GRE General Test, TOEFL, minimum GPA of 3.0. Application deadline: 2/1. Application fee: $35. *Expenses:* Tuition of $2450 per year full-time, $103 per credit part-time for state residents; $5766 per year full-time, $243 per credit part-time for nonresidents. Fees of $151 per year full-time, $10.45 per year (minimum) part-time. *Financial aid:* In 1990–91, 0 fellowships, 27 research assistantships, 21 teaching assistantships awarded. *Faculty research:* Astronomy, theoretical and observational astrophysics, paleontology, petrology, crystallography. Total annual research budget: $3.2-million. • Dr. Gilbert Hansen, Chairman, 516-632-8200.

University of California, Los Angeles, College of Letters and Science, Department of Earth and Space Sciences, Program in Geochemistry, Los Angeles, CA 90024. Program awards MS, PhD, C Phil. Matriculated students: 5 full-time (1 woman), 0 part-time. 9 applicants, 11% accepted. In 1990, 1 master's, 3 doctorates awarded. *Degree requirements:* For master's, thesis or comprehensive exams required, foreign language not required; for doctorate, dissertation, written and oral qualifying exams required, foreign language not required. *Entrance requirements:* For master's and doctorate, GRE General Test, GRE Subject Test. Application fee: $40. *Expenses:* Tuition of $0 for state residents; $7699 per year for nonresidents. Fees of $2907 per year. *Financial aid:* In 1990–91, 7 students received a total of $124,797 in aid awarded. 7 fellowships, 4 research assistantships, 4 teaching assistantships were awarded; full and partial tuition waivers, federal work-study, institutionally sponsored loans also available. Financial aid application deadline: 3/1. • Dr. Arthur Montana, Chair, Department of Earth and Space Sciences, 310-825-3880.

See full description on page 255.

University of California, Santa Barbara, College of Letters and Science, Department of Geological Sciences, Santa Barbara, CA 93106. Department offers programs in geological sciences (MA, PhD), geophysics (MS). Matriculated students: 52 full-time (16 women), 0 part-time; includes 4 foreign. 61 applicants, 49% accepted. In 1990, 3 master's, 7 doctorates awarded. *Degree requirements:* For master's, 1 foreign language, thesis or exam; for doctorate, variable foreign language requirement, dissertation. *Entrance requirements:* GRE, TOEFL (minimum score of 550 required). Application deadline: 2/10. Application fee: $40. *Expenses:* Tuition of $0 for state residents; $7699 per year for nonresidents. Fees of $2307 per year. *Financial aid:* Fellowships, research assistantships, teaching assistantships, full and partial tuition waivers, federal work-study, institutionally sponsored loans, and career-related internships or fieldwork available. Financial aid application deadline: 1/31. • Michael Fuller, Chair, 805-893-3508. Application contact: Leslie Buxton, Graduate Secretary, 805-893-3329.

See full description on page 257.

University of Cincinnati, McMicken College of Arts and Sciences, Department of Geology, Cincinnati, OH 45221. Department awards MS, PhD. Faculty: 11 full-time, 0 part-time. Matriculated students: 26 full-time (10 women), 16 part-time (8 women); includes 0 minority, 5 foreign. 32 applicants, 28% accepted. In 1990, 9 master's, 2 doctorates awarded. *Degree requirements:* For master's, thesis required, foreign language not required; for doctorate, 1 foreign language, dissertation. *Entrance requirements:* GRE General Test, GRE Subject Test. Application deadline: 2/15. Application fee: $20. *Tuition:* $131 per credit hour for state residents; $261 per credit hour for nonresidents. *Financial aid:* In 1990–91, 23 teaching assistantships awarded; fellowships, research assistantships, full tuition waivers also available. Aid available to part-time students. Average monthly stipend for a graduate assistantship: $583. Financial aid application deadline: 5/1. • Dr. I. Attila Kilinc, Acting Head, 513-556-5034.

See full description on page 261.

University of Georgia, College of Arts and Sciences, Department of Geology, Athens, GA 30602. Offerings include geochemistry (MS, PhD). Department faculty: 16 full-time (2 women), 0 part-time. *Degree requirements:* 1 foreign language (computer language can substitute), thesis/dissertation. *Entrance requirements:* GRE General Test. Application fee: $10. *Expenses:* Tuition of $598 per quarter full-time, $48 per quarter part-time for state residents; $1558 per quarter full-time, $144 per quarter part-time for nonresidents. Fees of $118 per quarter. • Dr. Michael F. Roden, Graduate Coordinator, 404-542-2416.

See full description on page 263.

University of Illinois at Chicago, College of Liberal Arts and Sciences, Department of Geological Sciences, Chicago, IL 60680. Offerings include geochemistry (MS, PhD). Department faculty: 11 full-time (2 women). *Degree requirements:* Thesis/dissertation. *Entrance requirements:* GRE General Test, TOEFL (minimum score of

550 required), minimum GPA of 3.75 (on a 5.0 scale). Application deadline: 7/5. Application fee: $20. *Expenses:* Tuition of $1369 per semester full-time, $521 per semester (minimum) part-time for state residents; $3840 per semester full-time, $1454 per semester (minimum) part-time for nonresidents. Fees of $458 per semester full-time, $398 per semester (minimum) part-time. • Norman Smith, Acting Head, 312-996-3153. Application contact: Kelvin Rodolfo, Graduate Director, 312-996-3154.

See full description on page 269.

University of Illinois at Urbana-Champaign, College of Liberal Arts and Sciences, Department of Geology, Champaign, IL 61820. Offerings include geochemistry (MS, PhD). Department faculty: 19 full-time (1 woman), 5 part-time (0 women). *Degree requirements:* Thesis/dissertation required, foreign language not required. *Entrance requirements:* GRE General Test, TOEFL. Application fee: $25. *Tuition:* $1838 per semester full-time, $708 per semester part-time for state residents; $4314 per semester full-time, $1673 per semester part-time for nonresidents. • Dr. R. James Kirkpatrick, Head, 217-333-3542. Application contact: Dr. Wang-Ping Chen, Graduate Admissions, 217-244-4065.

University of Michigan, College of Literature, Science, and the Arts, Department of Geological Sciences, Program in Oceanography: Marine Geology and Geochemistry, Ann Arbor, MI 48109. Program awards MS, PhD. *Degree requirements:* For master's, thesis required, foreign language not required; for doctorate, variable foreign language requirement, dissertation, preliminary exam. *Entrance requirements:* GRE General Test, GRE Subject Test. Application deadline: 2/1 (applications processed on a rolling basis). Application fee: $30. *Tuition:* $3255 per semester full-time, $352 per credit (minimum) part-time for state residents; $6803 per semester full-time, $746 per credit (minimum) part-time for nonresidents. *Financial aid:* Application deadline 3/15. • Henry Pollack, Chair, Department of Geological Sciences, 313-764-1435.

University of Nevada, Reno, Mackay School of Mines, Department of Geological Sciences, Reno, NV 89557. Department awards Geol E, MS, PhD. Programs offered in geochemistry (MS, PhD), geology (MS, PhD). Faculty: 21. Matriculated students: 53 (15 women). *Degree requirements:* For master's, thesis optional, foreign language not required; for doctorate, 1 foreign language, dissertation. *Entrance requirements:* For master's, GRE General Test, TOEFL, minimum GPA of 2.75; for doctorate, GRE General Test, TOEFL, minimum GPA of 3.0. Application deadline: 8/1 (priority date, applications processed on a rolling basis). Application fee: $20. *Expenses:* Tuition of $0 for state residents; $3600 per year full-time, $66 per credit hour part-time for nonresidents. Fees of $66 per credit hour. *Financial aid:* Research assistantships, teaching assistantships, full tuition waivers, institutionally sponsored loans available. Average monthly stipend for a graduate assistantship: $740. Financial aid applicants required to submit FAF. *Faculty research:* Hydrothermal ore deposits, metamorthic and igneous petrogenesis, sedimentary rock record of earth history, field and petrographic investigation of magnetism, rock fracture mechanics. Total annual research budget: $983,540. • Dr. Bob Watters, Chairman, 702-784-6050.

See full description on page 275.

University of Pittsburgh, Faculty of Arts and Sciences, Department of Geology and Planetary Science, Pittsburgh, PA 15260. Department awards MS, PhD. Part-time programs available. Faculty: 12 full-time (0 women), 1 (woman) part-time. Matriculated students: 17 full-time (5 women), 18 part-time (6 women); includes 4 minority (2 Asian American, 2 black American), 5 foreign. 59 applicants, 17% accepted. In 1990, 6 master's, 1 doctorate awarded. *Degree requirements:* Thesis/dissertation required, foreign language not required. *Entrance requirements:* GRE General Test, TOEFL. Application deadline: 8/1 (applications processed on a rolling basis). Application fee: $15 ($25 for foreign students). *Expenses:* Tuition of $2920 per semester full-time, $241 per credit part-time for state residents; $5840 per semester full-time, $482 per credit part-time for nonresidents. Fees of $156 per year. *Financial aid:* In 1990–91, 13 students received a total of $94,930 in aid awarded. 4 research assistantships (0 to first-year students), 9 teaching assistantships (3 to first-year students) were awarded; full and partial tuition waivers, federal work-study, institutionally sponsored loans, and career-related internships or fieldwork also available. Aid available to part-time students. Average monthly stipend for a graduate assistantship: $900. Financial aid application deadline: 3/1; applicants required to submit FAF. *Faculty research:* Plate tectonics as basement studies, sedimentology and image analysis, marine paleoecology and geoarchaeology, geophysics and paleomagnetics, planetary geology and meteoritics. Total annual research budget: $490,472. • Dr. Thomas H. Anderson, Chairman, 412-624-8780. Application contact: Dr. Walter L. Pilant, Graduate Adviser, 412-624-8870.

See full description on page 281.

University of Texas at Austin, Graduate School, College of Natural Sciences, Department of Geological Sciences, Austin, TX 78712. Department awards MA, PhD. Matriculated students: 179 full-time (41 women), 0 part-time; includes 5 minority (3 Asian American, 2 Hispanic American), 39 foreign. 90 applicants, 67% accepted. In 1990, 15 master's, 4 doctorates awarded. *Entrance requirements:* GRE. Application deadline: 2/1 (priority date, applications processed on a rolling basis). Application fee: $40 ($75 for foreign students). *Tuition:* $510.30 per semester for state residents; $1806 per semester for nonresidents. *Financial aid:* Fellowships, research assistantships, teaching assistantships available. Financial aid application deadline: 3/1. • Dr. Clark R. Wilson, Chairman, 512-471-5172. Application contact: Dr. William Carlson, Graduate Adviser, 512-471-6098.

See full description on page 285.

University of Texas at Dallas, School of Natural Sciences and Mathematics, Program in Geosciences, Richardson, TX 75083-0688. Program awards MS, PhD. Part-time and evening/weekend programs available. Faculty: 16 full-time (0 women), 7 part-time (0 women). Matriculated students: 44 full-time (10 women), 35 part-time (7 women); includes 7 minority (3 Asian American, 1 black American, 3 Hispanic American), 21 foreign. Average age 33. In 1990, 7 master's, 5 doctorates awarded. *Degree requirements:* Computer language, thesis/dissertation required, foreign language not required. *Entrance requirements:* GRE General Test (minimum combined score of 1000 required), TOEFL (minimum score of 550 required), minimum GPA of 3.0 in upper level course work in field. Application deadline: 7/15 (applications processed on a rolling basis). Application fee: $0 ($75 for foreign students). *Expenses:* Tuition of $360 per semester full-time, $100 per semester (minimum) part-time for state residents; $2196 per semester full-time, $122 per semester hour (minimum) part-time for nonresidents. Fees of $338 per semester full-time, $22 per hour (minimum) part-time. *Financial aid:* In 1990–91, 1 student received a total of $8175 in aid awarded. Fellowships, research assistantships, teaching assistantships, federal work-study, and career-related internships or fieldwork available. Aid available to part-time students. Financial aid application deadline: 11/1; applicants required to submit FAF. *Faculty research:* Geochemistry, structural geology-tectonics, sedimentology-stratigraphy, micropaleontology. • Dr. James L. Carter, Head, 214-690-2401.

See full description on page 287.

University of Toledo, College of Arts and Sciences, Department of Geology, Program in Geochemistry, Toledo, OH 43606. Program awards MS. *Degree requirements:* Thesis required, foreign language not required. *Entrance requirements:* GRE General Test. Application deadline: 9/8 (priority date). Application fee: $30. *Tuition:* $122.59 per credit hour for state residents; $193.40 per credit hour for nonresidents. *Financial aid:* Application deadline 4/1; applicants required to submit FAF. • Dr. James A. Harrell, Chairman, Department of Geology, 419-537-2193.

University of Tulsa, College of Engineering and Physical Sciences, Department of Geosciences, Tulsa, OK 74104. Offerings include geochemistry (MS, PhD). Department faculty: 7 full-time (1 woman), 1 part-time (0 women). *Degree requirements:* For master's, computer language, thesis or alternative required, foreign language not required; for doctorate, computer language, dissertation required, foreign language not required. *Entrance requirements:* For master's, GRE General Test, TOEFL; for doctorate, GRE General Test, GRE Subject Test, TOEFL. Application deadline: rolling. Application fee: $30. *Tuition:* $350 per credit hour. • Dr. Colin Barker, Chairperson, 918-631-3014.

Virginia Polytechnic Institute and State University, College of Arts and Sciences, Department of Geological Sciences, Blacksburg, VA 24061. Department offers programs in geological sciences (MS, PhD), geophysics (MS, PhD). Faculty: 19 full-time (0 women), 0 part-time. Matriculated students: 32 full-time (5 women), 4 part-time (2 women); includes 1 minority (Hispanic American), 5 foreign. 42 applicants, 50% accepted. In 1990, 7 master's, 9 doctorates awarded. *Degree requirements:* 1 foreign language, thesis/dissertation. *Entrance requirements:* GRE General Test, GRE Subject Test. Application deadline: 2/15 (priority date). Application fee: $10. *Tuition:* $1889 per semester full-time, $606 per credit hour part-time for state residents; $2627 per semester full-time, $853 per credit hour part-time for nonresidents. *Financial aid:* In 1990–91, 4 fellowships, 14 research assistantships, 22 teaching assistantships (11 to first-year students), 7 assistantships awarded; full tuition waivers and career-related internships or fieldwork also available. *Faculty research:* Paleocology, sedimentology, evolutionary theory, tectonics, structure. • Dr. Donald Bloss, Chairman, 703-231-6521.

Washington University, Graduate School of Arts and Sciences, Department of Earth and Planetary Sciences, St. Louis, MO 63130. Offerings include geochemistry (PhD). Terminal master's awarded for partial completion of doctoral program. *Degree requirements:* Dissertation. *Entrance requirements:* GRE General Test. Application deadline: 1/15 (priority date, applications processed on a rolling basis). Application fee: $0. *Tuition:* $15,960 per year full-time, $665 per credit hour part-time. • Dr. Larry A. Haskin, Chairman, 314-935-5610.

See full description on page 295.

Yale University, Graduate School of Arts and Sciences, Department of Geology and Geophysics, New Haven, CT 06520. Offerings include geochemistry (PhD). Terminal master's awarded for partial completion of doctoral program. Department faculty: 24 full-time (2 women), 1 part-time (0 women). *Degree requirements:* Dissertation. *Entrance requirements:* GRE General Test, GRE Subject Test. Application deadline: 1/2. Application fee: $45. *Tuition:* $15,160 per year full-time, $1895 per course part-time. • Dr. B. Saltzman, Chairman, 203-432-3114. Application contact: Susan Webb, Coordinator of Admissions, 203-432-2770.

See full description on page 299.

Geodetic Sciences

Columbia University, Graduate School of Arts and Sciences, Division of Natural Sciences, Department of Geological Sciences, New York, NY 10027. Offerings include geodetic sciences (MA, M Phil, PhD). Department faculty: 21 full-time, 19 part-time. *Degree requirements:* For master's, thesis or alternative, fieldwork, written exam required, foreign language not required; for doctorate, 1 foreign language, dissertation. *Entrance requirements:* GRE General Test, GRE Subject Test, TOEFL, major in natural or physical science. Application deadline: 1/5. Application fee: $50. *Expenses:* Tuition of $7836 per semester for state residents; $426 per credit for nonresidents. Fees of $148 per semester for state residents. • Dennis Hayes, Chair, 212-854-4525.

Iowa State University of Science and Technology, College of Engineering, Department of Civil and Construction Engineering, Ames, IA 50011. Offerings include geodesy and photogrammetry (MS, PhD). M Arch/MS jointly offered with Department of Architecture. *Degree requirements:* Foreign language not required. *Entrance requirements:* GRE General Test (foreign students), TOEFL (minimum score of 550 required). Application fee: $20 ($30 for foreign students). *Expenses:* Tuition of $1158 per semester full-time, $129 per credit part-time for state residents; $3340 per semester full-time, $372 per credit part-time for nonresidents. Fees of $10 per semester. • Dr. Lowell F. Greimann, Chair, 515-294-3532.

Ohio State University, College of Mathematical and Physical Sciences, Department of Geodetic Science and Surveying, Columbus, OH 43210. Offerings include geodetic science (MS, PhD). Department faculty: 9. *Degree requirements:* For master's, thesis optional, foreign language not required; for doctorate, dissertation required, foreign language not required. *Application deadline:* 8/15 (applications processed on a rolling basis). *Application fee:* $0 ($25 for foreign students). *Tuition:* $1213 per quarter full-time, $364 per course part-time for state residents; $3143 per quarter full-time, $943 per course part-time for nonresidents. • Ivan I. Mueller, Chairman, 614-292-2269.

SECTION 3: EARTH AND PLANETARY SCIENCES

Directories: Geodetic Sciences; Geology

Université Laval, Faculty of Forestry and Geodesy, Department of Geodesy, Sainte-Foy, PQ G1K 7P4, Canada. Department awards M Sc, PhD. Matriculated students: 23 full-time (5 women), 9 part-time (2 women). 36 applicants, 58% accepted. In 1990, 7 master's, 1 doctorate awarded. *Application deadline:* 3/1. *Application fee:* $15. *Expenses:* Tuition of $792 per year full-time for Canadian residents; $5914 per year full-time for nonresidents. Fees of $120 per year full-time. • Jacques Jobin, Director, 418-656-7182.

University of New Brunswick, Faculty of Engineering, Department of Surveying Engineering, Fredericton, NB E3B 5A3, Canada. Offerings include mapping, charting and geodesy (Diploma). *Application deadline:* 3/1 (priority date). *Expenses:* Tuition of $2100 per year. Fees of $45 per year. • Dr. J. D. McLaughlin, Chairperson, 506-453-4698. Application contact: Dr. R. Langley, Director of Graduate Studies, 506-453-4698.

Geology

Acadia University, Faculty of Pure and Applied Science, Department of Geology, Wolfville, NS B0P 1X0, Canada. Department awards M Sc. Faculty: 7 full-time (2 women), 0 part-time. Matriculated students: 4 full-time (0 women), 2 part-time (1 woman); includes 3 foreign. 2 applicants, 100% accepted. In 1990, 1 degree awarded (100% found work related to degree). *Degree requirements:* Thesis required, foreign language not required. *Application deadline:* 5/31 (priority date, applications processed on a rolling basis). *Expenses:* Tuition of $2405 per year full-time, $495 per course part-time. Fees of $1700 per year full-time, $340 per year part-time for nonresidents. *Financial aid:* In 1990–91, 2 students received a total of $15,000 in aid awarded. 2 fellowships (both to first-year students), 0 student assistantships were awarded; career-related internships or fieldwork also available. *Faculty research:* General, igneous, metamorphic, structural, and economic geology. Total annual research budget: $130,000. • Dr. J. Colwell, Head, 902-542-2201 Ext. 208.

Arizona State University, College of Liberal Arts and Sciences, Department of Geology, Tempe, AZ 85287. Department offers programs in geological engineering (MS, PhD), natural science (MNS). Matriculated students: 32 full-time (7 women), 30 part-time (9 women); includes 2 minority (both Hispanic American), 5 foreign. 97 applicants, 45% accepted. In 1990, 15 master's, 2 doctorates awarded. *Degree requirements:* For master's, thesis or alternative; for doctorate, dissertation. *Entrance requirements:* GRE. Application fee: $25. *Tuition:* $1528 per year full-time, $80 per hour part-time for state residents; $6934 per year full-time, $289 per hour part-time for nonresidents. *Faculty research:* Mechanics and deposits of volcanic eruptions, possible controls on global climate, electron microprobe studies of ore minerals. • Dr. Ronald Greeley, Chair, 602-965-5081.

Auburn University, College of Sciences and Mathematics, Department of Geology, Auburn University, AL 36849. Department awards MS. Part-time programs available. Faculty: 8 full-time (0 women), 0 part-time. Matriculated students: 2 full-time (1 woman), 14 part-time (2 women); includes 0 minority, 1 foreign. In 1990, 1 degree awarded (100% continued full-time study). *Degree requirements:* 1 foreign language (computer language can substitute), field camp. *Entrance requirements:* GRE General Test, GRE Subject Test. Application deadline: 9/1 (applications processed on a rolling basis). Application fee: $15. *Expenses:* Tuition of $1596 per year full-time, $44 per credit hour part-time for state residents; $4788 per year full-time, $132 per credit hour part-time for nonresidents. Fees of $92 per quarter for state residents; $276 per quarter for nonresidents. *Financial aid:* Research assistantships, teaching assistantships, federal work-study available. Aid available to part-time students. Financial aid application deadline: 3/15. • Dr. Robert B. Cook, Head, 205-844-4282. Application contact: Dr. Norman J. Doorenbos, Dean, Graduate School, 205-844-4700.

Announcement: The master's program offers a broad-based curriculum taking advantage of Auburn's location on the boundary between the Appalachian front and the Gulf coastal plain. Low student-faculty ratio results in small class size, close relationships between students and faculty, and relaxed and informal atmosphere. Research opportunities span all disciplines. Current research includes economic geology, geochemistry, and tectonics of the southern Appalachians; coastal-plain stratigraphy/sedimentology; and taphonomy and paleoecology of Phanerozoic megafloral and invertebrate assemblages. Current investigations are focused in San Salvador, Bahamas; eastern Borneo, Indonesia; Norway and eastern Europe; southern California; and the North American Precambrian craton. See Auburn's full description in the Graduate and Professional Programs (An Overview) volume of this series.

Ball State University, College of Sciences and Humanities, Department of Geology, 2000 University Avenue, Muncie, IN 47306. Department awards MS. Faculty: 6. Matriculated students: 7. In 1990, 1 degree awarded. *Degree requirements:* Foreign language not required. *Application fee:* $15. *Expenses:* Tuition of $1140 per semester (minimum) full-time for state residents; $2680 per semester (minimum) full-time for nonresidents. Fees of $6 per credit hour. *Financial aid:* In 1990–91, 4 teaching assistantships awarded; career-related internships or fieldwork also available. *Faculty research:* Environmental geology, geophysics, stratigraphy, micropaleontology, geomorphology. • Dr. Alan Samuelson, Chairman, 317-285-8270.

Announcement: Students benefit from instruction in techniques of "shallow-earth" geophysics, hydrogeology, computer modeling, micropaleontology, and fractal geometric analysis. The department houses a collection of core and cuttings from 120 wells. Active research is being performed in geohydrology, fluvial and glacial geomorphology, biostratigraphy, sedimentology, and till and bedrock fracture studies.

Baylor University, College of Arts and Sciences, Department of Geology, Waco, TX 76798. Department offers programs in earth science (MA), geology (MS, PhD). Matriculated students: 7 full-time (4 women), 26 part-time (3 women); includes 3 minority (1 black American, 2 Hispanic American), 1 foreign. In 1990, 8 master's awarded. *Degree requirements:* For master's, 1 foreign language, thesis. *Entrance requirements:* For master's, GRE General Test, GRE Subject Test; for doctorate, GRE. Application deadline: rolling. Application fee: $0. *Expenses:* Tuition of $4440 per year full-time, $185 per credit hour part-time. Fees of $510 per year full-time. *Financial aid:* Research assistantships, teaching assistantships, federal work-study, institutionally sponsored loans available. Financial aid applicants required to submit FAF. • Dr. H. H. Beaver, Chairman, 817-755-2361.

Boise State University, College of Arts and Sciences, Department of Geosciences, Boise, ID 83725. Department offers programs in applied geophysics (MS), geology (MS). Offered jointly with Idaho State University. Part-time programs available. Faculty: 10 full-time (0 women), 10 part-time (2 women). Matriculated students: 7 full-time (1 woman), 2 part-time (0 women). In 1990, 1 degree awarded. *Degree requirements:* Thesis. *Entrance requirements:* GRE General Test, minimum GPA of 2.75. Application deadline: 8/1 (priority date). Application fee: $15. *Expenses:* Tuition of $0 for state residents; $2000 per year full-time, for nonresidents. Fees of $1654 per year full-time, $83 per credit hour part-time. *Financial aid:* In 1990–91, 6 research assistantships (3 to first-year students), 2 grants (0 to first-year students) awarded; partial tuition waivers, federal work-study, institutionally sponsored loans, and career-related internships or fieldwork also available. Financial aid application deadline: 4/1. *Faculty research:* Seismology, geothermal aquifers, sedimentation, tectonism seismo/acoustic propagation. • Dr. Paul Donaldson, Chairman, 202-385-1631.

Boston College, Graduate School of Arts and Sciences, Department of Geology and Geophysics, Chestnut Hill, MA 02167. Department offers programs in geology and geophysics (MS, MST), oceanography (MS, MST). *Degree requirements:* 1 foreign language, thesis. *Entrance requirements:* GRE General Test, GRE Subject Test. *Tuition:* $412 per credit hour.

Boston University, Graduate School, Department of Geology, Boston, MA 02215. Department awards MA, PhD. Faculty: 7 full-time (0 women), 4 part-time (0 women). Matriculated students: 15 full-time (10 women), 8 part-time (6 women); includes 3 minority (2 Hispanic American, 1 Native American), 4 foreign. Average age 32. 30 applicants, 33% accepted. In 1990, 3 master's awarded (100% found work related to degree); 2 doctorates awarded (100% entered university research/teaching). Terminal master's awarded for partial completion of doctoral program. *Degree requirements:* 1 foreign language, thesis/dissertation. *Entrance requirements:* GRE General Test, TOEFL (minimum score of 550 required). Application deadline: 7/1 (applications processed on a rolling basis). Application fee: $45. *Expenses:* Tuition of $15,950 per year full-time, $498 per credit part-time. Fees of $120 per year full-time, $35 per semester part-time. *Financial aid:* In 1990–91, $82,273 in aid awarded. 0 fellowships, 2 research assistantships (0 to first-year students), 11 teaching assistantships (3 to first-year students) were awarded; federal work-study also available. Aid available to part-time students. Financial aid application deadline: 1/15. *Faculty research:* Animal/sediment interaction, microbial geology, sedimentology, coastal geology, hydrogeology. Total annual research budget: $177,995. • Christopher Baldwin, Chairman, 617-353-2532.

Bowling Green State University, College of Arts and Sciences, Department of Geology, Bowling Green, OH 43403. Department awards MS. Part-time programs available. Faculty: 7 full-time (1 woman), 0 part-time. Matriculated students: 15 full-time (8 women), 4 part-time (2 women); includes 0 minority, 2 foreign. Average age 26. 13 applicants, 62% accepted. In 1990, 11 degrees awarded. *Degree requirements:* Thesis or alternative required, foreign language not required. *Entrance requirements:* GRE General Test, TOEFL (minimum score of 550 required). Application deadline: 8/27. Application fee: $10. *Tuition:* $185 per credit hour for state residents; $357 per credit hour for nonresidents. *Financial aid:* In 1990–91, 14 teaching assistantships (8 to first-year students), 1 non-teaching assistantshi (to a first-year student) awarded; fellowships, full tuition waivers, institutionally sponsored loans, and career-related internships or fieldwork also available. Financial aid application deadline: 4/15; applicants required to submit FAF. *Faculty research:* Economic geochemistry, basin analysis, paleontology, geophysics, structure, carbonates. Total annual research budget: $119,000. • Dr. Charles F. Kahle, Chair, 419-372-2886. Application contact: Dr. Charles Onash, Graduate Adviser, 419-372-2886.

See full description on page 217.

Brigham Young University, College of Physical and Mathematical Sciences, Department of Geology, Provo, UT 84602. Department offers programs in earth science teaching (MA), geology (MS). Faculty: 14 full-time (0 women), 2 part-time (0 women). Matriculated students: 21 full-time (1 woman), 0 part-time; includes 1 minority (Hispanic American), 3 foreign. Average age 25. 9 applicants, 56% accepted. In 1990, 4 degrees awarded (75% found work related to degree, 25% continued full-time study). *Degree requirements:* Thesis. *Entrance requirements:* GRE General Test (score in 50th percentile or higher required), minimum GPA of 3.0 during previous 60 credit hours. Application deadline: 3/1 (applications processed on a rolling basis). Application fee: $30. *Tuition:* $1170 per semester (minimum) full-time, $130 per credit hour (minimum) part-time. *Financial aid:* In 1990–91, $60,000 in aid awarded. Fellowships, research assistantships, teaching assistantships, partial tuition waivers, institutionally sponsored loans, and career-related internships or fieldwork available. *Faculty research:* Regional tectonics, hydrogeochemistry, crystal chemistry and crystallography, stratigraphy, environmental geophysics. • Dr. Dana T. Griffen, Chairman, 801-378-3919. Application contact: Dr. Bart J. Kowallis, Graduate Coordinator, 801-378-2467.

Brock University, Faculty of Mathematics and Science, Department of Geological Sciences, St. Catharines, ON L2S 3A1, Canada. Department awards M Sc. Part-time programs available. Faculty: 9 full-time (0 women), 1 part-time (0 women). Matriculated students: 5 full-time, 1 part-time; includes 1 foreign. Average age 24. In 1990, 1 degree awarded. *Degree requirements:* Thesis optional, foreign language not required. *Entrance requirements:* Honor's B Sc in geology. Application deadline: 6/30. Application fee: $0. *Tuition:* $2667 per year full-time, $360 per trimester part-time for Canadian residents; $11,010 per year full-time for nonresidents. *Financial aid:* In 1990–91, $10,000 in aid awarded. 3 research assistantships, 6 teaching assistantships (3 to first-year students) were awarded; federal work-study, institutionally sponsored loans, and career-related internships or fieldwork also

Directory: Geology

available. Financial aid application deadline: 4/30. *Faculty research:* Quaternary geology, petrology, crustal studies, hydrology, carbonate geochemistry. Total annual research budget: $60,000. • Dr. S. Haynes, Chair, 416-688-5550 Ext. 3525. Application contact: Dr. Frank Fueten, Graduate Adviser, 416-688-5550 Ext. 3526.

Brooklyn College of the City University of New York, Department of Geology, Bedford Avenue and Avenue H, Brooklyn, NY 11210. Department awards MA. Faculty: 8 full-time, 0 part-time. Matriculated students: 20 full-time (3 women), 11 part-time (1 woman); includes 8 foreign. Average age 25. 8 applicants, 100% accepted. In 1990, 3 degrees awarded (100% found work related to degree). *Degree requirements:* Computer language, qualifying exam, comprehensive exam required, foreign language and thesis not required. *Entrance requirements:* TOEFL, bachelor's degree in geology or equivalent, fieldwork. Application deadline: 4/1. Application fee: $30. *Expenses:* Tuition of $1202 per semester full-time for state residents; $2350 per semester full-time for nonresidents. Fees of $45 per semester full-time. *Financial aid:* In 1990–91, $20,000 in aid awarded. 4 fellowships, 3 research assistantships, 0 teaching assistantships were awarded; partial tuition waivers, institutionally sponsored loans, and career-related internships or fieldwork also available. *Faculty research:* Geochemistry, petrology, tectonophysics, paleobiology, carbonate sedimentology. Total annual research budget: $225,000. • Dr. John Chamberlain Jr., Chairperson, 718-780-5416.

Brown University, Department of Geological Sciences, Box 1846, Providence, RI 02912. Department awards AM, Sc M, PhD. Faculty: 18 full-time, 1 part-time. Matriculated students: 44 full-time (15 women); includes 8 foreign. 76 applicants, 25% accepted. In 1990, 9 master's, 10 doctorates awarded. *Degree requirements:* For master's, foreign language and thesis not required; for doctorate, dissertation, preliminary exam required, foreign language not required. Application deadline: 1/2. Application fee: $45. *Expenses:* Tuition of $16,256 per year full-time, $2032 per course part-time. Fees of $372 per year. *Financial aid:* In 1990–91, $700,061 in aid awarded. 33 research assistantships, 10 teaching assistantships were awarded; assistantships, full and partial tuition waivers also available. Financial aid application deadline: 1/2; applicants required to submit GAPSFAS. *Faculty research:* Geochemistry, mineral kinetics, igneous and metamorphic petrology, tectonophysics including geophysics and structural geology, climate dynamics. Total annual research budget: $4-million. • Application contact: Richard Yund, Graduate Representative, 401-863-3065.

Bryn Mawr College, Graduate School of Arts and Sciences, Department of Geology, Bryn Mawr, PA 19010. Department awards MA, PhD. Faculty: 5. Matriculated students: 1 (woman) full-time, 5 part-time (4 women). 2 applicants, 100% accepted. In 1990, 0 master's, 0 doctorates awarded. *Degree requirements:* For master's, 1 foreign language, thesis; for doctorate, 2 foreign languages, dissertation. *Entrance requirements:* GRE General Test. Application deadline: 8/20. Application fee: $35. *Tuition:* $13,900 per year full-time, $2350 per course part-time. *Financial aid:* In 1990–91, 0 fellowships, 0 research assistantships, 3 teaching assistantships awarded; federal work-study, institutionally sponsored loans also available. Aid available to part-time students. Financial aid application deadline: 2/1; applicants required to submit GAPSFAS. • Dr. Bruce Saunders, Chairman, 215-526-5114. Application contact: Patricia Saukewitsch, Administrative Coordinator, 215-526-5072.

California Institute of Technology, Division of Geological and Planetary Sciences, Pasadena, CA 91125. Division offers programs in cosmochemistry (PhD), geobiology (PhD), geochemistry (MS, PhD), geology (MS, PhD), geophysics (MS, PhD), planetary science (MS, PhD). Faculty: 28 full-time (0 women), 0 part-time. Matriculated students: 88 full-time (29 women), 0 part-time; includes 12 minority (all Asian American), 23 foreign. Average age 27. In 1990, 12 master's awarded (100% continued full-time study); 10 doctorates awarded. *Entrance requirements:* GRE General Test, GRE Subject Test. Application deadline: 1/15. Application fee: $0. *Expenses:* Tuition of $14,100 per year. Fees of $8 per quarter. *Financial aid:* Fellowships, research assistantships, teaching assistantships available. • Dr. David J. Stevenson, Chairman, 818-356-6108.

See full description on page 219.

California State University, Bakersfield, School of Arts and Sciences, Program in Geology, 9001 Stockdale Highway, Bakersfield, CA 93311. Program awards MS. Matriculated students: 0 full-time, 10 part-time (1 woman); includes 1 minority (Hispanic American). In 1990, 1 degree awarded. *Degree requirements:* Thesis. *Expenses:* Tuition of $360 per semester full-time, $228 per semester part-time. Fees of $164 per unit for nonresidents. • Dr. Robert M. Negrini, Graduate Coordinator, 805-664-2185.

California State University, Chico, College of Natural Sciences, Department of Geology and Physical Sciences, Chico, CA 95929. Department offers program in physical science (MS), including chemistry, earth sciences, geology, hydrology and hydrogeology, mathematics, physics. Faculty: 13 full-time (2 women), 6 part-time (3 women). Matriculated students: 5 full-time (1 woman), 2 part-time (1 woman); includes 1 minority (black American), 0 foreign. Average age 35. 6 applicants, 83% accepted. In 1990, 2 degrees awarded. *Degree requirements:* Thesis required, foreign language not required. *Entrance requirements:* GRE General Test. Application deadline: 4/1 (applications processed on a rolling basis). Application fee: $55. *Expenses:* Tuition of $548 per semester full-time, $350 per semester part-time. Fees of $246 per unit for nonresidents. *Financial aid:* In 1990–91, 1 fellowship, 0 teaching assistantships awarded. • Dr. Howard Stensrud, Chair, 916-898-5262. Application contact: Dr. K. R. Gina Rothe, Graduate Coordinator, 916-898-6269.

California State University, Fresno, Division of Graduate Studies and Research, School of Natural Sciences, Department of Geology, 5241 North Maple Avenue, Fresno, CA 93710. Department awards MS. Faculty: 9 full-time, 4 part-time. Matriculated students: 10. *Degree requirements:* Thesis required, foreign language not required. *Entrance requirements:* GRE General Test (minimum score of 450 on verbal or 430 on quantitative required), undergraduate geology degree. Application fee: $45. *Tuition:* $1098 per year full-time, $4485 per year part-time. *Financial aid:* Teaching assistantships available. • Arthur H. Barabas, Chairman, 209-278-3086. Application contact: Robert Merrill, 209-278-3086.

Announcement: Master's programs in applied geology (comprising hydrogeology, water quality and geochemistry, engineering geology, and metallic ore deposits), sedimentary and structural geology, petrology, and volcanology. Applied geology courses, offered evenings, are geared to working geologists. Good local job opportunities for graduate students. Teaching assistantships available to qualified applicants.

California State University, Hayward, School of Science, Department of Geological Sciences, Hayward, CA 94542. Department offers programs in geology (MS), including environmental geology; marine sciences (MS), including marine sciences. Evening/weekend programs available. Faculty: 5. Matriculated students: 0 full-time, 15 part-time (3 women). In 1990, 3 degrees awarded. *Degree requirements:* Thesis or comprehensive exam required, foreign language not required. *Entrance requirements:* GRE, minimum GPA of 2.75 in field, 2.5 overall. Application deadline: 4/19 (priority date, applications processed on a rolling basis). Application fee: $55. *Expenses:* Tuition of $0 for state residents; $137 per unit for nonresidents. Fees of $895 per year full-time, $188 per quarter part-time. *Financial aid:* Federal work-study, institutionally sponsored loans, and career-related internships or fieldwork available. Aid available to part-time students. Financial aid application deadline: 3/1. • Dr. Sue E. Hirschfeld, Chair, 415-881-3486. Application contact: Dr. Robert Trinchero, Acting Associate Vice President, Admissions and Enrollment, 415-881-3828.

California State University, Long Beach, School of Natural Sciences, Department of Geological Sciences, 1250 Bellflower Boulevard, Long Beach, CA 90840. Department awards MS. Offered jointly with California State University, Northridge and California State University, Los Angeles. Matriculated students: 5 full-time (1 woman), 12 part-time (4 women); includes 1 minority (Asian American), 0 foreign. Average age 31. 7 applicants, 86% accepted. In 1990, 4 degrees awarded. *Degree requirements:* Thesis required, foreign language not required. *Entrance requirements:* GRE General Test. Application deadline: 8/1 (applications processed on a rolling basis). Application fee: $55. *Expenses:* Tuition of $0 for state residents; $246 per unit for nonresidents. Fees of $1120 per year full-time, $724 per year part-time. Financial aid application deadline 3/2; applicants required to submit FAF. • Dr. Stanley C. Finney, Chair, 310-985-4809.

California State University, Los Angeles, School of Natural and Social Sciences, Department of Geology, Los Angeles, CA 90032. Department awards MS. Offered jointly with California State University Northridge. Part-time and evening/weekend programs available. Faculty: 6 full-time, 10 part-time. Matriculated students: 3 full-time (2 women), 48 part-time (13 women); includes 3 minority (1 Asian American, 2 black American), 0 foreign. In 1990, 8 degrees awarded. *Degree requirements:* Thesis or comprehensive exam. *Entrance requirements:* TOEFL (minimum score of 550 required). Application deadline: 8/7 (applications processed on a rolling basis). Application fee: $55. *Expenses:* Tuition of $0 for state residents; $164 per unit for nonresidents. Fees of $1046 per year full-time, $650 per semester (minimum) part-time. *Financial aid:* Federal work-study available. Aid available to part-time students. Financial aid application deadline: 3/1; applicants required to submit FAF. • Dr. Gary Novak, Chair, 213-343-2400.

California State University, Northridge, School of Science and Mathematics, Department of Geological Sciences, Northridge, CA 91330. Department awards MS. Offered jointly with California State University, Los Angeles or Long Beach. Part-time and evening/weekend programs available. Matriculated students: 1 full-time (0 women), 16 part-time (2 women). Average age 31. 9 applicants, 78% accepted. In 1990, 4 degrees awarded. *Degree requirements:* Computer language, thesis required, foreign language not required. *Entrance requirements:* GRE General Test, minimum GPA of 2.75. Application deadline: 11/30. Application fee: $55. *Expenses:* Tuition of $0 for state residents; $205 per unit for nonresidents. Fees of $1128 per year full-time, $366 per semester part-time. *Financial aid:* Research assistantships, teaching assistantships, federal work-study available. *Faculty research:* Petrology of California Miocene volcanics, sedimentology of California Miocene formations, Eocene gastropods, structure of White/Inyo Mountains, seismology of Californian and Mexican earthquakes. Total annual research budget: $23,780. • Dr. Peter Weingand, Chairman, 818-885-3541. Application contact: Dr. A. Eugene Fritsche, Graduate Adviser, 818-885-3541.

Carleton University, Faculty of Science, Ottawa-Carleton Geoscience Centre, Ottawa, ON K1S 5B6, Canada. Center awards M Sc, PhD. M Sc and PhD offered jointly with the University of Ottawa. Evening/weekend programs available. Faculty: 56 full-time. Matriculated students: 78 full-time, 24 part-time; includes 10 foreign. Average age 23. 53 applicants, 26% accepted. In 1990, 18 master's, 9 doctorates awarded. *Degree requirements:* For master's, thesis; for doctorate, dissertation, comprehensive exam. *Entrance requirements:* For master's, honor's degree in science; for doctorate, M Sc. Application deadline: 7/1. Application fee: $10. *Tuition:* $985 per semester full-time, $284 per semester part-time for Canadian residents; $3939 per semester full-time, $1171 per semester part-time for nonresidents. *Financial aid:* In 1990–91, $446,653 in aid awarded. 72 fellowships (14 to first-year students), 64 research assistantships (18 to first-year students), 78 teaching assistantships (18 to first-year students) were awarded. Financial aid application deadline: 3/1. *Faculty research:* Precambrian geology, mineral resources, structural geology, geodynamics, Northern studies. Total annual research budget: $1.4-million. • Keiko Hattori, Director, 613-564-8169. Application contact: Helene De Gouffe, Secretary, 613-564-6834.

Case Western Reserve University, Department of Geological Sciences, Cleveland, OH 44106. Department awards MS, PhD. Part-time programs available. Faculty: 6 full-time (0 women), 0 part-time. Matriculated students: 0 full-time, 3 part-time (2 women); includes 0 minority, 1 foreign. Average age 29. 3 applicants, 0% accepted. In 1990, 1 master's awarded (100% found work related to degree); 0 doctorates awarded. Terminal master's awarded for partial completion of doctoral program. *Degree requirements:* For master's, thesis or alternative required, foreign language not required; for doctorate, dissertation required, foreign language not required. *Entrance requirements:* GRE General Test, GRE Subject Test, TOEFL (minimum score of 550 required). Application deadline: 3/31 (priority date, applications processed on a rolling basis). Application fee: $25. *Expenses:* Tuition of $13,600 per year full-time, $567 per credit part-time. Fees of $320 per year. *Financial aid:* In 1990–91, 3 students received a total of $21,936 in aid awarded. 2 research assistantships (0 to first-year students), 1 teaching assistantship (to a first-year student) were awarded; federal work-study also available. Average monthly stipend for a graduate assistantship: $750. Financial aid application deadline: 3/31; applicants required to submit FAF. *Faculty research:* Geochemistry, environmental studies and benthic ecology, regional geology and geochronology, sedimentology, paleoecology. Total annual research budget: $460,000. • Steven Stanley, Chairman, 216-368-3690. Application contact: Gerald Matisoff, Graduate Committee, 216-368-3677.

Centenary College of Louisiana, Department of Geology, Shreveport, LA 71134. Department awards MS. Part-time and evening/weekend programs available. Faculty: 4 full-time, 1 part-time. In 1990, 0 degrees awarded. *Degree requirements:* Foreign language not required. Application fee: $20. *Tuition:* $525 per course. • Dr. Austin Sartin, Chairman, 318-869-5234.

City College of the City University of New York, Graduate School, College of Liberal Arts and Science, Division of Science, Department of Earth and Planetary Sciences, Convent Avenue at 138th Street, New York, NY 10031. Department offers programs in earth and environmental science (PhD), geology (MA). PhD offered through the Graduate School and University Center of the City University of New York. Matriculated students: 0. 4 applicants, 25% accepted. In 1990, 2 master's awarded. *Degree requirements:* For master's, thesis, comprehensive exam required, foreign language not required. *Entrance requirements:* For master's, TOEFL (minimum score of 300 required), appropriate bachelor's degree. Application deadline: 6/1 (priority date, applications processed on a rolling basis). Application fee: $30. *Expenses:* Tuition of $2204 per year full-time, $95 per credit part-time for state residents; $4700 per year full-time, $199 per credit part-time for nonresidents. Fees of $15 per semester. *Financial aid:* Fellowships and career-related internships or fieldwork available. *Faculty research:* Water resources, high-temperature geochemistry, sedimentary basin analysis, tectonics. • Dr. Dennis Weiss, Chairman, 212-650-6984. Application contact: O. Lehn Franke, Adviser, 212-650-6984.

Clemson University, College of Sciences, Department of Earth Sciences, Clemson, SC 29634. Department offers program in geology (MS). Faculty: 4 full-time (0 women), 3 part-time (1 woman). *Entrance requirements:* GRE General Test (minimum combined score of 1500 required, minimum GPA of 3.0 in last 2 years. Application deadline: 6/1. Application fee: $25. *Expenses:* Tuition of $102 per credit hour. Fees of $80 per semester full-time. *Faculty research:* Ground water geology, envrionmental geology, geochemistry, structural geology, stratigraphy. Total annual research budget: $45,000. • R. D. Warner, Graduate Coordinator, 803-656-5023.

Colgate University, Department of Geology, Hamilton, NY 13346. Department awards MA. Part-time programs available. Faculty: 2 full-time (0 women), 0 part-time. Matriculated students: 0. Average age 30. 0 applicants. In 1990, 0 degrees awarded. *Degree requirements:* Thesis required, foreign language not required. *Entrance requirements:* GRE General Test. Application fee: $40. *Tuition:* $1779 per course. *Financial aid:* Research assistantships, partial tuition waivers, institutionally sponsored loans available. Financial aid application deadline: 2/1; applicants required to submit FAF. *Faculty research:* Geochemistry, clay mineralogy, sedimentology, sedimentary petrology. Total annual research budget: $125,000. • Dr. Art Goldstein, Chairman, 315-824-7203.

Colorado School of Mines, Department of Geology and Geological Engineering, Golden, CO 80401. Department offers programs in geological engineering (ME, MS, PhD), geology (MS, PhD). Faculty: 19 full-time (3 women), 18 part-time (0 women). Matriculated students: 71 full-time (6 women), 68 part-time (27 women); includes 26 foreign. 70 applicants, 63% accepted. In 1990, 20 master's, 5 doctorates awarded. *Degree requirements:* For master's, thesis required (for some programs), foreign language not required; for doctorate, 1 foreign language, dissertation, comprehensive exam. *Entrance requirements:* GRE General Test, GRE Subject Test, minimum GPA of 3.0. Application deadline: 4/15 (priority date, applications processed on a rolling basis). Application fee: $15 ($25 for foreign students). *Expenses:* Tuition of $3178 per year full-time, $124 per semester hour part-time for state residents; $10,304 per year full-time, $344 per semester hour part-time for nonresidents. Fees of $374 per year full-time. *Financial aid:* In 1990–91, 15 fellowships, 6 research assistantships, 13 teaching assistantships awarded. *Faculty research:* Mineral exploration, stratigraphy, hydrogeology. Total annual research budget: $827,026. • Dr. Gregory S. Holden, Head, 303-273-3800.

Colorado State University, College of Forestry and Natural Resources, Department of Earth Resources, Program in Geology, Fort Collins, CO 80523. Program awards MS. Part-time programs available. Faculty: 14 full-time (0 women), 0 part-time. Matriculated students: 14 full-time (2 women), 12 part-time (2 women); includes 1 minority (Asian American), 4 foreign. Average age 27. 40 applicants, 40% accepted. In 1990, 9 degrees awarded. *Degree requirements:* Computer language, thesis required, foreign language not required. *Entrance requirements:* GRE General Test (minimum score of 600 on each section required), TOEFL (minimum score of 550 required), minimum GPA of 3.0. Application deadline: 4/1 (priority date, applications processed on a rolling basis). Application fee: $30. *Tuition:* $1322 per semester full-time for state residents; $3673 per semester full-time for nonresidents. *Financial aid:* In 1990–91, 0 fellowships, 0 research assistantships, 14 teaching assistantships (1 to a first-year student), 1 traineeship (0 to first-year students) awarded; federal work-study also available. Average monthly stipend for a graduate assistantship: $698-750. Financial aid application deadline: 2/28. *Faculty research:* Economic geology, fluvial geomorphology, sedimentology, mineralogy, structural geology. • Dr. Donald O. Doehring, Head, Department of Earth Resources, 303-491-5661. Application contact: Barbara Holtz, 301-491-5662.

Columbia University, Graduate School of Arts and Sciences, Division of Natural Sciences, Department of Geological Sciences, New York, NY 10027. Department offers programs in geochemistry (MA, M Phil, PhD), geodetic sciences (MA, M Phil, PhD), geophysics (MA, M Phil, PhD), oceanography (MA, M Phil, PhD). Faculty: 21 full-time, 19 part-time. Matriculated students: 100 full-time (35 women), 0 part-time; includes 2 minority (both Hispanic American), 19 foreign. Average age 27. 100 applicants, 42% accepted. In 1990, 19 master's, 12 doctorates awarded. *Degree requirements:* For master's, thesis or alternative, fieldwork, written exam required, foreign language not required; for doctorate, 1 foreign language, dissertation. *Entrance requirements:* GRE General Test, GRE Subject Test, TOEFL, major in natural or physical science. Application deadline: 1/5. Application fee: $50. *Expenses:* Tuition of $7836 per semester for state residents; $426 per credit for nonresidents. Fees of $148 per semester for state residents. *Financial aid:* In 1990–91, $1.34-million in aid awarded. 10 fellowships (0 to first-year students), 65 teaching assistantships (11 to first-year students) were awarded; federal work-study, institutionally sponsored loans also available. Financial aid application deadline: 1/5; applicants required to submit GAPSFAS. *Faculty research:* Structural geology and stratigraphy, petrology, paleontology, rare gas, isotope and aqueous geochemistry. • Dennis Hayes, Chair, 212-854-4525.

Cornell University, Graduate Fields of Arts and Sciences, Field of Geological Sciences, Ithaca, NY 14853. Field offers programs in economic geology (M Eng, MS, PhD), engineering geology (M Eng, MS, PhD), general geology (M Eng, MS, PhD), geobiology (M Eng, MS, PhD), geochemistry and isotope geology (M Eng, MS, PhD), geohydrology (M Eng, MS, PhD), geomorphology (M Eng, MS, PhD), geophysics (M Eng, MS, PhD), geotectonics (M Eng, MS, PhD), mineralogy (M Eng, MS, PhD), paleontology (M Eng, MS, PhD), petroleum geology (M Eng, MS, PhD), petrology (M Eng, MS, PhD), planetary geology (M Eng, MS, PhD), Precambrian geology (M Eng, MS, PhD), quarternary geology (M Eng), quaternary geology (MS, PhD), rock mechanics (M Eng, MS, PhD), sedimentology (M Eng, MS, PhD). Faculty: 31 full-time, 0 part-time. Matriculated students: 36 full-time (5 women), 0 part-time; includes 0 minority, 9 foreign. 108 applicants, 16% accepted. In 1990, 3 master's, 14 doctorates awarded. Terminal master's awarded for partial completion of doctoral program. *Degree requirements:* For master's, thesis required, foreign language not required; for doctorate, 1 foreign language, dissertation. *Entrance requirements:* GRE General Test, TOEFL. Application deadline: 1/10. Application fee: $55. *Expenses:* Tuition of $16,170 per year. Fees of $28 per year. *Financial aid:* In 1990–91, 27 students received aid. 1 fellowship (to a first-year student), 22 research assistantships (4 to first-year students), 2 teaching assistantships (1 to a first-year student), 2 graduate assistantships (0 to first-year students) awarded; full and partial tuition waivers, federal work-study, institutionally sponsored loans, and career-related internships or fieldwork also available. Financial aid application deadline: 1/10; applicants required to submit GAPSFAS. *Faculty research:* Geophysics and mineral physics; structural geology and geotectonics;, geochemistry and petrology; planetology and geomorphology; paleontology and geohydrology. • William A. Bassett, Graduate Faculty Representative, 607-255-7502.

Announcement: Research and study include geotectonics and structural geology, seismology, seismic reflection profiling, structure of faults, physical properties and mechanical states of earth materials over a wide range of temperature and pressure, isotopic and trace-element studies, petrology and geochemistry, mineral resources, ecology and evolution of fossil organisms, geomorphology, recent crustal movements, sediments of orogenic zones, petroleum geology, planetary geology, mantle convection and geomechanics, geohydrology. Special emphasis placed on relations of these to plate tectonics and continental evolution. Contact Professor W. A. Bassett, Snee Hall, 607-255-7502.

Dalhousie University, College of Arts and Science, Faculty of Science, Department of Geology, Halifax, NS B3H 4H6, Canada. Department awards M Sc, PhD. Part-time programs available. Faculty: 13 full-time, 2 part-time. Matriculated students: 28 full-time, 5 part-time. 50 applicants, 30% accepted. In 1990, 6 master's, 1 doctorate awarded. *Degree requirements:* 1 foreign language, thesis/dissertation. *Entrance requirements:* For master's, TOEFL (minimum score of 550 required); for doctorate, TOEFL (minimum score of 550 required), M Sc. Application deadline: 7/15 (applications processed on a rolling basis). Application fee: $20. *Tuition:* $2594 per year full-time for Canadian residents; $4294 per year full-time for nonresidents. *Financial aid:* In 1990–91, $207,936 in aid awarded. Fellowships and career-related internships or fieldwork available. *Faculty research:* Marine geology and geophysics, Appalachian geology, metallurgy. • Dr. P. J. C. Ryall, Chair, 902-494-2358.

Dartmouth College, School of Arts and Sciences, Department of Earth Sciences, Hanover, NH 03755. Department offers program in geology (MS, PhD). Faculty: 9 full-time (1 woman), 1 part-time (0 women). Matriculated students: 21 full-time (6 women), 0 part-time; includes 2 minority (1 black American, 1 Hispanic American), 3 foreign. 44 applicants, 27% accepted. In 1990, 5 master's awarded (20% entered university research/teaching, 60% found other work related to degree, 20% continued full-time study); 1 doctorate awarded (100% found work related to degree). Terminal master's awarded for partial completion of doctoral program. *Degree requirements:* For master's, thesis; for doctorate, variable foreign language requirement, dissertation. *Entrance requirements:* GRE General Test, GRE Subject Test. Application deadline: 3/1. *Expenses:* Tuition of $16,230 per year. Fees of $650 per year. *Financial aid:* In 1990–91, $307,380 in aid awarded. 11 fellowships (7 to first-year students), 5 research assistantships were awarded; full tuition waivers, federal work-study, institutionally sponsored loans, and career-related internships or fieldwork also available. *Total annual research budget:* $586,988. • Charles Drake, Chairman, 603-646-2373.

Announcement: The Dartmouth graduate program offers both MS and PhD degree programs in geology. Active research, generally combining field and laboratory/theoretical work, covers petrology, isotope geochemistry, geophysics/tectonic modeling, stratigraphy, clay mineralogy, hydrogeology, mineral deposits, remote sensing, and history of geology. The department has about 20 graduate students, divided evenly between MS and PhD candidates. All graduate students receive tuition fellowships and full stipend.

Duke University, Graduate School, Department of Biological Anthropology and Anatomy, Durham, NC 27706. Offerings include gross anatomy and physical anthropology (PhD), with options in comparative morphology of human and non-human primates, primate social behavior, vertebrate paleontology. Department faculty: 13 full-time, 2 part-time. *Degree requirements:* 1 foreign language, dissertation. *Entrance requirements:* GRE General Test. Application deadline: 1/31. Application fee: $50. *Expenses:* Tuition of $8640 per year full-time, $360 per unit part-time. Fees of $1356 per year full-time. • Kathleen Smith, Director of Graduate Studies, 919-684-3887.

Duke University, Graduate School, Department of Geology, Durham, NC 27706. Department awards MS, PhD. Faculty: 11 full-time, 4 part-time. Matriculated students: 29 full-time (6 women), 3 part-time (1 woman); includes 0 minority, 9 foreign. 20 applicants, 85% accepted. In 1990, 6 master's, 1 doctorate awarded. *Degree requirements:* Thesis/dissertation. *Entrance requirements:* GRE General Test. Application deadline: 1/31. Application fee: $50. *Expenses:* Tuition of $8640 per year full-time, $360 per unit part-time. Fees of $1356 per year full-time. *Financial aid:* In 1990–91, $81,202 in aid awarded. Federal work-study available. Financial aid application deadline: 1/31; applicants required to submit GAPSFAS. • Paul Baker, Director of Graduate Studies, 919-684-5847.

Announcement: Geology department offers instruction and research opportunities in the areas of continental margin, rift-lake, and deep-sea sedimentation; clastic and carbonate facies analysis; sediment dynamics; igneous petrology and geochemistry; midocean-ridge basalt geochemistry; structure and development of transform faults, rift basins, spreading centers, passive margins; isotope geochemistry; carbonate diagenesis; coastal-zone management; paleoecology; paleoceanography; marine micropaleontology; remote sensing; earthquake seismology; computer and mathematical applications to geology; seismic stratigraphy; and geohydrology.

East Carolina University, College of Arts and Sciences, Department of Geology, Greenville, NC 27858-4353. Department awards MS. Part-time programs available. Faculty: 7. Matriculated students: 7 full-time (3 women), 7 part-time (0 women); includes 0 minority, 1 foreign. 9 applicants, 100% accepted. In 1990, 4 degrees awarded. *Degree requirements:* 1 foreign language (computer language can substitute), thesis, comprehensive exams. *Entrance requirements:* GRE General Test, TOEFL. Application deadline: 6/1 (priority date, applications processed on a rolling basis). Application fee: $25. *Tuition:* $627 per semester for state residents; $3154 per semester for nonresidents. *Financial aid:* In 1990–91, $70,000 in aid awarded. Research assistantships, teaching assistantships available. Aid available to part-time students. Financial aid application deadline: 6/1. • Dr. Stanley Riggs, Director of Graduate Studies, 919-757-6379. Application contact: Paul D. Tschetler, Assistant Dean, 919-757-6012.

Eastern Kentucky University, College of Natural and Mathematical Sciences, Department of Geology, Richmond, KY 40475. Department offers programs in geology (MS), hydrogeology (MS), mining (MS), petroleum (MS). Part-time programs available. Matriculated students: 1 full-time (0 women), 4 part-time (2 women); includes 0 minority, 0 foreign. Average age 25. In 1990, 10 degrees awarded. *Degree requirements:* Thesis. *Entrance requirements:* GRE General Test, minimum GPA of 2.5. Application fee: $0. *Tuition:* $1440 per year full-time, $88 per credit hour part-time for state residents; $4320 per year full-time, $248 per credit hour part-time for nonresidents. *Financial aid:* Research assistantships, teaching assistantships, federal work-study available. Aid available to part-time students. • Dr. Gary Kuhnhenn, Chair, 606-622-1273.

Eastern Washington University, College of Science, Mathematics and Technology, Department of Geology, Cheney, WA 99004. Department awards MS. Faculty: 9 full-time (1 woman), 0 part-time. 13 applicants, 46% accepted. In 1990, 5 degrees awarded. *Degree requirements:* Thesis, comprehensive oral exam. *Entrance requirements:* GRE General Test, minimum GPA of 3.0. Application deadline: rolling. Application fee: $25. *Tuition:* $2700 per year full-time, $180 per credit (minimum) part-time for state residents; $8187 per year full-time, $546 per credit (minimum) part-time for nonresidents. *Financial aid:* In 1990–91, 3 research assistantships, 3 teaching assistantships awarded; federal work-study, institutionally sponsored loans, and career-related internships or fieldwork also available. Financial aid applicants required to submit FAF. • Dr. Eugene Kiver, Chair, 509-359-2286. Application contact: Dr. William Steele, Graduate Director, 509-359-7494.

Florida Atlantic University, College of Science, Department of Geology, Boca Raton, FL 33431. Department awards MS. Faculty: 8 full-time (1 woman), 3 part-time (2 women). Matriculated students: 10 full-time (5 women), 4 part-time (2 women). *Degree requirements:* 1 foreign language, thesis. *Entrance requirements:* GRE General Test (minimum combined score of 1000 required), minimum GPA of 3.0. Application fee: $15. *Tuition:* $89.28 per credit hour for state residents; $291 per credit hour for nonresidents. *Faculty research:* Water paleontology, beach erosion, stratigraphy, hydrogeology. • Roy Lemon, Chair, 407-367-3311.

Florida International University, College of Arts and Sciences, Department of Geology, Miami, FL 33199. Department awards MS. Part-time and evening/weekend programs available. *Entrance requirements:* TOEFL. Application deadline: 4/1. Application fee: $15. *Tuition:* $462 per semester (minimum) full-time, $38.50 per credit part-time for state residents; $2217 per semester (minimum) full-time, $185 per credit part-time for nonresidents.

Florida State University, College of Arts and Sciences, Department of Geology, Tallahassee, FL 32306. Department offers programs in geology (MS, PhD), geophysical fluid dynamics (PhD). Faculty: 16 full-time, 0 part-time. Matriculated students: 11 full-time (1 woman), 22 part-time (5 women); includes 1 minority (Asian American), 3 foreign. Average age 25. 20 applicants, 20% accepted. In 1990, 3 master's awarded (33% entered university research/teaching, 67% found other work related to degree); 4 doctorates awarded (75% entered university research/teaching, 25% found other work related to degree). *Degree requirements:* For master's, thesis, departmental qualifying exam required, foreign language not required; for doctorate, 1 foreign language, dissertation, departmental qualifying exam. *Entrance requirements:* GRE General Test (minimum combined score of 1000 required), GRE Subject Test, minimum GPA of 3.0. Application deadline: 7/19 (applications processed on a rolling basis). Application fee: $15. *Tuition:* $76.29 per credit hour for state residents; $238 per credit hour for nonresidents. *Financial aid:* In 1990–91, 11 students received a total of $20,699 in aid awarded. 1 fellowship (to a first-year student), 1 research assistantship (0 to first-year students), 10 teaching assistantships (1 to a first-year student) were awarded; career-related internships or fieldwork also available. Average monthly stipend for a graduate assistantship: $700. Financial aid application deadline: 2/15. *Faculty research:* Igneous and metamorphic petrology of Southern Appalachians, invertebrate marine paleoecology, uranium series applications to geochronology. Total annual research budget: $100,000. • Dr. James B. Cowart, Chairman, 904-644-5860.

Fort Hays State University, College of Arts and Sciences, Department of Earth Science, Hays, KS 67601. Department offers program in geology (MS). Part-time programs available. Faculty: 6 full-time (0 women), 0 part-time. Matriculated students: 5 full-time (0 women), 4 part-time (1 woman); includes 0 minority, 0 foreign. Average age 27. 8 applicants, 75% accepted. In 1990, 4 degrees awarded. *Degree requirements:* Thesis required, foreign language not required. *Application deadline:* 7/1 (priority date, applications processed on a rolling basis). *Application fee:* $0. *Tuition:* $58.75 per credit for state residents; $140 per credit for nonresidents. *Financial aid:* In 1990–91, $16,378 in aid awarded. 1 research assistantship, 8 teaching assistantships were awarded; institutionally sponsored loans and career-related internships or fieldwork also available. Aid available to part-time students. Financial aid applicants required to submit FAF. *Faculty research:* Isotope dating of volcanic rock, chemical analysis of volcanic rock, water resources. • Dr. Michael Nelson, Chair, 913-628-5389.

George Washington University, Graduate School of Arts and Sciences, Department of Geology, Washington, DC 20052. Offerings include geology (MS, PhD). Terminal master's awarded for partial completion of doctoral program. Department faculty: 5 full-time (1 woman), 1 (woman) part-time. *Degree requirements:* For master's, thesis or alternative, comprehensive exam; for doctorate, dissertation, general exam. *Entrance requirements:* For master's, GRE General Test, bachelor's degree in field, minimum GPA of 3.0; for doctorate, GRE General Test, minimum GPA of 3.0. Application deadline: 7/1. Application fee: $45. *Expenses:* Tuition of $490 per semester hour. Fees of $215 per year full-time, $125.60 per year (minimum) part-time. • Dr. George C. Stephens, Chair, 202-994-6190.

Georgia State University, College of Arts and Sciences, Department of Geology, Atlanta, GA 30303. Department offers programs in geology (MS), hydrogeology (Certificate). Part-time and evening/weekend programs available. Faculty: 7 full-time (0 women), 2 part-time (0 women). Matriculated students: 5 full-time (3 women), 19 part-time (7 women); includes 0 minority, 3 foreign. 25 applicants, 32% accepted. In 1990, 3 master's, 3 Certificates awarded. *Degree requirements:* For master's, 1 foreign language (computer language can substitute), thesis, comprehensive exam; for Certificate, foreign language not required. *Entrance requirements:* For master's, GRE General Test, TOEFL (minimum score of 550 required), minimum GPA of 3.0; for Certificate, TOEFL (minimum score of 500 required), minimum GPA of 3.0. Application deadline: 7/15. Application fee: $10. *Expenses:* Tuition of $38 per quarter hour for state residents; $130 per quarter hour for nonresidents. Fees of $58 per quarter. *Financial aid:* In 1990–91, $42,000 in aid awarded. 5 research assistantships (2 to first-year students), 7 teaching assistantships (3 to first-year students) were awarded; partial tuition waivers, federal work-study, institutionally sponsored loans, and career-related internships or fieldwork also available. Aid available to part-time students. Financial aid application deadline: 7/15. *Faculty research:* Petrology; sedimentology; structural geology; groundwater chemistry, resources, and protection; coastal resources. Total annual research budget: $285,000. • Dr. Vernon J. Henry, Jr., Chair, 404-651-2272. Application contact: Dr. Seth Rose, Director of Graduate Studies, 404-651-2272.

Announcement: Most traditional geology study areas are available. The department offers specialized programs in hydrogeology or environmental geology and operates a well-equipped marine lab on Skidaway Island. The department also works closely with the nearby Georgia Geologic Survey and the U.S. Geologic Survey. Students may enter the program at the beginning of any quarter, and financial support is available for qualified students. Georgia State University is an equal opportunity/affirmative action institution.

Harvard University, Graduate School of Arts and Sciences, Department of Earth and Planetary Sciences, Cambridge, MA 02138. Department awards AM, PhD. Faculty: 12 full-time (0 women), 2 part-time (1 woman). Matriculated students: 31 full-time (7 women), 0 part-time; includes 0 minority, 10 foreign. In 1990, 2 master's awarded (100% continued full-time study); 6 doctorates awarded (33% entered university research/teaching, 67% continued full-time study). Terminal master's awarded for partial completion of doctoral program. *Degree requirements:* For master's, 1 foreign language, thesis or alternative; for doctorate, 1 foreign language, dissertation. *Entrance requirements:* GRE General Test, GRE Subject Test. Application deadline: 1/2. Application fee: $60. *Expenses:* Tuition of $14,860 per year. Fees of $550 per year. *Financial aid:* In 1990–91, $180,233 in aid awarded. 17 fellowships (9 to first-year students), 31 research assistantships (5 to first-year students), 24 teaching assistantships (9 to first-year students) were awarded; federal work-study, institutionally sponsored loans, and career-related internships or fieldwork also available. Financial aid application deadline: 1/2; applicants required to submit GAPSFAS. *Faculty research:* Economic geography, geochemistry, geophysics, mineralogy, crystallography. • Dr. Rick O'Connell, Director of Graduate Studies, 617-495-2351. Application contact: Office of Admissions and Financial Aid, 617-495-5315.

See full description on page 221.

Idaho State University, College of Arts and Sciences, Department of Geology, Pocatello, ID 83209. Department offers programs in geology (MS), natural science (MNS). Faculty: 5 full-time, 1 part-time. Matriculated students: 12 full-time (2 women), 9 part-time (3 women); includes 1 minority (Asian American), 0 foreign. In 1990, 4 degrees awarded. *Degree requirements:* Thesis required, foreign language not required. *Application deadline:* 8/1 (priority date). *Application fee:* $10. *Expenses:* Tuition of $0 full-time, $1060 per semester part-time for state residents; $0 for nonresidents. Fees of $809 per semester full-time, $72.50 per credit part-time. *Financial aid:* In 1990–91, 4 teaching assistantships (2 to first-year students) awarded. • Dr. Paul K. Link, Chairman, 208-236-3846.

Indiana State University, College of Arts and Sciences, Department of Geography and Geology, Program in Geology, Terre Haute, IN 47809. Program awards MA. Matriculated students: 3 full-time (0 women), 1 part-time (0 women); includes 0 minority, 1 foreign. 3 applicants, 33% accepted. In 1990, 1 degree awarded. *Entrance requirements:* GRE General Test. Application deadline: rolling. Application fee: $10 ($20 for foreign students). *Tuition:* $90 per hour for state residents; $199 per hour for nonresidents. *Financial aid:* Teaching assistantships available. Financial aid application deadline: 3/1. • Dr. William Dando, Chairperson, Department of Geography and Geology, 812-237-2261.

Indiana University Bloomington, College of Arts and Sciences, Department of Geological Sciences, Bloomington, IN 47405. Department awards MAT, MS, PhD, MSES/MS. Programs offered in earth science education (MAT), geochemistry (MS, PhD), geophysics (MS, PhD), regional analysis and planning (PhD). Part-time programs available. Faculty: 24 full-time (1 woman), 6 part-time (0 women). Matriculated students: 56 full-time (11 women), 33 part-time (7 women); includes 2 minority (both Asian American), 12 foreign. In 1990, 19 master's awarded (63% found work related to degree, 37% continued full-time study); 1 doctorate awarded (100% found work related to degree). *Degree requirements:* For master's, 1 foreign language, computer language; for doctorate, dissertation. *Entrance requirements:* GRE General Test. Application deadline: 4/15. Application fee: $25 ($35 for foreign students). *Tuition:* $99.85 per credit hour for state residents; $288 per credit hour for nonresidents. *Financial aid:* In 1990–91, $169,855 in aid awarded. 3 fellowships (all to first-year students), 23 teaching assistantships (11 to first-year students) were awarded; full tuition waivers, federal work-study, institutionally sponsored loans, and career-related internships or fieldwork also available. Financial aid application deadline: 2/15; applicants required to submit FAF. *Total annual research budget:* $250,000. • Lee J. Suttner, Chairman, 812-855-5581. Application contact: David G. Towell, Chairman, 812-855-7214.

See full description on page 223.

Indiana University–Purdue University at Indianapolis, School of Science, Department of Geology, Indianapolis, IN 46205. Department awards MS. Part-time and evening/weekend programs available. Faculty: 5 full-time (0 women), 4 part-time (0 women). Matriculated students: 2 full-time (1 woman), 17 part-time (2 women). Average age 34. 0 applicants. In 1990, 0 degrees awarded. *Degree requirements:* Computer language, thesis (for some programs) required, foreign language not required. *Application fee:* $20. *Tuition:* $100 per credit for state residents; $288 per credit for nonresidents. *Financial aid:* In 1990–91, 6 fellowships (0 to first-year students), 1 teaching assistantship (to a first-year student), 2 lecturers (0 to first-year students) awarded. *Faculty research:* Hazardous waste disposal, solid waste disposal, fracture permeability, groundwater contamination. • Arthur Mirsky, Chair, 317-274-7484. Application contact: Pascal de Caprariis, 317-274-7484.

Iowa State University of Science and Technology, College of Liberal Arts and Sciences, Department of Geological and Atmospheric Sciences, Ames, IA 50011. Department offers programs in earth science (MS, PhD), geology (MS, PhD), meteorology (MS, PhD). Matriculated students: 17 full-time (2 women), 11 part-time (1 woman); includes 1 minority (Asian American), 15 foreign. In 1990, 3 master's, 4 doctorates awarded. *Degree requirements:* For master's, thesis optional; for doctorate, dissertation. *Application fee:* $20 ($30 for foreign students). *Expenses:* Tuition of $1158 per semester full-time, $129 per credit part-time for state residents; $3340 per semester full-time, $372 per credit part-time for nonresidents. Fees of $10 per semester. *Financial aid:* In 1990–91, 1 fellowship (0 to first-year students), 19 research assistantships (6 to first-year students), 19 teaching assistantships (10 to first-year students), 9 scholarships (3 to first-year students) awarded. • Dr. Karl E. Seifert, Chair, 515-294-4478.

See full description on page 225.

Johns Hopkins University, School of Arts and Sciences, Department of Earth and Planetary Sciences, Program in Geology, Baltimore, MD 21218. Program awards MA, PhD. Faculty: 9 full-time (0 women). Matriculated students: 16. Average age 24. Terminal master's awarded for partial completion of doctoral program. *Degree requirements:* For doctorate, 1 foreign language, dissertation. *Entrance requirements:* GRE General Test. Application deadline: 2/1. Application fee: $40. *Expenses:* Tuition of $15,500 per year full-time, $1550 per course part-time. Fees of $400 per year full-time. *Financial aid:* Federal work-study, institutionally sponsored loans available. Financial aid application deadline: 3/14; applicants required to submit FAF. • Bruce Marsh, Chairman, Department of Earth and Planetary Sciences, 301-338-7135.

Kansas State University, College of Arts and Sciences, Department of Geology, Manhattan, KS 66506. Department awards MS, PhD. PhD offered jointly with the University of Kansas. Faculty: 12 full-time (0 women), 0 part-time. Matriculated students: 6 full-time (0 women), 4 part-time (0 women); includes 0 foreign. 9 applicants, 56% accepted. In 1990, 7 master's, 0 doctorates awarded. *Degree requirements:* For master's, thesis required, foreign language not required. *Entrance*

Directory: Geology

requirements: For master's, GRE General Test, GRE Subject Test, TOEFL. Application deadline: 3/15. *Expenses:* Tuition of $1668 per year full-time, $51 per credit hour part-time for state residents; $5382 per year full-time, $142 per credit hour part-time for nonresidents. Fees of $305 per year full-time, $64.50 per semester part-time. *Financial aid:* In 1990–91, 2 research assistantships, 6 teaching assistantships awarded; scholarships also available. Financial aid application deadline: 3/15. • Joseph L. Graf Jr., Head, 913-532-6724.

Kent State University, Department of Geology, Kent, OH 44242. Department awards MS, PhD. Faculty: 15 full-time, 0 part-time. Matriculated students: 30 full-time (9 women), 26 part-time (9 women). 19 applicants, 42% accepted. In 1990, 13 master's, 3 doctorates awarded. *Degree requirements:* For master's, thesis required, foreign language not required; for doctorate, 1 foreign language, dissertation. *Entrance requirements:* For doctorate, GRE General Test, GRE Subject Test. Application deadline: 7/12 (applications processed on a rolling basis). Application fee: $25. *Tuition:* $1601 per semester full-time, $133.75 per hour part-time for state residents; $3101 per semester full-time, $258.75 per hour part-time for nonresidents. *Financial aid:* Fellowships, research assistantships, teaching assistantships, full tuition waivers, federal work-study available. Financial aid application deadline: 2/1. *Faculty research:* Applied geology (ground water, surface water, engineering geology, petroleum geology, exploration geology, geophysics). • Dr. Richard A. Heimlich, Chairman, 216-672-2680. Application contact: Dr. Ernest H. Carlson, Graduate Coordinator, 216-672-3778.

Lakehead University, Faculty of Arts and Science, Department of Geology, Thunder Bay, ON P7B 5E1, Canada. Department awards M Sc. Part-time and evening/weekend programs available. Faculty: 7 full-time (1 woman), 0 part-time. Matriculated students: 6 full-time (2 women), 3 part-time (1 woman); includes 0 foreign. 18 applicants, 28% accepted. In 1990, 5 degrees awarded. *Degree requirements:* Thesis. *Entrance requirements:* TOEFL (minimum score of 550 required). Application fee: $0. *Expenses:* Tuition of $1969 per year full-time, $527 per year part-time for Canadian residents; $3712 per year full-time, $2618 per year part-time for nonresidents. Fees of $2735 per year full-time, $100 per year part-time for Canadian residents; $11,136 per year full-time, $100 per year part-time for nonresidents. *Financial aid:* In 1990–91, 6 students received a total of $47,732 in aid awarded. 6 teaching assistantships (2 to first-year students) were awarded; fellowships, entrance awards and research stipends also available. Average monthly stipend for a graduate assistantship: $680. Financial aid application deadline: 3/30. *Faculty research:* Structural geology, petrology-kimber lites, earth resources and geochemical cycles, genesis of banded iron formations. Total annual research budget: $139,000. • Dr. Graham J. Borradaile, Chairman, 807-343-8328.

Laurentian University, Programme in Geology, Sudbury, ON P3E 2C6, Canada. Program awards M Sc. Part-time programs available. Faculty: 9 full-time (0 women), 0 part-time. Matriculated students: 3 full-time (1 woman), 4 part-time (0 women); includes 3 foreign. 10 applicants, 40% accepted. In 1990, 1 degree awarded. *Degree requirements:* Thesis or alternative required, foreign language not required. *Entrance requirements:* Honor's bachelor's with second class or better. *Tuition:* $1926 per year full-time, $364 per course part-time for Canadian residents; $11,138 per year full-time for nonresidents. *Financial aid:* In 1990–91, $87,780 in aid awarded. 2 fellowships (both to first-year students), 5 teaching assistantships were awarded; research assistantships, partial tuition waivers, institutionally sponsored loans also available. *Faculty research:* Pre-Cambrian minerals, Archean greenstone belts, Huronian, early Paleozoic biostratigraphy-paleontology. Total annual research budget: $190,103. • Dr. A. E. Beswick, Chairman, 705-675-1151 Ext. 2265. Application contact: Admissions Department, 705-675-1151 Ext. 3915.

Lehigh University, College of Arts and Sciences, Department of Geological Sciences, Bethlehem, PA 18015. Department awards MS, PhD. Faculty: 9 full-time (0 women), 0 part-time. Matriculated students: 13 full-time (2 women), 2 part-time (0 women); includes 4 foreign. 26 applicants, 69% accepted. In 1990, 2 master's, 2 doctorates awarded. *Degree requirements:* For master's, thesis required, foreign language not required; for doctorate, dissertation, language at the discretion of the PhD committee. *Entrance requirements:* GRE, TOEFL. Application deadline: rolling. Application fee: $40. *Tuition:* $15,650 per year full-time, $655 per credit hour part-time. *Financial aid:* Fellowships, research assistantships, teaching assistantships available. • Dr. Bobb Carson, Chairman, 215-758-3660. Application contact: Dr. Carl Moses, Graduate Program Coordinator, 215-758-4907.

Loma Linda University, Department of Paleontology, Loma Linda, CA 92350. Department awards MS. Part-time programs available. Matriculated students: 2 full-time (1 woman), 4 part-time (3 women); includes 2 minority (1 Asian American, 1 Hispanic American), 0 foreign. 2 applicants, 100% accepted. In 1990, 1 degree awarded. *Degree requirements:* Thesis. *Entrance requirements:* GRE General Test (minimum combined score of 1500 required). Application fee: $30. *Tuition:* $248 per unit. *Financial aid:* In 1990–91, $15,000 in aid awarded. 2 fellowships (1 to a first-year student) were awarded; partial tuition waivers, federal work-study also available. Aid available to part-time students. Average monthly stipend for a graduate assistantship: $700. Financial aid applicants required to submit FAF. • Dr. Paul Buchheim, Coordinator, 714-824-4530.

Louisiana State University and Agricultural and Mechanical College, College of Basic Sciences, Department of Geology and Geophysics, Baton Rouge, LA 70803. Department awards MS, PhD. Faculty: 30 full-time (1 woman), 2 part-time (1 woman). Matriculated students: 42 full-time (12 women), 14 part-time (4 women); includes 2 minority (both Asian American), 19 foreign. Average age 32. 47 applicants, 32% accepted. In 1990, 6 master's, 7 doctorates awarded. Terminal master's awarded for partial completion of doctoral program. *Degree requirements:* Thesis/dissertation required, foreign language not required. *Entrance requirements:* GRE General Test (minimum combined score of 1000 required), TOEFL (minimum score of 525 required, 550 for assistantships), minimum GPA of 3.0. Application deadline: 7/1 (priority date, applications processed on a rolling basis). Application fee: $25. *Tuition:* $1020 per semester full-time for state residents; $2620 per semester full-time for nonresidents. *Financial aid:* In 1990–91, 42 students received a total of $440,884 in aid awarded. 7 fellowships (1 to a first-year student), 10 research assistantships (1 to a first-year student), 24 teaching assistantships (2 to first-year students) were awarded; federal work-study, institutionally sponsored loans, and career-related internships or fieldwork also available. Average monthly stipend for a graduate assistantship: $910. Financial aid application deadline: 3/15. *Faculty research:* Geophysics, geochemistry of sediments, isotope geochemistry, igneous and metamorphic petrology, micropaleontology. Total annual research budget: $1.46-million. • Dr. Joseph Hazel, Chair, 504-388-3354. Application contact: Elizabeth Holt, Graduate Coordinator, 504-388-3415.

See full description on page 227.

Massachusetts Institute of Technology, School of Science, Department of Earth, Atmospheric, and Planetary Sciences, Cambridge, MA 02139. Offerings include geology (SM, PhD, Sc D). Terminal master's awarded for partial completion of doctoral program. Department faculty: 37 full-time (3 women), 0 part-time. *Degree requirements:* For master's, thesis required, foreign language not required; for doctorate, dissertation, general exam required, foreign language not required. *Entrance requirements:* GRE General Test, GRE Subject Test. Application deadline: 1/15. Application fee: $45. *Tuition:* $16,900 per year. • Dr. Thomas H. Jordan, Chairman, 617-253-3382. Application contact: Anita Killian, Graduate Administrator, 617-253-3381.

McGill University, Faculty of Graduate Studies and Research, Department of Geological Sciences, Montreal, PQ H3A 2A7. Department awards M Sc, M Sc A, PhD, Diploma. Faculty: 17 full-time (0 women), 0 part-time. *Degree requirements:* For master's, thesis (M Sc); for Diploma, thesis not required. *Tuition:* $698.50 per semester full-time, $46.57 per credit part-time for Canadian residents; $3480 per semester full-time, $234 per semester part-time for nonresidents. *Financial aid:* Fellowships, teaching assistantships available. • D. Francis, Chairman, 514-398-6768. Application contact: R. Doig, Chairman, Graduate Admissions, 514-398-4884.

See full description on page 229.

McGill University, Faculty of Graduate Studies and Research, Department of Meteorology, Montreal, PQ H3A 2T5, Canada. Offerings include physical oceanography (M Sc, PhD). Terminal master's awarded for partial completion of doctoral program. Department faculty: 12 full-time (0 women), 0 part-time. *Degree requirements:* Thesis/dissertation required, foreign language not required. *Entrance requirements:* For master's, GRE General Test, minimum GPA of 3.0; for doctorate, GRE, master's degree in meteorology or related field. Application deadline: 7/1 (priority date, applications processed on a rolling basis). Application fee: $25. *Tuition:* $698.50 per semester full-time, $46.57 per credit part-time for Canadian residents; $3480 per semester full-time, $234 per semester part-time for nonresidents. • H. G. Leighton, Chairman, 514-398-3764.

McMaster University, Faculty of Science, Department of Geology, Hamilton, ON L8S 4L8, Canada. Department offers programs in geochemistry (PhD), geology (M Sc, PhD). *Degree requirements:* For master's, 1 foreign language, thesis; for doctorate, 1 foreign language, dissertation, comprehensive exam. *Expenses:* Tuition of $2250 per year full-time, $810 per semester part-time for Canadian residents; $10,340 per year full-time, $4050 per semester part-time for nonresidents. Fees of $76 per year full-time, $49 per semester part-time.

Memorial University of Newfoundland, School of Graduate Studies, Department of Earth Sciences, St. John's, NF A1C 5S7, Canada. Offerings include earth sciences (M Sc, PhD), with options in geology (M Sc, PhD), geophysics (M Sc, PhD). *Degree requirements:* For master's, thesis; for doctorate, dissertation, comprehensive exam.

Memphis State University, College of Arts and Sciences, Department of Geological Sciences, Memphis, TN 38152. Department offers programs in geology (MS), geophysics (MS). Faculty: 16 full-time (2 women), 1 part-time (0 women). Matriculated students: 7 full-time (2 women), 16 part-time (6 women); includes 0 minority, 10 foreign. In 1990, 4 degrees awarded. *Degree requirements:* Thesis, seminar presentation. *Entrance requirements:* GRE General Test, GRE Subject Test. Application deadline: 8/1 (applications processed on a rolling basis). Application fee: $5. *Expenses:* Tuition of $92 per credit hour for state residents; $239 per credit hour for nonresidents. Fees of $45 per year for state residents. *Financial aid:* In 1990–91, 0 research assistantships, 5 teaching assistantships awarded. *Faculty research:* Eocene and oligocene sediments, lab synthesis of dolomite, crystal structure of luenburgite, gregoryorite and lamprophyllite, Kramer evaporite ore. • Dr. Phili Deboo, Chairman, 901-678-2177. Application contact: Dr. Eugene S. Schweig, Coordinator of Graduate Studies, 901-678-2177.

Miami University, College of Arts and Sciences, Department of Geology, Oxford, OH 45056. Department awards MA, MS, PhD. Part-time programs available. Faculty: 11. Matriculated students: 20 full-time (9 women), 26 part-time (9 women); includes 0 minority, 2 foreign. 24 applicants, 67% accepted. In 1990, 10 master's, 3 doctorates awarded. *Degree requirements:* For master's, thesis (for some programs), final exam; for doctorate, 1 foreign language, dissertation, comprehensive exam, final exam. *Entrance requirements:* For master's, GRE General Test, GRE Subject Test, minimum undergraduate GPA of 2.75 or 3.0 during previous 2 years; for doctorate, GRE General Test, GRE Subject Test, minimum undergraduate GPA of 2.75 or 3.0 graduate. Application deadline: 3/1. Application fee: $30. *Expenses:* Tuition of $3196 per year full-time, $133 per semester (minimum) part-time for state residents; $7134 per year full-time, $312 per semester (minimum) part-time for nonresidents. Fees of $371 per year full-time, $165.50 per semester hour part-time. *Financial aid:* In 1990–91, $160,840 in aid awarded. Fellowships, research assistantships, teaching assistantships, full tuition waivers, federal work-study available. Financial aid application deadline: 3/1. • Maryellen Cameron, Chair, 513-529-3216. Application contact: Dr. Larry Mayer, Director of Graduate Study, 513-529-3224.

Announcement: Field research programs include modern carbonate sedimentology in Bahamas; structural geology, geochronology, and petrology in Adirondack Mountains and in Spitsbergen; tectonic geomorphology in southwestern US; clastic sedimentology in Spain; and volcanology in Cascade Range. Areas of laboratory research include geochemistry, isotope geochemistry, mineralogy/crystallography, tectonic geomorphology, and carbonate and clastic sedimentation. Equipment includes computer laboratory, mass spectrometer, DCP spectrometer, and 4-circle, single-crystal diffractometer. MA without thesis offered.

Michigan State University, College of Natural Science, Department of Geological Sciences, East Lansing, MI 48824-1115. Department awards MAT, MS, PhD. Faculty: 16 full-time, 0 part-time. Matriculated students: 28 full-time (5 women), 13 part-time (5 women); includes 1 minority, 4 foreign. 23 applicants, 40% accepted. In 1990, 6 master's, 2 doctorates awarded. *Degree requirements:* Thesis/dissertation. *Application deadline:* rolling. *Application fee:* $25 ($40 for foreign students). *Tuition:* $104.75 per credit for state residents; $211.75 per credit for nonresidents. *Financial aid:* In 1990–91, $180,782 in aid awarded. 9 fellowships (2 to first-year students), 3 research assistantships (0 to first-year students), 17 teaching assistantships (11 to first-year students) were awarded. • Thomas Vogel, Acting Chair, 517-355-4626.

Announcement: A program emphasizing the study of the water cycle and its interaction with the atmosphere, biosphere, and lithosphere is available. It addresses problems facing the modern environment and Earth's past and future by integrating an understanding of biological, chemical, and physical processes. Special strength is in stable-isotope geochemistry.

See full description on page 231.

Michigan Technological University, College of Engineering, Department of Geology and Geological Engineering, Program in Geology, Houghton, MI 49931. Program awards MS, PhD. Part-time programs available. Matriculated students: 22 full-time

(4 women), 0 part-time; includes 1 minority (Hispanic American), 4 foreign. Average age 29. 14 applicants, 64% accepted. In 1990, 1 master's, 1 doctorate awarded. *Degree requirements:* Thesis/dissertation required, foreign language not required. *Entrance requirements:* GRE, TOEFL (minimum score of 550 required). Application deadline: rolling. Application fee: $20. *Expenses:* Tuition of $852 per semester full-time, $71 per credit part-time for state residents; $2065 per semester full-time, $172 per credit part-time for nonresidents. Fees of $75 per semester. *Financial aid:* In 1990–91, 20 students received a total of $158,000 in aid awarded. 10 fellowships (2 to first-year students), 6 research assistantships, 4 teaching assistantships (2 to first-year students) were awarded; federal work-study and career-related internships or fieldwork also available. Aid available to part-time students. Average monthly stipend for a graduate assistantship: $800-933. Financial aid applicants required to submit FAF. *Faculty research:* Volcanology, aqueous geochemistry, hydrothermal systems, sedimentology, image processing and remote sensing. • Dr. William I. Rose, Head, Department of Geology and Geological Engineering, 906-487-2531. Application contact: Dr. Douglas McDowell, Graduate Adviser, 906-487-2531.

Mississippi State University, College of Arts and Sciences, Department of Geology and Geography, Mississippi State, MS 39762. Department offers program in geology (MS). Part-time programs available. Faculty: 9 full-time (0 women), 0 part-time. Matriculated students: 17 full-time (7 women), 4 part-time (1 woman); includes 0 minority, 2 foreign. Average age 28. 29 applicants, 17% accepted. In 1990, 1 degree awarded. *Degree requirements:* 1 foreign language, thesis. *Entrance requirements:* GRE General Test (minimum combined score of 800 required), GRE Subject Test, TOEFL (minimum score of 550 required), minimum GPA of 2.5. Application deadline: 3/15 (priority date, applications processed on a rolling basis). *Expenses:* Tuition of $891 per semester full-time, $99 per credit hour part-time for state residents; $1622 per semester full-time, $180 per credit hour part-time for nonresidents. Fees of $221 per semester full-time, $25 per semester (minimum) part-time. *Financial aid:* In 1990–91, 16 students received a total of $52,325 in aid awarded. 2 research assistantships (1 to a first-year student), 12 teaching assistantships (5 to first-year students) were awarded; partial tuition waivers, institutionally sponsored loans, and career-related internships or fieldwork also available. Average monthly stipend for a graduate assistantship: $600. Financial aid application deadline: 3/15. *Faculty research:* Climatology, micropaleontology, quaternary geology, sedimentology, stratigraphy. Total annual research budget: $90,000. • Dr. Charles Wax, Head, 601-325-3915. Application contact: Dr. Bruce Panuska, Graduate Coordinator, 601-325-3915.

Montana College of Mineral Science and Technology, Graduate School, Geological Engineering Program, Butte, MT 59701. Offerings include geoscience (MS), with options in geological engineering, geology, hydrogeological engineering, hydrogeology. Program faculty: 4 full-time (0 women), 3 part-time (0 women). *Degree requirements:* Thesis required, foreign language not required. *Entrance requirements:* GRE General Test, GRE Subject Test, TOEFL (minimum score of 525 required), minimum B average. Application deadline: 3/31 (priority date, applications processed on a rolling basis). Application fee: $20. *Tuition:* $524.25 per credit hour full-time, $310.50 per credit hour part-time for state residents; $1362 per credit hour full-time, $310.50 per credit hour part-time for nonresidents. • Dr. H. Peter Knudsen, Department Head, 406-496-4395. Application contact: Mary J. Seccombe, Graduate School Administrator, 406-496-4128.

Montclair State College, School of Mathematical and Natural Sciences, Department of Physics/Geoscience, Program in Geoscience, Upper Montclair, NJ 07043. Program awards MA. Part-time and evening/weekend programs available. Faculty: 10 full-time, 0 part-time. In 1990, 0 degrees awarded. *Degree requirements:* Thesis, comprehensive exam required, foreign language not required. *Entrance requirements:* GRE General Test. Application deadline: 7/1 (priority date, applications processed on a rolling basis). Application fee: $25. *Expenses:* Tuition of $130 per credit for state residents; $165 per credit for nonresidents. Fees of $276 per year. *Financial aid:* Application deadline 3/15. • Dr. Charles L. Hamilton, Adviser, 201-893-7273.

New Mexico Institute of Mining and Technology, Department of Geoscience, Program in Geology, Socorro, NM 87801. Program awards MS, PhD. Faculty: 3 full-time (0 women), 2 part-time (0 women). Matriculated students: 40 full-time (5 women), 0 part-time; includes 0 minority, 10 foreign. Average age 25. 34 applicants, 76% accepted. In 1990, 6 master's, 2 doctorates awarded. *Degree requirements:* For master's, thesis optional, foreign language not required; for doctorate, 1 foreign language, dissertation. *Entrance requirements:* For master's, GRE General Test, TOEFL (minimum score of 540 required); for doctorate, GRE General Test, GRE Subject Test, TOEFL (minimum score of 540 required). Application deadline: 6/1 (priority date, applications processed on a rolling basis). Application fee: $16. *Expenses:* Tuition of $617 per semester full-time, $51 per hour part-time for state residents; $2656 per semester full-time, $209 per hour part-time for nonresidents. Fees of $207 per semester. *Financial aid:* In 1990–91, 22 students received a total of $65,640 in aid awarded. 0 fellowships, 12 research assistantships (3 to first-year students), 10 teaching assistantships (5 to first-year students) were awarded; federal work-study, institutionally sponsored loans also available. Average monthly stipend for a graduate assistantship: $600. Financial aid application deadline: 3/15; applicants required to submit GAPSFAS or FAF. *Faculty research:* Mineralogy, petrology, sedimentology, stratigraphy, structure. Total annual research budget: $127,704. • Dr. David Norman, Coordinator, 505-835-5771. Application contact: Dr. J. A. Smoake, Dean, Graduate Studies, 505-835-5513.

Announcement: MS and PhD programs also available in geochemistry (volcanology, isotope geochemistry, geochemistry of ore deposits), geophysics (exploration geophysics, earthquake seismology, geothermics), and hydrology (groundwater transport and contamination, modeling, isotope hydrology). Courses, research, and support augmented by 14 adjuncts, most of whom are professionals with the on-campus NM Bureau of Mines and Mineral Resources. See full description of NMIMT in the Graduate and Professional Programs (An Overview) volume of this series.

North Carolina State University, College of Physical and Mathematical Sciences, Department of Marine, Earth, and Atmospheric Sciences, Raleigh, NC 27695. Department offers programs in geology (MS, PhD), geophysics (MS, PhD), meteorology (MS, PhD), oceanography (MS, PhD). Faculty: 35 full-time (2 women), 3 part-time (1 woman). Matriculated students: 74 full-time (15 women), 3 part-time (0 women); includes 2 minority (both Asian American), 23 foreign. Average age 24. 75 applicants, 49% accepted. In 1990, 18 master's, 6 doctorates awarded. Terminal master's awarded for partial completion of doctoral program. *Degree requirements:* For master's, thesis, final oral exam required, foreign language not required; for doctorate, dissertation, preliminary written and oral exams, final oral exams required, foreign language not required. *Entrance requirements:* GRE General Test, minimum GPA of 3.0. Application deadline: 5/1 (priority date, applications processed on a rolling basis). Application fee: $35. *Tuition:* $1138 per year for state residents; $5805 per year for nonresidents. *Financial aid:* In 1990–91, $383,267 in aid awarded. 0 fellowships, 33 research assistantships (10 to first-year students), 18 teaching assistantships (6 to first-year students) were awarded; institutionally sponsored loans also available. Average monthly stipend for a graduate assistantship: $750. Financial aid application deadline: 3/1. *Faculty research:* Boundary-layer and synoptic meteorology; physical, chemical, geological, and biological oceanography. Total annual research budget: $6.41-million. • Dr. Leonard J. Pietrafesa, Head, 919-737-3717. Application contact: G. S. Janowitz, Graduate Administrator, 919-737-7837.

See full description on page 317.

Northern Arizona University, College of Arts and Sciences, Department of Geology, Flagstaff, AZ 86011. Department offers programs in earth science (MAT, MS), geology (MAT, MS). *Degree requirements:* Thesis required, foreign language not required. *Entrance requirements:* GRE General Test, GRE Subject Test. Application deadline: 3/15 (priority date, applications processed on a rolling basis). Application fee: $0 ($25 for foreign students). *Expenses:* Tuition of $1528 per year full-time, $80 per credit part-time for state residents; $6180 per year full-time, $258 per credit part-time for nonresidents. Fees of $25 per semester. *Financial aid:* In 1990–91, 2 fellowships (both to first-year students), 9 research assistantships (3 to first-year students), 15 teaching assistantships (8 to first-year students) awarded; scholarships, full and partial tuition waivers, federal work-study, and career-related internships or fieldwork also available. Financial aid applicants required to submit FAF. *Faculty research:* Sedimentology, stratigraphy, paleontology, environmental geology, geochemistry. • Chair, 602-523-4561. Application contact: Program Adviser, 602-523-4561.

Northern Illinois University, College of Liberal Arts and Sciences, Department of Geology, De Kalb, IL 60115. Department awards MS, PhD. Part-time programs available. Faculty: 15 full-time (2 women), 1 part-time (0 women). Matriculated students: 27 full-time (3 women), 22 part-time (2 women); includes 4 minority (1 Asian American, 2 Hispanic American, 1 Native American), 14 foreign. Average age 30. 103 applicants, 21% accepted. In 1990, 12 master's, 0 doctorates awarded. *Degree requirements:* For master's, comprehensive exam required, thesis optional, foreign language not required; for doctorate, variable foreign language requirement, dissertation defense, internship, comprehensive exam. *Entrance requirements:* For master's, GRE General Test, TOEFL, bachelor's degree in related field, minimum GPA of 2.75; for doctorate, GRE General Test, TOEFL, bachelor's or master's degree in related field, minimum graduate GPA of 3.2, undergraduate 2.75. Application deadline: 6/1. Application fee: $0. *Tuition:* $1339 per semester full-time for state residents; $3163 per semester full-time for nonresidents. *Financial aid:* In 1990–91, 5 fellowships, 8 research assistantships, 17 teaching assistantships awarded; full tuition waivers, federal work-study, and career-related internships or fieldwork also available. Financial aid applicants required to submit FAF. *Faculty research:* Geochemistry, hydrogeology, petrology, sedimentology, micropaleontology, biostratigraphy. Total annual research budget: $769,700. • Dr. Ross Powell Jr., Chair, 815-753-1943. Application contact: Dr. James A. Walker, Director, Graduate Studies, 815-753-7936.

See full description on page 233.

Northwestern University, College of Arts and Sciences, Department of Geological Sciences, Evanston, IL 60208. Department awards MS, PhD. Part-time programs available. Faculty: 12 full-time (1 woman), 0 part-time. Matriculated students: 18 full-time (4 women), 0 part-time; includes 1 minority (black American), 6 foreign. 15 applicants, 73% accepted. In 1990, 3 master's, 9 doctorates awarded. Terminal master's awarded for partial completion of doctoral program. *Degree requirements:* For doctorate, dissertation. *Entrance requirements:* For doctorate, GRE General Test, TOEFL. Application deadline: 8/30. Application fee: $40 ($45 for foreign students). *Tuition:* $4665 per quarter full-time, $1704 per course part-time. *Financial aid:* In 1990–91, $86,804 in aid awarded. 4 fellowships, 10 research assistantships, 7 teaching assistantships were awarded; federal work-study, institutionally sponsored loans, and career-related internships or fieldwork also available. Financial aid application deadline: 1/15; applicants required to submit GAPSFAS. *Faculty research:* Geophysics, geochemistry, petrology/mineralogy, paleoclimate, stratigraphy/structure. Total annual research budget: $1-million. • Seth Stein, Chairman, 708-491-3238. Application contact: Richard Gordon, Graduate Adviser, 708-491-1170.

Ohio State University, College of Mathematical and Physical Sciences, Department of Geology, Columbus, OH 43210. Offerings include geology (MS, PhD), mineralogy (MS, PhD), paleontology (MS, PhD). Department faculty: 25. *Degree requirements:* For master's, thesis required, foreign language not required; for doctorate, 1 foreign language, dissertation. *Entrance requirements:* GRE General Test, GRE Subject Test. Application deadline: 8/15 (applications processed on a rolling basis). Application fee: $0 ($25 for foreign students). *Tuition:* $1213 per quarter full-time, $364 per course part-time for state residents; $3143 per quarter full-time, $943 per course part-time for nonresidents. • Peter Webb, Chairman, 614-292-2721.

Ohio University, Graduate Studies, College of Arts and Sciences, Department of Geological Sciences, Athens, OH 45701. Department awards MS. Faculty: 7 full-time (0 women), 0 part-time. Matriculated students: 23 full-time, 8 part-time; includes 2 minority (1 Asian American, 1 Hispanic American), 10 foreign. 34 applicants, 47% accepted. In 1990, 6 degrees awarded. *Degree requirements:* Thesis. *Application fee:* $25. *Tuition:* $1112 per quarter hour full-time, $138 per credit hour part-time for state residents; $2227 per quarter hour full-time, $277 per credit hour part-time for nonresidents. *Financial aid:* In 1990–91, 25 students received a total of $101,914 in aid awarded. 3 research assistantships (all to first-year students), 11 teaching assistantships (all to first-year students), 19 scholarships (all to first-year students) were awarded; full and partial tuition waivers, federal work-study, institutionally sponsored loans also available. Average monthly stipend for a graduate assistantship: $811. Financial aid application deadline: 3/15; applicants required to submit FAF. *Faculty research:* Environmental geology, hydrology, geophysics, glacial geology, structure and tectonics. Total annual research budget: $109,913. • Dr. Damian Nance, Chairman, 614-593-1101.

Oklahoma State University, College of Arts and Sciences, Department of Geology, Stillwater, OK 74078. Department awards MS. Faculty: 10 full-time, 0 part-time. Matriculated students: 41 (7 women); includes 0 minority, 2 foreign. In 1990, 12 degrees awarded. *Degree requirements:* Thesis required, foreign language not required. *Entrance requirements:* TOEFL (minimum score of 550 required). Application deadline: 7/1 (priority date). Application fee: $15. *Tuition:* $63.75 per credit for state residents; $138.25 per credit for nonresidents. *Financial aid:* In 1990–91, 9 research assistantships, 6 teaching assistantships awarded; partial tuition waivers, federal work-study, and career-related internships or fieldwork also available. Aid available to part-time students. Financial aid application deadline: 3/1; applicants required to submit GAPSFAS or FAF. *Faculty research:* Groundwater hydrology, petroleum geology. • Dr. Wayne A. PettyJohn, Head, 405-744-6358.

Directory: Geology

Old Dominion University, College of Sciences, Department of Geological Sciences, Norfolk, VA 23529. Department awards MS. Faculty: 10 full-time (1 woman), 0 part-time. Matriculated students: 11 full-time (2 women), 20 part-time (0 women); includes 0 minority, 1 foreign. Average age 25. 16 applicants, 100% accepted. In 1990, 5 degrees awarded (100% found work related to degree). *Degree requirements:* Comprehensive exams. *Entrance requirements:* GRE General Test, GRE Subject Test, TOEFL, minimum GPA of 3.0, interview. Application deadline: 7/1 (applications processed on a rolling basis). Application fee: $20. *Expenses:* Tuition of $148 per credit hour for state residents; $375 per credit hour for nonresidents. Fees of $64 per year full-time. *Financial aid:* In 1990–91, $8800 in aid awarded. 1 research assistantship, 1 tuition grant were awarded; teaching assistantships and career-related internships or fieldwork also available. • Dr. Randall Spencer, Chairperson, 804-683-4301. Application contact: Dr. Joseph Rule, Director, Graduate Program, 804-683-4301.

Announcement: Program strengthens fundamental areas of geology and places a strong emphasis on disciplines required for framework coastal-plain geological studies. Subjects such as groundwater and surface water, oil and gas exploration, industrial and agricultural applications, and exploration for natural resources are emphasized in various research programs. Close association with the USGS and Smithsonian Institution broaden and strengthen the department's expertise. Current areas of research include environmental geochemistry, geomorphology, geophysics, hydrogeology, micropaleontology, mineralogy, paleoecology, sedimentology, and structural geology. Flexibility allows a strong complement of courses in allied fields, such as oceanography, engineering, physics, chemistry, and business. Teaching assistantships available; application deadline of March 1.

Oregon State University, Graduate School, College of Science, Department of Geosciences, Corvallis, OR 97331. Department awards MA, MS, PhD. Part-time programs available. Faculty: 24 full-time (2 women), 12 part-time (0 women). In 1990, 29 master's, 2 doctorates awarded. Terminal master's awarded for partial completion of doctoral program. *Degree requirements:* For master's, variable foreign language requirement, thesis optional; for doctorate, 1 foreign language, dissertation. *Entrance requirements:* GRE General Test, GRE Subject Test, TOEFL (minimum score of 550 required), minimum GPA of 3.0 in last 90 hours. Application fee: $40. *Tuition:* $1140 per trimester full-time, $449 per year part-time for state residents; $1816 per trimester full-time, $674 per year part-time for nonresidents. *Financial aid:* Fellowships, research assistantships, teaching assistantships, federal work-study, institutionally sponsored loans available. Aid available to part-time students. Financial aid application deadline: 2/14; applicants required to submit GAPSFAS or FAF. • Dr. Cyrus W. Field, Chair. Application contact: Graduate Admissions Coordinator, 503-737-1201.

See full description on page 235.

Pennsylvania State University University Park Campus, College of Earth and Mineral Sciences, Department of Geosciences, University Park, PA 16802. Offerings include geology (MS, PhD). Department faculty: 39. *Entrance requirements:* GRE General Test, TOEFL. Application fee: $35. *Tuition:* $203 per credit for state residents; $403 per credit for nonresidents. • Dr. David Eggler, Chair, 814-865-6393.

Announcement: Penn State offers MS and PhD research opportunities in paleoclimatology, paleontology, palynology, coal geology, sedimentology, stratigraphy, structural geology and tectonophysics, rock mechanics, geomorphology, remote sensing, hydrogeology, and glaciology.

See full description on page 237.

Portland State University, College of Liberal Arts and Sciences, Department of Geology, Portland, OR 97207. Department awards MA, MS, PhD. PhD offered in conjunction with Interdisciplinary Program in Environmental Sciences and Resources. Part-time programs available. Faculty: 7 full-time (0 women), 6 part-time (1 woman). Matriculated students: 12 full-time (3 women), 6 part-time (2 women); includes 1 minority (Native American), 1 foreign. Average age 29. 11 applicants, 82% accepted. In 1990, 5 master's awarded. *Degree requirements:* For master's, computer language, thesis (for some programs), field comprehensive required, foreign language not required; for doctorate, computer language, dissertation, two years residence. *Entrance requirements:* For master's, TOEFL (minimum score of 550 required), minimum GPA of 3.0 in upper-division course work or 2.75 overall, undergraduate degree in geology or equivalent; for doctorate, GRE General Test, GRE Subject Test. Application deadline: 4/15 (priority date, applications processed on a rolling basis). Application fee: $40. *Tuition:* $1151 per trimester for state residents; $1827 per trimester for nonresidents. *Financial aid:* In 1990–91, 2 research assistantships (both to first-year students), 8 teaching assistantships (2 to first-year students) awarded; federal work-study, institutionally sponsored loans, and career-related internships or fieldwork also available. Aid available to part-time students. Average monthly stipend for a graduate assistantship: $435. Financial aid application deadline: 3/1. *Faculty research:* Volcanic stratigraphy, stratigraphy of Columbia River basalt, Neogene tectonics, economic geology, low and medium temperature alteration. Total annual research budget: $175,000. • Dr. Ansel Johnson, Head, 503-725-3022.

Portland State University, College of Liberal Arts and Sciences, Interdisciplinary Program in Environmental Sciences and Resources, Portland, OR 97207. Offers environmental sciences/biology (PhD), environmental sciences/chemistry (PhD), environmental sciences/civil engineering (PhD), environmental sciences/geology (PhD), environmental sciences/physics (PhD). Part-time programs available. Faculty: 1 (woman) full-time, 0 part-time. Matriculated students: 24 full-time (7 women), 8 part-time (2 women); includes 0 minority, 11 foreign. Average age 35. 13 applicants, 100% accepted. In 1990, 3 degrees awarded. *Degree requirements:* Variable foreign language requirement, computer language, dissertation, qualifying exam, oral exam. *Entrance requirements:* TOEFL (minimum score of 550 required), minimum GPA of 3.0 in upper-division course work or 2.75 overall. Application deadline: 7/12. Application fee: $40. *Tuition:* $1151 per trimester for state residents; $1827 per trimester for nonresidents. *Financial aid:* In 1990–91, 1 research assistantship (to a first-year student) awarded; federal work-study, institutionally sponsored loans also available. Aid available to part-time students. Average monthly stipend for a graduate assistantship: $578. Financial aid application deadline: 1/31. *Faculty research:* Aquatic biology and chemistry, atmospheric pollution, natural resources, ecology, biophysics. Total annual research budget: $1.5-million. • Dr. Pavel Smejtek, Director, 503-725-4980.

Announcement: PhD in environmental sciences and resources/geology is part of multidisciplinary program sponsored by Departments of Biology, Chemistry, Civil Engineering, Geology, and Physics. Geology thesis research is concerned with basic problems in geology that are environmentally relevant, including hazardous-waste disposal and recovery; quality, quantity, and location of groundwater supplies; land stability; earthquake hazards; chemical alterations in rock; and delineations of strata for purposes of identifying mineral resources and ground dislocation zones. Contact Dr. Pavel Smejtek, Program Director, 503-725-4980.

Princeton University, Department of Geological and Geophysical Sciences, Princeton, NJ 08544. Department offers programs in atmospheric and oceanic sciences (PhD), geological and geophysical sciences (PhD), water resources (PhD). Faculty: 16 full-time (3 women), 0 part-time. Matriculated students: 26 full-time (4 women), 0 part-time; includes 1 minority (Asian American), 12 foreign. 28 applicants, 54% accepted. In 1990, 11 doctorates awarded (45% entered university research/teaching, 55% found other work related to degree). Terminal master's awarded for partial completion of doctoral program. *Degree requirements:* For doctorate, 1 foreign language, dissertation. *Entrance requirements:* For doctorate, GRE General Test, GRE Subject Test. Application deadline: 1/8. Application fee: $45 ($50 for foreign students). *Tuition:* $16,670 per year. *Financial aid:* Fellowships, research assistantships, teaching assistantships, federal work-study, institutionally sponsored loans available. Financial aid application deadline: 1/8. • Gerta Keller, Director of Graduate Studies, 609-258-5807. Application contact: Michele Spreen, Director of Graduate Admissions, 609-258-3034.

Purdue University, School of Science, Department of Earth and Atmospheric Sciences, West Lafayette, IN 47907. Department awards MS, PhD. Faculty: 24 full-time, 1 part-time. Matriculated students: 74 full-time (10 women), 6 part-time (3 women); includes 4 minority (1 Asian American, 1 black American, 1 Hispanic American, 1 Native American), 27 foreign. Average age 28. 87 applicants, 23% accepted. In 1990, 19 master's, 6 doctorates awarded. *Degree requirements:* For master's, thesis required, foreign language not required; for doctorate, 1 foreign language, dissertation. *Entrance requirements:* GRE General Test. Application deadline: 6/1. Application fee: $25. *Tuition:* $1162 per semester full-time, $83.50 per credit hour part-time for state residents; $3720 per semester full-time, $244.50 per credit hour part-time for nonresidents. *Financial aid:* In 1990–91, 3 fellowships (2 to first-year students), 22 research assistantships (1 to a first-year student), 26 teaching assistantships (3 to first-year students) awarded. • Dr. E. M. Agee, Head, 317-494-0251.

Queens College of the City University of New York, Mathematics and Natural Sciences Division, Department of Geology, 65-30 Kissena Boulevard, Flushing, NY 11367. Department awards MA, MS Ed. MS Ed awarded through the School of Education. Part-time and evening/weekend programs available. Matriculated students: 0 full-time, 11 part-time (3 women). 10 applicants, 70% accepted. In 1990, 0 degrees awarded. *Degree requirements:* Thesis, comprehensive exam required, foreign language not required. *Entrance requirements:* GRE, TOEFL (minimum score of 550 required), previous course work in calculus, physics, and chemistry, minimum GPA of 3.0. Application deadline: 4/1 (applications processed on a rolling basis). Application fee: $35. *Tuition:* $2700 per year full-time, $45 per credit part-time for state residents; $4700 per year full-time, $199 per credit part-time for nonresidents. *Financial aid:* In 1990–91, 4 fellowships (0 to first-year students) awarded; partial tuition waivers, federal work-study, institutionally sponsored loans, and career-related internships or fieldwork also available. Aid available to part-time students. Financial aid application deadline: 4/1. *Faculty research:* Sedimentology/stratigraphy, paleontology, field petrology. • Dr. Allan Ludman, Chairperson, 718-997-3300. Application contact: Dr. Robert Finks, Graduate Adviser, 718-997-3300.

Queen's University at Kingston, Faculty of Arts and Sciences, Department of Geological Sciences, Kingston, ON K7L 3N6, Canada. Department awards M Sc, M Sc Eng, PhD. Part-time programs available. Matriculated students: 74 full-time, 23 part-time; includes 10 foreign. In 1990, 21 master's, 3 doctorates awarded. *Degree requirements:* For master's, thesis optional, foreign language not required; for doctorate, 1 foreign language, dissertation, comprehensive exam. *Entrance requirements:* TOEFL (minimum score of 550 required). Application deadline: 2/28 (priority date). Application fee: $35. *Tuition:* $2861 per year full-time, $426 per trimester part-time for Canadian residents; $10,613 per year full-time, $4998 per trimester part-time for nonresidents. *Financial aid:* Fellowships, research assistantships, teaching assistantships, institutionally sponsored loans available. Financial aid application deadline: 3/1. *Faculty research:* Mineralogy, petrology, structural geology, stratigraphy, sedimentology. • Dr. John Dixon, Head, 613-545-2000. Application contact: Dr. H. Helmstaedt, Graduate Coordinator, 613-545-6175.

Rensselaer Polytechnic Institute, School of Science, Department of Earth and Environmental Sciences, Troy, NY 12180. Department offers programs in geochemistry (MS, PhD), geophysics (MS, PhD), planetary geology (MS, PhD). Faculty: 8 full-time (0 women), 2 part-time (both women). Matriculated students: 23 full-time (5 women), 6 part-time (3 women); includes 0 minority, 4 foreign. 17 applicants, 65% accepted. In 1990, 8 master's, 1 doctorate awarded. *Degree requirements:* For master's, thesis optional; for doctorate, dissertation. *Entrance requirements:* GRE General Test, TOEFL. Application deadline: 2/1. Application fee: $30. *Expenses:* Tuition of $455 per credit hour. Fees of $195.57 per semester. *Financial aid:* In 1990–91, 21 students received a total of $115,000 in aid awarded. 4 fellowships (2 to first-year students), 13 research assistantships (2 to first-year students), 8 teaching assistantships (2 to first-year students) were awarded; partial tuition waivers and career-related internships or fieldwork also available. *Faculty research:* Seismology and tectonics, igneous and metamorphic petrology, geohydrology, fission tracks and thermal history, asteroids and early history of earth. Total annual research budget: $550,000. • E. B. Watson, Chairman, 518-276-6474. Application contact: M. Brian Bayly, Graduate Coordinator, 518-276-6494.

See full description on page 239.

Rice University, Wiess School of Natural Sciences, Department of Geology and Geophysics, Houston, TX 77251. Department offers programs in geochemistry (MA, PhD), geophysics (MA, PhD). Faculty: 15 full-time (0 women), 0 part-time. Matriculated students: 49 full-time (12 women), 1 part-time (0 women); includes 3 minority (2 Asian American, 1 black American), 2 foreign. *Degree requirements:* Thesis/dissertation required, foreign language not required. *Entrance requirements:* GRE General Test, GRE Subject Test, TOEFL (minimum score of 550 required), minimum GPA of 3.0. Application deadline: 3/1. Application fee: $0. *Expenses:* Tuition of $8300 per year full-time, $400 per credit hour part-time. Fees of $167 per year. *Financial aid:* Fellowships, research assistantships, full and partial tuition waivers available. Financial aid applicants required to submit FAF. *Faculty research:*

Stratigraphy sedimentation, igneous and metamorphic petrology, oceanography, structural geology, paleontology. • J. C. Stormer, Chairman, 713-527-4880.

Rutgers, The State University of New Jersey, Newark, Department of Geology, Newark, NJ 07102. Department offers programs in biostratigraphy (MS), geochemistry (MS), geomorphology (MS), marine geology (MS), mineralogy (MS), paleoecology (MS), paleontology (MS), petrology (MS), sedimentology (MS), stratigraphy (MS), structural geology (MS). Part-time and evening/weekend programs available. Faculty: 5 full-time (0 women), 6 part-time (2 women). Matriculated students: 4 full-time, 18 part-time; includes 6 minority (5 Asian American, 1 Native American), 5 foreign. Average age 29. 7 applicants, 71% accepted. In 1990, 2 degrees awarded. *Degree requirements:* Comprehensive exam required, thesis optional. *Entrance requirements:* GRE, minimum undergraduate B average. Application deadline: 6/1. Application fee: $35. *Tuition:* $2216 per semester full-time for state residents; $3248 per semester full-time for nonresidents. *Financial aid:* In 1990–91, 1 fellowship (to a first-year student), 5 teaching assistantships (1 to a first-year student) awarded; research assistantships, federal work-study, and career-related internships or fieldwork also available. Aid available to part-time students. Financial aid application deadline: 3/1; applicants required to submit GAPSFAS. *Faculty research:* Environmental geology, plate tectonics. • Dr. George Theokritoff, Director, 201-648-5100.

Rutgers, The State University of New Jersey, New Brunswick, Program in Geological Sciences, New Brunswick, NJ 08903. Program awards MS, PhD. Part-time programs available. Faculty: 10 full-time (1 woman), 0 part-time. Matriculated students: 15 full-time (5 women), 10 part-time (2 women); includes 0 minority, 3 foreign. 36 applicants, 31% accepted. In 1990, 4 master's awarded (100% found work related to degree); 1 doctorate awarded (100% found work related to degree). *Degree requirements:* For master's, thesis required, foreign language not required; for doctorate, 1 foreign language, dissertation. *Entrance requirements:* GRE General Test, GRE Subject Test. Application deadline: 7/1. Application fee: $35. *Expenses:* Tuition of $4432 per year full-time, $183 per credit part-time for state residents; $6496 per year full-time, $270 per credit part-time for nonresidents. Fees of $458 per year full-time, $117 per year part-time. *Financial aid:* In 1990–91, $12,000 in aid awarded. 2 fellowships (1 to a first-year student), 3 research assistantships (2 to first-year students), 7 teaching assistantships (5 to first-year students) were awarded. Financial aid application deadline: 3/1. *Faculty research:* Petrology, paleoecology, sedimentology, geophysics, geochemistry. Total annual research budget: $210,000. • Richard K. Olsson, Director, 908-932-3043.

San Diego State University, College of Sciences, Department of Geological Sciences, San Diego, CA 92182. Department awards MS. Faculty: 16 full-time, 0 part-time. Matriculated students: 4 full-time (2 women), 28 part-time (9 women); includes 2 minority (1 Asian American, 1 Hispanic American), 2 foreign. In 1990, 14 degrees awarded. *Degree requirements:* Foreign language and thesis not required. *Entrance requirements:* GRE General Test (minimum combined score of 1000 required), GRE Subject Test (minimum score of 650 required), bachelor's degree in related field. Application deadline: 8/1 (applications processed on a rolling basis). Application fee: $55. *Expenses:* Tuition of $0 for state residents; $189 per unit for nonresidents. Fees of $1974 per year full-time, $692 per year part-time for state residents; $1074 per year full-time, $692 per year part-time for nonresidents. *Financial aid:* In 1990–91, 5 fellowships awarded. *Faculty research:* Groundwater, sedimentation and tectonics, geochemistry, geochronology, paleontology. Total annual research budget: $350,000. • Dr. Michael Wallawender, Chair, 619-594-5586.

San Jose State University, School of Science, Department of Geology, San Jose, CA 95192. Department awards MS. Faculty: 14 full-time (2 women), 12 part-time (2 women). Matriculated students: 0 full-time, 19 part-time (3 women); includes 3 minority (1 Asian American, 2 Hispanic American), 2 foreign. Average age 32. 12 applicants, 83% accepted. In 1990, 8 degrees awarded. *Degree requirements:* Thesis required, foreign language not required. *Entrance requirements:* GRE. Application deadline: 6/1 (applications processed on a rolling basis). Application fee: $55. *Expenses:* Tuition of $0 for state residents; $246 per unit for nonresidents. Fees of $592 per year. *Financial aid:* Teaching assistantships, federal work-study available. • Dr. John Williams, Chair, 408-924-5050. Application contact: Dr. Scott Creely, Graduate Adviser, 408-924-5036.

South Dakota School of Mines and Technology, Department of Geology and Geological Engineering, Program in Geology, Rapid City, SD 57701. Program awards MS, PhD. Part-time programs available. Faculty: 3 full-time (0 women), 2 part-time (0 women). Matriculated students: 10 full-time (1 woman), 12 part-time (1 woman); includes 0 minority, 10 foreign. Average age 28. In 1990, 6 master's, 3 doctorates awarded. *Degree requirements:* For master's, thesis required, foreign language not required; for doctorate, dissertation. *Entrance requirements:* GRE General Test, GRE Subject Test, TOEFL (minimum score of 520 required). Application deadline: 7/15. Application fee: $15 ($45 for foreign students). *Expenses:* Tuition of $1474 per year full-time, $61.40 per credit hour part-time for state residents; $2917 per year full-time, $122 per credit hour part-time for nonresidents. Fees of $871 per year full-time, $132 per semester (minimum) part-time. *Financial aid:* In 1990–91, 17 students received aid. 4 fellowships, 12 research assistantships, 4 teaching assistantships awarded; federal work-study, institutionally sponsored loans also available. Aid available to part-time students. Average monthly stipend for a graduate assistantship: $300. Financial aid application deadline: 5/15. *Faculty research:* Mineral deposits (coal, precious metals, uranium, oil, gas), paleontology, geophysics. • Dr. W. M. Roggenthen, Head, Department of Geology and Geological Engineering, 605-394-2461. Application contact: Dr. Briant L. Davis, Dean, Graduate Division, 605-394-2493.

South Dakota School of Mines and Technology, Department of Geology and Geological Engineering, Program in Paleontology, Rapid City, SD 57701. Program awards MS. Part-time programs available. Faculty: 2 full-time (0 women), 1 (woman) part-time. Matriculated students: 2 full-time (1 woman), 2 part-time (1 woman); includes 0 minority, 0 foreign. Average age 28. In 1990, 1 degree awarded. *Degree requirements:* Thesis. *Entrance requirements:* GRE General Test, GRE Subject Test, TOEFL (minimum score of 520 required). Application deadline: 7/15. Application fee: $15 ($45 for foreign students). *Expenses:* Tuition of $1474 per year full-time, $61.40 per credit hour part-time for state residents; $2917 per year full-time, $122 per credit hour part-time for nonresidents. Fees of $871 per year full-time, $132 per semester (minimum) part-time. *Financial aid:* In 1990–91, 4 students received aid. 0 fellowships, 3 research assistantships (1 to a first-year student), 2 teaching assistantships (1 to a first-year student) awarded; federal work-study, institutionally sponsored loans also available. Aid available to part-time students. Average monthly stipend for a graduate assistantship: $300. Financial aid application deadline: 5/15. *Faculty research:* Holocene vertebrates associated with human occupations, cretaceous marine reptiles, Miosene vertebrates, Oligocene vertebrates, badlands of South Dakota. • Dr. Philip Bjork, Director of Museum of Geology, 605-394-2467.

Southern Illinois University at Carbondale, College of Science, Department of Geology, Carbondale, IL 62901. Department awards MS, PhD. Faculty: 17 full-time (1 woman), 0 part-time. Matriculated students: 26 full-time (6 women), 5 part-time (1 woman); includes 0 minority, 6 foreign. Average age 25. 50 applicants, 24% accepted. In 1990, 12 master's, 2 doctorates awarded. *Degree requirements:* For master's, thesis required, foreign language not required; for doctorate, 1 foreign language (computer language can substitute), dissertation. *Entrance requirements:* For master's, GRE, TOEFL (minimum score of 550 required), minimum GPA of 2.7; for doctorate, GRE General Test, TOEFL (minimum score of 550 required), minimum GPA of 3.25. Application deadline: 2/15 (priority date, applications processed on a rolling basis). Application fee: $0. *Expenses:* Tuition of $1638 per year full-time, $204.75 per semester hour part-time for state residents; $4914 per year full-time, $614.25 per semester hour part-time for nonresidents. Fees of $700 per year full-time, $216 per year part-time. *Financial aid:* In 1990–91, $202,536 in aid awarded. 5 fellowships (1 to a first-year student), 3 research assistantships (1 to a first-year student), 21 teaching assistantships (6 to first-year students) were awarded; full tuition waivers, federal work-study, institutionally sponsored loans also available. Aid available to part-time students. *Faculty research:* Geomorphology, hydrogeology, geophysics, sedimentology, coal petrology. Total annual research budget: $670,000. • Dr. John Utgaard, Chair, 618-453-3351.

Announcement: Strong, broad-based MS program and recently established PhD program are supported by excellent analytical laboratories for research in coal/organic petrology, organic geochemistry, SEM/EDS, and X-ray crystallography. Two drilling/coring rigs, a vibracorer, and a 40-foot by 8-foot flume for hydrogeology/geomorphology research; field data acquisition equipment and data processing facilities for research in geophysics; and facilities for research in ore deposits, structural geology, sedimentology, sedimentary petrology, paleontology, environments of deposition, computer mapping, and coal geology are available. PhD students are currently conducting research in coal geology/coal petrology, organic geochemistry, sedimentology, paleontology, environments of deposition, seismology, and geomorphology.

Southern Methodist University, Dedman College, Department of Geological Sciences, Dallas, TX 75275. Department offers programs in geochemistry (MS, PhD); geological sciences (MS, PhD); geophysics (MS, PhD), including applied geophysics (MS), geophysics (PhD); planetary studies (MS). Part-time programs available. Faculty: 12 full-time (1 woman), 0 part-time. Matriculated students: 6 full-time (2 women), 14 part-time (5 women); includes 0 minority, 7 foreign. Average age 29. 40 applicants, 23% accepted. In 1990, 7 master's, 1 doctorate awarded. *Degree requirements:* For master's, thesis, qualifying exam required, foreign language not required; for doctorate, 1 foreign language, dissertation, qualifying exam. *Entrance requirements:* GRE General Test (minimum combined score of 1200 required), minimum GPA of 3.0. Application deadline: 2/15. Application fee: $25. *Expenses:* Tuition of $435 per credit. Fees of $664 per semester for state residents; $56 per year for nonresidents. *Financial aid:* In 1990–91, $137,000 in aid awarded. 2 fellowships (1 to a first-year student), 5 research assistantships (1 to a first-year student), 13 teaching assistantships (5 to first-year students) were awarded; full and partial tuition waivers also available. Financial aid application deadline: 2/15. *Faculty research:* Sedimentology, geochemistry, igneous and metamorphic petrology, vertebrate paleontology, seismology. Total annual research budget: $2.5-million. • Dr. Michael J. Holdaway, Chairman, 214-692-2770. Application contact: Dr. Peter Scholle, Graduate Adviser, 214-692-4011.

See full description on page 241.

Stanford University, School of Earth Sciences, Department of Geology, Stanford, CA 94305. Department awards MS, PhD. Faculty: 12 (2 women). Matriculated students: 48 full-time (20 women), 0 part-time; includes 3 minority (1 Asian American, 2 Hispanic American), 12 foreign. Average age 28. 49 applicants, 33% accepted. In 1990, 2 master's, 4 doctorates awarded. Terminal master's awarded for partial completion of doctoral program. *Degree requirements:* Thesis/dissertation. *Entrance requirements:* GRE General Test, GRE Subject Test, TOEFL. Application deadline: 1/1. Application fee: $55. *Expenses:* Tuition of $15,102 per year. Fees of $28 per quarter. *Financial aid:* Fellowships, research assistantships, teaching assistantships, federal work-study, institutionally sponsored loans available. Financial aid application deadline: 1/1; applicants required to submit GAPSFAS. • Application contact: Graduate Admissions Coordinator, 415-723-2538.

See full description on page 243.

State University of New York at Albany, College of Science and Mathematics, Department of Geological Sciences, Albany, NY 12222. Department awards MS, PhD. Evening/weekend programs available. Faculty: 6 full-time (0 women), 0 part-time. Matriculated students: 14 full-time (4 women), 2 part-time (1 woman); includes 0 minority, 6 foreign. Average age 28. In 1990, 0 master's, 3 doctorates awarded. *Degree requirements:* For master's, 1 foreign language, thesis; for doctorate, 2 foreign languages, dissertation. *Entrance requirements:* GRE General Test, GRE Subject Test. *Expenses:* Tuition of $2450 per year full-time, $103 per credit part-time for state residents; $5765 per year full-time, $243 per credit part-time for nonresidents. Fees of $25 per year full-time, $0.85 per credit part-time. *Financial aid:* Minority assistantships available. *Faculty research:* Experimental microstructural geology; tectonics and geochemistry of ophiolites; collisional tectonics; experimental petrology; trace-element geochemistry. • Stephen DeLong, Chairman, 518-442-4466.

State University of New York at Binghamton, School of Arts and Sciences, Department of Geological Sciences and Environmental Studies, Binghamton, NY 13902-6000. Offerings include geological sciences (MA, PhD), with option in geophysics. Terminal master's awarded for partial completion of doctoral program. *Degree requirements:* For master's, thesis or alternative; for doctorate, variable foreign language requirement, dissertation, departmental qualifying exam. *Entrance requirements:* GRE General Test, GRE Subject Test, TOEFL. Application deadline: 4/15 (priority date). Application fee: $35. *Expenses:* Tuition of $2450 per year full-time, $102.50 per credit part-time for state residents; $5766 per year full-time, $242.50 per credit part-time for nonresidents. Fees of $77 per year full-time, $27.85 per semester (minimum) part-time. • Dr. Francis Wu, Chairperson, 607-777-2264.

State University of New York at Buffalo, Graduate School, Faculty of Natural Sciences and Mathematics, Department of Geology, Buffalo, NY 14260. Department awards MA, PhD. Part-time programs available. Faculty: 11 full-time (0 women), 1 part-time (0 women). Matriculated students: 27 full-time (6 women), 16 part-time (3 women); includes 1 minority (Asian American), 5 foreign. Average age 30. 34 applicants, 71% accepted. In 1990, 7 master's awarded (100% found work related to degree); 1 doctorate awarded (100% found work related to degree). *Degree requirements:* For master's, variable foreign language requirement, thesis, project or comprehensive exam; for doctorate, variable foreign language requirement, dissertation, oral and written exams, comprehensive exam. *Entrance requirements:* GRE General Test, GRE Subject Test (geology), TOEFL (minimum score of 550 required). Application deadline: 3/1 (priority date, applications processed on a rolling basis). Application fee: $35. *Expenses:* Tuition of $1600 per semester full-time, $134 per hour part-time for state residents; $3258 per semester full-time, $274 per hour part-time for nonresidents. Fees of $137 per semester full-time, $115 per semester (minimum) part-time. *Financial aid:* In 1990–91, 25 students received a total of $181,343 in aid awarded. 0 fellowships, 5 research assistantships (0 to first-year students), 12 teaching assistantships (3 to first-year students), 5 graduate assistantships were awarded; federal work-study also available. Average monthly stipend for a graduate assistantship: $718. Financial aid application deadline: 2/28; applicants required to submit FAF. *Faculty research:* Glaciology and paleoclimatology, geophysics, clay mineralogy and geochemistry, marine geology and tectonics, volcanology. Total annual research budget: $487,972. • Dr. Michael F. Sheridan Jr., Chairman, 716-636-6100. Application contact: Paul H. Reitan, Director of Graduate Studies, 716-636-3988.

State University of New York at Stony Brook, College of Arts and Sciences, Division of Physical Sciences and Mathematics, Department of Earth and Space Sciences, Stony Brook, NY 11794. Department offers programs in astronomical sciences (M Phil, MS, PhD), including astronomical sciences, astrophysics; geological sciences (M Phil, MS, PhD), including geochemistry, geology, geophysics. Faculty: 23 full-time, 0 part-time. Matriculated students: 48 full-time (14 women), 4 part-time (2 women); includes 1 minority (Asian American), 26 foreign. 116 applicants, 28% accepted. In 1990, 4 master's, 1 doctorate awarded. Terminal master's awarded for partial completion of doctoral program. *Degree requirements:* For master's, thesis or alternative required, foreign language not required; for doctorate, dissertation required, foreign language not required. *Entrance requirements:* GRE General Test, TOEFL, minimum GPA of 3.0. Application deadline: 2/1. Application fee: $35. *Expenses:* Tuition of $2450 per year full-time, $103 per credit part-time for state residents; $5766 per year full-time, $243 per credit part-time for nonresidents. Fees of $151 per year full-time, $10.45 per year (minimum) part-time. *Financial aid:* In 1990–91, 0 fellowships, 27 research assistantships, 21 teaching assistantships awarded. *Faculty research:* Astronomy, theoretical and observational astrophysics, paleontology, petrology, crystallography. Total annual research budget: $3.2-million. • Dr. Gilbert Hansen, Chairman, 516-632-8200.

State University of New York College at Fredonia, Geology Department, Fredonia, NY 14063. Department awards MS, MS Ed. Part-time and evening/weekend programs available. Faculty: 3 full-time (0 women), 0 part-time. Matriculated students: 0 full-time, 1 part-time (0 women); includes 0 minority, 0 foreign. 0 applicants. In 1990, 2 degrees awarded. *Degree requirements:* Thesis required, foreign language not required. *Application deadline:* 7/5. *Application fee:* $35. *Tuition:* $1600 per semester full-time, $134 per credit part-time for state residents; $3288 per semester full-time, $274 per credit part-time for nonresidents. *Financial aid:* In 1990–91, 0 research assistantships, 0 teaching assistantships awarded; full and partial tuition waivers and career-related internships or fieldwork also available. Aid available to part-time students. Financial aid application deadline: 3/15; applicants required to submit FAF. • Dr. Walther M. Barnard, Chairman, 716-673-3303.

State University of New York College at New Paltz, Faculty of Liberal Arts and Sciences, Program in Geological Sciences, New Paltz, NY 12561. Program awards MA, MAT, MS Ed. Faculty: 10. Matriculated students: 1 (woman) full-time, 1 part-time (0 women); includes 0 minority, 0 foreign. In 1990, 1 degree awarded. *Degree requirements:* Thesis, comprehensive exam. *Entrance requirements:* GRE General Test, minimum GPA of 3.0. Application deadline: 4/1 (priority date, applications processed on a rolling basis). Application fee: $35. *Tuition:* $1600 per semester full-time, $134 per credit part-time for state residents; $3258 per semester full-time, $274 per credit part-time for nonresidents. *Financial aid:* Research assistantships, teaching assistantships, federal work-study, institutionally sponsored loans available. Financial aid applicants required to submit FAF. • Dean Manos, Chairman, 914-257-3760.

Stephen F. Austin State University, School of Sciences and Mathematics, Department of Geology, Nacogdoches, TX 75962. Department awards MS, MSNS. Faculty: 6 full-time (0 women), 0 part-time. Matriculated students: 11 full-time (0 women), 26 part-time (7 women); includes 0 minority, 0 foreign. Average age 23. 12 applicants, 83% accepted. In 1990, 6 degrees awarded (0% entered university research/teaching, 83% found other work related to degree, 0% continued full-time study). *Degree requirements:* Comprehensive exam required, foreign language not required. *Entrance requirements:* GRE General Test (minimum combined score of 1000 required), minimum GPA of 2.5 overall, 2.8 in last half of major. Application deadline: 8/1. Application fee: $0 ($25 for foreign students). *Expenses:* Tuition of $18 per semester hour for state residents; $122 per semester hour for nonresidents. Fees of $14 per semester hour. *Financial aid:* In 1990–91, $5400 in aid awarded. 6 teaching assistantships (2 to first-year students), 0 assistantships were awarded; federal work-study also available. Financial aid application deadline: 6/1. *Faculty research:* Stratigraphy of Kaibab limestone, Utah; structure of Ouachita Mountains, Arkansas; groundwater chemistry of Carrizo Sand, Texas. Total annual research budget: $8000. • Dr. M. Carey Crocker, Chairman, 409-568-3701. Application contact: Dr. R. LaRell Nielson, Director of Graduate Program, 409-568-2248.

Sul Ross State University, School of Arts and Sciences, Department of Geology, Alpine, TX 79832. Department awards MS. Faculty: 2 full-time (0 women), 1 (woman) part-time. Matriculated students: 2 full-time (0 women), 6 part-time (0 women); includes 0 minority, 3 foreign. Average age 32. In 1990, 4 degrees awarded. *Degree requirements:* Thesis required, foreign language not required. *Entrance requirements:* GRE General Test, GRE Subject Test. *Tuition:* $473 per semester full-time for state residents; $1769 per semester full-time for nonresidents. *Financial aid:* In 1990–91, 6 students received a total of $20,274 in aid awarded. 5 research assistantships were awarded; federal work-study, institutionally sponsored loans, and career-related internships or fieldwork also available. Aid available to part-time students. Average monthly stipend for a graduate assistantship: $444. *Faculty research:* Stratigraphy, origin of mineral deposits, geochemistry, petrology of igneous and metamorphic rocks. • Dr. James Whitford-Stark, Chairman, 915-837-8259.

Announcement: Program stresses integrated field and laboratory research. The University is situated in an area of diverse and well-exposed geology. Research equipment includes XRF, EPMA, SEM, XRD, AA, CL, and NAA. Current faculty research in evaporites, petroleum geology, environmental geology, mineralogy, volcanology, trace-element geochemistry, ore deposits, paleontology, biostratigraphy, paleoecology, planetary geology, and remote sensing.

Syracuse University, College of Arts and Sciences, Department of Geology, Syracuse, NY 13244. Department offers programs in geology (MA, MS, PhD), hydrogeology (MS). Faculty: 12 full-time, 1 part-time. Matriculated students: 24 full-time (8 women), 0 part-time; includes 2 minority (1 black American, 1 Hispanic American), 3 foreign. Average age 31. 18 applicants, 78% accepted. In 1990, 6 master's, 3 doctorates awarded. *Degree requirements:* For master's, thesis (for some programs), 1 research tool; for doctorate, dissertation, 2 research tools. *Entrance requirements:* GRE General Test. Application fee: $40. *Expenses:* Tuition of $381 per credit. Fees of $289 per year full-time, $34.50 per semester part-time. *Financial aid:* In 1990–91, 4 research assistantships, 16 teaching assistantships awarded; fellowships, partial tuition waivers, federal work-study also available. Financial aid application deadline: 3/1; applicants required to submit FAF. • M. E. Bickford, Chairman, 315-443-2672.

Announcement: Department housed in Heroy Geology Laboratory with excellent geology library. Areas of research include paleontology, stratigraphy, sedimentology, petrology, hydrogeology, low-temperature and isotopic geochemistry, paleomagnetism, seismic reflection studies, marine geology, paleoclimatology, tectonics, and crustal evolution. Facilities include flume, X-ray spectrometer, mass spectrometer, plasma-emission spectrometer, and spinner magnetometer.

Temple University, College of Arts and Sciences, Department of Geology, Philadelphia, PA 19122. Department awards MA. Faculty: 9 full-time (2 women), 0 part-time. Matriculated students: 13 (3 women); includes 0 minority, 2 foreign. Average age 27. 23 applicants, 43% accepted. In 1990, 5 degrees awarded. *Degree requirements:* 1 foreign language, thesis. *Entrance requirements:* GRE General Test (minimum combined score of 1000 required), GRE Subject Test, minimum GPA of 2.8 overall, 3.0 during previous 2 years. Application deadline: 5/1. Application fee: $30. *Tuition:* $224 per credit for state residents; $283 per credit for nonresidents. *Financial aid:* In 1990–91, 4 research assistantships, 9 teaching assistantships awarded. • Dr. David Grandstaff, Chair, 215-787-8229. Application contact: Dr. Gene Ulmer, Co-Chair, Graduate Committee, 215-787-7171.

Texas A&I University, College of Arts and Sciences, Department of Geosciences, Kingsville, TX 78363. Department offers program in applied geology (MS). Part-time and evening/weekend programs available. Faculty: 7 full-time (1 woman), 0 part-time. Matriculated students: 0 full-time, 17 part-time (5 women); includes 4 minority (all Hispanic American), 1 foreign. Average age 26. 10 applicants, 50% accepted. In 1990, 1 degree awarded (100% found work related to degree). *Degree requirements:* Thesis or alternative, comprehensive exam required, foreign language not required. *Entrance requirements:* GRE General Test (minimum combined score of 800 required), TOEFL (minimum score of 500 required), minimum GPA of 3.0. Application deadline: 6/1 (applications processed on a rolling basis). Application fee: $0 ($25 for foreign students). *Expenses:* Tuition of $180 per semester full-time, $120 per semester part-time for state residents; $1152 per semester full-time, $768 per semester part-time for nonresidents. Fees of $149 per semester full-time, $101 per semester part-time. *Financial aid:* In 1990–91, $8000 in aid awarded. 2 fellowships (both to first-year students), 2 teaching assistantships (1 to a first-year student) were awarded; federal work-study, institutionally sponsored loans, and career-related internships or fieldwork also available. Aid available to part-time students. Financial aid application deadline: 5/15. *Faculty research:* Straitigraphy and sedimentology of modern coastal sediments, sandstone diagnosis, vertabrate paleontology, strctural geology. Total annual research budget: $300,000. • Dr. Jon A. Baskin, Graduate Coordinator, 512-595-3310.

Texas A&M University, College of Geosciences, Department of Geology, College Station, TX 77843. Department awards MS, PhD. Faculty: 33. Matriculated students: 116 full-time (35 women), 0 part-time; includes 2 minority (1 Asian American, 1 Hispanic American), 28 foreign. 51 applicants, 82% accepted. In 1990, 17 master's, 6 doctorates awarded. *Entrance requirements:* GRE General Test, TOEFL. Application deadline: 7/15. Application fee: $25 ($50 for foreign students). *Expenses:* Tuition of $100 per semester (minimum) for state residents; $128 per credit hour for nonresidents. Fees of $459 per year full-time, $252 per semester part-time. *Financial aid:* Fellowships, research assistantships, teaching assistantships available. • John H. Spang, Head. Application contact: Graduate Adviser, 409-845-2451.

Announcement: Texas A&M's geology department offers MS and PhD degrees; 33 faculty in hydrogeology, petroleum engineering, structural geology, paleontology, stratigraphy, geochemistry, and petrology. The 85,000-square-foot Halbouty Building, housing a new automated microprobe, is well equipped for graduate research. Assistantships and scholarships are plentiful. Contact Graduate Advisor, 409-845-2451.

Texas Christian University, Add Ran College of Arts and Sciences, Department of Geology, Fort Worth, TX 76129. Offerings include geology (MS). *Entrance requirements:* GRE. *Expenses:* Tuition of $244 per semester hour. Fees of $423 per semester full-time, $18.50 per semester hour part-time. • Dr. John Breyer, Chairperson, 817-921-7270.

Texas Tech University, Graduate School, College of Arts and Sciences, Department of Geosciences, Program in Geology, Lubbock, TX 79409. Program awards MS, PhD. Faculty: 10 full-time. Matriculated students: 46. In 1990, 7 master's, 3 doctorates awarded. *Degree requirements:* Variable foreign language requirement, thesis/dissertation. *Entrance requirements:* GRE General Test. Application deadline: 4/15 (priority date, applications processed on a rolling basis). Application fee: $0 ($50 for foreign students). *Tuition:* $494 per semester full-time, $20 per credit hour part-time for state residents; $1790 per semester full-time, $455 per credit hour part-time for nonresidents. *Faculty research:* High-temperature drilling muds, clay technology. • Dr. James E. Barrick, Adviser, 806-742-3102.

Tulane University, Department of Geology, New Orleans, LA 70118. Department awards MAT, MS, PhD. *Degree requirements:* For master's, 1 foreign language, thesis or alternative; for doctorate, dissertation. *Entrance requirements:* For master's, GRE General Test, TOEFL (minimum score of 600 required) or TSE (minimum score of 220 required), minimum B average in undergraduate course work. Application deadline: 7/1. Application fee: $35. *Expenses:* Tuition of $16,750 per year full-time, $931 per hour part-time. Fees of $230 per year full-time, $40 per hour part-time. *Financial aid:* Teaching assistantships, federal work-study, institutionally sponsored loans, and career-related internships or fieldwork available. Financial aid application deadline: 5/1; applicants required to submit GAPSFAS. • Dr. John McDowell, Chairman, 504-865-5198.

See full description on page 247.

Université de Montréal, Faculty of Arts and Sciences, Department of Geology, Montreal, PQ H3C 3J7, Canada. Department awards M Sc, PhD. Faculty: 11 full-time (0 women), 1 part-time (0 women). Matriculated students: 24 full-time (3 women), 9 part-time (4 women); includes 3 foreign. Average age 29. In 1990, 7 master's, 1 doctorate awarded. *Degree requirements:* For master's, thesis or alternative required, foreign language not required; for doctorate, 1 foreign language, dissertation. *Entrance requirements:* For doctorate, M Sc in geology or related field. Application deadline: 3/1. Application fee: $15. *Financial aid:* Fellowships, teaching assistantships, partial tuition waivers available. Financial aid application deadline: 4/15. *Faculty research:* Geochemistry, petrology, stratigraphy, tectonic, geomorphology. Total annual research budget: $835,890. • Pierre J. Lesperance, Chairman, 514-343-6821. Application contact: Walter Trzcienski, 514-343-5977.

Université Laval, Faculty of Sciences and Engineering, Department of Geology, Sainte-Foy, PQ G1K 7P4, Canada. Department awards M Sc, PhD. Matriculated students: 46 full-time (11 women), 10 part-time (3 women). 56 applicants, 68% accepted. In 1990, 5 master's, 1 doctorate awarded. *Application deadline:* 3/1. *Application fee:* $15. *Expenses:* Tuition of $792 per year full-time for Canadian residents; $5914 per year full-time for nonresidents. Fees of $120 per year full-time. *Faculty research:* Engineering, economics, regional geology. • Michel Rocheleau, Director, 418-656-7340.

University of Akron, Buchtel College of Arts and Sciences, Department of Geology, Akron, OH 44325. Department awards MS. Part-time programs available. Faculty: 9 full-time, 0 part-time. Matriculated students: 17 full-time (2 women), 3 part-time (1 woman); includes 0 minority, 4 foreign. Average age 25. 25 applicants, 80% accepted. In 1990, 6 degrees awarded. *Degree requirements:* Thesis required, foreign language not required. *Application deadline:* 3/1 (applications processed on a rolling basis). *Application fee:* $25. *Tuition:* $119.93 per credit hour for state residents; $210.93 per credit hour for nonresidents. *Financial aid:* Research assistantships, teaching assistantships, federal work-study available. *Faculty research:* Broad-range geology, petrology (sedimentary, igneous, metamorphic, and clay), geochemistry, geophysics. Total annual research budget: $80,000. • Dr. James Teeter, Head, 216-972-7631.

University of Alabama, College of Arts and Sciences, Department of Geology, Tuscaloosa, AL 35487-0132. Department awards MS, PhD. Part-time programs available. Faculty: 12 full-time (2 women), 2 part-time (0 women). Matriculated students: 17 full-time (5 women), 14 part-time (1 woman); includes 0 minority, 3 foreign. Average age 26. 16 applicants, 44% accepted. In 1990, 7 master's awarded (100% found work related to degree); 1 doctorate awarded (100% entered university research/teaching). *Degree requirements:* For master's, thesis required, foreign language not required; for doctorate, 1 foreign language (computer language can substitute), dissertation. *Entrance requirements:* For master's, GRE General Test (minimum combined score of 1500 on 3 sections required), GRE Subject Test, minimum GPA of 3.0; for doctorate, GRE General Test, GRE Subject Test. Application deadline: 7/6 (priority date, applications processed on a rolling basis). Application fee: $20. *Tuition:* $968 per year full-time, $82 per credit part-time for state residents; $2400 per year full-time, $218 per credit part-time for nonresidents. *Financial aid:* In 1990–91, 22 students received a total of $165,000 in aid awarded. 2 fellowships (0 to first-year students), 9 research assistantships (3 to first-year students), 11 teaching assistantships (5 to first-year students) were awarded; federal work-study, institutionally sponsored loans, and career-related internships or fieldwork also available. Average monthly stipend for a graduate assistantship: $835. Financial aid application deadline: 3/15. *Faculty research:* Paleontology, structure, tectonics, petrology, stratigraphy. Total annual research budget: $650,000. • D. Joe Benson, Chairperson, 205-348-1876. Application contact: C. Michael Lesher, Director of Graduate Studies, 205-348-5099.

University of Alaska Fairbanks, College of Natural Sciences, Department of Geology and Geophysics, Fairbanks, AK 99775. Department offers programs in geology (MS, PhD), geophysics (MS, PhD), geoscience (MAT). Faculty: 27 full-time (3 women), 0 part-time. Matriculated students: 29 full-time (9 women), 27 part-time (5 women); includes 1 minority (Native American), 8 foreign. Average age 28. 21 applicants, 71% accepted. In 1990, 15 master's, 1 doctorate awarded. *Degree requirements:* For master's, variable foreign language requirement, thesis; for doctorate, 1 foreign language, dissertation. *Entrance requirements:* GRE General Test, GRE Subject Test. Application deadline: 3/1. Application fee: $20. *Expenses:* Tuition of $1620 per year full-time, $90 per credit part-time for state residents; $3240 per year full-time, $180 per credit part-time for nonresidents. Fees of $464 per year full-time. *Financial aid:* In 1990–91, 6 teaching assistantships (2 to first-year students) awarded. Financial aid application deadline: 3/1. *Faculty research:* Glacial surging, Alaska as geologic fragments, natural zeolites. • Dr. S. E. Swanson, Head, 907-474-7565.

University of Alberta, Faculty of Graduate Studies and Research, Department of Geology, Edmonton, AB T6G 2J9, Canada. Department offers programs in exploration geology (Postgraduate Diploma), geology (M Sc, PhD). Matriculated students: 63 full-time, 10 part-time. *Application fee:* $0. *Expenses:* Tuition of $1495 per year full-time, $748 per year part-time for Canadian residents; $2243 per year full-time, $1121 per year part-time for nonresidents. Fees of $301 per year full-time, $118 per year part-time. • Dr. B. D. E. Chatterton, Chair, 403-492-3265.

University of Arizona, College of Arts and Sciences, Faculty of Science, Department of Geosciences, Tucson, AZ 85721. Department awards MS, PhD. Part-time programs available. Faculty: 31 full-time (3 women), 0 part-time. Matriculated students: 82 full-time (21 women), 40 part-time (12 women); includes 3 minority (all Hispanic American), 19 foreign. Average age 31. 191 applicants, 25% accepted. In 1990, 23 master's, 17 doctorates awarded. *Degree requirements:* For master's, thesis or alternative required, foreign language not required; for doctorate, 1 foreign language, dissertation. *Entrance requirements:* GRE General Test. Application deadline: 2/15 (applications processed on a rolling basis). Application fee: $25. *Expenses:* Tuition of $0 for state residents; $5406 per year full-time, $209 per credit hour part-time for nonresidents. Fees of $1528 per year full-time, $80 per credit hour part-time. *Financial aid:* In 1990–91, $503,412 in aid awarded. 1 fellowship, 79 research assistantships, 76 teaching assistantships, 12 scholarships were awarded; full tuition waivers, institutionally sponsored loans also available. Financial aid application deadline: 2/15; applicants required to submit FAF. *Faculty research:* Geology, geochemistry, geophysics. Total annual research budget: $1.881-million. • Dr. Clement C. Chase, Head, 602-621-4051. Application contact: Graduate Program Office, 602-621-6004.

See full description on page 249.

University of Arkansas, J. William Fulbright College of Arts and Sciences, Department of Geology, Fayetteville, AR 72701. Department awards MS. Faculty: 6 full-time (0 women), 0 part-time. Matriculated students: 6 full-time (2 women), 6 part-time (0 women); includes 0 minority, 0 foreign. In 1990, 7 degrees awarded. *Degree requirements:* Thesis required, foreign language not required. *Application fee:* $15. *Expenses:* Tuition of $2050 per year full-time, $103 per credit hour part-time for state residents; $4400 per year full-time, $220 per credit hour part-time for nonresidents. Fees of $50 per year full-time, $1.50 per credit hour part-time. *Financial aid:* In 1990–91, 11 teaching assistantships awarded; research assistantships also available. • Dr. Walter Manger, Chairperson, 501-575-3355.

University of British Columbia, Faculty of Science, Department of Geological Sciences, Vancouver, BC V6T 1Z1, Canada. Department offers programs in geological engineering (MA Sc, M Eng, PhD), geological sciences (MA Sc, M Eng, M Sc, PhD). *Degree requirements:* For master's, thesis; for doctorate, dissertation, comprehensive exam.

Announcement: Department's programmes are supervised by 18 full-time faculty members. Research areas include economic geology, petrology, geochemistry, mineralogy, experimental petrology and structure, geochronology, tectonics, coal, hydrogeology, computer modelling, paleontology, marine geology. Teaching assistantships available. Equipment: mass spectrometers, microprobe, XRD, XRF, AA, high-pressure apparatus, computers, machine shop. Contact W. K. Fletcher, Graduate Advisor, 604-822-2392.

University of Calgary, Faculty of Science, Department of Geology and Geophysics, Calgary, AB T2N 1N4, Canada. Department offers programs in geology (M Sc, PhD), geophysics (M Sc, PhD). Part-time programs available. Terminal master's awarded for partial completion of doctoral program. *Degree requirements:* Thesis/dissertation. *Entrance requirements:* For master's, TOEFL (minimum score of 550 required); for doctorate, TOEFL (minimum score of 550 required), B Sc or M Sc. Application fee: $25. *Tuition:* $1705 per year full-time, $427 per course part-time for Canadian residents; $3410 per year full-time, $854 per course part-time for nonresidents. *Faculty research:* Geochemistry, petrology, paleontology, stratigraphy, exploration and solid-earth geophysics.

See full description on page 253.

University of California at Berkeley, College of Letters and Science, Department of Geology and Geophysics, Berkeley, CA 94720. Department offers programs in geology (MA, MS, PhD), geophysics (MA, PhD). Faculty: 16. Matriculated students: 47 full-time, 0 part-time; includes 3 minority (all Hispanic American), 12 foreign. Terminal master's awarded for partial completion of doctoral program. *Degree requirements:* For master's, oral exam; for doctorate, dissertation, candidacy and comprehensive exams. *Entrance requirements:* GRE General Test, minimum GPA of 3.0. Application deadline: 2/11. Application fee: $40. *Expenses:* Tuition of $0. Fees of $1909 per year for state residents; $7825 per year for nonresidents. • Donald DePaolo, Chair.

University of California, Davis, Program in Geology, Davis, CA 95616. Program awards MS, PhD. Faculty: 20 full-time, 1 part-time. Matriculated students: 29. 60 applicants, 32% accepted. In 1990, 7 master's, 3 doctorates awarded. *Degree requirements:* Thesis/dissertation. *Entrance requirements:* GRE General Test, GRE Subject Test, minimum GPA of 3.0. Application deadline: 2/15. Application fee: $40. *Expenses:* Tuition of $0 for state residents; $7699 per year full-time, $3849 per year part-time for nonresidents. Fees of $2718 per year full-time, $1928 per year part-time. *Financial aid:* Fellowships, teaching assistantships available. • Graduate Adviser, 916-752-9100.

University of California, Los Angeles, College of Letters and Science, Department of Earth and Space Sciences, Program in Geology, Los Angeles, CA 90024. Program awards MS, PhD, C Phil. Matriculated students: 34 full-time (8 women), 0 part-time. 41 applicants, 34% accepted. In 1990, 3 master's, 6 doctorates, 2 C Phils awarded. *Degree requirements:* For master's, thesis or comprehensive exams required, foreign language not required; for doctorate, dissertation, written and oral qualifying exams required, foreign language not required. *Entrance requirements:* For master's and doctorate, GRE General Test. Application fee: $40. *Expenses:* Tuition of $0 for state residents; $7699 per year for nonresidents. Fees of $2907 per year. *Financial aid:* In 1990–91, 42 students received a total of $521,425 in aid awarded. 31 fellowships, 25 research assistantships, 29 teaching assistantships were awarded; full and partial tuition waivers, federal work-study, institutionally sponsored loans also available. Financial aid application deadline: 3/1. • Dr. Arthur Montana, Chair, Department of Earth and Space Sciences, 310-825-3880.

See full description on page 255.

University of California, Riverside, Graduate Division, College of Natural and Agricultural Sciences, Department of Earth Sciences, Program in Geological Sciences, Riverside, CA 92521. Program awards MS, PhD. Part-time programs available. Terminal master's awarded for partial completion of doctoral program. *Degree requirements:* For master's, comprehensive exams or thesis required, foreign language not required; for doctorate, 1 foreign language (computer language can substitute), dissertation, qualifying exams. *Entrance requirements:* GRE General Test, TOEFL (minimum score of 550 required). Application deadline: 6/1. Application fee: $40. *Tuition:* $950 per quarter full-time, $264 per quarter part-time for state residents; $3517 per quarter full-time, $1758 per quarter part-time for nonresidents. *Faculty research:* Seismic refraction, geothermics and ore genesis, orogenesis and tectonics, geohydrology, evolution and chronostratigraphy.

Announcement: The geological sciences program includes geology, geochemistry, geophysics, geomorphology, and paleontology. Research strengths include orogenesis and neotectonics, arid-lands geomorphology, stratigraphy and sedimentology, evolution and trace fossils, vertebrate biostratigraphy, mineral deposits, hydrothermal geochemistry, geothermal resources, stable-isotope geochemistry, groundwater resources and geohydrology, gravity, seismic refraction, heat flow, geoelectricity, and geomagnetism.

University of California, Santa Barbara, College of Letters and Science, Department of Geological Sciences, Santa Barbara, CA 93106. Department offers programs in geological sciences (MA, PhD), geophysics (MS). Matriculated students: 52 full-time (16 women), 0 part-time; includes 4 foreign. 61 applicants, 49% accepted. In 1990, 3 master's, 7 doctorates awarded. *Degree requirements:* For master's, 1 foreign language, thesis or exam; for doctorate, variable foreign language requirement, dissertation. *Entrance requirements:* GRE, TOEFL (minimum score of 550 required). Application deadline: 2/10. Application fee: $40. *Expenses:* Tuition of $0 for state residents; $7699 per year for nonresidents. Fees of $2307 per year. *Financial aid:* Fellowships, research assistantships, teaching assistantships, full and partial tuition waivers, federal work-study, institutionally sponsored loans, and career-related internships or fieldwork available. Financial aid application deadline: 1/31. • Michael Fuller, Chair, 805-893-3508. Application contact: Leslie Buxton, Graduate Secretary, 805-893-3329.

See full description on page 257.

University of Cincinnati, McMicken College of Arts and Sciences, Department of Geology, Cincinnati, OH 45221. Department awards MS, PhD. Faculty: 11 full-time, 0 part-time. Matriculated students: 26 full-time (10 women), 16 part-time (8 women); includes 0 minority, 5 foreign. 32 applicants, 28% accepted. In 1990, 9 master's, 2 doctorates awarded. *Degree requirements:* For master's, thesis required, foreign language not required; for doctorate, 1 foreign language, dissertation. *Entrance requirements:* GRE General Test, GRE Subject Test. Application deadline: 2/15. Application fee: $20. *Tuition:* $131 per credit hour for state residents; $261 per credit hour for nonresidents. *Financial aid:* In 1990–91, 23 teaching assistantships awarded; fellowships, research assistantships, full tuition waivers also

Directory: Geology

available. Aid available to part-time students. Average monthly stipend for a graduate assistantship: $583. Financial aid application deadline: 5/1. • Dr. I. Attila Kilinc, Acting Head, 513-556-5034.

See full description on page 261.

University of Colorado at Boulder, College of Arts and Sciences, Department of Geological Sciences, Boulder, CO 80309. Department awards MS, PhD. Faculty: 26 full-time (2 women). Matriculated students: 84 full-time (26 women), 22 part-time (2 women); includes 6 minority (2 Asian American, 3 Hispanic American, 1 Native American), 10 foreign. 103 applicants, 53% accepted. In 1990, 16 master's, 11 doctorates awarded. Terminal master's awarded for partial completion of doctoral program. *Degree requirements:* For master's, thesis or alternative, comprehensive exam required, foreign language not required; for doctorate, 1 foreign language, dissertation. *Entrance requirements:* GRE General Test. Application deadline: 3/1 (priority date, applications processed on a rolling basis). Application fee: $30 ($50 for foreign students). *Expenses:* Tuition of $2308 per year full-time, $387 per semester (minimum) part-time for state residents; $8730 per year full-time, $1455 per semester (minimum) part-time for nonresidents. Fees of $207 per semester full-time, $27.26 per semester (minimum) part-time. *Financial aid:* In 1990–91, $88,350 in aid awarded. 2 fellowships (0 to first-year students), 49 research assistantships (0 to first-year students), 21 teaching assistantships (0 to first-year students) were awarded; full tuition waivers also available. Financial aid application deadline: 3/1. *Total annual research budget:* $534,000. • Donald Runnells, Chairman, 303-492-8141.

University of Connecticut, College of Liberal Arts and Sciences, Field of Geology, Storrs, CT 06269. Field awards MS, PhD. Faculty: 12. Matriculated students: 11 full-time (2 women), 11 part-time (2 women); includes 1 minority (Native American), 3 foreign. Average age 29. 30 applicants, 33% accepted. In 1990, 1 master's, 0 doctorates awarded. Terminal master's awarded for partial completion of doctoral program. *Degree requirements:* For doctorate, dissertation. *Entrance requirements:* GRE General Test, TOEFL. Application deadline: 6/1 (priority date, applications processed on a rolling basis). Application fee: $25. *Expenses:* Tuition of $3428 per year full-time, $571 per course part-time for state residents; $8914 per year full-time, $1486 per course part-time for nonresidents. Fees of $636 per year full-time, $87 per course part-time. *Financial aid:* In 1990–91, $91,114 in aid awarded. 2 fellowships (0 to first-year students), 4 research assistantships (0 to first-year students), 6 teaching assistantships (2 to first-year students) were awarded. Financial aid application deadline: 2/15; applicants required to submit GAPSFAS or FAF. • Norman H. Gray, Head, 203-486-4434.

University of Delaware, College of Arts and Science, Department of Geology, Newark, DE 19716. Department awards MS, PhD. Part-time programs available. Faculty: 9 full-time (1 woman), 5 part-time (0 women). Matriculated students: 13 full-time (6 women), 14 part-time (5 women); includes 1 minority (Asian American), 7 foreign. 15 applicants, 67% accepted. In 1990, 5 master's awarded (20% continued full-time study); 2 doctorates awarded (100% entered university research/teaching). *Degree requirements:* For master's, thesis required, foreign language not required; for doctorate, 1 foreign language, dissertation, comprehensive exams. *Entrance requirements:* GRE General Test, GRE Subject Test. Application deadline: 7/1. Application fee: $40. *Tuition:* $179 per credit hour for state residents; $467 per credit hour for nonresidents. *Financial aid:* In 1990–91, 14 students received a total of $98,000 in aid awarded. 2 fellowships (1 to a first-year student), 1 research assistantship (to a first-year student), 10 teaching assistantships (6 to first-year students), 0 scholarships were awarded; full and partial tuition waivers also available. Average monthly stipend for a graduate assistantship: $749. *Faculty research:* Stratigraphy and sedimentology, micropaleontology, geochemistry, geophysics, coastal geology. Total annual research budget: $100,000. • Dr. Billy P. Glass, Chairman, 302-451-8229.

Announcement: The department has 13 laboratories equipped for research in aminostratigraphy, biostratigraphy, environmental geophysics, micropaleontology, coastal and marine geology and geophysics, geoarchaeology, geomorphology, mineralogy, organic geochemistry, paleoceanography, petrology, reflection seismology, sedimentology, stratigraphy, structure, taphonomy, tectonics. Major equipment includes X-ray diffractometer, liquid and gas chromatograph, SEM/EDS, drilling barge, multichannel seismograph, Sun Workstation.

University of Florida, College of Liberal Arts and Sciences, Department of Geology, Gainesville, FL 32611. Department awards MS, MST, PhD. Faculty: 22. Matriculated students: 21 full-time (5 women), 15 part-time (3 women); includes 2 minority (1 Asian American, 1 Hispanic American), 1 foreign. 12 applicants, 58% accepted. In 1990, 10 master's, 0 doctorates awarded. *Degree requirements:* For master's, thesis required, foreign language not required; for doctorate, 1 foreign language, dissertation. *Entrance requirements:* GRE General Test, minimum GPA of 3.0. Application deadline: 6/1 (priority date, applications processed on a rolling basis). Application fee: $15. *Tuition:* $87 per credit hour for state residents; $289 per credit hour for nonresidents. *Financial aid:* In 1990–91, $86,000 in aid awarded. 7 fellowships, 9 research assistantships, 10 teaching assistantships were awarded; federal work-study, institutionally sponsored loans also available. Aid available to part-time students. Financial aid application deadline: 3/1. *Faculty research:* Economic geology, engineering geology, environmental geology, paleontology, geochemistry. • Dr. Anthony Randazzo, Chairman, 904-392-2231.

Announcement: Advanced study and research in geochemistry, geomorphology, geophysics, industrial mineralogy, marine geology, paleontology, paleomagnetism, stratigraphy, sedimentary petrology, carbonate sedimentology, and hydrology. Lab facilities include equipment for paleomagnetics, geochronology, stable isotopes, X-ray spectroscopy, scanning electron microscopy, and a seismological station. Interdisciplinary approach to solving geologic problems with environmental implications. Field camp in Taos, New Mexico.

University of Georgia, College of Arts and Sciences, Department of Geology, Athens, GA 30602. Department offers programs in geochemistry (MS, PhD), geophysics (MS, PhD). Faculty: 16 full-time (2 women), 0 part-time. Matriculated students: 37 full-time (14 women), 11 part-time (2 women); includes 1 minority (Hispanic American), 11 foreign. 30 applicants, 57% accepted. In 1990, 7 master's, 3 doctorates awarded. *Degree requirements:* 1 foreign language (computer language can substitute), thesis/dissertation. *Entrance requirements:* GRE General Test. Application fee: $10. *Expenses:* Tuition of $598 per quarter full-time, $48 per quarter part-time for state residents; $1558 per quarter full-time, $144 per quarter part-time for nonresidents. Fees of $118 per quarter. *Financial aid:* Fellowships, research assistantships, teaching assistantships, assistantships available. • Dr. Michael F. Roden, Graduate Coordinator, 404-542-2416.

See full description on page 263.

University of Hawaii at Manoa, School of Ocean and Earth Science and Technology, Department of Geology and Geophysics, Honolulu, HI 96822. Department awards MS, PhD. Faculty: 42 (6 women). Matriculated students: 65 (20 women). *Degree requirements:* For master's, thesis; for doctorate, 1 foreign language, dissertation, comprehensive exams. *Entrance requirements:* GRE, TOEFL, minimum GPA of 3.0. Application deadline: 2/1. Application fee: $0. *Tuition:* $800 per semester full-time for state residents; $2405 per semester full-time for nonresidents. • Application contact: Application Inquiries, 808-956-8763.

Announcement: The department offers MS and PhD degrees in 6 fields: high-pressure geophysics and geochemistry; hydrogeology and engineering geology; marine geology and geophysics; planetary geosciences and remote sensing; seismology and solid-earth geophysics; and volcanology, petrology, and geochemistry. Tuition waivers and teaching and research assistantships are available to qualified students. Research centers include Hawaii Institute of Geophysics, University Marine Center, and Water Resources Research Center. The department has been incorporated into the newly established School of Ocean and Earth Science and Technology (SOEST), and a 7-story Pacific Ocean Science and Technology building is scheduled for completion in 1993.

See full description on page 265.

University of Houston, College of Natural Sciences and Mathematics, Department of Geosciences, 4800 Calhoun, Houston, TX 77004. Department offers programs in geophysics (MS, PhD), geosciences (MS, PhD). Part-time and evening/weekend programs available. *Degree requirements:* For master's, 1 foreign language; for doctorate, 1 foreign language, dissertation. *Entrance requirements:* GRE General Test, TOEFL. Application fee: $0. *Expenses:* Tuition of $30 per hour for state residents; $134 per hour for nonresidents. Fees of $240 per year full-time, $125 per year part-time. *Faculty research:* Isotope and basalt geochemistry, micropaleontology, plate and regional tectonics, mathematical modeling, meteoritics.

See full description on page 267.

University of Idaho, College of Graduate Studies, College of Mines and Earth Resources, Department of Geology and Geological Engineering, Moscow, ID 83843. Department offers programs in geological engineering (MS), geology (MS, PhD), geophysics (MS), hydrology (MS). Faculty: 16 full-time (2 women), 1 part-time (0 women). Matriculated students: 31 full-time (5 women), 52 part-time (8 women); includes 1 minority (Asian American), 6 foreign. In 1990, 22 master's, 5 doctorates awarded. *Degree requirements:* For doctorate, 1 foreign language, dissertation. *Entrance requirements:* For master's, minimum GPA of 2.8; for doctorate, minimum undergraduate GPA of 2.8, graduate GPA of 3.0. Application deadline: 8/1. Application fee: $20. *Expenses:* Tuition of $0 for state residents; $4146 per year for nonresidents. Fees of $818 per semester full-time, $82.75 per credit part-time. *Financial aid:* In 1990–91, 1 fellowship, 3 research assistantships (0 to first-year students), 11 teaching assistantships (0 to first-year students) awarded. Financial aid application deadline: 3/1. • Dr. Rolland R. Reid, Head, 208-885-6192.

Announcement: Baccalaureate and advanced degrees are awarded in geology, geological engineering, hydrogeology, and geophysics. Environmental themes are emphasized in studying many aspects of human activity requiring waste disposal and pollution control and abatement; mineral search and production; and manufacturing and construction. Sound basis in math-physics-chemistry required. Earth science teacher training also emphasized.

University of Illinois at Chicago, College of Liberal Arts and Sciences, Department of Geological Sciences, Chicago, IL 60680. Department offers programs in geochemistry (MS, PhD); geology (MS, PhD); geophysics (MS, PhD); geotechnical engineering and geosciences (PhD), including engineering; water resources (MS, PhD). Faculty: 11 full-time (2 women). Matriculated students: 20 full-time (4 women), 9 part-time (3 women); includes 0 minority, 16 foreign. 45 applicants, 20% accepted. In 1990, 6 master's, 1 doctorate awarded. *Degree requirements:* Thesis/dissertation. *Entrance requirements:* GRE General Test, TOEFL (minimum score of 550 required), minimum GPA of 3.75 (on a 5.0 scale). Application deadline: 7/5. Application fee: $20. *Expenses:* Tuition of $1369 per semester full-time, $521 per semester (minimum) part-time for state residents; $3840 per semester full-time, $1454 per semester (minimum) part-time for nonresidents. Fees of $458 per semester full-time, $398 per semester (minimum) part-time. *Financial aid:* In 1990–91, 1 fellowship, 1 research assistantship, 19 teaching assistantships awarded. • Norman Smith, Acting Head, 312-996-3153. Application contact: Kelvin Rodolfo, Graduate Director, 312-996-3154.

See full description on page 269.

University of Illinois at Urbana-Champaign, College of Liberal Arts and Sciences, Department of Geology, Champaign, IL 61820. Offerings include geology (MS, PhD). Department faculty: 19 full-time (1 woman), 5 part-time (0 women). *Degree requirements:* Thesis/dissertation required, foreign language not required. *Entrance requirements:* GRE General Test, TOEFL. Application fee: $25. *Tuition:* $1838 per semester full-time, $708 per semester part-time for state residents; $4314 per semester full-time, $1673 per semester part-time for nonresidents. • Dr. R. James Kirkpatrick, Head, 217-333-3542. Application contact: Dr. Wang-Ping Chen, Graduate Admissions, 217-244-4065.

University of Iowa, College of Liberal Arts, Department of Geology, Iowa City, IA 52242. Department awards MS, PhD. Faculty: 17 full-time, 0 part-time. Matriculated students: 22 full-time (6 women), 28 part-time (8 women); includes 3 minority (1 Asian American, 2 Hispanic American), 9 foreign. 33 applicants, 55% accepted. In 1990, 7 master's, 2 doctorates awarded. *Entrance requirements:* For master's, GRE General Test. Application fee: $20. *Expenses:* Tuition of $1158 per semester full-time, $387 per semester hour (minimum) part-time for state residents; $3372 per semester full-time, $387 per semester hour (minimum) part-time for nonresidents. Fees of $60 per semester (minimum). *Financial aid:* In 1990–91, 1 fellowship (to a first-year student), 19 research assistantships (3 to first-year students), 15 teaching assistantships (5 to first-year students) awarded. • Holmes Semken, Chair, 319-335-1820.

University of Kansas, College of Liberal Arts and Sciences, Department of Geology, Lawrence, KS 66045. Department awards MS, PhD. PhD offered jointly with Kansas State University. Faculty: 15 full-time (1 woman), 1 part-time (0 women). Matriculated students: 23 full-time (4 women), 27 part-time (8 women); includes 0 minority, 13 foreign. In 1990, 4 master's, 2 doctorates awarded. *Degree requirements:* For master's, thesis or alternative required, foreign language not required; for doctorate, dissertation required, foreign language not required. *Entrance requirements:* GRE General Test, GRE Subject Test, TOEFL (minimum score of 570 required). Application fee: $25. *Expenses:* Tuition of $1668 per year full-time, $56 per credit hour part-time for state residents; $5382 per year full-time, $179 per credit hour part-time for nonresidents. Fees of $338 per year full-time, $25 per credit hour part-time. *Financial aid:* Fellowships, research assistantships, teaching assistantships available. • Anthony Walton, Chairperson, 913-864-4974.

University of Kentucky, Graduate School Programs from the College of Arts and Sciences, Program in Geology, Lexington, KY 40506-0032. Program awards MS, PhD. Faculty: 17 full-time (1 woman), 0 part-time. Matriculated students: 13 full-time (4 women), 18 part-time (2 women); includes 0 minority, 6 foreign. 33 applicants, 33% accepted. In 1990, 8 master's, 5 doctorates awarded. *Degree requirements:* For master's, thesis, comprehensive exam required, foreign language not required; for doctorate, 1 foreign language, dissertation, comprehensive exam. *Entrance requirements:* For master's, GRE (verbal, quantitative, and analytical sections), minimum undergraduate GPA of 2.5; for doctorate, GRE (verbal, quantitative, and analytical sections), minimum graduate GPA of 3.0. Application deadline: 7/19 (applications processed on a rolling basis). Application fee: $20 ($25 for foreign students). *Tuition:* $1002 per semester full-time, $101 per credit hour part-time for state residents; $2782 per semester full-time, $299 per credit hour part-time for nonresidents. *Financial aid:* In 1990–91, 1 fellowship (to a first-year student), 7 research assistantships (8 to first-year students), 12 teaching assistantships (all to first-year students) awarded; federal work-study, institutionally sponsored loans also available. Aid available to part-time students. Financial aid applicants required to submit FAF. *Faculty research:* Geochemistry, carbonate hydrogeology, paleoecology, sedimentology, geophysics. • Dr. Ronald Street, Director of Graduate Studies, 606-257-4777. Application contact: Dr. Constance L. Wood, Associate Dean for Academic Administration, 606-257-4905.

University of Maine, Institute for Quaternary Studies, Orono, ME 04469. Institute awards MS. Part-time programs available. Faculty: 11 full-time (2 women), 0 part-time. Matriculated students: 8 full-time (4 women), 2 part-time (1 woman); includes 0 foreign. 6 applicants, 83% accepted. In 1990, 5 degrees awarded. *Degree requirements:* Thesis. *Entrance requirements:* TOEFL (minimum score of 550 required). Application deadline: 12/15 (priority date, applications processed on a rolling basis). Application fee: $25. *Tuition:* $100 per credit hour for state residents; $275 per credit hour for nonresidents. *Financial aid:* Research assistantships available. *Faculty research:* Geology, glacial geology, anthropology, climate. • Dr. George Denton, Director, 207-581-2190.

University of Maine, College of Sciences, Department of Geological Sciences, Orono, ME 04469. Department awards MS, PhD. Part-time programs available. Faculty: 21 full-time, 0 part-time. Matriculated students: 32 full-time (10 women), 0 part-time. In 1990, 3 master's awarded. *Degree requirements:* For master's, thesis required, foreign language not required; for doctorate, 1 foreign language, computer language, dissertation. *Entrance requirements:* For master's, GRE General Test, GRE Subject Test, TOEFL (minimum score of 550 required). Application deadline: 3/1. Application fee: $25. *Tuition:* $100 per credit hour for state residents; $275 per credit hour for nonresidents. *Financial aid:* Research assistantships, teaching assistantships, federal work-study, institutionally sponsored loans available. *Faculty research:* Appalachian bedrock geology, Quaternary studies, marine geology. Total annual research budget: $4.503-million. • Dr. Bradford A. Hall, Chairperson, 207-581-2151. Application contact: Daniel Belknap, Graduate Coordinator, 207-581-2159.

University of Manitoba, Faculty of Science, Department of Geological Sciences, Winnipeg, MB R3T 2N2, Canada. Offerings include geology (M Sc, PhD). *Degree requirements:* Thesis/dissertation. *Entrance requirements:* GRE Subject Test.

University of Maryland College Park, College of Computer, Mathematical and Physical Sciences, Department of Geology, College Park, MD 20742. Department awards MS, PhD. Faculty: 15 (4 women). Matriculated students: 7 full-time (3 women), 10 part-time (3 women); includes 1 minority (black American), 1 foreign. 15 applicants, 47% accepted. In 1990, 2 master's, 0 doctorates awarded. *Entrance requirements:* For master's, GRE General Test, minimum GPA of 3.0; for doctorate, GRE General Test. Application deadline: rolling. Application fee: $25. *Expenses:* Tuition of $143 per credit hour for state residents; $256 per credit hour for nonresidents. Fees of $171.50 per semester. *Financial aid:* In 1990–91, 5 fellowships (2 to first-year students), 1 research assistantship, 7 teaching assistantships awarded. • Dr. Michael Brown, Chairman, 301-405-4082.

See full description on page 271.

University of Massachusetts at Amherst, College of Arts and Sciences, Faculty of Natural Sciences and Mathematics, Department of Geology and Geography, Program in Geology, Amherst, MA 01003. Program awards MS, PhD. Matriculated students: 9 full-time (2 women), 35 part-time (13 women); includes 1 minority (Asian American), 4 foreign. Average age 30. 32 applicants, 63% accepted. In 1990, 8 master's, 7 doctorates awarded. *Degree requirements:* For master's, thesis optional, foreign language not required; for doctorate, 1 foreign language, dissertation. *Entrance requirements:* GRE General Test, GRE Subject Test. Application deadline: 2/1 (applications processed on a rolling basis). Application fee: $35. *Tuition:* $2568 per year full-time, $107 per credit part-time for state residents; $7920 per year full-time, $330 per credit part-time for nonresidents. *Financial aid:* Application deadline 3/1. • Dr. Richard Yuretich, Director, 413-545-0538.

University of Miami, Rosenstiel School of Marine and Atmospheric Science, Division of Marine Geology and Geophysics, Coral Gables, FL 33124. Division awards MA, MS, PhD. Faculty: 14 full-time (1 woman), 1 (woman) part-time. Matriculated students: 23 full-time (7 women), 0 part-time; includes 7 minority (3 Asian American, 4 Hispanic American), 6 foreign. Average age 27. 25 applicants, 20% accepted. In 1990, 3 master's, 3 doctorates awarded. Terminal master's awarded for partial completion of doctoral program. *Degree requirements:* For master's, thesis required (for some programs), foreign language not required; for doctorate, dissertation required, foreign language not required. *Entrance requirements:* GRE General Test, TOEFL (minimum score of 550 required). Application fee: $35. *Expenses:* Tuition of $567 per credit hour. Fees of $87 per semester full-time. *Financial aid:* In 1990–91, 2 fellowships (1 to a first-year student), 18 research assistantships (1 to a first-year student), 3 teaching assistantships awarded. *Faculty research:* Carbonate sedimentology, low-temperature geochemistry, marine geophysics, paleoceonography, deep sea petrology. Total annual research budget: $1-million. • Dr. Peter Swart, Head, 305-361-4103. Application contact: Dr. Larry Peterson, Graduate Adviser, 305-361-4692.

University of Michigan, College of Literature, Science, and the Arts, Department of Geological Sciences, Ann Arbor, MI 48109. Department offers programs in geology (MS, PhD), mineralogy (MS, PhD), oceanography: marine geology and geochemistry (MS, PhD). Faculty: 28. Matriculated students: 46 full-time (11 women), 0 part-time; includes 2 minority (1 Asian American, 1 Hispanic American), 12 foreign. 75 applicants, 56% accepted. In 1990, 11 master's, 10 doctorates awarded. Terminal master's awarded for partial completion of doctoral program. *Degree requirements:* For master's, thesis required, foreign language not required; for doctorate, variable foreign language requirement, dissertation, preliminary exam. *Entrance requirements:* GRE General Test, GRE Subject Test. Application deadline: 2/1 (applications processed on a rolling basis). Application fee: $30. *Tuition:* $3255 per semester full-time, $352 per credit (minimum) part-time for state residents; $6803 per semester full-time, $746 per credit (minimum) part-time for nonresidents. *Financial aid:* Career-related internships or fieldwork available. Financial aid application deadline: 3/15. • Henry Pollack, Chair, 313-764-1435.

University of Minnesota, Duluth, Graduate School, College of Science and Engineering, Department of Geology, Duluth, MN 55812. Department awards MS, PhD. PhD program offered in cooperation with the University of Minnesota–Twin Cities. Part-time programs available. Faculty: 6 full-time (0 women), 3 part-time (1 woman). Matriculated students: 9 full-time (1 woman), 0 part-time; includes 0 minority, 5 foreign. Average age 24. 15 applicants, 73% accepted. In 1990, 5 master's awarded (80% found work related to degree, 20% continued full-time study). *Degree requirements:* For master's, foreign language and thesis not required. *Entrance requirements:* For master's, GRE General Test. Application deadline: 7/15 (applications processed on a rolling basis). Application fee: $30. *Tuition:* $1184 per quarter full-time, $301 per credit (minimum) part-time for state residents; $2168 per quarter full-time, $602 per credit (minimum) part-time for nonresidents. *Financial aid:* In 1990–91, $87,516 in aid awarded. 2 research assistantships (1 to a first-year student), 8 teaching assistantships (1 to a first-year student) were awarded; full tuition waivers, institutionally sponsored loans, and career-related internships or fieldwork also available. Aid available to part-time students. Average monthly stipend for a graduate assistantship: $884. Financial aid applicants required to submit FAF. *Faculty research:* Precambrian geology, petrology, sedimentology, ore deposits. Total annual research budget: $144,360. • James Grant, Director of Graduate Study, 218-726-7218.

Announcement: The MS degree and a cooperative PhD are awarded. Both stress broad competence and field-based research. Almost all graduate students receive assistantships that include tuition scholarships. Duluth is well situated for living and for the study of geology. Specialties include Precambrian geology, glacial geology, economic geology, and geoarchaeology.

University of Minnesota, Twin Cities Campus, Institute of Technology, Department of Geology and Geophysics, Minneapolis, MN 55455. Department offers programs in geology (M Ed, MS, PhD), geophysics (MS, PhD). Faculty: 24. Matriculated students: 60. *Degree requirements:* For master's, thesis optional; for doctorate, dissertation. *Entrance requirements:* GRE General Test, TOEFL. *Expenses:* Tuition of $1084 per quarter full-time, $301 per credit part-time for state residents; $2168 per quarter full-time, $602 per credit part-time for nonresidents. Fees of $118 per quarter. • Peter Hudelston, Director of Graduate Studies, 612-624-1333.

See full description on page 273.

University of Mississippi, Graduate School, School of Engineering, Program in Geology, University, MS 38677. Program awards MS. Matriculated students: 3 full-time (1 woman), 1 part-time (0 women); includes 0 minority, 1 foreign. In 1990, 1 degree awarded. *Degree requirements:* Thesis required, foreign language not required. *Entrance requirements:* GRE General Test, minimum GPA of 3.0. Application deadline: 8/1. Application fee: $15 ($25 for foreign students). *Expenses:* Tuition of $1011 per semester full-time, $99 per semester part-time for state residents; $1842 per semester full-time, $180 per semester part-time for nonresidents. Fees of $219 per year full-time. • Dr. George D. Brunton, Chairman, 601-232-7499.

University of Missouri–Columbia, College of Arts and Sciences, Department of Geological Sciences, Columbia, MO 65211. Department awards MS, PhD. Faculty: 15 full-time, 0 part-time. Matriculated students: 11 full-time (1 woman), 14 part-time (3 women); includes 2 minority (1 Asian American, 1 black American), 2 foreign. In 1990, 6 master's, 0 doctorates awarded. *Degree requirements:* For master's, thesis required, foreign language not required; for doctorate, variable foreign language requirement, dissertation. *Entrance requirements:* GRE General Test, minimum GPA of 3.0. Application deadline: 8/1 (priority date, applications processed on a rolling basis). Application fee: $20 ($40 for foreign students). *Expenses:* Tuition of $89.90 per credit hour full-time, $98.35 per credit hour part-time for state residents; $244 per credit hour full-time, $252.45 per credit hour part-time for nonresidents. Fees of $123.55 per semester (minimum) full-time. • Dr. James H. Stitt, Director of Graduate Studies, 314-882-6388. Application contact: Gary L. Smith, Director of Admissions and Registrar, 314-882-7651.

University of Missouri–Kansas City, College of Arts and Sciences, Department of Urban Environmental Geology, Kansas City, MO 64110. Department awards MS. Part-time programs available. Faculty: 9 full-time, 1 part-time. Matriculated students: 2 full-time (0 women), 9 part-time (0 women); includes 0 minority, 2 foreign. In 1990, 5 degrees awarded. *Degree requirements:* Thesis required, foreign language not required. *Entrance requirements:* GRE, minimum GPA of 2.8. Application fee: $0. *Expenses:* Tuition of $2200 per year full-time, $92 per credit hour part-time for state residents; $5503 per year full-time, $229 per credit hour part-time for nonresidents. Fees of $122 per semester full-time, $9 per credit hour part-time. *Financial aid:* Full and partial tuition waivers, federal work-study, institutionally sponsored loans available. Aid available to part-time students. *Faculty research:* Origin of metalliferous shales of the midcontinent, fluid inclusion, economic geography of middle America. Total annual research budget: $22,027. • Edwin Goebel, Chairperson, 816-235-2983. Application contact: Dr. Paul Hilpman, Graduate Adviser, 816-235-2975.

University of Missouri–Rolla, School of Mines and Metallurgy, Department of Geology and Geophysics, Program in Geology, Rolla, MO 65401. Program awards MS, PhD. Part-time programs available. Faculty: 7 full-time (1 woman), 0 part-time. Average age 22. In 1990, 3 master's awarded (67% found work related to degree, 33% continued full-time study); 0 doctorates awarded. Terminal master's awarded for partial completion of doctoral program. *Degree requirements:* For master's, departmental qualifying exam required, foreign language not required; for doctorate, dissertation, departmental qualifying exam. *Entrance requirements:* GRE General Test (minimum combined score of 1100 required), GRE Subject Test. Application deadline: 7/1 (applications processed on a rolling basis). Application fee: $20 ($40 for foreign students). *Expenses:* Tuition of $2090 per year full-time, $87.10 per credit hour part-time for state residents; $5582 per year full-time, $232.60 per credit hour part-time for nonresidents. Fees of $349 per year full-time, $61.63 per semester (minimum) part-time. *Financial aid:* In 1990–91, $71,400 in aid awarded. 4 fellowships (0 to first-year students), 4 research assistantships (1 to a first-year student), 6 teaching assistantships (1 to a first-year student) were awarded; federal work-study, institutionally sponsored loans also available. Aid available to part-time students. Financial aid application deadline: 3/31. *Faculty research:* Remote sensing,

SECTION 3: EARTH AND PLANETARY SCIENCES

Directory: Geology

ore microscopy, x-ray analysis, petroleum geology, economic geology. • Richard D. Hagni, Chairman, Department of Geology and Geophysics, 314-341-4616.

Announcement: Recent highlights include the department's move into entirely new facilities of McNutt Hall, the acquisition of an infrared microscope to complement the department's unparalleled applied ore-microscopy laboratory, and the initiation of a new program in groundwater and environmental geology in cooperation with the geological engineering department.

University of Montana, College of Arts and Sciences, Department of Geology, Missoula, MT 59812. Department awards MS, PhD. *Degree requirements:* Thesis/dissertation required, foreign language not required. *Entrance requirements:* GRE General Test. Application deadline: 2/15. Application fee: $20. *Tuition:* $495 per quarter hour full-time for state residents; $1239 per quarter hour full-time for nonresidents.

University of Nebraska–Lincoln, College of Arts and Sciences, Department of Geology, Lincoln, NE 68588. Department awards MS, PhD. Faculty: 9 full-time (1 woman), 0 part-time. Matriculated students: 18 full-time (2 women), 24 part-time (4 women); includes 2 minority (both Asian American), 13 foreign. Average age 32. In 1990, 5 master's, 3 doctorates awarded. Terminal master's awarded for partial completion of doctoral program. *Degree requirements:* For master's, variable foreign language requirement, thesis, departmental qualifying exam; for doctorate, variable foreign language requirement, dissertation, comprehensive exams, departmental qualifying exam. *Entrance requirements:* GRE General Test, TOEFL (minimum score of 500 required). Application deadline: 5/1 (priority date, applications processed on a rolling basis). Application fee: $25. *Expenses:* Tuition of $75.75 per credit hour for state residents; $187.25 per credit hour for nonresidents. Fees of $161 per year full-time. *Financial aid:* Fellowships, research assistantships, teaching assistantships, federal work-study available. Aid available to part-time students. Financial aid application deadline: 2/15. • Dr. Eric Durrance, Chairperson, 402-472-2663.

University of Nevada, Las Vegas, College of Science and Mathematics, Department of Geoscience, Las Vegas, NV 89154. Department awards MS. Part-time programs available. Faculty: 9 full-time (1 woman), 5 part-time (0 women). Matriculated students: 8 full-time (2 women), 17 part-time (5 women); includes 0 minority, 0 foreign. 17 applicants, 47% accepted. In 1990, 6 degrees awarded. *Degree requirements:* Thesis, comprehensive exam required, foreign language not required. *Entrance requirements:* GRE Subject Test, minimum GPA of 2.75. Application deadline: 6/15. Application fee: $20. *Expenses:* Tuition of $66 per credit. Fees of $1800 per semester for nonresidents. *Financial aid:* In 1990–91, 7 research assistantships awarded; teaching assistantships also available. Financial aid application deadline: 3/1. • Dr. David Weide, Chair, 702-739-3262. Application contact: Graduate College Admissions Evaluator, 702-739-3320.

University of Nevada, Reno, Mackay School of Mines, Department of Geological Sciences, Reno, NV 89557. Department awards Geol E, MS, PhD. Programs offered in geochemistry (MS, PhD), geology (MS, PhD). Faculty: 21. Matriculated students: 53 (15 women). *Degree requirements:* For master's, thesis optional, foreign language not required; for doctorate, 1 foreign language, dissertation. *Entrance requirements:* For master's, GRE General Test, TOEFL, minimum GPA of 2.75; for doctorate, GRE General Test, TOEFL, minimum GPA of 3.0. Application deadline: 8/1 (priority date, applications processed on a rolling basis). Application fee: $20. *Expenses:* Tuition of $0 for state residents; $3600 per year full-time, $66 per credit hour part-time for nonresidents. Fees of $66 per credit hour. *Financial aid:* Research assistantships, teaching assistantships, full tuition waivers, institutionally sponsored loans available. Average monthly stipend for a graduate assistantship: $740. Financial aid applicants required to submit FAF. *Faculty research:* Hydrothermal ore deposits, metamorthic and igneous petrogenesis, sedimentary rock record of earth history, field and petrographic investigation of magnetism, rock fracture mechanics. Total annual research budget: $983,540. • Dr. Bob Watters, Chairman, 702-784-6050.

See full description on page 275.

University of Nevada, Reno, Mackay School of Mines, Program in Hydrology and Hydrogeology, Reno, NV 89557. Program awards MS, PhD. Part-time programs available. Faculty: 22 (1 woman). Matriculated students: 33 (7 women); includes 6 foreign. Average age 30. In 1990, 15 master's, 2 doctorates awarded. *Degree requirements:* For master's, thesis optional, foreign language not required; for doctorate, 1 foreign language, dissertation. *Entrance requirements:* For master's, GRE General Test, TOEFL, minimum GPA of 2.75; for doctorate, GRE General Test, TOEFL, minimum GPA of 3.0. Application deadline: 8/1 (priority date, applications processed on a rolling basis). Application fee: $20. *Expenses:* Tuition of $0 for state residents; $3600 per year full-time, $66 per credit hour part-time for nonresidents. Fees of $66 per credit hour. *Financial aid:* In 1990–91, 1 fellowship, 16 research assistantships, 0 teaching assistantships awarded; partial tuition waivers, institutionally sponsored loans, and career-related internships or fieldwork also available. Average monthly stipend for a graduate assistantship: $740. Financial aid applicants required to submit FAF. *Faculty research:* Hydrogeology, groundwater, water resources, surface water, soil science. • Dr. W. Berry Lyons, Head, 702-784-6465.

University of New Brunswick, Faculty of Science, Department of Geology, Fredericton, NB E3B 5A3, Canada. Department awards M Sc, PhD. *Degree requirements:* Thesis/dissertation. *Entrance requirements:* TOEFL, minimum GPA of 3.0. Application deadline: 3/1 (priority date). *Expenses:* Tuition of $2100 per year. Fees of $45 per year. *Financial aid:* Research assistantships, teaching assistantships available. • Dr. H. W. Van Der Poll, Chairperson, 506-453-4803. Application contact: Dr. P. Stringer, Director of Graduate Studies, 506-453-4804.

University of New Hampshire, College of Engineering and Physical Sciences, Department of Earth Sciences, Durham, NH 03824. Offerings include earth sciences geology (MS, PhD). Department faculty: 26 full-time. *Degree requirements:* For master's, thesis required, foreign language not required; for doctorate, 1 foreign language, dissertation. *Entrance requirements:* GRE General Test. Application deadline: 7/1 (priority date, applications processed on a rolling basis). Application fee: $25. *Tuition:* $1645 per semester full-time, $183 per credit hour part-time for state residents; $4920 per semester full-time, $547 per credit hour part-time for nonresidents. • Dr. S. Lawrence Dingman, Chairperson, 603-862-1718. Application contact: Dr. Francis Birch, 603-862-1718.

University of New Mexico, College of Arts and Sciences, Department of Geology, Albuquerque, NM 87131. Department awards MS, PhD. Faculty: 12 full-time (0 women), 9 part-time (2 women). Terminal master's awarded for partial completion of doctoral program. *Degree requirements:* For master's, computer language, thesis required, foreign language not required; for doctorate, 1 foreign language, computer language, dissertation. *Entrance requirements:* For master's, GRE General Test; for doctorate, GRE General Test, GRE Subject Test. Application deadline: 1/31. Application fee: $25. *Expenses:* Tuition of $467 per semester (minimum) full-time, $67.50 per credit hour part-time for state residents; $1549 per semester (minimum) full-time, $67.50 per credit hour part-time for nonresidents. Fees of $16 per semester. *Financial aid:* Fellowships, research assistantships, teaching assistantships, federal work-study, institutionally sponsored loans, and career-related internships or fieldwork available. Aid available to part-time students. Financial aid application deadline: 1/31. *Faculty research:* Geochemistry, meteoritics, tectonics. • Stephen G. Wells, Chairman, 505-277-4204.

See full description on page 277.

University of New Orleans, College of Sciences, Department of Geology and Geophysics, New Orleans, LA 70148. Department offers programs in geology (MS), geophysics (MS). Evening/weekend programs available. Faculty: 16 full-time (1 woman), 2 part-time (1 woman). Matriculated students: 23 full-time, 29 part-time; includes 3 minority (all black American), 6 foreign. Average age 24. 30 applicants, 33% accepted. In 1990, 15 degrees awarded. *Degree requirements:* Thesis required, foreign language not required. *Entrance requirements:* GRE General Test. Application deadline: 7/1 (priority date, applications processed on a rolling basis). Application fee: $20. *Tuition:* $962 per quarter hour full-time for state residents; $2308 per quarter hour full-time for nonresidents. *Financial aid:* In 1990–91, $163,000 in aid awarded. 2 fellowships (1 to a first-year student), 1 research assistantship (0 to first-year students), 15 teaching assistantships (7 to first-year students) were awarded; federal work-study, institutionally sponsored loans, and career-related internships or fieldwork also available. *Faculty research:* Continental margin structure and seismology, burial diagenesis of siliclastic sediments, tectonics at convergent plate margins, continental shelf sediment stability, early diagenesis of carbonates. Total annual research budget: $150,000. • William C. Craig, Chairman, 504-286-6793. Application contact: William H. Busch, Graduate Coordinator, 504-286-7230.

University of North Carolina at Chapel Hill, College of Arts and Sciences, Department of Geology, Chapel Hill, NC 27599. Department awards MA, MS, PhD. Faculty: 20 full-time, 0 part-time. Matriculated students: 37 full-time (8 women), 0 part-time; includes 1 minority (Hispanic American), 5 foreign. 45 applicants, 31% accepted. In 1990, 9 master's, 1 doctorate awarded. *Degree requirements:* For master's, thesis, comprehensive exam required, foreign language not required; for doctorate, 1 foreign language, dissertation, comprehensive exam. *Entrance requirements:* GRE General Test (minimum combined score of 1000 required), GRE Subject Test, minimum GPA of 3.0. Application deadline: 2/15. Application fee: $35. *Tuition:* $621 per semester full-time for state residents; $3555 per semester full-time for nonresidents. *Financial aid:* In 1990–91, 2 fellowships, 15 research assistantships, 16 teaching assistantships awarded. • Paul Geoffrey Feiss, Chairman, 919-962-0693.

University of North Carolina at Wilmington, College of Arts and Sciences, Department of Earth Sciences, Wilmington, NC 28403. Department offers program in geology (MS). Faculty: 4 full-time (0 women), 0 part-time. Matriculated students: 3 full-time (1 woman), 6 part-time (3 women); includes 0 minority, 0 foreign. Average age 31. 6 applicants, 100% accepted. In 1990, 0 degrees awarded. *Degree requirements:* Thesis, written and oral comprehensive exams. *Entrance requirements:* GRE General Test (minimum combined score of 1000 required), GRE Subject Test (geology), minimum B average in undergraduate major and basic courses for prerequisite to geology. Application deadline: 7/1 (applications processed on a rolling basis). Application fee: $15. *Tuition:* $651 per semester full-time for state residents; $3178 per semester full-time for nonresidents. *Financial aid:* In 1990–91, $5000 in aid awarded. 5 teaching assistantships were awarded; federal work-study and career-related internships or fieldwork also available. Aid available to part-time students. Financial aid application deadline: 3/15; applicants required to submit FAF. • Dr. Richard Laws, Chairman, 919-395-3736. Application contact: Dr. Eric G. Bolen, Dean of Graduate School, 919-395-3135.

University of North Dakota, School of Engineering and Mines, Department of Geology, Grand Forks, ND 58202. Department awards MS, PhD. Faculty: 11 full-time (1 woman), 0 part-time. Matriculated students: 11 full-time (3 women), 12 part-time (1 woman). 13 applicants, 100% accepted. In 1990, 4 master's, 2 doctorates awarded. *Degree requirements:* For master's, thesis; for doctorate, 1 foreign language (computer language can substitute), dissertation. *Entrance requirements:* GRE General Test, GRE Subject Test, TOEFL (minimum score of 550 required), minimum GPA of 3.0. Application deadline: 3/15 (priority date, applications processed on a rolling basis). Application fee: $20. *Tuition:* $2250 per year full-time, $94 per semester hour part-time for state residents; $5616 per year full-time, $234 per semester hour part-time for nonresidents. *Financial aid:* In 1990–91, 7 research assistantships, 10 teaching assistantships awarded; fellowships, full and partial tuition waivers, federal work-study, institutionally sponsored loans, and career-related internships or fieldwork also available. Financial aid application deadline: 3/15. • Dr. F. Karner, Chair, 701-777-2811.

University of Oklahoma, College of Geosciences, School of Geology and Geophysics, Program in Geology, Norman, OK 73019. Program awards MS, PhD. Faculty: 12 full-time (0 women), 1 part-time (0 women). Matriculated students: 36; includes 0 minority. Average age 25. In 1990, 13 master's awarded (8% found work related to degree, 92% continued full-time study). *Degree requirements:* For master's, computer language, thesis, comprehensive exam required, foreign language not required; for doctorate, 1 foreign language, computer language, dissertation, general exam. *Entrance requirements:* For master's, GRE General Test, TOEFL (minimum score of 550 required), bachelor's degree in field; for doctorate, GRE General Test, TOEFL (minimum score of 550 required). Application deadline: 2/1 (priority date, applications processed on a rolling basis). Application fee: $10. *Expenses:* Tuition of $63 per credit hour for state residents; $192 per credit hour for nonresidents. Fees of $67.50 per semester. *Financial aid:* In 1990–91, $193,270 in aid awarded. 16 research assistantships, 11 teaching assistantships were awarded; partial tuition waivers, federal work-study, institutionally sponsored loans also available. Financial aid application deadline: 2/1. *Faculty research:* Petroleum geology, stratigraphy, sedimentology, geochemistry, mineralogy/petrology. • Dr. Charles Gilbert, Director, School of Geology and Geophysics, 405-325-3253.

See full description on page 279.

University of Oregon, Graduate School, College of Arts and Sciences, Department of Geological Sciences, Eugene, OR 97403. Department awards MA, MS, PhD. Faculty: 14 full-time (1 woman), 2 part-time (0 women). Matriculated students: 30 full-time (8 women), 9 part-time (1 woman); includes 4 minority (2 Asian American, 2 Hispanic American), 5 foreign. Average age 26. 79 applicants, 11% accepted. In 1990, 3 master's awarded (67% found work related to degree, 33% continued full-time study); 4 doctorates awarded (25% entered university research/teaching, 75% found other work related to degree). *Degree requirements:* For master's, 1 foreign language (computer language can substitute), thesis; for doctorate, dissertation. *Entrance requirements:* GRE General Test, GRE Subject Test. Application deadline: 8/15. Application fee: $40. *Tuition:* $1171 per quarter full-time, $247 per credit part-time for state residents; $1980 per quarter full-time, $336 per credit part-time for nonresidents. *Financial aid:* In 1990–91, $120,000 in aid awarded. 0 fellowships, 9 research assistantships (0 to first-year students), 16 teaching assistantships (4 to first-year students) were awarded; federal work-study and career-related internships or fieldwork also available. *Faculty research:* Petrology, geochemistry, structural geology, sedimentology/paleontology, geophysics. Total annual research budget: $800,000. • Dr. Jack M. Rice, Chairman, 503-346-4586.

Announcement: Graduate study in the University of Oregon Department of Geological Sciences is open to students in geology, chemistry, physics, and biology. Areas of active research include biostratigraphy, economic geology, experimental petrology, geochemistry, geophysics, igneous petrology, metamorphic petrology, mineral spectroscopy, paleobotany, paleontology, palepedology, sedimentary petrology, seismology, structural geology, tectonics, and neotectonics.

University of Ottawa, Faculty of Science, Department of Geology, Ottawa, ON K1N 6N5, Canada. Department awards M Sc, PhD. Offered jointly with Carleton University. *Degree requirements:* Thesis/dissertation required, foreign language not required. *Entrance requirements:* For master's, honors bachelor's degree or equivalent, minimum B average; for doctorate, minimum B+ average. Application deadline: 3/1. Application fee: $10. *Faculty research:* Pre-Cambrian geology, structural geology and geodynamics, Arctic studies, resource geology.

University of Pennsylvania, School of Arts and Sciences, Graduate Group in Geology, Philadelphia, PA 19104. Group awards MS, PhD. Part-time programs available. Faculty: 8 full-time (0 women), 2 part-time (1 woman). Matriculated students: 10 full-time (3 women), 3 part-time (0 women); includes 1 minority (Asian American), 3 foreign. 19 applicants, 37% accepted. In 1990, 4 master's, 3 doctorates awarded. *Degree requirements:* 1 foreign language, thesis/dissertation. *Entrance requirements:* GRE General Test. Application fee: $40. *Expenses:* Tuition of $15,619 per year full-time, $1978 per course part-time. Fees of $965 per year full-time, $112 per course part-time. *Financial aid:* Fellowships, research assistantships, teaching assistantships, federal work-study, and career-related internships or fieldwork available. Financial aid application deadline: 2/1. *Faculty research:* Isotope geochemistry, regional tectonics, environmental geology, metamorphic and igneous petrology, paleontology. • Dr. Hermann W. Pfefferkorn, Chairman, 215-898-5156.

Announcement: Three areas of concentration: environmental geology currently includes biogeochemistry of forested ecosystems and industrial contamination of aquifers; in tectonics, fission track and isotopic analyses elucidate regional geologic histories; in paleobiology, evolutionary characteristics of flora and fauna are analyzed, together with the processes of their preservation as fossils.

University of Pittsburgh, Faculty of Arts and Sciences, Department of Geology and Planetary Science, Pittsburgh, PA 15260. Department awards MS, PhD. Part-time programs available. Faculty: 12 full-time (0 women), 1 (woman) part-time. Matriculated students: 17 full-time (5 women), 18 part-time (6 women); includes 4 minority (2 Asian American, 2 black American), 5 foreign. 59 applicants, 17% accepted. In 1990, 6 master's, 1 doctorate awarded. *Degree requirements:* Thesis/dissertation required, foreign language not required. *Entrance requirements:* GRE General Test, TOEFL. Application deadline: 8/1 (applications processed on a rolling basis). Application fee: $15 ($25 for foreign students). *Expenses:* Tuition of $2920 per semester full-time, $241 per credit part-time for state residents; $5840 per semester full-time, $482 per credit part-time for nonresidents. Fees of $156 per year. *Financial aid:* In 1990–91, 13 students received a total of $94,930 in aid awarded. 4 research assistantships (0 to first-year students), 9 teaching assistantships (3 to first-year students) were awarded; full and partial tuition waivers, federal work-study, institutionally sponsored loans, and career-related internships or fieldwork also available. Aid available to part-time students. Average monthly stipend for a graduate assistantship: $900. Financial aid application deadline: 3/1; applicants required to submit FAF. *Faculty research:* Plate tectonics as basement studies, sedimentology and image analysis, marine paleoecology and geoarchaeology, geophysics and paleomagnetics, planetary geology and meteoritics. Total annual research budget: $490,472. • Dr. Thomas H. Anderson, Chairman, 412-624-8780. Application contact: Dr. Walter L. Pilant, Graduate Adviser, 412-624-8870.

See full description on page 281.

University of Regina, Faculty of Graduate Studies and Research, Faculty of Science, Department of Geology, Regina, SK S4S 0A2, Canada. Department awards M Sc. Faculty: 9 full-time, 0 part-time. Matriculated students: 19. In 1990, 4 degrees awarded. *Degree requirements:* Thesis required, foreign language not required. *Application deadline:* 7/2 (applications processed on a rolling basis). *Tuition:* $1500 per year full-time, $242 per year (minimum) part-time. • Dr. L. Vigrass, Head, 306-585-4571.

University of Rhode Island, College of Arts and Sciences, Department of Geology, Kingston, RI 02881. Department awards MS. *Degree requirements:* Computer language required, thesis optional. *Application deadline:* 4/15. *Application fee:* $25. *Expenses:* Tuition of $2575 per year full-time, $120 per credit hour part-time for state residents; $5900 per year full-time, $274 per credit hour part-time for nonresidents. Fees of $696 per year full-time.

University of Rochester, College of Arts and Science, Department of Geological Sciences, Rochester, NY 14627-0001. Department offers program in geology (MS, PhD). Faculty: 8 full-time, 2 part-time. Matriculated students: 24 full-time (9 women), 0 part-time; includes 0 minority, 9 foreign. 64 applicants, 13% accepted. In 1990, 7 master's, 4 doctorates awarded. Terminal master's awarded for partial completion of doctoral program. *Degree requirements:* Thesis/dissertation required, foreign language not required. *Entrance requirements:* GRE General Test. Application deadline: 2/15 (priority date). Application fee: $25. *Expenses:* Tuition of $473 per credit hour. Fees of $243 per year full-time. *Financial aid:* Fellowships, research assistantships, teaching assistantships, and career-related internships or fieldwork available. Financial aid application deadline: 2/15. *Faculty research:* Paleontology, geochemistry, petrology of igneous and metamorphic rocks and mineral deposits, structural geology, geophysics. • Asish Basu, Chair, 716-275-5713.

Announcement: Department of Geological Sciences offers broad range of advanced studies in isotope geochemistry, petrology, sedimentology, structural geology, tectonics, vertebrate and invertebrate paleontology, paleoecology, and environmental studies. Excellent analytical facilities, including thermal ionization, noble gas, accelerator mass spectrometers, and computational support, offer students outstanding laboratory and field-oriented research opportunities. Teaching and research assistantship and fellowship support available.

University of Saskatchewan, Colleges of Arts and Sciences and Engineering, Department of Geological Sciences, Saskatoon, SK S7N 0W0, Canada. Department awards M Sc, PhD. *Degree requirements:* Thesis/dissertation. *Entrance requirements:* TOEFL. Application fee: $0.

See full description on page 283.

University of South Carolina, Graduate School, College of Science and Mathematics, Department of Geological Sciences, Columbia, SC 29208. Department awards MS, PhD. Faculty: 24 full-time (1 woman), 0 part-time. Matriculated students: 80 full-time (30 women), 10 part-time; includes 3 minority (2 black American, 1 Hispanic American), 15 foreign. 90 applicants, 22% accepted. In 1990, 15 master's awarded (60% found work related to degree, 40% continued full-time study); 9 doctorates awarded. Terminal master's awarded for partial completion of doctoral program. *Degree requirements:* For master's, thesis required, foreign language not required; for doctorate, dissertation, published paper required, foreign language not required. *Entrance requirements:* GRE General Test. Application deadline: 3/1 (priority date, applications processed on a rolling basis). Application fee: $25. *Tuition:* $1404 per semester full-time. *Financial aid:* In 1990–91, 40 research assistantships (10 to first-year students), 28 teaching assistantships (10 to first-year students) awarded; federal work-study and career-related internships or fieldwork also available. Financial aid application deadline: 3/1. *Faculty research:* Sedimentary geology, geochemistry, tectonics, marine geology, petrology. • Dr. Robert Thunell Jr., Chairman, 803-777-7593. Application contact: Dr. Pradeep Talwani, Director of Graduate Studies, 803-777-4519.

University of Southern California, Graduate School, College of Letters, Arts and Sciences, Division of Natural Sciences and Mathematics, Department of Geological Sciences, Los Angeles, CA 90089. Department awards MS, PhD. Part-time programs available. Faculty: 17 (1 woman). Matriculated students: 64 full-time (17 women), 8 part-time (4 women); includes 6 minority (5 Asian American, 1 Hispanic American), 27 foreign. Average age 26. 70 applicants, 44% accepted. In 1990, 11 master's, 3 doctorates awarded. *Degree requirements:* For doctorate, dissertation. *Entrance requirements:* GRE General Test. Application deadline: 7/1 (priority date). Application fee: $50. *Expenses:* Tuition of $12,120 per year full-time, $505 per unit part-time. Fees of $280 per year. *Financial aid:* 54 students received aid. Fellowships, research assistantships, teaching assistantships, federal work-study, institutionally sponsored loans available. Average monthly stipend for a graduate assistantship: $878. Financial aid application deadline: 3/1. • Dr. Thomas L. Henyey, Chair, 213-740-3832. Application contact: Rene Kirby, Academic Administrator, 213-740-6106.

Announcement: Specialties in biostratigraphy, chemical oceanography, earthquake prediction and hazards, exploration geophysics, heat flow, geomorphology, glacial geology, igneous petrology, metamorphic petrology, invertebrate paleontology, isotope geochemistry, stable isotope geochemistry, marine geochemistry and geophysics, marine geology–sedimentology, micropaleontology, paleoecology-paleobiology, paleoceanography, physical oceanography, rock magnetism, rock mechanics, stratigraphy, basin analysis, sedimentary petrology, coastal sedimentology, seismology and regional seismicity, statistical applications to geology, structural geology–regional tectonics, tectonophysics, structural petrology.

University of Southern Mississippi, College of Science and Technology, Department of Geology, Hattiesburg, MS 39406. Department awards MS. Part-time programs available. Faculty: 7 full-time (1 woman), 0 part-time. Matriculated students: 7 full-time (1 woman), 7 part-time (4 women); includes 0 minority. In 1990, 1 degree awarded (100% continued full-time study). *Degree requirements:* Computer language, thesis required, foreign language not required. *Entrance requirements:* GRE General Test (minimum combined score of 1000 required), BS in geology, minimum GPA of 2.75. Application deadline: 8/9 (priority date, applications processed on a rolling basis). Application fee: $0 ($25 for foreign students). *Expenses:* Tuition of $968 per semester full-time, $93 per semester hour part-time. Fees of $12 per semester part-time for state residents; $591 per year full-time, $12 per semester part-time for nonresidents. *Financial aid:* In 1990–91, 9 students received a total of $45,000 in aid awarded. 1 research assistantship, 6 teaching assistantships (all to first-year students) were awarded; fellowships, full tuition waivers, federal work-study, and career-related internships or fieldwork also available. Financial aid application deadline: 3/15. *Faculty research:* Volcanic rocks and associated minerals, marine stratigraphy and seismology, hydrology, micropaleontology, isotopegeology. Total annual research budget: $140,000. • Dr. Daniel A. Sundeen, Chairman, 601-266-4532.

University of South Florida, College of Arts and Sciences, Department of Geology, Tampa, FL 33620. Department offers programs in geology (MS), hydrogeology (MS, Adv C). Part-time programs available. Faculty: 10 full-time (1 woman). Matriculated students: 25 full-time (9 women), 21 part-time (3 women); includes 3 minority (1 black American, 2 Hispanic American), 4 foreign. Average age 30. 25 applicants, 68% accepted. In 1990, 9 master's awarded. *Degree requirements:* For master's, thesis (for some programs). *Entrance requirements:* For master's, GRE General Test (minimum combined score of 1000 required), minimum GPA of 3.0 in last two years. Application fee: $15. *Tuition:* $79.40 per credit hour for state residents; $241.33 per credit hour for nonresidents. *Financial aid:* In 1990–91, 23 students received a total of $73,740 in aid awarded. 1 fellowship, 11 research assistantships, 10 teaching assistantships were awarded. *Faculty research:* Coastal geology, environmental geology and hydrogeology, paleontology and volcanology. Total annual research budget: $485,000. • Mark T. Stewart, Graduate Admissions, 813-974-2236.

University of Southwestern Louisiana, College of Engineering, Department of Geology, Lafayette, LA 70504. Department awards MS. Part-time programs available. Faculty: 6 full-time (0 women), 0 part-time. Matriculated students: 2 full-time (1 woman), 22 part-time (4 women); includes 1 minority (Hispanic American), 0 foreign. Average age 27. 7 applicants, 71% accepted. In 1990, 6 degrees awarded. *Degree requirements:* Thesis required, foreign language not required. *Entrance requirements:* GRE General Test. Application deadline: 8/15. Application fee: $5. *Tuition:* $1560 per year full-time, $228 per credit (minimum) part-time for state residents; $3310 per year full-time, $228 per credit (minimum) part-time for nonresidents. *Financial aid:* In 1990–91, $44,500 in aid awarded. 9 teaching assistantships were awarded; full and partial tuition waivers, federal work-study also available. Aid available to part-time students. Financial aid application deadline: 5/1. *Faculty research:* Subsurface geology and geomorphology of the Gulf Coast, carbonate petrology, geochemistry of peat, isotope geochemistry of volcanic rocks. • Dr. Daniel Tucker, Head, 318-231-5353. Application contact: Dr. Richard Birdseye, Graduate Coordinator, 318-231-5352.

University of Tennessee, Knoxville, College of Liberal Arts, Department of Geological Sciences, Knoxville, TN 37996. Department offers program in geology (MS, PhD). Part-time programs available. Faculty: 17 (2 women). Matriculated students: 35 full-time (5 women), 12 part-time (4 women); includes 1 minority (Asian American), 9 foreign. 31 applicants, 55% accepted. In 1990, 2 master's, 2 doctorates awarded. *Degree requirements:* For master's, thesis required, foreign language not required; for doctorate, 1 foreign language, dissertation. *Entrance requirements:* GRE General Test, GRE Subject Test, TOEFL (minimum score of 525 required), minimum GPA of 2.5. Application deadline: 2/1 (priority date, applications processed on a rolling basis). Application fee: $15. *Tuition:* $1086 per semester full-time, $142 per credit

Directory: Geology

hour part-time for state residents; $2768 per semester full-time, $308 per credit hour part-time for nonresidents. *Financial aid:* In 1990–91, 0 fellowships, 8 research assistantships, 26 teaching assistantships, 0 assistantships awarded; federal work-study, institutionally sponsored loans also available. Financial aid application deadline: 2/1; applicants required to submit FAF. • Dr. Hap McSween, Head, 615-974-2366.

University of Texas at Arlington, College of Science, Department of Geology, Arlington, TX 76019. Department awards MS. Matriculated students: 6 full-time (4 women), 31 part-time (9 women); includes 1 minority (Hispanic American), 0 foreign. 3 applicants, 67% accepted. In 1990, 7 degrees awarded. *Degree requirements:* 1 foreign language, thesis. *Entrance requirements:* GRE General Test, GRE Subject Test. Application deadline: rolling. Application fee: $25. *Tuition:* $40 per hour for state residents; $148 per hour for nonresidents. *Financial aid:* In 1990–91, 2 research assistantships, 4 teaching assistantships awarded. *Total annual research budget:* $43,086. • Dr. Brooks Ellwood, Acting Chair, 817-273-2987. Application contact: Dr. Burke Burkart, Graduate Adviser, 817-273-2987.

University of Texas at Austin, Graduate School, College of Natural Sciences, Department of Geological Sciences, Austin, TX 78712. Department awards MA, PhD. Matriculated students: 179 full-time (41 women), 0 part-time; includes 5 minority (3 Asian American, 2 Hispanic American), 39 foreign. 90 applicants, 67% accepted. In 1990, 15 master's, 4 doctorates awarded. *Entrance requirements:* GRE. Application deadline: 2/1 (priority date, applications processed on a rolling basis). Application fee: $40 ($75 for foreign students). *Tuition:* $510.30 per semester for state residents; $1806 per semester for nonresidents. *Financial aid:* Fellowships, research assistantships, teaching assistantships available. Financial aid application deadline: 3/1. • Dr. Clark R. Wilson, Chairman, 512-471-5172. Application contact: Dr. William Carlson, Graduate Adviser, 512-471-6098.

See full description on page 285.

University of Texas at Dallas, School of Natural Sciences and Mathematics, Program in Geosciences, Richardson, TX 75083-0688. Program awards MS, PhD. Part-time and evening/weekend programs available. Faculty: 16 full-time (0 women), 7 part-time (0 women). Matriculated students: 44 full-time (10 women), 35 part-time (7 women); includes 7 minority (3 Asian American, 1 black American, 3 Hispanic American), 21 foreign. Average age 33. In 1990, 7 master's, 5 doctorates awarded. *Degree requirements:* Computer language, thesis/dissertation required, foreign language not required. *Entrance requirements:* GRE General Test (minimum combined score of 1000 required), TOEFL (minimum score of 550 required), minimum GPA of 3.0 in upper level course work in field. Application deadline: 7/15 (applications processed on a rolling basis). Application fee: $0 ($75 for foreign students). *Expenses:* Tuition of $360 per semester full-time, $100 per semester (minimum) part-time for state residents; $2196 per semester full-time, $122 per semester hour (minimum) part-time for nonresidents. Fees of $338 per semester full-time, $22 per hour (minimum) part-time. *Financial aid:* In 1990–91, 1 student received a total of $8175 in aid awarded. Fellowships, research assistantships, teaching assistantships, federal work-study, and career-related internships or fieldwork available. Aid available to part-time students. Financial aid application deadline: 11/1; applicants required to submit FAF. *Faculty research:* Geochemistry, structural geology-tectonics, sedimentology-stratigraphy, micropaleontology. • Dr. James L. Carter, Head, 214-690-2401.

See full description on page 287.

University of Texas at El Paso, College of Science, Department of Geological Sciences, 500 West University Avenue, El Paso, TX 79968. Department offers programs in geological sciences (MS, PhD), geophysics (MS). Matriculated students: 33 full-time (8 women), 8 part-time (1 woman); includes 3 minority (all Hispanic American), 19 foreign. In 1990, 4 master's, 1 doctorate awarded. *Application deadline:* 7/1 (priority date, applications processed on a rolling basis). *Application fee:* $0 ($50 for foreign students). *Expenses:* Tuition of $360 per semester full-time, $100 per semester (minimum) part-time for state residents; $2304 per semester full-time, $128 per credit hour (minimum) part-time for nonresidents. Fees of $137 per semester full-time, $28.50 per semester (minimum) part-time. *Financial aid:* Application deadline 3/1. • Randy Keller, Chair, 915-747-5501. Application contact: Diana Guerrero, Admissions Office, 915-747-5576.

University of Texas at San Antonio, College of Sciences and Engineering, Division of Earth and Physical Sciences, San Antonio, TX 78285. Offerings include geology (MS). *Entrance requirements:* GRE. Application deadline: 7/1 (applications processed on a rolling basis). Application fee: $20. *Expenses:* Tuition of $100 per semester hour (minimum) for state residents; $128 per semester hour (minimum) for nonresidents. Fees of $48 per semester hour (minimum). • Dr. Robert K. Smith, Director, 512-691-4455.

Announcement: Graduate study leads to the MS degree with following emphases: hydrogeology (fluid properties and flow of groundwater and hydrogeochemical models as they pertain to water-rock interaction at various temperatures and pressures) and applied geology (standard graduate geology program plus environmental and engineering geology). Out-of-state residents who are employed at least half-time as teaching or research assistants pay in-state tuition.

University of Texas of the Permian Basin, Graduate School, Division of Science and Engineering, Program in Geology, Odessa, TX 79762. Program awards MS. Matriculated students: 0 full-time, 13 part-time (2 women); includes 0 minority. Average age 34. In 1990, 5 degrees awarded. *Degree requirements:* Thesis or alternative required, foreign language not required. *Entrance requirements:* GRE General Test (minimum combined score of 1200 required). *Expenses:* Tuition of $480 per year full-time, $240 per year part-time for state residents; $2304 per year full-time, $1536 per year part-time for nonresidents. Fees of $243 per year full-time, $162 per year part-time. • Dr. Emilio Mutis-Duplat, Chairman, 915-367-2159.

University of Toledo, College of Arts and Sciences, Department of Geology, Toledo, OH 43606. Department offers programs in geochemistry (MS), geology (MS). Faculty: 8 full-time (0 women), 0 part-time. Matriculated students: 17 full-time (2 women), 10 part-time (2 women); includes 0 minority, 10 foreign. Average age 27. 41 applicants, 29% accepted. In 1990, 11 degrees awarded. *Degree requirements:* Thesis required, foreign language not required. *Entrance requirements:* GRE General Test. Application deadline: 9/8 (priority date). Application fee: $30. *Tuition:* $122.59 per credit hour for state residents; $193.40 per credit hour for nonresidents. *Financial aid:* In 1990–91, 2 research assistantships, 17 teaching assistantships awarded. Aid available to part-time students. Financial aid application deadline: 4/1; applicants required to submit FAF. *Faculty research:* Geohydrology, concrete mites. Total annual research budget: $72,047. • Dr. James A. Harrell, Chairman, 419-537-2193.

See full description on page 289.

University of Toronto, School of Graduate Studies, Physical Sciences Division, Department of Geology, Toronto, ON M5S 1A1, Canada. Department offers programs in earth sciences (MA Sc, M Sc, PhD), geophysics (M Sc, PhD). Faculty: 34. Matriculated students: 46 full-time (15 women), 9 part-time (3 women); includes 20 foreign. 92 applicants, 22% accepted. In 1990, 7 master's, 2 doctorates awarded. *Degree requirements:* Thesis/dissertation. *Application deadline:* 4/15. *Application fee:* $50. *Expenses:* Tuition of $2220 per year full-time, $666 per year part-time for Canadian residents; $10,198 per year full-time, $305.05 per year part-time for nonresidents. Fees of $277.56 per year full-time, $82.73 per year part-time. *Financial aid:* Application deadline 2/1. • J. Westgate, Chair, 416-978-3021.

See full description on page 291.

University of Tulsa, College of Engineering and Physical Sciences, Department of Geosciences, Tulsa, OK 74104. Offerings include geology (MS, PhD). Department faculty: 7 full-time (1 woman), 1 part-time (0 women). *Degree requirements:* For master's, computer language, thesis or alternative required, foreign language not required; for doctorate, computer language, dissertation required, foreign language not required. *Entrance requirements:* For master's, GRE General Test, TOEFL; for doctorate, GRE General Test, GRE Subject Test, TOEFL. Application deadline: rolling. Application fee: $30. *Tuition:* $350 per credit hour. • Dr. Colin Barker, Chairperson, 918-631-3014.

University of Utah, College of Mines and Earth Sciences, Department of Geology and Geophysics, Salt Lake City, UT 84112. Department offers programs in geological engineering (ME, MS, PhD), geology (MS, PhD), geophysics (MS, PhD). Faculty: 23 (1 woman). Matriculated students: 48 full-time (13 women), 12 part-time (1 woman); includes 0 minority, 21 foreign. Average age 31. 103 applicants, 30% accepted. In 1990, 8 master's, 0 doctorates awarded. Terminal master's awarded for partial completion of doctoral program. *Degree requirements:* Computer language, thesis/dissertation. *Entrance requirements:* GRE General Test, TOEFL, minimum GPA of 3.25. Application deadline: 8/1. Application fee: $25 ($50 for foreign students). *Tuition:* $195 per credit for state residents; $505 per credit for nonresidents. *Financial aid:* In 1990–91, $288,000 in aid awarded. 8 fellowships, 24 research assistantships, 15 teaching assistantships were awarded; stipends, institutionally sponsored loans also available. Financial aid application deadline: 3/1. *Faculty research:* Igneous, metamorphic, and sedimentary petrology; ore deposits; aqueous geochemistry; isotope geochemistry; heat flow. • Francis H. Brown, Chairman, 801-581-7162. Application contact: Peter H. Roth, Director of Graduate Studies, 801-581-7266.

University of Vermont, College of Arts and Sciences, Department of Geology, Burlington, VT 05405. Department awards MAT, MS, MST. Matriculated students: 14; includes 0 minority, 1 foreign. 18 applicants, 67% accepted. In 1990, 5 degrees awarded. *Degree requirements:* Thesis required, foreign language not required. *Entrance requirements:* GRE General Test, TOEFL (minimum score of 550 required). Application deadline: 4/1 (priority date, applications processed on a rolling basis). Application fee: $25. *Expenses:* Tuition of $206 per credit for state residents; $564 per credit for nonresidents. Fees of $150 per semester full-time. *Financial aid:* In 1990–91, 1 research assistantship, 8 teaching assistantships awarded. Financial aid application deadline: 3/1. *Faculty research:* Mineralogy, lake sediments, structural geology. • Dr. J. Hannah, Acting Chairperson, 802-656-3396. Application contact: Dr. B. Doolan, Coordinator, 802-656-3396.

University of Washington, College of Arts and Sciences, Department of Geological Sciences, Seattle, WA 98195. Department awards MS, PhD. Faculty: 19 full-time (1 woman), 2 part-time (0 women). Matriculated students: 40 full-time (10 women), 6 part-time (4 women); includes 1 minority (Asian American), 1 foreign. Average age 30. 110 applicants, 16% accepted. In 1990, 5 master's, 5 doctorates awarded. *Degree requirements:* For master's, thesis or alternative required, foreign language not required; for doctorate, dissertation required, foreign language not required. *Entrance requirements:* GRE, TOEFL (minimum score of 500 required), minimum GPA of 3.0. Application deadline: 2/15 (priority date, applications processed on a rolling basis). Application fee: $35. *Tuition:* $1129 per quarter full-time, $324 per credit (minimum) part-time for state residents; $2824 per quarter full-time, $809 per credit (minimum) part-time for nonresidents. *Financial aid:* In 1990–91, 36 students received a total of $333,867 in aid awarded. 12 fellowships (0 to first-year students), 11 research assistantships (3 to first-year students), 25 teaching assistantships (7 to first-year students) were awarded; federal work-study also available. Average monthly stipend for a graduate assistantship: $953. Financial aid application deadline: 3/1. *Faculty research:* Physics and chemistry of magmatic processes; quaternary geology and environmental change; surface processes and geomorphology; remote sensing; cordilleran tectonics. Total annual research budget: $1.5-million. • Darrel S. Cowan, Chairman, 206-543-1190.

University of Waterloo, Faculty of Science, Department of Earth Sciences, Waterloo, ON N2L 3G1, Canada. Department awards M Sc, PhD. Faculty: 28 full-time, 3 part-time. Matriculated students: 119 full-time (28 women), 19 part-time (2 women); includes 37 foreign. In 1990, 20 master's, 4 doctorates awarded. *Degree requirements:* Thesis/dissertation. *Entrance requirements:* For master's, TOEFL (minimum score of 530 required), honor's degree, minimum B average; for doctorate, TOEFL (minimum score of 530 required), master's degree. Application deadline: 8/1. Application fee: $25. *Expenses:* Tuition of $757 per year full-time, $530 per year part-time for Canadian residents; $3127 per year for nonresidents. Fees of $68 per year full-time, $17 per year part-time. *Financial aid:* In 1990–91, 261 students received aid. 180 research assistantships, 81 teaching assistantships awarded; institutionally sponsored loans and career-related internships or fieldwork also available. Average monthly stipend for a graduate assistantship: $1000. *Faculty research:* Hydrogeology, engineering geology, quaternary geology, geophysics, low temperature and isotope geochemistry. Total annual research budget: $5.3-million. • Dr. J. P. Greenhouse, Chairman, 519-885-1211 Ext. 3791. Application contact: S. Fischer, Graduate Secretary, 519-885-1211 Ext. 6870.

University of Western Ontario, Physical Sciences Division, Department of Geology, London, ON N6A 3K7, Canada. Department awards M Sc, PhD. *Degree requirements:* Thesis/dissertation required, foreign language not required. *Entrance requirements:* For master's, minimum B average, honors degree; for doctorate, honors B Sc, minimum B average during previous 2 years, or M Sc. *Tuition:* $1015 per year full-time, $1050 per year part-time for Canadian residents; $4207 per year for nonresidents.

University of Windsor, Faculty of Science, Department of Geology, Windsor, ON N9B 3P4, Canada. Department awards M Sc. Part-time programs available. Faculty: 11 full-time (0 women), 0 part-time. Matriculated students: 15 full-time, 0 part-time; includes 6 foreign. In 1990, 1 degree awarded. *Degree requirements:* Thesis. *Entrance requirements:* GRE, TOEFL, minimum B average. Application deadline: 7/1 (priority date, applications processed on a rolling basis). Application fee: $0. *Tuition:* $819.15 per semester full-time for Canadian residents; $3646 per semester full-time for nonresidents. *Financial aid:* Teaching assistantships available. Average monthly stipend for a graduate assistantship: $700. • Dr. William H. Blackburn, Head,

Directory: Geology

519-253-4232 Ext. 2486. Application contact: Admissions Officer, 519-253-4232 Ext. 2108.

University of Wisconsin–Madison, College of Letters and Science, Department of Geology and Geophysics, Madison, WI 53706. Department offers programs in geology (MS, PhD), geophysics (MS, PhD). Faculty: 22 full-time, 0 part-time. Matriculated students: 58 full-time (15 women), 5 part-time (1 woman); includes 3 minority (2 Asian American, 1 black American), 7 foreign. 89 applicants, 31% accepted. In 1990, 10 master's, 5 doctorates awarded. *Entrance requirements:* GRE General Test, GRE Subject Test. Application deadline: rolling. Application fee: $20. *Financial aid:* In 1990–91, 52 students received aid. 5 fellowships, 30 research assistantships, 16 teaching assistantships, 1 project assistantship awarded; institutionally sponsored loans also available. Financial aid application deadline: 1/15; applicants required to submit FAF. • Herbert F. Wang, Chairperson, 608-262-8960. Application contact: Joanne Nagy, Assistant Dean of the Graduate School, 608-262-2433.

University of Wisconsin–Milwaukee, College of Letters and Sciences, Department of Geosciences, Milwaukee, WI 53201. Department offers program in geological sciences (MS, PhD). Faculty: 17 full-time, 0 part-time. Matriculated students: 14 full-time (4 women), 30 part-time (6 women); includes 0 minority, 6 foreign. 39 applicants, 44% accepted. In 1990, 6 master's, 1 doctorate awarded. *Degree requirements:* For master's, thesis required, foreign language not required; for doctorate, 1 foreign language (computer language can substitute), dissertation. *Entrance requirements:* GRE General Test. Application deadline: 3/1 (priority date, applications processed on a rolling basis). Application fee: $20. *Financial aid:* In 1990–91, 3 research assistantships, 16 teaching assistantships, 1 project assistantship awarded. Financial aid application deadline: 4/15. • Douglas Cherkauer, Program Representative, 414-229-4562.

University of Wyoming, College of Arts and Sciences, Department of Geology and Geophysics, Laramie, WY 82071. Department offers programs in geology and geophysics (MS, PhD), geophysics (MS, PhD). Part-time programs available. Faculty: 20 full-time (1 woman), 3 part-time (1 woman). Matriculated students: 38 full-time (6 women), 35 part-time (10 women); includes 1 minority (Native American), 6 foreign. Average age 26. 136 applicants, 54% accepted. In 1990, 24 master's, 1 doctorate awarded. *Degree requirements:* Variable foreign language requirement, thesis/dissertation. *Entrance requirements:* For master's, GRE General Test (minimum combined score of 900 required), minimum GPA of 3.0; for doctorate, GRE General Test (minimum combined score of 1000 required), minimum GPA of 3.0. Application deadline: 1/31 (priority date, applications processed on a rolling basis). Application fee: $30. *Tuition:* $1554 per year full-time, $74.25 per credit hour part-time for state residents; $4358 per year full-time, $74.25 per credit hour part-time for nonresidents. *Financial aid:* In 1990–91, 49 students received a total of $429,872 in aid awarded. 17 fellowships, 15 research assistantships, 17 teaching assistantships (14 to first-year students) were awarded; federal work-study, institutionally sponsored loans, and career-related internships or fieldwork also available. Average monthly stipend for a graduate assistantship: $740. Financial aid application deadline: 1/31. *Faculty research:* Structural geology, diagenesis and hydrocarbon maturation, tectonics and sedimentation, igneous and metamorphic petrology, crustal studies. Total annual research budget: $5-million. • Dr. James R. Steidtman, Head, 307-766-3386. Application contact: Salina Renninger, Admissions Coordinator, 307-766-3389.

Utah State University, College of Science, Department of Geology, Logan, UT 84322. Offerings include geology (MS), with option in hydrogeology; geology-ecology (MS). Department faculty: 7 full-time (0 women), 0 part-time. *Degree requirements:* Thesis required, foreign language not required. *Entrance requirements:* GRE General Test (score in 40th percentile or higher required), minimum GPA of 3.0. Application deadline: 2/15 (priority date, applications processed on a rolling basis). Application fee: $25 ($30 for foreign students). *Tuition:* $426 per quarter (minimum) full-time, $184 per quarter (minimum) part-time for state residents; $1133 per quarter (minimum) full-time, $505 per quarter (minimum) part-time for nonresidents. • Dr. Don Fiesinger, Head, 801-750-1274.

Vanderbilt University, Department of Geology, Nashville, TN 37240. Department awards MS. Faculty: 8 full-time (2 women), 0 part-time. Matriculated students: 6 full-time (3 women), 2 part-time (1 woman); includes 0 minority, 0 foreign. Average age 24. 5 applicants, 100% accepted. In 1990, 7 degrees awarded. *Degree requirements:* Thesis required, foreign language not required. *Entrance requirements:* GRE General Test, GRE Subject Test. Application deadline: 1/15. Application fee: $25. *Expenses:* Tuition of $624 per semester hour. Fees of $196 per year. *Financial aid:* Fellowships, research assistantships, teaching assistantships, federal work-study, institutionally sponsored loans available. Financial aid application deadline: 1/15. • Leonard P. Alberstadt, Chairman, 615-322-2976. Application contact: Calvin F. Miller, Director of Graduate Studies, 615-322-2986.

Virginia Polytechnic Institute and State University, College of Arts and Sciences, Department of Geological Sciences, Blacksburg, VA 24061. Department offers programs in geological sciences (MS, PhD), geophysics (MS, PhD). Faculty: 19 full-time (0 women), 0 part-time. Matriculated students: 32 full-time (5 women), 4 part-time (2 women); includes 1 minority (Hispanic American), 5 foreign. 42 applicants, 50% accepted. In 1990, 7 master's, 9 doctorates awarded. *Degree requirements:* 1 foreign language, thesis/dissertation. *Entrance requirements:* GRE General Test, GRE Subject Test. Application deadline: 2/15 (priority date). Application fee: $10. *Tuition:* $1889 per semester full-time, $606 per credit hour part-time for state residents; $2627 per semester full-time, $853 per credit hour part-time for nonresidents. *Financial aid:* In 1990–91, 4 fellowships, 14 research assistantships, 22 teaching assistantships (11 to first-year students), 7 assistantships awarded; full tuition waivers and career-related internships or fieldwork also available. *Faculty research:* Paleoecology, sedimentology, evolutionary theory, tectonics, structure. • Dr. Donald Bloss, Chairman, 703-231-6521.

Virginia State University, School of Natural Sciences, Department of Geological Sciences, Petersburg, VA 23803. Department awards M Ed, MS. Faculty: 3 full-time (1 woman), 0 part-time. Matriculated students: 2 full-time (both women), 2 part-time (0 women). In 1990, 0 degrees awarded. *Degree requirements:* Thesis. *Entrance requirements:* GRE General Test. Application deadline: 8/15 (applications processed on a rolling basis). Application fee: $10. *Expenses:* Tuition of $1479 per semester full-time, $90 per credit hour part-time for state residents; $2879 per semester full-time, $215 per credit hour part-time for nonresidents. Fees of $25 per year full-time, $12 per year part-time. *Financial aid:* Application deadline 5/1; applicants required to submit FAF. • Dr. Constance Hill, Chairman, 804-524-5462. Application contact: Dr. Edgar A. Toppin, Dean, 804-524-5984.

Washington State University, College of Sciences and Arts, Division of Sciences, Department of Geology, Pullman, WA 99164. Department awards MS, PhD. Faculty: 12 full-time (0 women), 7 part-time (2 women). Matriculated students: 28 full-time (10 women), 11 part-time (1 woman); includes 1 minority (Hispanic American), 3 foreign. Average age 25. In 1990, 3 master's awarded; 3 doctorates awarded (33% entered university research/teaching, 67% found other work related to degree). *Degree requirements:* For master's, thesis (for some programs); for doctorate, 1 foreign language (computer language can substitute), dissertation. *Entrance requirements:* GRE General Test, minimum GPA of 3.0. Application deadline: 3/1 (priority date, applications processed on a rolling basis). Application fee: $25. *Tuition:* $1694 per semester full-time, $169 per credit hour part-time for state residents; $4236 per semester full-time, $424 per credit hour part-time for nonresidents. *Financial aid:* In 1990–91, 6 research assistantships, 21 teaching assistantships, 1 teaching associateship awarded; partial tuition waivers, federal work-study, institutionally sponsored loans, and career-related internships or fieldwork also available. Average monthly stipend for a graduate assistantship: $975. Financial aid application deadline: 4/1. *Faculty research:* Occurrence and foundation of economic minerals, geohydrology of the Pacific Northwest, geochemistry and petrology of plateau basalts. Total annual research budget: $223,000. • Dr. F. F. Foit, Chairman, 509-335-3009.

Washington University, Graduate School of Arts and Sciences, Department of Earth and Planetary Sciences, St. Louis, MO 63130. Offerings include geology (MA, PhD). Terminal master's awarded for partial completion of doctoral program. *Degree requirements:* Thesis/dissertation. *Entrance requirements:* GRE General Test. Application deadline: 1/15 (priority date, applications processed on a rolling basis). Application fee: $0. *Tuition:* $15,960 per year full-time, $665 per credit hour part-time. • Dr. Larry A. Haskin, Chairman, 314-935-5610.

See full description on page 295.

Wayne State University, College of Liberal Arts, Department of Geology, Detroit, MI 48202. Department awards MS. Faculty: 4 (0 women). Matriculated students: 0 full-time, 4 part-time (2 women); includes 0 foreign. Average age 30. 3 applicants, 67% accepted. In 1990, 1 degree awarded. *Degree requirements:* Thesis required, foreign language not required. Application deadline: 7/1. Application fee: $20 ($30 for foreign students). *Expenses:* Tuition of $119 per credit hour for state residents; $258 per credit hour for nonresidents. Fees of $50 per semester. *Faculty research:* Geologic history of Southern California, heavy metal contamination of soils, remote seismic thematic mapper asgeologic tool, determination of groundwater table using radar. • Dr. Robert B. Furlong, Chairman, 313-577-2506. Application contact: Dr. Luciano Ronca, Graduate Committee Chair, 313-577-2506.

West Chester University of Pennsylvania, College of Arts and Sciences, Department of Geology and Astronomy, West Chester, PA 19383. Department offers program in physical science (MA). Faculty: 0 full-time, 4 part-time. Matriculated students: 5 full-time (3 women), 20 part-time (7 women); includes 3 minority (2 Asian American, 1 black American), 0 foreign. Average age 32. 8 applicants, 75% accepted. In 1990, 4 degrees awarded. *Degree requirements:* Comprehensive exam required, foreign language and thesis not required. Application deadline: rolling. Application fee: $20. *Tuition:* $127 per credit for state residents; $160 per credit for nonresidents. *Financial aid:* In 1990–91, 1 research assistantship awarded. Aid available to part-time students. Financial aid application deadline: 4/1. • Dr. Allen Johnson, Chair, 215-436-2727. Application contact: Dr. Sandra Pritchard, Graduate Coordinator, 215-436-2721.

Western Michigan University, College of Arts and Sciences, Department of Geology, Kalamazoo, MI 49008. Department offers programs in earth science (MS), geology (MS). Matriculated students: 9 full-time (2 women), 19 part-time (6 women); includes 1 minority (black American), 2 foreign. 28 applicants, 64% accepted. In 1990, 7 degrees awarded. *Degree requirements:* Oral exam required, foreign language not required. *Entrance requirements:* GRE. Application deadline: 2/15 (priority date, applications processed on a rolling basis). Application fee: $25. *Expenses:* Tuition of $100 per credit hour for state residents; $244 per credit hour for nonresidents. Fees of $178 per semester full-time, $76 per semester part-time. *Financial aid:* Fellowships, research assistantships, teaching assistantships, federal work-study available. Financial aid application deadline: 2/15. • Dr. W. Thomas Straw, Chairperson, 616-387-5485. Application contact: Paula J. Boodt, Director, Graduate Student Services, 616-387-3570.

Western Washington University, College of Arts and Sciences, Department of Geology, Bellingham, WA 98225. Department awards MS. Faculty: 13 (12 women). Matriculated students: 19 full-time (4 women), 6 part-time (1 woman); includes 0 minority, 0 foreign. 25 applicants, 88% accepted. In 1990, 10 degrees awarded. *Degree requirements:* Thesis required, foreign language not required. *Entrance requirements:* GRE General Test, TOEFL (minimum score of 535 required), minimum GPA of 3.0 for last 60 semester hours or last 90 quarter hours. Application deadline: 6/1 (priority date, applications processed on a rolling basis). Application fee: $25. *Tuition:* $900 per quarter full-time, $90 per credit part-time for state residents; $2729 per quarter full-time, $273 per credit part-time for nonresidents. *Financial aid:* In 1990–91, 7 teaching assistantships awarded; partial tuition waivers, federal work-study, institutionally sponsored loans, and career-related internships or fieldwork also available. Aid available to part-time students. Financial aid application deadline: 3/31; applicants required to submit FAF. • Dr. Chris Suczek, Chairperson, 206-676-3582.

West Texas State University, College of Agriculture, Nursing, and Natural Sciences, Department of Biology and Geosciences, Program in Geology, Canyon, TX 79016. Program awards MS. Part-time programs available. *Degree requirements:* Thesis required, foreign language not required. *Entrance requirements:* GRE General Test, minimum GPA of 3.0. *Expenses:* Tuition of $600 per year full-time, $20 per hour part-time for state residents; $3840 per year full-time, $128 per hour part-time for nonresidents. Fees of $448 per year full-time, $15 per hour part-time.

West Virginia University, College of Arts and Sciences, Department of Geology and Geography, Program in Geology, Morgantown, WV 26506. Program awards MS, PhD. Part-time programs available. Matriculated students: 21 full-time (5 women), 5 part-time (0 women); includes 0 minority, 4 foreign. Average age 30. 45 applicants, 22% accepted. In 1990, 8 master's awarded; 1 doctorate awarded (100% entered university research/teaching). Terminal master's awarded for partial completion of doctoral program. *Degree requirements:* For master's, thesis required, foreign language not required; for doctorate, dissertation, comprehensive exam required, foreign language not required. *Entrance requirements:* GRE General Test, GRE Subject Test, TOEFL (minimum score of 550 required), minimum GPA of 2.5. Application deadline: rolling. Application fee: $25. *Expenses:* Tuition of $390 per year full-time for state residents; $1270 per year full-time for nonresidents. Fees of $1555 per year full-time for state residents; $3985 per year full-time for nonresidents. *Financial aid:* In 1990–91, $159,339 in aid awarded. 0 fellowships, 3 research assistantships, 18 teaching assistantships (9 to first-year students) were awarded; full and partial tuition waivers, federal work-study, institutionally sponsored loans, and career-related internships or fieldwork also available. Financial aid application deadline: 2/1; applicants required to submit FAF. *Faculty research:*

Directories: Geology; Geophysics

Geophysics, basin analysis, hydrogeology, environmental geology. • Dr. Alan Donaldson, Chairperson, Department of Geology and Geography, 304-293-5603.

Announcement: The department emphasizes basin analysis and the environmental geoscience area, with strengths in energy, hydrology, surficial geology, and geophysics. Associations with West Virginia Geological Survey, environmental agencies and industries, DOE agency, and energy-related industries enhance research expertise and opportunities. Field courses are in West Virginia, New England, Florida, and Europe. Seismic lab supports new geophysics program.

Wichita State University, Fairmount College of Liberal Arts and Sciences, Department of Geology, Wichita, KS 67208. Department awards MS. Part-time programs available. Faculty: 9 full-time (1 woman), 0 part-time. Matriculated students: 1 full-time (0 women), 10 part-time (2 women); includes 0 minority, 1 foreign. 7 applicants, 100% accepted. In 1990, 3 degrees awarded. *Degree requirements:* Thesis, comprehensive exam required, foreign language not required. *Entrance requirements:* GRE General Test (minimum combined score of 1000 required), TOEFL (minimum score of 550 required). Application deadline: 8/1 (priority date, applications processed on a rolling basis). Application fee: $0 ($25 for foreign students). *Expenses:* Tuition of $1590 per year full-time, $53 per credit part-time for state residents; $2574 per year full-time, $171.60 per credit part-time for nonresidents. Fees of $12.20 per credit. *Financial aid:* In 1990–91, $200,000 in aid awarded. 3 research assistantships (2 to first-year students), 3 teaching assistantships (2 to first-year students) were awarded; federal work-study, institutionally sponsored loans also available. Aid available to part-time students. Financial aid application deadline: 4/1. *Faculty research:* Midcontinent and Permian basin stratigraphy studies, recent sediments of Belize and Florida, image analysis of sediments and porosity. Total annual research budget: $50,000. • Dr. John Gries, Chairperson, 316-689-3140.

Wright State University, College of Science and Mathematics, Department of Geological Sciences, Dayton, OH 45435. Department offers programs in earth science education (MST), geological sciences (MS). Part-time programs available. Faculty: 11 full-time (1 woman), 2 part-time (0 women). Matriculated students: 36 full-time (6 women), 10 part-time (2 women); includes 1 minority (Hispanic American), 4 foreign. Average age 26. 36 applicants, 83% accepted. In 1990, 16 degrees awarded. *Degree requirements:* Computer language, thesis required, foreign language not required. *Application fee:* $25. *Tuition:* $3342 per year full-time, $106 per credit hour part-time for state residents; $5991 per year full-time, $190 per credit hour part-time for nonresidents. *Financial aid:* In 1990–91, 24 students received a total of $140,000 in aid awarded. 4 fellowships (all to first-year students), 7 research assistantships, 14 teaching assistantships (6 to first-year students), 7 graduate assistantships (4 to first-year students) were awarded; federal work-study also available. Average monthly stipend for a graduate assistantship: $500. *Faculty research:* Groundwater contamination; oil, gas, and coal exploration. Total annual research budget: $155,000. • Dr. Byron Kulander, Chair, 513-873-3455. Application contact: Deborah Hapner, Assistant to Chair, 513-873-3455.

See full description on page 297.

Yale University, Graduate School of Arts and Sciences, Department of Geology and Geophysics, New Haven, CT 06520. Department offers programs in geochemistry (PhD), geophysics (PhD), meteorology (PhD), mineralogy and crystallography (PhD), oceanography (PhD), paleoecology (PhD), paleontology and stratigraphy (PhD), petrology (PhD), structural geology (PhD). Faculty: 24 full-time (2 women), 1 part-time (0 women). Matriculated students: 44 full-time (5 women), 0 part-time; includes 0 minority, 25 foreign. 54 applicants, 30% accepted. In 1990, 5 master's, 10 doctorates awarded. Terminal master's awarded for partial completion of doctoral program. *Degree requirements:* For doctorate, dissertation. *Entrance requirements:* For doctorate, GRE General Test, GRE Subject Test. Application deadline: 1/2. Application fee: $45. *Tuition:* $15,160 per year full-time, $1895 per course part-time. • Dr. B. Saltzman, Chairman, 203-432-3114. Application contact: Susan Webb, Coordinator of Admissions, 203-432-2770.

Announcement: Department offers individualized programs of study leading to the doctorate (4 years). It welcomes applicants interested in earth sciences who have bachelor's or master's degree in biology, chemistry, engineering, mathematics, meteorology, or physics, as well as geology. Program has no required curriculum of credit courses but is designed to encourage development of individual interests under guidance of faculty advisory committee.

See full description on page 299.

Geophysics

Arizona State University, College of Liberal Arts and Sciences, Department of Physics, Tempe, AZ 85287. Offerings include geophysics (PhD). *Degree requirements:* Dissertation. *Entrance requirements:* GRE. Application fee: $25. *Tuition:* $1528 per year full-time, $80 per hour part-time for state residents; $6934 per year full-time, $289 per hour part-time for nonresidents. • Dr. John D. Dow, Chair, 602-965-3561.

Boise State University, College of Arts and Sciences, Department of Geosciences, Program in Applied Geophysics, Boise, ID 83725. Program awards MS. Part-time programs available. Faculty: 10 full-time (0 women), 10 part-time (2 women). Matriculated students: 7 full-time (1 woman), 2 part-time (0 women). Average age 25. In 1990, 1 degree awarded (100% found work related to degree). *Degree requirements:* Computer language, thesis. *Entrance requirements:* GRE, TOEFL (minimum score of 550 required), minimum GPA of 3.0. Application deadline: 8/1 (priority date). Application fee: $15. *Expenses:* Tuition of $0 for state residents; $2000 per year full-time, for nonresidents. Fees of $1654 per year full-time, $83 per credit hour part-time. *Financial aid:* In 1990–91, $40,000 in aid awarded. 2 research assistantships (both to first-year students), 4 grants (all to first-year students) were awarded; federal work-study, institutionally sponsored loans, and career-related internships or fieldwork also available. Financial aid application deadline: 4/1. *Faculty research:* Shallow seismic profile, seismic hazard, tectonics, hazardous waste disposal. Total annual research budget: $500,000. • Dr. John R. Pelton, Coordinator, 208-385-3640.

Boston College, Graduate School of Arts and Sciences, Department of Geology and Geophysics, Chestnut Hill, MA 02167. Department offers programs in geology and geophysics (MS, MST), oceanography (MS, MST). *Degree requirements:* 1 foreign language, thesis. *Entrance requirements:* GRE General Test, GRE Subject Test. *Tuition:* $412 per credit hour.

California Institute of Technology, Division of Geological and Planetary Sciences, Pasadena, CA 91125. Offerings include geophysics (MS, PhD). Division faculty: 28 full-time (0 women), 0 part-time. *Entrance requirements:* GRE General Test, GRE Subject Test. Application deadline: 1/15. Application fee: $0. *Expenses:* Tuition of $14,100 per year. Fees of $8 per quarter. • Dr. David J. Stevenson, Chairman, 818-356-6108.

See full description on page 219.

Colorado School of Mines, Department of Geophysics, Golden, CO 80401. Department offers programs in geophysical engineering (ME, MS, PhD), geophysics (MS, PhD). Faculty: 17 full-time (1 woman), 5 part-time (0 women). Matriculated students: 61 full-time (7 women), 16 part-time (1 woman); includes 45 foreign. 63 applicants, 67% accepted. In 1990, 13 master's, 5 doctorates awarded. *Degree requirements:* For master's, thesis required (for some programs), foreign language not required; for doctorate, dissertation required, foreign language not required. *Entrance requirements:* GRE General Test, minimum GPA of 3.0. Application deadline: 4/15 (priority date, applications processed on a rolling basis). Application fee: $15 ($25 for foreign students). *Expenses:* Tuition of $3178 per year full-time, $124 per semester hour part-time for state residents; $10,304 per year full-time, $344 per semester hour part-time for nonresidents. Fees of $374 per year full-time. *Financial aid:* In 1990–91, 23 fellowships, 19 research assistantships, 7 teaching assistantships awarded. *Faculty research:* Exploration techniques, modeling, integrated studies, case histories. Total annual research budget: $2.211-million. • Dr. Phillip Romig, Head, 303-273-3450.

Columbia University, Graduate School of Arts and Sciences, Division of Natural Sciences, Department of Geological Sciences, New York, NY 10027. Offerings include geophysics (MA, M Phil, PhD). Department faculty: 21 full-time, 19 part-time. *Degree requirements:* For master's, thesis or alternative, fieldwork, written exam required, foreign language not required; for doctorate, 1 foreign language, dissertation. *Entrance requirements:* GRE General Test, GRE Subject Test, TOEFL, major in natural or physical science. Application deadline: 1/5. Application fee: $50. *Expenses:* Tuition of $7836 per semester for state residents; $426 per credit for nonresidents. Fees of $148 per semester for state residents. • Dennis Hayes, Chair, 212-854-4525.

Emory University, Graduate School of Arts and Sciences, Department of Physics, Atlanta, GA 30322. Department offers programs in environmental sciences (MS); physics (MA, MS, PhD), including biophysics, radiological physics, solid-state physics. Faculty: 17. Matriculated students: 3 full-time (0 women), 2 part-time (0 women); includes 0 minority, 4 foreign. Average age 33. 0 applicants. In 1990, 0 master's, 1 doctorate awarded. Terminal master's awarded for partial completion of doctoral program. *Degree requirements:* For master's, thesis required, foreign language not required; for doctorate, dissertation, comprehensive exams required, foreign language not required. *Entrance requirements:* GRE General Test, TOEFL, minimum GPA of 3.0. Application deadline: 1/20 (priority date). Application fee: $35. *Expenses:* Tuition of $7370 per semester full-time, $642 per semester hour part-time. Fees of $160 per year full-time. *Financial aid:* In 1990–91, $109,180 in aid awarded. 4 fellowships (0 to first-year students), 0 teaching assistantships, 4 tuition scholarships (0 to first-year students) were awarded; partial tuition waivers also available. Average monthly stipend for a graduate assistantship: $1000. Financial aid application deadline: 2/1. *Faculty research:* Theory of semiconductors and superlattices; experimental laser optics and submillimeter spectroscopy theory; neural networks and stereoscopic vision, experimental studies of the structu re and function of metalloproteins. Total annual research budget: $953,952. • Dr. Krishan K. Bajaj, Chair, 404-727-6584. Application contact: Dr. Edmund Day, Director of Graduate Studies, 404-727-6584.

See full description on page 583.

Florida State University, College of Arts and Sciences, Departments of Geology, Mathematics, and Oceanography, Program in Geophysical Fluid Dynamics, Tallahassee, FL 32306. Program awards PhD. Faculty: 10 full-time (1 woman), 0 part-time. Matriculated students: 10 full-time (1 woman), 0 part-time; includes 10 foreign. Average age 30. 10 applicants, 30% accepted. In 1990, 1 degree awarded. *Degree requirements:* Computer language, dissertation required, foreign language not required. *Entrance requirements:* GRE General Test, TOEFL. Application deadline: 12/30. Application fee: $15. *Tuition:* $76.29 per credit hour for state residents; $238 per credit hour for nonresidents. *Financial aid:* Fellowships, research assistantships available. *Faculty research:* Hurricane dynamics, topography, convection, air-sea interaction, wave-mean flow interaction. Total annual research budget: $500,000. • Dr. Richard L. Pfeffer, Director, 904-644-5594.

Indiana University Bloomington, College of Arts and Sciences, Department of Geological Sciences, Program in Geophysics, Bloomington, IN 47405. Program awards MS, PhD. Faculty: 1 full-time. *Degree requirements:* For master's, 1 foreign language, computer language; for doctorate, 2 foreign languages (computer language can substitute for one), dissertation. *Entrance requirements:* GRE General Test. Application deadline: 4/15. Application fee: $25 ($35 for foreign students). *Tuition:* $99.85 per credit hour for state residents; $288 per credit hour for nonresidents. *Financial aid:* Application deadline 2/15. • Lee J. Suttner, Chairman, Department of Geological Sciences, 812-855-5581. Application contact: David G. Towell, Chairman, 812-855-7214.

See full description on page 223.

Johns Hopkins University, School of Arts and Sciences, Department of Earth and Planetary Sciences, Program in Geophysics, Baltimore, MD 21218. Program awards MA, PhD. Faculty: 3 full-time (0 women). Matriculated students: 10. Average age 26. Terminal master's awarded for partial completion of doctoral program. *Degree requirements:* For doctorate, 1 foreign language, dissertation. *Entrance requirements:* GRE General Test. Application deadline: 2/1. Application fee: $40.

Expenses: Tuition of $15,500 per year full-time, $1550 per course part-time. Fees of $400 per year full-time. *Financial aid:* Application deadline 3/14; applicants required to submit FAF. • Bruce Marsh, Chairman, Department of Earth and Planetary Sciences, 301-338-7135.

Louisiana State University and Agricultural and Mechanical College, College of Basic Sciences, Department of Geology and Geophysics, Baton Rouge, LA 70803. Department awards MS, PhD. Faculty: 30 full-time (1 woman), 2 part-time (1 woman). Matriculated students: 42 full-time (12 women), 14 part-time (4 women); includes 2 minority (both Asian American), 19 foreign. Average age 32. 47 applicants, 32% accepted. In 1990, 6 master's, 7 doctorates awarded. Terminal master's awarded for partial completion of doctoral program. *Degree requirements:* Thesis/dissertation required, foreign language not required. *Entrance requirements:* GRE General Test (minimum combined score of 1000 required), TOEFL (minimum score of 525 required, 550 for assistantships), minimum GPA of 3.0. Application deadline: 7/1 (priority date, applications processed on a rolling basis). Application fee: $25. *Tuition:* $1020 per semester full-time for state residents; $2620 per semester full-time for nonresidents. *Financial aid:* In 1990–91, 42 students received a total of $440,884 in aid awarded. 7 fellowships (1 to a first-year student), 10 research assistantships (1 to a first-year student), 24 teaching assistantships (2 to first-year students) were awarded; federal work-study, institutionally sponsored loans, and career-related internships or fieldwork also available. Average monthly stipend for a graduate assistantship: $910. Financial aid application deadline: 3/15. *Faculty research:* Geophysics, geochemistry of sediments, isotope geochemistry, igneous and metamorphic petrology, micropaleontology. Total annual research budget: $1.46-million. • Dr. Joseph Hazel, Chair, 504-388-3354. Application contact: Elizabeth Holt, Graduate Coordinator, 504-388-3415.

See full description on page 227.

Massachusetts Institute of Technology, School of Science, Department of Earth, Atmospheric, and Planetary Sciences, Cambridge, MA 02139. Offerings include geophysics (SM, PhD, Sc D). Terminal master's awarded for partial completion of doctoral program. Department faculty: 37 full-time (3 women), 0 part-time. *Degree requirements:* For master's, thesis required, foreign language not required; for doctorate, dissertation, general exam required, foreign language not required. *Entrance requirements:* GRE General Test, GRE Subject Test. Application deadline: 1/15. Application fee: $45. *Tuition:* $16,900 per year. • Dr. Thomas H. Jordan, Chairman, 617-253-3382. Application contact: Anita Killian, Graduate Administrator, 617-253-3381.

McGill University, Faculty of Graduate Studies and Research, Department of Geological Sciences, Montreal, PQ H3A 2A7. Department awards M Sc, M Sc A, PhD, Diploma. Faculty: 17 full-time (0 women), 0 part-time. *Degree requirements:* For master's, thesis (M Sc); for Diploma, thesis not required. *Tuition:* $698.50 per semester full-time, $46.57 per credit part-time for Canadian residents; $3480 per semester full-time, $234 per semester part-time for nonresidents. *Financial aid:* Fellowships, teaching assistantships available. • D. Francis, Chairman, 514-398-6768. Application contact: R. Doig, Chairman, Graduate Admissions, 514-398-4884.

See full description on page 229.

Memorial University of Newfoundland, School of Graduate Studies, Department of Earth Sciences, St. John's, NF A1C 5S7, Canada. Offerings include earth sciences (M Sc, PhD), with options in geology (M Sc, PhD), geophysics (M Sc, PhD). *Degree requirements:* For master's, thesis; for doctorate, dissertation, comprehensive exam.

Memphis State University, College of Arts and Sciences, Department of Geological Sciences, Memphis, TN 38152. Offerings include geophysics (MS). Department faculty: 16 full-time (2 women), 1 part-time (0 women). *Degree requirements:* Thesis, seminar presentation. *Entrance requirements:* GRE General Test, GRE Subject Test. Application deadline: 8/1 (applications processed on a rolling basis). Application fee: $5. *Expenses:* Tuition of $92 per credit hour for state residents; $239 per credit hour for nonresidents. Fees of $45 per year for state residents. • Dr. Phili Deboo, Chairman, 901-678-2177. Application contact: Dr. Eugene S. Schweig, Coordinator of Graduate Studies, 901-678-2177.

Michigan State University, College of Natural Science, Department of Geological Sciences, East Lansing, MI 48824-1115. Department awards MAT, MS, PhD. Faculty: 16 full-time, 0 part-time. Matriculated students: 28 full-time (5 women), 13 part-time (5 women); includes 1 minority, 4 foreign. 23 applicants, 40% accepted. In 1990, 6 master's, 2 doctorates awarded. *Degree requirements:* Thesis/dissertation. *Application deadline:* rolling. *Application fee:* $25 ($40 for foreign students). *Tuition:* $104.75 per credit for state residents; $211.75 per credit for nonresidents. *Financial aid:* In 1990–91, $180,782 in aid awarded. 9 fellowships (2 to first-year students), 3 research assistantships (0 to first-year students), 17 teaching assistantships (11 to first-year students) were awarded. • Thomas Vogel, Acting Chair, 517-355-4626.

See full description on page 231.

Michigan Technological University, College of Engineering, Department of Geology and Geological Engineering, Program in Geophysics, Houghton, MI 49931. Program awards MS. Part-time programs available. Matriculated students: 0. Average age 29. 11 applicants, 18% accepted. In 1990, 1 degree awarded. *Degree requirements:* Thesis required, foreign language not required. *Entrance requirements:* TOEFL (minimum score of 550 required). Application deadline: rolling. Application fee: $20. *Expenses:* Tuition of $852 per semester full-time, $71 per credit part-time for state residents; $2065 per semester full-time, $172 per credit part-time for nonresidents. Fees of $75 per semester. *Financial aid:* 0 students received aid. Fellowships, research assistantships, teaching assistantships, federal work-study, and career-related internships or fieldwork available. Aid available to part-time students. Average monthly stipend for a graduate assistantship: $800. Financial aid applicants required to submit FAF. *Faculty research:* Paleomagnetism/rock magnetism, magnetotellurics, applied geophysics, signal processing, electrical geophysics. • Dr. William I. Rose, Head, Department of Geology and Geological Engineering, 906-487-2531. Application contact: Dr. Douglas McDowell, Graduate Adviser, 906-487-2531.

New Mexico Institute of Mining and Technology, Department of Geoscience, Program in Geophysics, Socorro, NM 87801. Program awards MS, PhD. Faculty: 2 full-time (0 women), 0 part-time. Matriculated students: 7 full-time (1 woman), 0 part-time; includes 0 minority, 1 foreign. Average age 25. 20 applicants, 45% accepted. In 1990, 2 master's, 1 doctorate awarded. *Degree requirements:* For master's, thesis optional, foreign language not required; for doctorate, 1 foreign language, dissertation. *Entrance requirements:* For master's, GRE General Test, TOEFL (minimum score of 540 required); for doctorate, GRE General Test, GRE Subject Test, TOEFL (minimum score of 540 required). Application deadline: 6/1 (priority date, applications processed on a rolling basis). Application fee: $16. *Expenses:* Tuition of $617 per semester full-time, $51 per hour part-time for state residents; $2656 per semester full-time, $209 per hour part-time for nonresidents. Fees of $207 per semester. *Financial aid:* In 1990–91, 5 students received a total of $12,301 in aid awarded. 3 research assistantships (2 to first-year students), 2 teaching assistantships (1 to a first-year student) were awarded; federal work-study, institutionally sponsored loans also available. Average monthly stipend for a graduate assistantship: $600. Financial aid application deadline: 3/15; applicants required to submit GAPSFAS or FAF. *Faculty research:* Earthquake seismology, crustal exploration, tectonophysics, geothermics. Total annual research budget: $125,099. • Dr. Allan Sanford, Coordinator, 505-835-5212. Application contact: Dr. J. A. Smoake, Dean, Graduate Studies, 505-835-5513.

Announcement: MS and PhD programs also available in geology (structural geology, depositional environments, ore deposits, fluid inclusions), geochemistry (volcanology, isotope geochemistry, geochemistry of ore deposits), and hydrology (groundwater transport and contamination, modeling, isotope hydrology). Courses, research, and support augmented by 14 adjuncts, most of whom are professionals with the on-campus NM Bureau of Mines and Mineral Resources. See full description of NMIMT in the Graduate and Professional Programs (An Overview) volume of this series.

North Carolina State University, College of Physical and Mathematical Sciences, Department of Marine, Earth, and Atmospheric Sciences, Raleigh, NC 27695. Department offers programs in geology (MS, PhD), geophysics (MS, PhD), meteorology (MS, PhD), oceanography (MS, PhD). Faculty: 35 full-time (2 women), 3 part-time (1 woman). Matriculated students: 74 full-time (15 women), 3 part-time (0 women); includes 2 minority (both Asian American), 23 foreign. Average age 24. 75 applicants, 49% accepted. In 1990, 18 master's, 6 doctorates awarded. Terminal master's awarded for partial completion of doctoral program. *Degree requirements:* For master's, thesis, final oral exam required, foreign language not required; for doctorate, dissertation, preliminary written and oral exams, final oral exams required, foreign language not required. *Entrance requirements:* GRE General Test, minimum GPA of 3.0. Application deadline: 5/1 (priority date, applications processed on a rolling basis). Application fee: $35. *Tuition:* $1138 per year for state residents; $5805 per year for nonresidents. *Financial aid:* In 1990–91, $383,267 in aid awarded. 0 fellowships, 33 research assistantships (10 to first-year students), 18 teaching assistantships (6 to first-year students) were awarded; institutionally sponsored loans also available. Average monthly stipend for a graduate assistantship: $750. Financial aid application deadline: 3/1. *Faculty research:* Boundary-layer and synoptic meteorology; physical, chemical, geological, and biological oceanography. Total annual research budget: $6.41-million. • Dr. Leonard J. Pietrafesa, Head, 919-737-3717. Application contact: G. S. Janowitz, Graduate Administrator, 919-737-7837.

See full description on page 317.

Northern Illinois University, College of Liberal Arts and Sciences, Department of Geology, De Kalb, IL 60115. Department awards MS, PhD. Part-time programs available. Faculty: 15 full-time (2 women), 1 part-time (0 women). Matriculated students: 27 full-time (3 women), 22 part-time (2 women); includes 4 minority (1 Asian American, 2 Hispanic American, 1 Native American), 14 foreign. Average age 30. 103 applicants, 21% accepted. In 1990, 12 master's, 0 doctorates awarded. *Degree requirements:* For master's, comprehensive exam required, thesis optional, foreign language not required; for doctorate, variable foreign language requirement, dissertation defense, internship, comprehensive exam. *Entrance requirements:* For master's, GRE General Test, TOEFL, bachelor's degree in related field, minimum GPA of 2.75; for doctorate, GRE General Test, TOEFL, bachelor's or master's degree in related field, minimum graduate GPA of 3.2, undergraduate 2.75. Application deadline: 6/1. Application fee: $0. *Tuition:* $1339 per semester full-time for state residents; $3163 per semester full-time for nonresidents. *Financial aid:* In 1990–91, 5 fellowships, 8 research assistantships, 17 teaching assistantships awarded; full tuition waivers, federal work-study, and career-related internships or fieldwork also available. Financial aid applicants required to submit FAF. *Faculty research:* Geochemistry, hydrogeology, petrology, sedimentology, micropaleontology, biostratigraphy. Total annual research budget: $769,700. • Dr. Ross Powell Jr., Chair, 815-753-1943. Application contact: Dr. James A. Walker, Director, Graduate Studies, 815-753-7936.

See full description on page 233.

Oregon State University, Graduate School, College of Oceanography, Program in Geophysics, Corvallis, OR 97331. Program awards MA, MS, PhD. Part-time programs available. Faculty: 8 full-time (1 woman), 0 part-time. Matriculated students: 5 full-time (0 women), 0 part-time; includes 0 minority, 5 foreign. Average age 25. In 1990, 0 master's, 0 doctorates awarded. Terminal master's awarded for partial completion of doctoral program. *Degree requirements:* For master's, thesis optional, foreign language not required; for doctorate, dissertation required, foreign language not required. *Entrance requirements:* GRE General Test, TOEFL (minimum score of 550 required), minimum GPA of 3.0 in last 90 hours. Application deadline: 3/1. Application fee: $40. *Tuition:* $1140 per trimester full-time, $449 per year part-time for state residents; $1816 per trimester full-time, $674 per year part-time for nonresidents. *Financial aid:* In 1990–91, 11 research assistantships (2 to first-year students) awarded; fellowships, teaching assistantships, federal work-study, institutionally sponsored loans also available. Aid available to part-time students. Financial aid application deadline: 2/1; applicants required to submit FAF. *Faculty research:* Seismic waves; gravitational, geothermal, and electromagnetic fields; rock magnetism; paleomagnetism. • Dr. Douglas R. Caldwell, Director, 503-737-3504. Application contact: Dr. Jefferson J. Gonor, Head Adviser, 503-737-3504.

Oregon State University, Graduate School, College of Science, Department of Geosciences, Corvallis, OR 97331. Department awards MA, MS, PhD. Part-time programs available. Faculty: 24 full-time (2 women), 12 part-time (0 women). In 1990, 29 master's, 2 doctorates awarded. Terminal master's awarded for partial completion of doctoral program. *Degree requirements:* For master's, variable foreign language requirement, thesis optional; for doctorate, 1 foreign language, dissertation. *Entrance requirements:* GRE General Test, GRE Subject Test, TOEFL (minimum score of 550 required), minimum GPA of 3.0 in last 90 hours. Application fee: $40. *Tuition:* $1140 per trimester full-time, $449 per year part-time for state residents; $1816 per trimester full-time, $674 per year part-time for nonresidents. *Financial aid:* Fellowships, research assistantships, teaching assistantships, federal work-study, institutionally sponsored loans available. Aid available to part-time students. Financial aid application deadline: 2/14; applicants required to submit GAPSFAS or FAF. • Dr. Cyrus W. Field, Chair. Application contact: Graduate Admissions Coordinator, 503-737-1201.

See full description on page 235.

Pennsylvania State University University Park Campus, College of Earth and Mineral Sciences, Department of Geosciences, University Park, PA 16802. Offerings include geophysics (MS, PhD). Department faculty: 39. *Entrance requirements:* GRE General

Test, TOEFL. Application fee: $35. *Tuition:* $203 per credit for state residents; $403 per credit for nonresidents. • Dr. David Eggler, Chair, 814-865-6393.

Announcement: Penn State offers MS and PhD research opportunities in seismology, exploration geophysics, geophysical signal analysis, thermal-mechanical modeling, mineral physics and tectonophysics, potential fields (gravity, magnetic, electrical, thermal), and remote sensing.

See full description on page 237.

Princeton University, Department of Geological and Geophysical Sciences, Princeton, NJ 08544. Department offers programs in atmospheric and oceanic sciences (PhD), geological and geophysical sciences (PhD), water resources (PhD). Faculty: 16 full-time (3 women), 0 part-time. Matriculated students: 26 full-time (4 women), 0 part-time; includes 1 minority (Asian American), 12 foreign. 28 applicants, 54% accepted. In 1990, 11 doctorates awarded (45% entered university research/teaching, 55% found other work related to degree). Terminal master's awarded for partial completion of doctoral program. *Degree requirements:* For doctorate, 1 foreign language, dissertation. *Entrance requirements:* For doctorate, GRE General Test, GRE Subject Test. Application deadline: 1/8. Application fee: $45 ($50 for foreign students). *Tuition:* $16,670 per year. *Financial aid:* Fellowships, research assistantships, teaching assistantships, federal work-study, institutionally sponsored loans available. Financial aid application deadline: 1/8. • Gerta Keller, Director of Graduate Studies, 609-258-5807. Application contact: Michele Spreen, Director of Graduate Admissions, 609-258-3034.

Purdue University, School of Science, Department of Earth and Atmospheric Sciences, West Lafayette, IN 47907. Department awards MS, PhD. Faculty: 24 full-time, 1 part-time. Matriculated students: 74 full-time (10 women), 6 part-time (3 women); includes 4 minority (1 Asian American, 1 black American, 1 Hispanic American, 1 Native American), 27 foreign. Average age 28. 87 applicants, 23% accepted. In 1990, 19 master's, 6 doctorates awarded. *Degree requirements:* For master's, thesis required, foreign language not required; for doctorate, 1 foreign language, dissertation. *Entrance requirements:* GRE General Test. Application deadline: 6/1. Application fee: $25. *Tuition:* $1162 per semester full-time, $83.50 per credit hour part-time for state residents; $3720 per semester full-time, $244.50 per credit hour part-time for nonresidents. *Financial aid:* In 1990–91, 3 fellowships (2 to first-year students), 22 research assistantships (1 to a first-year student), 26 teaching assistantships (3 to first-year students) awarded. • Dr. E. M. Agee, Head, 317-494-0251.

Rensselaer Polytechnic Institute, School of Science, Department of Earth and Environmental Sciences, Troy, NY 12180. Offerings include geophysics (MS, PhD). Department faculty: 8 full-time (0 women), 2 part-time (both women). *Degree requirements:* For master's, thesis optional; for doctorate, dissertation. *Entrance requirements:* GRE General Test, TOEFL. Application deadline: 2/1. Application fee: $30. *Expenses:* Tuition of $455 per credit hour. Fees of $195.57 per semester. • E. B. Watson, Chairman, 518-276-6474. Application contact: M. Brian Bayly, Graduate Coordinator, 518-276-6494.

See full description on page 239.

Rice University, Wiess School of Natural Sciences, Department of Geology and Geophysics, Houston, TX 77251. Department offers programs in geochemistry (MA, PhD), geophysics (MA, PhD). Faculty: 15 full-time (0 women), 0 part-time. Matriculated students: 49 full-time (12 women), 1 part-time (0 women); includes 3 minority (2 Asian American, 1 black American), 2 foreign. *Degree requirements:* Thesis/dissertation required, foreign language not required. *Entrance requirements:* GRE General Test, GRE Subject Test, TOEFL (minimum score of 550 required), minimum GPA of 3.0. Application deadline: 3/1. Application fee: $0. *Expenses:* Tuition of $8300 per year full-time, $400 per credit hour part-time. Fees of $167 per year. *Financial aid:* Fellowships, research assistantships, full and partial tuition waivers available. Financial aid applicants required to submit FAF. *Faculty research:* Stratigraphy sedimentation, igneous and metamorphic petrology, oceanography, structural geology, paleontology. • J. C. Stormer, Chairman, 713-527-4880.

Saint Louis University, College of Arts and Sciences, Department of Earth and Atmospheric Sciences, Program in Geophysics, St. Louis, MO 63103. Program awards M Pr Gph, MS(R), PhD. Faculty: 6 full-time, 2 part-time. Matriculated students: 12 full-time (1 woman), 4 part-time (1 woman); includes 0 minority, 4 foreign. 4 applicants, 75% accepted. In 1990, 1 master's awarded. *Degree requirements:* For master's, computer language, comprehensive oral exam, thesis for MS(R) required, foreign language not required; for doctorate, computer language, dissertation, preliminary degree exams required, foreign language not required. *Entrance requirements:* GRE General Test. Application deadline: rolling. Application fee: $30. *Tuition:* $342 per credit hour. *Financial aid:* In 1990–91, 9 research assistantships awarded. Financial aid application deadline: 4/1. • Dr. Brian Mitchell, Director, 314-658-3131. Application contact: Dr. Robert J. Nikolai, Associate Dean of the Graduate School, 314-658-2240.

Southern Methodist University, Dedman College, Department of Geological Sciences, Program in Geophysics, Dallas, TX 75275. Offers applied geophysics (MS), geophysics (PhD). Part-time programs available. Matriculated students: 0 full-time, 9 part-time (2 women); includes 0 minority, 1 foreign. Average age 29. 11 applicants, 45% accepted. *Degree requirements:* For master's, thesis (for some programs), qualifying exam required, foreign language not required; for doctorate, 1 foreign language, dissertation, qualifying exam. *Entrance requirements:* GRE General Test (minimum combined score of 1200 required), minimum GPA of 3.0. Application deadline: 2/15. Application fee: $25. *Expenses:* Tuition of $435 per credit. Fees of $664 per semester for state residents; $56 per year for nonresidents. *Financial aid:* In 1990–91, 1 fellowship (0 to first-year students), 4 research assistantships (1 to a first-year student), 2 teaching assistantships (1 to a first-year student) awarded; full and partial tuition waivers also available. Financial aid application deadline: 2/15. *Faculty research:* Seismology. • Application contact: Dr. Peter Scholle, Graduate Adviser, 214-692-4011.

See full description on page 241.

Stanford University, School of Earth Sciences, Department of Geophysics, Stanford, CA 94305. Department awards MS, PhD. Faculty: 9 full-time (0 women), 0 part-time. Matriculated students: 52 full-time (8 women), 0 part-time; includes 2 minority (both Hispanic American), 25 foreign. Average age 29. 49 applicants, 37% accepted. In 1990, 17 master's, 6 doctorates awarded. Terminal master's awarded for partial completion of doctoral program. *Degree requirements:* For master's, foreign language and thesis not required; for doctorate, dissertation required, foreign language not required. *Entrance requirements:* GRE General Test, TOEFL. Application deadline: 1/1. Application fee: $55. *Expenses:* Tuition of $15,102 per year. Fees of $28 per quarter. *Financial aid:* Fellowships, research assistantships, teaching assistantships, institutionally sponsored loans available. Financial aid application deadline: 1/1; applicants required to submit GAPSFAS or FAF. • Application contact: Administrative Assistant, 415-723-3715.

State University of New York at Binghamton, School of Arts and Sciences, Department of Geological Sciences and Environmental Studies, Binghamton, NY 13902-6000. Department offers program in geological sciences (MA, PhD), including geophysics. Matriculated students: 22 full-time (5 women), 16 part-time (4 women); includes 1 minority (Hispanic American), 10 foreign. Average age 31. 66 applicants, 27% accepted. Terminal master's awarded for partial completion of doctoral program. *Degree requirements:* For master's, thesis or alternative; for doctorate, variable foreign language requirement, dissertation, departmental qualifying exam. *Entrance requirements:* GRE General Test, GRE Subject Test, TOEFL. Application deadline: 4/15 (priority date). Application fee: $35. *Expenses:* Tuition of $2450 per year full-time, $102.50 per credit part-time for state residents; $5766 per year full-time, $242.50 per credit part-time for nonresidents. Fees of $77 per year full-time, $27.85 per semester (minimum) part-time. *Financial aid:* In 1990–91, 25 students received a total of $196,875 in aid awarded. 1 fellowship, 9 research assistantships (3 to first-year students), 9 teaching assistantships (1 to a first-year student), 6 graduate assistantships (1 to a first-year student) were awarded; federal work-study, institutionally sponsored loans, and career-related internships or fieldwork also available. Aid available to part-time students. Average monthly stipend for a graduate assistantship: $788. Financial aid application deadline: 2/15. • Dr. Francis Wu, Chairperson, 607-777-2264.

State University of New York at Stony Brook, College of Arts and Sciences, Division of Physical Sciences and Mathematics, Department of Earth and Space Sciences, Stony Brook, NY 11794. Department offers programs in astronomical sciences (M Phil, MS, PhD), including astronomical sciences, astrophysics; geological sciences (M Phil, MS, PhD), including geochemistry, geology, geophysics. Faculty: 23 full-time, 0 part-time. Matriculated students: 48 full-time (14 women), 4 part-time (2 women); includes 1 minority (Asian American), 26 foreign. 116 applicants, 28% accepted. In 1990, 4 master's, 1 doctorate awarded. Terminal master's awarded for partial completion of doctoral program. *Degree requirements:* For master's, thesis or alternative required, foreign language not required; for doctorate, dissertation required, foreign language not required. *Entrance requirements:* GRE General Test, TOEFL, minimum GPA of 3.0. Application deadline: 2/1. Application fee: $35. *Expenses:* Tuition of $2450 per year full-time, $103 per credit part-time for state residents; $5766 per year full-time, $243 per credit part-time for nonresidents. Fees of $151 per year full-time, $10.45 per year (minimum) part-time. *Financial aid:* In 1990–91, 0 fellowships, 27 research assistantships, 21 teaching assistantships awarded. *Faculty research:* Astronomy, theoretical and observational astrophysics, paleontology, petrology, crystallography. Total annual research budget: $3.2-million. • Dr. Gilbert Hansen, Chairman, 516-632-8200.

Texas A&M University, College of Geosciences, Department of Geophysics, College Station, TX 77843. Department awards MS, PhD. Faculty: 19. Matriculated students: 64 full-time (8 women), 0 part-time; includes 7 minority (1 Asian American, 2 black American, 4 Hispanic American), 31 foreign. 26 applicants, 88% accepted. In 1990, 7 master's, 2 doctorates awarded. *Entrance requirements:* GRE General Test, TOEFL. Application deadline: 7/15. Application fee: $25 ($50 for foreign students). *Expenses:* Tuition of $100 per semester (minimum) for state residents; $128 per credit hour for nonresidents. Fees of $459 per year full-time, $252 per semester part-time. *Financial aid:* Fellowships, research assistantships, teaching assistantships available. • Joel S. Watkins, Head, 409-845-1371.

Announcement: The focus of research is primarily on studies of the solid earth in the general areas of tectonophysics, physical properties, seismology, paleomagnetism, geodynamics, and engineering geophysics. Instructional programs are tailored to each student's interests. The department has extensive computing facilities. Complete optical, physical-property, and seismic equipment is available.

See full description on page 245.

University of Alaska Fairbanks, College of Natural Sciences, Department of Geology and Geophysics, Fairbanks, AK 99775. Department offers programs in geology (MS, PhD), geophysics (MS, PhD), geoscience (MAT). Faculty: 27 full-time (3 women), 0 part-time. Matriculated students: 29 full-time (9 women), 27 part-time (5 women); includes 1 minority (Native American), 8 foreign. Average age 28. 21 applicants, 71% accepted. In 1990, 15 master's, 1 doctorate awarded. *Degree requirements:* For master's, variable foreign language requirement, thesis; for doctorate, 1 foreign language, dissertation. *Entrance requirements:* GRE General Test, GRE Subject Test. Application deadline: 3/1. Application fee: $20. *Expenses:* Tuition of $1620 per year full-time, $90 per credit part-time for state residents; $3240 per year full-time, $180 per credit part-time for nonresidents. Fees of $464 per year full-time. *Financial aid:* In 1990–91, 6 teaching assistantships (2 to first-year students) awarded. Financial aid application deadline: 3/1. *Faculty research:* Glacial surging, Alaska as geologic fragments, natural zeolites. • Dr. S. E. Swanson, Head, 907-474-7565.

University of Alberta, Faculty of Graduate Studies and Research, Department of Physics, Edmonton, AB T6G 2J9, Canada. Offerings include geophysics (M Sc, PhD). *Application fee:* $0. *Expenses:* Tuition of $1495 per year full-time, $748 per year part-time for Canadian residents; $2243 per year full-time, $1121 per year part-time for nonresidents. Fees of $301 per year full-time, $118 per year part-time. • Dr. H. R. Glyde, Chair, 403-492-5286.

University of British Columbia, Faculty of Science, Department of Geophysics and Astronomy, Vancouver, BC V6T 1Z1, Canada. Offerings include geophysics (MA Sc, M Sc, PhD). *Degree requirements:* For master's, thesis; for doctorate, dissertation, comprehensive exam.

University of Calgary, Faculty of Science, Department of Geology and Geophysics, Calgary, AB T2N 1N4, Canada. Department offers programs in geology (M Sc, PhD), geophysics (M Sc, PhD). Part-time programs available. Terminal master's awarded for partial completion of doctoral program. *Degree requirements:* Thesis/dissertation. *Entrance requirements:* For master's, TOEFL (minimum score of 550 required); for doctorate, TOEFL (minimum score of 550 required), B Sc or M Sc. Application fee: $25. *Tuition:* $1705 per year full-time, $427 per course part-time for Canadian residents; $3410 per year full-time, $854 per course part-time for nonresidents. *Faculty research:* Geochemistry, petrology, paleontology, stratigraphy, exploration and solid-earth geophysics.

See full description on page 253.

University of California at Berkeley, College of Letters and Science, Department of Geology and Geophysics, Berkeley, CA 94720. Department offers programs in geology (MA, MS, PhD), geophysics (MA, PhD). Faculty: 16. Matriculated students: 47 full-time, 0 part-time; includes 3 minority (all Hispanic American), 12 foreign. Terminal master's awarded for partial completion of doctoral program. *Degree requirements:* For master's, oral exam; for doctorate, dissertation, candidacy and comprehensive exams. *Entrance requirements:* GRE General Test, minimum GPA of 3.0. Application deadline: 2/11. Application fee: $40. *Expenses:* Tuition of $0. Fees of $1909 per year for state residents; $7825 per year for nonresidents. • Donald DePaolo, Chair.

Directory: Geophysics

University of California, Los Angeles, College of Letters and Science, Department of Earth and Space Sciences, Program in Geophysics and Space Physics, Los Angeles, CA 90024. Program awards MS, PhD. Matriculated students: 36 full-time (9 women), 0 part-time. 70 applicants, 26% accepted. In 1990, 1 master's, 5 doctorates awarded. *Degree requirements:* For master's, thesis or comprehensive exams required, foreign language not required; for doctorate, dissertation, written and oral qualifying exams required, foreign language not required. *Entrance requirements:* GRE General Test, GRE Subject Test. Application fee: $40. *Expenses:* Tuition of $0 for state residents; $7699 per year for nonresidents. Fees of $2907 per year. *Financial aid:* In 1990–91, 27 students received a total of $397,345 in aid awarded. 30 fellowships, 21 research assistantships, 21 teaching assistantships were awarded; full and partial tuition waivers, federal work-study, institutionally sponsored loans also available. Financial aid application deadline: 3/1. • Dr. Arthur Montana, Chair, Department of Earth and Space Sciences, 310-825-3880.

See full description on page 255.

University of California, Santa Barbara, College of Letters and Science, Department of Geological Sciences, Santa Barbara, CA 93106. Department offers programs in geological sciences (MA, PhD), geophysics (MS). Matriculated students: 52 full-time (16 women), 0 part-time; includes 4 foreign. 61 applicants, 49% accepted. In 1990, 3 master's, 7 doctorates awarded. *Degree requirements:* For master's, 1 foreign language, thesis or exam; for doctorate, variable foreign language requirement, dissertation. *Entrance requirements:* GRE, TOEFL (minimum score of 550 required). Application deadline: 2/10. Application fee: $40. *Expenses:* Tuition of $0 for state residents; $7699 per year for nonresidents. Fees of $2307 per year. *Financial aid:* Fellowships, research assistantships, teaching assistantships, full and partial tuition waivers, federal work-study, institutionally sponsored loans, and career-related internships or fieldwork available. Financial aid application deadline: 1/31. • Michael Fuller, Chair, 805-893-3508. Application contact: Leslie Buxton, Graduate Secretary, 805-893-3329.

See full description on page 257.

University of Chicago, Division of the Physical Sciences, Department of the Geophysical Sciences, Chicago, IL 60637. Department offers programs in atmospheric sciences (SM, PhD), earth sciences (SM, PhD), planetary and space sciences (SM, PhD). *Degree requirements:* For master's, thesis, seminar; for doctorate, 1 foreign language, dissertation, seminar. *Entrance requirements:* For master's, TOEFL; for doctorate, GRE, TOEFL (minimum score of 550 required). Application deadline: 1/6. Application fee: $45. *Expenses:* Tuition of $16,275 per year full-time, $8140 per year part-time. Fees of $356 per year. *Faculty research:* Mesoscale meteorology, climate dynamics, paleontology, geochemistry.

See full description on page 259.

University of Colorado at Boulder, College of Arts and Sciences, Department of Physics, Boulder, CO 80309. Offerings include geophysics (PhD). Terminal master's awarded for partial completion of doctoral program. Department faculty: 66 full-time (1 woman). *Degree requirements:* 1 foreign language, dissertation. *Entrance requirements:* GRE General Test, GRE Subject Test. Application deadline: 3/1 (priority date, applications processed on a rolling basis). Application fee: $30 ($50 for foreign students). *Expenses:* Tuition of $2308 per year full-time, $387 per semester (minimum) part-time for state residents; $8730 per year full-time, $1455 per semester (minimum) part-time for nonresidents. Fees of $207 per semester full-time, $27.26 per semester (minimum) part-time. • William O'Sullivan, Chairman, 303-492-8703.

University of Connecticut, College of Liberal Arts and Sciences, Field of Geophysics, Storrs, CT 06269. Field awards MS, PhD. Faculty: 2. Matriculated students: 5 full-time (0 women), 4 part-time (1 woman); includes 0 minority, 4 foreign. Average age 30. 29 applicants, 10% accepted. In 1990, 1 master's, 1 doctorate awarded. Terminal master's awarded for partial completion of doctoral program. *Degree requirements:* For doctorate, dissertation. *Entrance requirements:* GRE General Test, TOEFL. Application deadline: 6/1 (priority date, applications processed on a rolling basis). Application fee: $25. *Expenses:* Tuition of $3428 per year full-time, $571 per course part-time for state residents; $8914 per year full-time, $1486 per course part-time for nonresidents. Fees of $636 per year full-time, $87 per course part-time. *Financial aid:* In 1990–91, $29,370 in aid awarded. 0 fellowships, 2 research assistantships (0 to first-year students), 1 teaching assistantship (to a first-year student) were awarded. Financial aid application deadline: 2/15; applicants required to submit GAPSFAS or FAF. • Norman H. Gray, Head, 203-486-4434.

University of Georgia, College of Arts and Sciences, Department of Geology, Athens, GA 30602. Offerings include geophysics (MS, PhD). Department faculty: 16 full-time (2 women), 0 part-time. *Degree requirements:* 1 foreign language (computer language can substitute), thesis/dissertation. *Entrance requirements:* GRE General Test. Application fee: $10. *Expenses:* Tuition of $598 per quarter full-time, $48 per quarter part-time for state residents; $1558 per quarter full-time, $144 per quarter part-time for nonresidents. Fees of $118 per quarter. • Dr. Michael F. Roden, Graduate Coordinator, 404-542-2416.

See full description on page 263.

University of Hawaii at Manoa, School of Ocean and Earth Science and Technology, Department of Geology and Geophysics, Honolulu, HI 96822. Department awards MS, PhD. Faculty: 42 (6 women). Matriculated students: 65 (20 women). *Degree requirements:* For master's, thesis; for doctorate, 1 foreign language, dissertation, comprehensive exams. *Entrance requirements:* GRE, TOEFL, minimum GPA of 3.0. Application deadline: 2/1. Application fee: $0. *Tuition:* $800 per semester full-time for state residents; $2405 per semester full-time for nonresidents. • Application contact: Application Inquiries, 808-956-8763.

See full description on page 265.

University of Houston, College of Natural Sciences and Mathematics, Department of Geosciences, Program in Geophysics, 4800 Calhoun, Houston, TX 77004. Program awards MS, PhD. Part-time and evening/weekend programs available. *Degree requirements:* For master's, 1 foreign language, thesis (for some programs); for doctorate, 1 foreign language, dissertation. *Entrance requirements:* GRE General Test, TOEFL. Application fee: $0. *Expenses:* Tuition of $30 per hour for state residents; $134 per hour for nonresidents. Fees of $240 per year full-time, $125 per year part-time. *Faculty research:* Seismic modeling, exploration, paleomagnetics.

See full description on page 267.

University of Idaho, College of Graduate Studies, College of Mines and Earth Resources, Department of Geology and Geological Engineering, Program in Geophysics, Moscow, ID 83843. Program awards MS. Matriculated students: 2 full-time (0 women), 3 part-time (0 women); includes 0 minority, 0 foreign. In 1990, 0 degrees awarded. *Entrance requirements:* Minimum GPA of 2.8. Application deadline: 8/1. Application fee: $20. *Expenses:* Tuition of $0 for state residents; $4146 per year for nonresidents. Fees of $818 per semester full-time, $82.75 per credit part-time. *Financial aid:* Application deadline 3/1. • Dr. Rolland R. Reid, Head, Department of Geology and Geological Engineering, 208-885-6192.

University of Illinois at Chicago, College of Liberal Arts and Sciences, Department of Geological Sciences, Chicago, IL 60680. Offerings include geophysics (MS, PhD). Department faculty: 11 full-time (2 women). *Degree requirements:* Thesis/dissertation. *Entrance requirements:* GRE General Test, TOEFL (minimum score of 550 required), minimum GPA of 3.75 (on a 5.0 scale). Application deadline: 7/5. Application fee: $20. *Expenses:* Tuition of $1369 per semester full-time, $521 per semester (minimum) part-time for state residents; $3840 per semester full-time, $1454 per semester (minimum) part-time for nonresidents. Fees of $458 per semester full-time, $398 per semester (minimum) part-time. • Norman Smith, Acting Head, 312-996-3153. Application contact: Kelvin Rodolfo, Graduate Director, 312-996-3154.

See full description on page 269.

University of Illinois at Urbana-Champaign, College of Liberal Arts and Sciences, Department of Geology, Champaign, IL 61820. Offerings include geophysics (MS, PhD). Department faculty: 19 full-time (1 woman), 5 part-time (0 women). *Degree requirements:* Thesis/dissertation required, foreign language not required. *Entrance requirements:* GRE General Test, TOEFL. Application fee: $25. *Tuition:* $1838 per semester full-time, $708 per semester part-time for state residents; $4314 per semester full-time, $1673 per semester part-time for nonresidents. • Dr. R. James Kirkpatrick, Head, 217-333-3542. Application contact: Dr. Wang-Ping Chen, Graduate Admissions, 217-244-4065.

University of Kansas, College of Liberal Arts and Sciences, Department of Physics and Astronomy, Lawrence, KS 66045. Department offers programs in atmospheric science (MS), physics (MA, MS, PhD). Faculty: 25 full-time (1 woman), 2 part-time (0 women). Matriculated students: 38 full-time (6 women), 15 part-time (1 woman); includes 3 minority (1 Asian American, 2 Hispanic American), 20 foreign. In 1990, 4 master's, 3 doctorates awarded. *Degree requirements:* For master's, foreign language and thesis not required; for doctorate, computer language, dissertation. *Entrance requirements:* TOEFL (minimum score of 570 required). Application fee: $25. *Expenses:* Tuition of $1668 per year full-time, $56 per credit hour part-time for state residents; $5382 per year full-time, $179 per credit hour part-time for nonresidents. Fees of $338 per year full-time, $25 per credit hour part-time. *Financial aid:* Fellowships, research assistantships, teaching assistantships available. • Ray Ammar, Chairperson, 913-864-4626.

See full description on page 649.

University of Manitoba, Faculty of Science, Department of Geological Sciences, Winnipeg, MB R3T 2N2, Canada. Offerings include geophysics (M Sc, PhD). *Degree requirements:* Thesis/dissertation. *Entrance requirements:* GRE Subject Test.

University of Miami, Rosenstiel School of Marine and Atmospheric Science, Division of Marine Geology and Geophysics, Coral Gables, FL 33124. Division awards MA, MS, PhD. Faculty: 14 full-time (1 woman), 1 (woman) part-time. Matriculated students: 23 full-time (7 women), 0 part-time; includes 7 minority (3 Asian American, 4 Hispanic American), 6 foreign. Average age 27. 25 applicants, 20% accepted. In 1990, 3 master's, 3 doctorates awarded. Terminal master's awarded for partial completion of doctoral program. *Degree requirements:* For master's, thesis required (for some programs), foreign language not required; for doctorate, dissertation required, foreign language not required. *Entrance requirements:* GRE General Test, TOEFL (minimum score of 550 required). Application fee: $35. *Expenses:* Tuition of $567 per credit hour. Fees of $87 per semester full-time. *Financial aid:* In 1990–91, 2 fellowships (1 to a first-year student), 18 research assistantships (1 to a first-year student), 3 teaching assistantships awarded. *Faculty research:* Carbonate sedimentology, low-temperature geochemistry, marine geophysics, paleoceonography, deep sea petrology. Total annual research budget: $1-million. • Dr. Peter Swart, Head, 305-361-4103. Application contact: Dr. Larry Peterson, Graduate Adviser, 305-361-4692.

University of Minnesota, Twin Cities Campus, Institute of Technology, Department of Geology and Geophysics, Minneapolis, MN 55455. Department offers programs in geology (M Ed, MS, PhD), geophysics (MS, PhD). Faculty: 24. Matriculated students: 60. *Degree requirements:* For master's, thesis optional; for doctorate, dissertation. *Entrance requirements:* GRE General Test, TOEFL. *Expenses:* Tuition of $1084 per quarter full-time, $301 per credit part-time for state residents; $2168 per quarter full-time, $602 per credit part-time for nonresidents. Fees of $118 per quarter. • Peter Hudelston, Director of Graduate Studies, 612-624-1333.

See full description on page 273.

University of Missouri–Rolla, School of Mines and Metallurgy, Department of Geology and Geophysics, Program in Geophysics, Rolla, MO 65401. Program awards MS, PhD. Part-time programs available. Faculty: 3 full-time (0 women), 0 part-time. Average age 22. In 1990, 1 master's awarded (100% continued full-time study); 0 doctorates awarded. Terminal master's awarded for partial completion of doctoral program. *Degree requirements:* For master's, departmental qualifying exam required, foreign language not required; for doctorate, dissertation, departmental qualifying exam. *Entrance requirements:* GRE General Test (minimum combined score of 1100 required), GRE Subject Test. Application deadline: 7/1 (applications processed on a rolling basis). Application fee: $20 ($40 for foreign students). *Expenses:* Tuition of $2090 per year full-time, $87.10 per credit hour part-time for state residents; $5582 per year full-time, $232.60 per credit hour part-time for nonresidents. Fees of $349 per year full-time, $61.63 per semester (minimum) part-time. *Financial aid:* In 1990–91, $5100 in aid awarded. 1 teaching assistantship (0 to first-year students) was awarded; research assistantships, federal work-study, institutionally sponsored loans also available. Aid available to part-time students. Financial aid application deadline: 3/31. *Faculty research:* Digital filtering, modeling, field interpretation. • Richard D. Hagni, Chairman, Department of Geology and Geophysics, 314-341-4616.

University of New Orleans, College of Sciences, Department of Geology and Geophysics, New Orleans, LA 70148. Department offers programs in geology (MS), geophysics (MS). Evening/weekend programs available. Faculty: 16 full-time (1 woman), 2 part-time (1 woman). Matriculated students: 23 full-time, 29 part-time; includes 3 minority (all black American), 6 foreign. Average age 24. 30 applicants, 33% accepted. In 1990, 15 degrees awarded. *Degree requirements:* Thesis required, foreign language not required. *Entrance requirements:* GRE General Test. Application deadline: 7/1 (priority date, applications processed on a rolling basis). Application fee: $20. *Tuition:* $962 per quarter hour full-time for state residents; $2308 per quarter hour full-time for nonresidents. *Financial aid:* In 1990–91, $163,000 in aid awarded. 2 fellowships (1 to a first-year student), 1 research assistantship (0 to first-year students), 15 teaching assistantships (7 to first-year students) were awarded; federal work-study, institutionally sponsored loans, and career-related internships or fieldwork also available. *Faculty research:* Continental margin structure and seismology, burial diagenesis of siliclastic sediments, tectonics at convergent plate margins, continental shelf sediment stability, early diagenesis of carbonates. Total annual research budget: $150,000. • William C. Craig, Chairman, 504-286-6793. Application contact: William H. Busch, Graduate Coordinator, 504-286-7230.

Directory: Geophysics

University of Oklahoma, College of Geosciences, School of Geology and Geophysics, Program in Geophysics, Norman, OK 73019. Program awards MS. Faculty: 2 full-time (0 women), 0 part-time. Matriculated students: 11. Average age 25. In 1990, 3 degrees awarded (33% found work related to degree, 67% continued full-time study). *Degree requirements:* Computer language, thesis, comprehensive exam required, foreign language not required. *Entrance requirements:* GRE General Test, TOEFL (minimum score of 550 required), bachelor's degree in field. Application deadline: 2/1 (priority date, applications processed on a rolling basis). Application fee: $10. *Expenses:* Tuition of $63 per credit hour for state residents; $192 per credit hour for nonresidents. Fees of $67.50 per semester. *Financial aid:* In 1990–91, 2 research assistantships (0 to first-year students), 1 teaching assistantship (0 to first-year students) awarded; fellowships, partial tuition waivers, federal work-study, institutionally sponsored loans also available. Financial aid application deadline: 2/1. *Faculty research:* Technophysics and solid-earth geophysics, exploration geophysics, paleomagnetics. • Dr. Charles Gilbert, Director, School of Geology and Geophysics, 405-325-3253.

See full description on page 279.

University of Pittsburgh, Faculty of Arts and Sciences, Department of Geology and Planetary Science, Pittsburgh, PA 15260. Department awards MS, PhD. Part-time programs available. Faculty: 12 full-time (0 women), 1 (woman) part-time. Matriculated students: 17 full-time (5 women), 18 part-time (6 women); includes 4 minority (2 Asian American, 2 black American), 5 foreign. 59 applicants, 17% accepted. In 1990, 6 master's, 1 doctorate awarded. *Degree requirements:* Thesis/dissertation required, foreign language not required. *Entrance requirements:* GRE General Test, TOEFL. Application deadline: 8/1 (applications processed on a rolling basis). Application fee: $15 ($25 for foreign students). *Expenses:* Tuition of $2920 per semester full-time, $241 per credit part-time for state residents; $5840 per semester full-time, $482 per credit part-time for nonresidents. Fees of $156 per year. *Financial aid:* In 1990–91, 13 students received a total of $94,930 in aid awarded. 4 research assistantships (0 to first-year students), 9 teaching assistantships (3 to first-year students) were awarded; full and partial tuition waivers, federal work-study, institutionally sponsored loans, and career-related internships or fieldwork also available. Aid available to part-time students. Average monthly stipend for a graduate assistantship: $900. Financial aid application deadline: 3/1; applicants required to submit FAF. *Faculty research:* Plate tectonics as basement studies, sedimentology and image analysis, marine paleoecology and geoarchaeology, geophysics and paleomagnetics, planetary geology and meteoritics. Total annual research budget: $490,472. • Dr. Thomas H. Anderson, Chairman, 412-624-8780. Application contact: Dr. Walter L. Pilant, Graduate Adviser, 412-624-8870.

See full description on page 281.

University of Saskatchewan, Colleges of Arts and Sciences and Engineering, Department of Geological Sciences, Saskatoon, SK S7N 0W0, Canada. Department awards M Sc, PhD. *Degree requirements:* Thesis/dissertation. *Entrance requirements:* TOEFL. Application fee: $0.

See full description on page 283.

University of Texas at Austin, Graduate School, College of Natural Sciences, Department of Geological Sciences, Austin, TX 78712. Department awards MA, PhD. Matriculated students: 179 full-time (41 women), 0 part-time; includes 5 minority (3 Asian American, 2 Hispanic American), 39 foreign. 90 applicants, 67% accepted. In 1990, 15 master's, 4 doctorates awarded. *Entrance requirements:* GRE. Application deadline: 2/1 (priority date, applications processed on a rolling basis). Application fee: $40 ($75 for foreign students). *Tuition:* $510.30 per semester for state residents; $1806 per semester for nonresidents. *Financial aid:* Fellowships, research assistantships, teaching assistantships available. Financial aid application deadline: 3/1. • Dr. Clark R. Wilson, Chairman, 512-471-5172. Application contact: Dr. William Carlson, Graduate Adviser, 512-471-6098.

See full description on page 285.

University of Texas at Dallas, School of Natural Sciences and Mathematics, Program in Geosciences, Richardson, TX 75083-0688. Program awards MS, PhD. Part-time and evening/weekend programs available. Faculty: 16 full-time (0 women), 7 part-time (0 women). Matriculated students: 44 full-time (10 women), 35 part-time (7 women); includes 7 minority (3 Asian American, 1 black American, 3 Hispanic American), 21 foreign. Average age 33. In 1990, 7 master's, 5 doctorates awarded. *Degree requirements:* Computer language, thesis/dissertation required, foreign language not required. *Entrance requirements:* GRE General Test (minimum combined score of 1000 required), TOEFL (minimum score of 550 required), minimum GPA of 3.0 in upper level course work in field. Application deadline: 7/15 (applications processed on a rolling basis). Application fee: $0 ($75 for foreign students). *Expenses:* Tuition of $360 per semester full-time, $100 per semester (minimum) part-time for state residents; $2196 per semester full-time, $122 per semester hour (minimum) part-time for nonresidents. Fees of $338 per semester full-time, $22 per hour (minimum) part-time. *Financial aid:* In 1990–91, 1 student received a total of $8175 in aid awarded. Fellowships, research assistantships, teaching assistantships, federal work-study, and career-related internships or fieldwork available. Aid available to part-time students. Financial aid application deadline: 11/1; applicants required to submit FAF. *Faculty research:* Geochemistry, structural geology-tectonics, sedimentology-stratigraphy, micropaleontology. • Dr. James L. Carter, Head, 214-690-2401.

See full description on page 287.

University of Texas at El Paso, College of Science, Department of Geological Sciences, Program in Geophysics, 500 West University Avenue, El Paso, TX 79968. Program awards MS. *Application deadline:* 7/1 (priority date, applications processed on a rolling basis). *Application fee:* $0 ($50 for foreign students). *Expenses:* Tuition of $360 per semester full-time, $100 per semester (minimum) part-time for state residents; $2304 per semester full-time, $128 per credit hour (minimum) part-time for nonresidents. Fees of $137 per semester full-time, $28.50 per semester (minimum) part-time. *Financial aid:* Application deadline 3/1. • Application contact: Diana Guerrero, Admissions Office, 915-747-5576.

University of Toronto, School of Graduate Studies, Physical Sciences Division, Department of Geology, Program in Geophysics, Toronto, ON M5S 1A1, Canada. Program awards M Sc, PhD. Offered jointly with Department of Physics. *Degree requirements:* Thesis/dissertation. *Application deadline:* 4/15. *Application fee:* $50. *Expenses:* Tuition of $2220 per year full-time, $666 per year part-time for Canadian residents; $10,198 per year full-time, $305.05 per year part-time for nonresidents. Fees of $277.56 per year full-time, $82.73 per year part-time. *Financial aid:* Application deadline 2/1. • J. Westgate, Chair, Department of Geology, 416-978-3021.

See full description on page 291.

University of Tulsa, College of Engineering and Physical Sciences, Department of Geosciences, Tulsa, OK 74104. Offerings include geophysics (MS, PhD). Department faculty: 7 full-time (1 woman), 1 part-time (0 women). *Degree requirements:* For master's, computer language, thesis or alternative required, foreign language not required; for doctorate, computer language, dissertation required, foreign language not required. *Entrance requirements:* For master's, GRE General Test, TOEFL; for doctorate, GRE General Test, GRE Subject Test, TOEFL. Application deadline: rolling. Application fee: $30. *Tuition:* $350 per credit hour. • Dr. Colin Barker, Chairperson, 918-631-3014.

University of Utah, College of Mines and Earth Sciences, Department of Geology and Geophysics, Salt Lake City, UT 84112. Department offers programs in geological engineering (ME, MS, PhD), geology (MS, PhD), geophysics (MS, PhD). Faculty: 23 (1 woman). Matriculated students: 48 full-time (13 women), 12 part-time (1 woman); includes 0 minority, 21 foreign. Average age 31. 103 applicants, 30% accepted. In 1990, 8 master's, 0 doctorates awarded. Terminal master's awarded for partial completion of doctoral program. *Degree requirements:* Computer language, thesis/dissertation. *Entrance requirements:* GRE General Test, TOEFL, minimum GPA of 3.25. Application deadline: 8/1. Application fee: $25 ($50 for foreign students). *Tuition:* $195 per credit for state residents; $505 per credit for nonresidents. *Financial aid:* In 1990–91, $288,000 in aid awarded. 8 fellowships, 24 research assistantships, 15 teaching assistantships were awarded; stipends, institutionally sponsored loans also available. Financial aid application deadline: 3/1. *Faculty research:* Igneous, metamorphic, and sedimentary petrology; ore deposits; aqueous geochemistry; isotope geochemistry; heat flow. • Francis H. Brown, Chairman, 801-581-7162. Application contact: Peter H. Roth, Director of Graduate Studies, 801-581-7266.

University of Victoria, Faculty of Arts and Science, Department of Physics, Victoria, BC V8W 2Y2, Canada. Offerings include geophysics (M Sc, PhD). Department faculty: 21 full-time (0 women), 0 part-time. *Degree requirements:* Thesis/dissertation required, foreign language not required. *Application deadline:* 5/31 (priority date, applications processed on a rolling basis). *Application fee:* $20. *Expenses:* Tuition of $754 per semester. Fees of $23 per year. • Dr. L. P. Robertson, Chair, 604-721-7698. Application contact: Dr. J. T. Weaver, Graduate Adviser, 604-721-7768.

University of Washington, Interdisciplinary Programs and College of Arts and Sciences, Program in Geophysics, Seattle, WA 98195. Program awards MS, PhD. Part-time programs available. Faculty: 23 full-time (2 women). Matriculated students: 44 full-time (7 women), 0 part-time; includes 1 minority (Native American), 14 foreign. Average age 28. 92 applicants, 27% accepted. In 1990, 9 master's awarded; 2 doctorates awarded (100% entered university research/teaching). *Degree requirements:* For master's, thesis or alternative required, foreign language not required; for doctorate, dissertation or alternative, departmental qualifying exam required, foreign language not required. *Entrance requirements:* GRE General Test. Application deadline: 7/1. Application fee: $35. *Tuition:* $1129 per quarter full-time, $324 per credit (minimum) part-time for state residents; $2824 per quarter full-time, $809 per credit (minimum) part-time for nonresidents. *Financial aid:* In 1990–91, $367,650 in aid awarded. 3 fellowships (0 to first-year students), 28 research assistantships (5 to first-year students), 2 teaching assistantships (both to first-year students) were awarded; federal work-study also available. Financial aid application deadline: 2/15; applicants required to submit FAF. *Faculty research:* Volcanic seismology, seismic networks, fluid dynamics, geomagnetism, ionospheric and magnetospheric physics. Total annual research budget: $2.8-million. • Ronald T. Merrill, Chairman, 206-543-8020. Application contact: Rachel Munz, Academic Secretary, 206-543-8020.

See full description on page 293.

University of Western Ontario, Physical Sciences Division, Department of Geophysics, London, ON N6A 3K7, Canada. Department awards M Sc, PhD. *Degree requirements:* Thesis/dissertation required, foreign language not required. *Entrance requirements:* For master's, minimum B average; for doctorate, minimum B average, master's degree with thesis. *Tuition:* $1015 per year full-time, $1050 per year part-time for Canadian residents; $4207 per year for nonresidents.

University of Wisconsin–Madison, College of Letters and Science, Department of Geology and Geophysics, Madison, WI 53706. Department offers programs in geology (MS, PhD), geophysics (MS, PhD). Faculty: 22 full-time, 0 part-time. Matriculated students: 58 full-time (15 women), 5 part-time (1 woman); includes 3 minority (2 Asian American, 1 black American), 7 foreign. 89 applicants, 31% accepted. In 1990, 10 master's, 5 doctorates awarded. *Entrance requirements:* GRE General Test, GRE Subject Test. Application deadline: rolling. Application fee: $20. *Financial aid:* In 1990–91, 52 students received aid. 5 fellowships, 30 research assistantships, 16 teaching assistantships, 1 project assistantship awarded; institutionally sponsored loans also available. Financial aid application deadline: 1/15; applicants required to submit FAF. • Herbert F. Wang, Chairperson, 608-262-8960. Application contact: Joanne Nagy, Assistant Dean of the Graduate School, 608-262-2433.

University of Wisconsin–Oshkosh, College of Letters and Science, Department of Physics and Astronomy, Oshkosh, WI 54901. Offerings include physics (MS), with options in geophysics, instrumentation, magnetic resonance, physics education, solid state physics. Department faculty: 7 full-time (1 woman), 1 part-time (0 women). *Degree requirements:* Thesis required, foreign language not required. *Application deadline:* rolling. *Application fee:* $20. • Dr. Sandra Gade, Chair, 414-424-4433. Application contact: Dr. John Karl, Coordinator, 414-424-4432.

University of Wyoming, College of Arts and Sciences, Department of Geology and Geophysics, Laramie, WY 82071. Department offers programs in geology and geophysics (MS, PhD), geophysics (MS, PhD). Part-time programs available. Faculty: 20 full-time (1 woman), 3 part-time (1 woman). Matriculated students: 38 full-time (6 women), 35 part-time (10 women); includes 1 minority (Native American), 6 foreign. Average age 26. 136 applicants, 54% accepted. In 1990, 24 master's, 1 doctorate awarded. *Degree requirements:* Variable foreign language requirement, thesis/dissertation. *Entrance requirements:* For master's, GRE General Test (minimum combined score of 900 required), minimum GPA of 3.0; for doctorate, GRE General Test (minimum combined score of 1000 required), minimum GPA of 3.0. Application deadline: 1/31 (priority date, applications processed on a rolling basis). Application fee: $30. *Tuition:* $1554 per year full-time, $74.25 per credit hour part-time for state residents; $4358 per year full-time, $74.25 per credit hour part-time for nonresidents. *Financial aid:* In 1990–91, 49 students received a total of $429,872 in aid awarded. 17 fellowships, 15 research assistantships, 17 teaching assistantships (14 to first-year students) were awarded; federal work-study, institutionally sponsored loans, and career-related internships or fieldwork also available. Average monthly stipend for a graduate assistantship: $740. Financial aid application deadline: 1/31. *Faculty research:* Structural geology, diagenesis and hydrocarbon maturation, tectonics and sedimentation, igneous and metamorphic petrology, crustal studies. Total annual research budget: $5-million. • Dr. James R. Steidtman, Head, 307-766-3386. Application contact: Salina Renninger, Admissions Coordinator, 307-766-3389.

Virginia Polytechnic Institute and State University, College of Arts and Sciences, Department of Geological Sciences, Program in Geophysics, Blacksburg, VA 24061. Program awards MS, PhD. Faculty: 5 full-time (0 women), 0 part-time. Matriculated students: 8 full-time (2 women), 2 part-time (0 women); includes 0 minority, 2 foreign. 26 applicants, 46% accepted. In 1990, 2 master's, 0 doctorates awarded. *Degree requirements:* 1 foreign language, thesis/dissertation. *Entrance requirements:* GRE General Test, GRE Subject Test. Application deadline: 2/15 (priority date). Application fee: $10. *Tuition:* $1889 per semester full-time, $606 per credit hour part-time for state residents; $2627 per semester full-time, $853 per credit hour part-time for nonresidents. *Financial aid:* Fellowships, research assistantships, teaching assistantships, assistantships, full tuition waivers, and career-related internships or fieldwork available. *Faculty research:* Earthquake seismology, exploration seismology, reflection seismology, exploration geophysics, theoretical and observational seismology. • Dr. Donald Bloss, Chairman, Department of Geological Sciences, 703-231-6521.

Washington University, Graduate School of Arts and Sciences, Department of Earth and Planetary Sciences, St. Louis, MO 63130. Offerings include geophysics (PhD). Terminal master's awarded for partial completion of doctoral program. *Degree requirements:* Dissertation. *Entrance requirements:* GRE General Test. Application deadline: 1/15 (priority date, applications processed on a rolling basis). Application fee: $0. *Tuition:* $15,960 per year full-time, $665 per credit hour part-time. • Dr. Larry A. Haskin, Chairman, 314-935-5610.

See full description on page 295.

Yale University, Graduate School of Arts and Sciences, Department of Geology and Geophysics, New Haven, CT 06520. Department offers programs in geochemistry (PhD), geophysics (PhD), meterology (PhD), mineralogy and crystallography (PhD), oceanography (PhD), paleoecology (PhD), paleontology and stratigraphy (PhD), petrology (PhD), structural geology (PhD). Faculty: 24 full-time (2 women), 1 part-time (0 women). Matriculated students: 44 full-time (5 women), 0 part-time; includes 0 minority, 25 foreign. 54 applicants, 30% accepted. In 1990, 5 master's, 10 doctorates awarded. Terminal master's awarded for partial completion of doctoral program. *Degree requirements:* For doctorate, dissertation. *Entrance requirements:* For doctorate, GRE General Test, GRE Subject Test. Application deadline: 1/2. Application fee: $45. *Tuition:* $15,160 per year full-time, $1895 per course part-time. • Dr. B. Saltzman, Chairman, 203-432-3114. Application contact: Susan Webb, Coordinator of Admissions, 203-432-2770.

See full description on page 299.

Planetary Sciences

Arizona State University, College of Liberal Arts and Sciences, Department of Physics, Tempe, AZ 85287. Offerings include planetary sciences (PhD). *Degree requirements:* Dissertation. *Entrance requirements:* GRE. Application fee: $25. *Tuition:* $1528 per year full-time, $80 per hour part-time for state residents; $6934 per year full-time, $289 per hour part-time for nonresidents. • Dr. John D. Dow, Chair, 602-965-3561.

Brown University, Department of Geological Sciences, Box 1846, Providence, RI 02912. Department awards AM, Sc M, PhD. Faculty: 18 full-time, 1 part-time. Matriculated students: 44 full-time (15 women); includes 8 foreign. 76 applicants, 25% accepted. In 1990, 9 master's, 10 doctorates awarded. *Degree requirements:* For master's, foreign language and thesis not required; for doctorate, dissertation, preliminary exam required, foreign language not required. *Application deadline:* 1/2. *Application fee:* $45. *Expenses:* Tuition of $16,256 per year full-time, $2032 per course part-time. Fees of $372 per year. *Financial aid:* In 1990–91, $700,061 in aid awarded. 33 research assistantships, 10 teaching assistantships were awarded; assistantships, full and partial tuition waivers also available. Financial aid application deadline: 1/2; applicants required to submit GAPSFAS. *Faculty research:* Geochemistry, mineral kinetics, igenous and metamorphic petrology, tectonophysics including geophysics and structural geology, climate dynamics. Total annual research budget: $4-million. • Application contact: Richard Yund, Graduate Representative, 401-863-3065.

Announcement: Department awards MA, Sc M, PhD. Program in observational, analytical, experimental, and theoretical aspects of planetary surface and interior processes. Study of terrestrial planets, outer planet satellites, small bodies (comets, asteroids), and planetary materials. Faculty and students participate in many NASA programs and missions (e.g., Magellan, Mars Observer) and in cooperative programs with Soviet space-science institutions. Site of NASA Northeast Regional Planetary Data Center, National Space Grant College and Fellowship Program, and Reflectance Experiment Laboratory (RELAB). Graduate Representative: Richard Yund, 401-863-3065 or 863-3338.

California Institute of Technology, Division of Geological and Planetary Sciences, Pasadena, CA 91125. Offerings include planetary science (MS, PhD). Division faculty: 28 full-time (0 women), 0 part-time. *Entrance requirements:* GRE General Test, GRE Subject Test. Application deadline: 1/15. Application fee: $0. *Expenses:* Tuition of $14,100 per year. Fees of $8 per quarter. • Dr. David J. Stevenson, Chairman, 818-356-6108.

See full description on page 219.

Columbia University, Graduate School of Arts and Sciences, Division of Natural Sciences, Program in Atmospheric and Planetary Science, New York, NY 10027. Program awards M Phil, PhD. *Degree requirements:* For doctorate, variable foreign language requirement, dissertation. *Entrance requirements:* For doctorate, GRE General Test, GRE Subject Test, TOEFL, previous course work in mathematics and physics. Application deadline: 1/5. Application fee: $50. *Expenses:* Tuition of $7836 per semester for state residents; $426 per credit for nonresidents. Fees of $148 per semester for state residents. *Financial aid:* Application deadline 1/5. *Faculty research:* Climate, weather prediction. • Dr. Roger S. Bagnall, Dean, Graduate School of Arts and Sciences, 212-854-2861.

See full description on page 527.

Florida Institute of Technology, College of Science and Liberal Arts, Department of Physics and Space Sciences, Program in Space Science, Melbourne, FL 32901. Program awards MS, PhD. Part-time and evening/weekend programs available. Faculty: 11 full-time (0 women), 6 part-time (0 women). Matriculated students: 7 full-time (1 woman), 6 part-time (3 women); includes 4 foreign. 57 applicants, 33% accepted. In 1990, 12 master's, 0 doctorates awarded. *Degree requirements:* For master's, comprehensive exam required, thesis optional, foreign language not required; for doctorate, dissertation, comprehensive exam. *Entrance requirements:* For master's, minimum GPA of 3.0, proficiency in a computer language; for doctorate, minimum GPA of 3.2. Application deadline: rolling. Application fee: $35. *Tuition:* $234 per credit hour. *Financial aid:* Research assistantships, teaching assistantships, federal work-study available. Financial aid application deadline: 3/1. *Faculty research:* Remote sensing, planetary science, astronomical image processing. • Application contact: Carolyn P. Farrior, Director of Graduate Admissions, 407-768-8000 Ext. 8027.

Harvard University, Graduate School of Arts and Sciences, Department of Earth and Planetary Sciences, Cambridge, MA 02138. Department awards AM, PhD. Faculty: 12 full-time (0 women), 2 part-time (1 woman). Matriculated students: 31 full-time (7 women), 0 part-time; includes 0 minority, 10 foreign. In 1990, 2 master's awarded (100% continued full-time study); 6 doctorates awarded (33% entered university research/teaching, 67% continued full-time study). Terminal master's awarded for partial completion of doctoral program. *Degree requirements:* For master's, 1 foreign language, thesis or alternative; for doctorate, 1 foreign language, dissertation. *Entrance requirements:* GRE General Test, GRE Subject Test. Application deadline: 1/2. Application fee: $60. *Expenses:* Tuition of $14,860 per year. Fees of $550 per year. *Financial aid:* In 1990–91, $180,233 in aid awarded. 17 fellowships (9 to first-year students), 31 research assistantships (5 to first-year students), 24 teaching assistantships (9 to first-year students) were awarded; federal work-study, institutionally sponsored loans, and career-related internships or fieldwork also available. Financial aid application deadline: 1/2; applicants required to submit GAPSFAS. *Faculty research:* Economic geography, geochemistry, geophysics, minerology, crystallography. • Dr. Rick O'Connell, Director of Graduate Studies, 617-495-2351. Application contact: Office of Admissions and Financial Aid, 617-495-5315.

See full description on page 221.

Massachusetts Institute of Technology, School of Science, Department of Earth, Atmospheric, and Planetary Sciences, Cambridge, MA 02139. Department offers programs in geology (SM, PhD, Sc D), geophysics (SM, PhD, Sc D), meteorology and physical oceanography (SM, PhD, Sc D), planetary science (SM, PhD). Faculty: 37 full-time (3 women), 0 part-time. Matriculated students: 163 full-time (45 women), 0 part-time; includes 3 minority (1 Asian American, 1 black American, 1 Hispanic American), 54 foreign. Average age 24. 189 applicants, 15% accepted. In 1990, 12 master's awarded; 30 doctorates awarded (70% entered university research/teaching, 30% found other work related to degree). Terminal master's awarded for partial completion of doctoral program. *Degree requirements:* For master's, thesis required, foreign language not required; for doctorate, dissertation, general exam required, foreign language not required. *Entrance requirements:* GRE General Test, GRE Subject Test. Application deadline: 1/15. Application fee: $45. *Tuition:* $16,900 per year. *Financial aid:* In 1990–91, 20 fellowships (2 to first-year students), 93 research assistantships (17 to first-year students), 14 teaching assistantships (1 to a first-year student) awarded; institutionally sponsored loans also available. Financial aid application deadline: 1/15. *Faculty research:* Evolution of main features of the planetary system; origin, composition, structure, and state of the atmospheres, oceans surfaces, and interiors of planets; dynamics of planets and satellite motions. Total annual research budget: $12-million. • Dr. Thomas H. Jordan, Chairman, 617-253-3382. Application contact: Anita Kilian, Graduate Administrator, 617-253-3381.

Rensselaer Polytechnic Institute, School of Science, Department of Earth and Environmental Sciences, Program in Planetary Geology, Troy, NY 12180. Program awards MS, PhD. *Degree requirements:* For master's, thesis optional; for doctorate, dissertation. *Entrance requirements:* GRE General Test, TOEFL. Application deadline: 2/1. Application fee: $30. *Expenses:* Tuition of $455 per credit hour. Fees of $195.57 per semester. • Application contact: M. Brian Bayly, Graduate Coordinator, 518-276-6494.

See full description on page 239.

Southern Methodist University, Dedman College, Department of Geological Sciences, Program in Planetary Studies, Dallas, TX 75275. Program awards MS. *Degree requirements:* Thesis, qualifying exam required, foreign language not required. *Entrance requirements:* GRE General Test (minimum combined score of 1200 required), minimum GPA of 3.0. Application deadline: 2/15. Application fee: $25. *Expenses:* Tuition of $435 per credit. Fees of $664 per semester for state residents; $56 per year for nonresidents. *Financial aid:* Application deadline 2/15. • Dr. Michael J. Holdaway, Chairman, Department of Geological Sciences, 214-692-2770. Application contact: Dr. Peter Scholle, Graduate Adviser, 214-692-4011.

See full description on page 241.

State University of New York at Stony Brook, College of Arts and Sciences, Division of Physical Sciences and Mathematics, Department of Earth and Space Sciences, Program in Astronomical Sciences, Stony Brook, NY 11794. Offers astronomical sciences (M Phil, MS, PhD), astrophysics (M Phil, MS, PhD). Faculty: 8 full-time. Matriculated students: 14 full-time (3 women), 0 part-time. *Degree requirements:* For master's, thesis or alternative required, foreign language not required; for doctorate, dissertation required, foreign language not required. *Entrance requirements:* GRE General Test, TOEFL, minimum GPA of 3.0. Application deadline: 2/1. Application fee: $35. *Expenses:* Tuition of $2450 per year full-time, $103 per credit part-time for state residents; $5766 per year full-time, $243 per credit part-time for nonresidents. Fees of $151 per year full-time, $10.45 per year (minimum) part-time. *Financial aid:* Fellowships available. • Dr. Gilbert Hansen, Chairman, Department of Earth and Space Sciences, 516-632-8200.

University of Arizona, College of Arts and Sciences, Faculty of Science, Department of Planetary Sciences, Tucson, AZ 85721. Department awards MS, PhD. Faculty: 32 full-time (3 women), 0 part-time. Matriculated students: 23 full-time (9 women), 3 part-time (1 woman); includes 0 minority, 6 foreign. Average age 30. 56 applicants,

SECTION 3: EARTH AND PLANETARY SCIENCES

Directory: Planetary Sciences; Cross-Discipline Announcement

21% accepted. In 1990, 0 master's awarded; 6 doctorates awarded (100% found work related to degree). *Degree requirements:* For master's, thesis required (for some programs), foreign language not required; for doctorate, 1 foreign language, dissertation. *Entrance requirements:* GRE General Test, GRE Subject Test. Application deadline: 1/15 (applications processed on a rolling basis). Application fee: $25. *Expenses:* Tuition of $0 for state residents; $5406 per year full-time, $209 per credit hour part-time for nonresidents. Fees of $1528 per year full-time, $80 per credit hour part-time. *Financial aid:* In 1990–91, $485,000 in aid awarded. 14 fellowships, 30 research assistantships, 4 teaching assistantships, 6 scholarships were awarded; partial tuition waivers also available. Financial aid application deadline: 2/15; applicants required to submit FAF. *Faculty research:* Cosmochemistry, planetary geology, astronomy, space physics, planetary physics. Total annual research budget: $10-million. • Dr. Eugene H. Levy, Head, 602-621-6962. Application contact: Chair, Admissions Committee, 602-621-6963.

See full description on page 251.

University of California, Los Angeles, College of Letters and Science, Department of Earth and Space Sciences, Program in Geophysics and Space Physics, Los Angeles, CA 90024. Program awards MS, PhD. Matriculated students: 36 full-time (9 women), 0 part-time. 70 applicants, 26% accepted. In 1990, 1 master's, 5 doctorates awarded. *Degree requirements:* For master's, thesis or comprehensive exams required, foreign language not required; for doctorate, dissertation, written and oral qualifying exams required, foreign language not required. *Entrance requirements:* GRE General Test, GRE Subject Test. Application fee: $40. *Expenses:* Tuition of $0 for state residents; $7699 per year for nonresidents. Fees of $2907 per year. *Financial aid:* In 1990–91, 27 students received a total of $397,345 in aid awarded. 30 fellowships, 21 research assistantships, 21 teaching assistantships were awarded; full and partial tuition waivers, federal work-study, institutionally sponsored loans also available. Financial aid application deadline: 3/1. • Dr. Arthur Montana, Chair, Department of Earth and Space Sciences, 310-825-3880.

See full description on page 255.

University of Chicago, Division of the Physical Sciences, Department of the Geophysical Sciences, Chicago, IL 60637. Offerings include planetary and space sciences (SM, PhD). *Degree requirements:* For master's, thesis, seminar; for doctorate, 1 foreign language, dissertation, seminar. *Entrance requirements:* For master's, TOEFL; for doctorate, GRE, TOEFL (minimum score of 550 required). Application deadline: 1/6. Application fee: $45. *Expenses:* Tuition of $16,275 per year full-time, $8140 per year part-time. Fees of $356 per year.

See full description on page 259.

University of Pittsburgh, Faculty of Arts and Sciences, Department of Geology and Planetary Science, Pittsburgh, PA 15260. Department awards MS, PhD. Part-time programs available. Faculty: 12 full-time (0 women), 1 (woman) part-time. Matriculated students: 17 full-time (5 women), 18 part-time (6 women); includes 4 minority (2 Asian American, 2 black American), 5 foreign. 59 applicants, 17% accepted. In 1990, 6 master's, 1 doctorate awarded. *Degree requirements:* Thesis/dissertation required, foreign language not required. *Entrance requirements:* GRE General Test, TOEFL. Application deadline: 8/1 (applications processed on a rolling basis). Application fee: $15 ($25 for foreign students). *Expenses:* Tuition of $2920 per semester full-time, $241 per credit part-time for state residents; $5840 per semester full-time, $482 per credit part-time for nonresidents. Fees of $156 per year. *Financial aid:* In 1990–91, 13 students received a total of $94,930 in aid awarded. 4 research assistantships (0 to first-year students), 9 teaching assistantships (3 to first-year students) were awarded; full and partial tuition waivers, federal work-study, institutionally sponsored loans, and career-related internships or fieldwork also available. Aid available to part-time students. Average monthly stipend for a graduate assistantship: $900. Financial aid application deadline: 3/1; applicants required to submit FAF. *Faculty research:* Plate tectonics as basement studies, sedimentology and image analysis, marine paleoecology and geoarchaeology, geophysics and paleomagnetics, planetary geology and meteoritics. Total annual research budget: $490,472. • Dr. Thomas H. Anderson, Chairman, 412-624-8780. Application contact: Dr. Walter L. Pilant, Graduate Adviser, 412-624-8870.

See full description on page 281.

Utah State University, College of Science, Department of Physics, Logan, UT 84322. Offerings include physics (MS, PhD), with options in astronomy and astrophysics (MS), earth and planetary sciences (PhD), engineering physics (PhD), physics (MS). Terminal master's awarded for partial completion of doctoral program. Department faculty: 22 full-time (0 women), 3 part-time (0 women). *Degree requirements:* Dissertation required, foreign language not required. *Entrance requirements:* GRE General Test (score in 40th percentile or higher required), minimum GPA of 3.0. Application deadline: 7/15 (priority date, applications processed on a rolling basis). Application fee: $25 ($30 for foreign students). *Tuition:* $426 per quarter (minimum) full-time, $184 per quarter (minimum) part-time for state residents; $1133 per quarter (minimum) full-time, $505 per quarter (minimum) part-time for nonresidents. • Dr. W. John Raitt, Head, 801-750-2848. Application contact: O. Harry Otteson, Assistant Head, 801-750-2850.

Washington University, Graduate School of Arts and Sciences, Department of Earth and Planetary Sciences, St. Louis, MO 63130. Department offers programs in earth and planetary sciences (MA), geochemistry (PhD), geology (MA, PhD), geophysics (PhD), planetary sciences (PhD). Matriculated students: 32 full-time (7 women), 0 part-time; includes 0 minority, 7 foreign. In 1990, 3 master's, 3 doctorates awarded. Terminal master's awarded for partial completion of doctoral program. *Degree requirements:* Thesis/dissertation. *Entrance requirements:* GRE General Test. Application deadline: 1/15 (priority date, applications processed on a rolling basis). Application fee: $0. *Tuition:* $15,960 per year full-time, $665 per credit hour part-time. *Financial aid:* In 1990–91, 8 fellowships, 15 research assistantships, 9 teaching assistantships awarded; full and partial tuition waivers, federal work-study, institutionally sponsored loans also available. Financial aid application deadline: 1/15. • Dr. Larry A. Haskin, Chairman, 314-935-5610.

See full description on page 295.

York University, Faculty of Science, Program in Earth and Space Science, North York, ON M3J 1P3, Canada. Program awards M Sc, PhD. Part-time and evening/weekend programs available. Faculty: 37 full-time (1 woman), 13 part-time (0 women). Matriculated students: 35 full-time (10 women), 18 part-time (0 women); includes 8 foreign. 25 applicants, 64% accepted. In 1990, 8 master's, 2 doctorates awarded. *Degree requirements:* For master's, computer language required, thesis optional; for doctorate, variable foreign language requirement, computer language, dissertation. Application deadline: 2/15. Application fee: $35. *Tuition:* $2436 per year for Canadian residents; $11,480 per year for nonresidents. *Financial aid:* In 1990–91, $440,157 in aid awarded. 1 fellowship, 37 research assistantships, 24 teaching assistantships were awarded. • Dr. G. Jarvis, Director, 416-736-5247.

Cross-Discipline Announcement

University of Virginia, Graduate School of Arts and Sciences, Department of Environmental Sciences, Charlottesville, VA 22906.

Environmental Sciences is an established interdisciplinary program dealing with the analysis of atmospheric, hydrologic, geologic, and ecologic processes and the links among them. The faculty is an integral part of the growing community of global experts developing the capability to model and predict global-scale environmental changes. For further information, see the Environmental Sciences Section in Book 3 of this series. Contact the Director of Graduate Admissions, Clark Hall.

SECTION 3: EARTH AND PLANETARY SCIENCES

BOWLING GREEN STATE UNIVERSITY

Department of Geology

Programs of Study

The department offers graduate studies leading to the M.S. degree in geology. Programs of study include carbonate petrology, economic geology, environmental geology, geochemistry, geographic information systems, hydrology, paleontology, sedimentology/stratigraphy, and structural geology/tectonics. Students with backgrounds in geology or related areas are encouraged to apply.

Requirements for the Master of Science degree are 30 semester hours of course work and a thesis, including a formal thesis proposal and defense. These requirements are normally completed in two years.

Research Facilities

The Department of Geology has modern and extensive research facilities, including an atomic absorption spectrometer; X-ray diffraction and fluorescence spectrometers; a fluid inclusion stage; a scanning electron microscope; a cathodoluminescence microscope; research petrographic microscopes; geophysical equipment, including field and base station seismographs, magnetometers, and a radon sampler; complete sample preparation equipment, including automated thin section and polishing machines; an image analysis workstation; a rock mechanics laboratory; and field vehicles. In addition to campus mainframe IBM and VAX computers, the department has a Sun minicomputer and numerous microcomputers. Peripheral devices such as plotters, scanners, and digitizers are available for student use. A full-time technician provides assistance with equipment and instrumentation.

The department moved into totally renovated quarters in Overman Hall in the spring of 1991. The building provides ample classroom, laboratory, office, and storage space for faculty and students. This space, along with the research facilities, are among the most extensive of any master's degree program in the country.

Financial Aid

Teaching and research assistantships and externally funded fellowships are available on a competitive basis. The stipend for a half-time assistantship in 1991–92 is $6900. In addition to the stipend, a waiver of tuition and the nonresident fee are included. Assistantships for the summer are also available. Funds to support graduate research are available from the department, from the graduate school, and from faculty research grants.

Cost of Study

In 1991–92, tuition for Ohio residents is $1567 per semester. The nonresident fee is an additional $1675 per semester. A general fee of $258 is charged to all students.

Cost of Living

Off-campus living within walking distance of campus averages $275 per month for a furnished one-bedroom apartment with utilities included. Two-bedroom apartments rent for $350–$400 per month.

Student Group

The enrollment at Bowling Green is 18,000, which includes 2,200 graduate students. The Department of Geology has 20 graduate students, most of whom receive some form of financial support.

Location

Bowling Green, with a population of 27,000, is located in northwest Ohio, 20 miles from Toledo and 65 miles from Detroit. Being a county seat, the community offers a wide range of activities and services, and any not found in Bowling Green are available a short distance away in Toledo. Lake Erie is the focus of many recreational activities in the region.

The University

Bowling Green State University was established in 1910. Since that time, it has grown into a nationally recognized university with 730 full-time faculty members and 170 undergraduate, 69 master's, and 14 Ph.D. degree programs. The University combines the resources of a high-quality research institution with the friendliness of a small college. Graduate studies are supported by excellent facilities, including extensive library holdings, state-of-the-art laboratories, excellent computing facilities, and a network of nationally recognized research centers and institutes.

Applying

Applications for fall semester admission should be made before March 1 if financial aid is desired. A limited amount of financial aid is available for admission in the spring semester. The GRE General Test and Subject Test in geology are required for admission. Application forms for admission and financial aid may be obtained from the Department of Geology.

Correspondence and Information

Graduate Coordinator
Department of Geology, 190 Overman Hall
Bowling Green State University
Bowling Green, Ohio 43403
Telephone: 419-372-2886

Peterson's Guide to Graduate Programs in the Physical Sciences and Mathematics 1992

SECTION 3: EARTH AND PLANETARY SCIENCES

Bowling Green State University

THE FACULTY AND THEIR RESEARCH

James E. Evans, Assistant Professor; Ph.D., Washington (Seattle), 1987. Clastic sedimentology, including sedimentation and tectonics, basin analysis, petroleum geology, fluvial sedimentology, sediment transport, glacial geology, soils, geomorphology, and hydrology. Most recent research deals with tectonic and depositional history of nonmarine sedimentary basins in the Pacific northwest.

Jane L. Forsyth, Professor; Ph.D., Ohio State, 1956. Glacial geology of Ohio and New England, environmental geology, soils, hydrology, and geobotany. Most recent research deals with environmental problems in the Great Lakes region.

Joseph P. Frizado, Associate Professor; Ph.D., Northwestern, 1980. Diagenesis of sediments, computer applications in geology, geographical information systems, groundwater geochemistry. Most recent research deals with the use of GIS and DRASTIC to assess groundwater pollution potential in Ohio and the development and management of petrologic databases.

Richard D. Hoare, Professor; Ph.D., Missouri, 1957. Invertebrate paleontology of Mississippian and Pennsylvanian gastropods, Rostroconchia, Polyplacophora, and smaller foraminifera. Most recent research deals with Mississippian and Pennsylvanian ostrocodes and Mississippian bivalves of the Appalachian basin.

John A. Howe, Associate Professor; Ph.D., Nebraska, 1961. Vertebrate paleontology, including horses and other mammals. Most recent research deals with evolution of Pleistocene horses.

Charles F. Kahle, Professor; Ph.D., Kansas, 1962. Carbonate geology, including carbonate depositional and diagenetic environments, origin and significance of carbonate fabrics, karst and subaerial exposure, syntectonic sedimentation. Most recent research deals with effect of tectonism on deposition of carbonates of northwest Ohio.

Joseph J. Mancuso, Professor; Ph.D., Michigan State, 1960. Economic geology, including metallic ore deposits, Precambrian hydrocarbons, iron formations, and Precambrian geology of Great Lakes region. Most recent research deals with the origin of Precambrian pyrobitumen and Lake Superior iron formations.

Charles M. Onasch, Associate Professor; Ph.D., Penn State, 1977. Structural geology and geophysics, including strain and microstructures in sandstones, structural evolution of foreland fold and thrust belts, Appalachian geology, recurrent tectonics in the craton, engineering geology, and shallow geophysics. Most recent research deals with the relationship between strain and microstructures in quartz arenites.

Charles C. Rich, Professor; Ph.D., Harvard, 1960. Geomorphology, glacial geology of the Great Lakes region, environmental geology, surface and groundwater supplies in glacial deposits and karst regions, and geological education. Most recent research deals with environmental problems along the shoreline of Lake Erie.

Don C. Steinker, Professor; Ph.D., Berkeley, 1969. Invertebrate paleontology and stratigraphy, including ecology and paleoecology of foraminifera, Neogene micropaleontology, and stratigraphy. Most recent research deals with the reconstruction of Plio-Pleistocene paleoenvironments in Florida using microfossils.

Cooperating Faculty

Louis P. Fulcher, Professor; Ph.D., Virginia, 1972. Geophysics, potential field studies.
Charles Shirkey, Associate Professor; Ph.D., Ohio State, 1969. Geophysics, signal processing.

SECTION 3: EARTH AND PLANETARY SCIENCES

CALIFORNIA INSTITUTE OF TECHNOLOGY

Division of Geological and Planetary Sciences

Programs of Study Graduate programs in the Division of Geological and Planetary Sciences lead to the Ph.D. degree in geology, geophysics, geochemistry, geobiology, and planetary science. Students are not normally admitted to work toward the M.S. degree.

The Division is especially interested in students with sound and thorough training in chemistry, physics, mathematics, astronomy, or biology. Applicants with majors in these subjects and with a strong interest in earth and planetary sciences are given the same consideration for admission and appointment as geology majors.

Research Facilities Caltech research in the geological and planetary sciences ranges from the study of the center of the earth through its core, mantle, crust, oceans, and atmospheres to the surface of the moon, planetary atmospheres, and the rings of Saturn, Uranus, and Jupiter, using the disciplines of geology, physics, chemistry, and biology. The Division is housed in three well-equipped five-story buildings that provide office, classroom, and laboratory space for the staff and students. Special facilities and equipment for student research include electron microprobe and other analytical and X-ray facilities, scanning electron microscopes, superclean laboratories for preparation for isotopic analyses, an ion microprobe for isotopic and chemical analysis of microvolumes of complex geological specimens, mass spectrometers for volatile and nonvolatile elements, tunable far-infrared and infrared laser spectrometers, three laboratories for low- and high-pressure experimental petrology, a magnetic analysis laboratory, optical spectrometers, and microscopes. The Division ranks as a world leader in studies of faults and seismicity, basing research in part on data from a computerized network of 240 seismograph stations—the largest and most fully automated in the world. Other supporting facilities include a geological library, computers, machine and electronics shops, field vehicles, and extensive rock, mineral, and fossil collections. Graduate students have access to a wide variety of optical, infrared, and radio observational facilities; data from interplanetary spacecraft; meteorites; and Apollo lunar samples.

Financial Aid Each year the Division awards a number of graduate teaching and research assistantships. In 1991–92, they provide stipends of up to $13,332 for 20 hours per week during the academic year and the summer. The holder of such an appointment also receives a tuition scholarship and is permitted to carry a full academic program.

Also available are a number of fellowships that require no departmental duties. These provide tuition, an initial academic-year stipend of up to $9999 (plus supplemental assistance to raise the amount to the graduate assistant level), and, in some instances, a grant to cover research expenses. The Beno Gutenberg Fellowship, awarded to an outstanding first-year student in geophysics, provides a stipend of $9999 in addition to tuition.

Cost of Study Tuition for the academic year 1991–92 is $14,100, which includes medical insurance and the use of Institute facilities.

Cost of Living Unmarried students may live in one of the three graduate dormitories. In 1991–92, the cost is $288–$372 per month for the academic year. Four-bedroom, apartment-style accommodations cost $336 per month for the full year, not including utilities, in the single-student Catalina Complex I. In the Catalina Complex II and Complex III, the cost is $754 per month for married students and $378 each for single students for a two-bedroom apartment. One-bedroom apartments in Complex III rent for $634 per month. These costs include room only. Many students choose to live in apartments or rented rooms near campus. Meals are available at the campus dining hall.

Student Group Caltech has 615 male and 172 female undergraduate students and 1,000 graduate students, 201 of whom are women. The Division has ten undergraduate majors and 85 to 90 graduate students, nearly all of whom are receiving financial aid of some kind.

Location Numerous adjacent field areas near the campus are available for year-round field study and research in almost all aspects of geology. Ready access to the mountains, deserts, ocean, and Metropolitan Los Angeles provides a variety of recreational and cultural opportunities.

The Division The study of the earth sciences has evolved from a strongly descriptive treatment to more quantitative and analytical approaches. The Division of Geological and Planetary Sciences at Caltech, which was established in 1926, has built up a faculty, curriculum, and student body designed to be in the forefront of this evolution. The aim is to produce men and women who are well grounded in the basic sciences and who have an understanding of the great promise offered by a quantitative-analytical approach to research in geology. There is no wish to abandon classic geological methods; the goal is to blend the old with the new in what is believed to be the most promising approach for future advances in the earth sciences. The continuing heavy demand for men and women with this training—from universities, from research organizations, and from industry—assures an ongoing need.

Applying Complete applications for admission and financial aid should be received before January 15. Applicants will be notified of the results before April 1. North American applicants are required to submit GRE scores on the General Test and on the Subject Test in their field of undergraduate specialty. The GRE tests are strongly recommended for foreign applicants.

Prior to the first term of graduate work, the student must take placement examinations covering basic aspects of the earth sciences and elementary physics, mathematics, and chemistry.

Correspondence and Information
Division Chairman
Division of Geological and Planetary Sciences
California Institute of Technology
1201 East California Boulevard
Pasadena, California 91125

California Institute of Technology

THE FACULTY AND THEIR RESEARCH

Members of the faculty are listed below; in addition, 20 postdoctoral fellows are in residence each year. The supporting technical staff numbers about 60. To further augment the program of teaching and research, the Division invites each year a number of distinguished visiting professors, lecturers, and investigators, who spend from a week to a term or more in the Division.

Thomas J. Ahrens, Professor of Geophysics; Ph.D., RPI, 1962. Equations of state of earth material impact processes, earth stresses.

Arden L. Albee, Professor of Geology and Project Scientist for the Mars Observer; Ph.D., Harvard, 1957. Petrologic, mineralogic, and microprobe investigations; remote sensing of planetary surfaces.

Clarence R. Allen, Professor Emeritus of Geology and Geophysics; Ph.D., Caltech, 1954. Seismotectonics, seismic hazards evaluation.

Don L. Anderson, Eleanor and John R. McMillan Professor of Geophysics; Ph.D., Caltech, 1962. Elastic and anelastic properties of the deep interior of the earth by use of seismic body waves, surface waves and free oscillations.

Geoffrey A. Blake, Assistant Professor of Cosmochemistry and Planetary Science; Ph.D., Caltech, 1986. High-resolution laser spectroscopy of chemical and physical processes of importance in extreme environments, cluster and ultrafine particle catalysis, millimeter and sub-millimeter wave astrophysics.

Donald S. Burnett, Professor of Nuclear Geochemistry; Ph.D., Berkeley, 1963. Problems of nucleosynthesis, elemental abundances and chemical evolution of solar system, meteorite and lunar sample analyses and laboratory synthesis experiments.

Robert W. Clayton, Professor of Geophysics and Executive Officer, Geophysics; Ph.D., Stanford, 1981. Reflection seismology, numerical wave simulation, tomographic reconstructions, inverse scattering methods, tectonics.

Hermann Engelhardt, Senior Research Associate in Geophysics, Dr. rer. nat. in Physics, Munich Technical, 1964; Habilitation in Geophysics, Münster, 1984. Ice physics and glaciology: glacier dynamics; Antarctic ice sheet stability and global climate.

Samuel Epstein, William E. Leonhard Professor Emeritus of Geology; Ph.D., McGill, 1944. Geochemical processes: stable isotope records in nature.

Peter Goldreich, Lee A. DuBridge Professor of Astrophysics and Planetary Physics; Ph.D., Cornell, 1962. Astrophysics and planetary physics; principal interest in theoretical astrophysics and geophysics.

David G. Harkrider, Professor of Geophysics; Ph.D., Caltech, 1963. Excitation and coupling of surface waves in the earth, oceans, and atmosphere.

Donald V. Helmberger, Professor of Geophysics; Ph.D., California, San Diego, 1967. Seismic wave propagation and fine structures of the upper mantle and crust, as determined by P and S wave inversion; modeling tectonic release associated with explosions.

Ian D. Hutcheon, Senior Research Associate in Geochemistry; Ph.D., Berkeley, 1975. Origin and evolution of solid bodies in the solar system, with emphasis on the development and use of microanalytical instrumentation.

Andrew P. Ingersoll, Professor of Planetary Science and Executive Officer, Planetary Science; Ph.D., Harvard, 1966. Dynamic meteorology and climatology, spacecraft studies of the earth and other planets.

Barclay Kamb, Barbara and Stanley R. Rawn Jr. Professor of Geology and Geophysics; Ph.D., Caltech, 1956. Mechanics of glacier flow, with emphasis on basal sliding and surging; structures produced by rock flow and fracture in the earth.

Hiroo Kanamori, John E. and Hazel S. Smits Professor of Geophysics and Director of the Seismological Laboratory; Ph.D., Tokyo, 1964. Earthquake mechanisms and crustal deformation, crust and mantle structure, mechanisms of volcanic eruptions, application of seismology to engineering problems.

Joseph L. Kirschvink, Associate Professor of Geobiology; Ph.D., Princeton, 1979. Geobiology, paleomagnetism, magnetobiology.

Heinz A. Lowenstam, Professor Emeritus of Paleoecology; Ph.D., Chicago, 1939. Biomineralization of molluscs and sipunculida, evolution and biostratigraphy of mid-Triassic crinoids.

Duane O. Muhleman, Professor of Planetary Science; Ph.D., Harvard, 1963. Radio and radar astronomy of planetary surfaces and atmospheres.

Bruce C. Murray, Professor of Planetary Science; Ph.D., MIT, 1955. Surfaces of Earth, Moon, Mars, Mercury, and Venus; techniques and technology of deep space exploration.

Dimitri A. Papanastassiou, Senior Research Associate in Geochemistry; Ph.D., Caltech, 1970. Formation and evolution of meteorites and planets, high-precision mass spectrometry, nucleosynthesis.

Clair C. Patterson, Professor of Geochemistry; Ph.D., Caltech, 1969. Isotopic clean laboratory techniques to study biochemical processes carried out in lead-free sanctuaries.

George R. Rossman, Professor of Mineralogy; Ph.D., Caltech, 1971. Spectroscopic, chemical, and structural properties of minerals, which involve metal ion site occupancy, trace water, radiation effects, and X-ray amorphous materials.

Jason B. Saleeby, Professor of Geology; Ph.D., California, Santa Barbara, 1975. Tectonic and geochronological studies in orogenic terranes of western North America, emphasizing the paleogeographic development of the Pacific basin and margins.

Robert P. Sharp, Professor Emeritus of Geology ; Ph.D., Harvard, 1938. Glaciers and glaciation, arid region geomorphology, features of the Martian landscape.

Kerry E. Sieh, Professor of Geology; Ph.D., Stanford, 1977. Earthquake geology.

Leon T. Silver, W. M. Keck Foundation Professor for Resource Geology; Ph.D., Caltech, 1955. Petrology, tectonics, and applications of geology and isotope geochemistry to geochronology, crustal evolution, ore deposits, and comparative planetology.

David J. Stevenson, Professor of Planetary Science and Chairman of the Division; Ph.D., Cornell, 1976. Origin, evolution, and structure of planets.

Edward M. Stolper, William E. Leonhard Professor of Geology and Executive Officer, Geochemistry; Ph.D., Harvard, 1979. Petrology of meteoritic, lunar, and terrestrial igneous rocks; physical and chemical properties of liquid silicates.

Toshiro Tanimoto, Assistant Professor of Geophysics; Ph.D., Berkeley, 1982. Theoretical and observational seismology, earth structure.

Hugh P. Taylor Jr., Robert P. Sharp Professor of Geology and Executive Officer, Geology; Ph.D., Caltech, 1959. Oxygen and hydrogen isotope geochemistry and petrology of granitic batholiths, ophiolite complexes, and other igneous and metamorphic rocks and ore deposits.

Gerald J. Wasserburg, Crafoord Laureate and John D. McArthur Professor of Geology and Geophysics; Ph.D., Chicago, 1954. Application of chemical physics to the solution of fundamental problems in the origin and evolution of the earth, the planets, and the solar system.

James A. Westphal, Professor of Planetary Science and Principal Investigator for the Wide Field/Planetary Camera on the Hubble Space Telescope; B.S., Tulsa, 1954. Visible and UV detector development, spacecraft observations.

Peter J. Wyllie, Professor of Geology; Ph.D., St. Andrews (Scotland), 1958. Experimental petrology: origin of batholiths, andesites, kimberlites, carbonatites, and associated ore deposits.

Yuk L. Yung, Professor of Planetary Science; Ph.D., Harvard, 1974. Atmospheric chemistry, planetary atmospheres, evolution of atmospheres, global changes.

SECTION 3: EARTH AND PLANETARY SCIENCES

HARVARD UNIVERSITY

Department of Earth and Planetary Sciences

Programs of Study The department offers instruction and opportunities for research in a wide variety of fields within the broad scope of earth sciences. Requirements for admission are highly flexible, and adequate preparation in the related sciences of physics, chemistry, mathematics, and biology is considered as important as a background in geology. Those with undergraduate majors in another science are encouraged to apply. The master's degree is not a prerequisite for entering the Ph.D. program.

Graduate study leading to the A.M. and Ph.D. degrees is supervised by a faculty advisory committee made up in accordance with the student's aims and interests. The student and the committee choose a plan of courses and research that generally covers four specialized fields. Fields include economic geology, geochemistry, geophysics, mineralogy, oceanography, paleontology, petrology, sedimentology, stratigraphy, and structural geology. In addition, some of the fields may be in chemistry, physics, engineering, or biology. Special interdisciplinary programs may be arranged, as, for example, between economics and economic geology or between geochemistry and environmental chemistry. Under reciprocal arrangements, graduate students at Harvard may take and receive credit for courses given at the Massachusetts Institute of Technology in Cambridge. There are also opportunities for research and cross-registration for courses at the Woods Hole Oceanographic Institute, the Geophysical Laboratory of the Carnegie Institution of Washington, and other research organizations.

Research Facilities The department is located in Hoffman Laboratory, situated in the midst of the scientific area of the University. Physical facilities include the Geological and Mineralogical museums. The Hoffman Laboratory of Experimental Geology offers modern laboratory facilities for research in geophysics, geochemistry, and other experimental areas in earth sciences. Work in vertebrate and invertebrate paleontology is carried on in the Museum of Comparative Zoology, an integral part of the Museum complex. Harvard has one of the world's best research and reference collections in mineralogy, paleontology, and economic geology. Laboratories are fully equipped with modern instrumentation for teaching and research, including an automated Cameca-MBX electron microprobe, a scanning electron microscope, absorption and flame photometers, an emission spectrograph, mass spectrometers, and facilities for electrochemical and wet chemical analysis. Department facilities include a VG ISOMAS 54 thermal ionization mass spectrometer and an ultraclean laboratory for Nd, Sr, and Pb chemistry. Experimental facilities for X-ray diffraction, high-pressure mineral synthesis and phase equilibrium studies, and geophysics are extensive. Within the department, there is a computer system built around a 1-megabyte machine with a 512-megabyte disk memory, magnetic tape drives, plotting facilities, and other peripherals. A great variety of specialized equipment is available in other science departments.

Financial Aid Most graduate students in the department receive some financial support for both tuition and living expenses while studying for advanced degrees. Support may come from Harvard University; independent fellowships, such as the Shell Fellowship; teaching fellowships; research assistantships; or some combination thereof. In addition, some curatorial, library, and other part-time work is available. All students who need to do fieldwork are given special summer grants-in-aid from endowed funds.

Cost of Study Full-time tuition for the 1991–92 academic year is $16,002. This figure includes the cost of health insurance and covers the use of most facilities and services of the University. For the first two years, students pay full tuition. In the third and fourth years, they qualify for reduced tuition. After the fourth year, students pay only a facilities fee (while in residence) until they receive their degree. The department provides funding for tuition through the first four years.

Cost of Living Unmarried students may live at the Graduate Center dormitories of Harvard and Radcliffe College at costs ranging from $2150 to $3850 in 1991–92 (room only) for the academic year. Married graduate students may live in Peabody Terrace or other University-owned apartments. Many students choose to live in rooms or apartments in Cambridge or surrounding communities.

Student Group The department has from 35 to 50 graduate students and about 40 undergraduates from a wide variety of locations in North America and elsewhere. The Geology Club, more than sixty years old, organizes lectures, serves as a social organization, and participates with the faculty in discussions of matters of interest to the whole department.

Location The Cambridge-Boston area is one of the most concentrated centers of educational, intellectual, and cultural activity in the world. At the same time, there is ready access to the scenic and historic New England countryside, which is of diverse and abundant geologic interest. Recreational activities range from the full scope of indoor athletic activities provided by Harvard to sailing and skiing. The great concentration of educational organizations in the area makes the total student population quite large, and the services offered for the benefit of students are correspondingly numerous.

The University Harvard, the oldest institution of higher learning in the United States, offers an educational life covering the entire span of the fields of learning, old and new. Tradition is strong, but innovation and change are characteristic. Graduate students in the Department of Earth and Planetary Sciences mingle with students from other graduate departments, with undergraduates, and with students at the various professional schools. Perhaps the strongest drive in the University is the pursuit of excellence by students and faculty in a setting that is immensely varied, both physically and intellectually. A distinguished faculty is devoted to teaching and research in close contact with students.

Applying Completed applications for admission and scholarship aid should be submitted before January 7. Scores on the GRE General Test are required, and the examination should be taken before October 15.

Correspondence and Information
Professor Michael B. McElroy, Chairman
Department of Earth and Planetary Sciences
Hoffman Laboratory
Harvard University
20 Oxford Street
Cambridge, Massachusetts 02138
Telephone: 617-495-2351

SECTION 3: EARTH AND PLANETARY SCIENCES

Harvard University

THE FACULTY AND THEIR RESEARCH

Carl B. Agee, Assistant Professor of Experimental Petrology. Physics and chemistry of minerals and melts at high pressure, origin and evolution of planetary interiors.

James G. Anderson, Philip S. Weld Professor of Atmospheric Chemistry. Gas phase kinetics of free radicals, chemistry of planetary atmospheres, laser chemistry, stratospheric measurements by balloon and rocket.

Jeremy Bloxham, John L. Loeb Associate Professor of the Natural Sciences. Planetary magnetic fields, dynamo theory, structure and dynamics of the earth's core and lower mantle, inverse theory, mathematical geophysics.

Charles W. Burnham, Professor of Mineralogy. Mineralogy, crystallography, crystal chemistry of rock-forming minerals.

James N. Butler, Gordon McKay Professor of Applied Chemistry. Chemistry of natural and polluted waters, physical chemistry of interfaces, electrochemistry, oil pollution of the oceans.

Adam M. Dziewonski, Professor of Geology. Theoretical seismology, internal structure of the earth, seismic tomography, earthquake source mechanisms, geodynamics.

Goran A. Ekstrom, Assistant Professor of Geophysics. Seismology, forward and inverse problems of seismic source.

Brian F. Farrell, Associate Professor of Dynamic Meteorology.

Stephen J. Gould, Professor of Geology. Stochastic simulation of evolutionary patterns, the evolution of growth, Pleistocene and Recent evolution of the Bahamian land snail *Cerion*, the evolution of brain size, the relationship of ontogeny and phylogeny.

Heinrich D. Holland, Professor of Geochemistry. Chemistry of the ocean-atmosphere-crust system, chemistry of ore-forming solutions, origin of hydrothermal uranium deposits.

Stein B. Jacobsen, Professor of Geology. Isotope geology; petrology and geochemistry of oceanic and continental rifts, ultramafic rocks, granulites, eclogites, and chondritic meteorites.

Andrew H. Knoll, Professor of Biology. Paleontology and sedimentary geology of Precambrian terrains, evolution of vascular plants in geologic time.

Ursula B. Marvin, Lecturer in Geology. Mineralogy and paragenesis of lunar rocks and meteorites, origins and development of certain geological ideas.

James J. McCarthy, Alexander Agassiz Professor of Biological Oceanography. Biological oceanography, phytoplankton ecology, nitrogen nutrition of phytoplankton.

Michael B. McElroy, Abbott Lawrence Rotch Professor of Atmospheric Science and Chairman. Chemistry of the atmosphere and ocean, including interactions with the biosphere; evolution of planetary atmospheres.

Richard J. O'Connell, Professor of Geology. Elastic and flow properties of rocks and minerals, mantle flow, convection and plate tectonics.

Ulrich Petersen, Harry C. Dudley Professor of Economic Geology. Economic mineral deposits, particularly hydrothermal ore deposits; zoning in ore deposits; mineral policy, concession contracts, and taxation in developing countries.

James R. Rice, Gordon McKay Professor of Engineering Sciences and Geophysics. Earthquake source processes, laboratory rock friction and fault instability, fracture mechanics, materials science.

Allan R. Robinson, Gordon McKay Professor of Geophysical Fluid Dynamics. Physical oceanography and geophysical fluid dynamics: dynamics of oceanic motions (theory and observations), numerical models of ocean currents and circulations, design and interpretation of field experiments.

Jane Selverstone, John L. Loeb Associate Professor of Petrology. Metamorphic petrology, phase equilibria, and thermodynamics; application of petrologic techniques to interpretation of tectonic processes; fluid inclusion studies.

Raymond Siever, Professor of Geology. Sedimentary petrology and geochemistry: origin and evolution of atmosphere and the earth's crust; geochemical mass balance of seawater composition, present and past; distribution of oceanic sediment; origin of chert and other siliceous sediments.

Joann Stock, Associate Professor of Geophysics. Plate reconstructions, tectonics, structural geology, crustal stresses, marine geophysics.

James B. Thompson Jr., Professor of Mineralogy. Metamorphic petrology, thermodynamics, properties of rock-forming minerals, theory phase equilibrium, geology of the northern Appalachians.

Brian P. Wernicke, Professor of Geology. Structural geology and tectonics, with emphasis on field geological mapping; implications of Basin and Range extensional tectonics for passive margin development.

John A. Wood, Practicing Professor of Geology. Petrology of meteorites and lunar samples, origin and early evolution of the moon and planets.

Associated Faculty

Alfred W. Crompton, Professor of Biology. Vertebrate paleontology: origin of mammals and early dinosaurs.

Farish Jenkins, Professor of Biology. Early evolution of mammals during the Mesozoic period, with specific reference to the postcranial muscular skeletal system.

David R. Pilbeam, Professor of Anthropology. Biological anthropology, paleoanthropology of Africa and Asia.

SECTION 3: EARTH AND PLANETARY SCIENCES

INDIANA UNIVERSITY BLOOMINGTON

Department of Geological Sciences

Programs of Study

The Department of Geological Sciences at Indiana University offers training at the master's and doctoral levels in many fields of the geological sciences, including biogeochemistry, coal geology, economic geology, aqueous and isotope geochemistry, geomorphology, geophysics, glacial geology, hydrogeology, mineralogy, paleoecology, paleontology, petroleum geology, petrology, sedimentology, stratigraphy, structural geology, and tectonics. The programs are open to students with undergraduate degrees in geology, chemistry, engineering, mathematics, physics, and the biological sciences.

A minimum of three years of graduate work (90 hours), completion of a dissertation in the major area, proficiency in one language or a tool skill, and an outside minor are required for the Ph.D. degree. Three major examinations are part of the Ph.D. program: a preliminary examination usually taken during the third semester of graduate study, a qualifying examination taken after completion of the language or tool skill requirement, and a defense-of-thesis examination. The program of course work is flexible, depending on individual goals and background, but consists of a minimum of 35 hours of graduate course work.

The M.S. program consists of 30 hours of graduate work plus demonstrated proficiency in a foreign language, a tool skill, or a thesis. The thesis may include a maximum of 8 hours toward the 30-hour requirement.

Research Facilities

The spacious geology building contains many laboratories with such equipment as molecular and isotopic mass spectrometers, gas chromatographs, an electron microprobe, X-ray diffraction units, high-temperature experimental facilities, and atomic absorption and emission spectrometric equipment (including inductively coupled plasma). There is a seismograph station, Seismoline seismic modeling equipment, and magnetic tape analog field and playback equipment. Several microcomputer workstations are available in the building that are networked both locally in the department and with the University computer services, which includes a system of seven DEC VAX minicomputers for research. Mainframe research computing is also available on IBM 3030-120E and IBM 3030-300S computers. The University is linked to a national supercomputer center. The Geology Library (with 89,000 cataloged volumes, 50,000 uncataloged holdings, and 300,000 maps) and the University Library system are undergoing enhancement of their resources for research through on-line catalogs and databases. The department maintains a field station in southwest Montana that serves as a base for numerous graduate research projects.

Financial Aid

A substantial number of teaching assistantships as well as research assistantships and fellowships are available to qualified students. The stipends are competitive with those offered by other universities. The department and University also have some funds for expenses connected with dissertation research. A number of summer assistantships at the field station and on campus are also available.

Cost of Study

Tuition for 1990–91 was $93.30 per credit hour for Indiana residents and $266.60 for nonresidents. It is subject to increase for 1991–92. Teaching and research assistants and University Fellows ordinarily pay only a fee of $250 each semester for up to a maximum of 12 hours per semester.

Cost of Living

In 1990–91, housing for unmarried students in the Eigenmann Graduate Center ranged from $3238 to $3774 for the academic year, including meals and other fees. Rents for University married student housing ranged from $254 to $588 per month.

Student Group

Approximately 65 full-time graduate students are currently studying in the Department of Geological Sciences, about 75 percent of them receiving financial support. Fifty-one percent are working toward the Ph.D. degree. They come from all parts of the United States and from a number of foreign countries. The department also has approximately 35 undergraduate majors.

Location

Bloomington is located in the south-central part of the state in a scenic region of low but rugged relief. There are many recreational opportunities (fishing, boating, hiking, canoeing, cycling, camping, spelunking, skiing, and swimming) and several state parks and state and national forests in the immediate vicinity. Lake Monroe, the largest lake wholly within the state, is a major recreational site about 10 miles southeast of the city. The Bloomington area has a permanent population of about 50,000, with an economy based largely on the University and several electronics and electrical appliance firms. Indianapolis (population about 750,000) is 50 miles northeast and offers all the facilities of a large city. Cincinnati, Louisville, Chicago, and St. Louis are all just a few hours' drive away.

The University

Indiana University was founded as a small seminary in Bloomington in 1820. This makes it the second-oldest state-supported university in the Midwest. It is also one of the largest, with approximately 35,000 students on the Bloomington campus studying in practically all major fields of science, social science, humanities, and the arts. Despite its size, Indiana has retained much of the charm of a smaller campus, with open space and forested areas preserved near the center of the campus. The University is an important cultural center with a full and varied program of theater, opera, symphony, jazz, and pop concerts, performed by both professional touring companies and student groups. The University also has a well-balanced athletic program of both intramural and major intercollegiate sports.

Applying

Applications for admission to graduate study can be made at any time, but applications for financial assistance should be made before February 15 for consideration for the following fall semester. Two copies of an official transcript and three letters of reference must be supplied. The Graduate Record Examinations General Test results are required. The Subject Test in geology is optional.

Correspondence and Information

Chairman
Committee for Graduate Studies
Department of Geological Sciences
Indiana University
Bloomington, Indiana 47405
Telephone: 812-855-7214

SECTION 3: EARTH AND PLANETARY SCIENCES

Indiana University Bloomington

THE FACULTY AND THEIR RESEARCH INTERESTS

Lee J. Suttner, Professor, Chairman, and Director, Indiana Geologic Field Station; Ph.D., Wisconsin. Physical stratigraphy, sedimentology, sedimentary petrology.
Abhijit Basu, Professor; Ph.D., Indiana. Mineralogy and petrology, lunar geology.
Robert F. Blakely, Professor Emeritus; Ph.D., Indiana. Geophysics, computer applications.
Simon C. Brassell, Professor; Ph.D., Bristol (England). Organic geochemistry, molecular paleontology, petroleum geochemistry.
Donald D. Carr, Professor (part-time); Ph.D., Indiana. Industrial minerals, coal geology.
J. Robert Dodd, Professor; Ph.D., Caltech. Paleoecology, carbonate petrology, biogeochemistry, marine geology.
John B. Droste, Professor; Ph.D., Illinois. Sedimentary petrology, clay mineralogy, evaporites, environmental geology.
Donald E. Hattin, Professor; Ph.D., Kansas. Stratigraphy, paleoecology, Cretaceous sedimentology and biostratigraphy.
John M. Hayes, Professor; Ph.D., MIT. Organic geochemistry, lipid fractions in organisms, stable isotopes in carbon compounds.
Norman C. Hester, Professor; Ph.D., Cincinnati. Depositional environments, Carboniferous stratigraphy, coal and petroleum geology.
N. Gary Lane, Professor; Ph.D., Kansas. Paleontology, crinoids, Mississippian and Pennsylvanian faunas, Paleozoic geology, paleoecology.
Judson Mead, Professor Emeritus; Ph.D., MIT. Geophysics, rock magnetism, experimental seismology.
Haydn H. Murray, Professor; Ph.D., Illinois. Clay mineralogy, nonmetallic mineral deposits, economic geology, coal geology.
Peter J. Ortoleva, Professor; Ph.D., Cornell. Geochemistry, thermodynamics, kinetics, and transport mechanisms.
Carl B. Rexroad, Professor (part-time); Ph.D., Iowa. Conodonts, paleontology, Paleozoic stratigraphy.
Edward M. Ripley, Professor; Ph.D., Penn State. Metallic ore deposits, geochemical exploration, stable isotope and fluid inclusion geochemistry.
Albert J. Rudman, Professor; Ph.D., Indiana. Geophysics, gravity, magnetics.
Robert H. Shaver, Professor Emeritus; Ph.D., Illinois. Paleontology, ostracods, Silurian stratigraphy and paleontology.
Charles J. Vitaliano, Professor Emeritus; Ph.D., Columbia. Petrology, igneous and metamorphic geology, ignimbrites.
James G. Brophy, Associate Professor; Ph.D., Johns Hopkins. Igneous petrology, tectonophysics of magma ascension.
Jeremy Dunning, Associate Professor; Ph.D., North Carolina. Structural geology, tectonics, microseismology.
Gordon S. Fraser, Associate Professor (part-time); Ph.D., Illinois. Pleistocene and Holocene sedimentology.
Hendrik M. Haitjema, Associate Professor; Ph.D., Minnesota. Hydrology, groundwater flow modeling.
Brian D. Keith, Associate Professor (part-time); Ph.D., RPI. Carbonate depositional models and facies, carbonate petrology, and petroleum geology.
Noel C. Krothe, Associate Professor; Ph.D., Penn State. Hydrogeology, hydrochemistry and stable isotopes.
Enrique Merino, Associate Professor; Ph.D., Berkeley. Chemical sedimentology, solution chemistry, calorimetry.
Gregory Olyphant, Associate Professor; Ph.D., Iowa. Geomorphic processes in Alpine environments, Quaternary paleoclimatology.
Lawrence Onesti, Associate Professor; Ph.D., Wisconsin. Geomorphology, snow and ice hydrology.
Gary L. Pavlis, Associate Professor; Ph.D., Washington (Seattle). Seismology, location of earthquake events using inverse theory, fracture mechanics in relation to geophysical inverse-theory problems.
Lisa Pratt, Associate Professor; Ph.D., Princeton. Organic geochemistry, stable-isotope geochemistry, sedimentology.
David G. Towell, Associate Professor; Ph.D., MIT. Trace-element and isotope geochemistry.
Jeffrey R. White, Associate Professor; Ph.D., Syracuse. Aquatic chemistry, limnology, biogeochemistry.
Robert P. Wintsch, Associate Professor; Ph.D., Brown. Metamorphic, structural, and theoretical petrology, thermochronology.
Ned K. Bleuer, Assistant Professor (part-time); Ph.D., Wisconsin. Glacial geology.
Michael Hamburger, Assistant Professor; Ph.D., Cornell. Observational seismology, tectonics, earthquake prediction.
Andrea Koziol, Visiting Assistant Professor; Ph.D., Chicago. Thermodynamics of crystalline solutions, experimental mineralogy, field-based thermobarometry.
Vishnu Ranganathan, Assistant Professor; Ph.D., LSU. Hydrogeology, groundwater.
Michael Savarese, Assistant Professor; Ph.D., California, Davis. Paleontology, paleobiology, neontology, modern marine environments.
Alan S. Horowitz, Senior Scientist; Ph.D., Indiana. Paleontology, Mississippian faunas, quantitative methods.
Michael D. Dorais, Assistant Scientist; Ph.D., Georgia. Igneous petrology, alkaline magma genesis, mafic enclaves.
Bruce Douglas, Assistant Scientist; Ph.D., Princeton. Regional tectonics, field and experimental study of deformation mechanisms.

Indiana Geologic Field Station, Montana.

Indiana University Geology Building.

SECTION 3: EARTH AND PLANETARY SCIENCES

IOWA STATE UNIVERSITY

Department of Geological and Atmospheric Sciences

Programs of Study The department offers M.S. and Ph.D. degrees in geology and earth science with specializations in hydrogeology, environmental geology, geochemistry, petrology, mineralogy, structural geology/tectonics, stratigraphy/sedimentation, economic geology, glacial geology, and earth science education.

Graduate degree programs in geology include specific requirements, yet remain flexible enough to allow them to be formulated to suit each individual student. The earth science programs have different specified requirements to provide opportunities in nontypical areas of emphasis. The department participates in interdisciplinary Water Resources and Mineral Resources programs and is a major member department in the Iowa State Mining and Mineral Resources Research Institute. A geology/water resources major, emphasizing the geology and geochemistry of groundwater, can also be obtained in the department.

The M.S. program ordinarily requires two and a half years of study, particularly for those who hold assistantships. The degree involves preparation of a thesis. The Ph.D. degree normally requires six years beyond the bachelor's degree and is strongly research oriented.

Research Facilities The department has research laboratories with equipment and facilities that include X-ray diffraction and X-ray fluorescence units, fluid inclusion stages, an automated electron microprobe, a scanning electron microscope, ultraviolet and infrared spectrophotometers, a gas chromatograph, Logitech thin-section apparatus, low-termperature crystal growth chambers, a low-field torque magnetometer, a Bison magnetic susceptibility bridge, earth resistivity apparatus, proton precession magnetometers, portable seismic refraction equipment, and high-temperature and high-pressure experimental laboratories containing high-temperature furnaces, cold-seal Stellite and TZM pressure vessels, piston cyclinder apparatus, large-volume hydrothermal reaction vessels, and Griggs deformation apparatus. A network of personal computers is present within the department. Projects with greater computer requirements can access an NAS 9160 mainframe, four VAX-11/780s, and a VAX-11/785. Remote access to a Cray supercomputer is also available.

Equipment is available elsewhere on campus for ICP, MS/GC, and FT-IR spectroscopy; NMR spectroscopy; laser-Raman spectroscopy; and computerized single-crystal X-ray diffraction, and there are complete drafting and photographic facilities. The main campus library is excellent; it contains 4 million items, including 1.7 million books and bound serials, 2 million microforms, 115,000 maps, and 20,000 currently received journals.

Financial Aid The Department of Geological and Atmospheric Sciences offers nearly all students teaching and research assistantships with stipends that range from $8100 to $9000 for a nine-month half-time appointment. The teaching assistantship generally involves 20 hours of work per week in the form of preparation, teaching, and grading for laboratories in elementary classes. Some summer assistantships are also available. Applicants of high quality can be nominated by the department to receive a Graduate College Premium for Academic Excellence Award. This award can be for one to three years and equals half of resident tuition for each semester.

In addition, a Graduate College Scholarship Credit covering a portion of the resident fee is provided for each teaching and research assistant on full admission. The scholarships award $556.50 per semester ($310 in summer) for each student on a half-time assistantship and $278.25 per semester ($155 in summer) for each student on a quarter-time appointment or more (but less than a half-time appointment).

Cost of Study Full-time tuition and fees for the academic year 1990–91 were $1113 per semester and $620 per summer for Iowa residents and $3211 for nonresidents (who pay resident tuition if on an assistantship).

Cost of Living It is estimated that for a single student the average cost of room, board, books, and supplies for the academic year is approximately $4400.

Student Group There are approximately 26,000 students at Iowa State University; about 4,000 are graduate students. The department has 30 graduate students, of whom 11 are pursuing Ph.D. degrees. These students come from all parts of the United States and from several foreign countries.

Location Ames has population of 50,000 and is surrounded by some of the most fertile farmland in the United States. The city maintains twenty-one parks and 650 acres of streams, woods, and open meadows and offers outstanding facilities for indoor and outdoor recreational activities. The campus is approximately 30 miles north of Des Moines, the capital and largest city in Iowa (with a population of approximately 250,000). Chicago, Minneapolis, St. Louis, and Kansas City are all just a few hours' drive away.

The University Iowa State University, which encompasses over 1,000 acres, was founded in 1858 as the land-grant institution for the state of Iowa. The University is an outstanding cultural center and offers an excellent program of pop and blues concerts, symphony, theater, and plays, all performed by international, national, and local companies. The University offers a wide variety of major intercollegiate and intramural sports.

Applying Completed applications for admission and financial assistance, TOEFL scores for applicants whose native language is not English, three letters of recommendation, and official transcripts should be received before February 15 for admission in the following fall semester. The department does not require, but recommends, submission of GRE General Test scores and scores on the GRE Subject Test in geology.

Correspondence and Information
Professor Paul G. Spry
Graduate Applications Coordinator
Department of Geological and Atmospheric Sciences
253 Science Hall I
Iowa State University
Ames, Iowa 50011
Telephone: 515-294-4477 or 9637

SECTION 3: EARTH AND PLANETARY SCIENCES

Iowa State University

THE FACULTY AND THEIR RESEARCH

Robert D. Cody, Associate Professor; Ph.D., Colorado, 1968. Geochemical sedimentology, experimental growth of authigenic minerals, diagenesis, evaporites, environmental geochemistry.

Frederick P. DeLuca, Associate Professor; Ph.D., Oklahoma, 1970. Coordinator of Earth Science Education Program.

George R. Hallberg, Adjunct Professor; Ph.D., Iowa, 1975. Quaternary geology and geomorphology, stratigraphic, hydrologic, pedologic and engineering properties of the unconsolidated materials of Iowa.

Carl E. Jacobson, Associate Professor; Ph.D., UCLA, 1980. Structural geology, metamorphic petrology, and tectonics of western United States.

Matthew J. Kramer, Adjunct Assistant Professor; Ph.D., Iowa State, 1988. Experimental determination of the rheology of minerals, computer applications to geosciences.

John Lemish, Professor; Ph.D., Michigan, 1955. Economic geology, industrial materials, fossil fuels, metallic mineral deposits, geology of Iowa.

Richard Markuszewski, Adjunct Professor; Ph.D., Iowa State, 1976. Characterization and analyses of coal, development of chemical and physical methods for the removal of deleterious impurities in coal.

Bert E. Nordlie, Professor; Ph.D., Chicago, 1967. Volcanology, field and theoretical studies of magmatic gases, high-pressure–high-temperature experimental studies of gas-melt interactions.

Steven M. Richardson, Associate Professor; Ph.D., Harvard, 1975. Aqueous and environmental geochemistry, geochemistry of hydrothermal systems, geochemical history of planetary atmospheres and oceans.

Karl E. Seifert, Professor and Chairman; Ph.D., Wisconsin, 1963. Igneous petrology, rare-earth and trace-element geochemistry, single-crystal experimental deformation studies of feldspar.

William W. Simpkins, Assistant Professor; Ph.D., Wisconsin, 1989. Geohydrology, groundwater flow and geochemistry of glacial and bedrock aquifers and confining units, environmental geochemistry.

Paul G. Spry, Associate Professor; Ph.D., Toronto, 1984. Economic geology, metallic mineral deposits, hydrothermal experimental studies, mineralogy, fluid inclusion and stable isotope geochemistry.

Carl F. Vondra, Distinguished Professor; Ph.D., Nebraska, 1963. Cenozoic fluvial stratigraphy and sedimentation of North America, Asia, and Africa; geology of early man.

Kenneth E. Windom, Associate Professor; Ph.D., Penn State, 1976. Experimental and theoretical mineralogy and petrology, nature of the earth's crust and upper mantle.

Science Hall houses geological sciences on the Iowa State campus.

Students on an anticline at Iowa State University's geology field station in the Big Horn Mountains, Wyoming.

An automated ARL-SEMQ electron microprobe is utilized for graduate student research.

SECTION 3: EARTH AND PLANETARY SCIENCES

LOUISIANA STATE UNIVERSITY

School of Geoscience

Programs of Study

The School of Geoscience with its four components—the Department of Geology and Geophysics, Coastal Studies Institute, Basin Research Institute, and Louisiana Geological Survey (Division of Research)—offers an ideal opportunity to pursue a wide variety of graduate studies. Areas of specialization include stratigraphy, biostratigraphy and paleontology (micropaleontology, paleobiology), computer applications of geosciences (including well logs and chronostratigraphy), geochemistry (inorganic, isotope, fluids), geophysics and seismology, paleomagnetism, petrology (igneous, metamorphic, carbonates, clays), physical oceanography and remote sensing, soft rock geology (clastics and carbonates, ancient and modern depositional environments, processes, sea-level fluctuations), structure and tectonics, and volcanology. The number and scope of these areas is highly suitable for interdisciplinary approaches.

The graduate programs in geology and geophysics at LSU are a balanced combination of course work and independent research. The thesis or dissertation is a key requirement for the Master of Science or the Doctor of Philosophy degree. Graduate course work is generally selected by the student together with his or her major professor, with the approval of an advisory committee, from numerous offerings of the Department of Geology and Geophysics. Students may also select courses in related science departments or in the fields of mathematics, experimental statistics, and computer science. Incoming students are permitted to enter the Ph.D. program directly, but they are encouraged to begin their graduate careers in pursuit of an M.S. degree. Twenty-four semester hours of course work are required for the master's degree, and a total of 60 hours (including courses and dissertation research credit) are required for the doctorate. Students usually complete the M.S. degree within 2½ years and the Ph.D. degree in about 4 years of total graduate work. Active programs include fieldwork on six continents or at sea. Graduates of the department have been successful in academia, government, and industry; the petroleum industry has been the largest single employer of department graduates.

Research Facilities

The Department of Geology and Geophysics has all the equipment and facilities customarily used in teaching undergraduate and graduate courses in earth sciences and in conducting advanced research in specialized topics. Facilities and equipment include a scanning electron microscope with an EDS X-ray analyzer system; a VAX 8200 and a vector processing computer; a fully automated JEOL 733 Superprobe electron microprobe with computer-linked analytical programs; a modern computer facility, linked directly to the main campus computer, with flatbed plotter, digitizer, and support equipment; an automated X-ray powder diffractometer; an NOAA Satellite Earth Station (HRPT) and advanced image-processing system; an atomic absorption spectrophotometer; an inductively coupled plasma unit for chemical analysis; a differential thermal analysis unit; an infrared spectrometer; an optical spectrophotometer; a 12-inch mass spectrometer for heavy-isotope analysis, used at present for high-resolution analysis of barium in natural waters; a 6-inch mass spectrometer with gas-extraction lines for stable-isotope analysis; an X-ray radiography laboratory; a cathodoluminescence petrographic microscope; a gas chromatograph; and a laboratory with complete facilities for thin-section and polished-section preparations. This equipment is backed by standard laboratories. LSU operates research vessels and small boats from a central facility in Cocodrie.

Financial Aid

Financial assistance is available through scholarships, teaching assistantships, research assistantships, and fellowships. Most graduate students receive financial support. Assistantships require 16 to 20 hours of work per week and are supported by a stipend of a minimum of $8531 at the M.S. level and $10,151 at the Ph.D. level. The department usually administers about eight special fellowships, with stipends ranging from $6000 to $15,000. These are awarded on an annual basis and are renewable for up to four years. University student loans, work-study funds, and grants in support of research are also available. Funds donated by LSU alumni and the petroleum industry to the Geology Endowment also support student research and travel.

Cost of Study

Tuition and fees for full-time graduate students in 1991–92 are $1050 per semester for Louisiana residents and graduate assistants; nonresidents are expected to pay an additional $1600 per semester. Approximately $110 of those fees are not covered by financial aid and must be paid out-of-pocket.

Cost of Living

The University maintains two- and three-bedroom apartments for married students, which currently rent for $195 to $264 per month. Single students may live in dormitories for $391 to $681 per semester. A fifteen-meal-per-week board ticket costs approximately $435 for the semester. Off-campus housing generally costs $250 to $350 for monthly rent and utilities. Living costs in Baton Rouge are moderate.

Student Group

There are approximately 65 graduate students in the department; about 50 percent are pursuing the Ph.D. Most of them received a baccalaureate degree from a college or university outside Louisiana.

Location

Baton Rouge is the capital of Louisiana and a major port. The area has a mild, almost semitropical climate. The city is in the geographic center of southern Louisiana's main cultural and recreational attractions. Within 2 hours' driving time are the antebellum homes of the Mississippi Valley, the famous attractions of New Orleans, and the Louisiana bayou country, with an abundance of festivals and opportunities for recreation.

The University

Louisiana State University and Agricultural and Mechanical College is the oldest and largest institution in the LSU System. It exerts a major influence on the economic, social, and cultural life of the state through an extensive program of instruction, research, and service. The Board of Regents calls for LSU to continue to function as the comprehensive state university, with an emphasis on graduate, professional, and senior-division programs. Within this framework, study in geology and geophysics has developed as one of the most successful major programs on the campus.

Applying

Admission is subject to all rules specified in the current issue of the LSU *Graduate School Catalog*. Applicants must submit directly to the department a completed application form, two complete official transcripts of all undergraduate and graduate course work, three letters of recommendation, GRE General Test scores, and, for international students, TOEFL results and translated transcripts. Students are encouraged to contact any of the faculty members concerning scientific specifics for graduate study.

Correspondence and Information

Graduate Advisor
Department of Geology and Geophysics
Louisiana State University
Baton Rouge, Louisiana 70803-4101
Telephone: 504-388-3353

SECTION 3: EARTH AND PLANETARY SCIENCES

Louisiana State University

THE FACULTY AND THEIR RESEARCH

Charles E. Adams Jr., Professor; Ph.D., Florida State, 1980. Sedimentology, oceanography.
Paul H. Aharon, Professor; Ph.D., Australian National, 1980. Stable-isotope geochemistry and reef-carbonates.
Ajoy K. Baksi, Associate Professor; Ph.D., Toronto, 1970. Geochronology and isotope geochemistry.
Arnold H. Bouma, McCord Professor and Director of the School of Geoscience; Ph.D., Utrecht (Netherlands), 1961. Marine geology, with emphasis on submarine fans, interactions between shallow and deep water, and comparison of modern with ancient deep-sea fans.
Gary R. Byerly, Professor; Ph.D., Michigan State, 1974. Igneous and metamorphic petrology.
Lui-Heung Chan, Research Professor; Ph.D., Harvard, 1966. Chemical oceanography; geochemistry of rivers, estuaries, and oceans.
James M. Coleman, Boyd Professor and Executive Vice Chancellor; Ph.D., LSU, 1966. Deltaic and marine geology.
Roy K. Dokka, Professor; Ph.D., USC, 1980. Structural geology, tectonics, geochronology.
Barbara Dutrow, Assistant Research Professor; Ph.D., SMU, 1985. Experimental mineralogy and metamorphic petrology.
Ray E. Ferrell, Professor; Ph.D., Illinois at Urbana-Champaign, 1966. Geology and mineralogy of clays and clay minerals.
Charles G. Groat, Professor (Adjunct) and Director of the Louisiana Geological Survey; Ph.D., Texas at Austin, 1968. Economic geology; Gulf Coast energy, mineral, and water resources.
Jeffrey S. Hanor, Professor; Ph.D., Harvard, 1967. Geochemistry of sediments, natural waters, and low-temperature ore deposits.
George F. Hart, Professor and Director of Research, Louisiana Geological Survey; Ph.D., Sheffield (England), 1961. Biostratigraphy and petroleum geology.
Joseph E. Hazel, Campanile Professor and Department Chairman; Ph.D., LSU, 1963. Stratigraphic micropaleontology, stratigraphy, chronostratigraphic modeling.
Darrell Henry, Associate Professor; Ph.D., Wisconsin, 1981. Metamorphic petrology.
Shih-Ang Hsu, Professor; Ph.D., Texas at Austin, 1969. Coastal and marine meteorology.
Vindell Hsu, Assistant Professor; Ph.D., Hawaii, 1981. Seismology and paleomagnetism.
Oscar K. Huh, Professor; Ph.D., Penn State, 1968. Oceanography, with emphasis on remote sensing; sediment transport and field geology.
Masamichi Inoue, Assistant Professor; Ph.D., Texas A&M, 1982. Physical oceanography, with emphasis on numerical modeling and climate change.
Charles L. McCabe Jr., Associate Professor; Ph.D., Michigan, 1985. Paleomagnetism and tectonics.
Clyde H. Moore Jr., Professor; Ph.D., Texas, 1961. Sedimentology, petrology and geochemistry of carbonates.
Stephen P. Murray, Professor; Ph.D., Chicago, 1966. Physical oceanography, dynamic sedimentation.
Dag Nummedal, Professor; Ph.D., Illinois at Urbana-Champaign, 1974. Coastal and shallow marine sedimentary processes, planetary geology.
Jeffrey A. Nunn, Associate Professor; Ph.D., Northwestern, 1981. Theoretical tectonophysics and geodynamics.
Harry H. Roberts, Professor and Director of the Coastal Studies Institute; Ph.D., LSU, 1969. Physical processes in carbonate environments, deltaic sedimentation.
James E. Roche, Professor and Field Camp Director; Ph.D., Illinois at Urbana-Champaign, 1969. Carbonate geology, field geology.
Lawrence J. Rouse Jr., Associate Professor; Ph.D., LSU, 1972. Environmental remote sensing.
Andrew Schedl, Assistant Professor; Ph.D., Michigan, 1986. Structural geology.
Judith A. Schiebout, Adjunct Associate Professor and Director of the Geosciences Museum; Ph.D., Texas at Austin, 1973. Tertiary mammalian paleontology and paleoenvironments.
Barun K. Sen Gupta, Professor; Ph.D., Indian Institute of Technology, 1963. Foraminiferal ecology, paleoecology, biostratigraphy.
Willem A. van den Bold, Professor Emeritus; Ph.D., Utrecht (Netherlands), 1946. Micropaleontology of Ostracoda and Foraminifera, Cenozoic stratigraphy.
Stanley N. Williams, Associate Professor; Ph.D., Dartmouth, 1983. Volcanology, with emphasis on eruptive processes and forecasting; geology and tectonics of volcanic arcs; volcanic gas measurements.
William J. Wiseman Jr., Professor; Ph.D., Johns Hopkins, 1969. Shelf and estuarine dynamics and arctic coastal processes.

SECTION 3: EARTH AND PLANETARY SCIENCES

McGILL UNIVERSITY

Department of Geological Sciences
Graduate Program in Geology and Geophysics

Programs of Study

The department offers graduate programs leading to the Ph.D., M.Sc., Applied M.Sc. in Mineral Exploration, and Diploma degrees. The Ph.D. degree requires a thesis and a program of courses recommended by the thesis committee, but there is no formally specified course requirement. Students admitted to the Ph.D. program have normally completed an M.Sc. degree; however, exceptional students can be admitted to the Ph.D. program at the end of the first year in the M.Sc. program. The M.Sc. degree requires a thesis and 12 credits of graduate-level courses. The Applied M.Sc. in Mineral Exploration is designed to train professional geologists for careers in the mineral industry and consists of a two-year program of prescribed courses but does not require a thesis. The Diploma degree consists of a one-year program of courses (24 credits) at the graduate level and does not require a thesis.

Research Facilities

Research in the department is varied. The department is well equipped, especially in the areas of analytical and experimental geochemistry, with a Philips computer-automated X-ray spectrometer equipped with an automatic sample changer, a CAMECA computer-automated electron microprobe equipped with four WDS spectrometers and an EDS spectrometer with a germanium crystal, two Perkin-Elmer atomic absoption spectrometers equipped with a graphite furnace and a hydride generator, an automated high-pressure ion-diffusion chromatograph, an automated Siemens X-ray diffractometer, a fluid inclusion laboratory equipped for microthermometric work, a high-pressure piston cylinder apparatus for experimental petrology, a high-pressure experimental hydrothermal laboratory, and a cathodoluminescence system. Through active cooperation with other departments, the Université de Montréal, and the Université du Québec à Montréal, the department has access to scanning and transmission electron microscopes, a laser-Raman microprobe, a neutron activation laboratory for the analysis of rare earth elements, an inductively coupled plasma–mass spectrometer (ICP-MS) equipped with laser oblation for trace-element analysis, and two VG-sector mass spectrometers for radiogenic isotope analysis. The department has a variety of personal computers and workstations, many of which are connected by a department-wide network that is linked to the mainframe computers in the computer centre of the University. The University museums, the Redpath and McCord, are open to students. The Redpath Museum contains many geological collections and displays assembled by McGill geologists, paleontologists, and mineralogists.

Financial Aid

Students admitted to a thesis program are offered a minimum of $8000 during the academic session, half of which is earned by acting as teaching assistants in one undergraduate course per semester and half of which comes from the research funds of the thesis director. Many students receive additional support from their thesis directors during the summer session, bringing the typical financial support to approximately Can$12,000 per year. Students admitted to the applied mineral exploration program are eligible for Can$4000 in the form of teaching assistantships. In addition, the department has a large number of major and minor scholarships that are awarded on the basis of academic merit. The graduate faculty offers University scholarships, fellowships, and fee waivers for superior international students. Applicants are encouraged to apply early, and certainly before January to be considered for the University scholarships.

Cost of Study

The tuition fee for one academic year of resident study in a degree program is Can$891.43 (Can$4082.93). Students in nonthesis programs must register for a minimum of 12 credits per term in order to be considered full-time. Candidates requiring additional terms beyond the required period of residence must pay a tuition fee of Can$80 (Can$590) per term. A "special student" is not enrolled in a degree program, and tuition fees are assessed on a per-course basis.
Current tuition fees (including health insurance) for international students are shown in parentheses.

Cost of Living

In 1991–92, single students should allow about Can$10,000 per year for living expenses. Many graduate students live in apartments within a kilometre of the University. Accommodation for graduate students in the University residences is limited. The Student Services Office is prepared to help incoming students, particularly international students, to find accommodation. It publishes a list of apartments that are available each summer.

Student Group

In 1990–91, there were 60 students pursuing graduate studies in all areas of geology and geophysics. There are about 4,500 students enrolled in the Faculty of Graduate Studies and Research, and there is an active Graduate Students' Association.

Location

The McGill campus is within a few blocks of the cultural and commercial centre of Montreal. Although located in a downtown area, McGill lies on the flank of Mount Royal, whose summit is one of the largest and finest urban parks in Canada. The University is convenient to concert halls, theatres, art galleries, and museums. The city's sports teams have played leading roles in hockey and baseball. Both the Forum and the Olympic Stadium are easily accessible from the University by the Metro. Montreal is also the centre of winter sports in eastern Canada, with skiing areas within easy reach on three sides. Montreal's mixture of French, English, and many other cultures makes it a unique and always fascinating city.

The University and The Department

McGill has served both the Engligh and the French communities for over 150 years, but the language of teaching at the University is English. The Department of Geological Sciences occupies most of the Frank Dawson Adams Building on the east side of the lower campus. The building is connected directly to the engineering buildings, the Macdonald-Stewart Library of Physical Sciences and Engineering, the University Computing Centre, and the Map Library. The department occupied the building in the early fifties and expanded into additional space in the early seventies. Graduate students are provided with desks and study space in small offices throughout the building. Laboratory space is also made available to them.

Applying

Applicants must submit a formal application and have an official transcript of their academic record forwarded separately. Confidential letters of recommendation must be submitted from at least two referees. International students must obtain a student visa and demonstrate competence in English. Application for admission should be made several months before the desired time of admission; most students begin study in September.

Correspondence and Information

Graduate Admission Committee
Geological Sciences
McGill University
3450 University Street
Montreal, QC H3A 2A7
Canada
Telephone: 514-398-6767
Fax: 514-398-4680

McGill University

THE FACULTY AND THEIR RESEARCH

T. Ahmedali, Faculty Lecturer; M.Sc., Pakistan, 1962. Analytical geochemistry.

C. Alpers, Assistant Professor; Ph.D., Berkeley, 1986. Applied geochemistry, acid mine waters, weathering, stable isotopes in surficial environments.

J. Arkani-Hamed, Professor; Ph.D., MIT, 1969. Geophysical potential fields, dynamics of planetary interiors.

D. Baker, Assistant Professor; Ph.D., Penn State, 1985. High-temperature geochemistry and igneous petrology, experimental investigation of petrogenetic processes, structure and properties of silicate melts and glasses, physical and chemical controls on volcanic eruptions.

D. J. Crossley, Associate Professor; Ph.D., British Columbia, 1973. Global geodynamics and physics of Earth's interior.

B. d'Anglejan, Associate Professor; Ph.D., California, San Diego, 1965. Geological oceanography, sedimentology and geochemistry of estuaries—Quaternary studies.

R. Doig, Professor; Ph.D., McGill, 1964. Rb-Sr isotopic studies, Precambrian geology, measurement of natural radioactivity applied to geological problems.

D. Francis, Professor and Chair; Ph.D., MIT, 1974. Origin and evolution of basic magmas in the mantles of terrestrial planets.

R. Hesse, Professor; Ph.D., Munich, 1964. Sedimentology of modern and ancient continental margins (clastic sediments, diagenesis, marine geology, and plate tectonics).

A. J. Hynes, Associate Professor; Ph.D., Cambridge, 1972. Tectonics and structural geology; transpression in the Canadian Cordillera, origin of the Hudson Bay Arc, gravity features of sutures in the Canadian Shield, uplift of the Laurentides, paleomagnetism and plate motions.

O. G. Jensen, Professor; Ph.D., British Columbia, 1971. Seismology: tectonophysics, geophysical systems analyses.

W. H. MacLean, Professor; Ph.D., McGill, 1968. Economic geology; massive sulphides, alteration systems, mass changes, wallrock stratigraphy.

R. F. Martin, Professor; Ph.D., Stanford, 1969. Igneous petrology; nonorogenic and orogenic magmatism, alkali feldspars as indicators of magmatic and postmagmatic processes.

E. W. Mountjoy, Professor; Ph.D., Toronto, 1960. Sedimentation and diagenesis, acient and modern carbonates, Cordilleran structure and stratigraphy.

A. Mucci, Associate Professor; Ph.D., Miami (Florida), 1981. Low-temperature geochemistry and chemical oceanography, chemical thermodynamics and kinetics of solid-solution reactions in natural environments, early diagenesis of marine coastal and estuarine sediments.

C. W. Stearn, Professor; Ph.D., Yale, 1952. Paleontology, paleoecology, and stratigraphy.

A. E. Williams-Jones, Associate Professor; Ph.D., Queen's at Kingston, 1973. Economic geology, application of fluid inclusion, isotopic and experimental studies to the genesis of granitoid-related Sn–W–Mo–rare metal and epithermal Au-Ag deposits.

S. A. Wood, Associate Professor; Ph.D., Princeton, 1985. Theoretical and experimental geochemistry and its application to hydrothermal ore deposits.

The Macdonald-Stewart Library of Physical Sciences and Engineering and the Geological Sciences Building on the lower campus.

SECTION 3: EARTH AND PLANETARY SCIENCES

NORTHERN ILLINOIS UNIVERSITY

Department of Geology

Programs of Study

The department offers graduate studies leading to the M.S. and Ph.D. degrees in geology. Programs of study may emphasize geochemistry, geochronology, stratigraphy/sedimentation, paleontology/paleoecology, geophysics, petrology, structural geology, hydrogeology, mineralogy, environmental geology, glacial geology/sedimentology, geomorphology/Pleistocene geology, tectonics, or volcanology.

The Master of Science degree program ordinarily requires two years of study, particularly for those who hold assistantships. The M.S. degree is usually taken with a thesis option; however, the nonthesis option is also offered under some circumstances. The Ph.D. degree program normally requires four years beyond the master's degree or five years beyond the bachelor's degree.

Research Facilities

The Department of Geology has complete facilities for scanning electron microscopy, fully automated electron microprobe analysis, X-ray diffraction and fluorescence studies, differential thermal analysis, solid- and gas-source mass spectrometry, DC plasma spectrometry, atomic absorption spectrometry, cathodoluminescence, and particle-size, gravimetric, magnetic, seismic, and electrical geophysical studies. Laboratories and apparatus are available for sample and thin-section preparation, petrography, phase optical work, and photomicroscopy. The University's Amdahl VTA computer is accessible to any graduate student through four terminals in Davis Hall, and the department's Sun Workstation and computer network are available for hands-on research.

Financial Aid

Teaching and research assistantships include a waiver of tuition and a stipend of $945–$1000 per month in 1991–92. NSF graduate traineeships, grants-in-aid, University fellowships, and minority fellowships are available to qualified students. Traineeships and fellowships also carry a waiver of tuition.

Cost of Study

In 1991–92, tuition is $869 per semester for in-state students and $2607 for out-of-state students. In addition, all students are assessed $283.92 per semester in fees. Books and supplies cost about $300 per year. The graduation fee is $25. Tuition is waived for students with teaching or research appointments, as described above.

Cost of Living

Graduate students report that off-campus rooms and apartments are less expensive than University housing, and living costs average $400 per month for lodging, meals, and miscellaneous expenses. A one-bedroom unfurnished apartment with heat and water costs about $350 a month; about $25 monthly should be added for a furnished apartment.

Student Group

About 25,000 graduate and undergraduate students attend the University. Most of the 43 graduate students currently enrolled in the Department of Geology receive some sort of institutional support. Their research interests range, geographically, from Alaska to Antarctica.

Location

DeKalb and nearby Sycamore have about 45,000 permanent residents. Their economic bases are a number of medium-sized industries, agriculture, and the University. Shops in town are numerous and offer a wide range of goods and services. Most athletic, recreational, and cultural events and opportunities are connected with the University. Chicago, where cultural and recreational opportunities are almost unlimited, is less than 2 hours away. Summer and winter sports of many kinds are available in northern Illinois and nearby Wisconsin.

DeKalb is located in north-central Illinois on the south flank of the Kankakee Arch in an area of lower Paleozoic carbonate and clastic sedimentary rocks. The bedrock is mantled by Wisconsin glacial deposits. The campus is within easy reach of classic sedimentary, igneous, and metamorphic terranes.

The University

Northern Illinois University has been an institution of higher learning for over ninety years. It has expanded greatly in recent years, especially in its attention to and support of graduate work. The geology group has grown rapidly since its establishment as a department in 1968, with encouragement and support from the University. Space for the department was recently increased considerably, and its facilities have been completely remodeled. The University sponsors a full range of dramatic, musical, and artistic activities, both local and touring. The University believes in equal educational opportunity and awards scholastic support to qualified persons without regard to creed, color, sex, race, or national origin.

Applying

Application forms for admission and for assistantships may be obtained from either the department or the Graduate School but should be returned to the Graduate School. Applications must be completed by June 1 for the fall semester, November 1 for the spring semester, and April 1 for the summer term. However, to ensure consideration for assistantships or fellowships, applications should be submitted by March 1 for the fall semester. Assistantship funds are seldom available for new students entering in the spring or summer semester.

Correspondence and Information

Director of Graduate Studies
Department of Geology
Northern Illinois University
DeKalb, Illinois 60115
Telephone: 815-753-1943

Peterson's Guide to Graduate Programs in the Physical Sciences and Mathematics 1992

SECTION 3: EARTH AND PLANETARY SCIENCES

Northern Illinois University

THE FACULTY AND THEIR RESEARCH

Jonathan H. Berg, Professor; Ph.D., Massachusetts, 1976. Metamorphic and igneous petrology, anorthosites, Precambrian geology.
Colin J. Booth, Associate Professor; Ph.D., Penn State, 1984. Hydrogeology.
Elizabeth A. Burton, Assistant Professor; Ph.D., Washington (St. Louis), 1988. Carbonate geochemistry.
Philip J. Carpenter, Associate Professor; Ph.D., New Mexico Tech, 1984. Groundwater and engineering geophysics, earthquake seismology.
Keros Cartwright, Adjunct Professor; Ph.D., Illinois at Urbana-Champaign, 1973. Hydrogeology, geophysics.
Clarence J. Casella, Associate Professor; Ph.D., Columbia, 1962. Structural geology, granite petrology.
Donald M. Davidson Jr., Professor; Ph.D., Columbia, 1965. Analytical structure, tectonics, resource geology, remote sensing.
C. Patrick Ervin, Associate Professor; Ph.D., Wisconsin, 1972. Applied geophysics, potential fields, computer applications in the earth sciences.
Ruth I. Kalamarides, Associate Professor; Ph.D., Massachusetts, 1982. Mineralogy and stable-isotope geochemistry.
Ronald S. Kaufmann, Assistant Professor; Ph.D., Arizona, 1984. Groundwater and stable-isotope geochemistry.
Daniel E. Lawson, Adjunct Professor; Ph.D., Illinois at Urbana-Champaign, 1977. Glacial and periglacial sedimentology.
Hsin Yi Ling, Professor; Ph.D., Washington (St. Louis), 1963. Biostratigraphy, micropaleontology.
Paul Loubere, Associate Professor; Ph.D., Oregon State, 1980. Micropaleontology, paleoclimatology, isotope geochemistry.
Lyle D. McGinnis, Adjunct Professor; Ph.D., Illinois, 1965. Geophysics, seismology.
Carla W. Montgomery, Associate Professor; Ph.D., MIT, 1977. Isotope geochemistry, geochronology, Precambrian crustal evolution.
Michael G. Mudrey Jr., Adjunct Professor; Ph.D., Minnesota, 1973. Resource geology, igneous petrology.
I. Edgar Odom, Adjunct Professor; Ph.D., Illinois, 1963. Mineralogy, clay mineralogy, geochemistry and mineralogy of sediments.
Eugene C. Perry Jr., Professor; Ph.D., MIT, 1963. Stable light isotope geochemistry.
Ross Powell, Associate Professor and Chair; Ph.D., Ohio State, 1980. Stratigraphy and sedimentology of clastic rocks.
Paul R. Stoddard, Assistant Professor; Ph.D., Northwestern, 1989. Tectonics, plate kinematics, geophysics.
Jay A. Stravers, Assistant Professor; Ph.D., Colorado, 1986. Glacial geology and Quaternary stratigraphy.
James A. Walker, Associate Professor and Assistant Chair; Ph.D., Rutgers, 1982. Igneous petrology and volcanology.
Malcolm P. Weiss, Professor Emeritus; Ph.D., Minnesota, 1953. Stratigraphy, sedimentary petrology, reefs and carbonate rocks.

SECTION 3: EARTH AND PLANETARY SCIENCES

OREGON STATE UNIVERSITY

Department of Geosciences

Programs of Study Graduate programs leading to the M.S., M.A., and Ph.D. degrees are available in stratigraphy, sedimentation, paleontology, biostratigraphy, historical biogeography, structural geology, sedimentary and igneous petrology, economic geology, Quaternary geology, forest geomorphology, geochemistry, geophysics, neotectonics, and volcanology. In addition, programs in marine geology and soils are offered in conjunction with other departments. The M.S. and Ph.D. programs may be completed in an average of two and four years, respectively. Theses usually involve both laboratory work and fieldwork.

Research Facilities Facilities for instructional and research purposes within the Department of Geosciences (Geology Program) include equipped and operational sample preparation, mineral separation, and thin-section laboratories; IBM PC computers; a 30,000-psi triaxial deformation device; an X-ray diffractometer; an X-ray radiograph; student and research model binocular, ore, and petrographic microscopes; a luminoscope; a fluid inclusion heating-freezing stage; extraction lines for sulfur isotope analyses of sulfate and sulfide minerals and for oxygen and hydrogen isotope analyses of hydrous minerals and waters; and a Cameca SX-50 electron microprobe with digital image-processing capabilities, shared with the College of Oceanography. Acid-etching facilities, serial sectioning equipment, and photographic instruments are available for macrofossil and microfossil studies. A Silurian-Devonian worldwide brachiopod collection is also available. Elsewhere within the College of Science and the College of Oceanography are computers, an emission spectrograph, atomic absorption and mass spectrometers, a scanning electron microscope, a rock paleomagnetism laboratory, a K-Ar radiometric dating laboratory, and neutron activation facilities that are available to both faculty and graduate students.

Financial Aid Fellowships and traineeships sponsored by industry, foundations, and government agencies are available for graduate students. A number of teaching assistantships are awarded each year, with stipends that range from $2759 to $5518 for the nine-month academic year in 1991–92. Graduate students may also obtain employment within the institution under the work-study program. In addition, funds may be borrowed from the Office of Financial Aid on a deferred-payment plan. Most fieldwork for theses is supported by industry or by various research grants.

Cost of Study For 1991–92, tuition and fees for full-time graduate study are $912 per quarter for state residents and $1546 per quarter for nonresidents. However, the charge for all graduate students holding research and teaching assistantships, fellowships, and traineeships is $408 per year. U.S. citizens may establish Oregon residence after twelve months. (Figures are subject to change.)

Cost of Living Graduate students generally prefer to live in off-campus housing. Single students' costs in 1991–92 range from $190 to $350 per month for a room. There is limited housing for married students in University apartments with rents that range from $190 to $215 per month. Housing for unmarried students is available in the dormitories at a cost of $2835 to $3465 for room and board for the academic year. Information concerning on-campus and off-campus rentals may be obtained from the Department of Housing.

Student Group Approximately 16,000 students are enrolled at Oregon State University. Of this number, about 2,700 are graduate students, and the proportion of resident to nonresident students is about equal. The University imposes a quota on graduate students, so that admission is selective. The Department of Geosciences (Geology Program), which has about 45 graduate students, is within the College of Science, which constitutes the largest division of the University. Approximately 20 percent and 25 percent of the undergraduate and graduate student bodies, respectively, are enrolled in the College of Science.

Location Corvallis is a university-oriented community of approximately 43,000 people, located in the Willamette Valley 85 miles south of Portland between the Coast and Cascade ranges. Convenient access is provided by freeway and scheduled airline, bus, and train service. The local academic and scientific atmosphere is enhanced by research and development firms; agricultural, forest product, and metal industries; and federal research laboratories of the Bureau of Mines, the Forest Service, the Department of Agriculture, and the Water Resources Research Institute. Ocean beaches, lakes, rivers, desert, and rugged mountains are all within a 150-mile radius of Corvallis. This diversity provides unexcelled recreational opportunities that include boating, camping, fishing, hunting, mountain climbing, and skiing.

The Department Oregon State University was founded in 1868. The Department of Geosciences (Geology Program) was developed within the College of Science in 1932. It has 14 staff members and about 40 undergraduate majors in addition to its graduate students. The staff is complemented by visiting lecturers and professors. About 20 percent of the graduate students are in the Ph.D. program, and more than half receive financial support.

Applying Applicants for admission to the Graduate School must have a baccalaureate degree from an accredited college or university. In addition, they must show scholarship and other attributes indicative of their ability to complete the program of graduate study successfully. GRE scores on the General Test and on the Subject Test in geology are required. Applicants for financial aid should arrange to take the GRE examinations by January at the latest. Three letters of reference must be supplied by the applicant. An approved summer field camp providing at least 9 hours of credit is required for a graduate degree. There is no deadline for applications for admission to the Graduate School; however, applications for financial aid, together with GRE scores, must be received no later than February 14.

Correspondence and Information
Graduate Admissions Coordinator
Department of Geosciences
Wilkinson Hall 104
Oregon State University
Corvallis, Oregon 97331-5506
Telephone: 503-737-1201
Fax: 503-737-1200

SECTION 3: EARTH AND PLANETARY SCIENCES

Oregon State University

THE FACULTY AND THEIR RESEARCH

Allen F. Agnew, Courtesy Professor; Ph.D., Stanford, 1949. Engineering geology, hydrogeology, mineral resources.

Arthur J. Boucot, Courtesy Professor; Ph.D., Harvard, 1953. Local, regional, and worldwide taxonomic, ecologic, evolutionary, and biogeographic studies of Silurian and Devonian brachiopods and gastropods.

Peter U. Clark, Assistant Professor; Ph.D., Colorado, 1984. Quaternary geology and geomorphology, glacial sedimentation and erosion, paleoclimatology.

John H. Dilles, Assistant Professor; Ph.D., Stanford, 1984. Geology of ore deposits, with emphasis on the petrology, geochemistry, and tectonic setting of mineralized plutons in the western United States.

Cyrus W. Field, Professor and Chairman; Ph.D., Yale, 1961. Sulfur isotope geochemistry of mineral deposits; geology, geochemistry, mineralogy, and petrology of altered and mineralized plutons of the western cordillera.

Gordon E. Grant, Courtesy Instructor; Ph.D., Johns Hopkins, 1986. Fluvial geomorphology, mountain streams, Oregon Cascade and Coast ranges.

Anita L. Grunder, Assistant Professor; Ph.D., Stanford, 1986. Field, petrologic, and geochemical studies of continental volcanic rocks.

J. Granville Johnson, Professor; Ph.D., UCLA, 1964. Paleontology, biostratigraphy, and paleography of Paleozoic rocks of western and arctic North America.

Robert D. Lawrence, Associate Professor; Ph.D., Stanford, 1968. Structural geology and tectonics, mechanics of rock deformation, metamorphic structures, tectonics of the Pacific Northwest and Pakistan, remote sensing.

R. J. Lillie, Associate Professor; Ph.D., Cornell, 1984. Application of exploration geophysical techniques to large-scale tectonic problems.

Ellen J. Moore, Courtesy Research Associate; M.S., Oregon, 1950. Paleontology of West Coast Tertiary mollusks.

George W. Moore, Courtesy Professor; Ph.D., Yale, 1960. Tectonostratigraphic terranes; paleotectonics of the Circum-Pacific region.

Roger L. Nielsen, Courtesy Assistant Professor; Ph.D., SMU, 1983. Experimental and theoretical igneous petrology, trace-element geochemistry, X-ray microanalysis, computer simulation of open igneous systems processes.

Alan R. Niem, Associate Professor; Ph.D., Wisconsin–Madison, 1971. Sedimentology, sedimentary petrography, and stratigraphy of clastic rocks; geology of western Washington and Oregon.

Wendy A. Niem, Courtesy Research Associate; M.S., Oregon State, 1976. Basin analysis, interactive volcanic and sedimentary processes, volcaniclastic rocks, geomorphology.

Frederick J. Swanson, Courtesy Professor; Ph.D., Oregon, 1972. Forest geomorphology and sedimentology, Oregon Cascades.

William H. Taubeneck, Emeritus Professor; Ph.D., Columbia, 1955. Layering in igneous rocks, petrogenesis of granitic rocks, Miocene dike swarms of the western United States, regional tectonics of the Pacific Northwest, Columbia River basalt.

Edward M. Taylor, Associate Professor; Ph.D., Washington State, 1967. Mineralogy, X-ray diffraction and spectroscopy, volcanic petrology of the Oregon High Cascades and adjacent areas.

Robert S. Yeats, Professor; Ph.D., Washington (Seattle), 1958. Active faults of southern California, Pakistan, and Turkey; tectonics and geophysics of the Cascade Range of Washington, the Transverse ranges, and the offshore borderland of California.

Smith Rock, Crook County, Oregon: volcanic tuffs of the John Day Formation, with Pliocene intracanyon basalts exposed along the Crooked River.

Wallowa Mountains, Baker County, Oregon: Permian-Triassic country rocks intruded by the Cornucopia stock.

SECTION 3: EARTH AND PLANETARY SCIENCES

PENNSYLVANIA STATE UNIVERSITY
Department of Geosciences

Programs of Study

The Department of Geosciences offers M.S. and Ph.D. degree programs. A wide range of faculty interests (see reverse side of page) and exceptional laboratory and other support facilities provide an extensive variety of areas of specialization in which students may choose their course work and research topics. These areas include the newly established Earth System Science Center, which is an interdepartmental program directed toward a global, multidisciplinary view of the earth and its variability.

Research Facilities

The department maintains a variety of unexcelled modern facilities and equipment for research, including a Sun Microsystems computer network with direct access to the NSF's supercomputer network. Students have access to laboratories for research on the petrography and petrology of igneous, metamorphic, and sedimentary rocks, including coal and organic sediments; complete palynological processing and microscopy facilities; rock preparation and rock mechanics laboratories; high-temperature and high-pressure/high-temperature equipment for dry or hydrothermal experiments; mass spectrometers and ancillary equipment for isotope analysis; a seismic observatory, ultrasonic model and paleomagnetism laboratories, and field equipment for seismic, electrical, magnetic, and gravity surveys; facilities and data for remote sensing of earth resources; laboratories and field facilities for the study of the hydrogeology and geochemistry of natural waters; an X-ray laboratory for single-crystal and powder methods at low and high temperatures; and coastal marine laboratories in Virginia. The department and the Materials Characterization Laboratories are equipped for both classical methods of chemical analysis and modern instrumental methods, such as atomic absorption, emission, and absorption spectroscopy; electron microscopy and scanning transmission electron microscopy; automated X-ray diffractometry; and ion microprobe and automated electron microprobe analysis. The department has excellent collections of rocks, minerals, and ore and coal samples; paleontological, paleobotanical, and palynological collections; lunar photographs and data; and maps. Over 50,000 volumes related to earth sciences are housed in the library of the College of Earth and Mineral Sciences. A nuclear reactor and an IBM 3090/400E mainframe computer are available on campus. The department also maintains a doctoral research program in cooperation with the Geophysical Laboratory of the Carnegie Institution in Washington.

Financial Aid

Nearly all of the department's on-campus graduate students receive support from assistantships, fellowships, or traineeships. Half-time teaching or research assistantships of $4080 per semester, plus full tuition and limited medical coverage, were available to qualified applicants in 1990–91. One-quarter-time and three-quarter-time assistantships are awarded in special cases. Most research assistantships involve the study of problems appropriate for thesis research. Financial support for thesis research unsupported by grants is available through a special fund.

Cost of Study

For 1990–91, the self-supporting student who was a Pennsylvania resident paid $2225 per semester; nonresidents paid $4445 per semester.

Cost of Living

Living costs are moderate. In 1990–91, for single students, a room in the coeducational University residence halls and board were available for $1615 per semester. Privately owned housing can be found within walking distance of the campus. For married students, ample accommodations are available locally at a wide range of prices, and graduate student apartments on campus rented for $220 and $340 per month for one- and two-bedroom units, respectively.

Student Group

In the department there are approximately 115 graduate students. The total student enrollment at the main University Park campus is about 34,000.

Location

The University Park campus of the Pennsylvania State University is in the town of State College and in a metropolitan area of over 100,000 people in the center of the commonwealth. Located in a rural and scenic part of the Appalachian Mountains, the area is only 3 to 4 hours away from Washington, D.C., Pittsburgh, and Philadelphia. Varied cultural, educational, and athletic activities are available throughout the year.

The University and The Department

Founded in 1855, Penn State is the land-grant university of Pennsylvania. The University Park campus has 258 major buildings on 4,786 acres, of which 540 acres constitute the beautifully landscaped central campus. The College of Earth and Mineral Sciences, of which the department is a part, has approximately 130 faculty members and 450 graduate students in the earth sciences and closely related fields in the Departments of Geography, Materials Science and Engineering, Meteorology, Mineral Economics, and Mineral Engineering. The College occupies a complex of four buildings on the west side of the campus, with the Department of Geosciences housed primarily in the modern, air-conditioned Deike Building. Facilities for graduate student research are available within the department and in the Materials Characterization Labs, the Energy and Fuels Research Center, the Materials Research Lab, and the Earth System Science Center. The size of the faculty promotes close personal relationships between faculty and students. Each faculty member works with a research group averaging 2–4 students. In cooperation with a faculty adviser and committee, each student designs and pursues a course and research program tailored to his or her individual interests and needs.

Applying

The University offers two 15-week semesters and an 8-week summer session beginning approximately August 20, January 10, and June 10, respectively. Candidates may apply for admission in either fall or spring. Applications must be received by July for admission to the fall semester; if financial support is required, applications must be received by February 15. All correspondence regarding admission and financial aid should be sent to the address given below.

Correspondence and Information

Associate Head for the Geosciences Graduate Program
303 Deike Building
The Pennsylvania State University
University Park, Pennsylvania 16802
Telephone: 814-865-7394

Peterson's Guide to Graduate Programs in the Physical Sciences and Mathematics 1992

SECTION 3: EARTH AND PLANETARY SCIENCES

Pennsylvania State University

THE FACULTY AND THEIR RESEARCH

M. A. Arthur, Professor of Geosciences and Head of the Department; Ph.D., Princeton, 1979. Marine geology, stable isotope geochemistry, sedimentary geochemistry.
S. S. Alexander, Professor of Geophysics; Ph.D., Caltech, 1963. Seismology, time-series analysis, remote sensing.
R. B. Alley, Assistant Professor of Geosciences; Ph.D., Wisconsin–Madison, 1987. Glaciology, ice sheet stability, paleoclimates from ice cores.
H. L. Barnes, Professor of Geochemistry; Ph.D., Columbia, 1958. Ore deposits, hydrothermal fluids.
E. J. Barron, Professor of Geosciences and Director, Earth System Science Center; Ph.D., Miami (Florida), 1980. Earth system science, paleoclimatology.
S. L. Brantley, Associate Professor of Geosciences; Ph.D., Princeton, 1986. Aqueous geochemistry, geochemical kinetics.
R. J. Cuffey, Professor of Paleontology; Ph.D., Indiana, 1966. Paleontology, evolution, systematics, paleoecology, bryozoans, reefs.
A. Davis, Professor of Geology; Ph.D., Durham (England), 1965. Coal geology, coal petrology, paleogeology of coal deposits.
P. Deines, Professor of Geochemistry; Ph.D., Penn State, 1967. Isotope geochemistry.
W. Duke, Assistant Professor of Geology; Ph.D., McMaster, 1985. Clastic stratigraphy and sedimentology.
D. H. Eggler, Professor of Petrology and Associate Head, Graduate Programs; Ph.D., Colorado, 1967. Experimental mineralogy and petrology of the upper mantle.
T. Engelder, Professor of Geosciences; Ph.D., Texas A&M, 1973. Rock mechanics, structural geology.
D. M. Fisher, Assistant Professor of Geosciences; Ph.D., Brown, 1988. Regional tectonics, structural geology.
K. P. Furlong, Professor of Geosciences; Ph.D., Utah, 1981. Plate tectonics, thermal and mechanical evolution of the lithosphere.
T. W. Gardner, Professor of Geology; Ph.D., Cincinnati, 1978. Quantitative fluvial geomorphology.
D. P. Gold, Professor of Geology; Ph.D., McGill, 1963. Petrology, structural geology, remote sensing, economic geology.
E. K. Graham, Professor of Geophysics; Ph.D., Penn State, 1969. Experimental solid-state geophysics, planetary models.
R. J. Greenfield, Professor of Geophysics; Ph.D., MIT, 1965. Magnetic and electrical fields, seismology.
A. L. Guber, Professor of Geology; Ph.D., Illinois, 1962. Paleozoology, evolution, paleoecology.
J. F. Kasting, Associate Professor of Geosciences; Ph.D., Michigan, 1979. Atmospheric evolution, planetary atmospheres, paleoclimates.
D. M. Kerrick, Professor of Petrology; Ph.D., Berkeley, 1968. Metamorphic petrology.
L. R. Kump, Associate Professor of Geosciences; Ph.D., South Florida, 1986. Geochemical cycles, low-temperature sedimentary geochemistry.
C. A. Langston, Professor of Geophysics; Ph.D., Caltech, 1976. Seismology, inversion theory, tectonics.
P. M. Lavin, Professor of Geophysics; Ph.D., Penn State, 1962. Gravity and magnetic surveying, crustal tectonics.
J. N. Louie, Assistant Professor of Geosciences; Ph.D., Caltech, 1987. High-resolution seismology.
M. L. Machesky, Assistant Professor of Geochemistry; Ph.D., Wisconsin–Madison, 1986. Aqueous geochemistry.
S. J. Mackwell, Assistant Professor of Geosciences; Ph.D., Australian National, 1985. Ductile deformation.
H. Ohmoto, Professor of Geochemistry; Ph.D., Princeton, 1969. Stable isotopes, ore deposits.
R. R. Parizek, Professor of Geology; Ph.D., Illinois, 1961. Groundwater, glacial, and environmental geology.
A. W. Rose, Professor of Geochemistry; Ph.D., Caltech, 1958. Geochemical exploration, ore deposits, environmental geochemistry.
R. Roy, Professor of Solid-State Sciences; Ph.D., Penn State, 1948. Crystal chemistry, synthesis and stability of minerals.
R. F. Schmalz, Professor of Geology; Ph.D., Harvard, 1959. Oceanography, chemistry of sedimentation.
R. L. Slingerland, Professor of Geology; Ph.D., Penn State, 1977. Sedimentology, deterministic models.
D. K. Smith, Professor of Mineralogy; Ph.D., Minnesota, 1956. X-ray crystallography.
C. P. Thornton, Professor of Petrology and Associate Head, Undergraduate Programs; Ph.D., Yale, 1953. Volcanology, igneous petrology.
A. Traverse, Professor of Palynology; Ph.D., Harvard, 1951. Palynology of Paleozoic-Recent sediments.
B. Voight, Professor of Geology; Ph.D., Columbia, 1964. Engineering and structural geology, New England geology.
W. B. White, Professor of Geochemistry; Ph.D., Penn State, 1962. Carbonate groundwaters, spectroscopy.

Emeriti

T. F. Bates, Professor Emeritus of Mineralogy; Ph.D., Columbia, 1944. Clay mineralogy, public education in geosciences.
C. W. Burnham, Professor Emeritus of Geochemistry; Ph.D., Caltech, 1955. Experimental petrology, geochemistry of ore deposits.
J. C. Griffiths, Professor Emeritus of Petrography; Ph.D., Wales, 1936; Ph.D., London, 1940. Statistics in earth sciences.
B. F. Howell Jr., Professor Emeritus of Geophysics and Associate Dean Emeritus of the Graduate School; Ph.D., Caltech, 1949. Seismology, tectonics.
M. L. Keith, Professor Emeritus of Geochemistry; Ph.D., MIT, 1939. General geochemistry, geotectonics.
E. Osborn, Professor Emeritus of Geochemistry; Ph.D., Caltech, 1938. Igneous petrology.
R. Scholten, Professor Emeritus of Petroleum Geology; Ph.D., Michigan, 1950. Tectonics, Rocky Mountain geology, habitat of oil.
W. Spackman, Professor Emeritus of Paleobotany; Ph.D., Harvard, 1949. Paleobotany, coal petrology, modern organic sediments.
E. G. Williams, Professor Emeritus of Geology; Ph.D., Penn State, 1957. Carboniferous stratigraphy, sedimentation.
L. A. Wright, Professor Emeritus of Geology; Ph.D., Caltech, 1951. Industrial minerals, Great Basin geology, tectonics.

The Deike Building, which houses the Department of Geosciences.

SECTION 3: EARTH AND PLANETARY SCIENCES

RENSSELAER POLYTECHNIC INSTITUTE
Department of Earth and Environmental Sciences

Programs of Study

Graduate programs leading to the degrees of Master of Science and Doctor of Philosophy are offered in the fields of geochemistry, geophysics, surficial geology, hydrogeology, petrology, planetary science, structural geology, and tectonics. These programs recognize the interdependence of geology and other sciences. Students with undergraduate majors in geology have the opportunity and are generally expected to broaden their education, while those with undergraduate majors in biology, chemistry, engineering, mathematics, and physics are given the opportunity to complement their backgrounds with formal and reading courses in geology. The program for the master's degree usually includes a thesis of 6 to 9 credit hours, with exceptions made under special circumstances.

Research Facilities

The department is equipped for research in a number of fields. In addition to routine equipment for specimen preparation and sampling, the following are available: solid media high-pressure equipment, hydrothermal apparatus, a 5-spectrometer electron microprobe, a gamma-ray spectrometer, and an X-ray diffractometer. Seismology and neotectonic studies are supported by a seismograph station, a twelve-channel seismic system, a gravity meter, a magnetometer, a resistivity meter, and a network of Sun Workstations; petrology, fission-track studies, and asteroid spectrography are also supported by exceptional software. Essential geology is well represented, since classical areas of Proterozoic, Paleozoic, and Pleistocene geology are located close to campus.

Research is supported by such state-of-the-art facilities as the Folsom Library, whose automated retrieval systems link users to its own holdings as well as 5,600 libraries and 250 databases and the Voorhees Computing Center, which offers students exceptionally wide access to a diverse array of advanced-function workstations, microcomputers, and mainframes that are interconnected to each other and to national networks via the campus network.

Financial Aid

Financial aid is available in the form of fellowships, research or teaching assistantships, and scholarships. The stipend for assistantships ranges up to $8400 for the nine-month 1991–92 academic year. In addition, full tuition scholarships are granted. Additional compensation for research during the summer months is also available. Outstanding students may qualify for university-supported Rensselaer Scholar Fellowships, which carry a stipend of $13,500 and a full tuition scholarship. Low-interest, deferred-repayment graduate loans are also available to U.S. citizens with demonstrated need. Most enrolled students receive some kind of financial assistance.

Cost of Study

Tuition for 1991–92 is $455 per credit hour. Other fees amount to approximately $230 per semester. The application fee for admission is $30. Books and supplies cost about $930 per year.

Cost of Living

The cost of rooms for single students in residence halls or apartments varies from $2400 to $3800 for the 1991–92 academic year. A twenty-meal-per-week board plan costs $2500 for the academic year. Apartments for married students, with monthly rents of $360 to $590, are available on campus.

Student Group

There are about 4,500 undergraduate and 2,000 graduate students representing all fifty states and over sixty-five countries at Rensselaer. The Department of Earth and Environmental Sciences has about 25 graduate students and postdoctoral research fellows.

Location

Rensselaer is situated on a scenic 260-acre hillside campus in Troy, New York, across the Hudson River from the state capital of Albany. Troy's central Northeast location provides students with a supportive, active, medium-sized community in which to live and an easy commute to Boston, New York, and Montreal. The Capital Region has one of the largest concentrations of academic institutions in the United States. Sixty thousand students attend fourteen area colleges and benefit from shared activities and courses, as well as access to the Saratoga Performing Arts Center, the Empire Plaza, the Rensselaer Technology Park, and some of the country's finest outdoor recreation, including Lake George, Lake Placid, and the Adirondack, Catskill, Berkshire, and Vermont mountains.

The Institute

Founded in 1824 and the first American college to award degrees in engineering and science, Rensselaer Polytechnic Institute today is accredited by the Middle States Association of Colleges and Schools and is a private, nonsectarian, coeducational university. Rensselaer's five schools—Architecture, Engineering, Management, Science, and Humanities and Social Sciences—offer a total of eighty-seven graduate degrees in forty fields.

Applying

Admissions applications and all supporting credentials should be submitted well in advance of the preferred semester of entry to allow sufficient time for departmental review and processing. Graduate Record Examinations' General Test scores are required. Since the first departmental awards are made in February and March for the next full academic year, applicants requesting financial aid are encouraged to submit all required credentials by February 1 to ensure that they will receive consideration.

Correspondence and Information

For further information about graduate work:
Department of Earth and Environmental
  Sciences
Rensselaer Polytechnic Institute
Troy, New York 12180-3590
Telephone: 518-276-6475

For admissions information and applications:
Director of Graduate Admissions
Graduate Center
Rensselaer Polytechnic Institute
Troy, New York 12180-3590
Telephone: 518-276-6789

Peterson's Guide to Graduate Programs in the Physical Sciences and Mathematics 1992

SECTION 3: EARTH AND PLANETARY SCIENCES

Rensselaer Polytechnic Institute

THE FACULTY AND THEIR RESEARCH

Professors
M. B. Bayly, Ph.D., Chicago. Structural geology, rheological properties of earth materials.
G. M. Friedman (Emeritus), Ph.D., Columbia. Sedimentology.
M. J. Gaffey, Ph.D., MIT. Planetary science.
S. Katz (Emeritus), Ph.D., Columbia. Geophysics.
R. G. LaFleur, Ph.D., Rensselaer. Hydrogeology, glacial geology.
D. S. Miller, Ph.D., Columbia. Geochemistry, isotope geology, fission-track research.
S. W. Roecker, Ph.D., MIT. Geophysics, seismology, tectonics.
F. S. Spear, Ph.D., UCLA. Metamorphic petrology, tectonics.
E. B. Watson, Ph.D., MIT. Igneous petrology, high-temperature geochemistry.

Assistant Professor
R. McCaffrey, Ph.D., California, Santa Cruz. Marine geophysics, potential fields, tectonics and seismology.

Adjunct Faculty
Y. W. Isachsen, Ph.D., Cornell. Metamorphic petrology, Precambrian geology, field studies.
T. Morgan, Ph.D., Houston. Reflection seismology.

Research Assistant Professor
S. J. Gaffey, Ph.D., Hawaii. Carbonate petrology.

Research Associates
C. Ravenhurst, Ph.D., Dalhousie. Geochemistry, fission-track research.
M. Roden, Ph.D., Rensselaer. Geochemistry, isotope geology.

Postdoctoral Associates
T. Skulski, Ph.D., McGill. Igneous petrology, geochemistry.
T. R. Stanton, Ph.D., Arizona State. Igneous petrology, chemistry of silicate melts.

Research Scientists
D. Cherniak, Ph.D., SUNY at Albany. Geochemical kinetics.
D. A. Wark, Ph.D., Texas at Austin. Igneous petrology, geochemistry, microprobes.

SECTION 3: EARTH AND PLANETARY SCIENCES

SOUTHERN METHODIST UNIVERSITY

Department of Geological Sciences

Programs of Study The Department of Geological Sciences offers graduate programs leading to the Ph.D. and M.S. degrees with specializations in the areas of geophysics, igneous and metamorphic petrology, isotope geology, low-temperature geochemistry, sedimentology, and vertebrate paleontology. In addition, the department offers an M.S. in applied geophysics, which is a nonthesis program.

The Ph.D. program encompasses academic course work, including graduate core courses, and research training. After passing a preliminary oral examination on two original research propositions, Ph.D. candidates devote most of their academic effort to research and the preparation of a dissertation that must be defended publicly.

The M.S. program requires a minimum of 30 hours of course work, tailored to the requirements of each student, that includes graduate core courses and 6 hours of thesis work. Master's students must pass a preliminary oral examination on one proposition and complete and defend a thesis.

The M.S. in applied geophysics requires 33 hours of specified course work, an industry internship, a written report, an oral presentation, and a comprehensive written exam.

Research Facilities The department is housed in Heroy Hall, a modern research building devoted to geology, geophysics, anthropology, and statistics. Students in residence have office space and ready access to professors, staff, and other graduate students. The geology and geophysics library collection, housed separately, has grown, through generous endowment, to be one of the foremost in the country.

Analytical research facilities in the department include a fully automated JEOL 733 four-spectrometer electron microprobe and scanning electron microscope, two gas-source mass spectrometers (C-O-H-sample preparation), a modern X-ray diffractometer, and a multichannel gamma-ray spectrometer. In addition to a radiocarbon laboratory, fission-track and U-series disequilibrium (alphaspectrometry) facilities are currently active. Facilities for experimental petrology include hydrothermal furnaces. The SMU mainframe computer (an IBM 3081 D) is readily accessible to all graduate students. The SMU Computer Center also provides a diverse collection of microcomputers, a software library, graphics packages, and tutorial services for use by faculty members and students. The department has over a dozen personal computers that are accessible to students. The geophysics group operates a network of Sun Workstations for data analysis and modeling. The department also has well-equipped laboratories for thermal-conductivity measurements, heat-flow studies, and seismology (including modern digital acquisition systems for field measurements). The Shuler Museum of Paleontology houses a permanent working collection (primarily of vertebrates) that has complete fossil preparation capabilities. In addition to basic facilities for the production and polishing of petrographic thin sections, petrographic capabilities include reflected- and transmitted-light research scopes, CRT monitors, cathode luminescence, epifluorescence, and image processing.

Financial Aid Most graduate students in residence receive financial aid. All tuition and fees are paid for students on stipend support, which may take the form of research assistantships, endowed scholarships, or teaching assistantships. Stipends range from $12,000 to $15,000 per twelve months for 1991–92.

Cost of Study For 1991–92, tuition is $435 per semester hour, plus a student fee of $56 per semester hour.

Cost of Living Campus housing costs $1131 per semester per person (double occupancy). Board on campus is $1215 per semester (including tax). SMU-owned efficiency and one-, two-, and three-bedroom apartments are available; rent ranges from $180 to $555 per month (some apartments have utilities paid; they may be furnished or unfurnished). In addition, privately owned apartments are readily available in the University area.

Student Group Of the approximately 8,000 students enrolled at SMU, about 3,000 are graduate and professional students. The Department of Geological Sciences has a graduate student body of about 30, most of whom are pursuing Ph.D. degrees.

Location Dallas is the center of an attractive metropolitan area of 2 million people. It has fine parks, lakes, museums, theaters, orchestras, libraries, and churches. Both active and spectator sports fans find it exciting. Clean and progressive, the city continues to grow as a center of business and light industry. Dallas's electronics, aircraft manufacturing, industrial, and geological/geophysical research opportunities are outstanding.

The University Established in 1911 by the Methodist Church, Southern Methodist University is a nonsectarian institution dedicated to academic excellence and freedom of inquiry. Through its eight undergraduate, graduate, and professional colleges and schools, it is educating thousands of men and women for creative and useful lives. The University is centrally located on a spacious, handsome campus that adjoins a residential area.

In addition to the Department of Geological Sciences, other strong departments include those of anthropology, mathematics, statistics, chemistry, and biology within Dedman College and various departments in the School of Engineering and Applied Science.

Applying Completed applications for admission and financial aid should be received before February 15. Applications filed later than this date will be considered for aid if funds are available. Financial aid awards are based on merit. The minimum standards for admission are an overall GPA of 3.0 (on a 4.0 scale), a GPA in the major of 3.25, and a GRE General Test score of 1200.

Correspondence and Information
Graduate Committee
Department of Geological Sciences
Southern Methodist University
Dallas, Texas 75275-0395

Telephone: 214-692-2750
Fax: 214-692-4289

Peterson's Guide to Graduate Programs in the Physical Sciences and Mathematics 1992

Southern Methodist University

THE FACULTY AND THEIR RESEARCH

David D. Blackwell, Hamilton Professor; Ph.D., Harvard, 1967. Geothermal studies and their application to plate tectonics, especially of the western United States; energy resource estimates and geothermal exploration.

Patrick V. Brady, Assistant Professor; Ph.D., Northwestern, 1989. Fluid-rock interaction within the earth's crust, environmental chemistry, hydrothermal geochemistry.

James E. Brooks, Professor and President, Institute for the Study of Earth and Man; Ph.D., Washington (Seattle), 1954. Stratigraphy, geomorphology.

Michael A. Dungan, Associate Professor; Ph.D., Washington (Seattle), 1974. Igneous petrology and geochemistry, Cenozoic volcanism of the western United States (Rio Grande Rift and San Juan volcanic field), calc-alkaline volcanism of the Chilean Andes.

Robert T. Gregory, Associate Professor; Ph.D., Caltech, 1981. Stable-isotope geology and geochemistry, evolution of Earth's fluid envelope and lithosphere.

Vicki L. Hansen, Assistant Professor; Ph.D., UCLA, 1987. Structural geology and tectonics, with emphasis on ductile deformation; tectonic evolution of the Canadian-Alaskan Cordillera and Proterozoic evolution of the southwestern United States and Antarctica.

Eugene T. Herrin, Shuler-Foscue Professor; Ph.D., Harvard, 1958. Theoretical and applied seismology, solid earth properties, computer analysis of geophysical data.

Michael J. Holdaway, Professor and Chairman; Ph.D., Berkeley, 1962. Experimental petrology and crystal chemistry of pelitic minerals, metamorphic petrology of pelitic schists in New England and New Mexico.

Louis L. Jacobs, Associate Professor and Director, Shuler Museum of Paleontology; Ph.D., Arizona, 1977. Vertebrate paleontology, evolution.

Robert L. Laury, Professor and Director of Undergraduate Studies; Ph.D., Wisconsin, 1966. Sedimentology and sedimentary petrology, deposition and diagenesis of Paleozoic marine sediments, Cenozoic terrestrial depositional systems.

A. Lee McAlester, Professor; Ph.D., Yale, 1960. Marine ecology–paleoecology, evolutionary theory, Paleozoic geology, petroleum geology.

Roger J. Phillips, Matthews Professor and Graduate Committee Chairman; Ph.D., Berkeley, 1968. Potential fields, evolution and state of the interiors of the terrestrial planets, thermomechanical behavior of the lithosphere and its relationship to tectonics and to convection in the mantle.

Peter A. Scholle, Albritton Professor; Ph.D., Princeton, 1970. Carbonate depositional models and diagenesis, sedimentary petrology, arid region studies, cyclic sedimentation and secular variation.

Brian W. Stump, Associate Professor; Ph.D., Berkeley, 1979. Seismic wave propagation, inverse theory, earthquake and explosion source theory.

Adjunct Faculty

John F. Ferguson, Adjunct Associate Professor; Ph.D., SMU, 1981. Seismology.
John W. Goodge, Adjunct Assistant Professor; Ph.D., UCLA. Metamorphic petrology, structural geology.
Herbert Haas, Adjunct Professor and Director, Radiocarbon Laboratory; Ph.D., SMU, 1971. Radiocarbon dating.
Bonnie F. Jacobs, Adjunct Assistant Professor; Ph.D., Arizona, 1983. Palynology.
Shari A. Kelley, Adjunct Assistant Professor; Ph.D., SMU, 1984. Geothermal research.
Robert K. Suchecki, Adjunct Assistant Professor; Ph.D., Texas at Austin, 1980. Sedimentology and sedimentary petrology.

SECTION 3: EARTH AND PLANETARY SCIENCES

STANFORD UNIVERSITY
School of Earth Sciences
Department of Geology

Programs of Study

The Stanford Department of Geology offers programs leading to M.S. and Ph.D. degrees in a wide range of fields, including structural geology, tectonics, ore deposits, sedimentary geology, paleoceanography, volcanology, hard-rock and soft-rock petrology, and geochemistry (with a strong emphasis on the physics, chemistry, and surface chemistry of silicates, oxides, and sulphides. The size and the breadth of interest of the department's faculty provides extensive opportunities for field-based research projects as well as for experiments in the laboratory and on computers and for the most modern applications of instrumental analysis and spectroscopy to problems in the earth sciences. Students are encouraged to draw upon the skills of more than 1 research adviser in interdisciplinary projects. Complementary programs in fields such as hydrogeology, remote sensing, rock mechanics, geostatistics, petroleum geology, seismology, paleomagnetism, and oil and geothermal exploration and production are provided by the other departments of the School of Earth Sciences (Applied Earth Sciences, Geophysics, and Petroleum Engineering). The geology department maintains close connections with other departments at Stanford, including the Departments of Civil and Chemical Engineering, Chemistry, Biology, Applied Physics, and Material Science as well as with the Stanford Center for Materials Research, the Stanford Synchrotron Radiation Laboratory, and the Western Regional Headquarters of the U.S. Geological Survey.

Research Facilities

Laboratories, collections, offices, and the earth sciences library are located in several adjacent buildings close to the Departments of Chemistry, Biology, Engineering, and Physics and to the Center for Materials Research. Laboratory facilities in the geology department or associated with it include X-ray fluorescence and diffraction instruments, an electron microprobe with an image analysis system, scanning and transmission electron microscopes, solids nuclear magnetic resonance (NMR) and Mössbauer spectrometers, an ESCA laboratory and a scanning tunneling microscope for surface studies, thermal-analysis equipment, high-pressure and high-temperature apparatus for experimental petrology and the synthesis of minerals and magmas, a laser-heated Ar-Ar dating system, paleomagnetic equipment, a large biaxial deformation apparatus, gas chromatographs, and a gas chromatograph–mass spectrometer. Full-time staff technicians or research associates are in charge of many of these laboratories. A wide variety of interconnected microcomputers and mainframe computers are available. The Stanford Synchrotron Radiation Laboratory provides some of the best facilities in the world for studying the structure of materials by X-ray scattering and spectroscopy. A variety of facilities, including stable and radiogenic isotope laboratories, are available through ongoing collaborations with the U.S. Geological Survey, whose Western Regional Headquarters is located about 2 miles from the campus.

Financial Aid

Graduate students receive full support from combinations of departmental fellowships, teaching or research assistantships, and other sources; tuition is generally paid in full by the department. Special funds help support student-initiated research projects and summer fieldwork. Special fellowships are available from both the University and the department to encourage applications from qualified minority group members.

Cost of Study

The 1990–91 tuition for a three-quarter academic year was approximately $9000. This fee is, in general, paid by the fellowship awards that fund most students. (Students generally enroll for 9 units of credit per quarter.)

Cost of Living

Unmarried graduate student housing on campus is available at a range of costs from about $3000 to $5500 for nine months. Married student housing is also available on campus. After their first year, most graduate students find housing in nearby off-campus rental houses and apartments.

Student Group

The Department of Geology has about 50 graduate students, most of whom are enrolled in the Ph.D. program. About 40 percent of the geology graduate students are women, and about 25 percent are from foreign countries. Almost all students receive substantial financial aid. Recent graduates are employed in a variety of academic positions; in the oil and mining industries; in geotechnical, environmental, and high-tech materials companies; and at the U.S. Geological Survey and national laboratories.

Location

Stanford is located at the foot of the northern California coastal hills, near the lower end of San Francisco Bay and about 30 miles southeast of San Francisco. The small city of Palo Alto is adjacent to the campus. San Francisco and San Jose provide numerous cultural opportunities. The nearby Pacific Ocean beaches and Sierra Nevada mountains allow easy access to a wide range of outdoor activities ranging from sailing to hiking, skiing, and rock climbing as well as to very diverse geological environments.

The University and The School

The University enrolls about 6,000 undergraduate and 6,000 graduate students and is located on a large campus (about 10 square miles), much of which is natural open space: hills, oak parklands, and lakes. The tradition of the School of Earth Sciences goes back to the founding of the University, whose first professor (and later president) was a geologist. The Department of Geology is the largest part of the School. Its faculty and students have a very diverse range of interests and interact extensively with other members of the University community.

Applying

Applications are particularly welcome from students with strong backgrounds in chemistry, physics, mathematics, biology, or materials science, including students who have majored in one of these fields and who have strong interests in the earth sciences. Completed applications for admission should be submitted before January 1, although late applications are considered when possible. Scores on the GRE General Test are required. The department encourages prospective students to contact individual faculty members directly and to visit Stanford.

Correspondence and Information

Professor Jonathan F. Stebbins
Department of Geology
Stanford University
Stanford, California 94305-2115
Telephone: 415-723-1140

Peterson's Guide to Graduate Programs in the Physical Sciences and Mathematics 1992

SECTION 3: EARTH AND PLANETARY SCIENCES

Stanford University

THE FACULTY AND THEIR RESEARCH

Dennis K. Bird, Associate Professor of Geology. Theoretical and field studies of the geochemistry of hydrothermal systems, fracturing and mass transfer associated with igneous intrusions.
Gordon E. Brown Jr., Professor and Chairman of Geology. Inorganic geochemistry and mineralogy, structural studies of earth materials using X-ray diffraction and spectroscopic methods, geochemistry of mineral surfaces, mineralogy of pegmatites.
Robert G. Coleman, Professor of Geology. Petrogenesis and tectonics of accreted terranes, ophiolites, rift systems.
Marco T. Einaudi, Professor of Applied Earth Sciences and Geology. Field, petrologic, and geochemical studies of hydrothermal ore deposits.
W. Gary Ernst, Professor of Geology and Geophysics and Dean of the School. Field, analytical, and experimental phase-equilibrium studies of metamorphic terranes of the circum-Pacific, plate-tectonic implications.
Stephan A. Graham, Professor of Applied Earth Sciences and Geology. Field and subsurface investigations of sedimentary basins, petroleum geology.
Michael F. Hochella Jr., Associate Professor of Geology, Research. Mineral-water interface geochemistry: atomic structure, morphology, composition and reactivity of mineral surfaces; geochemical applications.
James C. Ingle Jr., Professor of Geology. Micropaleontology, marine stratigraphy, paleoceanography, particularly the Cenozoic history of the Pacific Ocean as recorded by continental margin sequences.
Simon Klemperer, Associate Professor of Geophysics. Reflection seismology: tectonic problems of the crust and upper mantle.
Juhn G. Liou, Professor of Geology. Field and experimental studies of metamorphic petrology, particularly low-T, high-P assemblages and zeolite paragenesis.
Donald R. Lowe, Professor of Geology. Sedimentology, sedimentary petrology, and Precambrian geology, with an emphasis on deep-sea sedimentation and processes on the Archean earth.
Gail Mahood, Associate Professor of Geology. Field and petrologic studies of silicic and intermediate composition magma systems, volcanology.
Michael O. McWilliams, Associate Professor of Geophysics and Geology. Kinematics of tectonic processes, paleomagnetic and isotopic age determinations in the study of crustal accretion and deformation.
Elizabeth L. Miller, Associate Professor of Geology. Structural geology and regional tectonics, field-based studies of physical processes during mountain building.
George A. Parks, Professor of Applied Earth Sciences and Geology. Geochemical processes involving mineral surfaces, experimental and theoretical low-T aqueous geochemistry.
David D. Pollard, Professor of Applied Earth Sciences and Geology. Field, laboratory, and theoretical studies of rock mechanics and structural geology; rock fracture and the transport of fluids in the crust.
Irwin Remson, Professor of Applied Earth Sciences and Geology. Environmental earth sciences and applied hydrogeology, including field studies and hydrologic simulations.
Norman H. Sleep, Professor of Geophysics and Geology. Physical processes of the earth's interior: thermal structure of spreading centers, hydrothermal circulation, influence of the thermal history of the earth on the evolution of life.
Jonathan F. Stebbins, Associate Professor of Geology. Experimental and theoretical studies of silicate melts, minerals, and igneous processes; NMR spectroscopy of silicate and oxide materials.
George A. Thompson, Professor of Geophysics and Geology. Crustal tectonics and seismic imaging, particularly in extending regions.

Consulting Faculty

Steven Bohlen, U.S. Geological Survey. Petrogenesis of metamorphic and igneous rocks, evolution of the continental crust, high-pressure experimental petrology.
Edward Clifton, U.S. Geological Survey. Near-shore clastic sedimentation, origin of bedforms and facies, deep-sea clastic sedimentation.
Brent Dalrymple, U.S. Geological Survey. Isotope geochemistry, radiometric dating.
Gerard Demaison, Chevron Overseas Petroleum, Inc. Integration of organic geochemistry, petroleum exploration, and basin evaluation: source-rock stratigraphy, basin mapping, modeling of oil and gas generation.
N. Timothy Hall, Geometrix. Neotectonics of California, geological hazards; engineering geology.
Thomas Hanks, U.S. Geological Survey. Neotectonics, earthquake seismology.
Keith A. Kvenvolden, U.S. Geological Survey. Organic and petroleum geochemistry.
Joseph Ruetz, Teck Resources, Inc. Photogeology, sedimentary and field geology.
Z. M. Zhang, Chinese Academy of Geological Sciences. Geology and tectonics of China.

Emeritus Faculty

Robert R. Compton. Structural geology, rock fabrics, Cenozoic deformation in the western United States.
Willam R. Evitt. Palynology and paleontology.
Konrad B. Krauskopf. Geochemistry and origin of ore deposits, petrology of granitic rocks; radioactive waste disposal.
Benjamin M. Page. Tectonics: geologic effects of plate interactions, especially in active continental margins; neotectonics.
Ernest I. Rich. Quaternary geology, geomorphology, remote sensing.
Tjeerd H. van Andel. Geoarchaeology, marine geology, paleoceanography.

Research Associates

Alex Blum, U.S. Geological Survey. Experimental, theoretical, and field studies of mineral dissolution and chemical weathering.
Nancy Breen. Geophysical and geological studies of plate boundaries, marine geology and tectonics.
Ian Farnan. NMR spectroscopy: structure and molecular motion in solids and liquids.
Julie Paque, Center for Materials Research. Mineralogy and petrology of meteorites, electron microprobe analysis and imaging.
Joel W. Sparks. Igneous petrology and basaltic volcanism, X-ray fluorescence analysis of rocks and minerals.
Glenn Waychunas, Center for Materials Research. X-ray crystallography, mineral spectroscopy, computer simulation of crystal structures.

TEXAS A&M UNIVERSITY

College of Geosciences
Department of Geophysics

Programs of Study

The department offers graduate programs that lead to the M.S. and Ph.D. degrees in both applied and theoretical geophysics. Research interests of the 20 participating faculty members vary widely but focus primarily on studies of the solid earth in the general areas of tectonophysics, geodynamics, and petroleum and engineering geophysics. Representative areas include experimental and theoretical geophysical rock mechanics, rheology, seismology, the earth's gravity and magnetic fields, electromagnetic and sonar probing of rock masses, and characterization and recovery of energy sources. Many research programs are carried out in collaboration with the Departments of Geology and Oceanography. An M.S. program in petroleum geophysics is available for students with little or no previous preparation in geophysics. Applicants with undergraduate degrees in mathematics, physics, geology, or engineering are encouraged to apply. Normally, completion of the master's program requires from two to three years; the Ph.D. program typically takes three to five years, depending on the student's background and goals. Both programs require the successful completion and defense of a thesis on an approved research topic. The Ph.D. program requires that a preliminary written and oral examination be passed prior to advancement to candidacy for the degree.

Research Facilities

The Department of Geophysics is housed in the M. T. Halbouty building, which also contains the geology department and the Earth Resources Institute. Departmental computing facilities include a VAX-11/780 computer with an array processor, a Landmark Graphics workstation, Sun Workstations, a GIS/LIS workstation, about fifty terminals, and printers, plotters, etc. The University computing facilities include Cray and IBM mainframe computers, several VAX computers, and a variety of printers and plotters. The Department of Geophysics also has a terminal with a direct line to a supercomputer at the Houston Area Research Center. Complete field and laboratory systems are available for multichannel digital recording and processing of seismic data; these systems include an analog computer and a two-dimensional seismic modeling laboratory for the study of elastic wave propagation. Other field exploration equipment includes a LaCoste-Romberg gravimeter and Sharpe flux-gate and Geometrics proton precession magnetometers as well as matched, optically pumped rubidium vapor magnetometers, an IDR induced polarization-resistivity system, and both long- and short-range radar and sonar probing equipment. The department and the adjacent Center for Tectonophysics employ a wide variety of high-pressure/high-temperature triaxial deformation apparatus for studies of the effects of important physical and chemical variables on mechanical properties and flow laws of rocks. Complete optical, X-ray, and electron microscope equipment is available for analyses of flow processes in experimentally and naturally deformed specimens. A state-of-the-art three-component cryogenic magnetometer is available for research in paleomagnetism and rock magnetism.

Financial Aid

Teaching and research programs are supported by an average of ten teaching and nonteaching assistantships (including Dean's Awards) from University funds provided by the state of Texas. In addition, several graduate fellowships funded by petroleum companies are available for especially outstanding students. The remainder of the students are usually supported as research assistants from funds provided to several faculty members by various research grants and contracts. Stipends range from $840 a month for beginning master's candidates to $1000 a month for advanced Ph.D. students. Except for fellowships, awards are normally made for a nine-month period, but some may be extended to twelve. There are also other kinds of aid available through the Student Financial Aid Office.

Cost of Study

For 1991–92, tuition for Texas residents is $20 per semester credit hour; students supported by assistantships and fellowships are regarded as residents for tuition purposes and need not pay the $128 per semester credit hour required of nonresidents. Other fees total approximately $185 per semester.

Cost of Living

The cost of living varies widely according to the type of accommodations sought, marital status, and other considerations. Room and board costs on campus for single students range from $275 to $400 per month. A limited number of University-owned apartments, both furnished and unfurnished, are available for married students. A large number of privately owned apartments are also available in the community. Information regarding housing and general living costs is available through the Housing Office.

Student Group

Texas A&M has approximately 40,000 students, who come from all states and many foreign countries. Nearly 6,800 are graduate students. The department has 65 graduate students and 25 undergraduate majors.

Location

The Bryan–College Station area is in south central Texas, about 90 miles north of Houston, 175 miles south of Dallas, and 95 miles east of Austin. The Gulf Coast, pine forests, and lakes are within a 3-hour drive. The population of the two adjacent cities, including students, totals more than 100,000. The community has excellent schools, churches (approximately twenty denominations), hospitals, theaters, parks, shopping malls, motels, and restaurants.

The University

Texas A&M University, founded as a land-grant college in 1876, is the state's oldest public institution of higher education. Its physical plant is currently valued at about $300-million. Areas of study include the sciences, arts and humanities, architecture, agriculture, engineering, veterinary medicine, business, education, and transportation. Although graduate training leading to the M.S. degree began in 1888 and the first doctorate was conferred in 1940, more than 75 percent of all master's and professional degrees and 85 percent of all doctorates have been conferred in the last two decades.

Applying

Inquiries regarding admission, financial aid, and graduate study in the Department of Geophysics should be addressed to the graduate adviser.

Correspondence and Information

Graduate Adviser
Department of Geophysics
Texas A&M University
College Station, Texas 77843-3114

Texas A&M University

THE FACULTY AND THEIR RESEARCH

Michele Caputo, Professor Emeritus of Geophysics; Doctor of Mathematics, Ferrara (Italy); Doctor of Physics, Bologna (Italy). Theoretical and earthquake seismology, gravitational potential field theory, theoretical geodesy, finite-strain determinations.

Richard L. Carlson, Professor; Associate Director of the Geodynamics Research Institute; Ph.D., Washington (Seattle). Elastic properties of rocks and minerals at in situ pressures, kinematic aspects of plate tectonics.

Neville L. Carter, Professor; Ph.D., UCLA. Experimental determinations of the rheology of rock-forming materials, analyses of deformation mechanisms and development of structures in naturally deformed rock systems.

Davis A. Fahlquist, Professor; Associate Dean of the College of Geosciences; Ph.D., MIT. Gravity and marine geophysics.

Andrew T. Fisher, Assistant Adjunct Professor; Ph.D., Miami (Florida). Physical properties.

Timothy J. G. Francis, Professor; Deputy Director, Ocean Drilling Program; Ph.D., Cambridge. Marine geophysics.

Anthony F. Gangi, Professor; Ph.D., UCLA. Theoretical geophysics and seismology, seismic exploration, filter theory.

John W. Handin, Distinguished Professor Emeritus; Director of the Earth Resources Institute; Research Associate in the Center for Tectonophysics; Ph.D., UCLA. Experimental rock deformation with applications to earthquake-hazards reduction, geothermal-energy recovery, underground storage, and underground structures.

Steven H. Harder, Visiting Assistant Professor; Ph.D., Texas at El Paso. Inversion of seismic velocities for the anisotropic elastic tensor.

Thomas W. C. Hilde, Professor; Director of the Geodynamics Research Institute; D.Sc., Tokyo. Marine geology and geophysics, plate tectonics, marine magnetics.

Earl R. Hoskins, Brockett Professor; Associate and Deputy Dean of the College of Geosciences; Leader of the Mineral Resources Program of the Earth Resources Institute; Ph.D., Australian National. Field and laboratory rock mechanics, development of mineral resources.

Andreas K. Kronenberg, Associate Professor; Ph.D., Brown. High-temperature deformation of rocks.

John M. Logan, Professor; Ph.D., Oklahoma. Experimental deformation of crustal rocks with special emphasis on fault systems and earthquake-hazards reduction.

Robert J. McCabe, Associate Professor; Ph.D., Tokyo. Paleomagnetism, plate reconstruction.

Frank Dale Morgan, Professor; Ph.D., MIT. Geoelectricity, rock physics, inverse problems, seismology.

Philip D. Rabinowitz, Professor; Director, Ocean Drilling Program; Ph.D., Columbia. Marine geophysics and plate reconstruction.

James E. Russell, Professor; Research Associate in the Center for Tectonophysics; Ph.D., Northwestern. Rock mechanics and computer simulation; design, construction, and utilization of structures in rocks.

William W. Sager, Associate Professor; Ph.D., Hawaii. Paleomagnetism, plate reconstructions.

Terry W. Spencer, Professor; Ph.D., Caltech. Seismic exploration, communication theory, acoustic properties of rocks.

Laura B. Stokking, Assistant Adjunct Professor; Ph.D., California, San Diego. Paleomagnetism.

Robert R. Unterberger, Professor Emeritus; Ph.D., Duke. Electric and magnetic properties of rocks, electromagnetic and sonar probing of rock masses.

Seiya Uyeda, Professor; D. B. Harris Chair in Geophysics; D.Sc., Tokyo. Global geophysics, geodynamics.

Joel S. Watkins, Earl F. Cook Professor of Geosciences; Head of the Department; Ph.D., Texas at Austin. Exploration geophysics, seismic stratigraphy.

Supporting Faculty, Center for Tectonophysics

Patrick A. Domenico, Harris Professor of Geology; Faculty Associate in the Center for Tectonophysics; Ph.D., Nevada. Mass and energy transport in porous media.

Melvin Friedman, Professor of Geology; Ph.D., Rice. Dynamic petrofabric and residual strain analysis, thermal cracking, borehole stability, frictional sliding and analysis of experimentally folded and faulted rock.

John H. Spang, Professor of Geology and Head of the Geology Department; Research Associate in the Center for Tectonophysics; Ph.D., Brown. Field studies of faults, using modern dynamic analysis, finite strain, and mathematical techniques; laboratory model studies in multilayered clays.

SECTION 3: EARTH AND PLANETARY SCIENCES

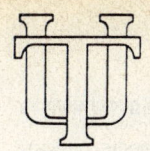

TULANE UNIVERSITY
Department of Geology

Programs of Study
The Department of Geology offers graduate programs leading to the degrees of Master of Science in broad areas of geology and paleontology and Doctor of Philosophy in geology and, in cooperation with the Department of Biology, a program leading to the degree of Doctor of Philosophy in paleontology. Two master's degree programs are available: the principal one requires 24 semester hours of graduate course work and successful completion, presentation, and defense of a thesis that reflects individual research accomplishments. A second, nonthesis, program requires 36 semester hours of course work and a significant research paper. The Ph.D. programs in both geology and paleontology each require a minimum of 48 semester hours of course work, oral and written examinations, and an original contribution in the form of a written dissertation suitable for publication in a learned journal. Areas of research in geology and paleontology include sedimentary geochemistry, theoretical geochemistry, sedimentary geology, carbonate petrology, environmental geology, igneous petrology, volcanology, Precambrian paleontology and sedimentary geology, molluscan paleontology, micropaleontology, and paleontology of primitive echinoderms. Special emphasis is given to geology of the Gulf Coast region, including Yucatan, and Latin America, especially Mexico.

Research Facilities
The department's research facilities, partially supported by a departmental endowment, include a scanning electron microscope with an energy-dispersive X-ray system, an X-ray fluorescence spectrometer, an electron microprobe, an X-ray diffractometer, a cathodoluminescence microscope, a DC plasma spectrometer, a wet chemistry laboratory, and a computer laboratory with microcomputers, a graphics laboratory, and terminals to the University Computer Center. In addition, single-crystal X-ray diffraction equipment, a transmission electron microscope with an energy-dispersive X-ray system, a high-resolution optical microscope, and other equipment are available in a coordinated instrumentation facility. Extensive collections in Tertiary Mollusca provide unusual research opportunities in paleontology.

Financial Aid
Graduate teaching and research assistantships are available to qualified students and provide nine-month stipends, including departmental supplements, ranging from $8500 to $9500. Two fellowships, each with a $12,000 stipend, are also offered by the department. All assistantships and fellowships are accompanied by a tax-free full tuition scholarship.

Cost of Study
Full-time tuition for 1991–92 is $8375 per semester plus a $115 student fee. Tuition on a part-time basis is $900 per credit hour plus fees.

Cost of Living
A limited amount of University housing is available for graduate students. Most graduate students choose to live off campus, where costs vary greatly depending on the type of accommodation selected. A cost-of-living figure of $750 per month is quoted to international graduate students for purposes of entry.

Student Group
Tulane currently enrolls 8,750 full-time and 2,320 part-time students. Of these, approximately 800 are registered in the Graduate School. In recent years, graduate students have come to Tulane from over 380 colleges and universities, from all fifty states, and from thirty-six foreign countries.

The department seeks to admit 5 to 7 students per year. There are currently 13 students in residence. Graduate students, in coordination with a member of the faculty and with departmental support, organize a program of speakers.

Location
Tulane's eleven colleges and schools, with the exception of the medical divisions, are located on 100 acres in a residential area of New Orleans. New Orleans' mild climate, many parks, and proximity to the Gulf Coast provide opportunities for a wide variety of outdoor activities. The city's many art galleries and museums offer regularly scheduled exhibits throughout the year. New Orleans is famous for its French Quarter, Mardi Gras, Creole cuisine, and jazz.

The University and The Department
Tulane is a private nonsectarian university offering a wide range of undergraduate, graduate, and professional courses of study for men and women. The University's history dates from 1834, when it was founded as the Medical College of Louisiana. Graduate work was first offered in 1883. In 1884, the University was organized under its present form of administration and renamed for Paul Tulane, a wealthy New Orleans merchant who endowed it generously. Tulane is a member of the American Association of Universities, a group of fifty-six major North American research universities. It is among the top twenty-five private universities in the amount of outside support received for research each year.

The Department of Geology is in the Liberal Arts and Sciences division of the University, which has strong programs in biology, chemistry, mathematics, and physics as well as in geology. Graduate students in geology are encouraged to enroll in appropriate courses in one or more of these disciplines. Cross-enrollment with the School of Engineering is also available, as are environmental courses offered by the School of Law.

Applying
The general deadlines for applying are July 1 for the fall semester and December 1 for the spring semester. For those requesting financial aid, however, the application deadline is February 1. Students entering in January are not likely to be able to obtain financial aid until the following fall semester. Students should write to the dean of the Graduate School for application forms. The Graduate School will not forward the application to the department for consideration for admission until all of the following documents, plus the $35 application fee, have been received: a completed application form, three completed recommendation forms, official transcripts of all undergraduate and graduate work, and official results of the Graduate Record Examinations General Test, taken within the past five years. International applicants for admission must present satisfactory evidence of competence in English by submitting a score of at least 220 on the TSE (Test of Spoken English) or, if this test is not available, a minimum score of 600 on the TOEFL. Admission is based on academic accomplishments and promise. Tulane is an affirmative action/equal employment opportunity institution.

Correspondence and Information
For application forms and admission:
Dean of the Graduate School
Tulane University
New Orleans, Louisiana 70118
Telephone: 504-865-5100

For specific information regarding programs:
Associate Chairman
Department of Geology
Tulane University
New Orleans, Louisiana 70118
Telephone: 504-865-5198

Peterson's Guide to Graduate Programs in the Physical Sciences and Mathematics 1992

Tulane University

THE FACULTY AND THEIR RESEARCH

George C. Flowers, Ph.D., Berkeley, 1979. Theoretical geochemistry, sedimentary geochemistry, and environmental geochemistry of estuarine sediments.

Ivan P. Gill, Ph.D., LSU, 1989. Carbonate petrology: deposition and diagenesis of modern and ancient carbonate sediments, geology and ecology of coral reef systems, hydrogeology, isotopic geochemistry, dolomitization, interaction of carbonate minerals with subsurface fluids.

Robert J. Horodyski, Ph.D., UCLA, 1965. Precambrian paleontology and sedimentology: early evolution of life; development and Precambrian evolution of prokaryotes; plant and animal protists, metaphytes, and metazoans; stromatolites. Proterozoic sedimentary deposits and geologic history of the western United States.

John P. McDowell, Ph.D., Johns Hopkins, 1968. Sedimentary petrology: Recent sedimentation and sedimentary processes, current structures, paleocurrent anlaysis; environmental geology.

Stephen A. Nelson, Ph.D., Berkeley, 1979. Igneous petrology: petrologic studies of volcanoes; relationships between volcanism and tectonism, particularly in Mexico; volcanic hazards studies; mechanisms of explosive volcanism; thermodynamic modeling of silicate systems; fluid mechanical processes in magmatic systems.

Ronald L. Parsley, Ph.D., Cincinnati, 1969. Paleontology: paleobiology, paleoecology, and evolution of lower Paleozoic primitive Echinodermata; Paleozoic faunas in general.

Hubert C. Skinner, Ph.D., Oklahoma, 1954. Micropaleontology and history of geology: systematic foraminiferal micropaleontology, coiling patterns and shell architecture, paleoecology of marine organisms, history of geology.

Emily H. Vokes, Ph.D., Tulane, 1967. Molluscan paleontology: taxonomy; biogeographic and stratigraphic relationships of Tertiary and Recent Gastropoda, particularly those of the western Atlantic and Caribbean region; geography and history.

Recent Publications

Flowers, G. C. Environmental sedimentology of the Pontchartrain Estuary. *Trans. Gulf Coast Assoc. Geological Soc.* 40:237–50, 1990.

Gill, I. P., D. K. Hubbard, P. P. McLaughlin, and C. H. Moore Jr. Sedimentological and tectonic evolution of Tertiary St. Croix. In *Terrestrial and Marine Geology of St. Croix*, Spec. Publ. no. 8, pp. 49–72, ed. D. K. Hubbard. U.S. Virgin Islands: West Indies Laboratory, 1989.

Horodyski, R. J., and C. Mankeiwicz. Possible late Proterozoic skeletal algae from the Pahrump Group, Kingston Range, southeastern California. *Am. J. Sci.* 290-A:149–69, 1990.

Hubbard, D. K., R. Burke, and I. P. Gill. Styles of accretion along a steep shelf-edge reef, St. Croix, U.S. Virgin Islands. *J. Sediment. Petrol.* 56:848–61, 1986.

Isphording, W. C., F. D. Imsand, and G. C. Flowers. Aging and sediment characteristics of northern Gulf of Mexico estuaries. *Trans. Gulf Coast Assoc. Geological Soc.* 39:387–401, 1989.

Nelson, S. A., and J. Hegre. Volcán Las Navajas, a Plio-Pleistocene trachyte/peralkaline rhyolite volcano in the northwestern Mexican Volcanic Belt. *Bull. Volcanology* 52:186–204, 1990.

Parsley, R. L. Aristocystites: A recumbent diploporid (Echinodermata) from the Middle and Upper Ordovician of Bohemia, CSSR. *J. Paleontol.* 64:278–93, 1990.

Parsley, R. L. Review of selected North American mitrate stylophorans (Homalozoa: Echinodermata). *Bull. Am. Paleontol.* 100:1–57, 1991.

Skinner, H. C. Modern paleoecological techniques: An evolution of the role of paleoecology in Gulf Coast exploration. In *Gulf Coast Geology*, vol. 3, *Biostratigraphy and Paleoecology of Gulf Coast Cenozoic Foraminifera*, pp. 53–78, 1982.

Verma, S. P., and S. A. Nelson. Isotopic and trace-element constraints on the origin and evolution of calc-alkaline and alkaline magmas in the northwestern portion of the Mexican Volcanic Belt. *J. Geophys. Res.* 94:4531–44, 1989.

Vokes, E. H. Two new species of *Chicoreus* subgenus *Siratus* (Gastropoda: Muricidae) from northeastern Brazil. *Nautilus* 103:124–30, 1990.

Vokes, E. H. Cenozoic Muricidae of the western Atlantic region. Part VII—*Murex* s.s., *Haustellum*, *Chicoreus*, and *Hexaplex*; additions and corrections: *Tulane Stud. Geol. Paleontol.* 23:1–96, 1990.

Walter, M., R. Du, and R. J. Horodyski. Coiled carbonaceous megafossils from the Middle Proterozoic of China and the USA. *Am. J. Sci.* 290-A: 133–48, 1990.

Recent Thesis Topics

"Depositional Environment and Diagenesis of Vicksburg Sandstones, Tabasco Field, Hidalgo County, Texas," Sydney A. Rasbury (1986).

"The Biostratigraphy and Paleoecology of Miocene Benthic Foraminiferida from the Salina Basin, State of Veracruz, Mexico," Betsy M. Strachan (1986).

"The Genus *Strombus* in Western Atlantic," Samuel C. Kindervater (1987).

"A Survey of Modern Peritidal Stromatolitic Mats of the Yucatan Peninsula and a Depositional Model of a Carbonate Tidal Flat in Rio Lagartos, Yucatan, Mexico," Jerry B. Pennington (1987).

"The Paleoecology of *Crepidula* (Gastropoda), James City Formation (Pleistocene), North Carolina," Diana M. Woods (1987).

"The Effect of Lithology, Diagenesis, and Low-Grade Metamorphism on the Ultrastructure and Surface Sculpture of Acritarchs from the Late Proterozoic Chuar Group, Grand Canyon, Arizona," Yvonne Halprin (1988).

"Provenance of Sandstones from the Belt Supergroup (Middle Proterozoic), Montana," Kathleen Kordesh (1988).

"Cyanobacterial Assemblages in Phanerozoic Cherts," Kenneth J. Tobin (1989).

"A Neotectonic Study Along the Rio Ameca—Implications of the Northern Boundary of the Jalisco Block of Western Mexico," Troy Rasbury (1990).

"Petrologic Study of the Volcanic Rocks in the Chiconquiaco–Palma Sola Area, Central Veracruz, Mexico," Manuel Lopez Infanzon (1991).

"The Stratigraphy and Volcanic History of Post Tertiary Volcanics, Tuxtla Volcanic Field, Veracruz, Mexico," Bentley K. Reinhardt (1991).

SECTION 3: EARTH AND PLANETARY SCIENCES

UNIVERSITY OF ARIZONA
Department of Geosciences

Programs of Study

The Department of Geosciences offers academic and research programs in field, laboratory, and theoretical studies leading to the M.S. and Ph.D. degrees in geosciences. Programs can be arranged for multidisciplinary work in such fields as planetary geology (with the Department of Planetary Sciences) and geoarchaeology (with the Department of Anthropology).

Applicants for the master's degree program must have completed the general equivalent of the baccalaureate degree with a major in geosciences or an allied discipline. Qualified students with a master's degree or a bachelor's degree may be accepted into the doctoral program.

The M.S. program requires a minimum of 30 units, 2 to 8 of which can be for work on a thesis or research report. A minimum of 48 units are required for the Ph.D., plus 18 for work on the dissertation.

A wide variety of study areas are available in the academic and research degree program. These include economic geology, geohydrology, geophysics, mineralogy-petrology-geochemistry, planetary geology, Quaternary-paleoenvironmental studies, stratigraphy-paleontology, and tectonics. Applicants are encouraged to identify faculty members with whom they may share research interests.

Research Facilities

In addition to conventional facilities and equipment, the department has modern research equipment to facilitate scientific investigations in the laboratory-intensive disciplines of isotope geochemistry, organic geochemistry, geophysics, economic geology, geochronology, mineralogy, paleomagnetism and rock magnetism, petrology, paleontology, and paleoenvironmental studies. Noteworthy among research equipment are a tandem accelerator mass spectrometer (TAMS); a VG isotopes mass spectrometer; a microprobe; X-ray diffraction and analytical facilities; a high-pressure–temperature laboratory; an SEM-EDS image analysis laboratory; a dedicated reflection seismology computer, Convex C1, for high-speed vector and scalar data processing; and petrography, mineralogy, and fluid inclusion laboratories.

The department has numerous in-house computer rooms and terminals linked to the University Computer Center's Convex C-240, IBM 3090, and DEC VAX 8650 systems. Above and beyond the advantages they derive from in-house laboratory research facilities, faculty and students benefit enormously from the exquisite geologic exposures that surround Tucson.

Financial Aid

Scholarships, fellowships, and teaching and research assistantships are available to qualified students. The assistantships are available on a quarter-time, third-time, and half-time basis. Pay scales, subject to change, ranged from $4232 to $10,086 for the academic year 1990–91. Students on teaching or research appointments are exempt from nonresident tuition but not from registration fees; however, the University makes a limited number of tuition and registration fee scholarships available to the department.

Many major mineral and fossil-fuel industrial companies annually conduct employment interviews with students in the department for summertime and permanent positions in the United States and abroad.

Cost of Study

Registration fees in 1991–92 range from $80 to $764 per semester according to the number of units taken. Nonresidents of Arizona pay an additional tuition fee, which ranges from $209 to $2703 per semester according to the number of units taken. Students on assistantship appointments have the tuition fee waived.

Cost of Living

Graduate students are generally responsible for their own living arrangements. Residence hall rates in University housing range from $1200 to $2000 per year in 1991–92. In addition, the University has a limited number of housing units for married students.

Student Group

The Department of Geosciences has 70 undergraduates and 120 M.S. and Ph.D. students. Graduate students, nearly all from out of state, come from all parts of the United States and several foreign countries.

Location

Tucson has a metropolitan population of over 600,000 and is located in the lower Santa Cruz River basin. It has a general elevation of 2,500 feet; surrounding mountains rise to about 9,000 feet. Situated in the center of Tucson, the University enjoys a warm, semiarid climate, but the mountain crests offer cool Alpine temperatures.

Tucson, rich in historical heritage, has numerous organizations sponsoring events in all of the fine arts. University programs also contribute greatly to cultural activities in the community.

The University

Authorized in 1885 by the 13th Territorial Legislature, the University of Arizona began construction of several buildings in 1886 on 40 acres of desert land donated by a local saloonkeeper and two gamblers. In 1891, the College of Agriculture, the College of Mines and Engineering, and the Agriculture Experiment Station opened their doors to students. Today, the 310-acre campus, with more than 122 buildings, serves over 31,000 students coming from all parts of the United States as well as 104 foreign countries. The University offers the bachelor's degree in 138 academic fields, the master's in 131, and the doctorate in 85. It also has over forty divisions of research and special interest.

Applying

The Department of Geosciences requires scores on the General Test of the Graduate Record Examinations, three letters of recommendation, and several other items in addition to the Graduate College application. Applications for the fall semester should be received before February 15. The Department of Geosciences accepts applications for fall term only.

Correspondence and Information

Graduate Student Coordinator
Department of Geosciences
Gould-Simpson Building
University of Arizona
Tucson, Arizona 85721
Telephone: 602-621-6004
Fax: 602-621-2672

University of Arizona

THE FACULTY AND THEIR RESEARCH

L. M. Anovitz, Assistant Professor; Ph.D., Michigan, 1987. Metamorphic/igneous petrology, mineralogy.
V. R. Baker, Regents Professor (joint appointment, Planetary Sciences); Ph.D., Colorado, 1971. Fluvial geomorphology.
S. L. Baldwin, Assistant Professor; Ph.D., SUNY at Albany, 1988. Thermochronology, $^{40}Ar/^{39}Ar$.
M. D. Barton, Associate Professor; Ph.D., Chicago, 1981. Geology, geochemistry.
S. L. Beck, Assistant Professor; Ph.D., Michigan, 1987. Seismology, tectonics.
W. B. Bull, Professor; Ph.D., Stanford, 1960. Tectonic and climatic geomorphology.
R. F. Butler, Professor; Ph.D., Stanford, 1972. Geophysics, paleomagnetics.
C. G. Chase, Professor and Head; Ph.D., California, San Diego (Scripps), 1970. Geophysics, geotectonics.
A. S. Cohen, Associate Professor; Ph.D., California, Davis, 1982. Stratigraphy, paleobiology.
P. J. Coney, Professor; Ph.D., New Mexico, 1964. Regional tectonics.
G. H. Davis, Professor and Vice Provost; Ph.D., Michigan, 1971. Structural geology.
O. K. Davis, Associate Professor; Ph.D., Minnesota, 1981. Quaternary paleoecology.
S. N. Davis, Professor (joint appointment, Hydrology and Water Resources); Ph.D., Yale, 1955. Groundwater, hydrogeology.
M J. Drake, Professor (joint appointment, Planetary Sciences); Ph.D., Oregon, 1972. Planetary geochemistry.
K. W. Flessa, Professor; Ph.D., Brown, 1972. Invertebrate paleontology, paleobiology.
J. Ganguly, Professor; Ph.D., Chicago, 1967. Petrology, geochemistry.
G. E. Gehrels, Assistant Professor; Ph.D., Caltech, 1986. Tectonics, geochronology.
J. M. Guilbert, Professor; Ph.D., Wisconsin, 1962. Mineral deposits, mineragraphy-petrology.
C. V. Haynes, Professor (joint appointment, Anthropology); Ph.D., Arizona, 1965. Geology of early man.
R. A. Johnson, Assistant Professor; Ph.D., Wyoming, 1984. Reflection seismology, crustal and lithospheric structure.
P. L. Kresan, Lecturer; M.S., Arizona, 1975. Geomorphology, environmental geology.
E. H. Lindsay Jr., Professor; Ph.D., Berkeley, 1967. Vertebrate paleontology.
A. Long, Professor; Ph.D., Arizona, 1966. Environmental isotope geochemistry, hydrogeochemistry, paleoclimatology.
E. J. McCullough Jr., Professor and Dean, Faculty of Science; Ph.D., Arizona, 1963. Environmental geology.
H. J. Melosh, Professor (joint appointment, Planetary Sciences); Ph.D., Caltech, 1972. Planetary geophysics.
B. S. Nagy, Professor; Ph.D., Penn State, 1953. Organic geochemistry.
D. L. Norton, Professor; Ph.D., California, Riverside, 1964. Mineral deposits, geochemistry.
J. T. Parrish, Associate Professor; Ph.D., California, Santa Cruz, 1979. Paleoclimatology.
P. J. Patchett, Professor (joint appointment, Arizona Research Laboratories); Ph.D., Edinburgh, 1976. Isotope geochemistry.
J. Quade, Assistant Professor; Ph.D., Utah, 1990. Soil geochemistry.
R. M. Richardson, Associate Professor; Ph.D., MIT, 1978. Solid-earth geophysics.
J. Ruiz, Associate Professor; Ph.D., Michigan, 1983. Geochemistry.
J. F. Schreiber Jr., Professor; Ph.D., Utah, 1958. Sedimentology, sedimentary petrography.
R. B. Singer, Associate Professor (joint appointment, Planetary Sciences); Ph.D., MIT, 1980. Planetary geology image processing.
E. A. Snow, Assistant Professor; Ph.D., Brown, 1987. Mineralogy, mineral kinetics.
S. R. Titley, Professor; Ph.D., Arizona, 1958. Mineral deposits, regional geology.
T. C. Wallace, Associate Professor; Ph.D., Caltech, 1983. Seismology.

Professors Emeriti
J. W. Anthony, Ph.D., Harvard, 1965. Mineralogy, crystallography.
P. E. Damon, Ph.D., Columbia, 1957. Geochronology, Cordilleran geology.
W. R. Dickinson, Ph.D., Stanford, 1958. Sedimentary geology, tectonics.
L. M. Gould, D.Sc., Michigan, 1925. Glaciology.
J. W. Harshbarger, Ph.D., Arizona, 1949. Hydrogeology.
G. O. W. Kremp, Dr.rer.nat., Posen (Germany), 1945. Palynology-paleobotany.
P. S. Martin, Ph.D., Michigan, 1956. Paleoecology.
T. L. Smiley, M.A., Arizona, 1949. Geochemistry, paleoclimatology.
J. S. Sumner, Ph.D., Wisconsin, 1955. Geophysical exploration.

Adjunct Professors
A. K. Behrensmeyer, Ph.D., Harvard, 1973. Vertebrate paleontology, stratigraphy.
A. A. Brant, Ph.D., Berlin, 1936. Geophysics.
C. P. Miller, Ph.D., Stanford, 1957. Mineral exploration.
E. L. Ohle, Ph.D., Harvard, 1950. Economic geology.
J. H. Stewart, Ph.D., Stanford, 1961. Stratigraphy, tectonics.
R. M. Turner, Ph.D., Washington State, 1954. Paleobiology.
T. R. Van Devender, Ph.D., Arizona, 1973. Geochronology, paleontology.
A. T. Wilson, Ph.D., Berkeley, 1954. Isotope geochemistry.
J. A. Wolfe, Ph.D., Berkeley, 1960. Paleontology.
K. L. Zonge, Ph.D., Arizona, 1972. Electrical engineering.

Visiting Professors
G. H. Davis, Professor at University of Vermont; Ph.D., Michigan, 1971. Structural geology.
F. Ortega-Gutierrez, Professor at Universidad Nacional Autónoma de México; Ph.D., Leeds (England), 1975. Regional tectonics.
T. Lawton, Assistant Professor at New Mexico State University; Ph.D., Arizona, 1983. Stratigraphy.
S. J. Reynolds, Associate Professor at Arizona State University; Ph.D., Arizona, 1982. Tectonics.

Research Scientists
C. J. Eastoe, Ph.D., Tasmania, 1979. Geochemistry, mineral deposits.
A. J. T. Jull, Ph.D., Bristol, 1972. Geochemistry.
R. Kra, M.A., NYU, 1979. Anthropology.
M. Shafiqullah, Ph.D., Carleton (Ottawa), 1970. Geochronology.

Adjunct Associate Research Scientists
J. L. Betancourt, Ph.D., Arizona, 1990. Climatic change, geomorphology.
R. H. Webb, Ph.D., Arizona, 1985. Geomorphology, climatic change.

Adjunct Assistant Research Scientists
P. A. Pearthree, M.S., Arizona, 1982. Geomorphology.
J. E. Spencer, Ph.D., MIT, 1981. Structural geology.

Research Associates
L. M. Cranwell, Ph.D., Auckland (New Zealand), 1959. Palynology.
J. A. White, Ph.D., Kansas, 1953. Vertebrate paleontology.

Visiting Scholar
A. Roe, Ph.D., Northwestern, 1938. Mineralogy, chemistry.

SECTION 3: EARTH AND PLANETARY SCIENCES

UNIVERSITY OF ARIZONA

Department of Planetary Sciences/Lunar and Planetary Laboratory

Program of Study

The graduate program prepares students for careers in solar system research. Recognizing that the study of the solar system is fundamentally an interdisciplinary enterprise, the department maintains faculty expertise in several important areas of planetary science. As a part of the Ph.D. program, the department requires its students to complete a minor consisting of at least 12 units in a scientific area relevant to planetary science. The purpose of the minor is to deepen the student's knowledge of a subject that will support his or her professional career. The minor may be fulfilled in another department or approved program of the University or in the Department of Planetary Sciences. The combination of core courses, minor requirements, and interaction with faculty/research personnel provides students with comprehensive training in modern planetary science. The program is oriented toward granting the Ph.D., although M.S. degrees are awarded in special circumstances.

Upon admission, a student is assigned an adviser in his or her general scientific area. At that time, the student may be appointed to a half-time research assistantship involving duties in a research project. Early course work consists primarily of core planetary sciences courses and courses in other departments as dictated by the choice of the minor. Students advance to Ph.D. candidacy by passing a written and oral preliminary examination after completing the required course work. The examination is normally taken 2 to 2½ years after matriculation. Students typically complete their dissertations and receive the Ph.D. 2 to 3 years later. Because of the low student-faculty ratio, students receive close supervision and guidance from their advisers and committee members. Possible dissertation areas include, but are not limited to, observational planetary astronomy; physics of the sun and interplanetary medium; observational, experimental, and theoretical studies of planetary atmospheres, surfaces, and interiors; studies of the interstellar medium and the origin of the solar system; and the geology and chemistry of the surfaces and interiors of solar system bodies. Graduates typically obtain employment with NASA and other national laboratories, universities, or industry.

Research Facilities

The Lunar and Planetary Laboratory (LPL), founded by Gerard P. Kuiper in 1960, is the primary research organization for solar system research and education at the University of Arizona. The department and laboratory function as a unit and are housed in the Gerard P. Kuiper Space Sciences Building on the campus. Neighboring facilities include the Tucson headquarters of the National Optical Astronomy Observatory (formerly known as Kitt Peak National Observatory), the National Radio Astronomy Observatory, the Steward Observatory, the Optical Sciences Center, the Flandrau Planetarium, and the Planetary Sciences Institute. The facilities of the University observatories are available to researchers in the LPL. These instruments include the multiple-mirror telescope (six 1.8-m mirrors) on Mt. Hopkins, the 2.3-m telescope on Kitt Peak, two 1.5-m telescopes and a 1-m telescope on Mt. Lemmon, and several smaller telescopes. The University is developing a new observatory on Mt. Graham. A variety of telescopes are planned, including an 8-m and a submillimeter telescope. For cosmochemical research, the LPL operates a scanning electron microscope and a Cameca XF50 microprobe, high-temperature furnaces and high-pressure equipment for rock-melting experiments, and a radiochemistry separation and gamma-ray spectrometry facility for neutron activation analysis. These facilities are used for studying meteorites, lunar samples, and terrestrial analogues. Also available are a well-equipped electronics shop, a machine shop, and darkroom facilities. The Space Imagery Center at the LPL is one of several regional facilities supported by NASA as a repository for images of planets and satellites obtained by spacecraft, as well as for topographical and geologic maps produced from such imagery. By agreement with NASA, this collection continues to be updated as material from current missions becomes available. LPL's Planetary Imaging Research Laboratory is a modern image processing facility for the analysis of planetary and astronomical data. The Laboratory maintains an extensive computer network, including central disk servers and numerous workstations. Various research groups maintain specialized computer systems for data taking and analysis and for theoretical computations. The Laboratory network is connected through campus facilities to national and international networks for communications and for interaction with remote computers. University central computing facilities include a variety of systems and network facilities and several Convex superminicomputers.

Financial Aid

Most planetary sciences graduate students receive graduate research assistantships for the academic year. These assistantships normally require 20 hours of work per week on a sponsored research project. For the nine-month academic year, such assistantships paid $9500. For students who passed their preliminary examination and were advanced to Ph.D. candidacy, the pay was increased to $10,800. In addition, most students work full-time (40 hours per week) on research projects during the summer term.

Cost of Study

For 1991–92, fees for Arizona residents taking 1–6 units are $80 per unit. For 7 or more units the cost is $764. Out-of-state students who do not have a research or teaching assistantship are also charged tuition.

Cost of Living

Typical costs for off-campus housing, food, and entertainment for a single graduate student total about $500–$750 per month. Several dormitories are available for single graduate students; Christopher City is a University-run apartment complex and is available for married and single-parent graduate students.

Student Group

In 1990–91, there were 35 graduate students enrolled in the planetary science program. Most come from undergraduate or M.S. programs in chemistry, physics, geology, or astronomy, and some have been employed for several years prior to entering graduate school. In addition, approximately 20 students from other departments conduct their Ph.D. dissertation research in LPL.

Location

Tucson is located in the Sonoran Desert, about 100 kilometers north of the Mexican border. The climate is dry and warm and is favorable for outdoor activities during the entire year. Hiking, mountain climbing, horseback riding, swimming, and tennis are popular year-round activities.

The University

The University is a state-supported institution with an enrollment of approximately 35,000. In addition to the planetary sciences program, major research efforts and graduate programs exist in chemistry, engineering, physics, astronomy, optical sciences, and geosciences. LPL interacts closely with these groups.

Applying

Completed application forms, three letters of reference, and GRE scores must be received by January 15 in order to receive full consideration. All applicants are required to submit GRE General Test scores as well as the Subject Test score in a physical science or other relevant area.

Correspondence and Information

Chairman, Admissions Committee
Department of Planetary Sciences
University of Arizona
Tucson, Arizona 85721
Telephone: 602-621-6963

Peterson's Guide to Graduate Programs in the Physical Sciences and Mathematics 1992

SECTION 3: EARTH AND PLANETARY SCIENCES

University of Arizona

THE FACULTY AND THEIR RESEARCH

Victor R. Baker, Professor; Ph.D., Colorado, 1971. Planetary geomorphology.
William V. Boynton, Professor; Ph.D., Carnegie-Mellon, 1971. Trace-element cosmochemistry and geochemistry, chemistry of planetary bodies by spacecraft remote sensing and in situ instrumentation, neutron activation analysis.
Lyle A. Broadfoot, Senior Research Scientist; *Voyager* Experiment Principal Investigator; Ph.D., Saskatchewan, 1963. Ultraviolet spectroscopy, planetary atmospheres, instrument development.
Michael J. Drake, Professor; Ph.D., Oregon, 1972. Experimental and theoretical geochemistry, petrology and geochemistry of lunar samples and meteorites.
Uwe Fink, Professor; Ph.D., Penn State, 1965. Infrared Fourier spectroscopy, planetary atmospheres.
Tom Gehrels, Professor; Ph.D., Chicago, 1956. Asteroid astronomy, survey and origin of the solar system.
Richard J. Greenberg, Professor; Ph.D., MIT, 1972. Celestial mechanics, studies of planetary accumulation, satellite and ring dynamics.
Jay B. Holberg, Associate Research Scientist; Ph.D., Berkeley, 1974. Far-UV spectra, interstellar medium, planetary rings and atmospheres.
Lon L. Hood, Associate Research Scientist; Ph.D., UCLA, 1979. Geophysics, space physics, solar-terrestrial physics.
William B. Hubbard, Professor; Ph.D., Berkeley, 1967. High-pressure physics, planetary interiors and atmospherics.
Donald M. Hunten, Professor; Ph.D., McGill, 1950. Earth and planetary atmospheres, composition, structure, aeronomy.
J. R. Jokipii, Professor; Ph.D., Caltech, 1965. Theoretical astrophysics, cosmic rays, solar wind, astrophysical plasmas.
Harold P. Larson, Research Professor; Ph.D., Purdue, 1967. High-resolution interferometry, infrared Fourier spectroscopy, planetary atmospheres.
Larry A. Lebofsky, Senior Research Scientist; Ph.D., MIT, 1974. Planetary astronomy.
Eugene H. Levy, Professor, Head of the Department, and Director of the Laboratory; Ph.D., Chicago, 1971. Theoretical astrophysics, generation and behavior of magnetic fields, solar physics, cosmic rays.
John S. Lewis, Professor; Ph.D., California, San Diego, 1968. Cosmochemistry, planetary atmospheres.
Jonathan I. Lunine, Associate Professor; Ph.D., Caltech, 1985. Theoretical planetary physics, condensed-matter studies, structure of planets.
Robert S. McMillan, Assistant Research Scientist; Ph.D., Texas, 1977. Doppler spectroscopy and asteroid detection survey.
H. Jay Melosh, Professor; Ph.D., Caltech, 1972. Planetary geophysics, planetary surfaces.
Carolyn C. Porco, Associate Professor; Ph.D., Caltech, 1983. Planetary rings, image processing.
George H. Rieke, Professor; Ph.D., Harvard, 1969. Gamma-ray, infrared astronomy, cosmic radiation.
Elizabeth Roemer, Professer; Ph.D., Berkeley, 1955. Comets, minor planets, astrometry.
Bill R. Sandel, Associate Research Scientist; Ph.D., Rice, 1972. Astrophysical plasmas, planetary atmospheres, airglow emissions, instrument development.
Donald E. Shemansky, Senior Research Scientist; Ph.D., Saskatchewan, 1965. Plasma physics, physical chemistry, atomic and molecular physics, planetary atmospheres.
Robert B. Singer, Associate Professor; Ph.D., MIT, 1980. Planetary geology and image processing, remote studies of surface composition.
Charles P. Sonett, Professor; Ph.D., UCLA, 1954. Planetary physics, solar wind.
Robert G. Strom, Professor; M.S., Stanford, 1957. Lunar and planetary surfaces, spacecraft imaging of planetary surfaces.
Timothy D. Swindle, Assistant Professor; Ph.D., Washington (St. Louis), 1986. Cosmochemistry, noble gas studies of meteorites.
William C. Tittemore, Assistant Professor; Ph.D., MIT, 1988. Solar system dynamics, chaotic dynamical systems.
Martin G. Tomasko, Research Professor; Ph.D., Princeton, 1969. Planetary atmospheres, radiative transfer theory.
Ann Vickery, Assistant Research Scientist; Ph.D., SUNY at Stony Brook, 1984. Impact cratering, computer modeling.
Roger V. Yelle, Assistant Research Scientist; Ph.D., Wisconsin–Madison, 1984. Planetary atmospheres, radiative transfer, atmospheric transport processes.

Space Sciences Building, which houses the Department of Planetary Sciences and the Lunar and Planetary Laboratory.

SECTION 3: EARTH AND PLANETARY SCIENCES

UNIVERSITY OF CALGARY

Department of Geology and Geophysics

Programmes of Study The department offers programmes leading to the M.Sc. and Ph.D. degrees in geology and geophysics. All programmes require a thesis. Research areas in geology include economic geology, geochemistry, igneous and metamorphic petrology, macropaleontology, micropaleontology, mineralogy, palynology, petroleum geology, quantitative geology, Quaternary geology and geomorphology, sedimentology, stratigraphy, and structural geology. Research areas in geophysics include exploration geophysics (seismic, potential, and electromagnetic methods), seismology, and geodynamics.

The M.Sc. and Ph.D. programmes are open to students with a previous degree in geology, geophysics, or a field related to geophysics (such as physics, mathematics, or engineering).

Research Facilities The department is well equipped for field and laboratory research in stratigraphy, paleontology, structure, sedimentation, geochemistry, petrography, and exploration geophysics. It has large paleontology and micropaleontology collections; the Ellis and Messina Catalogue of Foraminifera; automated X-ray diffraction instrumentation; Gandolfi and Guinier cameras; mobile chemical analysis equipment; cathodoluminescence instrumentation; hydrothermal synthesis equipment; a DFS III and Sercel 338HR seismic recording unit; a SQUID magnetometer; a seismic physical modelling facility; gravimetric instrumentation; potentiometric and electromagnetic prospecting equipment; an electromagnetic scale-model tank; a core tester; a rock press; a porometer and porosimeter; mercury-injection equipment for capillary-pressure measurement; core and micro flow testing equipment; gamma spectrometer and electromagnetic modelling equipment; and a Logitec thin-section grinder. A Cambridge 250 scanning electron microscope, equipped with an energy-dispersive X-ray unit, is available in the department. Stable isotope analysis facilities are available in the Department of Physics. In addition to the basic equipment for petrological studies, the department has an automated ARL-SEMQ electron microprobe, two wet chemistry laboratories and an automatic XRF laboratory, a Perkin-Elmer 5000 atomic absorption unit with graphite furnace, an advanced microscopy laboratory with a fluid-inclusion heating-freezing stage, a Nomarski interference microscope, a laser interferometer, and an image analyzer. A personal computer-based geological mapping laboratory, equipped with digitizing and plotting facilities, is available for support of field research and GIS development.

The department operates a Perkin-Elmer 3242 computer and peripherals for seismic data processing and computing and numerous IBM personal computers for graduate student use. The department also houses the data processing centre for the national LITHOPROBE project, which operates a CDC CYBER 835 mainframe computer and a MAP V array processor. An industry-sponsored seismic exploration research effort within the department (the CREWES project) operates an IBM 4381 mainframe computer, an IBM 3838 array processor, and a Landmark seismic interpretation workstation. Various standard geophysical software packages are available for use on the department's computers. The University operates a Bull DPS 8/70M mainframe computer, a CDC CYBER 860 mainframe computer, and a number of other minicomputers and microcomputers.

Research and study facilities are available in the Gallagher Library of Geology and Geophysics (in the Earth Sciences Building) and the main University Library.

Financial Aid A varying number of departmental graduate teaching assistantships and graduate research scholarships are available. A number of University scholarships and fellowships are also available. The starting stipend for a graduate teaching assistantship is $9760 for September through April. Such an assistantship also entitles the student to remission of programme fees. Graduate research scholarships, valued at $3670 each, are usually available for support for May through August.

Cost of Study In 1990–91, programme fees were $1270 per year, and general fees were $223.50 per year. Visa students are required to pay an additional tuition fee of one half the programme fee each year. Fees are subject to increase without notice.

Cost of Living The cost of living in Calgary is reasonable. Information regarding housing on campus may be obtained from the Assistant Manager, University Housing.

Student Group The total University enrolment is about 20,000 students, of whom about 2,500 are graduate students. There are typically 50–60 graduate students in the Department of Geology and Geophysics.

Location Calgary is located 80 kilometres east of the Rockies and has a population of 693,000. Skiing and hiking in the nearby mountains are popular activities. The city, host of the 1988 Winter Olympic Games, features superb athletic, cultural, and recreational facilities.

The University and The Department The campus is located in the northwest section of Calgary and has well-equipped facilities for teaching, research, continuing education, and recreation.

The Department of Geology and Geophysics is close to the Institute of Sedimentary and Petroleum Geology of the Geological Survey of Canada and to the Core Storage Centre of the Energy Resources Conservation Board.

Applying Students are urged to apply as early as possible. Most of the offers for fall admission are made between February and April. The deadline for applications for University scholarships is February 1.

Correspondence and Information Enquiries and applications for assistantships and scholarships should be addressed to:
Graduate Advisor
Department of Geology and Geophysics
University of Calgary
Calgary, Alberta T2N 1N4
Canada
Telephone: 403-220-5841

SECTION 3: EARTH AND PLANETARY SCIENCES

University of Calgary

THE FACULTY AND THEIR RESEARCH

R. J. Brown. Seismic anisotrophy, theory, and physical modelling; geophysics of salt and carbonate units; earthquake seismology; inductive coupling in induced polarization.

F. A. Cook. Seismic reflection, crustal structure and evolution, Precambrian geology.

K. Duckworth. Mining geophysics, model studies in electromagnetic interpretation, induced-polarization studies of sulphides.

E. D. Ghent. Geochemistry and petrology of metamorphic rocks, electron microprobe analysis, application of physical chemistry to petrology.

T. M. Gordon. Theoretical and numerical techniques applied to chemical thermodynamics and transport processes in geology, field studies of metamorphic rocks.

P. E. Gretener. Petrophysics, structural geology, geothermics. Current interests: role of fractures in oil reservoirs, lateral continuity of sedimentary lithological units, evidence for vertically stacked stress regimes.

R. L. Hall. Taxonomy, biostratigraphy, and biogeography of Jurassic ammonites in western Canada and the circum-Pacific.

C. M. Henderson. Taxonomy, biostratigraphy, and paleoecology of Upper Paleozoic conodonts; Paleozoic and Mesozoic stratigraphy and biostratigraphy.

L. V. Hills. Devonian miospores, megaspores; Carboniferous, Triassic, Jurassic, Cretaceous, and Tertiary miospores, taxonomy, biostratigraphy, and paleoecology; dinoflagellate and acritarch biostratigraphy, Pleistocene geology-palynology, stratigraphy; clastic sedimentology.

J. C. Hopkins. Sedimentology: petrography and facies analysis, modern and ancient platform-interior carbonates (Turks and Caicos Islands, western Canada), Mesozoic siliciclastic paleochannels (Alberta and Montana).

I. E. Hutcheon. Low-temperature inorganic geochemistry and diagenesis of clastic and carbonate rocks.

F. F. Krause. Sedimentology, sequence stratigraphy, and petroleum reservoir geology of Cretaceous and Carboniferous carbonate and siliciclastic rocks (Western Canada Sedimentary Basin), modern platform carbonates (Turks and Caicos Islands, British West Indies).

E. S. Krebes. Theoretical seismology, theoretical and computational studies of seismic wave propagation in anelastic media.

D. C. Lawton. Exploration seismology, multicomponent seismic techniques, seismic modelling; applications of integrated seismic, gravity, and magnetic methods in geological studies and in resource exploration.

A. A. Levinson. Exploration and environmental geochemistry; exploration for uranium, gold, base metal, and diamond deposits.

J. Nicholls. Igneous petrology, application of physics and thermodynamics to petrology, mineralogy, and crystallography.

A. E. Oldershaw. Genesis and diagenesis of carbonate sediments and sedimentary rocks, applications of the scanning electron microscope and X-ray energy dispersion analysis in sedimentary petrology; burial diagenesis and the evaluation of carbonate cements; reservoir characterization and behavior of siliciclastic and carbonate reservoir rocks under conditions of heavy-oil steam extraction; problems of secondary and tertiary recovery in conventional hydrocarbon reservoirs.

G. D. Osborn. Holocene/late Pleistocene glacial chronology and paleoclimatology, tephrostratigraphy, glacial geology; Tertiary stratigraphy and geomorphology of the North American plains; desert geomorphology.

D. R. M. Pattison. Metamorphism: formation of granulites, partial melting and granitoid genesis, fluid interactions in the crust, geothermometry and geobarometry, contact metamorphism.

P. S. Simony. Structural geology: polyphase folding and thrust faulting in Cordilleran core zone, structural analysis of plutons, Proterozoic stratigraphy.

R. J. Spencer. Evaporites, geochemistry and sedimentology.

D. A. Spratt. Structural geology, geometry, and mechanical origin of thrust faults and folds, Rocky Mountain foothills, and front ranges; strain analysis.

R. R. Stewart. Exploration seismology: vertical seismic profiling, multicomponent seismic analysis, tomography; Signal processing: median filters, inverse theory.

N. C. Wardlaw. Petroleum reservoir geology, petrophysics, multiphase fluid flow in porous media, petroleum reservoir production.

P. Wu. Geodynamics: rheology of the mantle, model deformations of the earth.

SECTION 3: EARTH AND PLANETARY SCIENCES

UNIVERSITY OF CALIFORNIA, LOS ANGELES

Department of Earth and Space Sciences

Programs of Study The Department of Earth and Space Sciences offers programs leading to the M.S. and Ph.D. degrees in geochemistry, geology, and geophysics and space physics. The program in geochemistry offers study in biogeochemistry, organic geochemistry, crystal chemistry, experimental petrology, isotopic studies of stable and radioactive elements, marine geochemistry, meteorite research, planetology, and lunar geochemistry. The program in geology includes geological physics, geomorphology, mineral deposits, mineralogy, nonrenewable natural resources, paleobiology, petrology, sedimentology, stratigraphy, structural geology, and tectonophysics. The program in geophysics and space physics offers study in geophysical fluid dynamics (turbulence, rotating systems, stability, and hydromagnetism), geophysics (seismology, gravity, thermal regime, geomagnetism, and tectonics), planetology (orbital dynamics, planetary interiors, surfaces and atmospheres, and solar-system origin), and space physics (magnetosphere, radiation belts, solar wind, magnetic fields, and cosmic rays). Many other areas of study are possible.

Research Facilities Research facilities in the earth, space, and planetary sciences at UCLA are in the geology, space sciences, and chemistry buildings, which are interconnected. There is very extensive equipment and support staff to assist faculty, students, and postdoctorals in their research. The geology building houses one of the largest collections of geoscience/space science literature and maps in the world, with complete literature-search capabilities; it is down the hall from the equally outstanding chemistry library. Specialized equipment available to all students includes a complete personal computer laboratory, two computer-automated electron microprobes, ICP and atomic-absorption spectrometers, a neutron-activation facility, X-ray diffractometers and XRF units, Mössbauer spectrometers, electron microscopes (SEM, TEM, STEM), numerous mass spectrometers and clean labs for stable and radioactive isotopes, extensive high-temperature and high-pressure facilities for experimental geochemistry and petrology, an outstanding planetary instrumentation laboratory, a modern planetary remote-sensing facility, complete field gear for exploration geophysics and GPS geodesy, and complete machine and electronics shops.

Financial Aid Most graduate students in the department receive some type of support. Research and teaching assistantships require 20 hours of work per week and carry a stipend of $1064 to $1494 per month. Fellowship support is also available, and some grants are provided to cover tuition and fees.

Cost of Study Fees for California residents are $2907 per year. Nonresidents pay an additional $7699. U.S. citizens normally are considered residents after one year.

Cost of Living A typical annual budget for a single student (exclusive of tuition) ranges from $11,000 to $13,000. This includes room and board, transportation, and academic and personal expenses. There is a dormitory on campus for single students, and the University owns apartment buildings that accommodate married students and single parents.

Student Group UCLA enrolls about 36,000 students, of whom approximately 11,000 are graduate students. The Department of Earth and Space Sciences has approximately 85 graduate students and 40 undergraduate majors. The Earth and Space Sciences Student Organization, which includes both graduate and undergraduate students, organizes weekly lectures, seminars, and social functions, as well as field trips.

Location Located in Westwood at the base of the Santa Monica Mountains near the Pacific Ocean, UCLA is in an area unmatched anywhere in the world in geologic diversity, with great exposure and access and a Mediterranean climate. It is proximal to all of the southern California centers for aerospace research and planetary exploration. Bounded by Beverly Hills, Bel Air, and Santa Monica, UCLA offers a pleasant and stimulating academic environment.

The University and The Department UCLA is the only top-ranked university in the country founded in the twentieth century. The Department of Earth and Space Sciences offers many diverse areas for graduate study, including geochemistry, geology, geophysics, paleontology, planetology, and space physics.

Applying Application may be made for the fall, winter, or spring quarter. Students needing financial support should apply in December of the preceding year. Applicants will be notified of the results before March 15. The GRE General Test is required.

Correspondence and Information
Graduate Adviser
Department of Earth and Space Sciences
University of California, Los Angeles
Los Angeles, California 90024-1567

Peterson's Guide to Graduate Programs in the Physical Sciences and Mathematics 1992

SECTION 3: EARTH AND PLANETARY SCIENCES

University of California, Los Angeles

THE FACULTY AND THEIR RESEARCH

Orson L. Anderson, Professor of Geophysics. Mineral physics, elastic properties and phase transitions, equations of state, planetary interiors, structure of Earth's core.
Peter Bird, Professor of Geophysics and Geology. Modeling of deformation of the lithosphere, Tertiary orogenies in the western United States, subduction.
Donald Carlisle, Professor Emeritus of Geology and Mineral Resources. Mineral deposits, environments of uranium deposition.
John M. Christie, Professor of Geology. Geology and structural analysis of metamorphic rocks, deformation of quartz and other minerals, transmission electron microscopy of minerals.
Paul J. Coleman, Professor of Geophysics and Space Physics. Physics of plasmas in space, magnetic fields of stars and planetary bodies, cosmic rays.
Jon P. Davidson, Assistant Professor of Geology. Geochemistry of volcanic rocks, particularly at convergent margins; crust-magma interaction; evolution of the continental crust; isotope studies of sedimentary provenance.
Paul M. Davis, Professor of Geophysics. Physical volcanology, tectomagnetism, seismology, earth tides, tectonic deformation, dislocation theory.
Wayne A. Dollase, Professor of Geology. Mineralogy, crystal chemistry, crystal structure analysis; relation of structure to properties; phase transformations; Mössbauer spectroscopy.
Clarence A. Hall Jr., Professor of Geology and Director, White Mountain Research Station, eastern California. Structural geology of the Coast Ranges and White Mountains in California.
T. Mark Harrison, Professor of Geochemistry. Application of isotopic techniques to analysis of tectonothermal events, geochemistry of noble gases, experimental and theoretical studies of the emplacement of granitoid magmas, diffusion theory, geochronology.
Robert E. Holzer, Professor Emeritus of Space Physics. Magnetospheric energy supply processes.
Raymond V. Ingersoll, Professor of Geology. Sedimentation and tectonics, basin analysis, stratigraphy, sandstone petrology.
David D. Jackson, Professor of Geophysics and Geology. Seismology, inverse problems and interpretation theory, crustal deformation, earthquake prediction, GPS geodesy.
Isaac R. Kaplan, Professor of Geology and Geochemistry. Biogeochemistry of Recent sediments, origin of petroleum; isotope geochemistry of terrestrial rocks, environmental geochemistry.
William M. Kaula, Professor of Geophysics. Planetary and orbital dynamics, planetary evolution, mantle convection, geodesy.
Margaret G. Kivelson, Professor of Space Physics. Physics of particles and fields in the magnetospheres of Earth and Jupiter, models of geomagnetic activity.
Helen Tappan Loeblich, Professor Emerita of Paleontology. Micropaleontology, systematics of foraminifera.
Craig E. Manning, Assistant Professor of Geology and Geochemistry. Metamorphic petrology and geochemistry, fluid-rock interaction, fluid production and migration in the crust, petroleum and ore deposits.
Charles Marshall, Assistant Professor of Paleontology. DNA-DNA hybridization, molecular evolution, phylogeny.
Robert L. McPherron, Professor of Space Physics and Geophysics. Physics of the magnetosphere, exploration geophysics, magnetic field measurements.
Arthur Montana, Professor of Geochemistry and Department Chairman. High-pressure and high-temperature geochemistry and petrology.
Clemens A. Nelson, Professor Emeritus of Geology. Cambrian and Precambrian stratigraphy, geology of eastern California and the Great Basin.
William I. Newman, Associate Professor of Planetary Physics, Astronomy, and Mathematics. Planetary physics, nonlinear dynamics and computation.
Gerhard Oertel, Professor Emeritus of Geology. Structural geology, mechanics of slaty cleavage, mesoscopic and X-ray fabrics.
David A. Paige, Assistant Professor of Planetary Science. Terrestrial planets; the surface, atmosphere, polar caps, and climate of Mars; remote sensing and spacecraft exploration.
Walter E. Reed, Associate Professor of Geology. Stratigraphy, sedimentology, and sedimentary petrology; development and sedimentation in circum-Arctic Paleozoic basins.
Mary R. Reid, Assistant Professor of Geochemistry. Geochemistry of the continents, volcanic arcs, ion microprobe related to crystal growth and kinetics.
John L. Rosenfeld, Professor Emeritus of Geology. Rates of metamorphic processes, Appalachian and Alpine geology.
Bruce N. Runnegar, Professor of Paleontology. Evolution of Mollusca, origin and early history of the Metazoa, molecular evolution.
Christopher T. Russell, Professor of Geophysics and Space Physics. Planetary magnetism, space plasma physics, planetary magnetospheres.
J. William Schopf, Professor of Paleobiology. Paleobiology-evolution of Precambrian organisms, evolutionary biology, atmospheric evolution.
Gerald Schubert, Professor of Geophysics and Planetary Physics. Geophysical fluid dynamics, planetary interiors, mantle convection, planetary atmospheres.
Ronald L. Shreve, Professor of Geology and Geophysics. Geomorphology, glaciology, geological physics, theoretical fluvial geomorphology, sediment transport.
John T. Wasson, Professor of Geochemistry and Chemistry. Cosmochemistry, origin and composition of meteorites, neutron activation, major impacts on the Earth.
An Yin, Assistant Professor of Geology. Structural geology, tectonics, mechanics of the continental crust, thrust systems and detachment fault systems.

SECTION 3: EARTH AND PLANETARY SCIENCES

UNIVERSITY OF CALIFORNIA, SANTA BARBARA

Department of Geological Sciences

Programs of Study	The department offers programs leading to the M.A. and Ph.D. degrees in geological sciences (including geophysics) and the M.S. degree in geophysics. The geological sciences program offers study in geochemistry, geomorphology and Quaternary geology, paleobiology, petrology, sedimentology and stratigraphy, structural geology and crustal evolution, and tectonics. The program in geophysics offers study in magnetism, marine geology and geophysics, and seismotectonics. The geophysics group places strong emphasis on geological applications of geophysics, fieldwork, and close interaction with the rest of the geological sciences faculty. The Marine Science Institute on campus coordinates interdisciplinary studies with other departments in five marine areas: marine/aquatic biology (Biological Sciences), marine geology and geophysics (Geology), physical oceanography/remote sensing (Geography), ocean management and policy (Economics), and ocean engineering (Mechanical and Environmental Engineering).
Students determine their program of study in consultation with their major professor and advisory committee in accordance with their special interests and needs. A departmental written comprehensive project and an oral examination on the project are required of all students. Master's students normally submit a thesis but may (with department approval) elect to obtain the degree by written examination. In addition to the comprehensive exam, Ph.D. students must also pass an oral examination before being advanced to candidacy and must present a dissertation defense prior to submission of the dissertation.
Research Facilities	Major equipment includes an electron microprobe and a scanning electron microscope, both with energy-dispersive analytical systems; X-ray diffraction and fluorescence apparatus; mass spectrometers for analyses of rare gases and radiogenic isotopes; an atomic absorption spectrometer; a magnetic shielded room with spinner and cryogenic magnetometers; a LaCoste-Romberg land gravity meter; Elsec proton magnetometers; various rheometric devices; and seismic refraction equipment. Also available are petrographic and biological optical equipment, a cathodoluminescence scope, photomicroscopes, and a variety of computers, including microcomputers, MicroVAX II workstations, and a VAX-11/750 with laser printers, a 36-inch color electrostatic plotter, several powerful color graphics workstations, and high-density tape drives. Graduate students also have access to all NSF supercomputer centers, plus network connections to almost all other major computing facilities in the United States. The Preston Cloud Research Laboratory is equipped with a scanning electron microscope, an electron microprobe, and a variety of optical equipment. There is a transmission electron microscope available in the Department of Biological Sciences. Graduate students have the opportunity to interact with faculty and professional researchers affiliated with the Institute for Crustal Studies (ICS), which is an organized research unit funded by various sources, including NSF, USGS, and various private sources. Available at ICS are a Sun minicomputer, two VAX workstations, an electronics lab, and other facilities. Supporting laboratories include those for mineral separation, wet chemical and thin-section work, and fossil preparation; there is also a photographic darkroom. Construction of specialized equipment is done by personnel in a machine shop, an electronics lab, and a woodworking shop. The Sciences and Engineering Library contains a large and growing collection of serials as well as a large collection of topographic and geologic maps of the world and a comprehensive collection of high-altitude aerial, ERTS, and LANDSAT imagery. The department has modern and sophisticated oceanographic instrumentation and equipment, including high-resolution profiling sensors and bottom-sampling gear. Through the Scripps Institution, ships and sophisticated sonar systems such as Deep Tow and Seabeam are available.
Financial Aid	Department support in the form of teaching and research assistantships, fellowships, and nonresident tuition fellowships are available and are awarded on a competitive basis.
Cost of Study	In 1990–91, fees were assessed at $558.66 per quarter, or $1676 for the academic year. Students who were not residents of the state of California were charged an additional $6416 for nonresident tuition for the year ($2138.66 per quarter). U.S. citizens may establish California residence after one year.
Cost of Living	The average cost of shared living arrangements is $425 per month for rent. On-campus housing, including married-student apartment complexes, is available through the Residential Contracts Office, 1501 Residential Services Building (telephone: 805-893-4021). Help in obtaining privately owned off-campus housing may be arranged for through the Community Housing Office (telephone: 805-893-4371). University-owned off-campus housing is available through the Office of Apartment Living (telephone: 805-893-4501).
Student Group	The department currently has 68 graduate students (22 women and 46 men) from colleges and universities across the United States and from various foreign countries. More than half are pursuing the Ph.D. degree.
Location	Located on the coast of southern California, Santa Barbara offers a wide diversity of cultural and recreational activities. UCSB is ideally located for tectonic, geophysical, petrogenic, stratigraphic, geomorphic, and biogeologic research in the western Cordillera and Pacific borderlands.
The University and The Department	UCSB is a major research institution with about 18,000 undergraduates, 2,000 graduate students, and 900 faculty members. The Department of Geological Sciences began offering graduate degrees in 1964 and is ranked among the top ten earth science departments in the country by the American Council of Learned Societies. The faculty has a wide breadth of scientific interests, and 4 are members of the National Academy of Sciences. An active speakers program brings weekly lecturers to campus. There are numerous opportunities for field trips and seagoing expeditions.
Applying	Application materials may be obtained from the graduate adviser. The application deadline for fall admission is February 10. Normally, graduate students are accepted for fall admission only. Those applying for fellowships should submit a completed application by January 31. For need-based financial aid such as grants and loans, separate application should be made to the Office of Financial Aid (telephone: 805-893-2432). GRE General Test scores are required for all applicants. The Subject Test in geology is not required but is recommended. Prospective students should take the GRE in October or December. Applicants from non-English-speaking countries are required to pass the TOEFL with a score of 550 or better.
Correspondence and Information	Graduate Advisor Department of Geological Sciences University of California Santa Barbara, California 93106 Telephone: 805-893-3329 800-323-9023 (toll-free)

SECTION 3: EARTH AND PLANETARY SCIENCES

University of California, Santa Barbara

THE FACULTY AND THEIR RESEARCH

Ralph J. Archuleta, Professor; Ph.D., California, San Diego, 1976. Earthquake source mechanics, analysis of strong motion data, numerical modeling of wave propagation and earthquake source processes, field measurements of local earthquakes.

Tanya M. Atwater, Professor; Ph.D., California, San Diego (Scripps), 1972. Tectonic behavior of plate boundaries, triple junctions, and microplates; detailed structure and evolution of oceanic spreading centers.

Stanley M. Awramik, Professor; Ph.D., Harvard, 1973. Biogeology, pre-Phanerozoic fossil record; fossil algae and invertebrate paleontology.

James R. Boles, Professor; Ph.D., Otago (New Zealand), 1972. Sedimentary petrology, low-temperature geochemistry, diagenesis of clastic sediments, low-grade metamorphism.

Cathy Busby-Spera, Associate Professor; Ph.D., Princeton, 1982. Paleogeographic reconstructions and depositional processes in sedimentary and volcanic terranes.

Michael DeNiro, Professor; Ph.D., Caltech, 1977. Stable isotopes and geobiology.

Richard V. Fisher, Professor; Ph.D., Washington (Seattle), 1957. Physical volcanology, stratigraphy and sedimentology of pyroclastic rocks.

Michael D. Fuller, Professor; Ph.D., Cambridge, 1961. Geophysics, rock magnetism, paleomagnetism.

Rachel M. Haymon, Lecturer; Ph.D., California, San Diego (Scripps), 1982. Marine hydrothermal geochemistry, volcanogenic massive sulfide deposits, sulfide mineral petrology, deposition of metalliferous sediments, volcanic and tectonic processes on and near oceanic spreading centers, evolution and hydrothermal processes of young seamounts, formation and emplacement of ophiolites.

Clifford A. Hopson, Professor; Ph.D., Johns Hopkins, 1955. Igneous and metamorphic petrology, volcanology, and tectonics; volcanism and plutonism in the Cascade Mountains; petrology and tectonic evolution of ophiolites.

Edward A. Keller, Professor; Ph.D., Purdue, 1973. Fluvial geomorphology, tectonic geomorphology, environmental geology. (Joint appointment with Environmental Studies Program)

James P. Kennett, Professor and Director, Marine Science Institute; Ph.D., Wellington (New Zealand), 1965. Paleoceanography, marine geology.

David W. Lea, Assistant Professor; Ph.D., MIT/Woods Hole, 1989. Chemical oceanography and paleoceanography.

John E. Lupton, Associate Adjunct Professor; Ph.D., Caltech, 1971. Isotope geochemistry, chemical oceanography.

Bruce P. Luyendyk, Professor and Director/Chairman, Institute of Crustal Studies; Ph.D., California, San Diego (Scripps), 1969. Application of geophysical observations, especially magnetism, to tectonic problems; midocean ridge structure and paleomagnetism of southern California.

Ken C. Macdonald, Professor; Ph.D., MIT/Woods Hole, 1975. Marine geophysics; applications of magnetism, seismology, geomorphology to plate tectonic problems, particularly tectonics of midocean ridges; transform faults and subduction zones.

James M. Mattinson, Professor and Chairman; Ph.D., California, Santa Barbara, 1970. Isotope geology, geochronology, and petrology of oceanic and continental igneous rocks.

Brian E. Patrick, Assistant Professor; Ph.D., Washington (Seattle), 1987. Structural geology, metamorphic petrology, regional tectonics and strain analysis.

William A. Prothero, Professor; Ph.D., California, San Diego, 1967. Ocean-bottom seismology, structure and seismicity of midocean ridges and the Califoria borderlands, seismic instrumentation, structure of the lithosphere.

Richard H. Sibson, Adjunct Professor; Ph.D., London, 1977. Structural geology, especially structures and slip processes of earthquake-generation faults; rock mechanics; field geology.

Frank J. Spera, Professor; Ph.D., Berkeley, 1977. Igneous petrology, magma transport phenomena.

Arthur G. Sylvester, Professor; Ph.D., UCLA, 1966. Structural petrology and structural geology, structure and neotectonics, wrench fault tectonics, petrology and emplacement of plutons.

Bruce H. Tiffney, Associate Professor; Ph.D., Harvard, 1977. Evolutionary biology, paleobotany.

William S. Wise, Professor; Ph.D., Johns Hopkins, 1961. Mineralogy and geochemistry; mineral paragenesis, especially of zeolites and basilicates; petrochemistry of volcanic rocks.

Andre R. Wyss, Assistant Professor; Ph.D., Columbia, 1989. Vertebrate paleontology.

Professors Emeriti

John C. Crowell, Ph.D., UCLA, 1947. Tectonics, sedimentation in relation to tectonics, paleoclimates, especially ancient glaciation.

Robert M. Norris, Ph.D., Scripps, 1951. Geomorphology, Quaternary geology.

George R. Tilton, Ph.D., Chicago, 1951. Isotope geology, particularly geochronology; mineral age determinations, using isotopic lead and strontium methods; the use of strontium and lead isotopes as natural tracers in geologic processes; meteorite and lunar studies.

George Tunell, Ph.D., Harvard, 1930. Geochemistry, mineralogy.

Donald W. Weaver, Ph.D., Berkeley, 1960. Stratigraphic paleontology and engineering geology, Cretaceous and Cenozoic history of depositional basins of the West Coast Ranges, geologic problems in land development and in the development of natural resources.

SECTION 3: EARTH AND PLANETARY SCIENCES

UNIVERSITY OF CHICAGO
Department of the Geophysical Sciences

Programs of Study

The department, which offers programs of study leading to the M.S. and the Ph.D. degrees, is dedicated to research and education in the sciences of the solid earth, the atmosphere and oceans, and earth's neighbors in the solar system. Faculty research areas include geology and geophysics, geochemistry, cosmochemistry, atmospheric radiation, paleontology, climate change, physical meteorology, and oceanography. More specific information on current research and recent dissertation titles is available upon request.

Applicants with undergraduate majors in physics, chemistry, mathematics, astronomy, geology, biology, or engineering are encouraged to apply. The department emphasizes an individual approach to graduate study. In the first year, the student and an advisory committee design a program to fit the student's interests, which frequently are interdisciplinary. Nine courses in three quarters of residence are required for the M.S. degree (which is not a requirement for the Ph.D.), in addition to a seminar on a research paper. At the end of the first year, students in the Ph.D. program select a field of concentration and a research sponsor. During the second year, the Ph.D. preliminary examination is taken, including both written and oral parts, and a research prospectus is submitted. The student then concentrates on dissertation research.

Research Facilities

The department is housed in the Henry Hinds Laboratory for the Geophysical Sciences, which includes general research facilities, classrooms, machine and wood shops, and offices for faculty, students, and staff. Special laboratories are available for hydrodynamics; thermal convection; wave dynamics; sediment transport; satellite radiation budget measurement; cloud physics; remote atmospheric probing; synoptic meteorology; glaciology; high-pressure and high-temperature experiments in mineralogy, petrology, and geophysics; fossil preparation; low-temperature geochemistry and sedimentology; trace-element analysis by neutron activation; and studies employing the following equipment: an electron microprobe, an ion microprobe, a scanning electron microscope, stable-isotope mass spectrometers, X-ray diffractometers, and single-crystal diffractometers. The department has numerous remote terminals to computing facilities on campus as well as to supercomputers at the National Center for Atmospheric Research, the National Science Foundation supercomputing centers, and parallel processors located at Argonne National Laboratory. Various research projects in the department also have computers, including Sun, IRIS, and IBM workstations, which are available for advanced graphics and image processing applications. The resources of Regenstein and Crerar libraries, each of which contains over 4 million volumes, are available within a block of the department.

Cooperative research links exist with the Field Museum of Natural History; the Argonne National Laboratory near Chicago; the NASA Goddard Space Flight Center; the Brookhaven National Laboratory in Long Island, New York; the Max Planck Institute for Chemistry at Mainz, Germany; various marine laboratories; and the National Center for Atmospheric Research in Boulder, Colorado. Recent geology and paleontology field trips to Baja California and the Death Valley region have been subsidized.

Financial Aid

Research assistantships, course assistantships, and fellowships are assigned on a competitive basis. Course assistantships require up to 20 hours of service per week and currently provide a nine-month stipend of $10,350 plus full tuition. Research assistantships offer equivalent compensation and are usually extended for the summer. The McCormick Fellowship has a twelve-month stipend supplement of $2000 per year for three years, and the Salisbury Fellowship has a stipend plus full tuition. Other forms of aid are available through the University loan program, the work-study program, and the employment services of the Personnel Office, which is also open to spouses. Currently, all students in the department receive some form of financial aid.

Cost of Study

Tuition costs for the academic year 1991–92 are $16,275. After completion of the residence requirement of nine quarters for the Ph.D., tuition is $11,340 for the academic year.

Cost of Living

Several housing options are available for graduate students, most within walking distance of the campus. The University has ten-month leases available on apartments for married or single students at rents of $460–$800 for furnished units and $400–$750 for unfurnished units; two- and three-bedroom apartments are available for married students with families. International House accommodates more than 500 women and men; current quarterly rates range from $818 to $1066 for single rooms and are $770 per person for double rooms; there is a dining room fee of $350 per quarter, which allows meals to be purchased at a reduced rate.

Student Group

Of the University's graduate enrollment of 5,821, about 50 men and women are students in the Department of the Geophysical Sciences. Most students in the department are working toward the Ph.D. degree. The University has an undergraduate enrollment of 3,460 students, 10 of whom are in the department.

Location

Hyde Park, a distinctive, politically independent, interracial community, is located 15 minutes from the Loop, the center of Chicago. All the cultural opportunities of a large city are available, including the renowned Chicago Symphony Orchestra, the Lyric Opera, the Art Institute, and various theaters, galleries, concert halls, and museums, and there are also many fine restaurants. Less than a mile from the University on Lake Michigan are parks, beaches, and harbors that border the eastern edge of the city.

The University

The University dates back to 1892, with a well-known tradition of free inquiry, scholarship, and intellectual leadership. It encompasses an undergraduate college and four graduate divisions: Physical Sciences, Biological Sciences, Social Sciences, and Humanities. There are also seven graduate and professional schools: Business, Law, Medicine, Graduate Library, Divinity, Social Service Administration, and Education.

Applying

Applications for admission and financial aid for the fall quarter should be submitted by February 1; after that date, the possibility of aid is reduced. Applications for financial aid are considered on the basis of scholastic merit, and all applicants are required to submit GRE scores. Foreign students must also submit TOEFL scores. Applications, transcripts, and three letters of recommendation should be sent directly to the department.

Correspondence and Information

Marilyn Bowie, Administrative Assistant for Student Affairs
Department of the Geophysical Sciences
University of Chicago
5734 South Ellis Avenue
Chicago, Illinois 60637

Peterson's Guide to Graduate Programs in the Physical Sciences and Mathematics 1992

University of Chicago

THE FACULTY AND THEIR RESEARCH

Alfred T. Anderson Jr., Professor; Ph.D., Princeton. Development of continents and oceans through volcanology.

Victor Barcilon, Professor; Ph.D., Harvard. Inverse problems, geophysical fluid dynamics.

Roscoe R. Braham Jr., Professor Emeritus; Ph.D., Chicago. Cloud physics and weather modification, inadvertent weather modification by urban pollution, lake snows.

Robert N. Clayton, Enrico Fermi Distinguished Service Professor; Ph.D., Caltech. Conditions in the early solar nebula; isotopic studies of lunar samples and terrestrial igneous, metamorphic, and sedimentary rocks.

Paul J. Crutzen, Visiting Professor; Ph.D., Stockholm. Atmospheric chemistry, biogeochemical cycles.

Leo J. Donner, Assistant Professor; Ph.D., Chicago. Roles of clouds and convection in the atmospheric general circulation.

John E. Frederick, Professor; Ph.D., Colorado. Atmospheric chemistry, radiation transfer, remote sensing.

T. Theodore Fujita, Professor Emeritus; D.Sci., Tokyo. Satellite and mesometeorology, structure of tornadoes and hurricanes.

Dave Fultz, Professor Emeritus; Ph.D., Chicago. Experimental geophysical fluid dynamics, thermal convection, rotating fluids.

Julian R. Goldsmith, Charles E. Merriam Distinguished Service Professor Emeritus; Ph.D., Chicago. Phase equilibria, crystal chemistry, experimental mineralogy and petrology.

Lawrence Grossman, Professor; Ph.D., Yale. Chemistry and mineralogy of meteorites, origin of the solar system and the planets, abundance and distribution of the elements in the cosmos.

Dion L. Heinz, Assistant Professor; Ph.D., Berkeley. Ultrahigh-pressure experimental geophysics, solid-state geophysics, mantle petrology.

David Jablonski, Professor; Ph.D., Yale. Invertebrate paleontology, evolutionary paleobiology.

Susan M. Kidwell, Associate Professor; Ph.D., Yale. Taphonomy and stratigraphy of marine sequences.

Ole J. Kleppa, Professor Emeritus; Dr.Techn., Norwegian Institute of Technology. Thermodynamic properties of fused salt mixtures, high-temperature calorimetry of refractory ionic and metallic systems, calorimetry above 1100°C.

Hsiao-Lan Kuo, Professor Emeritus; Ph.D., Chicago. Instability theory of atmospheric and oceanic disturbances, dynamics of monsoon circulation and mesoscale systems, radiative temperature changes in the atmosphere.

Michael C. LaBarbera, Associate Professor; Ph.D., Duke. Functional morphology and biomechanics of marine invertebrates.

Douglas R. MacAyeal, Associate Professor; Ph.D., Princeton. Antarctic Ice Sheet physics, polar ocean circulation.

Marc C. Monaghan, Assistant Professor; Ph.D., Yale. Radionuclides as natural tracers in water, soils, sediments, and volcanics.

Paul B. Moore, Professor; Ph.D., Chicago. Systematic classification of crystal structures, theory of ionic accretion, transition-metal solid-state inorganic chemistry, lone-pair electronic systems, chemistry of butterfly wings.

Robert C. Newton, Professor; Ph.D., UCLA. Experimental determination of pressure-temperature stabilities of the important rock-forming minerals.

Raymond Pierrehumbert, Professor; Ph.D., MIT. Dynamic meteorology, geophysical fluid dynamics.

George W. Platzman, Professor Emeritus; Ph.D., Chicago. Dynamic meteorology; dynamic oceanography, including waves and tides.

David M. Raup, Sewell Avery Distinguished Service Professor; Ph.D., Harvard. Mathematical models of evolution and extinction.

Frank M. Richter, Professor and Chairman; Ph.D., Chicago. Solid-earth geophysics, plate tectonics, numerical simultion of mantle convection, dynamical and chemical models for melt segregation.

J. John Sepkoski Jr., Professor; Ph.D., Harvard. Phanerozoic fossil diversity, mass extinction, Cambrian paleoecology, quantitative paleontology.

Joseph V. Smith, Louis Block Professor and Director, Center for Advanced Radiation Sources; Ph.D., Cambridge. Mineralogy of the upper mantle and lower crust, crystal structures of zeolites, feldspar minerals, X-ray fluorescence microprobe analysis of trace elements, mineralogy of achondritic meteorites, application of high-intensity and neutron sources to the biological and physical sciences.

Ramesh C. Srivastava, Professor; Ph.D., McGill. Radar observation of precipitating clouds and cloud systems, numerical modeling of convective clouds.

Stephen M. Wickham, Assistant Professor; Ph.D., Cambridge. Geochemistry and fluid dynamics of granites, high-temperature processes related to fluid flow and melting in the crust.

Alfred M. Ziegler, Professor; D.Phil., Oxford. Community paleoecology of the Silurian period, Phanerozoic paleogeography, paleoclimatology, and paleobiogeography.

SECTION 3: EARTH AND PLANETARY SCIENCES

UNIVERSITY OF CINCINNATI
Department of Geology

Programs of Study The Department of Geology offers graduate programs leading to the M.S. and Ph.D. degrees in the geological sciences. The major programs of teaching and research include clay mineralogy, geomorphology, glacial geology, hydrogeology, igneous petrology and geochemistry, invertebrate paleontology, metamorphic petrology, micropaleontology, mineralogy, paleoceanography, petroleum geology, sedimentology, and tectonics. The M.S. program, which requires a thesis, normally takes two years to complete. The Ph.D. program requires a dissertation and usually three years beyond the M.S. for completion. Interdisciplinary studies are possible in both degree programs with the help of extradepartmental members of the student's advisory committee. Environmental geology and paleontology are frequently pursued as interdisciplinary studies.

Research Facilities In 1987, the department moved into the new Geology-Physics Building, which provides modern facilities for teaching and research. Specialized research equipment housed in the department includes X-ray diffraction and X-ray fluorescence equipment; an electron microprobe; cathodoluminescence, high-temperature, high-pressure, and hydrothermal experimental equipment; a fluid inclusion heating/freezing stage; an atmosphere gas mixing apparatus; paleontological preparation equipment; a thin-sectioning laboratory; a complete mineral separation laboratory; and research polarizing microscopes and stereomicroscopes. A scanning electron microscope and high-resolution TEM are available on campus. The technical staff includes a thin-section technician, 2 electronics-computer technicians, and a draftsperson.

The Geology Library, which is housed adjacent to the new building, includes over 36,000 bound volumes, 20,000 microfiche, and a unique collection of 1,500 guidebooks. The Willis Meyer Map Library is an official depository for the U.S. Geological Survey and U.S. Bureau of Mines. The Geology Museum houses a large research, exhibit, and teaching collection of rocks, minerals, and invertebrate fossils.

Computing facilities in the department available to students include many IBM-compatible and Apple Macintosh microcomputers, plotters, digitizers, and printers. The University of Cincinnati Computing Center operates an Amdahl 470V/7A, an IBM 3081 D, an IBM 3033 N, a COMTEN 3690, two VAX-11/750s, and a VAX-11/780. An electrostatic plotter and a large plotter with extensive graphics software are available through the Computing Center.

Financial Aid Graduate assistantships and University Graduate Scholarships are awarded on the basis of academic merit. Students with assistantships receive a stipend of $7800 for the nine-month academic year in addition to a full waiver of tuition and fees. These students serve as teaching assistants in introductory geology and other departmental courses. This service amounts to about 15 hours per week. Students with University Graduate Scholarships receive a waiver of tuition only and are not required to serve as teaching assistants. Departmental fellowships from endowed funds (Caster, Fenneman, and Willis Meyer funds) and fellowships provided by grants from industry offer a stipend equal to the graduate assistantship, a waiver of tuition and fees, and, in some cases, a summer stipend. These awards carry no teaching duties and are usually made to advanced graduate students. M.S. students can receive up to two years of departmental support, and Ph.D. students can receive up to three years of departmental support beyond the M.S. Some students are supported as research assistants on grants made to individual faculty members. Currently, most full-time graduate students receive some form of financial aid, and most graduate assistants receive summer stipends. The cost of thesis research can usually be defrayed through departmental funds and summer fellowships from the University Research Council. The Research Council and the Department of Geology also provide support for student travel to attend meetings and present papers.

Cost of Study Tuition and fees for Ohio residents were $1472 per quarter ($4416 per year) in 1990–91. Out-of-state tuition and fees were $2896 per quarter ($8688 per year). Tuition and fees are waived for students receiving assistantships and fellowships.

Cost of Living Living expenses in the Cincinnati area are comparable to those in other midwestern urban areas. Most graduate students live in off-campus apartments. In 1990–91, on-campus housing was available as efficiency studio apartments (about $350 per month), one-bedroom apartments ($425 per month), and two-bedroom apartments ($515 per month). Dormitory room and board charges ranged from $1269 to $1316 per quarter.

Student Group Currently in the department, there are about 65 graduate students; approximately 50 students are in residence. Ph.D. students constitute a third of the total enrollment.

Location The University of Cincinnati is an urban campus situated on a hilltop about 2 miles from downtown Cincinnati. The Greater Cincinnati area, with a population of over a million, provides a wealth of cultural and recreational resources. Major attractions include music for every preference, ballet, opera, theater, museums, and the 1990 World Champion Cincinnati Reds baseball team.

The University The University is one of Ohio's two major comprehensive state universities, with a total enrollment of over 35,000 students in eighteen colleges, schools, and divisions. The University operates on a quarter system, with the autumn quarter beginning late in September and the spring quarter ending in early June. The University provides the rich cultural and recreational activities of a major university in the setting of one of America's most livable cities.

Applying Information and applications for admission and financial aid are available from the director of graduate studies in the Department of Geology. To apply, a student must submit a completed application, a statement of purpose, an official set of transcripts, three letters of recommendation, and scores on the GRE General Test and GRE Subject Test in geology or some other science. To qualify for admission, students should have an average that is above B in undergraduate courses and a year each of chemistry, physics, and calculus. Completed applications should be submitted by February 15 for admission in the autumn quarter. Inquiries about specific programs of study will be answered by the faculty members most knowledgeable about the area.

Correspondence and Information
Director of Graduate Studies
Department of Geology
University of Cincinnati
Cincinnati, Ohio 45221-0013
Telephone: 513-556-3732
Fax: 513-556-6931

Peterson's Guide to Graduate Programs in the Physical Sciences and Mathematics 1992

SECTION 3: EARTH AND PLANETARY SCIENCES

University of Cincinnati

THE FACULTY AND THEIR RESEARCH

Madeleine Briskin, Professor; Ph.D., Brown, 1972. Micropaleontology and oceanography.
Kees A. DeJong, Associate Professor; Ph.D., Utrecht (Netherlands), 1969. Tectonics.
Craig Dietsch, Assistant Professor; Ph.D., Yale, 1985. Metamorphic geology.
John E. Grover, Professor; Ph.D., Yale, 1972. Mineralogy and crystal chemistry.
Warren D. Huff, Professor; Ph.D., Cincinnati, 1963. Clay mineralogy.
Attila I. Kilinc, Professor and Head; Ph.D., Penn State, 1969. Experimental petrology and geochemistry.
Thomas V. Lowell, Assistant Professor; Ph.D., SUNY at Buffalo, 1986. Glacial and Quaternary geology.
J. Barry Maynard, Professor; Ph.D., Harvard, 1972. Sedimentary geochemistry.
David L. Meyer, Professor; Ph.D., Yale, 1971. Invertebrate paleobiology.
Arnold I. Miller, Assistant Professor; Ph.D., Chicago, 1986. Evolutionary paleontology and paleoecology.
David B. Nash, Associate Professor; Ph.D., Michigan, 1977. Geomorphology.
Paul Edwin Potter, Professor; Ph.D., Chicago, 1959. Sedimentology and sedimentary petrology.
Wayne A. Pryor, Professor; Ph.D., Rutgers, 1959. Sedimentation and stratigraphy.

SECTION 3: EARTH AND PLANETARY SCIENCES

UNIVERSITY OF GEORGIA

Department of Geology

Programs of Study

The department offers instruction and opportunities for research in archaeological geology, economic geology, geochemistry and isotope geology, geophysics, hydrogeology, mathematical geology, paleontology and micropaleontology, petrology, sedimentation, stratigraphy, structural geology, and other fields. Graduate programs leading to the M.S. and Ph.D. are individually arranged to fit each student and his or her background.

A minimum of one academic year of residence is required for the M.S. degree. A program of study consisting of 45 quarter hours of credit (including 5 quarter hours of thesis), a thesis, and an examination covering the thesis and course work are requirements for the M.S. degree.

For the Ph.D. degree, a minimum of three years of study beyond the bachelor's degree, a minimum of three continuous quarters of residence, a reading knowledge of one or two modern foreign languages at the discretion of the student's committee, a dissertation, and preliminary and final oral examinations are required.

Classes average fewer than 10 students for graduate-level courses. The student-faculty ratio is about 3:1.

Research Facilities

The geology department houses several well-equipped analytical and experimental facilities that have X-ray diffraction and fluorescence equipment, electron microscopes, an electron microprobe, stable and radiogenic isotope mass spectrometers, INAA counting facilities, and a C-14 dating laboratory. Other facilities include a micropaleontological laboratory, geochemical and geophysical laboratories, microcomputers for research and instruction, and laboratories for preparation of samples, optical determination, and photomicroscopy. The University's Marine Institute on Sapelo and Skidaway islands provides facilities for research in marine geology. Companion facilities include the University computer center, an electron microscopy laboratory, field vehicles, instrument shops, and rock, mineral, and fossil collections.

Financial Aid

For 1991–92, graduate, nonteaching, teaching, and research assistantships carry nine-month stipends of $7072–$7545 and remission of tuition. Prospective Ph.D. candidates are also eligible for Alumni Foundation fellowships. Recipients of these awards have minimal departmental duties, and the awards include payment of tuition. Marine geology students are eligible for the Levy-Logan Memorial award of $750.

Cost of Study

In 1991–92, the maximum cost of study for state residents is $692 per quarter; nonresidents are assessed tuition and fees of $1840 per quarter. Graduate assistants are not charged tuition.

Cost of Living

Housing for male students and dormitory accommodations for women are available on campus for about $430–$650 per quarter. A full range of University-owned housing for married students is offered for approximately $175–$230 a month, depending on the size and type of housing.

Student Group

The University of Georgia has an enrollment of over 28,000, including 5,000 graduate students. Students are enrolled from 50 states and more than 100 countries. At present, there are 50 graduate students enrolled in the geology department.

Location

Athens is located in the Piedmont, only a few hours' drive from the Appalachian and Blue Ridge mountains and the Atlantic and Gulf coastal plains. Excellent fishing, camping, golf, swimming, and various other outdoor sports are available either in Athens or a short distance away. Athens is 65 miles from Atlanta, providing the student with access to all of its recreational and shopping facilities. The cultural life in Athens centers on the University's offering of diverse and interesting programs. The climate is pleasant the year round.

The Department

The geology department was established as a separate department in 1961 and enrolled its first graduate students in 1962. However, geology courses have been taught at Georgia intermittently since 1823. The faculty members represent diverse geographic and geologic interests as well as different research specialties. Their graduate training was carried out at twenty different institutions. The resident staff is aided throughout the year by a program of visiting scientists.

Applying

Application forms for admission and financial assistance may be obtained by writing to the geology department, as indicated below. All prospective candidates must have a bachelor's degree from an accredited institution and are required to take the Graduate Record Examinations. Forms must be returned at least six weeks before the proposed registration date.

Correspondence and Information

Graduate Coordinator
Department of Geology
University of Georgia
Athens, Georgia 30602
Telephone: 404-542-2652

University of Georgia

THE FACULTY AND THEIR RESEARCH

Faculty members engage in both teaching and research. They have conducted field research throughout the Appalachian area as well as in Africa, Alaska, Brazil, Canada, Europe, Greenland, the Caribbean, California, and the Rocky Mountain states.

Gilles O. Allard, Professor Emeritus; Ph.D., Johns Hopkins, 1956. Economic geology, ore microscopy, exploration geochemistry.
Robert H. Carpenter, Adjunct Professor; Ph.D., Wisconsin, 1965. Economic geology, ore microscopy, exploration geochemistry.
Robert E. Carver, Professor; Ph.D., Missouri, 1961. Sedimentary petrology, industrial mineralogy, hydrogeology.
Douglas E. Crowe, Assistant Professor; Ph.D., Wisconsin, 1990. Economic geology, laser microprobe systems, stable isotope geochemistry.
R. David Dallmeyer, Professor; Ph.D., SUNY at Stony Brook, 1972. Structural geology, metamorphic petrology, geochronology.
John R. Ertel, Assistant Professor; Ph.D., Washington, (Seattle), 1985. Organic geochemistry.
Raymond Freeman-Lynde, Associate Professor; Ph.D., Columbia, 1981. Marine geology, carbonate sedimentology.
Robert W. Frey, Professor; Ph.D., Indiana, 1969. Invertebrate paleontology, paleoecology, ichnology.
Susan T. Goldstein, Assistant Professor; Ph.D., Berkeley, 1983. Micropaleontology, paleontology.
Elizabeth Gordon, Assistant Professor; Ph.D., SUNY at Binghamton, 1985. Stratigraphy, sedimentation.
Robert B. Hawman, Assistant Professor; Ph.D., Princeton, 1988. Seismology, geophysical data processing.
Willis B. Hayes, Research Associate and Lecturer; Ph.D., California, San Diego (Scripps), 1969. Oceanography, computer-assisted instruction.
V. J. Henry Jr., Adjunct Professor; Ph.D., Texas A&M, 1961. Marine geology.
Norman Herz, Professor and Acting Head; Ph.D., Johns Hopkins, 1950. Archaeological geology, petrology, economic geology, geochemistry.
Steven M. Holland, Assistant Professor; Ph.D., Chicago, 1990. Sequence stratigraphy, basin analysis, community paleoecology and evolution.
J. Hatten Howard III, Associate Professor; Ph.D., Stanford, 1969. Geochemistry, photogeology, geomorphology.
Vernon J. Hurst, Emeritus Research Professor; Ph.D., Johns Hopkins, 1954. Crystallography, experimental petrology, economic geology.
George S. Koch Jr., Professor; Ph.D., Harvard, 1955. Statistical geology, ore deposits.
Joshua Laerm, Adjunct Assistant Professor; Ph.D., Illinois, 1976. Vertebrate paleontology, evolution, functional morphology.
John E. Noakes, Professor; Ph.D., Texas A&M, 1962. Geochronology, geochemistry.
Alberto E. Patino-Douce, Assistant Professor; Ph.D., Oregon State, 1990. Experimental petrology, high-grade metamorphism, crustal magmatism.
L. Bruce Railsback, Assistant Professor; Ph.D., Illinois, 1989. Paleoceanography, carbonate petrology, sedimentary geochemistry.
Mark Rich, Professor Emeritus; Ph.D., Illinois, 1959. Stratigraphy, sedimentary petrology.
Michael F. Roden, Associate Professor; Ph.D., MIT, 1982. Igneous petrology, trace-element and radiogenic isotope geochemistry, mineralogy.
Barbara L. Ruff, Lecturer; M.S., Georgia, 1975. Vertebrate paleontology.
David B. Wenner, Associate Professor; Ph.D., Caltech, 1971. Stable isotope geochemistry.
James A. Whitney, Professor; Ph.D., Stanford, 1972. Igneous petrology, geochemistry, economic geology.

SECTION 3: EARTH AND PLANETARY SCIENCES

UNIVERSITY OF HAWAII AT MANOA

School of Ocean and Earth Science and Technology
Department of Geology and Geophysics

Programs of Study

The Department of Geology and Geophysics has been incorporated into the newly established School of Ocean and Earth Science and Technology (SOEST). The department offers programs of research and study leading to the M.S. and Ph.D. degrees in high-pressure geophysics and geochemistry; hydrogeology and engineering geology; marine geology and geophysics; planetary geosciences and remote sensing; seismology and solid-earth geophysics; and volcanology, petrology, and geochemistry. Because of the University's location on an island in the middle of the Pacific Ocean, programs of study and research related to the Pacific Basin and its islands and margin are emphasized.

Entering students normally have an undergraduate degree in geology or geophysics or in related sciences such as physics, chemistry, mathematics, engineering, or biology. A minimum of one year of calculus, physics, and chemistry is required of all students prior to admittance.

The M.S. program requires a minimum of 30 semester credit hours, including at least 24 credits of formal course work and 12 credits of thesis research. M.S. candidates must complete and defend a research thesis. A nonthesis option is available only in certain special cases.

The Ph.D. program has no formal course work requirements other than courses necessary to prepare students for the qualifying examination (waived for students entering with an M.S. in earth science fields), language exam, comprehensive exam, and dissertation work. Ph.D. candidates must demonstrate reading competence in one foreign language, pass a comprehensive exam, and then complete and defend a dissertation.

Research Facilities

The department is housed in the Hawaii Institute of Geophysics (HIG), which conducts multidisciplinary research in marine, earth, and planetary geosciences. Laboratory facilities are available for high-pressure research, geochemistry, stable and radioactive X-ray fluorescence and diffraction, electron microprobe and scanning electron microscopy analysis, atomic adsorption and inductively coupled plasma spectrophotometry, mineralogy, petrology, micropaleontology, sedimentology and paleomagnetism, earthquake and marine refraction and reflection seismology, ocean-bottom and downhole seismometry, gravity, bathymetry, telescope and remote-sensing field observations, computer image and spectrum processing, and instrument development in planetary and remote-sensing science.

SOEST operates the deep-sea research vessel *Moana Wave;* the coastal vessel *Kila;* two research submersibles, the *Makalii* (400 meters) and the *Pisces V* (2,000 meters); a sidescan sonar and bathymetric mapping system; and the University Marine Center at Snug Harbor in Honolulu. SOEST maintains an engineering support facility; machine and electronics shops; a library; core and rock storage; a thin-section lab; scientific data archives; a network of computers, including an Alliant FX/8 superminicomputer; and graphics, photography, publications, and other support facilities. A seven-story Pacific Ocean Science and Technology building is scheduled for completion in 1993.

The Water Resources Research Center, located adjacent to HIG, conducts cooperative research in hydrology with the department and maintains research facilities for water quality analysis, hydrologic and hydraulic studies, soil and rock testing, pump test analysis, geophysical well logging, and stream gauging.

Financial Aid

Teaching and research assistantships, which are available to all qualified students, provide free tuition and stipends ranging from approximately $12,000 to $22,000 per year.

Cost of Study

For 1991–92, graduate tuition is $2405 per semester for nonresidents and $800 per semester for state residents.

Cost of Living

The cost of housing in Honolulu is quite high. Most students share apartments near the campus. Housing and meals generally amount to $700–$900 per month.

Student Group

About 20,000 students study at the Manoa campus, and the Department of Geology and Geophysics normally has 65–75 graduate students in residence. About one third of the students are women, and 15 percent are from foreign countries.

Location

Hawaii, perceived by many as the paradise of the Pacific, abounds with experiences in multicultural living. Its capital, Honolulu, is a modern, cosmopolitan, tropical metropolis of 800,000 people. Recreational and cultural features include the ocean and mountains, parks, spectator and participant sports, the Honolulu Symphony, the Academy of Arts, the Bishop Museum, and several theatrical and musical groups. Social life tends to be informal.

The University

The University of Hawaii was founded in 1907 as a land-grant institution; it now operates as both a sea-grant and a land-grant university. The Manoa campus, where all graduate programs are conducted, is located on some 300 acres in Manoa Valley, about 5 kilometers from downtown Honolulu and 3 kilometers from Waikiki.

Applying

Application forms for admission and financial aid are available from the department. The deadline for application is February 1 for the fall semester and September 1 for the spring semester. Deadlines for international applications are January 15 and August 1, respectively. Scores on the GRE General Test and an appropriate Subject Test are required of all applicants; international students must also submit scores on the TOEFL.

Correspondence and Information

Graduate Admissions
Department of Geology and Geophysics
University of Hawaii at Manoa
2525 Correa Road
Honolulu, Hawaii 96822

Peterson's Guide to Graduate Programs in the Physical Sciences and Mathematics 1992

SECTION 3: EARTH AND PLANETARY SCIENCES

University of Hawaii at Manoa

THE FACULTY AND THEIR RESEARCH

R. Batiza, Ph.D., California, San Diego (Scripps). Igneous petrology, volcanology, marine geology.
J. F. Bell, Ph.D., Hawaii. Planetary geosciences.
D. Bercovici, Ph.D., UCLA. Geodynamics, geophysical fluid dynamics.
E. Berg, Dr.rer.nat., Saarlandes (Germany). Seismology.
P. A. Cooper, Ph.D., Hawaii. Seismology, marine tectonics.
F. K. Duennebier, Ph.D., Hawaii. Seismology, marine geophysics.
A. I. El-Kadi, Ph.D., Cornell. Hydrogeology, engineering geology.
P. F. Fan, Ph.D., UCLA. Mineralogy of sediments, geology of Asia.
F. P. Fanale, Ph.D., Columbia. Planetary sciences, geochemistry.
C. H. Fletcher, Ph.D., Delaware. Marine geology, coastal processes.
L. N. Frazer, Ph.D., Princeton. Theoretical seismology.
G. J. Fryer, Ph.D., Hawaii. Seismology, marine geophysics.
P. B. Fryer, Ph.D., Hawaii. Marine geology, petrology, planetary geosciences.
M. O. Garcia, Ph.D., UCLA. Igneous petrology, volcanology.
C. R. Glenn, Ph.D., Rhode Island. Marine geology and geochemistry.
B. R. Hawke, Ph.D., Brown. Planetary geosciences.
C. E. Helsley, Ph.D., Princeton. Paleomagnetism, marine geophysics.
J. L. Karsten, Ph.D., Washington. Igneous petrology, marine geology.
K. Keil, Dr.rer.nat., Mainz. Planetary geosciences, meteorites.
P. D. Lee, M.S., Hawaii. Hawaiian geology.
P. G. Lucey, Ph.D., Hawaii. Planetary geosciences.
J. J. Mahoney, Ph.D., California, San Diego (Scripps). Geochemistry.
M. H. Manghnani, Ph.D., Montana State. High-pressure mineralogy.
T. B. McCord, Ph.D., Caltech. Planetary sciences and remote sensing.
L. C. Ming, Ph.D., Rochester. High-pressure mineralogy, crystallography.
R. Moberly, Ph.D., Princeton. Marine geology, sedimentology.
G. F. Moore, Ph.D., Cornell. Marine geology and seismology.
P. J. Mouginis-Mark, Ph.D., Lancaster (England). Planetary geosciences, remote sensing.
K. A. Pankiwskyj, Ph.D., Harvard. Mineralogy, petrology.
F. L. Peterson, Ph.D., Stanford. Hydrogeology, engineering geology.
B. N. Popp, Ph.D., Indiana. Paleontology, paleoceanography.
S. Postawko, Ph.D., Michigan. Planetary geosciences, atmospheres.
C. B. Raleigh, Ph.D., UCLA. Tectonics, rock mechanics.
J. M. Resig, Dr.rer.nat., Kiel (Germany). Marine geology, micropaleontology, foraminifera.
E. Scott, Ph.D., Cambridge. Planetary geosciences.
S. Self, Ph.D., Imperial College (London). Volcanology.
J. M. Sinton, Ph.D., Otago (New Zealand). Igneous petrology, marine geology.
B. Taylor, Ph.D., Columbia. Tectonics, marine geology and geophysics.
G. P. L. Walker, Ph.D., Leeds (England). Physical volcanology, volcanic hazards.
S. Zisk, Ph.D., Stanford. Planetary geosciences.

Other Graduate Faculty
E. H. DeCarlo, Ph.D., Hawaii. Marine geochemistry.
A. S. Furumoto, Ph.D., Saint Louis. Seismology, geophysics.
R. N. Hey, Ph.D., California, San Diego (Scripps). Marine and plate tectonics.
B. H. Keating, Ph.D., Texas at Dallas. Marine geology and geophysics; paleomagnetism.
L. W. Kroenke, Ph.D., Hawaii. Marine geology and geophysics.
B. R. Lienert, Ph.D., Texas at Dallas. Seismology, computer applications.
F. T. Mackenzie, Ph.D., Lehigh. Geochemistry.
S. K. Sharma, Ph.D., Indian Institute of Technology (New Delhi). Experimental petrology.
A. N. Shor, Ph.D., Harvard. Marine geology, seafloor mapping.
D. M. Thomas, Ph.D., Hawaii. Geochemistry, geothermal systems.
J. S. Tribble, Ph.D., Northwestern. Marine geochemistry.
D. A. Walker, Ph.D., Hawaii. Seismology.
R. H. Wilkens, Ph.D., Washington (Seattle). Rock properties, borehole geophysics.
L. Wilson, Ph.D., London. Theoretical volcanology.

Affiliate Graduate Faculty
A. L. Clark, Ph.D., Idaho. Economic geology, resources.
J. Dorian, Ph.D., Hawaii. Mineral and economic geology, resources.
J. C. Gradie, Ph.D., Arizona. Planetary geosciences.
D. M. Hussong, Ph.D., Hawaii. Marine geology, seafloor mapping.
C. J. Johnson, Ph.D., Penn State. Mineral economics, resources.
J. N. Kellogg, Ph.D., Princeton. Marine tectonics.
J. P. Lockwood, Ph.D., Princeton. Modern volcanic processes.
G. H. Sutton, Ph.D., Columbia. Seismology, marine geophysics.
C. I. Voss, Ph.D., Princeton. Hydrogeology.
T. L. Wright, Ph.D., Johns Hopkins. Volcanology.

Emeritus Faculty
D. Cox, Ph.D., Harvard. Geophysics.
S. H. Laurila, Ph.D., Finland Institute of Technology. Geodesy, electronic surveying.
J. C. Rose, Ph.D., Wisconsin. Geophysics.

SECTION 3: EARTH AND PLANETARY SCIENCES

UNIVERSITY OF HOUSTON

Department of Geosciences

Programs of Study
The Department of Geosciences offers graduate programs leading to the M.S. and Ph.D. degrees in geology and geophysics. Candidates for the M.S. degree may elect a thesis or a nonthesis option. The thesis option requires the completion of a minimum of 24 hours of graduate course work with a grade point average of 3.0 or higher and the successful completion, presentation, and defense of a thesis that reflects individual research accomplishments. All supported students must select this option. The nonthesis option is designed for students employed in the geoscience community who study part-time. It requires the completion of selected graduate-level course work (36 hours for geology and 30 hours for geophysics) with a minimum grade point average of 3.0 and the successful completion of a comprehensive examination. Students selecting the nonthesis option are required to complete and defend a research project. Ph.D. students must pass an oral dissertation proposal/comprehensive examination and demonstrate a reading knowledge of one foreign language in order to be admitted to candidacy. One year of full-time residency is required. The independent research that constitutes the dissertation is defended in an oral examination.

Research Facilities
Research in the geosciences at the University of Houston can be carried out within the Department of Geosciences and the Allied Geophysical Laboratories (AGL), which are closely coordinated. The AGL is an organization that carries out the exploration geophysics research program of the University and educates geophysicists, computer scientists, and electrical engineers. Research facilities in the geosciences are located on the second and third floors and in the basement of the Science and Research I Building and in the building of the AGL. Arrangements can be made to use additional facilities on campus or nearby. Equipment available includes facilities for powder and single-crystal X-ray diffraction analysis; chemical analysis of rocks and mineral separates (atomic absorption spectrophotometer, X-ray fluorescence, ICP, electron microprobe); stable isotope geochemistry (Nuclide 3-inch and Finnigan MAT Delta-E mass spectrometers); general geochemistry; petrography (including grain-size analysis, cathode luminescence, fluorescence, and research petrographic microscopes); preparation and characterization of macro- and micropaleontological specimens; rock mechanics (triaxial cells and strain measuring equipment); mathematical/computer geology; planetology; paleomagnetism (magnetically shielded room, cryogenic magnetometer, and thermal and alternating field demagnetization); and potential field studies (magnetometers and gravity meters). The Allied Geophysical Laboratories facilities for exploration geophysics include those of the Seismic Acoustic Laboratory (model-building capability, solid-modeling system, and modeling tank with PDP-11/44 computer); Image Processing Laboratory for seismic modeling and interpretation (graphics systems: Adage 3000 color, Evans & Sutherland PS-3000 color, Genisco SpaceGraphy 3D, and Lexidata Lex 90); Research Computation Laboratory (VAX-11/780 with links to CDC CYBER 205 and CRAY-1 supercomputers); and Field Research Laboratory, a joint venture with Curtin University (48-channel DFS IV system). To support these facilities, the department maintains a complete photographic area, terminals for access to the University's AS-9000 and VAX computers, a number of Macintosh microcomputers, thin section and sample preparation labs, a map room and drafting area, and field vehicles and boats. Access to a machine shop and to transmission and scanning electron microscopes with EDAX can be arranged.

Financial Aid
Outstanding students may receive financial support packages worth up to $13,300. Packages include a teaching assistantship, fellowship, or research assistantship (the department supports approximately 20 students in these positions with stipends that begin at $7650 for the M.S. and $8100 for the Ph.D. for nine months), relief from out-of-state tuition (approximately $2700), and additional scholarship support for selected students of $3000 for the first year (renewable on a competitive basis). Limited summer support is also available through the department.

Cost of Study
In 1991–92, tuition and fees for Texas residents are approximately $690 per semester for full-time study. Nonresident students who receive no University support pay approximately $2250 per semester.

Cost of Living
Space is available in residence halls for students; application should be made early. No on-campus housing is available for married students. Entering students may contact the University's apartment referral service.

Student Group
The University enrollment is approximately 30,000, including about 7,000 graduate students. The Department of Geosciences has about 30 undergraduate majors and 130 graduate students, with the latter divided between the geology and geophysics options. About 30 graduate students attend full-time; the remainder work in local industry and pursue their degrees in the evening.

Location
The Houston metropolitan area has the largest resident population of geologists and geophysicists in the world. In the local area are Rice University, the Johnson Spacecraft Center, and many research companies. There are part-time and full-time employment opportunities for students and their spouses.

The University and The Department
The University, founded in 1927, became a four-year institution in 1934 and a state-supported university in 1963. The department is part of the College of Natural Sciences and Mathematics, which also offers programs in chemistry, the biophysical sciences, physics, mathematics, and computer science.

Applying
A GPA of 3.0 or greater for the last 60 semester hours of work completed plus a minimum GRE General Test score of 1600 are required for unconditional admission to the M.S. program. A student with a GPA of less than 3.0 but greater than 2.6 for the last 60 hours may receive conditional admission if the GRE score is greater than 1700. To apply for the Ph.D. program, completion of an M.S. degree or 30 hours of graduate credit and submission of GRE General Test and Subject Test scores, three letters of academic recommendation, transcripts, and curriculum vitae are required. All students whose native language is not English must score 550 or better on the TOEFL examination. Applications for admission and financial aid may be obtained from the Department of Geosciences. Completed applications, transcripts of all previous college work, three letters of recommendation, and GRE and TOEFL scores should be submitted to the department by the end of February for fellowship and assistantship awards for the following fall and by the end of September for awards for the following spring.

Correspondence and Information
Graduate Advisor
Department of Geosciences
University of Houston
Houston, Texas 77204-5503
Telephone: 713-749-1803
Fax: 713-748-7906

SECTION 3: EARTH AND PLANETARY SCIENCES

University of Houston

THE FACULTY AND THEIR RESEARCH

Kevin Burke, Professor; Ph.D., London. Plate tectonics, regional geology, Caribbean evolution, rift systems.
John Butler, Professor; Ph.D., Ohio State. Statistical applications, data analysis, mineralogy, crystal chemistry, petrology; effects of closure on a data matrix; software for undergraduate geosciences.
Regina M. Capuano, Assistant Professor; Ph.D., Arizona. Hydrogeology, geochemistry, hydrology.
John Casey, Associate Professor; Ph.D., SUNY at Albany. Tectonic and petrologic studies of ophiolite complexes in Newfoundland and Turkey and of mid-ocean ridge systems, geochemistry of siliciclastic sediments in orogenic belts.
Henry Chafetz, Professor; Ph.D., Texas at Austin. Deposition and diagenesis of carbonates, travertines, and bacterially induced precipitation of carbonate.
Peter Copeland, Assistant Professor; Ph.D., SUNY at Albany. Isotope geochemistry, thermal history of orogenic belts.
William Dupré, Associate Professor; Ph.D., Stanford. Geomorphology and sedimentology of Quaternary depositional systems in coastal regions; geologic modeling of enhanced oil recovery.
Ian Evans, Associate Professor; Ph.D., Texas A&M. Palaeomagnetic studies in Turkey, northern Mexico, and Newfoundland; paleoecology of Pennsylvanian faunas.
Stuart Hall, Associate Professor; Ph.D., Newcastle. Magnetic anomalies in the Caribbean, Gulf of Mexico, and the Australian-Antarctic discordance; palaeomagnetic studies in North America and Turkey.
Petr Jakes, Visiting Professor; Ph.D., Australian National. Petrology of granulites, nature of the lower crust, geochemistry of island arcs, planetary studies.
Elbert King, Professor; Ph.D., Harvard. Petrography and petrology of meteoritic chondrules, geology of Mars and the Moon, inclusions in carbonaceous chondites.
Timothy Kusky, Visiting Assistant Professor; Ph.D., Johns Hopkins. Field and structural geology.
James Lawrence, Associate Professor; Ph.D., Caltech. Stable isotope studies of precipitation in storms, circulation in Texas estuarines, and alterations in the oceanic crust.
Rosalie Maddocks, Professor; Ph.D., Kansas. Taxonomy, ecology, and evolution of Ostracoda; general marine micropaleontology.
John McDonald, Professor; Ph.D., SMU. Seismic signals in anisotropic media, laboratory modeling of seismic data, cross-hole tomography, enhanced oil recovery.
Ann McGuire, Research Assistant Professor; Ph.D., Stanford. Petrology of lower crustal and mantle-derived nodules, electron probe microanalysis.
Carl Norman, Associate Professor; Ph.D., Ohio State. Active faults of the Gulf Coastal region, time-dependent deformation of rock.
Arch Reid, Professor and Chairman of the Department; Ph.D., Pittsburgh. Petrology of achondrite meteorites, melting and deformation associated with terrestrial impacts, mineral chemistry.
Robert Sheriff, Professor; Ph.D., Ohio State. Seismic stratigraphy, reservoir geophysics.
Alex Woronow, Associate Professor; Ph.D., Harvard. Carbonate diagenesis in reef environments, analysis and interpretation of compositional data.
Sandra J. Wyld, Visiting Assistant Professor; Ph.D., Stanford. Tectonics, geologic evolution of convergent margins.
Hua-Wei Zhou, Assistant Professor; Ph.D., Caltech. Seismotectonics of plate boundaries, mantle seismic structures, geophysical data processing.

Professors Emeriti

Margaret S. Bishop, Ph.D., Michigan. Earth science education.
Max F. Carman Jr., Professor; Ph.D., UCLA. Igneous petrology, optical mineralogy, feldspar systematics.
Paul Fan, Ph.D., Iowa. Geochemistry, petroleum geology.
DeWitt C. Van Siclen, Ph.D., Princeton. Petroleum geology, structural geology.

Adjunct Faculty

David H. Carlson, Ph.D., Case Western Reserve. Mathematics.
Gerald H. F. Gardner, Ph.D., Princeton. Mathematical physics.
William P. Gore, B.S., Alabama. Mechanical engineering.
Peter F. Morse, Ph.D., Utah. Mathematics.
Illika Noponen, Ph.D., Helsinki. Geophysics.
Steven Schafersman, Ph.D., Rice. Geology.
Robert Tatham, Ph.D., Columbia. Geophysics.

SECTION 3: EARTH AND PLANETARY SCIENCES

UNIVERSITY OF ILLINOIS AT CHICAGO

Department of Geological Sciences

Programs of Study

The department offers graduate programs leading to the M.S. degree in the areas of glacial geology, geomorphology, environmental geology, sedimentology, paleontology, marine geology, geophysics, mineralogy, crystallography, petrology, and geochemistry. A Ph.D. program is offered jointly with the Department of Civil Engineering, Mechanics, and Metallurgy. Student research can include fieldwork, laboratory studies, theoretical investigations, or some combination of these. Students with geology majors or with strong backgrounds in other fields of science are encouraged to apply.

The M.S. degree program, including course work, generally takes two years to complete. A thesis is required and must be publicly presented. There is an oral thesis exam.

The Ph.D. requires an oral preliminary exam, a dissertation, and an oral defense of the dissertation.

Research Facilities

The offices and research facilities of the Department of Geological Sciences are housed in the modern Science and Engineering South Building, while undergraduate teaching laboratories are in the adjacent Science and Engineering Laboratories. These buildings also house the excellent science library, the Computer Center, and other campus science and engineering departments. The department maintains fully equipped laboratories for paleontology, mineralogy, petrology, X-ray crystallography, rock preparation, sedimentology, experimental petrology, and geomorphology. Specific equipment includes single-crystal and powder X-ray diffraction equipment, an automated electron microprobe, a scanning electron microscope, an atomic absorption unit, one-atmosphere furnaces, an internally heated pressure vessel, piston-cylinder apparatus, direct-current plasma apparatus, recirculating flumes, and a digitizer. The department also has a drafting room and several darkrooms. The department makes extensive use of the University Computer Center, which currently supports an advanced IBM 3081 D computer. In addition to a number of video terminals, the department has numerous smaller computers. Faculty and students have access to the facilities and collections of the Field Museum of Natural History.

Financial Aid

Financial support for qualified graduate students is available in the forms of fellowships and teaching and research assistantships. Work-study and separate tuition and fee waivers are also available. In 1991–92, assistantships provide approximately $8000 for the academic year plus tuition and fee waivers.

Cost of Study

Graduate tuition and fees for the 1991–92 academic year are $3640 for Illinois residents and $8300 for nonresidents.

Cost of Living

On-campus housing is available in either apartments or dormitories. Costs for these apartments range from $3792 to $4396 for an 11½-month contract. Dormitory expenses, which include a meal plan, are $4600 to $5200 for the academic year. In addition, off-campus apartments and rooms are available for rents ranging upward from $250 per month.

Student Group

UIC enrolls approximately 25,000 students, nearly one fifth of whom are at the graduate level. Students come from throughout the United States and overseas, as well as from Illinois. The department enrolls 60 undergraduates and 35 graduate students.

Location

UIC is located in a pleasant neighborhood on Chicago's Near West Side, five minutes by public transit from the Loop. The neighborhood contains Jane Addams Hull House, two historic-landmark residential areas, and the West Side Medical Center District, the world's largest concentration of health-care facilities. Residents have easy access to the city's numerous cultural and social amenities, including world-renowned museums and fine restaurants. Chicago's many educational institutions contribute to a lively intellectual atmosphere.

The University

The University of Illinois at Chicago is one of the two campuses of the University of Illinois. It was formed in 1982 by the merger of the preexisting Chicago Circle (1965) and Medical Center (1896) campuses and constitutes the largest public university in the Chicago area. The combined campuses support extensive research and teaching facilities, many of which are at the forefront of their fields. More than 1.3 million volumes are housed in the University's library and specialized collections. UIC offers bachelor's degrees in ninety-seven subjects, master's degrees in seventy-nine areas, and doctorates in more than forty-four specialties.

Applying

Applications for admission and financial aid are available from the department. Completed applications, including a personal statement, three letters of reference, transcripts, and scores on the GRE General Test (quantitative, verbal, and analytical), should be submitted by January 15 for fall-quarter admission. Although it is possible to begin study at any quarter, most students enter in the fall and receive most of the available financial aid. The GRE should be taken well in advance of application to ensure that the scores are available. Applicants should have their scores sent directly to the department by ETS. A minimum grade point average of 4.0 on a 5.0 scale is usually required for admission.

Correspondence and Information

Dr. Kelvin S. Rodolfo
Director of Graduate Studies
Department of Geological Sciences
University of Illinois at Chicago
Box 4348
Chicago, Illinois 60680
Telephone: 312-996-3154

Peterson's Guide to Graduate Programs in the Physical Sciences and Mathematics 1992

SECTION 3: EARTH AND PLANETARY SCIENCES

University of Illinois at Chicago

THE FACULTY AND THEIR RESEARCH

John R. Bolt, Adjunct Associate Professor; Ph.D., Chicago, 1968. Vertebrate paleontology, morphology and adaptations of Paleozoic amphibians.

Robert E. DeMar, Professor; Ph.D., Chicago, 1961. Vertebrate paleontology, functional morphology, evolution of dentitions in reptiles and mammals.

Martin F. J. Flower, Professor; Ph.D., Manchester, 1971. Petrology and geochemistry, magma genesis, geochemistry of the mantle, magmatic differentiation in the early Earth.

Stephen Guggenheim, Professor; Ph.D., Wisconsin–Madison, 1976. Mineralogy and crystallography, silicate mineralogy, crystal chemistry and stability of layer silicates.

August F. Koster van Groos, Professor; Ph.D., Leiden (Netherlands), 1966. Experimental petrology and geochemistry, origin of crust and carbonatites, liquid immiscibility, high-pressure differential thermal analysis.

Roy E. Plotnick, Associate Professor; Ph.D., Chicago, 1983. Invertebrate paleontology, paleoecology, functional morphology and taphonomy of arthropods, mathematical and statistical models.

Kelvin S. Rodolfo, Professor; Ph.D., USC, 1967. Marine geology and sedimentation; volcanoclastic sedimentation and tectonics of island arcs; geology of Southeast Asia, with emphasis on the Philippines.

Norman D. Smith, Professor; Ph.D., Brown, 1967. Sedimentology and stratigraphy; fluvial, lacustrine, and glacier-related sedimentology; placer deposits.

Carol Ann Stein, Assistant Professor; Ph.D., Columbia, 1984. Marine geophysics and seismology, plate tectonics.

SECTION 3: EARTH AND PLANETARY SCIENCES

UNIVERSITY OF MARYLAND COLLEGE PARK

Department of Geology

Programs of Study

The Department of Geology offers M.S. and Ph.D. degree programs in two general areas of concentration: lithospheric processes and environmental and surface studies. The lithospheric process group focuses on a wide range of fundamental problems in crustal formation and evolution and employs theoretical, field, and experimental techniques. The environmental and surface studies group addresses problems in geohydrology, geomorphology, and environmental geochemistry and geology. More specific faculty research interests are described on the reverse of this page. Within these groups, there are opportunities for collaborative research with area institutions, including the Smithsonian Institution, the United States Geological Survey, the National Aeronautical and Space Administration, and the National Institute of Standards and Technology.

Research Facilities

The department maintains a variety of modern facilities and equipment for research, including a Sun Microsystem computer network with direct access to supercomputer facilities; laboratories for research on the petrography and petrology of igneous, metamorphic, and sedimentary rocks; a Cue 3 color image analysis system; a Fluid, Inc. stage for fluid inclusion analysis; research microscopes with instrumentation for the measurement of reflectance; rock preparation laboratories; high-temperature and high-pressure/high-temperature equipment for dry or hydrothermal experiments; mass spectrometers and ancillary equipment for isotope analysis; laboratories and field facilities for the study of the hydrogeology and geochemistry of natural waters; flame and graphite furnace atomic absorption equipment; a JEOL 840 electron microprobe; and automated X-ray diffractometry apparatus. Analytical scanning and transmission electron microscopy are available on campus for geological research.

Financial Aid

Half-time teaching assistantships, research assistantships, and graduate school fellowships, with stipends ranging between $9500 and $14,000 plus tuition and medical coverage, were available to qualified applications in 1990–91.

Cost of Study

In fall 1991, the self-supporting student who is a Maryland resident pays $143 per credit hour plus mandatory fees of $140. Out-of-state students pay $256 per credit hour plus the mandatory fees.

Cost of Living

Some University graduate student housing is available. Costs range from $4500 per year for a one-bedroom apartment to $5000 per year for a two-bedroom apartment. Slightly more expensive off-campus housing in College Park is plentiful. Single rooms are available for $250–$350 per month. Estimated expenses for single students are $2000 for board, $1300 for transportation, and $1500 for personal expenses. Estimated expenses for married students include a higher figure for board.

Student Group

The Department of Geology has approximately 25 graduate students. The total graduate student enrollment on the College Park campus (full-time and part-time) in the fall of 1990 is 8,962; the undergraduate enrollment is 26,863. Most students graduating from the program with an M.S. go on to a Ph.D. program or find employment in the federal government or in the private sector in the Washington, D.C., metropolitan area.

Location

The College Park campus of the University of Maryland is located in the town of College Park, a suburb of Washington, D.C. A wealth of cultural, educational, and athletics activities are located in the metropolitan area. The Metro (the Washington area subway) will soon connect College Park with downtown Washington.

The University and The Department

The College Park campus has 280 major buildings on 1,300 acres. The department is part of the College of Computer, Mathematical and Physical Sciences, which has 1,556 undergraduate students, 834 graduate students, and a faculty of 486. The department was established in 1973 and its graduate program began in 1982. A strong sense of collegiality and cooperative spirit characterizes the department. Faculty and students work on many problems, but focus primarily on structural, isotopic, and petrologic investigations of tectonic and metamorphic processes; mechanisms of sediment transport; surface, near-surface, and deep-crustal fluid flow; and laboratory, isotopic and field studies of magmatic and ore-forming processes. Although the department is divided into two groups, research crosses group boundaries. For more information about the research programs, students can contact any of the faculty members in the department.

Applying

Candidates may apply for admission in either fall or spring. Applications must be received by May 1 for admission to the fall semester. If financial support is required, applications must be received by February 15. Scores on the Graduate Record Examinations General Test are required for consideration for admission; the Graduate Record Examinations Subject Test in geology is recommended but not required.

Correspondence and Information

Chair, Graduate Committee
Department of Geology
University of Maryland
College Park, Maryland 20742
Telephone: 301-405-4079
            301-405-4365

Peterson's Guide to Graduate Programs in the Physical Sciences and Mathematics 1992

SECTION 3: EARTH AND PLANETARY SCIENCES

University of Maryland College Park

THE FACULTY AND THEIR RESEARCH

Michael Brown, Professor and Chairman; Ph.D., Keele (England), 1975. Petrology of metamorphic rocks, the petrogenesis of high-temperature metamorphic belts, and the petrogenesis of migmatites and anatectic granites; the petrogenesis of calc-alkaline plutonic rocks; the Cadomian orogen of northwest France; the displacement history and tectonic significance of the Atacama fault system in north Chile. (Telephone: 301-405-4080)

Philip A. Candela, Associate Professor; Ph.D., Harvard, 1982. Thermodynamics and mass transfer dynamics of magmatic-hydrothermal systems; experimental studies of the distribution of ore metals and related substances in high-temperature, multicomponent, polyphase systems; field studies of granitic rocks in the western United States, with special reference to the behavior of halogens and ore metals in magmatic systems; modeling of low-temperature and high-temperature aqueous systems; geology of the Appalachian Piedmont. (Telephone: 301-405-2783)

Luke L. Y. Chang, Professor; Ph.D., Chicago, 1963. Structural and stability relations in mineral systems; mineralogy of Pb-Ag-Cu-Sb-Bi-Sn sulfides, selenides, and tellurides; process mineralogy of industrial materials. (Telephone: 301-405-4086)

David Harding, Assistant Research Scientist; Ph.D., Cornell, 1988. Tectonic evolution of the East African rift and Afar depression in Ethiopia, structural analysis of the high-temperature deformation history of oceanic lithosphere as recorded in ophiolites, quantitative application of multispectral remote-sensing data to geologic mapping, analysis of terrain morphometrics and geomorphic processes from digital topographic data, acquisition of topographic data via laser altimetry. (Telephone: 301-405-4083)

Eirik J. Krogstad, Assistant Professor; Ph.D., SUNY at Stony Brook, 1988. Formation and evolution of the continental crust in the Late Archean of south India and the Paleozoic of the Appalachian orogen; isotopic tracer, major, and trace element petrogenetic studies; sediment provenance studies applied to tectonics; U-Pb geochronology. (Telephone: 301-405-4088)

Eileen L. McLellan, Associate Professor; Ph.D., Cambridge, 1983. Field and lab studies of structure and petrology of metamorphic belts, on scales from microscopic to regional; petrogenesis of migmatites; applications of remote sensing to regional tectonics; fluid inclusion studies; geology and public policy with special interests in resource issues. (Telephone: 301-405-4087)

Karen L. Prestegaard, Associate Professor; Ph.D., Berkeley, 1982. Sediment transport and depositional processes in mountain gravel-bed streams; mechanisms of streamflow generation and their variations with watershed scale, geology, and land use; hydrologic behavior of frozen ground; hydrologic and erosional consequences of storm type and hydrologic consequences of climate change; hydrology of coastal and riparian wetlands. (Telephone: 301-405-6982)

Robert W. Ridky, Associate Professor; Ph.D., Syracuse, 1973. Landform analysis and interpretation of glacial dynamics from Quaternary age landforms; geomorphic analysis of drainage basin development; geologic amelioration techniques for isolation of hazardous-waste repositories from surface/groundwater systems; analysis of process dependent shapes (landform scale to particle size); geologic education, teacher training, and development of instructional material for public education. (Telephone: 301-405-4090)

Antonio V. Segovia, Associate Professor; Ph.D., Penn State, 1963. Identification of recent deformation of the earth's crust by means of geomophologic analysis and remote-sensing image interpretation, effects of fracture zones on soil development and groundwater circulation. (Telephone: 301-405-6979)

Peter B. Stifel, Associate Professor; Ph.D., Utah, 1964. Paleontology, paleobiology, and biostratigraphy of mid-Atlantic Phanerozoic rocks; ontogenetic studies; mulluscanpaleoecology, Tertiary otoliths, and trace fossil studies. (Telephone: 301-405-4078)

Richard J. Walker, Assistant Professor; Ph.D., SUNY at Stony Brook, 1984. Geochemical evolution of the earth's crust and mantle; origin and evolution of early solar system materials, including iron meteorites and chondrites; petrogenesis of granites and granitic pegmatites; petrogenesis of ore systems (Sudbury, Canada; Stillwater, United States; Pechenga, USSR); geochemistry of Re and Os. (Telephone: 301-405-4089)

Nicholas B. Woodward, Assistant Research Scientist; Ph.D., Johns Hopkins, 1981. Fold and thrust belts; tectonics of the central Rocky Mountains, southern Appalachians, and Tasman orogenic belt in southeastern Australia; mechanics of folding and thrusting. (Telephone: 301-405-4084)

Ann G. Wylie, Associate Professor; Ph.D., Columbia, 1972. Economic geology of Appalachian metal and industrial deposits; mineralogy and human health, with particular interest in minerals identified as carcinogens, such as asbestos and crystalline silica; ore microscopy and the study of ore minerals as petrogenetic indicators; geology and tectonic history of the central Appalachian Piedmont. (Telephone: 301-405-4079)

E-An Zen, Adjunct Professor; Ph.D., Harvard, 1955. Phase equilibria in metamorphic and sedimentary rocks; igneous petrology; tectonics of the Appalachians and northern Rockies. (Telephone: 301-405-4081)

Associated Nonresident Faculty

James F. Luhr, Adjunct Associate Professor; Ph.D., Berkeley, 1980. Igneous petrology, volcanology. (Smithsonian Institution)

Sorena S. Sorenson, Adjunct Assistant Professor; Ph.D., UCLA, 1984. Metamorphic petrology. (Smithsonian Institution)

Yuming Zhang, Assistant Research Scientist; Ph.D., Maryland, 1988. Mineralogy, geochemistry. (National Institute of Standards and Technology)

UNIVERSITY OF MINNESOTA
Department of Geology and Geophysics

Programs of Study

Programs leading toward the degrees of Master of Science and Doctor of Philosophy are conducted in two disciplines: geology and geophysics. M.S. programs may be completed with or without a thesis. Ph.D. programs may be interdepartmental. Courses are chosen in consultation with an adviser in the student's field of interest. Graduate class sizes normally range from 5 to 15.

An M.Ed. program in geology is offered for secondary school science teachers. Of the required 45 credits, 27 must be in geology, 9 in education, and the remainder in related science and mathematics courses.

Research Facilities

There are several X-ray generators; powder and single-crystal cameras; diffractometers; two solid-source mass spectrometers, including a Finnigan 262, a two-stage instrument, and a Finnigan Delta-E gas mass spectrometer; carbonate and silicate extraction lines; a clean chemistry laboratory for trace-element and isotopic studies; a luminoscope; hydrothermal and mineral synthesis laboratories; water chemistry laboratories equipped for DC-plasma emission spectrometry, ion chromatography, gas chromatography, and carbon analysis; pollen and diatom laboratories; geophysical equipment for seismic studies; gravity meters; magnetometers; electrical resistivity meters; computer graphics workstations (including CRAY X-MP and IBM 3090-600J supercomputers of the Minnesota Supercomputer Institute); a scanning electron microscope and a high-resolution analytical transmission electron microscope; a fluid-inclusion laboratory; a rock magnetism laboratory; a high-pressure diamond anvil laboratory equipped for ultrahigh-pressure X-ray diffraction and optical studies; and a rock deformation laboratory with high-temperature and high-pressure testing apparatus. Cold-room facilities are available in the department for glaciological studies, and the department has access to a well-equipped glaciological field camp in northern Sweden. A comprehensive collection of geological books and journals is housed in the Institute of Technology Library.

Financial Aid

A number of half-time research assistantships and teaching assistantships, carrying nine-month stipends up to about $8500, are awarded each year. A student with a half-time assistantship automatically receives a full tuition scholarship for up to 15 credits per quarter. Several fellowships are also awarded by the department and by the Graduate School. Graduate theses involving geological, hydrogeologic, mineral deposits, and geophysical problems pertinent to the state of Minnesota are sometimes supported by the Minnesota Geological Survey or by other service and research centers in the University or the Twin Cities area. Students receiving fellowship support may also receive partial or full tuition scholarships.

Cost of Study

Tuition for 1990–91 was $1030 per quarter for Minnesota residents, $2060 per quarter for nonresidents, and $300 after completion of the Ph.D. preliminary examinations. All students were assessed a student services fee of approximately $120 per quarter, which included prepaid general outpatient health care. Hospitalization insurance is required and was available at approximately $130 per quarter for a single policy. Tuition for 1991–92 is expected to increase.

Cost of Living

Dormitory room and board rates vary from about $800 per quarter for small double rooms to about $950 per quarter for large single rooms and suites. Many rooming houses and apartments exist close to campus. Sharing of rooms and flexibility in eating arrangements make expenses variable.

Student Group

The University of Minnesota has approximately 47,300 students, including about 7,900 graduate students. The Department of Geology and Geophysics has about 50 undergraduates and about 60 graduate students. Graduate students come from many areas of the United States and several foreign countries; most receive some form of financial support. Recent graduates have been employed by universities, petroleum companies, the U.S. Geological Survey, and environmental agencies.

Location

The Twin Cities metropolitan area consists of about 2 million people. It is a center for several electronics and computer firms as well as for machinery and grain industries. Cultural and recreational facilities abound: symphony, ballet, opera; numerous theaters including the famed Tyrone Guthrie Theatre; professional baseball, basketball, hockey, and football; many lakes and parks with public facilities for swimming and boating; skating rinks, small ski slopes, and cross-country ski trails for winter activities. The climate ranges from -30°C in winter to 35°C in summer as extremes, with a 1-m average snowfall in winter, and about 700 mm of rain per year.

The Department

The department was one of the first in the University and celebrated its centennial in 1972. The first Ph.D. at Minnesota was in geology and was awarded in 1897. The department houses large collections of minerals, rocks, and fossils; machine and electronics shops; and sample preparation laboratories.

Applying

Prospective graduate students should complete an application for admission and, if desired, an application for financial aid. These may be obtained from the Graduate School Admissions Office and normally should be returned by January 15, together with all supporting material. Financial awards are made in mid-March. Quarters commence in late September, early January, and late March. Entrance exams are not required, although a B average is necessary for admission. The GRE General Test is required. A TOEFL score is required of all international applicants.

A strong background in the primary sciences is highly recommended for admission to graduate work. A bachelor's degree in geology or geophysics is not required for matriculation as a candidate for advanced degrees in these fields, however, and students holding degrees in other sciences or engineering are encouraged to apply for graduate study. All graduate students are required to have, or to obtain early in their graduate program, a background in geology and cognate sciences equivalent to that of the University's graduating seniors.

Correspondence and Information

Director of Graduate Studies
Department of Geology and Geophysics
310 Pillsbury Drive SE
University of Minnesota
Minneapolis, Minnesota 55455
Telephone: 612-624-1333

SECTION 3: EARTH AND PLANETARY SCIENCES

University of Minnesota

THE FACULTY AND THEIR RESEARCH

E. Calvin Alexander Jr., Ph.D., Missouri–Rolla. Isotope geochemistry, geohydrology, geochronology, ground water pollution and environmental geology, karst hydrology and geomorphology, cosmochronology, rare-gas isotopic studies.

Subir K. Banerjee, Ph.D., Sc.D., Cambridge; Director, Institute for Rock Magnetism. Geophysics: geomagnetism and paleomagnetism, secular variation and reversal of the earth's magnetic field as recorded in rocks and sediments, theory and applications of rock magnetism, magnetic proxy records of paleoenvironmental and paleoclimatic changes, sources of magnetic anomalies, history of medieval Islamic science.

Val Chandler, Ph.D., Purdue; Geophysicist, Minnesota Geological Survey. Gravity and magnetic interpretation, Precambrian studies, structure of continental crust.

R. Lawrence Edwards, Ph.D., Caltech. Isotope geochemistry and geochronology; use of uranium series and long-lived isotopic tracers and chronometers to examine climatic, petrologic, oceanographic, and seismic processes.

Daniel Engstrom, Ph.D., Minnesota. Quaternary paleolimnology, limochronology, and lacustrine geochemistry; paleohydrology of saline lakes; lacustrine records of environmental pollution; neoglacial lake ontogeny in Glacier Bay, Alaska.

Priscilla C. Grew, Ph.D., Berkeley; Director, Minnesota Geological Survey. Geology and public affairs, environmental geology, natural resources management, science and regulatory policy studies.

Roger LeB. Hooke, Ph.D., Caltech. Geomorphology and glaciology: fieldwork in Sweden on glacier dynamics with particular emphasis on processes at the glacier-bed interface, ice deformation, and processes of formation of glacial landforms; fieldwork in southern California on processes on alluvial fans.

Peter J. Hudleston, Ph.D., Imperial College (London); Head of the School of Earth Sciences. Structural geology and tectonics: mechanics and geometry of folds, experimental and numerical modeling, development of foliation and crystallographic fabric, structures in glacial ice, regional structural analysis.

Emi Ito, Ph.D., Chicago. Isotope geochemistry: oxygen, carbon, and hydrogen isotopic tracer methods applied to paleoclimate and paleoecology studies; problems of crust-mantle cycling and island arc volcanism; studies of rock-water interaction related to metamorphism and tectonism.

Robert G. Johnson, Ph.D., Iowa State. Paleoclimatology and Pleistocene climate change mechanisms, Milankovitch orbital effects on glacial and interglacial climates, Barbados.

Shun-ichiro Karato, Ph.D., Tokyo. Mineral and rock physics, geophysics: plastic deformation of minerals and rocks, seismic anisotropy, mantle rheology and dynamics.

Kerry R. Kelts, Ph.D., Zurich; Director, Limnological Research Center. Limnogeology, past global change, geochemical sedimentology, Quaternary studies, marine and limnetic paleoenvironments.

Karen L. Kleinspehn, Ph.D., Princeton. Clastic sedimentology, basin analysis, tectonics and sedimentation, dynamics of strike-slip fault zones, fission-track dating, fieldwork in British Columbia and Spitsbergen.

David L. Kohlstedt, Ph.D., Illinois at Urbana-Champaign. Geophysics; high-temperature, high-pressure flow of rocks and minerals; point defects, dislocations, and grain boundaries in silicates; kinetic processes in rocks; melt migration in mantle rocks; diffusion of water and hydrogen in minerals.

Sally Gregory Kohlstedt, Ph.D., Illinois at Urbana-Champaign. History of science, nineteenth-century natural history, American scientific institutions, women in science, science in culture and education, with special reference to museums and the nature study movement.

Glenn B. Morey, Ph.D., Minnesota; Associate Director/Chief Geologist, Minnesota Geological Survey. Precambrian geology; geologic mapping, stratigraphy, and sedimentology; geology of Lake Superior region; iron and manganese deposits; history of geology.

Bruce M. Moskowitz, Ph.D., Minnesota; Associate Director, Institute for Rock Magnetism. Geophysics: rock magnetism and paleomagnetism, magnetism of fine particles, theoretical and experimental study of magnetic domains, biomagnetism.

V. Rama Murthy, Ph.D., Yale. Mantle geochemistry and petrology, isotope geology, the early history of the Earth, chemical differentiation processes.

Christopher Paola, Ph.D., MIT/Woods Hole. Sedimentology, bedform dynamics, dynamics of river and lake sediments, quantitative basin analysis.

Hans-Olaf Pfannkuch, Dr.Ing., Paris. Groundwater geology: watershed studies, water resources systems, flow of fluids in porous media, structure of porous media, dispersion in porous systems; mass transport in aquifers, especially hazardous and toxic wastes and hydrogeologic implications for environmental geology; hydrogeologic vulnerability and risk assessment.

William E. Seyfried Jr., Ph.D., USC. Aqueous geochemistry, experimental geochemistry, geochemistry of hydrothermal ore deposits.

Joseph Shapiro, Ph.D., Yale; Associate Director, Limnological Research Center. Biological and chemical limnology, pollution limnology, lake restoration, lake management.

Robert E. Sloan, Ph.D., Chicago. Paleontology: paleontology, stratigraphy of latest Cretaceous and earliest Tertiary rocks of the western United States and the Ordovician of the Upper Mississippi Valley.

David L. Southwick, Ph.D., Johns Hopkins; Assistant Director and Research Associate, Minnesota Geological Survey. Precambrian geology; geologic mapping, structural and petrologic studies in Precambrian terranes of Minnesota; Precambrian tectonics.

James H. Stout, Ph.D., Harvard; Director of Graduate Studies. Phase petrology, mineral chemistry, and geology of igneous and metamorphic rocks in Norway, Colorado, and Hawaii; theory of multisystem nets and their application to mantle petrology and melting; high-pressure phase equilibria and PVT equations of state using diamond anvil and energy-dispersive diffraction techniques.

Frederick M. Swain (Emeritus), Ph.D., Kansas. Micropaleontology, biostratigraphy, and organic geochemistry; Mesozoic microfaunas of Gulf and Atlantic regions; Cenozoic nonmarine microfaunas of Great Basin; carbohydrate geochemistry.

Christian P. Teyssier, Ph.D., Monash (Australia). Structural geology, tectonics, microtectonic and petrofabric analyses, rock and lithospheric mechanics. Current field areas are central Australia, western Australia, Spitsbergen (Norway), Mexico, and the western United States.

Matt S. Walton (Emeritus), Ph.D., Columbia; former Director, Minnesota Geological Survey. Fundamental problems of Minnesota geology and geologic factors related to regional planning and natural resources utilization, dynamic relationships between geologic processes and activities of humans.

Barbara J. Wanamaker, Ph.D., Princeton. Mineral and rock physics and geophysics: attenuation, electrical conductivity, point defects and diffusion, partially rotten rocks.

Paul W. Weiblen, Ph.D., Minnesota; Director, Space Science Center. Crustal evolution based on field and petrologic studies of the Precambrian volcanic and gabbroic rocks of northern Minnesota, Archean gneisses of the Minnesota River Valley, and lunar samples and meteorites; quantitative analysis with electron and heavy-ion beam techniques.

Herbert E. Wright Jr. (Emeritus), Ph.D., Harvard. Pleistocene studies, limnology, geomorphology: glacial geology, paleolimnology, paleoecology, and vegetation history in North America, Europe, and the Andes.

David A. Yuen, Ph.D., UCLA; Fellow, Minnesota Supercomputer Institute. Geophysics: geophysical fluid dynamics, mantle convection, inferences of mantle viscosity, magma processes, viscoelasticity, mathematical geophysics, supercomputing and graphics.

Tibor Zoltai (Emeritus), Ph.D., MIT. Determination and comparative study of the crystal and crystal surface structures of minerals, systematics of minerals and crystal structures, physical properties and environmental effects of mineral particles (like asbestos) in air pollution.

SECTION 3: EARTH AND PLANETARY SCIENCES

UNIVERSITY OF NEVADA, RENO

Mackay School of Mines
Department of Geological Sciences

Programs of Study

The Department of Geological Sciences, part of the Mackay School of Mines, offers the Master of Science degree in geological engineering and geochemistry and both master's- and doctoral-level programs in geology and related earth sciences, geophysics, hydrology, and hydrogeology. One of the greatest strengths of the graduate program is that the School is located in a structurally and lithologically varied region. The area offers opportunities for a wide variety of petrologic, structural, stratigraphic, and economic studies of rocks, ranging in age from Precambrian to Holocene. Seismic, volcanic, and geothermal phenomena are abundant, as are many research opportunities in rock, soil, and snow mechanics; groundwater; earthquake engineering; remote sensing; environmental geology; geostatistics; and engineering geology.

A major research strength of the department is the close relationship, via joint faculty appointments, with other state agencies housed in the School of Mines. These agencies include the Nevada Bureau of Mines and Geology, the Seismological Laboratory, and the Center for Neotectonic Studies. Additionally, courses are taught by faculty from the Desert Research Institute in groundwater and environmental areas. The United States Geological Survey (USGS) maintains a satellite office at the School in the area of mineral development.

Research Facilities

Mackay School of Mines occupies four buildings with 150,000 square feet of floor space. Over seventy well-equipped laboratories are available for research. More than $7-million is allocated for the remodeling of the historic Mackay Mines Building in 1991–92. The department has excellent research facilities and laboratories in the areas of remote sensing, rock and soil mechanics, seismology, geochemistry, petrology, paleomagnetism, ore volcanic chemistry, and hydrology/hydrogeology. A complete suite of analytical equipment is available for students' research. Students should phone the department chair for an updated list.

The University's library contains 1 million volumes and subscribes to 6,000 periodicals. The Mines Library has approximately 10 percent of the University's holdings. Computer search capabilities and interlibrary loan agreements facilitate the locating and retrieving of reference materials from around the world.

A broadband computer network provides access to NSF supercomputers and thirteen University of Nevada multiuser systems. Specialized computer systems are also available for research involving CAD/CAM, control systems, data acquisition and analysis, image processing, microprocessors, neuronetworks, power systems, signal processing, and seismographic data analysis. Several personal computer labs containing ten to fifty units each are also available to students for general use.

Financial Aid

Fellowships, teaching assistantships, and research assistantships are available. Both the college and the School attempt to support as many graduate students as possible. Teaching assistantships are part-time (20 hours per week) and pay a minimum of $8000 per academic year plus tuition. Research assistantships can be for a full twelve months—research assistants work part-time (20 hours per week) during the academic year and full-time (40 hours per week) during the summer. Two $15,000 fellowships are awarded each year to outstanding students by a new NASA space grant to UNR.

Cost of Study

For 1991–92, the registration fee for graduate-level courses is $66 per credit. Students who are not residents of Nevada are also charged tuition of $1800 per semester.

Cost of Living

Accommodations (including meals) in the campus dormitories cost $3720 per academic year. Married student housing costs $260 per month. Transportation and personal expenses typically cost $2700 per year.

Student Group

Twelve thousand students are enrolled at the University of Nevada, Reno. In spring 1991, enrollment in the School of Mines was 201 undergraduates and 182 graduate students.

Location

The main campus is located on 200 acres of rolling hills north of Reno's business district. The majestic Sierra Nevada and the vast high deserts of Nevada provide boundless recreational opportunities for skiing, backpacking, fishing, and other wilderness experiences.

The University

The University of Nevada, Reno, has a proud history and a bright future. Today, its intimate atmosphere and excellent programs provide an academic environment of high quality.

Applying

To apply for graduate standing, geology students must have an appropriate bachelor's degree (or the equivalent) from an accredited college, have an undergraduate GPA of at least 2.75 (or 3.0 in the last two years), and file an application with the admissions and records office. Applications for graduate standing must be approved by the appropriate department chairman, college dean, and graduate school dean. GRE scores are required. International students must have TOEFL scores of at least 500 to be considered for admission. Admission applications and all credentials must be received at least three weeks prior to anticipated registration. Three letters of recommendation must also be sent directly to the department.

Correspondence and Information

Department of Geological Sciences/168
Mackay School of Mines
University of Nevada, Reno
Reno, Nevada 89557-0138
Telephone: 702-784-6050

Peterson's Guide to Graduate Programs in the Physical Sciences and Mathematics 1992

SECTION 3: EARTH AND PLANETARY SCIENCES

University of Nevada, Reno

REPRESENTATIVE FACULTY MEMBERS AND THEIR RESEARCH

Robert J. Watters, Chairman of the Department; Ph.D., London; PE. Geological engineering, slope stability, applied rock mechanics, acoustical emissions.

John G. Anderson, Ph.D., Columbia. Seismic hazards, strong earthquake ground motions.
James R. Brune, Ph.D., Columbia. Source theory, earth structure, geophysical engineering.
James R. Carr, Ph.D., Arizona; PE. Mathematical geology, aerospace remote sensing, geological engineering.
James R. Firby, Ph.D., Berkeley. Invertebrate paleontology, micropaleontology.
Don Helm, Ph.D., Berkeley. Groundwater and hydraulics.
Malcolm J. Hibbard, Ph.D., Washington (Seattle). Magma mixing, granite genesis.
Liang-chi Hsu, Ph.D., UCLA. Geochemistry of ore deposits, hydrothermal phase relationships.
Elizabeth Jacobson, Ph.D., Arizona. Geostatistics and inverse theory in hydrology.
Roger Jacobson, Ph.D., Penn State. Geochemistry of groundwater.
Robert Karlin, Ph.D., Arizona State. Geophysics, paleomagnetism, marine geology, sediment geochemistry.
Lawrence T. Larson, Ph.D., Wisconsin; PG. Economic geology, geology of ore deposits, exploration geochemistry.
W. Berry Lyons, Ph.D., Connecticut. Low-temperature geochemistry, hydrogeology.
Donald C. Noble, Ph.D., Stanford. Volcanogenic ore deposits, elemental and isotope geochemistry, tectonics.
Richard A. Schultz, Ph.D., Purdue. Tectonics, rock mechanics, geological engineering, planetary science.
Richard Schweickert, Ph.D., Stanford. Structural geology and tectonics, stratigraphy.
James V. Taranik, Ph.D., Colorado School of Mines. Applied geophysics, aerospace remote sensing.
James H. Trexler, Ph.D., Washington (Seattle). Sedimentology, stratigraphy.
Steven Wheatcraft, Ph.D., Hawaii. Groundwater contaminant transport, numerical modeling.
Steven Wesnousky, Ph.D., Columbia. Neotectonics, seismology.

SECTION 3: EARTH AND PLANETARY SCIENCES

UNIVERSITY OF NEW MEXICO

Department of Geology

Programs of Study
The Department of Geology offers programs leading to the M.S. and Ph.D. degrees. The M.S. program requires a minimum of 24 credit hours and a thesis. The Ph.D. program requires a minimum of 24 credit hours beyond the master's degree program and a dissertation. Students will generally be expected to complete the M.S. before being admitted to the Ph.D. program. In both programs, the distribution of courses is flexible and will be designed to fit the needs and interests of the student. A minimum of one foreign language is required for the Ph.D. degree; there is no foreign language requirement for the M.S.

Programs are available leading to M.S. and Ph.D. degrees with emphasis in crystallography, geochemistry, geomorphology, geopedology and soil chemistry, geophysics, hydrogeology, meteoritics, mineralogy, paleoecology, paleomagnetism, paleontology, petrology, Quaternary geology, sedimentology, stable-isotope geochemistry, stratigraphy, structural geology, and tectonics. The full scope of areas of specialization is indicated by the research interests of the faculty (listed on the reverse of this page). The Institute of Meteoritics offers exceptional opportunities for students interested in meteoritics and planetary studies.

Research Facilities
The Department of Geology occupies a modern four-story building. In addition to conventional instructional facilities and equipment, the department has the following instrumental facilities and associated laboratories: automated JEOL 733 superprobe; Hitachi 450 scanning electron microscope with EDX system; instrumental neutron activation analysis system; JEOL 2000FX analytical electron microscope; powder and single-crystal X-ray diffraction equipment; Scintag automated powder X-ray diffractometer; high-temperature crystal synthesis facilities; 12-inch, 90° solid-source mass spectrometer for Rb-Sr work; 3-inch, 60° mass spectrometer; Finnigan Delta-E gas-source isotope ratio mass spectrometer; VG Plasma Quad PQ2 ICP-MS; three atomic absorption spectrophotometers; two 3-SQUID-axis, horizontal access, superconducting rock magnetometers with automated alternating field and thermal demagnetization capability; spinner magnetometer; conventional demagnetization and rock magnetism instruments; proton precession magnetometer; gravity meter; portable seismograph; and shallow resistivity unit. A Cameca ion microprobe will be on campus in spring 1992. Laboratories for petrography, polished and thin-section preparation, and analytical chemistry are well equipped. Computer facilities include an IBM 3081K mainframe networked with a VAXcluster consisting of a VAX 8650, a VAX-11/785, a VAX-11/780, two MicroVAX IIs, and a Sequent Balance 8000. Departmental computational facilities include a VAXstation II/GPX running VMS with DECnet access to the VAXcluster and a large number of microcomputers (both MS-DOS and Macintosh), many of which are hardwired into the campuswide broadband network for mainframe access. There are 6 technicians associated with the various laboratories noted above. Facilities at the Los Alamos Scientific Laboratory, Sandia National Laboratories, and the Albuquerque Seismological Laboratory are also available.

The Institute of Meteoritics is an integral part of the Department of Geology and has an extensive collection of planetary materials for research and teaching.

Financial Aid
Financial assistance is available to eligible students in the form of graduate assistantships (teaching and research) paying a minimum of $2200 and a maximum of $7200 for nine months in 1991–92. A number of University and departmental fellowships and scholarships are also available. The Leon T. Silver and Vincent C. Kelley Fellowships provide a monthly stipend of $1400 for nine or twelve months for personal expenses and up to $3000 per year for travel and research expenses. The Caswell Silver Foundation pays all tuition and University fees for holders of these fellowships.

Cost of Study
In 1991–92, tuition fees for full-time graduate students (12 or more credit hours) are $842 for New Mexico residents and $2828 for nonresidents. For students carrying fewer than 12 credit hours, the fees for residents and nonresidents are approximately $71.50 and $237 per credit hour, respectively.

Cost of Living
Full-time students can expect expenses of about $4000 (residents) or $5600 (nonresidents) per semester. These figures include tuition and are based on dormitory residence including board.

Student Group
The University of New Mexico has about 24,000 students, of whom approximately 3,500 are graduate students. Sixty undergraduate majors and 60 graduate students are enrolled in geology programs. Many geology graduate students receive financial aid.

Location
The University of New Mexico is located in Albuquerque, a city of 500,000 inhabitants and the major business, light industry, and cultural center of the state. The city lies along the Rio Grande at an average elevation of 5,300 feet; it enjoys a dry climate with distinct seasonal, though not extreme, temperatures. Nearby mountains and deserts provide a wide variety of opportunities for outdoor recreational activities, including skiing, boating, and camping. Numerous theaters, museums, and other resources in Albuquerque and nearby Santa Fe provide a wide range of cultural activities. Albuquerque enjoys a tradition in the arts that is unusual for a city of its size; contemporary crafts as well as traditional Indian and New Mexican crafts are noteworthy.

The University
The University of New Mexico opened in 1892. Since then it has grown to be the largest institution of higher learning in the state in both undergraduate and graduate programs. The University libraries have more than 1 million volumes. A number of public museums are maintained, particularly in geology, anthropology, and art. A University Program Series provides a wide range of cultural offerings during the year. There is a modern Student Union building and an International Students' Center. The presence in the state of a number of research organizations (Los Alamos Scientific Laboratory, Sandia National Laboratory, Holloman and Kirtland Air Force bases, and the Lovelace Foundation) has enhanced the scientific and engineering climate at the University. Geology has had an important and growing role in this broad scientific climate.

Applying
Students holding a bachelor's degree in geology are invited to apply for admission to the graduate program. Students holding degrees in other science areas may also be admitted with the understanding that they will be required to make up certain deficiencies in basic geology courses. Applications for admission should be received by the department by January 31 (financial aid) or May 1 (admission only) for the fall semester and November 1 (admission only) for the spring semester.

Correspondence and Information
Barry S. Kues, Chairman
Department of Geology
University of New Mexico
Albuquerque, New Mexico 87131-1116

SECTION 3: EARTH AND PLANETARY SCIENCES

University of New Mexico

THE FACULTY AND THEIR RESEARCH

Gary D. Acton, Caswell Silver Research Professor; Ph.D., Northwestern. Geodynamics and paleomagnetism.
Roger Y. Anderson, Professor; Ph.D., Stanford. Paleoclimatology, palynology; seasonal laminations and processes; reconstruction of limnic and evaporitic environments.
Douglas G. Brookins, Professor; Ph.D., MIT. Isotope geochemistry; geochronology; petrogenesis of carbonatites and kimberlites; geochemistry of uranium deposits; geochemistry of radioactive waste disposal sites and related studies.
Michael E. Campana, Associate Professor; Ph.D., Arizona. Hydrogeology, geothermal systems, arid land hydrology, groundwater and geologic processes.
Laura J. Crossey, Assistant Professor; Ph.D., Wyoming. Sedimentary geochemistry, low-temperature geochemistry.
Maya Elrick, Assistant Professor; Ph.D., Virginia Tech. Carbonate sedimentology, sequence stratigraphy.
Wolfgang E. Elston, Professor; Ph.D., Columbia. Volcanology, economic geology, planetology.
Rodney C. Ewing, Professor; Ph.D., Stanford. Mineralogy and crystallography.
John W. Geissman, Associate Professor; Ph.D., Michigan. Geophysics, paleomagnetics.
Jeffrey A. Grambling, Professor; Ph.D., Princeton. Metamorphic petrology, silicate mineralogy, Precambrian geology.
Stephen P. Huestis, Associate Professor; Ph.D., California, San Diego. Geophysics, geophysical inverse theory, heat flow, gravity and magnetics.
Karl E. Karlstrom, Associate Professor; Ph.D., Wyoming. Structural geology, evolution of continental lithosphere, Precambrian geology.
Cornelis Klein, Professor; Ph.D., Harvard. Mineralogy, crystallography, Precambrian geology.
Albert M. Kudo, Professor; Ph.D., California, San Diego. Igneous petrology, volcanology, and geochemistry; phase equilibria.
Barry S. Kues, Associate Professor and Chairman; Ph.D., Indiana. Invertebrate and vertebrate paleontology, paleoecology, and biostratigraphy.
Leslie D. McFadden, Associate Professor; Ph.D., Arizona. Geopedology, geomorphology, soil chemistry, tectonic and climatic geomorphology.
James J. Papike, Professor and Director of the Institute of Meteoritics; Ph.D., Minnesota. Crystal chemistry, volcanology, comparative planetology.
Lee A. Woodward, Professor; Ph.D., Washington (Seattle). Structure, regional tectonics, and mineral exploration; tectonics of Rocky Mountain foreland and overthrust zone of Cordilleran foldbelt.
Crayton J. Yapp, Associate Professor; Ph.D., Caltech. Stable-isotope geochemistry.

Professors Emeriti

J. Paul Fitzsimmons, Professor Emeritus; Ph.D., Washington (Seattle). Optical mineralogy, petrography, regional geology.
Stuart A. Northrop, Research Professor Emeritus; Ph.D., Yale. History of New Mexico earthquakes; history of New Mexico mineral occurrences.
Sherman A. Wengerd, Professor Emeritus; Ph.D., Harvard. Petroleum and groundwater exploration, stratigraphic analysis, geomorphology, and photointerpretation.

Adjunct Faculty and Research Staff

M. Susan Barger, Adjunct Associate Professor; Ph.D., Penn State. Material science of art objects.
Adrian J. Brearley, Research Scientist; Ph.D., Manchester. Meteoritics, cosmochemistry, mineralogy and petrology of metamorphic rocks, electron microscopy.
William F. Chambers, Adjunct Associate Professor; Ph.D., Duke. Electron microprobe analysis.
Ernest S. Gladney, Adjunct Professor; Ph.D., Maryland. Nuclear, environmental, and analytical chemistry.
Charles D. Harrington, Adjunct Associate Professor; Ph.D., Indiana. Geomorphology and Quaternary geochronology.
Rhian H. Jones, Research Scientist; Ph.D., Manchester. Meteoritics, cosmochemistry, experimental petrology, mineralogy.
Steven J. Lambert, Adjunct Professor; Ph.D., Caltech. Geochemistry.
Spencer G. Lucas, Adjunct Assistant Professor; Ph.D., Yale. Vertebrate paleontology, Mesozoic-Cenozoic stratigraphy.
Christopher K. Mawer, Adjunct Assistant Professor; Ph.D., Monash (Australia). Structural geology.
Horton E. Newsom, Research Scientist, Institute of Meteoritics; Ph.D., Arizona. Evolution of the moon, petrology, mineralogy, chemistry.
Frank V. Perry, Research Scientist; Ph.D., UCLA. Igneous petrology, isotope geochemistry.
Harald Poths, Research Associate; Ph.D., Mainz. Stable-isotope geochemistry.
Fransiscus J. M. Rietmeijer, Research Scientist; Ph.D., Utrecht (Netherlands). Cosmochemistry, solar system evolution, petrology.
Harrison "Jack" H. Schmitt, Adjunct Professor; Ph.D., Harvard. Astrogeology.
Charles K. Shearer Jr., Research Scientist; Ph.D., Massachusetts. Igneous petrology, with emphasis on terrestrial granite systems and lunar volcanic systems.
John W. Shomaker, Adjunct Associate Professor; M.S., New Mexico. Coal geology.
Gary A. Smith, Staff Scientist-Curator; Ph.D., Oregon. Sedimentology, physical volcanology.
Carol L. Stein, Adjunct Assistant Professor; Ph.D., Harvard. Low-temperature geochemistry.
Daniel B. Stephens, Adjunct Professor; Ph.D., Arizona. Subsurface hydrology.
Lu-Min Wang, Senior Research Associate; Ph.D., Wisconsin–Madison. Electron microscopy.

SECTION 3: EARTH AND PLANETARY SCIENCES

UNIVERSITY OF OKLAHOMA

School of Geology and Geophysics

Programs of Study	The School offers the M.S. in geology, M.S. in geophysics, and Ph.D. in geology. The Ph.D. in geology may be obtained with an emphasis in geophysics. M.S. programs are 30-hour thesis programs; the Ph.D. requires a dissertation and 60 hours of course work in addition to the master's degree work. Each student must take the General Examination. The principal fields of research activity are basin analysis, carbonate geochemistry, depositional environments, hydrothermal ore deposits, igneous petrology, low-temperature geochemistry, metamorphic petrology, paleoecology, paleomagnetics, paleontology, petroleum geology, petroleum and organic geochemistry, sedimentary geochemistry, sedimentary petrology, sedimentology, stable-isotope geochemistry, structural geology, and tectonophysics. Research programs in geophysics include exploration geophysics, fluid and heat transfer, geomechanics, geomagnetism, and seismic stratigraphy.
Research Facilities	The School's research facilities include the Texaco X-Ray Lab, which contains Rigaku automated XRD and XRF units, and a $3.5-million Geoscience Computing Network (GCN), comprising a VAX 6520, a VAX-11/785, MicroVAXs, Sun and Silicon Graphics workstations, graphics imaging and visualization systems, a microcomputer laboratory, seismic data processing hardware, and a scientific staff of 8 full-time employees. The GCN supports 2-D and 3-D seismic data processing, numerical modeling, and image processing. The School also has a new Cameca SX-50 electron microprobe with five wavelength-dispersive X-ray spectrometers and Kevex Delta-class EDS; a high-pressure rock mechanics lab with rock deformation and petrophysical capabilities; a structural models lab and a photoelastic lab with an 18-inch polarscope and 3-D oven; a paleomagnetism lab with a magnetically shielded room (0.5 percent of the earth's ambient field), containing a cryogenic three-axis magnetometer, alternating field and thermal demagnetization equipment, and a vibrating sample magnetometer/electromagnet; an organic and petroleum geochemistry lab with seven gas chromatographs and two HPLC systems, a CHN analyzer, a Finnigan ion trap detector system, a Finnigan TSQ 70 mass spectrometer and associated data system; a scanning UV spectrophotometer system, an IR spectrometer, two Chemical Data System pyrolysis units and a PYRAN-INCOS 50 pyrolysis unit; a stable-isotope lab with a Finnigan Delta E isotope ratio mass spectrometer and a Micromass Model 602C spectrometer; a geochemistry lab with an atomic absorption unit, as well as neutron activation equipment; a lab for fluid inclusion microthermometry and data analysis; and a modern experimental facility for mineral synthesis and phase equilibrium studies to 1200°C, 400 MPa. There is a Subsurface Interface Radar to obtain high-resolution shallow profiles. Equipment is available in the School for rock thin sectioning and photography. Other analytical facilities on campus include an electron microscopy facility with a new JEOL 880 SEM and JEOL 2000FX 200-kV TEM, which share a Kevex Delta-class EDS system; a Van de Graaff–PIXE ion probe; Raman, NMR, and FT-IR spectroscopy equipment; and an ICP laboratory. The Oklahoma Geological Survey provides a core and sample library, a chemical rock analysis lab, a coal analysis laboratory, a cartographic section, a geological information systems group, and a geophysical observatory near Tulsa, Oklahoma, which monitors seismic activity and the earth's magnetic and gravity fields. The Geoscience Library contains approximately 85,000 volumes and over 100,000 maps.
Financial Aid	The School offers approximately fifteen teaching and seventeen research graduate assistantships with stipends of $9000 and up for nine months plus a waiver of out-of-state tuition in 1991–92. In addition, several privately supported scholarships are awarded annually, and research appointments associated with externally funded grants are normally available. Research support is also provided by the Oklahoma Geological Survey, the Energy Resources Institute, the Mining and Minerals Resource Research Institute, and the energy industry.
Cost of Study	In 1991–92, resident tuition for graduate students is $58.20 per credit hour; nonresidents pay $187.10 per credit hour. Students pay special fees of $65 per semester plus an activities fee of $5.15 per credit hour. Teaching and research assistants pay resident tuition. All nonresident applicants pay a $10 application fee.
Cost of Living	In 1991–92, room and board for single students in University housing range from $1408 to $1973 per semester. University-owned apartments for married students can be rented for $225–$422 per month, including utilities. Rates for other apartments in town are comparable but do not include utilities. Bus service is available on campus, as well as to apartments off campus.
Student Group	As the country's oldest petroleum geology school, the School has one of the country's largest alumni groups. About 55 graduate students are enrolled, most from out of state.
Location	Norman, Oklahoma, has the relaxed atmosphere of a medium-sized university town but offers access to the metropolitan advantages of nearby Oklahoma City, including an international airport, baseball, soccer, and ballet. The Arbuckle, Ouachita, and Wichita mountains are within a few hours' drive.
The University	The first legislature of the Territory of Oklahoma founded the University in 1890 on 40 acres. The main campus has since grown to more than 2,700 acres. The School of Geology and Geophysics moved in 1986 to the University's Energy Center, a research and teaching complex that brings together the energy-related programs on campus and provides a national focal point for energy research. The School is part of the College of Geosciences, which includes the School of Meteorology and Department of Geography. Interdisciplinary work with other departments in the College and in other Colleges is encouraged. The University has strong programs in physics, mathematics, chemistry, biology, computer sciences, and engineering.
Applying	The application deadlines are February 1 for the fall semester with financial aid, May 1 for the fall semester without financial aid, and September 1 for the spring semester either with or without financial aid. Graduate College requirements are an undergraduate degree (or the equivalent) with a minimum 3.0 grade point average (4.0 scale), GRE General Test scores, and three letters of recommendation. International students whose native language is not English must have a TOEFL score of at least 550.
Correspondence and Information	Donna S. Montgomery School of Geology and Geophysics Sarkeys Energy Center, 810D University of Oklahoma Norman, Oklahoma 73019-0628 Telephone: 405-325-3255 Fax: 405-325-3140

SECTION 3: EARTH AND PLANETARY SCIENCES

University of Oklahoma

THE FACULTY AND THEIR RESEARCH

Professors
Harvey Blatt, Ph.D., UCLA. Mineralogy, petrology, and geochemistry of sandstones and shales.
James M. Forgotson, Ph.D., Northwestern. Petroleum geology, stratigraphy.
M. Charles Gilbert, Director, School of Geology and Geophysics; Ph.D., UCLA. Igneous and metamorphic petrology, experimental geochemistry, physical geology for engineers, environmental geology.
Charles Harper, Ph.D., Caltech. Invertebrate paleontology, paleoecology, quantitative biostratigraphy.
R. Paul Philp, Klabzuba Professor; Ph.D., Sydney. Organic geochemistry, petroleum geochemistry, biomarkers.
Ze'ev Reches, Kerr-McGee Professor; Ph.D., Stanford. Structural geology.
David Stearns, Monnett Professor of Energy Resources; Ph.D., Texas A&M. Structural geology, tectonophysics.

Associate Professors
Judson Ahern, Ph.D., Cornell. Geomechanics: thermal and mechanical modeling of the lithosphere, sedimentary basins, magma migration.
Douglas Elmore, Ph.D., Michigan. Paleomagnetism of sedimentary rocks, dating of diagenetic events using paleomagnetism, sedimentology, depositional systems.
Michael Engel, Ph.D., Arizona. Organic geochemistry, origin and diagenetic history of organic matter, detection of amino acids, oil migration.
David London, Ph.D., Arizona State. Magmatic and hydrothermal evolution of pegmatites and granites, economic petrology, metamorphic rocks.
Barry Weaver, Ph.D., Birmingham (England). Igneous petrology, geochemistry of trace and rare earth elements.
Roger A. Young, Ph.D., Toronto. Exploration geophysics, crustal studies.

Assistant Professors
David Deming, Ph.D., Utah. Geophysics, heat flow and fluid flow.
Thomas Dewers, Ph.D., Indiana. Low-temperature geochemistry.
John Pigott, Ph.D., Northwestern. Two-D and 3-D reflection geophysics, quantitative basin analysis and sedimentary geochemistry.

Professors Emeriti
Robert DuBois, Ph.D., Washington (Seattle). Geomagnetism, magnetic studies of secular variation of direction and intensity.
Patrick Sutherland, Ph.D., Cambridge. Stratigraphy, biostratigraphy and paleoecology.
Leonard Wilson, Ph.D., Wisconsin. Palynology, paleobotany.

SECTION 3: EARTH AND PLANETARY SCIENCES

UNIVERSITY OF PITTSBURGH

Department of Geology and Planetary Science

Programs of Study

The department offers programs of study in geochemistry, geology, geophysics, and planetary physics leading to the M.S. and Ph.D. degrees in geology and planetary science. Particular fields of interest are archaelogical geology, coal geology, engineering geology, environmental geology, extraterrestrial materials, igneous and metamorphic petrology, invertebrate and vertebrate paleontology, isotope geochemistry, paleoecology, paleomagnetism, physics of rock magnetism, planetary surfaces, Precambrian geology, regional geology, seismology, stratigraphy, and tectonics.

Eighteen credits of formal course work are required for the M.S. and 36 credits for the Ph.D. There is little restriction on the courses that may be taken, because the requirement is designed to broaden the students' background at the intermediate level and to allow them to begin their specialization. Candidates for the Ph.D. take a preliminary examination early in their program to demonstrate competence in their major area and to determine those areas that need strengthening. A comprehensive examination is taken by both M.S. and Ph.D. candidates at about the time independent research is begun. Degrees are awarded after successful defense of a thesis.

Research Facilities

The research programs in geochemistry, geology, geophysics, and planetary science are supported by a number of laboratories for the analysis of rocks and minerals as well as by cooperative work with the Department of Chemistry and the Carnegie Museum of Natural History. Methods applied include wet and dry chemical analysis, mineral separation, X-ray diffraction, X-ray fluorescence, scanning electron microscope analysis, and a black-and-white video digitizing system for image analysis. Two mass spectrometers are available for research problems involving isotope and nuclear geochemistry. A variety of X-ray and UV ion sources, together with associated high-vacuum equipment and spectrophotometers, are available for studies of radiation effects. The geophysical research programs are supported by well-equipped laboratories in paleomagnetism, seismology, and planetary science. The paleomagnetism laboratory contains a wide variety of equipment for rock-magnetic and paleomagnetic research. For planetary studies the department has an extensive collection of planetary images taken by spacecraft. Additional analytical instrumentation is available through the Surface Science Center on campus.

The University of Pittsburgh maintains state-of-the-art computing facilities linked by a high-speed fiber-optic network. A departmental computing lab contains DOS and Macintosh computers, laser printing, and facilities for digitizing, image analysis, and map-making. The University maintains a cluster of DEC 8800 and 5000 mainframes and is a host for the Pittsburgh Supercomputer Center, which operates a CRAY Y-MP supercomputer. Finally, more than 500 networked DOS and Macintosh microcomputers and workstations are available for undergraduate and graduate student use at public sites on campus.

Financial Aid

Financial aid is available. In 1991–92, stipends for two terms of half-time duties are $8490–$8920 for teaching assistantships and teaching fellowships and $7000–$9200 for research fellowships. Third-term financial aid is also available. All teaching assistantships, teaching fellowships, and research fellowships include a scholarship covering full tuition and fees.

Cost of Study

Tuition for 1991–92 for full-time students is $3776 per term for Pennsylvania residents and $5996 per term for nonresidents. Tuition is $241 per credit for part-time students who are state residents; nonresident part-time students pay $482 per credit hour.

Cost of Living

Although limited dormitory space is available, most students live in rooms or apartments in Oakland and pay, typically, $230–$350 per month for housing and $180 per month for meals.

Student Group

The University of Pittsburgh enrolls 22,500 students, including 6,900 students in graduate and professional programs. The graduate body in the Department of Geology and Planetary Science in the 1990–91 academic year consisted of 50 students, of whom 14 were women.

Location

Pittsburgh's famous Golden Triangle is enclosed by the Allegheny and Monongahela rivers, which meet at the Point in downtown Pittsburgh to form the Ohio River. Pittsburgh has enjoyed a renaissance and is now one of America's cleanest cities. It provides outstanding offerings in the arts, sports, and recreation. The University is located about 3 miles east of downtown Pittsburgh in the city's cultural center. Adjacent to the campus or nearby are Carnegie Mellon University; the Carnegie Museum and Institute; Schenley Park; Heinz Hall, the home of the famous Pittsburgh Symphony Orchestra; the Pittsburgh Playhouse; and Three Rivers Stadium. Professional sports teams include the Steelers (football), Pirates (baseball), and Penguins (ice hockey). Several city and county parks provide a full range of recreational activities, including swimming, ice-skating, and skiing. Several ski resorts are located about 50 miles from the campus in the Allegheny Mountains, which provide ample opportunities for many kinds of outdoor recreation.

The University

The University of Pittsburgh is one of the oldest institutions west of the Allegheny Mountains. Although privately endowed and controlled, the University has become state-related to permit lower tuition rates for Pennsylvania residents and to provide a steady source of funds for all of its programs.

Applying

Applications for September admission and assistantships should be made prior to April 1. However, special cases may be considered at a later date if there are cancellations. A B.S. degree in one of the physical sciences or engineering is the most common preparation for graduate study. The GRE General Test is required, and an appropriate GRE Subject Test is strongly recommended.

Correspondence and Information

Admissions Committee
Department of Geology and Planetary Science
321 Old Engineering Hall
University of Pittsburgh
Pittsburgh, Pennsylvania 15260
Telephone: 412-624-8780

Peterson's Guide to Graduate Programs in the Physical Sciences and Mathematics 1992

University of Pittsburgh

THE FACULTY AND THEIR RESEARCH

Thomas H. Anderson, Professor and Chairman of Geology and Planetary Science; Ph.D., Texas at Austin, 1969. Structural geology, geochronology, evolution of mountain belts.

James M. Adovasio, Professor and Chairman of Anthropology and Director of the Cultural Resources Management Program; Ph.D., Utah, 1970. Prehistory; archaeological method and theory; prehistoric technology and material culture analysis; geoarchaeology of North America, Mesoamenca, and the Near East.

Anthony D. Barnosky, Adjunct Assistant Professor of Vertebrate Paleontology; Ph.D., Washington (Seattle), 1983. Neogene mammals.

Michael Bikerman, Associate Professor of Geology; Ph.D., Arizona, 1965. Geochronology.

Uwe Brand, Adjunct Professor of Geochemistry; Ph.D., Ottawa, 1979. Geochemistry, paleontology.

David K. Brezinski, Adjunct Assistant Professor of Geology; Ph.D., Pittsburgh, 1986. Invertebrate paleontology and stratigraphy.

John L. Carter, Adjunct Professor of Paleontology; Ph.D., Cincinnati, 1966. Invertebrate paleontology.

William A. Cassidy, Professor of Geology; Ph.D., Penn State, 1961. Extraterrestrial materials.

Alvin J. Cohen, Professor Emeritus; Ph.D., Illinois, 1949. Radiation damage in minerals and glasses.

Mary R. Dawson, Adjunct Professor of Paleontology; Ph.D., Kansas, 1957. Fossil mammals.

Maurice Deul, Adjunct Professor of Geology; M.S., Colorado, 1947. Coal geology.

Jack Donahue, Professor of Geology; Ph.D., Columbia, 1967. Sedimentology and depositional environments.

George D. Gatewood, Professor of Physics and Astronomy; Ph.D., Pittsburgh, 1972. Extraterrestrial geology.

Bruce W. Hapke, Professor of Planetary Science; Ph.D., Cornell, 1962. Planetary surface studies and remote sensing.

William P. Harbert, Assistant Professor of Geophysics; Ph.D., Stanford, 1987. Geophysics.

Chiao-Min Hsieh, Professor Emeritus of Geography; Ph.D., Syracuse, 1953. Geography.

Stephen K. Kennedy, Research Assistant Professor of Geology; Ph.D., South Carolina, 1982. Sedimentology.

James E. King, Adjunct Professor of Geology; Ph.D., Arizona, 1972. Palynology.

Leonard Krishtalka, Adjunct Associate Professor of Vertebrate Paleontology; Ph.D., Texas Tech, 1975. Zoology, vertebrate paleontology.

Edward G. Lidiak, Professor of Geology; Ph.D., Rice, 1963. Igneous and metamorphic petrology, Precambrian geology.

Walter L. Pilant, Associate Professor of Geophysics; Ph.D., UCLA, 1960. Seismology, signal processing.

Harold B. Rollins, Professor of Paleontology; Ph.D., Columbia, 1967. Invertebrate paleontology, paleoecology.

Victor A. Schmidt, Professor of Geophysics; Ph.D., Carnegie Tech, 1966. Rock magnetism, paleomagnetism.

A. G. Sharkey, Research Professor of Mass Spectrometry; M.S., Case Institute of Technology, 1943. Mass spectrometer instrumentation and coal science.

Ellis Strick, Associate Professor Emeritus of Geophysics; Ph.D., Purdue, 1950. Seismology, physics, nuclear physics.

Eugene W. Sucov, Volunteer Senior Research Associate; Ph.D., NYU, 1959. Analytical geochemistry.

Cathy L. Whitlock, Adjunct Associate Professor of Geology; Ph.D., Washington (Seattle), 1983. Palynology.

SECTION 3: EARTH AND PLANETARY SCIENCES

UNIVERSITY OF SASKATCHEWAN

Department of Geological Sciences
Graduate Program in Geology, Geophysics and Geological Engineering

Programs of Study	The College of Graduate Studies and Research at the University of Saskatchewan offers programs through the Department of Geological Sciences leading to the Postgraduate Diploma and the M.Sc. and Ph.D. degrees. Areas of study include economic geology; geochemistry; geological engineering; mineralogy and petrology; paleontology including micropaleontology, paleobotany, and palynology; sedimentology and stratigraphy; seismology and other aspects of geophysics; and structural geology. The department maintains close ties and pursues cooperative research with several other departments in science and engineering and with government agencies and exploration companies. Students entering M.Sc. and Ph.D. programs must complete a thesis and a program of course work approved by the college. The normal course of study is two years for the M.Sc., three years for the Ph.D. The Postgraduate Diploma requires the completion of a group of courses assigned by the department and approved by the college, and it may be completed in one year.
Research Facilities	The department moved into a large, well-equipped new building in 1986. Equipment includes a Philips PW1450 automatic X-ray fluorescence spectrometer and facilities for sample preparation; a Perkin-Elmer 2380 atomic absorption spectrometer; a Rigaku rotation-anode X-ray diffractometer; a JEOL 8600 microprobe with three WDS, EDS, and TN-8500 image analysis systems; a Perkin-Elmer Sciex Elan 5000 ICP-MS with PE Sciex 320 Laser Sampler; fully equipped stable-isotope laboratories with two Finnigan MAT and one VG mass spectrometers; a radiogenic-isotope laboratory, including a clean laboratory and a Finnigan MAT 261 mass spectrometer with five multicollectors; a USGS fluid-inclusion microscope stage; a rock mechanics laboratory equipped to determine a wide range of rock physical properties; several photomicroscopes; and a fully equipped darkroom. Computer facilities include a VAX-11/785 system with two array processors (FPS 5205 and FPS 100), a VAXstation 3100, a VAXstation 3200, a SPARCstation 2, a Raster Technology ONE/80S graphic workstation, an AST Premium 386 PC, several color graphics CRTs, a microcomputer laboratory with twelve networked AT&T PC 6300 workstations, 22-inch and 11-inch Versatec electrostatic plotters, and Tektronix 4696 and HP 7475 color plotters. Geophysics equipment includes a DFSV forty-eight-channel seismograph, a Sercel 338HR forty-eight-channel seismograph, a Bison Geoprobe twenty-four-channel seismograph, a mobile geophysical laboratory for field surveys, two gravity meters, two Omni Plus magnetometer vlf instruments, and a well-equipped electronics testing laboratory. Primary software systems include the Aurora seismic processing system, Geoquest's Interactive Exploration System, Sierra geophysical and geological modelling portfolios, Zycor's Z-MAP+ mapping system, VISTA, and the UNIRAS graphic software suite. A new seismic observatory houses earthquake and earth-strain recording equipment. The department has extensive micropaleontological collections for comparative study. There is an excellent department library with on-line access to nationwide data and information banks.
Financial Aid	Canadian applicants may apply for a number of government- and industry-sponsored scholarships. In 1991–92, all applicants are eligible for University of Saskatchewan Scholarships, which are awarded on the basis of academic merit and pay approximately $11,250 per year for M.Sc. students and $15,000 per year for Ph.D. students. Most students who receive financial aid are expected to do a limited amount of laboratory instruction, for which they may earn up to about $4200 per year. Some graduate students are paid from research funds or contract funds of individual faculty members.
Cost of Study	The annual tuition fees are $2288 in 1991–92. The annual fee is paid only once for the M.Sc. and the Postgraduate Diploma, three times for the Ph.D. without the M.Sc., and twice for the Ph.D. after completion of the M.Sc.
Cost of Living	For the academic year 1991–92, single students should allow about $9000 per year for living expenses in Saskatoon. There is a limited amount of University housing.
Student Group	In 1991–92, there are 50 students pursuing research in all areas of geology and geophysics. There are about 1,300 students enrolled in the College of Graduate Studies and Research, and there is an active Graduate Students' Association.
Location	Saskatoon has a dry, sunny, continental climate. There is good access to camping, fishing, and skiing areas. With 179,000 inhabitants, Saskatoon is a small but culturally alive city. It has two professional theatre groups, a symphony orchestra, an established art gallery, and museums dedicated to western Canadian culture. Medical insurance in Saskatchewan is provided by the provincial government to all residents at no direct cost. Saskatchewan is one of Canada's prairie provinces. Geologically, it can be divided into two regions of more or less equal area. In the north, the rocks of the Precambrian Shield contain deposits of uranium, gold, zinc, and other metalliferous ores; and in the south, a southerly-thickening wedge of Phanerozoic deposits extends into the northern part of the Williston Basin. Much of Saskatchewan, particularly in the south, is blanketed by Pleistocene glacial deposits.
The University and The Department	The University of Saskatchewan stands on the banks of the South Saskatchewan River, close to downtown Saskatoon. The University is the focus of many cultural events, and modern, well-equipped sports facilities are available both on and off campus. The government of Saskatchewan maintains a provincial geological survey, including a core laboratory, in Regina, the provincial capital. The Department works cooperatively with the provincial survey and with the geology division of the Saskatoon-based Saskatchewan Research Council.
Applying	Applicants must submit a formal application and must have official transcripts of their undergraduate record forwarded separately. Confidential letters of recommendation must be submitted from three referees. Foreign students must obtain a student visa and demonstrate competence in English. Application for admission should be made several months before the desired time of admission; most students begin study in September. Application forms and additional information may be obtained by writing to the address below.
Correspondence and Information	Chairman Graduate Admissions Committee Department of Geological Sciences University of Saskatchewan Saskatoon, Saskatchewan S7N 0W0 Canada Telephone: 306-966-5683 Fax: 306-966-8593

University of Saskatchewan

THE FACULTY AND THEIR RESEARCH

J. F. Basinger, Associate Professor; Ph.D., Alberta, 1979. Paleobotany, Cretaceous and Tertiary plants of western and arctic Canada.
W. K. Braun, Professor; Ph.D., Tübingen (Germany), 1958. Devonian and Jurassic micropaleontology and biostratigraphy.
L. C. Coleman, Professor; Ph.D., Princeton, 1955. Mineralogy and geochemistry of igneous rocks.
D. J. Gendzwill, Professor; Ph.D., Saskatchewan, 1969. Geophysics applied to engineering and exploration, induced seismicity, high-resolution seismic surveys, electromagnetic sounding, gravity and magnetic interpretation.
Z. Hajnal, Professor; Ph.D., Manitoba, 1970. Seismic data processing, crustal seismology, experimental and theoretical signal modelling.
H. E. Hendry, Professor and Head; Ph.D., Edinburgh, 1970. Clastic sedimentology, deposition of conglomerates, diagenesis.
R. W. Kerrich, George J. McLeod Professor; Ph.D., London, 1975. Geochemistry, mechanics of rock deformation, fluid-rock interactions, concentration of precious metals.
W. O. Kupsch, Professor Emeritus; Ph.D., Michigan, 1950. Quaternary geology, terrain analysis, northern nonrenewable resource development, history of western and northern Canadian geological explorations.
T. K. Kyser, Professor; Ph.D., Berkeley, 1980. Aqueous geochemistry, igneous and metamorphic petrology, isotope geochemistry.
F. F. Langford, Professor; Ph.D., Princeton, 1960. Economic geology, uranium deposits, potash deposits.
J. B. Merriam, Associate Professor; Ph.D., York, 1976. Geodynamics, Earth tides, rotation of the Earth.
E. G. Nisbet, Professor; Ph.D., Cambridge, 1974. Early history of the Earth, development of the continents and oceans, origin and early distribution of life, geology of northern Saskatchewan and western Ontario, geology of Zimbabwe.
B. R. Pratt, Assistant Professor; Ph.D., Toronto, 1989. Paleontology, carbonate sedimentology of early Paleozoic rocks.
M. J. Reeves, Professor; Ph.D., Durham, 1971. Flow of fluids in porous media, computer applications in geological sciences.
R. W. Renaut, Associate Professor; Ph.D., London, 1982. Evaporites, nonmarine carbonates, lacustrine sedimentation, rift-valley sedimentation, Quaternary geology of interior British Columbia.
W. A. S. Sarjeant, Professor; Ph.D., Sheffield, 1959; D.Sc., Nottingham, 1972. Morphology, classification, and biostratigraphy of dinoflagellate cysts and acritarchs; terrestrial paleoichnology and paleoecology; history and bibliography of geology.
M. R. Stauffer, Professor; Ph.D., Australian National, 1965. Structural geology, tectonic rock fabrics, Canadian Shield geology.
D. Stead, Assistant Professor; Ph.D., Nottingham, 1984. Geotechnique, rock mechanics, slope stability.

Research Associates
C. M. R. Fowler, Ph.D., Cambridge, 1977. Crustal seismology, heat flow in the Earth, origin of sedimentary basins.
S. Fowler, Ph.D., Saskatchewan, 1985. Lower cretaceous foraminifera of arctic Canada.
B. Pandit, Ph.D., Toronto, 1971. Physical properties of rocks, seismic modelling.

The Geological Sciences Building.

SECTION 3: EARTH AND PLANETARY SCIENCES

UNIVERSITY OF TEXAS AT AUSTIN

Department of Geological Sciences

Programs of Study	The department offers graduate programs leading to the M.A. and Ph.D. degrees in geological sciences, including economic geology, environmental geology, geochemistry, geochronology, geophysics, hydrogeology, invertebrate and vertebrate paleontology, micropaleontology, mineralogy and petrology, stratigraphy and sedimentation, and structural geology and tectonics.
	The M.A. requires 24 semester hours of course work, a knowledge of one foreign language, and a thesis that represents individual research. Programs for the Ph.D. are individually designed with the advice of each student's faculty supervisor and committee. Students may pursue the Ph.D. immediately after receiving the bachelor's degree. Ph.D. students must complete and defend a research proposal prior to candidacy.
Research Facilities	The department occupies six floors of a modern building. Among the facilities are laboratories for organic and inorganic chemical analysis, including an automated electron microprobe equipped with wavelength- and energy-dispersive modes of analysis; a scanning electron microscope with energy-dispersive analysis; an inductively coupled plasma spectrometer; four mass spectrometers (including a multicollector solid-source instrument) for analysis of light and heavy isotopes; ultraclean laboratories; an automated X-ray diffractometer; an X-ray precession camera; X-ray fluorescence equipment; an atomic absorption spectrometer; and an HPLC instrument. The department also houses a fission-track thermochronology laboratory, a fluid-inclusion laboratory, apparatus for hydrothermal and controlled-atmosphere experiments, a paleomagnetic laboratory with a superconducting magnetometer, an extensive petrographic microscope laboratory with cathodoluminescence and heating/freezing stages, and laboratories for the preparation and study of microfossils and fossil vertebrates and invertebrates. Research collections include over 165,000 cataloged specimens of fossil vertebrates (especially strong in the Permian, Triassic, and Tertiary of Texas), plus modern vertebrate skeletal material, and extensive holdings of fossil invertebrates (especially strong in the Late Paleozoic, Cretaceous, and Tertiary of Texas). Collections of core and sample materials are housed in and curated by the Core Research Center of the Bureau of Economic Geology.
	Field geophysical equipment includes a gravimeter, fluxgate and proton precession magnetometers, a sledgehammer seismograph, a high-resolution marine-sparker profiling system, and marine seismic reflection and ocean-bottom seismometer systems. Two Trimble 4000 SST GPS receivers are available for high-precision geodetic studies. The department and the Institute for Geophysics maintain a Sun-4 computer system with seismic-reflection processing software and color and vector graphics devices. Access to other campus computers (IBM, CRAY Y-MP 8/864) is available through terminals in the geology building and through Ethernet. The CRAY Y-MP also serves the department's seismic and processing needs. A departmental microcomputer laboratory contains Macintosh and IBM computers and peripheral devices.
	A geological sciences library of over 86,000 volumes in the geology building contains the comprehensive combined holdings of the Department of Geological Sciences and the Bureau of Economic Geology. The Walter Geology Library also has more than 40,000 geologic and topographic maps.
Financial Aid	Approximately forty teaching assistantships, carrying a nine-month stipend of $7800–$9000 in 1991–92 for 20 hours of employment per week, are awarded each year. Approximately twenty to twenty-five research assistantships with comparable stipends are awarded each year through the Bureau of Economic Geology, which functions as the state geological survey. Fellowships and traineeships sponsored by industry, foundations, government agencies, and the University are also available. The Institute for Geophysics provides several fellowships each year as well as numerous research assistantships.
Cost of Study	Tuition for Texas residents is $18 per semester hour in 1991–92, plus required fees of approximately $165. Out-of-state residents pay $122 per semester hour, plus required fees. Students holding assistantships pay the Texas resident rate. Some fellowships waive the out-of-state portion of tuition. Tuition and required fees may be waived for some fellowship holders.
Cost of Living	Graduate students generally prefer to live off campus. Single students may, however, live in a graduate section of one of the University dormitories at a contract rate beginning at $3300 for room and meals for the 1991–92 academic year. Contracts for room only are also available. Off-campus apartments, furnished and unfurnished, may be rented for $250–$600 per month. Shuttle-bus service is available.
Student Group	The department has approximately 130 undergraduate majors and 180 graduate students. About 40 percent of the graduate students are seeking the master's degree; the remainder are Ph.D. students.
Location	Austin is the state capital, with a population of about 500,000. It is centrally located in the state, along the Balcones fault zone that separates the Gulf Coastal Province from the Edwards Plateau that overlies the craton. The many geologic features within an hour's drive range from Tertiary sediments of the Gulf to Precambrian basement rocks of the Llano Uplift.
The University and The Department	The University of Texas at Austin, founded in 1881, is a state institution endowed and maintained by legislative grants. The grounds consist of the original 40-acre campus and numerous sites acquired by gift and purchase. The University owns the 393-acre Balcones Research Center, 12 kilometers north of the campus, and a number of other properties throughout Texas.
	The department is in the College of Natural Sciences, which has strong programs in biology, botany, chemistry, computer sciences, mathematics, physics, and zoology as well as geological sciences. The College of Engineering provides opportunities for cross-disciplinary research in geological sciences.
Applying	Application forms for admission to the Graduate School and for financial aid are available from the department. For fall registration, the admissions application should be completed and returned no later than February 1; the application deadline for the spring semester is October 1. Late applications will be considered only if enrollment limits have not been reached. Classes for the fall semester begin the last week in August.
Correspondence and Information	Graduate Advisor Department of Geological Sciences University of Texas at Austin P.O. Box 7909 Austin, Texas 78713-7909

Peterson's Guide to Graduate Programs in the Physical Sciences and Mathematics 1992

SECTION 3: EARTH AND PLANETARY SCIENCES

University of Texas at Austin

THE FACULTY AND THEIR RESEARCH

Professors
Milo M. Backus, Ph.D., MIT, 1956. Seismic exploration, seismic data processing.
Daniel S. Barker, Ph.D., Princeton, 1961. Igneous petrology, geochemistry, volcanology.
Robert E. Boyer, Dean, College of Natural Sciences; Ph.D., Michigan, 1959. Structural geology, remote sensing, earth science education.
Richard Buffler, Ph.D., Berkeley, 1967. Marine geology and geophysics, seismic stratigraphy.
William D. Carlson, Ph.D., UCLA, 1980. Experimental petrology, metamorphic petrology, kinetics.
Ian W. D. Dalziel, Ph.D., Edinburgh, 1963. Regional tectonics, structural geology.
William L. Fisher, Director, Bureau of Economic Geology; Ph.D., Kansas, 1961. Energy and mineral resources.
William E. Galloway, Ph.D., Texas at Austin, 1971. Clastic depositional systems, sequence stratigraphy, energy resource geology.
Gary Kocurek, Ph.D., Wisconsin, 1980. Sedimentology: depositional environments, eolian processes.
J. Richard Kyle, Ph.D., Western Ontario, 1977. Ore deposits geology, fluid inclusion studies, mineral exploration.
Lynton S. Land, Ph.D., Lehigh, 1966. Isotope geochemistry, diagenesis, low-temperature aqueous geochemistry.
Leon E. Long, Ph.D., Columbia, 1959. Geochemistry, isotope studies.
Ernest L. Lundelius Jr., Ph.D., Chicago, 1954. Vertebrate paleontology, Pleistocene faunas.
Toshimatsu Matsumoto, Ph.D., Tokyo, 1961. Earthquake seismology, geophysics.
Arthur E. Maxwell, Director, Institute for Geophysics; Ph.D., Scripps, 1959. Marine geophysics, oceanography.
Earle F. McBride, Ph.D., Johns Hopkins, 1960. Sedimentary processes and sedimentary petrology.
Sharon Mosher, Ph.D., Illinois, 1978. Deformation mechanisms, pressure solution, strain analysis, metamorphic terranes.
William R. Muehlberger, Ph.D., Caltech, 1954. Tectonics.
Yosio Nakamura, Ph.D., Penn State, 1963. Geophysics, lunar and planetary physics.
Amos Salvador, Ph.D., Stanford, 1950. Stratigraphy, geology of the Gulf of Mexico Basin and Caribbean, petroleum geology.
John M. Sharp, Ph.D., Illinois, 1974. Hydrogeology, groundwater geology, relation of groundwater to ore genesis and hydrocarbon migration.
Douglas Smith, Ph.D., Caltech, 1969. Igneous and metamorphic petrology, mantle processes, geochemistry.
James Sprinkle, Ph.D., Harvard, 1971. Primitive echinoderms, blastoids, Paleozoic paleontology.
Paul L. Stoffa, Ph.D., Columbia, 1974. Acquisition and processing of multichannel seismic data.
Willem C. J. van Rensburg, Ph.D., Wisconsin, 1965. International minerals and energy policy issues, coal.
Clark R. Wilson, Chairman, Department of Geological Sciences; Ph.D., California, San Diego (Scripps), 1975. Geodesy, seismology.

Associate Professors
Mark Cloos, Ph.D., UCLA, 1981. Structural geology and tectonics.
Martin B. Lagoe, Ph.D., Stanford, 1982. Micropaleontology, stratigraphy of continental margins.
Timothy Rowe, Ph.D., Berkeley, 1986. Vertebrate paleontology and development, tetrapod evolution, computer imaging.

Assistant Professors
Jay L. Banner, Ph.D., SUNY at Stony Brook, 1986. Carbonates, fluid-rock interaction, isotope geochemistry.
Philip C. Bennett, Ph.D., Syracuse, 1988. Hydrogeology, low-temperature aqueous geochemistry.
Stephen P. Grand, Ph.D., Caltech, 1986. Seismology of the earth's interior.
Michelle A. Kominz, Ph.D., Columbia, 1986. Geodynamics.
Nicholas W. Walker, Ph.D., California, Santa Barbara, 1986. Regional tectonics, isotope geology, crustal evolution.

Lecturers, Research Scientists, and Research Associates
Wulf Gose, Ph.D., SMU, 1970. Paleomagnetism, structural history of Central America, magnetostratigraphy.
Mark Helper, Ph.D., Texas at Austin, 1985. Metamorphic petrology, tectonics.
Fred W. McDowell, Ph.D., Columbia, 1966. Geochemistry, geochronology.
Sally Sutton, Ph.D., Cincinnati, 1987. Low-grade metamorphism, clay mineralogy, deformation.

Professors Emeriti
Virgil E. Barnes, Ph.D., Wisconsin, 1930. Stratigraphy, geologic mapping, tektites.
L. Frank Brown Jr., Ph.D., Wisconsin, 1955. Stratigraphy, depositional systems, environmental geology.
Fred M. Bullard, Ph.D., Michigan, 1929. Volcanology.
Stephen E. Clabaugh, Ph.D., Harvard, 1950. Metamorphic petrology and volcanic rocks.
Ronald K. DeFord, M.S., Colorado School of Mines, 1922. Stratigraphy, history of geology.
Samuel P. Ellison Jr., Ph.D., Missouri, 1940. Energy resource geology, micropaleontology.
Peter T. Flawn, Ph.D., Yale, 1951. Economic geology, geology and public affairs.
Robert L. Folk, Ph.D., Penn State, 1952. Petrology of sediments, sandstones, and limestones.
F. Earl Ingerson, Ph.D., Yale, 1934. Geochemistry.
Edward C. Jonas, Ph.D., Illinois, 1954. Clay minerals, pyroclastic sediments and uranium deposits.
Wann Langston Jr., Research Scientist, Texas Memorial Museum; Ph.D., Berkeley, 1952. Paleontology of lower vertebrates.
John C. Maxwell, Ph.D., Princeton, 1946. Regional tectonics.
John A. Wilson, Ph.D., Michigan, 1941. Vertebrate biostratigraphy.
Keith Young, Ph.D., Wisconsin, 1948. Mesozoic stratigraphy and paleontology, geology of the environment of man.

SECTION 3: EARTH AND PLANETARY SCIENCES

UNIVERSITY OF TEXAS AT DALLAS

Programs in Geosciences

Programs of Study
The Programs in Geosciences lead to the M.S. and Ph.D. degrees in a variety of areas, with particular emphasis in geophysics, seismology, geochemistry-petrology, mineral resources, clastic and carbonate sedimentology, structural geology, tectonics, stratigraphy, and micropaleontology. In addition to completing courses in the major field of study, students are required to complete two to four courses in secondary fields, a computer programming course, and a thesis or dissertation. The qualifying oral examination is based on a written research proposal describing the intended project. The examination must be taken by the third (M.S.) or fourth (Ph.D.) semester of study or upon completion of 30 (M.S.) or 42 (Ph.D.) hours.

Research Facilities
Laboratories are well equipped for research in experimental petrology, geochemistry, geochronology, geohydrology, geophysics, micropaleontology, mineral resources, sedimentology, seismology, stratigraphy, structural geology, and tectonics. Available are four automated mass spectrometers; two clean laboratories; an automated X-ray diffractometer; an atomic absorption spectrophotometer with graphite furnace capabilities; an automated scanning electron microprobe with energy dispersive and cathodoluminescence capabilities; a heating and cooling stage for fluid inclusion studies; a scanning electron microscope; four solid-media piston cylinder devices for studies up to 1800°C and 50 kb; two solid-media MA-8 apparatus capable of attaining pressures up to 30 GPa and temperatures in excess of 2000°C; three 1-atmosphere furnaces for studies up to 1800°C at controlled oxygen fugacities; a CNS analyzer; a gas chromatograph; an amino acid analyzer; a cathodoluminoscope; a Coste-Romberg Model G gravimeter; a Nikon theodolite and data collector; a Zeiss Total Station electronic distance meter and theodolite; two 16-channel 4000 SST Trimble dual frequency geodetic receivers; a dual-channel Trimble Pathfinder II Global Positioning System (GPS) receiver; an array of recording three-component magnetic variometers; a multiuser virtual memory vector computer system (Convex C-1); numerous Sun Workstations; an Intel iPSC/860; a 24-channel floating point seismic acquisition system; a Geometrics G806 magnetometer; a Zone GP-12 digital electrical recording receiver; several IBM-compatible PCs, including one with a 48-inch GTCO digitizing table; a microfilm library of WWSSN records; a solid-confining-medium Griggs-type press for high temperature-pressure experiments; a large liquid-confining-medium press for low temperature and pore pressure experiments; and an Instron materials testing system (Model 1127). There is also direct network access to the University's IBM 4381, NCUBE, and Multimax computers, to two Silicon Graphics workstations, and to a CRAY Y-MP. A core collection provides samples for subsurface projects. The Geological Information Library contains almost 5 million well logs and other valuable geological information.

Financial Aid
Teaching and research assistantships are awarded each year. Stipends range from $700 per month for first-year students to $800–$1000 per month for Ph.D. students, plus insurance benefits and waiver of out-of-state tuition. Assistantship holders are required to register for 12 hours each semester. The Anton L. Hales Fellowship in Geophysics, awarded each year to an exceptional incoming Ph.D. applicant, carries an annual stipend of $12,000, insurance benefits, waiver of in-state tuition and fees for each semester and the summer session, and a $500 travel allowance to attend geology/geophysics meetings. University Merit Fellowships of $15,000–$21,800 are awarded annually to exceptional applicants seeking the doctoral degree. The Amoco Fellowship in Geophysics is awarded to a superior student (who must be a U.S. citizen) seeking the master's degree in geophysics. Various scholarships are available to assist outstanding students nearing completion of their thesis who have extraordinary expenses. These awards are made to students pursuing the M.S. or Ph.D. degrees on a full-time basis in some field of geosciences. Nonsupported students are eligible for work-study.

Cost of Study
In-state tuition and fees total $427 per semester in 1991–92. Additional fees are assessed for field trips.

Cost of Living
Approximately 200 garden apartments are located on campus. Most students live in apartments in the nearby communities; the rates for these vary, depending on the location and type of accommodation; they begin at about $430 to $450 per month.

Student Group
The program has about 55 full-time and about 35 part-time graduate students. Many local geologists, who are not seeking advanced degrees, also enroll in courses. Because of the large number of geologists in the Greater Dallas area, frequent opportunities exist for interaction between students and professional geologists through courses and the weekly program lecture series. Upon graduation, M.S. students generally pursue careers in the fossil-fuel industries. Ph.D. students select similar careers as well as opportunities in academic or government organizations.

Location
The campus is located in Richardson, a largely residential suburb of Dallas. Numerous outdoor recreational opportunities exist in the area because of the many lakes and parks within a few miles of the city. Major-league spectator sports are football, baseball, soccer, and basketball. A variety of scenic areas within Texas and the surrounding states are easily accessible by car. Dallas is a dynamic city, and the community and the campus offer many cultural activities for all tastes.

The University
The University was created in 1969 as a component of the University of Texas System. One of the purposes of UT-Dallas is to continue the high-level scientific research that had existed at the Southwest Center for Advanced Studies, the private research institution that was the nucleus of the new campus. The University is a graduate and undergraduate institution with a strong emphasis on research, graduate education, and science. About half of the 8,600 students are graduate students.

Applying
Requirements for admission include an undergraduate degree in geology (equivalent to the program's B.S. degree) and strength in other sciences, although individuals with degrees in other sciences are encouraged to apply when their background and interests are compatible (e.g., a math or physics degree for specialization in geophysics). Complete applications for admission and financial aid should be received before February 1, but later applications will be considered. GRE General Test scores are required, and the examination should be taken early in the preceding fall.

Correspondence and Information
Head, Programs in Geosciences
University of Texas at Dallas
P.O. Box 830688/FO 2.1
Richardson, Texas 75083-0688
Telephone: 214-690-2401
Fax: 214-690-2537

Peterson's Guide to Graduate Programs in the Physical Sciences and Mathematics 1992

University of Texas at Dallas

THE FACULTY AND THEIR RESEARCH

Carlos L. V. Aiken, Professor. Gravity and potential field techniques applied to resource exploration and to the properties of the crust and mantle.

James L. Carter, Associate Professor and Head of the Programs in Geosciences. Geochemistry and geochemical exploration, metallic mineral resources, geochemistry of mantle and lower crystal xenoliths, lunar studies.

David E. Dunn, Professor and Dean of the School of Natural Sciences and Mathematics. Experimental rock mechanics, especially brittle fracture mechanisms; structural studies in orogenic belts.

John F. Ferguson, Associate Professor. Seismology, linear inverse methods, inversion theory, structure of the Earth.

Anton L. Hales, Professor Emeritus. Seismology of the crust and mantle, earthquake seismology, physics of the interior of the Earth.

John D. Humphrey, Associate Professor. Carbonate diagenesis, carbonate sedimentology, stable isotope geochemistry.

Mark Landisman, Professor. Geophysics of the crust and mantle, seismology, electrical methods, geothermal studies.

Jose F. Longoria, Associate Professor. Micropaleontology (Mesozoic calpionellids and foraminifera), regional stratigraphy, carbonate microfacies analysis, geology of Mexico.

William I. Manton, Professor. Isotope geochemistry, geochronology, heavy metals in the environment.

George A. McMechan, Endowed Professor in Geosciences and Director of the Center for Lithospheric Studies. Reflection and refraction seismology, synthetic seismogram modeling, wave field imaging, tomography, numerical methods.

Richard M. Mitterer, Professor. Carbonate geochemistry, amino acid diagenesis, organic and sedimentary geochemistry.

Kent C. Nielsen, Associate Professor. Structural geology, experimental rock deformation, field geology, tectonics.

Emile A. Pessagno Jr., Professor. Micropaleontology and biostratigraphy, Mesozoic radiolaria and planktonic foraminifera.

Dean C. Presnall, Professor. Theoretical and experimental igneous petrology, geochemistry of ultramafic nodules and oceanic basalts.

Robert H. Rutford, Professor and President of the University of Texas at Dallas. Glaciology, Antarctic geology.

Kristian Soegaard, Associate Professor. Clastic depositional systems, sedimentary petrology, basin analysis and tectonics, Precambrian geology.

Robert J. Stern, Associate Professor. Geochronology and igneous petrology, especially as applied to problems in modern and ancient crustal evolution; geology of the Mideast; development of island arcs.

SECTION 3: EARTH AND PLANETARY SCIENCES

UNIVERSITY OF TOLEDO

Department of Geology

Programs of Study

The Department of Geology offers the degrees of Master of Science and Master of Science and Education in geology. Requirements for the degrees include a minimum of 45 quarter hours, with no more than 9 of these hours credited to thesis research. A minimum of 18 quarter hours must be in geology courses. Students must also successfully defend their thesis before the faculty. Students who have not taken a summer field course or its equivalent are required to complete a field course during the first summer session after enrollment. No foreign language is required.

The program is designed with a flexibility that accommodates applicants with an undergraduate degree in geology or the allied fields of chemistry, biology, physics, mathematics, engineering, and science education. Emphasis is placed on the interdisciplinary applied aspects of the earth sciences. Special areas of strength include hydrogeology, structural geology, stratigraphy, sedimentary petrology, coal petrology, igneous petrology, seismology, invertebrate paleontology, and crystal chemistry. The student and his or her adviser are given the opportunity to tailor a program of study designed to fit the student's interest and needs.

Research Facilities

The department occupies approximately 31,000 square feet of modern laboratory, lecture, and office space in three buildings on the main campus. Laboratory space contains up-to-date facilities for XRD, XRF, AA, and ICP analyses; electron microscopy (TEM and SEM); petrographic studies; coal characterization; and porosimetry analyses. Other facilities include a processing, storage, and data retrieval center for drill cores and cuttings; a complete machine shop; a computer station with mainframe terminals and microcomputers; a computer graphics laboratory; a geology library; and sample preparation facilities. Adjunct facilities include a campus computer center and a campus library with several million volumes and a large map collection.

Financial Aid

About twenty teaching and research assistantships are available for incoming students. These normally cover tuition for six consecutive quarters (summer excluded) and provide a stipend of at least $7000 per academic year (nine months). A student receiving a stipend is expected to carry a full-time course load (12 or more credit hours) and to spend up to 20 hours a week in laboratory instruction and/or other departmental duties. Support beyond six quarters may be provided in special cases. Research assistants are not required to teach or perform other departmental duties.

Cost of Study

Tuition in 1991–92 is $1042 per quarter (12–16 credit hours) for Ohio residents and $2127 per quarter for nonresidents.

Cost of Living

There is no University housing available for graduate students or married couples; however, the director of housing keeps a current update of off-campus housing available for single and married students. The listings offer housing within walking distance of the University or accessible to it via University or city bus service. A typical rent for one-bedroom apartments is about $350 per month.

Student Group

The University of Toledo now has an enrollment of over 25,000. The Graduate School has about 3,200 students enrolled in more than forty master's and doctoral programs in seven colleges. There were about 30 full- and part-time graduate students enrolled in the Department of Geology in 1990–91.

Location

The city of Toledo, in which the University's 400-acre campus is located, has a metropolitan population of 450,000. The city is situated on the western end of Lake Erie and has become a major Great Lakes seaport as well as the third-largest rail center in the United States. There are twenty-seven major industrial employers in Toledo, including three major glass manufacturers, which has led to the city's title of "Glass Capital of the World." Toledo is within easy driving distance of Detroit (1 hour to the north), Cleveland (2 hours to the east), and Columbus (3 hours to the southeast).

Toledo is the home of the famed Toledo Museum of Art and has one of the finest zoos in the nation. It supports its own symphony orchestra and theatrical and musical organizations. It is also the site of the Medical College of Ohio.

The University

The University of Toledo was founded in 1872 as a municipally supported institution. In 1967 the University joined the state system of higher education. The faculty now numbers about 1,300. Many of the students enrolled are commuters, and the University provides a strong program of adult and continuing education.

Applying

Prospective students should apply as early as possible. All available assistantships for the fall quarter are awarded shortly after April 1, and applications for them must be received before that date. Requests from students entering at times other than the fall quarter should be received at least thirty days in advance of enrollment. Those students interested in a research assistantship should inquire about the available research areas. Requests for information and application forms should be directed to the Department of Geology or to the Dean of the Graduate School. Applicants must meet the minimum standards for admission set by the Graduate School (i.e., an overall GPA of 2.7 or higher on a 4.0 scale). Applicants requesting financial aid must have earned a GPA of at least 3.0 in geology and achieved a combined score of 950 or better on the verbal and quantitative portions of the GRE General Test. Foreign applicants seeking financial aid must have a TOEFL score of at least 600.

Correspondence and Information

Dr. James A. Harrell, Chairman
Department of Geology
University of Toledo
Toledo, Ohio 43606-3390
Telephone: 419-537-2009
Fax: 419-537-2193

University of Toledo

THE FACULTY AND THEIR RESEARCH

James A. Harrell, Associate Professor and Chairman; Ph.D., Cincinnati, 1983. Sedimentary petrology, environmental geology, computer and statistical applications, geoarchaeology.

Stuart L. Dean, Professor; Ph.D., West Virginia, 1966. Structural geology, tectonics, structural petrology.

Craig B. Hatfield, Professor; Ph.D., Indiana, 1964. Stratigraphy.

Lon C. Ruedisili, Professor; Ph.D., Wisconsin, 1968. Hydrogeology, energy resources, environmental geology.

Mark J. Camp, Associate Professor and Director, Subsurface Data Center; Ph.D., Ohio State, 1974. Paleontology, paleoecology.

Michael W. Phillips, Associate Professor; Ph.D., Virginia Tech, 1972. Mineralogy, crystal chemistry.

Donald J. Stierman, Associate Professor; Ph.D., Stanford, 1977. Applied geophysics, seismology.

Vernon M. Brown, Assistant Professor; Ph.D., Missouri–Rolla, 1983. Igneous and metamorphic petrology, ore microscopy.

David Dollimore, Adjunct Professor; Ph.D., London, 1952. Geochemistry, mineral processing.

Gordon A. Parker, Adjunct Professor; Ph.D., Wayne State, 1966. Analytical geochemistry.

Margaret A. Kitchen, Adjunct Associate Professor; M.A., Michigan, 1946; J.D., Toledo, 1952. General geology.

Ronald E. Gallagher, Adjunct Assistant Professor; M.A., Toledo, 1978. Hydrology, engineering geology.

Ronald H. Hall, Adjunct Assistant Professor; M.S., Penn State, 1956. Analytical geochemistry.

William A. Kneller, Professor Emeritus; Ph.D., Michigan, 1964. Coal geology, organic petrology, economic geology of industrial rocks and minerals, concrete petrology.

The University of Toledo main campus.

SECTION 3: EARTH AND PLANETARY SCIENCES

UNIVERSITY OF TORONTO

Programmes in the Earth Sciences

Programmes of Study	Master's and Doctor of Philosophy degrees are offered in geology, geophysics (Department of Physics), metallurgy and materials science, geotechnical engineering (Geotechnical Section of the Department of Civil Engineering), geography, and interdisciplinary studies. A brochure is available on request.
Research Facilities	The Department of Geology is well equipped with various laboratories for studies in X-ray fluorescence analysis, atomic absorption spectroscopy, inductively coupled plasma spectroscopy, radiochemical and instrumental neutron activation analysis, ultrasensitive ion microprobe spectroscopy, electron microprobe analysis, electron microscopy, fluid inclusion research, isotope geochemistry including Sr-Rb and Nd-Sm studies, scanning electron microscopy, mass spectrometry, zircon geochronology, magnetotellurics, electrical properties, minicomputer and microcomputer automation of instruments, rock deformation, data and information processing, X-ray diffraction, palynology, invertebrate paleontology, applied geophysics, and high-sensitivity magnetics; for studies in Quaternary geology, hydrogeology, paleomagnetism, and high-pressure physical properties; and for the experimental study of chemical reactions involving both silicate and sulfide minerals at high pressures and temperatures.
Geophysics can support research work in field studies of magnetotellurics, seismic studies, field gravity studies, electromagnetic modelling, electromagnetic surveys, and geochronology.
The Department of Metallurgy and Materials Science has a wide range of equipment for high-temperature microradiography, stereomicroradiography, and mineral-processing research.
Soil and rock mechanics laboratories, an air-photo-interpretation laboratory, and a foundation laboratory can be found in the Geotechnical Section.
The Department of Geography has a remote-sensing laboratory at Erindale, a physical sedimentology laboratory at Scarborough, and major items of laboratory equipment, including an atomic absorption spectrophotometer, a viscosimeter, microscopes, scanning stereoscopes, and a 20-foot tilting, recirculating flume with appurtenances.
Financial Aid	All graduate students are eligible for financial support. Most students who are proceeding by thesis and who are Canadian citizens or have landed-immigrant status receive support at competitive levels through scholarships and/or teaching and research assistantships. For the academic session that runs from September 1991 through April 1992, inclusive, the minimum stipend is $8770. Students can usually obtain additional support for the summer months through employment in government or industry or through additional research assistantships. Qualified international students who are not landed immigrants are eligible for teaching assistantships and a limited number of scholarship awards. An additional award to cover their higher fees is usually available.
Cost of Study	Tuition and incidental fees in 1991–92 are $2477.56 per academic year for Canadians and permanent residents and $10,455.56 for international students. Fees are subject to change without notice.
Cost of Living	The cost of living in Toronto is high. In 1991–92, it is estimated that the maintenance cost for a graduate student is $12,000 (including lodging, meals, and books) for an eight-month period. There are many apartments and boarding houses available in the residential areas surrounding the University campus. Information regarding accommodations may be obtained from the University Housing Service, 214 College Street.
Student Group	The University of Toronto has a staff of approximately 4,500 teachers and an enrolment of 47,000 students, including more than 8,690 graduate students. There are about 175 graduate students enrolled in all the earth sciences programmes.
Location	Toronto, the capital of the province of Ontario and a city of more than 2 million people, is located on the north shore of Lake Ontario. The largest campus of the University, the St. George campus, occupies about 160 acres in the downtown core and is surrounded by residential, mercantile, and financial districts. The other two campuses, Scarborough and Erindale, are 30 kilometres east and west, respectively, of St. George. Toronto's cultural centre—the Royal Alexandra Theatre, O'Keefe Centre, Art Gallery, and Royal Ontario Museum—is only a few minutes from the downtown campus via public transit.
Within reach of Toronto by bus or car are golf courses, ski resorts, and the lakes, hiking trails, and canoe routes of Georgian Bay, Muskoka, and Algonquin Park.
The University	The University of Toronto is the largest university in Canada. It consists of the Faculty of Arts and Science with its seven colleges, the Faculty of Engineering, Erindale College, Scarborough College, seventeen professional faculties and schools, the School of Graduate Studies, Woodsworth College (offering part-time degree programmes), the School of Continuing Studies, and a number of large academic divisions that support the work of teaching and research, as well as "service" divisions and administrative departments.
Applying	Admission requirements in geology are flexible, and each application is individually considered, but, in general, a 75 percent average in work for the bachelor's degree is advised as a minimum standard (approximately equivalent to B+ or upper second class). Graduates of degree programmes in sciences other than geology are encouraged to apply.
Students wishing to pursue extensive studies within one of the participating interdisciplinary departments must be prepared to meet the minimum prerequisites for courses and graduate supervision specified by those departments.
Students interested in graduate work in any of the earth science areas should write directly to the graduate secretary of their preferred department. Students should also feel free to write to individual faculty members in their own field of interest.
Correspondence and Information	Graduate Admissions Officer
Department of Geology
Earth Sciences Centre
University of Toronto
Toronto, Ontario M5S 3B1
Canada

Peterson's Guide to Graduate Programs in the Physical Sciences and Mathematics 1992

SECTION 3: EARTH AND PLANETARY SCIENCES

University of Toronto

THE FACULTY AND THEIR RESEARCH
Geology
G. M. Anderson: geochemistry. R. C. Bailey: geomagnetism. A. R. Cruden: structural geology. N. Eyles: Quaternary geology/sedimentology. J. J. Fawcett: metamorphic and igneous petrology. J. Gittens: igneous and metamorphic petrology. M. P. Gorton: analytical geochemistry. C. J. Hale: geophysics. H. C. Halls: paleomagnetics. G. S. Henderson: mineralogy. K. W. F. Howard: hydrogeology. D. Kobluk: invertebrate paleontology. T. E. Krogh: geochronology, Rb-Sr, U-Pb, zircons. J. Mandarino: mineralogy. J. H. McAndrews: Quaternary biogeography. A. D. Miall: stratigraphy/sedimentology. A. J. Naldrett: mineral deposits geology, petrology of mafic and ultramafic rocks. G. Norris: stratigraphic palynology. J. G. Patterson: Precambrian stratigraphy tectonics. P.-Y. F. Robin: mineral physics. J. C. Rucklidge: X-ray analysis of minerals. D. J. Schulze: igneous petrology. W. M. Schwerdtner: Precambrian shield structure. S. D. Scott: marine geology, massive sulfide deposits. E. T. C. Spooner: economic geology, fluid inclusions, industrial minerals. J. C. Van Loon: environmental geochemistry, atomic absorption. P. H. von Bitter: paleontology. J. A. Westgate: Quaternary geology, tephrochronology. F. J. Wicks: mineralogy.

Geophysics
R. C. Bailey: geomagnetism, magnetotellurics. D. J. Dunlop: rock magnetism, paleomagnetism. R. N. Edwards: electrical conductivity structure. R. M. Farquhar: lead isotopes, geochronology. G. F. West: electromagnetics, regional geophysics. D. York: geochronology, potassium-argon.

Metallurgy and Materials Science
S. A. Argyropoulos: process metallurgy, applications of microprocessors. K. T. Aust: corrosion, interfacial phenomena. R. A. Bergman: production of known-ferrous metals. W. J. Bratina: fracture mechanics, fatigues. G. S. Dobby: minerals processing. S. N. Flengas: thermal dynamics of fused salt systems. U. M. Franklin: archaeometry, ancient materials. J. D. Labers: electrometallurgy. J. M. Lee: biomaterials, polymers. R. T. McAndrew: hydrometallurgy. A. McLean: iron and steelmaking. W. A. Miller: alloy microstructures, interfaces in materials. T. H. North: welding engineering. A. R. Perrin: computer-aided engineering, microalloyed steels, electronic material. R. M. Pilliar: biomaterials. B. Ramaswami: mechanical properties, electron microscopy. J. W. Rutter: solidification. I. D. Summerville: steelmaking. S. Thorpe: corrosion, amorphous materials, electronic packaging materials. J. M. Toguri: nonferrous metallurgy. G. C. Weatherly: physical metallurgy, welding, electron microscopy, interfaces.

Geotechnical Engineering
A. M. Crawford: rock mechanics. J. Curran: geotechniques. T. C. Kenney: soil properties. E. I. Robinsky: foundation. J. Ting: soil mechanics.

Geography
R. Bryan: environmental geomorphology. A. Davis: paleoecology. B. Greenwood: coastal geomorphology. S. Luk: geomorphology. D. S. Munro: microclimatology. A. Price: meteorology.

RESEARCH ACTIVITIES
Chemistry and physics of Earth: mineral deposits, petrology, physical geochemistry, isotope geochemistry, mineralogy, materials science.
Age and evolution of Earth: paleontology, stratigraphy and sedimentation, Precambrian geology, isotope geochemistry.
Structure and dynamics of Earth: structural geology, plate tectonics, solid-earth geophysics, marine geology, geophysics.
Environmental and engineering geology: environmental geochemistry, geomorphology, remote sensing, geotechnical engineering, hydrogeology.
Minerals exploration: mineral deposits geology, exploration geochemistry, exploration geophysics.

SECTION 3: EARTH AND PLANETARY SCIENCES

UNIVERSITY OF WASHINGTON
Geophysics Program

Programs of Study

Geophysics is the study of the earth's physical makeup, behavior, and planetary environment. Based directly on physical laws and encompassing many different mathematical and observational methods, geophysics examines the complex nature of the solid, liquid, and gaseous components of the earth and its environment. Current research also focuses increasingly on the interactions of these components, for example, the climate system. Modern geophysics is thus an interdisciplinary science, not a separate and isolated area of knowledge.

All students are required to take a core curriculum of courses on the earth's interior, the oceans, the atmosphere, and the solar-terrestrial environment. Beyond this sequence, students select a course of study from offerings in the Geophysics Program and the participating Departments of Astronomy, Atmospheric Sciences, Civil Engineering, Electrical Engineering, Geological Sciences, Oceanography, and Physics. Because of the diversity of disciplines represented in geophysics, there are few formal course requirements, and it is the responsibility of the student and his or her advisory committee to plan a course of study that is appropriate as a background for thesis research. The principal areas of activity in the program today are seismology, geomagnetism, high-pressure experimental geophysics, glaciology, geophysical fluid dynamics, space physics, and aeronomy.

At the end of the first year of study, the student takes a qualifying examination that emphasizes fundamental preparation in the physical sciences, material from the introductory courses, and overall competence in the geophysical disciplines. During the second year, M.S. students either prepare a thesis or satisfy the nonthesis option by preparing a manuscript suitable for publication. Ph.D. students pursue a course of study selected in conjunction with the advisory committee and prepare a dissertation proposal for presentation at a general examination. Work then proceeds on the dissertation until it is ready to be presented at the final examination. The Ph.D. program typically requires five years.

Research Facilities

Laboratory facilities are available for a wide range of experimental work. They include a cryogenic magnetometer for rock magnetism; a large convection cell for studying flow in porous media; a high-pressure/temperature laboratory, including diamond anvil cells for studying rock and mineral properties; cold room facilities for glaciological research; geophysical fluid laboratories; a space physics laboratory for preparing balloon and satellite experiments; a laboratory for the development of optical instrumentation; and assorted electronics and machine shop facilities. Computer systems are used for various types of on-line experiments and for modeling calculations. A statewide telemetered seismograph system with extensive support facilities is available, as are major facilities for field experiments, including ocean-bottom seismographs, seismic profiling equipment, magnetometers, gravimeters for both crustal deformation and conventional surveys, portable seismic networks, and equipment for geomagnetic induction and glaciological studies. In addition, the research facilities of all participating departments are available to students through their thesis supervisor.

Financial Aid

Graduate student support is typically provided by research assistantships. A limited number of teaching assistantship appointments are also made available to geophysics students by the Departments of Geological Sciences and Physics. All assistantship appointments require 20 hours of service per week and provide a beginning monthly income of approximately $940. The student's interests are matched as closely as possible with fields of available support. Changes can be made after the first year, if appropriate.

Cost of Study

Tuition for the 1991–92 academic year is approximately $1129 per quarter for in-state students and $2824 per quarter for out-of-state students. However, students holding assistantships are required to pay only a minor portion (approximately $110 per quarter) of these tuition costs.

Cost of Living

Housing for single students is available on campus through the Housing Assignment Office. In 1991–92, on-campus costs are about $420 per month, including meals. A limited number of married student units may also be available. Off-campus housing arrangements are typically $200–$300 per person per month for group-living rentals.

Student Group

Geophysics usually has about 40 students, most of whom are working toward a Ph.D. Backgrounds are diverse, but an entering student normally has the equivalent of a B.S. in physics, with some special interest, course work, or experience in one of the geophysical sciences. Graduates at both the M.S. and Ph.D. levels have found excellent employment opportunities during the past decade, and this is expected to continue.

Location

Seattle's University District is both environmentally attractive and culturally interesting. It lies between Lake Washington and Puget Sound at the northern end of downtown Seattle. All the cultural advantages of a large city are available, while some of the country's most spectacular natural environments, from mountains to ocean beaches, can be found nearby.

The University

The University, founded in 1861, is the oldest state-assisted institution of higher education on the Pacific Coast. The campus is large, with a total enrollment of nearly 33,000 and a national reputation for quality in graduate research and education. The Geophysics Program is administered by the College of Arts and Sciences.

Applying

Applications for admission and for financial support should be sent to the graduate program adviser (address below). The deadline for financial aid applications is February 15. Admission for the fall quarter can be requested as late as July 1, but earlier application is strongly recommended. Scores on the General Test of the Graduate Record Examinations are required for consideration for both admission and financial aid. Assistantship applications, statements of purpose, and three letters of recommendation should be sent directly to the Geophysics Program. Graduate School applications, transcripts, and test scores should be sent to Graduate Admissions, AD-10.

Correspondence and Information

Dr. John R. Booker
Graduate Program Adviser
Geophysics Program, AK-50
University of Washington
Seattle, Washington 98195
Telephone: 206-543-8020

University of Washington

THE FACULTY AND THEIR RESEARCH

Marcia B. Baker, Professor of Geophysics and Atmospheric Sciences; Ph.D., Washington (Seattle), 1971. Atmospheric geophysics.
John R. Booker, Professor of Geophysics and Atmospheric Sciences; Ph.D., California, San Diego, 1968. Geomagnetic induction, magnetotellurics, inverse theory, geophysical fluid dynamics.
J. Michael Brown, Associate Professor of Geophysics and Geological Sciences; Ph.D., Minnesota, 1980. Experimental and theoretical mineral and rock physics.
Robert A. Charlson, Professor of Atmospheric Science, Environmental Studies, Chemistry and Geophysics; Ph.D., Washington (Seattle), 1964. Atmospheric chemistry, tropospheric aerosols.
Kenneth C. Creager, Assistant Professor of Geophysics; Ph.D., California, San Diego, 1984. Seismology and geophysical inverse theory.
Robert S. Crosson, Professor of Geophysics and Geological Sciences; Ph.D., Stanford, 1966. Seismology, structure and tectonics, earthquake hazards.
John T. Ely, Research Associate Professor; Ph.D., Washington (Seattle), 1969. Cosmic rays, magnetospheric physics.
Gonzalo Hernandez, Research Professor; Ph.D., Rochester, 1962. Optical interference phenomena, remote sensing of atmospheres.
Robert H. Holzworth, Associate Professor of Geophysics and Physics; Ph.D., Berkeley, 1977. Experimental space plasma physics, atmospheric and magnetospheric electric fields, thunderstorm electrification.
David Jay, Research Assistant Professor; Ph.D., Washington (Seattle), 1987. Fluid mechanics, estuarine and coastal oceanography, wave dynamics and stratified flows.
Conway B. Leovy, Professor of Atmospheric Sciences, Geophysics, and Astronomy; Ph.D., MIT, 1963. Dynamics and radiative transfer of the stratosphere and mesosphere, the atmospheres of other planets.
Brian T. R. Lewis, Professor of Geophysics and Oceanography; Ph.D., Wisconsin, 1970. Processes of lithosphere generation and consumption using seismology, gravity, and magnetics.
Stephen D. Malone, Research Professor; Ph.D., Nevada, 1972. Seismicity of Mt. St. Helens, the Cascade volcanoes, and eastern Washington; computer applications in seismic network analysis.
Gary Maykut, Research Professor; Ph.D., Washington (Seattle), 1969. Sea-air-ice interaction in the polar oceans.
Michael McCarthy, Research Assistant Professor; Ph.D., Washington (Seattle), 1988. Solar wind and magnetospheric physics.
James A. Mercer, Research Assistant Professor; Ph.D., Washington (Seattle), 1983. Seismic acoustic studies, oceanacoustic tomography.
Ronald T. Merrill, Professor of Geophysics, Geological Sciences, and Oceanography; Ph.D., Berkeley, 1967. Geomagnetism, geophysics of solids, rock magnetism.
George K. Parks, Professor of Geophysics, Atmospheric Sciences, and Physics; Ph.D., Berkeley, 1966. Particles and waves in auroral, magnetospheric, and interplanetary space plasma phenomena.
Anthony Qamar, Research Associate Professor; Ph.D., Berkeley, 1971. Earthquakes associated with volcanoes and glaciers, earth-structure and earthquake hazards.
Charles F. Raymond, Professor of Geophysics, Quaternary Research, and Geological Sciences; Ph.D., Caltech, 1969. Glaciology, glacier and ice sheet dynamics.
Stewart W. Smith, Professor of Geophysics and Geological Sciences; Ph.D., Caltech, 1961. Seismology, earthquake risk, seismotectonics.
Yosiko Sato Sorensen, Research Associate Professor; Ph.D., Tokyo, 1973. High-pressure experiments in geophysics, physics and chemistry of minerals.
Norbert Untersteiner, Professor of Atmospheric Sciences and Geophysics; Ph.D., Innsbruck, 1950. Glaciology, arctic sea ice.
Edwin D. Waddington, Research Associate Professor; Ph.D., British Columbia, 1981. Glacier and ice sheet modeling, interpretation of ice sheet stratigraphy.
Stephen G. Warren, Associate Professor of Atmospheric Sciences, Geophysics, and Quaternary Research; Ph.D., Harvard, 1973. Radiation and climate, glaciology.
Scott Werden, Assistant Professor; Ph.D., Washington (Seattle), 1988. Solar terrestrial interactions and magnetospheric physics.

The eruption of Mount St. Helens, May 18, 1980.

Students and faculty prepare launch of stratospheric payload for electrical field measurements.

SECTION 3: EARTH AND PLANETARY SCIENCES

WASHINGTON UNIVERSITY

Department of Earth and Planetary Sciences

Programs of Study

The Department of Earth and Planetary Sciences offers programs leading to M.A. and Ph.D. degrees, with specialization in planetary sciences, geodynamics, evolution of the continental crust, and fluid-rock interactions.

A minimum of 72 semester hours is required for the Ph.D., including core courses in three of the following areas: geochemistry, geodynamics, petrology, and tectonics and sedimentation. Candidates must also pass an in-depth examination in the chosen area and an oral defense of the dissertation research. The requirements for the M.A. include good performance in required course work, completion of 39 semester hours of graduate courses, preparation of a thesis, and satisfactory performance in a final oral defense of the thesis.

Interdisciplinary studies are facilitated by the McDonnell Center for the Space Sciences, a University-wide center that involves the faculties of the Departments of Earth and Planetary Sciences, Physics, Chemistry, and Engineering.

Research Facilities

Housed in its own building, Wilson Hall, the department contains numerous collections; laboratory and office space; an excellent earth science library with more than 35,000 volumes; an automated electron microprobe; a thin-section preparation laboratory; facilities for neutron activation analysis, atomic absorption spectrophotometry, and X-ray diffraction and fluorescence studies; two laser Raman microprobes; an array of research microscopes; a twelve-channel exploration seismograph; a gravimeter; advanced spectropolarimetric imaging and mapping instruments for remote sensing (e.g., Fabry-Perot interferometer spectrometers); a complete collection of lunar and planetary photographic and cartographic products and extensive collections of digital planetary geology and geophysics data. The department is home to NASA's Planetary Data System Geosciences Node and to the Regional Planetary Image Facility. The node is responsible for curation and distribution of NASA's planetary geological and geophysical data. The department has a variety of computing systems available, including a suite of networked workstations for imaging processing, reduction and analysis of seismic data, and general purpose computing. The systems are on the campus network, Internet, and the Space Physics Analysis Network. In addition, the University operates an IBM 360/65. A modern laboratory building is under construction, with completion expected in early 1993.

Within the McDonnell Center for the Space Sciences, facilities include an ion microprobe (SIMS), a scanning electron microscope with microprobe attachment, rare-gas spectrometers, a solid-source mass spectrometer, a computer system, a nuclear-particle track laboratory, and equipment for mineral separation, thin-section preparation, and laser spectroscopy. The usual complement of research facilities can be found in the Departments of Chemistry and Physics.

Field studies take place near the University and in a variety of remote locations. Studies are currently under way in Quebec, Ontario, Arkansas, Minnesota, Greenland, Egypt, and the Sudan.

Financial Aid

Each year the department awards a number of teaching and research assistantships, with 1991–92 stipends of $9360 plus tuition for the academic year. Fifteen hours per week are usually devoted to duties associated with the appointment. Full or partial fellowship and scholarship stipends range from tuition remission to $13,160 plus tuition remission. Comparable stipends are available for summer support.

Cost of Study

The cost of graduate study at Washington University is comparable to that at other institutions of its type and caliber. In 1991–92, tuition for full-time students is $7975 per semester. Graduate students enrolled for fewer than 12 units pay $665 per unit.

Cost of Living

Moderately priced rental units may be obtained near the University at average monthly rates of $400 (one bedroom), $475 (two bedrooms), and $575 (three bedrooms).

Student Group

Almost half of the 8,400 students at the University are graduate and professional school students. The faculty has more than 1,200 full-time members. The department has 33 graduate students, 7 of whom are women.

Location

The community surrounding the University is both residential and commercial. Entertainment is varied: music lovers can enjoy the St. Louis Symphony, the St. Louis Philharmonic, and many local jazz, blues, rock, and dance clubs; theatergoers can attend performances of several repertory groups and musicals at the summer Municipal Opera. Forest Park, within walking distance of the campus, contains a golf course, bicycle paths, horseback-riding facilities, a fine zoo, the St. Louis Science Center, and the St. Louis Art Museum.

The University

Washington University was founded in 1853 as a private, coeducational institution. In 1904, the University moved to its present 168-acre hilltop campus bordering on the 1,430 acres of Forest Park. Undergraduate programs are offered in arts and sciences, engineering, business, and fine arts, and graduate programs are offered in all major fields of human inquiry. Eighteen Nobel Prize recipients have done all or part of their distinguished work at Washington University. Faculty members serve on editorial boards of more than 250 professional and scholarly journals; nineteen are members of the National Academy of Sciences. In a National Science Foundation report for fiscal 1987, Washington University ranked tenth among private universities in federal support for research and development. The University ranked seventh among private institutions receiving Department of Health and Human Services funding and eleventh among all universities. Washington University ranks among the top thirty schools in the nation in enrolling National Merit Scholars. Since 1976, Washington University has placed first or second in the William Lowell Putnam Mathematical Competition seven times, among the top ten twelve times, and eleventh once. Graduates often receive such prestigious graduate study awards as Fulbright, Marshall, Beinecke, and Truman scholarships and Mellon, Putnam, National Science Foundation, and NASA graduate fellowships.

Applying

To ensure consideration of a student for admission in September, a completed application, transcript, financial statement, Graduate Record Examinations scores, and three letters of recommendation must be received by January 15. Applicants should have had courses in calculus, chemistry, and physics and should have specialized in earth sciences, engineering, physics, biology, chemistry, or mathematics.

Correspondence and Information

Raymond E. Arvidson, Chairman
Department of Earth and Planetary Sciences
Washington University
One Brookings Drive
St. Louis, Missouri 63130-4899
Telephone: 314-889-5610

Peterson's Guide to Graduate Programs in the Physical Sciences and Mathematics 1992

Washington University

THE FACULTY AND THEIR RESEARCH

Raymond E. Arvidson, Professor and Chairman; Ph.D., Brown, 1974. Remote sensing of planetary surfaces, focused on retrieval of compositional and textural information; origin and evolution of planetary surfaces; image processing and database management.

Professors

Ghislaine Crozaz, Ph.D., Libre (Brussels), 1967. Cosmochemistry: petrogenesis of meteorites and lunar and terrestrial rocks, measuring in situ abundances of rare earths and other trace elements by secondary-ion mass spectrometry (ion microprobe).

Robert F. Dymek, Ph.D., Caltech, 1977. Field and chemical petrology of metamorphic and igneous rocks applied to problems of Precambrian crustal evolution, petrogenesis of anorthosites and related rocks, trace-element behavior in metamorphism, electron microprobe analysis, mineral chemistry and phase petrology.

Larry A. Haskin, Ralph E. Morrow Distinguished University Professor, Ph.D., Kansas, 1960. Geochemistry: abundances and behavior of rare earths and other trace elements; neutron activation analysis of layered igneous complexes, lavas, and sediments and of lunar igneous rocks, breccias, and soils; mathematical modeling of trace-element behavior during igneous processes; determination of trace-element behavior in synthetic silicates by use of electron paramagnetic resonance, electrochemistry, and electron microprobe techniques.

Harold L. Levin, Ph.D., Washington (St. Louis), 1954. Paleontology and stratigraphy: geowriting, evolutionary trends among Cenozoic calcareous phytoplankton, utilization of coccolithophorids in chronostratigraphic zonation and correlation, study of ultrastructure of calcareous nannoplankton.

Jill Dill Pasteris, Ph.D., Yale, 1980. Ore petrology and geochemistry: magmatic sulfide deposits, fluid-rock interaction, fluid inclusions, laser Raman spectroscopy.

Frank A. Podosek, Ph.D., Berkeley, 1969. Studies of isotopic geochemistry using thermal ionization and noble-gas mass spectrometry, with application to cosmochemistry and terrestrial geochemistry; meteorite chronology, origin of the solar system, pre-solar isotopic structures, isotopic systematics of volcanic rocks and hydrothermal ore deposits.

William Hayden Smith, Ph.D., Princeton, 1966. Spectropolarimetric mapping and imaging of the Earth and solar system objects, searches for external planets, laboratory spectroscopic studies of atmospheres and atmospheric-surface interaction processes (e.g., weathering).

Associate Professors

Ian J. Duncan, Ph.D., British Columbia, 1982. Structural geology: tectonophysics, structure of shear zones in basement terrains, field studies, strain analysis, Cordilleran structure, finite-element modeling, application of remote sensing data to structural problems, tectonics of Venus.

M. Bruce Fegley, Ph.D., MIT, 1980. Planetary science and cosmochemistry, with an emphasis on experimental and theoretical studies of chemical processes in the early solar system, on planetary surfaces, and in planetary atmospheres.

William B. McKinnon, Ph.D., Caltech, 1981. Geophysics and planetary science: physics of impact cratering; planetary tectonics; thermal and bombardment history, structure, and evolution of icy satellites and Pluto; radio astronomy.

Douglas A. Wiens, Ph.D., Northwestern, 1985. Seismology and geophysics: intraplate seismicity and stress in the lithosphere, seismotectonics of convergent and divergent plate boundaries, structure of the earth's interior.

Assistant Professors

Everett L. Shock, Ph.D., Berkeley, 1987. Geochemistry: theoretical organic geochemistry, hydrothermal stability of organic compounds, transport and deposition of petroleum, organic synthesis in carbonaceous meteorites, origin of life, mass transfer in geochemical processes, supercritical aqueous solution chemistry, magmatic/hydrothermal systems.

Michael E. Wysession, Ph.D., Northwestern, 1991. Geophysics and planetary structure: dynamics and composition of Earth's mantle and core from seismic waves; locations and recurrence of earthquakes.

Research Professor

Ernst Zinner, Ph.D., Washington (St. Louis), 1972. Study of extraterrestrial materials by microanalytical techniques, especially SIMS; nucleosynthesis in stars and the formation of the solar system. (Joint appointment with Physics)

Research Associate Professors

Thomas Bernatowicz, Ph.D., Washington (St. Louis), 1980. Planetary degassing, physical chemistry of noble-gas interactions with materials of geochemical interest, noble-gas isotopic fractionation, problems of grain formation in stellar atmospheres and laboratory study of interstellar grains. (Joint appointment with Physics)

Randall L. Korotev, Ph.D., Wisconsin–Madison, 1976. Geochemistry: processes of formation and development of the lunar regolith and crust, procedures for chemical analysis of geologic materials by neutron activation.

Senior Research Scientists

Russell Colson, Ph.D., Tennessee, 1986. Effects of variations in intensive parameters on partitioning of trace elements between phases; experimental determination of partition coefficients and modeling of trace-element partitioning; effects of composition, fracturing, and shock-induced dislocations on diffusion rates of cations in common silicates; study of diffusion rates in plagioclase and glass.

Edward A. Guinness, Ph.D., Washington (St. Louis), 1980. Planetary geology and sedimentology: remote sensing of planetary surfaces using reflected visible/infrared and emitted infrared radiation.

Mohamed Sultan, Ph.D., Washington (St. Louis), 1984. Petrology: study of the assembly and postaccretionary tectonic history of the Arabian-Nubian Shield, applying remote sensing, field, geochronologic and isotopic, and laboratory techniques.

Cooperating Faculty and Senior Research Staff

Charles Hohenberg, Professor of Physics; Ph.D., Berkeley, 1968. Planetary science: rare-gas mass spectrometry, particularly of extraterrestrial materials, applied to problems of extinct isotopes and the formation of the elements and solar system.

Robert M. Walker, McDonnell Professor of Physics and Director, McDonnell Center for the Space Sciences; Ph.D., Yale, 1954. Planetary science: nuclear track studies bearing on the history of the solar system and the nature of energetic radiations in space; infrared and isotopic studies of interplanetary dust and meteorites related to the nature of comets, protostars, and formation of the elements.

SECTION 3: EARTH AND PLANETARY SCIENCES

WRIGHT STATE UNIVERSITY
Department of Geological Sciences

Programs of Study
The Department of Geological Sciences offers programs in geochemistry, geophysics, hydrogeology, petroleum geology, petrology, sedimentary geology, and structural geology leading to the Master of Science (M.S.) in geological sciences. A Master of Science in Teaching (M.S.T.) program is also available. The programs include course work (minimum 36 quarter credits) and independent research (minimum 9 quarter credits). A thesis, evaluated by a faculty advisory committee, is required, and field data acquisition and hands-on experience in use of state-of-the-art equipment are stressed. Current thesis work is concentrated in Midwest cratonic basins, the Appalachians, and the Rocky Mountains. Employment opportunities for M.S. graduates have been outstanding. Major oil and hydrogeological companies interview directly in the department. In 1986, the department was recognized as a center of excellence in Ohio and received an Academic Challenge Grant for budget augmentation. The hydrogeology program at Wright State has been ranked, by an independent study, in a tie for seventh place among the best hydrogeology programs in the United States and Canada.

Research Facilities
The Department of Geological Sciences is housed primarily in the Brehm Laboratory Building. Department facilities include twelve teaching and research laboratories and a wide variety of specialized equipment. A geology field equipment building was completed in 1986. The facilities of the geophysics program include a 96-channel truck-mounted seismic reflection system, a Geophysical Work Station for seismic modeling, a Sun SPARCstation for seismic data processing, three gravity meters (LaCoste-Romberg and Worden), a magnetic gradiometer system, a ground-penetrating radar system, a resistivity meter, and a portable seismic refraction system. For the hydrogeology/hydrogeochemistry programs, the department has a Perkin-Elmer atomic absorption spectrophotometer (Model 3030), a Dionex ion chromatograph, a light isotope preparation facility, X-ray spectrographic and fluorescence equipment, an electron microprobe, an extensive collection of software for hydrogeological and hydrogeochemical modeling, and extensive equipment for field sampling, well drilling, and well logging. Students and faculty members in the petrology/sedimentary geology program may use Zeiss petrographic microscopes (including a Universal model), a cathodoluminescence system and ultraviolet luminescence, a morphometer facility, complete Logitech thin-section and rock preparation facilities, a sedimentological laboratory, photographic equipment and a darkroom, mineral separation facilities, and facilities for chemical analyses of rock. Computer laboratories within the department provide access to PCs, Sun Workstations, and the University mainframe network, which includes supercomputer capabilities. All programs are backed by a fleet of twenty department vehicles, including three drilling rigs. The library, in which geological sciences holdings have recently been significantly upgraded, contains more than 345,000 bound volumes, 700,400 microforms, 157,000 U.S. and Ohio government documents, 3,948 periodical subscriptions, and a national map depository with approximately 56,000 geological and topographical maps from all over the United States.

Financial Aid
Several industry-funded research fellowships, nine graduate teaching assistantships, and six graduate assistantships, all carrying fee remissions, are awarded. Additional research assistantships connected with supported research projects, summer opportunities on research projects, and research contracts are also available. Tutoring positions in the nonresidential IRIS program are another opportunity for graduate students.

Cost of Study
Quarterly fees in 1991–92 are $1114 for full-time (11 to 18 credit hours) students who are Ohio residents and $1997 for nonresidents. Charges per credit hour for students taking fewer than 11 or more than 18 hours are $106 for residents and $190 for nonresidents.

Cost of Living
For a limited number of students who want the convenience of living on or near the campus, the University provides one residence hall and apartment complexes. Most graduate students obtain privately owned housing near campus. Comfortable and attractive apartments shared by 2 students typically cost $250–$300. The overall cost of living in the Dayton area is relatively low.

Student Group
The University enrolls more than 17,000 students, of whom 3,000 are graduate and professional students. Students regularly participate in meetings of national and regional professional societies such as SEG, AAPG, GSA, and OAS. Wright State University has more than 100 programs of study leading to nine different baccalaureate degrees and 30 programs of graduate and professional study. The graduate program in geological sciences currently enrolls 48 students from over fifteen states and several foreign countries.

Location
Wright State University is located in a suburban area near Dayton, Ohio, a city of industrial importance that has a metropolitan population of more than 830,000. Dayton offers cultural and recreational facilities and has many public and private schools and educational facilities. Dayton is sometimes referred to as the "Birthplace of Aviation," a reference to native Daytonians Orville and Wilbur Wright and their accomplishments in powered flight. Wright Patterson Air Force Base is a major center for government research and development. Dayton is relatively close to the Appalachian Mountains, the Great Lakes, Kentucky caves, and interesting geological outcrops.

The University
Wright State is an exciting and expanding university with facilities for academics, support programs, recreation, and arts activities. Modern architecture facilitates access for all individuals.

Applying
Admission to graduate programs in the department is competitive. The faculty for each program determines eligibility and acceptance based upon the academic records, experience, and recommendation received for each applicant. If a student is undecided on a program concentration area, the Graduate Studies Committee of the department determines admission. Admission is related to program capacity, and students with excellent records may not be admitted if a program enrollment limit has been reached. For this reason, early application is encouraged. The annual review cycle begins in January, when applications are received and documentation verified. In February, program faculty members review applications and make recommendations to the Graduate Studies Committee.

Correspondence and Information
For more information, including admission requirements and a graduate course catalog, prospective students should contact:
Deborah Hapner
Department of Geological Sciences
Wright State University
Dayton, Ohio 45435
Telephone: 513-873-3455

Peterson's Guide to Graduate Programs in the Physical Sciences and Mathematics 1992

SECTION 3: EARTH AND PLANETARY SCIENCES

Wright State University

THE FACULTY AND THEIR RESEARCH

Cindy Carney, Assistant Professor; Ph.D., West Virginia. Carbonate petrology: carbonate sedimentology and petrology, interpretation of ancient depositional environments, mixed carbonate-clastic units, recent carbonate environments, carbonate diagenesis, all with special emphasis on deep burial stages.

Songlin Cheng, Assistant Professor; Ph.D., Arizona. Hydrogeochemistry: isotope hydrology/geochemistry, hydrogeochemical modeling and contaminant hydrogeology.

David Dominic, Assistant Professor; Ph.D., West Virginia. Clastic sedimentology: fluvial depositional environments, quantitative interpretation of sedimentary structures and textures, Pennsylvanian depositional history of the central Appalachian basin, controls of alluvial stratigraphy.

Bryan Gregor, Professor; D.Sc., Utrecht (Netherlands). Historical geochemistry, sedimentology: modeling the rock cycle, global inventory of sediments by age and lithology, estimating the main reservoirs (sedimentary, crystalline) and the fluxes between and within them.

Kenneth Kramer, Associate Professor; Ph.D., Florida State. Geochemistry: the effect of pH on element distribution between solid solutions and aqueous solutions; chemical evolution of groundwater within selected environments.

Byron Kulander, Professor; Ph.D., West Virginia. Regional geology, structural geology, geophysics: mechanism, style, and chronology of deformation in the central and southern Appalachian Plateau and Valley and Ridge structural provinces; gravity and magnetics of the central and southern Appalachians and Appalachian basement tectonics; interpretation and formation mechanism of regional rock fracture geometry; application of fractography to core and outcrop fracture investigations; detailed remapping of the Valley and Ridge structural province in West Virginia.

Paul Pushkar, Professor; Ph.D., California, San Diego. Isotope geochemistry, igneous petrology, mineral deposits, field geology: geology, geochemistry, and geochronology of volcanic rocks; field mapping of various geological and geochemical processes; Tertiary volcanic rocks in Gravelly Range, Montana; geochemistry of volcanic rocks in the Central Antilles; petrologic and geochemical studies in the Canadian Shield; strontium isotope studies of oil field brines.

Benjamin Richard, Professor; Ph.D., Indiana. Geophysics, hydrogeology, field geology, structural geology: use of gravity measurements and seismic refraction and reflection techniques to map buried bedrock topography and to aid in the definition of subsurface coal seams and small hydrocarbon traps.

Robert W. Ritzi, Assistant Professor; Ph.D., Arizona. Hydrogeology: analytical and numerical modeling (and inverse modeling) methodologies for flow and mass transport, geostatistics, hydrologic systems analysis, nonanthropogenic variance in groundwater pressure and chemistry, environments for hazardous-waste isolation, groundwater buffering of acid precipitation, island hydrogeology, hydrogeology of riparian systems.

Ronald Schmidt, Professor; Ph.D., Cincinnati. Hydrogeology, environmental geology, engineering geology: water resource planning and management, especially groundwater resources of islands; drilling and sample methodology; slope stability; mapping exploration methodology.

Karel Toman, Professor; Dr.Sc., Prague Technical; C.Sc., Czechoslovak Academy of Sciences. Crystallography: crystal structure, crystal imperfections, structure of solid solutions, seismic data enhancement, methodology of crystal structure determination, applied X-ray spectroscopy, determination of heavy metals in groundwater by X-ray fluorescence, data processing in reflection seismology.

Raphael Unrug, Professor; Ph.D., D.Sc., Jagiellonian (Krakow). Tectonics, basin analysis, sedimentology, remote sensing: basin analysis and sedimentology in the Blue Ridge Mountains and Central and Southern Appalachians; tectonic evolution of the eastern margin of the Laurentian craton; structural controls of drainage patterns in Ohio; regional tectonics of Africa.

Paul Wolfe, Professor; Ph.D., Case Western Reserve. Exploration geophysics: integrated geophysical techniques for petroleum exploration and for solving groundwater and environmental problems, subsurface cavity detection, deep crustal structure determination with seismic reflection.

SECTION 3: EARTH AND PLANETARY SCIENCES

YALE UNIVERSITY
Department of Geology and Geophysics

Programs of Study	The Department of Geology and Geophysics offers instruction and opportunities for research leading to the Ph.D. degree in a broad range of earth sciences. Fields represented in the department include geochemistry, geophysics, sedimentology, stratigraphy, structural geology, petrology, mineralogy, economic geology, paleoecology, vertebrate and invertebrate paleontology, meteorology, and oceanography. Requirements for admission are flexible, but adequate preparation in the related sciences of physics, chemistry, mathematics, and biology is important for graduate study in geology and geophysics. Those with majors in another science are encouraged to apply. At least three full years of residence are required for the Ph.D. Full tuition is charged to Ph.D. candidates through four years or until the dissertation is submitted, whichever occurs first. The Ph.D. program includes course work during the first two years, a comprehensive qualifying examination in the area of specialty at the end of the third term, research on a dissertation topic, and the completion and oral defense of the dissertation. Normal time for completion is five to six years. All entering graduate students are advised by a faculty committee selected by the student and the director of graduate studies in accordance with the student's interests and objectives. A thesis research committee (which may include specialists from other institutions, as well as department faculty) is appointed, when a dissertation topic is approved, to supervise the dissertation research and evaluate the dissertation and its defense.
Research Facilities	The department is located in the modern Kline Geology Laboratory adjacent to Yale's Peabody Museum of Natural History and the Bingham Oceanographic Laboratory. This facility occupies approximately 100,000 square feet and contains laboratories for research in geophysical fluid dynamics, geophysics, geochemistry, rock mechanics, metamorphic petrology, igneous petrology, and benthic marine ecology and sedimentology. Laboratories are fully equipped with modern microscopes, instrumentation, and extensive computer facilities, including a state-of-the-art electron microprobe, mass spectrometers, an X-ray diffractometer, radiometric dating facilities, an AA/AE spectrophotometer, and standard geophysical apparatus such as seismographs and magnetometers. Computational facilities include three Sun SPARCstations II, three Sun IPC computers, eight MicroVAX workstations, two UNIX-based Silicon Graphics Superworkstations, numerous IBM and Macintosh computers, and network connections to parallel-processing computers elsewhere on campus and supercomputers off campus. Also available in the Peabody Museum are a morphometrics laboratory and one of the world's most important collections of fossil vertebrates, invertebrates, and minerals. The Yale Computer Center and the Kline Science Library are nearby. The Geology Library, containing more than 110,000 volumes, is housed within the Kline Geology Laboratory, as is the map collection, which consists of more than 180,000 items. A great variety of other specialized equipment (such as the Van de Graaff accelerator in the physics department) is available in other science departments. The Peabody marine field station is located on Long Island Sound, 30 minutes away.
Financial Aid	Most graduate students in the department receive financial support for both tuition and living expenses through the four-year period during which full tuition is charged. Ph.D. candidates may receive support from Yale University, independent fellowships, teaching assistantships, research assistantships, or some combination of these. In addition, some curatorial, library, or other part-time work is available.
Cost of Study	Full-time tuition for the 1991–92 academic year is $15,160. After four years of study, students are charged a registration fee of $100 per term. Payment of full tuition automatically enrolls the student in the University health plan. An additional $375 is charged for hospitalization coverage. Spouses and dependents may be enrolled in the health plan for an additional fee.
Cost of Living	Rooms are available to single students in the Hall of Graduate Studies and Helen Hadley Hall for $3042 to $3954 for the academic year. Additional rooms for single men are available in small residence halls in other nearby locations. The University maintains a number of housing units for married students, ranging from efficiency apartments ($510–$525 per month) to one-, two-, and three-bedroom apartments ($575–$740 per month). Additional accommodations are available in privately owned housing in the Greater New Haven area.
Student Group	Approximately 45 full-time graduate students are currently studying in the department, and nearly all of them receive some type of financial support. They come from all parts of North America and elsewhere. There are approximately 20 undergraduate majors. The Dana Club (the graduate student club named after James Dwight Dana) sponsors student activities and social events and consults with the faculty.
Location	New Haven is located on the north shore of Long Island Sound, approximately 90 miles from New York City and 150 miles from Boston. Recreational opportunities include various sports, boating, sailing, swimming, fishing, hiking, cycling, and (a few hours to the north) skiing. New Haven and Hartford (40 miles away) feature frequent sports events, concerts, theater, ballet, and films.
The University	Yale was founded in 1701 and began offering graduate education in 1847. The Graduate School is one of twelve schools constituting the University. Yale was the first university in North America to award the Ph.D. degree, conferring three in 1861. The University community consists of approximately 5,000 undergraduates, 6,000 students in the various graduate and professional schools, 1,500 faculty members, and hundreds of supporting research staff members. Among the facilities for research and study, beyond those of the department, are the University Library of over 6 million volumes, the Beinecke Rare Book and Manuscript Library, the Becton Engineering and Applied Science Center, the Arthur W. Wright Nuclear Structure Laboratory, the Peabody Museum of Natural History, and the many laboratories in other science departments, the School of Medicine, and the School of Forestry and Environmental Studies. Interdepartmental and interdisciplinary study is encouraged throughout the University.
Applying	Inquiries should be made directly to the Department of Geology and Geophysics, Office of the Director of Graduate Studies. Applications for admission should be submitted by January 2 to the Office of Admissions, Graduate School, Box 1504A Yale Station, New Haven, Connecticut 06520. Scores on the General Test of the Graduate Record Examinations are required, and the examination should be taken before December; applicants are encouraged but not required to submit GRE Subject Test scores.
Correspondence and Information	Professor Antonio C. Lasaga Director of Graduate Studies Department of Geology and Geophysics Box 6666 Yale University New Haven, Connecticut 06511

SECTION 3: EARTH AND PLANETARY SCIENCES

Yale University

THE FACULTY AND THEIR RESEARCH

Jay J. Ague, Assistant Professor of Geology and Geophysics. Igneous petrology, economic geology, genesis of granitic batholiths and associated ore deposits.

Robert A. Berner, Alan M. Bateman Professor of Geology and Geophysics. Kinetics of mineral-solution interactions and chemical diagenesis of sediments.

Edward W. Bolton, Assistant Professor of Geology and Geophysics. Geophysical fluid dynamics.

Mark T. Brandon, Associate Professor of Geology and Geophysics. Structural geology, tectonics.

Leo W. Buss, Professor of Biology and Geology and Geophysics. Evolutionary biology and paleobiology.

Robert B. Gordon, Professor of Geophysics and Applied Mechanics. Mechanics of earth materials and surficial processes, utilization of earth resources.

Leo J. Hickey, Professor of Geology and Geophysics and Biology. Paleobotany.

Antonio C. Lasaga, Professor of Geology and Geophysics and Director of Graduate Studies, Department of Geology and Geophysics. Theoretical geochemistry and petrology.

Jonathan M. Lees, Assistant Professor of Geology and Geophysics. Crustal seismology.

John H. Ostrom, Professor of Geology. Functional morphology; evolution and paleoecology of ancient reptiles, particularly archosaurs.

Jeffrey J. Park, Associate Professor of Geology and Geophysics. Seismology, theoretical geophysics and geologic time-series analysis.

Neil Ribe, Associate Professor of Geology and Geophysics. Geodesy and mantle processes.

Danny M. Rye, Professor of Geology and Geophysics. Isotope geochemistry related to ore deposits, metamorphic rocks, and paleoenvironments.

Barry Saltzman, Professor of Geophysics and Chairman, Department of Geology and Geophysics. Dynamical meteorology, theory of climate.

Adolf Seilacher, Adjunct Professor of Geology and Geophysics. Paleozoic trace fossils, taphonomy and morphology.

Brian J. Skinner, Eugene Higgins Professor of Geology and Geophysics. Geochemistry of mineral deposits, sulfide mineralogy.

Ronald B. Smith, Professor of Geology and Geophysics and Mechanical Engineering. Geophysical fluid dynamics, dynamical meteorology, theoretical structural geology.

Karl K. Turekian, Silliman Professor of Geology and Geophysics. Geochemistry of oceanic and earth-surface processes and planetary evolution.

J. Rimas Vaisnys, Professor of Electrical Engineering and Geophysics. Solid-state geophysics.

George Veronis, Henry Barnard Davis Professor of Geophysics and Applied Mathematics. Physical oceanography, geophysical fluid dynamics.

Elisabeth S. Vrba, Professor of Geology and Geophysics and Biology. Evolutionary paleobiology.

Kuo-Yen Wei, Assistant Professor of Geology and Geophysics. Micropaleontology, paleoceanography.

Professors Emeriti in Residence

Sydney P. Clark Jr., Sidney J. Weinberg Professor of Geophysics Emeritus. Constitution of the earth's interior, heat flow and the thermal state of the earth.

John Rodgers, Silliman Professor of Geology Emeritus. Structural and regional geology.

Karl M. Waage, Alan M. Bateman Professor of Geology Emeritus. Invertebrate paleontology and stratigraphy, Mesozoic paleoenvironments.

Horace Winchell, Professor of Mineralogy and Crystallography Emeritus. Physical properties of minerals, systematic mineralogy, computer applications, amphiboles and pyroxenes.

Associates

William C. Graustein, Research Scientist. Atmospheric and surficial geochemistry.

William J. Pegram, Associate Research Scientist. Osmium isotope geochemistry.

Mikhail Verbitsky, Associate Research Scientist. Glaciology.

Section 4
Marine Sciences/Oceanography

This section contains a directory of institutions offering graduate work in marine sciences/oceanography, followed by two-page entries submitted by institutions that chose to prepare detailed program descriptions. Additional information about programs listed in the directory but not augmented by a two-page entry may be obtained by writing directly to the dean of a graduate school or chair of a department at the address given in the directory.

For programs offering related work, see also in this book: Chemistry; Earth and Planetary Sciences; Meteorology and Atmospheric Sciences; and Physics; in Book 3: Biology and Biomedical Sciences; Ecology, Environmental Biology, and Evolutionary Biology; and Zoology (Marine Biology); and in Book 5: Civil and Environmental Engineering; Engineering and Applied Sciences; and Ocean Engineering.

CONTENTS

Program Directory 302

Announcements

Florida Institute of Technology	302
Nova University	303
Old Dominion University	303
State University of New York at Stony Brook	303

Cross-Discipline Announcements

Florida Institute of Technology	305
McGill University	305

Full Descriptions

College of William and Mary	307
Dalhousie University	309
Florida State University	311
Louisiana State University	313
Massachusetts Institute of Technology/Woods Hole Oceanographic Institute	315
North Carolina State University	317
State University of New York at Stony Brook	319
University of California, San Diego	321
University of Connecticut	323
University of Hawaii at Manoa	325
University of Puerto Rico, Mayagüez Campus	327
University of Rhode Island	329
University of South Florida	331
University of Texas at Austin	333

See also:

Yale University—Geology and Geophysics	299

FIELD DEFINITION

In an effort to broaden prospective students' understanding of the discipline of marine sciences/oceanography, an educator in the field has provided the following statement.

Marine Sciences/Oceanography

Oceanography, viewed broadly, is the study of the processes, both physical and biological, occurring in the sea. The field is traditionally divided into five disciplines: physical oceanography, marine chemistry, geochemistry, marine geology, and marine biology or biological oceanography. The boundary lines between the fields are overlapping, and much of oceanography is truly interdisciplinary in nature.

Physical oceanography deals with mechanisms of energy transfer through the sea and across its boundaries and with physical interactions within the seas. Included are air-sea interactions and climatology, the general circulation of the oceans, fluctuations of currents, properties of waves, the thermodynamic description of the sea as a nonequilibrium system, and the optical and acoustic properties of the sea.

Marine chemistry and geochemistry deal with chemical and isotopic processes operating within the marine environment and with the interactions of the components of seawater with the atmosphere, with sediments and the underlying oceanic crust, with ocean circulation, and with organisms. Research in marine chemistry relates strongly to all of the disciplines of oceanography and includes biochemistry as well as inorganic and organic chemistry.

Marine geology is concerned with the origin, nature, and history of ocean basins and their shores and sediments and with the geological processes that affect them. Studies include tectonics and volcanism, geomorphology of the ocean crust and margins, sedimentation, stratigraphy and paleontology, petrology, geochemistry, and geophysics.

Biological oceanography is concerned with the interactions of populations of marine organisms with one another and with the environment. Studies range from population dynamics of single species and investigations of the structure of communities and ecosystems to primary productivity and the physiological, behavioral, and biochemical adaptations of organisms.

The area of applied ocean sciences is concerned with the application of engineering techniques to the solution of basic and applied research problems in the sea. Its goal is to produce oceanographers who are knowledgeable of modern engineering and engineers who know about the sea.

Oceanography is a synthetic graduate subject. Students who wish to pursue graduate study in one of the fields of oceanography should acquire a strong background in one or more of the basic sciences, mathematics, or engineering.

Graduates with degrees in one of the disciplines of oceanography have found jobs in a variety of areas—universities, federal and state government laboratories and agencies, and private industry.

Richard Rosenblatt, Professor
Graduate Department
Scripps Institution of Oceanography
University of California, San Diego

Marine Sciences/Oceanography

Boston College, Graduate School of Arts and Sciences, Department of Geology and Geophysics, Chestnut Hill, MA 02167. Offerings include oceanography (MS, MST). *Degree requirements:* 1 foreign language, thesis. *Entrance requirements:* GRE General Test, GRE Subject Test. *Tuition:* $412 per credit hour.

California State University, Hayward, School of Science, Departments of Biological Sciences and Geological Sciences, Moss Landing Marine Laboratory, Hayward, CA 94542. Offers program in marine sciences (MS). Faculty: 9. Matriculated students: 1 full-time (0 women), 3 part-time (2 women). In 1990, 1 degree awarded. *Degree requirements:* Thesis required, foreign language not required. *Entrance requirements:* GRE Subject Test, minimum GPA of 3.0 in biology or 2.75 in geology. Application deadline: 4/19 (priority date, applications processed on a rolling basis). Application fee: $55. *Expenses:* Tuition of $0 for state residents; $137 per unit for nonresidents. Fees of $895 per year full-time, $188 per quarter part-time. *Financial aid:* Application deadline 3/1. • John H. Martin, Director, 408-633-3304. Application contact: Dr. Robert Trinchero, Acting Associate Vice President, Admissions and Enrollment, 415-881-3828.

California State University, Sacramento, School of Arts and Sciences, Department of Biological Sciences, Sacramento, CA 95819. Offerings include marine science (MS). *Application fee:* $45. *Expenses:* Tuition of $0 for state residents; $246 per unit for nonresidents. Fees of $530 per semester full-time, $332 per semester part-time.

College of William and Mary, School of Marine Science/Virginia Institute of Marine Science, Williamsburg, VA 23185. School awards MA, PhD. Faculty: 54 full-time (4 women), 7 part-time (1 woman). Matriculated students: 68 full-time (20 women), 42 part-time (15 women); includes 2 minority (1 Asian American, 1 black American), 23 foreign. Average age 29. 102 applicants, 59% accepted. In 1990, 15 master's, 8 doctorates awarded. Terminal master's awarded for partial completion of doctoral program. *Degree requirements:* For master's, 1 foreign language, thesis, comprehensive exam, defense; for doctorate, 1 foreign language, dissertation, comprehensive exam, qualifying exam, defense. *Entrance requirements:* For master's, GRE, TOEFL, appropriate bachelor's degree; for doctorate, GRE, TOEFL, appropriate bachelor's, master's degree. Application deadline: 2/15. Application fee: $20. *Expenses:* Tuition of $2240 per year full-time, $120 per credit hour part-time for state residents; $8960 per year full-time, $320 per credit hour part-time for nonresidents. Fees of $1490 per year full-time. *Financial aid:* In 1990–91, $800,000 in aid awarded. 2 fellowships (1 to a first-year student), 97 research assistantships (26 to first-year students), 2 teaching assistantships, 7 tuition fellowships (5 to first-year students) were awarded; federal work-study and career-related internships or fieldwork also available. Aid available to part-time students. Financial aid application deadline: 2/15; applicants required to submit FAF. *Faculty research:* Physical, biological, geological, and chemical oceanography; marine fisheries science; resource management. Total annual research budget: $13-million. • Dr. Frank O. Perkins, Dean, 804-642-7102. Application contact: Dean of Graduate Studies, 804-642-7105.

See full description on page 307.

Columbia University, Graduate School of Arts and Sciences, Division of Natural Sciences, Department of Geological Sciences, New York, NY 10027. Offerings include oceanography (MA, M Phil, PhD). Department faculty: 21 full-time, 19 part-time. *Degree requirements:* For master's, thesis or alternative, fieldwork, written exam required, foreign language not required; for doctorate, 1 foreign language, dissertation. *Entrance requirements:* GRE General Test, GRE Subject Test, TOEFL, major in natural or physical science. Application deadline: 1/5. Application fee: $50. *Expenses:* Tuition of $7836 per semester for state residents; $426 per credit for nonresidents. Fees of $148 per semester for state residents. • Dennis Hayes, Chair, 212-854-4525.

Dalhousie University, College of Arts and Science, Faculty of Science, Department of Oceanography, Halifax, NS B3H 4H6, Canada. Department awards M Sc, PhD. Faculty: 20 full-time (0 women), 0 part-time. Matriculated students: 58 full-time (15 women), 3 part-time (1 woman); includes 31 foreign. 113 applicants, 17% accepted. In 1990, 4 master's, 5 doctorates awarded. *Degree requirements:* Thesis/dissertation required, foreign language not required. *Application deadline:* 5/1 (applications processed on a rolling basis). *Application fee:* $20. *Tuition:* $2594 per year full-time for Canadian residents; $4294 per year full-time for nonresidents. *Financial aid:* In 1990–91, 29 students received a total of $450,667 in aid awarded. Fellowships available. Financial aid application deadline: 5/1. *Faculty research:* Biological and physical oceanography, chemical and geological oceanography, atmospheric sciences. • Dr. E. L. Mills, Chairman, 902-494-3557.

See full description on page 309.

Florida Institute of Technology, College of Engineering, Department of Oceanography and Ocean Engineering, Melbourne, FL 32901. Department offers programs in environmental science (MS, PhD); ocean engineering (MS, PhD); oceanography (MS, PhD), including biological oceanography (MS, PhD), chemical oceanography (MS, PhD), coastal zone management (MS), geoscience oceanography (MS), physical oceanography (MS, PhD). Part-time programs available. Faculty: 13 full-time (0 women), 3 part-time (1 woman). Matriculated students: 37 full-time (9 women), 23 part-time (7 women); includes 1 minority (black American). 101 applicants, 50% accepted. In 1990, 22 master's, 4 doctorates awarded. *Degree requirements:* For master's, foreign language not required; for doctorate, dissertation, comprehensive exam, departmental qualifying exam. *Entrance requirements:* For master's, minimum GPA of 3.0; for doctorate, GRE General Test, minimum GPA of 3.2. Application deadline: rolling. Application fee: $35. *Tuition:* $234 per credit hour. *Financial aid:* Research assistantships, teaching assistantships, federal work-study available. Financial aid application deadline: 3/1. *Faculty research:* Marine geophysics, taxonomy and ecology of marine phytoplankton (particularly dinoflagellates), coastal processes. • Dr. N. T. Stephens, Head, 407-768-8000 Ext. 8096. Application contact: Carolyn P. Farrior, Director of Graduate Admissions, 407-768-8000 Ext. 8027.

Announcement: Oceanography graduate students emphasize biological, chemical, geological, or physical oceanography or coastal-zone management. Coastal-zone management students complete internship rather than thesis. Research opportunities: benthic community ecology, marine fisheries, benthic/pelagic relations, phytoplankton ecology, toxic dinoflagellates; trace-metal geochemistry and pollution, trace organic analysis, toxic substances/marine pollution, artificial reefs; sea-level change, coastal/shelf processes, sediment transport/morphodynamics, numerical modeling of inlet hydrodynamics, surf-zone processes, wave mechanics, wave generation, wave modeling, air-sea interaction. Research facilities: well-equipped laboratories on campus, oceanfront Indian River Marine Science Research Center, estuarine and oceangoing research vessels. Unique location on Florida's east coast, with direct access to Indian River Lagoon and Atlantic Ocean from University facilities.

Florida State University, College of Arts and Sciences, Department of Oceanography, Tallahassee, FL 32306. Department offers programs in geophysical fluid dynamics (PhD), oceanography (MS, PhD). Faculty: 17 full-time (2 women), 0 part-time. Matriculated students: 39 full-time (15 women), 3 part-time (1 woman); includes 0 minority, 22 foreign. Average age 28. 50 applicants, 16% accepted. In 1990, 6 master's awarded (67% found work related to degree, 33% continued full-time study); 5 doctorates awarded (40% entered university research/teaching, 60% found other work related to degree). Terminal master's awarded for partial completion of doctoral program. *Degree requirements:* Thesis/dissertation required, foreign language not required. *Entrance requirements:* GRE General Test (minimum combined score of 1100 required). Application deadline: 1/1 (priority date, applications processed on a rolling basis). Application fee: $15. *Tuition:* $76.29 per credit hour for state residents; $238 per credit hour for nonresidents. *Financial aid:* In 1990–91, 0 fellowships, 28 research assistantships (5 to first-year students), 3 teaching assistantships (1 to a first-year student) awarded. *Faculty research:* Trace metals in seawater, currents and waves, modeling, benthic ecology. Total annual research budget: $1.54-million. • Dr. Ya Hsueh III, Chairman, 904-644-6700. Application contact: Academic Coordinator, 904-644-6700.

See full description on page 311.

Johns Hopkins University, School of Arts and Sciences, Department of Earth and Planetary Sciences, Program in Oceanography, Baltimore, MD 21218. Program awards MA, PhD. Faculty: 5 full-time (0 women). Matriculated students: 15. Terminal master's awarded for partial completion of doctoral program. *Degree requirements:* For doctorate, 1 foreign language, dissertation. *Entrance requirements:* GRE General Test. Application deadline: 2/1. Application fee: $40. *Expenses:* Tuition of $15,500 per year full-time, $1550 per course part-time. Fees of $400 per year full-time. *Financial aid:* Federal work-study, institutionally sponsored loans available. Financial aid application deadline: 3/14; applicants required to submit FAF. • Bruce Marsh, Chairman, Department of Earth and Planetary Sciences, 301-338-7135.

Louisiana State University and Agricultural and Mechanical College, College of Basic Sciences, Department of Oceanography and Coastal Sciences, Baton Rouge, LA 70803. Department awards MS, PhD. Faculty: 14 full-time (1 woman), 3 part-time (1 woman). Matriculated students: 30 full-time (7 women), 13 part-time (3 women); includes 0 minority, 9 foreign. Average age 31. 27 applicants, 41% accepted. In 1990, 1 master's awarded (67% found work related to degree, 33% continued full-time study); 5 doctorates awarded (12% entered university research/teaching, 88% found other work related to degree). *Degree requirements:* For master's, thesis required (for some programs), foreign language not required; for doctorate, 1 foreign language, dissertation. *Entrance requirements:* For master's, GRE General Test, TOEFL, minimum GPA of 3.0; for doctorate, GRE General Test, TOEFL, MA or MS, minimum GPA of 3.0. Application deadline: 7/1 (priority date, applications processed on a rolling basis). Application fee: $25. *Tuition:* $1020 per semester full-time for state residents; $2620 per semester full-time for nonresidents. *Financial aid:* In 1990–91, 32 students received a total of $357,290 in aid awarded. 4 fellowships (2 to first-year students), 20 research assistantships (5 to first-year students) were awarded; federal work-study, institutionally sponsored loans also available. Average monthly stipend for a graduate assistantship: $942. *Faculty research:* Management and development of estuarine and coastal areas and resources; physical, chemical, geological, and biological research. • Dr. R. Eugene Turner, Chair, 504-388-6308.

See full description on page 313.

Massachusetts Institute of Technology, School of Science, Department of Biology, Program in Biological Oceanography, Cambridge, MA 02139. Program awards PhD, Sc D. *Entrance requirements:* GRE General Test, GRE Subject Test (biology). *Tuition:* $16,900 per year. • Dr. Richard O. Hynes, Chairman, Department of Biology, 617-253-4701. Application contact: Hannah D. Roberts, Graduate Administrator, 617-253-4738.

Massachusetts Institute of Technology, School of Science, Department of Earth, Atmospheric, and Planetary Sciences, Center for Meteorology and Physical Oceanography, Cambridge, MA 02139. Center awards SM, PhD, Sc D. Faculty: 12 full-time (1 woman), 0 part-time. Matriculated students: 62 full-time (16 women), 0 part-time; includes 1 minority (black American), 28 foreign. Average age 24. 61 applicants, 39% accepted. In 1990, 6 master's awarded (67% found work related to degree, 33% continued full-time study); 7 doctorates awarded. Terminal master's awarded for partial completion of doctoral program. *Degree requirements:* Thesis/dissertation required, foreign language not required. *Entrance requirements:* GRE General Test, GRE Subject Test. Application deadline: 1/15. Application fee: $45. *Tuition:* $16,900 per year. *Financial aid:* In 1990–91, 10 fellowships (2 to first-year students), 32 research assistantships (6 to first-year students), 2 teaching assistantships (0 to first-year students) awarded; institutionally sponsored loans also available. Financial aid application deadline: 1/15. *Faculty research:* Origin, composition, structure, and state of atmospheres and oceans. • Dr. Kerry Emanuel, Head, 617-253-2281.

Massachusetts Institute of Technology/Woods Hole Oceanographic Institution, Program in Oceanography/Oceanographic Engineering, Cambridge, MA 02139. Program awards MS, PhD, Sc D, Eng. Offerings include biological oceanography (PhD, Sc D), chemical oceanography (PhD, Sc D), marine geochemistry (PhD, Sc D), marine geology (PhD, Sc D), oceanographic engineering (MS, PhD, Sc D), physical oceanography (PhD, Sc D). Faculty: 152 full-time, 0 part-time. Matriculated students: 121 full-time (40 women), 0 part-time; includes 40 foreign. 133 applicants, 29% accepted. *Degree requirements:* For master's, thesis (for some programs); for doctorate, dissertation. *Entrance requirements:* For master's, GRE General Test; for doctorate, GRE General Test, GRE Subject Test. Application deadline: 1/15 (priority date). Application fee: $45. *Financial aid:* Research assistantships, teaching assistantships available. • Dr. Sallie W. Chisholm, Director, 617-253-1771. Application contact: Ronni Schwatz, Administrator, 617-253-7544.

See full description on page 315.

McGill University, Faculty of Graduate Studies and Research, Department of Meteorology, Montreal, PQ H3A 2T5, Canada. Department offers programs in meteorology (M Sc, PhD), physical oceanography (M Sc, PhD). Faculty: 12 full-time (0 women), 0 part-time. Matriculated students: 43 full-time (8 women), 0 part-time; includes 15 foreign. Average age 24. 454 applicants, 24% accepted. In 1990, 6 master's awarded (50% found work related to degree, 50% continued full-time study); 0 doctorates awarded. Terminal master's awarded for partial completion of doctoral program. *Degree requirements:* Thesis/dissertation required, foreign language not required. *Entrance requirements:* For master's, GRE General Test, minimum GPA of 3.0; for doctorate, GRE, master's degree in meteorology or related field. Application deadline: 7/1 (priority date, applications processed on a rolling basis). Application fee: $25. *Tuition:* $698.50 per semester full-time, $46.57 per credit part-time for Canadian residents; $3480 per semester full-time, $234 per semester part-time for nonresidents. *Financial aid:* In 1990–91, 40 students received a total of $365,000 in aid awarded. 1 fellowship (0 to first-year students), 30 research assistantships (9 to first-year students), 10 teaching assistantships (1 to a first-year student) were awarded. Average monthly stipend for a graduate assistantship: $1000. Financial aid application deadline: 7/1. *Faculty research:* Atmospheric radiation, cloud physics, climate research, dynamics, radar. Total annual research budget: $1.6-million. • H. G. Leighton, Chairman, 514-398-3764.

Memorial University of Newfoundland, School of Graduate Studies, Department of Physics, St. John's, NF A1C 5S7, Canada. Offerings include physical oceanography (M Sc); physics (M Sc, PhD), with options in condensed matter physics (PhD), molecular physics (PhD), physical oceanography (PhD). *Degree requirements:* Thesis/dissertation.

Naval Postgraduate School, Department of Oceanography, Monterey, CA 93943. Department awards MS, PhD. Program only open to commissioned officers of the United States and friendly nations and selected United States federal civilian employees. *Degree requirements:* Thesis/dissertation. *Tuition:* $0.

North Carolina State University, College of Physical and Mathematical Sciences, Department of Marine, Earth, and Atmospheric Sciences, Raleigh, NC 27695. Department offers programs in geology (MS, PhD), geophysics (MS, PhD), meteorology (MS, PhD), oceanography (MS, PhD). Faculty: 35 full-time (2 women), 3 part-time (1 woman). Matriculated students: 74 full-time (15 women), 3 part-time (0 women); includes 2 minority (both Asian American), 23 foreign. Average age 24. 75 applicants, 49% accepted. In 1990, 18 master's, 6 doctorates awarded. Terminal master's awarded for partial completion of doctoral program. *Degree requirements:* For master's, thesis, final oral exam required, foreign language not required; for doctorate, dissertation, preliminary written and oral exams, final oral exams required, foreign language not required. *Entrance requirements:* GRE General Test, minimum GPA of 3.0. Application deadline: 5/1 (priority date, applications processed on a rolling basis). Application fee: $35. *Tuition:* $1138 per year for state residents; $5805 per year for nonresidents. *Financial aid:* In 1990–91, $383,267 in aid awarded. 0 fellowships, 33 research assistantships (10 to first-year students), 18 teaching assistantships (6 to first-year students) were awarded; institutionally sponsored loans also available. Average monthly stipend for a graduate assistantship: $750. Financial aid application deadline: 3/1. *Faculty research:* Boundary-layer and synoptic meteorology; physical, chemical, geological, and biological oceanography. Total annual research budget: $6.41-million. • Dr. Leonard J. Pietrafesa, Head, 919-737-3717. Application contact: G. S. Janowitz, Graduate Administrator, 919-737-7837.

See full description on page 317.

Nova University, Oceanographic Center and Institute of Marine and Coastal Studies, Dania, FL 33004. Center offers programs in coastal-zone management (MS), marine biology (MS), oceanography (PhD). Part-time and evening/weekend programs available. Faculty: 7 full-time, 5 part-time. Matriculated students: 22 full-time (8 women), 25 part-time (10 women); includes 5 minority (3 Asian American, 2 black American), 5 foreign. Average age 32. In 1990, 5 master's, 2 doctorates awarded. *Degree requirements:* For master's, thesis optional, foreign language not required; for doctorate, dissertation, departmental qualifying exam. *Entrance requirements:* For master's, GRE General Test (minimum combined score of 1000 required); for doctorate, GRE General Test (minimum combined score of 1100 required), master's degree. Application deadline: rolling. Application fee: $30. *Tuition:* $250 per credit. *Financial aid:* Partial tuition waivers, federal work-study, and career-related internships or fieldwork available. Aid available to part-time students. Financial aid applicants required to submit FAF. *Faculty research:* Physical, geological, and chemical biological oceanography. Total annual research budget: $950,000. • Dr. Julian McCreary, Director, 305-920-1909. Application contact: Dr. Richard Dodge, Associate Director, 305-920-1909.

Announcement: MS degree program in marine biology and coastal-zone management. Requirements for thesis option are 30 course credits and 6 thesis credits; for nonthesis option, 36 course credits and 3 term-paper credits. Evening classes; 4 terms per year; 2 courses in each specialty per term. Joint MS program (marine biology and coastal-zone management) as well as PhD program in oceanography for advanced students (previous master's degree).

Old Dominion University, College of Sciences, Department of Oceanography, Norfolk, VA 23529. Department awards MS, PhD. Part-time and evening/weekend programs available. Faculty: 18 full-time (2 women), 0 part-time. Matriculated students: 45 full-time (16 women), 17 part-time (7 women); includes 4 minority (1 Asian American, 1 black American, 2 Hispanic American), 23 foreign. Average age 30. 32 applicants, 78% accepted. In 1990, 5 master's awarded (100% found work related to degree); 2 doctorates awarded (100% found work related to degree). Terminal master's awarded for partial completion of doctoral program. *Degree requirements:* For master's, thesis or alternative, 10 days of ship time required, foreign language not required; for doctorate, 2 foreign languages (computer language can substitute for one), dissertation, 10 days of ship time. *Entrance requirements:* For master's, GRE General Test, GRE Subject Test, TOEFL, minimum QPA of 3.0 in major, minimum QPA of 2.5 overall; for doctorate, GRE General Test (minimum combined score of 1000 required), GRE Subject Test (minimum score of 600 required), TOEFL. Application deadline: 7/1 (applications processed on a rolling basis). Application fee: $20. *Expenses:* Tuition of $148 per credit hour for state residents; $375 per credit hour for nonresidents. Fees of $64 per year full-time. *Financial aid:* In 1990–91, $36,800 in aid awarded. 5 research assistantships, 1 tuition grant were awarded; fellowships, teaching assistantships, federal work-study, institutionally sponsored loans, and career-related internships or fieldwork also available. Financial aid applicants required to submit FAF. *Faculty research:* Fisheries management, inorganic/organic marine chemistry, sediment transport, shelf dynamics, ocean circulation modeling. Total annual research budget: $1.2-million. • Dr. William M. Dunstan, Chairman, 804-683-4285. Application contact: Graduate Program Directors, 804-683-4285.

Announcement: Offers MS and PhD in biological, chemical, geological, physical oceanography. Department interested in students with strong mathematics and science backgrounds to apply their talents to global and regional research programs, such as ocean's role in climate or marine biogeochemical cycles. Interdisciplinary research opportunities exist in coastal and open-ocean areas. Contact Graduate Program Director, 804-683-4285.

Oregon State University, Graduate School, College of Oceanography, Corvallis, OR 97331. College offers programs in geophysics (MA, MS, PhD), marine resource management (MA, MS), oceanography (MA, MS, PhD). Part-time programs available. Faculty: 67 full-time (3 women), 7 part-time (0 women). Matriculated students: 70 full-time (22 women), 4 part-time (1 woman); includes 2 minority (1 Asian American, 1 Hispanic American), 35 foreign. 131 applicants, 24% accepted. In 1990, 23 master's, 0 doctorates awarded. Terminal master's awarded for partial completion of doctoral program. *Degree requirements:* For master's, thesis optional, foreign language not required; for doctorate, dissertation required, foreign language not required. *Entrance requirements:* GRE General Test, TOEFL (minimum score of 550 required), minimum GPA of 3.0 in last 90 hours. Application deadline: 3/1. Application fee: $40. *Tuition:* $1140 per trimester full-time, $449 per year part-time for state residents; $1816 per trimester full-time, $674 per year part-time for nonresidents. *Financial aid:* In 1990–91, 1 fellowship, 58 research assistantships (13 to first-year students), 8 teaching assistantships awarded; federal work-study, institutionally sponsored loans, and career-related internships or fieldwork also available. Aid available to part-time students. Financial aid application deadline: 2/1; applicants required to submit FAF. *Faculty research:* Air-sea interactions; biological, chemical, geological, and physical oceanography. Total annual research budget: $13.4-million. • Dr. Douglas R. Caldwell, Dean, 503-737-3504. Application contact: Dr. Jefferson J. Gonor, Head Adviser, 503-737-3504.

Princeton University, Department of Geological and Geophysical Sciences, Program in Atmospheric and Oceanic Sciences, Princeton, NJ 08544. Program awards PhD. Faculty: 14 full-time (0 women), 0 part-time. Matriculated students: 16 full-time (5 women), 0 part-time; includes 0 minority, 11 foreign. 24 applicants, 38% accepted. In 1990, 3 doctorates awarded (33% entered university research/teaching, 0% found other work related to degree, 0% continued full-time study). Terminal master's awarded for partial completion of doctoral program. *Degree requirements:* For doctorate, 1 foreign language, dissertation. *Entrance requirements:* For doctorate, GRE General Test, GRE Subject Test. Application deadline: 1/8. Application fee: $45 ($50 for foreign students). *Tuition:* $16,670 per year. *Financial aid:* Fellowships, research assistantships, federal work-study, institutionally sponsored loans available. Financial aid application deadline: 1/8. *Faculty research:* Climate dynamics, middle atmosphere dynamics and chemistry, oceanic circulation, marine geochemistry, numerical modeling. • S. George H. Philander, Director of Graduate Studies, 609-258-6571. Application contact: Michele Spreen, Director of Graduate Admissions, 609-258-3034.

Royal Roads Military College, Graduate Program in Oceanography and Acoustics, FMO, Victoria, BC V0S 1B0, Canada. College awards M Sc. Part-time programs available. Faculty: 9 full-time (0 women), 1 part-time (0 women). Matriculated students: 7 full-time (0 women), 1 part-time (0 women); includes 0 minority, 0 foreign. Average age 28. 12 applicants, 25% accepted. In 1990, 5 degrees awarded (100% found work related to degree). *Degree requirements:* Thesis required, foreign language not required. *Application deadline:* 6/15. *Tuition:* $0. *Financial aid:* Federal work-study available. *Faculty research:* Mesoscale processes, coastal dynamics, acoustic oceanography, internal waves, surface gravity waves. Total annual research budget: $300,000. • Dr. David P. Krauel, Dean of Graduate Studies, 604-363-4580.

San Jose State University, School of Science, Program in Marine Science, San Jose, CA 95192. Program awards MS. Faculty: 8 full-time (0 women), 14 part-time (0 women). Matriculated students: 2 full-time (1 woman), 35 part-time (16 women); includes 3 minority (2 Asian American, 1 Hispanic American), 1 foreign. Average age 27. 13 applicants, 77% accepted. In 1990, 6 degrees awarded. *Degree requirements:* Thesis, qualifying exam required, foreign language not required. *Entrance requirements:* GRE. Application deadline: 6/1 (applications processed on a rolling basis). Application fee: $55. *Expenses:* Tuition of $0 for state residents; $246 per unit for nonresidents. Fees of $592 per year. *Financial aid:* In 1990–91, $101,199 in aid awarded. Teaching assistantships and career-related internships or fieldwork available. Aid available to part-time students. *Faculty research:* Physical oceanography, marine geology, ecology, ichthyology, invertebrate zoology. • Dr. John Martin, Director, 408-633-3304. Application contact: Dr. Howard Shellhammer, Graduate Adviser, 408-924-4897.

State University of New York at Stony Brook, Marine Sciences Research Center, Stony Brook, NY 11794. Center offers programs in coastal oceanography (PhD), marine environmental sciences (MS). Evening/weekend programs available. Faculty: 31 full-time, 23 part-time. Matriculated students: 99 full-time (36 women), 19 part-time (8 women); includes 8 minority (5 Asian American, 3 Hispanic American), 43 foreign. 97 applicants, 58% accepted. In 1990, 12 master's awarded; 10 doctorates awarded (100% entered university research/teaching). *Degree requirements:* For master's, thesis, written comprehensive exam required, foreign language not required; for doctorate, 1 foreign language, dissertation, written comprehensive exam. *Entrance requirements:* For master's, GRE General Test, TOEFL (minimum score of 550 required), minimum GPA of 3.0; for doctorate, GRE General Test, TOEFL (minimum score of 550 required), minimum graduate GPA of 3.0. Application deadline: 2/1. Application fee: $35. *Expenses:* Tuition of $2450 per year full-time, $103 per credit part-time for state residents; $5766 per year full-time, $243 per credit part-time for nonresidents. Fees of $151 per year full-time, $10.45 per year (minimum) part-time. *Financial aid:* In 1990–91, 10 fellowships, 51 research assistantships, 37 teaching assistantships awarded; career-related internships or fieldwork also available. *Faculty research:* Coastal studies, coastal zone management, fisheries and aquaculture, marine environment. Total annual research budget: $3.74-million. • Dr. Jerry Schubel, Director, 516-632-8700.

Announcement: The Marine Sciences Research Center is the oceanographic center for the SUNY system. Emphasis is placed on coastal oceanography, with comprehensive programs in biological, chemical, geological, and physical oceanography; in fishery biology; in mariculture; in coastal-zone management; and in fishery management. MSRC has extensive facilities, and support is available for study at other SUNY campuses.

See full description on page 319.

Texas A&M University, College of Geosciences, Department of Oceanography, College Station, TX 77843. Department awards MS, PhD. Faculty: 46. Matriculated students: 87 full-time (23 women), 0 part-time; includes 3 minority (1 Asian American, 2 Hispanic American), 33 foreign. 41 applicants, 80% accepted. In 1990, 7 master's, 9 doctorates awarded. *Entrance requirements:* GRE General Test, TOEFL. Application deadline: 7/15. Application fee: $25 ($50 for foreign students). *Expenses:* Tuition of $100 per semester (minimum) for state residents; $128 per credit hour for nonresidents. Fees of $459 per year full-time, $252 per semester part-time. *Financial aid:* Fellowships, research assistantships, teaching assistantships available. • Gilbert Rowe, Head, 409-845-7211.

Université du Québec à Rimouski, Program in Maritime Studies, Rimouski, PQ G5L 3A1, Canada. Program awards Diploma. Part-time programs available. Matriculated students: 8 full-time (1 woman), 3 part-time (0 women); includes 10 foreign. 36 applicants, 78% accepted. In 1990, 7 degrees awarded. *Degree requirements:* Thesis not required. *Entrance requirements:* Appropriate bachelor's degree, proficiency in French. Application deadline: 5/1 (priority date). Application fee: $15. *Expenses:* Tuition of $555 per trimester full-time, $37 per credit part-time for Canadian residents; $3480 per trimester full-time, $234 per credit part-time for nonresidents. Fees of $57.50 per trimester full-time, $19.50 per trimester part-time. *Financial aid:* Fellowships, research assistantships, teaching assistantships available. • Michel LaChance, Director, 418-724-1544. Application contact: Conrad Lavoie, Office of Admissions, 418-724-1433.

Directory: Marine Sciences/Oceanography

Université du Québec à Rimouski, Program in Oceanography, Rimouski, PQ G5L 3A1, Canada. Program awards M Sc, PhD. Part-time programs available. Matriculated students: 30 full-time (11 women), 31 part-time (11 women); includes 15 foreign. 35 applicants, 74% accepted. In 1990, 11 master's, 0 doctorates awarded. *Degree requirements:* Thesis/dissertation. *Entrance requirements:* For master's, appropriate bachelor's degree, proficiency in French; for doctorate, appropriate master's degree, proficiency in French. Application deadline: 5/1 (priority date). Application fee: $15. *Expenses:* Tuition of $555 per trimester full-time, $37 per credit part-time for Canadian residents; $3480 per trimester full-time, $234 per credit part-time for nonresidents. Fees of $57.50 per trimester full-time, $19.50 per trimester part-time. *Financial aid:* Fellowships, research assistantships, teaching assistantships available. • Gaston Desrosiers, Director, 418-724-1765. Application contact: Conrad Lavoie, Office of Admissions, 418-724-1433.

University of Alaska Fairbanks, School of Fisheries and Ocean Sciences, Graduate Program in Marine Sciences and Limnology, Fairbanks, AK 99775. Offerings include oceanography (MS, PhD). Program faculty: 9 full-time (1 woman), 2 part-time (0 women). *Entrance requirements:* For master's, GRE. Application fee: $30. *Expenses:* Tuition of $1620 per year full-time, $90 per credit part-time for state residents; $3240 per year full-time, $180 per credit part-time for nonresidents. Fees of $464 per year full-time. • William S. Reeburgh, Head, 907-474-7289.

University of British Columbia, Faculty of Science, Department of Oceanography, Vancouver, BC V6T 1Z1, Canada. Department awards M Sc, PhD. Part-time programs available. *Degree requirements:* For master's, thesis; for doctorate, variable foreign language requirement, dissertation, oral comprehensive exam. *Entrance requirements:* TOEFL, minimum GPA of 3.0. Application deadline: 6/1. Application fee: $0.

University of California, San Diego, Scripps Institution of Oceanography, 9500 Gilman Drive, La Jolla, CA 92093. Institution offers programs in applied ocean science (MS, PhD), biological oceanography (MS, PhD), geochemistry and marine chemistry (MS, PhD), marine biology (MS, PhD), physical oceanography and geological sciences (MS, PhD). Faculty: 73. Matriculated students: 183 (57 women); includes 59 foreign. 273 applicants, 26% accepted. In 1990, 14 master's, 29 doctorates awarded. Terminal master's awarded for partial completion of doctoral program. *Entrance requirements:* GRE General Test, GRE Subject Test (marine biology). Application fee: $40. *Expenses:* Tuition of $0 for state residents; $7699 per year full-time, $1283 per quarter part-time for nonresidents. Fees of $2798 per year full-time, $669 per quarter part-time. *Financial aid:* Fellowships, research assistantships available. • J. Freeman Gilbert, Chair, 619-534-3208. Application contact: Betty Stover, Graduate Coordinator, 619-534-3206.

See full description on page 321.

University of California, San Diego, Scripps Institution of Oceanography and Departments of Applied Mechanics and Engineering Sciences and Electrical and Computer Engineering, Program in Applied Ocean Science, 9500 Gilman Drive, La Jolla, CA 92093. Program awards MS, PhD. *Degree requirements:* For master's, foreign language not required; for doctorate, variable foreign language requirement, dissertation. *Entrance requirements:* GRE General Test. Application fee: $40. *Expenses:* Tuition of $0 for state residents; $7699 per year full-time, $1283 per quarter part-time for nonresidents. Fees of $2798 per year full-time, $669 per quarter part-time. • J. Freeman Gilbert, Chair, Scripps Institution of Oceanography, 619-534-3208. Application contact: Betty Stover, Graduate Coordinator, 619-534-3206.

University of California, Santa Cruz, Division of Natural Sciences, Program in Marine Sciences, Santa Cruz, CA 95064. Program awards MS. Faculty: 12 full-time (3 women), 0 part-time. Matriculated students: 23 full-time (15 women), 0 part-time; includes 3 minority (1 Asian American, 1 black American, 1 Hispanic American), 2 foreign. 47 applicants, 32% accepted. In 1990, 4 degrees awarded. *Degree requirements:* Thesis. *Entrance requirements:* GRE General Test, GRE Subject Test. Application deadline: 1/1. Application fee: $40. *Expenses:* Tuition of $0 for state residents; $7699 per year for nonresidents. Fees of $3021 per year. *Financial aid:* In 1990–91, 0 fellowships, 5 research assistantships, 16 teaching assistantships awarded; federal work-study, institutionally sponsored loans, and career-related internships or fieldwork also available. Average monthly stipend for a graduate assistantship: $1335. Financial aid applicants required to submit GAPSFAS. *Faculty research:* Oceanography, biology of higher marine vertebrates, ecology of coastal zone, marine geology. • Dr. Kenneth Bruland, Chairperson, 408-459-4730.

University of Connecticut, College of Liberal Arts and Sciences, Field of Marine Sciences, Storrs, CT 06269. Field offers program in oceanography (MS, PhD). Faculty: 15. Matriculated students: 18 full-time (5 women), 3 part-time (all women); includes 0 minority, 5 foreign. Average age 30. 12 applicants, 10% accepted. In 1990, 4 master's, 0 doctorates awarded. Terminal master's awarded for partial completion of doctoral program. *Degree requirements:* For doctorate, dissertation. *Entrance requirements:* GRE General Test, GRE Subject Test, TOEFL. Application deadline: 6/1 (priority date, applications processed on a rolling basis). Application fee: $30. *Expenses:* Tuition of $3428 per year full-time, $571 per course part-time for state residents; $8914 per year full-time, $1486 per course part-time for nonresidents. Fees of $636 per year full-time, $87 per course part-time. *Financial aid:* In 1990–91, $170,175 in aid awarded. 1 fellowship (to a first-year student), 16 research assistantships (3 to first-year students), 1 teaching assistantship (0 to first-year students) were awarded. Financial aid application deadline: 2/15; applicants required to submit GAPSFAS or FAF. • Robert B. Whitlatch, Head, 203-445-3467.

See full description on page 323.

University of Delaware, College of Marine Studies, Newark, DE 19716. College awards MA, MMP, MS, PhD. Part-time programs available. Faculty: 44 full-time, 0 part-time. Matriculated students: 66 full-time (21 women), 39 part-time (13 women); includes 1 minority (black American), 31 foreign. Average age 25. In 1990, 8 master's, 9 doctorates awarded. *Degree requirements:* Thesis/dissertation required, foreign language not required. *Entrance requirements:* GRE General Test, GRE Subject Test. Application deadline: 7/1. Application fee: $40. *Tuition:* $179 per credit hour for state residents; $467 per credit hour for nonresidents. *Financial aid:* In 1990–91, 7 fellowships (3 to first-year students), 65 research assistantships (30 to first-year students), 2 teaching assistantships (0 to first-year students) awarded; full and partial tuition waivers, federal work-study, and career-related internships or fieldwork also available. Financial aid application deadline: 3/1. *Faculty research:* Marine biology and biochemistry, oceanography, applied ocean science, marine policy. • Dr. Carolyn A. Thoroughgood, Dean, 302-451-2841. Application contact: Dixie Goldner, Academic Affairs Coordinator, 302-645-4213.

University of Hawaii at Manoa, School of Ocean and Earth Science and Technology, Department of Oceanography, Honolulu, HI 96822. Department awards MS, PhD. *Entrance requirements:* GRE. *Tuition:* $800 per semester full-time for state residents; $2405 per semester full-time for nonresidents.

See full description on page 325.

University of Maine, College of Sciences, Department of Oceanography, Orono, ME 04469. Department awards MS, PhD. Part-time programs available. Faculty: 12 full-time (0 women), 0 part-time. Matriculated students: 9 full-time (3 women), 2 part-time (1 woman); includes 0 minority, 1 foreign. In 1990, 1 master's awarded. *Degree requirements:* Thesis/dissertation required, foreign language not required. *Entrance requirements:* GRE General Test, GRE Subject Test, TOEFL (minimum score of 550 required). Application deadline: 2/1. Application fee: $25. *Tuition:* $100 per credit hour for state residents; $275 per credit hour for nonresidents. *Financial aid:* In 1990–91, 5 research assistantships (2 to first-year students) awarded; fellowships, full and partial tuition waivers, federal work-study, and career-related internships or fieldwork also available. Aid available to part-time students. Financial aid application deadline: 2/1. *Faculty research:* Benthic ecology, paleoceanography, coastal processes, microbial ecology, crustacean systematics. • Dr. Lawrence Mayer, Head, 207-581-1437. Application contact: Dr. Detmar Schnitker, Graduate Coordinator, 207-563-3146.

University of Maryland Eastern Shore, Department of Natural Sciences, Program in Marine, Estuarine, and Environmental Sciences, Princess Anne, MD 21853. Program awards MS, PhD. Part-time programs available. Faculty: 6 full-time, 1 part-time. Matriculated students: 17 full-time (7 women), 17 part-time (8 women); includes 4 minority (1 Asian American, 2 black American, 1 Hispanic American), 10 foreign. 54 applicants, 98% accepted. In 1990, 5 master's, 2 doctorates awarded. *Degree requirements:* For master's, thesis required, foreign language not required; for doctorate, dissertation. *Entrance requirements:* For master's, GRE or MAT, TOEFL (minimum score of 550 required), minimum GPA of 3.0; for doctorate, TOEFL (minimum score of 550 required). Application deadline: 3/15 (priority date, applications processed on a rolling basis). Application fee: $20. *Tuition:* $114 per credit for state residents; $203 per credit for nonresidents. *Financial aid:* 7 students received aid. Fellowships, research assistantships, teaching assistantships, grants, and career-related internships or fieldwork available. Average monthly stipend for a graduate assistantship: $950. Financial aid application deadline: 3/15. • Dr. Steve Rebach, Coordinator, 301-651-2200 Ext. 325.

University of Maryland Graduate School, Baltimore, Graduate School, Program in Marine-Estuarine-Environmental Sciences, Baltimore, MD 21228. Program awards MS, PhD, JD/MS. Faculty: 16 full-time, 0 part-time. Matriculated students: 0 full-time, 8 part-time (5 women); includes 1 minority (black American), 0 foreign. 10 applicants, 10% accepted. In 1990, 1 master's awarded. *Degree requirements:* For master's, foreign language and thesis not required; for doctorate, dissertation required, foreign language not required. *Entrance requirements:* GRE General Test, GRE Subject Test, TOEFL, minimum GPA of 3.0. Application deadline: 7/1. Application fee: $25. *Tuition:* $134 per credit for state residents; $245 per credit for nonresidents. *Financial aid:* Fellowships, research assistantships, teaching assistantships, and career-related internships or fieldwork available. *Faculty research:* Research on Chesapeake Bay, marine population research. • Dr. Thomas Cronin, Coordinator, 301-455-3449. Application contact: Dr. Robert Nauman, Coordinator, 301-328-7538.

University of Miami, Rosenstiel School of Marine and Atmospheric Science, Coral Gables, FL 33124. School awards MA, MS, MSOE, PhD. Faculty: 72 (9 women). Matriculated students: 126 full-time (37 women), 7 part-time (2 women); includes 27 minority (7 Asian American, 2 black American, 18 Hispanic American), 44 foreign. Average age 30. 252 applicants, 27% accepted. In 1990, 16 master's, 15 doctorates awarded. Terminal master's awarded for partial completion of doctoral program. *Degree requirements:* For doctorate, variable foreign language requirement, dissertation. *Entrance requirements:* GRE General Test, TOEFL (minimum score of 550 required). Application fee: $35. *Expenses:* Tuition of $567 per credit hour. Fees of $87 per semester full-time. *Financial aid:* Fellowships, research assistantships, teaching assistantships available. • Dr. Bruce Rosendahl, Dean, 305-361-4630. Application contact: Dr. Frank Millero, Associate Dean, 305-361-4155.

University of Michigan, College of Engineering, Department of Atmospheric, Oceanic, and Space Sciences, Program in Oceanography: Physical, Ann Arbor, MI 48109. Program awards MS, PhD. *Degree requirements:* For doctorate, dissertation. *Entrance requirements:* GRE. Application deadline: 2/1 (applications processed on a rolling basis). Application fee: $30. *Tuition:* $3506 per semester full-time, $352 per credit (minimum) part-time for state residents; $7140 per semester full-time, $757 per credit (minimum) part-time for nonresidents. *Financial aid:* Application deadline 3/15. • Stanley Jacobs, Chair, 313-764-3335.

University of New Hampshire, College of Engineering and Physical Sciences, Department of Earth Sciences, Durham, NH 03824. Offerings include earth sciences oceanography (MS, PhD). Department faculty: 26 full-time. *Degree requirements:* For master's, thesis required, foreign language not required; for doctorate, 1 foreign language, dissertation. *Entrance requirements:* GRE General Test. Application deadline: 7/1 (priority date, applications processed on a rolling basis). Application fee: $25. *Tuition:* $1645 per semester full-time, $183 per credit hour part-time for state residents; $4920 per semester full-time, $547 per credit hour part-time for nonresidents. • Dr. S. Lawrence Dingman, Chairperson, 603-862-1718. Application contact: Dr. Francis Birch, 603-862-1718.

University of North Carolina at Chapel Hill, College of Arts and Sciences, Curriculum in Marine Sciences, Chapel Hill, NC 27599. Curriculum awards MS, PhD. Faculty: 17 full-time, 0 part-time. Matriculated students: 19 full-time (6 women), 0 part-time; includes 0 minority, 2 foreign. 31 applicants, 32% accepted. In 1990, 3 master's, 4 doctorates awarded. *Degree requirements:* For master's, thesis, comprehensive exam required, foreign language not required; for doctorate, 1 foreign language, dissertation, comprehensive exam. *Entrance requirements:* GRE General Test (minimum combined score of 1000 required), GRE Subject Test, minimum GPA of 3.0. Application deadline: 2/15. Application fee: $35. *Tuition:* $621 per semester full-time for state residents; $3555 per semester full-time for nonresidents. *Financial aid:* In 1990–91, 0 fellowships, 15 research assistantships, 3 teaching assistantships awarded. • Dr. Dirk Frankenberg, Chairman, 919-962-1252.

University of North Carolina at Wilmington, College of Arts and Sciences, Department of Biological Sciences, Wilmington, NC 28403. Offerings include biological oceanography (MS). Department faculty: 18 full-time (1 woman), 0 part-time. *Degree requirements:* 1 foreign language (computer language can substitute), thesis, written and oral comprehensive exams. *Entrance requirements:* GRE General Test (minimum combined score of 1000 required), GRE Subject Test, minimum B average in undergraduate major. Application deadline: 7/1 (applications processed on a rolling basis). Application fee: $15. *Tuition:* $651 per semester full-time for state residents; $3178 per semester full-time for nonresidents. • Dr. Ronald K. Sizemore, Chairman, 919-395-3470. Application contact: Dr. Eric G. Bolen, Dean of Graduate School, 919-395-3135.

University of Puerto Rico, Mayagüez Campus, College of Arts and Sciences, Department of Marine Sciences, Mayagüez, PR 00709. Department awards MS, PhD. Faculty: 19 full-time (0 women), 0 part-time. Matriculated students: 60 full-time (18 women), 0 part-time; includes 39 minority (all Hispanic American), 21 foreign. 20 applicants, 45% accepted. In 1990, 10 master's, 3 doctorates awarded. *Degree requirements:* For master's, 1 foreign language, thesis; for doctorate, 1 foreign language, dissertation, qualifying exams. *Entrance requirements:* For doctorate, GRE. Application deadline: 10/15. Application fee: $15. *Expenses:* Tuition of $45 per credit for commonwealth residents; $3000 per semester for nonresidents. Fees of $344 per semester. *Financial aid:* In 1990–91, 23 research assistantships (2 to first-year students), 1 teaching assistantship (0 to first-year students) awarded; federal work-study, institutionally sponsored loans also available. *Faculty research:* Marine botany, ecology, chemistry, and parasitology; fishery; physical oceanography; ichthyology; aquaculture. • Dr. Manuel Hernandez-Avila, Director, 809-832-4040 Ext. 3839.

See full description on page 327.

University of Rhode Island, Graduate School of Oceanography, Kingston, RI 02881. School awards MS, PhD. *Application deadline:* 4/15. *Application fee:* $25. *Expenses:* Tuition of $2575 per year full-time, $120 per credit hour part-time for state residents; $5900 per year full-time, $274 per credit hour part-time for nonresidents. Fees of $696 per year full-time.

See full description on page 329.

University of San Diego, College of Arts and Sciences, Department of Marine Science and Ocean Studies, San Diego, CA 92110-2492. Department offers programs in marine science (MA), ocean studies (MA). Faculty: 3 full-time (0 women), 0 part-time. Matriculated students: 4 full-time, 13 part-time. Average age 25. 15 applicants, 80% accepted. In 1990, 0 degrees awarded. *Entrance requirements:* TOEFL (minimum score of 600 required). Application fee: $35. *Expenses:* Tuition of $400 per unit. Fees of $25 per semester. *Financial aid:* Fellowships, federal work-study, institutionally sponsored loans, and career-related internships or fieldwork available. Aid available to part-time students. Financial aid application deadline: 5/1. • Dr. Lou Burnett, Chair, 619-260-4464. Application contact: Maureen Phalen, Director of Graduate Admissions, 619-260-4524.

University of South Carolina, Graduate School, College of Science and Mathematics, Marine Science Program, Columbia, SC 29208. Program awards MS, PhD. Faculty: 26 full-time (3 women), 0 part-time. Matriculated students: 27 full-time (8 women), 2 part-time (1 woman); includes 2 minority (both black American), 6 foreign. Average age 29. 45 applicants, 22% accepted. In 1990, 7 master's awarded (14% entered university research/teaching, 72% found other work related to degree, 14% continued full-time study); 3 doctorates awarded (67% entered university research/teaching, 33% found other work related to degree). *Degree requirements:* For master's, thesis required, foreign language not required; for doctorate, 1 foreign language, dissertation. *Entrance requirements:* GRE. Application deadline: 5/1 (applications processed on a rolling basis). Application fee: $25. *Tuition:* $1404 per semester full-time. *Financial aid:* In 1990–91, $130,000 in aid awarded. 3 fellowships (0 to first-year students), 10 research assistantships (1 to a first-year student), 14 teaching assistantships (9 to first-year students) were awarded; federal work-study, institutionally sponsored loans, and career-related internships or fieldwork also available. Financial aid application deadline: 5/1. *Faculty research:* Benthic ecology, geochemistry, estuarine ecology, micropaleontology, aquaculture. • Dr. Bruce Coull, Director, 803-777-2692.

University of Southern Mississippi, College of Science and Technology, Department of Marine Science, Hattiesburg, MS 39406. Department awards MS. Part-time programs available. Faculty: 8 full-time (1 woman), 23 part-time (2 women). *Degree requirements:* Computer language, thesis. *Entrance requirements:* GRE General Test (minimum combined score of 1000 required), minimum GPA of 3.0. Application deadline: 8/9 (priority date, applications processed on a rolling basis). Application fee: $0 ($25 for foreign students). *Expenses:* Tuition of $968 per semester full-time, $93 per semester hour part-time. Fees of $12 per semester part-time for state residents; $591 per year full-time, $12 per semester part-time for nonresidents. *Financial aid:* Research assistantships, institutionally sponsored loans available. Financial aid application deadline: 3/15. *Faculty research:* Chemical, biological, physical, and geological marine science; remote sensing. Total annual research budget: $432,781. • Application contact: Dr. George A. Knauer, Director, 601-688-3177.

University of South Florida, College of Arts and Sciences, Department of Marine Science, Tampa, FL 33620. Department offers programs in marine biology (MS, PhD), marine science (MS, PhD), oceanography (MS, PhD). Program in Oceanography offered jointly with Florida State University. Part-time and evening/weekend programs available. Faculty: 21. Matriculated students: 42 full-time (15 women), 44 part-time (11 women); includes 3 minority (1 Asian American, 1 black American, 1 Hispanic American), 16 foreign. Average age 30. 69 applicants, 55% accepted. In 1990, 13 master's, 4 doctorates awarded. *Degree requirements:* For master's, thesis; for doctorate, 2 foreign languages (computer language can substitute for one), dissertation. *Entrance requirements:* For master's, GRE General Test (minimum combined score of 1000 required), minimum GPA of 3.0 (in last 60 credit hours). Application fee: $15. *Tuition:* $79.40 per credit hour for state residents; $241.33 per credit hour for nonresidents. *Financial aid:* In 1990–91, 22 students received a total of $32,511 in aid awarded. • Peter Betzer, Chairperson, 813-893-9130. Application contact: Edward VanVleet, Coordinator, 813-893-9130.

See full description on page 331.

University of Texas at Austin, Graduate School, College of Natural Sciences, Department of Marine Science, Austin, TX 78712. Department awards MA, PhD. Matriculated students: 10 full-time (2 women), 0 part-time; includes 0 minority, 1 foreign. 18 applicants, 33% accepted. In 1990, 2 master's awarded. *Application deadline:* 2/1 (priority date, applications processed on a rolling basis). *Application fee:* $40 ($75 for foreign students). *Tuition:* $510.30 per semester for state residents; $1806 per semester for nonresidents. *Financial aid:* Application deadline 3/1. • Dr. Robert S. Jones, Chairman, 512-749-6730. Application contact: Dr. Patrick Parker, Graduate Adviser, 512-749-6730.

See full description on page 333.

University of Washington, College of Ocean and Fishery Sciences, School of Marine Affairs, Seattle, WA 98195. School awards MMA, MMA/MAIS. Program offered in marine affairs (MMA). Part-time programs available. Faculty: 9 full-time (0 women), 0 part-time. Matriculated students: 61 full-time (33 women), 0 part-time; includes 4 minority (1 Asian American, 2 black American, 1 Native American), 3 foreign. Average age 30. 61 applicants, 57% accepted. In 1990, 11 degrees awarded (100% found work related to degree). *Degree requirements:* Thesis required, foreign language not required. *Entrance requirements:* GRE General Test, TOEFL, minimum GPA of 3.0. Application deadline: 2/1 (priority date). Application fee: $35. *Tuition:* $1129 per quarter full-time, $324 per credit (minimum) part-time for state residents; $2824 per quarter full-time, $809 per credit (minimum) part-time for nonresidents. *Financial aid:* In 1990–91, 2 students received a total of $37,000 in aid awarded. 0 fellowships, 10 research assistantships (5 to first-year students), 0 teaching assistantships were awarded; full tuition waivers, federal work-study, institutionally sponsored loans, and career-related internships or fieldwork also available. Aid available to part-time students. Average monthly stipend for a graduate assistantship: $940. Financial aid application deadline: 3/1; applicants required to submit FAF. *Faculty research:* Marine pollution, port authorities, fisheries management, global climate change, marine environmental protection. Total annual research budget: $314,763. • Edward L. Miles, Director, 206-543-7004. Application contact: Marc L. Miller, Graduate Program Coordinator, 206-543-7004.

University of Washington, College of Ocean and Fishery Sciences, School of Oceanography, Seattle, WA 98195. School offers programs in biological oceanography (MS, PhD), chemical oceanography (MS, PhD), marine geology and geophysics (MS, PhD), physical oceanography (MS, PhD). Part-time programs available. Faculty: 51 full-time (5 women), 0 part-time. Matriculated students: 86 full-time (36 women), 5 part-time (1 woman); includes 5 minority (4 Asian American, 1 Hispanic American), 10 foreign. Average age 29. 138 applicants, 20% accepted. In 1990, 7 master's awarded; 12 doctorates awarded (100% entered university research/teaching). *Degree requirements:* For master's, research project required, foreign language and thesis not required; for doctorate, dissertation required, foreign language not required. *Entrance requirements:* GRE General Test. Application deadline: 2/1 (priority date). Application fee: $35. *Tuition:* $1129 per quarter full-time, $324 per credit (minimum) part-time for state residents; $2824 per quarter full-time, $809 per credit (minimum) part-time for nonresidents. *Financial aid:* In 1990–91, 65 students received a total of $1.227-million in aid awarded. 5 fellowships (all to first-year students), 52 research assistantships (11 to first-year students), 8 teaching assistantships (0 to first-year students) were awarded. Aid available to part-time students. Average monthly stipend for a graduate assistantship: $1035. Financial aid application deadline: 2/1. *Total annual research budget:* $8.76-million. • Arthur R. M. Nowell, Director, 206-543-5039. Application contact: Della Rogers, Student Services Coordinator, 206-543-5039.

University of Wisconsin–Madison, Program in Oceanography and Limnology, Madison, WI 53706. Program awards MS, PhD. Matriculated students: 16 full-time (8 women), 1 part-time (0 women); includes 0 minority, 5 foreign. 17 applicants, 24% accepted. In 1990, 0 master's, 3 doctorates awarded. *Application deadline:* rolling. *Application fee:* $20. *Financial aid:* In 1990–91, 13 students received aid. 0 fellowships, 9 research assistantships, 4 teaching assistantships awarded; institutionally sponsored loans also available. Financial aid application deadline: 1/15; applicants required to submit FAF. • C. J. Bowser, Chairperson, 608-263-3264. Application contact: Joanne Nagy, Assistant Dean of the Graduate School, 608-262-2433.

Western Connecticut State University, School of Arts and Sciences, Department of Biological and Environmental Sciences, Program in Oceanography and Limnology, Danbury, CT 06810. Program awards MA. Part-time and evening/weekend programs available. Faculty: 5 full-time, 0 part-time. Matriculated students: 1 full-time, 16 part-time. In 1990, 0 degrees awarded. *Degree requirements:* Thesis, comprehensive exam required, foreign language not required. *Entrance requirements:* Minimum GPA of 2.5. Application deadline: 8/1 (priority date). Application fee: $20. *Expenses:* Tuition of $860 per semester full-time, $115 per credit hour part-time for state residents; $2395 per semester full-time, $115 per credit hour part-time for nonresidents. Fees of $488 per semester full-time, $5 per semester part-time. *Financial aid:* In 1990–91, 2 fellowships (both to first-year students) awarded; federal work-study and career-related internships or fieldwork also available. Aid available to part-time students. Average monthly stipend for a graduate assistantship: $266. Financial aid application deadline: 5/1; applicants required to submit GAPSFAS or FAF. • Dr. K. Kanungo, Head, Department of Biological and Environmental Sciences, 203-797-4207.

Yale University, Graduate School of Arts and Sciences, Department of Geology and Geophysics, New Haven, CT 06520. Offerings include oceanography (PhD). Terminal master's awarded for partial completion of doctoral program. Department faculty: 24 full-time (2 women), 1 part-time (0 women). *Degree requirements:* Dissertation. *Entrance requirements:* GRE General Test, GRE Subject Test. Application deadline: 1/2. Application fee: $45. *Tuition:* $15,160 per year full-time, $1895 per course part-time. • Dr. B. Saltzman, Chairman, 203-432-3114. Application contact: Susan Webb, Coordinator of Admissions, 203-432-2770.

See full description on page 299.

Cross-Discipline Announcements

Florida Institute of Technology, College of Science and Liberal Arts, Department of Biological Sciences, Melbourne, FL 32901.

MS and PhD in marine biology with research emphases in mollusks, echinoderms, coral reef fishes, and manatees and community studies of lagoonal, mangrove, and reef systems. Applications range from physiological and ecological systematics to biochemical and molecular biological studies in these areas.

McGill University, Faculty of Graduate Studies and Research, Department of Meteorology, Montreal, PQ H3A 2T5, Canada.

The department offers opportunities for study and research at the M Sc and PhD levels in physical oceanography. Research carried out includes large-scale ocean circulation, isopycnal diffusion, interannual and decadal fluctuations in air-ice-sea interactions, under-ice dynamics of river plumes, coastal processes, and biological-physical interactions. For further information, see Announcement in Meteorology and Atmospheric Sciences Section.

SECTION 4: MARINE SCIENCES/OCEANOGRAPHY

COLLEGE OF WILLIAM AND MARY

Virginia Institute of Marine Science
School of Marine Science

Programs of Study The School of Marine Science offers the Master of Arts and the Doctor of Philosophy degrees in marine science. Majors in biological oceanography, physical oceanography, chemistry and toxicology, geological oceanography, marine fisheries oceanography, and marine resource management (marine affairs) are available at both levels. A program in marine education is offered in cooperation with the School of Education.

The M.A. degree requires 30 semester hours of graduate work completed with a grade of B or better, with at least 12 hours in advanced courses and 1–6 hours of thesis credit; successful completion of a comprehensive examination and defense of the thesis; and reading knowledge of one foreign language. The Ph.D. requires three years of study beyond the baccalaureate, a reading knowledge of one foreign language, successful completion of a course of study determined by the student's committee, a comprehensive examination, a qualifying examination, and a dissertation defense. Students working on either degree must spend one academic year in residence (at least 12 hours each semester). Upon completion of residence requirements, students who are on campus and utilizing facilities are required to register for 3 hours per semester (1 hour during each summer session). The program is interdisciplinary, and all students are required to take core courses in biological, chemical, physical, and geological oceanography and statistics. Students interested in careers in government and industry are encouraged to take courses in environmental law and marine affairs.

The primary orientation of the faculty is toward estuarine, coastal, slope, and tidal freshwater environments. Many of the faculty members are actively engaged in applied research of direct concern to industry and management agencies. Students often find that their assistantship duties and/or research topics bring them into close contact with industry and management agencies at state, federal, and regional levels.

There is ample opportunity for students to engage in field-oriented thesis and dissertation research, and all students are encouraged to participate in field studies to gain experience in research at sea.

Research Facilities The main research campus of the School of Marine Science/Virginia Institute of Marine Science is located at Gloucester Point on the York River. There is a branch laboratory at Wachapreague on Virginia's Eastern Shore that is ideally suited for research on the barrier island and lagoon systems. The Institute is well equipped with modern laboratory and seagoing instrumentation. Computer support is provided by a Prime 9955 Model II on campus and many microprocessors. The Institute is equipped with running salt water. Major equipment includes transmission and scanning electron microscopes, a gas chromatograph/mass spectrometer, computer-interfaced gas chromatographs, an atomic absorption spectrometer, computer-assisted image analysis, sidescan sonar, a multifrequency echo sounder, and a hydraulic flume. The Institute has several vessels for estuarine and coastal operations and an aircraft for remote-sensing research.

Financial Aid Financial aid in the form of research assistantships, teaching assistantships, workships, and fellowships is available to a large percentage of the student body. Twelve-month assistantships for 1991–92 are $9800 for M.A. candidates and $10,200 for Ph.D. candidates. A limited number of hourly workships are also available. Students on assistantships are eligible for consideration for resident tuition. The majority of assistantships are funded through grants and contracts. Every effort is made to continue funding once a student has been assigned to an assistantship, but support is not guaranteed throughout a student's tenure. Many students are successful in obtaining their own financial support from federal and private fellowship programs.

Cost of Study The 1991–92 cost of full-time study (9 or more hours) is $1865 for Virginia residents and $5225 for nonresidents per semester. Hourly fees are $120 per hour for residents and $320 per hour for nonresidents.

Cost of Living No student housing is available in Gloucester Point. Apartments and housing in the area rent for $325 or more per month. The need for an automobile is strongly emphasized. A list of apartments and houses in the area is available from the graduate dean's office.

Student Group The School enrolls about 120 students; 10 to 15 percent of the students are from foreign countries. The majority of students have a biological orientation. About 40 percent are women. Recruiting initiatives in the coming years will be focused on increasing enrollment in the physical, geological, and chemical sciences and increasing the number of minority students in all disciplines. In recent years, graduating students have entered primarily the governmental or industrial sector. Over 90 percent of the School's graduates hold positions in which they use their graduate training.

Location Gloucester Point offers easy access by automobile to the urban regions surrounding Hampton Roads. Williamsburg, the site of the main campus of William and Mary, is 20 minutes away. The urban areas and the College support a vigorous cultural program, including theater, opera, ballet, and art. Two large coliseums with full programs are within an hour's drive.

The University and The School The College of William and Mary, the second-oldest university in the United States, will celebrate its 300th anniversary in 1993. It is a state-supported institution that prides itself on a dedication to excellence in its undergraduate program and a few select graduate programs.

The School of Marine Science of the College of William and Mary draws its faculty and academic support facilities from the Virginia Institute of Marine Science (VIMS), a research laboratory of the College whose origins go back to 1940. The dean of the School is also the director of VIMS.

Applying Applications close on February 15 for the following academic year. The majority of students are admitted for the fall semester; students are admitted to the spring semester only under exceptional circumstances. Applications must be submitted on an appropriate form obtainable from the College of William and Mary and must include appropriate transcripts, scores on the General Test and a Subject Test of the Graduate Record Examinations, and three letters of recommendation. Personal interviews are encouraged and may be arranged by advance appointment.

Correspondence and Information
Dean of Graduate Studies
School of Marine Science
Virginia Institute of Marine Science
College of William and Mary
P.O. Box 1346
Gloucester Point, Virginia 23062
Telephone: 804-642-7106 or 7105

Peterson's Guide to Graduate Programs in the Physical Sciences and Mathematics 1992

SECTION 4: MARINE SCIENCES/OCEANOGRAPHY

College of William and Mary

THE FACULTY AND THEIR RESEARCH

Frank O. Perkins, Ph.D., Professor and Dean. Diseases of marine organisms, cell biology of fungi and protozoa.
Henry Aceto Jr., Ph.D., Acting Dean of Graduate Studies. In vitro tumor radiobiology, nuclear magnetic resonance spectroscopy of biological systems.
Kenneth L. Webb, Ph.D., Chancellor Professor. Energy flow and nutrient cycling in marine environments, physiology of marine organisms, image analysis.

Professors

Herbert M. Austin, Ph.D. Fisheries oceanography, physical environmental parameters and interannual variability abundance of living marine resources.
Michael E. Bender, Ph.D. Ecology, pollution, toxicology.
Rudolf H. Bieri, Dr.rer.nat. Instrumental organic analysis in particular polynuclear aromatic hydrocarbon metabolites; environmental chemistry; noble gases in the marine environment, atmosphere, and meteorites.
John D. Boon III, Ph.D. Geological oceanography, mathematical geology, tidal hydraulics.
Robert J. Byrne, Ph.D. Geological oceanography, sediment transport, marine resource management.
Michael Castagna, M.S. Shellfish biology, aquaculture of shellfish, hatchery design and management.
Mark E. Chittenden Jr., Ph.D. Ecology and population dynamics of marine fishes.
William D. DuPaul, Ph.D. Fisheries science, marine advisory services, commercial fisheries development.
George C. Grant, Ph.D. Taxonomy and ecology of marine and estuarine zooplankton.
William J. Hargis Jr., Ph.D. Marine affairs, pathobiology, biological oceanography.
Robert J. Huggett, Ph.D. Chemical oceanography, environmental chemistry.
Albert Y. Kuo, Ph.D. Physical oceanography, estuarine hydrodynamics, estuarine circulation and water quality, turbulence, mathematical modeling.
Joseph G. Loesch, Ph.D. Estuarine and coastal fisheries, biostatistics, population dynamics and fish ecology.
Maurice P. Lynch, Ph.D. Marine ecology, physiology, information systems, marine resource management.
William G. MacIntyre, Ph.D. Chemical oceanography, groundwater contaminant transport theory, organic solute transport solubility.
Roger L. Mann, Ph.D. Physiology, biochemistry, and ecology of marine bivalve molluscs; larval behavior; ecology.
John A. Musick, Ph.D. Ecology of marine fishes, sea turtles, and marine mammals; systematics and ecology of marine fishes.
Bruce J. Neilson, Ph.D.; PE. Nonpoint-source pollution, enrichment of estuaries, water-quality management.
Maynard M. Nichols, Ph.D. Geological oceanography, sedimentology.
Morris H. Roberts Jr., Ph.D. Aquatic toxicology, culture of marine invertebrates.
Gene M. Silberhorn, Ph.D. Coastal and wetlands plant ecology and systematics, coastal-zone management, submerged aquatic vegetation reproductive life cycles.
N. Bartlett Theberge, LL.M. Ocean and coastal law, marine affairs, marine resource management.
Richard L. Wetzel, Ph.D. Estuarine and coastal ecology, energetics, systems ecology.
L. Donelson Wright, Ph.D. Sediment transport, benthic boundary layer and morphodynamics of coastal zone.

Associate Professors

John M. Brubaker, Ph.D. Physical oceanography, estuarine hydrodynamics, shallow-water fronts.
Eugene M. Burreson, Ph.D. Parasitology; ecology, pathology, and immunology of marine parasitic protozoa; systematics and zoogeography of marine leeches.
Fu Lin Chu, Ph.D. Nutrition-biochemistry studies of molluscs, diseases and immunopathology of shellfish, oyster and algal culture.
Robert J. Diaz, Ph.D. Benthic ecology; population dynamics, energetics, and quantitative analyses of benthic communities; sediment resuspension and accumulation rates; invertebrate taxonomy.
David A. Evans, D.Phil. Applied mathematics and physics, numerical analysis.
Leonard W. Haas, Ph.D. Heterotrophic and autotrophic microplankton ecology, estuarine plankton.
John M. Hamrick, Ph.D. Estuarine and coastal physical oceanography, computational fluid dynamics.
Carl H. Hershner, Ph.D. Estuarine and coastal ecology, marine resource management.
Howard I. Kator, Ph.D. Microbiology, ecology of estuarine and marine bacteria, biodegradation processes, sanitary microbiology of estuarine waters.
Robert J. Orth, Ph.D. Biology and ecology of submerged aquatic vegetation, benthic communities.
Evon P. Ruzecki, Ph.D. Physical oceanography, estuarine circulation, continental shelf studies.
Michael E. Sieracki, Ph.D. Marine microbial plankton ecology, digital image analysis, marine bacterial and protistan biology.
Beverly A. Weeks, Ph.D. Immunotoxicology of marine organisms.

Assistant Professors

Bruce J. Barber, Ph.D. Physiopathology and reproduction of bivalve molluscs, oyster biology, fisheries science.
Thomas A. Barnard Jr., M.A. Wetlands management, wetlands ecology, marine resource management.
James A. Colvocoresses, Ph.D. Ecology and recruitment dynamics of marine and estuarine fishes.
Rebecca M. Dickhut, Ph.D. Chemical fate and transport in aquatic environments.
John E. Graves, Ph.D. Fishery genetics.
Robert C. Hale, Ph.D. Fate and effects of pollutants in the marine environment.
Carl H. Hobbs III, M.S. Geological oceanography.
James E. Kirkley, Ph.D. Resource economics, production, trade, resource management, quantitative methods and bioeconomics.
Romuald N. Lipcius, Ph.D. Ecology, behavior and population dynamics of crustaceans, benthic predator-prey and recruitment dynamics, biostatistics, quantitative ecology.
Jerome P.-Y. Maa, Ph.D. Geological oceanography, sediment transport.
John E. Olney, M.A. Ichthyology, systematics, taxonomy and early life history of marine fish, taxonomy and ecology of marine zooplankton.
Linda C. Schaffner, Ph.D. Biological oceanography, invertebrate ecology and behavior (general and blue crabs), animal-sediment interactions, effects of disturbance on marine ecosystems.
Peter A. Van Veld, Ph.D. Biochemical responses of marine organisms to pollutants, detoxification enzymes.

SECTION 4: MARINE SCIENCES/OCEANOGRAPHY

DALHOUSIE UNIVERSITY

Department of Oceanography

Program of Study

The Department of Oceanography offers a program leading to M.Sc. and Ph.D. degrees in oceanography. The program is designed to provide students with a general knowledge of the field as a whole and with advanced training in one of the specialties. Introductory and advanced courses are available in atmospheric sciences, physical and chemical oceanography, marine geology and geophysics, and biological oceanography. Students also commonly take courses at the graduate level in appropriate basic sciences. However, the main emphasis is on thesis research rather than formal course work, particularly after the first year. Ph.D. candidates take a written comprehensive examination, normally during the first year, and all students must make an oral thesis defence. Mastery of shipboard oceanographic techniques is an integral part of the program.

The University term is from mid-September to mid-April, but scholarships are granted on a year-round basis and students are expected to be in residence during the summer to carry on their thesis research.

Research Facilities

The department's facilities within the modern Life Sciences Centre include a large running-seawater system with environmentally controlled rooms, high-pressure facilities for simulation of deep-sea conditions, and large tanks for fish studies. Seagoing facilities are supplied by federal oceanographic ships. The department is equipped with the collecting gear and laboratory instrumentation required for modern oceanographic research. Library facilities are centralized in a science library nearby. In addition to numerous terminals to the University's VAX 8800 computer, the department has a variety of microcomputers and desktop calculators with plotters.

Financial Aid

Virtually all students not receiving external scholarships are supported by Dalhousie Fellowships of about $13,000 in 1991–92. Some larger scholarships are available for outstanding students. Applicants are urged to seek external scholarships that may be available to them.

Cost of Study

In 1991–92, the cost of tuition is $2594 per year for one to three years (depending upon the degree level) and $569 for students who are completing their thesis. Foreign students pay an additional $1700 per year, but supplementary scholarships may be offered.

Cost of Living

The cost of living in Halifax is higher than in the eastern United States. In general, scholarship support is adequate for an unmarried student. Married students find that their spouses need to get a job to supplement their income. Students may qualify for provincially supported medical and hospital services. University housing is limited, and most students live off campus, although finding good housing at a reasonable cost is one of the most troublesome problems that graduate students have.

Student Group

There are currently about 50 graduate students in the Department of Oceanography; the student-faculty ratio is 4:1.

Location

Halifax, a city of about 150,000 with a metropolitan population of 300,000, is built on a peninsula that is surrounded by arms of the sea. Dalhousie is situated in a residential area in the southwestern part of the peninsula, 0.6 kilometres from the water. Halifax is a mixture of the old and the new—high-rise apartments, modern office buildings, fine residential areas, and some historic neighborhoods. There is an unusual amount of park area. Museums, several repertory theatres, and a symphony orchestra supplement the variety of cultural activities sponsored by the University. By large-city standards the pace of life is leisurely. Halifax is surrounded by rugged granite country, largely forested and with numerous lakes. Wild country a half-hour drive from the centre of the city provides ample opportunity for outdoor activities.

The University

Dalhousie, founded in 1818, has a current enrolment of 11,000 students. In addition to the undergraduate and graduate faculties of arts and sciences, there are schools of law, medicine, dentistry, and other professions. The Life Sciences Centre houses the Departments of Oceanography, Biology, Geology, and Psychology. There is student representation on the Senate and Board of Governors and in most academic departments. The Dalhousie Association of Graduate Students has a year-round social facility.

Applying

Requests for application forms for admission and scholarship support should be directed to the registrar at Dalhousie University. In order to be considered for scholarships, students must submit applications for fall admission before February 1. There are no midyear admissions. Admission is granted primarily on the basis of the undergraduate record and letters of recommendation; scores on the General Test of the Graduate Record Examinations or the equivalent are required if available. If English is not their first language, students must demonstrate a knowledge of English by taking the TOEFL. A bachelor's degree in mathematics or in one of the basic sciences is ordinarily required for admission, although consideration is also given to applicants with a background in engineering. Non-Canadian students normally enter Canada on a student visa with a Dalhousie Fellowship. For more specific details, applicants should refer to the international student information letter provided by the Dalhousie Association of Graduate Students.

Correspondence and Information

The Graduate Co-ordinator
Department of Oceanography
Dalhousie University
Halifax, Nova Scotia B3H 4J1
Canada
Telephone: 902-424-3557

Dalhousie University

THE FACULTY AND THEIR RESEARCH

C. Beaumont, Professor; Ph.D., Dalhousie. Tidal loading and ocean tides, earth rheology.
B. P. Boudreau, Assistant Professor; Ph.D., Yale. Diagenesis, sediment-water interactions, mathematical modelling.
A. J. Bowen, Professor; Ph.D., California, San Diego (Scripps). Physical oceanography, nearshore dynamics, sediment transport.
C. M. Boyd, Professor; Ph.D., California, San Diego (Scripps). Zooplankton distribution, behaviour, and sensory physiology.
P. Chylek, Professor; Ph.D., California, Riverside. Radiation physics; radiative transfer in theory, in the laboratory, and in the atmosphere.
R. C. Courtney, Adjunct Professor; Ph.D., Cambridge; Atlantic Geoscience Centre, Bedford Institute of Oceanography. Marine geophysics, high-resolution seismics, geodynamics.
R. O. Fournier, Professor; Ph.D., Rhode Island. Biological oceanography, phytoplankton ecology and systematics.
K. T. Frank, Adjunct Professor; Ph.D., Toledo; Bedford Institute of Oceanography. Population dynamics, stock assessment, fisheries oceanography, recruitment.
J. Grant, Assistant Professor; Ph.D., South Carolina. Benthic ecology, shellfisheries.
B. T. Hargrave, Adjunct Professor; Ph.D., British Columbia; Bedford Institute of Oceanography. Benthic ecology.
O. Hertzman, Assistant Professor; Ph.D., Washington (Seattle). Mesoscale meteorology, radar meteorology, dynamics of marine storms.
D. A. Huntley, Adjunct Professor; Ph.D., Bristol; Institute of Marine Studies, Plymouth Polytechnic Southwest. Physical oceanography, nearshore processes, waves and turbulence.
W. Hyde, Assistant Professor; Ph.D., Toronto. Climate modelling, paleoclimatology.
B. D. Johnson, Assistant Professor; Ph.D., Dalhousie. Chemical oceanography, physical chemistry of seawater.
D. Kelley, Assistant Professor; Ph.D., Dalhousie. Ocean mixing, deep convection, double diffusion, arctic mixing.
G. B. Lesins, Assistant Professor; Ph.D., Toronto. Climate thermodynamics, cloud and precipitation physics.
M. R. Lewis, Associate Professor; Ph.D., Dalhousie. Biological oceanography, marine ecosystem modelling.
K. E. Louden, Associate Professor; Ph.D., MIT/Woods Hole. Marine geophysics.
L. A. Mayer, Associate Professor; Ph.D., California, San Diego (Scripps). Marine geology.
E. L. Mills, Professor; Ph.D., Yale. Benthic ecology, history of oceanography.
R. M. Moore, Professor; Ph.D., Southampton. Chemical oceanography, trace-metal distribution and behaviour in marine and estuarine environments.
S. Pearre Jr., Adjunct Professor; Ph.D., Dalhousie. Zooplankton ecology, predator-prey relationships.
D. J. W. Piper, Adjunct Professor; Ph.D., Cambridge; Atlantic Geoscience Center, Bedford Institute of Oceanography. Marine sedimentology.
B. R. Ruddick, Associate Professor; Ph.D., MIT/Woods Hole. Physical oceanography.
C. T. Taggart, Assistant Professor; Ph.D., McGill. Fisheries oceanography, early life-history phenomena, biological/physical interactions.
K. R. Thompson, Assistant Professor; Ph.D., Liverpool. Physical oceanography/climatology.
P. J. Wangersky, Professor; Ph.D., Yale. Chemical oceanography, biochemistry of seawater, biogeochemistry.

Research Associates

The research associates in the listing below have primary appointments in other Dalhousie departments or in neighbouring institutes and help with the teaching program and with thesis supervision in oceanography.

R. G. Ackman, Ph.D., London. Technical University of Nova Scotia.
N. E. Balch, Ph.D., Dalhousie. Manager, Dalhousie Aquatron.
J. S. Craigie, Ph.D., Queen's at Kingston. Atlantic Regional Laboratory, National Research Council.
F. W. Dobson, Ph.D., British Columbia. Atlantic Oceanographic Laboratory, Bedford Institute of Oceanography.
R. W. Doyle, Ph.D., Yale. Department of Biology, Dalhousie University.
J. A. Elliott, Ph.D., British Columbia. Atlantic Oceanographic Laboratory, Bedford Institute of Oceanography.
W. G. Harrison, Ph.D., North Carolina State. Bedford Institute of Oceanography.
W. D. Jamieson, Ph.D., Cambridge. Atlantic Research Laboratory, National Research Council.
C. Keen, Ph.D., Cambridge. Regional Reconnaissance, Bedford Institute of Oceanography. M. J. Keen, Ph.D., Cambridge. Atlantic Geoscience Centre, Bedford Institute of Oceanography.
S. R. Kerr, Ph.D., Dalhousie. Bedford Institute of Oceanography.
D. Lefaivre, Ph.D., Laval. Fisheries and Oceans, Government of Canada.
J. W. Loder, Ph.D., Dalhousie. Department of Fisheries and Oceans, Bedford Institute of Oceanography.
J. A. Novitsky, Ph.D., Oregon State. Department of Biology, Dalhousie University.
N. S. Oakey, Ph.D., McMaster. Ocean Circulation Division, Bedford Institute of Oceanography.
B. D. Petrie, Ph.D., Dalhousie. Atlantic Oceanographic Laboratory, Bedford Institute of Oceanography.
T. Platt, Ph.D., Dalhousie. Bedford Institute of Oceanography.
S. Sathyendranath, Ph.D., Marie Curie-Sklodowska (Poland). Bedford Institute of Oceanography; Oceanography, Dalhousie University.
M. Sinclair, Ph.D., California, San Diego (Scripps). Fisheries and Oceans, Government of Canada.
J. N. Smith, Ph.D., Toronto. Atlantic Oceanographic Laboratory, Bedford Institute of Oceanography.
P. C. Smith, Ph.D., MIT/Woods Hole. Coastal Oceanography, Bedford Institute of Oceanography.
R. L. Stephenson, Ph.D., Canterbury. Fisheries and Oceans, Government of Canada.
G. S. Stockmal, Ph.D., Brown. Atlantic Geoscience Centre, Bedford Institute of Oceanography.
B. J. Topliss, Ph.D., University College, North Wales. Coastal Oceanography, Bedford Institute of Oceanography.
D. G. Wright, Ph.D., British Columbia. Atlantic Oceanographic Laboratory, Bedford Institute of Oceanography.

SECTION 4: MARINE SCIENCES/OCEANOGRAPHY

FLORIDA STATE UNIVERSITY

Department of Oceanography

Programs of Study

A graduate program in oceanography has existed at Florida State University since 1949, first in an interdisciplinary institute and since 1966 in a department within the College of Arts and Sciences. The Department of Oceanography, which offers both the M.S. and Ph.D. degrees in oceanography with specializations in physical, biological, or chemical oceanography, is the center for marine studies at the University. Additional marine and environmental research is conducted by the Departments of Biological Sciences, Chemistry, Geology, Mathematics, Meteorology, Physics, and Statistics, as well as the Geophysical Fluid Dynamics Institute and the Institute of Molecular Biophysics. Both formal and informal cooperative efforts between these science departments and the Department of Oceanography have flourished for years.

The M.S. degree program requires the completion of 33 semester hours of course work and a thesis covering an original research topic. Students pursuing the Ph.D. degree must complete 18 semester hours of formal course work beyond the master's degree course requirements and perform original research leading to a dissertation that makes a contribution to the science of oceanography.

The first year of graduate study is generally concerned with required course work and examinations. A supervisory committee, chosen for the individual student, directs the examinations and supervises the student's progress. Under its direction, the student begins thesis research as soon as possible—no later than the fourth semester in residence. There is no foreign language requirement for either the M.S. or the Ph.D.

Research Facilities

Oceanography department headquarters, offices, and laboratories are located in the Oceanography-Statistics Building in the science area of the campus. Some of the laboratories currently in operation are for water quality analysis, organic geochemistry, trace-element analysis, radiochemistry, microbial ecology, phytoplankton ecology, numerical modeling, and fluid dynamics. The department also has the benefit of a fully equipped machine shop and a current-meter facility using standard current meters with modern state-of-the-art instruments. The Florida State University Marine Laboratory on the Gulf of Mexico is located at Turkey Point near Carrabelle, about 45 miles southwest of Tallahassee. The R/V *Bellows*, a 65-foot research vessel, is shared by FSU and other campuses of the State University System of Florida.

Departmental facilities are augmented by those in other FSU departments and institutes, such as the Van de Graaff accelerator in the physics department, the Antarctic Marine Geology Research Facility and Core Library in the geology department, the Geophysical Fluid Dynamics Institute laboratories, the Electron Microscopy Laboratory in the biological sciences department, the Statistical Consulting Center in the statistics department, and the CRAY Y-MP in the Supercomputer Computations Research Institute.

The research activities of the faculty are heavily supported by federal funding, and these programs involve fieldwork, often at sea, all over the world. Faculty and students have worked aboard a great many of the major research vessels of the U.S. fleet; because this kind of active collaboration works so well, it has not seemed necessary for the University to have its own major research vessel.

Financial Aid

Fellowships and teaching and research assistantships are available on a competitive basis. University fellowships pay $10,000 per academic year. Research and teaching assistantships range from $7585 to $9066 for half-time work for the nine-month academic year. In addition, out-of-state tuition waivers are available for assistantship and fellowship recipients. Financial aid is also available for the summer term. Currently, most of the full-time students in the department are receiving financial assistance.

Cost of Study

Tuition for 1990–91 was $76.29 per credit hour for Florida residents; out-of-state students paid an additional $161.93 per credit hour. A slight increase in these rates is expected for 1991–92. The normal course load is 9 credit hours per semester for research and teaching assistants.

Cost of Living

A double room for single students in the graduate dormitories on campus costs about $185 per month (including utilities and local telephone). Married student housing at Alumni Village costs $175–$285 per month. There are many off-campus apartment complexes in Tallahassee, with rents beginning at about $260 per month.

Student Group

The department currently has 43 graduate students enrolled in the program (28 M.S., 15 Ph.D.). Of these, 40 are full-time. The students come from all areas of the country, with the greatest number representing the Northeast, South, and Midwest. Sixteen of the students are women. During the last five years, twenty-nine M.S. and sixteen Ph.D. degrees were awarded in oceanography. Graduates have taken positions in federal and state agencies, universities, and private companies.

Location

Florida State University is located in Tallahassee, the state capital. Although it is among the nation's fastest-growing cities, Tallahassee has managed to preserve its natural beauty. The northern Florida location has a landscape and climate that are substantially different from those of southern Florida. Heavy forest covers much of the area, with the giant live oak being the chief tree of the clay hills. Five large lakes surrounding Tallahassee, as well as the nearby Gulf of Mexico, offer numerous recreational opportunities. Life in Tallahassee has been described as a combination of the ambience of traditional southern living with the bustle of a modern capital city.

The University

Florida State University is a public coeducational institution founded in 1857. Current enrollment is over 28,000. The University has great diversity in its cultural offerings and is rich in traditions. It has outstanding science departments and excellent schools and departments in such varied areas as law, music, theater, and religion.

Applying

Applications should be submitted as early as possible in the academic year prior to anticipated enrollment. The deadline for applications for fall semester enrollment is in February for international students and in July for domestic students. Each prospective candidate must have a bachelor's degree, with a major pertinent to the student's chosen specialty area in oceanography. Minimum undergraduate preparation must include differential and integral calculus and one year of college chemistry and physics. Most students need more advanced mathematics. A minimum undergraduate GPA of 3.0 and a minimum GRE General Test score of 1100 are required. The average undergraduate GPA and the average GRE score of currently enrolled students are 3.45 and 1174, respectively.

Correspondence and Information

Admissions Committee
Department of Oceanography
Florida State University
Tallahassee, Florida 32306-3048

Telephone: 904-644-6700
Fax: 904-644-2581

Florida State University

THE FACULTY AND THEIR RESEARCH

These faculty members are important in the academic program of students in the Department of Oceanography. This list includes faculty from other departments who interact regularly with the department's faculty and students.

Lawrence G. Abele, Professor of Biological Sciences and Chairman; Ph.D., Miami (Florida). Ecology, community biology, systematics of decapod crustaceans.

James E. Bauer, Assistant Professor of Oceanography; Ph.D., Maryland. Isotope geochemistry of oceanic carbon and nitrogen cycles, biogeochemistry of benthic and pelagic food webs.

Werner A. Baum, Professor of Meteorology and Dean Emeritus; Ph.D., Chicago. Descriptive climatology.

Steven L. Blumsack, Associate Professor of Mathematics; Ph.D., MIT. Theory of rotating fluids.

William C. Burnett, Professor of Oceanography; Ph.D., Hawaii. Uranium-series isotopes and geochemistry of authigenic minerals of the seafloor, elemental composition of suspended material from estuaries and the deep ocean, environmental studies.

Jeffrey P. Chanton, Assistant Professor of Oceanography; Ph.D., North Carolina at Chapel Hill. Major element cycling, light stable isotopes, methane production and transport, coastal biogeochemical processes.

Allan J. Clarke, Professor of Oceanography; Ph.D., Cambridge. Climate dynamics, coastal oceanography, equatorial dynamics, tides.

William K. Dewar, Associate Professor of Oceanography; Ph.D., MIT/Woods Hole. Gulf Stream ring dynamics, general circulation theory, intermediate- and large-scale interaction, mixed layer processes.

E. Imre Friedmann, Professor of Biological Sciences; Ph.D., Vienna. Microorganisms in extreme environments, biology of marine algae.

Penny J. Gilmer, Associate Professor of Chemistry; Ph.D., Berkeley. Biochemical nature of cell-cell recognition.

William F. Herrnkind, Professor of Biological Sciences; Ph.D., Miami (Florida). Behavior and migration of marine animals.

Louis N. Howard, Professor of Mathematics; Ph.D., Princeton. Theory of rotating and stratified flows, hydrodynamic stability, bifurcation theory, geophysical fluid dynamics, chemical waves, biological oscillations. Member, National Academy of Sciences.

Ya Hsueh, Professor of Oceanography and Chairman; Ph.D., Johns Hopkins. Variabilities in sea level, coastal currents, water density in continental shelf waters.

Christopher Hunter, Professor of Mathematics; Ph.D., Cambridge. Dynamics of fluids and stellar systems.

Richard L. Iverson, Professor of Oceanography; Ph.D., Oregon State. Physiology and ecology of marine phytoplankton.

Russell H. Johnsen, Professor of Chemistry; Ph.D., Wisconsin–Madison. High-energy radiation absorption, kinetics of radical decay.

Charles L. Jordan, Professor of Meteorology; Ph.D., Chicago. Synoptic meteorology.

Michael Kasha, Professor of Chemistry, Institute of Molecular Biophysics; Ph.D., Berkeley. Molecular electronic spectroscopy, molecular quantum mechanics. Member, National Academy of Sciences.

Ruby E. Krishnamurti, Professor of Oceanography; Ph.D., UCLA. Ocean circulation, atmospheric convection, bioconvection, stability and transition to turbulence. Fellow, American Meteorological Society.

William M. Landing, Assistant Professor of Oceanography; Ph.D., California, Santa Cruz. Biogeochemistry of trace elements in the oceans, with emphasis on the effects of biological and inorganic processes on dissolved/particulate fractionation.

Paul A. La Rock, Professor of Oceanography; Ph.D., RPI. Microbial ecology, sulfate reduction in anoxic marine basins, microbial growth in marine environments, ocean frontal systems.

Noel E. La Seur, Professor of Meteorology; Ph.D., Chicago. Meteorology of hurricanes over oceans. Former Director, National Hurricane Research Laboratories.

Fred A. Leysieffer, Professor of Statistics; Ph.D., Michigan. Environmental statistics, sample surveys, stochastic processes.

Robley J. Light, Professor of Chemistry; Ph.D., Duke. Biosynthesis, metabolism, and structure of lipids and related compounds.

Robert J. Livingston, Professor of Biological Sciences; Ph.D., Miami (Florida). Estuarine ecology, aquatic pollution biology.

Nancy H. Marcus, Associate Professor of Oceanography; Ph.D., Yale. Population biology and genetics of marine zooplankton dormancy, photoperiodism, biological rhythms.

Richard N. Mariscal, Professor of Biological Sciences; Ph.D., Berkeley. Marine biology; behavior, physiology, and ultrastructure of invertebrates.

Duane A. Meeter, Professor of Statistics; Ph.D., Wisconsin–Madison. Ecological statistics, computer applications.

John W. Nelson, Professor of Physics; Ph.D., Texas at Austin. Application of particle accelerators to environmental research.

Doron Nof, Professor of Oceanography; Ph.D., Wisconsin–Madison. Fluid motions in the ocean, dynamics of equatorial outflows and formation of eddies, geostrophic adjustment in sea straits and estuaries, generation of oceanfronts.

James J. O'Brien, Professor of Meteorology and Oceanography; Ph.D., Texas A&M. Modeling of coastal upwelling and equatorial circulation, upper oceanfronts, climate scale fluctuations. Fellow, American Meteorological Society; Recipient, Sverdrup Gold Medal in Air-Sea Interactions.

John K. Osmond, Professor of Geology; Ph.D., Wisconsin–Madison. Uranium-series isotopes.

Paul C. Ragland, Professor of Geology; Ph.D., Rice. Petrology, geochemistry.

Daniel Simberloff, Professor of Biological Sciences; Ph.D., Harvard. Population and community ecology.

Steven Stage, Associate Professor of Meteorology; Ph.D., Washington (Seattle). Air-sea interaction, turbulence.

Melvin E. Stern, Professor of Oceanography; Ph.D., MIT. Theory of ocean circulation, salt fingers.

Donald R. Strong Jr., Professor of Biological Sciences; Ph.D., Oregon. Population and marine ecology, plant/insect ecology.

David W. Stuart, Professor and Chairman of Meteorology; Ph.D., UCLA. Mesoscale observational studies of the meteorology of upwelling regions, global weather, air-sea interactions. Principal Scientist, Field Experiments on El Niño.

Wilton Sturges III, Professor of Oceanography; Ph.D., Johns Hopkins. Ocean currents.

William F. Tanner, Professor of Geology; Ph.D., Oklahoma. Fluid dynamics, paleoceanography, sedimentology, structural geology.

David Thistle, Professor of Oceanography; Ph.D., California, San Diego (Scripps). Ecology of sediment communities, meiofauna ecology, deep-sea biology, crustacean systematics.

Thomas J. Vickers, Professor of Chemistry; Ph.D., Florida. Spectroscopic techniques for chemical analysis of trace elements relating to human health.

Georges L. Weatherly, Professor of Oceanography; Ph.D., Nova. Deep-ocean circulation and near-bottom currents.

John W. Winchester, Professor of Oceanography; Ph.D., MIT. Atmospheric chemistry, trace-element and aerosol-particle analysis.

Sherwood W. Wise, Professor of Geology; Ph.D., Illinois. Micropaleontology, marine geology, diagenesis of pelagic sediments, biomineralization.

SECTION 4: MARINE SCIENCES/OCEANOGRAPHY

LOUISIANA STATE UNIVERSITY

Department of Oceanography and Coastal Sciences

Programs of Study The Department of Oceanography and Coastal Sciences (DOCS) is in the College of Basic Sciences. The DOCS faculty members hold joint appointments in DOCS and the following research institutes: the Coastal Studies Institute, the Coastal Ecology Institute, the Coastal Fisheries Institute, and the Wetland Biogeochemistry Institute. Students usually conduct research through one of these institutes. The department offers M.S. and Ph.D. degree programs that are multidisciplinary and sufficiently flexible to meet special interests of individual students. Emphasis is given to understanding and applying knowledge about the physical, chemical, ecological, geological, and meteorological aspects of those coastal environments usually identified as shallow-water, nearshore, estuarine, and wetland. The M.S. program includes a minimum of 24 semester hours of graduate-level courses and 6 hours of thesis research. All students must take one course from each of the following four major core areas: biological, chemical, geological, and physical science. Each student, in conjunction with an advisory committee, develops a course of study appropriate to his or her primary interest. A 36-hour nonthesis degree option is also available. Students pursuing the Ph.D. must develop a program that includes at least 48 hours of course work beyond the baccalaureate, at least one course from each of the core areas, a minor (usually 12 hours) in another department, research leading to a dissertation, and reading knowledge of an appropriate language other than English. Concurrent degree programs are available with the Department of Experimental Statistics (M.Ap.Stat./M.S.) and through an interdepartmental program in public administration (M.P.A./M.S.). Students generally complete an M.S. degree in two to three years and a Ph.D. in four to five years.

Research Facilities Since field studies are the major focus, a fleet of small craft for inshore and coastal work is available. Other facilities include excellent equipment for the measurement of physical processes in the field; diverse equipment for chemical, geological, and biological sampling; and shops for making and maintaining this equipment. Summer field courses and research facilities are available at the Louisiana Universities Marine Consortium at Cocodrie and the Gulf Coast Research Laboratory in Ocean Springs, Mississippi. Extensive aerial photograph and map collections, an image-processing laboratory, nuclear and computer science facilities, fully equipped chemical laboratories, biological and sedimentological laboratories, greenhouses, growth chambers, and library facilities are available.

Financial Aid Graduate Research Assistantships beginning at $11,871 per year are available. Because student research is closely tied to ongoing research, students can expect to devote about 20 hours per week to assistantship duties. Currently, most of the full-time students in the department receive some form of financial support. The Graduate School offers a limited number of fellowships to outstanding Ph.D. students with undergraduate GPAs of at least 3.5 and combined GRE scores (verbal and quantitative) of at least 1250. Fellowships carry a stipend of $15,000 per year.

Cost of Study Full-time student fees for 1991–92 are $1020 per semester for state residents and $2620 per semester for out-of-state students. These rates are subject to increase. For fee purposes, research assistants are considered residents for the period of appointment.

Cost of Living Dormitory rates for 1991–92 are $480–$1170 per semester. Rental rates for University apartments for married students for 1991–92 are $220–$295 per month. Off-campus rates are approximately $200–$245 per month for one-bedroom apartments.

Student Group The department currently has 12 M.S. and 31 Ph.D. students from all parts of America and several other countries. During the last five years, twenty-eight M.S. and twenty-seven Ph.D. degrees were awarded.

Location LSU is located in Baton Rouge, the capital of Louisiana, an inland port, and a major petrochemical center. New Orleans is 80 miles to the southeast. To the south is Cajun country, with its rich cultural heritage and many bayous, marshes, and lakes that offer fishing, hunting, boating, and other recreation.

The University LSU occupies more than 2,000 acres bordering the Mississippi River. The Baton Rouge campus is the oldest and largest institution in the LSU System. Its instructional programs include about 250 curricula leading to undergraduate, graduate, and professional degrees. LSU has been designated by the Louisiana Board of Regents as the state's only comprehensive university. It is recognized nationally as a Research I university and is one of twenty-five universities in the country designated as both a land-grant and a sea-grant institution. Enrollment is 24,000 undergraduates and over 4,000 graduate and professional students. LSU has approximately 1,275 faculty members.

Applying Students with strong baccalaureate or graduate degrees in natural or engineering sciences and competence in math through first-year calculus are encouraged to apply. Those admitted without these prerequisites may be required to take additional undergraduate courses. Before a qualified student is admitted, a faculty member must agree to advise the student and to supervise his or her program. Prospective students are encouraged to correspond with faculty members who have compatible research interests.
Minimum requirements for regular admission to the LSU Graduate School are a bachelor's degree from an accredited U.S. institution or the equivalent from a foreign institution; a GPA of at least 3.0 on a 4.0 scale on all undergraduate and graduate work completed; acceptable scores on the GRE and for international students, on the TOEFL; and satisfactory academic standing at the last institution attended. Advanced study application forms, official GRE scores and TOEFL scores (if applicable), official transcripts of all previous undergraduate and graduate study, and a $25 nonrefundable fee should be submitted to the LSU Office of Graduate Admissions. Three letters of recommendation should be sent directly to the chairman of the department. Application deadlines are May 1 for the summer term, July 1 for the fall semester, and December 1 for the spring semester. There is a late application fee of $25. Applicants for fall are strongly advised to complete applications by the previous February to be eligible for assistantships, which are generally awarded in March.

Correspondence and Information
Graduate Advisor
Department of Oceanography and Coastal Sciences
Louisiana State University
Baton Rouge, Louisiana 70803-7505
Telephone: 504-388-6308

Louisiana State University

THE FACULTY AND THEIR RESEARCH

Charles E. Adams Jr., Professor; Ph.D., Florida State, 1980. Physical oceanography, boundary-layer processes, sediment transport.
Donald M. Baltz, Associate Professor; Ph.D., California, Davis, 1980. Fish ecology, life history, and habitat selection; ecology of marine vertebrates; life history strategy.
Robert S. Carney, Associate Professor; Ph.D., Oregon State, 1977. Biological oceanography, benthic ecology, research administration.
James M. Coleman, Boyd Professor and Executive Vice Chancellor; Ph.D., LSU, 1966. Deltaic sedimentation, riverine processes, continental shelf sediments.
Richard E. Condrey, Associate Professor; Ph.D., Washington (Seattle), 1981. Population dynamics, fisheries.
John W. Day Jr., Professor; Ph.D., North Carolina State, 1971. Estuarine ecology, systems ecology, coastal management.
Robert P. Gambrell, Professor; Ph.D., North Carolina State, 1974. Environmental chemistry of soils and sediment-water systems.
James G. Gosselink, Professor Emeritus; Ph.D., Rutgers, 1959. Wetland vegetation processes, plant ecophysiology, systems ecology.
S. A. Hsu, Professor; Ph.D., Texas, 1969. Coastal and marine meteorology, air-sea interactions.
Oscar K. Huh, Professor; Ph.D., Penn State, 1968. Satellite and airplane remote sensing, field geology.
Masamichi Inoue, Assistant Professor; Ph.D., Texas A&M, 1982. Numerical modeling of ocean circulation, climate changes.
Irving A. Mendelssohn, Professor; Ph.D., North Carolina State, 1974. Plant physiological ecology, wetland plant ecology, barrier island vegetation.
Samuel P. Meyers, Professor; Ph.D., Columbia, 1957. Marine microbial ecology, ocean food resources, aquaculture nutrition.
Stephen P. Murray, Professor; Ph.D., Chicago, 1966. Shallow-water physical oceanography, tidal hydrodynamics.
William H. Patrick Jr., Boyd Professor; Ph.D., LSU, 1954. Sediment chemistry, nutrient cycling in wetlands, environmental chemistry of soils.
James H. Power, Assistant Professor; Ph.D., Maine, 1982. Fisheries oceanography, larval fish ecology, marine ecology.
Harry H. Roberts, Professor; Ph.D., LSU, 1969. Marine geology, carbonate sedimentology and reef processes.
Lawrence J. Rouse Jr., Associate Professor; Ph.D., LSU, 1969. Remote sensing, physical oceanography.
Richard F. Shaw, Associate Professor; Ph.D., Maine, 1981. Ichthyoplankton ecology and dynamics, transport and recruitment mechanisms.
R. Eugene Turner, Professor; Ph.D., Georgia, 1974. Wetland ecology, biological oceanography.
Jack R. Van Lopik, Professor; Ph.D., LSU, 1955. Research administration, coastal zone management, resource policy.
Flora C. Wang, Professor; Ph.D., Stanford, 1963. Resources planning and management, wetland hydrology and coastal hydrodynamics, mathematical modeling and computer simulation.
Charles A. Wilson, Associate Professor; Ph.D., South Carolina, 1984. Fishery development, fisheries biology, artificial reef development.
William J. Wiseman Jr., Professor; Ph.D., Johns Hopkins, 1969. Shelf and estuarine dynamics, arctic coastal processes, geophysical signal processing.

Associated Faculty

Michael J. Dagg, Adjunct Professor (Louisiana Universities Marine Consortium); Ph.D., Washington (Seattle), 1975. Biological oceanography, food chain dynamics.
Quay Dortch, Adjunct Assistant Professor (Louisiana Universities Marine Consortium); Ph.D., Washington (Seattle), 1980. Phytoplankton ecology, nitrogen cycling, biochemical indicators.
James P. Geaghan, Associate Professor of Experimental Statistics; Ph.D., North Carolina State, 1980. Population dynamics and trend analysis.
Charles W. Lindau, Associate Professor; Ph.D., Texas A&M, 1980. Stable isotopes, nitrogen transformations, sediment chemistry.
Nancy N. Rabalais, Adjunct Assistant Professor (Louisiana Universities Marine Consortium); Ph.D., Texas, 1983. Continental shelf ecosystems, benthic ecology, biological oceanography.

SECTION 4: MARINE SCIENCES/OCEANOGRAPHY

MASSACHUSETTS INSTITUTE OF TECHNOLOGY/ WOODS HOLE OCEANOGRAPHIC INSTITUTION

Joint Program in Oceanography/Oceanographic Engineering

Program of Study

The Massachusetts Institute of Technology (MIT) and the Woods Hole Oceanographic Institution (WHOI) offer joint doctoral and professional degrees in oceanography and oceanographic engineering. The Joint Program leads to a single degree awarded by both institutions. Graduate study in oceanography encompasses virtually all of the basic sciences as they apply to the marine environment: physics, chemistry, geology, geophysics, and biology. Oceanographic engineering allows for concentration in the major engineering fields of civil, mechanical, electrical, and ocean engineering. The graduate programs are administered by joint MIT/WHOI committees drawn from the faculty and staff of both institutions. The Joint Program involves several departments at MIT: Earth, Atmospheric, and Planetary Sciences and Biology in the School of Science; and Civil Engineering, Electrical Engineering and Computer Science, Mechanical Engineering, and Ocean Engineering in the School of Engineering. WHOI departments are Physical Oceanography, Biology, Chemistry, Geology and Geophysics, and Applied Ocean Physics and Engineering. Upon admission, students register in the appropriate MIT department and at WHOI simultaneously and are assigned academic advisers at each institution. The usual steps to a doctoral degree are entering the program the summer preceding the first academic year and working in a laboratory at WHOI, following an individually designed program in preparation for a general (qualifying) examination to be taken before the third year, submitting a dissertation of significant orginal theoretical or experimental research, and conducting a public oral defense of the thesis. The normal time it takes to achieve the doctoral degree is about five years from the bachelor's degree. Students entering with a master's degree in the field may need less time. There are few formal course requirements, but each student is expected to become familiar with the principal areas of oceanography in addition to demonstrating a thorough knowledge of at least one major field. Subjects, seminars, and opportunities for research participation are offered at both MIT and WHOI. Courses and seminars are supplemented by cross-registration privileges with Harvard, Brown, and the Boston University Marine Program. Students also have the opportunity to participate in oceanographic cruises during graduate study.

Research Facilities

A broad spectrum of equipment and facilities are available. The wide-ranging deep-sea research vessels at WHOI include *Oceanus*, *Atlantis II*, and *Knorr*. In addition, the deep-diving submersible *ALVIN*, which is carried on *Atlantis II*, is operated by WHOI, as are several smaller coastal vessels. Both MIT and WHOI utilize the latest developments in computer technology, from personal computers to large multiuser access systems. A microwave link between MIT and WHOI provides interactive video-voice transmission for classes and a high-speed data link for research. Broad-based engineering design and support shop facilities (machining, electronics) are available at both MIT and WHOI. There are over twenty libraries at MIT containing more than 2 million volumes. Cooperative arrangements with other libraries in the Boston area provide students with access to substantial research collections. WHOI library facilities are shared with the Marine Biological Laboratory and are supplemented by collections of the Northeast Fisheries Center and the U.S. Geological Survey, all located in Woods Hole.

Financial Aid

Research assistantships are available to most entering graduate students in the Joint Program and are usually awarded on a full-year basis. Such awards, as well as a few special fellowships, cover full tuition and provide a stipend adjusted annually to current living expenses.

Cost of Study

Because tuition and stipend are usually paid for students in good standing, the main costs to the student are for medical insurance, books, and supplies.

Cost of Living

Place of residence is determined by the student's selected program of study and research interests. Graduate students traditionally live off campus at both MIT and WHOI, although there is some graduate housing at both campuses. Housing and living costs tend to be expensive in both the Cambridge and Woods Hole areas, although reasonable housing is available. Estimated twelve-month living costs in 1990–91 were $29,300 for a single graduate student and $35,300 for a married student.

Student Group

There are approximately 125 graduate students registered in the Joint Program in any given year. Among the five disciplines, they are divided as follows: physical oceanography (35), marine geology and geophysics (20), biological oceanography (20), chemical oceanography (15), and oceanographic engineering (35).

Location

MIT's 146-acre campus extends more than a mile along the Cambridge side of the Charles River, overlooking downtown Boston. Metropolitan Boston offers diverse recreational and cultural opportunities. New England beaches and mountains are within easy reach. WHOI is in the small fishing village of Woods Hole, about 80 miles southeast of Boston and 20 miles west of Hyannis.

The University and The Institution

MIT is an independent, coeducational, privately endowed university. It is broadly organized into five academic schools: Architecture and Planning, Engineering, Humanities and Social Sciences, Management, and Science. Within these schools there are twenty-two academic departments. Total enrollment is approximately 9,500 divided almost evenly between undergraduate and graduate students. The MIT faculty numbers approximately 1,000 with a total teaching staff of 1,860. WHOI is one of the largest independent, unaffiliated oceanographic institutions and research fleet operators in the world. There is a staff of approximately 900 scientists, engineers, technicians, research vessel crews, and support personnel.

Applying

Application for admission to the Joint Program is made on the MIT graduate school application forms. Complete application files, including college transcripts, three letters of recommendation, test scores, and the application fee should be filed no later than January 15 for admission beginning in June or September. The General Test and one Subject Test of the GRE are required. The TOEFL is required of all international students whose schooling has not been predominantly in English.

Admission is offered on a competitive basis to those who appear most likely to benefit from the Joint Program. Notification of admission decisions are sent out prior to March 31.

Correspondence and Information

Dean of Graduate Studies
Education Office
Woods Hole Oceanographic Institution
Woods Hole, Massachusetts 02543
Telephone: 508-457-2000 Ext. 2200

or

MIT/WHOI Joint Program Office
54-911
Massachusetts Institute of Technology
Cambridge, Massachusetts 02139
Telephone: 617-253-7544

SECTION 4: MARINE SCIENCES/OCEANOGRAPHY

Massachusetts Institute of Technology/Woods Hole Oceanographic Institution

THE DEPARTMENTS AND THEIR RESEARCH

BIOLOGICAL OCEANOGRAPHY

David H. Marks, Ph.D., Chairman, Civil Engineering Department, MIT.
Richard O. Hynes, Ph.D., Chairman, Biology Department, MIT.
Peter H. Wiebe, Ph.D., Chairman, Biology Department, WHOI.

At WHOI: Phytoplankton and zooplankton ecology; population biology; the kinetics of nutrient assimilation; natural history and biology of oceanic fishes; biological effects of hydrocarbons and marine pollutants; molecular toxicology in marine species; theoretical and experimental population ecology; estuarine and salt marsh ecology and biogeochemistry; ecology of deep-sea and coastal benthos; ecology and physiology of gelatinous zooplankton; microbial ecology and biochemistry; development and reproductive biology of marine invertebrates; physiology of marine borers; microbial and invertebrate symbioses; communication systems of cetaceans.

At MIT: Molecular biology of cells and differentiation; the synthesis, structure, and shape of macromolecules; cellular and molecular immunology; various aspects of gene expression; toxicology; nutrition.

CHEMICAL OCEANOGRAPHY

David H. Marks, Ph.D., Chairman, Civil Engineering Department, MIT.
Thomas H. Jordan, Ph.D., Chairman, Department of Earth, Atmospheric, and Planetary Sciences, MIT.
Frederick L. Sayles, Ph.D., Chairman, Chemistry Department, WHOI.

Water column geochemistry: natural radionuclides, trace metals, rare earth elements, lipids, environmental quality, pigments, organic nitrogen cycle, volatiles, stable carbon oxygen isotopes, radiocarbon, helium/tritium, noble gases and artificial radionuclides.
Sedimentary geochemistry: major elements, radionuclides, stable isotopes, radiocarbon, petroleum, paleoceanographic applications.
Seawater-basalt interactions: major and trace elements, stable isotopes, chemistry of hydrothermal solutions.
Atmospheric chemistry: carbon dioxide cycle, photochemistry, inorganic nitrogen cycle, trace gases, aerosol organic geochemistry, sea/air exchange.

MARINE GEOLOGY AND GEOPHYSICS

Thomas H. Jordan, Ph.D., Chairman, Department of Earth, Atmospheric, and Planetary Sciences, MIT.
G. Michael Purdy, Ph.D., Chairman, Geology and Geophysics Department, WHOI.

Micropaleontological biostratigraphy; planktonic and benthonic foraminifera; paleoceanography; paleobiogeography; benthic boundary-layer processes; paleocirculation and paleoecology; igneous petrology and volcanic processes; crustal structure and tectonics; marine magnetic anomalies; heat flow of the ocean floor; upper mantle petrology; seismic stratigraphy; fractionation processes of stable isotopes and stable isotope stratigraphy; metamorphosis of high-strain zones; gravity; observational and theoretical reflection and refraction seismology; earthquake seismology; relative and absolute plate motions; coastal processes; marine sedimentation.

OCEANOGRAPHIC ENGINEERING

David H. Marks, Ph.D., Chairman, Civil Engineering Department, MIT.
Paul L. Penfield Jr., Ph.D., Chairman, Electrical Engineering and Computer Sciences Department, MIT.
David N. Wormley, Ph.D., Chairman, Mechanical Engineering Department, MIT.
T. Francis Ogilvie, Ph.D., Chairman, Ocean Engineering Department, MIT.
Albert J. Williams III, Ph.D., Chairman, Applied Ocean Physics and Engineering Department, WHOI.

Optical instrumentation, laser velocimetry; volcanic, tectonic, and hydrothermal processes; deep submergence systems (imaging, control, robotics); underwater acoustics (acoustic tomography, scattering, remote sensing, Arctic acoustics, bottom acoustics propagation through the ocean interior and sediments, array design); buoy and mooring engineering; ocean instrumentation; signal processing theory; fluid dynamics, sediment transport, nearshore processes, bottom boundary layer and mixed layer dynamics; turbulence, wave prediction, numerical modeling; seismic profiling; data acquisition and communication systems, microprocessor-based instrumentation; fiber optics; sonar systems; march ecology; ship design; offshore structures; material science; groundwater flow.

PHYSICAL OCEANOGRAPHY

Thomas H. Jordan, Ph.D., Chairman, Department of Earth, Atmospheric, and Planetary Sciences, MIT.
James R. Luyten, Ph.D., Chairman, Physical Oceanography Department, WHOI.

General circulation: distribution of tracer fields, models of idealized gyres, abyssal circulation, heat transport.
Air-sea interaction: water mass transportation, upper ocean response to atmospheric forcing, equatorial ocean circulation.
Shelf dynamics: coastal upwelling and fronts, coastal-trapped waves, deep ocean-shelf exchange.
Mesoscale processes: Gulf Stream Rings, oceanic fronts, barotropic and baroclinic instability, eddy-mean interactions.
Small-scale processes: double diffusion, intrusion, internal waves, convection.

SECTION 4: MARINE SCIENCES/OCEANOGRAPHY

NORTH CAROLINA STATE UNIVERSITY
Department of Marine, Earth and Atmospheric Sciences

Programs of Study

The department offers M.S. and Ph.D. degrees in oceanography, meteorology, geology, and geophysics.

In oceanography, students specialize in biological, chemical, geological, or physical oceanography. In the biological area, research topics are in benthic, plankton, or invertebrate physiological ecology. Research in the chemical area exists in the study of organic and inorganic processes in estuarine, coastal, and deep-sea environments. Emphasis in geological oceanography is in sedimentology and micropaleontology. In physical oceanography, research topics include the study of the dynamics of estuarine, shelf-slope, and deep-sea waters.

In meteorology, research topics exist in modeling and parameterizing the planetary atmospheric boundary layer, in physically and theoretically modeling dispersion over complex terrain, in air-sea interaction, in atmospheric chemistry, and in climate dynamics. Other research areas are those of cloud-aerosol interaction, cloud chemistry and acid rain deposition, plant-atmosphere interaction, severe localized storm systems, and mesoscale phenomena and processes related to East Coast fronts and cyclones.

In earth science, research topics are in the areas of hard-rock and soft-rock geology, tectonics, sedimentary geochemistry, pure and applied geophysics, and interdisciplinary research in groundwater and remote sensing. In hard-rock geology the emphasis is on igneous and metamorphic petrology. Soft-rock studies span both recent and ancient detrital deposits, including economic deposits, with field-based studies of facies relationships and associated depositional environments.

The M.S. degree program requires 30 semester credit hours of course work, a research thesis, and a final oral examination. A nonthesis option is available to students on leave from government or industry. The Ph.D. program requires at least 18 credit hours beyond the M.S. degree, as well as a thesis, preliminary written and oral examinations, and a dissertation defense.

Research Facilities

Jordan Hall, the department's home, is a modern structure dedicated to research in natural resources, which has been specially designed to accommodate department research laboratories. Modern facilities currently exist in all program areas. Students have access to the million-volume D. H. Hill Library and the University Computing Center IBM 3090 and CRAY Y-MP computers. In addition, the department operates a VAX-8700 computer and several microcomputers. Other specialized departmental equipment includes a Technavia WEFAX satellite receiver, a Finnigan MAT 251 Ratio Mass Spectrometer, a McIDAS workstation, an electron microprobe, an X-ray diffractometer, and an atomic absorption spectrometer. Elsewhere on campus, students have access to electron microscopes, ion microprobes, and a nuclear reactor for neutron activation analyses. Students also have access to the EPA Fluid Modeling Facility tow tank and wind tunnel in nearby Research Triangle Park. The department is a member of the Duke/UNC consortium, which operates the 131-foot R/V *Cape Hatteras*, a vessel used for both educational and research cruises. Students have also used facilities at the Duke University Marine Laboratory and the National Marine Fisheries Laboratory, both at Pivers Island, North Carolina. The department is a member of the University Corporation for Atmospheric Research, which provides access to the computing and observational systems of the National Center for Atmospheric Research.

Financial Aid

A number of teaching and research assistantships are available on a competitive basis. The stipends, for 20 hours of service per week, range in 1991–92 from $689 to $916 per month on a nine-month basis for teaching assistants and a twelve-month basis for research assistants. A limited number of out-of-state tuition waivers are available to first-year students on assistantships. These offset the difference between the out-of-state and in-state tuition rates, which is currently $4962 per year for a full course load.

Cost of Study

Tuition and fees for 1991–92 for a full course load of 9 or more credits are $575 per semester for in-state students and $3056 per semester for out-of-state students. U.S. citizens may be able to establish North Carolina residence after one year and then be eligible for in-state tuition rates.

Cost of Living

The University has graduate dormitory rooms that cost about $1100 per semester. Married student housing is available at King Village for about $230 per month for a one-bedroom apartment. Off-campus housing is available starting at about $250 per month.

Student Group

University enrollment reached 26,683 in fall 1990, with an undergraduate enrollment of 18,294, a graduate enrollment of 4,142, and a continuing-education enrollment of 4,247. The department has 100 undergraduate majors and 96 graduate students, with 23 in marine science, 23 in earth science, and 50 in atmospheric science. Seventeen of the graduate students are women.

Location

Raleigh, a modern growing city with a population of over 200,000 in a metropolitan area of over 600,000, is situated in rolling terrain in the eastern Piedmont near the upper Coastal Plain. Raleigh, the state capital, is one vertex of the Research Triangle area, with Durham and Chapel Hill the other vertices. Numerous colleges and industrial and government laboratories are located in the Triangle area, which each year also attracts some of the world's foremost symphony orchestras and ballet companies. Located within a 3-hour drive of the campus are both the seashore and the mountains, which offer many opportunities for skiing, hiking, swimming, boating, and fishing.

The University

North Carolina State University, the state's land-grant and chief technological institution, recently celebrated its centennial year. A graduate faculty of 1,400 and more than 100 major buildings are located on the 623-acre main campus. The 780-acre Centennial Campus, which has just been acquired adjacent to the main campus, ensures room for future expansion. The University is organized into nine colleges plus the Graduate School. The department is one of six in the College of Physical and Mathematical Sciences.

Applying

For fall admission, the completed application form, transcripts, recommendation forms, GRE scores, and TOEFL scores (for foreign students) should be received no later than March 1 to ensure full consideration for assistantship support. Applications for summer and spring admission are also considered, but assistantship support is less likely.

Correspondence and Information

Graduate Administrator
Department of Marine, Earth and Atmospheric Sciences
North Carolina State University
Box 8208
Raleigh, North Carolina 27695-8208
Telephone: 919-737-7837

Peterson's Guide to Graduate Programs in the Physical Sciences and Mathematics 1992

North Carolina State University

THE FACULTY AND THEIR RESEARCH

Charles E. Anderson, Professor; Ph.D., MIT, 1960. Severe weather, cloud dynamics.
Viney Aneja, Research Associate Professor; Ph.D., North Carolina State, 1977. Atmospheric chemistry.
S. Pal Arya, Professor; Ph.D., Colorado State, 1968. Micrometeorology, atmospheric turbulence and diffusion, air-sea interaction.
Michael G. Bevis, Associate Professor; Ph.D., Cornell, 1982. Geophysics, tectonics, tectonophysics.
Neal E. Blair, Associate Professor; Ph.D., Stanford, 1980. Chemical oceanography, biogeochemistry, organic geochemistry.
Roscoe R. Braham, Scholar in Residence; Ph.D., Chicago, 1951. Cloud physics, thunderstorms, weather modification.
Steven Businger, Assistant Professor; Ph.D., Washington (Seattle), 1986. Synoptic/dynamic meteorology.
Victor V. Cavaroc, Professor; Ph.D., LSU, 1969. Sedimentary petrology/petrography, lithostratigraphy, coal stratigraphy.
David E. Checkley, Research Assistant Professor; Ph.D., California, San Diego, 1978. Zooplankton and larval fish ecology, nitrogen cycling, effects of storms on larval fish nutrition.
Jerry M. Davis, Professor; Ph.D., Ohio State, 1971. Agricultural meteorology, climatology, statistical meteorology, planetary boundary layer.
David J. DeMaster, Professor; Ph.D., Yale, 1979. Marine geochemistry and radio chemistry in the nearshore and deep-sea environments.
Ronald V. Fodor, Professor; Ph.D., New Mexico, 1972. Igneous petrology, volcanoes, meteorites.
James Hibbard, Assistant Professor; Ph.D., Cornell, 1988. Structural geology.
Thomas S. Hopkins, Visiting Professor; Ph.D., Washington (Seattle), 1971. Physical oceanography.
Gerald S. Janowitz, Professor and Graduate Administrator; Ph.D., Johns Hopkins, 1967. Geophysical fluid mechanics, continental shelf and ocean circulation.
Daniel L. Kamykowski, Professor; Ph.D., California, San Diego, 1973. Effects of physical factors on phytoplankton behavior and physiology, global plant nutrient distributions.
Michael M. Kimberley, Associate Professor; Ph.D., Princeton, 1974. Sedimentary geochemistry, sedimentary ore deposits, chemistry of natural and polluted water.
Charles E. Knowles, Associate Professor; Ph.D., Texas A&M, 1970. Estuarine and coastal processes, surface gravity wave measurements.
Elana L. Leithold, Assistant Professor; Ph.D., Washington (Seattle), 1987. Nearshore and shelf sedimentation and stratigraphy, sediment transport.
Lisa A. Levin, Associate Professor; Ph.D., California, San Diego, 1982. Benthic ecology, invertebrate life histories and dispersal, deep-sea community structure.
Yuh Lang Lin, Assistant Professor; Ph.D., Yale, 1984. Modeling of mesoscale atmospheric dynamics.
Thomas F. Malone, University Distinguished Scholar; Sc.D., MIT, 1946. Meteorology.
John M. Morrison, Associate Professor; Ph.D., Texas A&M, 1977. Descriptive physical oceanography, general ocean circulation, air-sea interaction/climatic problems.
Leonard J. Pietrafesa, Professor and Head; Ph.D., Washington (Seattle), 1973. Estuarine and continental margin physical processes, seismology.
Sethu S. Raman, Professor; Ph.D., Colorado State, 1972. Air-sea interactions, boundary layer meteorology and air pollution.
Allen J. Riordan, Associate Professor; Ph.D., Wisconsin, 1977. Satellite meteorology, Antarctic meteorology.
Vin K. Saxena, Professor; Ph.D., Rajasthan, 1969. Cloud physics, acid precipitation, weather modification.
Frederick Semazzi, Associate Professor; Ph.D., Nairobi, 1983. Climate dynamics.
Ping Tung Shaw, Assistant Professor; Ph.D., Woods Hole/MIT, 1982. Shelf-slope physical oceanography and Lagrangian analysis.
William J. Showers, Associate Professor; Ph.D., Hawaii, 1982. Stable-isotope geochemistry, paleoceanography, micropaleontology, environmental monitoring, geoarchaeology.
Stephen Snyder, Assistant Professor; Ph.D., South Florida, 1991. Geological oceanography.
J. Alexander Speer, Assistant Professor; Ph.D., Virginia Tech, 1976. Mineralogy, igneous and metamorphic petrology.
Edward F. Stoddard, Associate Professor and Associate Head; Ph.D., UCLA, 1976. Metamorphic petrology, silicate mineralogy, Piedmont geology.
Gerald F. Watson, Associate Professor; Ph.D., Florida State, 1971. Synoptic meteorology, weather analysis and forecasting.
Charles W. Welby, Professor; Ph.D., MIT, 1952. Hydrogeology/groundwater pollution, environmental geology, land-use planning, invertebrate paleontology, petroleum exploration, shoreline erosion/coastal processes.
Donna L. Wolcott, Visiting Associate Professor; Ph.D., Berkeley, 1972. Physiological ecology of terrestrial crabs.
Thomas G. Wolcott, Professor; Ph.D., Berkeley, 1971. Physiological ecology of marine invertebrates, biotelemetry.

Jordan Hall, home of the department.

SECTION 4: MARINE SCIENCES/OCEANOGRAPHY

STATE UNIVERSITY OF NEW YORK AT STONY BROOK

Marine Sciences Research Center
Programs in Marine Environmental Sciences and Coastal Oceanography

Programs of Study

The master's degree program in marine environmental sciences is designed to prepare students for effective careers in research, management, environmental protection, and resource development of the coastal zone. It consists of a rigorous interdisciplinary curriculum. Students must successfully complete an approved course of study of 30 graduate credits, including core courses in biological, chemical, geological, and physical oceanography. A research thesis is required. All requirements for the M.S. must be completed within three years of admission.

The doctoral program in coastal oceanography prepares students to formulate and attack coastal oceanographic problems effectively on theoretical and applied levels through interdisciplinary training in biological, chemical, geological, and physical oceanography. A student's entire program is carefully tailored to his or her needs. At the core of each program is an apprenticeship to one or several key faculty members. Students advance to candidacy after completion of course work, a departmental examination, demonstration of proficiency in a foreign language, participation in oceanographic research cruises, one year in residence, one semester of teaching experience, and an oral qualifying examination. All requirements for the Ph.D. must be satisfied within five years after completing 24 hours of graduate courses in the Stony Brook department or program in which the student is to receive the degree.

The faculty of the Center conducts a broad range of basic and applied research in all facets of coastal oceanography throughout much of the world. The focus of the programs is on the coastal ocean, and support comes from many federal and state agencies as well as from private foundations. Students at both the master's and doctoral degree levels are active participants in all phases of the Center's research.

Research Facilities

The Marine Sciences Research Center (MSRC) has 7,969 square meters of laboratory and office space that is well equipped for most analyses and includes excellent computing facilities. The equipment and facilities on campus, at other SUNY units, and at Brookhaven National Laboratory and Cold Spring Harbor Laboratory are available by arrangement. The Center has a laboratory with aquarium facilities at the Flax Pond salt marsh, approximately 7 kilometers from campus. The 1-square-kilometer salt marsh is managed by the Center. The Center operates a 20-meter research vessel, the R/V *Onrust*, outfitted for oceanographic sampling and also maintains a fleet of small boats.

Financial Aid

Various forms of financial aid are available through the Center. Graduate teaching and research assistantships carry stipends of up to $8850 for the academic year 1991–92 and $3885 for the summer. Sea Grant traineeships provide $12,500 for the calendar year 1992. Inter-Campus Exchange fellowships carry a variable stipend, and Graduate Council fellowships provide $10,000 per academic year in 1991–92. All assistantships and fellowships carry a waiver of tuition. Work-study programs are available through the Financial Aids Office. Federal and state loans are also available.

Cost of Study

For 1991–92, New York State residents pay tuition of $1225 per semester full-time or $102.50 per credit part-time. Tuition for out-of-state residents is $2883 per semester full-time or $242.50 per credit part-time. A University fee of $12.50 per semester and an activity fee of $18 per semester are also charged.

Cost of Living

In 1991–92, living costs excluding tuition and fees total approximately $10,350 for twelve months if the student is single and lives on campus. Furnished apartments for graduate students are available. Apartments have one, two, or three bedrooms; each bedroom can accommodate 2 persons. Costs range from $372 to $816 per month per apartment (depending on the number of occupants per apartment) and include all utilities except telephone. Costs are subject to change without notice. Rental is for twelve months. There are optional board plans available on campus. A number of private homes, rooms, and apartments are also available off campus at a wide range of rents.

Student Group

There are 63 students enrolled in the M.S. program and 47 in the Ph.D. program. Students come from all parts of the United States and throughout the world. The student body includes 39 women and 46 married students. Nearly all of the full-time students receive financial aid.

To date, 275 students have received M.S. degrees and 45 have received Ph.D. degrees. Virtually all are employed in oceanographic research institutions; various federal, state, and local environmental protection and management units; and industry. Employment prospects continue to look favorable.

Location

Stony Brook is located 96 kilometers east of Manhattan on the wooded North Shore of Long Island, within a few kilometers of picturesque villages, harbors, and beaches. The cultural, scientific, and commercial resources of New York City are readily accessible by road and rail.

The Center

The Marine Sciences Research Center is the oceanographic research center of the State University of New York. Situated on Long Island, it is ideally located for studies of diverse coastal environments, including estuaries, lagoons, salt marshes, barrier islands, and continental shelf waters. The Center offers the only SUNY graduate degree programs in marine environmental sciences and coastal oceanography.

Two features that distinguish the Marine Sciences Research Center from other leading oceanographic institutions are its clear focus on the coastal ocean and the effectiveness with which the Center's staff members have translated the results of their research and that of others into forms directly applicable to the resolution of important societal problems.

Applying

Admission requires a B.S. in basic science with introductory courses in other sciences (biology, chemistry, physics, and geology) and mathematics through calculus (B average). The General Test of the Graduate Record Examinations is required. Application deadlines are March 1 for the fall and October 1 for the spring. Application materials are available from the Graduate Programs Office.

The State University of New York at Stony Brook is an affirmative action/equal opportunity educator and employer.

Correspondence and Information

Graduate Programs Office
Marine Sciences Research Center
State University of New York at Stony Brook
Stony Brook, New York 11794
Telephone: 516-632-8681

Peterson's Guide to Graduate Programs in the Physical Sciences and Mathematics 1992

SECTION 4: MARINE SCIENCES/OCEANOGRAPHY

State University of New York at Stony Brook

THE FACULTY AND THEIR RESEARCH

Josephine Y. Aller, Associate Research Professor; Ph.D., USC, 1975. Marine benthic ecology, invertebrate zoology, marine microbiology, biogeochemistry.

Robert C. Aller, Professor; Ph.D., Yale, 1977. Marine geochemistry, marine animal–sediment relations.

Henry J. Bokuniewicz, Professor; Ph.D., Yale, 1976. Nearshore transport processes, coastal sedimentation, marine geophysics.

Malcolm J. Bowman, Professor; Ph.D., Saskatchewan, 1971. Coastal ocean and estuarine dynamics.

Vincent T. Breslin, Assistant Research Professor; Ph.D., FIT, 1986. Stabilized waste interaction with the marine environment, metal leachability from particulate and stabilized combustion residues, trace metal geochemistry.

V. Monica Bricelj, Assistant Professor; Ph.D., SUNY at Stony Brook, 1984. Molluscan physiological ecology, benthic ecology.

Bruce J. Brownawell, Assistant Professor; Ph.D., MIT/Woods Hole, 1986. Biogeochemistry of organic pollutants in seawater and groundwater.

Edward J. Carpenter, Professor; Ph.D., North Carolina State, 1969. Nitrogen cycling among plankton and ambient seawater, phytoplankton and zooplankton ecology.

Harry H. Carter, Professor Emeritus; M.S., Scripps, 1948. Estuarine and coastal dynamics, turbulent diffusion.

Robert M. Cerrato, Assistant Professor; Ph.D., Yale, 1980. Benthic ecology, population and community dynamics.

J. Kirk Cochran, Professor; Ph.D., Yale, 1979. Marine geochemistry, use of radionuclides as geochemical tracers, diagenesis of marine sediments.

David O. Conover, Associate Professor; Ph.D., Massachusetts, 1981. Ecology of fish, fisheries biology.

Elizabeth M. Cosper, Associate Research Professor; Ph.D., CUNY, City College, 1981. Phytoplankton physiology and ecology, resistance of microalgae to pollutants.

Robert K. Cowen, Assistant Professor; Ph.D., California, San Diego (Scripps), 1985. Fishery oceanography, nearshore fish populations, fish ecology.

Nicholas S. Fisher, Professor; Ph.D., SUNY at Stony Brook, 1974. Marine phytoplankton physiology and ecology, biochemistry of metals, marine pollution.

Roger D. Flood, Associate Professor; Ph.D., MIT, 1978. Marine geology, sediment dynamics, continental margin sedimentation.

Valrie A. Gerard, Associate Professor; Ph.D., California, Santa Cruz, 1976. Marine macrophyte ecology and physiology.

Herbert Herman, Professor (Department of Engineering and Applied Sciences with joint appointment in MSRC); Ph.D., Northwestern, 1961. Ocean engineering, undersea vehicles, marine materials.

Richard K. Koehn, Professor (Department of Ecology and Evolution with joint appointment in MSRC) and Dean of the Division of Biological Sciences; Ph.D., Arizona, 1967. Evolutionary genetics of natural populations and evolution of physiological variation, using marine bivalves and mice.

Lee E. Koppelman, Professor; Ph.D., Cornell, 1970. Coastal zone management, planning, and policy studies.

Cindy Lee, Professor; Ph.D., California, San Diego (Scripps), 1975. Marine geochemistry of organic compounds, organic and inorganic nitrogen-cycle biochemistry.

Darcy J. Lonsdale, Assistant Professor; Ph.D., Maryland, 1979. Zooplankton ecology, with special interest in physiology; life history studies.

Glenn R. Lopez, Associate Professor; Ph.D., SUNY at Stony Brook, 1976. Benthic ecology, animal-sediment interactions.

Kamazima Lwiza, Assistant Research Professor; Ph.D., North Wales, 1991. Coastal ocean circulation, tides and tidal fronts, mixing.

James E. Mackin, Associate Professor; Ph.D., Chicago, 1983. Geochemistry of suspended sediment/solution interactions.

John L. McHugh, Professor Emeritus; Ph.D., UCLA, 1950. Fishery management, fishery oceanography; whales and whaling.

William J. Meyers, Associate Professor (Department of Earth and Space Sciences with joint appointment in MSRC); Ph.D., Rice, 1973. Carbonates, sedimentology.

Steven Morgan, Assistant Professor; Ph.D., Maryland College Park, 1986. Larval ecology and evolution of life histories.

Charles Nittrouer, Professor; Ph.D., Washington (Seattle), 1978. Geological oceanography, continental margin sedimentation.

Akira Okubo, Professor; Ph.D., Johns Hopkins, 1963. Oceanic diffusion, animal dispersal, mathematical ecology.

Hartmut Peters, Assistant Professor; Ph.D., Kiel (Germany), 1981. Turbulence and mixing, internal waves, equatorial circulation, coastal and estuarine circulation.

Donald W. Pritchard, Professor Emeritus; Ph.D., Scripps, 1951. Estuarine and coastal dynamics, coastal zone management.

Sheldon Reaven, Associate Professor (Department of Technology and Society with joint appointment in MSRC); Ph.D., Berkeley, 1975. Energy and environmental problems and issues, especially waste management.

Frank J. Roethel, Lecturer; Ph.D., SUNY at Stony Brook, 1982. Environmental chemistry, behavior of coal waste in the environment; solution chemistry.

J. R. Schubel, Professor; Ph.D., Johns Hopkins, 1968. Coastal sedimentation, suspended sediment transport, coastal zone management.

Mary I. Scranton, Associate Professor; Ph.D., MIT, 1977. Marine geochemistry, biological-chemical interactions in seawater.

Lawrence B. Slobodkin, Professor (Department of Ecology and Evolution with joint appointment in MSRC); Ph.D., Yale, 1951. Theoretical ecology, marine ecology.

R. Lawrence Swanson, Adjunct Professor and Director of the Waste Management Institute; Ph.D., Oregon State, 1971. Recycling and reuse of waste materials, waste management, waste disposal.

Gordon Taylor, Assistant Professor; Ph.D., USC, 1983. Marine microbial ecology and biogeochemistry.

Dong Ping Wang, Professor; Ph.D., Miami (Florida), 1975. Coastal ocean dynamics.

Franklin F. Y. Wang, Professor; Ph.D., Illinois, 1956. Ocean engineering, ocean structures, energy.

Peter K. Weyl, Professor; Ph.D., Chicago, 1953. Coastal zone planning, physical oceanography, paleoceanography.

Robert E. Wilson, Associate Professor; Ph.D., Johns Hopkins, 1973. Estuarine and coastal ocean dynamics.

Peter M. J. Woodhead, Research Professor; B.S., Durham (England), 1953. Behavior and physiology of fish, coral reef ecology, ocean energy conversion systems.

Charles F. Wurster, Associate Professor; Ph.D., Stanford, 1957. Effects of chlorinated hydrocarbons on phytoplankton communities.

Jeannette Yen, Assistant Professor; Ph.D., Washington (Seattle), 1982. Marine zooplankton ecology, predator-prey interactions, sensory perception and lipid metabolism of copepods.

Jonathan Zehr, Assistant Research Professor; Ph.D., California, Davis, 1985. Aquatic microbiology.

UNIVERSITY OF CALIFORNIA, SAN DIEGO

Department of the Scripps Institution of Oceanography

Programs of Study

The Department of the Scripps Institution of Oceanography offers programs of study leading to Ph.D. degrees in earth sciences, marine biology, and oceanography. A student's work is normally concentrated in one of several curricular programs within the department: biological oceanography, geochemistry and marine chemistry, marine biology, geological sciences, geophysics, physical oceanography, and applied ocean science. The interdisciplinary nature of research in marine and earth sciences is emphasized; students are encouraged to take courses in several programs and from various UCSD departments and to select research projects of an interdisciplinary character. Programs of study vary widely among the curricular groups, but generally first-year students are expected to enroll in core courses that cover physical, geological, chemical, and biological oceanography and in other courses recommended by the student's guidance committee or faculty adviser.

Doctoral candidates are normally required to take a department examination not later than the second year of study. This is primarily an oral exam, although written parts may be included. The student is required to demonstrate quantitative and analytical comprehension of the required subject material. After the student has passed the department examination and has completed an appropriate period of additional study, a doctoral committee is appointed to determine the student's qualifications for independent research. The oral qualifying examination is usually taken in the third year. It should be noted that the nature of this qualifying exam varies between curricular groups. Candidates for the Ph.D. degree are required to submit a dissertation and pass a final examination in which the thesis is publicly defended.

Research Facilities

Special resources and facilities include the SIO Library, which houses more than 189,000 volumes; radio station WWD, operated by the U.S. National Marine Fisheries Service; the 1,090-foot-long Scripps pier, which houses apparatus for a number of serial oceanographic observations and is used as a landing place for skiffs; the saltwater system, which provides clean seawater to biological laboratories and the aquarium; an underwater area for research and for collecting offshore from the institution; deep-sea sediment cores from several thousand widely scattered localities and original echograms and underway geophysical data along several hundred thousand miles of ships' tracks in the oceans of the world; an oceanographic data archive of some half a million bathythermograph observations; electron microprobe laboratories; nine mass spectrometers; the Piñon Flat Geophysical Observatory and the Anza Seismographic Array; several thousand samples of seawater from the world's oceans; a high-resolution scanning electron microscope laboratory; the Scripps fish collection of more than 2 million specimens of some 3,200 species of marine fish; and the large oceanographic collection of benthic invertebrates, plankton, rocks, and cores. There is access to the San Diego Supercomputer Center and a shipboard computer group with computers on several of the larger ships in the fleet as well as another computer on the SIO campus. The institution operates four ships specially fitted for oceanographic research and two research platforms.

Financial Aid

The majority of students are supported by research assistantships funded through the grants and contracts of the faculty and research staff. Twelve-month research assistantships for 1991–92 carry stipends ranging from $12,400 to $15,500 and include tuition and fee remission. Each year, various fellowships and scholarships within the department are available on a competitive basis, as are teaching fifteen to twenty assistantships.

Cost of Study

In 1990–91, the registration fees for California residents were $2041.50 and, for nonresidents, tuition and fees were $7957.50. Tuition and fees are remitted for research assistants. Registration fees include health insurance.

Cost of Living

There are a limited number of University apartments available. In 1990–91, rents ranged from $340 per month for a studio to $579 per month for a three-bedroom family apartment. Most students live in apartments or share houses nearby, but rents are considerably higher near the Scripps campus.

Student Group

There are approximately 180 graduate students enrolled in all disciplines offered by the department. They come from all areas of the United States, and about 25 percent of the student population are from foreign countries. Each year, 25–30 students graduate. Most of them remain in a university environment, becoming postdoctoral students and assistant professors or accepting research positions.

Location

The Scripps Institution of Oceanography is located on the coastline 15 miles north of downtown San Diego and just a few miles from the main University of California, San Diego, campus.

The University

San Diego is one of the newest campuses of the University of California System and has become one of the major research institutions in the United States, ranking in the top ten in the nation in the amount of federal dollars it receives for scientific research. In fall 1990, the total enrollment at UCSD, including undergraduate, graduate, and medical school students, was over 17,800.

Applying

Candidates for admission should have a bachelor's or master's degree in one of the physical, biological, or earth sciences. In some cases, a degree in mathematics or engineering science is accepted. Scores on the General Test of the Graduate Record Examinations are required and, for marine biology only, the Subject Test in biology. Students whose native language is not English are required to submit scores on the TOEFL. Test scores should be sent to the department. Application materials are available from the department. Students are admitted for the fall quarter only. The deadline for application is January 15.

Correspondence and Information

Admissions Office, 0208
Department of the Scripps Institution of Oceanography
University of California, San Diego
La Jolla, California 92093-0208
Telephone: 619-534-3206

SECTION 4: MARINE SCIENCES/OCEANOGRAPHY

University of California, San Diego

THE FACULTY AND THEIR RESEARCH

Freeman Gilbert, Professor and Chairman. Theoretical geodynamics and related problems.
Duncan C. Agnew, Professor. Crustal deformation measurement and interpretation, tides, low-frequency seismology.
Laurence Armi, Professor. Oceanic turbulence, boundary layers and intrusions.
Gustaf Arrhenius, Professor. Structure and composition of ocean minerals.
Farooq Azam, Professor in Residence. Marine microbial ecology.
George E. Backus, Professor. Theoretical geophysics, continuum mechanics, fluid dynamics.
Jeffrey L. Bada, Professor. Biogeochemistry, organic reaction mechanisms.
Douglas H. Bartlett, Assistant Professor. Marine bacterial genetics, molecular biology.
Wolfgang H. Berger, Professor. Biogenous sediments in the deep sea, history of the oceans.
Michael J. Buckingham, Professor. Ocean acoustics, acoustical oceanography, underwater acoustic imaging.
Paterno R. Castillo, Assistant Professor. Petrology and isotope geochemistry of oceanic igneous rocks.
Charles S. Cox, Professor. Turbulence, internal waves, air-sea interactions.
Harmon Craig, Professor. Marine and atmospheric chemistry, isotopic geochemistry.
Joseph R. Curray, Professor. Sediments, structure, and geological history of continental margins.
Russ E. Davis, Professor. Fluid dynamics and oceanic processes.
Paul K. Dayton, Professor. Near-shore community ecology.
Richard B. Deriso, Associate Adjunct Professor. Population dynamics, fisheries science, quantitative methods.
Leroy M. Dorman, Professor. Studies of the earth's crust and upper mantle using refraction seismology.
James T. Enright, Professor. Animal behavior, with particular emphasis on rhythmicity.
D. John Faulkner, Professor. Natural product chemistry, organic chemistry.
Horst Felbeck, Associate Professor. Physiology of metabolism, biochemistry of hydrothermal-vent animals.
William H. Fenical, Professor in Residence. Organic chemistry and marine natural products.
Edward A. Frieman, Professor and Director of the Institution. Plasma physics.
Carl H. Gibson, Professor. Turbulent mixing and geophysical fluid mechanics.
Joris M. Gieskes, Professor. Physical chemistry of seawater.
Edward D. Goldberg, Professor. Marine chemistry, geochronology, marine sediments.
Robert T. Guza, Professor. Physical oceanography, especially surf-zone dynamics.
James T. Hawkins, Professor. Studies of igneous and metamorphic rocks of the ocean basins.
Margo G. Haygood, Assistant Professor. Physiology of marine bacteria, molecular biology of symbiosis.
Walter F. Heiligenberg, Professor. Behavioral physiology.
Myrl C. Hendershott, Professor. Deep-sea tides, internal and planetary waves.
Timothy D. Herbert, Assistant Professor. Paleoceanography, paleoclimatology.
Robert R. Hessler, Professor. Deep-sea ecology and geography.
John A. Hildebrand, Assistant Professor. Deep-sea instrumentation, ocean bottom gravimetry.
William S. Hodgkiss, Associate Professor. Signal processing as applied to underwater acoustics.
Nicholas D. Holland, Professor. Invertebrate zoology, cell biology.
Douglas L. Inman, Professor. Waves, currents, and the transportation of sediments.
Miriam Kastner, Professor. Geochemistry of deep-sea sediments.
Charles D. Keeling, Professor. Geochemistry, chemical transport and exchange in natural environments.
Devendra Lal, Professor. Nuclear geophysics.
Ralph A. Lewin, Professor. Experimental phycology and microbiology.
Peter F. Lonsdale, Professor. Abyssal circulation, sedimentation, geomorphology.
J. Douglas Macdougall, Professor. Isotope studies as tracers of mantle and crustal chemical evolution.
T. Guy Masters, Associate Professor. Low-frequency seismology, structure of the earth.
John A. McGowan, Professor. Biological oceanography, oceanic zoogeography, zooplankton ecology.
Jean-Bernard Minster, Professor. Seismology, plate tectonics, tectonophysics.
Michael M. Mullin, Professor. Plankton ecology, especially feeding and energy relationships of zoolankton in the pelagic environment.
William A. Newman, Professor. Biogeography of benthic faunas.
Pearn P. Niiler, Professor. General circulation of the oceans, ocean mixed-layer physics.
John A. Orcutt, Professor. Synthetic and ocean bottom seismology.
Robert L. Parker, Professor. Geophysical inverse theory, geomagnetism.
W. Jason Phipps Morgan, Associate Professor. Marine geophysics, geodynamics, tectonophysics.
Robert Pinkel, Professor. Upper-ocean physical oceanography, internal waves, acoustics.
V. Ramanathan, Professor. Climate dynamics, atmospheric radiation, satellite measurements.
Joseph L. Reid, Professor. Physical oceanography: formation and circulation of water masses of the world ocean.
Dean H. Roemmich, Associate Professor. Physical oceanography, general circulation.
Richard H. Rosenblatt, Professor. Systematics, distribution, and ecology of fishes.
Richard L. Salmon, Professor. Theory and simulation of large-scale ocean flow.
David T. Sandwell, Associate Professor. Marine geophysics, satellite geodesy, remote sensing.
John G. Sclater, Professor. Heat flow, plate tectonics, behavior of aggregates.
Robert E. Shadwick, Assistant Professor. Biomechanical adaptations in marine organisms.
Richard C. J. Somerville, Professor. Air-sea interactions, climate modeling.
Arthur J. Spivack, Assistant Professor. Isotope geochemistry, marine chemistry.
George Sugihara, Associate Professor. Mathematical ecology.
Lynne D. Talley, Associate Professor. General circulation of the oceans, instability of geophysical flows.
Lisa Tauxe, Associate Professor. Paleomagnetism, sedimentary rock magnetism.
Victor D. Vacquier, Professor. Developmental and cell biology, invertebrate reproduction.
Charles W. Van Atta, Professor. Turbulent flow, geophysical fluid mechanics.
Martin Wahlen, Professor. Isotope geochemistry, global trace-gas cycles, trace gases in ice cores.
Kenneth M. Watson, Professor. Hydrodynamics, statistical mechanics.
Ray F. Weiss, Professor. Geochemistry of oceanic and atmospheric trace gases.
Clinton D. Winant, Professor. Physical oceanography, especially dynamics of near-shore processes.
Edward L Winterer, Professor. Stratigraphy and petrology of modern and ancient sediments.
William R. Young, Associate Professor. Theory and simulation of large-scale ocean circulation.

SECTION 4: MARINE SCIENCES/OCEANOGRAPHY

UNIVERSITY OF CONNECTICUT

Department of Marine Sciences and Marine Sciences Institute

Programs of Study   The University of Connecticut offers programs of research and study in the fields of chemical and physical oceanography, marine biology, marine geology, and geophysics that lead to M.S. and Ph.D. degrees. An M.S. program in ocean engineering is offered in cooperation with the School of Engineering.

Research Facilities   The Department of Marine Sciences and Marine Sciences Institute waterside laboratories are located at Avery Point, near New London, on a 73-acre point of land extending into Long Island Sound. Additional laboratories are at Noank, also a coastal site.

At Avery Point, the Institute occupies 60,000 square feet of space in several buildings, which contain (1) biological laboratories for conducting physiological studies of marine organisms, analyzing benthic biota, and conducting histological investigations; (2) chemistry laboratories equipped with a class-100 ultraclean room for trace-level chemical analyses, infrared plasma-emission and UV-visible spectrophotometers, several atomic absorption spectrometers, and gas and high-performance liquid chromatographs with associated equipment for the extraction and concentration of trace organic constituents, as well as thin-layer chromatographic and electrophoresis apparatus; (3) geology laboratories equipped with a diamond rock saw, instruments for determining sediment particle sizes, and sediment transport facilities that include a tilting/recirculating flume; (4) a geochronology laboratory; (5) wet laboratories with running seawater, tanks, and water tables; and (6) a darkroom with excellent black-and-white and color photographic equipment.

At Noank, there are additional marine biological research laboratories that have running seawater tables and animal-holding capabilities. The Institute's vessel operations are centered there.

Support facilities offered by the Institute include electronics, machine, and carpentry shops staffed by skilled technicians for fabrication and maintenance of equipment. Field equipment consists of dredges, specialized corers, underwater cameras, plankton nets, noncontaminating water-sampling equipment for trace analyses, a towed magnetometer, a seismic air gun and sparkers, and other specialized instrumentation. The Institute operates the 65-foot R/V *UConn*, especially fitted for heavy lift work; the 34-foot *Libinia*, used primarily for biological and diving support; and a fleet of small craft for inshore and nearshore work. The Institute's dive master supervises a fully equipped diver locker to support scuba work in the full range of research opportunities.

Within the Institute, the Connecticut Sea Grant Institutional Program sponsors marine research and carries out extension education through its Marine Advisory Service. The Northeastern Center of the National Underseas Research Program, National Oceanic and Atmospheric Administration, sponsors research using manned and unmanned diving vehicles. Also sponsored is the development of gear and instrumentation for underwater experimentation, robotics, and related subjects.

The Institute's marine library has a collection of more than 6,000 volumes and 120 periodicals staffed by a full-time librarian. This library offers quick turnaround service with the main library. Several CMS terminals are linked to the University's IBM 3081 mainframe computer. A number of microcomputers and minicomputers are used for controlling laboratory equipment and data acquisition and processing.

Collocated with the Institute at Avery Point are the U.S. Coast Guard's Research and Development Center and field-oriented education program, Project Oceanology. Both are available for research collaboration. The Mystic Marinelife Aquarium also offers opportunities for collaborative research, especially on marine mammals.

Financial Aid   Financial aid in the form of graduate research assistantships (beginning in 1991–92 at approximately $9700 for nine months) and fellowships is awarded on the basis of merit. Need-based financial aid is also available. All assistantships and fellowships carry a tuition waiver and health benefits.

Cost of Study   During 1991–92, Connecticut residents pay tuition of $1714 per semester full-time and $190 per credit part-time. Students eligible for the New England Regional Student Program pay tuition of $2571 per semester if registered for 9 or more credits and $286 per credit part-time. Tuition for out-of-state residents is $4457 per semester full-time and $495 per credit part-time. A University fee of $26 per credit up to a maximum of $308 per semester and an activity fee of $10 are also charged.

Cost of Living   Graduate students at the Institute traditionally live on the Connecticut shore rather than on campus at Storrs. However, some students may prefer to spend their first year on the main campus while completing formal course work. Reasonably priced housing is available in the many shore communities and in the Groton–New London metropolitan area.

Student Group   There are usually about 25 graduate students enrolled in the M.S. and Ph.D. programs in oceanography and approximately 20 in ocean engineering. A familylike atmosphere prevails at the Institute, with very close student-faculty interactions. About half of the students are married, and a few come from abroad.

Location   The Department of Marine Sciences and Marine Sciences Institute is located in Groton on the shore of eastern Long Island Sound at the mouth of the Thames River. The Groton–New London region is the center of marine-related activities in Connecticut and is the "submarine capital of the world." Located here are the U.S. Naval Submarine Base (New London), General Dynamics' Electric Boat Division (submarine manufacture), the U.S. Coast Guard Academy, and many related industries and research laboratories. The Institute's southeastern Connecticut location allows quick access to the excellent cultural activities on campus at Storrs and to the cultural and professional sports activities in Hartford, Boston, and New York City. Appalachian Mountain wilderness areas are about 3 to 5 hours away.

The Institute   The Marine Sciences Institute, established in 1968, is the only major institution for marine science research and education in Connecticut.

Applying   Admission requires a B.S. in a basic science with introductory courses in other sciences and mathematics through the calculus level. The General Test of the Graduate Record Examinations is also required. The TOEFL is required for international students whose native language is not English. Application deadlines are June 1 for the fall and November 1 for the spring. Application materials are available from the department chairman. The University of Connecticut is an Equal Opportunity/Affirmative Action educator.

Correspondence and Information
Chairman
Department of Marine Sciences
University of Connecticut
Groton, Connecticut 06340
Telephone: 203-445-3466
Fax: 203-445-3484

Peterson's Guide to Graduate Programs in the Physical Sciences and Mathematics 1992

SECTION 4: MARINE SCIENCES/OCEANOGRAPHY

University of Connecticut

THE FACULTY AND THEIR RESEARCH

Walter F. Bohlen, Professor of Marine Science; Ph.D., MIT (Woods Hole). Physical oceanography, nearshore sediment dynamics.
John D. Buck, Professor of Marine Science; Ph.D., Miami (Florida). Marine microbiology, aquatic mycology, microbial ecology.
William D. Chappel, Professor of Biological Sciences (jointly appointed with Marine Sciences); Ph.D., Stanford. Postural mechanisms in invertebrate nervous systems, crustacean behavior.
John C. Cooke, Associate Professor of Biological Sciences (jointly appointed with Marine Sciences); Ph.D., Georgia. Morphogenesis of Ascomycetes, natural products from marine fungi.
Richard A. Cooper, Professor of Marine Science and Director of the National Undersea Research Program; Ph.D., Rhode Island. Fisheries biology, continental shelf ecology.
Hans G. Dam, Assistant Professor of Marine Science; Ph.D., SUNY at Stony Brook. Feeding ecology and physiology of pelagic copepods, role of physical processes in the distribution of plankton.
John J. Dowling, Associate Professor of Marine Science; Ph.D., Saint Louis. Marine and solid-earth geophysics.
William F. Fitzgerald, Professor of Marine Science; Ph.D., MIT (Woods Hole). Chemical oceanography, marine inorganic and atmospheric chemistry.
Larry Frankel, Professor of Geological Sciences (jointly appointed with Marine Sciences); Ph.D., Nebraska. Marine and estuarine organism-sediment relationships.
Robert G. Jeffers, Professor of Mechanical Engineering and Head of Department of Mechanical Engineering; Ph.D., Pennsylvania. Shell theory and solid mechanics.
Hans Laufer, Professor of Biological Sciences (jointly appointed with Marine Sciences); Ph.D., Cornell. Developmental biology, metamorphism and regeneration in Crustacea.
Douglas S. Lee, Assistant Professor-in-Residence; Ph.D., Michigan State. Marine and freshwater fisheries ecology.
Richard P. Long, Professor of Civil Engineering and Head of Department of Civil Engineering; Ph.D., RPI. Soil mechanics and foundations, consolidation and settlement.
Edward C. Monahan, Professor of Marine Science and Director of the Connecticut Sea Grant Program; Ph.D., MIT; D.Sc., Ireland. Physical oceanography/air-sea interaction.
James O'Donnell, Assistant Professor of Marine Science; Ph.D., Delaware. Physical oceanography, circulation modeling, frontal dynamics.
Mark A. Powell, Assistant Professor of Marine Science; Ph.D., California, San Diego. Environmental physiology, biochemical evolution.
Peter H. Rich, Associate Professor of Biological Sciences (jointly appointed with Marine Sciences); Ph.D., Michigan State. Carbon cycling in aquatic ecosystems, ecosystem theory.
Donald F. Squires, Professor of Marine Science; Ph.D., Cornell. Marine policy and marine resource management.
Thomas Torgersen, Associate Professor of Marine Science; Ph.D., Columbia. Aqueous process geochemistry, geochemistry of rare gases.
Barbara L. Welsh, Associate Professor of Marine Science; Ph.D., Rhode Island. Marine systems ecology.
Robert B. Whitlatch, Professor of Marine Science and Chairman of the Department; Ph.D., Chicago. Benthic ecology, population and community dynamics, feeding and trophic dynamics.
Charles Yarish, Professor of Ecology and Evolutionary Biology (jointly appointed with Marine Sciences); Ph.D., Rutgers. Ecology and physiology of marine macroalgae.

SECTION 4: MARINE SCIENCES/OCEANOGRAPHY

UNIVERSITY OF HAWAII AT MANOA

School of Ocean and Earth Science and Technology
Department of Oceanography

Programs of Study
The Department of Oceanography offers master's and doctoral programs in marine biology and in physical, chemical, geological, and biological oceanography. Applicants must have had intensive, rigorous training in one of the basic sciences or engineering. Completion of mathematical training—including calculus through first-order ordinary differential equations—and a year each of physics and chemistry are required of all applicants, regardless of major.

Students pursuing a degree program must take the core courses in physical, geological, and chemical oceanography. For students in nonbiological areas the core sequence is completed by taking biological oceanography. Students specializing in biology complete the core sequence by taking three of the following courses: biological oceanography, marine microplankton ecology, ecology of pelagic marine animals, and benthic biological oceanography. Students may be admitted to the M.S. program upon successful completion of the core courses. To be admitted to the Ph.D. program, a student must receive a recommendation from a Ph.D. qualifying committee. All students must demonstrate competence in digital computing and the use of computerized databases, must accumulate (or have accumulated) at least one month of field experience, and must take at least one graduate seminar course in oceanography. The M.S. program requires a minimum of 36 credit hours, including 24 credit hours of course work and 12 credit hours of thesis research. Candidates for the Ph.D. must pass a comprehensive examination and a final oral examination in defense of the dissertation. They must also qualify in one foreign language. All requirements for the M.S. and Ph.D. degrees should be completed within three and six years of admission, respectively.

Research Facilities
The University operates four oceangoing research vessels from its marine operations center within the state's main port, Honolulu Harbor. These R/Vs are the *Moana Wave*, the *Kila*, the *S. P. Lee*, and the *John Wesley Powell*. The latter two are operated for the U.S. Geological Survey. Several smaller nearshore vessels are operated by the Hawaii Institute of Geophysics and the Hawaii Institute of Marine Biology. Two research submersibles are operated under the aegis of Hawaii Undersea Research Laboratory; the *Makalii* (1,200 feet maximum diving depth) and the *Pisces V* (6,000 feet maximum diving depth) are available to all University of Hawaii researchers on a proposal-competitive basis.

Computing facilities include an Alliant FX/8 vector computer, a growing network of about twenty-five Sun Workstations, a VAX 750/VMS, and graphic peripherals, including laser and Versatec printers. The computers are interconnected via the campuswide research network. An IBM 3081 and VAX 8000s running Ultrix and VMS are available at the University of Hawaii Computing Center. Computers at other institutions nationwide and NSF supercomputers are accessible through a 128-kbps link to the U.S. national research Internet.

Precision instruments include mass spectrometers, a flow cytometer, a CHN analyzer, gas and liquid chromatographs, a scanning electron microscope with an energy-dispersive X-ray fluorescence microelemental analyzer, electron microprobe equipment, an indirectly coupled plasma/atomic emission spectrometer, an atomic absorption system with a graphite atomizer, a liquid scintillation counting system, α and γ spectrometers, and a nutrient autoanalyzer.

Financial Aid
The department has four state-funded teaching assistantships. A variable number of research assistantships are federally funded through individual faculty members' research grants and contracts. All assistantships carry a waiver of tuition. In addition, several scholarships are available to oceanography students. For example, the ARCS foundation offers three $1500 scholarships in 1991–92.

Cost of Study
For 1990–91, Hawaii state residents paid tuition of $730 per semester full-time or $61 per credit part-time. Tuition for nonresidents was $2190 per semester full-time or $183 per credit part-time.

Cost of Living
Estimated living costs are about $7000 to $8000 for a single student living off campus. Semester rental rates per person for on-campus residential facilities range from $602 to $1024; priority of assignment is given to Hawaii state residents.

Student Group
Sixty-two graduate students were enrolled in 1990–91. About 25 percent of the students are women, and 25 percent are from foreign countries. Most of them obtain financial support from various sources. A representative of the group participates in regular departmental meetings.

Location
The principal campus of the University is located on some 300 acres in Manoa Valley, a residential section close to the heart of Metropolitan Honolulu. Students may enjoy the advantages of a mild tropical climate the year round, easy access to the beaches, and the colorful multiracial cultural life.

The Department
The oceanography department occupies the six-story Marine Science Building on campus. This building, designed with input from the faculty, represents very significant assets in the form of space and scientific facilities available to the program.

The Department of Oceanography is part of the University's School of Ocean and Earth Science and Technology. Besides the Department of Oceanography, the School includes the Departments of Geology and Geophysics, Meteorology, and Ocean Engineering as well as the Hawaii Institute of Geophysics, the Hawaii Institute of Marine Biology, the Hawaii Natural Energy Institute, the Sea Grant College Program, the Hawaii Undersea Research Laboratory, and the Joint Institute for Marine and Atmospheric Research. The many resources of these institutions are available to faculty and students of the department. There is also interaction with marine scientists and research programs in departments outside the School, such as the Departments of Botany, Microbiology, and Zoology.

Applying
Application forms for admission and financial aid are available from the department. The deadline for application is February 1 for the fall semester and September 1 for the spring semester. The General Test of the Graduate Record Examinations is required. International students must take the TOEFL.

Correspondence and Information
Chairman, Department of Oceanography
University of Hawaii
1000 Pope Road
Honolulu, Hawaii 96822

Peterson's Guide to Graduate Programs in the Physical Sciences and Mathematics 1992

SECTION 4: MARINE SCIENCES/OCEANOGRAPHY

University of Hawaii at Manoa

THE FACULTY AND THEIR RESEARCH

Biological Oceanography
Robert R. Bidigare, Associate Professor; Ph.D., Texas A&M. Biooptical oceanography, pigment biochemistry, biogeochemical cycling.
Lisa Campbell, Assistant Professor; Ph.D., SUNY at Stony Brook. Microbial and phytoplankton ecology, population dynamics, immunochemistry and flow cytometry.
Thomas A. Clarke, Associate Professor; Ph.D., California, San Diego (Scripps). Marine fish ecology.
Fred C. Dobbs, Assistant Professor; Ph.D., Florida State. Microbial ecology, symbiosis, benthic biogeochemistry.
Richard W. Grigg, Professor; Ph.D., California, San Diego (Scripps). Coral reef ecology, paleoceanography, fisheries management.
Jed Hirota, Associate Professor; Ph.D., California, San Diego (Scripps). Plankton/zooplankton, fisheries, oceanic productivity and nutrient cycles.
David M. Karl, Professor; Ph.D., California, San Diego (Scripps). Microbiological oceanography, oceanic productivity, biogeochemical fluxes.
Michael R. Landry, Associate Professor; Ph.D., Washington (Seattle). Zooplankton ecology, population dynamics, marine ecosystem modeling.
Edward A. Laws, Professor and Chairman; Ph.D., Harvard. Phytoplankton ecology, aquatic pollution, aquaculture.
Craig R. Smith, Associate Professor; Ph.D., California, San Diego (Scripps). Benthic biological oceanography, bioturbation, deep-sea carbon flux.
Christopher D. Winn, Assistant Professor; Ph.D., Hawaii. Marine microbiology, organic chemistry, biogeochemistry.
Richard E. Young, Professor; Ph.D., Miami (Florida). Ecology of midwater animals, especially cephalopod mollusks.

Chemical Oceanography
Marlin J. Atkinson, Associate Professor; Ph.D., Hawaii. Coral reef biogeochemistry, solid-state sensor technology.
Keith E. Chave, Professor; Ph.D., Chicago. Marine geochemistry, coral reefs.
Antony D. Clarke, Ph.D., Washington (Seattle). Marine aerosols, biogeochemical cycles, precipitation chemistry.
James P. Cowen, Ph.D., California, Santa Cruz. Deep-sea hydrothermal vent biogeochemistry, microbial geochemistry, particle aggregation.
Eric H. DeCarlo, Associate Professor; Ph.D., Hawaii. Geochemistry of marine mineral deposits, hydrothermal processes.
Yuan-Hui Li, Professor; Ph.D., Columbia. Marine geochemistry, marine pollution studies.
Christopher I. Measures, Associate Professor; Ph.D., Southampton (England). Trace-element geochemistry, hydrothermal systems, elemental mass balance.
Francis J. Sansone, Associate Professor; Ph.D., North Carolina at Chapel Hill. Marine organic geochemistry, lava-seawater interactions, anaerobic reef processes.
Stephen V. Smith, Professor; Ph.D., Hawaii. Mass balance in ecosystems, dynamics of calcification and community metabolism.

Geological Oceanography
Patricia Fryer, Associate Geologist; Ph.D., Hawaii. Marine geology, petrology, tectonics.
Richard N. Hey, Professor; Ph.D., Princeton. Plate tectonics.
Fred T. Mackenzie, Professor; Ph.D., Lehigh. Geochemistry, sedimentology, greenhouse effect, biogeochemical cycles and global environmental change.
Alexander Malahoff, Professor; Ph.D., Hawaii. Geological and geophysical oceanography, submarine volcanism, hydrothermal and mineral formation processes.
Gary M. McMurtry, Associate Professor; Ph.D., Hawaii. Marine sediment geochemistry, marine mineral formation and resources, submarine hydrothermal processes, radiochemistry.
Michael J. Mottl, Associate Professor; Ph.D., Harvard. Submarine hydrothermal processes, geochemical cycles, seawater-seafloor chemical interactions.
Jane S. Tribble, Associate Professor; Ph.D., Northwestern. Sedimentary geochemistry and diagenesis, sedimentation and diagenesis at accretionary plate margins.

Physical Oceanography
Eric Firing, Associate Professor; Ph.D., MIT. Equatorial circulation, physical oceanographic technology.
Pierre J. Flament, Assistant Professor; Ph.D., California, San Diego (Scripps). Dynamics of the surface layer, mesoscale structures, remote sensing.
Roger Lukas, Professor; Ph.D., Hawaii. Equatorial circulation, air-sea interaction and climate.
Lorenz Magaard, Professor; Ph.D., Kiel (Germany). Ocean waves, ocean turbulence.
Gary T. Mitchum, Assistant Professor; Ph.D., Florida State. Tropical ocean dynamics.
Dennis W. Moore, Professor and Director, Joint Institute for Marine and Atmospheric Research; Ph.D., Harvard. Equatorial oceanography, geophysical fluid dynamics.
Peter K. Muller, Associate Professor; Ph.D., Hamburg (Germany). Ocean circulation, large-scale ocean-atmosphere interaction, sea level, climate.

Affiliate Graduate Faculty
John E. Bardach, Ph.D., Wisconsin. Aquaculture, fisheries, law of the sea.
Paul K. Bienfang, Ph.D., Hawaii. Phytoplankton ecology.
George W. Boehlert, Ph.D., California, San Diego (Scripps). Ecology of early life history stages of marine fishes.
Lewis M. Rothstein, Ph.D., Hawaii. Physical oceanography, analytical modeling of equatorial ocean dynamics.
Roland Wollast, Ph.D., Libre (Brussels). Geochemical modeling, chemistry of coastal sediments, mineral weathering.
Alfred H. Woodcock, D.Sc., LIU. Sea salt particles in marine air, sea fogs and marine aerosols, air-sea interaction.

Cooperating Graduate Faculty
Louis Herman, Ph.D., Penn State. Behavior and ecology of marine mammals, animal cognition.
Richard L. Radtke, Ph.D., South Carolina. Population dynamics, fish biology, calcification.

SECTION 4: MARINE SCIENCES/OCEANOGRAPHY

UNIVERSITY OF PUERTO RICO, MAYAGÜEZ

Department of Marine Sciences

Programs of Study
The Department of Marine Sciences on the Mayagüez campus of the University of Puerto Rico is a graduate department offering a program leading to Master of Science and Doctor of Philosophy degrees in marine sciences. The aim is to train marine scientists for careers in teaching, research, and development. Students specialize in biological, physical, chemical, or geological oceanography; fisheries biology; aquaculture; or one of the areas within marine biology. They also gain an overall appreciation of marine sciences through core course requirements and electives. Much of the teaching and research is carried out at the marine station, 22 miles south of Mayagüez, but students are able to elect courses in other departments and to use facilities at the computer center and the Research Development Center.

A minimum of 35 semester hours of credit in approved graduate courses is required for the M.S. degree; 72 for the Ph.D. Courses in the Department of Marine Sciences are taught in English and Spanish. Because Puerto Rico has a Spanish culture and the University is bilingual, candidates are expected to gain a functional knowledge of Spanish as well as English before finishing their degree, if they are not already proficient in both languages. Further requirements for the M.S. are residence of at least one academic year, passing a departmental examination, completing a satisfactory thesis, and passing a comprehensive final examination; for the Ph.D., residence of at least two years, passing a qualifying examination, subsequently passing a comprehensive examination, completing a satisfactory thesis, and passing a final examination in defense of the thesis.

Research Facilities
Teaching and research facilities, complete with standard and sophisticated equipment, exist both on campus and at the field station. A department library specializing in marine science publications is located in the campus building. The field station is on 18-acre Magueyes Island within a protected embayment off La Parguera, 22 miles from Mayagüez. In addition to classroom and laboratory facilities, the station has indoor and outdoor aquaria and tanks with running seawater; a museum containing reference collections of fish, invertebrates, and algae; a 51-foot Thompson trawler; the 72-foot R/V *Isla Magueyes* trawler; a 35-foot diesel Downeast; and a number of smaller boats. In conjunction with the underwater research and training activities, the department operates a hyperbaric chamber. Research facilities for warmwater aquaculture include some 8 acres of earthen ponds, two hatcheries and numerous concrete tanks, plastic pools, and aquaria facilities with running water available for controlled environmental studies.

Financial Aid
The department is able to provide assistantships, which include a tuition waiver, and scholarships for a few students each year. These averaged $400 per month in 1990–91. Students in need of financial help should indicate this when applying for admission. Assistantships are assigned according to merit; they are not available to new students in the first semester.

Cost of Study
Residents carrying a full program (9 to 12 credits) paid $55 per credit hour in 1990–91. General fees of approximately $65 per semester were added to these costs. Resident status may be established in one year. Nonresidents, including aliens, paid $3500 per year except for students from U.S. institutions having reciprocal tuition-reduction agreements with Puerto Rico (a list of states is available on request).

Cost of Living
Apartments and houses can be found in Mayagüez and San Germán for $175 to $250 per month and single rooms for less. Single students in the department usually share apartments.

Student Group
Total enrollment at the Mayagüez campus of the University of Puerto Rico is about 9,500. The department's enrollment is about 65 graduate students.

Location
Mayagüez, the third-largest city in Puerto Rico, has a population of 100,000. It is a seaport on the west coast of the island. The economy of the city centers largely on shipping, commercial fishing, light industry, and the University. San Germán, where many University people live, is 10 miles south of Mayagüez. The main campus of Inter-American University is located there. An increasing number of concerts, art exhibits, and other cultural activities are arranged in Mayagüez and San Germán, although San Juan, a 3-hour drive from Mayagüez, remains the center for music (with, for instance, the yearly Casals Festival), drama, art, and other activities sponsored by various civic entities and the Institute of Puerto Rican Culture. Many of these events are held at the Center for Fine Arts. In addition, repertory theaters are becoming very popular in the metropolitan area.

The University and The Department
The University of Puerto Rico, Mayagüez, had its beginning in 1911 as the College of Agriculture, an extension of the University in Río Piedras. In 1912 the name was changed to the College of Agriculture and Mechanic Arts. Following a general reform of the University in 1942, the college became a regular campus of the University under a vice-chancellor and, in 1966, an autonomous campus with its own chancellor.

The main marine sciences program of the University of Puerto Rico is in Mayagüez. It began in 1954 with the establishment of the Institute of Marine Biology. A master's degree program in marine biology was initiated in 1963. In 1968 the institute became the Department of Marine Sciences, a graduate department, with its own academic staff. The doctoral program began in 1972. Research remains an important function of the department.

Applying
Application forms are available from the Graduate School. An undergraduate science degree is required. The applicant should have had at least basic courses in biology, chemistry, physics, geology, and mathematics through calculus. An engineering degree may be acceptable in some circumstances. Applications should be submitted before April 15 for the fall semester and before October 15 for the spring semester.

Correspondence and Information
Director, Graduate School
Box 5000
University of Puerto Rico
Mayagüez, Puerto Rico 00709-5000

Peterson's Guide to Graduate Programs in the Physical Sciences and Mathematics 1992

SECTION 4: MARINE SCIENCES/OCEANOGRAPHY

University of Puerto Rico, Mayagüez

THE FACULTY AND THEIR RESEARCH

Luis R. Almodóvar, Professor; Ph.D. (botany), Florida State. Taxonomy, morphology, physiology, and ecology of marine algae.
Dallas E. Alston, Associate Professor; Ph.D. (invertebrate aquaculture), Auburn. Culture of invertebrate organisms.
Nilda E. Aponte, Assistant Investigator; Ph.D. (marine botany), Puerto Rico, Mayagüez. Taxonomy, morphology and life history of marine algae.
Richard S. Appeldoorn, Professor; Ph.D. (fisheries biology), Rhode Island. Fisheries biology.
David L. Ballantine, Professor; Ph.D. (marine botany), Puerto Rico, Mayagüez. Taxonomy and ecology of marine algae.
Yury Belyayev, Instructor and Director of Underwater Activities; M.S. (physics), Karkov (USSR). Diving and submersible operations.
Jorge E. Corredor, Professor; Ph.D. (marine chemistry), Miami (Florida). Chemical oceanography, pollution.
Ricardo Cortés Maldonado, Associate Professor; M.S. (aquaculture), Puerto Rico, Mayagüez. Aquaculture.
Bertha D. Cutress, Assistant Investigator; M.S. (zoology), Oregon State. Biology of marine coelenterates, cidaroid systematics and paleontology.
Jorge R. Garcia Sais, Assistant Investigator; Ph.D. candidate (biological oceanography), Rhode Island. Zooplankton ecology.
Juan G. González Lagoa, Professor and Associate Dean of the Faculty of Arts and Sciences; Ph.D. (planktology), Rhode Island. Biological oceanography, planktology.
Dannie A. Hensley, Professor; Ph.D. (ichthyology), South Florida. Systematics and ecology of fishes.
Manuel L. Hernández-Avila, Professor and Director of the University of Puerto Rico Sea Grant Program; Ph.D. (physical oceanography), LSU. Coastal physical oceanography.
John Kubaryk, Associate Professor and Acting Director of the Department; Ph.D. (seafood technology), Auburn. Seafood technology, aquaculture.
José M. López, Associate Professor; Ph.D. (environmental chemistry), Texas. Water pollution control.
Aurelio Mercado Irizarry, Assistant Professor; Ph.D. candidate (physical oceanography), Miami (Florida). Geophysical fluid dynamics.
Jack Morelock, Professor; Ph.D. (oceanography), Texas A&M. Sediments, beach and littoral studies, reef sediments and marine terraces, geophysical surveys.
Mark Reigle, Director of Marine Operations; B.S. (engineering), Virginia Tech. Oceanographic instrument design.
Douglas Y. Shapiro, Professor; M.D., Case Western Reserve; Ph.D. (animal behavior), Cambridge. Marine animal behavior.
Thomas R. Tosteson, Professor; Ph.D. (physiology), Pennsylvania. Marine physiology and pharmacology.
Ernest H. Williams, Professor; Ph.D. (fisheries), Auburn. Systematics and culture of parasites of fishes.
Amos Winter, Associate Professor; Ph.D. (paleoceanography), Hebrew (Jerusalem). Paleoceanography.
Paul Yoshioka, Associate Professor; Ph.D. (marine ecology), California, San Diego. Marine ecology.
Baqar R. Zaidi, Assistant Professor; Ph.D. (marine physiology), Puerto Rico, Mayagüez. Marine physiology.

Field station at Magueyes Island, La Parguera, Puerto Rico.

Graduate student documenting underwater observations for thesis research.

SECTION 4: MARINE SCIENCES/OCEANOGRAPHY

UNIVERSITY OF RHODE ISLAND

Graduate School of Oceanography

Programs of Study	The Graduate School of Oceanography offers the M.S. and the Ph.D. degrees in oceanography with specializations in biological, chemical, geological, and physical oceanography. The M.S. degree requires the completion of 30 credit hours of course work and thesis research beyond the baccalaureate and the writing and defense of a thesis. The Ph.D. degree requires the completion of 72 credit hours of course work and research beyond the baccalaureate (this may include 30 credits taken for the master's degree) and the writing and defense of a dissertation. The master's thesis and the doctoral dissertation must be publishable work that is a contribution to the field of oceanography. Students are required to complete a program of core courses. Although courses in this program vary with the chosen area of specialization, all students must take at least two oceanography courses outside their specialization and must participate each semester in the oceanography seminar program. A program of study is planned by students in consultation with their major professor and their program committee. On completion of course work, Ph.D. students must pass a written and an oral comprehensive examination. There is no general language requirement, but the individual student's major professor may require the demonstration of ability in one or more foreign languages. No formal courses are offered during the summer, but students are advised to register for research credits or for individual study with a member of the faculty. All students are expected to participate in a regular oceanic research cruise sometime during their residence. Students are encouraged to complete the M.S. requirement in two years and the Ph.D. requirements in five years but cannot exceed four years for the M.S. or seven years for the Ph.D.
Research Facilities	Located on the Narragansett Bay campus, the Graduate School of Oceanography is ideally situated for marine studies on the shore of the West Passage of Narragansett Bay. A public service building, dining center, research aquarium, marine ecosystems laboratory, marine geomechanics laboratory, and 100-foot tow tank are among the fifteen major University buildings on the Bay campus. Construction of a new laboratory building for the study of atmospheric chemistry begins in fall 1991. There are docking facilities for the University's Research Vessel *Endeavor*, a 177-foot oceanographic ship capable of working in all parts of the world's oceans. There is also a marine science library of more than 28,000 volumes. The research activities of the School require an extensive and specialized collection of scientific and technical equipment and services, including a research reactor, a scanning electron microscope, a transmission electron microscope, a microprobe, three mass spectrometers, extensive computing facilities, specialized sample collections, and mechanical and electronics shops.
Financial Aid	There are a limited number of graduate assistantships available to incoming students. They are provided by the University and by sponsored research grants and contracts. Students also find employment at the Kingston campus of the University or at one of the federal, state, or University agencies at the Narragansett Bay campus. In 1991–92, assistantships pay up to $8350 for the academic year and require up to 20 hours of work per week. Full-time summer assistants are paid up to $7422 for sixteen weeks. (These figures are subject to change.) A few fellowships are available from the University. Students holding graduate assistantships and fellowships do not pay tuition fees for the academic year.
Cost of Study	For 1991–92, student fees are as follows: Rhode Island residents registered for full-time study pay tuition of $2166 per year; students eligible for the New England Regional Student Program pay $3240 per year; out-of-state students pay tuition of $4958 per year. Tuition includes other fees collected annually such as the registration fee of $20, student union fee of $126, graduate student assessment of $20, and health service fee of $227. (These figures are subject to change.)
Cost of Living	Graduate students traditionally live off campus, although there are a limited number of apartments available on the Kingston campus. A single student sharing a house can expect to pay approximately $300 per month for rent and $200 per month for food and other living expenses.
Student Group	As of January 1991, there were 131 students enrolled in the oceanography programs, 52 as M.S. and 79 as Ph.D. students. There were 56 students in the biological oceanography specialization, 25 in the chemical, 26 in the geological, and 24 in the physical. There were 52 women and 79 men, and 20 percent of the enrollment was made up of foreign students.
Location	The Graduate School of Oceanography is located on the 165-acre Narragansett Bay campus in the southern part of the state. It is a rural setting on the shore of the scenic West Passage of Narragansett Bay. The area is unsurpassed for its beaches and water sports, including swimming, fishing, and sailing. The University is situated in the Northeast Corridor between New York City and Boston and has the cultural advantages of being 30 miles from Providence, 70 miles from Boston, and 160 miles from New York City.
The School	The University of Rhode Island has two campuses: the main campus in Kingston and the Bay campus in Narragansett. The Graduate School of Oceanography shares the Bay campus with other divisions of the University, the state, and the federal government. The campus is home to the University's Division of Marine Resources, an important link between the academic world and the professional marine community, as well as home to some of the facilities of the Department of Ocean Engineering. The state's 2-megawatt research reactor provides an outstanding research tool for the campus community. The Environmental Protection Agency maintains a laboratory on campus, as does NOAA's National Marine Fisheries Service. Although students in the School spend most of their time on the Bay campus, they also use the facilities and resources of the main library on the Kingston campus. The combination of University, state, and federal programs contributes to the lively and enriching intellectual environment.
Applying	Applicants should have majored in a basic science and have had introductory courses in biology, chemistry, geology, physics, and mathematics through calculus. An official transcript from each college attended, a statement of research interests and career objectives, three letters of recommendation, and scores on the GRE General Test and an appropriate GRE Subject Test are required. Applicants wanting to enroll in September should file their application by the preceding April 15.
Correspondence and Information	For program information and application materials: Academic Affairs Office Graduate School of Oceanography University of Rhode Island Narragansett, Rhode Island 02882-1197	For application information and status of application: Graduate Admissions Office University of Rhode Island Kingston, Rhode Island 02881-0807

Peterson's Guide to Graduate Programs in the Physical Sciences and Mathematics 1992

SECTION 4: MARINE SCIENCES/OCEANOGRAPHY

University of Rhode Island

THE FACULTY AND THEIR RESEARCH

Michael L. Bender, Professor; Ph.D., Columbia, 1970. Geochemistry.
Steven Carey, Assistant Professor; Ph.D., Rhode Island, 1983. Volcanology, marine volcaniclastic sedimentation.
Peter Cornillon, Associate Professor; Ph.D., Cornell, 1973. Ocean engineering, remote-sensing oceanography.
Robert S. Detrick Jr., Professor; Ph.D., MIT/Woods Hole, 1978. Marine geophysics: oceanic crustal structure, tectonics of midocean ridges and transform faults.
Steven D'Hondt, Assistant Professor; Ph.D., Princeton, 1989. Marine micropaleontology, plankton paleobiology, paleoceanography.
Robert A. Duce, Professor; Ph.D., MIT, 1964. Sea-air chemical exchange, atmospheric chemistry.
Ann G. Durbin, Associate Professor; Ph.D., Rhode Island, 1976. Fish and zooplankton physiology, ecology, trophic relationships.
Edward G. Durbin, Associate Professor; Ph.D., Rhode Island, 1976. Marine planktonic food chains, zooplankton, fish ecology.
Paul J. Fox, Professor; Ph.D., Columbia, 1972. Plate tectonics, oceanic crustal structure.
Paul E. Hargraves, Professor; Ph.D., William and Mary, 1968. Phytoplankton systematics, morphology, biogeography.
Brian Heikes, Assistant Professor; Ph.D., Michigan, 1984. Atmospheric chemistry, atmospheric oxidants, heterogeneous processes.
Barry J. Huebert, Professor; Ph.D., Northwestern, 1970. Atmospheric chemistry, sampling methodology.
H. Perry Jeffries, Professor; Ph.D., Rutgers, 1959. Structure of inshore communities.
Dana R. Kester, Professor; Ph.D., Oregon State, 1969. Physical chemistry of seawater.
Christopher Kincaid, Assistant Professor; Ph.D., Johns Hopkins, 1989. Solid earth geophysics.
John King, Assistant Professor; Ph.D., Minnesota, 1983. Paleomagnetism, rock magnetism, palynology.
John A. Knauss, Professor; Ph.D., California, San Diego (Scripps), 1959. Ocean circulation, marine policy.
Roger L. Larson, Professor; Ph.D., California, San Diego (Scripps), 1970. Marine geophysics, plate tectonics.
Margaret Leinen, Professor; Ph.D., Rhode Island, 1979. Sediment geochemistry and mineralogy, paleoceanography.
John T. Merrill, Associate Professor; Ph.D., Colorado, 1976. Atmospheric transport, atmospheric waves.
Scott W. Nixon, Professor; Ph.D., North Carolina, 1969. Estuarine and wetland ecosystems.
Candace A. Oviatt, Professor; Ph.D., Rhode Island, 1967. Marine ecology.
Michael E. Q. Pilson, Professor; Ph.D., California, San Diego (Scripps), 1964. Chemistry of seawater, experimental biogeochemistry and ecology.
James G. Quinn, Professor; Ph.D., Connecticut, 1967. Marine organic chemistry.
Kenneth A. Rahn, Professor; Ph.D., Michigan, 1971. Atmospheric chemistry in the Arctic and midlatitudes.
Hans Thomas Rossby, Professor; Ph.D., MIT, 1966. Ocean circulation and instrumentation.
Lewis M. Rothstein, Associate Professor; Ph.D., Hawaii, 1983. Geophysical fluid dynamics.
Jean-Guy Schilling, Professor; Ph.D., MIT, 1966. Volcanology, isotope and trace-element geochemistry, mantle heterogeneities and dynamics.
John M. Sieburth, Professor; Ph.D., Minnesota, 1954. Marine microbial ecology.
Haraldur Sigurdsson, Professor; Ph.D., Durham (England), 1970. Petrology of igneous rocks, ocean ridge and island arc volcanism.
Theodore J. Smayda, Professor; Dr.Philos., Oslo, 1967. Phytoplankton ecology and physiology.
Elijah Swift, Professor; Ph.D., Johns Hopkins, 1967. Marine phytoplankton, oceanic bioluminescence.
Robert C. Tyce, Associate Professor; Ph.D., California, San Diego (Scripps), 1976. Ocean engineering, acoustics of sound in water, seafloor reflectivity.
D. Randolph Watts, Associate Professor; Ph.D., Cornell, 1973. Dynamics and variability of ocean currents and water masses.
Mark Wimbush, Associate Professor; Ph.D., California, San Diego (Scripps), 1969. Turbulence, tides and waves.
Howard E. Winn, Professor; Ph.D., Michigan, 1955. Animal behavior, bioacoustics orientation, fishes and whales.
Karen Wishner, Associate Professor; Ph.D., California, San Diego (Scripps), 1979. Marine zooplankton ecology, deep-sea biology.
James A. Yoder, Associate Professor; Ph.D., Rhode Island, 1978. Phytoplankton systems ecology and remote sensing.

SECTION 4: MARINE SCIENCES/OCEANOGRAPHY

UNIVERSITY OF SOUTH FLORIDA

Department of Marine Science

Programs of Study	The St. Petersburg Campus of the University of South Florida offers M.S. and Ph.D. degrees with specializations in biological, chemical, geological, and physical oceanography. Students are strongly encouraged to take interdisciplinary programs encompassing two or more of the four basic disciplines. Both M.S. and Ph.D. candidates are required to complete a core course program covering the four basic oceanographic disciplines. Requirements for the M.S. degree include 32 credit hours of course and research work, plus defense of a thesis that makes an original contribution to oceanography. Ph.D. candidates are required to demonstrate proficiency in two computer techniques or foreign languages and successfully complete a written and oral comprehensive examination. After completing 90 credit hours of course and research work, they must defend a dissertation that represents a publishable contribution to marine science. Students are encouraged to participate in oceanographic research cruises during the course of their enrollment. Over the past five years, the department's students and faculty have conducted research in the Pacific, Atlantic, Indian, and Antarctic oceans; the Gulf of Mexico; and the Norwegian, Arabian, and Bering seas as part of a variety of national and international programs supported by the National Science Foundation and other federal agencies.
Research Facilities	The Department of Marine Science is located at Bayboro Harbor, Port of St. Petersburg. The harbor has immediate access to Tampa Bay and can accommodate any ship in the fleet of U.S. oceanographic vessels. Bayboro Harbor is home port to the R/V *Suncoaster* (110 feet) and the R/V *Bellows* (71 feet), the principal vessels operated by the Florida Institute of Oceanography (FIO) for the entire state university system. FIO is located within the principal building (82,000 square feet) of the Department of Marine Science. The department's research facilities are adjacent to the Florida Marine Research Institute, the research arm of the Florida Department of Natural Resources. In 1988, the St. Petersburg Campus became the site of the Center for Coastal Geology of the United States Geological Survey. The department's specialized laboratories include those for trace-metal analysis, physical chemistry, water quality, organic and isotope geochemistry, optical oceanography, satellite imagery and numerical modeling, sedimentology, micropaleontology, physiology, benthic ecology, microbiology, ichthyology, planktology, and geophysics. The department has a large flume facility for interdisciplinary boundary-layer studies. Major items of equipment include an ISI (DS-130) scanning electron microscope, a Finnigan MAT-250 isotope ratio mass spectrometer, high-resolution gas chromatographs, a combined gas chromatograph–mass spectrometer, flame and graphite furnace atomic absorption spectrometers, multichannel autoanalyzer systems, UV-visible scanning spectrophotometers, an EG&G uniboom high-resolution continuous seismic reflection profiling system, an EG&G side-scan sonar system, X-ray diffraction systems, and a Mössbauer spectroscopy system. The department has its own marine science library collection on the St. Petersburg Campus as well as access to the University's Tampa Campus facilities.
Financial Aid	Approximately fifteen state-supported assistantships are available each year for beginning students. In addition, approximately ten fellowships are available from endowments. After their first year, students are expected to pursue research on grant-supported projects. Assistantships pay in the range of $10,000 to $12,000 for twelve months. A number of out-of-state tuition waivers are available for first-year students.
Cost of Study	For 1991–92, tuition is $78 per semester hour for state residents and $243 per semester hour for nonresident students. Out-of-state students who are American citizens may establish Florida residence after one year if certain criteria are met.
Cost of Living	There are no on-campus accommodations in St. Petersburg, but students easily find apartments or houses near the campus. The rents range between $200 and $450 per month, and single students often share in rental expenses. The average total living expenses for students are currently about $450 per month.
Student Group	For the 1991–92 academic year, there are over 100 active graduate students in the marine science program. More than half are working toward a Ph.D. degree. Approximately 45 students are in biological oceanography, 20 in chemical oceanography, 25 in geological oceanography, and 10 in physical oceanography. More than half are married, and over a third are women. All program graduates have obtained positions in their field or have gone on to other universities for their Ph.D.'s.
Location	The department is situated just a few blocks from downtown St. Petersburg and the Bayfront Center Arena-Theater (where major performing groups entertain), museums (including the Salvadore Dali Museum), Al Lang Baseball Stadium (where major-league spring training games are held), Albert Whitted Airport (a general aviation airport servicing the USF Flying Club), numerous restaurants, department stores, shops, the St. Petersburg Yacht Club, and marinas. The Gulf Coast beaches are within a 20-minute drive of the campus. Because the marine environment is extremely important in this area—both from a commercial and a leisure-activity point of view—the department receives considerable attention and support from the community.
The University and The Department	The University of South Florida was founded in 1956. It was the first major state university in America to be planned and built entirely in this century, and it is the second-largest public university in Florida. The Department of Marine Science, established in 1967, is a department in the College of Arts and Sciences and is the only department located entirely on the St. Petersburg Campus. The main campus in Tampa, with its wealth of supportive and cultural activities, is just 35 miles away via Interstate Highway 275. Construction on a new 50,000-square-foot marine science building is slated to begin in 1992. With this addition, the department's faculty is expected to expand to include approximately 30 members by 1995.
Applying	Applications should be submitted by June 30 for the fall semester and November 1 for the spring semester. (The deadlines for international applicants are May 1 for the fall and September 1 for the spring semesters.) The financial aid deadlines are February 15 and October 1, respectively. Minimum requirements for admission are an undergraduate major in biology, chemistry, engineering, geology, math, or physics; an upper-level GPA of 3.0; and a GRE General Test (verbal and quantitative sections) score of 1100. Applicants are also expected to have successfully completed 1 year of calculus.
Correspondence and Information	Chairman Department of Marine Science University of South Florida St. Petersburg, Florida 33701 Telephone: 813-893-9130

Peterson's Guide to Graduate Programs in the Physical Sciences and Mathematics 1992

SECTION 4: MARINE SCIENCES/OCEANOGRAPHY

University of South Florida

THE FACULTY AND THEIR RESEARCH

Peter R. Betzer, Professor and Chairman; Ph.D., Rhode Island, 1971. Chemical oceanography, chemical tracers, pollutant transfer, particle fluxes, role of organisms in modifying chemistry of seawater.

Norman J. Blake, Professor; Ph.D., Rhode Island, 1972. Ecology and physiology of marine invertebrates, inshore environmental ecology and pollution, reproductive physiology of mollusks and crustaceans.

John C. Briggs, Professor; Ph.D., Stanford, 1952. Systematics and behavior of marine fishes, marine zoogeography.

Robert H. Byrne, Professor; Ph.D., Rhode Island, 1974. Chemical oceanography, physical chemistry of seawater, ionic interactions, marine surface chemistry, oceanic CO_2 system chemistry.

Kendall L. Carder, Professor; Ph.D., Oregon State, 1970. Physical oceanography, ocean optics, suspended-particle dynamics, instrument development, ocean remote sensing.

John S. Compton, Assistant Professor; Ph.D., Harvard, 1986. Diagenesis of marine sediments, low-temperature and sedimentary geochemistry, sedimentary petrology, paleoceanography.

Larry J. Doyle, Professor; Ph.D., USC, 1973. Marine geology, sedimentology, sediments and sedimentary processes of the continental margins.

Kent A. Fanning, Professor; Ph.D., Rhode Island, 1973. Chemical oceanography, pore-water geochemistry, nutrients in the ocean, marine radiochemistry.

Boris Galperin, Associate Professor; Ph.D., Technion (Israel), 1982. Physical oceanography, boundary layers, turbulence, renormalization group theory, numerical modeling of oceanic circulation.

Giselher R. Gust, Professor; Ph.D., Kiel (Germany), 1975. Physical oceanography, benthic boundary-layer flows, geochemical fluxes across the sediment-water interface, biological-hydrodynamical interactions, instrument developments.

Pamela Hallock-Muller, Professor; Ph.D., Hawaii, 1977. Micropaleontology, paleoceanography, carbonate sedimentology, coral reef ecology.

Albert C. Hine, Professor; Ph.D., South Carolina, 1975. Carbonate sedimentology, coastal sedimentary processes, geological oceanography, sequence stratigraphy.

Thomas L. Hopkins, Professor; Ph.D., Florida State, 1974. Biological oceanography, marine plankton and micronekton ecology, oceanic food webs.

Mark E. Luther, Associate Professor; Ph.D., North Carolina at Chapel Hill, 1982. Physical oceanography, numerical modeling of ocean circulation, equatorial dynamics, air-sea interaction, climate variability, estuarine circulation.

Frank E. Müller-Karger, Assistant Professor; Ph.D., Maryland, 1988. Marine, estuarine, and environmental science; biological oceanography; remote sensing; nutrient cycles.

David F. Naar, Assistant Professor; Ph.D., California, San Diego (Scripps), 1990. Marine geophysics, plate tectonics, marine tectonics, midocean ridge processes, physical modeling using molten wax.

John H. Paul, Professor; Ph.D., Miami (Florida), 1980. Marine microbiology and genetics, gene transfer mechanisms.

William M. Sackett, Distinguished Professor; Ph.D., Washington (St. Louis), 1958. Marine organic and isotope geochemistry.

Joseph J. Torres, Professor; Ph.D., California, Santa Barbara, 1980. Biological oceanography, deep-sea biology, bioenergetics of pelagic animals, comparative physiology.

Edward S. Van Vleet, Professor; Ph.D., Rhode Island, 1978. Chemical oceanography, organic geochemistry, molecular biomarkers, hydrocarbon pollution.

Gabriel A. Vargo, Associate Professor; Ph.D., Rhode Island, 1976. Biological oceanography; phytoplankton ecology, physiology, and nutrient dynamics.

John J. Walsh, Distinguished Professor; Ph.D., Miami (Florida), 1969. Continental shelf ecosystems, systems analysis of marine food webs, global carbon and nitrogen cycles.

Robert W. Weisberg, Professor; Ph.D., Rhode Island, 1975. Physical oceanography, equatorial ocean dynamics, estuarine and nearshore circulation studies.

Raymond R. Wilson Jr., Assistant Professor; Ph.D., California, San Diego (Scripps), 1984. Ichthyology, deep-sea ecology, fisheries biology.

Adjunct Faculty

Thomas G. Bailey, Assistant Professor; Ph.D., California, Santa Barbara, 1984. Physiology and ecology of deep-sea fishes and invertebrates.

Ronald C. Baird, Associate Professor; Ph.D., Harvard, 1969. Systematics, ecology, and behavior of fishes; biology of oceanic mesopelagic fishes; energetics and the organization of pelagic marine ecosystems.

Theresa M. Bert, Assistant Professor; Ph.D., Yale, 1985. Evolution, systematics, population biology, physiology, genetics of fish and shellfish.

Roy E. Crabtree, Assistant Professor; Ph.D., William and Mary, 1984. Ecology, physiology, and early life history of gamefish; ichthyology.

Richard A. Davis Jr., Distinguished Professor; Ph.D., Illinois, 1964. Dynamics and sediments of beach and barrier island systems.

Robert B. Halley (U.S. Geological Survey), Professor; Ph.D., SUNY at Stony Brook, 1974. Carbonate sedimentation, chemistry and diagenesis, stratigraphy, coastal sedimentation, coral reefs, paleoclimate records and climate variability.

Gary W. Litman, Professor; Ph.D., Minnesota, 1972. Molecular genetics, evolution, immunology, developmental regulation.

R. Edmond Matheson, Assistant Professor; Ph.D., Texas A&M, 1983. Ecology and population biology of estuarine fish, ichthyology.

Anne Meylan, Assistant Professor; Ph.D., Florida, 1984. Ecology, migrations, and evolutionary history of marine turtles; biology of demosponges.

Robert G. Muller, Assistant Professor; Ph.D., Hawaii, 1976. Fisheries biology, population dynamics, modeling of exploited populations, fisheries statistics.

Esther C. Peters, Assistant Professor; Ph.D., Rhode Island, 1985. Comparative histopathology, coral biology, invertebrate oncology.

John E. Reynolds III, Associate Professor; Ph.D., Miami (Florida), 1980. Marine mammals: population dynamics, management, and functional anatomy.

Gary E. Rodrick, Associate Professor; Ph.D., Oklahoma, 1971. Medical malacology, comparative physiology and immunology of invertebrates, biochemistry of mitochondrial enzymes and nucleic acids, parasite metabolism.

Asbury H. Sallenger Jr. (U.S. Geological Survey), Professor; Ph.D., Virginia, 1975. Nearshore sedimentary and wave processes, coastal erosion, sediment transport.

Eugene A. Shinn (U.S. Geological Survey), Professor; H.C., Kensington, 1987. Carbonate diagenesis, tidal flat deposition, reef development, coral reef ecology and geology.

Albert C. Smith, Professor; Ph.D., California, Irvine, 1967; M.D., Hawaii, 1975. Marine biology and medicine, pathology, immunology and serology, molecular phylogeny and systematics, population biology.

Karen A. Steidinger, Ph.D., South Florida, 1979. Dinoflagellates, red tides, ultrastructure of unicells, cytology.

Yves Tardy, Distinguished Professor; Ph.D., Louis Pasteur (Strasbourg), 1969. Geochemical thermodynamics, mineral-solution equilibria, global chemical cycles, weathering and erosion.

Roland F. Wollast, Distinguished Professor; Ph.D., Libre (Brussels), 1960. Geochemical equilibria and kinetics, global cycling of elements.

SECTION 4: MARINE SCIENCES/OCEANOGRAPHY

UNIVERSITY OF TEXAS AT AUSTIN
Department of Marine Science

Programs of Study

The Department of Marine Science offers research opportunities and course work leading to the M.A. and Ph.D. degrees in marine science. Graduate students usually begin their academic program with course work on the Austin campus and move to the University of Texas Marine Science Institute at Port Aransas for specialized advanced courses and thesis or dissertation research. Core courses are required in several subdisciplines, including marine biology/ecology, marine chemistry, marine geology, and physical oceanography. Areas of research available in Port Aransas include physiology and ecology of marine organisms, biological oceanography, geochemistry, marine environmental quality, coastal processes, and mariculture. Students in marine minerals and marine mining research complete course work at Austin but engage in fieldwork elsewhere.

Research Facilities

The Marine Science Institute at Port Aransas is located near Corpus Christi and provides opportunities to study living organisms in the laboratory and under field conditions. A wide variety of environments are readily accessible, such as the pass connecting Corpus Christi Bay with the Gulf of Mexico, the continental shelf, and many bays and estuaries, including brackish estuaries and the hypersaline Laguna Madre. Facilities available include outside open and covered seawater tanks, a pier lab with running seawater, a reference collection of most of the plants and animals of the area, and controlled-environment chambers. Vessels include the R/V *Longhorn*, a 105-foot research vessel with navigation and laboratory capabilities for most research projects; the R/V *Katy*, a 54-foot boat with dredge and trawl equipment for collection of specimens; and several smaller boats. Specialized features include a remote terminal linked with the computation center in Austin. Laboratories are equipped to study animal physiology, bacterial and algal physiology, bacterial and algal ecology, fish ecology, marine phycology, mariculture, sea grass ecology and physiology, and geochemistry. Research is under way in benthic ecology, biological oceanography, fish behavior, invertebrate biology, phytoplankton ecology, and taxonomy of marine organisms. The Institute also provides teaching facilities in Port Aransas, including upper-division and graduate course offerings during the summer. Facilities are available on the Austin campus for research in marine sedimentology and in marine mineral deposits, including genesis, exploration, and recovery.

Financial Aid

Research and teaching assistantships are available through graduate advisers or the department chairman. E. J. Lund Fellowships and Scholarships for research at the Marine Science Institute are awarded annually.

Cost of Study

In 1991–92, tuition and required fees for Texas residents and any students holding an assistantship are approximately $300 per semester for 9 credit hours. Nonresident tuition and required fees for 9 credit hours total $1250 per semester.

Cost of Living

In Port Aransas in 1991–92, furnished University apartments are available for students at approximately $150–$200 per month plus utilities. Non-University housing off-campus costs approximately $300–$400 per month. For the Department of Marine Science's Summer Program, dormitory and dining facilities are also available to registered students.

In Austin, University dormitories and apartments, furnished and unfurnished, are available. Rooms and apartments near the campus are available year-round. Shuttle bus service is available.

Student Group

The enrollment of the University of Texas at Austin is over 50,000, including approximately 11,500 graduate students. The College of Natural Sciences has about 7,250 students. An average of 10–20 graduate students reside at the Marine Science Institute in Port Aransas.

Location

Austin is the state capital, with a population of approximately 350,000. Cultural events sponsored by the University are abundant, and there are many recreational facilities available. Port Aransas is a small coastal town approximately 200 miles south of Austin, where the Gulf of Mexico and surrounding bays and estuaries provide excellent boating, fishing, and swimming.

The University

The University of Texas at Austin was founded in 1883 and is part of the University of Texas System. The Department of Marine Science is in the College of Natural Sciences.

Applying

Prospective students must apply to both the director of admissions of the Graduate School and the Graduate Studies Committee of the Department of Marine Science in order to be considered for admission to the department. Application forms may be obtained from the Graduate School and from the department office. Only admission applications completed by February 1 can be considered for fellowship or teaching assistantship awards.

Correspondence and Information

Graduate Adviser
Marine Science Institute
P.O. Box 1267
Port Aransas, Texas 78373
Telephone: 512-749-6721

Chairman
Department of Marine Science
University of Texas at Austin
Port Aransas, Texas 78373

SECTION 4: MARINE SCIENCES/OCEANOGRAPHY

University of Texas at Austin

THE FACULTY AND THEIR RESEARCH

E. William Behrens, Associate Professor, Austin; Ph.D., Rice, 1963. Marine geology, sedimentology, continental slopes.

Ronald H. Benner, Assistant Professor, Port Aransas; Ph.D., Georgia, 1984. Microbial ecology and biogeochemistry, microbial utilization of dissolved and particulate organic matter, lignin degradation, role of bacteria in food webs, nutrient cycling.

Edward J. Buskey, Assistant Professor, Port Aransas; Ph.D., Rhode Island, 1983. Marine plankton ecology, sensory perception and behavior of marine organisms, bioluminescence of marine organisms.

James N. Cameron, Professor, Port Aransas; Ph.D., Texas at Austin, 1969. Gas transfer in gills, regulation of CO_2 and pH, dynamics of circulation in lower vertebrates and invertebrates, ion exchange mechanisms.

Kenneth H. Dunton, Assistant Professor, Port Aransas; Ph.D., Alaska Fairbanks, 1985. Physiological ecology of marine algae and seagrasses, in situ production in marine macrophytes, trophic relations in seagrass and algal communities, photosynthetic performance in arctic and antarctic macroalgae.

Lee A. Fuiman, Assistant Professor, Port Aransas; Ph.D., Michigan, 1983. Ichthyology, especially biology and ecology of fish larvae; behavior, ecology, organismal biology, evolution, systematics of fishes.

Robert S. Jones, Professor and Chairman, Port Aransas; Ph.D., Hawaii, 1967. Ecology of marine fishes, with emphasis on habitat requirements, trophic interactions, and nondestructive censusing techniques.

Paul A. Montagna, Assistant Professor, Port Aransas; Ph.D., South Carolina, 1983. Marine benthic ecology, animal-sediment interactions, invertebrate-microbial trophic dynamics, biogeochemistry, ecosystems ecology, modeling, biostatistics.

J. Robert Moore, Professor, Austin; Ph.D., Wales, 1964. Geological oceanography, marine minerals exploration and recovery, marine sediments, marine mining.

Carl H. Oppenheimer, Professor, Austin; Ph.D., UCLA (Scripps Institution of Oceanography), 1951. Ecology, microbiology, and geochemistry; hydrocarbon microbiology, geomicrobiology, nutrient cycles and coastal-zone management for estuarine environments.

Patrick L. Parker, Professor, Port Aransas; Ph.D., Arkansas, 1960. Marine chemistry and geochemistry; stable carbon, oxygen, and nitrogen isotope ratio variations in geochemical, natural, and aquaculture food webs; organic chemistry of marine waters and sediments, especially lipids; environmental chemistry of petroleum.

Curtis A. Suttle, Assistant Professor, Port Aransas; Ph.D., British Columbia, 1987. Nutrient cycling in aquatic food webs, nutrient competition and metabolism in phytoplankton and bacteria, indigenous marine viruses.

Peter Thomas, Professor, Port Aransas; Ph.D., Leicester (England), 1977. Fish reproductive physiology, purification and molecular actions of hormones, environmental endocrinology, applications of endocrinology to fish culture, biochemical and environmental toxicology of marine fishes, especially reproduction.

Terry E. Whitledge, Associate Professor, Port Aransas; Ph.D., Washington (Seattle), 1973. Marine chemistry, nutrients, instrument development, Bering Sea chemistry.

Section 5
Mathematical Sciences

This section contains directories of institutions offering graduate work in applied mathematics, biometrics, biostatistics, mathematics, and statistics, followed by two-page entries submitted by institutions that chose to prepare detailed program descriptions. Additional information about programs listed in the directories but not augmented by a two-page entry may be obtained by writing directly to the dean of a graduate school or chair of a department at the address given in the directory.

For programs offering work in related fields, see all other areas in this book; in Book 2, see: Economics; Library Science; and Psychology; in Book 3: Agricultural Sciences; Biology and Biomedical Sciences; Biophysics; Genetics and Developmental Biology; Natural Resource Sciences; and Pharmacology and Toxicology; in Book 5: Biomedical Engineering; Computer and Information Sciences; Electrical and Power Engineering; Engineering and Applied Sciences; and Industrial/Management Engineering, Operations Research, and Systems Engineering; and in Book 6: Business Administration and Management and Public and Community Health.

CONTENTS

Program Directories

Applied Mathematics	338
Biometrics	345
Biostatistics	347
Mathematics	350
Statistics	375

Announcements

Brandeis University	351
Brown University	351
California State University, Fresno	351
California State University, Fullerton	352
California State University, Los Angeles	352
Case Western Reserve University	347
Claremont Graduate School	353
Clemson University	
Mathematical Sciences	353
Statistics	376
College of William and Mary	353
Cornell University	
Applied Mathematics	338
Biometry	345
Drexel University	347
Emory University	354
Florida International University	355
Georgia Institute of Technology	355
Harvard University	377
Kent State University	357
Louisiana State University Medical Center	345
Loyola University Chicago	357
Miami University	358
New Jersey Institute of Technology	340
North Carolina State University	
Biomathematics	346
Statistics	378
Pennsylvania State University	379
Princeton University	379
Purdue University	379
Rensselaer Polytechnic Institute	341
Rutgers, The State University of New Jersey, New Brunswick	361
South Dakota State University	362
State University of New York College at Brockport	363
Temple University	364

University of Alabama at Birmingham	
Biomathematics	346
Biostatistics	348
University of California at Berkeley	381
University of California, Davis	381
University of California, Santa Barbara	381
University of California, Santa Cruz	366
University of Cincinnati	366
University of Denver	367
University of Florida	367
University of Illinois at Chicago	343, 367, 382
University of Minnesota, Duluth	343
University of Minnesota, Twin Cities Campus	349
University of North Carolina at Chapel Hill	
Biostatistics	349
Mathematics	369
University of North Florida	369
University of Rhode Island	384
University of Southern California	346
University of Texas at Dallas	371
University of Virginia	372

Cross-Discipline Announcements

Case Western Reserve University	386
Ohio State University	386
Southern Illinois University at Carbondale	386
University of Arizona	386
University of California at Berkeley	386
University of California, Los Angeles	386
University of Washington	386
Virginia Commonwealth University	386

Full Descriptions

American University	387
Arizona State University	389
Brown University	
Applied Mathematics	391
Mathematics	393
California Institute of Technology	395
Carnegie Mellon University	
Mathematics	397
Statistics	399
Colorado State University	401
Columbia University	403
Cornell University	405
Dartmouth College	407
Drexel University	409
Emory University	
Biostatistics	411
Mathematics and Computer Science	413
George Washington University	415
Howard University	417
Johns Hopkins University	
Mathematical Sciences	419
Mathematics	421
Lehigh University	423
Medical University of South Carolina	425
Michigan State University	427
Michigan Technological University	429
Mount Sinai School of Medicine of the City University of New York	431
New York University	433
Northwestern University	
Applied Mathematics	435
Mathematics	437
Ohio State University	439
Old Dominion University	441
Oregon State University	
Mathematics	443
Statistics	445

Peterson's Guide to Graduate Programs in the Physical Sciences and Mathematics 1992

SECTION 5: MATHEMATICAL SCIENCES

Contents; Field Definitions

Purdue University	
Mathematics	447
Statistics	449
Rensselaer Polytechnic Institute	
Decision Sciences and Engineering Systems	451
Mathematical Sciences	453
Simon Fraser University	455
State University of New York at Stony Brook	457
Temple University	459
University of Arizona	461
University of California, Los Angeles	
Biomathematics	463
Biostatistics	465
University of California, Riverside	467
University of Delaware	469
University of Florida	471
University of Georgia	
Mathematics	473
Statistics	475
University of Houston	477
University of Illinois at Chicago	479
University of Illinois at Urbana-Champaign	481
University of Iowa	483
University of Kentucky	485
University of Maryland Graduate School, Baltimore	487
University of Notre Dame	489
University of Rochester	491
University of Southern California	493
University of South Florida	495
University of Southwestern Louisiana	497
University of Utah	499
University of Washington	
Applied Mathematics	501
Biostatistics	503
University of Wyoming	505
Utah State University	507
Vanderbilt University	509
Virginia Polytechnic Institute and State University	
Mathematics	511
Statistics	513
Wesleyan University	515
Yale University	517

See also:

Institute of Paper Science and Technology—Chemistry	109

FIELD DEFINITIONS

In an effort to broaden prospective students' understanding of disciplines in mathematical sciences, educators in the field have submitted the following statements on applied mathematics, biometrics, biostatistics, mathematics, and statistics.

Applied Mathematics

The American tradition of mathematical education neglected the study of applied mathematics as a full-fledged academic discipline until the time of the Second World War, when the exigencies of wartime research drew attention to a shortage of American-trained applied mathematicians. The first U.S. universities to introduce systematic programs in this field were New York University under Richard Courant and Brown University under William Prager, scholars who were, through their backgrounds, able to draw on the long-standing European tradition of teaching pure and applied mathematics in an integrated manner.

The training of applied mathematicians has two fairly distinct aspects: they have to provide themselves with a broad arsenal of mathematical methods and techniques that will be essential to them in their professional life, and they have to study one or more application areas in order to learn firsthand how mathematics is used in actual, real-life problems—whether in a physical, biological, or social science; in engineering; or in industry and government. In their study of mathematics they will naturally gravitate toward the more applicable subdisciplines, and in applications, toward those currently posing mathematical challenge and promise. The latter have changed over the years. "Classical" applied mathematics principally addressed itself to the mathematical problems posed by the mechanics of particles, of rigid and deformable bodies, and of fluids and also by electromagnetic theory.

Although there are still many outstanding and challenging problems in these "classical" fields—for instance, those concerning turbulent fluid flow and the mechanics of unorthodox materials—the purview of applied mathematics has been considerably broadened since, perhaps, the middle 1950s. This is partly due to advances in mathematics itself, many stimulated by applications, but also to the fact that we can now profitably attack more complex and less idealized problems than before; the tool of computer simulation has been added to our armory. Large-scale scientific computation has in fact become an important subfield of applied mathematics, yielding challenging analytical and algorithmic problems. In addition, the emergence of supercomputers—both vector machines and massively parallel processors—is providing new challenges to applied mathematicians. These computers require different mathematical methods for their efficient utilization from those implemented on orthodox architectures and require numerical algorithms of a different sort for the solution of large-scale problems in meteorology, oil exploration, operations research and optimization, economic forecasting, aircraft and spacecraft design, nuclear technology, etc. In addition to the physical sciences, the medical, biological, and social sciences now provide a host of stimulating applications for mathematical analysis. With this has come a widening of job opportunities for graduates. No longer do they largely enter the academic profession. Industry, commerce, and government have come to appreciate the analytical and computational skills of Ph.D. graduates trained in the better applied mathematics departments.

Walter Freiberger, Professor and Associate Chairman
Division of Applied Mathematics
Brown University

Biometrics

The field of biometrics is concerned with the application of quantitative methodology to biological or, more generally, scientific problems concerning natural phenomena. In scope, biometrics can refer to such diverse fields as medicine, meteorology, genetics, agriculture, sociology, chemistry, epidemiology, physics, and demography. One specialty is medical information sciences, relating to medical information processing by computers.

Biometrics is a natural field of further study for the student with strength in both mathematics and biology and the desire to utilize his or her background in both areas and in a professional field where employment opportunities remain good.

In complexity, the practice of biometrics can range from the simple calculation of an average to the derivation of original quantitative theory or the development of sophisticated computer software. A student's background should generally include matrix algebra and analysis (calculus and differential equations), chemistry, general biology, genetics, physiology, and exposure to computers.

Richard H. Jones, Professor
Department of Preventive Medicine and Biometrics
University of Colorado School of Medicine

Biostatistics

Biostatistics is an integral part of biomedical and health research. Statistical models of observed phenomena and randomization in the design of experiments form the basis for inductive reasoning. Generalizations from such studies involve statistics in planning, analysis, and interpretation. The specialized statistical approaches for variables and investigations that are biomedical, epidemiologic, and health-service related constitute the field of biostatistics. Biostatistics has developed as a separate field of statistics not just because it is an intrinsic part of life science but because of society's concern over the quality of the environment, the development of new medical treatments, the provision of health services, the prevention of disease, and the assurance of the efficacy and safety of drug regimens.

Graduate study in biostatistics at the master's level prepares students for careers in the application of statistical methods to the design and analysis of biomedical and clinical research studies and the planning and evaluation of epidemiologic investigations and health services programs. In addition, doctoral study provides the opportunity for research on biostatistical methodology and greater responsibility and leadership in collaborative research. The demand for biostatistics graduates at both the master's and doctoral levels is very high, and it is anticipated that the demand will continue to grow.

The understanding and development of biostatistical methodology requires a background in mathematics, at least through calculus and linear algebra. Thus the primary preparation for graduate study in biostatistics is in mathematics. Course work in statistics and a cognate area related to health, but not necessarily biology itself, is needed at the undergraduate level mainly to stimulate interest in the application of mathematics and statistics to the health sciences and to provide the motivation for graduate study and a career in biostatistics.

Richard G. Cornell, Professor
Department of Biostatistics
University of Michigan

Mathematics

Mathematics is the broadest of all scientific fields, ranging over an enormous and ever-growing variety of subjects, styles, and applications. Yet a common structure subtly links the various branches, and the unity of mathematics is triumphantly demonstrated over and over again by discoveries of the unsuspected relevance of results and methods of one field to another one, seemingly far removed. No mathematician understands more than a fraction of the body of mathematics, except on a superficial level; no single department, even among the largest ones, can adequately teach all of it.

The origins of mathematics lie in concrete problems outside of mathematics; yet most mathematics develops as an autonomous subject, and its branches are usually taught as such. This line of development has been crowned with many recent successes, in topology (classification of 4 manifolds), in number theory (rational solution of algebraic equations), in algebra (classification of finite simple groups), in differential geometry, in several complex variables, etc. There is also a strong recent trend back to external sources. The study of dynamical systems is flourishing, spurred by recent startling discoveries (new classes of completely integrable systems and their stability, solitons, universal features of chaotic behavior). Fluid dynamics has undergone a similar recrudescence (strange attractors and other aspects of turbulence, propagation of shock waves). Mathematics has made new contacts with physics, in particular in statistical mechanics, quantum field theories, and plasma dynamics; relatively new applied branches are beginning to emerge, such as mathematical physiology, biology, and ecology; and attempts are being made to mathematicize some of the social sciences. Much, though not all, of this resurgence of interest in applied topics has been stimulated by the availability of powerful computer systems, which has made numerical modeling possible for purposes of design and to discover new phenomena. Computing is becoming an increasingly powerful tool in pure mathematics as well; furthermore, the problem of utilizing computers efficiently and the study of the structure of computer programs are emerging as separate disciplines on the border between mathematics, logic, and computer science.

In view of this diversity, no two departments of mathematics are alike. Prospective graduate students should try to match their interests with those of the department where they plan to study; they should make sure that they are evaluating departments on the basis of up-to-date information.

There are today good positions available for teaching and research in departments of mathematics. There are also good opportunities for employment to do research in industrial or national laboratories; previous acquaintance with the spirit and some of the substance of applied mathematics is good preparation for such positions, where mathematicians often work in collaboration with physicists, engineers, or even biologists.

Mathematics is a highly technical subject; advanced graduate training is partly an apprenticeship, and the Ph.D. dissertation is partly a demonstration of the candidate's proficiency in using the tools of the trade. However, the motive for doing mathematics is not to overcome technical difficulties but rather to gain understanding.

Peter D. Lax
Professor of Mathematics
Courant Institute of Mathematical Sciences
New York University

Statistics

Statistics deals with the collection, description, and analysis of data. The field is broad with respect to the theory-applications spectrum and also in terms of the diversity of fields of applications. Statistics is widely applied in the social, biological, chemical, and management sciences. Research topics in statistics reflect the breadth of the field in both of the above dimensions. Graduate work typically requires a strong undergraduate math background; a background in computing is increasingly important.

Employment opportunities for statisticians at the M.S. and Ph.D. levels are excellent and promise to be so for the foreseeable future. Opportunities exist in universities, federal and state government agencies, and industry. Notable examples include federal health agencies such as NIH and FDA; the National Bureau of Standards; the biopharmaceutical, chemical, automotive, and communications industries; and private consulting firms.

William E. Strawderman
Professor and Chairman
Department of Statistics
Rutgers University, New Brunswick

Applied Mathematics

Air Force Institute of Technology, School of Engineering, Department of Mathematics and Computer Science, Wright-Patterson AFB, OH 45433. Offerings include applied mathematics (MS, PhD). Department faculty: 19 full-time (2 women), 0 part-time. *Degree requirements:* Computer language, thesis/dissertation required, foreign language not required. *Entrance requirements:* For master's, GRE General Test (minimum combined score of 1100 required), minimum GPA of 3.0, must be military officer, DOD civilian, or government employee; for doctorate, Gre General Test (minimum combined score of 1200 required), minimum GPA of 3.5, must be military officer, DOD civilian, or government employee. *Tuition:* $0. • Dr. Vadim Komkov, Head, 513-255-3098.

American University, College of Arts and Sciences, Department of Mathematics and Statistics, Program in Applied Mathematics, Washington, DC 20016. Program awards MA. Part-time and evening/weekend programs available. Faculty: 16 full-time (4 women), 2 part-time (0 women). Matriculated students: 0. 14 applicants, 71% accepted. In 1990, 0 degrees awarded. *Degree requirements:* 1 foreign language required (computer language can substitute), thesis optional. *Application deadline:* 2/15. *Application fee:* $50. *Expenses:* Tuition of $475 per semester hour. Fees of $20 per semester. *Financial aid:* Fellowships, teaching assistantships, federal work-study, institutionally sponsored loans, and career-related internships or fieldwork available. Aid available to part-time students. Financial aid application deadline: 2/15; applicants required to submit FAF. • Dr. Robert W. Jernigan, Chair, Department of Mathematics and Statistics, 202-885-3120.

Arizona State University, College of Liberal Arts and Sciences, Department of Mathematics, Tempe, AZ 85287. Offerings include applied mathematics (MA, PhD). *Degree requirements:* For doctorate, 1 foreign language, dissertation. *Entrance requirements:* For doctorate, GRE. Application fee: $25. *Tuition:* $1528 per year full-time, $80 per hour part-time for state residents; $6934 per year full-time, $289 per hour part-time for nonresidents. • Dr. William Trotter, Chair, 602-965-3951.

See full description on page 389.

Bowling Green State University, College of Arts and Sciences, Department of Mathematics and Statistics, Bowling Green, OH 43403. Offerings include applied mathematics (MA). Department faculty: 27 full-time (1 woman), 0 part-time. *Application deadline:* 8/7. *Application fee:* $10. *Tuition:* $185 per credit hour for state residents; $357 per credit hour for nonresidents. • Dr. Hassoon Al-Amiri, Chair, 419-372-2636. Application contact: Dr. James Albert, Graduate Coordinator, 419-372-7456.

Brown University, Division of Applied Mathematics, Providence, RI 02912. Division awards Sc M, PhD. Faculty: 24 full-time (1 woman), 0 part-time. Matriculated students: 63 full-time (13 women), 6 part-time (1 woman); includes 8 minority (7 Asian American, 1 black American), 32 foreign. 90 applicants, 16% accepted. In 1990, 11 master's, 14 doctorates awarded. *Degree requirements:* For master's, thesis or alternative required, foreign language not required; for doctorate, 1 foreign language, dissertation, oral exam. *Entrance requirements:* GRE General Test. Application deadline: 1/2. Application fee: $40. *Expenses:* Tuition of $16,256 per year full-time, $2032 per course part-time. Fees of $372 per year. *Financial aid:* In 1990–91, $1.081-million in aid awarded. 9 fellowships (4 to first-year students), 24 research assistantships (2 to first-year students), 14 teaching assistantships (0 to first-year students), 1 assistantship (0 to first-year students) were awarded; full and partial tuition waivers, federal work-study, institutionally sponsored loans also available. Financial aid application deadline: 1/2; applicants required to submit GAPSFAS. • Harold J. Kushner, Chairman, 401-863-2353. Application contact: Allen Pipkin, Graduate Representative, 401-863-2354.

See full description on page 391.

California Institute of Technology, Division of Engineering and Applied Science, Option in Applied Mathematics, Pasadena, CA 91125. Option awards PhD. Faculty: 6 full-time (0 women), 0 part-time. Matriculated students: 17 full-time (1 woman), 0 part-time; includes 0 minority, 8 foreign. 70 applicants, 9% accepted. In 1990, 5 doctorates awarded. Terminal master's awarded for partial completion of doctoral program. *Degree requirements:* For doctorate, dissertation required, foreign language not required. *Entrance requirements:* For doctorate, GRE Subject Test. Application deadline: 1/15. Application fee: $0. *Expenses:* Tuition of $14,100 per year. Fees of $8 per quarter. *Faculty research:* Theoretical and computational fluid mechanics, numerical analysis, ordinary and partial differential equations, linear and nonlinear wave propagation, perturbation and asymptotic methods. • Executive Officer, 818-356-4560. Application contact: Donald S. Cohen, Option Representative, 818-356-4560.

See full description on page 395.

California State Polytechnic University, Pomona, College of Science, Department of Mathematics, Pomona, CA 91768. Offerings include applied mathematics (MS). *Degree requirements:* Thesis or alternative. *Application fee:* $55. *Expenses:* Tuition of $0 for state residents; $137 per unit for nonresidents. Fees of $880 per year full-time, $182 per quarter (minimum) part-time. • Dr. Scott Sportsman, Coordinator, 714-869-3478.

California State University, Hayward, School of Science, Department of Mathematics and Computer Science, Hayward, CA 94542. Offerings include applied mathematics (MS). Department faculty: 32. *Degree requirements:* Thesis or comprehensive exam required, foreign language not required. *Entrance requirements:* Minimum GPA of 3.0 in field. Application deadline: 4/19 (priority date, applications processed on a rolling basis). Application fee: $55. *Expenses:* Tuition of $0 for state residents; $137 per unit for nonresidents. Fees of $895 per year full-time, $188 per quarter part-time. • Dr. Edward Keller, Chair, 415-881-3414. Application contact: Dr. Robert Trinchero, Acting Associate Vice President, Admissions and Enrollment, 415-881-3828.

California State University, Long Beach, School of Humanities, Department of Mathematics, 1250 Bellflower Boulevard, Long Beach, CA 90840. Offerings include applied mathematics (MA). *Degree requirements:* Thesis or comprehensive exam required, foreign language not required. *Application deadline:* 8/1 (applications processed on a rolling basis). *Application fee:* $55. *Expenses:* Tuition of $0 for state residents; $246 per unit for nonresidents. Fees of $1120 per year full-time, $724 per year part-time. • Dr. Samuel G. Councilman, Chair, 310-985-4721.

California State University, Los Angeles, School of Natural and Social Sciences, Department of Mathematics and Computer Science, Los Angeles, CA 90032. Offerings include mathematics (MS), with options in applied mathematics, mathematics. Department faculty: 26 full-time, 44 part-time. *Degree requirements:* Thesis or comprehensive exam required, foreign language not required. *Entrance requirements:* TOEFL (minimum score of 550 required), previous course work in mathematics. Application deadline: 8/7 (applications processed on a rolling basis). Application fee: $55. *Expenses:* Tuition of $0 for state residents; $164 per unit for nonresidents. Fees of $1046 per year full-time, $650 per semester (minimum) part-time. • Dr. Marshall Cates, Chair, 213-343-2150.

Carnegie Mellon University, Mellon College of Science, Department of Mathematics, Pittsburgh, PA 15213. Offerings include applied mathematics (MS). Department faculty: 32 full-time (2 women), 0 part-time. *Degree requirements:* Foreign language and thesis not required. *Entrance requirements:* GRE General Test, TOEFL. Application deadline: 2/10. Application fee: $0. *Expenses:* Tuition of $15,250 per year full-time, $212 per unit part-time. Fees of $80 per year. • Dr. William O. Williams, Head, 412-268-2545.

See full description on page 397.

Case Western Reserve University, Department of Mathematics and Statistics, Cleveland, OH 44106. Offerings include applied mathematics (MS). Department faculty: 22 full-time, 3 part-time. *Degree requirements:* Thesis not required. *Entrance requirements:* GRE General Test, GRE Subject Test, TOEFL (minimum score of 550 required). Application deadline: 6/26 (applications processed on a rolling basis). Application fee: $25. *Expenses:* Tuition of $13,600 per year full-time, $567 per credit part-time. Fees of $320 per year. • C. A. Cullis, Chairman, 216-368-2880. Application contact: Program Coordinator, 216-368-2918.

Central Missouri State University, College of Arts and Sciences, Department of Mathematics and Computer Science, Warrensburg, MO 64093. Offerings include applied mathematics (MS). Department faculty: 23 (7 women). *Application deadline:* 6/30 (priority date, applications processed on a rolling basis). *Application fee:* $0. *Tuition:* $1968 per year full-time, $82 per hour part-time for state residents; $3744 per year full-time, $156 per hour part-time for nonresidents. • Dr. Edward W. Davenport, Chair, 816-429-4931.

Claremont Graduate School, Department of Mathematics, Claremont, CA 91711. Offerings include applied mathematics (MA, MS). Department faculty: 5 full-time (0 women), 32 part-time (4 women). *Degree requirements:* Foreign language and thesis not required. *Entrance requirements:* GRE General Test. Application deadline: 2/15 (priority date, applications processed on a rolling basis). Application fee: $40. *Expenses:* Tuition of $13,900 per year full-time, $620 per unit part-time. Fees of $55 per semester. • William Lucas, Chairman, 714-621-8080.

Clark Atlanta University, School of Arts and Sciences, Department of Mathematical Sciences, Atlanta, GA 30314. Offerings include applied mathematics (MS). *Degree requirements:* 1 foreign language (computer language can substitute), thesis. *Entrance requirements:* GRE General Test, minimum GPA of 2.5. Application deadline: 4/1 (applications processed on a rolling basis). Application fee: $40. *Expenses:* Tuition of $4860 per year. Fees of $300 per year. • Dr. Abdulelim Shebezz, Chairman, 404-880-8540. Application contact: Peggy Wade, Marketing and Recruitment, 404-880-8427.

Clemson University, College of Sciences, Department of Mathematical Sciences, Clemson, SC 29634. Offerings include applied and core mathematics (MS, PhD). Department faculty: 37 full-time (4 women), 0 part-time. *Degree requirements:* For master's, computer language, final project required, thesis optional, foreign language not required; for doctorate, computer language, dissertation, qualifying exams required, foreign language not required. *Entrance requirements:* GRE General Test. Application deadline: 6/1 (priority date). Application fee: $25. *Expenses:* Tuition of $102 per credit hour. Fees of $80 per semester full-time. • J. P. Jarvis, Head, 803-656-3434.

Cleveland State University, College of Arts and Sciences, Department of Mathematics, Program in Applied Mathematics, Cleveland, OH 44115. Program awards MS. Part-time programs available. Faculty: 17 full-time (1 woman), 0 part-time. Matriculated students: 0 full-time, 10 part-time (4 women); includes 1 minority (Asian American), 0 foreign. Average age 29. 10 applicants, 80% accepted. In 1990, 8 degrees awarded. *Degree requirements:* Foreign language and thesis not required. *Application deadline:* 9/1 (priority date, applications processed on a rolling basis). *Application fee:* $0. *Tuition:* $90 per credit for state residents; $180 per credit for nonresidents. *Financial aid:* In 1990–91, $12,270 in aid awarded. 2 teaching assistantships (0 to first-year students) were awarded; federal work-study, institutionally sponsored loans also available. • Dr. John Chao, Chairperson, 216-687-4680. Application contact: Ann Melville, Administrative Secretary, 216-523-7162.

Colorado School of Mines, Department of Mathematical and Computer Sciences, Golden, CO 80401. Department offers program in applied mathematics (MS, PhD). Faculty: 19 full-time (5 women), 5 part-time (3 women). Matriculated students: 36 full-time (7 women), 21 part-time (7 women); includes 13 foreign. 57 applicants, 86% accepted. In 1990, 10 master's, 1 doctorate awarded. *Degree requirements:* For master's, thesis required, foreign language not required; for doctorate, dissertation, written and oral comprehensive exams required, foreign language not required. *Entrance requirements:* GRE General Test, minimum GPA of 3.0. Application deadline: 4/15 (priority date, applications processed on a rolling basis). Application fee: $15 ($25 for foreign students). *Expenses:* Tuition of $3178 per year full-time, $124 per semester hour part-time for state residents; $10,304 per year full-time, $344 per semester hour part-time for nonresidents. Fees of $374 per year full-time. *Financial aid:* In 1990–91, 2 fellowships, 11 research assistantships, 7 teaching assistantships awarded. *Faculty research:* Classical applied mathematics, operations research, applied statistics, computer science, numerical computation. Total annual research budget: $749,073. • Dr. Ardel Boes, Head, 303-273-3860.

Columbia University, School of Engineering and Applied Science, Department of Applied Physics, Program in Applied Physics, New York, NY 10027. Offerings include applied physics (MS, Eng Sc D, PhD), with options in applied mathematics (PhD), plasma physics (MS, Eng Sc D, PhD), quantum electronics (MS, Eng Sc D, PhD), solid state physics (MS, Eng Sc D, PhD). Terminal master's awarded for partial completion of doctoral program. Program faculty: 9 full-time (0 women), 6 part-time (1 woman). *Application deadline:* 2/1. *Application fee:* $45. *Tuition:* $15,520 per year full-time, $518 per credit part-time. • Dr. Gerald A. Navratil, Chairman, Department of Applied Physics, 212-854-4457.

Concordia University, Faculty of Arts and Science, Department of Mathematics and Statistics, Montreal, PQ H3G 1M8, Canada. Offerings include applied mathematics (MA, M Sc). *Application fee:* $15. *Expenses:* Tuition of $10 per credit for Canadian residents; $195 per credit for nonresidents. Fees of $223 per year full-time, $38.85 per year part-time for Canadian residents; $118 per year full-time, $35.35 per year part-time for nonresidents. • Dr. W. Byers, Chair, 514-848-3232.

Cornell University, Graduate Fields of Arts and Sciences, Center for Applied Mathematics, Ithaca, NY 14853. Center awards PhD. Faculty: 79 full-time, 0 part-time. Matriculated students: 36 full-time (6 women), 0 part-time; includes 1 minority (Hispanic American), 18 foreign. 202 applicants, 5% accepted. In 1990, 5 master's, 5 doctorates awarded. Terminal master's awarded for partial completion of doctoral program. *Degree requirements:* For doctorate, 1 foreign language, dissertation. *Entrance requirements:* For doctorate, GRE General Test, TOEFL. Application deadline: 1/10. Application fee: $55. *Expenses:* Tuition of $16,170 per year. Fees of $28 per year. *Financial aid:* In 1990–91, 29 students received aid. 3 fellowships

(0 to first-year students), 11 research assistantships (0 to first-year students), 15 teaching assistantships (4 to first-year students) awarded; full and partial tuition waivers, federal work-study, institutionally sponsored loans also available. Financial aid application deadline: 1/10; applicants required to submit GAPSFAS. *Faculty research:* Applied mathematics; discrete and numerical mathematics; information and control theory; mechanics and dynamics; probability and statistics. • John Guckenheimer, Graduate Faculty Representative, 607-255-4335. Application contact: Robert Brashear, Director of Admissions, 607-255-4884.

Announcement: The Center for Applied Mathematics is an interdepartmental program with over 70 faculty members. Students may pursue studies over a broad range of the mathematical sciences and are admitted to the field from a variety of educational backgrounds with strong mathematical components. Students are normally awarded fellowships or teaching or research assistantships.

Cornell University, Graduate Fields of Engineering, Field of Operations Research, Ithaca, NY 14853. Offerings include applied probability and statistics (MS, PhD), mathematical programming (MS, PhD). Terminal master's awarded for partial completion of doctoral program. Faculty: 31 full-time, 0 part-time. *Degree requirements:* For doctorate, 1 foreign language, dissertation, oral exam. *Entrance requirements:* For doctorate, GRE General Test, TOEFL. Application deadline: 1/10. Application fee: $55. *Expenses:* Tuition of $16,170 per year. Fees of $28 per year. • Robert Bland, Graduate Faculty Representative, 607-255-5088. Application contact: Robert Brashear, Director of Admissions, 607-255-4884.

DePaul University, College of Liberal Arts and Sciences, Department of Mathematical Sciences, Program in Applied Mathematics, Chicago, IL 60604. Program awards MS. Matriculated students: 17 full-time (8 women), 22 part-time (7 women); includes 6 minority (3 Asian American, 3 black American), 0 foreign. 35 applicants, 66% accepted. In 1990, 3 degrees awarded. *Degree requirements:* Computer language, comprehensive exam required, foreign language and thesis not required. *Entrance requirements:* GRE. Application deadline: rolling. Application fee: $20. *Expenses:* Tuition of $215 per quarter hour. Fees of $10 per quarter hour (minimum). • Dr. Constantine Georgakis, Chairman, Department of Mathematical Sciences, 312-362-6769.

East Carolina University, College of Arts and Sciences, Department of Mathematics, Greenville, NC 27858-4353. Offerings include applied mathematics (MA). Department faculty: 19. *Application deadline:* 6/1. *Application fee:* $25. *Tuition:* $627 per semester for state residents; $3154 per semester for nonresidents. • Dr. Robert Hursey, Director of Graduate Studies, 919-757-6461. Application contact: Paul D. Tschetler, Assistant Dean, 919-757-6012.

École Polytechnique de Montréal, Department of Mathematics, Montreal, PQ H3C 3A7, Canada. Offerings include mathematical method in CA engineering (M Eng, M Sc A, PhD). Department faculty: 14 full-time (2 women), 4 part-time (0 women). *Degree requirements:* 1 foreign language, computer language, thesis/dissertation. *Entrance requirements:* For master's, minimum GPA of 2.75; for doctorate, minimum GPA of 3.0. Application deadline: 4/1. Application fee: $15. *Tuition:* $1400 per year full-time, $52 per credit part-time for Canadian residents; $7200 per year full-time, $290 per credit part-time for nonresidents. • Jacques Gauvin, Chairman, 514-340-4824.

Florida Institute of Technology, College of Engineering, Department of Applied Mathematics, Melbourne, FL 32901. Department awards MS, PhD. Part-time programs available. Faculty: 13 full-time (3 women), 6 part-time (1 woman). Matriculated students: 11 full-time (4 women), 11 part-time (6 women); includes 1 minority (Asian American), 18 foreign. 28 applicants, 29% accepted. In 1990, 3 master's, 0 doctorates awarded. *Degree requirements:* For master's, computer language, comprehensive exam required, thesis optional, foreign language not required; for doctorate, dissertation, comprehensive exam required, foreign language not required. *Entrance requirements:* For master's, proficiency in FORTRAN, minimum GPA of 3.0; for doctorate, minimum GPA of 3.2. Application fee: $35. *Tuition:* $234 per credit hour. *Financial aid:* Research assistantships, teaching assistantships, federal work-study available. Financial aid application deadline: 3/1. *Faculty research:* Methods of nonlinear analysis, spectral theory of operators, reaction diffusion equations, mathematical modeling. • Dr. V. Lakshmikantham, Head, 407-768-8000 Ext 8091. Application contact: Carolyn P. Farrior, Director of Graduate Admissions, 407-768-8000 Ext. 8027.

Florida State University, College of Arts and Sciences, Department of Mathematics, Program in Applied Mathematics, Tallahassee, FL 32306. Program awards MA, MS, PhD. Part-time programs available. Faculty: 11 full-time (0 women), 0 part-time. Matriculated students: 32 full-time (2 women), 0 part-time; includes 0 minority, 23 foreign. Average age 25. 57 applicants, 56% accepted. In 1990, 6 master's awarded (33% found work related to degree, 67% continued full-time study); 2 doctorates awarded (100% entered university research/teaching). Terminal master's awarded for partial completion of doctoral program. *Degree requirements:* For master's, thesis optional, foreign language not required; for doctorate, 2 foreign languages, computer language, dissertation. *Entrance requirements:* For master's, GRE General Test (minimum score of 650 on quantitative section, 400 on verbal required), minimum GPA of 3.0; for doctorate, minimum GPA of 3.0. Application deadline: 11/1 (priority date, applications processed on a rolling basis). Application fee: $15. *Tuition:* $76.29 per credit hour for state residents; $238 per credit hour for nonresidents. *Financial aid:* In 1990–91, 27 students received aid. 0 fellowships, 5 research assistantships (0 to first-year students), 22 teaching assistantships (7 to first-year students) awarded; institutionally sponsored loans also available. Average monthly stipend for a graduate assistantship: $1065. Financial aid application deadline: 11/1; applicants required to submit FAF. *Faculty research:* Fluid dynamics, computational methods, partial differential equations, numerical analysis. Total annual research budget: $500,000. • Dr. Christopher Hunter, Director, 904-644-6467. Application contact: James E. Miller, Academic Programs Coordinator, 904-644-5868.

George Washington University, Graduate School of Arts and Sciences, Department of Mathematics, Program in Applied Mathematics, Washington, DC 20052. Program awards MA, MS. Part-time and evening/weekend programs available. Matriculated students: 1 (woman) full-time, 3 part-time (0 women); includes 1 minority (Asian American), 2 foreign. Average age 31. 7 applicants, 57% accepted. In 1990, 3 degrees awarded. *Degree requirements:* Thesis or alternative, comprehensive exam required, foreign language not required. *Entrance requirements:* GRE General Test, GRE Subject Test, minimum GPA of 3.0. Application deadline: 7/1. Application fee: $45. *Expenses:* Tuition of $490 per semester hour. Fees of $215 per year full-time, $125.60 per year (minimum) part-time. *Financial aid:* In 1990–91, Fellowships, teaching assistantships available. Financial aid application deadline: 2/15. • Dr. Irving Katz, Chair, Department of Mathematics, 202-994-6235.

Harvard University, Graduate School of Arts and Sciences, Division of Applied Sciences, Cambridge, MA 02138. Offerings include applied mathematics (ME, SM, PhD). Terminal master's awarded for partial completion of doctoral program. Division faculty: 58 full-time, 0 part-time. *Degree requirements:* For master's, thesis not required; for doctorate, dissertation. *Entrance requirements:* GRE General Test, GRE Subject Test. Application deadline: 1/2. Application fee: $60. *Expenses:* Tuition of $14,860 per year. Fees of $550 per year. • Dr. Paul C. Martin Jr., Dean, 617-495-2833. Application contact: Office of Admissions and Financial Aid, 617-495-5315.

Hofstra University, College of Liberal Arts and Sciences, Division of Natural Sciences, Mathematics, Engineering, and Computer Science, Department of Mathematics, Hempstead, NY 11550. Department offers program in applied mathematics (MA, MS). Faculty: 9 full-time (2 women), 0 part-time. Matriculated students: 5 full-time (2 women), 7 part-time (3 women). 14 applicants, 86% accepted. In 1990, 3 degrees awarded. *Degree requirements:* Thesis or alternative, proficiency in computer programming, comprehensive exam or oral defense of master's thesis. *Entrance requirements:* Bachelor's degree in mathematics or related field. Application deadline: rolling. Application fee: $25 ($50 for foreign students). *Expenses:* Tuition of $3816 per semester (minimum) full-time, $318 per semester hour part-time. Fees of $140 per semester. *Financial aid:* In 1990–91, 6 students received a total of $21,400 in aid awarded. 6 fellowships (3 to first-year students), 4 teaching assistantships (2 to first-year students), 4 tutoring assistantships (2 to first-year students) were awarded. *Faculty research:* Teacher training, multiparameter stochastic applications, mathematical ecology, dynamic systems, shockwaves. Total annual research budget: $59,275. • Dr. Robert J. Bumerot, Chairperson, 516-463-5571. Application contact: Harry Curtis, Director of Graduate Admissions, 516-463-6707.

Howard University, Graduate School of Arts and Sciences, Department of Mathematics, 2400 Sixth Street, NW, Washington, DC 20059. Offerings include applied mathematics (MS, PhD). *Degree requirements:* For master's, computer language, thesis or alternative, comprehensive exam required, foreign language not required; for doctorate, 2 foreign languages (computer language can substitute for one), dissertation, comprehensive exam. *Entrance requirements:* For master's, minimum GPA of 3.0. Application deadline: 4/1. Application fee: $25. *Expenses:* Tuition of $6100 per year full-time, $339 per credit hour part-time. Fees of $555 per year full-time, $245 per semester part-time.

See full description on page 417.

Hunter College of the City University of New York, Division of Sciences and Mathematics, Department of Mathematics, 695 Park Avenue, New York, NY 10021. Offerings include applied mathematics (MA). *Degree requirements:* 1 foreign language required, thesis not required. *Entrance requirements:* GRE General Test, TOEFL (minimum score of 550 required). Application deadline: 4/1. Application fee: $30. *Expenses:* Tuition of $2204 per year full-time, $95 per credit part-time for state residents; $4700 per year full-time, $199 per credit part-time for nonresidents. Fees of $15 per semester. • Tom Jambois, Chairperson, 212-772-5300. Application contact: Office of Admissions, 212-772-4490.

Illinois Institute of Technology, Lewis College of Sciences and Letters, Department of Mathematics, Chicago, IL 60616. Offerings include applied mathematics (MS). Department faculty: 13 full-time (3 women), 1 part-time (0 women). *Degree requirements:* Thesis (for some programs), comprehensive exam required, foreign language not required. *Entrance requirements:* TOEFL (minimum score of 500 required). Application deadline: 7/1 (applications processed on a rolling basis). Application fee: $30. *Expenses:* Tuition of $13,070 per year full-time, $435 per credit hour part-time. Fees of $20 per semester (minimum) full-time, $1 per credit hour part-time. • Dr. Maurice J. Frank, Chairman, 312-567-3162.

Indiana University Bloomington, College of Arts and Sciences, Department of Mathematics, Bloomington, IN 47405. Offerings include applied mathematics–numerical analysis (MA, PhD). Terminal master's awarded for partial completion of doctoral program. Department faculty: 50 full-time (1 woman), 0 part-time. *Degree requirements:* For doctorate, 2 foreign languages, dissertation. *Entrance requirements:* For doctorate, GRE General Test, GRE Subject Test. Application deadline: 3/1. Application fee: $25 ($35 for foreign students). *Tuition:* $99.85 per credit hour for state residents; $288 per credit hour for nonresidents. • Allan L. Edmonds, Chair, 812-855-2200. Application contact: Richard C. Bradley, Director of Graduate Studies, 812-855-2645.

Indiana University–Purdue University at Fort Wayne, Department of Mathematical Sciences, Fort Wayne, IN 46805-1499. Offerings include applied mathematics (MS). Department faculty: 13 full-time (2 women), 0 part-time. *Degree requirements:* Foreign language and thesis not required. *Entrance requirements:* Minimum GPA of 3.0, undergraduate major or minor in mathematics. Application deadline: 5/15 (priority date, applications processed on a rolling basis). Application fee: $25. *Tuition:* $78.05 per credit hour for state residents; $174.75 per credit hour for nonresidents. • Douglas Townsend, Chair, 219-481-6235.

Indiana University–Purdue University at Indianapolis, School of Science, Department of Mathematical Sciences, Indianapolis, IN 46205. Offerings include applied mathematics (MS, PhD). Terminal master's awarded for partial completion of doctoral program. Department faculty: 21 full-time (1 woman), 0 part-time. *Degree requirements:* For master's, foreign language and thesis not required; for doctorate, 2 foreign languages, dissertation. *Entrance requirements:* TOEFL (minimum score of 570 required). Application deadline: 2/1. Application fee: $20. *Tuition:* $100 per credit hour for state residents; $288 per credit hour for nonresidents. • Bartholomew S. Ng, Chair, 317-274-6918. Application contact: Chair, Graduate Committee, 317-274-6925.

Iowa State University of Science and Technology, College of Liberal Arts and Sciences, Department of Mathematics, Ames, IA 50011. Offerings include applied mathematics (MS, PhD). *Degree requirements:* For doctorate, 2 foreign languages, dissertation, comprehensive written exam. *Application fee:* $20 ($30 for foreign students). *Expenses:* Tuition of $1158 per semester full-time, $129 per credit part-time for state residents; $3340 per semester full-time, $372 per credit part-time for nonresidents. Fees of $10 per semester. • Dr. Howard A. Levine, Chair, 515-294-2175.

Johns Hopkins University, G. W. C. Whiting School of Engineering, Department of Mathematical Sciences, Baltimore, MD 21218. Offerings include applied mathematics (MA, MSE, PhD). Terminal master's awarded for partial completion of doctoral program. Department faculty: 11 full-time (0 women), 2 part-time (1 woman). *Degree requirements:* For master's, computer language, thesis (for some programs); for doctorate, 1 foreign language, computer language, dissertation. *Entrance requirements:* GRE General Test, GRE Subject Test. Application deadline: 2/15 (priority date, applications processed on a rolling basis). Application fee: $45. *Expenses:* Tuition of $15,500 per year full-time, $1550 per course part-time. Fees of $400 per year full-time. • Dr. John C. Wierman, Chairman, 301-338-7195.

See full description on page 419.

Directory: Applied Mathematics

Lehigh University, College of Arts and Sciences, Department of Mathematics, Bethlehem, PA 18015. Offerings include applied mathematics (MS, PhD). Department faculty: 23 full-time, 0 part-time. *Degree requirements:* For master's, comprehensive exam required, foreign language and thesis not required; for doctorate, 1 foreign language, dissertation, comprehensive and qualifying exams. *Entrance requirements:* TOEFL, minimum GPA of 3.0. Application deadline: rolling. Application fee: $40. *Tuition:* $15,650 per year full-time, $655 per credit hour part-time. • Dr. Andrew K. Snyder, Chairman, 215-758-3730. Application contact: Dr. Bruce Dodson, Graduate Program Coordinator.

See full description on page 423.

Long Island University, C. W. Post Campus, College of Arts and Sciences, Department of Mathematics, Brookville, NY 11548. Department offers programs in applied mathematics (MS), including classical mathematics, computer mathematics; mathematics for secondary school teachers (MS). Part-time and evening/weekend programs available. Faculty: 17 full-time (4 women). Matriculated students: 25 full-time (12 women), 25 part-time (15 women); includes 7 foreign. Average age 25. 20 applicants, 75% accepted. In 1990, 12 degrees awarded. *Degree requirements:* Thesis or alternative required, foreign language not required. *Application fee:* $30. *Expenses:* Tuition of $310 per credit. Fees of $235 per semester full-time, $90 per semester part-time. *Financial aid:* In 1990–91, 10 graduate assistantships (5 to first-year students) awarded; full and partial tuition waivers also available. *Faculty research:* Numerical analysis, matrix algebra, computer graphics, functional geometry. • Dr. Neo Cleopa, Chairman, 516-299-2447.

Massachusetts Institute of Technology, School of Science, Department of Mathematics, Cambridge, MA 02139. Offerings include applied mathematics (PhD, Sc D). Terminal master's awarded for partial completion of doctoral program. Department faculty: 50 full-time (0 women), 0 part-time. *Degree requirements:* 1 foreign language, dissertation, oral exam. *Entrance requirements:* GRE General Test, GRE Subject Test. Application deadline: 1/15 (applications processed on a rolling basis). Application fee: $45. *Tuition:* $16,900 per year. • David J. Benney, Head, 617-253-6976. Application contact: Phyllis Ruby, Graduate Administrator, 617-253-2689.

Memphis State University, College of Arts and Sciences, Department of Mathematical Sciences, Memphis, TN 38152. Offerings include applied mathematics (MS). Department faculty: 28 full-time (3 women), 0 part-time. *Degree requirements:* Comprehensive exams required, thesis not required. *Entrance requirements:* MAT (minimum score of 27 required) or GRE General Test (minimum combined score of 800 required), TOEFL (minimum score of 550 required), minimum GPA of 2.5. Application deadline: 8/1 (applications processed on a rolling basis). Application fee: $5. *Expenses:* Tuition of $92 per credit hour for state residents; $239 per credit hour for nonresidents. Fees of $45 per year for state residents. • Dr. Ralph Faudree, Chairman, 901-678-2482. Application contact: Dr. R. H. Schlep, Coordinator, Graduate Studies, 901-678-2495.

Michigan State University, College of Natural Science, Department of Mathematics, East Lansing, MI 48824. Offerings include applied mathematics (MS, PhD). Department faculty: 64 full-time (6 women), 0 part-time. *Degree requirements:* For doctorate, 2 foreign languages, dissertation. *Application deadline:* rolling. *Application fee:* $25 ($40 for foreign students). *Tuition:* $104.75 per credit for state residents; $211.75 per credit for nonresidents. • Dr. Jacob Plotkin, Acting Chairperson, 517-353-8484.

See full description on page 427.

Montclair State College, School of Mathematical and Natural Sciences, Department of Mathematics and Computer Science, Programs in Mathematics, Concentration in Pure and Applied Mathematics, Upper Montclair, NJ 07043. Concentration awards MA. Part-time and evening/weekend programs available. Faculty: 33 full-time, 0 part-time. In 1990, 1 degree awarded. *Degree requirements:* Written comprehensive exam required, foreign language and thesis not required. *Entrance requirements:* GRE General Test, minimum GPA of 2.67. Application deadline: 7/1 (priority date, applications processed on a rolling basis). Application fee: $25. *Expenses:* Tuition of $130 per credit for state residents; $165 per credit for nonresidents. Fees of $276 per year. *Financial aid:* Application deadline 3/15. • Application contact: Dr. Helen Roberts, Graduate Adviser, 201-893-7262.

New Jersey Institute of Technology, Department of Mathematics, Newark, NJ 07102. Department offers program in applied mathematics (MS). Evening/weekend programs available. Faculty: 32 full-time (4 women), 5 part-time (1 woman). Matriculated students: 17 full-time (3 women), 14 part-time (5 women); includes 3 minority (2 Asian American, 1 Hispanic American), 18 foreign. 70 applicants, 80% accepted. In 1990, 10 degrees awarded. *Degree requirements:* Foreign language and thesis not required. *Entrance requirements:* GRE General Test (minimum score of 450 on verbal section, 600 on mathematics, 550 on analytical required). Application deadline: 6/5 (priority date, applications processed on a rolling basis). Application fee: $30. *Tuition:* $2585 per semester full-time, $253 per credit part-time for state residents; $3864 per semester full-time, $349 per credit part-time for nonresidents. *Financial aid:* Application deadline 2/1. *Faculty research:* Applied mathematics, computational methods, probability and statistics. • John Tavantzis, Acting Chairman, 201-596-3493. Application contact: Petra Theodos, Director of Graduate Admissions, 201-596-3460.

Announcement: MS program emphasizes research in applied mathematics. Areas of specialization include wave propagation, fluid mechanics, scientific computation, analysis, dynamical systems, and probability and statistics. Extensive computing capabilities exist on campus and within the department. Applicants should have a bachelor's degree in mathematics, the sciences, or engineering. Teaching and research assistantships available.

New Mexico State University, College of Arts and Sciences, Department of Mathematical Sciences, Las Cruces, NM 88003. Department awards MS, PhD. Part-time programs available. Faculty: 28 full-time, 0 part-time. Matriculated students: 25 full-time (9 women), 15 part-time (2 women); includes 25 foreign. In 1990, 3 master's, 6 doctorates awarded. *Entrance requirements:* GRE General Test. Application deadline: 7/1. Application fee: $10. *Tuition:* $1608 per year full-time, $67 per credit hour part-time for state residents; $5304 per year full-time, $221 per credit hour part-time for nonresidents. *Financial aid:* In 1990–91, 25 teaching assistantships awarded; fellowships, research assistantships also available. *Faculty research:* Algebraic k-theory, harmonic analysis, functional analysis, algebraic topology. • Dr. Carol Walker, Head, 505-646-3901.

Nicholls State University, College of Arts and Sciences, Department of Mathematics, Thibodaux, LA 70310. Offerings include applied mathematics (MS). Department faculty: 0 full-time, 6 part-time (1 woman). *Degree requirements:* Foreign language and thesis not required. *Entrance requirements:* GRE General Test. Application fee: $10. *Expenses:* Tuition of $1345 per year full-time, $337 per semester part-time for state residents; $3145 per year full-time, $337 per semester part-time for nonresidents. Fees of $246 per year full-time, $99 per semester part-time. • Dr. Donald M. Bardwell, Head, 504-448-4380.

North Carolina Agricultural and Technical State University, Graduate School, School of Education, Department of Curriculum and Instruction, Program in Mathematics Education, Greensboro, NC 27411. Offerings include applied mathematics (MS). Program faculty: 6 full-time (1 woman), 0 part-time. *Degree requirements:* Thesis or alternative, qualifying exam, comprehensive exam required, foreign language not required. *Entrance requirements:* GRE General Test, minimum GPA of 3.0. Application deadline: 6/1 (priority date, applications processed on a rolling basis). Application fee: $15. *Tuition:* $614 per semester full-time for state residents; $3141 per semester full-time for nonresidents. • Dr. Wendell P. Jones, Chairman, Department of Mathematics and Computer Science, 919-334-7822.

North Carolina State University, College of Physical and Mathematical Sciences, Department of Mathematics, Raleigh, NC 27695. Offerings include applied mathematics (MS, PhD). Terminal master's awarded for partial completion of doctoral program. Department faculty: 56 full-time (3 women), 4 part-time (0 women). *Degree requirements:* For master's, computer language required, thesis optional, foreign language not required; for doctorate, 1 foreign language, dissertation. *Entrance requirements:* GRE General Test, GRE Subject Test. Application deadline: 4/1 (priority date, applications processed on a rolling basis). Application fee: $35. *Tuition:* $1138 per year for state residents; $5805 per year for nonresidents. • Dr. Robert H. Martin, Head, 919-737-3798. Application contact: John E. Franke, Administrator, 919-737-2382.

Northwestern University, College of Arts and Sciences, Department of Mathematics, Evanston, IL 60208. Offerings include applied mathematics (MA, MS, PhD). Terminal master's awarded for partial completion of doctoral program. Department faculty: 33 full-time (3 women), 1 part-time (0 women). *Degree requirements:* For master's, preliminary exam required, foreign language and thesis not required; for doctorate, dissertation, preliminary exam required, foreign language not required. *Entrance requirements:* GRE General Test. Application deadline: 8/30. Application fee: $40 ($45 for foreign students). *Tuition:* $4665 per quarter full-time, $1704 per course part-time. • Stewart Priddy, Chairman, 708-491-8035. Application contact: Clark Robinson, Chairman, Graduate Committee, 708-491-8035.

See full description on page 437.

Northwestern University, Robert R. McCormick School of Engineering and Applied Science, Department of Applied Mathematics, Evanston, IL 60208. Department awards MS, PhD. Part-time programs available. Faculty: 11 full-time (0 women), 0 part-time. Matriculated students: 33 full-time (11 women), 0 part-time; includes 2 minority (1 black American, 1 Hispanic American), 8 foreign. 54 applicants, 74% accepted. In 1990, 6 master's, 4 doctorates awarded. Terminal master's awarded for partial completion of doctoral program. *Degree requirements:* For master's, foreign language and thesis not required; for doctorate, dissertation required, foreign language not required. *Entrance requirements:* GRE General Test. Application deadline: 8/30. Application fee: $40 ($45 for foreign students). *Tuition:* $4665 per quarter full-time, $1704 per course part-time. *Financial aid:* In 1990–91, $214,746 in aid awarded. 9 fellowships, 8 research assistantships, 5 teaching assistantships were awarded; federal work-study, institutionally sponsored loans, and career-related internships or fieldwork also available. Financial aid application deadline: 1/15; applicants required to submit GAPSFAS. *Faculty research:* Asympotics, bifurcation and stability, combustion, wave propagation, solid and fluid mechanics. Total annual research budget: $931,665. • Dr. Stephen H. Davis, Chairman, 708-491-3345. Application contact: Recruitment Officer, 708-491-5586.

See full description on page 435.

Oakland University, College of Arts and Sciences, Department of Mathematical Sciences, Program in Industrial Applied Mathematics, Rochester, MI 48309. Program awards MS. Part-time and evening/weekend programs available. *Degree requirements:* Foreign language and thesis not required. *Entrance requirements:* Minimum GPA of 3.0 for unconditional admission. Application deadline: 7/15. Application fee: $25. *Expenses:* Tuition of $122 per credit hour for state residents; $270 per credit hour for nonresidents. Fees of $170 per year. • Application contact: Dr. James H. McKay, Graduate Coordinator, 313-370-3430.

Old Dominion University, College of Sciences, Department of Mathematics and Statistics, Programs in Computational and Applied Mathematics, Norfolk, VA 23529. Offerings in applied mathematics (MS, PhD), statistics (MS, PhD). Part-time and evening/weekend programs available. Faculty: 24 full-time (0 women), 0 part-time. Matriculated students: 25 full-time (10 women), 23 part-time (11 women); includes 7 minority (1 Asian American, 4 black American, 2 Hispanic American), 19 foreign. Average age 27. 36 applicants, 86% accepted. In 1990, 6 master's awarded (17% entered university research/teaching, 83% found other work related to degree); 9 doctorates awarded (100% entered university research/teaching). Terminal master's awarded for partial completion of doctoral program. *Degree requirements:* For master's, comprehensive exam required, foreign language and thesis not required; for doctorate, dissertation, candidacy exam required, foreign language not required. *Entrance requirements:* For master's, GRE General Test, GRE Subject Test, TOEFL, minimum QPA of 3.0 in major, minimum QPA of 2.5 overall; for doctorate, GRE General Test (minimum combined score of 1000 required), GRE Subject Test (minimum score of 600 required), TOEFL. Application deadline: 7/1 (applications processed on a rolling basis). Application fee: $20. *Expenses:* Tuition of $148 per credit hour for state residents; $375 per credit hour for nonresidents. Fees of $64 per year full-time. *Financial aid:* In 1990–91, $28,800 in aid awarded. 3 research assistantships, 2 teaching assistantships, 1 tuition grant were awarded; fellowships also available. *Faculty research:* Numerical analysis, computational and applied mathematics, integral equations, continuum mechanics. Total annual research budget: $240,000. • Dr. John J. Swetits, Director, 804-683-3911.

See full description on page 441.

Pennsylvania State University University Park Campus, College of Science, Department of Mathematics, University Park, PA 16802. Offerings include mathematics (MA, M Ed, D Ed, PhD), with option in applied mathematics (MA, PhD). Department faculty: 52. *Application fee:* $35. *Tuition:* $203 per credit for state residents; $403 per credit for nonresidents. • Dr. Richard H. Herman, Head, 814-865-7527. Application contact: Joel Anderson, Professor, 814-865-1123.

Polytechnic University, Brooklyn Campus, Division of Arts and Sciences, Department of Mathematics, 333 Jay Street, Brooklyn, NY 11201. Offerings include industrial and applied mathematics (MS, PhD). *Tuition:* $6000 per semester.

Princeton University, Departments of Mathematics and Physics, Program in Applied and Computational Mathematics, Princeton, NJ 08544. Program awards PhD. Matriculated students: 15 full-time (2 women), 0 part-time; includes 0 minority, 4 foreign. 36 applicants, 28% accepted. In 1990, 6 doctorates awarded (67% entered university research/teaching, 33% found other work related to degree). Terminal master's awarded for partial completion of doctoral program. *Degree requirements:* For doctorate, 1 foreign language, dissertation. *Entrance requirements:* For doctorate, GRE General Test, GRE Subject Test. Application deadline: 1/8. Application fee: $45 ($50 for foreign students). *Tuition:* $16,670 per year. *Financial aid:* Fellowships, research assistantships, teaching assistantships, federal work-study, institutionally sponsored loans available. Financial aid application deadline: 1/8. • Steven Orszag, Director of Graduate Studies, 609-258-4262. Application contact: Michele Spreen, Director of Graduate Admissions, 609-258-3034.

Princeton University, School of Engineering and Applied Science, Department of Chemical Engineering, Princeton, NJ 08544. Offerings include applied and computational mathematics (PhD). Program in polymer sciences and materials offered in conjunction with the Department of Chemistry. Terminal master's awarded for partial completion of doctoral program. Department faculty: 14 full-time (0 women), 0 part-time. *Degree requirements:* Dissertation, general exam. *Entrance requirements:* GRE General Test, GRE Subject Test, TOEFL. Application deadline: 1/8. Application fee: $45 ($50 for foreign students). *Tuition:* $16,670 per year. • Jay B. Benziger, Director of Graduate Studies, 609-258-4581. Application contact: Michele Spreen, Director of Graduate Admissions, 609-258-3034.

Purdue University Calumet, School of Liberal Arts and Sciences, Department of Mathematical Sciences, Hammond, IN 46323. Department offers program in applied mathematics (MS). *Degree requirements:* Foreign language and thesis not required. *Entrance requirements:* TOEFL. Application fee: $25. *Tuition:* $83.50 per credit hour for state residents; $189.90 per credit hour for nonresidents. • Dr. C. M. Murphy, Head, 219-989-2273.

Rensselaer Polytechnic Institute, School of Science, Department of Mathematical Sciences, Program in Applied Mathematics, Troy, NY 12180. Program awards MS, PhD. Faculty: 24 full-time (5 women), 1 (woman) part-time. Matriculated students: 16 full-time (5 women), 0 part-time; includes 0 minority, 0 foreign. 33 applicants, 64% accepted. In 1990, 9 master's awarded. *Degree requirements:* For master's, thesis optional, foreign language not required. *Entrance requirements:* For master's, GRE General Test, TOEFL. Application deadline: 2/1. Application fee: $30. *Expenses:* Tuition of $455 per credit hour. Fees of $195.57 per semester. *Financial aid:* Fellowships, research assistantships, teaching assistantships, and career-related internships or fieldwork available. • Joyce McLaughlin, Graduate Coordinator, 518-276-6349.

Announcement: At the MS level, the program is designed to prepare students to become practicing applied mathematicians in industry or government. Stresses the construction, analysis, and evaluation of mathematical models of real-world problems and emphasizes related areas of mathematics. Interaction with industry is featured regularly through visiting lecturers and research projects sponsored by industry. At the PhD level, the program interprets applied mathematics in the broadest possible sense. Areas of emphasis include physical mathematics and modeling, differential equations, analysis, scientific computation, mathematical programming and operations research, and applied geometry. Development of expertise in at least one field of application is stressed.

See full description on page 453.

Rice University, George R. Brown School of Engineering, Department of Mathematical Sciences, Houston, TX 77251. Department awards MA, MAM Sc, PhD. Part-time programs available. Faculty: 10 full-time (1 woman), 3 part-time (0 women). Matriculated students: 34 full-time (8 women), 2 part-time (1 woman). *Degree requirements:* For master's, thesis required (for some programs), foreign language not required; for doctorate, 1 foreign language, dissertation. *Entrance requirements:* GRE General Test, TOEFL (minimum score of 550 required), minimum GPA of 3.0. Application deadline: 3/1. Application fee: $0. *Expenses:* Tuition of $8300 per year full-time, $400 per credit hour part-time. Fees of $167 per year. *Financial aid:* Fellowships, research assistantships, full and partial tuition waivers available. Financial aid applicants required to submit FAF. *Faculty research:* Statistics, operations research, computer science, game theory, numerical analysis. • Dr. John E. Dennis, Chairman, 713-527-4805.

Roosevelt University, College of Arts and Sciences, Department of Mathematical Sciences, Chicago, IL 60605. Offerings include applied mathematics (MS). *Degree requirements:* Foreign language and thesis not required. *Application deadline:* 8/1 (priority date, applications processed on a rolling basis). *Application fee:* $20 ($30 for foreign students). *Expenses:* Tuition of $302 per semester hour. Fees of $12 per year full-time, $6 per semester part-time. • Jimmie Lee Johnson, Chairman, 312-341-3772. Application contact: Joan R. Ritter, Coordinator of Graduate Admissions, 312-341-3612.

Rutgers, The State University of New Jersey, New Brunswick, Program in Mathematics, New Brunswick, NJ 08903. Offerings include applied mathematics (MS, PhD). Terminal master's awarded for partial completion of doctoral program. Program faculty: 86 full-time (10 women), 2 part-time (0 women). *Degree requirements:* For master's, foreign language and thesis not required; for doctorate, 1 foreign language, dissertation. *Entrance requirements:* GRE General Test, GRE Subject Test. Application deadline: 5/1. Application fee: $35. *Expenses:* Tuition of $4432 per year full-time, $183 per credit part-time for state residents; $6496 per year full-time, $270 per credit part-time for nonresidents. Fees of $458 per year full-time, $117 per year part-time. • Jose Barros-Neto, Director, 908-932-3921.

St. John's University, Graduate School of Arts and Sciences, Department of Mathematics and Computer Science, Jamaica, NY 11439. Offerings include applied mathematics (MA). Department faculty: 21 full-time (4 women), 0 part-time. *Degree requirements:* Variable foreign language requirement. *Entrance requirements:* GRE General Test. Application deadline: rolling. Application fee: $20. *Expenses:* Tuition of $297 per credit. Fees of $130 per year full-time, $65 per semester part-time. • Dr. Edward Miranda, Chairman, 718-990-6161.

San Diego State University, College of Sciences, Department of Mathematical Sciences, Program in Applied Mathematics, San Diego, CA 92182. Program awards MS. Matriculated students: 1 full-time (0 women), 17 part-time (5 women); includes 3 minority (1 Asian American, 2 Hispanic American), 2 foreign. In 1990, 8 degrees awarded. *Degree requirements:* Comprehensive exam required, thesis not required. *Entrance requirements:* GRE General Test (minimum combined score of 950 required). Application deadline: 8/1 (applications processed on a rolling basis). Application fee: $55. *Expenses:* Tuition of $0 for state residents; $189 per unit for nonresidents. Fees of $1974 per year full-time, $692 per year part-time for state residents; $1074 per year full-time, $692 per year part-time for nonresidents. • Dr. Joseph Mahaffy, Adviser, 619-594-6191.

Santa Clara University, School of Engineering, Department of Applied Mathematics, Santa Clara, CA 95053. Department awards MSAM. Part-time and evening/weekend programs available. Matriculated students: 0 full-time, 20 part-time (4 women); includes 3 minority (2 Asian American, 1 black American), 0 foreign. Average age 29. 8 applicants, 88% accepted. In 1990, 1 degree awarded. *Degree requirements:* Thesis or alternative required, foreign language not required. *Entrance requirements:* GRE, TOEFL, TSE, minimum GPA of 2.75. Application deadline: 6/15. Application fee: $25. *Fees:* $48 per quarter. *Financial aid:* In 1990–91, 0 research assistantships, 0 teaching assistantships awarded; federal work-study also available. Aid available to part-time students. Financial aid application deadline: 2/1; applicants required to submit FAF. • Dr. George Fegan, Chairman, 408-554-4181. Application contact: Dr. James White, Director of Graduate Admissions, 408-554-4600.

Simon Fraser University, Faculty of Science, Department of Mathematics and Statistics, Burnaby, BC V5A 1S6, Canada. Offerings include applied mathematics (M Sc, PhD). Department faculty: 35 (2 women). *Degree requirements:* For master's, thesis; for doctorate, dissertation, comprehensive exams. *Entrance requirements:* GRE Subject Test, TOEFL (minimum score of 600 required). Application fee: $0. *Expenses:* Tuition of $612 per trimester full-time, $306 per trimester part-time. Fees of $68 per trimester full-time, $34 per trimester part-time. • Dr. A. R. Freedman, Chairman, 604-291-3378.

See full description on page 455.

Southern Methodist University, Dedman College, Department of Mathematics, Dallas, TX 75275. Offerings include applied mathematics (MS). Department faculty: 16 full-time (1 woman), 0 part-time. *Application deadline:* 6/30. *Application fee:* $25. *Expenses:* Tuition of $435 per credit. Fees of $664 per semester for state residents; $56 per year for nonresidents. • Dr. Ian Gladwell, Chairman, 214-692-2506. Application contact: Dr. Richard Haberman, Graduate Adviser, 214-692-2506.

State University of New York at Stony Brook, College of Engineering and Applied Sciences, Department of Applied Mathematics and Statistics, Stony Brook, NY 11794. Department offers programs in applied mathematics (MS, PhD), operations research (MS, PhD), statistics (MS, PhD). Matriculated students: 80 full-time (37 women), 38 part-time (19 women); includes 18 minority (15 Asian American, 2 black American, 1 Hispanic American), 48 foreign. 218 applicants, 28% accepted. *Degree requirements:* For master's, thesis or alternative required, foreign language not required; for doctorate, 1 foreign language, dissertation, comprehensive exams. *Entrance requirements:* GRE General Test, TOEFL. Application deadline: 2/1. Application fee: $35. *Expenses:* Tuition of $2450 per year full-time, $103 per credit part-time for state residents; $5766 per year full-time, $243 per credit part-time for nonresidents. Fees of $151 per year full-time, $10.45 per year (minimum) part-time. *Financial aid:* In 1990–91, 8 fellowships, 15 research assistantships, 27 teaching assistantships awarded. *Faculty research:* Biostatistics, combinatorial analysis, differential equations, modeling. Total annual research budget: $312,578. • Dr. J. Glimm, Chairman, 516-632-8360.

Stevens Institute of Technology, Department of Pure and Applied Mathematics, Hoboken, NJ 07030. Department awards MS, PhD. Part-time and evening/weekend programs available. Terminal master's awarded for partial completion of doctoral program. *Degree requirements:* For master's, computer language required, thesis optional, foreign language not required; for doctorate, variable foreign language requirement, computer language, dissertation. *Entrance requirements:* GRE. Application fee: $25. *Tuition:* $4850 per semester full-time, $485 per credit part-time. *Faculty research:* Dynamical systems, numerical analysis, graph theory, algebraic geometry, applied mathematics.

Technical University of Nova Scotia, Faculty of Engineering, Department of Applied Mathematics, Halifax, NS B3J 2X4, Canada. Department awards M Sc, PhD. Faculty: 7 full-time (0 women), 2 part-time (0 women). Matriculated students: 5 full-time (1 woman), 2 part-time (1 woman). 5 applicants, 20% accepted. In 1990, 2 master's awarded. *Degree requirements:* Thesis/dissertation required, foreign language not required. *Application deadline:* rolling. *Expenses:* Tuition of $1240 per year full-time, $384 per course part-time. Fees of $2000 per year for nonresidents. • Dr. S. N. Sarwal, Head, 902-420-7500.

Temple University, Ambler Campus, College of Arts and Sciences, Program in Applied Mathematics, Ambler, PA 19002-3999. Program awards MA.

Texas A&M University, College of Science, Department of Mathematics, College Station, TX 77843. Department awards MS, PhD. PhD offered jointly with Stephen F. Austin University. Faculty: 62. Matriculated students: 83 full-time (21 women), 0 part-time; includes 6 minority (4 Asian American, 1 black American, 1 Hispanic American), 26 foreign. 81 applicants, 59% accepted. In 1990, 11 master's, 2 doctorates awarded. *Entrance requirements:* GRE General Test, TOEFL. Application deadline: 7/15. Application fee: $25 ($50 for foreign students). *Expenses:* Tuition of $100 per semester (minimum) for state residents; $128 per credit hour for nonresidents. Fees of $459 per year full-time, $252 per semester part-time. *Financial aid:* Fellowships, research assistantships, teaching assistantships available. • Carl Pearcy, Head, 409-845-3261.

Tulane University, Department of Mathematics, Concentration in Applied Mathematics, New Orleans, LA 70118. Concentration awards MS. *Degree requirements:* Thesis or alternative required, foreign language not required. *Entrance requirements:* GRE General Test, GRE Subject Test, TOEFL (minimum score of 600 required) or TSE (minimum score of 220 required), minimum B average in undergraduate course work. Application deadline: 7/1. Application fee: $35. *Expenses:* Tuition of $16,750 per year full-time, $931 per hour part-time. Fees of $230 per year full-time, $40 per hour part-time. *Financial aid:* Teaching assistantships available. Financial aid application deadline: 5/1; applicants required to submit GAPSFAS. • Dr. Steven Rosencrans, Chairman, Department of Mathematics, 504-865-5727.

University of Akron, Buchtel College of Arts and Sciences, Department of Mathematical Sciences and Statistics, Program in Mathematics, Akron, OH 44325. Offerings include applied mathematics (MS). Program faculty: 20 full-time (1 woman), 0 part-time. *Degree requirements:* Thesis optional, foreign language not required. *Entrance requirements:* Minimum GPA of 2.75. Application deadline: 3/1 (applications processed on a rolling basis). Application fee: $25. *Tuition:* $119.93 per credit hour for state residents; $210.93 per credit hour for nonresidents. • Dr. William H. Beyer, Head, Department of Mathematical Sciences and Statistics, 216-972-7401.

University of Alabama at Birmingham, Graduate School, School of Natural Sciences and Mathematics, Department of Mathematics, Birmingham, AL 35294. Offerings include applied mathematics (PhD). Department faculty: 0 part-time. *Application deadline:* rolling. *Application fee:* $25 ($50 for foreign students). *Tuition:* $66 per quarter for state residents; $132 per quarter for nonresidents. • Dr. Lex G. Oversteegen, Chairman, 205-934-2154.

University of Alabama in Huntsville, College of Science, Department of Mathematics, Huntsville, AL 35899. Department offers programs in applied mathematics (PhD), mathematics (MA, MS). PhD offered jointly with University of Alabama-Tuscaloosa and University of Alabama-Birmingham. Part-time programs available. Faculty: 19 full-time (2 women), 2 part-time (0 women). Matriculated students: 23 full-time, 18 part-time; includes 1 minority (Asian American), 3 foreign. 46 applicants, 83%

SECTION 6: MATHEMATICAL SCIENCES

Directory: Applied Mathematics

accepted. In 1990, 5 master's, 1 doctorate awarded. Terminal master's awarded for partial completion of doctoral program. *Degree requirements:* For master's, thesis or alternative, oral and written exams required, foreign language not required; for doctorate, 1 foreign language, dissertation, oral and written exams. *Entrance requirements:* For master's, GRE General Test (minimum combined score of 1500 on 3 sections required), minimum GPA of 3.0. Application deadline: 5/16 (priority date, applications processed on a rolling basis). Application fee: $20. *Tuition:* $2500 per year full-time, $1250 per year part-time for state residents; $5000 per year full-time, $2500 per year part-time for nonresidents. *Financial aid:* In 1990–91, 21 students received a total of $48,767 in aid awarded. 0 fellowships, 7 research assistantships, 14 teaching assistantships were awarded; full and partial tuition waivers, federal work-study, institutionally sponsored loans, and career-related internships or fieldwork also available. Aid available to part-time students. Average monthly stipend for a graduate assistantship: $720. Financial aid application deadline: 3/1; applicants required to submit FAF. *Faculty research:* Statistical modeling, stochastic processes, combinatorial matrix theory. Total annual research budget: $231,275. • Dr. Peter Gibson, Chairman, 205-895-6470.

University of Arizona, College of Arts and Sciences, Faculty of Science, Department of Mathematics, Program in Applied Mathematics, Tucson, AZ 85721. Committee awards MS, PhD. Faculty: 92 full-time (1 woman), 0 part-time. Matriculated students: 68 full-time (19 women), 13 part-time (3 women); includes 5 minority (1 Asian American, 3 Hispanic American, 1 Native American), 36 foreign. Average age 29. 145 applicants, 23% accepted. In 1990, 14 master's awarded (14% entered university research/teaching, 35% found other work related to degree, 50% continued full-time study); 8 doctorates awarded (60% entered university research/teaching). Terminal master's awarded for partial completion of doctoral program. *Degree requirements:* For master's, computer language required, foreign language and thesis not required; for doctorate, 1 foreign language, computer language, dissertation. *Entrance requirements:* GRE. Application deadline: 6/15 (applications processed on a rolling basis). Application fee: $0. *Expenses:* Tuition of $0 for state residents; $5406 per year full-time, $209 per credit hour part-time for nonresidents. Fees of $1528 per year full-time, $80 per credit hour part-time. *Financial aid:* In 1990–91, 65 students received a total of $125,268 in aid awarded. 18 fellowships, 21 research assistantships, 11 teaching assistantships were awarded; scholarships, institutionally sponsored loans also available. Financial aid application deadline: 3/1; applicants required to submit FAF. *Faculty research:* Dynamical systems and chaos, fluid mechanics, partial differential equations, biological applications, optimization and control. Total annual research budget: $163,383. • Dr. T. W. Secomb, Acting Chairman, 602-621-4664. Application contact: Graduate Commitee, 602-621-2016.

See full description on page 461.

University of Arkansas at Little Rock, College of Sciences and Engineering Technology, Department of Mathematics and Statistics, Little Rock, AR 72204. Department offers program in applied mathematics (MS). Part-time and evening/weekend programs available. Faculty: 5 full-time (1 woman), 0 part-time. Matriculated students: 3 full-time (1 woman), 10 part-time (3 women); includes 3 minority (1 Asian American, 2 black American), 0 foreign. Average age 30. 8 applicants, 25% accepted. In 1990, 5 degrees awarded (100% found work related to degree). *Degree requirements:* Computer language, comprehensive exams required, foreign language and thesis not required. *Application deadline:* 8/1 (priority date). *Application fee:* $25 ($30 for foreign students). *Expenses:* Tuition of $2224 per year full-time, $104 per credit hour part-time for state residents; $4774 per year full-time, $223 per credit hour part-time for nonresidents. Fees of $40 per semester. *Financial aid:* In 1990–91, 6 students received a total of $36,000 in aid awarded. 4 research assistantships (1 to a first-year student), 3 teaching assistantships (all to first-year students) were awarded; institutionally sponsored loans also available. Aid available to part-time students. Average monthly stipend for a graduate assistantship: $600. Financial aid application deadline: 8/1. *Faculty research:* Mathematical statistics, numerical analysis. • Daniel B. McCallum, Chairperson, 501-569-3252. Application contact: Dr. Alan Johnson, Graduate Coordinator, 501-569-3252.

University of British Columbia, Institute of Applied Mathematics, Vancouver, BC V6T 1Z1, Canada. Institute awards M Sc, PhD. *Degree requirements:* For master's, thesis; for doctorate, 1 foreign language, dissertation, comprehensive exam. *Entrance requirements:* For doctorate, master's degree. *Faculty research:* Applied analysis, statistics, optimization.

University of California at Berkeley, College of Letters and Science, Department of Mathematics, Program in Applied Mathematics, Berkeley, CA 94720. Program awards PhD. Matriculated students: 19 full-time, 0 part-time; includes 5 minority (4 Asian American, 1 Hispanic American), 5 foreign. 0 applicants. *Degree requirements:* Dissertation. *Entrance requirements:* GRE General Test, minimum GPA of 3.0. Application deadline: 2/11. Application fee: $40. *Expenses:* Tuition of $0. Fees of $1909 per year for state residents; $7825 per year for nonresidents. • Dr. Alberto Grunbaum, Chair, Department of Mathematics, 415-642-4129.

University of California, Davis, Program in Applied Mathematics, Davis, CA 95616. Program awards MS, PhD. Matriculated students: 22. 31 applicants, 48% accepted. In 1990, 13 master's, 3 doctorates awarded. *Entrance requirements:* For master's, GRE General Test, GRE Subject Test, minimum GPA of 3.0; for doctorate, GRE General Test, GRE Subject Test, master's degree, minimum GPA of 3.0. Application deadline: 2/15. Application fee: $40. *Expenses:* Tuition of $0 for state residents; $7699 per year full-time, $3849 per year part-time for nonresidents. Fees of $2718 per year full-time, $1928 per year part-time. *Financial aid:* Fellowships, research assistantships available. • Graduate Adviser, 916-752-8131.

University of California, Los Angeles, School of Medicine and Graduate Division, Graduate Programs in Medicine, Department of Biomathematics, Los Angeles, CA 90024. Department awards MS, PhD, MD/PhD. Faculty: 6. Matriculated students: 10 full-time (3 women), 0 part-time; includes 1 minority (0 Asian American, 0 black American, 0 Hispanic American, 0 Native American), 5 foreign. 24 applicants, 29% accepted. In 1990, 4 master's, 0 doctorates awarded. *Degree requirements:* For master's, thesis or comprehensive exam required, foreign language not required; for doctorate, dissertation, written and oral qualifying exams required, foreign language not required. *Entrance requirements:* GRE General Test, GRE Subject Test. Application fee: $40. *Expenses:* Tuition of $0 for state residents; $7699 per year for nonresidents. Fees of $2907 per year. *Financial aid:* In 1990–91, 13 students received a total of $188,161 in aid awarded. 10 fellowships, 3 research assistantships, 4 teaching assistantships were awarded; full and partial tuition waivers, federal work-study, institutionally sponsored loans also available. Financial aid application deadline: 3/1. • Dr. Kenneth L. Lange, Chair, 310-825-5800.

See full description on page 463.

University of California, Riverside, Graduate Division, College of Natural and Agricultural Sciences, Department of Mathematics, Riverside, CA 92521. Department awards MA, MS, PhD. Part-time programs available. Terminal master's awarded for partial completion of doctoral program. *Degree requirements:* For master's, comprehensive exams required, foreign language and thesis not required; for doctorate, 1 foreign language, dissertation, qualifying exams. *Entrance requirements:* GRE General Test, TOEFL (minimum score of 550 required). Application deadline: 6/1. Application fee: $40. *Tuition:* $950 per quarter full-time, $264 per quarter part-time for state residents; $3517 per quarter full-time, $1758 per quarter part-time for nonresidents. *Faculty research:* Topology, applied math, analysis, probability.

See full description on page 467.

University of California, San Diego, Department of Mathematics, 9500 Gilman Drive, La Jolla, CA 92093. Offerings include applied mathematics (MA). Department faculty: 54. *Application fee:* $40. *Expenses:* Tuition of $0 for state residents; $7699 per year full-time, $1283 per quarter part-time for nonresidents. Fees of $2798 per year full-time, $669 per quarter part-time. • Harold Stark, Chair, 619-534-3594. Application contact: Lois Stewart, Graduate Coordinator, 619-534-6887.

University of California, Santa Barbara, College of Letters and Science, Department of Mathematics, Program in Applied Mathematics, Santa Barbara, CA 93106. Program awards MA. Matriculated students: 9 full-time (1 woman), 0 part-time; includes 2 foreign. 14 applicants, 100% accepted. In 1990, 3 degrees awarded. *Degree requirements:* Thesis or alternative required, foreign language not required. *Entrance requirements:* GRE General Test, GRE Subject Test, TOEFL (minimum score of 550 required). Application deadline: 5/1. Application fee: $40. *Expenses:* Tuition of $0 for state residents; $7699 per year for nonresidents. Fees of $2307 per year. *Financial aid:* Application deadline 1/31. • Thomas Sideris, Adviser, 805-893-2179. Application contact: Laurie Theobald, Graduate Secretary, 805-893-8192.

University of Central Oklahoma, College of Mathematics and Science, Department of Mathematics, Edmond, OK 73034. Offerings include applied mathematical sciences (MS). Department faculty: 9 full-time (1 woman), 0 part-time. *Degree requirements:* Computer language, thesis required, foreign language not required. *Application deadline:* 8/17. *Application fee:* $0. *Expenses:* Tuition of $46.85 per credit hour for state residents; $123 per credit hour for nonresidents. Fees of $3 per credit hour. • Dr. Donald Boyce, Chairperson, 405-341-2980 Ext. 5012.

University of Chicago, Division of the Physical Sciences, Department of Mathematics, Program in Applied Mathematics, Chicago, IL 60637. Program awards SM, PhD. *Application deadline:* 1/6. *Application fee:* $45. *Expenses:* Tuition of $16,275 per year full-time, $8140 per year part-time. Fees of $356 per year.

University of Cincinnati, McMicken College of Arts and Sciences, Department of Mathematics, Cincinnati, OH 45221. Offerings include applied mathematics (MS, PhD). Department faculty: 29 full-time, 0 part-time. *Degree requirements:* For doctorate, 1 foreign language, dissertation. *Entrance requirements:* For doctorate, GRE General Test, GRE Subject Test. Application deadline: 2/1. Application fee: $20. *Tuition:* $131 per credit hour for state residents; $261 per credit hour for nonresidents. • Dr. C. David Minda, Head, 513-556-4052.

University of Colorado at Boulder, College of Arts and Sciences, Department of Mathematics, Boulder, CO 80309. Offerings include applied mathematics (MS, PhD). Terminal master's awarded for partial completion of doctoral program. Department faculty: 45 full-time (1 woman). *Degree requirements:* For doctorate, 2 foreign languages, dissertation, comprehensive exam. *Application deadline:* 3/1 (priority date, applications processed on a rolling basis). *Application fee:* $30 ($50 for foreign students). *Expenses:* Tuition of $2308 per year full-time, $387 per semester (minimum) part-time for state residents; $8730 per year full-time, $1455 per semester (minimum) part-time for nonresidents. Fees of $207 per semester full-time, $27.26 per semester (minimum) part-time. • Arlan Ramsay, Chairman, 303-492-7256.

University of Colorado at Colorado Springs, College of Engineering and Applied Science, Department of Mathematics, Colorado Springs, CO 80933. Department offers program in applied mathematics (MS). Part-time and evening/weekend programs available. Faculty: 9 full-time (1 woman), 2 part-time (1 woman). Matriculated students: 5 full-time (2 women), 5 part-time (1 woman); includes 0 minority, 0 foreign. Average age 32. 2 applicants, 100% accepted. In 1990, 2 degrees awarded (100% found work related to degree). *Degree requirements:* Thesis required, foreign language not required. *Entrance requirements:* Minimum GPA of 3.0. Application deadline: 6/15. Application fee: $30 ($50 for foreign students). *Tuition:* $91 per credit hour for state residents; $276 per credit hour for nonresidents. *Financial aid:* In 1990–91, 4 teaching assistantships awarded. Financial aid application deadline: 5/1. *Faculty research:* Abelion groups and non-commutative rings, hormone analysis and computer vision, probability and mathematical physics, stochastic dynamics, probability models. Total annual research budget: $18,765. • Dr. Kulumani M. Rangaswamy, Chairman, 719-593-3311. Application contact: Gene Abrams, Graduate Adviser, 719-593-3182.

University of Colorado at Denver, College of Liberal Arts and Sciences, Program in Mathematics, Denver, CO 80217. Offerings include applied mathematics (MS, PhD). Program faculty: 23 full-time (4 women), 0 part-time. *Degree requirements:* For master's, thesis or alternative; for doctorate, 1 foreign language, dissertation. *Application deadline:* 6/1 (applications processed on a rolling basis). *Application fee:* $30 ($50 for foreign students). *Expenses:* Tuition of $1185 per semester full-time, $142 per semester hour part-time for state residents; $3969 per semester full-time, $476 per semester hour part-time for nonresidents. Fees of $103 per semester. • William Briggs, Chair, 303-556-8442. Application contact: Debbie Wangerin, Staff Assistant, 303-556-2341.

University of Dayton, College of Arts and Sciences, Department of Mathematics, Dayton, OH 45469. Offerings include applied mathematical systems (MS). Department faculty: 8 full-time (1 woman), 0 part-time. *Degree requirements:* Foreign language not required. *Entrance requirements:* GRE General Test, minimum GPA of 3.0 in undergraduate major. Application fee: $20. *Expenses:* Tuition of $262 per semester hour. Fees of $20 per semester. • Dr. Thomas E. Gantner, Chairman, 513-229-2511.

University of Delaware, College of Arts and Science, Department of Mathematical Sciences, Newark, DE 19716. Offerings include applied mathematics (MA, MA Sc, MS, PhD). Terminal master's awarded for partial completion of doctoral program. Department faculty: 40. *Degree requirements:* For master's, thesis (for some programs), written exam required, foreign language not required; for doctorate, 2 foreign languages, dissertation, qualifying exam. *Entrance requirements:* GRE General Test. Application deadline: 7/1. Application fee: $40. *Tuition:* $179 per credit hour for state residents; $467 per credit hour for nonresidents. • Dr. Ivar Stakgold, Chairman, 302-451-2651. Application contact: Pam Haverland, Graduate Secretary, 302-451-2654.

See full description on page 469.

University of Denver, Graduate Studies, Faculty of Natural Sciences, Mathematics and Engineering, Department of Mathematics and Computer Science, Denver, CO 80208. Offerings include applied mathematics (MA, MS). Department faculty: 14

full-time (1 woman), 4 part-time (2 women). *Degree requirements:* Foreign language, computer language or laboratory experience required, thesis not required. *Entrance requirements:* GRE General Test, GRE Subject Test, TOEFL (minimum score of 550 required). Application deadline: rolling. Application fee: $30. *Tuition:* $12,852 per year full-time, $357 per credit hour part-time. • Dr. James LaVita, Chairperson, 303-871-3344. Application contact: Rick Bell, Graduate Adviser, 303-871-2453.

University of Florida, College of Liberal Arts and Sciences, Department of Mathematics, Gainesville, FL 32611. Department offers programs in applied mathematics (MS, PhD), mathematics (MA, MS, PhD), mathematics teaching (MAT, MST). Faculty: 50 full-time. Matriculated students: 80 full-time; includes 28 foreign. *Degree requirements:* For master's, written exam (MS, MA), written exam or comprehensive written exam (MAT, MST); for doctorate, 1 foreign language, dissertation, qualifying exam. *Entrance requirements:* GRE General Test, TOEFL. Application deadline: 2/1 (priority date, applications processed on a rolling basis). Application fee: $15. *Tuition:* $87 per credit hour for state residents; $289 per credit hour for nonresidents. *Financial aid:* In 1990–91, 65 teaching assistantships awarded; fellowships also available. Financial aid application deadline: 2/1. *Faculty research:* Group theory, probability theory, logic, differential geometry and mathematical physics, topology and dynamical systems. • Dr. David A. Drake, Chairman, 904-392-0281.

See full description on page 471.

University of Georgia, College of Arts and Sciences, Department of Computer Science, Athens, GA 30602. Offerings include applied mathematical science (MAMS). Department faculty: 13 full-time (0 women), 0 part-time. *Application fee:* $10. *Expenses:* Tuition of $598 per quarter full-time, $48 per quarter part-time for state residents; $1558 per quarter full-time, $144 per quarter part-time for nonresidents. Fees of $118 per quarter. • Dr. Thiab R. Taha, Graduate Coordinator, 404-542-3477. Application contact: Dr. Walter D. Potter, Graduate Coordinator, 404-542-0361.

University of Georgia, College of Arts and Sciences, Department of Mathematics, Athens, GA 30602. Department awards MA, MAMS, PhD. Faculty: 35 full-time (2 women), 0 part-time. Matriculated students: 40 full-time (19 women), 3 part-time (0 women); includes 5 minority (1 Asian American, 4 black American), 19 foreign. 121 applicants, 21% accepted. In 1990, 12 master's, 2 doctorates awarded. *Degree requirements:* For master's, 1 foreign language, technical report (MAMS); for doctorate, 1 foreign language, dissertation. *Entrance requirements:* GRE General Test. Application fee: $10. *Expenses:* Tuition of $598 per quarter full-time, $48 per quarter part-time for state residents; $1558 per quarter full-time, $144 per quarter part-time for nonresidents. Fees of $118 per quarter. *Financial aid:* Fellowships, research assistantships, teaching assistantships, assistantships available. • Dr. Dhandapani Kannan, Graduate Coordinator, 404-542-2609.

See full description on page 473.

University of Georgia, College of Arts and Sciences, Department of Statistics, Athens, GA 30602. Offerings include applied mathematics (MAMS). Department faculty: 16 full-time (2 women), 0 part-time. *Application fee:* $10. *Expenses:* Tuition of $598 per quarter full-time, $48 per quarter part-time for state residents; $1558 per quarter full-time, $144 per quarter part-time for nonresidents. Fees of $118 per quarter. • Dr. Kermit Hutcheson, Graduate Coordinator, 404-542-8232.

See full description on page 475.

University of Guelph, College of Physical Science, Department of Mathematics and Statistics, Guelph, ON N1G 2W1, Canada. Offerings include applied mathematics (PhD). Department faculty: 28. *Entrance requirements:* Minimum B average. *Expenses:* Tuition of $898 per semester full-time, $450 per semester part-time for Canadian residents; $4053 per semester full-time, $2185 per semester part-time for nonresidents. Fees of $543 per semester full-time, $450 per semester part-time for Canadian residents; $2278 per semester full-time, $2185 per semester part-time for nonresidents. • Dr. Langford, Chair, 519-824-4120 Ext. 2155. Application contact: Dr. Carter, Graduate Coordinator, 519-824-4120 Ext. 3569.

University of Houston, College of Natural Sciences and Mathematics, Department of Mathematics, 4800 Calhoun, Houston, TX 77004. Offerings include applied mathematics (MS). *Degree requirements:* Thesis optional, foreign language not required. *Entrance requirements:* GRE General Test (minimum combined score of 900 required), TOEFL (minimum score of 550 required). Application deadline: 7/7. *Expenses:* Tuition of $30 per hour for state residents; $134 per hour for nonresidents. Fees of $240 per year full-time, $125 per year part-time.

See full description on page 477.

University of Illinois at Chicago, College of Liberal Arts and Sciences, Department of Mathematics, Statistics, and Computer Science, Program in Applied Mathematics, Chicago, IL 60680. Program awards MS, DA, PhD. *Degree requirements:* For master's, foreign language and thesis not required; for doctorate, 1 foreign language, dissertation. *Entrance requirements:* GRE General Test, TOEFL (minimum score of 550 required), minimum GPA of 3.75 (on a 5.0 scale). Application deadline: 7/5. Application fee: $20. *Expenses:* Tuition of $1369 per semester full-time, $521 per semester (minimum) part-time for state residents; $3840 per semester full-time, $1454 per semester (minimum) part-time for nonresidents. Fees of $458 per semester full-time, $398 per semester (minimum) part-time. • Application contact: Jeff E. Lewis, Director of Graduate Studies, 312-996-3041.

Announcement: Research/teaching areas include applied analysis, asymptotic analysis, elasticity, fluid dynamics, plasmadynamics, scattering and propagation theory, stability theory, stochastic population models, mathematical biophysics, statistical mechanics, scientific computation, queuing theory, singular perturbations. Students in mathematical sciences option elect science-related sequences such as mathematical physics. Students in the applications-oriented mathematics option elect sequences in analysis for applications.

See full description on page 479.

University of Illinois at Urbana-Champaign, College of Liberal Arts and Sciences, Department of Mathematics, Champaign, IL 61820. Offerings include applied mathematics (MS). Department faculty: 98 full-time, 4 part-time. *Degree requirements:* Foreign language and thesis not required. *Entrance requirements:* Minimum GPA of 4.0 on a 5.0 scale. Application fee: $25. *Tuition:* $1838 per semester full-time, $708 per semester part-time for state residents; $4314 per semester full-time, $1673 per semester part-time for nonresidents. • Ward Henson, Head, 217-333-3350.

University of Illinois at Urbana-Champaign, College of Engineering, Department of Theoretical and Applied Mechanics, Concentration in Applied Mathematics, Champaign, IL 61820. Concentration awards MS, PhD. *Degree requirements:* For master's, thesis or alternative required, foreign language not required; for doctorate, dissertation required, foreign language not required. *Application fee:* $25. *Tuition:* $1838 per semester full-time, $708 per semester part-time for state residents; $4314 per semester full-time, $1673 per semester part-time for nonresidents. *Financial aid:* Application deadline 2/15. • Frank J. Rizzo, Head, Department of Theoretical and Applied Mechanics, 217-333-2322. Application contact: Dr. Su Su Wang, Director, 217-333-1835.

University of Iowa, College of Liberal Arts, Program in Applied Mathematical Sciences, Iowa City, IA 52242. Program awards PhD. Faculty: 15 full-time, 0 part-time. Matriculated students: 9 full-time (2 women), 8 part-time (2 women); includes 1 minority (Asian American), 11 foreign. 28 applicants, 43% accepted. In 1990, 2 degrees awarded. *Application fee:* $20. *Expenses:* Tuition of $1158 per semester full-time, $387 per semester hour (minimum) part-time for state residents; $3372 per semester full-time, $387 per semester hour (minimum) part-time for nonresidents. Fees of $60 per semester (minimum). *Financial aid:* In 1990–91, 0 fellowships, 2 research assistantships, 15 teaching assistantships (2 to first-year students) awarded. • Herbert Hethcote, Chair, 319-335-0790.

University of Kansas, College of Liberal Arts and Sciences, Department of Mathematics, Program in Applied Mathematics and Statistics, Lawrence, KS 66045. Program awards MA, PhD. *Degree requirements:* For master's, thesis or alternative required, foreign language not required; for doctorate, 2 foreign languages, dissertation. *Entrance requirements:* TOEFL (minimum score of 570 required). Application fee: $25. *Expenses:* Tuition of $1668 per year full-time, $56 per credit hour part-time for state residents; $5382 per year full-time, $179 per credit hour part-time for nonresidents. Fees of $338 per year full-time, $25 per credit hour part-time. *Financial aid:* Fellowships, research assistantships, teaching assistantships, institutionally sponsored loans available. Aid available to part-time students. Financial aid application deadline: 2/1. • Dr. Charles Himmelberg, Chairperson, Department of Mathematics, 913-864-3651. Application contact: Dr. Saul Stahl, Graduate Director, 913-864-4324.

University of Louisville, Speed Scientific School, Department of Engineering Mathematics and Computer Science, Louisville, KY 40292. Offerings include applied mathematics (MS). Department faculty: 8 full-time (1 woman), 0 part-time. *Application deadline:* rolling. *Expenses:* Tuition of $1780 per year full-time, $99 per credit hour part-time for state residents; $5340 per year full-time, $297 per credit hour part-time for nonresidents. Fees of $60 per semester full-time, $12.50 per semester (minimum) part-time. • Dr. Khaled A. Kamel, Chair, 502-588-6304.

University of Manitoba, Faculty of Science, Department of Applied Mathematics, Winnipeg, MB R3T 2N2, Canada. Department awards M Sc.

University of Maryland College Park, College of Computer, Mathematical and Physical Sciences, Department of Mathematics, Applied Mathematics Program, College Park, MD 20742. Program awards MA, PhD. Matriculated students: 58 full-time (13 women), 33 part-time (5 women); includes 10 minority (8 Asian American, 2 Hispanic American), 35 foreign. 176 applicants, 40% accepted. In 1990, 4 master's, 4 doctorates awarded. *Degree requirements:* For master's, thesis or comprehensive exams required, foreign language not required; for doctorate, 2 foreign languages, dissertation. *Entrance requirements:* For master's, minimum GPA of 3.0. Application deadline: rolling. Application fee: $25. *Expenses:* Tuition of $143 per credit hour for state residents; $256 per credit hour for nonresidents. Fees of $171.50 per semester. *Financial aid:* In 1990–91, 11 fellowships (6 to first-year students) awarded. • Dr. Daniel Sweet, Director, 301-405-5062.

University of Maryland Graduate School, Baltimore, Graduate School, Department of Mathematics and Statistics, Program in Applied Mathematics, Baltimore, MD 21228. Offers applied mathematics (MS, PhD), including applied analysis (MS), control and communications (MS), operations research (MS), scientific computing and modeling (MS). Evening/weekend programs available. Matriculated students: 25 full-time (6 women), 36 part-time (7 women); includes 2 minority (1 Asian American, 1 black American), 14 foreign. 87 applicants, 31% accepted. In 1990, 7 master's, 1 doctorate awarded. *Degree requirements:* For master's, foreign language and thesis not required; for doctorate, dissertation required, foreign language not required. *Entrance requirements:* GRE General Test, GRE Subject Test, TOEFL, minimum GPA of 3.0. Application deadline: 7/1. Application fee: $25. *Tuition:* $134 per credit for state residents; $245 per credit for nonresidents. *Financial aid:* In 1990–91, 3 fellowships, 3 research assistantships, 18 teaching assistantships awarded. *Faculty research:* Numerical analysis, computers. Total annual research budget: $21,000. • Application contact: Dr. Thomas Mathew, Coordinator, 301-455-2418.

See full description on page 487.

University of Massachusetts at Amherst, College of Arts and Sciences, Faculty of Natural Sciences and Mathematics, Department of Mathematics and Statistics, Amherst, MA 01003. Offerings include applied mathematics (MA). Department faculty: 49 full-time (2 women), 0 part-time. *Degree requirements:* Foreign language and thesis not required. *Entrance requirements:* GRE General Test. Application deadline: 3/1 (applications processed on a rolling basis). Application fee: $35. *Tuition:* $2568 per year full-time, $107 per credit part-time for state residents; $7920 per year full-time, $330 per credit part-time for nonresidents. • Dr. Mei Ku, Director, 413-545-2282. Application contact: Dr. Wallace S. Martindale III, Chair, Admissions Committee, 413-545-0984.

University of Massachusetts, Lowell, College of Arts and Sciences, Department of Mathematics, 1 University Avenue, Lowell, MA 01854. Offerings include applied mathematics (MS). *Degree requirements:* Foreign language and thesis not required. *Entrance requirements:* GRE General Test. Application deadline: 4/1. *Expenses:* Tuition of $87 per credit hour for state residents; $271 per credit hour for nonresidents. Fees of $114 per credit hour.

University of Minnesota, Duluth, Graduate School, College of Science and Engineering, Department of Mathematics and Statistics, Duluth, MN 55812. Department offers program in applied and computational mathematics (MS). Part-time programs available. Faculty: 18 full-time (2 women), 0 part-time. Matriculated students: 23 full-time (9 women), 5 part-time (1 woman); includes 13 minority (12 Asian American, 1 black American), 14 foreign. Average age 24. 207 applicants, 6% accepted. In 1990, 3 degrees awarded (33% entered university research/teaching, 33% found other work related to degree, 34% continued full-time study). *Degree requirements:* Thesis or alternative required, foreign language not required. *Entrance requirements:* GRE, TOEFL. Application deadline: 7/15 (applications processed on a rolling basis). Application fee: $30. *Tuition:* $1184 per quarter full-time, $301 per credit (minimum) part-time for state residents; $2168 per quarter full-time, $602 per credit (minimum) part-time for nonresidents. *Financial aid:* In 1990–91, 26 students received a total of $206,856 in aid awarded. 2 research assistantships, 24 teaching assistantships were awarded; fellowships, institutionally sponsored loans, and career-related internships or fieldwork also available. Average monthly stipend for a graduate assistantship: $884. Financial aid application deadline: 3/15; applicants required to submit FAF. *Faculty research:* Numerical methods, dynamical systems, graph theory, applied statistics, control theory. Total annual research budget: $154,518. • Dr. Harlan W. Stech, Director of Graduate Studies, 218-726-8272.

Announcement: Program in Applied and Computational Mathematics emphasizes the role of mathematical modeling in industry, business, and science. Computing facilities include various personal computers, VAX, Sun, Encore (multiprocessor), and Cray systems. Faculty research includes graph theory, combinatorics, scientific computation, differential equations, control, probability, and statistics.

University of Missouri–Columbia, College of Arts and Sciences, Department of Mathematics, Program in Applied Mathematics, Columbia, MO 65211. Program awards MS. Matriculated students: 5 full-time (2 women), 0 part-time; includes 1 minority (black American), 0 foreign. In 1990, 1 degree awarded. *Degree requirements:* Thesis. *Entrance requirements:* GRE General Test, minimum GPA of 3.0. Application deadline: 8/1 (priority date, applications processed on a rolling basis). Application fee: $20 ($40 for foreign students). *Expenses:* Tuition of $89.90 per credit hour full-time, $98.35 per credit hour part-time for state residents; $244 per credit hour full-time, $252.45 per credit hour part-time for nonresidents. Fees of $123.55 per semester (minimum) full-time. • Application contact: Gary L. Smith, Director of Admissions and Registrar, 314-882-7651.

University of Missouri–Rolla, College of Arts and Sciences, Department of Mathematics and Statistics, Program in Applied Mathematics, Rolla, MO 65401. Program awards MS. Faculty: 22 full-time (2 women), 1 part-time (0 women). Matriculated students: 18 full-time (10 women), 0 part-time; includes 3 minority (all Asian American), 1 foreign. Average age 26. 15 applicants, 93% accepted. In 1990, 5 degrees awarded. *Degree requirements:* Thesis or alternative required, foreign language not required. *Entrance requirements:* GRE General Test, GRE Subject Test. Application deadline: 7/1 (applications processed on a rolling basis). Application fee: $20 ($40 for foreign students). *Expenses:* Tuition of $2090 per year full-time, $87.10 per credit hour part-time for state residents; $5582 per year full-time, $232.60 per credit hour part-time for nonresidents. Fees of $349 per year full-time, $61.63 per semester (minimum) part-time. *Financial aid:* In 1990–91, 35 students received a total of $281,000 in aid awarded. 1 research assistantship was awarded; teaching assistantships also available. Average monthly stipend for a graduate assistantship: $1133. *Faculty research:* Analysis, differential equations, statistics. Total annual research budget: $155,000. • Application contact: Troy Hicks, 314-341-4654.

University of Montana, College of Arts and Sciences, Department of Mathematical Sciences, Missoula, MT 59812. Offerings include applied mathematics (MA, PhD). Terminal master's awarded for partial completion of doctoral program. *Degree requirements:* For doctorate, 2 foreign languages (computer language can substitute for one), dissertation. *Entrance requirements:* For doctorate, GRE General Test. Application deadline: 9/15. Application fee: $20. *Tuition:* $495 per quarter hour full-time for state residents; $1239 per quarter hour full-time for nonresidents.

University of Nebraska–Lincoln, College of Arts and Sciences, Department of Mathematics and Statistics, Program in Applied Mathematics, Lincoln, NE 68588. Program awards MS. Faculty: 2 full-time (0 women), 0 part-time. Matriculated students: 32 full-time (13 women), 6 part-time (3 women); includes 2 minority (both black American), 29 foreign. Average age 29. In 1990, 13 degrees awarded. *Degree requirements:* Foreign language not required. *Entrance requirements:* GRE General Test, TOEFL (minimum score of 500 required). Application deadline: 5/1 (priority date, applications processed on a rolling basis). Application fee: $25. *Expenses:* Tuition of $75.75 per credit hour for state residents; $187.25 per credit hour for nonresidents. Fees of $161 per year full-time. *Financial aid:* Fellowships, teaching assistantships, federal work-study available. Aid available to part-time students. Financial aid application deadline: 2/15. • Dr. Colin Ramsay, Director, 402-472-2698.

University of Nevada, Las Vegas, College of Science and Mathematics, Department of Mathematical Sciences, Las Vegas, NV 89154. Offerings include applied mathematics (MS). Department faculty: 16 full-time (0 women), 0 part-time. *Degree requirements:* Oral exam required, thesis optional, foreign language not required. *Entrance requirements:* Minimum GPA of 2.75 overall, 3.0 during previous 2 years. Application deadline: 6/15. Application fee: $20. *Expenses:* Tuition of $66 per credit. Fees of $1800 per semester for nonresidents. • Dr. Peter Shive, Chairman, 702-739-3567. Application contact: Graduate College Admissions Evaluator, 702-739-3320.

University of Pittsburgh, Faculty of Arts and Sciences, Department of Mathematics and Statistics, Program in Applied Mathematics, Pittsburgh, PA 15260. Program awards MA, MS. Part-time programs available. Faculty: 46 full-time (5 women), 26 part-time (13 women). Matriculated students: 1 full-time (0 women), 1 part-time (0 women); includes 0 minority, 1 foreign. In 1990, 1 degree awarded. *Degree requirements:* Thesis (for some programs), comprehensive exam, preliminary exam required, foreign language not required. *Entrance requirements:* GRE General Test, TOEFL, minimum QPA of 3.0. Application deadline: 3/15 (priority date, applications processed on a rolling basis). Application fee: $15 ($25 for foreign students). *Expenses:* Tuition of $2920 per semester full-time, $241 per credit part-time for state residents; $5840 per semester full-time, $482 per credit part-time for nonresidents. Fees of $156 per year. *Financial aid:* 65 students received aid. Federal work-study, institutionally sponsored loans available. Aid available to part-time students. Average monthly stipend for a graduate assistantship: $1050. Financial aid application deadline: 3/1; applicants required to submit FAF. *Faculty research:* Numerical analysis, fluid mechanics, biomathematics, scientific computing, differential equations. • Application contact: Thomas A. Metzger, Director of Graduate Studies, 412-624-8343.

University of Rhode Island, College of Business Administration, Kingston, RI 02881. Offerings include applied mathematics (PhD). *Degree requirements:* Foreign language and dissertation not required. *Entrance requirements:* GMAT. Application fee: $25. *Expenses:* Tuition of $2575 per year full-time, $120 per credit hour part-time for state residents; $5900 per year full-time, $274 per credit hour part-time for nonresidents. Fees of $696 per year full-time.

University of Southern California, Graduate School, College of Letters, Arts and Sciences, Division of Natural Sciences and Mathematics, Department of Mathematics, Program in Applied Mathematics, Los Angeles, CA 90089. Program awards MA, MS, PhD. Matriculated students: 51 full-time (12 women), 9 part-time (2 women); includes 4 minority (3 Asian American, 1 black American), 42 foreign. Average age 26. 111 applicants, 69% accepted. In 1990, 21 master's, 4 doctorates awarded. *Degree requirements:* For doctorate, dissertation. *Entrance requirements:* GRE General Test. Application deadline: 7/1 (priority date). Application fee: $50. *Expenses:* Tuition of $12,120 per year full-time, $505 per unit part-time. Fees of $280 per year. *Financial aid:* 51 students received aid. Fellowships, research assistantships, teaching assistantships, federal work-study, institutionally sponsored loans available. Average monthly stipend for a graduate assistantship: $878. Financial aid application deadline: 3/1. • Dr. Gary Rosen, Director, 213-740-2446.

See full description on page 493.

University of South Florida, College of Arts and Sciences, Department of Mathematics, Program in Applied Mathematics, Tampa, FL 33620. Program awards PhD. Part-time and evening/weekend programs available. Matriculated students: 1 full-time (0 women), 0 part-time; includes 0 minority. Average age 24. 0 applicants. In 1990, 0 degrees awarded. *Degree requirements:* 2 foreign languages (computer language can substitute for one), dissertation. *Application fee:* $15. *Tuition:* $79.40 per credit hour for state residents; $241.33 per credit hour for nonresidents. *Financial aid:* In 1990–91, 1 student received a total of $7939 in aid awarded. • Edwin Clark, Co-Coordinator, 813-974-3838.

University of Tennessee Space Institute, Program in Applied Mathematics, Tullahoma, TN 37388. Program awards MS. Faculty: 3 full-time (0 women), 0 part-time. Matriculated students: 2 full-time (0 women), 1 part-time (0 women); includes 0 foreign. In 1990, 1 degree awarded. *Degree requirements:* Thesis required (for some programs), foreign language not required. *Application fee:* $15. *Tuition:* $135 per credit hour for state residents; $296 per credit hour for nonresidents. *Financial aid:* Research assistantships available. Aid available to part-time students. • Dr. K. C. Reddy, Degree Program Chairman, 615-455-0631 Ext.318. Application contact: Dr. Edwin M. Gleason, Assistant Dean for Admissions and Student Affairs, 615-455-0631 Ext. 472.

University of Texas at Dallas, School of Natural Sciences and Mathematics, Program in Mathematical Sciences, Richardson, TX 75083-0688. Offerings include applied mathematics (MS, PhD). Program faculty: 11 full-time (0 women), 18 part-time (3 women). *Degree requirements:* For master's, thesis optional, foreign language not required; for doctorate, dissertation required, foreign language not required. *Entrance requirements:* For master's, GRE General Test (minimum combined score of 1050 required), TOEFL (minimum score of 550 required), minimum GPA of 3.0 in upper level course work in field, secondary certificate in mathematics or computer science (MAT only); for doctorate, GRE General Test (minimum combined score of 1300 required), TOEFL (minimum score of 550 required), minimum GPA of 3.5 in upper level course work in field. Application deadline: 7/15 (applications processed on a rolling basis). Application fee: $0 ($75 for foreign students). *Expenses:* Tuition of $360 per semester full-time, $100 per semester (minimum) part-time for state residents; $2196 per semester full-time, $122 per semester hour (minimum) part-time for nonresidents. Fees of $338 per semester full-time, $22 per hour (minimum) part-time. • Dr. John Van Ness, Head, 214-690-2161.

University of Toledo, College of Arts and Sciences, Department of Mathematics, Program in Applied Mathematics, Toledo, OH 43606. Program awards MS. Matriculated students: 0. *Degree requirements:* Foreign language and thesis not required. *Entrance requirements:* GRE General Test, GRE Subject Test. Application deadline: 9/8 (priority date). Application fee: $30. *Tuition:* $122.59 per credit hour for state residents; $193.40 per credit hour for nonresidents. *Financial aid:* Application deadline 4/1; applicants required to submit FAF. • Dr. Harvey Wolff, Chairman, Department of Mathematics, 419-537-2568.

University of Tulsa, College of Engineering and Physical Sciences, Department of Mathematical and Computer Sciences, Tulsa, OK 74104. Offerings include applied mathematics (MS). Department faculty: 12 full-time (1 woman), 0 part-time. *Degree requirements:* Computer language, thesis or alternative required, foreign language not required. *Entrance requirements:* GRE General Test, TOEFL. Application deadline: rolling. Application fee: $30. *Tuition:* $350 per credit hour. • Dr. William Coberly, Chairperson, 918-631-3119. Application contact: Dr. Roger Wainwright, Adviser, 918-631-3143.

University of Victoria, Faculty of Arts and Science, Department of Mathematics, Victoria, BC V8W 2Y2, Canada. Offerings include applied mathematics (MA, M Sc). Department faculty: 24 full-time (1 woman), 0 part-time. *Degree requirements:* Thesis (for some programs), oral exam required, foreign language not required. *Entrance requirements:* Honors degree in mathematics. Application deadline: 5/31 (priority date, applications processed on a rolling basis). Application fee: $20. *Expenses:* Tuition of $754 per semester. Fees of $23 per year. • Dr. D. J. Leeming, Chair, 604-721-7436. Application contact: Dr. J. Phillips, Graduate Adviser, 604-721-7450.

University of Virginia, School of Engineering and Applied Science, Department of Applied Mathematics, Charlottesville, VA 22906. Department awards M Ap Ma, MS, PhD. Faculty: 10 full-time (3 women), 1 (woman) part-time. Matriculated students: 32 full-time (13 women), 0 part-time; includes 3 minority (all black American), 6 foreign. Average age 25. 33 applicants, 48% accepted. In 1990, 8 master's, 4 doctorates awarded. *Degree requirements:* For master's, foreign language and thesis not required; for doctorate, dissertation, comprehensive exam. *Entrance requirements:* For master's, GRE General Test. Application deadline: 8/1. Application fee: $40. *Expenses:* Tuition of $2740 per year full-time, $904 per year (minimum) part-time for state residents; $8950 per year full-time, $2960 per year (minimum) part-time for nonresidents. Fees of $586 per year full-time, $342 per year part-time. *Faculty research:* Numerical analysis, computational mechanics, probability and statistics. • James G. Simmonds, Chairman, 804-924-7201. Application contact: Ralph A. Lowry, Associate Dean, 804-924-3050.

University of Washington, College of Arts and Sciences, Department of Applied Mathematics, Seattle, WA 98195. Department awards MS, PhD. Part-time programs available. *Degree requirements:* For master's, computer language required, thesis optional, foreign language not required; for doctorate, 1 foreign language, computer language, dissertation. *Application deadline:* 7/1. *Application fee:* $35. *Tuition:* $1129 per quarter full-time, $324 per credit (minimum) part-time for state residents; $2824 per quarter full-time, $809 per credit (minimum) part-time for nonresidents. *Faculty research:* Mathematical modeling for physical, biological, social, and engineering sciences; development of mathematical methods for analysis, including perturbation, asymptotic, transform, vocational, and numerical methods.

See full description on page 501.

University of Waterloo, Faculty of Mathematics, Department of Applied Mathematics, Waterloo, ON N2L 3G1, Canada. Department awards M Math, M Phil, PhD. Part-time programs available. Faculty: 24 full-time, 0 part-time. Matriculated students: 34 full-time (10 women), 0 part-time; includes 7 foreign. 51 applicants, 31% accepted. In 1990, 5 master's, 0 doctorates awarded. *Degree requirements:* Thesis/dissertation required, foreign language not required. *Entrance requirements:* For master's, TOEFL (minimum score of 550 required), honor's degree in field, minimum B average; for doctorate, TOEFL (minimum score of 550 required), master's degree. Application deadline: 4/15. Application fee: $25. *Expenses:* Tuition of $757 per year full-time, $530 per year part-time for Canadian residents; $3127 per year for nonresidents. Fees of $68 per year full-time, $17 per year part-time. *Financial aid:* In 1990–91, 9 research assistantships, 35 teaching assistantships awarded; career-related internships or fieldwork also available. Average monthly stipend for a graduate assistantship: $530. *Faculty research:* Differential equations, quantum theory, statistical mechanics, fluid mechanics relativity, applied differential geometry. • Dr. F. O. Goodman, Chair, 519-885-1211 Ext. 2887. Application contact: Dr. E. Vescay, Graduate Officer, 519-885-1211 Ext. 4602.

University of Western Ontario, Physical Sciences Division, Department of Applied Mathematics, London, ON N6A 3K7, Canada. Department awards M Sc, PhD. *Degree requirements:* For master's, thesis or alternative required, foreign language not required; for doctorate, dissertation required, foreign language not required. *Entrance requirements:* Minimum B average. *Tuition:* $1015 per year full-time, $1050 per year part-time for Canadian residents; $4207 per year for nonresidents.

University of Wyoming, College of Arts and Sciences, Department of Mathematics, Laramie, WY 82071. Department awards MA, MAT, MS, MST, PhD. Faculty: 30 full-time (3 women), 0 part-time. Matriculated students: 27 full-time (10 women), 9 part-time (3 women); includes 1 minority (Hispanic American), 12 foreign. In 1990, 3 master's, 0 doctorates awarded. *Entrance requirements:* GRE General Test, minimum GPA of 3.0. Application deadline: 6/1 (priority date, applications processed on a rolling basis). Application fee: $30. *Tuition:* $1554 per year full-time, $74.25 per credit hour part-time for state residents; $4358 per year full-time, $74.25 per credit hour part-time for nonresidents. *Financial aid:* Application deadline 4/15. • Dr. William Bridges, Head, 307-766-4221.

See full description on page 505.

Virginia Commonwealth University, College of Humanities and Sciences, Department of Mathematical Sciences, Richmond, VA 23284. Offerings include applied mathematics (MS). Department faculty: 38. *Degree requirements:* Foreign language not required. *Entrance requirements:* GRE General Test, GRE Subject Test. Application deadline: 7/1. Application fee: $20. *Expenses:* Tuition of $2770 per year full-time, $154 per hour part-time for state residents; $7550 per year full-time, $419 per hour part-time for nonresidents. Fees of $717 per year full-time, $25.50 per hour part-time. • Dr. William E. Haver, Chair, 804-367-1319. Application contact: Dr. James A. Wood, Director of Graduate Studies, 804-367-1301.

Virginia Polytechnic Institute and State University, College of Arts and Sciences, Department of Mathematics, Blacksburg, VA 24061. Offerings include applied mathematics (MS, PhD). Terminal master's awarded for partial completion of doctoral program. Department faculty: 55 full-time (5 women), 0 part-time. *Degree requirements:* For master's, thesis required (for some programs), foreign language not required; for doctorate, 1 foreign language, dissertation. *Application deadline:* 2/15 (priority date). *Application fee:* $10. *Tuition:* $1889 per semester full-time, $606 per credit hour part-time for state residents; $2627 per semester full-time, $853 per credit hour part-time for nonresidents. • Dr. C. Wayne Patty, Head, 703-231-6536. Application contact: E. L. Green, Graduate Administrator, 703-231-6536.

See full description on page 511.

Washington State University, College of Sciences and Arts, Division of Sciences, Department of Pure and Applied Mathematics, Pullman, WA 99164. Department offers program in pure and applied mathematics (MS, DA, PhD). Faculty: 31 full-time (3 women), 4 part-time (1 woman). Matriculated students: 41 full-time (13 women), 0 part-time; includes 2 minority (1 Asian American, 1 Native American), 13 foreign. In 1990, 9 master's awarded; 5 doctorates awarded (100% entered university research/teaching). *Degree requirements:* For master's, thesis or alternative, core exams required, foreign language not required; for doctorate, 2 foreign languages, dissertation, core exams. *Entrance requirements:* For master's, GRE General Test, GRE Subject Test, TOEFL, minimum GPA of 3.0; for doctorate, GRE Subject Test, TOEFL, minimum GPA of 3.0. Application deadline: 3/1 (priority date, applications processed on a rolling basis). Application fee: $25. *Tuition:* $1694 per semester full-time, $169 per credit hour part-time for state residents; $4236 per semester full-time, $424 per credit hour part-time for nonresidents. *Financial aid:* In 1990–91, 0 fellowships, 4 research assistantships, 36 teaching assistantships, 0 teaching associateships awarded; partial tuition waivers, federal work-study, institutionally sponsored loans, and career-related internships or fieldwork also available. Average monthly stipend for a graduate assistantship: $975. Financial aid application deadline: 4/1. *Faculty research:* Computational mathematics, operations research, modeling in the natural sciences, applied statistics. Total annual research budget: $395,000. • Dr. Michael J. Kallaher, Chairman, 509-335-8518.

Wayne State University, College of Liberal Arts, Department of Mathematics, Program in Applied Mathematics, Detroit, MI 48202. Program awards MA, PhD. Matriculated students: 0 full-time, 3 part-time (1 woman). 22 applicants, 36% accepted. In 1990, 0 master's, 0 doctorates awarded. *Degree requirements:* For doctorate, dissertation. *Application deadline:* 7/1. *Application fee:* $20 ($30 for foreign students). *Expenses:* Tuition of $119 per credit hour for state residents; $258 per credit hour for nonresidents. Fees of $50 per semester. • Dr. Pao liu Chou, Chairperson, Department of Mathematics, 313-577-2479. Application contact: Dr. Peter Malcolmson, Graduate Committee Chair, 313-577-2472.

Western Michigan University, College of Arts and Sciences, Department of Mathematics and Statistics, Program in Applied Mathematics, Kalamazoo, MI 49008. Program awards MS. Matriculated students: 1 (woman) full-time, 3 part-time (1 woman); includes 1 minority (black American), 0 foreign. 14 applicants, 29% accepted. In 1990, 3 degrees awarded. *Degree requirements:* Oral exams required, thesis not required. *Application deadline:* 2/15 (priority date, applications processed on a rolling basis). *Application fee:* $25. *Expenses:* Tuition of $100 per credit hour for state residents; $244 per credit hour for nonresidents. Fees of $178 per semester full-time, $76 per semester part-time. *Financial aid:* Fellowships, research assistantships, teaching assistantships, federal work-study available. Financial aid application deadline: 2/15. • Dr. Yousef Alavi, Chairperson, Department of Mathematics and Statistics, 616-387-4513. Application contact: Paula J. Boodt, Director, Graduate Student Services, 616-387-3570.

Wichita State University, Fairmount College of Liberal Arts and Sciences, Department of Mathematics, Wichita, KS 67208. Department offers programs in applied mathematics (PhD), mathematics (MS), statistics (MS). Part-time programs available. Faculty: 23 full-time (3 women), 0 part-time. Matriculated students: 30 full-time (6 women), 16 part-time (8 women); includes 4 minority (3 Asian American, 1 Hispanic American), 23 foreign. 32 applicants, 72% accepted. In 1990, 6 master's awarded. *Degree requirements:* For master's, comprehensive exam required, thesis optional, foreign language not required; for doctorate, dissertation. *Entrance requirements:* TOEFL (minimum score of 550 required). Application deadline: 8/1 (priority date, applications processed on a rolling basis). Application fee: $0 ($25 for foreign students). *Expenses:* Tuition of $1590 per year full-time, $53 per credit part-time for state residents; $2574 per year full-time, $171.60 per credit part-time for nonresidents. Fees of $12.20 per credit. *Financial aid:* In 1990–91, $175,000 in aid awarded. 23 teaching assistantships were awarded; federal work-study, institutionally sponsored loans also available. Financial aid application deadline: 4/1. *Faculty research:* Partial differential equations, combinatorics, ring theory, minimal surfaces, several complex variables. • Dr. Buma Fridman, Chairperson, 316-689-3160.

Worcester Polytechnic Institute, Department of Mathematical Science, Worcester, MA 01609. Department awards MM, MS. Faculty: 25 full-time (4 women), 0 part-time. Matriculated students: 8 full-time (1 woman), 1 (woman) part-time. 18 applicants, 72% accepted. In 1990, 5 degrees awarded (20% entered university research/teaching, 60% found other work related to degree, 20% continued full-time study). *Degree requirements:* Thesis optional. *Entrance requirements:* GRE, TOEFL (minimum score of 550 required). Application deadline: 2/15 (priority date, applications processed on a rolling basis). Application fee: $25. *Expenses:* Tuition of $460 per credit hour. Fees of $20 per year full-time. *Financial aid:* In 1990–91, 4 students received a total of $62,284 in aid awarded. 1 research assistantship, 4 teaching assistantships were awarded; fellowships also available. Average monthly stipend for a graduate assistantship: $866. Financial aid application deadline: 2/15. *Faculty research:* Computational and theoretical fluid mechanics, Bayesian methods, composite materials and optimal design, graph theory, time series analysis. Total annual research budget: $383,305. • Dr. S. Rankin, Head, 508-831-5316.

Wright State University, College of Science and Mathematics, Department of Mathematics and Statistics, Program in Applied Mathematics, Dayton, OH 45435. Program awards MS. Faculty: 25 full-time (1 woman), 0 part-time. Matriculated students: 1 full-time (0 women), 8 part-time (2 women). Average age 32. In 1990, 4 degrees awarded. *Degree requirements:* Comprehensive exams required, foreign language and thesis not required. *Tuition:* $3342 per year full-time, $106 per credit hour part-time for state residents; $5991 per year full-time, $190 per credit hour part-time for nonresidents. *Financial aid:* In 1990–91, 1 fellowship, 1 research assistantship, 2 teaching assistantships awarded. *Faculty research:* Control theory, ordinary differential equations, partial differential equations, numerical analysis. • Dr. David F. Miller, Graduate Adviser, 513-873-2068.

Yale University, Graduate School of Arts and Sciences, Program in Applied Mathematics, New Haven, CT 06520. Program awards PhD. Faculty: 11 full-time, 0 part-time. Matriculated students: 0. 11 applicants, 0% accepted. *Entrance requirements:* GRE General Test, TOEFL. Application deadline: 1/2. Application fee: $45. *Tuition:* $15,160 per year full-time, $1895 per course part-time. • G. Veronis, Chairman, 203-432-4176. Application contact: Susan Webb, Coordinator of Admissions, 203-432-2770.

Biometrics

Case Western Reserve University, Schools of Medicine and Graduate Studies, Graduate Programs in Medicine, Department of Epidemiology and Biostatistics, Cleveland, OH 44106. Offerings include biometry (MS, PhD). Terminal master's awarded for partial completion of doctoral program. Department faculty: 13 full-time (6 women), 7 part-time (2 women). *Degree requirements:* For master's, thesis required (for some programs), foreign language not required; for doctorate, dissertation required, foreign language not required. *Entrance requirements:* GRE General Test, GRE Subject Test, TOEFL (minimum score of 550 required). Application deadline: 6/1 (priority date, applications processed on a rolling basis). Application fee: $25. • Harold B. Houser, Chairman, 216-368-3195.

Cornell University, Graduate Fields of Agriculture and Life Sciences, Field of Biometry, Ithaca, NY 14853. Field awards MS, PhD. Faculty: 8 full-time, 0 part-time. Matriculated students: 8 full-time (3 women), 0 part-time; includes 0 minority, 4 foreign. 18 applicants, 17% accepted. In 1990, 3 master's, 2 doctorates awarded. Terminal master's awarded for partial completion of doctoral program. *Degree requirements:* For master's, thesis, comprehensive exam; for doctorate, dissertation, qualifying exam. *Entrance requirements:* GRE General Test, TOEFL. Application deadline: 2/1. Application fee: $55. *Expenses:* Tuition of $7440 per trimester full-time. Fees of $28 per year full-time. *Financial aid:* In 1990–91, 6 students received aid. 2 fellowships (0 to first-year students), 2 research assistantships (0 to first-year students), 4 teaching assistantships (2 to first-year students) awarded; full and partial tuition waivers, federal work-study, institutionally sponsored loans also available. Financial aid application deadline: 1/10; applicants required to submit GAPSFAS. *Faculty research:* Statistical theory and methods, experiment design, linear models, ecological statistics, epidemiology. • Shayle R. Searle, Graduate Faculty Representative, 607-255-1646. Application contact: Robert Brashear, Director of Admissions, 607-255-4884.

Announcement: The program stresses the connection between theory and applications. Students participate with faculty in statistical consulting. Computer hardware includes supercomputers, mainframes, workstations, and microcomputers, and a wide variety of statistical software is available. The program is closely related to the Field of Statistics, which is described in the Statistics Directory.

Louisiana State University Medical Center, School of Graduate Studies in New Orleans, Department of Biometry and Genetics, Program in Biometry, 433 Bolivar Street, New Orleans, LA 70112. Program awards MS, PhD. Part-time programs available. Faculty: 8 full-time (0 women), 0 part-time. Matriculated students: 6 full-time (3 women), 2 part-time (1 woman); includes 0 minority, 3 foreign. Average age 30. 17 applicants, 100% accepted. In 1990, 2 master's awarded (50% entered university research/teaching, 50% continued full-time study); 1 doctorate awarded (100% entered university research/teaching). Terminal master's awarded for partial completion of doctoral program. *Degree requirements:* Computer language, thesis/dissertation required, foreign language not required. *Entrance requirements:* GRE

Directory: Biometrics

General Test (minimum combined score of 1100 required), TOEFL (minimum score of 550 required). Application deadline: 4/1. Application fee: $30. *Financial aid:* In 1990–91, 6 students received a total of $60,000 in aid awarded. 1 fellowship (to a first-year student), 5 research assistantships (1 to a first-year student) were awarded. Average monthly stipend for a graduate assistantship: $750-1250. Financial aid application deadline: 4/1. *Faculty research:* Statistical methodology for data analysis, especially for human genetics. Total annual research budget: $700,000. • Application contact: Dr. Miguel Guzman, Graduate Coordinator, 504-568-7000.

Announcement: The department offers master's and doctoral training in basic statistical theory and biometrical applications. Areas of faculty expertise include maximum likelihood estimation, general linear models, multivariate methods, categorical data analysis, and Bayesian methods. The department has a strong research program of applied biometry in human quantitative genetics and genetic epidemiology.

Medical University of South Carolina, Department of Biostatistics, Epidemiology, and Systems Science, Charleston, SC 29425. Department offers programs in biometrics (MS, PhD), biostatistics (MS, PhD), epidemiology (MS, PhD). Faculty: 15 full-time (3 women), 2 part-time (1 woman). Matriculated students: 37 full-time (14 women), 0 part-time; includes 1 minority (black American), 13 foreign. Average age 25. 65 applicants, 51% accepted. In 1990, 1 master's, 3 doctorates awarded. Terminal master's awarded for partial completion of doctoral program. *Degree requirements:* For master's, thesis, minimum GPA of 3.0, research seminar; for doctorate, dissertation, teaching and research seminar, oral and written exams, minimum GPA of 3.0. *Entrance requirements:* GRE General Test (minimum combined score of 1650 on all 3 sections required; foreign students required to take advanced test in the area most relevant to their field of interest), interview. Application deadline: 3/1 (priority date, applications processed on a rolling basis). Application fee: $25. *Tuition:* $2475 per year full-time, $75 per credit hour part-time. *Financial aid:* In 1990–91, 20 fellowships (10 to first-year students) awarded; research assistantships, teaching assistantships, full and partial tuition waivers, federal work-study, institutionally sponsored loans, and career-related internships or fieldwork also available. Financial aid application deadline: 3/1. *Faculty research:* Statistical modeling, survival analysis, cardiovascualr epidemiology, biomathematics, biomedical computing. Total annual research budget: $811,240. • Dr. M. C. Miller III, Chairman, 803-792-2261. Application contact: Wanda Taylor, Director of Admissions, 803-792-3281.

See full description on page 425.

Mount Sinai School of Medicine of the City University of New York, Graduate School of Biological Sciences, Department of Biomathematical Sciences, New York, NY 10029. Department awards PhD, MD/PhD. Faculty: 13 full-time, 0 part-time. Matriculated students: 4 full-time (1 woman), 0 part-time. *Degree requirements:* Computer language, dissertation required, foreign language not required. *Entrance requirements:* GRE General Test, GRE Subject Test, TOEFL. Application deadline: 4/15. Application fee: $30. *Expenses:* Tuition of $2700 per year for state residents; $4700 per year for nonresidents. Fees of $17 per year. *Financial aid:* Tuition grants available. • Dr. Craig Benham, Acting Chairman, 212-241-5851.

See full description on page 431.

North Carolina State University, College of Physical and Mathematical Sciences, Program in Biomathematics, Raleigh, NC 27695. Offers biomathematics (M Biomath, MS, PhD), ecology (PhD). Part-time programs available. Faculty: 4 full-time (0 women), 11 part-time (2 women). Matriculated students: 24 full-time (10 women), 1 part-time (0 women); includes 3 minority (1 black American, 2 Hispanic American), 4 foreign. Average age 26. 19 applicants, 53% accepted. In 1990, 0 master's awarded; 2 doctorates awarded (100% entered university research/teaching). Terminal master's awarded for partial completion of doctoral program. *Degree requirements:* For master's, thesis or alternative; for doctorate, dissertation. *Entrance requirements:* GRE General Test, TOEFL. Application deadline: 5/1. Application fee: $35. *Tuition:* $1138 per year for state residents; $5805 per year for nonresidents. *Financial aid:* In 1990–91, $110,000 in aid awarded. 1 fellowship (0 to first-year students), 9 research assistantships (1 to a first-year student), 6 teaching assistantships (4 to first-year students) were awarded; career-related internships or fieldwork also available. Average monthly stipend for a graduate assistantship: $850. Financial aid application deadline: 3/1. *Faculty research:* Analysis of mathematical models, stochastic differential equations, physiological models, systems and decision analysis, ecology models. • R. E. Stinner, Director, 919-737-2271.

Announcement: Biomathematics Graduate Program, North Carolina State University, offers a rigorous education at MS and PhD levels. Program encompasses diversity of interests, focusing on development and use of approaches dealing with uncertainty in biological systems. Program has a history of cooperation with departments representing biological, mathematical, and computational sciences. Students concentrate in applied biology (environmental management, agricultural systems, applied ecology), theoretical biology (ecology, neurobiology, genetics, physiology), or modeling methodology (system-science theory, simulation, stochastic modeling, decision modeling, statistical theory). Address inquiries to Biomathematics Graduate Program, Campus Box 8203.

Oregon State University, Graduate School, College of Science, Department of Statistics, Corvallis, OR 97331. Offerings include biometry (MA, MS, PhD). Department faculty: 15 full-time (2 women), 3 part-time (0 women). *Degree requirements:* For doctorate, dissertation required, foreign language not required. *Entrance requirements:* For doctorate, TOEFL (minimum score of 550 required), minimum GPA of 3.0 in last 90 hours. Application deadline: 3/1. Application fee: $40. *Tuition:* $1140 per trimester full-time, $449 per year part-time for state residents; $1816 per trimester full-time, $674 per year part-time for nonresidents. • Dr. Justus F. Seely, Chair, 503-737-3366.

See full description on page 445.

State University of New York at Buffalo, Graduate School, Graduate Programs in Biomedical Sciences at Roswell Park, Department of Biometry at Roswell Park, Buffalo, NY 14263. Department awards MS. Part-time programs available. Faculty: 9 full-time (1 woman), 0 part-time. Matriculated students: 0 full-time, 3 part-time (2 women); includes 0 minority, 0 foreign. Average age 30. 5 applicants, 60% accepted. In 1990, 3 degrees awarded. *Degree requirements:* Thesis. *Entrance requirements:* GRE General Test, TOEFL (minimum score of 600 required). Application deadline: 6/1 (priority date, applications processed on a rolling basis). Application fee: $35. *Expenses:* Tuition of $1600 per semester full-time, $134 per hour part-time for state residents; $3258 per semester full-time, $274 per hour part-time for nonresidents. Fees of $137 per semester full-time, $115 per semester (minimum) part-time. *Financial aid:* In 1990–91, $6552 in aid awarded. 0 research assistantships were awarded; federal work-study, institutionally sponsored loans also available. Financial aid application deadline: 2/1; applicants required to submit FAF. *Total annual research budget:* $250,000. • Dr. Robert Rein, Chairman, 716-845-3072. Application contact: Dr. Okhee Suh, Director of Graduate Studies, 716-845-8651.

University of Alabama at Birmingham, Schools of Medicine and Dentistry, Graduate Programs in Basic Sciences, Department of Biostatistics and Biomathematics, Program in Biomathematics, Birmingham, AL 35294. Program awards MS, PhD. Faculty: 15 full-time (4 women), 0 part-time. Matriculated students: 8 full-time (3 women), 5 part-time (2 women); includes 1 black American, 5 foreign. Average age 32. *Degree requirements:* Variable foreign language requirement, thesis/dissertation, comprehensive exam. *Entrance requirements:* GRE General Test (minimum combined score of 1650 on 3 sections required), minimum GPA of 3.0. Application deadline: rolling. Application fee: $25 ($50 for foreign students). *Tuition:* $66 per quarter for state residents; $132 per quarter for nonresidents. *Financial aid:* In 1990–91, 4 fellowships awarded. *Faculty research:* Mathematical modeling of biological systems, computational mathematics, molecular basis of contractility, nonlinear optimization, fluorescence spectroscopy. • Dr. Jane B. Hazelrig, Director, 205-934-4905.

Announcement: The program provides a balance between theory and application, with the objective of producing research-oriented statistical scientists who can advance statistical theory and can interact effectively with scientists in other disciplines to advance knowledge in those fields. Faculty members are extensively involved in collaborative research projects and consultations with other members of the University community. This enables students to work with researchers from other scientific areas to obtain practical experience in data collection, management, analysis, and interpretation of experimental results.

University of California, Los Angeles, School of Medicine and Graduate Division, Graduate Programs in Medicine, Department of Biomathematics, Los Angeles, CA 90024. Department awards MS, PhD, MD/PhD. Faculty: 6. Matriculated students: 10 full-time (3 women), 0 part-time; includes 1 minority (0 Asian American, 0 black American, 0 Hispanic American, 0 Native American), 5 foreign. 24 applicants, 29% accepted. In 1990, 4 master's, 0 doctorates awarded. *Degree requirements:* For master's, thesis or comprehensive exam required, foreign language not required; for doctorate, dissertation, written and oral qualifying exams required, foreign language not required. *Entrance requirements:* GRE General Test, GRE Subject Test. Application fee: $40. *Expenses:* Tuition of $0 for state residents; $7699 per year for nonresidents. Fees of $2907 per year. *Financial aid:* In 1990–91, 13 students received a total of $188,161 in aid awarded. 10 fellowships, 3 research assistantships, 4 teaching assistantships were awarded; full and partial tuition waivers, federal work-study, institutionally sponsored loans also available. Financial aid application deadline: 3/1. • Dr. Kenneth L. Lange, Chair, 310-825-5800.

See full description on page 463.

University of Colorado Health Sciences Center, School of Medicine and Graduate School, Programs in Biological and Medical Sciences, Department of Preventive Medicine and Biometrics, Program in Biometrics, Denver, CO 80262. Program awards MS, PhD. Matriculated students: 9 full-time (4 women), 6 part-time (5 women); includes 1 minority (Native American), 3 foreign. 2 applicants, 0% accepted. In 1990, 2 master's, 2 doctorates awarded. *Degree requirements:* Thesis/dissertation required, foreign language not required. *Application deadline:* 3/1. *Application fee:* $30. *Tuition:* $8804 per year full-time for state residents; $36,521 per year full-time for nonresidents. • Dr. Richard Jones, Head, 303-270-7605. Application contact: Kathy Trejo, Secretary, 303-270-7605.

University of Minnesota, Twin Cities Campus, Program in Health Informatics, Minneapolis, MN 55455. Program awards MS, PhD. Part-time programs available. Faculty: 18 (4 women). Matriculated students: 21 full-time (9 women), 8 part-time (3 women); includes 12 foreign. Average age 31. 30 applicants, 50% accepted. In 1990, 5 master's, 2 doctorates awarded. *Degree requirements:* For master's, thesis or alternative, project paper; for doctorate, dissertation. *Entrance requirements:* For master's, GRE General Test, previous course work in calculus, linear algebra, life sciences, programming, biology; for doctorate, GRE General Test, previous course work in life sciences, programming, differential equations. Application deadline: 7/15 (applications processed on a rolling basis). Application fee: $25. *Expenses:* Tuition of $1084 per quarter full-time, $301 per credit part-time for state residents; $2168 per quarter full-time, $602 per credit part-time for nonresidents. Fees of $118 per quarter. *Financial aid:* In 1990–91, 2 teaching assistantships (both to first-year students) awarded; fellowships, research assistantships, federal work-study also available. Financial aid application deadline: 1/15. *Faculty research:* Medical decision making, physiological control systems, population studies, health services research, clinical information systems. • Dr. Stanley Finkelstein, Director, 612-625-8440.

University of North Carolina at Chapel Hill, School of Medicine and Graduate School, Graduate Programs in Medicine, Department of Biomedical Engineering, Chapel Hill, NC 27599. Department awards MS, PhD. Faculty: 15 full-time, 0 part-time. Matriculated students: 45 full-time (8 women), 0 part-time; includes 0 minority, 4 foreign. 89 applicants, 40% accepted. In 1990, 6 master's, 4 doctorates awarded. *Degree requirements:* For master's, thesis, comprehensive exam required, foreign language not required; for doctorate, 1 foreign language, dissertation, qualifying exam. *Entrance requirements:* GRE General Test (minimum combined score of 1000 required), minimum GPA of 3.0. Application deadline: 2/1. Application fee: $35. *Tuition:* $621 per semester full-time for state residents; $3555 per semester full-time for nonresidents. *Financial aid:* In 1990–91, 3 fellowships, 20 research assistantships awarded; graduate assistantships and career-related internships or fieldwork also available. • Dr. Richard N. Johnson, Chairman, 919-966-1175.

University of Southern California, School of Medicine and Graduate School, Graduate Programs in Medicine, Department of Preventive Medicine, Program in Biometry, Los Angeles, CA 90089. Offers applied biometry/epidemiology (MS), biometry (MS, PhD), epidemiology (PhD), preventive medicine (health behavior) (PhD). Faculty: 24 full-time (10 women), 0 part-time. Matriculated students: 32 full-time (18 women), 0 part-time; includes 4 minority (3 Asian American, 1 Native American), 11 foreign. Average age 26. 30 applicants, 53% accepted. In 1990, 8 master's awarded (63% entered university research/teaching, 25% found other work related to degree, 12% continued full-time study); 2 doctorates awarded (50% entered university research/teaching, 50% found other work related to degree). Terminal master's awarded for partial completion of doctoral program. *Degree requirements:* Thesis/dissertation required, foreign language not required. *Entrance requirements:* For master's, GRE General Test (minimum combined score of 1000 required), GRE Subject Test, TOEFL, minimum GPA of 3.0; for doctorate, GRE General Test (minimum combined score of 1250 required), GRE Subject Test, TOEFL, minimum GPA of 3.0. Application deadline: 3/1. Application fee: $50. *Financial aid:* In 1990–91, 19 students received a total of $197,506 in aid awarded. 2 fellowships (0 to first-year students), 4 research assistantships (0 to first-year students), 13 teaching assistantships (6 to first-year students) were awarded; federal work-study, institutionally sponsored loans, and career-related internships or fieldwork also available. Average monthly stipend for a graduate assistantship: $1081. Financial aid application deadline: 4/1; applicants required to submit FAF. *Faculty research:* Clinical trials in ophthalmology and cancer research, methods of analysis for epidemiological studies. Total annual research budget: $1.2-million. • Dr. Stanley P. Azen, Director, 213-342-1810.

Announcement: MS and PhD programs prepare students in applied statistics/epidemiology, with emphasis in health science applications. Research activities include clinical trials, survival analysis, health-care delivery systems, and cancer research. Many opportunities are available for half-time teaching/research assistantships ($14,000–$16,000). Interdisciplinary research is encouraged.

University of Texas Health Science Center at Houston, Graduate School of Biomedical Sciences, Studies in Biomathematics, Houston, TX 77225. Awards MS, PhD, MD/PhD. Faculty: 5 full-time (0 women), 0 part-time. Matriculated students: 0. 1 applicants, 0% accepted. In 1990, 0 master's, 0 doctorates awarded. Terminal master's awarded for partial completion of doctoral program. *Degree requirements:* Thesis/dissertation required, foreign language not required. *Entrance requirements:* GRE General Test, TOEFL (minimum score of 550 required). Application deadline: 3/1 (priority date, applications processed on a rolling basis). Application fee: $10. *Expenses:* Tuition of $20 per hour for state residents; $128 per hour for nonresidents. Fees of $220 per year full-time, $9.17 per hour part-time. *Financial aid:* Fellowships, research assistantships, institutionally sponsored loans available. Average monthly stipend for a graduate assistantship: $1050. Financial aid application deadline: 3/31; applicants required to submit FAF. • Dr. Stuart Zimmerman, Coordinator, 713-792-2600. Application contact: Debra M. Moss, Coordinator, Admissions and Alumni Affairs, 713-792-4582.

University of Wisconsin–Madison, College of Agricultural and Life Sciences, Biometry Program, Madison, WI 53706. Program awards MS. Faculty: 0 full-time, 23 part-time. Matriculated students: 3 full-time (1 woman), 1 (woman) part-time; includes 0 minority, 2 foreign. 0 applicants. In 1990, 0 degrees awarded. *Application deadline:* rolling. *Application fee:* $20. *Financial aid:* 0 students received aid. Institutionally sponsored loans available. Financial aid application deadline: 1/15; applicants required to submit FAF. • Erik V. Nordheim, Chairperson, 608-262-0326. Application contact: Joanne Nagy, Assistant Dean of the Graduate School, 608-262-2433.

Biostatistics

Boston University, Graduate School, Program in Biostatistics, Boston, MA 02215. Program awards MA, PhD. 11 applicants, 36% accepted. *Entrance requirements:* GRE General Test, TOEFL (minimum score of 550 required). Application deadline: 7/1 (applications processed on a rolling basis). Application fee: $45. *Expenses:* Tuition of $15,950 per year full-time, $498 per credit part-time. Fees of $120 per year full-time, $35 per semester part-time. *Financial aid:* Application deadline 1/15. • Ralph B. D'Agostino, Director, 617-353-2560.

Boston University, School of Medicine, School of Public Health, Epidemiology and Biostatistics Section, Boston, MA 02215. Offerings include biostatistics (MPH). *Degree requirements:* Foreign language and thesis not required. *Entrance requirements:* GRE General Test. Application deadline: 4/15. *Tuition:* $15,950 per year full-time, $498 per credit part-time. • Dr. Theodore Colton, Chief, 617-638-5172. Application contact: Barbara St. Onge, Director of Admissions, 617-638-5052.

Case Western Reserve University, Schools of Medicine and Graduate Studies, Graduate Programs in Medicine, Department of Epidemiology and Biostatistics, Cleveland, OH 44106. Department awards MS, PhD, MD/PhD. Programs offered in biometry (MS, PhD), biostatistics (MS, PhD), clinical decision analysis (MS, PhD), computer science data (MS, PhD), epidemiology (MS, PhD). Faculty: 13 full-time (6 women), 7 part-time (2 women). Matriculated students: 9 full-time (2 women), 19 part-time (11 women); includes 0 minority, 15 foreign. Average age 33. 16 applicants, 50% accepted. In 1990, 0 doctorates awarded. Terminal master's awarded for partial completion of doctoral program. *Degree requirements:* For master's, thesis required (for some programs), foreign language not required; for doctorate, dissertation required, foreign language not required. *Entrance requirements:* GRE General Test, GRE Subject Test, TOEFL (minimum score of 550 required). Application deadline: 6/1 (priority date, applications processed on a rolling basis). Application fee: $25. *Financial aid:* In 1990–91, 13 students received a total of $89,400 in aid awarded. Full and partial tuition waivers and career-related internships or fieldwork available. Aid available to part-time students. Financial aid application deadline: 4/1. *Faculty research:* AIDS epidemiology, cancer epidemiology nutritional epidemiology, biostatistical method, clinical epidemiology. Total annual research budget: $2.80-million. • Harold B. Houser, Chairman, 216-368-3195.

Announcement: The department encompasses a range of disciplines (biostatistics, epidemiology, clinical decision analysis, and computer science) that apply to acquisition, processing, storage, and analysis of biological and medical data. The data must be analyzed, inferences drawn and synthesized with existing knowledge, and causal relations established. The multidisciplinary faculty and collaborative arrangements with basic and clinical science departments provide a unique opportunity for students to develop an integrated approach to the measurement, management, analysis, and interpretation of data.

Columbia University, School of Public Health, Division of Biostatistics, New York, NY 10032. Division awards MPH, MS, Dr PH, PhD. In 1990, 2 master's, 0 doctorates awarded. *Degree requirements:* For master's, foreign language not required; for doctorate, dissertation required, foreign language not required. *Entrance requirements:* GRE General Test, GMAT, or MAT. Application deadline: 5/15 (priority date, applications processed on a rolling basis). Application fee: $50. *Tuition:* $498 per credit. • Dr. Joseph L. Fleiss, Head, 212-305-3932.

Drexel University, Center for Multidisciplinary Study and Research, Biomedical Engineering and Science Institute, 32nd and Chestnut Streets, Philadelphia, PA 19104. Offerings include biostatistics (MS). Institute faculty: 38 full-time (4 women), 20 part-time (2 women). *Degree requirements:* Thesis (for some programs). *Entrance requirements:* TOEFL (minimum score of 550 required), minimum GPA of 3.0. Application deadline: 8/23 (applications processed on a rolling basis). Application fee: $25. *Expenses:* Tuition of $345 per credit hour. Fees of $81 per quarter full-time, $43 per quarter part-time. • Dr. Dov Jaron, Director, 215-895-2215. Application contact: Vernon L. Newhouse, Graduate Adviser, 215-895-2225.

Announcement: Biostatistics is offered as a specialization within Drexel's biomedical science program. The specialization answers the demand for life scientists with advanced training in biostatistics and experimental design; it is also suited for students with prior training in mathematics or business. See profile description in the Biomedical Engineering Section in the Engineering and Applied Sciences volume of this series.

Drexel University, College of Arts and Sciences, Department of Mathematics and Computer Science, 32nd and Chestnut Streets, Philadelphia, PA 19104. Department offers programs in computer science (MS), mathematics (MS, PhD). Part-time programs available. Faculty: 26 full-time (2 women), 0 part-time. Matriculated students: 30 full-time, 85 part-time; includes 30 foreign. 183 applicants, 47% accepted. In 1990, 30 master's awarded (80% found work related to degree, 20% continued full-time study); 2 doctorates awarded (50% entered university research/teaching, 50% found other work related to degree). *Degree requirements:* For master's, thesis not required; for doctorate, 1 foreign language, dissertation. *Entrance requirements:* For master's, TOEFL (minimum score of 550 required), TSE (for teaching assistants), GRE. Application deadline: 8/23 (applications processed on a rolling basis). Application fee: $25. *Expenses:* Tuition of $345 per credit hour. Fees of $81 per quarter full-time, $43 per quarter part-time. *Financial aid:* In 1990–91, $400,000 in aid awarded. 37 teaching assistantships (14 to first-year students) were awarded; institutionally sponsored loans also available. Financial aid application deadline: 2/1. *Faculty research:* Probability and statistics, combinatorics special functions, functional analysis, parallel processing, computer a;gebra. Total annual research budget: $80,000. • Dr. James Pool, Head, 215-895-2668. Application contact: Keith Brooks, Director of Graduate Admissions, 215-895-6700.

See full description on page 409.

Emory University, Graduate School of Arts and Sciences, Divisions of Epidemiology and Biostatistics, Atlanta, GA 30322. Division offers programs in biostatistics (MS, PhD), quantitative epidemiology (PhD). PhD programs offered jointly with the School of Public Health. Part-time programs available. Faculty: 20 full-time (5 women), 0 part-time. Matriculated students: 12 full-time (6 women), 10 part-time (5 women); includes 0 minority, 3 foreign. 30 applicants, 27% accepted. In 1990, 1 master's awarded (100% continued full-time study); 2 doctorates awarded (50% entered university research/teaching, 50% found other work related to degree). Terminal master's awarded for partial completion of doctoral program. *Degree requirements:* For master's, computer language, thesis required, foreign language not required; for doctorate, computer language, dissertation, comprehensive exam required, foreign language not required. *Entrance requirements:* GRE General Test, TOEFL, minimum GPA of 3.0. Application deadline: 1/20 (priority date, applications processed on a rolling basis). Application fee: $35. *Expenses:* Tuition of $7370 per semester full-time, $642 per semester hour part-time. Fees of $160 per year full-time. *Financial aid:* In 1990–91, 15 students received a total of $121,500 in aid awarded. 7 fellowships (5 to first-year students), 13 tuition scholarships (6 to first-year students) were awarded; partial tuition waivers, institutionally sponsored loans, and career-related internships or fieldwork also available. Financial aid application deadline: 2/1. *Faculty research:* Epidemiologic study design and analysis, surgery methodology, research data management, epidemic modeling, categorical data analysis. Total annual research budget: $1.4-million. • Dr. Michael Kutnererg, Associate Dean for Academic Affairs, 404-727-7692. Application contact: Dr. Dollie Daniels, Academic Coordinator of Graduate Studies, 404-727-8712.

See full description on page 411.

Georgetown University, School of Medicine and Graduate School, Graduate Programs in Medicine, Division of Biostatistics and Epidemiology, Washington, DC 20057. Division awards MS. Faculty: 5 full-time, 5 part-time. Matriculated students: 0 full-time, 15 part-time (14 women); includes 3 minority (2 Asian American, 1 black American), 0 foreign. Average age 31. In 1990, 5 degrees awarded. *Degree requirements:* Foreign language and thesis not required. *Entrance requirements:* GRE General Test, previous course work in algebra. Application deadline: 7/15. Application fee: $40. *Tuition:* $12,768 per year full-time, $532 per year part-time. *Financial aid:* Career-related internships or fieldwork available. Financial aid applicants required to submit FAF. *Faculty research:* Occupation epidemiology, cancer. • Dr. Leonard Chiazze Jr., Director.

George Washington University, Graduate School of Arts and Sciences, Department of Statistics/Computer and Information Systems, Washington, DC 20052. Department offers programs in applied statistics (MS, PhD), including applied statistics (MS), statistics (PhD); mathematical statistics (MA); statistical computing (MS). Part-time and evening/weekend programs available. Faculty: 6 full-time (2 women), 4 part-time (1 woman). Matriculated students: 11 full-time (5 women), 21 part-time (8 women); includes 1 minority (black American), 8 foreign. Average age 35. 31 applicants, 71% accepted. In 1990, 3 master's, 2 doctorates awarded. Terminal master's awarded for partial completion of doctoral program. *Degree requirements:* For master's, comprehensive exam required, thesis not required; for doctorate, computer language, dissertation, general exam. *Entrance requirements:* GRE General Test, minimum GPA of 3.0. Application deadline: 7/1. Application fee: $45. *Expenses:* Tuition of $490 per semester hour. Fees of $215 per year full-time, $125.60 per year (minimum) part-time. *Financial aid:* In 1990–91, $111,945 in aid awarded. 10 fellowships (4 to first-year students), 7 teaching assistantships (4 to first-year students) were awarded; partial tuition waivers, federal work-study also available. Financial aid application deadline: 2/15. • Dr. Arthur Kirsch, Chair, 202-994-6356.

See full description on page 415.

Harvard University, School of Public Health, Department of Biostatistics, Boston, MA 02115. Department awards SM, DPH, SD. Part-time programs available. Faculty: 33 full-time (8 women), 10 part-time (4 women). Matriculated students: 52 (29

Directory: Biostatistics

women). In 1990, 7 master's, 8 doctorates awarded. *Degree requirements:* For doctorate, dissertation, oral and written qualifying exams. *Entrance requirements:* GRE General Test, TOEFL, prior training in mathematics and/or statistics. Application deadline: 2/1. Application fee: $50. *Tuition:* $14,790 per year. *Financial aid:* Applicants required to submit FAF. *Faculty research:* Statistical methodology, clinical trials, cancer and AIDS research, environmental and public health. • Dr. Nan Laird, Chair, 617-432-1056. Application contact: Jennifer Powell, 617-432-1056.

Johns Hopkins University, School of Hygiene and Public Health, Department of Biostatistics, Baltimore, MD 21205. Department awards MHS, Sc M, PhD. Faculty: 16 full-time, 3 part-time. Matriculated students: 32 (14 women); includes 5 minority (4 Asian American, 1 Hispanic American), 10 foreign. 49 applicants, 59% accepted. In 1990, 2 master's, 2 doctorates awarded. *Degree requirements:* For master's, thesis (for some programs); for doctorate, 1 foreign language, dissertation. *Entrance requirements:* GRE. Application deadline: 3/1 (priority date, applications processed on a rolling basis). Application fee: $40. *Tuition:* $15,500 per year full-time, $323 per credit part-time. *Financial aid:* Scholarships, stipends, traineeships, federal work-study, institutionally sponsored loans available. Aid available to part-time students. Financial aid application deadline: 4/15; applicants required to submit GAPSFAS or FAF. *Faculty research:* Linear models, environmental statistics, statistical methods, sampling techniques, time series analysis. Total annual research budget: $529,000. • Dr. Charles Rohde, Chair, 301-955-3539. Application contact: Carolyn Danner, Senior Administrator, 301-955-3716.

Johns Hopkins University, School of Hygiene and Public Health, Program in Public Health, Baltimore, MD 21205. Offerings include biostatistics (MPH). *Degree requirements:* Foreign language and thesis not required. *Entrance requirements:* TOEFL (minimum score of 550 required). Application deadline: 1/31 (priority date, applications processed on a rolling basis). Application fee: $40. *Tuition:* $15,500 per year full-time, $323 per credit part-time. • Dr. Haroutune Armenian, Director, 301-955-4095. Application contact: Lenora Davis, Administrator, 301-955-1291.

Loma Linda University, School of Medicine, Graduate Programs in Public Health, Programs in Biostatistics, Loma Linda, CA 92350. Programs award MPH, MSPH. Faculty: 4 full-time (1 woman), 1 (woman) part-time. Matriculated students: 3 full-time (1 woman), 2 part-time (1 woman); includes 1 minority (Asian American), 1 foreign. 3 applicants, 67% accepted. In 1990, 1 degree awarded. *Entrance requirements:* GMAT (MHA), MTELP (minimum score of 92 required) or TOEFL (minimum score of 600 required). Application fee: $25. *Tuition:* $248 per unit. • Dr. Robert Cruise, Chairman, 714-824-4590.

McGill University, Faculties of Medicine and Graduate Studies and Research, Graduate Programs in Medicine, Department of Epidemiology and Biostatistics, Montreal, PQ H3A 2T5, Canada. Offerings include epidemiology and biostatistics (M Sc, PhD, Diploma), with options in community health (M Sc), environmental health (M Sc), epidemiology (M Sc), health-care evaluation (M Sc), medical statistics (M Sc), occupational health (M Sc). *Degree requirements:* For master's, thesis optional, foreign language not required; for doctorate, dissertation required, foreign language not required. *Tuition:* $698.50 per semester full-time, $46.57 per credit part-time for Canadian residents; $3480 per semester full-time, $234 per credit part-time for nonresidents.

Medical College of Wisconsin, Biostatistics/Epidemiology Section, Milwaukee, WI 53226. Section awards MS. Faculty: 4 full-time (0 women), 0 part-time. Matriculated students: 2 full-time (1 woman), 7 part-time (2 women); includes 0 minority, 2 foreign. Average age 30. 5 applicants, 80% accepted. In 1990, 3 degrees awarded (100% entered university research/teaching). *Degree requirements:* Thesis required, foreign language not required. *Entrance requirements:* GRE. Application deadline: 2/15 (priority date, applications processed on a rolling basis). Application fee: $15. *Expenses:* Tuition of $5790 per year full-time, $325 per credit part-time. Fees of $105 per year full-time. *Financial aid:* In 1990–91, 9 students received a total of $45,218 in aid awarded. 2 fellowships (both to first-year students), 1 research assistantship were awarded; institutionally sponsored loans also available. Aid available to part-time students. Average monthly stipend for a graduate assistantship: $833. Financial aid application deadline: 2/15. *Faculty research:* Cancer epidemiology, health services. Total annual research budget: $1.5-million. • Dr. Alfred Rimm, Director, 414-257-8280.

Medical University of South Carolina, Department of Biostatistics, Epidemiology, and Systems Science, Charleston, SC 29425. Offerings include biostatistics (MS, PhD). Terminal master's awarded for partial completion of doctoral program. Department faculty: 15 full-time (3 women), 2 part-time (1 woman). *Degree requirements:* For master's, thesis, minimum GPA of 3.0, research seminar; for doctorate, dissertation, teaching and research seminar, oral and written exams, minimum GPA of 3.0. *Entrance requirements:* GRE General Test (minimum combined score of 1650 on all 3 sections required; foreign students required to take advanced test in the area most relevant to their field of interest), interview. Application deadline: 3/1 (priority date, applications processed on a rolling basis). Application fee: $25. *Tuition:* $2475 per year full-time, $75 per credit hour part-time. • Dr. M. C. Miller III, Chairman, 803-792-2261. Application contact: Wanda Taylor, Director of Admissions, 803-792-3281.

See full description on page 425.

New York Medical College, Graduate School of Health Sciences, Programs in Public and Community Health, Valhalla, NY 10595. Offerings include biostatistics (MPH, MS). *Degree requirements:* Computer language, thesis required, foreign language not required. *Entrance requirements:* TOEFL. Application fee: $25. *Tuition:* $300 per credit.

Ohio State University, College of Mathematical and Physical Sciences, Department of Statistics, Program in Biostatistics, Columbus, OH 43210. Program awards PhD. Matriculated students: 4 full-time (2 women), 1 part-time (0 women); includes 0 minority, 3 foreign. 14 applicants, 21% accepted. In 1990, 0 degrees awarded. *Degree requirements:* Dissertation required, foreign language not required. *Application deadline:* 8/15 (applications processed on a rolling basis). *Application fee:* $0 ($25 for foreign students). *Tuition:* $1213 per quarter full-time, $364 per course part-time for state residents; $3143 per quarter full-time, $943 per course part-time for nonresidents. *Financial aid:* Fellowships, research assistantships, teaching assistantships, federal work-study, institutionally sponsored loans available. Aid available to part-time students. Financial aid applicants required to submit FAF. • Prem Goel, Chairman, Department of Statistics, 614-292-8110.

Queen's University at Kingston, Faculty of Medicine and School of Graduate Studies and Research, Graduate Programs in Medicine, Department of Community Health and Epidemiology, Kingston, ON K7L 3N6, Canada. Offerings include biostatistics (M Sc). *Degree requirements:* Thesis required, foreign language not required. *Entrance requirements:* TOEFL (minimum score of 550 required). Application fee: $25. *Tuition:* $2861 per year full-time, $426 per trimester part-time for Canadian residents; $10,613 per year full-time, $4998 per trimester part-time for nonresidents.

San Diego State University, College of Health and Human Services, Graduate School of Public Health, San Diego, CA 92182. Offerings include epidemiology (MPH), with option in biostatistics. School faculty: 23 full-time (7 women), 9 part-time (4 women). *Application deadline:* 8/1 (applications processed on a rolling basis). *Application fee:* $55. *Expenses:* Tuition of $0 for state residents; $189 per unit for nonresidents. Fees of $1974 per year full-time, $692 per year part-time for state residents; $1074 per year full-time, $692 per year part-time for nonresidents. • Dr. F. Douglas Scutchfield, Director, 619-594-6317. Application contact: Brenda Fass-Holmes, Admissions Coordinator, 619-594-4492.

Tulane University, School of Public Health and Tropical Medicine, Department of Biostatistics, New Orleans, LA 70118. Department awards MS, MSPH, PhD, Sc D. *Entrance requirements:* GRE General Test, TOEFL. Application deadline: 4/15. Application fee: $40. *Expenses:* Tuition of $370 per credit hour. Fees of $110 per year. *Financial aid:* Application deadline 5/1; applicants required to submit GAPSFAS. • Dr. Virginia Ktsanes, Chairman, 504-588-5164.

University of Alabama at Birmingham, Schools of Medicine and Dentistry, Graduate Programs in Basic Sciences, Department of Biostatistics and Biomathematics, Program in Biostatistics, Birmingham, AL 35294. Program awards MS, PhD. Faculty: 15 full-time (4 women), 0 part-time. Matriculated students: 8 full-time (3 women), 5 part-time (2 women); includes 1 minority (black American), 5 foreign. Average age 32. 34 applicants, 12% accepted. In 1990, 2 master's, 0 doctorates awarded. *Degree requirements:* For master's, variable foreign language requirement, thesis; for doctorate, variable foreign language requirement, dissertation, comprehensive exam. *Entrance requirements:* For master's, GRE General Test (minimum combined score of 1500 on 3 sections required) or MAT, minimum GPA of 3.0; for doctorate, GRE General Test (minimum combined score of 1650 on 3 sections required) or MAT, minimum GPA of 3.0. Application deadline: rolling. Application fee: $25 ($50 for foreign students). *Tuition:* $66 per quarter for state residents; $132 per quarter for nonresidents. *Financial aid:* In 1990–91, 4 fellowships awarded. *Faculty research:* Survival analysis, clinical trials, biostatistical modeling, distribution theory, nonparametric statistics. • Dr. Edwin L. Bradley, Director, 205-934-4905.

Announcement: The program provides a balance between theory and application, with the objective of producing research-oriented statistical scientists who can advance statistical theory and can interact effectively with scientists in other disciplines to advance knowledge in those fields. Faculty members are extensively involved in collaborative research projects and consultations with other members of the University community. This enables students to work with researchers from other scientific areas to obtain practical experience in data collection, management, analysis, and interpretation of experimental results.

University of California at Berkeley, School of Public Health, Department of Biomedical and Environmental Health Sciences, Berkeley, CA 94720. Offerings include biostatistics (MA, MPH, Dr PH, PhD). Department faculty: 23. *Degree requirements:* For doctorate, dissertation. *Entrance requirements:* For doctorate, GRE General Test, TOEFL (minimum score of 550 required), minimum GPA of 3.0. Application deadline: 2/10 (applications processed on a rolling basis). Application fee: $40. *Expenses:* Tuition of $0. Fees of $2680 per year for state residents; $10,375 per year for nonresidents. • George Sensabaugh, Chair.

University of California at Berkeley, School of Public Health, Group in Biostatistics, Berkeley, CA 94720. Group awards MA, PhD. Faculty: 24. Matriculated students: 21 full-time, 0 part-time; includes 1 minority (Hispanic American), 6 foreign. 32 applicants, 38% accepted. *Degree requirements:* For doctorate, dissertation. *Entrance requirements:* GRE, minimum GPA of 3.0. Application deadline: 2/10 (applications processed on a rolling basis). Application fee: $40. *Expenses:* Tuition of $0. Fees of $2680 per year for state residents; $10,375 per year for nonresidents. • Dr. Nicholas P. Jewell, Chair, 415-642-7900.

University of California, Los Angeles, School of Public Health, Department of Biostatistics, Los Angeles, CA 90024. Department awards MS, PhD. Faculty: 6. Matriculated students: 40 full-time (18 women), 0 part-time; includes 8 minority, 21 foreign. 58 applicants, 74% accepted. In 1990, 9 master's, 3 doctorates awarded. *Degree requirements:* For master's, comprehensive exam required, foreign language and thesis not required; for doctorate, dissertation, qualifying exams required, foreign language not required. *Entrance requirements:* For master's, GRE General Test (minimum combined score of 1100 required), written screening exam; for doctorate, GRE General Test (minimum combined score of 1200 required), written screening exam. Application fee: $40. *Expenses:* Tuition of $0 for state residents; $7699 per year for nonresidents. Fees of $2907 per year. *Financial aid:* In 1990–91, 30 students received a total of $354,133 in aid awarded. 24 fellowships, 11 research assistantships, 8 teaching assistantships were awarded; full and partial tuition waivers, federal work-study, institutionally sponsored loans, and career-related internships or fieldwork also available. Financial aid application deadline: 3/1. • Dr. William G. Cumberland, Chair, 310-825-5250.

See full description on page 465.

University of Cincinnati, College of Medicine and Division of Graduate Studies and Research, Graduate Programs in Medicine, Department of Environmental Health, Cincinnati, OH 45221. Offerings include epidemiology and biostatistics (MS). Department faculty: 20 full-time, 0 part-time. *Degree requirements:* Thesis required, foreign language not required. *Entrance requirements:* GRE General Test, bachelor's degree in science. Application deadline: 2/15. Application fee: $20. *Tuition:* $131 per credit hour for state residents; $261 per credit hour for nonresidents. • Dr. Roy Albert, Head, 513-558-5701. Application contact: Dr. Dale R. Johnson, Director of Graduate Studies, 513-558-5703.

University of Hawaii at Manoa, College of Health Sciences and Social Welfare, School of Public Health, Honolulu, HI 96822. Offerings include biostatistics (MPH, MS), biostatistics-epidemiology (PhD). *Tuition:* $800 per semester full-time for state residents; $2405 per semester full-time for nonresidents.

University of Hawaii at Manoa, School of Public Health and Graduate Programs in Biomedical Sciences, Program in Biostatistics-Epidemiology, Honolulu, HI 96822. Program awards PhD. *Entrance requirements:* GRE General Test. *Tuition:* $800 per semester full-time for state residents; $2405 per semester full-time for nonresidents.

University of Illinois at Chicago, School of Public Health, Biostatistics Section, Chicago, IL 60680. Section awards MS, PhD. *Degree requirements:* For master's, thesis, field practicum required, foreign language not required; for doctorate, dissertation, internship, independent research required, foreign language not required. *Entrance requirements:* GRE General Test (minimum combined score of 1000 required), TOEFL (minimum score of 550 required), minimum GPA of 3.75 (on a 5.0 scale). Application deadline: 7/5. Application fee: $20. *Expenses:* Tuition of $1369 per semester full-time, $521 per semester (minimum) part-time for state residents; $3840 per semester full-time, $1454 per semester (minimum) part-time for nonresidents. Fees of $458 per semester full-time, $398 per semester (minimum) part-time. • Dr. Paul Levy, Chair, 312-996-6625. Application contact: Dr. Shafdeen Amuwo, Assistant Dean, 312-996-6625.

University of Illinois at Chicago, School of Public Health, Program in Epidemiology and Biostatistics, Chicago, IL 60680. Program awards MPH, MS, Dr PH, PhD. *Degree requirements:* For master's, thesis, field practicum required, foreign language not required; for doctorate, dissertation, internship, independent research required, foreign language not required. *Entrance requirements:* GRE General Test (minimum combined score of 1000 required), TOEFL (minimum score of 550 required), minimum GPA of 3.75 (on a 5.0 scale). Application deadline: 7/5. Application fee: $20. *Expenses:* Tuition of $1369 per semester full-time, $521 per semester (minimum) part-time for state residents; $3840 per semester full-time, $1454 per semester (minimum) part-time for nonresidents. Fees of $458 per semester full-time, $398 per semester (minimum) part-time. • Dr. Paul S. Levy, Director, 312-996-8860. Application contact: Dr. Shafdeen Amuwo, Assistant Dean, 312-996-6625.

University of Iowa, College of Medicine and Graduate College, Graduate Programs in Medicine, Department of Preventive Medicine and Environmental Health, Iowa City, IA 52242. Offerings include biostatistics (MS, PhD). MS/MA offered jointly with Program in Urban and Regional Planning. Terminal master's awarded for partial completion of doctoral program. Department faculty: 24 full-time (4 women), 8 part-time (0 women). *Degree requirements:* For master's, thesis optional, foreign language not required; for doctorate, dissertation required, foreign language not required. *Entrance requirements:* For master's, GRE General Test (minimum combined score of 1050 required), TOEFL (minimum score of 600 required), minimum GPA of 2.7; for doctorate, GRE General Test (minimum combined score of 1050 required), TOEFL (minimum score of 600 required), minimum GPA of 3.0. Application deadline: rolling. Application fee: $20 ($30 for foreign students). *Expenses:* Tuition of $1158 per semester full-time, $387 per semester hour (minimum) part-time for state residents; $3372 per semester full-time, $387 per semester hour (minimum) part-time for nonresidents. Fees of $60 per semester. • Dr. Robert B. Wallace, Chair, 319-335-9627. Application contact: Barbara Scott, Graduate Studies Coordinator, 319-335-8992.

University of Michigan, School of Public Health, Program in Biostatistics, Ann Arbor, MI 48109. Program awards MS, PhD. Offered through Horace H. Rackham School of Graduate Studies. Faculty: 11. Matriculated students: 47 full-time (28 women), 0 part-time; includes 4 minority (2 Asian American, 1 black American, 1 Hispanic American), 12 foreign. 71 applicants, 46% accepted. In 1990, 12 master's, 3 doctorates awarded. *Degree requirements:* For master's, foreign language and thesis not required; for doctorate, dissertation, preliminary exam required, foreign language not required. *Entrance requirements:* GRE General Test. Application deadline: 4/1. Application fee: $30. *Tuition:* $6546 per year for state residents; $13,680 per year for nonresidents. *Financial aid:* Fellowships, teaching assistantships available. *Faculty research:* Biostatistical methodology, biomedical research. • Richard Cornell, Chair, 313-764-5451.

University of Minnesota, Twin Cities Campus, School of Public Health, Major in Biostatistics, Minneapolis, MN 55455. Major awards MPH, MS, PhD. 58 applicants, 31% accepted. *Degree requirements:* For master's, foreign language not required; for doctorate, dissertation required, foreign language not required. *Entrance requirements:* For master's, GRE General Test (minimum combined score of 1500 on 3 sections required), minimum GPA of 3.0; for doctorate, GRE General Test (minimum combined score of 1500 on 3 sections required). Application fee: $30. *Expenses:* Tuition of $1084 per quarter full-time, $301 per credit part-time for state residents; $2168 per quarter full-time, $602 per credit part-time for nonresidents. Fees of $118 per quarter. *Financial aid:* Fellowships, research assistantships, teaching assistantships, federal work-study, institutionally sponsored loans available. • Dr. Thomas Louis, Head, 612-624-4655. Application contact: Marcus Kjelsburg, Admissions Committee Chair, 612-624-4655.

Announcement: With 12 full-time faculty, numerous sponsored research projects, and a large support staff, the department offers well-rounded and exciting master's and doctoral programs. Courses include biostatistical theory and methods, statistical computing, clinical trials, computing, epidemiology and demography, and offerings in Department of Statistics and other departments. Research and teaching assistantships are available.

University of North Carolina at Chapel Hill, School of Public Health, Department of Biostatistics, Chapel Hill, NC 27599. Department awards MPH, MS, MSPH, Dr PH, PhD. Faculty: 26 full-time, 0 part-time. Matriculated students: 110 full-time (73 women), 0 part-time; includes 9 minority (4 Asian American, 3 black American, 1 Hispanic American, 1 Native American), 17 foreign. 96 applicants, 63% accepted. In 1990, 22 master's, 8 doctorates awarded. *Degree requirements:* For master's, comprehensive exam required, foreign language not required; for doctorate, dissertation, comprehensive exam required, foreign language not required. *Entrance requirements:* GRE General Test (minimum combined score of 1000 required), minimum GPA of 3.0. Application deadline: 2/15. Application fee: $35. *Tuition:* $860 per semester full-time for state residents; $3555 per semester full-time for nonresidents. *Financial aid:* In 1990–91, 12 fellowships, 26 research assistantships, 2 teaching assistantships, 10 graduate assistantships awarded. • Dr. Barry H. Margolin, Chairman, 919-966-2483.

Announcement: Graduate students are from many parts of the world. They take theoretical and applied biostatistics courses and have opportunities for field observation and consulting experience. They also take nondepartmental courses in fields of biostatistical applications, such as cancer, cardiovascular disease, demography, the environment, and other health-related sciences.

University of Oklahoma Health Sciences Center, College of Public Health, Program in Biostatistics and Epidemiology, Oklahoma City, OK 73190. Offers biostatistics (MPH, MS, Dr PH, PhD), epidemiology (MPH, MS, Dr PH, PhD). Part-time programs available. Faculty: 7 full-time (2 women), 0 part-time. Matriculated students: 18 full-time (8 women), 33 part-time (20 women); includes 5 minority (1 black American, 3 Hispanic American, 1 Native American), 6 foreign. Average age 36. 19 applicants, 74% accepted. In 1990, 16 master's, 1 doctorate awarded. *Degree requirements:* For master's, computer language, thesis (for some programs) required, foreign language not required; for doctorate, computer language, dissertation required, foreign language not required. *Entrance requirements:* For master's, TOEFL (minimum score of 550 required); for doctorate, GRE, TOEFL (minimum score of 550 required). Application deadline: 7/1 (applications processed on a rolling basis). Application fee: $15. *Tuition:* $68 per credit hour for state residents; $206 per credit hour for nonresidents. *Financial aid:* Research assistantships, traineeships, institutionally sponsored loans, and career-related internships or fieldwork available. Aid available to part-time students. Financial aid application deadline: 5/1; applicants required to submit FAF. *Faculty research:* Statistical methodology, applied statistics, acute and chronic disease epidemiology. • Dr. Nabih Asai Jr., Interim Chair, 405-271-2229.

University of Pittsburgh, Graduate School of Public Health, Department of Biostatistics, Pittsburgh, PA 15260. Department offers programs in biostatistics (MPH, MS Hyg, DPH, PhD), human genetics (MS, MS Hyg, PhD). Part-time programs available. Faculty: 12 full-time (3 women), 0 part-time. Matriculated students: 24 full-time (11 women), 12 part-time (7 women); includes 4 minority (1 Asian American, 1 black American, 2 Hispanic American), 13 foreign. 40 applicants, 50% accepted. In 1990, 3 master's, 1 doctorate awarded. *Degree requirements:* For master's, thesis required, foreign language not required; for doctorate, 1 foreign language (computer language can substitute), dissertation. *Entrance requirements:* For master's, GRE General Test, TOEFL; for doctorate, TOEFL. Application deadline: 7/1 (priority date, applications processed on a rolling basis). Application fee: $20 ($30 for foreign students). *Tuition:* $3425 per semester full-time, $283 per credit part-time for state residents; $6850 per semester full-time, $566 per credit part-time for nonresidents. *Financial aid:* Fellowships, research assistantships, teaching assistantships, federal work-study, institutionally sponsored loans available. Aid available to part-time students. Financial aid applicants required to submit FAF. • Dr. Carol K. Redmond, Chairman, 412-624-3022.

University of Rochester, College of Arts and Science, Department of Statistics, Program in Medical Statistics, Rochester, NY 14627-0001. Program awards MS. *Degree requirements:* Foreign language and thesis not required. *Application deadline:* 2/15 (priority date). *Application fee:* $25. *Expenses:* Tuition of $473 per credit hour. Fees of $243 per year full-time. *Financial aid:* Application deadline 2/15. • Dr. David Oakes, Chair, Department of Statistics, 716-275-3644.

University of South Carolina, Graduate School, College of Health, School of Public Health, Department of Epidemiology/Biostatistics, Program in Biostatistics, Columbia, SC 29208. Program awards MPH, MSPH, Dr PH, PhD. Part-time programs available. Faculty: 6 full-time (1 woman), 0 part-time. Matriculated students: 10 full-time, 8 part-time; includes 3 minority (all Asian American), 7 foreign. Average age 26. In 1990, 2 master's, 1 doctorate awarded. *Degree requirements:* For master's, thesis, practicum (MPH) required, foreign language not required; for doctorate, dissertation. *Entrance requirements:* For master's, GRE; for doctorate, GRE General Test. Application deadline: rolling. Application fee: $25. *Tuition:* $1404 per semester full-time. *Financial aid:* Research assistantships, teaching assistantships, traineeships available. *Faculty research:* Bayesian methods, biometric modeling, non-linear regression, health survey methodology, measurement of health status. • Application contact: Dr. Francisco Sy, Graduate Director, 803-777-7353.

University of Utah, School of Medicine and Graduate School, Graduate Programs in Medicine, Programs in Public Health, Salt Lake City, UT 84112. Offerings include biostatistics (M Stat). Faculty: 10 full-time (4 women), 28 part-time (6 women). *Application deadline:* 2/28 (priority date). *Application fee:* $25 ($50 for foreign students). *Tuition:* $195 per credit for state residents; $505 per credit for nonresidents. • Dr. Charles C. Hughes, Director, 801-581-7234. Application contact: Dorothy Crockett, Executive Secretary, 801-581-7234.

University of Vermont, College of Engineering and Mathematics, Department of Mathematics and Statistics, Program in Biostatistics, Burlington, VT 05405. Program awards MS. Matriculated students: 1; includes 0 minority, 0 foreign. 4 applicants, 25% accepted. In 1990, 2 degrees awarded. *Degree requirements:* Thesis or alternative required, foreign language not required. *Entrance requirements:* GRE General Test, TOEFL (minimum score of 550 required). Application deadline: 4/1 (priority date, applications processed on a rolling basis). Application fee: $25. *Expenses:* Tuition of $206 per credit for state residents; $564 per credit for nonresidents. Fees of $150 per semester full-time. *Financial aid:* Fellowships, research assistantships, teaching assistantships available. Financial aid application deadline: 3/1. • Dr. L. Haugh, Director, 802-656-2940.

University of Washington, School of Public Health and Community Medicine, Program in Biostatistics, Seattle, WA 98195. Program awards MS, PhD. Faculty: 28 full-time (7 women), 2 part-time (both women). Matriculated students: 44 full-time (21 women), 8 part-time (6 women); includes 6 minority (4 Asian American, 2 Hispanic American), 15 foreign. Average age 31. 53 applicants, 49% accepted. In 1990, 8 master's awarded (0% entered university research/teaching, 75% found other work related to degree, 13% continued full-time study); 7 doctorates awarded (71% entered university research/teaching, 0% found other work related to degree, 14% continued full-time study). *Degree requirements:* Computer language, thesis/dissertation, departmental qualifying exams required, foreign language not required. *Entrance requirements:* GRE General Test, TOEFL, 2 years advanced calculus, 1 course linear algebra, 1 course probability. Application deadline: 4/15 (priority date, applications processed on a rolling basis). Application fee: $35. *Tuition:* $1129 per quarter full-time, $324 per credit (minimum) part-time for state residents; $2824 per quarter full-time, $809 per credit (minimum) part-time for nonresidents. *Financial aid:* In 1990–91, 45 students received a total of $514,105 in aid awarded. 2 fellowships (both to first-year students), 25 research assistantships (13 to first-year students), 5 teaching assistantships (0 to first-year students), 13 traineeships (4 to first-year students) were awarded; career-related internships or fieldwork also available. Average monthly stipend for a graduate assistantship: $955. Financial aid application deadline: 4/15. *Faculty research:* Statistical methods for survival data analysis, clinical trials, epidemiological case control and cohort studies, statistical genetics. Total annual research budget: $3.5-million. • Dr. Polly Feigl, Coordinator, 206-543-1044.

See full description on page 503.

University of Western Ontario, Biosciences Division, Department of Epidemiology and Biostatistics, London, ON N6A 3K7, Canada. Department awards M Sc, PhD. Part-time programs available. Faculty: 7 full-time (1 woman), 14 part-time (5 women). Matriculated students: 13 full-time (9 women), 15 part-time (8 women); includes 3 minority (2 Asian American, 1 black American), 0 foreign. Average age 35. 31 applicants, 26% accepted. In 1990, 3 master's awarded (33% found work related to degree, 67% continued full-time study); 1 doctorate awarded (100% found work related to degree). *Degree requirements:* For master's, thesis required, foreign language not required; for doctorate, comprehensive exam, thesis proposal defence required, foreign language not required. *Entrance requirements:* For master's, BA or B Sc honor's degree, minimum B+ GPA in last 10 courses; for doctorate, M Sc or equivalent, minimum B+ GPA in last 10 courses. Application deadline: 2/1. Application fee: $20. *Tuition:* $1015 per year for Canadian residents; $4207 per year for nonresidents. *Financial aid:* In 1990–91, 3 students received a total of $13,478 in aid awarded. Research assistantships, teaching assistantships, and career-related internships or fieldwork available. Average monthly stipend for a graduate assistantship: $888. Financial aid application deadline: 4/30. *Faculty research:* Chronic disease epidemiology, clinical epidemiology, biostatistics. • Allan P. Donner, Head.

Directories: Biostatistics; Mathematics

Virginia Commonwealth University, School of Graduate Studies and Medical College of Virginia-Professional Programs, School of Basic Health Sciences, Department of Biostatistics, Richmond, VA 23284. Department awards MS, PhD. Part-time programs available. Faculty: 10 full-time, 3 part-time. Matriculated students: 25 full-time (11 women), 13 part-time (5 women); includes 2 minority (both black American), 12 foreign. 45 applicants, 31% accepted. In 1990, 2 master's awarded (100% found work related to degree); 1 doctorate awarded (100% found work related to degree). Terminal master's awarded for partial completion of doctoral program. *Degree requirements:* Thesis/dissertation required, foreign language not required. *Entrance requirements:* For master's, GRE General Test, MCAT, or DAT; for doctorate, GRE General Test. Application fee: $20. *Expenses:* Tuition of $2770 per year full-time, $154 per hour part-time for state residents; $7550 per year full-time, $419 per hour part-time for nonresidents. Fees of $717 per year full-time, $25.50 per hour part-time. *Financial aid:* 38 students received aid. Fellowships, teaching assistantships, and career-related internships or fieldwork available. *Total annual research budget:* $836,264. • Dr. W. Hans Carter, Chair, 804-786-9824. Application contact: Dr. S. James Kilpatrick, Graduate Program Director, 804-786-9824.

Western Michigan University, College of Arts and Sciences, Department of Mathematics and Statistics, Program in Biostatistics, Kalamazoo, MI 49008. Program awards MS. Matriculated students: 1 (woman) full-time, 5 part-time (4 women); includes 0 minority, 1 foreign. 7 applicants, 14% accepted. In 1990, 3 degrees awarded. *Degree requirements:* Written exams required, thesis not required. *Application deadline:* 2/15 (priority date, applications processed on a rolling basis). *Application fee:* $25. *Expenses:* Tuition of $100 per credit hour for state residents; $244 per credit hour for nonresidents. Fees of $178 per semester full-time, $76 per semester part-time. *Financial aid:* Fellowships, research assistantships, teaching assistantships, federal work-study available. Financial aid application deadline: 2/15. • Application contact: Paula J. Boodt, Director, Graduate Student Services, 616-387-3570.

Yale University, School of Medicine, Department of Epidemiology and Public Health, Division of Biostatistics, New Haven, CT 06520. Division awards MPH, Dr PH, PhD. PhD offered through the Graduate School. Faculty: 5 full-time (1 woman), 2 part-time (0 women). Matriculated students: 2 full-time (1 woman), 2 part-time (1 woman). Terminal master's awarded for partial completion of doctoral program. *Degree requirements:* For master's, thesis required, foreign language not required; for doctorate, dissertation, residency period, comprehensive exams required, foreign language not required. *Entrance requirements:* For master's, GRE General Test, TOEFL, minimum B average in last 2 years, selected course work; for doctorate, GRE General Test (minimum combined score of 1200 required), TOEFL, MPH or equivalent and professional experience (Dr PH). Application deadline: 4/1 (priority date, applications processed on a rolling basis). Application fee: $45. *Tuition:* $13,450 per year. *Financial aid:* Scholarships, federal work-study, institutionally sponsored loans, and career-related internships or fieldwork available. Aid available to part-time students. Financial aid application deadline: 4/15; applicants required to submit GAPSFAS. *Faculty research:* Statistical and genetic epidemiology, dynamics of cancer incidence rates, clinical trials, clinical and population sampling, design and analysis of longitudinal surveys. • Theodore R. Holford, Professor, Public Health and Biostatistics, 203-785-2867. Application contact: Joan Stenner, Admissions Officer, 203-785-2844.

Mathematics

Acadia University, Faculty of Pure and Applied Science, Department of Mathematics, Wolfville, NS B0P 1X0, Canada. Department awards M Sc. Program temporarily suspended. Faculty: 12 full-time, 1 part-time. Matriculated students: 0. In 1990, 0 degrees awarded. *Degree requirements:* Thesis required, foreign language not required. *Expenses:* Tuition of $2405 per year full-time, $495 per course part-time. Fees of $1700 per year full-time, $340 per year part-time for nonresidents. • Dr. Paul Cabilio, Head, 902-542-2201 Ext. 382.

Adelphi University, Graduate School of Arts and Sciences, Department of Mathematics and Computer Science, Garden City, NY 11530. Department offers program in mathematics (MS, DA). Evening/weekend programs available. Matriculated students: 2 full-time (0 women), 27 part-time (12 women); includes 2 minority (both black American), 7 foreign. Average age 26. 49 applicants, 69% accepted. In 1990, 3 master's, 3 doctorates awarded. *Degree requirements:* For master's, foreign language not required; for doctorate, variable foreign language requirement, dissertation. *Application deadline:* rolling. *Application fee:* $50. *Expenses:* Tuition of $5300 per semester full-time, $315 per credit part-time. Fees of $159 per semester part-time. *Financial aid:* In 1990–91, $3600 in aid awarded. 6 teaching assistantships (all to first-year students), 14 grants (7 to first-year students) were awarded; full tuition waivers also available. • Dr. David Lubell, Chairperson, 516-877-4480.

Alabama State University, School of Graduate Studies, College of Arts and Sciences, Department of Mathematics and Physical Science, Montgomery, AL 36101. Department offers program in mathematics (MS, Ed S). Part-time programs available. Faculty: 3 full-time (0 women), 0 part-time. Matriculated students: 1 full-time (0 women), 5 part-time (2 women); includes 6 minority (1 Asian American, 5 black American). 3 applicants, 67% accepted. In 1990, 10 master's awarded. *Degree requirements:* For master's, 1 foreign language (computer language can substitute), thesis, comprehensive exam; for Ed S, thesis. *Entrance requirements:* For master's, GRE General Test, GRE Subject Test. Application deadline: 7/15. Application fee: $10. *Tuition:* $55 per credit hour for state residents; $110 per credit hour for nonresidents. *Financial aid:* In 1990–91, $9000 in aid awarded. 2 teaching assistantships (1 to a first-year student) were awarded. *Faculty research:* Shift dynamical systems, classification in automorphisms of subshift of finite type. • Dr. Wallace Maryland, Head.

American University, College of Arts and Sciences, Department of Mathematics and Statistics, Program in Mathematics, Washington, DC 20016. Program awards MA. Part-time and evening/weekend programs available. Faculty: 16 full-time (4 women), 2 part-time (0 women). Matriculated students: 2 full-time (1 woman), 1 part-time (0 women); includes 0 minority, 1 foreign. Average age 28. 15 applicants, 47% accepted. In 1990, 2 degrees awarded. *Degree requirements:* 1 foreign language required (computer language can substitute), thesis optional. *Application deadline:* 2/15. *Application fee:* $50. *Expenses:* Tuition of $475 per semester hour. Fees of $20 per semester. *Financial aid:* Fellowships, teaching assistantships, federal work-study, institutionally sponsored loans, and career-related internships or fieldwork available. Aid available to part-time students. Financial aid application deadline: 2/15; applicants required to submit FAF. • Dr. Robert W. Jernigan, Chair, Department of Mathematics and Statistics, 202-885-3120.

See full description on page 387.

Angelo State University, College of Sciences, Department of Mathematics, San Angelo, TX 76909. Department awards MS. Part-time and evening/weekend programs available. Faculty: 5 full-time (0 women), 0 part-time. Matriculated students: 4 full-time (3 women), 4 part-time (1 woman); includes 0 minority, 0 foreign. Average age 32. 7 applicants, 71% accepted. In 1990, 0 degrees awarded. *Degree requirements:* Comprehensive exam required, thesis optional, foreign language not required. *Entrance requirements:* GRE General Test, minimum GPA of 2.5. Application deadline: 8/15 (priority date, applications processed on a rolling basis). Application fee: $0 ($50 for foreign students). *Expenses:* Tuition of $20 per semester hour full-time, $20 per semester (minimum) part-time for state residents; $128 per semester full-time, $128 per credit hour part-time for nonresidents. Fees of $195 per semester full-time, $67.50 per semester hour part-time. *Financial aid:* In 1990–91, 0 teaching assistantships, 2 graduate assistantships awarded; fellowships, partial tuition waivers, federal work-study also available. Aid available to part-time students. Average monthly stipend for a graduate assistantship: $816. Financial aid application deadline: 8/1. • Dr. Johnny M. Bailey, Head, 915-942-2111.

Appalachian State University, College of Arts and Sciences, Department of Mathematics, Boone, NC 28608. Department awards MA. Part-time programs available. Faculty: 21 full-time (5 women), 0 part-time. Matriculated students: 9 full-time (7 women), 24 part-time (19 women); includes 1 minority (black American), 0 foreign. 15 applicants, 60% accepted. In 1990, 5 degrees awarded. *Degree requirements:* 1 foreign language (computer language can substitute), comprehensive exam. *Entrance requirements:* GRE General Test. Application deadline: 7/31 (priority date). Application fee: $15. *Tuition:* $598 per semester for state residents; $3125 per semester full-time, $6215 per semester part-time for nonresidents. *Financial aid:* Fellowships, research assistantships, teaching assistantships available. Financial aid application deadline: 7/31. • Dr. H. W. Paul, Chairman, 704-262-3050.

Arizona State University, College of Liberal Arts and Sciences, Department of Mathematics, Tempe, AZ 85287. Department offers programs in applied mathematics (MA, PhD), mathematics (MA, MNS, PhD), statistics (MA, PhD). Matriculated students: 30 full-time (6 women), 36 part-time (8 women); includes 3 minority (1 Asian American, 1 black American, 1 Hispanic American), 44 foreign. 170 applicants, 32% accepted. In 1990, 5 master's, 5 doctorates awarded. *Degree requirements:* For master's, thesis or alternative; for doctorate, 1 foreign language, dissertation. *Entrance requirements:* GRE. Application fee: $25. *Tuition:* $1528 per year full-time, $80 per hour part-time for state residents; $6934 per year full-time, $289 per hour part-time for nonresidents. *Faculty research:* Mathematical biology, ordinary and partial differential equations, calculus of variations. • Dr. William Trotter, Chair, 602-965-3951.

See full description on page 389.

Arkansas State University, College of Arts and Sciences, Department of Computer Science, Mathematics, and Physics, State University, AR 72467. Offerings include mathematics (MS, MSE). Department faculty: 32 full-time (7 women), 2 part-time (0 women). *Degree requirements:* 1 foreign language, comprehensive exam required, thesis not required. *Entrance requirements:* Appropriate bachelor's degree. Application deadline: 8/1 (priority date, applications processed on a rolling basis). Application fee: $0 ($25 for foreign students). *Tuition:* $790 per semester full-time, $67 per hour part-time for state residents; $1415 per semester full-time, $120 per hour part-time for nonresidents. • Dr. Jerry Linnstaedter, Chairman, 501-972-3090.

Auburn University, College of Sciences and Mathematics, Division of Mathematics, Department of Foundations, Analysis, and Topology, Auburn University, AL 36849. Department awards MACT, MAM, MPS, MS, PhD. Part-time programs available. Faculty: 27 full-time, 0 part-time. Matriculated students: 22 full-time (6 women), 11 part-time (2 women); includes 1 minority (Asian American), 10 foreign. *Degree requirements:* For master's, thesis (MS) required, foreign language not required; for doctorate, 2 foreign languages, dissertation, written and oral exam. *Entrance requirements:* For master's, GRE General Test; for doctorate, GRE General Test (minimum score of 400 on each section required), GRE Subject Test. Application deadline: 9/1 (applications processed on a rolling basis). Application fee: $15. *Expenses:* Tuition of $1596 per year full-time, $44 per credit hour part-time for state residents; $4788 per year full-time, $132 per credit hour part-time for nonresidents. Fees of $92 per quarter for state residents; $276 per quarter for nonresidents. *Financial aid:* In 1990–91, Fellowships, research assistantships, teaching assistantships, federal work-study available. Aid available to part-time students. Financial aid application deadline: 3/15. • Dr. George A. Kozlowski, Head, 205-844-4290. Application contact: Dr. Norman J. Doorenbos, Dean, Graduate School, 205-844-4700.

Ball State University, College of Sciences and Humanities, Department of Mathematical Sciences, 2000 University Avenue, Muncie, IN 47306. Department offers programs in actuarial science (MA), mathematical statistics (MA), mathematics (MA, MAE, MS). Faculty: 19 full-time, 0 part-time. Matriculated students: 28. In 1990, 13 degrees awarded. *Degree requirements:* Foreign language not required. *Application fee:* $15. *Expenses:* Tuition of $1140 per semester (minimum) full-time for state residents; $2680 per semester (minimum) full-time for nonresidents. Fees of $6 per credit hour. *Financial aid:* In 1990–91, 16 research assistantships (9 to first-year students) awarded. *Faculty research:* Differential equations. • Dr. Donald R. Whitaker, Chairman, 317-285-8640.

Baylor University, College of Arts and Sciences, Department of Mathematics, Waco, TX 76798. Department awards MA, MS. Matriculated students: 9 full-time (4 women), 3 part-time (2 women); includes 1 minority (Asian American), 1 foreign. In

1990, 2 degrees awarded. *Degree requirements:* Computer language, thesis (for some programs), final oral exam required, foreign language not required. *Entrance requirements:* GRE General Test. Application deadline: 5/1 (applications processed on a rolling basis). Application fee: $0. *Expenses:* Tuition of $4440 per year full-time, $185 per credit hour part-time. Fees of $510 per year full-time. *Financial aid:* Teaching assistantships, federal work-study, institutionally sponsored loans, and career-related internships or fieldwork available. Aid available to part-time students. Financial aid application deadline: 5/1; applicants required to submit FAF. *Faculty research:* Algebra, statistics, probability, applied mathematics. • Dr. Danny W. Turner, Director of Graduate Studies, 817-755-3561.

Boston College, Graduate School of Arts and Sciences, Department of Mathematics, Chestnut Hill, MA 02167. Department awards MA, MST. Part-time programs available. *Degree requirements:* Oral presentation required, thesis optional, foreign language not required. *Application deadline:* 3/15. *Tuition:* $412 per credit hour. *Faculty research:* Abstract algebra and number theory, topology, probability and statistics, computer science, analysis.

Boston University, Graduate School, Department of Mathematics, Boston, MA 02215. Department offers programs in mathematics (MA, PhD), statistics (MA, PhD). Faculty: 32 full-time (3 women), 5 part-time (1 woman). Matriculated students: 49 full-time (20 women), 8 part-time (2 women); includes 2 minority (1 Asian American, 1 black American), 12 foreign. Average age 30. 111 applicants, 28% accepted. In 1990, 3 master's, 6 doctorates awarded. Terminal master's awarded for partial completion of doctoral program. *Degree requirements:* For master's, 1 foreign language, comprehensive exam required, thesis not required; for doctorate, 1 foreign language, dissertation. *Entrance requirements:* GRE General Test, GRE Subject Test, TOEFL (minimum score of 550 required). Application deadline: 5/1 (priority date, applications processed on a rolling basis). Application fee: $45. *Expenses:* Tuition of $15,950 per year full-time, $498 per credit part-time. Fees of $120 per year full-time, $35 per semester part-time. *Financial aid:* In 1990–91, 25 students received a total of $237,500 in aid awarded. Fellowships, research assistantships, teaching assistantships, full tuition waivers available. Average monthly stipend for a graduate assistantship: $1055. Financial aid application deadline: 1/15. *Faculty research:* Algebraic and differential geometry, dynamical systems, mathematical biology, number theory, statistics and probability. • Ralph B. D'Agostino, Chairman, 617-353-2560.

Bowling Green State University, College of Arts and Sciences, Department of Mathematics and Statistics, Bowling Green, OH 43403. Offerings include mathematics (MA, MAT, PhD). Department faculty: 27 full-time (1 woman), 0 part-time. *Degree requirements:* For doctorate, 1 foreign language (computer language can substitute), dissertation. *Entrance requirements:* For doctorate, GRE General Test, TOEFL (minimum score of 600 required). Application deadline: 8/7. Application fee: $10. *Tuition:* $185 per credit hour for state residents; $357 per credit hour for nonresidents. • Dr. Hassoon Al-Amiri, Chair, 419-372-2636. Application contact: Dr. James Albert, Graduate Coordinator, 419-372-7456.

Brandeis University, Graduate School of Arts and Sciences, Program in Mathematics, Waltham, MA 02254. Program awards MA, PhD. Faculty: 22 full-time (1 woman), 0 part-time. Matriculated students: 45 full-time (8 women), 0 part-time; includes 0 minority, 31 foreign. Average age 25. 98 applicants, 12% accepted. In 1990, 1 master's awarded (100% continued full-time study); 3 doctorates awarded (100% entered university research/teaching). Terminal master's awarded for partial completion of doctoral program. *Degree requirements:* For master's, 1 foreign language; for doctorate, 2 foreign languages, dissertation. *Entrance requirements:* GRE General Test. Application deadline: 2/15 (priority date, applications processed on a rolling basis). Application fee: $50. *Tuition:* $16,085 per year full-time, $2015 per course part-time. *Financial aid:* In 1990–91, 31 research assistantships (5 to first-year students), 31 teaching assistantships (5 to first-year students) awarded. Average monthly stipend for a graduate assistantship: $989. Financial aid application deadline: 2/15; applicants required to submit GAPSFAS. *Faculty research:* Algebra, analysis, number theory, combinatorics, topology. Total annual research budget: $477,066. • Dr. Mark Adler, Chair, 617-736-3052. Application contact: David Eisenbud, Graduate Adviser, 617-736-3051.

Announcement: Program directed primarily toward PhD in pure mathematics. Benefits from informality, flexibility, and warmth of small department and from intellectual vigor of faculty well known for research accomplishments. Brandeis-Harvard-MIT Colloquium and many joint seminars provide opportunities for contact with other Boston-area mathematicians. Students normally receive full-tuition scholarship and teaching assistantship or fellowship. Contact Professor David Eisenbud, Graduate Advisor, 617-736-3051.

Brigham Young University, College of Physical and Mathematical Sciences, Department of Mathematics, Provo, UT 84602. Department offers programs in mathematics (MA, MS, PhD), mathematics education (MA). Part-time programs available. Faculty: 22 full-time (0 women), 0 part-time. Matriculated students: 17 full-time (3 women), 29 part-time (12 women); includes 2 minority (1 Asian American, 1 Native American), 13 foreign. Average age 25. In 1990, 7 master's, 0 doctorates awarded. Terminal master's awarded for partial completion of doctoral program. *Degree requirements:* For master's, written exams or thesis required, foreign language not required; for doctorate, 2 foreign languages, dissertation, qualifying exams. *Entrance requirements:* For master's, minimum GPA of 3.0 during previous 60 credit hours; for doctorate, GRE General Test, GRE Subject Test, TOEFL (minimum score of 600 required), undergraduate degree in mathematics or related field. Application deadline: 6/15. Application fee: $30. *Tuition:* $1170 per semester (minimum) full-time, $130 per credit hour (minimum) part-time. *Financial aid:* In 1990–91, 32 students received a total of $250,000 in aid awarded. 23 fellowships (8 to first-year students), 1 research assistantship, 9 teaching assistantships (7 to first-year students) were awarded; institutionally sponsored loans also available. Aid available to part-time students. Financial aid application deadline: 3/31. *Faculty research:* Algebraic geometry, linear algebra, numerical analysis, partial differential equations, number theory. • Dr. Donald Robinson, Chairman, 801-378-5868. Application contact: Louis J. Chatterley, Chairman, Graduate Admissions Committee, 801-378-3286.

Brooklyn College of the City University of New York, Graduate Program in Mathematics, Bedford Avenue and Avenue H, Brooklyn, NY 11210. Offers mathematics (MA), secondary mathematics education (MA). Part-time and evening/weekend programs available. Faculty: 15 full-time, 3 part-time. Matriculated students: 0 full-time, 40 part-time (20 women); includes 13 minority (5 Asian American, 5 black American, 3 Hispanic American), 8 foreign. Average age 30. 15 applicants, 87% accepted. In 1990, 3 degrees awarded. *Degree requirements:* Thesis or alternative, comprehensive exam (MA in mathematics) required, foreign language not required. *Entrance requirements:* TOEFL, minimum GPA of 3.0. Application deadline: 4/1. Application fee: $30. *Expenses:* Tuition of $1202 per semester full-time for state residents; $2350 per semester full-time for nonresidents. Fees of $45 per semester full-time. *Financial aid:* Institutionally sponsored loans available. • Dr. G. Shapiro, Chairperson, 718-780-5246. Application contact: K. Marathe, Graduate Deputy Chairman, 718-780-5246.

Brown University, Department of Mathematics, Providence, RI 02912. Department awards Sc M, PhD. Faculty: 21 full-time (2 women), 2 part-time (0 women). Matriculated students: 39 full-time (10 women), 0 part-time; includes 1 minority (Asian American), 27 foreign. In 1990, 4 master's awarded (100% continued full-time study); 3 doctorates awarded (100% entered university research/teaching). Terminal master's awarded for partial completion of doctoral program. *Degree requirements:* For master's, 1 foreign language required, thesis not required; for doctorate, 2 foreign languages, dissertation. *Application deadline:* 1/2. *Application fee:* $50. *Expenses:* Tuition of $16,256 per year full-time, $2032 per course part-time. Fees of $372 per year. *Financial aid:* Application deadline 1/2. *Faculty research:* Algebraic geometry, number theory, functional analysis, geometry, topology. • Walter A. Strauss, Chairman, 401-863-1867. Application contact: Bruno Harris, Graduate Representative, 401-863-1126.

Announcement: Program directed primarily toward PhD in pure mathematics and a career in research or teaching at the college level. The relatively small enrollment gives students the advantages of small classes and close contact with faculty members. Numerous lively seminars, weekly colloquium, and proximity to Boston-area universities provide a valuable experience for graduate students.

See full description on page 393.

Bryn Mawr College, Graduate School of Arts and Sciences, Department of Mathematics, Bryn Mawr, PA 19010. Department awards MA, PhD. Part-time programs available. Faculty: 5. Matriculated students: 3 full-time (all women), 5 part-time (all women); includes 1 foreign. 6 applicants, 17% accepted. In 1990, 2 master's, 0 doctorates awarded. *Degree requirements:* For master's, 1 foreign language, thesis; for doctorate, 2 foreign languages, dissertation. *Entrance requirements:* GRE General Test. Application deadline: 8/20. Application fee: $35. *Tuition:* $13,900 per year full-time, $2350 per course part-time. *Financial aid:* In 1990–91, 4 teaching assistantships, 1 assistantship awarded; fellowships, federal work-study, institutionally sponsored loans also available. Aid available to part-time students. Financial aid application deadline: 2/1; applicants required to submit GAPSFAS. • Dr. Rhonda Hughes, Chairman, 215-526-5351. Application contact: Patricia Saukewitsch, Administrative Coordinator, 215-526-5072.

Bucknell University, College of Arts and Sciences, Department of Mathematics, Lewisburg, PA 17837. Department awards MA, MS. Faculty: 13 full-time. Matriculated students: 3. *Degree requirements:* Foreign language and thesis not required. *Entrance requirements:* GRE General Test, GRE Subject Test, TOEFL (minimum score of 550 required), minimum GPA of 2.8. Application deadline: 6/1 (applications processed on a rolling basis). Application fee: $25. *Tuition:* $1800 per course. *Financial aid:* Assistantships available. Average monthly stipend for a graduate assistantship: $630. • Dr. David S. Ray, Head, 717-524-1343.

California Institute of Technology, Division of Physics, Mathematics and Astronomy, Department of Mathematics, Pasadena, CA 91125. Department awards PhD. Faculty: 16 full-time (0 women), 0 part-time. Matriculated students: 26 full-time (3 women), 0 part-time; includes 0 minority, 21 foreign. Average age 25. 108 applicants, 26% accepted. In 1990, 6 degrees awarded. *Degree requirements:* 1 foreign language, dissertation, candidacy and final exams. *Entrance requirements:* GRE General Test, GRE Subject Test, TOEFL. Application deadline: 1/15. Application fee: $0. *Tuition:* $14,100 per year. *Financial aid:* In 1990–91, 25 students received a total of $352,500 in aid awarded. 6 fellowships (all to first-year students), 0 research assistantships, 23 teaching assistantships (10 to first-year students) were awarded; federal work-study, institutionally sponsored loans also available. Financial aid application deadline: 1/15. *Faculty research:* Number theory, combinatorics, differential geometry, dynamical systems, finite groups. Total annual research budget: $530,000. • D. Wales, Executive Officer, 818-356-4332. Application contact: D. Ramakrishnan, Option Representative, 818-356-4348.

California Polytechnic State University, San Luis Obispo, School of Science and Mathematics, Department of Mathematics, San Luis Obispo, CA 93407. Department awards MS. Faculty: 42 full-time, 12 part-time. Matriculated students: 15 full-time (5 women), 0 part-time. In 1990, 4 degrees awarded. *Entrance requirements:* Minimum GPA of 3.0. Application fee: $55. *Financial aid:* Federal work-study and career-related internships or fieldwork available. Aid available to part-time students. • Dr. Thomas Hale, Chair, 805-546-2208.

California State Polytechnic University, Pomona, College of Science, Department of Mathematics, Pomona, CA 91768. Department offers programs in applied mathematics (MS), pure mathematics (MS). Matriculated students: 0 full-time, 31 part-time (9 women); includes 5 minority (all Asian American). 22 applicants, 64% accepted. In 1990, 3 degrees awarded. *Degree requirements:* Thesis or alternative. *Application fee:* $55. *Expenses:* Tuition of $0 for state residents; $137 per unit for nonresidents. Fees of $880 per year full-time, $182 per quarter (minimum) part-time. *Financial aid:* Application deadline 3/2. • Dr. Scott Sportsman, Coordinator, 714-869-3478.

California State University, Chico, College of Natural Sciences, Department of Geology and Physical Sciences, Program in Physical Science, Chico, CA 95929. Offerings include mathematics (MS). Program faculty: 5 full-time (1 woman), 5 part-time (3 women). *Degree requirements:* Thesis required, foreign language not required. *Entrance requirements:* GRE General Test. Application deadline: 4/1 (applications processed on a rolling basis). Application fee: $55. *Expenses:* Tuition of $548 per semester full-time, $350 per semester part-time. Fees of $246 per unit for nonresidents. • Dr. K. R. Gina Rothe, Graduate Coordinator, 916-898-6269.

California State University, Fresno, Division of Graduate Studies and Research, School of Natural Sciences, Department of Mathematics, 5241 North Maple Avenue, Fresno, CA 93710. Department awards MA, MS. Faculty: 20 full-time, 7 part-time. Matriculated students: 49. *Degree requirements:* Thesis or alternative required, foreign language not required. *Entrance requirements:* GRE General Test (minimum score of 450 on verbal or 430 on quantitative required). *Tuition:* $1098 per year full-time, $4485 per year part-time. • Robert F. Arnold, Chairman, 209-278-2992. Application contact: M. Kursheed Ali, Graduate Coordinator, 209-278-2992.

Announcement: Master's degree program is designed to provide preparation for work in industry, high school and junior college teaching, and advanced graduate study for a PhD in mathematics. Extensive computing facilities available, including microcomputer labs and several mainframes accessible by terminals located throughout the campus.

Directory: Mathematics

California State University, Fullerton, School of Natural Science and Mathematics, Department of Mathematics, Fullerton, CA 92634. Department offers programs in mathematics (MA), mathematics for secondary school teachers (MA). Part-time programs available. Faculty: 29 full-time (4 women), 35 part-time (16 women). Matriculated students: 1 full-time (0 women), 61 part-time (22 women); includes 14 minority (8 Asian American, 2 black American, 3 Hispanic American, 1 Native American), 4 foreign. Average age 31. 55 applicants, 82% accepted. In 1990, 15 degrees awarded. *Degree requirements:* Comprehensive exam or project required, foreign language not required. *Entrance requirements:* Minimum GPA of 2.5 in last 60 units, major in mathematics or related field. Application fee: $55. *Expenses:* Tuition of $0 for state residents; $246 per unit for nonresidents. Fees of $554 per semester full-time, $356 per unit part-time. *Financial aid:* In 1990–91, 0 state grants awarded; research assistantships, teaching assistantships, federal work-study, institutionally sponsored loans, and career-related internships or fieldwork also available. Aid available to part-time students. Financial aid application deadline: 3/1. • Dr. James Friel, Chair, 714-773-3631.

Announcement: The program in applied mathematics was designed in consultation with mathematicians and scientists from the industrial community. It is intended specifically for those who seek or currently hold positions involving mathematical applications. The course work emphasizes modern applied mathematics, modeling, problem solving, and computation. All required courses are offered in the evening and can be completed in 2 calendar years. The culminating experience is a project in which students work in teams on current problems under contract with local industrial firms. Teaching assistantships and research assistantships are available.

California State University, Hayward, School of Science, Department of Mathematics and Computer Science, Hayward, CA 94542. Department offers programs in applied mathematics (MS), computer science (MS), mathematics (MS). Faculty: 32. Matriculated students: 17 full-time (8 women), 66 part-time (29 women). In 1990, 9 degrees awarded. *Degree requirements:* Thesis or comprehensive exam required, foreign language not required. *Entrance requirements:* Minimum GPA of 3.0 in field. Application deadline: 4/19 (priority date, applications processed on a rolling basis). Application fee: $55. *Expenses:* Tuition of $0 for state residents; $137 per unit for nonresidents. Fees of $895 per year full-time, $188 per quarter part-time. *Financial aid:* Federal work-study, institutionally sponsored loans, and career-related internships or fieldwork available. Aid available to part-time students. Financial aid application deadline: 3/1. • Dr. Edward Keller, Chair, 415-881-3414. Application contact: Dr. Robert Trinchero, Acting Associate Vice President, Admissions and Enrollment, 415-881-3828.

California State University, Long Beach, School of Humanities, Department of Mathematics, 1250 Bellflower Boulevard, Long Beach, CA 90840. Department offers programs in applied mathematics (MA), mathematics (MA). Matriculated students: 13 full-time (1 woman), 39 part-time (13 women); includes 16 minority (10 Asian American, 3 black American, 2 Hispanic American, 1 Native American), 2 foreign. Average age 30. 28 applicants, 82% accepted. In 1990, 6 degrees awarded. *Degree requirements:* Thesis or comprehensive exam required, foreign language not required. *Application deadline:* 8/1 (applications processed on a rolling basis). *Application fee:* $55. *Expenses:* Tuition of $0 for state residents; $246 per unit for nonresidents. Fees of $1120 per year full-time, $724 per year part-time. *Financial aid:* Application deadline 3/2; applicants required to submit FAF. • Dr. Samuel G. Councilman, Chair, 310-985-4721.

California State University, Los Angeles, School of Natural and Social Sciences, Department of Mathematics and Computer Science, Los Angeles, CA 90032. Department offers program in mathematics (MS), including applied mathematics, mathematics. Part-time and evening/weekend programs available. Faculty: 26 full-time, 44 part-time. Matriculated students: 11 full-time (2 women), 49 part-time (15 women); includes 18 minority (10 Asian American, 6 black American, 2 Hispanic American), 9 foreign. In 1990, 7 degrees awarded. *Degree requirements:* Thesis or comprehensive exam required, foreign language not required. *Entrance requirements:* TOEFL (minimum score of 550 required), previous course work in mathematics. Application deadline: 8/7 (applications processed on a rolling basis). Application fee: $55. *Expenses:* Tuition of $0 for state residents; $164 per unit for nonresidents. Fees of $1046 per year full-time, $650 per semester (minimum) part-time. *Financial aid:* Teaching assistantships, federal work-study available. Aid available to part-time students. Financial aid application deadline: 3/1; applicants required to submit FAF. *Faculty research:* Group theory, functional analysis, convexity theory. • Dr. Marshall Cates, Chair, 213-343-2150.

Announcement: MS programs offered in both pure and applied mathematics. Electives may be chosen from a rapidly expanding list of computer science courses. Teaching assistantships available. The school operates year-round on the quarter system, and evening classes are offered for working students. Employment opportunities are numerous in local aerospace firms.

California State University, Northridge, School of Science and Mathematics, Department of Mathematics, Northridge, CA 91330. Department awards MS. Part-time and evening/weekend programs available. Matriculated students: 12 full-time (4 women), 54 part-time (23 women); includes 9 minority (4 Asian American, 1 black American, 4 Hispanic American), 3 foreign. Average age 32. 32 applicants, 81% accepted. In 1990, 7 degrees awarded. *Degree requirements:* Thesis required (for some programs), foreign language not required. *Application deadline:* 11/30. *Application fee:* $55. *Expenses:* Tuition of $0 for state residents; $205 per unit for nonresidents. Fees of $1128 per year full-time, $366 per semester part-time. *Financial aid:* Teaching assistantships, federal work-study, institutionally sponsored loans available. Aid available to part-time students. • Dr. Donald Potts, Chairman, 818-885-2721. Application contact: Dr. David Protas, Graduate Coordinator, 818-885-5079.

California State University, Sacramento, School of Arts and Sciences, Department of Mathematics, Sacramento, CA 95819. Department awards MA. Part-time programs available. *Degree requirements:* Thesis or alternative, writing proficiency exam required, foreign language not required. *Entrance requirements:* TOEFL (minimum score of 550 required), minimum GPA of 2.5 during previous 2 years, 3.0 in mathematics, BA in mathematics or equivalent. Application fee: $45. *Expenses:* Tuition of $0 for state residents; $246 per unit for nonresidents. Fees of $530 per semester full-time, $332 per semester part-time.

California University of Pennsylvania, School of Science and Technology, Department of Mathematics and Computer Sciences, 250 University Avenue, California, PA 15419. Department offers programs in computer science (M Ed), mathematics (MA, M Ed). Part-time and evening/weekend programs available. Faculty: 0 full-time, 7 part-time (2 women). Matriculated students: 18 full-time (5 women), 26 part-time (12 women); includes 1 minority (black American), 15 foreign. 34 applicants, 94% accepted. In 1990, 12 degrees awarded. *Degree requirements:* Comprehensive exam required, foreign language not required. *Entrance requirements:* MAT, TOEFL (minimum score of 550 required), teaching certificate (M Ed), minimum GPA of 2.5 or MAT (minimum score of 35 required). Application deadline: rolling. Application fee: $20. *Tuition:* $2550 per year full-time, $127 per credit part-time for state residents; $3156 per year full-time, $160 per credit part-time for nonresidents. *Financial aid:* In 1990–91, 19 research assistantships awarded. • Dr. Leonard Hauser, Head, 412-938-4078.

Carleton University, Faculty of Science, Ottawa-Carleton Institute for Graduate Studies and Research in Mathematics and Statistics, Ottawa, ON K1S 5B6, Canada. Institute awards M Sc, PhD. PhD and M Sc offered jointly with the University of Ottawa. Faculty: 57. Matriculated students: 47 full-time, 6 part-time; includes 14 foreign. In 1990, 12 master's, 4 doctorates awarded. *Degree requirements:* For master's, thesis optional; for doctorate, dissertation. *Entrance requirements:* For master's, TOEFL (minimum score of 550 required), honor's degree; for doctorate, TOEFL (minimum score of 550 required), master's degree. Application deadline: 7/1 (priority date, applications processed on a rolling basis). Application fee: $15. *Tuition:* $985 per semester full-time, $284 per semester part-time for Canadian residents; $3939 per semester full-time, $1171 per semester part-time for nonresidents. *Financial aid:* Application deadline 3/1. *Faculty research:* Pure mathematics, applied mathematics, probability and statistics. • G. Ivanoff, Director, 613-788-2155.

Carnegie Mellon University, Mellon College of Science, Department of Mathematics, Pittsburgh, PA 15213. Department offers programs in algorithms, combinatorics, and optimization (PhD); applied mathematics (MS); mathematics (MS, DA, PhD); mathematics finance (PhD); operations research (MS); pure and applied logic (PhD). PhD in pure and applied logic offered jointly with the School of Computer Science and the Department of Philosophy. Faculty: 32 full-time (2 women), 0 part-time. Matriculated students: 38 full-time (4 women), 3 part-time (0 women); includes 0 minority, 20 foreign. Average age 27. 200 applicants, 12% accepted. In 1990, 5 master's awarded (80% continued full-time study); 3 doctorates awarded (67% entered university research/teaching, 33% found other work related to degree). Terminal master's awarded for partial completion of doctoral program. *Degree requirements:* For master's, foreign language and thesis not required; for doctorate, dissertation required, foreign language not required. *Entrance requirements:* GRE General Test, TOEFL. Application deadline: 2/10. Application fee: $0. *Expenses:* Tuition of $15,250 per year full-time, $212 per unit part-time. Fees of $80 per year. *Financial aid:* In 1990–91, $694,010 in aid awarded. 5 research assistantships (0 to first-year students), 31 teaching assistantships (6 to first-year students), 1 tuition scholarship (0 to first-year students) were awarded. Financial aid application deadline: 7/1. *Faculty research:* Continuum mechanics, numerical analysis, differential equations, discrete mathematics, mathematical biology. Total annual research budget: $80,000. • Dr. William O. Williams, Head, 412-268-2545.

See full description on page 397.

Case Western Reserve University, Department of Mathematics and Statistics, Cleveland, OH 44106. Department offers programs in applied mathematics (MS), mathematics (MS, PhD). Part-time programs available. Faculty: 22 full-time, 3 part-time. Matriculated students: 8 full-time (2 women), 5 part-time (0 women); includes 0 minority, 7 foreign. Average age 33. 33 applicants, 48% accepted. In 1990, 3 master's, 0 doctorates awarded. Terminal master's awarded for partial completion of doctoral program. *Degree requirements:* For master's, thesis not required; for doctorate, 1 foreign language, dissertation. *Entrance requirements:* GRE General Test, GRE Subject Test, TOEFL (minimum score of 550 required). Application deadline: 6/26 (applications processed on a rolling basis). Application fee: $25. *Expenses:* Tuition of $13,600 per year full-time, $567 per credit part-time. Fees of $320 per year. *Financial aid:* In 1990–91, 7 students received a total of $114,030 in aid awarded. 7 teaching assistantships (2 to first-year students) were awarded. Average monthly stipend for a graduate assistantship: $1810. Financial aid application deadline: 4/15; applicants required to submit FAF. *Faculty research:* Dynamic systems, differential geometry, functional analysis, differential equations, probability theory. Total annual research budget: $280,000. • C. A. Cullis, Chairman, 216-368-2880. Application contact: Program Coordinator, 216-368-2918.

Catholic University of America, School of Arts and Sciences, Department of Mathematics, Washington, DC 20064. Department awards MA, MS, MTS, PhD. Part-time programs available. Faculty: 11 full-time (1 woman), 0 part-time. Matriculated students: 2 full-time (1 woman), 2 part-time (0 women); includes 1 minority (Asian American), 1 foreign. Average age 34. 11 applicants, 64% accepted. In 1990, 2 master's, 1 doctorate awarded. *Degree requirements:* For master's, thesis or alternative, comprehensive exam required, foreign language not required; for doctorate, 2 foreign languages (computer language can substitute for one), dissertation, comprehensive exam. *Entrance requirements:* GRE General Test, TOEFL. Application deadline: 8/1 (applications processed on a rolling basis). Application fee: $30. *Expenses:* Tuition of $11,626 per year full-time, $445 per credit hour part-time. Fees of $240 per year full-time, $90 per year part-time. *Financial aid:* In 1990–91, $17,000 in aid awarded. Partial tuition waivers, federal work-study, institutionally sponsored loans available. Financial aid application deadline: 2/15; applicants required to submit GAPSFAS or FAF. *Faculty research:* Functional analysis, optimal control, computer simulation, software systems, random processes. Total annual research budget: $130,000. • Dr. Gustav B. Hensel, Chair, 202-319-5221.

Central Connecticut State University, School of Arts and Sciences, Department of Mathematics and Computer Science, New Britain, CT 06050. Offerings include actuarial mathematics (MA), mathematics (MA, MS). Department faculty: 5 full-time (2 women), 1 (woman) part-time. *Degree requirements:* Thesis or alternative, comprehensive exam or special project. *Entrance requirements:* TOEFL (minimum score of 550 required for international students), minimum GPA of 2.7. Application deadline: 8/31. Application fee: $20. *Expenses:* Tuition of $1720 per year full-time, $115 per credit part-time for state residents; $4790 per year full-time, $115 per credit part-time for nonresidents. Fees of $1013 per year full-time, $33 per semester part-time for state residents; $1655 per year full-time, $33 per semester part-time for nonresidents. • Dr. William Driscoll, Chair, 203-827-7374.

Central Michigan University, College of Arts and Sciences, Department of Mathematics, Mount Pleasant, MI 48859. Department awards MA, MAT. Faculty: 33 full-time (6 women). In 1990, 7 degrees awarded. *Degree requirements:* Thesis or alternative required, foreign language not required. *Application deadline:* 7/15 (priority date, applications processed on a rolling basis). *Application fee:* $30. *Tuition:* $96.50 per credit for state residents; $210.50 per credit for nonresidents. *Financial aid:* In 1990–91, 14 teaching assistantships awarded; federal work-study also available. Financial aid applicants required to submit FAF. *Faculty research:* Combinatorics, approximation theory, operations theory, functional analysis, stochastic models. • Dr. Robert G. Clason, Chairperson, 517-774-3596.

Central Missouri State University, College of Arts and Sciences, Department of Mathematics and Computer Sciences, Warrensburg, MO 64093. Offerings include mathematics and computer sciences (MS). Department faculty: 23 (7 women).

Application deadline: 6/30 (priority date, applications processed on a rolling basis). *Application fee:* $0. *Tuition:* $1968 per year full-time, $82 per hour part-time for state residents; $3744 per year full-time, $156 per hour part-time for nonresidents. • Dr. Edward W. Davenport, Chair, 816-429-4931.

Central Washington University, College of Letters, Arts, and Sciences, Department of Mathematics, Ellensburg, WA 98926. Department awards MAT, MS. Faculty: 13 full-time (0 women), 0 part-time. Matriculated students: 9 full-time (4 women), 3 part-time (2 women); includes 0 minority, 0 foreign. 12 applicants, 92% accepted. In 1990, 5 degrees awarded. *Degree requirements:* Thesis or alternative required, foreign language not required. *Entrance requirements:* GRE General Test, GRE Subject Test, minimum GPA of 3.0. Application deadline: 4/1 (priority date, applications processed on a rolling basis). Application fee: $25. *Expenses:* Tuition of $90 per credit for state residents; $273 per credit for nonresidents. Fees of $25 per year. *Financial aid:* In 1990–91, 4 research assistantships (all to first-year students), 0 teaching assistantships awarded; federal work-study also available. Average monthly stipend for a graduate assistantship: $865. Financial aid application deadline: 2/15; applicants required to submit FAF. • Dr. Steven Hinthorne, Chairman, 509-963-2103. Application contact: Britta Jo Hammond, Administrative Assistant, Graduate Studies and Research, 509-963-3103.

Chicago State University, College of Arts and Sciences, Department of Mathematics, Chicago, IL 60628. Department awards MS. Faculty: 5 full-time (1 woman). Matriculated students: 21 (8 women). *Degree requirements:* Oral exam required, thesis optional, foreign language not required. *Entrance requirements:* Minimum GPA of 2.75. *Expenses:* Tuition of $1680 per year full-time, $70 per semester part-time for state residents; $5040 per year full-time, $210 per semester part-time for nonresidents. Fees of $92 per year. *Financial aid:* Research assistantships available. • Dr. D. Bunt, Chairperson, 312-995-2102.

City College of the City University of New York, Graduate School, College of Liberal Arts and Science, Division of Science, Department of Mathematics, Convent Avenue at 138th Street, New York, NY 10031. Department awards MA. Part-time programs available. Matriculated students: 5 full-time (0 women), 10 part-time (3 women). 25 applicants, 80% accepted. In 1990, 7 degrees awarded. *Degree requirements:* 1 foreign language required (computer language can substitute), thesis not required. *Entrance requirements:* TOEFL (minimum score of 300 required). Application deadline: 6/1 (priority date, applications processed on a rolling basis). Application fee: $30. *Expenses:* Tuition of $2204 per year full-time, $95 per credit part-time for state residents; $4700 per year full-time, $199 per credit part-time for nonresidents. Fees of $15 per semester. *Financial aid:* Teaching assistantships, federal work-study available. Aid available to part-time students. Financial aid application deadline: 5/1. *Faculty research:* Group theory, number theory, logic, statistics, computational geometry. • Jacob Barshay, Chairman, 212-650-5173. Application contact: K. Hrbacek, Adviser, 212-650-5101.

Claremont Graduate School, Department of Mathematics, Claremont, CA 91711. Department offers programs in applied mathematics (MA, MS), engineering mathematics (PhD), mathematics (MA, MS, PhD), operations research and statistics (MA, MS), scientific computing (MA, MS), systems and control theory (MA, MS). PhD (engineering mathematics) offered jointly with California State University. Part-time programs available. Faculty: 5 full-time (0 women), 32 part-time (4 women). Matriculated students: 7 full-time (3 women), 34 part-time (7 women); includes 8 minority (6 Asian American, 1 black American, 1 Hispanic American), 9 foreign. Average age 28. 26 applicants, 77% accepted. In 1990, 7 master's, 2 doctorates awarded. Terminal master's awarded for partial completion of doctoral program. *Degree requirements:* For master's, foreign language and thesis not required; for doctorate, 2 foreign languages (computer language can substitute for one), dissertation. *Entrance requirements:* GRE General Test. Application deadline: 2/15 (priority date, applications processed on a rolling basis). Application fee: $40. *Expenses:* Tuition of $13,900 per year full-time, $620 per unit part-time. Fees of $55 per semester. *Financial aid:* In 1990–91, 13 students received a total of $109,530 in aid awarded. 0 research assistantships were awarded; fellowships, full and partial tuition waivers, federal work-study, institutionally sponsored loans, and career-related internships or fieldwork also available. Aid available to part-time students. Financial aid application deadline: 2/15. *Faculty research:* Algebra, analysis, topology, applied mathematics, statistics. Total annual research budget: $400,000. • William Lucas, Chairman, 714-621-8080.

Announcement: Emphasis on applied mathematics, including modeling, numerical analysis, partial differential equations, operations research, statistics, game theory. Operates leading Mathematics Clinic with industry-sponsored research projects conducted by student teams. Concentrations include physical applied mathematics, scientific computing, systems and control theory, operations research and statistics, and pure mathematics.

Clarion University of Pennsylvania, College of Arts and Sciences, Program in Mathematics, Clarion, PA 16214. Program awards M Ed. Faculty: 0 full-time, 15 part-time (0 women). Matriculated students: 0. 0 applicants. In 1990, 0 degrees awarded. *Degree requirements:* Foreign language and thesis not required. *Entrance requirements:* Minimum QPA of 2.75. Application deadline: 8/1 (priority date, applications processed on a rolling basis). Application fee: $15. *Tuition:* $2178 per year full-time, $12 per credit part-time for state residents; $2598 per year full-time, $144 per credit part-time for nonresidents. *Financial aid:* Research assistantships available. Aid available to part-time students. Financial aid application deadline: 5/1. • Dr. Benjamin Freed, Chairman, 814-226-2592.

Clarkson University, School of Science, Department of Mathematics and Computer Science, Potsdam, NY 13699. Department offers programs in computer science (MS), mathematics (MS, PhD). Faculty: 16 full-time (3 women), 4 part-time (0 women). Matriculated students: 12 full-time (4 women), 1 (woman) part-time; includes 0 minority, 9 foreign. Average age 30. 168 applicants, 39% accepted. In 1990, 4 master's awarded; 3 doctorates awarded (100% entered university research/teaching). Terminal master's awarded for partial completion of doctoral program. *Degree requirements:* For master's, foreign language and thesis not required; for doctorate, dissertation, departmental qualifying exam required, foreign language not required. *Application fee:* $10. *Expenses:* Tuition of $446 per credit hour. Fees of $75 per semester. *Financial aid:* In 1990–91, 0 research assistantships, 10 teaching assistantships awarded. *Total annual research budget:* $413,259. • Dr. Athanassios Fokas, Chairman, 315-268-2395.

Clemson University, College of Sciences, Department of Mathematical Sciences, Clemson, SC 29634. Department offers programs in applied and core mathematics (MS, PhD), computational mathematics (MS, PhD), management science (PhD), operations research (MS, PhD), statistics (MS, PhD). Part-time programs available. Faculty: 37 full-time (4 women), 0 part-time. Matriculated students: 96 full-time (38 women), 0 part-time; includes 1 minority (Asian American), 19 foreign. Average age 29. 75 applicants, 87% accepted. In 1990, 20 master's awarded (10% entered university research/teaching, 60% found other work related to degree, 30% continued full-time study); 2 doctorates awarded (100% entered university research/teaching). *Degree requirements:* For master's, computer language, final project required, thesis optional, foreign language not required; for doctorate, computer language, dissertation, qualifying exams required, foreign language not required. *Entrance requirements:* GRE General Test. Application deadline: 6/1 (priority date). Application fee: $25. *Expenses:* Tuition of $102 per credit hour. Fees of $80 per semester full-time. *Financial aid:* In 1990–91, $600,000 in aid awarded. 4 fellowships (all to first-year students), 6 research assistantships (0 to first-year students), 69 teaching assistantships (17 to first-year students) were awarded. Average monthly stipend for a graduate assistantship: $785. Financial aid application deadline: 4/15. *Faculty research:* Applied and computational analysis, discrete mathematics and mathematical programming, statistics and probability. Total annual research budget: $482,000. • J. P. Jarvis, Head, 803-656-3434.

Announcement: Program in pure and classical applied mathematics leading to PhD and master's degrees offered by comprehensive Department of Mathematical Sciences. Faculty expertise includes discrete mathematics and analysis, in both applied and theoretical. Diverse research expertise and integrated graduate programs provide preparation for wide variety of student career goals.

Cleveland State University, College of Arts and Sciences, Department of Mathematics, Cleveland, OH 44115. Offerings include mathematics (MA, MS). Department faculty: 17 full-time (1 woman), 0 part-time. *Degree requirements:* Foreign language and thesis not required. *Application deadline:* 9/1 (priority date, applications processed on a rolling basis). *Application fee:* $0. *Tuition:* $90 per credit for state residents; $180 per credit for nonresidents. • Dr. John Chao, Chairman, 216-687-4680. Application contact: Ann Melville, Administrative Secretary, 216-523-7162.

College of William and Mary, Faculty of Arts and Sciences, Department of Mathematics, Williamsburg, VA 23185. Department offers program in mathematics and operations research (MA, MS). Faculty: 16 full-time, 8 part-time. Matriculated students: 14 full-time (8 women), 5 part-time (2 women); includes 1 minority (Asian American), 3 foreign. Average age 27. In 1990, 5 degrees awarded. *Degree requirements:* Comprehensive exam required, foreign language not required. *Entrance requirements:* Minimum GPA of 2.5. Application fee: $5. *Expenses:* Tuition of $2240 per year full-time, $120 per credit hour part-time for state residents; $8960 per year full-time, $320 per credit hour part-time for nonresidents. Fees of $1490 per year full-time. *Financial aid:* Fellowships available. • Dr. David P. Stanford, Chairman, 804-221-2000. Application contact: Dr. Sidney H. Lawrence, Graduate Director, 804-253-4481.

Announcement: The Department of Mathematics offers a program in operations research that emphasizes broad training in the field. Following an initial survey of the field and preparation in probability and statistics, students select from a variety of in-depth courses including linear, nonlinear, and integer programming; stochastic processes; decision theory; inventory theory; and simulation.

Colorado State University, College of Natural Sciences, Department of Mathematics, Fort Collins, CO 80523. Department awards MS, PhD. Faculty: 24 full-time (2 women), 0 part-time. Matriculated students: 48 full-time (14 women), 4 part-time (0 women); includes 0 minority, 12 foreign. Average age 24. 216 applicants, 18% accepted. In 1990, 14 master's, 1 doctorate awarded. Terminal master's awarded for partial completion of doctoral program. *Degree requirements:* For master's, foreign language and thesis not required; for doctorate, 1 foreign language, dissertation. *Entrance requirements:* GRE General Test, TOEFL, minimum GPA of 3.0. Application deadline: 4/1 (priority date, applications processed on a rolling basis). Application fee: $30. *Tuition:* $1322 per semester full-time for state residents; $3673 per semester full-time for nonresidents. *Financial aid:* In 1990–91, $400,000 in aid awarded. 2 fellowships (both to first-year students), 1 research assistantship (0 to first-year students), 41 teaching assistantships (9 to first-year students), 0 traineeships were awarded; federal work-study, institutionally sponsored loans, and career-related internships or fieldwork also available. Average monthly stipend for a graduate assistantship: $698-750. Financial aid application deadline: 4/15. *Faculty research:* Applied mathematics, numerical analysis, algebraic geometry, partial differential equations, combinatorics. Total annual research budget: $325,000. • Robert E. Gaines, Head, 303-491-7925. Application contact: Janice Lucero, Secretary to the Graduate Director, 303-491-7925.

See full description on page 401.

Columbia University, Graduate School of Arts and Sciences, Division of Natural Sciences, Department of Mathematics, New York, NY 10027. Department awards MA, M Phil, PhD. Faculty: 30 full-time. Matriculated students: 48 full-time (8 women), 0 part-time; includes 2 minority (both Asian American), 21 foreign. Average age 26. 51 applicants, 51% accepted. In 1990, 13 master's, 8 doctorates awarded. *Degree requirements:* For master's, written exam required, foreign language and thesis not required; for doctorate, 2 foreign languages, dissertation. *Entrance requirements:* GRE General Test, TOEFL, major in mathematics. Application deadline: 1/5. Application fee: $50. *Expenses:* Tuition of $7836 per semester for state residents; $426 per credit for nonresidents. Fees of $148 per semester for state residents. *Financial aid:* In 1990–91, $757,589 in aid awarded. 1 fellowship (0 to first-year students), 18 teaching assistantships (1 to a first-year student) were awarded; federal work-study, institutionally sponsored loans also available. Financial aid application deadline: 1/5; applicants required to submit GAPSFAS. *Faculty research:* Algebra, topology, analysis. • Patrick X. Gallagher, Chair, 212-854-4113.

See full description on page 403.

Concordia University, Faculty of Arts and Science, Department of Mathematics and Statistics, Montreal, PQ H3G 1M8, Canada. Offerings include mathematics (MA, M Sc). *Application fee:* $15. *Expenses:* Tuition of $10 per credit for Canadian residents; $195 per credit for nonresidents. Fees of $223 per year full-time, $38.85 per year part-time for Canadian residents; $118 per year full-time, $35.35 per year part-time for nonresidents. • Dr. W. Byers, Chair, 514-848-3232.

Cornell University, Graduate Fields of Arts and Sciences, Field of Mathematics, Ithaca, NY 14853. Field awards MA, MS, PhD. Faculty: 52 full-time, 0 part-time. Matriculated students: 61 full-time (9 women), 0 part-time; includes 4 minority (3 Asian American, 1 Hispanic American), 31 foreign. 354 applicants, 13% accepted. In 1990, 11 master's, 11 doctorates awarded. Terminal master's awarded for partial completion of doctoral program. *Degree requirements:* For master's, thesis, teaching experience required, foreign language not required; for doctorate, 1 foreign language, dissertation, teaching experience. *Entrance requirements:* GRE General Test, GRE Subject Test, TOEFL. Application deadline: 1/10. Application fee: $55. *Expenses:* Tuition of $16,170 per year. Fees of $28 per year. *Financial aid:* In 1990–91, 56 students received aid. 1 fellowship (to a first-year student), 5 research assistantships (0 to first-year students), 50 teaching assistantships (15 to first-year students)

Directory: Mathematics

awarded; full and partial tuition waivers, federal work-study, institutionally sponsored loans also available. Financial aid application deadline: 1/10; applicants required to submit GAPSFAS. • Richard Durrett, Graduate Faculty Representative, 607-255-6757. Application contact: Robert Brashear, Director of Admissions, 607-255-4884.

See full description on page 405.

Creighton University, College of Arts and Sciences, Department of Mathematics, Statistics, and Computer Science, Program in Mathematics and Statistics, Omaha, NE 68178. Program awards MS. Part-time programs available. Faculty: 8 full-time, 0 part-time. *Degree requirements:* 1 foreign language required (computer language can substitute), thesis optional. *Entrance requirements:* GRE General Test, TOEFL. Application deadline: 3/15. Application fee: $20. *Expenses:* Tuition of $272 per credit hour. Fees of $140 per semester full-time, $14 per semester part-time. • Dr. John Mordeson, Chairman, Department of Mathematics, Statistics, and Computer Science, 402-280-2478. Application contact: Dr. Michael Lawler, Dean, Graduate School, 402-280-2870.

Dalhousie University, College of Arts and Science, Faculty of Science, Department of Mathematics, Statistics, and Computing Science, Halifax, NS B3H 4H6, Canada. Offerings include mathematics (MA, M Sc, PhD). Department faculty: 36 full-time, 1 part-time. *Degree requirements:* For master's, thesis required, foreign language not required; for doctorate, 1 foreign language, dissertation. *Entrance requirements:* TOEFL (minimum score of 550 required). Application deadline: 7/15 (applications processed on a rolling basis). Application fee: $20. *Tuition:* $2594 per year full-time for Canadian residents; $4294 per year full-time for nonresidents. • Dr. R. P. Gupta, Chair, 902-494-2572.

Dartmouth College, School of Arts and Sciences, Department of Mathematics and Computer Science, Hanover, NH 03755. Department offers programs in computer science (PhD), mathematics (AM, PhD). Faculty: 31 full-time (3 women), 0 part-time. Matriculated students: 21 full-time (6 women), 0 part-time; includes 0 minority, 4 foreign. 97 applicants, 11% accepted. In 1990, 7 master's awarded (100% continued full-time study); 1 doctorate awarded. Terminal master's awarded for partial completion of doctoral program. *Degree requirements:* For master's, foreign language and thesis not required; for doctorate, 2 foreign languages, dissertation. *Entrance requirements:* GRE General Test, GRE Subject Test. Application deadline: 3/1. *Expenses:* Tuition of $16,230 per year. Fees of $650 per year. *Financial aid:* In 1990–91, $636,582 in aid awarded. 16 fellowships (4 to first-year students), 5 research assistantships were awarded; teaching assistantships, full and partial tuition waivers, institutionally sponsored loans also available. *Total annual research budget:* $209,217. • K. Bogart, Chairman, 603-646-2421.

See full description on page 407.

Drexel University, College of Arts and Sciences, Department of Mathematics and Computer Science, 32nd and Chestnut Streets, Philadelphia, PA 19104. Department offers programs in computer science (MS), mathematics (MS, PhD). Part-time programs available. Faculty: 26 full-time (2 women), 0 part-time. Matriculated students: 30 full-time, 85 part-time; includes 30 foreign. 183 applicants, 47% accepted. In 1990, 30 master's awarded (80% found work related to degree, 20% continued full-time study); 2 doctorates awarded (50% entered university research/teaching, 50% found other work related to degree). *Degree requirements:* For master's, thesis not required; for doctorate, 1 foreign language, dissertation. *Entrance requirements:* For master's, TOEFL (minimum score of 550 required), TSE (for teaching assistants), GRE. Application deadline: 8/23 (applications processed on a rolling basis). Application fee: $25. *Expenses:* Tuition of $345 per credit hour. Fees of $81 per quarter full-time, $43 per quarter part-time. *Financial aid:* In 1990–91, $400,000 in aid awarded. 37 teaching assistantships (14 to first-year students) were awarded; institutionally sponsored loans also available. Financial aid application deadline: 2/1. *Faculty research:* Probability and statistics, combinatorics special functions, functional analysis, parallel processing, computer a;gebra. Total annual research budget: $80,000. • Dr. James Pool, Head, 215-895-2668. Application contact: Keith Brooks, Director of Graduate Admissions, 215-895-6700.

See full description on page 409.

Duke University, Graduate School, Department of Mathematics, Durham, NC 27706. Department awards PhD. Faculty: 30 full-time, 1 part-time. Matriculated students: 38 full-time (5 women), 2 part-time (1 woman); includes 5 minority (2 black American, 2 Hispanic American, 1 Native American), 12 foreign. 40 applicants, 50% accepted. In 1990, 1 master's, 7 doctorates awarded. Terminal master's awarded for partial completion of doctoral program. *Degree requirements:* For doctorate, 2 foreign languages, dissertation. *Entrance requirements:* For doctorate, GRE General Test, GRE Subject Test. Application deadline: 1/31. Application fee: $50. *Expenses:* Tuition of $8640 per year full-time, $360 per unit part-time. Fees of $1356 per year full-time. *Financial aid:* In 1990–91, $243,925 in aid awarded. Federal work-study available. Financial aid application deadline: 1/31; applicants required to submit GAPSFAS. • Greg Lawler, Director of Graduate Studies, 919-684-2321.

East Carolina University, College of Arts and Sciences, Department of Mathematics, Greenville, NC 27858-4353. Department offers programs in applied mathematics (MA), mathematics (MA, MA Ed). Part-time and evening/weekend programs available. Faculty: 19. Matriculated students: 3 full-time (all women), 15 part-time (11 women); includes 0 minority, 2 foreign. 12 applicants, 75% accepted. In 1990, 10 degrees awarded. *Degree requirements:* 1 foreign language (computer language can substitute), thesis, comprehensive exams. *Entrance requirements:* GRE General Test or MAT, TOEFL. Application deadline: 6/1. Application fee: $25. *Tuition:* $627 per semester for state residents; $3154 per semester for nonresidents. *Financial aid:* In 1990–91, $50,000 in aid awarded. Research assistantships, teaching assistantships available. Financial aid application deadline: 6/1. • Dr. Robert Hursey, Director of Graduate Studies, 919-757-6461. Application contact: Paul D. Tschetter, Assistant Dean, 919-757-6012.

Eastern Illinois University, College of Liberal Arts and Sciences, Department of Mathematics, 600 Lincoln Avenue, Charleston, IL 61920-3099. Offerings include mathematics (MA). Department faculty: 30 full-time (6 women), 0 part-time. *Degree requirements:* Foreign language and thesis not required. *Entrance requirements:* GRE General Test. Application deadline: rolling. Application fee: $0. *Tuition:* $1220 per semester full-time for state residents; $2984 per semester full-time for nonresidents. • Dr. Ira Rosenholtz, Chair, 217-581-2028. Application contact: Dr. Max Gerling, Coordinator, 217-581-6281.

Eastern Kentucky University, College of Natural and Mathematical Sciences, Department of Mathematics, Statistics and Computer Science, Richmond, KY 40475. Department offers program in mathematical sciences (MS). Part-time programs available. Matriculated students: 9 full-time (4 women), 15 part-time (9 women); includes 3 minority (2 Asian American, 1 black American), 2 foreign. Average age 25. In 1990, 2 degrees awarded. *Degree requirements:* Thesis not required. *Entrance requirements:* GRE General Test, minimum GPA of 2.5. Application fee: $0. *Tuition:* $1440 per year full-time, $88 per credit hour part-time for state residents; $4320 per year full-time, $248 per credit hour part-time for nonresidents. *Financial aid:* Research assistantships, teaching assistantships, federal work-study available. Aid available to part-time students. *Faculty research:* Computer science, statistics. • Charles Franke, Chair, 606-622-5942.

Eastern Michigan University, College of Arts and Sciences, Department of Mathematics, Ypsilanti, MI 48197. Department awards MA. Evening/weekend programs available. Faculty: 27 full-time (5 women). Matriculated students: 3 full-time (0 women), 68 part-time (28 women); includes 18 minority (14 Asian American, 2 black American, 2 Native American), 34 foreign. In 1990, 13 degrees awarded. *Degree requirements:* Thesis optional, foreign language not required. *Application deadline:* 6/15 (applications processed on a rolling basis). *Application fee:* $25. *Expenses:* Tuition of $89.50 per credit hour for state residents; $212 per credit hour for nonresidents. Fees of $90.25 per semester. • Dr. Donald R. Lick, Head, 313-487-1444. Application contact: Kenneth Shiskowska, Coordinator, 313-487-1294.

Eastern New Mexico University, College of Liberal Arts and Sciences, Department of Mathematical Sciences, Portales, NM 88130. Department awards MA. Part-time programs available. Faculty: 7 full-time (1 woman), 2 part-time (1 woman). Matriculated students: 1 (woman) full-time, 11 part-time (2 women); includes 0 minority, 3 foreign. 16 applicants, 56% accepted. In 1990, 3 degrees awarded. *Degree requirements:* Thesis optional, foreign language not required. *Entrance requirements:* Minimum GPA of 2.5. Application deadline: rolling. Application fee: $10. *Tuition:* $711 per semester full-time, $59.25 per credit part-time for state residents; $2325 per semester full-time, $193.75 per credit part-time for nonresidents. *Financial aid:* In 1990–91, $36,300 in aid awarded. 2 fellowships (both to first-year students), 7 teaching assistantships (1 to a first-year student) were awarded; federal work-study and career-related internships or fieldwork also available. Aid available to part-time students. Financial aid application deadline: 4/1. *Faculty research:* Applied mathematics, graph theory. • Dr. George Baldwin, Graduate Coordinator, 505-562-2328.

Eastern Washington University, College of Science, Mathematics and Technology, Department of Mathematics, Cheney, WA 99004. Department awards M Ed, MS. Part-time programs available. Faculty: 13 full-time (1 woman), 0 part-time. 41 applicants, 56% accepted. In 1990, 6 degrees awarded. *Degree requirements:* Thesis or alternative, comprehensive oral exam. *Entrance requirements:* GRE General Test, departmental qualifying exam, minimum GPA of 3.0. Application deadline: rolling. Application fee: $25. *Tuition:* $2700 per year full-time, $180 per credit (minimum) part-time for state residents; $8187 per year full-time, $546 per credit (minimum) part-time for nonresidents. *Financial aid:* In 1990–91, 13 teaching assistantships awarded; federal work-study, institutionally sponsored loans also available. Financial aid applicants required to submit FAF. • Dr. Ronald Dalla, Chairman, 509-359-6225. Application contact: Dr. Yves Nievergelt, Adviser, 509-359-2219.

East Tennessee State University, College of Arts and Sciences, Department of Mathematics, Johnson City, TN 37614. Department awards MS. Part-time and evening/weekend programs available. Faculty: 12 full-time (2 women), 0 part-time. Matriculated students: 3 full-time (2 women), 8 part-time (7 women); includes 1 minority (Hispanic American), 1 foreign. Average age 25. 17 applicants, 76% accepted. In 1990, 1 degree awarded. *Degree requirements:* Thesis or alternative, oral and written comprehensive exams required, foreign language not required. *Entrance requirements:* GRE General Test. Application deadline: 7/15. Application fee: $5. *Expenses:* Tuition of $1619 per year full-time, $82 per semester hour part-time for state residents; $4492 per year full-time, $207 per semester hour part-time for nonresidents. Fees of $14 per year full-time, $2 per year part-time. *Financial aid:* In 1990–91, $10,000 in aid awarded. 2 teaching assistantships (both to first-year students) were awarded; research assistantships also available. Average monthly stipend for a graduate assistantship: $555. Financial aid application deadline: 3/1. *Faculty research:* Approximation of coal washability curve, nonassociative rings in terms of derivation. Total annual research budget: $150,000. • Dr. George Poole, Chairman, 615-929-4349. Application contact: Dr. Tae-Il Suh, Graduate Adviser, 615-929-6980.

East Texas State University, College of Arts and Sciences, Department of Mathematics, Commerce, TX 75429. Department awards MA, MS. Faculty: 7 full-time (0 women), 1 part-time (0 women). Matriculated students: 6 full-time (3 women), 18 part-time (12 women); includes 1 minority (Hispanic American), 1 foreign. In 1990, 4 degrees awarded. *Degree requirements:* Thesis (for some programs), comprehensive exam. *Entrance requirements:* GRE General Test. Application deadline: rolling. Application fee: $0 ($25 for foreign students). *Tuition:* $430 per semester full-time for state residents; $1726 per semester full-time for nonresidents. *Financial aid:* Research assistantships, teaching assistantships, federal work-study, institutionally sponsored loans available. Financial aid applicants required to submit GAPSFAS or FAF. • Dr. Stuart Anderson, Head, 903-886-5944.

Emory University, Graduate School of Arts and Sciences, Department of Mathematics and Computer Science, Atlanta, GA 30322. Department offers programs in mathematics (MA, MS, PhD), mathematics/computer science (MS), statistics/computer science (MS). Faculty: 23 full-time (2 women), 0 part-time. Matriculated students: 22 full-time (10 women), 7 part-time (5 women); includes 0 minority, 8 foreign. 49 applicants, 31% accepted. In 1990, 6 master's, 1 doctorate awarded. Terminal master's awarded for partial completion of doctoral program. *Degree requirements:* For master's, thesis required, foreign language not required; for doctorate, 1 foreign language, dissertation, comprehensive exams. *Entrance requirements:* GRE General Test, TOEFL, minimum GPA of 3.0. Application deadline: 1/20 (priority date, applications processed on a rolling basis). Application fee: $35. *Expenses:* Tuition of $7370 per semester full-time, $642 per semester hour part-time. Fees of $160 per year full-time. *Financial aid:* In 1990–91, 22 students received a total of $398,260 in aid awarded. 19 fellowships (8 to first-year students), 19 teaching assistantships (8 to first-year students), 25 tuition scholarships (10 to first-year students) were awarded; full and partial tuition waivers also available. Average monthly stipend for a graduate assistantship: $900. Financial aid application deadline: 2/1. *Faculty research:* Combinatorics, differential equations, universal algebra, mathematical physics. • Dr. Ron Gould, Chairman, 404-727-7924. Application contact: Dr. Dwight Duffus, Director of Graduate Studies, 404-727-7579.

Announcement: The department offers the PhD in mathematics and master's degrees in mathematics, mathematics/computer science, statistics/computer science. Mathematics research includes algebra, analysis, applied mathematics, combinatorics, differential equations, dynamical systems. Computer science research includes image analysis, networking, operating systems. Assistantships, fellowships, and tuition waivers are available.

See full description on page 413.

Emporia State University, School of Graduate Studies, College of Liberal Arts and Sciences, Division of Mathematics and Computer Science, Emporia, KS 66801-5087. Division offers program in mathematics (MS). Faculty: 9 full-time (2 women), 0 part-time. Matriculated students: 9 full-time (2 women), 7 part-time (3 women); includes 0 minority. 4 applicants, 50% accepted. In 1990, 3 degrees awarded. *Degree requirements:* Thesis or comprehensive exam required, foreign language not required. *Entrance requirements:* GRE General Test, TOEFL (minimum score of 550 required). Application deadline: 8/16 (priority date, applications processed on a rolling basis). Application fee: $0 ($50 for foreign students). *Tuition:* $858 per semester full-time, $62 per credit hour part-time for state residents; $2072 per semester full-time, $143 per credit hour part-time for nonresidents. *Financial aid:* In 1990–91, 1 fellowship, 0 research assistantships, 6 teaching assistantships awarded; federal work-study, institutionally sponsored loans, and career-related internships or fieldwork also available. Average monthly stipend for a graduate assistantship: $472. Financial aid application deadline: 3/15; applicants required to submit FAF. • Dr. Larry Scott, Chair, 316-343-5281.

Fairleigh Dickinson University, Florham-Madison Campus, Maxwell Becton College of Arts and Sciences, Department of Mathematics, Computer Science and Physics, 285 Madison Avenue, Madison, NJ 07940. Department offers program in mathematics (MS). Faculty: 9 full-time (1 woman), 7 part-time (1 woman). Matriculated students: 1 (woman) full-time, 1 (woman) part-time; includes 0 minority, 0 foreign. Average age 27. 0 applicants. In 1990, 2 degrees awarded. *Degree requirements:* Computer language required, foreign language and thesis not required. *Entrance requirements:* GRE General Test. Application deadline: rolling. Application fee: $35. *Expenses:* Tuition of $348 per credit. Fees of $205 per year full-time, $90 per year part-time. *Financial aid:* In 1990–91, 0 fellowships, 0 research assistantships, 1 teaching assistantship awarded. • Dr. Peter Falley, Chairperson, 201-593-8680.

Fairleigh Dickinson University, Teaneck-Hackensack Campus, College of Science and Engineering, Department of Mathematics and Computer Science, Program in Mathematics, 1000 River Road, Teaneck, NJ 07666. Program awards MS. Admissions temporarily suspended. Matriculated students: 1 full-time (0 women), 5 part-time (4 women); includes 0 minority, 2 foreign. Average age 29. In 1990, 1 degree awarded. *Degree requirements:* Computer language required, thesis optional, foreign language not required. *Expenses:* Tuition of $348 per credit. Fees of $205 per year full-time, $90 per year part-time. • Dr. William Anderson, Chairperson, Department of Mathematics and Computer Science, 201-692-2795.

Florida Atlantic University, College of Science, Department of Mathematics, Boca Raton, FL 33431. Department awards MS, MST, PhD. Part-time programs available. Faculty: 17 full-time (1 woman), 9 part-time (3 women). Matriculated students: 5 full-time (1 woman), 7 part-time (0 women); includes 4 minority (2 Asian American, 2 Hispanic American), 2 foreign. Average age 25. In 1990, 2 master's awarded. Terminal master's awarded for partial completion of doctoral program. *Degree requirements:* For master's, 1 foreign language, thesis. *Entrance requirements:* For master's, GRE General Test (minimum combined score of 1000 required), minimum GPA of 3.0. Application fee: $15. *Tuition:* $89.28 per credit hour for state residents; $291 per credit hour for nonresidents. *Financial aid:* In 1990–91, $90,000 in aid awarded. 8 teaching assistantships (5 to first-year students) were awarded; fellowships, federal work-study also available. *Faculty research:* Algebra, analysis, combinatorics, mathematical physics. • Dr. James Brewer, Chairman, 407-367-3340. Application contact: Dr. Julio Bastida, Graduate Advisor, 407-367-3340.

Florida International University, College of Arts and Sciences, Department of Mathematical Sciences, Miami, FL 33199. Department offers program in mathematical sciences (MS). Evening/weekend programs available. *Degree requirements:* Computer language, thesis required, foreign language not required. *Entrance requirements:* GRE General Test, TOEFL. Application deadline: 4/1. Application fee: $15. *Tuition:* $462 per semester (minimum) full-time, $38.50 per credit part-time for state residents; $2217 per semester (minimum) full-time, $185 per credit part-time for nonresidents.

Announcement: The department offers an MS in mathematical sciences, with a written thesis required. Some course work may be taken, with a choice of mathematics, computer science, statistics, or numerical analysis orientation. Teaching assistantships, including tuition waivers, are available to qualified candidates.

Florida State University, College of Arts and Sciences, Department of Mathematics, Tallahassee, FL 32306. Offerings include mathematical sciences (MA, MS), pure mathematics (MA, MS, PhD). Terminal master's awarded for partial completion of doctoral program. Department faculty: 34 full-time (1 woman), 2 part-time (0 women). *Degree requirements:* For master's, thesis optional, foreign language not required; for doctorate, 2 foreign languages, computer language, dissertation. *Entrance requirements:* GRE General Test (minimum score of 650 on quantitative section, 400 on verbal required), minimum GPA of 3.0. Application deadline: 11/1 (priority date, applications processed on a rolling basis). Application fee: $15. *Tuition:* $76.29 per credit hour for state residents; $238 per credit hour for nonresidents. • Dr. H. Frederick Kreimer, Chairperson, 904-644-2202. Application contact: James E. Miller, Academic Programs Coordinator, 904-644-5868.

Fordham University, Graduate School of Arts and Sciences, Department of Mathematics, New York, NY 10458. Department awards MA. Matriculated students: 7. In 1990, 6 degrees awarded. *Degree requirements:* 1 foreign language required, thesis not required. *Entrance requirements:* GRE General Test, GRE Subject Test. Application fee: $35. *Tuition:* $352 per credit. • Dr. Theodore Faticoni, Chairman, 212-579-2362.

Fort Hays State University, College of Arts and Sciences, Department of Mathematics and Computer Science, Hays, KS 67601. Department offers programs in computer science (MAT), mathematics (MAT). Faculty: 7 full-time (2 women), 0 part-time. Matriculated students: 1 full-time (0 women), 5 part-time (4 women); includes 2 minority (1 Asian American, 1 Hispanic American), 0 foreign. Average age 30. 2 applicants, 0% accepted. In 1990, 1 degree awarded. *Degree requirements:* Thesis or alternative required, foreign language not required. *Application deadline:* 7/1 (priority date, applications processed on a rolling basis). *Application fee:* $0. *Tuition:* $58.75 per credit for state residents; $140 per credit for nonresidents. *Financial aid:* Institutionally sponsored loans available. Financial aid applicants required to submit FAF. *Faculty research:* Applied mathematics, probability and statistics, discrete mathematics topology, role of visual ability, CSI computer-assisted instruction. • Dr. Ellen Veed, Chair, 913-628-4240.

Framingham State College, Division of Graduate and Continuing Education, Program in Mathematics, Framingham, MA 01701. Program awards M Ed. Faculty: 3 full-time, 0 part-time. Matriculated students: 1 full-time, 2 part-time. In 1990, 0 degrees awarded. *Entrance requirements:* GRE General Test, minimum GPA of 3.0. Application fee: $25. *Expenses:* Tuition of $345 per course for state residents; $405 per course for nonresidents. Fees of $65 per semester. • Tom Koshy, Chair, 508-626-4500.

George Mason University, College of Arts and Sciences, Department of Mathematical Sciences, Fairfax, VA 22030. Department awards MS. Evening/weekend programs available. Faculty: 3 full-time, 15 part-time. Matriculated students: 3 full-time (0 women), 25 part-time (8 women); includes 7 minority (4 Asian American, 1 black American, 1 Hispanic American, 1 Native American), 5 foreign. 19 applicants, 79% accepted. In 1990, 9 degrees awarded. *Degree requirements:* Thesis, comprehensive exam. *Entrance requirements:* Minimum GPA of 3.0 in last 60 hours of undergraduate study. Application deadline: 5/1. Application fee: $25. *Expenses:* Tuition of $1872 per year full-time, $78 per semester hour part-time for state residents; $6264 per year full-time, $261 per semester hour part-time for nonresidents. Fees of $1080 per year full-time, $45 per semester hour part-time. *Financial aid:* In 1990–91, 1 fellowship awarded; teaching assistantships and career-related internships or fieldwork also available. Financial aid application deadline: 3/1. • Dr. Stephen Saperstone, Chairman, 703-993-1460. Application contact: Dr. Timothy Sauer, Graduate Coordinator, 703-993-1471.

George Washington University, Graduate School of Arts and Sciences, Department of Mathematics, Program in Mathematics, Washington, DC 20052. Program awards MA, PhD. Part-time and evening/weekend programs available. Matriculated students: 5 full-time (3 women), 7 part-time (1 woman); includes 1 minority (Asian American), 2 foreign. Average age 31. 9 applicants, 100% accepted. In 1990, 1 doctorate awarded. Terminal master's awarded for partial completion of doctoral program. *Degree requirements:* For master's, thesis or alternative, comprehensive exam required, foreign language not required; for doctorate, 2 foreign languages, dissertation, general exam. *Entrance requirements:* GRE General Test, GRE Subject Test, minimum GPA of 3.0. Application deadline: 7/1. Application fee: $45. *Expenses:* Tuition of $490 per semester hour. Fees of $215 per year full-time, $125.60 per year (minimum) part-time. *Financial aid:* Fellowships, teaching assistantships available. Financial aid application deadline: 2/15. • Dr. Irving Katz, Chair, Department of Mathematics, 202-994-6235.

Georgia Institute of Technology, College of Sciences, School of Mathematics, Atlanta, GA 30332. School offers programs in mathematics (MS, PhD), statistics (MSS). Faculty: 44 full-time (3 women), 0 part-time. Matriculated students: 50 full-time (18 women), 14 part-time (3 women); includes 6 minority (2 Asian American, 3 black American, 1 Hispanic American), 21 foreign. Average age 24. 60 applicants, 50% accepted. In 1990, 19 master's awarded (16% entered university research/teaching, 42% found other work related to degree, 32% continued full-time study); 4 doctorates awarded (75% entered university research/teaching, 25% found other work related to degree). Terminal master's awarded for partial completion of doctoral program. *Degree requirements:* For master's, thesis or alternative required, foreign language not required; for doctorate, 1 foreign language, dissertation. *Entrance requirements:* GRE General Test, TOEFL (minimum score of 570 required), minimum GPA of 3.0. Application deadline: 8/1 (applications processed on a rolling basis). Application fee: $15. *Expenses:* Tuition of $574 per semester full-time, $48 per credit part-time for state residents; $1380 per semester full-time, $115 per credit part-time for nonresidents. Fees of $132 per semester full-time. *Financial aid:* In 1990–91, 50 students received a total of $400,000 in aid awarded. 1 fellowship (to a first-year student), 6 research assistantships (3 to first-year students), 44 teaching assistantships (15 to first-year students) were awarded; partial tuition waivers, federal work-study, institutionally sponsored loans, and career-related internships or fieldwork also available. Aid available to part-time students. Average monthly stipend for a graduate assistantship: $1000. Financial aid application deadline: 2/15. *Faculty research:* Partial differential equations, statistics and probability, combinatorics, fractals, dynamical systems. Total annual research budget: $476,643. • Dr. Shui-Nee Chow, Director, 404-894-2700. Application contact: Dr. William Green, Graduate Coordinator, 404-853-9203.

Announcement: Environment of the technological university has produced an emphasis on applied and applicable mathematics. Particular strengths include analysis, differential equations, discrete mathematics, dynamical systems, computer graphical mathematics, probability, statistics, scientific computing, mathematical physics. Extensive computer facilities for student use. Excellent library holdings enhance research. Most students supported as teaching or research assistants. Georgia Tech has a tradition of producing outstanding engineering and science students.

Georgian Court College, Graduate School, Lakewood, NJ 08701-2697. Offerings include mathematics (MA). School faculty: 5 full-time (2 women), 23 part-time (6 women). *Application deadline:* 8/25 (applications processed on a rolling basis). *Application fee:* $20. *Expenses:* Tuition of $215 per credit. Fees of $10 per year. • Sr. Mary Arthur Beal, Dean, 908-367-1717. Application contact: Sr. Margaret Mary Foley, Director of Graduate Records, 908-367-1717.

Georgia Southern University, School of Arts and Sciences, Department of Mathematics and Computer Science, Statesboro, GA 30460. Department offers program in mathematics (MS). Faculty: 4 full-time (0 women), 0 part-time. Matriculated students: 0. In 1990, 0 degrees awarded. *Degree requirements:* Terminal exam. *Entrance requirements:* GRE General Test (minimum combined score of 1000 required), minimum GPA of 3.0. Application fee: $0. *Expenses:* Tuition of $430 per quarter. Fees of $430 per quarter full-time, $126 per quarter part-time for state residents; $986 per quarter full-time, $430 per quarter part-time for nonresidents. *Financial aid:* Research assistantships, teaching assistantships, federal work-study, and career-related internships or fieldwork available. Aid available to part-time students. Financial aid application deadline: 4/15; applicants required to submit FAF. • Dr. Arthur Sparks, Head, 912-681-5390. Application contact: Dr. Charlene R. Black, Dean of the Graduate School, 912-681-5384.

Graduate School and University Center of the City University of New York, Program in Mathematics, New York, NY 10036. Program awards PhD. Faculty: 43 full-time (2 women), 0 part-time. Matriculated students: 80 full-time (15 women), 0 part-time; includes 8 minority (5 Asian American, 2 black American, 1 Hispanic American), 28 foreign. Average age 32. 59 applicants, 75% accepted. In 1990, 4 doctorates awarded (50% entered university research/teaching, 0% found other work related to degree, 0% continued full-time study). Terminal master's awarded for partial completion of doctoral program. *Degree requirements:* For doctorate, 2 foreign languages, dissertation. *Entrance requirements:* For doctorate, GRE General Test. Application deadline: 4/1. Application fee: $30. *Tuition:* $2204 per year full-time, $95 per credit part-time for state residents; $4700 per year full-time, $199 per credit part-time for nonresidents. *Financial aid:* In 1990–91, $165,000 in aid awarded. 24 fellowships, 6 research assistantships, 6 teaching assistantships were awarded; full and partial tuition waivers, federal work-study, institutionally sponsored loans, and career-related internships or fieldwork also available. Financial aid application deadline: 2/1. • Dr. Martin Moskowitz, Executive Officer, 212-642-2458.

Directory: Mathematics

Hampton University, Department of Mathematics, Hampton, VA 23668. Department awards MA, MS. *Entrance requirements:* GRE General Test (minimum score of 450 on verbal section required). Application deadline: 7/1. Application fee: $10. *Tuition:* $3050 per semester full-time, $155 per credit hour part-time.

Harvard University, Graduate School of Arts and Sciences, Department of Mathematics, Cambridge, MA 02138. Department awards AM, PhD. Faculty: 27 full-time. Matriculated students: 57 full-time (7 women), 0 part-time; includes 0 minority. 215 applicants, 11% accepted. In 1990, 6 master's, 11 doctorates awarded. Terminal master's awarded for partial completion of doctoral program. *Degree requirements:* For doctorate, 2 foreign languages, dissertation, qualifying exam. *Entrance requirements:* For doctorate, GRE General Test, GRE Subject Test. Application deadline: 1/2. Application fee: $60. *Expenses:* Tuition of $14,860 per year. Fees of $550 per year. *Financial aid:* In 1990–91, $304,686 in aid awarded. 46 fellowships (12 to first-year students), 13 research assistantships (1 to a first-year student), 14 teaching assistantships (12 to first-year students) were awarded; federal work-study, institutionally sponsored loans, and career-related internships or fieldwork also available. Financial aid application deadline: 1/2; applicants required to submit GAPSFAS. • Dr. Joe Harris, Director of Graduate Studies, 617-495-2170. Application contact: Office of Admissions and Financial Aid, 617-495-5315.

Harvey Mudd College, Graduate Program in Mathematics, Claremont, CA 91711. College awards MS. Offered jointly with Claremont Graduate School, MS available to Harvey Mudd undergraduates only. *Expenses:* Tuition of $13,480 per year. Fees of $385 per year full-time. • Dr. Henry Krieger, Chairman, 714-621-8023.

Howard University, Graduate School of Arts and Sciences, Department of Mathematics, 2400 Sixth Street, NW, Washington, DC 20059. Department offers programs in applied mathematics (MS, PhD), mathematics (MS, PhD). Part-time programs available. *Degree requirements:* For master's, computer language, thesis or alternative, comprehensive exam required, foreign language not required; for doctorate, 2 foreign languages (computer language can substitute for one), dissertation, comprehensive exam. *Entrance requirements:* For master's, minimum GPA of 3.0. Application deadline: 4/1. Application fee: $25. *Expenses:* Tuition of $6100 per year full-time, $339 per credit hour part-time. Fees of $555 per year full-time, $245 per semester part-time. *Faculty research:* An electrical lemma, a lifting theorem for subordinated invariant kernels, efficient algorithms for computing Hausdorff dimensions.

See full description on page 417.

Hunter College of the City University of New York, Division of Sciences and Mathematics, Department of Mathematics, 695 Park Avenue, New York, NY 10021. Department offers programs in applied mathematics (MA), pure mathematics (MA). *Degree requirements:* 1 foreign language required, thesis not required. *Entrance requirements:* GRE General Test, TOEFL (minimum score of 550 required). Application deadline: 4/1. Application fee: $30. *Expenses:* Tuition of $2204 per year full-time, $95 per credit part-time for state residents; $4700 per year full-time, $199 per credit part-time for nonresidents. Fees of $15 per semester. • Tom Jambois, Chairperson, 212-772-5300. Application contact: Office of Admissions, 212-772-4490.

Idaho State University, College of Arts and Sciences, Department of Mathematics, Pocatello, ID 83209. Department offers programs in mathematics (MS, DA), natural science (MNS). Faculty: 13 full-time, 1 part-time. Matriculated students: 14 full-time (4 women), 10 part-time (4 women); includes 5 minority (3 Asian American, 2 Hispanic American), 2 foreign. In 1990, 2 master's, 2 doctorates awarded. *Degree requirements:* For master's, foreign language and thesis not required; for doctorate, dissertation required, foreign language not required. *Entrance requirements:* GRE General Test, GRE Subject Test. Application deadline: 8/1 (priority date). Application fee: $10. *Expenses:* Tuition of $0 full-time, $1060 per semester part-time for state residents; $0 for nonresidents. Fees of $809 per semester full-time, $72.50 per credit part-time. *Financial aid:* In 1990–91, 9 teaching assistantships awarded; fellowships also available. • Dr. Richard D. Hill, Chairman, 208-236-3176.

Illinois Institute of Technology, Lewis College of Sciences and Letters, Department of Mathematics, Chicago, IL 60616. Offerings include mathematics (PhD). Terminal master's awarded for partial completion of doctoral program. Department faculty: 13 full-time (3 women), 1 part-time (0 women). *Degree requirements:* Dissertation, comprehensive exam required, foreign language not required. *Entrance requirements:* TOEFL (minimum score of 500 required). Application deadline: 7/1 (applications processed on a rolling basis). Application fee: $30. *Expenses:* Tuition of $13,070 per year full-time, $435 per credit hour part-time. Fees of $20 per semester (minimum) full-time, $1 per credit hour part-time. • Dr. Maurice J. Frank, Chairman, 312-567-3162.

Illinois State University, College of Arts and Sciences, Department of Mathematics, Normal, IL 61761. Department awards MA, MS, DA. Faculty: 23 full-time (4 women), 0 part-time. Matriculated students: 19 full-time (11 women), 33 part-time (24 women); includes 4 minority (3 Asian American, 1 Hispanic American). In 1990, 13 master's, 2 doctorates awarded. *Degree requirements:* For master's, thesis or alternative; for doctorate, variable foreign language requirement, dissertation, oral comprehensive exam, 2 terms of residency. *Entrance requirements:* For master's, GRE General Test, minimum GPA of 2.8 during last 60 hours; for doctorate, GRE General Test. Application deadline: rolling. Application fee: $0. *Expenses:* Tuition of $1824 per year full-time, $76 per hour part-time for state residents; $5472 per year full-time, $228 per hour part-time for nonresidents. Fees of $630 per year full-time, $21 per hour (minimum) part-time. *Financial aid:* In 1990–91, $209,926 in aid awarded. 23 fellowships, 4 research assistantships, 19 teaching assistantships were awarded. Financial aid application deadline: 4/1. *Total annual research budget:* $326,644. • Jane Swafford, Chairperson, 309-438-8781.

Indiana State University, College of Arts and Sciences, Department of Mathematics, Terre Haute, IN 47809. Department awards MA, MS. Faculty: 18 full-time (2 women), 2 part-time (0 women). Matriculated students: 7 full-time (2 women), 5 part-time (3 women); includes 0 minority, 4 foreign. 27 applicants, 56% accepted. In 1990, 5 degrees awarded. *Degree requirements:* Foreign language not required. *Application deadline:* rolling. *Application fee:* $10 ($20 for foreign students). *Tuition:* $90 per hour for state residents; $199 per hour for nonresidents. *Financial aid:* In 1990–91, $20,000 in aid awarded. 6 teaching assistantships (1 to a first-year student) were awarded; partial tuition waivers also available. Financial aid application deadline: 3/1. • Dr. Laurence Kunes, Acting Chairperson, 812-237-2130.

Indiana University of Pennsylvania, College of Natural Sciences and Mathematics, Department of Mathematics, Indiana, PA 15705. Department awards M Ed, MS. Part-time programs available. Faculty: 7 full-time (1 woman), 0 part-time. Matriculated students: 14 full-time (7 women), 8 part-time (all women); includes 0 minority, 5 foreign. 47 applicants, 60% accepted. In 1990, 16 degrees awarded. *Degree requirements:* Thesis optional, foreign language not required. *Entrance requirements:* GRE General Test, TOEFL (minimum score of 500 required). Application deadline: 7/1 (priority date, applications processed on a rolling basis). Application fee: $20. *Expenses:* Tuition of $1139 per semester full-time, $127 per credit part-time for state residents; $1442 per semester full-time, $160 per credit part-time for nonresidents. Fees of $169 per semester full-time. *Financial aid:* In 1990–91, 7 research assistantships awarded; federal work-study also available. Aid available to part-time students. Financial aid application deadline: 3/15. • Dr. Gary Buaiok, Chairperson, 412-357-2608. Application contact: Dr. Joseph Angelo, Director of Graduate Studies, 412-357-2600.

Indiana University–Purdue University at Fort Wayne, Department of Mathematical Sciences, Fort Wayne, IN 46805-1499. Department offers programs in applied mathematics (MS), mathematics (MS), operations research (MS). Part-time and evening/weekend programs available. Faculty: 13 full-time (2 women), 0 part-time. Matriculated students: 4 full-time (1 woman), 2 part-time (0 women); includes 1 minority (Asian American), 1 foreign. Average age 25. 5 applicants, 80% accepted. In 1990, 3 degrees awarded (33% entered university research/teaching, 34% found other work related to degree, 33% continued full-time study). *Degree requirements:* Foreign language and thesis not required. *Entrance requirements:* Minimum GPA of 3.0, undergraduate major or minor in mathematics. Application deadline: 5/15 (priority date, applications processed on a rolling basis). Application fee: $25. *Tuition:* $78.05 per credit hour for state residents; $174.75 per credit hour for nonresidents. *Financial aid:* In 1990–91, 4 students received a total of $20,000 in aid awarded. 4 teaching assistantships (3 to first-year students) were awarded. Average monthly stipend for a graduate assistantship: $500. Financial aid application deadline: 8/1. *Faculty research:* Graph theory, approximation theory, statistical design, mathematics education, ring theory. • Douglas Townsend, Chair, 219-481-6235.

Indiana University–Purdue University at Indianapolis, School of Science, Department of Mathematical Sciences, Indianapolis, IN 46205. Department offers programs in applied mathematics (MS, PhD), mathematics (MS, PhD). Part-time programs available. Faculty: 21 full-time (1 woman), 0 part-time. Matriculated students: 15 full-time (1 woman), 4 part-time (1 woman); includes 1 minority (black American), 12 foreign. Average age 25. 50 applicants, 22% accepted. In 1990, 8 master's, 0 doctorates awarded. Terminal master's awarded for partial completion of doctoral program. *Degree requirements:* For master's, foreign language and thesis not required; for doctorate, 2 foreign languages, dissertation. *Entrance requirements:* TOEFL (minimum score of 570 required). Application deadline: 2/1. Application fee: $20. *Tuition:* $100 per credit for state residents; $288 per credit for nonresidents. *Financial aid:* In 1990–91, 18 students received a total of $130,000 in aid awarded. 3 fellowships (1 to a first-year student), 15 teaching assistantships (8 to first-year students) were awarded; research assistantships, full and partial tuition waivers, federal work-study also available. Average monthly stipend for a graduate assistantship: $750. Financial aid application deadline: 3/1. *Faculty research:* Fluid dynamics, partial differential equations, functional analysis, topology, mathematical economics. Total annual research budget: $100,000. • Bartholomew S. Ng, Chair, 317-274-6918. Application contact: Chair, Graduate Committee, 317-274-6925.

Institute of Paper Science and Technology, Program in Chemistry, Atlanta, GA 30318. Offers paper science and technology (MS, PhD). Part-time programs available. Terminal master's awarded for partial completion of doctoral program. *Degree requirements:* For master's, industrial experience required, foreign language and thesis not required; for doctorate, dissertation required, foreign language not required. *Entrance requirements:* For master's, GRE. Application deadline: 5/15 (priority date). Application fee: $0. *Tuition:* $0 full-time, $225 per credit hour part-time.

See full description on page 109.

Institute of Paper Science and Technology, Program in Physics/Mathematics, Atlanta, GA 30318. Offers paper science and technology (MS, PhD). Part-time programs available. Terminal master's awarded for partial completion of doctoral program. *Degree requirements:* For master's, industrial experience required, foreign language and thesis not required; for doctorate, dissertation required, foreign language not required. *Entrance requirements:* For master's, GRE. Application deadline: 5/15 (priority date). Application fee: $0. *Tuition:* $0 full-time, $225 per credit hour part-time.

Iowa State University of Science and Technology, College of Liberal Arts and Sciences, Department of Mathematics, Ames, IA 50011. Department offers programs in applied mathematics (MS, PhD), mathematics (MS, MSM, PhD). Matriculated students: 26 full-time (5 women), 37 part-time (14 women); includes 1 minority (Asian American), 32 foreign. In 1990, 15 master's, 4 doctorates awarded. *Degree requirements:* For master's, comprehensive written exam required, thesis optional; for doctorate, 2 foreign languages, dissertation, comprehensive written exam. *Entrance requirements:* For master's, GRE General Test. Application fee: $20 ($30 for foreign students). *Expenses:* Tuition of $1158 per semester full-time, $129 per credit part-time for state residents; $3340 per semester full-time, $372 per credit part-time for nonresidents. Fees of $10 per semester. *Financial aid:* In 1990–91, 1 fellowship (to a first-year student), 6 research assistantships (0 to first-year students), 56 teaching assistantships (27 to first-year students), 7 scholarships (all to first-year students) awarded. • Dr. Howard A. Levine, Chair, 515-294-2175.

Jackson State University, School of Science and Technology, Department of Mathematics, Jackson, MS 39217. Department awards MS, MST, Ed S. Evening/weekend programs available. Faculty: 19 full-time (6 women), 18 part-time (9 women). Matriculated students: 9 full-time (5 women), 9 part-time (5 women); includes 16 minority (1 Asian American, 15 black American). In 1990, 3 master's, 0 Ed Ss awarded. *Degree requirements:* For master's, thesis (for some programs). *Entrance requirements:* For master's, GRE General Test. Application deadline: rolling. Application fee: $0. *Expenses:* Tuition of $1114 per semester full-time, $113 per hour part-time for state residents; $1142 per semester full-time, $113 per hour part-time for nonresidents. Fees of $730 per semester for nonresidents. *Financial aid:* Federal work-study available. Financial aid application deadline: 5/1; applicants required to submit FAF. • Dr. Joseph Colen, Chair, 601-968-2161.

Jacksonville State University, College of Letters and Sciences, Department of Mathematics, Jacksonville, AL 36265. Department awards MS. Faculty: 11 full-time (0 women), 0 part-time. Matriculated students: 0 full-time, 8 part-time (3 women); includes 0 minority. In 1990, 1 degree awarded. *Degree requirements:* Thesis optional. *Entrance requirements:* GRE General Test or MAT. Application deadline: rolling. Application fee: $15. *Tuition:* $685 per semester full-time, $69 per semester hour part-time. • Dr. William D. Carr, Dean, College of Graduate Studies and Continuing Education, 205-782-5329.

James Madison University, College of Letters and Sciences, Department of Mathematics and Computer Science, Harrisonburg, VA 22807. Department offers programs in computer science (MS), mathematics (MAT, MS). Part-time programs available. Faculty: 27 full-time (4 women), 16 part-time. Matriculated students: 9 full-time (3 women), 21 part-time (8 women); includes 1 minority (Asian American), 5 foreign. Average age 30. 29 applicants, 97% accepted. In 1990, 2 degrees awarded. *Degree requirements:* Thesis or alternative required, foreign language not required.

Entrance requirements: GRE General Test. Application deadline: 7/1 (priority date, applications processed on a rolling basis). Application fee: $20. *Tuition:* $2544 per year full-time, $106 per credit hour part-time for state residents; $6936 per year full-time, $289 per credit hour part-time for nonresidents. *Financial aid:* Fellowships, teaching assistantships, federal work-study available. Financial aid application deadline: 3/18; applicants required to submit FAF. • Dr. Diane Spresser, Head, 703-568-6184.

John Carroll University, Department of Mathematics, University Heights, OH 44118. Department awards MA, MS. Part-time and evening/weekend programs available. Faculty: 14 full-time (0 women), 0 part-time. Matriculated students: 6 full-time (2 women), 8 part-time (4 women); includes 0 minority, 3 foreign. Average age 26. In 1990, 3 degrees awarded (100% found work related to degree). *Degree requirements:* Research essay, comprehensive exam required, foreign language and thesis not required. *Application deadline:* 8/17 (priority date, applications processed on a rolling basis). *Application fee:* $20. *Tuition:* $300 per credit. *Financial aid:* In 1990–91, $48,000 in aid awarded. 6 teaching assistantships (3 to first-year students) were awarded; partial tuition waivers also available. *Faculty research:* Algebraic topology, algebra, differential geometry, combinatorics, lie groups. • Dr. Carl R. Spitznagel, Chairperson, 216-397-4351.

Johns Hopkins University, G. W. C. Whiting School of Engineering, Department of Mathematical Sciences, Baltimore, MD 21218. Department offers programs in applied mathematics (MA, MSE, PhD), operations research (MA, MSE, PhD), statistics (MA, MSE, PhD). Faculty: 11 full-time (0 women), 2 part-time (1 woman). Matriculated students: 31 full-time (8 women), 0 part-time; includes 0 minority, 12 foreign. Average age 28. 128 applicants, 39% accepted. In 1990, 5 master's awarded (40% found work related to degree, 60% continued full-time study); 2 doctorates awarded (100% entered university research/teaching). Terminal master's awarded for partial completion of doctoral program. *Degree requirements:* For master's, computer language, thesis (for some programs); for doctorate, 1 foreign language, computer language, dissertation. *Entrance requirements:* GRE General Test, GRE Subject Test. Application deadline: 2/15 (priority date, applications processed on a rolling basis). Application fee: $45. *Expenses:* Tuition of $15,500 per year full-time, $1550 per course part-time. Fees of $400 per year full-time. *Financial aid:* In 1990–91, $265,420 in aid awarded. 2 fellowships (both to first-year students), 3 research assistantships (0 to first-year students), 10 teaching assistantships (6 to first-year students), 0 grants were awarded; full and partial tuition waivers, federal work-study, institutionally sponsored loans also available. Average monthly stipend for a graduate assistantship: $1111. Financial aid application deadline: 2/15; applicants required to submit FAF. *Faculty research:* Graph theory, matrix analysis, probability, statistics, optimization. Total annual research budget: $233,000. • Dr. John C. Wierman, Chairman, 301-338-7195.

See full description on page 419.

Johns Hopkins University, School of Arts and Sciences, Department of Mathematics, Baltimore, MD 21218. Department awards MA, PhD. Faculty: 16 full-time (0 women), 2 part-time (0 women). Matriculated students: 29 full-time (4 women), 0 part-time; includes 1 minority (Hispanic American), 18 foreign. Average age 23. 54 applicants, 24% accepted. In 1990, 4 master's awarded; 4 doctorates awarded (100% entered university research/teaching). Terminal master's awarded for partial completion of doctoral program. *Degree requirements:* For master's, 1 foreign language, oral exam required, thesis not required; for doctorate, 2 foreign languages, dissertation, oral exams. *Entrance requirements:* GRE General Test, GRE Subject Test. Application deadline: 2/15. Application fee: $40. *Expenses:* Tuition of $15,500 per year full-time, $1550 per course part-time. Fees of $400 per year full-time. *Financial aid:* In 1990–91, $616,800 in aid awarded. 2 fellowships (0 to first-year students), 27 teaching assistantships (7 to first-year students) were awarded; partial tuition waivers, federal work-study, institutionally sponsored loans also available. Financial aid application deadline: 3/14; applicants required to submit FAF. *Faculty research:* Analysis, representation theory, number theory, algebraic topology, algebraic/differential geometry. • Dr. Bernard Shiffman, Chairman, 301-338-7399. Application contact: Dr. W. S. Wilson, Graduate Admissions Committee, 301-338-7399.

See full description on page 421.

Kansas State University, College of Arts and Sciences, Department of Mathematics, Manhattan, KS 66506. Department awards MS, PhD. Faculty: 23 full-time (0 women), 0 part-time. Matriculated students: 47 full-time (6 women), 0 part-time; includes 23 foreign. 26 applicants, 81% accepted. In 1990, 4 master's, 2 doctorates awarded. *Degree requirements:* For master's, thesis required, foreign language not required; for doctorate, 1 foreign language, dissertation. *Expenses:* Tuition of $1668 per year full-time, $51 per credit hour part-time for state residents; $5382 per year full-time, $142 per credit hour part-time for nonresidents. Fees of $305 per year full-time, $64.50 per semester part-time. *Financial aid:* Teaching assistantships available. • Louis Pigno, Head, 913-532-6750. Application contact: David Surowski, Graduate Coordinator, 913-532-6750.

Kent State University, Department of Mathematics, Kent, OH 44242. Department awards MA, MS, PhD. Faculty: 31 full-time, 0 part-time. Matriculated students: 69 full-time (12 women), 69 part-time (15 women). 155 applicants, 68% accepted. In 1990, 17 master's, 1 doctorate awarded. *Degree requirements:* For master's, thesis optional, foreign language not required; for doctorate, 1 foreign language, dissertation. *Entrance requirements:* GRE. Application deadline: 7/12 (applications processed on a rolling basis). Application fee: $25. *Tuition:* $1601 per semester full-time, $133.75 per hour part-time for state residents; $3101 per semester full-time, $258.75 per hour part-time for nonresidents. *Financial aid:* Fellowships, research assistantships, teaching assistantships, full tuition waivers, federal work-study available. Financial aid application deadline: 2/1. • Dr. O. P. Stackelberg, Chairman, 216-672-2430.

Announcement: Department offers master's programs in mathematics and computer science. PhD areas include numerical, functional, and complex analysis; ring and group theory; computer algebra; computer science; and applied mathematics. Special biomedical mathematics PhD program in conjunction with Northeastern Ohio Universities College of Medicine available for suitably prepared students. Contact O. P. Stackelberg, Chairman.

Kutztown University of Pennsylvania, Program in Mathematics, Kutztown, PA 19530. Program awards MA. Part-time and evening/weekend programs available. Faculty: 6 full-time (1 woman), 0 part-time. Matriculated students: 0 full-time, 8 part-time (3 women); includes 0 minority. Average age 31. In 1990, 3 degrees awarded (100% found work related to degree). *Degree requirements:* Thesis or comprehensive exam required, foreign language not required. *Entrance requirements:* GRE General Test or MAT or minimum GPA of 3.0. Application fee: $15. *Expenses:* Tuition of $1139 per semester full-time, $127 per credit hour part-time for state residents; $1442 per semester full-time, $160 per credit hour part-time for nonresidents. Fees of $36 per semester. *Financial aid:* Graduate assistantships available. *Faculty research:* Operations research. • Dr. Edward Evans, Chairperson, 215-683-4410.

Lakehead University, Faculty of Arts and Science, Department of Mathematical Sciences, Thunder Bay, ON P7B 5E1, Canada. Department offers programs in computer science (M Sc), mathematics (MA, M Sc). Part-time and evening/weekend programs available. Faculty: 17 full-time (0 women), 0 part-time. Matriculated students: 6 full-time (1 woman), 2 part-time (1 woman); includes 0 foreign. 114 applicants, 10% accepted. In 1990, 1 degree awarded. *Entrance requirements:* TOEFL (minimum score of 550 required). Application fee: $0. *Expenses:* Tuition of $1969 per year full-time, $527 per year part-time for Canadian residents; $3712 per year full-time, $2618 per year part-time for nonresidents. Fees of $2735 per year full-time, $100 per year part-time for Canadian residents; $11,136 per year full-time, $100 per year part-time for nonresidents. *Financial aid:* In 1990–91, 7 students received a total of $62,354 in aid awarded. 7 teaching assistantships (all to first-year students) were awarded; fellowships, entrance awards and research stipends also available. Average monthly stipend for a graduate assistantship: $680. Financial aid application deadline: 3/30. *Faculty research:* Classical analysis, combinatorics, algebra, lattice theory, parallel computing. Total annual research budget: $53,300. • Dr. Bill Eames, Chairman, 807-343-8689. Application contact: Dr. M. Hasegawa, Graduate Coordinator, 807-343-8310.

Lamar University–Beaumont, College of Engineering, Department of Mathematics, Beaumont, TX 77705. Department awards MS. Faculty: 7 full-time (1 woman), 0 part-time. Matriculated students: 5 full-time (1 woman), 1 part-time (0 women); includes 0 minority, 5 foreign. In 1990, 5 degrees awarded. *Degree requirements:* Foreign language and thesis not required. *Entrance requirements:* GRE General Test (minimum combined score of 900 required), TOEFL (minimum score of 500 required), minimum GPA of 2.5, or 3.0 in last 60 hours of undergraduate course work. Application deadline: rolling. Application fee: $0. *Expenses:* Tuition of $630 per year full-time, $378 per year part-time for state residents; $4350 per year full-time, $2610 per year part-time for nonresidents. Fees of $480 per year. *Financial aid:* In 1990–91, 5 teaching assistantships awarded. • Dr. John R. Cannon, Head, 409-880-8792.

Lehigh University, College of Arts and Sciences, Department of Mathematics, Bethlehem, PA 18015. Department offers programs in applied mathematics (MS, PhD), mathematics (MS, PhD). Faculty: 23 full-time, 0 part-time. Matriculated students: 42 full-time (16 women), 9 part-time (1 woman); includes 5 minority (1 Asian American, 3 black American, 1 Hispanic American), 13 foreign. 45 applicants, 22% accepted. In 1990, 10 master's, 2 doctorates awarded. *Degree requirements:* For master's, comprehensive exam required, foreign language and thesis not required; for doctorate, 1 foreign language, dissertation, comprehensive and qualifying exams. *Entrance requirements:* TOEFL, minimum GPA of 3.0. Application deadline: rolling. Application fee: $40. *Tuition:* $15,650 per year full-time, $655 per credit hour part-time. *Financial aid:* Fellowships available. • Dr. Andrew K. Snyder, Chairman, 215-758-3730. Application contact: Dr. Bruce Dodson, Graduate Program Coordinator.

See full description on page 423.

Lehman College of the City University of New York, Division of Natural and Social Sciences, Department of Mathematics and Computer Science, Bedford Park Boulevard West, Bronx, NY 10468. Offerings include mathematics (MA). Department faculty: 9 full-time (1 woman), 0 part-time. *Application deadline:* 4/15 (applications processed on a rolling basis). *Application fee:* $30. *Expenses:* Tuition of $2204 per year full-time, $95 per credit part-time for state residents; $4700 per year full-time, $199 per credit part-time for nonresidents. Fees of $45 per year full-time, $25 per year part-time. • Robert Feinerman, Chairperson, 212-960-8116. Application contact: Connor Lazarov, Adviser, 212-960-8787.

Long Island University, C. W. Post Campus, College of Arts and Sciences, Department of Mathematics, Brookville, NY 11548. Offerings include applied mathematics (MS), with options in classical mathematics, computer mathematics. Department faculty: 17 full-time (4 women). *Degree requirements:* Thesis or alternative required, foreign language not required. *Application fee:* $30. *Expenses:* Tuition of $310 per credit. Fees of $235 per semester full-time, $90 per semester part-time. • Dr. Neo Cleopa, Chairman, 516-299-2447.

Louisiana State University and Agricultural and Mechanical College, College of Arts and Sciences, Department of Mathematics, Baton Rouge, LA 70803. Department awards MA, MS, PhD. Faculty: 44 full-time (1 woman), 1 part-time (0 women). Matriculated students: 63 full-time (20 women), 4 part-time (1 woman); includes 6 minority (2 Asian American, 3 black American, 1 Hispanic American), 30 foreign. Average age 28. 900 applicants, 59% accepted. In 1990, 10 master's awarded; 4 doctorates awarded (100% entered university research/teaching). Terminal master's awarded for partial completion of doctoral program. *Degree requirements:* For master's, foreign language and thesis not required; for doctorate, 2 foreign languages, dissertation. *Entrance requirements:* GRE General Test, TOEFL (minimum score of 525 required), minimum GPA of 3.0. Application deadline: 7/1 (priority date, applications processed on a rolling basis). Application fee: $25. *Tuition:* $1020 per semester full-time for state residents; $2620 per semester full-time for nonresidents. *Financial aid:* In 1990–91, 61 students received a total of $697,347 in aid awarded. 11 fellowships (5 to first-year students), 0 research assistantships, 48 teaching assistantships (21 to first-year students) were awarded; full tuition waivers, institutionally sponsored loans also available. Average monthly stipend for a graduate assistantship: $1087. Financial aid application deadline: 7/1. *Faculty research:* Algebra, graph theory and combinatorics, algebraic topology, analysis and probability, topological algebra. Total annual research budget: $986,888. • Dr. J. D. Lawson, Chair, 504-388-1618. Application contact: Dr. Len Richardson, Director of Graduate Studies, 504-388-1568.

Louisiana Tech University, College of Arts and Sciences, Department of Mathematics and Statistics, Ruston, LA 71272. Department awards MS. Part-time programs available. Faculty: 12 full-time (4 women), 0 part-time. Matriculated students: 6 full-time (2 women), 0 part-time; includes 1 minority (Asian American), 2 foreign. In 1990, 5 degrees awarded. *Degree requirements:* Computer language, thesis or alternative required, foreign language not required. *Entrance requirements:* GRE General Test. Application deadline: 8/13. Application fee: $5. *Tuition:* $613 per quarter full-time, $184 per semester (minimum) part-time for state residents; $999 per quarter full-time, $184 per semester (minimum) part-time for nonresidents. *Financial aid:* Fellowships available. Financial aid application deadline: 2/1; applicants required to submit GAPSFAS. • Dr. John E. Maxfield, Acting Head, 318-257-2538.

Loyola University Chicago, Graduate School, Department of Mathematical Sciences, 820 North Michigan Avenue, Chicago, IL 60611. Department offers program in computer science (MS). Part-time and evening/weekend programs available. Faculty: 25 full-time (5 women), 0 part-time. Matriculated students: 44 full-time (20 women), 43 part-time (12 women); includes 7 minority (4 Asian American, 3 black American), 46 foreign. Average age 27. 143 applicants, 67% accepted. In 1990, 23 degrees awarded (13% entered university research/teaching, 83% found other work related to degree, 4% continued full-time study). *Degree requirements:* Computer language, written comprehensive exam required, foreign language and thesis not required. *Entrance requirements:* GRE or TOEFL (minimum score of 500 required), minimum B average. Application deadline: 8/1 (applications processed on a rolling basis). Application fee: $30. *Tuition:* $252 per credit hour. *Financial aid:* In 1990–91, 25 students received a total of $140,000 in aid awarded. 7 research assistantships (5 to first-year students), 9 teaching assistantships (4 to first-year students) were awarded; partial tuition waivers, federal work-study, institutionally sponsored loans, and career-related internships or fieldwork also available. Aid available to part-time students. Average monthly stipend for a graduate assistantship: $800. Financial aid application deadline: 2/1; applicants required to submit FAF. *Faculty research:* Graph theory, algebraic coding theory, algorithms, cryptography, artificial intelligence. Total annual research budget: $230,000. • Dr. W. Cary Huffman, Chair, 312-508-3558. Application contact: Dr. Alan Saleski, Graduate Director, 312-508-3577.

Announcement: The Department of Mathematical Sciences has developed a very successful graduate program leading toward an MS in computer science. The student is required to take 8 courses, including 4 specified core courses, and pass a written comprehensive examination. Most courses are offered during the evening or on Saturday, thus accommodating part-time students, who constitute roughly half of the student body. Faculty members are currently engaged in research in cryptography, algorithms, operations research, complexity theory, recursion theory, system performance evaluation, computer graphics, fractals, parallel processing, combinatorial analysis, and algebraic coding theory. Graduates have obtained outstanding positions in business, teaching, and research.

Maharishi International University, Department of Mathematics, Fairfield, IA 52556. Department awards MS. Faculty: 5 full-time (2 women), 3 part-time (0 women). Matriculated students: 4 full-time (0 women), 0 part-time; includes 0 minority, 1 foreign. Average age 29. In 1990, 0 degrees awarded. *Degree requirements:* Thesis or alternative required, foreign language not required. *Entrance requirements:* GRE General Test, TOEFL, minimum GPA of 3.0. Application deadline: 4/15 (priority date, applications processed on a rolling basis). Application fee: $40. *Expenses:* Tuition of $7950 per year full-time, $1990 per year part-time. Fees of $170 per year. *Financial aid:* Fellowships, federal work-study available. Financial aid application deadline: 4/30. • Dr. John Price, Head, 515-472-5031 Ext. 3434. Application contact: Harry Bright, Director of Admissions, 515-472-1166.

Mankato State University, College of Natural Sciences, Mathematics and Home Economics, Department of Mathematics, Statistics, and Astronomy, Program in Computers, South Road and Ellis Avenue, Mankato, MN 56002-8400. Offers mathematics: computer science (MS). Faculty: 10 full-time (0 women), 10 part-time (1 woman). Matriculated students: 47 full-time (16 women), 12 part-time (5 women); includes 2 minority (1 Asian American, 1 Hispanic American), 54 foreign. Average age 28. 87 applicants, 87% accepted. In 1990, 10 degrees awarded. *Degree requirements:* 1 foreign language, computer language, thesis or alternative, comprehensive exam. *Entrance requirements:* GRE General Test, GRE Subject Test, minimum GPA of 2.75 for last 2 years of undergraduate study. Application deadline: 2/3 (priority date, applications processed on a rolling basis). Application fee: $15. *Expenses:* Tuition of $52 per credit for state residents; $75 per credit for nonresidents. Fees of $6.50 per quarter. *Financial aid:* In 1990–91, $12,922 in aid awarded. Research assistantships, teaching assistantships, federal work-study, institutionally sponsored loans available. Aid available to part-time students. Financial aid application deadline: 7/1. • Dr. Don Henderson, Head, 507-389-2968.

Mankato State University, College of Natural Sciences, Mathematics and Home Economics, Department of Mathematics, Statistics, and Astronomy, Program in Mathematics, South Road and Ellis Avenue, Mankato, MN 56002-8400. Program awards MA, MS. Matriculated students: 14 full-time (2 women), 7 part-time (4 women); includes 2 minority (1 Asian American, 1 black American), 4 foreign. Average age 29. 33 applicants, 55% accepted. In 1990, 4 degrees awarded. *Degree requirements:* 1 foreign language, computer language, thesis or alternative, comprehensive exam. *Entrance requirements:* GRE General Test, GRE Subject Test, minimum GPA of 2.75 for last 2 years of undergraduate study. Application deadline: 2/3 (priority date, applications processed on a rolling basis). Application fee: $15. *Expenses:* Tuition of $52 per credit for state residents; $75 per credit for nonresidents. Fees of $6.50 per quarter. *Financial aid:* Teaching assistantships available. Financial aid application deadline: 7/1. • Dr. Larry Pearson, Head, 517-389-1453.

Marquette University, College of Arts and Sciences, Department of Mathematics, Statistics, and Computer Science, Milwaukee, WI 53233. Offerings include algebraic theory of semi-groups (PhD), bio-mathematical modeling (PhD), mathematics (MS). Terminal master's awarded for partial completion of doctoral program. Department faculty: 28 full-time (2 women), 0 part-time. *Degree requirements:* For master's, thesis or alternative, comprehensive exam required, foreign language not required; for doctorate, 2 foreign languages, dissertation, comprehensive exam. *Entrance requirements:* TOEFL (minimum score of 550 required). Application fee: $25. *Tuition:* $300 per credit. • Dr. Peter R. Jones, Chairman, 414-288-7573.

Marshall University, College of Science, Department of Mathematics, Huntington, WV 25755. Department awards MA. Faculty: 10 (3 women). Matriculated students: 4 full-time (0 women), 12 part-time (6 women); includes 1 minority (Asian American), 1 foreign. In 1990, 5 degrees awarded. *Degree requirements:* Thesis optional, foreign language not required. *Entrance requirements:* GRE General Test. Application fee: $0. *Tuition:* $857 per semester full-time, $492 per semester part-time for state residents; $2207 per semester full-time, $1392 per semester part-time for nonresidents. • Dr. Charles W. Peele, Chairperson, 304-696-6482. Application contact: Dr. James Harless, Director of Admissions, 304-696-3160.

Massachusetts Institute of Technology, School of Science, Department of Mathematics, Cambridge, MA 02139. Department offers programs in applied mathematics (PhD, Sc D), mathematics (PhD, Sc D). Faculty: 50 full-time (0 women), 0 part-time. Matriculated students: 126 full-time (15 women), 0 part-time; includes 7 minority (6 Asian American, 1 black American), 62 foreign. Average age 24. 288 applicants, 19% accepted. In 1990, 5 master's awarded (0% entered university research/teaching, 40% found other work related to degree, 0% continued full-time study); 19 doctorates awarded (95% entered university research/teaching, 5% found other work related to degree). Terminal master's awarded for partial completion of doctoral program. *Degree requirements:* For doctorate, 1 foreign language, dissertation, oral exam. *Entrance requirements:* For doctorate, GRE General Test, GRE Subject Test. Application deadline: 1/15 (applications processed on a rolling basis). Application fee: $45. *Tuition:* $16,900 per year. *Financial aid:* In 1990–91, 78 students received a total of $2.4-million in aid awarded. 4 fellowships (2 to first-year students), 26 research assistantships (1 to a first-year student), 48 teaching assistantships (16 to first-year students) were awarded; federal work-study, institutionally sponsored loans also available. Average monthly stipend for a graduate assistantship: $1085-1200. Financial aid application deadline: 1/15. *Faculty research:* Analysis, topology, algebraic geometry, logic, Lie theory. Total annual research budget: $2.8-million. • David J. Benney, Head, 617-253-6976. Application contact: Phyllis Ruby, Graduate Administrator, 617-253-2689.

McGill University, Faculty of Graduate Studies and Research, Department of Mathematics, Montreal, PQ H3A 2T5, Canada. Department awards MA, M Sc, PhD. Matriculated students: 57. 91 applicants, 44% accepted. In 1990, 3 master's, 6 doctorates awarded. *Degree requirements:* For master's, thesis; for doctorate, 2 foreign languages, dissertation. *Application deadline:* 4/1 (priority date, applications processed on a rolling basis). *Application fee:* $25. *Tuition:* $698.50 per semester full-time, $46.57 per credit part-time for Canadian residents; $3480 per semester full-time, $234 per semester part-time for nonresidents. • K. P. Russell, Chairman, 514-398-7373.

McMaster University, Faculty of Science, Department of Mathematics and Statistics, Hamilton, ON L8S 4L8, Canada. Department awards M Sc, PhD. *Degree requirements:* For master's, thesis or alternative required, foreign language not required; for doctorate, 2 foreign languages, dissertation, comprehensive exam. *Expenses:* Tuition of $2250 per year full-time, $810 per semester part-time for Canadian residents; $10,340 per year full-time, $4050 per semester part-time for nonresidents. Fees of $76 per year full-time, $49 per semester part-time.

McNeese State University, College of Science, Department of Mathematics, Computer Science, and Statistics, Lake Charles, LA 70609. Offerings include mathematics (MS). Department faculty: 10 full-time (2 women), 0 part-time. *Degree requirements:* Computer language, thesis or alternative, written exam required, foreign language not required. *Entrance requirements:* GRE General Test. Application deadline: 7/15 (priority date, applications processed on a rolling basis). Application fee: $10 ($25 for foreign students). *Tuition:* $808 per semester full-time, $254 per course (minimum) part-time for state residents; $1583 per semester full-time, $254 per course (minimum) part-time for nonresidents. • Dr. George F. Mead Jr., Head, 318-475-5788.

Memorial University of Newfoundland, School of Graduate Studies, Department of Mathematics and Statistics, St. John's, NF A1C 5S7, Canada. Department awards MAS, M Phil, M Sc, PhD. Programs offered in mathematics (M Phil, M Sc, PhD), statistics (M Phil, M Sc, PhD). *Degree requirements:* For master's, thesis not required; for doctorate, dissertation.

Memphis State University, College of Arts and Sciences, Department of Mathematical Sciences, Memphis, TN 38152. Department offers programs in applied mathematics (MS), applied statistics (PhD), college teaching of mathematics (PhD), computer sciences (MS), general mathematics (MS), statistics (MS). Faculty: 28 full-time (3 women), 0 part-time. Matriculated students: 66 full-time (25 women), 29 part-time (10 women); includes 8 minority (4 Asian American, 4 black American), 55 foreign. In 1990, 12 master's, 2 doctorates awarded. *Degree requirements:* For master's, comprehensive exams required, thesis not required; for doctorate, 1 foreign language, dissertation. *Entrance requirements:* For master's, MAT (minimum score of 27 required) or GRE General Test (minimum combined score of 800 required), TOEFL (minimum score of 550 required), minimum GPA of 2.5; for doctorate, GRE General Test (minimum combined score of 1000 required). Application deadline: 8/1 (applications processed on a rolling basis). Application fee: $5. *Expenses:* Tuition of $92 per credit hour for state residents; $239 per credit hour for nonresidents. Fees of $45 per year for state residents. *Financial aid:* In 1990–91, 5 research assistantships, 25 teaching assistantships awarded. *Faculty research:* Research in approximation theory, graph theory, neural networks, differential equations, operator theory. • Dr. Ralph Faudree, Chairman, 901-678-2482. Application contact: Dr. R. H. Schlep, Coordinator, Graduate Studies, 901-678-2495.

Miami University, College of Arts and Sciences, Department of Mathematics and Statistics, Program in Mathematics, Oxford, OH 45056. Offers mathematics (MA, MAT, MS), mathematics/operations research (MS), statistics (MS). Part-time programs available. Faculty: 24. Matriculated students: 14 full-time (6 women), 5 part-time (3 women); includes 0 minority, 4 foreign. 41 applicants, 56% accepted. In 1990, 11 degrees awarded. *Degree requirements:* Final exam required, thesis not required. *Entrance requirements:* Minimum undergraduate GPA of 2.75 or 3.0 during previous 2 years. Application deadline: 3/1. Application fee: $30. *Expenses:* Tuition of $3196 per year full-time, $133 per semester (minimum) part-time for state residents; $7134 per year full-time, $312 per semester (minimum) part-time for nonresidents. Fees of $371 per year full-time, $165.50 per semester hour part-time. *Financial aid:* In 1990–91, $131,534 in aid awarded. Fellowships, research assistantships, teaching assistantships, full tuition waivers, federal work-study available. Financial aid application deadline: 3/1. • Application contact: Robert Schaefer, Director of Graduate Studies, 513-529-3527.

Announcement: Programs are offered leading to the following degrees: MA in mathematics, MAT in mathematics, MS in statistics, MS in mathematics, and MS in mathematics with option in operations research. Among the features of the program are the Statistical Consulting Center and computing facilities in classrooms and Computing Lab.

Michigan State University, College of Natural Science, Department of Mathematics, East Lansing, MI 48824. Department offers programs in applied mathematics (MS, PhD), computational mathematics (MS), mathematics (MA, MAT, MS, PhD), mathematics education (PhD). Faculty: 64 full-time (6 women), 0 part-time. Matriculated students: 128 full-time (29 women), 23 part-time (11 women); includes 5 minority (2 Asian American, 3 black American), 100 foreign. In 1990, 21 master's, 6 doctorates awarded. *Degree requirements:* For master's, foreign language and thesis not required; for doctorate, 2 foreign languages, dissertation. *Application deadline:* rolling. *Application fee:* $25 ($40 for foreign students). *Tuition:* $104.75 per credit for state residents; $211.75 per credit for nonresidents. *Financial aid:* In 1990–91, 1 research assistantship, 138 teaching assistantships awarded. • Dr. Jacob Plotkin, Acting Chairperson, 517-353-8484.

See full description on page 427.

Michigan Technological University, College of Sciences and Arts, Department of Mathematical Sciences, Houghton, MI 49931. Department offers program in mathematics (MS). Part-time programs available. Faculty: 26 full-time (2 women), 2 part-time (0 women). Matriculated students: 14 full-time (6 women), 3 part-time (1 woman); includes 0 minority, 3 foreign. Average age 29. 72 applicants, 33% accepted. In 1990, 9 degrees awarded. *Degree requirements:* Thesis or project required, foreign language not required. *Entrance requirements:* TOEFL (minimum score of 520 required). Application deadline: rolling. Application fee: $20. *Expenses:* Tuition of $852 per semester full-time, $71 per credit part-time for state residents; $2065 per semester full-time, $172 per credit part-time for nonresidents. Fees of $75 per semester. *Financial aid:* In 1990–91, 15 students received a total of $108,000 in aid awarded. 0 fellowships, 2 research assistantships (0 to first-year students), 13 teaching assistantships (7 to first-year students) were awarded; federal work-study also available. Aid available to part-time students. Average monthly stipend for a graduate assistantship: $800. Financial aid application deadline: 4/1; applicants required to submit FAF. *Faculty research:* Algebraic structures, applied statistics, combinatories, fluids/diffenerntial equations, signal processing. Total annual research budget: $61,182. • Dr. Alphonse H. Baartmans, Head, 906-487-2068. Application contact: Dr. Clark Givens, Graduate Director, 906-487-2068.

See full description on page 429.

Middle Tennessee State University, School of Basic and Applied Sciences, Department of Mathematics, Murfreesboro, TN 37132. Department offers programs in mathematics (MS), mathematics education (MST). Faculty: 14 full-time (1 woman), 0 part-time. Matriculated students: 4 full-time (all women), 31 part-time (19 women); includes 1 minority (black American), 1 foreign. Average age 30. 17 applicants, 100% accepted. In 1990, 7 degrees awarded. *Degree requirements:* 1 foreign language, comprehensive exams required, thesis not required. *Application deadline:* 8/1 (priority date). *Application fee:* $5. *Expenses:* Tuition of $1760 per year full-time, $89 per semester hour part-time for state residents; $4964 per year full-time, $229 per semester hour part-time for nonresidents. Fees of $15 per year full-time, $5 per year part-time. *Financial aid:* In 1990–91, 5 teaching assistantships (1 to a first-year student) awarded. • Dr. Harold Spraker, Chairman, 615-898-2669.

Mississippi State University, College of Arts and Sciences, Department of Mathematics and Statistics, Mississippi State, MS 39762. Offerings include mathematical sciences (PhD), mathematics (MA, MS). Terminal master's awarded for partial completion of doctoral program. Department faculty: 25 full-time (2 women), 0 part-time. *Degree requirements:* For master's, thesis optional, foreign language not required; for doctorate, variable foreign language requirement, computer language, dissertation. *Entrance requirements:* For master's, TOEFL; for doctorate, GRE, TOEFL. Application deadline: 4/15 (priority date, applications processed on a rolling basis). *Expenses:* Tuition of $891 per semester full-time, $99 per credit hour part-time for state residents; $1622 per semester full-time, $180 per credit hour part-time for nonresidents. Fees of $221 per semester full-time, $25 per semester (minimum) part-time. • Dr. J. L. Solomon, Head, 601-325-3414. Application contact: Dr. Paul W. Spikes, Graduate Coordinator, 601-325-3414.

Montana State University, College of Letters and Science, Department of Mathematical Sciences, 901 West Garfield Street, Bozeman, MT 59717. Department offers programs in mathematics (MS, PhD), statistics (MS, PhD). Faculty: 23 full-time (1 woman), 0 part-time. Matriculated students: 37 full-time (9 women), 9 part-time (3 women); includes 1 minority (Asian American), 3 foreign. Average age 31. 26 applicants, 73% accepted. In 1990, 16 master's, 3 doctorates awarded. *Degree requirements:* For master's, computer language, thesis or alternative required, foreign language not required; for doctorate, variable foreign language requirement, computer language, dissertation. *Entrance requirements:* GRE General Test, TOEFL (minimum score of 550 required). Application deadline: 6/15 (applications processed on a rolling basis). Application fee: $20. *Tuition:* $1150 per year full-time, $74 per credit (minimum) part-time for state residents; $2824 per year full-time, $100 per credit (minimum) part-time for nonresidents. *Financial aid:* Teaching assistantships and career-related internships or fieldwork available. Financial aid application deadline: 3/1. *Faculty research:* Numerical analysis, dynamical systems, mathematics education, multivariate analysis, biostatistical methodology. • Dr. Kenneth J. Tiahrt, Head, 406-994-3601.

Montclair State College, School of Mathematical and Natural Sciences, Department of Mathematics and Computer Science, Programs in Mathematics, Upper Montclair, NJ 07043. Offerings in computer science (MA), mathematics education (MA), pure and applied mathematics (MA), statistics (MA). Part-time and evening/weekend programs available. Faculty: 33 full-time, 0 part-time. In 1990, 5 degrees awarded. *Degree requirements:* Written comprehensive exam required, foreign language and thesis not required. *Entrance requirements:* GRE General Test, minimum GPA of 2.67. Application deadline: 7/1 (priority date, applications processed on a rolling basis). Application fee: $25. *Expenses:* Tuition of $130 per credit for state residents; $165 per credit for nonresidents. Fees of $276 per year. *Financial aid:* Application deadline 3/15. • Dr. Helen Roberts, Adviser, 201-893-7262.

Morgan State University, College of Arts and Sciences, Department of Mathematics, Baltimore, MD 21239. Department awards MA. Part-time and evening/weekend programs available. Matriculated students: 1 full-time (0 women), 4 part-time (0 women); includes 1 minority (black American), 2 foreign. In 1990, 0 degrees awarded. *Degree requirements:* Thesis, comprehensive exam required, foreign language not required. *Application deadline:* 7/1 (applications processed on a rolling basis). *Application fee:* $0. *Expenses:* Tuition of $85 per credit hour for state residents; $95 per credit hour for nonresidents. Fees of $98 per semester. *Financial aid:* Application deadline 4/1. • Dr. Nathaniel Knox, Chairman, 301-444-3240.

Murray State University, College of Sciences, Department of Mathematics, Murray, KY 42071. Department awards MA, MAT, MS. Part-time programs available. Faculty: 11 full-time (0 women), 0 part-time. Matriculated students: 3 full-time (0 women), 6 part-time (4 women); includes 0 minority, 0 foreign. 4 applicants, 75% accepted. In 1990, 1 degree awarded. *Degree requirements:* Thesis required (for some programs), foreign language not required. *Entrance requirements:* GRE General Test, TOEFL (minimum score of 500 required). Application fee: $0 ($20 for foreign students). *Tuition:* $775 per semester full-time, $87 per credit hour part-time for state residents; $2215 per semester full-time, $246 per credit hour part-time for nonresidents. *Financial aid:* Research assistantships, teaching assistantships, federal work-study available. Average monthly stipend for a graduate assistantship: $335. Financial aid application deadline: 4/1; applicants required to submit GAPSFAS. • Dr. Donald Bennett, Chairman, 502-762-2311.

Naval Postgraduate School, Department of Mathematics, Monterey, CA 93943. Department awards MS, PhD. Program only open to commissioned officers of the United States and friendly nations and selected United States federal civilian employees. *Degree requirements:* Thesis/dissertation. *Tuition:* $0.

New Mexico Highlands University, School of Science and Technology, Department of Computer Science and Mathematics, Las Vegas, NM 87701. Department awards MS. Program being phased out; applicants no longer accepted. Matriculated students: 0 full-time, 1 part-time (0 women). In 1990, 0 degrees awarded. *Expenses:* Tuition of $1248 per year full-time, $52 per credit hour part-time for state residents; $4488 per year full-time, $187 per credit hour part-time for nonresidents. Fees of $20 per year. • Dr. Charles Searcy, Head, 505-425-7511.

New Mexico Institute of Mining and Technology, Department of Mathematics, Socorro, NM 87801. Department offers programs in mathematics (MS), operations research (MS). Faculty: 8 full-time (1 woman), 0 part-time. Matriculated students: 11 full-time (2 women), 0 part-time; includes 0 minority, 5 foreign. Average age 25. 30 applicants, 70% accepted. In 1990, 8 degrees awarded (100% found work related to degree). *Degree requirements:* Thesis optional, foreign language not required. *Entrance requirements:* GRE General Test, TOEFL (minimum score of 540 required). Application deadline: 6/1 (priority date, applications processed on a rolling basis). Application fee: $16. *Expenses:* Tuition of $617 per semester full-time, $51 per hour part-time for state residents; $2656 per semester full-time, $209 per hour part-time for nonresidents. Fees of $207 per semester. *Financial aid:* In 1990–91, 10 students received a total of $37,571 in aid awarded. 0 research assistantships, 10 teaching assistantships (5 to first-year students) were awarded; fellowships, federal work-study, institutionally sponsored loans also available. Average monthly stipend for a graduate assistantship: $600. Financial aid application deadline: 3/15; applicants required to submit GAPSFAS or FAF. *Faculty research:* Abstract algebra, applied analysis, probability and statistics, stochastic processes, combinatorics. Total annual research budget: $50,000. • Dr. Dave Arterburn, Chairman, 505-835-5110. Application contact: Dr. J. A. Smoake, Dean, Graduate Studies, 505-835-5513.

New Mexico State University, College of Arts and Sciences, Department of Mathematical Sciences, Las Cruces, NM 88003. Department awards MS, PhD. Part-time programs available. Faculty: 28 full-time, 0 part-time. Matriculated students: 25 full-time (9 women), 15 part-time (2 women); includes 25 foreign. In 1990, 3 master's, 6 doctorates awarded. *Entrance requirements:* GRE General Test. Application deadline: 7/1. Application fee: $10. *Tuition:* $1608 per year full-time, $67 per credit hour part-time for state residents; $5304 per year full-time, $221 per credit hour part-time for nonresidents. *Financial aid:* In 1990–91, 25 teaching assistantships awarded; fellowships, research assistantships also available. *Faculty research:* Algebraic k-theory, harmonic analysis, functional analysis, algebraic topology. • Dr. Carol Walker, Head, 505-646-3901.

New York University, Graduate School of Arts and Science, Courant Institute of Mathematical Sciences, Department of Mathematics, New York, NY 10012. Department offers programs in mathematics (MS, PhD), mathematics and statistics/operations research (MS). Part-time and evening/weekend programs available. Faculty: 51 full-time, 0 part-time. Matriculated students: 89 full-time, 75 part-time; includes 10 minority (9 Asian American, 1 Hispanic American), 49 foreign. 297 applicants, 64% accepted. In 1990, 30 master's, 24 doctorates awarded. *Degree requirements:* For master's, thesis or alternative required, foreign language not required; for doctorate, 1 foreign language, dissertation, written and oral exams. *Entrance requirements:* GRE General Test, GRE Subject Test, TOEFL. Application deadline: 1/15 (priority date, applications processed on a rolling basis). Application fee: $30. *Tuition:* $467 per credit. *Financial aid:* In 1990–91, $842,814 in aid awarded. 5 fellowships (0 to first-year students), 34 research assistantships (10 to first-year students), 22 teaching assistantships (6 to first-year students) were awarded; full and partial tuition waivers, federal work-study also available. Aid available to part-time students. Average monthly stipend for a graduate assistantship: $1000. Financial aid application deadline: 1/15; applicants required to submit GAPSFAS or FAF. *Faculty research:* Wave propagation, computational fluid dynamics, numerical analysis, magneto fluid dynamics. • Harold Weitzner, Chairman, 212-998-3297. Application contact: W. Stephen Childress, Director of Graduate Studies, 212-998-3000.

See full description on page 433.

Nicholls State University, College of Arts and Sciences, Department of Mathematics, Thibodaux, LA 70310. Department offers programs in applied mathematics (MS), mathematics (MS). Part-time and evening/weekend programs available. Faculty: 0 full-time, 6 part-time (1 woman). Matriculated students: 3 full-time (2 women), 3 part-time (all women); includes 0 minority, 0 foreign. In 1990, 5 degrees awarded (100% found work related to degree). *Degree requirements:* Foreign language and thesis not required. *Entrance requirements:* GRE General Test. Application fee: $10. *Expenses:* Tuition of $1345 per year full-time, $337 per semester part-time for state residents; $3145 per year full-time, $337 per semester part-time for nonresidents. Fees of $246 per year full-time, $99 per semester part-time. *Faculty research:* Operations research, statistics, numerical analysis. • Dr. Donald M. Bardwell, Head, 504-448-4380.

North Carolina Central University, Division of Academic Affairs, College of Arts and Sciences, Department of Mathematics, Durham, NC 27707. Department awards MS. Part-time and evening/weekend programs available. Faculty: 8 full-time (1 woman), 0 part-time. Matriculated students: 1 full-time (0 women), 10 part-time (8 women); includes 10 minority (all black American), 0 foreign. Average age 23. 2 applicants, 100% accepted. In 1990, 4 degrees awarded (75% found work related to degree, 25% continued full-time study). *Degree requirements:* 1 foreign language (computer language can substitute), thesis, comprehensive exam. *Entrance requirements:* Minimum cumulative GPA of 2.5, 3.0 in major. Application deadline: 8/1. Application fee: $15. *Financial aid:* Fellowships, teaching assistantships, federal work-study, institutionally sponsored loans available. Aid available to part-time students. Financial aid application deadline: 5/1; applicants required to submit GAPSFAS or FAF. *Faculty research:* Structure of algebras with radicals; algorithms for finding eigenvalues of large, randomly sparse, symmetric matrices; nonlinear methods of optimizations with applications. • Dr. William T. Fletcher, Chairperson, 919-560-6315. Application contact: Dr. Mary M. Townes, Dean, College of Arts and Sciences, 919-560-6368.

North Carolina State University, College of Physical and Mathematical Sciences, Department of Mathematics, Raleigh, NC 27695. Department offers programs in applied mathematics (MS, PhD), mathematics (MS, PhD). Part-time programs available. Faculty: 56 full-time (3 women), 4 part-time (0 women). Matriculated students: 82 full-time (21 women), 11 part-time (4 women); includes 8 minority (6 black American, 2 Hispanic American), 27 foreign. Average age 24. 121 applicants, 56% accepted. In 1990, 14 master's awarded (7% entered university research/teaching, 57% found other work related to degree, 36% continued full-time study); 9 doctorates awarded (89% entered university research/teaching, 11% found other work related to degree). Terminal master's awarded for partial completion of doctoral program. *Degree requirements:* For master's, computer language required, thesis optional, foreign language not required; for doctorate, 1 foreign language, dissertation. *Entrance requirements:* GRE General Test, GRE Subject Test. Application deadline: 4/1 (priority date, applications processed on a rolling basis). Application fee: $35. *Tuition:* $1138 per year for state residents; $5805 per year for nonresidents. *Financial aid:* In 1990–91, 75 students received a total of $750,000 in aid awarded. 17 fellowships (4 to first-year students), 7 research assistantships (1 to a first-year student), 70 teaching assistantships (25 to first-year students) were awarded; institutionally sponsored loans and career-related internships or fieldwork also available. Average monthly stipend for a graduate assistantship: $1100. Financial aid application deadline: 4/1. *Faculty research:* Ring theory, numerical analysis, differential equations, linear algebra. Total annual research budget: $528,804. • Dr. Robert H. Martin, Head, 919-737-3798. Application contact: John E. Franke, Administrator, 919-737-2382.

North Dakota State University, College of Science and Mathematics, Mathematics Department, Fargo, ND 58105. Department awards MS, PhD. Faculty: 13 full-time (0 women), 1 part-time (0 women). Matriculated students: 11 full-time (1 woman), 0 part-time; includes 0 minority, 2 foreign. Average age 28. 14 applicants, 93% accepted. In 1990, 2 master's awarded (100% entered university research/teaching). *Degree requirements:* For master's, thesis or alternative required, foreign language not required; for doctorate, 2 foreign languages (computer language can substitute for one), dissertation. *Entrance requirements:* GRE, TOEFL (minimum score of 600 required). Application fee: $20. *Tuition:* $1411 per year full-time, $52 per credit hour part-time for state residents; $3571 per year full-time, $132 per credit hour part-time for nonresidents. *Financial aid:* In 1990–91, $51,262 in aid awarded. 1 fellowship, 1 research assistantship (0 to first-year students), 6 teaching assistantships (2 to first-year students) were awarded; federal work-study, institutionally sponsored loans also available. Financial aid application deadline: 4/15. *Faculty research:* Differential equations, discrete mathematics, number theory, ergodic theory, applied mathematics. • Dr. James H. Olsen, Chair, 701-237-8183.

Northeastern Illinois University, College of Arts and Sciences, Department of Mathematics, Chicago, IL 60625. Department awards MA, MS. Evening/weekend programs available. Faculty: 13 full-time (3 women), 0 part-time. 23 applicants, 87% accepted. In 1990, 8 degrees awarded. *Degree requirements:* Thesis or alternative, comprehensive exam required, foreign language not required. *Application deadline:* rolling. *Application fee:* $0. *Expenses:* Tuition of $70 per credit hour for state residents; $210 per credit hour for nonresidents. Fees of $37 per credit hour. • Dr. Joo Koo, Chairperson, 312-583-4050 Ext. 731.

Northeastern University, Graduate School of Arts and Sciences, Department of Mathematics, Boston, MA 02115. Department awards MS, PhD. Faculty: 46 full-time (6 women), 0 part-time. Matriculated students: 60 full-time (15 women), 13 part-time (2 women); includes 3 minority (2 Asian American, 1 black American), 30 foreign. 105 applicants, 75% accepted. In 1990, 7 master's, 4 doctorates awarded. *Degree requirements:* For doctorate, dissertation. *Entrance requirements:* TOEFL. Application fee: $40. *Expenses:* Tuition of $10,260 per year full-time, $285 per quarter hour part-time. Fees of $293 per year full-time. *Financial aid:* In 1990–91, 1 research assistantship, 27 teaching assistantships (8 to first-year students), 14 tuition assistantships (2 to first-year students) awarded. *Faculty research:* Algebraic and differential geometry combinations, topology and probability. Total annual research budget: $346,246. • Margaret Cozzens, Chairperson, 617-437-2450.

Northeast Missouri State University, Division of Mathematics, Kirksville, MO 63501. Division awards MA. Faculty: 12 full-time (1 woman), 0 part-time. Matriculated students: 2 full-time, 0 part-time. In 1990, 3 degrees awarded. *Degree requirements:* Comprehensive exam required, foreign language and thesis not required. *Entrance requirements:* GRE General Test, GRE Subject Test. Application deadline: 6/15. *Tuition:* $1458 per year full-time, $81 per credit part-time for state residents; $2844 per year full-time, $158 per credit part-time for nonresidents. *Financial aid:* In 1990–91, 1 research assistantship, 1 teaching assistantship (to a first-year student) awarded; federal work-study also available. Financial aid application deadline: 3/15. • Dr. Lanny Morley, Head, 816-785-4547.

Northern Arizona University, College of Arts and Sciences, Department of Mathematics, Flagstaff, AZ 86011. Department awards MAT, MS. Part-time programs available. *Degree requirements:* Departmental qualifying exam required, foreign language and thesis not required. *Application deadline:* 3/15 (priority date, applications processed on a rolling basis). *Application fee:* $0 ($25 for foreign students). *Expenses:* Tuition of $1528 per year full-time, $80 per credit part-time for state residents; $6180 per year full-time, $258 per credit part-time for nonresidents. Fees of $25 per semester. *Financial aid:* In 1990–91, 7 teaching assistantships (4 to first-year students) awarded; full and partial tuition waivers, federal work-study also available. *Faculty research:* Topology, statistics, groups, ring theory, number theory. • Dr. John Hagood, Chairman, 602-523-3482.

Northern Illinois University, College of Liberal Arts and Sciences, Department of Mathematical Sciences, De Kalb, IL 60115. Department offers programs in applied probability and statistics (MS), mathematical sciences (PhD), mathematics (MS). Part-time programs available. Faculty: 53 full-time (9 women), 2 part-time (0 women). Matriculated students: 42 full-time (14 women), 22 part-time (15 women); includes 6 minority (2 Asian American, 3 black American, 1 Hispanic American), 18 foreign. Average age 30. 137 applicants, 31% accepted. In 1990, 11 master's, 0 doctorates awarded. Terminal master's awarded for partial completion of doctoral program. *Degree requirements:* For master's, comprehensive exam required, thesis optional, foreign language not required; for doctorate, 2 foreign languages, computer language, candidacy exam, dissertation defense. *Entrance requirements:* For master's, GRE General Test, TOEFL, minimum GPA of 2.75; for doctorate, GRE General Test, TOEFL, minimum graduate GPA of 3.2. Application deadline: 6/1. Application fee: $0. *Tuition:* $1339 per semester full-time for state residents; $3163 per semester full-time for nonresidents. *Financial aid:* In 1990–91, 1 research assistantship, 56 teaching assistantships awarded; fellowships, full tuition waivers, federal work-study, and career-related internships or fieldwork also available. Aid available to part-time students. Financial aid applicants required to submit FAF. *Total annual research budget:* $225,700. • Dr. William D. Blair, Chair, 815-753-0566. Application contact: Dr. Robert F. Wheeler, Director, Graduate Studies, 815-753-6772.

Northwestern University, College of Arts and Sciences, Department of Mathematics, Evanston, IL 60208. Department offers programs in applied mathematics (MA, MS, PhD), mathematics (MA, MS, PhD). Part-time programs available. Faculty: 33 full-time (3 women), 1 part-time (0 women). Matriculated students: 47 full-time (19 women), 0 part-time; includes 0 minority, 10 foreign. 108 applicants, 68% accepted. In 1990, 7 master's, 4 doctorates awarded. Terminal master's awarded for partial completion of doctoral program. *Degree requirements:* For master's, preliminary exam required, foreign language and thesis not required; for doctorate, dissertation, preliminary exam required, foreign language not required. *Entrance requirements:* GRE General Test. Application deadline: 8/30. Application fee: $40 ($45 for foreign students). *Tuition:* $4665 per quarter full-time, $1704 per course part-time. *Financial aid:* In 1990–91, $272,071 in aid awarded. 9 fellowships (all to first-year students), 0 research assistantships, 23 teaching assistantships were awarded; federal work-study, institutionally sponsored loans, and career-related internships or fieldwork also available. Financial aid application deadline: 1/15; applicants required to submit GAPSFAS. *Faculty research:* Topology, dynamical systems, partial differential equations, analysis, algebra. • Stewart Priddy, Chairman, 708-491-8035. Application contact: Clark Robinson, Chairman, Graduate Committee, 708-491-8035.

See full description on page 437.

Oakland University, College of Arts and Sciences, Department of Mathematical Sciences, Rochester, MI 48309. Offerings include mathematics (MA). *Application deadline:* 7/15. *Application fee:* $25. *Expenses:* Tuition of $122 per credit hour for state residents; $270 per credit hour for nonresidents. Fees of $170 per year. • Dr. Darryl Schmidt, Chair, 313-370-3430. Application contact: Dr. James H. McKay, Graduate Coordinator, 313-370-3430.

Ohio State University, College of Mathematical and Physical Sciences, Department of Mathematics, Columbus, OH 43210. Department awards MA, MS, PhD. Faculty: 105. Matriculated students: 174 full-time (29 women), 11 part-time (9 women); includes 10 minority (3 Asian American, 4 black American, 3 Hispanic American), 87 foreign. 72 applicants, 35% accepted. In 1990, 33 master's, 16 doctorates awarded. *Degree requirements:* For master's, thesis optional, foreign language not required; for doctorate, 2 foreign languages, dissertation. *Entrance requirements:* GRE General Test, GRE Subject Test. Application deadline: 8/15 (applications processed on a rolling basis). Application fee: $0 ($25 for foreign students). *Tuition:* $1213 per quarter full-time, $364 per course part-time for state residents; $3143 per quarter full-time, $943 per course part-time for nonresidents. *Financial aid:* Fellowships, research assistantships, teaching assistantships, administrative assistantships, federal work-study, institutionally sponsored loans available. Aid available to part-time students. Financial aid applicants required to submit FAF. • Dijen Ray-Chaudhuri, Chairman, 614-292-4975.

See full description on page 439.

Ohio University, Graduate Studies, College of Arts and Sciences, Department of Mathematics, Athens, OH 45701. Department awards MS, PhD. Faculty: 29 full-time (2 women), 5 part-time (3 women). Matriculated students: 95 full-time, 6 part-time; includes 1 minority (Asian American), 81 foreign. 172 applicants, 42% accepted. In 1990, 54 master's, 5 doctorates awarded. *Degree requirements:* For master's, thesis or alternative; for doctorate, 1 foreign language, dissertation, comprehensive exam. *Application fee:* $25. *Tuition:* $1112 per quarter hour full-time, $138 per credit hour part-time for state residents; $2227 per quarter hour full-time, $277 per credit hour part-time for nonresidents. *Financial aid:* In 1990–91, 44 students received a total of $281,300 in aid awarded. 4 fellowships, 36 teaching assistantships (11 to first-year students) were awarded; full and partial tuition waivers, federal work-study, institutionally sponsored loans also available. Average monthly stipend for a graduate assistantship: $870. Financial aid application deadline: 3/15; applicants required to submit FAF. • Dr. Shih Wen, Chairman, 614-593-1252. Application contact: Dr. M. S. K. Sastry, Graduate Chair, 614-593-1256.

Oklahoma State University, College of Arts and Sciences, Department of Mathematics, Stillwater, OK 74078. Department awards MS, PhD. Faculty: 22 full-time, 0 part-time. Matriculated students: 64 (20 women); includes 1 minority (Asian American), 24 foreign. In 1990, 6 master's, 0 doctorates awarded. *Degree requirements:* For master's, foreign language and thesis not required; for doctorate, 1 foreign language, dissertation. *Entrance requirements:* TOEFL (minimum score of 550 required). Application deadline: 7/1 (priority date). Application fee: $15. *Tuition:* $63.75 per credit for state residents; $138.25 per credit for nonresidents. *Financial aid:* In 1990–91, 4 research assistantships, 53 teaching assistantships awarded; partial tuition waivers, federal work-study, and career-related internships or fieldwork also available. Aid available to part-time students. Financial aid application deadline: 3/1; applicants required to submit GAPSFAS or FAF. • Dr. Joel Haack, Interim Head, 405-744-5688.

Old Dominion University, College of Sciences, Department of Mathematics and Statistics, Norfolk, VA 23529. Department offers program in computational and applied mathematics (MS, PhD), including applied mathematics, statistics. Part-time and evening/weekend programs available. Faculty: 24 full-time (0 women), 0 part-time. Matriculated students: 25 full-time (10 women), 23 part-time (11 women); includes 7 minority (1 Asian American, 4 black American, 2 Hispanic American), 19 foreign. Average age 27. 36 applicants, 86% accepted. In 1990, 6 master's awarded (17% entered university research/teaching, 83% found other work related to degree); 9 doctorates awarded (100% entered university research/teaching). Terminal master's awarded for partial completion of doctoral program. *Degree requirements:* For master's, comprehensive exam required, foreign language and thesis not required; for doctorate, dissertation, candidacy exam required, foreign language not required. *Entrance requirements:* For master's, GRE General Test, GRE Subject Test, TOEFL, minimum QPA of 3.0 in major, minimum QPA of 2.5 overall; for doctorate, GRE General Test (minimum combined score of 1000 required), GRE Subject Test (minimum score of 600 required), TOEFL. Application deadline: 7/1 (applications processed on a rolling basis). Application fee: $20. *Expenses:* Tuition of $148 per credit hour for state residents; $375 per credit hour for nonresidents. Fees of $64 per year full-time. *Financial aid:* In 1990–91, $28,800 in aid awarded. 3 research assistantships, 2 teaching assistantships were awarded; fellowships also available. *Faculty research:* Computational mathematics, numerical analysis, continuum mechanics, integral equations. Total annual research budget: $240,000. • Dr. John Tweed, Chairperson, 804-683-3882. Application contact: Dr. John J. Swetits, Director, 804-683-3911.

See full description on page 441.

Oregon State University, Graduate School, College of Science, Department of Mathematics, Corvallis, OR 97331. Department awards MA, MAIS, MS, PhD. Faculty: 35 full-time (2 women), 8 part-time (0 women). Matriculated students: 54 full-time (15 women), 0 part-time; includes 1 minority (Hispanic American), 19 foreign. In 1990, 8 master's awarded; 1 doctorate awarded (100% entered university research/teaching). Terminal master's awarded for partial completion of doctoral program. *Degree requirements:* For master's, variable foreign language requirement, thesis or alternative; for doctorate, 2 foreign languages, dissertation. *Entrance requirements:* For master's, TOEFL (minimum score of 550 required), minimum GPA of 3.0 in last 90 hours; for doctorate, TOEFL (minimum score of 550 required), minimum GPA pf 3.0 in last 90 hours. Application deadline: 3/1. Application fee: $40. *Tuition:* $1140 per trimester full-time, $449 per year part-time for state residents; $1816 per trimester full-time, $674 per year part-time for nonresidents. *Financial aid:* In 1990–91, $200,000 in aid awarded. 6 research assistantships (3 to first-year students), 45 teaching assistantships (20 to first-year students) were awarded; federal work-study, institutionally sponsored loans, and career-related internships or fieldwork also available. Financial aid application deadline: 2/1; applicants required to submit FAF. *Faculty research:* Analysis, geometry, topology, algebra, applied mathematics. • Dr. Francis J. Flaherty, Chair, 503-737-4686. Application contact: Dr. Robert M. Burton Jr., Director of Graduate Studies, 503-737-4686.

See full description on page 443.

Pennsylvania State University Great Valley Graduate Center, Graduate Studies and Continuing Education, Program in Mathematics, Malvern, PA 19355. Program awards M Ed. Matriculated students: 0. In 1990, 1 degree awarded. *Application fee:* $35. *Tuition:* $203 per credit for state residents; $403 per credit for nonresidents. • Dr. John Mulhern, Chair, 215-889-1300.

Pennsylvania State University University Park Campus, College of Science, Department of Mathematics, University Park, PA 16802. Department offers program in mathematics (MA, M Ed, D Ed, PhD), including applied mathematics (MA, PhD). Faculty: 52. Matriculated students: 80 full-time (20 women), 19 part-time (8 women). In 1990, 13 master's, 7 doctorates awarded. *Entrance requirements:* GRE General Test. Application fee: $35. *Tuition:* $203 per credit for state residents; $403 per credit for nonresidents. • Dr. Richard H. Herman, Head, 814-865-7527. Application contact: Joel Anderson, Professor, 814-865-1123.

Pittsburg State University, College of Arts and Sciences, Department of Mathematics, Pittsburg, KS 66762. Department awards MA, MS. Faculty: 5 full-time (0 women), 0 part-time. Matriculated students: 7 full-time (3 women), 11 part-time (3 women); includes 0 minority, 4 foreign. In 1990, 6 degrees awarded. *Degree* *requirements:* Thesis or alternative required, foreign language not required. *Application fee:* $35. *Tuition:* $1670 per year full-time, $54 per credit hour part-time for state residents; $4108 per year full-time, $125 per credit hour part-time for nonresidents. *Financial aid:* In 1990–91, 5 teaching assistantships (4 to first-year students) awarded; federal work-study and career-related internships or fieldwork also available. *Faculty research:* Operations research, numerical analysis, applied analysis, applied algebra. • Dr. Elwyn Davis, Chairman, 316-231-7000 Ext. 4401.

Plymouth State College of the University System of New Hampshire, Department of Mathematics, Plymouth, NH 03264. Department offers program in mathematics education (M Ed). Z applicants. *Degree requirements:* Comprehensive exam required, thesis optional, foreign language not required. *Entrance requirements:* MAT or GRE General Test. *Tuition:* $130 per credit for state residents; $143 per credit for nonresidents. • Dr. Bill Roberts, Chairman, 603-535-2433. Application contact: Richard Evans, Coordinator, 603-535z2487.

Polytechnic University, Brooklyn Campus, Division of Arts and Sciences, Department of Mathematics, 333 Jay Street, Brooklyn, NY 11201. Department offers programs in applied statistics (MS, PhD), industrial and applied mathematics (MS, PhD). Evening/weekend programs available. *Tuition:* $6000 per semester.

Portland State University, Systems Science Program, Portland, OR 97207. Offerings include systems science/mathematics (PhD). Program faculty: 3 full-time (0 women), 0 part-time. *Degree requirements:* Variable foreign language requirement, computer language, dissertation. *Entrance requirements:* GMAT (score in 75th percentile or higher required), GRE General Test (score in 75th percentile or higher required), TOEFL (minimum score of 575 required), minimum GPA of 3.0 (undergraduate). Application deadline: 4/1 (applications processed on a rolling basis). Application fee: $40. *Tuition:* $1151 per trimester for state residents; $1827 per trimester for nonresidents. • Dr. Roy Koch, Coordinator, 503-725-4960.

Portland State University, College of Liberal Arts and Sciences, Department of Mathematical Sciences, Portland, OR 97207. Department awards MA, MAT, MS, MST, PhD. PhD offered in conjunction with Systems Science Program. Part-time and evening/weekend programs available. Faculty: 29 full-time (7 women), 9 part-time (3 women). Matriculated students: 29 full-time (13 women), 30 part-time (16 women); includes 5 minority (all Asian American), 5 foreign. Average age 31. 58 applicants, 97% accepted. In 1990, 9 master's awarded. *Degree requirements:* For master's, variable foreign language requirement, thesis or alternative, exams; for doctorate, 2 foreign languages, dissertation, exams. *Entrance requirements:* For master's, TOEFL (minimum score of 550 required), minimum GPA of 3.0 in upper-division course work or 2.75 overall; for doctorate, GRE General Test. Application deadline: 7/12 (priority date, applications processed on a rolling basis). Application fee: $40. *Tuition:* $1151 per trimester for state residents; $1827 per trimester for nonresidents. *Financial aid:* In 1990–91, 24 teaching assistantships (14 to first-year students) awarded; federal work-study, institutionally sponsored loans also available. Aid available to part-time students. Average monthly stipend for a graduate assistantship: $610. Financial aid application deadline: 3/1. *Faculty research:* Cohomology of groups, law dimension topology and knot theory, control theory, statistical robustness, statistical distribution theory. • Dr. Bruce Jensen, Head, 503-725-3621. Application contact: Gavin Bjork, Graduate Adviser, 503-725-3621.

Prairie View A&M University, College of Arts and Sciences, Department of Mathematics, Prairie View, TX 77446. Department offers program in). Faculty: 9 full-time (3 women), 1 (woman) part-time. Matriculated students: 12 full-time (5 women), 17 part-time (9 women); includes 16 minority (all black American), 10 foreign. 15 applicants, 80% accepted. In 1990, 3 degrees awarded (67% entered university research/teaching). *Degree requirements:* Thesis required, foreign language not required. *Entrance requirements:* GRE General Test. Application deadline: rolling. Application fee: $0. *Expenses:* Tuition of $162 per semester full-time, $144 per semester part-time for state residents; $1098 per semester full-time, $976 per semester part-time for nonresidents. Fees of $133 per year full-time, $125 per year part-time. *Financial aid:* In 1990–91, $50,490 in aid awarded. 10 teaching assistantships (3 to first-year students) were awarded; federal work-study also available. Aid available to part-time students. Average monthly stipend for a graduate assistantship: $561. Financial aid applicants required to submit FAF. *Faculty research:* Approximation theory, differential equations, algebraic topology, number theory complex analysis, operations research. • Dr. Evelyn E. Thornton, Head, 409-857-2026. Application contact: Dr. G. A. Roberts, Graduate Adviser, 409-857-3807.

Princeton University, Department of Mathematics, Princeton, NJ 08544. Department offers programs in applied and computational mathematics (PhD), mathematical physics (PhD), mathematics (PhD). Faculty: 42 full-time (1 woman), 0 part-time. Matriculated students: 51 full-time (8 women), 0 part-time; includes 6 minority (4 Asian American, 2 Hispanic American), 27 foreign. 166 applicants, 17% accepted. In 1990, 10 doctorates awarded (90% entered university research/teaching). Terminal master's awarded for partial completion of doctoral program. *Degree requirements:* For doctorate, 2 foreign languages, dissertation. *Entrance requirements:* For doctorate, GRE General Test, GRE Subject Test. Application deadline: 1/8. Application fee: $45 ($50 for foreign students). *Tuition:* $16,670 per year. *Financial aid:* Fellowships, research assistantships, teaching assistantships, federal work-study, institutionally sponsored loans available. Financial aid application deadline: 1/8. • Nicholas M. Katz, Director of Graduate Studies, 609-258-4200. Application contact: Michele Spreen, Director of Graduate Admissions, 609-258-3034.

Purdue University, School of Science, Department of Mathematics, West Lafayette, IN 47907. Department awards MAT, MS, PhD. Faculty: 72 full-time (3 women), 9 part-time (2 women). Matriculated students: 177 full-time (50 women), 2 part-time (1 woman); includes 6 minority (3 Asian American, 1 black American, 2 Hispanic American), 128 foreign. 326 applicants, 28% accepted. In 1990, 29 master's, 9 doctorates awarded. Terminal master's awarded for partial completion of doctoral program. *Degree requirements:* For master's, foreign language and thesis not required; for doctorate, 2 foreign languages, dissertation, written and oral exams. *Entrance requirements:* TOEFL (minimum score of 550 required). Application deadline: 3/15. Application fee: $25. *Tuition:* $1162 per semester full-time, $83.50 per credit hour part-time for state residents; $3720 per semester full-time, $244.50 per credit hour part-time for nonresidents. *Financial aid:* In 1990–91, 10 fellowships (4 to first-year students), 10 research assistantships (0 to first-year students), 150 teaching assistantships (45 to first-year students) awarded. Financial aid application deadline: 3/15. *Faculty research:* Algebra, analysis, topology, differential equations, applied mathematics. • Dr. Joseph Lipman, Head, 317-494-1908. Application contact: Dr. L. M. Lipshitz, Graduate Committee Chairman, 317-494-1961.

See full description on page 447.

Queens College of the City University of New York, Mathematics and Natural Sciences Division, Department of Mathematics, 65-30 Kissena Boulevard, Flushing, NY 11367. Department awards MA, MS Ed. MS Ed awarded through the School of Education. Part-time and evening/weekend programs available. Matriculated students: 0 full-time, 17 part-time (5 women). 24 applicants, 54% accepted. In 1990, 5 degrees awarded. *Degree requirements:* Comprehensive exams required, foreign language and thesis not required. *Entrance requirements:* TOEFL (minimum score of 550 required), minimum GPA of 3.0. Application deadline: 4/1 (applications processed on a rolling basis). Application fee: $35. *Tuition:* $2700 per year full-time, $45 per credit part-time for state residents; $4700 per year full-time, $199 per credit part-time for nonresidents. *Financial aid:* In 1990–91, 1 fellowship (0 to first-year students) awarded; partial tuition waivers, federal work-study, institutionally sponsored loans, and career-related internships or fieldwork also available. Aid available to part-time students. Financial aid application deadline: 4/1. *Faculty research:* Topology, differential equations, combinatorics. • Dr. Gerald Roskes, Chairperson, 718-997-5800. Application contact: Dr. Nick Metas, Graduate Adviser, 718-997-5800.

Queen's University at Kingston, Faculty of Arts and Sciences, Department of Mathematics and Statistics, Kingston, ON K7L 3N6, Canada. Department offers programs in mathematics (MA, M Sc, PhD), statistics (MA, M Sc, PhD). Part-time programs available. Matriculated students: 32 full-time, 5 part-time; includes 10 foreign. In 1990, 7 master's, 3 doctorates awarded. *Degree requirements:* For master's, thesis required, foreign language not required; for doctorate, 2 foreign languages, dissertation, comprehensive exam. *Entrance requirements:* TOEFL (minimum score of 550 required). Application deadline: 2/28 (priority date). Application fee: $35. *Tuition:* $2861 per year full-time, $426 per trimester part-time for Canadian residents; $10,613 per year full-time, $4998 per trimester part-time for nonresidents. *Financial aid:* Fellowships, research assistantships, teaching assistantships, institutionally sponsored loans available. Financial aid application deadline: 3/1. *Faculty research:* Algebra, analysis, applied mathematics, combinatorics, differential geometry. • Dr. L. L. Campbell, Head, 613-545-6147. Application contact: Dr. O. A. Nielsen, Graduate Coordinator, 613-545-2397.

Rensselaer Polytechnic Institute, School of Science, Department of Mathematical Sciences, Troy, NY 12180. Department offers programs in applied mathematics (MS, PhD), mathematics (MS, PhD). Faculty: 24 full-time (5 women), 1 (woman) part-time. Matriculated students: 71 full-time (17 women), 0 part-time; includes 3 minority (all Asian American), 13 foreign. 136 applicants, 50% accepted. In 1990, 18 master's, 5 doctorates awarded. *Degree requirements:* For master's, thesis optional; for doctorate, 1 foreign language, dissertation. *Entrance requirements:* GRE General Test, TOEFL. Application deadline: 2/1. Application fee: $30. *Expenses:* Tuition of $455 per credit hour. Fees of $195.57 per semester. *Financial aid:* In 1990–91, 51 students received a total of $840,000 in aid awarded. 15 fellowships (10 to first-year students), 6 research assistantships (0 to first-year students), 29 teaching assistantships (11 to first-year students) were awarded; career-related internships or fieldwork also available. *Faculty research:* Physical mathematics, modeling, applied analysis, optimization, applied geometry. Total annual research budget: $1.3-million. • Joseph G. Ecker, Chairman, 518-276-6346. Application contact: Joyce McLaughlin, Graduate Coordinator, 518-276-6349.

See full description on page 453.

Rhode Island College, Faculty of Arts and Sciences, Department of Mathematics, Providence, RI 02908. Department awards MA, MAT, CAGS. Evening/weekend programs available. Faculty: 24 full-time (5 women), 0 part-time. Matriculated students: 6 full-time (0 women), 8 part-time (3 women); includes 0 minority, 0 foreign. In 1990, 4 master's awarded. *Degree requirements:* For master's, foreign language and thesis not required; for CAGS, thesis required, foreign language not required. *Entrance requirements:* For master's, GRE or MAT. Application deadline: 4/1 (applications processed on a rolling basis). Application fee: $25. *Expenses:* Tuition of $2400 per year full-time, $100 per credit part-time for state residents; $4848 per year full-time, $202 per credit part-time for nonresidents. Fees of $167 per year full-time, $35 per semester (minimum) part-time. *Financial aid:* Career-related internships or fieldwork available. • Dr. James Sedlock, Chair, 401-456-8038.

Rice University, Wiess School of Natural Sciences, Department of Mathematics, Houston, TX 77251. Department awards MA, PhD. Faculty: 14 full-time, 1 part-time. Matriculated students: 16 full-time (1 woman), 0 part-time; includes 1 minority (Asian American), 7 foreign. *Degree requirements:* For master's, 1 foreign language, thesis; for doctorate, 2 foreign languages, dissertation. *Entrance requirements:* For master's, GRE General Test, TOEFL (minimum score Of 550 required), minimum GPA of 3.0; for doctorate, GRE General Test (score in 70th percentile or higher required), minimum GPA of 3.0. Application deadline: 3/1. Application fee: $0. *Expenses:* Tuition of $8300 per year full-time, $400 per credit hour part-time. Fees of $167 per year. *Financial aid:* Fellowships, research assistantships, full and partial tuition waivers available. Financial aid applicants required to submit FAF. *Faculty research:* Geometry, topology. • John P. Hempel, Chairman, 713-527-4829.

Rivier College, Department of Computer Science and Mathematics, Nashua, NH 03060. Department offers programs in computer science (MS), information science (MS), mathematics (MS). Evening/weekend programs available. Faculty: 1 full-time, 11 part-time. Matriculated students: 14 full-time (6 women), 69 part-time (29 women); includes 8 minority (5 Asian American, 3 black American), 7 foreign. In 1990, 29 degrees awarded. *Degree requirements:* Foreign language and thesis not required. *Entrance requirements:* GRE Subject Test. Application fee: $25. *Expenses:* Tuition of $210 per credit. Fees of $20 per semester. • Dr. Stefan Ehrlich, Chairman, 603-888-1311.

Roosevelt University, College of Arts and Sciences, Department of Mathematical Sciences, Program in Mathematical Sciences, Chicago, IL 60605. Program awards MS. Part-time and evening/weekend programs available. *Degree requirements:* Foreign language and thesis not required. *Application deadline:* 8/1 (priority date, applications processed on a rolling basis). *Application fee:* $20 ($30 for foreign students). *Expenses:* Tuition of $302 per semester hour. Fees of $12 per year full-time, $6 per semester part-time. *Financial aid:* Application deadline 2/15. • Application contact: Joan R. Ritter, Coordinator of Graduate Admissions, 312-341-3612.

Rutgers, The State University of New Jersey, New Brunswick, Program in Mathematics, New Brunswick, NJ 08903. Offers applied mathematics (MS, PhD), pure mathematics (MS, PhD). Part-time and evening/weekend programs available. Faculty: 86 full-time (10 women), 2 part-time (0 women). Matriculated students: 104 full-time (25 women), 37 part-time (12 women); includes 9 minority (1 Asian American, 4 black American, 4 Hispanic American), 63 foreign. 220 applicants, 51% accepted. In 1990, 7 master's, 11 doctorates awarded. Terminal master's awarded for partial completion of doctoral program. *Degree requirements:* For master's, foreign language and thesis not required; for doctorate, 1 foreign language, dissertation. *Entrance requirements:* GRE General Test, GRE Subject Test. Application deadline: 5/1. Application fee: $35. *Expenses:* Tuition of $4432 per year full-time, $183 per credit part-time for state residents; $6496 per year full-time, $270 per credit part-time for nonresidents. Fees of $458 per year full-time, $117 per year part-time. *Financial aid:* In 1990–91, 20 fellowships (13 to first-year students), 12 research assistantships, 51 teaching assistantships (10 to first-year students) awarded. Financial aid application deadline: 3/1. *Faculty research:* Combinatorics, mathematical physics, Lie theory, algebraic and analytic number theory, non-linear analysis. Total annual research budget: $1.4-million. • Jose Barros-Neto, Director, 908-932-3921.

Announcement: There is no strict separation between pure and applied mathematics in graduate study at Rutgers. Notably active areas of study include combinatorics/discrete mathematics, mathematical physics, and control and systems theory in applied mathematics; nonlinear analysis, Lie theory, number theory, group theory, logic, and partial differential equations in pure mathematics. For the 2nd year in a row, the program has received a major grant from the US Department of Education. These grants support a large number of fellowships, with stipends from $12,000 to $17,000 (plus tuition). These fellowships add to other forms of support previously available. Applications from all qualified students are encouraged, especially women and minorities.

St. Cloud State University, College of Science and Technology, Department of Mathematics, St. Cloud, MN 56301-4498. Department awards MS. Faculty: 27 full-time (4 women), 0 part-time. Matriculated students: 1 (woman) full-time, 2 part-time (both women). 5 applicants. In 1990, 3 degrees awarded. *Degree requirements:* Thesis or alternative required, foreign language not required. *Entrance requirements:* GRE General Test, minimum GPA of 2.75. Application deadline: 6/1 (priority date, applications processed on a rolling basis). Application fee: $15. *Expenses:* Tuition of $57.20 per credit for state residents; $82.50 per credit for nonresidents. Fees of $158 per year. *Financial aid:* In 1990–91, 1 graduate assistantship awarded; federal work-study also available. Financial aid application deadline: 3/1; applicants required to submit FAF. • Dr. James Kepner, Chairperson, 612-255-3001. Application contact: Graduate Studies Office, 612-255-2113.

St. John's University, Graduate School of Arts and Sciences, Department of Mathematics and Computer Science, Jamaica, NY 11439. Department offers programs in algebra (MA), analysis (MA), applied mathematics (MA), computer science (MA), geometry-topology (MA), logic and foundations (MA), probability and statistics (MA). Faculty: 21 full-time (4 women), 0 part-time. Matriculated students: 4 full-time (1 woman), 11 part-time (4 women); includes 3 minority (all Asian American), 5 foreign. Average age 25. In 1990, 5 degrees awarded. *Degree requirements:* Variable foreign language requirement. *Entrance requirements:* GRE General Test. Application deadline: rolling. Application fee: $20. *Expenses:* Tuition of $297 per credit. Fees of $130 per year full-time, $65 per semester part-time. • Dr. Edward Miranda, Chairman, 718-990-6161.

Saint Louis University, College of Arts and Sciences, Department of Mathematics, St. Louis, MO 63103. Department awards MA, MA(R), PhD. Faculty: 14 full-time, 0 part-time. Matriculated students: 11 full-time (5 women), 3 part-time (0 women); includes 3 minority (all Asian American), 0 foreign. 10 applicants, 50% accepted. In 1990, 7 master's, 0 doctorates awarded. *Degree requirements:* For master's, thesis (for some programs), comprehensive oral exam; for doctorate, 1 foreign language, dissertation, preliminary degree exams. *Entrance requirements:* GRE General Test. Application deadline: rolling. Application fee: $30. *Tuition:* $342 per credit hour. *Financial aid:* In 1990–91, 11 teaching assistantships awarded. Financial aid application deadline: 4/1. *Faculty research:* Algebraic, geometric topology; differential geometry; group representation theory. • Dr. John Cantwell, Chairman, 314-658-2444. Application contact: Dr. Robert J. Nikolai, Associate Dean of the Graduate School, 314-658-2240.

Salem State College, Department of Mathematics, Salem, MA 01970. Department awards MAT, MS. Faculty: 0 full-time, 1 part-time (0 women). Matriculated students: 0 full-time, 6 part-time (2 women). In 1990, 2 degrees awarded. *Degree requirements:* Foreign language and thesis not required. *Entrance requirements:* GRE General Test. Application fee: $25. *Expenses:* Tuition of $90 per credit hour (minimum) for state residents; $110 per credit hour (minimum) for nonresidents. Fees of $45 per year. • Dr. Arthur J. Rosenthal, Coordinator, 508-741-6288.

Sam Houston State University, College of Arts and Sciences, Division of Mathematical and Information Sciences, Program in Mathematics, Huntsville, TX 77341. Program awards MA, M Ed, MS. Faculty: 10 full-time (0 women), 0 part-time. Matriculated students: 5 full-time (0 women), 14 part-time (9 women); includes 1 foreign. 50 applicants, 80% accepted. *Degree requirements:* Foreign language and thesis not required. *Entrance requirements:* GRE General Test (minimum combined score of 800 required), TOEFL (minimum score of 550 required), minimum B average in undergraduate mathematics course work. Application deadline: 4/30 (priority date, applications processed on a rolling basis). Application fee: $0. *Expenses:* Tuition of $432 per year full-time, $216 per year part-time for state residents; $2880 per year full-time, $1440 per year part-time for nonresidents. Fees of $364 per year full-time, $220 per year part-time. *Financial aid:* In 1990–91, 3 teaching assistantships (1 to a first-year student) awarded; institutionally sponsored loans also available. Aid available to part-time students. • Application contact: Dr. Ronald Stoltenberg, Graduate Coordinator, 409-294-1589.

San Diego State University, College of Sciences, Department of Mathematical Sciences, San Diego, CA 92182. Offerings include mathematics (MA). *Application deadline:* 8/1 (applications processed on a rolling basis). *Application fee:* $55. *Expenses:* Tuition of $0 for state residents; $189 per unit for nonresidents. Fees of $1974 per year full-time, $692 per year part-time for state residents; $1074 per year full-time, $692 per year part-time for nonresidents. • Dr. John Elwin, Chair, 619-594-6191.

San Francisco State University, School of Science, Department of Mathematics, San Francisco, CA 94132. Department awards MA. *Application deadline:* 11/30 (priority date, applications processed on a rolling basis). *Application fee:* $55. *Expenses:* Tuition of $0 for state residents; $250 per unit for nonresidents. Fees of $950 per year full-time, $350 per semester part-time for state residents; $1050 per year full-time, $350 per semester part-time for nonresidents. *Financial aid:* Application deadline 3/1. • Dr. Newman Fisher, Chair, 415-338-2251.

Sangamon State University, School of Liberal Arts and Sciences, Program in Mathematical Sciences, Springfield, IL 62794-9243. Offerings include mathematics (MA). Program faculty: 7 full-time (2 women), 1 (woman) part-time. *Degree requirements:* Foreign language and thesis not required. *Entrance requirements:* Proficiency in calculus and programming, BA in mathematics or computer science. Application fee: $0. *Expenses:* Tuition of $69.50 per credit. Fees of $42 per year. • Mary Kate Yntema, Director, 217-786-6770.

San Jose State University, School of Science, Department of Mathematics and Computer Science, San Jose, CA 95192. Offerings include mathematics (MA, MS). Department faculty: 51 full-time (5 women), 3 part-time (0 women). *Degree requirements:* Thesis (for some programs), comprehensive exam required, foreign language not required. *Entrance requirements:* GRE Subject Test. Application deadline: 6/1 (applications processed on a rolling basis). Application fee: $55. *Expenses:* Tuition of $0 for state residents; $246 per unit for nonresidents. Fees of $592 per year. • Dr. Veril Phillips, Chair, 408-924-5100. Application contact: Dr. Howard Swann, Graduate Adviser, 408-924-5156.

Seton Hall University, College of Arts and Sciences, Department of Mathematics, South Orange, NJ 07079. Department offers program in pure mathematics and operations research (MA, MS). Part-time and evening/weekend programs available. Faculty: 16 full-time (3 women), 0 part-time. Matriculated students: 5 full-time (4 women), 3 part-time (1 woman); includes 2 minority (1 Asian American, 1 Hispanic American). 12 applicants, 75% accepted. In 1990, 3 degrees awarded (100% found work related to degree). *Degree requirements:* 1 foreign language required (computer language can substitute), thesis not required. *Entrance requirements:* GRE General Test, GRE Subject Test. Application fee: $30. *Expenses:* Tuition of $346 per credit. Fees of $105 per semester. *Financial aid:* In 1990–91, 2 students received a total of $8400 in aid awarded. 2 teaching assistantships (0 to first-year students) were awarded. Average monthly stipend for a graduate assistantship: $420. Financial aid application deadline: 5/1; applicants required to submit FAF. • Dr. Daniel Gross, Chairman, 201-761-9466.

Shippensburg University of Pennsylvania, College of Arts and Sciences, Department of Mathematics, Shippensburg, PA 17257. Department awards M Ed, MS. Faculty: 2 full-time (1 woman), 0 part-time. Matriculated students: 10 full-time (0 women), 20 part-time (9 women); includes 0 minority, 1 foreign. 100% of applicants accepted. In 1990, 6 degrees awarded. *Degree requirements:* Foreign language and thesis not required. *Entrance requirements:* Minimum GPA of 2.5 or GRE General Test. Application fee: $20. *Expenses:* Tuition of $1314 per semester full-time, $146 per credit part-time for state residents; $1677 per semester full-time, $186 per credit part-time for nonresidents. Fees of $240 per year full-time, $58 per semester (minimum) part-time. *Financial aid:* Application deadline 3/1. • Dr. Howard Bell, Chairperson, 717-532-1431.

Simon Fraser University, Faculty of Science, Department of Mathematics and Statistics, Burnaby, BC V5A 1S6, Canada. Department offers programs in applied mathematics (M Sc, PhD), pure mathematics (M Sc, PhD), statistics (M Sc, PhD). Faculty: 35 (2 women). Matriculated students: 60 full-time (13 women), 0 part-time. In 1990, 7 master's, 5 doctorates awarded. *Degree requirements:* For master's, thesis; for doctorate, dissertation, comprehensive exams. *Entrance requirements:* GRE Subject Test, TOEFL (minimum score of 600 required). Application fee: $0. *Expenses:* Tuition of $612 per trimester full-time, $306 per trimester part-time. Fees of $68 per trimester full-time, $34 per trimester part-time. *Financial aid:* In 1990–91, 18 fellowships awarded; research assistantships, teaching assistantships also available. *Faculty research:* Semi-groups, lattice ordered groups, summability, functional analysis, graph theory. • Dr. A. R. Freedman, Chairman, 604-291-3378.
See full description on page 455.

South Dakota State University, College of Engineering, Department of Mathematics, Brookings, SD 57007. Department awards MS, MST. *Degree requirements:* Thesis optional, foreign language not required. *Tuition:* $553 per semester full-time, $61 per credit part-time for state residents; $1094 per semester full-time, $122 per credit part-time for nonresidents.

Announcement: MS in mathematics with thesis and nonthesis options. Faculty of 16, 9 with the doctorate. Areas include analysis, algebra, number theory, numerical analysis, topology, and statistics. Small program provides opportunity for close working relationship with faculty. Graduate assistants receive a stipend of $600–$900 per month and pay one third of resident tuition.

Southeast Missouri State University, Department of Mathematics, Cape Girardeau, MO 63701. Department awards MNS. Faculty: 11 full-time (1 woman), 0 part-time. Matriculated students: 20 full-time, 0 part-time; includes 0 minority, 2 foreign. Average age 26. In 1990, 2 degrees awarded. *Degree requirements:* Thesis or alternative. *Application deadline:* 3/1 (applications processed on a rolling basis). *Application fee:* $10. *Expenses:* Tuition of $1860 per year full-time, $75 per credit part-time for state residents; $3420 per year full-time, $140 per credit part-time for nonresidents. Fees of $25 per year. *Financial aid:* In 1990–91, 4 teaching assistantships (2 to first-year students) awarded. • Dr. Harold Hager, Chairperson.

Southern Illinois University at Carbondale, College of Science, Department of Mathematics, Carbondale, IL 62901. Department offers programs in mathematics (MA, MS, PhD), statistics (MS). PhD offered jointly with Southeast Missouri State University. Part-time programs available. Faculty: 37 full-time (3 women), 0 part-time. Matriculated students: 40 full-time (11 women), 12 part-time (4 women); includes 3 minority (1 Asian American, 1 black American, 1 Hispanic American), 20 foreign. Average age 26. 81 applicants, 30% accepted. In 1990, 7 master's, 1 doctorate awarded. *Degree requirements:* For master's, 1 foreign language (computer language can substitute), thesis or alternative; for doctorate, 2 foreign languages (computer language can substitute for one), dissertation. *Entrance requirements:* For master's, TOEFL (minimum score of 550 required), minimum GPA of 2.7; for doctorate, TOEFL (minimum score of 550 required), minimum GPA of 3.25. Application deadline: rolling. Application fee: $0. *Expenses:* Tuition of $1638 per year full-time, $204.75 per semester hour part-time for state residents; $4914 per year full-time, $614.25 per semester hour part-time for nonresidents. Fees of $700 per year full-time, $216 per year part-time. *Financial aid:* In 1990–91, $265,470 in aid awarded. 6 fellowships (3 to first-year students), 2 research assistantships (0 to first-year students), 23 teaching assistantships (9 to first-year students) were awarded; full tuition waivers, federal work-study, institutionally sponsored loans also available. Aid available to part-time students. *Faculty research:* Ordinary differential equations, partial differential equations, stochastic process, combinatorics. Total annual research budget: $163,345. • Ronald Kirk, Chairperson, 618-453-5302. Application contact: Dr. Marvin Zeman, Coordinator, 618-453-5302.

Southern Illinois University at Edwardsville, School of Sciences, Department of Mathematics and Statistics, Edwardsville, IL 62026. Department awards MS. Part-time programs available. Faculty: 27 full-time (5 women), 9 part-time (3 women). Matriculated students: 19 full-time (8 women), 34 part-time (10 women); includes 7 minority (2 Asian American, 2 black American, 2 Hispanic American, 1 Native American), 12 foreign. In 1990, 17 degrees awarded. *Degree requirements:* Computer language, thesis or alternative, final exam required, foreign language not required. *Entrance requirements:* Undergraduate major in field. Application fee: $0. *Expenses:* Tuition of $1566 per year full-time, $43.40 per credit hour part-time for state residents; $4698 per year full-time, $130.20 per credit hour part-time for nonresidents. Fees of $291.75 per year full-time, $27.35 per year (minimum) part-time. *Financial aid:* In 1990–91, 1 fellowship, 1 research assistantship, 21 teaching assistantships, 1 assistantship awarded; federal work-study, institutionally sponsored loans also available. Aid available to part-time students. • Dr. Chung-wu Ho, Chairman, 618-692-2385.

Southern Methodist University, Dedman College, Department of Mathematics, Dallas, TX 75275. Department offers programs in applied mathematics (MS), mathematical sciences (PhD), mathematics (MA). Part-time programs available. Faculty: 16 full-time (1 woman), 0 part-time. Matriculated students: 10 full-time (3 women), 10 part-time (2 women); includes 6 minority (3 Asian American, 2 black American, 1 Hispanic American), 3 foreign. Average age 27. 40 applicants, 25% accepted. In 1990, 6 master's, 2 doctorates awarded. *Degree requirements:* For master's, foreign language and thesis not required; for doctorate, dissertation required, foreign language not required. *Application deadline:* 6/30. *Application fee:* $25. *Expenses:* Tuition of $435 per credit. Fees of $664 per semester for state residents; $56 per year for nonresidents. *Financial aid:* In 1990–91, 1 research assistantship (0 to first-year students), 14 teaching assistantships (4 to first-year students) awarded. *Faculty research:* Numerical analysis, scientific computation, biomedical modeling, nonlinear waves, perturbation methods. Total annual research budget: $106,000. • Dr. Ian Gladwell, Chairman, 214-692-2506. Application contact: Dr. Richard Haberman, Graduate Adviser, 214-692-2506.

Southern University and Agricultural and Mechanical College, College of Sciences, Program in Mathematics, Baton Rouge, LA 70813. Program awards MS. Faculty: 11 full-time (3 women), 2 part-time (0 women). Matriculated students: 5 full-time (2 women), 7 part-time (4 women); includes 10 minority (all black American), 2 foreign. Average age 25. 11 applicants, 82% accepted. In 1990, 5 degrees awarded. *Entrance requirements:* GRE General Test or GMAT, TOEFL. Application deadline: 7/1. Application fee: $5. *Tuition:* $1594 per year full-time, $299 per semester part-time for state residents; $3010 per year full-time, $299 per semester part-time for nonresidents. • Dr. Lovenia CeConge, Chairperson, 504-771-5180.

Southwest Missouri State University, College of Science and Mathematics, Springfield, MO 65804. College awards MA, MS. Part-time programs available. Matriculated students: 39 full-time (17 women), 18 part-time (10 women); includes 3 minority (1 Asian American, 1 black American, 1 Hispanic American). In 1990, 18 degrees awarded. *Degree requirements:* Thesis or alternative, comprehensive exam required, foreign language not required. *Application deadline:* 8/7 (priority date, applications processed on a rolling basis). *Application fee:* $25. *Expenses:* Tuition of $67 per credit hour (minimum). Fees of $53 per year. *Financial aid:* Fellowships, research assistantships, teaching assistantships, federal work-study, and career-related internships or fieldwork available. Aid available to part-time students. • Jerry Berlin, Dean, 417-836-5249.

Southwest Texas State University, School of Science, Department of Mathematics, San Marcos, TX 78666. Department awards MA, M Ed, MS. Part-time programs available. Faculty: 24 full-time (0 women), 0 part-time. Matriculated students: 8 full-time (2 women), 18 part-time (9 women); includes 4 minority (3 Asian American, 1 Hispanic American), 2 foreign. Average age 30. 14 applicants, 71% accepted. In 1990, 6 degrees awarded. *Degree requirements:* Thesis (for some programs), comprehensive exam required, foreign language not required. *Entrance requirements:* GRE General Test (minimum combined score of 900 required), TOEFL (minimum score of 550 required), minimum GPA of 2.75 in last 60 hours. Application deadline: 7/15 (priority date, applications processed on a rolling basis). Application fee: $0 ($50 for foreign students). *Expenses:* Tuition of $180 per semester (minimum) full-time, $100 per semester (minimum) part-time for state residents; $1152 per semester (minimum) full-time, $128 per semester (minimum) part-time for nonresidents. Fees of $217 per semester (minimum) full-time, $73 per semester (minimum) part-time. *Financial aid:* In 1990–91, 28 teaching assistantships (16 to first-year students) awarded; federal work-study also available. Aid available to part-time students. Financial aid application deadline: 3/1. *Faculty research:* Differential equation, topology, number theory, evolution equation/epidemiology, mathematics education. Total annual research budget: $40,000. • Dr. Stanley G. Wayment Jr., Chair, 512-245-2551. Application contact: Dr. John A. Chatfield, Graduate Adviser, 512-245-2551.

Stanford University, School of Humanities and Sciences, Department of Mathematics, Stanford, CA 94305. Department awards MAT, MS, PhD. Faculty: 34 (0 women). Matriculated students: 62 full-time (11 women), 0 part-time; includes 10 minority (8 Asian American, 2 Hispanic American), 23 foreign. Average age 25. 149 applicants, 26% accepted. In 1990, 13 master's, 7 doctorates awarded. Terminal master's awarded for partial completion of doctoral program. *Degree requirements:* For master's, foreign language and thesis not required; for doctorate, 2 foreign languages, dissertation, oral exam. *Entrance requirements:* For master's, GRE, TOEFL; for doctorate, GRE General Test, GRE Subject Test, TOEFL. Application deadline: 1/1. Application fee: $55. *Expenses:* Tuition of $15,102 per year. Fees of $28 per quarter. *Financial aid:* Fellowships, research assistantships, teaching assistantships, federal work-study, institutionally sponsored loans available. Financial aid application deadline: 1/1; applicants required to submit GAPSFAS or FAF. • Application contact: Administrator, 415-723-2601.

State University of New York at Albany, College of Science and Mathematics, Department of Mathematics and Statistics, Albany, NY 12222. Department offers programs in actuarial sciences (MA), secondary teaching (MA), statistics (MA, PhD). Evening/weekend programs available. Faculty: 32 full-time (1 woman), 1 part-time (0 women). Matriculated students: 43 full-time (9 women), 23 part-time (7 women); includes 4 minority (1 Asian American, 1 black American, 2 Hispanic American), 10 foreign. Average age 25. In 1990, 20 master's, 3 doctorates awarded. *Degree requirements:* For master's, foreign language and thesis not required; for doctorate, 1 foreign language, dissertation. *Entrance requirements:* For master's, GRE; for doctorate, GRE General Test, GRE Subject Test. *Expenses:* Tuition of $2450 per year full-time, $103 per credit part-time for state residents; $5765 per year full-time, $243 per credit part-time for nonresidents. Fees of $25 per year full-time, $0.85 per credit part-time. *Financial aid:* In 1990–91, 1 fellowship, 29 teaching assistantships awarded; research assistantships, minority assistantships also available. • Timothy Lance, Chairman, 518-442-4602.

State University of New York at Binghamton, School of Arts and Sciences, Department of Mathematical Sciences, Binghamton, NY 13902-6000. Department offers program in mathematical sciences (MA, PhD), including computer science, probability and statistics. Matriculated students: 42 full-time (12 women), 19 part-time (4 women); includes 0 minority, 13 foreign. Average age 29. 79 applicants, 70% accepted. Terminal master's awarded for partial completion of doctoral program. *Degree requirements:* For master's, thesis or alternative; for doctorate, 2 foreign languages, dissertation. *Entrance requirements:* GRE General Test, GRE Subject Test, TOEFL. Application deadline: 4/15 (priority date). Application fee: $35. *Expenses:* Tuition of $2450 per year full-time, $102.50 per credit part-time for state residents; $5766 per year full-time, $242.50 per credit part-time for nonresidents. Fees of $77 per year full-time, $27.85 per semester (minimum) part-time. *Financial aid:* In 1990–91, 41 students received a total of $309,156 in aid awarded. 1 fellowship (to a first-year student), 1 research assistantship, 27 teaching assistantships (11 to first-year students), 12 graduate assistantships (3 to first-year students) were awarded; federal work-study, institutionally sponsored loans, and career-related internships or fieldwork also available. Aid available to part-time students. Average monthly stipend for a graduate assistantship: $754. Financial aid application deadline: 2/15. • Dr. David L. Hanson, Chairperson, 607-777-2147.

State University of New York at Buffalo, Graduate School, Faculty of Natural Sciences and Mathematics, Department of Mathematics, Buffalo, NY 14260. Department awards MA, PhD. Part-time programs available. Faculty: 41 full-time (2 women), 0 part-time. Matriculated students: 53 full-time (13 women), 27 part-time (4 women); includes 1 minority (Hispanic American), 43 foreign. Average age 28. 368 applicants, 18% accepted. In 1990, 13 master's awarded; 4 doctorates awarded (100% entered university research/teaching). Terminal master's awarded for partial completion of doctoral program. *Degree requirements:* For master's, thesis, project or comprehensive exam required, foreign language not required; for doctorate, 2 foreign languages, dissertation, exams. *Entrance requirements:* TOEFL (minimum score of 550 required). Application deadline: 2/1 (applications processed on a rolling basis). Application fee: $35. *Expenses:* Tuition of $1600 per semester full-time, $134 per hour part-time for state residents; $3258 per semester full-time, $274 per hour part-time for nonresidents. Fees of $137 per semester full-time, $115 per semester (minimum) part-time. *Financial aid:* In 1990–91, $556,977 in aid awarded. 0 research assistantships, 33 teaching assistantships (16 to first-year students), 18 graduate assistantships (5 to first-year students) were awarded; fellowships, federal work-study, institutionally sponsored loans also available. Financial aid application deadline: 2/1; applicants required to submit FAF. *Faculty research:* Applied mathematics, logic, analysis, algebra, topology. Total annual research budget: $271,300. • Dr. Lewis A. Coburn, Chairman, 716-831-2148. Application contact: Dr. Ching Chou, Director of Graduate Studies, 716-831-2146.

State University of New York at Stony Brook, College of Arts and Sciences, Division of Physical Sciences and Mathematics, Department of Mathematics, Stony Brook, NY 11794. Department awards MA, M Phil, PhD. Faculty: 40 full-time, 7 part-time. Matriculated students: 66 full-time (17 women), 21 part-time (10 women); includes 4 minority (3 Asian American, 1 Hispanic American), 36 foreign. 163 applicants, 36% accepted. In 1990, 20 master's awarded; 5 doctorates awarded (100% entered university research/teaching). *Degree requirements:* For master's, foreign language and thesis not required; for doctorate, 2 foreign languages, dissertation. *Entrance requirements:* GRE General Test, TOEFL. Application deadline: 2/1. Application fee: $35. *Expenses:* Tuition of $2450 per year full-time, $103 per credit part-time for state residents; $5766 per year full-time, $243 per credit part-time for nonresidents. Fees of $151 per year full-time, $10.45 per year (minimum) part-time. *Financial aid:* In 1990–91, 8 fellowships, 6 research assistantships, 48 teaching assistantships awarded. *Faculty research:* Real analysis, relativity and mathematical physics, complex analysis, topology, combinatorics. Total annual research budget: $633,108. • Blaine Lawson, Chairman, 516-632-8290.

See full description on page 457.

State University of New York College at Brockport, School of Letters and Sciences, Department of Mathematics, Brockport, NY 14420. Department offers program in mathematics (MA). Part-time and evening/weekend programs available. Faculty: 14 full-time (3 women), 10 part-time (6 women). Matriculated students: 11 full-time (6 women), 18 part-time (11 women); includes 0 minority, 1 foreign. Average age 29. 20 applicants, 90% accepted. In 1990, 2 degrees awarded (100% found work related to degree). *Degree requirements:* 1 foreign language (computer language can substitute), thesis or alternative, comprehensive exam. *Entrance requirements:* Minimum GPA of 3.0. Application fee: $35. *Expenses:* Tuition of $1225 per semester full-time, $103 per credit hour part-time for state residents; $2883 per semester full-time, $243 per credit hour part-time for nonresidents. Fees of $62.50 per semester full-time, $5.05 per credit hour part-time. *Financial aid:* In 1990–91, $15,000 in aid awarded. 3 teaching assistantships (all to first-year students) were awarded; federal work-study also available. Aid available to part-time students. Financial aid application deadline: 4/1; applicants required to submit FAF. *Faculty research:* Applications of computers, secondary school teacher retraining, complex analysis statistics, combinatorics. Total annual research budget: $400,000. • Dr. Norman Bloch, Chairperson, 716-395-2194.

Announcement: Flexible program leading to MA in mathematics. Three required courses in algebra, analysis, statistics. Elect additional 4 courses in mathematics or computer science. Remaining 3 courses may be chosen from other departments, including education. Several assistantships available, which include a stipend of $5000 and a full tuition waiver.

State University of New York College at Fredonia, Department of Mathematics, Fredonia, NY 14063. Department awards MA, MS Ed. Faculty: 5 full-time (0 women), 0 part-time. Matriculated students: 0 full-time, 2 part-time (both women); includes 0 minority, 0 foreign. 0 applicants. In 1990, 0 degrees awarded. *Degree requirements:* Thesis or alternative required, foreign language not required. *Application deadline:* 7/5. *Application fee:* $35. *Tuition:* $1600 per semester full-time, $134 per credit part-time for state residents; $3288 per semester full-time, $274 per credit part-time for nonresidents. *Financial aid:* In 1990–91, 0 research assistantships, 0 teaching assistantships awarded; full and partial tuition waivers also available. Aid available to part-time students. Financial aid application deadline: 3/15; applicants required to submit FAF. • Dr. Albert Polimemi, Chairman, 716-673-3243.

State University of New York College at New Paltz, Faculty of Liberal Arts and Sciences, Department of Mathematics and Computer Science, Program in Mathematics, New Paltz, NY 12561. Program awards MA, MS Ed. Matriculated students: 0 full-time, 3 part-time (2 women); includes 1 minority (Asian American), 0 foreign. In 1990, 2 degrees awarded. *Degree requirements:* Thesis (MA), comprehensive exam (MS Ed) required, foreign language not required. *Entrance requirements:* GRE General Test, minimum GPA of 3.0. Application deadline: 4/1 (priority date, applications processed on a rolling basis). Application fee: $35. *Tuition:* $1600 per semester full-time, $134 per credit part-time for state residents; $3258 per semester full-time, $274 per credit part-time for nonresidents. *Financial aid:* Teaching assistantships, full tuition waivers, federal work-study, institutionally sponsored loans available. Financial aid applicants required to submit FAF. • Lawrence Filkow, Chairman, Department of Mathematics and Computer Science, 914-257-3535. Application contact: H. P. Sankappanavar, Graduate Adviser, 914-257-3535.

State University of New York College at Potsdam, School of Liberal Studies, Department of Mathematics, Potsdam, NY 13676. Department awards MA. Part-time and evening/weekend programs available. Faculty: 5 full-time (1 woman), 1 part-time (0 women). Matriculated students: 0 full-time, 1 (woman) part-time; includes 1 minority (Asian American), 0 foreign. 3 applicants, 67% accepted. In 1990, 3 degrees awarded. *Degree requirements:* Comprehensive exam required, foreign language and thesis not required. *Entrance requirements:* Minimum GPA of 2.75 in last 60 hours of undergraduate course work. Application deadline: rolling. Application fee: $35. *Tuition:* $1600 per semester full-time, $134 per credit hour part-time for state residents; $3258 per semester full-time, $274 per credit hour part-time for nonresidents. *Financial aid:* Federal work-study available. Aid available to part-time students. Financial aid application deadline: 3/1; applicants required to submit GAPSFAS or FAF. • Dr. Vasily Cateforis, Chairperson, 315-267-2064. Application contact: Dr. Conrad A. Bautz, Dean of Professional and Graduate Studies, 315-267-2515.

Stephen F. Austin State University, School of Sciences and Mathematics, Department of Mathematics and Statistics, Nacogdoches, TX 75962. Department offers programs in mathematics (MS), mathematics education (MS), statistics (MS). Faculty: 14 full-time (1 woman), 0 part-time. Matriculated students: 11 full-time (4 women), 7 part-time (2 women); includes 1 minority (black American), 0 foreign. In 1990, 6 degrees awarded. *Degree requirements:* Comprehensive exam required, thesis optional, foreign language not required. *Entrance requirements:* GRE General Test, minimum GPA of 2.5 overall, 2.8 in last half of major. Application deadline: 8/1. Application fee: $0 ($25 for foreign students). *Expenses:* Tuition of $18 per semester hour for state residents; $122 per semester hour for nonresidents. Fees of $14 per semester hour. *Financial aid:* In 1990–91, 12 students received a total of $56,000 in aid awarded. 12 teaching assistantships (10 to first-year students) were awarded. Average monthly stipend for a graduate assistantship: $667. *Faculty research:* Kernel type estimators, fractal mappings, spline curve fitting, robust regression continua theory. • Dr. Jasper Adams, Chairman, 409-568-3805.

Stevens Institute of Technology, Department of Pure and Applied Mathematics, Hoboken, NJ 07030. Department awards MS, PhD. Part-time and evening/weekend programs available. Terminal master's awarded for partial completion of doctoral program. *Degree requirements:* For master's, computer language required, thesis optional, foreign language not required; for doctorate, variable foreign language requirement, computer language, dissertation. *Entrance requirements:* GRE. Application fee: $25. *Tuition:* $4850 per semester full-time, $485 per credit part-time. *Faculty research:* Dynamical systems, numerical analysis, graph theory, algebraic geometry, applied mathematics.

Syracuse University, College of Arts and Sciences, Department of Mathematics, Syracuse, NY 13244. Department awards MS, PhD. Faculty: 32 full-time, 3 part-time. Matriculated students: 49 full-time (18 women), 8 part-time (2 women); includes 6 minority (2 Asian American, 2 black American, 2 Hispanic American), 37 foreign. Average age 29. 80 applicants, 51% accepted. In 1990, 21 master's, 3 doctorates awarded. Terminal master's awarded for partial completion of doctoral program. *Degree requirements:* For master's, foreign language and thesis not required; for doctorate, 2 foreign languages, dissertation, qualifying exam. *Entrance requirements:* GRE General Test, GRE Subject Test, TOEFL. Application deadline: rolling. Application fee: $40. *Expenses:* Tuition of $381 per credit. Fees of $289 per year full-time, $34.50 per semester part-time. *Financial aid:* In 1990–91, 67 teaching assistantships awarded; fellowships, research assistantships, partial tuition waivers, federal work-study also available. Financial aid application deadline: 3/1; applicants required to submit FAF. *Faculty research:* Pure mathematics, numerical mathematics, computing statistics. • Daniel Waterman, Chair, 315-443-1472. Application contact: Howard Johnson, Graduate Program Director, 315-443-2373.

Temple University, College of Arts and Sciences, Department of Mathematics, Philadelphia, PA 19122. Department awards MA, PhD. Faculty: 49 full-time (5 women), 0 part-time. Matriculated students: 61 (19 women); includes 7 minority (3 Asian American, 3 black American, 1 Native American), 23 foreign. Average age 31. 178 applicants, 20% accepted. In 1990, 17 master's, 4 doctorates awarded. *Degree requirements:* For master's, written exam required, foreign language and thesis not required; for doctorate, 2 foreign languages, dissertation, written and oral exams. *Entrance requirements:* GRE General Test (minimum combined score of 1000 required), GRE Subject Test, minimum GPA of 2.8 overall, 3.0 during previous 2 years. Application deadline: 8/1. Application fee: $30. *Tuition:* $224 per credit for state residents; $283 per credit for nonresidents. *Financial aid:* In 1990–91, $525,000 in aid awarded. 7 fellowships (3 to first-year students), 35 teaching assistantships (9 to first-year students) were awarded; full and partial tuition waivers, institutionally sponsored loans also available. Financial aid application deadline: 8/1. • Dr. R. Srinivasan, Chair, 215-787-7840. Application contact: Dr. Bruce Conrad, Graduate Chair, 215-787-7287.

Announcement: Master's (thesis and nonthesis options) and doctoral programs in pure mathematics, applied and computational mathematics. Flexible tracks with highly favorable faculty-student ratio. Numerous student-faculty seminars. Assistantships and fellowships at competitive levels with nearly all students receiving support. Faculty research interests include most major fields, with particular strength in harmonic analysis, approximation theory, partial differential equations, number theory, algebra, combinatorics, probability, and statistics. Excellent departmental library and computing facilities. Proximity of several other major universities enhances opportunities for interaction and collaboration. Write to Graduate Chair, Box P, Department of Mathematics.

Tennessee State University, College of Arts and Sciences, Department of Physics, Mathematics and Computer Science, Nashville, TN 37209-1561. Department offers program in mathematics (MS). Part-time and evening/weekend programs available. Faculty: 9 full-time (2 women), 0 part-time. Matriculated students: 9 full-time (3 women), 13 part-time (4 women); includes 5 minority (all black American), 5 foreign. Average age 28. 25 applicants, 88% accepted. In 1990, 7 degrees awarded (14% entered university research/teaching, 14% found other work related to degree, 42% continued full-time study). *Degree requirements:* Computer language, thesis, comprehensive exam required, foreign language not required. *Entrance requirements:* GRE. Application deadline: 8/1 (priority date, applications processed on a rolling basis). Application fee: $5. *Tuition:* $1814 per year full-time, $92 per semester hour (minimum) part-time for state residents; $5018 per year full-time, $232 per semester hour (minimum) part-time for nonresidents. *Financial aid:* In 1990–91, 7 students received a total of $15,000 in aid awarded. 6 graduate assistantships (3 to first-year students) were awarded; federal work-study, institutionally sponsored loans also available. Average monthly stipend for a graduate assistantship: $400. Financial aid application deadline: 8/1; applicants required to submit FAF. *Faculty research:* Chaos theory, semi-coherent light scattering, lattices of topologies, hulls of semigroups, ramsey theory. Total annual research budget: $150,000. • Dr. Raymond Richardson, Head, 615-251-1575. Application contact: Dr. Sandra Scheick, Mathematics Coordinator, 615-251-1575.

Tennessee Technological University, College of Arts and Sciences, Department of Mathematics, Cookeville, TN 38505. Department awards MS. Part-time programs available. Faculty: 17 full-time (4 women), 0 part-time. Matriculated students: 4 full-time (3 women), 6 part-time (3 women); includes 1 minority (black American), 1 foreign. Average age 27. 17 applicants, 53% accepted. In 1990, 1 degree awarded. *Degree requirements:* Thesis required, foreign language not required. *Entrance requirements:* GRE General Test, TOEFL (minimum score of 525 required). Application deadline: 3/1 (priority date). Application fee: $5 ($30 for foreign students). *Tuition:* $2026 per year full-time, $102 per credit hour part-time for state residents; $5486 per year full-time, $253 per credit hour part-time for nonresidents. *Financial aid:* In 1990–91, 9 teaching assistantships (3 to first-year students) awarded. Average monthly stipend for a graduate assistantship: $473. Financial aid application deadline: 4/1. • Dr. Alice Mason, Chairperson, 615-372-3441. Application contact: Dr. Rebecca Quattlebaum, Interim Dean, 615-372-3233.

Texas A&I University, College of Arts and Sciences, Department of Mathematics, Kingsville, TX 78363. Department awards MS. Faculty: 11 full-time, 0 part-time. Matriculated students: 8 full-time (3 women), 11 part-time (6 women); includes 0 minority, 7 foreign. Average age 30. In 1990, 3 degrees awarded. *Degree requirements:* Thesis or alternative, comprehensive exam required. *Entrance requirements:* GRE General Test (minimum combined score of 800 required), TOEFL (minimum score of 500 required). Application deadline: 6/1 (applications processed on a rolling basis). Application fee: $0 ($25 for foreign students). *Expenses:* Tuition of $180 per semester full-time, $120 per semester part-time for state residents; $1152 per semester full-time, $768 per semester part-time for nonresidents. Fees of $149 per semester full-time, $101 per semester part-time. *Financial aid:* In 1990–91, 6 teaching assistantships (all to first-year students) awarded. Financial aid application deadline: 5/15. • Dr. Margaret Land, Chairman, 512-595-3517. Application contact: Dr. H. Wu, Graduate Coordinator, 512-595-3517.

Texas A&M University, College of Science, Department of Mathematics, College Station, TX 77843. Department awards MS, PhD. PhD offered jointly with Stephen F. Austin University. Faculty: 62. Matriculated students: 83 full-time (21 women), 0 part-time; includes 6 minority (4 Asian American, 1 black American, 1 Hispanic American), 26 foreign. 81 applicants, 59% accepted. In 1990, 11 master's, 2 doctorates awarded. *Entrance requirements:* GRE General Test, TOEFL. Application deadline: 7/15. Application fee: $25 ($50 for foreign students). *Expenses:* Tuition of $100 per semester (minimum) for state residents; $128 per credit hour for nonresidents. Fees of $459 per year full-time, $252 per semester part-time. *Financial aid:* Fellowships, research assistantships, teaching assistantships available. • Carl Pearcy, Head, 409-845-3261.

Texas Christian University, Add Ran College of Arts and Sciences, Department of Mathematics, Fort Worth, TX 76129. Department awards MS. In 1990, 0 degrees awarded. *Degree requirements:* 1 foreign language required, thesis optional. *Entrance requirements:* GRE. *Expenses:* Tuition of $244 per semester hour. Fees of $423 per semester full-time, $18.50 per semester hour part-time. • Dr. Robert Doran, Chairperson, 817-921-7335.

Texas Southern University, College of Arts and Sciences, Department of Mathematics, Houston, TX 77004. Department awards MA, MS. Faculty: 7 full-time (1 woman), 0 part-time. Matriculated students: 10 full-time (4 women), 12 part-time (3 women); includes 12 minority (2 Asian American, 10 black American). 14 applicants, 50% accepted. In 1990, 2 degrees awarded. *Degree requirements:* Thesis, comprehensive exam required, foreign language not required. *Entrance requirements:* GRE General Test, TOEFL, minimum GPA of 2.5. Application deadline: 8/1. *Financial aid:* In 1990–91, 8 students received a total of $36,000 in aid awarded. 2 teaching assistantships (0 to first-year students), 5 graduate assistantships (1 to a first-year student) were awarded; federal work-study, institutionally sponsored loans also available. Average monthly stipend for a graduate assistantship: $500. Financial aid application deadline: 5/1. *Faculty research:* Statistics, number theory, topology, differential equations, numerical analysis. • Dr. Robert Nehs, Acting Head, 713-527-7002.

Texas Tech University, Graduate School, College of Arts and Sciences, Department of Mathematics, Lubbock, TX 79409. Offerings include mathematics (MA, MS, PhD). Department faculty: 43. *Degree requirements:* For master's, foreign language not required; for doctorate, 1 foreign language, dissertation. *Entrance requirements:* GRE General Test. Application deadline: 4/15 (priority date, applications processed on a rolling basis). Application fee: $0 ($50 for foreign students). *Tuition:* $494 per semester full-time, $20 per credit hour part-time for state residents; $1790 per semester full-time, $455 per credit hour part-time for nonresidents. • Dr. Ron Anderson, Chairman, 806-742-2566.

Texas Woman's University, College of Arts and Sciences, Department of Mathematics and Computer Science, Denton, TX 76204-1925. Department offers program in mathematics (MA, MS, MSSE). Part-time and evening/weekend programs available. Faculty: 13 full-time (6 women), 2 part-time (both women). Matriculated students: 16 full-time (11 women), 69 part-time (58 women); includes 19 minority (3 Asian American, 11 black American, 5 Hispanic American), 10 foreign. In 1990, 10 degrees awarded. *Degree requirements:* 1 foreign language, thesis. *Entrance requirements:* Minimum GPA of 3.0. Application fee: $0. *Tuition:* $508 per semester full-time, $317 per semester part-time for state residents; $1804 per semester full-time, $955 per semester part-time for nonresidents. *Financial aid:* In 1990–91, 0 research assistantships, 9 teaching assistantships (3 to first-year students) awarded. Financial aid application deadline: 4/1. • Dr. Rose Marie Smith, Chair, 817-898-2166.

Trenton State College, Graduate Division, School of Arts and Sciences, Department of Mathematical Science, Trenton, NJ 08650-4700. Department awards MA, M Ed. Part-time and evening/weekend programs available. Faculty: 5 full-time, 0 part-time. Matriculated students: 2 full-time, 10 part-time. In 1990, 10 degrees awarded. *Degree requirements:* Comprehensive exam required, foreign language and thesis not required. *Entrance requirements:* GRE General Test, minimum GPA of 2.75 overall or 3.0 in field. Application deadline: 6/1 (applications processed on a rolling basis). Application fee: $40. *Tuition:* $151.60 per credit hour for state residents; $200.60 per credit hour for nonresidents. *Financial aid:* Application deadline 5/1. • Dr. David Boliver, Coordinator, 609-771-2269.

Trinity College, Department of Mathematics, Hartford, CT 06106. Department awards MS. Part-time and evening/weekend programs available. Faculty: 1 full-time (0 women), 0 part-time. Matriculated students: 0 full-time, 1 part-time (0 women); includes 0 minority, 0 foreign. Average age 32. 1 applicants, 100% accepted. In 1990, 2 degrees awarded. *Degree requirements:* Foreign language and thesis not required. *Entrance requirements:* Minimum GPA of 3.0. Application deadline: 4/1. Application fee: $35. *Expenses:* Tuition of $675 per course. Fees of $25 per year. *Financial aid:* Partial tuition waivers available. Aid available to part-time students. • Dr. E. Finlay Whittlesey, Graduate Adviser, 203-297-2290.

Tufts University, Graduate School of Arts and Sciences, Department of Mathematics, Medford, MA 02155. Department awards MS, PhD. Faculty: 20 full-time, 5 part-time. Matriculated students: 15 (4 women); includes 6 foreign. 77 applicants, 19% accepted. In 1990, 2 master's, 2 doctorates awarded. Terminal master's awarded for partial completion of doctoral program. *Degree requirements:* For master's, 1 foreign language, thesis; for doctorate, 2 foreign languages, dissertation. *Entrance requirements:* GRE General Test, GRE Subject Test, TOEFL (minimum score of 550 required). Application deadline: 2/15. Application fee: $50. *Expenses:* Tuition of $16,755 per year full-time, $2094 per course part-time. Fees of $885 per year. *Financial aid:* Teaching assistantships, full and partial tuition waivers, federal work-study available. Financial aid application deadline: 2/15; applicants required to submit GAPSFAS. *Faculty research:* Probability, dynamical systems, group theory, functional analysis, algebraic geometry. • Richard Weiss, Chair, 617-381-3234. Application contact: William Reynolds, 617-381-3234.

Tulane University, Department of Mathematics, New Orleans, LA 70118. Department offers programs in applied mathematics (MS), mathematics (MAT, MS, PhD), statistics (MS). *Degree requirements:* For doctorate, dissertation. *Entrance requirements:* For master's, GRE General Test, GRE Subject Test, TOEFL (minimum score of 600 required) or TSE (minimum score of 220 required), minimum B average in undergraduate course work; for doctorate, GRE General Test, GRE Subject Test, TOEFL (minimum score of 600 required) or TSE (minimum score of 220 required). Application deadline: 7/1. Application fee: $35. *Expenses:* Tuition of $16,750 per year full-time, $931 per hour part-time. Fees of $230 per year full-time, $40 per hour part-time. *Financial aid:* Fellowships, teaching assistantships, federal work-study, institutionally sponsored loans, and career-related internships or fieldwork available. Financial aid application deadline: 5/1; applicants required to submit GAPSFAS. • Dr. Steven Rosencrans, Chairman, 504-865-5727.

Université de Montréal, Faculty of Arts and Sciences, Department of Mathematics and Statistics, Montreal, PQ H3C 3J7, Canada. Department offers program in mathematics (M Sc, PhD). Faculty: 45 full-time (3 women), 0 part-time. Matriculated students: 97 full-time (20 women), 4 part-time (1 woman); includes 25 foreign. 56 applicants, 82% accepted. In 1990, 16 master's awarded (50% found work related to degree, 50% continued full-time study); 8 doctorates awarded (62% entered university research/teaching, 25% found other work related to degree, 12% continued full-time study). *Degree requirements:* Thesis/dissertation. *Entrance requirements:* Proficiency in French. Application deadline: 2/1. Application fee: $15. *Financial aid:* In 1990–91, $403,779 in aid awarded. 6 fellowships (0 to first-year students), 27 research assistantships (0 to first-year students), 17 teaching assistantships (0 to first-year students), 101 monitorships (46 to first-year students) were awarded. Financial aid application deadline: 4/1. *Faculty research:* Pure and applied mathematics, statistics, actuarial mathematics. • Pierre Berthiaume, Chairman, 514-343-6710.

Université de Sherbrooke, Faculty of Sciences, Department of Mathematics and Informatics, Sherbrooke, PQ J1K 2R1, Canada. Department awards M Sc, PhD. Faculty: 33 full-time, 0 part-time. Matriculated students: 53 full-time (13 women), 0 part-time. 47 applicants, 45% accepted. In 1990, 11 master's, 1 doctorate awarded. *Degree requirements:* Thesis/dissertation required, foreign language not required. *Entrance requirements:* For doctorate, master's degree. Application deadline: 6/30. Application fee: $15. *Expenses:* Tuition of $585 per trimester full-time, $43.34 per credit part-time for Canadian residents; $2900 per trimester full-time, $195 per credit part-time for nonresidents. Fees of $125 per trimester full-time, $7.75 per credit part-time for Canadian residents; $610 per year full-time, $7.50 per credit part-time for nonresidents. *Financial aid:* Fellowships, research assistantships, teaching assistantships available. *Faculty research:* Measure theory, differential equations, probability, statistics, error control codes. Total annual research budget: $350,000. • Jacques Dubois, Chairman, 819-821-7030. Application contact: Denis Fournier, Secretary, 819-821-7033.

Université du Québec à Montréal, Program in Mathematics, Montreal, PQ H3C 3P8, Canada. Program awards M Sc, PhD. Part-time programs available. Matriculated students: 32 full-time (10 women), 22 part-time (1 woman); includes 6 foreign. 44 applicants, 36% accepted. In 1990, 19 master's, 0 doctorates awarded. *Degree requirements:* Thesis/dissertation. *Entrance requirements:* For master's, proficiency in French, appropriate bachelor's degree; for doctorate, appropriate master's degree, proficiency in French. Application deadline: 5/1. Application fee: $15. *Expenses:* Tuition of $555 per trimester full-time, $37 per credit part-time for Canadian residents; $3480 per trimester full-time, $234 per credit part-time for nonresidents. Fees of $57.50 per trimester full-time, $19.50 per trimester part-time. *Financial aid:* Fellowships, research assistantships, teaching assistantships available. • Manzoor Ahmad, Director, 514-987-7092. Application contact: Lucille Boisselle-Roy, Admissions Officer, 514-987-3128.

Université Laval, Faculty of Sciences and Engineering, Department of Mathematics and Statistics, Sainte-Foy, PQ G1K 7P4, Canada. Offerings include mathematics (M Sc, PhD). *Application deadline:* 3/1. *Application fee:* $15. *Expenses:* Tuition of $792 per year full-time for Canadian residents; $5914 per year full-time for nonresidents. Fees of $120 per year full-time. • Robert Coté, Director, 418-656-2355.

University of Akron, Buchtel College of Arts and Sciences, Department of Mathematical Sciences and Statistics, Program in Mathematics, Akron, OH 44325. Offers applied mathematics (MS), mathematics (MS). Part-time and evening/weekend programs available. Faculty: 20 full-time (1 woman), 0 part-time. Matriculated students: 29 full-time (18 women), 10 part-time (2 women); includes 1 minority (black American), 18 foreign. Average age 26. 41 applicants, 85% accepted. In 1990, 3 degrees awarded. *Degree requirements:* Thesis optional, foreign language not required. *Entrance requirements:* Minimum GPA of 2.75. Application deadline: 3/1 (applications processed on a rolling basis). Application fee: $25. *Tuition:* $119.93 per credit hour for state residents; $210.93 per credit hour for nonresidents. *Financial aid:* In 1990–91, $4800 in aid awarded. 13 teaching assistantships (4 to first-year students) were awarded. Financial aid application deadline: 3/1. *Faculty research:* Topology analysis. Total annual research budget: $200,000. • Dr. William H. Beyer, Head, Department of Mathematical Sciences and Statistics, 216-972-7401.

University of Alabama, College of Arts and Sciences, Department of Mathematics, Tuscaloosa, AL 35487-0132. Department awards MA, PhD, Ed S. PhD offered with the University of Alabama at Birmingham and the University of Alabama at Huntsville. Faculty: 33 full-time (1 woman), 0 part-time. Matriculated students: 20 full-time (8 women), 0 part-time; includes 0 minority, 14 foreign. Average age 31. In 1990, 2 master's awarded (100% continued full-time study); 0 doctorates awarded; 0 Ed Ss awarded. *Degree requirements:* For master's, thesis or alternative required, foreign language not required; for doctorate, 2 foreign languages, dissertation, teaching experience; for Ed S, 2 foreign languages, thesis or alternative. *Entrance requirements:* GRE General Test (minimum combined score of 1100 required), TOEFL (minimum score of 550 required), minimum GPA of 3.0. Application deadline: 7/6 (applications processed on a rolling basis). Application fee: $20. *Tuition:* $968 per year full-time, $82 per credit part-time for state residents; $2400 per year full-time, $218 per credit part-time for nonresidents. *Financial aid:* In 1990–91, 20 students received a total of $32,200 in aid awarded. 0 fellowships, 28 teaching assistantships (11 to first-year students), 6 graders (3 to first-year students) were awarded; federal work-study also available. Average monthly stipend for a graduate assistantship: $700. Financial aid application deadline: 7/14. *Faculty research:* Applied mathematics, analysis, topology, algebra, differential equations. • Wei Shen Hsia, Chairperson, 205-348-5071. Application contact: Rita Reese, Administrative Specialist, Graduate Advisory Committee, 204-348-5074.

University of Alabama at Birmingham, Graduate School, School of Natural Sciences and Mathematics, Department of Mathematics, Birmingham, AL 35294. Department offers programs in applied mathematics (PhD), mathematics (MS). Faculty: 0 part-time. Matriculated students: 22 full-time (9 women), 3 part-time (all women); includes 3 minority (1 Asian American, 2 black American), 7 foreign. 53 applicants, 38% accepted. In 1990, 4 master's awarded. *Degree requirements:* For master's, variable foreign language requirement. *Entrance requirements:* For master's, GRE General Test or MAT. Application deadline: rolling. Application fee: $25 ($50 for foreign students). *Tuition:* $66 per quarter for state residents; $132 per quarter for nonresidents. *Financial aid:* In 1990–91, 2 fellowships awarded; career-related internships or fieldwork also available. *Faculty research:* Differential equations, topology. • Dr. Lex G. Oversteegen, Chairman, 205-934-2154.

University of Alabama in Huntsville, College of Science, Department of Mathematics, Huntsville, AL 35899. Department offers programs in applied mathematics (PhD), mathematics (MA, MS). PhD offered jointly with University of Alabama-Tuscaloosa and University of Alabama-Birmingham. Part-time programs available. Faculty: 19 full-time (2 women), 2 part-time (0 women). Matriculated students: 23 full-time, 18 part-time; includes 1 minority (Asian American), 3 foreign. 46 applicants, 83% accepted. In 1990, 5 master's, 1 doctorate awarded. Terminal master's awarded for partial completion of doctoral program. *Degree requirements:* For master's, thesis or alternative, oral and written exams required, foreign language not required; for doctorate, 1 foreign language, dissertation, oral and written exams. *Entrance requirements:* For master's, GRE General Test (minimum combined score of 1500 on 3 sections required), minimum GPA of 3.0. Application deadline: 5/16 (priority date, applications processed on a rolling basis). Application fee: $20. *Tuition:* $2500 per year full-time, $1250 per year part-time for state residents; $5000 per year full-time, $2500 per year part-time for nonresidents. *Financial aid:* In 1990–91, 21 students received a total of $48,767 in aid awarded. 0 fellowships, 7 research assistantships, 14 teaching assistantships were awarded; full and partial tuition waivers, federal work-study, institutionally sponsored loans, and career-related internships or fieldwork also available. Aid available to part-time students. Average monthly stipend for a graduate assistantship: $720. Financial aid application deadline: 3/1; applicants required to submit FAF. *Faculty research:* Statistical modeling, stochastic processes, combinatorial matrix theory. Total annual research budget: $231,275. • Dr. Peter Gibson, Chairman, 205-895-6470.

University of Alberta, Faculty of Graduate Studies and Research, Department of Mathematics, Edmonton, AB T6G 2J9, Canada. Department awards M Sc, PhD. Matriculated students: 46 full-time, 2 part-time. *Application fee:* $0. *Expenses:* Tuition of $1495 per year full-time, $748 per year part-time for Canadian residents; $2243 per year full-time, $1121 per year part-time for nonresidents. Fees of $301 per year full-time, $118 per year part-time. • Dr. R. D. Bercov, Chair, 403-492-3396.

University of Arizona, College of Arts and Sciences, Faculty of Science, Department of Mathematics, Tucson, AZ 85721. Department awards MA, M Ed, MS, PhD. Program offered in applied mathematics (MS, PhD). Part-time programs available. Faculty: 54 full-time (2 women), 2 part-time (0 women). Matriculated students: 76 full-time (15 women), 11 part-time (5 women); includes 3 minority (2 Asian American, 1 Hispanic American), 32 foreign. Average age 30. 133 applicants, 41% accepted. In 1990, 10 master's, 0 doctorates awarded. *Degree requirements:* For master's, computer language required, foreign language and thesis not required; for doctorate, 2 foreign languages (computer language can substitute for one), dissertation. *Entrance requirements:* GRE (strongly preferred). Application fee: $25. *Expenses:* Tuition of $0 for state residents; $5406 per year full-time, $209 per credit hour part-time for nonresidents. Fees of $1528 per year full-time, $80 per credit hour part-time. *Financial aid:* In 1990–91, $430,000 in aid awarded. 10 fellowships (0 to first-year students), 1 research assistantship, 50 teaching assistantships, 0 scholarships were awarded; full and partial tuition waivers also available. Financial aid application deadline: 3/5; applicants required to submit FAF. *Faculty research:* Algebra/number theory, computational science, dynamical systems, geometry, analysis. Total annual research budget: $1.3-million. • Dr. Alan Newell, Head, 602-621-2868. Application contact: Dr. W. M. Greenlee, Chairman, Graduate Committee, 602-621-2068.

See full description on page 461.

University of Arkansas, J. William Fulbright College of Arts and Sciences, Department of Mathematical Sciences, Fayetteville, AR 72701. Department offers programs in mathematics (MS, PhD), secondary mathematics (MA), statistics (MS). Faculty: 22 full-time (1 woman), 0 part-time. Matriculated students: 23 full-time (9 women), 19 part-time (6 women); includes 1 minority (Asian American), 12 foreign. In 1990, 8 master's, 1 doctorate awarded. *Degree requirements:* For master's, foreign language not required; for doctorate, 2 foreign languages, dissertation. *Application fee:* $15. *Expenses:* Tuition of $2050 per year full-time, $103 per credit hour part-time for state residents; $4400 per year full-time, $220 per credit hour part-time for nonresidents. Fees of $50 per year full-time, $1.50 per credit hour part-time. *Financial aid:* In 1990–91, 25 teaching assistantships awarded. • Dr. John Duncan, Chairman, 501-575-3351.

University of Bridgeport, College of Science and Engineering, Department of Mathematics, 380 University Avenue, Bridgeport, CT 06601. Department awards MS. Faculty: 5 full-time (2 women), 0 part-time. Matriculated students: 0. 2 applicants, 0% accepted. In 1990, 0 degrees awarded. *Degree requirements:* Thesis optional, foreign language not required. *Entrance requirements:* GRE General Test. Application deadline: rolling. Application fee: $30 ($50 for foreign students). *Expenses:* Tuition of $6010 per semester full-time, $310 per credit part-time for state residents; $6010 per semester full-time, $610 per credit part-time for nonresidents. Fees of $50 per year full-time, $25 per year part-time. *Financial aid:* Federal work-study, institutionally sponsored loans, and career-related internships or fieldwork available. Aid available to part-time students. Financial aid application deadline: 6/1; applicants required to submit FAF. • Dr. James V. Tucci, Chairman, 203-576-4271.

University of British Columbia, Faculty of Science, Department of Mathematics, Vancouver, BC V6T 1Z1, Canada. Department awards MA, M Sc, PhD. *Degree requirements:* For doctorate, dissertation, comprehensive exam.

University of Calgary, Faculty of Science, Department of Mathematics and Statistics, Calgary, AB T2N 1N4, Canada. Department awards M Sc, PhD. *Application fee:* $20. *Tuition:* $1705 per year full-time, $427 per course part-time for Canadian residents; $3410 per year full-time, $854 per course part-time for nonresidents.

University of California at Berkeley, College of Letters and Science, Department of Mathematics, Berkeley, CA 94720. Department offers programs in applied mathematics (PhD), mathematics (MA, PhD). Faculty: 80. Matriculated students: 252 full-time, 0 part-time; includes 32 minority (17 Asian American, 3 black American, 12 Hispanic American), 73 foreign. 392 applicants, 46% accepted. Terminal master's awarded for partial completion of doctoral program. *Degree requirements:* For doctorate, dissertation. *Entrance requirements:* GRE General Test, minimum GPA of 3.0. Application deadline: 2/11. Application fee: $40. *Expenses:*

Directory: Mathematics

Tuition of $0. Fees of $1909 per year for state residents; $7825 per year for nonresidents. • Dr. Alberto Grunbaum, Chair, 415-642-4129.

University of California, Irvine, School of Physical Sciences, Department of Mathematics, Irvine, CA 92717. Department awards MS, PhD. Matriculated students: 58 (19 women); includes 22 minority (17 Asian American, 3 black American, 2 Hispanic American), 5 foreign. 75 applicants, 45% accepted. In 1990, 5 master's, 1 doctorate awarded. *Degree requirements:* For master's, 1 foreign language (computer language can substitute); for doctorate, 2 foreign languages, computer language, dissertation. *Entrance requirements:* GRE General Test. Application deadline: 2/1 (applications processed on a rolling basis). Application fee: $40. *Expenses:* Tuition of $0 for state residents; $7699 per year full-time, $3850 per year part-time for nonresidents. Fees of $2930 per year full-time, $2139 per year part-time. *Financial aid:* Fellowships, research assistantships, teaching assistantships, and career-related internships or fieldwork available. *Faculty research:* Analysis, algebra, geometry and topology, probability and statistics, mathematical logic. • Ronald Stern, Chair, 714-856-5544. Application contact: Robert Doedens, Associate Dean, 714-856-6507.

University of California, Los Angeles, College of Letters and Science, Department of Mathematics, Los Angeles, CA 90024. Department awards MA, MAT, PhD, C Phil. Faculty: 66. Matriculated students: 220 full-time (61 women), 0 part-time; includes 41 minority, 80 foreign. 327 applicants, 78% accepted. In 1990, 36 master's, 11 doctorates, 11 C Phils awarded. *Degree requirements:* For master's, comprehensive exam required, foreign language and thesis not required; for doctorate, 2 foreign languages, dissertation, written qualifying exams. *Entrance requirements:* For master's, GRE General Test, GRE Subject Test, minimum GPA of 3.2 in mathematics; for doctorate, GRE General Test, GRE Subject Test, minimum GPA of 3.5 in mathematics. Application fee: $40. *Expenses:* Tuition of $0 for state residents; $7699 per year for nonresidents. Fees of $2907 per year. *Financial aid:* In 1990–91, 155 students received a total of $2.04-million in aid awarded. 104 fellowships, 65 research assistantships, 107 teaching assistantships were awarded; full and partial tuition waivers, federal work-study, institutionally sponsored loans also available. Financial aid application deadline: 3/1. • Dr. Alfred Hales, Chair, 310-825-4701.

University of California, Riverside, Graduate Division, College of Natural and Agricultural Sciences, Department of Mathematics, Riverside, CA 92521. Department awards MA, MS, PhD. Part-time programs available. Terminal master's awarded for partial completion of doctoral program. *Degree requirements:* For master's, comprehensive exams required, foreign language and thesis not required; for doctorate, 1 foreign language, dissertation, qualifying exams. *Entrance requirements:* GRE General Test, TOEFL (minimum score of 550 required). Application deadline: 6/1. Application fee: $40. *Tuition:* $950 per quarter full-time, $264 per quarter part-time for state residents; $3517 per quarter full-time, $1758 per quarter part-time for nonresidents. *Faculty research:* Topology, applied math, analysis, probability.

See full description on page 467.

University of California, San Diego, Department of Mathematics, 9500 Gilman Drive, La Jolla, CA 92093. Department offers programs in applied mathematics (MA), mathematics (MA, PhD), statistics (MS). Faculty: 54. Matriculated students: 107 (26 women); includes 22 foreign. 245 applicants, 23% accepted. In 1990, 18 master's, 9 doctorates awarded. *Degree requirements:* For master's, 1 foreign language required, thesis not required; for doctorate, 2 foreign languages, dissertation. *Entrance requirements:* GRE General Test, GRE Subject Test. Application fee: $40. *Expenses:* Tuition of $0 for state residents; $7699 per year full-time, $1283 per quarter part-time for nonresidents. Fees of $2798 per year full-time, $669 per quarter part-time. • Harold Stark, Chair, 619-534-3594. Application contact: Lois Stewart, Graduate Coordinator, 619-534-6887.

University of California, Santa Barbara, College of Letters and Science, Department of Mathematics, Santa Barbara, CA 93106. Department offers programs in applied mathematics (MA), mathematics (MA, PhD), statistics (PhD). PhD in statistics and applied probability offered in conjunction with the Statistics and Applied Probability Program. Matriculated students: 68 full-time (22 women), 0 part-time; includes 29 foreign. 103 applicants, 79% accepted. In 1990, 9 master's, 5 doctorates awarded. *Degree requirements:* For master's, thesis or alternative required, foreign language not required; for doctorate, 2 foreign languages, dissertation. *Entrance requirements:* GRE General Test, GRE Subject Test, TOEFL (minimum score of 550 required). Application deadline: 5/1. Application fee: $40. *Expenses:* Tuition of $0 for state residents; $7699 per year for nonresidents. Fees of $2307 per year. *Financial aid:* Fellowships, research assistantships, teaching assistantships, full and partial tuition waivers, federal work-study, institutionally sponsored loans available. Financial aid application deadline: 1/31. • Adil Yagub, Chair, 805-893-8340. Application contact: Laurie Theobald, Graduate Secretary, 805-893-8192.

University of California, Santa Cruz, Division of Natural Sciences, Department of Mathematics, Santa Cruz, CA 95064. Department awards MA, PhD. Faculty: 17 full-time (1 woman), 0 part-time. Matriculated students: 43 full-time, 1 part-time; includes 6 minority (4 Asian American, 1 black American, 1 Hispanic American), 11 foreign. 51 applicants, 47% accepted. In 1990, 0 master's, 2 doctorates awarded. Terminal master's awarded for partial completion of doctoral program. *Degree requirements:* For doctorate, 1 foreign language (computer language can substitute), dissertation, qualifying exam. *Entrance requirements:* For doctorate, GRE General Test, GRE Subject Test. Application deadline: 2/1. Application fee: $40. *Expenses:* Tuition of $0 for state residents; $7699 per year for nonresidents. Fees of $3021 per year. *Financial aid:* In 1990–91, 8 fellowships (5 to first-year students), 2 research assistantships (0 to first-year students), 27 teaching assistantships (8 to first-year students) awarded; federal work-study, institutionally sponsored loans also available. Average monthly stipend for a graduate assistantship: $1335. Financial aid applicants required to submit GAPSFAS. *Faculty research:* Algebra, analysis, geometry, applied mathematics. • Dr. Tudor Ratin, Chairperson, 408-459-2085.

Announcement: The Mathematics Department at UC, Santa Cruz, offers MA and PhD programs in pure, applied, and computational mathematics. Active research areas include dynamical systems, finite-group theory, geometric mechanics, global nonlinear analysis, Lie groups and Lie algebras, number theory and automorphic forms, operator theory and integral equations, ordinary and partial differential equations, Riemannian geometry, applied mathematics, computing and computer graphics, and combinatorics and graph theory. Teaching assistantships currently pay $12,015 per 9 months, with possibilities for summer teaching. For more information write to Professor Maria Schonbek, Vice Chair for Graduate Studies, Mathematics Department, University of California, Santa Cruz, CA 95064.

University of Central Florida, College of Arts and Sciences, Program in Mathematical Science, Orlando, FL 32816. Program awards MS. Part-time and evening/weekend programs available. Faculty: 19 full-time (0 women), 2 part-time (0 women). Matriculated students: 41 full-time (12 women), 13 part-time (5 women); includes 11 minority (7 Asian American, 4 black American), 11 foreign. 64 applicants, 64% accepted. In 1990, 11 degrees awarded. *Degree requirements:* Thesis or alternative required, foreign language not required. *Entrance requirements:* GRE General Test, minimum GPA of 3.0 in last 60 hours. Application deadline: 7/27. Application fee: $15. *Expenses:* Tuition of $81 per credit hour for state residents; $364 per credit hour for nonresidents. Fees of $50 per semester. *Financial aid:* In 1990–91, 1 student received a total of $8000 in aid awarded. 26 teaching assistantships (12 to first-year students) were awarded; partial tuition waivers, federal work-study, institutionally sponsored loans, and career-related internships or fieldwork also available. Aid available to part-time students. Average monthly stipend for a graduate assistantship: $800. *Faculty research:* Applied mathematics, analysis, approximation theory, graph theory, mathematical statistics. • Dr. L. Debnath, Chair, 407-823-6284. Application contact: Dr. Larry C. Andrews, Coordinator, 407-823-6284.

University of Chicago, Division of the Physical Sciences, Department of Mathematics, Chicago, IL 60637. Offerings include mathematics (SM, PhD). *Degree requirements:* For master's, 1 foreign language, oral exams; for doctorate, 2 foreign languages, dissertation. *Entrance requirements:* GRE General Test, GRE Subject Test, TOEFL. Application deadline: 1/6. Application fee: $45. *Expenses:* Tuition of $16,275 per year full-time, $8140 per year part-time. Fees of $356 per year.

University of Cincinnati, McMicken College of Arts and Sciences, Department of Mathematics, Cincinnati, OH 45221. Department offers programs in applied mathematics (MS, PhD), mathematics education (MAT), pure mathematics (MS, PhD), statistics (MS, PhD). Faculty: 29 full-time, 0 part-time. Matriculated students: 51 full-time (21 women), 17 part-time (8 women); includes 6 minority (5 Asian American, 1 black American), 39 foreign. 293 applicants, 8% accepted. In 1990, 28 master's, 3 doctorates awarded. *Degree requirements:* For master's, foreign language not required; for doctorate, 1 foreign language, dissertation. *Entrance requirements:* GRE General Test, GRE Subject Test. Application deadline: 2/1. Application fee: $20. *Tuition:* $131 per credit hour for state residents; $261 per credit hour for nonresidents. *Financial aid:* In 1990–91, 33 teaching assistantships awarded; fellowships, research assistantships, full tuition waivers also available. Aid available to part-time students. Average monthly stipend for a graduate assistantship: $683. Financial aid application deadline: 3/1. *Faculty research:* Algebra, analysis, differential equations, numerical analysis. • Dr. C. David Minda, Head, 513-556-4052.

Announcement: Nonthesis MS programs are offered in pure and applied mathematics and statistics. PhD can be pursued in several areas, including algebra, analysis, differential equations, numerical analysis, probability and statistics. The department has an excellent library and computing facilities. Assistantships and fellowships range from $8500–$9500, including tuition remission.

University of Colorado at Boulder, Interdisciplinary Program in Basic Sciences, Boulder, CO 80309. Offerings include mathematics (MBS). *Degree requirements:* Thesis or alternative, comprehensive exam required, foreign language not required. *Entrance requirements:* GRE General Test, GRE Subject Test. Application deadline: 3/1 (priority date, applications processed on a rolling basis). Application fee: $30 ($50 for foreign students). *Expenses:* Tuition of $2308 per year full-time, $387 per semester (minimum) part-time for state residents; $8730 per year full-time, $1455 per semester (minimum) part-time for nonresidents. Fees of $207 per semester full-time, $27.26 per semester (minimum) part-time. • Jim Wailes, Director, 303-492-8674.

University of Colorado at Boulder, College of Arts and Sciences, Department of Mathematics, Boulder, CO 80309. Department offers programs in applied mathematics (MS, PhD), mathematics (MA, PhD). Faculty: 45 full-time (1 woman). Matriculated students: 79 full-time (22 women), 8 part-time (2 women); includes 6 minority (3 Asian American, 1 black American, 2 Hispanic American), 16 foreign. 166 applicants, 57% accepted. In 1990, 12 master's, 5 doctorates awarded. Terminal master's awarded for partial completion of doctoral program. *Degree requirements:* For master's, thesis or alternative, comprehensive exam required, foreign language not required; for doctorate, 2 foreign languages, dissertation, comprehensive exam. *Application deadline:* 3/1 (priority date, applications processed on a rolling basis). *Application fee:* $30 ($50 for foreign students). *Expenses:* Tuition of $2308 per year full-time, $387 per semester (minimum) part-time for state residents; $8730 per year full-time, $1455 per semester (minimum) part-time for nonresidents. Fees of $207 per semester full-time, $27.26 per semester (minimum) part-time. *Financial aid:* In 1990–91, $328,500 in aid awarded. 25 fellowships (0 to first-year students), 3 research assistantships (0 to first-year students), 41 teaching assistantships (0 to first-year students) were awarded; full tuition waivers also available. Financial aid application deadline: 3/1. *Total annual research budget:* $290,000. • Arlan Ramsay, Chairman, 303-492-7256.

University of Colorado at Denver, College of Liberal Arts and Sciences, Program in Mathematics, Denver, CO 80217. Offers applied mathematics (MS, PhD). Part-time and evening/weekend programs available. Faculty: 23 full-time (4 women), 0 part-time. Matriculated students: 5 full-time, 61 part-time; includes 3 minority (2 Asian American, 1 Hispanic American), 6 foreign. In 1990, 7 master's awarded. *Degree requirements:* For master's, thesis or alternative; for doctorate, 1 foreign language, dissertation. *Application deadline:* 6/1 (applications processed on a rolling basis). *Application fee:* $30 ($50 for foreign students). *Expenses:* Tuition of $1185 per semester full-time, $142 per semester hour part-time for state residents; $3969 per semester full-time, $476 per semester hour part-time for nonresidents. Fees of $103 per semester. *Financial aid:* Fellowships, research assistantships, teaching assistantships, federal work-study available. Financial aid application deadline: 3/1. *Total annual research budget:* $1.059-million. • William Briggs, Chair, 303-556-8442. Application contact: Debbie Wangerin, Staff Assistant, 303-556-2341.

University of Connecticut, College of Liberal Arts and Sciences, Field of Mathematics, Storrs, CT 06269. Field awards MS, PhD. Faculty: 36. Matriculated students: 88 full-time (43 women), 8 part-time (4 women); includes 1 minority (Asian American), 69 foreign. Average age 29. 146 applicants, 44% accepted. In 1990, 12 master's, 6 doctorates awarded. Terminal master's awarded for partial completion of doctoral program. *Degree requirements:* For doctorate, dissertation. *Entrance requirements:* GRE General Test, GRE Subject Test. Application deadline: 6/1 (priority date, applications processed on a rolling basis). Application fee: $25. *Expenses:* Tuition of $3428 per year full-time, $571 per course part-time for state residents; $8914 per year full-time, $1486 per course part-time for nonresidents. Fees of $636 per year full-time, $87 per course part-time. *Financial aid:* In 1990–91, $398,026 in aid awarded. 5 fellowships (1 to a first-year student), 1 research assistantship (0 to first-year students), 43 teaching assistantships (8 to first-year students) were awarded. Financial aid application deadline: 2/15; applicants required to submit GAPSFAS or FAF. • Soon–Kyu Kim, Head, 203-486-3923.

University of Delaware, College of Arts and Science, Department of Mathematical Sciences, Newark, DE 19716. Department offers programs in applied mathematics (MA, MA Sc, MS, PhD), mathematics (MA, MA Sc, MS, PhD), statistics (MA, MA Sc, MS, PhD). Part-time programs available. Faculty: 40. Matriculated students: 51 full-time (12 women), 29 part-time (9 women); includes 3 minority (1 black American, 1 Hispanic American, 1 Native American), 21 foreign. Average age 25. 100 applicants, 12% accepted. In 1990, 11 master's, 2 doctorates awarded. Terminal master's awarded for partial completion of doctoral program. *Degree requirements:* For master's, thesis (for some programs), written exam required, foreign language not required; for doctorate, 2 foreign languages, dissertation, qualifying exam. *Entrance requirements:* GRE General Test. Application deadline: 7/1. Application fee: $40. *Tuition:* $179 per credit hour for state residents; $467 per credit hour for nonresidents. *Financial aid:* In 1990–91, 2 fellowships, 2 research assistantships (1 to a first-year student), 29 teaching assistantships (5 to first-year students) awarded; institutionally sponsored loans and career-related internships or fieldwork also available. Average monthly stipend for a graduate assistantship: $900. Financial aid application deadline: 3/1. *Faculty research:* Differential equations, complex analysis, numerical analysis, combinatorics, inverse problems. • Dr. Ivar Stakgold, Chairman, 302-451-2651. Application contact: Pam Haverland, Graduate Secretary, 302-451-2654.

See full description on page 469.

University of Denver, Graduate Studies, Faculty of Natural Sciences, Mathematics and Engineering, Department of Mathematics and Computer Science, Denver, CO 80208. Department offers programs in applied mathematics (MA, MS), computer science (MS), mathematics and computer science (PhD). Part-time programs available. Faculty: 14 full-time (1 woman), 4 part-time (2 women). Matriculated students: 64 (14 women); includes 8 minority (1 Asian American, 5 black American, 2 Hispanic American), 20 foreign. 70 applicants, 90% accepted. In 1990, 13 master's, 1 doctorate awarded. Terminal master's awarded for partial completion of doctoral program. *Degree requirements:* For master's, foreign language, computer language or laboratory experience required, thesis not required; for doctorate, 1 foreign language (computer language can substitute), dissertation, written and oral exams. *Entrance requirements:* GRE General Test, GRE Subject Test, TOEFL (minimum score of 550 required). Application deadline: rolling. Application fee: $30. *Tuition:* $12,852 per year full-time, $357 per credit hour part-time. *Financial aid:* In 1990–91, 28 students received a total of $303,880 in aid awarded. 18 teaching assistantships, 10 scholarships were awarded; fellowships, research assistantships, federal work-study, institutionally sponsored loans, and career-related internships or fieldwork also available. Aid available to part-time students. Financial aid application deadline: 3/1. • Dr. James LaVita, Chairperson, 303-871-3344. Application contact: Rick Bell, Graduate Adviser, 303-871-2453.

Announcement: Symbolic computations, mathematical physics, and functional analysis are focal points for graduate program in mathematics. Course offerings and research activities enhanced by the administrative coupling of mathematics and computer science in one department. PhD students expected to study mathematics and computer science with concentration in one. Teaching assistantships ($6500 plus tuition) available.

University of Detroit Mercy, College of Engineering and Science, Department of Mathematics, Detroit, MI 48221. Department offers programs in computer science (MACS), elementary mathematics education (MATM), junior high mathematics education (MATM), mathematics (MA), secondary mathematics education (MATM). Evening/weekend programs available. Faculty: 8 full-time (1 woman), 0 part-time. Matriculated students: 42 full-time (18 women), 57 part-time (34 women); includes 16 minority (2 Asian American, 14 black American), 59 foreign. Average age 29. In 1990, 71 degrees awarded. *Degree requirements:* Foreign language not required. *Entrance requirements:* Minimum GPA of 3.0. Application deadline: 8/1 (priority date, applications processed on a rolling basis). Application fee: $25 ($35 for foreign students). *Tuition:* $360 per credit hour. *Financial aid:* Fellowships and career-related internships or fieldwork available. • James Lanahan, Chairman, 313-927-1209.

University of Florida, College of Liberal Arts and Sciences, Department of Mathematics, Gainesville, FL 32611. Department offers programs in applied mathematics (MS, PhD), mathematics (MA, MS, PhD), mathematics teaching (MAT, MST). Faculty: 50 full-time. Matriculated students: 80 full-time; includes 28 foreign. *Degree requirements:* For master's, written exam (MS, MA), written exam or comprehensive written exam (MAT, MST); for doctorate, 1 foreign language, dissertation, qualifying exam. *Entrance requirements:* GRE General Test, TOEFL. Application deadline: 2/1 (priority date, applications processed on a rolling basis). Application fee: $15. *Tuition:* $87 per credit hour for state residents; $289 per credit hour for nonresidents. *Financial aid:* In 1990–91, 65 teaching assistantships awarded; fellowships also available. Financial aid application deadline: 2/1. *Faculty research:* Group theory, probability theory, logic, differential geometry and mathematical physics, topology and dynamical systems. • Dr. David A. Drake, Chairman, 904-392-0281.

Announcement: Full fellowships and assistantships are offered to qualified applicants. Teaching assistants received $9000 in the 1991–92 academic year, and nearly all tuition was waived. Oustanding applicants may also be awarded a $3000 supplemental fellowship. Supplemental summer teaching assistantships are available.

See full description on page 471.

University of Georgia, College of Arts and Sciences, Department of Mathematics, Athens, GA 30602. Department awards MA, MAMS, PhD. Faculty: 35 full-time (2 women), 0 part-time. Matriculated students: 40 full-time (19 women), 3 part-time (0 women); includes 5 minority (1 Asian American, 4 black American), 19 foreign. 121 applicants, 21% accepted. In 1990, 12 master's, 2 doctorates awarded. *Degree requirements:* For master's, 1 foreign language, technical report (MAMS); for doctorate, 1 foreign language, dissertation. *Entrance requirements:* GRE General Test. Application fee: $10. *Expenses:* Tuition of $598 per quarter full-time, $48 per quarter part-time for state residents; $1558 per quarter full-time, $144 per quarter part-time for nonresidents. Fees of $118 per quarter. *Financial aid:* Fellowships, research assistantships, teaching assistantships, assistantships available. • Dr. Dhandapani Kannan, Graduate Coordinator, 404-542-2609.

See full description on page 473.

University of Guelph, College of Physical Science, Department of Mathematics and Statistics, Guelph, ON N1G 2W1, Canada. Department offers programs in applied mathematics (PhD), applied statistics (PhD), mathematics and statistics (M Sc). Faculty: 28. Matriculated students: 29; includes 10 foreign. 67 applicants, 19% accepted. In 1990, 7 master's awarded. *Entrance requirements:* For master's, minimum B average during last two years; for doctorate, minimum B average. *Expenses:* Tuition of $898 per semester full-time, $450 per semester part-time for Canadian residents; $4053 per semester full-time, $2185 per semester part-time for nonresidents. Fees of $543 per semester full-time, $450 per semester part-time for Canadian residents; $2278 per semester full-time, $2185 per semester part-time for nonresidents. *Financial aid:* Fellowships, research assistantships, teaching assistantships available. *Faculty research:* Dynamical systems, mathematical biology, numerical analysis and reproductive research, linear and non-linear models, reliability and bioassay. • Dr. Langford, Chair, 519-824-4120 Ext. 2155. Application contact: Dr. Carter, Graduate Coordinator, 519-824-4120 Ext. 3569.

University of Hawaii at Manoa, College of Arts and Sciences, Department of Mathematics, Honolulu, HI 96822. Department awards MA, PhD. *Degree requirements:* For master's, comprehensive exams required, foreign language and thesis not required; for doctorate, 2 foreign languages (computer language can substitute for one), dissertation, comprehensive exams. *Entrance requirements:* GRE, TOEFL, minimum GPA of 3.0. *Tuition:* $800 per semester full-time for state residents; $2405 per semester full-time for nonresidents. *Faculty research:* Analysis, algebra, lattice theory, logic topology, differential geometry.

University of Houston, College of Natural Sciences and Mathematics, Department of Mathematics, 4800 Calhoun, Houston, TX 77004. Department offers programs in applied mathematics (MS), mathematics (MS, PhD). Part-time programs available. *Degree requirements:* For master's, thesis optional, foreign language not required; for doctorate, 2 foreign languages, dissertation. *Entrance requirements:* For master's, GRE General Test (minimum combined score of 900 required), TOEFL (minimum score of 550 required); for doctorate, MS in mathematics or equivalent. Application deadline: 7/7. *Expenses:* Tuition of $30 per hour for state residents; $134 per hour for nonresidents. Fees of $240 per year full-time, $125 per year part-time. *Faculty research:* Numerical analysis and scientific computing, ordinary and partial differential equations, topology, algebra.

See full description on page 477.

University of Houston–Clear Lake, School of Natural and Applied Sciences, Program in Mathematical Sciences, Houston, TX 77058. Program awards MS. Matriculated students: 3 full-time (1 woman), 41 part-time (18 women); includes 4 minority (3 Asian American, 1 Hispanic American), 2 foreign. In 1990, 3 degrees awarded. *Degree requirements:* Foreign language not required. *Entrance requirements:* GRE General Test. Application fee: $0. *Tuition:* $40 per credit hour for state residents; $134 per credit hour for nonresidents. *Financial aid:* Application deadline 5/1. • Dr. Carroll B. Lassiter, Director. Application contact: Dr. Eldon Husband, Associate Dean and Director of Student Affairs, 713-283-3710.

University of Idaho, College of Graduate Studies, College of Letters and Science, Department of Mathematics and Applied Statistics, Program in Mathematics, Moscow, ID 83843. Offers mathematics (M Nuc Sci, MS, PhD), mathematics education (MAT). Matriculated students: 17 full-time (7 women), 5 part-time (1 woman); includes 0 minority, 3 foreign. In 1990, 5 master's, 1 doctorate awarded. *Degree requirements:* For master's, foreign language and thesis not required; for doctorate, 2 foreign languages, dissertation. *Entrance requirements:* For master's, minimum GPA of 2.8; for doctorate, minimum undergraduate GPA of 2.8, graduate GPA of 3.0. Application deadline: 8/1. Application fee: $20. *Expenses:* Tuition of $0 for state residents; $4146 per year for nonresidents. Fees of $818 per semester full-time, $82.75 per credit part-time. *Financial aid:* Application deadline 3/1. *Faculty research:* Algebra, topology, analysis. • Dr. Clarence J. Potrate, Chair, Department of Mathematics and Applied Statistics, 208-885-6742.

University of Illinois at Chicago, College of Liberal Arts and Sciences, Department of Mathematics, Statistics, and Computer Science, Chicago, IL 60680. Department offers programs in applied mathematics (MS, DA, PhD), computer science (MS, DA, PhD), probability and statistics (MS, DA, PhD), pure mathematics (MS, DA, PhD), teaching of mathematics (MST). Faculty: 78 full-time (4 women). Matriculated students: 245; includes 11 minority (10 Asian American, 1 black American), 115 foreign. 384 applicants, 27% accepted. In 1990, 30 master's, 11 doctorates awarded. *Degree requirements:* For master's, comprehensive exam required, foreign language and thesis not required; for doctorate, 1 foreign language, dissertation. *Entrance requirements:* GRE General Test, TOEFL (minimum score of 550 required), minimum GPA of 3.75 (on a 5.0 scale). Application deadline: 7/5. Application fee: $20. *Expenses:* Tuition of $1369 per semester full-time, $521 per semester (minimum) part-time for state residents; $3840 per semester full-time, $1454 per semester (minimum) part-time for nonresidents. Fees of $458 per semester full-time, $398 per semester (minimum) part-time. *Financial aid:* In 1990–91, $800,000 in aid awarded. 6 fellowships, 12 research assistantships, 98 teaching assistantships were awarded; full tuition waivers also available. • John W. Wood, Head, 312-413-2153. Application contact: Jeff E. Lewis, Director of Graduate Studies, 312-996-3041.

Announcement: Master's and doctoral programs in pure mathematics, applied mathematics, computer science, and probability and statistics. In recent years, 141 MS, 58 MST, 2 DA, and 28 PhD degrees awarded. Currently 245 graduate students, including 98 teaching assistants, 6 fellows, and 12 research assistants. Departmental library and excellent computing facilities.

See full description on page 479.

University of Illinois at Urbana-Champaign, College of Liberal Arts and Sciences, Department of Mathematics, Champaign, IL 61820. Department offers programs in applied mathematics (MS), mathematics (MS, PhD), teaching of mathematics (MS). Faculty: 98 full-time, 4 part-time. Matriculated students: 172 full-time (41 women), 1 (woman) part-time; includes 16 minority (11 Asian American, 3 black American, 2 Hispanic American), 106 foreign. 210 applicants, 56% accepted. In 1990, 19 master's, 8 doctorates awarded. *Degree requirements:* For master's, foreign language and thesis not required; for doctorate, 2 foreign languages, dissertation. *Entrance requirements:* For master's, minimum GPA of 4.0 on a 5.0 scale. Application fee: $25. *Tuition:* $1838 per semester full-time, $708 per semester part-time for state residents; $4314 per semester full-time, $1673 per semester part-time for nonresidents. *Financial aid:* In 1990–91, 3 fellowships, 6 research assistantships, 157 teaching assistantships awarded. Financial aid application deadline: 2/15. • Ward Henson, Head, 217-333-3350.

University of Iowa, College of Liberal Arts, Department of Mathematics, Iowa City, IA 52242. Department awards MS, PhD. Faculty: 48 full-time, 0 part-time. Matriculated students: 44 full-time (17 women), 61 part-time (19 women); includes 8 minority (3 Asian American, 2 black American, 2 Hispanic American, 1 Native American), 64 foreign. 117 applicants, 70% accepted. In 1990, 18 master's, 6 doctorates awarded. *Application fee:* $20. *Expenses:* Tuition of $1158 per semester full-time, $387 per semester hour (minimum) part-time for state residents; $3372 per semester full-time, $387 per semester hour (minimum) part-time for nonresidents. Fees of $60 per semester (minimum). *Financial aid:* In 1990–91, 3 fellowships (all to first-year students), 5 research assistantships (0 to first-year students), 78 teaching assistantships (14 to first-year students) awarded. • William Kirk, Chair, 319-335-0714.

See full description on page 483.

University of Kansas, College of Liberal Arts and Sciences, Department of Mathematics, Lawrence, KS 66045. Department offers programs in applied mathematics and statistics (MA, PhD), mathematics (MA, PhD). Faculty: 31 full-time (2 women), 0 part-time. Matriculated students: 39 full-time (17 women), 16 part-time (6 women); includes 2 minority (1 Asian American, 1 Native American), 27 foreign. In 1990, 13 master's, 1 doctorate awarded. *Degree requirements:* For master's, thesis or alternative required, foreign language not required; for doctorate, 2 foreign languages, dissertation. *Entrance requirements:* TOEFL (minimum score of 570 required). Application fee: $25. *Expenses:* Tuition of $1668 per year full-time, $56 per credit hour part-time for state residents; $5382 per year full-time, $179 per credit hour part-time for nonresidents. Fees of $338 per year full-time, $25 per credit hour part-time. *Financial aid:* Fellowships, research assistantships, teaching assistantships, institutionally sponsored loans available. Aid available to part-time students. Financial aid application deadline: 2/1. • Dr. Charles Himmelberg, Chairperson, 913-864-3651. Application contact: Dr. Saul Stahl, Graduate Director, 913-864-4324.

University of Kentucky, Graduate School Programs from the College of Arts and Sciences, Program in Mathematics, Lexington, KY 40506-0032. Program awards MA, MS, PhD. Faculty: 38 full-time (1 woman), 0 part-time. Matriculated students: 45 full-time (9 women), 17 part-time (6 women); includes 2 minority (both Asian American), 16 foreign. 145 applicants, 24% accepted. In 1990, 5 master's, 2 doctorates awarded. *Degree requirements:* For master's, 1 foreign language, comprehensive exam required, thesis optional; for doctorate, 2 foreign languages, dissertation, comprehensive exam. *Entrance requirements:* For master's, GRE (verbal, quantitative, and analytical sections), minimum undergraduate GPA of 2.5; for doctorate, GRE (verbal, quantitative, and analytical sections), minimum graduate GPA of 3.0. Application deadline: 7/19 (applications processed on a rolling basis). Application fee: $20 ($25 for foreign students). *Tuition:* $1002 per semester full-time, $101 per credit hour part-time for state residents; $2782 per semester full-time, $299 per credit hour part-time for nonresidents. *Financial aid:* In 1990–91, 4 fellowships (2 to first-year students), 3 research assistantships, 50 teaching assistantships (15 to first-year students) awarded; federal work-study, institutionally sponsored loans also available. Aid available to part-time students. Financial aid applicants required to submit FAF. *Faculty research:* Algebraic, general and infinite-dimensional topology, group theory, ring theory, algebraic geometry. • Dr. Brauch Fugate, Director of Graduate Studies, 606-257-4781. Application contact: Dr. Constance L. Wood, Associate Dean for Academic Administration, 606-257-4905.

University of Louisville, College of Arts and Sciences, Department of Mathematics, Louisville, KY 40292. Department awards MA, MAT. Evening/weekend programs available. Faculty: 13 full-time (0 women), 0 part-time. Matriculated students: 17 full-time (8 women), 15 part-time (7 women); includes 2 minority (both Asian American), 1 foreign. In 1990, 9 degrees awarded. *Degree requirements:* Foreign language and thesis not required. *Entrance requirements:* GRE General Test (minimum combined score of 1150 required for MA), GRE General Test (minimum combined score of 1050 required). Application deadline: rolling. *Expenses:* Tuition of $1780 per year full-time, $99 per credit hour part-time for state residents; $5340 per year full-time, $297 per credit hour part-time for nonresidents. Fees of $60 per semester full-time, $12.50 per semester (minimum) part-time. • Dr. Robert McFadden, Chair, 502-588-6826.

University of Maine, College of Sciences, Department of Mathematics, Orono, ME 04469. Department awards MA. Faculty: 18 full-time (1 woman). Matriculated students: 12 full-time (4 women), 0 part-time; includes 5 foreign. In 1990, 2 degrees awarded. *Degree requirements:* Thesis optional, foreign language not required. *Entrance requirements:* GRE General Test, GRE Subject Test, TOEFL (minimum score of 550 required). Application deadline: 12/15 (priority date, applications processed on a rolling basis). Application fee: $25. *Tuition:* $100 per credit hour for state residents; $275 per credit hour for nonresidents. *Financial aid:* Teaching assistantships available. Financial aid application deadline: 3/1. • Dr. Grattan Murphy, Chair, 207-581-3902.

University of Manitoba, Faculty of Science, Department of Mathematics, Winnipeg, MB R3T 2N2, Canada. Department awards MA, M Sc, PhD. *Degree requirements:* For master's, 1 foreign language, thesis or alternative; for doctorate, 1 foreign language, dissertation.

University of Maryland College Park, College of Computer, Mathematical and Physical Sciences, Department of Mathematics, College Park, MD 20742. Department offers programs in applied mathematics (MA, PhD), mathematical statistics (MA, PhD), mathematics (MA, PhD). Faculty: 95 (5 women). Matriculated students: 145 full-time (38 women), 83 part-time (21 women); includes 25 minority (14 Asian American, 7 black American, 4 Hispanic American). 443 applicants, 50% accepted. In 1990, 15 master's, 17 doctorates awarded. *Degree requirements:* For master's, thesis or alternative required, foreign language not required; for doctorate, 2 foreign languages, dissertation. *Entrance requirements:* For master's, minimum GPA of 3.0. Application deadline: rolling. Application fee: $25. *Expenses:* Tuition of $143 per credit hour for state residents; $256 per credit hour for nonresidents. Fees of $171.50 per semester. *Financial aid:* In 1990–91, 27 fellowships, 7 research assistantships, 110 teaching assistantships awarded. • Dr. Nelson G. Markley, Chairman, 301-405-5048.

University of Massachusetts, Lowell, College of Arts and Sciences, Department of Mathematics, 1 University Avenue, Lowell, MA 01854. Offerings include mathematics (MS). *Degree requirements:* Foreign language and thesis not required. *Entrance requirements:* GRE General Test. Application deadline: 4/1. *Expenses:* Tuition of $87 per credit hour for state residents; $271 per credit hour for nonresidents. Fees of $114 per credit hour.

University of Miami, College of Arts and Sciences, Department of Mathematics and Computer Science, Coral Gables, FL 33124. Offerings include mathematics (MA, MS, DA, PhD). Terminal master's awarded for partial completion of doctoral program. Department faculty: 31 (2 women). *Degree requirements:* For master's, comprehensive exam or project required, foreign language and thesis not required; for doctorate, 1 foreign language, dissertation, qualifying exams. *Entrance requirements:* GRE General Test (minimum combined score of 1000 required), TOEFL (minimum score of 550 required), minimum GPA of 3.0. Application deadline: 7/1 (applications processed on a rolling basis). Application fee: $35. *Expenses:* Tuition of $567 per credit hour. Fees of $87 per semester full-time. • Dr. Alan Zame, Chairman, 305-284-2348. Application contact: Dr. Marvin Mielke, Graduate Adviser, 305-284-2348.

University of Michigan, College of Literature, Science, and the Arts, Department of Mathematics, Ann Arbor, MI 48109. Department awards AM, MS, PhD. Faculty: 63. Matriculated students: 140 full-time (37 women), 0 part-time; includes 9 minority (3 Asian American, 2 black American, 4 Hispanic American), 45 foreign. 355 applicants, 63% accepted. In 1990, 33 master's, 10 doctorates awarded. *Degree requirements:* For master's, foreign language and thesis not required; for doctorate, 2 foreign languages, dissertation, preliminary exam. *Entrance requirements:* GRE General Test, GRE Subject Test. Application deadline: 2/1 (applications processed on a rolling basis). Application fee: $30. *Tuition:* $3255 per semester full-time, $352 per credit (minimum) part-time for state residents; $6803 per semester full-time, $746 per credit (minimum) part-time for nonresidents. *Financial aid:* Fellowships, research assistantships, teaching assistantships available. Financial aid application deadline: 3/15. • Donald Lewis, Chair, 313-764-0335.

University of Minnesota, Twin Cities Campus, Institute of Technology, School of Mathematics, Minneapolis, MN 55455. School awards MA, MS, PhD. *Degree requirements:* For master's, thesis optional; for doctorate, dissertation. *Expenses:* Tuition of $1084 per quarter full-time, $301 per credit part-time for state residents; $2168 per quarter full-time, $602 per credit part-time for nonresidents. Fees of $118 per quarter. • Eugene Fabes, Head, 612-625-7575.

University of Mississippi, Graduate School, College of Liberal Arts, Department of Mathematics, University, MS 38677. Department awards MA, MS, PhD. Matriculated students: 22 full-time (8 women), 4 part-time (2 women); includes 1 minority (black American), 5 foreign. In 1990, 12 master's, 2 doctorates awarded. *Degree requirements:* For master's, thesis required (for some programs), foreign language not required; for doctorate, dissertation required, foreign language not required. *Entrance requirements:* For master's, GRE General Test, minimum GPA of 3.0; for doctorate, GRE General Test (minimum combined score of 900 required). Application deadline: 8/1. Application fee: $15 ($25 for foreign students). *Expenses:* Tuition of $1011 per semester full-time, $99 per semester part-time for state residents; $1842 per semester full-time, $180 per semester part-time for nonresidents. Fees of $219 per year full-time. *Financial aid:* Application deadline 3/1. • Dr. Eldon Miller, Acting Chairman, 601-232-7071.

University of Missouri–Columbia, College of Arts and Sciences, Department of Mathematics, Columbia, MO 65211. Department offers programs in applied mathematics (MS), mathematics (MA, MST, PhD). Faculty: 34 full-time, 1 part-time. Matriculated students: 35 full-time (10 women), 13 part-time (3 women); includes 1 minority (Asian American), 19 foreign. In 1990, 1 master's, 1 doctorate awarded. *Degree requirements:* For master's, foreign language and thesis not required; for doctorate, 2 foreign languages, dissertation. *Entrance requirements:* GRE General Test, minimum GPA of 3.0. Application deadline: 8/1 (priority date, applications processed on a rolling basis). Application fee: $20 ($40 for foreign students). *Expenses:* Tuition of $89.90 per credit hour full-time, $98.35 per credit hour part-time for state residents; $244 per credit hour full-time, $252.45 per credit hour part-time for nonresidents. Fees of $123.55 per semester (minimum) full-time. • Dr. Mark Ashbaugh, Director of Graduate Studies, 314-882-4558. Application contact: Gary L. Smith, Director of Admissions and Registrar, 314-882-7651.

University of Missouri–Kansas City, College of Arts and Sciences, Department of Mathematics, Kansas City, MO 64110. Department awards MA, MS, PhD. Part-time programs available. Faculty: 15 full-time (1 woman), 0 part-time. Matriculated students: 6 full-time (3 women), 13 part-time (4 women); includes 3 minority (1 Asian American, 1 black American, 1 Hispanic American), 2 foreign. In 1990, 2 master's, 1 doctorate awarded. *Degree requirements:* For master's, foreign language and thesis not required; for doctorate, 2 foreign languages, dissertation. *Entrance requirements:* GRE General Test or GMAT. Application fee: $0. *Expenses:* Tuition of $2200 per year full-time, $92 per credit hour part-time for state residents; $5503 per year full-time, $229 per credit hour part-time for nonresidents. Fees of $122 per semester full-time, $9 per credit hour part-time. *Financial aid:* Fellowships, teaching assistantships, full and partial tuition waivers, federal work-study, institutionally sponsored loans available. Aid available to part-time students. *Faculty research:* Classical real variables, matrix theory, ring theory, linear numerical analysis, point set topology. • Dr. Phil Barker, Chairperson, 816-235-2842.

University of Missouri–Rolla, College of Arts and Sciences, Department of Mathematics and Statistics, Rolla, MO 65401. Department offers programs in applied mathematics (MS), mathematics (PhD), statistics (MS, PhD). Faculty: 21 full-time (2 women), 4 part-time (0 women). Matriculated students: 26 full-time (14 women), 0 part-time; includes 4 minority (all Asian American), 2 foreign. Average age 26. 19 applicants, 95% accepted. In 1990, 5 master's awarded; 2 doctorates awarded (100% entered university research/teaching). Terminal master's awarded for partial completion of doctoral program. *Degree requirements:* For master's, thesis or alternative required, foreign language not required; for doctorate, 1 foreign language, dissertation. *Entrance requirements:* GRE General Test, GRE Subject Test. Application deadline: 7/1 (applications processed on a rolling basis). Application fee: $20 ($40 for foreign students). *Expenses:* Tuition of $2090 per year full-time, $87.10 per credit hour part-time for state residents; $5582 per year full-time, $232.60 per credit hour part-time for nonresidents. Fees of $349 per year full-time, $61.63 per semester (minimum) part-time. *Financial aid:* In 1990–91, 35 students received a total of $281,000 in aid awarded. 1 research assistantship (0 to first-year students) was awarded; teaching assistantships also available. Average monthly stipend for a graduate assistantship: $1133. *Faculty research:* Analysis, differential equations, topology, statistics. Total annual research budget: $155,000. • William T. Ingram, Chairman, 314-341-4641. Application contact: Troy Hicks, 314-341-4654.

University of Missouri–St. Louis, College of Arts and Sciences, Department of Mathematical Sciences, Normandy, MO 63121-4499. Department awards MA. Part-time and evening/weekend programs available. Faculty: 16 full-time (2 women), 0 part-time. Matriculated students: 7 full-time (5 women), 9 part-time (2 women); includes 3 Asian American, 0 black American, 3 foreign. Average age 32. 12 applicants, 83% accepted. In 1990, 1 degree awarded. *Degree requirements:* Foreign language and thesis not required. *Entrance requirements:* GRE General Test. Application deadline: 7/1 (applications processed on a rolling basis). Application fee: $0. *Expenses:* Tuition of $2157 per year full-time, $89.90 per credit hour part-time for state residents; $5856 per year full-time, $244 per credit hour part-time for nonresidents. Fees of $235 per year full-time, $9.80 per credit hour part-time. *Financial aid:* In 1990–91, $7100 in aid awarded. 2 teaching assistantships were awarded. *Faculty research:* Number theory, harmonic analysis, mathematics, physics, algebraic topology. Total annual research budget: $185,000. • Dr. Edward Andalafte, Chairman, 314-553-5741.

University of Montana, College of Arts and Sciences, Department of Mathematical Sciences, Missoula, MT 59812. Department offers programs in algebra (MA, PhD), analysis (PhD), applied mathematics (MA, PhD), mathematics (MAT), operations research (MA), statistics (MA, PhD). Terminal master's awarded for partial completion of doctoral program. *Degree requirements:* For master's, 1 foreign language, thesis (for some programs); for doctorate, 2 foreign languages (computer language can substitute for one), dissertation. *Entrance requirements:* GRE General Test. Application deadline: 9/15. Application fee: $20. *Tuition:* $495 per quarter hour full-time for state residents; $1239 per quarter hour full-time for nonresidents. *Faculty research:* Harmonic analysis, ring theory, nonparametric statistics, mathematical modeling, differential equations.

University of Nebraska at Omaha, College of Arts and Sciences, Department of Mathematics and Computer Science, Omaha, NE 68182. Offerings include mathematics (MA, MAT, MS). *Degree requirements:* Foreign language not required. *Entrance requirements:* GRE General Test. Application deadline: 7/15. Application fee: $25. *Expenses:* Tuition of $64.50 per credit hour for state residents; $154 per credit hour for nonresidents. Fees of $57.25 per semester.

University of Nebraska–Lincoln, College of Arts and Sciences, Department of Mathematics and Statistics, Lincoln, NE 68588. Department offers programs in applied mathematics (MS), mathematics (M Sc T, PhD), mathematics and statistics (MA, MS), mathematics education (MAT), statistics (PhD). Faculty: 29 full-time (1 woman), 0 part-time. Matriculated students: 94 full-time (36 women), 15 part-time (6 women); includes 4 minority (2 Asian American, 2 black American), 54 foreign. Average age 29. In 1990, 29 master's, 4 doctorates awarded. Terminal master's awarded for partial completion of doctoral program. *Degree requirements:* For master's, foreign language not required; for doctorate, dissertation, comprehensive exams. *Entrance requirements:* TOEFL (minimum score of 500 required). Application deadline: 5/1 (priority date, applications processed on a rolling basis). Application fee: $25. *Expenses:* Tuition of $75.75 per credit hour for state residents; $187.25 per credit hour for nonresidents. Fees of $161 per year full-time. *Financial aid:* Fellowships, research assistantships, teaching assistantships, federal work-study available. Aid available to part-time students. Financial aid application deadline: 2/15. • Dr. Jim Lewis, Chairperson, 402-472-3731.

University of Nevada, Las Vegas, College of Science and Mathematics, Department of Mathematical Sciences, Las Vegas, NV 89154. Department offers programs in applied mathematics (MS), mathematics (MS), pure mathematics (MS), statistics (MS). Part-time programs available. Faculty: 16 full-time (0 women), 0 part-time. Matriculated students: 15 (7 women). 8 applicants, 63% accepted. In 1990, 1 degree awarded. *Degree requirements:* Oral exam required, thesis optional, foreign language not required. *Entrance requirements:* Minimum GPA of 2.75 overall, 3.0 during previous 2 years. Application deadline: 6/15. Application fee: $20. *Expenses:* Tuition of $66 per credit. Fees of $1800 per semester for nonresidents. *Financial aid:* In 1990–91, 4 teaching assistantships awarded. Financial aid application deadline: 3/1. • Dr. Peter Shive, Chairman, 702-739-3567. Application contact: Graduate College Admissions Evaluator, 702-739-3320.

University of Nevada, Reno, College of Arts and Science, Department of Mathematics, Reno, NV 89557. Department awards MATM, MS. Faculty: 15 full-time (1 woman), 0 part-time. Matriculated students: 15. Average age 30. In 1990, 4 degrees awarded. *Degree requirements:* Thesis optional, foreign language not required. *Entrance requirements:* GRE General Test, GRE Subject Test, TOEFL, minimum GPA of 2.75. Application deadline: 8/1 (priority date, applications processed on a rolling basis). Application fee: $20. *Expenses:* Tuition of $0 for state residents; $3600 per year full-time, $66 per credit hour part-time for nonresidents. Fees of $66 per credit hour. *Financial aid:* Institutionally sponsored loans available. Financial aid applicants required to submit FAF. *Faculty research:* Operator algebra, non-linear systems, differential equations. • Dr. Robert N. Tompson, Chairman, 702-784-6773.

University of New Brunswick, Faculty of Arts, Department of Mathematics and Statistics, Fredericton, NB E3B 5A3, Canada. Department awards M Sc, PhD. *Degree requirements:* For master's, thesis or alternative; for doctorate, dissertation. *Entrance requirements:* TOEFL, minimum GPA of 3.0. Application deadline: 3/1 (priority date). *Expenses:* Tuition of $2100 per year. Fees of $45 per year. • Peter C. Kent, Dean, Faculty of Arts.

University of New Hampshire, College of Engineering and Physical Sciences, Department of Mathematics, Durham, NH 03824. Department offers programs in mathematics (MS, MST, PhD), mathematics education (PhD). Faculty: 23 full-time. Matriculated students: 8 full-time (2 women), 22 part-time (8 women); includes 7 foreign. 29 applicants, 41% accepted. In 1990, 16 master's, 1 doctorate awarded. Terminal master's awarded for partial completion of doctoral program. *Degree requirements:* For master's, foreign language and thesis not required; for doctorate, 2 foreign languages (computer language can substitute for one), dissertation. *Application deadline:* 7/1 (priority date, applications processed on a rolling basis). *Application fee:* $25. *Tuition:* $1645 per semester full-time, $183 per credit hour part-time for state residents; $4920 per semester full-time, $547 per credit hour part-time for nonresidents. *Financial aid:* In 1990–91, 0 fellowships, 23 teaching assistantships, 2 scholarships awarded; full and partial tuition waivers, federal work-study, and career-related internships or fieldwork also available. Aid available to part-time students. Financial aid application deadline: 2/15. *Faculty research:* Ring theory, group theory, operator theory, algebras, statistics, category theory. • Dr. Donald Van Osdol, Chairperson, 603-862-2320. Application contact: Dr. Donald Hadwin, 603-862-2320.

University of New Mexico, College of Arts and Sciences, Department of Mathematics and Statistics, Albuquerque, NM 87131. Department awards MA, PhD. Faculty: 26 full-time (2 women), 9 part-time (2 women). *Degree requirements:* For master's, foreign language and thesis not required; for doctorate, 1 foreign language, dissertation. *Application fee:* $25. *Expenses:* Tuition of $467 per semester (minimum) full-time, $67.50 per credit hour part-time for state residents; $1549 per semester (minimum) full-time, $67.50 per credit hour part-time for nonresidents. Fees of $16 per semester. *Financial aid:* Teaching assistantships and career-related internships or fieldwork available. *Faculty research:* Pure and applied mathematics, applied statistics. • Frank Gilfeather, Chairman, 505-277-4613.

University of New Orleans, College of Sciences, Department of Mathematics, New Orleans, LA 70148. Department awards MS. Part-time programs available. Faculty: 18 full-time (1 woman), 0 part-time. Matriculated students: 33 full-time (10 women), 9 part-time (4 women); includes 7 minority (3 Asian American, 3 black American, 1 Hispanic American), 14 foreign. Average age 22. In 1990, 9 degrees awarded (67% found work related to degree, 33% continued full-time study). *Degree requirements:* Thesis not required. *Application deadline:* 7/1 (priority date, applications processed on a rolling basis). *Application fee:* $20. *Tuition:* $962 per quarter hour full-time for state residents; $2308 per quarter hour full-time for nonresidents. *Financial aid:* In 1990–91, 30 teaching assistantships (8 to first-year students) awarded. *Faculty research:* Differential equations, combinatorics, statistics, complex analysis, algebra. • Dr. Carroll F. Blakemore, Chairman, 504-286-6331.

University of North Carolina at Chapel Hill, College of Arts and Sciences, Department of Mathematics, Chapel Hill, NC 27599. Department awards MA, MS, PhD. Faculty: 33 full-time, 0 part-time. 121 applicants, 40% accepted. In 1990, 9 master's, 2 doctorates awarded. *Degree requirements:* For master's, 1 foreign language (computer language can substitute), comprehensive exam; for doctorate, 2 foreign languages, dissertation, comprehensive exam. *Entrance requirements:* GRE General Test (minimum combined score of 1000 required), GRE Subject Test, minimum GPA of 3.0. Application deadline: 2/15. Application fee: $35. *Tuition:* $621 per semester full-time for state residents; $3555 per semester full-time for nonresidents. *Financial aid:* In 1990–91, 0 fellowships, 45 teaching assistantships awarded. • Dr. Sheldon E. Newhouse, Chairman, 919-962-9621.

Announcement: The department, with 33 faculty members and 45–50 graduate students, offers unusual opportunities for student-faculty interaction in master's and doctoral programs. Flexible MS program permits substantial course work in allied disciplines. The proximity of North Carolina State University, Duke, and the Research Triangle Park enhances opportunities in both programs.

University of North Carolina at Charlotte, College of Arts and Sciences, Department of Mathematics, Charlotte, NC 28223. Department awards MA, MS. Part-time and evening/weekend programs available. Faculty: 33 full-time (2 women), 0 part-time. Matriculated students: 2 full-time (1 woman), 28 part-time (15 women); includes 2 minority (both black American), 8 foreign. Average age 28. 20 applicants, 80% accepted. In 1990, 7 degrees awarded. *Degree requirements:* Foreign language and thesis not required. *Entrance requirements:* GRE General Test or MAT, minimum GPA of 2.5 overall, 3.0 in undergraduate major. Application deadline: 7/1. Application fee: $15. *Tuition:* $574.50 per semester full-time for state residents; $3105 per semester full-time for nonresidents. *Financial aid:* In 1990–91, 1 research assistantship, 17 teaching assistantships (2 to first-year students) awarded; federal work-study also available. Financial aid application deadline: 4/15; applicants required to submit FAF. *Faculty research:* Probability and statistics, analysis. • Dr. Joseph Quinn, Chairman, 704-547-4551. Application contact: Kathi M. Baucom, Director of Admissions, 704-547-2213.

University of North Carolina at Greensboro, College of Arts and Sciences, Department of Mathematics, Greensboro, NC 27412. Department awards MA, M Ed, CAS. Faculty: 14 full-time (1 woman), 1 (woman) part-time. Matriculated students: 12 (7 women); includes 0 minority. *Degree requirements:* For master's, thesis required, foreign language not required. *Entrance requirements:* For master's, for MA: GRE General Test, MAT; for M Ed: GRE General Test or MAT or NTE; for CAS, GRE General Test or MAT or NTE. Application fee: $35. *Tuition:* $751 per semester full-time for state residents; $3685 per semester full-time for nonresidents. *Financial aid:* Research assistantships, teaching assistantships available. • Dr. Paul Duvall, Head, 919-334-5836.

University of North Carolina at Wilmington, College of Arts and Sciences, Department of Mathematical Sciences, Wilmington, NC 28403. Department awards MS. Faculty: 9 full-time (0 women), 0 part-time. Matriculated students: 3 full-time (1 woman), 7 part-time (6 women); includes 0 minority, 3 foreign. Average age 31. 5 applicants, 100% accepted. In 1990, 0 degrees awarded. *Degree requirements:* Thesis, written and oral comprehensive exams. *Entrance requirements:* GRE General Test (minimum combined score of 1000 required), GRE Subject Test, minimum B average in undergraduate major. Application deadline: 7/1 (applications processed on a rolling basis). Application fee: $15. *Tuition:* $651 per semester full-time for state residents; $3178 per semester full-time for nonresidents. *Financial aid:* In 1990–91, $5000 in aid awarded. 7 teaching assistantships were awarded; federal work-study and career-related internships or fieldwork also available. Aid available to part-time students. Financial aid application deadline: 3/15; applicants required to submit FAF. • Dr. Douglas D. Smith, Chairman, 919-395-3290.

University of North Dakota, College of Arts and Sciences, Department of Mathematics, Grand Forks, ND 58202. Department awards MS. Part-time programs available. Faculty: 15 full-time (1 woman), 0 part-time. Matriculated students: 11 full-time (4 women), 2 part-time (1 woman). 9 applicants, 100% accepted. In 1990, 5 degrees awarded. *Degree requirements:* Thesis or alternative. *Entrance requirements:* TOEFL (minimum score of 550 required), minimum GPA of 3.0. Application deadline: 3/15 (priority date, applications processed on a rolling basis). Application fee: $20. *Tuition:* $2250 per year full-time, $94 per semester hour part-time for state residents; $5616 per year full-time, $234 per semester hour part-time for nonresidents. *Financial aid:* In 1990–91, 14 students received aid. 1 fellowship, 0 research assistantships, 13 teaching assistantships awarded; full and partial tuition waivers, federal work-study, institutionally sponsored loans also available. Average monthly stipend for a graduate assistantship: $720. Financial aid application deadline: 3/15. • Dr. James Rue, Chairperson, 701-777-2881.

University of Northern Colorado, College of Arts and Sciences, Department of Mathematics and Applied Statistics, Program in Mathematics, Greeley, CO 80639. Offers educational mathematics (PhD), mathematics (MA). Faculty: 6 full-time (1 woman), 0 part-time. Matriculated students: 14 full-time (5 women), 4 part-time (3 women); includes 0 minority, 4 foreign. 14 applicants, 93% accepted. In 1990, 3 master's, 0 doctorates awarded. *Degree requirements:* For master's, thesis or alternative, comprehensive exams. *Entrance requirements:* GRE General Test. Application deadline: rolling. Application fee: $30. *Expenses:* Tuition of $1900 per year full-time, $106 per credit hour part-time for state residents; $6078 per year full-time, $338 per credit hour part-time for nonresidents. Fees of $320 per year full-time, $18 per credit hour part-time. *Financial aid:* In 1990–91, 2 fellowships (1 to a first-year student), 7 teaching assistantships (5 to first-year students), 2 graduate assistantships (1 to a first-year student) awarded; research assistantships also available. Financial aid application deadline: 3/1. • Dr. Christopher Cotter, Head, 303-351-2744. Application contact: Dr. Igor Szozyrba, Coordinator, 303-351-2011.

University of Northern Iowa, College of Natural Sciences, Department of Mathematics and Computer Science, Cedar Falls, IA 50614. Department offers programs in computer science education (MA), mathematics (MA), mathematics for elementary and middle schools (MA). Part-time programs available. Faculty: 14 full-time (3 women), 0 part-time. Matriculated students: 8 full-time (4 women), 6 part-time (5 women); includes 0 minority, 2 foreign. Average age 33. 8 applicants, 75% accepted. In 1990, 4 degrees awarded. *Degree requirements:* Thesis or alternative required, foreign language not required. *Application deadline:* 8/1 (priority date, applications processed on a rolling basis). *Application fee:* $20. *Tuition:* $2192 per year full-time, $240 per hour part-time for state residents; $5492 per year full-time, $240 per hour part-time for nonresidents. *Financial aid:* In 1990–91, 6 scholarships awarded; full and partial tuition waivers, federal work-study, and career-related internships or fieldwork also available. Aid available to part-time students. Financial aid application deadline: 3/1; applicants required to submit FAF. • Dr. Philip J. East, Acting Head, 319-273-2631.

University of North Florida, College of Arts and Sciences, Department of Mathematics and Statistics, Jacksonville, FL 32216. Department offers programs in computer science (MA), mathematical sciences (MA), statistics (MA). Part-time and evening/weekend programs available. Faculty: 20 full-time (3 women), 0 part-time. Matriculated students: 0 full-time, 21 part-time (11 women); includes 5 minority. *Degree requirements:* Comprehensive exam required, thesis optional, foreign

Directory: Mathematics

language not required. *Entrance requirements:* GRE, TOEFL, minimum GPA of 3.0. Application fee: $15. *Financial aid:* In 1990–91, 7 teaching assistantships awarded. • Dr. Leonard J. Lipkin, Chairman, 904-646-2653. Application contact: Dr. William J. Wilson, Graduate Director, 904-646-2653.

Announcement: MA in mathematical sciences requires 32 semester credits, with tracks in each of the disciplines of mathematics, statistics, and computing. Building upon a common core of multidisciplinary courses, advanced courses in each track generally emphasize applicability of the subject. Microcomputers, an IBM 4341 mainframe, and a modern suburban campus provide an attractive academic atmosphere.

University of North Texas, College of Arts and Sciences, Department of Mathematics, Denton, TX 76203. Department awards MA, MS, PhD. Part-time programs available. Faculty: 20 full-time (1 woman), 0 part-time. Matriculated students: 57 full-time (15 women), 17 part-time (4 women); includes 1 minority (Hispanic American), 23 foreign. Average age 27. 49 applicants, 43% accepted. In 1990, 12 master's, 5 doctorates awarded. Terminal master's awarded for partial completion of doctoral program. *Degree requirements:* For master's, 1 foreign language; for doctorate, 2 foreign languages, dissertation. *Entrance requirements:* GRE General Test. Application deadline: 8/1. Application fee: $25. *Expenses:* Tuition of $40 per credit hour for state residents; $128 per credit hour for nonresidents. Fees of $298 per year full-time, $38 per year part-time. *Financial aid:* In 1990–91, 6 research assistantships (3 to first-year students), 44 teaching assistantships (12 to first-year students) awarded; federal work-study, institutionally sponsored loans also available. Financial aid application deadline: 6/1. *Faculty research:* Differential equations, descriptive set theory, combinatorics, functional analysis, algebra. Total annual research budget: $259,000. • Dr. John Ed Allen, Chair, 817-565-2155.

University of Notre Dame, College of Science, Department of Mathematics, Notre Dame, IN 46556. Department awards MS, PhD. Faculty: 38 full-time (3 women), 0 part-time. Matriculated students: 37 full-time (10 women), 0 part-time; includes 0 minority, 28 foreign. Average age 25. 134 applicants, 6% accepted. In 1990, 7 master's awarded (100% continued full-time study); 6 doctorates awarded (50% entered university research/teaching, 50% found other work related to degree). Terminal master's awarded for partial completion of doctoral program. *Degree requirements:* For master's, comprehensive exam; for doctorate, dissertation, qualifying exam. *Entrance requirements:* GRE, TOEFL. Application deadline: 2/15 (priority date). Application fee: $25. *Tuition:* $13,385 per year full-time, $744 per credit hour part-time. *Financial aid:* In 1990–91, $764,000 in aid awarded. 7 fellowships (all to first-year students), 31 teaching assistantships (all to first-year students) were awarded; full and partial tuition waivers also available. Financial aid application deadline: 2/15. *Faculty research:* Analysis, topology, geometry, logic. Total annual research budget: $386,000. • Dr. Andrew J. Sommese, Chairman, 219-239-7083. Application contact: Dr. Alexander Hahn, Director of Graduate Study, 219-239-7245.

See full description on page 489.

University of Oklahoma, College of Arts and Sciences, Department of Mathematics, Norman, OK 73019. Department awards MA, MS, PhD, MBA/MS. Part-time programs available. Faculty: 35 full-time, 3 part-time. Matriculated students: 40 full-time (10 women), 18 part-time (6 women); includes 7 minority (4 Asian American, 1 black American, 1 Hispanic American, 1 Native American), 20 foreign. 131 applicants, 60% accepted. In 1990, 9 master's, 1 doctorate awarded. Terminal master's awarded for partial completion of doctoral program. *Degree requirements:* For master's, comprehensive exam required, thesis optional, foreign language not required; for doctorate, 2 foreign languages, dissertation, qualifying exam. *Entrance requirements:* TOEFL (minimum score of 550 required), TSE (minimum score of 210 required). Application deadline: 6/1 (priority date, applications processed on a rolling basis). Application fee: $10. *Expenses:* Tuition of $63 per credit hour for state residents; $192 per credit hour for nonresidents. Fees of $67.50 per semester. *Financial aid:* In 1990–91, $278,000 in aid awarded. 1 fellowship (to a first-year student), 6 research assistantships (0 to first-year students) were awarded; teaching assistantships, partial tuition waivers, federal work-study also available. Aid available to part-time students. *Faculty research:* Algebra, analysis, topology, geometry, applied mathematics. • Dr. Andy Magid, Chairman, 405-325-6711. Application contact: Curtis McKnight, Graduate Liaison, 405-325-6711.

University of Oregon, Graduate School, College of Arts and Sciences, Department of Mathematics, Eugene, OR 97403. Department awards MA, MS, PhD. Part-time programs available. Faculty: 32 full-time (2 women), 0 part-time. Matriculated students: 66 full-time (20 women), 8 part-time (1 woman); includes 22 foreign. Average age 29. 108 applicants, 86% accepted. In 1990, 13 master's, 4 doctorates awarded. Terminal master's awarded for partial completion of doctoral program. *Degree requirements:* For master's, thesis not required; for doctorate, 2 foreign languages, dissertation. *Entrance requirements:* GRE General Test, TOEFL. Application deadline: 3/1. Application fee: $40. *Tuition:* $1171 per quarter full-time, $247 per credit part-time for state residents; $1980 per quarter full-time, $336 per credit part-time for nonresidents. *Financial aid:* In 1990–91, $365,752 in aid awarded. 1 fellowship, 48 teaching assistantships (12 to first-year students) were awarded; federal work-study also available. Aid available to part-time students. Financial aid application deadline: 3/1; applicants required to submit FAF. *Faculty research:* Algebra, topology, analysis geometry, numerical analysis, statistics. Total annual research budget: $419,000. • Dr. Frank W. Anderson, Head, 503-346-4705.

University of Ottawa, Faculty of Science, Department of Mathematics, Ottawa, ON K1N 6N5, Canada. Department awards M Sc, PhD. Offered jointly with Carleton University. *Degree requirements:* For master's, thesis optional, foreign language not required; for doctorate, 1 foreign language, dissertation. *Entrance requirements:* For master's, honors bachelor's degree or equivalent, minimum B average; for doctorate, minimum B+ average. Application deadline: 3/1. Application fee: $10. *Faculty research:* Pure mathematics, applied mathematics, probability and statistics.

University of Pennsylvania, School of Arts and Sciences, Graduate Group in Mathematics, Philadelphia, PA 19104. Group awards AM, PhD. Faculty: 34 full-time (0 women), 0 part-time. Matriculated students: 40 full-time (8 women), 5 part-time (2 women); includes 17 minority (all Asian American), 27 foreign. Average age 25. 126 applicants, 14% accepted. In 1990, 6 master's, 5 doctorates awarded. Terminal master's awarded for partial completion of doctoral program. *Degree requirements:* For master's, 1 foreign language, thesis or alternative; for doctorate, 2 foreign languages, dissertation. *Entrance requirements:* GRE General Test, GRE Subject Test, TOEFL. Application deadline: 2/1. Application fee: $40. *Expenses:* Tuition of $15,619 per year full-time, $1978 per course part-time. Fees of $965 per year full-time, $112 per course part-time. *Financial aid:* In 1990–91, 11 fellowships (3 to first-year students), 34 teaching assistantships (7 to first-year students) awarded; institutionally sponsored loans also available. Average monthly stipend for a graduate assistantship: $933. Financial aid application deadline: 2/1; applicants required to submit GAPSFAS. • Dr. Ted Chinburg, Chairperson, 215-898-4080.

University of Pittsburgh, Faculty of Arts and Sciences, Department of Mathematics and Statistics, Program in Mathematics, Pittsburgh, PA 15260. Program awards MA, MS, PhD. Part-time programs available. Faculty: 46 full-time (5 women), 26 part-time (13 women). Matriculated students: 70 full-time (17 women), 29 part-time (12 women); includes 9 minority (8 Asian American, 1 Hispanic American), 44 foreign. 325 applicants, 31% accepted. In 1990, 13 master's, 6 doctorates awarded. Terminal master's awarded for partial completion of doctoral program. *Degree requirements:* For master's, oral exam required, foreign language and thesis not required; for doctorate, 2 foreign languages (computer language can substitute for one), dissertation, comprehensive exam, preliminary exam. *Entrance requirements:* GRE General Test, TOEFL. Application deadline: 3/15 (priority date, applications processed on a rolling basis). Application fee: $15 ($25 for foreign students). *Expenses:* Tuition of $2920 per semester full-time, $241 per credit part-time for state residents; $5840 per semester full-time, $482 per credit part-time for nonresidents. Fees of $156 per year. *Financial aid:* In 1990–91, 65 students received a total of $650,000 in aid awarded. 3 fellowships (0 to first-year students), 7 research assistantships (0 to first-year students), 52 teaching assistantships (15 to first-year students) were awarded; federal work-study, institutionally sponsored loans also available. Aid available to part-time students. Average monthly stipend for a graduate assistantship: $1050. Financial aid application deadline: 3/1; applicants required to submit FAF. *Faculty research:* Analysis, topology, combinatorics, differential geometry, differential equations. Total annual research budget: $750,000. • Application contact: Thomas A. Metzger, Director of Graduate Studies, 412-624-8343.

University of Puerto Rico, Mayagüez Campus, College of Arts and Sciences, Department of Mathematics, Mayagüez, PR 00709. Department awards MS. Part-time programs available. Faculty: 35 full-time (11 women), 0 part-time. Matriculated students: 20 full-time (6 women), 0 part-time; includes 16 minority (all Hispanic American), 4 foreign. 8 applicants, 50% accepted. In 1990, 1 degree awarded (100% found work related to degree). *Degree requirements:* 1 foreign language, thesis. *Application deadline:* 10/15. *Application fee:* $15. *Expenses:* Tuition of $45 per credit for commonwealth residents; $3000 per semester for nonresidents. Fees of $344 per semester. *Financial aid:* In 1990–91, 1 research assistantship (0 to first-year students), 10 teaching assistantships (8 to first-year students) awarded; federal work-study, institutionally sponsored loans also available. *Faculty research:* Automata theory, applied mathematics, linear algebra, statistics, logic. • Julio Quintana, Director, 809-832-4040 Ext. 3848.

University of Puerto Rico, Río Piedras, Faculty of Natural Sciences, Department of Mathematics, Río Piedras, PR 00931. Department awards MA. Part-time and evening/weekend programs available. Faculty: 46 (14 women). Matriculated students: 29 full-time (23 women), 37 part-time (28 women); includes 8 foreign. In 1990, 6 degrees awarded. *Degree requirements:* Thesis, comprehensive exam, minimum GPA of 3.0 required, foreign language not required. *Entrance requirements:* Minimum GPA of 2.75. Application deadline: 2/1. Application fee: $45. *Expenses:* Tuition of $55 per credit hour for commonwealth residents; $55 per credit hour (minimum) for nonresidents. Fees of $286 per year. *Financial aid:* Fellowships, research assistantships, teaching assistantships, partial tuition waivers, federal work-study, institutionally sponsored loans available. Financial aid application deadline: 5/31. *Faculty research:* Investigation in database logistics, cryptograph systems, distribution and spectral theory, Boolean function, differential equations. • Dr. Jorge López, Chairperson, 809-764-0000 Ext. 4671.

University of Regina, Faculty of Graduate Studies and Research, Faculty of Science, Department of Mathematics and Statistics, Regina, SK S4S 0A2, Canada. Offerings include mathematics (MA, M Sc, PhD). Department faculty: 21 full-time, 0 part-time. *Degree requirements:* For master's, thesis required, foreign language not required; for doctorate, variable foreign language requirement, dissertation. *Application deadline:* 7/2 (applications processed on a rolling basis). *Tuition:* $1500 per year full-time, $242 per year (minimum) part-time. • Dr. R. J. Tompkins, Head, 306-585-4148.

University of Rhode Island, College of Arts and Sciences, Department of Mathematics, Kingston, RI 02881. Department awards MS, PhD. *Degree requirements:* For master's, thesis optional; for doctorate, 1 foreign language, dissertation. *Application deadline:* 4/15. *Application fee:* $25. *Expenses:* Tuition of $2575 per year full-time, $120 per credit hour part-time for state residents; $5900 per year full-time, $274 per credit hour part-time for nonresidents. Fees of $696 per year full-time.

University of Rochester, College of Arts and Science, Department of Mathematics, Rochester, NY 14627-0001. Department awards MA, MS, PhD. Faculty: 29 full-time, 3 part-time. Matriculated students: 38 full-time (13 women), 1 part-time (0 women); includes 1 minority (Hispanic American), 30 foreign. 102 applicants, 11% accepted. In 1990, 14 master's, 2 doctorates awarded. Terminal master's awarded for partial completion of doctoral program. *Degree requirements:* For master's, thesis required (for some programs), foreign language not required; for doctorate, 1 foreign language (computer language can substitute), dissertation, written and oral preliminary exams. *Entrance requirements:* For doctorate, GRE, TOEFL. Application deadline: 2/15 (priority date). Application fee: $25. *Expenses:* Tuition of $473 per credit hour. Fees of $243 per year full-time. *Financial aid:* Fellowships, teaching assistantships, federal work-study, institutionally sponsored loans, and career-related internships or fieldwork available. Financial aid application deadline: 2/15. *Faculty research:* Analysis, algebra, analytic number theory, topology, probability. • Samuel Gitler, Chair, 716-275-9422. Application contact: Chairman, Graduate Committee, 716-275-4411.

University of Saskatchewan, College of Arts and Sciences, Department of Mathematics, Saskatoon, SK S7N 0W0, Canada. Department awards MA, M Math, PhD. *Degree requirements:* Thesis/dissertation. *Entrance requirements:* TOEFL. Application fee: $0.

University of South Alabama, College of Arts and Sciences, Department of Mathematics, Mobile, AL 36688. Department awards MS. Part-time and evening/weekend programs available. Faculty: 16 full-time (2 women), 0 part-time. Matriculated students: 10 full-time (4 women), 3 part-time (0 women); includes 1 minority (black American), 7 foreign. 31 applicants, 39% accepted. In 1990, 2 degrees awarded. *Degree requirements:* Computer language, written comprehensive exam required, thesis not required. *Entrance requirements:* GRE, minimum B average. Application deadline: 9/1 (priority date, applications processed on a rolling basis). Application fee: $10. *Tuition:* $46 per credit hour. *Financial aid:* In 1990–91, 22 research assistantships awarded; fellowships also available. Aid available to part-time students. Financial aid application deadline: 4/1. *Total annual research budget:* $13,500. • Dr. Suzanne McGill, Chairperson, 205-460-6264.

University of South Carolina, Graduate School, College of Science and Mathematics, Department of Mathematics, Columbia, SC 29208. Department offers programs in mathematics (MA, MS, PhD), mathematics education (MAT, M Math). MAT offered in cooperation with the College of Education. Part-time programs available. Faculty: 36

full-time (1 woman), 0 part-time. Matriculated students: 92 (30 women); includes 37 foreign. 143 applicants, 48% accepted. In 1990, 14 master's awarded; 4 doctorates awarded (100% entered university research/teaching). Terminal master's awarded for partial completion of doctoral program. *Degree requirements:* For master's, thesis required, foreign language not required; for doctorate, 1 foreign language, computer language, dissertation. *Entrance requirements:* GRE General Test. Application deadline: 9/1 (priority date, applications processed on a rolling basis). Application fee: $25. *Tuition:* $1404 per semester full-time. *Financial aid:* In 1990–91, $450,000 in aid awarded. 1 fellowship (to a first-year student), 50 teaching assistantships (15 to first-year students) were awarded. Financial aid application deadline: 3/15. *Faculty research:* Applied mathematics, analysis, discrete mathematics, algebra, topology. • Dr. Colin Bennett Jr., Chairman, 803-777-4224. Application contact: Dr. Anton R. Schep, Graduate Director, 803-777-4226.

University of South Dakota, College of Arts and Sciences, Department of Mathematics, Vermillion, SD 57069-2390. Department awards MA, MNS. Faculty: 7 full-time (0 women), 1 part-time (0 women). Matriculated students: 11 full-time (4 women), 6 part-time (3 women); includes 1 minority (Asian American), 1 foreign. 9 applicants, 78% accepted. In 1990, 9 degrees awarded. *Degree requirements:* Foreign language not required. *Application fee:* $15. *Expenses:* Tuition of $61 per hour for state residents; $122 per hour for nonresidents. Fees of $24.41 per hour. *Financial aid:* Teaching assistantships available. Financial aid applicants required to submit GAPSFAS. • Dr. Alexander Mehaffey, Chairman, 605-677-5262. Application contact: Dr. Wallace Raab, Adviser, 605-677-5262.

University of Southern California, Graduate School, College of Letters, Arts and Sciences, Division of Natural Sciences and Mathematics, Department of Mathematics, Los Angeles, CA 90089. Department offers programs in applied mathematics (MA, MS, PhD), mathematics (MA, PhD), statistics (MS). Faculty: 33 (2 women). Matriculated students: 86 full-time (18 women), 17 part-time (5 women); includes 11 minority (8 Asian American, 2 black American, 1 Hispanic American), 63 foreign. Average age 27. 187 applicants, 64% accepted. In 1990, 29 master's, 6 doctorates awarded. *Degree requirements:* For doctorate, dissertation. *Entrance requirements:* GRE General Test. Application deadline: 7/1 (priority date). Application fee: $50. *Expenses:* Tuition of $12,120 per year full-time, $505 per unit part-time. Fees of $280 per year. *Financial aid:* 82 students received aid. Fellowships, research assistantships, teaching assistantships, federal work-study, institutionally sponsored loans available. Average monthly stipend for a graduate assistantship: $878. Financial aid application deadline: 3/1. • Dr. Robert Guralnick, Chairman, 213-740-1717.

See full description on page 493.

University of Southern Mississippi, College of Science and Technology, Department of Mathematics, Hattiesburg, MS 39406. Department awards MS. Part-time programs available. Faculty: 16 full-time, 0 part-time. Matriculated students: 11 full-time (3 women), 0 part-time; includes 0 minority, 6 foreign. In 1990, 2 degrees awarded. *Degree requirements:* Thesis or alternative, written/oral comprehensive exam required, foreign language not required. *Entrance requirements:* GRE General Test (minimum combined score of 1000 required), TOEFL (minimum score of 527 required), minimum GPA of 3.0 in mathematics, bachelors degree in mathematics or related field. Application deadline: 8/9 (priority date, applications processed on a rolling basis). Application fee: $0 ($25 for foreign students). *Expenses:* Tuition of $968 per semester full-time, $93 per semester hour part-time. Fees of $12 per semester part-time for state residents; $591 per year full-time, $12 per semester part-time for nonresidents. *Financial aid:* In 1990–91, 11 students received aid. 8 teaching assistantships (3 to first-year students) awarded; federal work-study, institutionally sponsored loans also available. Financial aid application deadline: 3/15. *Faculty research:* Algebra, mathematical physics, graph theory, combinatorics, probability. Total annual research budget: $77,840. • Dr. Wallace Pye, Chair, 601-266-4289.

University of South Florida, College of Arts and Sciences, Department of Mathematics, Tampa, FL 33620. Department offers programs in applied mathematics (PhD), mathematics (MA, PhD). Part-time and evening/weekend programs available. Faculty: 25. Matriculated students: 42 full-time (11 women), 18 part-time (6 women); includes 4 minority (2 Asian American, 1 Hispanic American, 1 Native American), 19 foreign. Average age 31. 171 applicants, 36% accepted. In 1990, 13 master's, 4 doctorates awarded. *Degree requirements:* For doctorate, 2 foreign languages (computer language can substitute for one), dissertation. *Entrance requirements:* For master's, GRE General Test (minimum combined score of 1000 required), minimum GPA of 3.0 (in last 60 credit hours). Application fee: $15. *Tuition:* $79.40 per credit hour for state residents; $241.33 per credit hour for nonresidents. *Financial aid:* In 1990–91, 35 students received a total of $89,456 in aid awarded. • Kenneth L. Pothoven, Chairperson, 813-974-2643.

See full description on page 495.

University of Southwestern Louisiana, College of Sciences, Department of Mathematics, Lafayette, LA 70504. Department awards MS, PhD. Faculty: 14 full-time (2 women), 0 part-time. Matriculated students: 25 full-time (8 women), 7 part-time (4 women); includes 1 minority (Asian American), 16 foreign. Average age 23. 10 applicants, 80% accepted. In 1990, 5 master's awarded (100% continued full-time study); 4 doctorates awarded (100% entered university research/teaching). Terminal master's awarded for partial completion of doctoral program. *Degree requirements:* For master's, thesis or alternative required, foreign language not required; for doctorate, 2 foreign languages (computer language can substitute for one), dissertation. *Entrance requirements:* For master's, GRE General Test (minimum combined score of 850 required); for doctorate, GRE General Test (minimum combined score of 1000 required). Application deadline: 8/15. Application fee: $5. *Tuition:* $1560 per year full-time, $228 per credit (minimum) part-time for state residents; $3310 per year full-time, $228 per credit (minimum) part-time for nonresidents. *Financial aid:* In 1990–91, $154,387 in aid awarded. 5 fellowships (3 to first-year students), 1 research assistantship (0 to first-year students), 13 teaching assistantships (4 to first-year students) were awarded; full tuition waivers also available. Financial aid application deadline: 3/1. *Faculty research:* Topology, algebra, applied mathematics, analysis. • Dr. Bradd Clark, Head, 318-231-6702. Application contact: Dr. C. Y. Chan, Graduate Coordinator, 318-231-5288.

See full description on page 497.

University of Tennessee, Knoxville, College of Liberal Arts, Department of Mathematics, Knoxville, TN 37996. Department offers programs in mathematical ecology (PhD), mathematics (M Math, MS, PhD). Part-time programs available. Faculty: 46 (3 women). Matriculated students: 47 full-time (12 women), 40 part-time (20 women); includes 4 minority (2 Asian American, 1 black American, 1 Hispanic American), 22 foreign. 107 applicants, 56% accepted. In 1990, 17 master's, 1 doctorate awarded. *Degree requirements:* For master's, thesis or alternative required, foreign language not required; for doctorate, 2 foreign languages, dissertation. *Entrance requirements:* TOEFL (minimum score of 525 required), minimum GPA of 2.5. Application deadline: 2/1 (priority date, applications processed on a rolling basis). Application fee: $15. *Tuition:* $1086 per semester full-time, $142 per credit hour part-time for state residents; $2768 per semester full-time, $308 per credit hour part-time for nonresidents. *Financial aid:* In 1990–91, 2 fellowships, 0 research assistantships, 53 teaching assistantships awarded; federal work-study, institutionally sponsored loans also available. Financial aid application deadline: 2/1; applicants required to submit FAF. • Dr. John B. Conway, Head, 615-974-2464.

University of Texas at Arlington, College of Science, Department of Mathematics, Arlington, TX 76019. Department offers programs in mathematical sciences (PhD), mathematics (MS). PhD offered jointly with the University of Texas at Dallas. Matriculated students: 39 full-time (15 women), 26 part-time (6 women); includes 12 minority (9 Asian American, 1 black American, 2 Hispanic American), 19 foreign. 90 applicants, 36% accepted. In 1990, 4 master's, 4 doctorates awarded. *Entrance requirements:* For master's, GRE General Test. Application deadline: rolling. Application fee: $25. *Tuition:* $40 per hour for state residents; $148 per hour for nonresidents. *Financial aid:* In 1990–91, 6 teaching assistantships awarded. *Total annual research budget:* $186,799. • Dr. George Fix, Chair, 817-273-3261. Application contact: Dr. A. Alan Gillespie, Graduate Adviser, 817-273-3261.

University of Texas at Austin, Graduate School, College of Natural Sciences, Department of Mathematics, Austin, TX 78712. Department offers programs in mathematics (MA, PhD), statistics (MS Stat). Matriculated students: 149 full-time (34 women), 0 part-time; includes 12 minority (9 Asian American, 3 Hispanic American), 41 foreign. 117 applicants, 83% accepted. In 1990, 12 master's, 14 doctorates awarded. *Entrance requirements:* GRE. Application deadline: 2/1 (priority date, applications processed on a rolling basis). Application fee: $40 ($75 for foreign students). *Tuition:* $510.30 per semester for state residents; $1806 per semester for nonresidents. *Financial aid:* Fellowships, teaching assistantships available. Financial aid application deadline: 3/1. • Dr. John D. Dollard, Chairman, 512-471-0117. Application contact: Dr. Ted Odell, Graduate Adviser, 512-471-7711.

University of Texas at Dallas, School of Natural Sciences and Mathematics, Program in Mathematical Sciences, Richardson, TX 75083-0688. Offers applied mathematics (MS, PhD), applied statistics (MS, PhD), pure mathematics (MS), theoretical statistics (MS, PhD). Part-time and evening/weekend programs available. Faculty: 11 full-time (0 women), 18 part-time (3 women). Matriculated students: 25 full-time (5 women), 43 part-time (18 women); includes 7 minority (2 Asian American, 3 black American, 2 Hispanic American), 18 foreign. Average age 33. In 1990, 7 master's, 1 doctorate awarded. *Degree requirements:* For master's, thesis optional, foreign language not required; for doctorate, dissertation required, foreign language not required. *Entrance requirements:* For master's, GRE General Test (minimum combined score of 1050 required), TOEFL (minimum score of 550 required), minimum GPA of 3.0 in upper level course work in field, secondary certificate in mathematics or computer science (MAT only); for doctorate, GRE General Test (minimum combined score of 1300 required), TOEFL (minimum score of 550 required), minimum GPA of 3.5 in upper level course work in field. Application deadline: 7/15 (applications processed on a rolling basis). Application fee: $0 ($75 for foreign students). *Expenses:* Tuition of $360 per semester full-time, $100 per semester (minimum) part-time for state residents; $2196 per semester full-time, $122 per semester hour (minimum) part-time for nonresidents. Fees of $338 per semester full-time, $22 per hour (minimum) part-time. *Financial aid:* 0 students received aid. Fellowships, research assistantships, teaching assistantships, federal work-study available. Aid available to part-time students. Financial aid application deadline: 11/1; applicants required to submit FAF. • Dr. John Van Ness, Head, 214-690-2161.

Announcement: PhD in theoretical and applied statistics and applied mathematics. MS available in these areas plus pure mathematics. Research interests include systems and control theory, signal processing, inverse scattering, linear models, time series, multivariate statistics, stochastic processes, robust statistics, nonparametric statistics, pattern recognition, relativity theory, differential equations, game theory, and risk assessment. Excellent computing facilities, including parallel machines.

University of Texas at El Paso, College of Science, Department of Mathematical Sciences, 500 West University Avenue, El Paso, TX 79968. Department awards MAT, MS. Matriculated students: 18 full-time (5 women), 7 part-time (4 women); includes 6 minority (2 Asian American, 4 Hispanic American), 14 foreign. In 1990, 7 degrees awarded. *Application deadline:* 7/1 (priority date, applications processed on a rolling basis). *Application fee:* $0 ($50 for foreign students). *Expenses:* Tuition of $360 per semester full-time, $100 per semester (minimum) part-time for state residents; $2304 per semester full-time, $128 per credit hour (minimum) part-time for nonresidents. Fees of $137 per semester full-time, $28.50 per semester (minimum) part-time. *Financial aid:* Application deadline 3/1. • Simon Bernau, Chair, 915-747-5761. Application contact: Diana Guerrero, Admissions Office, 915-747-5576.

University of Texas at San Antonio, College of Sciences and Engineering, Division of Mathematics, Computer Science and Statistics, San Antonio, TX 78285. Division offers programs in computer science (MS); mathematics (MS), including mathematics education, statistics. Part-time and evening/weekend programs available. Faculty: 23 full-time (4 women), 1 part-time (0 women). Matriculated students: 87 (25 women); includes 21 minority (4 Asian American, 17 Hispanic American), 13 foreign. 26 applicants, 69% accepted. In 1990, 9 degrees awarded. *Degree requirements:* Computer language, comprehensive exam required, foreign language and thesis not required. *Entrance requirements:* GRE General Test, TOEFL, minimum GPA of 3.0. Application deadline: 7/1 (applications processed on a rolling basis). Application fee: $20. *Expenses:* Tuition of $100 per semester hour (minimum) for state residents; $128 per semester hour (minimum) for nonresidents. Fees of $48 per semester hour (minimum). *Financial aid:* In 1990–91, 4 research assistantships (2 to first-year students), 20 teaching assistantships (12 to first-year students) awarded. Average monthly stipend for a graduate assistantship: $880. *Faculty research:* Computer applications to medicine, software engineering, computer graphics, differential equations, mathematical physics. • Shair Ahmad, Director, 512-691-4454.

University of Texas at Tyler, School of Sciences and Mathematics, Department of Mathematics and Computer Science, Program in Mathematics, Tyler, TX 75701. Offers interdisciplinary studies (MA, MS). Faculty: 4 full-time (1 woman), 0 part-time. Matriculated students: 9. In 1990, 1 degree awarded. *Degree requirements:* Comprehensive exam required, thesis not required. *Entrance requirements:* GRE General Test (minimum combined score of 1000 required). Application fee: $0. *Expenses:* Tuition of $760 per year full-time, $150 per course part-time for state residents; $3300 per year full-time, $410 per course part-time for nonresidents. Fees of $340 per year full-time, $30 per course part-time. *Financial aid:* Application deadline 7/1. *Faculty research:* Commutative algebra, differential equations, real analysis, probability and statistics, applied mathematics. • Dr. Robert Cranford, Chair, 903-566-7402. Application contact: Martha D. Wheat, Director of Admissions, 903-566-7201.

Directory: Mathematics

University of Texas–Pan American, College of Arts and Sciences, Department of Mathematics, Edinburg, TX 78539. Department awards MSIS. Part-time and evening/weekend programs available. Faculty: 4 full-time (1 woman), 0 part-time. Matriculated students: 8 full-time; includes 4 minority (all Hispanic American). In 1990, 0 degrees awarded. *Degree requirements:* Thesis or alternative, comprehensive exam. *Entrance requirements:* GRE General Test, minimum GPA of 3.0. Application fee: $0. *Tuition:* $438.40 per semester full-time for state residents; $1734 per semester full-time for nonresidents. *Financial aid:* Teaching assistantships, partial tuition waivers, federal work-study, institutionally sponsored loans available. Aid available to part-time students. Financial aid application deadline: 6/1. *Faculty research:* Boundary value problems in differential equations, training of public school teachers in methods of presenting mathematics. Total annual research budget: $40,000. • Dr. Joe Chance, Chair, 512-381-3452.

University of the District of Columbia, College of Physical Science, Engineering, and Technology, Department of Mathematics, 4200 Connecticut Avenue, NW, Washington, DC 20008. Department awards MST. Part-time and evening/weekend programs available. Matriculated students: 7. In 1990, 3 degrees awarded. *Degree requirements:* Comprehensive exam required, foreign language and thesis not required. *Entrance requirements:* GRE General Test, writing proficiency exam. Application deadline: 7/1 (priority date, applications processed on a rolling basis). Application fee: $10. *Tuition:* $632 per semester full-time for district residents; $1232 per semester full-time for nonresidents. • Dr. Jagyt Bakshi, Chairman, 202-282-3171. Application contact: LaHugh Bankston, Registrar and Director of Admissions, Graduate Studies and Research, 202-282-3578.

University of Toledo, College of Arts and Sciences, Department of Mathematics, Toledo, OH 43606. Offerings include mathematics (MA, MS Ed, PhD). Department faculty: 29 full-time (2 women), 0 part-time. *Degree requirements:* For doctorate, 2 foreign languages, dissertation. *Entrance requirements:* For doctorate, GRE General Test, GRE Subject Test. Application deadline: 9/8 (priority date). Application fee: $30. *Tuition:* $122.59 per credit hour for state residents; $193.40 per credit hour for nonresidents. • Dr. Harvey Wolff, Chairman, 419-537-2568.

University of Toronto, School of Graduate Studies, Physical Sciences Division, Department of Mathematics and Applied Mathematics, Toronto, ON M5S 1A1, Canada. Department awards M Sc, M Sc T, PhD. Faculty: 64. Matriculated students: 72 full-time (11 women), 4 part-time (1 woman); includes 22 foreign. 152 applicants, 47% accepted. In 1990, 17 master's, 4 doctorates awarded. *Degree requirements:* For master's, thesis optional; for doctorate, dissertation. *Application deadline:* 4/15. *Application fee:* $50. *Expenses:* Tuition of $2220 per year full-time, $666 per year part-time for Canadian residents; $10,198 per year full-time, $305.05 per year part-time for nonresidents. Fees of $277.56 per year full-time, $82.73 per year part-time. *Financial aid:* Application deadline 2/1. • J. Friedlander, Chair, 416-978-3320.

University of Tulsa, College of Engineering and Physical Sciences, Department of Mathematical and Computer Sciences, Tulsa, OK 74104. Offerings include mathematical and computer sciences (MS). Department faculty: 12 full-time (1 woman), 0 part-time. *Degree requirements:* Computer language, thesis or alternative required, foreign language not required. *Entrance requirements:* GRE General Test, TOEFL. Application deadline: rolling. Application fee: $30. *Tuition:* $350 per credit hour. • Dr. William Coberly, Chairperson, 918-631-3119. Application contact: Dr. Roger Wainwright, Adviser, 918-631-3143.

University of Utah, College of Science, Department of Mathematics, Salt Lake City, UT 84112. Department awards MA, M Phil, MS, M Stat, PhD. Part-time programs available. Faculty: 42 (2 women). Matriculated students: 61 full-time (14 women), 23 part-time (8 women); includes 1 minority (Hispanic American), 35 foreign. Average age 29. In 1990, 13 master's, 3 doctorates awarded. Terminal master's awarded for partial completion of doctoral program. *Degree requirements:* For master's, 1 foreign language, thesis or alternative, written or oral exam; for doctorate, 2 foreign languages, dissertation, written and oral exams. *Application deadline:* 3/15. *Application fee:* $25 ($50 for foreign students). *Tuition:* $195 per credit for state residents; $505 per credit for nonresidents. *Financial aid:* In 1990–91, 50 teaching assistantships awarded. Financial aid application deadline: 3/15. *Faculty research:* Algebraic geometry, differential geometry, scientific computing, topology, mathematical biology. • Klaus Schmitt, Chair, 801-581-6851. Application contact: Herbert Clemens, Director of Graduate Studies, 801-581-8005.

See full description on page 499.

University of Vermont, College of Engineering and Mathematics, Department of Mathematics and Statistics, Program in Mathematics, Burlington, VT 05405. Program awards MAT, MS, MST. Matriculated students: 15; includes 1 minority (Hispanic American), 0 foreign. 27 applicants, 63% accepted. In 1990, 5 degrees awarded. *Degree requirements:* Foreign language and thesis not required. *Entrance requirements:* GRE General Test, GRE Subject Test, TOEFL (minimum score of 550 required). Application deadline: 4/1 (priority date, applications processed on a rolling basis). Application fee: $25. *Expenses:* Tuition of $206 per credit for state residents; $564 per credit for nonresidents. Fees of $150 per semester full-time. *Financial aid:* In 1990–91, 1 research assistantship, 16 teaching assistantships awarded; fellowships also available. Financial aid application deadline: 3/1. • Application contact: Dr. D. Archdeacon, Coordinator, 802-656-2940.

University of Victoria, Faculty of Arts and Science, Department of Mathematics, Victoria, BC V8W 2Y2, Canada. Department offers programs in applied mathematics (MA, M Sc), industrial mathematics (MA), pure mathematics (MA, M Sc), statistics (MA, M Sc). Faculty: 24 full-time (1 woman), 0 part-time. Matriculated students: 10 full-time (2 women), 0 part-time; includes 4 foreign. Average age 29. 59 applicants, 39% accepted. In 1990, 2 degrees awarded. *Degree requirements:* Thesis (for some programs), oral exam required, foreign language not required. *Entrance requirements:* Honors degree in mathematics. Application deadline: 5/31 (priority date, applications processed on a rolling basis). Application fee: $20. *Expenses:* Tuition of $754 per semester. Fees of $23 per year. *Financial aid:* Fellowships, research assistantships, teaching assistantships, institutionally sponsored loans, and career-related internships or fieldwork available. *Faculty research:* Functional analysis, differential equations, approximation theory. • Dr. D. J. Leeming, Chair, 604-721-7436. Application contact: Dr. J. Phillips, Graduate Adviser, 604-721-7450.

University of Virginia, Graduate School of Arts and Sciences, Department of Mathematics, Charlottesville, VA 22906. Department awards MA, MAT, MS, PhD. Faculty: 31 full-time (1 woman), 2 part-time (both women). Matriculated students: 56 full-time (21 women), 4 part-time (1 woman); includes 0 minority, 9 foreign. Average age 27. 86 applicants, 83% accepted. In 1990, 9 master's, 5 doctorates awarded. *Degree requirements:* For master's, 1 foreign language, thesis; for doctorate, 2 foreign languages, dissertation. *Entrance requirements:* GRE General Test, GRE Subject Test. Application deadline: 7/15. Application fee: $40. *Expenses:* Tuition of $2740 per year full-time, $904 per year (minimum) part-time for state residents; $8950 per year full-time, $2960 per year (minimum) part-time for nonresidents. Fees of $586 per year full-time, $342 per year part-time. *Financial aid:* Application deadline 2/1. • Lawrence E. Thomas, Chairman, 804-924-4919. Application contact: William A. Elwood, Associate Dean, 804-924-7184.

Announcement: The department seeks to integrate graduate students into an active research community. Balanced program with doctoral advisers available for main areas of mathematics, mathematical physics, and statistics. Students awarded teaching fellowships or teaching assistant positions receive tuition fellowships to cover required tuition fees. Write for description of Graduate Program.

University of Washington, College of Arts and Sciences, Department of Mathematics, Seattle, WA 98195. Department awards MA, MS, PhD. Part-time programs available. Faculty: 64 full-time (4 women), 3 part-time (0 women). *Degree requirements:* For master's, thesis optional; for doctorate, 2 foreign languages. *Entrance requirements:* GRE, TOEFL (minimum score of 500 required), minimum GPA of 3.0. Application deadline: 7/1. Application fee: $35. *Tuition:* $1129 per quarter full-time, $324 per credit (minimum) part-time for state residents; $2824 per quarter full-time, $809 per credit (minimum) part-time for nonresidents. *Financial aid:* Fellowships, research assistantships, teaching assistantships available. Financial aid application deadline: 2/15. *Faculty research:* Algebraic topology, algebra, analysis, partial differential equations, optimization. • Edgar Lee Stout, Chairman, 206-543-1150. Application contact: Ronald S. Irving, Graduate Adviser, 206-543-1199.

University of Waterloo, Faculty of Mathematics, Department of Combinatorics/Optimization, Waterloo, ON N2L 3G1, Canada. Department awards M Math, M Phil, PhD. Part-time programs available. Faculty: 21 full-time, 0 part-time. Matriculated students: 27 full-time (5 women), 1 part-time (0 women); includes 12 foreign. 54 applicants, 24% accepted. In 1990, 10 master's, 6 doctorates awarded. *Degree requirements:* For master's, thesis or alternative required, foreign language not required; for doctorate, 1 foreign language, dissertation. *Entrance requirements:* For master's, TOEFL (minimum score of 550 required), honor's degree in field, minimum B average; for doctorate, TOEFL (minimum score of 550 required), master's degree. Application deadline: 4/15. Application fee: $25. *Expenses:* Tuition of $757 per year full-time, $530 per year part-time for Canadian residents; $3127 per year for nonresidents. Fees of $68 per year full-time, $17 per year part-time. *Financial aid:* In 1990–91, 12 research assistantships (21 to first-year students), 29 teaching assistantships (38 to first-year students) awarded. Average monthly stipend for a graduate assistantship: $530. *Faculty research:* Graph theory, continuous optimization, operations research. • Dr. I. Goulden, Chair, 519-885-1211 Ext. 3481. Application contact: Dr. H. Wolkowicz, Associate Chair, 519-885-1211 Ext. 4597.

University of Waterloo, Faculty of Mathematics, Department of Pure Mathematics, Waterloo, ON N2L 3G1, Canada. Department awards M Math, M Phil, PhD. Faculty: 23 full-time, 0 part-time. Matriculated students: 17 full-time (2 women), 0 part-time; includes 7 foreign. 46 applicants, 28% accepted. In 1990, 3 master's, 1 doctorate awarded. *Degree requirements:* For master's, thesis required, foreign language not required; for doctorate, 1 foreign language, dissertation. *Entrance requirements:* For master's, TOEFL (minimum score of 550 required), honor's degree in field, minimum B average; for doctorate, TOEFL (minimum score of 550 required), master's degree. Application deadline: 4/15. Application fee: $25. *Expenses:* Tuition of $757 per year full-time, $530 per year part-time for Canadian residents; $3127 per year for nonresidents. Fees of $68 per year full-time, $17 per year part-time. *Financial aid:* In 1990–91, 8 research assistantships (2 to first-year students), 20 teaching assistantships (3 to first-year students) awarded; scholarships also available. Average monthly stipend for a graduate assistantship: $530. Financial aid application deadline: 4/15. *Faculty research:* Algebra, topology, information theory, analysis. • Dr. F. Zorzitto, Chair, 519-885-1211 Ext. 4073. Application contact: K. Davidson, Graduate Officer, 519-885-1211 Ext. 4081.

University of Western Ontario, Physical Sciences Division, Department of Mathematics, London, ON N6A 3K7, Canada. Department awards MA, PhD. Terminal master's awarded for partial completion of doctoral program. *Degree requirements:* For master's, thesis or alternative required, foreign language not required; for doctorate, 1 foreign language, dissertation, qualifying exam. *Entrance requirements:* For master's, TOEFL (minimum score of 550 required), minimum B average, honors degree; for doctorate, TOEFL (minimum score of 550 required), master's degree. *Tuition:* $1015 per year full-time, $1050 per year part-time for Canadian residents; $4207 per year for nonresidents.

University of West Florida, College of Arts and Sciences, Department of Mathematics and Statistics, Pensacola, FL 32514-5750. Offerings include mathematics (MA). Department faculty: 11 full-time (10 women), 0 part-time. *Degree requirements:* Thesis optional, foreign language not required. *Entrance requirements:* GRE (minimum score of 1000 required), minimum GPA of 3.0. Application deadline: 7/19. Application fee: $15. *Tuition:* $86.38 per credit hour for state residents; $288.79 per credit hour for nonresidents. • Dr. D. Byrkit, Chairperson, 904-474-2291.

University of Windsor, Faculty of Science, Department of Mathematics and Statistics, Windsor, ON N9B 3P4, Canada. Department awards M Sc, PhD. Part-time programs available. Faculty: 22 full-time (1 woman), 0 part-time. Matriculated students: 21 full-time, 0 part-time; includes 15 foreign. In 1990, 3 master's, 4 doctorates awarded. *Degree requirements:* For master's, thesis (for some programs); for doctorate, dissertation. *Entrance requirements:* For master's, TOEFL (minimum score of 550 required), GRE, minimum B average. Application deadline: 7/1 (priority date, applications processed on a rolling basis). Application fee: $0. *Tuition:* $819.15 per semester full-time for Canadian residents; $3646 per semester full-time for nonresidents. *Financial aid:* Research assistantships, teaching assistantships available. Average monthly stipend for a graduate assistantship: $700. • Dr. Richard Caron, Head, 519-253-4232 Ext. 3017. Application contact: Admissions Officer, 519-253-4232 Ext. 2108.

University of Wisconsin–Madison, College of Letters and Science, Department of Mathematics, Madison, WI 53706. Department awards MA, PhD. Faculty: 64 full-time, 0 part-time. Matriculated students: 186 full-time (42 women), 34 part-time (5 women); includes 4 minority (2 Asian American, 2 black American), 37 foreign. 374 applicants, 72% accepted. In 1990, 28 master's, 15 doctorates awarded. *Application deadline:* rolling. *Application fee:* $20. *Financial aid:* In 1990–91, 166 students received aid. 13 fellowships, 11 research assistantships, 135 teaching assistantships, 7 project assistantships awarded; institutionally sponsored loans also available. Financial aid application deadline: 1/15; applicants required to submit FAF. • Simon Hellerstein, Chairperson, 608-263-8884. Application contact: Joanne Nagy, Assistant Dean of the Graduate School, 608-262-2433.

University of Wisconsin–Milwaukee, College of Letters and Sciences, Department of Mathematical Sciences, Milwaukee, WI 53201. Department offers program in mathematics (MS, PhD). Faculty: 31 full-time, 0 part-time. Matriculated students: 36 full-time (8 women), 30 part-time (9 women); includes 3 minority (1 Asian American, 1 black American, 1 Hispanic American), 29 foreign. 115 applicants, 38% accepted. In 1990, 15 master's, 5 doctorates awarded. *Degree requirements:* For master's, foreign language and thesis not required; for doctorate, 2 foreign languages, dissertation. *Application deadline:* 3/1 (priority date, applications processed on a rolling basis). *Application fee:* $20. *Financial aid:* In 1990–91, 7 fellowships, 0 research assistantships, 48 teaching assistantships awarded. Financial aid application deadline: 4/15. • Mark Tepley, Program Representative, 414-229-5110.

University of Wyoming, College of Arts and Sciences, Department of Mathematics, Laramie, WY 82071. Department awards MA, MAT, MS, MST, PhD. Faculty: 30 full-time (3 women), 0 part-time. Matriculated students: 27 full-time (10 women), 9 part-time (3 women); includes 1 minority (Hispanic American), 12 foreign. In 1990, 3 master's, 0 doctorates awarded. *Entrance requirements:* GRE General Test, minimum GPA of 3.0. Application deadline: 6/1 (priority date, applications processed on a rolling basis). Application fee: $30. *Tuition:* $1554 per year full-time, $74.25 per credit hour part-time for state residents; $4358 per year full-time, $74.25 per credit hour part-time for nonresidents. *Financial aid:* Application deadline 4/15. • Dr. William Bridges, Head, 307-766-4221.

See full description on page 505.

Utah State University, College of Science, Department of Mathematics and Statistics, Logan, UT 84322. Department offers programs in applied statistics (MS), mathematical sciences (PhD), mathematics (M Math, MS). Part-time programs available. Faculty: 24 full-time (3 women), 0 part-time. Matriculated students: 8 full-time (3 women), 28 part-time (16 women); includes 2 minority (1 Asian American, 1 Hispanic American), 17 foreign. Average age 24. 48 applicants, 48% accepted. In 1990, 15 master's awarded; 1 doctorate awarded (100% entered university research/teaching). Terminal master's awarded for partial completion of doctoral program. *Degree requirements:* For master's, qualifying exam required, foreign language and thesis not required; for doctorate, 1 foreign language, dissertation, comprehensive exams. *Entrance requirements:* GRE General Test (score in 40th percentile or higher required), TOEFL (minimum score of 550 required), minimum GPA of 3.0. Application deadline: 7/15 (priority date, applications processed on a rolling basis). Application fee: $25 ($30 for foreign students). *Tuition:* $426 per quarter (minimum) full-time, $184 per quarter (minimum) part-time for state residents; $1133 per quarter (minimum) full-time, $505 per quarter (minimum) part-time for nonresidents. *Financial aid:* In 1990–91, $144,000 in aid awarded. 3 fellowships (1 to a first-year student), 2 research assistantships (1 to a first-year student), 15 teaching assistantships (8 to first-year students) were awarded; partial tuition waivers, federal work-study also available. Aid available to part-time students. Average monthly stipend for a graduate assistantship: $1000. Financial aid application deadline: 4/1; applicants required to submit FAF. *Faculty research:* Differential geometry, differential equations, statistics, computational mathematics, probability. Total annual research budget: $75,000. • L. Duane Loveland, Head, 801-750-2809. Application contact: Ian Anderson, Graduate Chairman, 801-750-2818.

See full description on page 507.

Vanderbilt University, Department of Mathematics, Nashville, TN 37240. Department awards MA, MAT, MS, PhD. Faculty: 28 full-time (1 woman), 2 part-time (0 women). Matriculated students: 31 full-time (9 women), 0 part-time; includes 0 minority, 6 foreign. Average age 26. 42 applicants, 38% accepted. In 1990, 8 master's, 5 doctorates awarded. Terminal master's awarded for partial completion of doctoral program. *Degree requirements:* For master's, foreign language and thesis not required; for doctorate, 2 foreign languages, dissertation. *Entrance requirements:* GRE General Test, GRE Subject Test. Application deadline: 1/15. Application fee: $25. *Expenses:* Tuition of $624 per semester hour. Fees of $196 per year. *Financial aid:* Fellowships, teaching assistantships, federal work-study, institutionally sponsored loans available. Financial aid application deadline: 1/15. • Glen F. Webb, Chairman, 615-322-6672. Application contact: Matthew I. Gould, Director of Graduate Studies, 615-322-6672.

See full description on page 509.

Villanova University, Graduate School of Arts and Sciences, Department of Mathematical Sciences, Villanova, PA 19085. Offerings include mathematics (MA, MS). Department faculty: 30 full-time (5 women), 5 part-time (0 women). *Application deadline:* 8/1. *Application fee:* $25. *Expenses:* Tuition of $270 per credit. Fees of $28 per semester. • Dr. Frederick Hartmann, Chairperson, 215-645-4850.

Virginia Commonwealth University, College of Humanities and Sciences, Department of Mathematical Sciences, Richmond, VA 23284. Department offers programs in applied mathematics (MS), computer science (MS), mathematics (MS), operations research (MS), statistics (MS). Faculty: 38. Matriculated students: 17 full-time (8 women), 19 part-time (2 women); includes 3 minority (all Asian American), 4 foreign. 19 applicants, 42% accepted. In 1990, 13 degrees awarded. *Degree requirements:* Foreign language not required. *Entrance requirements:* GRE General Test, GRE Subject Test. Application deadline: 7/1. Application fee: $20. *Expenses:* Tuition of $2770 per year full-time, $154 per hour part-time for state residents; $7550 per year full-time, $419 per hour part-time for nonresidents. Fees of $717 per year full-time, $25.50 per hour part-time. *Financial aid:* 24 students received aid. Fellowships, research assistantships, teaching assistantships, federal work-study, institutionally sponsored loans available. Aid available to part-time students. Financial aid applicants required to submit FAF. *Total annual research budget:* $89,329. • Dr. William E. Haver, Chair, 804-367-1319. Application contact: Dr. James A. Wood, Director of Graduate Studies, 804-367-1301.

Virginia Polytechnic Institute and State University, College of Arts and Sciences, Department of Mathematics, Blacksburg, VA 24061. Department offers programs in applied mathematics (MS, PhD), mathematical physics (MS, PhD), pure mathematics (MS, PhD). Part-time programs available. Faculty: 55 full-time (5 women), 0 part-time. Matriculated students: 89 full-time (26 women), 4 part-time (1 woman); includes 2 minority (1 black American, 1 Native American), 45 foreign. 126 applicants, 56% accepted. In 1990, 21 master's, 9 doctorates awarded. Terminal master's awarded for partial completion of doctoral program. *Degree requirements:* For master's, thesis required (for some programs), foreign language not required; for doctorate, 1 foreign language, dissertation. *Application deadline:* 2/15 (priority date). *Application fee:* $10. *Tuition:* $1889 per semester full-time, $606 per credit hour part-time for state residents; $2627 per semester full-time, $853 per credit hour part-time for nonresidents. *Financial aid:* In 1990–91, 1 fellowship (to a first-year student), 7 research assistantships, 50 teaching assistantships (16 to first-year students), 23 assistantships (1 to a first-year student) awarded. *Faculty research:* Differential equations, operator theory, numerical analysis, algebra, control theory. • Dr. C. Wayne Patty, Head, 703-231-6536. Application contact: E. L. Green, Graduate Administrator, 703-231-6536.

See full description on page 511.

Virginia State University, School of Natural Sciences, Department of Mathematics, Petersburg, VA 23803. Department offers programs in mathematics (MS), mathematics education (M Ed). Faculty: 6 full-time (3 women), 0 part-time. Matriculated students: 5 full-time (2 women), 5 part-time (all women). In 1990, 1 degree awarded. *Degree requirements:* Thesis. *Application deadline:* 8/15 (applications processed on a rolling basis). *Application fee:* $10. *Expenses:* Tuition of $1479 per semester full-time, $90 per credit hour part-time for state residents; $2879 per semester full-time, $215 per credit hour part-time for nonresidents. Fees of $25 per year full-time, $12 per year part-time. *Financial aid:* Application deadline 5/1; applicants required to submit FAF. • Dr. Loretta M. Braxton, Chairman, 804-524-5920. Application contact: Dr. Edgar A. Toppin, Dean, 804-524-5984.

Wake Forest University, Department of Mathematics, Winston-Salem, NC 27109. Department awards MA. Part-time programs available. Faculty: 17 full-time (3 women), 1 (woman) part-time. Matriculated students: 10 full-time (6 women), 3 part-time (all women); includes 2 foreign. Average age 25. 17 applicants, 47% accepted. In 1990, 10 degrees awarded (100% found work related to degree). *Degree requirements:* 1 foreign language (computer language can substitute), thesis. *Entrance requirements:* GRE General Test, GRE Subject Test. Application deadline: 3/1. Application fee: $25. *Tuition:* $10,800 per year full-time, $280 per hour part-time. *Financial aid:* In 1990–91, 11 students received a total of $162,700 in aid awarded. 6 fellowships (2 to first-year students), 0 research assistantships, 4 teaching assistantships (3 to first-year students), 1 scholarship (0 to first-year students) were awarded. Aid available to part-time students. Average monthly stipend for a graduate assistantship: $800. Financial aid application deadline: 3/1; applicants required to submit FAF. *Faculty research:* Algebra, ring theory, topology, differential equations. • Dr. Richard Carmichael, Chairman, 919-759-5354.

Washington State University, College of Sciences and Arts, Division of Sciences, Department of Pure and Applied Mathematics, Pullman, WA 99164. Department offers program in pure and applied mathematics (MS, DA, PhD). Faculty: 31 full-time (3 women), 4 part-time (1 woman). Matriculated students: 41 full-time (13 women), 0 part-time; includes 2 minority (1 Asian American, 1 Native American), 13 foreign. In 1990, 9 master's awarded; 5 doctorates awarded (100% entered university research/teaching). *Degree requirements:* For master's, thesis or alternative, core exams required, foreign language not required; for doctorate, 2 foreign languages, dissertation, core exams. *Entrance requirements:* For master's, GRE General Test, GRE Subject Test, TOEFL, minimum GPA of 3.0; for doctorate, GRE Subject Test, TOEFL, minimum GPA of 3.0. Application deadline: 3/1 (priority date, applications processed on a rolling basis). Application fee: $25. *Tuition:* $1694 per semester full-time, $169 per credit hour part-time for state residents; $4236 per semester full-time, $424 per credit hour part-time for nonresidents. *Financial aid:* In 1990–91, 0 fellowships, 4 research assistantships, 36 teaching assistantships, 0 teaching associateships awarded; partial tuition waivers, federal work-study, institutionally sponsored loans, and career-related internships or fieldwork also available. Average monthly stipend for a graduate assistantship: $975. Financial aid application deadline: 4/1. *Faculty research:* Computational mathematics, operations research, modeling in the natural sciences, applied statistics. Total annual research budget: $395,000. • Dr. Michael J. Kallaher, Chairman, 509-335-8518.

Washington University, Graduate School of Arts and Sciences, Department of Mathematics, St. Louis, MO 63130. Department offers programs in mathematics (MA, PhD), mathematics education (MAT), statistics (MA, PhD). Part-time programs available. Matriculated students: 46 full-time (15 women), 2 part-time (0 women); includes 0 minority, 29 foreign. In 1990, 3 master's, 3 doctorates awarded. Terminal master's awarded for partial completion of doctoral program. *Degree requirements:* For master's, thesis or alternative; for doctorate, dissertation. *Entrance requirements:* GRE. Application deadline: 1/15 (priority date, applications processed on a rolling basis). Application fee: $0. *Tuition:* $15,960 per year full-time, $665 per credit hour part-time. *Financial aid:* In 1990–91, 8 fellowships, 5 research assistantships, 18 teaching assistantships awarded; full and partial tuition waivers, federal work-study, institutionally sponsored loans also available. Aid available to part-time students. Financial aid application deadline: 1/15. • Dr. Gary Jensen, Chairman, 314-935-6760.

Wayne State University, College of Liberal Arts, Department of Mathematics, Program in Mathematics, Detroit, MI 48202. Program awards MA, PhD. Matriculated students: 19 full-time (5 women), 31 part-time (6 women). 48 applicants, 31% accepted. In 1990, 11 master's, 1 doctorate awarded. *Degree requirements:* For master's, foreign language and thesis not required; for doctorate, 2 foreign languages, dissertation. *Application deadline:* 7/1. *Application fee:* $20 ($30 for foreign students). *Expenses:* Tuition of $119 per credit hour for state residents; $258 per credit hour for nonresidents. Fees of $50 per semester. • Dr. Pao liu Chou, Chairperson, Department of Mathematics, 313-577-2479. Application contact: Dr. Peter Malcolmson, Graduate Committee Chair, 313-577-2472.

Wesleyan University, Department of Mathematics, Middletown, CT 06459. Department awards MA, PhD. Faculty: 20 full-time (5 women), 0 part-time. Matriculated students: 18 full-time (8 women), 0 part-time; includes 9 foreign. Average age 28. 102 applicants, 3% accepted. In 1990, 3 master's awarded (100% found work related to degree); 1 doctorate awarded (100% entered university research/teaching). Terminal master's awarded for partial completion of doctoral program. *Degree requirements:* For master's, 1 foreign language, thesis; for doctorate, 2 foreign languages, dissertation. *Entrance requirements:* For master's, GRE General Test, GRE Subject Test; for doctorate, GRE Subject Test. Application deadline: 2/15. Application fee: $0. *Expenses:* Tuition of $2035 per course. Fees of $627 per year full-time. *Financial aid:* In 1990–91, 15 teaching assistantships (3 to first-year students) awarded; full and partial tuition waivers also available. Average monthly stipend for a graduate assistantship: $875. *Faculty research:* Topology, analysis. • Dr. Carol Wood, Chairman, 203-347-9411 Ext. 2648.

See full description on page 515.

West Chester University of Pennsylvania, College of Arts and Sciences, Department of Mathematics and Computer Science, West Chester, PA 19383. Department offers programs in computer science (MS), mathematics (MA, M Ed). Faculty: 0 full-time, 22 part-time. Matriculated students: 35 full-time (10 women), 54 part-time (25 women); includes 48 minority (46 Asian American, 1 black American, 1 Native American), 39 foreign. Average age 30. 82 applicants, 61% accepted. In 1990, 24 degrees awarded. *Degree requirements:* Comprehensive exam required, foreign language and thesis not required. *Entrance requirements:* GRE General Test (MA, M Ed). Application deadline: rolling. Application fee: $20. *Tuition:* $127 per credit for state residents; $160 per credit for nonresidents. *Financial aid:* In 1990–91, 4 research assistantships awarded. Aid available to part-time students. Financial aid application deadline: 4/1. • Dr. John Weaver, Chair, 215-436-2440. Application contact: Dr. Elaine Milito, Graduate Coordinator, 215-436-2595.

Western Carolina University, School of Arts and Sciences, Department of Mathematics and Computer Science, Cullowhee, NC 28723. Department awards MA Ed, MS. Part-time and evening/weekend programs available. Faculty: 11 (0 women). Matriculated students: 8 full-time (2 women), 11 part-time (7 women); includes 2 minority (1 Asian American, 1 Hispanic American), 5 foreign. 14 applicants, 86% accepted. In 1990, 13 degrees awarded. *Degree requirements:* Comprehensive exams required, foreign language not required. *Entrance requirements:* GRE General Test, GRE Subject Test (MS), NTE (MA Ed). Application

Directory: Mathematics

deadline: rolling. Application fee: $15. *Tuition:* $635 per semester full-time, $226 per course (minimum) part-time for state residents; $3162 per semester full-time, $1490 per course (minimum) part-time for nonresidents. *Financial aid:* In 1990–91, $37,652 in aid awarded. 5 research assistantships, 2 teaching assistantships were awarded; fellowships, federal work-study, institutionally sponsored loans also available. Financial aid application deadline: 3/15; applicants required to submit FAF. • Joseph B. Klerlein, Head, 704-227-7245. Application contact: Kathleen Owen, Assistant to the Dean, 704-227-7398.

Western Connecticut State University, School of Arts and Sciences, Department of Mathematics and Computer Science, Danbury, CT 06810. Offerings include mathematics and computer science (MA), theoretical mathematics (MA). Department faculty: 5 full-time, 0 part-time. *Degree requirements:* Thesis, comprehensive exam required, foreign language not required. *Entrance requirements:* Minimum GPA of 2.5. Application deadline: 8/1 (priority date). Application fee: $20. *Expenses:* Tuition of $860 per semester full-time, $115 per credit hour part-time for state residents; $2395 per semester full-time, $115 per credit hour part-time for nonresidents. Fees of $488 per semester full-time, $5 per semester part-time. • Dr. Bruce King, Chairman, 203-797-4086.

Western Illinois University, College of Arts and Sciences, Department of Mathematics, Macomb, IL 61455. Department awards MS. Part-time programs available. Faculty: 24 full-time (2 women), 0 part-time. Matriculated students: 13 full-time (2 women), 10 part-time (5 women); includes 0 minority, 9 foreign. Average age 32. 42 applicants, 38% accepted. In 1990, 5 degrees awarded. *Degree requirements:* Thesis or alternative required, foreign language not required. *Application fee:* $0. *Expenses:* Tuition of $870 per semester full-time, $520 per semester hour part-time for state residents; $2193 per semester full-time, $1402 per semester hour part-time for nonresidents. Fees of $537 per year full-time, $148 per year part-time. *Financial aid:* In 1990–91, $38,375 in aid awarded. 12 research assistantships were awarded. • Dr. MecKinley Scott, Chairperson, 309-298-1054. Application contact: Barbara Baily, Assistant to the Dean, 309-298-4806.

Western Kentucky University, Ogden College of Science, Technology, and Health, Department of Mathematics, Bowling Green, KY 42101. Department awards MA Ed, MS. Part-time programs available. Faculty: 14 full-time (3 women), 0 part-time. Matriculated students: 3 full-time (2 women), 12 part-time (10 women); includes 1 minority (black American), 0 foreign. 25 applicants, 52% accepted. In 1990, 3 degrees awarded. *Degree requirements:* Variable foreign language requirement, thesis or alternative, written exam. *Entrance requirements:* GRE General Test (minimum combined score of 1150 required). Application deadline: 8/1 (priority date, applications processed on a rolling basis). *Tuition:* $1580 per year full-time, $87 per credit part-time for state residents; $4460 per year full-time, $247 per credit part-time for nonresidents. *Financial aid:* In 1990–91, 2 service awards (both to first-year students) awarded; federal work-study, institutionally sponsored loans also available. Average monthly stipend for a graduate assistantship: $525. Financial aid application deadline: 4/1. • Dr. Robert Bueker, Head, 502-745-3651.

Western Michigan University, College of Arts and Sciences, Department of Mathematics and Statistics, Kalamazoo, MI 49008. Department offers programs in applied mathematics (MS); biostatistics (MS); computational mathematics (MS); graph theory and computer science (PhD); mathematics (MA, PhD), including mathematics (MA), mathematics education (MA, PhD); operations research (MS); statistics (MS, PhD). Faculty: 34 full-time. Matriculated students: 29 full-time (13 women), 60 part-time (29 women); includes 4 minority (2 Asian American, 1 black American, 1 Native American), 21 foreign. 102 applicants, 36% accepted. In 1990, 25 master's, 5 doctorates awarded. *Degree requirements:* For master's, thesis not required; for doctorate, 1 foreign language, dissertation. *Entrance requirements:* For doctorate, GRE General Test. Application deadline: 2/15 (priority date, applications processed on a rolling basis). Application fee: $25. *Expenses:* Tuition of $100 per credit hour for state residents; $244 per credit hour for nonresidents. Fees of $178 per semester full-time, $76 per semester part-time. *Financial aid:* In 1990–91, 5 fellowships awarded; research assistantships, teaching assistantships, federal work-study also available. Financial aid application deadline: 2/15. • Dr. Yousef Alavi, Chairperson, 616-387-4513. Application contact: Paula J. Boodt, Director, Graduate Student Services, 616-387-3570.

Western Oregon State College, School of Liberal Arts and Sciences, Division of Natural Sciences and Mathematics, Monmouth, OR 97361. Division offers program in interdisciplinary studies (MAIS, MSIS). Faculty: 25 full-time (4 women), 0 part-time. *Degree requirements:* Written exams required, foreign language and thesis not required. *Entrance requirements:* GRE General Test or MAT, minimum GPA of 3.0 in last 60 semester hours. Application deadline: 6/1 (priority date). Application fee: $35. *Tuition:* $1099 per trimester full-time for state residents; $1754 per trimester full-time for nonresidents. *Financial aid:* Research assistantships, teaching assistantships, federal work-study, and career-related internships or fieldwork available. Aid available to part-time students. Financial aid application deadline: 3/1; applicants required to submit FAF. *Faculty research:* Paleontology, gamma ray spectroscopy, mycology, animal physiology, plant physiological ecology. • Dr. J. Morris Johnson, Chair, 503-838-8207.

Western Washington University, College of Arts and Sciences, Department of Mathematics, Bellingham, WA 98225. Department awards MS. Faculty: 12 (1 woman). Matriculated students: 19 full-time (5 women), 1 part-time (0 women); includes 0 minority, 1 foreign. 15 applicants, 87% accepted. In 1990, 12 degrees awarded. *Degree requirements:* Thesis required (for some programs), foreign language not required. *Entrance requirements:* GRE General Test, TOEFL (minimum score of 535 required), minimum GPA of 3.0 for last 60 semester hours or last 90 quarter hours. Application deadline: 6/1 (priority date, applications processed on a rolling basis). Application fee: $25. *Tuition:* $900 per quarter full-time, $90 per credit part-time for state residents; $2729 per quarter full-time, $273 per credit part-time for nonresidents. *Financial aid:* In 1990–91, 5 teaching assistantships awarded; partial tuition waivers, federal work-study, institutionally sponsored loans also available. Aid available to part-time students. Financial aid application deadline: 3/31; applicants required to submit FAF. • Dr. Thomas T. Read, Chairperson, 206-676-3785. Application contact: Dr. John Reay, Graduate Program Adviser, 206-676-3467.

West Texas State University, College of Agriculture, Nursing, and Natural Sciences, Department of Mathematics and Physical Sciences, Program in Mathematics, Canyon, TX 79016. Program awards MS. Part-time programs available. Faculty: 0 part-time. Matriculated students: 23. In 1990, 0 degrees awarded. *Degree requirements:* Thesis or alternative required, foreign language not required. *Entrance requirements:* GRE General Test. Application deadline: rolling. *Expenses:* Tuition of $600 per year full-time, $20 per hour part-time for state residents; $3840 per year full-time, $128 per hour part-time for nonresidents. Fees of $448 per year full-time, $15 per hour part-time. • Dr. William Cooke, Head, 806-656-2540.

West Virginia University, College of Arts and Sciences, Department of Mathematics, Morgantown, WV 26506. Department awards MS, PhD. Part-time programs available. Faculty: 30 full-time (2 women), 14 part-time (8 women). Matriculated students: 23 full-time (7 women), 11 part-time (6 women); includes 0 minority, 10 foreign. Average age 28. 40 applicants, 85% accepted. In 1990, 11 master's, 0 doctorates awarded. Terminal master's awarded for partial completion of doctoral program. *Degree requirements:* For master's, foreign language and thesis not required; for doctorate, 1 foreign language, computer language, dissertation, comprehensive exam. *Entrance requirements:* For master's, TOEFL (minimum score of 550 required), minimum GPA of 2.5; for doctorate, GRE, TOEFL (minimum score of 550 required). Application deadline: 3/1 (priority date, applications processed on a rolling basis). Application fee: $25. *Expenses:* Tuition of $390 per year full-time for state residents; $1270 per year full-time for nonresidents. Fees of $1555 per year full-time for state residents; $3985 per year full-time for nonresidents. *Financial aid:* In 1990–91, $336,000 in aid awarded. 3 research assistantships (1 to a first-year student), 21 teaching assistantships (13 to first-year students) were awarded; full and partial tuition waivers, federal work-study, institutionally sponsored loans also available. Average monthly stipend for a graduate assistantship: $745. Financial aid application deadline: 2/1; applicants required to submit FAF. *Faculty research:* Combinatorics and graph theory, topology, differential equations, applied and computational mathematics. • Dr. James Lightbourne, Chairperson, 304-293-2011. Application contact: Dr. Ian Christie, Director of Graduate Studies, 304-293-2014.

Wichita State University, Fairmount College of Liberal Arts and Sciences, Department of Mathematics, Wichita, KS 67208. Department offers programs in applied mathematics (PhD), mathematics (MS), statistics (MS). Part-time programs available. Faculty: 23 full-time (3 women), 0 part-time. Matriculated students: 30 full-time (6 women), 16 part-time (8 women); includes 4 minority (3 Asian American, 1 Hispanic American), 23 foreign. 32 applicants, 72% accepted. In 1990, 6 master's awarded. *Degree requirements:* For master's, comprehensive exam required, thesis optional, foreign language not required; for doctorate, dissertation. *Entrance requirements:* TOEFL (minimum score of 550 required). Application deadline: 8/1 (priority date, applications processed on a rolling basis). Application fee: $0 ($25 for foreign students). *Expenses:* Tuition of $1590 per year full-time, $53 per credit part-time for state residents; $2574 per year full-time, $171.60 per credit part-time for nonresidents. Fees of $12.20 per credit. *Financial aid:* In 1990–91, $175,000 in aid awarded. 23 teaching assistantships were awarded; federal work-study, institutionally sponsored loans also available. Financial aid application deadline: 4/1. *Faculty research:* Partial differential equations, combinatorics, ring theory, minimal surfaces, several complex variables. • Dr. Buma Fridman, Chairperson, 316-689-3160.

Wilkes University, Department of Mathematics, Wilkes-Barre, PA 18766. Department awards MS, MS Ed. *Degree requirements:* Thesis or alternative required, foreign language not required. *Application fee:* $25.

Winthrop College, College of Arts and Sciences, Department of Mathematics, Rock Hill, SC 29733. Department awards M Math. Part-time programs available. Faculty: 8 full-time (1 woman), 0 part-time. Matriculated students: 1 full-time (0 women), 6 part-time (3 women); includes 0 minority, 0 foreign. Average age 33. In 1990, 6 degrees awarded. *Degree requirements:* Foreign language and thesis not required. *Entrance requirements:* GRE General Test or NTE, minimum GPA of 3.0. Application deadline: 7/15 (priority date, applications processed on a rolling basis). Application fee: $35. *Tuition:* $117 per credit hour for state residents; $210 per credit hour for nonresidents. *Financial aid:* Graduate assistantships graduate scholarships, federal work-study available. Aid available to part-time students. Financial aid application deadline: 2/1; applicants required to submit FAF. • Dr. Ronnie Goolsby, Chairman, 803-323-2175. Application contact: Sharon Johnson, Supervisor of the Graduate Office, 803-323-2204.

Wright State University, College of Science and Mathematics, Department of Mathematics and Statistics, Program in Mathematics, Dayton, OH 45435. Program awards MS. Faculty: 25 full-time (1 woman), 0 part-time. Matriculated students: 2 full-time (0 women), 2 part-time (1 woman). Average age 33. In 1990, 3 degrees awarded. *Degree requirements:* Comprehensive exams required, thesis optional, foreign language not required. *Tuition:* $3342 per year full-time, $106 per credit hour part-time for state residents; $5991 per year full-time, $190 per credit hour part-time for nonresidents. *Financial aid:* In 1990–91, 1 fellowship, 1 research assistantship, 2 teaching assistantships awarded. *Faculty research:* Analysis, combinatorics, geometry, graph theory, operator theory. • Dr. Joanne M. Dombrowski, Graduate Adviser, 513-873-3218.

Yale University, Graduate School of Arts and Sciences, Department of Mathematics, New Haven, CT 06520. Department awards MS, PhD. Faculty: 24 full-time (1 woman), 1 part-time (0 women). Matriculated students: 36 full-time (14 women), 1 part-time (0 women); includes 2 minority (1 Asian American, 1 black American), 19 foreign. 154 applicants, 20% accepted. In 1990, 3 master's, 12 doctorates awarded. Terminal master's awarded for partial completion of doctoral program. *Degree requirements:* For doctorate, 2 foreign languages, dissertation. *Entrance requirements:* For doctorate, GRE General Test. Application deadline: 1/2. Application fee: $45. *Tuition:* $15,160 per year full-time, $1895 per course part-time. • Tsuneu Tamagawa, Chairman, 203-432-4172. Application contact: Susan Webb, Coordinator of Admissions, 203-432-2770.

See full description on page 517.

York University, Faculty of Arts, Program in Mathematics and Statistics, North York, ON M3J 1P3, Canada. Program awards MA, PhD. Part-time programs available. Faculty: 45 full-time (9 women), 2 part-time (0 women). Matriculated students: 29 full-time (5 women), 9 part-time (5 women); includes 7 foreign. 136 applicants, 26% accepted. In 1990, 22 master's, 0 doctorates awarded. *Degree requirements:* For master's, thesis optional, foreign language not required; for doctorate, dissertation. *Application deadline:* 3/31. *Application fee:* $35. *Tuition:* $2436 per year for Canadian residents; $11,480 per year for nonresidents. *Financial aid:* In 1990–91, $343,171 in aid awarded. 7 fellowships, 39 research assistantships, 30 teaching assistantships were awarded. • Dr. G. O'Brien, Director, 416-736-2100.

Youngstown State University, College of Arts and Sciences, Department of Mathematics, Youngstown, OH 44555. Department awards MS. Part-time programs available. Faculty: 19 full-time (0 women), 1 part-time (0 women). Matriculated students: 3 full-time (1 woman), 32 part-time (16 women); includes 0 minority, 4 foreign. 11 applicants, 91% accepted. In 1990, 10 degrees awarded. *Degree requirements:* Computer language, comprehensive exam required, thesis optional, foreign language not required. *Entrance requirements:* Minimum GPA of 2.7 in mathematics and computer science. Application deadline: 8/15 (priority date, applications processed on a rolling basis). Application fee: $30. *Expenses:* Tuition of $1566 per year full-time, $58 per quarter hour part-time for state residents; $2808 per year full-time, $104 per quarter hour part-time for nonresidents. Fees of $432 per year full-time, $16 per quarter hour part-time. *Financial aid:* In 1990–91, 7 students received a total of $64,519 in aid awarded. 7 graduate assistantships (6 to first-year students) were awarded; federal work-study, institutionally sponsored loans also available. Aid available to part-time students. Average monthly stipend for a graduate assistantship: $833. Financial aid application deadline: 7/8. *Faculty research:* Computer-aided tools for multiple-valued logic circuits, parallel algorithms for Markov chains. • Dr. Albert J. Klein, Chair, 216-742-3302. Application contact: Dr. Sally M. Hotchkiss, Dean of Graduate Studies, 216-742-3091.

Statistics

American University, College of Arts and Sciences, Department of Mathematics and Statistics, Program in Statistics, Washington, DC 20016. Program awards MA, PhD. Part-time and evening/weekend programs available. Faculty: 16 full-time (4 women), 2 part-time (0 women). Matriculated students: 10 full-time (4 women), 13 part-time (7 women); includes 0 minority, 16 foreign. Average age 32. 45 applicants, 87% accepted. In 1990, 2 master's, 2 doctorates awarded. Terminal master's awarded for partial completion of doctoral program. *Degree requirements:* For master's, 1 foreign language required (computer language can substitute), thesis optional; for doctorate, 2 foreign languages (computer language can substitute for one), dissertation. *Application deadline:* 2/15. *Application fee:* $50. *Expenses:* Tuition of $475 per semester hour. Fees of $20 per semester. *Financial aid:* Fellowships, teaching assistantships, federal work-study, institutionally sponsored loans, and career-related internships or fieldwork available. Aid available to part-time students. Financial aid application deadline: 2/15; applicants required to submit FAF. *Faculty research:* Statistical computing; data analysis; random processes; environmental, meteorological and biological applications. • Dr. Robert W. Jernigan, Chair, Department of Mathematics and Statistics, 202-885-3120.

See full description on page 387.

Arizona State University, College of Liberal Arts and Sciences, Department of Mathematics, Tempe, AZ 85287. Offerings include statistics (MA, PhD). *Degree requirements:* For doctorate, 1 foreign language, dissertation. *Entrance requirements:* For doctorate, GRE. Application fee: $25. *Tuition:* $1528 per year full-time, $80 per hour part-time for state residents; $6934 per year full-time, $289 per hour part-time for nonresidents. • Dr. William Trotter, Chair, 602-965-3951.

See full description on page 389.

Arizona State University, College of Liberal Arts and Sciences, Interdisciplinary Program in Statistics, Tempe, AZ 85287. Program awards MS. 2 applicants, 0% accepted. *Application fee:* $25. *Tuition:* $1528 per year full-time, $80 per hour part-time for state residents; $6934 per year full-time, $289 per hour part-time for nonresidents. • Dr. Mike Driscoll, Director, 602-965-3951.

Auburn University, College of Sciences and Mathematics, Division of Mathematics, Department of Algebra, Combinatorics, and Analysis, Auburn University, AL 36849. Department awards MACT, MAM, MPS, MS, PhD. Part-time programs available. Faculty: 29 full-time, 0 part-time. Matriculated students: 31 full-time (18 women), 16 part-time (5 women); includes 1 minority (Asian American), 7 foreign. In 1990, 0 master's awarded. *Degree requirements:* For master's, thesis (MS); for doctorate, 1 foreign language, dissertation, written and oral exams. *Entrance requirements:* For master's, GRE General Test, GRE Subject Test, minimum B average in mathematics; for doctorate, GRE General Test (minimum score of 400 on each section required), GRE Subject Test, minimum B average in mathematics. Application deadline: 9/1 (applications processed on a rolling basis). Application fee: $15. *Expenses:* Tuition of $1596 per year full-time, $44 per credit hour part-time for state residents; $4788 per year full-time, $132 per credit hour part-time for nonresidents. Fees of $92 per quarter for state residents; $276 per quarter for nonresidents. *Financial aid:* In 1990–91, 24 teaching assistantships awarded; fellowships, research assistantships, federal work-study also available. Aid available to part-time students. Financial aid application deadline: 3/15. • Dr. James R. Wall, Head, 205-844-5111. Application contact: Dr. Norman J. Doorenbos, Dean, Graduate School, 205-844-4700.

Ball State University, College of Sciences and Humanities, Department of Mathematical Sciences, Program in Mathematical Statistics, 2000 University Avenue, Muncie, IN 47306. Program awards MA. Matriculated students: 4. In 1990, 1 degree awarded. *Degree requirements:* Foreign language not required. *Application fee:* $15. *Expenses:* Tuition of $1140 per semester (minimum) full-time for state residents; $2680 per semester (minimum) full-time for nonresidents. Fees of $6 per credit hour. *Faculty research:* Robust methods. • Mir Ali, Director, 317-285-8670.

Baruch College of the City University of New York, School of Business and Public Administration, Department of Statistics and Computer Information Systems, Program in Statistics, 17 Lexington Avenue, New York, NY 10010. Program awards MBA, MS. Part-time and evening/weekend programs available. Faculty: 11 full-time (2 women), 1 (woman) part-time. Matriculated students: 10 full-time (2 women), 35 part-time (11 women). Average age 27. 23 applicants, 52% accepted. In 1990, 2 degrees awarded. *Degree requirements:* Computer language required, foreign language not required. *Entrance requirements:* GMAT or GRE (for MS), TOEFL, TWE. Application deadline: 6/1. Application fee: $30. *Tuition:* $950 per semester full-time, $82 per credit part-time for state residents; $2350 per semester full-time, $199 per credit part-time for nonresidents. *Financial aid:* Fellowships, research assistantships, full and partial tuition waivers, federal work-study available. Aid available to part-time students. Financial aid application deadline: 5/15. • Dr. Israel Pressman, Chairman, Department of Statistics and Computer Information Systems, 212-447-3080.

Baylor University, College of Arts and Sciences, Department of Psychology, Program in Behavioral Statistics, Waco, TX 76798. Program awards PhD. Matriculated students: 6 full-time (5 women), 2 part-time (1 woman); includes 0 minority, 0 foreign. In 1990, 1 doctorate awarded. Terminal master's awarded for partial completion of doctoral program. *Degree requirements:* For doctorate, computer language, dissertation, comprehensive exam required, foreign language not required. *Entrance requirements:* For doctorate, GRE General Test. Application deadline: 2/15 (applications processed on a rolling basis). Application fee: $0. *Expenses:* Tuition of $4440 per year full-time, $185 per credit hour part-time. Fees of $510 per year full-time. *Financial aid:* Research assistantships, teaching assistantships, federal work-study, institutionally sponsored loans, and career-related internships or fieldwork available. Financial aid application deadline: 2/15; applicants required to submit FAF. *Faculty research:* Experimental design, Bayesian statistics, factor analysis, database management systems. • Dr. Roger Kirk, Director, 817-755-2961.

Boston University, Graduate School, Department of Mathematics, Boston, MA 02215. Offerings include statistics (MA, PhD). Terminal master's awarded for partial completion of doctoral program. Department faculty: 32 full-time (3 women), 5 part-time (1 woman). *Degree requirements:* For master's, 1 foreign language, comprehensive exam required, thesis not required; for doctorate, 1 foreign language, dissertation. *Entrance requirements:* GRE General Test, GRE Subject Test, TOEFL (minimum score of 550 required). Application deadline: 5/1 (priority date, applications processed on a rolling basis). Application fee: $45. *Expenses:* Tuition of $15,950 per year full-time, $498 per credit part-time. Fees of $120 per year full-time, $35 per semester part-time. • Ralph B. D'Agostino, Chairman, 617-353-2560.

Bowling Green State University, College of Arts and Sciences, Department of Mathematics and Statistics, Bowling Green, OH 43403. Offerings include applied statistics (MS), statistics (MA, MAT, PhD). Department faculty: 27 full-time (1 woman), 0 part-time. *Degree requirements:* For doctorate, 1 foreign language (computer language can substitute), dissertation. *Entrance requirements:* For doctorate, GRE General Test, TOEFL (minimum score of 600 required). Application deadline: 8/7. Application fee: $10. *Tuition:* $185 per credit hour for state residents; $357 per credit hour for nonresidents. • Dr. Hassoon Al-Amiri, Chair, 419-372-2636. Application contact: Dr. James Albert, Graduate Coordinator, 419-372-7456.

Bowling Green State University, College of Business Administration, Department of Applied Statistics and Operations Research, Bowling Green, OH 43403. Offerings include applied statistics (MS). Department faculty: 9 full-time (2 women), 0 part-time. *Degree requirements:* Comprehensive exam required, foreign language and thesis not required. *Entrance requirements:* GRE General Test, TOEFL (minimum score of 550 required). Application fee: $10. *Tuition:* $185 per credit hour for state residents; $357 per credit hour for nonresidents. • Dr. Wei Shih, Chair, 419-372-2363. Application contact: Ralph St. John, Graduate Coordinator, 419-372-8098.

Brigham Young University, College of Physical and Mathematical Sciences, Department of Statistics, Provo, UT 84602. Department awards MS. Faculty: 13 full-time (0 women), 2 part-time (0 women). Matriculated students: 27 full-time (7 women), 0 part-time; includes 3 minority (all Asian American), 5 foreign. Average age 27. 11 applicants, 91% accepted. In 1990, 10 degrees awarded. *Degree requirements:* Thesis required, foreign language not required. *Entrance requirements:* GRE General Test, GRE Subject Test (mathematics), minimum GPA of 3.0 during previous 60 credit hours. Application deadline: 6/15. Application fee: $30. *Tuition:* $1170 per semester (minimum) full-time, $130 per credit hour (minimum) part-time. *Financial aid:* In 1990–91, 20 students received a total of $15,000 in aid awarded. 19 research assistantships (10 to first-year students) were awarded; teaching assistantships and career-related internships or fieldwork also available. Average monthly stipend for a graduate assistantship: $860. Financial aid application deadline: 4/1. *Faculty research:* Software development, health care, sample survey, environmental impact of pesticides quality and productivity. Total annual research budget: $40,000. • Dr. Leland J. Hendrix, Chair, 801-378-4505.

California State University, Fullerton, School of Business Administration and Economics, Department of Management Science, Fullerton, CA 92634. Offerings include statistics (MS). Department faculty: 24 full-time (2 women), 16 part-time (2 women). *Application fee:* $55. *Expenses:* Tuition of $0 for state residents; $246 per unit for nonresidents. Fees of $554 per semester full-time, $356 per unit part-time. • Dr. Zvi Drezner, Chair, 714-773-2221. Application contact: Dr. Richard Stolz, Associate Dean, 714-773-2211.

California State University, Hayward, School of Science, Department of Statistics, Hayward, CA 94542. Department awards MS. Faculty: 8. Matriculated students: 4 full-time (2 women), 14 part-time (6 women). Average age 29. In 1990, 3 degrees awarded. *Degree requirements:* Comprehensive exam required, foreign language and thesis not required. *Entrance requirements:* Minimum GPA of 2.5 in previous 2 years. Application deadline: 4/19 (priority date, applications processed on a rolling basis). Application fee: $55. *Expenses:* Tuition of $0 for state residents; $137 per unit for nonresidents. Fees of $895 per year full-time, $188 per quarter part-time. *Financial aid:* Federal work-study, institutionally sponsored loans available. Aid available to part-time students. Financial aid application deadline: 3/1. • Dr. Michael Orkin, Chair, 415-881-3435. Application contact: Dr. Robert Trinchero, Acting Associate Vice President, Admissions and Enrollment, 415-881-3828.

California State University, Los Angeles, School of Business and Economics, Department of Economics and Statistics, Los Angeles, CA 90032. Department offers programs in analytical quantitative economics (MA), business economics (MA, MBA, MS), economics (MA). Part-time and evening/weekend programs available. Faculty: 18 full-time, 18 part-time. Matriculated students: 13 full-time (4 women), 28 part-time (10 women); includes 11 minority (4 Asian American, 4 black American, 3 Hispanic American), 15 foreign. In 1990, 8 degrees awarded. *Degree requirements:* Thesis or comprehensive exam required, foreign language not required. *Entrance requirements:* GMAT, TOEFL (minimum score of 550 required), minimum GPA of 2.5 during last 2 years. Application deadline: 6/30 (applications processed on a rolling basis). Application fee: $55. *Expenses:* Tuition of $0 for state residents; $164 per unit for nonresidents. Fees of $1046 per year full-time, $650 per semester (minimum) part-time. *Financial aid:* Federal work-study and career-related internships or fieldwork available. Aid available to part-time students. Financial aid application deadline: 3/1; applicants required to submit FAF. • Dr. Eduardo Ochoa, Chair, 213-343-2930.

Carleton University, Faculty of Science, Ottawa-Carleton Institute for Graduate Studies and Research in Mathematics and Statistics, Ottawa, ON K1S 5B6, Canada. Institute awards M Sc, PhD. PhD and M Sc offered jointly with the University of Ottawa. Faculty: 57. Matriculated students: 47 full-time, 6 part-time; includes 14 foreign. In 1990, 12 master's, 4 doctorates awarded. *Degree requirements:* For master's, thesis optional; for doctorate, dissertation. *Entrance requirements:* For master's, TOEFL (minimum score of 550 required), honor's degree; for doctorate, TOEFL (minimum score of 550 required), master's degree. Application deadline: 7/1 (priority date, applications processed on a rolling basis). Application fee: $15. *Tuition:* $985 per semester full-time, $284 per semester part-time for Canadian residents; $3939 per semester full-time, $1171 per semester part-time for nonresidents. *Financial aid:*

Directory: Statistics

Application deadline 3/1. *Faculty research:* Pure mathematics, applied mathematics, probability and statistics. • G. Ivanoff, Director, 613-788-2155.

Carnegie Mellon University, College of Humanities and Social Sciences, Department of Statistics, Pittsburgh, PA 15213. Department offers program in statistics (MS, PhD), including applied statistics (PhD), computational statistics (PhD), theoretical statistics (PhD). Faculty: 14 full-time (1 woman), 0 part-time. Matriculated students: 27 full-time (10 women), 4 part-time (0 women); includes 0 minority, 16 foreign. Average age 26. 159 applicants, 13% accepted. In 1990, 10 master's awarded (40% found work related to degree, 60% continued full-time study); 4 doctorates awarded (75% entered university research/teaching, 25% found other work related to degree). Terminal master's awarded for partial completion of doctoral program. *Degree requirements:* For master's, foreign language and thesis not required; for doctorate, dissertation, oral and written comprehensive exams. *Entrance requirements:* GRE General Test, TOEFL (minimum score of 550 required). Application deadline: 3/15. Application fee: $20. *Expenses:* Tuition of $15,250 per year (minimum) full-time, $212 per unit (minimum) part-time. Fees of $100 per year full-time. *Financial aid:* In 1990–91, 29 students received a total of $521,162 in aid awarded. 0 fellowships, 1 research assistantship (0 to first-year students), 26 teaching assistantships (9 to first-year students) were awarded; full and partial tuition waivers, federal work-study, institutionally sponsored loans, and career-related internships or fieldwork also available. Aid available to part-time students. Average monthly stipend for a graduate assistantship: $900. Financial aid application deadline: 3/15. *Faculty research:* Stochastic processes, Bayesian statistics, statistical computing, decision theory, psychiatric statistics. Total annual research budget: $1.14-million. • Dr. John P. Lehoczky, Head, 412-268-8725. Application contact: Dr. Robert Kass, Admissions Coordinator, 412-268-8723.
See full description on page 399.

Central Connecticut State University, School of Arts and Sciences, Department of Mathematics and Computer Science, New Britain, CT 06050. Offerings include statistics (MA). Department faculty: 5 full-time (2 women), 1 (woman) part-time. *Application deadline:* 8/31. *Application fee:* $20. *Expenses:* Tuition of $1720 per year full-time, $115 per credit part-time for state residents; $4790 per year full-time, $115 per credit part-time for nonresidents. Fees of $1013 per year full-time, $33 per semester part-time for state residents; $1655 per year full-time, $33 per semester part-time for nonresidents. • Dr. William Driscoll, Chair, 203-827-7374.

Claremont Graduate School, Department of Mathematics, Claremont, CA 91711. Offerings include operations research and statistics (MA, MS). Department faculty: 5 full-time (0 women), 32 part-time (4 women). *Degree requirements:* Foreign language and thesis not required. *Entrance requirements:* GRE General Test. Application deadline: 2/15 (priority date, applications processed on a rolling basis). Application fee: $40. *Expenses:* Tuition of $13,900 per year full-time, $620 per unit part-time. Fees of $55 per semester. • William Lucas, Chairman, 714-621-8080.

Clemson University, College of Sciences, Department of Mathematical Sciences, Program in Statistics, Clemson, SC 29634. Program awards MS, PhD. Faculty: 7 full-time (0 women), 0 part-time. Matriculated students: 12 full-time (5 women), 0 part-time; includes 2 minority (both Asian American), 5 foreign. Average age 29. 25 applicants, 64% accepted. In 1990, 8 master's awarded (63% found work related to degree, 37% continued full-time study); 1 doctorate awarded (100% entered university research/teaching). *Degree requirements:* For master's, computer language, final project required, thesis optional, foreign language not required; for doctorate, computer language, dissertation, qualifying exams required, foreign language not required. *Entrance requirements:* GRE General Test. Application deadline: 6/1. Application fee: $25. *Expenses:* Tuition of $102 per credit hour. Fees of $80 per semester full-time. *Financial aid:* In 1990–91, $150,000 in aid awarded. 1 fellowship (to a first-year student), 1 research assistantship (0 to first-year students), 12 teaching assistantships (4 to first-year students) were awarded. Average monthly stipend for a graduate assistantship: $800. Financial aid application deadline: 4/15. *Faculty research:* Data analysis, computational statistics, stochastics, mathematical statistics, quality control. Total annual research budget: $100,000. • R. D. Ringeisen, Head, 803-656-3434. Application contact: T. G. Proctor, Associate Head, 803-656-5234.

Announcement: Program in statistics leading to PhD and master's degrees offered by comprehensive Department of Mathematical Sciences. Faculty expertise includes interactive data analysis, sampling techniques, and mathematical statistics. Diverse research expertise and integrated graduate programs provide preparation for wide variety of student career goals.

Colorado State University, College of Natural Sciences, Department of Statistics, Fort Collins, CO 80523. Department awards MS, PhD. Faculty: 12 full-time (0 women), 4 part-time (0 women). Matriculated students: 29 full-time (8 women), 7 part-time (2 women); includes 3 minority (all Asian American), 16 foreign. Average age 28. 75 applicants, 56% accepted. In 1990, 5 master's, 3 doctorates awarded. *Degree requirements:* For master's, computer language, project, seminar required, thesis optional, foreign language not required; for doctorate, computer language, dissertation, candidacy exam, preliminary exam seminar required, foreign language not required. *Entrance requirements:* GRE General Test, TOEFL, minimum GPA of 3.0. Application deadline: 4/1 (priority date, applications processed on a rolling basis). Application fee: $30. *Tuition:* $1322 per semester full-time for state residents; $3673 per semester full-time for nonresidents. *Financial aid:* In 1990–91, $150,000 in aid awarded. 2 fellowships (both to first-year students), 2 research assistantships (1 to a first-year student), 19 teaching assistantships (4 to first-year students), 0 traineeships were awarded; federal work-study, institutionally sponsored loans, and career-related internships or fieldwork also available. Average monthly stipend for a graduate assistantship: $698-750. Financial aid application deadline: 3/15. *Faculty research:* Applied probability, linear models, experimental design, time-series analysis, statistical inference. Total annual research budget: $350,000. • Duane Boes, Chairman, 303-491-5269.

Columbia University, Graduate School of Arts and Sciences, Division of Natural Sciences, Department of Statistics, New York, NY 10027. Department awards MA, M Phil, PhD, MD/PhD. Part-time programs available. Faculty: 10 full-time, 1 part-time. Matriculated students: 20 full-time (7 women), 5 part-time (0 women); includes 1 minority (Asian American), 12 foreign. Average age 28. 25 applicants, 40% accepted. In 1990, 2 master's, 7 doctorates awarded. *Degree requirements:* For master's, foreign language and thesis not required; for doctorate, dissertation, M Phil required, foreign language not required. *Entrance requirements:* GRE General Test, GRE Subject Test, TOEFL. Application deadline: 1/5. Application fee: $50. *Expenses:* Tuition of $7836 per semester for state residents; $426 per credit for nonresidents. Fees of $148 per semester for state residents. *Financial aid:* In 1990–91, $249,771 in aid awarded. 7 teaching assistantships (0 to first-year students) were awarded; federal work-study, institutionally sponsored loans also available. Financial aid application deadline: 1/5; applicants required to submit GAPSFAS. • David Krantz, Chair, 212-854-3652.

Cornell University, Graduate Fields of Agriculture and Life Sciences, Field of Statistics, Ithaca, NY 14853. Field offers programs in biometry (MS, PhD), decision theory (MS, PhD), economic and social statistics (MS, PhD), engineering statistics (MS, PhD), experimental design (MS, PhD), mathematical statistics (MS, PhD), probability (MS, PhD), sampling (MS, PhD), statistical computing (MS, PhD), stochastic processes (MS, PhD). Faculty: 31 full-time, 0 part-time. Matriculated students: 16 full-time (7 women), 0 part-time; includes 2 minority (1 Asian American, 1 Hispanic American), 9 foreign. 111 applicants, 16% accepted. In 1990, 2 master's, 6 doctorates awarded. Terminal master's awarded for partial completion of doctoral program. *Degree requirements:* For master's, thesis required, foreign language not required; for doctorate, 1 foreign language, dissertation. *Entrance requirements:* GRE General Test, TOEFL. Application deadline: 1/10. Application fee: $55. *Expenses:* Tuition of $7440 per trimester full-time. Fees of $28 per year full-time. *Financial aid:* In 1990–91, 13 students received aid. 2 fellowships (1 to a first-year student), 1 research assistantship (0 to first-year students), 10 teaching assistantships (4 to first-year students) awarded; full and partial tuition waivers, federal work-study, institutionally sponsored loans also available. Financial aid application deadline: 1/10; applicants required to submit GAPSFAS. *Faculty research:* Decision theory; empirical Bayes analysis; statistical computing; non-linear and robust methods; stochastic processes. • George Casella, Graduate Faculty Representative, 607-255-8066. Application contact: Robert Brashear, Director of Admissions, 607-255-4884.

Creighton University, College of Arts and Sciences, Department of Mathematics, Statistics, and Computer Science, Program in Mathematics and Statistics, Omaha, NE 68178. Program awards MS. Part-time programs available. Faculty: 8 full-time, 0 part-time. *Degree requirements:* 1 foreign language required (computer language can substitute), thesis optional. *Entrance requirements:* GRE General Test, TOEFL. Application deadline: 3/15. Application fee: $20. *Expenses:* Tuition of $272 per credit hour. Fees of $140 per semester full-time, $14 per semester part-time. • Dr. John Mordeson, Chairman, Department of Mathematics, Statistics, and Computer Science, 402-280-2478. Application contact: Dr. Michael Lawler, Dean, Graduate School, 402-280-2870.

Dalhousie University, College of Arts and Science, Faculty of Science, Department of Mathematics, Statistics, and Computing Science, Halifax, NS B3H 4H6, Canada. Offerings include statistics (M Sc, PhD). Department faculty: 36 full-time, 1 part-time. *Degree requirements:* For doctorate, 1 foreign language, dissertation. *Entrance requirements:* For doctorate, TOEFL (minimum score of 550 required). Application deadline: 7/15 (applications processed on a rolling basis). Application fee: $20. *Tuition:* $2594 per year full-time for Canadian residents; $4294 per year full-time for nonresidents. • Dr. R. P. Gupta, Chair, 902-494-2572.

Duke University, Graduate School, Institute of Statistics and Decision Sciences, Durham, NC 27706. Institute awards PhD. Faculty: 9 full-time, 2 part-time. Matriculated students: 6 full-time (2 women), 0 part-time; includes 6 foreign. 15 applicants, 60% accepted. *Degree requirements:* Dissertation. *Entrance requirements:* GRE General Test. Application deadline: 1/31. Application fee: $50. *Expenses:* Tuition of $8640 per year full-time, $360 per unit part-time. Fees of $1356 per year full-time. *Financial aid:* In 1990–91, $54,534 in aid awarded. • Mike West, Director of Graduate Studies, 919-684-4210.

Emory University, Graduate School of Arts and Sciences, Department of Mathematics and Computer Science, Program in Statistics/Computer Science, Atlanta, GA 30322. Program awards MS. *Degree requirements:* Computer language, thesis required, foreign language not required. *Entrance requirements:* GRE General Test, TOEFL, minimum GPA of 3.0. Application deadline: 1/20 (priority date, applications processed on a rolling basis). Application fee: $35. *Expenses:* Tuition of $7370 per semester full-time, $642 per semester hour part-time. Fees of $160 per year full-time. *Financial aid:* Application deadline 2/1. • Dr. Ron Gould, Chairman, Department of Mathematics and Computer Science, 404-727-7924. Application contact: Dr. Dwight Duffus, Director of Graduate Studies, 404-727-7579.
See full description on page 413.

Florida State University, College of Arts and Sciences, Department of Statistics, Tallahassee, FL 32306. Department offers programs in applied statistics (MS, PhD), mathematical statistics (MS, PhD), operations research (PhD), probability (PhD). Part-time programs available. Faculty: 15 full-time (1 woman), 0 part-time. Matriculated students: 34 full-time (11 women), 4 part-time (3 women); includes 0 minority, 21 foreign. 249 applicants, 24% accepted. In 1990, 9 master's awarded (33% found work related to degree, 67% continued full-time study); 3 doctorates awarded (67% entered university research/teaching, 33% found other work related to degree). Terminal master's awarded for partial completion of doctoral program. *Degree requirements:* For master's, foreign language and thesis not required; for doctorate, dissertation, departmental qualifying exam required, foreign language not required. *Entrance requirements:* GRE General Test (minimum combined score of 1000 required), minimum GPA of 3.0. Application deadline: 2/15 (priority date, applications processed on a rolling basis). Application fee: $15. *Tuition:* $76.29 per credit hour for state residents; $238 per credit hour for nonresidents. *Financial aid:* In 1990–91, 23 students received a total of $190,000 in aid awarded. 0 fellowships, 3 research assistantships (0 to first-year students), 20 teaching assistantships (8 to first-year students) were awarded; federal work-study, institutionally sponsored loans also available. Aid available to part-time students. Average monthly stipend for a graduate assistantship: $800. Financial aid application deadline: 4/1; applicants required to submit FAF. *Faculty research:* Statistical inference, probability theory, biostatistics, reliability. Total annual research budget: $360,853. • Dr. Fred Leysieffer, Chairman, 904-644-3218.

George Mason University, School of Information Technology and Engineering, Department of Operations Research and Applied Statistics, Program in Statistical Science, Fairfax, VA 22030. Program awards MS. Matriculated students: 5 full-time (4 women), 7 part-time (1 woman); includes 1 minority (black American), 1 foreign. 8 applicants, 63% accepted. In 1990, 1 degree awarded. *Degree requirements:* Computer language, oral exam required, thesis optional. *Entrance requirements:* Previous course work in calculus, probability, matrix or linear algebra, microeconomics, scientific computing, minimum GPA of 3.0 in last 60 hours of undergraduate study. Application deadline: 5/1. Application fee: $25. *Expenses:* Tuition of $1872 per year full-time, $78 per semester hour part-time for state residents; $6264 per year full-time, $261 per semester hour part-time for nonresidents. Fees of $1080 per year full-time, $45 per semester hour part-time. *Financial aid:* Application deadline 3/1. • Dr. Carl M. Harris, Chairman, Department of Operations Research and Applied Statistics, 703-993-1687.

George Washington University, Graduate School of Arts and Sciences, Department of Statistics/Computer and Information Systems, Washington, DC 20052. Department offers programs in applied statistics (MS, PhD), including applied statistics (MS), statistics (PhD); mathematical statistics (MA); statistical computing (MS). Part-time and evening/weekend programs available. Faculty: 6 full-time (2 women), 4 part-time (1 woman). Matriculated students: 11 full-time (5 women), 21 part-time (8 women); includes 1 minority (black American), 8 foreign. Average age 35. 31 applicants, 71% accepted. In 1990, 3 master's, 2 doctorates awarded. Terminal master's awarded for partial completion of doctoral program. *Degree requirements:* For master's, comprehensive exam required, thesis not required; for doctorate, computer language, dissertation, general exam. *Entrance requirements:* GRE General Test, minimum GPA of 3.0. Application deadline: 7/1. Application fee: $45. *Expenses:* Tuition of $490 per semester hour. Fees of $215 per year full-time, $125.60 per year (minimum) part-time. *Financial aid:* In 1990–91, $111,945 in aid awarded. 10 fellowships (4 to first-year students), 7 teaching assistantships (4 to first-year students) were awarded; partial tuition waivers, federal work-study also available. Financial aid application deadline: 2/15. • Dr. Arthur Kirsch, Chair, 202-994-6356.

See full description on page 415.

Georgia Institute of Technology, Program in Industrial and Systems Engineering and Schools of Mathematics and Management, Program in Statistics, Atlanta, GA 30332. Program awards MSS. Part-time programs available. In 1990, 4 degrees awarded. *Degree requirements:* Foreign language not required. *Entrance requirements:* GRE General Test, TOEFL (minimum score of 550 required), minimum GPA of 2.7. Application deadline: 8/1. Application fee: $15. *Expenses:* Tuition of $574 per semester full-time, $48 per credit part-time for state residents; $1380 per semester full-time, $115 per credit part-time for nonresidents. Fees of $132 per semester full-time. *Financial aid:* Partial tuition waivers, federal work-study, institutionally sponsored loans, and career-related internships or fieldwork available. Aid available to part-time students. Financial aid application deadline: 2/15. *Faculty research:* Statistical control procedures, statistical modeling of transportation systems. • Dr. W. W. Hines, Associate Director, 404-894-4289.

Georgia State University, College of Arts and Sciences, Department of Mathematics and Computer Science, Atlanta, GA 30303. Department offers program in statistics (MA, MAT, MS). Part-time and evening/weekend programs available. Faculty: 23 full-time (4 women), 1 part-time (0 women). Matriculated students: 33 full-time (9 women), 29 part-time (12 women); includes 8 minority (4 Asian American, 2 black American, 2 Hispanic American), 16 foreign. Average age 29. 121 applicants, 22% accepted. In 1990, 13 degrees awarded. *Degree requirements:* 1 foreign language (computer language can substitute), thesis or alternative. *Entrance requirements:* GRE General Test, TOEFL (minimum score of 550 required), minimum GPA of 2.75. Application deadline: 7/15. Application fee: $10. *Expenses:* Tuition of $38 per quarter hour for state residents; $130 per quarter hour for nonresidents. Fees of $58 per quarter. *Financial aid:* In 1990–91, $10,200 in aid awarded. 18 research assistantships (1 to a first-year student), 3 teaching assistantships were awarded; full tuition waivers, federal work-study, institutionally sponsored loans, and career-related internships or fieldwork also available. Aid available to part-time students. Average monthly stipend for a graduate assistantship: $600. Financial aid applicants required to submit FAF. *Faculty research:* Linear algebra and graph theory, numerical and functional analysis, computer graphics and software engineering, applied statistics and probability, mathematics education. Total annual research budget: $166,000. • Dr. Fred A. Massey, Acting Chair, 404-651-2253. Application contact: Dr. George J. Davis, Director of Graduate Studies, 404-651-2253.

Harvard University, Graduate School of Arts and Sciences, Department of Statistics, Cambridge, MA 02138. Department awards AM, PhD. Faculty: 5 full-time, 3 part-time. Matriculated students: 19 full-time (4 women), 0 part-time; includes 0 minority. Average age 28. 45 applicants, 18% accepted. In 1990, 6 master's, 6 doctorates awarded. Terminal master's awarded for partial completion of doctoral program. *Degree requirements:* For master's, 1 foreign language required (computer language can substitute), thesis not required; for doctorate, 1 foreign language (computer language can substitute), dissertation, qualifying paper and exam. *Entrance requirements:* GRE General Test, GRE Subject Test. Application deadline: 1/2. Application fee: $60. *Expenses:* Tuition of $14,860 per year. Fees of $550 per year. *Financial aid:* In 1990–91, $80,717 in aid awarded. 6 fellowships (2 to first-year students), 14 research assistantships (1 to a first-year student), 9 teaching assistantships (2 to first-year students) were awarded; federal work-study, institutionally sponsored loans, and career-related internships or fieldwork also available. Financial aid application deadline: 1/2; applicants required to submit GAPSFAS. *Faculty research:* Interactive graphic analysis of multidimensional data, data analysis, modeling and inference, statistical modeling of U.S. economic time series. • Donald B. Rubin, Chairman, 617-495-5496. Application contact: Office of Admissions and Financial Aid, 617-495-5315.

Announcement: The Department of Statistics offers courses of study and independent research in theoretical and applied problems leading to both the AM and PhD degrees. It encourages applications from motivated students with good mathematical backgrounds. Although small in relation to some departments, the statistics department offers broad training, research, teaching, and consulting opportunities.

Illinois Institute of Technology, Lewis College of Sciences and Letters, Department of Mathematics, Chicago, IL 60616. Offerings include applied statistics (MS). Department faculty: 13 full-time (3 women), 1 part-time (0 women). *Degree requirements:* Thesis (for some programs), comprehensive exam required, foreign language not required. *Entrance requirements:* TOEFL (minimum score of 500 required). Application deadline: 7/1 (applications processed on a rolling basis). Application fee: $30. *Expenses:* Tuition of $13,070 per year full-time, $435 per credit hour part-time. Fees of $20 per semester (minimum) full-time, $1 per credit hour part-time. • Dr. Maurice J. Frank, Chairman, 312-567-3162.

Indiana University Bloomington, College of Arts and Sciences, Department of Mathematics, Bloomington, IN 47405. Offerings include probability-statistics (MA, PhD). Terminal master's awarded for partial completion of doctoral program. Department faculty: 50 full-time (1 woman), 0 part-time. *Degree requirements:* For doctorate, 2 foreign languages, dissertation. *Entrance requirements:* For doctorate, GRE General Test, GRE Subject Test. Application deadline: 3/1. Application fee: $25 ($35 for foreign students). *Tuition:* $99.85 per credit hour for state residents; $288 per credit hour for nonresidents. • Allan L. Edmonds, Chair, 812-855-2200. Application contact: Richard C. Bradley, Director of Graduate Studies, 812-855-2645.

Iowa State University of Science and Technology, College of Liberal Arts and Sciences, Department of Statistics, Ames, IA 50011. Department awards MS, PhD. Matriculated students: 90 full-time (27 women), 46 part-time (17 women); includes 4 minority (2 Asian American, 1 black American, 1 Hispanic American), 71 foreign. In 1990, 29 master's, 11 doctorates awarded. *Application fee:* $20 ($30 for foreign students). *Expenses:* Tuition of $1158 per semester full-time, $129 per credit part-time for state residents; $3340 per semester full-time, $372 per credit part-time for nonresidents. Fees of $10 per semester. *Financial aid:* In 1990–91, 4 fellowships (2 to first-year students), 26 research assistantships (10 to first-year students), 33 teaching assistantships (18 to first-year students), 19 scholarships (14 to first-year students) awarded. • Dr. Dean L. Isaacson, Head, 515-294-3440.

Johns Hopkins University, G. W. C. Whiting School of Engineering, Department of Mathematical Sciences, Baltimore, MD 21218. Offerings include statistics (MA, MSE, PhD). Terminal master's awarded for partial completion of doctoral program. Department faculty: 11 full-time (0 women), 2 part-time (1 woman). *Degree requirements:* For master's, computer language, thesis (for some programs); for doctorate, 1 foreign language, computer language, dissertation. *Entrance requirements:* GRE General Test, GRE Subject Test. Application deadline: 2/15 (priority date, applications processed on a rolling basis). Application fee: $45. *Expenses:* Tuition of $15,500 per year full-time, $1550 per course part-time. Fees of $400 per year full-time. • Dr. John C. Wierman, Chairman, 301-338-7195.

See full description on page 419.

Kansas State University, College of Arts and Sciences, Department of Statistics, Manhattan, KS 66506. Department awards MS, PhD. Faculty: 12 full-time (1 woman), 0 part-time. Matriculated students: 43 full-time (12 women), 0 part-time; includes 25 foreign. 46 applicants, 57% accepted. In 1990, 2 master's, 0 doctorates awarded. *Degree requirements:* For master's, thesis required, foreign language not required; for doctorate, 1 foreign language, dissertation. *Expenses:* Tuition of $1668 per year full-time, $51 per credit hour part-time for state residents; $5382 per year full-time, $142 per credit hour part-time for nonresidents. Fees of $305 per year full-time, $64.50 per semester part-time. *Financial aid:* In 1990–91, 5 research assistantships, 12 teaching assistantships awarded. • James Higgins, Head, 913-532-6883. Application contact: John Boyer, Graduate Coordinator, 913-532-6883.

Louisiana State University and Agricultural and Mechanical College, College of Agriculture, Department of Experimental Statistics, Baton Rouge, LA 70803. Department offers program in applied statistics (M Ap Stat). Part-time programs available. Faculty: 13 full-time (2 women), 0 part-time. Matriculated students: 94 full-time (5 women), 24 part-time (1 woman); includes 0 minority, 6 foreign. Average age 28. 9 applicants, 44% accepted. In 1990, 1 degree awarded (100% found work related to degree). *Degree requirements:* Project required, foreign language and thesis not required. *Entrance requirements:* GRE General Test (minimum combined score of 1000 required), minimum GPA of 3.0. Application deadline: 7/1 (priority date, applications processed on a rolling basis). Application fee: $25. *Tuition:* $1020 per semester full-time for state residents; $2620 per semester full-time for nonresidents. *Financial aid:* In 1990–91, 8 students received a total of $70,450 in aid awarded. 4 research assistantships (0 to first-year students), 3 teaching assistantships (1 to a first-year student), 0 assistantships were awarded; institutionally sponsored loans and career-related internships or fieldwork also available. Average monthly stipend for a graduate assistantship: $833. Financial aid application deadline: 4/1. *Faculty research:* Linear models, statistical computing, ecological statistics, applied statistics. Total annual research budget: $325,000. • Dr. Lynn R. LaMotte, Head, 504-388-8303.

Louisiana Tech University, College of Arts and Sciences, Department of Mathematics and Statistics, Ruston, LA 71272. Department awards MS. Part-time programs available. Faculty: 12 full-time (4 women), 0 part-time. Matriculated students: 6 full-time (2 women), 0 part-time; includes 1 minority (Asian American), 2 foreign. In 1990, 5 degrees awarded. *Degree requirements:* Computer language, thesis or alternative required, foreign language not required. *Entrance requirements:* GRE General Test. Application deadline: 8/13. Application fee: $5. *Tuition:* $613 per quarter full-time, $184 per semester (minimum) part-time for state residents; $999 per quarter full-time, $184 per semester (minimum) part-time for nonresidents. *Financial aid:* Fellowships available. Financial aid application deadline: 2/1; applicants required to submit GAPSFAS. • Dr. John E. Maxfield, Acting Head, 318-257-2538.

Mankato State University, College of Natural Sciences, Mathematics and Home Economics, Department of Mathematics, Statistics, and Astronomy, Program in Statistics, South Road and Ellis Avenue, Mankato, MN 56002-8400. Program awards MS. Matriculated students: 1 (woman) full-time, 1 (woman) part-time; includes 0 minority, 2 foreign. Average age 34. In 1990, 0 degrees awarded. *Degree requirements:* 1 foreign language, computer language, thesis or alternative. *Entrance requirements:* GRE General Test, GRE Subject Test, minimum GPA of 2.75 for last 2 years of undergraduate study. Application deadline: 2/3 (priority date, applications processed on a rolling basis). Application fee: $15. *Expenses:* Tuition of $52 per credit for state residents; $75 per credit for nonresidents. Fees of $6.50 per quarter. *Financial aid:* Application deadline 7/1. • Dr. Larry Pearson, Head, 507-389-1453.

Marquette University, College of Arts and Sciences, Department of Mathematics, Statistics, and Computer Science, Milwaukee, WI 53233. Offerings include statistics (MS). Department faculty: 28 full-time (2 women), 0 part-time. *Degree requirements:* Thesis or alternative, comprehensive exam required, foreign language not required. *Entrance requirements:* TOEFL (minimum score of 550 required). Application fee: $25. *Tuition:* $300 per credit. • Dr. Peter R. Jones, Chairman, 414-288-7573.

McMaster University, Program in Statistics, Hamilton, ON L8S 4L8, Canada. Offers applied statistics (M Sc), medical statistics (M Sc), statistical theory (M Sc). *Degree requirements:* Thesis or alternative required, foreign language not required. *Expenses:* Tuition of $2250 per year full-time, $810 per semester part-time for Canadian residents; $10,340 per year full-time, $4050 per semester part-time for nonresidents. Fees of $76 per year full-time, $49 per semester part-time.

McNeese State University, College of Science, Department of Mathematics, Computer Science, and Statistics, Lake Charles, LA 70609. Offerings include statistics (MS). Department faculty: 10 full-time (2 women), 0 part-time. *Degree requirements:* Computer language, thesis or alternative, written exam required, foreign language not required. *Entrance requirements:* GRE General Test. Application deadline: 7/15 (priority date, applications processed on a rolling basis). Application fee: $10 ($25 for foreign students). *Tuition:* $808 per semester full-time, $254 per course (minimum) part-time for state residents; $1583 per semester full-time, $254 per course (minimum) part-time for nonresidents. • Dr. George F. Mead Jr., Head, 318-475-5788.

Memorial University of Newfoundland, School of Graduate Studies, Department of Mathematics and Statistics, St. John's, NF A1C 5S7, Canada. Department awards MAS, M Phil, M Sc, PhD. Programs offered in mathematics (M Phil, M Sc, PhD), statistics (M Phil, M Sc, PhD). *Degree requirements:* For master's, thesis not required; for doctorate, dissertation.

Memphis State University, College of Arts and Sciences, Department of Mathematical Sciences, Memphis, TN 38152. Offerings include applied statistics (PhD), statistics (MS). Department faculty: 28 full-time (3 women), 0 part-time. *Degree requirements:* For master's, comprehensive exams required, thesis not required; for doctorate, 1 foreign language, dissertation. *Entrance requirements:* For master's, MAT (minimum score of 27 required) or GRE General Test (minimum combined score of 800 required), TOEFL (minimum score of 550 required), minimum GPA of 2.5; for doctorate, GRE General Test (minimum combined score of 1000 required). Application deadline: 8/1 (applications processed on a rolling basis). Application fee: $5. *Expenses:* Tuition of $92 per credit hour for state residents; $239 per credit hour for nonresidents. Fees of $45 per year for state residents.

Directory: Statistics

• Dr. Ralph Faudree, Chairman, 901-678-2482. Application contact: Dr. R. H. Schlep, Coordinator, Graduate Studies, 901-678-2495.

Miami University, College of Arts and Sciences, Department of Mathematics and Statistics, Program in Statistics, Oxford, OH 45056. Program awards MS. Part-time programs available. Faculty: 13. Matriculated students: 11 full-time (5 women), 6 part-time (3 women); includes 0 minority, 0 foreign. 19 applicants, 68% accepted. In 1990, 4 degrees awarded. *Degree requirements:* Final exam required, thesis not required. *Entrance requirements:* Minimum undergraduate GPA of 2.75 or 3.0 during previous 2 years. Application deadline: 3/1. Application fee: $30. *Expenses:* Tuition of $3196 per year full-time, $133 per semester (minimum) part-time for state residents; $7134 per year full-time, $312 per semester (minimum) part-time for nonresidents. Fees of $371 per year full-time, $165.50 per semester hour part-time. *Financial aid:* In 1990–91, $64,786 in aid awarded. Fellowships, research assistantships, teaching assistantships, full tuition waivers, federal work-study available. Financial aid application deadline: 3/1. • Application contact: Robert Schaefer, Director of Graduate Studies, 513-529-3527.

Michigan State University, College of Natural Science, Department of Statistics and Probability, East Lansing, MI 48824. Department offers programs in applied statistics (MS), computational statistics (MS), operations research-statistics (MS), statistics (MA, PhD). Part-time programs available. Faculty: 13 full-time (1 woman), 0 part-time. Matriculated students: 46 full-time (14 women), 29 part-time (13 women); includes 3 minority (1 Asian American, 2 black American), 48 foreign. In 1990, 17 master's, 7 doctorates awarded. Terminal master's awarded for partial completion of doctoral program. *Degree requirements:* For master's, foreign language and thesis not required; for doctorate, 1 foreign language, dissertation. *Application deadline:* rolling. *Application fee:* $25 ($40 for foreign students). *Tuition:* $104.75 per credit for state residents; $211.75 per credit for nonresidents. *Financial aid:* In 1990–91, 1 fellowship (to a first-year student), 3 research assistantships, 9 teaching assistantships (8 to first-year students) awarded. *Faculty research:* Weak convergence in statistical inference, stochastic approximation, nonparametrics, sequential procedures, stochastic processes. • Habib Salehi, Chairman, 517-353-3391. Application contact: James Hannan, Graduate Program Director, 517-355-9677.

Mississippi State University, College of Arts and Sciences, Department of Mathematics and Statistics, Mississippi State, MS 39762. Offerings include statistics (MA, MS). Department faculty: 25 full-time (2 women), 0 part-time. *Degree requirements:* Thesis optional, foreign language not required. *Entrance requirements:* TOEFL. Application deadline: 4/15 (priority date, applications processed on a rolling basis). *Expenses:* Tuition of $891 per semester full-time, $99 per credit hour part-time for state residents; $1622 per semester full-time, $180 per credit hour part-time for nonresidents. Fees of $221 per semester full-time, $25 per semester (minimum) part-time. • Dr. J. L. Solomon, Head, 601-325-3414. Application contact: Dr. Paul W. Spikes, Graduate Coordinator, 601-325-3414.

Montana State University, College of Letters and Science, Department of Mathematical Sciences, 901 West Garfield Street, Bozeman, MT 59717. Department offers programs in mathematics (MS, PhD), statistics (MS, PhD). Faculty: 23 full-time (1 woman), 0 part-time. Matriculated students: 37 full-time (9 women), 9 part-time (3 women); includes 1 minority (Asian American), 3 foreign. Average age 31. 26 applicants, 73% accepted. In 1990, 16 master's, 3 doctorates awarded. *Degree requirements:* For master's, computer language, thesis or alternative required, foreign language not required; for doctorate, variable foreign language requirement, computer language, dissertation. *Entrance requirements:* GRE General Test, TOEFL (minimum score of 550 required). Application deadline: 6/15 (applications processed on a rolling basis). Application fee: $20. *Tuition:* $1150 per year full-time, $74 per credit (minimum) part-time for state residents; $2824 per year full-time, $100 per credit (minimum) part-time for nonresidents. *Financial aid:* Teaching assistantships and career-related internships or fieldwork available. Financial aid application deadline: 3/1. *Faculty research:* Numerical analysis, dynamical systems, mathematics education, multivariate analysis, biostatistical methodology. • Dr. Kenneth J. Tiahrt, Head, 406-994-3601.

Montclair State College, School of Mathematical and Natural Sciences, Department of Mathematics and Computer Science, Programs in Mathematics, Concentration in Statistics, Upper Montclair, NJ 07043. Concentration awards MA. Part-time and evening/weekend programs available. Faculty: 33 full-time, 0 part-time. In 1990, 1 degree awarded. *Degree requirements:* Written comprehensive exam required, foreign language and thesis not required. *Entrance requirements:* GRE General Test, minimum GPA of 2.67. Application deadline: 7/1 (priority date, applications processed on a rolling basis). Application fee: $25. *Expenses:* Tuition of $130 per credit for state residents; $165 per credit for nonresidents. Fees of $276 per year. *Financial aid:* Application deadline 3/15. • Application contact: Dr. Helen Roberts, Graduate Adviser, 201-893-7262.

New Mexico State University, College of Agriculture and Home Economics, Department of Experimental Statistics, Las Cruces, NM 88003. Department awards MS. Faculty: 9 full-time, 0 part-time. Matriculated students: 10 full-time (4 women), 0 part-time. In 1990, 2 degrees awarded. *Degree requirements:* Computer language required, thesis optional. *Entrance requirements:* GRE General Test, 3 semesters of calculus. Application deadline: 7/1. Application fee: $10. *Tuition:* $1608 per year full-time, $67 per credit hour part-time for state residents; $5304 per year full-time, $221 per credit hour part-time for nonresidents. • Dr. Carroll Hall, Acting Head, 505-646-2936.

New Mexico State University, College of Arts and Sciences, Department of Mathematical Sciences, Las Cruces, NM 88003. Department awards MS, PhD. Part-time programs available. Faculty: 28 full-time, 0 part-time. Matriculated students: 25 full-time (9 women), 15 part-time (2 women); includes 25 foreign. In 1990, 3 master's, 6 doctorates awarded. *Entrance requirements:* GRE General Test. Application deadline: 7/1. Application fee: $10. *Tuition:* $1608 per year full-time, $67 per credit hour part-time for state residents; $5304 per year full-time, $221 per credit hour part-time for nonresidents. *Financial aid:* In 1990–91, 25 teaching assistantships awarded; fellowships, research assistantships also available. *Faculty research:* Algebraic k-theory, harmonic analysis, functional analysis, algebraic topology. • Dr. Carol Walker, Head, 505-646-3901.

New York University, Leonard N. Stern School of Business, Department of Statistics and Operations Research, New York, NY 10006. Department offers program in statistics and operations research (MBA, MS, PhD, APC). Faculty: 22 full-time (3 women), 28 part-time (0 women). Matriculated students: 18 full-time, 34 part-time. In 1990, 13 master's, 2 doctorates, 0 APCs awarded. *Degree requirements:* For master's, computer language, thesis or alternative required, foreign language not required; for doctorate, computer language, dissertation required, foreign language not required; for APC, foreign language and thesis not required. *Entrance requirements:* For master's and doctorate, GMAT. Application deadline: 4/15. Application fee: $60. *Expenses:* Tuition of $8000 per year full-time, $3200 per semester (minimum) part-time. Fees of $377 per semester full-time, $144 per semester (minimum) part-time. *Financial aid:* Application deadline 1/31. *Faculty research:* Time-series analysis, stochastic processes, sampling theory, integer programming, combinatorial optimization and nonlinear regression. Total annual research budget: $41,334. • Aaron Tenenbein, Chairman, 212-285-6180. Application contact: Jane de Vos, Chair, Admissions Committee, 212-285-6251.

North Carolina State University, College of Physical and Mathematical Sciences, Department of Statistics, Raleigh, NC 27695. Department awards MS, M Stat, PhD. Part-time programs available. Faculty: 35 full-time (4 women), 0 part-time. Matriculated students: 76 full-time (28 women), 9 part-time (2 women); includes 5 minority (1 Asian American, 3 black American, 1 Hispanic American), 33 foreign. 130 applicants, 62% accepted. In 1990, 10 master's awarded; 4 doctorates awarded (50% entered university research/teaching, 50% found other work related to degree). *Degree requirements:* For master's, thesis (for some programs), comprehensive exam, final oral exam required, foreign language not required; for doctorate, dissertation, written and oral preliminary exams, final oral and written exams required, foreign language not required. *Entrance requirements:* GRE General Test, TOEFL. Application deadline: 6/25. Application fee: $35. *Tuition:* $1138 per year for state residents; $5805 per year for nonresidents. *Financial aid:* In 1990–91, 66 students received a total of $614,500 in aid awarded. 3 fellowships (2 to first-year students), 25 research assistantships (6 to first-year students), 30 teaching assistantships (10 to first-year students) were awarded; career-related internships or fieldwork also available. Average monthly stipend for a graduate assistantship: $1000. Financial aid application deadline: 3/1. *Faculty research:* Time series, variance components, nonlinear models, nonparametric and robust inference, statistical computing. Total annual research budget: $2.25-million. • Dr. Daniel L. Solomon, Head, 919-737-2420. Application contact: Dr. Thomas M. Gerig, Director of Graduate Programs, 919-737-2528.

Announcement: NC State offers master's and PhD programs in statistics with applied orientation. Large faculty with varied research and consulting interests provides education that blends theory, methods, and practice. Substantive concentrations in industrial, environmental, and genetic statistics. Excellent study environment and computing facilities. Fellowships, teaching and research assistantships, and industry internships available.

North Carolina State University, Colleges of Humanities and Social Sciences and Agriculture and Life Sciences, Division of Economics and Business, Program in Management, Raleigh, NC 27695. Offerings include statistics (MS). Program faculty: 28 full-time (3 women), 1 part-time (0 women). *Degree requirements:* Computer language required, foreign language and thesis not required. *Entrance requirements:* GRE or GMAT, TOEFL (minimum score of 550 required), minimum undergraduate GPA of 3.0. Application deadline: 6/25 (applications processed on a rolling basis). Application fee: $35. *Tuition:* $1138 per year for state residents; $5805 per year for nonresidents. • J. C. Poindexter, Director, 919-737-7157. Application contact: Nancy Fisher, Graduate Adviser, 919-737-7157.

North Dakota State University, College of Science and Mathematics, Department of Statistics, Fargo, ND 58105. Department awards MS, PhD. Faculty: 5 full-time (1 woman), 0 part-time. Matriculated students: 15 full-time, 6 part-time. Average age 24. In 1990, 2 master's awarded. *Entrance requirements:* TOEFL (minimum score of 525 required). Application fee: $20. *Tuition:* $1411 per year full-time, $52 per credit hour part-time for state residents; $3571 per year full-time, $132 per credit hour part-time for nonresidents. *Financial aid:* Federal work-study, institutionally sponsored loans, and career-related internships or fieldwork available. Financial aid application deadline: 4/15. *Faculty research:* Linear models, nonparametric statistics, multivariate analysis, distribution theory, inference modeling. • Rhonda Magel, Acting Chair, 701-237-8171.

Northern Illinois University, College of Liberal Arts and Sciences, Department of Mathematical Sciences, Program in Applied Probability and Statistics, De Kalb, IL 60115. Program awards MS. Part-time programs available. Matriculated students: 11 full-time (3 women), 6 part-time (0 women); includes 0 minority, 5 foreign. Average age 30. 22 applicants, 41% accepted. In 1990, 4 degrees awarded. *Degree requirements:* Thesis or alternative, comprehensive exam required, foreign language not required. *Entrance requirements:* GRE General Test, TOEFL, minimum GPA of 2.75. Application deadline: 6/1. Application fee: $0. *Tuition:* $1339 per semester full-time for state residents; $3163 per semester full-time for nonresidents. *Financial aid:* Fellowships, research assistantships, teaching assistantships, full tuition waivers, federal work-study, and career-related internships or fieldwork available. Aid available to part-time students. Financial aid applicants required to submit FAF. *Total annual research budget:* $22,400. • Dr. Ibrahim Ahmad, Director, Division of Statistics, 815-753-6796.

Northwestern University, College of Arts and Sciences, Department of Statistics, Evanston, IL 60208. Department awards MS, PhD. Part-time programs available. Faculty: 3 full-time (0 women), 3 part-time (0 women). Matriculated students: 10 full-time (5 women), 3 part-time (1 woman); includes 0 minority, 3 foreign. 17 applicants, 59% accepted. In 1990, 3 master's, 1 doctorate awarded. Terminal master's awarded for partial completion of doctoral program. *Degree requirements:* For master's, final exam required, foreign language and thesis not required; for doctorate, dissertation, preliminary exam, final exam required, foreign language not required. *Entrance requirements:* GRE General Test. Application deadline: 8/30. Application fee: $40 ($45 for foreign students). *Tuition:* $4665 per quarter full-time, $1704 per course part-time. *Financial aid:* In 1990–91, $44,600 in aid awarded. 1 fellowship, 4 research assistantships, 2 teaching assistantships were awarded; federal work-study, institutionally sponsored loans also available. Financial aid application deadline: 1/15; applicants required to submit GAPSFAS. *Faculty research:* Qualitative data sampling, experimental design, survival analysis, linear models. Total annual research budget: $150,000. • Bruce D. Spencer, Chairman, 708-491-5810. Application contact: Ajit C. Tamhane, 708-491-3577.

Oakland University, College of Arts and Sciences, Department of Mathematical Sciences, Program in Applied Statistics, Rochester, MI 48309. Program awards MS. Part-time and evening/weekend programs available. *Degree requirements:* Foreign language and thesis not required. *Entrance requirements:* Minimum GPA of 3.0 for unconditional admission. Application deadline: 7/15. Application fee: $25. *Expenses:* Tuition of $122 per credit hour for state residents; $270 per credit hour for nonresidents. Fees of $170 per year. • Application contact: Dr. James H. McKay, Graduate Coordinator, 313-370-3430.

Oakland University, College of Arts and Sciences, Department of Mathematical Sciences, Program in Statistical Methods, Rochester, MI 48309. Program awards Certificate. *Application deadline:* 7/15. *Application fee:* $25. *Expenses:* Tuition of $122 per credit hour for state residents; $270 per credit hour for nonresidents. Fees of $170 per year. • Application contact: Dr. James H. McKay, Graduate Coordinator, 313-370-3430.

Ohio State University, College of Mathematical and Physical Sciences, Department of Statistics, Columbus, OH 43210. Offerings include statistics (M Appl Stat, MS, PhD). Department faculty: 20. *Degree requirements:* For master's, thesis optional, foreign language not required; for doctorate, dissertation required, foreign language not required. *Application deadline:* 8/15 (applications processed on a rolling basis). *Application fee:* $0 ($25 for foreign students). *Tuition:* $1213 per quarter full-time, $364 per course part-time for state residents; $3143 per quarter full-time, $943 per course part-time for nonresidents. • Prem Goel, Chairman, 614-292-8110.

Oklahoma State University, College of Arts and Sciences, Department of Statistics, Stillwater, OK 74078. Department awards MS, PhD. Faculty: 7 full-time, 0 part-time. Matriculated students: 26 (9 women); includes 1 minority (Asian American), 11 foreign. In 1990, 4 master's, 3 doctorates awarded. *Degree requirements:* For master's, foreign language not required; for doctorate, dissertation required, foreign language not required. *Entrance requirements:* GRE, TOEFL (minimum score of 550 required). Application deadline: 7/1 (priority date). Application fee: $15. *Tuition:* $63.75 per credit for state residents; $138.25 per credit for nonresidents. *Financial aid:* In 1990–91, 2 research assistantships, 12 teaching assistantships awarded; partial tuition waivers, federal work-study also available. Aid available to part-time students. Financial aid application deadline: 3/1; applicants required to submit GAPSFAS or FAF. • Dr. J. Leroy Folks, Head, 405-744-5684.

Old Dominion University, College of Sciences, Department of Mathematics and Statistics, Programs in Computational and Applied Mathematics, Norfolk, VA 23529. Offerings in applied mathematics (MS, PhD), statistics (MS, PhD). Part-time and evening/weekend programs available. Faculty: 24 full-time (0 women), 0 part-time. Matriculated students: 25 full-time (10 women), 23 part-time (11 women); includes 7 minority (1 Asian American, 4 black American, 2 Hispanic American), 19 foreign. Average age 27. 36 applicants, 86% accepted. In 1990, 6 master's awarded (17% entered university research/teaching, 83% found other work related to degree); 9 doctorates awarded (100% entered university research/teaching). Terminal master's awarded for partial completion of doctoral program. *Degree requirements:* For master's, comprehensive exam required, foreign language and thesis not required; for doctorate, dissertation, candidacy exam required, foreign language not required. *Entrance requirements:* For master's, GRE General Test, GRE Subject Test, TOEFL, minimum QPA of 3.0 in major, minimum QPA of 2.5 overall; for doctorate, GRE General Test (minimum combined score of 1000 required), GRE Subject Test (minimum score of 600 required), TOEFL. Application deadline: 7/1 (applications processed on a rolling basis). Application fee: $20. *Expenses:* Tuition of $148 per credit hour for state residents; $375 per credit hour for nonresidents. Fees of $64 per year full-time. *Financial aid:* In 1990–91, $28,800 in aid awarded. 3 research assistantships, 2 teaching assistantships, 1 tuition grant were awarded; fellowships also available. *Faculty research:* Numerical analysis, computational and applied mathematics, integral equations, continuum mechanics. Total annual research budget: $240,000. • Dr. John J. Swetits, Director, 804-683-3911.

See full description on page 441.

Oregon State University, Graduate School, College of Science, Department of Statistics, Corvallis, OR 97331. Department awards MA, M Agr, MAIS, MS, PhD. Programs offered in biometry (MA, MS, PhD), operations research (MA, MAIS, MS), statistics (MA, MS, PhD). Part-time programs available. Faculty: 15 full-time (2 women), 3 part-time (0 women). Matriculated students: 28 full-time (11 women), 4 part-time (1 woman); includes 1 minority (Asian American), 15 foreign. Average age 31. In 1990, 8 master's, 3 doctorates awarded. *Degree requirements:* For master's, foreign language and thesis not required; for doctorate, dissertation required, foreign language not required. *Entrance requirements:* TOEFL (minimum score of 550 required), minimum GPA of 3.0 in last 90 hours. Application deadline: 3/1. Application fee: $40. *Tuition:* $1140 per trimester full-time, $449 per year part-time for state residents; $1816 per trimester full-time, $674 per year part-time for nonresidents. *Financial aid:* In 1990–91, $163,426 in aid awarded. 14 research assistantships (1 to a first-year student), 20 teaching assistantships (7 to first-year students) were awarded; institutionally sponsored loans also available. Aid available to part-time students. Financial aid application deadline: 2/1; applicants required to submit FAF. *Faculty research:* Bayes fitting, robust estimation, methodology time data, general regression models, analysis of escapement methods. Total annual research budget: $222,593. • Dr. Justus F. Seely, Chair, 503-737-3366.

See full description on page 445.

Pennsylvania State University University Park Campus, College of Science, Department of Statistics, University Park, PA 16802. Department awards MA, MS, PhD. Faculty: 15. Matriculated students: 31 full-time (14 women), 14 part-time (6 women). In 1990, 9 master's, 6 doctorates awarded. *Entrance requirements:* GRE General Test. Application fee: $35. *Tuition:* $203 per credit for state residents; $403 per credit for nonresidents. • Dr. James L. Rosenberger, Acting Head, 814-865-1348.

Announcement: Department faculty evenly divided between theoretical and applied interests. This balance important to student success in finding rewarding jobs. Computers quite accessible; students learn several statistical packages (Minitab, SAS, BMDP, SPSS). Minitab, one of the most widely used packages, was developed in department. Most students supported by fellowships and teaching assistantships that carry stipends of $850–$950 per month and full tuition waiver.

Polytechnic University, Brooklyn Campus, Division of Arts and Sciences, Department of Mathematics, 333 Jay Street, Brooklyn, NY 11201. Offerings include applied statistics (MS, PhD). *Tuition:* $6000 per semester.

Princeton University, School of Engineering and Applied Science, Departments of Civil Engineering and Operations Research and Electrical Engineering, Program in Statistics and Operations Research, Princeton, NJ 08544. Program awards MSE, PhD. *Degree requirements:* For master's, 1 foreign language, thesis; for doctorate, 1 foreign language, dissertation, qualifying exam. *Entrance requirements:* For master's, GRE General Test, GRE Subject Test, baccalaureate degree in engineering or science; for doctorate, GRE General Test, GRE Subject Test. Application deadline: 1/8. Application fee: $45 ($50 for foreign students). *Tuition:* $16,670 per year. *Financial aid:* Fellowships, research assistantships, teaching assistantships available. Financial aid application deadline: 1/8. • Dr. Erhan Cinlar, Director, 609-258-5995. Application contact: Michele Spreen, Director of Graduate Admissions, 609-258-3034.

Announcement: The program draws upon diverse course offerings and research activities in several departments. Areas of research emphasis are in stochastic processes, data analysis, statistical inference for processes, combinatorial and probabilistic optimization, linear and nonlinear programming, large-scale networks and stochastic programming. For further information, see full description in the Operations Research Section in the Engineering and Applied Sciences volume of this series.

Purdue University, School of Science, Department of Statistics, West Lafayette, IN 47907. Department awards MS, PhD. Faculty: 20. Matriculated students: 43 full-time (13 women), 22 part-time (6 women); includes 2 minority (1 Asian American, 1 Native American), 44 foreign. 153 applicants, 11% accepted. In 1990, 11 master's, 5 doctorates awarded. *Degree requirements:* For master's, foreign language and thesis not required; for doctorate, 1 foreign language, dissertation. *Entrance requirements:* TOEFL (minimum score of 550 required). Application fee: $25. *Tuition:* $1162 per semester full-time, $83.50 per credit hour part-time for state residents; $3720 per semester full-time, $244.50 per credit hour part-time for nonresidents. *Financial aid:* In 1990–91, 2 fellowships (both to first-year students), 5 research assistantships (0 to first-year students), 46 teaching assistantships (6 to first-year students) awarded; institutionally sponsored loans also available. Financial aid application deadline: 4/15. • Dr. S. S. Gupta, Head, 317-494-6031.

Announcement: MS programs train students to use data in planning market analysis, decision making, and production control. Most MS graduates select industrial positions. PhD graduates pursue careers in industry or in university teaching and research.

See full description on page 449.

Queen's University at Kingston, Faculty of Arts and Sciences, Department of Mathematics and Statistics, Kingston, ON K7L 3N6, Canada. Department offers programs in mathematics (MA, M Sc, PhD), statistics (MA, M Sc, PhD). Part-time programs available. Matriculated students: 32 full-time, 5 part-time; includes 10 foreign. In 1990, 7 master's, 3 doctorates awarded. *Degree requirements:* For master's, thesis required, foreign language not required; for doctorate, 2 foreign languages, dissertation, comprehensive exam. *Entrance requirements:* TOEFL (minimum score of 550 required). Application deadline: 2/28 (priority date). Application fee: $35. *Tuition:* $2861 per year full-time, $426 per trimester part-time for Canadian residents; $10,613 per year full-time, $4998 per trimester part-time for nonresidents. *Financial aid:* Fellowships, research assistantships, teaching assistantships, institutionally sponsored loans available. Financial aid application deadline: 3/1. *Faculty research:* Algebra, analysis, applied mathematics, combinatorics, differential geometry. • Dr. L. L. Campbell, Head, 613-545-6147. Application contact: Dr. O. A. Nielsen, Graduate Coordinator, 613-545-2397.

Rensselaer Polytechnic Institute, Interdisciplinary Programs, Department of Decision Sciences and Engineering Systems, Troy, NY 12180. Department offers programs in industrial and management engineering (M Eng, MS); management systems (MS), including management systems; operations research and statistics (MS, PhD), including decision science and engineering systems (PhD), operations research and statistics (MS). Part-time programs available. Faculty: 17 full-time (0 women), 2 part-time (0 women). Matriculated students: 69 full-time (20 women), 14 part-time (2 women); includes 4 minority (1 black American, 2 Hispanic American, 1 Native American), 33 foreign. 176 applicants, 70% accepted. In 1990, 34 master's, 5 doctorates awarded. *Degree requirements:* For master's, thesis optional, foreign language not required; for doctorate, dissertation required, foreign language not required. *Entrance requirements:* For doctorate, GRE General Test, TOEFL. Application deadline: 2/1. Application fee: $30. *Expenses:* Tuition of $455 per credit hour. Fees of $195.57 per semester. *Financial aid:* in 1990–91, 56 students received a total of $658,155 in aid awarded. 4 fellowships (3 to first-year students), 24 research assistantships (8 to first-year students), 33 teaching assistantships (23 to first-year students) were awarded; full and partial tuition waivers and career-related internships or fieldwork also available. Financial aid application deadline: 2/1. *Faculty research:* Industrial engineering, manufacturing, management information systems, applied statistics, operations research. Total annual research budget: $2.7-million. • James M. Tien, Chair, 518-276-6486. Application contact: Madabhushi Raghavachi, Associate Chair, 518-276-2962.

See full description on page 451.

Rice University, George R. Brown School of Engineering, Department of Statistics, Houston, TX 77251. Department awards MA, M Stat, PhD. *Degree requirements:* For master's, thesis (MA); for doctorate, dissertation. *Entrance requirements:* GRE General Test, TOEFL (minimum score of 550 required), minimum GPA of 3.0. *Expenses:* Tuition of $8300 per year full-time, $400 per credit hour part-time. Fees of $167 per year. • D. W. Scott, Chairman.

Rochester Institute of Technology, College of Engineering, Center of Quality and Applied Statistics, Rochester, NY 14623. Center offers program in applied statistics (MS). Part-time and evening/weekend programs available. Matriculated students: 7 full-time (3 women), 76 part-time (27 women); includes 7 minority (3 Asian American, 2 black American, 2 Hispanic American), 1 foreign. 33 applicants, 85% accepted. In 1990, 40 degrees awarded. *Degree requirements:* Oral exam required, foreign language and thesis not required. *Entrance requirements:* TOEFL, previous course work in calculus, minimum GPA of 3.0. Application deadline: 3/1 (priority date, applications processed on a rolling basis). Application fee: $35. *Tuition:* $12,657 per year full-time, $359 per hour part-time. *Financial aid:* Research assistantships available. • Dr. Edward Schilling, Chairman, 716-475-6129.

Rutgers, The State University of New Jersey, New Brunswick, Program in Statistics, New Brunswick, NJ 08903. Offers quality and productivity management (MS), statistics (MS, PhD). Part-time programs available. Faculty: 19 full-time (2 women), 0 part-time. Matriculated students: 31 full-time (18 women), 55 part-time (26 women); includes 2 minority (both black American), 49 foreign. Average age 28. 95 applicants, 85% accepted. In 1990, 11 master's, 1 doctorate awarded. *Degree requirements:* For master's, exam, essay required, foreign language and thesis not required; for doctorate, 1 foreign language (computer language can substitute), dissertation, written and oral exams. *Entrance requirements:* GRE General Test. Application deadline: 8/1. Application fee: $35. *Expenses:* Tuition of $4432 per year full-time, $183 per credit part-time for state residents; $6496 per year full-time, $270 per credit part-time for nonresidents. Fees of $458 per year full-time, $117 per year part-time. *Financial aid:* In 1990–91, $150,000 in aid awarded. 6 fellowships (0 to first-year students), 2 research assistantships (0 to first-year students), 11 teaching assistantships (1 to a first-year student) were awarded; federal work-study and career-related internships or fieldwork also available. Financial aid application deadline: 3/1; applicants required to submit FAF. *Faculty research:* Probability, decision theory, sequential analysis, linear models, multivariate statistics. Total annual research budget: $400,000. • Harold Sackrowitz, Director, 908-932-2693.

St. John's University, Graduate School of Arts and Sciences, Department of Mathematics and Computer Science, Jamaica, NY 11439. Offerings include probability and statistics (MA). Department faculty: 21 full-time (4 women), 0 part-time. *Degree requirements:* Variable foreign language requirement. *Entrance requirements:* GRE General Test. Application deadline: rolling. Application fee: $20. *Expenses:* Tuition of $297 per credit. Fees of $130 per year full-time, $65 per semester part-time. • Dr. Edward Miranda, Chairman, 718-990-6161.

Directory: Statistics

San Diego State University, College of Sciences, Department of Mathematical Sciences, Program in Statistics, San Diego, CA 92182. Program awards MS. Faculty: 4 full-time, 3 part-time. Matriculated students: 1 full-time (0 women), 18 part-time (8 women); includes 2 minority (1 Asian American, 1 black American), 1 foreign. In 1990, 5 degrees awarded. *Degree requirements:* Comprehensive exam required, foreign language and thesis not required. *Entrance requirements:* GRE General Test (minimum combined score of 950 required). Application deadline: 8/1 (applications processed on a rolling basis). Application fee: $55. *Expenses:* Tuition of $0 for state residents; $189 per unit for nonresidents. Fees of $1974 per year full-time, $692 per year part-time for state residents; $1074 per year full-time, $692 per year part-time for nonresidents. *Faculty research:* Time-series analysis, statistical inference, experimental design, biostatistics (public health), nonparametric statistics. • Dr. C. J. Park, Adviser, 619-594-6191.

Sangamon State University, School of Liberal Arts and Sciences, Program in Mathematical Sciences, Springfield, IL 62794-9243. Offerings include statistics (MA). Program faculty: 7 full-time (2 women), 1 (woman) part-time. *Degree requirements:* Foreign language and thesis not required. *Entrance requirements:* Proficiency in calculus and programming, BA in mathematics or computer science. Application fee: $0. *Expenses:* Tuition of $69.50 per credit. Fees of $42 per year. • Mary Kate Yntema, Director, 217-786-6770.

Simon Fraser University, Faculty of Science, Department of Mathematics and Statistics, Burnaby, BC V5A 1S6, Canada. Department offers programs in applied mathematics (M Sc, PhD), pure mathematics (M Sc, PhD), statistics (M Sc, PhD). Faculty: 35 (2 women). Matriculated students: 60 full-time (13 women), 0 part-time. In 1990, 7 master's, 5 doctorates awarded. *Degree requirements:* For master's, thesis; for doctorate, dissertation, comprehensive exams. *Entrance requirements:* GRE Subject Test, TOEFL (minimum score of 600 required). Application fee: $0. *Expenses:* Tuition of $612 per trimester full-time, $306 per trimester part-time. Fees of $68 per trimester full-time, $34 per trimester part-time. *Financial aid:* In 1990–91, 18 fellowships awarded; research assistantships, teaching assistantships also available. *Faculty research:* Semi-groups, lattice ordered groups, summability, functional analysis, graph theory. • Dr. A. R. Freedman, Chairman, 604-291-3378.

See full description on page 455.

Southern Illinois University at Carbondale, College of Science, Department of Mathematics, Program in Statistics, Carbondale, IL 62901. Program awards MS. Faculty: 37 full-time (3 women), 0 part-time. Matriculated students: 3 full-time (0 women), 0 part-time; includes 0 minority, 0 foreign. Average age 24. 6 applicants, 50% accepted. In 1990, 0 degrees awarded. *Degree requirements:* 1 foreign language (computer language can substitute), thesis or alternative. *Entrance requirements:* TOEFL (minimum score of 550 required), minimum GPA of 2.7. Application deadline: rolling. Application fee: $0. *Expenses:* Tuition of $1638 per year full-time, $204.75 per semester hour part-time for state residents; $4914 per year full-time, $614.25 per semester hour part-time for nonresidents. Fees of $700 per year full-time, $216 per year part-time. *Financial aid:* In 1990–91, $30,118 in aid awarded. 0 fellowships, 1 research assistantship (0 to first-year students), 4 teaching assistantships (2 to first-year students) were awarded; full tuition waivers, federal work-study, institutionally sponsored loans also available. *Faculty research:* Statistical decision theory, linear models, sampling theory. • Ronald Kirk, Chairperson, Department of Mathematics, 618-453-5302. Application contact: Dr. Marvin Zeman, Coordinator, 618-453-5302.

Southern Illinois University at Edwardsville, School of Sciences, Department of Mathematics and Statistics, Edwardsville, IL 62026. Department awards MS. Part-time programs available. Faculty: 27 full-time (5 women), 9 part-time (3 women). Matriculated students: 19 full-time (8 women), 34 part-time (10 women); includes 7 minority (2 Asian American, 2 black American, 2 Hispanic American, 1 Native American), 12 foreign. In 1990, 17 degrees awarded. *Degree requirements:* Computer language, thesis or alternative, final exam required, foreign language not required. *Entrance requirements:* Undergraduate major in field. Application fee: $0. *Expenses:* Tuition of $1566 per year full-time, $43.40 per credit hour part-time for state residents; $4698 per year full-time, $130.20 per credit hour part-time for nonresidents. Fees of $291.75 per year full-time, $27.35 per year (minimum) part-time. *Financial aid:* In 1990–91, 1 fellowship, 1 research assistantship, 21 teaching assistantships, 1 assistantship awarded; federal work-study, institutionally sponsored loans also available. Aid available to part-time students. • Dr. Chung-wu Ho, Chairman, 618-692-2385.

Southern Methodist University, Dedman College, Department of Statistical Science, Dallas, TX 75275. Department awards MS, PhD. Faculty: 10 full-time (1 woman), 0 part-time. Matriculated students: 17 full-time (6 women), 10 part-time (4 women); includes 1 minority (Hispanic American), 11 foreign. Average age 28. 40 applicants, 80% accepted. In 1990, 6 master's, 1 doctorate awarded. Terminal master's awarded for partial completion of doctoral program. *Degree requirements:* For master's, written and oral exams required, foreign language and thesis not required; for doctorate, dissertation, written and oral exams required, foreign language not required. *Entrance requirements:* For master's, GRE, TOEFL (minimum score of 550 required). Application deadline: 4/30. Application fee: $25. *Expenses:* Tuition of $435 per credit. Fees of $664 per semester for state residents; $56 per year for nonresidents. *Financial aid:* In 1990–91, $220,000 in aid awarded. 15 teaching assistantships (6 to first-year students) were awarded; full tuition waivers also available. Financial aid application deadline: 4/30. *Faculty research:* Bayesian analysis, nonparametric, quality control, regression, time series. Total annual research budget: $115,000. • Dr. Wayne Woodward, Chairperson, 214-692-2270. Application contact: Dr. Campbell Read, Graduate Adviser, 214-692-2456.

Stanford University, School of Humanities and Sciences, Department of Statistics, Stanford, CA 94305. Department awards MS, PhD. Faculty: 16 (2 women). Matriculated students: 49 full-time (15 women), 0 part-time; includes 5 minority (all Asian American), 24 foreign. Average age 27. 49 applicants, 41% accepted. In 1990, 17 master's, 10 doctorates awarded. Terminal master's awarded for partial completion of doctoral program. *Degree requirements:* For master's, computer language required, foreign language not required; for doctorate, computer language, dissertation, oral exam required, foreign language not required. *Entrance requirements:* For master's, GRE, TOEFL; for doctorate, GRE General Test, GRE Subject Test, TOEFL. Application deadline: 1/1. Application fee: $55. *Expenses:* Tuition of $15,102 per year. Fees of $28 per quarter. *Financial aid:* Fellowships, research assistantships, teaching assistantships, institutionally sponsored loans available. Financial aid application deadline: 1/1; applicants required to submit GAPSFAS. • Application contact: Administrator, 415-723-2625.

State University of New York at Albany, College of Science and Mathematics, Department of Mathematics and Statistics, Albany, NY 12222. Department offers programs in actuarial sciences (MA), secondary teaching (MA), statistics (MA, PhD). Evening/weekend programs available. Faculty: 32 full-time (1 woman), 1 part-time (0 women). Matriculated students: 43 full-time (9 women), 23 part-time (7 women); includes 4 minority (1 Asian American, 1 black American, 2 Hispanic American), 10 foreign. Average age 25. In 1990, 20 master's, 3 doctorates awarded. *Degree requirements:* For master's, foreign language and thesis not required; for doctorate, 1 foreign language, dissertation. *Entrance requirements:* For master's, GRE; for doctorate, GRE General Test, GRE Subject Test. *Expenses:* Tuition of $2450 per year full-time, $103 per credit part-time for state residents; $5765 per year full-time, $243 per credit part-time for nonresidents. Fees of $25 per year full-time, $0.85 per credit part-time. *Financial aid:* In 1990–91, 1 fellowship, 29 teaching assistantships awarded; research assistantships, minority assistantships also available. • Timothy Lance, Chairman, 518-442-4602.

State University of New York at Binghamton, School of Arts and Sciences, Department of Mathematical Sciences, Binghamton, NY 13902-6000. Offerings include mathematical sciences (MA, PhD), with options in computer science, probability and statistics. Terminal master's awarded for partial completion of doctoral program. *Degree requirements:* For master's, thesis or alternative; for doctorate, 2 foreign languages, dissertation. *Entrance requirements:* GRE General Test, GRE Subject Test, TOEFL. Application deadline: 4/15 (priority date). Application fee: $35. *Expenses:* Tuition of $2450 per year full-time, $102.50 per credit part-time for state residents; $5766 per year full-time, $242.50 per credit part-time for nonresidents. Fees of $77 per year full-time, $27.85 per semester (minimum) part-time. • Dr. David L. Hanson, Chairperson, 607-777-2147.

State University of New York at Buffalo, Graduate School, School of Medicine, Graduate Programs in Medicine, Department of Statistics, Buffalo, NY 14260. Department awards MA, PhD. Faculty: 6 full-time (0 women), 1 part-time (0 women). Matriculated students: 19 full-time (7 women), 1 (woman) part-time; includes 5 minority (3 Asian American, 2 black American), 10 foreign. Average age 27. 95 applicants, 22% accepted. In 1990, 4 master's awarded (75% found work related to degree, 25% continued full-time study); 4 doctorates awarded (100% entered university research/teaching). Terminal master's awarded for partial completion of doctoral program. *Degree requirements:* For master's, thesis or alternative, project required, foreign language not required; for doctorate, dissertation required, foreign language not required. *Entrance requirements:* TOEFL (minimum score of 550 required). Application deadline: 2/15 (priority date, applications processed on a rolling basis). Application fee: $35. *Expenses:* Tuition of $1600 per semester full-time, $134 per hour part-time for state residents; $3258 per semester full-time, $274 per hour part-time for nonresidents. Fees of $137 per semester full-time, $115 per semester (minimum) part-time. *Financial aid:* In 1990–91, 12 students received a total of $993,352 in aid awarded. 2 fellowships (1 to a first-year student), 6 teaching assistantships (4 to first-year students), 4 graduate assistantships (2 to first-year students) were awarded; federal work-study, institutionally sponsored loans, and career-related internships or fieldwork also available. Average monthly stipend for a graduate assistantship: $793. Financial aid application deadline: 2/15; applicants required to submit FAF. *Faculty research:* Statistical interferences, biostatistics, stochastic processes, Bayesian statistics. Total annual research budget: $21,691. • Dr. Peter Enis, Chairman, 716-831-3690. Application contact: Dr. M. M. Desu, Director of Graduate Studies, 716-831-3690.

State University of New York at Stony Brook, College of Engineering and Applied Sciences, Department of Applied Mathematics and Statistics, Stony Brook, NY 11794. Department offers programs in applied mathematics (MS, PhD), operations research (MS, PhD), statistics (MS, PhD). Matriculated students: 80 full-time (37 women), 38 part-time (19 women); includes 18 minority (15 Asian American, 2 black American, 1 Hispanic American), 48 foreign. 218 applicants, 28% accepted. *Degree requirements:* For master's, thesis or alternative required, foreign language not required; for doctorate, 1 foreign language, dissertation, comprehensive exams. *Entrance requirements:* GRE General Test, TOEFL. Application deadline: 2/1. Application fee: $35. *Expenses:* Tuition of $2450 per year full-time, $103 per credit part-time for state residents; $5766 per year full-time, $243 per credit part-time for nonresidents. Fees of $151 per year full-time, $10.45 per year (minimum) part-time. *Financial aid:* In 1990–91, 8 fellowships, 15 research assistantships, 27 teaching assistantships awarded. *Faculty research:* Biostatistics, combinatorial analysis, differential equations, modeling. Total annual research budget: $312,578. • Dr. J. Glimm, Chairman, 516-632-8360.

Stephen F. Austin State University, School of Sciences and Mathematics, Department of Mathematics and Statistics, Nacogdoches, TX 75962. Department offers programs in mathematics (MS), mathematics education (MS), statistics (MS). Faculty: 14 full-time (1 woman), 0 part-time. Matriculated students: 11 full-time (4 women), 7 part-time (2 women); includes 1 minority (black American), 0 foreign. In 1990, 6 degrees awarded. *Degree requirements:* Comprehensive exam required, thesis optional, foreign language not required. *Entrance requirements:* GRE General Test, minimum GPA of 2.5 overall, 2.8 in last half of major. Application deadline: 8/1. Application fee: $0 ($25 for foreign students). *Expenses:* Tuition of $18 per semester hour for state residents; $122 per semester hour for nonresidents. Fees of $14 per semester hour. *Financial aid:* In 1990–91, 12 students received a total of $56,000 in aid awarded. 12 teaching assistantships (10 to first-year students) were awarded. Average monthly stipend for a graduate assistantship: $667. *Faculty research:* Kernel type estimators, fractal mappings, spline curve fitting, robust regression continua theory. • Dr. Jasper Adams, Chairman, 409-568-3805.

Teachers College, Columbia University, Graduate Faculty of Education, Division of Psychology and Education, Department of Measurement, Evaluation, and Statistics, New York, NY 10027. Offerings include educational measurement and evaluation (Ed M, MA, MS, Ed D, PhD), with options in applied statistics (MS), measurement and evaluation (Ed M, Ed D, PhD), psychology in education (MA). Department faculty: 5 full-time (2 women), 0 part-time. *Application deadline:* 7/15. *Application fee:* $35. *Expenses:* Tuition of $435 per credit. Fees of $100 per semester. • Richard M. Wolf, Chair, 212-678-3355. Application contact: Roland A. Hence, Office of Admissions, 212-678-3710.

Temple University, School of Business and Management, Department of Statistics, Philadelphia, PA 19122. Department awards MS, PhD. Faculty: 22 full-time (2 women), 0 part-time. Matriculated students: 45 (10 women); includes 1 minority (black American), 29 foreign. Average age 33. 81 applicants, 43% accepted. In 1990, 5 master's, 2 doctorates awarded. *Degree requirements:* For master's, comprehensive exam required, foreign language and thesis not required; for doctorate, dissertation, comprehensive exams, specialty exam required, foreign language not required. *Entrance requirements:* GRE General Test (minimum score of 550 on math section required), TOEFL (minimum score of 550 required), minimum GPA of 2.8 overall, 3.0 during previous 2 years. Application deadline: 7/1. Application fee: $30. *Tuition:* $224 per credit for state residents; $283 per credit for nonresidents. *Financial aid:* In 1990–91, 9 teaching assistantships awarded. • Dr. Damaraju Raghavarao, Chairperson, 215-787-8459. Application contact: Dr. Jagbir Singh, Graduate Chair, 215-787-5069.

See full description on page 459.

Temple University, Ambler Campus, School of Business and Management, Department of Statistics, Ambler, PA 19002-3999. Department awards MBA.

Texas A&M University, College of Science, Department of Statistics, College Station, TX 77843. Department awards MS, PhD. Faculty: 27. Matriculated students: 67 full-time (27 women), 0 part-time; includes 6 minority (2 Asian American, 4 Hispanic American), 32 foreign. 68 applicants, 74% accepted. In 1990, 7 master's, 2 doctorates awarded. *Entrance requirements:* GRE General Test, TOEFL. Application deadline: 7/15. Application fee: $25 ($50 for foreign students). *Expenses:* Tuition of $100 per semester (minimum) for state residents; $128 per credit hour for nonresidents. Fees of $459 per year full-time, $252 per semester part-time. *Financial aid:* Fellowships, research assistantships, teaching assistantships available. • H. S. Newton, Head, 409-845-3141.

Texas Tech University, Graduate School, College of Arts and Sciences, Department of Mathematics, Program in Statistics, Lubbock, TX 79409. Program awards MS. Faculty: 4. Matriculated students: 8. In 1990, 5 degrees awarded. *Degree requirements:* Foreign language not required. *Entrance requirements:* GRE General Test. Application deadline: 4/15 (priority date, applications processed on a rolling basis). Application fee: $0 ($50 for foreign students). *Tuition:* $494 per semester full-time, $20 per credit hour part-time for state residents; $1790 per semester full-time, $455 per credit hour part-time for nonresidents. • Dr. Ron Anderson, Chairman, Department of Mathematics, 806-742-2566.

Tulane University, Department of Mathematics, Concentration in Statistics, New Orleans, LA 70118. Concentration awards MS. *Entrance requirements:* GRE General Test, GRE Subject Test, TOEFL (minimum score of 600 required) or TSE (minimum score of 220 required), minimum B average in undergraduate course work. Application deadline: 7/1. Application fee: $35. *Expenses:* Tuition of $16,750 per year full-time, $931 per hour part-time. Fees of $230 per year full-time, $40 per hour part-time. *Financial aid:* Teaching assistantships available. Financial aid application deadline: 5/1; applicants required to submit GAPSFAS. • Dr. Arnold Levine, Chairman, 504-865-5727.

University of Akron, Buchtel College of Arts and Sciences, Department of Mathematical Sciences and Statistics, Program in Statistics, Akron, OH 44325. Program awards MS. Part-time and evening/weekend programs available. Faculty: 9 full-time (1 woman), 0 part-time. Matriculated students: 10 full-time (4 women), 5 part-time (2 women); includes 3 minority (1 Asian American, 2 black American), 6 foreign. Average age 25. 17 applicants, 88% accepted. In 1990, 7 degrees awarded. *Degree requirements:* Thesis optional, foreign language not required. *Entrance requirements:* Minimum GPA of 2.75. Application deadline: 3/1 (applications processed on a rolling basis). Application fee: $25. *Tuition:* $119.93 per credit hour for state residents; $210.93 per credit hour for nonresidents. *Financial aid:* In 1990–91, $5550 in aid awarded. 12 teaching assistantships (3 to first-year students) were awarded. Financial aid application deadline: 3/1. *Faculty research:* Experimental design, sampling biostatistics. • Dr. William H. Beyer, Head, Department of Mathematical Sciences and Statistics, 216-972-7401.

University of Alabama, The Manderson Graduate School of Business, Department of Management Science and Statistics, Program in Applied Statistics, Tuscaloosa, AL 35487-0132. Program awards PhD. Faculty: 8 full-time (1 woman), 1 part-time (0 women). In 1990, 2 degrees awarded (100% found work related to degree). *Degree requirements:* 1 foreign language, dissertation, comprehensive exam. *Entrance requirements:* GRE or GMAT, undergraduate major in related field. Application deadline: 7/6 (applications processed on a rolling basis). Application fee: $20. *Tuition:* $968 per year full-time, $82 per credit part-time for state residents; $2400 per year full-time, $218 per credit part-time for nonresidents. *Financial aid:* Fellowships, research assistantships, teaching assistantships available. *Faculty research:* Design of experiments, linear models, regression, nonparametric statistics. • Jean D. Gibbons, Chairman, 205-348-6085.

University of Alberta, Faculty of Graduate Studies and Research, Department of Statistics and Applied Probability, Edmonton, AB T6G 2J9, Canada. Department offers programs in applied probability (M Sc, PhD), statistics (M Sc, PhD), statistics and applied probability (M Sc, PhD). Matriculated students: 14 full-time, 1 part-time. *Application fee:* $0. *Expenses:* Tuition of $1495 per year full-time, $748 per year part-time for Canadian residents; $2243 per year full-time, $1121 per year part-time for nonresidents. Fees of $301 per year full-time, $118 per year part-time. • Dr. K. L. Mehra, Chair, 403-492-2052.

University of Arizona, College of Arts and Sciences, Faculty of Science, Department of Statistics, Tucson, AZ 85721. Department awards MS. Part-time programs available. Faculty: 4 full-time (2 women), 3 part-time (1 woman). Matriculated students: 3 full-time (2 women), 1 part-time (0 women); includes 0 minority, 2 foreign. Average age 26. 10 applicants, 40% accepted. In 1990, 4 degrees awarded. *Degree requirements:* Thesis or alternative required, foreign language not required. *Entrance requirements:* Minimum GPA of 3.0 in all mathematics course work, 3 semesters of calculus, 6 units of upper division mathematics. Application fee: $25. *Expenses:* Tuition of $0 for state residents; $5406 per year full-time, $209 per credit hour part-time for nonresidents. Fees of $1528 per year full-time, $80 per credit hour part-time. *Financial aid:* In 1990–91, $46,053 in aid awarded. 1 fellowship, 1 research assistantship, 4 teaching assistantships (0 to first-year students), 2 scholarships were awarded; partial tuition waivers and career-related internships or fieldwork also available. Financial aid application deadline: 7/1; applicants required to submit FAF. *Faculty research:* Clinical trials, stochastic processes, extreme value theory, spatial time series; statistical consulting. Total annual research budget: $35,295. • Dr. Yashawini Mittal, Head, 602-621-4158. Application contact: Dr. J. L. Denny, 602-621-6208.

University of Arkansas, J. William Fulbright College of Arts and Sciences, Department of Mathematical Sciences, Program in Statistics, Fayetteville, AR 72701. Program awards MS. Matriculated students: 6 full-time (2 women), 2 part-time (1 woman); includes 0 minority, 3 foreign. In 1990, 1 degree awarded. *Degree requirements:* Thesis required, foreign language not required. *Application fee:* $15. *Expenses:* Tuition of $2050 per year full-time, $103 per credit hour part-time for state residents; $4400 per year full-time, $220 per credit hour part-time for nonresidents. Fees of $50 per year full-time, $1.50 per credit hour part-time. *Financial aid:* Teaching assistantships available. • Dr. James E. Dunn, Chairman of Studies, 501-575-3351.

University of British Columbia, Faculty of Science, Department of Statistics, Vancouver, BC V6T 1Z1, Canada. Department awards M Sc, PhD. *Degree requirements:* For doctorate, 1 foreign language, dissertation, comprehensive exam.

University of Calgary, Faculty of Science, Department of Mathematics and Statistics, Calgary, AB T2N 1N4, Canada. Department awards M Sc, PhD. *Application fee:* $20. *Tuition:* $1705 per year full-time, $427 per course part-time for Canadian residents; $3410 per year full-time, $854 per course part-time for nonresidents.

University of California at Berkeley, College of Letters and Science, Department of Statistics, Berkeley, CA 94720. Department awards MA, PhD. Faculty: 26. Matriculated students: 55 full-time, 0 part-time; includes 9 minority (8 Asian American, 1 Hispanic American), 27 foreign. 104 applicants, 21% accepted. Terminal master's awarded for partial completion of doctoral program. *Degree requirements:* For doctorate, dissertation. *Entrance requirements:* GRE General Test, minimum GPA of 3.0. Application deadline: 2/11. Application fee: $40. *Expenses:* Tuition of $0. Fees of $1909 per year for state residents; $7825 per year for nonresidents. • Terence Speed, Chair.

Announcement: Department of Statistics offers diverse program including probability, theoretical statistics, statistical computing, data analysis, biostatistics, and statistical consulting. Department has the finest statistical computing facilities available anywhere and is involved in extensive computer-oriented research and instruction. PhD and master's programs available in statistics and biostatistics (jointly with Public Health).

University of California, Davis, Program in Statistics, Davis, CA 95616. Program awards MS, PhD. Faculty: 14 full-time, 0 part-time. Matriculated students: 23 full-time (7 women), 1 (woman) part-time; includes 3 minority (2 Asian American, 1 Hispanic American), 9 foreign. Average age 26. 39 applicants, 79% accepted. In 1990, 7 master's awarded (57% found work related to degree, 43% continued full-time study); 1 doctorate awarded (100% found work related to degree). *Degree requirements:* For master's, thesis not required; for doctorate, dissertation. *Entrance requirements:* GRE General Test, TOEFL, minimum GPA of 3.0. Application fee: $40. *Expenses:* Tuition of $0 for state residents; $7699 per year full-time, $3849 per year part-time for nonresidents. Fees of $2718 per year full-time, $1928 per year part-time. *Financial aid:* In 1990–91, 22 students received aid. 1 research assistantship, 13 teaching assistantships (7 to first-year students) awarded; fellowships, federal work-study, institutionally sponsored loans also available. Average monthly stipend for a graduate assistantship: $1335. *Faculty research:* Nonparametric analysis, time series analysis, biostatistics, curve estimation, reliability. Total annual research budget: $306,543. • Application contact: Francisco J. Samaniego, Graduate Adviser, 916-752-6492.

Announcement: The Division of Statistics has 14 faculty in the division and an additional 20 faculty from other disciplines, forming a Graduate Group in Statistics. This intercollege unit creates an atmosphere conducive to the development of interdisciplinary programs. The Statistical Laboratory, operated by the Division, provides training in consulting for students and services to campus clientele.

University of California, Los Angeles, School of Medicine and Graduate Division, Graduate Programs in Medicine, Department of Biomathematics, Los Angeles, CA 90024. Department awards MS, PhD, MD/PhD. Faculty: 6. Matriculated students: 10 full-time (3 women), 0 part-time; includes 1 minority (0 Asian American, 0 black American, 0 Hispanic American, 0 Native American), 5 foreign. 24 applicants, 29% accepted. In 1990, 4 master's, 0 doctorates awarded. *Degree requirements:* For master's, thesis or comprehensive exam required, foreign language not required; for doctorate, dissertation, written and oral qualifying exams required, foreign language not required. *Entrance requirements:* GRE General Test, GRE Subject Test. Application fee: $40. *Expenses:* Tuition of $0 for state residents; $7699 per year for nonresidents. Fees of $2907 per year. *Financial aid:* In 1990–91, 13 students received a total of $188,161 in aid awarded. 10 fellowships, 3 research assistantships, 4 teaching assistantships were awarded; full and partial tuition waivers, federal work-study, institutionally sponsored loans also available. Financial aid application deadline: 3/1. • Dr. Kenneth L. Lange, Chair, 310-825-5800.

See full description on page 463.

University of California, Riverside, Graduate Division, College of Natural and Agricultural Sciences, Department of Statistics, Riverside, CA 92521. Department offers programs in applied statistics (PhD), statistics (MS). Part-time programs available. *Degree requirements:* For master's, computer language, comprehensive exams required, foreign language and thesis not required; for doctorate, computer language, dissertation, qualifying exams, 3 quarters teaching experience required, foreign language not required. *Entrance requirements:* GRE General Test, TOEFL (minimum score of 550 required). Application deadline: 6/1. Application fee: $40. *Tuition:* $950 per quarter full-time, $264 per quarter part-time for state residents; $3517 per quarter full-time, $1758 per quarter part-time for nonresidents. *Faculty research:* Stochastic models, paired comparisons, linear models, discrete data analysis, Bayesian inference.

University of California, San Diego, Department of Mathematics, 9500 Gilman Drive, La Jolla, CA 92093. Offerings include statistics (MS). Department faculty: 54. *Application fee:* $40. *Expenses:* Tuition of $0 for state residents; $7699 per year full-time, $1283 per quarter part-time for nonresidents. Fees of $2798 per year full-time, $669 per quarter part-time. • Harold Stark, Chair, 619-534-3594. Application contact: Lois Stewart, Graduate Coordinator, 619-534-6887.

University of California, Santa Barbara, College of Letters and Science, Statistics and Applied Probability Program, Santa Barbara, CA 93106. Program awards MA, PhD. Matriculated students: 9 full-time (0 women), 0 part-time; includes 3 foreign. 14 applicants, 71% accepted. In 1990, 10 master's awarded. *Degree requirements:* For master's, thesis or alternative required, foreign language not required; for doctorate, dissertation required, foreign language not required. *Entrance requirements:* GRE, TOEFL (minimum score of 550 required). Application deadline: 5/1. Application fee: $40. *Expenses:* Tuition of $0 for state residents; $7699 per year for nonresidents. Fees of $2307 per year. *Financial aid:* Fellowships, research assistantships, teaching assistantships, full and partial tuition waivers, federal work-study, institutionally sponsored loans available. Financial aid application deadline: 1/31. • S. R. Jammalamadaka, Chairman, 805-893-3119. Application contact: Josianne Merminod, Graduate Secretary, 805-893-4857.

Announcement: Master's courses cover mathematical and applied statistics and operations research leading to high-level professional competence. Doctoral program is strongly research oriented in probability, stochastic processes, and statistics. Main research areas covered are theoretical and applied probability and mathematical and applied statistics. Some financial support available for students.

University of Chicago, Graduate School of Business, Chicago, IL 60637. Offerings include econometrics (MBA), statistics (MBA). School faculty: 99 full-time, 50 part-time. *Degree requirements:* Foreign language not required. *Entrance requirements:* GMAT or GRE General Test, TOEFL (minimum score of 550 required). Application deadline: 11/8 (priority date, applications processed on a rolling basis). Application fee: $75. *Expenses:* Tuition of $16,700 per year full-time, $1670 per course part-time. Fees of $240 per year. • John P. Gould, Dean, 312-702-7121. Application contact: Raymond A. Brown, Director of Admissions and Financial Aid, 312-702-7369.

Directory: Statistics

University of Chicago, Division of the Physical Sciences, Department of Statistics, Chicago, IL 60637. Department awards SM, PhD. Terminal master's awarded for partial completion of doctoral program. *Degree requirements:* For master's, thesis; for doctorate, 1 foreign language, dissertation. *Entrance requirements:* GRE General Test, GRE Subject Test, TOEFL. Application deadline: 1/6. Application fee: $45. *Expenses:* Tuition of $16,275 per year full-time, $8140 per year part-time. Fees of $356 per year.

University of Cincinnati, McMicken College of Arts and Sciences, Department of Mathematics, Cincinnati, OH 45221. Offerings include statistics (MS, PhD). Department faculty: 29 full-time, 0 part-time. *Degree requirements:* For doctorate, 1 foreign language, dissertation. *Entrance requirements:* For doctorate, GRE General Test, GRE Subject Test. Application deadline: 1/2. Application fee: $20. *Tuition:* $131 per credit hour for state residents; $261 per credit hour for nonresidents. • Dr. C. David Minda, Head, 513-556-4052.

University of Connecticut, College of Liberal Arts and Sciences, Field of Statistics, Storrs, CT 06269. Field awards MS, PhD. Faculty: 10. Matriculated students: 24 full-time (6 women), 7 part-time (2 women); includes 1 minority (black American), 23 foreign. Average age 30. 33 applicants, 70% accepted. In 1990, 9 master's, 1 doctorate awarded. Terminal master's awarded for partial completion of doctoral program. *Degree requirements:* For master's, foreign language not required; for doctorate, dissertation required, foreign language not required. *Entrance requirements:* GRE General Test, TOEFL. Application deadline: 6/1 (priority date, applications processed on a rolling basis). Application fee: $25. *Expenses:* Tuition of $3428 per year full-time, $571 per course part-time for state residents; $8914 per year full-time, $1486 per course part-time for nonresidents. Fees of $636 per year full-time, $87 per course part-time. *Financial aid:* In 1990–91, $123,117 in aid awarded. 6 fellowships (0 to first-year students), 3 research assistantships (0 to first-year students), 16 teaching assistantships (3 to first-year students) were awarded. Financial aid application deadline: 2/15; applicants required to submit GAPSFAS or FAF. • Nitis Mukhopadhyay, Head, 203-486-3414.

University of Delaware, College of Arts and Science, Department of Mathematical Sciences, Newark, DE 19716. Offerings include statistics (MA, MA Sc, MS, PhD). Terminal master's awarded for partial completion of doctoral program. Department faculty: 40. *Degree requirements:* For master's, thesis (for some programs), written exam required, foreign language not required; for doctorate, 2 foreign languages, dissertation, qualifying exam. *Entrance requirements:* GRE General Test. Application deadline: 7/1. Application fee: $40. *Tuition:* $179 per credit hour for state residents; $467 per credit hour for nonresidents. • Dr. Ivar Stakgold, Chairman, 302-451-2651. Application contact: Pam Haverland, Graduate Secretary, 302-451-2654.

See full description on page 469.

University of Florida, College of Liberal Arts and Sciences, Department of Statistics, Gainesville, FL 32611. Department awards MS Stat, M Stat, PhD. Faculty: 31 full-time (3 women), 0 part-time. Matriculated students: 44 full-time (18 women), 13 part-time (4 women); includes 3 minority (2 black American, 1 Hispanic American), 30 foreign. 74 applicants, 76% accepted. In 1990, 10 master's, 1 doctorate awarded. *Degree requirements:* For master's, variable foreign language requirement, computer language, comprehensive exam, final oral exam required, thesis not required; for doctorate, variable foreign language requirement, computer language, dissertation. *Entrance requirements:* GRE General Test, TOEFL (minimum score of 600 required), minimum GPA of 3.0. Application deadline: 6/1 (priority date, applications processed on a rolling basis). Application fee: $15. *Tuition:* $87 per credit hour for state residents; $289 per credit hour for nonresidents. *Financial aid:* In 1990–91, $320,000 in aid awarded. 3 fellowships (1 to a first-year student), 16 research assistantships (2 to first-year students), 24 teaching assistantships (9 to first-year students) were awarded. Financial aid application deadline: 2/1. • Dr. Ronald Randles, Chairman, 904-392-1941.

University of Georgia, College of Arts and Sciences, Department of Statistics, Athens, GA 30602. Department offers programs in applied mathematics (MAMS), statistics (MS, PhD). Faculty: 16 full-time (2 women), 0 part-time. Matriculated students: 32 full-time (7 women), 8 part-time (5 women); includes 4 minority (all black American), 25 foreign. 63 applicants, 35% accepted. In 1990, 20 master's, 3 doctorates awarded. *Degree requirements:* For master's, technical report (MAMS) required, foreign language not required; for doctorate, 1 foreign language (computer language can substitute), dissertation. *Entrance requirements:* GRE General Test. Application fee: $10. *Expenses:* Tuition of $598 per quarter full-time, $48 per quarter part-time for state residents; $1558 per quarter full-time, $144 per quarter part-time for nonresidents. Fees of $118 per quarter. *Financial aid:* Fellowships, research assistantships, teaching assistantships, assistantships available. • Dr. Kermit Hutcheson, Graduate Coordinator, 404-542-8232.

See full description on page 475.

University of Guelph, College of Physical Science, Department of Mathematics and Statistics, Guelph, ON N1G 2W1, Canada. Department offers programs in applied mathematics (PhD), applied statistics (PhD), mathematics and statistics (M Sc). Faculty: 28. Matriculated students: 29; includes 10 foreign. 67 applicants, 19% accepted. In 1990, 7 master's awarded. *Entrance requirements:* For master's, minimum B average during last two years; for doctorate, minimum B average. *Expenses:* Tuition of $898 per semester full-time, $450 per semester part-time for Canadian residents; $4053 per semester full-time, $2185 per semester part-time for nonresidents. Fees of $543 per semester full-time, $450 per semester part-time for Canadian residents; $2278 per semester full-time, $2185 per semester part-time for nonresidents. *Financial aid:* Fellowships, research assistantships, teaching assistantships available. *Faculty research:* Dynamical systems, mathematical biology, numerical analysis and reproductive research, linear and non-linear models, reliability and bioassay. • Dr. Langford, Chair, 519-824-4120 Ext. 2155. Application contact: Dr. Carter, Graduate Coordinator, 519-824-4120 Ext. 3569.

University of Houston, College of Business Administration, Program in Statistics and Operations Research, 4800 Calhoun, Houston, TX 77004. Program awards MBA, PhD. *Degree requirements:* For master's, computer language required, foreign language and thesis not required; for doctorate, computer language, dissertation, comprehensive exam required, foreign language not required. *Entrance requirements:* For master's, GMAT (minimum score of 530 required), minimum GPA of 3.0; for doctorate, GMAT (minimum score of 530 required), minimum GPA of 3.25. Application deadline: 6/5. Application fee: $25 ($75 for foreign students). *Expenses:* Tuition of $30 per hour for state residents; $134 per hour for nonresidents. Fees of $240 per year full-time, $125 per year part-time. *Financial aid:* Application deadline 4/1. • Dr. Everette Gardner, Chair, 713-749-3957. Application contact: Mary Gould, Director, Office of Student Services, 713-749-2893.

University of Idaho, College of Graduate Studies, College of Letters and Science, Department of Mathematics and Applied Statistics, Program in Applied Statistics, Moscow, ID 83843. Program awards MS. Matriculated students: 6 full-time (3 women), 2 part-time (0 women); includes 1 minority (Asian American), 3 foreign. In 1990, 4 degrees awarded. *Degree requirements:* Foreign language and thesis not required. *Entrance requirements:* Minimum GPA of 2.8. Application deadline: 8/1. Application fee: $20. *Expenses:* Tuition of $0 for state residents; $4146 per year for nonresidents. Fees of $818 per semester full-time, $82.75 per credit part-time. *Financial aid:* Application deadline 3/1. • Dr. Clarence J. Potrate, Chair, Department of Mathematics and Applied Statistics, 208-885-6742.

University of Illinois at Chicago, College of Liberal Arts and Sciences, Department of Mathematics, Statistics, and Computer Science, Program in Probability and Statistics, Chicago, IL 60680. Program awards MS, DA, PhD. *Degree requirements:* For master's, foreign language and thesis not required; for doctorate, 1 foreign language, dissertation. *Entrance requirements:* GRE General Test, TOEFL (minimum score of 550 required), minimum GPA of 3.75 (on a 5.0 scale). Application deadline: 7/5. Application fee: $20. *Expenses:* Tuition of $1369 per semester full-time, $521 per semester (minimum) part-time for state residents; $3840 per semester full-time, $1454 per semester (minimum) part-time for nonresidents. Fees of $458 per semester full-time, $398 per semester (minimum) part-time. • Application contact: Jeff E. Lewis, Director of Graduate Studies, 312-996-3041.

Announcement: Research/teaching areas include design of experiments, linear statistical models, multivariate analysis, time series, game theory, decision theory, probability theory, reliability theory, and stochastic processes. In the last 10 years, 15 students have completed doctoral degrees in probability and statistics, and about 4 students per year have completed master's degrees.

See full description on page 479.

University of Illinois at Urbana-Champaign, College of Liberal Arts and Sciences, Department of Statistics, Champaign, IL 61820. Department awards MS, PhD. Faculty: 14 full-time (0 women), 7 part-time (1 woman). Matriculated students: 33 full-time (13 women), 22 part-time (13 women); includes 3 minority (all Asian American), 44 foreign. Average age 25. 86 applicants, 78% accepted. In 1990, 4 master's awarded (25% found work related to degree, 50% continued full-time study); 5 doctorates awarded (60% entered university research/teaching, 40% found other work related to degree). Terminal master's awarded for partial completion of doctoral program. *Degree requirements:* Thesis/dissertation required, foreign language not required. *Entrance requirements:* For master's, TOEFL. *Tuition:* $1838 per semester full-time, $708 per semester part-time for state residents; $4314 per semester full-time, $1673 per semester part-time for nonresidents. *Financial aid:* In 1990–91, 1 fellowship, 12 research assistantships (1 to a first-year student), 12 teaching assistantships (2 to first-year students) awarded; full tuition waivers also available. Financial aid application deadline: 2/15. *Faculty research:* Statistical decision theory, sequential analysis, computer-aided stochastic modeling. • Jerome Sacks, Head, 217-333-2167.

See full description on page 481.

University of Iowa, College of Liberal Arts, Department of Statistics, Iowa City, IA 52242. Department awards MS, PhD. Faculty: 22 full-time, 0 part-time. Matriculated students: 71 full-time (23 women), 18 part-time (5 women); includes 4 minority (3 Asian American, 1 Hispanic American), 59 foreign. 174 applicants, 70% accepted. In 1990, 21 master's, 3 doctorates awarded. *Application fee:* $20. *Expenses:* Tuition of $1158 per semester full-time, $387 per semester hour (minimum) part-time for state residents; $3372 per semester full-time, $387 per semester hour (minimum) part-time for nonresidents. Fees of $60 per semester (minimum). *Financial aid:* In 1990–91, 1 fellowship (0 to first-year students), 7 research assistantships, 38 teaching assistantships (6 to first-year students) awarded. • Richard Dykstra Jr., Chair, 319-335-0712.

University of Kansas, College of Liberal Arts and Sciences, Department of Mathematics, Program in Applied Mathematics and Statistics, Lawrence, KS 66045. Program awards MA, PhD. *Degree requirements:* For master's, thesis or alternative required, foreign language not required; for doctorate, 2 foreign languages, dissertation. *Entrance requirements:* TOEFL (minimum score of 570 required). Application fee: $25. *Expenses:* Tuition of $1668 per year full-time, $56 per credit hour part-time for state residents; $5382 per year full-time, $179 per credit hour part-time for nonresidents. Fees of $338 per year full-time, $25 per credit hour part-time. *Financial aid:* Fellowships, research assistantships, teaching assistantships, institutionally sponsored loans available. Aid available to part-time students. Financial aid application deadline: 2/1. • Dr. Charles Himmelberg, Chairperson, Department of Mathematics, 913-864-3651. Application contact: Dr. Saul Stahl, Graduate Director, 913-864-4324.

University of Kentucky, Graduate School Programs from the College of Arts and Sciences, Program in Statistics, Lexington, KY 40506-0032. Program awards MS, PhD. Faculty: 13 full-time (1 woman), 0 part-time. Matriculated students: 19 full-time (6 women), 11 part-time (3 women); includes 4 minority (3 Asian American, 1 Hispanic American), 13 foreign. 120 applicants, 11% accepted. In 1990, 12 master's, 1 doctorate awarded. *Degree requirements:* For master's, comprehensive exam required, thesis optional, foreign language not required; for doctorate, 1 foreign language, dissertation, comprehensive exam. *Entrance requirements:* For master's, GRE (verbal, quantitative, and analytical sections), minimum undergraduate GPA of 2.5; for doctorate, GRE (verbal, quantitative, and analytical sections), minimum graduate GPA of 3.0. Application deadline: 7/19 (applications processed on a rolling basis). Application fee: $20 ($25 for foreign students). *Tuition:* $1002 per semester full-time, $101 per credit hour part-time for state residents; $2782 per semester full-time, $299 per credit hour part-time for nonresidents. *Financial aid:* In 1990–91, 0 fellowships, 0 research assistantships, 23 teaching assistantships (5 to first-year students) awarded; federal work-study, institutionally sponsored loans also available. Aid available to part-time students. Financial aid applicants required to submit FAF. *Faculty research:* Applied statistics, mathematical statistics, applied probability, storage models, queuing problems. • William Griffith, Director of Graduate Studies, 606-257-6906. Application contact: Dr. Constance L. Wood, Associate Dean for Academic Administration, 606-257-4905.

See full description on page 485.

University of Manitoba, Faculty of Science, Department of Statistics, Winnipeg, MB R3T 2N2, Canada. Department awards M Sc, PhD. *Degree requirements:* For master's, thesis or alternative required, foreign language not required; for doctorate, 1 foreign language, dissertation.

University of Maryland College Park, College of Computer, Mathematical and Physical Sciences, Department of Mathematics, Program in Mathematical Statistics, College Park, MD 20742. Program awards MA, PhD. Matriculated students: 18 full-time (6 women), 16 part-time (6 women); includes 8 minority (4 Asian American, 3 black American, 1 Hispanic American), 14 foreign. 70 applicants, 39% accepted. In 1990, 5 master's, 3 doctorates awarded. *Degree requirements:* For master's, thesis or comprehensive exams required, foreign language not required; for doctorate, 2 foreign languages, dissertation. *Entrance requirements:* For master's, minimum GPA of 3.0. Application deadline: rolling. Application fee: $25. *Expenses:* Tuition of $143 per credit hour for state residents; $256 per credit hour for nonresidents. Fees of

$171.50 per semester. *Financial aid:* In 1990–91, 5 fellowships (2 to first-year students) awarded. • Dr. Paul J. Smith, Director, 301-405-5061.

University of Maryland Graduate School, Baltimore, Graduate School, Department of Mathematics and Statistics, Program in Statistics, Baltimore, MD 21228. Program awards MS, PhD. Matriculated students: 9 full-time (4 women), 8 part-time (2 women); includes 0 minority, 7 foreign. 25 applicants, 28% accepted. In 1990, 1 master's, 1 doctorate awarded. *Degree requirements:* For master's, foreign language and thesis not required; for doctorate, dissertation required, foreign language not required. *Entrance requirements:* GRE General Test, GRE Subject Test, TOEFL, minimum GPA of 3.0. Application deadline: 7/1. Application fee: $25. *Tuition:* $134 per credit for state residents; $245 per credit for nonresidents. *Financial aid:* In 1990–91, 1 research assistantship awarded; fellowships, teaching assistantships also available. • Application contact: Dr. Thomas Mathew, Coordinator, 301-455-2418.

See full description on page 487.

University of Massachusetts at Amherst, College of Arts and Sciences, Faculty of Natural Sciences and Mathematics, Department of Mathematics and Statistics, Amherst, MA 01003. Department offers programs in applied mathematics (MA), statistics (MA, PhD). Faculty: 49 full-time (2 women), 0 part-time. Matriculated students: 54 full-time (17 women), 13 part-time (6 women); includes 0 minority, 25 foreign. Average age 27. 672 applicants, 57% accepted. In 1990, 16 master's, 5 doctorates awarded. Terminal master's awarded for partial completion of doctoral program. *Degree requirements:* For master's, foreign language and thesis not required; for doctorate, 2 foreign languages, dissertation. *Entrance requirements:* GRE General Test. Application deadline: 3/1 (applications processed on a rolling basis). Application fee: $35. *Tuition:* $2568 per year full-time, $107 per credit part-time for state residents; $7920 per year full-time, $330 per credit part-time for nonresidents. *Financial aid:* In 1990–91, 1 fellowship, 14 research assistantships, 69 teaching assistantships awarded; federal work-study also available. Aid available to part-time students. Financial aid application deadline: 3/1; applicants required to submit FAF. • Dr. Mei Ku, Director, 413-545-2282. Application contact: Dr. Wallace S. Martindale III, Chair, Admissions Committee, 413-545-0984.

University of Michigan, College of Literature, Science, and the Arts, Department of Statistics, Ann Arbor, MI 48109. Department offers programs in applied statistics (AM), statistics (AM, PhD). Faculty: 14 full-time. Matriculated students: 35 (16 women); includes 1 minority (Asian American), 23 foreign. 80 applicants, 34% accepted. In 1990, 12 master's, 3 doctorates awarded. *Degree requirements:* For master's, foreign language and thesis not required; for doctorate, 1 foreign language (computer language can substitute), dissertation, preliminary exam. *Entrance requirements:* For master's, GRE General Test (applied statistics), GRE Subject Test (statistics); for doctorate, GRE General Test, GRE Subject Test. Application deadline: 2/15 (applications processed on a rolling basis). Application fee: $30. *Tuition:* $3255 per semester full-time, $352 per credit (minimum) part-time for state residents; $6803 per semester full-time, $746 per credit (minimum) part-time for nonresidents. *Financial aid:* Fellowships, teaching assistantships available. Financial aid application deadline: 3/15. *Faculty research:* Sequential analysis, Bayesian statistics, multivariate analysis. Total annual research budget: $45,000. • Robb Muirhead, Chair, 313-763-3520.

University of Michigan, School of Public Health, Department of Environmental Health Sciences, Interdepartmental Program in Clinical Research Design and Statistical Analysis, Ann Arbor, MI 48109. Program awards MS. Part-time programs available. Faculty: 6. Matriculated students: 28 full-time (9 women), 0 part-time; includes 2 minority (1 Asian American, 1 black American), 1 foreign. 42 applicants, 90% accepted. In 1990, 27 degrees awarded. *Degree requirements:* Thesis required, foreign language not required. *Entrance requirements:* GRE General Test. Application fee: $30. *Tuition:* $6546 per year for state residents; $13,680 per year for nonresidents. • Richard A. Cornell, Chair, 313-764-5451.

University of Minnesota, Twin Cities Campus, College of Liberal Arts, School of Statistics, Minneapolis, MN 55455. School awards MS, PhD. *Expenses:* Tuition of $1084 per quarter full-time, $301 per credit part-time for state residents; $2168 per quarter full-time, $602 per credit part-time for nonresidents. Fees of $118 per quarter. • Seymour Geisser, Director, 612-625-8046.

University of Missouri–Columbia, College of Arts and Sciences, Department of Statistics, Columbia, MO 65211. Department awards MA, PhD. Faculty: 13 full-time, 1 part-time. Matriculated students: 21 full-time (6 women), 17 part-time (4 women); includes 2 minority (1 Asian American, 1 black American), 28 foreign. In 1990, 3 master's, 1 doctorate awarded. *Degree requirements:* For master's, thesis not required; for doctorate, dissertation. *Entrance requirements:* GRE General Test, minimum GPA of 3.0. Application deadline: 8/1 (priority date, applications processed on a rolling basis). Application fee: $20 ($40 for foreign students). *Expenses:* Tuition of $89.90 per credit hour full-time, $98.35 per credit hour part-time for state residents; $244 per credit hour full-time, $252.45 per credit hour part-time for nonresidents. Fees of $123.55 per semester (minimum) full-time. *Financial aid:* Research assistantships, teaching assistantships, full and partial tuition waivers available. • Dr. Farroll Wright, Director of Graduate Studies, 314-882-4075. Application contact: Gary L. Smith, Director of Admissions and Registrar, 314-882-7651.

University of Missouri–Rolla, College of Arts and Sciences, Department of Mathematics and Statistics, Rolla, MO 65401. Offerings include statistics (MS, PhD). Terminal master's awarded for partial completion of doctoral program. Department faculty: 21 full-time (2 women), 4 part-time (0 women). *Degree requirements:* For master's, thesis or alternative required, foreign language not required; for doctorate, 1 foreign language, dissertation. *Entrance requirements:* GRE General Test, GRE Subject Test. Application deadline: 7/1 (applications processed on a rolling basis). Application fee: $20 ($40 for foreign students). *Expenses:* Tuition of $2090 per year full-time, $87.10 per credit hour part-time for state residents; $5582 per year full-time, $232.60 per credit hour part-time for nonresidents. Fees of $349 per year full-time, $61.63 per semester (minimum) part-time. • William T. Ingram, Chairman, 314-341-4641. Application contact: Troy Hicks, 314-341-4654.

University of Montana, College of Arts and Sciences, Department of Mathematical Sciences, Missoula, MT 59812. Offerings include statistics (MA, PhD). Terminal master's awarded for partial completion of doctoral program. *Degree requirements:* For doctorate, 2 foreign languages (computer language can substitute for one), dissertation. *Entrance requirements:* For doctorate, GRE General Test. Application deadline: 9/15. Application fee: $20. *Tuition:* $495 per quarter hour full-time for state residents; $1239 per quarter hour full-time for nonresidents.

University of Nebraska–Lincoln, College of Arts and Sciences, Department of Mathematics and Statistics, Lincoln, NE 68588. Department offers programs in applied mathematics (MS), mathematics (M Sc T, PhD), mathematics and statistics (MA, MS), mathematics education (MAT), statistics (PhD). Faculty: 29 full-time (1 woman), 0 part-time. Matriculated students: 94 full-time (36 women), 15 part-time (6 women); includes 4 minority (2 Asian American, 2 black American), 54 foreign. Average age 29. In 1990, 29 master's, 4 doctorates awarded. Terminal master's awarded for partial completion of doctoral program. *Degree requirements:* For master's, foreign language not required; for doctorate, dissertation, comprehensive exams. *Entrance requirements:* TOEFL (minimum score of 500 required). Application deadline: 5/1 (priority date, applications processed on a rolling basis). Application fee: $25. *Expenses:* Tuition of $75.75 per credit hour for state residents; $187.25 per credit hour for nonresidents. Fees of $161 per year full-time. *Financial aid:* Fellowships, research assistantships, teaching assistantships, federal work-study available. Aid available to part-time students. Financial aid application deadline: 2/15. • Dr. Jim Lewis, Chairperson, 402-472-3731.

University of Nevada, Las Vegas, College of Science and Mathematics, Department of Mathematical Sciences, Las Vegas, NV 89154. Offerings include statistics (MS). Department faculty: 16 full-time (0 women), 0 part-time. *Degree requirements:* Oral exam required, thesis optional, foreign language not required. *Entrance requirements:* Minimum GPA of 2.75 overall, 3.0 during previous 2 years. Application deadline: 6/15. Application fee: $20. *Expenses:* Tuition of $66 per credit. Fees of $1800 per semester for nonresidents. • Dr. Peter Shive, Chairman, 702-739-3567. Application contact: Graduate College Admissions Evaluator, 702-739-3320.

University of New Brunswick, Faculty of Arts, Department of Mathematics and Statistics, Fredericton, NB E3B 5A3, Canada. Department awards M Sc, PhD. *Degree requirements:* For master's, thesis or alternative; for doctorate, dissertation. *Entrance requirements:* TOEFL, minimum GPA of 3.0. Application deadline: 3/1 (priority date). *Expenses:* Tuition of $2100 per year. Fees of $45 per year. • Peter C. Kent, Dean, Faculty of Arts.

University of New Mexico, College of Arts and Sciences, Department of Mathematics and Statistics, Albuquerque, NM 87131. Department awards MA, PhD. Faculty: 26 full-time (2 women), 9 part-time (2 women). *Degree requirements:* For master's, foreign language and thesis not required; for doctorate, 1 foreign language, dissertation. *Application fee:* $25. *Expenses:* Tuition of $467 per semester (minimum) full-time, $67.50 per credit hour part-time for state residents; $1549 per semester (minimum) full-time, $67.50 per credit hour part-time for nonresidents. Fees of $16 per semester. *Financial aid:* Teaching assistantships and career-related internships or fieldwork available. *Faculty research:* Pure and applied mathematics, applied statistics. • Frank Gilfeather, Chairman, 505-277-4613.

University of North Carolina at Chapel Hill, College of Arts and Sciences, Department of Statistics, Chapel Hill, NC 27599. Department awards MS, PhD. Faculty: 12 full-time, 0 part-time. Matriculated students: 30 full-time (10 women), 0 part-time; includes 0 minority, 22 foreign. 65 applicants, 32% accepted. In 1990, 7 master's, 7 doctorates awarded. *Degree requirements:* For master's, comprehensive exam, thesis or essay required, foreign language not required; for doctorate, dissertation, comprehensive exam required, foreign language not required. *Entrance requirements:* For master's, GRE General Test (minimum combined score of 1000 required), GRE Subject Test, TOEFL, minimum GPA of 3.0; for doctorate, GRE General Test (minimum combined score of 1000 required), GRE Subject Test, minimum GPA of 3.0. Application deadline: 2/15. Application fee: $35. *Tuition:* $621 per semester full-time for state residents; $3555 per semester full-time for nonresidents. *Financial aid:* In 1990–91, 2 fellowships, 4 research assistantships, 23 teaching assistantships awarded. • Stamatis Cambanis, Chairman, 919-962-2307.

University of Northern Colorado, College of Arts and Sciences, Department of Mathematics and Applied Statistics, Program in Applied Statistics, Greeley, CO 80639. Program awards MS, PhD. Faculty: 8 full-time (0 women), 0 part-time. Matriculated students: 15 full-time (5 women), 4 part-time (1 woman); includes 4 minority (2 Asian American, 1 black American, 1 Hispanic American), 7 foreign. 26 applicants, 58% accepted. In 1990, 18 master's, 3 doctorates awarded. *Degree requirements:* For master's, thesis or alternative, comprehensive exams; for doctorate, dissertation, comprehensive exams. *Entrance requirements:* GRE General Test. Application deadline: rolling. Application fee: $30. *Expenses:* Tuition of $1900 per year full-time, $106 per credit hour part-time for state residents; $6078 per year full-time, $338 per credit hour part-time for nonresidents. Fees of $320 per year full-time, $18 per credit hour part-time. *Financial aid:* In 1990–91, 2 fellowships, 4 teaching assistantships (1 to a first-year student), 6 graduate assistantships (3 to first-year students) awarded; research assistantships also available. Financial aid application deadline: 3/1. • Application contact: Dr. Donald Searles, Coordinator, 303-351-2055.

University of North Florida, College of Arts and Sciences, Department of Mathematics and Statistics, Jacksonville, FL 32216. Offerings include statistics (MA). Department faculty: 20 full-time (3 women), 0 part-time. *Degree requirements:* Comprehensive exam required, thesis optional, foreign language not required. *Entrance requirements:* GRE, TOEFL, minimum GPA of 3.0. Application fee: $15. • Dr. Leonard J. Lipkin, Chairman, 904-646-2653. Application contact: Dr. William J. Wilson, Graduate Director, 904-646-2653.

University of Pennsylvania, Wharton School, Statistics Department, Philadelphia, PA 19104. Department awards MBA, PhD. *Entrance requirements:* For master's, GMAT; for doctorate, GMAT or GRE. *Tuition:* $17,750 per year. • Dr. Paul Shaman, Chairman, 215-898-8222. Application contact: Dr. Donald F. Morrison, Coordinator, 215-898-8222.

University of Pittsburgh, Faculty of Arts and Sciences, Department of Mathematics and Statistics, Program in Statistics, Pittsburgh, PA 15260. Program awards MA, MS. Part-time programs available. Faculty: 46 full-time (5 women), 26 part-time (13 women). Matriculated students: 4 full-time (1 woman), 2 part-time (0 women); includes 0 minority, 2 foreign. In 1990, 7 degrees awarded. *Degree requirements:* Thesis (for some programs), comprehensive exam, preliminary exam required, foreign language not required. *Entrance requirements:* GRE General Test, TOEFL, minimum QPA of 3.0. Application deadline: 3/15 (priority date, applications processed on a rolling basis). Application fee: $15 ($25 for foreign students). *Expenses:* Tuition of $2920 per semester full-time, $241 per credit part-time for state residents; $5840 per semester full-time, $482 per credit part-time for nonresidents. Fees of $156 per year. *Financial aid:* 65 students received aid. Federal work-study, institutionally sponsored loans available. Aid available to part-time students. Average monthly stipend for a graduate assistantship: $1050. Financial aid application deadline: 3/1; applicants required to submit FAF. *Faculty research:* Reliability theory, multivariate analysis, time series. • Application contact: Thomas A. Metzger, Director of Graduate Studies, 412-624-8343.

University of Regina, Faculty of Graduate Studies and Research, Faculty of Science, Department of Mathematics and Statistics, Regina, SK S4S 0A2, Canada. Offerings include statistics (MA, M Sc). Department faculty: 21 full-time, 0 part-time. *Degree requirements:* Thesis required, foreign language not required. *Application deadline:* 7/2 (applications processed on a rolling basis). *Tuition:* $1500 per year full-time, $242 per year (minimum) part-time. • Dr. R. J. Tompkins, Head, 306-585-4148.

Directory: Statistics

University of Rhode Island, College of Arts and Sciences, Department of Computer Science and Statistics, Kingston, RI 02881. Department awards MS, PhD. *Degree requirements:* For master's, thesis optional; for doctorate, 1 foreign language, dissertation. *Entrance requirements:* For master's, GRE Subject Test. Application deadline: 4/15. Application fee: $25. *Expenses:* Tuition of $2575 per year full-time, $120 per credit hour part-time for state residents; $5900 per year full-time, $274 per credit hour part-time for nonresidents. Fees of $696 per year full-time.

Announcement: Department of Computer Science and Statistics offers MS and interdisciplinary PhD with statistics concentration. Research areas include experimental design, sampling, ecological statistics and biostatistics, statistical computation, multivariate analysis, nonparametric methods, analysis of variance. Academic-year assistantships begin at $7250 plus waiver of tuition and fees.

University of Rochester, College of Arts and Science, Department of Statistics, Rochester, NY 14627-0001. Department offers programs in medical statistics (MS), statistics (MA, PhD). Faculty: 8 full-time, 0 part-time. Matriculated students: 18 full-time (9 women), 1 (woman) part-time; includes 2 minority (1 Asian American, 1 black American), 13 foreign. 113 applicants, 18% accepted. In 1990, 9 master's, 4 doctorates awarded. Terminal master's awarded for partial completion of doctoral program. *Degree requirements:* For master's, foreign language and thesis not required; for doctorate, dissertation, qualifying exam required, foreign language not required. *Entrance requirements:* For doctorate, GRE, TOEFL. Application deadline: 2/15 (priority date). Application fee: $25. *Expenses:* Tuition of $473 per credit hour. Fees of $243 per year full-time. *Financial aid:* In 1990–91, 10 fellowships (5 to first-year students), 3 research assistantships (0 to first-year students), 2 teaching assistantships (0 to first-year students) awarded; full and partial tuition waivers and career-related internships or fieldwork also available. Financial aid application deadline: 2/15. *Faculty research:* Statistical theory, applied statistics and biostatistics, probability. • Dr. David Oakes, Chair, 716-275-3644.
See full description on page 491.

University of South Carolina, Graduate School, College of Science and Mathematics, Department of Statistics, Columbia, SC 29208. Department awards MS, PhD. Part-time programs available. Faculty: 9 full-time (1 woman), 0 part-time. Matriculated students: 19 full-time (12 women), 8 part-time (4 women); includes 1 minority (Native American), 7 foreign. Average age 25. 64 applicants, 42% accepted. In 1990, 4 master's awarded (100% found work related to degree); 1 doctorate awarded (100% entered university research/teaching). Terminal master's awarded for partial completion of doctoral program. *Degree requirements:* For master's, thesis required, foreign language not required; for doctorate, computer language, dissertation required, foreign language not required. *Entrance requirements:* GRE General Test (minimum combined score of 1050 required). Application deadline: 3/1 (priority date, applications processed on a rolling basis). Application fee: $25. *Tuition:* $1404 per semester full-time. *Financial aid:* In 1990–91, 15 students received a total of $113,800 in aid awarded. 3 research assistantships (0 to first-year students), 12 teaching assistantships (4 to first-year students) were awarded; career-related internships or fieldwork also available. Average monthly stipend for a graduate assistantship: $900. *Faculty research:* Reliability and quality control, multiple comparisons, ranking and selection, sequential analysis, applied probability. Total annual research budget: $87,399. • W. J. Padgett, Chairman, 803-777-5070. Application contact: John D. Spurrier, Graduate Director, 803-777-5072.

University of Southern California, Graduate School, College of Letters, Arts and Sciences, Division of Natural Sciences and Mathematics, Department of Mathematics, Program in Statistics, Los Angeles, CA 90089. Program awards MS. Matriculated students: 11 full-time (2 women), 1 (woman) part-time; includes 1 minority (Asian American), 7 foreign. Average age 30. 20 applicants, 65% accepted. In 1990, 1 degree awarded. *Degree requirements:* Thesis. *Entrance requirements:* GRE General Test. Application deadline: 7/1 (priority date). Application fee: $50. *Expenses:* Tuition of $12,120 per year full-time, $505 per unit part-time. Fees of $280 per year. *Financial aid:* 5 students received aid. Fellowships, research assistantships, teaching assistantships, federal work-study, institutionally sponsored loans available. Average monthly stipend for a graduate assistantship: $878. Financial aid application deadline: 3/1. • Dr. Simon Tavare, Director, 213-740-8766.
See full description on page 493.

University of Southern Maine, College of Arts and Science, Program in Statistics, Portland, ME 04103. Program awards MS. Faculty: 2 full-time (0 women), 2 part-time (0 women). Matriculated students: 12 full-time (4 women), 8 part-time (2 women); includes 0 minority, 4 foreign. 20 applicants, 65% accepted. In 1990, 0 degrees awarded. *Degree requirements:* Thesis optional, foreign language not required. *Entrance requirements:* GRE General Test. Application deadline: 8/1 (priority date). Application fee: $25. *Tuition:* $312 per course for state residents; $861 per course for nonresidents. *Financial aid:* In 1990–91, 2 research assistantships (both to first-year students) awarded. Financial aid applicants required to submit FAF. • Bhisham Gupta, Director, 207-780-4225. Application contact: Michelle Mondor, Administrative Assistant, 207-780-4386.

University of Southwestern Louisiana, College of Sciences, Department of Statistics, Lafayette, LA 70504. Department awards MS, PhD. Faculty: 4 full-time (0 women), 0 part-time. Matriculated students: 19 full-time (5 women), 5 part-time (2 women); includes 1 minority (black American), 13 foreign. 20 applicants, 50% accepted. In 1990, 4 master's awarded (50% found work related to degree, 50% continued full-time study); 5 doctorates awarded. Terminal master's awarded for partial completion of doctoral program. *Degree requirements:* For master's, thesis or alternative required, foreign language not required; for doctorate, dissertation. *Entrance requirements:* GRE General Test. Application deadline: 8/15. Application fee: $5. *Tuition:* $1560 per year full-time, $228 per credit (minimum) part-time for state residents; $3310 per year full-time, $228 per credit (minimum) part-time for nonresidents. *Financial aid:* In 1990–91, $146,389 in aid awarded. 4 fellowships (1 to a first-year student), 18 teaching assistantships (3 to first-year students) were awarded; full tuition waivers, federal work-study also available. Financial aid application deadline: 5/1. *Faculty research:* Quality control, inference, experimental design, reliability. • Dr. Thomas Boullion, Head, 318-231-6771.
See full description on page 497.

University of Tennessee, Knoxville, College of Business Administration, Department of Statistics, Knoxville, TN 37996. Department offers programs in industrial statistics (MS), statistics (MS). Part-time programs available. Faculty: 20 (3 women). Matriculated students: 18 full-time (12 women), 7 part-time (4 women); includes 0 minority, 4 foreign. 39 applicants, 28% accepted. In 1990, 10 degrees awarded. *Degree requirements:* Thesis or alternative required, foreign language not required. *Entrance requirements:* GMAT of GRE General Test, TOEFL (minimum score of 525 required), minimum GPA of 2.5. Application deadline: 2/1 (priority date, applications processed on a rolling basis). Application fee: $15. *Tuition:* $1086 per semester for state residents; $2768 per semester for nonresidents. *Financial aid:* In 1990–91, 0 fellowships, 0 teaching assistantships, 21 assistantships awarded; federal work-study, institutionally sponsored loans, and career-related internships or fieldwork also available. Financial aid application deadline: 3/1; applicants required to submit FAF. • Dr. David Sylwester, Head, 615-974-2556.

University of Tennessee, Knoxville, College of Business Administration, Program in Business Administration, Knoxville, TN 37996. Offerings include statistics (MBA). Program faculty: 55 (6 women). *Degree requirements:* Computer language, thesis or alternative required, foreign language not required. *Entrance requirements:* GMAT, TOEFL (minimum score of 550 required), minimum GPA of 2.5. Application deadline: 2/1 (applications processed on a rolling basis). Application fee: $15. *Tuition:* $1086 per semester for state residents; $2768 per semester for nonresidents. • Dr. Roger L. Jenkins, Associate Dean, 615-974-5033.

University of Texas at Austin, Graduate School, College of Natural Sciences, Department of Mathematics, Program in Statistics, Austin, TX 78712. Program awards MS Stat. Matriculated students: 21 full-time (13 women), 0 part-time; includes 1 minority (Asian American), 11 foreign. 15 applicants, 80% accepted. In 1990, 6 degrees awarded. *Entrance requirements:* GRE. Application deadline: 2/1 (priority date, applications processed on a rolling basis). Application fee: $40 ($75 for foreign students). *Tuition:* $510.30 per semester for state residents; $1806 per semester for nonresidents. *Financial aid:* Fellowships available. Financial aid application deadline: 3/1. • Application contact: Dr. Peter John, Graduate Adviser, 512-471-7154.

University of Texas at Dallas, School of Natural Sciences and Mathematics, Program in Mathematical Sciences, Richardson, TX 75083-0688. Offerings include applied statistics (MS, PhD), theoretical statistics (MS, PhD). Program faculty: 11 full-time (0 women), 18 part-time (3 women). *Degree requirements:* For master's, thesis optional, foreign language not required; for doctorate, dissertation required, foreign language not required. *Entrance requirements:* For master's, GRE General Test (minimum combined score of 1050 required), TOEFL (minimum score of 550 required), minimum GPA of 3.0 in upper level course work in field, secondary certificate in mathematics or computer science (MAT only); for doctorate, GRE General Test (minimum combined score of 1300 required), TOEFL (minimum score of 550 required), minimum GPA of 3.5 in upper level course work in field. Application deadline: 7/15 (applications processed on a rolling basis). Application fee: $0 ($75 for foreign students). *Expenses:* Tuition of $360 per semester full-time, $100 per semester (minimum) part-time for state residents; $2196 per semester full-time, $122 per semester hour (minimum) part-time for nonresidents. Fees of $338 per semester full-time, $22 per hour (minimum) part-time. • Dr. John Van Ness, Head, 214-690-2161.

University of Texas at San Antonio, College of Sciences and Engineering, Division of Mathematics, Computer Science and Statistics, San Antonio, TX 78285. Offerings include mathematics (MS), with options in mathematics education, statistics. Division faculty: 23 full-time (4 women), 1 part-time (0 women). *Degree requirements:* Computer language, comprehensive exam required, foreign language and thesis not required. *Entrance requirements:* GRE General Test, TOEFL, minimum GPA of 3.0. Application deadline: 7/1 (applications processed on a rolling basis). Application fee: $20. *Expenses:* Tuition of $100 per semester hour (minimum) for state residents; $128 per semester hour (minimum) for nonresidents. Fees of $48 per semester hour (minimum). • Shair Ahmad, Director, 512-691-4454.

University of Toledo, College of Arts and Sciences, Department of Mathematics, Program in Statistics, Toledo, OH 43606. Program awards MS. Evening/weekend programs available. Matriculated students: 0. In 1990, 0 degrees awarded. *Degree requirements:* Computer language required, foreign language and thesis not required. *Entrance requirements:* GRE General Test, GRE Subject Test. Application deadline: 9/8 (priority date). Application fee: $30. *Tuition:* $122.59 per credit hour for state residents; $193.40 per credit hour for nonresidents. *Financial aid:* Career-related internships or fieldwork available. Financial aid application deadline: 4/1; applicants required to submit FAF. • Dr. Harvey Wolff, Chairman, Department of Mathematics, 419-537-2568.

University of Toronto, School of Graduate Studies, Physical Sciences Division, Department of Statistics, Toronto, ON M5S 1A1, Canada. Department awards M Sc, PhD. Faculty: 15. Matriculated students: 28 full-time (5 women), 9 part-time (4 women); includes 9 foreign. 64 applicants, 56% accepted. In 1990, 4 master's, 3 doctorates awarded. *Degree requirements:* For master's, thesis not required; for doctorate, dissertation. *Application deadline:* 4/15. *Application fee:* $50. *Expenses:* Tuition of $2220 per year full-time, $666 per year part-time for Canadian residents; $10,198 per year full-time, $305.05 per year part-time for nonresidents. Fees of $277.56 per year full-time, $82.73 per year part-time. *Financial aid:* Application deadline 2/1. • M. S. Srivastava, Acting Chair, 416-978-4450.

University of Utah, Interdepartmental Program in Statistics, Salt Lake City, UT 84112. Program awards M Stat. Part-time programs available. Matriculated students: 3 full-time (all women), 7 part-time (0 women); includes 0 minority, 3 foreign. In 1990, 0 degrees awarded. *Degree requirements:* Comprehensive exam, projects required, foreign language and thesis not required. *Entrance requirements:* TOEFL, minimum GPA of 3.0, previous course work in calculus, matrix theory, statistics. Application fee: $25 ($50 for foreign students). *Tuition:* $195 per credit for state residents; $505 per credit for nonresidents. *Financial aid:* Career-related internships or fieldwork available. *Faculty research:* Biostatistics, management, economics, educational psychology, mathematics, psychology. • David Mason, Chair, University Statistics Committee, 801-581-7650.

University of Vermont, College of Engineering and Mathematics, Department of Mathematics and Statistics, Program in Statistics, Burlington, VT 05405. Program awards MS. Matriculated students: 9; includes 1 minority (black American), 1 foreign. 12 applicants, 100% accepted. In 1990, 5 degrees awarded. *Degree requirements:* Foreign language not required. *Entrance requirements:* GRE General Test, TOEFL (minimum score of 550 required). Application deadline: 4/1 (priority date, applications processed on a rolling basis). Application fee: $25. *Expenses:* Tuition of $206 per credit for state residents; $564 per credit for nonresidents. Fees of $150 per semester full-time. *Financial aid:* In 1990–91, 0 research assistantships awarded; fellowships, teaching assistantships also available. Financial aid application deadline: 3/1. *Faculty research:* Applied statistics. • Dr. L. Haugh, Director, 802-656-2940.

University of Victoria, Faculty of Arts and Science, Department of Mathematics, Victoria, BC V8W 2Y2, Canada. Offerings include statistics (MA, M Sc). Department faculty: 24 full-time (1 woman), 0 part-time. *Degree requirements:* Thesis (for some programs), oral exam required, foreign language not required. *Entrance requirements:* Honors degree in mathematics. Application deadline: 5/31 (priority date, applications processed on a rolling basis). Application fee: $20. *Expenses:* Tuition of $754 per semester. Fees of $23 per year. • Dr. D. J. Leeming, Chair, 604-721-7436. Application contact: Dr. J. Phillips, Graduate Adviser, 604-721-7450.

University of Washington, College of Arts and Sciences, Department of Statistics, Seattle, WA 98195. Department awards MS, PhD. Part-time programs available. Faculty: 14 full-time (2 women), 3 part-time (1 woman). Matriculated students: 30 full-time (9 women), 0 part-time; includes 1 minority (Asian American), 15 foreign. Average age 30. 71 applicants, 27% accepted. In 1990, 6 master's awarded; 2 doctorates awarded (100% entered university research/teaching). *Degree requirements:* For master's, computer language required, thesis optional, foreign language not required; for doctorate, 1 foreign language, computer language, dissertation. *Entrance requirements:* GRE General Test (minimum combined score of 1100 on verbal and quantitative sections required), TOEFL (minimum score of 580 required). Application deadline: 2/1 (priority date, applications processed on a rolling basis). Application fee: $35. *Tuition:* $1129 per quarter full-time, $324 per credit (minimum) part-time for state residents; $2824 per quarter full-time, $809 per credit (minimum) part-time for nonresidents. *Financial aid:* In 1990–91, 30 students received a total of $470,000 in aid awarded. 0 fellowships, 19 research assistantships (0 to first-year students), 9 teaching assistantships (7 to first-year students), 1 graduate staff assistants (0 to first-year students) were awarded; career-related internships or fieldwork also available. Average monthly stipend for a graduate assistantship: $970. *Faculty research:* Mathematical statistics, stochastic modeling, spatial statistics, statistical computing. Total annual research budget: $1-million. • Elizabeth A. Thompson, Chair, 206-685-0108. Application contact: Peter Guttorp, Graduate Program Coordinator, 206-685-7439.

University of Waterloo, Faculty of Mathematics, Department of Statistics and Actuarial Science, Waterloo, ON N2L 3G1, Canada. Department awards M Math, M Phil, PhD. Faculty: 35 full-time, 0 part-time. Matriculated students: 44 full-time (7 women), 8 part-time (1 woman); includes 21 foreign. 102 applicants, 51% accepted. In 1990, 16 master's, 3 doctorates awarded. *Degree requirements:* For master's, thesis or alternative required, foreign language not required; for doctorate, dissertation required, foreign language not required. *Entrance requirements:* For master's, TOEFL (minimum score of 550 required), honor's degree in field, minimum B average; for doctorate, TOEFL (minimum score of 550 required), master's degree. Application deadline: 4/15. Application fee: $25. *Expenses:* Tuition of $757 per year full-time, $530 per year part-time for Canadian residents; $3127 per year for nonresidents. Fees of $68 per year full-time, $17 per year part-time. *Financial aid:* In 1990–91, 22 research assistantships, 44 teaching assistantships, 20 scholarships awarded; career-related internships or fieldwork also available. Average monthly stipend for a graduate assistantship: $530. Financial aid application deadline: 4/15. *Faculty research:* Biometry, multivariate analysis, risk theory, inference, stochastic processes. • Dr. K. S. Brown, Chair, 519-885-1211 Ext. 4499. Application contact: D. Matthews, Graduate Officer, 519-885-1211 Ext. 4497.

University of Western Ontario, Physical Sciences Division, Department of Statistical and Actuarial Sciences, London, ON N6A 3K7, Canada. Department awards M Sc. *Tuition:* $1015 per year full-time, $1050 per year part-time for Canadian residents; $4207 per year for nonresidents.

University of West Florida, College of Arts and Sciences, Department of Mathematics and Statistics, Pensacola, FL 32514-5750. Offerings include statistics (MA). Department faculty: 11 full-time (10 women), 0 part-time. *Degree requirements:* Thesis optional, foreign language not required. *Entrance requirements:* GRE (minimum score of 1000 required), minimum GPA of 3.0. Application deadline: 7/19. Application fee: $15. *Tuition:* $86.38 per credit hour for state residents; $288.79 per credit hour for nonresidents. • Dr. D. Byrkit, Chairperson, 904-474-2291.

University of Wisconsin–Madison, School of Business, Madison, WI 53706. Offerings include business (MA, MBA, MS, PhD), with options in accounting, finance, investment, and banking (MBA), arts administration (MA), diversified (MBA, MS), finance, investment, and banking (MS), health care fiscal management (MBA, MS), health services administration (MA), information systems analysis and design (MBA, MS), international business (MBA, MS), management (MBA, MS), marketing (MBA, MS), program administration (MBA, MS), public management (MBA, MS), quantitative analysis (MBA, MS), real estate and urban land economics (MBA), real estate appraisal and investment analysis (MBA, MS), risk and insurance (MBA, MS), statistics (MBA, MS), transportation and public utilities (MBA, MS). School faculty: 92 full-time, 0 part-time. *Application deadline:* rolling. *Application fee:* $20. • James C. Hickman, Dean, 608-262-1553. Application contact: E. J. Blakely, Associate Dean, 608-262-1555.

University of Wisconsin–Madison, College of Letters and Science, Department of Statistics, Madison, WI 53706. Department awards MS, PhD. Faculty: 21 full-time, 0 part-time. Matriculated students: 73 full-time (18 women), 27 part-time (9 women); includes 5 minority (4 Asian American, 1 Hispanic American), 74 foreign. 225 applicants, 42% accepted. In 1990, 22 master's, 13 doctorates awarded. *Application deadline:* rolling. *Application fee:* $20. *Financial aid:* In 1990–91, 60 students received aid. 0 fellowships, 12 research assistantships, 36 teaching assistantships, 12 project assistantships, t awarded; institutionally sponsored loans also available. Financial aid application deadline: 1/15; applicants required to submit FAF. • Robert B. Miller, Chairperson, 608-262-1009. Application contact: Joanne Nagy, Assistant Dean of the Graduate School, 608-262-2433.

University of Wyoming, College of Arts and Sciences, Department of Statistics, Laramie, WY 82071. Department awards MS, PhD. Faculty: 8 full-time (0 women), 0 part-time. Matriculated students: 24 full-time (10 women), 0 part-time; includes 0 minority, 10 foreign. Average age 33. In 1990, 3 master's, 3 doctorates awarded. *Degree requirements:* Thesis/dissertation. *Entrance requirements:* For master's, GMAT, GRE General Test, minimum GPA of 3.0; for doctorate, GRE General Test, minimum GPA of 3.0. Application deadline: 4/5 (priority date, applications processed on a rolling basis). Application fee: $30. *Tuition:* $1554 per year full-time, $74.25 per credit hour part-time for state residents; $4358 per year full-time, $74.25 per credit hour part-time for nonresidents. *Financial aid:* In 1990–91, $126,047 in aid awarded. 9 research assistantships (3 to first-year students), 11 teaching assistantships (1 to a first-year student) were awarded; federal work-study, institutionally sponsored loans also available. Financial aid application deadline: 4/15. *Faculty research:* Valdez oil spill. Total annual research budget: $395,000. • Dr. Robert S. Cochran, Head, 307-766-4229.

Utah State University, College of Science, Department of Mathematics and Statistics, Logan, UT 84322. Department offers programs in applied statistics (MS), mathematical sciences (PhD), mathematics (M Math, MS). Part-time programs available. Faculty: 24 full-time (3 women), 0 part-time. Matriculated students: 8 full-time (3 women), 28 part-time (16 women); includes 2 minority (1 Asian American, 1 Hispanic American), 17 foreign. Average age 24. 48 applicants, 48% accepted. In 1990, 15 master's awarded; 1 doctorate awarded (100% entered university research/teaching). Terminal master's awarded for partial completion of doctoral program. *Degree requirements:* For master's, qualifying exam required, foreign language and thesis not required; for doctorate, 1 foreign language, dissertation, comprehensive exams. *Entrance requirements:* GRE General Test (score in 40th percentile or higher required), TOEFL (minimum score of 550 required), minimum GPA of 3.0. Application deadline: 7/15 (priority date, applications processed on a rolling basis). Application fee: $25 ($30 for foreign students). *Tuition:* $426 per quarter (minimum) full-time, $184 per quarter (minimum) part-time for state residents; $1133 per quarter (minimum) full-time, $505 per quarter (minimum) part-time for nonresidents. *Financial aid:* In 1990–91, $144,000 in aid awarded. 3 fellowships (1 to a first-year student), 2 research assistantships (1 to a first-year student), 15 teaching assistantships (8 to first-year students) were awarded; partial tuition waivers, federal work-study also available. Aid available to part-time students. Average monthly stipend for a graduate assistantship: $1000. Financial aid application deadline: 4/1; applicants required to submit FAF. *Faculty research:* Differential geometry, differential equations, statistics, computational mathematics, probability. Total annual research budget: $75,000. • L. Duane Loveland, Head, 801-750-2809. Application contact: Ian Anderson, Graduate Chairman, 801-750-2818.

See full description on page 507.

Villanova University, Graduate School of Arts and Sciences, Department of Mathematical Sciences, Program in Applied Statistics, Villanova, PA 19085. Program awards MS. Part-time and evening/weekend programs available. Faculty: 4 full-time (0 women), 2 part-time (0 women). Matriculated students: 1 full-time (0 women), 5 part-time (2 women); includes 2 foreign. Average age 25. 6 applicants, 83% accepted. In 1990, 3 degrees awarded (100% found work related to degree). *Degree requirements:* Comprehensive exam required, thesis optional, foreign language not required. *Entrance requirements:* Minimum GPA of 3.0. Application deadline: 8/1. Application fee: $25. *Expenses:* Tuition of $270 per credit. Fees of $28 per semester. *Financial aid:* In 1990–91, $18,000 in aid awarded. 2 teaching assistantships (1 to a first-year student) were awarded; federal work-study also available. Financial aid application deadline: 4/1. • Dr. Michael Levitan, Director, 215-645-4850.

Virginia Commonwealth University, College of Humanities and Sciences, Department of Mathematical Sciences, Richmond, VA 23284. Offerings include statistics (MS). Department faculty: 38. *Degree requirements:* Foreign language not required. *Entrance requirements:* GRE General Test, GRE Subject Test. Application deadline: 7/1. Application fee: $20. *Expenses:* Tuition of $2770 per year full-time, $154 per hour part-time for state residents; $7550 per year full-time, $419 per hour part-time for nonresidents. Fees of $717 per year full-time, $25.50 per hour part-time. • Dr. William E. Haver, Chair, 804-367-1319. Application contact: Dr. James A. Wood, Director of Graduate Studies, 804-367-1301.

Virginia Polytechnic Institute and State University, College of Arts and Sciences, Department of Statistics, Blacksburg, VA 24061. Department awards MS, PhD. Faculty: 20 full-time (4 women), 0 part-time. Matriculated students: 50 full-time (17 women), 7 part-time (2 women); includes 4 minority (3 Asian American, 1 black American), 19 foreign. 134 applicants, 68% accepted. In 1990, 19 master's, 3 doctorates awarded. *Degree requirements:* For master's, computer language, qualifying exam required, thesis optional, foreign language not required; for doctorate, computer language, dissertation, preliminary exam required, foreign language not required. *Application deadline:* 2/15 (priority date). *Application fee:* $10. *Tuition:* $1889 per semester full-time, $606 per credit hour part-time for state residents; $2627 per semester full-time, $853 per credit hour part-time for nonresidents. *Financial aid:* In 1990–91, $300,000 in aid awarded. 0 fellowships, 0 research assistantships, 13 teaching assistantships (0 to first-year students), 37 assistantships (21 to first-year students) were awarded; institutionally sponsored loans also available. Financial aid application deadline: 2/15. *Faculty research:* Design and sampling theory, computing and simulation, nonparametric, robust and multivariate methods, biostatistics quality. • Dr. Klaus H. Hinkelmann, Head, 703-231-5657. Application contact: Golde L. Holtzman, Graduate Program Administrator, 703-961-5630.

See full description on page 513.

Washington University, Graduate School of Arts and Sciences, Department of Mathematics, St. Louis, MO 63130. Offerings include statistics (MA, PhD). Terminal master's awarded for partial completion of doctoral program. *Degree requirements:* For doctorate, dissertation. *Entrance requirements:* For doctorate, GRE. Application deadline: 1/15 (priority date, applications processed on a rolling basis). Application fee: $0. *Tuition:* $15,960 per year full-time, $665 per credit hour part-time. • Dr. Gary Jensen, Chairman, 314-935-6760.

Wayne State University, College of Liberal Arts, Department of Mathematics, Program in Statistics, Detroit, MI 48202. Program awards MA, PhD. Matriculated students: 0 full-time, 5 part-time (0 women). 10 applicants, 20% accepted. In 1990, 2 master's, 0 doctorates awarded. *Degree requirements:* For master's, foreign language and thesis not required; for doctorate, dissertation. *Application deadline:* 7/1. *Application fee:* $20 ($30 for foreign students). *Expenses:* Tuition of $119 per credit hour for state residents; $258 per credit hour for nonresidents. Fees of $50 per semester. • Dr. Pao liu Chou, Chairperson, Department of Mathematics, 313-577-2479. Application contact: Dr. Peter Malcolmson, Graduate Committee Chair, 313-577-2472.

Western Michigan University, College of Arts and Sciences, Department of Mathematics and Statistics, Program in Statistics, Kalamazoo, MI 49008. Program awards MS, PhD. Matriculated students: 7 full-time (5 women), 18 part-time (6 women); includes 1 minority (Native American), 6 foreign. 19 applicants, 53% accepted. In 1990, 4 master's awarded. *Degree requirements:* For master's, oral exams required, thesis not required; for doctorate, 1 foreign language, dissertation. *Entrance requirements:* For doctorate, GRE General Test. Application deadline: 2/15 (priority date, applications processed on a rolling basis). Application fee: $25. *Expenses:* Tuition of $100 per credit hour for state residents; $244 per credit hour for nonresidents. Fees of $178 per semester full-time, $76 per semester part-time. *Financial aid:* Fellowships, research assistantships, teaching assistantships, federal work-study available. Financial aid application deadline: 2/15. • Application contact: Paula J. Boodt, Director, Graduate Student Services, 616-387-3570.

West Virginia University, College of Arts and Sciences, Department of Statistics and Computer Science, Program in Statistics, Morgantown, WV 26506. Program awards MS. Matriculated students: 9 full-time (3 women), 1 (woman) part-time; includes 0 minority, 5 foreign. Average age 26. 13 applicants, 54% accepted. In 1990, 7 degrees awarded (100% found work related to degree). *Degree requirements:* Computer language, thesis required, foreign language not required. *Entrance requirements:* TOEFL (minimum score of 550 required), minimum GPA of 3.0. Application deadline: 3/15 (priority date, applications processed on a rolling basis). Application fee: $25. *Expenses:* Tuition of $390 per year full-time for state residents; $1270 per year full-time for nonresidents. Fees of $1555 per year full-time for state residents; $3985 per year full-time for nonresidents. *Financial aid:* In 1990–91, $76,063 in aid awarded. 3 research assistantships (2 to first-year students), 7 teaching assistantships (4 to first-year students) were awarded; fellowships, full and partial tuition waivers, federal work-study, institutionally sponsored loans also available. Financial aid application deadline: 2/1; applicants required to submit FAF. *Faculty research:* Linear models, multivariate analysis, categorical data analysis,

SECTION 5: MATHEMATICAL SCIENCES

Directory: Statistics; Cross-Discipline Announcements

statistical computing, experimental design. Total annual research budget: $179,091. • Dr. Donald F. Butcher, Chairperson, 304-293-3601.

Wright State University, College of Science and Mathematics, Department of Mathematics and Statistics, Program in Statistics, Dayton, OH 45435. Program awards MS. Faculty: 7 full-time (1 woman), 0 part-time. Matriculated students: 6 full-time (5 women), 16 part-time (4 women). Average age 30. In 1990, 5 degrees awarded. *Degree requirements:* Computer language, comprehensive exams required, foreign language and thesis not required. *Tuition:* $3342 per year full-time, $106 per credit hour part-time for state residents; $5991 per year full-time, $190 per credit hour part-time for nonresidents. *Financial aid:* In 1990–91, 0 fellowships, 0 research assistantships, 3 teaching assistantships awarded. *Faculty research:* Contingency table analysis, reliability theory, stochastic processing, nonparametric statistics, design of experiments. • Dr. Harry J. Khamis, Graduate Adviser, 513-873-2433.

Yale University, Graduate School of Arts and Sciences, Department of Statistics, New Haven, CT 06520. Department awards MS, PhD. Faculty: 5 full-time (1 woman), 0 part-time. Matriculated students: 12 full-time (2 women), 2 part-time (1 woman); includes 1 minority (Asian American), 9 foreign. 17 applicants, 29% accepted. In 1990, 3 master's, 3 doctorates awarded. Terminal master's awarded for partial completion of doctoral program. *Degree requirements:* For master's, foreign language and thesis not required; for doctorate, dissertation required, foreign language not required. *Entrance requirements:* For doctorate, GRE General Test, GRE Subject Test. Application deadline: 1/2. Application fee: $45. *Tuition:* $15,160 per year full-time, $1895 per course part-time. • Dr. John Hartigan, Chairman, 203-432-0666. Application contact: Susan Webb, Coordinator of Admissions, 203-432-2770.

York University, Faculty of Arts, Program in Mathematics and Statistics, North York, ON M3J 1P3, Canada. Program awards MA, PhD. Part-time programs available. Faculty: 45 full-time (9 women), 2 part-time (0 women). Matriculated students: 29 full-time (5 women), 9 part-time (5 women); includes 7 foreign. 136 applicants, 26% accepted. In 1990, 22 master's, 0 doctorates awarded. *Degree requirements:* For master's, thesis optional, foreign language not required; for doctorate, dissertation. Application deadline: 3/31. Application fee: $35. *Tuition:* $2436 per year for Canadian residents; $11,480 per year for nonresidents. *Financial aid:* In 1990–91, $343,171 in aid awarded. 7 fellowships, 39 research assistantships, 30 teaching assistantships were awarded. • Dr. G. O'Brien, Director, 416-736-2100.

Cross-Discipline Announcements

Case Western Reserve University, Weatherhead School of Management and School of Graduate Studies, Department of Operations Research, Cleveland, OH 44106.

Operations research is concerned with the use of mathematical models and solutions to explain the behavior of complex man-machine systems, assist in system design, and improve current or planned operations. Faculty members and graduate research assistants currently have funded research grants and contracts with federal agencies and area businesses and industries.

Ohio State University, College of Mathematical and Physical Sciences, Department of Geodetic Science and Surveying, Columbus, OH 43210.

Geodetic Science uses advanced mathematical and statistical tools to solve problems related to Earth's gravity field, satellite orbits, Earth rotation, movements of Earth's crust, positioning by satellites for mapping, analysis of digital and photographic images (for mapping, transportation, archaeology, etc.), land information systems. Own world-class laboratories. Master's and PhD.

Southern Illinois University at Carbondale, Molecular Science Program, Carbondale, IL 62901.

Faculty are from sciences, engineering, and medicine. Research areas include molecular biology, biophysics, applied physics, chemical physics, catalysis, engineering science, material science, applied mathematics, and others. Several choices for prelims. University fellowships and accelerated entry available to outstanding students. See statistical data in Materials Sciences Section in the Engineering and Applied Sciences volume of this series.

University of Arizona, College of Arts and Sciences, Faculty of Science, Department of Ecology and Evolutionary Biology, Tucson, AZ 85721.

Ecology and evolutionary biology study offers exciting opportunities in applied mathematics. Department areas of interest include chaos in biological systems, biological clocks, multivariate statistical descriptions of organisms, reconstructing phylogenetic trees, evolution of genetic systems, DNA sequence analysis, and stochastic modeling. Contact Dr. Richard Strauss, Director of Graduate Studies, 602-621-1165.

University of California at Berkeley, Walter A. Haas School of Business, Berkeley, CA 94720.

The PhD program offers students interested in applied mathematics/statistics excellent graduate student support and research and placement opportunities, especially in accounting, finance, management science, and marketing. Doctoral course work in a second department (e.g., economics or computer science) is required. See full description in the Business Administration Section in the Business, Education, Health, and Law volume of this series.

University of California, Los Angeles, School of Public Health, Department of Biostatistics, Los Angeles, CA 90024.

Department offers MPH, Dr PH, MS, and PhD degrees, balancing theory and practice. Many applications arise in the biomedical, physical, and social sciences. See 2-page description in Mathematical Sciences Section.

University of Washington, School of Public Health and Community Medicine, Program in Biostatistics, Seattle, WA 98195.

The Graduate Program in Biostatistics offers MS and PhD degrees in quantitative methods applied to the medical and biological sciences. Research specialties include clinical trials, survival data analysis, and statistical computing. Students with undergraduate degrees in mathematics, statistics, or biology are encouraged to apply.

Virginia Commonwealth University, School of Graduate Studies and Medical College of Virginia-Professional Programs, School of Basic Health Sciences, Department of Human Genetics, Richmond, VA 23284.

Human quantitative genetics applies mathematical and biostatistical models to problems in the genetics of human populations. Genetic linkage and segregation analyses of disease are increasingly important tools in biomedical research. Linear structural equation modeling of physical and behavioral traits is used to study their development and biological and cultural inheritance.

SECTION 5: MATHEMATICAL SCIENCES

AMERICAN UNIVERSITY

Department of Mathematics and Statistics

Programs of Study

The Department of Mathematics and Statistics offers master's degrees in statistics, mathematics, and statistical computing and Ph.D. degrees in statistics and mathematics education.

Candidates for the M.S. in statistical computing must pass a comprehensive examination based on the core courses in computer science, statistics, and numerical analysis. Of the 36 semester hours required for the degree, 6 must be in research-level courses. Candidates for the M.A. in statistics may choose one of two tracks: mathematical statistics or applied statistics. The comprehensive examination is given in two parts (the composition of the exam depends on the track). Students must demonstrate a knowledge of French, German, Russian, or a computer language. A thesis or a paper based on independent study is also required. To complete the degree, students must earn 30 semester hours, including 6 for a thesis or 3 for an independent project. Candidates for the M.A. in mathematics must pass one comprehensive examination based largely on real and complex variables and linear and modern algebra. A knowledge of French, German, Russian, or an approved computer language is required. Students must complete 30 semester hours, including 6 for a thesis or 3 for an independent project or seminar.

Ph.D. candidates in statistics must pass a qualifying examination based on master's degree work. This exam must be taken during the first 24 semester hours of the program. Three written comprehensive examinations are required: one in advanced mathematical statistics and two chosen from probability, theory of sampling, linear estimation, multivariate analysis, decision theory, and statistical computing. An oral examination is required as well. The program consists of 72 semester hours, of which 30 may be credited for the student's master's program. Twelve hours are for dissertation research; there is an oral defense of the dissertation. Two tools of research are required. Candidates for the Ph.D. in mathematics education should have a master's degree in a mathematical science or in education. Four comprehensive examinations are required: a written exam in mathematics for mathematics education, a written exam in mathematics education, an oral examination given by the department, and an approved exam in a related field. An oral defense of the dissertation is required; also required are two tools of research—one in statistics and the other, a knowledge of French, German, Russian, Spanish, or an approved computer language. At least 42 semester hours of approved graduate work beyond the master's degree are required.

Research Facilities

The University Computer Center has a 3090 IBM mainframe with MVS/ESA batch and the CMS interactive systems. In addition, there are ten public computer laboratories on campus. The Advanced Technology Laboratory (ATL) provides equipment and material for research and instructional purposes. ATL facilities include an IBM InfoWindow system, a Kurzweil intelligent optical scanner, an optical mark reader, a color plotter, Macintosh computers, a NeXT workstation, a film recorder, and high-quality laser printers. There are numerous major cluster terminal locations, a social science computer research laboratory, and dozens of individual terminals. Statistical software packages such as SYSTAT, SPSS, SAS, and Statpak are available. The laboratory has Sun-3/60 Workstations, an IBM PC/AT, and PS/2, Macintosh, and other microcomputers networked to an AT&T 3B2 computer. Graphics terminals and plotters are also available. This lab's software includes SAS, SYSTAT, ISP, Pascal, FORTRAN, and C. The Intelligent Systems Laboratory has several expert system tools, including Nexpert and CLIPS.

The Bender Library and Learning Resources Center houses over 550,000 volumes and 3,500 periodicals, as well as microform collections and a nonprint media center. In addition, more than fourteen indexes in compact disk format are searchable using library microcomputers. ALADIN, American University's on-line collections catalog, also contains the catalogs of the other seven members of the Washington Research Library Consortium (WRLC). Graduate students have borrowing privileges at all WRLC libraries. Dozens of other research collections in the Washington, D.C., metropolitan area are easily accessible.

Financial Aid

Several types of aid are available to full-time students, including assistantships and fellowships. The Department of Mathematics and Statistics participates in the Patricia Roberts Harris Graduate Fellowship Program funded through the U.S. Department of Education. The Harris Fellowship Program provides financial assistance to specific higher education institutions that, in turn, recruit members of minority groups and women, who are underrepresented in specific academic and professional fields. These students are awarded a stipend and an educational allowance. Part-time work is also available as are loans and deferred-payment programs.

Cost of Study

Graduate tuition for 1991–92 is $475 per semester credit hour. Special fees include those charged for application, comprehensive examinations, and thesis processing.

Cost of Living

Housing costs in the area are comparable with those in most other major metropolitan areas; rates vary with distance from campus and the extent to which facilities are shared. Although many graduate students live off campus, the University provides a number of graduate accommodations.

Student Group

The Department of Mathematics and Statistics has approximately 50 graduate students.

Location

The University has an 85-acre campus in a residential area of northwest Washington from which public transportation offers quick access to downtown Washington.

The University

The American University was founded as a Methodist institution, chartered by Congress in 1893, and intended originally for graduate study only. The strong faculty-student relationships that are fostered in the department create an ideal learning environment.

Correspondence and Information

Dr. Robert W. Jernigan, Chairperson
Department of Mathematics and Statistics—PG
205 Clark Hall
American University
4400 Massachusetts Avenue, NW
Washington, D.C. 20016
Telephone: 202-885-3120

SECTION 5: MATHEMATICAL SCIENCES

American University

THE FACULTY AND THEIR RESEARCH

Austin M. Barron, Associate Professor and Associate Dean, College of Arts and Sciences; Ph.D., Purdue. Decision theory, nonparametric statistics.
Stephen D. Casey, Assistant Professor; Ph.D., Maryland. Harmonic analysis, complex analysis, mathematical modeling, differential and integral geometry.
I-Lok Chang, Associate Professor; Ph.D., Cornell. Complex variables, numerical analysis.
David S. Crosby, Professor; Ph.D., Arizona. Probability, statistical operator theory.
Nancy Flournoy, Associate Professor; Ph.D., Washington (Seattle). Biostatistics, stochastic processes.
Mary W. Gray, Professor; Ph.D., Kansas; J.D., American. Algebra, applied statistics, computer law.
Michael Greene, Associate Professor; Ph.D., Carnegie Mellon. Statistical computing, simulation, applications of statistics.
Jeffrey Hakim, Assistant Professor; Ph.D., Columbia. Number theory, harmonic analysis.
Stephen Hillis, Assistant Professor; Ph.D., Iowa. Linear estimation, M—estimation.
Robert W. Jernigan, Professor and Chairperson; Ph.D., South Florida. Time series, statistical computing.
Basil P. Korin, Professor; Ph.D., George Washington. Multivariate analysis.
John P. Nolan, Assistant Professor; Ph.D., Virginia. Probability.
Scott Parker, Associate Professor; Ph.D., Columbia. Scaling, nonmetric methods, applications to social science research.
Hannah Sandler, Assistant Professor; Ph.D., Columbia. Hyperbolic geometry and complex analysis.
Steven H. Schot, Professor; Ph.D., Maryland. Partial differential equations, singular integral equations, fluid dynamics.
Virginia Stallings-Roberts, Assistant Professor; Ph.D., Southern Mississippi. Mathematics education.
Ferdinand T. Wang, Assistant Professor; Ph.D., Rice. Nonparametric regression, density estimation.

Recent Faculty Publications

Casey, S. D. Analysis of fractal and Pereto-Levy sets: Theory and application. *Fourth Eur. Frequency Time Forum*, 205–11, 1990.
Flournoy, N., et al. Busulfan, cyclophosphamide and fractionated total body irradiation as a preparatory regimen for marrow transplantation in patients with advanced hematological malignancies: A Phase I study. *Bone Marrow Transplantation* 4:617–23, 1989.
Gray, M. W. and L. Hayden. A successful intervention program for high ability minority students. *Sch. Sci. Mathematics* 90:323–33, 1990.
Gray, M. W. Advanced calculus of murder (book review). *Math. Intelligencer* 12:77–79, 1990.
Hillis, S. and D. E. Furst. Immunosuppression with chlorambucil versus placebo for scleroderma. *Arthritis Rheum.* 32:584–93, 1989.
Jernigan, R. W. and P. J. Munson. A cubic spline extension of the Durbin-Watson Test. *Biometrika* 76, 1989.
Nolan, J. P. Local nondeterminism and local times for stable processes. *Probability Theory and Related Fields* 82:387–410, 1989.
Nolan, J. P., S. Cambanis, and J. Rosinski. On the oscillation of infinitely divisible processes. *Stochastic Processes and Their Applications* 35:87–97, 1990.

SECTION 5: MATHEMATICAL SCIENCES

ARIZONA STATE UNIVERSITY
Department of Mathematics

Programs of Study The Department of Mathematics offers graduate study leading to the Ph.D. and the M.A. degrees in most areas of mathematics.

The Ph.D. program is intended for the student with superior mathematical ability and emphasizes the development of creative scholarship and breadth and depth in background knowledge. A doctoral student must take certain advanced courses during his or her first year and take two comprehensive examinations within fourteen months of entering the program. Each student must pass three written examinations and one oral examination. In general, a reading knowledge of either French, German, or Russian is also required. After the examinations are passed, a dissertation that constitutes an original contribution to the discipline must be written.

The aim of the master's program is to offer students with a bachelor's degree in mathematics or related fields an opportunity to broaden their knowledge by undertaking course work at the graduate level. A master's degree student must pass three written examinations and complete a thesis. Precise requirements differ as to the option chosen (General, Applied, Statistics and Probability, Computational).

The department also offers a Master of Science degree with a major in statistics and a Master of Natural Science degree for interdisciplinary study, for which a thesis is not required.

Research Facilities There are 58 faculty members in the department, and their research interests cover most aspects of mathematics, from algebraic number theory to partial differential equations. The Daniel E. Noble Science Library, a member of the Center for Research Libraries, has a very good holding of mathematical texts and journals; its operations are fully computerized to aid in literature searches.

The Computing Services facilities are offered at no charge to the University community for use in academic pursuits. The Computer Accounts Office provides instructional, research, and individual computer accounts that are used to access Computing Services' equipment. The department has an advanced computing facility built around a network of three Titan class superminicomputers, which feature state-of-the-art graphics and have impressive computational power. For symbolic manipulation, there is also a cluster of workstations using a Titan as a file server. The University has also just installed two supercomputers (Cray and IBM), putting ASU among the few universities that have this computing capability.

Financial Aid Financial assistance for graduate students is available through teaching and research assistantships, fellowships, and a limited number of privately sponsored scholarships. Currently, 46 teaching assistantships are assigned each semester. Research assistantships are usually reserved for advanced students. A small number of fellowships are available for award at the discretion of the department. Teaching and research assistants must enroll for a minimum of 6 semester hours of graduate credit. They receive scholarships covering nonresident tuition but must pay the usual registration fee each semester. A Ph.D. student may expect to be supported for up to five years, a master's degree student for up to five semesters.

Cost of Study Arizona residents registered for 7 or more semester hours are considered full-time students for fee payment purposes and pay registration and tuition fees of $760 per semester in 1991-92. Out-of-state students pay an additional $289 per credit hour. Part-time students (6 hours or fewer) pay $80 per credit hour (in-state students) or $289 per credit hour (out-of-state students).

Cost of Living The room and board charges for University housing, available for graduate students, range between $1359 and $1450 per semester. A large number of privately owned apartments are available in the community. Students living off campus typically pay between $400 and $550 per month for food and rent. The University Residence Life Office can supply additional information.

Student Group The University currently has over 1,800 faculty members and 43,000 students, of whom 12,000 are graduate students—the largest graduate enrollment at a single university in the United States. There are 76 students enrolled in the graduate programs offered by the Department of Mathematics.

Location The state of Arizona is famous worldwide for its great natural beauty, from Monument Valley and the Grand Canyon in the north to the Sonoran desert in the south. The University is centrally located in Tempe, one of several neighboring cities occupying the wide expanse of the Salt River Valley (known locally as the Valley of the Sun). The population of Tempe is close to 150,000, and the city combines the advantages of a moderate-sized, university-oriented community with the cultural and technical resources of a major metropolitan area. Phoenix is one of the fastest-growing cities in the country, and recent economic development has brought major electronics and aerospace industries to the area, often to the mutual benefit of the company and the University.

The University Arizona State University was founded in 1885 as a training college for teachers in what was then the sparsely populated Arizona Territory. Following the rapid growth of the Phoenix metropolitan area, the institution became Arizona State College in 1945 and Arizona State University in 1958.

The University is noted for its attractive campus. There are extensive recreational facilities, and the University offers both a rich cultural heritage and a diverse student population.

Applying Application forms for admission and financial support, as well as additional information about the department and its programs, can be obtained from the address given below. All applicants must submit a completed application form and transcripts of academic records, together with Graduate Record Examinations General Test scores, to the Graduate College of the University. They must also submit to the graduate secretary of the Department of Mathematics at least three letters of recommendation from persons who are familiar with their qualifications to pursue graduate study in mathematics. All application materials must be received by March 15; appointments for teaching assistantships for the following year are made in late March, and late applicants will be considered only if there are vacancies.

Correspondence and Information
Dr. Andrew Bremner
Director of Graduate Studies
Department of Mathematics
Arizona State University
Tempe, Arizona 85287

Arizona State University

THE FACULTY AND THEIR RESEARCH

Bruce A. Anderson, Professor; Ph.D., Iowa, 1966. Combinatorics, graph theory.
D. Armbruster, Assistant Professor; Ph.D., Tübingen (Germany), 1984. Nonlinear partial differential equations, infinite dimensional dynamical systems.
Steven Baer, Assistant Professor; Ph.D., Illinois at Chicago, 1984. Bifurcation analysis, numerical methods, neurobiology.
Hélène Barcelo, Assistant Professor; Ph.D., California, San Diego, 1988. Lie algebras.
Douglas Blount, Assistant Professor; Ph.D., Wisconsin–Madison, 1987. Probability theory.
Andrew Bremner, Professor; Ph.D., Cambridge, 1978. Number theory.
Joaquin Bustoz, Professor; Ph.D., Arizona State, 1967. Classical analysis.
Michael Driscoll, Associate Professor; Ph.D., Arizona, 1971. Statistics.
A. Eden, Assistant Professor; Ph.D., Indiana, 1988. Differential equations, applied analysis.
Genghua Fan, Assistant Professor; Ph.D., Waterloo, 1988. Graph theory.
Frank D. Farmer, Associate Professor; Ph.D., Washington (Seattle), 1970. Combinatorial algebraic topology.
Alan Feldstein, Professor; Ph.D., UCLA, 1964. Numerical analysis, computer science, applied mathematics.
Myron Goldstein, Professor; Ph.D., UCLA, 1963. Riemann surfaces, approximation, function and potential theory.
Edward E. Grace, Professor; Ph.D., North Carolina, 1956. Point set topology.
Matthew J. Hassett, Professor; Ph.D., Rutgers, 1966. Mathematics of finance.
Jon Helton, Professor; Ph.D., Texas at Austin, 1970. Analysis, summability, application of mathematics to the biological sciences.
Domingo Herrero, Professor; Ph.D., Chicago, 1970. Operator theory.
Glenn Hurlbert, Assistant Professor; Ph.D., Rutgers, 1990. Discrete mathematics.
Edwin Ihrig, Professor; Ph.D., Toronto, 1974. General relativity.
Zdzislaw Jackiewicz, Professor; Ph.D., Gdansk (Poland), 1980. Numerical analysis.
Ronald Jacobowitz, Professor; Ph.D., Princeton, 1960. Algebra, biomathematics.
John Jones, Assistant Professor; Ph.D., Harvard, 1987. Arithmetic geometry, Iwasawa theory.
Kevin Kadell, Associate Professor; Ph.D., Penn State, 1979. Algebraic combinatorics, classical analysis.
Matthias Kawski, Assistant Professor; Ph.D., Colorado at Boulder, 1986. Control theory.
John Kelly, Professor; Ph.D., MIT, 1948. Combinatorics, number theory.
Henry Kierstead, Professor; Ph.D., California, San Diego, 1979. Discrete mathematics.
E. Kostelich, Assistant Professor; Ph.D., Maryland, 1985. Differential equations, dynamical systems.
Yang Kuang, Assistant Professor; Ph.D., Alberta, 1988. Dynamical systems, mathematical biology.
Hendrik Kuiper, Associate Professor; Ph.D., Wisconsin–Madison, 1971. Partial differential equations, numerical analysis.
Lynn Kurtz, Associate Professor; Ph.D., Utah, 1964. Functional analysis.
Philip Leonard, Professor; Ph.D., Penn State, 1968. Algebra, number theory.
Sharon Lohr, Assistant Professor; Ph.D., Wisconsin–Madison, 1987. Statistics.
Joan McCarter, Assistant Professor; M.A., Arizona, 1958. Applied mathematics.
John McDonald, Professor; Ph.D., Rutgers, 1969. Convexity, complex analysis.
Hans Mittelmann, Professor; Ph.D., Darmstadt (Germany), 1973. Numerical analysis.
J. Douglas Moore, Associate Professor; Ph.D., Syracuse, 1969. Algebra, automata theory.
Basil Nicolaenko, Professor; Ph.D., Michigan, 1968. Nonlinear partial differential equations, infinite dimensional dynamical systems.
E. Petrie, Assistant Professor; Ph.D., Cornell, 1989. Complex analytic geometry, number theory.
John Quigg, Associate Professor; Ph.D., Drexel, 1979. Operator algebras.
Rosemary Renaut, Assistant Professor; Ph.D., Cambridge, 1984. Numerical analysis.
Christian Ringhofer, Professor; Ph.D., Vienna, 1981. Numerical analysis.
Nevin Savage, Professor; Ph.D., UCLA, 1956. Analysis.
Thomas Sherman, Professor; Ph.D., Utah, 1963. Ordinary differential equations.
Hal Smith, Professor; Ph.D., Iowa, 1976. Differential equations, dynamical systems, mathematical biology.
Harvey Smith, Professor; Ph.D., Pennsylvania, 1964. Functional analysis, mathematical modeling.
Lehi Smith, Professor; Ed.D., Stanford, 1959. Mathematics education, teacher education.
John Spielberg, Assistant Professor; Ph.D., Berkeley, 1985. Operator algebras.
Donald Stewart, Associate Professor; Ph.D., Tennessee, 1963. Special functions, topology.
Alvin Swimmer, Associate Professor; Ph.D., Berkeley, 1963. Geometry, tensor analysis, Grassman algebras.
Betty Tang, Assistant Professor; Ph.D., USC, 1983. Mathematical biology.
Tom Taylor, Associate Professor; Ph.D., Harvard, 1983. Nonlinear control theory, stochastic processes, filtering, dynamical systems.
Horst Thieme, Associate Professor; Ph.D., Münster (Germany), 1976. Differential equations, mathematical modeling.
William Trotter, Professor; Ph.D., Alabama, 1969. Combinatorics, graph theory.
Alan Wang, Professor; Ph.D., UCLA, 1965. Applied mathematics.
Cecilia Wang, Professor; Ph.D., UCLA, 1970. Complex analysis.
Neil Weiss, Professor; Ph.D., UCLA, 1970. Probability theory, time-series analysis.
Bruno Welfert, Assistant Professor; Ph.D., California, San Diego, 1990. Numerical analysis.
Dennis L. Young, Professor; Ph.D., Purdue, 1970. Statistics.

SECTION 5: MATHEMATICAL SCIENCES

BROWN UNIVERSITY
Division of Applied Mathematics

Programs of Study

The Division of Applied Mathematics offers graduate programs leading to the Ph.D. and Sc.M. degrees.

The emphasis of the Ph.D. program is on both thesis research and obtaining a solid foundation for future work. Course programs are designed to suit each individual's needs. Admission to Ph.D. candidacy is based on a preliminary examination designed individually for each student in light of his or her interests. Research interests of the faculty can be gauged from the list on the reverse of this page and include partial differential equations and dynamical systems, control theory (including stochastic control), probability and statistics, numerical analysis and scientific computation, continuum and fluid mechanics, computer vision, image reconstruction and speech recognition, and pattern theory. A wide spectrum of graduate courses is offered, reflecting the broad interests of the faculty in the different areas of applied mathematics. Relevant courses are also offered by the Departments of Mathematics, Physics, Computer Sciences, Economics, Geological Sciences, Linguistics, and Psychology and the Divisions of Engineering and of Biology and Medicine.

The Sc.M. program does not require a thesis, and students with sound preparation usually complete it in a year.

Research Facilities

The University's science library houses an outstanding collection in mathematics and its applications. The Computer Center is equipped with an IBM 3090. Faculty members from the Division are affiliated with the Center for Scientific Computation, which maintains a high-speed link with NSF-sponsored supercomputer centers. In addition, the Division houses a local center of microcomputers and associated graphics equipment. Also available in the Division are powerful scientific workstations with supporting image- and signal-processing hardware. All computer systems on campus are interconnected by a high-speed network.

Financial Aid

Fellowships, scholarships, and research and teaching assistantships, which cover tuition and living expenses, are available for qualified full-time graduate students. Summer support can usually be arranged.

Cost of Study

Tuition fees for full-time students are $16,256 per year in 1991–92. Teaching assistants and research assistants are not charged for tuition.

Cost of Living

The cost of living in Providence is somewhat lower than the national average. Housing for graduate students in the Graduate Center is available at $3200 for the 1991–92 academic year.

Student Group

Brown University has approximately 5,600 undergraduates and 1,300 graduate students. The Division of Applied Mathematics has about 60 full-time graduate students, of whom about 50 receive financial support from the University. A number of other students hold outside fellowships.

Location

Brown University is located on a hill overlooking Providence, the capital of Rhode Island and one of America's oldest cities. The proximity of Providence to the excellent beaches and ocean ports of Rhode Island and Massachusetts provides considerable recreational opportunities. Numerous nearby ski facilities are available for winter recreation. The libraries, theaters, museums, and historic sites in Providence and Newport offer an abundance of cultural resources. In addition, Providence is only an hour from Boston and 4 hours from New York City by auto or train.

The University and The Division

Brown University was founded in 1764 in Warren, Rhode Island, as Rhode Island College. It is the seventh-oldest college in America and the third-oldest in New England. In 1770, the College was moved to College Hill, high above the city of Providence, where it has remained ever since. The name was changed to Brown University in 1804 in honor of Nicholas Brown, son of one of the founders of the College. The University awarded its first Doctor of Philosophy degree in 1889. The University attracts many distinguished lecturers both in the sciences and in the arts. Brown is a member of the Ivy League and participates in all intercollegiate sports.

Brown has the oldest tradition and one of the strongest programs in applied mathematics of all universities in the country. Based on a wartime program instituted in 1942, the Division of Applied Mathematics at Brown was established in 1946 as a center of graduate education and fundamental research. It includes several research centers, has cooperative programs with many other universities, and attracts many scientific visitors.

Applying

Applications for admission to the Graduate School and for financial aid may be obtained by writing directly to the Admissions Office, Graduate School. The bulletin of the Graduate School may also be requested from the Admissions Office.

Correspondence and Information

Professor Harold J. Kushner, Chairman
Division of Applied Mathematics
Brown University
Providence, Rhode Island 02912

Peterson's Guide to Graduate Programs in the Physical Sciences and Mathematics 1992

SECTION 5: MATHEMATICAL SCIENCES

Brown University

THE FACULTY AND THEIR RESEARCH

Harvey Thomas Banks, Professor of Applied Mathematics; Ph.D., Purdue. Control theory, mathematical biology, functional differential equations.
Frederic Edward Bisshopp, Professor of Applied Mathematics; Ph.D., Chicago. Asymptotics, nonlinear wave propagation.
Constantine Michael Dafermos, Professor of Applied Mathematics and Alumni-Alumnae University Professor; Ph.D., Johns Hopkins. Continuum mechanics, differential equations.
Philip Jacob Davis, Professor of Applied Mathematics; Ph.D., Harvard. Numerical analysis, approximation theory.
Paul G. Dupuis, Assistant Professor of Applied Mathematics; Ph.D., Brown. Stochastic control and probability theory.
Peter Lawrence Falb, Professor of Applied Mathematics; Ph.D., Harvard. Control and stability theory.
Paul F. Fischer, Assistant Professor of Applied Mathematics; Ph.D., MIT. Numerical analysis, scientific computation.
Wendell Helms Fleming, Professor of Applied Mathematics and Mathematics; Ph.D., Wisconsin. Stochastic differential equations, stochastic control theory.
Walter Frederick Freiberger, Professor of Applied Mathematics; Ph.D., Cambridge. Statistics, biostatistics.
Stuart Geman, Professor of Applied Mathematics; Ph.D., MIT. Image processing, computer vision, stochastic processes.
Basilis Gidas, Professor of Applied Mathematics; Ph.D., Michigan. Mathematical physics, nonlinear partial differential equations, artificial intelligence.
David Gottlieb, Professor of Applied Mathematics; Ph.D., Tel-Aviv. Numerical methods, scientific computation.
Ulf Grenander, Professor of Probability and Statistics and L. Herbert Ballou University Professor; Ph.D., Stockholm. Probability and statistics, pattern theory.
Din-Yu Hsieh, Professor of Applied Mathematics; Ph.D., Caltech. Fluid mechanics, mathematical physics.
Kazufumi Ito, Associate Professor of Applied Mathematics; D.Sc., Washington (St. Louis). Control and inverse problems for distributed systems, numerical methods, functional differential equations.
Christopher K. R. T. Jones, Professor of Applied Mathematics; Ph.D., Wisconsin. Differential equations, dynamical systems.
Herbert Kolsky, Professor of Applied Physics (Research); Ph.D., Sc.D., London. Solid mechanics, applied physics.
Harold Joseph Kushner, Professor of Applied Mathematics and Engineering and Chairman of the Division; Ph.D., Wisconsin. Stochastic control and stability, operations research.
John Mallet-Paret, Professor of Applied Mathematics; Ph.D., Minnesota. Ordinary and functional differential equations.
Martin Maxey, Associate Professor of Applied Mathematics and Engineering; Ph.D., Cambridge. Fluid dynamics, stability and turbulence.
Donald Ernest McClure, Professor of Applied Mathematics; Ph.D., Brown. Pattern analysis, image processing, mathematical statistics.
Allen Compere Pipkin, Professor of Applied Mathematics; Ph.D., Brown. Continuum mechanics.
Chi-Wang Shu, Assistant Professor of Applied Mathematics; Ph.D., UCLA. Numerical methods for conservation laws and other methods for computational fluid dynamics.
Lawrence Sirovich, Professor of Applied Mathematics; Ph.D., Johns Hopkins. Gasdynamics, perturbation methods, mathematical biology.
Panagiotis Souganidis, Professor of Applied Mathematics; Ph.D., Wisconsin. Nonlinear partial differential equations, differential games, control theory.
Walter Strauss, Professor of Mathematics and Applied Mathematics; Ph.D., MIT. Scattering theory, partial differential equations.
Chau-Hsing Su, Professor of Applied Mathematics; Ph.D., Princeton. Fluid mechanics, mathematical physics.

SECTION 5: MATHEMATICAL SCIENCES

BROWN UNIVERSITY

Department of Mathematics

Program of Study

The graduate program offered by the Department of Mathematics is primarily intended to lead to the degree of Doctor of Philosophy and a career in research or teaching at the college level. The relatively small enrollment in the department gives students the advantages of small classes and close contact with faculty members.

Students are admitted to Ph.D. candidacy upon demonstrating promise for research and a knowledge of four basic areas: algebra, complex analysis, real analysis, and topology. This is achieved either by satisfactorily completing appropriate courses in these areas or by passing diagnostic examinations and is normally accomplished in the first two years. Students then begin their research under the supervision of an adviser and continue to broaden and deepen their knowledge of mathematics by attending advanced courses and seminars, which are often offered in the following areas: algebra, algebraic geometry, algebraic number theory, algebraic topology, differential geometry, differential topology, functional analysis, partial differential equations, probability theory, Riemann surfaces, and several complex variables. Students may also take courses offered by the Division of Applied Mathematics and the Department of Computer Science. Numerous lively seminars and a weekly colloquium at which distinguished visiting mathematicians give lectures on their current research provide a valuable experience for graduate students.

Students must also demonstrate an ability to read mathematics in two of the following three foreign languages: French, German, and Russian.

The Ph.D. program, which culminates with the completion of a dissertation, is normally completed within five years. The dissertation is a report of original research and is usually published in a scientific journal.

Research Facilities

The Sciences Library has an excellent mathematics collection—one of the best in the country—containing 25,000 volumes and holding subscriptions to 650 journals in mathematics and mathematical sciences.

Among various computing facilities, the department has several Macintosh and IBM computers available for research use by graduate students. There are two computer rooms.

Financial Aid

The standard practice is to support qualified students in the Ph.D. program for five years, primarily through teaching assistantships. There are several fellowships for entering students and research assistantships for advanced students. Fellowships and assistantships pay a stipend plus tuition. For 1991–92, the stipend is in the range of $8500–$9500.

Cost of Study

Tuition and fees for the academic year 1991–92 are $16,628.

Cost of Living

The room charges for a single person in the Graduate Center are $3200 for the academic year 1991–92. Rents are rather low for shared local apartments.

Student Group

Brown University has about 5,500 undergraduates and 1,450 graduate students (including those in the Medical Program). The graduate program in mathematics has 41 students in 1991–92. A total of twenty-five Ph.D.'s have been awarded in the last five years.

Location

Providence, founded by Roger Williams in 1636, has a population of 165,000. The University is located on the East Side, which is a quiet residential section of the city. Most faculty members and graduate students live on the East Side, often within walking distance of the campus.

Rhode Island has a 400-mile shoreline. Newport is less than an hour's drive from Brown, and Cape Cod is only 2 hours away. The Greater Boston area, noted for its cultural activities, is 1 hour's drive away. New Haven and New York City are less than 2 hours and 4 hours away, respectively.

The University

Brown University was founded in 1764 in Warren, Rhode Island, as Rhode Island College—the seventh-oldest college in America and the third-oldest in New England. It moved to Providence in 1770, and its name was changed to Brown University in 1804. The University's first Ph.D. degree was awarded in 1889, and the Department of Mathematics granted its first Ph.D. in 1929. Most Ph.D. graduates in mathematics have gone on to academic positions; a substantial number have been active in research jobs in government and industry and in research firms.

Applying

Applicants should have a good background in undergraduate mathematics, regardless of their major. Applicants should submit GRE scores (General Test and Subject Test in math), and international students whose native language is not English must submit TOEFL scores. Students who have a background in advanced mathematics will find the program quite flexible. Visits and direct communication with the graduate representative are strongly recommended.

January 2 is the deadline for receiving application material and letters of recommendation. Admission and financial aid awards are usually announced by the middle of March.

Correspondence and Information

For further information about graduate work:
Graduate Representative
Department of Mathematics
Box 1917
Brown University
Providence, Rhode Island 02912
Telephone: 401-863-2708

For application forms and general information:
Graduate School
Box 1867
Brown University
Providence, Rhode Island 02912
Telephone: 401-863-2698

SECTION 5: MATHEMATICAL SCIENCES

Brown University

THE FACULTY AND THEIR RESEARCH
Robert D. M. Accola, Ph.D., Harvard, 1958. Riemann surfaces.
Thomas F. Banchoff, Ph.D., Berkeley, 1964. Geometry and topology.
    Cusps of Gauss Mappings. Pitman, 1982. (With Gaffney and McCrory)
    Linear Algebra Through Geometry. Springer, 1983. (With Wermer)
Andrew Browder, Ph.D., MIT, 1961. Functional analysis.
    Introduction to Function Algebras. Addison-Wesley, 1969.
Brian J. Cole, Ph.D., Yale, 1968. Functional analysis.
Walter Craig, Ph.D., NYU (Courant), 1981. Nonlinear, partial differential equations.
Herbert Federer (Emeritus), Ph.D., Berkeley, 1944. Geometric measure theory.
    Geometric Measure Theory. Springer, 1969.
Wendell H. Fleming, Ph.D., Wisconsin–Madison, 1951. Applied probability.
    Deterministic and Stochastic Optimal Control. Springer, 1975. (With Rishel)
    Functions of Several Variables, 2nd ed. Springer, 1976.
Thomas G. Goodwillie, Ph.D., Princeton, 1982. Topology.
Bruno Harris, Ph.D., Yale, 1956. Algebraic topology.
Jeffrey Hoffstein, Ph.D., MIT, 1978. Analytic number theory and automorphic forms.
Eva Kallin, Ph.D., Berkeley, 1963. Function algebras.
Nicolaos Kapouleas, Ph.D., Berkeley, 1988. Differential geometry and partial differential equations.
Kang-Tae Kim, Ph.D., UCLA, 1988. Several complex variables.
Alan Landman, Ph.D., Berkeley, 1967. Algebraic geometry.
Stephen Lichtenbaum, Ph.D., Harvard, 1964. Number theory.
Paul Lockhart, Ph.D., Columbia, 1990. Number theory.
Jonathan D. Lubin, Ph.D., Harvard, 1963. Algebraic geometry, number theory.
Randolph McCarthy III, Ph.D., Cornell, 1990. K-theory and cyclic homology.
Katsumi Nomizu, Ph.D., Chicago, 1953; Sc.D., Nagoya (Japan), 1955. Differential geometry.
    Lie Groups and Differential Geometry. Mathematical Society of Japan, 1956.
    Foundations of Differential Geometry, vols. I and II. Wiley, 1963 and 1969. (With Kobayashi)
    Fundamentals of Linear Algebra, 2nd ed. Chelsea Publishing, 1979.
Jill C. Pipher, Ph.D., UCLA, 1985. Partial differential equations.
Michael I. Rosen, Ph.D., Princeton, 1963. Algebraic numbers and functions.
    Elements of Number Theory. Bogden & Quigley, 1972. (With Ireland)
    A Classical Introduction to Modern Number Theory. Springer, 1982. (With Ireland)
Joseph Silverman, Ph.D., Harvard, 1982. Number theory.
    The Arithmetic of Elliptic Curves. Springer, 1986.
Frank M. Stewart (Emeritus), Ph.D., Harvard, 1947. Mathematical biology.
    Introduction to Linear Algebra. Van Nostrand, 1963.
Walter A. Strauss, Ph.D., MIT, 1962. Partial differential equations.
John Wermer, Ph.D., Harvard, 1951. Complex analysis.
    Potential Theory. Springer, 1974.
    Banach Algebras and Several Complex Variables, 2nd ed. Springer, 1976.
    Linear Algebra Through Geometry. Springer, 1983. (With Banchoff)

RESEARCH COURSES AND SEMINARS
In addition to regularly offered graduate courses—algebra, real function theory, complex function theory, topology, differential geometry, algebraic geometry, probability, and partial differential equations—the department has in recent years offered the following advanced courses, which initiate students into special research areas:

Topics in algebraic number theory
Topology of complex varieties
Geometry of discrete groups
Fourier series and group representations
Several complex variables
Topics in differential geometry
Topics in differential topology
Topics in theory of submanifolds
Geometry and analysis on manifolds
Topics in classical mechanics
Compactifications of locally symmetric spaces
Symplectic geometry and theta series
The arithmetic of elliptic curves
Diophantine geometry
Introduction to complex manifolds
Topics in dynamical systems
Topics in compact Riemann surfaces
Introduction to pseudodifferential and Fourier integral operators

Many seminars are conducted in the areas of analysis, algebraic geometry, number theory, topology and geometry, and differential geometry, as well as departmental colloquia and Graduate Student Seminar. Proximity to the Greater Boston area provides opportunities to participate in many outside mathematical activities.

SECTION 5: MATHEMATICAL SCIENCES

CALIFORNIA INSTITUTE OF TECHNOLOGY
Applied Mathematics Program

Program of Study

An interdisciplinary program of study in applied mathematics that leads to the Ph.D. degree is offered by the Institute. In addition to various basic and advanced courses taught by the applied mathematics faculty, broad selections are available in mathematics, physics, engineering, and other areas. Students are expected to become proficient in some special physical or nonmathematical field. A subject minor in applied computation is offered jointly with the computer science option.

The Institute requirements for a Ph.D. specify three academic years of study after the bachelor's degree. The first year usually consists of a full program of four or five courses. An oral candidacy examination is given at the end of the first term of the second year to assess the student's readiness for research. After this the emphasis is on research and thesis work.

Research Facilities

The Institute library facilities are excellent and are supplemented by more specialized departmental libraries that cater to specific needs. Remote consoles for easy on-line access to various computers are located throughout the campus. Microcomputers are also available. The applied mathematics group has access to supercomputers at UC San Diego's National Computing Center, to various national laboratories, and to a variety of hypercubes on the campus. In addition, the applied mathematics group has a local network of Sun, IRIS, and DEC workstations supporting scientific computation, algebraic manipulation programs, and other scientific software and computing languages. The local network is connected to campuswide and national networks. A state-of-the-art parallel supercomputer rated at up to 40 gigaflops was installed in April 1991. Access to conventional supercomputers, such as Cray supercomputers, is available through JPL.

Strong interactions with applied mathematics and computing programs at Rice University, Los Alamos National Laboratory, and Argonne National Laboratory exist as a result of the NSF Center for Research in Parallel Computing, of which Caltech is a major component.

Financial Aid

Teaching or research assistantships are available for qualified students. The assistantships include tuition scholarships and carry cash stipends ranging from $8667 per academic year for first-year graduate students to $13,212 for advanced students in 1991–92. The duties involved vary from grading to teaching to research, but they permit the holder to carry a full schedule of graduate study. A number of assistantships are available for research in the summer months.

In addition, Institute fellowships in applied mathematics are awarded; they carry stipends at least comparable to those mentioned above. Additional sources of support common at Caltech include NSF fellowships and traineeships and NDEA fellowships.

Cost of Study

Tuition for 1991–92 is $14,100 for the academic year, including health fees.

Cost of Living

Graduate student dormitories provide rooms for unmarried students at costs ranging from $3712 to $4860 for the academic year 1991–92. Meals are available at the campus dining hall.

Married students usually find attractive accommodations in one of the apartment buildings located in the vicinity of the campus. The Housing Office also maintains and rents single-family houses and apartments at costs considerably below competitive market value. Early application for these is advised, as demand exceeds supply.

Student Group

Caltech has approximately 850 undergraduate students and about 1,025 graduate students, a small number of whom are women. There are about 20 graduate students in applied mathematics, nearly all of whom are receiving financial support of some kind. Admission to graduate study in applied mathematics is highly competitive; only about 6 to 8 new students can be admitted each year.

Location

Pasadena is a city of approximately 140,000 inhabitants, located just northeast of Los Angeles at the foot of the San Gabriel Mountains. The city contains both residential and light industrial areas. Caltech is located in the center of a residential area but is within a few blocks of shopping facilities. Most of the cultural activities of the large metropolitan area may be reached in an hour or less by car.

Recreational facilities abound. Hiking, camping, bicycling, and (occasionally) skiing are immediately at hand in the mountains; the Pacific Ocean is an hour or two away by car; and on a weekend any part of southern California can be explored.

The Institute

Throughout its history, Caltech has maintained a small, select student body and a faculty that is unusually active in research. The staff currently numbers about 600, including postdoctoral fellows; of this number more than 195 are full professors.

There is a strong emphasis on fundamental studies in science and engineering, with a minimum of specialization. There is also considerable emphasis on humanistic studies. The humanities have always played an essential role in the undergraduate curriculum, and undergraduates are permitted to major in these areas.

There are approximately 17,000 Institute alumni scattered all over the world, many eminent in their fields of science and engineering. Twenty of them have received Nobel Prizes.

Musical and dramatic events are presented on campus throughout the year.

Applying

Students are admitted only at the beginning of the fall term, and applications should be received by January 15. Only full-time Ph.D. candidates are accepted. All applicants are urged to take the mathematics Subject Test of the Graduate Record Examinations. Application forms for admission and financial assistance may be obtained from the Graduate Office.

Correspondence and Information

Dean of Graduate Studies
California Institute of Technology
Pasadena, California 91125

SECTION 5: MATHEMATICAL SCIENCES

California Institute of Technology

THE FACULTY AND THEIR RESEARCH

The faculty members most concerned with teaching, advising, and supervising the research of students in applied mathematics are listed below.

Kirk Brattkus, von Karman Instructor of Applied Mathematics; Ph.D., Northwestern, 1988. Pattern formation, multiphase dynamics.
Thomas K. Caughey, Professor of Applied Mechanics; Ph.D., Caltech, 1954. Dynamics, system theory and control, stochastic processes.
K. Mani Chandy, Professor of Computer Science; Ph.D., MIT, 1969. Parallel programs, concurrency, algorithms, performance modeling.
Robert W. Clayton, Assistant Professor of Exploration Geophysics; Ph.D., Stanford, 1981. Geophysical applications of finite-difference wave modeling, inverse scattering methods, tomographic reconstructions.
Donald S. Cohen, Professor of Applied Mathematics; Ph.D., NYU (Courant), 1962. Perturbation theory, bifurcation theory, diffusive systems.
Noel R. Corngold, Professor of Applied Physics; Ph.D., Harvard, 1954. Statistical mechanics; theory of particle transport in gases, liquids, and plasmas.
Joel N. Franklin, Professor of Applied Mathematics; Ph.D., Stanford, 1953. Mathematical programming, ill-posed problems, numerical analysis, stochastic processes, computer algorithms.
John J. Hopfield, Roscoe G. Dickinson Professor of Chemistry and Biology; Ph.D., Cornell, 1958. Collective computation.
Herbert B. Keller, Professor of Applied Mathematics; Ph.D., NYU (Courant), 1954. Numerical analysis, bifurcation theory, large-scale scientific computing.
James K. Knowles, Professor of Applied Mechanics; Ph.D., MIT, 1957. Mathematical problems in solid mechanics.
Daniel I. Meiron, Associate Professor of Applied Mathematics; Sc.D., MIT, 1981. Computational physics, large-scale scientific computing.
Ellen Randall, von Karman Instructor of Applied Mathematics; Ph.D., MIT, 1991. Fluid dynamics: linear and nonlinear waves, compressible flow, stability theory, dynamical systems.
Philip G. Saffman, Professor of Applied Mathematics; Ph.D., Cambridge, 1956. Theoretical and computational fluid mechanics.
Charles L. Seitz, Professor of Computer Science; Ph.D., MIT, 1971. Computer architecture and design, VLSI systems, self-timing systems, concurrency, switching theory, graphics and image systems.
Stephen Taylor, Instructor in Computer Science; Ph.D., Weizmann (Israel), 1989. Programming languages for parallelization, parallel programs, concurrency, algorithms, performance modeling.
Eric Van de Velde, Senior Research Fellow in Applied Mathematics; Ph.D., NYU (Courant), 1986. Concurrent scientific computing.
Gerald B. Whitham, Charles Lee Powell Professor of Applied Mathematics; Ph.D., Manchester, 1953. Fluid mechanics, wave propagation, nonlinear problems.
Stephen R. Wiggins, Assistant Professor of Applied Mechanics; Ph.D., Cornell, 1985. Nonlinear dynamical systems, chaotic phenomena.

RESEARCH ACTIVITIES

Because mathematics is applied by most scientists in almost all disciplines at Caltech, significant research by applied mathematics students is encouraged in all areas. Areas in which research has been particularly strong include fluid dynamics, magnetohydrodynamics, kinetic theory, elasticity, dynamics and celestial mechanics, numerical analysis, ordinary and partial differential equations, integral equations, linear and nonlinear wave propagation, water waves, bifurcation theory, perturbation and asymptotic methods, stability theory, computational fluid mechanics, stochastic processes, variational methods, large-scale scientific computing, applications of parallel processing, and other related branches of analysis. The department has many visitors, from senior scientists to postdoctoral research fellows, who specialize in these and other topics for varying periods.

SECTION 5: MATHEMATICAL SCIENCES

CARNEGIE MELLON UNIVERSITY

Mellon College of Science
Department of Mathematics

Programs of Study

The Department of Mathematics offers programs leading to the degrees of Master of Science in mathematics and applied mathematics; Ph.D. and D.A. in mathematics; Ph.D. in algorithms, combinatorics, and optimization; Ph.D. in pure and applied logic; and Ph.D. in mathematical finance.

The master's degree programs involve rather strict course requirements and a comprehensive examination; a master's thesis or a research project is optional. The usual time taken in residence to complete the master's is two years, although unusually well qualified students can meet all requirements within one year.

The first requirement for the doctoral degrees is completion of a set of core courses, followed by passage of the comprehensive examination. After this, the emphasis is on preparation for research. After passing a qualifying examination that measures competence in the proposed research area, the student begins serious research. The differences between Ph.D. and D.A. theses is that the Ph.D. thesis must consist of original publishable research in depth; the D.A. thesis is a scholarly publishable work that need not report new results.

The research of the department is mainly specialized in several areas of applied mathematics, in algebra, and in logic. Greater breadth in course offerings and in research areas is available through cooperation with the Department of Comptuer Science, the Department of Statistics, and the operations research and finance groups in the Graduate School of Industrial Administration. The Department of Mathematics houses the Center for Nonlinear Analysis, which is funded by the Army Research Office to perform basic research on mathematical issues arising in such areas as materials science, stochastic modeling, and continuum mechanics.

Research Facilities

The University offers outstanding computational facilities, including easy access to powerful networked workstations and to the CRAY Y-MP of the Pittsburgh Supercomputer Center. In addition, the department is well equipped with workstations.

Financial Aid

The most common mode of support for students is through teaching assistantships, with approximately 30 students so supported. Advanced students may obtain research assistantships. The number available varies, but is on the order of ten. Both forms of assistantship include a tuition waiver and a stipend competitive with other institutions. Some first-year students receive fellowships from the Center for Nonlinear Analysis.

Cost of Study

Tuition and fees for the academic year 1991–92 are $15,250.

Cost of Living

Pittsburgh provides an attractive and reasonably priced living environment. On-campus housing is limited, but the Off-Campus Housing Office assists students in finding suitable accommodations. Most graduate students choose to live in nearby rooms and apartments, which are readily available.

Student Group

CMU has about 4,200 undergraduates and 1,100 graduate students. The Department of Mathematics graduate program includes 45 students in 1991–92. A total of twenty-six Ph.D.'s and forty-nine master's degrees have been awarded in the past five years.

Location

Pittsburgh is in a large metropolitan area with a population of 2.3 million people. It has been rated by the *Rand-McNally Places Rated Almanac* as the nation's most livable city. Pittsburgh is the headquarters for fifteen Fortune 500 corporations, and there is a large concentration of research laboratories in the area. Carnegie Mellon is located in Oakland, the cultural and civic center of Pittsburgh. The campus covers 90 acres and adjoins Schenley Park, the largest city park. The city's cultural and recreational opportunities are truly outstanding.

The University

Carnegie Mellon is the result of a 1967 merger of the Carnegie Institute of Technology, founded by Andrew Carnegie and known primarily for preparing engineers, and the Mellon Institute, founded by A. W. and R. B. Mellon, which carried out pure and applied research in conjunction with local industry. By carefully selecting areas of research concentration and encouraging interdisciplinary programs, Carnegie Mellon maintains a research program that rivals those of universities many times its size. Thus, the University gives graduate students both the advantages of a wide range of opportunities for research and the comparative intimacy of a small university. In addition to the seven colleges—the Carnegie Institute of Technology, the Mellon College of Science, the College of Fine Arts, the College of the Humanities and Social Sciences, the School of Urban and Public Affairs, the Graduate School of Industrial Administration, and the School of Computer Science—the interdisciplinary centers on campus, including the Center for Nonlinear Analysis, the Robotics Institute, the Engineering Design Research Center, and the Magnetic Materials Research Group, further enrich the opportunities for research collaboration. Since several of these colleges and research centers draw heavily upon mathematics, it is an excellent environment for mathematical research. The research activities of the University are complemented by those of the University Teaching Center, whose programs are open to graduate students and can be helpful to the beginning instructor.

Applying

Completed applications and credentials for graduate study in mathematics should be submitted by February 10 for decision by mid-April. However, admission decisions are made on a continuous basis, and applications are considered at any time. In addition to the application form, transcripts from all college-level institutions attended, three letters of recommendation, and an official report of the applicant's scores on the General Test and the Subject Test in mathematics of the Graduate Record Examinations are required. A full description of procedures and programs is given in the booklet *Carnegie Mellon Graduate Studies in Mathematics*, which will be sent on request.

Correspondence and Information

Graduate Applications Committee
Department of Mathematics
Carnegie Mellon University
Pittsburgh, Pennsylvania 15213
Telephone: 412-268-2545

SECTION 5: MATHEMATICAL SCIENCES

Carnegie Mellon University

THE FACULTY AND THEIR RESEARCH

Michael H. Albert, Assistant Professor; D.Phil., Oxford, 1984. Mathematical logic, category theory, universal algebra, model theory.
Peter B. Andrews, Professor; Ph.D., Princeton, 1964. Mathematical logic, automated theorem proving.
Egon Balas, Professor; D.Sc.Ec., Brussels, 1967; D.U., Paris, 1968. Disjunctive programming, combinatorial optimization, algorithmic methods for optimization.
Charles V. Coffman, Professor; Ph.D., Johns Hopkins, 1963. Ordinary and partial differential equations, functional analysis.
Gerard P. Cornuejols, Professor; Ph.D., Cornell, 1974. Integer and combinatorial programming.
Richard J. Duffin, University Professor (Professor Emeritus); Ph.D., Illinois, 1935. Optimization of systems or devices.
Irene M. Fonseca, Assistant Professor; Ph.D., Minnesota, 1985. Partial differential equations, elastic crystals, phase transformations.
Alan M. Frieze, Professor; Ph.D., London, 1975. Combinatorial optimization, random graphs, computational complexity.
Rami Grossberg, Assistant Professor; Ph.D., Hebrew (Jerusalem), 1986. Mathematical logic, model theory, set theory.
Morton E. Gurtin, Professor; Ph.D., Brown, 1961. Continuum mechanics, population dynamics, variational calculus, partial differential equations.
William J. Hrusa, Associate Professor; Ph.D., Brown, 1982. Partial differential equations, integrodifferential equations.
Ravindran Kannan, Associate Professor; Ph.D., Cornell, 1979. Algorithms, complexity, operations research, mathematics of computation.
David Kinderlehrer, Professor; Ph.D., Berkeley, 1968. Nonlinear partial differential equations, calculus of variations, liquid crystal theory.
Ignace I. Kolodner, Professor Emeritus; Ph.D., NYU, 1950. Nonlinear problems in analysis, differential and integral equations.
John P. Lehoczky, Professor; Ph.D., Stanford, 1969. Stochastic control, with applications to finance manufacturing and queueing.
Ling Ma, Visiting Assistant Professor; Ph.D., Minnesota, 1991. Numerical and analytic study of variational problems, numerical study of moving phase boundaries.
Richard C. MacCamy, Professor; Ph.D., Berkeley, 1956. Integral equations, integrodifferential equations, nonlinear and degenerate models, partial differential equations.
Victor J. Mizel, Professor; Ph.D., MIT, 1955. Calculus of variations, stochastic control problems, hereditary phenomena.
R. A. Nicolaides, Professor; Ph.D., London, 1972. Numerical solution of partial differential equations, computational fluid dynamics, phase transitions.
Walter Noll, Professor; Ph.D., Indiana, 1954. Axiomatic foundations of physical theories, continuum mechanics and thermodynamics, theory of relativity.
David R. Owen, Professor; Ph.D., Brown, 1968. Plastic and viscoelastic materials, foundations of classical thermodynamics.
Roger N. Pederson, Professor; Ph.D., Minnesota, 1957. Partial differential equations, complex analysis.
Jack Schaeffer, Associate Professor; Ph.D., Indiana, 1983. Nonlinear partial differential equations, Poisson-Vlasov and Maxwell-Vlasov systems.
Juan Jorge Schäffer, Professor; Dr.sc.techn., Swiss Federal Institute of Technology, 1956; Dr.phil., Zürich, 1956. Functional analysis, ordinary and functional differential equations.
Dana S. Scott, University Professor of Computer Science and Mathematical Logic and Philosophy; Ph.D., Princeton, 1958. Logic, model theory, automata, modal and intuitionistic logic, semantics of programming languages.
Robert F. Sekerka, Professor of Physics and Mathematics; Ph.D., Harvard, 1965. Morphological stability of phase transitions.
Steven S. Shreve, Professor; Ph.D., Illinois, 1977. Optimization in the presence of uncertainty, control of diffusion processes.
H. Mete Soner, Associate Professor; Ph.D., Brown, 1986. Stochastic partial differential equations, optimal control and nonlinear partial differential equations.
Richard Statman, Professor; Ph.D., Stanford, 1974. Computation, symbolic computation, lambda calculus and combinatory algebra.
Gabriella Tarantello, Assistant Professor; Ph.D., NYU (Courant), 1986. Nonlinear ordinary and partial differential equations.
Luc C. Tartar, Professor; Ph.D., Paris VI (Curie), 1971. Nonlinear functional analysis, nonlinear partial differential equations.
Gerald L. Thompson, IBM Professor of Systems and Operations Research; Ph.D., Michigan, 1953. Mathematical programming, combinatorial optimization, large-scale linear programming, large-scale network algorithms.
Russell C. Walker, Senior Lecturer and Assistant Department Head; D.A., Carnegie Mellon, 1972. Mathematical exposition.
Noel J. Walkington, Assistant Professor, Ph.D., Texas, 1988. Numerical analysis, nonlinear partial differential equations, acoustic equations.
Ross D. Willard, Assistant Professor, Ph.D., Waterloo, 1989. Universal algebra, lattice theory, computational complexity.
William O. Williams, Professor and Head; Ph.D., Brown, 1967. General laws of mechanics and thermodynamics, phase transitions, viscoplasticity, biological mechanics.
Oswald Wyler, Professor; Dr.sc.math., Swiss Federal Institute of Technology, 1950. Theory of categories, categorical logic, quasitopoi.

Doctoral Theses Since 1987

On Matroid Intersection.
A Higher Order Vortex Method for Two- and Three-Dimensional Spaces.
A Two-Dimensional Eddy Current Problem.
Fractional Order Volterra Equations in Hilbert Spaces.
On a Class of Models for Heat Flow in Materials with Memory.
Analysis of a Ladyzhenskaya Model for Incompressible Viscous Flow and Its Finite Element Approximations.
On Non-Equilibrium Phase Transitions in Mixtures with Interfacial Structure.
On One-Dimensional Nonlinear Thermoelasticity with Second Sound: Existence of Globally Smooth Solutions.
Semi-Linear Programming and the Unidimensional Similarities Problem.
On Balanced and Perfect Matrices.
An Eigenvalue Problem for the Mean Curvature Operator in Non-Radial Domains.
Interface Problems in Elastodynamics.
Existence, Uniqueness, and Finite Element Approximation of Solutions of the Equations of Stationary, Incompressible MHD.
Tchebychev Nets on Two-Dimensional Riemannian Manifolds.
Studies on Set Covering, Set Partitioning, and Vehicle Routine Problems.
On Local Existence and Formation of Singularities in Nonlinear Thermoelasticity.
Lavrentiev Phenomenon in Nonlinear Elasticity.
Analysis and Finite Element Approximation of Some Optimal Control Problems Associated with the Navier-Stokes Equations.
A Duality Approach to a Stochastic Consumption/Portfolio Decision Problem in a Continuous Time Market with Short-Selling Prohibition.
A Mathematical Theory of Cold Drawing.
Covolume Techniques for Anistropic Media/Application of Spectral Methods to a Cahn-Hilliard Model of Phase Transition.
A Computational Study of the Set Covering Problem.
Mobile Phase Boundaries in Elastic Media.

SECTION 5: MATHEMATICAL SCIENCES

CARNEGIE MELLON UNIVERSITY
Department of Statistics

Programs of Study

While the Department of Statistics is most widely recognized for its research in applied and theoretical Bayesian statistics and Bayesian decision theory, substantial contributions have also been made in the areas of the analysis of discrete data, computational statistics, federal statistical programs, the foundations of statistics, legal statistics, psychiatric statistics, stochastic modeling, and time-series analysis. In addition, the department has recently established a center for statistical research in quality improvement. The Department of Statistics at Carnegie Mellon offers an emphasis on both application and theory, outstanding computing facilities that aid training in computational statistics, and a moderate size and congenial environment to foster close working relationships between students and faculty.

The graduate programs are designed to train students for positions in industry, government, and academic institutions. The interdisciplinary nature of the programs brings together faculty and students interested in statistical applications in engineering, science, social science, and management, as well as in statistical theory. Each student may follow a course of study determined by individual interests.

The master's degree program trains students in applied statistics by imparting knowledge of the theory and practice of statistics. Requirements are satisfactory completion of 96 course units and a written comprehensive examination. There is no thesis requirement. Students complete the program in one or two years, depending on their previous preparation.

The Ph.D. program is structured to prepare students for careers in university teaching and research and for industrial and government positions that involve consulting and research in new statistical methods. Doctoral candidates first complete the requirements for the M.S. in statistics. They then typically complete another year of courses in probability and statistics. A written Ph.D. comprehensive examination and an oral thesis proposal presentation and defense are required. Proficiency in the use of the computer is required. There are no foreign language requirements.

The Ph.D. program in psychiatry and mental health statistics is sponsored jointly with the Department of Psychiatry, School of Medicine, University of Pittsburgh. After two years of graduate training in statistics, students take courses in biostatistics and epidemiology at the University of Pittsburgh. Students become involved in research projects at the Western Psychiatric Institute and Clinic, which is the specialty psychiatric hospital of the University Health Center of Pittsburgh. Qualified candidates receive a prestigious National Research Service Award, which covers all tuition and fees and provides a regular stipend.

Research Facilities

The computational resources available to students at Carnegie Mellon are unsurpassed and are a major strength of the program. Statistics operates its own computer facilities, which provide students with experience using state-of-the-art equipment. Facilities include more than fifty workstations interconnected by a department Ethernet, which in turn is connected to the University and worldwide networks. The workstations are, for the most part, DECstation 5000s and DECstation 3100s; there are also several personal computers, terminals, and laser printers. The department has a graphics laboratory with equipment for producing computer-animated videotapes and computer-generated color photographs.

Financial Aid

The department attempts to provide financial aid for as many of its students, both master's and Ph.D. candidates, as possible. Tuition scholarships are usually granted in conjunction with graduate assistantships, which currently offer a stipend of $8325 for nine months in return for duties as teaching or research assistants. Students who receive both tuition scholarships and graduate assistantships are expected to maintain a full course load and devote effort primarily to their studies and assigned duties. These duties require not more than 12 hours per week. Exceptionally well qualified candidates may qualify for a fellowship that pays tuition and a stipend and requires no assistantship duties.

Cost of Study

The tuition fee for full-time graduate students in 1991–92 is $15,250 per academic year.

Cost of Living

Pittsburgh has attractive, reasonably priced neighborhoods where students can live comfortably.

Student Group

Carnegie Mellon University has 4,502 undergraduate and 2,892 graduate students. The teaching faculty numbers approximately 650. During 1990–91, there were 27 full-time students in the graduate program, of whom 21 were working toward a Ph.D. degree.

Location

Located in a metropolitan area of over 2 million people, Pittsburgh is the headquarters of many of the nation's largest corporations. There is an unusually large concentration of research laboratories in the area. Carnegie Mellon is located in Oakland, the cultural center of the city. The campus is within walking distance of museums and libraries and is close to the many cultural and sports activities of the city.

The University

One of the leading universities in the country, Carnegie Mellon has long been devoted to liberal professional education. Four colleges—the Carnegie Institute of Technology, the College of Fine Arts, the College of Humanities and Social Sciences, and the Mellon College of Science—offer both undergraduate and graduate programs. The Graduate School of Industrial Administration, the School of Urban and Public Affairs, and the School of Computer Science offer graduate programs only.

Applying

There is no strict deadline; however, students who want to begin study in September are strongly encouraged to send in their applications by January 15. Graduate students in statistics have diverse backgrounds, with typical preparation being an undergraduate program in mathematics or in engineering, science, economics, or management. A course in probability and statistics at the level of DeGroot's *Probability and Statistics* is highly desirable, but excellence and promise always balance a lack of formal preparation. The General Test of the Graduate Record Examinations is required of all applicants. Foreign applicants are also required to take the TOEFL.

Correspondence and Information

Department of Statistics
Carnegie Mellon University
Pittsburgh, Pennsylvania 15213-3890

Peterson's Guide to Graduate Programs in the Physical Sciences and Mathematics 1992

SECTION 5: MATHEMATICAL SCIENCES

Carnegie Mellon University

THE FACULTY AND THEIR RESEARCH

David L. Banks, Assistant Professor of Statistics; Ph.D., Virginia Tech, 1984. Nonparametric inference, Bayesian inference.
  Patterns of oppression: A statistical analysis of human rights data. *J. Am. Stat. Assoc.*, 1989. Histospline smoothing the Bayesian bootstrap. *Biometrika*, 1988. Bootstrapping II, *Encyclopedia of Statistical Sciences*, 1988.

Ngai Hang Chan, Associate Professor of Statistics; Ph.D., Maryland, 1985. Time-series analysis, asymptotics, parametric inference.
  Limiting distributions of least squares estimates of unstable autoregressive processes. *Ann. Stat.*, 1988. With Wei. On the parameter inference for nearly nonstationary time series. *J. Am. Stat. Assoc.*, 1988. Asymptotic inference for unstable autoregressive time series with drifts. *J. Stat. Plann. Inference*, 1989.

George T. Duncan, Professor of Statistics (primary appointment in School of Urban and Public Affairs); Ph.D., Minnesota, 1970. Applied statistics, decision theory, public policy.
  Entering access to data while protecting confidentiality: Prospects for the future. *Stat. Sci.*, 1991. With Pearson. The risk of disclosure for microdata. *J. Business Econ. Stat.*, 1989. With Lambert. The role of mandates in third-party intervention. *Negotiation J.*, 1988. With Kaufman.

William F. Eddy, Professor of Statistics; Ph.D., Yale, 1976. Computational methods, graphical statistics, geometric probability.
  Cross-disciplinary research in the statistical sciences. *Stat. Sci.*, 1990. With the IMS Panel on Cross-Disciplinary Research in the Statistical Sciences. Random number generators for parallel processors. *J. Comput. Appl. Math.*, 1990. Asynchronous iteration. *Computer Science and Statistics: Proceedings of the 20th Symposium on the Interface*, 1989. With Schervish. Determining properties of minimal spanning trees by local sampling. *Computer Science and Statistics: Proceedings of the 20th Symposium on the Interface*, 1989. With McIntosh. An introduction to color systems. *Stat. Comput. Stat. Graphics*, 1990.

Michael D. Escobar, Assistant Professor of Statistics; Ph.D., Yale, 1988. Empirical Bayes methods, statistical consulting, biostatistics.
  Generalized binomial models to examine the historical control assumption in active control studies. *The Statistician*, 1989. With Makuch and Stephens. EEG coherence of prefrontal areas in normal and schizophrenic males during perceptual activation. *J. Neuropsychiatry Clin. Neurosci.*, 1990. With Hoffman, et al. Estimating normal means with a Dirichlet process prior. Presented at the Joint Statistical Meetings, Washington, D.C., 1990. Prevalence of reading disability in boys and girls: Results of Connecticut longitudinal study. *J. Am. Med. Assoc.*, 1990. With E. E. Shaywitz, B. A. Shaywitz, and Fletcher.

Stephen E. Fienberg, Maurice Falk Professor of Statistics and Social Science (joint appointment in Department of Social and Decision Sciences) and Dean of the College of Humanities and Social Sciences; Ph.D., Harvard, 1968. Categorical data, criminal justice statistics, federal statistics, multivariate data analysis, statistical inference.
  The Analysis of Cross-Classified Categorical Data, 2nd ed. Cambridge: MIT Press, 1980. *The Evolving Role of Statistical Assessment as Evidence in the Courts.* Edited NRC Report. New York: Springer-Verlag, 1989. *A Statistical Model: Frederick Mosteller's Contributions to Statistics, Science, and Public Policy.* New York: Springer-Verlag, 1990. Edited with Hoaglin, Kruskal, and Tanur.

Joel B. Greenhouse, Associate Professor of Statistics; Ph.D., Michigan, 1982. Methodology, biostatistics, psychiatric statistics.
  Selection models and the file-drawer problem. *Stat. Sci.*, 1988. With Iyengar. An introduction to survival analysis: Statistical methods for analysis of clinical trial data. *J. Consult. Clin. Psychol.*, 1989. With Stangl and Bromberg. On some applications of Bayesian methods in cancer clinical trials. *Stat. Med.*, 1991.

Brian Junker, Assistant Professor of Statistics; Ph.D., Illinois, 1988. Large-sample theory, associated random variables, psychometric models.
  Essential independence and ability estimation for polytomous items. *Psychometrika*, in press. Exploratory statistical methods with applications in psychiatry. *Research in Psychiatry: Issues, Strategies and Methods*, in press. With Greenhouse.

Joseph B. Kadane, Leonard J. Savage Professor of Statistics and Social Science (joint appointment in Department of Social and Decision Sciences); Ph.D., Stanford, 1966. Statistical inference, econometrics, statistical methods in social sciences, sequential problems.
  Adjusting the 1980 census of housing and population. *J. Am. Stat. Assoc.*, 1989. With Ericksen and Tukey. A version was published by the House of Representatives, Committee on Post Office and Civil Service, Subcommittee on Census and Population, Record of the Hearing of August 17, 1987. Fully exponential Laplace's approximations to expectations and variances of non-positive functions. *J. Am. Stat. Assoc.*, 1989. With Tierney and Kass. A statistical analysis of adverse impact of employer decisions. *J. Am. Stat. Assoc.*, 1990. Randomization in a Bayesian perspective. *J. Stat. Plann. Inference*, 1990.

Robert E. Kass, Associate Professor of Statistics; Ph.D., Chicago, 1980. Bayesian inference, asymptotics, differential geometry in statistics.
  The geometry of asymptotic inference. *Stat. Sci.*, 1989. Asymptotics in Bayesian computation. *Bayesian Statistics 3*, 1988. With Kadane and Tierney. Approximate Bayesian inference in conditionally independent hierarchical models (parametric empirical Bayes models). *J. Am. Stat. Assoc.*, 1989. With Steffey.

John P. Lehoczky, Professor of Statistics and Head; Ph.D., Stanford, 1969. Applied stochastic processes with applications in real-time computer and communication systems, economics and finance, stochastic control theory, inference for stochastic processes.
  Statistical analysis of 24-hour blood pressure monitoring data. *Stat. Med.*, 1988. With Marler and Jacob. Existence and uniqueness of a multi-agent equilibrium in a stochastic, dynamic consumption/investment model. *Math. Operations Res.*, 1990. With Karatzas and Shreve. Superposition of independent renewal processes. *Adv. Appl. Probab.*, 1991. With Lam.

Michael M. Meyer, Adjunct Associate Professor of Statistics; Ph.D., Minnesota, 1981. Statistical computing, categorical data analysis.
  Statistical analysis of multiple sociometric relationships. *J. Am. Stat. Assoc.*, 1985. With Fienberg and Wasserman.

Mark J. Schervish, Professor of Statistics; Ph.D., Illinois, 1979. Statistical computing, foundations of statistics, multivariate analysis.
  A general method for comparing probability assessors. *Ann. Stat.*, 1989. Applications of parallel computation to statistical inference. *JASA*, 1988.

Teddy Seidenfeld, Professor of Philosophy and Statistics (primary appointment in Department of Philosophy); Ph.D., Columbia, 1976. Foundations of statistical inference and decision theory.
  Decision theory without "independence" or without "ordering," what is the difference? *Econ. Philos.*, 1988. An approach to consensus and certainty with increasing evidence. *J. Stat. Plann. Inference*, in press. With Schervish.

Norma Terrin, Assistant Professor of Statistics; Ph.D., Boston University, 1990. Limit theorems, long-range dependence, self-similar processes, time series.
  A noncentral limit theorem for quadratic forms of Gaussian stationary sequences. *J. Theor. Probab.*, 1990. With Taqqu. Convergence in distribution of sums of bivariate Appell polynomials with long-range dependence. *Probab. Theor. Related Fields*, in press. With Taqqu.

Isabella Verdinelli, Visiting Associate Professor; M.Sc., University College, London, 1976. Bayesian design of experiments, reliability and designs, outliers.
  Advances in Bayesian experimental design. Invited paper at the Fourth Valencia International Meeting on Bayesian Statistics, 1991. Bayesian designs for maximizing information and outcome. *JASA*, in press. With Kadane. Target attainment and experimental designs, a Bayesian approach. In *Optimal Design and Analysis of Experiments*, eds. Y. Dodge et al. New York: North Holland, 1988. With Wynn.

Larry A. Wasserman, Assistant Professor of Statistics; Ph.D., Toronto, 1988. Robust Bayesian inference, upper and lower probabilities.
  A robust Bayesian interpretation of likelihood regions. *Ann. Stat.*, 1989. Bayes' theorem for Choquet capacities. *Ann. Stat.*, 1990. With Kadane. Prior envelopes based on belief function. *Ann. Stat.*, 1990.

SECTION 5: MATHEMATICAL SCIENCES

COLORADO STATE UNIVERSITY

Department of Mathematics

Programs of Study

The Department of Mathematics offers programs leading to the degrees of M.S. and Ph.D.

The M.S. degree requires 33 semester credits and usually takes two years for a student on an assistantship. The student must write a master's paper or a thesis of original content. No foreign language is required. Programs with up to half of the credits in a related area of specialization, such as atmospheric science, computer science, economics, engineering, optimization, physics, or statistics, are also available with advisers doing research in these areas. There is a final oral examination in all cases.

The Ph.D. degree usually takes four or five years beyond the bachelor's degree. To enter the program, a student must have demonstrated a definite aptitude for mathematics (for example, by completing an M.S. degree in mathematics with a good record in course work). Requirements for the doctorate include a reading knowledge of one foreign language (chosen from French, German, and Russian) and a dissertation based on independent publishable research. A qualifying examination covering basic areas of mathematics is required; it is normally taken near the end of the student's first year at Colorado State University. The Ph.D. preliminary examination, administered by the student's Ph.D. committee, is a written and/or oral examination taken near the end of the course work and at the beginning of the work on the dissertation. All dissertations must be successfully defended in an open final oral examination.

All courses of study are planned individually by students and their advisers in consultation with the students' graduate committees; programs are designed to allow a broad interdisciplinary range.

Research Facilities

The library subscribes to about 275 research journals in mathematics and related areas. Students at Colorado State have easy access to two MicroVAX II minicomputers, a Sequent multiprocessor super minicomputer, a visualization lab, and several networks connected to most of the U.S. supercomputers. The University has an outstanding Department of Statistics, and the Department of Mathematics works closely with it.

Financial Aid

A number of assistantships are available for qualified graduate students.

In 1991–92, stipends for teaching assistantships provide $9600–$10,600 for nine months. In addition, some summer assistantships are available for continuing students. The typical graduate assistant teaches a single small section of first- or second-semester calculus, though other assignments are available. The department currently provides supplemental support for between 10 and 15 graduate students each summer. The level of stipend depends on the qualifications and seniority of the student and the type of work assigned.

Research assistantships are occasionally available for advanced students to work with professors on research grants. Stipends are comparable to those for teaching assistantships.

Cost of Study

Teaching assistantships include payment of tuition on behalf of the student. Tuition payments for research assistantships must be negotiated with the principal investigator. Fees are paid by all students and totaled $244.40 per semester for the 1990–91 academic year. For those not on appointments, tuition and fees in 1990–91 for the nine-month academic year were $2486.80 for in-state and $6850.80 for out-of-state students. (The State Board of Agriculture reserves the right to change the schedule of tuition and fees at any time. It is anticipated that the costs for 1991–92 will increase by 5 percent.)

Cost of Living

In 1990–91, the cost for room and board, books and classroom supplies, incidental personal expenses, and health insurance was about $7245 for nine months. Furnished housing for married students was available on campus at $342–$368 per month, which included utilities, telephone, and cable. Single graduate housing is also available. For more information students should call the Office of Housing and Food Services at 491-6511.

Student Group

Colorado State University has an enrollment of approximately 20,000 regular on-campus students, including approximately 3,000 graduate students. The Department of Mathematics has about 55 graduate students.

Location

Fort Collins, a city of about 90,000, is located on the eastern slope of the Rocky Mountains, 65 miles north of Denver and about 40 miles from Rocky Mountain National Park. The mountains offer excellent opportunities for outdoor recreation, such as hiking, fishing, skiing, hunting, and camping. The climate, with clear air, an abundance of sunny days, and cool summer nights, is one of the most attractive features of the area. Fort Collins and the University offer a wide variety of cultural and other activities.

The University

The University was founded in 1870 as the Agricultural College of Colorado. In 1957, it became Colorado State University, and it has grown rapidly in recent years. At present, one emphasis of the University is the development of upper-level and graduate programs while maintaining the first- and second-year undergraduate enrollment at a stable level.

Applying

Completed applications are accepted anytime, but it is best if they arrive in the Department of Mathematics by April 15. A $30 nonrefundable application fee must accompany the admission application. Interested students should write to the department for application forms.

Correspondence and Information

Janice Lucero
Secretary to the Graduate Director
Department of Mathematics
Colorado State University
Fort Collins, Colorado 80523
Telephone: 303-491-7925
Fax: 303-491-2161

Peterson's Guide to Graduate Programs in the Physical Sciences and Mathematics 1992

SECTION 5: MATHEMATICAL SCIENCES

Colorado State University

THE FACULTY AND THEIR RESEARCH

Eugene Allgower, Professor; Ph.D., IIT, 1964. Numerical analysis, nonlinear systems.
W. E. Brumley, Associate Professor; Ph.D., Purdue, 1967. Differential equations, numerical analysis.
Duane J. Clow, Associate Professor; Ph.D., Colorado State, 1968. Biomathematics.
Richard B. Darst, Professor; Ph.D., LSU, 1960. Measure theory, integration theory, probability, real and functional analysis.
E. R. Deal, Associate Professor; Ph.D., Michigan, 1962. Functional analysis.
Frank R. DeMeyer, Professor; Ph.D., Oregon, 1965. Ring theory, algebraic geometry.
Paul C. DuChateau, Professor; Ph.D., Purdue, 1970. Applied mathematics, partial differential equations.
Jeanne Duflot, Associate Professor; Ph.D., MIT, 1980. Algebraic topology.
H. Howard Frisinger, Professor; Dr.Ed. (mathematics), Michigan, 1964. History of science, geometry.
Robert Gaines, Professor; Ph.D., Colorado, 1967. Ordinary differential equations, mathematical models of economic growth, optimal control.
Kurt Georg, Associate Professor; Ph.D., Bonn, 1968. Numerical analysis.
Darel Hardy, Associate Professor; Ph.D., New Mexico State, 1967. Algebra, semigroups.
Michael Kirby, Assistant Professor; Ph.D., Brown, 1988. Applied mathematics.
Kenneth F. Klopfenstein, Associate Professor; Ph.D., Purdue, 1967. Hilbert spaces of analytic functions, operator theory.
Nicholas Krier, Associate Professor; Ph.D., Ohio State, 1969. Geometry, combinatorics.
Steven Landsburg, Associate Professor; Ph.D., Chicago, 1979. Algebraic K-theory, commutative algebra.
Bernard Levinger, Associate Professor; Ph.D., NYU, 1960. Matrix theory, numerical analysis, group theory.
Robert A. Liebler, Professor; Ph.D., Michigan, 1970. Finite-group theory, geometry, combinatorics.
John Locker, Professor; Ph.D., Michigan, 1965. Functional analysis of differential equations.
Arne Magnus, Professor; Ph.D., Washington (St. Louis), 1953. Complex analysis, continued fractions, rational approximation.
Bennet Manvel, Professor; Ph.D., Michigan, 1970. Graph theory, combinatorics.
Kelly McArthur, Assistant Professor; Ph.D., Montana State, 1987. Numerical analysis.
Rick Miranda, Associate Professor; Ph.D., MIT, 1979. Algebraic geometry.
Richard P. Osborne, Professor; Ph.D., Michigan State, 1965. Combinatorial group theory, recursive functions, geometric topology.
Richard Painter, Professor; Ph.D., North Carolina at Chapel Hill, 1963. Algebra, linear algebra.
Aubrey B. Poore, Professor; Ph.D., Caltech, 1972. Applied mathematics, numerical analysis, optimization and control.
P. M. Prenter, Professor; Ph.D., Oregon State, 1965. Variational methods, numerical analysis, operator approximations.
Jaya N. Srivastava, Professor; Ph.D., North Carolina at Chapel Hill, 1962. Combinatorial mathematics, matrix theory, statistics.
Gerald D. Taylor, Professor; Ph.D., Michigan, 1965. Approximation theory, numerical analysis.
James W. Thomas, Professor; Ph.D., Arizona, 1967. Applied mathematics, functional analysis, bifurcation theory, nonlinear water waves, numerical simulation of fluid flows.
Jaak Vilms, Associate Professor; Ph.D., Columbia, 1967. Differential geometry, topology, tensor analysis.
David W. Zachmann, Professor; Ph.D., Arizona, 1970. Applied mathematics, soil physics.

Skiing facilities can be found within an hour of Colorado State's campus.

SECTION 5: MATHEMATICAL SCIENCES

COLUMBIA UNIVERSITY

Graduate School of Arts and Sciences
Department of Mathematics

Programs of Study

Graduate training in mathematics at Columbia University is an intensive course of study leading to the Ph.D. degree. It is designed for the full-time student planning a career in research and teaching at the university level or in basic research in a nonacademic setting.

Admission is limited and selective. Applicants should present an undergraduate major in mathematics from a college with strong mathematics offerings. A reading knowledge of two languages (French, German, or Russian) is recommended but not required at the time of admission. Graduating seniors are given preference. The department does not admit part-time students or those who have the master's degree as their principal objective.

In the first year, students must pass written qualifying examinations in the areas of the basic first-year courses—analysis, complex variables, algebra, and differential manifolds and topology. Most of the formal course work is completed in the second year, when an oral examination in two selected topics must be passed. Also required is a reading knowledge of two languages, chosen from French, German, and Russian. The third and fourth years are devoted to seminars and the preparation of a dissertation.

An important part of the training of the graduate student is participation in the department's undergraduate educational work. This usually takes the form of assistant teaching during six semesters of the four-year period.

There are allied graduate programs available in mathematical statistics and in computer science.

Research Facilities

The mathematics and statistics departments are housed in a comfortable building containing an excellent Mathematics Library, a lounge for tea and conversation, and numerous seminar and lecture rooms.

A wide range of computing facilities is available on campus. In addition, the mathematics department has a computer available exclusively to mathematics faculty and graduate students.

Financial Aid

The department has a broad fellowship program designed to enable qualified students to achieve the Ph.D. degree in the shortest practicable time. Each student admitted to this program is appointed a fellow in the Department of Mathematics for the duration of his or her doctoral candidacy, up to a total of four years. A fellow receives a stipend of $9945 for the 1991–92 nine-month academic year and is exempt from payment of tuition and fees.

A fellow in the Department of Mathematics may hold a fellowship from a source outside Columbia University. When not prohibited by the terms of the outside fellowship, the University supplements the outside stipend to bring it up to the level of the University fellowship. Candidates for admission are urged to apply for fellowships for which they are eligible (National Science Foundation, New York State Regents, etc.).

Cost of Study

All students admitted become fellows in the department and are exempt from fees, as explained above.

Cost of Living

Students in the program have managed to live comfortably in the University neighborhood on their fellowship stipends.

Student Group

The graduate program in mathematics usually has an enrollment of 40–50 students. Normally, 8–12 students enter each year. While students come from all over the world, they have always been socially as well as scientifically cohesive and mutually supportive.

Location

New York City is America's major center of culture. Columbia University's remarkably pleasant and sheltered campus, near the Hudson River and Riverside Park, is situated within 20 minutes of Lincoln Center, Broadway theaters, Greenwich Village, and major museums. Most department members live within a short walk of the University.

The University

Since receiving its charter from King George II in 1754, Columbia University has played an eminent role in American education. In addition to its various faculties and professional schools (such as Engineering, Law, and Medicine), the University has close ties with nearby museums, schools of music and theology, the United Nations, and the city government.

Applying

The application deadline is mid-January; however, applicants of unusual merit are considered beyond the application deadline. Applicants who expect to be in the New York vicinity are encouraged to arrange a department visit and interview.

Correspondence and Information

For information on the department and program:
Chairman
Department of Mathematics
Columbia University
New York, New York 10027
Telephone: 212-854-4112

For applications:
Graduate School of Arts and Sciences
Office of Student Affairs
107 Low Memorial Library
Columbia University
New York, New York 10027
Telephone: 212-854-3808

SECTION 5: MATHEMATICAL SCIENCES

Columbia University

THE FACULTY AND THEIR RESEARCH

Hyman Bass, Professor; Ph.D., Chicago, 1959. Algebraic K theory, group theory.
David A. Bayer, Ritt Assistant Professor; Ph.D., Harvard, 1982. Algebraic geometry.
Joan Birman, Professor; Ph.D., NYU, 1968. Low-dimensional topology, knot theory.
Huai-Dong Cao, Ritt Assistant Professor; Ph.D., Princeton, 1986. Differential geometry, topology, nonlinear partial differential equations.
Fred Diamond, Ritt Assistant Professor; Ph.D., Princeton, 1988. Number theory.
F. Thomas Farrell, Professor; Ph.D., Yale, 1967. Algebraic topology.
Benji Fisher, Ritt Assistant Professor; Ph.D., Princeton, 1990. Number theory.
Sidney Frankel, Ritt Assistant Professor; Ph.D., Stanford, 1986. Differential geometry, several complex variables, mathematical physics, partial differential equations.
Robert Friedman, Professor; Ph.D., Harvard, 1981. Algebraic geometry.
Patrick X. Gallagher, Professor and Chairman; Ph.D., Princeton, 1959. Analytic number theory, group theory.
Dorian Goldfeld, Professor; Ph.D., Columbia, 1969. Number theory.
Hervé Jacquet, Professor; Dr.Sci.Math., Paris, 1967. Representation theory, automorphic functions.
Troels Jorgensen, Professor; Cand.Scient., Copenhagen, 1970. Hyperbolic geometry, complex analysis.
Masatake Kuranishi, Professor; Ph.D., Nagoya (Japan), 1952. Partial differential equations.
Xiao-Song Lin, Ritt Assistant Professor; Ph.D., California, San Diego, 1988. Topology.
Yuri I. Manin, Eilenberg Visiting Professor; Ph.D., Steklov Mathematics Institute (Moscow), 1963. Number theory, mathematical physics, algebraic geometry.
Boris Moishezon, Professor; Ph.D., USSR Academy of Sciences, 1962. Algebraic geometry.
Ngaiming Mok, Professor; Ph.D., Stanford, 1980. Differential geometry.
John W. Morgan, Professor; Ph.D., Rice, 1969. Geometric topology, manifold theory.
Joseph Oesterle, Visiting Professor; Ph.D., Paris XI (South), 1984. Number theory.
Duong H. Phong, Professor; Ph.D., Princeton, 1977. Analysis.
Jonathan Pila, Ritt Assistant Professor; Ph.D., Stanford, 1988. Computational number theory.
Henry C. Pinkham, Professor; Ph.D., Harvard, 1974. Algebraic geometry.
David Rana, Ritt Assistant Professor; Ph.D., Princeton, 1987. Mathematical physics.
Daniel Rockmore, Ritt Assistant Professor; Ph.D., Harvard, 1989. Efficient computer algorithms, group representations.
Xiaochun Rong, Ritt Assistant Professor; Ph.D., SUNY at Stony Brook, 1990. Differential geometry.
Roberto Silvotti, Ritt Assistant Professor; Ph.D., Zurich, 1990. Mathematical physics.
Ki-Seng Tan, Ritt Assistant Professor; Ph.D., Harvard, 1990. Group representation theory.
Andrew Winkler, Ritt Assistant Professor; Ph.D., NYU (Courant), 1987. Automorphic forms.
Peter Woit, Ritt Assistant Professor; Ph.D., Princeton, 1985. Mathematical physics, topology.
Siye Wu, Ritt Assistant Professor; Ph.D., MIT, 1990. Mathematical physics.
George Zettler, Ritt Assistant Professor; Ph.D., Maryland, 1983. Analysis, algebraic geometry.

SECTION 5: MATHEMATICAL SCIENCES

CORNELL UNIVERSITY
Field of Mathematics

Program of Study

The graduate program in the Field of Mathematics at Cornell leads to the Ph.D. degree. It takes most students about five years of graduate study to complete the program. Normally only students applying for candidacy for the Ph.D. degree are considered.

One of the attractive features of the program is the broad range of interests of the faculty. Besides researchers in the usual areas of algebra, analysis, geometry, and topology, the Field has outstanding groups of people in the areas of mathematic logic, partial differential equations, and probability and statistics. The Field also maintains close ties with distinguished sister fields and graduate programs in the Fields of Applied Mathematics, Computer Science, Operations Research, and Statistics. Interdisciplinary study flourishes at Cornell.

Students must pass a test of reading ability in mathematical German or Russian. Reading familiarity with mathematical French is taken for granted. Cornell provides ample facilities for language study.

Research Facilities

The total number of volumes in the Cornell library system is nearly 5 million. The Mathematics Library, housed in the Field of Mathematics, contains all the important mathematical journals and a very wide selection of books on mathematics and related subjects.

The Field of Mathematics maintains its own minicomputer facility of networked Sun Workstations with gateways to campuswide, national, and international networks.

Financial Aid

Everyone who applies for admission to the Graduate School is automatically considered for financial assistance from Cornell. For many years the Field has been able to provide financial support, through a teaching assistantship, fellowship, or graduate research assistantship, to every graduate student who is making satisfactory progess toward the Ph.D. degree, and it expects to continue this practice. Assistantship stipends range between $9070 and $9370 in 1991–92. It is usually possible to supplement a University fellowship by working as a part-time teaching fellow. Fellowships paying full tuition are normally also provided.

Cost of Study

Tuition for the 1991–92 academic year is $16,170. After six semesters in residence, tuition is reduced to approximately $4500 per academic year if the student has passed the admission-to-candidacy examination and completed all course requirements.

Cost of Living

For the 1991–92 academic year, graduate student dormitories provide rooms for unmarried students at $3484 for a regular single and $3786 for a large single. Student families usually find accommodations in one of the University's three apartment complexes at approximately $407 per month on a twelve-month contract.

Student Group

The Cornell Graduate School has an enrollment of around 4,450 students, and the graduate faculty has about 1,600 members. Cornell differs from many other graduate schools in that approximately 98 percent of the students are full-time degree candidates, with the majority in programs leading to the Ph.D. degree. There are at present approximately 60 graduate students from many parts of the world in the Field of Mathematics. Since the Field has over 40 faculty members, it is possible to pay a great deal of attention to the individual needs of each student.

Location

Ithaca, New York, is a small town in the heart of the Finger Lakes region. It offers the cultural activities of a large university and the diversions of a rural environment. Facilities for skiing, sailing, camping, hiking, soaring, and other sports are close at hand.

The University

Graduate study at Cornell is an exceptional experience. The University has managed to foster excellence in research without forsaking the ideals of liberal education. In many ways, the cohesiveness and rigor of the mathematics graduate program are a reflection of the Cornell tradition. The campus is widely acknowledged as one of the most beautiful in the country. Cornell's size and diversity guarantee a rich cultural life.

Applying

To be considered for admission to the graduate Field of Mathematics, a student must have completed the work for an undergraduate major in mathematics. That work should have included a rigorous course in advanced calculus and real variable theory that will serve as an introduction to measure theory. Students should also have some familiarity with applications of advanced calculus and should have had courses in linear algebra and modern abstract algebra at an advanced level.

There is no deadline for the receipt of applications for admission in the fall term; however, students wanting to be considered for fellowships must have their application completed by January 10. All applicants should take the GRE General Test and Subject Test in mathematics. International students for whom English is not the native language must submit TOEFL scores.

Correspondence and Information

For program information:
Graduate Faculty Representative
Field of Mathematics
White Hall
Cornell University
Ithaca, New York 14853-7901
Telephone: 607-255-6757

For applications and financial aid forms:
Dean
Graduate School
Sage Graduate Center
Cornell University
Ithaca, New York 14853-6201
Telephone: 607-255-4884

Peterson's Guide to Graduate Programs in the Physical Sciences and Mathematics 1992

SECTION 5: MATHEMATICAL SCIENCES

Cornell University

THE FACULTY AND THEIR RESEARCH

Dan Barbasch, Professor. Lie theory.
Israel Berstein, Professor. Algebraic and differential topology.
Louis Billera, Professor. Algebraic methods for combinatorial problems arising in geometry.
James H. Bramble, Professor. Numerical solutions of partial differential equations.
Kenneth S. Brown, Professor. Algebra, topology, group theory.
Lawrence D. Brown, Professor. Statistics.
Stephen U. Chase, Professor. Algebra, algebraic number theory, homological algebra.
Marshall M. Cohen, Professor. Topology, geometric (combinational) group theory.
Robert Connelly, Professor. Geometry, rigidity, topology.
R. Keith Dennis, Professor and Chairman. Commutative and noncommutative algebra, algebraic K-theory.
Richard Durrett, Professor. Probability.
Eugene B. Dynkin, Professor. Probability theory.
Clifford J. Earle, Professor. Complex variables, Teichmuller spaces.
Roger H. Farrell, Professor. Mathematical statistics, measure theory.
Leonard Gross, Professor. Functional analysis, constructive quantum field theory.
John Guckenheimer, Professor. Dynamical systems.
Allen Hatcher, Professor. Geometric topology.
David W. Henderson, Professor. Geometry, geometric topology, mathematics education.
Philip J. Holmes, Professor. Nonlinear mechanics, dynamical systems, bifurcation theory.
John H. Hubbard, Professor. Analysis, differential equations, differential geometry.
Jiunn T. Hwang, Professor. Statistics, confidence set theory.
Peter J. Kahn, Professor. Algebraic and differential topology, cobordism and homotopy type of manifolds.
Harry Kesten, Professor. Probability theory, limit theorems, percolation theory.
Dexter Kozen, Professor. Theory of computation, logic and semantics of programs.
Richard Liu, Assistant Professor. Statistics.
G. Roger Livesay, Professor. Differential topology, group actions.
Michael D. Morley, Professor and Associate Chairman. Mathematical logic, model theory.
Anil Nerode, Professor and Director of the Mathematical Sciences Institute. Mathematical logic, recursive functions, theoretical computer science.
Lawrence E. Payne, Professor. Partial differential equations, ill-posed and nonstandard problems.
Richard Platek, Associate Professor. Mathematical logic, recursion theory, set theory, computer science.
James Renegar, Associate Professor. Interior methods for linear programming, design of algorithms.
Oscar S. Rothaus, Professor. Several complex variables, combinatorics.
Alfred H. Schatz, Professor. Numerical solutions of partial differential equations.
Shankar Sen, Professor. Algebraic number theory.
Richard D. Shore, Professor. Mathematical logic, recursion theory, set theory.
John Smillie, Associate Professor. Dynamical systems.
Birgit E. Speh, Professor. Lie groups.
Frank L. Spitzer, Professor. Probability theory.
Michael E. Stillman, Assistant Professor. Algebraic geometry, computational algebra.
Robert S. Strichartz, Professor. Harmonic analysis, partial differential equations.
Bernd Sturmfels, Assistant Professor. Computational algebra, discrete geometry, combinatorics.
Moss E. Sweedler, Professor. Algebra, computer algebra, phenomena, algebraic cohomologies.
Karen Vogtmann, Associate Professor. Topology, cohomology of groups.
Lars B. Wahlbin, Professor. Numerical solutions of partial differential equations.
James E. West, Professor. Geometric topology, infinite-dimensional topology.

SECTION 5: MATHEMATICAL SCIENCES

DARTMOUTH COLLEGE

Department of Mathematics and Computer Science

Program of Study

The Dartmouth Ph.D. program in mathematics is designed to develop mathematicians highly qualified for both teaching and research at the college or university level or for research in the mathematical sciences in industry or government. Students earn a master's degree as part of becoming a candidate for the Ph.D. degree but should not apply to study only for a master's degree.

During the first six terms (eighteen months) of residence, the student develops a strong basic knowledge of algebra, analysis, topology, and a fourth area of mathematics chosen by the student. Areas recently chosen for this fourth area include combinatorics, computer science, geometry, logic, number theory, probability, and statistics. Rather than using traditional qualifying exams, the department requires that 2 faculty members certify that the student knows the material on the departmental syllabus in each of the four areas. This certification may be based on a formal oral exam, course work, informal discussions, supervised independent study, seminar presentations, informal oral exams, or any means that seems appropriate. Students and faculty usually find a formal oral exam to be the most efficient route to certification.

After completion of at least eight graduate courses and certification, students are awarded the master's degree and, subject to departmental approval, are admitted to candidacy for the Ph.D. degree. This normally occurs during the second year of graduate study. After admission to candidacy, the student chooses a thesis adviser and thesis area and begins in-depth study of the chosen area. Normally, the thesis is completed during the fourth year of graduate study, although it is not uncommon to use a fifth year. The typical thesis consists of publishable original work. Areas recently chosen for thesis research include algebra, analysis, applied mathematics, combinatorics, computer science, geometry, number theory, set theory, and topology. Students continue taking courses according to their interests and demonstrate competence in two foreign languages while doing their thesis research.

Dartmouth is committed to helping its graduate students develop as teachers by providing examples of effective teaching in the graduate courses, by instruction in the graduate teaching seminar, and by provision of carefully chosen opportunities to gain real teaching experience. These opportunities begin as tutorial or discussion leader positions for courses taught by senior faculty. They culminate in the third and fourth years after completion of the teaching seminar in the opportunity to teach one course for one term each year. The first of these courses is a section of a multisection course supervised by a senior faculty member, and the second of these courses is chosen to fit the interests and needs of the students and the deparment.

Research Facilities

In Bradley Hall, the department has well-equipped offices and a lounge for graduate students, a research library with an outstanding collection of journals and advanced books, faculty offices, classrooms, lounges, and computer facilities. The campus computing network provides connections to VAX (UNIX and VMS), IBM, Honeywell, and other computers including a superminicomputer, all on campus, and to supercomputer facilities off campus. The department has its own computer laboratories, where it has a network of Sun and Xerox workstations for research and a network of Macintosh computers for research and teaching. Other computers available for research include VAX computers operating under UNIX and VMS, IBM mainframe supercomputers, and numerous microcomputers. In addition, students may purchase one Macintosh much cheaper and other personal computers at a discount through the College.

Financial Aid

Students admitted to the program receive a full tuition scholarship and the Dartmouth College teaching fellowship, of which the stipend in 1991–92 is $971 per month. This stipend continues for twelve months per year through the fourth year of graduate study and may be renewable for a fifth year as well. Duties involve 8–12 hours per week for two terms in each of the first two years and about 20 hours per week for one term in each of the last two years.

Cost of Study

With the exception of textbooks, all costs of study are covered by the scholarship.

Cost of Living

Students find that $971 per month suffices comfortably for living in College housing, renting local apartments, or sharing a rented house with other students. A married student whose spouse does not work or hold a similar fellowship can maintain a Spartan life in College-owned married student housing.

Student Group

Dartmouth attracts and admits students from colleges and universities of all types. About 25 percent are women, the percentage of married students has varied from 5 to 35, 1 or 2 often are not recent graduates, and 1 or 2 often are not from North America. The department has 22 graduate students, and it offers an effective placement program for its Ph.D. graduates. Recipients of the Ph.D. degree from Dartmouth have found employment at a broad cross section of academic institutions, including Ivy League institutions, major state universities, and outstanding four-year liberal arts colleges. Some are now working as research mathematicians in industry and government as well.

Location

Dartmouth is in a small town that has an unusual metropolitan flavor. There are adequate shopping facilities but no large cities nearby. Hiking, boating, fishing, swimming, ski touring, and Alpine skiing are all available in the immediate area. A car is a pleasant luxury, but many students find it unnecessary.

The College

Dartmouth has about 4,000 undergraduate students, who are among the most talented and best motivated in the nation. There are under 1,000 graduate students in the College (arts and sciences faculty) and in associated professional schools in engineering, medicine, and business. With a faculty–graduate student ratio higher than 1:1, the department is a friendly place where student-faculty interaction in both academic and social activities is encouraged.

Applying

Application forms are available from the department. Applicants should send to the address below a completed application form, an undergraduate transcript, and three letters of recommendation that describe their mathematical background and ability, estimate their potential as teachers, and compare them with a peer group of the recommender's choice. Applicants must take both the General Test and Subject Test of the Graduate Record Examinations and have the scores sent to the department. All sections of the TOEFL, including the Test of Spoken English, are required of applicants whose native language is not English. Applicants whose files are complete by March 1 receive first consideration.

Correspondence and Information

Graduate Admissions Committee Chair
Ph.D. Program in Mathematics
Department of Mathematics and Computer Science
Bradley Hall
Dartmouth College
Hanover, New Hampshire 03755
Telephone: 603-646-2415

SECTION 5: MATHEMATICAL SCIENCES

Dartmouth College

THE FACULTY AND THEIR RESEARCH

Professors

Martin Arkowitz, Ph.D., Cornell, 1960. Algebraic topology and differential geometry. Provides thesis supervision in these areas.

James E. Baumgartner, Ph.D., Berkeley, 1970. Set theory, general topology, and mathematical logic. Currently working with combinatorial set theory, theory of forcing, reflection principles obtained from large cardinals, and consistency questions in general topology. Provides thesis supervision in all areas listed above.

Thomas Bickel, Ph.D., Michigan, 1965. Finite groups, permutation groups, and representation theory. Provides thesis supervision in group theory and related topics in algebra and combinatorial mathematics.

Kenneth P. Bogart, Department Chair; Ph.D., Caltech, 1968. Combinatorial mathematics and algebra and their applications. Current research in ordered sets, the theory of generating functions, algebraic coding theory, and database design. Provides thesis supervision in algebra, combinatorics, and their applications.

Edward M. Brown, Ph.D., MIT, 1963. Topology and hyperbolic geometry of three-dimensional manifolds; knot theory, current research in topological and geometric classification of noncompact three-manifolds. Provides thesis supervision in areas above.

Richard H. Crowell, Ph.D., Princeton, 1955. Topology and algebra related to knot theory. Recent work in congruences and imbeddings of groupoids and on covering groups (generalization of derived module of a homomorphism, for example, the Alexander module of knot theory). Provides thesis supervision in related areas of algebra and topology.

Robert L. (Scot) Drysdale III, Ph.D. (computer science), Stanford, 1979. Design and analysis of algorithms. Current interest in computational geometry, algorithms, graph theoretic algorithms, and sorting networks. Provides thesis supervision in algorithms.

Carolyn Gordon, Ph.D., Washington (St. Louis), 1979. Geometry. Provides thesis supervision in differential geometry.

Donald B. Johnson, Ph.D. (computer science), Cornell, 1973. Algorithms and complexity. Provides thesis supervision in computer science.

Donald L. Kreider, Ph.D., MIT, 1959. Logic and computer science. Provides thesis supervision in logic and related areas.

Thomas E. Kurtz, Ph.D. (statistics), Princeton, 1956. Statistics and computer science (especially programming languages and software design). Current interests include development of new American National Standard for the BASIC programming language.

Charles Dwight Lahr, Ph.D., Syracuse, 1971. Analysis, especially functional analysis. Provides thesis supervision in functional analysis.

John Lamperti, Ph.D., Caltech, 1957. Theory of probability, especially stochastic processes. Current interests in statistical theory and practice and the social implications of science and technology. Provides thesis supervision in areas above.

Fillia Makedon, Ph.D. (computer science), Northwestern, 1982. Parallel algorithms, VLSI theory.

Robert Z. Norman, Ph.D., Michigan, 1954; Miller Prize, 1954. Combinatorial mathematics and applications of mathematics in political science, psychology, and sociology. Current work in combinatorial problems in Diophantine equations and in measurement and decision processes in politics and sociology. Provides thesis supervision in combinatorial mathematics and its applications.

Reese T. Prosser, Ph.D., Berkeley, 1955. Analysis, geometry, and physics. Provides thesis supervision in analysis and its applications in geometry and physics.

William E. Slesnick, M.A., Oxford, 1954. Mathematical education.

Ernst Snapper, Ph.D., Princeton, 1941; National Science Foundation Postdoctoral Fellow, Harvard, 1953–54; Allendorfer Prize, 1980. Algebra, geometry, and the philosophy of mathematics. Provides thesis supervision in all these areas.

J. Laurie Snell, Ph.D., Illinois, 1954. Probability theory and its applications, interaction of probability and computing. Current interest in the relation between electrical network theory and Markov chains. Provides thesis supervision in probability theory and its applications.

Richard E. Williamson, Ph.D., Pennsylvania, 1955. Classical analysis, especially integral transforms, potential theory, and control theory. Provides thesis supervision in these areas.

Associate Professors

Dorothy W. Andreoli, Ph.D., California, San Diego, 1982. Number theory, especially analytic number theory. Provides thesis supervision in number theory.

Marcia Groszek, Ph.D., Harvard, 1981. Logic. Provides thesis supervision in logic.

Thomas R. Shemanske, Ph.D., Rochester, 1979. Number theory and modular forms. Currently interested in Hilbert/Siegel modular forms and theta series. Provides thesis supervision in number theory and related areas of mathematics.

David L. Webb, Ph.D., Cornell, 1983. Algebraic K theory. Provides thesis supervision in algebra.

Dana P. Williams, Ph.D., Berkeley, 1979. Analysis. Provides thesis supervision in analysis.

Assistant Professors

Samuel W. Bent, Ph.D. (computer science), Stanford, 1982. Design and analysis of algorithms. Provides thesis supervision in computer science.

Matthew A. Bishop, Ph.D. (computer science), Purdue, 1984. Computer security. Provides thesis supervision in computer science.

James R. Driscoll, Ph.D. (computer science), Carnegie Mellon, 1986. Permutation groups, graph embeddings, persistent data structures, geometric algorithms. Provides thesis supervision in computer science.

Dennis Healy Jr., Ph.D., California, San Diego, 1986. Analysis. Provides thesis supervision in analysis.

Daniel Rockmore, Ph.D., Harvard, 1989. Analysis and representation theory.

Peter Sandon, Ph.D. (computer science), Wisconsin–Madison, 1987. Computer vision, machine learning, fine-grained parallel networks. Provides thesis supervision in computer science.

Clifford J. Walinsky, Ph.D. (computer science), Oregon Graduate Center, 1987. Programming language theory. Provides thesis supervision in computer science.

John Wesley Young Research Instructors

The JWY Research Instructorship is a two-year visiting position; the people involved and their fields thus change from year to year.

James Cummings, Ph.D., Cambridge, 1988. Logic and set theory.

Garth Isaak, Ph.D., Rutgers, 1990. Combinatorics and operations research.

Marcelo E. Laca, Ph.D., Berkeley, 1989. Analysis.

Timothy E. Olson, Ph.D., Auburn, 1991. Analysis, approximation theory.

Visiting and Adjunct Faculty

Richard E. Brown, Lecturer in Computer Science and Manager of Special Projects and Computing Services, Kiewit Computation Center; M.S., Pennsylvania, 1980.

Denis Devlin, Adjunct Assistant Professor of Mathematics and Computer Science and Director of Computing, Department of Mathematics and Computer Science; Ph.D., Dartmouth, 1980.

John Finn, Lecturer in Mathematics and Computer Science; M.A., Dartmouth, 1978.

Anne Schwartz, Visiting Assistant Professor of Mathematics; Ph.D., California, San Diego, 1989.

SECTION 5: MATHEMATICAL SCIENCES

DREXEL UNIVERSITY

College of Arts and Sciences
Chemistry, Mathematics and Computer Science, and Physics and Atmospheric Science Programs

Programs of Study

Drexel University's College of Arts and Sciences offers distinctive graduate programs in chemistry, mathematics and computer science, and physics and atmospheric science. Studies in all three areas lead to M.S. and Ph.D. degrees. Both full-time and part-time programs are available; full-time students can usually complete the M.S. in two years.

Advanced study in the Department of Chemistry can be conducted in analytical, inorganic, organic, physical, or polymer chemistry. Research areas within these major branches include atmospheric, biological, environmental, organic, mechanistic, theoretical, and synthetic chemistry.

In the Department of Mathematics and Computer Science, the M.S. and the Ph.D. are available in applied mathematics, probability and statistics, and computer science. Courses in mathematics and statistics emphasize applications, computational mathematics, and theory. Computer algebra systems and parallel or distributed computing are emphasized in computer science. The Ph.D. is awarded for academic achievement and the proven ability to carry out independent research, rather than for the completion of a prescribed course of study. The equivalent of four or five years beyond the B.S. degree is usually necessary for the Ph.D.; candidates must write and defend an original dissertation.

In the Department of Physics and Atmospheric Science, the requirement for the M.S. degree is 45 quarter credits in an approved program; there is no thesis or foreign language requirement. The Ph.D. degree requires 90 quarter credits beyond the bachelor's degree. The successful Ph.D. candidate must pass a written and oral candidacy examination, satisfy a residence requirement, perform original research, and write and defend a thesis. The Ph.D. degree must be completed within seven years.

In addition to the programs outlined here, the College of Arts and Sciences offers studies in bioscience and biotechnology, environmental science, and nutrition science/food science, as well as teacher preparation in the sciences and mathematics.

Research Facilities

The Drexel library houses over 473,000 volumes, including more than 120,000 in the science-technology section. Within each department, state-of-the-art equipment and laboratories are available for student use, including a broad range of instrumentation in the chemistry department, numerous computing resources in mathematics and computer science, and extensive facilities for laboratory research in all aspects of physics and atmospheric science.

Financial Aid

Teaching and research assistantships are available in chemistry, mathematics, and physics; job requirements vary according to department. Stipends range from $8000 to $11,000 per calendar year, and many assistantships include tuition remission. Information on general loan programs may be obtained from the Financial Aid Office.

Cost of Study

In 1991–92, tuition is $395 per credit hour. The general University fee is $81 per term for full-time students and $43 per term for part-time students.

Cost of Living

For a full calendar year (four terms), living expenses for a single student, excluding tuition, are estimated at $9000. This sum includes housing, meals, books, travel, and miscellaneous expenses.

Student Group

The University has a total enrollment of about 12,400 students, including 2,900 at the graduate level. The Department of Chemistry enrolls 50 to 60 graduate students, and the Department of Mathematics and Computer Science and the Department of Physics and Atmospheric Science each enroll between 40 and 50.

Location

As a part of the University City area of west Philadelphia, Drexel is conveniently located within minutes of downtown Philadelphia, a great cultural, educational, and industrial center. New York City and Washington, D.C., are easily reached by train, bus, or car.

The University

Founded in 1891, Drexel University is a private institution offering undergraduate and graduate programs in business and administration, design arts, engineering, humanities and social sciences, information studies, and science. The University operates on a year-round calendar, with four terms per year.

Applying

Graduate students may apply with the intention of enrolling in any of Drexel's four terms (these begin in January, March, June, and September; application deadlines vary accordingly). Transcripts and letters of recommendation are required. The GRE General Test is recommended for all applicants to the College of Science. For assistantship consideration, students must submit their application by February 1.

Correspondence and Information

Keith T. Brooks, Director
Office of Graduate Admissions
Drexel University
Philadelphia, Pennsylvania 19104
Telephone: 215-895-6700

Peterson's Guide to Graduate Programs in the Physical Sciences and Mathematics 1992

SECTION 5: MATHEMATICAL SCIENCES

Drexel University

THE TENURED FACULTY AND THEIR RESEARCH

Department of Chemistry
Anthony W. Addison, Professor; Ph.D., Kent (England), 1971: inorganic chemistry. Alan R. Bandy, Professor; Ph.D., Florida, 1968: atmospheric and analytical chemistry. Franklin A. Davis, George S. Sasin Professor of Organic Chemistry; Ph.D., Syracuse, 1966: organic chemistry. James P. Friend, Hanson Professor of Atmospheric Chemistry; Ph.D., Columbia, 1956: atmospheric chemistry. Robert O. Hutchins, Professor and Department Head; Ph.D., Purdue, 1967: organic chemistry. Jack G. Kay, Professor; Ph.D., Kansas, 1960: physical and inorganic chemistry. Frederick R. Longo, Professor; Ph.D., Pennsylvania, 1962: physical chemistry. Amar Nath, Professor; Ph.D., Moscow, 1961: physical chemistry. Kevin Owens, Assistant Professor; Ph.D., Indiana, 1989: analytical chemistry. Louis L. Pytlewski, Professor; Ph.D., Pennsylvania, 1960: inorganic chemistry. Carey M. Rosenthal, Associate Professor; Ph.D., Harvard. 1969: chemical physics. Allan L. Smith, Professor; Ph.D., MIT, 1965: physical chemistry. Sally Solomon, Associate Professor; Ph.D., Pennsylvania, 1966: physical chemistry, chemical education. Irwin H. Suffet, P. Walter Purdom Professor of Environmental Chemistry; Ph.D., Rutgers, 1968: analytical and environmental chemistry. Peter A. Wade, Associate Professor; Ph.D., Purdue, 1973: organic chemistry. Yen Wei, Associate Professor; Ph.D., CUNY, Staten Island, 1986: polymer chemistry.

Department of Mathematics and Computer Science
Loren N. Argabright, Professor; Ph.D., Washington (Seattle), 1963: functional analysis, wavelets, abstract harmonic analysis and the theory of group representations. Robert P. Boyer, Associate Professor; Ph.D., Pennsylvania, 1978: functional analysis, C* algebras and the theory of group representations. Robert C. Busby, Professor; Ph.D., Pennsylvania, 1969: functional analysis, C* algebras and group representations, wavelets, computer science. Robin Carr, Assistant Professor; Ph.D., Toronto, 1985: mathematical physics, symbolic algebra, combinatorics. Bruce W. Char, Associate Professor; Ph.D., Berkeley, 1980: symbolic mathematical computation; algorithms and systems for computer algebra, automatic scientific programming, parallel and distributed computation. William M. Y. Goh, Assistant Professor; Ph.D., Ohio State, 1987: number theory, approximation theory and special functions, combinatorics, asymptotic analysis. Herman Gollwitzer, Associate Professor; Ph.D., Minnesota, 1967: applied mathematics, differential equations, data analysis, computer science. William J. Gordon, Professor; Ph.D., Brown, 1965: numerical analysis, multivariate interpolation and approximation, numerical solution of partial differential equations, computer graphics. Nira Herrmann, Associate Professor; Ph.D., Stanford, 1976: mathematical and applied statistics, early decision problems, expert systems in statistics, computer science, multivariate analysis, biostatistics. Jeremy R. Johnson, Assistant Professor; Ph.D., Ohio State, 1991: computer algebra, parallel computation, algorithmic algebraic number theory. Bernard Kolman, Professor; Ph.D., Pennsylvania, 1965: Lie algebras; theory, applications, and computational techniques; operations research. David Magagnosc, Assistant Professor; Ph.D., Dartmouth, 1987: computer science, combinatorial optimization, networks, combinatorial and parallel algorithms. Charles J. Mode, Professor; Ph.D., California, Davis, 1956: probability and statistics, biostatistics, epidemiology, mathematical demography, data analysis, computer-intensive methods. Marci A. Perlstadt, Associate Professor; Ph.D., Berkeley, 1978: applied mathematics, special functions, numerical analysis of function reconstruction, signal processing, combinatorics. James C. T. Pool, Professor and Department Head; Ph.D., Iowa, 1963: mathematical physics, scientific computing, mathematical software and supercomputing. Jeffrey L. Popyack, Associate Professor; Ph.D., Virginia, 1982: operations research, discrete and stochastic optimization, computational methods for Markov decision processes, artificial intelligence. Chris Rorres, Professor; Ph.D., NYU (Courant), 1969: applied mathematics, scattering theory, mathematical modeling in biological sciences, solar collector systems. Eric Schmutz, Assistant Professor; Ph.D., Pennsylvania, 1988: discrete mathematics, combinatorics, number theory, graph theory. Justin Smith, Associate Professor; Ph.D., NYU (Courant), 1976: computer science; parallel algorithms, algebraic topology, homotopy theory. Jet Wimp, Professor; Ph.D., Edinburgh, 1968: applied mathematics; special functions, approximation theory, numerical techniques, asymptotic analysis. Doron Zeilberger, Professor; Ph.D., Weizmann (Israel), 1976: combinatorial mathematics, enumerative combinatorics, partial difference equations, special functions. Stanley Zietz, Associate Professor; Ph.D., Berkeley, 1977: biomathematics, modeling of cellular modification, population dynamics, optimal control theory applied to medicine and biology, enzyme kinetics.

Department of Physics and Atmospheric Science
Shyamalendu M. Bose, Professor; Ph.D., Maryland, 1967: theory of surfaces and interfaces, disordered systems, electronic and X-ray spectroscopy of solids, superconductivity. Joan Centrella, Associate Professor; Ph.D., Cambridge, 1980: large-scale computations of galactic structures, numerical modeling of the universe. Ariel Cohen, Professor; Ph.D., Hebrew (Jerusalem), 1970: atmospheric science, lider measurements, electromagnetic scattering for nonspherical particles. Leonard D. Cohen, Professor; Ph.D., Pennsylvania, 1959: electromagnetic scattering of random-size particles, atmospheric physics. N. John DiNardo, Assistant Professor; Ph.D., Pennsylvania, 1982: surfaces and interfaces of semiconductors and metals; scanning tunneling, photoemission, and electron-scattering spectroscopies. Da Hsuan Feng, Professor; Ph.D., Minnesota, 1972: development of symmetry-dictated truncations of the spherical shell model for collective states in nuclei, electromagnetic properties of nuclei, high-spin spectroscopy, properties of nuclei far from B-stability, nuclear astrophysics, quantum chaos. Frank A. Ferrone, Associate Professor; Ph.D., Princeton, 1974: experimental and theoretical protein dynamics, kinetics of biological self-assembly. Leonard X. Finegold, Professor; Ph.D., London, 1959: biological physics, phase transitions in biomembranes. Robert Gilmore, Professor; Ph.D., MIT, 1967: applications of compact and noncompact Lie algebras to problems in nuclear, atomic, and molecular physics; nonlinear dynamics and chaos; laser instabilities. Richard D. Haracz, Professor and Acting Head; Ph.D., Wayne State, 1964: electromagnetic scattering of irregular objects, atmospheric physics. Frederick B. House, Professor and Director, Earth Radiation Budget Data Management Facility; Ph.D., Wisconsin, 1965: satellite meteorology, earth energy budget. Carl W. Kreitzberg, Professor and Director, Data Processing and Modeling Center; Ph.D., Washington (Seattle), 1963: atmospheric science, mesoscale computational weather prediction, dynamical data assimilation, air quality and modeling on scales of 10 km to 1000 km. Charles E. Lane, Assistant Professor; Ph.D., Caltech, 1987: experimental nuclear physics, experimental tests of invariance principles and conservation laws, magnetic monopoles and high-energy cosmic neutrinos, solar neutrinos and neutrino oscillations. Donald C. Larson, Professor and Director, Center for Insulation Studies; Ph.D., Harvard, 1962: optical waveguides in semiconductors, fiber-optical sensors, heat transfer. Teck-Kah Lim, Professor; Ph.D., Adelaide (Australia), 1968: structures and dynamics of small nuclear and molecular systems, spin-polarized quantum systems, physics in two dimensions, computer-aided physics education. Arthur E. Lord, Professor; Ph.D., Columbia, 1964: physical, mechanical, and transport properties of geosynthetic material; centrifuge modeling of geosynthetic/soil structures. James A. McCray, Professor; Ph.D., Caltech, 1962: pulsed-laser and synchroton radiation (Brookhaven and DESY), kinetic studies of the molecular mechanism of muscle contraction and neural transmitter function, laser photolysis of caged compounds, development of laser temperature-jump instrumentation systems. Stephen L. McMillan, Assistant Professor; Ph.D., Harvard, 1983: stellar dynamics, large-scale computations of stellar systems. Irvin A. Miller, Associate Professor; Ph.D., Temple, 1968: physics education. Lorenzo M. Narducci, Francis K. Davis Professor; Ph.D., Milan, 1964: laser physics, quantum optics, nonlinear dynamical systems, spatial patterns. Donald J. Perkey, Associate Professor; Ph.D., Penn State, 1976: atmospheric science, atmospheric circulations, mesoscale numerical weather prediction, cyclogenesis/frontogenesis, effects of boundary-layer and latent-heat processes. Richard I. Steinberg, Associate Professor; Ph.D., Yale, 1969: Experimental tests of invariance principles and conservation laws, experimental search for magnetic monopoles and high-energy cosmic neutrinos, solar neutrinos and neutrino oscillations. Somdev Tyagi, Associate Professor; Ph.D., Brigham Young, 1976: high-temperature superconductivity, magnetic properties of thin sputtered films of amorphous metallic allies, fiber-optical sensors. Michel Vallieres, Associate Professor; Ph.D., Pennsylvania, 1972: large-scale (supercomputer) calculations of nuclear and quark structures, computer architecture for nuclear physics problems. Jian-Min Yuan, Associate Professor; Ph.D., Chicago, 1973: nonlinear dynamics and chaos for atomic and molecular systems, group theoretical studies of scattering processes.

SECTION 5: MATHEMATICAL SCIENCES

EMORY UNIVERSITY
Graduate School of Arts and Sciences
School of Public Health
Division of Biostatistics

Programs of Study The Division of Bistatistics at Emory University School of Public Health, in conjunction with the Graduate School of Arts and Sciences, offers programs of study leading to the Master of Science and Doctor of Philosophy degrees in biostatistics. The M.S./Ph.D. curriculum is designed for individuals with a strong background and interest in the biological and mathematical sciences. M.S./Ph.D. graduates have pursued a variety of career options in academia as well as in public and private research organizations.

The Division of Biostatistics also offers an M.P.H. degree through the School of Public Health. The M.P.H. curriculum is designed for practicing health professionals as well as those who wish to enter the health field. M.P.H. graduates are qualified for administrative, clinical, and research positions in health agencies.

Students interested in studying mathematical models for infectious and chronic diseases can take advantage of the division's close working relationship with the neighboring Centers for Disease Control by participating in collaborative research projects. Those interested in developing methodology for cancer epidemiology can pursue research opportunities with the National Cancer Institute–supported Surveillance, Epidemiology and End Results (SEER) Program at Emory University or with the American Cancer Society, whose national headquarters are located near the University. Research opportunities are available in other divisions of the School, the Carter Center, the Georgia Department of Human Resources, Morehouse School of Medicine, and the five large teaching hospitals affiliated with Emory. These resources, as well as others in the clinical and basic science divisions of the Emory University School of Medicine, provide students with a wide range of areas in which to study and conduct research.

Biostatistics students are required to complete a core curriculum that consists of graduate courses in biostatistics and epidemiology. However, advanced course work and research are tailored to the experience, training, area of concentration, and degree objective of each student. The M.S. and M.P.H. programs usually include two to three semesters of course work and generally take 1½ to 2 years to complete. The Ph.D. degree program normally requires four calendar years to complete, including approximately five semesters of course work. Some qualified health professionals, such as those with an M.P.H. or a doctoral-level degree, may be able to satisfy the Ph.D. requirements in a shorter period of time.

Research Facilities The Division of Biostatistics conducts research in biostatistics, epidemic modeling, categorical data analysis, complex sample survey methods, and methods for infectious disease epidemiology. The School of Public Health is equipped with state-of-the-art computers and numerous microcomputers. Mainframe computers are accessible to the School through high-speed telecommunications lines. Research laboratories are housed at the Centers for Disease Control. Health sciences libraries are conveniently located at Emory University, the American Cancer Society headquarters, and the Centers for Disease Control.

Financial Aid Financial aid programs are available through the Emory Office of Financial Aid. Graduate School fellowships, including full tuition coverage, are available.

Cost of Study Tuition for 1991–92 is $7290 per semester for full-time students or $608 per credit hour. The student activity and athletic fees total $80 per semester. The cost of books and supplies averages $650 per year.

Cost of Living Living expenses for a single person are estimated at $12,000 per year. Information regarding University and off-campus housing may be obtained from the Housing Office of Emory University.

Student Group Emory University has a total enrollment of about 9,300 students. Enrollments in the various schools of the University are restricted in order to maintain a favorable balance between resources, faculty members, and students. There are about 5,400 students in the undergraduate college and 3,900 in the eight graduate and professional schools. The student body represents all areas of the United States and many foreign countries.

Location The Atlanta metropolitan area has a population of nearly 3 million. It is the major academic center in the Southeast: there are eight major universities in the metropolitan area. Atlanta is green the year round, with numerous parks and a temperate climate. Professional, athletic, cultural, and recreational activities are available throughout the year. Atlanta is one of the leading convention centers in the United States, and the city is served by the busiest airport in the world. It is the site of the 1996 Olympics.

The University and The School In recent years, Emory University has been ranked among the twenty-five most distinguished centers for higher education in the United States. The heavily wooded 550-acre campus features a blend of traditional and contemporary architecture. A main corridor through the campus incorporates the expanding health sciences complex with the Centers for Disease Control and the American Cancer Society. Within a short drive from the main campus are a variety of affiliated resources, such as the Georgia Mental Health Institute, the Georgia Department of Human Resources, the Carter Center, and Grady Memorial Hospital.

The Emory University School of Public Health was officially constituted in September of 1990 and is the newest School of Public Health in the United States and Canada. The establishment of the School of Public Health was made possible because of a strong community health-delivery base in the metropolitan area and because of Emory's flexibility in developing mutually beneficial institutional affiliations. The School of Public Health has five academic divisions—Biostatistics, Epidemiology, Health Policy and Management, Environmental/Occupational Health, and Behavioral Sciences and Health Education—each of which offers an M.P.H. degree. The Center for International Health is an interdisciplinary program of the School of Public Health. The Epidemiology Division offers a Ph.D. through the Graduate School of Arts and Sciences.

Applying Information and application forms for admission to the M.S./Ph.D. program and financial aid may be obtained from the Graduate School of Arts and Sciences of Emory University, Atlanta, Georgia 30322. Applicants for an M.S. or Ph.D. in biostatistics should have a strong undergraduate background in biology and mathematics, including calculus. Requirements for admission include satisfactory completion of a baccalaureate degree, satisfactory completion of at least two semesters of calculus, and satisfactory performance on the GRE General Test or MCAT. International students whose schooling has not been in English must submit a TOEFL score. Similar information on the M.P.H. degree can be obtained from the Office of Admissions, School of Public Health, 1599 Clifton Road, NE, Atlanta, Georgia 30329.

Correspondence and Information
Dollie Daniels
Emory University School of Public Health
1599 Clifton Road, NE
Atlanta, Georgia 30329
Telephone: 404-727-8712

SECTION 5: MATHEMATICAL SCIENCES

Emory University

THE FACULTY AND THEIR RESEARCH

Donna J. Brogan, Professor and Director of Graduate Studies; Ph.D., Iowa State, 1967. Sample survey design and analysis, epidemiology of hypertension, renal disease and breast cancer.
W. Scott Clark, Assistant Professor; Ph.D., Emory, 1990. Models of infectious diseases, biostatistical consulting.
Michael J. Haber, Associate Professor; Ph.D., Hebrew (Jerusalem), 1976. Statistical theory, categorical data analysis, models of infectious diseases.
E. C. Hall, Professor; Ph.D., North Carolina, 1966. Biostatistical consulting, statistical methodology.
M. Elizabeth Halloran, Assistant Professor; M.D., Berlin, 1983; D.Sc., Harvard, 1989. Infectious diseases, epidemic modeling, vaccine efficacy, population dynamics, vector-borne diseases.
Andrzej S. Kosinski, Assistant Professor; Ph.D., Washington (Seattle), 1990. Linear models, clinical trials.
Michael H. Kutner, Professor and Director, Division of Biostatistics; Ph.D., Texas A&M, 1971. Linear models, variance components, experimental design.
Lillian S. Lin, Assistant Professor; Ph.D., Washington (Seattle), 1990. Clinical trials.
Ira M. Longini, Associate Professor; Ph.D., Minnesota, 1977. Stochastic processes, models for infectious diseases.

Jointly Appointed Faculty
Stephanie L. Sherman, Assistant Professor of Pediatrics; Ph.D., Indiana, 1981. Population and medical genetics.

Adjunct Faculty
John M. Karon, Mathematical Statistician, Division of HIV/AIDS of the Center for Infectious Diseases, Centers for Disease Control; Ph.D., Stanford, 1968. Epidemiologic methods, HIV/AIDS modeling.
Glen A. Satten, Mathematical Statistician, Division of HIV/AIDS of the Center for Infectious Diseases, Centers for Disease Control; Ph.D., Harvard, 1985. Stochastic processes, HIV/AIDS modeling.
Donna F. Stroup, Assistant Director for the Science, Epidemiology Program Office, Centers for Disease Control; Ph.D., Princeton, 1980. Stopping rules for stochastic approximations, disease surveillance.

Associate Faculty
Brooke D. Fielding, M.S., North Carolina, 1987. Statistical methodology.
Elaine W. Flagg, M.S., Georgia, 1985. Sample survey conduct and analysis, nutrition and cancer epidemiology.
Michael J. Lynn, M.S., Mississippi State, 1976. Data management, clinical trials, statistical applications in ophthalmic research.
Azhar Nizam, M.S., South Carolina, 1987. Multiple comparisons, statistical education.
Frank Schipani, M.A., UCLA, 1966, M.Ln., Emory, 1979. Information systems, database administration.

SECTION 5: MATHEMATICAL SCIENCES

EMORY UNIVERSITY

Department of Mathematics and Computer Science

Programs of Study The department offers programs leading to the Ph.D. in mathematics and the M.A. and M.S. in mathematics, mathematics/computer science, and statistics/computer science. Students working toward the Ph.D. in mathematics can concurrently earn any of the master's degrees. Graduates who choose to combine the Ph.D. and the master's degree in mathematics/computer science have found this combination particularly useful when seeking academic employment.

Core courses in all programs are offered regularly. In addition, students in each program may choose from a variety of elective courses. Students take basic and advanced classes, and the faculty directs weekly seminars in all the major research areas. Students are encouraged to participate in seminars from their first year. Frequent visitors also contribute to a lively research atmosphere.

The faculty members work actively in all major research areas of mathematics but are especially productive in algebra, analysis, applied mathematics, combinatorics, and computer science research areas, including operating systems, networking, artificial intelligence, and image analysis.

The requirements for the Ph.D. include 48 credit hours beyond the master's degree (including four semesters in residence), doctoral examinations, reading knowledge of a foreign language (chosen from French, German, and Russian and satisfied by an acceptable score on an ETS examination or translation of a research paper), and a dissertation.

Requirements for master's programs include at least 28 credit hours and a thesis.

Research Facilities Emory's mathematics and computer science library has an up-to-date collection of well over 10,000 volumes, including over 200 journal subscriptions. Students in the program also use the department's computer lab. The lab includes a Sun SPARCserver 490 serving a network of Sun SPARCstations and Macintosh II A/UX workstations. There is a network connection to the University mainframes as well as to the Internet and BITNET networks.

Financial Aid The department offers teaching assistantships, fellowships, and tuition waivers. Awards carry tuition waivers and stipends of between $8500 and $11,200 for the 1991–92 academic year. Assistants typically teach one class per semester. Summer support is sometimes available.

The University also offers a number of Woodruff Fellowships. In 1991–92, these fellowships carry a $12,000 stipend and a tuition waiver for three academic years. The department may nominate 5 students to compete for these awards. The deadline for application is February 15. Departmental aid decisions also begin at this time. Late applications are considered to the extent that funds remain available.

Cost of Study In 1991–92, tuition for full-time graduate students is $14,580 for the academic year or $2432 for a 4-credit class.

Cost of Living Private rooms and apartments are available within the community at a wide variety of costs. The University also provides limited housing for graduate students. Interested students are urged to apply early.

Student Group Emory has a total enrollment of about 9,000 students. Enrollment in the various schools of the University is restricted to maintain the most favorable use possible of facilities and resources. There are about 3,500 students in the undergraduate college and about 5,500 students in the eight graduate and professional schools. The department has about 25 full-time students.

Location Atlanta is the capital of Georgia and is the leading business and cultural center of the Southeast. With a metropolitan population of over 2 million, Atlanta is a growing, dynamic, and progressive city. It offers a wide variety of cultural and recreational activities, including those of the Atlanta Memorial Arts Center, the Civic Center, two major sports complexes, and numerous public and private organizations.

The University Emory is a coeducational, privately owned university with a heritage of scholarship and educational excellence. The main campus is located on 550 wooded acres in a quiet suburban neighborhood, about 6 miles northeast of Atlanta. The University includes schools of medicine, law, nursing, dentistry, business administration, and theology, as well as graduate and undergraduate schools.

Applying Applications for admission and financial aid should be received by February 1; awards are announced by March 15. Scores on the General Test of the Graduate Record Examinations are required. International students whose native language is not English can provide scores on the Test of English as a Foreign Language (TOEFL) in addition to GRE scores. An official transcript in duplicate of the applicant's record at each college and graduate school attended must also be sent, along with three letters of recommendation. A nonrefundable fee of $25 must accompany each application.

Correspondence and Information

For further information:
Director of Graduate Studies
Department of Mathematics
   and Computer Science
Emory University
Atlanta, Georgia 30322
Telephone: 404-727-7580

For application forms:
Dean
Graduate School
Emory University
Atlanta, Georgia 30322
Telephone: 404-727-6028

SECTION 5: MATHEMATICAL SCIENCES

Emory University

THE FACULTY AND THEIR RESEARCH

Chang Mo Bang, Associate Professor; Ph.D., Vanderbilt. Abelian groups, modules, combinatorics, artificial intelligence.
Steve Batterson, Associate Professor; Ph.D., Northwestern. Geometric analysis, dynamical systems.
Shun Yan Cheung, Assistant Professor; Ph.D., Georgia Tech. Distributed systems, networking, computer science.
Dwight Duffus, Professor; Ph.D., Calgary. Combinatorics, lattice theory.
David A. Ford, Associate Professor; Ph.D., Utah. Mathematical economics, analysis, computer science.
Ronald J. Gould, Professor and Chairman; Ph.D., Western Michigan. Graph theory, combinatorics, computer science.
William S. Mahavier, Professor; Ph.D., Texas at Austin. General topology, computer science.
Kenneth Mandelberg, Associate Professor; Ph.D., Cornell. Operating systems, networking, image analysis, commutative algebra.
Mary Francis Neff, Associate Professor; Ph.D., Florida. Lattice theory, universal algebra.
Vladimir I. Oliker, Professor; Ph.D., Leningrad. Partial differential equations, differential geometry, applied mathematics.
Franz J. Pedit, Assistant Professor; Ph.D., Innsbruck. Geometric analysis and mathematical physics.
Victoria Powers, Assistant Professor; Ph.D., Cornell. Quadratic forms, algebra.
Vojtech Rödl, Professor; Ph.D., Karlova (Prague). Combinatorics, algebra, topology.
Robert L. Roth, Associate Professor; Ph.D., Ohio State. Combinatorics.
Kunimochi Sakamoto, Assistant Professor; Ph.D., Brown. Dynamical systems and applied mathematics.
Rudolf Schmid, Associate Professor; Ph.D., Zurich. Global analysis, mathematical physics.
V. S. Sunderam, Assistant Professor; Ph.D., Kent (England). Computer networks, software engineering, operating systems.
Philip C. Tonne, Associate Professor; Ph.D., North Carolina at Chapel Hill. Analysis.
Paul Waltman, Professor; Ph.D., Missouri–Columbia. Differential equations, mathematical biology.
Bevan K Youse, Associate Professor; M.S., Georgia. Analysis, mathematics education.

SECTION 5: MATHEMATICAL SCIENCES

GEORGE WASHINGTON UNIVERSITY

Department of Statistics/Computer and Information Systems

Programs of Study

The Department of Statistics/Computer and Information Systems offers graduate programs leading to the M.A., M.S., and Ph.D. degrees in statistics. The master's programs provide advanced training in modern data-analytic techniques and develop a conceptual understanding of the foundations and methodology of statistics. The master's degree can serve either as a professional degree for those intending to work in government or industry or as preparation for the doctoral degree. The Ph.D. program aims to prepare students for university teaching and research and for government and industrial positions that involve consulting and the development of new methodology.

The degrees of Master of Arts in mathematical statistics, Master of Science in applied statistics, and Master of Science in statistical computing require a minimum of 30 semester hours without a thesis. A concentration in biostatistics is available in the M.S. in applied statistics. All master's degree candidates must pass a written master's comprehensive examination.

Candidates for the Ph.D. in statistics customarily begin by fulfilling the requirements for a master's degree. Students are then required to take four core courses in probability, distribution theory, and inference and to pass a written qualifying examination based on this core. After further course work is taken, an oral thesis proposal presentation is required. Demonstration of computer proficiency is also required. There is no foreign language requirement. Students must complete a dissertation that demonstrates capability to carry out original research.

Part-time study toward either master's or Ph.D. degrees is possible. Students enrolled part-time are required to maintain a course load of 6 hours per semester.

Research Facilities

The Gelman Library has an extensive collection of statistical publications, including complete files of many statistical journals. In addition, George Washington University students have access to the collected holdings of the Washington Research Library Consortium, which comprises eight local universities. The department maintains a small library, which contains reference books and files of a number of leading journals. The University Computer Center, located in the modern Academic Center, maintains an IBM 4381 with excellent supporting hardware, an IBM 4341, and a large software collection, including the statistical packages SAS, SPSS, GLIM, and BMDP. The Computer Center also maintains forty IBM personal computers. The departmental computing laboratory is equipped with IBM personal computers linked to the main computer and laser printers. The Biostatistics Center, administered by the department, serves as the coordinating center for several large-scale clinical trials. The center provides statistical design, data management, and statistical analysis for these and other medical studies. The center maintains a dedicated IBM 4381 computer with extensive supporting hardware and software. Several faculty members hold joint appointments with the department and the center, and many research projects involving both faculty and graduate students arise directly from activities of the center.

Financial Aid

A number of graduate teaching assistantships and research assistantships are available. University fellowships are also available with these appointments. Teaching assistants receive 18 semester hours of tuition credit per year in addition to a salary and a stipend. Their duties include assisting faculty with undergraduate instruction or statistical consultation at the Computer Information and Resource Center. Research assistants generally work with faculty at the Biostatistics Center. Student loan programs are also available.

Cost of Study

The tuition fee for graduate study is $490 per credit hour for 1991–92.

Cost of Living

Many graduate students live in nearby rooms and apartments. The University's location on a main subway line gives quick access to campus from lower-cost areas in Virginia and Maryland as well as the District of Columbia.

Student Group

There are approximately 40 students enrolled in the department's graduate programs, of whom about two thirds are part-time students. Currently, 18 students are working toward a Ph.D. degree.

Location

The University is located in downtown Washington, D.C., a city rich in architectural and historical treasures. The campus is 10 minutes' walk from the John F. Kennedy Center for the Performing Arts and equally close to the White House. A multitude of museums, libraries, restaurants, and cultural and historic attractions are easily accessible from the University. Nearby recreational opportunities include Rock Creek Park, the C&O Canal, the Chesapeake Bay, and the Blue Ridge Mountains.

Washington is the center of statistical activity for the federal government. The Washington Statistical Society is the largest chapter of the American Statistical Association and offers a wide variety of seminars and short courses. The presence of the federal statistical community enriches both the teaching and the research activities of the department.

The University and The Department

George Washington University is a private, nonsectarian institution. Beginning as the Columbian College in 1821, it has awarded graduate degrees since the 1880s. A comprehensive university, it offers professional training in law, medicine, engineering, business, and education in addition to a full range of undergraduate and graduate studies. The Department of Statistics/Computer and Information Systems, founded in 1935 as the Department of Statistics, is the oldest statistics department in an arts and sciences college in the United States. The department includes specialists in both statistics and computer science and offers undergraduate programs in both subjects.

Applying

Applications for financial assistance should be received by February 15. For admission without financial aid, deadlines are July 1 for the fall semester and November 1 for the spring semester. For admission to all graduate programs the Graduate Record Examinations General Test is required. International applicants for financial aid are required to take the TOEFL and the Test of Spoken English.

Master's program applicants should have a background that includes multivariate calculus and linear algebra. A course in regression analysis is also required but may be taken (without master's credit) during the first semester of graduate study. Applicants to the master's program in statistical computing should have taken a course in data structures.

Correspondence and Information

For application forms and a *University Bulletin:*
Graduate School of Arts and Sciences
George Washington University
Washington, D.C. 20052

For information about the department:
Chairman
Department of Statistics/C&IS
George Washington University
Washington, D.C. 20052

Peterson's Guide to Graduate Programs in the Physical Sciences and Mathematics 1992

George Washington University

THE FACULTY AND THEIR RESEARCH

Joseph L. Gastwirth, Professor of Statistics and Economics; Ph.D., Columbia, 1963. Nonparametric and robust statistical inference; applications in law, economics, and public policy.
  Statistical Reasoning in Law and Public Policy. Orlando: Academic Press, 1988.
  Bayesian analysis of screening data: application to AIDS in blood donors. *Can. J. Stat.*, 1991. With Johnson and Renau.
  Statistical properties of measures of between-group income differentials. *J. Econometrics*, 1989. With Nayak and Wang.

Samuel W. Greenhouse, Professor Emeritus of Statistics; Ph.D., George Washington, 1959. Biostatistics, clinical trials, epidemiologic methods, statistical inference.
  The comparability of longitudinal studies and the role of clinical trials in normal aging research. *Exp. Gerontol.*, 1986.
  Estimating a common relative risk: Application in equal employment. *J. Am. Stat. Assoc.*, 1986. With Gastwirth.

David A. Grier, Assistant Professor of Statistics; Ph.D., Washington (Seattle), 1986. Expert systems, numerical methods, symbolic computation with uncertainty, connectionist artificial intelligence, Monte Carlo algorithms, parallelism.
  Solving non-standard least squares problems. In *Computing and Graphics in Statistics*, eds. A. Buja and P. Tukey, 1991.
  Monte Carlo and supercomputers. *Chance*, 1988.
  Bootstrap and variance reduction. In *Proc. Winter Simulation Conference*, 1988.

Catherine B. Hurley, Assistant Professor of Statistics; Ph.D., Washington (Seattle), 1987. Statistical computing, graphics, computing environments for data analysis, exploratory multivariate methods.
  Analyzing high-dimensional data with motion graphics. *SIAM J. Sci. Stat. Comput.*, 1990. With Buja.
  A software model for statistical graphics. In *Computing and Graphics in Statistics*, eds. A. Buja and P. Tukey, 1991. With Oldford.

Arthur D. Kirsch, Professor of Statistics and Psychology and Chair; Ph.D., Purdue, 1957. Survey research, experimental design, statistics in law and public policy.
  Trends in religious institutions: Findings from a national survey. In *Spring Research Forum Independent Sector*, St. Louis, 1988.
  An empirical study of selected cost measures of clinical engineering services. *J. Clin. Eng.*, 1988. With Ibrahim.

John M. Lachin, Professor of Statistics and Director of the Biostatistics Center; Sc.D., Pittsburgh, 1972. Statistical methods for clinical trials, repeated measures with incomplete observations, sample size evaluation, and randomization.
  Estimates and tests in the analysis of multiple non-independent 2 x 2 tables with partially missing observations. *Biometrics*, 1988. With Wei.
  Analysis of recurrent events: Nonparametric methods for random-interval count data. *J. Am. Stat. Assoc.*, 1988. With Thall.
  Statistical properties of randomization procedures in clinical trials. *Controlled Clinical Trials*, 1988.

Gordon Lan, Professor of Statistics; Ph.D., Columbia, 1974. Biostatistics, clinical trials.
  Linear rank tests for survival data: Equivalence of two formulations. *Am. Statistician*, 1990. With Wittes.
  Implementation of group sequential logrank tests in a maximum duration trial. *Biometrics*, 1990. With Lachin.

Hubert Lilliefors, Professor of Statistics; Ph.D., George Washington, 1964. Simulations, goodness of fit tests.
  Old simulation results and when to stop. *Am. Stat.*, 1987.
  It's time to stop. In *Computer Science and Statistics, Proceedings of the 19th Symposium on the Interface*, 1988.
  Kolmogorov-Smirnov and Anderson-Darling statistics with parameters estimated by linear combinations of order statistics and some robustness properties. In *Colloquia Mathematica Societatis Janos*, Bolyai, 1987.

Hosam M. Mahmoud, Associate Professor of Statistics; Ph.D., Ohio State, 1983. Analysis of algorithms, design of combinatorial algorithms, random structures.
  On the joint distribution of the insertion path length and the number of comparisons in search trees. *Discrete Applied Math*, 1988. With Pittel.
  Analysis of the space of search trees under the random insertion algorithm. *J. Algorithms*, 1989. With Pittel.
  Limiting distributions for path lengths in recursive trees. *Probl. Eng. Inf. Sci.*, 1991.

Tapan K. Nayak, Associate Professor of Statistics; Ph.D., Pittsburgh, 1983. Software reliability, diversity analysis, inference.
  Estimation of location and scale parameters using generalized Pitman nearness criterion. *J. Stat. Plann. Inference*, 1990.
  Estimating the number of component processes of a superimposed process. *Biometrika*, 1991.
  The use of diversity analysis to assess the relative influence of factors affecting the income distribution. *J. Bus. Econ. Stat.*, 1989. With Gastwirth.

Nozer D. Singpurwalla, Professor of Operations Research and of Statistics; Ph.D., NYU, 1968. Reliability theory (hardware and software), quality control, warranty analysis, Bayesian inference and decision theory, Kalman filtering and control theory.
  On the evidence needed to reach agreement between adversaries. *J. Am. Stat. Assoc.*, 1991. With Lindley.
  A Bayesian analysis of some nonhomogeneous autoregressive processes. *J. Roy. Stat. Soc. B*, 1991. With Soyer.
  Determining an optimal time interval for testing and debugging software. *IEEE Trans. Software Eng.*, 1991.

Robert T. Smythe, Professor of Statistics; Ph.D., Stanford, 1969. Biostatistics, stochastic processes, probability.
  Interval estimation with restricted randomization rules. *Biometrika*, 1989. With Wei.
  The use of historical control information in long-term carcinogenicity bioassays. In *Statistical Methods in Toxicological Research*, eds. Franklin and D. Krewski. New York: Gordon and Breach, 1991.
  On the distribution of leaves in rooted subtrees of recursive trees. *Ann. Appl. Probl.*, 1991. With Mahmoud.

Blaza Toman, Assistant Professor of Statistics; Ph.D., Ohio State, 1987. Optimal design of experiments, Bayesian statistics.
  Bayesian optimal experimental design for treatment-control comparisons in the presence of two-way heterogeneity, *J. Stat. Plann. Inference*, 1990. With Notz.

SECTION 5: MATHEMATICAL SCIENCES

HOWARD UNIVERSITY
Department of Mathematics

Programs of Study

The department offers programs leading to the M.S. and Ph.D. degrees in both pure and applied areas of mathematics. Current research areas include functional analysis, harmonic analysis, differential topology and geometry, several complex variables, semitopological groups, general topology, algebraic number theory, differential equations (ODE and PDE), dynamical systems, combinatorics, approximation theory, and fluid mechanics.

For the M.S. degree, a thesis is optional. Students electing the thesis are required to complete 30 course credits (including a maximum of 6 graduate credits for work on the thesis). Students electing the nonthesis option must complete 36 hours of course work and pass a qualifying examination.

The Ph.D. program requires a minimum of 60 graduate credits beyond the bachelor's degree, or a minimum of 36 graduate credits beyond the master's degree in mathematics; a doctoral dissertation for which 12 graduate credits will be given; two years of full-time study in the Graduate School of Howard University; and two approved foreign languages.

Research Facilities

The department is located in the Academic Support Building B on the main campus. On-campus research facilities include the University Computer Center; the University Library System comprising Founders Library for graduate students, the Undergraduate Library, and eleven branches in various schools, colleges, and departments. The mathematics collection of books and journals at the Founders Library is considerable and ranks among the best in the Consortium of Universities of the Washington Metropolitan Area. Facilities in the consortium are available to students.

Financial Aid

Prospective students may apply for fellowships, graduate assistantships, tuition scholarships, and loans. Graduate assistantships pay a stipend of $8000, and the recipients are entitled to remission of tuition. Inquiries about the Patricia Roberts Harris and the Dorothy Danforth Compton fellowships, which provide stipends ranging from $10,000 to $16,000 and cover tuition and fees, and graduate assistantships should be sent to the director of graduate studies at the address given at the bottom of the page. Inquiries concerning loans and tuition scholarships, as well as those pertaining to application procedures, should be directed to the Office of Financial Aid and Student Employment (telephone: 202-806-2800).

Cost of Study

Tuition for full-time students is $3312.50 a semester for 9–15 hours; for part-time students it is $368 a credit hour for 1–8 hours (1991–92 figures). Other fixed costs for graduate students include the enrollment fee, $150 (a one-time charge that is payable on acceptance and is not refundable); the health fee, $148 per semester; the student activities fee, $37.50 per semester; the University fee, $125 per semester for full-time students and $100 for part-time students; and the graduation fee of $100. A fee of $60 is charged for the binding of three copies of the thesis or dissertation.

Cost of Living

Generally, graduate students are expected to arrange their own housing in the community. However, the newly opened Howard Plaza Towers provide graduate students with an attractive, competitively priced option for on-campus housing. For more information about these accommodations as well as information about off-campus housing, students should contact the Office of Student Housing at 202-806-6131. Requests for assistance should be made well in advance of the school year for which the housing is needed. The estimated total cost of living (including direct educational expenses) for one year is about $14,000 for unmarried students and about $16,000 for married students.

Student Group

The average number of graduate students enrolled in programs leading to the various advanced degrees offered through the Graduate School of Arts and Sciences has been around 1,100 over the last few years. The graduate enrollment includes students from twenty-nine states, two U.S. possessions, and fifty-eight foreign countries.

Location

Washington, D.C., is a large, beautiful urban community with a distinct cosmopolitan flavor. It is an ideal location for many fields of study because of the resource materials available at the Library of Congress, federal agencies, the consortium of universities, and the many museums in the area.

The University

Howard University is a fully accredited, nonsectarian institution incorporated by an Act of Congress and approved by President Andrew Johnson in 1867. Although a private institution, Howard receives partial financial support from the federal government. While always open to students and faculty without regard to race, creed, sex, or national origin, Howard University has a special commitment to providing academic opportunities for those who have traditionally been denied access to higher education for reasons other than their intellectual abilities. The University's record in carrying out this commitment is widely known. The main campus, site of the Graduate School and many of the University's other eighteen schools and colleges, occupies 75 acres in northwest Washington. The West Campus, also located in northwest Washington, houses one of the professional schools, the Howard University Press, and several institutes. The University also owns a 108-acre site at nearby Beltsville, Maryland, that is to be developed as a campus for research in the life sciences and training in veterinary medicine.

Applying

Application forms for admission and assistantships may be obtained from the department at the address given below. Completed forms must be accurate and accompanied by a nonrefundable application fee of $25 in the form of a certified check or money order, payable to Howard University. All applicants are required to submit scores on the GRE General Test (scores not more than five years old) as part of their application. Applications for financial assistance must be filed by April 1 for the fall semester or November 1 for the spring semester.

Correspondence and Information

Director, Graduate Studies
Department of Mathematics
Howard University
Washington, D.C. 20059

SECTION 5: MATHEMATICAL SCIENCES

Howard University

THE FACULTY AND THEIR RESEARCH

Adeniran Adeboye, Ph.D., Johns Hopkins. Differential geometry, global physics.
Sanjay Arora, Ph.D., Colorado. Statistical design, probability.
Richard Bourgin, Ph.D., Washington (Seattle). Functional analysis, computational geometry.
Colette Calmelet, Ph.D., Vanderbilt. Applied mathematics, fluid mechanics.
Gerald Chachere, Ph.D., Berkeley. Differential geometry, computational geometry.
Senhuei Chen, Ph.D., Maryland. Partial differential equations, applied mathematics.
James A. Donaldson, Ph.D., Illinois. Partial differential equations, applied mathematics.
Seyoum Getu, Ph.D., Missouri. Nonassociative algebras, combinatorics.
Adnan Haider, Ph.D., American. Mathematical statistics.
Isom Herron, Ph.D., Johns Hopkins. Applied mathematics, fluid dynamics.
Neil Hindman, Ph.D., Wesleyan. Topological semigroups, combinatorics.
Fern Hunt, Ph.D., NYU. Mathematical biology, dynamical systems.
Olufemi Idowu, Ph.D., Northeastern. Algebraic topology, actuarial studies.
Abdulcadir Issa, Ph.D., Howard. Harmonic analysis.
David James, Ph.D., Chicago. Differential topology, compact transformation groups.
James Joseph, M.S., Howard. Point set topology.
Rodney Kerby, Ph.D., Maryland. Harmonic analysis.
Myung Kwack, Ph.D., Berkeley. Differential manifolds, several complex variables, topology.
Joshua Leslie, Ph.D., Paris IV (Sorbonne). Algebraic topology, infinite dimensional Lie algebras, dynamical systems.
Clement Lutterodt, Ph.D., Birmingham (England). Approximation theory, several complex variables.
Amir Maleki, Ph.D., George Washington. Operator theory.
Walter Miller, Ph.D., CUNY Graduate Center. Dynamical systems.
Bhamini Nayar, Ph.D., Delhi. Topology.
Paul Peart, Ph.D., West Indies. Optimization theory, combinatorics, approximation theory.
Francois Ramaroson, Ph.D., Johns Hopkins. Number theory, algebra.
Louise Raphael, Ph.D., Catholic University. Elliptic eigenfunction expansions, summability theory, approximation theory.
Cora Sadosky, Ph.D., Chicago. Harmonic analysis.
Louis Shapiro, Ph.D., Maryland. Combinatorics, group theory.
Duraiswamy Sundararaman, Ph.D., Columbia. Algebraic, complex geometry, supermanifold.
Daniel Williams, Ph.D., Berkeley. Partial differential equations, numerical analysis.
Wen-jin Woan, Ph.D., Illinois. Group theory, combinatorics.
Abdul-Aziz Yakubu, Ph.D., North Carolina State. Dynamical systems.

SECTION 5: MATHEMATICAL SCIENCES

JOHNS HOPKINS UNIVERSITY
Department of Mathematical Sciences

Programs of Study

The Department of Mathematical Sciences offers programs in statistics/probability/stochastic processes and in operations research/optimization/decision science leading to the M.A., M.S.E., and Ph.D. degrees. These programs are supported by a strong curriculum in computational and applied mathematics, including numerical analysis, linear algebra, and graph theory. A graduate program may emphasize one of the primary areas or may be more diversified in the mathematical sciences. Research specializations for the doctoral dissertation may be selected from an area represented in the department or may be interdisciplinary and involve allied faculty in other departments.

Fields represented in the department include probability, stochastic processes, mathematical statistics, statistical inference, applied statistics, operations research, continuous and discrete optimization, numerical optimization, computer modeling, game theory, numerical analysis, matrix analysis, graph theory, and combinatorics. Closely allied departments include Biostatistics, which offers programs involving the application of statistics in the life sciences; Health Policy and Management, which offers programs involving the application of operations research methodology to health systems; and Geography and Environmental Engineering, which offers programs involving the application of operations research to public-sector planning and policy analysis. Close liaisons are also maintained with the Departments of Computer Science and Mathematics, the Johns Hopkins Applied Physics Laboratory, and the Chesapeake Biological Laboratory.

Master's degree programs require students to take eight to ten 1-semester graduate courses in a coherent program. Doctoral degree programs include, in addition, a program of original research and its clear exposition in a written dissertation worthy of publication as a significant contribution to knowledge.

Research Facilities

The University's Milton S. Eisenhower Library, on the Homewood campus, one of the nation's foremost research facilities, has over 2 million volumes, general stack access, a highly qualified staff, capability for computerized literature searches, ample photocopying facilities, and study carrels for graduate students. The facilities of the University Computing Center, with its IBM 3081 time-sharing system, and the Engineering Computing Facility, with its DEC VAX 8600, AT&T 3B4000, and graphics laboratory, are available to students for research and instruction. Terminals for access are available in the department as well as elsewhere in its building, Maryland Hall. The department maintains a variety of personal computers, a computing room with terminals and printers, a small reference collection of books and journals, and a lounge for use by faculty and graduate students. Office space is provided for full-time resident graduate students.

Financial Aid

Full tuition and an academic-year stipend are awarded to nearly all Ph.D. candidates. The stipend level depends on the source, which may be a fellowship, teaching assistantship, or research assistantship. The minimum nine-month stipend for 1991–92 is $10,530. In addition, summer employment opportunities are usually available through the University or in the Baltimore-Washington area.

Cost of Study

Tuition for 1991–92 is $15,500; first-year students also pay a matriculation fee of $400.

Cost of Living

Nearby off-campus accommodations, ranging from simple efficiencies to spacious apartments with French windows and marble steps, are available in five University-owned security apartment buildings. Current rents, including furnishings and utilities, vary from $235 to $450 per month for single students and from $460 to $560 for married students. Other off-campus housing is also available nearby. General assistance is provided by the University's Housing Office. The campus cafeterias and other neighborhood eating establishments offer a variety of culinary attractions at reasonable prices.

Student Group

The University enrolls approximately 2,400 undergraduates and 850 graduate students in the School of Arts and Sciences and the School of Engineering. There are 45–50 undergraduates, 2–5 master's candidates, and 20–25 Ph.D. candidates in the Department of Mathematical Sciences. Among the graduate students, about one third come from foreign countries and about one third are women. Formal graduate classes at Hopkins are often quite small, and students and their research advisers usually have close working relationships.

Location

The attractive 140-acre Homewood campus is located in a pleasant residential neighborhood in the northern section of Baltimore, a historic port city. Adjacent to campus is the Baltimore Museum of Art. Within convenient distances are the Peabody Conservatory of Music, the Walters Art Gallery, the Baltimore Civic Opera Company, several legitimate theaters, Memorial Stadium (home of the Orioles), Pimlico race track (site of the Preakness), Druid Hill Park, the Pennsylvania railway station, and the downtown Inner Harbor with its scientific exhibitions and festival activities.

The University and The Department

Privately endowed, the Johns Hopkins University was founded in 1876 as a graduate and research institution offering collegiate preparation. The faculty members seek a balance between their commitment to scholarship and research and their commitment to teaching. In addition to the School of Arts and Sciences and the School of Engineering at the Homewood campus, the University has several other divisions. In East Baltimore, contiguous with the renowned Johns Hopkins Hospital, are the School of Medicine and the School of Hygiene and Public Health. (The University provides a free shuttle between the Homewood and East Baltimore campuses.) The School of Advanced International Studies is located in Washington, D.C.; this school also has a center for foreign studies in Bologna, Italy. The Johns Hopkins Applied Physics Laboratory, noted for contributions to applied sciences in a variety of fields, lies midway between Baltimore and Washington. The Peabody Institute, a leading professional school of music affiliated with Johns Hopkins, is located near the Homewood campus. The Homewood campus is also the location of the scientific research facility for the Space Telescope, an instrument placed in orbit around the earth in 1990. The Athletic Center, on the Homewood campus, is available to students and faculty seven days a week.

Applying

Application materials for admission and financial assistance are available from the department, along with further information on programs and facilities. Completed applications, letters of recommendation, transcripts, and GRE scores (the General Test plus the Subject Test in mathematics) are due by February 15 for initial decisions. International students whose native language is not English must also submit TOEFL scores at this time.

Correspondence and Information

Chairman, Graduate Admissions Committee
Department of Mathematical Sciences
Johns Hopkins University
Baltimore, Maryland 21218
Telephone: 301-338-7195

Peterson's Guide to Graduate Programs in the Physical Sciences and Mathematics 1992

SECTION 5: MATHEMATICAL SCIENCES

Johns Hopkins University

THE FACULTY AND THEIR RESEARCH

Members of the primary department faculty are listed below. In addition, several jointly appointed or visiting faculty members join the department each year.

James A. Fill, Associate Professor; Ph.D., Chicago, 1980. Probability, stochastic processes.
Alan J. Goldman, Professor; Ph.D., Princeton, 1956. Operations research, game theory, optimization, graph theory.
Shih-Ping Han, Professor; Ph.D., Wisconsin, 1974. Parallel optimization, mathematical programming.
Roger A. Horn, Professor; Ph.D., Stanford, 1967. Analysis, complex variables, matrix analysis.
Alan F. Karr, Professor; Ph.D., Northwestern, 1973. Stochastic processes, probability, image analysis.
Daniel Q. Naiman, Associate Professor; Ph.D., Illinois at Urbana-Champaign, 1982. Statistics, probability.
Jong-Shi Pang, Professor; Ph.D., Stanford, 1976. Mathematical programming, network equilibrium, parallel optimization, linear complementarity.
Edward R. Scheinerman, Associate Professor; Ph.D., Princeton, 1984. Graph theory, combinatorics.
Michael H. Schneider, Assistant Professor; Ph.D., Northwestern, 1984. Network optimization, nonlinear optimization, equilibrium analysis.
Robert J. Serfling, Professor; Ph.D., North Carolina, 1967. Probability, statistics, asymptotic theory.
John C. Wierman, Professor and Chair; Ph.D., Washington (Seattle), 1976. Probability, stochastic processes, statistics, random graphs.
Colin Wu, Assistant Professor; Ph.D., Berkeley, 1990. Statistics, semiparametric models, and robust statistics.

RESEARCH ACTIVITIES

Recent research projects of the department faculty and graduate students include:

Problems in stochastic games and discrete dynamic programming.
Network programming models of ecosystem development.
Inference and state estimation for stochastic point processes.
Partially ordered sets.
Simultaneous statistical inference.
Regression and analysis of variance models.
Nonlinear methods in multivariate analysis.
Numerical methods in nonlinear programming.
Probabilistic analysis of algorithms.
Critical values in bond, site, and mixed percolation models.
Optimal and equitable location of obnoxious facilities.
Large sample theory for general statistical inference.
Probability theory for random fields and Markov processes.
Numerical analysis and algorithms.
Singular value inequalities useful in control theory.
Combinatorial games.
Optimal maneuvers and schedules.
Mathematical models of precipitation.
The linear complementarity problem.
Finite-dimensional variational inequalities.
Theory of intersection graphs.
Robust estimation of Rayleigh distribution parameters.
Rates of convergence of Markov chains.
Applied statistics.
Mathematical models of health care.
Random graphs.
Survival data analysis with time-dependent covariate effects.
Algorithms for network optimization.
Generalized U-statistics and L-statistics.

ANNUAL MATHEMATICAL SCIENCES LECTURE SERIES

The Department of Mathematical Sciences annually hosts a weeklong, research-level lecture series during the summer. A distinguished mathematical scientist delivers the lectures, and the conference is attended by advanced graduate students and researchers. The conference is sponsored by the department, the Johns Hopkins University Press, and federal agencies. Recent speakers and topics have been Ulf Grenander (statistical analysis of patterns and images), 1991; W. T. Trotter (partially ordered sets), 1990; Arthur P. Veinott Jr. (lattice programming), 1989; Robert C. Thompson (matrix spectral inequalities), 1988; Richard Karp (probabilistic analysis of algorithms), 1987; Peter C. Fishburn (nonlinear utility theory), 1986; Charles R. Johnson (combinatorial aspects of matrix theory), 1985; A. N. Shiryayev (inference for diffusion processes), 1984; Peter J. Bickel (adaptive statistical inference), 1983; Ralph L. Disney (queuing networks and applications), 1982; Darwin Klingman (network flows), 1981.

SELECTED RECENT PAPERS

Fill, J. A. (with P. Diaconis). Examples for the theory of strong stationary duality with countable state spaces. *Probability in the Engineering and Informational Sciences* 4:157–80, 1990.
Goldman, A. J. (with R. H. Byrd and M. Heller). Recognizing unbounded integer programs. *Op. Res.* 35:140–42, 1987.
Horn, R. A. (with Y. P. Hong). A characterization of unitary congruence. *Linear and Multilinear Algebra* 25:105–19, 1989.
Naiman, D. Q. Conservative confidence bands in curvilinear regression. *Ann. Statist.* 14:896–906, 1986.
Pang, J.-S. Inexact Newton methods for the nonlinear complementarity problem. *Math. Program.* 36:54–71, 1986.
Scheinerman, E. R. (with N. Alon and M. Katchalski). Not all graphs are segment T-graphs. *European J. Combinatorics* 11:7–13, 1990.
Schneider, M. H. Matrix scaling, entropy minimization, and conjugate duality (I): Existence conditions. *Linear Algebra and Its Applications* 114/115:785–813, 1989.
Serfling, R. J. (with W. Hardle and P. Janssen). Strong uniform consistency rates for estimators of conditional functionals. *Ann. Stat.* 16:1428–49, 1988.
Wierman, J. C. (with T. Luczak). Counterexamples in AB percolation. *J. Phys.* A22:185–91, 1989.

BOOKS BY FACULTY MEMBERS

Horn, R. A., and C. R. Johnson. *Matrix Analysis.* New York: Cambridge University Press, 1985. *Topics in Matrix Analysis.* New York: Cambridge University Press, 1988.
Karr, A. F. *Point Processes and Their Statistical Inference.* New York: Marcel Dekker, 1986.
Serfling, R. J. *Approximation Theorems of Mathematical Statistics.* New York: Wiley, 1980.
Wierman, J. C., and R. T. Smythe. *First-passage Percolation on the Square Lattice.* New York: Springer, 1978.

JOHNS HOPKINS UNIVERSITY

Department of Mathematics
Mathematics Graduate Program

Program of Study

The Mathematics Graduate Program at Johns Hopkins prepares students for research and teaching in mathematics. Both the Master of Arts and the Doctor of Philosophy degrees are offered. Areas of research include algebraic geometry, algebraic groups, algebraic number theory, differential geometry, mathematical physics, partial differential equations, representation theory, several complex variables, and topology. These areas can be supplemented with courses in computer science, mechanics, probability, theoretical physics, etc., offered by other departments.

Study in the program is centered on seven basic graduate courses: algebraic geometry, algebraic topology, Lie groups and Lie algebras, number theory, real variables, Riemannian geometry, and several complex variables. These courses are intended to inform students of current developments in these areas and to prepare students for research study in their chosen field. This phase of study generally lasts one or two years and leads to the M.A. degree. One of these courses is required for the M.A. degree and three for the Ph.D. degree. A reading knowledge of French, German, or Russian is also required for either degree.

Some entering students may need to take one or more of the fundamental courses (introduction to advanced algebra, methods of complex analysis, complex function theory, point set topology) as additional preparation for graduate courses. It is expected, however, that those students will take at least one of the basic seven courses in their first year.

The main requirement for the Ph.D. degree is the dissertation. Work on the dissertation is generally begun in the third year under a faculty adviser. Students may continue to attend courses while working on the dissertation. The Ph.D. program lasts four or five years for most students. The final examination is the dissertation defense.

Research Facilities

The University's Milton S. Eisenhower Library has an extensive collection of mathematical books and journals, and the stacks are open to students. The department also has a small, useful reference library. Students share departmental offices, and places to study can also be reserved in the University library. Students have access to the department's Sun computer as well as to the University's three mainframes. Computer terminals are located in the department as well as in the University Academic Computing Center in the same building.

Financial Aid

Most students admitted to the Ph.D. program receive full tuition fellowships and teaching assistantships. The salary for teaching assistants in 1991–92 is at least $9500 for nine months. As long as satisfactory progress is made, a student can expect to receive support for at least four years. Some summer research support is available for those students who are working on their dissertations.

Exceptional applicants are considered for the George E. Owen Fellowship, which includes full tuition, a stipend, and no teaching duties for the first year.

Cost of Study

For 1991–92, tuition is $15,500. Fees for health insurance (required for students without other insurance coverage) are $800. The cost of books is approximately $400. There is also a one-time matriculation fee, currently $400.

Cost of Living

An estimated cost of living for one year, including housing and personal expenses, is $8000. The University suggests adding $2500 for each dependent. Housing accommodations for single and married students are available through University Housing. Most students, however, live in the numerous privately owned apartments within walking distance of the campus.

Student Group

There are currently 34 graduate students in the Department of Mathematics, 25 men and 9 women. Four or 5 students receive their Ph.D.'s each year and find varied employment in and out of the academic world.

Location

The Johns Hopkins University's Homewood campus is located in a pleasant residential area of Baltimore, about 15 minutes from downtown and the Inner Harbor. There are found shops, restaurants, the National Aquarium, the Maryland Science Center, and the Pier 6 Concert Pavilion. The downtown area is the locale of much of Baltimore's nightlife as well as Meyerhoff Symphony Hall, the Morris Mechanic Theatre, Center Stage, the Walters Art Gallery, and the new Orioles stadium.

Washington, D.C., is about 1 hour from Hopkins and has many national museums and facilities.

The Department

Johns Hopkins was founded in 1876 as the first graduate school in the country, with the mathematician J. J. Sylvester as one of the original faculty members. Since then, the department has maintained a mathematics faculty of international eminence in a small setting. The Department of Mathematics currently has 18 full-time faculty members; personal, one-on-one interaction exists between students and faculty.

Mathematicians from around the country are invited to speak at the weekly colloquia held by the department. The department also sponsors picnics and other social events for faculty and students. The Japan–U.S. Mathematics Institute (JAMI) has its home in the department. Each year, JAMI invites at least 4 outstanding Japanese mathematicians to visit, usually for a semester. Weekly seminars are organized by these visiting professors. A JAMI workshop or conference is also usually held once a year. The department also runs the *American Journal of Mathematics*, the oldest scientific journal in the country.

Applying

Admission to the Department of Mathematics is based on academic records, letters of recommendation, and GRE scores. Students whose native language is not English must also submit a TOEFL score.

The application deadline for fall 1992 admission is February 14, 1992. (There is no spring admission.) Applications can be obtained by writing to the address below.

Correspondence and Information

Graduate Admissions Committee
Department of Mathematics
Johns Hopkins University
Baltimore, Maryland 21218
Telephone: 410-516-7411
Fax: 410-516-5549
Electronic mail: grad@chow.mat.jhu.edu

SECTION 5: MATHEMATICAL SCIENCES

Johns Hopkins University

THE FACULTY AND THEIR RESEARCH

John M. Boardman, Professor; Ph.D., Cambridge, 1965. Algebraic and differential topology.
Wei-Liang Chow, Professor Emeritus. Algebra, algebraic geometry, complex varieties.
Rui-tao Dong, Assistant Professor; Ph.D., San Diego, 1990. Partial differential equations.
Jun-ichi Igusa, Professor; Ph.D., Kyoto, 1953. Algebra, algebraic geometry, modular functions, number theory.
Kevin Keating, Assistant Professor; Ph.D., Harvard, 1987. Number theory.
George Kempf, Professor; Ph.D., Columbia, 1970. Algebraic geometry.
Jean-Pierre Meyer, Professor; Ph.D., Cornell, 1954. Algebraic topology, category theory.
Jack Morava, Professor; Ph.D., Rice, 1969. Algebraic topology, mathematical physics.
Takashi Ono, Professor; Ph.D., Nagoya, 1958. Algebra, number theory, algebraic groups.
Hal Sadofsky, Assistant Professor; Ph.D., MIT, 1990. Algebraic topology.
Joseph H. Sampson, Professor Emeritus. Differential geometry, global analysis, algebraic geometry.
Joseph A. Shalika, Professor; Ph.D., Johns Hopkins, 1966. Algebraic groups and representations, number theory.
Bernard Shiffman, Professor and Chair; Ph.D., Berkeley, 1968. Several complex variables, differential geometry.
Joel Spruck, Professor; Ph.D., Stanford, 1971. Partial differential equations, geometric analysis, mathematical physics.
W. Stephen Wilson, Professor; Ph.D., MIT, 1972. Algebraic topology.
Steven Zelditch, Associate Professor; Ph.D., Berkeley, 1981. Partial differential equations.
Steven Zucker, Professor; Ph.D., Princeton, 1974. Differential geometry, algebraic geometry.
Maciej Zworski, Assistant Professor; Ph.D., MIT, 1989. Partial differential equations, scattering theory.

Books by Faculty Members

J.-i. Igusa. *Theta Functions.* Springer-Verlag, 1972.
   Forms of Higher Degree. Springer-Verlag, 1978.
G. R. Kempf. *Complex Abelian Varieties.* Springer-Verlag, 1990.
   Algebraic Varieties. In press.
G. D. Mostow, J. H. Sampson, and J. P. Meyer. *Fundamental Structures of Algebra.* McGraw-Hill, 1963.
T. Ono. *An Introduction to Algebraic Number Theory.* Plenum, 1990.
B. Shiffman and A. J. Sommese. *Vanishing Theorems on Complex Manifolds.* Birkhauser, 1985.

SECTION 5: MATHEMATICAL SCIENCES

LEHIGH UNIVERSITY

Department of Mathematics

Programs of Study

The Department of Mathematics offers programs of study leading to both the M.S. and Ph.D. degrees in mathematics. In addition, the Division of Applied Mathematics within the department administers a program leading to the M.S. and Ph.D. degrees in applied mathematics. Graduate study in mathematics covers a broad spectrum of pure mathematics, applied mathematics, and statistics. Normally, a student desiring a master's degree will study for two years, and a student desiring a doctorate will study for four years or more, frequently obtaining an M.S. on the way to the Ph.D.

All entering graduate students take a comprehensive examination during their first year of graduate work. A total of 30 credit hours is needed to complete the requirements for the M.S. degree, and up to 6 of these hours may be used for research and the writing of a thesis.

To be admitted to candidacy for a doctoral degree, the student who has passed the comprehensive examination must take a qualifying examination that consists of three separate examinations chosen from seven areas. Doctoral students must also take a general examination and a language examination and complete and publicly defend a dissertation.

Research Facilities

Over 21,000 volumes in the mathematical sciences, including monographs and bound periodicals, are housed in the new E. W. Fairchild-Martindale Library. The University libraries contain over 850,000 volumes and subscribe to approximately 8,800 serials. The fully automated libraries have a public-access on-line catalog and circulation system. On-line searches of external databases, as well as traditional information services, are available.

The resources of the libraries are augmented by memberships in the following consortia: the Center for Research Libraries, the Lehigh Valley Association of Independent Colleges, OCLC (Online Computer Library Center), and PALINET (Pennsylvania Area Library Network).

Computing resources available to all faculty and staff include the VAX 8530 and CYBER 850 mainframes for academic computing; the Network Server, an IBM 4381 mainframe dedicated to providing information services, such as electronic mail and electronic forms; and ASA, Automated System Access, which is the on-line card catalog system of the University library system. Approximately 250 microcomputers are available to the University community.

The principal programming languages used are FORTRAN, COBOL, Pascal, C++, APL, Reduce, and Maple. The University has acquired a campuswide license for Maple. All full-time faculty, staff, and students may obtain personal copies of the microcomputer versions.

Financial Aid

The mathematics department will appoint 23 teaching assistants in 1991–92; the stipend is between $9300 and $9600 for the academic year, and tuition is covered. The University awards about forty-five fellowships in various amounts and forty tuition scholarships that are available to students in all departments by competition.

Cost of Study

The tuition for a full-time student in 1991–92 is $15,500. Teaching assistants receive a tuition waiver for 9 credit hours each semester. Fellows receive a tuition waiver for 9 credit hours each semester.

Cost of Living

The cost of living in the Lehigh Valley is moderate. In addition to the University's apartments in the Saucon Valley, numerous privately owned apartments near the University are available to students.

Student Group

The University has approximately 4,400 undergraduates and 1,900 graduate students. The mathematics department has about 40 full-time students enrolled in its graduate programs. The department has been awarding graduate degrees for more than fifty years; more than 225 M.S. and 110 Ph.D. degrees in mathematics have been awarded in the past twenty-five years.

Location

Bethlehem is 60 miles north of Philadelphia, 90 miles southwest of New York City, and 180 miles northeast of Washington, D.C. Founded in 1742 by Moravians seeking religious freedom, Bethlehem has a rich cultural heritage. The Department of Mathematics shares in this history by occupying Christmas-Saucon Hall, part of which was originally built over a century ago to serve as a Moravian church. Dozens of historic buildings and locales have been remarkably preserved and are in current use, giving the community a charming Colonial atmosphere. The Lehigh Valley is also an important commercial and industrial center.

The University

Lehigh University is an independent, nondenominational, coeducational university. Founded in 1865 as a predominantly technical four-year institution, the University now has four major units: the College of Engineering and Applied Science, College of Business and Economics, College of Arts and Science, and College of Education. Eleven interdisciplinary research centers and nine institutes complement the research efforts of the four units.

Most of the University's more than 120 buildings are located on a 900-acre wooded campus on the north slope and top of South Mountain. On an additional 600 acres in Saucon Valley, on the south side of South Mountain, there are a field house, a convocation center, playing fields, and garden apartments for graduate students.

The *Journal of Differential Geometry* is published by the University.

Applying

Application forms for admission and financial aid may be obtained from the graduate adviser. Applicants requesting financial aid should apply by February 1. The TOEFL is required of all students whose native language is not English, and the TSE is required for teaching assistants from non-English-speaking countries. The GRE General Test and Subject Test in mathematics are recommended for all applicants.

Correspondence and Information

Graduate Advisor
Department of Mathematics, Building 14
Lehigh University
Bethlehem, Pennsylvania 18015
Telephone: 215-758-3730

Peterson's Guide to Graduate Programs in the Physical Sciences and Mathematics 1992

Lehigh University

THE FACULTY AND THEIR RESEARCH

Members of the department are on the editorial boards of *American Mathematical Monthly*, *Arab Journal of Mathematics*, *Bulletin of the Institute of Mathematics (Academia Sinica)*, *Journal of Differential Geometry*, and *Sequential Analysis*.

Edward F. Assmus Jr., Professor of Mathematics; Ph.D., Harvard, 1958. Algebra, combinatorial theory.
Donald M. Davis, Professor of Mathematics; Ph.D., Stanford, 1972. Algebraic topology, homotopy theory.
Vladimir Dobric, Assistant Professor of Mathematics; Ph.D., Zagreb (Yugoslavia), 1985. Analysis, measure theory, probability.
Bruce A. Dodson, Associate Professor of Mathematics; Ph.D., SUNY at Stony Brook, 1976. Algebra, arithmetic geometry.
Dominic G. B. Edelen, Professor of Mathematics; Ph.D., Johns Hopkins, 1956. Applied mathematics, differential geometry, mathematical physics.
Bennett Eisenberg, Professor of Mathematics; Ph.D., MIT, 1968. Probability, mathematical statistics.
Bhaskar K. Ghosh, Professor of Mathematics; Ph.D., London, 1959. Statistics.
Wei-Min Huang, Associate Professor of Mathematics; Ph.D., Rochester, 1982. Statistics, probability.
David L. Johnson, Associate Professor of Mathematics; Ph.D., MIT, 1977. Differential geometry, algebraic geometry.
Samir A. Khabbaz, Professor of Mathematics; Ph.D., Kansas, 1960. Algebra, topology, singularities.
Jerry P. King, Professor of Mathematics; Ph.D., Kentucky, 1962. Complex analysis, summability.
Gregory T. McAllister, Professor of Mathematics; Ph.D., Berkeley, 1962. Optimal control, numerical solution of differential equations.
George E. McCluskey Jr., Professor of Mathematics and Astronomy; Ph.D., Pennsylvania, 1965. Binary stars, space astronomy.
Clifford S. Queen, Associate Professor of Mathematics; Ph.D., Ohio State, 1969. Algebra, number theory.
Eric P. Salathe, Professor of Mathematics; Ph.D., Brown, 1965. Applied mathematics, physiological transport phenomena.
Murray Schechter, Professor of Mathematics; Ph.D., NYU, 1964. Numerical analysis, mathematical programming.
Penny Smith, Associate Professor of Mathematics; Ph.D., Polytechnic of New York, 1978. Partial differential equations.
Andrew K. Snyder, Professor and Chairperson of the Department of Mathematics; Ph.D., Lehigh, 1965. Functional analysis.
Lee J. Stanley, Associate Professor of Mathematics; Ph.D., Berkeley, 1977. Set theory, mathematical logic.
Gilbert A. Stengle, Professor of Mathematics; Ph.D., Wisconsin–Madison, 1961. Classical analysis, differential equations.
Susan Szczepanski, Associate Professor of Mathematics; Ph.D., Rutgers, 1980. Algebraic topology, geometric topology.
Ramamirtham Venkataraman, Associate Professor of Mathematics; Ph.D., Brown, 1968. Applied mathematics, fluid mechanics.
Albert Wilansky, University Distinguished Professor of Mathematics; Ph.D., Brown, 1947. Functional analysis, summability.
Joseph E. Yukich, Associate Professor of Mathematics; Ph.D., MIT, 1982. Probability, analysis.

Professors Emeriti

Theodore Hailperin, Ph.D., Cornell, 1943.
Chuan-Chih Hsiung, Ph.D., Michigan State, 1948.
A. Everett Pitcher, Ph.D., Harvard, 1935.

RESEARCH COURSES AND SEMINARS

The department regularly offers graduate courses in algebra, real analysis, complex analysis, topology, differential geometry, functional analysis, differential equations, numerical analysis, probability, and statistics. In addition, advanced seminars are offered in research areas of current faculty interest. Seminars for 1990–91 included:

Probability
Statistics
Set theory
Differential geometry
Algebraic topology
Functional analysis
Applied mathematics

The department conducts the prestigious Pitcher Lecture series, an annual three-day series of lectures given by a distinguished mathematician and directed at graduate students. In addition, the department offers a weekly colloquium featuring outside speakers and an annual Geometry/Topology Festival. Proximity to Philadelphia, New York, and Princeton provides ample opportunity to participate in outside mathematical activities.

SECTION 5: MATHEMATICAL SCIENCES

MEDICAL UNIVERSITY OF SOUTH CAROLINA

College of Graduate Studies
Department of Biostatistics, Epidemiology, and Systems Science

Programs of Study

The Department of Biostatistics, Epidemiology, and Systems Science offers programs of study leading to the Master of Science and Doctor of Philosophy degrees. Students may specialize in one of four areas: biostatistics, biomedical computing and systems science, epidemiology, or health-care systems science. The goal of these programs is to produce biometrists who can function as contributing members of biomedical research teams as well as original researchers.

Requirements for the master's degree include a minimum of 24 semester hours of credit—at least 15 hours in the major department and 4 hours outside the department. In addition to course credits, at least 9 hours must be earned in research and preparation of the thesis.

Requirements for the Ph.D. degree include a minimum of 60 semester hours of credit beyond the bachelor's degree, 45 of which must be in course credits exclusive of research. Of these 45 hours, at least 12 hours must be taken outside the department. Completion of a satisfactory dissertation is considered the most important requirement for the Ph.D.

Research Facilities

The department has laboratories for student research in microprocessing, image analysis, and artificial intelligence. The department has several Sun Workstations and direct access to a VAX-11/785. A variety of smaller computers (e.g., Macintosh and IBM) are also available within the department. The department maintains close ties with research laboratories in anatomy, physiology, nuclear medicine, radiology, pathology, and other basic and clinical science departments.

Financial Aid

Graduate fellowships are granted to qualified Ph.D. applicants on a competitive basis. The 1991–92 stipend is $10,200 per year. Additional fellowships are available to exceptional students.

Cost of Study

In 1991–92, tuition fees for the College of Graduate Studies are $2800 per year for both state residents and out-of-state students. Students with fellowships and traineeships pay reduced fees. Comprehensive health care is offered.

Cost of Living

Accommodations may be readily found in the surrounding community, with costs varying according to individual needs.

Student Group

Although the majority of the students at MUSC are native South Carolinians, most of the students in the Department of Biostatistics, Epidemiology, and Systems Science are from out of state. In 1991–92, there are about 40 graduate students in the department.

Location

Charleston is located on a peninsula between the Ashley and Cooper rivers and has one of the best harbors on the eastern seaboard. Numerous islands, all of which are popular resort areas, provide scenic charm and opportunities for recreation, including golf, tennis, surfing, fishing, and sailing.

The University

The Medical University of South Carolina was chartered in 1824 as a school of medicine. As the need has arisen, colleges have been organized for other divisions of the health sciences. The present College of Graduate Studies, evolving from a program of graduate education in the basic medical sciences organized in 1949, attained full divisional status in 1965. Its purpose is to further the objectives of the Medical University by offering qualified students opportunities for advanced learning and research and to promote the extension of knowledge in various areas.

The Medical University of South Carolina is situated on a 45-acre complex that also includes seven hospitals, one of which is the Medical University Hospital. There are six colleges in the University—Medicine, Nursing, Dental Medicine, Pharmacy, Graduate Studies, and Health Related Professions.

Applying

Students may apply at any time but are usually admitted to begin work in August. Candidates for admission are expected to present evidence of knowledge in mathematics and in an area related to biomedical research. These prerequisites, however, may be satisfied by noncredit courses taken during a probationary admission period. The General Test of the Graduate Record Examinations is required, and foreign applicants are required to take the GRE Subject Test in the area most relevant to their field of interest as well as the TOEFL. GRE scores more than ten years old are not acceptable.

Correspondence and Information

Dr. Eberhard O. Voit
Department of Biostatistics, Epidemiology, and Systems Science
Medical University of South Carolina
171 Ashley Avenue
Charleston, South Carolina 29425
Telephone: 803-792-2261

Peterson's Guide to Graduate Programs in the Physical Sciences and Mathematics 1992

SECTION 5: MATHEMATICAL SCIENCES

Medical University of South Carolina

THE FACULTY AND THEIR RESEARCH

Professors

M. Clinton Miller III, Chairman; Ph.D., Oklahoma Health Sciences Center, 1961. Experimental design, health-care systems evaluation, epidemiological models, medical decision models, operations research, quality control.

John B. Dunbar, Professor; D.M.D., Alabama at Birmingham, 1953; Dr.P.H., Tulane, 1963. Epidemiology of hypertension, atherosclerosis, coronary heart disease.

Alan J. Gross, Ph.D., North Carolina, 1962. Life tables, survival and reliability theory, stochastic processes, outlier theory, sample size determination.

Julian E. Keil, Dr.P.H., North Carolina, 1975. Environmental, occupational, and cardiovascular epidemiology; demography.

Chan F. Lam, Ph.D., Clemson, 1970. Signal and image processing, modeling and simulation, optimization, systems analysis, biomedical engineering, medical expert systems.

Associate Professors

Hurshell H. Hunt, Ph.D., Oklahoma State, 1968. Sampling, probability, distribution theory, Bayesian statistics, estimation, outlier theory.

Rebecca G. Knapp, Ph.D., Medical University of South Carolina, 1984. Biostatistics, evaluation methodology.

Philip F. Rust, Ph.D., Berkeley, 1976. Markov processes, stochastic processes, nonparametrics, categorical analysis, genetics, epidemiology.

Eberhard O. Voit, Ph.D., Cologne, 1981. Biomathematics, nonlinear modeling, regenerative growth, biological models simulation, predator-prey systems, S-systems.

Assistant Professors

Alexander A. Georgiev, Ph.D., Wroclaw Technical (Poland), 1984. Nonlinear systems analysis, nonparametric procedures in signal and image processing, computer vision.

James K. Dias, Ph.D., Medical University of South Carolina, 1973. Applications of statistics to the environment, marine sciences, and biology; health services research.

Daniel T. Lackland, Dr.P.H., Pittsburgh, 1990. Cardiovascular, cancer, and environmental epidemiology; survey methodology; epidemiology of hypertension and behavioral risk factors.

June Stevens, Ph.D., Cornell, 1986. Nutritional epidemiology, assessment of adiposity, determinants of energy intake, dietary fiber.

Zhen Zhang, Ph.D., Pittsburgh, 1987. Image and signal processing, artificial intelligence.

Instructors

Paul M. Darden, M.D., Houston, 1981. Epidemiology of vaccine preventable diseases, evaluation of health-care practice.

Ammasi Periasamy, Ph.D., Indian Institute of Technology, 1983. Biomedical image processing, medical optics.

Charles M. Robinson, M.S.T., American, 1969. Medical, hospital, and business computing.

Susan E. Sutherland, M.S., Medical University of South Carolina, 1984. Biostatistical consulting, statistical computing.

Dual/Joint Appointments

Thomas C. Hulsey, Research Associate; Ph.D., Johns Hopkins, 1987. Epidemiology of maternal and child health, family planning.

David L. Sisco, Instructor; M.S., Medical University of South Carolina, 1984. Information systems, medical databases and microcomputing.

Adjunct Faculty

Loren Cobb, Ph.D., Cornell, 1973. Artificial intelligence, nonlinear models, catastrophe theory, biomathematics.

Richard F. Daniels, Assistant Professor; Ph.D., Rutgers, 1981. Biological systems modeling, quantitative ecology, genetic response modeling, management decision models.

Bonnie P. Dumas, Instructor; M.S., Medical University of South Carolina, 1978. Forestry biometrics, statistical computing and information systems.

Emma L. Frazier, Ph.D., Medical University of South Carolina, 1988. Artificial intelligence, statistical computing, epidemiological research design.

Michael L. Geis, Associate Professor; Ph.D., MIT, 1970. Linguistics, natural language, artificial intelligence.

Gary W. Harrison, Associate Professor; Ph.D., Michigan State, 1975. Mathematical ecology, dynamic systems, stability and dynamic models.

C. Boyd Loadholt, Professor; Ph.D., Virginia Tech, 1968. Experimental design, statistical genetics, demography, epidemiology.

SECTION 5: MATHEMATICAL SCIENCES

MICHIGAN STATE UNIVERSITY
Department of Mathematics

Programs of Study

The department offers graduate work leading to the degrees of Master of Arts, Master of Arts for Teachers, Master of Computation Mathematics, Master of Science, Master of Science in Applied Mathematics, and Doctor of Philosophy in mathematics and applied mathematics. Doctoral candidates may pursue study and research in the areas of algebra, analysis, applied mathematics, combinatorics and graph theory, geometry, logic, and topology. (The faculty research activities section on the reverse of this page gives the specific research topics.) Students may take courses outside mathematics, for example, statistics or computer science. The department has an arrangement through which a student may receive a master's degree in mathematics and in a related field or may pursue a master's degree in an outside area while completing the doctorate in mathematics.

The master's programs require 45 quarter credits of approved course work and also require the student to pass a certifying exam covering a part of this course work. The M.A.T. degree requires the candidate to have completed the requirements for a teaching certificate. Normally, the master's program takes two years (three for a dual master's). The Ph.D. program requires students to pass a qualifying examination in basic mathematics, pass the comprehensive examination, show competence in two foreign languages, and write a thesis that is acceptable to the faculty. The qualifying examination can be waived for students with strong backgrounds.

Research Facilities

The Department of Mathematics is housed in Wells Hall, a modern air-conditioned office and classroom building. The Department of Statistics and Probability is also in Wells Hall. Assistants are usually assigned 3 or 4 per office.

The Vernon G. Grove Mathematics Library is housed in Wells Hall. It contains about 37,200 volumes and subscribes to more than 470 mathematical and statistical periodicals from all over the world. The University has IBM 3090, VAX 8650, and Convex C-220 computers that are available to faculty and students. The Department of Mathematics has a Sun SPARCstation laboratory dedicated to research use by faculty and graduate students. The department also has a room containing Macintoshes and other PCs that is open to students.

Financial Aid

The department usually employs about 110 teaching assistants each year. Their duties, including classroom work, paper grading, and student consultation, normally do not exceed 15 hours per week. In the 1990–91 academic year, the stipend for teaching assistants was approximately $9000–$10,500, depending on the student's qualifications and experience. The University offers fellowships, which, in conjunction with a small amount of teaching duties, pay $18,000 per year and carry a full tuition waiver. The assistantships carry a tuition waiver for 6 credits per term.

Cost of Study

The 1990–91 tuition was $97.75 per credit hour for Michigan residents and $197.75 per credit hour for out-of-state students. All assistants pay in-state tuition. The normal class load is 6–9 credits per quarter.

Cost of Living

Unmarried students may stay in the graduate dormitory. Currently, the cost of these accommodations is $718 per quarter, including the food plan. A limited number of apartments in housing for married students are also available for unmarried students. The cost of married student apartments is currently $252–$276 per month. Off-campus housing is also available.

Student Group

The University has a total enrollment of about 43,000, including 6,700 graduate students. There are about 150 graduate students in mathematics.

Location

The University occupies approximately 2,000 beautifully landscaped acres in East Lansing. East Lansing is a residential community, and the University is near its business district. East Lansing is adjacent to Lansing, the state capital. The University is about a 90-minute drive from Detroit.

Recreational facilities, both summer and winter, including many lakes and ski resorts, are nearby.

The University and The Department

Michigan State University, founded in 1855, is one of the leading institutions of higher education in the country. It ranks among the fifteen largest in the nation. All of the facilities of a large university are available to Michigan State's graduate students, including many cultural and sports events.

The Department of Mathematics has an important role in the academic life of the University. This is shown by the number of undergraduates taking mathematics courses and also by the number of graduate students from outside the department taking mathematics courses. On the undergraduate level, the mathematics teams entered by Michigan State University in the Putnam Competition have placed first on three occasions. On one occasion, members of the team placed first and second. Recent master's degree recipients have found teaching positions in high schools and junior colleges as well as employment in industry. Most of the department's Ph.D. recipients have gone into college or university teaching, although some in applied mathematics have taken jobs in industry.

Applying

Applicants for admission should have at least 15 quarter credits of mathematics beyond calculus. Applicants for financial aid must take the GRE Subject Test in mathematics. Students are admitted to the graduate program at any time. However, most assistantships start in September, and decisions on these are usually made in February. Therefore, all material, including GRE scores, should reach the department by February 1.

In line with its policy of encouraging people from all sectors of society to pursue graduate studies in mathematics, the department especially welcomes applications from women and members of minority groups.

Correspondence and Information

Chairperson
   or
Director of Graduate Studies
Department of Mathematics
Michigan State University
East Lansing, Michigan 48824-1027
Telephone: 517-353-4650

Peterson's Guide to Graduate Programs in the Physical Sciences and Mathematics 1992

Michigan State University

THE FACULTY AND AREAS OF RESEARCH

Algebra
W. C. Brown, Professor; Ph.D., Northwestern. J. I. Hall, Professor; D.Phil., Oxford. M. D. Hestenes, Professor; Ph.D., Michigan. K. K. Hickin, Associate Professor; Ph.D., Michigan State. E. C. Ingraham, Professor; Ph.D., Oregon. J. W. Kerr, Associate Professor; Ph.D., California, San Diego. W. E. Kuan, Professor; Ph.D., Berkeley. U. Meierfrankenfeld, Associate Professor; Ph.D., Bielefeld (Germany). R. E. Phillips, Professor; Ph.D., Kansas. C. Rotthaus, Professor; Ph.D., Münster (Germany). S. E. Schuur, Assistant Professor; Ph.D., Michigan. I. Sinha, Professor; Ph.D., Wisconsin–Madison. L. M. Sonneborn, Professor; Ph.D., Caltech. M. L. Tomber, Professor; Ph.D., Pennsylvania. C. Tsai, Professor; Ph.D., IIT. B. Ulrich, Professor; Ph.D., Saarland (Germany). D. L. Winter, Professor; Ph.D., Carnegie Tech.

Analysis
G. D. Anderson, Professor; Ph.D., Michigan. S. Axler, Professor; Ph.D., Berkeley. S. V. Dragosh, Associate Professor; Ph.D., Wisconsin–Milwaukee. C. C. Ganser, Associate Professor; Ph.D., Wisconsin–Madison. J. C. Kurtz, Professor; Ph.D., Utah. P. K. Lamm, Associate Professor; Ph.D., Brown. P. A. Lappan, Professor; Notre Dame. J. J. Masterson, Professor; Ph.D., Purdue. T. L. McCoy, Professor; Ph.D., Wisconsin–Madison. W. H. Ow, Professor; Ph.D., UCLA. W. C. Ramey, Associate Professor; Ph.D., Wisconsin–Madison. H. Salehi, Professor; Ph.D., Indiana. J. D. Schuur, Professor; Ph.D., Michigan. J. H. Shapiro, Professor; Ph.D., Michigan. W. T. Sledd, Professor; Ph.D., Kentucky. A. Volberg, Professor; Ph.D., Leningrad. C. E. Weil, Professor; Ph.D., Purdue. P. K. Wong, Professor; Ph.D., Carnegie Tech.

Applied Mathematics
C. Chiu, Assistant Professor; Ph.D., Carnegie Mellon. B. Drachman, Professor; Ph.D., Brown. Q. Du, Assistant Professor; Ph.D., Carnegie Mellon. D. R. Dunninger, Professor; Ph.D., Maryland. D. W. Hall, Professor; Ph.D., Princeton. R. O. Hill, Associate Professor; Ph.D., Northwestern. N. L. Hills, Professor; Ph.D., NYU. F. Hoppensteadt, Professor; Ph.D., Wisconsin–Madison. P. K. Lamm, Associate Professor; Ph.D., Brown. T. Y. Li, Professor; Ph.D., Maryland. C. Y. Lo, Professor; Ph.D., Maryland. C. R. MacCluer, Professor; Ph.D., Michigan. M. Miklavcic, Associate Professor; Ph.D., Virginia Tech. J. D. Schuur, Professor; Ph.D., Michigan. L. M. Sonneborn, Professor; Ph.D., Caltech. V. P. Sreedharan, Professor; Ph.D., Carnegie Tech. C. Y. Wang, Professor; Ph.D., MIT. M. J. Winter, Professor; Ph.D., Carnegie Tech. P. K. Wong, Professor; Ph.D., Carnegie Tech. D. Yen, Professor; Ph.D., NYU. V. Zeidan, Associate Professor; Ph.D., British Columbia.

Combinatorics and Graph Theory
J. I. Hall, Professor; D.Phil., Oxford. E. M. Palmer, Professor; Ph.D., Michigan. B. E. Sagan, Associate Professor; Ph.D., MIT.

Differential Geometry
D. E. Blair, Professor; Ph.D., Illinois. B. Y. Chen, Professor; Ph.D., Notre Dame. W. E. Kuan, Professor; Ph.D., Berkeley. G. D. Ludden, Professor; Ph.D., Notre Dame. T. H. Parker, Associate Professor; Ph.D., Stanford. B. Ulrich, Professor; Ph.D., Saarland (Germany). J. G. Wolfson, Associate Professor; Ph.D., Berkeley.

Logic
K. K. Hickin, Associate Professor; Ph.D., Michigan State. J. M. Plotkin, Professor; Ph.D., Cornell.

Mathematics Education
W. H. Fitzgerald, Professor; Ph.D., Michigan. G. Lappan, Professor; Ed.D., Georgia. J. J. Masterson, Professor; Ph.D., Purdue. S. L. Senk, Associate Professor; Ph.D., Chicago. I. E. Vance, Professor; Ph.D., Michigan. M. J. Winter, Professor; Ph.D., Carnegie Tech.

Topology
S. Akbulut, Professor; Ph.D., Berkeley. H. S. Davis, Associate Professor; Ph.D., Illinois. R. A. Fintushel, Professor; Ph.D., SUNY at Binghamton. M. Fuchs, Professor; Ph.D., Saarland (Germany). K. W. Kwun, Professor; Ph.D., Michigan. J. D. McCarthy, Associate Professor; Ph.D., Columbia. D. A. Moran, Professor; Ph.D., Illinois. R. C. O'Neill, Associate Professor; Ph.D., Purdue. C. L. Seebeck, Professor; Ph.D., Florida State.

RESEARCH ACTIVITIES

Algebra. Finite groups, infinite groups, linear groups, group representation, group rings, coding theory, noncommutative rings, commutative algebra and algebraic geometry, logic, recursive functions.

Analysis. Differentiation and integration theory (real variable), boundary behavior and cluster sets of functions in the unit disk, HP spaces, operator theory, quasi-conformal mappings, Riemann surfaces and several complex variables, harmonic analysis.

Applied Mathematics. Ordinary and partial differential equations, solid and fluid mechanics, dynamical systems, operations research, applied functional analysis, numerical analysis, information and control theory, applied probability, biomathematics.

Combinatorics and Graph Theory. Enumeration, algebraic combinatorics, random graphs, algorithms, extremal graph theory, finite geometry, designs.

Differential Geometry. Complex Koehler, symplectic, and Riemannian geometry; geometric partial differential equations and Yang-Mills theory; symmetric spaces; submanifold theory; variational problems.

Topology. Low-dimensional manifolds, Teichmuller theory, topology of algebraic varieties, gauge theory, classifying spaces, topology of quantum field theory.

SECTION 5: MATHEMATICAL SCIENCES

MICHIGAN TECHNOLOGICAL UNIVERSITY

Department of Mathematical Sciences

Program of Study The department offers the M.S. degree in mathematics. The objective is to train students for careers in teaching, industry, government, or the national laboratories or to prepare students for advanced work and research leading to the doctoral degree in mathematics or related areas. The department offers three options with corresponding course work requirements: applied mathematics (ordinary and partial differential equations, approximate analysis, and numerical analysis), pure mathematics (topology, algebra, geometry, and real or complex analysis), and statistics (mathematical statistics, stochastic processes, linear models, multivariate analysis, and data analysis). The Graduate School imposes additional requirements. They include 45 quarter credit hours of course work and/or research work. There are three options for completing the degree requirements: course work, research thesis, or independent study project. Most students choose the third option. Each student must pass an oral examination. In the first option, the examination covers only the course work, while in the second and third options it generally covers the thesis or project material. The typical student with the equivalent of a bachelor's degree in mathematics from MTU takes about two years to complete the degree requirements for the master's degree. During the first five quarters, students concentrate on course work, while the last quarter or two are generally devoted more to independent study under the supervision of a faculty member. Thesis problems have recently included theoretical and computational fluid mechanics, possibly in connection with the Fluids Research Oriented Group (FROG); numerical solutions of boundary value problems; Householder matrices and applications; and the effect of nonnormality of $\bar{x}$ and R charts.

Research Facilities The Michigan Tech library contains 710,000 volumes and subscribes to 12,000 serials and periodicals, many in the mathematical sciences. The library has open stacks, microfilm, graduate student and faculty carrels, interlibrary loan privileges, photocopying facilities, and computerized bibliographic search services.
Academic Computing Services maintains computer facilities, including IBM 4381 R14 running under VM/SP HPO level 4.2, with three remote batch stations and various terminal rooms that access the system; three microcomputer laboratories that house IBM and Apple microcomputers with associated hardware and software; one of the most complete computer graphics design laboratories in the nation; two Sequents and an Intel Hypercube in the Center for Experimental Computation; Sequent; Celerity; two VAX-11/750 systems; an FPS-164 array processor; a Simult 2010 six-node parallel processing computer; an Ardent 2 CPU graphics supercomputer; and links to BITNET, MERIT, and Internet. The department's equipment includes several Sun, IBM, and Apple machines. Departmental faculty and graduate students have access to research groups on campus, including FROG, Institute for Condensed Matter Studies, Institute for Wood Research, Institute of Materials Processing, and Keweenaw Research Center. Available to departmental faculty and students are Ph.D. programs in other departments, such as mechanical engineering/engineering mechanics, chemistry and chemical engineering, structural engineering, sensing and signal processing, biology, forestry, and applied physics.

Financial Aid The majority of mathematics graduate students are supported by graduate teaching assistantships (GTA), which are considered half-time teaching. Duties include teaching one or possibly two classes per quarter. The stipend is $7200 for the 1991–92 academic year (nine months); tuition is paid by the University. Graduate research assistantships may be available.

Cost of Study In 1990–91, quarterly tuition for graduate students was $76.75 per credit hour (for 1–11 credit hours) for Michigan residents and $172.75 per credit hour for others. (Tuition is waived for those with a GTA.) There was a matriculation fee of $20 per quarter and a Memorial Union Expansion fee of $15 per quarter.

Cost of Living University residence hall accommodations, including a nineteen-meal plan, range from $3288 for double occupancy to $3764 for single occupancy. The University maintains one-, two-, and three-bedroom apartments; rates are $275 per month for one bedroom, $305 for two bedrooms, and $370 for three bedrooms. Off-campus apartments and rooms are available and are generally considered to be inexpensive compared with the rest of the nation. Other costs of living are generally low to moderate.

Student Group During 1990–91, there were 20 students (including 6 women), 14 of whom were graduate teaching assistants and 1 of whom graduated with an M.S. degree. For 1991–92, the department plans to enroll 20 students and have available seventeen graduate teaching assistantships. (The department expects to be able to maintain that level or grow slowly.) Recent graduates have gone on to Ph.D. programs elsewhere, been hired to teach at the community college level, or been hired by industry.

Location The University is located in Houghton, Michigan. Proximity to Lake Superior provides moderation of temperatures in summer and winter and an abundance of snowfall (over 200 inches per year on average). Autumn and spring are cool and colorful. The remoteness imposed by geography is mitigated by the beauty of the countryside, which is relatively unspoiled and free of pollution. There are abundant opportunities for outdoor recreational activities during all seasons.

The University Michigan Technological University was founded in 1885 as a school of mining and metallurgical engineering. It has expanded its programs to include nearly all areas of science and engineering. It has an excellent reputation in engineering and science education at the undergraduate and graduate levels, and it excels in several spheres of science and technology. The University operates an indoor ice arena, an eighteen-hole golf course, 8 kilometers of Nordic ski trails, its own ski hill with a chair lift, and an indoor tennis center. Cultural attractions are sponsored by the University and surrounding community.

Applying Application materials should be obtained from the department and submitted to the Graduate School. It is assumed that applicants will have a bachelor's degree in mathematics. Students with a background in another field will be considered for admission if there is evidence to indicate likely success in the M.S. program. Evaluations for admission are based on undergraduate performance, both overall and in mathematics, and letters of recommendation. Submission of GRE test scores is recommended. TOEFL scores are required for all foreign applicants whose native language is not English. There are no fixed deadlines for receipt of applications. However, evaluation of applications for financial aid begins in late March.

Correspondence and Information
Anant P. Godbole
Director of Graduate Studies
Department of Mathematical Sciences
Michigan Technological University
Houghton, Michigan 49931
Telephone: 906-487-2884

Peterson's Guide to Graduate Programs in the Physical Sciences and Mathematics 1992

Michigan Technological University

THE FACULTY AND THEIR RESEARCH

Alphonse H. Baartmans, Professor and Department Head; Ph.D. (mathematics), Michigan State, 1967. Combinatorics, design theory, algebra.
John P. Beckwith, Associate Professor; Ph.D. (mathematical statistics), Wayne State, 1970. Statistics.
Barbara S. Bertram, Assistant Professor; Ph.D. (applied mathematics), New Mexico, 1987. Singular integral equations, numerical analysis.
Phyllis O. Boutilier, Associate Professor; M.S. (mathematics), Michigan Tech, 1968. Linear algebra, mathematics transition (high school to college in science and engineering), placement and course design.
Thomas D. Drummer, Assistant Professor; Ph.D. (statistics), Wyoming, 1985. Statistical ecology, model-based sampling, applications of statistics to wildlife management.
Lee Erlebach, Assistant Professor; Ph.D. (harmonic analysis), Washington (Seattle), 1969. Discrete mathematics, topological groups, game theory.
William P. Francis, Associate Professor; Ph.D. (physics), Cornell, 1969. Applied mathematics, cosmology, field theories.
Michael J. Gilpin, Associate Professor; Ph.D. (algebra), Oregon, 1972. Combinatorics, discrete mathematics.
Beverly J. Gimmestad, Associate Professor; Ph.D. (mathematics education), Colorado at Boulder, 1976. Mathematical problem-solving strategies, instructional technology, spatial visualization.
Clark Givens, Associate Professor; M.S. (physics), Michigan, 1960. Applied linear algebra, signal processing, differential geometry, mathematical physics.
Anant P. Godbole, Assistant Professor; Ph.D. (probability and statistics), Michigan State, 1984. Strong limit theorems of probability, applied probability.
Mangalam R. Gopal, Associate Professor; Ph.D. (mathematics), Michigan, 1967. Extremal problems in the class of univalent functions.
Sidney W. Graham, Associate Professor; Ph.D. (mathematics), Michigan, 1977. Analytic number theory.
Konrad J. Heuvers, Professor; Ph.D. (mathematics), Ohio State, 1969. Functional equations, linear algebra, combinatorics, group theory.
Darrell L. Hicks, Professor; Ph.D. (mathematics), New Mexico, 1969. Mathematical and computer modeling, material dynamics, parallel algorithms and advanced architectures, numerical analysis..
John W. Hilgers, Associate Professor; Ph.D. (mathematics), Wisconsin–Madison, 1973. Integral equations, functional analysis, signal processing, EM-wave generation and propagation, astrophysics and cosmology.
Glenn R. Ierley, Associate Professor; Ph.D. (applied mathematics), MIT, 1982. Geophysical fluid dynamics, hydrodynamic stability theory.
Robert W. Kolkka, Assistant Professor; Ph.D. (applied mechanics), Lehigh, 1982. Bifurcation and stability theory, viscoelasticity, non-Newtonian fluid mechanics, nonlinear elastic buckling analysis, constitutive equations.
Donald L. Kreher, Associate Professor; Ph.D. (mathematics and computer science), Nebraska, 1984. Computational combinatorics, coding theory, algorithms, cryptography.
Kenneth Kuttler, Assistant Professor; Ph.D. (mathematics), Texas, 1981. Partial differential equations.
Gilbert N. Lewis, Associate Professor; Ph.D. (applied mathematics), Wisconsin–Milwaukee, 1976. Asymptotics, singular perturbations, numerical solutions of ODE boundary value problems.
Daniel S. Moak, Associate Professor; Ph.D. (mathematics), Wisconsin–Madison, 1979. Special functions, functional equations.
Russell M. Reid, Associate Professor; Ph.D. (applied mathematics), Wisconsin–Madison, 1979. Control theory of distributed systems.
Otto G. Ruehr, Professor; Ph.D. (applied mathematics), Michigan, 1963. Classical analysis, problem solving, continued fractions, asymptotics, integral transforms.
Robert J. Spahn, Associate Professor; Ph.D. (mathematical physics), Michigan State, 1963. Applied mathematics, wave propagation.
Allan A. Struthers, Assistant Professor; Ph.D. (mathematics), Carnegie Mellon, 1991. Applied mathematics, continuum mechanics.
Vladimir D. Tonchev, Associate Professor; Ph.D. (mathematics), Bulgarian Academy of Sciences, 1987. Combinatorics, coding theory.
Peter C. Wollan, Assistant Professor; Ph.D. (statistics), Iowa, 1985. Inequality-constrained estimation and testing, mathematical statistics, statistical computing.

SECTION 5: MATHEMATICAL SCIENCES

MOUNT SINAI SCHOOL OF MEDICINE
of the City University of New York

Department of Biomathematical Sciences

Program of Study

The Department of Biomathematical Sciences seeks to train future leaders of theoretical research in the quantitative biology of the twenty-first century. Students who enter this program will have extraordinary opportunities to develop careers at the forefront of the biological revolution. Areas of faculty expertise span the biomathematical sciences, with emphasis on macromolecular structure and function, physiology, epidemiology, and statistics.

All entering students take a core curriculum that includes molecular and cellular biology, mathematical methods and modeling, numerical and computational methods, and statistics. In the first and second years, students are oriented to the research opportunities available in the department and the School by participating on a rotating basis in circumscribed research projects with each of several faculty members. First-year and second-year students also participate in a seminar program and a journal club. In addition to the core curriculum, students choose advanced courses in accord with their interests. These may include appropriate courses given at any branch of the City University of New York, at the Courant Institute of Mathematical Sciences of New York University, or at other graduate institutions in Manhattan. As students become increasingly involved in the research program, each develops a dissertation proposal. After passing a written proficiency examination in biomathematical sciences and giving a satisfactory oral presentation of the dissertation proposal, the student devotes full time to research in close consultation with his or her chosen adviser. The progress of each student is overseen at all stages by a committee of 3 advisers, who ensure that the program is carefully tailored to individual needs and interests. The usual time needed to complete all requirements for a Ph.D. degree is four to five years.

The department's extensive links with other basic science and clinical units in the medical school provide ample opportunities to develop collaborative research projects with experimental groups. In addition, the department maintains close ties with other institutions in the New York area that have active research programs in mathematical biology. These include the Courant Institute of Mathematical Sciences of New York University, the Computer Science Department of CUNY, the Applied Mathematics Department at SUNY at Stony Brook, and the Cold Spring Harbor Laboratory.

A combined M.D./Ph.D. program also is available.

Research Facilities

The department is located on the thirteenth floor of the Annenberg Building at the Mount Sinai School of Medicine. Its modern and well-equipped computing resources include a Convex C220 parallel vectorizing computer, a VAX 4000 computer, Hewlett-Packard and Sun workstations, and several graphics workstations. Extensive calculations also can be performed on the IBM 3090/400 located at the CUNY computer center. All local computing resources are interconnected through a high-speed network. Access to remote systems is available via BITNET, Internet, and Nysernet.

Financial Aid

Students are supported to a common stipend level using Mount Sinai and CUNY fellowships, graduate assistantships and traineeships, and departmental research grant funds. In the 1991–92 year, students receive tuition, a stipend of $14,000 per annum, and a book/travel allowance of $350 per annum.

Cost of Study

The cost of study usually is met by financial aid awards, as described above.

Cost of Living

Student housing is available for all students at a wide range of costs. Housing is available in dormitories and other buildings owned by the School of Medicine.

Student Group

The biomathematical sciences subarea was recently added to the doctoral program at Mount Sinai. The first students to enter are in their third year of study. A low student-faculty ratio is maintained to preserve a productive and congenial atmosphere in which students receive close faculty guidance.

Location

The Mount Sinai School of Medicine derives many advantages from its location adjoining Central Park on the Upper East Side of Manhattan. Several important medical and scientific institutions are nearby, including the New York Academy of Medicine and the New York Academy of Sciences. Students can attend meetings, symposia, and conferences at these centers. Students also have access to several major specialty libraries in the area. The recreational and cultural opportunities available in New York City are without peer. Discounted tickets and free passes to many theater, concert, and dance performances, as well as to other cultural events, are available to students through the Mount Sinai Entertainment Office.

The School

The School of Medicine was founded in 1963 as part of the Mount Sinai Medical Center, a major medical research center. The biomedical sciences doctoral program is affiliated with the City University of New York and its Graduate Center. The biomathematical sciences subarea was added to this doctoral program in 1988.

Applying

The department seeks to attract students with strong research potential. Because of the interdisciplinary nature of the program, students with a wide variety of majors are encouraged to apply. Academic background should include sound training in mathematics and/or computer science, plus at least one year each of physics, biology, and chemistry and a course in organic chemistry. In some cases, prerequisites can be made up after admission.

If an applicant is interested in biomathematics but would like to explore other areas before committing to a program, Mount Sinai offers a flexible entry option.

All applicants to the doctoral program must submit Graduate Record Examinations scores, TOEFL scores if their native language is not English, official transcripts of undergraduate and any previous graduate work, and letters of recommendation. Application is made directly to the Graduate School of the Mount Sinai School of Medicine.

Correspondence and Information

Director of Graduate Studies
Department of Biomathematical Sciences
Box 1023
Mount Sinai School of Medicine
1 Gustave Levy Place
New York, New York 10029

Mount Sinai School of Medicine

THE FACULTY AND THEIR RESEARCH

Craig J. Benham, Professor and Acting Chairman; Ph.D. (mathematics), Princeton, 1972. Topological properties of biomolecules and their relationships to energetics, structure, and function, with emphasis on nucleic acids.

Agnes Berger, Professorial Lecturer; Ph.D. (mathematics), Budapest, 1939. Statistical methods for studying chronic diseases, late-onset genetic diseases.

Carol Bodian, Assistant Professor; Dr.P.H. (biostatistics), Columbia, 1983. Biostatistics for clinical research, cancer epidemiology.

David Greenberg, Associate Professor; Ph.D. (physiology and biophysics), Washington (St. Louis), 1976. Population genetics, human genetics.

Warren Hirsch, Professorial Lecturer; Ph.D. (mathematics), NYU, 1952. Epidemiology of infectious diseases, theoretical immunology, mathematical physiology of the kidney.

Michael Lacker, Associate Professor; M.D., Ph.D. (physiology), NYU, 1977. Mathematical physiology, reproductive endocrinology, developmental biology.

Tai-Shing Lau, Assistant Professor; Ph.D. (statistics), Purdue, 1983. Epidemiology and screening tests, meta-analysis, Bayesian estimation, optimal design, mathematical physiology.

John Mandeli, Assistant Professor; Ph.D. (biostatistics), Cornell, 1978. Design and analysis of clinical experiments, combinatorial problems in experimental design.

George Nemethy, Professor; Ph.D. (chemistry), Cornell, 1962. Theory of protein structures and interactions, physical chemistry of aqueous solutions.

Henry Sacks, Associate Professor; M.D., Ph.D. (microbiology), SUNY at Albany, 1971. Clinical epidemiology, including meta-analysis, decision analysis, clinical trials methodology, and AIDS epidemiology.

Harold Scheraga, Adjunct Professor; Ph.D. (physical chemistry), Duke, 1946. Theory of protein conformation and the mechanisms of folding, chemistry of blood clotting, structure of water and dilute aqueous solutions.

Istvan Sugar, Associate Professor; Ph.D. (biophysics), Eötvös Loránd (Budapest), 1973. Theory of magnetic resonance, theoretical membrane biophysics.

Sylvan Wallenstein, Associate Professor; Ph.D. (statistics), Rutgers, 1971. Biostatistics, with applications to epidemiology, clustering, clinical trials, and experimental design.

SECTION 5: MATHEMATICAL SCIENCES

NEW YORK UNIVERSITY

Courant Institute of Mathematical Sciences
Department of Mathematics

Program of Study

The graduate program offers balanced training in mathematical analysis and the applications of mathematics in the broadest sense. It includes computational applied mathematics and cross relations with computer science. The program of study leads to the M.S. and Ph.D. degrees. It is possible to earn both degrees by part-time study, but most students in the Ph.D. program are full-time.

Several options exist for obtaining the M.S. degree, including career-oriented programs for students who do not plan to engage in graduate study in mathematics beyond the master's degree. One of these is conducted jointly with the Graduate School of Business Administration. The M.S. can be completed in the equivalent of three or four terms of full-time study.

Doctoral students obtain the M.S. degree as they fulfill the requirements for the Ph.D. degree. A student must accumulate 72 course and research points for the Ph.D., but no specific courses are required. From the outset, students (in consultation with their advisers) select courses that prepare them for the degree examinations and for research. The Ph.D. program also involves independent reading of mathematical literature, active participation in a seminar, written and oral preliminary examinations, and preparation and defense of a doctoral dissertation. Candidates for the degree must demonstrate a reading knowledge of French, German, or Russian. Students are encouraged to complete the Ph.D. in four years of full-time study.

The department occupies a leading position in applied mathematics and mathematical analysis, especially in ordinary and partial differential equations, probability theory, and mathematical physics. In applied mathematics, the program encompasses many research activities not found in most mathematics departments: wave propagation, computational fluid dynamics, numerical analysis, magnetofluiddynamics, combustion theory, statistical mechanics, climatology, geophysics, mathematical biology, and aerodynamics. Some of the applied research, directed by the faculty of the Courant Institute, is conducted in connection with major federally sponsored basic research projects involving large staffs of postdoctoral scientists.

The department is successful in helping its Ph.D. graduates find positions in first-rate university departments and in nonacademic employment.

Research Facilities

An outstanding mathematical sciences library and a large-scale computer installation are housed together with the other facilities of the Courant Institute in Warren Weaver Hall. The library collection consists of more than 55,000 volumes. The library subscribes to more than 550 journals devoted to the mathematical sciences. The Courant Institute possesses a large array of computer facilities. Students involved in certain research projects also have network access to the largest computers in the United States.

Financial Aid

Fellowships and assistantships are awarded to students who intend to engage in full-time Ph.D. study; they cover tuition and, in 1991–92, provide a stipend of $13,000 for the nine-month academic year. Some summer positions dealing with Courant Institute research projects are available to assistants with computational skills. Because the department is unable to support all of its full-time students, applicants should apply for other support as well. U.S. students with a strong interest in applications should note that Hertz Foundation Fellowships are tenable at the Courant Institute. Federally funded low-interest loans are available to qualified U.S. citizens on the basis of need.

Cost of Study

In 1991–92, tuition and fees are calculated at $467 per point. A full-time program of study normally consists of 24 points per year (four 3-point courses each term). A limited deferred tuition payment plan is in effect.

Cost of Living

University housing for graduate students is limited. It consists mainly of shared studio apartments in buildings adjacent to Warren Weaver Hall and shared suites in residence halls within walking distance of the University. University housing rents in 1991–92 range from $504 to $682 per month.

Student Group

In 1990–91, the department had approximately 170 graduate students. More than half were full-time students.

Location

New York City is a world capital for art, music, and drama and for the financial and communications industries. NYU is located at Washington Square in Greenwich Village, just north of SoHo, in a residential neighborhood consisting of apartments, lofts, art galleries, theaters, restaurants, and shops.

The University and The Institute

New York University, founded in 1831, enrolls 46,000 students and is one of the major private universities in the country. Its various schools offer a wide range of undergraduate, graduate, and professional degrees. Among its internationally known divisions is the Courant Institute of Mathematical Sciences. Named for its founder, Richard Courant, the Institute combines research in the mathematical sciences with advanced training at the graduate and postdoctoral levels. Its activities are supported by the University, government, industry, and private foundations and individuals. The graduate program in mathematics is conducted by the faculty of the Courant Institute. The mathematics department ranks among leading departments in the country and is the only highly distinguished department to have made applications so much a central concern of its programs. Nine members of the Courant Institute faculty are members of the National Academy of Sciences.

Applying

The graduate program is open to students with strong mathematical interests, regardless of their undergraduate major. They are expected to have a knowledge of the elements of mathematical analysis. Applications for admission are evaluated throughout the year, but a major annual review of applications for financial aid occurs in February, and most awards for the succeeding academic year are made by early March. Financial aid applications must include GRE scores on both the General and Subject tests and must be received by December 15.

Correspondence and Information

For program and financial aid information:
Fellowship Committee
Courant Institute
New York University
251 Mercer Street
New York, New York 10012
Telephone: 212-998-3256

For application forms and a Graduate School bulletin:
Office of Admissions and Financial Aid
Graduate School of Arts and Science
New York University
6 Washington Square North
New York, New York 10003
Telephone: 212-998-8050

SECTION 5: MATHEMATICAL SCIENCES

New York University

THE FACULTY AND THEIR RESEARCH

Professors
Jack Bazer, Ph.D. Partial differential equations, applied mathematics.
Jerome Berkowitz, Ph.D. Differential equations, magnetofluiddynamics.
Simeon M. Berman, Ph.D. Probability theory and stochastic processes.
Sylvain Cappell, Ph.D. Algebraic and geometric topology.
Jeff Cheeger, Ph.D. Differential geometry and its connections with topology and analysis.
W. Stephen Childress, Ph.D. Fluid dynamics, applied mathematics.
Demetrios Christodoulou, Ph.D. General relativity, nonlinear partial differential equations, differential geometry.
Martin Davis, Ph.D. Mathematical logic, theory of computation, logic programming.
Percy A. Deift, Ph.D. Mathematical physics, scattering theory.
Monroe Donsker, Ph.D. Probability theory, stochastic processes.
Harold M. Edwards, Ph.D. Number theory, history of mathematics.
Paul R. Garabedian, Ph.D. Computational fluid dynamics, partial differential equations.
Jonathan Goodman, Ph.D. Numerical analysis, fluid dynamics, partial differential equations.
Frederick P. Greenleaf, Ph.D. Functional analysis, Lie groups, invariant partial differential equations.
Eliezer Hameiri, Ph.D. Magnetofluiddynamics, applied mathematics.
Melvin Hausner, Ph.D. Nonstandard analysis, geometry, combinatorics.
Samuel N. Karp, Ph.D. Applied mathematics, electromagnetic theory.
Robert V. Kohn, Ph.D. Elasticity, composite materials, optimal control, formation of singularities.
Anneli Lax, Ph.D. Mathematical education and exposition.
Peter D. Lax, Ph.D. Fluid dynamics, partial differential equations, computing.
Fang-Hua Lin, Ph.D. Partial differential equations, geometric measure theory.
Henry P. McKean, Ph.D. Partial differential equations, Riemann surfaces.
Cathleen S. Morawetz, Ph.D. Transonic flow, scattering theory.
Charles M. Newman, Ph.D. Statistical mechanics, probability theory, mathematical ecology.
Louis Nirenberg, Ph.D. Partial differential equations, differential geometry, complex analysis.
Albert B. J. Novikoff, Ph.D. Analysis, history of mathematics, pedagogy.
George Papanicolaou, Ph.D. Stochastic differential equations, applied mathematics.
Jerome Percus, Ph.D. Mathematical physics, mathematical biology.
Charles S. Peskin, Ph.D. Physiology, fluid dynamics, numerical methods.
Richard M. Pollack, Ph.D. Discrete and computational geometry.
Jacob T. Schwartz, Ph.D. Robotics and computer vision, computer design, language design, compiler optimization, nonnumerical computations, operating systems.
Harold N. Shapiro, Ph.D. Number theory, algebra, combinatorics.
Jalal Shatah, Ph.D. Partial differential equations, analysis.
Joel H. Spencer, Ph.D. Algebra, combinatorics, probabilistic methods, theory of algorithms.
Gang Tian, Ph.D. Differential geometry, Kahler geometry, analysis on manifold, complex analysis.
Lu Ting, Ph.D. Applied mathematics, fluid mechanics.
Srinivasa S. R. Varadhan, Ph.D. Probability theory, stochastic processes, partial differential equations.
Harold Weitzner, Ph.D. Plasma physics, fluid dynamics, differential equations.
Olof Widlund, Phil.Dr. Numerical analysis, partial differential equations, parallel computing.

Associate Professors
Marco M. Avellaneda, Ph.D. Partial differential equations, homogenization, turbulence theory.
Malcolm Goldman, Ph.D. Probability and statistics, functional analysis.
Graeme Milton, Ph.D. Theoretical physics, composite materials.
Daniel Tranchina, Ph.D. Vision, neural coding, mathematical modeling.

Assistant Professors
Thomas Buttke, Ph.D. Computational physics, fluid dynamics, vortex dynamics.
Leslie Greengard, Ph.D. Scientific computing, fast algorithms, potential theory.
Thomas Hou, Ph.D. Numerical analysis, computational fluid dynamics, homogenization.
Tamar Schlick, Ph.D. Mathematical biology, numerical analysis, computational chemistry.
Zhouping Xin, Ph.D. Hyperbolic equations, nonlinear conservation laws.
Horng-Tzer Yau, Ph.D. Partial differential equations, statistical mechanics.

Affiliated Professor
Alan Sokal, Ph.D.

Courant Institute Instructors
Yi Fang, Ph.D.
Changfeng Gui, Ph.D.
Phillippe LeFloch, Ph.D.
Sijue Wu, Ph.D.

SECTION 5: MATHEMATICAL SCIENCES

NORTHWESTERN UNIVERSITY

Program in Applied Mathematics

Programs of Study

Graduate programs leading to the M.S. and Ph.D. degrees in applied mathematics are offered by the Department of Engineering Sciences and Applied Mathematics. Qualified students with backgrounds in engineering, mathematics, or natural science are eligible for admission to these programs. Study plans are drawn up to meet the needs of the individual student; they encompass courses in mathematical methods and in one or more fields of science or engineering, where significant applications of mathematics are made. These programs prepare students for careers in fields in which the uses of mathematical tools play essential roles.

A student can obtain the M.S. degree in one academic year of full-time study. This entails the successful completion of an approved program of courses, followed by an examination relative to the work. No thesis is required for the M.S. degree. The Ph.D. program takes a minimum of three years beyond the B.S. degree. For the Ph.D. degree, the student must achieve a distinguished record in an approved program of courses and pass both a preliminary and a qualifying examination in the general research area to be followed in the doctoral dissertation. A final examination on the doctoral dissertation is required upon its completion.

Research Facilities

The Department of Engineering Sciences and Applied Mathematics is part of the McCormick School of Engineering and Applied Science, which includes computer sciences, materials science, and the traditional engineering departments. A Science-Engineering Library has an extensive collection of journals, books, and other reference works. The Vogelback Computing Center is also nearby.

Financial Aid

Students who have been accepted for graduate study toward the Ph.D. in applied mathematics are eligible for various forms of financial support. University scholarships and fellowships as well as teaching and research assistantships cover tuition costs and provide a monthly stipend for living expenses. Some students may also qualify for a University loan. All full-time students are entitled to the benefits of the University Health Service as well as hospitalization and surgical insurance coverage.

Cost of Study

Tuition totals $13,995 for the academic year 1991–92, with varying rates applicable during the summer quarter.

Cost of Living

A variety of housing is available in the area. In addition to University-owned dwellings, there are apartments and rooms to be rented in the surrounding community. Costs vary according to the type of housing and the location. General living costs are slightly above the national average.

Student Group

Graduate students in the applied mathematics program come from all parts of the United States and from various foreign countries. There are about 35 students in the program.

Location

The main campus of the University is located in the residential suburb of Evanston, immediately north of Chicago. It occupies some 170 acres, partly bounded by a mile of Lake Michigan shoreline. This location provides the combined advantages of a lovely suburban community and the many cultural opportunities of Metropolitan Chicago. The University maintains its own beach for use by students, faculty, and staff.

The University

Northwestern University was founded in 1851 and is the only privately supported school in the Big Ten. It is a coeducational institution that offers a full range of educational opportunities. Research awards at the University exceeded $100-million last year.

Applying

Northwestern University operates on the quarter system. It is usual practice for new students to enter at the beginning of the fall quarter. Admissions applications that request financial aid should be completed by late January.

Correspondence and Information

Director of Applied Mathematics
Department of Engineering Sciences and Applied Mathematics
McCormick School of Engineering and Applied Science
Northwestern University
Evanston, Illinois 60208-3125

Peterson's Guide to Graduate Programs in the Physical Sciences and Mathematics 1992

SECTION 5: MATHEMATICAL SCIENCES

Northwestern University

THE FACULTY AND THEIR RESEARCH

Jan D. Achenbach, Ph.D., Stanford. Theoretical and applied mechanics.
Erwin H. Bareiss, Ph.D., Zurich. Numerical analysis and code development.
Alvin Bayliss, Ph.D., NYU (Courant). Numerical analysis, scientific computations.
Andrew J. Bernoff, Ph.D., Cambridge. Fluid dynamics, chemical instabilities, pattern selection.
Stephen H. Davis, Ph.D., RPI. Hydrodynamic stability, interfacial phenomena, phase-change phenomena.
Thomas Erneux, Ph.D., Free University of Brussels. Chemical instabilities and quantum optics.
William L. Kath, Ph.D., Caltech. Wave propagation and optics.
Moshe Matalon, Ph.D., Cornell. Asymptotic methods, combustion, fluid dynamics.
Bernard J. Matkowsky, Ph.D., NYU (Courant). Bifurcation and stability, combustion, stochastic differential equations.
Michael J. Miksis, Ph.D., NYU (Courant). Multiphase flow and free boundary problems.
Toshio Mura, Ph.D., Tokyo. Micromechanics.
W. Edward Olmstead, Ph.D., Northwestern. Reaction-diffusion theory and integral equations.
Edward L. Reiss, Ph.D., NYU. Bifurcation analysis, elastic buckling and hydrodynamic stability.
Hermann Riecke, Ph.D., Bayreuth (Germany). Theoretical physics, phase dynamics in fluid flows.
M. Grae Worster, Ph.D., Cambridge. Fluid dynamics, convection heat and mass transfer.
Tai Te Wu, Ph.D., Harvard. Protein structure and bacterial evolution.

Visiting Faculty (1991–92) and Their Affiliations
D. S. Ahluwalia, Courant Institute, New York University.
M. Booty, Southern Methodist University.
M. Garbey, University of Valenciennes (France).
L. M. Hocking, University College, London.
S. W. Joo, University of Michigan.
M. Klosek-Dygas, University of Wisconsin–Milwaukee.
C. Knessl, University of Illinois at Chicago.
S. Linz, Saarbruecken University (Germany).
Z. Schuss, Tel-Aviv University.
G. Shkadinskaya, Institute of Chemical Physics, USSR Academy of Sciences.
K.G. Shkadinsky, Institute of Structural Macrokinetics, USSR Academy of Sciences.
C. Tier, University of Illinois at Chicago Circle.
L. Ting, Courant Institute, New York University.
A. Umantsev, Central Scientific Research Institute of Ferrous Metallurgy (USSR).
Vitaly A. Volpert, Institute of Chemical Physics, USSR Academy of Sciences.
Vladimir A. Volpert, Institute of Structural Macrokinetics, USSR Academy of Sciences.

SECTION 5: MATHEMATICAL SCIENCES

NORTHWESTERN UNIVERSITY
Department of Mathematics

Programs of Study

Graduate programs leading to the M.S. and Ph.D. degrees are offered by the Department of Mathematics. The program of study is quite flexible and offers a wide range of options in pure and applied mathematics. There are strong faculty research interests in the areas of algebra, algebraic topology, analysis, applied mathematics, differential geometry, dynamical systems, ergodic theory, partial differential equations, and probability. Students are guided in their studies and research by distinguished and internationally known mathematicians in all these fields. The program prepares students for careers in teaching and basic research, as well as in industrial and commercial fields where mathematics plays a prominent role.

A student can obtain the M.S. degree in one academic year of full-time study. This entails the successful completion of an approved program of courses, followed by an examination based on the work. No thesis is required for the M.S. degree. The Ph.D. program requires a minimum of three years beyond the B.S. degree. For the Ph.D. degree, the student must achieve a distinguished record in an approved program of courses and pass a qualifying examination based on that work and the general research area to be followed in the doctoral dissertation. A final examination is required upon completion of the dissertation.

Research Facilities

The Department of Mathematics has its own building, Lunt Hall, which contains offices for faculty and students, as well as classrooms and conference rooms. The mathematics library is also located in Lunt Hall and contains a complete collection of most mathematical journals and texts. Faculty and students have complete access to the library 24 hours a day. The facilities of Vogelback Computing Center, ten Sun Workstations, and several microcomputers are readily available for research by faculty and students.

Financial Aid

Students who have been accepted for graduate study in mathematics are eligible for various forms of financial support. University scholarships and fellowships, as well as teaching and research assistantships, cover tuition costs and provide a monthly stipend for living expenses. Some students may also qualify for a University loan. All full-time students are entitled to the benefits of the University Health Services, as well as to hospitalization and surgical insurance coverage.

Cost of Study

Tuition totals $13,995 for the 1991–92 academic year.

Cost of Living

A variety of housing is available in the area. In addition to University-owned graduate student housing, there are apartments and rooms for rent in the surrounding community. Costs vary according to the type of housing and the location.

Student Group

Students in the mathematics program come from all parts of the United States and from various foreign countries. There are 43 students in the graduate program in 1991–92.

Location

The main campus of the University is located in the residential suburb of Evanston, immediately north of Chicago. It occupies some 170 acres, including a mile of Lake Michigan shorefront. This location provides the combined advantages of a lovely suburban community and the many cultural opportunities of Metropolitan Chicago. The University maintains its own beach to be used by faculty, students, and staff.

The University

Northwestern University was founded in 1851 and is the only privately supported school in the Big Ten Conference. It is a coeducational institution that offers a full range of educational opportunities. Northwestern University has world-famous theater and music departments, which present theatrical and concert programs for the University community. In addition, many outside groups and speakers are brought to the campus for special events.

Applying

Northwestern University operates on the quarter system. While students may begin graduate study in any quarter, it is standard practice to enter at the beginning of the fall quarter. Admission applications that request financial aid should be completed by January 15. Other applications may be submitted by August 15.

Correspondence and Information

Application forms, the *Graduate Study in Mathematics* booklet, and additional information can be obtained by writing to the address below.

Graduate Program
Department of Mathematics
Lunt Hall
Northwestern University
2033 Sheridan Road
Evanston, Illinois 60208

Peterson's Guide to Graduate Programs in the Physical Sciences and Mathematics 1992

SECTION 5: MATHEMATICAL SCIENCES

Northwestern University

THE FACULTY AND THEIR RESEARCH

Donald G. Austin. Probability.
Michael G. Barratt. Topology, algebraic and geometric homological algebra.
Alexandra Bellow. Probability and analysis.
Ralph P. Boas (Emeritus). Fourier series and integrals, complex analysis, inequalities.
Keith Burns. Ergodic theory, differential geometry.
Allen Devinatz. Functional analysis, differential equations.
Emmanuele DiBenedetto. Partial differential equations, analysis.
Meyer Dwass (Emeritus). Statistics, probability.
Leonard Evens. Algebra, number theory, group theory, computers in mathematics.
Stephen D. Fisher. Analysis, complex analysis.
John Franks. Dynamical systems.
Eric Friedlander. Algebraic geometry, algebraic topology.
George Gasper. Analysis, Fourier analysis, special functions.
Colin C. Graham. Analysis, Fourier analysis, functional analysis.
Elton Pei Hsu. Probability and analysis.
C. Ionescu-Tulcea. Analysis.
Joseph W. Jerome. Analysis, constructive function theory, numerical functional analysis.
Daniel S. Kahn. Algebraic topology.
Joshua Leslie. Lie groups, global analysis.
Mark E. Mahowald. Algebraic topology.
Eben Matlis. Algebra, commutative rings.
Kenneth R. Mount. Algebraic geometry, mathematical economics.
Mark A. Pinsky. Probability, stochastic differential equations.
Stewart B. Priddy. Algebraic topology, group cohomology.
Clark Robinson. Dynamical systems.
Donald G. Saari. Analysis, differential equations, celestial mechanics.
Judith D. Sally. Algebra, commutative ring theory.
Michael R. Stein. Algebra, algebraic K-theory.
Neil Trudinger. Partial differential equations.
Robert Welland. Analysis.
Sandy Zabell. Probability, statistics.
Daniel Zelinsky. Algebra, rings and modules.

An aerial view of the Northwestern campus.

Northwestern University

SECTION 5: MATHEMATICAL SCIENCES

OHIO STATE UNIVERSITY

Department of Mathematics

Programs of Study

The Department of Mathematics offers programs leading to the M.A., M.S., and Ph.D. degrees. Courses of study are available in all of the principal branches of mathematics: algebra, analysis, applied mathematics, combinatorics, geometry, group theory, logic, probability, and topology.

Students pursuing a Master of Science degree may specialize in pure mathematics, applied mathematics, actuarial science, or interdisciplinary studies. The interdisciplinary studies concentration encompasses, for example, a program that is offered jointly by the Departments of Mathematics and Computer Science and leads to two M.S. degrees, one from each department. A program of study leading to the Master of Arts degree provides a mathematics-content degree specifically designed for teachers of secondary mathematics.

Research Facilities

The Ohio State University libraries have more than 3.3 million volumes and are served by a campuswide automated circulation system. In addition to the Main Library, there are twenty-four departmental libraries. The Department of Mathematics library serves graduate research in mathematics, statistics, and geodetic sciences. It houses more than 31,000 volumes and currently receives 450 serials. Students have ready access to microcomputers and the University mainframes. The Academic Computing Center is a research and service facility that serves all departments of the University, offering short-term seminars and noncredit courses directed to the needs of faculty, staff, and graduate students.

Financial Aid

Complete financial assistance, in the form of fellowships and/or teaching associateships, is offered to nearly every mathematics graduate student attending Ohio State. Both forms of support carry a stipend plus total waiver of all regular tuition and fees. In 1990–91, stipends ranged from $920 to $1034 per month. Teaching associateships of 5 hours a week involve 15 hours per week of classroom contact and preparation. (TAs in the M.A. program may teach 10 hours per week for $1840 per month.) Each new graduate student at Ohio State is also encouraged to accept an initial Head Start Summer Fellowship, which carries a stipend of $2000 in 1991. Many continuing students receive summer quarter support in the form of either fellowships or research associate or teaching associate appointments.

Cost of Study

In 1991–92, tuition for full-time graduate study is $1213 per quarter (Ohio residents) or $3143 (nonresidents).

Cost of Living

Convenient housing is located on and near the Columbus campus. Two dormitories house graduate students exclusively, at costs ranging from $215 to $265 per person per month. Ohio State maintains an apartment complex for married students; monthly rents range from $300 for a one-bedroom unit to $355 for a two-bedroom unit, including gas and water. The Off-Campus Student Center keeps files of available off-campus housing as a free service.

Student Group

The mathematics department has approximately 200 graduate students, representing twenty states and twenty-eight countries. Each year some 15 students earn the Ph.D. degree, while 20–25 earn the M.S. degree and 10–15 the M.A. degree. Most Ph.D. recipients seek and find employment in academic institutions, whereas M.S. recipients are employed for the most part by business, industry, or government.

Location

Columbus is centrally located in the eastern half of the country, 316 miles from Chicago, 555 miles from New York City, 560 miles from Atlanta, and 404 miles from St. Louis. The Columbus metropolitan area has approximately 1.25 million residents. It is the only major population center in Ohio to have gained residents since 1970. Columbus has a wide variety of fine restaurants and cultural activities. The city supports its own orchestra and ballet and is home to one of the premier zoos in the nation. Temperatures average 73.5°F in August and 30.1°F in January. The average annual precipitation is 36.29 inches.

The University

The Ohio State University, comprising seventeen colleges, nine schools, four regional campuses, and the Graduate School, is the principal center for graduate and professional study in Ohio and one of the leading institutions of higher education in the United States. Approximately 55,000 students are enrolled in the University, nearly 10,000 pursuing a graduate or professional degree. Each year the University attracts large numbers of visiting scholars (about 60 in mathematics) who contribute to the intellectual vigor of the Ohio State community. Many Ohio State faculty members play important roles as consultants to federal and state government bodies and to private enterprise. The mathematics department is home to the International Mathematics Research Institute at the Ohio State University.

Applying

Candidates for admission to the Ohio State Graduate School must file an application form (application for a fellowship and/or a teaching associateship may be made on the same form) and all college transcripts with the Ohio State Admissions Office. In addition, applicants must forward to the address below three letters of recommendation, a brief autobiography, and scores on the GRE Subject Test in mathematics. Scores on the GRE General Test are required of all fellowship nominees.

GRE scores are not required of applicants to the Master of Arts program who have a B average or better. Applicants from non-English-speaking countries must take the TOEFL. Applications for fellowships must be completed before February 1.

Correspondence and Information

For application forms or information, students should write to:
Dijen Ray-Chandhuri, Chairman
Department of Mathematics
Ohio State University
231 West 18th Avenue
Columbus, Ohio 43210-1174
Telephone: 614-292-6274
Fax: 614-292-1479
Electronic mail: dijen@function.mps.ohio-state.edu

Peterson's Guide to Graduate Programs in the Physical Sciences and Mathematics 1992

SECTION 5: MATHEMATICAL SCIENCES

Ohio State University

THE FACULTY AND THEIR RESEARCH

Professors
Harry Allen, Ph.D., Yale: algebra. Avner Ash, Ph.D., Harvard: arithmetic groups. Bogdan Baishanski, Ph.D., Belgrade: analysis. Gregory Baker, Ph.D., Caltech: numerical analysis. Ranko Bojanic, Ph.D., Serbian Academy of Science: approximation theory, numerical analysis. Robert Brown, Ph.D., Chicago: algebra. Dan Burghelea, Ph.D., Romanian Academy: topology. Francis Carroll, Ph.D., Purdue: analysis. Ruth Charney, Ph.D., Princeton: algebraic topology. Michael Davis, Ph.D., Princeton: topology. William Davis, Ph.D., Case Tech: functional analysis. David Dean, Ph.D., Illinois: functional analysis. Frank Demana, Ph.D., Michigan State: group theory. Thomas Dowling, Ph.D., North Carolina: combinatorics. Alexander Dynin, Ph.D., Moscow: applied mathematics, partial differential equations. Gerald Edgar, Ph.D., Harvard: functional analysis, measure theory. Joseph Ferrar, Ph.D., Yale: algebra. Gregory Forest, Ph.D., Arizona: nonlinear partial differential equations, wave theory. Harvey Friedman, Ph.D., MIT: philosophy of mathematics. Ulrich Gerlach, Ph.D., Princeton: theoretical physics, quantum field theory. Henry Glover, Ph.D., Michigan: topology. Robert Gold, Ph.D., MIT: algebraic number theory. David Goss, Ph.D., Harvard: number theory. Koichiro Harada, Ph.D., Tokyo: group theory. John Hsia, Ph.D., MIT: number theory (quadratic forms). John Philip Huneke, Ph.D., Wesleyan: topology. Joan Leitzel, Ph.D., Indiana: mathematics curriculum and instruction. Manohar Madan, Ph.D., Göttingen (Germany): algebraic numbers, function theory. Stephen Milne, Ph.D., Caltech: combinatorics. Guido Mislin, Ph.D., Swiss Federal Institute of Technology: topology. Boris Mityagin, Ph.D., Moscow: functional analysis. Henri Moscovici, Ph.D., Bucharest: harmonic analysis, representation theory. Walter Neumann, Ph.D., Bonn: topology. Paul Nevai, Ph.D., Szeged (Hungary): approximation theory, orthogonal polynomials. Boris Pittel, Ph.D., Leningrad: analysis of combinatorial algorithms. Steve Rallis, Ph.D., MIT: harmonic analysis, representation theory. Dijen Ray-Chaudhuri, Ph.D., North Carolina: combinatorics. Neil Robertson, Ph.D., Waterloo: graph theory. Joseph Rosenblatt, Ph.D., Washington (Seattle): harmonic analysis, measure theory. Karl Rubin, Ph.D., Harvard: number theory. Ernst Ruh, Ph.D., Brown: differential geometry. Surinder Sehgal, Ph.D., Notre Dame: group/representation theory, group rings. Daniel Shapiro, Ph.D., Berkeley: quadratic forms. Warren Sinnott, Ph.D., Stanford: algebraic number theory. Ronald Solomon, Ph.D., Yale: theory of finite groups. Robert Stanton, Ph.D., Cornell: harmonic analysis on Lie groups. Louis Sucheston, Ph.D., Wayne State: ergodic theory, probability. Michel Talagrand, Ph.D., Paris: functional analysis. Bert Waits, Ph.D., Ohio State: math education. Alan Woods, Ph.D., Manchester: geometry of numbers, diophantine approximation. Bostwick Wyman, Ph.D., Berkeley: system/control theory.

Professors Emeriti
Henry Colson, Ph.D., Brown: applied mathematics. Arnold Ross, Ph.D., Chicago: number theory. Charles Saltzer, Ph.D., Minnesota: integral equations. Hans Zassenhaus, Ph.D., Hamburg: finite group theory.

Associate Professors
Vitaly Bergelson, Ph.D., Hebrew (Jerusalem): ergodic theory/probability. Timothy Carlson, Ph.D., Minnesota: logic, foundations and combinatorics. Luis Casian, Ph.D., MIT: representation theory. Andrzej Derdzinski, Ph.D., Wroclaw (Poland): differential geometry. Zita Divis, Ph.D., Heidelberg: partial differential equations, approximation theory. Neil Falkner, Ph.D., British Columbia: probability theory. Zbigniew Fiedorowicz, Ph.D., Chicago: topology. Yuval Flicker, Ph.D., Cambridge: representation theory. Matthew Foreman, Ph.D., Berkeley: logic. Thomas Kappeler, Ph.D., Brown: topology. Yuji Kodama, Ph.D., Clarkson: differential equations, mathematical physics. James Leitzel, Ph.D., Indiana: mathematics curriculum and instruction. Yung-Chen Lu, Ph.D., Berkeley: topology, algebraic geometry. George Majda, Ph.D., NYU (Courant): applied mathematics. William McWorter, Ph.D., Ohio State: group theory. Leroy Meyers, Ph.D., Illinois at Chicago: number theory. Edward Overman, Ph.D., Arizona: applied mathematics. Alayne Parson, Ph.D., Illinois at Chicago: number theory, complex analysis. Paul Ponomarev, Ph.D., Yale: number theory. Thomas Ralley, Ph.D., Illinois: mathematics of computation. John Riedl, Ph.D., Notre Dame: mathematics curriculum and instruction. John Scheick, Ph.D., Syracuse: analysis, numerical analysis. James Schultz, Ph.D., Ohio State: mathematics curriculum and instruction. Thomas Schwartzbauer, Ph.D., Minnesota: probability. Alice Silverberg, Ph.D., Princeton: number theory. Saleh Tanveer, Ph.D., Caltech: applied mathematics. David Terman, Ph.D., Minnesota: applied mathematics. Monique Vuilleumier, Ph.D., Geneva: analysis. Herbert Walum, Ph.D., Colorado: classical number theory. Sia Wong, Ph.D., Monash (Australia): group theory. Joseph Zilber, Ph.D., Harvard: topology. Shimshon Zimering, Ph.D., Brussels: real analysis, nuclear physics.

Assistant Professors
Anthony Bloch, Ph.D., Harvard: applied mathematics. Randall Dougherty, Ph.D., Berkeley: logic. Roy Joshua, Ph.D., Northwestern: applied mathematics. Peter March, Ph.D., Minnesota: probability. Crichton Ogle, Ph.D., Brandeis: topology. Akos Seress, Ph.D., Ohio State: combinatorics. Michael Shapiro, Ph.D., Michigan: topology. Amarendra Sinha, Ph.D., Columbia: applied mathematics. James Turner, Ph.D., Carnegie Mellon: applied mathematics.

Research Instructors and Instructors:
David Ginzburg, Research Instructor; Ph.D., Tel Aviv: representation theory. Thomas Ward, Research Instructor; Ph.D., Warwick (England): ergodic theory.
Tomas Erdelyi, Instructor; Ph.D., South Carolina: approximation theory. Patrick McDonald, Instructor; Ph.D., MIT: differential geometry. André Nachbin, Instructor; Ph.D., NYU (Courant): applied mathematics. Conrad Plaut, Instructor; Ph.D., Maryland: differential geometry. Alan Reid, Instructor; Ph.D., Aberdeen (Scotland): topology.

SECTION 5: MATHEMATICAL SCIENCES

OLD DOMINION UNIVERSITY

Department of Mathematics and Statistics

Programs of Study The department offers a graduate program (entitled Computational and Applied Mathematics) leading to the M.S. and Ph.D. degrees, with options in applied mathematics and statistics.

Master's students must complete a total of 36 credit hours of course work. Up to 9 of these credits may be chosen from a field of application in which the student applies analytical and numerical techniques to another discipline. The program is flexible and may vary considerably, depending on the student's interests and career goals. A master's thesis is not required.

The Ph.D. program requires a minimum of 24 credit hours beyond the master's degree and exclusive of doctoral dissertation work. A foreign language is not required.

In applied mathematics, students choose from courses in ordinary and partial differential equations, biomathematics, complex and real analysis, optimization techniques, numerical analysis, continuum mechanics, integral equations, transform methods, numerical fluid dynamics, tensor analysis, calculus of variations, and singular perturbation methods.

The statistics curriculum offers courses in sampling theory, design and analysis of experiments, regression and analysis of variance, stochastic models, nonparametric methods, reliability and life testing, linear models, multivariate analysis, time-series analysis, statistical inference, and probability theory. All students in the statistics option are required to complete a modeling project.

Research Facilities ODU's computer center operates a CDC CYBER 930-931 and an IBM 3090 with VM/XA-SP and MVS/SP operating systems. The department also owns several IBM microcomputers for student and faculty use and also has a Sun network.

The University library holdings of more than a million items are accessible through a computer output microfiche catalog located in each academic department. The library has a comprehensive collection of periodicals and books in support of the graduate curriculum and research in mathematics, statistics, and related areas. The library offers computer-assisted searches of over 100 indexing and abstracting services. Interlibrary loan services are conducted on-line and are available to graduate students.

Financial Aid Departmental graduate assistantships offer stipends ranging up to $8000 in 1991–92. Nonresidents of Virginia who hold assistantships pay resident tuition. Tuition is waived for doctoral students who have a master's degree and are graduate assistants. In addition, a number of teaching and research positions are available for financial support of graduate assistants during the summer months.

There are occasional opportunities for graduate students to participate in joint research projects at NASA Langley Research Center. NASA Fellowships carry stipends of $12,000 and provide for payment of tuition, fees, and some additional expenses.

Cost of Study Tuition for the academic year 1991–92 is $127 per semester hour for Virginia residents and $313 per semester hour for nonresidents. All holders of assistantships qualify for resident status.

Cost of Living Off-campus apartments are available, starting at about $300 per month for a furnished one-bedroom apartment. The area also offers opportunities for shared housing arrangements.

Student Group The University enrolls approximately 15,000 students; graduate students account for a quarter of the enrollment. There are about 50 graduate students in applied mathematics and statistics.

Location Norfolk, in eastern Virginia, is among the most rapidly growing urban areas along the Atlantic seaboard. The city is located in one of the most favored climatic regions of the United States. Proximity to the ocean and Gulf Stream moderates the summer's heat, and winter temperatures usually approximate those in northern Florida.

The University's 146-acre campus lies within the seven-city Hampton Roads metropolitan area that is nationally known for its historical, recreational, cultural, educational, and military facilities. The many points of interest include Virginia Beach; the Chesapeake Bay; the historic towns of Williamsburg, Jamestown, and Yorktown; commercial and naval shipping and shipbuilding facilities; and the NASA-Langley Research Center. The area supports a full schedule of cultural events, including theater, opera, ballet, dance, art exhibitions, symphony and popular music concerts, and lecture series.

The University Old Dominion University is a coeducational, state-supported public institution. ODU is organized into six academic colleges: Business Administration, Arts and Letters, Education, Engineering, Sciences, and Health Sciences. The University currently offers forty-seven master's programs and thirteen doctoral programs.

Applying Applicants to the master's program should have a bachelor's degree in either mathematics, statistics, computer science, or an application area with a strong mathematics component (such as physics or engineering). Undergraduate mathematics preparation should include course work in linear algebra, advanced calculus, differential equations, probability, and numerical methods.

Applicants to the Ph.D. program are normally students who have earned a master's degree in mathematics or statistics or a related application area.

Correspondence and Information
Graduate Program Director
Department of Mathematics and Statistics
Old Dominion University
Norfolk, Virginia 23529
Telephone: 804-683-3887

Old Dominion University

THE FACULTY AND THEIR RESEARCH

Applied Mathematics
John Adam, Ph.D., London. Mathematical biology, astrophysical magnetohydrodynamics, mathematical modeling.
Przemck Bogacki, Ph.D., SMU. Numerical solution of initial value problems for ordinary differential equations, including implementation for parallel processing.
Charlie H. Cooke, Ph.D., North Carolina State. Differential equations, numerical fluid mechanics, finite-element methods.
J. Mark Dorrepaal, Ph.D., Toronto. Stokes flow, transform methods, asymptotic analysis.
John H. Heinbockel, Ph.D., North Carolina State. Differential equations, systems analysis–optimization, numerical methods, mathematical modeling, integral transforms, tensor analysis, solar energy.
Jiashi Hou, Ph.D., RPI. Solid and fluid mechanics with application to the modeling of human joints.
Fang Hu, Ph.D., Florida State. Computational fluids with an emphasis in turbulent mixing.
Thomas L. Jackson, Ph.D., RPI. Combustion theory, asymptotic analysis.
Hideaki Kaneko, Ph.D., Clemson. Numerical solution of integral equations, fixed-point theory, approximation theory.
John E. Kroll, Ph.D., Yale. Geophysical fluid dynamics, oceanography.
David G. Lasseigne, Ph.D., Northwestern. Combustion theory, application of integral equations.
Wu Li, Ph.D., Penn State. Approximation theory, optimization theory.
Gordon Melrose, Ph.D., Old Dominion. Singular integral equations, special functions, fracture mechanics.
Richard D. Noren, Ph.D., Virginia Tech. Volterra integral equations.
John Swetits, Ph.D., Lehigh. Approximation theory, functional analysis.
John Tweed, Ph.D., Glasgow. Integral equations, transform techniques, elasticity and fracture mechanics.
Stanley E. Weinstein, Ph.D., Michigan State. Numerical analysis and approximation.
Philip R. Wohl, Ph.D., Cornell. Asymptotic analysis, differential equations, fluid dynamics, mathematical biology.

Statistics
N. Rao Chaganty, Ph.D., Florida State. Applied probability, reliability.
Ram C. Dahiya, Ph.D., Wisconsin. Statistical methodology and inference, biostatistics, applied statistics.
Michael J. Doviak, Ph.D., Florida. Applied statistics.
Larry D. Lee, Ph.D., Missouri. Statistical inference, reliability.
John P. Morgan, Ph.D., North Carolina. Design of experiments.
Dayanand N. Naik, Ph.D., Pittsburgh. Linear models.

OREGON STATE UNIVERSITY

College of Science
Department of Mathematics

Programs of Study

Graduate programs are offered leading to the M.A., M.S., and Ph.D. degrees. Courses of study are available in all of the principal branches of mathematics: algebra, analysis, applied mathematics, geometry, probability, and topology. In addition, there is a program in mathematics education leading to an M.A. or an M.S. degree with a teaching emphasis and a career program featuring an occupational internship leading to an M.A. or M.S. for students planning government or industrial careers in mathematics. There are separate Departments of Computer Science and Statistics, and joint work with the faculty from these departments is encouraged.

Research Facilities

The main library maintains an excellent mathematics collection, subscribing to more than 200 journals of pure and applied mathematics. The Department of Mathematics has a network of Sun Workstations for research computing. Numerous microcomputers are also attached to the network. The Sun system has various software packages for numerical computation, symbolic computation, and graphics. The departmental network is connected to the central campus network. This provides direct access to computer systems in other departments, to the campus computer center, and to Internet, which provides electronic mail service and direct access to remote facilities, such as the NSF supercomputer network.

Financial Aid

About ten to fifteen new graduate teaching assistantships are available each year carrying stipends of $7300–$8000 in 1991–92 for teaching one course per term and participating in related activities.

Cost of Study

In 1991–92, Oregon resident graduate students pay $912 per term and nonresident graduate students pay $1546 per term in tuition and fees. For graduate assistants, these amounts are reduced to $128 per term. Tuition is subject to change.

Cost of Living

Expenditures vary widely, but $4000–$5000 per year might be a reasonable average. The University has a few apartments for rent to married students at modest monthly rates. Student housing is plentiful and reasonably priced in the Corvallis area.

Student Group

There are about 50 graduate mathematics students, with about 45 holding graduate teaching assistantships. Each year, about a dozen students complete an advanced degree in mathematics. Graduate students come from all areas of the United States and several foreign countries.

Location

Oregon State University is located in Corvallis, a city of 41,000 that lies south of Portland in the gently rolling Willamette Valley. Seventy miles to the east, the Cascade Mountains invite campers, hikers, climbers, and skiers. Beachcombing, crabbing, clam digging, and salmon fishing are frequent temptations for those visiting the Oregon Coast, 50 miles to the west. The climate in Corvallis is mild, with winter lows ranging in the high 20s and summer highs in the low 90s. International artists are drawn to Oregon State during the year through the Music Association, the Friends of Chamber Music, and the OSU Foundation Center. Various regional activities in art, music, and drama are promoted by the Corvallis Arts Center and other local groups.

The University

Oregon State University was founded in 1868 and granted its first graduate degrees in 1876. It has been a land-grant university from its beginning and a sea-grant university since 1971. Graduate work and research are being carried on in all major areas of physical science and mathematics.

Applying

Applications for admission to the Graduate School should be made to the Office of Admissions, Oregon State University. Applications for graduate teaching assistantships should be made to the Department of Mathematics. There is no definite closing date for applications for assistantships, but offers are made about February 15 and March 15 each year. Sometimes there are a few additional openings later in the spring or summer. Oregon State University supports equal educational opportunity without regard to sex, race, handicap, national origin, marital status, or religion.

Correspondence and Information

Graduate Committee
Department of Mathematics
Oregon State University
Corvallis, Oregon 97331–4605
Telephone: 503-737-4686

SECTION 5: MATHEMATICAL SCIENCES

Oregon State University

THE FACULTY AND THEIR RESEARCH

The following mathematics faculty members supervise programs in the fields indicated.

C. S. Ballantine, Ph.D., Stanford, 1959. Matrix theory.
R. M. Burton Jr., Ph.D., Stanford, 1977. Probability, ergodic theory, dynamical systems.
L. K. Chen, Ph.D., Chicago, 1986. Harmonic analysis, singular integral equations.
J. Davis, Ph.D., Wisconsin, 1966. Numerical analysis, approximation theory, nonlinear problems.
T. P. Dick, Ph.D., New Hampshire, 1984. Mathematics education.
T. Dray, Ph.D., Berkeley, 1981. Differential geometry, relativity.
A. Faridani, Ph.D., Universität Münster, 1988. Applied mathematics, numerical analysis.
B. I. Fein, Ph.D., Oregon, 1965. Algebraic number theory, division algebras, group representation.
D. V. Finch, Ph.D., MIT, 1977. Singularity of mappings, partial differential equations, tomography.
F. J. Flaherty, Ph.D., Berkeley, 1965. Differential geometry, differential topology, general relativity, gauge theories.
M. E. Flahive, Ph.D., Ohio State, 1976. Number theory, algebra.
D. J. Garity, Ph.D., Wisconsin, 1980. Geometric topology.
R. B. Guenther, Ph.D., Colorado, 1964. Applied mathematics.
R. L. Higdon, Ph.D., Stanford, 1981. Partial differential equations, numerical analysis.
J. W. Lee, Ph.D., Stanford, 1969. Differential and integral equations, approximation theory.
L. F. Murphy, Ph.D., Carnegie-Mellon, 1980. Applied mathematics, biomathematics, modeling.
G. L. Musser, Ph.D., Miami (Florida), 1970. Mathematics education.
M. N. L. Narasimhan, Ph.D., Kharagpur (India), 1958. Continuum mechanics, fracture mechanics, turbulence.
S. M. Newberger, Ph.D., MIT, 1964. Applications of analysis.
M. E. Ossiander, Ph.D., Washington (Seattle), 1985. Probability.
H. R. Parks, Ph.D., Princeton, 1974. Geometric measure theory, minimal hypersurfaces.
B. E. Petersen, Ph.D., MIT, 1968. Partial differential equations, pseudo differential operators.
P. J. Pohjanpelto, Ph.D., Minnesota, 1989. Symmetries of differential equations, Lie groups, applied mathematics.
R. O. Robson, Ph.D., Stanford, 1981. Real algebraic geometry, commutative algebra.
R. M. Schori, Ph.D., Iowa, 1964. Infinite dimensional topology.
J. M. Shaughnessy, Ph.D., Michigan State, 1976. Mathematics education, Van Hiele levels, misconceptions of probability and statistics.
J. W. Smith, Ph.D., Columbia, 1957. Algebraic and differential topology.
D. C. Solmon, Ph.D., Oregon State, 1974. Image reconstruction, transform theory.
E. A. Thomann, Ph.D., Berkeley, 1985. Partial differential equations and applications.
E. Waymire, Ph.D., Arizona, 1976. Probability, mathematical physics, geophysics.
J. A. C. Weideman, Ph.D., Orange Free State (South Africa), 1986. Applied mathematics, numerical analysis.
H. L. Wilson, Ph.D., Illinois, 1966. Mathematics education.

Kidder Hall, location of the Department of Mathematics.

Kerr Library, the main library, is located opposite Kidder Hall.

SECTION 5: MATHEMATICAL SCIENCES

OREGON STATE UNIVERSITY

College of Science
Department of Statistics

Programs of Study The department offers graduate work leading to the M.S. and Ph.D. degrees. Major fields are applied statistics, biometry, mathematical statistics, and operations research (M.S. only). Students can concentrate on theory or applications, and programs can be tailored to emphasize such areas of interest as ecology, engineering, forestry, mathematics, or oceanography.

The M.S. degree is designed to prepare a candidate for a career in industry or government or for further study at the Ph.D. level. An M.S. candidate chooses two of the three basic sequences in applied statistics, mathematical statistics, and operations research and must pass comprehensive examinations covering the two areas selected. Each candidate must gain some consulting and teaching experience. A thesis or research paper is optional. Prerequisites for admission are multivariable calculus, linear algebra, and an undergraduate sequence in mathematical statistics, but some applicants without the usual prerequisites may be admitted on a provisional basis. A student can normally complete the M.S. program in four to six quarters.

To enter the Ph.D. program, a student must have the equivalent of an M.S. degree and must pass the department's M.S. comprehensive examinations. A Ph.D. candidate takes one or two years of advanced course work. Current research areas in the department include analysis of enumerative data, nonparametric statistics, asymptotics, experimental design, generalized regression models, linear model theory, reliability theory, sampling methodology, statistical and systems ecology, survival analysis, and wildlife surveys. A student can normally complete the Ph.D. degree program in two to four years.

Research Facilities The department is located in Kidder Hall, which is adjacent to the University library and the University Computing Services Center. In addition to these University facilities, the department maintains its own library and its own personal computers. The departmental library contains approximately 700 volumes of research and reference works and subscribes to most of the major journals in statistics and operations research. The departmental computer system consists of an expanding network of Sun Workstations and personal computers.

Financial Aid In 1991–92, assistantships carry stipends ranging from $2800 to $8100 per academic year, depending on the qualifications of the applicant and the source of funds. Depending upon the terms of the appointment, graduate assistants are required to spend 8 to 20 hours per week assisting in teaching or departmental research. Tuition is waived for graduate assistants, but fees of approximately $140 per term must be paid. A limited number of Oregon Merit Awards are also available. They provide $2000 to be used toward tuition for first-year graduate students not receiving assistantships.

Cost of Study In 1991–92, tuition and fees for graduate students are approximately $1030 per term for residents and approximately $1700 per term for nonresidents. Oregon residency can be established within a year or less. Foreign students must obtain financial clearance from the Office of International Education.

Cost of Living Single students may obtain room and board in University residence halls for approximately $3175 per academic year. Residence hall housing is also available during the summer term. The University maintains a number of furnished apartments for married students. Rents for off-campus housing vary considerably but are reasonable considered by national standards. Additional information is available from the University's Department of Housing.

Student Group The Department of Statistics, which has an active student statistical association, typically has about 50 graduate students. In the past three years, the average number of degrees awarded annually has been ten M.S. and three Ph.D. degrees. Enrollment in the University is about 16,000 students, of whom approximately 2,800 are graduate students.

Location The city of Corvallis, an attractive community with a population of 42,000, is situated in the Willamette Valley between Portland and Eugene. An hour to the west is the Pacific Ocean; an hour to the east, the Cascade Mountains. The region offers numerous recreational opportunities, including camping, fishing, hiking, hunting, and skiing. A variety of cultural activities are offered in Corvallis, Eugene, and Portland, and the famous Oregon Shakespearean Festival is in Ashland. The area has a temperate climate with generally mild winters and sunny summers with moderate temperatures.

The University and The Department Oregon State offers a variety of programs in its various schools and colleges: Agricultural Sciences, Business, Education, Engineering, Forestry, Health and Physical Education, Home Economics, Liberal Arts, Oceanography, Pharmacy, Science, and Veterinary Medicine. The University started as a land-grant institution in 1868 and in 1971 was also designated a sea-grant institution.

A statistical consulting and computer service was initiated at Oregon State in 1947, and ten years later the Department of Statistics was established. The first M.S. degree was awarded in 1962 and the first Ph.D. degree in 1969. The University has an active Survey Research Center, established in 1973.

Applying Application forms for graduate study at OSU are available from the Department of Statistics and through the Office of Admissions. Two copies of the completed form, a $40 nonrefundable application fee, official (sealed) transcripts of work completed at other institutions, and a letter indicating the applicant's particular field of interest should be sent to the Office of Admissions. A third copy of the completed application form, three letters of reference, additional transcripts, GRE scores (optional), and a copy of the applicant's letter of interest should be sent to the Department of Statistics.

Applications for admission to the Department of Statistics are considered at any time, but applicants desiring financial aid must apply by March 1.

Correspondence and Information
Director of Graduate Studies
Department of Statistics
Oregon State University
Kidder Hall 44
Corvallis, Oregon 97331-4606

SECTION 5: MATHEMATICAL SCIENCES

Oregon State University

THE FACULTY AND THEIR RESEARCH

Jeffrey L. Arthur, Associate Professor; Ph.D., Purdue, 1977. Operations research.
Helen Berg, Assistant Professor; M.S., Oregon State, 1973. Sampling, applied statistics.
David S. Birkes, Associate Professor; Ph.D., Washington (Seattle), 1969. Linear models, mathematical statistics.
H. Daniel Brunk, Professor Emeritus; Ph.D., Rice, 1944. Fellow of IMS and ASA; elected member of ISI. Probability, mathematical statistics.
David A. Butler, Professor; Ph.D., Stanford, 1975. Operations research, reliability analysis.
Lyle D. Calvin, Professor Emeritus; Ph.D., North Carolina State, 1953. Fellow of ASA and AAAS; elected member of ISI. Sampling methods, experimental design.
G. David Faulkenberry, Professor; Ph.D., Oklahoma State, 1965. Inference, sample surveys.
Robert G. Mason, Professor; Ph.D., Stanford, 1962. Survey methodology.
Allan H. Murphy, Professor; Ph.D., Michigan, 1974. Bayesian statistics, decision analysis, applications in meteorology and climatology.
W. Scott Overton, Professor; Ph.D., North Carolina State, 1964. Statistical ecology, systems ecology.
Cliff Pereira, Research Associate; Ph.D., Oregon State, 1985. Linear models, applications of statistics in biology.
Dawn Peters, Assistant Professor; Ph.D., Florida, 1988. Nonparametric statistics, asymptotics.
Roger G. Petersen, Professor Emeritus; Ph.D., North Carolina State, 1954. Applied statistics, experimental design.
Donald A. Pierce, Professor; Ph.D., Oklahoma State, 1965. Fellow of IMS and ASA; elected member of ISI. Theory of inference, applied statistics.
Fred L. Ramsey, Professor; Ph.D., Iowa State, 1964. Elected member of ISI. Statistical ecology, time series, stochastic processes.
Kenneth E. Rowe, Professor; Ph.D., Iowa State, 1966. Statistical computing, design and analysis of experiments.
Daniel W. Schafer, Associate Professor; Ph.D., Chicago, 1982. Applied statistics, regression analysis, generalized linear models.
Justus F. Seely, Professor and Chairman; Ph.D., Iowa State, 1969. Fellow of IMS and ASA; elected member of ISI. Linear models.
David R. Thomas, Professor; Ph.D., Iowa State, 1965. Nonparametric statistics, applied statistics.
N. Scott Urquhart, Professor; Ph.D., Colorado State, 1965. Applied and environmental statistics.
Edward C. Waymire, Associate Professor; Ph.D., Arizona, 1976. Probability, stochastic processes.

SECTION 5: MATHEMATICAL SCIENCES

PURDUE UNIVERSITY

Department of Mathematics

Programs of Study

The Department of Mathematics offers programs leading to the degrees of Master of Science and Doctor of Philosophy. There are several programs leading to the Master of Science degree, some of which prepare the student to seek nonacademic employment; others prepare the student to continue to the Ph.D. degree.

The master's degree program requires 30 hours of course work. There are no required oral or written examinations, and a thesis is not required. A student with a half-time teaching assistantship normally takes two years to complete the master's degree program.

Among the requirements for the Ph.D. are a minimum of 42 hours of graduate work, reading knowledge in two foreign languages, passing a written qualifying examination and oral specialty examinations, writing a thesis, and passing a final oral examination based on the thesis. A student with a half-time teaching assistantship would require a minimum of four years to complete the Ph.D. program, and most students spend five or six years in the program.

Research Facilities

The Mathematics Library, located in the Mathematical Sciences Building, features an outstanding collection of research journals and reference materials in pure and applied mathematics. The department maintains a network of over twenty-five Sun Workstations, several high performance graphics workstations, and equipment for high-quality graphics output. Supported software includes TEX, MACSYMA, Maple, Mathematica, and MATLAB. University facilities for research computing include CDC CYBER 205 and ETA-10 supercomputers.

Financial Aid

Beginning graduate students who intend to work toward the Ph.D. degree and who supply application material by February 1 will be considered for a fellowship. Purdue University Fellowships provide a tax-free, nine- or twelve-month stipend of $1167 per month. This fellowship is renewable for a second year. The department also makes nominations for the Doctoral Fellowship Program of Black and Other Ethnic Minorities, which provides a tax-free, nine- or twelve-month stipend of $1000 per month for up to three years, with a research allowance totaling $1200 available during the second and third years. Final selection of these fellows from the list of nominees is made by University committees.

A number of graduate teaching assistantships are available with stipends ranging from $8500 to $12,500 per academic year (1991–92), depending on duties. Half-time positions are available in the summer for assistants who perform satisfactorily in course work and assistantship duties.

For advanced students, research fellowships are available for both the summer and the academic year.

Cost of Study

Students holding any of the fellowships or assistantships described above receive remission of tuition and fees (in 1991–92, $1162 for Indiana residents and $3720 for nonresidents per semester), except for $205 each semester and $95 for the summer session.

Cost of Living

Unmarried students may live in one of the graduate dormitories. The minimum cost is $6.90 a day for room only. The University operates 1,300 apartments for married students. Monthly rates range from $255 to $347, depending on the type of accommodations. These rates normally include heat, gas, water, and electricity, but not telephone.

Student Group

Purdue has approximately 28,000 undergraduate students and 5,000 graduate students. There are about 175 graduate students in mathematics, nearly all of whom receive financial support.

Location

Purdue's main campus is located on 650 acres in the city of West Lafayette, Indiana, across the Wabash River from Lafayette. The population of the two cities exceeds 65,000. Amtrak and major bus lines serve the Greater Lafayette area. Two airlines, American Eagle to Chicago-O'Hare and US Air Express to Dayton, Ohio, operate out of the Purdue Airport. Purdue is 60 miles northwest of Indianapolis, the state capital, and 126 miles southeast of Chicago. The West Lafayette campus offers a wide variety of cultural and recreational opportunities for graduate students and their families.

The University

During the past decade, Purdue has ranked high among American universities in awarding the Doctor of Philosophy degree. The Graduate School is nationally recognized as being in the top category for the competence of its faculty and the quality of its graduates in the broad life science area, engineering, and basic physical sciences and mathematics.

Applying

Application forms may be obtained by writing to the Mathematics Graduate Office. Completed applications for assistantships beginning in the fall semester should be received before March 15. Completed applications for fellowships should be received by February 1. Applicants should arrange to take the General Test and the Subject Test in mathematics of the Graduate Record Examinations so that scores are received by the department before February 1. A minimum TOEFL score of 550 is required for all applicants whose native language is not English. Purdue does not discriminate against qualified handicapped persons in any of its programs or activities.

Purdue is an Equal Opportunity/Equal Access University.

Correspondence and Information

Graduate Committee Chairman
Department of Mathematics
Mathematical Sciences Building
Purdue University
West Lafayette, Indiana 47907
Telephone: 317-494-1961

Peterson's Guide to Graduate Programs in the Physical Sciences and Mathematics 1992

Purdue University

THE FACULTY AND THEIR RESEARCH

I. Aberbach, Visiting Assistant Professor; Ph.D., Michigan 1990. Commutative algebra.
S. Abhyankar, Professor; Ph.D., Harvard, 1955. Algebraic geometry.
D. Arapura, Associate Professor; Ph.D., Columbia, 1985. Algebraic geometry.
A. Arnold, Research Assistant Professor; Ph.D., Berlin Technical, 1990. Applied mathematics.
L. Avramov, Professor; Ph.D., Moscow, 1975. Commutative algebra.
R. Bañuelos, Associate Professor; Ph.D., UCLA, 1984. Probability, harmonic analysis.
P. Bauman, Associate Professor; Ph.D., Minnesota, 1982. Partial differential equations, real analysis.
J. Becker, Professor; Ph.D., Michigan, 1964. Algebraic topology.
S. Bell, Professor; Ph.D., MIT, 1980. Several complex variables.
M. Benjamin, Associate Professor; Ph.D., Moscow, 1971. Group representations.
L. Berkovitz, Professor; Ph.D., Chicago, 1951. Control theory.
L. de Branges, Professor; Ph.D., Cornell, 1957. Functional analysis.
J. Brown, Associate Professor; Ph.D., Michigan, 1979. Complex variables.
L. Brown, Professor; Ph.D., Harvard, 1968. Operator theory.
N. Carlson, Assistant Professor; Ph.D., Berkeley, 1991. Applied mathematics.
D. Catlin, Professor; Ph.D., Princeton, 1978. Several complex variables.
C. Cowen, Professor; Ph.D., Berkeley, 1976. Operator theory.
B. Davis, Professor; Ph.D., Illinois, 1968. Probability.
H. Donnelly, Professor; Ph.D., Berkeley, 1974. Differential geometry.
J. Douglas, Professor; Ph.D., Rice, 1952. Computational modeling, numerical analysis.
D. Drasin, Professor; Ph.D., Cornell, 1966. Complex variables.
M. Drazin, Professor; Ph.D., Cambridge, 1953. Algebra.
E. Dubinsky, Professor; Ph.D., Michigan, 1962. Mathematics education.
A. Eremenko, Visiting Professor; Ph.D., Rostov (USSR), 1979. Analysis.
E. Farjoun, Visiting Professor; Ph.D., MIT, 1971. Topology.
R. Gambill, Professor; Ph.D., Purdue, 1954. Differential equations.
N. Garofalo, Associate Professor; Ph.D., Minnesota, 1987. Partial differential equations.
W. Gautschi, Professor; Ph.D., Basel (Switzerland), 1953. Numerical analysis.
D. Goldberg, Research Assistant Professor; Ph.D., Maryland, 1991. Representation theory.
D. Gottlieb, Professor; Ph.D., UCLA, 1962. Algebraic topology.
W. Heinzer, Professor; Ph.D., Florida, 1966. Commutative algebra.
C. Huneke, Professor; Ph.D., Yale, 1978. Commutative algebra.
R. Hunt, Professor; Ph.D., Washington (St. Louis), 1965. Harmonic analysis.
S. Lalley, Associate Professor; Ph.D., Stanford, 1980. Probability.
L. Lempert, Professor; Ph.D., Eötvös Loránd Tudományegyetem (Budapest), 1979. Several complex variables.
J. Lillo, Professor; Ph.D., Princeton, 1957. Differential equations.
J. Lipman, Professor; Ph.D., Harvard, 1965. Algebraic geometry.
L. Lipshitz, Professor; Ph.D., Princeton, 1972. Logic.
B. Lucier, Professor; Ph.D., Chicago, 1981. Numerical analysis.
R. Lynch, Professor; Ph.D., Harvard, 1963. Numerical analysis.
M. Marson, Research Assistant Professor; Ph.D., California, San Diego, 1990. Several complex variables.
F. Milner, Associate Professor; Ph.D., Chicago, 1983. Numerical analysis.
T. Moh, Professor; Ph.D., Purdue, 1969. Algebraic geometry.
T. Mullikin, Professor; Ph.D., Harvard, 1958. Applied mathematics.
C. Neugebauer, Professor; Ph.D., Ohio State, 1954. Harmonic analysis.
F. Nier, Visiting Assistant Professor; Ph.D., Polytechnic (Paris), 1991. Applied mathematics.
R. Penney, Professor; Ph.D., MIT, 1971. Group representations.
D. Phillips, Associate Professor; Ph.D., Minnesota, 1981. Partial differential equations.
J. Price, Professor; Ph.D., Pennsylvania, 1956. Orthogonal expansions.
P. Protter, Professor; Ph.D., California, San Diego, 1975. Probability.
C. Putnam, Professor; Ph.D., Johns Hopkins, 1948. Operators in Hilbert space.
M. Ramachandran, Research Assistant Professor; Ph.D., Illinois at Chicago, 1990. Differential geometry.
J. Roberts, Visiting Associate Professor; Ph.D., Houston, 1976. Applied mathematics.
Z. Robinson, Research Assistant Professor; Ph.D., Harvard, 1990. Real and p-adic subanalytic geometry, logic.
J. Rubin, Professor; Ph.D., Stanford, 1955. Logic.
A. Sa Barreto, Research Assistant Professor; Ph.D., MIT, 1988. Partial differential equations.
R. Schultz, Professor; Ph.D., Chicago, 1968. Algebraic topology.
F. Shahidi, Professor; Ph.D., Johns Hopkins, 1975. Automorphic forms.
Z. Shen, Assistant Professor; Ph.D., Chicago, 1989. Partial differential equations.
W.-X. Shi, Assistant Professor; Ph.D., Harvard, 1990. Differential geometry.
J. Smith, Assistant Professor; Ph.D., MIT, 1981. Topology.
J. Thurber, Professor; Ph.D., NYU, 1961. Applied mathematics.
Y. Tong, Professor; Ph.D., Johns Hopkins, 1970. Complex manifolds.
A. Tumanov, Visiting Associate Professor; Ph.D., Moscow, 1980. Several complex variables.
J. Wang, Professor; Ph.D., Cornell, 1966. Lie groups.
S. Weingram, Associate Professor; Ph.D., Princeton, 1962. Algebraic topology.
A. Weitsman, Professor; Ph.D., Syracuse, 1968. Complex variables.
V. Weston, Professor; Ph.D., Toronto, 1956. Applied mathematics, inverse problems.
C. Wilkerson, Professor; Ph.D., Rice, 1970. Algebraic topology.
Y. Xu, Research Assistant Professor; Ph.D., Illinois at Chicago, 1990. Algebraic geometry.
E. Zachmanoglou, Professor; Ph.D., Berkeley, 1962. Partial differential equations.
R. Zink, Professor; Ph.D., Minnesota, 1953. Functional analysis.

SECTION 5: MATHEMATICAL SCIENCES

PURDUE UNIVERSITY

Department of Statistics

Programs of Study

Master of Science degrees in applied statistics or in statistics and computer science are available for students interested in careers as statisticians or operations analysts in industry and government. These degree programs do not require a thesis and are usually completed in two years. Students in the programs are encouraged to participate in the department's consulting service.

The Doctor of Philosophy program in statistics prepares students for careers in university teaching or in government or industrial research. Students entering the program spend four semesters acquiring a basic background in probability and mathematical statistics and take general examinations on these subjects. More specialized study follows with the thesis research, which usually begins in the third year. This research may be concentrated in any area of statistics or probability in which a faculty member is interested. Students also have the opportunity to gain experience in applied statistics through participation in statistical consulting. Completion of the Ph.D. program normally requires three to five years.

Research Facilities

The Department of Statistics is housed in the Mathematical Sciences Building along with the Department of Mathematics. This modern building contains faculty and graduate student offices, seminar rooms, classrooms, the Mathematical Sciences Library, and the Computing Center. The Center for Statistical Decision Sciences, affiliated with the Department of Statistics, provides research and training opportunities—both theoretical and interdisciplinary—in statistical sciences. The library contains 39,000 volumes and subscribes to 540 journals in mathematics and related fields.

All graduate student offices have terminals that provide access to the statistics department's UNIX-based computers, to the Purdue University Computing Center's IBM 3083 and CYBER 205 supercomputer, and to other computers worldwide via ARPANET and BITNET. The department also has Sun Workstations, graphics terminals, pen plotters, laser printers, and microcomputers.

Financial Aid

In 1991–92, most students have half-time assistantships that carry stipends of $800–$900 per month for ten months, plus reduction of tuition and fees to $200 per semester. Purdue University fellowships are available to entering students on a competitive basis; the stipends are $12,000 for one year, and tuition and fees are reduced to $200 per semester. The appointments may be renewed for one year for students with exceptionally good records. A limited number of research assistantships are available for students who have begun their Ph.D. research.

Many students' spouses are employed in the Lafayette area. Competition is keen for certain professional work, particularly teaching.

Cost of Study

In 1991–92, tuition is $2280 per year for Indiana residents and $7170 per year for out-of-state students. Tuition for assistants and fellows is $400 per year. Graduation fees are $10 for the M.S. and $15 for the Ph.D.

Cost of Living

Dormitory rooms in University-supervised graduate residences cost $207–$273 per month for 1991–92, and meals in the campus cafeterias cost $215–$290 per month. The cost of one- or two-bedroom furnished apartments for married students ranges from $227 to $367 per month. Off-campus rates are comparable.

Student Group

Purdue University has about 6,000 graduate students. The Department of Statistics has about 60 students working toward the Ph.D. and 15 enrolled in the applied M.S. programs. There are, typically, 12–15 female graduate students and about 50 international students. About two thirds of the American students begin graduate work immediately after the B.S., one sixth enter directly after the M.S., and one sixth have some work experience.

Location

Purdue University is the principal institution in West Lafayette, Indiana, which has a population of 20,000. In the larger Lafayette–West Lafayette area (population, 110,000), there are several other large organizations. Although Lafayette is an industrial town, it retains the atmosphere of an agricultural county seat. However, there are shopping malls, an airport, restaurants of note, two hospitals, and two large municipal parks. Community groups present plays, operettas, and musical programs several times each year, and the University offers musical, dramatic, and athletic events that may be attended by the general public. Several state parks are close to Lafayette.

The University and The Department

Purdue University, the land-grant college of Indiana, has achieved international recognition in the areas of agriculture, engineering, and science. Since 1960 the University has greatly expanded its efforts in the humanities, but it remains a predominantly technical and scientific institution. The West Lafayette campus of Purdue enrolls about 36,000 students. Purdue University is an Equal Opportunity/Affirmative Action employer.

The Department of Statistics is one of the seven departments that constitute Purdue's School of Science. The department's faculty interacts with other faculty members and graduate students through teaching, research, and a statistical consulting service.

Applying

Application materials and information may be obtained from the department. An applicant's mathematical training should include linear algebra and advanced calculus; probability and mathematical statistics are desirable.

Applications for Purdue University fellowships should be received before February 1. Other applications are considered at any time, although the availability of assistantships becomes limited after mid-April.

Correspondence and Information

Department of Statistics
532 Mathematical Sciences Building
Purdue University
West Lafayette, Indiana 47907
Telephone: 317-494-6030

Purdue University

THE FACULTY AND THEIR RESEARCH

James O. Berger, Ph.D., Cornell, 1974. Decision theory, mathematical statistics, Bayesian analysis.
Mary Ellen Bock, Ph.D., Illinois at Urbana-Champaign, 1974. Decision theory, econometrics.
Bertrand S. Clarke, Ph.D., Illinois at Urbana-Champaign, 1989. Statistical inference, information theory, modeling biological systems.
Anirban DasGupta, Ph.D., Indian Statistical Institute, 1983. Decision theory, mathematical statistics.
Burgess Davis, Ph.D., Illinois at Urbana-Champaign, 1968. Probability theory, mathematical analysis.
Holger Dette, Ph.D., Hannover, 1989. Optical designs, nonparametric tests. (Visiting)
Jayanta K. Ghosh, Ph.D., Calcutta, 1964. Decision theory, Bayesian analysis. (Visiting)
Chong Gu, Ph.D., Wisconsin, 1989. Statistical computing, interactive splines, time series.
Shanti S. Gupta, Ph.D., North Carolina, 1956. Selection and ranking, decision theory, order statistics, reliability.
Siu L. Hui, Ph.D., Yale, 1979. Biostatistics. (Adjunct)
Thomas Kuczek, Ph.D., Purdue, 1980. Design of experiments, stochastic processes.
Steven P. Lalley, Ph.D., Stanford, 1980. Sequential analysis, decision theory, probability.
Wei-Liem Loh, Ph.D., Stanford, 1988. Multivariate analysis, decision theory.
George P. McCabe, Ph.D., Columbia, 1970. Regression analysis, sequential analysis.
Vincent Melfi, Ph.D., Michigan, 1991. Sequential analysis, renewal theory.
David S. Moore, Ph.D., Cornell, 1967. Large-sample theory, tests of fit.
Louis R. Pericchi, Ph.D., London, 1991. Decision theory, subjective Bayesian statistics. (Visiting)
Dimitris N. Politis, Ph.D., Stanford, 1990. Time series, resampling methods.
Philip Protter, Ph.D., California, San Diego, 1975. Probability theory.
Herman Rubin, Ph.D., Chicago, 1948. Mathematical statistics, probability theory, numerical methods.
Myra L. Samuels, Ph.D., Berkeley, 1969. Applied probability, biostatistics. (Visiting)
Stephen M. Samuels, Ph.D., Stanford, 1964. Probability theory, optimal stopping, expert systems.
Thomas Sellke, Ph.D., Stanford, 1982. Sequential analysis, probability theory.
William J. Studden, Ph.D., Stanford, 1962. Optimal designs, Tchebysheff systems.

RECENT PUBLICATIONS

Monographs and Books

J. O. Berger. *Statistical Decision Theory and Bayesian Analysis*. New York: Springer-Verlag, 1986.
J. O. Berger and R. Wolpert. *The Likelihood Principle: A Review and Generalizations* (Institute of Mathematical Statistics Monograph Series). Hayward, Calif.: Institute of Mathematical Statistics, 1984.
M. E. Bock and G. G. Judge. *The Statistical Implications of Pretest and Stein-rule Estimators in Econometrics*, vol. 25, *Series of Studies in Mathematical and Managerial Economics*. Amsterdam: North Holland Publishing Company, 1979.
S. S. Gupta and J. O. Berger, eds. *Statistical Decision Theory and Related Topics IV*, vols. 1 and 2. New York: Springer-Verlag, 1987.
S. S. Gupta and D. Y. Huang. *Multiple Statistical Decision Theory: Recent Developments* (in the Lecture Notes in Statistics series). New York: Springer-Verlag, 1981.
S. S. Gupta and S. Panchapakesan. *Multiple Decision Procedures: Theory and Methodology of Selecting and Ranking Populations*. New York: John Wiley and Sons, 1979.
D. S. Moore. *Statistics: Concepts and Controversies*. San Francisco: W. H. Freeman and Co., 1979.
D. S. Moore and G. P. McCabe. *Introduction to the Practice of Statistics*. New York: W. H. Freeman and Co., 1989.
M. L. Samuels. *Statistics for the Life Sciences*. San Francisco: Dellen Publishers, 1989.

Research Reports

S. S. Gupta and Y. Liao. Subset Selection Procedures for the Binomial Models Based on Some Priors and Some Applications.
S. P. Lalley. Finite Range Random Walk on Free Groups and Homogeneous Trees.
B. Liseo. Elimination of Nuisance Parameters with Reference Noninformative Priors.
A. D. Barbour, L. H. Y. Chen, and W.-L. Loh. Compound Poisson Approximation for Nonnegative Random Variables Via Stein's Method.
S. S. Gupta and S. N. Hande. On Some Nonparametric Selection Procedures. (Revised and updated version of 90-26C.)
C. Gu and G. Wahba. Semiparametric Anova with Tensor Product Thin Plate Splines.
S. S. Gupta and S. N. Hande. On Selecting a Population Close to a Control: A Nonparametric Approach.
M. Zen and A. DasGupta. Estimating a Binomial Parameter: Is Robust Bayes Real Bayes?
A. DasGupta. Bounds on Asymptotic Relative Efficiencies of Robust Estimates of Locations for Random Contaminations.
T. G. Kurtz and P. Protter. Wong-Zakai Corrections Random Evolutions and Simulation Schemes for SDE's.
W. L. Loh. Stein's Method and Multinomial Approximation.
D. N. Politis and J. P. Romano. The Stationary Bootstrap.
J. O. Berger and W. Jeffreys. The Application of Robust Bayesian Analysis to Hypothesis Testing and Occam's Razor.
H. Dette. On General Equations for Orthogonal Polynomials.
M. J. Bayarri and M. H. DeGroot. Difficulties and Ambiguities in the Definition of a Likelihood Function.
D. N. Politis and J. P. Romano. A Circular Block-Resampling Procedure for Stationary Data.
T. G. Kurtz and P. Protter. Characterizing the Weak Convergence of Stochastic Integrals.
P. Müller. A Generic Approach to Posterior Integration and Gibbs Sampling.
B. S. Clarke and B. W. Junker. Inference from the Product of Marginals of a Dependent Likelihood.
H. Dette and W. J. Studden. On a New Characterization of the Classical Orthogonal Polynomials.
M. J. Bayarri and M. H. DeGroot. A "Bad" View at Weighted Distributions and Selection Models.
D. N. Politis. Applications of Resampling Schemes for Stationary Time Series.
S. S. Gupta and S. N. Hande. Single-Stage Bayes and Empirical Bayes Rules for Ranking Multinomial Events.
J. O. Berger and J. M. Bernardo. On the Development of the Reference Prior Method.
H. Dette. Some Generalizations of Elfings Theorem.
D. N. Politis. Non-parametric Maximum Entropy.
S. P. Lalley and D. Gatzouras. Hausdorff and Box Dimensions of Certain Self-Affine Fractals.
C. Gu and C. Qiu. Smoothing Spline Density Estimation: Theory.
D. Duffie and P. Protter. From Discrete to Continuous Time Finance: Weak Convergence of the Financial Gain Process. (Revised and updated version of 89-02.)
M. J. Bayarri and M. H. DeGroot. The Analysis of Published Significant Results.
T. G. Kurtz, E. Paradoux, and P. Protter. Stratonovich Stochastic Differential Equations Driven by General Semimartingales.
H. Dette. On a Mixture of the D- and D_1-Optimality Criterion in Polynomial Regression.
F. Pukelsheim and W. J. Studden. E-Optimal Designs for Polynomial Regression.
T. C. Liang. On Empirical Bayes Test Procedures for Uniform Distributions.
T. C. Liang. Convergence Rates for Empirical Bayes Estimation of the Scale Parameter in a Pareto Distribution.

SECTION 5: MATHEMATICAL SCIENCES

RENSSELAER POLYTECHNIC INSTITUTE

Department of Decision Sciences and Engineering Systems

Programs of Study

Rensselaer recognized the need for a new endeavor in decision sciences by forming an unusual interschool department, Decision Sciences and Engineering Systems (DSES), involving the Schools of Engineering, Management, and Science. The objectives of the department are to conduct research that leads to a better understanding of how information technology and quantitative analysis and modeling can support individuals, groups, and organizations in problem solving and decision making and to prepare engineers and managers to design, develop, and implement complex decision support systems. In order to accomplish these objectives, knowledge from disciplines such as systems and industrial engineering, statistics, probability, operations research, artificial intelligence, computer science, and economics must be extended and integrated. The department combines programs in industrial and management engineering, management systems, and operations research and statistics.

The graduate industrial and management engineering curriculum offers two programs that focus on the use of analytical techniques to solve complex problems. Courses give students practice in such techniques as computer simulation and mathematical modeling. Nine-credit-hour concentrations are offered in health systems, management information systems, manufacturing systems, and operations research and statistics. The Master of Engineering program requires 30–36 credit hours, including six required courses and four electives. Among the required courses is a 6-credit-hour master's project. The Master of Science program requires 30–36 credit hours, including six required courses and a statistics elective.

The Master of Science program in the operations research and statistics curriculum focuses on mathematical modeling techniques that are applicable to a wide range of problems connected with physical, economic, social, and biological systems. Students learn the theories and methodologies that underlie a range of analytical and optimization approaches. The program requires 30–36 credit hours, including four core courses, an advanced modeling course, and courses in computer simulation techniques, probability and statistics, and mathematical programming. Students may elect to devote 6 credit hours to a master's thesis or a research project.

The graduate management systems curriculum is designed to prepare researchers and professionals to manage technological systems and to employ technology to study and resolve management problems. The Master of Science program requires 48 credit hours, including fourteen required courses and two electives.

The Doctor of Philosophy program in decision sciences and engineering systems requires 90 credit hours of graduate studies with specialization either in industrial and management engineering, in management systems, or in operations research and statistics. Within these broad disciplines, areas of concentration are offered in applied probability; applied statistics; management information systems; mathematical programming; and production, inventory, and logistics. In addition to the course work, the program includes a qualifying examination, an area examination, and candidacy and dissertation requirements.

Research Facilities

The DSES department is located in the George M. Low Center for Industrial Innovation, which also houses the Centers for Integrated Electronics and for Manufacturing Productivity and Technology Transfer and the Rensselaer Design Research Center. This proximity provides a symbiotic environment conducive to excellence in research in the decision sciences. Other resources include an advanced computer graphics facility, an automated manufacturing laboratory, and an artificial intelligence facility.

Financial Aid

Financial aid is available in the form of teaching assistantships and scholarships. The stipend for assistantships ranges up to $8800 for the 1991–92 academic year. Tuition waivers are usually also granted. Outstanding students may qualify for university-supported Rensselaer Scholar Fellowships, which carry a stipend of $13,500 and a full tuition scholarship. Low-interest, deferred-repayment graduate loans are also available for U.S. citizens with demonstrated need.

Cost of Study

Tuition for 1991–92 is $455 per credit hour. Other fees amount to approximately $230 per semester. The application fee for admission is $30. Books and supplies cost about $930 per year.

Cost of Living

The cost of rooms for single students in residence halls or apartments varies from $2400 to $3800 for the 1991–92 academic year. A twenty-meal-per-week board plan costs $2500 for the academic year. Apartments for married students, with monthly rates ranging from $360 to $590, are available on campus.

Student Group

There are about 4,500 undergraduates and 2,000 graduate students, representing fifty states and more than sixty-five countries at Rensselaer.

Location

Rensselaer is situated on a scenic 260-acre hillside campus in Troy, New York, across the Hudson River from the state capital of Albany. Troy's central Northeast location provides students with a supportive, active, medium-sized community in which to live and an easy commute to Boston, New York, and Montreal. The Capital Region has one of the largest concentrations of academic institutions in the United States. Sixty thousand students attend fourteen area colleges and benefit from shared activities and courses, as well as access to the Saratoga Performing Arts Center, the Empire Plaza, the Rensselaer Technology Park, and some of the country's finest outdoor recreation, including Lake George, Lake Placid, and the Adirondack, Catskill, Berkshire, and Vermont mountains.

The Institute

Founded in 1824 and the first American college to award degrees in engineering and science, Rensselaer Polytechnic Institute today is a private, nonsectarian, coeducational university, accredited by the Middle States Association of Colleges and Schools. Rensselaer's five schools—Architecture, Engineering, Management, Science, and Humanities and Social Sciences—offer a total of eighty-seven graduate degrees in forty fields.

Applying

Admissions applications and all supporting credentials should be submitted well in advance of the preferred semester of entry. GRE General Test scores are required. Since the first departmental awards are made in February and March for the next full academic year, applicants requesting financial aid are encouraged to submit all required credentials by February 1 to ensure consideration.

Correspondence and Information

For further information about graduate work:
Administrative Secretary (Admissions)
Department of Decision Sciences and
 Engineering Systems
Rensselaer Polytechnic Institute
Troy, New York 12180-3590
Telephone: 518-276-6681

For applications and admissions information:
Director of Graduate Admissions
Graduate Center
Rensselaer Polytechnic Institute
Troy, New York 12180-3590
Telephone: 518-276-6789

SECTION 5: MATHEMATICAL SCIENCES

Rensselaer Polytechnic Institute

THE FACULTY AND THEIR RESEARCH

D. Berg, Institute Professor; Ph.D., Yale. Management of technological organizations, innovation, policy, robotics, policy issues of research and development in the service sector.

J. G. Ecker, Professor; Ph.D., Michigan. Mathematical programming, multiobjective programming, geometric programming, mathematical programming applications, ellipsoid algorithms.

J. Haddock, Associate Professor; Ph.D., Purdue. Simulation manufacturing/production, inventory control systems.

C. Hsu, Associate Professor; Ph.D., Ohio State. Large-scale information systems, database and knowledge-based systems, computerized manufacturing, probabilistic programming and scheduling.

R. P. Leifer, Associate Professor; Ph.D., Wisconsin. Organizational impacts of computerized information systems, organizational design.

G. F. List, Associate Professor; Ph.D., Pennsylvania. Real-time control of transportation network operations; multiobjective routing, scheduling, and fleet sizing; operations planning; hazardous materials logistics.

C. J. Malmborg, Associate Professor; Ph.D., Georgia Tech. Applied mathematical modeling, logistical and manufacturing systems, decision theory.

J. Mittenthal, Assistant Professor; Ph.D., Michigan. Scheduling, mathematical modeling, stochastic programming.

R. M. O'Keefe, Associate Professor; Ph.D., Southampton. Expert systems, simulation.

M. Raghavachari, Professor and Associate Chairman; Ph.D., Berkeley. Statistical inference, quality control, multivariate methods, scheduling problems.

W. Reitman, Professor; Ph.D., Michigan. Effective management, management of technology, management information systems, artificial intelligence, expert systems.

G. Runger, Assistant Professor; Ph.D., Minnesota. Productivity, quality control, design or reliability test plans, statistical modeling.

D. P. Schneider, Associate Professor; Ph.D., Florida. Mathematical modeling, health systems analysis, computer simulation modeling.

G. R. Simons, Professor, Manufacturing Center; Ph.D., Rensselaer. Project planning and control, manufacturing systems.

P. Sullo, Associate Professor; Ph.D., Florida State. Reliability, life testing, quality assurance, biostatistics, policy and risk analysis.

M. Terrab, Assistant Professor; Ph.D., MIT. Transportation networks and logistics, air traffic control, stochastic modeling, large-scale optimization.

J. M. Tien, Professor and Chairman; Ph.D., MIT. Systems modeling, queuing theory, public policy and decision analysis, computer performance evaluation, information systems, expert systems.

W. A. Wallace, Professor; Ph.D., Rensselaer. Public management and systems, decision support systems, expert systems.

J. W. Wilkinson, Professor; Ph.D., North Carolina. Applied statistics, design of experiments, regression modeling, data analysis, statistical quality control.

T. R. Willemain, Associate Professor; Ph.D., MIT. Probabilistic modeling, data analysis, forecasting.

Research and Adjunct Faculty

W. J. Foley, Research Assistant Professor; Ph.D., Rensselaer. Computer simulation modeling, health applications of operations research, health-care policy analysis.

M. Grabowski, Research Assistant Professor; Ph.D., Rensselaer. Management information systems, expert systems.

L. E. Hanifin, Research Professor; D.Eng., Detroit. Manufacturing, strategy, teamwork on the manufacturing floor, modeling of manufacturing facilities and systems, automated assembly systems.

M. Kupferschmid, Adjunct Associate Professor; Ph.D., Rensselaer; PE. Mathematical programming, algorithm performance evaluation, engineering applications.

Affiliated Faculty

A. A. Desrochers, Associate Professor; Ph.D., Purdue. Nonlinear systems, robotics, control of automated manufacturing systems.

F. Dicesare, Associate Professor; Ph.D., Carnegie-Mellon. Mathematical modeling, information systems, microprocessor applications.

W. R. Franklin, Associate Professor; Ph.D., Harvard. Computational geometry, graphics, CAD, cartography, parallel algorithms, large databases, expert systems.

D. H. Goldenberg, Associate Professor; Ph.D., Florida. Corporation finance, investment.

D. A. Grivas, Associate Professor; Ph.D., Purdue. Geostochastics, applications of probability to the description of soil behavior, statistical analysis of soil parameters, numerical methods in geotechnical engineering, reliability analysis of soil structures and foundations.

R. P. LeMay, Associate Professor; Ph.D., Iowa. Manufacturing planning and control.

C. Lemke, Ford Foundation Professor Emeritus; Ph.D., Carnegie Tech. Mathematical programming, complementarity.

J. E. Mitchell, Assistant Professor; Ph.D., Cornell. Mathematical programming, combinatorics, nonlinear programming.

J. R. Norsworthy, Professor; Ph.D., Virginia. Economics of productivity, productivity measurements, industrial economics.

A. S. Paulson, Professor; Ph.D., Virginia Tech. Risk management, financial models, multivariate statistics, time series and forecasting, survival data analysis.

D. S. Rebne, Assistant Professor; Ph.D., UCLA. Industrial relations of technological change, industrial relations of higher education, human resource management in professional occupations, social action theory, labor productivity.

D. Sandhu, Assistant Professor; Ph.D., Toronto. Stochastic models in operations research, complex queuing networks and applications to communication and manufacturing systems.

G. N. Saridis, Professor; Ph.D., Purdue. Intelligent control systems, pattern recognition, computer systems, robotics, protheses.

J. U. Turner, Assistant Professor; Ph.D., Rensselaer. Solid modeling, computer graphics, computer-aided design.

SECTION 5: MATHEMATICAL SCIENCES

RENSSELAER POLYTECHNIC INSTITUTE
Department of Mathematical Sciences

Programs of Study The Department of Mathematical Sciences offers graduate studies leading to the Master of Science and the Doctor of Philosophy. The program of study may be selected to emphasize either mathematics or applied mathematics.

The interests of most faculty members center on applied mathematics and analysis. Applied mathematics is interpreted in the broadest possible sense, with particular emphasis on mathematical modeling; methods of applied mathematics, differential equations, and functional analysis; applications of mathematics in the physical sciences, biological sciences, and engineering; applied geometry; numerical analysis and computer science; and mathematical programming and operations research.

While specific course requirements are minimal, every doctoral program in mathematics includes advanced courses in several fundamental areas of mathematics. In addition, doctoral candidates must pass a written preliminary examination covering basic topics in three areas; demonstrate a reading knowledge of French, German, or Russian; pass an oral candidacy examination in their field of proposed research and related areas of mathematics; and present an acceptable thesis containing the results of their own research. The Ph.D. program is normally completed within about four years, and the M.S. program in two years or less.

Research Facilities In addition to the mathematics collection in the main library, there is also a department reference collection. The department operates, jointly with Computer Science, a computational facility that houses a network of Sun Workstations, two Sequent multiprocessor computers, a hypercube, and a MasPar computer for parallel computation. Graduate students are expected to utilize the computer facilities for performance of course assignments and for computation necessary to their research.

Research is supported by such state-of-the-art facilities as the Folsom Library, whose automated retrieval systems link users to its own holdings as well as 5,600 libraries and 250 databases and the Voorhees Computing Center, which offers students exceptionally wide access to a diverse array of advanced-function workstations, microcomputers, and mainframes that are interconnected to each other and to national networks via the campus network.

Financial Aid Rensselaer Scholar Fellowships, competitive across the university, provide stipends of $13,500. These fellowships also include a full tuition and fees scholarship. Graduate assistantships, involving teaching, research, or some combination of these, provide stipends ranging up to $9700 for the academic year, plus tuition for up to 21 credits per academic year. Mathematical Sciences Fellowships, funded by an award from the U.S. Department of Education, will be available through the 1992–93 academic year. These fellowships consist of a $12,000 stipend and a tuition and fees scholarship. Special research fellowships are available within the department for highly qualified graduate students to enable them to work in selected areas. These range up to $10,000 for the academic year. Additional summer support is also available.

Full or partial tuition scholarships, low-interest loans, and part-time work are also available.

Cost of Study Tuition for 1991–92 is $455 per credit hour. Other fees amount to approximately $230 per semester. The application fee for admission is $30. Books and supplies cost about $930 per year.

Cost of Living The cost of rooms for single students in residence halls or apartments varies from $2400 to $3800 for the 1991–92 academic year. A twenty-meal-per-week board plan costs $2500 for the academic year. Apartments for married students, with monthly rents of $360 to $590, are available on campus.

Student Group There are about 4,500 undergraduates and 2,000 graduate students representing all fifty states and over sixty-five countries at Rensselaer.

Location Rensselaer is situated on a scenic 260-acre hillside campus in Troy, New York, across the Hudson River from the state capital of Albany. Troy's central Northeast location provides students with a supportive, active, medium-sized community in which to live and an easy commute to Boston, New York, and Montreal. The Capital Region has one of the largest concentrations of academic institutions in the United States. Sixty thousand students attend fourteen area colleges and benefit from shared activities and courses, as well as access to the Saratoga Performing Arts Center, the Empire Plaza, the Rensselaer Technology Park, and some of the country's finest outdoor recreation, including Lake George, Lake Placid, and the Adirondack, Catskill, Berkshire, and Vermont mountains.

The Institute Founded in 1824 and the first American college to award degrees in engineering and science, Rensselaer Polytechnic Institute today is accredited by the Middle States Association of Colleges and Schools and is a private, nonsectarian, coeducational university. Rensselaer's five schools—Architecture, Engineering, Management, Science, and Humanities and Social Sciences offer a total of eighty-seven graduate degrees in forty fields.

Applying Students with quite varied backgrounds are welcomed into the program. While the majority of the entering graduate students have an undergraduate major in mathematics, the department is pleased to accept students who have majored in such other areas as physics, engineering, management, and computer science and who are particularly concerned with the applications of mathematics to these fields. The GRE is required.

Admissions applications and all supporting credentials should be submitted well in advance of the preferred semester of entry to allow sufficient time for departmental review and processing. Since the first departmental awards are made in February and March for the next full academic year, applicants requesting financial aid are encouraged to submit all required credentials by February 1 to ensure consideration. Entrance in the spring term is possible.

Correspondence and Information

For further information about graduate work:
Department of Mathematical Sciences
Graduate Admissions Committee
Rensselaer Polytechnic Institute
Troy, New York 12180-3590
Telephone: 518-276-6894

For application and admissions information:
Director of Graduate Admissions
Graduate Center
Rensselaer Polytechnic Institute
Troy, New York 12180-3590
Telephone: 518-276-6789

SECTION 5: MATHEMATICAL SCIENCES

Rensselaer Polytechnic Institute

THE FACULTY AND THEIR RESEARCH

Professors
W. E. Boyce. Applied mathematics, random differential equations.
J. D. Cole. Applied mathematics, fluid mechanics, transonic flow, physiological modeling.
D. A. Drew. Applied mathematics, fluid mechanics, multiphase flow, semiconductor device modeling.
J. G. Ecker. Mathematical programming and operations research.
B. A. Fleishman (Emeritus). Free boundary problems, precollege education.
G. J. Habetler. Functional analysis, numerical analysis.
G. H. Handelman (Emeritus). Elasticity, biological applications.
M. H. Holmes. Applied mathematics, mathematical physiology.
D. Isaacson. Mathematical physics, numerical analysis, biomedical computer imaging.
M. J. Jacobson (Emeritus). Applied mathematics, sound transmission, signal processing.
A. K. Kapila. Applied mathematics, reactive fluid mechanics, combustion.
C. E. Lemke (Emeritus). Mathematical programming and operations research.
E. H. Luchins. Mathematics education, math psychology, history and philosophy of math.
H. W. McLaughlin II. Geometric modeling.
J. R. McLaughlin. Differential equations, inverse problems, optimization.
R. E. O'Malley. Asymptotic analysis and singular perturbation methods.
L. A. Rubenfeld. Applied mathematics, mathematics and science education.
W. L. Siegmann. Applied mathematics, wave propagation.

Associate Professors
E. K. Boyce. Analysis.
M. Cheney. Applied mathematics, analysis, differential equations, mathematical physics, inverse problems.
M. Levi. Dynamical systems, applied analysis, nonlinear ordinary differential equations.
V. Roytburd. Applied mathematics, analysis, partial differential equations.

Assistant Professors
S. Cole. Applied mathematics, asymptotic analysis, fluid dynamics.
G. Kovacic. Applied dynamical systems, Hamiltonian dynamics, bifurcations and perturbation theory.
C. C. Lim. Applied mathematics, vortex dynamics, dynamical systems, applied combinatorics.
J. E. Mitchell. Mathematical programming, combinatorial optimization, operations research.
J. Peters. Applied geometry, computer-aided design, numerical analysis.
J. Pimbley. Applied mathematics, semiconductor device modeling.
B. R. Piper. Computer-aided geometric design, applied geometry.
D. Schwendeman. Scientific computing, wave propagation.

JOINT APPOINTMENTS WITH COMPUTER SCIENCE

Professors
J. E. Flaherty. Applied mathematics, numerical analysis.
E. H. Rogers. VLSI architectures, computer applications.

SECTION 5: MATHEMATICAL SCIENCES

SIMON FRASER UNIVERSITY

Department of Mathematics and Statistics

Programs of Study

The department offers programs leading to the M.Sc. and Ph.D. degrees in applied mathematics, pure mathematics, and statistics. Some areas of particular strength are continuum mechanics, differential equations, numerical analysis, population dynamics, and relativity in applied mathematics; discrete mathematics, graph theory, history of mathematics, mathematical logic, probability, and real analysis in pure mathematics; and applied statistics, Bayesian inference, biometrics, goodness-of-fit, large-sample theory, linear functional models, multivariate estimation, robustness, statistics of directions, and stochastic processes in statistics.

With the exception of the project option in statistics mentioned below, the requirements for an M.Sc. are 20 semester hours of course work and the writing and successful defence of a thesis. The M.Sc. program in statistics (project option) requires 28 semester hours of course work, two semesters' participation in the Statistical Consulting Service, and the completion and successful defence of a project involving statistical analysis and computing. In principle, it is possible to complete an M.Sc. in twelve months, but many students take twenty months.

Candidates for the Ph.D. are required to complete 28 semester hours of course work (or 8 if the candidate has an acceptable master's degree), pass a general examination, and write and defend a thesis embodying the results of significant original research. The course work in all cases involves the study of four subareas of mathematics and/or statistics. The general examination normally consists of written examinations in three subjects, the choice of which for a particular candidate will be subject to approval by the candidate's supervisory committee and the Graduate Studies Committee of the department. The time taken to complete a Ph.D. varies widely: candidates entering with a master's should aim to complete the degree in three years, while those entering with a bachelor's should allow four years.

Research Facilities

The University library, 3 minutes' walk from the department, is a large, modern facility, which houses almost a million print volumes and 800,000 microforms. There are 14,000 volumes and 212 periodicals whose main focus is mathematics, applied mathematics, or statistics. The collection is constantly supplemented. Students may also draw on the considerable library resources of the University of British Columbia as well as interlibrary loan.

Computing resources available to researchers include the University's mainframe—an IBM 3081 GX. For numerically intensive applications, the University provides a high-performance parallel processing Silicon Graphics Power Series computer. The Graphics Center operates a VAX-11/750 and an IIS-70/F4 display processor. Within the department there is a UNIX-based network of twelve Sun SPARCstation 1's. The department also maintains a network of Macintosh computers and some stand-alone MC-DOS machines.

Financial Aid

Teaching assistantships are the principal source of income for many students. The semester salary for a work load of 5 base units is $4803 for M.Sc. students and $5673 for Ph.D. students. In the summer semester, when fewer teaching assistantships are available, graduate students are often employed as research assistants at the same semester salaries. In 1991–92, more than a third of the department's students hold graduate fellowships ($3850 for M.Sc. and $4450 for Ph.D.) during at least one semester. The fellowships are offered each year on a competitive basis. Outstanding students are encouraged to apply for a University entrance scholarship ($15,000 per annum).

Cost of Study

The fee unit is currently $588. The minimum for the master's degree is 6 fee units, unless the degree is completed within sixteen months, in which case a reduction applies. The minimum for the Ph.D. is 8 fee units. Students pay 1 fee unit per semester until the minimum fee requirement for their programs has been paid and ½ fee unit per semester thereafter.

Cost of Living

Burnaby and the surrounding municipalities provide an attractive and reasonably priced living environment. On-campus housing is limited, but the Housing Office carries many listings of suitable rooms and apartments.

Student Group

The enrollment is 18,568, of whom some 1,750 are graduate students. The graduate enrollment in the department is about 60; these students come from ten different countries.

Location

Simon Fraser University is located atop 400-metre-high Burnaby Mountain, 16 kilometres east of downtown Vancouver. The 360-degree view includes the city, its deepwater harbour and bordering mountains, the farmland of the Fraser Valley, and the Fraser River as it flows to the Strait of Georgia.

The University

Simon Fraser University was established by the province of British Columbia in 1963. The Vancouver firm of Erikson and Massey won the design competition and produced the breathtaking and award-winning structures that compose the main campus. The University opened its doors in September 1965.

Applying

Applicants should submit evidence of completion of a bachelor's degree from a recognized institution, with a GPA of 3.0 or equivalent; GRE General Test scores (waived for Canadian applicants); and TOEFL scores for those whose native language is not English. They should also arrange for three letters of reference to be sent to the address below. Women in particular are encouraged to apply.

Correspondence and Information

Graduate Admissions Committee
Department of Mathematics and Statistics
Simon Fraser University
Burnaby, British Columbia V5A 1S6
Canada
Telephone: 604-291-3801

Simon Fraser University

THE FACULTY AND THEIR RESEARCH

B. R. Alspach, Professor; Ph.D., California, Santa Barbara, 1966. Graph theory, discrete mathematics.
J. L. Berggren, Professor; Ph.D., Washington (Seattle), 1966. History of mathematics, algebra.
G. N. Bojadziev, Professor; Ph.D., Sofia (Bulgaria), 1957. Differential equations, population mechanics.
T. C. Brown, Professor; Ph.D., Washington (St. Louis), 1964. Algebra, combinatorics.
A. Das, Professor; Ph.D., Ireland (Dublin), 1961. Relativity, quantum mechanics.
C. Dean, Assistant Professor; Ph.D., Waterloo, 1988. Discrete and lifetime data, extra-Poisson variation.
D. Eaves, Associate Professor; Ph.D., Washington (Seattle), 1966. Biometrics, generalized linear modeling, theory of inference.
A. R. Freedman, Associate Professor; Ph.D., Oregon State, 1965. Combinatorial number theory and analysis.
H. Gerber, Assistant Professor; Ph.D., Penn State, 1965. Mathematical logic, mathematics education.
L. Goddyn, Assistant Professor; Ph.D., Waterloo, 1989. Combinatorics, discrete optimization.
G. A. C. Graham, Professor; Ph.D., Glasgow, 1966. Viscoelastic moving boundary value problems.
R. Harrop, Professor; Ph.D., Cambridge, 1953. Medical computing, mathematical logic.
K. Heinrich, Professor; Ph.D., Newcastle (Australia), 1979. Combinatorial designs, graph theory.
P. Hell, Professor; Ph.D., Montreal, 1972. Computational discrete mathematics.
C. Kim, Associate Professor; Ph.D., Stanford, 1965. Analysis, probability.
A. H. Lachlan, Professor; Ph.D., Cambridge, 1964. Mathematical logic.
R. W. Lardner, Professor; Ph.D., Cambridge, 1963. Applied and computational mathematics.
R. A. Lockhart, Associate Professor; Ph.D., Berkeley, 1979. Goodness-of-fit testing, inference on stochastic processes, large-sample theory.
A. Mekler, Professor; Ph.D., Stanford, 1976. Mathematical logic.
E. Pechlaner, Associate Professor; Ph.D., Vienna, 1965. Relativity.
N. R. Reilly, Professor; Ph.D., Glasgow, 1965. Algebra.
R. Routledge, Associate Professor; Ph.D., Dalhousie, 1975. Biometrics, estimating the sizes of animal populations.
R. D. Russell, Professor; Ph.D., New Mexico, 1971. Numerical analysis.
D. Ryeburn, Assistant Professor; Ph.D., Ohio State, 1962. General topology.
J. J. Sember, Associate Professor; Ph.D., New Mexico State, 1967. Functional analysis and summability.
D. Sharma, Associate Professor; Ph.D., Illinois, 1963. Elasticity, applied mathematics.
C. Y. Shen, Professor; Ph.D., Oregon State, 1967. Applied analysis, scientific computing.
E. M. Shoemaker, Professor; Ph,.D., Carnegie Tech, 1955. Glaciology.
M. Singh, Professor; Ph.D., Brown, 1965. Applied mathematics, continuum mechanics.
M. A. Stephens, Professor; Ph.D., Toronto, 1962. Goodness-of-fit testing, directional data.
T. Swartz, Assistant Professor; Ph.D., Toronto, 1986. Statistical computing, theory of inference.
T. Tang, Assistant Professor; Ph.D., Leeds (England), 1989. Applied mathematics.
S. K. Thomason, Professor; Ph.D., Cornell, 1966. Mathematical logic.
B. S. Thomson, Professor; Ph.D., Waterloo, 1968. Real analysis.
M. Trummer, Associate Professor; Dr.sc.math., Swiss Federal Institute of Technology, 1983. Numerical analysis.
C. Villegas, Professor; Ing.Ind., Uruguay, 1953. Foundations of Bayesian inference.
K. L. Weldon, Associate Professor; Ph.D., Stanford, 1969. Cross-sectional sampling, statistical consulting.

SECTION 5: MATHEMATICAL SCIENCES

STATE UNIVERSITY OF NEW YORK AT STONY BROOK

Departments of Mathematics and of Applied Mathematics and Statistics

Programs of Study

The Department of Mathematics and the Department of Applied Mathematics and Statistics offer programs leading to the Master of Arts and Doctor of Philosophy degrees for over 150 full-time graduate students. The programs prepare students for careers in research, teaching, and industry. Recent master's degree recipients have found employment in community college and secondary school teaching and in all sectors of industry. Recent Ph.D.'s have taken academic positions at such institutions as Massachusetts Institute of Technology, the University of Pennsylvania, New York University, the University of California at Berkeley, the Institute for Advanced Study (Princeton), Bowdoin College, and Colgate University, as well as nonacademic positions at organizations such as Bell Labs, IBM, Los Alamos Scientific Laboratory, Chase Manhattan Bank, and Eli Lilly Pharmaceuticals.

Special strengths in applied mathematics and statistics include computational fluid dynamics, operations research, applied statistics, numerical analysis, and game theory. Special strengths in mathematics include geometry, topology, complex analysis, operator theory, and dynamical systems. All students receive a broad range of basic and advanced training in mathematics or applied mathematics.

The Ph.D. program consists of an individually designed selection of advanced courses followed by a program of research leading to a dissertation. Students are required to pass a written comprehensive examination and an oral preliminary examination. The master's program focuses on specific mathematical sciences skills needed for a career as a teacher or an industrial mathematician.

All students receive personal attention and advising at all levels. There are numerous seminars and colloquia, featuring distinguished guest speakers and Stony Brook faculty members, that present both accelerated background knowledge and ongoing research. The Institute for Mathematical Sciences (IMS) has many joint activities and shares many faculty members with the mathematics department. The Institute for Mathematical Modeling (IMM) is home to the federally funded Center of Excellence in the Mathematics of Nonlinear Systems as well as a major Department of Energy research program in computational fluid dynamics. In addition, the two departments have close ties and joint activities with many other units on campus, including the Institute for Theoretical Physics, Harriman College of Management, and the computer science department.

Research Facilities

The two departments are housed together in the Mathematics Building. Each graduate student shares an office with one or more other students. The mathematics-physics library, located in an adjoining building, contains 50,000 books and subscribes to about 500 journals. The departments have about fifty Sun Workstations, and most are available to graduate students. There is also a 32-node Intel iPSC/860 Hypercube parallel supercomputer along with graphics workstations in the Institute for Mathematical Modeling.

Financial Aid

Almost all full-time students are supported by assistantships or fellowships carrying stipends ranging from $9000 to $12,500 plus full tuition for the 1991–92 academic year. There are several fellowships available to outstanding applicants who are U.S. nationals; these include the Department of Education GAANN Fellowships, the C. Burghardt Turner Fellowships for minority students, and the Graduate Council Fellowships. There are also unrestricted departmental fellowships. Some of the more advanced students are supported by faculty research grants.

Cost of Study

In 1991–92, tuition for full-time study is $2950 for New York State residents and $6265 for nonresidents. Additional fees total approximately $300 per year.

Cost of Living

In 1991–92, a single student living on campus requires about $10,000–$11,000 to cover normal living expenses for twelve months.

Student Group

The University enrolls about 16,000 students, including 5,000 graduate students. The number of full-time graduate students in applied mathematics and statistics is about 85 and in mathematics about 70.

Location

Situated on Long Island's North Shore, 50 miles east of New York City, Stony Brook enjoys the best of two worlds—the wide range of cultural and intellectual activities available about an hour and a half away in the city and the lush, rolling countryside and beaches of the North Shore, where one can find eighteenth-century homes close to modern shopping centers. From the adjacent sailing center of Port Jefferson, New England is 80 minutes away via ferry to Connecticut. The 1,100-acre campus is in a wooded area 3 miles from Long Island Sound. There are public beaches for swimming and for all kinds of boating within a few miles of the campus. The state and national parklands and the sandy Atlantic beaches on the South Shore of Long Island are only 45 minutes away by automobile.

The University

The State University of New York at Stony Brook is one of four University Centers in the State University of New York System. In addition to its outstanding departments in mathematics and the sciences, Stony Brook has many excellent programs in the humanities, the social sciences, and the fine arts. Of particular interest to many is the University's Fine Arts Center, which hosts a steady stream of concerts, recitals, and theatrical performances by visiting artists, faculty members, and students.

Applying

Applications for admission are welcome at all times. Transcripts, GRE General Test scores, and three letters of recommendation are required. Completed applications should be received by March 1 for fall admission and by February 1 for financial aid.

Correspondence and Information

Director of the Graduate Program
Department of Mathematics
State University of New York at Stony Brook
Stony Brook, New York 11794-3651
Telephone: 516-632-8282

or

Director of the Graduate Program
Department of Applied Mathematics and Statistics
State University of New York at Stony Brook
Stony Brook, New York 11794-3600
Telephone: 516-632-8360

SECTION 5: MATHEMATICAL SCIENCES

State University of New York at Stony Brook

THE FACULTY AND THEIR RESEARCH

Faculty in Mathematics
Michael Anderson, Associate Professor; Ph.D., Berkeley. Differential geometry.
William Barcus, Professor; D.Phil., Oxford. Algebraic topology.
Benjamin Bielefeld, Assistant Professor; Ph.D., Cornell. Dynamical systems.
Emil Bifet, Assistant Professor; Ph.D., Chicago. Algebraic geometry.
Christopher Bishop, Assistant Professor; Ph.D., Chicago. Analysis.
Jeff Cheeger, Distinguished Professor; Ph.D., Princeton. Differential geometry.
Ronald G. Douglas, Leading Professor and Vice-Provost for Undergraduate Studies; Ph.D., LSU. Operator theory, functional analysis.
David Ebin, Professor; Ph.D., MIT. Global analysis.
William Fox, Associate Professor; Ph.D., Michigan. Complex analysis.
Lenore Frank, Lecturer; M.S., Yeshiva. Mathematics education.
Daryl Geller, Associate Professor; Ph.D., Princeton. Analysis.
James Glimm, Distinguished Professor and Director, IMM; Ph.D., Columbia. Computational fluid dynamics, conservation laws.
Christophe Gole, Assistant Professor; Ph.D., Boston University. Dynamics.
Detlef Gromoll, Professor; Ph.D., Bonn. Differential geometry.
C. Denson Hill, Professor and Director, Undergraduate Program; Ph.D., NYU. Partial differential equations, several complex variables.
Yunping Jiang, Assistant Professor; Ph.D., CUNY Graduate Center. Dynamics.
Lowell Jones, Professor; Ph.D., Yale. Topology.
Anthony Knapp, Professor; Ph.D., Princeton. Lie groups.
Irwin Kra, Leading Professor and Dean, Division of Physical Sciences and Mathematics; Ph.D., Columbia. Complex analysis.
Paul Kumpel, Professor; Ph.D., Brown. Algebraic topology.
Henry Laufer, Professor; Ph.D., Princeton. Several complex variables.
H. Blaine Lawson Jr., Leading Professor; Ph.D., Stanford. Differential geometry, topology.
Claude LeBrun, Associate Professor; D.Phil., Oxford. Complex analysis, mathematical physics.
William Lister, Professor; Ph.D., Yale. Algebra.
Mikhail Lyubich, Associate Professor; Ph.D., Tashkent State (USSR). Dynamics.
Bernard Maskit, Professor; Ph.D., NYU. Complex analysis, Kleinian groups.
Hisayosi Matumoto, Assistant Professor; Ph.D., MIT. Lie groups.
Dusa McDuff, Professor and Chair; Ph.D., Cambridge. Operator theory, topology.
Marie-Louise Michelsohn, Professor and Director, Graduate Program; Ph.D., Chicago. Differential geometry.
John W. Milnor, Distinguished Professor and Director, IMS; Ph.D., Princeton. Topology, geometry, dynamical systems.
Anthony Phillips, Professor; Ph.D., Princeton. Differential topology.
Joel Pincus, Professor; Ph.D., NYU. Operator theory, integral equations.
Bradley James Plohr, Associate Professor; Ph.D., Princeton. Computational fluid dynamics.
Chih-Han Sah, Professor; Ph.D., Princeton. Algebra.
Vadim Schechtman, Professor; Ph.D., Moscow State. Algebra.
Ralf Spatzier, Assistant Professor; Ph.D., Warwick (England). Dynamical systems, differential geometry.
Scott Sutherland, Assistant Professor; Ph.D., Boston University. Dynamical systems.
Grzegorz Swiatek, Assistant Professor; Ph.D., Warsaw. Dynamical systems.
Peter Szusz, Professor; Ph.D., Budapest. Analytic number theory.
Folkert M. Tangerman, Assistant Professor; Ph.D., Boston University. Dynamical systems, differential geometry.
Nicolae Teleman, Professor; Ph.D., MIT. Differential geometry.
Gang Tian, Associate Professor; Ph.D., Harvard. Differential geometry.
Eugene Vinegrad, Lecturer. Mathematics education.
Wen-xiang Wang, Assistant Professor; Ph.D., Princeton. Complex differential geometry.
Eugene Zaustinsky, Professor; Ph.D., USC. Differential geometry.

Faculty in Applied Mathematics and Statistics
Esther Arkin, Assistant Professor; Ph.D., Stanford. Combinatorial optimization, computational geometry.
Laurence Baxter, Associate Professor; Ph.D., University College, London. Applied probability, reliability theory.
Edward Beltrami, Professor; Ph.D., Adelphi. Nonlinear models, stochastic models.
Hung Chen, Associate Professor; Ph.D., Berkeley. Nonparametric statistics.
Yung Ming Chen, Professor; Ph.D., Maryland. Numerical analysis.
Yuefan Deng, Assistant Professor; Ph.D., Columbia. Computational fluid dynamics, parallel computing.
Daniel Dicker, Professor; Ph.D., Columbia. Porous flow problems.
Vaclav Dolezal, Professor; Ph.D., Czechoslovak Academy of Sciences. Mathematical systems theory.
Pradeep Dubey, Professor; Ph.D., Cornell. Game theory, mathematical economics.
Eugene Feinberg, Professor of Management; Ph.D., Moscow State. Applied probability.
Stephen Finch, Associate Professor; Ph.D., Princeton. Applied statistics.
James Glimm, Distinguished Professor, Chair, and Director, IMM; Ph.D., Columbia. Computational fluid dynamics, conservation laws, mathematical physics.
John Grove, Assistant Professor; Ph.D., Ohio State. Computational fluid dynamics.
Woo Jong Kim, Professor and Director, Graduate Program; Ph.D., Carnegie Mellon. Ordinary differential equations.
Brent Lindquist, Associate Professor; Ph.D., Cornell. Computational fluid dynamics, reservoir modeling.
Nancy Mendell, Associate Professor; Ph.D., North Carolina. Applied statistics.
Jean Pierre Mertens, Professor; Ph.D., Louvain (Belgium). Game theory, mathematical economics.
Joseph Mitchell, Associate Professor; Ph.D., Stanford. Computational geometry.
Abraham Neyman, Professor; Ph.D., Hebrew (Jerusalem). Game theory, mathematical economics.
Akiri Okubo, Professor of Marine Science; Ph.D., Johns Hopkins. Nonlinear phenomena.
Bradley Plohr, Associate Professor; Ph.D., Princeton. Computational fluid dynamics.
Steven Skiena, Assistant Professor of Computer Science; Ph.D., Illinois. Combinatorial algorithms.
Darko Skorin-Kapov, Assistant Professor of Management; Ph.D., British Columbia. Integer programming.
Jadranka Skorin-Kapov, Assistant Professor of Management, Ph.D., British Columbia. Mathematical programming.
Matthew Sobel, Professor; Ph.D., Stanford. Stochastic models, production management.
Ram Srivastav, Professor; Dc.S., Glasgow. Numerical analysis, integral equations.
Michael Taksar, Professor; Ph.D., Cornell. Stochastic processes, control theory.
Reginald Tewarson, Professor; Ph.D., Boston University. Numerical analysis, biomathematics.
Alan Tucker, Distinguished Teaching Professor; Ph.D., Stanford. Combinatorial optimization.
Qiqing Yu, Assistant Professor; Ph.D., UCLA. Decision theory, biostatistics.
Qiang Zhang, Assistant Professor; Ph.D., NYU. Computational fluid dynamics, conservation laws.
Kevin Zumbrun, Assistant Professor; Ph.D., NYU. Computational fluid dynamics.

SECTION 5: MATHEMATICAL SCIENCES

TEMPLE UNIVERSITY
of the Commonwealth System of Higher Education
Department of Statistics
Statistics and Biostatistics Programs

Programs of Study
The Department of Statistics at Temple University offers graduate programs leading to the M.S. and Ph.D. degrees in statistics. The aim of the department's graduate programs is to provide broad training for statisticians and biostatisticians. Graduates have been placed in academic, business, and government positions. The department also offers the Ph.D. in business administration with a major in statistics.

The course requirement for the master's degree consists of 30 semester hours, of which 12 hours are core courses and the remainder electives. Core courses are provided for students with no previous statistical knowledge, but students with a background in statistics may waive part or all of the core courses by taking placement exams. A written core exam must be passed prior to the awarding of the degree. There are no thesis or foreign language requirements for the master's degree. Students are required to acquire proficiency in the use of computers.

In addition to fulfilling the requirements for the master's degree, Ph.D. students are expected to take a number of advanced courses, pass Ph.D. qualifying examinations, and write a dissertation. There is reasonable flexibility in tailoring a program of study to a student's interests. After passing the Ph.D. qualifying exams, the student works with a faculty adviser, who assists the student in planning a program of study and guides his or her research. The Ph.D. degree requires a minimum of one year of full-time residence; the remainder of the program may be completed on a part-time basis. Proficiency in the use of computers is required. Advanced students are encouraged to consult in the Data Analysis Laboratory of the Department of Statistics. There is no foreign language requirement for the Ph.D.

The department conducts seminars weekly during the academic year. In addition, the Philadelphia chapter of the American Statistical Association, with over 400 members, holds monthly meetings.

Research Facilities
The University Computer Center has IBM 4381, CDC CYBER 860, and FPS parallel-processor mainframe computing systems. The Department of Statistics operates its own UNIX-based VAX-11/750 superminicomputer, two Sun-3 graphics workstations, an IMAGEN 8/300 graphics laser printer for mathematical typesetting, and IBM PCs and terminals connected to the University's network. Additional terminals and PCs are available throughout the campus. All graduate students have free use of these facilities. Major statistical packages and mathematical software are available.

The Samuel Paley Library contains an extensive collection of books in the fields of biometrics, statistics, and operations research. Journals and periodicals are housed in the Mathematics, Statistics, and Computer Science Library, located in the Mathematics–Computer Activity Building. The statistics department's reading room in the department's suite in Speakman Hall holds recent journal issues, reference books, and reprints.

The Department of Statistics has arranged joint programs with statisticians at several of the Philadelphia-area hospitals, medical research centers, and pharmaceutical companies for the supervision and practical training of advanced graduate students. The department's Data Analysis Laboratory (DAL) is a joint faculty–graduate student activity that provides design, analysis, and other statistical consulting services for all academic and administrative units at Temple University. The DAL also provides similar services to outside agencies on a contract or grant basis. In addition, the laboratory serves as a training vehicle for both students and faculty in the application of statistical theory to the solution of real problems.

Financial Aid
A number of research assistantships, fellowships, and tuition scholarships are available for qualified applicants. Assistantships provide a stipend plus tuition. Advanced graduate students are considered for part-time teaching positions in the department. Several loan programs also are available.

Cost of Study
Temple is a state-related university. Tuition for Pennsylvania residents is $205 per credit in 1991–92. Out-of-state tuition is $258 per credit. All tuition rates are subject to change.

Cost of Living
Many full-time graduate students live in nearby Cooney Hall or Yorktown Apartments. Others live in apartments or rooms in various parts of the city, since the University's campuses can be conveniently reached by public transportation. Suitable apartments can be found for about $300 per month.

Student Group
The Department of Statistics is of moderate size, and each student receives individual attention. During 1990–91, there were more than 65 graduate students, some of whom were part-time.

Location
Philadelphia, the nation's fifth-largest city, offers a multitude of educational, cultural, and recreational opportunities. Its several universities, libraries, museums, world-renowned orchestra, and resident opera, ballet, and theater groups are readily accessible to area residents. In addition, ocean and mountain recreational areas and the rich resources of New York City lie within a 100-mile radius of the city.

The University
Temple University was founded in 1884 and became a state-related institution in 1965. It offers bachelor's degrees in 100 fields, master's degrees in 100 areas, and doctorates in 75 specialties. The faculty numbers nearly 3,000. The main campus is 1½ miles from the center of Philadelphia, and the Temple University Health Science center is about 1½ miles to the north on Broad Street. The center includes the Schools of Medicine, Dentistry, and Pharmacy; the College of Allied Health Professions; and the University Hospital. The University is dedicated to working with the community and providing outstanding educational opportunities for all students, including those who wish to study part-time. Graduate courses in the Department of Statistics are taught in the afternoon and evening to permit statisticians working in the area to pursue advanced degrees. Shopping, dining, and entertainment facilities are close to the University.

Applying
Application for admission can be made for either semester, although applicants for financial aid should apply as early as possible. All applicants must hold at least a bachelor's degree and must take the GRE General Test. Foreign students must take the Test of English as a Foreign Language (TOEFL). Part-time students are encouraged to apply. Both the statistics and the biostatistics programs require a knowledge of advanced calculus and matrix algebra. Although admission is not denied on the basis of a deficient mathematical background, students are expected to remove any deficiencies by the end of the first year.

Correspondence and Information
Dr. Sanat K. Sarkar, Director of Graduate Study
Department of Statistics
Temple University
Philadelphia, Pennsylvania 19122
Telephone: 215-787-6878
Fax: 215-787-5698

Peterson's Guide to Graduate Programs in the Physical Sciences and Mathematics 1992

Temple University

THE FACULTY AND THEIR RESEARCH

Zhidong Bai, Associate Professor; Ph.D., University of Science and Technology of China, 1982. Probability theory, asymptotic theory, multirate analysis. Fellow, the Third World Academia of Sciences. Elected member, International Statistical Institute. Associate editor, *Statistica Sinica*.

Luisa T. Fernholz, Assistant Professor; Ph.D., Rutgers, 1979. Asymptotic theory, robustness, statistical functionals. Author, *Von Mises Calculus for Statistical Functionals*. In *Lecture Notes in Statistics*. New York: Springer-Verlag, 1983.

Richard M. Heiberger, Associate Professor; Ph.D., Harvard, 1972. Linear models, experimental design, statistical computing, robust estimation. Author, *Computation for the Analysis of Designed Experiments*. New York: Wiley, 1989.

Burt S. Holland, Associate Professor; Ph.D., North Carolina State, 1970. Simultaneous inference, linear models, experimental design, econometric modeling.

Francis Hsuan, Associate Professor; Ph.D., Cornell, 1974. Statistical decision theory, statistical computing, linear models. Author, *[2]-Inverses and Their Statistical Applications*. In *Lecture Notes in Statistics*. New York: Springer-Verlag, 1988.

Boris Iglewicz, Professor; Ph.D., Virginia Tech, 1967. Robust and exploratory procedures, clinical trials, quality control, sequential analysis, survey sampling. Elected member, International Statistical Institute. Coauthor, *Introduction to Mathematical Reasoning*. New York: Macmillan, 1973. Associate editor, *Journal of Quality Technology*.

Alan J. Izenman, Associate Professor and Director, Statistical Computing Laboratory; Member, Center for Advanced Computational Science; Ph.D., Berkeley, 1972. Multivariate analysis, time-series analysis, nonparametric density estimation, welfare quality control, statistical computing and graphics. Associate editor, *Journal of the American Statistical Association*.

Fanny Yuen-Ching Ki, Assistant Professor; Ph.D., Wisconsin–Madison, 1986. Decision theory, multivariate analysis, nonparametric regression, Poisson regression.

Dirk F. Moore, Assistant Professor; Ph.D., Washington (Seattle), 1985. Statistical methods in AIDS and teratology; generalized linear models.

Milton Parnes, Associate Professor; Ph.D., Wayne State, 1968. Survival analysis, theory of gambling, probability theory.

Damaraju Raghavarao, Laura H. Carnell Professor and Chairman; Acting Director, Data Analysis Laboratory; Ph.D., Bombay, 1961. Experimental design, survey sampling, combinatorics, multivariate analysis. Fellow, Institute of Mathematical Statistics. Fellow, American Statistical Association. Elected member, International Statistical Institute. Associate editor, *Journal of Statistical Planning and Inference*, *Communications in Statistics*, and *Utilitas Mathematica*. Author, *Constructions and Combinatorial Problems in Design of Experiments*. New York: Wiley, 1971; Dover, 1988. *Matrix Theory*. India: Oxford and International Book House, 1972. *Statistical Techniques in Agriculture and Biological Research*. India: Oxford and International Book House, 1983. *Exploring Statistics*. New York: Dekker, 1988. Coauthor, *Sample Size Methodology*. San Diego: Academic Press.

Sanat K. Sarkar, Associate Professor and Director of Graduate Study; Ph.D., Calcutta, 1982. Multivariate analysis, probability inequalities, regression analysis, contingency tables.

Jagbir Singh, Professor; Ph.D., Florida State, 1967. Linear models, applied probability, survival analysis. Elected member, International Statistical Institute. Coauthor, *Statistical Methods in Food and Consumer Research*. Orlando, Fla.: Academic Press, 1985. Editorial Board member, *Journal of Sensory Studies*.

Woollcott K. Smith, Associate Professor; Ph.D., Johns Hopkins, 1969. Stochastic processes, statistical ecology, atmospheric research, survey design. Coeditor, *Ecological Diversity in Theory and Practice*. In *International Statistical Ecology Program*. Md.: International Cooperative Publishing House, 1979.

Marcus J. Sobel, Assistant Professor; Ph.D., Berkeley, 1983. Bayes, empirical Bayes, cross-validation, ranking and selection, statistical inference.

Mohamed Tahir, Assistant Professor; Ph.D., Michigan, 1987. Sequential analysis, decision theory.

William W. S. Wei, Professor; Ph.D., Wisconsin–Madison, 1974. Time-series analysis, regression asymptotic theory, forecasting, applications of statistics in economics and business. Associate editor, *Journal of Forecasting*. Editorial Board member, *Advances in Quantitative Analysis of Finance and Accounting*. Author, *Time Series Analysis: Univariate and Multivariate Methods*. Redwood City, Calif.: Addison-Wesley, 1989.

Adjunct and Visiting Faculty

In addition to the regular faculty listed above, there are adjunct and visiting faculty members from other educational institutions, corporations, and government agencies. The following members are scheduled for 1990–91.

Parthasarathy Bagchi, Adjunct Assistant Professor; Ph.D., Toronto, 1987. Bayesian inference.

Kalimuthu Krishnamoorthy, Adjunct Assistant Professor; Ph.D., IIT, Kanpur (India), 1984. Multivariate analysis.

Hameed Bayo Lawal, Adjunct Associate Professor; Ph.D., Essex (England), 1981. Categorical data analysis.

Outside Speakman Hall, location of the Department of Statistics.

SECTION 5: MATHEMATICAL SCIENCES

UNIVERSITY OF ARIZONA

Department of Mathematics and Program in Applied Mathematics

Programs of Study

The Department of Mathematics at the University of Arizona offers a broad spectrum of graduate courses and seminars in algebra, analysis, applied mathematics, geometry, mathematical physics, probability, and statistics that lead to the degrees of Master of Arts, Master of Science, Master of Education, and Doctor of Philosophy with majors in mathematics.

In addition, the interdisciplinary Program in Applied Mathematics offers courses of study leading to the degrees of Master of Science and Doctor of Philosophy with majors in applied mathematics.

The M.A. in mathematics is designed for students who wish to combine mathematics with some related field outside the Department of Mathematics, while the M.S. in mathematics is intended for students who wish to take primarily mathematics courses. Both master's degrees can serve as a basis for further study toward a Ph.D. degree. Highly flexible programs in pure, applied, and applications-oriented mathematics are offered in the Ph.D. program in mathematics. Completion of course work in major and minor fields and a thesis presenting the student's original research are required.

Students entering the Program in Applied Mathematics take a one-year sequence of mathematics courses tailored to the needs of applied mathematics. Beyond this, the program offers great flexibility. Both M.S. and Ph.D. candidates are required to complete a certain number of courses outside the mathematics department. Ph.D. students complete a dissertation embodying original research under the direction of a member of the program faculty. These faculty are 89 in number and have appointments in more than twenty departments of the University. The highly interdisciplinary faculty membership is a distinguishing feature of the Program in Applied Mathematics.

In both mathematics and applied mathematics, a computer programming examination must be passed, and in addition the Ph.D. degrees require a reading knowledge of French, German, or Russian. Both Ph.D. programs may normally be completed in four to five years.

Research Facilities

The Department of Mathematics occupies a seven-story building in which most seminars and graduate courses are held. The department operates an extensive computer system, including a network of Sun and IRIS Workstations, which is available to graduate students in both mathematics and applied mathematics. The mathematics reading room contains some 5,000 books and displays the latest issues of mathematical research journals.

The University Library has extensive collections in most fields, including mathematics and applied mathematics. The University Computer Center has a Convex superminicomputer, an IBM 3090 with vector processor, and several VAX computers.

Financial Aid

Teaching assistantships are available for qualified graduate students. The stipend in 1991–92 is $9179 and up for teaching 4 class hours. Associate teaching assistantships are available for students with a master's degree or the equivalent; the stipend ranges upward from $9948. Out-of-state tuition is waived, but a registration fee of $764 is required for each semester. Fellowships of $10,000 for the academic year are available on a highly competitive basis. Outstanding applications are also considered for fellowship funding with a stipend of $12,000 for the fiscal year.

Cost of Study

Full-time students who are Arizona residents are required to pay registration fees of $764 per semester in 1991–92. Full-time out-of-state students pay registration fees plus tuition of up to $2889 each semester. Part-time students pay proportionately.

Cost of Living

University-maintained off-campus housing is available for married students. Single graduate students may live in the dormitories. Meals are available on campus. Excluding registration/tuition fees, the minimum cost for an academic year for a student living on campus has been estimated to be $6116 for the current year.

Student Group

The University of Arizona has more than 35,000 students, including more than 7,000 graduate students. Total enrollment in the mathematics and applied mathematics graduate programs is about 170, of whom over one fifth are women. Most students receive financial aid. Many recent graduates hold positions in universities, national research laboratories, or industry.

Location

The Tucson metropolitan area has a population of 600,000. It is located in a valley surrounded by mountain ranges, some of which rise 7,000 feet above the city. The Mexican border is 65 miles south of the University. The climate and surroundings lend themselves to outdoor recreational activities. Classical music concerts, theater, skiing, and mountain hiking are all readily available.

The University and The Department

The state-supported University of Arizona, founded in 1885, is located in the center of Tucson. The Department of Mathematics has a professional staff of about 60, who maintain a high level of research activity.

Applying

Application should be made to only one of the two programs at any one time. Forms are available from the addresses below. GRE scores should be submitted to the appropriate program. International students must ensure that their complete applications for admission and financial aid are received by March 1. Applications from U.S. students can be accepted somewhat later.

Correspondence and Information

Graduate Committee
Department of Mathematics
University of Arizona
Tucson, Arizona 85721
Telephone: 602-621-2068

Graduate Committee
Program in Applied Mathematics
University of Arizona
Tucson, Arizona 85721
Telephone 602-621-2016 or 4664

SECTION 5: MATHEMATICAL SCIENCES

University of Arizona

THE FACULTY
DEPARTMENT OF MATHEMATICS
Professors. Clark T. Benson, John D. Brillhart, M. S. Cheema, James R. Clay, J. M. Cushing, John L. Denny, William Faris, Hermann Flaschka, W. M. Greenlee, Helmut Groemer, Larry C. Grove, George L. Lamb Jr., Peter Li, David O. Lomen, John S. Lomont, David Lovelock, Henry B. Mann (emeritus), Warren L. May, Donald E. Myers, Alan Newell, Charles Newman, R. S. Pierce, Alwyn Scott, Moshed Shaked, A. H. Steinbrenner, Elias Toubassi, William Velez, Stephen Willoughby.
Associate Professors. William E. Conway, Carl L. DeVito, Nicholas Ercolani, David Gay, Oma Hamara, Sheldon Kamienny, Thomas Kennedy, Theodore Laetsch, C. David Levermore, Daniel J. Madden, William McCallum, Timothy W. Secomb, F. W. Stevenson, Richard B. Thompson, Maciej Wojtkowski, Bruce Wood, Arthur L. Wright, Lai Sang Young.
Assistant Professors. Bruce Bayly, Moysey Brio, Kwok Chow, Paul Fan, Luc Haine, Brenton LeMesurier, Robert Maier, Doug Pickrell, W. Raskind, M. Rychlik, Yon-Quan Yin.
Lecturers. Robert C. Dillon, John L. Leonard, Stephen D. Tellman.

Research Activities
Algebra and Number Theory. Algebraic function fields, algebraic number theory, arithmetical algebraic geometry, associative algebras, Boolean algebra and lattice theory, combinatorics, Galois theory, geometry of numbers, infinite Abelian groups, number theory and computational algorithms, representations of finite groups, ring and near-ring theory, theory of equations, theory of partitions, universal algebra.
Fluid Dynamics. Aerodynamics, computational fluid mechanics, hydrodynamics, soil mechanics.
Geometry. Symplectic and Poisson manifolds, Abelian surfaces, algebraic geometry of dynamical systems, differential geometry and analysis on manifolds.
Mathematical Biology and Chemistry. Mathematical ecology, nonlinear diffusion, population dynamics.
Mathematical Physics. Nonlinear waves and solitons, quantum mechanics and operator theory, quantum optics, relativity and classical field theory, statistical mechanics and quantum field theory.
Nonlinear and Computational Analysis. Approximation theory, bifurcation theory, calculus of variations, convexity, eigenvalue problems, smooth ergodic theory, integral equations, linear and nonlinear functional analysis, numerical methods, ordinary and partial differential equations, singular perturbations. Special emphasis is placed on the nonlinear phenomena encountered in physical and engineering structures.
Probability and Statistics. Probability, random matrices, statistics, stochastic processes.

PROGRAM IN APPLIED MATHEMATICS
Key to abbreviations: A.M.E.—Aerospace and Mechanical Engineering; E.C.E.—Electrical and Computer Engineering; E.E.B.—Ecology and Evolutionary Biology; L.P.L.—Lunar and Planetary Laboratory; M.I.S.—Management Information Systems; N.E.E.—Nuclear and Energy Engineering; S.I.E.—Systems and Industrial Engineering; S.W.S.—Soil and Water Science.

Dynamical systems and chaos, including applications in optics, population biology, and nonlinear waves. H. Flaschka (Mathematics), G. L. Lamb Jr. (Mathematics and Optical Science), P. Meystre (Optical Science), J. V. Moloney (Mathematics), A. C. Newell (Mathematics), W. M. Schaffer (E.E.B.), A. C. Scott (Mathematics), M. P. Wojtkowski (Mathematics), L. S. Young (Mathematics).
Fluid mechanics, hydrodynamic stability, transition to turbulence. T. F. Balsa (A.M.E.), B. J. Bayly (Mathematics), K. W. Chow (Mathematics), H. F. Fasel (A.M.E.), K. Y. Fung (A.M.E.), J. C. Heinrich (A.M.E.), E. J. Kerschen (A.M.E.).
Analysis and numerical methods for partial differential equations. M. Brio (Mathematics), N. M. Ercolani (Mathematics), W. M. Greenlee (Mathematics), B. LeMesurier (Mathematics), C. D. Levermore (Mathematics), H. Rund (Applied Mathematics).
Biological applications, including population dynamics and genetics, medical imaging, flow and transport in living tissue. H. H. Barrett (Radiology), J. M. Cushing (Mathematics), W. J. Dallas (Radiology), J. F. Gross (Chemical Engineering), J. O. Kessler (Physics), R. E. Michod (E.E.B.), R. B. Roemer (A.M.E.), M. L. Rosenzweig (E.E.B.), T. W. Secomb (Physiology), T. Triffet (Engineering Mechanics), J. B. Walsh (E.E.B.), A. Winfree (E.E.B.).
Statistical mechanics, solid-state physics, quantum mechanics. P. A. Carruthers (Physics), W. G. Faris (Mathematics), T. G. Kennedy (Mathematics), S. W. Koch (Physics), W. E. Lamb Jr. (Optical Sciences/Physics), S. Mazumdar (Physics), C. W. Newman (Mathematics), J. N. Palmer (Mathematics), A. N. Patrascioiu (Physics), D. L. Stein (Physics).
Astronomy, astrophysics, and planetary science. W. D. Arnett (Physics), A. S. Burrows (Physics), W. B. Hubbard (Planetary Science and L.P.L.), J. R. Jokipii (Astronomy and Planetary Science), E. H. Levy (Planetary Science and L.P.L.), J. I. Lunine (Planetary Science), H. J. Melosh (L.P.L.), W. C. Tittemore (Planetary Science).
Mathematical programming and optimization, control systems, operations research. R. G. Askin (S.I.E), F. E. Cellier (E.C.E.), J. B. Goldberg (S.I.E.), A. M. Law (M.I.S.), S. Sen (S.I.E.), M. K. Sundareshan (E.C.E.), T. L. Vincent (A.M.E.).
Computer science, image and signal processing. P. J. Downey (Computer Science), B. R. Hunt (E.C.E.), U. Manber (Computer Science), E. W. Myers (Computer Science), R. Schowengerdt (Arid Lands/Electrical Engineering).
Applied probability and statistics. D. E. Myers (Mathematics), M. F. Neuts (S.I.E.), M. Shaked, (Mathematics), A. L. Wright (Mathematics).
Nuclear physics, nuclear engineering. B. R. Barrett (Physics), W. Filippone (N.E.E.), B. D. Ganapol (N.E.E.), D. L. Hetrick (N.E.E.), R. L. Morse (Applied Mathematics).
Electromagnetics and electronics. A. C. Cangellaris (E.C.E.), D. G. Dudley (E.C.E.), O. A. Palusinski (E.C.E.), M. N. Szilagyi (E.C.E.), J. R. Wait (E.C.E. and Geosciences), R. N. Ziolkowski (E.C.E.).
Multibody mechanics, solid mechanics and materials science. A. Arabyan (A.M.E.), C. L. Chan (A.M.E.), A. Chandra (A.M.E.), C. S. Desai (Civil Engineering), E. Madenci (A.M.E.).
Geosciences, soil sciences, and hydrology. D. O. Lomen (Mathematics), S. P. Neuman (Hydrology), R. M. Richardson (Geosciences), T. C. Wallace (Geosciences), A. W. Warrick (S.W.S.).
Social and behavioral sciences. R. L. Hamblin (Sociology), V. L. Smith (Economics), M. E. Sobel (Sociology).

SECTION 5: MATHEMATICAL SCIENCES

UNIVERSITY OF CALIFORNIA, LOS ANGELES

School of Medicine
Department of Biomathematics

Programs of Study

The Department of Biomathematics offers a graduate program leading to the Master of Science and Doctor of Philosophy degrees in biomathematics. The goal of the doctoral program is to train creative, fully independent investigators who can initiate research in both applied mathematics and their chosen biomedical specialty. The department's orientation is away from abstract modeling and toward theoretical and applied research vital to the advancement of current biomedical frontiers. This is reflected in a curriculum providing doctoral-level competence in a biomedical specialty; substantial training in applied mathematics, statistics, and computing; and appropriate biomathematics courses and research experience. A low student-faculty ratio permits close and frequent contact between students and faculty throughout the training and research years.

Entering students come from a variety of backgrounds in mathematics, biology, the physical sciences, and computer science. Some of the students are enrolled in the UCLA M.D./Ph.D. program. Doctoral students generally use the first two years to take the core sequence and electives in biomathematics, to broaden their backgrounds in biology and mathematics, and to begin directed individual study or research. Comprehensive examinations in biomathematics are taken after this period, generally followed by the choice of a major field and dissertation area. Individualized programs permit students to select graduate courses in applied mathematics, biomathematics, and statistics appropriate to their area of research and to choose among diverse biomedical specialties. At present, approved fields of special emphasis for which courses of study and qualifying examinations have been developed include genetics, physiology, neurosciences, pharmacology, and immunology. Other major fields can be added to the list by petition. The expected time for completion of the Ph.D. degree is 5 to 5½ years.

The master's program is used primarily as a step to further graduate work in biomathematics, but it can also be adapted to the needs of researchers desiring supplemental biomathematical training or of individuals wishing to provide methodologic support to biomedical researchers. The M.S. program requires at least five graduate biomathematics courses and either a thesis or comprehensive examination plan. The master's degree can be completed in one or two years.

Research Facilities

The department is situated in the Center for the Health Sciences, close to UCLA's rich research and educational resources in the School of Medicine and in the Departments of Mathematics, Biology, Computer Science, Engineering, Chemistry, and Physics. The department has for many years housed multidisciplinary research programs comprising innovative modeling, statistical, and computing methods directed to many areas of biomedical research. It was the original home of the BMDP statistical programs and has an active consulting clinic for biomedical researchers. Computers within the department and in graduate student offices include IBM PS/2's, VMS and Ultrix workstations, and Macintosh II's, with good software support and networking. Nearby are terminals to the campus's IBM 3090 mainframe computer and extensive software resources. Students may apply for time on the San Diego Supercomputer Center CRAY X-MP as part of a UCLA block grant. The Biomedical Library is one of the finest libraries of its kind in the country, and nearby are the Engineering and Mathematical Sciences Library and other subject libraries of the renowned nineteen-branch University Library. The department maintains a small library with selected titles in mathematical biology and statistics.

Financial Aid

The department maintains its own NIH Systems and Integrative Biomathematics Training Grant for eligible students. Tuition, fees, some supplies, and travel to scientific meetings are paid for by the grant. Each trainee also receives a yearly cost of living allowance, which is set at $8800 in 1991–92. Supplementation is also possible from unrestricted funds, a teaching associate position, research assistantships, consulting, UCLA fellowships, and other merit-based funds. These additional sources of support and nonresident tuition waivers are available to exceptionally well qualified international students as well.

Cost of Study

The 1991–92 registration and other fees are estimated at $2940 per year and nonresident tuition fees at $7700 per year. Domestic students may attain residency after one year.

Cost of Living

Besides the cost of study, the estimated cost of living varies from $9900 for a single student living in the graduate residence hall to $11,200 for a single student in off-campus housing.

Student Group

Currently about one dozen graduate students are enrolled in the department's program. About a third of the students are international. Five students are on the NIH predoctoral training grant, and most students are receiving financial support or are employed on campus in the area of their research or both. Many graduates hold tenure-track appointments at leading universities and research appointments at the National Institutes of Health and in industry.

Location

UCLA's 411 acres are cradled in rolling green hills just 5 miles inland from the ocean, in one of the most attractive areas of southern California. The campus is bordered on the north by the protected wilderness of the Santa Monica Mountains and at its southern gate by Westwood Village, one of the entertainment magnets of Los Angeles.

The University

UCLA is one of America's most prestigious and influential public universities, serving over 33,000 students. The Department of Biomathematics is one of ten basic science departments in the School of Medicine. The medical school, regarded by many to be among the best in the nation, is situated on the south side of the UCLA campus, just adjacent to the Life Sciences Building and the Court of Sciences.

Applying

Most students enter in the fall quarter, but applications for winter or spring quarter entry are considered. However, it is advantageous for candidates applying for financial support to initiate the application by the middle of January for decisions for the following fall. The department prefers that applicants for direct admission to the doctoral program submit scores on the General Test of the Graduate Record Examinations and on one GRE Subject Test of the student's choice. Inquiries are welcome from students early in their undergraduate training.

Correspondence and Information

Admissions Committee Chair
Department of Biomathematics
UCLA School of Medicine
Los Angeles, California 90024-1766
Telephone: 213-206-1748
Fax: 213-825-8685
E-mail: JSneyd@biomath.medsch.ucla.edu

Peterson's Guide to Graduate Programs in the Physical Sciences and Mathematics 1992

SECTION 5: MATHEMATICAL SCIENCES

University of California, Los Angeles

THE FACULTY AND THEIR RESEARCH

A. A. Afifi, Professor of Biostatistics and Biomathematics; Ph.D. (statistics), Berkeley, 1965. Theory of applied statistical methods, with applications to biomedical and public health problems.

Edward C. DeLand, Adjunct Professor of Anesthesiology and Biomathematics; Ph.D. (mathematics), UCLA, 1956. Kinetic and steady-state fluid and electrolyte systems, biochemical modeling, computer-assisted instruction, expert systems, biological control, patient monitoring.

Wilfrid J. Dixon, Professor of Biomathematics, Biostatistics, and Psychiatry (Emeritus); Ph.D. (statistics), Princeton, 1944. Statistical computation, statistical theory, biological applications, data analysis, psychiatric research.

Janet D. Elashoff, Adjunct Professor of Biomathematics; Ph.D. (statistics), Harvard, 1966. Design and analysis of clinical trials and evaluations of robustness of standard statistical techniques.

* Robert M. Elashoff, Professor of Biomathematics and Biostatistics; Ph.D. (statistics), Harvard, 1963. Markov renewal models in survival analysis, random coefficient regression models.

Eli Engel, Adjunct Assistant Professor of Biomathematics; M.D., Buffalo, 1951; Ph.D. (physiology), UCLA, 1975. Mechanisms for acid neutralization in gastric mucus, facilitated transport of oxygen, theory of intracellular microelectrodes.

Alan B. Forsythe, Adjunct Professor of Biomathematics and Dentistry; Ph.D. (biometry), Yale, 1967. Methods development in robust regression and hypothesis testing, design and analysis of clinical and epidemiological studies, computer systems design.

Karim F. Hirji, Adjunct Assistant Professor of Biomathematics; Ph.D. (biostatistics), Harvard, 1986. Exact statistical methods for categorical data.

* Sung-Cheng (Henry) Huang, Professor in Residence of Radiological Sciences and Biomathematics; D.Sc. (electrical engineering), Washington (St. Louis), 1973. Positron emission computed tomography and physiological modeling.

Donald J. Jenden, Professor of Pharmacology and Biomathematics; M.B.,B.S. (pharmacology and therapeutics), Westminster (London), 1950. Pharmacokinetic modeling, chemical pharmacology, analysis of GC/MS data, neuropharmacology.

* Robert I. Jennrich, Professor of Mathematics, Biomathematics, and Biostatistics; Ph.D. (mathematics), UCLA, 1960. Statistical methodology, computational algorithms, nonlinear regression, factor analysis, compartment analysis.

* Elliot M. Landaw, Associate Professor of Biomathematics; M.D., Chicago, 1972; Ph.D. (biomathematics), UCLA, 1980. Identifiability and optimal experiment design for compartmental models; nonlinear regression; modeling/estimation applications in pharmacokinetics, ligand-receptor analysis, transport, and pediatrics.

* Kenneth L. Lange, Professor of Biomathematics and Chair; Ph.D. (mathematics), MIT, 1971. Statistical and mathematical methods for human genetics and population growth, image reconstruction algorithms.

* Roderick J. A. Little, Professor of Biomathematics and Vice Chair; Ph.D. (statistics), Imperial College (London), 1974. Statistical analysis with missing data, survey research methods, Bayesian modeling.

* Carol M. Newton, Professor of Biomathematics and Radiation Oncology; Ph.D. (physics and mathematics), Stanford, 1956; M.D., Chicago, 1960. Simulation; cellular models for hematopoiesis, cancer treatment strategies, optimization; interactive graphics for modeling; model-based exploration of complex data structures (pedigrees).

* Arthur Peskoff, Adjunct Professor of Biomathematics and Physiology; Ph.D. (electrical engineering), MIT, 1960. Mathematical biophysics, electrodiffusion theory for ion channels, reaction-diffusion in gastric mucus, electric potential theory in cells and syncytia.

* Michael E. Phelps, Jennifer Jones Simon Professor, Professor of Biomathematics, and Chief, Laboratory of Nuclear Medicine; Ph.D. (nuclear chemistry), Washington (St. Louis), 1970. Positron emission tomography (PET), tracer kinetic modeling of biochemical and pharmacokinetic processes.

James Sneyd, Assistant Professor of Biomathematics; Ph.D. (mathematics), NYU, 1989. Differential equations and deterministic modeling in the biological sciences; biological control processes, oscillations, and spatial structure; phototransduction and light adaptation; principles of self-organization in biological systems.

M. Anne Spence, Professor in Residence of Psychiatry and Biomathematics; Ph.D. (genetics), Hawaii, 1969. Human genetics: mathematical analyses of family data, including segregation, linkage, and pedigree analyses.

---

* Recent chairs/members of biomathematics doctoral committees.

SECTION 5: MATHEMATICAL SCIENCES

UNIVERSITY OF CALIFORNIA, LOS ANGELES

School of Public Health
Department of Biostatistics

Programs of Study The Department of Biostatistics offers programs leading to M.P.H., M.S., Dr.P.H., and Ph.D. degrees. The M.P.H. and M.S. degrees are typically two-year programs but can be completed in one year by well-prepared students. The M.P.H. degree emphasizes public health, exposing students to important areas of health research. The M.S. degree is the appropriate choice for students planning to continue on to doctoral-level training. The Dr.P.H. program prepares biostatisticians skilled in applying statistical methods to problems in the health sciences. The Ph.D. degree trains biostatisticians to solve problems in the health sciences and develop biostatistical methodology. The doctoral degrees require two years of academic residence at UCLA. The time to complete the degree varies, but the usual amount of time is four years of post-master's study. Computing skills are essential tools of biostatisticians; every degree program integrates substantial training in statistical packages and health data management.

Faculty members participate in research projects in areas such as cancer, AIDS, gerontology, genetics, immunology, dentistry, medical imaging, mental health, and air pollution. Students work with the faculty as research associates during their training. This practical experience often results in coauthored publications before graduation and makes the graduates highly attractive to future employers.

Research Facilities The School of Public Health is located in the UCLA Center for the Health Sciences and has close relationships with the Schools of Medicine, Nursing, and Dentistry. The School of Public Health has extensive research laboratories and facilities available for faculty and student use. Its Microcomputer Instructional Center offers computers that are used by faculty and students in teaching, learning, and research. Students also have access to the resources of the campus library system, which includes the Engineering and Mathematical Sciences Library and the Louise Darling Biomedical Library.

Financial Aid The department holds a training grant in Biostatistics Training for AIDS Research that supports excellent doctoral students who concentrate in this area. This grant pays a stipend plus tuition and fees. UCLA is a major center for AIDS research and is one of the few departments with such training opportunities. Other support for outstanding students includes nonresident tuition waivers and campus fellowship funds. Some federal public health traineeships are available to support U.S. citizens and permanent residents. Through the Health Career Opportunity Program, the University has special scholarship funds to support minority students who have high potential for graduate study. Teaching assistantships that pay up to $12,000 for the three quarters are also offered. These usually require at least 1 year of statistics study. Most students hold assistantships involving research in medicine or health in the Center for the Health Sciences.

Cost of Study For 1991-92, full-time annual graduate fees are $2313 for California residents. Nonresidents must also pay an additional out-of-state tuition of $7700. Mandatory health insurance is $594 per year and is usually paid by UCLA employers if the student works more than 10 hours per week.

Cost of Living Accommodations in the residence halls and a board plan (nineteen meals per week) cost about $5000 for three quarters. There is one graduate dormitory, Hershey Hall, that is located about 200 yards from the School. Space in the graduate dormitory can be reserved for outstanding students. One-bedroom apartments within 3 miles of campus cost about $800 per month; apartments farther away usually cost $100–$150 less. Married student housing is available, but students must apply early to be considered.

Student Group About 10 students are admitted to the doctoral programs and about 20 students enter the master's programs each year. Students come from many countries and states and range in age from 22 to 40 years. Graduates have found work in teaching, research, and consulting in medicine, public health, the life sciences, survey research, and computer science. The field of biostatistics has grown tremendously in recent years, and many employers insist on biostatistical input for their research and marketing efforts. Federal regulations require biostatistical planning and analysis in medical and drug research. The department has a superior record of training graduate students, with over 300 master's and 100 doctoral degrees awarded since 1963.

Location Los Angeles is one of the leading cities of the world and the major U.S. city of the Pacific Rim. It has outstanding theater, music, sports, and other entertainment facilities, as well as excellent restaurants and museums. The UCLA campus is a significant cultural component of the city and offers many low-cost opportunities for students. Both mountain and beach recreation are within 1 hour of the campus.

The University and The Department Academically, UCLA is ranked among the leading universities in the country. A distinguished faculty, an extensive library system, excellent research laboratories, and a variety of centers and institutes provide graduate students with many opportunities. UCLA's program of music, dance, theater, films, and lectures offers student productions and performances and public lectures by nationally and internationally known artists and scholars. The campus also provides extensive recreational facilities.

The Department of Biostatistics was founded in 1960 and is widely recognized as one of the best departments in the country. The first Ph.D. was awarded in 1963. Faculty members of the department have served as officers of major statistical societies and have been recognized by important statistical societies.

Applying Typical applicants hold a bachelor's degree in mathematics or in the biological, physical, or social sciences. Although an interest in a field of application is desirable, advanced training is not necessary for admission. Applicants to the M.P.H. program must have at least 1 year of calculus; applicants to the M.S. program must have at least 2 years of calculus. Admission to either of the doctoral programs usually requires the equivalent of an M.S. in biostatistics, statistics, or mathematics. Students with outstanding undergraduate records may be admitted directly to a doctoral program. Although admission is possible in any quarter, it is recommended that students start in the fall quarter, which begins at the end of September. Students who wish to be considered for University fellowships must have their applications completed by January 1. Students who wish to be considered for departmental traineeships, teaching assistantships, or nonresident waivers must have their applications completed by March 1. Applications must be completed with three letters of recommendation, all transcripts, and GRE scores. Students from foreign countries are encouraged to apply early to ensure adequate time for the receipt of their materials.

Correspondence and Information
Chair, Admissions Committee
Department of Biostatistics
UCLA School of Public Health
Los Angeles, California 90024-1772
Telephone: 213-825-5250

Peterson's Guide to Graduate Programs in the Physical Sciences and Mathematics 1992

SECTION 5: MATHEMATICAL SCIENCES

University of California, Los Angeles

THE FACULTY AND THEIR RESEARCH

Professors
A. Afifi, Dean of the School of Public Health; Ph.D., Berkeley. Survival analysis, multivariate analysis, health needs of underserved populations.
P. Chang, Ph.D., Minnesota. Regression analysis, goodness-of-fit for generalized linear models, applications.
R. Elashoff, Ph.D., Harvard. Survival analysis, measurement error models, risk analysis in carcinogenic bioassay. (Joint appointment with Biomathematics)
D. Guthrie, Ph.D., Stanford. Applications in mental retardation and child psychiatry, statistical computing. (Joint appointment with Psychiatry and Biobehavioral Sciences)
R. Jennrich, Ph.D., UCLA. Statistical computing, statistical software, algorithms. (Joint appointment with Mathematics)
P. Lachenbruch, Ph.D., UCLA. Robustness, discriminant analysis, statistical computing, applications to geriatrics, injury prevention, psychiatry.

Associate Professors
W. Cumberland, Ph.D., Johns Hopkins. Finite-population sampling, small-domain estimation, stochastic modeling, applications to cancer, immunology, health insurance.
V. Flack, Ph.D., Berkeley. Variable selection algorithms, evaluation of model fit, applications to dentistry, malnutrition in children.
J. Taylor, Ph.D., Berkeley. Data transformations, robustness, statistical problems in AIDS research and radiation research. (Joint appointment with Radiation Oncology)

Assistant Professors
D. Dabrowska, Ph.D., Berkeley. Inference in nonparametric and semiparametric models, survival analysis, data transformations, analysis of bone marrow transplant data.
N. Schenker, Ph.D., Chicago. Missing data—nonresponse, census undercount, Bayesian methods for logistic regression.
R. Weiss, Ph.D., Minnesota. Diagnostics, Bayesian data analysis, longitudinal data and graphics.
W. Wong, Ph.D., Minnesota. Optimal design of experiments, linear models.

Adjunct Faculty and Lecturers
D. Gjertson, Ph.D., UCLA. Statistical genetics, measurement errors.
J. Lee, Ph.D., UCLA. Survival analysis, statistical computing, statistical graphics, applications to dental research.
M. Lee, Ph.D., UCLA. Applications to pharmaceutical problems.
J. Sayre, Dr.P.H., UCLA. Computational statistics and database management, clinical trials, statistical methodology in medical diagnostic systems.

SECTION 5: MATHEMATICAL SCIENCES

UNIVERSITY OF CALIFORNIA, RIVERSIDE

Department of Mathematics
Programs in Mathematics

Programs of Study The department offers the degrees of Master of Arts, Master of Science, and Doctor of Philosophy in mathematics and the Master of Science in applied mathematics.

Research areas include algebraic geometry, commutative algebra, Lie algebras, differential equations, differential geometry, functional analysis, approximation theory, topology, order theory, combinatorics, numerical analysis, and probability and statistics. (A Ph.D. in applied statistics is available through the Department of Statistics.)

In addition to completing a thesis, students in the doctoral program must demonstrate a reading knowledge of French, German, Italian, or Russian and pass written qualifying examinations in four of the following areas: algebra, applied mathematics, complex analysis, geometry/topology, and real analysis.

The master's degree programs are usually completed within two years, although it is possible for a well-prepared student to earn a master's degree in one year. For the M.A. or M.S. in mathematics or applied mathematics, the student must pass qualifying examinations in two areas.

Research Facilities The campus library (a 3-minute walk from the Department of Mathematics) maintains extensive holdings of mathematics books and journals, including back issues. Current issues of journals are kept in a department reading room. Publications not available locally can be obtained through interlibrary loan from the Berkeley and Los Angeles campuses of the University of California.

The department owns a Sun-4/490 server, two Sun-4/260 servers, over fifty Sun Workstations (twenty-three SPARCstation 4/60s, two color 3/110s, three color 3/60s, thirteen monochrome 3/60s, and fifteen 3/50s), and many ordinary terminals. All machines run UNIX and are networked together over Ethernet. Students also have access to the University's general computing facilities, including an 8820/6310 VAXcluster running VMS, an IBM 4341, and a microcomputer lab, and networked access to computing systems at other branches of the University of California, including a CRAY Y-MP supercomputer on the San Diego campus.

Financial Aid Fellowships are awarded by the Graduate Division on a competitive basis, with stipends ranging from $5500 to $12,500 for the nine-month academic year. These awards include payment of all assessed registration fees. The department offers teaching assistantships, community teaching fellowships, and research assistantships. A half-time appointment as a teaching assistant or community teaching fellow carries a stipend of $11,466 for the 1991–92 academic year. A limited number of tuition waivers are available to cover the additional tuition fee for out-of-state students.

Cost of Study For 1991–92, California residents pay approximately $2820 a year in fees. Nonresidents are charged an additional tuition fee of $10,518. These amounts are subject to change.

Cost of Living Riverside offers graduate students one of the lowest costs of living of all the cities with a UC campus. Room and board in residence halls cost $5420 for the 1991–92 academic year. The University owns 268 houses that are available to married students and single students with children and rent from $340 to $375 per month in 1991–92. Rents for the 150 apartments and 92 suites available for single students range from $147 to $493 per month. Abundant off-campus housing is available within walking distance of the campus.

Student Group The department currently enrolls about 45 graduate students, who come from all sections of the United States. Most are supported by research or teaching assistantships or fellowships. About two thirds of the mathematics graduate students are enrolled in the Ph.D. program. The campus has about 8,700 students, of whom over 1,550 are graduate students.

Location The Riverside campus, consisting of more than 1,000 acres, is located 3 miles east of the center of Riverside in the shelter of the Box Springs Mountains. A community of more than 170,000 people, Riverside has excellent recreational facilities, a symphony orchestra, an opera association, a community theater, an art center, and several other colleges. Within a 60-mile radius are the mountains, the desert, the ocean, and Metropolitan Los Angeles. The average year-round maximum temperature is 79 degrees. The region is semiarid, with relatively low rainfall; consequently, students spend much of their leisure time out of doors.

The University and The Department The Riverside campus of the University of California began as a Citrus Experimental Station in 1907. In 1954 the College of Letters and Science opened for classes, and in 1959 Riverside became a general campus. The department began the graduate program in mathematics in 1961.

Applying Applications for admission are rated in terms of academic scholarship and preparation and the applicant's score on the General Test of the Graduate Record Examinations. Application forms and more detailed information may be obtained from the Department of Mathematics or from the Graduate Division.

To receive full consideration for financial support, applications, together with a $40 application fee, should be received by February 1. Later applications will be considered if any support is still available.

Correspondence and Information
Graduate Secretary
Department of Mathematics
University of California, Riverside
Riverside, California 92521-0135
Telephone: 714-787-3114

Peterson's Guide to Graduate Programs in the Physical Sciences and Mathematics 1992

SECTION 5: MATHEMATICAL SCIENCES

University of California, Riverside

THE FACULTY AND THEIR RESEARCH

John Baez, Mathematical physics.
Theodore J. Barth. Analytic functions.
Richard E. Block. Lie theory.
Bruce L. Chalmers. Approximation theory.
Mei-Chu Chang. Algebraic geometry.
Vyjayanthi Chari. Lie theory.
John E. de Pillis. Numerical linear algebra.
Le Baron O. Ferguson. Approximation theory.
Gerhard Gierz. Order theory.
Neil E. Gretsky. Functional analysis, mathematical economics.
Lawrence H. Harper. Combinatorial theory.
George Kempf. Algebraic geometry.
Michel Lapidus. Analysis, fractal geometry.
Frederic T. Metcalf. Applied mathematics, approximation theory.
J. Keith Oddson. Applied mathematics.
Ivan Penkov. Lie theory.
Yat Sun Poon. Differential geometry.
Mihai Putinar. Operator theory.
Ziv Ran. Algebraic geometry.
Malempati M. Rao. Probability theory, functional analysis.
Louis J. Ratliff Jr. Commutative algebra.
David E. Rush. Commutative algebra.
Victor L. Shapiro. Harmonic analysis, partial differential equations.
James D. Stafney. Analysis.
Albert R. Stralka. Order theory.
John Walsh. Topology.
Bun Wong. Several complex variables, differential geometry.

SECTION 5: MATHEMATICAL SCIENCES

UNIVERSITY OF DELAWARE

Department of Mathematical Sciences

Programs of Study	The Department of Mathematical Sciences offers master's and Ph.D. programs in mathematics, applied mathematics, and statistics. Students receive instruction in a broad range of courses and may specialize in many areas of mathematics and statistics. Strong departmental research groups exist in applied mathematics, partial differential equations, combinatorics and computer algebra, complex analysis, probability and stochastic processes, statistics, and numerical analysis. Master's programs normally require two years for completion, while the Ph.D. usually takes four to five years. Internship programs with industry and government are available in statistics.
Research Facilities	The University libraries contain 2 million volumes and documents and subscribe to 22,500 periodicals and serials. The University library belongs to the Association of Research Libraries.
	The Academic Computing Services facility provides time-sharing access to an IBM 3090-300 and two Sun-4/480 minicomputers. More than 1,100 terminals are distributed around the campus in addition to Sun Workstations and PCs. The mathematical sciences department has its own network of Sun Workstations, a Sun Workstation classroom, and a microcomputer laboratory.
	The department fosters an active research environment, with numerous seminars and colloquia and many national and international visitors. In the most recent (1983) national assessment, the Ph.D. program in mathematics was ranked among the five most improved programs in the country.
Financial Aid	Graduate assistantships and fellowships are available on a competitive basis. Teaching assistantships in 1991–92 range from $10,945 to $11,925 plus tuition remission for nine months, which includes two semesters and a winter session. Additional summer session stipends are also often available. At present, most full-time students receive some financial support. Some research assistantships are also available.
Cost of Study	Course fees for full-time students in 1991–92 are $2890 per academic year for residents of Delaware and $7680 per academic year for out-of-state students. Fees for the summer sessions and for part-time students are $161 per credit for Delaware residents and $427 per credit for nonresidents. The graduation fee is $30 for the master's degree and $65 for the Ph.D.
Cost of Living	While prices vary widely throughout the area, average monthly rent for a one-bedroom apartment is $410 plus utilities.
Student Group	There are approximately 50 full-time and 28 part-time graduate students in the Department of Mathematical Sciences. About one quarter of these are international students.
Location	The University is located in Newark, Delaware, a pleasant college community of about 30,000 people. Newark is 14 miles southwest of Wilmington, halfway between Philadelphia and Baltimore. It offers the advantages of a small community yet is within easy driving distance of Philadelphia, New York, Baltimore, and Washington, D.C. It is also close to the recreational areas on the Atlantic Ocean and Chesapeake Bay.
The University	The University of Delaware grew out of a small academy founded in 1743. It has been a degree-granting institution since 1834. In 1867, an act of the Delaware General Assembly made the University a part of the nationwide system of land-grant colleges and universities. Delaware College and the Women's College, an affiliate, were combined under the name of the University of Delaware in 1921. In 1950, the Graduate College was organized to administer the existing graduate programs and to develop new ones. In the past fifteen years, the University has greatly expanded the scope of its educational endeavors. In 1976, the University was named a sea-grant college. In 1985, it was one of the first six universities to receive an NSF grant for an Engineering Research Center.
Applying	Application forms may be obtained from the address below. Completed applications, including letters of recommendation, a $40 application fee, GRE General Test scores, and transcripts of previous work, should be submitted as early as possible but no later than March 1 to be considered for financial aid for the fall semester.
Correspondence and Information	Professor Thomas S. Angell Coordinator of Graduate Studies Department of Mathematical Sciences University of Delaware Newark, Delaware 19716 Telephone: 302-451-2654

Peterson's Guide to Graduate Programs in the Physical Sciences and Mathematics 1992

SECTION 5: MATHEMATICAL SCIENCES

University of Delaware

THE FACULTY AND THEIR RESEARCH

Thomas S. Angell, Professor; Ph.D., Michigan. Optimal control theory, differential equations.
Ronald D. Baker, Professor; Ph.D., Ohio State. Combinatorics, algebra.
Willard E. Baxter, Professor; Ph.D., Pennsylvania. Algebra.
David P. Bellamy, Professor; Ph.D., Michigan State. Topology.
John G. Bergman, Associate Professor and Associate Chairman; Ph.D., Illinois at Urbana-Champaign. Functional analysis, probability.
David L. Colton, Professor; Ph.D., D.Sc., Edinburgh. Partial differential equations, integral equations.
L. Pamela Cook-Ioannidis, Professor; Ph.D., Cornell. Applied mathematics, perturbation theory, transonic flow.
Richard J. Crouse, Associate Professor; Ph.D., Delaware. Mathematics education.
Gary L. Ebert, Professor; Ph.D., Wisconsin–Madison. Combinatorics.
Paul P. Eggermont, Associate Professor; Ph.D., SUNY at Buffalo. Numerical analysis, integral equations, image reconstruction.
Robert P. Gilbert, Unidel Chair Professor; Ph.D., Carnegie Mellon. Integral and differential equations, function theory.
David J. Hallenbeck, Professor; Ph.D., SUNY at Albany. Function theory.
Joseph S. Hemmeter, Assistant Professor; Ph.D., Ohio State. Combinatorics, graph theory.
George C. Hsiao, Professor; Ph.D., Carnegie Mellon. Differential and integral equations, perturbation theory, fluid dynamics.
Jinsoo Hwang, Assistant Professor; Ph.D., Purdue. Categorical data analysis, survival analysis.
Judy A. Kennedy, Associate Professor; Ph.D., Auburn. Topology.
Ralph E. Kleinman, Professor; Ph.D., Delft University of Technology. Integral equations, boundary-value problems, electromagnetic theory.
Vincent N. LaRiccia, Associate Professor; Ph.D., Texas A&M. Mathematical statistics.
Felix Lazebnik, Assistant Professor; Ph.D., Pennsylvania. Graph theory, combinatorics, algebra.
Yuk J. Leung, Associate Professor; Ph.D., Michigan. Function theory.
Richard J. Libera, Professor; Ph.D., Rutgers. Function theory.
Albert E. Livingston, Professor; Ph.D., Rutgers. Function theory.
Walter K. Mallory, Assistant Professor; Ph.D., Rutgers. Mathematical logic, fuzzy set theory.
David M. Mason, Professor; Ph.D., Washington (Seattle). Probability, statistics.
Peter Monk, Associate Professor; Ph.D., Rutgers. Numerical analysis.
M. Zuhair Nashed, Professor; Ph.D., Michigan. Numerical analysis, functional analysis, integral equations, optimization and approximation theory.
David Olagunju, Assistant Professor; Ph.D., Northwestern. Applied mathematics.
Georgia B. Pyrros, Instructor; M.S., McMaster. Nuclear physics.
Rakesh, Assistant Professor; Ph.D., Cornell. Partial differential equations.
Lidia Rejtö, Associate Professor; Ph.D., Eotvos Loránd (Budapest). Probability and statistics.
David P. Roselle, Professor and President of the University; Ph.D., Duke. Combinatorics.
Lillian M. Russell, Instructor; M.S., Delaware. Statistics.
Fadil Santosa, Associate Professor; Ph.D., Illinois at Urbana-Champaign. Analysis and numerical modeling in inverse problems and wave propagation.
Gilberto Schleiniger, Assistant Professor; Ph.D., UCLA. Scientific computing, numerical analysis.
John H. Schuenemeyer, Professor; Ph.D., Georgia. Multivariate analysis, applied statistics.
Clifford W. Sloyer, Professor; Ph.D., Lehigh. Topology, mathematics education.
Ivar Stakgold, Professor and Chairman; Ph.D., Harvard. Nonlinear boundary-value problems.
Robert M. Stark, Professor; Ph.D., Delaware. Applied probability, operations research, civil engineering systems.
Howard M. Taylor, Professor; Ph.D., Stanford. Probability, stochastic modeling.
Henry B. Tingey, Professor; Ph.D., Minnesota. Biometry, applied statistics.
Richard J. Weinacht, Professor; Ph.D., Maryland. Partial differential equations.
Ronald H. Wenger, Associate Professor and Director, Mathematical Sciences Teaching and Learning Center; Ph.D., Michigan State. Algebra, mathematics education.
Shangyou Zhang, Assistant Professor; Ph.D., Penn State. Numerical analysis and scientific computation.

Visitors 1991–92
Peter Hähner, Ph.D., Göttingen (Germany). Numerical analysis and applied mathematics.
Krzysztof Samotij, Ph.D., Wroclaw Technical (Poland). Complex analysis.
Boris Vainberg, Ph.D., Moscow. Asymptotic methods, scattering theory.
Lin Wei, Ph.D., Zhongshan (China). Function theory.

Joint Appointments with Other Departments
Morris W. Brooks, Ph.D., Harvard. Computer-based instruction.
Bobby F. Caviness, Ph.D., Carnegie Mellon. Computer algebra.
Kathleen Hollowell, Ed.D., Boston University. Mathematics education.
William B. Moody, Ed.D., Maryland. Mathematics education.
Richard S. Sacher, Ph.D., Stanford. Scientific computing, operations research.
David Saunders, Ph.D., Wisconsin–Madison. Computer algebra.
Leonard W. Schwartz, Ph.D., Stanford. Fluid mechanics.

Adjunct Faculty and Their Affiliations
Steven P. Bailey, DuPont Company. Design of experiments, Bayesian inference.
Spencer Free. Consultant in biostatistics.
Alan Jeffrey, University of Newcastle-upon-Tyne. Wave propagation.
Rainer Kress, University of Göttingen. Integral equations, scattering theory.
James M. Lucas, DuPont Company. Response surface methodology and quality control.
Donald W. Marquardt, DuPont Company. Administration of consulting, nonlinear estimation, biased estimation, spectrum estimation.
Charles G. Pfeifer, DuPont Company. Group testing, applied statistics.
Darryl Pregibon, Bell Laboratories. Expert systems, regression.
Gary Roach, University of Strathclyde. Operator theory, scattering theory.
Ronald Snee, DuPont Company. Quality management, statistical quality control, design of experiments.
Malcolm Taylor. U.S. Army Ballistics Research Laboratory.
Dana Ullery, DuPont Company. Applied statistics.

SECTION 5: MATHEMATICAL SCIENCES

UNIVERSITY OF FLORIDA
Department of Mathematics

Programs of Study The Department of Mathematics offers programs of study leading to the degrees of Master of Arts, Master of Science, Master of Arts in Teaching, and Doctor of Philosophy. There are opportunities for concentrated study in a number of specific areas of pure and applied mathematics at both the master's and doctoral levels.

The requirements for the master's degree include 32 semester hours of course work and a comprehensive written examination. A thesis is not required. Two master's programs are available, one in pure mathematics and one in applied mathematics. A student normally takes two years to complete either program. At least 36 semester hours of graduate course work in specified advanced graduate courses are required for the Ph.D. degree, out of a total of 90 semester hours. The student must pass a written and oral comprehensive preliminary examination to become a candidate for the degree. Students must pass one reading knowledge examination in French, German, or Russian.

The dissertation is an important part of the doctoral program in mathematics. The topic for the dissertation may be chosen from a number of areas of current research in pure and applied mathematics. A minimum of four years is required to complete the Ph.D. program.

The Department of Mathematics interacts closely with the Departments of Statistics, Computer and Information Sciences, Industrial and Systems Engineering, and Engineering Sciences and participates actively in the programs of the interdisciplinary Center for Mathematical Systems Theory, the Center for Applied Mathematics, and the Institute for Fundamental Theory.

Research Facilities The holdings of the libraries at the University of Florida number over 2 million cataloged items. A comprehensive collection of mathematics books and journals is housed in the University's new Central Science Library.

Faculty members and graduate students have unlimited access to a network of Sun computers in the mathematics department. The University of Florida houses the central facilities of the Northeast Regional Data Center (NERDC) of the State University System. The Northeast Regional Data Center has an IBM 3090-200 (with a vector processor) and an IBM 3081D that can be used for research. An extensive collection of software provides the faculty and students with database management systems, major statistical packages, libraries of scientific and mathematical routines (including IMSL, SAS, SPSSX, and the HARWELL library), graphics programs, plotting software, minicomputer and microcomputer support, and special-purpose languages.

Another facility available to the faculty and students is the Center for Instructional and Research Computing Activities (CIRCA), which houses a VAXcluster (a VAX 8600 and two VAX 780s). The center also offers a variety of computing services, including consulting, programming and analysis, database design and implementation, statistical analysis, and data entry services and provides interactive terminals as well as local and remote terminal access to both NERDC and CIRCA computers.

Financial Aid A limited number of graduate fellowships with a stipend of $12,000 for nine months are available. Special fellowships for minority students are also offered.

Teaching assistantships, which carry a basic nine-month stipend of $9000 in 1991–92, are available. Outstanding students may also receive a $3000 supplemental fellowship. Advanced graduate students receive a $1000 supplement to their basic stipend upon being admitted to candidacy for the Ph.D. degree. In addition, several annual departmental supplementary fellowship awards of $1000–$4000 are available to continuing graduate students.

Summer support in the form of teaching assistantships and fellowships is usually available to continuing graduate students.

Cost of Study Fees for 1990–91 were $76.78 per credit hour for Florida residents and $238.71 per credit hour for nonresidents. Nearly all tuition fees are waived for holders of teaching assistantships and fellowships. Microfilming and binding fees (for theses and dissertations) are $60.

Cost of Living Housing for single and married students is available both on and off campus. In 1990–91, dormitory room costs ranged from $463 to $1097 per semester, and apartment rents for single and married students ranged from $182 to $318 per month. Off-campus, privately owned housing is also available at reasonable rates.

Student Group The University has a student body of approximately 35,000, including 7,800 graduate students. There are about 80 graduate students in the mathematics department, drawn from all parts of the United States and from several foreign countries. Recent graduates have been successful in finding employment in government laboratories, private industry, and academia.

Location Gainesville, a community of 122,000, is located in north-central Florida, 50 miles from the Gulf of Mexico and 60 miles from the Atlantic Ocean. Proximity to springs, lakes, beaches, and parks offers opportunities for camping, swimming, fishing, boating, and hiking. Gainesville offers a variety of cultural activities, including two theater groups, a civic ballet, an orchestra, and several museums.

The University Founded in 1853 as Florida's first university, the University of Florida is now one of three universities in the country that offer such a wide scope of professional fields on a single campus. Nationally, it ranks in the top fifty colleges and universities in the amount of federal funds received, which attests to the high quality of its faculty and research programs. The diversity and scope of campus programs enable students to fulfill their intellectual and professional objectives.

Applying Graduate study may begin in any semester; the application deadline is four months prior to the start of the fall semester and two months prior to the start of the spring and summer semesters. Fellowship applicants must begin study in the fall semester and must have submitted complete applications, including officially reported Graduate Record Examinations scores, by early February of the previous winter. Assistantships are usually awarded for the fall semester, and the financial aid decision is made the previous spring. Applicants must submit their up-to-date transcripts and current Graduate Record Examinations scores to the registrar as early as possible.

Correspondence and Information
Graduate Selection Committee
Department of Mathematics
University of Florida
Gainesville, Florida 32611
Telephone: 904-392-0281

Peterson's Guide to Graduate Programs in the Physical Sciences and Mathematics 1992

SECTION 5: MATHEMATICAL SCIENCES

University of Florida

THE FACULTY AND THEIR RESEARCH

Graduate Research Professor
R. E. Kalman, D.Sci., Columbia, 1957. Mathematical system theory and control theory.

Professors
K. Alladi, Ph.D., UCLA, 1978. Number theory.
A. R. Bednarek, Ph.D., Buffalo, 1961. Relation algebras and combinatorial theory.
L. S. Block, Ph.D., Northwestern, 1973. Dynamical systems.
B. L. Brechner, Ph.D., LSU, 1964. Topology.
J. K. Brooks, Ph.D., Ohio State, 1964. Measure and integration theory.
D. A. Cenzer, Ph.D., Michigan, 1972. Logic, recursion theory.
N. Dinculeanu, Ph.D., Bucharest, 1951. Vector measures, integration theory, stochastic processes, functional analysis.
D. A. Drake, Ph.D., Syracuse, 1967. Finite geometries and combinatorial theory.
B. Edwards, Ph.D., Dartmouth, 1976. Numerical analysis.
P. E. Ehrlich, Ph.D., SUNY at Stony Brook, 1974. Riemannian and pseudo-Riemannian differential geometry.
G. Emch, Ph.D., Geneva, 1963. Mathematical physics.
A. Fathi, Ph.D., Paris-Sud, 1980. Dynamical systems, topology.
J. Glover, Ph.D., California, San Diego, 1978. Probability theory and stochastic processes.
W. W. Hager, Ph.D., MIT, 1974. Numerical analysis, optimization, optimal control.
C. Ho, Ph.D., Chicago, 1972. Group theory, finite geometrics.
J. E. Keesling, Ph.D., Miami (Florida), 1968. Topology, applied mathematics, dynamical systems.
J. R. Klauder, Ph.D., Princeton, 1959. Mathematical physics.
J. A. Larson, Ph.D., Dartmouth, 1972. Combinatorial set theory.
J. Martinez, Ph.D., Tulane, 1969. Partially ordered algebraic structures, topology, commutative algebra.
W. J. Mitchell, Ph.D., Berkeley, 1970. Logic, set theory.
C. N. Nelson, Ph.D., Maryland, 1962. Mathematics education.
V. M. Popov, Ph.D., Bucharest, 1968. Stability theory of differential equations.
Z. R. Pop-Stojanovic, Ph.D., Belgrade, 1964. Probability theory and stochastic processes.
M. Rao, Ph.D., Tata Institute (India), 1963. Probability theory and stochastic processes.
G. Robinson, D.Phil., Oxford, 1981. Algebra, including group representations.
A. Turull, Ph.D., Chicago, 1982. Group representations.
A. K. Varma, Ph.D., Alberta, 1964. Approximation theory.
N. L. White, Ph.D., Harvard, 1972. Combinatorial theory, symbolic algebra.
D. C. Wilson, Ph.D., Rutgers, 1969. Topology, image processing.

Associate Professors
P. Bacon, Ph.D., Tennessee, 1964. Topology.
R. M. Crew, Ph.D., Princeton, 1981. Algebraic geometry, number theory.
D. Groisser, Ph.D., Harvard, 1983. Differential geometry, mathematical physics.
M. P. Hale Jr., Ph.D., Illinois, 1969. Finite groups and geometries.
J. L. King, Ph.D., Stanford. Dynamical systems, ergodic theory.
B. Mair, Ph.D., McGill, 1983. Partial differential equations.
T. O. Moore, Ph.D., Missouri, 1955. Topology.
S. A. Saxon, Ph.D., Florida State, 1969. Functional analysis.
L. C. Shen, Ph.D., Wisconsin, 1981. Complex analysis.
K. Sigmon, Ph.D., Florida, 1966. Topological algebra, numerical analysis, parallel computing, algebraic topology.
R. Smith, Ph.D., Penn State, 1979. Theory of computation, program verification.
C. Stark, Ph.D., Michigan, 1982. Algebraic topology.
S. Summers, Ph.D., Harvard, 1979. Mathematical physics, theory of operator algebras.
A. Vince, Ph.D., Michigan, 1981. Graph theory, combinatorics.
H. Volklein, Ph.D., Erlangen (West Germany), 1983. Geometry, Galois theory.
T. Walsh, Ph.D., Chicago, 1969. Singular integrals.

Assistant Professors
Y. Chen, Ph.D., Fudan (China). Partial differential equations.
L. Flaminio, Ph.D., Stanford, 1985. Dynamical systems.
F. G. Garvan, Ph.D., Penn State, 1986. Number theory.
S. A. McCullough, Ph.D., California, San Diego, 1987. Functional analysis.
P. L. Robinson, Ph.D., Warwick, 1984. Symplectic geometry, differential geometry, functional analysis.
P. K. W. Sin, D.Phil., Oxford, 1986. Algebra, including group representations.
C. Y. Xu, Ph.D., MIT. Partial differential equations.

SECTION 5: MATHEMATICAL SCIENCES

UNIVERSITY OF GEORGIA

Department of Mathematics

Programs of Study	The Department of Mathematics offers programs leading to the degrees of Master of Arts, Master of Applied Mathematical Science, and Doctor of Philosophy. The M.A. requires 55 quarter hours of course work and comprehensive exams or 40 quarter hours of course work and a thesis. Both options for the M.A. degree require a reading knowledge of one foreign language. The Master of Applied Mathematical Science (M.A.M.S.) degree requires 55 quarter hours of courses, with comprehensive exams at the end of the first year and a substantial technical report based on a research project carried out during the second year of the program. Students normally take two years to complete either the M.A. or the M.A.M.S. program. The Ph.D. degree program has a residence requirement of three full years of study beyond the bachelor's degree. Three written preliminary exams and an oral exam must be passed, normally by the end of the third year. Reading knowledge of two foreign languages is required. The final degree requirement is the Ph.D. dissertation. Students usually take five or six years to complete the Ph.D. program. Areas of research that are particularly strong at Georgia are Lie groups and harmonic analysis, operator theory, number theory, algebra, algebraic geometry, topology, applied mathematics, and stochastic equations.
Research Facilities	The Department of Mathematics is located in the Boyd Graduate Studies Research Center, which also houses the University's science library. The library's current mathematics collection is excellent. It includes the American Mathematical Society library, which was donated to the University in the 1960s and which contains a comprehensive collection of nineteenth- and early-twentieth-century mathematics books and journals. The department has a computer laboratory with more than 200 personal computers, which are available to graduate students on a priority basis. The department also has several terminals that can be used by graduate students to access the University's IBM 3090 mainframe computer and the department's Sun minicomputer. In addition, personal computers and terminals at various campus sites are available for general student use. The University's CDC CYBER 205 supercomputer can also be utilized by graduate students working on specific projects directed by research professors.
Financial Aid	The department offers graduate assistantships, which typically range from $7073 to $13,413 for nine months in 1991–92. Some assistantships are also available for the summer quarter at $2357 to $2515. The department competes for University-wide nonteaching graduate assistantships and minority graduate assistantships. Typical duties of a first-year graduate assistant are 10–13 hours a week of grading written work and conducting problem sessions. More advanced assistants may teach some precalculus courses.
Cost of Study	In 1991–92, students with assistantships pay a $25 matriculation fee plus $115 in transportation, athletic, and health fees per quarter. Students without assistantships pay a total of approximately $650 a quarter if they are residents of Georgia or approximately $1700 if they are from out of state. Graduating master's degree students pay a $15 graduation fee. Graduating Ph.D. students pay a $35 graduation fee and a $45 microfilm fee.
Cost of Living	The estimated cost of living for a single student, including room and board, books and supplies, and personal expenses, is about $1400 per quarter or about $4200 for the nine-month academic year. The University provides graduate dormitory housing at $410 per quarter and married student housing at $135–$180 per month for one- and two-bedroom furnished apartments, including the costs of water and cable television. Off-campus housing is available in a wide range of prices that start at about $250 per month for a two-bedroom apartment.
Student Group	There are more than 28,000 students at the University of Georgia, including over 5,000 graduate students. The Department of Mathematics has about 55 graduate students, of whom about 45 percent are from foreign countries, 40 percent are women, and 12 percent are members of minority groups. Almost all full-time graduate students in the department receive financial support.
Location	Athens is a friendly college town of about 75,000, located in rural northeast Georgia about 65 miles from Atlanta. The Appalachian Mountains are a 2-hour drive to the north, and the seacoast is 5 hours to the southeast. Athens is the recreational and cultural center of northeast Georgia, with a lively community of local artists and musicians. Popular annual local events include the Athens Criterium (a bicycle race), the Athens Human Rights Festival, the North Georgia Folk Festival, and the Athens New Jazz Festival.
The University	The University of Georgia was chartered in 1785. It is part of the University System of Georgia, which also includes the Georgia Institute of Technology and Georgia State University in Atlanta and the Medical College of Georgia in Augusta. The Graduate School was organized in 1910. The resources of the Department of Mathematics are augmented by the University's Departments of Computer Science, Statistics, and Mathematics Education, as well as by the Advanced Computational Methods Center. Several mathematics faculty members act as consultants for departments in the Division of Biological Sciences, a preeminent research unit of the University.
Applying	It is possible to begin graduate work in any quarter, but the Department of Mathematics begins all of its full-year courses in the fall, and it is considered inadvisable to start at any other time. Each applicant is required to submit GRE General Test Scores, original transcripts of all college work, and three letters of recommendation. Applicants whose native language is not English must score at least 550 on the TOEFL and are strongly advised to take the TSE (Test of Spoken English). Application forms can be obtained from the director of graduate admissions. The deadline for applications for graduate assistantships in the Department of Mathematics for the fall quarter of 1991 is February 15, 1992. Decisions on assistantships are made in late February, and applicants are notified immediately thereafter. Late applications are considered subject to the availability of support.
Correspondence and Information	Dr. Dhandapani Kannan Graduate Coordinator Department of Mathematics University of Georgia Athens, Georgia 30602 Telephone: 404-542-2211	Mary Ann Keller Director of Graduate Admissions Graduate Studies Research Center University of Georgia Athens, Georgia 30602 Telephone: 404-542-1787

Peterson's Guide to Graduate Programs in the Physical Sciences and Mathematics 1992

SECTION 5: MATHEMATICAL SCIENCES

University of Georgia

THE FACULTY AND THEIR RESEARCH

Malcolm R. Adams, Assistant Professor; Ph.D., MIT, 1982. Global analysis, mathematical physics.
George Adomian, Professor; Ph.D., UCLA, 1963. Nonlinear stochastic systems theory.
Edward A. Azoff, Professor; Ph.D., Michigan, 1972. Operator theory.
Sybilla K. Beckman, Assistant Professor; Ph.D., Pennsylvania, 1986. Arithmetic algebraic geometry.
Brian D. Boe, Assistant Professor; Ph.D., Yale, 1982. Representation theory of Lie groups and Lie algebras.
Richard H. Bouldin, Professor; Ph.D., Virginia, 1968. Functional analysis.
Thomas R. Brahana, Professor; Ph.D., Michigan, 1954. Lie algebras, group theory.
James C. Cantrell, Professor; Ph.D., Tennessee, 1961. Topology.
Yulin Cao; Ph.D., Brown, 1989. Differential equations.
Jon F. Carlson, Professor; Ph.D., Virginia, 1967. Representations of group algebras.
Leonard Chastkofsky, Associate Professor; Ph.D., Yale, 1978. Representations of Chevalley groups.
Kevin F. Clancey, Professor; Ph.D., Purdue, 1969. Analysis.
Douglas N. Clark, Professor; Ph.D., Johns Hopkins, 1967. Analysis.
C. H. Edwards, Professor; Ph.D., Tennessee, 1960. History and applications of mathematics.
David A. Edwards, Associate Professor; Ph.D., Columbia, 1971. Topology, mathematical physics.
Joseph H. G. Fu, Assistant Professor; Ph.D., MIT, 1984. Geometric measure theory, integral geometry.
David E. Galewski, Associate Professor; Ph.D., Michigan State, 1969. Topology of manifolds.
Thomas C. Gard, Associate Professor; Ph.D., Tennessee, 1974. Differential equations, mathematical biology.
Elliot C. Gootman, Professor; Ph.D., MIT, 1970. Operator algebras, group representations.
John A. Gosselin, Associate Professor; Ph.D., Purdue, 1972. Harmonic analysis, singular integrals.
Andrew J. Granville, Assistant Professor; Ph.D., Queen's at Kingston, 1987. Analytic number theory.
Nathan Habegger, Associate Professor; Ph.D., Geneva, 1981. Topology.
John G. Hollingsworth, Professor; Ph.D., Rice, 1967. Geometric topology.
J. G. Horne, Professor; Ph.D., Tulane, 1956. Applied linear algebra.
Kenneth D. Johnson, Professor; Ph.D., Berkeley, 1968. Lie groups and harmonic analysis.
Dhandapani Kannan, Professor; Ph.D., Wayne State, 1970. Stochastic processes, applied mathematics, probability theory.
William A. Kazez, Assistant Professor; Ph.D., Cornell, 1982. Low-dimensional topology.
Ray A. Kunze, Professor and Head of the Department; Ph.D., Chicago, 1957. Analysis on homogeneous spaces.
Michel L. Lapidus, Associate Professor; Ph.D., Paris VI (Curie), 1980. Functional analysis, mathematical physics.
Frank G. Lether, Professor; Ph.D., Utah, 1969. Numerical analysis.
Helmut Maier, Assistant Professor; Ph.D., Minnesota, 1981. Number theory.
Gordana Matic, Assistant Professor; Ph.D., Utah, 1986. Topology.
Stephen H. McCleary, Professor; Ph.D., Wisconsin, 1967. Ordered groups.
Clinton G. McCrory, Professor; Ph.D., Brandeis, 1972. Topology and geometry of singularities.
David E. Penney, Associate Professor; Ph.D., Tulane, 1965. Number theory and applications.
Carl Pomerance, Professor; Ph.D., Harvard, 1972. Number theory.
Peter M. Rice, Professor; Ph.D., Florida State, 1963. Game theory, modeling in the social sciences.
Mitchell J. Rothstein, Assistant Professor; Ph.D., UCLA, 1984. Mathematical physics, infinite-dimensional geometry.
Robert S. Rumely, Associate Professor; Ph.D., Princeton, 1978. Number theory and logic.
John Marshall Saade, Assistant Professor; Ph.D., Emory, 1966. General algebraic structures.
Theodore Shifrin, Associate Professor; Ph.D., Berkeley, 1979. Differential and algebraic geometry.
Roy C. Smith, Associate Professor; Ph.D., Utah, 1977. Algebraic geometry.
Robert Varley, Associate Professor; Ph.D., North Carolina, 1977. Algebraic geometry.
Paul R. Wenston, Associate Professor; Ph.D., Carnegie-Mellon, 1974. Partial differential equations, numerical analysis.
Mladen Victor Wickerhauser, Assistant Professor; Ph.D., Yale, 1985. Harmonic analysis.
Miljenko Zabcic, Assistant Professor; Ph.D., Utah, 1987. Representation theory, complex geometry.

SECTION 5: MATHEMATICAL SCIENCES

UNIVERSITY OF GEORGIA

Department of Statistics

Programs of Study

The Department of Statistics offers graduate degree programs leading to the Master of Science in statistics, Master of Applied Mathematical Sciences, and Doctor of Philosophy in statistics.

The Doctor of Philosophy degree in statistics is a research degree designed to prepare students to work on the frontiers in the discipline of statistics. This program prepares students for careers in research and teaching as well as for leadership roles in industry, business, and government, where statistical methodology and unusual statistical applications are required.

The Master of Science requires 40 quarter hours of graduate-level courses (of which at least 15 hours must be in 800-level courses) and a thesis. A nonthesis M.S. degree program requires 55 quarter hours of graduate-level courses (of which at least 15 hours must be in 800-level courses). Specialization may be in mathematical statistics or applied statistics.

The Master of Applied Mathematical Sciences (M.A.M.S.) requires 55 hours of graduate-level courses and one or two technical reports (in lieu of a thesis) prepared either as part of a regular course or in a special seminar. Specialization is possible in applied mathematics, operations research, or computer science and statistical analysis. The program is designed for students who seek broad training in applied quantitative methods as preparation for professional employment in business, government, or industry.

Research Facilities

The department has its own computer system of eleven Sun Workstations and thrity supermicrocomputers. Two labs containing Sun Workstations, ATs, and PRO/380 microcomputers provide the user with the versatility of in-house computing or connection to the University Computer Center's computer systems. Experience in statistical and computational consulting can be acquired by students through the Statistical Consulting Service, operated by the department. Experience with actual problems can be gained through the cooperative program that exists between the department and several University and research institutes and organizations.

Located nearby is the University Computer Center. It houses an IBM 3090 mainframe, VAX-11/780 minicomputer, CDC 850 mainframe, CDC 845 mainframe, and ETA-10 supercomputer. There are eighteen computer sites containing terminals, microcomputers, and printers, all of which connect to the Computer Center's systems. Also nearby is the Science Library, which has a very strong collection in the mathematical sciences and subscribes to all the major statistics journals.

Financial Aid

Teaching and research assistantships are available for qualified applicants. The 1991–92 stipends are $7072 and $7545 for master's and doctoral students, respectively. Graduate students are also eligible for a number of National Science Foundation grants and University graduate assistantships awarded in the Graduate School on a University-wide, competitive basis. A limited number of out-of-state tuition waivers are also available. Graduate students achieve teaching experience through a structured program of teaching assignments, in which initial assignments of a laboratory-assistance nature lead to actual teaching responsibilities.

Cost of Study

In 1991–92, the rate for state residents for 12 hours or more is $692 per quarter; the nonresident rate for 12 hours or more is $1840. Tuition fees are prorated for fewer than 12 hours. In addition, there is a $118 student activity fee. Tuition fees for students on assistantships are waived except for a $25 activity fee.

Cost of Living

Housing is available for married students in apartments on campus. There are also residence halls reserved exclusively for graduate students. Off-campus housing and apartments are also available. A realistic budget for a single person living on campus would be slightly more than $1100 per quarter in 1991–92.

Student Group

The total University enrollment for the fall of 1991 is estimated to be 28,500 students, including 7,000 graduate students. The graduate student body in all degree programs in the Department of Statistics consists of about 50 students.

Location

The city of Athens is a typical college town with a population of approximately 50,000. Athens is about 65 miles from Atlanta, and it is only a few hours' drive from the mountains to the north and the Atlantic coast to the east.

The University

The University of Georgia, the first chartered state university in the United States, was founded in 1785. It contains thirteen schools and colleges. The campus covers 3,500 acres and has its own transit system.

Applying

Admission is possible in all quarters, but fall admission is preferable. Applicants seeking financial assistance should apply before February 15. Applications must include a completed application form, three letters of recommendation, official transcripts, and GRE General Test scores.

Correspondence and Information

Graduate Coordinator
Department of Statistics
University of Georgia
Athens, Georgia 30602
Telephone: 404-542-5232

SECTION 5: MATHEMATICAL SCIENCES

University of Georgia

THE FACULTY AND THEIR RESEARCH

Jon E. Anderson, Assistant Professor; Ph.D., Minnesota, 1991. Survival analysis, biostatistics, survey sampling.
Rolf E. Bargmann, Professor Emeritus; Ph.D., North Carolina, 1957. Multivariate analysis, statistical computation.
Ishwar V. Basawa, Professor; Ph.D., Sheffield, 1970. Inference in stochastic processes, time series, asymptotics.
Lynne Billard, Professor; Ph.D., New South Wales, 1969. Stochastic processes, sequential analysis, time series.
Ralph A. Bradley, Research Professor; Ph.D., North Carolina, 1949. Design of experiments, nonparametric statistics, multivariate analysis.
Manas K. Chattopadhyay, Assistant Professor; Ph.D., Minnesota, 1990. Bayesian nonparametrics, bandit problems.
Hubert J. Chen, Professor; Ph.D., Rochester, 1974. Ranking and selection, multiple-comparison procedures.
A. Clifford Cohen, Professor Emeritus; Ph.D., Michigan, 1941. Truncated and censored sampling, statistical quality control.
Gauri Sankar Datta, Assistant Professor; Ph.D., Florida, 1990. Statistical inference, linear models, sampling theory.
Somnath Datta, Assistant Professor; Ph.D., Michigan State, 1988. Compound decision, Empirical Bayes, bootstrap.
Christine A. Franklin, Temporary Instructor; M.A., North Carolina, 1980. Statistical educaton, computing.
Patrick Homblé, Assistant Professor; Ph.D., Iowa State, 1990. Stochastic dynamical systems.
Kermit Hutcheson, Associate Professor and Assistant Head; Ph.D., Virginia Tech, 1970. Ecological diversity, multinomial-type distribution.
Sun Young Hwang, Temporary Assistant Professor; M.S., Seoul (Korea), 1985. Time series, inference in stochastic processes, multivariate analysis.
Carl F. Kossack, Professor Emeritus; Ph.D., Michigan, 1939. Classification techniques, sample survey.
Nancy I. Lyons, Associate Professor; Ph.D., North Carolina State, 1975. Inference in ecological models, sample survey.
William P. McCormick, Associate Professor; Ph.D., Stanford, 1978. Extreme value theory, Gaussian processes.
David M. Nickerson, Associate Professor; Ph.D., Florida, 1985. Sequential and shrinkage estimation.
Stephen L. Rathbun, Assistant Professor; Ph.D., Iowa State, 1990. Spatial statistics, ecological statistics, inference in stochastic processes.
Jaxk H. Reeves, Associate Professor; Ph.D., Berkeley, 1982. Statistical demography, ranking and selection, sampling.
Leonard R. Shenton, Professor Emeritus; Ph.D., Edinburgh, 1940. Estimation and asymptotic series.
Daphne L. Smith, Assistant Professor; Ph.D., MIT, 1985. Empirical probability distributions and Gaussian processes.
Tharuvai N. Sriram, Assistant Professor; Ph.D., Michigan State, 1986. Sequential estimation.
Robert L. Taylor, Professor and Head; Ph.D., Florida State, 1971. Laws of large numbers, stochastic convergence and density estimation.

SECTION 5: MATHEMATICAL SCIENCES

UNIVERSITY OF HOUSTON

Department of Mathematics

Programs of Study

The Department of Mathematics offers programs leading to the Doctor of Philosophy and Master of Science degrees in mathematics and the Master of Science in Applied Mathematics. Candidates for the M.S. degree may elect either the thesis or nonthesis option. The thesis option requires a minimum of 30 hours of course work as well as the successful completion and defense of a thesis. The nonthesis option requires 36 hours and the completion of a tutorial project. In both options, at least 15 hours must be taken in core courses in analysis, algebra, and topology. Candidates for the Ph.D. degree must demonstrate reading knowledge of two foreign languages and pass a comprehensive preliminary examination in mathematics. Ph.D. candidates must also present and orally defend a dissertation representing significant original research. At least one year of continuous full-time residence at the University of Houston, after the attainment of the equivalent of the M.S. degree, is required for the Ph.D.

The degree of Master of Science in Applied Mathematics requires a minimum of 36 hours of course work and the completion of a tutorial project. Eighteen hours must be taken in required courses in applied analysis, numerical analysis, and probability and statistics. Admission to the degree program requires a baccalaureate degree in a field related to mathematics and a substantial background in undergraduate mathematics.

Research Facilities

The department occupies the entire sixth floor of the Philip G. Hoffman Building. The department has its own computing facilities, including numerous workstations, a computer server, X-window terminals, and two eight-processor parallel computers. The University Computing Center also provides a diverse computing environment, including VAXclusters running VMS and UNIX multiprocessors. Network access to worldwide computing facilities is excellent. The University library contains over a million volumes plus large quantities of material in microform. Journal subscriptions in mathematical sciences number over 300. The library facilities of other Houston-area universities are available through reciprocal access arrangements.

Financial Aid

Teaching fellowships are available to most full-time graduate students in mathematics. In 1991–92, stipends range from $8500 to $8800 for nine months. Support for summer study is usually available. The department also offers a small number of instructorships, paying $12,500 for nine months, to advanced graduate students of proven teaching ability. Financial aid in the form of work-study programs and student loans is also available. Most fellowships are awarded in the early spring for the following academic year; however, late applications are considered if positions are still available.

Cost of Study

Tuition and fees for full-time graduate students are $34 per semester hour for Texas residents and $138 per semester hour for nonresidents in 1991–92. Nonresident students holding teaching fellowships or instructorships pay the Texas-resident tuition and fees.

Cost of Living

Dormitory rates for single students, including room and board, range from $3075 to $3925 per academic year (fall and spring). Limited University-owned housing for single graduate students is available at $260–$295 per month. The University has new on-campus housing (Cambridge Oaks) for married students, seniors, and graduates, with rents ranging from $345 to $800 per month. Off-campus apartment rates range upward from $250 per month.

Student Group

Enrollment at the University is approximately 31,700 students, including 6,100 graduate students. The Department of Mathematics has 75 graduate students, of whom 40 attend full-time. Approximately half of the graduate students in the department are pursuing the Ph.D. degree.

Location

Houston is the nation's fourth-largest city and the largest city in the South. Opportunities for cultural and recreational activities are plentiful. The city has many parks, theaters, and museums as well as a symphony orchestra and opera and ballet companies. The area has several collegiate and professional teams in baseball, football, basketball, and other sports. Houston is a center for medical, aerospace, and geophysical research and offers many opportunities for employment in highly technical fields.

The University

The University of Houston was founded in 1927 and became a state-supported university in 1963. It is now the third-largest university in Texas. The campus is located 3 miles from the downtown area of Houston and is easily accessible by automobile or public transportation. The College of Natural Sciences and Mathematics and the Department of Mathematics are committed to high standards of research and teaching in all areas of mathematics, particularly in those areas closely connected to current progress in science and technology.

Applying

Consideration for admission will be given to students in good standing who expect to receive a baccalaureate degree in mathematics or a closely related area prior to entrance. A GPA of 3.0 for the last 60 hours and a combined GRE General Test score of at least 1000 are required. International students from non-English-speaking countries must score at least 550 on the TOEFL. International students applying for a teaching fellowship must also score at least 220 on the TSE. Application material should be submitted directly to the Department of Mathematics three months prior to the date of proposed entrance. Teaching fellowship applicants should complete their files by March 15 to ensure full consideration.

Correspondence and Information

Director of Graduate Studies
Department of Mathematics
University of Houston
4800 Calhoun Road
Houston, Texas 77204-3476
Telephone: 713-749-4827

Peterson's Guide to Graduate Programs in the Physical Sciences and Mathematics 1992

SECTION 5: MATHEMATICAL SCIENCES

University of Houston

THE FACULTY AND THEIR RESEARCH

Neal R. Amundson, Cullen Professor; Ph.D., Minnesota, 1945. Applied mathematics.
James F. G. Auchmuty, Professor; Ph.D., Chicago, 1970. Applied mathematics, partial differential equations.
Joseph G. Baldwin, Associate Professor; Dr.rer.nat., Göttingen (Germany), 1963. Probability.
David Bao, Assistant Professor; Ph.D., Berkeley, 1983. Mathematical physics.
David P. Blecher, Assistant Professor; Ph.D., Edinburgh, 1988. Operator algebras and operator theory.
Dennison R. Brown, Professor; Ph.D., LSU, 1963. Topological semigroups.
Richard D. Byrd, Professor; Ph.D., Tulane, 1966. Retractible groups, ordered algebraic systems.
John R. Cannon, Adjunct Professor; Ph.D., Rice, 1962. Partial differential equations.
S. S. Chern, Distinguished Visiting Professor; Ph.D., Hamburg, 1936. Differential geometry.
Howard Cook, Professor; Ph.D., Texas, 1962. Point-set topology.
Edward Dean, Assistant Professor; Ph.D., Rice, 1985. Numerical analysis.
Henry P. Decell, Professor; Ph.D., LSU, 1963. Applied mathematics.
Garret J. Etgen, Professor and Chairman of the Department; Ph.D., North Carolina at Chapel Hill, 1964. Ordinary differential equations.
Siemion Fajtlowicz, Professor; Ph.D., Wroclaw (Poland), 1967. Graph theory, universal algebra.
William E. Fitzgibbon, Professor; Ph.D., Vanderbilt, 1972. Partial differential equations, nonlinear analysis.
Michael Friedberg, Professor; Ph.D., LSU, 1965. Topological algebra, topological semigroups, uniquely divisible semigroups.
John Froelich, Assistant Professor; Ph.D., Iowa, 1984. Operator theory.
Roland Glowinski, Cullen Professor; Thèse d'État, Paris, 1970. Numerical analysis, applied mathematics.
Martin Golubitsky, Professor; Ph.D., MIT, 1970. Bifurcation theory, singularity theory.
John T. Hardy, Associate Professor; Ph.D., LSU, 1965. Number theory.
Jutta Hausen, Professor; Ph.D., Frankfurt, 1967. Abelian groups and module theory.
Shanyu Ji, Assistant Professor; Ph.D., Johns Hopkins, 1988. Complex analysis, differential geometry.
Gordon G. Johnson, Professor; Ph.D., Tennessee, 1964. Analysis.
Johnny A. Johnson, Professor; Ph.D., California, Riverside, 1968. Structural properties of algebraic systems.
Klaus Kaiser, Professor; Ph.D., Bonn, 1966. Model theory.
Barbara L. Keyfitz, Professor; Ph.D., NYU, 1970. Nonlinear partial differential equations, applied mathematics.
Andrew S. Lelek, Professor; Ph.D., Wroclaw (Poland), 1959. Set-theoretic topology, continua theory.
Justin T. Lloyd, Associate Professor; Ph.D., Tulane, 1964. Algebra.
Ian Melbourne, Assistant Professor; Ph.D., Warwick (England), 1987. Bifurcation theory, singularity theory.
Christopher B. Murray, Assistant Professor; Ph.D., Texas, 1964. Analysis, statistics.
Matthew J. O'Malley, Professor; Ph.D., Florida State, 1967. Algebra.
Vern I. Paulsen, Professor; Ph.D., Michigan, 1977. Operator theory, C^* algebras.
Burnis C. Peters, Associate Professor; Ph.D., Texas A&M, 1973. Mathematical statistics, multivariate analysis.
Richard Sanders, Associate Professor; Ph.D., UCLA, 1981. Numerical solutions of partial differential equations.
Joseph A. Schatz, Associate Professor; Ph.D., Brown, 1952. Applied mathematics.
Ridgway Scott, Professor; Ph.D., MIT, 1973. Numerical solution of partial differential equations, scientific computation.
Richard D. Sinkhorn, Professor; Ph.D., Wisconsin, 1962. Linear algebra and matrix theory, doubly stochastic matrices, linear inequalities.
John M. Slye, Associate Professor; Ph.D., Texas, 1953. Point-set topology.
James W. Stepp, Professor; Ph.D., Kentucky, 1968. Topological semigroups.
Charles T. Tucker, Associate Professor; Ph.D., Texas, 1966. Functional analysis.
Arnold R. Vobach, Associate Professor; Ph.D., LSU, 1963. Topology and algebra.
David H. Wagner, Associate Professor; Ph.D., Michigan, 1980. Nonlinear partial differential equations.
Philip W. Walker, Associate Professor; Ph.D., Georgia, 1969. Ordinary differential equations.
Lewis T. Wheeler, Professor; Ph.D., Caltech, 1969. Applied mathematics, partial differential equations.
Mary F. Wheeler, Affiliated Senior Scientist; Ph.D., Rice, 1971. Numerical solution of partial differential equations.
Clifton T. Whyburn, Associate Professor; Ph.D., North Carolina at Chapel Hill, 1964. Analytic number theory.
Tiee-Jian Wu, Associate Professor; Ph.D., Indiana, 1982. Nonparametric statistics.
James N. Younglove, Professor; Ph.D., Texas, 1958. Point-set topology.

The University campus: the Department of Mathematics is in the building on the right, and downtown Houston is in the background.

SECTION 5: MATHEMATICAL SCIENCES

UNIVERSITY OF ILLINOIS AT CHICAGO

Department of Mathematics, Statistics, and Computer Science

Programs of Study

The department offers programs leading to the M.S., M.S.T., Ph.D., and D.A. degrees in pure mathematics, mathematical computer science, applied mathematics, and probability and statistics. Research and teaching areas of the current faculty are listed on the reverse side of this page.

Students in the M.S. program specialize in pure mathematics, applied mathematics, computer science, or probability and statistics. A thesis is not required for the master's degree. The programs are designed to prepare the student for further academic work, teaching, or industrial employment.

The M.S.T. program is arranged on an individual basis and is designed to strengthen the preparation and background of secondary school teachers. A candidate must earn 48 hours of graduate credit. A candidate who is teaching can complete the program through evening courses and summer courses. No thesis is required.

A student continuing for a Ph.D. must pass two written qualifying exams and also complete a minor. If both written exams are within the computer science area, the minor must be in another area (pure mathematics, applied mathematics, or probability and statistics). A student must demonstrate reading proficiency in French, German, or Russian. The Ph.D. dissertation is a significant original contribution to mathematical research.

The Doctor of Arts program is designed to educate mathematics instructors for teaching in two- and four-year colleges. The program includes study and research in the methodology and techniques necessary for successfully teaching college mathematics; a dissertation is required.

Research Facilities

The Mathematics Library, located in the department, is part of the University library system. The mathematics collection numbers well over 20,000 volumes and 240 journals and is constantly augmented. The University's library and special collections comprise more than 1.5 million volumes, and students may also draw upon the considerable library resources of the Urbana campus and those of several nearby universities. A statistical laboratory is available in the department.

The University Computer Center provides excellent facilities, including VM/CMS, a powerful interactive system that supports a full-screen editor, electronic mail, and many other programming facilities; MVS, a powerful batch computing system; and an Express system for running small batch jobs quickly. A large collection of software is available on both CMS and MVS. There is a Laboratory for Advanced Computing in the department.

Financial Aid

Teaching assistantships and fellowships are offered. For 1991–92, the salary is $8500 for nine months. Tuition is waived for all teaching assistants. Fees of approximately $510 per year must be paid by the student. Assistantship duties require 4–6 contact hours per week. Some summer aid is available. Half of all graduate students receive financial aid. An increasing number of research assistantships and fellowships are available annually. Citizens and permanent residents are eligible for National Needs Fellowships funded by the Department of Education.

Cost of Study

Tuition and fees total $1827 per semester for state residents and $4298 per semester for others in 1991–92. Tuition and fees are waived for teaching assistants; a limited number of tuition-and-fee waivers are awarded to others.

Cost of Living

In addition to UIC residence halls, apartments and private rooms are available both near the campus and throughout the city at a wide range of rents. The campus is easily accessible by public transportation.

Student Group

UIC enrolls approximately 25,000 students from throughout Illinois, the United States, and the world. Nearly one fifth of the total enrollment is at the graduate level. There are approximately 250 graduate students in mathematics, of whom one fourth are in the doctoral program.

Location

UIC is located on the Near West Side, 5 minutes by public transit from Chicago's downtown center, the Loop. It is part of a neighborhood that includes Jane Addams's Hull House and two historic landmark residential areas. In addition to the campus's recreation facilities and cultural and entertainment programs, the city offers a wealth of concert halls, theaters, movies, galleries, museums, parks, restaurants serving a variety of ethnic foods, and a lakefront setting. Several other distinguished educational institutions help to generate a wealth of mathematical activity.

The University

The University of Illinois at Chicago is the largest institution of higher learning in the Chicago area and one of the top 100 research universities in the United States. UIC offers bachelor's degrees in ninety-eight fields, master's degrees in seventy-nine, and doctorates in forty-six.

Applying

Applicants are required to take the General Test of the GRE; the Subject Test in mathematics is recommended also. Applications for admission and teaching assistantships should be received by February 1. To be considered for a fellowship, students should make sure that their applications are received by January 1. The department requires a minimum grade average of B in postcalculus mathematics and three letters of recommendation. Students seeking admission to the doctoral programs from other graduate institutions must complete work equivalent to the M.S. program, and they may be required to pass an exam in pure mathematics to fully satisfy admission requirements. It is possible to begin study in any quarter.

Correspondence and Information

Professor Jeff E. Lewis, Director of Graduate Studies
Department of Mathematics, Statistics, and Computer Science (Mail Code 249)
University of Illinois at Chicago
P.O. Box 4348
Chicago, Illinois 60680-4348
Telephone: 312-996-3041

Peterson's Guide to Graduate Programs in the Physical Sciences and Mathematics 1992

SECTION 5: MATHEMATICAL SCIENCES

University of Illinois at Chicago

THE FACULTY AND THEIR RESEARCH

Algebra
A. O. L. Atkin, Ph.D., Cambridge, 1952. Modular forms, number theory.
R. Terry Czerwinski, Ph.D., Michigan, 1965. Projective planes, finite geometries.
Paul Fong, Ph.D., Harvard, 1959. Group theory, representation theory of finite groups.
David A. Foulser, Ph.D., Michigan, 1963. Projective planes, finite geometries.
Richard G. Larson, Ph.D., Chicago, 1965. Hopf algebras, application of computers to algebra, algorithms.
David E. Radford, Ph.D., North Carolina at Chapel Hill, 1970. Hopf algebras, algebraic groups.
Mark A. Ronan, Ph.D., Oregon, 1978. Buildings, geometries of finite groups.
Fredrick L. Smith, Ph.D., Ohio State, 1972. Group theory.
Stephen D. Smith, D.Phil., Oxford, 1973. Finite groups, representation theory.
Bhama Srinivasan, Ph.D., Manchester, 1960. Representation theory of finite and algebraic groups.
Geremy Teitelbaum, Ph.D., Harvard, 1986. Number theory.

Analysis
Herbert J. Alexander, Ph.D., Berkeley, 1968. Several complex variables, function algebras.
Calixto P. Calderon, Ph.D., Buenos Aires, 1969. Harmonic analysis, differentiation theory.
Shmuel Friedland, D.Sc., Technion (Israel), 1971. Matrix theory and its applications.
Melvin L. Heard, Ph.D., Purdue, 1967. Integrodifferential equations.
Pei Hsu, Ph.D., Stanford, 1984. Probability, Brownian motion.
Jeff E. Lewis, Ph.D., Rice, 1966. Partial differential equations, microlocal analysis.
Charles S. C. Lin, Ph.D., Berkeley, 1967. Operator theory, perturbation theory, functional analysis.
Howard A. Masur, Ph.D., Minnesota, 1974. Quasiconformal mappings, Teichmuller spaces.
James W. Moeller, Ph.D., NYU, 1961. Functional analysis, probability.
Yoram Sagher, Ph.D., Chicago, 1967. Harmonic analysis, interpolation theory.
Zbigniew Slodkowski, D.Sc., Warsaw, 1981. Several complex variables.
David S. Tartakoff, Ph.D., Berkeley, 1969. Partial differential equations, several complex variables.

Applied Mathematics
Eugene M. Barston, Ph.D., Stanford, 1964. Stability theory, fluid dynamics, plasma dynamics.
Neil E. Berger, Ph.D., NYU, 1968. Applied elasticity, fluid dynamics.
Susan Friedlander, Ph.D., Princeton, 1972. Geophysical and fluid dynamics.
Floyd B. Hanson, Ph.D., Brown, 1968. Numerical methods, asymptotic methods, stochastic bioeconomics.
Charles Knessl, Ph.D., Northwestern, 1986. Stochastic models, perturbation methods, queuing theory.
Alexander Lifshits, Ph.D., Moscow State, 1982. Mathematical physics.
G. V. Ramanathan, Ph.D., Princeton, 1965. Statistical mechanics, critical phenomena.
Charles Tier, Ph.D., NYU, 1976. Analysis of stochastic models, queuing theory.
Victor Twersky, Ph.D., NYU, 1950. Multiple scattering and propagation of waves, biophysical applications.

Computer Science and Combinatorics
Janet S. Beissinger, Ph.D., Pennsylvania, 1981. Combinatorics, algorithms.
Robert Grossman, Ph.D., Princeton, 1985. Applications of computers to analysis, symbolic computation.
Norman T. Hamilton, Ph.D., Chicago, 1955. Computer mathematics, ancient and medieval astronomy.
Jeffrey S. Leon, Ph.D., Caltech, 1971. Computer methods in group theory and combinatorics, algorithms.
Wolfgang Maass, Dr.rer.nat., Munich, 1974. Complexity, lower bound arguments, logic, recursion theory.
Glenn Manacher, Ph.D., Carnegie Tech, 1961. Algorithms, complexity, computer language design.
Uri N. Peled, Ph.D., Waterloo, 1976. Optimization, combinatorial algorithms, computational complexity.
Vera Pless, Ph.D., Northwestern, 1957. Coding theory, combinatorics.
Gyorgy Turan, Ph.D., Attila József (Hungary), 1982. Complexity theory, logic, combinatorics.

Geometry and Topology
A. K. Bousfield, Ph.D., MIT, 1966. Algebraic topology, homotopy theory.
Marc Culler, Ph.D., Berkeley, 1978. Low-dimensional topology, group theory.
Lawrence Ein, Ph.D., Berkeley, 1981. Algebraic geometry.
Henri Gillet, Ph.D., Harvard, 1978. Algebraic K-theory, algebraic geometry.
Brayton I. Gray, Ph.D., Chicago, 1965. Homotopy theory, cobordism theory.
V. K. A. M. Gugenheim, D.Phil., Oxford, 1952. Algebraic topology.
James L. Heitsch, Ph.D., Chicago, 1971. Differential topology, theory of foliations.
Steven Hurder, Ph.D., Illinois at Urbana-Champaign, 1980. Differential topology, theory of foliations.
Louis Kauffman, Ph.D., Princeton, 1972. Differential topology, knot theory of singularities.
Anatoly S. Libgober, Ph.D., Tel-Aviv, 1977. Topology of varieties, theory of singularities.
Peter Shalen, Ph.D., Harvard, 1972. Low-dimensional topology, group theory.
Nicholas Shepherd-Barron, Ph.D., Warwick (England), 1981. Algebraic geometry.
Martin C. Tangora, Ph.D., Northwestern, 1966. Algebraic topology, homotopy theory.
Philip Wagreich, Ph.D., Columbia, 1966. Algebraic geometry, discrete groups, mathematics education.
John W. Wood, Ph.D., Berkeley, 1968. Differential topology, topology of varieties.
Stephen S.-T. Yau, Ph.D., SUNY at Stony Brook, 1976. Complex geometry, singularities of complex algebraic varieties.

Logic and Universal Algebra
John T. Baldwin, Ph.D., Simon Fraser, 1971. Model theory, universal algebra.
Joel D. Berman, Ph.D., Washington (Seattle), 1970. Lattice theory, universal algebra.
Willem J. Blok, Ph.D., Amsterdam, 1976. Algebraic logic, universal algebra, nonclassical logics.
William A. Howard, Ph.D., Chicago, 1956. Foundations of mathematics, proof theory.
David Marker, Ph.D., Yale, 1983. Model theory, models of arithmetic.

Mathematics Education
Steven L. Jordan, Ph.D., Berkeley, 1970. Education, computer graphics, computational geometry.
David A. Page, M.A., Illinois, 1950. Elementary mathematics education.
A. I. Weinzweig, Ph.D., Harvard, 1957. Teaching and learning of mathematics, microcomputers in education.

Probability and Statistics
Tomasz Bielecki, Ph.D., Warsaw, 1987. Stochastic analysis, control theory.
Emad El-Neweihi, Ph.D., Florida State, 1973. Reliability theory, probability, stochastic processes.
Nasrollah Etemadi, Ph.D., Minnesota, 1974. Probability theory, stochastic processes.
Moon-Chall Han, Ph.D., UCLA, 1987. Biostatistics.
A. Hedayat, Ph.D., Cornell, 1969. Optimal designs, sampling theory, linear models, discrete optimization.
Dibyen Majumdar, Ph.D., Indian Statistical Institute, 1981. Optimal designs, linear models.
Klaus J. Miescke, Dr.rer.nat., Heidelberg, 1972. Statistics, decision theory, selection procedures.
T. E. S. Raghavan, Ph.D., Indian Statistical Institute, 1966. Game theory, optimization methods in matrices, statistics.

SECTION 5: MATHEMATICAL SCIENCES

UNIVERSITY OF ILLINOIS AT URBANA–CHAMPAIGN
Department of Statistics

Programs of Study

Master of Science degrees in statistics as well as applied master's degrees with specializations in biostatistics, psychometrics and behavioral sciences, and statistical genetics are available for students interested in careers as statisticians in industry or government. These degree programs do not require a thesis and are usually completed in two years. Master's students are encouraged to participate in the consulting service that is staffed by the department.

The Doctor of Philosophy program in statistics prepares students for careers in university teaching and research or in government or industrial research. Students entering the program spend two years acquiring a basic background in mathematical and applied statistics and in probability and taking general written examinations in these subjects. More-specialized study follows, leading into thesis research, which usually begins in the third year. The research may be concentrated in any pure or applied area of statistics or probability in which a faculty member has expertise and is interested. Ph.D. students are encouraged to participate in the statistical consulting service that is staffed by the department.

Research in the Department of Statistics ranges from pure to applied. The department's faculty have a long and distinguished history of theoretical research in probability and statistics. Far less well known, but just as important, is the effort that faculty and students direct toward applied research into topics such as robust statistics, density estimation, nonparametric regression, information theory, spatial statistics, computational statistics, and reliability and even into substantive area research such as social science and biological science stochastic modeling, medical statistics, and environmental statistics. Much applied research is heavily computational.

Research Facilities

Outstanding research facilities make the departmental environment a stimulating one for graduate study and research. The University Library is the largest public university library in the nation. The easily accessible Mathematics and Statistics Library has a superb collection of journals and books relevant to graduate study and research in statistics. Computing facilities at Illinois are among the nation's best. The department has a well-maintained UNIX-based network system of workstations and also has easy access to various mainframes and supercomputers.

Financial Aid

Departmental half-time assistantships are available on a competitive basis. In 1991–92 these carry stipends of $1044–$1088 per month with additional summer support possible. Tuition and fees are waived for students with at least quarter-time assistantship support, except for $582 of annual fees. Advanced students often receive research assistantships with duties consisting of research, typically closely related to their thesis topics. A limited number of University fellowships are available also. Some tuition and fee waivers, loans, and work-study grants are also available. Further, research assistantships from other academic units are often available for students with statistical and/or computer expertise.

Cost of Study

In 1991–92 graduate tuition and fees for full-time study total $3676 for state residents and $8628 for nonresidents.

Cost of Living

A variety of campus and community housing is available at rents from $250 to $450 per month per person. Food and entertainment costs in the community are moderate by American standards.

Student Group

In 1990–91 the graduate and professional population constituted 9,321 of the University's total enrollment of 35,766 students. Most graduate students are full-time; 36 percent are women, and 28 percent are international students. The Department of Statistics has about 35 students working toward the Ph.D. and about 20 working toward a master's degree.

Location

Urbana and Champaign are twin cities 135 miles south of Chicago, with a combined population of 100,000. They are currently served by Amtrak twice daily, by four airlines, and by two bus lines. The cities lie at the intersection of Interstate Highways 57, 72, and 74.

The University

Founded in 1867, UIUC is one of the nation's major comprehensive public universities. Excellence in engineering and the physical sciences is balanced by outstanding social science, humanities, fine arts, business, agriculture, and education programs. The University supports a wide array of cultural, intellectual, and recreational activities.

Applying

A baccalaureate or equivalent degree and a grade point average of 4.0 (A = 5.0) for the last 60 hours of undergraduate and all previous graduate work are required. The application deadline for financial aid is February 15. The application for fellowships and assistantships is part of the application for admission to graduate study.

Correspondence and Information

For program information:
Graduate Admissions
Department of Statistics
101 Illini Hall
University of Illinois at Urbana-Champaign
725 South Wright Street
Champaign, Illinois 61820
Telephone: 217-333-2167

For general information:
Graduate College
202 Coble Hall
University of Illinois at Urbana-Champaign
801 South Wright Street
Champaign, Illinois 61820
Telephone: 217-333-0035

Peterson's Guide to Graduate Programs in the Physical Sciences and Mathematics 1992

SECTION 5: MATHEMATICAL SCIENCES

University of Illinois at Urbana-Champaign

THE FACULTY AND THEIR RESEARCH

Juha Alho, Assistant Professor (joint with Environmental Studies); Ph.D., Northwestern, 1983. Demography, population forecasting, environmental risk assessment.

Andrew Barron, Associate Professor (joint with Electrical and Computer Engineering); Ph.D., Stanford, 1985. Information theory, probability limit theorems, statistical inference, complexity.

Robert E. Bohrer, Professor; Ph.D., North Carolina, 1965. Simultaneous inference in linear models, time-series analysis, statistical computing.

Donald L. Burkholder, Professor (joint with Mathematics); Ph.D., North Carolina, 1955. Probability theory and its applications to other areas of analysis.

Dennis D. Cox, Professor; Ph.D., Washington (Seattle), 1980. Robust and nonparametric statistics, asymptotics, time-series analysis, computational statistics, curve estimation.

Joseph L. Doob, Professor Emeritus (joint with Mathematics); Ph.D., Harvard, 1932. Probability and potential theory.

Lawrence J. Hubert, Professor (joint with Psychology); Ph.D., Stanford, 1971. Combinatorial data analysis, nonparametric statistics, models of data representation.

Peter B. Imrey, Professor (joint with Medical Science Information); Ph.D., North Carolina, 1972. Analysis of multivariate categorical data, general linear models, sample survey methodology, quantitative epidemiology, diet and cancer, health hazard analysis, screening, clinical trials.

Kumar Joag-dev, Professor; Ph.D., Berkeley, 1962. Nonparametric statistics, multivariate inequalities, reliability.

Frank B. Knight, Professor Emeritus (joint with Mathematics); Ph.D., Princeton, 1959. Probability theory, particularly Markov processes.

John I. Marden, Professor; Ph.D., Chicago, 1978. Decision theory related to hypothesis testing, multivariate analysis, ranking data, nonparametrics.

Adam T. Martinsek, Associate Professor; Ph.D., Columbia, 1981. Sequential analysis, adaptive methods, density estimation.

Ditlev Monrad, Associate Professor (joint with Mathematics); Ph.D., Berkeley, 1976. Stochastic processes, random fields.

Walter Philipp, Professor (joint with Mathematics); Ph.D., Vienna, 1960. Probability limit theorems, probabilistic number theory.

Stephen L. Portnoy, Professor (joint with Life Science); Ph.D., Stanford, 1969. Decision theory, robust estimation, large-sample theory, modeling and analysis of biological systems.

Jerome Sacks, Professor; Ph.D., Cornell, 1956. Model robust statistical inference, experimental design, and computational statistics.

Douglas G. Simpson, Associate Professor (joint with Environmental Studies); Ph.D., North Carolina, 1985. Robust statistical inference, multivariate data analysis, applied statistics.

William F. Stout, Professor (joint with Mathematics); Ph.D., Purdue, 1967. Psychometrics, especially probability modeling and statistical analysis of standardized test data.

Stanley Wasserman, Professor (joint with Psychology); Ph.D., Harvard, 1977. Applied statistics, psychology and sociology.

Robert A. Wijsman, Professor Emeritus; Ph.D., Berkeley, 1952. Multivariate analysis, invariant tests, simultaneous inference, sequential analysis, estimation theory.

Wei Wu, Assistant Professor; Ph.D., North Carolina, 1990. Regression, filtering, and bootstrapping.

Zhiliang Ying, Assistant Professor; Ph.D., Columbia, 1987. Survival analysis, recursive estimation and adaptive control, large-sample theory.

SECTION 5: MATHEMATICAL SCIENCES

UNIVERSITY OF IOWA
Department of Mathematics

Programs of Study

The department offers programs leading to the M.S. and Ph.D. degrees.

Three programs lead to an M.S. degree in mathematics: one emphasizes pure mathematics, one is designed for secondary school teachers, and one emphasizes applied mathematics. The M.S. degree requires 30 semester hours of graduate course work in mathematics (although some courses in statistics or computer science may be substituted), along with written examinations appropriate to the program option.

The Ph.D. program requires 72 semester hours of graduate course work and is designed to be completed within approximately five years of the bachelor's degree. In addition to the Ph.D. thesis, degree candidates must pass written comprehensive examinations in three areas (e.g., algebra, analysis, logic, topology, partial differential equations) and demonstrate reading proficiency in French, German, or Russian.

Faculty members in the department are engaged in research in many areas, including algebra, analysis, differential geometry, logic and foundations, mathematical biology, mathematics education, numerical analysis, operator theory/mathematical physics, partial differential equations, and topology.

Research Facilities

The Department of Mathematics shares MacLean Hall with the Departments of Computer Science and Statistics. The Mathematical Sciences Library, conveniently located in MacLean Hall, serves all three departments. The library has extensive holdings and subscribes to all major research journals. Excellent computing facilities are available for students working in relevant areas of research.

Financial Aid

The department supports approximately 90 students, with 80 holding graduate teaching assistantships. These positions carry a stipend that ranges from $10,000 to $11,150 in 1991–92. University fellowships and tuition supplements are available for especially well qualified entering students. Some research assistantships are available for advanced students. Approximately 90 percent of all graduate students in the department receive some form of financial support.

Cost of Study

For students taking 9 or more semester hours, the 1991–92 graduate tuition and fees are $1158 per semester for Iowa residents and $3372 per semester for nonresidents. Students holding teaching or research assistantships are charged only resident tuition and fees.

Cost of Living

Ample private and University housing in a variety of price ranges is available in the immediate campus area, as well as in areas served by the excellent city and University bus systems. The cost of living is typical of a Midwestern town.

Student Group

The Department of Mathematics has over 100 graduate students, and the total University enrollment is currently stable at approximately 28,000 students.

Location

The University of Iowa is located in Iowa City, a community of 55,000 inhabitants, located in rolling countryside and situated along the scenic Iowa River. Iowa City is 240 miles west of Chicago, and several major airlines serve the nearby Cedar Rapids airport. The area offers a wide range of recreational and cultural activities; the University and the city sponsor performances by nationally renowned concert artists and theatrical, ballet, and opera companies. Excellent facilities for recreational sports and student theater and music performances provide low-cost opportunities.

The University

Founded in 1847 as Iowa's first public institution of higher education, the University has become a major intellectual and cultural center for the state of Iowa. It has won international recognition for its wealth of achievements in the arts, sciences, and humanities. Iowa was the first public university in the nation to admit men and women on an equal basis and the first institution to accept creative work in theater, writing, music, and art as theses for advanced degrees.

Applying

Students seeking to apply should request the appropriate forms from the Director of Graduate Admissions, University of Iowa. To receive full consideration for financial support, completed applications should be received in the Office of Graduate Admissions by March 1.

Correspondence and Information

Associate Chair/Director of Graduate Programs
Department of Mathematics
University of Iowa
Iowa City, Iowa 52242
Telephone: 319-335-0694

Peterson's Guide to Graduate Programs in the Physical Sciences and Mathematics 1992

SECTION 5: MATHEMATICAL SCIENCES

University of Iowa

THE FACULTY AND THEIR RESEARCH

D. Anderson, Professor; Ph.D., Chicago, 1974. Commutative algebra.
K. Atkinson, Professor; Ph.D., Wisconsin, 1966. Numerical analysis.
R. Baker, Assistant Professor; Ph.D., Berkeley, 1987. Functional analysis.
T. Branson, Professor; Ph.D., MIT, 1979. Differential geometry.
N. Cac, Professor; Ph.D., Cambridge, 1965. Partial differential equations, functional analysis, topology, vector spaces.
V. Camillo, Professor; Ph.D., Rutgers, 1969. Noncommutative ring theory.
H.-I. Choi, Associate Professor; Ph.D., Berkeley, 1982. Differential geometry.
R. Curto, Professor and Associate Chair for Graduate Studies; Ph.D., SUNY at Stony Brook, 1978. Functional analysis, multivariable operator theory.
O. Durumeric, Associate Professor; Ph.D., SUNY at Stony Brook, 1982. Differential geometry.
C. Frohman, Associate Professor; Ph.D., Indiana, 1984. Topology of manifolds.
K. Fuller, Professor; Ph.D., Oregon, 1967. Rings, modules, representation theory.
J. Gatica, Professor; Ph.D., Iowa, 1972. Biomathematics, differential equations, fixed-point theory.
M. Geraghty, Associate Professor; Ph.D., Notre Dame, 1959. Geometry.
F. Goodman, Professor; Ph.D., Berkeley, 1979. Functional analysis.
H. Hethcote, Professor; Ph.D., Michigan, 1968. Differential equations, mathematical biology.
E. Johnson, Professor; Ph.D., California, Riverside, 1966. Lattice theory, commutative algebra.
N. Johnson, Professor; Ph.D., Washington State, 1968. Finite geometry, combinatorics.
P. Jorgensen, Professor; Ph.D., Aarhus (Denmark), 1973. Operator theory, functional analysis, harmonic analysis, mathematical physics.
S. Khurana, Professor; Ph.D., Illinois, 1968. Functional analysis, measure theory, general topology.
W. Kirk, Professor; Ph.D., Missouri, 1962. Nonlinear functional analysis.
E. Kleinfeld, Professor; Ph.D., Wisconsin, 1951. Algebra, foundations of geometry.
M. Kleinfeld, Associate Professor; Ph.D., Syracuse, 1965. Nonassociative rings.
F. Kosier, Professor; Ph.D., Michigan State, 1960. Nonassociative rings.
P. Kutzko, Professor; Ph.D., Wisconsin, 1972. Number theory.
J. Lediaev, Associate Professor; Ph.D., California, Riverside, 1967. Applied math, axiomatic evolutionary frameworks.
B.-L. Lin, Professor; Ph.D., Northwestern, 1963. Banach space theory.
E. Madison, Professor; Ph.D., Illinois, 1966. Mathematical logic.
D. Manderscheid, Associate Professor; Ph.D., Yale, 1981. Representation theory.
P. Muhly, Professor; Ph.D., Michigan, 1969. Functional analysis.
G. Nelson, Professor; Ph.D., Case Western Reserve, 1968. Algebraic model theory, logical and computable aspects of reduced powers.
R. Oehmke, Professor; Ph.D., Chicago, 1954. Nonassociative algebra and its application, semigroup theory.
K. O'Hara, Assistant Professor; Ph.D., Berkeley, 1984. Combinatorics.
G. Paulik, Assistant Professor; Ph.D., Indiana, 1985. Partial differential equations.
F. Potra, Professor; Ph.D., Bucharest, 1980. Numerical analysis, functional analysis.
R. Randell, Professor and Chairman; Ph.D., Wisconsin, 1973. Differential topology.
D. Roseman, Associate Professor; Ph.D., Michigan, 1968. Topology.
H. Schoen, Professor; Ph.D., Ohio State, 1971. Math education.
F. Schulz, Professor; Dr.Sci., Göttingen (Germany), 1979. Partial differential equations.
W. Seaman, Associate Professor; Ph.D., Massachusetts, 1983. Differential geometry.
J. Simon, Professor; Ph.D., Wisconsin, 1969. Topology, low-dimensional manifolds.
G. Ströhmer, Associate Professor; Dr.Sci., Göttingen (Germany), 1978. Calculus of variations, partial differential equations, fluid dynamics.
K. Stroyan, Professor; Ph.D., Caltech, 1971. Infinitesimal analysis.
T. Ton-That, Professor; Ph.D., California, Irvine, 1974. Harmonic analysis, math physics.
E. Venturino, Assistant Professor; Ph.D., SUNY at Stony Brook, 1984. Numerical analysis.
Y. Ye, Assistant Professor; Ph.D., Columbia, 1986. Representation theory.
M. Zweng, Professor; Ph.D., Wisconsin, 1963. Verbal problem solving.

A scenic lookout on campus.

MacLean Hall, which houses the Department of Mathematics.

The Iowa River and a view of the Art Building.

SECTION 5: MATHEMATICAL SCIENCES

UNIVERSITY OF KENTUCKY
Department of Statistics

Programs of Study
The department offers programs of study leading to the degrees of Master of Science (with or without a thesis) and Doctor of Philosophy. All graduate students are required to take courses in each of the three basic areas of statistical inference, probability and stochastic processes, and linear models and design. The Ph.D. program also includes course work in advanced inference and probability and in other selected areas, according to the student's needs and goals. The Ph.D. degree prepares the student for university teaching and research, as well as consulting and research roles in industry and government, and usually requires four to five years of study. Current areas of research interest are listed with the faculty on the reverse of this page.

The department cooperates closely with the Departments of Computer Science and Mathematics. Joint graduate degrees between Statistics and Computer Science are possible. The University is expanding its research capability in areas associated with statistics, such as biological and engineering systems, econometrics, operations research, and quantitative genetics. Students with undergraduate majors in mathematics, statistics, the physical sciences, or an applied field are urged to apply for graduate work in statistics; they are expected to have a good background in mathematics, including advanced calculus and matrix algebra.

Written comprehensive examinations must be passed for all degrees. A reading knowledge of one foreign language is required for the doctorate but not for the master's degree.

Research Facilities
Offices and a library for the mathematical sciences are located in a modern, air-conditioned office tower building. The library contains a very comprehensive collection of books and journals in statistics, mathematics, and related areas. The University Computing Center provides computing facilities for faculty and students through its IBM 3090E and Prime computers. In addition, a VAX research computer and numerous microcomputers are available to the mathematical sciences departments.

Financial Aid
The University and the Department of Statistics offer a variety of fellowships and assistantships. The department has teaching, research, and consulting assistantships and fellowships that pay $980–$1000 per month in 1991–92. These assistantships involve teaching undergraduate statistics courses, working with faculty members on research projects, or doing statistical consulting with researchers in various colleges, such as Agriculture and Medicine, and in other divisions of the University. Some three-quarter-time and full-time appointments are available for advanced graduate students.

Cost of Study
Registration fees per semester are $890 for residents of Kentucky and $2670 for nonresidents in 1991–92. Part-time fees are $63 per semester hour for residents and $188 per semester hour for nonresidents. Graduate students holding fellowships or assistantships receive tuition scholarships to cover both in-state and out-of-state fees.

Cost of Living
University apartments for graduate and married students are available ($265–$450 per month in 1990–91).

Student Group
The University had a total on-campus enrollment of 23,000 students in 1990–91, of whom about 19 percent were graduate students. The Department of Statistics had approximately 25 students.

Location
Metropolitan Lexington is in the heart of the Bluegrass country and has a population of 240,000. The area is widely known for its white-fenced horse farms and the world's largest burley tobacco market. Lexington and its surrounding areas are rich in historic landmarks and have an unspoiled natural beauty.

The city is 80 miles south of Cincinnati, 75 miles east of Louisville, and centrally located in relation to numerous very well developed state parks. Varied sports and recreational facilities and one of the South's leading medical and hospital centers make Lexington a very pleasant community in which to study and live. In addition, the Civic Center, which opened in 1976, provides a 23,000-seat arena for the University basketball team and supplies outstanding facilities for national and international conventions and meetings.

Lexington is conveniently located at the intersection of Interstates 75 and 64 and at the terminus of the Bluegrass Parkway. Nearby Blue Grass Field offers service from Comair, Delta, Piedmont, United, and USAir airlines.

The University
The University was founded in 1865, offered its first graduate work in 1870, and in the past fifteen years has experienced exceptional growth and improvement. It is the land-grant institution of Kentucky and consists of the Colleges of Agriculture, Allied Health Professions, Architecture, Arts and Sciences, Business and Economics, Communications, Dentistry, Education, Engineering, Fine Arts, Home Economics, Law, Library and Information Science, Medicine, Nursing, Pharmacy, and Social Work, as well as the Patterson School of Diplomacy and International Commerce and the Graduate School. Doctoral programs are available in forty areas of study, including mathematics. A variety of cultural and sporting events are presented throughout the year. Plays, concerts, films, and lectures provide a many-faceted cultural program. The University also has an outstanding basketball team and a rapidly improving football program.

Applying
Prospective graduate students should submit two official transcripts of all previous college work and a written application at least two months before anticipated entrance. Applications for fellowships and assistantships must be submitted on official forms no later than March 15. Applications are considered to the extent that funds are available, and no further applications are processed after May 1. All fellowship applicants must also submit scores on the General Test of the Graduate Record Examinations.

Correspondence and Information
Further information regarding the statistics graduate program, graduate fellowships, and assistantships, as well as the necessary application materials, can be obtained from:

Director of Graduate Studies
Department of Statistics
University of Kentucky
Lexington, Kentucky 40506
Telephone: 606-257-6902
Fax: 606-258-1973
Electronic mail: moses@ms.uky.edu

Peterson's Guide to Graduate Programs in the Physical Sciences and Mathematics 1992

SECTION 5: MATHEMATICAL SCIENCES

University of Kentucky

THE FACULTY AND THEIR RESEARCH

D. K. Aaron, Assistant Professor (joint appointment with Animal Sciences); Ph.D., Oklahoma State, 1984. Variance components, beef cattle breeding.
W. Alexander, Assistant Professor; Ph.D., Texas A&M, 1989. Statistical computing, nonparametrics.
D. M. Allen, Professor and Chairman; Ph.D., North Carolina State, 1968. Statistical computing, linear and nonlinear models, prediction.
R. L. Anderson, Professor Emeritus; Ph.D., Iowa State, 1941. Variance components, linear models, consulting in pharmaceutical industry with design of experiments and analysis of data for clinical trials.
V. P. Bhapkar, Professor; Ph.D., North Carolina, 1959. Categorical data, nonparametric methods, multivariate analysis.
S. M. Butler, Assistant Professor (joint appointment with Sanders Brown Center on Aging); Ph.D., California, Santa Barbara, 1989. Analysis of repeated measurements, empirical Bayes methods, stochastic processes, survival analysis.
P. L. Cornelius, Professor (joint appointment with Agronomy); Ph.D., Illinois, 1972. Statistical genetics, experimental design.
Z. Govindarajulu, Professor; Ph.D., Minnesota, 1961. Nonparametric inference, sequential analysis, large-sample theory.
W. Griffith, Associate Professor; Ph.D., Pittsburgh, 1979. Reliability theory, applied probability.
R. L. Kryscio, Professor and Director, Mathematical Sciences Consulting Laboratory; Ph.D., SUNY at Buffalo, 1971. Stochastic processes, epidemic theory.
H. E. McKean, Professor; Ph.D., Purdue, 1958. Epidemiology, biostatistics, experimental design.
W. S. Rayens, Assistant Professor; Ph.D., Duke, 1986. Linear discriminant analysis, pattern recognition.
C. Srinivasan, Associate Professor; Ph.D., Indian Statistical Institute, 1979. Statistical decision theory, efficiency of estimators.
R. Stockbridge, Assistant Professor; Ph.D., Wisconsin, 1987. Probability, stochastic control theory.
A. J. Stromberg, Assistant Professor; Ph.D., North Carolina at Chapel Hill, 1989. Computational statistics, nonlinear regression.
K. J. Utikal, Assistant Professor; Ph.D., Florida State, 1987. Survival analysis, inference for stochastic processes.
C. L. Wood, Associate Professor; Ph.D., Florida State, 1975. Limit theorems, nonparametric procedures, goodness-of-fit tests.
M. Zhou, Assistant Professor; Ph.D., Columbia, 1986. Survival analysis, nonparametrics, sequential clinical trials.

Patterson Office Tower houses the Department of Statistics.

The Bluegrass region of Kentucky.

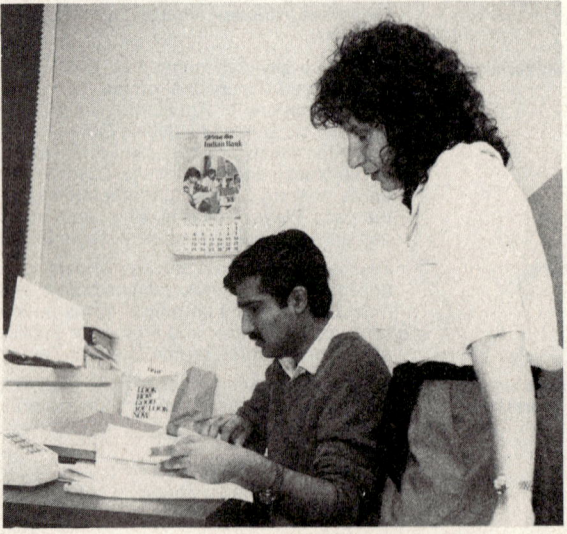

Graduate students working on a consulting project.

SECTION 5: MATHEMATICAL SCIENCES

UNIVERSITY OF MARYLAND GRADUATE SCHOOL, BALTIMORE
Programs in Applied Mathematics and Statistics

Programs of Study The Department of Mathematics and Statistics at UMBC offers the M.S. and Ph.D. in applied mathematics and statistics. Emphasis is placed on (1) applied and numerical analysis, (2) systems theory and operations research, and (3), through a separate program, statistics/probability/stochastic processes.

A master's program begins with a broad foundation of introductory core course work, then proceeds to specialized courses and/or a master's thesis. Students pursuing the doctorate continue advanced study and undertake dissertation research, with specialization in any of the departmental fields or in an interdisciplinary area.

The research programs of the faculty cover a broad spectrum of both classical and modern applied mathematics and different branches of statistics. Particular emphasis is given to the following areas: functional and applied functional analysis, function theory, number theory, partial differential equations, fluid dynamics, control theory, spectral estimation, signal processing, dynamical systems and chaos, numerical analysis, optimization algorithms, game theory, stochastic programming, multivariate analysis, statistical inference, probability theory, linear statistical models, statistical decision theory, time series analysis, general Markov processes, and combinatorics.

It is the policy of the department to provide the graduate students with a study/research environment that is most conducive to the attainment of their educational and career objectives. Consequently, emphasis is placed on close individual advising and on promoting friendly and informal interactions between students and faculty. Students are invited to participate in the department's colloquium and seminar series, as well as in occasional social functions.

Research Facilities The library of the University's Baltimore County campus has over 430,000 books, and its collection is growing rapidly. It carries over 140 mathematical and statistical journals. In addition, graduate students have access to other large and first-rate libraries in the Baltimore-Washington area.

The Consulting Center for Statistics within the department provides graduate students with opportunities for practical statistical experience. Computer facilities include two clustered VAX 8600s (VMS operating system), a VAX 785, and a VAX 750 (UNIX operating system). Department facilities include five DEC 5000 workstations, five DEC 3100 workstations, and a four-processor Stardent 3000, all running UNIX and capable of supporting high-level graphics. The department also has in-house VMS capability on two MicroVAXes.

Financial Aid Teaching assistants receive academic-year stipends of $9200 and up, plus remission of tuition; research assistants receive comparable stipends. University Merit Fellowships and two departmental Aziz fellowships of $10,000 (plus tuition) for the academic year are awarded to doctoral candidates on a competitive basis. Government loans and work-study opportunities are also available. In addition, fellowships and grants-in-aid are available for students from minority groups.

Cost of Study Tuition for graduate courses is $141 per credit hour for Maryland residents and $252 per credit hour for nonresidents in 1991–92. The application fee is $25.

Cost of Living Housing in campus apartments for graduate students costs approximately $1186 per semester, plus utilities. Off-campus housing is available at somewhat higher rates.

Student Group The University of Maryland Graduate School, Baltimore, has 1,358 graduate students. The Department of Mathematics and Statistics has about 80 graduate students, of whom 25 are full-time receiving full financial support. The graduate student population is expected to increase considerably in the near future.

Location The Graduate School is located at the hub of one of the greatest concentrations of health-care institutions, research facilities, government agencies, and professional associations in the nation. Both of the University's Baltimore campuses are within easy reach of such resources as the Library of Congress, National Institutes of Health, National Library of Medicine, and Smithsonian Institution.

Outside the metropolitan area, and just minutes from the campus, is the Maryland countryside. The state offers a great variety of recreational and leisure activities in its many fine national and state parks, from the Catoctin Mountains in western Maryland to the Assateague Island National Seashore on the Eastern Shore, all within a pleasant drive of the campus. Historic Annapolis, the state capital, is only a short drive away.

The University The 1984 merger of graduate education and research facilities for the University of Maryland's Baltimore and Baltimore County campuses represented a milestone in graduate education in Maryland. The linkage broadens the scope of graduate offerings in the region, enhances the collective research base, and makes possible collaborative efforts that cross disciplines in which each campus has particular strengths.

The University of Maryland Baltimore County campus was established in 1966 so that undergraduate and graduate students in Metropolitan Baltimore could benefit from the presence of a major public research university and share in the rich tradition of academic excellence that is the mark of the Maryland system.

In addition to offering the traditional arts, humanities, and sciences curricula, plus new programs in computer science, engineering, and operations analysis, UMBC has introduced the nation's first graduate major in emergency health services and the first curriculum in genetic engineering (applied molecular biology).

Applying In addition to submitting the completed application form and official college transcripts, applicants are requested to have three letters of recommendation sent on their behalf. Scores on the General Test of the Graduate Record Examinations are required. International students must earn a score of 550 or higher on the Test of English as a Foreign Language (TOEFL).

Correspondence and Information
Dr. Thomas Mathew, Director of Graduate Programs
Department of Mathematics and Statistics
University of Maryland Baltimore County
5401 Wilkens Avenue
Catonsville, Maryland 21228
Telephone: 301-455-2418 or 2410

Peterson's Guide to Graduate Programs in the Physical Sciences and Mathematics 1992

SECTION 5: MATHEMATICAL SCIENCES

University of Maryland Graduate School, Baltimore

THE FACULTY AND THEIR RESEARCH

Thomas E. Armstrong, Professor; Ph.D., Princeton, 1973. Functional analysis and measure theory; probability, mathematical economics.
Abdul K. Aziz, Professor Emeritus; Ph.D., Maryland, 1958. Functional and numerical analysis, control theory for partial differential equations.
Nam P. Bhatia, Professor; Dr.rer.nat., Dresden, 1961. Dynamical and semidynamical systems, stability theory, evolutionary processes and chaos.
William Coleman, Associate Professor; Ph.D., SUNY at Buffalo, 1982. Applied statistics, statistical computing, topology.
Immaculata Jacinta Curiel, Assistant Professor; Ph.D., Nijmegen (Netherlands), 1988. Combinatorial game theory, operations research.
John F. Dillon, Visiting Professor; Ph.D., Maryland, 1974. Combinatorics and coding theory.
Jerzy A. Filar, Professor; Ph.D., Illinois at Chicago, 1980. Game theory, operations research, mathematical programming.
M. S. Gowda, Associate Professor; Ph.D., Wisconsin, 1982. Function theory in C^n, operator theory, mathematical programming.
James M. Greenberg, Professor and Chairman; Ph.D., Brown, 1966. Applied mathematics, modeling and computing.
Fred Gross, Professor; Ph.D., UCLA, 1962. Functional equations, complex function theory, meromorphic functions.
Soren S. Jensen, Assistant Professor; Ph.D., Maryland, 1985. Numerical analysis.
Jacob Kogan, Assistant Professor; Ph.D., Weizmann (Israel), 1985. Calculus of variations, optimal control theory, optimization.
James T. Lo, Professor; Ph.D., USC, 1969. Optimal filtering, stochastic control, signal detection, system analysis.
Yen-Mow Lynn, Professor; Ph.D., Caltech, 1961. Fluid dynamics, mathematical physics.
Thomas Mathew, Associate Professor; Ph.D., Indian Statistical Institute, 1984. Inference in linear models and variance component models, design of experiments.
Peter C. Matthews, Associate Professor; Ph.D., Stanford, 1986. Group theory in probability and statistics, survival analysis.
N. K. Nagaraj, Assistant Professor; Ph.D., Iowa State, 1986. Time series analysis, econometrics.
Arthur O. Pittenger, Professor and Dean of Arts and Sciences; Ph.D., Stanford, 1967. General Markov processes, probability theory.
Richard C. Roberts, Professor; Ph.D., Brown, 1949. Numerical analysis.
Rouben Rostamian, Professor; Ph.D., Brown, 1977. Qualitative theory of partial differential equations, nonlinear and degenerate equations, stability, boundary-value problems, elasticity, modeling.
Andrew L. Rukhin, Professor; Ph.D., Steklov Mathematical Institute (USSR), 1970. Decision theory, estimation theory, mathematical statistics.
Christoph Schwab, Assistant Professor; Ph.D., Maryland College Park, 1989. Applied functional analysis, boundary value problems, partial differential equations.
Thomas I. Seidman, Professor; Ph.D., NYU, 1959. Control theory, nonlinear partial differential equations, inverse problems.
Bimal K. Sinha, Professor; Ph.D., Calcutta, 1973. Multivariate analysis, statistical inference, linear models, decision theory, robustness and asymptotic theory.
Manil Suri, Associate Professor; Ph.D., Carnegie-Mellon, 1983. Numerical analysis, partial differential equations.
Marc Teboulle, Associate Professor; D.Sc., Technion (Israel), 1985. Nonlinear optimization, stochastic programming, mathematical modeling in engineering and economics.
Yin Zhang, Assistant Professor; Ph.D., SUNY at Stony Brook, 1987. Numerical analysis, optimization, scientific computing.

SECTION 5: MATHEMATICAL SCIENCES

UNIVERSITY OF NOTRE DAME

Graduate Studies in Mathematics

Program of Study

The purpose of the graduate program in mathematics is to give students the opportunity to develop into educated and creative mathematicians. In most instances, the doctoral program starts with two years of basic training, including supervised teaching experience and introductory and advanced course work in the fundamentals of algebra, analysis, geometry, and topology. This is followed by thesis work done in close association with one or more members of the faculty. Limited enrollment and the presence of several active groups of research mathematicians provide thesis opportunities in many areas of algebra, algebraic geometry, complex analysis, differential geometry, logic, partial differential equations, and topology.

Research Facilities

Every effort is made to enable students to avail themselves of the opportunities provided by the excellent mathematics faculty at Notre Dame. The Department of Mathematics has its own building with all modern facilities, including a comprehensive research library of 27,000 volumes that subscribes to 250 current journals. All graduate students have comfortable offices (two to an office). Students are ensured a stimulating and challenging intellectual experience.

Financial Aid

In 1991–92, all new students receive a stipend of up to $10,200 and have no teaching duties the first year. Then they become teaching assistants with a stipend of over $9400 and begin the three stages of supervised teaching provided by the department. A teaching assistant usually starts by doing tutorial work in freshman and sophomore calculus courses (4 classroom hours per week); this is followed by a variety of duties in advanced undergraduate courses; the final, lecturing stage involves independent teaching in the classroom. All doctoral students in mathematics also receive a full tuition fellowship. Support is available for citizens and noncitizens.

Cost of Study

Virtually all graduate students in mathematics are supported by fellowships or assistantships, which include tuition scholarships.

Cost of Living

University housing includes rooms for single men, town houses for single women, and two-bedroom apartments for married students at rents ranging from $300 to $500 per month in 1991–92. Comfortable and attractive off-campus rooms normally cost between $200 and $400 per month. Other expenses are lower than in most metropolitan areas.

Student Group

The carefully selected men and women who make up the student body of the University come from every state in the Union and sixty-six foreign countries. There are about 35 graduate students working for their doctorate in mathematics. The faculty-student ratio is almost 1:1.

Location

The University is just north of South Bend, a pleasant Midwestern city with a population of about 110,000. The Notre Dame campus is exceptionally rich in active cultural programs, and the wide variety of cultural, educational, and recreational facilities of Chicago and Lake Michigan are less than 2 hours away by car.

The University

Founded in 1842, the University of Notre Dame has a 1,250-acre campus. Much of the campus is heavily wooded, and two delightful lakes lie entirely within it. Total enrollment is about 9,000; approximately one fifth of these are graduate students. The University is proud of its tradition as a Catholic university with a profound commitment to intellectual freedom in every area of contemporary thought. The students and faculty represent a rich diversity of religious, racial, and ethnic backgrounds.

Applying

All applicants are urged to take the General Test and the Subject Test in mathematics of the Graduate Record Examinations. Application for these tests should be made to Educational Testing Service in Princeton, New Jersey 08541, or at 1947 Center Street, Berkeley, California 94704. While there is no final application date, completed forms should be filed before February 15. All applicants are considered without regard to race, sex, or religious affiliation.

Correspondence and Information

Professor Andrew Sommese, Chairman
Department of Mathematics
University of Notre Dame
P.O. Box 398
Notre Dame, Indiana 46556
Telephone: 219-239-7245
Fax: 219-239-6579

Peterson's Guide to Graduate Programs in the Physical Sciences and Mathematics 1992

University of Notre Dame

THE FACULTY AND THEIR RESEARCH

Algebra
Matthew Dyer. Finite groups.
Kenneth Grant. Algebraic number theory, homological algebra.
Alex Hahn. Algebraic number theory, quadratic forms, classical groups.
George Kolettis. Abelian groups, homological algebra.
Timothy O'Meara. Algebraic number theory, integral linear groups, quadratic forms. (Currently Provost of the University.)
Kok-Wee Phan. Finite groups.
Barth Pollak. Quadratic forms, orthogonal groups, composition algebras.
Warren Wong. Theory of finite groups and their representations.

Algebraic and Analytic Geometry
Mario Borelli. Algebraic geometry, computer graphics.
Juan Migliore. Algebraic geometry.
Dennis Snow. Complex group actions.
Andrew Sommese. Algebraic geometry.

Analysis
Alex Himonas. Partial differential equations.
Mei-Chi Shaw. Partial differential equations.
Nancy Stanton. Partial differential equations, several complex variables.
Wilhelm Stoll. Several complex variables, complex manifolds.
Pit-Mann Wong. Several complex variables.

Applied Mathematics
Mark Alber. Dynamical systems.
Bei Hu. Nonlinear partial differential equations.
Joachim Rosenthal. Geometric control theory.

Geometry
Abraham Goetz. Differential geometry.
Alan Howard. Complex manifolds.
Cecil Mast. Differential geometry, relativity.
Brian Smyth. Differential geometry.
Gudlaugur Thorbergsson. Differential geometry.
Frederico Xavier. Differential geometry.

Logic
Steven Buechler. Model theory.
Julia Knight. Recursion theory.
Anand Pillay. Model theory.

Topology
Francis Connolly. Differential and algebraic topology.
John Derwent. Differential and algebraic topology.
William G. Dwyer. Algebraic topology.
Stephan Stolz. Algebraic topology.
Laurence Taylor. Geometric and algebraic topology.
Dariusz Wilczynski. Algebraic topology.
Bruce Williams. Geometric topology, homotopy theory.

UNIVERSITY OF ROCHESTER
Department of Statistics

Programs of Study

The Department of Statistics offers graduate programs leading to the M.A. and Ph.D. degrees. A joint M.A. program in mathematics and statistics and an M.S. program in medical statistics are also available. The department interprets the term "statistics" very broadly, with specialization available in theoretical and applied probability, statistical theory and analysis, and biostatistics. The department cooperates closely with the Division of Biostatistics of the School of Medicine and Dentistry, the Departments of Mathematics and Computer Science, and various programs in substantive areas. Students have opportunities for supervised teaching and statistical consulting experience. The department gives individual attention to each student through intensive advising, extensive small seminars, and research collaboration. Several of the Ph.D. students have had as many as three publications based on research done in collaboration with various faculty members in statistics, management, and medicine. Ph.D. program graduates have found employment at Bowman Gray School of Medicine, Carnegie Mellon, Case Western Reserve, Harvard, Johns Hopkins, Lehigh, Rochester, RIT, Rutgers, ten state universities, and numerous industrial concerns. M.A. and M.S. graduates are in various academic programs and in industrial, government, research, and consulting positions.

The M.S. program in medical statistics is primarily intended for students who wish to follow careers in the pharmaceutical industry or other health-related professions. The M.S. degree requires one or two years of study, depending upon preparation, and includes an applied project.

Entering M.A. and Ph.D. students need undergraduate preparation in mathematics, including mathematical analysis (advanced calculus) and linear algebra and a year of probability and statistics. The prerequisites for the M.S. program in medical statistics are somewhat less stringent. Minor deficiencies can be made up after matriculation. Normally, doctoral students are initially considered M.A. candidates; this nonthesis degree can be completed in three or four semesters or, in some cases, in one calendar year. The degree prepares a student for doctoral studies or for professional employment. Ph.D. studies consist of additional specialized courses and seminars and supervised research leading to a dissertation. There is no foreign language requirement.

Research Facilities

The department is located in the eleven-story Ray P. Hylan Building. Computing facilities include terminals connected to the University's VAX and IBM computers and the department's own Sun Workstations. Students and faculty also have access to a well-equipped computer-graphics laboratory. The nearby Carlson Library holds an excellent collection of books and journals in statistics and related areas. Office space is available for teaching and research assistants.

Financial Aid

In 1992–93, a well-qualified beginning graduate student will receive up to $9500 for the academic year. Teaching assistants receive $2500 for 12–15 hours of service per week; additional fellowship support from $5500 to $7000 is available. A teaching assistant's duties may include teaching two recitation sections, assisting on consulting projects, computer programming, or grading tests and papers. Advanced students may be supported as research assistants or instructors. Some partial stipends are offered, and some summer stipends are also available.

Cost of Study

In 1991–92, tuition is $15,150 per year but is waived for students receiving support. A student health fee covers basic medical care and hospitalization for the calendar year through the University Health Service.

Cost of Living

Apartments are available in four University-operated facilities near the campus. In 1991–92, a single person pays $250 to $350 per month, depending on the type of accommodation desired. The cost for a married couple without children is $400; for a married couple with children, between $450 and $550. The Office of the Housing Coordinator maintains a file of private rooms and apartments that are available in the area.

Student Group

In 1990–91, there were 17 full-time graduate students in the department from the United States, Europe, and Asia. Several had completed master's degrees at Rochester or elsewhere.

Location

Located on the south shore of Lake Ontario, a short drive from the Finger Lakes area, Rochester is a cultural center for upstate New York. The more than 700,000 residents of the Rochester metropolitan area enjoy the year-round activities of the University-owned Memorial Art Gallery, the cinematic and photographic offerings of George Eastman House, and the events at the University's Eastman Theatre, where concerts are given by the Rochester Philharmonic Orchestra and by students and faculty of the University's Eastman School of Music. Known as the photographic and optical capital of the world, Rochester is the home of many highly technological industries. Sailing, skiing, and camping facilities abound.

The University

The University of Rochester, founded in 1850, is a medium-sized coeducational private university. It has developed into a major center for graduate education in recent times. A survey conducted by the American Council on Education gave the highest possible rating to the University's graduate programs in nine of the thirteen fields evaluated. There are 5,000 undergraduates and 3,000 full-time graduate students. The main campus is along the Genesee River at the edge of the city, where a riverside path for bicyclists and strollers and a launching dock for canoeists are available. The medical campus is adjacent to the main campus, and the Eastman School of Music is downtown.

Applying

Applications requesting financial support should be submitted by February 15. GRE General Test scores are required; students from non-English-speaking countries must also submit TOEFL scores.

Correspondence and Information

Department of Statistics
University of Rochester
Rochester, New York 14627
Telephone: 716-275-3644

SECTION 5: MATHEMATICAL SCIENCES

University of Rochester

THE FACULTY AND THEIR RESEARCH

Several of the faculty members listed below have joint appointments in the Division of Biostatistics of the School of Medicine and Dentistry or the Graduate School of Management. Other faculty members at the University have related interests and cooperate closely with the department's programs. Among them are a number with appointments in mathematics, management, economics, education, engineering, philosophy, physics, political science, preventive medicine, psychology, and sociology.

Professors
K. Ruben Gabriel, Ph.D., Hebrew (Jerusalem). Data analysis, multivariate analysis, graphics.
W. J. Hall, Ph.D., North Carolina. Statistical theory, sequential analysis, robust inference.
G. S. Mudholkar, Ph.D., North Carolina. Statistical theory, multivariate analysis.
David Oakes, Chair; Ph.D., London. Survival analysis, medical statistics, semiparametric inference.
P. S. R. S. Rao, Ph.D., Harvard. Sampling theory, variance components, empirical Bayes.
Martin A. Tanner, Director, Biostatistics; Ph.D., Chicago. Bayesian inference, missing data, imputation, survival analysis.

Assistant Professors
Steven P. Ellis, Ph.D., Berkeley. Foundations of data analysis, applied statistics.
Michael McDermott, Ph.D., Rochester. Order-restricted inference.
Richard Raubertas, Ph.D., Wisconsin–Madison. Biostatistics.

Research projects, many federally funded, in which the department's faculty members are currently major participants include:
Multidimensional Slit-Scan Detection of Bladder Cancer.
Isolation of Human Fetal Cells from Maternal Blood.
Multicenter Study of Silent Myocardial Ischemia.
Approximate Likelihood and Multivariate Survival Analysis.
Technological Assessment Using Clinical/Economic Analysis.
Multicenter Diltiazem Post-Infarction Trial.
Clustering and Trends in Disease Incidence Data.
Analysis of Weather Modification Experiments.
Deprenyl and Tocopherol Anti-Oxidative Therapy of Parkinsonism.
Goodness-of-fit Tests and Isotones.
Statistical Analysis of Multiple Event Time Data.
Topics in Semiparametric Inference, with Special Reference to Survival Analysis.
A New Approach to Order-Restricted Inference.
Nonparametric Analysis of Censored Data.
Correlates of Alzheimer's Disease.

The Ray P. Hylan Building, which houses the Departments of Mathematics and Statistics, provides classrooms; seminar, conference, and computing rooms; and an eleventh-floor commons room with two balconies.

SECTION 5: MATHEMATICAL SCIENCES

UNIVERSITY OF SOUTHERN CALIFORNIA

Department of Mathematics

Programs of Study

The department offers degree programs leading to the Master of Arts in mathematics and applied mathematics, the Master of Science in applied mathematics and statistics, and the Doctor of Philosophy in mathematics and applied mathematics.

Introductory graduate courses are offered in algebra, real analysis, complex analysis, numerical analysis, linear and nonlinear functional analysis, algebraic topology, combinatorics, differential geometry, ordinary differential equations, partial differential equations, control theory, applied mathematics, probability, and statistics. Advanced courses in these and related areas, leading to research topics, are also offered.

The Ph.D. degree can normally be completed within four years. Candidates must take a written prequalifying exam in algebra or analysis by the end of their second semester and two written qaulifying exams, as well as an oral exam, by the end of their fifth semester. Once these exams have been successfully completed, independent research is begun under the direction of a senior faculty member. The student must write an acceptable dissertation that exhibits original and independent research. A reading knowledge of two languages, other than English, in which there is a significant body of research in mathematics is required.

The M.A. degree program normally requires three semesters of graduate study; it is completed either by passing written comprehensive examinations or by writing a thesis. In both the M.S. in applied mathematics and the M.S. in statistics, the degree is completed by carrying out a practicum (research project) in a selected area. The M.S. in statistics also has statistical consulting as a requirement.

Research Facilities

All graduate assistants are provided with office space in the mathematics building. Seaver Science Library contains more than 200,000 books and current subscriptions to about 350 journals of mathematical interest (3,000 journals in all fields). Facilities of the UCLA and Caltech libraries are also available through interlibrary loan, as well as the facilities of the Research Libraries Group, a corporation of major universities and research institutions of which USC is a member. A variety of computing facilities is available for academic and research purposes on the main campus and in the department. Also within the Department of Mathematics is the Center for Applied Mathematical Sciences, an interdisciplinary research group with interests in the application of mathematics to problems in engineering and the sciences.

Financial Aid

Almost all of the department's graduate students are supported by assistantships or fellowships for the duration of their study. Especially well qualified students receive an additional Dean's Fellowship ($4600 in 1991–92). About thirty teaching assistantships are available for new graduate students each year. These assistantships carry a stipend of $9784–$10,842 for the 1991–92 academic year (depending on seniority), tuition remission for up to 12 units per semester during the academic year (9 units is a normal load), and tuition remission for 12 units the following summer. In addition, many summer teaching assistantships are available at stipends of $2446–$2710. A C. W. Trigg fellowship is available for the applicant best demonstrating mathematical problem-solving ability, and several summer research assistantships are available for first-year Ph.D. students who have completed the Ph.D. screening procedure. A limited number of University fellowships, research assistantships, and tuition fellowships are also available. California residents are strongly urged to apply for California State Graduate Fellowships, even if they expect to receive other financial aid.

Cost of Study

For 1991–92, the tuition for full-time students is $510 per unit, and fees (student health service, etc.) are $120 per semester. A one-time publication fee of $113 is charged for microfilming and binding the thesis.

Cost of Living

In 1991–92, monthly rates for graduate student family housing vary between $546–$650 for a one-bedroom apartment and $761 for a two-bedroom apartment. Monthly rates for single graduate student housing vary between $334 for a two-bedroom suite (2 students per unit) to $568 for a one-bedroom single. There are also many privately owned apartments available in the area.

Student Group

There are currently about 100 graduate students in the department (working with about 40 faculty members); approximately 20 percent are women, and 60 percent are international students. The total University enrollment is about 30,000, of whom about 14,000 are graduate students.

Location

The main campus of USC is located near the center of Los Angeles and its varied cultural attractions. It takes only a few minutes to reach the Music Center, the Art Museum, Dodger Stadium, and Hollywood. The campus is within an hour's drive of mountains, deserts, and beaches and within only a few hours' drive of Mexico or the Sierra Nevada. Shuttle-bus service is available to and from the campuses of UCLA and Caltech.

The University

Founded in 1880, USC is the oldest major independent, coeducational, nonsectarian university in the West. There are 105 buildings on its 150 urban acres. The University's Idyllwild School of Music and the Arts is located on a 205-acre campus in the San Jacinto Mountains. More graduate and professional degrees than bachelor's degrees have been awarded by USC each year since 1960.

Applying

Although applications for admission and financial aid are considered at all times, they should be submitted by March 1 to ensure consideration for the following academic year. Scores on the General Test of the Graduate Record Examinations (GRE), as well as transcripts and three letters of recommendation, are mandatory requirements for admission to all programs. Scores on the GRE Subject Test in mathematics are required for admission to the Ph.D. program. International applicants who are applying for a teaching assistantship are advised to take the TSE and TOEFL exams. Applicants to the Ph.D. and M.A. programs in mathematics must have taken full-year sequences at the undergraduate level in real analysis and algebra; admission in special standing may be feasible while remedying deficiencies, but this delays progress. Applicants for the M.S. in applied mathematics and statistics programs should have a substantial undergraduate background in mathematics, which includes one semester of real analysis or advanced calculus and one semester of linear algebra.

Correspondence and Information

Graduate Vice-Chair
Department of Mathematics
University of Southern California
Los Angeles, California 90089-1113
Telephone: 213-740-2400

SECTION 5: MATHEMATICAL SCIENCES

University of Southern California

THE FACULTY AND THEIR RESEARCH

Kenneth Alexander, Associate Professor; Ph.D., MIT. Probability, statistics.
Henry A. Antosiewicz, Professor; Ph.D., Vienna. Differential equations and applications to finance.
Richard Arratia, Associate Professor; Ph.D., Wisconsin–Madison. Probability, combinatorics.
H. Thomas Banks, Professor; Ph.D., Purdue. Applied mathematics.
Peter Baxendale, Professor; Ph.D., Warwick (England). Probability.
Edward K. Blum, Professor; Ph.D., Columbia. Numerical analysis, mathematical neuroscience.
Francis Bonahon, Professor; Ph.D., Paris. Geometry, topology.
Robert Brooks, Professor; Ph.D., Harvard. Geometry, topology.
Ronald E. Bruck, Professor; Ph.D., Chicago. Nonlinear functional analysis.
Richard S. Bucy, Professor; Ph.D., Berkeley. Control theory, probability.
Zhiqiang Cai, Assistant Professor; Ph.D., Colorado. Numerical analysis.
Dennis R. Estes, Professor; Ph.D., LSU. Algebra, number theory.
Larry Goldstein, Associate Professor; Ph.D., California, San Diego. Statistics.
Solomon Golomb, Professor; Ph.D., Harvard. Combinatorial analysis, number theory.
Louis Gordon, Professor; Ph.D., Stanford. Statistics.
Robert M. Guralnick, Professor; Ph.D., UCLA. Group theory, representation theory.
Eugene Gutkin, Associate Professor; Ph.D., Brandeis. Dynamical systems, mathematical physics.
William A. Harris Jr., Professor; Ph.D., Minnesota. Differential and difference equations.
Nicolai T. A. Haydn, Assistant Professor; Ph.D., Warwick (England). Ergodic theory, dynamical systems.
Kazufumi Ito, Associate Professor; Ph.D., Washington (St. Louis). Applied mathematics, control theory.
Russell Johnson, Professor; Ph.D., Minnesota. Differential equations, dynamical systems.
Sheldon Kamienny, Professor; Ph.D., Harvard. Algebraic geometry, number theory.
Charles P. Lanski, Professor; Ph.D., Chicago. Noncommutative ring theory.
Susan Montgomery, Professor; Ph.D., Chicago. Noncommutative rings, Hopf algebra.
William Navidi, Assistant Professor; Ph.D., Berkeley. Statistics.
Claudia Neuhauser, Assistant Professor; Ph.D., Cornell. Probability.
Robert Penner, Associate Professor; Ph.D., MIT. Geometry, topology.
Wlodek Proskurowski, Associate Professor; Ph.D., Royal Institute of Technology (Stockholm). Numerical analysis, scientific computing.
Simeon Reich, Professor; D.Sc., Technion (Israel). Nonlinear functional analysis.
Gary Rosen, Associate Professor; Ph.D., Brown. Applied mathematics.
Boris Rozovskii, Professor; Ph.D., Moscow State. Applied mathematics, stochastic partial differential equations.
Robert J. Sacker, Professor; Ph.D., NYU. Differential equations and dynamical systems.
Alan Schumitzky, Professor; Ph.D., Cornell. Applied mathematics.
Mario Taboada, Assistant Professor; Ph.D., Minnesota. Differential equations.
Simon Tavaré, Professor; Ph.D., Sheffield (England). Applied probability, statistics, molecular evolution.
Nikolaus Vonessen, Assistant Professor; Ph.D., MIT. Noncommutative ring and invariant theory.
Zdenek Vorel, Professor; Ph.D., Czechoslovak Academy of Sciences. Ordinary and functional differential equations.
Chunming Wang, Assistant Professor; Ph.D., Brown. Control theory.
Michael S. Waterman, Professor; Ph.D., Michigan State. Applied mathematics, statistics, molecular biology.
Joseph Watkins, Assistant Professor; Ph.D., Wisconsin–Madison. Probability.
Paul Yang, Professor; Ph.D., Berkeley. Differential geometry.

Emeritus Professors

Herbert Busemann. Geometry
Theodore E. Harris. Probability.
Donald E. Hyers. Integral equations, functional analysis.
Gerhard Tintner. Mathematical economics.
B. Andreas Troesch. Numerical analysis, applied mathematics.
Paul A. White. Mathematical education.
Albert L. Whiteman. Combinatorial analysis.

SECTION 5: MATHEMATICAL SCIENCES

UNIVERSITY OF SOUTH FLORIDA

Department of Mathematics

Programs of Study

The department offers programs in mathematics leading to the M.A. and Ph.D. degrees. Interdisciplinary programs with the Colleges of Engineering and Medicine and with the Department of Physics are available at the Ph.D. level. At the master's level, the department offers the choice of a thesis or nonthesis program.

It usually takes one to two years to earn a master's degree and an additional two to three years for the Ph.D. degree. Master's degree candidates must complete at least 30 semester hours of graduate work and either successfully defend a master's thesis or pass one of the qualifying examinations at the master's level, which may be selected from the areas of algebra, real analysis, topology, or mathematical statistics. Students must demonstrate a reading knowledge of Chinese, French, German, or Russian. A computer science project may be substituted for the language.

Candidates for the Ph.D. degree must complete 90 hours of graduate work and pass four qualifying exams at the Ph.D. level. Ph.D. students begin their graduate study with a sequence of courses taken in preparation for their qualifying examinations, which may be selected from the areas of algebra, applied statistical methods, complex analysis, differential equations, probability theory, real analysis, statistics, theoretical computer science, and topology. These examinations are usually taken at the end of the first year and during the second year. Students must also demonstrate a reading knowledge of two of the languages mentioned above. Finally, a dissertation that demonstrates the student's ability to do original independent research must be submitted and defended.

Research Facilities

The University's library contains over 10,000 volumes and 168 journal titles devoted to mathematics. The Computer Research Center operates an IBM 3033, IBM 3081, and an IBM 3090-300E vector processor with on-line terminals. Other computers are operated by various departments or divisions on campus. The Department of Mathematics operates a network of Suns running UNIX and attached to the Internet.

Financial Aid

Several types of financial aid are available. Teaching assistantships offer stipends for two semesters that range from $9000 to $10,500 for duties not to exceed 8 contact hours. The duties involve teaching lower-level courses, leading recitation sections of large lecture-type courses, assisting faculty in their teaching, or working in the mathematics tutorial lab. The teaching assistantships are renewable, depending upon satisfactory teaching performance and academic record. About fifteen teaching assistantships were available during the academic year 1990–91. University Fellowships provide a stipend for two semesters that is $7000. No duties are involved. Out-of-state-tuition waivers are available for up to 9 hours per semester to qualified graduate students. A limited number of teaching assistantships are available during the summer term. The Center for Mathematical Services often employs faculty and qualified graduate students as research assistants with summer support and provides an opportunity to interact with business and industry.

Completed applications for assistantships should be received by March 15. Assistantships for the fall are announced by April 30. Deadlines for the other types of financial aid vary.

Cost of Study

Tuition for graduate courses is approximately $80 per credit hour for in-state students and $241 per credit hour for out-of-state students in 1991–92. (Costs are subject to change.)

Cost of Living

Some on-campus housing is available for single graduate students. There is off-campus housing near the campus, with one-bedroom apartments renting for $250–$350 per month. The cost of living is lower in Tampa than in most other cities.

Student Group

The University enrollment is more than 31,000, including about 5,000 graduate students. There are approximately 70 graduate students in mathematics, drawn from all parts of the United States and several foreign countries.

Location

The University of South Florida's large, modern campus is located on the northeast edge of Tampa, just 2 miles east of I-275. The rapidly growing Tampa Bay area has several historical and cultural institutions, including the Ringling museums and the Florida Symphony. Theatrical productions and concerts are given on a regular basis. The warm, sunny climate allows one to take advantage of the abundance of outdoor recreational facilities. Tampa is less than a 2-hour drive from Disney World and Cypress Gardens.

The University

The University of South Florida, founded in 1956, is part of the state university system. The campus consists of modern buildings with many classrooms and offices opening onto courtyards and open-air halls. The University comprises ten colleges, including the College of Engineering, College of Medicine, College of Public Health, and New College, a nationally recognized innovative liberal arts college located in Sarasota.

Applying

Completed applications for admission, including official copies of transcripts and GRE General Test scores, must be received by October 28 for the spring semester beginning in January; by March 11 for the summer semester beginning in May; and by June 24 for the fall semester beginning in August. International applicants must submit their applications for admission at least six months prior to the date of desired enrollment. All applicants must have earned a B average or better in all work attempted while an upper-division student and must have a quantitative plus verbal score of at least 1000 on the General Test of the GRE, with at least 650 on the quantitative section. Applicants whose native language is not English must score at least 550 on the Test of English as a Foreign Language.

In order for an application for graduate study to be complete, all applicants must submit to the Office of Admissions of the University a completed admission form, an official copy of their transcripts, and an official copy of their scores on the GRE General Test. All applicants interested in financial aid must submit to the Graduate Admissions Committee, Department of Mathematics, three letters of recommendation and a completed Application for Teaching Assistantship Form.

Correspondence and Information

Graduate Admissions Committee
Department of Mathematics
University of South Florida
Tampa, Florida 33620-5700
Telephone: 813-974-5329

Peterson's Guide to Graduate Programs in the Physical Sciences and Mathematics 1992

SECTION 5: MATHEMATICAL SCIENCES

University of South Florida

THE FACULTY AND THEIR RESEARCH

W. Edwin Clark, Professor; Ph.D., Tulane, 1964. Algebraic, geometric, and combinatoric problems arising from the study of error control codes: arithmetic codes, cyclic codes, bit and byte error control codes; semigroups, rings, associative algebras.

R. W. R. Darling, Associate Professor; Ph.D., Warwick (England), 1982. Stochastic flows, stochastic calculus on manifolds, probability models in climatology, multiattribute decision problems.

A. W. Goodman, Distinguished Service Professor; Ph.D., Columbia, 1947. Geometric function theory, graph theory.

Mourad E. H. Ismail, Professor; Ph.D., Alberta, 1974. Applications of special functions and orthogonal polynomials: applications to infinite divisibility and Bessel processes (in probability), enumeration problems (in combinatorics), spectral analysis (in functional analysis and mathematical physics), how eigenvalues of parameter-dependent systems vary with the parameters.

A. G. Kartsatos, Professor; Ph.D., Athens, 1969. Differential equations in Banach spaces, nonlinear analysis involving accretive and monotone operators in Banach spaces.

Joseph J. Liang, Professor; Ph.D., Ohio State, 1969. Algebraic number theory, finite fields, arithmetic coding theory, mathematical aspects of pattern recognition, neural networks and fractal geometry.

Y. F. Lin, Professor; Ph.D., Florida, 1964. Topological semigroups and areas of general topology concerning the impact of functions on spaces, almost continuous mappings and Baire spaces.

M. N. Manougian, Professor; Ph.D., Texas at Austin, 1968. Perron integration for the approximate solution of nonlinear ordinary and partial differential equations, oscillation theory.

Gregory L. McColm, Assistant Professor; Ph.D., UCLA, 1986. Abstract recursion, expressibility, and complexity theory; combinatorics and theoretical biology.

Arunava Mukherjea, Professor; Ph.D., Wayne State, 1967. Probability measures and random walks on semigroups, groups, and matrices; nonhomogeneous Markov chains; products of random matrices; random mappings; stochastic automata.

R. Kent Nagle, Associate Professor; Ph.D., Michigan, 1975. Existence and properties of solutions to nonlinear differential equations—in particular, equations with nonlinearities that involve derivatives of the unknown function, nonlinear third order equations, and systems of nonlinear ordinary differential equations.

Ralph Oberste-Vorth, Assistant Professor; Ph.D., Cornell, 1987. Dynamical systems, particularly the iteration of complex analytic maps in one and several variables; complex analysis; fractal geometry.

Mary E. Parrott, Associate Professor; Ph.D., Memphis State, 1979. Existence, approximation, and stability of solutions to delay differential equations and applications to population dynamics.

John Pedersen, Assistant Professor; Ph.D., Emory, 1984. Decision problems in universal algebra, term-rewriting systems, cellular automata, computational aspects of combinatorics, algebra, and logic.

Kenneth L. Pothoven, Associate Professor and Chairman; Ph.D., Western Michigan, 1969. Measure theory, functional analysis, and nonlinear differential equations: specifically, boundary value problems involving nonlinearities dependent on derivatives and methods of nonlinear analysis for generating existence results.

K. M. Ramachandran, Assistant Professor; Ph.D., Brown, 1987. Optimal and nearly optimal control of wideband noise–driven systems, using weak convergence and martingale methods; computational aspects of control problems; asymptotic behaviors of stochastic systems; learning systems.

A. N. V. Rao, Professor; Ph.D., Rhode Island, 1971. Reliability analysis, stochastic modeling, stochastic control systems.

J. S. Ratti, Professor; Ph.D., Wayne State, 1966. Real analysis (summability and functional equations), complex analysis (polynomials and univalent functions), graph theory (algebraic aspects).

E. B. Saff, Graduate Research Professor; Ph.D., Maryland, 1968. Approximation and interpolation by polynomials and rational functions, orthogonal polynomials, and their asymptotics, asymptotics of zero distributions of sequences of polynomials, analysis of numerical methods, complex function theory, Padé approximants, convergence theory for rational approximants.

Boris Shekhtman, Associate Professor; Ph.D., Kent State, 1980. Approximation theory, abstract and classical analysis.

W. Richard Stark, Associate Professor; Ph.D., Wisconsin–Madison, 1975. Parallel computing and distributed processes—especially algebraic representations of these issues, mathematical logic.

Ernest Thieleker, Associate Professor; Ph.D., Chicago, 1968. Representations of Lie groups and Lie algebras, applications to quantum systems with symmetry.

Vilmos Totik, Professor; Ph.D., Attila József (Hungary), 1979. Approximation theory, interpolation spaces, potential theory, orthogonal polynomials.

Chris P. Tsokos, Professor; Ph.D., Connecticut, 1968. Linear and nonlinear statistical models for health sciences, operations research problems, and economic systems; frequentist and Bayesian reliability analysis and sensitivity modeling; forecasting models for stationary and nonstationary time series analysis; differential stochastic control systems.

Carol A. Williams, Professor; Ph.D., Yale, 1967. Development of algorithms to generate literal solutions of equations of dynamical systems by computer, algebraic and geometric studies of normalizations and simplifications of Hamiltonian systems; development of lunar and planetary theories.

You Yuncheng, Assistant Professor; Ph.D., Minnesota, 1988. Differential equations, control theory, dynamic systems.

SECTION 5: MATHEMATICAL SCIENCES

UNIVERSITY OF SOUTHWESTERN LOUISIANA

Departments of Mathematics and Statistics

Programs of Study

The Department of Mathematics offers graduate programs leading to the degrees of Master of Science and Doctor of Philosophy. Complete programs at both levels are offered in the research areas of ring theory, near rings, universal algebra, low-dimensional algebra, continua theory, topological groups, functional analysis, differential equations, partial differential equations, numerical analysis, and applied mathematics.

Graduate faculty members in the Department of Mathematics are active scholars whose research is regularly published in professional journals. Class sizes are small. This allows close faculty-student relationships. The two-year master's program involves either 24 hours and a thesis or 36 hours and a written exam. The five-year doctoral program includes a written exam covering basic graduate-level courses and an oral exam in the area of specialization. A reading knowledge of either two foreign languages or one foreign language and a knowledge of either computer science or statistics are required.

The Department of Statistics offers programs of study leading to the Master of Science and Doctor of Philosophy degrees. The curriculum is a balanced combination of theoretical and applied statistics courses preparing students for immediate industrial or academic employment. In addition to pursuing formal course work, faculty and graduate students participate in consulting in a wide variety of fields. Knowledge of a high-level computer language and proficiency with statistical packages satisfy the foreign language requirement.

Research Facilities

The departments' research is supported by the University's outstanding computer facilities, including the IBM 3090/200 mainframe. The departments also maintain a networked 386 PC laboratory. Software supported by the two departments includes Mathematica, GAUSS, Minitab, Quattro Pro, Ellpack, SAS, and BMDP. Dupre Library, which recently implemented the IBM-Dobis integrated library system, houses a large collection of volumes and series titles in the areas of mathematics, statistics, and allied fields.

Financial Aid

Master's and doctoral assistantships offered by the departments carry a nine-month stipend and a waiver of tuition and most fees. Research grants also provide assistantship funding. University fellowships available to outstanding students are awarded at the master's and doctoral levels. Renewable doctoral fellowships funded by the University and Louisiana Board of Regents offer twelve-month stipends of $12,000 to $15,000. Information on loans may be obtained from the Financial Aid Office.

Cost of Study

In spring 1991, tuition and fees for Louisiana residents totaled $780 per semester; nonresidents paid $1725. Tuition and fees are subject to change without notice.

Cost of Living

The cost of room and board (fifteen meals per week) for single students living in residence halls in 1990–91 was $1245 per semester. A limited number of rooms in the University Conference Center rent for $750 per semester. Married student apartments were available at $230 per month. The community of Lafayette provides a wide selection of privately owned housing convenient to the campus.

Student Group

Total University enrollment is approximately 15,000; 1,400 are graduate students. In fall 1990, approximately 55 percent of the graduate students were women. A large percentage of the full-time student body receives some type of financial assistance.

Location

The University is located in Lafayette, the central city of the geographic area known as Acadiana, which is populated by French-speaking descendants of the exiled Acadians of Nova Scotia and is characterized by a joie de vivre that has given it an international reputation. The city, with its population of over 95,000, provides many recreational and cultural opportunities. Interstate 10 offers easy access to New Orleans and Houston.

The University and The Departments

The University was established by legislative act in 1898 as the Southwestern Louisiana Industrial Institute; in 1960 it became the University of Southwestern Louisiana. A rapidly developing center of research and graduate study, the 1,366-acre campus reflects the beauty and culture of the area.

The Ph.D. degree has been awarded by the departments since 1968. Statistics became a separate department in 1985. The University is an institutional member of the American Mathematical Society, the Society for Industrial and Applied Mathematics, and the American Statistical Association.

Applying

Basic requirements for admission include an earned bachelor's degree from an accredited institution; satisfactory performance in undergraduate and, if applicable, graduate work; submission of satisfactory scores on the GRE; three letters of recommendation; and acceptance by a department. International applicants must provide satisfactory TOEFL scores and may be required to take ESOL courses upon their arrival on campus. The application fee is $5 for U.S. citizens and $15 for non–U.S. citizens.

Correspondence and Information

For mathematics program information:
C. Y. Chan
University of Southwestern Louisiana
P.O. Box 41010
Lafayette, Louisiana 70504
Telephone: 318-231-5288

For statistics program information:
T. L. Boullion
University of Southwestern Louisiana
P.O. Box 41006
Lafayette, Louisiana 70504
Telephone: 318-231-6771

Peterson's Guide to Graduate Programs in the Physical Sciences and Mathematics 1992

SECTION 5: MATHEMATICAL SCIENCES

University of Southwestern Louisiana

THE FACULTY AND THEIR RESEARCH

Mathematics
Patricia Beaulieu, Assistant Professor; Ph.D., LSU, 1991. Algebra.
Gary Birkenmeier, Professor; Ph.D., Wisconsin–Milwaukee, 1975. Algebra.
Chiu Yeung Chan, Professor; Ph.D., Toronto, 1969. Partial differential equations.
Bradd Clark, Professor; Ph.D., Wyoming, 1976. Topology.
Keng Deng, Assistant Professor; Ph.D., Iowa State, 1990. Differential equations.
Carroll Guillory, Associate Professor; Ph.D., Berkeley, 1978. Analysis.
Henry Heatherly, Professor; Ph.D., Texas A&M, 1968. Algebra.
Awad Iskander, Professor; Ph.D., Moscow, 1965. Algebra.
Baker Kearfott, Associate Professor; Ph.D., Utah, 1977. Numerical analysis.
Steve Ligh, Professor; Ph.D., Texas A&M, 1969. Algebra.
Victor Schneider, Professor; Ph.D., Massachusetts, 1970. Topology.
Robert Sidman, Professor; Ph.D., RPI, 1968. Applied mathematics.
A. S. Vatsala, Associate Professor; Ph.D., Indian Institute of Technology, 1973. Differential equations.
Thelma West, Assistant Professor; Ph.D., Houston, 1986. Topology.
Wilbur Whitten, Professor; Ph.D., Pittsburgh, 1961. Topology.

Statistics
Charles Anderson, Associate Professor; Ph.D., SMU, 1969. Probability, decision theory.
Calvin Berry, Assistant Professor; Ph.D., Cornell, 1985. Bayes inference, decision theory.
Thomas L. Boullion, Professor; Ph.D., Texas, 1966. Linear model time series.
Nabendu Pal, Assistant Professor; Ph.D., Maryland, 1989. Decision theory, biostatistics.
Thomas Rizzuto, Assistant Professor; Ph.D., Texas Tech, 1973. Operations research, multivariate analysis.
John W. Seaman Jr., Assistant Professor; Ph.D., Texas at Dallas, 1983. Inference, probability theory.

SECTION 5: MATHEMATICAL SCIENCES

UNIVERSITY OF UTAH
Department of Mathematics

Programs of Study

The Department of Mathematics offers programs leading to the degrees of Doctor of Philosophy, Master of Arts, and Master of Science in mathematics.

The master's degree requires 45 hours of course work beyond certain basic prerequisites. The candidate for the M.A. degree must satisfy the standard proficiency requirement in one foreign language; a further requirement is an expository thesis of good quality or an approved three-quarter graduate course sequence.

The doctoral degree carries a minimum course requirement designed to prepare the student to pass a written preliminary examination in the basic fields of mathematics. An oral examination, with emphasis on the candidate's area of specialization, is also required. For the Ph.D., the department has adopted a minimum language requirement of standard proficiency in two foreign languages or advanced proficiency in one, chosen from French, German, and Russian. A dissertation describing independent and original work is required. The Department of Mathematics stresses excellence in research. A master's degree is not a requirement for the Ph.D.

The University also awards a Master of Philosophy degree that requires the same qualifications for admission and scholarly achievement as the Doctor of Philosophy degree; however, a dissertation is not required.

Research Facilities

The mathematics building contains a research library consisting of over 5,000 books and 205 journals. In addition, there is a general library that contains many more mathematics books and journals with articles of mathematical interest. There are extensive interactive computing and computer graphics facilities available in the department.

Financial Aid

Approximately 65 percent of the mathematics graduate students are supported by fellowships. There are teaching fellowships that grant from $9900 to $10,400 plus state-resident tuition and fees. In most cases, nonresident tuition is also waived. Application for University research fellowships of $4500 can be made through the Research Committee Office, 120 Park, at the University.

The normal teaching load for a teaching assistant (M.S. candidate) is the equivalent of two 4-hour sections during one quarter and one 4-hour section during two quarters. The normal teaching load for a teaching fellow (Ph.D. candidate) is the equivalent of one 4-hour section during each quarter. Summer teaching assignments for an additional stipend are available.

Cost of Study

For 1991–92, tuition is $611.50 per quarter for Utah residents and $1725 per quarter for nonresidents (12 credit hours). (Tuition rates may change without notice.) All resident tuition fees are waived for teaching assistants and fellows, and, in most cases, nonresident tuition is also waived. However, there is a computer fee all students must pay.

Cost of Living

A wide variety of housing is offered by the University on or near the campus. The cost of a single room is $1885 per academic year and board is about $800 per quarter for a single student. University Village is operated by the University for married students. One-, two-, and three-bedroom apartments range from $230 to $570 per month, including heat, hot water, electricity, range, and refrigerator. (These rates may change without notice.) There is a waiting period of about six months for the University's married student housing. Privately owned housing near the campus is also available.

Student Group

The University's total enrollment is currently over 24,000. The Department of Mathematics has 80 graduate students, 52 of whom receive financial support.

Location

The Salt Lake City metropolitan area has a population of about a million and is the cultural, economic, and educational center of the Intermountain West. The Utah Symphony and Ballet West are located in Salt Lake City. The Salt Palace is the home of professional hockey and basketball teams. Climate and geography combine in the Salt Lake environs to provide ideal conditions for outdoor sports. Some of the world's best skiing is available less than an hour's drive from the University campus.

The University and The Department

The University of Utah is a state-supported coeducational institution. Founded in 1850, it is the oldest state university west of the Missouri River.

In the last five years, the Department of Mathematics has awarded 57 Ph.D. degrees. In recent years, the Graduate School has been awarding about 180 doctoral degrees per year. The University faculty has approximately 3,300 members.

Applying

Admission to graduate status requires that students hold a bachelor's degree or its equivalent and that they show promise for success in graduate work. Applicants are urged to take the mathematics Subject Test of the Graduate Record Examinations.

Students are normally admitted at the beginning of the autumn term. It is desirable that applications for teaching fellowships, as well as for other financial grants, be submitted as early as possible. All applications received before March 1 will automatically be considered for fellowships.

Correspondence and Information

Graduate Fellowship Committee
Attention: Graduate Secretary
Department of Mathematics
University of Utah
Salt Lake City, Utah 84112

SECTION 5: MATHEMATICAL SCIENCES

University of Utah

THE FACULTY AND THEIR RESEARCH

Professors
P. W. Alfeld, Ph.D., Dundee (Scotland), 1977. Numerical analysis.
R. M. Brooks, Ph.D., LSU, 1963. Topological algebras.
C. E. Burgess, Ph.D., Texas, 1951. Topology.
J. A. Carlson, Ph.D., Princeton, 1971. Algebraic geometry.
C. H. Clemens, Ph.D., Berkeley, 1966. Algebraic geometry.
W. J. Coles, Ph.D., Duke, 1954. Ordinary differential equations.
E. A. Davis, Ph.D., Berkeley, 1951. Mathematical economics, teacher training.
S. N. Ethier, Ph.D., Wisconsin–Madison, 1975. Probability and statistics.
P. C. Fife, Ph.D., NYU, 1959. Applied mathematics.
E. S. Folias, Ph.D., Caltech, 1963. Applied mathematics, elasticity.
S. M. Gersten, Ph.D., Cambridge, 1965. Algebra.
L. C. Glaser, Ph.D., Wisconsin–Madison, 1964. Geometric topology.
K. G. Goodearl, Ph.D., Washington (Seattle), 1971. Algebra.
F. I. Gross, Ph.D., Caltech, 1964. Algebra.
G. B. Gustafson, Ph.D., Arizona State, 1968. Ordinary differential equations.
H. Hecht, Ph.D., Columbia, 1974. Lie groups.
J. P. Keener, Ph.D., Caltech, 1972. Applied mathematics.
J. Kollar, Brandeis, 1984. Algebraic geometry.
J. D. Mason, Ph.D., California, Riverside, 1968. Probability.
D. Milicic, Ph.D., Zagreb (Yugoslavia), 1973. Lie groups.
H. G. Othmer, Ph.D., Minnesota, 1969. Applied mathematics.
P. C. Roberts, Ph.D., McGill, 1974. Commutative algebra, algebraic geometry.
H. Rossi, Ph.D., MIT, 1960. Complex analysis.
T. B. Rushing, Ph.D., Georgia, 1968. Topology.
K. Schmitt, Ph.D., Nebraska, 1967. Differential equations.
J. L. Taylor, Ph.D., LSU, 1964. Abstract analysis.
D. Toledo, Ph.D., Cornell, 1972. Algebraic and differential geometry.
A. E. Treibergs, Ph.D., Stanford, 1980. Differential geometry.
P. C. Trombi, Ph.D., Illinois at Urbana-Champaign, 1970. Lie groups.
D. H. Tucker, Ph.D., Texas, 1958. Differential equations, functional analysis.
C. H. Wilcox, Ph.D., Harvard, 1955. Applied mathematics, scattering theory.
D. W. Willett, Ph.D., Caltech, 1963. Differential equations.
J. H. Wolfe, Ph.D., Harvard, 1948. Geometric integration theory.
B. K. Zimmermann-Huisgen, Munich, 1974. Algebra.

Emeritus Professors
R. E. Chamberlin, Ph.D., Harvard, 1950. Topology, numerical analysis.
W. R. Scott, Ph.D., Ohio State, 1947. Algebra.

Visiting Faculty
A. Cherkaev, Ph.D., Leningrad Polytechnic Institute, 1988. Applied math. (W, S 1991–92)
S. Ivanov, Ph.D., Moscow, 1988. Group theory. (A, W, S 1991–92)
J. Keesling, Ph.D., Miami (Florida), 1968. Topology. (A, W, S 1991–92)
J. McKernan, Ph.D., Harvard, 1990. Algebraic geometry. (A, W, S 1991–92)
S. Mori, Ph.D., Kyoto (Japan), 1978. Algebraic geometry. (A 1990–91)
A. Panfilov, Ph.D., Institute of Biological Physics, Moscow, 1983. Applied mathematics. (A, W, S 1990–91)
A. Rosenberg, Ph.D., Moscow, 1973. Algebra. (A, W, S 1991–92)
S. A. Stromme, D.Phil., Oslo (Norway), 1983. Algebraic geometry. (A, W, S 1991–92)
S. Zheng, Ph.D., Fudan-China. PDE. (A 1991–92)

Associate Professors
A. L. Fogelson, Ph.D., NYU, 1982. Computational fluids.
K. M. Golden, Ph.D., NYU, 1984. Applied math.
L. Horvath, Ph.D., Szeged (Hungary), 1982. Probability, statistics.
N. J. Korevaar, Ph.D., Stanford, 1981. Partial differential equations.

Assistant Professors
P. J. Braam, D.Phil., Oxford, 1987. Topology.
P. K. Maini, D.Phil., Oxford, 1986. Mathematical biology.
R. Palais, Ph.D., Berkeley, 1986. Applied mathematics.
N. Smale, Ph.D., Berkeley, 1987. Differential geometry.
J. Zhu, Ph.D., Courant, 1989. Applied and computational math, computational fluid dynamics.

Adjunct Professors
J. E. Anderson, M.Phil., Utah, 1970: mathematics education. N. Beebe, Ph.D., Florida, 1972: numerical analysis. D. D. Clark, Ed.D., Brigham Young, 1974: secondary foundations and instruction. E. Cohen, Ph.D., Syracuse, 1974: computer-aided geometric design. M. J. Egger, Ph.D., Stanford, 1979: statistics. H. G. Ehrbar, Ph.D., Munich, 1969: probability. S. Foresti, Ph.D., Pavia (Italy), 1987: scientific computing. D. Henderson, Ph.D., Utah, 1961: scientific computing. C. Johnson, Ph.D., Utah, 1990: cardiovascular research. L. G. Lewis, Ph.D., Indiana, 1969: complex analysis. M. Pernice, Ph.D., Colorado, 1986: scientific computing. J. C. Reading, Ph.D., Stanford, 1970: statistics. A. D. Roberts, Ph.D., McGill, 1972: analysis.

Instructors
V. A. Alexeev, Ph.D., Moscow, 1990: algebraic geometry. K. M. Boucher, Ph.D., Michigan, 1990: geometric topology. P. Burchard, Ph.D., Chicago, 1989: representation theory. D. H. Dang, Ph.D., Ho Chi Minh City, 1990: nonlinear analysis, differential equations. L. Y. Fong, Ph.D., Brandeis, 1991: algebraic geometry. J. S. Huang, Ph.D., MIT, 1989: Lie groups. J. M. Johnson, Ph.D., Brandeis, 1989: automorphic forms. S. Keel, Ph.D., Chicago, 1989: algebraic geometry. S. M. Krone, Ph.D., Massachusetts, 1990: probability. M. J. Lai, Ph.D., Texas A&M, 1989: numerical analysis. E. S. Letzter, Ph.D., Washington (Seattle), 1988: ring theory. A. Marini, Ph.D., Chicago, 1990: differential equations and math physics. A. Morlet, Ph.D., Caltech, 1990: numerical analysis.

UNIVERSITY OF WASHINGTON
Department of Applied Mathematics

Program of Study

The Department of Applied Mathematics offers an independent, interdisciplinary graduate degree program involving training in mathematics as well as significant study in at least one outside field. Graduate work in applied mathematics leading to M.S. and Ph.D. degrees consists of broad training in mathematical techniques that have been found generally useful in applications, in-depth exposure to at least one field of application, and the opportunity to explore certain specialized aspects of applied mathematics.

At present, the principal areas of study in applied mathematics include applied linear algebra, complex variables, control and estimation theory, numerical analysis, optimization theory and mathematical programming, ordinary differential equations, partial differential equations, perturbation and approximation techniques, probability and statistics, and special functions and approximations. An extensive range of appropriate outside fields has been identified, including all branches of engineering, the physical and geophysical sciences, the biological sciences, computer science, economics and management science, and certain areas of medical science. Nontraditional fields of application may be approved by the Department of Applied Mathematics where appropriate.

Each individual program of study is designed by the student in consultation with and with the approval of a supervisory committee. In addition to study in appropriate courses, close collaboration between student and faculty members in research is essential, and each student works under the supervision of a faculty member to develop the techniques and insight necessary for successful research.

Students pursuing the Master of Science may select either a thesis or a nonthesis program. Entry into the Ph.D. program is determined by performance on the Qualifying Examination. Advancement to doctoral candidacy is determined by performance on the General Examination, set by the Supervisory Committee. A satisfactory dissertation presentation and defense (the Final Examination) is required for completion of the Ph.D. degree.

Research Facilities

An extensive library system includes separate specialized libraries in science, mathematics, engineering, physics, oceanography, chemistry, medicine, and other areas. A departmental network of DEC/RISC graphics workstations and file servers with links to additional centralized campus computing resources is available. Supported software on departmental systems includes a symbolic mathematics package (Mathematica), an interactive linear algebra environment (MATLAB), FORTRAN callable subroutine libraries (NAG, SLATEC, CMLIB, Colsys), interactive and program callable graphics packages (SPlus, AVS, Tec Plot, NCAR), desktop publishing software (TeX, LaTeX, FRAME), and numerous utility programs. Easy access to all major academic computer networks (NSFNET, BITNET, CSnet, etc.) is provided, permitting access to remote supercomputer sites nationwide.

Financial Aid

A limited number of teaching and research assistantships are available and are granted on a merit basis. The minimum stipend is $940 per month in 1991–92. The tuition for holders of assistantships is waived except for approximately $125 per quarter in fees. In addition, students are encouraged to apply for various loans, fellowships, and scholarships that are available through the Financial Aid Office and the Graduate School.

Cost of Study

Tuition and fees for the academic year 1991–92 are $1129 per quarter for Washington residents and $2824 per quarter for nonresidents. Holders of assistantships pay only approximately $125 in fees per quarter. Costs may be subject to change.

Cost of Living

The cost of room and board on campus for the academic year is about $3510. Married student housing ranges from $3360 to $5400 per calendar year. The area offers shared housing, starting at about $240 per month, and apartments, starting at about $280 per month.

Student Group

During the 1991–92 academic year, of the 39 graduate students in the program, 72 percent were men, and 28 percent were international students. Eighty-two percent of the students were awarded assistantships, and 13 percent had their own fellowship/scholarship funds or had part-time positions in industry.

Location

Seattle, the largest city in Washington, is situated between Puget Sound and Lake Washington, with the Olympic Mountains to the west and the Cascade range to the east. It combines the cultural attributes of an urban setting, such as good restaurants, opera, ballet, and symphonies, with immediately available outdoor activities, such as boating, skiing, and hiking. Seattle has professional basketball, football, and baseball teams as well as amateur leagues in all sports. City parks are located in many scenic areas of the city, and bus transportation is good.

The University

The University of Washington was founded in 1861 and is the oldest state-assisted institution of higher education on the Pacific Coast. There are 33,238 students studying in a variety of fields, including arts and sciences, business administration, dentistry, education, engineering, fisheries, forest resources, law, medicine, nursing, pharmacy, and public health and community medicine. The Graduate School has approximately 7,700 students and 2,300 faculty members. Comprehensive intercollegiate and intramural athletic programs are offered, plus a range of musical and cultural programs. The beautiful green campus encompasses 680 acres and is bordered on the east by Lake Washington and on the south by Lake Union. The University of Washington is one of the top research universities in the United States. Its faculty consistently receives more research grants than any other public university in the country.

Applying

Application forms may be obtained from the Department of Applied Mathematics or from Graduate Admissions, AD-10, at the University of Washington. Although the deadline for admission in the autumn quarter is July 1, applications should be completed by February 1 to be competitive for financial assistance. The GRE General Test is required, and three letters of recommendation and a personal statement must be submitted directly to the department in support of an application. In addition, international students must take the Test of English as a Foreign Language (TOEFL).

Correspondence and Information

Department of Applied Mathematics, FS-20
University of Washington
Seattle, Washington 98195
Telephone: 206-543-5493

SECTION 5: MATHEMATICAL SCIENCES

University of Washington

THE FACULTY AND THEIR RESEARCH

Loyce M. Adams, Associate Professor; Ph.D., Virginia, 1982. Numerical analysis, scientific computing. Current research: Multilevel methods for linear systems of equations arising from elliptic partial differential equations, eigenvalue problems, design and implementation of algorithms for parallel computers.

Christopher S. Bretherton, Associate Professor; Ph.D., MIT, 1984. Fluid dynamics. Current research: Mixing processes, dynamics of moist convection, complex behavior and disorder in geophysical systems, applications to mesoscale meteorology, bifurcation.

William O. Criminale Jr., Professor; Ph.D., Johns Hopkins, 1960. Fluid dynamics, nonlinear mechanics. Current research: Large-scale oscillations in turbulent flows, particle dynamics, mixing.

Jirair Kevorkian, Professor; Ph.D., Caltech, 1961. Partial differential equations, perturbation theory. Current research: Development and application of perturbation techniques for analysis of problems in nonlinear wave propagation, other topics of current interest in theoretical mechanics.

Mark Kot, Assistant Professor; Ph.D., Arizona, 1987. Mathematical ecology, nonlinear dynamics, difference equations. Current research: Chaotic population dynamics, discrete-time growth-dispersal models, traveling waves.

Randall LeVeque, Associate Professor; Ph.D., Stanford, 1982. Numerical analysis. Current research: Analysis of numerical methods for differential equations, particularly those arising in fluid mechanics; mathematical problems in physiology.

James D. Murray, Professor; D.Sc., Oxford, 1968. Mathematical biology, particularly embryology, ecology, and epidemiology. Current research: Generation of biological pattern and form, wound healing, spatial spread of disease, spatial aspects in the release of genetically engineered organisms for biological control of pests.

Robert E. O'Malley, Professor and Chair; Ph.D., Stanford, 1966. Singular perturbations. Current research: Singular perturbations asymptotic methods, differential equations and control theory.

Carl E. Pearson, Professor; Ph.D., Brown, 1949. Fluid dynamics, numerical analysis. Current research: Streamline methods in compressible gas flow, applications of singular integral equations, optimization, theory of games, electromagnetic waves.

R. Tyrrell Rockafellar, Professor; Ph.D., Harvard, 1963. Optimization problems and associated mathematical analysis, in particular nonlinear programming, network programming, and optimal control. Current research: Structured large-scale convex programming models in stochastic optimization and optimal control, especially solution techniques based on duality and piecewise linear-quadratic approximations; new developments in nonsmooth analysis with applications to optimality conditions in nonlinear programming and optimal control, including sensitivity to perturbations; Hamilton-Jacobi theory; monotropic optimization.

Ka Kit Tung, Professor; Ph.D., Harvard, 1977. Atmospheric sciences, geophysical fluid mechanics. Current research: Nonlinear dynamics in the atmosphere, tracer transport in the stratosphere, modeling of atmosphere composition, barotropic and baroclinic instabilities, large-amplitude waves.

Frederic Y. M. Wan, Professor and Associate Dean, College of Arts and Sciences; Ph.D., MIT, 1965. Asymptotic, variational, and stochastic methods for differential equations and their applications to problems in elastic solids, resource economics, and biomechanics. Current research: Elastic and viscoelastic theories of beams, plates, and shells; asymptotic solutions of nonlinear boundary value problems; optimal control problems in urban land economics and forestry economics.

Adjunct Faculty

Marcia B. Baker, Professor; Ph.D., Washington (St. Louis), 1971. Cloud physics. Current research: Dynamics and microphysics in layer and convective clouds.

Marshall Baker, Professor (Physics); Ph.D., Harvard, 1958. Theoretical physics, with emphasis on elementary particle theory and relativistic quantum field theory. Current research: Long-distance QCD and the nature and physics of quark confinement.

Bruce H. Faaland, Professor (Quantitative Methods); Ph.D., Stanford, 1971. Integer and network programming, dynamic programming. Current research: Models that maximize yield in selecting sawmill patterns to cut raw logs into lumber, portfolio selection models, optimal design of B-trees for secondary computer-based storage systems, linear time algorithms for specially structured linear programs (these programs arise in advertising and production problems), selection of the optimal size of a crew to achieve a balanced assembly line, coordination of job shop operations.

Akira Ishimaru, Professor (Electrical Engineering); Ph.D., Washington (Seattle), 1958. Wave propagation and scattering. Current research: Wave propagation and scattering in dense geophysical media; multiple scattering effects on transmission through the atmosphere; statistical study of electromagnetic interference; ultrasound imaging of biological medium; space-to-ground microwave and millimeter-wave communications; underwater acoustics; advanced radar technology; electromagnetic, optical, and acoustic waves in random media; imaging and inverse problems.

George Kosály, Professor (Mechanical Engineering); Ph.D., Eötvös Loránd (Budapest), 1974. Theoretical physics, with emphasis on application to random phenomena; inelastic scattering of electrons and neutrons in gases, liquids, and crystals; linear neutron transport in nuclear power reactors; turbulence theory. Current research: Mixing of inert and reacting scalars in turbulent flow.

James J. Riley, Professor (Mechanical Engineering); Ph.D., Johns Hopkins, 1971. Fluid mechanics, especially transition and turbulence, and geophysical fluid mechanics. Current research: Stability, transition, and turbulence in jets and mixing layers; turbulent, chemically reacting flows; turbulent, density-stratified flows; nonlinear waves and quasi-horizontal vortices in rotating, stably stratified flows; particle motion in turbulent flows.

Duane W. Storti, Associate Professor (Mechanical Engineering); Ph.D., Cornell, 1984. Dynamical systems and nonlinear ordinary differential equations, especially applications of perturbation theory and bifurcation theory to nonlinear oscillators arising in mechanical and biological systems. Current research: Study of phase locking in systems of coupled limit-cycle oscillators, flow-induced oscillations, lattice models of dynamic systems.

Juris Vagners, Professor; Ph.D., Stanford, 1967. Control theory, optimization. Current research: Applied optimal control and estimation, specifically as applied to problems in the aviation industry; biomedical engineering; development of real-time algorithms for system identification, adaptive control, performance seeking control; development of inertial navigation algorithms.

Harry H. Yeh, Associate Professor (Civil Engineering); Ph.D., Berkeley, 1983. Fluid mechanics. Current research: Nonlinear edge waves, sheet flows, bores (broken waves), gravity currents.

Affiliate Faculty

Robert M. Miura, Professor (Mathematics, British Columbia); Ph.D., Princeton, 1966. Nonlinear wave propagation, mathematical neurobiology, differential-difference equations, partial differential equations. Current research: Modeling and analysis of linear and nonlinear phenomena in excitable media, mainly those arising in neurobiology; rotating waves in excitable media; singular perturbation analyses of boundary-value problems for linear and nonlinear differential-difference equations; nonlinear diffusive and dispersive equations.

John L. Nazareth, Professor (Washington State University, Pullman, Washington); Ph.D., Berkeley, 1973. Computer science. Current research: Operations research and optimization, numerical analysis, computational linear algebra, pratical implementation of algorithms, development of mathematical programming models and decision support systems.

Erik W. Pearson, Assistant Professor (Battelle Pacific Northwest Laboratories, Richland, Washington); Ph.D., Harvard, 1983. Computational chemistry, scientific computing. Current research: Parallel computational methods for differential equations, chemical kinetics and nonlinear dynamics.

Kraig B. Winters, Applied Mathematician (Applied Physics Laboratory, University of Washington, Seattle, Washington); Ph.D., Washington (Seattle), 1989. Applied mathematics. Current research: Scientific computation, wave propagation, fluid dynamics, inverse problems.

UNIVERSITY OF WASHINGTON

Graduate Program in Biostatistics

Programs of Study

The Graduate Program in Biostatistics offers Master of Science and Doctor of Philosophy degrees in quantitative methods applied to the medical and biological sciences. The Master of Science degree takes an average of three years and the Doctor of Philosophy degree an average of six years to complete.

The goal of the graduate program is to equip students to develop and apply the quantitative techniques of mathematics, statistics, and computing appropriate to medicine and biology. Because of the faculty's involvement in a diversity of statistical applications, students in the graduate program receive an education of high quality. Students are recruited from undergraduate programs in mathematics, statistics, and biology and are selected on the basis of outstanding quantitative ability.

Research Facilities

The extensive University of Washington library system includes separate specialized libraries in natural sciences, mathematics, fisheries/oceanography, chemistry, and medicine. The Health Sciences Library is located in the same building as the Department of Biostatistics and has a research-level collection in biostatistics. The Department of Biostatistics maintains a computer lab for student use. Resources include IBM-RT and Sun UNIX Workstations; MS-DOS computers running a wide variety of statistical software; network capabilities with TCP/IP Ethernet, Internet, and BITNET; and an electronic mail system. Free use of campus mainframes and the Health Sciences Library PC/MAC II Lab is also available.

Financial Aid

A limited number of teaching assistantships, research assistantships, and traineeships (fellowships) are available for student support during the academic year. In 1990–91, TAs and RAs received $886–$1086 per month, plus tuition remission. Fees amounting to about $110 per quarter were not covered by assistantships. Traineeships are restricted to U.S. citizens and permanent residents. Trainees received a stipend of $708 per month, plus remission of tuition and fees.

Cost of Study

Tuition and fees for the academic year 1991–92 are $1129 per quarter for Washington residents and $2824 per quarter for nonresidents. Holders of assistantships pay approximately $110 in fees per quarter.

Cost of Living

The cost of room and board on campus for the academic year is approximately $2800. Married students' housing ranges from $3900 to $5700 per calendar year. The area offers shared housing, starting at about $200 per month, and apartments start at about $400 per month.

Student Group

During the 1990–91 academic year, there were 11 M.S. and 39 Ph.D. students in the graduate program. Half of these 50 students were women, and 15 came from abroad. Thirteen students were supported by traineeships, and most of the remainder, by research or teaching assistantships.

Location

Seattle, the largest city in Washington, is situated between Puget Sound and Lake Washington, with the Olympic Mountains to the west and the Cascade range to the east. The city combines the cultural attributes of an urban setting (for example, good restaurants, opera, ballet, and symphonies) with outdoor activities, such as boating, skiing, and hiking. Seattle has professional basketball, football, and baseball teams as well as amateur leagues in all sports. City parks are located in all areas of the city, and bus transportation is good.

The University

The University of Washington was founded in 1861 and is the oldest state-assisted institution of higher education on the Pacific Coast. The University enrolls approximately 33,000 students in a variety of fields, including arts and sciences, business administration, dentistry, education, engineering, fisheries, forest resources, law, medicine, nursing, pharmacy, and public health and community medicine. The Graduate School has approximately 8,900 students and 2,100 faculty members. Comprehensive intercollegiate and intramural athletic programs are offered, as is a range of musical and cultural programs. The beautiful green campus encompasses 680 acres and is bordered on the east by Lake Washington and on the south by Lake Union. The University of Washington is ranked in the top quartile of universities in the United States and is an outstanding center of academic excellence in the Northwest.

Applying

Further information and application materials can be obtained from the student services counselor of the Graduate Program in Biostatistics. Applications are welcomed from students with an undergraduate or master's degree in mathematics, statistics, or a biological field. Applications for the fall term should be submitted by April 15 of the preceding academic year.

The biostatistics program is particularly interested in recruiting members of minority groups currently underrepresented in biomedical science.

Correspondence and Information

Dr. Polly Feigl, Graduate Program Coordinator
Graduate Program in Biostatistics, SC-32
University of Washington
Seattle, Washington 98195
Telephone: 206-543-1044

SECTION 5: MATHEMATICAL SCIENCES

University of Washington

THE FACULTY AND THEIR RESEARCH

Professors

B. Bruce Bare, Department of Biostatistics and the Center for Quantitative Science in Forestry, Fisheries, and Wildlife; Ph.D., Purdue, 1969. Biometry, forest management, operations research, multiobjective programming.

Norman E. Breslow, Department of Biostatistics; Ph.D., Stanford, 1967. Clinical trials, epidemiology, survival and categorical data, environmental health.

John Crowley, Department of Biostatistics; Ph.D., Washington (Seattle), 1973. Survival analysis: cancer clinical trials.

Kathryn Davis, Department of Biostatistics; Ph.D., Washington (Seattle), 1974. Density estimation, cardiovascular data analysis, clinical trials.

Timothy DeRouen, Department of Biostatistics; Ph.D., Virginia Tech, 1971. Applications in dentistry and the epidemiology of sexually transmitted diseases.

Paula Diehr, Department of Biostatistics; Ph.D., UCLA, 1971. Application of statistics to health services research, small-area analysis, multiple regression.

Polly Feigl, Department of Biostatistics; Ph.D., Minnesota, 1961. Application of statistics to cancer control research, prevention studies, and large-scale patient data systems.

Lloyd D. Fisher, Department of Biostatistics; Ph.D., Dartmouth, 1966. Cardiovascular data analysis, clinical trials, multivariate statistics, longitudinal data analysis.

Thomas R. Fleming, Department of Biostatistics; Ph.D., Maryland, 1976. Survival analysis, cancer clinical trials, AIDS research, sequential analysis.

David Ford, Department of Biostatistics and the Center for Quantitative Science in Forestry, Fisheries, and Wildlife; Ph.D., University College, London, 1963. Modeling plant response to the environment, interpreting ecological processes from spatial patterns, statistical inference for complex models, measurement theory for ecological processes, time-series analysis.

Vincent F. Gallucci, Department of Biostatistics and the Center for Quantitative Science in Forestry, Fisheries, and Wildlife; Ph.D., North Carolina State, 1971. Application of stochastic processes and differential equations to the biological sciences, population dynamics, management of harvested populations.

James Grizzle, Research Professor, Department of Biostatistics; Ph.D., North Carolina State, 1960. Clinical trials, cancer prevention studies.

Alfred P. Hallstrom, Research Professor, Department of Biostatistics; Ph.D., Brown, 1975. Clinical trial methodologies, especially in cardiovascular (chronic) applications and emergency services applications.

Richard A. Kronmal, Department of Biostatistics; Ph.D., UCLA, 1964. Nonparametric density estimation, computer algorithms, cardiovascular data analysis, clinical trials.

Donald C. Martin, Department of Biostatistics; Ph.D., Florida State, 1968. Statistical computing, design of statistical systems, classification methods, approximations for probability functions, signal processing, randomization tests.

Suresh Moolgavkar, Departments of Epidemiology and Biostatistics; Ph.D., Johns Hopkins, 1973. Stochastic models for carcinogenesis, statistical inference in epidemiologic studies.

Edward Perrin, Department of Health Sciences; Ph.D., Stanford, 1960. Health information, systems stochastic modeling, research methodology.

Arthur V. Peterson, Department of Biostatistics; Ph.D., Stanford, 1975. Survival data methodology, competing risks, design of medical studies, methodologic issues of prevention studies, random number generation.

Ross L. Prentice, Department of Biostatistics; Ph.D., Toronto, 1970. Failure time analysis, disease prevention trials, epidemiologic methods, dietary lactose and disease.

Elizabeth Thompson, Department of Statistics; Ph.D., 1974, Sc.D., 1988, Cambridge. Statistics of genetics, population genetics, evolution.

Patricia Wahl, Department of Biostatistics; Ph.D., Washington (Seattle), 1971. Multivariate statistical techniques, especially regression analysis applied to cardiovascular data.

Jon A. Wellner, Department of Statistics; Ph.D., Washington (Seattle), 1975. Empirical processes, semiparametric models, asymptotic efficiency, survival analysis, martingales.

Associate Professors

Jacqueline K. Benedetti, Research Associate Professor, Department of Biostatistics; Ph.D., Washington (Seattle), 1974. Statistical methodology in infectious disease research.

Brent A. Blumenstein, Research Associate Professor, Department of Biostatistics; Ph.D., Emory, 1974. Clinical trials, categorical data analysis, research data management.

Loveday Conquest, Center for Quantitative Science in Forestry, Fisheries, and Wildlife; Ph.D., Washington (Seattle), 1975. Sampling and experimental design for environmental monitoring studies, especially spatial and temporal aspects; statistical methods for timber, fish, and wildlife studies; analysis of aquatic toxicity data and environmental monitoring data; estimation of animal abundance; general biometry.

Stephanie Green, Department of Biostatistics; Ph.D., Wisconsin, 1979. Longitudinal data analysis, clinical trials, cancer research.

Kenneth Kopecky, Research Associate Professor, Department of Biostatistics; Ph.D., Oregon, 1977. Clinical trials design and analysis, survival data analysis, epidemiologic methodology, goodness of fit, biomedical and cancer-related applications.

Barbara McKnight, Department of Biostatistics; Ph.D., Wisconsin, 1981. Statistical applications in animal carcinogenesis testing, epidemiology, survival analysis and competing risks.

Finbarr O'Sullivan, Department of Statistics; Ph.D., Wisconsin, 1983. Inverse problems, statistical computing.

Steven G. Self, Department of Biostatistics; Ph.D., Washington (Seattle), 1981. Longitudinal data analysis, survival time models, cancer prevention and screening trials.

Nancy R. Temkin, Department of Biostatistics; Ph.D., SUNY at Buffalo, 1976. Clinical trials, recovery models, statistical modeling of epileptic phenomena, survival analysis.

Assistant Professors

Kevin C. Cain, Department of Biostatistics; Ph.D., Harvard, 1982. Cardiovascular and health services applications, measurement error models.

Danyu Lin, Department of Biostatistics; Ph.D., Michigan, 1989. Survival analysis, clinical trials methodology, AIDS research, liver transplantation, environmental health.

T. E. Raghunathan, Department of Biostatistics; Ph.D., Harvard, 1987. Analysis of incomplete data, Bayes and empirical Bayes analysis, survey sampling, longitudinal data analysis.

Mark D. Thornquist, Department of Biostatistics; Ph.D., Wisconsin, 1985. Ordinal response, repeated measures data, categorical response.

Ellen M. Wijsman, Research Assistant Professor, Department of Biostatistics; Ph.D., Wisconsin, 1981. Statistical genetics, population genetics, genetic epidemiology.

SECTION 5: MATHEMATICAL SCIENCES

UNIVERSITY OF WYOMING

Department of Mathematics

Programs of Study

The Department of Mathematics at the University of Wyoming offers graduate studies for M.S., M.S. in Teaching (M.S.T.), and Ph.D. degrees in mathematics. The department also offers special concentration in combinatorial studies and applied mathematics (applied analysis and scientific computation) and a joint Ph.D. in mathematics and computer science.

All entering graduate students are required to take a placement exam at the start of the academic year to guide their initial graduate work. The M.S. program requires 30 credit hours of course work and the successful completion of a qualifying exam (written and oral) together with a master's paper or thesis. The M.S.T. program is primarily for teachers who have certification and experience in primary or secondary education.

The Ph.D. degree is offered in both pure and applied areas in mathematics and is intended for students with superior mathematical ability and research interests. The Ph.D. program requires each student to pass a written qualifying examination testing the student's breadth of knowledge related to proposed research. A reading knowledge of French, German, or Russian is also required. To earn a Ph.D., a student must write and defend an original dissertation containing significant, publishable results. Students pursuing the joint degree in mathematics and computer science must develop an interdisciplinary program of study with a committee of faculty members from both departments.

Research Facilities

There are currently 27 faculty members and several visiting professors. The library has an excellent holding of mathematical books and journals, and its operation is fully computerized under the CARL system.

The deparment maintains active research ties with the other science and engineering departments at the University. Many members in the department are integrally involved in the Institute for Scientific Computation, the Wyoming Center for Energy Research, and the Enhanced Oil Recovery Institute.

The computing facilities at the University of Wyoming are excellent. Through the Institute for Scientific Computation, the department has networked access to Alliant FX/8 and Ardent Titan parallel-architecture computers, a VAX-11/785, an IRIS 4D/70GT graphics workstation, and other Sun Workstations and peripherals. The department uses off-site supercomputers and advanced-architecture machines via West Net, a network linking Rocky Mountain institutions to the nationwide NSFNet.

Financial Aid

There are approximately 30 graduate assistantships and several research fellowships for each academic year. New teaching assistants normally teach two sections of a precalculus course (6–8 hours) and receive a stipend of up to $8500 plus full tuition and fees. A Ph.D. student may expect to be supported for five to six years with satisfactory academic progress. Summer support is available on the basis of academic year performance.

Cost of Study

The tuition fee for full time (12–20 hours) for the fall 1991 semester is $777 for residents and $2179 for nonresidents. Part-time students are charged on a per-credit basis.

Cost of Living

The University maintains housing for both married students and single students. The cost of housing for married students with a single-bedroom apartment is $183 a month plus utilities, and a two-bedroom apartment is $223 a month plus utilities. The residence hall charges for room and board are $1630 for a double room and $1968 for a single room. Off-campus housing can be located through the University Housing Office or through local advertisements.

Student Group

The 1991 spring enrollment at the University was approximately 10,500 students, of whom 10 percent were graduate students. There are currently 35 graduate assistants in the Department of Mathematics.

Location

The University of Wyoming is located in Laramie, a city of 25,000 inhabitants, perched 7,200 feet above sea level in a scenic valley in the Snowy Range Mountains of the Rockies. Laramie offers outstanding year-round access to outdoor recreation and is 2½ hours by car from Denver, Colorado.

The University

The University of Wyoming was founded in 1886 and is the only four-year university in Wyoming. The Graduate School was inaugurated in 1946. The campus is attractive and beautifully landscaped. The University offers a rich cultural heritage and a student population of diverse international cultures.

Applying

Application forms for admission and financial aid as well as any additional information can be obtained from the address given below. All application material including the GRE scores must be received by March 1. Late application for graduate assistantships will be considered only if there are vacancies.

Correspondence and Information

Graduate Committee
Department of Mathematics
202 Ross Hall
University of Wyoming
Laramie, Wyoming 82071
Telephone: 307-766-4221

SECTION 5: MATHEMATICAL SCIENCES

University of Wyoming

THE FACULTY AND THEIR RESEARCH

Myron Allen, Ph.D., Princeton. Numerical analysis, partial differential equations.
Leonard Asimow, Ph.D., Washington (Seattle). Functional analysis, optimization.
William Bridges, Ph.D., Caltech. Combinatorics, matrix theory.
Robert Buschman, Ph.D., Colorado. Special functions, integral transforms.
Benito Chen, Ph.D., Caltech. Differential equations, applied mathematics.
Shue-Sum Chow, Ph.D., Australian National. Numerical analysis, partial differential equations.
Richard Ewing, Ph.D., Texas. Numerical analysis, partial differential equations.
George Gastl, Ph.D., Wisconsin. Computer languages, topology.
John George, Ph.D., Alabama. Mathematical modeling, stability theory.
Sylvia Hobart, Ph.D., Michigan. Combinatorics, algebra.
Syed Husain, Ph.D., Purdue. Functional analysis, Fourier analysis.
Lynne Ipiña, Ph.D., NYU. Fluid dynamics, partial differential equations.
Eli Isaacson, Ph.D., NYU. Differential equations, numerical analysis.
Farhad Jafari, Ph.D., Wisconsin. Functional analysis, harmonic analysis.
Terry Jenkins, Ph.D., Nebraska. Ring theory, radical theory.
Raytcho Lazarov, Ph.D., Moscow. Numerical analysis, partial differential equations.
Joseph Martin, Ph.D., Iowa. Geometric topology.
Eric Moorhouse, Ph.D., Toronto. Combinatorics.
A. Duane Porter, Ph.D., Colorado. Matrix theory, field theory.
Ben Roth, Ph.D., Dartmouth. Analysis, rigidity theory.
John Rowland, Ph.D., Penn State. Numerical analysis, approximation theory.
Virindra Sehgal, Ph.D., Wayne State. Functional analysis, fixed point theory.
Bryan Shader, Ph.D., Wisconsin. Combinatorics.
Leslie Shader, Ph.D., Colorado. Algebra, combinatorics.
Shagi-Di Shih, Ph.D., Maryland. Singular perturbations.
Raymond Smithson, Ph.D., Oregon. Topology, relation theory.
Junping Wang, Ph.D., Chicago. Numerical analysis, partial differential equations.

SECTION 5: MATHEMATICAL SCIENCES

UTAH STATE UNIVERSITY
Department of Mathematics and Statistics

Programs of Study   The Department of Mathematics and Statistics offers four graduate degree programs: the Ph.D. in mathematical sciences, the M.S. in mathematics, the M.S. in applied statistics, and the Master of Mathematics.

Each master's degree program requires approximately two years of graduate work. The Master of Mathematics is a nonthesis program intended primarily for the classroom teacher at the secondary school and junior college levels, and a secondary school teaching certificate is an additional requirement for this degree. The M.S. programs offer both thesis and nonthesis options with considerable flexibility in course selection, thus allowing both pure and applied emphases within the same framework. All master's degrees require that the student successfully complete a written qualifying examination.

The Ph.D. in mathematical sciences program offers three directions for graduate study: the Pure and Applied Mathematics Option, the Interdisciplinary Option, and the College Teaching Option. The Pure and Applied Mathematics Option is a traditional doctoral program in mathematics offering broad training in modern mathematics and requiring a dissertation that represents a significant and original contribution to mathematics research in a chosen area of specialization. The Interdisciplinary Option combines advanced training in mathematics, advanced training in another chosen area (e.g., physics, engineering, economics), and a dissertation involving advanced mathematics and representing significant original research in the interdisciplinary area. The College Teaching Option is designed to prepare students to teach mathematics in two- and four-year colleges and universities. This option emphasizes broad training in graduate-level mathematics, requires a teaching internship, and allows for an expository dissertation. All three options require successful completion of comprehensive examinations and one-year college-level proficiency in a foreign language.

Research Facilities   The library subscribes to most major research journals in mathematics. Other books and journals are available through interlibrary loan. Computing facilities are good and are readily available.

Financial Aid   Teaching fellowships and assistantships are available to qualified students. Teaching fellowships are for doctoral-level students who already have a master's degree; the fellowships pay $10,000 per year; teaching assistantships are for master's-level students and pay $9000 per year. Duties for both amount to full responsibility for teaching one course per quarter. In addition, several foreign students are supported on graduate assistantships paying $5000–$6000 per year with paper-grading duties. All fellowships and assistantships carry a waiver of nonresident tuition.

Cost of Study   For the fall quarter of 1991, state-resident tuition and registration fees for most graduate mathematics students vary from $270 to $405 per quarter, depending on the course load. Student insurance for a single student is $91 per quarter. Tuition and fees are subject to change for any quarter.

Cost of Living   A variety of housing is available. On-campus room and board for a single student are about $860 per quarter. The University offers married student housing, the cost of which (including utilities) ranges between $215 and $300 per month, depending on the number of bedrooms, kind of furnishings, and so on. There is a six-month waiting list for the University's married student housing. Roughly comparable privately owned housing is available near campus.

Student Group   In 1990–91, the University's total enrollment was 13,719. There were 2,538 graduate students. The department had 40 graduate students, most of whom received financial support.

Location   Surrounded by the Wasatch Range of the Rocky Mountains, Cache Valley is the home of Utah State University and provides a spectacular setting. Logan, the center of Cache Valley activity, is the fifth-largest city in Utah, with a population of 32,000.

The outdoor recreation activities for which opportunities are provided by the beautiful mountain setting include hiking, camping, backpacking, downhill skiing, cross-country skiing, waterskiing, hunting, fishing, snowmobiling, and horseback riding. Proceeding east up lovely Logan Canyon, one comes to the Cache National Forest boundary within 1 mile of Logan, arrives at Beaver Mountain Ski Resort within 28 miles, and reaches beautiful Bear Lake within 40 miles.

The University   Utah State University, founded in 1888 as a land-grant institution, has grown into a major modern university that offers degree programs through eight colleges and forty-five departments. Some of the programs and facilities of interest on campus are the Exceptional Child Center, the Water Research Laboratory, the Electro-Dynamics Laboratory, the Center for Atmospheric and Space Studies, and the Chase Fine Arts Center, which includes theaters, a concert hall, and an arts museum.

Applying   Admission to the School of Graduate Studies requires a bachelor's degree or equivalent and demonstrated aptitude in mathematics or a related field. The GRE General Test is required of all applicants, and the TOEFL is required of all international students. Applications are accepted throughout the year. Students applying for financial aid should do so as early as possible and no later than April 1 when applying for the following academic year.

Correspondence and Information

For program information:
Graduate Chairman
Department of Mathematics and Statistics
Utah State University
Logan, Utah 84322-3900
Telephone: 801-750-2809

For application information:
Dean
School of Graduate Studies
Utah State University
Logan, Utah 84322-0900
Telephone: 801-750-1189

SECTION 5: MATHEMATICAL SCIENCES

Utah State University

THE FACULTY AND THEIR RESEARCH

Professors
I. M. Anderson, Ph.D., Arizona, 1976. Differential geometry, global analysis.
L. B. Beasley, Ph.D., British Columbia, 1969. Matrix theory, linear algebra, group theory, combinatorics.
R. V. Canfield, Ph.D., Wyoming, 1975. Statistics, extreme value theory.
L. O. Cannon, Ph.D., Utah, 1965. Topology, algebra.
L. D. Loveland, Ph.D., Utah, 1965. Geometric topology.
J. R. Ridenhour, Ph.D., Arizona State, 1971. Differential equations.
D. V. Sisson, Ph.D., Iowa State, 1962. Statistics.
R. C. Thompson, Ph.D., Utah, 1973. Differential equations.
H. F. Walker, Ph.D., NYU, 1970. Applied mathematics, partial differential equations, statistical pattern recognition.
M. P. Windham, Ph.D., Rice, 1970. Numerical cluster analysis and classification.

Associate Professors
A. H. Bringhurst, M.S., Utah State, 1965. Mathematics education.
C. S. Coray, Ph.D., Utah, 1973. Numerical analysis.
E. R. Heal, Ph.D., Utah, 1971. Analysis, statistics.
L. L. Littlejohn, Ph.D., Penn State, 1981. Differential equations, special functions.
R. Schaaf, Ph.D., Heidelberg, 1981. Partial differential equations.
D. L. Turner, Ph.D., Colorado State, 1975. Statistics, linear models.
E. E. Underwood, M.A., Illinois, 1961. Matrix theory, linear algebra.
S. C. Williams, Ph.D., North Texas State, 1983. Measure theory, analysis.

Assistant Professors
D. Coster, Ph.D., Berkeley, 1986. Optimal design, computational statistics.
A. Cutler, Ph.D., Berkeley, 1988. Statistics, statistical computing.
D. R. Cutler, Ph.D., Berkeley, 1988. Statistics, experimental design.
K. Hestir, Ph.D., Berkeley, 1986. Probability theory, stochastic processes.
J. V. Koebbe, Ph.D., Wyoming, 1988. Applied mathematics, computational fluid dynamics.
J. Powell, Ph.D., Arizona, 1990. Applied mathematics.
K. Turner, Ph.D., Rice, 1987. Numerical analysis.
Z. Wang, Ph.D., Academia Sinica (China), 1986. Partial differential equations.

Aerial view of campus, looking east up Logan Canyon.

Eastern view from Taggart Student Center patio.

Western view of Old Main Building with Wellsville Mountains in background.

SECTION 5: MATHEMATICS

VANDERBILT UNIVERSITY
Department of Mathematics

Programs of Study

The department offers the Master of Arts, Master of Science, and Doctor of Philosophy degrees in mathematics. The Master of Arts in Teaching is also available.

The M.A. and M.S. degrees require completion of 24 hours of course work and a thesis or completion of 36 hours of course work. In the 36-hour program, up to 12 hours may be taken in related fields such as computer science, economics, or physics. The 36-hour degree program is flexible and is particularly suited to meet the needs of students preparing for careers in industry, actuarial work, and government. The M.A.T. degree requires 36 hours of course work, including at least 18 hours in mathematics. Students take the professional education courses necessary for certification in the state in which they wish to teach.

After completing the core curriculum, Ph.D. candidates begin in-depth study in their area of concentration, undertaking seminar work and writing a dissertation. The qualifying process has been specifically designed to accelerate progress toward research. In addition to an examination in the area of competence, candidates write an expository paper of some magnitude that usually leads to a dissertation topic. Candidates for the Ph.D. must demonstrate a reading knowledge of French, German, or Russian.

With a faculty-student ratio greater than 1:1, graduate students receive a great deal of individual attention. Research opportunities are available in many areas of algebra, analysis, applied mathematics, differential equations, graph theory, and topology.

Research Facilities

The Jean and Alexander Heard Library is one of the important research libraries in the South, with more than 1.8 million volumes in eight divisions. It contains over 17,000 mathematics volumes and approximately 250 mathematics journals.

The University's computer facilities include two VAX 8800s, an IBM 4361, and multiple minicomputers and microcomputers. The department has its own DECsystem 3100 computer running the Ultrix operating system, which is freely available for use by graduate students. Graduate students also have access to numerous microcomputers.

Financial Aid

Most graduate students receive financial support in the form of teaching assistantships, which currently provide a stipend of $10,000 for nine months plus a tuition waiver. Some applicants are awarded fellowships with a stipend over and above the regular departmental award: the Harold Stirling Vanderbilt Scholarships and the Graduate Select Scholarships currently provide additional stipends of $3000 for a total of $13,000 for nine months. Applicants may also receive University Graduate Fellowships, which currently carry a stipend of $12,000 for nine months plus a tuition waiver and are free of duties. A student holding one of these awards may also be awarded a Colowick Scholarship paying an additional $2000. Through a grant from the Danforth Foundation, the Graduate School also awards several Compton Fellowships (stipends of $11,000) to black graduate students who plan teaching and research careers in arts and science disciplines. All stipends are expected to increase for 1992–93.

Cost of Study

Virtually all graduate students in mathematics are supported by assistantships or fellowships, which include tuition waivers. For others, tuition is $624 per semester hour.

Cost of Living

Although ample privately owned rental accommodations are available within walking distance of the campus, graduate students may apply for housing in University apartments. Accommodations for single and married students range from $228 to $635 per month. The University maintains several cafeterias on campus. Health service benefits and major medical insurance coverage for the student are provided.

Student Group

University enrollment is about 9,000, including 1,300 students in the Graduate School. There are 31 graduate students in mathematics, including 13 women. Almost all of the mathematics graduate students are pursuing the Ph.D. degree.

Location

Vanderbilt is near the center of Nashville, a metropolitan area of half a million people in the heart of middle Tennessee. The rolling hills and waterways of the region have made Nashville a mecca for lovers of the outdoors. Nashville is the cultural, commercial, and financial center of the mid-South.

The University

Vanderbilt is an independent, privately supported university with a fine liberal arts college and a full range of graduate and professional programs. Founded in 1873, the University now includes the College of Arts and Science, the Graduate School, and eight professional schools. The University is a member of the Association of American Universities and is accredited by the Southern Association of Colleges and Schools.

Applying

Applicants should submit complete credentials to the Graduate School prior to January 15 to assure themselves of consideration. Late applications will be considered if positions are still available. Prospective applicants are asked to note that GRE scores are required for all applicants, and TOEFL scores are required for all applicants from outside the United States whose native language is not English.

Correspondence and Information

Matthew Gould
Director of Graduate Studies
Department of Mathematics
Vanderbilt University
Nashville, Tennessee 37235
Telephone: 615-322-6659

Peterson's Guide to Graduate Programs in the Physical Sciences and Mathematics 1992

Vanderbilt University

THE FACULTY AND THEIR RESEARCH

John F. Ahner, Associate Professor; Ph.D., Delaware, 1972. Integral equation methods in scattering and potential theory, fractional calculus and mixed boundary value problems, elasticity theory.
Richard F. Arenstorf, Professor; Ph.D., Mainz (Germany), 1956. Analytic number theory, automorphic functions, zeta functions, celestial mechanics, Hamiltonian dynamical systems.
Philip S. Crooke, Professor; Ph.D., Cornell, 1970. Applied mathematics: mathematical modeling, differential equations, mathematical modeling in medicine.
Mark N. Ellingham, Assistant Professor; Ph.D., Waterloo, 1986. Graph theory, including reconstruction problems, coloring problems, and existence problems for paths and cycles.
Richard R. Goldberg, Professor; Ph.D., Harvard, 1956. Harmonic analysis, integral transforms.
Matthew Gould, Professor; Ph.D., Penn State, 1967. Universal algebra, semigroup theory, lattice theory.
Douglas P. Hardin, Assistant Professor; Georgia Tech, 1985. Fractal geometry, ergodic theory, dynamical systems, mathematical biology.
Robert L. Hemminger, Professor; Ph.D., Michigan State, 1963. Graph theory and combinatorics.
C. Bruce Hughes, Associate Professor; Ph.D., Kentucky, 1981. Geometric and algebraic topology, manifold theory, controlled topology, stratified spaces.
Bjarni Jònsson, Distinguished Professor of Mathematics; Ph.D., Berkeley, 1946. Universal algebra and lattice theory.
Charles Kahane, Professor; Ph.D., NYU, 1962. Integral equations, asymptotic behavior of reaction diffusion systems.
Keith A. Kearnes, Assistant Professor; Ph.D., Berkeley, 1988. Universal algebra.
John A. Kelingos, Associate Professor; Ph.D., Michigan, 1963. Boundary behavior of analytic and harmonic functions with Schwartz and Beurling distributional boundary values.
Denise E. Kirschner, Assistant Professor; Ph.D., Tulane, 1991. Analysis.
Richard J. Larsen, Associate Professor; Ph.D., Rutgers, New Brunswick, 1970. Probability and statistics.
Peter B. Massopust, Assistant Professor; Ph.D., Georgia Tech, 1986. Applied mathematics.
Charles K. Megibben, Professor; Ph.D., Auburn, 1963. Abelian groups, set-theoretical algebra, theory of rings.
Michael L. Mihalik, Associate Professor; Ph.D., SUNY at Binghamton, 1979. Algebraic topology, low dimensional topology, geometric group theory.
Michael D. Plummer, Professor; Ph.D., Michigan, 1966. Graph theory and combinatorics.
Efstratios Prassidis, Assistant Professor; Notre Dame, 1989. High dimensional topology.
John G. Ratcliffe, Associate Professor; Ph.D., Michigan, 1977. Combinatorial group theory, three-dimensional manifolds, hyperbolic geometry.
Eric Schechter, Associate Professor; Ph.D., Chicago, 1978. Nonlinear initial value problems, functional analysis, models of set theory.
Larry L. Schumaker, Stevenson Professor of Mathematics; Ph.D., Stanford, 1966. Approximation theory, spline theory, computer-aided design.
Peter Takác, Assistant Professor, Ph.D., Minnesota, Twin Cities, 1986. Systems of reaction-diffusion equations, singular perturbation problems, strongly monotonic dynamical systems.
Steven T. Tschantz, Assistant Professor; Ph.D., Berkeley, 1983. Logic, universal algebra, geometric group theory, computer algebra.
Constantine Tsinakis, Associate Professor; Ph.D., Berkeley, 1979. Lattice theory, universal algebra, theoretical computer science.
Patrick J. Van Fleet, Assistant Professor; Ph.D., Southern Illinois, 1991. Approximation theory.
Glenn F. Webb, Professor and Chairman; Ph.D., Emory, 1968. Mathematical biology, population dynamics, models of tumor growth, differential equations.
Horace E. Williams, Professor; Ph.D., George Peabody, 1962. Elementary mathematics education.
Daoxing Xia, Professor; Ph.D., Jijian (China), 1952. Operator theory and its application.

Downtown Nashville viewed from Vanderbilt campus.

The Mathematics Building at Vanderbilt University.

SECTION 5: MATHEMATICAL SCIENCES

VIRGINIA POLYTECHNIC INSTITUTE AND STATE UNIVERSITY
Department of Mathematics

Programs of Study

The department offers programs of study in pure and applied mathematics and in mathematical physics leading to the M.S. and Ph.D. degrees. These programs prepare students for careers in teaching, research, industry, and government service.

The M.S. degree requires 30 semester hours of graduate credit and successful performance on a comprehensive final exam. Candidates for the degree may choose the thesis, the nonthesis, or the interdisciplinary option, depending on their interests and career objectives. Students on a teaching assistantship normally complete the degree in two years.

The Ph.D. degree program emphasizes a strong foundation in fundamental areas of mathematics as well as specialized study in selected research areas. Requirements include at least 27 semester hours of upper-level graduate course work (in addition to the prerequisite lower-level courses), a thesis, proficiency in one foreign language approved by the student's committee, successful performance on three preliminary exams, and a final oral exam and thesis defense. Students generally need an additional two to three years beyond the M.S. to earn the Ph.D.

Research Facilities

The University's Carol M. Newman Library has extensive holdings in all areas of mathematics and related scientific fields. The department has a computer network consisting of UNIX systems, a VMS VAXcluster, and personal computers. The UNIX systems include a Sun SPARCstation 2GX, a Sun-3/80, a VAX 3800, a VAXstation 2000, a DECstation 2100, and a Silicon Graphics 4D25TG. The VMS VAXcluster includes a VAX 3600, a VAX 3200, and several VAXstations. The department's network is part of a University network with access to an IBM 3090-300E/VF and 3084 complex running VM/CMS/MVS/AIX and a VAX 8800 running VMS. The University's network also provides access to Internet and BITNET.

The Interdisciplinary Center for Applied Mathematics (ICAM) promotes and facilitates interdisciplinary research and education in applied mathematics at Virginia Tech. ICAM provides a wide range of research and educational programs emphasizing interaction between engineers, mathematicians, and scientists.

Financial Aid

The University offers graduate teaching and research assistantships and graduate project assistantships (for more advanced students working on specific research projects). For the 1991–92 academic year, graduate teaching and research assistantships offer monthly stipends ranging from $1113 to $1181. Graduate project assistantships pay monthly stipends of $1441 to $1509. Full and partial tuition waivers for exceptionally well qualified students are also available. Continuing students often receive aid for the summer months.

Cost of Study

The 1991–92 tuition and fees for full-time graduate students are $1889 per semester.

Cost of Living

On-campus housing for graduate students is very limited. However, there are numerous apartment complexes, town houses, and rooms for rent within walking or biking distance of campus, and a town bus system provides transportation directly to campus. (The cost of this service is included in the student fees.)

Student Group

The mathematics department has approximately 90 graduate students, of whom one third are women. Most are from out of state; about 40 percent are international students. About 80 percent have assistantships, and about 60 percent are in the Ph.D. program. Graduates at both the M.S. and Ph.D. levels have been very successful in obtaining teaching and research positions in colleges and universities and positions in business, industry, and the government.

Location

Virginia Tech is located in Blacksburg, a town of about 30,000, situated on a plateau between the Blue Ridge and Allegheny mountains in southwest Virginia. The area is noted for its beauty and high quality of life. Even though the climate is generally mild the year round, the region enjoys four seasonal changes. Opportunities for outdoor activities such as hiking, canoeing, bicycling, and skiing abound. The city of Roanoke (population 100,000) is only 40 miles away by interstate highway.

The University and The Department

Virginia Tech is Virginia's senior land-grant university. It has an enrollment of over 22,000, including almost 4,000 graduate students. The University has been known for many years as a center of science and engineering but more recently has emerged as an internationally recognized comprehensive institution of higher learning. The mathematics department, in particular, is increasingly gaining recognition for the high quality of its research and graduate programs. A poll published in the April 1983 issue of *Notices of the American Mathematical Society* rated the department's graduate programs within the top 14 (out of 114 programs) in the country in terms of improvement over a five-year period.

Applying

Application forms for admission and financial aid are supplied upon request by the mathematics department and the Graduate School Admissions Office. Applicants are encouraged to take the GRE General Test and Subject Test in mathematics, and those whose native language is not English must submit TOEFL scores. While applications are accepted at any time during the year, decisions on admission and financial aid are customarily made in January and February for the following fall term. Consequently, completed applications should reach the department by February 1 to receive the fullest consideration.

Correspondence and Information

Chairman, Graduate Admissions Committee
Department of Mathematics
Virginia Polytechnic Institute and State University
Blacksburg, Virginia 24061
Telephone: 703-231-6536

SECTION 5: MATHEMATICAL SCIENCES

Virginia Polytechnic Institute and State University

THE FACULTY AND THEIR RESEARCH

University Distinguished Professor
F. S. Quinn, Ph.D., Princeton. Topology of manifolds.

Professors
J. T. Arnold, Ph.D., Florida State. Commutative rings.
C. E. Aull, Ph.D., Colorado. General topology.
J. A. Ball, Ph.D., Virginia. Operator theory, systems theory.
M. B. Boisen, Ph.D., Nebraska. Mathematical crystallography, commutative rings.
E. A. Brown, Ph.D., LSU. Number theory, history of mathematics.
J. A. Burns, Ph.D., Oklahoma. Applied mathematics, control theory.
D. R. Farkas, Ph.D., Chicago. Ring theory.
C. D. Feustel, Ph.D., Dartmouth. Algorithms.
P. Fletcher, Ph.D., North Carolina. General topology.
E. L. Green, Ph.D., Brandeis. Representation theory of rings and algebras.
W. Greenberg, Ph.D., Harvard. Mathematical physics, statistical mechanics.
M. D. Gunzburger, Ph.D., NYU. Numerical analysis, computational mechanics.
G. A. Hagedorn, Ph.D., Princeton. Mathematical physics.
K. D. Hannsgen, Ph.D., Wisconsin. Volterra equations, control theory.
T. L. Herdman, Ph.D., Oklahoma. Applied mathematics, functional differential equations.
J. R. Holub, Ph.D., LSU. Functional analysis.
L. W. Johnson, Ph.D., Michigan State. Numerical analysis.
M. Klaus, Ph.D., Zurich. Mathematical physics.
W. E. Kohler, Ph.D., RPI. Applied mathematics.
R. A. McCoy, Ph.D., Iowa State. General topology.
R. F. Olin, Ph.D., Indiana. Operator theory, functional analysis.
C. J. Parry, Ph.D., Michigan State. Number theory.
C. W. Patty, Ph.D., Georgia. Topology.
C. L. Prather, Ph.D., Northwestern. Complex analysis.
M. Renardy, Dr.rer.nat., Stuttgart. Nonlinear partial differential equations, fluid mechanics.
R. D. Riess, Ph.D., Iowa State. Numerical analysis.
D. L. Russell, Ph.D., Minnesota. Ordinary differential equations, partial differential equations, systems theory.
J. K. Shaw, Ph.D., Kentucky. Complex analysis, differential equations.
R. L. Snider, Ph.D., Miami (Florida). Ring theory.
J. E. Thomson, Ph.D., North Carolina. Operator theory.
R. L. Wheeler, Ph.D., Wisconsin. Integrodifferential equations, control theory.

Associate Professors
C. A. Beattie, Ph.D., Johns Hopkins. Functional analysis and numerical analysis.
R. S. Crittenden, Ph.D., North Carolina. Ring theory.
M. V. Day, Ph.D., Colorado. Stochastic processes, probability.
W. J. Floyd, Ph.D., Princeton. Topology and geometric group theory.
H. L. Johnson, Ph.D., Minnesota. Applied mathematics, partial differential equations.
J. U. Kim, Ph.D., Brown. Nonlinear partial differential equations.
J. W. Layman, Ph.D., Virginia. Applied mathematics, interpolation theory.
P. A. Linnell, Ph.D., Cambridge. Group rings.
M. A. Murray, Ph.D., Yale. Harmonic analysis.
C. C. Oehring, Ph.D., Tennessee. Analysis.
J. Peterson, Ph.D., Tennessee. Numerical analysis.
B. E. Reed, Ph.D., Georgia. Nonassociative algebras.
Y. Renardy, Ph.D., Western Australia. Fluid mechanics.
J. F. Rossi, Ph.D., Hawaii. Complex analysis.
J. E. Shockley, Ph.D., North Carolina. Number theory.
J. C. Smith, Ph.D., Duke. General topology.

Assistant Professors
P. E. Haskell, Ph.D., Brown. Index theory.
T. Lin, Ph.D., Wyoming. Numerical analysis, large-scale numerical simulation.
R. C. Rogers, Ph.D., Maryland. Partial differential equations, continuum mechanics.
J. K. Washenberger, Ph.D., Iowa State. Analysis, functional analysis.

SECTION 5: MATHEMATICAL SCIENCES

VIRGINIA POLYTECHNIC INSTITUTE AND STATE UNIVERSITY
Department of Statistics

Programs of Study

The Department of Statistics offers thesis and nonthesis Master of Science degrees and a Doctor of Philosophy degree. Both M.S. programs require 30 semester hours of credit. The nonthesis program requires 26 semester hours of course work within the department. In the thesis program there is no fixed departmental credit hour requirement, and the thesis may count for 6 to 10 semester hours of credit. The master's degree can be obtained in sixteen months of graduate study. First-year core courses for both programs include Probability and Distribution Theory, Statistical Inference, Linear Models Theory, Applied Statistics, and Experimental Design and Analysis. Additional courses may be taken in statistics and mathematics or in approved areas of application. Each student participates in statistical consulting activities for at least one semester. For the Ph.D. program, students must complete a minimum of 90 credit hours of graduate study, including at least 54 semester hours of course work. Core courses (beyond the first-year core courses) are Measure and Probability and Advanced Statistical Inference. Students are expected to complete at least two semesters of courses in two areas of concentration, chosen in conjunction with the Advisory Committee. A field of application may be selected in place of one area of concentration. Each student participates for three semesters in specialized professional training in statistical consulting and/or teaching. Students with no previous graduate training in statistics can expect to complete the Ph.D. program in four to five years.

Research Facilities

The department has excellent facilities for classwork, consulting, and research. The University's modern computer center has an IBM 3084 processor complex, an IBM 3090 supercomputer-grade processor complex, and excellent supporting hardware. The extensive software collection includes BMDP, SPSSX, MINITAB, MPS, and SAS statistical packages. The departmental computing laboratory is equipped with terminals to the University Computing Center and a number of microcomputers. The University library has an extensive collection of statistical publications, including complete files of most statistical journals. The department maintains a small library containing reference books, recent issues of major journals, and a large reprint file. The department operates the University's Statistical Consulting Center, through which the faculty and students work with members of other departments in various research activities.

Financial Aid

Graduate assistantships, with stipends for the 1991–92 academic year of $1000 to $1078 per month for nine months, are available to highly qualified applicants. Responsibilities include 20 hours per week of grading, consulting, teaching, and/or special assignments, and graduate assistants must carry between 9 and 12 credits per semester. Instructional fee scholarships are provided for exceptional students, and some financial aid is available in the summer.

Cost of Study

The 1991–92 instructional fee (tuition) for full-time graduate students is $1665 per semester for Virginia residents and for nonresidents who hold graduate assistantships paying more than $2000 per year. Nonresidents who do not hold such assistantships pay an additional $738 per semester. All students pay a comprehensive fee of $224 per semester that includes health services, student activities, athletics, and bus service.

Cost of Living

Privately owned housing (both rooms and apartments) is available in Blacksburg and the surrounding area.

Student Group

The current graduate enrollment in the department is approximately 50. Students have previously studied at forty different universities in nine countries. Graduates of the department have never experienced difficulty in obtaining excellent academic, industrial, and government positions. Salaries for professional statisticians are appreciably higher than those offered after similar training in most other fields of science.

Location

The University is located in Blacksburg, a town of almost 30,000 people in scenic southwestern Virginia. The 2,300-acre campus lies on a plain between the Blue Ridge and Allegheny mountains. The area is noted for its beauty and recreational opportunities. Boating, swimming, camping, and fishing facilities are available at nearby Claytor Lake State Park. Also nearby are Mountain Lake, the Blue Ridge Parkway, the Roanoke and Shenandoah valleys, Jefferson National Forest, and the Appalachian Trail. Roanoke, 38 miles to the east, is easily reached via four-lane highways. Commercial air service is provided through Roanoke. There is direct bus service to Blacksburg.

The University and The Department

Founded in 1872, Virginia Tech, with approximately 23,000 students, now has the largest resident enrollment in the state. Master's degree programs are offered in sixty fields and Ph.D. programs in forty.

The Department of Statistics is one of the oldest in the nation. The Statistical Laboratory was organized in 1948, and the department was established in the following year. The department's reputation for significant research in modern statistical theory and methodology is supported by an impressive list of publications and research grants. With a faculty–graduate student ratio of approximately 1:2, there are ample opportunities for individual attention. An undergraduate curriculum is also offered.

Applying

Admission to the Graduate School normally requires a minimum cumulative grade point average of 2.75 (on a 4.0 scale) for the equivalent of the last two years of undergraduate study. Exceptions may be made upon recommendation of the department; substantial evidence of ability to succeed in graduate work must be presented. Prospective applicants are urged to take the GRE General Test. The department encourages applications from students in fields other than mathematics and statistics, although mathematical training at least through advanced calculus and matrix algebra is desirable. Matrix algebra may be taken after enrollment, but advanced calculus should be completed before the start of the first academic year. Applications for admission should be forwarded to the Graduate School office. A complete application consists of an application form, two official and up-to-date transcripts of the student's undergraduate and graduate records, and at least three letters of recommendation, usually from former professors. A $10 application fee is also required. Students desiring to apply for financial aid should so indicate to the dean of the Graduate School when applying for admission. Applications for financial aid should be received before January 31; awards are announced in mid-March. Later applications will be considered if vacancies occur or new positions are made available.

Correspondence and Information

For application forms and a graduate catalog:
Graduate School
Virginia Polytechnic Institute
  and State University
Blacksburg, Virginia 24061-0325
Telephone: 703-231-6691

For more information about the department:
Dr. Golde I. Holtzman, Graduate Program
  Administrator
Department of Statistics
Virginia Polytechnic Institute
  and State University
Blacksburg, Virginia 24061-0439
Telephone: 703-231-5630

Peterson's Guide to Graduate Programs in the Physical Sciences and Mathematics 1992

Virginia Polytechnic Institute and State University

THE FACULTY AND THEIR RESEARCH

Jesse C. Arnold, Professor; Ph.D., Florida State, 1967. Teaching interests: nonparametric statistics, inference, statistics for biological and health sciences. Research areas: estimation, sampling, epidemiology, nutrition. Other activities: consulting.

Jeffrey B. Birch, Associate Professor; Ph.D., Washington (Seattle), 1977. Teaching interests: statistical methods, regression, bioassay, exploratory and robust analysis. Research areas: robust procedures, Monte Carlo methods, regression analysis. Other activities: consulting in biostatistics; associate editor, *Biometrics*.

Clint W. Coakley, Assistant Professor; Ph.D., Penn State, 1991. Teaching interests: nonparametric methods. Research areas: nonparametric methods, robust regression.

Whitfield Cobb, Associate Professor Emeritus; Ph.D., North Carolina, 1959.

Robert V. Foutz, Professor; Ph.D., Ohio State, 1974. Teaching interests: time series, statistical inference. Research areas: time series, large sample theory, statistical inference. Other activities: consulting.

I. J. Good, University Distinguished Professor of Statistics and Adjunct Professor of Philosophy; Ph.D., Cambridge, 1941. Research areas: foundations and applications of probability and statistics, including probability estimation for both continuous and categorical data. Other activities: contributing editor, *Journal of Statistical Computation and Simulation* and *Journal of Statistical Planning and Inference*.

Boyd Harshbarger, Professor Emeritus; Ph.D., George Washington, 1943. Archivist for the Virginia Academy of Science.

Klaus H. Hinkelmann, Professor and Department Head; Ph.D., Iowa State, 1963. Teaching interests: experimental design, linear models, analysis of variance, genetic statistics, statistical methods. Research areas: experimental design, genetic statistics. Other activities: consulting; interdepartmental genetics program; editor, *Biometrics*.

Golde I. Holtzman, Associate Professor; Ph.D., North Carolina State, 1980. Teaching interests: biostatistics. Research areas: biomathematics, mathematical ecology, population dynamics. Other activities: Graduate Program Administrator.

Donald R. Jensen, Professor; Ph.D., Iowa State, 1962. Teaching interests: probability and distribution theory, mathematical statistics, multivariate analysis. Research areas: multivariate analysis, large sample theory, simultaneous inference, process control.

Richard G. Krutchkoff, Professor; Ph.D., Columbia, 1964. Teaching interests: biometry, inference, stochastic simulation, decision theory, stochastic modeling. Research areas: environmental statistics, stochastic simulation, Bayesian decision theory. Other activities: editor, *Journal of Statistical Computation and Simulation*; consultant on environmental problems.

Marvin Lentner, Professor; Ph.D., Kansas State, 1967. Teaching interests: statistical methods, experimental design. Research areas: computer applications. Other activities: consultant, College of Agriculture and Life Sciences.

Jerry E. Mann, Associate Professor; Ph.D., Southwestern Louisiana, 1973. Teaching interest: theoretical statistics. Research areas: modeling, sequential decision theory. Other activities: consulting.

Ann M. McGuirk, Assistant Professor of Agricultural Economics and Statistics; Ph.D., Cornell, 1988. Teaching interests: econometrics.

Raymond H. Myers, Professor; Ph.D., Virginia Tech, 1963. Teaching interests: linear models, response surface analysis, experimental design, regression, engineering statistics. Research areas: experimental design and analysis; response surface analysis. Other activities: associate editor, *Technometrics*; director, Statistical Consulting Center.

Panickos N. Palettas, Assistant Professor; Ph.D., Ohio State, 1988. Teaching interests: statistical computing. Research areas: reliability theory, statistical computing.

Walter R. Pirie, Associate Professor; Ph.D., Florida State, 1970. Teaching interests: nonparametric theory and methods, limit theory of statistics, probability and mathematical statistics, log-linear models. Research areas: nonparametrics, limit theory, estimation. Other activities: statistical consulting for on-campus research, Undergraduate Coordinator; associate editor, *STATS*.

Marion R. Reynolds Jr., Professor of Statistics and Forestry; Ph.D., Stanford 1972. Teaching interests: sequential analysis, quality control, statistical inference, probability and distribution theory, engineering statistics. Research areas: sequential analysis, statistical process control, mathematical modeling in natural resource problems. Other activities: consultant, Department of Forestry and Wildlife.

Robert S. Schulman, Associate Professor; Ph.D., North Carolina at Chapel Hill, 1974. Teaching interests: statistics for social sciences, applied statistics. Research areas: test theory, psychometric methods. Other activities: consultant for research in social sciences.

Eric P. Smith, Associate Professor; Ph.D., Washington (Seattle), 1982. Teaching interests: statistical methods, biometry, multivariate methods. Research areas: multivariate analysis, multivariate graphics, biological sampling, modeling. Other activities: consultant for Anaerobe Laboratory and Center for Environmental Studies.

George R. Terrell, Assistant Professor; Ph.D., Rice, 1978. Teaching interests: mathematical statistics, probability, statistical computing. Research areas: nonparametric density estimation, multivariate nonparametric methods, projection pursuit methods. Other activities: consulting.

Keying Ye, Assistant Professor; Ph.D., Purdue, 1990. Teaching interests: mathematical statistics, Bayesian statistics. Research interests: Bayesian inference, statistical decision theory, sequential analysis.

C. C. Garvin Visiting Endowed Professors 1991–92

Nozer D. Singpurwalla; Professor, George Washington Unversity; Ph.D., NYU; 1968.

Donald Michie; Chief Scientist, Turing Institute; D.Phil., 1953, D.Sc., 1971, Oxford.

Selected Publications

Reynolds, M., R. Amin, and J. C. Arnold. Cusum charts with variable sampling intervals (with discussion). *Technometrics* 32:371–96, 1990.

Agard, D. B., and J. B. Birch. Robust inferential procedures applied to regression. *Proc. Stat. Comp. Section*, ASA, 1990.

Koons, B. K., and R. V. Foutz. Estimating moving average parameters in the presence of measurement error. *Comm. Stat. Theory Meth.* 19:3179–87, 1990.

Good, I. J. Speculations concerning the future of statistics. *J. Stat. Plan. Inference* 25:441–66, 1990.

Singh, M., and K. H. Hinkelmann. On generation of efficient partial diallel crosses plans. *Biom. J.* 32:177–87, 1990.

Jensen, D. R., and D. E. Ramirez. Dispersion-diminishing transformations. *Comm. Stat. Theory Meth.* 19:3259–66, 1990.

Krutchkoff, R. G. Comparisons using simulation. *J. Stat. Comp. Simul.* 36:47–48, 1990.

Vining, G. G., and R. H. Myers. Combining Taguchi and response surface philosophies: A dual response problem. *J. Qual. Techn.* 22:38–45, 1990.

Palettas, P. N. Validity assessment of predictive models. *Proc. Stat. Graphics Section*, ASA, 1990.

Marx, B. D., and E. P. Smith. Principal component regression for generalized linear models. *Biometrika* 77:23–31, 1990.

Terrell, G. R. The maximal smoothing principle in density estimation. *J. Amer. Stat. Assoc.* 85:470–77, 1990.

Recent Dissertations

"Robust Inference Procedures Applied to Regression," David B. Agard (1990).

"X Control Charts in the Presence of Correlation," Jai W. Baik (1991).

"An Examination of Outliers and Interaction in a Nonreplicated Two-way Table," Barbara R. Kuzmak (1990).

"Analysis of Multispecies Microcosm Experiments," Donald E. Mercante (1990).

"Estimation of Group Delay in the Presence of Short Data Records," Philip J. Ramsey (1989).

"Estimating the Hausdorff Dimension," Russell L. Reeve (1990).

"Effective Design Augmentation for Prediction," Michael A. Rozum (1990).

SECTION 5: MATHEMATICAL SCIENCES

WESLEYAN UNIVERSITY

Department of Mathematics

Program of Study

The Department of Mathematics offers a program of courses and research leading to the degrees of Master of Arts and Doctor of Philosophy.

The Ph.D. degree demands breadth of knowledge, intensive specialization in one field, original contribution to that field, and expository skill. Three first-year courses are designed to provide a strong foundation in algebra, analysis, and topology. Written preliminary examinations in these three areas are normally taken in the middle of the second year. During the second year, the student continues with a variety of courses, sampling areas of possible concentration. By the start of the third year, the student chooses a specialty and begins research work under the guidance of a thesis adviser. Also required is the ability to read mathematics in at least two of the following languages: French, German, and Russian. The usual time required for completion of all requirements for a Ph.D., including the dissertation, is four to five years.

After passing the preliminary examinations, most Ph.D. candidates teach one course per term, typically a small section (fewer than 20 students) of first-year calculus or linear algebra.

The M.A. degree is designed to ensure basic knowledge and the capacity for sustained scholarly study; requirements are six semester courses at the graduate level and the writing and oral presentation of a thesis. The thesis requires (at least) independent search and study of the literature.

Students are also involved in a variety of departmental activities, including seminars and colloquia. The small size of the program, with more faculty members than graduate students, contributes to an atmosphere of informality and accessibility.

Historically, the emphasis at Wesleyan has been in pure mathematics, and most Wesleyan Ph.D.'s in mathematics have chosen academic careers. In computer science, particularly its theoretical aspects, recent additions have been made in the faculty and in the program of study. In this area, several assistantships are available, and the department encourages all of its students to avail themselves of growing opportunities.

Research Facilities

The department is housed in the Science Center, where all graduate students and faculty members have offices. Computer facilities (currently, two networked DECSYSTEM-20s, a VAX, and various microprocessor networks) are located on the fifth floor and are available for both learning and research purposes. The Science Library collection has about 120,000 volumes, with extensive mathematics holdings; there are over 200 subscriptions to mathematics journals, and approximately 60 new mathematics books arrive each month. The proximity of students and faculty and the daily gatherings at teatime are also key elements of the research environment.

Financial Aid

Each applicant for admission is automatically considered for appointment to an assistantship. For the 1991–92 academic year, this carries a stipend of $8370, plus a dependency allowance when appropriate. All students in good standing are given financial support. Approximately $2790 more is usually available for the student who wishes to remain on campus to study during the summer. Tuition and health fees are waived by the University.

Cost of Study

The only academic costs to the student are books and other educational materials.

Cost of Living

The University provides some subsidized housing and assists in finding private housing. The 1991–92 academic-year cost of a single student's housing (a private room in a 2- or 4-person house, with common kitchen and living area) is $3000.

Student Group

The number of graduate students in mathematics ranges from 9 to 14, with an entering class of 3 to 5 each year. There have always been both male and female students, graduates of small colleges and large universities, and U.S. and foreign students, including, in recent years, students from China, Ethiopia, Greece, Ireland, Israel, Mexico, Pakistan, Sri Lanka, and Yugoslavia.

All of the department's recent Ph.D. recipients have obtained academic employment. Some of these have subsequently taken positions as industrial mathematicians.

Location

Middletown, Connecticut, is a small city of 40,000 on the Connecticut River, about 19 miles southeast of Hartford and 25 miles northeast of New Haven, midway between New York and Boston. The University provides many cultural and recreational opportunities, supplemented by those in the countryside and in larger cities nearby. Several members of the mathematical community are actively involved in sports, including distance running, handball, hiking, squash, table tennis, volleyball, and cycling.

The University

Founded in 1831, Wesleyan is an independent coeducational institution of liberal arts and sciences, with Ph.D. programs in biology, chemistry, ethnomusicology, mathematics, and physics and master's programs in a number of departments. Current enrollments show about 2,800 undergraduates and 145 graduate students.

Applying

No specific courses are required for admission, but it is expected that the equivalent of an undergraduate major in mathematics will have been completed. The complete application consists of the application form, transcripts of all previous academic work at or beyond the college level, letters of recommendation from three college instructors familiar with the applicant's mathematical ability and performance, and GRE scores (if available). Applications should be submitted by February 15 in order to receive adequate consideration, but requests for admission from outstanding candidates are welcome at any time. A visit to the campus is strongly recommended for its value in determining the suitability of the program for the applicant.

Correspondence and Information

Graduate Education Committee
Department of Mathematics
Wesleyan University
Middletown, Connecticut 06459
Telephone: 203-347-9411 Ext. 2398

Peterson's Guide to Graduate Programs in the Physical Sciences and Mathematics 1992

SECTION 5: MATHEMATICAL SCIENCES

Wesleyan University

THE FACULTY AND THEIR RESEARCH

Professors
W. Wistar Comfort, Ph.D., Washington (Seattle). Point-set topology, ultrafilters, set theory, topological groups.
Ethan M. Coven, Ph.D., Yale. Dynamical systems.
Anthony W. Hager, Ph.D., Penn State. Lattice-ordered algebraic structures, general and categorical topology.
F. E. J. Linton, Ph.D., Columbia. Categorical algebra, functorial semantics, topoi.
William L. Reddy, Ph.D., Syracuse. Topological dynamics, branched coverings.
James D. Reid, Ph.D., Washington (Seattle). Abelian groups, module theory.
Lewis C. Robertson, Ph.D., UCLA. Lie groups, topological groups, representation theory.
Carol Wood, Ph.D., Yale. Mathematical logic, applications of model theory to algebra.

Associate Professors
Adam Fieldsteel, Ph.D., Berkeley. Ergodic theory.
Rae Michael Shortt, Ph.D., MIT. Probability, descriptive theory of sets.

Assistant Professors
Karen Collins, Ph.D., MIT. Combinatorics.
Philip H. Scowcroft, Ph.D., Cornell. Foundations of mathematics, model-theoretic algebra.
Donna M. Testerman, Ph.D., Oregon. Algebraic groups, finite groups of Lie type, representation theory.

Associate Professor of Computer Science
Michael Rice, Ph.D., Wesleyan. Parallel computing, formal specification methods.

Assistant Professors of Computer Science
Daniel Dougherty Jr., Ph.D., Maryland. Logic, theoretical computer science.
Sorin Istrail, Ph.D., Bucharest. Logic of programming, models of computation, complexity.

Visitors During the Academic Year 1991–92
Alexander Blokh, Ph.D., SUNY at Stony Brook. Dynamical systems.
Jan van Mill, Ph.D., Free University, Amsterdam. Set-theoretic topology, geometric topology.

Faculty-student conferences, daily gatherings at teatime, and discussions in graduate students' offices are key ingredients of the research environment in the Department of Mathematics.

SECTION 5: MATHEMATICAL SCIENCES

YALE UNIVERSITY
Department of Mathematics

Program of Study

The main object of the graduate program in mathematics is to train mathematicians who intend to make mathematical research their life work. The interests of the faculty members of the Yale Department of Mathematics cover the major fields of contemporary mathematics. In addition to the traditional triad of algebra, analysis, and topology, the department has significant strength in the interdisciplinary areas of Lie theory and mathematical physics.

Basic courses are given each year in algebra, real and complex analysis, and topology, and a wide variety of more advanced courses and seminars are offered, both by regular members of the department and by visiting mathematicians. Courses in many allied fields are also available at the University. Since the dominant type of employment for research mathematicians is university teaching, the program provides opportunities for students to gain teaching experience during their course of study.

The program offers the Ph.D. degree; M.Phil. and M.S. degrees subsidiary to the Ph.D. program; and a special (terminal) M.S. Requirements for the Ph.D. include: (1) completing eight 1-term graduate courses; (2) passing a comprehensive qualifying examination (commonly during the third term); (3) demonstrating reading competence in two of the languages French, German, and Russian; (4) completing a dissertation that definitely advances the subject it treats; and (5) participating in the instruction of undergraduates. Students interested in computer programming are encouraged to learn computer languages and systems. The dissertation must be the result of substantially independent work done under the guidance of a faculty adviser, who is chosen by the student at roughly the time of the comprehensive examination. First- and second-year students have very light teaching duties. Third- and fourth-year students teach one section of calculus or an equivalent alternative. The normal time needed to complete the Ph.D. program is four years.

Research Facilities

The department has a superb mathematics library, housed in the department and accessible to faculty and students 24 hours a day. The department also has a local network of microcomputer workstations.

In addition to the department's resources, the facilities of the University Library, the Computer Center, and the computer science department are available.

Financial Aid

The University has a fellowship program that is intended to foster progress toward the Ph.D. University Fellowships, which include a tuition waiver and an academic-year living stipend that ranges up to $10,000 in 1991–92, are available to the majority of students. A student making normal progress toward the Ph.D. can expect support to continue for four years. Some money for summer living support is also provided, according to the availability of funds.

Cost of Study

Tuition, including bond, health fee, and infirmary fee, is $15,160 for 1991–92. (This is waived for recipients of University Fellowships as explained above.) In addition, there is a $430 assessment for hospitalization coverage. This may be waived for students who can document adequate alternative coverage.

Cost of Living

The estimated average living cost for a single student is $10,530. This includes room and board, transportation, and academic and personal expenses. The estimated budget is $18,180 for a married couple and increases by approximately $7000 for each dependent. Students may live in University dormitories, but most live in apartments near the University.

Student Group

The average class size is 10 students, and the total of the mathematics graduate student group is about 50. This is a relatively close-knit group within a total graduate student population at Yale of over 2,000.

Location

Yale University is located adjacent to downtown New Haven, Connecticut. Yale and New Haven offer a broad range of cultural and recreational activities: two resident theaters, as well as student drama; several concert series; many film series; and major collegiate sports, intramural sports, and gymnasium facilities. The easily accessible Connecticut countryside supports a rich variety of outdoor activities, including skiing in winter. Travel time to New York by car or train is approximately 90 minutes, and Boston is about 2½ hours away by car.

The Department

The department is best described as informal and friendly. Teas are served daily at 4 p.m., affording students easy contact with faculty members and fellow students. There are departmental picnics in the fall and spring, and there is a Christmas party.

The department is located in Dunham Laboratory (DL), 10 Hillhouse Avenue, 4th floor, and in Leet Oliver Memorial Hall (LOM), 12 Hillhouse Avenue. DL contains faculty, administrative, and graduate students offices and the computer center. LOM contains classrooms, faculty offices, and the mathematics research library of the department.

The department has two special lecture series, the Frank J. Hahn Lectures and the Whittemore Lectures, each normally presented once a year. These lectures provide a forum for surveys of topics of outstanding interest in contemporary research. The Mathematics Colloquium is held weekly. It is devoted to reports on current developments in mathematical research by leading mathematicians from this country and abroad.

Applying

Students are selected for admission strictly according to their prospects for completing the Ph.D. program with distinction. Factors that weigh most heavily in the selection process are (1) the quality of the applicant's previous training in mathematics, and (2) letters of recommendation from qualified persons who are in a position to estimate the applicant's potential as a research mathematician.

Applicants are required to take the GRE General Test and Subject Test in mathematics. Applications for admission and financial aid are available from the Graduate Admissions Office, Box 1504A Yale Station. The *Bulletin* of the Graduate School is available upon request and payment of $3. Applications must be received complete by January 2.

Correspondence and Information

Professor George B. Seligman
Director of Graduate Studies
Department of Mathematics
Yale University
Box 2155 Yale Station
New Haven, Connecticut 06520

SECTION 5: MATHEMATICAL SCIENCES

Yale University

THE FACULTY AND THEIR RESEARCH

Professors

Richard W. Beals, Ph.D., Yale, 1964. Differential operators, partial differential equations, real and complex function theory, scattering theory.

Ronald R. Coifman, Ph.D., Geneva, 1965. Nonlinear analysis, scattering theory, real and complex analysis, singular integrals.

Walter Feit, Ph.D., Michigan, 1954. Finite groups.

Igor B. Frenkel, Ph.D., Yale, 1980. Infinite-dimensional algebras, representation theory, applications of Lie theory, mathematical physics.

Howard Garland, Ph.D., Berkeley, 1964. Algebraic groups, Lie algebras, structure theory, infinite dimensional algebras, representation theory, discrete and arithmetic groups, applications of Lie theory, mathematical physics.

Roger E. Howe, Ph.D., Berkeley, 1969. Representation theory, automorphic forms, harmonic analysis, invariant theory.

Peter W. Jones, Ph.D., UCLA, 1978. Real and complex function theory, singular integrals, potential theory, Fourier and complex analysis.

Serge Lang, Ph.D., Princeton, 1951. Algebraic number theory, algebraic geometry, Diophantine problems, automorphic forms.

Ronnie Lee, Ph.D., Michigan, 1968. Algebraic topology, differential topology, transformation groups, K theory.

Benoit B. Mandelbrot, D.Sc., Paris, 1952. Fractals, random processes and sets, applications.

Gregory A. Margulis, Ph.D., Moscow, 1970. Lie group theory, ergodic theory, number theory, network theory and dynamics.

William S. Massey, Ph.D., Princeton, 1948. Algebraic topology, differential topology, homotopy theory, theory of fiber bundles.

Vincent Moncrief, Ph.D., Maryland, 1972. Relativity, mathematical physics.

George D. Mostow, Ph.D., Harvard, 1948. Algebraic groups, Lie algebras, structure theory, representation theory, discrete and arithmetic groups.

Ilya I. Piatetski-Shapiro, D.Sc., Steklov Institute of Mathematics (USSR), 1958. Representation theory, automorphic forms, L-functions.

Vladimir Rokhlin, Ph.D., Rice, 1983. Numerical scattering theory, elliptic partial differential equations, numerical solution of integral equations.

George B. Seligman, Ph.D., Yale, 1954. Lie algebras and generalizations, algebraic groups, Lie algebras, structure theory, representation theory.

Robert H. Szczarba, Ph.D., Chicago, 1960. Algebraic topology, differential topology, homotopy theory, theory of fiber bundles.

Tsuneo Tamagawa, D.Sc., Tokyo, 1954. Algebraic number theory, algebraic geometry, Diophantine problems, algebraic groups, Lie algebras, structure theory, representation theory, automorphic forms, L-functions.

Gregg J. Zuckerman, Ph.D., Princeton, 1975. Representation theory, applications of Lie theory, mathematical physics.

Harkness Tower is one of the architectural landmarks of Yale University.

The mathematics department is located in Leet Oliver Memorial Hall on Hillhouse Avenue.

Sterling Memorial Library, the main library at Yale, contains approximately 4 million volumes.

Section 6
Meteorology and Atmospheric Sciences

This section contains a directory of institutions offering graduate work in meteorology and atmospheric sciences, followed by two-page entries submitted by institutions that chose to prepare detailed program descriptions. Additional information about programs listed in the directory but not augmented by a two-page entry may be obtained by writing directly to the dean of a graduate school or chair of a department at the address given in the directory.

For programs offering related work, see also in this book: Astronomy and Astrophysics; Earth and Planetary Sciences; Marine Sciences/Oceanography; and Physics; in Book 3: Biology and Biomedical Sciences and Biophysics; and in Book 5: Civil and Environmental Engineering; Engineering and Applied Sciences; and Mechanical Engineering, Mechanics, and Aerospace/Aeronautical Engineering.

CONTENTS

Program Directory   520

Announcements
McGill University   520
South Dakota School of Mines and Technology   521
University of California, Davis   522

Cross-Discipline Announcement
University of Chicago   523

Full Descriptions
Colorado State University   525
Columbia University   527
Pennsylvania State University   529
State University of New York at Albany   531
State University of New York at Stony Brook   533
University of Maryland College Park   535

See also:
Clemson University—Physics and Astronomy   581
Drexel University—Chemistry, Mathematics and Computer
 Science, and Physics and Atmospheric Sciences   409
North Carolina State University—Marine, Earth, and
 Atmospheric Sciences   317
University of Chicago—Geophysical Sciences   259
University of Kansas—Physics and Astronomy   649
Yale University—Geology and Geophysics   299

FIELD DEFINITION

In an effort to broaden prospective students' understanding of meteorology and atmospheric sciences, an educator in the field has provided the following statement.

Meteorology and Atmospheric Sciences

The field of meteorology or atmospheric science is a relative newcomer as an academic discipline. Most graduate programs in meteorology/atmospheric science were established during the 1950s and 1960s. During that time period, the discipline experienced rapid growth, which has continued into the present era.

Many advances in meteorology/atmospheric science can be attributed to breakthroughs in related areas of mathematics, physics, engineering, and technology. Supercomputers, weather radars, meteorological satellites, and the latest in other remote-sensing technologies have been applied not only to scientific inquiry but also to hour-to-hour and minute-to-minute operational weather forecasting. Some meteorological research is devoted to the observation, numerical simulation, and prediction of weather systems such as hurricanes and typhoons, cyclones, severe storms, and heavy snow events. Other meteorologists/atmospheric scientists perform research in such diverse areas as atmospheric dynamics, atmospheric chemistry, climate modeling and climate change, turbulence, precipitation physics, and solar-terrestrial interactions. Interdisciplinary research efforts focus on national initiatives in earth system science, global change, acid rain, the "ozone hole," and several other areas of concern. Historically, meteorologists have been at the forefront in the use of all sizes of computers and peripheral devices for data handling, modeling, and the graphical imaging of results. Supercomputers and the latest supercomputing techniques are an integral part of the science of meteorology.

Meteorological research and graduate education span a continuum from highly theoretical to more applied projects and courses. Graduate students are recruited and usually receive financial support to tackle these tough scientific problems under the guidance of a faculty member with a research grant. Graduate student research projects might focus on the derivation of a new mathematical theory to describe the behavior of hurricanes, the application of a remote-sensing technique to severe storm and tornado detection, the complex mechanisms that produce acid precipitation, or the effects of climate variability on wheat production.

Graduate students in meteorology or atmospheric science come from a variety of undergraduate backgrounds. Some have B.S. degrees in meteorology or atmospheric science. Others come with undergraduate majors in mathematics, physics, chemistry, computer science, or one of the engineering disciplines. A thorough grounding in differential equations and physics is the most important requirement for admission to a graduate program in meteorology/atmospheric science.

Graduate degree recipients in meteorology/atmospheric science have traditionally sought employment within the federal government or the academic community. However, a growing number have found employment within the private sector. A recent study showed that 30 percent of the master's degree and 17 percent of the Ph.D. recipients in meteorology are now entering promising career paths in private companies.

James F. Kimpel, Dean
College of Geosciences
University of Oklahoma

Meteorology and Atmospheric Sciences

Clemson University, College of Sciences, Department of Physics and Astronomy, Clemson, SC 29634. Department offers program in physics (MS, PhD). Faculty: 23 full-time (0 women), 0 part-time. Matriculated students: 45 full-time (4 women), 0 part-time; includes 2 minority (1 black American, 1 Hispanic American), 10 foreign. Average age 27. 92 applicants, 26% accepted. In 1990, 4 master's awarded (25% entered university research/teaching, 50% found other work related to degree, 25% continued full-time study); 3 doctorates awarded (67% entered university research/teaching, 33% found other work related to degree). Terminal master's awarded for partial completion of doctoral program. *Degree requirements:* For master's, thesis required (for some programs), foreign language not required; for doctorate, dissertation required, foreign language not required. *Entrance requirements:* GRE General Test, TOEFL. Application deadline: 6/1 (priority date, applications processed on a rolling basis). Application fee: $25. *Expenses:* Tuition of $102 per credit hour. Fees of $80 per semester full-time. *Financial aid:* In 1990–91, 40 students received a total of $421,400 in aid awarded. 2 fellowships (both to first-year students), 15 research assistantships (1 to a first-year student), 23 teaching assistantships (13 to first-year students) were awarded. Average monthly stipend for a graduate assistantship: $877. Financial aid application deadline: 6/1. *Faculty research:* Astrophysics, atmosphere physics, condensed matter, radiation physics, solid state physics. Total annual research budget: $457,126. • Dr. P. J. McNulty, Head, 803-656-3416. Application contact: J. A. Gilreath, Graduate Student Adviser, 803-656-3416.

See full description on page 581.

Colorado State University, College of Engineering, Department of Atmospheric Science, Fort Collins, CO 80523. Department awards MS, PhD. PhD offered jointly with South Dakota School of Mines and Technology. Faculty: 14 full-time (0 women), 0 part-time. Matriculated students: 66 full-time (8 women), 7 part-time (0 women); includes 4 minority (1 Asian American, 1 black American, 2 Hispanic American), 18 foreign. Average age 28. 121 applicants, 26% accepted. In 1990, 17 master's, 5 doctorates awarded. *Degree requirements:* Thesis/dissertation required, foreign language not required. *Entrance requirements:* For master's, GRE General Test, TOEFL (minimum score of 550 required), minimum of 1 year of physics, minimum GPA of 3.0; for doctorate, GRE General Test, TOEFL, minimum GPA of 3.0. Application deadline: 4/1 (priority date, applications processed on a rolling basis). Application fee: $30. *Tuition:* $1322 per semester full-time for state residents; $3673 per semester full-time for nonresidents. *Financial aid:* In 1990–91, 0 fellowships, 50 research assistantships (12 to first-year students), 4 teaching assistantships (0 to first-year students), 0 traineeships awarded. Average monthly stipend for a graduate assistantship: $698-750. *Faculty research:* Global circulation and climate, atmospheric radiation and remote sensing, tropical and marine meteorology, mesoscale modeling and analysis. Total annual research budget: $4.1-million. • Dr. Stephen K. Cox, Head, 303-491-8594. Application contact: Dr. Thomas McKee, Student Counselor, 303-491-8545.

See full description on page 525.

Columbia University, Graduate School of Arts and Sciences, Division of Natural Sciences, Program in Atmospheric and Planetary Science, New York, NY 10027. Program awards M Phil, PhD. *Degree requirements:* For doctorate, variable foreign language requirement, dissertation. *Entrance requirements:* For doctorate, GRE General Test, GRE Subject Test, TOEFL, previous course work in mathematics and physics. Application deadline: 1/5. Application fee: $50. *Expenses:* Tuition of $7836 per semester for state residents; $426 per credit for nonresidents. Fees of $148 per semester for state residents. *Financial aid:* Application deadline 1/5. *Faculty research:* Climate, weather prediction. • Dr. Roger S. Bagnall, Dean, Graduate School of Arts and Sciences, 212-854-2861.

See full description on page 527.

Cornell University, Graduate Fields of Agriculture and Life Sciences, Field of Soil, Crop, and Atmospheric Sciences, Ithaca, NY 14853. Field offers programs in atmospheric science (MS), atmospheric sciences (MPS, PhD), field crop science (MPS, MS, PhD), remote sensing (MPS, MS, PhD), seed technology (MPS, MS, PhD), soil science (MPS, MS, PhD). Faculty: 45 full-time, 0 part-time. Matriculated students: 56 full-time (13 women), 0 part-time; includes 4 minority (2 black American, 2 Hispanic American), 30 foreign. 6 applicants, 33% accepted. In 1990, 5 master's, 10 doctorates awarded. Terminal master's awarded for partial completion of doctoral program. *Degree requirements:* Thesis/dissertation required, foreign language not required. *Entrance requirements:* GRE General Test, TOEFL. Application deadline: 1/10. Application fee: $55. *Expenses:* Tuition of $7440 per trimester full-time. Fees of $28 per year full-time. *Financial aid:* In 1990–91, 29 students received aid. 1 fellowship (0 to first-year students), 25 research assistantships (3 to first-year students), 3 teaching assistantships (0 to first-year students) awarded; full and partial tuition waivers, federal work-study, institutionally sponsored loans also available. Financial aid application deadline: 1/10; applicants required to submit GAPSFAS. *Faculty research:* Field crop science; soil science; atmospheric science; remote sensing; seed technology. • John H. Peverly, Graduate Faculty Representative, 607-255-5408. Application contact: Robert Brashear, Director of Admissions, 607-255-4884.

Creighton University, College of Arts and Sciences, Program in Atmospheric Science, Omaha, NE 68178. Program awards MS. Faculty: 3 full-time, 10 part-time. Matriculated students: 1 full-time (0 women), 6 part-time (0 women). 3 applicants, 100% accepted. *Degree requirements:* Computer language, thesis. *Entrance requirements:* GRE General Test, TOEFL. Application deadline: 3/15. Application fee: $20. *Expenses:* Tuition of $272 per credit hour. Fees of $140 per semester full-time, $14 per semester part-time. • Dr. Arthur Douglas, Director, 402-280-2420. Application contact: Dr. Michael Lawler, Dean, Graduate School, 402-280-2870.

Drexel University, College of Arts and Sciences, Department of Physics and Atmospheric Science, Program in Atmospheric Science, 32nd and Chestnut Streets, Philadelphia, PA 19104. Program awards MS, PhD. Faculty: 4 full-time (0 women), 0 part-time. Terminal master's awarded for partial completion of doctoral program. *Degree requirements:* For master's, foreign language and thesis not required; for doctorate, dissertation required, foreign language not required. *Entrance requirements:* GRE, TOEFL (minimum score of 550 required). Application deadline: 8/23 (applications processed on a rolling basis). Application fee: $25. *Expenses:* Tuition of $345 per credit hour. Fees of $81 per quarter full-time, $43 per quarter part-time. *Financial aid:* Research assistantships, teaching assistantships available. Financial aid application deadline: 2/1. *Faculty research:* Numerical weather prediction, mesoscale meteorology, earth energy radiation budget. • Application contact: Keith Brooks, Director of Graduate Admissions, 215-895-6700.

See full description on page 409.

Florida State University, College of Arts and Sciences, Department of Meteorology, Tallahassee, FL 32306. Department awards MS, PhD. Faculty: 21 full-time (1 woman), 0 part-time. Matriculated students: 63 full-time (10 women), 6 part-time (0 women); includes 4 minority (2 black American, 2 Hispanic American), 14 foreign. Average age 28. 88 applicants, 42% accepted. In 1990, 10 master's awarded (90% found work related to degree, 10% continued full-time study); 5 doctorates awarded (20% entered university research/teaching, 80% found other work related to degree). Terminal master's awarded for partial completion of doctoral program. *Degree requirements:* For master's, foreign language and thesis not required; for doctorate, dissertation required, foreign language not required. *Entrance requirements:* GRE General Test (minimum combined score of 1000 required) or minimum GPA of 3.0. Application fee: $15. *Tuition:* $76.29 per credit hour for state residents; $238 per credit hour for nonresidents. *Financial aid:* In 1990–91, 62 students received aid. 8 fellowships (2 to first-year students), 29 research assistantships (4 to first-year students), 3 teaching assistantships (0 to first-year students), 0 traineeships awarded; federal work-study and career-related internships or fieldwork also available. Average monthly stipend for a graduate assistantship: $850. Financial aid application deadline: 4/1. *Faculty research:* Physical, dynamic, and synoptic meteorology, climate. Total annual research budget: $3.5-million. • Dr. David Stuart, Chairman, 904-644-6205. Application contact: Anna R. Nelson, Student Affairs Coordinator, 904-644-8582.

Massachusetts Institute of Technology, School of Science, Department of Earth, Atmospheric, and Planetary Sciences, Center for Meteorology and Physical Oceanography, Cambridge, MA 02139. Center awards SM, PhD, Sc D. Faculty: 12 full-time (1 woman), 0 part-time. Matriculated students: 62 full-time (16 women), 0 part-time; includes 1 minority (black American), 28 foreign. Average age 24. 61 applicants, 39% accepted. In 1990, 6 master's awarded (67% found work related to degree, 33% continued full-time study); 7 doctorates awarded. Terminal master's awarded for partial completion of doctoral program. *Degree requirements:* Thesis/dissertation required, foreign language not required. *Entrance requirements:* GRE General Test, GRE Subject Test. Application deadline: 1/15. Application fee: $45. *Tuition:* $16,900 per year. *Financial aid:* In 1990–91, 10 fellowships (2 to first-year students), 32 research assistantships (6 to first-year students), 2 teaching assistantships (0 to first-year students) awarded; institutionally sponsored loans also available. Financial aid application deadline: 1/15. *Faculty research:* Origin, composition, structure, and state of atmospheres and oceans. • Dr. Kerry Emanuel, Head, 617-253-2281.

McGill University, Faculty of Graduate Studies and Research, Department of Meteorology, Montreal, PQ H3A 2T5, Canada. Department offers programs in meteorology (M Sc, PhD), physical oceanography (M Sc, PhD). Faculty: 12 full-time (0 women), 0 part-time. Matriculated students: 43 full-time (8 women), 0 part-time; includes 15 foreign. Average age 24. 454 applicants, 24% accepted. In 1990, 6 master's awarded (50% found work related to degree, 50% continued full-time study); 0 doctorates awarded. Terminal master's awarded for partial completion of doctoral program. *Degree requirements:* Thesis/dissertation required, foreign language not required. *Entrance requirements:* For master's, GRE General Test, minimum GPA of 3.0; for doctorate, GRE, master's degree in meteorology or related field. Application deadline: 7/1 (priority date, applications processed on a rolling basis). Application fee: $25. *Tuition:* $698.50 per semester full-time, $46.57 per credit part-time for Canadian residents; $3480 per semester full-time, $234 per semester part-time for nonresidents. *Financial aid:* In 1990–91, 40 students received a total of $365,000 in aid awarded. 1 fellowship (0 to first-year students), 30 research assistantships (9 to first-year students), 10 teaching assistantships (1 to a first-year student) were awarded. Average monthly stipend for a graduate assistantship: $1000. Financial aid application deadline: 7/1. *Faculty research:* Atmospheric radiation, cloud physics, climate research, dynamics, radar. Total annual research budget: $1.6-million. • H. G. Leighton, Chairman, 514-398-3764.

Announcement: Graduate research topics include radar meteorology, cloud physics, satellite data analysis, climate research, large-scale dynamics, the numerical simulation of cumulus convection, mesoscale computer modeling, physical oceanography, and air-ice-ocean interactions. Students have access to a wide range of computers and radar and modern synoptics laboratories.

Naval Postgraduate School, Department of Meteorology, Monterey, CA 93943. Department awards MS, PhD. Program only open to commissioned officers of the United States and friendly nations and selected United States federal civilian employees. *Degree requirements:* Thesis/dissertation. *Tuition:* $0.

New Mexico Institute of Mining and Technology, Department of Physics, Socorro, NM 87801. Offerings include physics (MS, PhD), with options in astrophysics (MS, PhD), atmospheric physics (MS, PhD), instrumentation (MS), mathematical physics (PhD). Department faculty: 15 full-time (1 woman), 1 part-time (0 women). *Degree requirements:* For master's, thesis required, foreign language not required; for doctorate, 1 foreign language, dissertation. *Entrance requirements:* For master's, GRE General Test, TOEFL (minimum score of 540 required); for doctorate, GRE General Test, GRE Subject Test, TOEFL (minimum score of 540 required). Application deadline: 6/1 (priority date, applications processed on a rolling basis). Application fee: $16. *Expenses:* Tuition of $617 per semester full-time, $51 per hour part-time for state residents; $2656 per semester full-time, $209 per hour part-time for nonresidents. Fees of $207 per semester. • Dr. David Raymond, Chairman, 505-835-5610. Application contact: Dr. J. A. Smoake, Dean, Graduate Studies, 505-835-5513.

New York University, Graduate School of Arts and Science, Department of Applied Science, New York, NY 10011. Offerings include energy science; atmospheric science (MS). Department faculty: 6 full-time, 3 part-time. *Degree requirements:* Thesis or project required, foreign language not required. *Entrance requirements:* TOEFL, minimum GPA of 3.0. Application deadline: 1/15 (priority date, applications processed on a rolling basis). Application fee: $30. *Tuition:* $467 per credit. • Dr. Martin I. Hoffert, Chairman, 212-998-8995. Application contact: Gabriel Miller, Director of Graduate Studies, 212-998-8995.

North Carolina State University, College of Physical and Mathematical Sciences, Department of Marine, Earth, and Atmospheric Sciences, Raleigh, NC 27695. Department offers programs in geology (MS, PhD), geophysics (MS, PhD), meteorology (MS, PhD), oceanography (MS, PhD). Faculty: 35 full-time (2 women), 3 part-time (1 woman). Matriculated students: 74 full-time (15 women), 3 part-time (0 women); includes 2 minority (both Asian American), 23 foreign. Average age 24. 75 applicants, 49% accepted. In 1990, 18 master's, 6 doctorates awarded. Terminal master's awarded for partial completion of doctoral program. *Degree requirements:* For master's, thesis, final oral exam required, foreign language not required; for doctorate, dissertation, preliminary written and oral exams, final oral exams required, foreign language not required. *Entrance requirements:* GRE General Test, minimum GPA of 3.0. Application deadline: 5/1 (priority date, applications processed on a rolling basis). Application fee: $35. *Tuition:* $1138 per year for state residents; $5805 per year for nonresidents. *Financial aid:* In 1990–91, $383,267 in aid awarded. 0 fellowships, 33 research assistantships (10 to first-year students), 18 teaching assistantships (6 to first-year students) were awarded; institutionally sponsored loans also available. Average monthly stipend for a graduate assistantship: $750. Financial aid application deadline: 3/1. *Faculty research:* Boundary-layer and synoptic meteorology; physical, chemical, geological, and biological oceanography.

Total annual research budget: $6.41-million. • Dr. Leonard J. Pietrafesa, Head, 919-737-3717. Application contact: G. S. Janowitz, Graduate Administrator, 919-737-7837.

See full description on page 317.

Ohio State University, College of Engineering, Program in Atmospheric Sciences, Columbus, OH 43210. Program awards MS, PhD. Faculty: 1. Matriculated students: 6 full-time (0 women), 0 part-time; includes 0 minority, 4 foreign. 35 applicants, 23% accepted. In 1990, 0 master's, 0 doctorates awarded. *Degree requirements:* Computer language, thesis/dissertation required, foreign language not required. *Entrance requirements:* GRE General Test, minimum GPA of 3.0. Application deadline: 8/15 (applications processed on a rolling basis). Application fee: $0 ($25 for foreign students). *Tuition:* $1213 per quarter full-time, $364 per course part-time for state residents; $3143 per quarter full-time, $943 per course part-time for nonresidents. *Financial aid:* Fellowships, research assistantships, teaching assistantships, federal work-study, institutionally sponsored loans available. Aid available to part-time students. Financial aid applicants required to submit FAF. *Faculty research:* Climatology, aeronomy, solar-terrestrial physics, air environment. • John N. Rayner, Graduate Studies Committee Chair, 614-292-2514.

Oregon State University, Graduate School, College of Science, Department of Atmospheric Sciences, Corvallis, OR 97331. Department awards MA, MAIS, MS, PhD. Faculty: 9 full-time (0 women), 0 part-time. Matriculated students: 16 full-time (1 woman), 1 part-time (0 women); includes 0 minority, 10 foreign. Average age 30. In 1990, 3 master's awarded (100% found work related to degree); 1 doctorate awarded (100% entered university research/teaching). *Degree requirements:* For master's, variable foreign language requirement, thesis, qualifying exams; for doctorate, dissertation, qualifying exams required, foreign language not required. *Entrance requirements:* GRE General Test, TOEFL (minimum score of 550 required), minimum GPA of 3.0 in last 90 hours. Application deadline: 3/1. Application fee: $40. *Tuition:* $1140 per trimester full-time, $449 per year part-time for state residents; $1816 per trimester full-time, $674 per year part-time for nonresidents. *Financial aid:* In 1990–91, $90,490 in aid awarded. 13 research assistantships (3 to first-year students), 1 teaching assistantship (0 to first-year students) were awarded; institutionally sponsored loans also available. Aid available to part-time students. Financial aid application deadline: 2/1; applicants required to submit FAF. *Faculty research:* Planetary atmospheres, boundary layer dynamics, climate, statistical meteorology, satellite meteorology. Total annual research budget: $867,011. • Dr. Steven K. Esbensen, Chair, 503-737-4557.

Pennsylvania State University University Park Campus, College of Earth and Mineral Sciences, Department of Meteorology, University Park, PA 16802. Department awards MS, PhD. Faculty: 26. Matriculated students: 72 full-time (18 women), 3 part-time (0 women). In 1990, 13 master's, 11 doctorates awarded. *Entrance requirements:* GRE General Test. Application fee: $35. *Tuition:* $203 per credit for state residents; $403 per credit for nonresidents. • Bruce Albrecht, Acting Head, 814-865-0478. Application contact: Peter Bannon, 814-863-1309.

See full description on page 529.

Princeton University, Department of Geological and Geophysical Sciences, Program in Atmospheric and Oceanic Sciences, Princeton, NJ 08544. Program awards PhD. Faculty: 14 full-time (0 women), 0 part-time. Matriculated students: 16 full-time (5 women), 0 part-time; includes 0 minority, 11 foreign. 24 applicants, 38% accepted. In 1990, 3 doctorates awarded (33% entered university research/teaching, 0% found other work related to degree, 0% continued full-time study). Terminal master's awarded for partial completion of doctoral program. *Degree requirements:* For doctorate, 1 foreign language, dissertation. *Entrance requirements:* For doctorate, GRE General Test, GRE Subject Test. Application deadline: 1/8. Application fee: $45 ($50 for foreign students). *Tuition:* $16,670 per year. *Financial aid:* Fellowships, research assistantships, federal work-study, institutionally sponsored loans available. Financial aid application deadline: 1/8. *Faculty research:* Climate dynamics, middle atmosphere dynamics and chemistry, oceanic circulation, marine geochemistry, numerical modeling. • S. George H. Philander, Director of Graduate Studies, 609-258-6571. Application contact: Michele Spreen, Director of Graduate Admissions, 609-258-3034.

Purdue University, School of Science, Department of Earth and Atmospheric Sciences, West Lafayette, IN 47907. Department awards MS, PhD. Faculty: 24 full-time, 1 part-time. Matriculated students: 74 full-time (10 women), 6 part-time (3 women); includes 4 minority (1 Asian American, 1 black American, 1 Hispanic American, 1 Native American), 27 foreign. Average age 28. 87 applicants, 23% accepted. In 1990, 19 master's, 6 doctorates awarded. *Degree requirements:* For master's, thesis required, foreign language not required; for doctorate, 1 foreign language, dissertation. *Entrance requirements:* GRE General Test. Application deadline: 6/1. Application fee: $25. *Tuition:* $1162 per semester full-time, $83.50 per credit hour part-time for state residents; $3720 per semester full-time, $244.50 per credit hour part-time for nonresidents. *Financial aid:* In 1990–91, 3 fellowships (2 to first-year students), 22 research assistantships (1 to a first-year student), 26 teaching assistantships (3 to first-year students) awarded. • Dr. E. M. Agee, Head, 317-494-0251.

Rutgers, The State University of New Jersey, New Brunswick, Program in Meteorology, New Brunswick, NJ 08903. Offers air pollution meteorology (MS), air-sea interactions (MS), applied climatology (MS), biological and agricultural meteorology (MS). Interdisciplinary PhD option available. Part-time and evening/weekend programs available. Faculty: 9 full-time (0 women), 0 part-time. Matriculated students: 4 full-time (0 women), 14 part-time (1 woman); includes 3 minority (all Asian American), 3 foreign. Average age 24. 14 applicants, 79% accepted. In 1990, 2 degrees awarded (100% found work related to degree). *Degree requirements:* Thesis required, foreign language not required. *Entrance requirements:* GRE General Test. Application deadline: 8/15. Application fee: $35. *Expenses:* Tuition of $4432 per year full-time, $183 per credit part-time for state residents; $6496 per year full-time, $270 per credit part-time for nonresidents. Fees of $458 per year full-time, $117 per year part-time. *Financial aid:* In 1990–91, 7 students received a total of $40,000 in aid awarded. 3 research assistantships (2 to first-year students), 1 teaching assistantship (0 to first-year students) were awarded; federal work-study and career-related internships or fieldwork also available. Average monthly stipend for a graduate assistantship: $850. Financial aid application deadline: 3/1. *Faculty research:* Climatology, weather forecasting, air pollution, agricultural meteorology. Total annual research budget: $200,000. • Nathan M. Reiss, Director, 908-932-9387.

Saint Louis University, College of Arts and Sciences, Department of Earth and Atmospheric Sciences, Program in Atmospheric Science, St. Louis, MO 63103. Program awards M Pr Met, MS(R), PhD. Faculty: 6 full-time, 1 part-time. Matriculated students: 9 full-time (1 woman), 10 part-time (0 women); includes 0 minority, 4 foreign. 14 applicants, 79% accepted. In 1990, 3 master's, 0 doctorates awarded. *Degree requirements:* For master's, computer language, comprehensive oral exam, thesis for MS(R) required, foreign language not required; for doctorate, computer language, dissertation, preliminary degree exams required, foreign language not required. *Entrance requirements:* GRE General Test. Application deadline: rolling. Application fee: $30. *Tuition:* $342 per credit hour. *Financial aid:* In 1990–91, 1 research assistantship, 3 teaching assistantships awarded. Financial aid application deadline: 4/1. • Dr. G. V. Rao, Director, 314-658-3116. Application contact: Dr. Robert J. Nikolai, Associate Dean of the Graduate School, 314-658-2240.

San Jose State University, School of Science, Department of Meteorology, San Jose, CA 95192. Department awards MS. Faculty: 5 full-time (2 women), 4 part-time (0 women). Matriculated students: 0 full-time, 4 part-time (0 women); includes 0 minority, 0 foreign. Average age 28. 4 applicants, 50% accepted. In 1990, 3 degrees awarded. *Degree requirements:* Thesis or alternative required, foreign language not required. *Entrance requirements:* GRE. Application deadline: 6/1 (applications processed on a rolling basis). Application fee: $55. *Expenses:* Tuition of $0 for state residents; $246 per unit for nonresidents. Fees of $592 per year. • Dr. Peter Lester, Chair, 408-924-5200. Application contact: Dr. Robert Bornstein, Graduate Adviser, 408-924-5200.

South Dakota School of Mines and Technology, Department of Meteorology, Rapid City, SD 57701. Department offers program in atmospheric sciences (MS, PhD). PhD offered jointly with Colorado State University. Part-time programs available. Faculty: 7 full-time (0 women), 2 part-time (0 women). Matriculated students: 16 full-time (4 women), 0 part-time; includes 0 minority, 8 foreign. Average age 28. In 1990, 10 master's awarded. *Degree requirements:* For master's, thesis required, foreign language not required. *Entrance requirements:* For master's, TOEFL (minimum score of 520 required). Application deadline: 7/15. Application fee: $15 ($45 for foreign students). *Expenses:* Tuition of $1474 per year full-time, $61.40 per credit hour part-time for state residents; $2917 per year full-time, $122 per credit hour part-time for nonresidents. Fees of $871 per year full-time, $132 per semester (minimum) part-time. *Financial aid:* In 1990–91, 15 students received a total of $43,358 in aid awarded. 0 fellowships, 22 research assistantships, 0 teaching assistantships were awarded; federal work-study, institutionally sponsored loans also available. Aid available to part-time students. Average monthly stipend for a graduate assistantship: $300. Financial aid application deadline: 5/15. *Faculty research:* Hailstorms observations and numerical modeling; microbursts, lightning, and acid rain; air pollution; radiative transfer. Total annual research budget: $1.5-million. • Dr. H. D. Orville, Head, 605-394-2291. Application contact: Dr. Briant L. Davis, Dean, Graduate Division, 605-394-2493.

Announcement: The MS program in atmospheric sciences (co-op PhD with CSU) emphasizes physical meteorology; topics such as radar meteorology, cloud and precipitation physics (including hail and cloud modification), atmospheric electricity, air pollution and atmospheric radiation. Also high-quality education is offered in dynamic meteorology and forecasting, mesoscale meteorology, climatology, and satellite imagery.

State University of New York at Albany, College of Science and Mathematics, Department of Atmospheric Science, Albany, NY 12222. Department awards MS, PhD. Evening/weekend programs available. Faculty: 10 full-time (0 women), 0 part-time. Matriculated students: 32 full-time (5 women), 6 part-time (0 women); includes 1 minority (black American), 13 foreign. Average age 25. In 1990, 5 master's, 3 doctorates awarded. *Degree requirements:* For master's, 1 foreign language, thesis, comprehensive exam; for doctorate, 2 foreign languages, dissertation, comprehensive and oral exams. *Entrance requirements:* GRE General Test. Application fee: $35. *Expenses:* Tuition of $2450 per year full-time, $103 per credit part-time for state residents; $5765 per year full-time, $243 per credit part-time for nonresidents. Fees of $25 per year full-time, $0.85 per credit part-time. *Financial aid:* Fellowships, research assistantships, teaching assistantships, minority assistantships available. • Jon Scott, Chairman, 518-442-4556.

See full description on page 531.

State University of New York at Stony Brook, Division of Physical Sciences and Mathematics and College of Engineering and Applied Sciences, Institute for Terrestrial and Planetary Atmospheres, Stony Brook, NY 11794. Institute awards PhD. *Entrance requirements:* GRE General Test, TOEFL. Application deadline: 3/1. Application fee: $35. *Expenses:* Tuition of $2450 per year full-time, $103 per credit part-time for state residents; $5766 per year full-time, $243 per credit part-time for nonresidents. Fees of $151 per year full-time, $10.45 per year (minimum) part-time. *Financial aid:* Fellowships, assistantships, summer research assistantships available. • Marvin A. Geller, Director, 516-632-6170.

See full description on page 533.

Texas A&M University, College of Geosciences, Department of Meteorology, College Station, TX 77843. Department awards MS, PhD. Faculty: 12. Matriculated students: 57 full-time (8 women), 0 part-time; includes 2 minority (1 black American, 1 Hispanic American), 16 foreign. 35 applicants, 94% accepted. In 1990, 8 master's, 2 doctorates awarded. *Degree requirements:* Thesis/dissertation. *Entrance requirements:* GRE General Test, TOEFL. Application deadline: 7/15. Application fee: $25 ($50 for foreign students). *Expenses:* Tuition of $100 per semester (minimum) for state residents; $128 per credit hour for nonresidents. Fees of $459 per year full-time, $252 per semester part-time. *Financial aid:* Fellowships, research assistantships, teaching assistantships available. • Edward J. Zipser, Head, 409-845-7688.

Texas Tech University, Graduate School, College of Arts and Sciences, Department of Geosciences, Programs in Atmospheric Sciences, Lubbock, TX 79409. Programs award MS, PhD. Faculty: 5. Matriculated students: 16. In 1990, 3 master's awarded. *Degree requirements:* Thesis/dissertation. *Application deadline:* 4/15 (priority date, applications processed on a rolling basis). *Application fee:* $0 ($50 for foreign students). *Tuition:* $494 per semester full-time, $20 per credit hour part-time for state residents; $1790 per semester full-time, $455 per credit hour part-time for nonresidents. • Dr. James E. Barrick, Graduate Adviser, 806-742-3102.

Université du Québec à Montréal, Programs in Atmospheric Sciences and Meteorology, Montreal, PQ H3C 3P8, Canada. Programs award M Sc, Diploma. Part-time programs available. Matriculated students: 23 full-time (4 women), 6 part-time (2 women); includes 3 foreign. 30 applicants, 50% accepted. In 1990, 9 master's, 4 Diplomas awarded. *Degree requirements:* For master's, thesis required, foreign language not required; for Diploma, thesis not required. *Entrance requirements:* Proficiency in French, appropriate bachelor's degree. Application deadline: 5/1. Application fee: $15. *Expenses:* Tuition of $555 per trimester full-time, $37 per credit part-time for Canadian residents; $3480 per trimester full-time, $234 per credit part-time for nonresidents. Fees of $57.50 per trimester full-time, $19.50 per trimester part-time. *Financial aid:* Fellowships, research assistantships, teaching assistantships available. • Peter Zwack, Director, 514-987-3312. Application contact: Lucille Boisselle-Roy, Admissions Officer, 514-987-3128.

Directory: Meteorology and Atmospheric Sciences

University of Alaska Fairbanks, College of Natural Sciences, Department of Physics, Fairbanks, AK 99775. Offerings include atmospheric science (MS, PhD). Terminal master's awarded for partial completion of doctoral program. Department faculty: 25 full-time (1 woman), 0 part-time. *Degree requirements:* For master's, thesis; for doctorate, 1 foreign language (computer language can substitute), dissertation. *Entrance requirements:* For master's, GRE General Test, GRE Subject Test (minimum score of 550 required); for doctorate, GRE General Test, GRE Subject Test. Application deadline: 2/15. Application fee: $20. *Expenses:* Tuition of $1620 per year full-time, $90 per credit part-time for state residents; $3240 per year full-time, $180 per credit part-time for nonresidents. Fees of $464 per year full-time. • Dr. John Morack, Head, 907-474-7339.

University of Arizona, College of Arts and Sciences, Faculty of Science, Department of Atmospheric Sciences, Tucson, AZ 85721. Department awards MS, PhD. Faculty: 10 full-time (0 women), 2 part-time (0 women). Matriculated students: 22 full-time (2 women), 7 part-time (0 women); includes 2 minority (both Asian American), 3 foreign. Average age 30. 39 applicants, 46% accepted. In 1990, 2 master's, 3 doctorates awarded. *Degree requirements:* For master's, computer language, thesis or alternative required, foreign language not required; for doctorate, computer language, dissertation required, foreign language not required. *Application deadline:* 4/15 (applications processed on a rolling basis). *Application fee:* $25. *Expenses:* Tuition of $0 for state residents; $5406 per year full-time, $209 per credit hour part-time for nonresidents. Fees of $1528 per year full-time, $80 per credit hour part-time. *Financial aid:* In 1990–91, $156,182 in aid awarded. 1 fellowship (0 to first-year students), 7 research assistantships, 9 teaching assistantships, 2 scholarships were awarded; full tuition waivers also available. Financial aid applicants required to submit FAF. *Faculty research:* Climate dynamics, radiative transfer and remote sensing, atmospheric chemistry, atmosphere dynamics, atmospheric electricity. Total annual research budget: $375,000. • Dr. E. Philip Krider, Head, 602-621-6831. Application contact: Nancy Emptage, Secretary, 602-621-6831.

University of California, Davis, Program in Atmospheric Science, Davis, CA 95616. Program awards MS, PhD. Faculty: 8. Matriculated students: 28. 30 applicants, 47% accepted. In 1990, 7 master's, 1 doctorate awarded. *Degree requirements:* For master's, thesis or comprehensive exam; for doctorate, dissertation, 3 part qualifying exam. *Entrance requirements:* GRE General Test, TOEFL (minimum score of 550 required), minimum GPA of 3.0. Application fee: $40. *Expenses:* Tuition of $0 for state residents; $7699 per year full-time, $3849 per year part-time for nonresidents. Fees of $2718 per year full-time, $1928 per year part-time. *Financial aid:* Fellowships, research assistantships, teaching assistantships available. *Faculty research:* Air pollution/air quality, biometeorology, boundary-layer meteorology, climate dynamics, mesoscale modeling. • Graduate Adviser, 916-752-1669.

Announcement: Research areas: air-quality and boundary-layer meteorology, including ground and aircraft field programs, theoretical, numerical simulations; mesoscale meteorology, focusing mainly on numerical modeling; biometeorology, including observations, modeling of turbulent flow/transfer in plant canopies; climate dynamics, including seasonal climate problems, fluid dynamics, NWP. Affiliated groups: applied mathematics, ecology, and hydrology programs and National Institute for Global Environmental Change.

University of California, Los Angeles, College of Letters and Science, Department of Atmospheric Sciences, Los Angeles, CA 90024. Department awards MS, PhD, C Phil. Faculty: 12. Matriculated students: 46 full-time (15 women), 0 part-time; includes 3 minority, 28 foreign. 53 applicants, 64% accepted. In 1990, 9 master's, 5 doctorates, 5 C Phils awarded. *Degree requirements:* For master's, thesis or comprehensive exam required, foreign language not required; for doctorate, dissertation, comprehensive exam, oral qualifying exam required, foreign language not required. *Application fee:* $40. *Expenses:* Tuition of $0 for state residents; $7699 per year for nonresidents. Fees of $2907 per year. *Financial aid:* In 1990–91, 46 students received a total of $578,087 in aid awarded. 45 fellowships, 23 research assistantships, 6 teaching assistantships were awarded; full and partial tuition waivers, federal work-study, institutionally sponsored loans also available. Financial aid application deadline: 3/1. • Dr. Michael Ghil, Chair, 310-825-1217.

University of Chicago, Division of the Physical Sciences, Department of the Geophysical Sciences, Chicago, IL 60637. Offerings include atmospheric sciences (SM, PhD). *Degree requirements:* For master's, thesis, seminar; for doctorate, 1 foreign language, dissertation, seminar. *Entrance requirements:* For master's, TOEFL; for doctorate, GRE, TOEFL (minimum score of 550 required). Application deadline: 1/6. Application fee: $45. *Expenses:* Tuition of $16,275 per year full-time, $8140 per year part-time. Fees of $356 per year.

See full description on page 259.

University of Colorado at Boulder, College of Arts and Sciences, Department of Astrophysical, Planetary, and Atmospheric Sciences, Boulder, CO 80309. Offerings include astrophysical and geophysical fluid dynamics (MS, PhD). Terminal master's awarded for partial completion of doctoral program. Department faculty: 72 full-time (4 women). *Degree requirements:* For master's, thesis or alternative, comprehensive exam required, foreign language not required; for doctorate, 1 foreign language, dissertation. *Entrance requirements:* GRE General Test, GRE Subject Test. Application deadline: 3/1 (priority date, applications processed on a rolling basis). Application fee: $30 ($50 for foreign students). *Expenses:* Tuition of $2308 per year full-time, $387 per semester (minimum) part-time for state residents; $8730 per year full-time, $1455 per semester (minimum) part-time for nonresidents. Fees of $207 per semester full-time, $27.26 per semester (minimum) part-time. • Ellen Zweibel, Chairman, 303-492-8915.

University of Delaware, College of Arts and Science, Department of Geography, Program in Climatology, Newark, DE 19716. Program awards PhD. Faculty: 10 full-time (2 women), 0 part-time. Matriculated students: 8 full-time (5 women), 4 part-time (2 women); includes 1 black American, 0 Native American, 4 foreign. 8 applicants, 50% accepted. In 1990, 1 degree awarded. *Degree requirements:* Computer language, dissertation required, foreign language not required. *Entrance requirements:* GRE General Test. Application deadline: 2/1. Application fee: $40. *Tuition:* $179 per credit hour for state residents; $467 per credit hour for nonresidents. *Financial aid:* In 1990–91, $74,270 in aid awarded. 3 fellowships (1 to a first-year student), 2 research assistantships, 3 teaching assistantships (1 to a first-year student) were awarded. *Faculty research:* Physical and applied climatology, water budgets, socioeconomic value of weather, glaciology. • Dr. Cort J. Willmott, Chair, Department of Geography, 302-451-2294.

University of Hawaii at Manoa, School of Ocean and Earth Science and Technology, Department of Meteorology, Honolulu, HI 96822. Department awards MS, PhD. *Tuition:* $800 per semester full-time for state residents; $2405 per semester full-time for nonresidents.

University of Kansas, College of Liberal Arts and Sciences, Department of Physics and Astronomy, Lawrence, KS 66045. Department offers programs in atmospheric science (MS), physics (MA, MS, PhD). Faculty: 25 full-time (1 woman), 2 part-time (0 women). Matriculated students: 38 full-time (6 women), 15 part-time (1 woman); includes 3 minority (1 Asian American, 2 Hispanic American), 20 foreign. In 1990, 4 master's, 3 doctorates awarded. *Degree requirements:* For master's, foreign language and thesis not required; for doctorate, computer language, dissertation. *Entrance requirements:* TOEFL (minimum score of 570 required). Application fee: $25. *Expenses:* Tuition of $1668 per year full-time, $56 per credit hour part-time for state residents; $5382 per year full-time, $179 per credit hour part-time for nonresidents. Fees of $338 per year full-time, $25 per credit hour part-time. *Financial aid:* Fellowships, research assistantships, teaching assistantships available. • Ray Ammar, Chairperson, 913-864-4626.

See full description on page 649.

University of Maryland College Park, College of Computer, Mathematical and Physical Sciences, Department of Meteorology, College Park, MD 20742. Department awards MS, PhD. Faculty: 19 (1 woman). Matriculated students: 40 full-time (8 women), 11 part-time (3 women); includes 3 minority (2 Asian American, 1 black American), 28 foreign. 87 applicants, 51% accepted. In 1990, 6 master's, 1 doctorate awarded. *Entrance requirements:* For master's, minimum GPA of 3.0. Application deadline: rolling. Application fee: $25. *Expenses:* Tuition of $143 per credit hour for state residents; $256 per credit hour for nonresidents. Fees of $171.50 per semester. *Financial aid:* In 1990–91, 8 fellowships (5 to first-year students), 32 research assistantships, 0 teaching assistantships awarded. • Dr. Robert Hudson, Acting Chairman, 301-405-5391.

See full description on page 535.

University of Miami, Rosenstiel School of Marine and Atmospheric Science, Division of Meteorology and Physical Oceanography, Coral Gables, FL 33124. Division offers programs in atmospheric science (MA, MS, PhD), physical oceanography (MA, MS, PhD). Faculty: 13 (2 women). Matriculated students: 23 full-time (11 women), 1 part-time (0 women); includes 4 minority (1 black American, 3 Hispanic American), 14 foreign. Average age 32. In 1990, 6 master's awarded; 10 doctorates awarded (100% found work related to degree). Terminal master's awarded for partial completion of doctoral program. *Degree requirements:* For master's, thesis optional, foreign language not required; for doctorate, dissertation required, foreign language not required. *Entrance requirements:* GRE General Test, TOEFL (minimum score of 550 required). Application fee: $35. *Expenses:* Tuition of $567 per credit hour. Fees of $87 per semester full-time. *Financial aid:* In 1990–91, 1 fellowship awarded; research assistantships, teaching assistantships also available. *Faculty research:* Observation and theoretical modeling of oceanic and atmospheric circulations, climate theory, remote sensing. Total annual research budget: $2.5-million. • Dr. Otis Brown, Head, 305-350-7257.

University of Michigan, College of Engineering, Department of Atmospheric, Oceanic, and Space Sciences, Program in Atmospheric and Space Sciences, Ann Arbor, MI 48109. Program awards MS, PhD. *Degree requirements:* For doctorate, dissertation. *Application deadline:* 2/1 (applications processed on a rolling basis). *Application fee:* $30. *Tuition:* $3506 per semester full-time, $352 per credit (minimum) part-time for state residents; $7140 per semester full-time, $757 per credit (minimum) part-time for nonresidents. *Financial aid:* Application deadline 3/15. • Suskil Atreya, Chair, 313-764-3335.

University of Missouri–Columbia, College of Agriculture, Department of Atmospheric Science, Columbia, MO 65211. Department awards MS, PhD. Faculty: 5 full-time, 0 part-time. Matriculated students: 12 full-time (1 woman), 9 part-time (0 women); includes 1 minority (black American), 14 foreign. In 1990, 2 master's, 0 doctorates awarded. Terminal master's awarded for partial completion of doctoral program. *Degree requirements:* Thesis/dissertation required, foreign language not required. *Entrance requirements:* GRE General Test, minimum GPA of 3.0. Application deadline: 8/1 (priority date, applications processed on a rolling basis). Application fee: $20 ($40 for foreign students). *Expenses:* Tuition of $89.90 per credit hour full-time, $98.35 per credit hour part-time for state residents; $244 per credit hour full-time, $252.45 per credit hour part-time for nonresidents. Fees of $123.55 per semester (minimum) full-time. • Dr. Wayne L. Decker, Director of Graduate Studies, 314-882-6591. Application contact: Gary L. Smith, Director of Admissions and Registrar, 314-882-7651.

University of Oklahoma, College of Geosciences, School of Meteorology, Norman, OK 73019. School awards MS Metr, PhD. Part-time programs available. Faculty: 11 full-time (0 women), 0 part-time. Matriculated students: 77; includes 0 minority. Average age 27. In 1990, 10 master's awarded; 1 doctorate awarded (100% entered university research/teaching). *Degree requirements:* For master's, thesis or alternative, comprehensive exam required, foreign language not required; for doctorate, 1 foreign language (computer language can substitute), dissertation, departmental qualifying exam. *Entrance requirements:* For master's, GRE, TOEFL (minimum score of 550 required), bachelor's degree in field or related area; for doctorate, GRE, TOEFL (minimum score of 550 required). Application deadline: 2/1 (priority date, applications processed on a rolling basis). Application fee: $10. *Expenses:* Tuition of $63 per credit hour for state residents; $192 per credit hour for nonresidents. Fees of $67.50 per semester. *Financial aid:* In 1990–91, $300,000 in aid awarded. 0 fellowships, 25 research assistantships (8 to first-year students), 10 teaching assistantships (5 to first-year students) were awarded; federal work-study, institutionally sponsored loans, and career-related internships or fieldwork also available. Aid available to part-time students. Financial aid application deadline: 2/1. *Faculty research:* Atmospheric dynamics, cloud physics, climatology, synoptic and mesometeorology. Total annual research budget: $1.310-million. • Dr. Claude E. Duchon, Director, 405-325-6561.

University of Utah, College of Mines and Earth Sciences, Department of Meteorology, Salt Lake City, UT 84112. Department awards MS, PhD. Faculty: 9 full-time (1 woman), 0 part-time. Matriculated students: 13 full-time (3 women), 0 part-time; includes 1 minority (Asian American), 5 foreign. Average age 30. 16 applicants, 31% accepted. In 1990, 2 master's, 3 doctorates awarded. *Degree requirements:* For master's, thesis optional, foreign language not required; for doctorate, dissertation required, foreign language not required. *Entrance requirements:* GRE General Test, TOEFL, minimum GPA of 3.0. Application deadline: 8/1. Application fee: $25 ($50 for foreign students). *Tuition:* $195 per credit for state residents; $505 per credit for nonresidents. *Financial aid:* In 1990–91, $90,000 in aid awarded. 10 research assistantships, 0 teaching assistantships were awarded. Financial aid application deadline: 3/1; applicants required to submit FAF. *Faculty research:* Micrometeorology, cloud physics, air pollution, dynamical processes, satellite meteorology. • Dr. J. E. Geisler, Chairman, 801-581-6136.

University of Washington, College of Arts and Sciences, Department of Atmospheric Sciences, Seattle, WA 98195. Department awards MS, PhD. Part-time programs available. Faculty: 17 full-time (1 woman), 12 part-time (1 woman). Matriculated students: 63 full-time (20 women), 6 part-time (2 women); includes 4 minority (3

Asian American, 1 Hispanic American), 10 foreign. Average age 24. 98 applicants, 21% accepted. In 1990, 7 master's awarded (57% found work related to degree, 43% continued full-time study); 7 doctorates awarded (29% entered university research/teaching, 71% found other work related to degree). *Degree requirements:* For master's, thesis required, foreign language not required; for doctorate, dissertation, qualifying exam required, foreign language not required. *Entrance requirements:* GRE General Test. Application deadline: 3/1 (priority date, applications processed on a rolling basis). Application fee: $35. *Tuition:* $1129 per quarter full-time, $324 per credit (minimum) part-time for state residents; $2824 per quarter full-time, $809 per credit (minimum) part-time for nonresidents. *Financial aid:* In 1990–91, 0 fellowships, 54 research assistantships (14 to first-year students), 3 teaching assistantships (0 to first-year students) awarded. Average monthly stipend for a graduate assistantship: $940. Financial aid application deadline: 3/1. *Faculty research:* Climate change, synoptic and mesoscale meterology, atmospheric chemistry, cloud physics, dynamics of the atmosphere. Total annual research budget: $4.7-million. • Norbert Untersteiner, Chairman, 206-543-4250. Application contact: Kathryn Stout, Academic Counselor, 206-543-6471.

University of Wisconsin–Madison, College of Letters and Science, Department of Meteorology, Madison, WI 53706. Department awards MS, PhD. Faculty: 16 full-time, 0 part-time. Matriculated students: 66 full-time (16 women), 4 part-time (1 woman); includes 2 minority (1 Asian American, 1 Hispanic American), 25 foreign. 109 applicants, 28% accepted. In 1990, 13 master's, 3 doctorates awarded. *Application deadline:* rolling. Application fee: $20. *Financial aid:* In 1990–91, 52 students received aid. 4 fellowships, 41 research assistantships, 5 teaching assistantships, 2 project assistantships awarded; institutionally sponsored loans also available. Financial aid application deadline: 1/15; applicants required to submit FAF. • John E. Kutzbach, Graduate Chairperson, 608-262-2827. Application contact: Joanne Nagy, Assistant Dean of the Graduate School, 608-262-2433.

University of Wyoming, College of Engineering, Department of Atmospheric Science, Laramie, WY 82071. Department awards MS, PhD. Faculty: 9 full-time (0 women), 0 part-time. Matriculated students: 10 full-time (2 women), 1 (woman) part-time; includes 4 foreign. Average age 28. 30 applicants, 13% accepted. In 1990, 3 master's awarded (67% entered university research/teaching, 33% found other work related to degree); 1 doctorate awarded (100% entered university research/teaching). *Degree requirements:* Thesis/dissertation required, foreign language not required. *Entrance requirements:* GRE General Test, TOEFL, minimum GPA of 3.0. Application deadline: 4/15 (priority date, applications processed on a rolling basis). Application fee: $30. *Tuition:* $1554 per year full-time, $74.25 per credit hour part-time for state residents; $4358 per year full-time, $74.25 per credit hour part-time for nonresidents. *Financial aid:* In 1990–91, 11 students received a total of $113,000 in aid awarded. 11 research assistantships were awarded; federal work-study, institutionally sponsored loans, and career-related internships or fieldwork also available. Average monthly stipend for a graduate assistantship: $742. Financial aid application deadline: 4/15. *Faculty research:* Cloud and precipitation processes, mesoscale dynamics, weather modification, winter storms, aircraft instrumentation. Total annual research budget: $1.5-million. • Dr. John D. Marwitz, Head, 307-766-3246.

Yale University, Graduate School of Arts and Sciences, Department of Geology and Geophysics, Program in Meteorology, New Haven, CT 06520. Program awards PhD. *Degree requirements:* Dissertation. *Entrance requirements:* GRE General Test, GRE Subject Test. Application deadline: 1/2. Application fee: $45. *Tuition:* $15,160 per year full-time, $1895 per course part-time. • Dr. B. Saltzman, Chairman, Department of Geology and Geophysics, 203-432-3114. Application contact: Susan Webb, Coordinator of Admissions, 203-432-2770.

See full description on page 299.

Cross-Discipline Announcement

University of Chicago, Division of the Physical Sciences, Department of the Geophysical Sciences, Chicago, IL 60637.

The department offers graduate-level course work and research in the atmospheric sciences leading to the MS and PhD degrees. Current research areas include cloud physics, greenhouse effect and global warming, atmospheric chemistry and radiation, remote sensing, and fluid dynamics.

SECTION 6: METEOROLOGY AND ATMOSPHERIC SCIENCES

COLORADO STATE UNIVERSITY
Department of Atmospheric Science

Programs of Study

The Department of Atmospheric Science offers programs leading to the M.S. and Ph.D. degrees. Full-time students typically complete the M.S. in four semesters and the Ph.D. in four years beyond the B.S. degree. Students working part-time normally require two years for the M.S. and four to six years beyond the B.S. for a Ph.D.

Active academic and research programs are offered in the fields of dynamic meteorology; general circulation and climate modeling; tropical meteorology and tropical cyclones; mesoscale meteorology; weather modification; cloud and precipitation physics; cumulus convection and cloud dynamics; atmospheric chemistry and air pollution; satellite meteorology; airborne meteorological instrumentation; theoretical and dynamic meteorology; atmospheric radiation; remote sensing; global, regional, and local climatology; and mountain meteorology.

Research Facilities

The Department of Atmospheric Science is located on the CSU Foothills campus. The three-story Atmospheric Science Building and several smaller buildings provide over 30,000 square feet of office and research space. Academic and research facilities include a geosynchronous satellite receiving station; a digital image processing system for display and analysis of satellite data; a weather laboratory; UNIFAX; optical and infrared radiometers; a cloud simulation laboratory; an extensive field observation system including a 10-cm Doppler radar; a UHF Doppler wind profiler; rawinsonde and tethered balloon systems; an infrared interferometer and spectral radiometers; multiple surface energy budget stations; towers; and electronics and machine shops. Extensive computer resources are available ranging from PCs to superworkstations capable of running complex atmospheric models. Convenient access is provided to mainframe computers and supercomputers throughout the United States through ARPANET and NSFNET.

The Cooperative Institute for Research in the Atmosphere (CIRA) has been formed between CSU and the National Oceanic and Atmospheric Administration (NOAA). CIRA research programs are typically interdisciplinary and involve students and faculty from the Department of Atmospheric Science and other CSU departments as well as NOAA scientists.

Financial Aid

Aid for some qualified students is available in the form of research assistantships. For the 1990–91 academic year, the half-time stipend for assistantships was $907–$937 per month for M.S. students and $967–$1027 per month for Ph.D. students, including a tuition waiver. Graduate assistants pay only a special registration and fees assessment, which totaled $244.40 per semester for the 1990–91 academic year. Full-time summer employment is also generally available. Research assistants are assigned to a research project consistent with their area of major interest.

Cost of Study

During 1990–91, tuition was $999 per semester for Colorado residents and $3181 per semester for out-of-state residents. Fees are $244.40 per semester. As noted above, tuition is paid by research grants for those holding graduate research assistantships. These figures are subject to change.

Cost of Living

Accommodations for single graduate students are available. Within the city of Fort Collins, rent ranges from $250 to $350 per month for buffet apartments to $600 or more for a furnished house. There is limited campus housing for married students. In 1990–91, a two-bedroom unit rented for $342–$368 per month, and a three-bedroom unit rented for $411 per month, including utilities except telephone service. These figures are subject to change.

Student Group

The total enrollment at Colorado State University is approximately 18,100 regular, on-campus students, of whom about 2,400 are graduate students. About 90 students are enrolled in the graduate program in atmospheric science.

Location

Fort Collins, a college community with a population of approximately 95,000, is located 65 miles north of Denver. The rolling green foothills west of the city are accentuated by a backdrop of the Rocky Mountains. The climate is mild throughout the year, with a summer temperature ranging from an average high of 82 degrees to an average low of 52 degrees; the winter temperature ranges from an average high of 41 degrees to an average low of 13 degrees. Over 55 inches of snow fall in an average winter season while the summers have frequent thunderstorms. Numerous recreational and cultural activities are available. Fort Collins offers symphony and community concert series, a fine arts festival, a city-owned museum, an olympic-size natatorium/ice rink complex, art league exhibits, and theater. In addition, visiting campus speakers are frequently invited to lecture on a variety of subjects, ranging from engineering science to poetry. Horsetooth Reservoir (on the western edge of Fort Collins), Cache la Poudre River Canyon, and Red Feather lakes, as well as Rocky Mountain National Park, are nearby summer recreation centers for swimming, fishing, boating, and hiking. Cross-country and downhill skiing areas are within a 1½-hour drive to the west.

The University

CSU, founded in 1870 as a land-grant college, now has three campuses. The main campus, covering more than 400 acres, is located about half a mile south of the Fort Collins business district; the Foothills research campus of 1,700 acres is located 4 miles west of the main campus; and the Pingree Park campus is located high in the mountains 55 miles west of Fort Collins.

Applying

Students applying for admission to the atmospheric science program should have a B.S. degree in meteorology, physics, geophysics, mathematics, chemistry, or engineering. Further information and application forms may be obtained from the Department of Atmospheric Science. Completed forms should be filed six months before the date on which the applicant wishes to be admitted. Applications for financial aid should be filed by March 1 for consideration for the fall semester.

Correspondence and Information

Dr. Stephen K. Cox, Head
Department of Atmospheric Science
Colorado State University
Fort Collins, Colorado 80523
Telephone: 303-491-8360

SECTION 6: METEOROLOGY AND ATMOSPHERIC SCIENCES

Colorado State University

THE FACULTY AND THEIR RESEARCH

Professors
William R. Cotton, Ph.D., Penn State, 1970. Numerical modeling, cloud physics and dynamics, mesoscale meteorology.
Stephen K. Cox, Head of the Department; Ph.D., Wisconsin, 1967. Radiation physics, general circulation.
Lewis O. Grant, M.S., Caltech, 1948. Precipitation physics, weather modification, mountain weather, hydrometeorology.
William M. Gray, Ph.D., Chicago, 1964. Tropical meteorology, atmospheric vortices, cumulus convection.
Richard H. Johnson, Ph.D., Washington (Seattle), 1975. Atmospheric convection, boundary-layer meteorology, synoptic and mesoscale meteorology.
Thomas B. McKee, Colorado State Climatologist; Ph.D., Colorado State, 1972. Climatology, atmospheric physics.
Roger A. Pielke, Ph.D., Penn State, 1973. Mesoscale modeling, weather forecasting, air-quality modeling.
Wayne H. Schubert, Ph.D., UCLA, 1973. Dynamic and theoretical meteorology, parameterization of cumulus convection, planetary circulations.
Graeme L. Stephens, Ph.D., Melbourne, 1977. Radiation theory, radiative parameterization, cloud/climate studies.
Thomas H. Vonder Haar, Ph.D., Wisconsin, 1968. Satellite meteorology, radiation physics, global climate.

Associate Professors
David Randall, Ph.D., UCLA, 1976. General circulation modeling, planetary boundary layer, convective cloud dynamics.
Steven Rutledge, Ph.D., Oregon State, 1983. Radar and mesoscale meteorology, precipitation physics, atmospheric electricity.
Peter C. Sinclair, Ph.D., Arizona, 1965. Severe storms, cumulus dynamics, thunderstorm modification, meteorological instrumentation.

The CSU tethered balloon boundary-layer profiler being flown near Vail, Colorado.

The Colorado State University Weather Station No. 053005, featuring both research and operational weather observations since 1889.

The atmospheric science graduate education and research facilities nestled in the foothills of the Rockies.

SECTION 6: METEOROLOGY AND ATMOSPHERIC SCIENCES

COLUMBIA UNIVERSITY/ NASA GODDARD SPACE FLIGHT CENTER'S INSTITUTE FOR SPACE STUDIES

Atmospheric and Planetary Science Program

Program of Study

The Departments of Applied Physics, Astronomy, Geological Sciences, and Physics jointly offer a graduate program in atmospheric and planetary science leading to the Ph.D. degree. Four to six years are generally required to complete the Ph.D., including the earning of M.A. and M.Phil. degrees. Applicants should have a strong background in physics and mathematics, including advanced undergraduate courses in mechanics, electromagnetism, advanced calculus, and differential equations.

The program is conducted in cooperation with the NASA Goddard Space Flight Center's Institute for Space Studies, which is adjacent to Columbia University. Members of the Institute hold adjunct faculty appointments, offer courses, and supervise the research of graduate students in the program. The Institute holds colloquia and scientific conferences in which the University community participates. Opportunities for visiting scientists to conduct research at the Institute are provided by postdoctoral research programs administered by the National Academy of Sciences–National Research Council and Columbia and supported by NASA.

Research at the Institute focuses on broad studies of natural and anthropogenic global changes that affect the planet's habitability. Areas of study include global climate, biogeochemical cycles, earth observations, planetary atmospheres, and related interdisciplinary studies. The global climate program involves basic research on climatic variations and climate processes, including the development of one-, two-, and three-dimensional numerical models to study the climatic effects of increasing carbon dioxide and other trace gases, aerosols, solar variability, and changing surface conditions. Biogeochemical cycles research utilizes three-dimensional models to study the distribution of trace gases in the troposphere and stratosphere and to examine the role of the biosphere in the global carbon cycle. The earth observations program entails research in the retrieval of cloud, aerosol, and surface radiative properties from global satellite radiance data to further understanding of their effects on climate. The planetary atmospheres program includes comparative modeling of radiative transfer and dynamics applied to Venus, Titan, and the Jovian planets; participation in spacecraft experiments; and analysis of ground-based observations. Interdisciplinary research includes studies of turbulence and solar system formation, evolution of solar-type stars, and quantum chemistry.

Research Facilities

The Institute operates a modern general-purpose scientific computing facility consisting of Amdahl 5870 and IBM 4381 computers, four IBM RISC 6000 workstations, and peripheral equipment. An IBM 7350 video display system permits interactive processing, display, and analysis of satellite imagery and other digital data. Daily weather maps, analyses, and satellite images are received from the National Weather Service via a DIFAX printer and VGA software graphics. The Institute conducts the Cloud Photopolarimeter experiment on the *Pioneer Venus Orbiter* spacecraft and the Photopolarimeter Radiometer experiment on the *Galileo Jupiter Orbiter* spacecraft. The Institute is the Global Processing Center for the International Satellite Cloud Climatology Project, a decade-long observational program involving data from five geostationary and two polar-orbiting Earth satellites. Institute scientists are developing the Earth Observing Scanning Polarimeter for flight on the first Earth Observing System satellite. Institute personnel frequently collaborate with scientists at the Goddard Space Flight Center in Greenbelt, Maryland. Close research ties also exist with the Lamont-Doherty Geological Observatory of Columbia University, especially in the areas of geochemistry, oceanography, and paleoclimate studies. All facilities, including the Institute's library containing approximately 17,000 volumes, are made available to students in the program.

Financial Aid

Research assistantships are available to most students in the program. Graduate assistantships in 1991–92 carry a twelve-month stipend of $1155 per month and include a tuition waiver and payment of fees.

Cost of Study

Tuition and fees for 1991–92 are estimated at $16,472. As noted above, tuition and fees are paid for graduate students holding research assistantships.

Cost of Living

Limited on-campus housing is available for single and married graduate students. Rates range from $3230 for a single room on a 250-day contract to $3790 for a double room on a 350-day contract. Studios, suites, and one-bedroom apartments for married students, rated on a 350-day contract, range from $4010 to $6675. Most students live off campus, many of them in apartments owned and operated by the University within a few blocks of the campus.

Student Group

Of the 18,500 students at Columbia, 3,500 are students in the Graduate School of Arts and Sciences. In 1990–91, there were over 40 Columbia students at the Institute for Space Studies, 15 of whom were Ph.D. candidates in the Atmospheric and Planetary Science Program. There were also 34 University and 4 National Academy of Sciences research associates in residence at the Institute.

Location

Columbia University is located in the Morningside Heights section of Manhattan in New York City. New York's climate is moderate, with average maximum and minimum temperatures of 85 and 69 degrees in July and 40 and 28 in January. New York is one of the top cultural centers in the United States and, as such, provides unrivaled opportunities for attending concerts, operas, and plays and for visiting world-renowned art, scientific, and historical museums. Student discount tickets for many musical and dramatic performances are available in the Graduate Student Lounge. A comprehensive network of public transportation alleviates the need for keeping an automobile in the city. The superb beaches of Long Island, including the Fire Island National Seashore, are within easy driving distance, as are the numerous ski slopes, state parks, and other mountain recreational areas of upstate New York and southern New England.

The University

Columbia University, founded in 1754 by royal charter of King George II of England, is a member of the Ivy League. It is the oldest institution of higher learning in New York State and the fifth-oldest in the United States. It consists of sixteen separate schools and colleges with over 1,700 full-time faculty members.

Applying

To enter the program, an application must be submitted to one of the four participating departments. Completed forms should be received by January 15 from students applying for September admission.

Correspondence and Information

Dr. Anthony D. Del Genio
Atmospheric and Planetary Science Program
Armstrong Hall—GISS
Columbia University
2880 Broadway
New York, New York 10025
Telephone: 212-678-5588

Peterson's Guide to Graduate Programs in the Physical Sciences and Mathematics 1992

SECTION 6: METEOROLOGY AND ATMOSPHERIC SCIENCES

Columbia University/NASA Goddard Space Flight Center's Institute for Space Studies

THE INSTITUTE STAFF AND THEIR RESEARCH

Michael Allison, Ph.D., Rice, 1982. Planetary atmospheric dynamics, remote sensing meteorology of Jupiter and the outer planets.

Vittorio M. Canuto, Ph.D., Turin (Italy), 1960. Theory of fully developed turbulence, analytical models for large-scale turbulence and their applications to geophysics and astrophysics.

Barbara E. Carlson, Ph.D., SUNY at Stony Brook, 1984. Radiative transfer in planetary atmospheres, remote sensing and cloud modeling of Earth and Jovian planets.

Anthony D. Del Genio, Ph.D., UCLA, 1978. Dynamics of planetary atmospheres, parameterization of clouds and cumulus convection, climate change, general circulation.

Inez Y. Fung, Sc.D., MIT, 1977. Carbon cycle modeling and remote sensing, dynamic meteorology and oceanography, air-sea interactions.

Vivien Gornitz, Ph.D., Columbia, 1969. Planetary geology, remote sensing of Earth resources, sea-level changes, impact of anthropogenic land-cover changes on climate.

Sheldon Green, Ph.D., Harvard, 1971. Theoretical chemical physics: quantum chemistry of molecular structures and properties; scattering calculations of molecular collisions; radioastronomical, atmospheric, and remote sensing applications.

James E. Hansen, Head of the Institute for Space Studies; Ph.D., Iowa, 1967. Remote sensing of Earth and planetary atmospheres, global modeling of climate processes and climate sensitivity.

Andrew A. Lacis, Ph.D., Iowa, 1970. Radiative transfer, climate modeling, remote sensing of Earth and planetary atmospheres.

Dorothy M. Peteet, Ph.D., NYU, 1983. Paleoclimatology, palynology, ecology, botany.

Michael J. Prather, Ph.D., Yale, 1975. Atmospheric chemistry, global distribution and budgets of trace gases, evolution of Earth and planetary atmospheres.

David H. Rind, Ph.D., Columbia, 1976. Atmospheric and climate dynamics, stratospheric modeling and remote sensing.

Cynthia Rosenzweig, Ph.D., Massachusetts at Amherst, 1991. Parameterization of ground hydrology and biosphere, impacts of climate change on agriculture.

William B. Rossow, Ph.D., Cornell, 1976. Planetary atmospheres and climate, cloud physics and climatology, general circulation.

Gary L. Russell, Ph.D., Columbia, 1976. Numerical methods, general circulation modeling.

Richard B. Stothers, Ph.D., Harvard, 1964. Astronomy, climatology, geophysics, solar physics, history of science.

Larry D. Travis, Associate Chief of the Institute for Space Studies; Ph.D., Penn State, 1971. Remote sensing of Earth and planetary atmospheres, radiative transfer, numerical modeling.

The Institute hosts conferences and workshops that bring scientists together to discuss relevant dynamics, radiation, and chemistry issues. Conferences and workshops on Jovian atmospheres, clouds in climate, and vegetation-climate interactions have been held recently.

Fossil pollen and spores obtained by coring swamp sediments are used by Institute scientists to document ancient climate changes.

Graduate students can use the Institute's IBM 7350 video display system to process and analyze visible and infrared data from earth-orbiting satellites and to view the results of global climate model simulations.

SECTION 6: METEOROLOGY AND ATMOSPHERIC SCIENCES

PENNSYLVANIA STATE UNIVERSITY
Department of Meteorology

Programs of Study

The Department of Meteorology at Penn State is one of the oldest and largest in the country. It offers individually tailored graduate programs of academic study leading to the M.S. and Ph.D. degrees. The diverse teaching and research interests of the faculty and the varied research activities of students encompass the full range of specialty areas in meteorology and the atmospheric sciences.

Faculty advisers work closely with each graduate student to assist in developing a scholarly approach to, and understanding of, both fundamental and advanced material. The department's goal is to develop within its students an appreciation of the multitude of applied environmental problems that require meteorological knowledge for their solution and of the diversity of available knowledge in the sciences that can be used to solve contemporary meteorological problems.

Areas of research emphasized in the department are dynamic and physical meteorology: the theory of large-scale and mesoscale motions; synoptic and numerical weather analysis and prediction; tropical meteorology; climate theory and modeling; atmospheric turbulence; the theory and practice of direct and indirect atmospheric measurements and sounding; and the theory and laboratory study of atmospheric precipitation mechanisms, atmospheric chemistry, and radiative transfer.

Interdisciplinary programs in applied mathematics, fluid mechanics, air pollution, biometeorology, and earth systems science are also encouraged by the graduate faculty. Students preferring to terminate their graduate studies with the M.S. degree may select a nonthesis option emphasizing applied meteorology.

Research Facilities

The department's excellent research facilities include Doppler acoustic sounders, VHF and UHF Doppler wind profilers, microwave radiometers, and a comprehensive set of micrometeorological and air-chemistry- and air-pollution-related instruments. Several minicomputer systems are employed for synoptic studies and field measurement programs. Both the University's IBM 3090 Model 600 computer and the Cray computers at NCAR are used for developing and running boundary layer, mesoscale, and climate simulation models. The weather observatory includes a full complement of weather data services, radar, and digital satellite picture-receiving equipment.

Financial Aid

The department offers half-time graduate assistantships that paid $12,000–$14,500 per year in addition to tuition in 1990–91. Students may also compete for a variety of tuition-free fellowships administered by the University. Most assistantships and fellowships are government sponsored. Numerous opportunities exist for students' spouses who seek employment.

Cost of Study

Tuition fees for full-time study in 1990–91 were $2225 per semester for Pennsylvania residents and $4445 per semester for nonresidents. No tuition is paid by graduate assistants.

Cost of Living

University dormitories and private rooms are available to single students, and inexpensive University apartments are available to married students. Off-campus apartments are also available. Application for graduate housing should be made as soon as possible.

Student Group

The average graduate enrollment in the department is about 80; two thirds are M.S. candidates, and one third are Ph.D. candidates. Students from every state of the Union and many foreign countries have been in residence. With few exceptions, graduate students receive financial aid.

Location

The University is located in the town of State College, 1,200 feet above sea level in the green, rolling hills of the Alleghenies, close to the geographic center of Pennsylvania. State College has a population of about 30,000. The area has excellent fishing and hunting sites, lakes for boating and sailing, ski resorts, and many other recreational facilities. Most area residents are connected either with the University or with local research and development companies.

The University

The main campus of the Pennsylvania State University is located on a tract of 4,500 acres, only one tenth of which is occupied by buildings. With its large open areas, malls lined with old trees, well-kept greens, woods, and flower gardens, it is a beautiful campus. The enrollment at the main campus exceeds 30,000 students, and there are more than 2,200 faculty members. Students have access to facilities for nearly all summer and winter sports. Art exhibits, opera, ballet, concerts, and student musicals are presented throughout the year, adding to the rich cultural life of the area.

Applying

The academic graduate major is open to all students with a B.S. degree in science or engineering, in addition to those students who have majored in meteorology. Requirements include mathematics at least through differential equations and at least one year of college physics.

Penn State operates on a semester system, and applications are processed throughout the year. Prospective students should, however, submit their applications as early as possible, preferably with a lead time of six months or more, especially if they wish to compete for one of the government-sponsored fellowships or traineeships. Offers for financial assistance are made at the earliest possible moment. Admission is granted by the Graduate School of the University with the advice of the department.

Correspondence and Information

Graduate Admissions Officer
Department of Meteorology
503 Walker Building
Pennsylvania State University
University Park, Pennsylvania 16802
Telephone: 814-865-0478

Peterson's Guide to Graduate Programs in the Physical Sciences and Mathematics 1992

SECTION 6: METEOROLOGY AND ATMOSPHERIC SCIENCES

Pennsylvania State University

THE FACULTY AND THEIR RESEARCH

William M. Frank, Professor of Meteorology and Department Head; Ph.D., Colorado State. Cumulus parameterization, tropical meteorology.

Thomas P. Ackerman, Associate Professor of Meteorology; Ph.D., Washington (Seattle). Atmospheric physics, radiative transfer, climate theory.

Bruce A. Albrecht, Associate Professor of Meteorology; Ph.D., Colorado State. Marine boundary layers, cloud-climate interactions, remote sensing of clouds.

Peter R. Bannon, Associate Professor of Meteorology; Ph.D., Colorado at Boulder. Airflow over and around mountains, cyclogenesis, frontal formation and propagation, monsoon meteorology.

Alfred K. Blackadar, Professor Emeritus of Meteorology; Ph.D., NYU. Atmospheric planetary boundary layer.

Craig F. Bohren, Professor of Meteorology; Ph.D., Arizona. Atmospheric radiation, meteorological optics.

William H. Brune, Associate Professor of Meteorology; Ph.D., Johns Hopkins. Atmospheric chemistry, phenomena of the middle atmosphere, atmospheric and laboratory chemistry measurements.

John J. Cahir, Professor of Meteorology and Associate Dean for Resident Instruction; Ph.D., Penn State. Dynamic and synoptic meteorology.

Toby N. Carlson, Professor of Meteorology; Ph.D., Imperial College (London). Boundary-layer/biosphere modeling.

Carl R. Chelius, Assistant Professor of Meteorology; D.Ed., Penn State. Mesoscale phenomena, particularly mesoscale convective complexes.

John H. E. Clark, Associate Professor of Meteorology; Ph.D., Florida State. Mesoscale and synoptic scale dynamics.

Judith A. Curry, Associate Professor of Meteorology; Ph.D., Chicago. Polar climate and meteorology, atmospheric physics, remote sensing.

John A. Dutton, Professor of Meteorology and Dean of the College of Earth and Mineral Sciences; Ph.D., Wisconsin. Dynamics, theoretical meteorology, and atmospheric energetics.

Gregory S. Forbes, Associate Professor of Meteorology; Ph.D., Chicago. Mesoanalysis, forecasting severe storms, operational meteorology.

Alistair B. Fraser, Professor of Meteorology; Ph.D., Imperial College (London). Atmospheric optics, cloud physics and dynamics.

J. Michael Fritsch, Professor of Meteorology; Ph.D., Colorado State. Numerical modeling, synoptic and mesoscale meteorology, operational meteorology.

Charles L. Hosler, Professor of Meteorology, Senior Vice President for Research, and Dean of the Graduate School; Ph.D., Penn State. Cloud physics and dynamics, inadvertent weather modification.

James F. Kasting, Associate Professor; Ph.D., Michigan. Atmospheric chemistry, atmospheric thermodynamics, biogeochemical cycles.

Dennis Lamb, Associate Professor of Meteorology; Ph.D., Washington (Seattle). Atmospheric chemistry and cloud physics.

Hans Neuberger, Professor Emeritus of Meteorology; D.Sc., Hamburg. Physical meteorology, biometeorology.

John J. Olivero, Professor of Meteorology; Ph.D., Michigan. Aeronomy, atmospheric photochemistry, microwave measurements.

Rosa G. de Pena, Professor Emerita of Meteorology; Ph.D., Buenos Aires. Atmospheric chemistry and cloud physics.

Nelson L. Seaman, Assistant Professor of Meteorology; Ph.D., Penn State. Mesoscale modeling, objective analysis techniques, finite differencing processes.

Hampton N. Shirer, Associate Professor of Meteorology; Ph.D., Penn State. Theoretical meteorology, nonlinear dynamics, chaos, atmospheric waves, spectral modeling, convection, general circulation.

Dennis W. Thomson, Professor of Meteorology; Ph.D., Wisconsin. Indirect and direct atmospheric measurements, physical and marine meteorology.

Thomas T. Warner, Associate Professor of Meteorology; Ph.D., Penn State. Dynamics, dynamic meteorology, mesoscale dynamics and circulations, numerical weather predictions, physical climatology, climate modeling.

Peter J. Webster, Professor of Meteorology; Ph.D., MIT. Tropical meteorology, dynamics and climate theory.

John C. Wyngaard, Professor of Meteorology; Ph.D., Penn State. Turbulence, boundary-layer meteorology, instrumentation.

George S. Young, Assistant Professor of Meteorology; Ph.D., Colorado State. Observational and diagnostic studies of turbulence and mesoscale weather systems.

Theoretical studies of chaos in the atmosphere are being conducted.

The 94-GHz Doppler cloud radar.

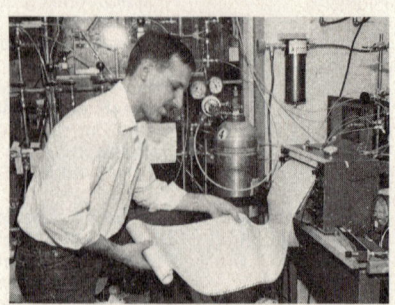

An atmospheric chemistry laboratory.

SECTION 6: METEOROLOGY AND ATMOSPHERIC SCIENCES

STATE UNIVERSITY OF NEW YORK AT ALBANY

Department of Atmospheric Science

Programs of Study The Department of Atmospheric Science offers programs of study leading to master's and doctoral degrees. The areas of active research include cloud and precipitation physics, solar energy meteorology, aerosol physics, dynamical and theoretical meteorology, synoptic-scale and mesoscale meteorology, numerical weather prediction, tropical meteorology, micrometeorology, atmospheric electricity, upper atmosphere physics, atmospheric chemistry, and physical limnology.

Admission to graduate study generally requires a bachelor's degree in a natural science or mathematics. With the exception of those with clearly superior academic records, students are expected to obtain a master's degree before beginning doctoral study.

Research Facilities The department's research facilities include the National Lightning Detection Network, an interactive computer/video weather data-handling system called McIDAS, a GOES satellite image recorder, a high-speed printer that outputs all National Weather Service surface and upper-air reports, and a field station at Whiteface Mountain in the Adirondacks. All these facilities are described in detail on the reverse side of this page.

Financial Aid Fellowships and assistantships in the amounts of $10,500 to $17,000 for fall 1991 through summer 1992, inclusive, are available to students admitted to graduate study. Tuition scholarships are available for students who are receiving financial support.

Cost of Study Resident tuition in 1991–92 is expected to be $1225 per term, and nonresident tuition is expected to be $2450 per term. The cost of books is estimated at $250 per term. Master's theses costs come to about $25; doctoral dissertations, $75.

Cost of Living Residence hall facilities plus two meals per day are $1728 per term. There is no University campus housing for married students. Off-campus accommodations include apartments renting for about $300 per month.

Student Group During the 1990–91 academic year there were 40 full-time students in the graduate programs.

Location The Albany area has abundant cultural and recreational opportunities. Located nearby are the Saratoga Performing Arts Center, summer home of the New York City Ballet and the Philadelphia Orchestra; Tanglewood, summer home of the Boston Symphony Orchestra; Jacob's Pillow, which features prominent artists in ballet and other forms of dance; a full range of theatrical activities and concerts at the Palace Theater in Albany, Proctor's Theater in Schenectady, Troy Music Hall, and Cohoes Music Hall; and the New York State Cultural Education Center, which includes the State Museum, Convention Center, and Performing Arts Center at the Nelson A. Rockefeller Empire State Plaza in downtown Albany.

The area also lends itself naturally to winter and summer recreational activities. It is surrounded by the Berkshire, Adirondack, Helderberg, and Catskill mountains. Historical and vacation spots within leisurely driving distance include New York City, Boston, Cape Cod, Connecticut, Vermont, Montreal, Lake George, Lake Placid, and Lake Champlain.

The University The State University of New York at Albany, the oldest unit and one of four University Centers of the statewide system, acknowledges the three traditional obligations of the University—teaching, research, and service to its community. The campus, designed by Edward Durell Stone, contains fourteen academic buildings within a common platform, all connected by a continuous roof and an enclosed below-level corridor.

Applying Applications for general admission are handled through the Office of Graduate Admissions. However, persons requiring information or applications for assistantships should write directly to the departmental chair as noted below. Applicants are required to submit scores on the General Test of the Graduate Record Examinations.

Correspondence and Information

For admission:
Office of Graduate Admissions
State University of New York at Albany
1400 Washington Avenue
Albany, New York 12222
Telephone: 518-442-3980

For applications:
Chair, Department of Atmospheric Science
State University of New York at Albany
1400 Washington Avenue
Albany, New York 12222
Telephone: 518-442-4556

Peterson's Guide to Graduate Programs in the Physical Sciences and Mathematics 1992

SECTION 6: METEOROLOGY AND ATMOSPHERIC SCIENCES

State University of New York at Albany

THE FACULTY AND THEIR RESEARCH

Lance F. Bosart, Professor of Atmospheric Science; Ph.D., MIT, 1969. Synoptic and mesoscale meteorology.

Marx Brook, Visiting Professor at Atmospheric Science; Ph.D., UCLA, 1953. Atmospheric electricity.

*Julius S. Chang, Research Professor of Atmospheric Science; Ph.D., SUNY at Stony Brook, 1971. Atmospheric chemistry, numerical modeling.

Ulrich H. Czapski, Associate Professor of Atmospheric Science; Ph.D., Hamburg, 1953. Turbulence and convection.

Kenneth L. Demerjian, Professor of Atmospheric Science and Director, Atmospheric Sciences Research Center; Ph.D., Ohio State, 1973. Atmospheric chemistry.

*David R. Fitzjarrald, Research Professor of Atmospheric Science; Ph.D., Virginia, 1980. Boundary-layer meteorology.

Harry L. Hamilton, Associate Professor of Atmospheric Science; Ph.D., Wisconsin–Madison, 1965. Micrometeorology, solar energy meteorology.

*Lee C. Harrison, Research Professor of Atmospheric Science; Ph.D., Washington (Seattle), 1982. Aerosol physics.

Vincent P. Idone, Research Professor of Atmospheric Science; Ph.D., SUNY at Albany, 1982. Atmospheric electricity.

Robert G. Keesee, Associate Professor of Atmospheric Science; Ph.D., Colorado, 1979. Atmospheric chemistry.

Daniel Keyser, Associate Professor of Atmospheric Science; Ph.D., Penn State, 1981. Synoptic-dynamic and mesoscale meteorology.

Jai S. Kim, Professor of Atmospheric Science; Ph.D., Saskatchewan, 1958. Aurora and airglow, solar-terrestrial relations.

David Knight, Assistant Professor of Atmospheric Science; Ph.D., Washington (Seattle), 1987. Mesoscale meteorology, numerical modeling.

*G. Garland Lala, Research Professor of Atmospheric Science; Ph.D., SUNY at Albany, 1972. Cloud and precipitation physics.

Michael G. Landin, Adjunct Assistant Professor of Atmospheric Science; M.S., SUNY at Albany, 1982. Synoptic meteorology.

Arthur Z. Loesch, Associate Professor of Atmospheric Science; Ph.D., Chicago, 1973. Geophysical fluid dynamics.

Volker A. Mohnen, Professor of Atmospheric Science; Ph.D., Munich, 1966. Air pollution, aerosol physics.

John E. Molinari, Associate Professor of Atmospheric Science; Ph.D., Florida State, 1979. Numerical weather prediction, tropical meteorology.

S. T. Rao, Adjunct Professor of Atmospheric Science; Ph.D., SUNY at Albany, 1973. Atmospheric turbulence and dispersion.

*James J. Schwab, Research Professor of Atmospheric Science; Ph.D., Harvard, 1983. Atmospheric chemistry.

Jon T. Scott, Associate Professor of Atmospheric Science and Chair, Department of Atmospheric Science; Ph.D., Wisconsin–Madison, 1963. Bioclimatology, physical oceanography.

*Bernard Vonnegut, Distinguished Professor of Atmospheric Science; Ph.D., MIT, 1939. Atmospheric electricity.

*Christopher J. Walcek, Research Professor of Atmospheric Science; Ph.D., UCLA, 1983. Cloud physics, cloud chemistry.

*Wei-Chyung Wang, Research Professor of Atmospheric Science; D.Eng.Sc., Columbia, 1973. Global climatic change.

SPECIAL ACADEMIC AND RESEARCH FACILITIES

The department is home to the National Lightning Detection Network. The network is configured to record cloud-to-ground lightning discharges, peak current, flash polarity, and the number of return strokes in each discharge. Data coverage is national, and almost all cloud-to-ground lightning discharges across the continental United States are monitored in real time. A sophisticated IBM-based computer system is used to process and archive the lightning data. A variety of research projects in atmospheric electricity and mesometeorology, based upon data from the lightning detection network, are open to students.

An active synoptic-dynamic and mesometeorology research and teaching program is supported by a fully equipped map room, a synoptic laboratory, and interactive computer systems. Domestic and international surface and upper-air weather observations, together with GOES–East and West and METEOSAT satellite observations (visible, infrared, and water-vapor images over the Pacific and Atlantic oceans, North and South America, and Europe and Asia), can be accessed and analyzed through the departmental McIDAS (which accesses data worldwide and allows their analysis along with displays of satellite video images), UNIDATA (hardware and software systems for the analysis and display of geophysical data, sponsored by a consortium of UCAR universities), and GEMPAK/GEMPLT (a NASA interactive computer system for the analysis and display of weather data). National Weather Service conventional data and facsimile maps are also available through a DIFAX map plotter and weather data printer. Diagnostic and prognostic research investigations can also be conducted through the University VAX and IBM mainframes or by remote link to the NCAR computers. Year-round forecasting is open to students to help them apply their theoretical knowledge to real-world situations.

The Atmospheric Sciences Research Center (ASRC) is a University-wide research center affiliated with the SUNY Albany campus. The center has close ties with the Department of Atmospheric Science; most of the researchers in ASRC hold joint appointments with the department. The center offers unusual research opportunities in atmospheric sciences, with special emphasis on atmospheric chemistry, air pollution and aerosol physics, planetary boundary-layer processes, cloud physics and fog research, and solar energy–related topics. ASRC maintains a specialized field station at Whiteface Mountain, which is located in the midst of the Adirondack Mountains' high peak area. The Whiteface Mountain research observatory provides routine monitoring instrumentation for numerous atmospheric chemical and geophysical parameters known to be active in atmospheric chemical interactions. Sophisticated instruments enable a variety of chemical compounds to be measured in the gas, aqueous (cloud or precipitation), and aerosol phases, together with a detailed meteorological characterization of the chemical environment.

In order to understand the interactions between chemical and meteorological processes in the troposphere, several sophisticated theoretical models of tropospheric chemistry have been developed at ASRC. These models integrate the known physical, chemical, meteorological, and biological processes that control the chemical composition of the atmosphere. One of these models, the Regional Acid Deposition Model (RADM), is being used by the U.S. Environmental Protection Agency to assist in the formulation of a strategy to mitigate the effects of acidic deposition (acid rain). These models provide a useful tool for the interpretation and analysis of chemical measurements and are continuously being evaluated and refined using atmospheric trace species measurements.

Primary responsibility with the Atmospheric Sciences Research Center.

SECTION 6: METEOROLOGY AND ATMOSPHERIC SCIENCES

STATE UNIVERSITY OF NEW YORK AT STONY BROOK
Institute for Terrestrial and Planetary Atmospheres

Program of Study
The Institute for Terrestrial and Planetary Atmospheres coordinates an interdepartmental teaching and research program for students interested in the physics, chemistry, and dynamics of the atmospheres of Earth and other planets. The Institute maintains vigorous activities in a broad range of modern atmospheric research. Activities are pursued in several Stony Brook departments affiliated with the Institute. Approximately five years is generally required to complete the Ph.D. after receiving the bachelor's degree.

Each graduate student's program of study begins with a development of the fundamental principles of atmospheric sciences through course work. At the same time, students are encouraged to join an ongoing research activity in which they participate with an increasing degree of responsibility. Completion of the degree program thus entails a thorough understanding of the principles of atmospheric science and their application to significant problems.

Research is done on a wide range of problems relating to terrestrial and planetary atmospheres. Research is carried out in state-of-the-art laboratories, by ground-based remote sensing at various locations around the earth, by analysis of conventional and satellite data, and by the development and analysis of theoretical models. Research is being performed to better understand past climate changes as well as to predict the future climate. Comprehensive data sets are analyzed as are the results of comprehensive three-dimensional climate models together with results of simplified models. Cloud-radiative effects on climate are of particular interest and are investigated in models and by analyzing satellite data from the Earth Radiation Budget Experiment. Satellite data, models, and conventional data are also analyzed to better understand the influences of latent heat release in the tropics on global climate. This is being done in preparation for analysis of satellite data to come from NASA's Tropical Rain Measurement Mission, which is planned for launching in 1997. There is extensive activity aimed at better understanding the physical basis for predictions of the size and timing of future greenhouse warming.

Atmospheric chemistry is another area of emphasis. Experimental research has been carried out for nearly a decade using state-of-the-art remote-sensing equipment to measure stratospheric ozone and those chemicals that catalyze its destruction. With the launch of NASA's Upper Atmospheric Research Satellite in late 1991, the Institute will analyze daily global measurements of stratospheric energetics, chemistry, and dynamics. The Institute is involved in continuing activity in the modeling of global tropospheric chemistry. Several faculty members are investigators on NASA's Earth Observation System, which is the most comprehensive planned international investigation of the global climate system. There is also research activity on the physics, chemistry, and evolution of terrestrial and planetary thermospheres-ionospheres, including those of Mars, Venus, and the outer planets and their satellites. Research is also carried out in a state-of-the-art infrared spectroscopy laboratory to determine those molecular parameters, such as line shape and strength, that are needed for atmospheric heating calculations as well as for remote sensing.

Research Facilities
The Institute computer facilities include one VAX 6310, one VAX 6510 with vector processing capabilities, nine workstations, printers, graphics terminals, and hard-copy plotters. An IBM 3080 is also available at the University Computer Center. The spectroscopy laboratories are equipped with infrared (grating) spectrometers, low-temperature absorption cells, a tunable diode laser spectrometer, and a high-resolution Fourier-transform spectrometer. Students have access to millimeter-wave remote-sensing equipment, specially developed at Stony Brook, and to data from the Infrared Telescope Facility, Pioneer Venus, and the Earth Radiation Budget Experiment.

Financial Aid
Assistantships and fellowships provide a stipend of $8000–$9500 for the 1991–92 academic year. Tuition is waived for all supported students. Summer research assistantships with stipends of up to $3800 are available.

Cost of Study
The tuition fee for the 1991–92 academic year is $2450 for residents of New York State and $5766 for out-of-state students. Miscellaneous fees, such as insurance and activity fees, total approximately $300. Tuition is waived for teaching and research assistants.

Cost of Living
In 1991–92, estimated living costs are approximately $600–$800 per month for single students living on campus. Off-campus rentals in communities surrounding the campus are also popular with graduate students.

Student Group
At any given time, 15 to 20 graduate students are engaged in research in atmospheric sciences in collaboration with the faculty members shown on the reverse side of this page.

Location
Stony Brook is located about 60 miles east of Manhattan on the wooded North Shore of Long Island, convenient to New York City's cultural life and Suffolk County's tranquil, recreational countryside and seashores. Long Island's hundreds of miles of magnificent coastline attract many swimming, boating, and fishing enthusiasts from around the world.

The University
Established thirty years ago as New York's comprehensive State University Center for Long Island and Metropolitan New York, Stony Brook offers excellent programs in a broad spectrum of academic subjects. The University conducts major research and public service projects. Over the past decade, externally funded support for Stony Brook's research programs has grown faster than that of any other university in the United States and now exceeds $70-million per year. The University's internationally renowned faculty members teach courses from the undergraduate to the doctoral level to more than 16,000 students. More than 100 undergraduate and graduate departmental and interdisciplinary majors are offered. Extensive resources and expert support services help foster intellectual and personal growth.

Applying
Students applying for graduate study should hold a B.S. degree in such fields as physics, chemistry, mathematics, engineering, or atmospheric science. Applicants may write for additional information about admission and financial aid to a graduate program faculty member whose research is of primary interest to them or write to the Institute director. Applications should be received by March 1 for September admission.

Correspondence and Information
Professor Marvin A. Geller, Director
Institute for Terrestrial and Planetary Atmospheres
State University of New York at Stony Brook
Stony Brook, New York 11794-3800
Telephone: 516-632-6170 or 8009
Fax: 516-632-6251

SECTION 6: METEOROLOGY AND ATMOSPHERIC SCIENCES

State University of New York at Stony Brook

THE FACULTY AND THEIR RESEARCH

Robert D. Cess, Leading Professor of Atmospheric Sciences, Department of Mechanical Engineering; Ph.D., Pittsburgh, 1959. Radiative transfer and climate modeling; greenhouse effect; nuclear winter theory; atmospheric structures of Mars, Saturn, and Jupiter.

Robert G. Currie, Research Associate Professor of Atmospheric Sciences, Department of Mechanical Engineering; Ph.D., UCLA, 1965. Climatology of droughts, variations in sea level, geomagnetic field.

Robert L. deZafra, Professor of Physics, Department of Physics; Ph.D., Maryland, 1958. Monitoring and detection of trace gases in the terrestrial stratosphere, changes in the ozone layer, remote-sensing instrumentation.

Jane L. Fox, Associate Professor of Atmospheric Sciences, Department of Mechanical Engineering; Ph.D., Harvard, 1978. Aeronomy of Earth and other planets, chemical and thermal structures of thermospheres and ionospheres, airglow and aurora, atmospheric evolution.

Marvin A. Geller, Professor and Director of Institute for Terrestrial and Planetary Atmospheres; Ph.D., MIT, 1969. Atmospheric dynamics, stratosphere dynamics, ozone behavior.

Sultan Hameed, Professor of Atmospheric Sciences, Department of Mechanical Engineering; Ph.D., Manchester, 1968. Atmospheric chemistry, urban air pollution, climate change, droughts.

Kenji Takano, Research Assistant Professor of Atmospheric Sciences; Ph.D., Tokyo, 1976. Numerical modeling of the atmosphere.

Prasad Varanasi, Professor of Atmospheric Sciences, Department of Mechanical Engineering; Ph.D., California, San Diego, 1967. Infrared spectroscopic measurements in support of NASA's space missions, atmospheric remote sensing, greenhouse effect and climate research molecular physics at low temperatures.

Minghua Zhang, Assistant Professor of Atmospheric Sciences; Ph.D., Academia Sinica (China), 1987. Atmospheric dynamics and climate modeling.

State University of New York at Stony Brook

UNIVERSITY OF MARYLAND COLLEGE PARK

Department of Meteorology

Programs of Study

The Department of Meteorology offers graduate study leading to the M.S. and Ph.D. degrees. An extremely vigorous research program is maintained. As an integral part of their graduate education, students are expected to take an active role in one of a variety of areas, such as biosphere-atmosphere interactions, chemistry, climate theory and modeling, cloud studies, fluid dynamics, general circulation, numerical weather prediction, oceanography, pollution, radiation, remote sensing, and turbulence and diffusion. Interdisciplinary programs in areas such as chemical physics and chemistry are encouraged, as are collaborative endeavors with neighboring institutions such as the NASA Goddard Space Flight Center, National Oceanic and Atmospheric Administration (NOAA), and National Institute of Standards and Technology (NIST).

Under a newly revised curriculum, students wishing to earn a Ph.D. are required to demonstrate an understanding of the basics of meteorology by passing (usually in the second year) a written comprehensive examination in the areas of general meteorology, dynamics, radiation, and a specialty area. To show the ability to plan and conduct original research, Ph.D. students must pass an oral defense of a research proposal and later, a final defense of the dissertation. Master's students must pass the comprehensive exam at a lower level and prepare a scholarly paper.

Research Facilities

Students and faculty in the department enjoy access to extensive research facilities, including several mainframe computers. The department operates DEC and Apollo workstation networks and maintains an IBM 4381 mainframe. Up-to-date networking capabilities provide easy access to all machines on the networks. All facilities allow state-of-the-art graphics visualization. Access to Cray supercomputers is provided by satellite links to the San Diego Supercomputer Center and the National Center for Atmospheric Research in Boulder and by high-speed land links to the NASA Goddard Space Flight Center and the National Meteorological Center. In addition, the University IBM 3081, Unisys 1100/92, Digital VAX-11/785 with FPS N64/30 array processor, and various smaller systems, as well as the Amdahls at Goddard, are available for research and classwork. The meteorology department has installed a UNIDATA computer graphics animation system that ingests, manages, and displays current weather map, satellite, and radar data in color for research and instruction. The department also has a solar radiation monitoring station and an instrumented weather station (an NOAA cooperative observing station). For air chemistry, two sets of state-of-the-art sensors have been developed. One is used in a mobile laboratory for surface measurements and the second for airborne experiments on aircraft such as the NOAA Hurricane Hunters.

Financial Aid

The Department of Meteorology offers graduate research assistantships to qualified students; scholarships and fellowships are also available. Typically, several incoming students are awarded prestigious University fellowships, and every year the department awards the Louis Allen Memorial Scholarship.

Cost of Study

Tuition for 1990–91 was $143 per credit hour for Maryland residents and $256 per hour for nonresidents; the stipend given to graduate research assistants and fellows includes an additional increment to cover the full cost of tuition.

Cost of Living

A limited amount of space in University dormitories and apartments is available to students. Apartments off campus are plentiful; rent decreases dramatically with distance from downtown Washington.

Student Group

The department's typical graduate student enrollment is about 50, with members drawn from across the United States and many countries. Nearly all students receive financial aid; the few who do not are generally employed by one of the many nearby government agencies.

Location

Situated in the Maryland suburbs of Washington, D.C., the University is in an ideal location for interaction with the large scientific community in the area. Cooperative research agreements have been made with NOAA, NASA, and NIST, and government scientists often collaborate with students on research projects. Nearby Washington (the White House is less than 10 miles away) offers a truly international atmosphere and a rich variety of cultural and recreational opportunities, such as the Kennedy Center and the Smithsonian Institution. The University lies between the Blue Ridge Mountains (about 50 miles, or 80 kilometers, to the west) and Chesapeake Bay (about 35 miles, or 55 kilometers, to the east). Summers are warm and sometimes humid, and the winters are mild; especially pleasant weather prevails in the spring and fall. The coldest weather occurs in late January and early February, with an average daily maximum temperature of 7°C (45°F) and an average daily minimum of −2°C (28°F). The warmest time is late July, when daily high temperatures commonly exceed 30°C (86°F). Sunny weather is common the year round, with most precipitation falling in showers; thunderstorms occur on about one out of every ten days in the summer.

The University

The University of Maryland is a land-grant university, originating in 1807. The entire University system includes about 2,300 faculty members and 12,000 graduate students. The College Park campus, where the Department of Meteorology is located, is the system's flagship institution and is undergoing a period of sustained growth.

Applying

Students from a wide variety of backgrounds are encouraged to apply. A bachelor's degree in meteorology; in a science such as biology, chemistry, oceanography, mathematics, or physics; or in engineering is required. The application deadlines for admission with financial aid are February 1 for the fall semester and September 15 for the spring semester.

Correspondence and Information

Chair, Admissions Committee
Department of Meteorology
University of Maryland
College Park, Maryland 20742-2425
Telephone: 301-405-5392

SECTION 6: METEOROLOGY AND ATMOSPHERIC SCIENCES

University of Maryland College Park

THE FACULTY AND THEIR RESEARCH

Ferdinand Baer, Professor; Ph.D., Chicago, 1961. Numerical weather prediction.
James Carton, Associate Professor; Ph.D., Princeton, 1983. Tropical ocean modeling and data assimilation.
Russell R. Dickerson, Associate Professor; Ph.D., Michigan, 1980. Atmospheric chemistry and air pollution, effects of clouds on photochemistry.
Robert G. Ellingson, Professor and Director of the Cooperative Institute for Climate Studies; Ph.D., Florida State, 1972. Atmospheric radiation and remote sensing.
Alan J. Faller, Research Professor Emeritus; Sc.D., MIT, 1957. Geophysical fluid dynamics.
Robert D. Hudson, Professor and Chairman; Ph.D., Reading, 1959. Atmospheric chemistry, stratospheric physics, satellite measurements.
James L. Kinter III, Assistant Research Scientist; Ph.D., Princeton, 1984. Atmospheric modeling and general circulation.
Istvan Laszlo, Assistant Research Scientist; Ph.D., Loránd Eötvös (Hungary), 1984. Remote sensing, surface radiation, atmospheric physics.
Sumant Nigam, Assistant Research Scientist; Ph.D., Princeton, 1984. Large-scale atmospheric dynamics and tropical meteorology.
Rachel T. Pinker, Associate Professor; Ph.D., Maryland, 1976. Land-atmosphere interactions and remote sensing.
Eugene Rasmusson, Senior Research Scientist; Ph.D., MIT, 1966. Diagnostic studies of short-term climate variability.
Alan Robock, Associate Professor; Ph.D., MIT, 1977. Climate modeling, climate theory, climatic data analysis, nuclear winter, volcanoes and climate.
Edwin K. Schneider, Senior Research Scientist; Ph.D., Harvard, 1976. General circulation of the atmosphere and boundary-layer processes.
Piers J. Sellers, Adjunct Professor; Ph.D., Leeds, 1981. Air-land modeling and atmosphere-biosphere interactions.
Jagadish Shukla, Professor and Director of the Center for Ocean-Land-Atmosphere Interactions; Sc.D., MIT, 1976. Climate modeling, predictability, monsoon dynamics.
David Straus, Associate Research Scientist; Ph.D., Cornell, 1977. General circulation of the atmosphere, predictability, atmospheric dynamics.
Owen E. Thompson, Professor; Ph.D., Missouri, 1966. Satellite meteorology and remote sensing, use of microcomputers in data acquisition.
Huug van den Dool, Associate Research Scientist; Ph.D., Utrecht (Netherlands), 1975. Climate variability, climate modeling, long-range weather prediction.
Anandu D. Vernekar, Professor; Ph.D., Michigan, 1966. Climate modeling and general circulation of the atmosphere, statistics in atmospheric science.

Graphic depiction of cumulus formation.

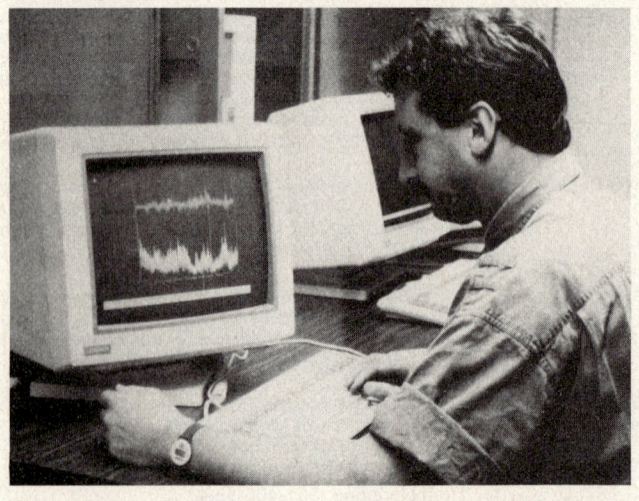

The Inner Harbor in nearby Baltimore.

Research associate working at an Apollo computer workstation.

Section 7
Physics

This section contains directories of institutions offering graduate work in acoustics, applied physics, mathematical physics, optical sciences, physics, and plasma physics, followed by two-page entries submitted by institutions that chose to prepare detailed program descriptions. Additional information about programs listed in the directories but not augmented by a two-page entry may be obtained by writing directly to the dean of a graduate school or chair of a department at the address given in the directory.

For programs offering related work, see all other areas in this book; in Book 3, see: Biology and Biomedical Sciences and Biophysics; in Book 5: Electrical and Power Engineering; Engineering and Applied Sciences; Engineering Physics; Materials Sciences and Engineering; Mechanical Engineering, Mechanics, and Aerospace/Aeronautical Engineering; and Nuclear Engineering; and in Book 6: Allied Health Professions and Optometry and Vision Sciences.

CONTENTS

Program Directories

Acoustics	541
Applied Physics	541
Mathematical Physics	544
Optical Sciences	544
Physics	545
Plasma Physics	566

Announcements

Brooklyn College of the City University of New York	546
California State University, Fresno	546
City College of the City University of New York	547
Clarkson University	547
Clark University	547
Colorado School of Mines	547
East Carolina University	548
East Texas State University	548
Florida International University	549
Florida State University	549
Georgia Institute of Technology	549
Indiana University Bloomington	550
New Jersey Institute of Technology	542
Northern Illinois University	552
Old Dominion University	553
Oregon State University	553
Pennsylvania State University	553
Portland State University	553
Rensselaer Polytechnic Institute	554
San Francisco State University	555
University of Alabama	557
University of Alabama at Birmingham	557
University of Alaska Fairbanks	557
University of Colorado at Boulder	558
University of Maryland College Park	560
University of Maryland Graduate School, Baltimore	560
University of Miami	560
University of Minnesota, Duluth	560
University of Missouri-Columbia	560
University of Montana	561
University of Nebraska-Lincoln	561
University of Nevada, Las Vegas	561
University of New Mexico	561
University of North Carolina at Chapel Hill	562
University of North Texas	562
University of Texas at Arlington	563
University of Washington	564
Wright State University	566
Yale University	543

Cross-Discipline Announcements

Columbia University	567
Georgetown University	567
Harvard University	567
Johns Hopkins University	567
Louisiana State University	567
Northwestern University	567
University of Kentucky	567
University of Maryland College Park	567
University of Michigan	567
University of Rochester	567
University of Virginia	568
University of Washington	
Bioengineering	568
Materials Science and Engineering	568
University of Wisconsin-Madison	568
Virginia Commonwealth University	568

Full Descriptions

Auburn University	569
Boston University	571
Brandeis University	573
Brown University	575
California Institute of Technology	577
California State University, Los Angeles	579
Clemson University	581
Emory University	583
Florida State University	585
Harvard University	587
Indiana University Bloomington	589
Johns Hopkins University	591
Kent State University	593
Louisiana State University	595
Michigan State University	597
New Mexico State University	599
Northeastern University	601
Northwestern University	603
Ohio University	605
Oregon State University	607
Pennsylvania State University	609
Rice University	611
Rutgers, The State University of New Jersey, New Brunswick	613
State University of New York at Buffalo	615
State University of New York at Stony Brook	617
Stevens Institute of Technology	619
Syracuse University	621
Temple University	623
Texas A&M University	625
Tufts University	627
University of Akron	629
University of Arizona	
Optics	631
Physics	633
University of California, Los Angeles	635
University of California, Riverside	637
University of Cincinnati	639
University of Connecticut	641
University of Florida	643
University of Georgia	645
University of Houston	647
University of Kansas	649
University of Massachusetts Lowell	651
University of Michigan	653
University of Notre Dame	655
University of Pennsylvania	657

Peterson's Guide to Graduate Programs in the Physical Sciences and Mathematics 1992

SECTION 7: PHYSICS

Contents; Field Definitions

University of Pittsburgh	659
University of Puerto Rico, Río Piedras	661
University of Rochester	
Optics	663
Physics and Astronomy	665
University of South Carolina	667
University of Tennessee, Knoxville	669
University of Texas at Austin	671
University of Virginia	673

See also:

Drexel University—Chemistry, Mathematics and Computer Science, and Physics and Atmospheric Sciences	409
Institute of Paper Science and Technology—Chemistry	109
Virginia Polytechnic Institute and State University—Mathematics	511

FIELD DEFINITIONS

In an effort to broaden prospective students' understanding of disciplines in physics, educators in the field have submitted the following statements on acoustics, applied physics, mathematical physics, optical sciences, physics, and plasma physics.

Acoustics

Acoustics is the science of sound. It contains a large number of subfields, many of which are interdisciplinary. Some of the subfields are architectural, musical, noise control, physical, physiological, psychological, speech communication, ultrasonic, and underwater acoustics. Several of these branches of acoustics are concerned with how we produce and perceive sound and how we use it to communicate information and emotions. Others use sound to learn more about our environment, sensing the properties of crystals, parts of the human body, or objects in the ocean. A graduate degree in acoustics can lead to employment as a research scientist in an industrial, government, or university setting.

The appropriate preparation for graduate study in acoustics depends on the branch(es) of acoustics one desires to study and on how one desires to contribute to that field. A solid undergraduate physics program is recommended, with special attention to classical mechanics, the wave equation and its application to problems involving boundary conditions, and Fourier and Laplace transforms. Electrical engineering courses are helpful for work in signal processing, and experience with computers is useful. Some knowledge of mechanical engineering is particularly suitable for design of transducers and control of noise and vibration.

Gary L. Gibian
Department of Physics
American University
and Senior Scientist
Planning Systems, Inc.

Applied Physics

Applied physics is not a well-defined branch of physics, such as solid-state or nuclear physics; rather, it is an umbrella designation for a group of fields, akin to physics, with a fascinating diversity. In fact, most fields of human intellectual activity have received contributions from physics in one form or another, and frequently a separate branch of applied physics arose out of the need of a certain field. Biophysics, geophysics, electrophysics, medical physics, and others are well known, yet there are also contributions of physics to archaeology and history, to the fine arts, and to forensic sciences. Best known, of course, are those applied physics areas that are related to engineering disciplines, e.g., electronics, semiconductor devices, microwaves, computer technology, nuclear fission, aerospace, and optics and lasers.

New areas of applied physics appear as our knowledge of physics advances. Sometimes these new fields emerge slowly, sometimes quickly. We are told that Wilhelm Roentgen's discovery of X rays was followed, within days, by the first radiological examination of a bone fracture in a local hospital, marking the appearance of a new field of applied physics, now called radiology. On the other hand, almost half a century passed between Lord Rutherford's basic discovery of α scattering by nuclei and the application of this technique to the study of surfaces and of thin films.

What is the best preparation for a career in applied physics? A thorough familiarity with basic physics is almost always necessary. Usually this knowledge is acquired by taking graduate courses in quantum mechanics, classical mechanics, electricity and magnetism, thermodynamics, and mathematical methods of physics. Some specialized courses should also be taken; their selection is indicated by the applied physics branch selected by the student. These courses may or may not be in physics. For example, a student interested in semiconductor devices may wish to take solid-state physics and electrical engineering courses. A flair for experimental work, for data analysis, and for data reporting (technical writing) would be an asset. Skill courses such as FORTRAN programming may have been taken as part of the undergraduate curriculum.

Students considering applied physics as a career should be cautioned: If you wish to succeed in applied physics, you must be able to work professionally with nonphysicists. You, as a physicist, are interested in the basic laws of nature; their interests may be quite different. You must learn to understand and to respect their motives. If you can do this, and if your academic preparation is adequate, then you will find that a career in applied physics is both stimulating and rewarding.

Gunter H. R. Kegel, Professor
Department of Physics and Applied Physics
College of Arts and Sciences
University of Lowell

Mathematical Physics

Mathematical physics is a specialty that encompasses activities ranging from the construction of apparatus to abstract theorizing. For example, the design of a machine to study plasma fusion employs many mathematical physicists in the study of the electromagnetic fields in a particular geometry, how the plasma interacts because of the fields and conductor geometry, and how practical an eventual configuration design is for applications. On a more abstract level, mathematical physicists study models of physical theories that are analytically tractable. Into these models more and more "reality" is injected to improve the description of the model. The aim here is to demonstrate the mathematical soundness or existence of a particular model.

Between these two extremes and mixed with them, there is an area of theorizing that is closely connected with mathematical forms of expression. General relativity—a highly abstruse field of study—proceeds from very few physical assumptions to very far reaching conclusions through analysis of the mathematical properties of the theory. The existence of black holes is a subject stimulated by the study of the singular solutions to the Einstein field equations. More recently, the study of nonlinear waves has received a great impetus from a few simple but startling discoveries. Computer analyses led to analytical studies of nonlinear field theories. These analytical studies, in turn, stimulated a significant amount of thinking about the physical systems that the theories described. The result of this thinking is that the generality of certain physical theories, especially quantum field theory, is recognizably much greater than had been anticipated. This is one of the best examples of a situation in physics where a theory and the language in which it is expressed are developed simultaneously.

As computers have become more sophisticated and powerful, physicists have found greater application for them in experiments as well as theoretical analysis. This, in turn, has stimulated the growth of a new subdiscipline, computational physics. This field is currently in an emergent state. Journals

SECTION 7: PHYSICS

Field Definitions

have already appeared that are devoted to exposition of new ideas, such as algorithms to make effective use of supercomputers, numerical simulation of experiments, and proofs of mathematical theorems.

Many areas of mathematical physics have become traditional. One of the more important of these, because of its connections with quantum theory, is group theory. Contemporary studies of this important area focus on nonlinear representations of groups, especially Lie groups, and their related algebras. This subject is applied broadly in the physical sciences and is the topic of regular international conferences.

*R. Philip B. Burt, Professor
Department of Physics and Astronomy
Clemson University*

Optical Sciences

Optics, the study of light, is both an important branch of physics and an exciting field in engineering. In optics, basic scientific discoveries are being made each year, and each year optical physicists and engineering scientists are making useful new applications combining lasers, electronics, and computers. This renaissance in optics started with the application of communications theory, an engineering science, to optics and continued with the invention of the hologram and the laser. Today, optical scientists and engineers are working on electronic printing and imaging, guided wave optics, new lasers, optics in robotics, phase conjugation, laser radar, laser fusion, laser and X-ray lithography, laser optical disks, optical interconnections, neural optical computers, and new optoelectronic materials.

Optoelectronics, an important branch of the optical sciences, is emerging rapidly as new hybrid systems are discovered. There are three main aspects to optoelectronics: optical materials, novel devices, and systems. Optoelectronics is stimulated in its growth as new materials are perfected that are being tailor-made for novel devices, which are key technology enablers for integrated optics, integrated circuits, and nonlinear optics. Extensive industrial growth is projected for optoelectronics and the other optical sciences during the next twenty years.

Optics is a science that greatly enriches other areas of science and technology. Optical techniques are used in physics, chemistry, and biology laboratories for measuring lengths, illuminating samples, forming images, and detecting emitted radiation. Optical communications is a challenging, rapid-growth discipline giving rise to marked improvements in data transmission. Optical data acquisition is a key aspect of many space programs and ranges from the detection of environmental pollutants to the discovery of the characteristics of the Martian landscape. Lasers are now used in everyday life, in such applications as laying drainpipe, providing automatic scanners for supermarket checkout systems, and automating industrial process control. As the telephone industry switches from conventional copper cable to fiber optics, it will require the services of many more engineering scientists in optics. Advanced optical technology now includes lasers, fast electrooptic and acoustooptic modulators, low-loss fibers, wideband detectors, integrated circuits, and holographic and optical disc recording.

Optics is a field with far-ranging appeal. Whether students aspire to become theoreticians, experimentalists, engineering scientists, or design engineers, the field of optics is broad enough to provide a challenging career and a rewarding profession. For example, students with a bachelor's degree in physics, mathematics, computer science, or any discipline of engineering can obtain an M.S. in optics with one year of graduate education and be well prepared to enter the field at a highly professional level. Students whose career goal is to conduct basic research in the optical sciences are well advised to continue their studies through the Ph.D. Whether or not the graduating Ph.D. wishes to be known professionally as an optical physicist or an optical engineer is largely a matter of choice depending on which area of specialization he or she elects. In the optical sciences it is important for the graduate student to have rigorous core courses in the principles of optics, including electromagnetic theory, quantum mechanics, solid-state physics, and mathematics. The industrial community expects a Ph.D. recipient to be highly competent in each of three specialized areas: optical systems and instrumentation—conception and design of telescopes, lens groupings for imaging, information processing, or energy transfer; fields and waves, including coherency—propagation and statistical optics; and quantum optics—interaction of light and matter, and device physics (lasers, detectors, and active elements).

*Nicholas George, Professor
Institute of Optics
Associate Dean for Research
University of Rochester*

Physics

The goal of physics is to discover the fundamental laws of nature. "What man knows about inanimate nature is physics—or, rather, the most lasting and universal things that he knows make up physics" (*Physics in Perspective,* National Academy of Sciences, 1972). Physics is an experimental science. Experiments range in scale from those performed using very large accelerators at national laboratories to those carried out in small laboratories on college campuses. There is also a large effort in theoretical physics aimed at understanding past experiments and predicting the results of future experiments.

Physics naturally divides itself into subfields: astrophysics and relativity, plasma and fluid physics, acoustics, optics, condensed-matter (solid-state) physics, atomic and molecular physics, nuclear physics, and elementary particle physics. Although this division is useful, it is not unusual for a single experiment to overlap several subfields, e.g., optical techniques used to study atomic transitions in a plasma.

The goal of graduate study in physics is to prepare for a research career in an academic institution, an industrial research laboratory, or a government research center. Employment prospects are, in general, quite good, though there is a shortage of permanent positions at academic institutions. Students who have completed graduate work in physics are in great demand by industrial employers. This strong demand is due in part to the physicist's ability to bring his or her research training to bear on a very broad spectrum of problems. The national need to solve technical problems in such diverse areas as environmental protection, energy conservation, and national defense promises a bright future for employment in this field.

*Thomas A. Griffy
Professor of Physics
University of Texas at Austin*

Plasma Physics

Plasma is the state of matter at temperatures so high that an appreciable fraction of the molecules and atoms are dissociated into ions and electrons. Stellar and interstellar matter is mostly in the plasma state, as is matter in the magnetosphere, the ionosphere, flames, chemical and nuclear explosions, and electrical discharges. In addition, a number of current high-technology areas involve matter in the plasma state, including gas lasers, free-electron lasers, certain microwave amplifiers, and plasma deposition and etching operations. Also, very significantly, the working fuel in a controlled thermonuclear reactor would be in the plasma state, and it has been largely the study and development of this "fusion reactor" concept for large-scale electric power generation that has led to our extensive current knowledge of plasmas. Fusion power would utilize, in a socially and ecologically safe way, the almost limitless nuclear energy that is potentially extractable from the world's abundant resources of the light elements, particularly hydrogen and lithium. The kindling temperature for the very slow nuclear burning envisioned for a fusion reactor is around

Peterson's Guide to Graduate Programs in the Physical Sciences and Mathematics 1992

SECTION 7: PHYSICS

Field Definitions

100,000,000°C, many times hotter than the center of the sun, a fact that explains why the reacting fuel is necessarily in the plasma state.

In terms of gross properties, plasma differs from ordinary (un-ionized) gas by its very high electrical and thermal conductivity and by the emission of electromagnetic radiation such as microwaves, light, bremsstrahlung, and even X rays. At high temperatures, interparticle collisions, which dominate ordinary gas behavior, become very infrequent, and mean free paths many times longer than the apparatus dimensions are commonly achieved in laboratory experiments. Plasmas may therefore exhibit fascinating "memory" effects and can respond in dramatic ways to electric and magnetic fields.

Plasma physics is not a specialized discipline but broadly spans modern experimental and theoretical physics and technology. On the academic front, the study and the teaching of plasma physics invoke the blending of knowledge from electricity and magnetism, atomic physics, statistical mechanics and kinetic theory, computer science, and applied mathematics. Graduate curricula in plasma physics typically include the canonical graduate physics courses in quantum mechanics, electromagnetic theory, and statistical mechanics, together with advanced work in magnetohydrodynamics, plasma waves and instabilities, irreversible processes, and nonlinear interactions. Research laboratories with sophisticated apparatus, diagnostics, and data-handling facilities are associated with almost every institution offering graduate plasma programs and offer students invaluable training in current experimental research using state-of-the-art techniques.

Because of the manifold possibilities for interaction among its different aspects, plasma physics provides extraordinarily diverse, fertile, and still largely unexplored ground for fundamental theoretical and experimental research. Plasma physics is itself a large field and in many of its aspects is still a young field with challenging applications in astrophysics, space physics, electronics, materials sciences, and fusion research—where success in our understanding will have enormous impact on reaching a satisfactory long-range solution of the world's energy problems.

Thomas H. Stix, Associate Chairman
Department of Astrophysical Sciences
Princeton University

Acoustics

Catholic University of America, School of Engineering and Architecture, Department of Mechanical Engineering, Program in Acoustics, Washington, DC 20064. Program awards MME, MS Engr, PhD. Faculty: 1 full-time (0 women), 3 part-time (0 women). Matriculated students: 2 full-time (0 women), 24 part-time (1 woman); includes 2 foreign. 3 applicants. *Degree requirements:* For master's, comprehensive exam required, thesis optional, foreign language not required; for doctorate, dissertation, comprehensive exam, oral exam required, foreign language not required. *Entrance requirements:* For master's, minimum GPA of 3.0; for doctorate, minimum GPA of 3.5. Application deadline: rolling. Application fee: $30. *Expenses:* Tuition of $11,726 per year full-time, $445 per credit part-time. Fees of $150 per year. *Financial aid:* Application deadline 2/1. • Dr. John Gilheany, Director, 202-319-5170.

George Washington University, School of Engineering and Applied Science, Department of Civil, Mechanical, and Environmental Engineering, Program in Mechanical Engineering, Concentration in Aeroacoustics, Washington, DC 20052. Concentration awards MS, D Sc, Ap Sci, Engr. Matriculated students: 1 full-time (0 women), 2 part-time (0 women); includes 0 minority, 0 foreign. Average age 57. 0 applicants. In 1990, 3 master's awarded. *Degree requirements:* For master's, thesis or alternative, comprehensive exam required, foreign language not required; for doctorate, computer language, dissertation, qualifying exam, final exam required, foreign language not required; for other advanced degree, minimum GPA of 3.0 required, foreign language and thesis not required. *Entrance requirements:* For master's, TOEFL or George Washington University English as a Foreign Language Test, minimum GPA of 3.0, appropriate bachelor's degree; for doctorate, TOEFL or George Washington University English as a Foreign Language Test, appropriate master's degree, minimum GPA of 3.4; for other advanced degree, TOEFL or George Washington University English as a Foreign Language Test, appropriate master's degree, minimum GPA of 3.0. Application deadline: 8/1 (applications processed on a rolling basis). Application fee: $45. *Expenses:* Tuition of $490 per semester hour. Fees of $223.06 per year full-time, $13.06 per year part-time. *Financial aid:* Fellowships, research assistantships, teaching assistantships, and career-related internships or fieldwork available. Financial aid application deadline: 4/1. • Dr. Ali M. Kiper, Chair, Department of Civil, Mechanical, and Environmental Engineering, 202-994-6749.

Pennsylvania State University University Park Campus, Intercollege Graduate Programs and College of Engineering, Intercollege Graduate Program in Acoustics, University Park, PA 16802. Program awards M Eng, MS, PhD. Faculty: 43. Matriculated students: 27 full-time (1 woman), 32 part-time (4 women). In 1990, 8 master's, 3 doctorates awarded. *Degree requirements:* For master's, foreign language not required; for doctorate, dissertation. *Application fee:* $35. *Tuition:* $203 per credit for state residents; $403 per credit for nonresidents. • Dr. Jiri Tichy, Head, 814-865-6364.

Royal Roads Military College, Graduate Program in Oceanography and Acoustics, FMO, Victoria, BC V0S 1B0, Canada. College awards M Sc. Part-time programs available. Faculty: 9 full-time (0 women), 1 part-time (0 women). Matriculated students: 7 full-time (0 women), 1 part-time (0 women); includes 0 minority, 0 foreign. Average age 28. 12 applicants, 25% accepted. In 1990, 5 degrees awarded (100% found work related to degree). *Degree requirements:* Thesis required, foreign language not required. *Application deadline:* 6/15. *Tuition:* $0. *Financial aid:* Federal work-study available. *Faculty research:* Mesoscale processes, coastal dynamics, acoustic oceanography, internal waves, surface gravity waves. Total annual research budget: $300,000. • Dr. David P. Krauel, Dean of Graduate Studies, 604-363-4580.

University of Houston, Cullen College of Engineering, Program in Acoustics, 4800 Calhoun, Houston, TX 77004. Program awards MS, PhD. Part-time and evening/weekend programs available. Faculty: 3 full-time (0 women), 1 part-time (0 women). Matriculated students: 2 full-time (1 woman), 0 part-time; includes 0 minority, 1 foreign. Average age 26. 6 applicants, 17% accepted. In 1990, 1 master's awarded (100% found work related to degree); 0 doctorates awarded. Terminal master's awarded for partial completion of doctoral program. *Degree requirements:* For master's, foreign language not required; for doctorate, dissertation, qualifying exam required, foreign language not required. *Entrance requirements:* GRE General Test. Application deadline: 7/7 (priority date, applications processed on a rolling basis). Application fee: $0 ($75 for foreign students). *Expenses:* Tuition of $30 per hour for state residents; $134 per hour for nonresidents. Fees of $240 per year full-time, $125 per year part-time. *Financial aid:* Fellowships, research assistantships, federal work-study available. Financial aid application deadline: 4/1. *Faculty research:* Ultrasonics, jet noise, acoustic systems. • Dr. R. D. Finch, Director, 713-749-2437.

Applied Physics

Alabama Agricultural and Mechanical University, School of Arts and Sciences, Department of Physics, PO Box 285, Normal, AL 35762. Offerings include applied physics (MS, PhD). Department faculty: 9 full-time (0 women), 0 part-time. *Degree requirements:* For master's, thesis optional, foreign language not required; for doctorate, computer language, dissertation required, foreign language not required. *Entrance requirements:* For master's, GRE, BS in physics or electrical engineering; for doctorate, GRE General Test (minimum combined score of 1000 required). Application deadline: 3/1 (priority date, applications processed on a rolling basis). Application fee: $15. *Tuition:* $79 per credit hour for state residents; $158 per credit hour for nonresidents. • Dr. Jai-Ching Wang, Chairman, 205-851-5305.

Appalachian State University, College of Arts and Sciences, Department of Physics, Boone, NC 28608. Offerings include applied physics (MS). Department faculty: 11 full-time (0 women), 0 part-time. *Degree requirements:* Thesis, comprehensive exam. *Entrance requirements:* GRE General Test. Application deadline: 7/31 (priority date). Application fee: $15. *Tuition:* $598 per semester for state residents; $3125 per semester full-time, $6215 per semester part-time for nonresidents. • Karl Mamola, Chairperson, 704-262-3090.

Arizona State University, College of Liberal Arts and Sciences, Department of Physics, Program in Applied Physics, Tempe, AZ 85287. Program awards PhD. *Degree requirements:* Dissertation, written and comprehensive exams, final oral exam, departmental qualifying exam required, foreign language not required. *Entrance requirements:* GRE. Application fee: $25. *Tuition:* $1528 per year full-time, $80 per hour part-time for state residents; $6934 per year full-time, $289 per hour part-time for nonresidents. • Dr. John D. Dow, Chair, Department of Physics, 602-965-3561.

Boston University, Graduate School, Department of Physics, Boston, MA 02215. Offerings include applied physics (PhD). PhD in applied physics offered through the Division of Applied Sciences and Engineering. Terminal master's awarded for partial completion of doctoral program. *Degree requirements:* 1 foreign language, dissertation. *Entrance requirements:* GRE General Test, GRE Subject Test, TOEFL (minimum score of 550 required). Application deadline: 4/1. Application fee: $45. *Expenses:* Tuition of $15,950 per year full-time, $498 per credit part-time. Fees of $120 per year full-time, $35 per semester part-time. • Lawrence R. Sulak, Chairman, 617-353-2623. Application contact: Beverly Pacheco, Admissions Coordinator, 617-353-2623.

See full description on page 571.

Brooklyn College of the City University of New York, Department of Physics, Bedford Avenue and Avenue H, Brooklyn, NY 11210. Offerings include applied physics (MA). Department faculty: 25 full-time, 0 part-time. *Degree requirements:* Comprehensive exam required, foreign language not required. *Entrance requirements:* GRE, TOEFL. Application deadline: 4/1. Application fee: $30. *Expenses:* Tuition of $1202 per semester full-time for state residents; $2350 per semester full-time for nonresidents. Fees of $45 per semester full-time. • Dr. Louis Celenza, Chairperson, 718-780-5418.

California Institute of Technology, Division of Engineering and Applied Science, Option in Applied Physics, Pasadena, CA 91125. Option offers programs in applied physics (MS, PhD), plasma physics (MS, PhD). Faculty: 10 full-time (0 women), 0 part-time. Matriculated students: 65 full-time (2 women), 0 part-time; includes 2 minority, 23 foreign. 129 applicants, 8% accepted. In 1990, 5 master's, 9 doctorates awarded. *Degree requirements:* For master's, foreign language and thesis not required; for doctorate, dissertation required, foreign language not required. *Application fee:* $0. *Expenses:* Tuition of $14,100 per year. Fees of $8 per quarter. *Financial aid:* Research assistantships, teaching assistantships available. *Faculty research:* Solid-state electronics, quantum electronics, plasmas, linear and nonlinear laser optics, electromagnetic theory. • Dr. Noel Corngold, Head, Steering Committee, 818-356-4129. Application contact: Dr. Paul Bellan, Option Representative, 818-356-4827.

Carnegie Mellon University, Mellon College of Science, Department of Physics, Pittsburgh, PA 15213. Offerings include applied physics (PhD). Terminal master's awarded for partial completion of doctoral program. Department faculty: 34 full-time (1 woman), 0 part-time. *Degree requirements:* Dissertation required, foreign language not required. *Entrance requirements:* GRE General Test. Application deadline: 4/15. Application fee: $0. *Expenses:* Tuition of $15,250 per year full-time, $212 per unit part-time. Fees of $80 per year. • Dr. Robert W. Kraemer, Head, 412-268-2740. Application contact: Ned S. Vanderven, 412-268-2766.

Case Western Reserve University, Graduate Programs in Engineering, Department of Electrical Engineering and Applied Physics, Program in Applied Physics, Cleveland, OH 44106. Program awards MS, PhD. *Degree requirements:* For master's, thesis or alternative required, foreign language not required; for doctorate, dissertation required, foreign language not required. *Entrance requirements:* TOEFL (minimum score of 550 required). Application fee: $25. *Expenses:* Tuition of $13,600 per year full-time, $567 per credit part-time. Fees of $320 per year. *Financial aid:* Applicants required to submit FAF. *Faculty research:* Electromagnetic wave propagation, applied superconductivity, solid-state device physics and materials. • Sheldon Gruber, Chairman, Department of Electrical Engineering and Applied Physics, 216-368-4089.

Colorado School of Mines, Department of Physics, Program in Applied Physics, Golden, CO 80401. Program awards PhD. Faculty: 13 full-time, 4 part-time. Matriculated students: 16 full-time (0 women), 6 part-time (0 women). *Degree requirements:* Dissertation, written and oral comprehensive exams required, foreign language not required. *Entrance requirements:* GRE General Test, minimum GPA of 3.0, BS in physics. Application deadline: 3/1. Application fee: $15 ($25 for foreign students). *Expenses:* Tuition of $3178 per year full-time, $124 per semester hour part-time for state residents; $10,304 per year full-time, $344 per semester hour part-time for nonresidents. Fees of $374 per year full-time. • Dr. John U. Trefny, Head, Department of Physics, 303-273-3830. Application contact: Dr. James McNeil, 303-273-3844.

Columbia University, School of Engineering and Applied Science, Department of Applied Physics, Program in Applied Physics, New York, NY 10027. Offers applied physics (MS, Eng Sc D, PhD), including applied mathematics (PhD), plasma physics (MS, Eng Sc D, PhD), quantum electronics (MS, Eng Sc D, PhD), solid state physics (MS, Eng Sc D, PhD). Part-time programs available. Faculty: 9 full-time (0 women), 6 part-time (1 woman). Matriculated students: 65 full-time (9 women), 7 part-time (2 women); includes 4 minority (all Asian American), 40 foreign. In 1990, 16 master's awarded (31% found work related to degree, 69% continued full-time study); 8 doctorates awarded (37% entered university research/teaching, 63% found other work related to degree). Terminal master's awarded for partial completion of doctoral program. *Degree requirements:* For master's, foreign language and thesis not required; for doctorate, dissertation, qualifying exam required, foreign language not required. *Entrance requirements:* GRE General Test, GRE Subject Test, TOEFL. Application deadline: 2/1. Application fee: $45. *Tuition:* $15,520 per year full-time, $518 per credit part-time. *Financial aid:* In 1990–91, 57 students received a total of $500,000 in aid awarded. 3 fellowships (1 to a first-year student), 48 research assistantships (7 to first-year students), 4 teaching assistantships (all to first-year students), 2 assistantships (1 to a first-year student) were awarded; federal work-study also available. Average monthly stipend for a graduate assistantship: $1083. Financial aid application deadline: 2/1; applicants required to submit GAPSFAS. *Faculty research:* Controlled fusion and spare plasma physics, free electron lasers, laser diagnostics, quantum electronics of solids, large-scale scientific computation. Total annual research budget: $2.66-million. • Dr. Gerald A. Navratil, Chairman, Department of Applied Physics, 212-854-4457.

Cornell University, Graduate Fields of Engineering, Graduate Programs in the Physical Sciences and Mathematics, Field of Applied Physics, Ithaca, NY 14853. Field awards PhD. Faculty: 55 full-time, 0 part-time. Matriculated students: 85 full-time (10 women), 0 part-time; includes 1 minority (Asian American), 25 foreign. 200

Directory: Applied Physics

applicants, 12% accepted. In 1990, 12 degrees awarded. *Degree requirements:* Dissertation, oral exam, written exam required, foreign language not required. *Entrance requirements:* GRE General Test, TOEFL. Application deadline: 1/10 (priority date). Application fee: $55. *Expenses:* Tuition of $16,170 per year. Fees of $28 per year. *Financial aid:* Fellowships, research assistantships, teaching assistantships, federal work-study, institutionally sponsored loans available. Financial aid application deadline: 1/10; applicants required to submit GAPSFAS. *Faculty research:* Optics; condensed matter physics and device research; plasma and astrophysics; biological physics; electron, ion and x-ray microscopy and spectroscopy. • Bruce Kusse, Graduate Faculty Representative, 607-255-0642. Application contact: Robert Brashear, Director of Admissions, 607-255-4884.

DePaul University, College of Liberal Arts and Sciences, Department of Physics, Programs in Physics, Chicago, IL 60604. Offerings include applied physics (MS). *Degree requirements:* Written and oral exams required, thesis optional, foreign language not required. *Application deadline:* rolling. *Application fee:* $20. *Expenses:* Tuition of $215 per quarter hour. Fees of $10 per quarter hour (minimum). • Dr. Donald Vanostenburg, Chairman, Department of Physics, 312-362-8659.

East Carolina University, College of Arts and Sciences, Department of Physics, Program in Applied Physics, Greenville, NC 27858-4353. Program awards MP. *Degree requirements:* 1 foreign language (computer language can substitute), comprehensive exam. *Entrance requirements:* GRE General Test, TOEFL. Application deadline: 6/1 (priority date, applications processed on a rolling basis). Application fee: $25. *Tuition:* $627 per semester for state residents; $3154 per semester for nonresidents. *Financial aid:* Available to part-time students. Financial aid application deadline: 6/1. • Dr. Carl Adler, Chairman, Department of Physics, 919-757-6739. Application contact: Dr. James Joyce, Director of Graduate Studies, 919-757-6483.

George Mason University, College of Arts and Sciences, Department of Physics, Fairfax, VA 22030. Department offers program in applied physics (MS). Faculty: 11 (3 women). Matriculated students: 1 full-time (0 women), 28 part-time (5 women); includes 1 minority (Asian American), 1 foreign. 13 applicants, 85% accepted. In 1990, 2 degrees awarded. *Degree requirements:* Thesis or alternative, comprehensive exam. *Entrance requirements:* Minimum GPA of 3.0 in last 60 hours of undergraduate study. Application deadline: 5/1. Application fee: $25. *Expenses:* Tuition of $1872 per year full-time, $78 per semester hour part-time for state residents; $6264 per year full-time, $261 per semester hour part-time for nonresidents. Fees of $1080 per year full-time, $45 per semester hour part-time. *Financial aid:* Teaching assistantships available. Financial aid application deadline: 3/1. • Dr. Menas Kafatos, Chairman, 703-993-1280. Application contact: Dr. John Evans, Graduate Adviser, 703-993-1285.

Georgia Institute of Technology, College of Sciences, School of Physics, Atlanta, GA 30332. Offerings include applied physics (MSAP). School faculty: 28 full-time (2 women), 0 part-time. *Application deadline:* 1/15 (priority date, applications processed on a rolling basis). *Application fee:* $5. *Expenses:* Tuition of $574 per semester full-time, $48 per credit part-time for state residents; $1380 per semester full-time, $115 per credit part-time for nonresidents. Fees of $132 per semester full-time. • Dr. Henry S. Valk, Acting Director, 404-894-5200.

Harvard University, Graduate School of Arts and Sciences, Division of Applied Sciences, Cambridge, MA 02138. Offerings include applied physics (ME, SM, PhD); medical engineering/medical physics (PhD), with options in applied physics, engineering sciences, physics. Terminal master's awarded for partial completion of doctoral program. Division faculty: 58 full-time, 0 part-time. *Degree requirements:* For master's, thesis not required; for doctorate, dissertation. *Entrance requirements:* GRE General Test, GRE Subject Test. Application deadline: 1/2. Application fee: $60. *Expenses:* Tuition of $14,860 per year. Fees of $550 per year. • Dr. Paul C. Martin Jr., Dean, 617-495-2833. Application contact: Office of Admissions and Financial Aid, 617-495-5315.

Harvard University, Divisions of Health Sciences and Technology and Applied Sciences and Department of Physics, Program in Medical Engineering/Medical Physics, Cambridge, MA 02138. Offerings include applied physics (PhD). Program offered jointly with Massachusetts Institute of Technology. *Degree requirements:* Dissertation, oral and written qualifying exams. *Application deadline:* 1/15 (applications processed on a rolling basis). *Tuition:* $15,410 per year. • Thomas McMahon, Head, 617-495-5854. Application contact: Graduate School of Arts and Sciences, Office of Admissions and Financial Aid, 617-495-5315.

Laurentian University, Programme in Applied Physics, Sudbury, ON P3E 2C6, Canada. Program awards M Sc. Part-time programs available. Faculty: 9 full-time (0 women), 0 part-time. Matriculated students: 6 full-time (1 woman), 6 part-time (0 women); includes 0 foreign. 8 applicants, 63% accepted. In 1990, 3 degrees awarded. *Degree requirements:* Thesis or alternative required, foreign language not required. *Entrance requirements:* Honors bachelor's with second class or better. *Tuition:* $1926 per year full-time, $364 per course part-time for Canadian residents; $11,138 per year full-time for nonresidents. *Financial aid:* In 1990–91, $71,545 in aid awarded. 3 fellowships (2 to first-year students), 6 teaching assistantships were awarded; partial tuition waivers, institutionally sponsored loans also available. *Faculty research:* Acoustics, fine particles, solid state, theoretical. Total annual research budget: $245,880. • Dr. G. A. Rubin, Chairman, 705-675-1151 Ext. 2221. Application contact: Admissions Department, 705-675-1151 Ext. 3915.

Michigan Technological University, College of Sciences and Arts, Department of Physics, Houghton, MI 49931. Offerings include applied physics (PhD). PhD offered jointly with Department of Metallurgy. Department faculty: 16 full-time (1 woman), 0 part-time. *Degree requirements:* Dissertation required, foreign language not required. *Entrance requirements:* TOEFL (minimum score of 570 required), BS in physics or related area, minimum GPA of 3.0. Application deadline: rolling. Application fee: $20. *Expenses:* Tuition of $852 per semester full-time, $71 per credit part-time for state residents; $2065 per semester full-time, $172 per credit part-time for nonresidents. Fees of $75 per semester. • Dr. Bryan H. Suits, Head, 906-487-2086. Application contact: Dr. Robert Weidman, Chairman, Graduate Committee, 906-487-2126.

New Jersey Institute of Technology, Department of Physics, Newark, NJ 07102. Department offers program in applied physics (MS). Faculty: 19 full-time (0 women), 0 part-time. Matriculated students: 8 full-time (2 women), 0 part-time; includes 1 minority (Hispanic American), 6 foreign. 20 applicants, 80% accepted. In 1990, 2 degrees awarded. *Degree requirements:* Thesis required, foreign language not required. *Entrance requirements:* GRE General Test (minimum score of 450 on verbal section, 600 on mathematics, 550 on analytical required). Application deadline: 6/5 (priority date, applications processed on a rolling basis). Application fee: $30. *Tuition:* $2585 per semester full-time, $253 per credit part-time for state residents; $3864 per semester full-time, $349 per credit part-time for nonresidents. *Faculty research:* Surface microstructures, molecular beam epitaxy, ion implantation, computational physics, sensors and actuators. • Dr. Leon J. Buteau, Acting Chairperson, 201-596-3561. Application contact: Petra Theodos, Director of Graduate Admissions, 201-596-3460.

Announcement: The master's program emphasizes research in applied physics. Areas of study of study include solid-state devices, chemical vapor deposition, molecular-beam-epitaxy growth of III-V materials, microelectronics, infrared spectroscopy, modulation spectroscopy, and Rutherford backscattering/channeling materials. Extensive laboratory and computing facilities exist on campus. Applicants should have a bachelor's degree in physics, applied physics, or engineering. Teaching and research assistantships are available.

Northern Illinois University, College of Liberal Arts and Sciences, Department of Physics, De Kalb, IL 60115. Offerings include applied physics (MS). Department faculty: 19 full-time (2 women), 3 part-time. *Degree requirements:* Thesis, comprehensive exam required, foreign language not required. *Entrance requirements:* GRE General Test, TOEFL. Application deadline: 6/1. Application fee: $0. *Tuition:* $1339 per semester full-time for state residents; $3163 per semester full-time for nonresidents. • Richard Preston, Chair, 815-753-6470.

Old Dominion University, College of Sciences, Department of Physics, Program in Applied Physics, Norfolk, VA 23529. Program awards PhD. Faculty: 13 full-time (0 women), 3 part-time (0 women). Matriculated students: 26 full-time (2 women), 4 part-time (1 woman); includes 3 minority (1 Asian American, 2 black American), 10 foreign. Average age 25. 20 applicants, 80% accepted. In 1990, 2 doctorates awarded. Terminal master's awarded for partial completion of doctoral program. *Degree requirements:* For doctorate, dissertation, comprehensive exam required, foreign language not required. *Entrance requirements:* For doctorate, GRE General Test (minimum combined score of 1000 required), GRE Subject Test (minimum score of 600 required), TOEFL, minimum GPA of 3.0. Application deadline: 7/1 (applications processed on a rolling basis). Application fee: $20. *Expenses:* Tuition of $148 per credit hour for state residents; $375 per credit hour for nonresidents. Fees of $64 per year full-time. *Financial aid:* In 1990–91, $28,000 in aid awarded. 3 research assistantships (2 to first-year students), 1 teaching assistantship were awarded; fellowships, federal work-study, and career-related internships or fieldwork also available. *Faculty research:* Atomic physics, nuclear physics, gamma ray optics, condensed matter physics, radiation damage, plasma physics, mossbauer physics. Total annual research budget: $700,000. • Dr. James L. Cox Jr., Chairman, Department of Physics, 804-683-3468. Application contact: Dr. Govind Khandelwal, Coordinator, 804-683-3468.

Oregon Graduate Institute of Science and Technology, Department of Applied Physics and Electrical Engineering, Beaverton, OR 97006. Department offers programs in applied physics (MS, PhD), electrical engineering (MS, PhD). MS in applied physics offered in conjunction with Pacific University and Reed College. Part-time programs available. Faculty: 9 full-time (0 women), 29 part-time (0 women). Matriculated students: 29 full-time (4 women), 6 part-time (0 women); includes 0 minority, 17 foreign. Average age 29. 80 applicants, 30% accepted. In 1990, 5 master's awarded (40% found work related to degree, 60% continued full-time study); 5 doctorates awarded (40% entered university research/teaching, 60% found other work related to degree). Terminal master's awarded for partial completion of doctoral program. *Degree requirements:* For master's, thesis optional, foreign language not required; for doctorate, dissertation, comprehensive exam, oral defense required, foreign language not required. *Entrance requirements:* GRE General Test, GRE Subject Test, TOEFL (minimum score of 550 required). Application deadline: 3/1 (priority date, applications processed on a rolling basis). Application fee: $40. *Expenses:* Tuition of $9600 per year full-time, $240 per credit part-time. Fees of $60 per quarter. *Financial aid:* In 1990–91, 28 students received a total of $445,000 in aid awarded. 0 fellowships, 28 research assistantships (9 to first-year students) were awarded; institutionally sponsored loans also available. Average monthly stipend for a graduate assistantship: $833. Financial aid application deadline: 3/1. *Faculty research:* Semiconductor materials, microwave circuits, atmospheric optics, surface physics, electron and ion optics. Total annual research budget: $633,450. • Dr. James J. Huntzicker, Acting Chairman, 503-690-1072. Application contact: Margaret B. Day, Director of Admissions and Records, 503-690-1028.

Pittsburg State University, College of Arts and Sciences, Department of Physics, Pittsburg, KS 66762. Offerings include applied physics (MS). Department faculty: 4 full-time (0 women), 0 part-time. *Degree requirements:* Thesis or alternative required, foreign language not required. *Application fee:* $35. *Tuition:* $1670 per year full-time, $54 per credit hour part-time for state residents; $4108 per year full-time, $125 per credit hour part-time for nonresidents. • Dr. Orville Brill, Chairman, 316-231-7000 Ext. 4390.

Polytechnic University, Farmingdale Campus, Division of Engineering, School of Electrical Engineering and Computer Science, Route 110, Farmingdale, NY 11735. Offerings include electrophysics (MS, PhD). *Degree requirements:* For master's, computer language. *Application fee:* $40.

Princeton University, School of Engineering and Applied Science, Department of Mechanical and Aerospace Engineering, Princeton, NJ 08544. Offerings include applied physics and materials sciences (MSE, PhD). Terminal master's awarded for partial completion of doctoral program. Department faculty: 26 full-time (0 women), 0 part-time. *Degree requirements:* 1 foreign language, thesis/dissertation. *Entrance requirements:* GRE General Test, GRE Subject Test. Application deadline: 1/8. Application fee: $45 ($50 for foreign students). *Tuition:* $16,670 per year. • Harvey Lam, Director of Graduate Studies, 609-258-4683. Application contact: Michele Spreen, Director of Graduate Admissions, 609-258-3034.

Rensselaer Polytechnic Institute, School of Science, Department of Physics, Troy, NY 12180. Offerings include applied physics (MS, PhD). Department faculty: 28 full-time (1 woman), 7 part-time (2 women). *Degree requirements:* Thesis/dissertation required, foreign language not required. *Entrance requirements:* GRE General Test, GRE Subject Test, TOEFL (minimum score of 575 required). Application deadline: 2/1. Application fee: $30. *Expenses:* Tuition of $455 per credit hour. Fees of $195.57 per semester. • Timothy Hayes, Chair, 518-276-6419.

Rutgers, The State University of New Jersey, New Brunswick, Program in Radiation Science, New Brunswick, NJ 08903. Offerings include nuclear physics (MS). Program faculty: 8 full-time (0 women), 0 part-time. *Degree requirements:* Essay or thesis, exam required, foreign language not required. *Entrance requirements:* GRE General Test, previous course work in chemistry, physics, biology, and calculus. Application deadline: 8/1. Application fee: $35. *Expenses:* Tuition of $4432 per year full-time, $183 per credit part-time for state residents; $6496 per year full-time, $270 per credit part-time for nonresidents. Fees of $458 per year full-time, $117 per year part-time. • Francis J. Haughey, Director, 908-932-2551.

Directory: Applied Physics

Stanford University, School of Humanities and Sciences, Department of Applied Physics, Stanford, CA 94305. Department awards MS, PhD. Faculty: 18 (0 women). Matriculated students: 119 full-time (18 women), 0 part-time; includes 19 minority (15 Asian American, 3 Hispanic American, 1 Native American), 33 foreign. Average age 26. 224 applicants, 20% accepted. In 1990, 22 master's, 21 doctorates awarded. Terminal master's awarded for partial completion of doctoral program. *Degree requirements:* For master's, thesis optional, foreign language not required; for doctorate, dissertation required, foreign language not required. *Entrance requirements:* GRE General Test, GRE Subject Test, TOEFL. Application deadline: 1/1. Application fee: $55. *Expenses:* Tuition of $15,102 per year. Fees of $28 per quarter. *Financial aid:* Fellowships, research assistantships, federal work-study, institutionally sponsored loans available. Financial aid application deadline: 1/1; applicants required to submit GAPSFAS. • Application contact: Graduate Admissions Coordinator, 415-723-4027.

State University of New York at Binghamton, School of Arts and Sciences, Department of Physics, Applied Physics, and Astronomy, Binghamton, NY 13902-6000. Offerings include applied physics (MS). *Application deadline:* 4/15 (priority date). *Application fee:* $35. *Expenses:* Tuition of $2450 per year full-time, $102.50 per credit part-time for state residents; $5766 per year full-time, $242.50 per credit part-time for nonresidents. Fees of $77 per year full-time, $27.85 per semester (minimum) part-time. • Dr. Robert L. Pompi, Chairperson, 607-777-2217.

State University of New York at Buffalo, Graduate School, School of Engineering and Applied Sciences, Department of Electrical and Computer Engineering, Buffalo, NY 14260. Offerings include applied physics (ME, MS, PhD). Terminal master's awarded for partial completion of doctoral program. Department faculty: 33 full-time (0 women), 11 part-time (0 women). *Degree requirements:* For master's, project, exam required, foreign language not required; for doctorate, dissertation required, foreign language not required. *Entrance requirements:* GRE General Test, TOEFL (minimum score of 550 required). Application deadline: 4/1 (applications processed on a rolling basis). Application fee: $35. *Expenses:* Tuition of $1600 per semester full-time, $134 per hour part-time for state residents; $3258 per semester full-time, $274 per hour part-time for nonresidents. Fees of $137 per semester full-time, $115 per semester (minimum) part-time. • Dr. Wayne Anderson, Chairman, 716-636-2422. Application contact: Dr. Raj K. Kaul, Director of Graduate Studies, 716-636-2427.

Texas Tech University, Graduate School, College of Arts and Sciences, Department of Physics and Engineering Physics, Lubbock, TX 79409. Offerings include applied physics (MS, PhD). Department faculty: 22. *Degree requirements:* Variable foreign language requirement, thesis/dissertation. *Entrance requirements:* GRE General Test. Application deadline: 4/15 (priority date, applications processed on a rolling basis). Application fee: $0 ($50 for foreign students). *Tuition:* $494 per semester full-time, $20 per credit hour part-time for state residents; $1790 per semester full-time, $455 per credit hour part-time for nonresidents. • Dr. Walter Borst, Chairman, 806-742-3767.

University of Central Oklahoma, College of Mathematics and Science, Department of Industrial and Applied Physics, Edmond, OK 73034. Department awards MS. Part-time and evening/weekend programs available. Faculty: 7 full-time (0 women), 0 part-time. Matriculated students: 1 full-time (0 women), 13 part-time (1 woman); includes 3 minority (1 Asian American, 1 black American, 1 Native American), 3 foreign. Average age 35. 5 applicants, 100% accepted. In 1990, 7 degrees awarded. *Degree requirements:* Thesis optional, foreign language not required. *Application deadline:* 8/17. *Application fee:* $0. *Expenses:* Tuition of $46.85 per credit hour for state residents; $123 per credit hour for nonresidents. Fees of $3 per credit hour. *Financial aid:* In 1990–91, $8600 in aid awarded. 2 teaching assistantships (1 to a first-year student) were awarded; graduate assistantships also available. *Faculty research:* Acoustics, solid-state physics/optical properties, molecular dynamics, nuclear physics, crystallography. Total annual research budget: $42,700. • Dr. John P. King, Chairperson, 405-341-2980 Ext. 5468.

University of Colorado at Boulder, College of Arts and Sciences, Department of Physics, Boulder, CO 80309. Offerings include applied physics (PhD). Terminal master's awarded for partial completion of doctoral program. Department faculty: 66 full-time (1 woman). *Degree requirements:* 1 foreign language, dissertation. *Entrance requirements:* GRE General Test, GRE Subject Test. Application deadline: 3/1 (priority date, applications processed on a rolling basis). Application fee: $30 ($50 for foreign students). *Expenses:* Tuition of $2308 per year full-time, $387 per semester (minimum) part-time for state residents; $8730 per year full-time, $1455 per semester (minimum) part-time for nonresidents. Fees of $207 per semester full-time, $27.26 per semester (minimum) part-time. • William O'Sullivan, Chairman, 303-492-8703.

University of Maryland Graduate School, Baltimore, Graduate School, Department of Physics, Baltimore, MD 21228. Department awards MS, PhD. Program offered in applied physics (MS). Faculty: 12 full-time (0 women), 1 part-time (0 women). Matriculated students: 12 full-time (2 women), 3 part-time (2 women); includes 0 minority, 6 foreign. Average age 25. 40 applicants, 28% accepted. In 1990, 7 master's awarded (15% found work related to degree, 85% continued full-time study). *Degree requirements:* For master's, thesis optional, foreign language not required. *Entrance requirements:* For master's, GRE General Test, GRE Subject Test, TOEFL, minimum GPA of 3.0. Application deadline: 3/1 (priority date, applications processed on a rolling basis). Application fee: $25. *Tuition:* $134 per credit for state residents; $245 per credit for nonresidents. *Financial aid:* In 1990–91, 12 students received a total of $120,000 in aid awarded. 1 fellowship (0 to first-year students), 2 research assistantships (1 to a first-year student), 9 teaching assistantships (all to first-year students) were awarded. Average monthly stipend for a graduate assistantship: $1100. Financial aid application deadline: 3/1. *Faculty research:* Physics of solids, spectroscopy, nonlinear and quantum optics, atmospheric research, materials science. • Dr. Geoffrey Summers, Chairman, 301-455-2513. Application contact: Dr. Robert Reno, Coordinator, 301-455-2530.

University of Massachusetts at Boston, College of Arts and Sciences, Program in Applied Physics, Boston, MA 02125. Program awards MS. Matriculated students: 7 full-time (1 woman), 11 part-time (3 women). *Degree requirements:* Thesis or alternative, comprehensive exams. *Entrance requirements:* Minimum GPA of 2.75. Application deadline: 6/1 (applications processed on a rolling basis). Application fee: $20. *Expenses:* Tuition of $2568 per year full-time, $107 per credit part-time for state residents; $7920 per year full-time, $330 per credit part-time for nonresidents. Fees of $767 per year full-time, $339 per year part-time. *Financial aid:* In 1990–91, 4 research assistantships (3 to first-year students), 8 teaching assistantships (3 to first-year students) awarded. • Dr. Leonard Catz, Director, 617-287-6050. Application contact: Lisa Lavely, Director of Graduate Admissions, 617-287-6400.

University of Massachusetts, Lowell, College of Arts and Sciences, Department of Physics and Applied Physics, 1 University Avenue, Lowell, MA 01854. Department offers programs in applied physics (PhD), computational physics (PhD), energy engineering (PhD), physics (MS), radiological sciences and protection (MS, PhD). Terminal master's awarded for partial completion of doctoral program. *Degree requirements:* For master's, thesis required, foreign language not required; for doctorate, 2 foreign languages (computer language can substitute for one), dissertation. *Entrance requirements:* GRE General Test. Application deadline: 4/1. *Expenses:* Tuition of $87 per credit hour for state residents; $271 per credit hour for nonresidents. Fees of $114 per credit hour.

See full description on page 651.

University of Michigan, College of Literature, Science, and the Arts, Program in Applied Physics, Ann Arbor, MI 48109. Program awards PhD. Faculty: 6. Matriculated students: 28 full-time (4 women), 0 part-time; includes 1 minority (black American), 13 foreign. 66 applicants, 27% accepted. In 1990, 0 degrees awarded. *Degree requirements:* Dissertation, preliminary exam. *Entrance requirements:* GRE General Test, GRE Subject Test. Application deadline: 2/1 (applications processed on a rolling basis). Application fee: $30. *Tuition:* $3255 per semester full-time, $352 per credit (minimum) part-time for state residents; $6803 per semester full-time, $746 per credit (minimum) part-time for nonresidents. *Financial aid:* Fellowships, research assistantships available. Financial aid application deadline: 3/15. • Roy Clarke, Director, 313-764-4466.

University of New Orleans, College of Sciences, Department of Physics, New Orleans, LA 70148. Offerings include applied physics (MS). Department faculty: 9 full-time (1 woman). *Degree requirements:* Thesis required (for some programs), foreign language not required. *Entrance requirements:* GRE General Test (minimum combined score of 1000 required), TOEFL (minimum score of 500 required). Application deadline: 7/1 (priority date, applications processed on a rolling basis). Application fee: $20. *Tuition:* $962 per quarter hour full-time for state residents; $2308 per quarter hour full-time for nonresidents. • Milton D. Slaughter, Chair, 504-286-6341. Application contact: Graduate Coordinator, 504-286-6343.

University of Southwestern Louisiana, College of Sciences, Department of Physics, Lafayette, LA 70504. Offerings include applied physics (MS). Department faculty: 6 full-time (0 women), 0 part-time. *Degree requirements:* Thesis required, foreign language not required. *Entrance requirements:* GRE General Test. Application deadline: 8/15. Application fee: $5. *Tuition:* $1560 per year full-time, $228 per credit (minimum) part-time for state residents; $3310 per year full-time, $228 per credit (minimum) part-time for nonresidents. • Dr. Davy Bernard, Head, 318-231-6691. Application contact: Dr. L. Dwynn Lafleur, Graduate Coordinator, 318-231-6696.

University of Washington, College of Arts and Sciences, Department of Physics, Seattle, WA 98195. Department awards MS, PhD. Part-time and evening/weekend programs available. Faculty: 56 full-time (2 women). Matriculated students: 125 full-time (15 women), 68 part-time (9 women); includes 6 minority (2 Asian American, 2 black American, 2 Hispanic American), 34 foreign. Average age 30. 370 applicants, 22% accepted. In 1990, 24 master's, 14 doctorates awarded. *Degree requirements:* For master's, foreign language and thesis not required; for doctorate, dissertation required, foreign language not required. *Entrance requirements:* For doctorate, GRE General Test, GRE Subject Test, TOEFL. Application deadline: 2/15 (priority date, applications processed on a rolling basis). Application fee: $35. *Tuition:* $1129 per quarter full-time, $324 per credit (minimum) part-time for state residents; $2824 per quarter full-time, $809 per credit (minimum) part-time for nonresidents. *Financial aid:* In 1990–91, $1.2-million in aid awarded. 2 fellowships (both to first-year students), 4 research assistantships (all to first-year students) were awarded; teaching assistantships, federal work-study also available. Average monthly stipend for a graduate assistantship: $950. Financial aid application deadline: 2/15. *Faculty research:* Atomic, molecular, and condensed-matter physics; elementary particles; nuclear physics; astrophysics. Total annual research budget: $7.554-million. • Mark McDermott, Chairman, 206-543-2770. Application contact: Graduate Program Assistant, 206-543-2770.

Virginia Commonwealth University, College of Humanities and Sciences, Department of Physics, Richmond, VA 23284. Offerings include applied physics (MS). Department faculty: 8 full-time, 0 part-time. *Degree requirements:* Thesis required, foreign language not required. *Entrance requirements:* GRE. Application deadline: 8/1. Application fee: $20. *Expenses:* Tuition of $2770 per year full-time, $154 per hour part-time for state residents; $7550 per year full-time, $419 per hour part-time for nonresidents. Fees of $717 per year full-time, $25.50 per hour part-time. • Dr. Peru Jena, Chair, 804-367-1313.

Yale University, Graduate School of Arts and Sciences, Council of Engineering and Applied Science, Program of Applied Physics, New Haven, CT 06520. Program awards MS, PhD. Faculty: 13 full-time (1 woman), 0 part-time. Matriculated students: 32 full-time (4 women), 4 part-time (0 women); includes 0 minority, 11 foreign. 65 applicants, 17% accepted. In 1990, 2 master's, 3 doctorates awarded. Terminal master's awarded for partial completion of doctoral program. *Degree requirements:* For master's, foreign language and thesis not required; for doctorate, dissertation, area exam required, foreign language not required. *Entrance requirements:* GRE General Test, GRE Subject Test, TOEFL. Application deadline: 1/2. Application fee: $45. *Tuition:* $15,160 per year full-time, $1895 per course part-time. *Financial aid:* In 1990–91, 15 fellowships awarded. *Faculty research:* Condensed-matter, plasma, laser, and molecular physics. Total annual research budget: $1.9-million. • Werner Wolf, Chairman, 203-432-2210. Application contact: Mary Lally, 203-432-4250.

Announcement: Applied physics at Yale University encompasses numerous areas of theoretical and experimental condensed-matter, plasma, laser, and molecular physics. Specific programs include localization, surface science, microlithography, optical properties of micro-objects, magnetic materials and phase transitions, atomic collision theory, chaos, and nonequilibrium and transport properties of plasma.

Mathematical Physics

Indiana University Bloomington, College of Arts and Sciences, Department of Physics, Bloomington, IN 47405. Department awards MAT, MS, PhD. Part-time programs available. Faculty: 47 full-time (2 women), 1 part-time (0 women). Matriculated students: 88 full-time (13 women), 0 part-time; includes 2 minority (both Hispanic American), 47 foreign. Average age 29. 169 applicants, 23% accepted. In 1990, 27 master's awarded (0% entered university research/teaching, 0% found other work related to degree, 77% continued full-time study); 6 doctorates awarded (50% entered university research/teaching, 50% found other work related to degree). Terminal master's awarded for partial completion of doctoral program. *Degree requirements:* For master's, qualifying exam required, foreign language and thesis not required; for doctorate, dissertation, qualifying exam required, foreign language not required. *Entrance requirements:* GRE, TOEFL (minimum score of 550 required). Application deadline: 2/1. Application fee: $25 ($35 for foreign students). *Tuition:* $99.85 per credit hour for state residents; $288 per credit hour for nonresidents. *Financial aid:* In 1990–91, 8 fellowships (all to first-year students), 53 research assistantships (21 to first-year students), 28 teaching assistantships (10 to first-year students) awarded; federal work-study and career-related internships or fieldwork also available. Aid available to part-time students. Financial aid application deadline: 2/1. *Faculty research:* Condensed matter, elementary particles, nuclear physics, astrophysics. Total annual research budget: $11.5-million. • Dr. J. Timothy Londergan, Chair, 812-855-1247. Application contact: June Dizer, Academic Services Coordinator, 812-855-3973.

See full description on page 589.

Princeton University, Departments of Physics and Mathematics, Program in Mathematical Physics, Princeton, NJ 08544. Program awards PhD. *Degree requirements:* 1 foreign language, dissertation. *Entrance requirements:* GRE General Test, GRE Subject Test. Application deadline: 1/8. Application fee: $45 ($50 for foreign students). *Tuition:* $16,670 per year. *Financial aid:* Federal work-study, institutionally sponsored loans available. Financial aid application deadline: 1/8. • Application contact: Michele Spreen, Director of Graduate Admissions, 609-258-3034.

University of Colorado at Boulder, College of Arts and Sciences, Department of Physics, Boulder, CO 80309. Offerings include mathematical physics (PhD). Terminal master's awarded for partial completion of doctoral program. Department faculty: 66 full-time (1 woman). *Degree requirements:* 1 foreign language, dissertation. *Entrance requirements:* GRE General Test, GRE Subject Test. Application deadline: 3/1 (priority date, applications processed on a rolling basis). Application fee: $30 ($50 for foreign students). *Expenses:* Tuition of $2308 per year full-time, $387 per semester (minimum) part-time for state residents; $8730 per year full-time, $1455 per semester (minimum) part-time for nonresidents. Fees of $207 per semester full-time, $27.26 per semester (minimum) part-time. • William O'Sullivan, Chairman, 303-492-8703.

Virginia Polytechnic Institute and State University, College of Arts and Sciences, Department of Mathematics, Blacksburg, VA 24061. Offerings include mathematical physics (MS, PhD). Terminal master's awarded for partial completion of doctoral program. Department faculty: 55 full-time (5 women), 0 part-time. *Degree requirements:* For master's, thesis required (for some programs), foreign language not required; for doctorate, 1 foreign language, dissertation. *Application deadline:* 2/15 (priority date). *Application fee:* $10. *Tuition:* $1889 per semester full-time, $606 per credit hour part-time for state residents; $2627 per semester full-time, $853 per credit hour part-time for nonresidents. • Dr. C. Wayne Patty, Head, 703-231-6536. Application contact: E. L. Green, Graduate Administrator, 703-231-6536.

See full description on page 511.

Optical Sciences

Alabama Agricultural and Mechanical University, School of Arts and Sciences, Department of Physics, PO Box 285, Normal, AL 35762. Offerings include optics (MS, PhD). Department faculty: 9 full-time (0 women), 0 part-time. *Degree requirements:* For master's, thesis optional, foreign language not required; for doctorate, computer language, dissertation required, foreign language not required. *Entrance requirements:* For master's, GRE, BS in physics or electrical engineering; for doctorate, GRE General Test (minimum combined score of 1000 required). Application deadline: 3/1 (priority date, applications processed on a rolling basis). Application fee: $15. *Tuition:* $79 per credit hour for state residents; $158 per credit hour for nonresidents. • Dr. Jai-Ching Wang, Chairman, 205-851-5305.

Cleveland State University, College of Arts and Sciences, Department of Physics, Cleveland, OH 44115. Offerings include applied optics (MS). Department faculty: 7 full-time (0 women), 0 part-time. *Degree requirements:* Computer language required, foreign language and thesis not required. *Application deadline:* 9/1 (priority date, applications processed on a rolling basis). *Application fee:* $0. *Tuition:* $90 per credit for state residents; $180 per credit for nonresidents. • Dr. Francis Stephenson, Chairman, 216-687-2425. Application contact: Dr. Jack Soules, Director, 216-687-3517.

École Polytechnique de Montréal, Department of Engineering Physics, Montreal, PQ H3C 3A7, Canada. Offerings include optical engineering (M Eng, M Sc A, PhD). Department faculty: 11 full-time (0 women), 9 part-time (0 women). *Degree requirements:* 1 foreign language, computer language, thesis/dissertation. *Entrance requirements:* For master's, minimum GPA of 2.75; for doctorate, minimum GPA of 3.0. Application deadline: 4/1. Application fee: $15. *Tuition:* $1400 per year full-time, $52 per credit part-time for Canadian residents; $7200 per year full-time, $290 per credit part-time for nonresidents. • Guy Faucher, Chairman, 514-340-4768.

Ohio State University, College of Optometry and Graduate School, Graduate Program in Optometry, Program in Physiological Optics, Columbus, OH 43210. Program awards MS, PhD. Faculty: 14 full-time (3 women), 0 part-time. Matriculated students: 7 full-time (1 woman), 0 part-time; includes 1 minority (black American), 0 foreign. 9 applicants, 44% accepted. In 1990, 3 master's awarded (100% found work related to degree); 1 doctorate awarded. *Degree requirements:* For master's, thesis required, foreign language not required; for doctorate, 2 foreign languages (computer language can substitute for one), dissertation. *Entrance requirements:* For master's, OD; for doctorate, MS and OD. *Tuition:* $1213 per quarter full-time, $364 per course part-time for state residents; $3143 per quarter full-time, $943 per course part-time for nonresidents. *Financial aid:* In 1990–91, 4 teaching assistantships awarded; fellowships also available. • Dr. Richard Hill, Dean, College of Optometry, 614-292-5043.

Rose-Hulman Institute of Technology, Program in Applied Optics, Terre Haute, IN 47803. Program awards MS. *Degree requirements:* Computer language, thesis required, foreign language not required. *Entrance requirements:* GRE, TOEFL (minimum score of 550 required), minimum GPA of 3.0. Application fee: $0.

University of Arizona, Graduate Committee on Optical Sciences, Tucson, AZ 85721. Committee awards MS, PhD. Part-time programs available. Faculty: 48 full-time (2 women), 3 part-time (1 woman). Matriculated students: 121 full-time (21 women), 34 part-time (4 women); includes 18 minority (11 Asian American, 7 Hispanic American), 26 foreign. Average age 28. 194 applicants, 18% accepted. In 1990, 40 master's, 18 doctorates awarded. *Degree requirements:* For master's, thesis (for some programs), exam required, foreign language not required; for doctorate, dissertation, oral and written exams required, foreign language not required. *Entrance requirements:* GRE General Test, GRE Subject Test. Application deadline: 3/1 (applications processed on a rolling basis). Application fee: $25. *Expenses:* Tuition of $0 for state residents; $5406 per year full-time, $209 per credit hour part-time for nonresidents. Fees of $1528 per year full-time, $80 per credit hour part-time. *Financial aid:* In 1990–91, $137,333 in aid awarded. 8 fellowships, 85 research assistantships, 8 teaching assistantships, 3 scholarships were awarded. Financial aid applicants required to submit FAF. *Faculty research:* Medical optics, medical imaging, optical data storage, optical bistability, nonlinear optical effects. Total annual research budget: $7.3-million. • Robert R. Shannon, Director, 602-621-6997. Application contact: Dr. Jack D. Gaskill, Associate Director, Academic Affairs, 602-621-4111.

See full description on page 631.

University of Dayton, School of Engineering, Program in Electro-Optics, Dayton, OH 45469. Program awards MSEO. Part-time and evening/weekend programs available. Faculty: 1 full-time (0 women), 16 part-time (0 women). Matriculated students: 29 full-time (4 women), 13 part-time (1 woman); includes 3 foreign. In 1990, 10 degrees awarded (80% found work related to degree, 20% continued full-time study). *Degree requirements:* Thesis optional. *Entrance requirements:* TOEFL. Application deadline: 8/1. Application fee: $20 ($50 for foreign students). *Expenses:* Tuition of $262 per semester hour. Fees of $20 per semester. *Financial aid:* In 1990–91, 19 research assistantships, 3 teaching assistantships awarded; institutionally sponsored loans also available. *Faculty research:* Optical information processing, interferometry, computational optics, holography, laser diagnostics. Total annual research budget: $500,000. • Dr. Mohammad A. Karim, Director, 513-229-3611. Application contact: Dr. Gary A. Thiele, Associate Dean/Director, 513-229-2241.

University of Houston–Clear Lake, School of Natural and Applied Sciences, Houston, TX 77058. Offerings include electrooptical technology (MS). *Application fee:* $0. *Tuition:* $40 per credit hour for state residents; $134 per credit hour for nonresidents. • E. T. Dickerson, Dean, 713-283-3703. Application contact: Dr. Eldon Husband, Associate Dean and Director of Student Affairs, 713-283-3710.

University of Massachusetts, Lowell, College of Arts and Sciences, Department of Physics and Applied Physics, 1 University Avenue, Lowell, MA 01854. Department offers programs in applied physics (PhD), computational physics (PhD), energy engineering (PhD), physics (MS), radiological sciences and protection (MS, PhD). Terminal master's awarded for partial completion of doctoral program. *Degree requirements:* For master's, thesis required, foreign language not required; for doctorate, 2 foreign languages (computer language can substitute for one), dissertation. *Entrance requirements:* GRE General Test. Application deadline: 4/1. *Expenses:* Tuition of $87 per credit hour for state residents; $271 per credit hour for nonresidents. Fees of $114 per credit hour.

See full description on page 651.

University of Rochester, College of Engineering and Applied Science, Institute of Optics, Rochester, NY 14627-0001. Institute awards MS, PhD. Faculty: 14 full-time, 7 part-time. Matriculated students: 134 full-time (21 women), 16 part-time (3 women); includes 6 minority (3 Asian American, 1 black American, 2 Hispanic American), 21 foreign. 217 applicants, 23% accepted. In 1990, 27 master's, 13 doctorates awarded. *Degree requirements:* For master's, comprehensive exam required, foreign language and thesis not required; for doctorate, dissertation, preliminary and qualifying exams required, foreign language not required. *Entrance requirements:* For doctorate, GRE. Application deadline: 2/15 (priority date). Application fee: $25. *Expenses:* Tuition of $473 per credit hour. Fees of $243 per year full-time. *Financial aid:* Fellowships, research assistantships, teaching assistantships, full and partial tuition waivers available. Financial aid application deadline: 2/15. *Faculty research:* Physical optics, quantum optics, electrooptics, wave guides, laser systems and holography. • Duncan T. Moore, Chair, 716-275-7764.

See full description on page 663.

Physics

Adelphi University, Graduate School of Arts and Sciences, Department of Physics, Garden City, NY 11530. Department awards MS. Part-time and evening/weekend programs available. Matriculated students: 4 full-time (2 women), 8 part-time (3 women); includes 3 minority (2 Asian American, 1 Hispanic American), 2 foreign. Average age 30. 48 applicants, 60% accepted. In 1990, 4 degrees awarded. *Degree requirements:* Thesis or alternative required, foreign language not required. *Application deadline:* 8/30. *Application fee:* $50. *Expenses:* Tuition of $5300 per semester full-time, $315 per credit part-time. Fees of $159 per semester part-time. *Financial aid:* In 1990–91, $3300 in aid awarded. 0 fellowships, 2 research assistantships (1 to a first-year student), 2 teaching assistantships (both to first-year students) were awarded; partial tuition waivers also available. Financial aid application deadline: 8/30. *Faculty research:* General relativity, solid-state physics, atomic physics, optics. Total annual research budget: $200,000. • Dr. John Dooher, Chairman, 516-877-4883.

Alabama Agricultural and Mechanical University, School of Arts and Sciences, Department of Physics, PO Box 285, Normal, AL 35762. Department offers programs in applied physics (MS, PhD), materials science (MS, PhD), optics (MS, PhD). Faculty: 9 full-time (0 women), 0 part-time. Matriculated students: 31 full-time (6 women), 7 part-time (1 woman); includes 16 black American, 8 foreign. Average age 30. 9 applicants, 56% accepted. In 1990, 65 master's awarded (40% entered university research/teaching, 20% found other work related to degree, 40% continued full-time study); 1 doctorate awarded (100% entered university research/teaching). *Degree requirements:* For master's, thesis optional, foreign language not required; for doctorate, computer language, dissertation required, foreign language not required. *Entrance requirements:* For master's, GRE, BS in physics or electrical engineering; for doctorate, GRE General Test (minimum combined score of 1000 required). Application deadline: 3/1 (priority date, applications processed on a rolling basis). Application fee: $15. *Tuition:* $79 per credit hour for state residents; $158 per credit hour for nonresidents. *Financial aid:* In 1990–91, 25 students received aid. 2 fellowships (1 to a first-year student), 18 research assistantships (5 to first-year students), 5 teaching assistantships (2 to first-year students) awarded; career-related internships or fieldwork also available. Average monthly stipend for a graduate assistantship: $800. Financial aid application deadline: 8/1. *Faculty research:* Optics/lasers, nonlinear optics, materials science, crystal growth modeling. Total annual research budget: $4.2-million. • Dr. Jai-Ching Wang, Chairman, 205-851-5305.

American University, College of Arts and Sciences, Department of Physics, Washington, DC 20016. Department awards MS, PhD. Part-time and evening/weekend programs available. Faculty: 11 full-time (0 women), 2 part-time (0 women). Matriculated students: 15 full-time (4 women), 16 part-time (3 women); includes 5 minority (3 Asian American, 1 black American, 1 Hispanic American), 6 foreign. Average age 29. 61 applicants, 74% accepted. In 1990, 4 master's awarded (75% continued full-time study); 4 doctorates awarded (25% entered university research/teaching). Terminal master's awarded for partial completion of doctoral program. *Degree requirements:* For master's, thesis or alternative, comprehensive exams; for doctorate, 1 foreign language (computer language can substitute), dissertation, comprehensive exams. *Application deadline:* 2/15 (priority date, applications processed on a rolling basis). *Application fee:* $50. *Expenses:* Tuition of $475 per semester hour. Fees of $20 per semester. *Financial aid:* In 1990–91, 12 students received a total of $196,040 in aid awarded. 11 fellowships (3 to first-year students), 1 teaching assistantship (to a first-year student) were awarded; research assistantships, partial tuition waivers, federal work-study, institutionally sponsored loans, and career-related internships or fieldwork also available. Aid available to part-time students. Average monthly stipend for a graduate assistantship: $750. Financial aid application deadline: 2/15; applicants required to submit FAF. *Faculty research:* Nuclear particle experimental physics, solid-state physics, quantum electronics, critical phenomena, intense fields. • Dr. Romeo Segnan, Chair, 202-885-2750.

Andrews University, School of Graduate Studies, College of Arts and Sciences, Department of Physics, Berrien Springs, MI 49104. Department awards MS. *Application deadline:* rolling. *Application fee:* $40. *Expenses:* Tuition of $200 per quarter hour. Fees of $25 per year. • Robert E. Kingman, Chairman, 616-471-3431.

Appalachian State University, College of Arts and Sciences, Department of Physics, Boone, NC 28608. Department offers program in applied physics (MS). Faculty: 11 full-time (0 women), 0 part-time. Matriculated students: 6 full-time (2 women), 1 (woman) part-time; includes 1 minority (black American), 1 foreign. 19 applicants, 26% accepted. In 1990, 1 degree awarded. *Degree requirements:* Thesis, comprehensive exam. *Entrance requirements:* GRE General Test. Application deadline: 7/31 (priority date). Application fee: $15. *Tuition:* $598 per semester for state residents; $3125 per semester full-time, $6215 per semester part-time for nonresidents. *Financial aid:* Application deadline 7/31. • Karl Mamola, Chairperson, 704-262-3090.

Arizona State University, College of Liberal Arts and Sciences, Department of Physics, Tempe, AZ 85287. Department offers programs in applied physics (PhD), astronomy (PhD), biophysics (PhD), geophysics (PhD), interdisciplinary physics (MS), physics (MNS), physics teaching (MS), planetary sciences (PhD), pure physics (MS, PhD), science education (MNS), technical physics (MS). Matriculated students: 72 full-time (10 women), 18 part-time (3 women); includes 6 minority (5 Asian American, 1 Hispanic American), 39 foreign. 230 applicants, 12% accepted. In 1990, 5 master's, 10 doctorates awarded. *Degree requirements:* For master's, thesis, written and oral exams; for doctorate, dissertation. *Entrance requirements:* GRE. Application fee: $25. *Tuition:* $1528 per year full-time, $80 per hour part-time for state residents; $6934 per year full-time, $289 per hour part-time for nonresidents. *Faculty research:* Electromagnetic interaction of hadrons, investigation of tripartition fission, and beta activity of various elements formed in fission processes; phase transitions in solids. • Dr. John D. Dow, Chair, 602-965-3561.

Auburn University, College of Sciences and Mathematics, Department of Physics, Auburn University, AL 36849. Department awards MS, PhD. Part-time programs available. Faculty: 16 full-time, 0 part-time. Matriculated students: 16 full-time (2 women), 28 part-time (4 women); includes 1 minority (Native American), 23 foreign. In 1990, 7 master's, 0 doctorates awarded. *Degree requirements:* For master's, foreign language not required; for doctorate, 1 foreign language, dissertation, written and oral exams. *Entrance requirements:* GRE General Test, GRE Subject Test. Application deadline: 9/1 (applications processed on a rolling basis). Application fee: $15. *Expenses:* Tuition of $1596 per year full-time, $44 per credit hour part-time for state residents; $4788 per year full-time, $132 per credit hour part-time for nonresidents. Fees of $92 per quarter for state residents; $276 per quarter for nonresidents. *Financial aid:* Research assistantships, teaching assistantships, federal work-study available. Aid available to part-time students. Financial aid application deadline: 4/15. *Faculty research:* Atomic and molecular physics, biophysics, plasma physics, mathematical physics, nuclear physics. • Dr. Joe D. Perez, Head, 205-844-4264. Application contact: Dr. Norman J. Doorenbos, Dean of the Graduate School, 205-844-4700.

See full description on page 569.

Ball State University, College of Sciences and Humanities, Department of Physics and Astronomy, 2000 University Avenue, Muncie, IN 47306. Department offers program in physics (MA, MS). Faculty: 13. Matriculated students: 8. In 1990, 2 degrees awarded. *Degree requirements:* Foreign language not required. *Application fee:* $15. *Expenses:* Tuition of $1140 per semester (minimum) full-time for state residents; $2680 per semester (minimum) full-time for nonresidents. Fees of $6 per credit hour. *Financial aid:* In 1990–91, 6 research assistantships awarded. • Paul Errington, Head, 317-285-8860.

Baylor University, College of Arts and Sciences, Department of Physics, Waco, TX 76798. Department awards MA, MS, PhD. Matriculated students: 6 full-time (1 woman), 16 part-time (1 woman); includes 10 minority (7 Asian American, 2 black American, 1 Hispanic American), 10 foreign. In 1990, 3 master's, 2 doctorates awarded. *Degree requirements:* For master's, thesis or alternative required, foreign language not required; for doctorate, 1 foreign language (computer language can substitute), dissertation. *Entrance requirements:* GRE General Test. Application deadline: rolling. Application fee: $0. *Expenses:* Tuition of $4440 per year full-time, $185 per credit hour part-time. Fees of $510 per year full-time. *Financial aid:* Fellowships, teaching assistantships, federal work-study, institutionally sponsored loans available. Financial aid applicants required to submit FAF. • Dr. Darden Powers, Director of Graduate Studies, 817-755-2511.

Bishop's University, Division of Natural Sciences, Department of Physics, Lennoxville, PQ J1M 1Z7, Canada. Department awards M Sc. *Application fee:* $25. *Tuition:* $883 per semester for Canadian residents; $3910 per semester for nonresidents. • Dr. L. Nelson, Chairperson, 819-822-9600 Ext. 372.

Boston College, Graduate School of Arts and Sciences, Department of Physics, Chestnut Hill, MA 02167. Department awards MS, MST, PhD. Terminal master's awarded for partial completion of doctoral program. *Degree requirements:* For master's, thesis required (for some programs), foreign language not required; for doctorate, dissertation required, foreign language not required. *Entrance requirements:* GRE General Test, GRE Subject Test. Application deadline: 5/15. *Tuition:* $412 per credit hour. *Faculty research:* Atmospheric/space physics, astrophysics, atomic and molecular physics, fusion and plasmas, solid state physics.

Boston University, Graduate School, Department of Physics, Boston, MA 02215. Department offers programs in applied physics (PhD), astronomy and physics (MA, PhD), cellular biophysics (PhD), physics (MA, PhD). PhD in applied physics offered through the Division of Applied Sciences and Engineering. Matriculated students: 99 full-time (18 women), 2 part-time (0 women); includes 3 minority (1 Asian American, 1 black American, 1 Hispanic American), 50 foreign. Average age 28. 376 applicants, 13% accepted. In 1990, 10 master's, 6 doctorates awarded. Terminal master's awarded for partial completion of doctoral program. *Degree requirements:* For master's, 1 foreign language, thesis, comprehensive exam; for doctorate, 1 foreign language, dissertation. *Entrance requirements:* GRE General Test, GRE Subject Test, TOEFL (minimum score of 550 required). Application deadline: 4/1. Application fee: $45. *Expenses:* Tuition of $15,950 per year full-time, $498 per credit part-time. Fees of $120 per year full-time, $35 per semester part-time. *Financial aid:* Fellowships, research assistantships, teaching assistantships available. Financial aid application deadline: 1/15. *Faculty research:* Experimental and theoretical condensed matter, intermediate energy, nuclear physics, membrane biophysics, elemntal particle physics. • Lawrence R. Sulak, Chairman, 617-353-2623. Application contact: Beverly Pacheco, Admissions Coordinator, 617-353-2623.

See full description on page 571.

Brandeis University, Graduate School of Arts and Sciences, Program in Physics, Waltham, MA 02254. Program awards PhD. Faculty: 23 full-time (1 woman), 0 part-time. *Entrance requirements:* GRE General Test. Application deadline: 3/1 (priority date). Application fee: $50. *Tuition:* $16,085 per year full-time, $2015 per course part-time. *Financial aid:* Fellowships, research assistantships, teaching assistantships, full tuition waivers available. Financial aid application deadline: 3/1; applicants required to submit GAPSFAS. *Faculty research:* Theoretical physics, experimental physics. • Dr. John F. C. Wardle III, Chair.

See full description on page 573.

Bridgewater State College, Division of Natural Sciences and Mathematics, Department of Physics, Bridgewater, MA 02324. Department awards MAT, CAGS. *Entrance requirements:* For master's, GRE General Test. Application deadline: 3/1. Application fee: $25. *Tuition:* $1446 per semester for state residents; $3303 per semester for nonresidents. • Marilyn W. Barry, Graduate Dean, Graduate School, 508-697-1300.

Brigham Young University, College of Physical and Mathematical Sciences, Department of Physics and Astronomy, Provo, UT 84602. Department offers programs in physics (MS, PhD), physics and astronomy (PhD). Faculty: 30 full-time (0 women), 1 part-time (0 women). Matriculated students: 40 full-time (2 women), 5 part-time (0 women); includes 0 minority, 17 foreign. Average age 26. 70 applicants, 13% accepted. In 1990, 4 master's, 2 doctorates awarded. Terminal master's awarded for partial completion of doctoral program. *Degree requirements:* For master's, thesis; for doctorate, 1 foreign language, computer language, dissertation. *Entrance requirements:* For master's, GRE Subject Test (score in 40th percentile or higher required), minimum GPA of 3.0 during previous 60 credit hours; for doctorate, GRE Subject Test (score in 60th percentile or higher required). Application deadline: 2/15. Application fee: $30. *Tuition:* $1170 per semester (minimum) full-time, $130 per credit hour (minimum) part-time. *Financial aid:* In 1990–91, $388,000 in aid awarded. 5 fellowships (0 to first-year students), 15 research assistantships, 13 teaching assistantships (7 to first-year students) were awarded; institutionally sponsored loans also available. Aid available to part-time students. Financial aid application deadline: 2/15. *Faculty research:* Acoustics; astrophysics; atomic, nuclear, plasma, solid-state, and theoretical physics. • Dr. Daniel L. Decker, Chairman, 801-378-4361. Application contact: Dr. Dorian M. Hatch, Graduate Coordinator, 801-378-2427.

Brock University, Faculty of Mathematics and Science, Department of Physics, St. Catharines, ON L2S 3A1, Canada. Department awards M Sc. Faculty: 8 full-time, 0 part-time. Matriculated students: 7 full-time, 2 part-time; includes 7 foreign. Average age 26. 25 applicants, 0% accepted. In 1990, 3 degrees awarded. *Degree requirements:* Thesis optional, foreign language not required. *Entrance requirements:* Honor's B Sc. Application fee: $0. *Tuition:* $2667 per year full-time, $360 per trimester part-time for Canadian residents; $11,010 per year full-time for nonresidents. *Financial aid:* In 1990–91, $15,000 in aid awarded. 5 research assistantships (3 to first-year students) were awarded. *Faculty research:* Solid-state physics. Total annual research budget: $137,000. • Dr. J. Moore, Chair, 416-688-5550 Ext. 3414. Application contact: Dr. B. Mitrovic, Graduate Studies Officer, 416j-688-5550 Ext. 3415.

SECTION 7: PHYSICS

Directory: Physics

Brooklyn College of the City University of New York, Department of Physics, Bedford Avenue and Avenue H, Brooklyn, NY 11210. Department offers programs in applied physics (MA), physics (MA, PhD). Part-time programs available. Faculty: 25 full-time, 0 part-time. Matriculated students: 28 full-time (6 women), 0 part-time; includes 2 minority (1 Asian American, 1 black American), 26 foreign. 30 applicants, 20% accepted. In 1990, 2 master's awarded (100% continued full-time study). Terminal master's awarded for partial completion of doctoral program. *Degree requirements:* For master's, comprehensive exam required, foreign language not required; for doctorate, dissertation, comprehensive exam required, foreign language not required. *Entrance requirements:* GRE, TOEFL. Application deadline: 4/1. Application fee: $30. *Expenses:* Tuition of $1202 per semester full-time for state residents; $2350 per semester full-time for nonresidents. Fees of $45 per semester full-time. *Financial aid:* In 1990–91, $360,000 in aid awarded. 1 fellowship (to a first-year student), 12 research assistantships (all to first-year students), 13 teaching assistantships (all to first-year students) were awarded; full and partial tuition waivers also available. *Faculty research:* Experimental nuclear and condensed-matter and applied physics; theoretical, nuclear, condensed-matter, and atomic physics. Total annual research budget: $1-million. • Dr. Louis Celenza, Chairperson, 718-780-5418.

Announcement: All course work and research leading to PhD may be pursued on Brooklyn College campus. Research facilities include 3.75-MV Dynamitron particle accelerator and extremely well-equipped solid-state and applied physics laboratories, including UHV surface science facility and photoreflectance, Raman scattering, molecular-beam epitaxy, and electrochemical facilities. Major areas of research include experimental: applied, low-energy nuclear, condensed-matter physics; theoretical: applied, atomic, nuclear, solid-state physics. Teaching and/or research assistantships available at approximately $15,000 per year.

Brown University, Department of Physics, Providence, RI 02912. Department awards Sc M, PhD. Faculty: 27 full-time (1 woman), 0 part-time. Matriculated students: 77 full-time (7 women), 10 part-time (1 woman); includes 13 minority (9 Asian American, 2 black American, 2 Hispanic American), 48 foreign. 133 applicants, 17% accepted. In 1990, 27 master's, 10 doctorates awarded. *Degree requirements:* For master's, foreign language and thesis not required; for doctorate, dissertation, qualifying and oral exams required, foreign language not required. *Application deadline:* 1/2. *Application fee:* $40. *Expenses:* Tuition of $16,256 per year full-time, $2032 per course part-time. Fees of $372 per year. *Financial aid:* In 1990–91, $2.211-million in aid awarded. 6 fellowships (3 to first-year students), 45 research assistantships (0 to first-year students), 26 teaching assistantships (9 to first-year students), 2 assistantships (0 to first-year students) were awarded; full and partial tuition waivers, institutionally sponsored loans also available. Financial aid application deadline: 1/2; applicants required to submit GAPSFAS. • Robert E. Lanou Jr., Chairman, 401-863-2644. Application contact: Chung-I Tan, Graduate Representative, 401-863-2683.

See full description on page 575.

Bryn Mawr College, Graduate School of Arts and Sciences, Department of Physics, Bryn Mawr, PA 19010. Department awards MA, PhD. Faculty: 6. Matriculated students: 2 full-time (1 woman), 5 part-time (2 women); includes 2 foreign. 11 applicants, 45% accepted. In 1990, 1 master's, 3 doctorates awarded. *Degree requirements:* 1 foreign language, thesis/dissertation. *Entrance requirements:* GRE General Test, GRE Subject Test. Application deadline: 8/20. Application fee: $35. *Tuition:* $13,900 per year full-time, $2350 per course part-time. *Financial aid:* In 1990–91, 3 teaching assistantships awarded; fellowships, research assistantships, federal work-study, institutionally sponsored loans also available. Aid available to part-time students. Financial aid application deadline: 2/1; applicants required to submit GAPSFAS. • Dr. Peter Beckmann, Chairman, 215-526-5359. Application contact: Patricia Saukewitsch, Administrative Coordinator, 215-526-5072.

California Institute of Technology, Division of Physics, Mathematics and Astronomy, Department of Physics, Pasadena, CA 91125. Department awards PhD. Faculty: 45 full-time (1 woman), 0 part-time. Matriculated students: 147 full-time (26 women), 0 part-time; includes 1 minority (Asian American), 71 foreign. Average age 25. 389 applicants, 17% accepted. In 1990, 18 doctorates awarded. Terminal master's awarded for partial completion of doctoral program. *Degree requirements:* For doctorate, dissertation, candidacy and final exams required, foreign language not required. *Entrance requirements:* For doctorate, GRE General Test, GRE Subject Test, TOEFL. Application deadline: 1/15. Application fee: $0. *Tuition:* $14,100 per year. *Financial aid:* In 1990–91, 147 students received a total of $3.2-million in aid awarded. 4 fellowships (all to first-year students), 81 research assistantships (15 to first-year students), 57 teaching assistantships (8 to first-year students), 16 outside awards (5 to first-year students) were awarded; federal work-study, institutionally sponsored loans also available. Financial aid application deadline: 1/15. *Faculty research:* High-energy physics, nuclear physics, condensed-matter physics, theoretical physics and astrophysics, gravity physics. Total annual research budget: $18.3-million. • Steven C. Frautschi, Executive Officer, 818-356-6689. Application contact: Donna Driscoll, Graduate Secretary, 818-356-4244.

See full description on page 577.

California State University, Chico, College of Natural Sciences, Department of Geology and Physical Sciences, Program in Physical Science, Chico, CA 95929. Offerings include physics (MS). Program faculty: 5 full-time (1 woman), 5 part-time (3 women). *Degree requirements:* Thesis required, foreign language not required. *Entrance requirements:* GRE General Test. Application deadline: 4/1 (applications processed on a rolling basis). Application fee: $55. *Expenses:* Tuition of $548 per semester full-time, $350 per semester part-time. Fees of $246 per unit for nonresidents. • Dr. K. R. Gina Rothe, Graduate Coordinator, 916-898-6269.

California State University, Fresno, Division of Graduate Studies and Research, School of Natural Sciences, Department of Physics, 5241 North Maple Avenue, Fresno, CA 93710. Department awards MA, MS. Faculty: 10 full-time, 5 part-time. Matriculated students: 19. *Degree requirements:* Thesis or alternative required, foreign language not required. *Entrance requirements:* GRE General Test. *Tuition:* $1098 per year full-time, $4485 per year part-time. • Brandt Kehoe, Chairman, 209-278-2371. Application contact: Michael Zender, 209-278-2371.

Announcement: Programs leading to MS (goal: solid background in physics for further graduate study or industry) and MA (goal: high school teaching). Faculty involved in radiation, magnetics, thin films, chaos theory, lasers, semiconductors, Raman spectroscopy, computer calculations, applied physics. Assistantships approximately $10,000. Near Yosemite and Sequoia national parks.

California State University, Long Beach, School of Natural Sciences, Department of Physics and Astronomy, 1250 Bellflower Boulevard, Long Beach, CA 90840. Department offers programs in metals physics (MS), physics (MA). Matriculated students: 10 full-time (0 women), 45 part-time (4 women); includes 8 minority (4 Asian American, 1 black American, 3 Hispanic American), 3 foreign. Average age 27. 28 applicants, 75% accepted. In 1990, 6 degrees awarded. *Degree requirements:* Thesis or comprehensive exam required, foreign language not required. *Application deadline:* 8/1 (applications processed on a rolling basis). *Application fee:* $55. *Expenses:* Tuition of $0 for state residents; $246 per unit for nonresidents. Fees of $1120 per year full-time, $724 per year part-time. *Financial aid:* Application deadline 3/2; applicants required to submit FAF. • Dr. Bruce L. Scott, Chair, 310-985-4924.

California State University, Los Angeles, School of Natural and Social Sciences, Department of Physics and Astronomy, Los Angeles, CA 90032. Department awards MS. Part-time and evening/weekend programs available. Faculty: 11 full-time, 20 part-time. Matriculated students: 0 full-time, 24 part-time (6 women); includes 6 minority (5 Asian American, 1 Hispanic American), 9 foreign. 5 applicants. In 1990, 8 degrees awarded. *Degree requirements:* Thesis or comprehensive exam required, foreign language not required. *Entrance requirements:* TOEFL (minimum score of 550 required). Application deadline: 8/7 (applications processed on a rolling basis). Application fee: $55. *Expenses:* Tuition of $0 for state residents; $164 per unit for nonresidents. Fees of $1046 per year full-time, $650 per semester (minimum) part-time. *Financial aid:* Federal work-study available. Aid available to part-time students. Financial aid application deadline: 3/1; applicants required to submit FAF. *Faculty research:* Intermediate energy, nuclear physics, condensed-matter physics, biophysics. • Dr. Fleur Yano, Chair, 213-343-2100.

See full description on page 579.

California State University, Northridge, School of Science and Mathematics, Department of Physics and Astronomy, Northridge, CA 91330. Department offers program in physics (MS). Part-time and evening/weekend programs available. Matriculated students: 7 full-time (1 woman), 41 part-time (3 women). Average age 30. 27 applicants, 85% accepted. In 1990, 15 degrees awarded. *Degree requirements:* Thesis optional, foreign language not required. *Entrance requirements:* Minimum GPA of 3.0 or GRE General Test. Application deadline: 11/30. Application fee: $55. *Expenses:* Tuition of $0 for state residents; $205 per unit for nonresidents. Fees of $1128 per year full-time, $366 per semester part-time. *Financial aid:* Teaching assistantships available. • Dr. Giovan Natale, Chairman, 818-885-2775. Application contact: Dr. Robert Romagnoli, Graduate Coordinator, 818-885-2778.

Carleton University, Faculty of Science, Ottawa-Carleton Institute for Physics, Ottawa, ON K1S 5B6, Canada. Institute awards M Sc, PhD. M Sc and PhD offered jointly with the University of Ottawa. Faculty: 39. Matriculated students: 42 full-time, 3 part-time; includes 7 foreign. In 1990, 9 master's, 1 doctorate awarded. *Degree requirements:* For master's, thesis or alternative; for doctorate, dissertation, comprehensive exam. *Entrance requirements:* For master's, TOEFL (minimum score of 550 required), honor's degree in science; for doctorate, TOEFL (minimum score of 550 required), M Sc. Application deadline: 7/1 (priority date, applications processed on a rolling basis). Application fee: $15. *Tuition:* $985 per semester full-time, $284 per semester part-time for Canadian residents; $3939 per semester full-time, $1171 per semester part-time for nonresidents. *Financial aid:* Application deadline 3/1. *Faculty research:* Experimental and theoretical elementary particle physics, medical physics. • Brian Hird, Director, 613-788-4376.

Carnegie Mellon University, Mellon College of Science, Department of Physics, Pittsburgh, PA 15213. Department offers programs in applied physics (PhD), physics (MS, PhD). Faculty: 34 full-time (1 woman), 0 part-time. Matriculated students: 60 full-time (6 women), 3 part-time (0 women); includes 30 foreign. Average age 27. 480 applicants, 8% accepted. In 1990, 8 master's awarded (37% found work related to degree, 63% continued full-time study); 8 doctorates awarded (87% entered university research/teaching, 13% found other work related to degree). Terminal master's awarded for partial completion of doctoral program. *Degree requirements:* For master's, foreign language not required; for doctorate, dissertation required, foreign language not required. *Entrance requirements:* GRE General Test. Application deadline: 4/15. Application fee: $0. *Expenses:* Tuition of $15,250 per year full-time, $212 per unit part-time. Fees of $80 per year. *Financial aid:* In 1990–91, $826,000 in aid awarded. 0 fellowships, 41 research assistantships (0 to first-year students), 22 teaching assistantships (12 to first-year students) were awarded. *Faculty research:* Nuclear, particle, and condensed-matter physics; biophysics. Total annual research budget: $3.2-million. • Dr. Robert W. Kraemer, Head, 412-268-2740. Application contact: Ned S. Vanderven, 412-268-2766.

Case Western Reserve University, Department of Physics, Cleveland, OH 44106. Department awards MS, PhD. Faculty: 24 full-time, 0 part-time. Matriculated students: 18 full-time (5 women), 26 part-time (1 woman); includes 1 minority (Hispanic American), 19 foreign. 134 applicants, 30% accepted. In 1990, 5 master's, 4 doctorates awarded. *Degree requirements:* For doctorate, dissertation, departmental qualifying exam required, foreign language not required. *Entrance requirements:* TOEFL (minimum score of 550 required). Application fee: $25. *Expenses:* Tuition of $13,600 per year full-time, $567 per credit part-time. Fees of $320 per year. *Financial aid:* Research assistantships, teaching assistantships available. Financial aid application deadline: 3/1. *Faculty research:* Experimental and theoretical condensed-matter physics, intermediate and high-energy physics, surface and thin-film physics, acoustics. • William L. Gordon, Chairman, 216-368-4000. Application contact: T. G. Eck, Chairman, Graduate Admissions Committee, 216-368-4022.

Catholic University of America, School of Arts and Sciences, Department of Physics, Washington, DC 20064. Department awards MS, PhD. Part-time programs available. Faculty: 16 full-time (1 woman), 4 part-time (1 woman). Matriculated students: 27 full-time (5 women), 41 part-time (5 women); includes 5 minority (2 Asian American, 2 black American, 1 Hispanic American), 27 foreign. 200 applicants, 23% accepted. In 1990, 10 master's, 5 doctorates awarded. Terminal master's awarded for partial completion of doctoral program. *Degree requirements:* For master's, thesis or alternative, comprehensive exam required, foreign language not required; for doctorate, 1 foreign language, dissertation, comprehensive exam. *Entrance requirements:* For master's, GRE General Test or Doppelt Math Reasoning Test, TOEFL; for doctorate, GRE General Test or Doppelt Math Reasoning Test. Application deadline: 8/1 (applications processed on a rolling basis). Application fee: $30. *Expenses:* Tuition of $11,626 per year full-time, $445 per credit hour part-time. Fees of $240 per year full-time, $90 per year part-time. *Financial aid:* In 1990–91, $70,000 in aid awarded. 10 fellowships (4 to first-year students), 16 research assistantships (3 to first-year students), 6 teaching assistantships (1 to a first-year student) were awarded; full and partial tuition waivers, federal work-study, and career-related internships or fieldwork also available. Financial aid application deadline: 2/15; applicants required to submit GAPSFAS. *Faculty research:* Condensed-matter physics, intermediate energy physics, acoustics, astrophysics, biophysics. Total annual research budget: $3.0-million. • Dr. James G. Brennan, Chair, 202-319-5315.

Central Connecticut State University, School of Arts and Sciences, Department of Physics and Earth Science, New Britain, CT 06050. Department offers programs in earth science (MS), general science (MS), physics (MS). Faculty: 3 full-time (1 woman), 1 (woman) part-time. Matriculated students: 5 full-time (2 women), 18 part-time (10 women); includes 0 minority, 0 foreign. 10 applicants, 50% accepted. In 1990, 7 degrees awarded. *Degree requirements:* Thesis or alternative, comprehensive exam. *Entrance requirements:* TOEFL (minimum score of 550 required for international students), minimum GPA of 2.7. Application deadline: 8/31. Application fee: $20. *Expenses:* Tuition of $1720 per year full-time, $115 per credit part-time for state residents; $4790 per year full-time, $115 per credit part-time for nonresidents. Fees of $1013 per year full-time, $33 per semester part-time for state residents; $1655 per year full-time, $33 per semester part-time for nonresidents. *Financial aid:* Application deadline 3/15; applicants required to submit FAF. • Dr. Ali Antar, Chair, 203-827-7228.

Central Michigan University, College of Arts and Sciences, Department of Physics, Mount Pleasant, MI 48859. Department awards MS. Faculty: 13 full-time (0 women). In 1990, 2 degrees awarded. *Degree requirements:* Thesis required, foreign language not required. *Application deadline:* 7/15 (priority date, applications processed on a rolling basis). *Application fee:* $30. *Tuition:* $96.50 per credit for state residents; $210.50 per credit for nonresidents. *Financial aid:* In 1990–91, 7 teaching assistantships awarded; federal work-study and career-related internships or fieldwork also available. Financial aid applicants required to submit FAF. *Faculty research:* Polymer physics, laser spectroscopy, stellar acretion disks, nuclear physics, glassy thin-films. • Dr. Stanley Hirschi, Chairperson, 517-774-3321.

City College of the City University of New York, Graduate School, College of Liberal Arts and Science, Division of Science, Department of Physics, Convent Avenue at 138th Street, New York, NY 10031. Department awards MA, PhD. PhD offered through the Graduate School and University Center of the City University of New York. Matriculated students: 6 full-time (0 women), 3 part-time (0 women). 18 applicants, 22% accepted. In 1990, 3 master's awarded. Terminal master's awarded for partial completion of doctoral program. *Degree requirements:* For master's, comprehensive exam required, foreign language not required; for doctorate, dissertation required, foreign language not required. *Entrance requirements:* For master's, TOEFL (minimum score of 300 required); for doctorate, GRE. Application deadline: 6/1 (priority date, applications processed on a rolling basis). Application fee: $30. *Expenses:* Tuition of $2204 per year full-time, $95 per credit part-time for state residents; $4700 per year full-time, $199 per credit part-time for nonresidents. Fees of $15 per semester. *Financial aid:* Fellowships available. • Philip Baumel, Chairperson, 212-650-6832. Application contact: Victor Chung, Adviser, 212-650-6832.

Announcement: Particularly strong programs leading to MA and PhD in elementary particle physics, solid-state and condensed-matter physics, laser physics and quantum optics, atomic and molecular physics, astrophysics, fluid dynamics, and biophysics. Extensive laboratory and computational facilities. High faculty-student ratio. Financial support available at a level of $9500 plus tuition.

Clark Atlanta University, School of Arts and Sciences, Department of Physics, Atlanta, GA 30314. Department awards MS. Part-time programs available. Matriculated students: 13 (4 women). *Degree requirements:* 1 foreign language (computer language can substitute), thesis. *Entrance requirements:* GRE General Test, minimum GPA of 2.5. Application deadline: 4/1 (applications processed on a rolling basis). Application fee: $40. *Expenses:* Tuition of $4860 per year. Fees of $300 per year. *Faculty research:* Fusion energy, investigations of nonlinear differential equations, difference schemes, collisions in dense plasma. • Dr. Charles Brown, Chairman, 404-880-8798. Application contact: Peggy Wade, Marketing and Recruitment, 404-880-8427.

Clarkson University, School of Science, Department of Physics, Potsdam, NY 13699. Department awards MS, PhD. Faculty: 11 full-time (0 women), 0 part-time. Matriculated students: 16 full-time (4 women), 0 part-time; includes 0 minority, 8 foreign. Average age 33. 111 applicants, 18% accepted. In 1990, 5 master's awarded; 5 doctorates awarded (100% entered university research/teaching). *Degree requirements:* For master's, foreign language and thesis not required; for doctorate, dissertation, departmental qualifying exam required, foreign language not required. *Application fee:* $10. *Expenses:* Tuition of $446 per credit hour. Fees of $75 per semester. *Financial aid:* In 1990–91, 6 research assistantships, 8 teaching assistantships awarded. *Faculty research:* Solid-state and particle physics, non-linear and statistical physics, radiation physics. Total annual research budget: $348,769. • Dr. Lawrence S. Schulman, Chairman, 315-268-2396.

Announcement: Graduate students at Clarkson are engaged in a range of experimental and theoretical research, much of it sponsored by government and industry. Areas of current interest include statistical physics; critical phenomena; nonlinear physics and solitons; path integration; quantum foundations; astrophysics; and solid-state, fluid, surface, and medical physics.

Clark University, Department of Physics, Worcester, MA 01610. Department awards MA, PhD. Part-time programs available. Faculty: 6 full-time (0 women). Matriculated students: 13 full-time (4 women); includes 6 foreign. Terminal master's awarded for partial completion of doctoral program. *Degree requirements:* For master's, computer language, thesis or alternative required, foreign language not required; for doctorate, 1 foreign language, computer language, dissertation. *Application deadline:* 3/1 (priority date, applications processed on a rolling basis). *Application fee:* $40. *Tuition:* $15,000 per year full-time, $1875 per course part-time. *Financial aid:* Research assistantships, teaching assistantships, full and partial tuition waivers, federal work-study available. Financial aid application deadline: 4/1; applicants required to submit GAPSFAS. *Faculty research:* Condensed-matter physics, statistical and thermal physics, magnetic properties of materials, computer simulation. Total annual research budget: $418,000. • Dr. Christopher Landee, Chair, 508-793-7169.

Announcement: High-quality, small-scale, flexible PhD program emphasizing condensed-matter physics and early student research participation. Apply by March 1 for optimum consideration; each application evaluated individually. Application fee, $40 (waived if financial need). Contact Harvey Gould, 508-793-7485, hgould@clarku.BITNET.

Clemson University, College of Sciences, Department of Physics and Astronomy, Clemson, SC 29634. Department offers program in physics (MS, PhD). Faculty: 23 full-time (0 women), 0 part-time. Matriculated students: 45 full-time (4 women), 0 part-time; includes 2 minority (1 black American, 1 Hispanic American), 10 foreign. Average age 27. 92 applicants, 26% accepted. In 1990, 4 master's awarded (25% entered university research/teaching, 50% found other work related to degree, 25% continued full-time study); 3 doctorates awarded (67% entered university research/teaching, 33% found other work related to degree). Terminal master's awarded for partial completion of doctoral program. *Degree requirements:* For master's, thesis required (for some programs), foreign language not required; for doctorate, dissertation required, foreign language not required. *Entrance requirements:* GRE General Test, TOEFL. Application deadline: 6/1 (priority date, applications processed on a rolling basis). Application fee: $25. *Expenses:* Tuition of $102 per credit hour. Fees of $80 per semester full-time. *Financial aid:* In 1990–91, 40 students received a total of $421,400 in aid awarded. 2 fellowships (both to first-year students), 15 research assistantships (1 to a first-year student), 23 teaching assistantships (13 to first-year students) were awarded. Average monthly stipend for a graduate assistantship: $877. Financial aid application deadline: 6/1. *Faculty research:* Astrophysics, atmosphere physics, condensed matter, radiation physics, solid state physics. Total annual research budget: $457,126. • Dr. P. J. McNulty, Head, 803-656-3416. Application contact: J. A. Gilreath, Graduate Student Adviser, 803-656-3416.

See full description on page 581.

College of William and Mary, Faculty of Arts and Sciences, Department of Physics, Williamsburg, VA 23185. Department awards MA, MS, PhD. Faculty: 24 full-time, 0 part-time. Matriculated students: 47 full-time (6 women), 0 part-time; includes 1 minority (Hispanic American), 13 foreign. Average age 27. In 1990, 14 master's, 6 doctorates awarded. Terminal master's awarded for partial completion of doctoral program. *Degree requirements:* For master's, comprehensive exam, qualifying exam required, foreign language and thesis not required; for doctorate, dissertation, comprehensive exam, qualifying exam, oral exam required, foreign language not required. *Entrance requirements:* GRE General Test, GRE Subject Test, minimum GPA of 2.5. Application fee: $5. *Expenses:* Tuition of $2240 per year full-time, $120 per credit hour part-time for state residents; $8960 per year full-time, $320 per credit hour part-time for nonresidents. Fees of $1490 per year full-time. *Financial aid:* Fellowships, research assistantships, teaching assistantships, and career-related internships or fieldwork available. *Faculty research:* Nuclear, particle, solid-state, atomic, and plasma physics. Total annual research budget: $2.02-million. • Dr. Robert E. Welsh, Chairman, 804-221-3505. Application contact: Dr. Morton Eckhause, Graduate Director, 804-253-4471.

Colorado School of Mines, Department of Physics, Golden, CO 80401. Department offers programs in applied physics (PhD), physics (MS). Faculty: 13 full-time (0 women), 4 part-time (0 women). Matriculated students: 26 full-time (2 women), 9 part-time (1 woman); includes 22 foreign. 80 applicants, 66% accepted. In 1990, 4 master's, 0 doctorates awarded. *Degree requirements:* For master's, thesis required, foreign language not required; for doctorate, dissertation, written and oral comprehensive exams required, foreign language not required. *Entrance requirements:* GRE General Test, GRE Subject Test, TOEFL, minimum GPA of 3.0, BS in physics. Application deadline: 3/1. Application fee: $25 ($25 for foreign students). *Expenses:* Tuition of $3178 per year full-time, $124 per semester hour part-time for state residents; $10,304 per year full-time, $344 per semester hour part-time for nonresidents. Fees of $374 per year full-time. *Financial aid:* In 1990–91, 2 fellowships, 16 research assistantships, 7 teaching assistantships awarded. *Faculty research:* Solid-state and surface physics; nuclear physics; solar energy; particulates; optics. Total annual research budget: $1.538-million. • Dr. John U. Trefny, Head, 303-273-3830. Application contact: Dr. James McNeil, 303-273-3844.

Announcement: The physics department of the Colorado School of Mines, with an annual research budget of more than $1-million, supports graduate studies in condensed matter (amorphous materials, high-Tc superconductors, surfaces and interfaces), solar energy, lasers and quantum optics, low-temperature physics, experimental and theoretical nuclear physics, and geophysics. The physics department participates in interdisciplinary research through the School's Material Science Program. The department collaborates extensively with the nearby US Solar Energy Research Institute.

Colorado State University, College of Natural Sciences, Department of Physics, Fort Collins, CO 80523. Department awards MS, PhD. Faculty: 15 full-time (1 woman), 0 part-time. Matriculated students: 31 full-time (7 women), 3 part-time (0 women); includes 2 minority (1 black American, 1 Hispanic American), 13 foreign. Average age 26. 241 applicants, 8% accepted. In 1990, 10 master's, 1 doctorate awarded. Terminal master's awarded for partial completion of doctoral program. *Degree requirements:* For master's, thesis required (for some programs), foreign language not required; for doctorate, dissertation required, foreign language not required. *Entrance requirements:* GRE General Test, TOEFL (minimum score of 550 required), minimum GPA of 3.0. Application deadline: 4/1 (priority date, applications processed on a rolling basis). Application fee: $30. *Tuition:* $1322 per semester full-time for state residents; $3673 per semester full-time for nonresidents. *Financial aid:* In 1990–91, $250,000 in aid awarded. 1 fellowship (to a first-year student), 5 research assistantships (0 to first-year students), 25 teaching assistantships (9 to first-year students), 0 traineeships were awarded; federal work-study also available. Average monthly stipend for a graduate assistantship: $698-750. *Faculty research:* Experimental condensed matter physics, laser spectroscopy, optics, theoretical condensed matter physics, statistical mechanics. Total annual research budget: $2-million. • James R. Sites, Interim Chair, 303-491-6206. Application contact: Richard E. Eykholt Jr., Chairman, Graduate Admissions Committee, 303-491-7366.

Columbia University, Graduate School of Arts and Sciences, Division of Natural Sciences, Department of Physics, New York, NY 10027. Department awards MA, M Phil, PhD. Faculty: 23 full-time, 5 part-time. Matriculated students: 101 full-time (10 women), 0 part-time; includes 10 minority (9 Asian American, 1 black American), 40 foreign. Average age 26. 111 applicants, 42% accepted. In 1990, 21 master's, 13 doctorates awarded. *Degree requirements:* For master's, foreign language and thesis not required; for doctorate, dissertation required, foreign language not required. *Entrance requirements:* GRE General Test, GRE Subject Test, TOEFL, 3 years of course work in physics. Application deadline: 1/5. Application fee: $50. *Expenses:* Tuition of $7836 per semester for state residents; $426 per credit for nonresidents. Fees of $148 per semester for state residents. *Financial aid:* In 1990–91, $1.42-million in aid awarded. 3 fellowships (0 to first-year students), 65 teaching assistantships (2 to first-year students) were awarded; federal work-study, institutionally sponsored loans also available. Aid available to part-time students. Financial aid application deadline: 1/5; applicants required to submit GAPSFAS. *Faculty research:* Theoretical physics; astrophysics; low-, medium-, and high-energy physics. • Frank Sciulli, Chair, 212-854-3347.

Concordia University, Faculty of Arts and Science, Department of Physics, Montreal, PQ H3G 1M8, Canada. Department awards M Sc, PhD. Matriculated students: 11 full-time (1 woman), 2 part-time (0 women); includes 5 foreign. In 1990, 2 master's, 2 doctorates awarded. *Degree requirements:* For master's, thesis or alternative required, foreign language not required; for doctorate, dissertation, comprehensive exam required, foreign language not required. *Application fee:* $15. *Expenses:* Tuition of $10 per credit for Canadian residents; $195 per credit for nonresidents. Fees of $223 per year full-time, $38.85 per year part-time for Canadian residents; $118 per year full-time, $35.35 per year part-time for nonresidents. *Faculty research:* Fundamental physics, applied physics. Total annual research budget: $30,000. • Dr. B. Frank, Chair, 514-848-3271.

Directory: Physics

Cornell University, Graduate Fields of Arts and Sciences, Field of Physics, Ithaca, NY 14853. Field offers programs in experimental physics (MS, PhD), physics (MS, PhD), theoretical physics (MS, PhD). Faculty: 58 full-time, 0 part-time. Matriculated students: 194 full-time (34 women), 0 part-time; includes 16 minority (12 Asian American, 1 black American, 2 Hispanic American, 1 Native American), 55 foreign. 465 applicants, 20% accepted. In 1990, 23 master's, 35 doctorates awarded. Terminal master's awarded for partial completion of doctoral program. *Degree requirements:* Thesis/dissertation required, foreign language not required. *Entrance requirements:* For doctorate, TOEFL. Application deadline: 1/10. Application fee: $55. *Expenses:* Tuition of $16,170 per year. Fees of $28 per year. *Financial aid:* In 1990–91, 160 students received aid. 6 fellowships (1 to a first-year student), 100 research assistantships (2 to first-year students), 54 teaching assistantships (25 to first-year students) awarded; full and partial tuition waivers, federal work-study, institutionally sponsored loans, and career-related internships or fieldwork also available. Financial aid application deadline: 1/10; applicants required to submit GAPSFAS. *Faculty research:* Experimental condensed matter physics, theoretical condensed matter physics, experimental high energy particle physics, particle and field theory physics, theoretical astrophysics. • Toichiro Kinoshita, Graduate Faculty Representative, 607-255-6016. Application contact: Robert Brashear, Director of Admissions, 607-255-4884.

Creighton University, College of Arts and Sciences, Program in Physics, Omaha, NE 68178. Program awards MS. Faculty: 3 full-time. Matriculated students: 6 full-time (0 women), 4 part-time (0 women); includes 1 minority (black American), 2 foreign. 9 applicants, 56% accepted. *Degree requirements:* 1 foreign language, thesis or alternative. *Entrance requirements:* GRE General Test, GRE Subject Test, TOEFL. Application deadline: 3/15. Application fee: $20. *Expenses:* Tuition of $272 per credit hour. Fees of $140 per semester full-time, $14 per semester part-time. • Dr. Sam Cipolla, Director, 402-280-2835. Application contact: Dr. Michael Lawler, Dean, Graduate School, 402-280-2870.

Dalhousie University, College of Arts and Science, Faculty of Science, Department of Physics, Halifax, NS B3H 4H6, Canada. Department awards M Sc, PhD. Faculty: 17 full-time, 2 part-time. Matriculated students: 25 full-time (3 women), 1 part-time (0 women). 183 applicants, 10% accepted. In 1990, 4 master's, 1 doctorate awarded. *Degree requirements:* Thesis/dissertation required, foreign language not required. *Entrance requirements:* TOEFL (minimum score of 550 required). Application deadline: 7/15 (applications processed on a rolling basis). Application fee: $20. *Tuition:* $2594 per year full-time for Canadian residents; $4294 per year full-time for nonresidents. *Financial aid:* In 1990–91, $181,100 in aid awarded. Fellowships available. *Faculty research:* Applied, experimental, and solid-state physics. • Dr. A. M. Simpson, Chairman, 902-494-2339. Application contact: Dr. R. A. Dunlap, Graduate Coordinator.

Dartmouth College, School of Arts and Sciences, Department of Physics and Astronomy, Hanover, NH 03755. Department awards AM, PhD. Faculty: 14 full-time (1 woman), 0 part-time. Matriculated students: 30 full-time (5 women), 0 part-time; includes 1 minority (Asian American), 9 foreign. 191 applicants, 7% accepted. In 1990, 2 master's awarded (50% found work related to degree, 50% continued full-time study); 3 doctorates awarded. Terminal master's awarded for partial completion of doctoral program. *Degree requirements:* For master's, thesis required, foreign language not required; for doctorate, variable foreign language requirement, dissertation. *Entrance requirements:* GRE General Test, GRE Subject Test. Application deadline: 3/1. *Expenses:* Tuition of $16,230 per year. Fees of $650 per year. *Financial aid:* In 1990–91, $828,554 in aid awarded. 16 fellowships (6 to first-year students), 7 research assistantships were awarded; full tuition waivers, institutionally sponsored loans also available. Financial aid applicants required to submit GAPSFAS or FAF. *Total annual research budget:* $773,867. • Joseph Harris, Chairman, 603-646-2359.

Delaware State College, Department of Physics, Dover, DE 19901. Department offers programs in physics (MS), physics teaching (MS). Part-time and evening/weekend programs available. Faculty: 0 full-time, 3 part-time (1 woman). Matriculated students: 0 full-time, 7 part-time (2 women); includes 1 foreign. 4 applicants, 50% accepted. In 1990, 3 degrees awarded (33% entered university research/teaching, 33% found other work related to degree, 33% continued full-time study). *Degree requirements:* Foreign language and thesis not required. *Entrance requirements:* Minimum GPA of 2.75 overall, 3.0 in major. Application deadline: 6/30. Application fee: $10. *Tuition:* $95 per credit for state residents; $166 per credit for nonresidents. *Financial aid:* Full and partial tuition waivers, federal work-study, institutionally sponsored loans available. Aid available to part-time students. Financial aid applicants required to submit FAF. *Faculty research:* Thermal properties of solids, nuclear physics, radiation damage in solids. • Dr. Ehsan Helmy, Chairperson, 302-739-5158.

DePaul University, College of Liberal Arts and Sciences, Department of Physics, Programs in Physics, Chicago, IL 60604. Offerings in applied physics (MS), physics (MS). Part-time and evening/weekend programs available. Matriculated students: 10 full-time (1 woman), 11 part-time (3 women); includes 6 minority (4 Asian American, 1 black American, 1 Hispanic American), 2 foreign. In 1990, 0 degrees awarded. *Degree requirements:* Written and oral exams required, thesis optional, foreign language not required. *Application deadline:* rolling. *Application fee:* $20. *Expenses:* Tuition of $215 per quarter hour. Fees of $10 per quarter hour (minimum). *Financial aid:* Research assistantships available. • Dr. Donald Vanostenburg, Chairman, Department of Physics, 312-362-8659.

Drake University, College of Arts and Sciences, Department of Physics and Astronomy, Des Moines, IA 50311. Department offers programs in physical science (MA), physics (MS). Faculty: Matriculated students: 4 full-time (2 women), 0 part-time; includes 0 minority, 0 foreign. 6 applicants, 17% accepted. In 1990, 0 degrees awarded. *Degree requirements:* Thesis or alternative required, foreign language not required. *Entrance requirements:* GRE General Test or MAT. Application fee: $25. *Tuition:* $11,280 per year full-time, $365 per hour part-time. *Financial aid:* In 1990–91, $22,500 in aid awarded. 3 teaching assistantships (1 to a first-year student) were awarded. *Total annual research budget:* $20,000. • Dr. Larry Staunton, Chairperson, 515-271-3141. Application contact: Dr. Klaus Bartschat, Graduate Coordinator, 515-271-3750.

Drexel University, College of Arts and Sciences, Department of Physics and Atmospheric Science, Program in Physics, 32nd and Chestnut Streets, Philadelphia, PA 19104. Program awards MS, PhD. Faculty: 31 full-time (1 woman), 0 part-time. Terminal master's awarded for partial completion of doctoral program. *Degree requirements:* For master's, foreign language and thesis not required; for doctorate, dissertation required, foreign language not required. *Entrance requirements:* For master's, GRE, TOEFL (minimum score of 550 required); for doctorate, GRE Subject Test, TOEFL (minimum score of 550 required). Application deadline: 8/23 (applications processed on a rolling basis). Application fee: $25. *Expenses:* Tuition of $345 per credit hour. Fees of $81 per quarter full-time, $43 per quarter part-time. *Financial aid:* Research assistantships, teaching assistantships available. Financial aid application deadline: 2/1. *Faculty research:* Nuclear structure, biophysics, numerical astrophysics, quantum optics, surface physics. • Application contact: Keith Brooks, Director of Graduate Admissions, 215-895-6700.

See full description on page 409.

Duke University, Graduate School, Department of Physics, Durham, NC 27706. Department awards PhD. Faculty: 24 full-time, 9 part-time. Matriculated students: 71 full-time (5 women), 1 part-time (0 women); includes 1 minority (black American), 20 foreign. 76 applicants, 58% accepted. In 1990, 9 master's, 8 doctorates awarded. Terminal master's awarded for partial completion of doctoral program. *Degree requirements:* For doctorate, dissertation. *Entrance requirements:* For doctorate, GRE General Test, GRE Subject Test. Application deadline: 1/31. Application fee: $50. *Expenses:* Tuition of $8640 per year full-time, $360 per unit part-time. Fees of $1356 per year full-time. *Financial aid:* In 1990–91, $249,390 in aid awarded. Federal work-study available. Financial aid application deadline: 1/31; applicants required to submit GAPSFAS. • Eric Gerbst, Director of Graduate Studies, 919-684-8210.

East Carolina University, College of Arts and Sciences, Department of Physics, Greenville, NC 27858-4353. Department offers programs in 2-year college teaching (MP), applied physics (MP), medical physics (MP), physics (MS). Part-time programs available. Faculty: 10. Matriculated students: 11 full-time (4 women), 2 part-time (0 women); includes 1 minority (Asian American), 4 foreign. 18 applicants, 39% accepted. In 1990, 4 degrees awarded. *Degree requirements:* 1 foreign language (computer language can substitute), comprehensive exam, thesis (MS). *Entrance requirements:* GRE General Test, TOEFL. Application deadline: 6/1. Application fee: $25. *Tuition:* $627 per semester for state residents; $3154 per semester for nonresidents. *Financial aid:* In 1990–91, $59,000 in aid awarded. Research assistantships, teaching assistantships, federal work-study available. Aid available to part-time students. Financial aid application deadline: 6/1. • Dr. Carl Adler, Chairman, 919-757-6739. Application contact: Dr. James Joyce, Director of Graduate Studies, 919-757-6483.

Announcement: MS in physics and MP programs in applied physics, medical physics, and 2-year college teaching. Research covers theoretical and experimental atomic physics (2-MV tandem accelerator), astronomy, solar energy, medical physics (facilities and adjunct physics faculty at ECU Medical School), and instrument/computer systems. Virtually all students receive teaching/research assistantships (up to $10,400), summer research support (up to $2660), and out-of-state tuition waivers.

Eastern Kentucky University, College of Natural and Mathematical Sciences, Department of Physics and Astronomy, Richmond, KY 40475. Department offers program in physics (MS). Part-time programs available. Matriculated students: 3 full-time (0 women), 2 part-time (0 women); includes 3 minority (2 Asian American, 1 black American), 2 foreign. Average age 25. In 1990, 2 degrees awarded. *Degree requirements:* Thesis optional. *Entrance requirements:* GRE, minimum GPA of 2.5. Application fee: $0. *Tuition:* $1440 per year full-time, $88 per credit hour part-time for state residents; $4320 per year full-time, $248 per credit hour part-time for nonresidents. *Financial aid:* Research assistantships, teaching assistantships, federal work-study available. Aid available to part-time students. • Dr. Jerry Faughn, Chair, 606-622-1521.

Eastern Michigan University, College of Arts and Sciences, Department of Physics and Astronomy, Ypsilanti, MI 48197. Department offers programs in general science (MS), physics (MS), physics education (MS). Evening/weekend programs available. Faculty: 11 full-time (3 women). Matriculated students: 1 (woman) full-time, 47 part-time (19 women); includes 10 minority (7 Asian American, 2 black American, 1 Hispanic American), 16 foreign. In 1990, 15 degrees awarded. *Degree requirements:* Foreign language not required. *Application deadline:* 6/15 (applications processed on a rolling basis). *Application fee:* $25. *Expenses:* Tuition of $89.50 per credit hour for state residents; $212 per credit hour for nonresidents. Fees of $90.25 per semester. • Dr. Daniel Trochet, Head, 313-487-4144.

East Texas State University, College of Arts and Sciences, Department of Physics, Commerce, TX 75429. Department awards MS. Faculty: 6 full-time (0 women), 0 part-time. Matriculated students: 11 full-time (2 women), 8 part-time (0 women); includes 4 minority (1 black American, 3 Hispanic American), 3 foreign. In 1990, 3 degrees awarded. *Degree requirements:* Thesis (for some programs), comprehensive exam. *Entrance requirements:* GRE General Test. Application deadline: rolling. Application fee: $0 ($25 for foreign students). *Tuition:* $430 per semester full-time for state residents; $1726 per semester full-time for nonresidents. *Financial aid:* Research assistantships, teaching assistantships, federal work-study, institutionally sponsored loans available. Financial aid applicants required to submit GAPSFAS or FAF. • Dr. Ben Doughty, Head, 903-886-5478.

Announcement: MS in physics, minor program in microcomputers/electronics. Research opportunities include surface physics, neutron physics, speech and image processing, satellite mapping, and computational physics. Major equipment includes shielded neutron facility, AEAPS and XPS spectrometers, high-resolution graphics equipment, and excellent computational facilities. Full and partial teaching assistantships available.

Emory University, Graduate School of Arts and Sciences, Department of Physics, Atlanta, GA 30322. Department offers programs in environmental sciences (MS); physics (MA, MS, PhD), including biophysics, radiological physics, solid-state physics. Faculty: 17. Matriculated students: 3 full-time (0 women), 2 part-time (0 women); includes 0 minority, 4 foreign. Average age 33. 0 applicants. In 1990, 0 master's, 1 doctorate awarded. Terminal master's awarded for partial completion of doctoral program. *Degree requirements:* For master's, thesis required, foreign language not required; for doctorate, dissertation, comprehensive exams required, foreign language not required. *Entrance requirements:* GRE General Test, TOEFL, minimum GPA of 3.0. Application deadline: 1/20 (priority date). Application fee: $35. *Expenses:* Tuition of $7370 per semester full-time, $642 per semester hour part-time. Fees of $160 per year full-time. *Financial aid:* In 1990–91, $109,180 in aid awarded. 4 fellowships (0 to first-year students), 0 teaching assistantships, 4 tuition scholarships (0 to first-year students) were awarded; partial tuition waivers also available. Average monthly stipend for a graduate assistantship: $1000. Financial aid application deadline: 2/1. *Faculty research:* Theory of semiconductors and superlattices; experimental laser optics and submillimeter spectroscopy theory; neural networks and stereoscopic vision, experimental studies of the structu re and function of metalloproteins. Total annual research budget: $953,952. • Dr. Krishan K. Bajaj, Chair, 404-727-6584. Application contact: Dr. Edmund Day, Director of Graduate Studies, 404-727-6584.

See full description on page 583.

Emporia State University, School of Graduate Studies, College of Liberal Arts and Sciences, Division of Physical Sciences, Emporia, KS 66801-5087. Division offers programs in chemistry (MS), earth science (MS), physics (MS). Faculty: 14 full-time (0 women), 1 part-time (0 women). Matriculated students: 7 full-time (1 woman), 8 part-time (2 women); includes 0 minority. 9 applicants, 100% accepted. In 1990, 13 degrees awarded. *Degree requirements:* Thesis or comprehensive exam required, foreign language not required. *Entrance requirements:* GRE General Test, TOEFL (minimum score of 550 required). Application deadline: 8/16 (priority date, applications processed on a rolling basis). Application fee: $0 ($50 for foreign students). *Tuition:* $858 per semester full-time, $62 per credit hour part-time for state residents; $2072 per semester full-time, $143 per credit hour part-time for nonresidents. *Financial aid:* In 1990–91, 1 research assistantship (0 to first-year students), 7 teaching assistantships awarded; federal work-study, institutionally sponsored loans also available. Financial aid application deadline: 3/15; applicants required to submit FAF. *Faculty research:* Bredigite, larnite, and dicalcium silicates-Marble Canyon. • DeWayne Backhus, Chair, 316-343-5472.

Fisk University, Department of Physics, Nashville, TN 37208. Department awards MA. Faculty: 3 full-time (0 women), 1 part-time (0 women). In 1990, 2 degrees awarded (100% continued full-time study). *Degree requirements:* Thesis, comprehensive exam required, foreign language not required. *Entrance requirements:* GRE, minimum GPA of 3.0. *Tuition:* $4950 per year full-time, $206 per credit hour part-time. • Dr. John Springer, Director, 615-329-8605.

Florida Atlantic University, College of Science, Department of Physics, Boca Raton, FL 33431. Department awards MS, MST, PhD. Part-time programs available. Faculty: 13 full-time (0 women), 2 part-time (0 women). Matriculated students: 8 full-time (3 women), 3 part-time (0 women). Average age 22. In 1990, 3 master's awarded (100% continued full-time study). *Degree requirements:* For master's, 1 foreign language, thesis. *Entrance requirements:* For master's, GRE General Test (minimum combined score of 1100 required), minimum GPA of 3.0. Application fee: $15. *Tuition:* $89.28 per credit hour for state residents; $291 per credit hour for nonresidents. *Financial aid:* In 1990–91, 5 research assistantships (all to first-year students), 8 teaching assistantships (4 to first-year students) awarded; fellowships, partial tuition waivers, federal work-study also available. *Faculty research:* Upper atmosphere, astrophysics, spectroscopy, mathematical physics, theory of metals. • Dr. Bjorn Lamborn, Chairman, 407-393-3380.

Florida Institute of Technology, College of Science and Liberal Arts, Department of Physics and Space Sciences, Melbourne, FL 32901. Department offers programs in physics (MS, PhD), space science (MS, PhD). Part-time and evening/weekend programs available. Faculty: 11 full-time (0 women), 6 part-time (0 women). Matriculated students: 16 full-time (3 women), 8 part-time (4 women); includes 13 foreign. 62 applicants, 47% accepted. In 1990, 17 master's, 0 doctorates awarded. *Degree requirements:* For master's, comprehensive exam required, thesis optional, foreign language not required; for doctorate, dissertation, comprehensive exam required, foreign language not required. *Entrance requirements:* For master's, minimum GPA of 3.0, proficiency in a computer language; for doctorate, minimum GPA of 3.2. Application deadline: rolling. Application fee: $35. *Tuition:* $234 per credit hour. *Financial aid:* Research assistantships, teaching assistantships, federal work-study available. Financial aid application deadline: 3/1. *Faculty research:* Electron emission from solids, fiber optics, terrestrial geomagnetism. • Dr. James Patterson, Head, 407-768-8000 Ext. 8098. Application contact: Carolyn P. Farrior, Director of Graduate Admissions, 407-768-8000 Ext. 8027.

Florida International University, College of Arts and Sciences, Department of Physics, Miami, FL 33199. Department awards MS. *Degree requirements:* 1 foreign language, thesis. *Entrance requirements:* GRE General Test, TOEFL. Application deadline: 4/1. Application fee: $15. *Tuition:* $462 per semester (minimum) full-time, $38.50 per credit part-time for state residents; $2217 per semester (minimum) full-time, $185 per credit part-time for nonresidents. *Faculty research:* Molecular collision processes (molecular beams), medium-energy theory, biophysical optics.

Announcement: The Department of Physics offers a rigorous program leading to the Master of Science. It consists of 30 semester hours of course work and 15 hours of research in 1 of the following areas: experimental molecular and atomic-beam physics; theoretical nuclear, medium-energy, and particle physics; experimental or theoretical solid-state physics; experimental or theoretical biophysics; astronomy. For further information, contact the Graduate Advisor, 305-348-2605.

Florida State University, College of Arts and Sciences, Department of Physics, Tallahassee, FL 32306. Department offers programs in chemical physics (MS, PhD), physics (MS, PhD). Faculty: 45 full-time (0 women), 0 part-time. Matriculated students: 83 full-time (8 women), 1 part-time (0 women); includes 9 minority (5 black American, 4 Hispanic American), 32 foreign. Average age 24. 379 applicants, 5% accepted. In 1990, 5 master's, 9 doctorates awarded. Terminal master's awarded for partial completion of doctoral program. *Degree requirements:* For master's, thesis required (for some programs), foreign language not required; for doctorate, dissertation required, foreign language not required. *Entrance requirements:* GRE General Test (minimum combined score of 1100 required), minimum GPA of 3.0. Application deadline: 2/15. Application fee: $15. *Tuition:* $76.29 per credit hour for state residents; $238 per credit hour for nonresidents. *Financial aid:* In 1990–91, 82 students received a total of $750,000 in aid awarded. 14 fellowships (5 to first-year students), 56 research assistantships (0 to first-year students), 27 teaching assistantships (19 to first-year students) were awarded; federal work-study and career-related internships or fieldwork also available. Aid available to part-time students. Average monthly stipend for a graduate assistantship: $1060. Financial aid application deadline: 2/15. *Faculty research:* Nuclear, high-energy, condensed-matter, and computational physics. Total annual research budget: $3.97-million. • Dr. Donald Robson, Chairman, 904-644-2867. Application contact: Dr. Dan Kimel, Director, Physics Graduate Programs, 904-644-3437.

Announcement: Extensive research opportunities exist in theoretical and experimental physics in the areas of atomic, condensed-matter, high-energy, and nuclear physics. Additional opportunities exist in the recently established Supercomputer Computations Research Institute, the MARTECH Center for Materials Science, and the National High-Magnetic Field Laboratory. Each full-time graduate student has an assistantship.

See full description on page 585.

Fort Hays State University, College of Arts and Sciences, Department of Physics, Hays, KS 67601. Department offers program in physical science (MS). Faculty: 4 full-time (0 women), 0 part-time. Matriculated students: 0 full-time, 2 part-time (both women); includes 0 minority, 0 foreign. Average age 30. 6 applicants, 67% accepted. In 1990, 1 degree awarded. *Application deadline:* 7/1 (priority date, applications processed on a rolling basis). *Application fee:* $0. *Tuition:* $58.75 per credit for state residents; $140 per credit for nonresidents. *Faculty research:* Chemisorbed thermionic cathode surfaces. • Dr. Kwo-Sun Chu, Chairman, 913-628-5844.

George Washington University, Graduate School of Arts and Sciences, Department of Physics, Washington, DC 20052. Department awards MA, PhD. Part-time and evening/weekend programs available. Faculty: 9 full-time (0 women), 2 part-time (0 women). Matriculated students: 6 full-time (1 woman), 5 part-time (0 women); includes 0 minority, 6 foreign. Average age 30. 6 applicants, 33% accepted. In 1990, 0 master's, 0 doctorates awarded. Terminal master's awarded for partial completion of doctoral program. *Degree requirements:* For master's, computer language, thesis or alternative, comprehensive exam; for doctorate, dissertation, general exam. *Entrance requirements:* GRE General Test, GRE Subject Test, minimum GPA of 3.0. Application deadline: 7/1. Application fee: $45. *Expenses:* Tuition of $490 per semester hour. Fees of $215 per year full-time, $125.60 per year (minimum) part-time. *Financial aid:* In 1990–91, $104,215 in aid awarded. 8 fellowships (0 to first-year students), 6 teaching assistantships (0 to first-year students) were awarded; full and partial tuition waivers, federal work-study also available. Financial aid application deadline: 2/15. • Dr. Donald R. Lehman, Chair, 202-994-6275.

Georgia Institute of Technology, College of Sciences, School of Physics, Atlanta, GA 30332. School offers programs in applied physics (MSAP), physics (MS, PhD). Part-time programs available. Faculty: 28 full-time (2 women), 0 part-time. Matriculated students: 87 full-time (17 women), 12 part-time (3 women); includes 10 minority (4 Asian American, 4 black American, 2 Hispanic American), 32 foreign. 142 applicants, 37% accepted. In 1990, 21 master's, 4 doctorates awarded. Terminal master's awarded for partial completion of doctoral program. *Degree requirements:* For master's, foreign language not required; for doctorate, dissertation, comprehensive exam required, foreign language not required. *Entrance requirements:* For master's, TOEFL (minimum score of 550 required), minimum GPA of 3.0; for doctorate, TOEFL (minimum score of 570 required), minimum GPA of 3.4. Application deadline: 1/15 (priority date, applications processed on a rolling basis). Application fee: $5. *Expenses:* Tuition of $574 per semester full-time, $48 per credit part-time for state residents; $1380 per semester full-time, $115 per credit part-time for nonresidents. Fees of $132 per semester full-time. *Financial aid:* In 1990–91, 75 students received a total of $497,124 in aid awarded. 44 research assistantships, 27 teaching assistantships were awarded; fellowships, partial tuition waivers, federal work-study, institutionally sponsored loans, and career-related internships or fieldwork also available. Aid available to part-time students. Average monthly stipend for a graduate assistantship: $875-1200. Financial aid application deadline: 1/15. *Faculty research:* Atomic and molecular physics, biophysics, lasers and optics, computer interfacing, acoustics. Total annual research budget: $2.1-million. • Dr. Henry S. Valk, Acting Director, 404-894-5200.

Announcement: Presidential Fellowships may be awarded to particularly well qualified students and add $4000 per annum to the standard assistantship stipend for a total of $14,500. Fees for students with assistantships are reduced to $151 per quarter.

Graduate School and University Center of the City University of New York, Program in Physics, New York, NY 10036. Program awards PhD. Faculty: 112 full-time (5 women), 0 part-time. Matriculated students: 150 full-time (27 women), 1 part-time (0 women); includes 11 minority (6 Asian American, 1 black American, 4 Hispanic American), 123 foreign. Average age 30. 245 applicants, 32% accepted. In 1990, 12 doctorates awarded (50% entered university research/teaching, 17% found other work related to degree, 8% continued full-time study). Terminal master's awarded for partial completion of doctoral program. *Degree requirements:* For doctorate, dissertation required, foreign language not required. *Entrance requirements:* For doctorate, GRE General Test. Application deadline: 4/1. Application fee: $30. *Tuition:* $2204 per year full-time, $95 per credit part-time for state residents; $4700 per year full-time, $199 per credit part-time for nonresidents. *Financial aid:* In 1990–91, $1.76-million in aid awarded. 24 fellowships, 4 research assistantships, 78 teaching assistantships were awarded; full and partial tuition waivers, federal work-study, institutionally sponsored loans, and career-related internships or fieldwork also available. Financial aid application deadline: 2/1. *Faculty research:* Condensed matter, particle, nuclear, and atomic physics. • Dr. Joseph Kreiger, Executive Officer, 212-642-2454.

Hampton University, Department of Physics, Hampton, VA 23668. Department awards MS. Part-time and evening/weekend programs available. *Degree requirements:* Thesis optional, foreign language not required. *Entrance requirements:* GRE General Test (minimum score of 450 on verbal section required). Application deadline: 7/1. Application fee: $10. *Tuition:* $3050 per semester full-time, $155 per credit hour part-time. *Faculty research:* Laser optics, remote sensing.

Harvard University, Graduate School of Arts and Sciences, Committee on Chemical Physics, Cambridge, MA 02138. Offerings include physics (AM). Faculty: 21 full-time, 0 part-time. *Expenses:* Tuition of $14,860 per year. Fees of $550 per year. • Dr. Donald Ciappenelli, Director of Graduate Studies, 617-495-4076. Application contact: Office of Admissions and Financial Aid, 617-495-5315.

Harvard University, Graduate School of Arts and Sciences, Department of Physics, Cambridge, MA 02138. Department offers programs in experimental physics (AM, PhD); medical engineering/medical physics (PhD), including applied physics, engineering sciences, physics; theoretical physics (AM, PhD). Faculty: 22 full-time, 14 part-time. Matriculated students: 116 full-time (13 women), 0 part-time; includes 2 minority (both Hispanic American). 318 applicants, 13% accepted. In 1990, 23 master's, 15 doctorates awarded. *Degree requirements:* For doctorate, dissertation required, foreign language not required. *Entrance requirements:* For doctorate, GRE General Test, GRE Subject Test. Application deadline: 1/2. Application fee: $60. *Expenses:* Tuition of $14,860 per year. Fees of $550 per year. *Financial aid:* In 1990–91, $524,193 in aid awarded. 61 fellowships (28 to first-year students), 62 research assistantships (10 to first-year students), 49 teaching assistantships (20 to first-year students) were awarded; federal work-study, institutionally sponsored loans, and career-related internships or fieldwork also available. Financial aid application deadline: 1/2; applicants required to submit GAPSFAS. *Faculty research:* Particle physics, condensed-matter physics, atomic physics. Total annual research budget: $3.5-million. • Dr. B. I. Halperin, Chairman, 617-495-2872. Application contact: Office of Admissions and Financial Aid, 617-495-5315.

See full description on page 587.

Harvard University, Divisions of Health Sciences and Technology and Applied Sciences and Department of Physics, Program in Medical Engineering/Medical Physics, Cambridge, MA 02138. Offerings include physics (PhD). Program offered jointly with Massachusetts Institute of Technology. *Degree requirements:* Dissertation, oral and written qualifying exams. *Application deadline:* 1/15 (applications processed on a rolling basis). *Tuition:* $15,410 per year. • Thomas McMahon, Head, 617-495-5854. Application contact: Graduate School of Arts and Sciences, Office of Admissions and Financial Aid, 617-495-5315.

SECTION 7: PHYSICS

Directory: Physics

Howard University, Graduate School of Arts and Sciences, Department of Physics and Astronomy, 2400 Sixth Street, NW, Washington, DC 20059. Department offers program in physics (MS, PhD). Part-time programs available. *Degree requirements:* For master's, thesis, comprehensive exam required, foreign language not required; for doctorate, dissertation, departmental qualifying exam, comprehensive exam required, foreign language not required. *Entrance requirements:* For master's, bachelor's degree in physics or related field, minimum GPA of 2.7; for doctorate, master's degree in physics or related field. Application deadline: 4/1. Application fee: $25. *Expenses:* Tuition of $6100 per year full-time, $339 per credit hour part-time. Fees of $555 per year full-time, $245 per semester part-time. *Faculty research:* Magnetic, optical, and structure studies of optical material and magnetic alloys; optimal materials and magnetic alloys.

Hunter College of the City University of New York, Division of Sciences and Mathematics, Department of Physics, 695 Park Avenue, New York, NY 10021. Department awards MA. Faculty: 14 full-time, 0 part-time. Matriculated students: 1 (woman) full-time, 0 part-time. 27 applicants, 33% accepted. In 1990, 2 degrees awarded. *Degree requirements:* Thesis or comprehensive exam required, foreign language not required. *Entrance requirements:* GRE General Test, TOEFL (minimum score of 550 required), minimum of 36 credits in physics and mathematics. Application deadline: 4/1. Application fee: $30. *Expenses:* Tuition of $2204 per year full-time, $95 per credit part-time for state residents; $4700 per year full-time, $199 per credit part-time for nonresidents. Fees of $15 per semester. *Financial aid:* Research assistantships, teaching assistantships, grants available. • Steve Greenbaum, Chair, 212-772-5248. Application contact: Office of Admissions, 212-772-4490.

Idaho State University, College of Arts and Sciences, Department of Physics, Pocatello, ID 83209. Department offers programs in natural science (MNS), physics (MS). Faculty: 5 full-time, 1 part-time. Matriculated students: 12 full-time (2 women), 10 part-time (2 women); includes 1 minority (Asian American), 9 foreign. In 1990, 3 degrees awarded. *Degree requirements:* Thesis required, foreign language not required. *Application deadline:* 8/1 (priority date). *Application fee:* $10. *Expenses:* Tuition of $0 full-time, $1060 per semester part-time for state residents; $0 for nonresidents. Fees of $809 per semester full-time, $72.50 per credit part-time. *Financial aid:* In 1990–91, 6 teaching assistantships awarded. • Dr. J. Frank Harmon, Chairman, 208-236-3150.

Illinois Institute of Technology, Lewis College of Sciences and Letters, Department of Physics, Chicago, IL 60616. Department offers programs in medical physics (MS, PhD), physics (MS, PhD). MS, PhD (medical physics) offered jointly with University of Health Sciences, Chicago Medical School. Part-time programs available. Faculty: 13 full-time (0 women), 0 part-time. Matriculated students: 24 full-time (6 women), 8 part-time (0 women); includes 3 minority (1 black American, 2 Hispanic American), 12 foreign. 55 applicants, 87% accepted. In 1990, 3 master's, 2 doctorates awarded. Terminal master's awarded for partial completion of doctoral program. *Degree requirements:* For master's, thesis (for some programs), comprehensive exam required, foreign language not required; for doctorate, dissertation, qualifying comprehensive exam required, foreign language not required. *Entrance requirements:* TOEFL (minimum score of 500 required). Application deadline: 7/1 (applications processed on a rolling basis). Application fee: $30. *Expenses:* Tuition of $13,070 per year full-time, $435 per credit hour part-time. Fees of $20 per semester (minimum) full-time, $1 per credit hour part-time. *Financial aid:* In 1990–91, 0 fellowships, 11 research assistantships, 15 teaching assistantships, 2 graduate assistantships awarded; federal work-study, institutionally sponsored loans also available. Financial aid application deadline: 3/1; applicants required to submit FAF. *Faculty research:* Magnetism, x-ray scattering, experimental and theoretical particle physics, solid state physics, radiation biophysics,. Total annual research budget: $780,000. • Dr. Porter Johnson, Chairman, 312-567-3375.

Indiana State University, College of Arts and Sciences, Department of Physics, Terre Haute, IN 47809. Department awards MA, MS. Part-time programs available. Faculty: 8 full-time (0 women), 0 part-time. Matriculated students: 7 full-time (2 women), 3 part-time (0 women); includes 0 minority, 6 foreign. Average age 24. 28 applicants, 64% accepted. In 1990, 6 degrees awarded. *Degree requirements:* Thesis required (for some programs), foreign language not required. *Entrance requirements:* TOEFL. Application deadline: rolling. Application fee: $10 ($20 for foreign students). *Tuition:* $90 per hour for state residents; $199 per hour for nonresidents. *Financial aid:* In 1990–91, $15,300 in aid awarded. 5 teaching assistantships (1 to a first-year student) were awarded; research assistantships, partial tuition waivers, federal work-study also available. Aid available to part-time students. Financial aid application deadline: 3/1. *Faculty research:* Astrophysics, image processing, holography, solid state, fluorescent spectroscopy. Total annual research budget: $40,000. • Dr. John Swez, Chairperson, 812-237-2064.

Indiana University Bloomington, College of Arts and Sciences, Department of Physics, Bloomington, IN 47405. Department awards MAT, MS, PhD. Part-time programs available. Faculty: 47 full-time (2 women), 1 part-time (0 women). Matriculated students: 88 full-time (13 women), 0 part-time; includes 2 minority (both Hispanic American), 47 foreign. Average age 29. 169 applicants, 23% accepted. In 1990, 27 master's awarded (0% entered university research/teaching, 0% found other work related to degree, 77% continued full-time study); 6 doctorates awarded (50% entered university research/teaching, 50% found other work related to degree). Terminal master's awarded for partial completion of doctoral program. *Degree requirements:* For master's, qualifying exam required, foreign language and thesis not required; for doctorate, dissertation, qualifying exam required, foreign language not required. *Entrance requirements:* GRE, TOEFL (minimum score of 550 required). Application deadline: 2/1. Application fee: $25 ($35 for foreign students). *Tuition:* $99.85 per credit hour for state residents; $288 per credit hour for nonresidents. *Financial aid:* In 1990–91, 8 fellowships (all to first-year students), 53 research assistantships (21 to first-year students), 28 teaching assistantships (10 to first-year students) awarded; federal work-study and career-related internships or fieldwork also available. Aid available to part-time students. Financial aid application deadline: 2/1. *Faculty research:* Condensed matter, elementary particles, nuclear physics, astrophysics. Total annual research budget: $11.5-million. • Dr. J. Timothy Londergan, Chair, 812-855-1247. Application contact: June Dizer, Academic Services Coordinator, 812-855-3973.

Announcement: Programs in theoretical and experimental nuclear, elementary particle, condensed-matter, and atomic physics; mathematical physics; accelerator physics; astrophysics; and chemical physics. Cyclotron facility; electron-cooled storage ring; VAX, IBM 3090, and Sun computers; low-temperature facilities; nuclear, optical, and NMR spectroscopy laboratories. Financial support available.

See full description on page 589.

Indiana University of Pennsylvania, College of Natural Sciences and Mathematics, Department of Physics, Indiana, PA 15705. Department awards MA, MS. Part-time programs available. Faculty: 6 full-time (0 women), 0 part-time. Matriculated students: 11 full-time (2 women), 1 part-time (0 women); includes 3 minority (all black American), 2 foreign. 64 applicants, 44% accepted. In 1990, 8 degrees awarded. *Degree requirements:* Thesis required (for some programs), foreign language not required. *Entrance requirements:* GRE General Test, TOEFL (minimum score of 500 required). Application deadline: 7/1 (priority date, applications processed on a rolling basis). Application fee: $20. *Expenses:* Tuition of $1139 per semester full-time, $127 per credit part-time for state residents; $1442 per semester full-time, $160 per credit part-time for nonresidents. Fees of $169 per semester full-time. *Financial aid:* In 1990–91, 11 research assistantships awarded; federal work-study also available. Aid available to part-time students. Financial aid application deadline: 3/15. • Dr. John Fox, Chairperson, 412-357-2371. Application contact: Dr. Kenneth Schwartzman, Graduate Coordinator, 412-357-2192.

Indiana University–Purdue University at Indianapolis, School of Science, Department of Physics, Indianapolis, IN 46205. Department offers programs in biological physics (PhD), physics (MS). Part-time and evening/weekend programs available. Faculty: 12 full-time, 0 part-time. Matriculated students: 10 full-time (3 women), 3 part-time (2 women); includes 6 minority (all Asian American), 6 foreign. Average age 30. 40 applicants, 10% accepted. In 1990, 2 master's awarded (100% continued full-time study); 1 doctorate awarded (100% entered university research/teaching). Terminal master's awarded for partial completion of doctoral program. *Degree requirements:* For master's, thesis optional, foreign language not required; for doctorate, dissertation required, foreign language not required. *Entrance requirements:* TOEFL (minimum score of 550 required). Application deadline: 3/1 (applications processed on a rolling basis). Application fee: $20. *Tuition:* $100 per credit for state residents; $288 per credit for nonresidents. *Financial aid:* In 1990–91, 11 students received a total of $92,000 in aid awarded. 2 fellowships (0 to first-year students), 2 research assistantships, 4 teaching assistantships (3 to first-year students) were awarded; full and partial tuition waivers, federal work-study also available. Aid available to part-time students. Average monthly stipend for a graduate assistantship: $730. Financial aid application deadline: 2/1. *Faculty research:* Magnetic resonance, photosynthesis, optical communications, biophysics. Total annual research budget: $300,000. • B. D. Nageswara Rao, Chair, 317-274-6901. Application contact: F. W. Kleinhans, Chair, Graduate Committee, 317-274-6901.

Institute of Paper Science and Technology, Program in Chemistry, Atlanta, GA 30318. Offers paper science and technology (MS, PhD). Part-time programs available. Terminal master's awarded for partial completion of doctoral program. *Degree requirements:* For master's, industrial experience required, foreign language and thesis not required; for doctorate, dissertation required, foreign language not required. *Entrance requirements:* For master's, GRE. Application deadline: 5/15 (priority date). Application fee: $0. *Tuition:* $0 full-time, $225 per credit hour part-time.

See full description on page 109.

Institute of Paper Science and Technology, Program in Physics/Mathematics, Atlanta, GA 30318. Offers paper science and technology (MS, PhD). Part-time programs available. Terminal master's awarded for partial completion of doctoral program. *Degree requirements:* For master's, industrial experience required, foreign language and thesis not required; for doctorate, dissertation required, foreign language not required. *Entrance requirements:* For master's, GRE. Application deadline: 5/15 (priority date). Application fee: $0. *Tuition:* $0 full-time, $225 per credit hour part-time.

John Carroll University, Department of Physics, University Heights, OH 44118. Department awards MS. Part-time programs available. Faculty: 7 full-time (1 woman), 0 part-time. Matriculated students: 12 full-time, 3 part-time; includes 0 minority, 7 foreign. Average age 23. In 1990, 4 degrees awarded (25% entered university research/teaching, 75% found other work related to degree). *Degree requirements:* Comprehensive exam, thesis or essay required, foreign language not required. *Entrance requirements:* Undergraduate degree in physics or electrical engineering. Application deadline: 8/17 (priority date, applications processed on a rolling basis). Application fee: $0. *Tuition:* $300 per credit. *Financial aid:* In 1990–91, 7 students received a total of $56,000 in aid awarded. 7 teaching assistantships (5 to first-year students) were awarded. *Faculty research:* Fiber optics, ultrasonics, complex fluids, light scattering, superconductivity. Total annual research budget: $220,000. • Dr. Klaus Fritsch, Chairperson, 216-397-4301.

Johns Hopkins University, School of Arts and Sciences, Department of Physics and Astronomy, Baltimore, MD 21218. Department offers programs in astronomy (MA, PhD), physics (MA, PhD). Faculty: 27 full-time (3 women), 10 part-time (0 women). Matriculated students: 94 full-time (7 women), 0 part-time; includes 3 minority (1 Asian American, 2 Hispanic American), 30 foreign. Average age 25. 157 applicants, 38% accepted. In 1990, 15 master's awarded; 11 doctorates awarded (100% entered university research/teaching). Terminal master's awarded for partial completion of doctoral program. *Degree requirements:* For doctorate, 1 foreign language, dissertation, comprehensive exam. *Entrance requirements:* For master's, GRE General Test; for doctorate, GRE General Test, GRE Subject Test, TOEFL. Application deadline: 2/15. Application fee: $40. *Expenses:* Tuition of $15,500 per year full-time, $1550 per course part-time. Fees of $400 per year full-time. *Financial aid:* In 1990–91, $1.485-million in aid awarded. 93 fellowships (22 to first-year students), 56 research assistantships (3 to first-year students), 37 teaching assistantships (19 to first-year students) were awarded; full and partial tuition waivers, federal work-study, institutionally sponsored loans also available. Financial aid application deadline: 3/14; applicants required to submit FAF. *Faculty research:* Experimental, theoretical, atomic, plasma, particle, nuclear, and condensed matter physics. Total annual research budget: $4.5-million. • Dr. James C. Walker Jr., Chairman, 301-338-7347. Application contact: Barbara A. Staicer, Graduate Admissions Administrator, 301-338-7347.

See full description on page 591.

Kansas State University, College of Arts and Sciences, Department of Physics, Manhattan, KS 66506. Department awards MS, PhD. Faculty: 24 full-time (1 woman), 0 part-time. Matriculated students: 40 full-time (3 women), 4 part-time (1 woman); includes 29 minority (all Asian American), 0 foreign. 134 applicants, 7% accepted. In 1990, 2 master's, 4 doctorates awarded. *Degree requirements:* For master's, thesis required, foreign language not required; for doctorate, dissertation. *Expenses:* Tuition of $1668 per year full-time, $51 per credit hour part-time for state residents; $5382 per year full-time, $142 per credit hour part-time for nonresidents. Fees of $305 per year full-time, $64.50 per semester part-time. *Financial aid:* In 1990–91, 11 teaching assistantships awarded; research assistantships also available. • James Legg, Head, 913-532-6786.

Kent State University, Department of Physics, Kent, OH 44242. Department awards MA, MS, PhD. Faculty: 24 full-time, 0 part-time. Matriculated students: 66 full-time (9 women), 6 part-time (0 women); includes 30 foreign. 83 applicants, 36% accepted. In 1990, 7 master's, 7 doctorates awarded. *Degree requirements:* For master's, thesis optional, foreign language not required; for doctorate, computer language, dissertation. *Application deadline:* 7/12 (applications processed on a rolling basis). *Application fee:* $25. *Tuition:* $1601 per semester full-time, $133.75 per hour part-time for state residents; $3101 per semester full-time, $258.75 per hour part-time for nonresidents. *Financial aid:* Fellowships, research assistantships, teaching assistantships, full tuition waivers, federal work-study available. Financial aid application deadline: 2/1. • Dr. Stanley Christensen, Chairman, 216-672-2880.

See full description on page 593.

Lakehead University, Faculty of Arts and Science, Department of Physics, Thunder Bay, ON P7B 5E1, Canada. Department awards M Sc. Faculty: 7 full-time (1 woman), 0 part-time. Matriculated students: 2 full-time (0 women), 0 part-time; includes 1 foreign. 17 applicants, 0% accepted. In 1990, 0 degrees awarded. *Degree requirements:* Thesis required, foreign language not required. *Entrance requirements:* TOEFL (minimum score of 550 required). Application fee: $0. *Expenses:* Tuition of $1969 per year full-time, $527 per year part-time for Canadian residents; $3712 per year full-time, $2618 per year part-time for nonresidents. Fees of $2735 per year full-time, $100 per year part-time for Canadian residents; $11,136 per year full-time, $100 per year part-time for nonresidents. *Financial aid:* In 1990–91, 2 students received a total of $11,183 in aid awarded. 2 teaching assistantships (0 to first-year students) were awarded; entrance awards and research stipends also available. Average monthly stipend for a graduate assistantship: $680. Financial aid application deadline: 3/30. *Faculty research:* Semiconductors, nuclear physics, condensed matter physics, dialectic studies of disordered solids, phase transitions in phospholipid bilayers. Total annual research budget: $43,000. • Dr. Margaret H. Hawton, Chairman, 807-343-8633. Application contact: Dr. V. V. Paranjape, Graduate Coordinator, 807-343-8258.

Lehigh University, College of Arts and Sciences, Department of Physics, Bethlehem, PA 18015. Department awards MS, PhD. Faculty: 19 full-time, 4 part-time. Matriculated students: 53 full-time (12 women), 4 part-time (2 women); includes 3 minority (2 black American, 1 Hispanic American), 21 foreign. 115 applicants, 7% accepted. In 1990, 9 master's, 2 doctorates awarded. Terminal master's awarded for partial completion of doctoral program. *Degree requirements:* For master's, research project required, foreign language and thesis not required; for doctorate, dissertation, exam required, foreign language not required. *Entrance requirements:* For master's, TOEFL; for doctorate, GRE General Test, TOEFL. Application deadline: rolling. Application fee: $40. *Tuition:* $15,650 per year full-time, $655 per credit hour part-time. *Financial aid:* Fellowships, research assistantships, teaching assistantships, federal work-study, institutionally sponsored loans available. Financial aid application deadline: 3/15. *Faculty research:* Solid-state physics, fluids and plasmas, atomic physics, high-energy physics. • Dr. Beall Fowler, Acting Chairman, 215-758-3931. Application contact: Dr. Gary Borse, Graduate Coordinator.

Louisiana State University and Agricultural and Mechanical College, College of Basic Sciences, Department of Physics and Astronomy, Baton Rouge, LA 70803. Department offers programs in astronomy (PhD), astrophysics (PhD), physics (MS, PhD). Faculty: 35 full-time (0 women), 0 part-time. Matriculated students: 65 full-time (15 women), 3 part-time (0 women); includes 3 minority (2 Asian American, 1 black American), 51 foreign. Average age 28. 100 applicants, 41% accepted. In 1990, 8 master's awarded (100% continued full-time study); 10 doctorates awarded (70% entered university research/teaching, 30% found other work related to degree). Terminal master's awarded for partial completion of doctoral program. *Degree requirements:* For master's, thesis or alternative required, foreign language not required; for doctorate, dissertation required, foreign language not required. *Entrance requirements:* For master's, GRE General Test, TOEFL (minimum score of 525 required for admission, 560 required for assistantships), minimum GPA of 3.0; for doctorate, GRE General Test, TOEFL (minimum score of 525 required), minimum GPA of 3.0. Application deadline: 7/1 (priority date, applications processed on a rolling basis). Application fee: $25. *Tuition:* $1020 per semester full-time for state residents; $2620 per semester full-time for nonresidents. *Financial aid:* In 1990–91, 64 students received a total of $757,068 in aid awarded. 3 fellowships (1 to a first-year student), 14 research assistantships (0 to first-year students), 46 teaching assistantships (11 to first-year students) were awarded; institutionally sponsored loans also available. Average monthly stipend for a graduate assistantship: $968. Financial aid application deadline: 3/15. *Faculty research:* Experimental atomic, nuclear, particle, cosmic-ray, low-temperature, and condensed-matter physics; theoretical atomic, nuclear, particle, and condensed-matter physics. • Dr. Jerry P. Draayer, Chair, 504-388-2261. Application contact: Dr. Michael Cherry, Chair, Assistantship Committee, 504-388-1194.

See full description on page 595.

Louisiana Tech University, College of Arts and Sciences, Department of Physics, Ruston, LA 71272. Department awards MS. Part-time programs available. Faculty: 5 full-time (0 women), 0 part-time. Matriculated students: 5 full-time (0 women), 0 part-time; includes 0 minority, 2 foreign. In 1990, 3 degrees awarded. *Degree requirements:* Computer language, thesis or alternative required, foreign language not required. *Entrance requirements:* GRE General Test. Application deadline: 8/13. Application fee: $5. *Tuition:* $613 per quarter full-time, $184 per semester (minimum) part-time for state residents; $999 per quarter full-time, $184 per semester (minimum) part-time for nonresidents. *Financial aid:* Fellowships available. Financial aid application deadline: 2/1; applicants required to submit GAPSFAS. • Dr. Richard L. Gibbs, Head, 318-257-4358.

Maharishi International University, Program in Physics, Fairfield, IA 52556. Program awards MS, PhD. Program admits applicants every other year. Faculty: 7 full-time (0 women), 0 part-time. Matriculated students: 12 full-time (0 women), 0 part-time; includes 0 minority, 7 foreign. Average age 34. 0 applicants. In 1990, 1 master's, 0 doctorates awarded. Terminal master's awarded for partial completion of doctoral program. *Degree requirements:* For master's, master's seminar presentation required, foreign language and thesis not required; for doctorate, dissertation, qualifying exam, comprehensive exam required, foreign language not required. *Entrance requirements:* GRE General Test, GRE Subject Test, TOEFL (minimum score of 550 required), bachelor's degree in physics or mathematics, minimum GPA of 3.0. Application deadline: 4/15 (priority date, applications processed on a rolling basis). Application fee: $40. *Expenses:* Tuition of $7950 per year full-time, $1990 per year part-time. Fees of $170 per year. *Financial aid:* In 1990–91, 15 teaching assistantships (0 to first-year students) awarded; federal work-study also available. Financial aid application deadline: 4/30; applicants required to submit FAF. *Faculty research:* Particle physics and cosmology, critical phenomena and phase transitions, quantum optics, nonlinear dynamical systems theory. Total annual research budget: $72,800. • Dr. John Hagelin, Chairman, 515-472-1162. Application contact: Harry Bright, Director of Admissions, 515-472-1166.

Mankato State University, School of Physics, Engineering and Technology, Department of Physics, South Road and Ellis Avenue, Mankato, MN 56002-8400. Department awards MA, MS. Faculty: 8 full-time (0 women), 0 part-time. Matriculated students: 7 full-time (1 woman), 5 part-time (0 women); includes 0 minority, 9 foreign. Average age 27. 19 applicants, 79% accepted. In 1990, 2 degrees awarded. *Degree requirements:* 1 foreign language, thesis or alternative, comprehensive exam. *Entrance requirements:* GRE General Test, minimum GPA of 2.75 for last 2 years of undergraduate study. Application deadline: 2/3 (priority date, applications processed on a rolling basis). Application fee: $15. *Expenses:* Tuition of $52 per credit for state residents; $75 per credit for nonresidents. Fees of $6.50 per quarter. *Financial aid:* In 1990–91, 12 students received a total of $28,386 in aid awarded. 12 teaching assistantships (5 to first-year students) were awarded; federal work-study also available. Aid available to part-time students. Average monthly stipend for a graduate assistantship: $262. Financial aid application deadline: 7/1. • Dr. Sandford Schuster, Chairperson, 507-389-1521. Application contact: Dr. Forrest Glick, Graduate Coordinator, 507-389-1521.

Marquette University, College of Arts and Sciences, Department of Physics, Milwaukee, WI 53233. Offerings include lasers and optics (MS), solid-state physics (MS), theoretical physics (MS). Department faculty: 10 full-time (0 women), 0 part-time. *Degree requirements:* Thesis or alternative, comprehensive exam required, foreign language not required. *Entrance requirements:* TOEFL (minimum score of 550 required). Application fee: $25. *Tuition:* $300 per credit. • Dr. Kenneth S. Mendelson, Chairman, 414-288-7247. Application contact: Dr. Larry Browning, Director of Graduate Studies, 414-288-1605.

Marshall University, College of Science, Department of Physical Science and Physics, Huntington, WV 25755. Offerings include physical science (MS). Department faculty: 9 (1 woman). *Degree requirements:* Thesis optional, foreign language not required. *Entrance requirements:* GRE General Test. Application fee: $0. *Tuition:* $857 per semester full-time, $492 per semester part-time for state residents; $2207 per semester full-time, $1392 per semester part-time for nonresidents. • Dr. Wesley L. Shanholtzer, Chairman, 304-696-6738. Application contact: Dr. James Harless, Director of Admissions, 304-696-3160.

Massachusetts Institute of Technology, School of Science, Department of Physics, Cambridge, MA 02139. Department awards SM, PhD, Sc D. Faculty: 85 full-time (4 women), 0 part-time. Matriculated students: 314 full-time (35 women), 0 part-time; includes 27 minority (16 Asian American, 4 black American, 6 Hispanic American, 1 Native American), 123 foreign. 626 applicants, 29% accepted. In 1990, 8 master's, 38 doctorates awarded. *Degree requirements:* Thesis/dissertation required, foreign language not required. *Entrance requirements:* GRE General Test, GRE Subject Test. Application deadline: 1/15. Application fee: $45. *Tuition:* $16,900 per year. *Financial aid:* In 1990–91, 263 students received a total of $3.63-million in aid awarded. 9 fellowships (3 to first-year students), 220 research assistantships (30 to first-year students), 34 teaching assistantships (10 to first-year students) were awarded. Average monthly stipend for a graduate assistantship: $1150. Financial aid application deadline: 1/15. • Robert Birgeneau, Head, 617-253-4801. Application contact: Peggy Berkovitz, Graduate Office Administrator, 617-253-4851.

McGill University, Faculty of Graduate Studies and Research, Department of Physics, Montreal, PQ H3A 2T5, Canada. Department awards M Sc, PhD. Faculty: 38 full-time (0 women), 0 part-time. Matriculated students: 96 full-time (9 women), 0 part-time; includes 0 minority, 32 foreign. Average age 25. 60 applicants, 67% accepted. In 1990, 12 master's, 9 doctorates awarded. Terminal master's awarded for partial completion of doctoral program. *Degree requirements:* Thesis/dissertation. *Application deadline:* 2/15. *Application fee:* $25. *Tuition:* $698.50 per semester full-time, $46.57 per credit part-time for Canadian residents; $3480 per semester full-time, $234 per semester part-time for nonresidents. *Financial aid:* Fellowships, research assistantships, teaching assistantships available. Financial aid application deadline: 2/15. *Faculty research:* High-energy, condensed-matter, nuclear, atmospheric, and medical physics. Total annual research budget: $3.13-million. • M. J. Zuckermann, Chairman, Graduate Studies Committee, 514-398-6485. Application contact: P. Domingues, Graduate Studies Secretary, 514-398-6485.

McMaster University, Faculty of Science, Department of Physics, Hamilton, ON L8S 4L8, Canada. Department offers programs in chemical physics (M Sc, PhD), health and radiation physics (M Sc), physics (PhD). *Degree requirements:* For master's, thesis or alternative required, foreign language not required; for doctorate, dissertation, comprehensive exam required, foreign language not required. *Expenses:* Tuition of $2250 per year full-time, $810 per semester part-time for Canadian residents; $10,340 per year full-time, $4050 per semester part-time for nonresidents. Fees of $76 per year full-time, $49 per semester part-time.

Memorial University of Newfoundland, School of Graduate Studies, Department of Physics, St. John's, NF A1C 5S7, Canada. Department offers programs in physical oceanography (M Sc); physics (M Sc, PhD), including condensed matter physics (PhD), molecular physics (PhD), physical oceanography (PhD). *Degree requirements:* Thesis/dissertation.

Memphis State University, College of Arts and Sciences, Department of Physics, Memphis, TN 38152. Department awards MS. Faculty: 9 full-time (0 women), 1 part-time (0 women). Matriculated students: 5 full-time (1 woman), 10 part-time (0 women); includes 3 minority (all black American), 2 foreign. In 1990, 7 degrees awarded. *Degree requirements:* Thesis or alternative, comprehensive exam. *Entrance requirements:* GRE or MAT, 20 undergraduate hours in physics. Application deadline: 8/1 (applications processed on a rolling basis). Application fee: $5. *Expenses:* Tuition of $92 per credit hour for state residents; $239 per credit hour for nonresidents. Fees of $45 per year for state residents. *Financial aid:* In 1990–91, 2 research assistantships, 6 teaching assistantships awarded. *Faculty research:* High technology superconduction, electrochemical systems, Burnett coefficients and diffusion, colofusion, metal-electrolyte interface. • Dr. Donald Franceschetti, Chairman, 901-678-2620. Application contact: Dr. Robert R. Marchini, Coordinator, Graduate Studies, 901-678-2620.

Miami University, College of Arts and Sciences, Department of Physics, Oxford, OH 45056. Department awards MA, MAT, MS. Part-time programs available. Faculty: 16. Matriculated students: 15 full-time (5 women), 9 part-time (4 women); includes 0 minority, 3 foreign. 47 applicants, 49% accepted. In 1990, 5 degrees awarded. *Degree requirements:* Final exam. *Entrance requirements:* Minimum undergraduate GPA of 2.75 or 3.0 during previous 2 years. Application deadline: 3/1. Application fee: $30. *Expenses:* Tuition of $3196 per year full-time, $133 per semester (minimum) part-time for state residents; $7134 per year full-time, $312 per semester (minimum) part-time for nonresidents. Fees of $371 per year full-time, $165.50 per semester hour part-time. *Financial aid:* In 1990–91, $106,380 in aid awarded. Fellowships, research assistantships, teaching assistantships, full tuition waivers, federal work-study available. Financial aid application deadline: 3/1. • Paul D. Scholten, Director of Graduate Study, 513-529-5636.

Michigan State University, College of Natural Science, Department of Physics and Astronomy, East Lansing, MI 48824. Department offers program in physics (MAT, MS, PhD). Faculty: 41 full-time (1 woman), 0 part-time. Matriculated students: 170 full-time (20 women), 17 part-time (7 women); includes 4 minority (1 Asian American, 2 black American, 1 Hispanic American), 88 foreign. In 1990, 36 master's, 9 doctorates awarded. Terminal master's awarded for partial completion

Directory: Physics

of doctoral program. *Degree requirements:* For master's, thesis or alternative required, foreign language not required; for doctorate, dissertation required, foreign language not required. *Application deadline:* 3/15 (applications processed on a rolling basis). *Application fee:* $25 ($40 for foreign students). *Tuition:* $104.75 per credit for state residents; $211.75 per credit for nonresidents. *Financial aid:* In 1990–91, 1 fellowship, 65 research assistantships, 41 teaching assistantships awarded. *Faculty research:* Nuclear and accelerator physics, high-energy physics, condensed-matter physics, astronomy and astrophysics. Total annual research budget: $23-million. • Dr. Gerard Crawley, Chairperson, 517-353-8662.

See full description on page 597.

Michigan Technological University, College of Sciences and Arts, Department of Physics, Houghton, MI 49931. Offerings include physics (MS). Department faculty: 16 full-time (1 woman), 0 part-time. *Degree requirements:* Thesis required (for some programs), foreign language not required. *Entrance requirements:* TOEFL (minimum score of 570 required), BS in physics or related area, minimum GPA of 3.0. Application deadline: rolling. Application fee: $20. *Expenses:* Tuition of $852 per semester full-time, $71 per credit part-time for state residents; $2065 per semester full-time, $172 per credit part-time for nonresidents. Fees of $75 per semester. • Dr. Bryan H. Suits, Head, 906-487-2086. Application contact: Dr. Robert Weidman, Chairman, Graduate Committee, 906-487-2126.

Mississippi State University, College of Arts and Sciences, Department of Physics and Astronomy, Mississippi State, MS 39762. Offerings include physics (MS). Department faculty: 13 full-time (1 woman), 0 part-time. *Degree requirements:* Thesis required, foreign language not required. *Entrance requirements:* GRE General Test, TOEFL (minimum score of 540 required), TSE (minimum score of 200 required). Application deadline: 3/1 (priority date, applications processed on a rolling basis). *Expenses:* Tuition of $891 per semester full-time, $99 per credit hour part-time for state residents; $1622 per semester full-time, $180 per credit hour part-time for nonresidents. Fees of $221 per semester full-time, $25 per semester (minimum) part-time. • Dr. Joe L. Ferguson, Head, 601-325-2806. Application contact: Dr. David Monts, 601-325-2806.

Montana State University, College of Letters and Science, Department of Physics, 901 West Garfield Street, Bozeman, MT 59717. Department awards MS, PhD. Faculty: 14 full-time (0 women), 0 part-time. Matriculated students: 58 full-time (8 women), 6 part-time (2 women); includes 11 minority (10 Asian American, 1 Hispanic American), 29 foreign. Average age 28. 74 applicants, 39% accepted. In 1990, 8 master's, 4 doctorates awarded. Terminal master's awarded for partial completion of doctoral program. *Degree requirements:* For master's, thesis or alternative required, foreign language not required; for doctorate, dissertation required, foreign language not required. *Entrance requirements:* GRE General Test, TOEFL (minimum score of 550 required). Application deadline: 6/15 (applications processed on a rolling basis). Application fee: $20. *Tuition:* $1150 per year full-time, $74 per credit (minimum) part-time for state residents; $2824 per year full-time, $100 per credit (minimum) part-time for nonresidents. *Financial aid:* Research assistantships, teaching assistantships available. Financial aid application deadline: 3/1. *Faculty research:* Surface science, general relativity/astrophysics, superconductivity, polymers and liquid crystals, laser spectroscopy. Total annual research budget: $800,000. • Dr. John C. Hermanson, Head, 406-994-3614.

Murray State University, College of Sciences, Department of Physics, Murray, KY 42071. Department awards MAT, MS. Part-time programs available. Faculty: 9 full-time (0 women), 0 part-time. Matriculated students: 2 full-time (0 women), 2 part-time (1 woman); includes 1 minority (black American), 0 foreign. 4 applicants, 75% accepted. In 1990, 0 degrees awarded. *Degree requirements:* Thesis required (for some programs), foreign language not required. *Entrance requirements:* GRE General Test, TOEFL (minimum score of 500 required). Application fee: $0 ($20 for foreign students). *Tuition:* $775 per semester full-time, $87 per credit hour part-time for state residents; $2215 per semester full-time, $246 per credit hour part-time for nonresidents. *Financial aid:* Research assistantships, teaching assistantships, federal work-study available. Average monthly stipend for a graduate assistantship: $335. Financial aid application deadline: 4/1; applicants required to submit GAPSFAS. • Dr. Robert C. Etherton, Chairman, 502-762-2993.

Naval Postgraduate School, Department of Physics, Monterey, CA 93943. Department offers programs in engineering acoustics (MS, PhD), engineering science (MS), physics (MS, PhD). Program only open to commissioned officers of the United States and friendly nations and selected United States federal civilian employees. Part-time programs available. Terminal master's awarded for partial completion of doctoral program. *Degree requirements:* Thesis/dissertation required, foreign language not required. *Tuition:* $0.

New Mexico Institute of Mining and Technology, Department of Physics, Socorro, NM 87801. Department offers program in physics (MS, PhD), including astrophysics (MS, PhD), atmospheric physics (MS, PhD), instrumentation (MS), mathematical physics (PhD). Faculty: 15 full-time (1 woman), 1 part-time (0 women). Matriculated students: 25 full-time (5 women), 0 part-time; includes 0 minority, 17 foreign. Average age 25. 60 applicants, 25% accepted. In 1990, 3 master's, 0 doctorates awarded. *Degree requirements:* For master's, thesis required, foreign language not required; for doctorate, 1 foreign language, dissertation. *Entrance requirements:* For master's, GRE General Test, TOEFL (minimum score of 540 required); for doctorate, GRE General Test, GRE Subject Test, TOEFL (minimum score of 540 required). Application deadline: 6/1 (priority date, applications processed on a rolling basis). Application fee: $16. *Expenses:* Tuition of $617 per semester full-time, $51 per hour part-time for state residents; $2656 per semester full-time, $209 per hour part-time for nonresidents. Fees of $207 per semester. *Financial aid:* In 1990–91, 24 students received a total of $67,194 in aid awarded. 15 research assistantships (4 to first-year students), 9 teaching assistantships (7 to first-year students) were awarded; fellowships, federal work-study, institutionally sponsored loans also available. Average monthly stipend for a graduate assistantship: $600. Financial aid applicants required to submit GAPSFAS or FAF. *Faculty research:* Cloud physics, stellar and extragalactic processes. Total annual research budget: $2.8-million. • Dr. David Raymond, Chairman, 505-835-5610. Application contact: Dr. J. A. Smoake, Dean, Graduate Studies, 505-835-5513.

New Mexico State University, College of Arts and Sciences, Department of Physics, Las Cruces, NM 88003. Department awards MS, PhD. Faculty: 16 full-time, 0 part-time. Matriculated students: 37 full-time (2 women), 13 part-time (1 woman). In 1990, 3 master's, 2 doctorates awarded. *Degree requirements:* For master's, thesis or alternative; for doctorate, dissertation required, foreign language not required. *Entrance requirements:* GRE General Test, minimum GPA of 3.0. Application deadline: 7/1. Application fee: $10. *Tuition:* $1608 per year full-time, $67 per credit hour part-time for state residents; $5304 per year full-time, $221 per credit hour part-time for nonresidents. *Financial aid:* Fellowships, research assistantships, teaching assistantships, federal work-study available. Aid available to part-time students. Financial aid applicants required to submit GAPSFAS or FAF. *Faculty research:* Optics; spectroscopy; geophysics; aerosol, nuclear, and particle physics; physics theory. Total annual research budget: $750,000. • Dr. George H. Goedecke, Head, 505-646-3832. Application contact: Dr. James Ni, Graduate Adviser, 505-646-1920.

See full description on page 599.

New York University, Graduate School of Arts and Science, Department of Physics, New York, NY 10011. Department awards MS, PhD. Part-time programs available. Faculty: 27 full-time, 1 part-time. Matriculated students: 59 full-time, 4 part-time; includes 0 minority, 46 foreign. Average age 26. 165 applicants, 44% accepted. In 1990, 6 master's awarded (50% found work related to degree, 50% continued full-time study); 10 doctorates awarded (80% entered university research/teaching, 20% found other work related to degree). Terminal master's awarded for partial completion of doctoral program. *Degree requirements:* For master's, thesis or alternative required, foreign language not required; for doctorate, 1 foreign language, dissertation, research seminar, teaching experience. *Entrance requirements:* For master's, GRE Subject Test, GRE General Test, TOEFL, bachelor's degree in physics; for doctorate, GRE Subject Test, GRE General Test, TOEFL. Application deadline: 4/15 (priority date, applications processed on a rolling basis). Application fee: $30. *Tuition:* $467 per credit. *Financial aid:* In 1990–91, $900,707 in aid awarded. 10 fellowships (7 to first-year students), 15 research assistantships (0 to first-year students), 30 teaching assistantships (6 to first-year students) were awarded; full and partial tuition waivers, federal work-study, institutionally sponsored loans also available. Average monthly stipend for a graduate assistantship: $1000. Financial aid application deadline: 1/15; applicants required to submit GAPSFAS or FAF. *Faculty research:* Atomic physics, elementary particles and fields, astrophysics, condensed-matter physics, neuromagnetism. Total annual research budget: $1.4-million. • Henry Stroke, Chairman, 212-998-7710. Application contact: Leonard Rosenberg, Director of Graduate Studies, 212-998-7736.

North Carolina State University, College of Physical and Mathematical Sciences, Department of Physics, Raleigh, NC 27695. Department awards MS, PhD. Part-time programs available. Faculty: 34 full-time (2 women), 0 part-time. Matriculated students: 75 full-time (12 women), 1 part-time (0 women); includes 5 minority (1 Asian American, 4 black American), 21 foreign. 85 applicants, 41% accepted. In 1990, 5 master's awarded; 9 doctorates awarded (33% entered university research/teaching, 67% found other work related to degree). Terminal master's awarded for partial completion of doctoral program. *Degree requirements:* Thesis/dissertation. *Entrance requirements:* GRE General Test, GRE Subject Test. Application deadline: 5/1 (priority date, applications processed on a rolling basis). Application fee: $35. *Tuition:* $1138 per year for state residents; $5805 per year for nonresidents. *Financial aid:* In 1990–91, $610,000 in aid awarded. 9 fellowships (2 to first-year students), 28 research assistantships, 38 teaching assistantships (14 to first-year students) were awarded; institutionally sponsored loans also available. Average monthly stipend for a graduate assistantship: $1000. Financial aid application deadline: 5/1. *Faculty research:* Condensed-matter physics, atomic physics, nuclear physics, plasma physics, astrophysics. Total annual research budget: $10.15-million. • Dr. Richard R. Patty, Head, 919-737-2521. Application contact: G. E. Mitchell, Graduate Administrator, 919-737-2521.

North Dakota State University, College of Science and Mathematics, Department of Physics, Fargo, ND 58105. Department awards MS, PhD. Faculty: 8 full-time (0 women), 3 part-time (0 women). Matriculated students: 10 full-time (1 woman), 0 part-time; includes 9 foreign. Average age 28. 55 applicants, 42% accepted. In 1990, 4 master's, 1 doctorate awarded. Terminal master's awarded for partial completion of doctoral program. *Degree requirements:* Thesis/dissertation required, foreign language not required. *Entrance requirements:* TOEFL (minimum score of 570 required). Application deadline: 4/15 (priority date). Application fee: $20. *Tuition:* $1411 per year full-time, $52 per credit hour part-time for state residents; $3571 per year full-time, $132 per credit hour part-time for nonresidents. *Financial aid:* In 1990–91, $36,000 in aid awarded. 2 research assistantships (0 to first-year students), 6 teaching assistantships (2 to first-year students) were awarded; federal work-study, institutionally sponsored loans, and career-related internships or fieldwork also available. Aid available to part-time students. Financial aid application deadline: 4/15. *Faculty research:* Biophysics, solid-state physics, surface physics, many-body physics, medical physics. Total annual research budget: $170,000. • Dr. Charles A. Sawicki, Chair, 701-237-8974. Application contact: Dr. Craig Rottman, Graduate Advisory Committee, 701-237-8974.

Northeastern Illinois University, College of Arts and Sciences, Department of Physics, Chicago, IL 60625. Department awards MS. Evening/weekend programs available. Faculty: 6 full-time (0 women), 0 part-time. 4 applicants, 50% accepted. In 1990, 3 degrees awarded. *Degree requirements:* Thesis or alternative, comprehensive exam required, foreign language not required. *Application deadline:* rolling. *Application fee:* $0. *Expenses:* Tuition of $70 per credit hour for state residents; $210 per credit hour for nonresidents. Fees of $37 per credit hour. • Dr. Charles Nissim-Sabat, Chairperson, 312-583-4050 Ext. 746.

Northeastern University, Graduate School of Arts and Sciences, Department of Physics, Boston, MA 02115. Department awards MS, PhD. Part-time programs available. Faculty: 33 full-time (2 women), 0 part-time. Matriculated students: 62 full-time (11 women), 11 part-time (1 woman); includes 1 minority (Asian American), 48 foreign. Average age 24. 387 applicants, 4% accepted. In 1990, 20 master's, 13 doctorates awarded. Terminal master's awarded for partial completion of doctoral program. *Degree requirements:* For master's, thesis optional, foreign language not required; for doctorate, dissertation required, foreign language not required. *Entrance requirements:* TOEFL (minimum score of 550 required). Application deadline: 3/1. Application fee: $40. *Expenses:* Tuition of $10,260 per year full-time, $285 per quarter hour part-time. Fees of $293 per year full-time. *Financial aid:* In 1990–91, $532,000 in aid awarded. 1 research assistantship (0 to first-year students), 35 teaching assistantships (12 to first-year students), 4 tuition assistantships (3 to first-year students) were awarded; federal work-study also available. Financial aid application deadline: 3/1. *Faculty research:* High-energy theory and experimentation, astrophysics, biophysics, condensed-matter theory and experimentation. • Stephen Reucroft, Chairman, 617-437-2902. Application contact: Clive Perry, Graduate Coordinator, 617-437-2913.

See full description on page 601.

Northeast Louisiana University, College of Pure and Applied Sciences, Department in Physics, Monroe, LA 71209. Department awards MS. Matriculated students: 1 (woman) full-time, 1 part-time (0 women); includes 0 minority, 2 foreign. In 1990, 3 degrees awarded. *Degree requirements:* Thesis. *Application deadline:* 7/1. *Application fee:* $5. *Tuition:* $816 per semester for state residents; $1608 per semester for nonresidents. • Dr. John Myers Jr., Head, 318-342-1937.

Northern Illinois University, College of Liberal Arts and Sciences, Department of Physics, De Kalb, IL 60115. Department offers programs in applied physics (MS), basic physics (MS), physics teaching (MS). Part-time programs available. Faculty: 19 full-time (2 women), 3 part-time. Matriculated students: 30 full-time (6 women), 4 part-time (3 women); includes 14 foreign. Average age 27. 116 applicants, 31%

accepted. In 1990, 11 degrees awarded. *Degree requirements:* Thesis, comprehensive exam required, foreign language not required. *Entrance requirements:* GRE General Test, TOEFL. Application deadline: 6/1. Application fee: $0. *Tuition:* $1339 per semester full-time for state residents; $3163 per semester full-time for nonresidents. *Financial aid:* In 1990–91, 9 research assistantships, 19 teaching assistantships awarded; fellowships, federal work-study also available. Financial aid application deadline: 3/1; applicants required to submit FAF. *Total annual research budget:* $824,800. • Richard Preston, Chair, 815-753-6470.

Announcement: Experimental MS thesis opportunities in Mössbauer spectroscopy, magnetic susceptibility, crystallography, acoustics and vibrations, and solar energy conversion (on campus); and in high-energy physics, superconductivity, and surface and thin-film physics (at nearby Fermilab, Argonne, and UW Synchrotron Radiation Center). Theoretical research on liquid metals, cooperative phenomena, and elementary particles. Contact Richard Preston, Chair, 815-753-6470.

Northwestern University, College of Arts and Sciences, Department of Physics and Astronomy, Evanston, IL 60208. Department offers programs in astronomy (MS, PhD), astrophysics (MS, PhD), physics (MS, PhD). Faculty: 34 full-time (2 women), 0 part-time. Matriculated students: 87 full-time (17 women), 0 part-time; includes 2 minority (both Asian American), 49 foreign. 160 applicants, 56% accepted. In 1990, 6 master's, 13 doctorates awarded. Terminal master's awarded for partial completion of doctoral program. *Degree requirements:* For master's, thesis or alternative, comprehensive exam required, foreign language not required; for doctorate, dissertation, qualifying exam, oral exam required, foreign language not required. *Entrance requirements:* For master's, GRE General Test; for doctorate, GRE General Test, GRE Subject Test (physics). Application deadline: 8/30. Application fee: $40 ($45 for foreign students). *Tuition:* $4665 per quarter full-time, $1704 per course part-time. *Financial aid:* In 1990–91, $491,669 in aid awarded. 13 fellowships, 46 research assistantships, 25 teaching assistantships were awarded; federal work-study, institutionally sponsored loans, and career-related internships or fieldwork also available. Financial aid application deadline: 1/15; applicants required to submit GAPSFAS. *Faculty research:* Condensed matter, high-energy physics, nuclear physics, elementary particle and fields, astrophysics. Total annual research budget: $5.2-million. • Dr. William P. Halpern, Chairperson, 708-491-5468. Application contact: David Taylor, Assistant Chairperson, 708-491-2053.

See full description on page 603.

Oakland University, College of Arts and Sciences, Department of Physics, Rochester, MI 48309. Offerings include medical physics (PhD), physics (MS). *Degree requirements:* For master's, foreign language and thesis not required; for doctorate, dissertation. *Entrance requirements:* For master's, minimum GPA of 3.0 for unconditional admission; for doctorate, GRE Subject Test. Application deadline: 7/15. Application fee: $25. *Expenses:* Tuition of $122 per credit hour for state residents; $270 per credit hour for nonresidents. Fees of $170 per year. • Dr. Norman Tepley, Chair, 313-370-3410. Application contact: Abraham Liboff, 313-370-3410.

Ohio State University, College of Mathematical and Physical Sciences, Department of Physics, Columbus, OH 43210. Department awards MS, PhD. Faculty: 47. Matriculated students: 146 full-time (18 women), 3 part-time (0 women); includes 2 minority (both Asian American), 65 foreign. 128 applicants, 20% accepted. In 1990, 12 master's, 17 doctorates awarded. *Degree requirements:* For master's, thesis optional, foreign language not required; for doctorate, dissertation required, foreign language not required. *Entrance requirements:* GRE General Test, GRE Subject Test. Application deadline: 8/15 (applications processed on a rolling basis). Application fee: $0 ($25 for foreign students). *Tuition:* $1213 per quarter full-time, $364 per course part-time for state residents; $3143 per quarter full-time, $943 per course part-time for nonresidents. *Financial aid:* Fellowships, research assistantships, teaching assistantships, federal work-study, institutionally sponsored loans available. Aid available to part-time students. Financial aid applicants required to submit FAF. • Frank DeLucia, Chairman, 614-292-5713.

Ohio University, Graduate Studies, College of Arts and Sciences, Department of Physics, Athens, OH 45701. Department awards MS, PhD. Faculty: 22 full-time (1 woman), 0 part-time. Matriculated students: 58 full-time (7 women), 0 part-time; includes 2 minority (1 Asian American, 1 black American), 39 foreign. Average age 27. 168 applicants, 25% accepted. In 1990, 5 master's awarded; 2 doctorates awarded (100% found work related to degree). Terminal master's awarded for partial completion of doctoral program. *Degree requirements:* For master's, thesis or alternative required, foreign language not required; for doctorate, 1 foreign language, dissertation, written and oral comprehensive exams. *Application fee:* $25. *Tuition:* $1112 per quarter hour full-time, $138 per credit hour part-time for state residents; $2227 per quarter hour full-time, $277 per credit hour part-time for nonresidents. *Financial aid:* In 1990–91, 30 students received a total of $330,000 in aid awarded. 1 fellowship (to a first-year student), 3 research assistantships (1 to a first-year student), 26 teaching assistantships (9 to first-year students) were awarded; full tuition waivers, federal work-study, institutionally sponsored loans also available. Average monthly stipend for a graduate assistantship: $1000. Financial aid application deadline: 3/15; applicants required to submit FAF. *Faculty research:* Nuclear physics, condensed-matter physics, statistical physics. Total annual research budget: $985,000. • Dr. Louis E. Wright, Chair, 614-593-1713. Application contact: Dr. Darrell Huwe, Graduate Chair, 614-593-1730.

See full description on page 605.

Oklahoma State University, College of Arts and Sciences, Department of Physics, Stillwater, OK 74078. Department awards MS, PhD. Faculty: 17 full-time, 0 part-time. Matriculated students: 54 (5 women); includes 6 minority (2 Asian American, 3 Hispanic American, 1 Native American), 14 foreign. In 1990, 7 master's, 3 doctorates awarded. *Degree requirements:* For master's, foreign language not required; for doctorate, dissertation required, foreign language not required. *Entrance requirements:* TOEFL (minimum score of 550 required). Application deadline: 7/1 (priority date). Application fee: $15. *Tuition:* $63.75 per credit for state residents; $138.25 per credit for nonresidents. *Financial aid:* In 1990–91, 22 research assistantships, 27 teaching assistantships awarded; partial tuition waivers, federal work-study also available. Aid available to part-time students. Financial aid application deadline: 3/1; applicants required to submit GAPSFAS or FAF. • Dr. H. Larry Scott Jr., Head, 405-744-5796.

Old Dominion University, College of Sciences, Department of Physics, Norfolk, VA 23529. Department offers programs in applied physics (PhD), physics (MS). Faculty: 13 full-time (0 women), 3 part-time (0 women). Matriculated students: 28 full-time (2 women), 6 part-time (1 woman); includes 3 minority (1 Asian American, 2 black American), 10 foreign. Average age 25. 20 applicants, 80% accepted. In 1990, 6 master's, 2 doctorates awarded. Terminal master's awarded for partial completion of doctoral program. *Degree requirements:* For master's, thesis or alternative, comprehensive exam required, foreign language not required; for doctorate, dissertation, comprehensive exam required, foreign language not required. *Entrance requirements:* For master's, TOEFL, minimum GPA of 3.0 in field, BS in physics or related field; for doctorate, GRE General Test (minimum combined score of 1000 required), GRE Subject Test (minimum score of 600 required), TOEFL, minimum GPA of 3.0. Application deadline: 7/1 (applications processed on a rolling basis). Application fee: $20. *Expenses:* Tuition of $148 per credit hour for state residents; $375 per credit hour for nonresidents. Fees of $64 per year full-time. *Financial aid:* In 1990–91, $28,800 in aid awarded. 3 research assistantships, 1 teaching assistantship, 1 tuition grant were awarded; fellowships, federal work-study, and career-related internships or fieldwork also available. *Faculty research:* Atomic physics, nuclear physics, gamma ray optics, condensed matter physics, radiation damage, plasma physics, mossbauer physics. Total annual research budget: $700,000. • Dr. James L. Cox Jr., Chairman, 804-683-3468. Application contact: Dr. Govind Khandelwal, Coordinator, 804-683-3468.

Announcement: Active research areas: atomic and molecular, condensed-matter, laser, Mössbauer, nuclear, radiation, and plasma physics; gamma-ray optics. Facilities at ODU, as well as at nearby CEBAF and NASA Langley Research Center. Excellent University computing facilities. Grants from NSF, NASA, and NRL; fellowships and assistantships, with tuition waiver, available.

Oregon State University, Graduate School, College of Science, Department of Physics, Corvallis, OR 97331. Department awards MA, MAIS, MS, PhD. Part-time programs available. Faculty: 18 full-time (2 women), 0 part-time. Matriculated students: 51 full-time (10 women), 1 (woman) part-time; includes 2 minority (1 Asian American, 1 black American), 25 foreign. Average age 27. In 1990, 17 master's awarded; 2 doctorates awarded (100% entered university research/teaching). Terminal master's awarded for partial completion of doctoral program. *Degree requirements:* For master's, qualifying exams required, foreign language and thesis not required; for doctorate, dissertation, qualifying exams required, foreign language not required. *Entrance requirements:* TOEFL (minimum score of 550 required), minimum GPA of 3.0 in last 90 hours. Application deadline: 3/1. Application fee: $40. *Tuition:* $1140 per trimester full-time, $449 per year part-time for state residents; $1816 per trimester full-time, $674 per year part-time for nonresidents. *Financial aid:* In 1990–91, $342,019 in aid awarded. 9 fellowships (5 to first-year students), 32 research assistantships (5 to first-year students), 30 teaching assistantships (15 to first-year students) were awarded; federal work-study, institutionally sponsored loans also available. Aid available to part-time students. Financial aid application deadline: 2/1; applicants required to submit FAF. *Faculty research:* Atomic nuclear optics and solid-state physics, computational physics, nonlinear optics, laser spectroscopy, exotic matter. Total annual research budget: $592,285. • Dr. Kenneth S. Krane, Chair, 503-737-4631.

Announcement: Advanced degrees offered in experimental or theoretical atomic, optical, solid-state, nuclear, and particle physics. Facilities include nonlinear optics, atomic beams, diode laser, NMR, superconductivity, and thin-film laboratories. Nuclear and particle experiments performed at LAMPF, TRIUMF, and Oak Ridge. Supercomputers, superminis, and workstations support theoretical studies.

See full description on page 607.

Pennsylvania State University University Park Campus, College of Science, Department of Physics, University Park, PA 16802. Department awards M Ed, MS, D Ed, PhD. Faculty: 35. Matriculated students: 70 full-time (14 women), 8 part-time (0 women). In 1990, 8 master's, 18 doctorates awarded. *Entrance requirements:* GRE General Test. Application fee: $35. *Tuition:* $203 per credit for state residents; $403 per credit for nonresidents. • Dr. Howard Grotch, Head, 814-865-7533.

Announcement: Experimental and theoretical research opportunities in acoustics, atomic-molecular-optical physics, condensed-matter physics, elementary-particle physics, low-temperature physics, and surface physics. In 1990–91, teaching assistantships were set at $9320, plus tuition waiver. Research assistantships also available with comparable stipend. Summer support available. All students are automatically considered for additional supplementary fellowship ranging from $2000–$3500.

See full description on page 609.

Pittsburg State University, College of Arts and Sciences, Department of Physics, Pittsburg, KS 66762. Offerings include physics (MS), professional physics (MS). Department faculty: 4 full-time (0 women), 0 part-time. *Degree requirements:* Thesis or alternative required, foreign language not required. *Application fee:* $35. *Tuition:* $1670 per year full-time, $54 per credit hour part-time for state residents; $4108 per year full-time, $125 per credit hour part-time for nonresidents. • Dr. Orville Brill, Chairman, 316-231-7000 Ext. 4390.

Polytechnic University, Brooklyn Campus, Division of Arts and Sciences, Department of Physics, 333 Jay Street, Brooklyn, NY 11201. Department awards MS, PhD. Evening/weekend programs available. *Tuition:* $6000 per semester.

Portland State University, College of Liberal Arts and Sciences, Department of Physics, Portland, OR 97207. Department awards MA, MS, PhD. PhD offered in conjunction with Interdisciplinary Program in Environmental Sciences and Resources. Part-time programs available. Faculty: 12 full-time (0 women), 1 part-time (0 women). Matriculated students: 7 full-time (0 women), 5 part-time (2 women); includes 0 minority, 4 foreign. Average age 28. 6 applicants, 100% accepted. In 1990, 1 master's awarded. *Degree requirements:* For master's, variable foreign language requirement, thesis, oral exam. *Entrance requirements:* For master's, TOEFL (minimum score of 550 required), minimum GPA of 3.0 in upper-division course work or 2.75 overall. Application deadline: 7/12. Application fee: $40. *Tuition:* $1151 per trimester for state residents; $1827 per trimester for nonresidents. *Financial aid:* In 1990–91, 2 research assistantships (0 to first-year students), 4 teaching assistantships (1 to a first-year student) awarded; federal work-study, institutionally sponsored loans, and career-related internships or fieldwork also available. Aid available to part-time students. Average monthly stipend for a graduate assistantship: $725. Financial aid application deadline: 3/1. *Faculty research:* Statistical physics, membrane biophysics, low-temperature physics, electron microscopy, materials science. Total annual research budget: $400,000. • Dr. Mark Gurevitch, Acting Head, 503-725-3812.

Portland State University, College of Liberal Arts and Sciences, Interdisciplinary Program in Environmental Sciences and Resources, Portland, OR 97207. Offers environmental sciences/biology (PhD), environmental sciences/chemistry (PhD), environmental sciences/civil engineering (PhD), environmental sciences/geology (PhD), environmental sciences/physics (PhD). Part-time programs available. Faculty: 1 (woman) full-time, 0 part-time. Matriculated students: 24 full-time (7 women), 8 part-time (2 women); includes 0 minority, 11 foreign. Average age 35. 13 applicants, 100% accepted. In 1990, 3 degrees awarded. *Degree requirements:* Variable foreign language requirement, computer language, dissertation, qualifying exam, oral exam. *Entrance requirements:* TOEFL (minimum score of 550 required), minimum GPA of

SECTION 7: PHYSICS

Directory: Physics

3.0 in upper-division course work or 2.75 overall. Application deadline: 7/12. Application fee: $40. *Tuition:* $1151 per trimester for state residents; $1827 per trimester for nonresidents. *Financial aid:* In 1990–91, 1 research assistantship (to a first-year student) awarded; federal work-study, institutionally sponsored loans also available. Aid available to part-time students. Average monthly stipend for a graduate assistantship: $578. Financial aid application deadline: 1/31. *Faculty research:* Aquatic biology and chemistry, atmospheric pollution, natural resources, ecology, biophysics. Total annual research budget: $1.5-million. • Dr. Pavel Smejtek, Director, 503-725-4980.

Announcement: The PhD in environmental sciences and resources/physics is part of a multidisciplinary program sponsored by the Departments of Biology, Chemistry, Civil Engineering, Geology, and Physics. Physics thesis research is concerned with basic problems in physics that are environmentally relevant, including low-temperature physics, finite-time thermodynamics, membrane biophysics, electron optics, environmental sensors, and materials science. Contact Dr. Pavel Smejtek, Program Director, 503-725-4980.

Princeton University, Department of Physics, Princeton, NJ 08544. Department offers programs in applied and computational mathematics (PhD), mathematical physics (PhD), physics (PhD), physics and chemical physics (PhD). Faculty: 44 full-time (1 woman), 0 part-time. Matriculated students: 116 full-time (10 women), 0 part-time; includes 7 minority (all Asian American), 37 foreign. 355 applicants, 18% accepted. In 1990, 14 doctorates awarded (71% entered university research/teaching, 29% found other work related to degree). Terminal master's awarded for partial completion of doctoral program. *Degree requirements:* For doctorate, 1 foreign language, dissertation. *Entrance requirements:* For doctorate, GRE General Test, GRE Subject Test. Application deadline: 1/8. Application fee: $45 ($50 for foreign students). *Tuition:* $16,670 per year. *Financial aid:* Fellowships, research assistantships, teaching assistantships, federal work-study, institutionally sponsored loans available. Financial aid application deadline: 1/8. • Kirk McDonald, Director of Graduate Studies, 609-258-4403. Application contact: Michele Spreen, Director of Graduate Admissions, 609-258-3034.

Purdue University, School of Science, Department of Physics, West Lafayette, IN 47907. Department awards MS, PhD. Part-time programs available. Faculty: 58 full-time (0 women), 0 part-time. Matriculated students: 110 full-time (18 women), 25 part-time (6 women); includes 7 minority (3 Asian American, 1 black American, 3 Hispanic American), 55 foreign. Average age 25. 147 applicants, 44% accepted. In 1990, 30 master's, 12 doctorates awarded. Terminal master's awarded for partial completion of doctoral program. *Degree requirements:* For master's, foreign language and thesis not required; for doctorate, dissertation required, foreign language not required. *Entrance requirements:* TOEFL (minimum score of 550 required). Application deadline: 3/30. Application fee: $25. *Tuition:* $1162 per semester full-time, $83.50 per credit hour part-time for state residents; $3720 per semester full-time, $244.50 per credit hour part-time for nonresidents. *Financial aid:* In 1990–91, $1-million in aid awarded. 18 fellowships (8 to first-year students), 62 research assistantships (0 to first-year students), 88 teaching assistantships (25 to first-year students) were awarded; institutionally sponsored loans also available. Financial aid application deadline: 3/30. *Faculty research:* Solid-state, elementary particle, and nuclear physics; biological physics; acoustics; astrophysics. Total annual research budget: $4-million. • Dr. Arnold Tubis, Head, 317-494-3000.

Queens College of the City University of New York, Mathematics and Natural Sciences Division, Department of Physics, 65-30 Kissena Boulevard, Flushing, NY 11367. Department awards MA, MS Ed. MS Ed awarded through the School of Education. Part-time and evening/weekend programs available. Matriculated students: 4 full-time (0 women), 2 part-time (0 women). 21 applicants, 33% accepted. In 1990, 5 degrees awarded. *Degree requirements:* Comprehensive exam required, foreign language and thesis not required. *Entrance requirements:* TOEFL (minimum score of 550 required), previous course work in calculus, minimum GPA of 3.0. Application deadline: 4/1 (applications processed on a rolling basis). Application fee: $35. *Tuition:* $2700 per year full-time, $45 per credit part-time for state residents; $4700 per year full-time, $199 per credit part-time for nonresidents. *Financial aid:* Fellowships, partial tuition waivers, federal work-study, institutionally sponsored loans, and career-related internships or fieldwork available. Aid available to part-time students. Financial aid application deadline: 4/1. *Faculty research:* Solid-state physics, low temperature physics, elementary particles and fields. • Dr. Kenneth Rafanelli, Chairperson, 718-997-3350. Application contact: Dr. Marion J. Dickey, Graduate Adviser, 718-997-3350.

Queen's University at Kingston, Faculty of Arts and Sciences, Department of Physics, Kingston, ON K7L 3N6, Canada. Department awards MA, M Sc, M Sc Eng, PhD. MA offered jointly with Trent University. Part-time programs available. Matriculated students: 59 full-time, 4 part-time; includes 5 foreign. In 1990, 23 master's, 7 doctorates awarded. *Degree requirements:* For master's, thesis required, foreign language not required; for doctorate, dissertation, comprehensive exam required, foreign language not required. *Entrance requirements:* TOEFL (minimum score of 550 required). Application deadline: 2/28 (priority date). Application fee: $35. *Tuition:* $2861 per year full-time, $426 per trimester part-time for Canadian residents; $10,613 per year full-time, $4998 per trimester part-time for nonresidents. *Financial aid:* Fellowships, research assistantships, teaching assistantships, institutionally sponsored loans available. Financial aid application deadline: 3/1. *Faculty research:* Experimental nuclear physics, experimental solid-state physics, applied solid-state and nuclear physics, theoretical physics, astronomy and astrophysics. • Dr. D. R. Taylor, Head, 613-545-2706. Application contact: Dr. P. J. Scanlon, Graduate Coordinator, 613-545-2707.

Rensselaer Polytechnic Institute, School of Science, Department of Physics, Troy, NY 12180. Department offers programs in applied physics (MS, PhD), physics (MS, PhD). Faculty: 28 full-time (1 woman), 7 part-time (2 women). Matriculated students: 94 full-time (15 women), 2 part-time (0 women); includes 11 minority (9 Asian American, 2 Hispanic American), 43 foreign. 169 applicants, 58% accepted. In 1990, 17 master's, 9 doctorates awarded. *Degree requirements:* Thesis/dissertation required, foreign language not required. *Entrance requirements:* GRE General Test, GRE Subject Test, TOEFL (minimum score of 575 required). Application deadline: 2/1. Application fee: $30. *Expenses:* Tuition of $455 per credit hour. Fees of $195.57 per semester. *Financial aid:* In 1990–91, 73 students received a total of $1-million in aid awarded. 4 fellowships (0 to first-year students), 33 research assistantships (1 to a first-year student), 36 teaching assistantships (14 to first-year students) were awarded; career-related internships or fieldwork also available. *Faculty research:* Astrophysics, educational development, biophysics, optics, condensed matter physics. Total annual research budget: $1.6-million. • Timothy Hayes, Chair, 518-276-6419.

Announcement: Areas include astrophysics, condensed matter, surface phenomena, optical physics and optical properties of materials, biophysics, intermediate-energy physics, and physics education. Interdisciplinary research opportunities in materials physics, optoelectronics, and electronic-device physics are pursued in Rensselaer centers or with other departments. Astrophysics and nuclear groups use observatories and accelerator facilities worldwide. See full description in the Graduate and Professional Programs (An Overview) volume of this series.

Rice University, Wiess School of Natural Sciences, Department of Physics, Houston, TX 77251. Department awards MA, PhD. Faculty: 17 full-time (1 woman), 0 part-time. Matriculated students: 46 full-time (0 women), 0 part-time; includes 2 minority (both Asian American), 24 foreign. *Degree requirements:* Thesis/dissertation required, foreign language not required. *Entrance requirements:* For master's, GRE General Test, TOEFL (minimum score 0f 550 required), minimum GPA of 3.0; for doctorate, GRE General Test (score in 70th percentile or higher required), minimum GPA of 3.0. Application deadline: 3/1. Application fee: $0. *Expenses:* Tuition of $8300 per year full-time, $400 per credit hour part-time. Fees of $167 per year. *Financial aid:* Fellowships, research assistantships, full and partial tuition waivers available. Financial aid applicants required to submit FAF. *Faculty research:* Atomic, solid-state, and molecular physics; biophysics; medium- and high-energy physics. • Billy E. Bonner, Chairman, 713-527-4938.

See full description on page 611.

Rice University, Wiess School of Natural Sciences, Department of Space Physics and Astronomy, Houston, TX 77251. Department offers program in space physics and astronomy (MS, PhD). Faculty: 14 full-time, 0 part-time. Matriculated students: 41 full-time (8 women), 0 part-time; includes 2 minority (both Asian American), 8 foreign. *Degree requirements:* Thesis/dissertation required, foreign language not required. *Entrance requirements:* For master's, GRE General Test, TOEFL (minimum score 0f 550 required), minimum GPA of 3.0; for doctorate, GRE General Test (score in 70th percentile or higher required), minimum GPA of 3.0. Application deadline: 3/1. Application fee: $0. *Expenses:* Tuition of $8300 per year full-time, $400 per credit hour part-time. Fees of $167 per year. *Financial aid:* Fellowships, research assistantships, full and partial tuition waivers available. Financial aid applicants required to submit FAF. *Faculty research:* Magnetospheric physics, planetary atmospheres, astrophysics. • Alexander J. Dessler, Chairman, 713-527-4939.

Rutgers, The State University of New Jersey, New Brunswick, Program in Physics, New Brunswick, NJ 08903. Offers astrophysics (MS, PhD), condensed matter physics (MS, PhD), elementary particle physics (MS, PhD), intermediate energy nuclear physics (MS, PhD), nuclear physics (MS, PhD), physics (MST), theoretical physics (MS, PhD). Part-time programs available. Faculty: 58 full-time (4 women), 0 part-time. Matriculated students: 95 full-time (9 women), 9 part-time (2 women); includes 6 minority (2 Asian American, 2 black American, 2 Hispanic American), 61 foreign. Average age 25. 225 applicants, 14% accepted. In 1990, 8 master's, 4 doctorates awarded. Terminal master's awarded for partial completion of doctoral program. *Degree requirements:* For master's, thesis optional, foreign language not required; for doctorate, dissertation required, foreign language not required. *Entrance requirements:* GRE General Test, GRE Subject Test. Application deadline: 6/1. Application fee: $35. *Expenses:* Tuition of $4432 per year full-time, $183 per credit part-time for state residents; $6496 per year full-time, $270 per credit part-time for nonresidents. Fees of $458 per year full-time, $117 per year part-time. *Financial aid:* In 1990–91, 90 students received aid. 17 fellowships (4 to first-year students), 30 research assistantships (3 to first-year students), 37 teaching assistantships (4 to first-year students) awarded. Average monthly stipend for a graduate assistantship: $1000. Financial aid application deadline: 3/15. *Total annual research budget:* $3-million. • David Harrington, Director, 908-932-3892.

See full description on page 613.

St. Bonaventure University, School of Arts and Sciences, Department of Physics, St. Bonaventure, NY 14778. Department awards MS. Part-time programs available. Faculty: 5 full-time (0 women), 0 part-time. Matriculated students: 1 full-time (0 women), 1 part-time (0 women); includes 0 minority, 0 foreign. Average age 26. 4 applicants, 100% accepted. In 1990, 3 degrees awarded. *Degree requirements:* 1 foreign language (computer language can substitute), written and oral comprehensive exams required, thesis optional. *Entrance requirements:* GRE General Test, GRE Subject Test, TOEFL (minimum score of 550 required). Application deadline: rolling. Application fee: $25. *Tuition:* $275 per credit. *Financial aid:* In 1990–91, 2 students received a total of $22,548 in aid awarded. 2 teaching assistantships (1 to a first-year student) were awarded; federal work-study also available. Average monthly stipend for a graduate assistantship: $555. Financial aid applicants required to submit FAF. *Faculty research:* Elementary particle physics; ultrasonics; applied energy research; chemical, computational, and solid-state physics. • Dr. John Neeson, Head, 716-375-2516.

St. Francis Xavier University, Program in Physics, Antigonish, NS B2G 1C0, Canada. Program awards M Sc. Faculty: 7 full-time. Matriculated students: 0. 2 applicants, 0% accepted. In 1990, 1 degree awarded. *Degree requirements:* Thesis required, foreign language not required. *Application deadline:* 8/1 (priority date, applications processed on a rolling basis). *Application fee:* $25. *Tuition:* $400 per course. • Dr. Y. N. Joshi, Chair, 902-867-3977.

Sam Houston State University, College of Arts and Sciences, Department of Physics, Huntsville, TX 77341. Department awards MS. Faculty: 8 full-time (0 women), 2 part-time (0 women). Matriculated students: 21 full-time (5 women), 0 part-time; includes 18 foreign. Average age 23. 90 applicants, 23% accepted. In 1990, 4 degrees awarded (100% continued full-time study). *Degree requirements:* 1 foreign language (computer language can substitute), thesis. *Entrance requirements:* GRE General Test (minimum combined score of 800 required), TOEFL (minimum score of 550 required). Application deadline: 3/15. Application fee: $0. *Expenses:* Tuition of $432 per year full-time, $216 per year part-time for state residents; $2880 per year full-time, $1440 per year part-time for nonresidents. Fees of $364 per year full-time, $220 per year part-time. *Financial aid:* In 1990–91, 23 students received a total of $62,400 in aid awarded. 8 research assistantships (3 to first-year students), 7 teaching assistantships (5 to first-year students) were awarded; institutionally sponsored loans also available. Average monthly stipend for a graduate assistantship: $600. *Faculty research:* Solid-state physics, optical properties of semiconductors, EPR of semiconductors, meteorites, ion sources. Total annual research budget: $150,000. • Dr. Russell L. Palma, Chair, 409-294-1600.

San Diego State University, College of Sciences, Department of Physics, San Diego, CA 92182. Offerings include physics (MA, MS). Department faculty: 15 full-time, 3 part-time. *Degree requirements:* Variable foreign language requirement, oral exam or thesis. *Entrance requirements:* GRE General Test (minimum combined score of 950 required). Application deadline: 8/1 (applications processed on a rolling basis).

Application fee: $55. *Expenses:* Tuition of $0 for state residents; $189 per unit for nonresidents. Fees of $1974 per year full-time, $692 per year part-time for state residents; $1074 per year full-time, $692 per year part-time for nonresidents. • Dr. Roger Lilly, Chair, 619-594-6240.

San Francisco State University, School of Science, Department of Physics and Astronomy, San Francisco, CA 94132. Department offers program in physics and astrophysics (MS). *Degree requirements:* Thesis or alternative. *Application deadline:* 11/30 (priority date, applications processed on a rolling basis). *Application fee:* $55. *Expenses:* Tuition of $0 for state residents; $250 per unit for nonresidents. Fees of $950 per year full-time, $350 per semester part-time for state residents; $1050 per year full-time, $350 per semester part-time for nonresidents. *Financial aid:* Research assistantships, teaching assistantships available. Financial aid application deadline: 3/1. *Faculty research:* Quark search, thin-films, dark matter detection, search for planetary systems, low temperature. • Dr. Gerald Fisher, Chair, 415-338-1659. Application contact: Dr. Oliver Johns, Graduate Coordinator, 415-338-1691.

Announcement: A hands-on, faculty-supervised research experience (the department focus) supplements standard MS curriculum and allows students to successfully compete for entrance into PhD programs or to obtain high-level industrial positions. Most faculty have joint appointments or affiliations with neighboring institutions, including LBL, UCB, Stanford, SLAC, Lick Observatory. Ample grant support provided by NSF, DOE, and others.

San Jose State University, School of Science, Department of Physics, San Jose, CA 95192. Department awards MS. Part-time and evening/weekend programs available. Faculty: 18 full-time (1 woman), 0 part-time. Matriculated students: 9 full-time (2 women), 56 part-time (7 women); includes 21 minority (14 Asian American, 1 black American, 6 Hispanic American), 8 foreign. Average age 28. 52 applicants, 75% accepted. In 1990, 14 degrees awarded. *Degree requirements:* Thesis optional, foreign language not required. *Entrance requirements:* GRE. Application deadline: 6/1 (applications processed on a rolling basis). Application fee: $55. *Expenses:* Tuition of $0 for state residents; $246 per unit for nonresidents. Fees of $592 per year. *Financial aid:* In 1990–91, 7 teaching assistantships (3 to first-year students) awarded; federal work-study, institutionally sponsored loans, and career-related internships or fieldwork also available. Aid available to part-time students. Financial aid application deadline: 3/1; applicants required to submit FAF. *Faculty research:* Astrophysics, atmospheric physics, elementary particles, dislocation theory, general relativity. • Dr. Donald Strandburg, Chair, 408-924-5210. Application contact: Dr. Franklin Muirhead, Graduate Adviser, 408-924-5258.

Simon Fraser University, Faculty of Science, Department of Physics, Burnaby, BC V5A 1S6, Canada. Department offers programs in biophysics (M Sc, PhD), chemical physics (M Sc, PhD), physics (M Sc, PhD). Faculty: 22 full-time (0 women), 0 part-time. Matriculated students: 49 full-time (9 women), 0 part-time. In 1990, 7 master's, 2 doctorates awarded. *Degree requirements:* 1 foreign language, thesis/dissertation. *Application fee:* $0. *Expenses:* Tuition of $612 per trimester full-time, $306 per trimester part-time. Fees of $68 per trimester full-time, $34 per trimester part-time. *Financial aid:* In 1990–91, 16 fellowships awarded; research assistantships, teaching assistantships also available. *Faculty research:* Solid-state physics, magnetism, energy research, superconductivity. • M. Plischke, Chairman, 604-291-3154.

South Dakota School of Mines and Technology, Department of Physics, Rapid City, SD 57701. Department awards MS. Part-time programs available. Faculty: 6 full-time (1 woman), 0 part-time. Matriculated students: 5 full-time (1 woman), 4 part-time (2 women); includes 0 minority, 8 foreign. Average age 28. In 1990, 2 degrees awarded. *Degree requirements:* Thesis optional, foreign language not required. *Entrance requirements:* TOEFL (minimum score of 520 required). Application deadline: 7/15. Application fee: $15 ($45 for foreign students). *Expenses:* Tuition of $1474 per year full-time, $61.40 per credit hour part-time for state residents; $2917 per year full-time, $122 per credit hour part-time for nonresidents. Fees of $871 per year full-time, $132 per semester (minimum) part-time. *Financial aid:* In 1990–91, 9 students received a total of $20,537 in aid awarded. 0 fellowships, 3 research assistantships, 12 teaching assistantships were awarded; federal work-study, institutionally sponsored loans also available. Aid available to part-time students. Average monthly stipend for a graduate assistantship: $300. Financial aid application deadline: 5/15. *Faculty research:* Solid-state physics, transport properties, applied physics, interfacial physics, high temperature superconductors. Total annual research budget: $75,213. • Dr. T. Ashworth, Head, 605-394-2361. Application contact: Dr. Briant L. Davis, Dean, Graduate Division, 605-394-2493.

South Dakota State University, College of Engineering, Department of Physics, Brookings, SD 57007. Department awards MS, MST. *Degree requirements:* Thesis required, foreign language not required. *Tuition:* $553 per semester full-time, $61 per credit part-time for state residents; $1094 per semester full-time, $122 per credit part-time for nonresidents. *Faculty research:* Neutron activation analysis, remote sensing, electron spin resonance.

Southern Illinois University at Carbondale, College of Science, Department of Physics, Carbondale, IL 62901. Department awards MS. Faculty: 14 full-time (0 women), 0 part-time. Matriculated students: 9 full-time (2 women), 0 part-time; includes 0 minority, 3 foreign. 33 applicants, 21% accepted. In 1990, 3 degrees awarded. *Degree requirements:* 1 foreign language (computer language can substitute), thesis. *Entrance requirements:* TOEFL (minimum score of 550 required), minimum GPA of 2.7. Application deadline: rolling. Application fee: $0. *Expenses:* Tuition of $1638 per year full-time, $204.75 per semester hour part-time for state residents; $4914 per year full-time, $614.25 per semester hour part-time for nonresidents. Fees of $700 per year full-time, $216 per year part-time. *Financial aid:* In 1990–91, $80,652 in aid awarded. 0 fellowships, 2 research assistantships (1 to a first-year student), 7 teaching assistantships (3 to first-year students) were awarded; full tuition waivers, federal work-study, institutionally sponsored loans, and career-related internships or fieldwork also available. Financial aid application deadline: 2/15. *Faculty research:* Atomic, molecular, nuclear, and mathematical physics; statistical mechanics; solid-state and low-temperature physics; material science; applied physics. Total annual research budget: $731,886. • Frank C. Sanders, Chairperson, 618-453-2643.

Southern Illinois University at Edwardsville, School of Sciences, Department of Physics, Edwardsville, IL 62026. Department awards MS. Faculty: 10 full-time (0 women), 1 part-time (0 women). Matriculated students: 5 full-time (1 woman), 8 part-time (0 women); includes 0 minority, 7 foreign. In 1990, 5 degrees awarded. *Degree requirements:* Final exam required, foreign language not required. *Application fee:* $0. *Expenses:* Tuition of $1566 per year full-time, $43.40 per credit hour part-time for state residents; $4698 per year full-time, $130.20 per credit hour part-time for nonresidents. Fees of $291.75 per year full-time, $27.35 per year (minimum) part-time. *Financial aid:* In 1990–91, 2 research assistantships, 8 teaching assistantships, 2 assistantships awarded; fellowships, federal work-study, institutionally sponsored loans also available. Aid available to part-time students. • Dr. P. N. Swamy, Chairman, 618-692-2472.

Southwest Texas State University, School of Science, Department of Physics, San Marcos, TX 78666. Department awards MA, MS. Part-time programs available. Faculty: 5 full-time (0 women), 0 part-time. Matriculated students: 3 full-time (0 women), 6 part-time (0 women); includes 0 minority, 1 foreign. Average age 33. 4 applicants, 25% accepted. In 1990, 1 degree awarded. *Degree requirements:* Thesis (for some programs), comprehensive exam required, foreign language not required. *Entrance requirements:* GRE General Test (minimum combined score of 900 required), TOEFL (minimum score of 550 required), minimum GPA of 2.75 in last 60 hours. Application deadline: 7/15 (priority date, applications processed on a rolling basis). Application fee: $0 ($50 for foreign students). *Expenses:* Tuition of $180 per semester (minimum) full-time, $100 per semester (minimum) part-time for state residents; $1152 per semester (minimum) full-time, $128 per semester (minimum) part-time for nonresidents. Fees of $217 per semester (minimum) full-time, $73 per semester (minimum) part-time. *Financial aid:* Teaching assistantships and career-related internships or fieldwork available. Aid available to part-time students. Financial aid application deadline: 3/1. *Faculty research:* High temperature super conductors, historical astronomy, general relativity. • Dr. James R. Crawford, Chair, 512-245-2131. Application contact: Dr. William R. Jackson, Graduate Adviser, 512-245-2131.

Stanford University, School of Humanities and Sciences, Department of Physics, Stanford, CA 94305. Department awards PhD. Faculty: 51 (1 woman). Matriculated students: 125 full-time (16 women), 0 part-time; includes 26 minority (12 Asian American, 6 black American, 7 Hispanic American, 1 Native American), 33 foreign. Average age 25. 325 applicants, 18% accepted. In 1990, 26 doctorates awarded. Terminal master's awarded for partial completion of doctoral program. *Degree requirements:* For doctorate, dissertation, oral exam required, foreign language not required. *Entrance requirements:* For doctorate, GRE General Test, GRE Subject Test, TOEFL. Application deadline: 1/1. Application fee: $55. *Expenses:* Tuition of $15,102 per year. Fees of $28 per quarter. *Financial aid:* Fellowships, research assistantships, teaching assistantships, institutionally sponsored loans available. Financial aid application deadline: 1/1; applicants required to submit GAPSFAS. • Application contact: Graduate Administrator, 415-723-4346.

State University of New York at Albany, College of Science and Mathematics, Department of Physics, Albany, NY 12222. Department awards MS, PhD. Evening/weekend programs available. Faculty: 21 full-time (2 women), 0 part-time. Matriculated students: 72 full-time (22 women), 8 part-time (2 women); includes 6 minority (3 black American, 3 Hispanic American), 52 foreign. Average age 27. 93 applicants, 40% accepted. In 1990, 8 master's, 6 doctorates awarded. *Degree requirements:* For master's, 1 foreign language (computer language can substitute); for doctorate, 1 foreign language (computer language can substitute), dissertation. *Application deadline:* 6/1. *Application fee:* $35. *Expenses:* Tuition of $2450 per year full-time, $103 per credit part-time for state residents; $5765 per year full-time, $243 per credit part-time for nonresidents. Fees of $25 per year full-time, $0.85 per credit part-time. *Financial aid:* In 1990–91, $8000 in aid awarded. 1 fellowship, 23 teaching assistantships (7 to first-year students) were awarded; research assistantships, minority assistantships also available. Financial aid application deadline: 3/15. *Faculty research:* Condensed matter physics, high energy physics, applied physics, electronic materials. Total annual research budget: $1-million. • William A. Lanford, Chairman, 518-442-4500.

State University of New York at Binghamton, School of Arts and Sciences, Department of Physics, Applied Physics, and Astronomy, Binghamton, NY 13902-6000. Department offers programs in applied physics (MS), physics (MA, MS). Matriculated students: 9 full-time (1 woman), 7 part-time (1 woman); includes 0 minority, 5 foreign. Average age 27. 50 applicants, 72% accepted. *Degree requirements:* Thesis or alternative. *Entrance requirements:* GRE General Test, GRE Subject Test, TOEFL. Application deadline: 4/15 (priority date). Application fee: $35. *Expenses:* Tuition of $2450 per year full-time, $102.50 per credit part-time for state residents; $5766 per year full-time, $242.50 per credit part-time for nonresidents. Fees of $77 per year full-time, $27.85 per semester (minimum) part-time. *Financial aid:* In 1990–91, 6 students received a total of $44,500 in aid awarded. 0 fellowships, 1 research assistantship (to a first-year student), 2 teaching assistantships, 3 graduate assistantships (1 to a first-year student) were awarded; federal work-study, institutionally sponsored loans, and career-related internships or fieldwork also available. Aid available to part-time students. Average monthly stipend for a graduate assistantship: $742. Financial aid application deadline: 2/15. • Dr. Robert L. Pompi, Chairperson, 607-777-2217.

State University of New York at Buffalo, Graduate School, Faculty of Natural Sciences and Mathematics, Department of Physics, Buffalo, NY 14260. Department awards MA, PhD. Part-time programs available. Faculty: 25 full-time (0 women), 4 part-time (0 women). Matriculated students: 89 full-time (11 women), 4 part-time (2 women); includes 2 minority (both Asian American), 71 foreign. Average age 27. 166 applicants, 24% accepted. In 1990, 13 master's awarded (23% found work related to degree, 77% continued full-time study); 3 doctorates awarded (66% entered university research/teaching, 34% found other work related to degree). Terminal master's awarded for partial completion of doctoral program. *Degree requirements:* For master's, thesis or project required, foreign language not required; for doctorate, dissertation, departmental qualifying exams required, foreign language not required. *Entrance requirements:* GRE Subject Test, TOEFL (minimum score of 550 required). Application deadline: 5/1 (applications processed on a rolling basis). Application fee: $35. *Expenses:* Tuition of $1600 per semester full-time, $134 per hour part-time for state residents; $3258 per semester full-time, $274 per hour part-time for nonresidents. Fees of $137 per semester full-time, $115 per semester (minimum) part-time. *Financial aid:* In 1990–91, 60 students received a total of $550,863 in aid awarded. 2 fellowships (0 to first-year students), 15 research assistantships (1 to a first-year student), 36 teaching assistantships (13 to first-year students), 1 graduate assistantship (0 to first-year students) were awarded; federal work-study, institutionally sponsored loans also available. Average monthly stipend for a graduate assistantship: $820. Financial aid application deadline: 3/1; applicants required to submit FAF. *Faculty research:* Condensed-matter physics, laser physics (experimental), atomic and nuclear physics (theoretical), solid-state physics (experimental). Total annual research budget: $1.159-million. • Dr. Bruce McCombe, Chairman, 716-636-2017. Application contact: Dr. Moit-Lal Rustgi, Director of Graduate Studies, 716-636-2536.

See full description on page 615.

State University of New York at Stony Brook, College of Arts and Sciences, Division of Physical Sciences and Mathematics, Department of Physics, Stony Brook, NY 11794. Department awards MA, M Phil, PhD. Faculty: 59 full-time, 6 part-time. Matriculated students: 168 full-time (24 women), 16 part-time (3 women); includes 12 minority (10 Asian American, 1 black American, 1 Hispanic American), 100 foreign. 534 applicants, 20% accepted. In 1990, 25 master's, 21 doctorates awarded. *Degree requirements:* For master's, foreign language and thesis not

Directory: Physics

required; for doctorate, 1 foreign language, dissertation. *Entrance requirements:* GRE General Test, TOEFL. Application deadline: 2/1. Application fee: $35. *Expenses:* Tuition of $2450 per year full-time, $103 per credit part-time for state residents; $5766 per year full-time, $243 per credit part-time for nonresidents. Fees of $151 per year full-time, $10.45 per year (minimum) part-time. *Financial aid:* In 1990–91, 14 fellowships, 92 research assistantships, 62 teaching assistantships awarded. *Faculty research:* Theoretical, experimental, high-energy, nuclear, and solid-state physics. Total annual research budget: $10.95-million. • Dr. Gene Sprouse, Chairman, 516-632-8110.

See full description on page 617.

State University of New York College at Buffalo, Faculty of Natural and Social Sciences, Department of Physics, Buffalo, NY 14222. Department awards MA, MS Ed. Part-time and evening/weekend programs available. Faculty: 5 full-time (0 women), 1 part-time (0 women). Matriculated students: 0. 0 applicants. In 1990, 0 degrees awarded. *Degree requirements:* Thesis or alternative, project required, foreign language not required. *Application deadline:* 5/1. *Application fee:* $35. *Tuition:* $2450 per year full-time, $103 per credit part-time for state residents; $5765 per year full-time, $243 per credit part-time for nonresidents. *Financial aid:* Federal work-study available. Aid available to part-time students. Financial aid application deadline: 3/1. • Dr. James Wells, Chairperson, 716-878-6731.

State University of New York College at Fredonia, Department of Physics, Fredonia, NY 14063. Department awards MS, MS Ed. Part-time and evening/weekend programs available. Matriculated students: 0. 0 applicants. In 1990, 0 degrees awarded. *Degree requirements:* Thesis or alternative required, foreign language not required. *Application deadline:* 7/5. *Application fee:* $35. *Tuition:* $1600 per semester full-time, $134 per credit part-time for state residents; $3288 per semester full-time, $274 per credit part-time for nonresidents. *Financial aid:* Research assistantships, teaching assistantships, full and partial tuition waivers available. Aid available to part-time students. Financial aid application deadline: 3/15; applicants required to submit FAF. • Dr. Michael Grasso, Chairman, 716-673-3301.

State University of New York College at New Paltz, Faculty of Liberal Arts and Sciences, Program in Physics, New Paltz, NY 12561. Program awards MA, MS Ed. Faculty: 9. Matriculated students: 1 full-time (0 women), 3 part-time (1 woman); includes 0 minority, 2 foreign. In 1990, 2 degrees awarded. *Degree requirements:* Thesis (for some programs), comprehensive exam. *Entrance requirements:* GRE General Test, minimum GPA of 3.0. Application deadline: 4/1 (priority date, applications processed on a rolling basis). Application fee: $35. *Tuition:* $1600 per semester full-time, $134 per credit part-time for state residents; $3258 per semester full-time, $274 per credit part-time for nonresidents. *Financial aid:* Research assistantships, teaching assistantships, federal work-study, institutionally sponsored loans available. Financial aid applicants required to submit FAF. • Richard Veghte, Chairman, 914-257-3740.

Stephen F. Austin State University, School of Sciences and Mathematics, Department of Physics and Astronomy, Nacogdoches, TX 75962. Department offers program in physics (MS). Part-time programs available. Faculty: 6 full-time, 1 part-time. Matriculated students: 5 full-time (2 women), 4 part-time (0 women); includes 2 minority (1 black American, 1 Hispanic American), 1 foreign. Average age 30. 2 applicants, 100% accepted. In 1990, 5 degrees awarded (60% found work related to degree, 40% continued full-time study). *Degree requirements:* Comprehensive exam required, foreign language and thesis not required. *Entrance requirements:* GRE General Test (minimum combined score of 1000 required). Application deadline: 8/1. Application fee: $0 ($25 for foreign students). *Expenses:* Tuition of $18 per semester hour for state residents; $122 per semester hour for nonresidents. Fees of $14 per semester hour. *Financial aid:* In 1990–91, $21,000 in aid awarded. 4 teaching assistantships (2 to first-year students) were awarded; institutionally sponsored loans also available. Average monthly stipend for a graduate assistantship: $666. Financial aid application deadline: 6/1. *Faculty research:* Low-temperature physics, x-ray spectroscopy and metallic glasses, infrared spectroscopy. Total annual research budget: $30,000. • Dr. Harry D. Downing, Chairman, 409-568-3001.

Stevens Institute of Technology, Department of Physics and Engineering Physics, Hoboken, NJ 07030. Department awards ME, MS, PhD. Programs offered in physics (PhD), physics and engineering physics (MS). Part-time and evening/weekend programs available. Terminal master's awarded for partial completion of doctoral program. *Degree requirements:* For master's, thesis optional, foreign language not required; for doctorate, variable foreign language requirement, dissertation. *Entrance requirements:* GRE. Application fee: $25. *Tuition:* $4850 per semester full-time, $485 per credit part-time. *Faculty research:* Optics, biophysics, fusion and plasmas, low-temperature physics, surface physics.

See full description on page 619.

Syracuse University, College of Arts and Sciences, Department of Physics, Syracuse, NY 13244. Department offers programs in biophysics (PhD), physics (MS, PhD). Faculty: 22 full-time, 1 part-time. Matriculated students: 54 full-time (10 women), 0 part-time; includes 1 minority (Asian American), 39 foreign. Average age 29. 149 applicants, 14% accepted. In 1990, 5 master's, 12 doctorates awarded. Terminal master's awarded for partial completion of doctoral program. *Degree requirements:* For master's, thesis or alternative required, foreign language not required; for doctorate, dissertation required, foreign language not required. *Entrance requirements:* GRE General Test, GRE Subject Test, TOEFL. Application deadline: rolling. Application fee: $40. *Expenses:* Tuition of $381 per credit. Fees of $289 per year full-time, $34.50 per semester part-time. *Financial aid:* In 1990–91, 22 research assistantships, 21 teaching assistantships awarded; fellowships, partial tuition waivers, federal work-study also available. Financial aid application deadline: 3/1; applicants required to submit FAF. • Ed Lipson, Graduate Program Director, 315-443-3901.

See full description on page 621.

Temple University, College of Arts and Sciences, Department of Physics, Philadelphia, PA 19122. Department awards MA, PhD. Faculty: 27 full-time (0 women), 0 part-time. Matriculated students: 40 (8 women); includes 3 minority (1 Asian American, 1 black American, 1 Hispanic American), 28 foreign. Average age 30. 17 applicants, 94% accepted. In 1990, 1 master's, 7 doctorates awarded. *Degree requirements:* For master's, thesis or alternative, comprehensive exam required, foreign language not required; for doctorate, dissertation required, foreign language not required. *Entrance requirements:* GRE General Test (minimum combined score of 1000 required), GRE Subject Test, minimum GPA of 2.8 overall, 3.0 during previous 2 years. Application deadline: 7/1. Application fee: $30. *Tuition:* $224 per credit for state residents; $283 per credit for nonresidents. *Financial aid:* In 1990–91, 1 fellowship, 19 research assistantships, 18 teaching assistantships awarded. • Dr. Robert Intemann, Chair, 215-787-7696. Application contact: Dr. Leonard Auerbach, Graduate Chair of Admissions and Awards, 215-787-7654.

See full description on page 623.

Texas A&M University, College of Science, Department of Physics, College Station, TX 77843. Department awards MS, PhD. Faculty: 43. Matriculated students: 145 full-time (19 women), 0 part-time; includes 8 minority (3 Asian American, 3 black American, 1 Hispanic American, 1 Native American), 101 foreign. 94 applicants, 64% accepted. In 1990, 15 master's, 4 doctorates awarded. *Entrance requirements:* GRE General Test, TOEFL. Application deadline: 7/15. Application fee: $25 ($50 for foreign students). *Expenses:* Tuition of $100 per semester (minimum) for state residents; $128 per credit hour for nonresidents. Fees of $459 per year full-time, $252 per semester part-time. *Financial aid:* Fellowships, research assistantships, teaching assistantships available. • Richard Arnowitt, Head, 409-845-7717.

See full description on page 625.

Texas Christian University, Add Ran College of Arts and Sciences, Department of Physics, Fort Worth, TX 76129. Department awards MA, MS, PhD. Matriculated students: 15. In 1990, 4 doctorates awarded. *Degree requirements:* For master's, thesis optional; for doctorate, dissertation. *Entrance requirements:* GRE. *Expenses:* Tuition of $244 per semester hour. Fees of $423 per semester full-time, $18.50 per semester hour part-time. *Financial aid:* Fellowships, teaching assistantships available. • Dr. Richard Lysiak, Chairperson, 817-921-7375.

Texas Tech University, Graduate School, College of Arts and Sciences, Department of Physics and Engineering Physics, Lubbock, TX 79409. Offerings include physics and engineering physics (MS, PhD). Department faculty: 22. *Degree requirements:* Variable foreign language requirement, thesis/dissertation. *Entrance requirements:* GRE General Test. Application deadline: 4/15 (priority date, applications processed on a rolling basis). Application fee: $0 ($50 for foreign students). *Tuition:* $494 per semester full-time, $20 per credit hour part-time for state residents; $1790 per semester full-time, $455 per credit hour part-time for nonresidents. • Dr. Walter Borst, Chairman, 806-742-3767.

Trent University, Program in Freshwater Science, Department of Physics, Peterborough, ON K9J 7B8, Canada. Department awards M Sc. Part-time programs available. Matriculated students: 0 full-time. *Degree requirements:* Thesis required, foreign language not required. *Application deadline:* 2/15 (priority date, applications processed on a rolling basis). *Application fee:* $0. *Expenses:* Tuition of $2326 per year full-time, $1163 per year part-time for Canadian residents; $9712 per year full-time, $1163 per year part-time for nonresidents. Fees of $225 per year full-time, $94 per year part-time. *Financial aid:* Fellowships, research assistantships, teaching assistantships available. *Faculty research:* Radiation physics, chemical physics. • Dr. A. J. Slavin, Chairman, 705-748-1289. Application contact: Dr. W. F. J. Evans, Director, Freshwater Science Program, 705-748-1622.

Tufts University, Graduate School of Arts and Sciences, Department of Physics and Astronomy, Medford, MA 02155. Department offers programs in astronomy (MS, PhD), physics (MS, PhD). Faculty: 21 full-time, 0 part-time. Matriculated students: 44 (5 women); includes 15 foreign. 52 applicants, 33% accepted. In 1990, 3 master's, 10 doctorates awarded. Terminal master's awarded for partial completion of doctoral program. *Degree requirements:* Thesis/dissertation required, foreign language not required. *Entrance requirements:* GRE General Test, GRE Subject Test, TOEFL (minimum score of 550 required). Application deadline: 2/15. Application fee: $50. *Expenses:* Tuition of $16,755 per year full-time, $2094 per course part-time. Fees of $885 per year. *Financial aid:* Research assistantships, teaching assistantships, federal work-study available. Financial aid application deadline: 2/15; applicants required to submit GAPSFAS. *Faculty research:* Search for nuclear decay, neutrino interactions, cosmology, superconductivity, protein structure. • David Weaver, Chair, 617-381-3515. Application contact: Anthony Mann, 617-381-3219.

See full description on page 627.

Tulane University, Department of Physics, New Orleans, LA 70118. Department awards MAT, MS, PhD. *Degree requirements:* For master's, thesis or alternative; for doctorate, dissertation. *Entrance requirements:* For master's, GRE General Test, TOEFL (minimum score of 600 required) or TSE (minimum score of 220 required), minimum B average in undergraduate course work; for doctorate, GRE General Test, TOEFL (minimum score of 600 required) or TSE (minimum score of 220 required). Application deadline: 7/1. Application fee: $35. *Expenses:* Tuition of $16,750 per year full-time, $931 per hour part-time. Fees of $230 per year full-time, $40 per hour part-time. *Financial aid:* Fellowships, teaching assistantships, federal work-study, institutionally sponsored loans, and career-related internships or fieldwork available. Financial aid application deadline: 5/1; applicants required to submit GAPSFAS. • G. Rosensteel, Chairman, 504-865-5520.

Université de Moncton, Faculty of Graduate Studies and Research, Faculty of Science, Department of Physics, Moncton, NB E1A 3E9, Canada. Department awards M Sc. Faculty: 23 full-time (1 woman), 1 part-time (0 women). Matriculated students: 6 full-time (1 woman), 0 part-time. Average age 24. 12 applicants, 17% accepted. In 1990, 1 degree awarded (100% found work related to degree). *Degree requirements:* Thesis required, foreign language not required. *Application deadline:* 6/1. *Expenses:* Tuition of $1915 per trimester full-time, $222 per course part-time. Fees of $108 per trimester full-time. *Financial aid:* In 1990–91, $20,000 in aid awarded. Fellowships, research assistantships, teaching assistantships, institutionally sponsored loans available. Financial aid application deadline: 6/30. *Faculty research:* Thin films, optical properties, electrochomism solar selective surfaces. Total annual research budget: $75,000. • Dr. Fernand Girouard, Director, 506-858-4339.

Université de Montréal, Faculty of Arts and Sciences, Department of Physics, Montreal, PQ H3C 3J7, Canada. Department awards M Sc, PhD. *Financial aid:* Fellowships, research assistantships, teaching assistantships available. *Faculty research:* Astronomy; biophysics; solid-state, plasma, and nuclear physics. • Jean-Robert Derome, Chairman, 514-343-6669.

Université de Sherbrooke, Faculty of Sciences, Department of Physics, Sherbrooke, PQ J1K 2R1, Canada. Department awards M Sc, PhD. Faculty: 14 full-time (0 women), 0 part-time. Matriculated students: 25 full-time (2 women), 0 part-time. 22 applicants, 41% accepted. In 1990, 8 master's, 4 doctorates awarded. *Degree requirements:* Thesis/dissertation required, foreign language not required. *Entrance requirements:* For doctorate, master's degree. Application deadline: 6/30. Application fee: $15. *Expenses:* Tuition of $585 per trimester full-time, $43.34 per credit part-time for Canadian residents; $2900 per trimester full-time, $195 per credit part-time for nonresidents. Fees of $125 per trimester full-time, $7.75 per credit part-time for Canadian residents; $610 per year full-time, $7.50 per credit part-time for nonresidents. *Financial aid:* In 1990–91, $150,000 in aid awarded. 25 fellowships, 15 teaching assistantships were awarded; research assistantships also available. *Faculty research:* Solid-state physics. Total annual research budget: $1.6-million. • Dr. Marcel Aubin, Chairman, 819-821-7055.

Université du Québec à Trois-Rivières, Program in Physics, Trois-Rivières, PQ G9A 5H7, Canada. Program awards M Sc. Part-time programs available. Matriculated students: 4 full-time (0 women), 7 part-time (0 women); includes 4 foreign. 16 applicants, 38% accepted. In 1990, 0 degrees awarded. *Degree requirements:*

SECTION 7: PHYSICS

Directory: Physics

Thesis. *Entrance requirements:* Appropriate bachelor's degree, proficiency in French. Application deadline: 4/1. Application fee: $15. *Expenses:* Tuition of $555 per trimester full-time, $37 per credit part-time for Canadian residents; $3480 per trimester full-time, $234 per credit part-time for nonresidents. Fees of $57.50 per trimester full-time, $19.50 per trimester part-time. *Financial aid:* Fellowships, research assistantships, teaching assistantships available. • Jean-Marie St-Arnaud, Director, 819-376-5107. Application contact: Michel Potvin, Office of Admissions, 819-376-5045.

Université Laval, Faculty of Sciences and Engineering, Department of Physics, Sainte-Foy, PQ G1K 7P4, Canada. Department awards M Sc, PhD. Matriculated students: 93 full-time (10 women), 10 part-time (2 women). 68 applicants, 38% accepted. In 1990, 11 master's, 7 doctorates awarded. *Application deadline:* 3/1. *Application fee:* $15. *Expenses:* Tuition of $792 per year full-time for Canadian residents; $5914 per year full-time for nonresidents. Fees of $120 per year full-time. • Gabriel Bédard, Director, 415-656-2152.

University of Akron, Buchtel College of Arts and Sciences, Department of Physics, Akron, OH 44325. Department awards MS. Part-time and evening/weekend programs available. Faculty: 10 full-time (0 women), 3 part-time (0 women). Matriculated students: 24 full-time (9 women), 4 part-time (0 women); includes 2 minority (both black American), 18 foreign. 47 applicants, 87% accepted. In 1990, 6 degrees awarded. *Degree requirements:* Thesis or alternative required, foreign language not required. *Application deadline:* 8/15 (applications processed on a rolling basis). *Application fee:* $25. *Tuition:* $119.93 per credit hour for state residents; $210.93 per credit hour for nonresidents. *Financial aid:* In 1990–91, 13 research assistantships (6 to first-year students), 9 teaching assistantships (4 to first-year students) awarded; full tuition waivers also available. *Faculty research:* Polymer physics, statistical physics, NMR, electron tunneling, solid-state physics. Total annual research budget: $200,000. • Dr. Roger Creel, Head, 216-972-7078. Application contact: Dr. Frank Griffin, 216-972-7138.

See full description on page 629.

University of Alabama, College of Arts and Sciences, Department of Physics and Astronomy, Tuscaloosa, AL 35487-0132. Department offers program in physics (MS, PhD). Faculty: 24 full-time (0 women), 0 part-time. Matriculated students: 27 full-time (6 women), 1 part-time (0 women); includes 0 minority, 21 foreign. Average age 29. 31 applicants, 52% accepted. In 1990, 4 master's awarded (100% continued full-time study); 3 doctorates awarded (33% entered university research/teaching, 33% found other work related to degree, 33% continued full-time study). *Degree requirements:* For master's, oral exam required, thesis optional, foreign language not required; for doctorate, 1 foreign language (computer language can substitute), dissertation, oral and written exams. *Entrance requirements:* For master's, GRE General Test (minimum combined score of 1500 on 3 wections required, TOEFL (minimum score of 550 required), minimum GPA of 3.0; for doctorate, GRE General Test (minimum combined score of 1500 on3 sections required), TOEFL (minimum score of 550 required), minimum GPA of 3.0. *Application deadline:* 7/6 (applications processed on a rolling basis). *Application fee:* $20. *Tuition:* $968 per year full-time, $82 per credit part-time for state residents; $2400 per year full-time, $218 per credit part-time for nonresidents. *Financial aid:* In 1990–91, $190,000 in aid awarded. 8 research assistantships (3 to first-year students), 16 teaching assistantships (6 to first-year students) were awarded; career-related internships or fieldwork also available. Financial aid application deadline: 4/1. *Faculty research:* Condensed matter, high-energy physics, optics, molecular spectroscopy, astrophysics. Total annual research budget: $713,000. • Dr. Philip W. Coulter, Chairman, 205-348-5050.

Announcement: Research areas include high-energy physics, with participation in experiments at Fermilab and CERN, in detector development for the SSC, and in theory and phenomenology of fundamental interactions; theoretical and experimental condensed-matter physics, with applications to materials science and information technology; astrophysics, with emphasis in extragalactic astronomy; molecular spectroscopy; and optics.

University of Alabama at Birmingham, Graduate School, School of Natural Sciences and Mathematics, Department of Physics, Birmingham, AL 35294. Department awards MS, PhD. Matriculated students: 28 full-time (5 women), 2 part-time (0 women); includes 4 minority (2 Asian American, 2 black American), 18 foreign. 184 applicants, 22% accepted. In 1990, 0 master's, 1 doctorate awarded. *Degree requirements:* For master's, 1 foreign language; for doctorate, 1 foreign language, dissertation. *Entrance requirements:* GRE General Test or MAT. *Application deadline:* rolling. *Application fee:* $25 ($50 for foreign students). *Tuition:* $66 per quarter for state residents, $132 per quarter for nonresidents. *Financial aid:* In 1990–91, 3 fellowships awarded. • Dr. David L. Shealy, Chairman, 205-934-4736.

Announcement: MS and PhD programs available. Research activities include optics (X-ray and laser optical design, nonlinear optics, integrated optics), biophysics (Raman/fluorescence spectroscopy and microscopy, dynamic light scattering, contractile protein structure), nuclear magnetic resonance, solid-state physics/materials science, theoretical physics, relativity, laboratory astrophysics, electron spin resonance, and Mössbauer spectroscopy (magnetic and extraterrestrial materials).

University of Alabama in Huntsville, College of Science, Department of Physics, Huntsville, AL 35899. Department awards MS, PhD. Part-time programs available. Faculty: 26 full-time (1 woman), 13 part-time (0 women). Matriculated students: 56 full-time, 32 part-time; includes 4 minority (2 Asian American, 2 Hispanic American), 22 foreign. 79 applicants, 70% accepted. In 1990, 16 master's, 8 doctorates awarded. Terminal master's awarded for partial completion of doctoral program. *Degree requirements:* For master's, thesis or alternative, oral and written exams required, foreign language not required; for doctorate, dissertation, oral and written exams required, foreign language not required. *Entrance requirements:* For master's, GRE General Test (minimum combined score of 1500 on 3 sections required), GRE Subject Test, minimum GPA of 3.0; for doctorate, GRE General Test (minimum combined score of 1500 on 3 sections required), GRE Subject Test. *Application deadline:* 5/16 (priority date, applications processed on a rolling basis). *Application fee:* $20. *Tuition:* $2500 per year full-time, $1250 per year part-time for state residents; $5000 per year full-time, $2500 per year part-time for nonresidents. *Financial aid:* In 1990–91, 31 students received a total of $153,068 in aid awarded. 3 fellowships, 31 research assistantships, 27 teaching assistantships were awarded; full and partial tuition waivers, federal work-study, institutionally sponsored loans, and career-related internships or fieldwork also available. Aid available to part-time students. Average monthly stipend for a graduate assistantship: $750. Financial aid application deadline: 3/1; applicants required to submit GAPSFAS or FAF. *Total annual research budget:* $1.945-million. • Dr. Gordon Emslie, Acting Chairman, 205-895-6276.

University of Alaska Fairbanks, College of Natural Sciences, Department of Physics, Fairbanks, AK 99775. Department offers programs in atmospheric science (MS, PhD), physics (MS, PhD), space physics (MS, PhD). Part-time programs available. Faculty: 25 full-time (1 woman), 0 part-time. Matriculated students: 38 full-time (5 women), 0 part-time; includes 2 minority (both Asian American), 21 foreign. Average age 27. 33 applicants, 30% accepted. In 1990, 1 master's awarded (100% continued full-time study); 2 doctorates awarded (100% found work related to degree). Terminal master's awarded for partial completion of doctoral program. *Degree requirements:* For master's, thesis; for doctorate, 1 foreign language (computer language can substitute), dissertation. *Entrance requirements:* For master's, GRE General Test, GRE Subject Test (minimum score of 550 required); for doctorate, GRE General Test, GRE Subject Test. *Application deadline:* 2/15. *Application fee:* $20. *Expenses:* Tuition of $1620 per year full-time, $90 per credit part-time for state residents; $3240 per year full-time, $180 per credit part-time for nonresidents. Fees of $464 per year full-time. *Financial aid:* In 1990–91, 38 students received a total of $631,940 in aid awarded. 30 research assistantships (4 to first-year students), 8 teaching assistantships (6 to first-year students) were awarded. Average monthly stipend for a graduate assistantship: $985. Financial aid application deadline: 2/15. *Faculty research:* Laser physics; condensed matter, space plasma physics; auroral and ionospheric physics; physical and polar meteorology. • Dr. John Morack, Head, 907-474-7339.

Announcement: Department graduate students are generally employed as teaching assistants within the department or as research assistants at the Geophysical Institute, where the majority of research is conducted. Active research pursued by department faculty and members of the Geophysical Institute includes laser physics, condensed-matter physics, space plasma physics, ionospheric and radio physics, solar-terrestrial relations, electron impact phenomena, optical spectroscopy and interferometry of the aurora and airglow, physical meteorology, and polar meteorology.

University of Alberta, Faculty of Graduate Studies and Research, Department of Physics, Edmonton, AB T6G 2J9, Canada. Department offers programs in engineering physics (M Sc, PhD), geophysics (M Sc, PhD), physics (M Sc, PhD). Matriculated students: 59 full-time, 2 part-time. *Application fee:* $0. *Expenses:* Tuition of $1495 per year full-time, $748 per year part-time for Canadian residents; $2243 per year full-time, $1121 per year part-time for nonresidents. Fees of $301 per year full-time, $118 per year part-time. • Dr. H. R. Glyde, Chair, 403-492-5286.

University of Arizona, College of Arts and Sciences, Faculty of Science, Department of Physics, Tucson, AZ 85721. Department awards M Ed, MS, PhD. Part-time programs available. Faculty: 44 full-time (2 women), 0 part-time. Matriculated students: 109 full-time (12 women), 12 part-time (1 woman); includes 9 minority (6 Asian American, 2 Hispanic American, 1 Native American), 54 foreign. Average age 30. 240 applicants, 16% accepted. In 1990, 7 master's, 6 doctorates awarded. Terminal master's awarded for partial completion of doctoral program. *Degree requirements:* For master's, thesis optional, foreign language not required; for doctorate, 1 foreign language, dissertation. *Entrance requirements:* GRE General Test, GRE Subject Test, minimum GPA of 3.0. *Application deadline:* 2/1 (applications processed on a rolling basis). *Application fee:* $25. *Expenses:* Tuition of $0 for state residents, $5406 per year full-time, $209 per credit hour part-time for nonresidents. Fees of $1528 per year full-time, $80 per credit hour part-time. *Financial aid:* In 1990–91, $900,000 in aid awarded. 20 fellowships, 40 research assistantships, 40 teaching assistantships, 14 scholarships were awarded; full tuition waivers, federal work-study also available. Aid available to part-time students. Financial aid application deadline: 5/1; applicants required to submit FAF. *Faculty research:* Astrophysics, high energy, condensed matter, atomic and molecular, optics. Total annual research budget: $2.5-million. • Dr. Peter Carruthers, Head, 602-621-6801. Application contact: Dr. Robert L. Thews, Associate Head, 602-621-2453.

See full description on page 633.

University of Arkansas, J. William Fulbright College of Arts and Sciences, Department of Physics, Fayetteville, AR 72701. Department awards MA, MS, PhD. Faculty: 12 full-time (1 woman), 0 part-time. Matriculated students: 12 full-time (3 women), 15 part-time (0 women); includes 1 minority (Asian American), 12 foreign. In 1990, 5 master's, 5 doctorates awarded. *Degree requirements:* Thesis/dissertation required, foreign language not required. *Application fee:* $15. *Expenses:* Tuition of $2050 per year full-time, $103 per credit hour part-time for state residents; $4400 per year full-time, $220 per credit hour part-time for nonresidents. Fees of $50 per year full-time, $1.50 per credit hour part-time. *Financial aid:* In 1990–91, 20 teaching assistantships awarded; research assistantships also available. *Total annual research budget:* $600,000. • Dr. Raj Gupta, Chairperson, 501-575-2506.

University of Bridgeport, College of Science and Engineering, Department of Physics, 380 University Avenue, Bridgeport, CT 06601. Department awards MS. Faculty: 4 full-time (1 woman), 1 part-time (0 women). Matriculated students: 0 full-time, 9 part-time (4 women); includes 0 minority, 4 foreign. 7 applicants, 57% accepted. In 1990, 3 degrees awarded. *Degree requirements:* Thesis optional, foreign language not required. *Entrance requirements:* GRE General Test. *Application deadline:* rolling. *Application fee:* $30 ($50 for foreign students). *Expenses:* Tuition of $6010 per semester full-time, $310 per credit part-time for state residents; $6010 per semester full-time, $610 per credit part-time for nonresidents. Fees of $50 per year full-time, $25 per year part-time. *Financial aid:* Federal work-study, institutionally sponsored loans, and career-related internships or fieldwork available. Aid available to part-time students. Financial aid application deadline: 6/1; applicants required to submit FAF. • Dr. James V. Tucci, Chairman, 203-576-4271.

University of British Columbia, Faculty of Science, Department of Physics, Vancouver, BC V6T 1Z1, Canada. Offerings include physics (M Sc, PhD). *Degree requirements:* For doctorate, dissertation, comprehensive exam. *Entrance requirements:* For doctorate, master's degree.

University of Calgary, Faculty of Science, Department of Physics and Astronomy, Calgary, AB T2N 1N4, Canada. Department awards M Sc, PhD. Part-time programs available. *Degree requirements:* Thesis/dissertation. *Entrance requirements:* GRE General Test, GRE Subject Test, TOEFL (minimum score of 550 required). *Application deadline:* 5/31. *Application fee:* $25. *Tuition:* $1705 per year full-time, $427 per course part-time for Canadian residents; $3410 per year full-time, $854 per course part-time for nonresidents. *Faculty research:* Astronomy and astrophysics, EPR, mass spectrometry, atmospheric physics, space physics.

University of California at Berkeley, College of Letters and Science, Department of Physics, Berkeley, CA 94720. Department awards MA, PhD. Faculty: 65. Matriculated students: 258 full-time, 0 part-time; includes 29 minority (18 Asian American, 2 black American, 9 Hispanic American), 30 foreign. 517 applicants, 25% accepted. Terminal master's awarded for partial completion of doctoral program. *Degree requirements:* For doctorate, dissertation. *Entrance requirements:* GRE General Test, minimum GPA of 3.0. *Application deadline:* 1/7. *Application fee:* $40.

SECTION 7: PHYSICS

Directory: Physics

Expenses: Tuition of $0. Fees of $1909 per year for state residents; $7825 per year for nonresidents. • P. Buford Price, Chair.

University of California, Davis, Program in Physics, Davis, CA 95616. Program awards MS, PhD. Faculty: 32. Matriculated students: 99. 120 applicants, 30% accepted. In 1990, 4 master's, 8 doctorates awarded. *Degree requirements:* Thesis/dissertation. *Entrance requirements:* GRE General Test, GRE Subject Test, minimum GPA of 3.0. Application deadline: 1/15. Application fee: $40. *Expenses:* Tuition of $0 for state residents; $7699 per year full-time, $3849 per year part-time for nonresidents. Fees of $2718 per year full-time, $1928 per year part-time. *Financial aid:* Fellowships, research assistantships, teaching assistantships available. • Graduate Adviser, 916-752-1501.

University of California, Irvine, School of Physical Sciences, Department of Physics, Irvine, CA 92717. Department awards MS, PhD. Matriculated students: 91 (11 women); includes 10 minority (5 Asian American, 5 Hispanic American), 26 foreign. 184 applicants, 25% accepted. In 1990, 12 master's, 9 doctorates awarded. *Degree requirements:* For master's, foreign language not required; for doctorate, dissertation required, foreign language not required. *Entrance requirements:* GRE General Test, GRE Subject Test. Application deadline: 2/1 (applications processed on a rolling basis). Application fee: $40. *Expenses:* Tuition of $0 for state residents; $7699 per year full-time, $3850 per year part-time for nonresidents. Fees of $2930 per year full-time, $2139 per year part-time. *Financial aid:* Fellowships, research assistantships, teaching assistantships, and career-related internships or fieldwork available. *Faculty research:* High energy physics, condensed matter physics, low temparature physics, plasma physics, astrophysics. • Walter Bron, Chair, 714-856-5438. Application contact: Robert Doedens, Associate Dean, 714-856-6507.

University of California, Los Angeles, College of Letters and Science, Department of Physics, Los Angeles, CA 90024. Department awards MAT, MS, PhD. Faculty: 45. Matriculated students: 172 full-time (14 women), 0 part-time; includes 27 minority, 53 foreign. 315 applicants, 27% accepted. In 1990, 20 master's, 18 doctorates awarded. *Degree requirements:* For master's, thesis or comprehensive exam required, foreign language not required; for doctorate, dissertation, written and oral comprehensive exams required, foreign language not required. *Entrance requirements:* GRE Subject Test. Application fee: $40. *Expenses:* Tuition of $0 for state residents; $7699 per year for nonresidents. Fees of $2907 per year. *Financial aid:* In 1990–91, 179 students received a total of $2.97-million in aid awarded. 162 fellowships, 131 research assistantships, 88 teaching assistantships were awarded; full and partial tuition waivers, federal work-study, institutionally sponsored loans also available. Financial aid application deadline: 3/1. • Dr. Roberto Peccei, Chair, 310-825-3224.

See full description on page 635.

University of California, Riverside, Graduate Division, College of Natural and Agricultural Sciences, Department of Physics, Riverside, CA 92521. Department awards MA, MS, PhD. Part-time programs available. Terminal master's awarded for partial completion of doctoral program. *Degree requirements:* For master's, comprehensive exams or thesis required, foreign language not required; for doctorate, dissertation, qualifying exams required, foreign language not required. *Entrance requirements:* GRE General Test, GRE Subject Test, TOEFL (minimum score of 550 required). Application deadline: 6/1. Application fee: $40. *Tuition:* $950 per quarter full-time, $264 per quarter part-time for state residents; $3517 per quarter full-time, $1758 per quarter part-time for nonresidents. *Faculty research:* Experimental high energy physics, relativistic heavy ion collisions, experimental high temperature superconductivity, disordered and amorphous systems, astrophysics and space physics.

See full description on page 637.

University of California, San Diego, Department of Physics, 9500 Gilman Drive, La Jolla, CA 92093. Department offers programs in biophysics (MS, PhD), physics (MS, PhD). Faculty: 47. Matriculated students: 135 (13 women); includes 47 foreign. 401 applicants, 24% accepted. In 1990, 16 master's, 16 doctorates awarded. *Degree requirements:* For master's, foreign language not required; for doctorate, dissertation. *Entrance requirements:* GRE General Test, GRE Subject Test. Application fee: $40. *Expenses:* Tuition of $0 for state residents; $7699 per year full-time, $1283 per quarter part-time for nonresidents. Fees of $2798 per year full-time, $669 per quarter part-time. *Total annual research budget:* $3.15-million. • Roger Dashen, Chair, 619-534-3292. Application contact: Debra Bomar, Graduate Coordinator, 619-534-3293.

University of California, Santa Barbara, College of Letters and Science, Department of Physics, Santa Barbara, CA 93106. Department awards PhD. Matriculated students: 114 full-time (9 women), 0 part-time; includes 18 foreign. 260 applicants, 35% accepted. In 1990, 2 master's, 7 doctorates awarded. Terminal master's awarded for partial completion of doctoral program. *Degree requirements:* For doctorate, dissertation required, foreign language not required. *Entrance requirements:* For doctorate, GRE General Test, GRE Subject Test, TOEFL (minimum score of 550 required). Application deadline: 5/1. Application fee: $40. *Expenses:* Tuition of $0 for state residents; $7699 per year for nonresidents. Fees of $2307 per year. *Financial aid:* Fellowships, research assistantships, teaching assistantships, full and partial tuition waivers, federal work-study, institutionally sponsored loans, and career-related internships or fieldwork available. Financial aid application deadline: 1/31. • Raymond S. Sawyer, Chair, 805-893-3495. Application contact: Suzanne Toso, Graduate Secretary, 805-893-4646.

University of California, Santa Cruz, Division of Natural Sciences, Program in Physics, Santa Cruz, CA 95064. Program awards MS, PhD. Faculty: 22 full-time (1 woman), 0 part-time. Matriculated students: 45 full-time (8 women), 0 part-time; includes 4 minority (1 Asian American, 3 Hispanic American), 14 foreign. 97 applicants, 37% accepted. In 1990, 2 master's, 3 doctorates awarded. *Degree requirements:* For master's, thesis; for doctorate, dissertation, qualifying exam. *Entrance requirements:* For doctorate, GRE General Test, GRE Subject Test. Application deadline: 1/16. Application fee: $40. *Expenses:* Tuition of $0 for state residents; $7699 per year for nonresidents. Fees of $3021 per year. *Financial aid:* In 1990–91, 4 fellowships, 21 research assistantships, 60 teaching assistantships awarded; federal work-study, institutionally sponsored loans, and career-related internships or fieldwork also available. Average monthly stipend for a graduate assistantship: $1335. Financial aid applicants required to submit GAPSFAS. *Faculty research:* Theoretical and experimental high-energy physics, theoretical and experimental solid-state physics, critical phenomena, theoretical fluid dynamics, experimental biophysics. • Dr. Bruce Rosenbloom, Chair, 408-459-2329.

University of Central Florida, College of Arts and Sciences, Program in Physics, Orlando, FL 32816. Program awards MS, PhD. Part-time and evening/weekend programs available. Faculty: 18 full-time (1 woman), 0 part-time. Matriculated students: 7 full-time (2 women), 21 part-time (4 women); includes 3 minority (1 Asian American, 1 black American, 1 Hispanic American), 7 foreign. In 1990, 3 master's awarded (100% found work related to degree); 0 doctorates awarded. *Degree requirements:* For master's, thesis required, foreign language not required; for doctorate, dissertation, qualifying exam, candidate exam. *Entrance requirements:* For master's, GRE General Test, minimum GPA of 3.0 in last 60 hours. Application fee: $15. *Expenses:* Tuition of $81 per credit hour for state residents; $364 per credit hour for nonresidents. Fees of $50 per semester. *Financial aid:* In 1990–91, 3 research assistantships, 5 teaching assistantships awarded; federal work-study, institutionally sponsored loans, and career-related internships or fieldwork also available. Aid available to part-time students. *Faculty research:* Atomic-molecular physics, condensed-matter physics, biophysics of proteins, laser physics. • Dr. S. K. Bose, Chair, 407-823-2325. Application contact: Dr. Denise Caldwell, Graduate Coordinator, 407-823-2325.

University of Chicago, Division of the Physical Sciences, Department of Physics, Chicago, IL 60637. Department awards SM, PhD. Terminal master's awarded for partial completion of doctoral program. *Degree requirements:* For master's, foreign language and thesis not required; for doctorate, dissertation required, foreign language not required. *Entrance requirements:* GRE General Test, GRE Subject Test, TOEFL. Application deadline: 1/6. Application fee: $45. *Expenses:* Tuition of $16,275 per year full-time, $8140 per year part-time. Fees of $356 per year. *Faculty research:* Astrophysics, particle physics, condensed matter physics, statistical physics, electron and ion microscopy.

University of Cincinnati, McMicken College of Arts and Sciences, Department of Physics, Cincinnati, OH 45221. Department awards MS, PhD. Faculty: 18 full-time, 0 part-time. Matriculated students: 56 full-time (8 women), 7 part-time (1 woman); includes 7 minority (1 Asian American, 5 black American, 1 Hispanic American), 42 foreign. 330 applicants, 5% accepted. In 1990, 13 master's, 6 doctorates awarded. *Degree requirements:* Thesis/dissertation. *Entrance requirements:* GRE General Test, GRE Subject Test. Application deadline: 2/15. Application fee: $20. *Tuition:* $131 per credit hour for state residents; $261 per credit hour for nonresidents. *Financial aid:* In 1990–91, 43 teaching assistantships awarded; fellowships, research assistantships, full tuition waivers also available. Aid available to part-time students. Average monthly stipend for a graduate assistantship: $750. Financial aid application deadline: 5/1. • Dr. Richard Newrock, Head, 513-556-0501.

See full description on page 639.

University of Colorado at Boulder, College of Arts and Sciences, Department of Physics, Boulder, CO 80309. Department offers programs in applied physics (PhD), chemical physics (PhD), geophysics (PhD), mathematical physics (PhD), physics (MS, PhD). Faculty: 66 full-time (1 woman). Matriculated students: 163 full-time (17 women), 10 part-time (1 woman); includes 6 minority (2 Asian American, 4 Hispanic American), 39 foreign. 313 applicants, 14% accepted. In 1990, 12 master's, 18 doctorates awarded. Terminal master's awarded for partial completion of doctoral program. *Degree requirements:* For master's, thesis or alternative, comprehensive exam required, foreign language not required; for doctorate, 1 foreign language, dissertation. *Entrance requirements:* GRE General Test, GRE Subject Test. Application deadline: 3/1 (priority date, applications processed on a rolling basis). Application fee: $30 ($50 for foreign students). *Expenses:* Tuition of $2308 per year full-time, $387 per semester (minimum) part-time for state residents; $8730 per year full-time, $1455 per semester (minimum) part-time for nonresidents. Fees of $207 per semester full-time, $27.26 per semester (minimum) part-time. *Financial aid:* In 1990–91, $243,400 in aid awarded. 3 fellowships (0 to first-year students), 110 research assistantships (0 to first-year students), 26 teaching assistantships (4 to first-year students) were awarded; full tuition waivers also available. Financial aid application deadline: 3/1. *Total annual research budget:* $3.6-million. • William O'Sullivan, Chairman, 303-492-8703.

Announcement: The Department of Physics offers MS and PhD graduate programs. Research areas include atomic and molecular, mathematical and theoretical, condensed-matter, elementary particle, and nuclear physics and geophysics. The size and scope of the department are greatly enhanced by close ties with numerous associated or neighboring laboratories and institutes.

University of Colorado at Colorado Springs, College of Letters, Arts and Sciences, Department of Physics, Colorado Springs, CO 80933. Department awards MBS, MS. Part-time programs available. Faculty: 7 full-time (1 woman), 3 part-time (0 women). Matriculated students: 3 full-time (1 woman), 8 part-time (4 women); includes 0 minority, 1 foreign. Average age 30. 6 applicants, 50% accepted. In 1990, 2 degrees awarded (50% found work related to degree, 50% continued full-time study). *Degree requirements:* Foreign language not required. *Entrance requirements:* GRE or minimum GPA of 2.75. Application deadline: 6/15 (applications processed on a rolling basis). Application fee: $35 ($50 for foreign students). *Tuition:* $91 per credit hour for state residents; $276 per credit hour for nonresidents. *Financial aid:* In 1990–91, 2 students received a total of $26,000 in aid awarded. 2 research assistantships (0 to first-year students), 1 teaching assistantship (0 to first-year students) were awarded; federal work-study and career-related internships or fieldwork also available. Financial aid application deadline: 5/1. *Faculty research:* Solid state/condensed matter physics, surface science, electron spectroscopies nonlinear physics. Total annual research budget: $136,500. • Dr. James F. Burkhart, Chairman, 719-593-3214. Application contact: Dr. Robert Camley, Graduate Adviser, 719-593-3512.

University of Connecticut, College of Liberal Arts and Sciences, Field of Physics, Storrs, CT 06269. Field awards MS, PhD. Faculty: 34. Matriculated students: 44 full-time (7 women), 15 part-time (1 woman); includes 3 minority (1 Asian American, 1 black American, 1 Hispanic American), 18 foreign. Average age 29. 112 applicants, 28% accepted. In 1990, 13 master's, 8 doctorates awarded. Terminal master's awarded for partial completion of doctoral program. *Degree requirements:* For doctorate, dissertation. *Entrance requirements:* GRE General Test, GRE Subject Test. Application deadline: 6/1 (priority date, applications processed on a rolling basis). Application fee: $25. *Expenses:* Tuition of $3428 per year full-time, $571 per course part-time for state residents; $8914 per year full-time, $1486 per course part-time for nonresidents. Fees of $636 per year full-time, $87 per course part-time. *Financial aid:* In 1990–91, $420,417 in aid awarded. 24 fellowships (2 to first-year students), 9 research assistantships (2 to first-year students), 35 teaching assistantships (8 to first-year students) were awarded. Financial aid application deadline: 2/15; applicants required to submit GAPSFAS or FAF. • Ralph H. Bartram, Head, 203-486-4924.

See full description on page 641.

University of Delaware, College of Arts and Science, Department of Physics and Astronomy, Newark, DE 19716. Department awards MS, PhD. Part-time programs available. Faculty: 24 full-time (1 woman), 0 part-time. Matriculated students: 63 full-time (8 women), 1 part-time (0 women); includes 1 minority (Asian American), 37 foreign. 400 applicants, 6% accepted. In 1990, 3 master's, 2 doctorates awarded. Terminal master's awarded for partial completion of doctoral program. *Degree requirements:* Thesis/dissertation required, foreign language not required. *Entrance requirements:* GRE General Test. Application deadline: 7/1. Application fee: $40. *Tuition:* $179 per credit hour for state residents; $467 per credit hour for nonresidents. *Financial aid:* In 1990–91, $500,000 in aid awarded. 5 fellowships (1 to a first-year student), 28 research assistantships (2 to first-year students), 30 teaching assistantships (8 to first-year students) were awarded. *Total annual research budget:* $1-million. • Dr. James B. Mehl, Chair, 302-451-2661.

University of Denver, Graduate Studies, Faculty of Natural Sciences, Mathematics and Engineering, Department of Physics, Denver, CO 80208. Department awards MS, PhD. Part-time programs available. Faculty: 4 full-time (0 women), 3 part-time (0 women). Matriculated students: 16 (4 women); includes 0 minority, 7 foreign. 16 applicants, 100% accepted. In 1990, 5 master's, 2 doctorates awarded. *Degree requirements:* For master's, thesis optional, foreign language not required; for doctorate, dissertation required, foreign language not required. *Entrance requirements:* GRE General Test, GRE Subject Test, TOEFL (minimum score of 570 required), minimum GPA of 3.0. Application deadline: rolling. Application fee: $30. *Tuition:* $12,852 per year full-time, $357 per credit hour part-time. *Financial aid:* In 1990–91, 7 students received a total of $92,370 in aid awarded. 5 teaching assistantships, 2 scholarships were awarded; research assistantships, federal work-study, institutionally sponsored loans, and career-related internships or fieldwork also available. Aid available to part-time students. Average monthly stipend for a graduate assistantship: $872. Financial aid application deadline: 3/1. • Dr. Herschel Neumann, Chair, 303-871-2238.

University of Florida, College of Liberal Arts and Sciences, Department of Physics, Gainesville, FL 32611. Department awards MS, MST, PhD. PhD offered jointly with Florida Atlantic University. Faculty: 56. Matriculated students: 72 full-time (10 women), 1 part-time (0 women); includes 10 minority (1 Asian American, 4 black American, 5 Hispanic American), 29 foreign. 124 applicants, 24% accepted. In 1990, 4 master's, 7 doctorates awarded. *Degree requirements:* For master's, variable foreign language requirement; for doctorate, 1 foreign language, dissertation. *Entrance requirements:* GRE General Test, minimum GPA of 3.0. Application deadline: 6/1 (priority date, applications processed on a rolling basis). Application fee: $15. *Tuition:* $87 per credit hour for state residents; $289 per credit hour for nonresidents. *Financial aid:* In 1990–91, 4 fellowships, 50 research assistantships, 17 teaching assistantships awarded. *Faculty research:* Astrophysics, condensed matter physics, elementary particle physics, statistical mechanics, quantum theory. Total annual research budget: $4.99-million. • Dr. Neil Sullivan Jr., Chairman, 904-392-0521.

See full description on page 643.

University of Georgia, College of Arts and Sciences, Department of Physics and Astronomy, Athens, GA 30602. Department offers program in physics (MS, PhD). Faculty: 30 full-time (0 women), 0 part-time. Matriculated students: 34 full-time (4 women), 2 part-time (0 women); includes 1 minority (Native American), 19 foreign. 148 applicants, 11% accepted. In 1990, 4 master's, 3 doctorates awarded. *Degree requirements:* For master's, thesis required, foreign language not required; for doctorate, 1 foreign language (computer language can substitute), dissertation. *Entrance requirements:* GRE General Test. Application fee: $10. *Expenses:* Tuition of $598 per quarter full-time, $48 per quarter part-time for state residents; $1558 per quarter full-time, $144 per quarter part-time for nonresidents. Fees of $118 per quarter. *Financial aid:* Fellowships, research assistantships, teaching assistantships, assistantships available. • Dr. Alan K. Edwards, Graduate Coordinator, 404-542-2891.

See full description on page 645.

University of Guelph, Guelph-Waterloo Centre for Graduate Work in Physics, Guelph, ON N1G 2W1, Canada. Center awards M Sc, PhD. *Entrance requirements:* For master's, minimum B average during last two years; for doctorate, minimum B average. *Expenses:* Tuition of $898 per semester full-time, $450 per semester part-time for Canadian residents; $4053 per semester full-time, $2185 per semester part-time for nonresidents. Fees of $543 per semester full-time, $450 per semester part-time for Canadian residents; $2278 per semester full-time, $2185 per semester part-time for nonresidents. *Financial aid:* Fellowships, research assistantships, teaching assistantships available. • Dr. Hallett, Director, 519-824-4120 Ext. 3989.

University of Guelph, College of Physical Science, Department of Physics, Guelph, ON N1G 2W1, Canada. Department awards M Sc, PhD. Program offered jointly with University of Waterloo. Faculty: 66. Matriculated students: 31; includes 7 foreign. *Entrance requirements:* For master's, GRE General Test, minimum B average during last two years; for doctorate, minimum B average. *Expenses:* Tuition of $898 per semester full-time, $450 per semester part-time for Canadian residents; $4053 per semester full-time, $2185 per semester part-time for nonresidents. Fees of $543 per semester full-time, $450 per semester part-time for Canadian residents; $2278 per semester full-time, $2185 per semester part-time for nonresidents. *Financial aid:* Fellowships, research assistantships, teaching assistantships, and career-related internships or fieldwork available. *Faculty research:* Condensed matter physics, atomic and subatomic physics, molecular physics, theoretical physics, astrophysics. • R. W. Ollerhead, Chair, 519-824-4120 Ext. 3771. Application contact: Dr. Brooks, Graduate Coordinator, 519-824-4120 Ext. 3991.

University of Hawaii at Manoa, College of Arts and Sciences, Department of Physics and Astronomy, Honolulu, HI 96822. Department awards MS, PhD. *Degree requirements:* For master's, qualifying exam or thesis; for doctorate, dissertation, master's degree, qualifying exam, oral comprehensive exam. *Entrance requirements:* GRE General Test, GRE Subject Test. *Tuition:* $800 per semester full-time for state residents; $2405 per semester full-time for nonresidents.

University of Houston, College of Natural Sciences and Mathematics, Department of Physics, 4800 Calhoun, Houston, TX 77004. Department awards MS, PhD. *Degree requirements:* For master's, thesis. *Entrance requirements:* GRE General Test. Application fee: $0. *Expenses:* Tuition of $30 per hour for state residents; $134 per hour for nonresidents. Fees of $240 per year full-time, $125 per year part-time.

See full description on page 647.

University of Houston–Clear Lake, School of Natural and Applied Sciences, Houston, TX 77058. Offerings include physical sciences (MS). *Application fee:* $0. *Tuition:* $40 per credit hour for state residents; $134 per credit hour for nonresidents. • E. T. Dickerson, Dean, 713-283-3703. Application contact: Dr. Eldon Husband, Associate Dean and Director of Student Affairs, 713-283-3710.

University of Idaho, College of Graduate Studies, College of Letters and Science, Department of Physics, Moscow, ID 83843. Department offers programs in physics (M Nat Sci, M Nuc Sci, MS, PhD), physics education (MAT). Faculty: 11 full-time (0 women), 1 (woman) part-time. Matriculated students: 13 full-time (3 women), 5 part-time (1 woman); includes 0 minority, 14 foreign. In 1990, 3 master's, 0 doctorates awarded. *Degree requirements:* Thesis/dissertation required, foreign language not required. *Entrance requirements:* For master's, GRE, minimum GPA of 2.8; for doctorate, GRE, minimum undergraduate GPA of 2.8, 3.0 graduate. Application deadline: 8/1. Application fee: $20. *Expenses:* Tuition of $0 for state residents; $4146 per year for nonresidents. Fees of $818 per semester full-time, $82.75 per credit part-time. *Financial aid:* In 1990–91, 5 research assistantships, 7 teaching assistantships (0 to first-year students) awarded. Financial aid application deadline: 3/1. • Dr. Henry Willmes, Acting Chair, 208-885-6380.

University of Illinois at Chicago, College of Liberal Arts and Sciences, Department of Physics, Chicago, IL 60680. Department awards MS, PhD. Faculty: 34 full-time. Matriculated students: 59 full-time (11 women), 24 part-time (5 women); includes 8 minority (5 Asian American, 2 black American, 1 Hispanic American), 47 foreign. 209 applicants, 23% accepted. In 1990, 6 master's, 3 doctorates awarded. Terminal master's awarded for partial completion of doctoral program. *Degree requirements:* For master's, foreign language and thesis not required; for doctorate, dissertation required, foreign language not required. *Entrance requirements:* GRE General Test, TOEFL (minimum score of 550 required), minimum GPA of 4.0 on a 5.0 scale. Application deadline: 7/5. Application fee: $20. *Expenses:* Tuition of $1369 per semester full-time, $521 per semester (minimum) part-time for state residents; $3840 per semester full-time, $1454 per semester (minimum) part-time for nonresidents. Fees of $458 per semester full-time, $398 per semester (minimum) part-time. *Financial aid:* In 1990–91, 3 fellowships, 1 research assistantship, 32 teaching assistantships awarded. *Faculty research:* High-energy, laser, and solid-state physics. • Paul Raccah, Head, 312-996-3400. Application contact: James S. Kouvel, Director of Graduate Studies, 312-996-5348.

University of Illinois at Urbana-Champaign, College of Engineering, Department of Physics, Champaign, IL 61820. Department awards MS, PhD. Faculty: 99 full-time, 0 part-time. Matriculated students: 303 full-time (30 women), 2 part-time (0 women); includes 19 minority (13 Asian American, 1 black American, 5 Hispanic American), 60 foreign. 376 applicants, 14% accepted. In 1990, 44 master's, 38 doctorates awarded. *Degree requirements:* For master's, foreign language and thesis not required; for doctorate, dissertation, departmental qualifying exam required, foreign language not required. *Entrance requirements:* For master's, minimum GPA of 4.0 on a 5.0 scale. Application fee: $25. *Tuition:* $1838 per semester full-time, $708 per semester part-time for state residents; $4314 per semester full-time, $1673 per semester part-time for nonresidents. *Financial aid:* In 1990–91, 23 fellowships, 178 research assistantships, 98 teaching assistantships awarded. Financial aid application deadline: 2/15. • Ansel C. Anderson, Head, 217-333-3761.

University of Iowa, College of Liberal Arts, Department of Physics and Astronomy, Program in Physics, Iowa City, IA 52242. Program awards MS, PhD. Matriculated students: 35 full-time (3 women), 44 part-time (3 women); includes 0 minority, 42 foreign. 123 applicants, 44% accepted. In 1990, 5 master's, 7 doctorates awarded. *Application fee:* $20. *Expenses:* Tuition of $1158 per semester full-time, $387 per semester hour (minimum) part-time for state residents; $3372 per semester full-time, $387 per semester hour (minimum) part-time for nonresidents. Fees of $60 per semester (minimum). *Financial aid:* In 1990–91, 7 fellowships (2 to first-year students), 39 research assistantships (3 to first-year students), 25 teaching assistantships (4 to first-year students) awarded. • Dwight Nicholson, Chair, Department of Physics and Astronomy, 319-335-1686.

University of Kansas, College of Liberal Arts and Sciences, Department of Physics and Astronomy, Lawrence, KS 66045. Department offers programs in atmospheric science (MS), physics (MA, MS, PhD). Faculty: 25 full-time (1 woman), 2 part-time (0 women). Matriculated students: 38 full-time (6 women), 15 part-time (1 woman); includes 3 minority (1 Asian American, 2 Hispanic American), 20 foreign. In 1990, 4 master's, 3 doctorates awarded. *Degree requirements:* For master's, foreign language and thesis not required; for doctorate, computer language, dissertation. *Entrance requirements:* TOEFL (minimum score of 570 required). Application fee: $25. *Expenses:* Tuition of $1668 per year full-time, $56 per credit hour part-time for state residents; $5382 per year full-time, $179 per credit hour part-time for nonresidents. Fees of $338 per year full-time, $25 per credit hour part-time. *Financial aid:* Fellowships, research assistantships, teaching assistantships available. • Ray Ammar, Chairperson, 913-864-4626.

See full description on page 649.

University of Kentucky, Graduate School Programs from the College of Arts and Sciences, Program in Physics and Astronomy, Lexington, KY 40506-0032. Program awards MS, PhD. Faculty: 27 full-time (0 women), 0 part-time. Matriculated students: 27 full-time (6 women), 16 part-time (4 women); includes 2 minority (1 Asian American, 1 Hispanic American), 26 foreign. 180 applicants, 7% accepted. In 1990, 4 master's, 2 doctorates awarded. *Degree requirements:* For master's, 1 foreign language, comprehensive exam required, thesis optional; for doctorate, dissertation, comprehensive exam. *Entrance requirements:* For master's, GRE (verbal, quantitative, and analytical sections), minimum undergraduate GPA of 2.5; for doctorate, GRE (verbal, quantitative, and analytical sections), minimum graduate GPA of 3.0. Application deadline: 7/19 (applications processed on a rolling basis). Application fee: $20 ($25 for foreign students). *Tuition:* $1002 per semester full-time, $101 per credit hour part-time for state residents; $2782 per semester full-time, $299 per credit hour part-time for nonresidents. *Financial aid:* In 1990–91, 1 fellowship (2 to first-year students), 17 research assistantships, 26 teaching assistantships (4 to first-year students) awarded; federal work-study, institutionally sponsored loans also available. Aid available to part-time students. Financial aid applicants required to submit FAF. *Faculty research:* Interstellar cloud dynamics, interstellar magnetic fields, formation of interstellar molecules, circumstellar masers, maser excitation. • Jesse Weil, Director of Graduate Studies, 606-257-3997. Application contact: Dr. Constance L. Wood, Associate Dean for Academic Administration, 606-257-4905.

University of Louisville, College of Arts and Sciences, Department of Physics, Louisville, KY 40292. Department awards MAT, MS. Faculty: 15 full-time (0 women), 0 part-time. Matriculated students: 10 full-time (0 women), 3 part-time (0 women); includes 1 minority (Asian American), 3 foreign. In 1990, 5 degrees awarded. *Degree requirements:* Thesis. *Entrance requirements:* GRE. Application deadline: rolling. *Expenses:* Tuition of $1780 per year full-time, $99 per credit hour part-time for state residents; $5340 per year full-time, $297 per credit hour part-time for nonresidents. Fees of $60 per semester full-time, $12.50 per semester (minimum) part-time. • Dr. Joseph S. Chalmers, Chairperson, 502-588-6787.

University of Maine, College of Sciences, Department of Physics and Astronomy, Orono, ME 04469. Department offers programs in engineering physics (M Eng), physics and astronomy (MS, PhD). Faculty: 17 full-time (1 woman), 1 part-time (0 women). Matriculated students: 26 full-time (3 women), 2 part-time (1 woman); includes 0 minority, 11 foreign. Average age 29. 42 applicants. In 1990, 2 master's awarded (100% found work related to degree); 4 doctorates awarded (100% entered university research/teaching). Terminal master's awarded for partial completion of doctoral program. *Degree requirements:* Thesis/dissertation required, foreign language not required. *Entrance requirements:* For master's, GRE General Test, GRE Subject Test, TOEFL (minimum score of 550 required). Application deadline: 12/15 (priority date, applications processed on a rolling basis). Application fee: $25. *Tuition:* $100 per credit hour for state residents; $275 per credit hour for nonresidents. *Financial aid:* In 1990–91, $230,000 in aid awarded. 2 fellowships (0

Directory: Physics

to first-year students), 9 research assistantships (3 to first-year students), 12 teaching assistantships (5 to first-year students) were awarded. *Faculty research:* Solid-state physics, fluids, biophysics, plasma physics, surface physics. Total annual research budget: $800,000. • Dr. Charles W. Smith, Chairperson, 207-581-1015. Application contact: Dr. Gerald S. Harmon, Graduate Coordinator, 207-581-1016.

University of Manitoba, Faculty of Science, Department of Physics, Winnipeg, MB R3T 2N2, Canada. Department awards M Sc, PhD. *Degree requirements:* For master's, thesis required, foreign language not required; for doctorate, 1 foreign language, dissertation.

University of Maryland College Park, College of Computer, Mathematical and Physical Sciences, Department of Physics and Astronomy, Program in Physics, College Park, MD 20742. Program awards MS, PhD. Faculty: 84 (1 woman). Matriculated students: 190 full-time (20 women), 73 part-time (13 women); includes 8 minority (6 Asian American, 1 black American, 1 Hispanic American), 129 foreign. 428 applicants, 37% accepted. In 1990, 26 master's, 28 doctorates awarded. *Degree requirements:* For master's, thesis or alternative required, foreign language not required; for doctorate, variable foreign language requirement, dissertation. *Entrance requirements:* For master's, GRE Subject Test, minimum GPA of 3.0; for doctorate, GRE Subject Test. Application deadline: rolling. Application fee: $25. *Expenses:* Tuition of $143 per credit hour for state residents; $256 per credit hour for nonresidents. Fees of $171.50 per semester. *Financial aid:* In 1990–91, 8 fellowships (5 to first-year students) awarded; research assistantships, teaching assistantships also available. • Application contact: Jean Clement, 301-405-5982.

Announcement: Department of Physics and Astronomy, one of the nation's major recipients of research grants, has extensive facilities for graduate study and research in space, plasma, elementary particles, condensed-matter, and nuclear physics, as well as in general relativity, accelerator design, quantum electronics, cosmic rays, and superconductivity. Contact Jean Clement, 301-405-5982.

University of Maryland Graduate School, Baltimore, Graduate School, Department of Physics, Baltimore, MD 21228. Department awards MS, PhD. Program offered in applied physics (MS). Faculty: 12 full-time (0 women), 1 part-time (0 women). Matriculated students: 12 full-time (2 women), 3 part-time (2 women); includes 0 minority, 6 foreign. Average age 25. 40 applicants, 28% accepted. In 1990, 7 master's awarded (15% found work related to degree, 85% continued full-time study). *Degree requirements:* For master's, thesis optional, foreign language not required. *Entrance requirements:* For master's, GRE General Test, GRE Subject Test, TOEFL, minimum GPA of 3.0. Application deadline: 3/1 (priority date, applications processed on a rolling basis). Application fee: $25. *Tuition:* $134 per credit for state residents; $245 per credit for nonresidents. *Financial aid:* In 1990–91, 12 students received a total of $120,000 in aid awarded. 1 fellowship (0 to first-year students), 2 research assistantships (1 to a first-year student), 9 teaching assistantships (all to first-year students) were awarded. Average monthly stipend for a graduate assistantship: $1100. Financial aid application deadline: 3/1. *Faculty research:* Physics of solids, spectroscopy, nonlinear and quantum optics, atmospheric research, materials science. • Dr. Geoffrey Summers, Chairman, 301-455-2513. Application contact: Dr. Robert Reno, Coordinator, 301-455-2530.

Announcement: The Department of Physics at the University of Maryland Graduate School, Baltimore, awards an MS degree in applied physics, with concentrations in optics or materials. In the optics concentration, graduate research opportunities exist in nonlinear optics and laser physics, infrared molecular spectroscopy, speckle interferometry, and theoretical quantum optics. In the materials track, research opportunities exist in X-ray diffraction, scanning electron microscopy, deep-level transient spectroscopy (DLTS), Mössbauer and gamma-ray perturbed angular correlation spectroscopy, positron annihilation, and the quantum theory of condensed matter. Materials studies include optical crystals, semiconductors, polymers, ceramics, and crystalline and amorphous metals and alloys.

University of Massachusetts at Amherst, College of Arts and Sciences, Faculty of Natural Sciences and Mathematics, Department of Physics and Astronomy, Program in Physics, Amherst, MA 01003. Program awards MS, PhD. Part-time programs available. Faculty: 44 full-time (2 women), 0 part-time. Matriculated students: 10 full-time (1 woman), 55 part-time (9 women); includes 1 minority (Asian American), 29 foreign. Average age 28. 137 applicants, 32% accepted. In 1990, 11 master's, 5 doctorates awarded. Terminal master's awarded for partial completion of doctoral program. *Degree requirements:* For master's, foreign language and thesis not required; for doctorate, dissertation required, foreign language not required. *Entrance requirements:* GRE General Test, GRE Subject Test. Application deadline: 3/1 (applications processed on a rolling basis). Application fee: $35. *Tuition:* $2568 per year full-time, $107 per credit part-time for state residents; $7920 per year full-time, $330 per credit part-time for nonresidents. *Financial aid:* In 1990–91, 1 fellowship, 48 research assistantships, 53 teaching assistantships awarded; federal work-study also available. Aid available to part-time students. Financial aid application deadline: 3/1; applicants required to submit FAF. • Dr. James F. Walker, Director, 413-545-1310. Application contact: Dr. Robert V. Krotkov, Chair, Admissions Committee, 413-545-2191.

University of Massachusetts Dartmouth, Graduate School, College of Arts and Sciences, Department of Physics, North Dartmouth, MA 02747. Department awards MS. Part-time programs available. Faculty: 5 full-time (0 women), 0 part-time. Matriculated students: 23 full-time (5 women), 2 part-time (0 women); includes 0 minority, 24 foreign. 29 applicants, 97% accepted. In 1990, 5 degrees awarded. *Degree requirements:* Thesis (for some programs). *Entrance requirements:* GRE General Test, GRE Subject Test, TOEFL. Application deadline: 4/20 (priority date, applications processed on a rolling basis). *Expenses:* Tuition of $1368 per semester full-time, $7688 per credit part-time for state residents; $4388 per semester full-time, $244 per credit part-time for nonresidents. Fees of $771 per year full-time, $59 per year part-time. *Financial aid:* In 1990–91, $88,900 in aid awarded. 22 research assistantships, 18 teaching assistantships were awarded. Financial aid application deadline: 5/1. *Faculty research:* Elementary particles, astrophysics. Total annual research budget: $190,000. • Dr. Wolfhard Kern, Director, 508-999-8356. Application contact: Carol A. Novo, Graduate Admissions Office, 508-999-8604.

University of Massachusetts, Lowell, College of Arts and Sciences, Department of Physics and Applied Physics, 1 University Avenue, Lowell, MA 01854. Department offers programs in applied physics (PhD), computational physics (PhD), energy engineering (PhD), physics (MS), radiological sciences and protection (MS, PhD). Terminal master's awarded for partial completion of doctoral program. *Degree requirements:* For master's, thesis required, foreign language not required; for doctorate, 2 foreign languages (computer language can substitute for one), dissertation. *Entrance requirements:* GRE General Test. Application deadline: 4/1. *Expenses:* Tuition of $87 per credit hour for state residents; $271 per credit hour for nonresidents. Fees of $114 per credit hour.

See full description on page 651.

University of Miami, College of Arts and Sciences, Department of Physics, Coral Gables, FL 33124. Department awards MS, DA, PhD. Faculty: 16 (0 women). Matriculated students: 20 full-time (4 women), 1 part-time (0 women); includes 2 minority (both Hispanic American), 15 foreign. Average age 34. 90 applicants, 6% accepted. In 1990, 1 master's, 1 doctorate awarded. Terminal master's awarded for partial completion of doctoral program. *Degree requirements:* For master's, foreign language and thesis not required; for doctorate, dissertation required, foreign language not required. *Entrance requirements:* GRE General Test, GRE Subject Test, TOEFL (minimum score of 560 required). Application deadline: 3/1 (priority date, applications processed on a rolling basis). Application fee: $35. *Expenses:* Tuition of $567 per credit hour. Fees of $87 per semester full-time. *Financial aid:* In 1990–91, 19 students received a total of $185,000 in aid awarded. 1 fellowship (to a first-year student), 4 research assistantships, 14 teaching assistantships (5 to first-year students) were awarded. *Faculty research:* High-energy theory, marine and atmospheric optics, plasma physics, solid-state physics. Total annual research budget: $700,000. • Dr. George Alexandrakis, Chairman, 305-284-2323. Application contact: Dr. Manuel A. Huerta, Adviser, 305-284-2323 Ext. 8.

Announcement: Department of Physics offers tuition-exempt research and teaching assistantships per 9 months. Additional summer support is available. Major research areas with active theoretical and experimental programs are high-energy physics, plasma physics, condensed-matter physics, optical physics, and satellite oceanography. The department is housed in a new facility.

University of Michigan, College of Literature, Science, and the Arts, Department of Physics, Ann Arbor, MI 48109. Department awards MS, PhD. Faculty: 54 full-time, 0 part-time. Matriculated students: 165 full-time (23 women), 0 part-time; includes 9 minority (5 Asian American, 1 black American, 3 Hispanic American), 68 foreign. 387 applicants, 50% accepted. In 1990, 13 master's, 10 doctorates awarded. *Degree requirements:* For master's, foreign language and thesis not required; for doctorate, 1 foreign language, dissertation, preliminary exam. *Entrance requirements:* GRE General Test. Application deadline: 2/1 (applications processed on a rolling basis). Application fee: $30. *Tuition:* $3255 per semester full-time, $352 per credit (minimum) part-time for state residents; $6803 per semester full-time, $746 per credit (minimum) part-time for nonresidents. *Financial aid:* Fellowships, research assistantships, teaching assistantships available. Financial aid application deadline: 3/15. *Faculty research:* Elementary particle, solid-state, atomic and molecular physics (theoretical and experimental). • Homer A. Neal, Chair, 313-936-0654.

See full description on page 653.

University of Minnesota, Duluth, Graduate School, College of Science and Engineering, Department of Physics, Duluth, MN 55812. Department awards MS. Part-time programs available. Faculty: 6 full-time (0 women), 1 part-time (0 women). Matriculated students: 8 full-time (0 women), 1 (woman) part-time; includes 0 minority, 1 foreign. Average age 32. 27 applicants, 37% accepted. In 1990, 5 degrees awarded (20% entered university research/teaching, 20% found other work related to degree, 60% continued full-time study). *Degree requirements:* Thesis optional, foreign language not required. *Application deadline:* 7/15 (applications processed on a rolling basis). *Application fee:* $30. *Tuition:* $1184 per quarter full-time, $301 per credit (minimum) part-time for state residents; $2168 per quarter full-time, $602 per credit (minimum) part-time for nonresidents. *Financial aid:* In 1990–91, 8 students received a total of $68,000 in aid awarded. 0 research assistantships, 8 teaching assistantships (4 to first-year students) were awarded; institutionally sponsored loans also available. Aid available to part-time students. Average monthly stipend for a graduate assistantship: $885. Financial aid application deadline: 2/15. *Faculty research:* Computer modeling, solid-state physics, theoretical physics, surface phenomena. Total annual research budget: $99,000. • B. R. Casserberg, Director of Graduate Study, 218-726-8247.

Announcement: MS in physics: concentrations in computer modeling, optics, and atomic, solid-state, and theoretical physics. Thesis research projects provide students with practical experience in numerical modeling, instrumentation, electronics, and microprocessor interfacing. Recent research has expanded into general relativity, quark model calculations, vacuum-tunneling microscopy, and photoreflectance solid-state measurements.

University of Minnesota, Twin Cities Campus, Institute of Technology, School of Physics and Astronomy, Minneapolis, MN 55455. School offers programs in astronomy (MS, PhD), physics (MS, PhD). *Degree requirements:* Thesis/dissertation. *Entrance requirements:* GRE General Test, GRE Subject Test. *Expenses:* Tuition of $1084 per quarter full-time, $301 per credit part-time for state residents; $2168 per quarter full-time, $602 per credit part-time for nonresidents. Fees of $118 per quarter. • Marvin Marshak, Head, 612-624-6062.

University of Mississippi, Graduate School, College of Liberal Arts, Department of Physics and Astronomy, University, MS 38677. Department offers program in physics (MA, MS, PhD). Matriculated students: 30 full-time (2 women), 2 part-time (1 woman); includes 1 minority (black American), 11 foreign. In 1990, 2 master's, 2 doctorates awarded. *Degree requirements:* For master's, thesis required (for some programs), foreign language not required; for doctorate, dissertation required, foreign language not required. *Entrance requirements:* For master's, GRE General Test, minimum GPA of 3.0; for doctorate, GRE General Test (minimum combined score of 900 required). Application deadline: 8/1. Application fee: $15 ($25 for foreign students). *Expenses:* Tuition of $1011 per semester full-time, $99 per semester part-time for state residents; $1842 per semester full-time, $180 per semester part-time for nonresidents. Fees of $219 per year full-time. *Financial aid:* Application deadline 3/1. • Dr. James Reidy, Chairman, 601-232-7046.

University of Missouri–Columbia, College of Arts and Sciences, Department of Physics, Columbia, MO 65211. Department awards MS, PhD. Faculty: 19 full-time, 0 part-time. Matriculated students: 23 full-time (0 women), 14 part-time (3 women); includes 1 minority (Asian American), 15 foreign. In 1990, 1 master's, 5 doctorates awarded. Terminal master's awarded for partial completion of doctoral program. *Degree requirements:* For master's, foreign language and thesis not required; for doctorate, 1 foreign language, dissertation. *Entrance requirements:* GRE General Test, minimum GPA of 3.0. Application deadline: 8/1 (priority date, applications processed on a rolling basis). Application fee: $20 ($40 for foreign students). *Expenses:* Tuition of $89.90 per credit hour full-time, $98.35 per credit hour part-time for state residents; $244 per credit hour full-time, $252.45 per credit hour part-time for nonresidents. Fees of $123.55 per semester (minimum) full-time. • Brian DeFacio, Director of Graduate Studies, 314-882-7024.

Announcement: Emphases include X-ray and neutron diffraction, neutron interferometry, surface and interface physics, gravitation, plasma physics, elementary particles and fields, and astrophysics. Outstanding facilities: 10-megawatt reactor for studying solids with elastic and inelastic neutron scattering as well as intense sources for Mössbauer spectroscopy. UHV facilities for high-resolution LEED, Auger, IR spectroscopy, and surface X-ray scattering. State-of-the-art laser lab for reflection and adsorption Raman studies of liquids and solids at high pressures. Total annual research budget $1-million. Financial: teaching appointments for 9 months are $9200 and include full tuition waiver. Five summer research appointments of $3000 are offered to incoming students. Six additional summer research fellowships of $2000 are offered annually. Contact Professor Brian Defacio, Director of Graduate Studies.

University of Missouri–Kansas City, College of Arts and Sciences, Department of Physics, Kansas City, MO 64110. Department awards MS. Part-time programs available. Faculty: 10 full-time (0 women), 1 part-time (0 women). Matriculated students: 12 full-time (4 women), 11 part-time (2 women); includes 0 minority, 9 foreign. In 1990, 5 degrees awarded. *Degree requirements:* Foreign language and thesis not required. *Entrance requirements:* GRE. Application fee: $0. *Expenses:* Tuition of $2200 per year full-time, $92 per credit hour part-time for state residents; $5503 per year full-time, $229 per credit hour part-time for nonresidents. Fees of $122 per semester full-time, $9 per credit hour part-time. *Financial aid:* Fellowships, research assistantships, teaching assistantships, full and partial tuition waivers, federal work-study, institutionally sponsored loans available. Aid available to part-time students. *Faculty research:* Condensed matter theory and experiment, optical properties of materials, relativity and quantum theory, scanning tunneling microscopy, chemical physics. Total annual research budget: $430,328. • Dr. James Phillips, Chairperson, 816-235-2501.

University of Missouri–Rolla, College of Arts and Sciences, Department of Physics, Rolla, MO 65401. Department awards MS, PhD. Faculty: 21 full-time (1 woman), 5 part-time (0 women). Matriculated students: 32 full-time (2 women), 0 part-time; includes 5 minority (3 Asian American, 1 Hispanic American, 1 Native American), 8 foreign. Average age 28. 9 applicants, 100% accepted. In 1990, 2 master's awarded (100% continued full-time study); 4 doctorates awarded (50% entered university research/teaching, 50% found other work related to degree). Terminal master's awarded for partial completion of doctoral program. *Degree requirements:* For master's, thesis optional, foreign language not required; for doctorate, 1 foreign language, dissertation. *Entrance requirements:* GRE General Test (minimum combined score of 1100 required, TOEFL (minimum score of 570 required). Application deadline: 7/1 (applications processed on a rolling basis). Application fee: $20 ($40 for foreign students). *Expenses:* Tuition of $2090 per year full-time, $87.10 per credit hour part-time for state residents; $5582 per year full-time, $232.60 per credit hour part-time for nonresidents. Fees of $349 per year full-time, $61.63 per semester (minimum) part-time. *Financial aid:* In 1990–91, 27 students received a total of $247,326 in aid awarded. 14 research assistantships (0 to first-year students), 13 teaching assistantships (5 to first-year students) were awarded; federal work-study, institutionally sponsored loans also available. Average monthly stipend for a graduate assistantship: $1127. Financial aid application deadline: 7/1. *Faculty research:* Solid-state physics, atomic and molecular physics, cloud physics. Total annual research budget: $660,161. • Dr. Ralph W. Alexander Jr., Chairman, 314-341-4702.

University of Missouri–St. Louis, College of Arts and Sciences, Department of Physics, Normandy, MO 63121-4499. Department awards MS, PhD. PhD offered jointly with University of Missouri-Rolla. Part-time and evening/weekend programs available. Faculty: 12 (0 women). Matriculated students: 11 full-time (2 women), 4 part-time (0 women); includes 5 minority (all Asian American), 7 foreign. Terminal master's awarded for partial completion of doctoral program. *Degree requirements:* For master's, foreign language and thesis not required; for doctorate, dissertation required, foreign language not required. *Entrance requirements:* For master's, GRE General Test; for doctorate, GRE General Test, GRE Subject Test. Application deadline: 7/1 (applications processed on a rolling basis). Application fee: $0. *Expenses:* Tuition of $2157 per year full-time, $89.90 per credit hour part-time for state residents; $5856 per year full-time, $244 per credit hour part-time for nonresidents. Fees of $235 per year full-time, $9.80 per credit hour part-time. *Financial aid:* In 1990–91, $9000 in aid awarded. 2 fellowships (1 to a first-year student), 2 research assistantships (0 to first-year students), 8 teaching assistantships (5 to first-year students) were awarded. *Faculty research:* Astronomy, solid-state physics, atomic physics, plasma physics, non-linear dynamics. • Dr. Bernard Feldman, Chairman, 314-553-5931. Application contact: Dr. T. P. Cheng, Graduate Coordinator, 314-553-5931.

University of Montana, College of Arts and Sciences, Department of Physics and Astronomy, Missoula, MT 59812. Department offers program in physics (MA, MS). *Degree requirements:* 1 foreign language, thesis. *Entrance requirements:* GRE General Test. Application deadline: 9/15. Application fee: $20. *Tuition:* $495 per quarter hour full-time for state residents; $1239 per quarter hour full-time for nonresidents.

Announcement: MA and MS programs of research are offered in ion optics, low- and medium-energy nuclear physics, computational physics, astronomy, astrophysics, fluid mechanics, plasma physics, and fluid dynamics. Facilities include access to excellent computing equipment, a solar telescope, and a 16-inch photometer-equipped telescope located at a nearby mountaintop observatory.

University of Nebraska–Lincoln, College of Arts and Sciences, Department of Physics and Astronomy, Lincoln, NE 68588. Department offers program in physics (MS, PhD). Faculty: 25 full-time (0 women), 1 part-time (0 women). Matriculated students: 30 full-time (4 women), 11 part-time (2 women); includes 4 minority (all Asian American), 20 foreign. Average age 29. In 1990, 4 master's, 4 doctorates awarded. Terminal master's awarded for partial completion of doctoral program. *Degree requirements:* For master's, foreign language and thesis not required; for doctorate, dissertation, comprehensive exams. *Entrance requirements:* GRE General Test, TOEFL (minimum score of 550 required). Application deadline: 5/1 (priority date, applications processed on a rolling basis). Application fee: $25. *Expenses:* Tuition of $75.75 per credit hour for state residents; $187.25 per credit hour for nonresidents. Fees of $161 per year full-time. *Financial aid:* Fellowships, research assistantships, teaching assistantships, federal work-study available. Aid available to part-time students. Financial aid application deadline: 2/15. • Dr. Anthony F. Starace, Chairperson, 402-472-2770.

Announcement: MS and PhD offered with full program of graduate courses. Research opportunities in experimental and theoretical atomic physics, experimental and theoretical condensed-matter physics, materials science, astronomy and astrophysics, elementary particles and general relativity, charged-particle interactions with matter, and archaeometry (MS only). Teaching assistantships and fellowships with tuition waiver available.

University of Nevada, Las Vegas, College of Science and Mathematics, Department of Physics, Las Vegas, NV 89154. Department awards MS, PhD. Part-time programs available. Faculty: 8 full-time (1 woman), 1 part-time (0 women). Matriculated students: 10 (3 women); includes 0 minority, 0 foreign. 21 applicants, 38% accepted. In 1990, 3 master's awarded. *Degree requirements:* For master's, thesis (for some programs), oral exam required, foreign language not required. *Entrance requirements:* For master's, GRE Subject Test, minimum GPA of 2.75 overall, 3.0 during previous 2 years. Application deadline: 6/15. Application fee: $20. *Expenses:* Tuition of $66 per credit. Fees of $1800 per semester for nonresidents. *Financial aid:* In 1990–91, 7 research assistantships awarded; teaching assistantships also available. Financial aid application deadline: 3/1. • Dr. James Selser, Chairman, 702-739-3563. Application contact: Graduate College Admissions Evaluator, 702-739-3320.

Announcement: Programs leading to MS and PhD degrees. Research includes experimental and theoretical laser physics, astronomy/astrophysics, and condensed-matter theory. Facilities include 4 laser laboratories, an automated telescope, observing time on the Hubble Space Telescope, and the UNLV Cray supercomputer. Financial aid: $7500/academic year, often with additional summer support.

University of Nevada, Reno, College of Arts and Science, Department of Physics, Reno, NV 89557. Department awards MS, PhD. Faculty: 8 (0 women). Matriculated students: 34 (6 women); includes 18 foreign. Average age 25. 50 applicants, 4% accepted. In 1990, 0 master's, 2 doctorates awarded. Terminal master's awarded for partial completion of doctoral program. *Degree requirements:* For master's, thesis optional, foreign language not required; for doctorate, 1 foreign language, dissertation. *Entrance requirements:* For master's, GRE General Test, GRE Subject Test, TOEFL, minimum GPA of 2.75; for doctorate, GRE General Test, GRE Subject Test, TOEFL. Application deadline: 8/1 (priority date, applications processed on a rolling basis). Application fee: $20. *Expenses:* Tuition of $0 for state residents; $3600 per year full-time, $66 per credit hour part-time for nonresidents. Fees of $66 per credit hour. *Financial aid:* In 1990–91, 15 research assistantships, 11 teaching assistantships awarded; federal work-study, institutionally sponsored loans also available. Average monthly stipend for a graduate assistantship: $740. Financial aid applicants required to submit FAF. *Faculty research:* Atomic and molecular physics. Total annual research budget: $43,120. • Dr. Neil Moore, Chairman, 702-784-6792.

University of New Brunswick, Faculty of Science, Department of Physics, Fredericton, NB E3B 5A3, Canada. Department awards M Sc, PhD. *Degree requirements:* Thesis/dissertation. *Entrance requirements:* TOEFL, minimum GPA of 3.0. Application deadline: 3/1 (priority date). *Expenses:* Tuition of $2100 per year. Fees of $45 per year. *Financial aid:* Research assistantships, teaching assistantships available. • Dr. G. R. Demille, Chairperson, 506-453-4723. Application contact: Dr. C. Linton, Director of Graduate Studies, 506-453-4723.

University of New Hampshire, College of Engineering and Physical Sciences, Department of Physics, Durham, NH 03824. Department awards MS, MST, PhD. Faculty: 26 full-time, 0 part-time. Matriculated students: 28 full-time (6 women), 9 part-time (1 woman); includes 9 foreign. 71 applicants, 25% accepted. In 1990, 8 master's, 3 doctorates awarded. Terminal master's awarded for partial completion of doctoral program. *Degree requirements:* For master's, foreign language and thesis not required; for doctorate, dissertation required, foreign language not required. *Application deadline:* 7/1 (priority date, applications processed on a rolling basis). *Application fee:* $25. *Tuition:* $1645 per semester full-time, $183 per credit hour part-time for state residents; $4920 per semester full-time, $547 per credit hour part-time for nonresidents. *Financial aid:* In 1990–91, 16 research assistantships, 16 teaching assistantships, 1 scholarship awarded; full and partial tuition waivers, federal work-study, and career-related internships or fieldwork also available. Aid available to part-time students. Financial aid application deadline: 2/15. *Faculty research:* Astrophysics and space physics, nuclear physics, atomic and molecular physics, nonlinear dynamical systems. • Dr. John R. Calarco, Chairperson, 603-862-1950. Application contact: Dr. Harvey Shepard, 603-862-1950.

University of New Mexico, College of Arts and Sciences, Department of Physics and Astronomy, Albuquerque, NM 87131. Department offers programs in optical sciences (MS, PhD), physics (MS, PhD). Faculty: 28 full-time (1 woman), 10 part-time (0 women). *Entrance requirements:* GRE General Test. Application fee: $25. *Expenses:* Tuition of $467 per semester (minimum) full-time, $67.50 per credit hour part-time for state residents; $1549 per semester (minimum) full-time, $67.50 per credit hour part-time for nonresidents. Fees of $16 per semester. *Financial aid:* Fellowships, research assistantships, teaching assistantships available. • Daniel Finley, Chairman, 505-277-2616.

Announcement: Research concentrations include abberation theory, astrophysics, biophysics, cosmic radiation, experimental surface physics, general relativity, infrared astronomy, laser physics, medical physics, molecular physics, theoretical nuclear physics, optical testing, experimental and theoretical particle physics, quantum optics, radio astronomy, scattering theory, solar energy, space physics, theoretical condensed matter. Research work is conducted on campus in newly remodeled labs of the department or at the University-based Center for High Technology Materials. Research is also possible at neighboring national facilities, such as Sandia National Laboratories, Phillips Laboratory, Los Alamos National Laboratories, and the Very Large Array Radio Observatory. Research sponsored by DOE, NSF, NASA, NIH, AFOSR, ONR. Twenty-six tuition-exempt, 9-month teaching assistantships with stipends of $7500 to $8000 are available. A few stipends up to $10,000 are awarded to exceptional students. Research assistantships available for advanced students.

University of New Orleans, College of Sciences, Department of Physics, New Orleans, LA 70148. Department offers programs in applied physics (MS), physics (MS). Part-time and evening/weekend programs available. Faculty: 9 full-time (1 woman). Matriculated students: 9 full-time (1 woman), 10 part-time (4 women); includes 1 minority (Asian American), 5 foreign. 6 applicants, 83% accepted. In 1990, 2 degrees awarded (100% found work related to degree). *Degree requirements:* Thesis required (for some programs), foreign language not required. *Entrance requirements:*

Directory: Physics

GRE General Test (minimum combined score of 1000 required), TOEFL (minimum score of 500 required). Application deadline: 7/1 (priority date, applications processed on a rolling basis). Application fee: $20. *Tuition:* $962 per quarter hour full-time for state residents; $2308 per quarter hour full-time for nonresidents. *Financial aid:* In 1990–91, 14 students received aid. 3 research assistantships (1 to a first-year student), 8 teaching assistantships (4 to first-year students) awarded; career-related internships or fieldwork also available. Average monthly stipend for a graduate assistantship: $1000. *Faculty research:* Underwater acoustics, applied electromagnetics, experimental atomic beams, digital signal processing, astrophysics. • Milton D. Slaughter, Chair, 504-286-6341. Application contact: Graduate Coordinator, 504-286-6343.

University of North Carolina at Chapel Hill, College of Arts and Sciences, Department of Physics and Astronomy, Chapel Hill, NC 27599. Department offers program in astronomy and astrophysics (MS, PhD). Faculty: 34 full-time, 2 part-time. *Degree requirements:* For master's, comprehensive exam required, foreign language not required; for doctorate, dissertation, comprehensive exam. *Entrance requirements:* GRE General Test (minimum combined score of 1000 required), minimum GPA of 3.0. Application deadline: 2/1. Application fee: $35. *Tuition:* $621 per semester full-time for state residents; $3555 per semester full-time for nonresidents. *Financial aid:* In 1990–91, 4 fellowships, 22 research assistantships, 33 teaching assistantships awarded. • Dr. Thomas B. Clegg, Chairman, 919-962-3016.

Announcement: Active research areas: astronomy and astrophysics; atomic, molecular, chemical physics; computational physics; condensed matter; microelectronics; field and particle theory; gravitation and relativity; nuclear and atomic physics. Facilities at UNC and Microelectronics Center of North Carolina, Triangle Universities Nuclear Laboratory, and North Carolina Supercomputing Center. 1990–91 salaries were $992–$1097 per month for teaching assistants and $914–$977 per month for research assistants.

University of North Carolina at Greensboro, College of Arts and Sciences, Department of Physics, Greensboro, NC 27412. Department awards M Ed, MS. Faculty: 7 full-time (0 women), 0 part-time. Matriculated students: 11 (2 women); includes 1 minority (black American), 0 foreign. In 1990, 1 degree awarded. *Degree requirements:* Foreign language not required. *Entrance requirements:* For MS: GRE General Test (preferred), MAT; for M Ed: GRE General Test or MAT or NTE. Application fee: $35. *Tuition:* $751 per semester full-time for state residents; $3685 per semester full-time for nonresidents. *Financial aid:* Research assistantships, teaching assistantships available. • Dr. Francis J. McCormack, Head, 919-334-5844.

University of North Dakota, College of Arts and Sciences, Department of Physics, Grand Forks, ND 58202. Department awards MS, PhD. Faculty: 8 full-time (0 women), 0 part-time. Matriculated students: 7 full-time (2 women), 4 part-time (0 women). 5 applicants, 100% accepted. In 1990, 3 master's, 0 doctorates awarded. *Degree requirements:* For master's, thesis; for doctorate, 1 foreign language, dissertation. *Entrance requirements:* For master's, TOEFL (minimum score of 550 required), minimum GPA of 3.0; for doctorate, GRE General Test, GRE Subject Test, TOEFL (minimum score of 550 required), minimum GPA of 3.0. Application deadline: 3/15 (priority date, applications processed on a rolling basis). Application fee: $20. *Tuition:* $2250 per year full-time, $94 per semester hour part-time for state residents; $5616 per year full-time, $234 per semester hour part-time for nonresidents. *Financial aid:* In 1990–91, 800 students received aid. 1 research assistantship, 7 teaching assistantships awarded; fellowships, full and partial tuition waivers, federal work-study, institutionally sponsored loans also available. Financial aid application deadline: 3/15. • B. S. Rao, Chairperson, 701-777-2911.

University of North Texas, College of Arts and Sciences, Department of Physics, Denton, TX 76203. Department awards MA, MS, PhD. Faculty: 20 full-time, 0 part-time. Matriculated students: 56 full-time (9 women), 14 part-time (0 women); includes 3 minority (1 Asian American, 2 Hispanic American), 50 foreign. Average age 26. 150 applicants, 25% accepted. In 1990, 8 master's, 5 doctorates awarded. Terminal master's awarded for partial completion of doctoral program. *Degree requirements:* For master's, thesis or alternative, comprehensive exam required, foreign language not required; for doctorate, 1 foreign language (computer language can substitute), dissertation, comprehensive exam. *Entrance requirements:* GRE General Test. Application deadline: 8/1. Application fee: $25. *Expenses:* Tuition of $40 per credit hour for state residents; $128 per credit hour for nonresidents. Fees of $298 per year full-time, $38 per year part-time. *Financial aid:* In 1990–91, $360,000 in aid awarded. 4 fellowships (0 to first-year students), 15 research assistantships (2 to first-year students), 38 teaching assistantships (2 to first-year students) were awarded; federal work-study, institutionally sponsored loans, and career-related internships or fieldwork also available. Financial aid application deadline: 4/1. *Faculty research:* Accelerator-based atomic physics, solid-state physics, molecular spectroscopy, magnetic resonance, astrophysics. Total annual research budget: $2-million. • Dr. Bruce J. West, Chair, 817-565-2630. Application contact: Dr. William Deering, Graduate Adviser, 817-565-2630.

Announcement: MS and PhD offered with full program of graduate courses in experimental physics: atomic, NMR, condensed-matter, molecular beams, and materials characterization; theoretical physics: all of the above, as well as nonlinear dynamics systems theory, biophysics, and nonequilibrium statistical mechanics. Teaching assistantships and fellowships with tuition waiver available.

University of Notre Dame, College of Science, Department of Physics, Notre Dame, IN 46556. Department awards MS, PhD. Faculty: 36 full-time (4 women), 0 part-time. Matriculated students: 83 full-time (9 women), 1 (woman) part-time; includes 1 minority (Asian American), 41 foreign. Average age 24. 215 applicants, 13% accepted. In 1990, 5 master's, 9 doctorates awarded. Terminal master's awarded for partial completion of doctoral program. *Degree requirements:* For master's, thesis or alternative; for doctorate, dissertation. *Entrance requirements:* GRE, TOEFL. Application deadline: 3/1 (priority date, applications processed on a rolling basis). Application fee: $25. *Tuition:* $13,385 per year full-time, $744 per credit hour part-time. *Financial aid:* In 1990–91, $1.23-million in aid awarded. 5 fellowships (2 to first-year students), 41 research assistantships (2 to first-year students), 37 teaching assistantships (13 to first-year students) were awarded; full and partial tuition waivers also available. Financial aid application deadline: 2/15. *Faculty research:* Elementary particle, nuclear, atomic, condensed-state physics; astrophysics. Total annual research budget: $3.647-million. • Dr. Gerald L. Jones, Chairman, 219-239-6386.

See full description on page 655.

University of Oklahoma, College of Arts and Sciences, Department of Physics and Astronomy, Program in Physics, Norman, OK 73019. Program awards M Nat Sci, MS, PhD. Part-time programs available. Terminal master's awarded for partial completion of doctoral program. *Degree requirements:* For master's, thesis or alternative, departmental qualifying exam required, foreign language not required; for doctorate, dissertation, written and oral exams, departmental qualifying exam, comprehensive exam required, foreign language not required. *Entrance requirements:* For master's, GRE General Test, GRE Subject Test, TOEFL (minimum score of 550 required), previous course work in physics; for doctorate, GRE General Test, GRE Subject Test, TOEFL (minimum score of 550 required). Application deadline: 4/1 (priority date, applications processed on a rolling basis). Application fee: $10. *Expenses:* Tuition of $63 per credit hour for state residents; $192 per credit hour for nonresidents. Fees of $67.50 per semester. *Financial aid:* In 1990–91, $23,200 in aid awarded. 14 research assistantships, 15 teaching assistantships (1 to a first-year student) were awarded; federal work-study, institutionally sponsored loans also available. Aid available to part-time students. Financial aid application deadline: 3/1. *Faculty research:* Atomic and molecular physics, experimental high-energy physics, experimental solid-state physics, atmospheric science. • Application contact: Dr. Robert F. Petry, Chair, Graduate Studies Committee, 405-325-3961.

University of Oklahoma, College of Arts and Sciences, Department of Physics and Astronomy, Program of Engineering Physics, Norman, OK 73019. Program awards M Nat Sci, MS, PhD. Part-time programs available. Faculty: 26 full-time (2 women), 0 part-time. Matriculated students: 46 full-time (8 women), 2 part-time (0 women); includes 1 minority, 19 foreign. Average age 32. 170 applicants, 34% accepted. In 1990, 4 master's awarded (100% continued full-time study); 4 doctorates awarded (100% entered university research/teaching). Terminal master's awarded for partial completion of doctoral program. *Degree requirements:* For master's, thesis or alternative, department qualifying exam required, foreign language not required; for doctorate, dissertation, department qualifying exam, comprehensive exam required, foreign language not required. *Entrance requirements:* GRE General Test, TOEFL (minimum score of 550 required), GRE Subject Test (physics), previous course work in physics. Application deadline: 3/1 (priority date, applications processed on a rolling basis). Application fee: $10. *Expenses:* Tuition of $63 per credit hour for state residents; $192 per credit hour for nonresidents. Fees of $67.50 per semester. *Financial aid:* In 1990–91, $156,000 in aid awarded. Fellowships, teaching assistantships available. *Faculty research:* Atomic and molecular physics, high energy physics, condensed matter physics, astrophysics, applied physics. • Application contact: Dr. Kimball Milton, Chair, Graduate Studies Committee, 405-325-3961.

University of Oregon, Graduate School, College of Arts and Sciences, Department of Physics, Eugene, OR 97403. Department awards MA, MS, PhD. Faculty: 29 full-time (0 women), 0 part-time. Matriculated students: 149 full-time (15 women), 13 part-time (0 women); includes 2 minority (1 Asian American, 1 Hispanic American), 98 foreign. 275 applicants, 59% accepted. In 1990, 19 master's, 12 doctorates awarded. Terminal master's awarded for partial completion of doctoral program. *Degree requirements:* For master's, foreign language and thesis not required; for doctorate, dissertation required, foreign language not required. *Entrance requirements:* GRE General Test, TOEFL (minimum score of 500 required). Application deadline: 8/15. Application fee: $40. *Tuition:* $1171 per quarter full-time, $247 per credit part-time for state residents; $1980 per quarter full-time, $336 per credit part-time for nonresidents. *Financial aid:* In 1990–91, $656,000 in aid awarded. 50 research assistantships (0 to first-year students), 30 teaching assistantships (13 to first-year students), 12 traineeships (0 to first-year students) were awarded; federal work-study, institutionally sponsored loans also available. Financial aid application deadline: 3/1; applicants required to submit FAF. *Faculty research:* Solid-state and chemical physics, optical physics, elementary particle physics, astrophysics, atomic and molecular physics. Total annual research budget: $3.4-million. • Dr. David McDaniels, Head, 503-346-4751.

University of Ottawa, Faculty of Science, Department of Physics, Ottawa, ON K1N 6N5, Canada. Department awards M Sc, PhD. Offered jointly with Carleton University. *Degree requirements:* Thesis/dissertation required, foreign language not required. *Entrance requirements:* For master's, honours bachelor's degree or equivalent, minimum B average; for doctorate, minimum B+ average. Application deadline: 3/1. Application fee: $10. *Faculty research:* Low-temperature, solid-state, ion, and high-energy physics.

University of Pennsylvania, School of Arts and Sciences, Graduate Group in Physics, Philadelphia, PA 19104. Group awards MS, PhD. Faculty: 41 (1 woman). Matriculated students: 102 full-time (5 women), 3 part-time (0 women). 250 applicants, 33% accepted. In 1990, 3 master's, 6 doctorates awarded. Terminal master's awarded for partial completion of doctoral program. *Degree requirements:* For master's, thesis or alternative required, foreign language not required; for doctorate, dissertation required, foreign language not required. *Entrance requirements:* GRE General Test, GRE Subject Test, TOEFL. Application deadline: 2/1. Application fee: $40. *Expenses:* Tuition of $15,619 per year full-time, $1978 per course part-time. Fees of $965 per year full-time, $112 per course part-time. *Financial aid:* Application deadline 2/1. • Dr. Terry Fortune, Chairperson, 215-898-3125.

See full description on page 657.

University of Pittsburgh, Faculty of Arts and Sciences, Department of Crystallography, Pittsburgh, PA 15260. Department awards MS, PhD. Part-time programs available. Faculty: 7 full-time (1 woman), 1 part-time (0 women). Matriculated students: 14 full-time (3 women), 0 part-time; includes 1 minority (Asian American), 12 foreign. 10 applicants, 20% accepted. In 1990, 0 master's, 1 doctorate awarded. Terminal master's awarded for partial completion of doctoral program. *Degree requirements:* For master's, thesis required, foreign language not required; for doctorate, 1 foreign language, dissertation. *Entrance requirements:* For master's, GRE General Test, TOEFL, minimum QPA of 3.0; for doctorate, GRE General Test, TOEFL. Application deadline: 8/1 (priority date, applications processed on a rolling basis). Application fee: $15 ($25 for foreign students). *Expenses:* Tuition of $2920 per semester full-time, $241 per credit part-time for state residents; $5840 per semester full-time, $482 per credit part-time for nonresidents. Fees of $156 per year. *Financial aid:* In 1990–91, 11 students received a total of $115,500 in aid awarded. 1 fellowship, 11 research assistantships (2 to first-year students) were awarded; federal work-study, institutionally sponsored loans also available. Aid available to part-time students. Average monthly stipend for a graduate assistantship: $875. Financial aid application deadline: 8/1; applicants required to submit FAF. *Faculty research:* Crystal structures of organic molecules, protein crystallography, crystallographic computing, charge density studies, theoretical calculations. • Dr. Bryan M. Craven, Chairman, 412-624-9317.

University of Pittsburgh, Faculty of Arts and Sciences, Department of Physics and Astronomy, Program in Physics, Pittsburgh, PA 15260. Program awards MS, PhD. Faculty: 45 full-time (1 woman), 2 part-time (0 women). Matriculated students: 73 full-time (12 women), 9 part-time (2 women); includes 0 minority, 51 foreign. 425 applicants, 5% accepted. In 1990, 12 master's, 13 doctorates awarded. *Degree requirements:* For master's, thesis or alternative required, foreign language not required; for doctorate, dissertation required, foreign language not required. *Entrance requirements:* For master's, GRE General Test, GRE Subject Test, TOEFL, minimum QPA of 3.0; for doctorate, GRE General Test, GRE General Test, TOEFL. Application deadline: 1/30 (priority date). Application fee: $15 ($25 for foreign students).

Expenses: Tuition of $2920 per semester full-time, $241 per credit part-time for state residents; $5840 per semester full-time, $482 per credit part-time for nonresidents. Fees of $156 per year. *Financial aid:* In 1990–91, 75 students received a total of $592,900 in aid awarded. 4 fellowships (0 to first-year students), 45 research assistantships (0 to first-year students), 26 teaching assistantships (21 to first-year students) were awarded; federal work-study, institutionally sponsored loans also available. Aid available to part-time students. Average monthly stipend for a graduate assistantship: $988. Financial aid application deadline: 1/30; applicants required to submit FAF. *Faculty research:* Atomic and atmospheric physics, intermediate and high energy physics (theory and experiment), general relativity (theory), condensed matter physics(theory and experiment), astronomy. Total annual research budget: $5.132-million. • Application contact: Raymond S. Willey, Admissions Chairman, 412-624-9041.

See full description on page 659.

University of Puerto Rico, Mayagüez Campus, College of Arts and Sciences, Department of Physics, Mayagüez, PR 00709. Department awards MS. Part-time programs available. Faculty: 12 full-time (2 women), 0 part-time. Matriculated students: 16 full-time (2 women), 1 part-time (0 women); includes 10 minority (all Hispanic American), 7 foreign. 9 applicants, 56% accepted. In 1990, 4 degrees awarded. *Degree requirements:* Thesis required, foreign language not required. *Application deadline:* 10/15. *Application fee:* $15. *Expenses:* Tuition of $45 per credit for commonwealth residents; $3000 per semester for nonresidents. Fees of $344 per semester. *Financial aid:* In 1990–91, 0 research assistantships awarded; teaching assistantships, federal work-study, institutionally sponsored loans also available. *Faculty research:* Atomic and molecular physics, nuclear physics, nonlinear thermostatics, fluid dynamics, molecular spectroscopy. • Dr. Rubén Méndez, Director, 809-832-4040 Ext. 3844.

University of Puerto Rico, Río Piedras, Faculty of Natural Sciences, Department of Physics, Río Piedras, PR 00931. Department offers programs in applied physics (MS), chemical physics (PhD), physics (MS). Part-time and evening/weekend programs available. Faculty: 15 (1 woman). Matriculated students: 13 full-time (8 women), 23 part-time (13 women); includes 11 foreign. In 1990, 2 master's, 0 doctorates awarded. *Degree requirements:* Thesis/dissertation, comprehensive exam required, foreign language not required. *Entrance requirements:* For master's, GRE, minimum GPA of 2.5; for doctorate, GRE, minimum GPA of 2.5, master's degree. Application deadline: 2/1. Application fee: $45. *Expenses:* Tuition of $55 per credit hour for commonwealth residents; $55 per credit hour (minimum) for nonresidents. Fees of $286 per year. *Financial aid:* Fellowships, research assistantships, teaching assistantships, partial tuition waivers, federal work-study, institutionally sponsored loans available. Financial aid application deadline: 5/31. *Faculty research:* Low frequency radio studies of active extragalactic sources, structural studies of cylindrical polyelectrolyte solutions, spectroscopy of small clusters of hydrogen, light scattering and infrared spectroscopy of solids, spontaneously broken global symmetrics in supersymmetric theories. • Dr. Alfredo Torruellas, Chairman, 809-764-0000 Ext. 4746.

See full description on page 661.

University of Regina, Faculty of Graduate Studies and Research, Faculty of Science, Department of Physics, Regina, SK S4S 0A2, Canada. Department awards M Sc, PhD. Faculty: 8 full-time, 0 part-time. Matriculated students: 5 (0 women). In 1990, 0 master's, 3 doctorates awarded. *Degree requirements:* For master's, thesis required, foreign language not required; for doctorate, variable foreign language requirement, dissertation. *Application deadline:* 7/2 (applications processed on a rolling basis). *Tuition:* $1500 per year full-time, $242 per year (minimum) part-time. • Dr. I. Naqvi, Head, 306-585-4149.

University of Rhode Island, College of Arts and Sciences, Department of Physics, Kingston, RI 02881. Department awards MS, PhD. *Application deadline:* 4/15. *Application fee:* $25. *Expenses:* Tuition of $2575 per year full-time, $120 per credit hour part-time for state residents; $5900 per year full-time, $274 per credit hour part-time for nonresidents. Fees of $696 per year full-time.

University of Rochester, College of Arts and Science, Department of Physics and Astronomy, Rochester, NY 14627-0001. Department awards MA, MS, PhD. Programs offered in astronomy (PhD), physics (MA, MS, PhD). Part-time programs available. Faculty: 37 full-time, 0 part-time. Matriculated students: 121 full-time (16 women), 1 part-time (0 women); includes 2 minority (both Native American), 63 foreign. 543 applicants, 11% accepted. In 1990, 21 master's, 18 doctorates awarded. Terminal master's awarded for partial completion of doctoral program. *Degree requirements:* For master's, thesis (for some programs), comprehensive exam required, foreign language not required; for doctorate, dissertation, comprehensive exam required, foreign language not required. *Entrance requirements:* For doctorate, GRE. Application deadline: 2/15 (priority date). Application fee: $25. *Expenses:* Tuition of $473 per credit hour. Fees of $243 per year full-time. *Financial aid:* Fellowships, research assistantships, teaching assistantships available. Financial aid application deadline: 2/15. *Faculty research:* Condensed matter, biophysics, quantum optics, astrophysics, observational astronomy. • Paul Slattery, Chair, 716-275-4344.

See full description on page 665.

University of Saskatchewan, College of Arts and Sciences, Department of Physics, Saskatoon, SK S7N 0W0, Canada. Department awards M Sc, PhD. *Degree requirements:* Thesis/dissertation. *Entrance requirements:* TOEFL. Application fee: $0.

University of South Carolina, Graduate School, College of Science and Mathematics, Department of Physics and Astronomy, Columbia, SC 29208. Department awards IMA, MAT, MS, PhD. IMA and MAT offered in cooperation with the College of Education. Part-time programs available. Faculty: 22 full-time (0 women), 2 part-time (0 women). Matriculated students: 33 full-time (6 women), 0 part-time; includes 0 minority, 11 foreign. Average age 29. 42 applicants, 10% accepted. In 1990, 1 master's awarded (100% continued full-time study); 4 doctorates awarded (100% entered university research/teaching). Terminal master's awarded for partial completion of doctoral program. *Degree requirements:* For master's, thesis required, foreign language not required; for doctorate, 1 foreign language, dissertation. *Entrance requirements:* GRE General Test. Application deadline: 8/1 (priority date, applications processed on a rolling basis). Application fee: $25. *Tuition:* $1404 per semester full-time. *Financial aid:* In 1990–91, 1 fellowship (0 to first-year students), 7 research assistantships (0 to first-year students), 24 teaching assistantships (5 to first-year students) awarded; federal work-study also available. Aid available to part-time students. *Faculty research:* Mechanics, electron spin resonance, magnetism, intermediate energy nuclear physics, high-energy physics. • Dr. Frank Avignone, Chairman, 803-777-4983. Application contact: H. A. Farach, Director of Graduate Studies, 803-777-6407.

See full description on page 667.

University of Southern California, Graduate School, College of Letters, Arts and Sciences, Division of Natural Sciences and Mathematics, Department of Physics, Los Angeles, CA 90089. Department awards MA, MS, PhD. Faculty: 21 (0 women). Matriculated students: 57 full-time (9 women), 5 part-time (1 woman); includes 2 minority (1 Asian American, 1 Hispanic American), 49 foreign. Average age 28. 93 applicants, 40% accepted. In 1990, 3 master's, 5 doctorates awarded. *Degree requirements:* For doctorate, dissertation. *Entrance requirements:* GRE General Test. Application deadline: 7/1 (priority date). Application fee: $50. *Expenses:* Tuition of $12,120 per year full-time, $505 per unit part-time. Fees of $280 per year. *Financial aid:* 54 students received aid. Fellowships, research assistantships, teaching assistantships, federal work-study, institutionally sponsored loans available. Average monthly stipend for a graduate assistantship: $878. Financial aid application deadline: 3/1. • Dr. Peter Lambropoulos, Chairman, 213-740-1108.

University of Southern Mississippi, College of Science and Technology, Department of Physics and Astronomy, Hattiesburg, MS 39406. Department awards MS. Faculty: 6 full-time, 2 part-time. *Entrance requirements:* GRE General Test (minimum combined score of 1000 required), minimum GPA of 2.75. Application deadline: 8/9 (priority date, applications processed on a rolling basis). Application fee: $0 ($25 for foreign students). *Expenses:* Tuition of $968 per semester full-time, $93 per semester hour part-time. Fees of $12 per semester part-time for state residents; $591 per year full-time, $12 per semester part-time for nonresidents. *Financial aid:* Application deadline 3/15. *Faculty research:* Polymers, atomic physics, fluid mechanics, liquid crystals, refractory materials. • Dr. Roger Hester, Chairman, 601-266-4934.

University of South Florida, College of Arts and Sciences, Department of Physics, Tampa, FL 33620. Department offers programs in engineering science/physics (PhD), physics (MA, MS). PhD administered through the College of Engineering. Part-time programs available. Faculty: 11 full-time (0 women), 0 part-time. Matriculated students: 13 full-time (0 women), 2 part-time (0 women); includes 2 minority (both Asian American), 4 foreign. Average age 29. 19 applicants, 53% accepted. In 1990, 10 master's, 0 doctorates awarded. Terminal master's awarded for partial completion of doctoral program. *Degree requirements:* For master's, thesis optional, foreign language not required; for doctorate, 2 foreign languages (computer language can substitute for one), dissertation. *Entrance requirements:* For master's, GRE General Test (minimum combined score of 1000 required), minimum GPA of 3.0 for the last 2 years. Application deadline: 6/6. Application fee: $15. *Tuition:* $79.40 per credit hour for state residents; $241.33 per credit hour for nonresidents. *Financial aid:* In 1990–91, 11 students received a total of $19,359 in aid awarded. Fellowships, research assistantships, teaching assistantships, federal work-study, institutionally sponsored loans available. *Faculty research:* Laser, medical, and solid-state physics. Total annual research budget: $600,000. • S. Sundaram, Chairperson, 813-974-2871. Application contact: H. R. Brooker, Coordinator, 813-974-2871.

University of Southwestern Louisiana, College of Sciences, Department of Physics, Lafayette, LA 70504. Department offers programs in applied physics (MS), physics (MS). Part-time programs available. Faculty: 6 full-time (0 women), 0 part-time. Matriculated students: 5 full-time (1 woman), 2 part-time (0 women); includes 0 minority, 7 foreign. Average age 24. 15 applicants, 60% accepted. In 1990, 1 degree awarded. *Degree requirements:* Thesis required, foreign language not required. *Entrance requirements:* GRE General Test. Application deadline: 8/15. Application fee: $5. *Tuition:* $1560 per year full-time, $228 per credit (minimum) part-time for state residents; $3310 per year full-time, $228 per credit (minimum) part-time for nonresidents. *Financial aid:* In 1990–91, $21,000 in aid awarded. 2 research assistantships, 2 teaching assistantships were awarded; federal work-study also available. Financial aid application deadline: 5/1. *Faculty research:* Environmental physics, geophysics, astrophysics, acoustics, atomic physics. Total annual research budget: $150,000. • Dr. Davy Bernard, Head, 318-231-6691. Application contact: Dr. L. Dwynn Lafleur, Graduate Coordinator, 318-231-6696.

University of Tennessee, Knoxville, College of Liberal Arts, Department of Physics and Astronomy, Knoxville, TN 37996. Department offers program in physics (MS, PhD). Part-time programs available. Faculty: 58 (4 women). Matriculated students: 75 full-time (14 women), 29 part-time (4 women); includes 3 minority (2 Asian American, 1 black American), 42 foreign. 201 applicants, 34% accepted. In 1990, 8 master's, 3 doctorates awarded. *Degree requirements:* For master's, thesis or alternative required, foreign language not required; for doctorate, 1 foreign language, dissertation. *Entrance requirements:* TOEFL (minimum score of 525 required), minimum GPA of 2.5. Application deadline: 2/1 (priority date, applications processed on a rolling basis). Application fee: $15. *Tuition:* $1086 per semester full-time, $142 per credit hour part-time for state residents; $2768 per semester full-time, $308 per credit hour part-time for nonresidents. *Financial aid:* In 1990–91, 1 fellowship, 38 research assistantships, 28 teaching assistantships awarded; federal work-study, institutionally sponsored loans also available. Financial aid application deadline: 2/1; applicants required to submit FAF. • Dr. William Bugg, Head, 615-974-3342.

See full description on page 669.

University of Tennessee Space Institute, Program in Physics, Tullahoma, TN 37388. Program awards MS, PhD. Faculty: 4 full-time (0 women), 2 part-time (0 women). Matriculated students: 15 full-time (2 women), 8 part-time (2 women); includes 4 foreign. In 1990, 1 master's, 0 doctorates awarded. *Degree requirements:* For master's, thesis required (for some programs), foreign language not required; for doctorate, 1 foreign language, dissertation. *Entrance requirements:* GRE General Test, GRE Subject Test. Application fee: $15. *Tuition:* $135 per credit hour for state residents; $296 per credit hour for nonresidents. *Financial aid:* In 1990–91, 1 fellowship awarded; research assistantships also available. Aid available to part-time students. • Dr. A. A. Mason, Degree Program Chairman, 615-455-0631 Ext. 466. Application contact: Dr. Edwin M. Gleason, Assistant Dean for Admissions and Student Affairs, 615-455-0631 Ext. 472.

University of Texas at Arlington, College of Science, Department of Physics, Arlington, TX 76019. Department offers programs in physics (MS, D Sc, PhD), radiological physics (MS). Faculty: 18. Matriculated students: 28 full-time (6 women), 10 part-time (2 women); includes 1 minority (Asian American), 16 foreign. 50 applicants, 20% accepted. In 1990, 1 doctorate awarded. *Degree requirements:* For master's, thesis optional, foreign language not required; for doctorate, dissertation, comprehensive exam. *Entrance requirements:* GRE General Test, TOEFL. Application deadline: rolling. Application fee: $25. *Tuition:* $40 per hour for state residents; $148 per hour for nonresidents. *Financial aid:* In 1990–91, 10 research assistantships, 16

Directory: Physics

teaching assistantships awarded. • Dr. R. N. West, Chairman, 817-273-2266. Application contact: Dr. Asok K. Ray, Graduate Admissions, 817-273-2503.

Announcement: Vigorous interdisciplinary research programs are being carried out in both experimental and theoretical physics. Current experimental research areas: experimental solid-state physics; high-energy physics; surface physics; positron studies of solids and surfaces; electron paramagnetic resonance; laser optics; inelastic and elastic scattering of protons and deuterons; thin-film targets; instrumentation; scanning tunneling microscopy. Current theoretical research areas: solid-state theory; energy-band theory; optical and magnetic properties of solids; clusters; chemisorption; surface physics; positron studies; superconductivity; alloy theory; quarks.

University of Texas at Austin, Graduate School, College of Natural Sciences, Department of Physics, Austin, TX 78712. Department awards MA, PhD. Matriculated students: 294 full-time (23 women), 0 part-time; includes 13 minority (9 Asian American, 4 Hispanic American), 105 foreign. 296 applicants, 47% accepted. In 1990, 11 master's, 32 doctorates awarded. *Entrance requirements:* GRE. Application deadline: 2/1 (priority date, applications processed on a rolling basis). Application fee: $40 ($75 for foreign students). *Tuition:* $510.30 per semester for state residents; $1806 per semester for nonresidents. *Financial aid:* Fellowships, research assistantships, teaching assistantships available. Financial aid application deadline: 3/1. • Dr. Austin Gleeson, Chairman, 512-471-1152. Application contact: Dr. Thomas Griffy, Graduate Adviser, 512-471-1053.

See full description on page 671.

University of Texas at Dallas, School of Natural Sciences and Mathematics, Program in Physics, Richardson, TX 75083-0688. Program awards MS, PhD. Part-time and evening/weekend programs available. Faculty: 14 full-time (0 women), 4 part-time (1 woman). Matriculated students: 54 full-time (5 women), 49 part-time (9 women); includes 10 minority (8 Asian American, 1 black American, 1 Hispanic American), 14 foreign. Average age 31. In 1990, 19 master's, 4 doctorates awarded. *Degree requirements:* For master's, industrial internship required, thesis optional, foreign language not required; for doctorate, dissertation, publishable paper required, foreign language not required. *Entrance requirements:* GRE General Test (minimum combined score of 1100 or minimum score of 700 on math section required), TOEFL (minimum score of 550 required), minimum GPA of 3.0 in upper level course work in field. Application deadline: 7/15 (applications processed on a rolling basis). Application fee: $0 ($75 for foreign students). *Expenses:* Tuition of $360 per semester full-time, $100 per semester (minimum) part-time for state residents; $2196 per semester full-time, $122 per semester hour (minimum) part-time for nonresidents. Fees of $338 per semester full-time, $22 per hour (minimum) part-time. *Financial aid:* In 1990–91, 3 students received a total of $15,436 in aid awarded. Fellowships, research assistantships, teaching assistantships, federal work-study, and career-related internships or fieldwork available. Aid available to part-time students. Financial aid application deadline: 11/1; applicants required to submit FAF. *Faculty research:* Atomic, molecular, atmospheric, chemical, solid state and space physics; optics and quantum electronics; relativity and astropysics; high energy particles. • Dr. John J. Hoffman, Head, 214-690-2846.

University of Texas at El Paso, College of Science, Department of Physics, 500 West University Avenue, El Paso, TX 79968. Department awards MS. Matriculated students: 13 full-time (4 women), 3 part-time (0 women); includes 4 minority (1 Asian American, 3 Hispanic American), 8 foreign. In 1990, 4 degrees awarded. *Application deadline:* 7/1 (priority date, applications processed on a rolling basis). *Application fee:* $0 ($50 for foreign students). *Expenses:* Tuition of $360 per semester full-time, $100 per semester (minimum) part-time for state residents; $2304 per semester full-time, $128 per credit hour (minimum) part-time for nonresidents. Fees of $137 per semester full-time, $28.50 per semester (minimum) part-time. *Financial aid:* Application deadline 3/1. • Clarence Cooper, Chair, 915-747-5715. Application contact: Diana Guerrero, Admissions Office, 915-747-5576.

University of the Pacific, Department of Physics, Stockton, CA 95211. Department awards MS. Faculty: 7 full-time (0 women), 0 part-time. Matriculated students: 1 (woman) full-time, 0 part-time; includes 1 foreign. Average age 23. In 1990, 0 degrees awarded. *Degree requirements:* Thesis required, foreign language not required. *Entrance requirements:* GRE General Test, GRE Subject Test. Application deadline: 5/1 (applications processed on a rolling basis). Application fee: $30. *Expenses:* Tuition of $14,160 per year full-time, $485 per unit part-time. Fees of $616 per unit. *Financial aid:* In 1990–91, $28,128 in aid awarded. 1 teaching assistantship (to a first-year student) was awarded. Aid available to part-time students. Financial aid application deadline: 3/1; applicants required to submit FAF. • Dr. Richard Perry, Chairman, 209-946-2220.

University of Toledo, College of Arts and Sciences, Department of Physics and Astronomy, Toledo, OH 43606. Offerings include physics (MS, MS Ed, PhD). Department faculty: 19 full-time (1 woman), 0 part-time. *Degree requirements:* For master's, thesis required, foreign language not required; for doctorate, 1 foreign language, dissertation, departmental qualifying exam. *Entrance requirements:* For master's, GRE General Test, GRE Subject Test. Application deadline: 9/8 (priority date). Application fee: $30. *Tuition:* $122.59 per credit hour for state residents; $193.40 per credit hour for nonresidents. • Dr. Philip James, Chairman, 419-537-2276.

University of Toronto, School of Graduate Studies, Physical Sciences Division, Department of Physics, Toronto, ON M5S 1A1, Canada. Department awards M Sc, PhD. Faculty: 65. Matriculated students: 147 full-time (16 women), 21 part-time (4 women); includes 35 foreign. 203 applicants, 41% accepted. In 1990, 48 master's, 16 doctorates awarded. *Degree requirements:* For master's, thesis optional; for doctorate, dissertation. *Application deadline:* 4/15. *Application fee:* $50. *Expenses:* Tuition of $2220 per year full-time, $666 per year part-time for Canadian residents; $10,198 per year full-time, $305.05 per year part-time for nonresidents. Fees of $277.56 per year full-time, $82.73 per year part-time. *Financial aid:* Application deadline 2/1. • M. B. Walker, Chair, 416-978-5205.

University of Utah, College of Science, Department of Physics, Salt Lake City, UT 84112. Department offers programs in chemical physics (PhD), physics (MA, M Phil, MS, PhD). Part-time programs available. Faculty: 28 (1 woman). Matriculated students: 42 full-time (4 women), 43 part-time (4 women); includes 1 minority (Asian American), 48 foreign. Average age 29. In 1990, 19 master's, 6 doctorates awarded. Terminal master's awarded for partial completion of doctoral program. *Degree requirements:* For master's, 1 foreign language, thesis or alternative, teaching experience; for doctorate, dissertation, departmental qualifying exam required, foreign language not required. *Entrance requirements:* For master's, GRE General Test, GRE Subject Test, minimum GPA of 3.0; for doctorate, GRE Subject Test, minimum GPA of 3.0. Application deadline: 8/1. Application fee: $25 ($50 for foreign students). *Tuition:* $195 per credit for state residents; $505 per credit for nonresidents. *Financial aid:* In 1990–91, 40 teaching assistantships awarded; fellowships, research assistantships, federal work-study, institutionally sponsored loans also available. Financial aid application deadline: 3/31. *Faculty research:* High-energy physics, crystal growth, low-temperature physics, cosmic ray physics, solid-state physics. • Craig Taylor, Chairman, 801-581-6901. Application contact: Carleton Detar, Director of Graduate Studies, 801-581-7115.

University of Utah, College of Science, Departments of Chemistry and Physics, Interdepartmental Program in Chemical Physics, Salt Lake City, UT 84112. Program awards PhD. *Degree requirements:* Dissertation, exams required, foreign language not required. *Application fee:* $25 ($50 for foreign students). *Tuition:* $195 per credit for state residents; $505 per credit for nonresidents. • C. H. Wang, Chair, 801-581-8445.

University of Vermont, College of Arts and Sciences, Department of Physics, Burlington, VT 05405. Department offers programs in engineering physics (MS), physical sciences (MST), physics (MAT, MS). Matriculated students: 10; includes 0 minority, 0 foreign. 21 applicants, 38% accepted. In 1990, 0 degrees awarded. *Degree requirements:* Computer language required, foreign language not required. *Entrance requirements:* GRE General Test, GRE Subject Test, TOEFL (minimum score of 550 required). Application deadline: 4/1 (priority date, applications processed on a rolling basis). Application fee: $25. *Expenses:* Tuition of $206 per credit for state residents; $564 per credit for nonresidents. Fees of $150 per semester full-time. *Financial aid:* In 1990–91, 1 fellowship, 0 research assistantships, 8 teaching assistantships awarded. Financial aid application deadline: 3/1. • Dr. David Smith, Chairperson, 802-656-2644. Application contact: L. Scarfone, Coordinator, 802-656-2644.

University of Victoria, Faculty of Arts and Science, Department of Physics, Victoria, BC V8W 2Y2, Canada. Department offers programs in astronomy and astrophysics (M Sc, PhD), condensed-matter physics (M Sc, PhD), geophysics (M Sc, PhD), nuclear and particle studies (M Sc, PhD), physics of fluids (M Sc, PhD), theoretical physics (M Sc, PhD). Faculty: 21 full-time (0 women), 0 part-time. Matriculated students: 50 full-time (9 women), 1 part-time (0 women); includes 20 foreign. Average age 29. 253 applicants, 13% accepted. In 1990, 3 master's, 5 doctorates awarded. *Degree requirements:* Thesis/dissertation required, foreign language not required. *Application deadline:* 5/31 (priority date, applications processed on a rolling basis). *Application fee:* $20. *Expenses:* Tuition of $754 per semester. Fees of $23 per year. *Financial aid:* Fellowships, research assistantships, teaching assistantships, institutionally sponsored loans, and career-related internships or fieldwork available. *Faculty research:* Geomagnetism, biophysics, nuclear magnetic resonance. • Dr. L. P. Robertson, Chair, 604-721-7698. Application contact: Dr. J. T. Weaver, Graduate Adviser, 604-721-7768.

University of Virginia, Graduate School of Arts and Sciences, Department of Physics, Charlottesville, VA 22906. Department awards MA, MAT, MS, PhD. Faculty: 39 full-time (1 woman), 0 part-time. Matriculated students: 84 full-time (8 women), 1 part-time (0 women); includes 1 minority (Hispanic American), 34 foreign. Average age 26. 108 applicants, 52% accepted. In 1990, 4 master's, 12 doctorates awarded. *Degree requirements:* Thesis/dissertation required, foreign language not required. *Entrance requirements:* GRE General Test, GRE Subject Test. Application deadline: 7/15. Application fee: $40. *Expenses:* Tuition of $2740 per year full-time, $904 per year (minimum) part-time for state residents; $8950 per year full-time, $2960 per year (minimum) part-time for nonresidents. Fees of $586 per year full-time, $342 per year part-time. *Financial aid:* Application deadline 2/1. • Michael Fowler, Chairman, 804-924-3781. Application contact: William A. Elwood, Associate Dean, 804-924-7184.

See full description on page 673.

University of Washington, College of Arts and Sciences, Department of Physics, Seattle, WA 98195. Department awards MS, PhD. Part-time and evening/weekend programs available. Faculty: 56 full-time (2 women). Matriculated students: 125 full-time (15 women), 68 part-time (9 women); includes 6 minority (2 Asian American, 2 black American, 2 Hispanic American), 34 foreign. Average age 30. 370 applicants, 22% accepted. In 1990, 24 master's, 14 doctorates awarded. *Degree requirements:* For master's, foreign language and thesis not required; for doctorate, dissertation required, foreign language not required. *Entrance requirements:* For doctorate, GRE General Test, GRE Subject Test, TOEFL. Application deadline: 2/15 (priority date, applications processed on a rolling basis). Application fee: $35. *Tuition:* $1129 per quarter full-time, $324 per credit (minimum) part-time for state residents; $2824 per quarter full-time, $809 per credit (minimum) part-time for nonresidents. *Financial aid:* In 1990–91, $1.2-million in aid awarded. 2 fellowships (both to first-year students), 4 research assistantships (all to first-year students) were awarded; teaching assistantships, federal work-study also available. Average monthly stipend for a graduate assistantship: $950. Financial aid application deadline: 2/15. *Faculty research:* Atomic, molecular, and condensed-matter physics; elementary particles; nuclear physics; astrophysics. Total annual research budget: $7.554-million. • Mark McDermott, Chairman, 206-543-2770. Application contact: Graduate Program Assistant, 206-543-2770.

Announcement: MS and PhD programs in theoretical and experimental particle, nuclear, condensed-matter, surface, and atomic physics; astrophysics; general relativity; and physics education. Facilities: condensed-matter, atomic, high-energy experiment, nuclear laboratories; Nuclear Theory Institute; departmental computer. Contact Graduate Admissions at above address.

University of Waterloo, Faculty of Science, Guelph-Waterloo Program in Physics, Waterloo, ON N2L 3G1, Canada. Program awards M Sc, PhD. Faculty: 36 full-time, 1 part-time. Matriculated students: 43 full-time (4 women), 3 part-time (0 women); includes 15 foreign. 87 applicants, 24% accepted. In 1990, 12 master's, 4 doctorates awarded. *Degree requirements:* Thesis/dissertation. *Entrance requirements:* For master's, TOEFL (minimum score of 500 required), honor's degree, minimum B average; for doctorate, TOEFL (minimum score of 500 required), master's degree. Application fee: $25. *Expenses:* Tuition of $757 per year full-time, $530 per year part-time for Canadian residents; $3127 per year for nonresidents. Fees of $68 per year full-time, $17 per year part-time. *Financial aid:* Research assistantships, teaching assistantships available. • Dr. F. R. Hallett, Director, 519-824-4120 Ext. 2263. Application contact: K. Harris, Administrative Assistant, 519-824-4420 Ext. 2263.

University of Western Ontario, Physical Sciences Division, Department of Physics, London, ON N6A 3K7, Canada. Department awards M Sc, PhD. *Tuition:* $1015 per year full-time, $1050 per year part-time for Canadian residents; $4207 per year for nonresidents.

University of Windsor, Faculty of Science, Department of Physics, Windsor, ON N9B 3P4, Canada. Department awards M Sc, PhD. Part-time programs available. Faculty: 15 full-time (0 women), 0 part-time. Matriculated students: 17 full-time, 0 part-time; includes 5 foreign. In 1990, 1 master's, 1 doctorate awarded. *Degree requirements:* For master's, thesis (for some programs); for doctorate, dissertation. *Entrance* *requirements:* For master's, GRE, TOEFL, minimum B average. Application deadline: 7/1 (priority date, applications processed on a rolling basis). Application fee: $0. *Tuition:* $819.15 per semester full-time for Canadian residents; $3646 per semester full-time for nonresidents. *Financial aid:* Research assistantships, teaching assistantships available. Average monthly stipend for a graduate assistantship: $700. • Dr. Mordechai Schlesinger, Head, 519-253-4232 Ext. 2647. Application contact: Admissions Officer, 519-253-4232 Ext. 2108.

University of Wisconsin–Madison, College of Letters and Science, Department of Physics, Madison, WI 53706. Department awards MS, PhD. Faculty: 46 full-time, 0 part-time. Matriculated students: 174 full-time (20 women), 26 part-time (2 women); includes 9 minority (5 Asian American, 1 black American, 2 Hispanic American, 1 Native American), 65 foreign. 552 applicants, 15% accepted. In 1990, 11 master's, 24 doctorates awarded. *Application deadline:* rolling. *Application fee:* $20. *Financial aid:* In 1990–91, 180 students received aid. 24 fellowships, 114 research assistantships, 39 teaching assistantships, 3 project assistantships awarded; institutionally sponsored loans also available. Financial aid application deadline: 1/15; applicants required to submit FAF. • Martin Olsson, Chairperson, 608-262-9678. Application contact: Joanne Nagy, Assistant Dean of the Graduate School, 608-262-2433.

University of Wisconsin–Milwaukee, College of Letters and Sciences, Department of Physics, Milwaukee, WI 53201. Department awards MS, PhD. Faculty: 21 full-time, 0 part-time. Matriculated students: 19 full-time (2 women), 22 part-time (6 women); includes 1 minority (Asian American), 19 foreign. 42 applicants, 43% accepted. In 1990, 6 master's, 6 doctorates awarded. *Degree requirements:* For master's, thesis or alternative required, foreign language not required; for doctorate, 1 foreign language, dissertation. *Entrance requirements:* GRE General Test. Application deadline: 3/1 (priority date, applications processed on a rolling basis). Application fee: $20. *Financial aid:* In 1990–91, 6 fellowships, 10 research assistantships, 19 teaching assistantships, 0 project assistantships awarded. • Moises Levy, Program Representative, 414-229-4541.

University of Wisconsin–Oshkosh, College of Letters and Science, Department of Physics and Astronomy, Oshkosh, WI 54901. Department offers program in physics (MS), including geophysics, instrumentation, magnetic resonance, physics education, solid state physics. Faculty: 7 full-time (1 woman), 1 part-time (0 women). Matriculated students: 0 full-time, 1 part-time (0 women); includes 0 minority, 0 foreign. Average age 27. 1 applicants, 0% accepted. In 1990, 3 degrees awarded. *Degree requirements:* Thesis required, foreign language not required. *Application deadline:* rolling. *Application fee:* $20. *Financial aid:* In 1990–91, 3 graduate assistantships (1 to a first-year student) awarded; full and partial tuition waivers, federal work-study, institutionally sponsored loans, and career-related internships or fieldwork also available. Financial aid application deadline: 3/15. *Faculty research:* Digital signal processing, magnetic resonance, geophysics. Total annual research budget: $90,000. • Dr. Sandra Gade, Chair, 414-424-4433. Application contact: Dr. John Karl, Coordinator, 414-424-4432.

University of Wyoming, College of Arts and Sciences, Department of Physics and Astronomy, Laramie, WY 82071. Department awards MS, MST, PhD. Faculty: 19 full-time (0 women), 1 part-time (0 women). Matriculated students: 24 full-time (5 women), 7 part-time (1 woman); includes 2 foreign. Average age 28. 150 applicants, 3% accepted. In 1990, 3 master's, 0 doctorates awarded. *Degree requirements:* Foreign language not required. *Entrance requirements:* GRE General Test, GRE Subject Test, minimum GPA of 3.0. Application deadline: 6/1 (priority date, applications processed on a rolling basis). Application fee: $30. *Tuition:* $1554 per year full-time, $74.25 per credit hour part-time for state residents; $4358 per year full-time, $74.25 per credit hour part-time for nonresidents. *Financial aid:* In 1990–91, 24 students received a total of $160,272 in aid awarded. 8 research assistantships (0 to first-year students), 15 teaching assistantships (4 to first-year students) were awarded; institutionally sponsored loans also available. Average monthly stipend for a graduate assistantship: $753. Financial aid application deadline: 4/15. *Total annual research budget:* $1.553-million. • Dr. Glen Rebka, Head, 307-766-6150.

Utah State University, College of Science, Department of Physics, Logan, UT 84322. Department offers program in physics (MS, PhD), including astronomy and astrophysics (MS), earth and planetary sciences (PhD), engineering physics (PhD), physics (MS). Part-time programs available. Faculty: 22 full-time (0 women), 3 part-time (0 women). Matriculated students: 5 full-time (1 woman), 39 part-time (9 women). Average age 25. 105 applicants, 57% accepted. In 1990, 4 master's awarded (100% found work related to degree); 3 doctorates awarded. Terminal master's awarded for partial completion of doctoral program. *Degree requirements:* Thesis/dissertation required, foreign language not required. *Entrance requirements:* GRE General Test (score in 40th percentile or higher required), minimum GPA of 3.0. Application deadline: 7/15 (priority date, applications processed on a rolling basis). Application fee: $25 ($30 for foreign students). *Tuition:* $426 per quarter (minimum) full-time, $184 per quarter (minimum) part-time for state residents; $1133 per quarter (minimum) full-time, $505 per quarter (minimum) part-time for nonresidents. *Financial aid:* In 1990–91, $200,000 in aid awarded. 1 fellowship (to a first-year student), 23 research assistantships (0 to first-year students), 7 teaching assistantships (4 to first-year students) were awarded; partial tuition waivers, federal work-study, institutionally sponsored loans also available. Aid available to part-time students. Financial aid application deadline: 4/1; applicants required to submit FAF. *Faculty research:* Upper-atmosphere physics, relativity, particle physics, medium energy nuclear physics, surface physics. Total annual research budget: $7.5-million. • Dr. W. John Raitt, Head, 801-750-2848. Application contact: O. Harry Otteson, Assistant Head, 801-750-2850.

Vanderbilt University, Department of Physics, Nashville, TN 37240. Offerings include physics (MA, MS, PhD). Department faculty: 25 full-time (1 woman), 12 part-time (2 women). *Degree requirements:* For master's, thesis; for doctorate, 1 foreign language, dissertation. *Entrance requirements:* For master's, GRE General Test; for doctorate, GRE General Test, GRE Subject Test. Application deadline: 1/15. Application fee: $25. *Expenses:* Tuition of $624 per semester hour. Fees of $196 per year. • W. T. Pinkston, Chairman, 615-322-2828. Application contact: Volker E. Oberacker, Director of Graduate Studies, 615-322-5035.

Virginia Commonwealth University, College of Humanities and Sciences, Department of Physics, Richmond, VA 23284. Department offers programs in applied physics (MS), physics (MS). Part-time programs available. Faculty: 8 full-time, 0 part-time. Matriculated students: 7 full-time (1 woman), 6 part-time (0 women); includes 3 minority (all black American), 2 foreign. 20 applicants, 20% accepted. In 1990, 3 degrees awarded. *Degree requirements:* Thesis required, foreign language not required. *Entrance requirements:* GRE. Application deadline: 8/1. Application fee: $20. *Expenses:* Tuition of $2770 per year full-time, $154 per hour part-time for state residents; $7550 per year full-time, $419 per hour part-time for nonresidents. Fees of $717 per year full-time, $25.50 per hour part-time. *Financial aid:* In 1990–91, 4 students received aid. 0 fellowships, 0 teaching assistantships awarded; full and partial tuition waivers, federal work-study, institutionally sponsored loans also available. Aid available to part-time students. Financial aid applicants required to submit FAF. *Faculty research:* Condensed matter theory and experimentation, electronic instrumentation, relativity. Total annual research budget: $110,006. • Dr. Peru Jena, Chair, 804-367-1313.

Virginia Polytechnic Institute and State University, College of Arts and Sciences, Department of Physics, Blacksburg, VA 24061. Department awards MS, PhD. Matriculated students: 46 full-time (4 women), 3 part-time (1 woman); includes 0 minority, 31 foreign. 89 applicants, 38% accepted. In 1990, 12 master's, 4 doctorates awarded. *Application deadline:* 2/15 (priority date). *Application fee:* $10. *Tuition:* $1889 per semester full-time, $606 per credit hour part-time for state residents; $2627 per semester full-time, $853 per credit hour part-time for nonresidents. *Financial aid:* In 1990–91, 17 research assistantships, 13 teaching assistantships, 13 assistantships awarded; fellowships also available. • Dr. Thomas E. Gilmer, Head, 703-231-6544.

Virginia State University, School of Natural Sciences, Department of Physics, Petersburg, VA 23803. Department awards MS. Faculty: 2 full-time (0 women), 0 part-time. Matriculated students: 7 full-time (2 women), 3 part-time (1 woman). In 1990, 2 degrees awarded. *Degree requirements:* 1 foreign language, thesis. *Entrance requirements:* GRE General Test. Application deadline: 8/15 (applications processed on a rolling basis). Application fee: $10. *Expenses:* Tuition of $1479 per semester full-time, $90 per credit hour part-time for state residents; $2879 per semester full-time, $215 per credit hour part-time for nonresidents. Fees of $25 per year full-time, $12 per year part-time. *Financial aid:* Application deadline 5/1; applicants required to submit FAF. • Dr. James C. Davenport, Chairman, 804-524-5913. Application contact: Dr. Edgar A. Toppin, Dean, 804-524-5984.

Wake Forest University, Department of Physics, Winston-Salem, NC 27109. Department awards MS, PhD. Part-time programs available. Faculty: 11 full-time (2 women), 0 part-time. Matriculated students: 17 full-time (3 women), 2 part-time (0 women); includes 4 foreign. Average age 27. 45 applicants, 53% accepted. In 1990, 2 master's awarded (100% continued full-time study). *Degree requirements:* 1 foreign language (computer language can substitute), thesis/dissertation. *Entrance requirements:* For master's, GRE General Test, GRE Subject Test, TOEFL. Application deadline: 3/1. Application fee: $25. *Tuition:* $10,800 per year full-time, $280 per hour part-time. *Financial aid:* In 1990–91, 17 students received a total of $228,900 in aid awarded. 0 fellowships, 0 research assistantships, 8 teaching assistantships (7 to first-year students), 9 scholarships (4 to first-year students) were awarded. Aid available to part-time students. Average monthly stipend for a graduate assistantship: $800. Financial aid application deadline: 3/1; applicants required to submit FAF. • Dr. Howard Shields, Chairman, 919-759-5337.

Washington State University, College of Sciences and Arts, Division of Sciences, Department of Physics, Pullman, WA 99164. Department offers programs in chemical physics (PhD), physics (MS, PhD). Faculty: 21 full-time (1 woman), 3 part-time (0 women). Matriculated students: 39 full-time (3 women), 5 part-time (1 woman); includes 4 minority (3 Asian American, 1 black American), 20 foreign. In 1990, 5 master's awarded; 5 doctorates awarded (100% entered university research/teaching). *Degree requirements:* For doctorate, dissertation. *Entrance requirements:* GRE General Test, minimum GPA of 3.0. Application deadline: 3/1 (priority date, applications processed on a rolling basis). Application fee: $25. *Tuition:* $1694 per semester full-time, $169 per credit hour part-time for state residents; $4236 per semester full-time, $424 per credit hour part-time for nonresidents. *Financial aid:* In 1990–91, 0 fellowships, 8 research assistantships, 18 teaching assistantships, 6 teaching associateships awarded; partial tuition waivers, federal work-study, institutionally sponsored loans also available. Average monthly stipend for a graduate assistantship: $975. Financial aid application deadline: 4/1. *Total annual research budget:* $1.08-million. • Dr. Michael Miller, Chair, 509-335-9531. Application contact: Dr. Y. Gupta, 509-335-3140.

Washington University, Graduate School of Arts and Sciences, Department of Physics, St. Louis, MO 63130. Department awards MA, PhD. Part-time programs available. Matriculated students: 70 full-time (11 women), 3 part-time (0 women); includes 2 minority (1 Asian American, 1 black American), 18 foreign. In 1990, 11 master's, 11 doctorates awarded. Terminal master's awarded for partial completion of doctoral program. *Degree requirements:* For master's, thesis or alternative; for doctorate, dissertation. *Entrance requirements:* GRE. Application deadline: 1/15 (priority date, applications processed on a rolling basis). Application fee: $0. *Tuition:* $15,960 per year full-time, $665 per credit hour part-time. *Financial aid:* In 1990–91, 10 fellowships, 35 research assistantships, 20 teaching assistantships awarded; full and partial tuition waivers, federal work-study, institutionally sponsored loans also available. Aid available to part-time students. Financial aid application deadline: 1/15. • Dr. Clifford Will, Chairperson, 314-935-6276.

Wayne State University, College of Liberal Arts, Department of Physics and Astronomy, Detroit, MI 48202. Department offers program in physics (MA, MS, PhD). Faculty: 27 full-time (2 women), 0 part-time. Matriculated students: 33 full-time (6 women), 21 part-time (3 women). 110 applicants, 25% accepted. In 1990, 3 master's, 2 doctorates awarded. *Degree requirements:* For master's, foreign language not required; for doctorate, dissertation. *Application deadline:* 7/1. *Application fee:* $20 ($30 for foreign students). *Expenses:* Tuition of $119 per credit hour for state residents; $258 per credit hour for nonresidents. Fees of $50 per semester. *Faculty research:* Thermal wave imaging, quantum theory of fields, HgCdTe, positron and atomic scattering. • Dr. David Fradkin, Chairperson, 313-577-2720.

Wesleyan University, Department of Physics, Middletown, CT 06459. Department awards MA, PhD. Faculty: 10 full-time (0 women), 0 part-time. Matriculated students: 13 full-time (3 women), 0 part-time; includes 6 minority, 6 foreign. Average age 25. 35 applicants, 14% accepted. In 1990, 1 master's, 0 doctorates awarded. Terminal master's awarded for partial completion of doctoral program. *Degree requirements:* Thesis/dissertation. *Entrance requirements:* For master's, GRE General Test, GRE Subject Test; for doctorate, GRE Subject Test. Application deadline: 3/1. Application fee: $0. *Expenses:* Tuition of $2035 per course. Fees of $627 per year full-time. *Financial aid:* In 1990–91, 13 teaching assistantships (2 to first-year students) awarded; full tuition waivers, institutionally sponsored loans also available. Average monthly stipend for a graduate assistantship: $875. *Faculty research:* Low-temperature physics, magnetic resonance, atomic collisions, laser spectroscopy, surface physics. Total annual research budget: $80,000. • Dr. Brian Stewart, Head, 203-347-9411 Ext. 3106.

Western Carolina University, School of Arts and Sciences, Department of Chemistry and Physics, Cullowhee, NC 28723. Department awards MA Ed, MS. Part-time and evening/weekend programs available. Faculty: 10 (0 women). Matriculated students: 5 full-time (2 women), 3 part-time (2 women); includes 1 minority (black American), 0 foreign. 8 applicants, 75% accepted. In 1990, 2 degrees awarded. *Degree requirements:* Variable foreign language requirement, thesis, comprehensive exams. *Entrance requirements:* GRE General Test, GRE Subject Test (for MS), NTE (for

Directories: Physics; Plasma Physics

MA Ed). Application deadline: rolling. Application fee: $15. *Tuition:* $635 per semester full-time, $226 per course (minimum) part-time for state residents; $3162 per semester full-time, $1490 per course (minimum) part-time for nonresidents. *Financial aid:* In 1990–91, $24,750 in aid awarded. 0 fellowships, 1 research assistantship, 5 teaching assistantships were awarded; federal work-study, institutionally sponsored loans also available. Financial aid application deadline: 3/15; applicants required to submit FAF. • F. Glenn Liming, Head, 704-227-7260. Application contact: Kathleen Owen, Assistant to the Dean, 704-227-7398.

Western Illinois University, College of Arts and Sciences, Department of Physics, Macomb, IL 61455. Department awards MS. Part-time programs available. Faculty: 8 full-time (0 women), 0 part-time. Matriculated students: 14 full-time (3 women), 3 part-time (1 woman); includes 0 minority, 14 foreign. Average age 26. 52 applicants, 25% accepted. In 1990, 6 degrees awarded. *Degree requirements:* Thesis or alternative required, foreign language not required. *Application fee:* $0. *Expenses:* Tuition of $870 per semester full-time, $520 per semester hour part-time for state residents; $2193 per semester full-time, $1402 per semester hour part-time for nonresidents. Fees of $537 per year full-time, $148 per year part-time. *Financial aid:* In 1990–91, $37,735 in aid awarded. 15 research assistantships were awarded. • Dr. Harold Hart, Chairperson, 309-298-1596. Application contact: Barbara Baily, Assistant to the Dean, 309-298-4806.

Western Michigan University, College of Arts and Sciences, Department of Physics, Kalamazoo, MI 49008. Department awards MA, PhD. Matriculated students: 4 full-time (2 women), 17 part-time (7 women); includes 0 minority, 16 foreign. 71 applicants, 30% accepted. In 1990, 2 master's awarded. *Degree requirements:* For master's, thesis required, foreign language not required; for doctorate, dissertation. *Entrance requirements:* For doctorate, GRE General Test. Application deadline: 2/15 (priority date, applications processed on a rolling basis). Application fee: $25. *Expenses:* Tuition of $100 per credit hour for state residents; $244 per credit hour for nonresidents. Fees of $178 per semester full-time, $76 per semester part-time. *Financial aid:* Fellowships, research assistantships, teaching assistantships, federal work-study available. Financial aid application deadline: 2/15. • Dr. John Tanis, Chairperson, 616-387-4940. Application contact: Paula J. Boodt, Director, Graduate Student Services, 616-387-3570.

West Virginia University, College of Arts and Sciences, Department of Physics, Morgantown, WV 26506. Department awards MS, PhD. Part-time programs available. Faculty: 26 full-time (3 women), 1 (woman) part-time. Matriculated students: 25 full-time (4 women), 2 part-time (1 woman); includes 0 minority, 19 foreign. Average age 27. 72 applicants, 13% accepted. In 1990, 3 master's awarded (67% found work related to degree, 33% continued full-time study); 4 doctorates awarded. Terminal master's awarded for partial completion of doctoral program. *Degree requirements:* For master's, thesis required, foreign language not required; for doctorate, 1 foreign language (computer language can substitute), dissertation, comprehensive exam. *Entrance requirements:* GRE General Test, GRE Subject Test, TOEFL (minimum score of 550 required), minimum GPA of 2.5. Application deadline: 3/1 (priority date, applications processed on a rolling basis). Application fee: $25. *Expenses:* Tuition of $390 per year full-time for state residents; $1270 per year full-time for nonresidents. Fees of $1555 per year full-time for state residents; $3985 per year full-time for nonresidents. *Financial aid:* In 1990–91, $207,390 in aid awarded. 7 research assistantships, 28 teaching assistantships were awarded; fellowships, full and partial tuition waivers, federal work-study, institutionally sponsored loans also available. Financial aid application deadline: 2/1; applicants required to submit FAF. *Faculty research:* Condensed matter and surface physics; magnetism; electronic structure of materials; high energy theory; experimental plasma, theoretical aerosol, and astrophysics. • Dr. Larry E. Halliburton, Chairperson, 304-293-3422. Application contact: Dr. Richard P. Treat, Graduate Adviser, 304-293-3422.

Wichita State University, Fairmount College of Liberal Arts and Sciences, Department of Physics, Wichita, KS 67208. Department awards MS. Part-time programs available. Faculty: 9 full-time (0 women), 0 part-time. Matriculated students: 7 full-time (2 women), 6 part-time (0 women); includes 0 minority, 10 foreign. Average age 25. 11 applicants, 55% accepted. In 1990, 7 degrees awarded. *Degree requirements:* Qualifying exam required, thesis optional. *Entrance requirements:* TOEFL (minimum score of 550 required). Application deadline: 8/1 (priority date, applications processed on a rolling basis). Application fee: $0 ($25 for foreign students). *Expenses:* Tuition of $1590 per year full-time, $53 per credit part-time for state residents; $2574 per year full-time, $171.60 per credit part-time for nonresidents. Fees of $12.20 per credit. *Financial aid:* In 1990–91, $59,000 in aid awarded. 3 research assistantships (all to first-year students), 9 teaching assistantships (4 to first-year students), 1 graduate assistantship (to a first-year student) were awarded; federal work-study, institutionally sponsored loans also available. Financial aid application deadline: 4/1. *Faculty research:* Solid-state theory, astrophysics. • Dr. David Alexander, Chairperson, 316-689-3190.

Wilkes University, Department of Physics, Wilkes-Barre, PA 18766. Department awards MS, MS Ed. *Degree requirements:* Thesis or alternative required, foreign language not required. *Application fee:* $25.

Worcester Polytechnic Institute, Department of Physics, Worcester, MA 01609. Department awards MS, PhD. Faculty: 15 full-time (0 women), 2 part-time (0 women). Matriculated students: 12 full-time (2 women), 2 part-time (1 woman); includes 0 minority, 11 foreign. Average age 33. 113 applicants, 2% accepted. In 1990, 0 master's awarded; 2 doctorates awarded (0% entered university research/teaching, 0% found other work related to degree, 50% continued full-time study). *Degree requirements:* Thesis/dissertation required, foreign language not required. *Entrance requirements:* TOEFL (minimum score of 550 required). Application deadline: 2/15. Application fee: $25. *Expenses:* Tuition of $460 per credit hour. Fees of $20 per year full-time. *Financial aid:* In 1990–91, 10 students received a total of $78,000 in aid awarded. 3 research assistantships (0 to first-year students), 7 teaching assistantships (0 to first-year students) were awarded; institutionally sponsored loans also available. Average monthly stipend for a graduate assistantship: $866. Financial aid application deadline: 2/15. *Faculty research:* Nuclear and solid-state physics, quantum and nonlinear optics, photoacoustic and optical spectroscopy, amorphous materials, chemical physics. Total annual research budget: $243,000. • Dr. Stephen N. Jasperson, Head, 508-831-5392.

Wright State University, College of Science and Mathematics, Department of Physics, Dayton, OH 45435. Department offers programs in physics (MS), physics education (MST). Part-time and evening/weekend programs available. Faculty: 11 full-time (0 women), 3 part-time (0 women). Matriculated students: 13 full-time (5 women), 9 part-time (1 woman); includes 2 minority (both Asian American), 7 foreign. Average age 28. 25 applicants, 44% accepted. In 1990, 6 degrees awarded. *Degree requirements:* Computer language, thesis required, foreign language not required. *Application deadline:* 3/1 (priority date, applications processed on a rolling basis). *Tuition:* $3342 per year full-time, $106 per credit hour part-time for state residents; $5991 per year full-time, $190 per credit hour part-time for nonresidents. *Financial aid:* In 1990–91, 11 students received a total of $150,100 in aid awarded. 0 fellowships, 2 research assistantships (both to first-year students), 7 teaching assistantships (4 to first-year students), 1 graduate assistantship were awarded; full and partial tuition waivers, federal work-study, institutionally sponsored loans also available. Average monthly stipend for a graduate assistantship: $667. Financial aid application deadline: 3/1; applicants required to submit FAF. *Faculty research:* Solid-state, semiconductors, metal alloys, atomic physics, plasma. Total annual research budget: $780,800. • Dr. Merrill L. Andrews, Chair, 513-873-2954.

Announcement: Solid State Electronics Research Center houses 2-MeV electron Van de Graaff accelerator, 400-keV ion Van de Graaff with RBS/channeling, 120-keV ion implanter, Polaron DLTS, high-speed internal friction/Young's modulus apparatus, positron annihilation angular correlation apparatus; Atomic Physics Lab has 2-meter Czerny Turner vacuum spectrometer, plasma mirror machine, plasma deposition system.

Yale University, Graduate School of Arts and Sciences, Department of Physics, New Haven, CT 06520. Department awards PhD. Faculty: 34 full-time (1 woman), 0 part-time. Matriculated students: 97 full-time (11 women), 1 part-time (0 women); includes 2 minority (both Asian American), 48 foreign. 187 applicants, 25% accepted. In 1990, 22 master's, 13 doctorates awarded. Terminal master's awarded for partial completion of doctoral program. *Degree requirements:* For doctorate, 1 foreign language, dissertation. *Entrance requirements:* For doctorate, GRE General Test, GRE Subject Test. Application deadline: 1/2. Application fee: $45. *Tuition:* $15,160 per year full-time, $1895 per course part-time. • Michael Zeller, Chairman, 203-432-3600. Application contact: Susan Webb, Coordinator of Admissions, 203-432-2770.

York University, Faculty of Science, Program in Physics and Astronomy, North York, ON M3J 1P3, Canada. Program awards M Sc, PhD. Part-time and evening/weekend programs available. Faculty: 36 full-time (2 women), 8 part-time (1 woman). Matriculated students: 37 full-time (5 women), 5 part-time (0 women); includes 3 foreign. 141 applicants, 9% accepted. In 1990, 2 master's, 3 doctorates awarded. *Degree requirements:* For master's, thesis optional, foreign language not required; for doctorate, dissertation required, foreign language not required. *Application fee:* $35. *Tuition:* $2436 per year for Canadian residents; $11,480 per year for nonresidents. *Financial aid:* In 1990–91, $531,339 in aid awarded. 2 fellowships, 46 research assistantships, 40 teaching assistantships were awarded. • Dr. R. P. McEachran, Director, 416-736-2100.

Plasma Physics

California Institute of Technology, Division of Engineering and Applied Science, Option in Applied Physics, Pasadena, CA 91125. Offerings include plasma physics (MS, PhD). Faculty: 10 full-time (0 women), 0 part-time. *Degree requirements:* For master's, foreign language and thesis not required; for doctorate, dissertation required, foreign language not required. *Application fee:* $0. *Expenses:* Tuition of $14,100 per year. Fees of $8 per quarter. • Dr. Noel Corngold, Head, Steering Committee, 818-356-4129. Application contact: Dr. Paul Bellan, Option Representative, 818-356-4827.

Columbia University, School of Engineering and Applied Science, Department of Applied Physics, Program in Applied Physics, New York, NY 10027. Offerings include applied physics (MS, Eng Sc D, PhD), with options in applied mathematics (PhD), plasma physics (MS, Eng Sc D, PhD), quantum electronics (MS, Eng Sc D, PhD), solid state physics (MS, Eng Sc D, PhD). Terminal master's awarded for partial completion of doctoral program. Program faculty: 9 full-time (0 women), 6 part-time (1 woman). *Degree requirements:* For master's, foreign language and thesis not required; for doctorate, dissertation, qualifying exam required, foreign language not required. *Entrance requirements:* GRE General Test, GRE Subject Test, TOEFL. Application deadline: 2/1. Application fee: $45. *Tuition:* $15,520 per year full-time, $518 per credit part-time. • Dr. Gerald A. Navratil, Chairman, Department of Applied Physics, 212-854-4457.

Emory University, Graduate School of Arts and Sciences, Department of Physics, Atlanta, GA 30322. Department offers programs in environmental sciences (MS); physics (MA, MS, PhD), including biophysics, radiological physics, solid-state physics. Faculty: 17. Matriculated students: 3 full-time (0 women), 2 part-time (0 women); includes 0 minority, 4 foreign. Average age 33. 0 applicants. In 1990, 0 master's, 1 doctorate awarded. Terminal master's awarded for partial completion of doctoral program. *Degree requirements:* For master's, thesis required, foreign language not required; for doctorate, dissertation, comprehensive exams required, foreign language not required. *Entrance requirements:* GRE General Test, TOEFL, minimum GPA of 3.0. Application deadline: 1/20 (priority date). Application fee: $35. *Expenses:* Tuition of $7370 per semester full-time, $642 per semester hour part-time. Fees of $160 per year full-time. *Financial aid:* In 1990–91, $109,180 in aid awarded. 4 fellowships (0 to first-year students), 0 teaching assistantships, 4 tuition scholarships (0 to first-year students) were awarded; partial tuition waivers also available. Average monthly stipend for a graduate assistantship: $1000. Financial aid application deadline: 2/1. *Faculty research:* Theory of semiconductors and superlattices; experimental laser optics and submillimeter spectroscopy theory; neural networks and stereoscopic vision, experimental studies of the structu re and function of metalloproteins. Total annual research budget: $953,952. • Dr. Krishan

SECTION 7: PHYSICS

Directory: Plasma Physics; Cross-Discipline Announcements

K. Bajaj, Chair, 404-727-6584. Application contact: Dr. Edmund Day, Director of Graduate Studies, 404-727-6584.
See full description on page 583.

Massachusetts Institute of Technology, School of Engineering, Department of Nuclear Engineering, Cambridge, MA 02139. Offerings include applied plasma physics (SM). Department faculty: 22 full-time (0 women), 0 part-time. *Degree requirements:* Thesis. *Application deadline:* 1/15. *Application fee:* $45. *Tuition:* $16,900 per year. • Dr. Mujid S. Kazimi, Head, 617-253-3801. Application contact: Clare Egan, Graduate Office Administrator, 617-253-3814.

Princeton University, Department of Astrophysical Sciences, Program in Plasma Physics, Princeton, NJ 08544. Program awards PhD. Terminal master's awarded for partial completion of doctoral program. *Degree requirements:* For doctorate, dissertation required, foreign language not required. *Entrance requirements:* For doctorate, GRE General Test, GRE Subject Test. Application deadline: 1/8. Application fee: $45 ($50 for foreign students). *Tuition:* $16,670 per year. *Financial aid:* Fellowships, research assistantships, teaching assistantships, federal work-study, institutionally sponsored loans available. Financial aid application deadline: 1/8. *Faculty research:* Magnetic fusion energy research, plasma physics, x-ray laser studies. • Nathaniel Fisch, Director of Graduate Studies, 609-243-2489. Application contact: Michele Spreen, Director of Graduate Admissions, 609-258-3034.

Rensselaer Polytechnic Institute, School of Engineering, Department of Electrical, Computer, and Systems Engineering, Program in Plasma Physics, Troy, NY 12180. Program awards M Eng, MS, D Eng, PhD. *Degree requirements:* Thesis/dissertation required, foreign language not required. *Entrance requirements:* GRE General Test, GRE Subject Test, TOEFL. Application deadline: 2/1. Application fee: $30. *Expenses:* Tuition of $455 per credit hour. Fees of $195.57 per semester. *Financial aid:* Application deadline 2/1. • R. L. Hickok, Graduate Coordinator, 518-276-6390. Application contact: Barbara Kochanin, Manager of Graduate Admissions and Financial Aid, 518-276-2719.

University of Colorado at Boulder, College of Arts and Sciences, Department of Astrophysical, Planetary, and Atmospheric Sciences, Boulder, CO 80309. Offerings include plasma physics (MS, PhD). Terminal master's awarded for partial completion of doctoral program. Department faculty: 72 full-time (4 women). *Degree requirements:* For master's, thesis or alternative, comprehensive exam required, foreign language not required; for doctorate, 1 foreign language, dissertation. *Entrance requirements:* GRE General Test, GRE Subject Test. Application deadline: 3/1 (priority date, applications processed on a rolling basis). Application fee: $30 ($50 for foreign students). *Expenses:* Tuition of $2308 per year full-time, $387 per semester (minimum) part-time full-time, $8730 per year full-time, $1455 per semester (minimum) part-time for nonresidents. Fees of $207 per semester full-time, $27.26 per semester (minimum) part-time. • Ellen Zweibel, Chairman, 303-492-8915.

University of Maryland College Park, College of Computer, Mathematical and Physical Sciences, Department of Physics and Astronomy, College Park, MD 20742. Offerings include plasma physics (PhD). Department faculty: 105 (2 women). *Entrance requirements:* GRE Subject Test. Application deadline: rolling. Application fee: $25. *Expenses:* Tuition of $143 per credit hour for state residents; $256 per credit hour for nonresidents. Fees of $171.50 per semester. • Dr. Derik Boyd, Chairman, 301-405-5982.

University of Minnesota, Twin Cities Campus, Institute of Technology, Department of Electrical Engineering, Minneapolis, MN 55455. Offerings include plasma physics (MEE, MSEE, PhD). *Expenses:* Tuition of $1084 per quarter full-time, $301 per credit part-time for state residents; $2168 per quarter full-time, $602 per credit part-time for nonresidents. Fees of $118 per quarter. • Mostafa Kaveh, Head, 612-625-0720.

Yale University, Graduate School of Arts and Sciences, Council of Engineering and Applied Science, New Haven, CT 06520. Offerings include plasma physics (MS, PhD). Terminal master's awarded for partial completion of doctoral program. Faculty: 43 full-time, 0 part-time. *Degree requirements:* For doctorate, dissertation, exam required, foreign language not required. *Entrance requirements:* For doctorate, GRE General Test, GRE Subject Test, TOEFL. Application deadline: 1/2. Application fee: $45. *Tuition:* $15,160 per year full-time, $1895 per course part-time. • Gary Haller, Chairman, 203-432-4220. Application contact: Susan Webb, Coordinator of Admissions, 203-432-2770.

Cross-Discipline Announcements

Columbia University, Graduate School of Arts and Sciences, Division of Natural Sciences, Department of Biochemistry and Molecular Biophysics, New York, NY 10032.

The research activities of the Department of Biochemistry and Molecular Biophysics encompass a variety of topics of interest to physics majors, including X-ray diffraction and NMR studies of biological macromolecules, computer studies of proteins and nucleic acids and the structure and function of membrane receptors. See full description in the Biophysics Section in the Biological and Agricultural Sciences volume of this series.

Georgetown University, College of Arts and Sciences, Department of Radiation Science, Washington, DC 20057.

MS program in health physics offers primarily evening courses and has a variety of internships in local federal facilities and institutions. Research areas include solid-state dosimetry and medical health physics. Degree requirements: qualifying examination and thesis. Application contact: James Rodgers, PhD, Director, 202-687-2212. See full description in Section 14 of Book 6 of this series.

Harvard University, Graduate School of Arts and Sciences, Committee on Biophysics, Cambridge, MA 02138.

The Committee on Higher Degrees in Biophysics at Harvard offers students with backgrounds in physics and chemistry graduate training in diverse areas of biophysical research with faculty from Biochemistry, Molecular, Cellular and Developmental Biology; Chemistry; Applied Physics; and the Medical Sciences Departments. Please see full description in the Biological and Agricultural Sciences volume of this series.

Johns Hopkins University, School of Medicine, Graduate Programs in Medicine, Program in Molecular Biophysics, Baltimore, MD 21218.

IPMB (The Inter-Campus Program in Molecular Biophysics) is staffed by about 35 faculty with interests in molecular biophysics. It offers special opportunities to applicants trained in the physical sciences or mathematics for graduate study in areas such as protein crystallography, NMR and ESR, thermodynamics, statistical mechanics, computer modeling, biophysical chemistry, and biochemistry. It emphasizes studies on macromolecules, or on interacting assemblies of macromolecules, for which a combination of approaches—molecular genetics and structural studies, for example—may be necessary for real progress. Collaborative projects between faculty are encouraged. For information, contact IPMB Office, 301-338-5197, fax 301-338-5199.

Louisiana State University and Agricultural and Mechanical College, College of Engineering, Department of Electrical and Computer Engineering, Baton Rouge, LA 70803.

A flexible curriculum featuring the theory, design, fabrication, and characterization of semiconductor devices is emphasized. The Solid State Laboratory has class-100 clean room for device processing. The new DoE-funded Center for Advanced Microstructures and Devices (CAMD) is available. Students may enter the program with a BS in engineering or another science discipline, such as physics or chemistry.

Northwestern University, Robert R. McCormick School of Engineering and Applied Science, Department of Chemical Engineering, Evanston, IL 60208.

MS and PhD programs. Physics-related research areas include: experimental and computational fluid mechanics; heat and mass transfer; electrical, optical, and structural properties of polymers; surface characterization of catalysts and supports; chaos and pattern formation; turbulence; electronic materials; and applications of solid-state NMR. Financial aid is available. See full description in Book 5, Chemical Engineering Section.

University of Kentucky, Graduate School Programs from the College of Allied Health, Program in Radiation Sciences, Lexington, KY 40506-0032.

Two-year track leading to MS in Radiological Medical Physics includes several months of intensive clinical and laboratory training in therapy and diagnostic imaging physics. Equipment includes 6- and 18-MeV linacs, Co-60 units, and MRI, SPECT, CT, and simulator imaging systems. About 20 students in division, 12 in medical physics. Campus visits welcomed. Call 606-233-6350 for information and application materials.

University of Maryland College Park, College of Computer, Mathematical and Physical Sciences, Department of Physics and Astronomy, Program in Chemical Physics, College Park, MD 20742.

The Chemical Physics Program offers MS and PhD degrees in chemical physics. Students wishing to pursue a career requiring in-depth knowledge of both physics and chemistry and whose needs cannot reasonably be accommodated within a single department are encouraged to apply. See full description in Chemistry Section.

University of Michigan, College of Engineering, Department of Materials Science and Engineering, Ann Arbor, MI 48109.

Interdisciplinary curriculum leads to MS and PhD degrees in materials science and engineering for physics students interested in condensed-matter physics, metals, ceramics, polymers, composites, and other engineering materials. Research assistantships and fellowships available on a competitive basis. See full description of the department in the Engineering and Applied Sciences volume of this series.

University of Rochester, College of Engineering and Applied Science, Department of Electrical Engineering, Rochester, NY 14627-0001.

Electrical engineering graduate study features flexible degree programs, interdisciplinary research, close faculty-student cooperation, numerous sponsored projects, and excellent computing resources and specialized facilities. Research areas: medical imaging, ultrasonics, robotics, signal and image processing, electromagnetic fields, superconducting electronics, semiconductor devices, electrooptics, high-speed electronics, and VLSI.

Peterson's Guide to Graduate Programs in the Physical Sciences and Mathematics 1992

SECTION 7: PHYSICS

Cross-Discipline Announcements

University of Virginia, Graduate School of Arts and Sciences, Interdisciplinary Program in Biophysics, Charlottesville, VA 22906.

The Interdisciplinary Program in Biophysics at the University of Virginia offers training and research opportunities with 45 faculty in the Schools of Graduate Arts and Sciences, Engineering, and Medicine. Macromolecular structure and physical biochemistry, membrane biophysics, and radiological physics are areas of specific research strength. All students are financially supported.

University of Washington, College of Engineering and Graduate Programs in Medicine, Center for Bioengineering, Seattle, WA 98195.

The Center for Bioengineering is an interdisciplinary graduate program, integrating engineering and the physical sciences with biology and medicine. Research areas include bioinstrumentation, biomechanics, sensors, biosystems, simulation of bioprocesses, biomaterials, imaging, and molecular bioengineering. Students with degrees in engineering, computer science, or the physical sciences are encouraged to apply.

University of Washington, College of Engineering, Department of Materials Science and Engineering, Seattle, WA 98195.

Programs in Department of Materials Science and Engineering apply condensed-matter physics to electronic and optical materials, ceramics, metals, and polymers. Current thrusts include superconducting ceramics, molecular-beam epitaxy in III–V compounds, colloidal ceramics, electron microscopy and imaging physics, and structure/property relationships for ceramic/polymer/metal composites. See the department's full description in the Materials Sciences and Engineering Section in the Engineering and Applied Sciences volume of this series.

University of Wisconsin–Madison, Institute for Molecular Virology, Program in Biophysics, Madison, WI 53706.

The Biophysics Program offers research that revolves around the application of physical techniques to biological problems. It is an interdisciplinary program that includes 38 faculty members from 9 departments, laboratories, and institutes. See full description in the Biological and Agricultural Sciences volume of this series.

Virginia Commonwealth University, School of Graduate Studies and Medical College of Virginia-Professional Programs, School of Basic Health Sciences, Program in Biomedical Engineering, Richmond, VA 23284.

The Biomedical Engineering Program offers tracks leading to the MS degree. Students may pursue course work in signal processing, medical imaging, bioinstrumentation, and molecular modeling and graphics. See full description in Book 5, Engineering and Applied Sciences.

SECTION 7: PHYSICS

AUBURN UNIVERSITY

Department of Physics

Programs of Study

Programs of study are offered that lead to Ph.D. and M.S. degrees in physics. The M.S. program includes both a thesis and a nonthesis option. The primary areas of research are condensed-matter physics, plasma physics, atomic physics, and space physics. Students in the doctoral program must pass a qualifying examination. The exam covers introductory graduate course material and is typically completed after no more than two years of course work.

The student-faculty ratio is low, which allows close interaction between students and professors. Auburn University's Department of Physics takes pride in preparing students for meaningful careers in teaching and research.

Research Facilities

The Department of Physics has state-of-the-art facilities for conducting research at the forefront of science. The Semiconductor Physics Laboratory contains facilities for epitaxial growth (VPE and LPE) and characterization of III-V semiconductor materials and contacts. The Surface Science Laboratory houses equipment for complete analysis of surfaces and interfaces (AES, XPS, ISS, and SEM). The Accelerator Laboratory uses a 3-Mv Dynamitron to apply the techniques of Rutherford backscattering (RBS) and light ion channeling to probe the structure of thin films and interfaces. A new magnetic confinement fusion device, the Compact Auburn Torsatron, was recently constructed to test theories of plasma optimization and transport. Also available are an ellipsometer, a laser-initiated vacuum spark device, and abundant data acquisition equipment. Links with supercomputers operated by the Department of Energy and NASA are in place. Collaborative research projects involving personnel at major national laboratories are available.

Financial Aid

Teaching and research assistantships are available with stipends of $10,600 per year. These stipends usually include summer support if the student wishes. Students receive an increase in stipend upon passage of the Ph.D. qualifying examination.

Cost of Study

Auburn operates on the quarter system. The base registration fee is $85 per quarter, with an additional charge of $40 per quarter hour. Until passage of the qualifying examination, most students take 10 hours per quarter. All students under assistantship are eligible to pay in-state tuition.

Cost of Living

Very affordable housing is available both on and off campus. For married students, the University maintains a complex of one- and two-bedroom apartments with separate outside entrances. Among the features are all-electric kitchens, furnished living rooms, dining rooms, and a bedroom. Current monthly rents range from $245 to $325. There are also many sources of affordable off-campus housing located nearby.

Student Group

The current enrollment at Auburn University is approximately 22,000 students, representing fifty states and fifty foreign countries. About 2,000 of these are graduate students. The physics department currently has about 45 graduate students, all of whom receive financial support. Recent graduates are successfully pursuing careers in academia and in industrial and government laboratories.

Location

Auburn University is located in Auburn, Alabama, in the eastern section of the state. Montgomery, the state capital, is about 60 miles east and Atlanta, Georgia, about 120 miles southwest. The University is close to the beaches of southern Alabama and northern Florida.

The University

The University community enjoys the advantages of security, open space, and clean air afforded by this residential city, surrounded by woodland and farms. The campus is distinguished by its architecture, lawns and flowers, trees, and playing fields. The University, chartered in 1856, consists of fourteen undergraduate schools and colleges, the Graduate School, and the Extension Division. As both a land-grant and a space-grant university, Auburn is dedicated to the service of the people of the state, nation, and world.

Applying

Application forms for admission and financial assistance may be obtained by request at the address given below. The department requires that applicants take the GRE General Test and the Subject Test in physics. Students whose native language is not English are required to provide a minimum score of 550 on the TOEFL. There is no fixed application deadline.

Correspondence and Information

Chairperson, Graduate Admission Committee
Physics Department
Auburn University
Auburn, Alabama 36849
Telephone: 205-844-4264

Peterson's Guide to Graduate Programs in the Physical Sciences and Mathematics 1992

Auburn University

THE FACULTY AND THEIR RESEARCH

William L. Alford, Professor; Ph.D., Caltech, 1953. Nuclear physics.
Raymond F. Askew, Professor; Ph.D., Virginia, 1960. Plasma physics.
Peter A. Barnes, Walter Professor; Ph.D., Simon Fraser, 1969. Condensed-matter physics.
Michael J. Bozack, Assistant Professor; Ph.D., Oregon Graduate Center, 1985. Surface physics.
Howard E. Carr, Professor Emeritus; Ph.D., Virginia, 1941. Atomic physics.
An-Ban Chen, Professor; Ph.D., William and Mary, 1971. Condensed-matter theory.
Eugene J. Clothiaux, Professor; Ph.D., New Mexico State, 1963. Plasma physics and spectroscopy.
John R. Cooper, Associate Professor; Ph.D., Auburn, 1970. Nuclear physics.
Albert T. Fromhold Jr., Professor; Ph.D., Cornell, 1961. Condensed-matter physics.
Junichiro Fukai, Associate Professor; Ph.D., Tennessee, 1972. Plasma physics.
Rex F. Gandy, Associate Professor; Ph.D., Texas, 1981. Plasma physics.
James D. Hanson, Associate Professor; Ph.D., Maryland, 1982. Plasma physics.
Satoshi Hinata, Professor; Ph.D., Illinois, 1973. Space physics.
Earl T. Kinzer Jr., Associate Professor; Ph.D., Virginia, 1961. Lattice dynamics.
Stephen F. Knowlton, Assistant Professor; Ph.D., MIT, 1984. Plasma and space physics.
Paul Latimer, Professor; Ph.D., Illinois, 1956. Biophysics: Light scattering.
Eugene Oks, Senior Research Fellow; Ph.D., Academy of Sciences of the USSR, Institute of General Physics, 1985. Plasma physics.
J. D. Perez, Professor and Department Head; Ph.D., Maryland, 1968. Space and plasma physics.
Michael S. Pindzola, Professor; Ph.D., Virginia, 1975. Atomic and molecular physics.
Vadim Shuets, Research Associate; Ph.D., Academy of Sciences of the USSR, Landau Institute of Theoretical Physics, 1981. Plasma physics.
Marllin L. Simon, Associate Professor; Ph.D., Missouri, 1972. Physics education.
D. Gary Swanson, Alumni Professor; Ph.D., Caltech, 1963. Plasma physics.
G. Donald Thaxton, Professor Emeritus; Ph.D., North Carolina, 1964. Condensed-matter physics.
Charlotte R. Ward, Associate Professor; Ph.D., Purdue, 1956. Physics education.
Jean-Marie Wersinger, Associate Professor; Ph.D., Federal Institute of Technology of Lausanne, 1977. Plasma physics.
John R. Williams, Associate Professor; Ph.D., North Carolina State, 1974. Nuclear and condensed-matter physics.

SECTION 7: PHYSICS

BOSTON UNIVERSITY
Department of Physics

Programs of Study

The Department of Physics offers programs leading to the Ph.D. and M.A. degrees in physics and in astronomy and physics and, through the Division of Applied Sciences and Engineering, to a Ph.D. degree in applied physics. The time it takes to obtain a Ph.D. degree is approximately 5½ years, although students have obtained their degree in as short a time as 2 years and as long as 8.

The master's degree requires the completion of eight semester courses, passed with a grade of B– or better; evidence of having successfully completed advanced courses in French, German, or Russian or passed the departmental language exam; and the achievement of a passing grade on the departmental comprehensive exam or the completion of a master's thesis. Each student must satisfy a residency requirement of a minimum of two consecutive semesters of full-time graduate study at Boston University.

The doctorate requires the completion of eight semester courses beyond the master's degree passed with a grade of B– or better; passing of the departmental language exam in French, Russian, or German; an honors grade on the departmental comprehensive exam; passing of an oral exam; and the completion of a dissertation and a dissertation defense. The dissertation must exhibit an original contribution to the field. Each student must satisfy a residency requirement of a minimum of two consecutive semesters of full-time graduate study at Boston University.

In addition to the programs mentioned above, a Ph.D. program in cellular biophysics and an interdisciplinary Ph.D. program are also available.

Research Facilities

The Department of Physics is located at 590 and 616 Commonwealth Avenue and is part of the new Science and Engineering Complex of Boston University. The complex houses well-equipped research laboratories in solid-state surface physics, low temperature and magnetism, nuclear physics, cosmic ray and high-energy physics, biophysics, and polymer studies. Research is supported by a 10,000-square-foot scientific instrument facility, a high bay assembly area, an electronics design facility, and excellent computational facilities consisting of many microcomputers and minicomputers that are located in the various laboratories and are networked to the departmental VAX 8600 and to an IBM 370/168. A recently obtained Connection Machine CM-2 is capable of performance up to 2 gigaflops on physics-related applications. Other facilities used by the Boston University faculty and students in their research are the Francis Bitter National Magnet Laboratory, the MIT-Bates linear accelerator, the Brookhaven National Laboratory's light source and particle accelerators, Stanford's linear accelerator, the National Balloon Launch Facility, the Lawrence Berkeley Laboratory's Bevelac, the Irvine-Michigan-Brookhaven underground detector facility, and the space shuttle long-duration-exposure facility. The Science Library, part of the Science and Engineering Complex, houses an excellent physics collection in addition to many of the latest periodicals in the field.

Financial Aid

Financial aid is available for qualified students in the form of teaching fellowships, research assistantships, and University fellowships, as well as many national fellowships such as that sponsored by the NSF. The amount of the aid varies from $23,000 to $25,000, including tuition, in 1991–92.

Cost of Study

For 1991–92, tuition costs for a normal two-semester load for graduate students (five semester courses) are $18,070, including fees. Books and other expenses might run an additional $300.

Cost of Living

There is limited graduate housing available on the Boston University campus at approximately $6500 per year for room and board. However, students generally rent apartments in the Boston area. The cost of apartments varies widely, depending on the area.

Student Group

Currently, the department has approximately 90 graduate students engaged in work toward the Ph.D. and M.A. degrees, and it prides itself on the close contact maintained between students and faculty.

Location

Boston University is located in Boston, Massachusetts, which is a major metropolitan center of cultural, scholarly, scientific, and technological activity. Besides Boston University, there are many major academic institutions in the area. Facilities of the local universities, if needed, are easily accessible. Colloquium programs in the area give graduate students a special opportunity to learn the current status of physics research.

The University and The Department

Boston University is a private urban university with a faculty of 2,500 members and a student population of 18,000. The University consists of fifteen schools and colleges.

The Department of Physics is part of the College of Liberal Arts and the Graduate School. In the recent past, the department has experienced significant growth. The department was the first to move into the Science and Engineering Complex. The Graduate Student Association is represented on the faculty Graduate Committee and has a major input in policies concerning graduate affairs in the department.

Applying

The application deadline for fall admission is April 1, although February 15 is suggested for applicants seeking financial aid. For admission to the graduate programs, a bachelor's degree in physics or astronomy is required, with no minimum undergraduate GPA specified. Scores on the Subject Test in physics of the GRE are required; scores on the General Test of the GRE are helpful but not required. The acceptable score for admission is dependent on the applicant's overall record; there is no specified minimum. Students from non-English-speaking countries are required to demonstrate proficiency in English by earning a minimum acceptable score of 550 on the Test of English as a Foreign Language (TOEFL).

Correspondence and Information

Chair, Graduate Admissions Committee
Department of Physics
Boston University
590 Commonwealth Avenue
Boston, Massachusetts 02215
Telephone: 617-353-2623

Peterson's Guide to Graduate Programs in the Physical Sciences and Mathematics 1992

Boston University

THE FACULTY AND THEIR RESEARCH

Professors
Steven Paul Ahlen, Ph.D., Berkeley, 1976. Experimental astrophysics, heavy-ion physics, monopole and quark searches.
Edward C. Booth, Ph.D., Johns Hopkins, 1955. Intermediate-energy particle physics.
Kenneth Brecher, joint appointment with Department of Astronomy; Ph.D., MIT, 1969. Theoretical astrophysics, relativity, cosmology.
James Brooks, Ph.D., Oregon, 1973. Low-temperature physics, magnetism.
Bernard Chasan, Ph.D., Cornell, 1961. Biophysics.
Robert S. Cohen, Ph.D., Yale, 1948. Philosophical and historical foundations of physics.
Ernesto Corinaldesi, Ph.D., Manchester (England), 1951. Quantum mechanics.
Alvaro DeRújula, joint appointment with CERN; Ph.D., Madrid, 1968. Theoretical particle physics, phenomenology.
Dean S. Edmonds Jr., Emeritus; Ph.D., MIT, 1958. Electronics and instrumentation.
Wolfgang Franzen, Emeritus; Ph.D., Pennsylvania, 1949. Atomic physics, surface physics.
Sheldon Glashow, Distinguished Physicist and Research Scholar (Harvard); Ph.D., Harvard, 1958. Theoretical particle physics.
William S. Hellman, Ph.D., Syracuse, 1961. Elementary particle theory.
William Klein, Ph.D., Temple, 1972. Condensed-matter theory.
Frank Krienen, Professor of Engineering and Applied Physics; I.R., Amsterdam, 1947. Experimental particle and accelerator physics, muon g-2.
Kenneth D. Lane, Ph.D., Johns Hopkins, 1970. Theoretical high-energy physics.
Lee Makowski, Ph.D., MIT. Electrical engineering, biophysics.
James Miller, Ph.D., Carnegie-Mellon, 1974. Intermediate- and high-energy experimental physics, muon g-2, CP violation.
Theodore Moustakis, faculty appointment in College of Engineering; Senior Research Associate in Department of Physics; Ph.D., Columbia, 1974. Synthetic novel materials.
Claudio Rebbi, Ph.D., Turin (Italy), 1967. Theoretical physics, lattice quantum chromodynamics, computational physics.
Sidney Redner, Ph.D., MIT, 1977. Statistical and polymer physics.
B. Lee Roberts, Ph.D., William and Mary, 1974. Intermediate- and high-energy experimental physics, muon g-2, CP violation.
James Rohlf, Ph.D., Caltech, 1980. Experimental particle physics, hadron collider physics.
Kenneth Rothschild, Ph.D., MIT, 1973. Biophysics, molecular electronics, physics of vision.
Abner Shimony, Ph.D. (philosophy), Yale, 1953; Ph.D. (physics), Princeton, 1962. Philosophical and historical foundations of physics, theoretical quantum mechanics.
William J. Skocpol, Ph.D., Harvard, 1974. Experimental condensed-matter physics.
John Stachel, Curator of Einstein papers in the United States; Ph.D., Stevens, 1952. General relativity, foundations of relativistic space-time theories.
H. Eugene Stanley, Ph.D., Harvard, 1967. Phase transitions, scaling, polymer physics, fractals and chaos.
Lawrence R. Sulak, Ph.D., Princeton, 1970. Experimental particle physics, proton decay, monopoles, muon g-2, neutrinos.
François Vannucci, joint appointment with University of Paris; Ph.D., Paris XIII (Nord), 1974. Experimental particle physics, neutrino oscillation studies.
Charles R. Willis, Ph.D., Syracuse, 1957. Theory of interaction of radiation with matter, statistical physics.
J. Scott Whitaker, Ph.D., Berkeley, 1976. Experimental colliding-beam physics, supersymmetric particle searches.
George O. Zimmerman, Ph.D., Yale, 1963. Low-temperature physics, magnetism.

Associate Professors
Rama Bansil, Ph.D., Rochester, 1974. Biophysics, polymers.
Maged El-Batanouny, Ph.D., California, Davis, 1978. Surface physics, solitons.
So-Young Pi, DOE Outstanding Junior Investigator; Ph.D., SUNY at Stony Brook, 1974. Theoretical field theory, elementary particles.
James L. Stone, Ph.D., Michigan, 1976. Experimental particle physics and astrophysics, neutrinos, proton decay, monopole studies.

Assistant Professors
Assa Auerbach, Ph.D., SUNY at Stony Brook, 1985. Theoretical condensed-matter physics.
James J. Beatty, Research Assistant Professor; Ph.D., Chicago, 1986. Experimental particle astrophysics.
R. Sekhar Chivukula, Ph.D., Harvard, 1987. Elementary particle theory.
Andrew G. Cohen, Ph.D., Harvard, 1986. Mathematical physics, high-energy theory.
Bennett B. Goldberg, Visiting Assistant Professor; Ph.D., Brown, 1987. Condensed-matter physics.
Karl Ludwig, Ph.D., Stanford, 1986. Experimental condensed-matter physics.
Ganpathy Murthy, Ph.D., Yale, 1987. Condensed-matter theory.
Ryan M. Rohm, Ph.D., Princeton, 1985. Theoretical particle physics.
Robert Wilson, Ph.D., Purdue, 1983. Experimental colliding-beam physics, supersymmetric particle searches.
William Worstell, Ph.D., Harvard, 1986. Experimental particle physics, monopole searches, muon g-2.

Lecturer
Thomas C. Hayes, J.D., Harvard, 1969. Electronics.

Research Associates
Preben Alstrom, Ph.D., Copenhagen, 1986. Statistical mechanics.
Esther Bullitt, Ph.D., Brandeis, 1988. Biophysics.
Zheming Cheng, Ph.D., Michigan, 1987. Statistical mechanics.
Antonio Coniglio, Laurea in physics, Naples, 1962. Statistical mechanics.
Steve Dye, Ph.D., Hawaii, 1988. Experimental particle physics.
Suzhou Huang, Ph.D., MIT, 1988. Theoretical particle physics.
Abdo Ibrahim, Ph.D., Northeastern, 1985. Low-temperature physics, magnetism.
Anthony Johnson, D.Phil., Oxford, 1984. Experimental colliding-beam physics.
Brond Larson, Ph.D., Harvard, 1987. Theoretical condensed-matter physics.
Theodore Lawry, Ph.D., Berkeley, 1985. Experimental particle physics, CP violation.
Paul Mankiewich. Experimental condensed-matter physics, electron microscopy.
Alexandru A. Marin, Ph.D., Central Institute for Physics (Bucharest), 1977. Experimental astrophysics, heavy-ion physics, monopole and quark searches.
Eric Myers, Ph.D., Yale, 1984. Computational physics with supercomputers.
Raman Nambudripad, Ph.D., Indian Institute of Science (Bangalore), 1984. Biophysics X-ray spectroscopy.
Jean Potvin, Ph.D., Colorado at Boulder, 1985. Experimental particle physics, theoretical particle physics, computational physics.
James Shank, Ph.D., Berkeley, 1988. Experimental particle physics.
Jakob Sidenius, Ph.D., Copenhagen, 1988. Theoretical high-enery physics.
T. Somasundaram, Ph.D., Indian Institute of Science (Bangalore), 1987. Molecular biophysics.
Bing Zhou, Ph.D., MIT, 1986. High-energy particle astrophysics.

SECTION 7: PHYSICS

BRANDEIS UNIVERSITY
Department of Physics

Program of Study

The Department of Physics offers a full-time program leading to the Ph.D. degree. As a rule, only candidates for the doctoral degree are accepted; the master's degree may be awarded upon evidence of appropriate proficiency in classical and modern physics.

To qualify for the Ph.D. degree, a student must ordinarily complete at least two full years in residence and nine semester courses and submit a research dissertation. The student must pass a qualifying examination and an advanced examination in the research subject in order to be admitted to candidacy. Course requirements are determined by the Graduate Committee, in consultation with the student. Previous educational experience may be taken into account.

Students may pursue basic research in theoretical physics, including quantum theory of fields, elementary particles, classical and quantum gravity, quantum statistical mechanics, quantum theory of the solid state, critical phenomena, and phase transitions, and in experimental physics, including high-energy physics, atomic and molecular physics, solid state, liquid crystals, light scattering, positron physics, radio astronomy, biophysical structure analysis, and surface physics.

The department maintains a very favorable student-faculty ratio and encourages close interaction between students and faculty. The department contributes significantly to the wide range of colloquia and seminars in the Greater Boston area, to which students have access.

Research Facilities

The department occupies a modern multistory building, which is part of the University's Science Center. Research and teaching laboratory facilities, faculty and graduate student offices, conference and teaching rooms, and student shop facilities are housed within the Physics Building. The Science Library, the University shop facilities, and the Computer Center are in contiguous buildings.

In addition to the specialized research computers in the department's laboratories, the Computer Center provides a VAX 8650 with appropriate peripherals and several microcomputer clusters (Macintosh and IBM PC) around campus for student use.

The department maintains a 24-inch reflector telescope atop the building. Off campus, major collaborative research projects are carried out at Fermilab, Superconducting Super Collider, Brookhaven, and several national radio astronomy observatories.

Financial Aid

Financial aid is available in the form of teaching assistantships, research assistantships, scholarships, and fellowships. All of these awards normally cover the full cost of tuition and provide a stipend as well. Teaching and research assistantships usually require 12–15 hours of work per week. It is expected that all graduate students do some undergraduate teaching during the course of their studies.

Cost of Study

Tuition for full-time students in 1991–92 is $16,085 per year, or $8042.50 per term. (Full tuition scholarships are available for most students.) Tuition for part-time students is prorated.

Cost of Living

Students generally live off campus in apartments or rented rooms. Rents in 1991–92 range from about $300 to $500 per month for a student sharing an apartment or from about $500 to $800 per month for an apartment for a couple.

Student Group

The Department of Physics has approximately 50 full-time graduate students enrolled; nearly all receive financial support of some kind. These graduate students have come to Brandeis from about twenty U.S. undergraduate institutions and eight foreign nations. Nearly all degree recipients have followed research careers, either in academic institutions or in industrial laboratories.

Location

Waltham is a suburb of Boston; the University is located just within the Route 128 perimeter. The Greater Boston area has a well-deserved reputation as one of the leading scientific, recreational, and cultural centers of the world. Other major universities, museums, theaters, conservatories, and libraries in Boston and in Cambridge are within a 30-minute drive. Beaches and ski areas can be reached in less than an hour. The mountains to the north and Cape Cod are within easy driving distance.

The University

Brandeis is a small, private, nonprofit research university that was founded in 1948. The large, verdant residential campus houses about two thirds of the undergraduate population of 2,800 students. There are about 650 graduate students currently enrolled in more than twenty advanced degree programs. The faculty numbers about 350. In 1990–91, annual government support for research carried out at the University was $29.5-million.

Applying

Applications should be submitted in duplicate as early as possible; application forms and catalog materials may be requested from the dean of the Graduate School. The application fee is $25. Applicants for financial aid must file a GAPSFAS form.

Correspondence and Information

Chairman, Graduate Admissions Committee
Department of Physics
Brandeis University
Waltham, Massachusetts 02254-9110
Telephone: 617-736-2835

SECTION 7: PHYSICS

Brandeis University

THE FACULTY AND THEIR RESEARCH

The faculty of the Department of Physics consists of 23 members, whose research specializations fall into six broad areas: astrophysics, atomic physics, biophysics, condensed-matter physics, elementary particle physics, and field theory/gravity/supergravity. The members of the faculty include 2 members of the National Academy of Sciences, 4 Fellows of the American Academy of Arts and Sciences, 5 Fellows of the American Physical Society, 5 former Guggenheim Fellows, 5 former Alfred P. Sloan Fellows, 3 members who among them have held six Fulbright awards, and 1 holder of an honorary D.Sc. from the University of Stockholm. All members of the department are active in research and teaching. As a matter of policy, graduate students are included on research projects whenever possible.

Professors
Laurence F. Abbott, Ph.D., Brandeis, 1977. Computational neuroscience.
James R. Bensinger, Ph.D., Wisconsin–Madison, 1970. Experimental high-energy physics.
Stephan Berko, William R. Kenan Jr. Professor of Physics; Ph.D., Virginia, 1953. Experimental solid-state and positron physics.
Karl F. Canter, Ph.D., Wayne State, 1970. Experimental solid-state and positron physics.
Donald L. D. Caspar, Ph.D., Yale, 1955. Biophysics.
Stanley A. Deser, Enid and Nate Ancell Professor of Physics; Ph.D., Harvard, 1953. Gravity, field theory.
Jack S. Goldstein, Ph.D., Cornell, 1953. Astrophysics, science and public policy.
Marcus T. Grisaru, Ph.D., Princeton, 1958. Field theory, supergravity.
Peter Heller, Ph.D., Harvard, 1963. Statistical physics.
Lawrence E. Kirsch, Ph.D., Rutgers, 1964. Experimental high-energy physics.
Robert B. Meyer, Ph.D., Harvard, 1970. Liquid crystals and complex fluids.
Hugh N. Pendleton, Ph.D., Carnegie Tech, 1961. Mathematical physics.
Alfred G. Redfield, Ph.D., Illinois, 1953. Nuclear magnetic resonance.
David H. Roberts, Ph.D., Stanford, 1973. Extragalactic astronomy.
Howard J. Schnitzer, Ph.D., Rochester, 1960. High-energy theory.
Silvan S. Schweber, Ph.D., Princeton, 1952. History of science.
John F. C. Wardle, Ph.D., Manchester, 1969. Radio astronomy.

Associate Professors
Craig A. Blocker, Ph.D., Berkeley, 1980. Experimental high-energy physics.
Eric S. Jensen, Ph.D., Cornell, 1982. Experimental solid-state physics.
Robert V. Lange, Ph.D., Harvard, 1963. Educational software.
Hermann F. Wellenstein, Ph.D., Texas at Austin, 1971. Experimental atomic physics.

Assistant Professors
Bulbul Chakraborty, Ph.D., SUNY at Stony Brook, 1979. Condensed-matter theory.
Seth Fraden, Ph.D., Brandeis, 1987. Liquid crystals, complex fluids, and multiple scattering.

Postdoctoral Research Associates
Steven Behrends, Ph.D., Rochester, 1987. Experimental high-energy physics.
Michelle Bourdeau, Ph.D., Syracuse, 1990. Field theory.
Leslie Brown, Ph.D. candidate, Brandeis. Radio astronomy.
Tian Yu Cao, Ph.D., Cambridge, 1987. Quantum field theory.
Avram Cohen, Ph.D., Technion (Israel), 1991. Condensed-matter theory.
John D. Cunningham, Ph.D., Notre Dame, 1991. Experimental high-energy physics.
Gustav Delius, Ph.D., SUNY at Stony Brook, 1990. Field theory.
Marialuisa Frau, Ph.D., Turin (Italy), 1985. Field theory.
Christopher Halkides, Ph.D., Wisconsin, 1990. Nuclear magnetic resonance.
Thomas B. Kepler, Ph.D., Brandeis, 1989. Computational neuroscience.
James G. McCarthy, Ph.D., Rockefeller, 1985. Field theory.
Ann-Frances Miller, Ph.D., Yale, 1989. Nuclear magnetic resonance.
Harold Riggs, Ph.D., Chicago, 1989. Field theory.

SECTION 7: PHYSICS

BROWN UNIVERSITY
Department of Physics

Program of Study

The department offers a program of study designed primarily for students seeking the degree of Doctor of Philosophy in physics. The Master of Science in physics is also offered.

Requirements for the Ph.D. are (1) a core of nine semester courses at the graduate level in physics and mathematics; (2) some specialization, normally three semester courses, in advanced graduate study; (3) a qualifying examination and a preliminary examination in general physics; (4) a thesis describing the results of independent research; and (5) an oral examination based on the thesis and research. Students normally complete their formal course work and begin research after two years. All graduate students are encouraged to remain on campus during the summer for study and research.

Research Facilities

The physics department is housed in the seven-story Barus and Holley Building, built in 1965. Nearby are the University computing laboratory, with two IBM computers, linked to the Barus and Holley Building, and a fourteen-story science library that houses the University's excellent collection. High-energy facilities at the Stanford Linear Accelerator Center, at Brookhaven National Laboratory, and at Fermi National Accelerator Laboratory are used by Brown research groups.

Financial Aid

Financial aid includes University fellowships, which cover full tuition and include a stipend sufficient for living costs, and NSF graduate fellowships, which are awarded directly by the granting organization. The most common form of aid is an assistantship—teaching, research, or a combination—which allows three-quarter-time study. Stipends for assistantships cover tuition and defray some living costs. Assistantships carry a total equivalent value of $22,364 for the nine-month academic year 1991–92. Almost all graduate students in the department receive some type of financial aid.

Cost of Study

Tuition is $16,256 for the academic year 1991–92, and $12,192 for assistants studying three-quarter time. After the equivalent of three full-time years, the student pays only a registration fee of $975.36 per semester while completing his or her work.

Cost of Living

The University's Graduate Center provides housing for both men and women at moderate cost. Students also live in apartments and rented rooms in the residential area surrounding the University. Meals are available at campus dining halls.

Student Group

There are 5,562 undergraduate students and 1,539 graduate students at Brown University. There are some 75 graduate students in physics.

Location

The Brown University campus is a few minutes' walk from the center of the city of Providence, the capital of Rhode Island. The metropolitan area has a population of half a million. It is an active commercial center and port and has theaters, concert and symphony halls, and museums. It is on main highway, rail, bus, and air routes. Boston is an hour away by car and 45 minutes by train. New York City is about 3½ hours away by car, train, or bus and 40 minutes by plane. Rhode Island has an extensive coastline, and there are many excellent beaches. Newport is a half-hour drive away. The rest of New England is also within easy driving distance: Cape Cod, 1½ hours; the Berkshires, 2 hours; the White Mountains of New Hampshire, 4½ hours; and the Maine coast, a little over 2 hours.

The University

Founded in 1764, Brown University first awarded advanced degrees in 1888. Instruction and research in the sciences are carried on within the framework of a strong and comprehensive program in the liberal arts. The University has consciously maintained a small, select student body and a faculty that is active in research and enthusiastic in teaching. A low student-faculty ratio permits small classes and individual attention to each student.

Applying

Completed applications for the following fall are due in the Office of the Dean of the Graduate School by January 2. Applications received after this date will be considered, although much of the available financial aid is committed by mid-March. It is strongly recommended that applicants take the Graduate Record Examinations. Applicants whose native language is not English must submit scores on the Test of English as a Foreign Language (TOEFL).

Correspondence and Information

Professor R. E. Lanou, Chairman
Department of Physics
Brown University
Providence, Rhode Island 02912
Telephone: 401-863-2644

Peterson's Guide to Graduate Programs in the Physical Sciences and Mathematics 1992

SECTION 7: PHYSICS

Brown University

THE FACULTY AND AREAS OF RESEARCH

J. C. Baird, R. T. Beyer, R. H. Brandenberger, P. J. Bray, L. Chang, L. N. Cooper, D. Cutts, C. Elbaum, P. J. Estrup, S. Fallieros, F. Fang, D. Feldman, H. M. Fried, H. J. Gerritsen, M. Glicksman, G. S. Guralnik, J. S. Hoftun, A. Houghton, A. Jevicki, K. Kang, H. Kolsky, J. M. Kosterlitz, R. E. Lanou, N. M. Lawandy, F. S. Levin, H. J. Maris, J. B. Marston, D. R. Maxson, P. F. Mende, A. V. Nurmikko, R. Partridge, R. A. Peck, R. A. Pelcovits, G. L. Petersen, G. M. Seidel, A. M. Shapiro, P. J. Stiles, C.-I. Tan, D. M. Targan, J. Tauc, P. T. Timbie, J. H. Weiner, P. J. Westervelt, M. Widgoff, A. O. Williams, G. Xiao, S. C. Ying.

RESEARCH ACTIVITIES

Astrophysics. A new program of experimental and theoretical work has been inaugurated in close cooperation with both the high-energy and condensed-matter programs. Experiments to search for anisotropy in cosmic microwave background radiation, using new types of sensitive infrared detectors, are under way. Theoretical work focuses on explaining the large-scale structure of the universe in the inflationary universe and cosmic string models.

Atomic Physics. Experimental studies of optically oriented atoms are being conducted in the areas of measurements of lifetimes and hyperfine structure for excited atomic states, determination of collision cross sections, and measurements of coherence effects and the range of atomic interactions.

Electronic, Thermal, Optical, Magnetic, and Low-Temperature Properties of Condensed Matter. Research activities include spin-spin and spin-phonon interactions in magnetic systems; low-temperature electronic and thermal properties of solids, particularly metals, through measurements of heat capacity, thermal conductivity, magnetic susceptibility, and resonance techniques; optical studies of the energy-band structure of solids; magnetoconductive, optical, and mechanical properties of amorphous metals and semiconductors; electron-electron interactions in two-dimensional electron plasmas at low temperatures; electronic and magnetic properties of artificial superlattices, submicron structures and ultrafine particles; the quantum Hall effect; physics and application of high-temperature superconductors; nonlinear optical phenomena and plasmadynamics studies in semiconductors, using nanosecond and picosecond laser pulses; holographic studies; studies of one-dimensional systems; properties of liquid and solid He^4 and He^3, including wave propagation, crystal growth, deformation manifestations of macroscopic quantum features, elementary excitations and their interactions, and thermal and dielectric behavior; properties of supercooled liquid H$_2$; pattern formation; and onset and evolution of convection in fluids.

Experimental High-Energy Physics. The properties of elementary particles and their interactions are being investigated through a wide range of experiments. These include studies of the electroweak and strong interactions at the highest available energy using the D0 detector at the newly commissioned Tevatron collider at Fermilab; studies of neutrinos and muons from astrophysical sources in the new underground laboratory at Gran Sasso, Italy; studies of neutrino and muon interactions at the Tevatron II (fixed target); and an effort to search for dark matter and understand the solar neutrino problem using a novel liquid helium detector. Data from these experiments are analyzed at Brown using a MicroVAX farm and the University's IBM 3090 computer. Brown is also active in planning and development for experiments at the Superconducting Supercollider (SSC).

Neural Networks. Biological mechanisms of learning and memory storage are studied. Theoretical research focuses on synaptic formation and modification. Resulting theories (mean-field, relaxation, spin glass) are compared to experimentally obtained information on visual cortical neurons. Artificial neural networks and their applications are studied.

Physical Acoustics. Experimental studies include finite-amplitude ultrasonic propagation in liquids and ultrasonic attenuation and velocity measurements in liquids. Theoretical studies include nonlinear acoustics.

Positive-Ion Reactions and Scattering. A 350-KV positive-ion accelerator has recently been moved into the Barus and Holley Building. The machine is suitable for experiments in atomic, nuclear, and surface physics and is now being used for nuclear reaction cross-section measurements.

Structure of Solids & Crystal Defects. Current investigations include NMR studies of structure and bonding of crystalline and glassy solids; nuclear quadrupole resonance (NQR) studies of electronic distributions in molecules of biological importance and in inorganic solids; the study of crystal defects by means of ultrasonic techniques and X-ray diffraction; the interaction of ultrasonic waves with charge carriers; nonlinear effects in wave propagation and the contribution of defects to the anharmonic behavior of solids; low-temperature studies of the effects of crystal defects on thermal and electrical resistivity; and study of crystalline, optical, and electrical properties of ternary III V semiconductor layers.

Surface Physics. Research in this area centers on the fundamental structural, electronic, and chemical properties of metal and semiconductor surfaces and films. Current topics include chemisorption phenomena, surface band structure, reconstruction, and two-dimensional phase transitions. The experiments employ low-energy electron diffraction, Auger and photoelectron spectroscopy, electron energy loss spectroscopy (EELS), and related techniques.

Theoretical Condensed-Matter Physics. Research problems currently under investigation include the quantum mechanical many-body problem; the properties of models relevant to high-T$_c$ superconductors and heavy fermion systems, e.g., the Anderson and Hubbard models; the study of the elementary excitations and the optical and transport properties of artificially structured materials; the effect of disorder in solids; phase transitions, including the development and application of renormalization group methods to the study of critical phenomena; topological order in two-dimensional systems; Josephson junction arrays in a magnetic field—classical and quantum aspects; surface reconstruction; microscopic theory of surface diffusion and vibrational relaxation; stability and growth of strained epitaxial layers; and liquid crystals.

Theoretical High-Energy Physics. Current activities include studies in quantum field theory, quantum chromodynamics, functional formulations and path-integral techniques, nonperturbative methods in field theory, solitons, monopoles, spontaneous symmetry breaking, lattice field theories, renormalization group, field theoretic approaches to condensed matter, gauge theories of weak and electromagnetic interactions, grand unification theory and phenomenology, flavor dynamics and mixings, cosmology (in particular inflationary universe models and cosmic strings), topological analytic S-matrix theory, phenomenology of scattering and production processes, the quantum theory of gravitation, supersymmetry, supergravity, superstrings, and geometrical aspects of gauge fields.

Theoretical Nuclear Physics. Research is done in areas of low- and intermediate-energy nuclear physics, including elementary excitations, nuclear reaction mechanisms, electromagnetic properties of nuclear states, and reactions of elementary particles with nuclei. Current topics include two-photon processes in nuclei, form factors, sum rules, relativistic features, intrinsic excitations of nucleons, many-body scattering theory, the three- and four-nucleon problems, and few-body models of nuclear reactions.

A view of the Graduate Center.

A researcher studying acoustic attenuation in the low-temperature lab.

The Barus and Holley Building.

SECTION 7: PHYSICS

CALIFORNIA INSTITUTE OF TECHNOLOGY

Department of Physics

Program of Study

The department offers a program of study leading to the degree of Doctor of Philosophy in physics. A Master of Science degree may be awarded upon completion of a one-year program of courses. Students are not normally admitted to work toward the M.S. degree in physics unless they are also working for a Ph.D. Requirements for the Ph.D. include passing a written candidacy examination, typically taken in the first or second year, covering basic material in physics; completing 54 units (equivalent to 12 semester hours) of advanced electives in physics; writing a thesis that describes the results of independent research; and passing a final oral examination based on this thesis and research.

A minor is not required, but a student may elect to pursue one. Furthermore, there are no language requirements for a Ph.D. in physics, but mastery of one or more foreign languages may be advantageous.

Some students admitted into the physics department find their research interests are best matched to Caltech faculty members outside the physics department. The two areas in which this is most common are condensed-matter physics, which is studied in both the applied physics and physics departments at Caltech, and astronomy. Students should apply directly to the applied physics or astronomy department if their interests are primarily in these areas.

Research Facilities

There are three electrostatic accelerators, which operate at 1, 6, and 12 MeV. A new 6-MeV tandem accelerator features high-intensity, high-transmission efficiency and superb energy and position resolution. Equipment is also available for precision investigations of nuclear gamma and beta rays. Equipment for experiments in high-energy physics is designed and constructed, and the resulting data analyzed, in Caltech laboratories. Other experimental facilities include one 130-foot and two 90-foot movable and steerable radio antennas with a 10-meter millimeter dish, a low-temperature laboratory, and facilities for cosmic-ray research on the ground, in balloons, and in spacecraft. The facilities of the nearby Jet Propulsion Laboratory (JPL) are often used. The Mount Palomar observatory is an outstanding facility as is the new 10-meter Keck telescope on Mauna Kea in Hawaii.

Financial Aid

Fellowships that require no duties are available. A typical stipend is $12,600 for the 1991–92 academic year and summer, plus tuition. In addition, the Richard P. Feynman Fellowship in theoretical physics and the Robert A. Millikan Fellowship in experimental physics are awarded. Each of these carries a stipend of $13,650.

Graduate teaching assistants are assigned teaching duties requiring a total of 15 hours per week during the academic year. At present, the stipend ranges from $8667 to $9909, depending on experience and the duties assigned. Additional financial support is also available for graduate teaching assistants and fellowship recipients who wish to continue their research during the summer months.

Graduate research assistantships are awarded to students who have acquired enough laboratory or shop skills to be employed for work on a research project. Graduate research assistants serve on various research projects for 15 hours per week during the academic year and for 30 hours per week during the summer. Currently, their stipend ranges from $13,650 to $14,850 for a calendar year, depending on experience.

Cost of Study

Tuition in 1991–92 is $14,100 for the academic year. There is a general deposit of $25. Books and supplies vary in cost; a typical figure might be $500 per year. Currently, all students receive a special award to cover tuition.

Cost of Living

The cost of accommodations in the graduate dormitories for single students ranges from $3449 to $3624 for the academic year 1991–92. Costs in the Catalina graduate apartment complex range from $4028 to $4528 for single students and $7608 to $9060 for married students; this covers rent for 12 months; utilities are extra. Housing is also available in the residential areas that surround the campus.

Student Group

Caltech has approximately 1,000 undergraduate students and about 1,000 graduate students, a small number of whom are women. There are about 150 graduate students in physics, all of whom are receiving financial support of some kind. Admission to graduate study in physics is highly competitive; about 30 new students are admitted each year.

There are more than 16,000 Institute alumni around the world, many of whom are eminent in their fields of science and engineering. Twenty-one alumni have received Nobel Prizes.

Location

Pasadena, a city of approximately 125,000 inhabitants, is located just northeast of Los Angeles at the foot of the San Gabriel Mountains. The city contains both residential and light industrial areas; Caltech is located in the center of a residential area but is within several blocks of shopping facilities. Most of the cultural activities of the large metropolitan area may be reached in an hour or less by car.

There are numerous recreational facilities in the vicinity. Hiking, camping, and occasionally skiing are immediately at hand in the mountains. The Pacific Ocean is less than an hour away by car, and any part of southern California can be explored on a weekend.

The Institute

Throughout its history, Caltech has maintained a small, select student body and a faculty that is unusually active in research. The staff currently numbers about 900, including postdoctoral fellows; of this number, more than 250 are full professors.

There is a strong emphasis on fundamental studies in science and engineering with a minimum of specialization. There is also considerable emphasis on humanistic studies. Humanities have always played an essential role in the undergraduate curriculum, and undergraduates can major in these areas.

Musical and dramatic events are presented on campus throughout the year.

Applying

Students are admitted only in September, and applications should be received by January 15. It is strongly advised that applicants take the GRE General Test and physics Subject Test by the October test date. Application materials may be obtained from the Institute Graduate Office, mail code 02-31.

Correspondence and Information

Donna Driscoll
Physics Graduate Coordinator
Mail Code 103-33
California Institute of Technology
Pasadena, California 91125

SECTION 7: PHYSICS

California Institute of Technology

THE FACULTY AND THEIR RESEARCH

B. C. Barish, Ph.D., California, 1962. Experimental high-energy physics, monopole searches, e^+e^- collisions.
C. A. Barnes, Ph.D., Cambridge, 1950. Nuclear physics, nuclear astrophysics.
R. D. Blandford, Ph.D., Cambridge, 1974. Astrophysical plasmas, pulsars, active galactic nuclei.
F. H. Boehm, Ph.D., Zurich, 1951. Nuclear physics, mesic X rays, time reversal symmetry in nuclei, neutrino interactions.
M. C. Cross, Ph.D., Cambridge, 1975. Condensed-matter theory, nonequilibrium physics.
R. P. W. Drever, Ph.D., Glasgow, 1958. Experimental gravitation, gravitational wave detection.
B. W. Filippone, Ph.D., Chicago, 1982. Experimental nuclear physics, nuclear astrophysics.
S. C. Frautschi, Ph.D., Stanford, 1958. Theoretical physics.
M. Gell-Mann, Ph.D., MIT, 1950. Symmetries and current algebra.
R. Gomez, Ph.D., MIT, 1956. Experimental high-energy physics, strong interactions at high energy.
D. L. Goodstein, Ph.D., Washington (Seattle), 1965. Two-dimensional matter, surfaces, phase transitions and interfaces, ballistic phonons, high-temperature superconductivity.
D. Hitlin, Ph.D., Columbia, 1968. Experimental high-energy physics.
R. W. Kavanagh, Ph.D., Caltech, 1956. Experimental nuclear physics, nuclear astrophysics.
H. J. Kimble, Ph.D., Rochester, 1978. Experimental quantum optics.
S. E. Koonin, Ph.D., MIT, 1975. Theoretical nuclear and many-body physics.
K. G. Libbrecht, Ph.D., Princeton, 1984. Solar phenomena, solar and stellar acoustic oscillations, solar oblateness.
R. D. McKeown, Ph.D., Princeton, 1979. Experimental nuclear physics, electromagnetic structure of nuclear matter and nucleons at high-momentum transfer, development of optically pumped polarized ^{3}He targets.
R. A. Mewaldt, Ph.D., Washington (St. Louis), 1971. Experimental galactic, solar, and interplanetary cosmic rays.
R. P. Mount, Ph.D., Cambridge, 1975. Experimental high-energy physics.
G. Neugebauer, Ph.D., Caltech, 1960. Observational infrared astronomy.
H. Newman, Sc.D., MIT, 1973. Experimental high-energy physics.
C. W. Peck, Ph.D., Caltech, 1964. Experimental high-energy physics, electron-positron colliding beams, monopole searches.
T. Phillips, D.Phil., Oxford, 1964. Submillimeter-wave astronomy.
E. S. Phinney, Ph.D., Cambridge, 1983. Active galactic nuclei, high-energy and relativistic astrophysics.
J. Pine, Ph.D., Cornell, 1956. Biophysics, experimental physics.
H. D. Politzer, Ph.D., Harvard, 1974. Theoretical physics.
F. C. Porter, Ph.D., California, 1977. Experimental high-energy physics, e^+e^- collisions.
J. Preskill, Ph.D., Harvard, 1980. Elementary particle theory.
T. Prince, Ph.D., Chicago, 1978. Gamma-ray astronomy, ground-based optical interferometry, concurrent computing.
F. J. Raab, Ph.D., SUNY at Stony Brook, 1980. Experimental gravitation, gravitational wave detection.
J. H. Schwarz, Ph.D., California, 1966. Theoretical high-energy physics, especially superstring theory.
B. Simon, Ph.D., Princeton, 1970. Mathematical physics.
T. Soifer, Ph.D., Cornell, 1972. Infrared astronomy.
E. C. Stone, Ph.D., Chicago, 1963. Experimental galactic and solar cosmic rays, magnetosphere physics.
R. Stroynowski, Ph.D., Geneva, 1973. High-energy particle physics.
K. S. Thorne, Ph.D., Princeton, 1965. Relativistic astrophysics, gravitational physics.
T. A. Tombrello, Ph.D., Rice, 1961. Application of ion beam techniques to material science and surface science, modeling of geophysical processes, radiation damage.
P. Vogel, Ph.D., Joint Institute for Nuclear Research (USSR), 1966. Nuclear theory, mesic X rays, neutrinos, fission.
R. E. Vogt, Ph.D., Chicago, 1961. Experimental astrophysics, experimental gravitation, gravitational wave detection.
P. B. Weichman, Ph.D., Cornell, 1986. Condensed-matter theory, phase transitions, critical phenomena.
A. J. Weinstein, Ph.D., Harvard, 1983. Experimental high-energy physics.
W. Whaling, Ph.D., Rice, 1949. Application of nuclear physics techniques to atomic spectroscopy, transition probabilities.
M. B. Wise, Ph.D., Stanford, 1980. Theoretical particle physics, cosmology.
N. C. Yeh, Ph.D., MIT, 1988. Experimental condensed-matter theory, superconductivity, magnetism, phase transitions.
F. Zachariasen, Ph.D., Chicago, 1957. Theory of high-energy physics.
H. Zirin, Ph.D., Harvard, 1953. Solar phenomena; flares, stellar spectroscopy, solar radio astronomy.
J. Zmuidzinas, Ph.D., California, 1987. Submillimeter astronomy and instrumentation.

RESEARCH ACTIVITIES

Biophysics. A small group is working in neurobiology. **Condensed Matter and Low-Temperature Physics.** Experimental work based on cryogenic techniques ranges from studies of the fundamental nature of superfluidity, phase transitions, critical phenomena, and dimensional crossovers in such materials as superconductors and thin films to the development of unique electronic devices from superconductivity. Theoretical work includes investigating phase transitions, critical phenomena, quantum fluids and solids, and nonequilibrium phenomena, including spatial structure, chaos, and turbulence. **Experimental Gravitation.** A new 40-meter interferometer for the detection of gravitational waves is now in operation, and design work on a multikilometer-long detector is in progress. **Experimental High-Energy Physics.** Experiments in elementary particle physics are carried out with accelerators at SLAC, CERN, DESY, and Fermilab. A large, partially operational detector to search for magnetic monopoles is under construction at Gran Sasso underground lab in Italy. A phenomenology group is concentrating on QCD in lepton and high-transverse momentum processes and on design and planning for the superconducting supercollider. **Kellogg Radiation Laboratory.** Studies of the structure and interactions of nuclei currently include experiments in the few-MeV energy range, carried out with Caltech's in-house tandem electrostatic accelerators. Experiments in the GeV range are carried out at SLAC and Los Alamos. **Nuclear Physics at Low and Intermediate Energies.** Properties of nuclei and elementary particles, particularly neutrinos, are studied. Experiments on neutrino oscillations and double beta decay are designed to help in the understanding of neutrino mixing and neutrino mass. The experimental program is complemented by theoretical studies of nuclear structure and particle properties. **Optical, Infrared, and Radio Astronomy.** Astrophysical observations are carried out at Mount Palomar and Big Bear Lake. Caltech has been a major participant in a recent survey of the infrared sky conducted by IRAS. Star formation, interstellar gas, galaxies, and quasars are studied using the 10-meter telescope on Mauna Kea in Hawaii. Far-infrared observations are made from NASA's Kuiper Airborne Observatory. **Quantum Optics.** Studies include quantum measurement, quantum dynamics of dissipative systems, and squeezed-state and cavity QED experiments. **Space Physics.** The astrophysical aspects of cosmic radiation are investigated with detectors flown in balloons and in spacecraft. Observational and theoretical studies of magnetic fields, velocity fields, and active regions on the sun are carried out. **Theoretical Physics.** Superstring theory, a promising candidate for unified theory of all elementary particles and fundamental forces, is emphasized. Other areas include phenomenology, quantum chromodynamics, and the electroweak theory. Early universe cosmology, quantum fluctuations in the topology of space-time, general relativity, and black-hole physics are investigated. Theoretical studies also include nuclear physics, mathematical physics, and computational physics, such as parallel processors and neural networks.

SECTION 7: PHYSICS

CALIFORNIA STATE UNIVERSITY, LOS ANGELES

Department of Physics and Astronomy
Program in Physics

Program of Study

The Master of Science program in physics at California State University, Los Angeles, offers an opportunity for graduate study with particular attention given to individual instruction and research supervision. Students undertaking graduate study receive excellent preparation for further work at the doctoral level or for careers in related fields or industry. Graduates of the program are regularly admitted to Ph.D. programs at various campuses of the University of California or at other leading institutions.

To undertake graduate study leading to a master's degree in physics, a student must have completed the equivalent of an undergraduate major in physics. Students with bachelor's degrees in other disciplines or with significant deficiencies in their undergraduate preparation should expect to complete some course requirements in addition to those of the M.S. program. Requirements for the master's degree include 45 quarter units of approved courses, of which at least 32 must be completed in residence at this University. The thesis option involves 27–36 units of formal course work and 9–18 units of guided theoretical or laboratory work and independent study.

All candidates for the degree must pass an examination administered by the department. For students electing the research option, the examination is an oral one based on the thesis; for students electing the course option, it is a comprehensive examination, which may be supplemented by an oral portion at the option of the examining committee.

The University operates on a year-round quarter system.

Research Facilities

The department occupies three floors of the Physical Science Building, which has facilities for experimental research in nuclear physics, solid-state physics, cryogenics, atomic physics, modern optics, and biophysics. Other departmental equipment includes several Sun Workstations for condensed-matter theory, a microcomputer lab, a complete X-ray diffraction unit, an electron microscope, a tunable dye laser, a vacuum double-beam spectrometer, cryogenic systems, a high-speed digital averager, electron and nuclear paramagnetic resonance apparatus, and an atomic resonance spectrometer. In addition, there are glass, machine, and electronics shops, all staffed by trained personnel. The University has access to optical and radio astronomy facilities.

Financial Aid

Approximately five graduate assistantships are available. Appointments are based on the applicant's previous record and letters of recommendation. Graduate assistants usually assist in four laboratory classes per week. The stipend for 1990–91 was $2592 per academic quarter.

Cost of Study

Fees in 1990–91 for a California resident averaged about $295 per quarter. For nonresidents (including foreign students), there was an additional fee of $137 per unit.

Cost of Living

There are apartments in a very attractive student housing complex that accommodates more than 1,000 students. There is also additional housing available close to the campus at rates comparable to those in most metropolitan areas. General living costs are also comparable, on the average, to those in other urban areas of the country.

Student Group

The department has approximately 70 students majoring in physics; about one third of these are graduate students.

Location

California State University, Los Angeles, is located six miles east of downtown Los Angeles, the nation's second-largest metropolitan area. The campus is bordered on two sides by the San Bernardino and Long Beach freeways and is only a short distance away from attractive residential communities and major cultural activities. Mountain and beach recreation areas are within easy reach of the campus. Several major universities are nearby, including the University of Southern California, the University of California, Los Angeles, and the California Institute of Technology. The region has numerous industrial laboratories spanning all areas of research interest.

The University

California State University, Los Angeles, established in 1947 as Los Angeles State College, has grown into a major urban center of higher education. The more than 20,000 students, from diverse cultural and national backgrounds, constitute an unusually well motivated and mature group. The University has several unusual educational programs connected with its urban focus.

Applying

An appropriate baccalaureate degree from an accredited institution is required for admission to unclassified graduate standing. Application deadlines and forms may be obtained from the Admissions Office. All applicants who did not earn their bachelor's degree from an institution where English was the language of instruction must score at least 550 on the TOEFL examination before they can be admitted.

Correspondence and Information

For admission:
Admissions Office
California State University, Los Angeles
5151 State University Drive
Los Angeles, California 90032

For additional information about programs and assistantships:
Fleur B. Yano, Chair
Department of Physics and Astronomy
California State University, Los Angeles
5151 State University Drive
Los Angeles, California 90032

Peterson's Guide to Graduate Programs in the Physical Sciences and Mathematics 1992

SECTION 7: PHYSICS

California State University, Los Angeles

THE FACULTY AND THEIR RESEARCH

Radi A. Al-Jishi, Ph.D., MIT, 1982. Theoretical condensed-matter physics.
Konrad A. Aniol, Ph.D., Australian National, 1977. Experimental medium-energy nuclear physics, muon-catalyzed fusion.
Roland L. Carpenter, Ph.D., UCLA, 1966. Relativistic astrophysics, extragalactic astronomy.
Robert H. Carr, Ph.D., Iowa State, 1963. Computer-assisted instruction.
Berken Chang, Ph.D., Berkeley, 1967. Experimental atomic physics and modern optics.
Harold L. Cohen, Ph.D., Rutgers, 1968. Theoretical and mathematical physics.
Charles C. Coleman, Ph.D., UCLA, 1968. Solid-state physics, modulation spectroscopy, ion implantation, crystal growth, thin films, bioluminescence, junctions, intercalation.
Martin B. Epstein, Ph.D., Maryland, 1967. Experimental nuclear physics, few-nucleon problems.
Perry S. Ganas, Ph.D., Sydney, 1968. Theoretical nuclear physics, nuclear shell model and electron atom scattering.
David T. Gregorich, Ph.D., California, Riverside, 1968. Infrared astronomy, theoretical physics.
Demetrius J. Margaziotis, Ph.D., UCLA, 1966. Experimental medium-energy nuclear physics, few-nucleon problems.
Milan Mijic, Ph.D., Caltech, 1987. Theoretical study of quantum and inflationary cosmology.
Edward H. Rezayi, Ph.D., Stanford, 1979. Condensed-matter theory, statistical mechanics.
José Rodriguez, Ph.D., Illinois, 1987. Theoretical condensed-matter physics.
Frieda A. Stahl, Ph.D., Claremont Graduate School, 1968. Low-temperature, solid-state physics.
William A. Taylor, Ph.D., California, Riverside, 1966. Experimental solid-state physics, thermodynamic studies of metals and alloys, cryogenics, optical studies in solids.
John C. Woolum, Ph.D., Cornell, 1965. Neurobiology, circadian rhythms.
Fleur B. Yano, Ph.D., Rochester, 1966. Theoretical nuclear physics, nuclear structure theory, application of group theory methods to nuclear physics, medium-energy theory, heavy-ion collisions.

Aerial view of the campus. The Physical Sciences Building is on the far left.

SECTION 7: PHYSICS

CLEMSON UNIVERSITY

Programs in Physics and Astronomy

Programs of Study

Clemson University offers programs leading to the M.S. and Ph.D. degrees in physics through the Department of Physics and Astronomy. Master's and doctoral programs in astronomy and astrophysics, atmospheric physics, and biophysics are also available; the degree awarded is in physics.

Requirements for the Ph.D. degree include the passing of a qualifying exam on graduate-level physics and an oral exam on the subject of the dissertation. There are no formal course requirements. A program of study is worked out with the student's advisory committee and the graduate advisers.

Requirements for the M.S. degree include the completion of 30 semester hours of credit for research and courses and the passing of an oral exam on the subject of the thesis. A nonthesis option, requiring 36 semester hours of credit, is also available.

Research Facilities

The Department of Physics and Astronomy is housed in a modern, completely air-conditioned building containing 64,000 square feet of teaching and research space. Major equipment and facilities include a computer-controlled microdensitometer, a Van de Graaff accelerator for electrons, electron spin resonance spectrometers equipped with a microcomputer, a superconducting solenoid for studies in magnetic fields up to 120,000 gauss, a computer-graphics laboratory, low-temperature apparatus for studies at temperatures down to 10K, electrostatically shielded rooms, complete electronics and machine shops, and microcomputers that are used in both teaching and research laboratories. Students and faculty members have direct access by terminals to the University computer center's mainframe computers—HDS AS/EX-80 and four VAX's (6000-410, 8650, 8810, 8820). Clemson is connected to BITNET, TELENET, SURANET, and ARPANET. The department has a number of workstations. Students have access to major astronomical observatories and radar facilities for atmospheric studies both in the United States and around the world.

Financial Aid

For students with a B.S. degree, University assistantships start at $10,200 in 1991–92. For students with an M.S. degree, the stipend is $10,500. Graduate assistants' duties usually involve two or three elementary laboratories each week as well as tutoring or grading in elementary courses. These stipends may be supplemented by fellowships for highly qualified students.

The University employs students' spouses, as do many local businesses and schools.

Cost of Study

The total cost of study for assistants is about $325 per semester and $200 for summer school in 1991–92, for in-state and out-of-state students alike.

Cost of Living

Dormitories range in cost from $650 to $825 per semester for double occupancy in 1991–92. All are air conditioned, and most include the cost of a room telephone. University apartments range in cost from $845 to $1030 per semester for modern duplex units; the cost of utilities varies. Many privately owned apartments are available. Costs vary considerably.

Food at University dining facilities in 1991–92 costs $619 per semester for three meals a day five days a week and $719 for seven days a week. Optional University health care is available at $80 per semester.

Student Group

There are approximately 45 graduate students and 35 undergraduate students in the department.

Location

Clemson University is located in Clemson, South Carolina, a small university town (population 11,000) in northwestern South Carolina in the foothills of the Blue Ridge Mountains. It is midway between Charlotte, North Carolina, and Atlanta, Georgia, and is about 10 miles north of Interstate Highway 85. In addition to the normal University activities, there are extensive opportunities for outdoor recreation. For example, Lake Hartwell with its 1,000-mile shoreline and beautifully clear water is just west of the campus. While shopping in Clemson is limited to rather small stores, Greenville and Anderson are within a few minutes' drive and offer more extensive shopping and entertainment.

The University

Clemson University is a fully accredited, state-supported land-grant university. The main campus is situated on a 1,400-acre site, part of which was once the John C. Calhoun plantation. The campus is surrounded by 21,000 acres of agricultural research land and bordered by Lake Hartwell. Enrollment is about 16,400, including about 3,500 graduate students. Clemson offers seventy-eight undergraduate and fifty-five graduate curricula.

Applying

Prospective students should write to the address below for the necessary application forms. International students should apply directly to the Graduate School for a special self-managed application package. Applications are normally due by March 15 but are considered at any time. The General Test of the Graduate Record Examinations is required for admission.

Correspondence and Information

J. A. Gilreath
Graduate Student Recruiter
Department of Physics and Astronomy
Clemson University
Clemson, South Carolina 29634-1911
Telephone: 803-656-3416

Peterson's Guide to Graduate Programs in the Physical Sciences and Mathematics 1992

Clemson University

THE FACULTY AND THEIR RESEARCH
Professors: P. B. Burt; D. D. Clayton; L. M. Duncan; H. W. Graben; F. J. Keller; L. L. Larcom; M. F. Larsen; A. L. Laskar; J. R. Manson; P. J. McNulty, Head; M. G. Miller; J. R. Ray; M. D. Sherrill; R. C. Turner; C. W. Ulbrich. **Associate Professors:** T. F. Collins; P. J. Flower; J. A. Gilreath; J. R. Letaw; J. W. Meriwether; P. A. Steiner; G. X. Tessema. **Assistant Professors:** D. H. Hartmann; M. D. Leising; B. S. Meyer. **Research Professors:** L. S. Miller; M. V. Nevitt; M. J. Skove; E. P. Stillwell.

RESEARCH AREAS

Astronomy and Astrophysics
Gamma-ray astronomy; origin of the solar system and of meteorites; nucleosynthesis. (D. D. Clayton, M.D. Leising)
Stellar structure and evolution of stars; color-magnitude diagrams of galactic and extragalactic star clusters. (P. J. Flower)
Gamma-ray astronomy; gamma-ray bursts; galactic structure; nucleosynthesis. (D. H. Hartmann)
Astrophysics; cosmic-ray propagation in space. (J. R. Letaw)
Nuclear astrophysics; stellar collapse; cosmology. (B. S. Meyer)

Atmospheric Physics
Upper-atmospheric phenomena involving active radio-wave experiments in space plasmas, radar studies of upper-atmospheric dynamics and the near-earth environment, and nonlinear phenomena in the ionosphere. (L. M. Duncan)
Vertical profiling of atmospheric winds with MST radars; applications in turbulence theory and dynamic meteorology. (M. F. Larsen)
Studies of atmospheric dynamics and optical emissions using Fabry-Perot interferometers and lidar instrumentation. (J. W. Meriwether)
Millimeter-wave-length radar techniques for remote sensing of geophysical atmospheric and oceanographic parameters. (L. S. Miller)
Physics of precipitation-forming processes in clouds and methods of remotely sensing precipitation parameters. (C. W. Ulbrich)

Biophysics
Structure and spectra of biological molecules and the interactions among these; DNA damage and its repair; effects of nonionizing radiation on biological systems. (L. L. Larcom)

Computational Physics
Interfacing of computers with experimental and theoretical programs; computer-assisted instruction. (J. A. Gilreath, M. G. Miller)
Modeling radiation environments in space and the effects on microstructures. (P. J. McNulty)
Computer simulation studies of solids and glasses, elastic properties, structural phase transformations, free-energy difference calculations, and epitaxial growth studies. (J. R. Ray)

Solid-State Physics
Interaction of radiation with matter; microdosimetry; elementary particle theory; interaction of electromagnetic radiation with atoms and molecules; soft errors and microelectronic circuits. (P. J. McNulty)
Quasi-one-dimensional metals: stress effects on charge density wave systems; high T_c superconductors; electron noise at transitions. (M. J. Skove, E. P. Stillwell, G. X. Tessema)
Dynamics of charge density wave systems; electrical and magnetic field effect on charge density wave systems. (G. X. Tessema)
Macroscopic quantum systems: superconductivity and Josephson tunneling. (M. D. Sherrill)
Defect properties and their transport in ionic and superionic solids: migration of crystalline defects as related to diffusion, electrical conductivity, and dielectric properties. (A. L. Laskar)
Defect properties of ionic crystals: the influence of defects on ionic crystal properties, such as electrical conductivity and polarization phenomena. (M. G. Miller)
Electron paramagnetic resonance: degradation, point defects, and free radicals in ionic crystals and synthetic polymers. (R. C. Turner)

Theoretical Physics
Quantum theory and quantum field theory: persistent interactions and nonperturbative interactions. (P. B. Burt)
Mathematical physics: construction of intrinsically nonlinear solutions of nonlinear field equations. (P. B. Burt)
Statistical mechanics; ensemble theories and fluctuations. (H. W. Graben, J. R. Ray)
Surface physics and solid-state theory: information about the nature of surface structure and surface interactions, obtained by studying the interactions of atomic and molecular beams impinging upon solid surfaces. (J. R. Manson)
Computer simulation studies of solids and glasses, elastic properties, structural phase transformations, free-energy difference calculations, and epitaxial growth studies. (J. R. Ray)

SECTION 7: PHYSICS

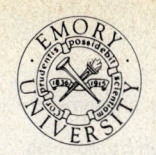

EMORY UNIVERSITY
Department of Physics

Programs of Study

The Department of Physics is currently undergoing a major expansion, providing outstanding opportunities for close interaction between graduate students and faculty. The Department offers programs leading to the Master of Arts, Master of Science, and Doctor of Philosophy degrees in physics. Current research activities in the department include experimental and theoretical condensed-matter physics and biophysics. The experimental condensed-matter physics programs include research activities in far-infrared and Raman studies of superlattices and superconductors, dynamic light-scattering investigations of complex fluids, and studies of pattern formation and fluid flow in porous media. Theoretical condensed-matter physics research includes calculations of electronic properties of semiconductors and solid-state devices, studies of nonequilibrium growth phenomena, pattern formation, fractals, surfaces and interfaces, neural networks, and spin glasses. The biophysics programs are focused on structural and functional studies of metalloenzymes, lipids and DNA molecules, and the statistical theory of steroscopic vision. The department's very low student-faculty ratio guarantees high levels of personal interaction. Regular colloquia and seminars provide additional stimulation. Advanced graduate students participate regularly in major national and regional physics meetings and are encouraged to present papers on these occasions.

The Graduate School requirements for the master's degrees include 24 semester hours of course, seminar, or research credit and a general examination and/or a thesis. These requirements take an average of two to three semesters to complete. The requirements for the Ph.D. include the master's degree or the equivalent, full residence (12 semester hours) for at least four semesters beyond the master's level, a general doctoral examination, and a doctoral dissertation. During the full-residence period, 48 semester hours of credit must be accumulated. Of these, 24 hours must be in courses, directed study, or seminars, and 8 hours must be in areas outside physics.

Research Facilities

The Department of Physics is located in the new Rollins Research Center, a state-of-the-art building designed for interdisciplinary scientific research. The departmental research facilities include EPR, Mössbauer, ENDOR, Raman, far-infrared, photoluminescence, and picosecond spectrometers; a SQUID susceptometer; a TEA CO_2 laser; fluorescence, flash photolysis, IR, UV, and visible spectrophotometers; and quasi-elastic and static light-scattering facilities. Other facilities include an electronics shop, a machine shop, and a materials preparation laboratory. The department operates an internal computer network, connected both to the University Computer Center's IBM 3090, Sun-4, and VAX computers and to external networks, including BITNET and Internet. Graduate students have unlimited access to these facilities. The library facilities include a 75,000-volume science library, a chemistry library, and a health sciences library that subscribes to all of the major U.S. and foreign physics, chemistry, mathematics, and biophysics journals. Emory University is a sponsoring member of the UNISOR facility at Oak Ridge, and the department also has access to the infrared beam at the Brookhaven National Laboratory Synchrotron light source.

Financial Aid

Graduate students in the department receive full funding, including a full tuition waiver and a combined fellowship and assistantship. First-year students do not teach. Second-year teaching assistants are assigned teaching duties that require a light load of 10–12 hours of teaching per week. Research assistantships and fellowships are available, allowing graduate students beyond the second year to carry out full-time research.

Cost of Study

Tuition for the 1991–92 academic year is $14,685. Admitted graduate students are granted full tuition waivers and combined fellowships and yearly $12,000 assistantships that cover expenses while they are earning their degree.

Cost of Living

Atlanta's cost of living ranks among the lowest of the nation's metropolitan areas. Housing in University apartments ranges from $280 to $560 per month. In addition, students have a wide range of options in rooms and apartments in the residential neighborhoods surrounding the campus.

Student Group

Emory has a total enrollment of about 9,000 students. Enrollment in the various schools of the University is restricted to maintain the most favorable use possible of facilities and resources. There are about 3,500 students in the undergraduate college and about 5,500 students in the eight graduate and professional schools.

Location

Atlanta is a metropolitan area with a population of more than 2 million. It offers many cultural and recreational opportunities, including the Atlanta Symphony Orchestra, a resident opera company, a resident repertory theater, a ballet company, and major-league sports. Emory is located in the northeastern part of the city in a residential area, within a short drive or bus ride of the downtown area. Within an hour's drive are the Blue Ridge Mountains and several large lakes that offer opportunities for camping, hiking, swimming, and boating.

The University and The Department

Emory University is a private university with a national reputation for scholarly and educational excellence. Expansion at Emory accelerated after 1980, when the Robert W. Woodruff gift boosted Emory's endowment, now ninth in the nation. The Emory University Graduate School has awarded advanced degrees since 1919, and many of its graduates occupy positions of leadership in education and research. Over 100 Ph.D. degrees were awarded last year. The physics department has a tradition of emphasizing personal contact among faculty and students, as well as offering high-quality training in scholarship and research.

Applying

Applications should be submitted as early as possible. Applications for financial aid and scores on the GRE General Test and Subject Test in physics are to be submitted by February 15. International students whose native language is not English must provide scores on the Test of English as a Foreign Language (TOEFL) in addition to the GRE scores. Awards are generally made by April 1. Emory University does not discriminate on the basis of race, color, religion, sex, national origin, handicap, age, or veteran status.

Correspondence and Information

Director of Graduate Studies
Department of Physics
Emory University
Atlanta, Georgia 30322
Telephone: 404-727-6584

SECTION 7: PHYSICS

Emory University

THE FACULTY AND THEIR RESEARCH

Scott R. Anderson, Assistant Professor; Ph.D., Chicago, 1987. Computational condensed-matter physics: critical phenomena, nonequilibrium growth phenomena.

Krishan K. Bajaj, Professor and Chair of the Department; Ph.D., Purdue, 1966. Theoretical solid-state physics: electronic properties of semiconductors and superlattices, solid-state devices.

Robert L. W. Chen, Professor; Ph.D., Syracuse, 1960. Theoretical physics, quantum mechanics.

Robert N. Coleman, Lecturer; M.S., Emory, 1974. Radioecology.

Edmund P. Day, Associate Professor; Ph.D., Stanford, 1973. Experimental biophysics: magnetic susceptibility measurement of enzymes, proteins, and molecular magnets.

Raymond C. DuVarney, Associate Professor and Associate Chair; Ph.D., Clark, 1968. Experimental solid-state physics: muon spin resonance.

Robert L. Eisner, Adjunct Assistant Professor; Ph.D., Purdue, 1968. Nuclear medicine.

Fereydoon Family, Samuel Candler Dobbs Professor of Condensed Matter Physics; Ph.D., Clark, 1974. Theoretical condensed-matter physics: nonequilibrium growth phenomena, pattern formation, fractals, surface and interface physics.

Peter Fong, Professor; Ph.D., Chicago, 1953. Theoretical nuclear physics, molecular biophysics, geophysics.

Ernest V. Garcia, Adjunct Associate Professor; Ph.D., Miami (Florida), 1974. Medical imaging.

H. George E. Hentschel, Associate Professor; Ph.D., Cambridge, 1978. Theoretical condensed-matter and statistical physics: neural networks, spin glasses, nonlinear dynamics, chaos and turbulence.

Boi Hanh Huynh, Professor; Ph.D., Columbia, 1974. Experimental biophysics: Mössbauer, ENDOR, and EPR studies of metalloenzymes.

Herbert M. Lindsay, Assistant Professor; Ph.D., UCLA, 1988. Experimental condensed-matter physics: static and dynamic light scattering, complex fluids.

John A. Malko, Adjunct Assistant Professor; Ph.D., Ohio, 1970. Nuclear medicine.

Sidney Perkowitz, Charles Howard Candler Professor of Condensed Matter Physics; Ph.D., Pennsylvania, 1967. Experimental condensed-matter physics: Raman, far-infrared, and photoluminescence spectroscopy of semiconductor superlattices and superconductors.

P. Venugopala Rao, Associate Professor; Ph.D., Oregon, 1964. Experimental nuclear and atomic physics.

Robert H. Rohrer, Professor Emeritus; Ph.D., Duke, 1954. Experimental radiological physics.

Richard M. Williamon, Adjunct Assistant Professor; Ph.D., Florida, 1972. Astronomy.

Kwok To Yue, Assistant Professor; Ph.D., Illinois, 1982. Experimental biophysics: Raman, optical studies of metalloenzymes, proteins, DNA, drug interaction and lipids.

RESEARCH INTERESTS

Theoretical

Condensed-matter physics: Nonequilibrium growth phenomena, pattern formation, fractals, spin glasses, nonlinear dynamics, physics of surfaces and interfaces. Anderson, Family, Hentschel, 4 postdoctoral associates.

Neural networks: Statistical theory of stereoscopic vision. Hentschel, 1 postdoctoral associate.

Solid-state physics: Theory of semiconductors and superlattices, electronic devices. Bajaj, 4 postdoctoral associates.

Experimental

Biophysics: Structural and functional studies of metalloenzymes, lipids, DNA, and drug interactions using spectroscopic methods, including Mössbauer, EPR, ENDOR, Raman, magnetic susceptibility, absorption, fluorescence, and flash photolysis. Day, Huynh, Yue, 6 postdoctoral associates.

Condensed-matter physics: Static and quasi-elastic light scattering, structure and growth of aggregates, colloids, complex fluids, colloidal crystals, flow in porous media. Lindsay.

Laser optics: Submillimeter spectroscopy. Perkowitz.

Solid-state physics: Raman, far-infrared, photoluminescence, and picosecond spectroscopy of semiconductor superlattices and superconductors, EPR, ENDOR, and muon spin resonance. DuVarney, Perkowitz, 2 postdoctoral associates.

Low-temperature facility in the Mössbauer research lab.

SECTION 7: PHYSICS

FLORIDA STATE UNIVERSITY

Department of Physics

Programs of Study

Programs of study are offered that lead to the M.S. and Ph.D. degrees. Students are required to take a diagnostic exam covering undergraduate material upon entering the graduate program and to pass a proficiency examination by the beginning of their second year of residence. For the M.S., either a thesis program or a course-work-only program may be chosen. The Ph.D. examination includes a written portion covering electrodynamics and quantum, classical, and statistical mechanics. Within one year of passing the written examination, an oral examination covering the student's prospective research must be passed. The only formal course requirements for the Ph.D. are three courses from a select list of advanced courses. All students are required to teach two sections of a beginning physics laboratory.

Students are encouraged to begin research early and must provide evidence of progress in research by the end of their second year in residence. The department offers extensive research opportunities in theoretical, experimental, pure, and applied physics in the areas of computational physics, atomic physics, materials science, high-energy particle physics, medium- and low-energy nuclear physics, condensed-matter physics, aerosol physics, and chemical physics.

Research at FSU offers close collaboration between experimentalists and theorists and between students and faculty. Extensive electronics and machine shops, as well as the availability of computers ranging from micros to supercomputers, enable almost any research topic to be pursued.

Research Facilities

The department occupies three adjacent buildings: an eight-story Physics Research Building, a Nuclear Research Building, and an undergraduate physics classroom and laboratory building. The experimental facilities include a 9.5-MV Super FN Tandem Van de Graaff accelerator with superconducting post accelerator; 3- and 4-MV Van de Graaffs; a detector development laboratory for high-energy particle detectors; high-resolution Fourier-transform IR spectrometers; facilities for ion implantation; liquid helium temperature research facilities; UHV facilities (including surface characterization, molecular beam epitaxy, and surface analysis by He atom scattering); facilities for high- and low-temperature superconductivity, small-angle and standard X-ray diffractometry, scanning electron and tunneling microscopy, image analysis, quasi-elastic light scattering, polarized electron energy loss spectroscopy, thick- and thin-film preparation, and high magnetic field studies; a unique aerosol physics–electron irradiation system; and the new National High Magnetic Field Laboratory. In addition to using in-house facilities, those engaged in ongoing experiments use accelerator and other research equipment at Fermilab, Brookhaven, SLAC, Los Alamos, CEBAF, Oak Ridge, and CERN. Computational facilities at the University include a CRAY Y-MP 432 supercomputer and a 64K-processor Connection Machine supercomputer. A VAX 3400 cluster (with thirty individual CPUs), numerous minicomputers, and workstations are present throughout the department. Extensive networking facilities provide access to computers on and off campus.

Financial Aid

The department offers teaching and research assistantships and fellowships. The fellowships include several that are designed to help develop promising young minority physicists. The assistantship stipend is $13,558 for twelve months, with a workload equivalent to 6 contact hours in an elementary laboratory. In general, summer assistantships are provided for all students, and students are teaching assistants during their first academic year but are supported by research assistantships after that year.

Cost of Study

Tuition and fees that must be paid by assistants and fellows who are Florida residents are $843 for twelve months in 1991–92. The additional charge for out-of-state tuition is normally waived for assistants and fellows.

Cost of Living

Apartments and houses are readily available in Tallahassee. A typical one-bedroom unfurnished apartment within walking distance of the physics building rents for $350 per month. For 1991–92, the University has married student housing with rents ranging from $210 per month for a one-bedroom apartment to $345 per month for a three-bedroom apartment. National surveys show that the cost of living in Tallahassee is 10 to 15 percent lower than that in most areas of the United States.

Student Group

Florida State University is a comprehensive university with a total of 28,000 students, of whom 4,500 are graduate or professional students. The physics department has about 85 graduate students. The average time for achieving a Ph.D. for students entering with a B.S. degree in physics is about six years.

Location

Tallahassee is the capital city of the state of Florida. Its population is about 180,000. Many employment opportunities exist for students' spouses in Tallahassee. Students can live in relatively rural surroundings and still be only 20 minutes from the University. Extensive sports facilities and active city leagues exist in the city. Because of the mild winter climate, people in this region tend to be outdoor oriented. Numerous sinkholes provide enjoyable nontraditional swimming areas. The Gulf of Mexico is about 30 miles from campus.

The University and The Department

The presentations of the Schools of Fine Arts and Music provide cultural opportunities that are usually available only in much larger cities. The University Symphony, the Flying High Circus, and other theater and music groups give students the opportunity to participate in many activities in addition to their physics studies. FSU has active programs in intercollegiate and intramural sports.

Recent major additions in the FSU Science Center have been the development of the Supercomputer Computations Research Institute (SCRI), the completion of the P. A. M. Dirac Science Library, and the start of construction on the new National High Magnetic Field Laboratory. Besides the teaching faculty at the Department of Physics (listed on the back of this page), there are 11 full-time research faculty members at SCRI. The lab will house the world's highest field magnets, making FSU one of the top centers for magnetic research.

Applying

Assistantship decisions are based on a student's transcript, GRE General Test scores, and three letters of reference. The deadline for completed applications to be on file with the physics department is February 15, 1992.

Correspondence and Information

Professor Vasken Hagopian
Graduate Physics Program
Department of Physics
Florida State University
Tallahassee, Florida 32306-3016
Telephone: 904-644-4473

Peterson's Guide to Graduate Programs in the Physical Sciences and Mathematics 1992

SECTION 7: PHYSICS

Florida State University

THE FACULTY AND THEIR RESEARCH

John Albright, Professor; Ph.D., Wisconsin–Madison, 1964. Theoretical physics: mathematical physics.
Howard Baer, Assistant Professor; Ph.D., Wisconsin–Madison, 1984. Theoretical physics: elementary particle theory.
Bernd Berg, Professor; Ph.D., Berlin, 1977. Theoretical physics: statistical mechanics, lattice gauge theory, quantum gravity, computational physics.
Scott D. Berry, Assistant Professor; Ph.D., Tennessee, 1985. Experimental physics: clusters, surface physics, magnetic and oxide superlattices.
Paul Cottle, Associate Professor; Ph.D., Yale, 1986. Experimental physics: heavy-ion nuclear physics.
Jack Crow, Professor; Ph.D., Rochester, 1967. Experimental physics: correlated electron systems, high-T_c superconductors and heavy fermions.
Lawrence C. Dennis, Professor; Ph.D., Virginia, 1979. Experimental physics: heavy-ion nuclear physics, electron scattering.
Edward Desloge, Professor; Ph.D., Saint Louis, 1957. Theoretical physics: thermal and statistical physics, classical and relativistic dynamics.
Dennis Duke, Associate Professor; Ph.D., Iowa State, 1974. Theoretical physics: elementary particle physics, computational physics.
Steve Edwards, Professor; Ph.D., Johns Hopkins, 1960. Theoretical physics: low-energy nuclear physics.
Neil R. Fletcher, Professor; Ph.D., Duke, 1961. Experimental physics: low-energy nuclear physics.
John D. Fox, Professor; Ph.D., Illinois, 1960. Experimental physics: superconducting accelerator development, heavy-ion physics.
Vasken Hagopian, Professor; Ph.D., Pennsylvania, 1963. Experimental physics: elementary particle physics.
Robert Hunt, Professor; Ph.D., Michigan, 1963. Experimental physics: infrared spectroscopy, molecular structure.
Kirby Kemper, Professor; Ph.D., Indiana, 1968. Experimental physics: polarization studies of nuclear reactions.
J. Daniel Kimel, Professor; Ph.D., Wisconsin–Madison, 1966. Theoretical physics: elementary particle physics, computational physics.
Robert Kromhout, Professor; Ph.D., Illinois, 1952. Theoretical physics: intermolecular interactions, phase transitions, cognition.
Joseph E. Lannutti, Professor; Ph.D., Berkeley, 1957. Experimental physics: elementary particle physics.
David Levinthal, Associate Professor; Ph.D., Columbia, 1980. Experimental physics: experimental particle physics.
David M. Lind, Assistant Professor; Ph.D., Rice, 1986. Experimental physics: surfaces, thin films, magnetic properties of solids, magnetic and oxide superlattices.
Efstratios Manousakis, Associate Professor; Ph.D., Illinois at Urbana-Champaign, 1985. Theoretical physics: condensed-matter physics, many-body theory, superfluidity, superconductivity.
William G. Moulton, Professor; Ph.D., Illinois, 1952. Experimental physics: low-temperature solid-state physics.
J. William Nelson, Professor; Ph.D., Texas, 1959. Experimental physics: nuclear methods in quantitative analysis.
H. K. Ng, Assistant Professor; Ph.D., McMaster, 1984. Experimental condensed-matter physics: far-infrared spectroscopy, superconductivity, quantum size effects in selected clusters.
Joseph F. Owens, Professor and Chairman of the Department; Ph.D., Tufts, 1973. Theoretical physics: elementary particle theory.
Fred L. Petrovich, Professor; Ph.D., Michigan State, 1970. Theoretical physics: nuclear structure; reaction theory.
John Philpott, Professor; D.Phil., Oxford, 1963. Theoretical physics: low-energy nuclear physics.
Hans S. Plendl, Professor; Ph.D., Yale, 1958. Experimental physics: medium-energy nuclear and particle physics.
Per Arne Rikvold, Associate Professor; Ph.D., Temple, 1983. Theoretical physics: condensed-matter physics, surface and interface science, computational physics.
Mark A. Riley, Assistant Professor; Ph.D., Liverpool, 1985. Experimental physics: nuclear structure physics.
Don Robson, Professor; Ph.D., Melbourne, 1963. Theoretical physics: interface between high-energy and nuclear theory.
Pedro Schlottmann, Professor; Ph.D., Munich Technical, 1973. Theoretical physics: condensed-matter physics, heavy fermions, magnetism, high-T superconductivity.
Shahid A. Shaheen, Assistant Professor; Ph.D., Ruhr-Bochum, 1985. Experimental physics: superconductivity, magnetism, materials science.
Raymond Sheline, Professor; Ph.D., Berkeley, 1949. Experimental physics: tests of nuclear models.
Neil Shelton, Professor; Ph.D., Florida State, 1962. Experimental physics: electron scattering on atoms, atomic collision calculations.
James G. Skofronick, Professor; Ph.D., Wisconsin–Madison, 1964. Experimental physics: surface physics.
Samuel L. Tabor, Professor; Ph.D., Stanford, 1972. Experimental physics: high-spin states in nuclei.
Louis R. Testardi, Professor; Ph.D., Pennsylvania, 1960. Experimental physics: structural instabilities, modulated structures, materials science.
David Van Winkle, Associate Professor; Ph.D., Colorado, 1984. Experimental physics: liquid crystals, colloids, macromolecules.
Horst Wahl, Professor; Ph.D., Vienna, 1969. Experimental physics: particle physics.
Yung-Li Wang, Professor; Ph.D., Pennsylvania, 1966. Theoretical physics: solid-state theory, computational physics.
Anthony G. Williams, Assistant Professor; Ph.D., Flinders University of South Australia, 1985. Theoretical physics: theoretical nuclear and particle physics, computational physics.
W. John Womersley, Assistant Professor; D.Phil., Oxford, 1986. Experimental physics: elementary particle physics.

RESEARCH ACTIVITIES

Theoretical
Condensed Matter. Many-body theory of magnetism, magnetic properties of solids. High-temperature superconductivity, adsorption, phase transitions, numerical simulations, heavy fermions.
Elementary Particles and Fields. Strong and electroweak interaction phenomenology in high-energy particle physics. Lattice gauge theory, numerical simulations, computational quantum gravity.
Mechanics. Classical mechanics.
Nuclear Physics. Direct reaction theory. Nuclear shell-model calculations. Nuclear structure studies via direct and resonance reactions at low and intermediate energies. Mean field studies. Quark dynamics in nuclei.
Relativity. Gravitation theory.
Other Theoretical/Mathematics. The many-body problem and mathematical physics.

Experimental
Atomic and Molecular Physics. Electron scattering by atoms and molecules. Infrared studies of gases of planetary atmospheres. Radiation effects.
Condensed-Matter Physics/Materials Science. Liquid crystals, cluster effects, magnetic properties, surface physics, modulated structures, spectroscopy.
Elementary Particles and Fields. Hadron spectroscopy, strong and electroweak interactions in high-energy particle physics, detector development and simulation.
Nuclear Physics. Resonance phenomena in heavy-ion reactions and fusion cross-section measurements. Heavy-ion reactions and reaction mechanisms, using polarized alkali beams. Fragmentation studies. Properties of nuclear systems at high angular momentum and extreme shapes. Electro- and photo-production of hypernuclei and hyperons, nuclear octupole excitations.
Other Experimental. Medium-energy hadron interactions with nuclei. Electrostatic precipitator development. Proton-induced X-ray emission for trace-element detection. Mu-mesonic atom studies.
Solid-State/Low-Temperature and Superconductivity. Low- and high-temperature superconductivity.
Surface physics. He-surface scattering, clusters, electron spectroscopies.

SECTION 7: PHYSICS

HARVARD UNIVERSITY
Department of Physics

Program of Study

The Department of Physics offers a program of graduate study leading to the Ph.D. degree in physics. The primary areas of experimental and theoretical research in the physics department are high-energy particle physics, atomic and molecular physics, physics of solids and fluids, astrophysics, certain aspects of nuclear physics, quantum field theory, statistical mechanics, mathematical physics, quantum optics, and relativity. The department is closely linked with the Division of Applied Sciences, which has an extensive program in theoretical and experimental studies of the properties of crystalline and disordered solids; the division also studies nonlinear optics and light scattering, earth and planetary physics, computer science, and applied mathematics.

The first year and a half of graduate study is normally spent on lecture courses. In the second year, students are expected to pass an oral examination on a subject of their choice and choose a field and adviser for their Ph.D. work. The requirements for the Ph.D. degree are demonstration of competence (usually through a year course in each field) in three fields of physics, satisfactory performance on a preliminary oral examination, and a Ph.D. dissertation based on independent scholarly research, which, upon conclusion, is defended in an oral examination before a Ph.D. committee. With normal preparation, students can usually complete the requirements for the Ph.D. degree in four to six years. The research interests of the faculty members in the physics department are listed on the reverse of this page.

A limited number of openings for postdoctoral research, with or without a stipend, are available each year to qualified applicants without regard to race, color, sex, or creed. Inquiries should be addressed to individual professors, under whose sponsorship these appointments are made.

Research Facilities

The facilities of the Department of Physics are concentrated in several buildings. Lyman Laboratory and Jefferson Laboratory form the center of departmental activity. These two buildings contain facilities for atomic physics experiments with fast atomic beams, apparatus for trapping and studying individual electrons and ions, equipment for producing nuclear and atomic polarization, superconducting magnets, lasers, dilution refrigerators for attaining very low temperatures, equipment for high-pressure studies, and equipment for optical and ultrasonic measurements. Additional facilities for the study of solid-state physics, laser physics, and materials science are located in Gordon McKay Laboratory. These include high-resolution electron microscopes, low-temperature facilities, a clean room for fabricating submicrometer structures, an MeV heavy-ion accelerator, high-resolution X-ray facilities, and materials preparation and characterization equipment. Studies of condensed-matter systems using synchrotron radiation are carried out at Brookhaven National Laboratory. Current projects in particle physics are being carried out at the Fermi National Accelerator Laboratory; at the European Center for Nuclear Research (CERN) in Geneva, Switzerland; at the Cornell Wilson Synchrotron Laboratory; and at KEK laboratory in Japan. Apparatus for these projects is built and data from these experiments are analyzed in part at the High Energy Physics Laboratory at Harvard. A VAXcluster computer facility, located at the High Energy Physics Laboratory, is accessible from terminals in Lyman Laboratory.

Financial Aid

The Department of Physics has generally provided full tuition and fees for all graduate students who are not supported in full by outside scholarships. Living stipends are provided via scholarships, teaching fellowships, and research assistantships. Summer support is also included and is available in the form of either a teaching fellowship or a research assistantship.

Cost of Study

Tuition and fees are provided for all graduate students as described above.

Cost of Living

There are a wide variety of dormitory rooms for single students, with costs ranging from $2350 (for a small single room) to $4150 (for a two-room suite) per academic year. These figures do not include meals.

Married students and single graduate students may apply for apartments in graduate student housing or other University-owned apartments. The monthly cost is $490–$736 for a one-room studio apartment, $624–$1062 for a one-bedroom apartment, $789–$1521 for a two-bedroom apartment, and $1141 and up for a three-bedroom apartment. There are also many privately owned accommodations nearby and within commuting distance.

Student Group

The Graduate School of Arts and Sciences has an enrollment of 2,858. About 150 men and women are pursuing Ph.D. research in the physics laboratories with physics department faculty. Students come from all parts of the United States, and about one fourth are from foreign nations.

Location

Cambridge, Massachusetts, is a city of 85,000, adjacent to Boston and its cultural benefits, yet suburban in nature. All of New England is within driving distance—the mountains of New Hampshire and Vermont with camping and skiing, the beaches and woodlands of Maine, and the seashore and seaports of Massachusetts, as well as the great array of colleges and universities spread across all six states. Cambridge itself is a scientific and intellectual center teeming with activity in all areas of creativity and study.

The University

Harvard College was established in 1636, and its charter, which still guides the University, was granted in 1650. Today, Harvard University, with its network of graduate schools, occupies a noteworthy position in the academic world, and the Department of Physics offers an educational program in keeping with the University's long-standing record of achievement.

Applying

Men and women who are completing a bachelor's degree or the equivalent should write to the Admissions Office of the Graduate School of Arts and Sciences for application material and to the Department of Physics for additional information on the program. Completed application forms and all supporting material should be returned to the Admissions Office by January 2.

Correspondence and Information

Information on the program:
Chairman
Department of Physics
Jefferson 364
Harvard University
Cambridge, Massachusetts 02138

Application forms for admission:
Admissions Office
Graduate School of Arts and Sciences
Harvard University
8 Garden Street, 2nd Floor
Cambridge, Massachusetts 02138

Peterson's Guide to Graduate Programs in the Physical Sciences and Mathematics 1992

SECTION 7: PHYSICS

Harvard University

THE FACULTY AND THEIR RESEARCH

Paul G. Bamberg, Ph.D., Director of Science Instruction in Continuing Education and Senior Lecturer on Physics.
Michael Bershadsky, Ph.D., Assistant Professor of Physics. Elementary particle theory.
George Brandenburg, Ph.D., Senior Research Fellow and Director of the High Energy Physics Laboratory.
Eric Carlson, Ph.D., Assistant Professor of Physics. Elementary particle theory.
Timothy Chupp, Ph.D., Associate Professor of Physics. Tests of fundamental symmetries in nuclear and atomic physics.
Sidney R. Coleman, Ph.D., Donner Professor of Science. Quantum field theory, relativity.
Henry Ehrenreich, Ph.D., Clowes Professor of Science (joint appointment with the Division of Applied Sciences). Theoretical condensed-matter physics.
Gary Feldman, Ph.D., Professor of Physics. Experimental high-energy physics.
Daniel S. Fisher, Ph.D., Professor of Physics. Statistical physics, condensed-matter theory.
Melissa Franklin, Ph.D., Associate Professor of Physics. Experimental high-energy physics.
Gerald Gabrielse, Ph.D., Professor of Physics. Atomic physics, experimental elementary particle physics.
Howard Georgi, Ph.D., Professor of Physics and Chairman. Field theory, elementary particle physics.
Sheldon L. Glashow, Ph.D., Higgins Professor of Physics. Theoretical elementary particle physics.
Roy J. Glauber, Ph.D., Mallinckrodt Professor of Physics. Elementary particle theory, high-energy nuclear physics, quantum optics, statistical mechanics.
Jene A. Golovchenko, Ph.D., Gordon McKay Professor of Applied Physics (joint appointment with the Division of Applied Sciences). Solid-state physics.
Benjamin Grinstein, Ph.D., Associate Professor of Physics. Theoretical elementary particle physics.
Bertrand I. Halperin, Ph.D., Mallinckrodt Professor of Physics. Solid-state physics, statistical theory.
Gerald Holton, Ph.D., Mallinckrodt Professor of Physics and Professor of the History of Science (joint appointment with the History of Science Department). Experimental molecular physics at high pressures, history of nineteenth- and twentieth-century physics.
Paul Horowitz, Ph.D., Professor of Physics. Experimental astrophysics, search for extraterrestrial intelligence.
Arthur M. Jaffe, Ph.D., Landon T. Clay Professor of Mathematics and Theoretical Science (joint appointment with the Department of Mathematics). Field theory, statistical physics, and connections between physics and mathematics.
Efthimios Kaxiras, Ph.D., Assistant Professor of Physics and of Applied Physics on the Gordon McKay Endowment Condensed-matter theory.
Kay Kinoshita, Ph.D., Associate Professor of Physics. Experimental high-energy physics.
Margaret E. Law, Ph.D., Director of the Physics Laboratories.
Andrzej Lesniewski, Ph.D., Assistant Professor of Physics. Mathematical physics.
Paul C. Martin, Ph.D., John H. Van Vleck Professor of Pure and Applied Physics and Dean of the Division of Applied Sciences (joint appointment with the Division of Applied Sciences). Statistical physics, condensed-matter theory.
Eric Mazur, Ph.D., Professor of Physics and of Applied Physics on the Gordon McKay Endowment (joint appointment with the Division of Applied Sciences). Quantum optics, molecular physics.
Sanjib Mishra, Ph.D., Assistant Professor of Physics. Experimental high-energy physics.
David R. Nelson, Ph.D., Professor of Physics. Statistical physics, solid-state theory.
Costas D. Papaliolios, Ph.D., Professor of Physics. Experimental astrophysics, atmospheric physics.
William Paul, Ph.D., Gordon McKay Professor of Applied Physics and of Physics (joint appointment with the Division of Applied Sciences). Experimental condensed-matter physics, amorphous semiconductors.
Peter S. Pershan, Ph.D., Gordon McKay Professor of Applied Physics and of Physics (joint appointment with the Division of Applied Sciences). Experimental condensed-matter physics, synchrotron radiation studies of properties of matter at interfaces and surfaces.
Francis M. Pipkin, Ph.D., Frank B. Baird Jr. Professor of Science. Experimental elementary particle physics, atomic and molecular physics.
Mara Prentiss, Ph.D., Associate Professor of Physics. Experimental atomic physics.
William H. Press, Ph.D., Professor of Astronomy and of Physics (joint appointment with the Department of Astronomy). Cosmology, theoretical astrophysics, computational physics.
Irwin I. Shapiro, Ph.D., Professor of Astronomy and of Physics (joint appointment with the Department of Astronomy). Radar and radio astronomy.
Isaac F. Silvera, Ph.D., T. D. Cabot Professor of Natural Sciences. Low-temperature physics of quantum fluids and solids, ultrahigh-pressure physics.
Karl Strauch, Ph.D., George Vasmer Leverett Professor of Physics. Experimental high-energy particle physics.
Michael Tinkham, Ph.D., Gordon McKay Professor of Applied Physics and Rumford Professor of Physics (joint appointment with the Division of Applied Sciences). Superconductivity, condensed-matter physics.
Cumrun Vafa, Ph.D., Professor of Physics. Elementary particle theory.
Robert M. Westervelt, Ph.D., Gordon McKay Professor of Applied Physics and of Physics (joint appointment with the Division of Applied Sciences). Experimental condensed-matter physics.
Richard Wilson, D.Phil., Mallinckrodt Professor of Physics. Experimental nuclear physics, elementary particle physics, energy-related environmental and medical physics.
Tai T. Wu, Ph.D., Gordon McKay Professor of Applied Physics and of Physics (joint appointment with the Division of Applied Sciences). Theoretical elementary particle physics, electromagnetic theory.
Hitoshi Yamamoto, Ph.D., Assistant Professor of Physics. Experimental high-energy physics.

Professors Emeriti
Kenneth T. Bainbridge, Ph.D., George Vasmer Leverett Professor of Physics.
Nicolaas Bloembergen, Ph.D., Gerhard Gade University Professor. Quantum optics, atomic and molecular physics, condensed-matter physics.
Robert V. Pound, M.A., Mallinckrodt Professor of Physics. Experimental physics.
Edward M. Purcell, Ph.D., Gerhard Gade University Professor.
Norman F. Ramsey, Ph.D., Higgins Professor of Physics. Experimental physics.

SECTION 7: PHYSICS

INDIANA UNIVERSITY BLOOMINGTON

Department of Physics

Programs of Study

The department offers work leading to the M.S., M.A.T., and Ph.D. degrees in nuclear physics, accelerator physics, fundamental particle physics, condensed-matter physics, and theoretical physics. Interdisciplinary programs leading to a Ph.D. degree in astrophysics, chemical physics, and mathematical physics are also offered.

A candidate for the master's degree in physics must complete 30 credit hours of graduate work (including a minimum of 20 hours in physics) and pass a written comprehensive examination.

Requirements for the M.A.T. are selected to prepare future teachers of high school physics and to acquaint in-service teachers with modern developments in physics. At least 8 hours of undergraduate physics must be completed before admission, and 20 hours must be taken within the department. An additional 16 hours should be chosen from mathematics, astronomy, chemistry, or graduate education courses.

The requirements for the Ph.D. include a minimum of 90 hours of graduate credit made up of course work, supervised reading, and research. Course work, while essential, is not the main criterion for obtaining the Ph.D. degree. The candidate must demonstrate an ability to do research by carrying out an investigation and presenting a publishable thesis and must show a broad grasp of the field of physics. An official qualifying examination is required by the Graduate School. It must be taken at least eight months before the degree is to be given. The final oral examination is conducted by the candidate's doctoral committee and consists of questions on the major and minor fields of work, as well as on the thesis.

Research Facilities

The Department of Physics is housed in Swain Hall. The Swain Hall library includes 80,000 volumes and more than 900 periodicals, 200 specifically in physics. The Indiana University Cyclotron Facility is one of the few medium-energy laboratories in the United States. The variable-energy, separated-magnet, isochronous cyclotron is the only accelerator of its type in the world providing beams of H, He, and Li in the medium-energy range with high precision and good intensity. A light-ion storage ring with internal targets has recently begun operation. It electron-cools the injected beams from the existing cyclotron and increases their energy through synchrotron acceleration by a factor of about 3 for most ions. The Nuclear Theory Center enhances and supports the commitment to nuclear physics at Indiana University through research in medium-energy nuclear reactions, relativistic nuclear structure theory, and subhadronic models of nuclei. The high-energy-physics group performs experiments at Brookhaven in Upton, New York; Fermilab in Batavia, Illinois; and SLAC, the electron positron colliding facility in Stanford, California. The high-energy laboratories contain wire-winding machines for proportional and drift chambers, electronics development facilities, two LSI-11 on-line computers for testing, a fully equipped VAX 6240 computer, and three VAXstations. The high-energy-astrophysics group performs balloon flight experiments and is part of a monopole search, MACRO, in Gran Sasso, Italy. Plans are being made for doing research on the Space Station. The condensed-matter experimental facilities include a high-vacuum multisource sputtering system for the production of metallic multilayers; a high-resolution, temperature-controlled X-ray powder diffractometer; three ultrahigh-vacuum surface analysis systems with LEED, Auger, and electron energy loss spectrometers as well as a scanning tunneling microscope (STM); crystal-growing facilities; and stabilized lasers. Low-temperature facilities in the condensed-matter group include several cryostats, a helium liquefier, and a dilution refrigerator. Microwave generation and detection equipment allows electron transport measurements from DC to 20 GHz. Computational facilities for condensed-matter experimentation and theory include a DEC PDP-11/34, a Sun Microsystems local network, and various workstations. University computing facilities include a DEC VAXcluster, an IBM 3090, and a visualization center that includes an Ardent and other hardware. Public clusters feature PCs, Macintoshes, NeXTs, and Suns. The theorists in the department are served by a network of Suns.

Financial Aid

Teaching assistantships carried stipends of $10,500 for the 1991–92 academic year of ten months. Teaching fellowships for outstanding students provided a teaching assistantship stipend with an additional award of up to $3000. Research assistantships are open to graduate students with previous research experience and to those with special talents. These carried stipends of $10,500 for a twelve-month period.

Cost of Study

In 1990–91, fees per credit hour for in-state graduate students were $93.30; for out-of-state graduate students, $266.60. Teaching and research assistants ordinarily pay only a fee of approximately $330 per semester.

Cost of Living

University housing and meals for single graduate students (men and women) were available for $3221–$3757 for the 1990–91 academic year. Married students chose from efficiency apartments and one-, two-, and three-bedroom apartments ranging in cost from $270–$588 per month. Both furnished and unfurnished units are available in University housing. All prices include utilities, except telephone.

Student Group

Indiana University is a large institution, with over 33,000 students enrolled at the Bloomington campus, including 6,075 graduate and professional school students. In fall 1990, there were 87 graduate students in physics, most of whom receive full financial support. Graduates hold jobs throughout the world in universities, colleges, and high schools as well as in government and industrial laboratories.

Location

Bloomington is located in the picturesque rolling hills of southern Indiana, 50 miles south of Indianapolis, the state capital. It is close to five state parks, two state forests, and Lake Monroe, the state's largest lake, where recreational activities include boating, camping, fishing, hiking, picnicking, swimming, and waterskiing.

The University

Indiana University is the oldest state university west of the Alleghenies. It was founded in 1820 and has been a pioneer in higher education in the Midwest. It is widely recognized for the beauty of its campus and for the diversity and high quality of its graduate programs in the arts, humanities, and sciences. The campus provides numerous facilities for all types of indoor and outdoor sports. The School of Music presents concerts and opera. Lectures, dramatic and musical Broadway productions, ballet, drama, and concerts are presented by the Auditorium and the University Theatre.

Applying

The deadline for assistantship and fellowship applications for the fall semester is February 1. For further information, students should write to one of the addresses or call one of the numbers given below.

Correspondence and Information

Dr. J. Timothy Londergan, Chairperson
Department of Physics
Indiana University
Bloomington, Indiana 47405-4201
Telephone: 812-855-1247

or

Graduate Admissions Committee
Department of Physics
Indiana University
Bloomington, Indiana 47405-4201
Telephone: 812-855-3973

Peterson's Guide to Graduate Programs in the Physical Sciences and Mathematics 1992

SECTION 7: PHYSICS

Indiana University Bloomington

THE FACULTY AND THEIR RESEARCH

Professors
J. Timothy Londergan, Chairperson; D.Phil., Oxford, 1969. Theoretical physics, nuclear theory.
Lawrence M. Langer, Professor Emeritus; Ph.D., NYU, 1938. Nuclear physics: radioactivity, nuclear spectroscopy.
Martin E. Rickey, Professor Emeritus; Ph.D., Washington (Seattle), 1958. Nuclear physics: cyclotron design; physics of music.
Roger G. Newton, Distinguished Professor; Ph.D., Harvard, 1953. Theoretical and mathematical physics: scattering theory.
Robert E. Pollock, Distinguished Professor; Ph.D., Princeton, 1963. Nuclear physics: nuclear reactions, cyclotron design.
Ethan D. Alyea, Ph.D., Caltech, 1962. Astrophysics (experimental).
Andrew D. Bacher, Ph.D., Caltech, 1967. Intermediate-energy nuclear physics (experimental).
Robert D. Bent, Ph.D., Rice, 1954. Experimental nuclear physics: nuclear structure, reactions, astrophysics.
Bennet B. Brabson, Ph.D., MIT, 1966. Elementary particle physics (experimental).
John M. Cameron, Ph.D., UCLA, 1967. Nuclear physics (experimental).
John L. Challifour, Ph.D., Cambridge, 1963. Theoretical physics, mathematical physics.
Ray R. Crittenden, Ph.D., Wisconsin, 1960. Elementary particle physics (experimental).
Alex R. Dzierba, Ph.D., Notre Dame, 1969. Elementary particle physics (experimental).
Steven M. Girvin, Ph.D., Princeton, 1977. Condensed-matter theory.
Charles Goodman, Ph.D., Rochester, 1959. Nuclear physics (experimental).
Richard R. Hake, Ph.D., Illinois, 1955. Condensed-matter and low-temperature physics.
Gail G. Hanson, Ph.D., MIT, 1973. Elementary particle physics (experimental).
Richard M. Heinz, Ph.D., Michigan, 1964. Astrophysics (experimental).
Archibald W. Hendry, Ph.D., Glasgow, 1962. Theoretical physics: elementary particles.
Charles J. Horowitz, Ph.D., Stanford, 1981. Nuclear theory.
Larry L. Kesmodel, Ph.D., Texas, 1974. Condensed-matter physics (experimental).
S. Y. Lee, Ph.D., SUNY at Stony Brook, 1972. Accelerator physics.
Andrew A. Lenard, Ph.D., Iowa, 1953. Theoretical physics, mathematical physics.
Don B. Lichtenberg, Ph.D., Illinois, 1955. Elementary particle physics (theory).
Allan H. MacDonald, Ph.D., Toronto, 1978. Condensed-matter theory.
Malcolm Macfarlane, Ph.D., Rochester, 1959. Nuclear theory.
Hugh J. Martin, Ph.D., Caltech, 1956. Elementary particle physics (experimental).
Hans Otto Meyer, Ph.D., Basel, 1970. Nuclear physics (experimental).
Daniel W. Miller, Ph.D., Wisconsin, 1951. Nuclear physics (experimental): nuclear reactions.
Hermann Nann, Ph.D., Frankfurt, 1967. Intermediate-energy nuclear physics (experimental).
Harold Ogren, Ph.D., Cornell, 1970. Elementary particle physics (experimental).
William L. Schaich, Ph.D., Cornell, 1970. Condensed-matter theory.
Peter Schwandt, Ph.D., Wisconsin, 1967. Nuclear physics (experimental).
Brian D. Serot, Ph.D., Stanford, 1979. Nuclear theory.
P. Paul Singh, Ph.D., British Columbia, 1960. Nuclear physics: nuclear reactions, nuclear spectroscopy.
James C. Swihart, Ph.D., Purdue, 1955. Condensed-matter theory.
Steven E. Vigdor, Ph.D., Wisconsin, 1973. Nuclear physics (experimental).
George E. Walker, Ph.D., Case Tech, 1966. Nuclear theory.
John G. Wills, Ph.D., Washington (Seattle), 1963. Theoretical nuclear physics: intermediate energy.
Andrej Zieminski, Ph.D., Warsaw, 1971. Elementary particle physics (experimental).

Associate Professors
Leslie C. Bland, Ph.D., Pennsylvania, 1983. Nuclear physics (experimental).
Steven A. Gottlieb, Ph.D., Princeton, 1978. Theoretical physics.
V. Alan Kostelecky, Ph.D., Yale, 1982. Theoretical physics.
Fred M. Lurie, Ph.D., Illinois, 1963. Condensed-matter physics: nuclear magnetic resonance.
Catherine Olmer, Ph.D., Yale, 1976. Intermediate-energy nuclear physics (experimental).

Assistant Professors
David V. Baxter, Ph.D., Caltech, 1984. Condensed-matter physics (experimental).
John P. Carini, Ph.D., Chicago, 1988. Condensed-matter physics (experimental).
James A. Musser, Ph.D., Berkeley, 1984. Astrophysics (experimental).
John J. Szymanski, Ph.D., Carnegie Mellon, 1987. Nuclear physics (experimental).
Scott W. Wissink, Ph.D., Stanford, 1986. Nuclear physics (experimental).

Allan MacDonald and Steven Girvin, condensed-matter theorists, discuss the physics of high-temperature superconductors subjected to strong magnetic fields. Some of their other interests include the quantum Hall effect and phase transitions in quantum and classical systems.

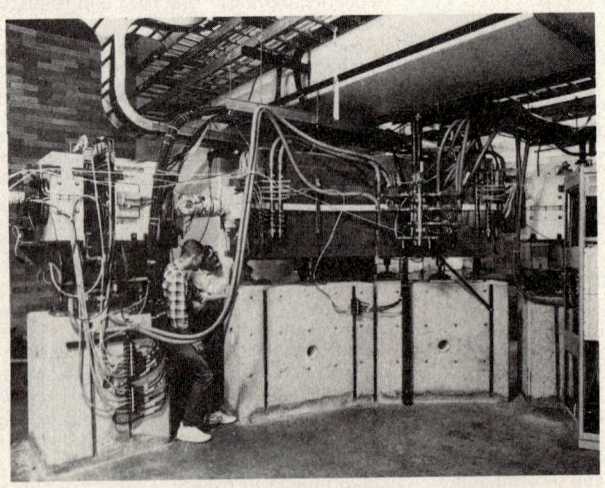

Professor Peter Schwandt, nuclear experimentalist, and a student work on the cooler ring at the Indiana University Cyclotron Facility.

SECTION 7: PHYSICS

JOHNS HOPKINS UNIVERSITY
Henry A. Rowland Department of Physics and Astronomy

Program of Study
The department offers a broad program for graduate and postdoctoral study in physics and astronomy, for which intermediate, advanced, and specialized courses in various fields are offered. These courses and the research that a student is encouraged to begin as soon as possible form the basis of the Ph.D. program. Considerable flexibility is available in the student's program, which is geared to individual needs on advice from staff advisers. In addition to required courses, candidates take written and oral qualifying examinations. The written examinations, covering intermediate-level material, are offered in two parts, during January and May, and are ordinarily passed within a year. These exams are followed by an intermediate-level oral examination in modern physics in the second year. There is also a reading examination in a foreign language with a scientific literature. A comprehensive oral examination is taken prior to the beginning of full-time research on the thesis topic (usually at the end of the second or the beginning of the third year). After completion of the student's research, there is an oral defense of the thesis. During residence, some teaching is required. Only those students who expect to complete the Ph.D. are admitted.

Research Facilities
The high-energy-physics group has facilities for constructing the electronics and detectors needed in modern experiments and also has independent computing capabilities that allow full analyses of massive amounts of data. Nuclear physics equipment includes facilities for relativistic heavy-ion collision studies and studies of nuclear interactions utilizing muons and pions with cryogenic targets. Facilities for solid-state physics include systems for molecular beam epitaxy, He^3-He^4 dilution refrigerator, high-rate sputtering, ultrahigh-vacuum thin-film deposition, automatic X-ray diffraction, scanning electron microscopy, X-ray fluorescence, LEED/Auger spectroscopy, SQUID and vibrating-sample magnetometry, ferromagnetic resonance, magnetooptics, and Mössbauer spectroscopy. For atomic and molecular physics, facilities include high-resolution and very sensitive spectrometers for measurements of infrared to ultraviolet wavelengths, a high-precision X-ray spectrometer, extensive spectroscopic facilities, and lasers. The astrophysics group maintains a calibration and test facility for testing instrumentation for rocket and space flights. Computer facilities in the department include a large number of VAX, Sun, and Apollo workstations as well as a VAX-11/780, an Apollo DN10000, and a Stardent Titan superminicomputer. These machines support a wide range of functions, including data reduction, image processing, simulation of physical processes, and creation of full-color video. All are networked to universities, national laboratories, and supercomputer facilities throughout the world. The Johns Hopkins University is the home of the Space Telescope Science Institute. Facilities at the following laboratories and observatories are also frequently used: Brookhaven National Laboratory, Stanford Linear Accelerator Center, Fermi National Accelerator Laboratory, the University's own Applied Physics Laboratory, National Bureau of Standards, Lawrence Berkeley Laboratory, Francis Bitter National Magnet Laboratory, Lawrence Livermore National Laboratory, the White Sands Missile Range, Kitt Peak National Observatory, Cerro Tololo Interamerican Observatory, the Very Large Array of the National Radio Astronomy Observatory, the Las Campanas Observatory, and NASA's Goddard Space Flight Center.

Financial Aid
Tuition fellowships are usually awarded to all full-time Ph.D. candidates. There are various fellowships. Nonservice University fellowships and teaching assistantships offer a minimum of $9500 (plus full tuition remission) for the nine-month academic year in 1991–92. Summer research assistantships are available at approximately $2300. Holders of teaching assistantships must assist in teaching general physics and introductory courses. This experience is useful for students interested in a college teaching career. In addition to the teaching assistantships, there are research assistantships available for upperclass graduate students that pay approximately $1000 per month (plus full tuition remission). These assistantships are awarded on the basis of experience, merit, and academic performance. (These awards are not usually given to first-year students unless they have special experience.) Loans and work-study arrangements are available from the Office of Student Financial Services.

Cost of Study
Tuition is $15,500 for the 1991–92 academic year. Books and supplies vary in cost but come to approximately $800. The average thesis cost is about $350 (typing, duplicating, and binding).

Cost of Living
The University owns four apartment buildings adjacent to the campus. In 1991–92, rates for rooms and apartments vary from $250 to $800 per month. A campus housing office assists students in finding rooms and apartments in the surrounding residential area. Meals are available in the campus cafeteria.

Student Group
The University's Homewood Campus (the Schools of Arts and Sciences and of Engineering) has approximately 3,050 undergraduates and 1,300 graduate students. There are approximately 110 graduate students in physics and astronomy, all of whom are receiving financial support of some kind. Admission to graduate study in the department is highly competitive; an average of 25 new students are admitted each year, the majority enrolling directly from college.

Location
Located in the northern section of Baltimore, the University is adjacent to one of the finest residential areas of the city, while most of the cultural activities of the large metropolitan area are but minutes away.

The University
The concept of graduate study came into being in America with the founding of the Johns Hopkins University in 1876. From the beginning, the hallmark of the University has been one of creative scholarship.

Applying
Requirements for admission after completion of the bachelor's or master's degree are transcripts of previous academic work, letters of recommendation, and GRE scores, including the General Test and the Subject Test in physics. Foreign students whose native language is not English must submit their scores on the Test of English as a Foreign Language. Students are admitted only in September. Applications and all supporting materials must be received by January 15. The application fee is $40, but it is waived for students with either financial need or foreign exchange problems. Application materials may be obtained from the Henry A. Rowland Department of Physics and Astronomy, Graduate Admissions, Bloomberg Center.

Correspondence and Information
Graduate Admissions
The Henry A. Rowland Department of Physics and Astronomy
Bloomberg Center
Johns Hopkins University
Homewood Campus
Baltimore, Maryland 21218
Telephone: 301-338-7347

SECTION 7: PHYSICS

Johns Hopkins University

THE FACULTY AND THEIR RESEARCH

Lloyd Armstrong Jr., Professor and Dean, School of Arts and Sciences; Ph.D., Berkeley, 1966. Atomic theory, quantum optics, group theory, electron-atom scattering.
Jonathan A. Bagger, Professor and Director, Theoretical Interdisciplinary Physics and Astrophysics Center; Ph.D., Princeton, 1983. Theoretical high-energy physics.
Bruce A. Barnett, Professor; Ph.D., Maryland, 1970. High-energy physics.
Barry J. Blumenfeld, Professor; Ph.D., Columbia, 1974. High-energy physics.
Collin Broholm, Assistant Professor; Ph.D., Copenhagen, 1988. Experimental condensed-matter physics.
Robert Brown, Research Professor (Space Telescope Science Institute); Ph.D., Harvard, 1971. Atomic physics, planetary atmospheres and environments, high-resolution spectroscopy.
Eric Chaisson, Adjunct Professor; Ph.D., Harvard, 1972. Astrophysics, cosmic evolution.
Chia-Ling Chien, Professor; Ph.D., Carnegie-Mellon, 1972. Condensed-matter physics, artificially structured solids.
Chih-Yung Chien, Professor; Ph.D., Yale, 1966. High-energy physics.
Arthur F. Davidsen, Professor; Ph.D., Berkeley, 1975. Astronomy, astrophysics.
Gabor Domokos, Professor; Ph.D., Dubna, 1963. Algebraic approaches to elementary particle physics, symmetries with application to field theories of particles, strong interactions at high energies.
S. Michael Fall, Research Professor (Space Telescope Science Institute); D.Phil., Oxford, 1976. Theoretical astrophysics, galaxy formation, structure of galaxies, cosmology.
William G. Fastie, Adjunct Research Professor. Planetary atmospheres.
Gordon Feldman, Professor; Ph.D., Birmingham, 1953. Quantum field theory, theory of elementary particles.
Paul D. Feldman, Professor; Ph.D., Columbia, 1964. Astrophysics, spectroscopy, space physics, planetary and cometary atmospheres.
Peter H. Fisher, Assistant Professor; Ph.D., Caltech, 1988. High-energy physics, particle detector development.
Holland Ford, Professor; Ph.D., Wisconsin, 1970. Stellar dynamics, stellar populations, active galactic nuclei, astronomical instrumentation.
Thomas Fulton, Professor; Ph.D., Harvard, 1954. Quantum electrodynamics, atomic theory, high-energy particle physics.
Riccardo Giacconi, Professor (Director, Space Telescope Science Institute); Ph.D., Milan, 1954. Astrophysics.
Timothy Heckman, Professor; Ph.D., Washington (Seattle), 1978. Astrophysics, active galaxies and quasars.
Richard C. Henry, Professor; Ph.D., Princeton, 1967. Astronomy, astrophysics.
Brian R. Judd, Professor; D.Phil., Oxford, 1955. Theoretical atomic and molecular physics, group theory, solid-state theory.
Chung W. Kim, Professor; Ph.D., Indiana, 1963. Nuclear theory, elementary particle theory, cosmology.
Belita Koiller, Visiting Professor; Ph.D., Berkeley, 1975. Theoretical condensed-matter physics.
Susan Kövesi-Domokos, Professor; Ph.D., Budapest, 1963. High-energy theory, algebraic theory of quarks, supersymmetry.
Julian H. Krolik, Associate Professor; Ph.D., Berkeley, 1977. Theoretical astrophysics.
Yung Keun Lee, Professor; Ph.D., Columbia, 1961. Nuclear physics.
Leon Madansky, Professor; Ph.D., Michigan, 1948. Nuclear physics, fundamental particles.
H. Warren Moos, Professor and Director, Center for Astrophysical Sciences; Ph.D., Michigan, 1962. Astrophysics, spectroscopy, plasma physics.
David A. Neufeld, Assistant Professor; Ph.D., Harvard, 1987. Theoretical astrophysics, form of dust grains, galactic theory.
Colin A. Norman, Professor (Chief of Academic Affairs, Space Telescope Science Institute); D.Phil., Oxford, 1973. Theoretical astrophysics.
Aihud Pevsner, Professor; Ph.D., Columbia, 1954. High-energy physics.
James Pringle, Research Professor (Space Telescope Science Institute) and Senior Member, Institute of Astronomy, Cambridge; Ph.D., Cambridge, 1973. Physics of accretion discs, binary stars, galactic nuclei, molecular clouds, prostars.
George T. Rado, Research Professor; Ph.D., MIT, 1943. Condensed-matter physics, theoretical and experimental magnetism.
Daniel H. Reich, Assistant Professor; Ph.D., Chicago, 1988. Experimental condensed-matter physics.
Mark O. Robbins, Associate Professor; Ph.D., Berkeley, 1983. Theoretical condensed-matter physics.
Dan Schechtman, Visiting Professor; Ph.D., International School for Electron Microscopy (Italy), 1972. Condensed-matter physics; quasi-crystals.
Sunil Sinha, Adjunct Professor; Ph.D., Cambridge, 1964. Condensed-matter physics.
Darrell F. Strobel, Professor (with primary appointment in Earth and Planetary Sciences); Ph.D., Harvard, 1969. Planetary atmospheres and astrophysics.
Alexander S. Szalay, Professor; Ph.D., Eötvös Lorand (Budapest), 1975. Theoretical astrophysics, galaxy formation.
Zlatko Tesanovic, Associate Professor; Ph.D., Minnesota, 1985. Theoretical condensed-matter physics.
Israel Vagner, Visiting Associate Professor; Ph.D., A. F. Ioffe Physicotechnical Institute (Leningrad), 1972. Theoretical condensed-matter physics.
J. C. Walker, Professor and Chair; Ph.D., Princeton, 1961. Condensed-matter physics, thin films and surfaces, nuclear physics.
Rosemary F. G. Wyse, Associate Professor; Ph.D., Cambridge, 1982. Astrophysics: galaxy formation and evolution.

RESEARCH ACTIVITIES

Astrophysics. Research in astrophysics spans a broad range of observational and theoretical programs. Observational programs include the use of ground-based optical and radio telescopes, analysis of archival data from previous space experiments, new research with existing satellites and sounding rockets, and space experiments. There is extensive laboratory work on detectors and instrument development for ultraviolet and optical astronomy. Research is concentrated in the following areas of astrophysics: cosmology, active galactic nuclei and quasars, galaxies and galaxy dynamics, stellar populations, the interstellar medium, comets and planetary atmospheres, and diffuse ultraviolet background studies.

Atomic Physics. Research in this area includes theoretical work on the electronic structure of atoms and molecules.

High-Energy Physics. Programs involve the study of strong, electromagnetic, and weak interactions. Experiments currently in progress are being performed at Brookhaven National Laboratory and, those using large magnetic spectrometers, at the Stanford Linear Collider, the Tevatron pp⁻ collider at Fermilab, and the LEP storage ring at CERN in Switzerland. Facilities for the construction and testing of particle detectors and associated electronics are available.

Plasma Spectroscopy. Extreme ultraviolet and soft X-ray diagnostic instrumentation is used to study high-temperature plasma devices used in controlled thermonuclear research.

Relativistic Heavy-Ion and Medium-Energy Nuclear Physics. The heavy-ion physics program includes the study of strange baryon and lepton pair production in relativistic heavy-ion collisons at the Lawrence Berkeley and Brookhaven National laboratories.

Solid-State Physics. Research programs involve studies of very thin magnetic films, interfaces and surfaces, amorphous materials, conducting, superconducting, and magnetic properties of artificially structured materials, nanocrystals of metals and alloys, and high-T_c superconductors. Techniques involve SQUID magnetometry, X-ray diffraction, Mössbauer spectroscopy, DC and AC conductivity, LEED and Auger spectroscopies, ferromagnetic resonance, and vibrating-sample magnetometry.

Theoretical Physics. Areas of current research include elementary particles, condensed-matter physics, quantum molecular and atomic structure, quantum electrodynamics, quantum optics, and astrophysics. Members of the theory group specializing in different areas maintain close contact with each other and with the experimental groups.

SECTION 7: PHYSICS

KENT STATE UNIVERSITY
Department of Physics

Programs of Study

The Department of Physics offers a diverse program of graduate study and research leading to the Master of Arts, Master of Science, and Doctor of Philosophy degrees in physics. Major areas of research are adsorption studies, critical phenomena, electron paramagnetic resonance studies, electrooptics, liquid-crystal physics, low-temperature physics, medium-energy nuclear physics, molecular physics, neutron-scattering studies, nonlinear optics, nuclear magnetic resonance studies, nuclear theory, quantum Monte Carlo simulations, quasi-elastic light scattering, relativistic heavy-ion physics, strongly correlated systems, superconductivity, and X-ray scattering. An interdisciplinary graduate program in chemical physics is offered jointly with the chemistry department.

A student typically takes core courses during the first two years of study. The M.A. and M.S. degrees require 32 semester hours of courses and proficiency in a computer language. The M.S. degree requires a thesis. The Ph.D. degree requires 90 semester hours of courses, seminars, and research work beyond the bachelor's degree or 60 semester hours beyond the master's degree. Doctoral students normally will be expected to pass the candidacy examination by the start of the third year. Proficiency in a computer language is required.

Research Facilities

The Department of Physics is located in Smith Hall and has extensive facilities for condensed-matter research, including NMR and EPR equipment, microcalorimetry equipment, laser-light-scattering equipment, a dilution refrigerator for millikelvin research, a scanning tunneling microscope, and a high-brilliance rotating-anode X-ray source. The experimental nuclear physics group has an extensive pool of state-of-the-art fast electronics, power supplies, and a transportable data-acquisition computer. Of special note are the large-volume, ultrafast neutron detectors and neutron polarimeters developed by the Kent faculty and their students.

The Center for Nuclear Research (CNR) has the mission to support, enhance, and promote academic activities in the nuclear physics program of the physics department. The objectives of the CNR are to enhance visibility of the nuclear physics program on a national and international level, to attract high-quality graduate students, to strengthen the academic environment through a visiting scholars program in nuclear physics, and to coordinate and strengthen collaborations with scientists at institutions throughout the world.

The Liquid Crystal Institute (LCI) at Kent State University is a strong academic center of liquid-crystal research and the only institute of its kind in the United States. In 1990 the National Science Foundation awarded the LCI a major grant to establish the Science and Technology Center for Research in Advanced Liquid Crystalline Optical Materials (ALCOM). Seven physics faculty members are principal investigators in the ALCOM center.

Financial Aid

In 1990–91 stipends for first-time regular graduate appointments were $8350 for the nine-month academic year plus a full tuition scholarship. Continuing appointments carry a stipend of $8900. Outstanding applicants may also receive an additional Provost's Fellowship of $2000. Half-time appointments are available that carry the same tuition scholarship as full-time appointments but have only half the stipend and half the service load. It is possible to enter the program at midyear. Qualified students generally receive support during the summer. Final decisions regarding stipends for the following academic year are usually made during spring or summer. Full-time research assistantships may be offered to students on completion of course work, usually at the end of the second academic year. Holders of teaching assistantships, teaching fellowships, and research assistantships receive a full tuition waiver.

Cost of Study

In 1990–91 annual tuition was $3202 for in-state residents pursuing full-time study and $6202 for out-of-state residents pursuing full-time study.

Cost of Living

Various housing options are available for graduate students. Rooms are available for single graduate students in some of the campus dormitories housing third- and fourth-year undergraduates. Current costs per month are $236.50 for a single, $219.75 for a double, and $275.25 for a deluxe single. One- and two-bedroom furnished apartments for married students are available on campus in the University-owned Allerton Apartments. Current costs per month are $299 for 1 bedroom, $307 for 1½ bedrooms, and $316 for 2 bedrooms. Additionally, a variety of reasonably priced rental housing can be found in the Kent area. The Campus Bus Service provides a transportation network for the Kent campus and links the campus with shopping centers and residential neighborhoods in nearby communities; this service is free to all Kent students.

Student Group

The total number of students studying on the Kent campus is about 24,000, including about 4,500 graduate students. The physics department has about 70 graduate students at present with more than twelve foreign countries represented.

Location

Kent is a city of about 30,000 located in northeastern Ohio. The Appalachian foothills to the east and Lake Erie to the north are within an hour's drive. Downtown Cleveland with the Cleveland Symphony, major-league sports, and world-class museums is an hour's drive also. The Akron-Canton, Youngstown, and Cleveland metropolitan areas are less than an hour's drive, providing job opportunities for spouses.

The University and The Department

Kent State offers degree programs ranging from undergraduate degrees in creative and performing arts to advanced graduate degrees in the sciences. Kent State is a strong research institution, ranking in the top dozen or so universities in income from patents, and has a research library of more than 1.6 million volumes.

Besides its physical facilities, Kent State's resources include thousands of acres of wildlife refuges, marshlands, and bogs preserved as learning laboratories and resources for the future.

The physics department has 20 regular faculty members (all with Ph.D.'s), who receive about $1.7-million per year in research support from federal and state agencies on more than twenty different major grants and contracts, which provide research assistantship support for more than 25 graduate students.

Applying

Application forms for admission and financial assistance may be obtained by writing to the Graduate College or to the Department of Physics. The deadline for fall admission is June 1. Graduate study may be initiated during any term, including the summer. Early application is recommended for consideration for financial aid.

Correspondence and Information

Dr. Stanley H. Christensen, Chairman
Department of Physics
Kent State University
Kent, Ohio 44242
Telephone: 216-672-2880

Peterson's Guide to Graduate Programs in the Physical Sciences and Mathematics 1992

SECTION 7: PHYSICS

Kent State University

THE FACULTY AND THEIR RESEARCH

Professors
David W. Allender, Ph.D., Illinois, 1975. Theoretical physics of condensed matter, superconductivity theory, liquid crystals and membrane models.
Bryon D. Anderson, Ph.D., Case Western Reserve, 1972. Medium-energy nuclear physics.
Stanley H. Christensen, Chairman of the Department; Ph.D., Cornell, 1963. Electron paramagnetic resonance, physics education.
J. William Doane, Ph.D., Missouri, 1965. Nuclear magnetic resonance and liquid crystals.
Edward Gelerinter, Ph.D., Cornell, 1966. Electron paramagnetic resonance and optical studies of liquid crystals and artificial membranes.
Wilbert N. Hubin, Ph.D., Illinois, 1969. Computer hardware and physics education.
David L. Johnson, Ph.D., Iowa State, 1966. Low-temperature and liquid-crystal physics, superconductivity.
Richard Madey, Ph.D., Berkeley, 1952. Medium-energy nuclear physics, transport of fluids through porous media.
David S. Moroi, Ph.D., Johns Hopkins, 1959. Theory of particles with internal degrees of freedom, interaction of intense laser beams with matter, liquid-crystal theory.
Alfred Saupe, Ph.D., Freiburg, 1958. Molecular structure and physical properties of liquid crystals.
Nathan Spielberg, Ph.D., Ohio State, 1952. X-ray physics, structure of liquid crystals.
Peter C. Tandy, Ph.D., Flinders (Australia), 1973. Nuclear reaction theory, multiparticle-scattering theory.
John W. Watson, Ph.D., Maryland, 1970. Medium-energy nuclear physics.

Associate Professors
George Fai, Ph.D., Eötvös Lorand (Budapest), 1974. Theoretical nuclear physics, relativistic nuclear collisions.
Satyendra Kumar, Ph.D., Illinois at Urbana-Champaign, 1981. Liquid-crystal physics, superconductivity, liquid-crystal electrooptical effects.
Michael A. Lee, Ph.D., Northwestern, 1977. Condensed-matter theory.
D. Mark Manley, Ph.D., Wyoming, 1981. Medium-energy nuclear physics.

Assistant Professors
Daniele Finotello, Ph.D., SUNY at Buffalo, 1985. Low-temperature and liquid-crystal physics, superconductivity.
Declan Keane, Ph.D., Ireland, 1981. Relativistic nuclear collisions.
Khandker F. Quader, Ph.D., SUNY at Stony Brook, 1983. Theoretical condensed-matter physics, superconductivity, strongly correlated systems, low-temperature physics.

SECTION 7: PHYSICS

LOUISIANA STATE UNIVERSITY

Department of Physics and Astronomy

Programs of Study
The department offers studies leading to the Master of Natural Science (M.N.S.), Master of Science (M.S.), and Doctor of Philosophy (Ph.D.) degrees. For the M.N.S., 36 hours of graduate courses are required. This degree program provides depth in science subjects, as well as the breadth in physics that is required of teachers in junior and senior high schools. The M.S. degree requires 30 hours of graduate work with a thesis or 36 hours without a thesis. Formal requirements for the Ph.D. degree include the Ph.D. qualifying examination; 12 hours of advanced graduate courses beyond the core level; the Ph.D. general examination; publication of research results; and the final examination. The Ph.D. qualifying examination consists of the GRE Subject Test in physics. The minimum qualifying score of 650 must be obtained within the first year of graduate study (international students must take this exam prior to admission). The Ph.D. general examination is offered twice each year. It is comprehensive and is based primarily on graduate physics; students should approach it as the central formal examination of their graduate career. Those with normal preparation must take the general examination within two years of their entrance into the department. The final examination is an oral defense of the thesis.

Research Facilities
LSU has fully staffed machine, electronics, and glassblowing shops, as well as a drafting facility. A full-time staff member operates a liquid-helium facility for the entire University. Terminals connect to LSU's computer center, which houses an IBM 3090 Model 600E, a VAX 8800, an FPS-500, and two FPS-264 Array processors. Via BITNET, SPAN, and SURAnet, LSU is networked to many other national and international computing centers. For research on problems that demand parallel processing, the department is building a Parallel Computing Facility whose focus is an 8K-node parallel processor. A cluster of Sun Workstations for graphics-intensive applications and a VAXcluster for the experimental groups are also supported for research. A campuswide Ethernet connects departmental computers with all other major computers on campus. Two professional computer programmers are staff members of the department. Research resources include a dilution refrigerator–high magnetic field (18 tesla) apparatus; a 1.2-GeV electron synchrotron (CAMD) for materials science, surface physics, and X-ray lithography applications; and laser optics and crystal growing laboratories. The experimental groups are involved in high-energy experiments at LAMPF, TRISTAN, and HERA; SSC-detector development; nuclear physics measurements at Oak Ridge, CERN, and Berkeley; condensed-matter experiments at NIST and Brookhaven; cosmic-ray and neutrino studies at IMB, Baksan, and balloon-launching sites from Canada to the Antarctic; and gravity-wave observations. Astronomers are involved in observations at Kitt Peak and in Chile and with the Hubble Space Telescope, IUE, GRO, and CRRES satellites.

Financial Aid
The department provides about forty-five teaching and research assistantships. Stipends for fall 1991 are approximately $12,000 for the calendar year (twelve months). The University offers Graduate School, Board of Regents, and NASA Space Grant Fellowships with yearly stipends of $16,000 to $18,000. To apply for financial aid, applicants should attach a statement to their application for admission indicating that they wish to be considered for an assistantship or a fellowship. Fellowship applicants must also report their score on the GRE Subject Test in physics and prepare a statement of their intellectual and professional goals. Research assistantships are usually available to students after their first year. There are currently over twenty grant-funded research assistantship positions for both thesis and manpower work; this number is expected to increase in the future.

Cost of Study
For 1990–91, fees were $1020 per semester for Louisiana residents and $2620 per semester for nonresidents. For the summer, the fees were $452 and $1252, respectively.

Cost of Living
The cost of room and board in dormitories runs from $632 to $1275 per semester in 1991–92. There are also 600 married student apartments that rent for $201–$273 per month. In addition, many moderately priced apartments are available near the campus. The cost of living in Baton Rouge is moderate.

Student Group
The physics graduate student body is geographically diverse; students come from all parts of the United States and the world. The 1991–92 entering class of 23 was selected from over 300 applicants, and the enrollment for the fall is expected to be about 80.

Location
Baton Rouge, located on the Mississippi River, is the capital of Louisiana. It has many of the recreational opportunities of a big city, while remaining a residential community of about 400,000 people.

The University
Louisiana State University, established over a hundred years ago, is a statewide system of higher education, with headquarters on the main campus in Baton Rouge. Other campuses are located in Alexandria, Eunice, New Orleans, and Shreveport. On the Baton Rouge campus, the student enrollment is about 24,000, of whom about 4,500 are graduate students. The bachelor's degree is awarded in 125 major fields. Master's degrees are awarded in 56 departmental fields and the Ph.D. in 45.

Applying
Students should request an application packet from the department. All application materials should be sent to the department at the address below to be forwarded to the appropriate offices. Students must submit an Application for Admission for Advanced Studies; their score on the GRE General Test (a score on the GRE Subject Test in physics is recommended for U.S. students and required for international students); two official transcripts of all college and university course work, with a minimum GPA of 3.0 required for all undergraduate and graduate work (A = 4.0); three letters of reference from people acquainted with the student's academic ability; and a nonrefundable application fee of $25. Students whose native language is not English must submit a score of at least 600 on the TOEFL. U.S. applicants must submit all credentials at least sixty days before the beginning of the fall semester and thirty days before the beginning of the spring semester or summer session; International students must submit all credentials at least ninety days before the beginning of any term. Application deadlines are February 15 for fellowships and March 15 for assistantships. Students are advised to apply well before these deadlines but are not discouraged from applying for assistantships after the deadline or for openings for the spring semester.

Correspondence and Information
Department of Physics and Astronomy
Attention: AGS
Louisiana State University
Baton Rouge, Louisiana 70803
Telephone: 504-388-1194

Peterson's Guide to Graduate Programs in the Physical Sciences and Mathematics 1992

SECTION 7: PHYSICS

Louisiana State University

THE FACULTY AND THEIR RESEARCH

Astronomy and Astrophysics
Ganesar Chanmugam, Professor; Ph.D., Brandeis, 1970. Magnetic degenerate stars, cataclysmic variables.
John S. Drilling, Professor; Ph.D., Case Western Reserve, 1967. Objective-prism surveys.
Juhan Frank, Associate Professor; Ph.D., Cambridge, 1978. Accretion in close binaries and active galactic nuclei.
Detlev Koester, Professor; Ph.D., Kiel (Germany), 1981. White dwarfs, stellar atmospheres, stellar evolution.
Arlo U. Landolt, Professor; Ph.D., Indiana, 1962. Stellar photometry.
Paul D. Lee, Associate Professor; Ph.D., Illinois at Urbana-Champaign, 1968. High-dispersion spectroscopy, planetary and galactic nebulae, spectroscopic binaries.
Charles Perry, Professor; Ph.D., Berkeley, 1965. Interstellar medium optical polarization.
Joel Tohline, Professor; Ph.D., California, Santa Cruz, 1978. Star formation, galaxy dynamics.

Atomic/Condensed-Matter Theory
Nathan Brener, Associate Professor–Research; Ph.D., LSU, 1971. Energy band theory, electronic structure of molecules.
Dana Browne, Assistant Professor; Ph.D., Stanford, 1981. Transport properties of very small devices, macroscopic quantum coherence effects, exotic forms of superconductivity.
Joseph Callaway, Boyd Professor; Ph.D., Princeton, 1956. Energy band theory, multichannel scattering theory, ferromagnetism.
Y. K. Ho, Associate Professor–Research; Ph.D., Western Ontario, 1975. Interactions of electrons and positrons with atoms.
Rajiv Kalia, Professor; Ph.D., Northwestern, 1976. Computational solid-state physics.
Robert F. O'Connell, Boyd Professor; Ph.D., Notre Dame, 1962. Inversion layers at insulator-semiconductor interface, Wigner distribution, two-dimensional degenerate electron gas.
A. R. P. Rau, Professor; Ph.D., Chicago, 1970. Atoms in electric and magnetic fields, threshold laws, mathematical physics.
Swaraj S. Tayal, Assistant Professor–Research; Ph.D., Roorkee (India), 1981. Electron scattering from atoms and ions, photoionization of atoms.
Priya Vashishta, FPS Chaired Professor of Computational Methods; Ph.D., Indian Institute of Technology, 1967. Materials science, molecular dynamics, computational physics.

Condensed-Matter/Solid-State/Atomic Experiment
Philip W. Adams, Assistant Professor; Ph.D., Rutgers, 1986. Transport properties in two-dimensional systems, transport properties in quench condensed superconducting films.
Roy G. Goodrich, Professor; Ph.D., California, Riverside, 1965. Electrical conduction in metals, magnitude and anisotropies in electron scattering.
Herbert Piller, Associate Professor; Ph.D., Vienna, 1954. Optical properties of crystalline and amorphous solids, Fabry-Perot cavity, Faraday rotation.
Volker Saile, Professor; Ph.D., Munich, 1976. Electronic structure of atomic and molecular systems.
Roger Stockbauer, Professor; Ph.D., Chicago, 1973. Surface science, electronic properties of materials.
Doyle Temple, Assistant Professor; Ph.D., MIT, 1988. Single-crystal growth, wave-mixing spectroscopy, nonlinear optics.
Nadim H. Zebouni, Associate Professor; Ph.D., LSU, 1961. Superconductivity, transport properties, Chevrel phase, magnetism.

Elementary Particle Theory
Lai-Him Chan, Professor; Ph.D., Harvard, 1966. Quark phenomenology, low-energy hadron dynamics, chiral symmetry and α model, derivative expansion, effective action expansion.
Richard W. Haymaker, Professor; Ph.D., Berkeley, 1967. Dynamical symmetry breaking, lattice gauge theories.

Experimental General Relativity
William O. Hamilton, Professor; Ph.D., Stanford, 1963. Gravitational radiation instrumentation, cryogenic antenna to detect supernova collapses, cavity accelerometer detector, superconducting oscillator.
Warren W. Johnson, Associate Professor–Research; Ph.D., Rutgers, 1974. Gravitational radiation detectors, Josephson devices, parametric transducers and quantum non-demolition.

Fluid Mechanics
R. G. Hussey, Professor; Ph.D., LSU, 1962. Low Reynolds number, boundary effects on axial motion.

High-Energy Astrophysics and Space Physics
Michael L. Cherry, Associate Professor; Ph.D., Chicago, 1978. Neutrinos, cosmic rays, high-energy-particle astrophysics.
T. Gregory Guzik, Assistant Professor–Research; Ph.D., Chicago, 1980. Solar flares, particle interactions, accelerator experiments, cosmic rays.
Richard W. Huggett, Professor; Ph.D., Indiana, 1957. Cosmic rays.
Robert Svoboda, Assistant Professor; Ph.D., Hawaii, 1985. Neutrino physics, proton decay, high-energy-particle astrophysics.
John P. Wefel, Professor; Ph.D., Washington (St. Louis), 1971. Astrophysics—experimental and theoretical; galactic cosmic radiation, solar energetic particles.

High-Energy Experiment
Ali R. Fazely, Assistant Professor–Research; Ph.D., Kent State, 1982. Neutrino physics at LANL, Solenoidal Detector Collaboration at SSC.
Richard L. Imlay, Professor; Ph.D., Princeton, 1967. Neutrino oscillations at Los Alamos National Laboratory (LANL), e^+e^- collisions at the TRISTAN Storage Ring in Japan.
Roger McNeil, Assistant Professor; Ph.D., California, Davis, 1986. e^+e^- collisions at the TRISTAN Storage Ring in Japan.
William J. Metcalf, Professor; Ph.D., Caltech, 1974. Neutrino oscillations at LANL, e^+e^- collisions at the TRISTAN Storage Ring in Japan.

Nuclear Experiment
Paul N. Kirk, Professor; Ph.D., MIT, 1969. Nuclear and quark matter, quark condensation.
Edward F. Zganjar, Professor; Ph.D., Vanderbilt, 1966. Nuclei far from stability, heavy-ion accelerator, nuclear spectroscopy.

Nuclear Theory
Jerry P. Draayer, Professor and Chairman; Ph.D., Iowa State, 1968. Shell model, statistical spectroscopy, group theory.

SECTION 7: PHYSICS

MICHIGAN STATE UNIVERSITY

Department of Physics and Astronomy

Programs of Study M.S. and Ph.D. degrees are offered with specializations in accelerator physics, chemical physics, elementary particle theory, experimental particle physics, low-temperature physics, many-body theory, nuclear physics, solid-state physics, and observational and theoretical astrophysics. The quarter system is followed. The ratio of faculty members to graduate students is about 1:2, and formal class sizes range from 5 to 25.

Research Facilities Research facilities include two superconducting cyclotrons, K500 and K1200, each injected by ECR ion sources and associated apparatus, including a large 92-inch scattering chamber; a magnetic spectrometer; an isotope separating beam line; a recoil mass separator; 4-pi charged-particle detectors; a high-energy gamma ray detector array; neutron and charged-particle detector hadoscopes; a number of data acquisition and analysis computers, VAX 8650 and 8530 computers, and a considerable number of MicroVAX workstations; light-scattering apparatus, including visible and (unique) infrared laser sources, two Raman spectrophotometers, and a Brillouin spectrometer; diffuse and discrete X-ray diffraction apparatus employing X-ray cameras, a four-circle diffractometer, and a 12-k-watt rotating anode X-ray source; high-pressure apparatus (1–300-k bar); an analytical electron microscope laboratory, containing a unique field-emission scanning transmission electron microscope (FESTEM) equipped for high-resolution imaging and quantitative diffraction, X-ray energy-dispersive analysis, electron energy loss spectroscopy, digital beam control and mapping, and temperature-dependent studies; a scanning electron microscope (SEM) capable of atomic resolution over the temperature range 4K–300K; photolithographic and electron-beam lithographic facilities for device fabrication with 100-nm resolution; cryogenic facilities, including three helium-3 refrigerators and three helium-3/helium-4 mixing refrigerators; apparatus for cooling by adiabatic demagnetization; two (55-kG and 105-kG) superconducting magnets; four electromagnets; an automated SQUID magnetometer; an electron spin resonance laboratory; an ultrahigh-vacuum four-gun sputtering system; a liquid argon calorimeter test station for the D=zero experiment at Fermilab and detector development for the Superconducting Super Collider; a high-energy physics laboratory, which is a state-of-the-art electronics design facility where detectors for experiments are developed, tested, and constructed; and numerous minicomputers and microcomputers in all research areas. Important off-campus facilities in the high-energy area are at the accelerator at Fermilab in Batavia, Illinois; at Brookhaven National Laboratory, Long Island, New York; CERN, Geneva, Switzerland, where experiments are currently being carried out by MSU faculty members and students; and the proposed Superconducting Super Collider in Waxahachie, Texas. MSU faculty and students are participating in studies and workshops in which the aim is to design an experiment to operate at this frontier facility. A department library has up-to-date collections of books and journals. Current research programs are described in materials available from the department. The astronomy faculty makes use of the facilities on campus and of observatories at Kitt Peak (Arizona), WIRO (Wyoming), Mounts Wilson and Palomar (California), Siding Springs (Australia), and Cerro Tololo and Las Campanas (Chile).

Financial Aid Half-time graduate assistantship stipends began at $10,800 for the 1990–91 academic year. Summer assistantships are readily available. Assistants spend up to 20 hours a week on their duties. In-class contact hours for teaching assistants range from 6 to 8 hours for recitation and laboratory classes. The normal course load for assistants is 8 to 10 credit hours. The duties of research assistants are commonly in the general area in which the Ph.D. thesis will be written. Fellowships and scholarships are also available.

Cost of Study Tuition for 1990–91 was $98 per credit hour for Michigan residents. Teaching and research assistants pay in-state rates and receive a tuition waiver of 6 credits per term. Out-of-state tuition was $198 per credit hour. All students paid a registration fee of $98 per term.

Cost of Living Single rooms in Owen Hall, the graduate residence center, rented for $952 per quarter and double rooms for $819 per student per quarter in 1990–91. This cost included $155 per term of meal credit. Food may be obtained from several campus cafeterias and local restaurants. The University owns and operates more than 2,000 one- and two-bedroom apartments to help meet the housing needs of married students. These rent for $281 and $309 per month, respectively, and include all utilities, essential furniture, and a private telephone. Privately owned off-campus rooms and apartments are also available.

Student Group The on-campus enrollment at Michigan State University for the 1990 fall term was approximately 42,000, including about 8,000 graduate and professional students. There were 143 physics graduate students, including 79 Ph.D. candidates, and there were 19 postdoctoral research associates.

Location East Lansing is a residential city adjacent to the Michigan State University campus and close to Lansing, the state capital. Many opportunities for cultural and social development are offered by the University and neighboring civic groups. Examples include the Wharton Center for the Performing Arts, the Kresge Art Center, the University Museum, and the Lecture-Concert, World Travel, and Foreign Film series.

The University and The Department Michigan State University, one of the oldest land-grant colleges, was founded in 1855 for the purpose of furthering the interests of agriculture and the mechanic arts. From this modest beginning it has grown to become one of America's largest universities, with many educational innovations to its credit. Through its 14 colleges and more than 100 departments, it offers 200 different programs leading to undergraduate and graduate degrees. The Department of Physics and Astronomy has grown rapidly in size and quality in recent years. This growth was facilitated and accelerated by a large, long-term grant to the University by the National Science Foundation under its University Science Development Plan for the development of "centers of excellence."

Applying Application forms may be obtained by writing to the Office of Admissions and Scholarships. Applications for admission and supporting documents should be received at least one month prior to the first enrollment together with a $25 application fee. Applicants should request that registrars of colleges previously attended send transcripts directly to this office. Applications for a graduate assistantship, fellowship, or scholarship should reach the office no later than six months prior to the first anticipated enrollment. Acceptance of graduate students is decided by a departmental committee maintained for this purpose.

Correspondence and Information
Professor Julius S. Kovacs, Associate Chairperson
Department of Physics and Astronomy
Michigan State University
East Lansing, Michigan 48824-1116
Telephone: 517-355-9276
Fax: 517-355-6661
BITNET: KOVACS@MSUPA

SECTION 7: PHYSICS

Michigan State University

THE FACULTY

Physics
Maris A. Abolins, Professor; Ph.D., California, San Diego, 1965.
Sam M. Austin, University Distinguished Professor and Director, National Superconducting Cyclotron Laboratory; Ph.D., Wisconsin–Madison, 1960.
Jack Bass, Professor; Ph.D., Illinois, 1964.
Wolfgang Bauer, Assistant Professor; Ph.D., Giessen (Germany), 1987.
Walter Benenson, Professor and Associate Director, National Superconducting Cyclotron Laboratory; Ph.D., Wisconsin–Madison, 1962.
George Bertsch, John A. Hannah Professor; Ph.D., Princeton, 1965.
Martin Berz, Associate Professor; Ph.D., Giessen (Germany), 1986.
Norman Birge, Assistant Professor; Ph.D., Chicago, 1986.
Henry G. Blosser, University Distinguished Professor; Ph.D., Virginia, 1954.
Jerzy Borysowicz, Professor; Ph.D., Institute for Nuclear Research (Warsaw), 1965.
Raymond L. Brock, Associate Professor; Ph.D., Carnegie-Mellon, 1980.
Carl M. Bromberg, Professor; Ph.D., Rochester, 1974.
B. Alex Brown, Professor; Ph.D., SUNY at Stony Brook, 1974.
Aurel Bulgac, Assistant Professor; Ph.D., Leningrad Nuclear Physics Institute, 1977.
Edward H. Carlson, Professor; Ph.D., Johns Hopkins, 1959.
Jerry A. Cowen, Professor; Ph.D., Michigan State, 1954.
Gerard M. Crawley, Professor and Chairperson; Ph.D., Princeton, 1965.
Pawel Danielewicz, Visiting Associate Professor; Ph.D., Warsaw, 1981.
Michael A. Dubson, Assistant Professor; Ph.D., Cornell, 1984.
Phillip M. Duxbury, Assistant Professor; Ph.D., New South Wales, 1983.
Carl L. Foiles, Professor; Ph.D., Arizona, 1964.
Aaron Galonsky, Professor; Ph.D., Wisconsin–Madison, 1954.
C. Konrad Gelbke, University Distinguished Professor; Ph.D., Heidelberg, 1972.
Brage Golding, Professor; Ph.D., MIT, 1966.
Morton M. Gordon, Professor; Ph.D., Washington (St. Louis), 1950.
Michael J. Harrison, Professor; Ph.D., Chicago, 1960.
William M. Hartmann, Professor; D.Phil., Oxford, 1965.
Jack Hetherington, Professor; Ph.D., Illinois, 1962.
Joey W. Huston, Assistant Professor; Ph.D., Rochester, 1982.
Thomas Kaplan, Professor; Ph.D., Pennsylvania, 1954.
Edwin Kashy, Professor; Ph.D., Rice, 1959.
Julius S. Kovacs, Professor and Associate Chairperson; Ph.D., Indiana, 1955.
James T. Linnemann, Associate Professor; Ph.D., Cornell, 1978.
William G. Lynch, Associate Professor; Ph.D., Washington (Seattle), 1980.
S. D. Mahanti, Professor; Ph.D., California, Riverside, 1968.
Hugh McManus, Emeritus Professor; Ph.D., Birmingham, 1947.
Jerry A. Nolen, Professor; Ph.D., Princeton, 1965.
Paul M. Parker, Professor; Ph.D., Ohio State, 1958.
Gerald L. Pollack, Professor; Ph.D., Caltech, 1962.
Bernard G. Pope, Professor; Ph.D., Columbia, 1971.
William Pratt, Professor; Ph.D., Minnesota, 1969.
Jon Pumplin, Professor; Ph.D., Michigan, 1968.
Wayne Repko, Professor; Ph.D., Wayne State, 1967.
Peter A. Schroeder, Professor; Ph.D., Bristol, 1955.
David K. Scott, Professor and Provost; D.Phil., Oxford, 1967.
Peter S. Signell, Professor; Ph.D., Rochester, 1958.
Daniel R. Stump, Associate Professor; Ph.D., MIT, 1976.
Michael Thoennessen, Assistant Professor; Ph.D., SUNY at Stony Brook, 1988.
Michael F. Thorpe, Professor; D.Phil., Oxford, 1968.
Roger C. Tobin, Assistant Professor; Ph.D., Berkeley, 1985.
David Tomanek, Assistant Professor; Ph.D., Berlin, 1983.
Hendrik J. Weerts, Associate Professor; Ph.D., Aachen (Germany), 1981.
Gary D. Westfall, Associate Professor; Ph.D., Texas at Austin, 1975.

Astronomy
Timothy C. Beers, Assistant Professor; Ph.D., Harvard, 1983.
Jeffrey R. Kuhn, Associate Professor; Ph.D., Princeton, 1981.
Albert P. Linnell, Professor; Ph.D., Harvard, 1958.
Edwin D. Loh, Associate Professor; Ph.D., Princeton, 1977.
Susan M. Simkin, Professor; Ph.D., Wisconsin–Madison, 1967.
Horace Smith, Associate Professor; Ph.D., Yale, 1980.
Robert F. Stein, Professor; Ph.D., Columbia, 1966.

SECTION 7: PHYSICS

NEW MEXICO STATE UNIVERSITY

Department of Physics

Programs of Study

The Department of Physics offers programs in many areas of specialization leading to the M.S. and Ph.D. degrees. Each graduate student has a faculty adviser who helps design an individual course of study and research. Individually tailored interdisciplinary programs are also available. The average graduate class size of 15 students facilitates close student-faculty interaction.

Students entering with a bachelor's degree usually need two years to complete requirements for an M.S., or between five and six years to complete requirements for a Ph.D. Proficiency in computer and writing skills has replaced the traditional foreign language requirement.

The principal experimental research areas in the department are laser spectroscopy, atomic and molecular physics, aerosol physics (through the NMSU Center for Atmospheric Sciences), nuclear and particle physics, X-ray spectroscopy, and geophysics. Major areas of theoretical research are nonlinear dynamics, linear and nonlinear optics, relativity and astrophysics, nuclear and particle physics, scattering theory, and condensed-matter physics. Currently, the department has $2-million in external funding for research.

Summer and/or academic-year placements are available for qualified students at Los Alamos National Laboratory; Sandia National Laboratories; the U.S. Army Atmospheric Sciences Laboratory and Nuclear Effects Laboratory at nearby White Sands Missile Range (WSMR); Lawrence Livermore National Laboratory; the Paul Scherrer Institute near Zurich, Switzerland; the NASA-Johnson Space Flight Center; the University's Physical Science Laboratory; and local offices of several private companies engaged in optics research. These placements often lead to employment after graduation.

Research Facilities

The department is housed in a 60,000-square-foot building. Computational physics is carried out on the department's VAX-11/750, two Sun SPARCstations 1+ computers, two VAXstation 3100s, and numerous personal computers; the University's VAX 8530; Sequent Symmetry S27 parallel processor; Sun-4 shared computer system; and two IBM 3081 D mainframes or via Internet on supercomputers at Los Alamos and other locations. The nuclear and particle physics group conducts experiments at the Los Alamos Meson Physics Facility and accelerators at Vancouver, Canada, and the Paul Scherrer Institute. Major research equipment in the building includes seismometers, magnetometers, and gravity meters; CO_2, Nd:yag, helium-neon, argon-ion, dye, and other laser systems; state-of-the-art optical detectors and related fast electronics; and a double-crystal X-ray spectrometer.

Financial Aid

Fellowships, assistantships, and laboratory internships are offered. During 1991–94, ten Department of Education fellowships are available. These carry an annual stipend of $10,000 plus tuition, fees, and an allowance for books. Fellowships from the Nuclear Effects Directorate at WSMR and the State of New Mexico are also available. For 1991–92, assistantship stipends are $9197 and $9397 for students entering with bachelor's and master's degrees, respectively. Highly qualified fellows and assistants may receive additional stipends. Laboratory internships usually include summer employment, which pays $1000 or more.

Cost of Study

For New Mexico residents, 1991–92 tuition and required fees are $71.50 per credit hour; nonresidents pay $225.50 per credit hour. Entering graduate assistants and fellows qualify for resident rates and may establish New Mexico residence during their first year.

Cost of Living

The University has dormitory facilities for single students and single-family and quad units for married students. Single student University rentals cost $550 to $760 per semester. Married student housing is about $300 per month. Many privately owned houses and apartments are available near campus, starting at about $250 per month. In general, the cost of living in the Las Cruces area is lower than that in a large city.

Student Group

In 1990–91, the physics department enrolled 43 undergraduate majors and 56 graduate students; of the latter, 14 are foreign nationals, 2 are women, and 6 others are members of minority groups.

Location

Las Cruces is a city of 70,000 people, located at an elevation of 3,900 feet in the historic Mesilla Valley of the Rio Grande. The city is 50 miles north of El Paso, Texas, and Juarez, Mexico. Just east of town are the rugged but beautiful Organ Mountains, a rock-climbers' paradise; somewhat farther to the northeast are White Sands Missile Range, White Sands National Monument, and, within 100 miles, two ski areas located in the Lincoln National Forest. The Gila Wilderness, the largest in the contiguous forty-eight states, lies 100 miles to the northwest. The climate is mild during the winter (often sunny and 65°F) and hot but dry during the summer. All buildings and dwellings are cooled by refrigerated air or by inexpensive evaporative cooling.

The University and The Department

The University is the land-grant public institution of higher education of New Mexico. More than 14,000 students are enrolled at NMSU in seventy major areas. Of these, about 25 percent are members of minority groups and 50 percent are women. More than 2,000 graduate students are enrolled in the forty-two master's and twenty-four doctoral programs. An exceptionally large campus with complete recreational and sports facilities helps to establish and maintain the University's relaxed and friendly atmosphere.

NMSU is ranked as a Research I University by the Carnegie Foundation. With total research contracts in excess of $150-million and more than $55-million this year in federal research support, NMSU ranks among the top thirty public universities receiving federal funds and third among universities receiving NASA funds. NMSU is also the administrative headquarters of the National Physical Sciences Consortium, a group of the major national laboratories and universities formed in 1988 that provides graduate fellowships in the physical sciences for minority and women students.

The physics department is one of twenty-two academic departments in the College of Arts and Sciences, the largest of the University's six colleges. The department has awarded about 100 Ph.D. degrees and more than 200 M.S. degrees since beginning its graduate programs in 1960.

Applying

Applications for admission and financial assistance are accepted at any time but should preferably be received prior to March 31 for the fall semester and November 15 for the spring semester. The GRE General Test is required of all students, and scores should be submitted at the time of application, if possible. Foreign students whose native tongue is not English must score at least 500 on the TOEFL.

Correspondence and Information

Graduate Admissions Committee
Department of Physics
New Mexico State University
Las Cruces, New Mexico 88003-0001

Telephone: 505-646-3831
Fax: 505-646-1934

Peterson's Guide to Graduate Programs in the Physical Sciences and Mathematics 1992

SECTION 7: PHYSICS

New Mexico State University

THE FACULTY AND THEIR RESEARCH

Robert L. Armstrong, Professor; Ph.D., Johns Hopkins, 1970. Optics, laser spectroscopy, atmospheric physics.
Douglas R. Brown, Adjunct Professor; Ph.D., Colorado, 1975. Aerosol physics.
Charles Bruce, Adjunct Professor; Ph.D., New Mexico State, 1970. Aerosol physics.
George R. Burleson, Professor; Ph.D., Stanford, 1960. Experimental nuclear and particle physics.
Alex F. Burr, Professor; Ph.D., Johns Hopkins, 1966. X-ray and condensed-matter physics, physics education.
Tuan Wu Chen, Professor; Ph.D., Syracuse, 1966. Elementary particle theory, scattering theory.
Horace H. Coburn, College Professor; Ph.D., Ohio State, 1943. Optics, physics education.
Sidney A. Coon, Associate Professor; Ph.D., Maryland, 1972. Nuclear and particle theory.
Harold A. Daw, College Professor; Ph.D., Utah, 1956. Optics, laser physics, physics education.
William R. Gibbs, College Professor; Ph.D., Rice, 1961. Nuclear theory.
James B. Gillespie, Adjunct Professor; Ph.D., New Mexico State, 1982. Atmospheric optics.
George H. Goedecke, Professor and Department Head; Ph.D., RPI, 1961. Relativity, scattering theory, optics, condensed-matter theory.
Michael E. Goggin, Adjunct Assistant Professor; Ph.D., Arkansas, 1988. Nonlinear dynamics.
Thomas M. Hearn, Assistant Professor; Ph.D., Caltech, 1985. Geophysics, geotomography, seismology.
Paul Higbie, Adjunct Professor; Ph.D., MIT., 1968. Space and plasma physics.
Richard L. Ingraham, Research Professor; Ph.D., Harvard, 1952. Astrophysics, nonlinear dynamics, relativity, field theory, condensed-matter theory.
James D. Klett, Adjunct Professor; Ph.D., UCLA, 1968. Atmospheric optics.
Gary S. Kyle, Associate Professor; Ph.D., Minnesota, 1979. Experimental nuclear and particle physics.
Robert J. Liefeld, Professor; Ph.D., Ohio State, 1959. X-ray and condensed-matter physics, physics education.
John Meason, Adjunct Professor; Ph.D., Arkansas, 1965. Nuclear physics.
August Miller, Emeritus Professor; Ph.D., New Mexico State, 1961. Molecular physics, atmospheric optics.
Paul Nachman, Assistant Professor; Ph.D., Chicago, 1978. Experimental optical, atomic, and molecular physics.
James F. Ni, Associate Professor; Ph.D., Cornell, 1984. Geophysics, geodesy, seismology, remote sensing.
J. David Pendleton, Adjunct Professor; Ph.D., Tennessee, Knoxville, 1982. Atmospheric physics.
Ronald Pinnick, Adjunct Professor; Ph.D., Wyoming, 1972. Atmospheric optics, laser physics.
Budh Ram, Professor; Ph.D., Colorado, 1963. Elementary particle theory.
Jean Robillard, Adjunct Professor; D.Sc., Paris IV (Sorbonne), 1974. Solid-state physics.
Stephen L. Salyards, College Assistant Professor; Ph.D., Caltech, 1988. Geophysics, paleomagnetics.
Thorsten F. Stromberg, Associate Professor; Ph.D., Iowa State, 1965. Low-temperature physics, optical properties of aerosols, physics education.
Robert A. Sutherland, Adjunct Professor; Ph.D., Missouri, 1972. Atmospheric physics.
Alan Van Heuvelen, Professor; Ph.D., Colorado, 1964. Biophysics, physics education.

Postdoctoral Research Associates
Sanjoy Mukhopadhyay, Ph.D., Northwestern, 1989. Experimental nuclear and particle physics.
Mohini Rawool-Sullivan, Ph.D., New Mexico State, 1988. Nuclear and particle physics.
Jing-Gang Xie, Ph.D., California, San Diego, 1989. Laser physics.

CURRENT RESEARCH ACTIVITIES

Theoretical
Astrophysics (Ingraham, Ram): Equations of state in extremely strong magnetic fields, quantum theory of the early universe.
Atmospheric Physics (Armstrong, Goedecke, Klett, Miller): Infrared and millimeter-wave propagation through aerosols, nonlinear interactions between intense laser beams and aerosols, radiative transfer, climatology.
Atomic and Molecular Physics (Armstrong, Burr): Line broadening mechanisms, rotational relaxation processes, X-ray atomic energy levels.
Condensed-Matter Physics (Goedecke, Ingraham): Quantum Hall effect, superconductivity, electronic properties of thin films.
Electromagnetism and Electrodynamics (Chen, Goedecke, Miller): Propagation of e-m waves in random media, classical descriptions of radiation reaction, scattering of e-m waves by aerosol particles, stochastic electrodynamics, foundations of quantum mechanics.
Elementary Particles and Fields (Chen, Coon, Goedecke, Ingraham, Ram): High-energy particle scattering and reactions, finite quantum field theory, conformal relativity, relativistic quark models, gauge theory, QCD, magnetic monopoles, mass spectrum of hadrons.
Geophysics (Hearn, Ni): Geotomography, inversion methods, tectonophysics.
Nonlinear Dynamics (Goggin, Ingraham): Stability, bifurcation theory, chaos.
Nuclear Physics (Chen, Coon, Gibbs, Goedecke): Pion-nucleus scattering, multiple scattering mechanisms, Eikonal-type approximations, large-angle scattering theories.
Optics (Chen, Goedecke, Miller): Scattering by particles of arbitrary shape and structure, multiple scattering and propagation, nonlinear effects, laser-induced plasmas.
Statistical Physics (Goedecke, Goggin, Ingraham): Applications of stability theory of nonlinear maps and coupled nonlinear differential equations, stochastic electrodynamics.

Experimental
Atmospheric Optics and Aerosol Physics (Armstrong, Bruce, Gillespie, Pinnick, Stromberg, Xie): Absorption and scattering of IR and mm waves by natural and artificial aerosols, laser-induced heating and vaporization of aerosols.
Atomic and Molecular Physics (Armstrong, Burr, Liefeld, Nachman, Stromberg, Van Heuvelen, Xie): Raman scattering, Rayleigh-Brillouin scattering, Doppler limited IR absorption, multiphoton spectroscopy, soft X-ray emission and absorption spectroscopy, X-ray line and continuum isochromats, Auger and photoelectron spectroscopy.
Biophysics (Van Heuvelen): Medical and environmental applications of physics, blood glucose monitoring.
Condensed Matter (Burr, Liefeld, Robillard): Density of states above Fermi level by X-ray isochromats and absorption spectra, density of states below Fermi level by X-ray emission and electron spectra, surface physics.
Geophysics (Hearn, Ni, Salyards): Magnetic, gravity, seismic, and resistivity studies in geothermal areas of New Mexico, Arizona, and west Texas; global paleomagnetics; geotomography; crustal movements; neotectonics.
Nuclear and Elementary Particle Physics (Burleson, Kyle, Meason, Mukhopadhyay, Rawool-Sullivan): Pion-nucleus and nucleon-nucleon interactions, studies of nucleon structure.
Optics (Armstrong, Daw, Nachman, Pinnick, Xie): Laser Raman spectroscopy, laser absorption and emission spectroscopy, laser-induced plasma spectroscopy.
Physics Education (Burr, Daw, Liefeld, Stromberg, Van Heuvelen): Development of new teaching methods for engineering physics, development of computer-assisted laboratories and classroom instruction.

NORTHEASTERN UNIVERSITY
Department of Physics

Programs of Study

The department offers a full-time program leading to the Ph.D. and full-time and part-time evening programs leading to the M.S. Requirements for the Ph.D. include 62 quarter hours of course work, a written qualifying examination, a thesis describing the results of independent research, and a final oral examination. Students may pursue basic research in elementary particle physics, condensed-matter physics, and molecular biophysics or in interdisciplinary areas such as materials science, chemical physics, biophysics, and applied engineering physics. They also may carry out cooperative research at technologically advanced industrial, governmental, and national and international laboratories and at medical research institutions in the Boston area. Requirements for the M.S. are 42 quarter hours of credit, up to 12 of which may be transfer credit, if approved, and up to 9 may be taken from a selection of senior-level undergraduate physics courses. Subject to approval, graduate courses in other science and engineering fields may be taken for up to 12 quarter hours. The department offers alternative M.S. options with concentrations in optics and instrumentation that require a thesis. There is no language requirement for either degree.

Research Facilities

The department is located in a modern building that houses research laboratories, machine shops, an electronics shop, conference and seminar rooms, and faculty and graduate student offices. Adjacent to this center is a technologically sophisticated central library with extensive print, database, and media collections supporting graduate studies. The department has numerous MicroVAX and personal computers dedicated to specific research programs. Also available are a VAXcluster with seven workstations, Cray and connection-machine supercomputers, and a PC cluster facility. All computers are connected to a dual VAX 6420 installation—a campuswide Ethernet with BITNET national and international file-transfer and electronic mail capabilities. A broad range of experimental facilities and spectroscopic instrumentation is on hand. Instrumentation includes Fourier transform interferometers; infrared, photoluminescence, laser excitation, and Raman spectrometers; and submillimeter, microwave, radio-frequency, electron spin resonance, and dielectric systems. Ancillary equipment exists for pressure, temperature, and electric and magnetic field studies. Superconducting quantum interference devices, quench condensation of metals onto substrates at ultra-low temperatures, and cryogenic apparatus are used to study normal and high-temperature superconductors. Surface physics techniques include scanning tunneling microscopy; scanning electron microscopy; X-ray diffraction; metallography; and ultraviolet, visible, and infrared reflectance procedures. Laboratories devoted to multilayer deposition systems, crystal growth, fabrication, sample preparation, and polishing facilities are available. High-energy and other experiments are carried out at national and international laboratories.

Financial Aid

Northeastern awards financial aid through the Perkins Loan, College Work-Study, and Stafford Loan programs and through a limited number of minority fellowships and Martin Luther King, Jr. Scholarships. The Graduate School offers teaching, research, and administrative assistantships that include tuition remission and a stipend (currently $11,900 for four quarters) and require a maximum of 20 hours of work per week. A few tuition assistantships provide partial or full tuition remission and require a maximum of 10 hours of work per week.

Cost of Study

Tuition for the 1991–92 academic year is $285 per quarter hour. Books and supplies cost about $300 per year. Tuition charges are made for Ph.D. thesis and continuation. Other charges include the Student Center fee and health and accident insurance fee, which are required of all full-time students.

Cost of Living

On-campus living expenses are estimated at $630 per month; on-campus housing is available on a limited basis to new students. In addition, the Office of Residential Life provides lists of apartments, rooms, and potential roommates. Off-campus living costs are estimated at $1000 per month. A public transportation system serves the Greater Boston area, with convenient subway, bus, and commuter rail services.

Student Group

Approximately 36,000 students are enrolled at the University, representing a wide variety of academic, professional, geographic, and cultural backgrounds. The Graduate School of Arts and Sciences has approximately 900 students, 65 percent of whom attend on a full-time basis. In the fall of 1991, the department expects to have approximately 80 full-time students, of whom 85 percent will receive some form of financial support. About 20 additional students are enrolled in the part-time, evening M.S. program. The department awards roughly six Ph.D. degrees and eight M.S. degrees per year. Most graduates have continued to pursue research careers, either in academic institutions as postdoctoral fellows or in industrial, medical, or government laboratories.

Location

The University is located in the Back Bay section of Boston, close to the Museum of Fine Arts, the New England Conservatory of Music, Symphony Hall, and historic Copley Square. Greater Boston is home to more universities and research facilities than any other area in the world. It is a place where the past is appreciated, the present enjoyed, and the future anticipated.

The University and The Department

Northeastern is a privately endowed nonsectarian institution of higher learning that is dedicated to the pursuit of excellence in graduate research and scholarship. Since its founding almost a century ago, it has grown into one of the largest private universities in the country, with a national reputation in cooperative and graduate education. The Department offers opportunities for students to work on a wide range of groundbreaking research programs with an internationally recognized faculty whose goal is to provide an effective education to students with varied backgrounds.

Applying

Although there is no absolute deadline for applying, completed applications should be received by February 15 to secure priority consideration for September acceptance, especially if financial assistance is sought. Scores on the GRE General Test and Subject Test in Physics are highly desirable. The latter is given considerable weight in the admissions and assistantship awarding process when the number of applicants is high. For international students, a minimum TOEFL score of 525 is required for admission, 550 if a teaching assistantship is sought.

Correspondence and Information

Graduate Director (Admissions)
Department of Physics
Northeastern University
Boston, Massachusetts 02115
Telephone: 617-437-2902
Fax: 617-437-2943

SECTION 7: PHYSICS

Northeastern University

THE FACULTY AND THEIR RESEARCH

Professors
Stephen Reucroft, Chairperson; Ph.D., Liverpool, 1969. High-energy experimental physics.
Ronald Aaron, Ph.D., Pennsylvania, 1961. Medical physics.
Petros Argyres, Ph.D., Berkeley, 1954. Condensed-matter theory.
Arun Bansil, Ph.D., Harvard, 1974. Condensed-matter theory.
Paul M. Champion, Ph.D., Illinois at Urbana-Champaign, 1975. Molecular biophysics.
Alan H. Cromer, Ph.D., Cornell, 1960. Biophysics and education.
William L. Faissler, Ph.D., Harvard, 1967. High-energy experimental physics.
Marvin H. Friedman, Ph.D., Illinois at Urbana-Champaign, 1952. High-energy theory.
David A. Garelick, Ph.D., MIT, 1963. High-energy experimental physics.
Michael J. Glaubman, Ph.D., Illinois at Urbana-Champaign, 1953. High-energy experimental physics.
Hyman Goldberg, Ph.D., MIT, 1963. Particle theory.
Walter Hauser, Ph.D., MIT, 1950. Education.
Jorge V. Jose, D.Sc., National of Mexico, 1976. Condensed-matter theory.
Robert P. Lowndes, Dean of Arts and Sciences; Ph.D., London, 1966. Condensed-matter experimental physics.
Bertram J. Malenka, Ph.D., Harvard, 1951. Particle theory.
Robert S. Markiewicz, Ph.D., Berkeley, 1975. Condensed-matter experimental physics.
Pran Nath, Ph.D., Stanford, 1964. Particle theory.
Clive H. Perry, Graduate Director; Ph.D., London, 1960. Condensed-matter experimental physics.
Eugene J. Salatan, Emeritus; Ph.D., Princeton, 1962. High-energy experimental physics.
Carl A. Shiffman, D.Phil., Oxford, 1956. Condensed-matter experimental physics.
Jeffrey B. Sokoloff, Ph.D., MIT, 1967. Condensed-matter theory.
Yogendra N. Srivastava, Ph.D., Indiana, 1964. Particle theory.
Michael T. Vaughn, Ph.D., Purdue, 1960. Particle theory.
Eberhard von Goeler, Ph.D., Illinois at Urbana-Champaign, 1961. High-energy experimental physics.
Allan Widom, Ph.D., Cornell, 1967. Condensed-matter theory.
Fa-Yueh Wu, Ph.D., Washington (St. Louis), 1963. Condensed-matter theory.

Associate Professors
George O. Alverson, Ph.D., Illinois at Urbana-Champaign, 1979. High-energy experimental physics.
Jacqueline Krim, Ph.D., Washington (Seattle), 1984. Condensed-matter experimental physics.
Marie E. Machacek, Ph.D., Iowa, 1973. Particle theory.

Assistant Professors
Narendra Jaggi, Ph.D., Bombay, 1982. Condensed-matter experimental physics.
Alain Karma, Ph.D., California, Santa Barbara, 1986. Condensed-matter theory.
Ian Leedom, Ph.D., Purdue, 1982. High-energy experimental physics.
Russell LoBrutto, Ph.D., SUNY at Buffalo, 1984. Molecular biophysics.
Srinivas Sridhar, Ph.D., Caltech, 1983. Condensed-matter experimental physics.
Tomasz Taylor, Ph.D., Warsaw, 1981. Particle theory.

Research Associates
Stefan Cordes, Ph.D., Princeton, 1986. Particle theory.
Stanislaw Kaprzyk, Ph.D., Stanislaw Staszic (Poland), 1981. Condensed-matter theory.
Matti Lindroos, Ph.D., Tampere Tech (Finland), 1979. Condensed-matter theory.
Christopher B. Lirakis, Ph.D., Northeastern, 1990. High-energy experimental physics.
Jorge H. Moromisato, Ph.D., Northeastern, 1971. High-energy experimental physics.
Michael O'Connor, Ph.D., Illinois at Urbana-Champaign, 1990. Particle theory.
Zbignlew Ryzak, Ph.D., MIT, 1987. Particle theory.
J. Timothy Sage, Ph.D., Illinois at Urbana-Champaign, 1986. Molecular biophysics.
Lucas Taylor, Ph.D., London, 1988. High-energy experimental physics.
Andrew Wells, Ph.D., MIT, 1989. Molecular biophysics.
Takahiro Yasuda, Ph.D., Tsukuba (Japan), 1983. High-energy experimental physics.

RESEARCH ACTIVITIES

Experimental Condensed-Matter Physics. Research activities focus on high-temperature superconductors (HTSC), including flux-lattice melting; Josephson-junction arrays; low-field HTSC magnets; linear and nonlinear electrodynamics of HTSCs; electromagnetic response of HTSCs at far infrared, microwave, and radio frequencies; the growth and characterization of new HTSC ceramics and single crystals; and the factors limiting critical currents. Surface studies concentrate on surface growth behavior and sliding friction measurements of atomic monolayer films. Other areas under investigation are electromagnetic and quantum chaos; electrorheological fluids; and Raman, FTIR, photoluminescence of multiple quantum well, and superlattice and single interface structures as a function of electric and magnetic fields, pressure, and temperature.

Experimental High-Energy Physics. The group is one of the principal participants in the design of a detector for the Superconducting Super Collider (SSC) in Texas. Subgroups are participating in the study of collisions of electrons and positrons up to 200 GeV at LEP I and LEP II at CERN, in a study of prompt photon production in π nucleus collisions (the role of gluons inside the pion) at Fermilab, in a measurement of the spin structure functions of neutrons and protons, in a particle astrophysics experiment (the Large Volume Detector) in the Gran Sasso tunnel (Italy), and in a search for exotic particles produced by 275-GeV electrons at Fermilab.

Molecular Biophysics. The group probes the structure and function of macromolecules, metalloproteins, and protein complexes. Specific research areas include electron transport, enzyme catalysis, and ligand binding and protein dynamics, using quasi-elastic scattering, transient absorption spectroscopy, Raman and fluorescence spectroscopy, Electron Spin-Echo Envelope Modulation, and Electron Paramagnetic Resonance.

Theoretical Condensed-Matter Physics. Research topics include transport theory, quantum chaos, Fermi liquid theory, charge density waves, and dense dipolar suspensions; and theory of Josephson junctions, catalytic properties of alloys, transport in nanostructures, structural phase transitions in DNA, nanotribology (atomic-level friction), electronic structure of disordered materials, magnetism, ferrites, Fermiology of HTSCs, exact and rigorous results in statistical mechanics, localization and percolation in order-disorder phase transitions, positron annihilation and photoemission spectroscopy, and pattern formation from solidification, hydrodynamic stabilities, and chemical waves.

Theoretical Elementary Particle Physics. Fundamental research includes the study of unified models based on superstrings; supersymmetric phenomenology, unified gauge theories in the TeV range, and precision calculations within and beyond the Standard Model; particle physics in the early universe; proton stability and neutrino masses; electroweak anomaly in the observed asymmetry of the baryon number, gravitational theory and quantum gravity, and computer simulations of topological structures in field theory; and finite temperature effects in quantum chromodynamics.

SECTION 7: PHYSICS

NORTHWESTERN UNIVERSITY
Department of Physics and Astronomy

Program of Study

The department offers a comprehensive program of course work and research leading to the degree of Doctor of Philosophy. A Master of Science degree may be obtained upon completion of a one-year program of courses; however, students are normally admitted only if they intend to complete the Ph.D. program and exhibit the ability to successfully do so. A candidate for the Ph.D. degree is required to complete 9 quarters of full-time registration (normally three years, but credits may be transferred from other graduate institutions), pass the departmental qualifying examination, prepare a thesis that presents the results of original research, and pass an oral examination based on this thesis.

The emphasis in the first year of graduate study is on giving the student a solid, fundamental background in mathematics, quantum mechanics, and electrodynamics. In the second year, students take the Ph.D. qualifying examination and continue with advanced course work in contemporary physics or astrophysics. All required course work is normally completed within the first two years, after which research with an adviser may begin.

Weekly departmental colloquia and topical seminars provide interaction and intellectual stimulation both within the department and with the wider scientific community.

Research Facilities

Resources are available for state-of-the-art research into a wide spectrum of physics and astrophysics topics. The library houses a full collection of reference books, journals, and reprints. The Vogelback Computing Center contains an IBM 4381, which can be used to connect with major computing centers across the nation, as well as a large number of microcomputers. Physics itself owns several superminicomputer systems. Technical support for the experimental programs is provided by a professionally staffed machine shop and a separate shop for do-it-yourself projects. Electronic equipment and construction facilities, including a printed-circuit laboratory, are also available. Solid-state facilities include equipment for electrical, magnetic, and thermal measurements at low temperatures, NMR spectrometers, thin-film evaporators, equipment for X-ray scattering studies, and crystal growth and orientation facilities. Experimental programs in nuclear and high-energy physics are supported by on-site data-analysis and detector-construction facilities. Both the Fermi National Accelerator Laboratory and the Argonne National Laboratory are within commuting distance of the Evanston campus and are used extensively by Northwestern's faculty and students.

Financial Aid

All applicants for admission are automatically considered for financial aid as teaching assistants. Students receiving assistantships are given a nine-month stipend ($9324 in 1991–92) plus full tuition. Teaching assistants spend 7 to 8 contact hours per week in laboratory teaching or other equivalent duties and approximately another 3 to 4 hours per week in preparation. Completed applications and Graduate Record Examinations scores must be submitted by February 1 to be considered for financial support. Students normally receive financial support during the summer.

Cost of Study

Tuition in 1991–92 is expected to be $14,000 for the three-quarter academic year. (Full tuition remission is offered with teaching assistantships, as noted above.) Books and supplies vary in cost, but a typical figure is $425 per year.

Cost of Living

In 1991, rental rates in the Evanston area ranged from $500 to $800 per month for a one-bedroom apartment and $750 to $1300 per month for a three-bedroom apartment, depending upon location and amenities. Single rooms in shared houses ranged from $300 to $550 per month. Campus dining facilities are available.

Student Group

There are currently 90 graduate students in physics and astronomy, and almost all receive full financial support. The undergraduate enrollment in the University is 7,300, with approximately equal numbers of men and women. The graduate enrollment on the Evanston campus of Northwestern is 4,000.

Location

The Evanston campus of Northwestern University stretches for a mile along the western shore of Lake Michigan. Evanston is the first suburb north of Chicago and is one of the most pleasant residential towns of the area. It has an excellent shopping center within walking distance of the campus and four lakefront parks with sandy beaches and picnic areas. Chicago, with its wide variety of shopping facilities, cultural activities, and entertainment, is easily reached by the elevated railroad running close to the campus. A wide variety of activities in the form of sports events, plays, concerts, and public lectures are an integral part of life at Northwestern University.

The University

Northwestern's geographical location enables the student to profit from the cultural advantages of a large city, the recreational opportunities of the local environment, and the quieter pace of town life when on campus. There is ample housing within a 20-minute walk of the campus.

Applying

Except in special cases, students are admitted only in September. Completed applications for the forthcoming year should be received by February 1. The General Test of the Graduate Record Examinations and the Subject Test in physics are required and should be taken early enough so that test scores are available at this time. TOEFL scores are required of candidates from non-English-speaking countries.

Correspondence and Information

Chairman, Graduate Admissions Committee
Department of Physics and Astronomy
Northwestern University
Evanston, Illinois 60208
Telephone: 708-491-3685

Peterson's Guide to Graduate Programs in the Physical Sciences and Mathematics 1992

SECTION 7: PHYSICS

Northwestern University

THE FACULTY AND THEIR RESEARCH

Bernard M. Abraham, Research Professor; Ph.D., Chicago, 1948. Experimental condensed-matter physics.
Paul R. Auvil, Associate Professor; Ph.D., Stanford, 1963. Theoretical particle physics.
John D. R. Bahng, Associate Professor; Ph.D., Wisconsin, 1957. Astronomy.
Martin H. Bailyn, Professor; Ph.D., Harvard, 1956. Theoretical condensed-matter physics and astrophysics.
Martin M. Block, Professor; Ph.D., Columbia, 1951. Experimental particle physics.
Eric A. Braaten, Associate Professor; Ph.D., Wisconsin, 1981. Theoretical particle physics.
Laurie M. Brown, Professor; Ph.D., Cornell, 1950. History of modern physics.
David A. Buchholz, Associate Professor; Ph.D., Pennsylvania, 1972. Experimental particle physics.
William Buscombe, Professor Emeritus; Ph.D., Princeton, 1950. Observational astrophysics.
Darwin Chang, Assistant Professor; Ph.D., Carnegie-Mellon, 1983. Theoretical particle physics.
Pulak Dutta, Associate Professor; Ph.D., Chicago, 1980. Experimental condensed-matter physics.
Donald E. Ellis, Professor; Ph.D., MIT, 1966. Theoretical condensed-matter physics.
Arthur J. Freeman, Professor; Ph.D., MIT, 1956. Theoretical condensed-matter physics.
Anupam Garg, Assistant Professor; Ph.D., Cornell, 1983. Condensed-matter theory.
Bruno Gobbi, Professor; Ph.D., Swiss Federal Institute of Technology, 1963. Experimental particle physics.
David Grabelsky, Research Assistant Professor; Ph.D., Columbia, 1985. Observational astrophysics.
William P. Halperin, Professor and Chairman; Ph.D., Cornell, 1974. Experimental condensed-matter physics.
Roderick L. Hines, Professor; Ph.D., Michigan, 1954. Experimental condensed-matter physics.
Joseph Keren, Associate Professor; Ph.D., Columbia, 1963. Experimental particle physics.
John B. Ketterson, Professor; Ph.D., Chicago, 1962. Experimental condensed-matter physics.
Liu Liu, Professor; Ph.D., Chicago, 1961. Theoretical condensed-matter physics.
David M. Meyer, Assistant Professor; Ph.D., UCLA, 1984. Observational astrophysics.
Donald H. Miller, Professor; Ph.D., Harvard, 1951. Experimental particle physics.
Robert J. Oakes, Professor; Ph.D., Minnesota, 1962. Theoretical particle physics.
John Peoples, Professor; Ph.D., Columbia, 1966. Experimental particle physics.
Jerome L. Rosen, Professor; Ph.D., Columbia, 1959. Experimental particle physics.
James A. Sauls, Associate Professor; Ph.D., SUNY at Stony Brook, 1980. Theoretical condensed-matter physics.
Heidi Schellman, Assistant Professor; Ph.D., Berkeley, 1984. Experimental particle physics.
Robert A. Schluter, Professor; Ph.D., Chicago, 1954. Experimental particle physics.
Arthur G. Schmidt, Senior Lecturer; Ph.D., Notre Dame, 1974. Nuclear physics.
Ralph E. Segel, Professor; Ph.D., Johns Hopkins, 1955. Nuclear physics.
Kamal K. Seth, Professor; Ph.D., Pittsburgh, 1957. Nuclear physics.
Ronald E. Taam, Professor; Ph.D., Columbia, 1973. Theoretical astrophysics.
David Taylor, Lecturer and Assistant Chairman; Ph.D., Maryland, 1983. Experimental surface science.
Melville P. Ulmer, Professor; Ph.D., Wisconsin, 1970. Observational astrophysics.
George K. Wong, Professor; Ph.D., Berkeley, 1974. Experimental condensed-matter physics.
Horace P. Yuen, Professor; Ph.D., MIT, 1970. Theoretical quantum optics.
Farhad Yusef-Zadeh, Assistant Professor; Ph.D., Columbia, 1986. Observational astrophysics.

RESEARCH ACTIVITIES

Astronomy and Astrophysics. The observational astrophysics program applies X-ray, radio, optical, and gamma-ray astronomy to the study of neutron stars, black holes, clusters of galaxies, and the galactic center. High-resolution optical spectroscopy is being used to study the interstellar medium, primordial gas clouds between quasars, and the temperature of the cosmic microwave background. There is a laboratory for producing and evaluating the X-ray mirrors used in X-ray astronomy studies. Theoretical work includes the investigation of stellar structure and evolution, stellar hydrodynamics, high-energy astrophysics, and general relativity. Educational facilities at the Lindheimer Observatory include 40-inch and 16-inch Boller and Chivens Cassegrain reflectors, as well as associated scanners and spectrographs. Other facilities include a historic 18½-inch Alvan Clark refractor at the Dearborn Observatory.

Experimental Condensed-Matter Physics. Major areas of interest include high-temperature superconductivity; superconducting artificial superlattices; magnetic, metallic, and semiconducting superlattices produced by MBE; structure and thermodynamics of monomolecular organic films; structure, magnetic properties, and nonlinear optical behavior of Langmuir-Blodgett films; magnetic and electrical properties of highly one-dimensional molecular crystals; quantum-size effects in metallic clusters and catalysts; collective excitations in superfluid He-3 at ultralow temperatures; physical characteristics of porous media, ceramics, and sandstones; and atomic and ionic diffusion in superionic conductors. There are laboratory facilities for measurements using X-ray, optical, ultrasonic, very low temperature, magnetic, and magnetic-resonance techniques; for production of thin organic films; molecular-beam epitaxial growth of semiconductors and metal films; and electron microscopy.

Experimental Nuclear Physics. An extensive program of research in medium-energy nuclear and particle physics is pursued at national accelerator facilities at Argonne, Los Alamos, Fermilab, SLAC, MIT, and Indiana University. These include the study of proton- and pion-induced reactions at the 800-MeV high-intensity linear accelerator at the Los Alamos Scientific Laboratory, antiproton-induced reactions at Fermilab, electron-induced reactions at the 1-GeV linac-storage ring at the Bates Laboratory of MIT and the 25-GeV linac at SLAC, and proton- and deuteron-induced reactions at the cyclotron at Indiana University.

Experimental Particle Physics. The elementary particles and their interactions are extensively studied, using the most advanced detection equipment. This equipment includes scintillation and Cerenkov counters, multiwire proportional and drift chambers, fast processing electronics, and on-line computers. Various experiments are being carried out at the Berkeley Bevelac and at the 800-GeV proton accelerator and the 1.6 TeV proton and antiproton collider at the Fermi National Accelerator Laboratory, which is within an hour's drive of the Evanston campus.

Theoretical Condensed-Matter Physics. Members of the condensed-matter theory group are actively studying the electronic structure of matter, many-body theory, and statistical physics. Current research in electronic structure includes calculations of the band structures and Fermi surfaces of metals and alloys and the magnetic properties of transition metals. Considerable research effort is devoted to calculations of the electronic properties of the high-temperature oxide superconductors, as well as the formulation and testing of models for the superconductivity mechanism. The many-body theorists are working on the theory of nonequilibrium superconductivity and superfluidity in quantum liquids (He-3 and He-4), as well as developing models for the effective interactions between elementary excitations in strongly interacting systems. There is also research in the area of statistical mechanics of phase transitions in systems exhibiting novel-broken symmetries (e.g., liquid crystals, exotic superconductors) and of transitions in reduced spare dimensions (e.g., monomolecular films, membranes, magnetic chains).

Theoretical Particle Physics. Fundamental research is being carried out with the aim of understanding the intrinsic properties and interactions of elementary particles. Implications of the standard $SU(3) \times SU(2) \times U(1)$ gauge theory of the strong, electromagnetic, and weak interactions are being explored using various techniques, including short-distance expansion, the large N limit, and perturbation theory. Investigations range from fundamental studies of the origin of CP nonconservation to more general phenomenological calculations based on effective low-energy, nonlinear Lagrangians.

SECTION 7: PHYSICS

OHIO UNIVERSITY

Department of Physics

Programs of Study

The Department of Physics offers graduate study and research programs leading to the Master of Arts, Master of Science, and Doctor of Philosophy degrees. The program of study emphasizes individual needs and interests in addition to essential general requirements of the discipline. Major areas of current research are experimental and theoretical nuclear and intermediate-energy physics, experimental condensed-matter and surface physics, theoretical condensed-matter and statistical physics, nonlinear dynamics and chaos, acoustics, planetary physics, special and general relativity, relativistic astrophysics, and science history and biography.

A student typically takes core courses during the first two years and becomes familiar with the professors in the department and their research. The Ph.D. comprehensive examination is taken in the spring of the second year. Students can usually retake the examination one year later if necessary. Master's degrees require completion of 45 graduate credits in physics and approved electives and have both thesis and nonthesis options. The Ph.D. requirements include passing the comprehensive exam, writing a dissertation, and orally defending the dissertation.

Research Facilities

The physics department occupies two wings of Clippinger Laboratories, a modern, well-equipped research building; the Edwards Accelerator Building, which contains Ohio University's 9-MeV high-intensity tandem accelerator; and the Surface Science Research Laboratory, which is isolated from mechanical and electrical disturbances. Specialized facilities for measuring structural, thermal, transport, optical, and magnetic properties of condensed matter are available. In addition to research computers in laboratories, students have free high-speed interactive access to minicomputers, three Ohio University mainframe computers (two IBM 4381s and one VAX 6440), and the CRAY Y-MP at the Ohio Supercomputer Center.

Financial Aid

Financial aid is available in the form of graduate assistantships (GAs), teaching assistantships (TAs), and research assistantships (RAs). All cover the full cost of tuition plus a stipend from which a quarterly fee of about $209 must be paid by the student. Current stipend levels for GAs are $9000 for the academic year plus $2000 for the summer; for TAs, $10,000 for the academic year plus $2000 for the summer. The stipend levels for RAs are set by the research grant holders but are at or above the level of the TA stipends. Both GAs and TAs require approximately 15 hours per week of laboratory and/or teaching duties. Special assistantships through the Condensed Matter and Surface Science (CMSS) program and tuition-only scholarships are also available.

Cost of Study

Tuition and fees are $1110 per quarter for Ohio residents and $2225 per quarter for out-of-state students. Tuition and fees for part-time students are prorated.

Cost of Living

On-campus rooms for single students range from $492 to $732 per quarter, while married student apartments cost from $295 to $495 per month. A number of off-campus apartments and rooms are available at various costs.

Student Group

About 18,000 students study on the main campus of the University, and about 2,500 of these are graduate students. The graduate student enrollment in the physics department ranges from 50 to 60.

Location

Athens is a city of about 25,000, situated in the rolling Appalachian foothills of southeastern Ohio. The surrounding landscape consists of wooded hills rising about the Hocking River valley, and the area offers many outdoor recreational opportunities. Eight state parks lie within easy driving distance of the campus and are popular spots for relaxation. The outstanding intellectual and cultural activities sponsored by this diverse university community are pleasantly blended in Athens with a lively tradition of music and crafts.

The University and The Department

Ohio University, founded in 1804 and the oldest institution of higher education in the Northwest Territory, is a comprehensive university with a wide range of graduate and undergraduate programs. The Ph.D. program in physics began in 1959, and more than 150 doctoral degrees have been awarded. Currently, the department has 23 regular faculty members, and additional part-time faculty and postdoctoral fellows. Sponsored research in the department amounts to over $1-million per year and comes from NSF, DOE, the Electric Power Research Institute, Battelle, and the state of Ohio.

Applying

Application forms for admission and for financial assistance may be obtained by writing to the Office of Graduate Students Services or to the Department of Physics. The deadline for assistantship and scholarship applications is April 1. Further information can be obtained by writing to one of the addresses given below.

Correspondence and Information

Professor Darrell O. Huwe
Graduate Appointments Committee
Department of Physics
Ohio University
Athens, Ohio 45701
Telephone: 614-593-1720

or

Professor Louis E. Wright
Chair, Department of Physics
Ohio University
Athens, Ohio 45701
Telephone: 614-593-1713

Peterson's Guide to Graduate Programs in the Physical Sciences and Mathematics 1992

SECTION 7: PHYSICS

Ohio University

THE FACULTY AND THEIR RESEARCH

Distinguished Professors
Raymond O. Lane, Ph.D., Iowa State, 1953. Nuclear physics.
Jacobo Rapaport, Ph.D., MIT, 1963. Nuclear and intermediate-energy physics.

Professors
Ernst Breitenberger, Dr.phil., Vienna, 1950; Ph.D., Cambridge, 1956. Theoretical physics, science history and biography.
Ronald L. Cappelletti, Ph.D., Illinois, 1966. Solid-state physics, superconductivity, mixed conductors, glasses, magnetism.
Charles C. Chen, Ph.D., Maryland, 1962. Solid-state theory, statistical mechanics.
James P. Dilley, Ph.D., Syracuse, 1963. Planetary physics.
Roger W. Finlay, Ph.D., Johns Hopkins, 1962. Nuclear and intermediate-energy physics.
Steven M. Grimes, Ph.D., Wisconsin–Madison, 1968. Nuclear physics.
Earle R. Hunt, Ph.D., Rutgers, 1962. Solid-state physics, magnetic resonance, surface physics.
Davis S. Onley, D.Phil., Oxford, 1960. Nuclear theory, electrodynamics.
Roger W. Rollins, Ph.D., Cornell, 1967. Solid-state physics, superconductivity, chaotic systems.
Edward R. Sanford, Ph.D., Iowa State, 1959. Solid-state physics.
Folden B. Stumpf, Ph.D., IIT, 1956. Acoustics, ultrasonics.
Louis E. Wright, Ph.D., Duke, 1966; Chair of the Department. Nuclear theory, electrodynamics, intermediate-energy theory.
Seung S. Yun, Ph.D., Brown, 1964. Physical acoustics, ultrasonics.

Associate Professors
Charles E. Brient, Ph.D., Texas at Austin, 1963. Nuclear physics, surface physics.
Darrell O. Huwe, Ph.D., Berkeley, 1964. High-energy physics.
David C. Ingram, Ph.D., Salford (England), 1980. Ion implantation, surface physics.
Sergio E. Ulloa, Ph.D., SUNY at Buffalo, 1984. Theoretical condensed-matter physics.

Assistant Professors
Charlotte Elster, Ph.D., Bonn, 1986. Nuclear and intermediate-energy theory.
Kenneth H. Hicks, Ph.D., Colorado, 1984. Nuclear and intermediate-energy physics.
Martin E. Kordesch, Ph.D., Case Western Reserve, 1984. Surface physics.
Prasun K. Kundu, Ph.D., Rochester, 1981. Relativistic cosmology, general relativity.

RESEARCH ACTIVITIES

Acoustics. Experimental studies of semiconductors, glasses, liquid mixtures, polymer solutions, and particle suspensions, using ultrasonics.

Condensed-Matter and Surface Science. Current projects include study and fabrication of high-temperature superconductors, optical properties of amorphous materials, scanning tunneling microscope (STM) images of graphite intercalation and dichalcogenide compounds, fabrication of diamond and diamondlike films and their characterization via novel electron microscopes, and theoretical study of electronic states in semiconductor superlattices. Some of these projects in physics are being pursued as an interdisciplinary effort with the Department of Chemistry and several engineering departments through the CMSS program.

Nonlinear Systems and Chaos. Investigations in this exciting new area of physics are being carried out from both theoretical and experimental points of view. Recent projects include experimental study of universality at the transition from quasi-periodicity to chaos in a coupled-diode resonator, numerical studies of fractal basin boundaries, and study of models that show chaotic behavior and simulate various phenomena such as environmentally assisted fracture and surface chemical reactions. Solitons, resonances, and stability problems are also being studied.

Nuclear and Intermediate Energy. Theoretical and experimental investigations of phenomena on the femtometer (10 to the −15 meter) length scale play a large role in this program. Included are conventional nuclear physics at the larger end of the scale and quark effects in nuclei at the smaller part of the scale. Various probes are used in these investigations, including electrons, photons, protons, neutrons, and heavier ions. Experimental work is performed at Ohio University's 9-MeV tandem and at intermediate-energy facilities at Indiana University, Los Alamos National Laboratory, and TRIUMF. Theoretical studies of nuclear structure, few-body systems, meson photoproduction, and electron scattering are under way, and a number of investigations are carried out in anticipation of the new electron scattering facility CEBAF.

Planetary Science. Topics under investigation are the evolution of atmospheres, effects of large collisions, and the origin of satellite systems.

Theoretical Physics. In the field of gravitational physics, research is concentrated in two distinct areas—classical general relativity and quantum field theory in curved space-time—with investigations of new, physically interesting exact solutions of the field equations and quantum effects near a black hole. Another general area of interest is what might be termed "fundamentals of physics," in which topics such as special relativity, generalizations of Maxwell's theory, the concept of identity, and uncertainty relations have been examined.

OREGON STATE UNIVERSITY

Department of Physics

Programs of Study

Oregon State University offers courses and research experience leading to the M.S. and Ph.D. degrees in physics. Thesis research may be pursued in experimental or theoretical atomic, nuclear, optical, particle, or solid-state physics. Interdisciplinary programs offer joint degrees with the Department of Electrical and Computer Engineering and the Center for Advanced Materials Research.

The master's degree is awarded following the satisfactory completion of 45 term (quarter) hours of credit and satisfactory performance on written and oral comprehensive examinations. A thesis is optional, and there is no foreign language requirement.

The Ph.D. degree is granted primarily for attainment and proven ability in research. A core curriculum of physics graduate courses is required, and the program must include a minimum of 36 term (quarter) hours in residence and the equivalent of at least three years of full-time graduate study beyond the bachelor's degree. For advancement to candidacy the student must demonstrate a broad working knowledge of physics by achieving a sufficiently high standard of performance in a written comprehensive examination and an oral examination. The most important part of the Ph.D. program is an original investigation in the field of the candidate's choice, under the direction of a member of the physics faculty. This investigation forms the basis for the student's doctoral thesis. The final requirement is an oral defense of the thesis.

Research Facilities

An excellent modern building houses the physics department's offices, shops, and research facilities. A solid-state laboratory for the study of magnetic properties includes a Faraday susceptibility apparatus, a pulsed NMR spectrometer, research electromagnets, minicomputers, and sample-preparation facilities. Properties of ceramics and high-temperature superconductors are studied using subnanosecond perturbed angular correlation, Mössbauer-effect, and positron-annihilation spectrometers. A new laboratory has been constructed for the study of superconducting thin films. Laser cooling and atom trapping studies are being done using diode lasers. The scattering of atoms in highly excited Rydberg states is studied by a pulsed-beam time-of-flight spectrometer with a field-ionization detector. A picosecond laser system in a new modern optics laboratory is the basis for research in ultrafast dynamics and nonlinear optical phenomena. Lasers and UHV equipment are used in studies of the structure and dynamics of thin films and interfaces. The nuclear physics group is running experiments at the Los Alamos and TRIUMF meson facilities and at the Oak Ridge heavy-ion facility. Much of the nuclear instrumentation is assembled on campus, including scintillation counters, multiwire proportional chambers, targets, and scattering chambers. Ultralow-temperature nuclear alignment experiments are under way in other laboratories of the physics department. The department maintains a machine shop and an electronics shop, which are well equipped and competently staffed. Three departmental superminicomputers, a link to a Cray system, and several workstation clusters support computational activities in theoretical nuclear and solid-state physics.

Financial Aid

The majority of the graduate students in physics receive financial support through teaching or research assistantships and fellowships. Assistants receive $8100 to $9000 for the 1991–92 academic year and are exempt from tuition. Several teaching fellowships are available; these pay up to $9000. Qualified applicants are automatically considered for these fellowships. Summer appointments are also available.

Cost of Study

Full-time annual tuition in 1991–92 is $3087 for Oregon residents and $4737 for out-of-state residents. However, tuition is waived for students receiving assistantships and teaching fellowships. Incidental and health service fees are approximately $135 per term.

Cost of Living

The University maintains a special dormitory for students over 21 years of age, with an annual charge in 1991–92 of $2590–$3689 for board and room, and cooperative living units costing approximately $2000 per year. Married students' housing is available at $135 per month for single-bedroom apartments, $185 for two-bedroom units, and $275 for three-bedroom units, including stove, refrigerator, and minimal furnishings. Many privately owned apartments and houses are also available for rent.

Student Group

Approximately 13,500 undergraduate and 2,600 graduate students are enrolled on the campus. In physics there are about 70 graduate students, 25 percent of whom are women.

Location

Corvallis is a university community of 41,570, located in an attractive natural setting on the Willamette River, 80 miles south of Portland. About an hour's drive to the west, through the scenic Coast Range, lies the Pacific Ocean, with hundreds of miles of open unspoiled coast. About the same distance to the east are the Cascade Mountains, which provide first-rate fishing, skiing, and hiking amid superb scenery.

The University and The Department

A congenial academic atmosphere and an accessible faculty contribute to the appeal of the graduate physics program at Oregon State University. The department maintains a vigorous colloquium program in which well-known physicists present lectures on current research. Students are also invited to participate in topical seminars, offered regularly in each of the major research areas, for the discussion of research results and for studies of specialized subjects at an advanced level.

Founded in 1868, Oregon State University has earned an excellent reputation for its graduate and undergraduate programs in the sciences. The Department of Physics, strengthened by recent additions to the faculty, is actively seeking applications from able students.

Excellent facilities for sports and recreation are available on campus. Nearby University forestlands offer miles of trails for jogging and exploring.

Applying

February 1 is the usual deadline for financial aid applications, but later awards are often possible. Oregon State University is an equal opportunity employer and welcomes students of all racial and ethnic backgrounds.

Correspondence and Information

Dr. Kenneth S. Krane, Chairman
Department of Physics
Weniger Hall 301
Oregon State University
Corvallis, Oregon 97331-6507
Telephone: 503-737-4631

SECTION 7: PHYSICS

Oregon State University

THE FACULTY AND THEIR RESEARCH

Atomic, Molecular, and Optical Physics
Charles W. Drake, Professor; Ph.D., Yale, 1958. Beam-foil spectroscopy, fundamental symmetries.
Clifford E. Fairchild, Professor; Ph.D., Washington (Seattle), 1962. Electrooptics, laser optics.
Peter R. Fontana, Professor; Ph.D., Yale, 1960. Radiative processes, theoretical atomic physics, optical information processing.
William M. Hetherington III, Associate Professor; Ph.D., Stanford, 1977. Nonlinear optics, chemical physics, surface physics.
Carl A. Kocher, Associate Professor; Ph.D., Berkeley, 1967. Highly excited states, atomic collisions, surface physics.
David H. McIntyre, Assistant Professor; Ph.D., Stanford, 1987. Laser spectroscopy, diode lasers, laser cooling.

Condensed-Matter Physics
John A. Gardner, Professor; Ph.D., Illinois, 1966. Hyperfine studies (NMR, Mössbauer, PAC) of ceramics, semiconductors, and superconductors; high-temperature and high-pressure phenomena.
David J. Griffiths, Professor; Ph.D., British Columbia, 1965. Hypervelocity impact in under-dense media.
Henri J. F. Jansen, Associate Professor; Ph.D., Groningen (Netherlands), 1981. Theoretical solid-state physics.
Janet Tate, Assistant Professor; Ph.D., Stanford, 1987. High-temperature superconductors, thin films.
William W. Warren, Professor; Ph.D., Washington (St. Louis), 1965. Solid-state physics, nuclear magnetic resonance.
Allen L. Wasserman, Professor; Ph.D., Iowa State, 1963. Theoretical solid-state physics.

Nuclear and Particle Physics
Kenneth S. Krane, Professor and Chairman; Ph.D., Purdue, 1970. Nuclear spectroscopy, nuclear orientation, heavy-ion reactions, angular correlations.
Rubin H. Landau, Professor; Ph.D., Illinois, 1970. Scattering theory, few-body problems, computational physics.
Victor A. Madsen, Professor; Ph.D., Washington (Seattle), 1961. Nuclear inelastic scattering theory, charge exchange reactions.
Corinne A. Manogue, Assistant Professor; Ph.D., Texas, 1984. Superstring theory, field theory in curved space-time.
Philip J. Siemens, Professor; Ph.D., Cornell, 1970. Nuclear theory, exotic nuclear matter.
Albert W. Stetz, Professor; Ph.D., Berkeley, 1968. Intermediate-energy tests of symmetry principles.
L. Wayne Swenson, Professor; Ph.D., MIT, 1960. Interactions of pions and protons with nuclei.

Weniger Hall, the physics building.

Recreational opportunities abound in the Cascades.

An atomic physics laboratory, with Dr. Carl Kocher and a graduate student.

SECTION 7: PHYSICS

PENNSYLVANIA STATE UNIVERSITY

Eberly College of Science
Department of Physics

Program of Study

The graduate program in physics is intended primarily for those studying for the Ph.D. degree. The main areas of departmental research, both experimental and theoretical, are in condensed matter, elementary particles and fields, and atomic-molecular-optical physics. The first year of graduate study is spent mainly on course work in mathematical methods, classical mechanics, electromagnetism, quantum mechanics, and statistical mechanics and in graduate laboratory. Formal course work is augmented by a large and diverse colloquium and seminar program featuring physicists primarily from other universities and laboratories but also from within the department. At the end of the first year a candidacy exam based on the first-year course material is given. The second year consists of elective advanced course work directed toward specialization together with entry into research. A research area, adviser, and Ph.D. committee are chosen by the end of the second year. An oral comprehensive exam is administered by the committee before the end of the first semester of the third year. The third and subsequent years are devoted primarily to research. Ultimately, a written thesis embodying the student's research is presented to the committee and defended by the student in an oral examination. The total duration of graduate study is variable but generally takes around 5½ years. Graduate study leading to the M.S. degree is also possible with two options available: a thesis submitted to the research adviser and to the Graduate School or a departmental paper submitted only to the research adviser. Either option takes about two years, with the first year devoted to course work and the second year to research.

Research Facilities

The Department of Physics is located in Davey Laboratory and in the adjoining Osmond Laboratory, with additional research facilities available at the Materials Research Laboratory located elsewhere on campus. Extensive state-of-the-art research equipment exists at these facilities, including apparatus for thin-film preparation by sputtering and molecular-beam evaporation; photoelectron spectrometers; numerous pulsed and continuous lasers; several cryostats operating between 77 K and 5m K; atomic-scale microscopes including STM, FIM, and FEM; LEED apparatus; in situ UHV ellipsometry apparatus; and atomic beam scattering apparatus. Extensive materials preparation and characterization facilities exist at the Materials Research Laboratory. Additional research in condensed-matter physics is also carried out at the Intense Pulsed Neutron Source at Argonne National Laboratory. The experimental high-energy physics program is centered in the Laboratory for Elementary Particle Science located in the basement of Osmond Laboratory. High-energy experiments are carried out at various national and international facilities: Brookhaven National Laboratory, Fermilab, and DESY in Hamburg, Germany. The high-energy group is also a member of the Solenoidal Detector Collaboration (SDC) for the Superconducting Super Collider. An extensive Physical Sciences Library is located in Davey Lab on one of the three floors occupied by the physics department. Computer facilities include the University's Center for Academic Computing featuring an IBM ES 3090-600S supercomputer, the departmental Sun and VAX systems, and numerous SPARCstations and PC-level computers.

Financial Aid

All students accepted for graduate study in the Ph.D. program are given an assistantship, generally a teaching assistantship. The teaching assistantship stipend is currently $9320 per academic year plus a tuition waiver. Teaching assistantship duties involve approximately 5 contact hours per week plus preparation. Summer support is offered for each of two summers after which the student is expected to obtain summer support in the form of a research assistantship from his or her research mentor. Limited summer teaching assistantship support is available. Full research assistantship support is available after the first or second year. Stipends are comparable with the teaching assistantship stipend. All students are automatically considered for supplementary fellowships ranging from $2000 to $3500. Entry-level M.S. students are not guaranteed support.

Cost of Study

Tuition rates for 1991–92 are $4450 for full-time in-state students and $8890 for full-time out-of-state students. Graduate tuition for part-time students is $187 per credit in-state and $370 per credit out-of-state.

Cost of Living

There is on-campus graduate housing for single students ($900–$1125 per semester) and for married students ($210–$320 per month, availability uncertain). Most students live in off-campus apartments with rents ranging from $350 to $500 per month for single occupancy. Basic living expenses are estimated at $6000–$7000 per year for a single student. Students on assistantship can obtain health insurance for $81 per year.

Student Group

There are currently about 38,780 total and 6,440 graduate students at Penn State with 80 graduate students in physics. Approximately half of the physics graduate students are international, and 15 percent are women. A typical incoming class consists of 15 students.

Location

Penn State is located at University Park in the middle of Pennsylvania. The mountainous, rural setting of the very pleasant surrounding countryside provides ample opportunity for outdoor recreation. The adjacent community of State College and contiguous townships have a population of about 75,000. Major cities are within several hours' driving time: Pittsburgh (3 hours), Philadelphia and Baltimore (3½ hours), Washington (4 hours), and New York (5 hours).

The University

Penn State is a land-grant institution founded in 1855. In addition to the University Park campus there are eighteen commonwealth campuses throughout Pennsylvania, which primarily offer the first two years of undergraduate education in a local community setting.

Applying

Requests for an application package (consisting of a brochure, financial assistantship application form, letter of recommendation forms, and explanatory information) should be sent directly to the Department of Physics. Both the General Test and the physics Subject Test of the GRE are required by the department. Average GRE scores for domestic students in the current incoming class are verbal, 86th percentile; quantitative, 89th percentile; analytical, 89th percentile; and physics, 62nd percentile. The average college GPA is 3.4. There is no deadline, but it is preferred that applications be received before February 15. At a later stage, formal application must be made also to the Graduate School for a fee of $35.

Correspondence and Information

Chairman, Committee on Graduate Admissions in Physics
104 Davey Laboratory
The Pennsylvania State University
University Park, Pennsylvania 16802

Telephone: 814-865-7534 (departmental secretary for graduate admissions)

Peterson's Guide to Graduate Programs in the Physical Sciences and Mathematics 1992

Pennsylvania State University

THE FACULTY AND THEIR RESEARCH
J. Annett, J. Banavar, G. R. Barsch, M. H. W. Chan, M. W. Cole, J. Collins, R. Collins, P. H. Cutler, R. Diehl, W. Ernst, T. E. Feuchtwang, G. N. Fleming, R. Graetzer, H. Grotch, M. Gunaydin, S. F. Heppelmann, R. M. Herman, E. Kazes, B. R. F. Kendall, J. S. Lannin, O. Lechtenfeld, S. Liang, J. D. Maynard, B. Y. Oh, J. Pliva, R. W. Robinett, G. A. Smith, P. Sokol, T. T. Tsong, K. Vedam, J. J. Whitmore, R. F. Willis.

Several faculty members at the various commonwealth campuses are also members of the Graduate Faculty and offer further opportunities for research. In addition there are currently 10 postdoctoral research associates and a senior scientist involved in departmental research.

EXPERIMENTAL RESEARCH ACTIVITIES
Acoustics. Sound propagation in superfluid helium, acoustical holography, nonlinear acoustics. (Maynard)

Atomic-Molecular-Optical Physics. High-resolution infrared spectroscopy, laser and microwave spectroscopy; molecular-beam techniques; nonlinear optics. (Ernst, Pliva)

Biophysics. Cellular mechanisms for repair of damaged DNA. (Graetzer)

Elementary-Particle and High-Energy Physics. Studies of the properties of quarks and gluons in hadronic interactions, precise measurements of masses and widths of charmonium states, low-energy antiproton annihilation in nucleons and nuclei, studies of the electroweak current by high-energy scattering, studies of nuclear color transparency with hard hadronic elastic collisions. (Heppelmann, Oh, Smith, Whitmore)

Low-Temperature Physics. Sound modes in liquid helium, critical phenomenon, phase transitions of adsorbed molecular films, superfluidity in reduced dimensionality and at millikelvin temperatures, wave propagation in random media, momentum distributions in quantum liquids and solids using deep inelastic neutron scattering, highly polarized liquid helium 3, thermodynamic properties of high-T_c superconductors and quasicrystals. (Chan, Maynard, Sokol)

Materials Physics. Elastic, dielectric, and optical properties of solids at high pressure; mechanical and optical properties of metal matrix composites; electrical and optical properties and X-ray diffraction of thin films, surfaces, and interfaces; thin-film sputtering phenomena; relaxation and amorphous semiconductors; field-ion microscopy; study of metals, piezoelectric, and ferroelectric materials; structural phase transitions in martensitic materials. (Barsch, R. Collins, Lannin, Tsong, Vedam, Willis)

Solid-State and Surface Physics. Raman spectroscopy; transport properties; band structure of transition metals; superlattice structures; phonons in clusters, amorphous and liquid semiconductors; investigation of adsorption-desorption phenomenon; interaction between surface atoms, random walk of atoms on surfaces, and structural and composition analysis of surface layers with the atom-probe microscope and the atom-probe field-ion microscope; diffraction of atomic beams by surfaces; photoemission; optical properties; inelastic neutron scattering; high-resolution low-energy electron diffraction; surface defects; structural studies of adsorbed overlayers; phase transitions of adsorbates; liquid-metal ionization; laser–solid surface interactions; scanning tunneling microscopy; magnetic and high-T_c superconducting thin films. (Barsch, R. Collins, Diehl, Lannin, Tsong, Vedam, Willis)

Vacuum Physics. Absolute pressure measurements at high vacuum, ion optics, new types of nonmagnetic mass spectrometers, electron-induced desorption, properties of spacecraft materials, levitation and ion trapping in vacuum. (Kendall)

THEORETICAL RESEARCH ACTIVITIES
Acoustics. Computational techniques for boundary value problems. (Maynard)

Atomic-Molecular-Optical Physics. Application of quantum electrodynamics to fine and hyperfine structure of simple systems; quantum mechanical coherence; nonlinear optics; interaction of atoms, molecules, and condensed matter with electromagnetic fields; liquid crystals. (Grotch, Herman)

Condensed-Matter Physics. Structural phase transitions, pattern formation, and shape-memory effect in martensitic materials; tunneling phenomena in normal metals and superconductors; electronic properties of surfaces and interfaces; contact and transport phenomena in submicron systems; lattice dynamics of metals; liquid metals and alloys; theory of the scanning tunneling microscope; field emission of electrons; photoemission; ion emission from solid and liquid metals; dynamical theory of particle-surface interaction; adsorption of atoms on surfaces; two-dimensional phase transitions; statistical mechanics; Hubbard model; high-T_c superconductivity; Monte Carlo simulation in hydrodynamics, solid-liquid-vapor systems, spin glasses, magnetic systems, and porous media. (Annett, Banavar, Barsch, Cole, Cutler, Feuchtwang, Kazes, Liang)

Elementary Particles and Fields. Quantum chromodynamics, quantum electrodynamics, electroweak interactions, collider and supercollider phenomenology, supersymmetry, superstring theory, conformal field theory. (J. Collins, Grotch, Gunaydin, Kazes, Lechtenfeld, Robinett)

Foundational Physics. Conceptual foundations of relativistic quantum theory. (Fleming)

SECTION 7: PHYSICS

RICE UNIVERSITY
Department of Physics

Program of Study

Rice University offers a program leading to the M.A. and the Ph.D. degrees in physics. The program consists of formal courses and original research conducted under the guidance of a faculty adviser. During the first academic year, the student concentrates on foundation course work. A thesis research area is chosen during the second semester. Instead of a written qualifying or comprehensive exam, each student is required to complete a master's thesis or a publishable research report and successfully defend it in an oral examination. This defense, plus at least a B average in graduate physics courses, is required for the Ph.D. candidacy. After the first year, research occupies the majority of the student's time and is conducted full-time during summers. Students usually complete the Ph.D. program in an average of five years. There is no foreign language requirement.

Major areas of research include experimental and theoretical programs in atomic and molecular physics, biophysics, condensed-matter and surface physics, and nuclear and particle physics. Interdisciplinary work in many areas of applied physics can be undertaken with faculty members from other departments. Some recent examples are in the fields of chemical physics, materials science, surface physics, and space physics.

Research Facilities

Research labs on campus are located in the Physics Building, the T. W. Bonner Nuclear Laboratory Building, and the Space Physics Building. Faculty and students participate in experiments at national and international laboratories such as CERN, FNAL, BNL, LAMPF, SLAC, SIN, TRIUMF, ORNL, and CEBAF. Detectors and apparatus for high-energy physics are designed and constructed in the Bonner Nuclear Laboratory. The data analysis is done using a cluster of six VAXstations and two new, powerful IBM RISC workstations.

Atomic, molecular, surface, and condensed-matter experiments are done in numerous well-equipped physics research laboratories housed in the Space Physics Building. The equipment includes two ultrahigh-vacuum chambers equipped with LEED/Auger/RHEED diagnostics, thin-film evaporators, polarized-beam sources, and energy- and angle-resolved electron polarization measurement capabilities. An optically pumped flowing helium afterglow apparatus forms the basis of the Rice polarized electron source. There are three high-vacuum chambers dedicated to atomic/molecular beam studies. These chambers are equipped with beam sources, position-sensitive particle detectors, and state-of-the-art laser systems. There is a UHV system equipped for ion-surface scattering experiments to probe surface magnetism. The laser cooling group employs tunable ring dye lasers and state-of-the-art electrooptic devices to cool atoms in a thermal beam. Frequency-stabilized tunable laser diodes are used for spectroscopic analysis. In addition, there are scanning tunneling microscopes that possess subatomic resolution. All of the above systems have computer-based data acquisition, control, and analysis systems.

An Itel AS/9000 mainframe computer is linked to terminals in the physics department as well as to other laboratories. Numerous Macintoshes and IBM PCs are available to students. Through the central computer center, there is access to BITNET, ARPANET, and DECnet/HEPnet. Rice is one of the primary centers on the new NSFNET high-speed backbone system. Students also have access to the SX-2 supercomputer located at the Houston Area Research Center. Rice's Fondren Library contains an excellent science collection.

Financial Aid

Most students receive financial aid in the form of a stipend. For the academic year 1991–92, this is $11,000 for twelve months for M.A. candidates and $11,500 for Ph.D. candidates. Currently, all students receive a tuition waiver.

Cost of Study

Tuition for 1991–92 is $8300; however, most students receive tuition waivers. Health insurance and other fees amount to $555 per year.

Cost of Living

Graduate student housing is available adjacent to the campus at the Graduate House. Monthly rents range from $240 to $335 for singles and $150 to $185 per person for doubles. Inexpensive off-campus housing is available in the area.

Student Group

There are currently 55 graduate students in the Department of Physics. Of these, approximately 40 percent are foreign students. About 10–12 students are admitted each year. Past experience has shown that about 50 percent of an entering class successfully complete the Ph.D. program. The student-faculty ratio is slightly over 2:1. Employment opportunities have been good for the department's graduates; they typically find positions in academic institutions, national laboratories, industrial laboratories, or entrepreneurial enterprises.

Location

Set in a 300-acre academic park, the campus is near dynamic downtown Houston, but it is insulated by quiet residential neighborhoods and tall hedges. The campus is adjacent to the Texas Medical Center, where world-famous labs, hospitals, and medical schools offer added opportunities for collaborative research. Hermann Park, one of the nation's largest urban parks, featuring hundreds of acres of picnic grounds, a zoo, a golf course, and an outdoor theater offering operas, plays, and concerts, is nearby. Also within easy walking distance are the Museum of Fine Arts, the Contemporary Arts Museum, and the Museum of Natural Science. Opportunities are plentiful for boating, sailing, freshwater or deep-sea fishing, and ocean swimming and surfing.

The University

Rice University is a private, coeducational, nondenominational university founded in 1891 from the estate of William Marsh Rice. The University was opened in 1912 as Rice Institute; the name was changed in 1960. The University has faculties of liberal arts, science, engineering, administration, and music. Current enrollment is approximately 2,700 undergraduates and 1,300 graduate students. The academic year consists of two semesters of sixteen weeks each. No graduate classes are taught during the summer, but graduate research activity rolls into high gear during this time.

Applying

Application materials for admission may be obtained by writing to the address below. Completed applications, together with transcripts and four letters of recommendation, should be sent to the same address by February 1. The department requires official scores from the Graduate Record Examinations, including the physics Subject Test. International students for whom English is not the native language must submit an official score from the TOEFL.

Correspondence and Information

Chairman of the Graduate Committee
Department of Physics
Rice University
Houston, Texas 77251-1892
Telephone: 713-285-5313
Fax: 713-527-9033

SECTION 7: PHYSICS

Rice University

THE FACULTY AND THEIR RESEARCH

S. D. Baker, Professor; Ph.D., Yale, 1963. Experimental physics.
B. E. Bonner, Professor and Chairman of the Department; Ph.D., Rice, 1965. Experimental nuclear and elementary particle physics.
D. C. Chang, Adjunct Associate Professor of Biophysics; Ph.D., Rice, 1970. Experimental biophysics.
M. D. Corcoran, Associate Professor; Ph.D., Indiana, 1977. Experimental high-energy physics.
S. A. Dodds, Associate Professor; Ph.D., Cornell, 1975. Experimental solid-state physics, muon spin rotation.
I. M. Duck, Professor; Ph.D., Caltech, 1961. Theoretical nuclear physics, elementary particles and fields.
F. B. Dunning, Joint Professor of Physics and of Space Physics and Astronomy; Ph.D., University College, London, 1969. Experimental atomic and molecular physics, physics of surfaces.
T. L. Estle, Professor; Ph.D., Illinois, 1957. Experimental solid-state physics, muon spin rotation.
J. P. Hannon, Professor; Ph.D., Rice, 1967. Theoretical solid-state physics, coherent gamma-ray optics, cooperative phenomena.
C. F. Hazlewood, Adjunct Professor; Ph.D., Tennessee, Memphis, 1962. Experimental biophysics.
H. W. Huang, Professor; Ph.D., Cornell, 1967. Theoretical statistical physics; theoretical biophysics; X-ray, neutron, and optical spectroscopies on membrane biophysics.
R. G. Hulet, Assistant Professor; Ph.D., MIT, 1984. Experimental atomic and molecular physics, laser cooling and atom trapping.
M. Kimura, Adjunct Associate Professor; Ph.D., Alberta, 1981. Theoretical chemical physics.
N. F. Lane, Professor; Ph.D., Oklahoma, 1964. Theoretical atomic and molecular physics.
F. C. Michel, Joint Professor of Physics and of Space Physics and Astronomy; Ph.D., Caltech, 1962. Theoretical space physics and astronomy.
H. E. Miettinen, Associate Professor; Ph.D., Michigan, 1973. Experimental elementary particle physics.
G. S. Mutchler, Professor; Ph.D., MIT, 1966. Experimental nuclear and particle physics.
P. Nordlander, Assistant Professor; Ph.D., Chalmers (Sweden), 1985. Condensed-matter theory, electronic properties of surfaces.
C. Rau, Professor; Ph.D., Munich Technical, 1970; Dr.habil., Munich, 1980. Surface science, magnetism.
J. B. Roberts Jr., Professor; Ph.D., Pennsylvania, 1969. Experimental elementary particle physics.
H. E. Rorschach Jr., Professor; Ph.D., MIT, 1952. Experimental solid-state physics, biophysics (NMR and neutron scattering).
R. E. Smalley, Joint Professor of Physics and Chemistry; Ph.D., Princeton, 1973. Experimental molecular (cluster) physics.
R. F. Stebbings, Joint Professor of Physics and of Space Physics and Astronomy; Ph.D., London, 1956. Experimental atomic and molecular physics.
P. M. Stevenson, Associate Professor; Ph.D., Imperial College (London), 1979. Quantum field theory, elementary particle physics.
G. T. Trammell, Professor; Ph.D., Cornell, 1950. Theoretical physics.
G. K. Walters, Professor; Ph.D., Duke, 1956. Experimental atomic and molecular physics, physics of surfaces.

RESEARCH ACTIVITIES

Atomic and Molecular Physics. A diverse theoretical and experimental program exists in atomic and molecular scattering processes. Work is done on charge and excitation transfer reactions and in the collisions and spectroscopy of high-Rydberg atoms. A new laser cooling laboratory for exploring the behavior of ultracold atoms has recently been established at Rice. This is a new and largely unexplored domain of physics that may exhibit dramatic and previously unobserved quantum statistical effects.

New methods and techniques in the atomic physics of electron spin have yielded (as a spinoff!) the world's most intense source of polarized electrons.

Biophysics. The techniques of nuclear magnetic resonance and neutron scattering are used to study the properties of water in cells and its interactions with the cellular macromolecules, particularly the proteins. There is good reason to believe that the protein motion may be quite different in solutions as compared with crystals. This work has as its aim to understand this in detail.

Ion channels, the basic unit of bioelectricity, are studied, using ultraviolet circular dichroism, neutron scattering, and X-ray scattering. The ultimate goal of this study is nothing less than to understand the structure-function relations of the basic units in the nervous system.

Nuclear and Particle Physics. The High Energy and Medium Energy groups work closely together on a whole range of experiments. New understanding of how particles are produced will come from an experiment using the world's highest energy (200-GeV) polarized proton beam at Fermilab. The production of hadron jets by 500-GeV photon scattering from quarks and gluons residing in protons and nuclei will show whether the standard model (QCD) can explain these fundamental processes. A search for a new state of matter, the quark gluon plasma, is being carried out in antiproton-nucleus and heavy ion–nucleus collisions at the Brookhaven AGS. A future experiment will measure the electromagnetic decays of excited hyperons at CEBAF in order to ascertain their quark wave functions. In order to answer the question of just what carries the spin of hadrons—the quarks, the gluons, or the sea partons—an experiment will be done to measure the internal spin structure of protons and neutrons by deep inelastic muon scattering at CERN.

Theoretical research concentrates on nonperturbative methods in quantum field theory, using the Gaussian effective potential method elaborated by Rice theorists. Other areas of study are models of quark confinement and the structure of the nucleon.

Surface Physics and Condensed-Matter Physics. A powerful array of new spectroscopies, based on spin-polarized electron- and atomic-beam techniques, have been invented and developed for probing surface electronic, geometric, and magnetic structure. These spectroscopies include spin-polarized low-energy electron diffraction (SPLEED), secondary electron spectroscopy (SES), spin-polarized metastable deexcitation spectroscopy (SPMDS), and electron capture spectroscopy (ECS). Bulk magnetic properties are studied, using muon spin rotation.

Theoretical studies of the interaction of radiation with matter have led to several new concepts such as the X-ray Resonance Exchange Scattering, which probes the static and fluctuating magnetic moments in rare earths, and Nuclear Bragg scattering, which can be used to achieve highly monochromatic long-coherence-length X rays from synchrotron radiation.

SECTION 7: PHYSICS

RUTGERS, THE STATE UNIVERSITY OF NEW JERSEY, NEW BRUNSWICK

Department of Physics and Astronomy

Programs of Study The Department of Physics and Astronomy offers programs leading to the Ph.D., M.S., M.S.T. (Master of Science for Teachers), and M.Phil. (Master of Philosophy) degrees in physics. The program for the Ph.D. degree involves an appropriate combination of research and course work, including a few required courses. Most students should expect to obtain the Ph.D. degree in about five years. The qualifying examinations, including both written and oral sections, are normally completed by the beginning of the second year. No foreign languages are required.

The M.S. requires a minimum of 30 credit hours, of which 6 can be devoted to research. In addition to passing an oral examination, the candidate must present either a critical essay or a thesis on some research problem. Two years are normally required to complete the M.S. program. While almost all of the currently enrolled graduate students are in the Ph.D. program, the master's programs provide attractive and useful alternatives to students who wish to complete their advanced education more quickly.

Research Facilities Major research efforts are devoted to elementary particle physics, astrophysics, condensed-matter physics, and nuclear and atomic physics. Experimental facilities are located in a modern, fully equipped three-story research laboratory and include extensive equipment for low-temperature research, a new surface modification center, several VAX computers, and VAX and Sun workstations. Access to supercomputers is also available. Intermediate- and high-energy experiments are done at Fermi National Accelerator Laboratory, the Los Alamos Meson Physics Facility, Stanford Linear Accelerator Center, and KEK in Japan. Rutgers astrophysicists use the observatory facilities at Kitt Peak, Cerro-Tololo and Arecibo observatories, as well as data from the IUE and Einstein satellites.

Financial Aid Virtually all students receive financial support from the department. First-year students are generally awarded teaching assistantships with stipends of at least $9400 for the academic year, a waiver of tuition, and comprehensive health insurance. More advanced students generally have research assistantships, with a stipend of $10,810 to $11,040 for the calendar year and a waiver of tuition. Outstanding applicants may be eligible for fellowships of up to $12,000 for the calendar year. Summer jobs for teaching assistants sometimes are available with a research group, and, in some cases, research assistants may supplement their income with limited teaching assignments.

Cost of Study The tuition for 1990–91 was $4066 per year for New Jersey residents and $5960 per year for nonresidents and was waived for students with assistantships or internships. In 1990–91, fees were $350 per year.

Cost of Living Assistantships provide modest but adequate support for students. In 1990–91, the cost of rooms in University housing for single students ranged from about $2365 per academic year in a dormitory to about $2905 in a University apartment shared with three other students. The cost of University apartments for married students ranged from $301 to $455 per month.

Student Group About 95 full-time graduate students are currently enrolled in the department; almost all of these are supported by assistantships or fellowships. Students come from a number of foreign countries as well as from all parts of the United States.

Location The department is located in Rutgers' Science Center in Piscataway, a pleasant suburban community about 10 minutes from urban New Brunswick in central New Jersey. Rutgers is about 35 miles from New York City, 40 miles from the New Jersey ocean beaches, and 16 miles from Princeton. Academic life in New Brunswick is enriched by lecture series, films, and, in particular, an extensive program of high-quality musical events. Athletic and other recreational facilities are also available.

The University and The Department Rutgers was founded in 1766 and is now the State University of New Jersey. There are more than 48,000 students on six campuses. About 33,300 of these students, including about 4,500 graduate students, are in New Brunswick and Piscataway. The Department of Physics and Astronomy has grown significantly in recent years, with new faculty in condensed-matter theory, string theory, and surface science. The department currently receives research support from outside sources in excess of $4-million per year.

Applying All necessary forms for admission to the Graduate School and for appointment as an assistant or fellow may be obtained by writing to the address below. Applications are accepted until about July 15; applicants for financial aid are encouraged to apply before March 1. The GRE General Test and the Subject Test in physics are required. The Test of English as a Foreign Language (TOEFL) is required of students whose native language is not English.

Correspondence and Information
Dr. David Harrington
Physics Graduate Program Director
Rutgers University
P.O. Box 849
Piscataway, New Jersey 08855-0849
Telephone: 201-932-2502

SECTION 7: PHYSICS

Rutgers, The State University of New Jersey, New Brunswick

THE FACULTY AND THEIR RESEARCH

Professors
Allen B. Robbins, Chairperson; Ph.D., Yale, 1956. Experimental nuclear physics.
David Harrington, Associate Chairperson; Ph.D., Carnegie Tech, 1961. Theoretical nuclear physics.
Elihu Abrahams, Ph.D., Berkeley, 1952. Theoretical condensed-matter physics.
Natan Andrei, Ph.D., Princeton, 1979. Theoretical elementary particle/condensed-matter physics.
Thomas Banks, Ph.D., MIT, 1973. Theoretical elementary particle physics.
John B. Bronzan, Ph.D., Princeton, 1963. Theoretical elementary particle physics.
Herman Y. Carr, Ph.D., Harvard, 1953. Experimental condensed-matter physics.
Thomas Devlin, Ph.D., Berkeley, 1961. Experimental elementary particle physics.
Glennys Farrar, Ph.D., Princeton, 1970. Theoretical elementary particle physics.
Daniel Friedan, Ph.D., Berkeley, 1980. Theoretical elementary particle physics.
Charles Glashausser, Ph.D., Princeton, 1966. Experimental nuclear physics.
Gerald A. Goldin, Ph.D., Princeton, 1969. Quantum theory.
Torgny Gustafsson, D.Sc., Chalmers University of Technology (Sweden), 1973. Experimental condensed-matter physics.
George K. Horton, Ph.D., Birmingham (England), 1949. Theoretical condensed-matter physics.
Shirley Jackson, Ph.D., MIT, 1973. Theoretical condensed-matter physics.
Haruo Kojima, Ph.D., UCLA, 1972. Experimental condensed-matter physics.
Noemie B. Koller, Ph.D., Columbia, 1958. Experimental nuclear physics.
Theodore K. Kruse, Ph.D., Columbia, 1959. Experimental nuclear physics.
Antti Kupiainen, Ph.D., Princeton, 1979. Statistical mechanics theory, math physics.
David C. Langreth, Ph.D., Illinois, 1964. Theoretical condensed-matter physics.
Paul L. Leath, Provost; Ph.D., Missouri–Columbia, 1966. Theoretical condensed-matter physics.
Joel Lebowitz, Ph.D., Syracuse, 1956. Statistical mechanics theory, math physics.
Peter Lindenfeld, Ph.D., Columbia, 1954. Experimental condensed-matter physics.
Claud Lovelace, B.S., Cape Town (South Africa), 1954. Theoretical elementary particle physics.
Theodore Madey, Director, Surface Modification and Interface Dynamics Center; Ph.D., Notre Dame, 1963. Experimental surface science physics.
Aram Mekjian, Ph.D., Maryland, 1968. Theoretical nuclear physics.
Joe H. Pifer, Undergraduate Coordinator; Ph.D., Illinois, 1966. Experimental condensed-matter physics.
Richard J. Plano, Ph.D., Chicago, 1956. Experimental elementary particle physics.
T. Alexander Pond, Executive Vice President and Chief Academic Officer; Ph.D., Princeton, 1953. Experimental nuclear physics.
Ronald Rockmore, Ph.D., Columbia, 1957. Theoretical nuclear physics.
Joseph Sak, Ph.D., Institute of Solid-State Physics (Czechoslovakia), 1968. Theoretical condensed-matter physics.
Felix R. Sannes, Ph.D., McGill, 1968. Experimental elementary particle physics.
Robert Schommer, Professor; Ph.D., Washington (Seattle), 1977. Experimental astrophysics.
Nathan Seiberg, Ph.D., Weizmann (Israel), 1980. Theoretical elementary particle physics.
Joel Shapiro, Ph.D., Cornell, 1967. Theoretical elementary particle physics.
Stephen H. Shenker, Ph.D., Cornell, 1980. Theoretical elementary particle physics.
George Sigel, Ph.D., Georgetown, 1968. Fiber optics.
Michael Stephen, D.Phil., Oxford, 1955. Theoretical condensed-matter physics.
Georges M. Temmer, Ph.D., Berkeley, 1949. Experimental nuclear physics.
Terence Watts, Ph.D., Yale, 1963. Experimental elementary particle physics.
Larry Zamick, Ph.D., MIT, 1961. Theoretical nuclear physics.
Alexander Zamoldchikov, Ph.D., Institute of Theoretical and Experimental Physics (Moscow), 1978. Theoretical elementary physics.
Harold S. Zapolsky, Ph.D., Cornell, 1962. Theoretical astrophysics.

Associate Professors
Jolie Cizewski, Ph.D., SUNY at Stony Brook, 1978. Experimental nuclear physics.
Piers Coleman, Ph.D., Princeton, 1984. Theoretical condensed-matter physics.
Mark Croft, Ph.D., Rochester, 1977. Experimental condensed-matter physics.
Mohan S. Kalelkar, Ph.D., Columbia, 1975. Experimental elementary particle physics.
Willem M. Kloet, Ph.D., Utrecht (Netherlands), 1973. Theoretical nuclear physics.
B. Gabriel Kotliar, Ph.D., Princeton, 1983. Theoretical condensed-matter physics.
Terry A. Matilsky, Ph.D., Princeton, 1971. Experimental astrophysics.
Herbert Neuberger, Ph.D., Tel-Aviv, 1979. Theoretical elementary particle physics.
Ronald Ransome, Ph.D., Texas at Austin, 1981. Experimental nuclear physics.
Andrei E. Ruckenstein, Ph.D., Cornell, 1984. Theoretical condensed-matter physics.
Stephen R. Schnetzer, Ph.D., Berkeley, 1981. Experimental elementary particle physics.
Jeremy Sellwood, Ph.D., Manchester (England), 1977. Theoretical astrophysics.
Gordon Thomson, Ph.D., Harvard, 1972. Experimental elementary particle physics.
David Vanderbilt, Ph.D., MIT, 1981. Theoretical condensed-matter physics.
Theodore B. Williams, Ph.D., Caltech, 1974. Experimental astrophysics.

Assistant Professors
Eva Y. Andrei, Ph.D., Rutgers, 1980. Experimental condensed-matter physics.
Robert Bartynski, Ph.D., Pennsylvania, 1986. Experimental condensed-matter physics.
Michael Douglas, Ph.D., Caltech, 1988. Theoretical elementary particle physics.
Ronald Gilman, Ph.D., Pennsylvania, 1985. Experimental nuclear physics.
Lev Ioffe, Ph.D., London Institute for Theoretical Physics, 1985. Theoretical condensed-matter physics.
David R. Merritt, Ph.D., Princeton, 1982. Theoretical astrophysics.
Carlton Pryor, Ph.D., Harvard, 1982. Experimental astrophysics.

SECTION 7: PHYSICS

STATE UNIVERSITY OF NEW YORK AT BUFFALO
Department of Physics and Astronomy
Physics Program

Programs of Study

The Department of Physics and Astronomy offers programs in physics leading to the M.A. and Ph.D. degrees and provides opportunities for postdoctoral research.

The M.A. program is individually planned by the student along with his or her adviser and approved by the Graduate School. The student can choose either a thesis or nonthesis program leading to a master's degree in physics. A minimum of 30 credit hours is required in either option. The student is also required to pass Part 1 of the Qualifying Examination.

The student qualifies as a Ph.D. candidate by passing Parts 1 and 2 of the Qualifying Examination and demonstrating satisfactory performance in graduate-level courses, with an average grade of B or better. When a student has qualified, he or she selects an area of specialization. A faculty committee is then formed to advise the student in planning a course of study and research leading to the degree. A thesis, embodying the results of the student's research and judged by the department to be of publishable quality, must be produced. The Ph.D. program requires a minimum of 72 credit hours at the graduate level. Graduate students and faculty attend weekly colloquia, informal seminars on specialized topics, and special lecture series on current topics.

Research Facilities

The department offers opportunities for graduate study and research in condensed-matter physics, including semiconductors and magnetism; superconductivity; low-temperature physics; light scattering; atomic and molecular physics; elementary particles; statistical physics; solid-state theory; nuclear physics; general relativity; and the philosophy of physics. Excellent electronics shop, machine shop, and cryogenic facilities are available. The Department of Physics and Astronomy is housed in a modern building equipped with major new research facilities on the University's Amherst Campus.

An IBM 3084 mainframe and several DEC VAX computers with on-campus terminals, a reactor at the Western New York Nuclear Research Center, and biophysical facilities at Roswell Park are used by members of the department and their students. Faculty members and graduate students participate in interdisciplinary research as part of the Center for Electronic and Electro-optic Materials and the New York State Institute on Superconductivity.

Facilities that aid in research are maintained within the department, including a research and preprint library, minicomputers, a Sun Workstation, and terminals to access the University computers. The University also has a 2.3-million-volume library.

Financial Aid

Assistantships pay up to $8500 for the 1991–92 academic year and normally include exemption from tuition. Graduate School fellowships are available for superior candidates. Summer teaching assistantships are also available. The graduate teaching assistants supervise laboratory and recitation sections on a part-time basis and register for 9 hours of graduate study. Application for scholarships, loans, and similar forms of financial assistance should be made to the Graduate School or the Office of Financial Aid. More than 90 percent of the graduate students in the department receive some type of financial support.

Cost of Study

Tuition for 1991–92 for full-time graduate students is $1600 per semester for New York State residents and $3258 per semester for nonresidents. This charge is normally waived for assistantship holders. Tuition rates are subject to change without prior notice. Student fees are approximately $110 per semester. Health insurance is $502 per year for international students, $385 per year for domestic students.

Cost of Living

On-campus graduate housing is available; off-campus housing is an alternative. Costs vary, depending upon size, accommodations, and services available. Average yearly expenditure is estimated to be $8135.

Student Group

In 1990–91, the department had 89 full-time graduate students, of whom 65 were enrolled in the Ph.D. program and 24 in the master's program. The students come from all parts of the United States and many other countries. In 1989–90, the department awarded three doctoral degrees and thirteen master's degrees. Graduates find employment in government and private institutions and in college teaching and research.

Location

Buffalo is the second-largest city in New York State. The metropolitan area has a population of 1.2 million and is a commercial center for western New York. It provides all city conveniences and yet is surrounded by many beautiful parks including the Niagara Falls National Park. Buffalo is approximately 100 miles from Toronto and 15 miles from Niagara Falls. The city is located next to Lake Erie and the Niagara River, thus providing unusual opportunities for water sports. Downhill skiing areas are only an hour's drive from the city. Contrary to the common impression, Buffalo enjoys a climate similar to that of other large northeastern cities.

Many cultural opportunities within the Buffalo metropolitan area enhance the educational experience. The Albright-Knox Art Gallery has an impressive collection of modern art, and the Historical Museum and Buffalo Museum of Science are delightful to visit. The Buffalo Philharmonic Orchestra and the University's own School of Music and its Chamber Music Society provide regularly scheduled concerts and recitals. Buffalo's theater district offers an opportunity to enjoy live performances.

Buffalo is a major sports city well known for the Buffalo Sabres ice hockey, Buffalo Bills football, and Buffalo Bisons baseball teams.

The University

The State University of New York at Buffalo was founded in 1846 as the University of Buffalo. It is the largest single unit and most comprehensive graduate center of the State University of New York System, having a total of 27,000 students. The department is located on the Amherst Campus, a $1-billion site on a tract of 1,200 acres located approximately 3 miles north of the city.

Applying

Applications for admission to graduate study should be filed at least one month prior to the opening of the term for which admission is sought. Students should apply to the department by March 1 for teaching assistantships and January 15 for fellowships.

Correspondence and Information

Director of Graduate Studies
Department of Physics and Astronomy
237 Fronczak Hall—Amherst Campus
State University of New York at Buffalo
Buffalo, New York 14260
Telephone: 716-636-2017
Fax: 716-636-2507

SECTION 7: PHYSICS

State University of New York at Buffalo

THE FACULTY AND THEIR RESEARCH

Bruce D. McCombe, Professor and Chairman; Ph.D., Brown. Experimental condensed-matter physics.
Eric W. Beth, Assistant Professor Emeritus; Ph.D., Vienna. Quantum mechanics.
David J. Bishop, Adjunct Professor; Ph.D., Cornell. Experimental solid-state physics.
Stephen G. Bishop, Adjunct Professor; Ph.D., Brown. Experimental solid-state physics.
Lyle B. Borst, Professor Emeritus; Ph.D., Chicago. Neutron physics.
Gilbert O. Brink, Professor; Ph.D., Berkeley. Experimental atomic and molecular physics.
Daniel G. Caldi, Associate Professor; Ph.D., Rockefeller. Theoretical high-energy physics.
Michael G. Fuda, Professor; Ph.D., RPI. Theoretical nuclear physics.
Shigeji Fujita, Professor; Ph.D., Maryland. Statistical mechanics, many-body problems, transport coefficients.
Francis M. Gasparini, Professor; Ph.D., Minnesota. Experimental low-temperature physics.
Robert I. Gayley, Associate Professor; Ph.D., Rutgers. Superconductivity, atmospheric physics.
Thomas F. George, Professor; Ph.D., Yale. Theoretical chemical physics.
Henry Goldberg, Assistant Professor; Ph.D., Maryland. Theoretical solid-state physics.
Richard Gonsalves, Assistant Professor; Ph.D., Columbia. Theoretical high-energy physics.
Juergen Heberle, Associate Professor Emeritus; Ph.D., Columbia. Classical electrodynamics, Mössbauer effect.
John T. Ho, Professor; Ph.D., MIT. Experimental condensed-matter physics and biophysics.
Richard J. Howard, Associate Professor; Ph.D., Ohio State. Experimental optics.
Robert P. Hurst, Professor; Ph.D., Texas at Austin. Theoretical atomic physics.
Akira Isihara, Professor Emeritus; D.Sc., Tokyo. Theoretical solid-state and low-temperature physics, statistical mechanics, many-body theory.
Piyare L. Jain, Professor; Ph.D., Michigan State. Experimental elementary-particle and relativistic heavy-ion physics, biophysics.
Yi-Han Kao, Professor; Ph.D., Columbia. Experimental solid-state physics, materials physics, low-temperature physics.
Andrzej Krol, Research Assistant Professor; Ph.D., Warsaw Technical. Experimental solid-state physics, materials physics.
Yung-Chang Lee, Associate Professor; Ph.D., Maryland. Condensed-matter physics, many-body theory.
Duo-Liang Lin, Associate Professor; Ph.D., Ohio State. Theoretical nuclear and solid-state physics.
Michael J. Naughton, Assistant Professor; Ph.D., Boston University. Experimental condensed-matter physics.
Athos Petrou, Associate Professor; Ph.D., Purdue. Experimental solid-state physics.
Michael Ram, Associate Professor; Ph.D., Columbia. Theoretical physics and atmospheric physics.
Jonathan F. Reichert, Professor; Ph.D., Washington (St. Louis). Experimental solid-state physics.
Jan P. Roalsvig, Associate Professor; Ph.D., Saskatchewan. Experimental nuclear physics.
Moti L. Rustgi, Professor; Ph.D., LSU. Theoretical nuclear and atomic physics, medical physics.
Mendel Sachs, Professor; Ph.D., UCLA. General relativity, elementary particles, astrophysics, philosophy of physics.
Bernard A. Weinstein, Professor; Ph.D., Brown. Experimental condensed-matter physics.
Ta-You Wu, Professor Emeritus; Ph.D., Michigan. Theoretical physics.

Determining the crystal structure of a magnetic alloy with an X-ray diffractometer.

A student and a faculty member consult on an ESR experiment in Advanced Laboratory.

Fronczak Hall, Department of Physics.

SECTION 7: PHYSICS

STATE UNIVERSITY OF NEW YORK AT STONY BROOK

Department of Physics and Institute for Theoretical Physics

Programs of Study The Department of Physics, together with the Institute for Theoretical Physics, offers M.A. and Ph.D. degrees. For the M.A., the student must complete 30 approved graduate credits and pass an examination. For advancement to candidacy for the Ph.D., a student must have one year of residence and pass the comprehensive examination. To be granted the Ph.D. degree, a candidate must complete a thesis, pass a thesis examination, and have two semesters of teaching experience as a graduate teaching assistant. The Master of Arts in Teaching degree requires two semesters of course work and one semester of a supervised internship experience teaching physics in a secondary school. The Master of Scientific Instrumentation requires a written and/or oral examination, one semester of teaching as a graduate teaching assistant, a major and a minor project, and course work.

The 58 faculty members teach a rich curriculum of courses that feature many special topics of current interest. Course requirements are kept to a minimum to allow the student to set up a flexible program. Students are encouraged to participate in research as soon as possible and to begin their thesis research no later than the end of their second year. As one of the top graduate and research departments in the nation, the Stony Brook physics department strives to make a graduate education in physics as intellectually stimulating and educationally rewarding as possible. The wide research program supported by its large faculty offers exciting opportunities for every interested and qualified student.

Research Facilities The department has a large base of research facilities installed in its Graduate Physics Building and is a principal user of the major high-energy and solid-state physics facilities at Brookhaven National Laboratory, located only 20 miles away. A number of institutes dedicated to specific fields of research are associated with the department and housed in the building. The Institute for Theoretical Physics, under the directorship of C. N. Yang, is dedicated to research in fundamental theory, such as string theory, supersymmetry, and statistical mechanics. The Nuclear Theory Institute is working on the theory of hadronic matter and nuclear astrophysics. The Stony Brook Radiation Laboratory supports experimental research in nuclear and high-energy physics. The nuclear physics group operates the superconducting linear accelerator for nuclear physics research on campus and uses the AGS accelerator at Brookhaven National Laboratory for research on relativistic heavy-ion physics. The high-energy group uses the large D0 detector at the collider at Fermilab for experiments in the TeV region. The Institute for Interface Phenomena concentrates on research in device-oriented solid-state physics based on superconductors and semiconductors and on the new field of single-electron devices. The Institute for Terrestrial and Planetary Atmospheres offers a program in atmospheric physics. In addition, there is a strong effort in solid-state physics research, some of it using an in-house 17-tesla magnet and dilution refrigerator, an array of He refrigerators, and the UV and X-ray rings of the National Synchrotron Light Source at Brookhaven. The latter is also the base for a growing effort in X-ray microscopy and holography. Atomic physics programs focus on cooling and trapping atoms and on quantum chaos.

Financial Aid Tuition is waived for most full-time assistants. Teaching and research asssistantships carry stipends of $9300 to $10,000 for nine months in 1991–92. Summer research funding ($2500–$3000) is available.

Cost of Study In 1991–92, tuition for the academic year for 9 credits of full-time graduate study is $1845 for New York State residents and $4365 for nonresidents. Tuition for 12 credits of full-time study is $2450 for residents and $5766 for nonresidents. Fees are subject to change without notice.

Cost of Living Room and board charges for students living on campus are approximately $3000 per academic year; however, most students elect to live off campus.

Student Group There are 190 full-time graduate students in the department. Most receive assistantship, traineeship, or fellowship stipends. There are about 16,000 students at Stony Brook, including approximately 5,200 graduate students. Students come from many foreign countries and from all sections of the United States.

Location Stony Brook is located on the historic North Shore of Long Island, approximately 50 miles east of New York City. The University enjoys advantageous proximity to the cultural, scientific, and industrial resources of the nation's largest city.

The University The State University of New York at Stony Brook is one of sixty-four campuses of a statewide system. Stony Brook, along with three other campuses, is designated a university center with a mandate to provide undergraduate and graduate programs through the Ph.D. in the humanities, sciences, social sciences, and certain specialized areas. At present, the Stony Brook campus includes the College of Arts and Sciences, the College of Engineering and Applied Sciences, the College of Urban and Policy Sciences, the School for Continuing Education, and the five schools of the Health Sciences Center.

Applying Applications for admission and financial support for the following fall semester should be received as early as possible during the winter, but not later than March 1. Late applications are reviewed only as space considerations permit. Minimum admission requirements include a B average (3.0 on a 4.0 scale) in physics and mathematics, as well as an overall 3.0 average. The application fee is $35.

Correspondence and Information
Professor Harold J. Metcalf, Director of the Graduate Program
Department of Physics
State University of New York at Stony Brook
Stony Brook, New York 11794-3800
Telephone: 516-632-8080

SECTION 7: PHYSICS

State University of New York at Stony Brook

THE FACULTY AND THEIR RESEARCH

Professors
Gene D. Sprouse, Chairman of the Department; Ph.D., Stanford. Experimental nuclear structure.
Philip B. Allen, Ph.D., Berkeley. Theoretical solid-state physics.
Nandor L. Balazs, Ph.D., Amsterdam. Theoretical physics: statistical mechanics, general relativity.
Peter Braun-Munzinger, Ph.D., Heidelberg. Experimental nuclear physics: relativistic heavy ions.
*Gerald E. Brown, Ph.D., Yale. Theoretical nuclear physics.
Robert L. deZafra, Ph.D., Maryland. Experimental atomic physics; optical pumping and double-resonance quantum electronics.
Roderich Engelmann, Ph.D., Heidelberg. Experimental high-energy physics.
Guido Finocchiaro, Ph.D., Catania (Italy). Experimental high-energy physics.
David B. Fossan, Ph.D., Wisconsin–Madison. Experimental nuclear physics.
*Alfred S. Goldhaber, Ph.D., Princeton. Theoretical physics, nuclear theory, particle physics.
Paul D. Grannis, Ph.D., Berkeley. Experimental high-energy physics.
Michael Gurvitch, Director of the Institute for Interface Phenomena; Ph.D., SUNY at Stony Brook. Experimental solid-state physics.
Andrew D. Jackson, Ph.D., Princeton. Nuclear theory.
Peter B. Kahn, Ph.D., Northwestern. Theoretical physics: nonlinear systems, statistical properties of spectra.
Janos Kirz, Ph.D., Berkeley. X-ray microscopy and holography.
Peter M. Koch, Ph.D., Yale. Atomic physics, synchrotron radiation.
Vladimir E. Korepin, Ph.D., Leningrad State. Theoretical statistical mechanics.
T. T. S. Kuo, Ph.D., Pittsburgh. Nuclear theory.
Linwood L. Lee Jr., Ph.D., Yale. Experimental nuclear physics.
Juliet Lee-Franzini, Ph.D., Columbia. Experimental high-energy physics.
James Lukens, Ph.D., California, San Diego. Experimental solid-state physics.
John H. Marburger III, President; Ph.D., Stanford. Theoretical laser physics.
Michael Marx, Ph.D., MIT. Experimental high-energy physics.
Robert L. McCarthy, Director of the Undergraduate Program; Ph.D., Berkeley. Experimental high-energy physics.
*Barry M. McCoy, Ph.D., Harvard. Theoretical statistical mechanics.
Robert L. McGrath, Ph.D., Iowa. Experimental nuclear physics.
Harold J. Metcalf, Director of the Graduate Program; Ph.D., Brown. Atomic physics, laser cooling, laser physics.
Laszlo Mihaly, Ph.D., Eotvos Lorand (Budapest). Experimental solid-state physics.
Herbert R. Muether, Ph.D., Princeton. Experimental nuclear physics.
Robert Nathans, Professor of Physics and Engineering and Director of the Institute for Pattern Recognition; Ph.D., Pennsylvania. Solid state, energy problems.
*Hwa-Tung Nieh, Ph.D., Harvard. Theoretical high-energy physics.
Peter Paul, Ph.D., Freiburg. Experimental nuclear physics: accelerator physics.
*Martin Rocek, Ph.D., Harvard. Theoretical physics.
*Robert Shrock, Ph.D., Princeton. Theoretical physics.
Edward Shuryak, Ph.D., Novosibirsk (USSR). Nuclear theory.
Warren Siegel, Ph.D., Berkeley. Theoretical physics.
Henry B. Silsbee, Ph.D., Harvard. Experimental physics: molecular and atomic beams, magnetic resonance.
*John Smith, Ph.D., Edinburgh. Theoretical high-energy physics.
*George Sterman, Ph.D., Maryland. Theoretical physics.
Arnold A. Strassenburg, Ph.D., Caltech. Curriculum development.
Clifford E. Swartz, Ph.D., Rochester. School curriculum revision.
*Peter van Nieuwenhuizen, Ph.D., Utrecht (Netherlands). Theoretical physics: quantum field theory, general relativity.
*William I. Weisberger, Ph.D., MIT. Theoretical physics.
*C. N. Yang, Einstein Professor of Physics and Director of the Institute for Theoretical Physics; Ph.D., Chicago. Theoretical physics: field theory, statistical mechanics, particle physics.

Associate Professors
Erlend H. Graf, Ph.D., Cornell. Experimental low-temperature physics.
Richard Mould, Ph.D., Yale. Theoretical physics: general relativity, quantum theory of measurements.
Michael Rijssenbeek, Ph.D., Amsterdam. Experimental high-energy physics.
Johanna Stachel, Ph.D., Mainz. Experimental nuclear physics: relativistic heavy ions.
Peter W. Stephens, Ph.D., MIT. Experimental solid-state physics.

Assistant Professors
Thomas Hemmick, Ph.D., Rochester. Experimental nuclear physics: relativistic heavy ions.
Vladimir J. Goldman, Ph.D., Maryland. Experimental solid-state physics.
Chris Jacobsen, Ph.D., SUNY at Stony Brook. X-ray microscopy and holography.
Jainendra Jain, Ph.D., SUNY at Stony Brook. Theoretical solid-state physics.
Chang Kee Jung, Ph.D., Indiana. Experimental high-energy physics.
Mohammad Mohammadi, Ph.D., Wisconsin. Experimental high-energy physics.
Luis Orozco, Ph.D., Texas. Atomic physics.
Ismail Zahed, Ph.D., MIT. Theoretical nuclear physics.

Research Assistant Professors
Siyuan Han, Ph.D., Iowa State. Experimental solid-state physics.
Madappa Prakash, Ph.D., Bombay. Theoretical nuclear physics.
Jac Verbaarschot, Ph.D., Utrecht (Netherlands). Theoretical nuclear physics.

Adjunct Professors
Ilan Ben-Zvi, Ph.D., Weizmann (Israel). Accelerator physics.
Marvin Geller, Director of the Institute for Terrestrial and Planetary Atmospheres; Ph.D., MIT. Atmospheric sciences.
*Robert Palmer, Ph.D., Imperial College (London). Accelerator physics.

Visiting Faculty
Dimitri Averin, Assistant Professor; Ph.D., Moscow State. Solid-state physics.
Konstantin Likharev, Professor; Ph.D., Moscow State. Solid-state physics.

---

*Members of the Institute for Theoretical Physics.

SECTION 7: PHYSICS

STEVENS INSTITUTE OF TECHNOLOGY

Department of Physics and Engineering Physics

Programs of Study

The department offers M.S. degrees in physics and engineering physics and the Ph.D. degree in physics. To earn an M.S. degree, a student must complete a specific program of courses. For advancement to candidacy for the Ph.D., a student must pass a preliminary qualifying examination that has a written and an oral part. To earn the Ph.D., a student must complete a specific program of courses, complete a thesis, and successfully defend the thesis. M.S. programs require a minimum of 30 credits; a thesis is optional. The Ph.D. program requires a minimum of 90 credits, of which at least 30 are thesis credits. There is a one-year residency requirement for the Ph.D.

A student with an assistantship might expect to complete the M.S. degree in two years; four more years are required for the Ph.D. Students can expect to work closely with their thesis advisers.

The Department of Physics and Engineering Physics carries out research in the following areas: astrophysics, computer applications, elementary particle physics, fluid dynamics, general relativity, nonlinear dynamics, plasma physics, surface physics, and solid-state physics.

Research Facilities

There are four experimental research laboratories under the supervision of department faculty members. These laboratories are the Plasma Physics Laboratory, the Plasma and Surface Physics Laboratory, the Laser-Optics Laboratory, and the Laser Spectroscopy Laboratory. Equipment in the Plasma Physics Laboratory includes two plasma focus machines, two fast oscilloscopes, neutron and X-ray scintillators, infrared detectors, schlieren photography equipment, a neodymium glass laser, and X-ray pinhole cameras. Major equipment in the Plasma and Surface Physics Laboratory includes plasma instability apparatus (e.g., coils producing uniform magnetic fields); a microwave spectrum analyzer; surface studies apparatus (e.g., an ultrahigh vacuum system, an Auger electron spectrometer, and an ion-scattering spectrometer); apparatus for ion scattering and sputtering studies (e.g., a magnetic mass spectrometer); plasma etcher apparatus for thermionic emission from ionic crystals; furnaces; thin-film apparatus for evaporation and sputtering; a scanning secondary ion mass spectrometer; two electron energy analyzers, and a high-pressure facility for 130,000 psi, among other equipment. A partial list of the major equipment in the Laser-Optics Laboratory includes Raman apparatus (e.g., an argon-ion laser and a spex monochrometer); Brillouin-scattering apparatus (e.g., a krypton-ion laser and a Fabry-Perot interferometer); a fluorescence ultraviolet spectrometer; and laser frequency-modulation spectroscopy apparatus. The facilities of the Laser Spectroscopy Laboratory include computer-controlled data acquisition equipment.

The department has two machine shops and employs a professional machinist.

Financial Aid

Teaching and research assistantships carry stipends of $7515–$8460 in 1991–92 and include full tuition remission. A very limited number of fellowships are available.

Cost of Study

Tuition is $485 per credit in 1991–92. All students pay a $60 enrollment fee, which is not covered by financial aid.

Cost of Living

Accommodations in single-student housing cost $1155–$1385 per semester; married student housing costs are $520 per month on a twelve-month basis (1991–92 rates).

Student Group

The Institute enrolls about 1,500 undergraduate students, about 75 percent of whom live on campus. There are about 2,000 graduate students. Most graduate students attend part-time; approximately 10 percent live on campus. In 1991–92, approximately 35 master's students, of whom one half are full-time, and about 35 doctoral students, of whom three quarters are full-time, are enrolled in the department.

Location

The Stevens campus overlooks the Hudson River, directly across from midtown Manhattan. Located in Hoboken, New Jersey, a small city occupying about 1 square mile, the 55-acre campus contains academic buildings, recreational facilities, and areas covered by grass and trees.

The Institute

Stevens was founded by and named after a distinguished family of engineers who were involved in the early 1800s in the design and development of steamboats, locomotives, the T rail, ironclads, and yachts, among other projects. The Institute opened its doors in 1870, offering a program in engineering firmly grounded in scientific principles. The mission of Stevens Institute of Technology is to sustain a community of individuals who are dedicated to the achievement of excellence and who share a vision related to engineering, management, and applied and pure science.

Applying

In order to apply for admission to a graduate program, the candidate must have a bachelor's degree in science or engineering and a GPA of at least 3.0. The GRE General Test and Subject Test in physics are recommended but not required, and there are no firm minimum scores. For students from non-English-speaking countries, a TOEFL score of at least 500 is required. Applications for admission and financial aid should be received by February 15; however, the Institute processes applications received after this date. The application fee is $35. Students should address applications to Dr. Timothy R. Hart, Graduate School.

Correspondence and Information

Professor Earl L. Koller
Chairman of the Graduate Committee and Department Head
Department of Physics and Engineering Physics
Stevens Institute of Technology
Castle Point Station
Hoboken, New Jersey 07030
Telephone: 201-216-5665 or 5663

SECTION 7: PHYSICS

Stevens Institute of Technology

THE FACULTY AND THEIR RESEARCH

Professors
James L. Anderson, Ph.D., Syracuse, 1952. Relativistic astrophysics, gravitational radiation.
Jeremy Bernstein, Ph.D., Harvard, 1955. Elementary particles and cosmology.
E. Byerly Brucker, Ph.D., Johns Hopkins, 1959. Experimental high-energy physics, optics.
Wayne E. Carr, Ph.D., Illinois, 1967. Plasma physics, electron and positive ion beams, computational physics.
Timothy R. Hart, Ph.D., MIT, 1970. Laser Raman spectroscopy of solid-state surfaces, including magnetic and semiconductor materials; infrared and laser homodyne-scattering applications in biophysics.
Norman J. Horing, Ph.D., Harvard, 1964. Quantum many-particle theory, solid-state physics, surface physics, high-magnetic field effects.
Earl L. Koller, Head of the Department of Physics and Engineering Physics; Ph.D., Columbia, 1959. Experimental particle physics, bubble chambers and hybrid systems.
Erich E. Kundhardt, Ph.D., Polytechnic, 1976. Nonequilibrium effects in condensed matter.
Franklin Pollock, Ph.D., Columbia, 1952. Energy policy, electron-scattering theory.
Bernard Rosen, Ph.D., NYU, 1959. Computational plasma physics.
Harold Salwen, Ph.D., Columbia, 1956. Fluid dynamics, kinetic theory, quantum mechanics.
Ralph Schiller, Ph.D., Syracuse, 1952. Biophysics of microorganisms.
George Schmidt, Ph.D., Hungarian Academy of Sciences, 1956. Plasma physics, thermonuclear fusion, nonlinear dynamics.
Milos Seidl, D.Sc., Prague Technical, 1962. Plasma physics, physical electronics.
Snowden Taylor, Ph.D., Columbia, 1959. Experimental high-energy particle physics, technical education for disadvantaged students and members of minority groups.
George J. Yevick, Sc.D., MIT, 1947. Foundations of quantum mechanics and the many-body problem.

Associate Professors
Arthur Layzer, Ph.D., Columbia, 1960. Computer science, language problems of the hearing-impaired, field theoretical solid-state and low-temperature physics.
Edward A. Whittaker, Ph.D., Columbia, 1982. Laser techniques.

Assistant Professors
Swapan K. Gayen, Ph.D., Connecticut, 1984. Laser spectroscopy of solids, tunable solid-state lasers.
Veneta G. Tsoukala, Ph.D., RPI, 1988. Condensed-matter physics, experimental light scattering and optical studies.

Adjunct Professors
Marvin Hutt, M.S., NYU, 1987. President and founder of Optics, Coating, Research. Applied optics and optical design.
Jay Mancini, Ph.D., Virginia Tech, 1982. Solid-state physics.
Steven Silverman, Ph.D., Stevens, 1980. Surface physics.
James Supplee, Ph.D., Texas at Dallas, 1979. Fluid dynamics, kinetic theory, quantum mechanics.
Robert Webb, M.S., NYU, 1966. Member of technical staff, Engineering Research Center (AT&T Bell Laboratories). Nonlinear optics.

Research Professor
Vittorio Nardi, Ph.D., Rome, 1954. Ion and electron beams, nuclear fusion plasmas, pulsed-power technology.

SECTION 7: PHYSICS

SYRACUSE UNIVERSITY
Department of Physics

Programs of Study
The department offers the M.S. and Ph.D. degrees in physics. The M.S. is usually obtained after two years of study. The student must also either complete a master's thesis or pass a qualifying examination. The Ph.D. is awarded to students who complete course requirements, pass a two-part qualifying examination, complete a written dissertation based upon original research, and pass a dissertation defense examination. A typical program of study takes about five years to complete. The Graduate Biophysics Program and the Solid-State Science and Technology Program are affiliated with the physics department. A student may transfer to one of these programs after being admitted as a graduate student in physics or may apply for direct admission to either program. Research in experimental particle physics uses the accelerator facilities at Cornell, Brookhaven, MIT, and CERN and emphasizes heavy quark production and charm meson, B meson decays, electroweak parameters, and quark gluon plasmas. Theoretical particle physics focuses especially on grand unified theories, supersymmetry, monopoles, quantum chromodynamics, and Weinberg-Salam theory. Condensed-matter physics involves experimental work in amorphous and crystalline semiconductors, surface physics, thin films, and nuclear spin-polarized solids; theoretical work in phase transitions, chaotic systems, high-field transport, and physics of liquids. Relativity and field theory include classical general relativity, quantum gravity, relativistic quantum mechanics, gauge theories, and experimental gravitation. Biophysics includes sensory information processing in single-celled organisms, and medical imaging, using antiprotons. Syracuse has a strong computational physics program centered on the use of advanced parallel computers and the development of new algorithms to solve problems in physics. Areas of interest are condensed-matter physics, high-energy physics, and astrophysics.

Research Facilties
The department occupies a modern six-story building that houses research laboratories, the physics library, offices, classrooms, teaching laboratories, a mechanical shop, an electronics shop, and computing facilities, including a departmental cluster based on VAX series computers. Other facilities include the Northeast Parallel Architectures Center, which includes two Connection Machines, a Multimax 320 and 520, and an Alliant FX/80. Analysis of recently acquired data from elementary particle physics experiments is carried out locally. Syracuse University houses a national facility for multi-element nuclear magnetic resonance, which is operated by the chemistry department. Off-campus facilities include the CLEO detector at the Cornell electron storage ring, a beam line at the 1-GeV synchrotron radiation facility in Stoughton, Wisconsin, and the spin-polarized electron spectrometer at MIT. Faculty members maintain on-campus laboratories in astrophysics and gravitation, low temperature, thin-film preparation, surface analysis, electron microscopy, optical and laser spectroscopy, electron spin resonance and nuclear magnetic resonance, X-ray diffraction, chromatography and electrophoresis, time-lapse video recording, and action spectroscopy.

Financial Aid
In 1991–92, teaching assistantships pay up to $9235 and include up to 24 hours of tuition scholarship. Teaching assistants spend up to 20 hours per week teaching laboratory or recitation classes, grading, and preparation. Research assistantships of approximately $9235 per academic year are generally offered to students who have completed one or more years of study in the department and include up to 24 hours of tuition scholarship. Summer research and teaching assistantships are also available.

University Graduate Fellowships that require no duties are available. The physics department nominates candidates, including both new and currently enrolled students; the University Fellowship Committee, upon the recommendation of a faculty advisory group, selects the University Fellows. Fellowships consist of an academic-year award of $8775 plus a full tuition scholarship. Where appropriate, these fellowships are awarded for more than one year and include, typically, a year of fellowship, followed by a year of teaching or research support, followed by a third year of fellowship support.

Cost of Study
In 1991–92, tuition for full-time graduate students is $9144 (this charge is normally waived for graduate assistants) and $381 per credit for part-time students. Student fees are approximately $275 per year.

Cost of Living
On-campus single student housing is available at approximately $3840 for the academic year. On-campus married student housing costs $420–$530 per month, including utilities. Off-campus housing is available within walking distance of the campus. Free shuttle-bus service is available within the University complex and to local shopping centers.

Student Group
There are approximately 65 graduate students in physics, nearly all of whom receive financial aid. Students come from all over the United States and many foreign countries. Admission to graduate study is competitive. Graduates find employment in government and private institutions and in college teaching and research.

Location
The Syracuse metropolitan area, with a population exceeding 500,000, has a strong and varied industrial and service economy. At the geographical center of New York State, Syracuse has excellent highway and air connections to New York City, Boston, Montreal, Toronto, Philadelphia, and Washington. Recreation opportunities are provided by the Adirondack Mountains, the Thousand Islands, the Finger Lakes, Lake Ontario, and the Mohawk River and Hudson River valleys. The city's cultural attractions include the Everson Museum of Art (designed by I. M. Pei); a professional repertory theater, the Syracuse Stage; several amateur theaters; and regular series of symphony concerts, ballets, and operas in the three-theater Civic Center. On campus, the 50,000-seat Carrier Dome is the site of concerts, sports events, and commencement.

The University
Syracuse University, founded in 1870, is an independent, privately endowed university with an international reputation. There are 16,000 students enrolled, 12,000 of whom are undergraduates. The 200-acre campus features a main quadrangle surrounded by academic buildings, with residential facilities nearby.

Applying
Catalogs and application forms may be obtained upon request from the address below. Students are normally admitted in September, and completed applications should be received by January 1 for applicants who wish to be considered for a fellowship or by February 1 for those who wish to be considered for other financial aid. Applicants should take the GRE physics Subject Test in time for scores to arrive by February 15. TOEFL scores (550 is normally the cutoff score) are required of foreign students whose native language is not English.

Correspondence and Information
Graduate Secretary
Syracuse University
Department of Physics
201 Physics Building
Syracuse, New York 13244-1130
Telephone: 315-443-3901

SECTION 7: PHYSICS

Syracuse University

THE FACULTY AND THEIR RESEARCH

Marina Artuso, Research Assistant Professor; Ph.D., Northwestern, 1986. Elementary particles and fields.
Abhay Ashtekar, Distinguished Professor; Ph.D., Chicago, 1974. General relativity, elementary particles and fields.
A. P. Balachandran, Professor; Ph.D., Madras (India), 1962. Elementary particles and fields.
Ofer Biham, Assistant Professor; Ph.D., Weizmann (Israel), 1988. Condensed-matter theory.
Mark Bowick, Assistant Professor; Ph.D., Caltech, 1983. Elementary particles and fields.
Nicola Cabibbo, Distinguished Visiting Professor; Ph.D., Rome, 1958. Elementary particles and fields.
Peter Dowben, Associate Professor; Ph.D., Cambridge, 1981. Experimental condensed-matter physics.
Kenneth Foster, Associate Professor; Ph.D., Caltech, 1972. Biological physics.
Geoffrey Fox, Professor; Ph.D., Cambridge, 1967. Computational physics.
Wojtek Furmanski, Research Professor; Ph.D., Jagiellonian (Krakow), 1977. Computational physics.
Joshua Goldberg, Professor; Ph.D., Syracuse, 1952. General relativity.
Marvin Goldberg, Professor and Chair; Ph.D., Syracuse, 1965. Elementary particles and fields.
Arnold Honig, Professor; Ph.D., Columbia, 1953. Magnetic resonance, condensed matter.
Nahmin Horwitz, Professor; Ph.D., Minnesota, 1954. Elementary particles and fields.
Theodore Kalogeropoulos, Professor; Ph.D., Berkeley, 1959. Biological physics, elementary particles and fields.
Harvey Kaplan, Professor; Ph.D., Berkeley, 1952. Condensed matter.
Edward Lipson, Professor and Associate Chair; Ph.D., Caltech, 1971. Biological physics.
M. Cristina Marchetti, Assistant Professor; Ph.D., Florida, 1982. Condensed matter.
Allen Miller, Associate Professor; Ph.D., Rutgers, 1960. Condensed matter.
Giancarlo Moneti, Professor; Ph.D., Rome, 1954. Elementary particles and fields.
Roger Penrose, Distinguished Visiting Professor; Ph.D., Cambridge, 1957. Relativity.
Carl Rosenzweig, Professor; Ph.D., Harvard, 1972. Elementary particles and fields.
Jureepan Saranak, Research Assistant Professor; Ph.D., CUNY, Mount Sinai, 1981. Biophysics.
Peter Saulson, Associate Professor; Ph.D., Princeton, 1981. Relativity, astrophysics.
Joseph Schechter, Professor; Ph.D., Rochester, 1965. Elementary particles and fields.
Eric Schiff, Associate Professor; Ph.D., Cornell, 1979. Condensed matter.
Tomasz Skwarnicki, Assistant Professor; Ph.D., Hamburg, 1986. Elementary particles and fields.
Lee Smolin, Professor; Ph.D., Harvard, 1979. General relativity.
Rafael Sorkin, Associate Professor; Ph.D., Caltech, 1974. General relativity.
Paul Souder, Professor; Ph.D., Princeton, 1971. Elementary particles and fields.
Sheldon Stone, Professor; Ph.D., Rochester, 1972. Elementary particles and fields.
Gianfranco Vidali, Associate Professor; Ph.D., Penn State, 1982. Condensed matter.
Richard Vook, Professor; Ph.D., Illinois, 1957. Materials and condensed matter.
Kameshwar Wali, Professor; Ph.D., Wisconsin, 1959. Elementary particles and fields.
Volker Weiss, Professor; Ph.D., Syracuse, 1957. Condensed matter.
Marcel Wellner, Professor; Ph.D., Princeton, 1958. Elementary particles and fields.

SECTION 7: PHYSICS

TEMPLE UNIVERSITY
of the Commonwealth System of Higher Education

Department of Physics

Programs of Study

The department offers the M.A. and Ph.D. degrees. There is an evening as well as a daytime M.A. program; both require 24 semester hours of credit. Normally, required courses for the M.A. degree encompass 18 hours; the other 6 semester hours are used for thesis research or for additional courses. The student must also pass the M.A. qualifying examination in physics.

No specific number of graduate credits is required for the Ph.D. degree, but an approved program of graduate courses must be satisfactorily completed. A dissertation and dissertation examination are required. An M.A. degree is not necessary for the Ph.D. degree; however, the M.A. examination must be taken within a year after entering the Ph.D. program. The comprehensive Ph.D. examination in physics is taken at the beginning of the third year. There is a one-year residence requirement for the Ph.D. degree.

Students whose native language is not English must pass an examination in spoken and written English. There is no other language requirement for either the Ph.D. or the M.A. degree.

Each full-time graduate student is given a desk in one of several student offices. Lecturers from other institutions describe their research activities at a weekly colloquium and a weekly seminar, and informal discussions with members of the faculty are frequent.

Research Facilities

The department is housed in completely air-conditioned Barton Hall, which has lecture halls, offices, classrooms, and laboratories. The Physics Reading Room contains frequently used journals and books; several thousand additional volumes are located in the Paley Library across the street from Barton Hall. There are a fully staffed machine shop, a student shop, and a materials preparation facility.

The University computer center is equipped with a CDC CYBER 860 and an IBM 3081 with an extensive network of terminals that is ideally suited for scientific work and is available for student use. A new supercomputer with parallel processors is also available. The elementary particle physics group has a number of computers that it uses for on-line data acquisition and off-line data analysis.

The research laboratories are conducting a variety of studies on many-body effects in crystals; on the properties of surface and interfaces; on laser spectroscopy; on low-temperature properties of rare earth metals and alloys, including valence fluctuations and heavy fermion behavior; on superconductivity and magnetic resonance; and on elementary particles. The department also uses outside facilities, including the Los Alamos Meson Physics Facility, Brookhaven National Laboratory, the Stanford Linear Accelerator Center, and the Francis Bitter National Magnet Laboratory.

Theoretical work is being conducted in such areas as elementary particles and their interactions, special and general theory of relativity, statistical mechanics, and solid-state theory.

Financial Aid

Aid is available to qualified full-time students in the form of assistantships, traineeships, and fellowships funded by the University and various extramural agencies. All forms of financial aid include a stipend plus tuition. Tuition scholarships are also available to qualified students. The specific type of aid offered to a particular student will depend on the student's qualifications and program of study. Summer support for qualified students consisting of stipends plus tuition is also normally available. Current stipends are $12,000 for the calendar year. For students with grant-supported research assistantships, the stipend is much higher.

Cost of Study

The annual tuition for full-time graduate study in 1991–92 is $205 per credit hour for residents of Pennsylvania and $258 per credit hour for nonresidents. Minimal fees are charged for various services, such as microfilming theses.

Cost of Living

Room and board costs for students living on campus are approximately $2200 per year in 1991–92. University-sponsored apartments, both furnished and unfurnished, are also available on the edge of the campus.

Student Group

The department has approximately 40 full-time graduate students, nearly all of whom are supported by assistantships or fellowships.

Location

Philadelphia is the fifth-largest city in the country, with a metropolitan population of over 2 million. It offers a varied menu of cultural attractions. The city has a world-renowned symphony orchestra, a ballet company, two professional opera companies, and a chamber music society. Besides attracting touring plays, Philadelphia has a professional repertory theater and many amateur troupes. All sports and forms of recreation are easily accessible. The city is world famous for its historic shrines and parks and for the eighteenth-century charm that is carefully maintained in the oldest section. The climate is temperate, with an average winter temperature of 33 degrees and an average summer temperature of 75 degrees.

The University

The development of Temple University has been in line with the ideal of "educational opportunity for the able and deserving student of limited means." With a rich heritage of social purpose, Temple seeks to provide the opportunity for high-quality education without regard to a student's race, creed, or station in life. Affiliation with the Commonwealth System of Higher Education undergirds Temple's character as a public institution.

Applying

All application material, both for admission and for financial awards, should be received by early March for admission in the fall semester. Notification regarding admission and the awarding of an assistantship is made as soon as the application has been screened.

Correspondence and Information

For program information and all applications:
Chairman, Admissions and Awards Committee
Department of Physics
Temple University
Philadelphia, Pennsylvania 19122
Telephone: 215-787-7654

For general information on graduate programs:
Dean
Graduate School
Temple University
Philadelphia, Pennsylvania 19122

Temple University

THE FACULTY AND THEIR RESEARCH

Condensed-Matter Physics
M. Bloch, Associate Professor; Ph.D., Hebrew (Jerusalem), 1980. Surface physics.
J. Karra, Associate Professor; Ph.D., Rutgers, 1964. Nuclear magnetic resonance, electron paramagnetic resonance, double resonance, pulsed magnetic resonance techniques.
C. L. Lin, Assistant Professor; Ph.D., Temple, 1985. Heavy fermions, crystal fields, valence fluctuations, the Kondo effect, high-temperature superconductivity.
T. Mihalisin, Professor; Ph.D., Rochester, 1967. Crystal fields, valence fluctuations and the Kondo effect in magnetic systems.
R. Tahir-Kheli, Professor and Chairman; D.Phil., Oxford, 1962. Theory of magnetism, randomly disordered systems.

Educational Development Physics
The teaching of physics, creation of innovative teaching methods; use of mass media.
L. Dubeck, Professor; Ph.D., Rutgers, 1965. Development, publication, and testing of pre-college science materials.
R. B. Weinberg, Professor; Ph.D., Columbia, 1963.

Elementary Particle Physics and Cosmology
L. B. Auerbach, Professor; Ph.D., Berkeley, 1962. Experimental particle physics; investigations of the properties of fundamental particles at Los Alamos Meson Physics Facility, Brookhaven National Laboratory, and CERN.
Z. Dziembowski, Associate Professor; Ph.D., Warsaw, 1975. Theoretical particle physics.
J. Franklin, Professor; Ph.D., Illinois, 1956. Theoretical particle physics; quark and parton theory, S-matrix theory.
Y. K. Ha, Visiting Assistant Professor; Ph.D., Yale, 1982. Quantum field theory.
V. L. Highland, Professor; Ph.D., Cornell, 1963. Experimental particle physics, investigations of the properties of fundamental particles at Los Alamos Meson Physics Facility and Brookhaven National Laboratory.
C. J. Martoff, Associate Professor; Ph.D., Berkeley, 1980. Experimental particle physics: investigation of weak interactions and dynamics of nuclei and particles, development of particle detectors for the study of "dark matter" using superconductivity.
W. K. McFarlane, Professor; Ph.D., Birmingham, 1964. Experimental particle physics; investigations of the properties of fundamental particles at Los Alamos Meson Physics Facility, Brookhaven National Laboratory, and CERN.
D. E. Neville, Professor; Ph.D., Chicago, 1962. Theoretical particle physics; symmetries and quark models, quantum gravity.

Optics
M. Lyyra, Associate Professor; Ph.D., Stockholm, 1984. Laser spectroscopy.

Relativity
P. Havas, Professor Emeritus; Ph.D., Columbia, 1944. Special and general relativity, elementary particle physics, mathematical physics.

Statistical Physics
T. Burkhardt, Professor; Ph.D., Stanford, 1967. Statistical mechanics and many-body theory.
D. Forster, Professor; Ph.D., Harvard, 1969. Statistical mechanics and many-body theory.
E. Gawlinski, Assistant Professor; Ph.D., Boston University, 1983. Statistical mechanics and computational physics.
S. Y. Larsen, Professor; Ph.D., Columbia, 1962. Quantum statistical physics, few-body problem, hyperspherical harmonics, molecular physics, chemical reactions.
G. Neofotistos, Visiting Assistant Professor; Ph.D., Temple, 1991. Statistical mechanics.

Theoretical Atomic Physics
R. L. Intemann, Professor; Ph.D., Stevens, 1964. Atomic physics, inner-shell processes.

Barton Hall, the physics building.

The Elementary Particle Physics Laboratory.

SECTION 7: PHYSICS

TEXAS A&M UNIVERSITY
Department of Physics

Programs of Study

The Department of Physics offers graduate studies leading to the degrees of Master of Science and Doctor of Philosophy. Major areas of research include atomic and molecular physics, optics and light scattering, low-temperature and condensed-matter physics, nuclear physics, high-energy physics, and accelerator physics, with both theoretical and experimental programs. An experimental program in medical physics is also available.

Both thesis and nonthesis options are available for the M.S. The nonthesis option may be satisfied by successfully completing 36 hours of approved course work, including a 6-hour research project and comprehensive written and oral exams. The thesis option requires 32 hours of approved course work, including research, a research thesis, and an oral exam. The M.S. can be completed within two years. Qualified students may proceed directly to the Ph.D. program without first obtaining the M.S.

Students in the doctoral program must pass a Ph.D. qualifying exam, usually given at the end of the third semester of graduate study. The exam covers the material in introductory graduate courses. The Ph.D. program is tailored to the individual student's needs and normally includes several specialty or research-oriented courses in the student's field of research as well as introductory graduate courses and research. There is no language requirement. Most students take about five years to complete the requirements for the Ph.D.

Research Facilities

The physics building provides many well-equipped laboratories for experimental research. Optics and atomic research equipment includes CW narrow-bandwidth ring dye laser systems; excimer-, Nd:yag-, and flashlamp-pumped pulsed dye laser systems; and a 100-kV ion accelerator. Equipment for low-temperature studies includes several superconducting solenoids ranging up to 140 kG, four dilution refrigerator cryostats for studies to 0.003K, and a CTI 1400 helium liquefier. For nanostructure device fabrication, there are two molecular beam epitaxy machines operating in an 800-square-foot clean room, along with an electron-beam writer for nanolithography. For materials research, there are a clean high-vacuum thin-film evaporation facility and extensive facilities for rapid quenching of alloys. The department has large, well-equipped machine and electronics shops, a VAX computer with access to a Silicon Graphics Attached Processor, a number of workstations of various kinds, and access to a University-owned CRAY Y-MP/116 supercomputer. Physics faculty and students have major programs at the Cyclotron Institute, which contains a K500 superconducting cyclotron and a powerful electron cyclotron resonance external ion source. Experimental equipment includes a K-400 magnetic spectrometer, a momentum achromat recoil mass spectrometer, a diproton spectrometer, a 4π neutron calorimeter and a 57-element BaF_2 array. The faculty members of the physics department also play a major role at the Texas Accelerator Center, a laboratory for advanced accelerator research and development, located at The Woodlands, Texas. Its functions include design, construction, and testing of superconducting magnets for the Superconducting Super Collider (SSC) and development of new RF accelerating structures for linac colliders. Facilities include a 200W liquid He refrigerator, magnet fabrication and test facilities, and an RF measurements laboratory.

Financial Aid

The research programs of the department are supported by University funds and grants from the federal government and private sponsors. Teaching or research fellowships and assistantships support about 140 graduate students. Research fellowships and assistantships are provided by the Office of Graduate Studies; government agencies, such as the NSF; research grants given to faculty members by private foundations, such as the Robert A. Welch Foundation; and industrial sponsors. In 1991–92, stipends begin at $950 per month for twelve months. Job opportunities for spouses are quite plentiful.

Cost of Study

In 1990–91, annual graduate tuition for Texas residents was $450 for full-time students or $18 per semester hour for part-time students. Annual tuition for nonresidents was $2928 for a full-time student and $122 per semester hour for part-time. Nonresident students on assistantships or fellowships pay resident tuition. Other fees totaled approximately $325 for a full-time student.

Cost of Living

A limited number of University-owned apartments, both furnished and unfurnished, are available for students at $175–$276 per month plus electricity. Further information is available from University Owned Apartments, University Mail Service, M.S. 3365. A large number of private apartments are also available in the community. For an information booklet, students can call 409-845-2264 or write to Off-Campus Center, Department of Student Affairs.

Student Group

Texas A&M University has an enrollment of approximately 41,170, including about 7,320 graduate students.

Location

The Bryan–College Station area is located in south-central Texas, approximately 95 miles from Houston and 175 miles from Dallas. With a population of nearly 155,200, including the Texas A&M student body, the two cities constitute the largest urban area in this part of the state. Community facilities include excellent public and private schools, churches representing approximately twenty denominations, hospitals, theaters, and various shopping centers.

The University

Founded in 1876 as a land-grant college under the Morrill Act, Texas A&M University began graduate training leading to the Master of Science in 1888. The first doctoral degree was conferred by the University in 1940. In recent years, the Office of Graduate Studies has grown quite rapidly, from 500 students in 1957 to its present enrollment of 7,320. More than 75 percent of all master's and professional degrees conferred by Texas A&M University and more than 85 percent of its doctorates have been granted in the past twenty years.

Applying

Inquiries regarding admission to graduate studies should be addressed to the Department of Physics. Inquiries about facilities for advanced studies, research, and requirements for graduate work in physics should also be addressed to the department. Applications for admission should be filed no later than four weeks prior to the opening of the semester. Students seeking admission with financial aid should send copies of all application materials by March 1 preceding the academic year for which the awards are sought. An application for financial aid will be considered independently of the admission decision.

Correspondence and Information

For admission:
Director of Admissions
Texas A&M University
College Station, Texas 77843

For departmental information or aid:
Department of Physics
Texas A&M University
College Station, Texas 77843
Telephone: 409-845-7717

SECTION 7: PHYSICS

Texas A&M University

THE FACULTY AND THEIR RESEARCH

Professors
Thomas W. Adair III, Ph.D., Texas A&M, 1965. Magnetic effects in gases and single crystals.
Roland E. Allen, Ph.D., Texas at Austin, 1968. Condensed-matter theory.
R. L. Arnowitt, Department Head; Ph.D., Harvard, 1953. High-energy theory.
William H. Bassichis, Ph.D., Case Institute of Technology, 1963. Physics education.
Ronald A. Bryan, Ph.D., Rochester, 1960. Theoretical nuclear and particle physics.
David A. Church, Ph.D., Washington (Seattle), 1969. Stored-ion collisions and spectroscopy, polarization spectroscopy, multicharged ions.
Robert B. Clark, Ph.D., Yale, 1968. Theoretical particle physics.
Thomas M. Cormier, Ph.D., MIT, 1974. High angular momentum in nuclear structure physics.
Michael Duff, Ph.D., Imperial College (London), 1972. Elementary particle theory.
Nelson M. Duller Jr., Ph.D., Rice, 1953. Cosmic-ray muons, high-energy particle detectors, cosmic-ray anisotropies, magnetic resonances.
David J. Ernst, Ph.D., MIT, 1970. Theoretical nuclear physics.
A. Lewis Ford, Ph.D., Texas at Austin, 1972. Ion-atom collision theory.
Edward S. Fry, Ph.D., Michigan, 1969. Atomic spectroscopy, electron-atom scattering, light-particle scattering.
Joe S. Ham, Ph.D., Chicago, 1954. High polymer physics, electrical fluctuation phenomena, composite materials.
John C. Hiebert, Ph.D., Yale, 1964. Fast neutrons, neutron polarization studies, nuclear reactions with polarization transfer.
Chia-Ren Hu, Ph.D., Maryland, 1968. Theory of superconductivity, superfluid ^{3}He, quantum Hall effect, and electromagnetic scattering.
F. R. Huson, Ph.D., Berkeley, 1964. Experimental elementary particle physics, pp colliding beams at very high energy, new particle searches, accelerator physics, superconducting magnet development.
Marko V. Jaric, Ph.D., CUNY, City College, 1978. Theoretical investigations of quasicrystal structure, stability, diffraction, elasticity, and growth.
George W. Kattawar, Ph.D., Texas A&M, 1963. Electromagnetic scattering, atmospheric and oceanic physics.
Robert A. Kenefick, Ph.D., Florida State, 1962. High-energy atomic collision processes, ion sources, ion traps.
Wiley P. Kirk, Ph.D., SUNY at Stony Brook, 1970. Condensed-matter physics, nanostructure, low-temperature physics.
Che-Ming Ko, Ph.D., SUNY at Stony Brook, 1973. Theoretical nuclear physics.
Eckhard Krotscheck, Ph.D., Cologne (Germany), 1974. Theoretical nuclear physics.
John A. McIntyre, Ph.D., Princeton, 1950. Gamma-ray camera for nuclear medicine.
Peter M. McIntyre, Ph.D., Chicago, 1972. Experimental elementary particle physics, pp colliding beams at Fermilab, supercollider detector development, microwave power devices.
Dimitri Nanopoulos, Ph.D., Sussex (England), 1973. High-energy theory.
Donald G. Naugle, Ph.D., Texas A&M, 1965. Superconductivity, thin metal films, metallic glasses, amorphous materials.
Lee C. Northcliffe Jr., Ph.D., Wisconsin, 1957. Fast neutrons, neutron polarization studies, nuclear reactions with polarization transfer.
John F. Reading, Ph.D., Birmingham (England), 1964. Scattering theory, inner-shell ionization of atoms.
Wayne M. Saslow, Ph.D., California, Irvine, 1968. Electromagnetic properties of solids, magnetic disorder, superfluidity.
Hans A. Schuessler, Ph.D., Heidelberg (Germany), 1964. Spectroscopy of free ions, ion storage, laser spectroscopy.
Roger A. Smith, Ph.D., Stanford, 1973. Many-body theory, nuclear matter, neutron star matter, condensed helium.
Robert E. Tribble, Ph.D., Princeton, 1973. Mass measurement of exotic nuclei, nuclear reactions, weak interactions.
Robert C. Webb, Ph.D., Princeton, 1972. Electronic counter experiments in high-energy particle physics, production of new particle states, symmetry properties of high-energy collisions, colliding beams experiments.
Dave H. Youngblood, Ph.D., Rice, 1965. Nuclear structure and reactions, giant multipole resonances.

Associate Professors
Siu Ah Chin, Ph.D., Stanford, 1975. Theoretical nuclear physics, high-density matter, lattice calculations, Monte Carlo methods.
Carl A. Gagliardi, Ph.D., Princeton, 1982. Weak interactions, beta decay, neutron scattering.
Christopher N. Pope, Ph.D., Cambridge, 1980. Elementary particle theory.
Ergin Sezgin, Ph.D., SUNY at Stony Brook, 1980. Supersymmetry and supergravity; theory of strings, membranes, and other extended objects; unification of gravity with other interactions; infinite dimensional symmetries and their physical applications.

Assistant Professors
Glen Agnolet, Ph.D., Cornell, 1983. Experimental low-temperature physics.
Teruki Kamon, Ph.D., Tsukuba (Japan), 1986. Proton, antiproton colliding beams at Fermilab; development of detectors at SSC.
Joseph H. Ross Jr., Ph.D., Illinois, 1986. Experimental condensed matter, materials research.
Michael B. Weimer, Ph.D., Caltech, 1986. Experimental surface science and low-temperature solid-state physics.
James T. White, Ph.D., California, San Diego, 1985. Experimental high-energy physics, Dϕ collider experiment at Fermilab and Solenoid Detector experiment at the SSC.

TUFTS UNIVERSITY

Department of Physics and Astronomy

Programs of Study

The Department of Physics and Astronomy offers programs of study leading to both M.S. and Ph.D. degrees for students interested in pursuing a wide variety of careers, including both teaching and research. Because there is a combined faculty of 23, many areas of physics and astronomy are available for study, and close faculty-student contact permits a deeper and more complete involvement in studies and research.

Candidates for the master's degree must complete eight graduate-level courses with grades of B– or better in their approved program; submission of a thesis is optional. While two semesters of residence are required, there are no language or examination requirements. The option of including research courses is also available. Each candidate is aided and advised by his or her own special committee of faculty members. Entering students should be familiar with intermediate-level physics and with mathematics through the level of calculus.

Ph.D. candidates need not fulfill a specific number of graduate courses but must demonstrate their proficiency in fundamental physics, quantum mechanics, their doctoral field, and a suitable fourth field. This may be done through course work or examinations. A preliminary oral examination is required on a subject determined by the candidate's committee, and, upon its successful completion, the committee will advise the student on further research work. Three academic years of study are required, at least one of which must be in residence; a dissertation and dissertation examination will conclude the Ph.D. program.

The principal areas of research in the department are elementary particle physics and condensed-matter physics; theoretical and experimental studies are conducted in both areas. Research in radio astronomy, medical physics, theoretical molecular biophysics, and general relativity and cosmology is also undertaken by individual department members.

Research Facilities

Experimental work is carried out at Tufts as well as at national institutions such as the Arecibo Laboratory (National Astronomy and Ionospheric Center), Argonne National Laboratory, Brookhaven National Laboratory, Fermi National Accelerator Laboratory, Haystack Observatory, Francis Bitter National Magnet Laboratory, Jet Propulsion Laboratories, National Radio Astronomy Observatories (Green Bank, West Virginia, and Socorro, New Mexico), Stanford Linear Accelerator Center, and Soudan II Underground Laboratory.

Financial Aid

Both teaching and research assistantships are available, as are tuition scholarships. Stipends are intended to provide financial support adequate for basic living costs in the area. Summer research and teaching appointments are available to qualified students.

Cost of Study

Tuition for the 1991–92 academic year is $16,755. Teaching and research assistants are charged only half this amount, which is usually covered by a tuition scholarship. For part-time students, the tuition is $2095 per course. There is a $50 application fee.

Cost of Living

Although no on-campus housing is provided, the local area contains many private apartment facilities that customarily accommodate several persons, thus rendering living costs quite economical. Local transportation is modestly priced and readily accessible.

Student Group

There are about 50 graduate students enrolled in the physics program, about half of whom conduct research during any given year. In 1990–91, four Ph.D. and five M.S. degrees were awarded.

Location

The Boston area offers an unusual combination of historical, cultural, and educational experiences and opportunities. Tufts University, on a small wooded campus in the suburb of Medford, is in a convenient location to take advantage of all of them.

The University

The University dates back to 1850 and has a proud heritage of both undergraduate and graduate achievement. There are about 6,500 students and 300 faculty members overall. The moderate size of the student body and good student-faculty ratio foster an informal family environment not found in many larger institutions. There are many active organizations on campus, as well as a variety of local activities available for student participation.

Applying

There is no deadline, but it is recommended that applications for fall admission be received by March 15. Relevant materials may be obtained from the Department of Physics and Astronomy, and applications should be sent to the Graduate School of Tufts University. Applicants will be considered on their individual merits for any and all available positions.

Correspondence and Information

Director of Graduate Studies
Department of Physics and Astronomy
Tufts University
Medford, Massachusetts 02155

SECTION 7: PHYSICS

Tufts University

THE FACULTY AND THEIR RESEARCH

David L. Weaver, Chairman; Ph.D., Iowa State, 1963. Molecular biophysics.

Allan M. Cormack, M.Sc., Cape Town, 1945; Nobel laureate, 1979. Experimental particle physics, image formation.
Allen E. Everett, Ph.D., Harvard, 1960. Theoretical particle physics.
Lawrence H. Ford, Ph.D., Princeton, 1974. General relativity, quantum field theory.
Gary R. Goldstein, Ph.D., Chicago, 1968. Theoretical particle physics.
Robert P. Guertin, Ph.D., Rochester, 1968. Experimental solid-state physics.
Leon Gunther, Ph.D., MIT, 1964. Theoretical solid-state physics.
Tomas Kafka, Ph.D., SUNY at Stony Brook, 1974. Experimental particle physics.
Kenneth R. Lang, Ph.D., Stanford, 1969. Astrophysics, radio astronomy.
W. Anthony Mann, Ph.D., Massachusetts, 1970. Experimental particle physics.
Kathryn A. McCarthy, Ph.D., Harvard, 1957. Experimental solid-state physics.
Richard H. Milburn, Ph.D., Harvard, 1954. Experimental particle physics.
George S. Mumford, Ph.D., Virginia, 1954. Astronomy, astrophysics.
Austin Napier, Ph.D., MIT, 1978. Experimental particle physics.
William P. Oliver, Ph.D., Berkeley, 1969. Experimental particle physics.
Howard H. Sample, Ph.D., Iowa State, 1966. Experimental solid-state physics.
J. Schneps, Ph.D., Wisconsin–Madison, 1956. Experimental particle physics.
Yaakov Shapira, Ph.D., MIT, 1964. Experimental solid-state physics.
Krzysztof Sliwa, Ph.D., Jagiellonian (Poland), 1980. Experimental particle physics.
Alexander Vilenkin, Ph.D., SUNY at Buffalo, 1977. Theory of condensed matter, general relativity, quantum field theory.
Robert F. Willson, Ph.D., Tufts, 1980. Radio astronomy, astrophysics.

SECTION 7: PHYSICS

UNIVERSITY OF AKRON
Department of Physics

Programs of Study

The department offers a program of graduate study leading to the degree of Master of Science in physics. In addition, an interdisciplinary program with the Department of Polymer Science enables physics students to pursue a course of study leading to the degree of Doctor of Philosophy in polymer science, with specialization in polymer physics.

Graduate study programs are quite flexible and are designed around the particular interests and needs of individual students. The requirements for the M.S. degree in physics are a minimum of 30 semester credits with a quality point average of at least 3.0 (B) and successful completion of a comprehensive examination in physics. A thesis, constituting the report of an original research study, is optional. A period of two years of full-time study is generally required for completion of the M.S. degree. Physics graduate course sequences normally begin in the fall semester and continue through the spring semester; the summer session is generally devoted to University research projects. Thus, while students may in principle begin graduate study at any time of the year, it is generally advisable for them to enter in September.

Research Facilities

The Department of Physics is housed in a modern, air-conditioned building with good facilities and space for research and instruction. The laboratories include experimental facilities for electron tunneling spectroscopy; electron tunneling microscopy; pulsed and continuous-wave NMR; and Shubnikov-deHaas measurements. Experimental projects in progress include studies in surface physics and thin films, diffusion measurements and NMR in polymers, solid-state physics, computer-assisted instruction, and nonequilibrium physics, Hamiltonian systems, and nonlinear dynamics. Theoretical projects in progress include critical phenomena and phase transitions, renormalization group theory, supersymmetry, polymer physics, and solid-state physics. Studies of physical properties of polymeric materials use the extensive facilities of the Department of Polymer Science and the Institute of Polymer Science.

The University operates several large computers (including IBM 3090-II Vector, IBM 4381, and DEC VAX computers). The Department of Physics operates about ten UNIX workstation computers and more than ten MS-DOS desktop computers dedicated to research. In addition, there are approximately forty-five Apple computers (Macintosh, Apple II) that support course work and teachings.

Financial Aid

Teaching and research assistantships are available with 1991–92 stipends ranging from $6800 to $9000 for the academic year, plus remission of tuition and fees. Summer employment in connection with University research projects is frequently possible.

Cost of Study

Graduate fees for 1991–92 are $97.15 per credit for Ohio residents and $177.01 per credit for nonresidents. All students pay a general service fee of $8.86 per credit to a maximum of $114.23 per semester.

Cost of Living

Housing for single students is available in University-operated residence halls on campus; room and board cost about $3400 for the nine-month 1991–92 academic year. In addition, there is ample private housing in apartments and rooms in the vicinity of the University. Food and living costs in Akron are about average for most cities its size in the United States.

Student Group

There are approximately 28,900 students, including more than 3,400 graduate and law students, enrolled at the University. The number is about equally divided between full-time (day) students and part-time (evening) students. The student body includes about 420 international students from sixty nations.

At present, there are approximately 25 full-time and 10 part-time graduate students in physics. Most of the full-time students receive financial support through the department.

Location

The University of Akron is located in the center of Akron, close to downtown stores, hotels, and restaurants. In addition to the many distinguished visiting speakers and special programs regularly brought to the University campus, there are numerous local cultural activities. The Akron Symphony is quite good, and there are several first-rate community theater groups. Blossom Center, the summer home of the Cleveland Orchestra, is located in the wooded Cuyahoga Valley, immediately to the north of Akron. Cleveland is only 35 miles away. The city of Akron has a population of about 220,000. Its technological importance is due primarily to its concentration of major medical and polymer-related business offices and research centers. Akron is conveniently served by both the Cleveland and Akron-Canton airports. Because of nearby interstate highways, the University campus is only 2 hours away by car from Columbus, Ohio, and Pittsburgh, Pennsylvania. Many outdoor recreational facilities, including municipal, state, and national parks, are within easy driving distance.

The University

Founded in 1870 as Buchtel College, the University of Akron became a municipal university in 1913 and a state university in 1967. It has grown rapidly in recent years. At present, the University has more than seventy-five modern buildings on a 162-acre campus.

Applying

The University of Akron operates on the semester system. Applications for admission to the Graduate School and for a graduate assistantship should be submitted as early as possible, preferably before March 1. The admission application fee is $25.

Qualified students from other countries are most welcome at the University. Students whose native language is not English must demonstrate proficiency in English by scoring 550 or better on the TOEFL. The TOEFL should be taken preferably during the fall preceding the desired date of admission.

Since course content and grading standards vary markedly from university to university, it is recommended that applicants supply GRE scores (General Test and Subject Test in physics) in order to help establish their qualifications for graduate assistantships. All international applicants should supply GRE General Test scores prior to admission.

Requests for application forms, as well as specific questions, should be addressed to the chairman of the Graduate Admissions Committee.

Correspondence and Information

Chairman, Graduate Admissions Committee
Department of Physics
University of Akron
Akron, Ohio 44325
Telephone: 216-972-7078

SECTION 7: PHYSICS

University of Akron

THE FACULTY AND THEIR RESEARCH

Philip R. Baldwin, Assistant Professor of Physics; Ph.D., Illinois at Urbana-Champaign, 1987. Nonequilibrium physics, polymer physics, coherent structures, liquid crystals, Hamiltonian systems. (Theory)

Harry T. Chu, Professor of Physics; Ph.D., SUNY at Stony Brook, 1969. Solid-state physics: Shubnikov-deHaas effect in bismuth-antimony alloys, quantum size effect, electron tunneling. (Theory and experiment)

Roger B. Creel, Professor of Physics; Ph.D., Iowa State, 1969. Solid-state chemical and polymer physics: NMR spectroscopy and electron tunneling spectroscopy. (Experiment)

Alan N. Gent, H. A. Morton Professor of Polymer Engineering and Physics; Ph.D., London, 1955. Polymer physics: deformation and fracture processes, crystallization, adhesion. (Experiment)

C. Frank Griffin, Professor of Physics; Ph.D., Ohio State, 1964. Solid-state physics: NMR spectroscopy, physics education research, computer-assisted instruction. (Experiment)

P. D. Gujrati, Professor of Physics and Polymer Science; Ph.D., Columbia, 1979. Theoretical physics: phase transitions and critical phenomena, polymer physics, combinatorics and graph theory, renormalization group and field theory. (Theory)

Peter N. Henriksen, Professor of Physics; Ph.D., Georgia, 1968. Atomic, molecular, and solid-state physics: electron tunneling, surface physics. (Experiment)

Robert R. Mallik, Assistant Professor of Physics; Ph.D., Leicester Polytechnic, 1985. Low-temperature physics, surface physics, electron tunneling. (Experiment)

Ronald E. Schneider, Associate Professor of Physics; Ph.D., Western Reserve, 1964. Nuclear physics, astrophysics. (Theory)

Ernst D. von Meerwall, Distinguished Professor of Physics; Ph.D., Northwestern, 1969. Condensed-matter physics, mainly polymers: NMR, diffusion structure-property relations, numerical methods, computations. (Experiment)

Informal discussion.

Surface physics laboratory.

NMR laboratory.

SECTION 7: PHYSICS

UNIVERSITY OF ARIZONA

Optical Sciences Center
Graduate Study in Optics

Programs of Study

The Optical Sciences Center, dedicated in 1970, is a graduate center for research and teaching. Qualified applicants holding undergraduate degrees in engineering, mathematics, or physics are admitted to programs that lead to the Master of Science or Doctor of Philosophy. The requirements for the M.S. degree are the successful completion of 32 units of graduate credit if the thesis option is selected or 35 units of graduate credit if the nonthesis option is selected and, in either option, a final examination. The requirements for the Ph.D. degree include the successful completion of a minimum of six semesters of full-time graduate study. An oral qualifying examination and a combined written/oral preliminary examination must also be passed. In addition, a suitable dissertation must be submitted and successfully defended at the final examination. There is no foreign language requirement for students enrolled in the Ph.D. program in optical sciences. The course load for any of these programs varies from student to student, but 9 to 12 units per semester are considered to be a normal load.

Current active research areas include coherent optics; holography; image processing; infrared techniques; interferometry; laser physics; medical optics; modulation spectroscopy; optical data storage; optical design, fabrication, and testing; optical properties of materials; quantum optics; remote sensing; solar energy conversion; and thin films.

Interdisciplinary programs currently involve the Departments of Astronomy, Atmospheric Sciences, Civil Engineering and Engineering Mechanics, Electrical and Computer Engineering, Microbiology, Pathology, Physics, Planetary Sciences, and Radiology.

Research Facilities

The Center houses laboratories and equipment used to conduct research in a number of areas of optics; included are CVD and vacuum deposition thin-film facilities, darkrooms, an electronics shop, an infrared laboratory, an instrument shop, a massive-optics shop, a small-optics shop, a student/faculty machine shop, and a teaching laboratory. In addition, an ECLIPSE minicomputer, a PDS microdensitometer, and a remote computer terminal are available for use in research projects. A reading room is housed in the Center, and study desks are available for most students.

Financial Aid

Financial assistance is often available through graduate assistantships in research and teaching, which pay stipends between $8700 and $10,350 in 1991–92 for ten months of half-time service. The nonresident tuition fee is waived for holders of these assistantships. A few special industry-sponsored fellowships are available. A number of students are supported for specified periods by the particular industrial concern or government agency that employs them. In addition, Graduate Tuition Scholarships are often available for nonresidents.

Cost of Study

The registration fee for 7 or more units for both state residents and nonresidents is $764 per semester in 1991–92. For fewer than 7 units, the fee is $80 per unit. Graduate students in absentia who are registered only for work on the thesis or dissertation also pay $80 per unit.

In addition to the appropriate registration fee, nonresidents are assessed a nonresident tuition fee, which ranges from $2022 per semester for 7 units to $3467 per semester for 12 or more units. For fewer than 7 units, the nonresident tuition fee is waived.

Cost of Living

Costs are generally comparable to those in a moderately large city. There are a limited number of University-operated apartments for married students at varying rents. However, there is generally a waiting period of one year to eighteen months. Most students rent privately owned rooms, apartments, or houses throughout the city, with costs a little higher than those at the University.

Student Group

A total of 153 graduate students were enrolled in degree programs at the Optical Sciences Center during the fall semester of 1990.

Location

The University of Arizona is situated in the southern Arizona city of Tucson, a richly historic town that has grown into a bustling metropolis of approximately 560,000 people and is a winter vacation center for visitors from all parts of the United States. There are many cultural, social, and recreational activities on and off campus. The elevation of 2,400 feet, the dry air, and the abundant sunshine all help provide one of the most healthful climates in the world. For a change in scenery, the pine forests of the Santa Catalina Mountains, at an elevation of 9,200 feet, are an hour's drive away; Nogales, in Sonora, Mexico, is 66 miles to the south.

This area is one of the most active regions of astronomical observation in the United States: Kitt Peak National Observatory, Steward Observatory, the Lunar and Planetary Laboratory, and the Smithsonian Astrophysical Observatory all have facilities near Tucson.

The University

The University of Arizona, which opened in 1891, is a member of several recognized educational organizations, including the National Association of State Universities and Land Grant Colleges. It is accredited by the North Central Association of Colleges and Schools and a number of other accrediting organizations. Today the University is the most complete university in the region and enjoys an outstanding reputation for research, teaching, and service. The fall 1989 enrollment at the University of Arizona was 35,735, of whom 7,253 were graduate students.

Applying

An application for admission and two complete sets of transcripts of all previous university work must be sent to the dean of the Graduate College. In addition, one complete set of transcripts, at least two letters of recommendation, and test scores on the General Test and one Subject Test (engineering, mathematics, or physics) of the Graduate Record Examinations must be submitted to the associate director for academic affairs in the Optical Sciences Center. All required application materials must be received no later than March 1; however, early application is advised because the enrollment is limited. Applications for assistantships and Graduate Tuition Scholarships should be sent directly to the associate director. There is no absolute deadline for these financial aid applications, but early application is advised.

Correspondence and Information

Dr. Jack D. Gaskill
Associate Director
Optical Sciences Center
University of Arizona
Tucson, Arizona 85721
Telephone: 602-621-4111

SECTION 7: PHYSICS

University of Arizona

THE FACULTY AND THEIR RESEARCH

Robert R. Shannon, Professor and Director; M.S., Rochester, 1957. Optical lens design, fabrication, testing, computer applications.
J. Roger Angel, Professor; D.Phil., Oxford, 1965; joint appointment in Astronomy. Optics manufacturing.
George H. Atkinson, Professor; Ph.D., Indiana, 1971; joint appointment in Chemistry. Laser system application, laser spectroscopy.
H. Brad Barber, Assistant Research Professor; Ph.D., Arizona, 1976; joint appointment in Nuclear Medicine. Medical instrumentation, gamma-ray imaging, medical probes.
Harrison H. Barrett, Professor; Ph.D., Harvard, 1968; joint appointment in Radiology. Medical optics, coded-aperture imaging, holographic techniques.
Peter H. Bartels, Professor; Ph.D., Göttingen (Germany), 1954; joint appointment in Pathology. Image analysis, pattern recognition.
James J. Burke, Professor; Ph.D., Arizona, 1972. Image processing, optical waveguides, integrated optics.
Katherine Creath, Assistant Professor; Ph.D., Arizona, 1985. Interferometry, optical metrology, holographic and speckle nondestructive testing, microscopy.
William J. Dallas, Associate Professor; Ph.D., California, San Diego, 1973; joint appointment in Radiology. Medical image acquisition, processing, and display.
Eustace L. Dereniak, Associate Professor; Ph.D., Arizona, 1976. Medical optics, radiology, infrared systems and devices.
Charles M. Falco, Professor; Ph.D., California, Irvine, 1974; joint appointment in Physics. Low-temperature metals, superconductivity.
Peter A. Franken, Professor; Ph.D., Columbia, 1952; joint appointment in Physics. Lasers, spectroscopy, nonlinear optics.
B. Roy Frieden, Professor; Ph.D., Rochester, 1966. Image enhancement and restoration, information theory, mathematical and statistical analysis.
Kenneth F. Galloway, Professor; Ph.D., South Carolina, 1966; joint appointment in Electrical and Computer Engineering. Microelectronics, solid-state devices.
Jack D. Gaskill, Professor and Associate Director; Ph.D., Stanford, 1968; joint appointment in Electrical and Computer Engineering. Coherent optics, holography, optical data processing, medical optics.
Hyatt M. Gibbs, Professor; Ph.D., Berkeley, 1965. Quantum optics, laser physics, optical bistability.
Arthur Gmitro, Associate Professor; Ph.D., Arizona, 1982; joint appointment in Radiology. Magnetic resonance imaging, optical computing, neural networks, laser angioplasty.
John E. Greivenkamp, Associate Professor; Ph.D., Arizona, 1980. Interferometry, optical fabrication and testing, optics of electronic imaging systems.
Stephen F. Jacobs, Professor; Ph.D., Johns Hopkins, 1956. Infrared detectors, lasers, optical communications, interferometry.
Michael Ray Jacobson, Associate Research Professor; Ph.D., Cornell, 1977; joint appointment with Material Science and Engineering. Optical thin films: deposition and characterization, optical materials, optical measurements, solar energy conversion.
Stephan W. Koch, Professor; Ph.D., Frankfurt, 1979; joint appointment in Physics. Theoretical studies of optical nonlinearities in laser-excited semiconductors, theory of optical bistability in semiconductors.
Raymond K. Kostuk, Assistant Professor; Ph.D., Stanford, 1986; joint appointment in Electrical and Computer Engineering. Holography, optical interconnects, optical data storage, electrooptics.
George L. Lamb Jr., Professor; Ph.D., MIT, 1958; joint appointment in Mathematics. Various problems in classical wave propagation, primarily with application to acoustics and vibration.
Willis E. Lamb Jr., Professor; Ph.D., Berkeley, 1938; joint appointment in Physics. Laser physics, atomic structure.
H. Angus Macleod, Professor; D.Tech., Council for National Academic Awards (Scotland), 1979. Design and properties of thin films, narrow-band filters, optical waveguides.
Masud Mansuripur, Associate Professor; Ph.D., Stanford, 1981. Optical data storage, magnetooptics; optics of polarized light in systems of high numerical aperture, magnetic and magnetooptical properties of thin solid films.
Arvind S. Marathay, Professor; Ph.D., Boston University, 1963. Coherence theory, partial polarization, optical spatial filtering.
Aden B. Meinel, Professor Emeritus; Ph.D., Berkeley, 1949; joint appointment in Astronomy. Astrophysics, spectroscopy, optical design, instrumentation, utilization of solar energy.
Pierre Meystre, Professor; Ph.D., Swiss Federal Institute of Technology, 1974. Theoretical analysis of fully microscopic systems of one or few atoms in interaction with a single mode of the electromagnetic field.
Tom D. Milster, Assistant Professor; Ph.D., Arizona, 1986. Optical data storage, optical testing, microoptics, statistical optics.
James M. Palmer, Research Associate and Adjunct Lecturer; Ph.D., Arizona, 1975. Radiometric calibration, solar radiation and atmospheric radiometry, optical radiation detectors.
Robert E. Parks, Adjunct Lecturer; M.A., Williams, 1966; Staff Scientist, Steward Observatory. Computer-controlled machining of aspheric surfaces, optical testing technology.
Nasser Peyghambarian, Associate Professor; Ph.D., Indiana, 1982. Laser spectroscopy of solids, nonlinear optical properties of semiconductors, optical bistability, quantum optics.
Ralph M. Richard, Professor; Ph.D., Purdue, 1961; joint appointment in Civil Engineering and Engineering Mechanics. Deformations due to mechanical and thermal loads, wavefront error analysis.
Hans Roehrig, Associate Research Professor; Ph.D., Giessen (Germany), 1964; joint appointment in Radiology. Medical imaging, image evaluation, photoelectronic imaging devices.
Murray Sargent III, Professor; Ph.D., Yale, 1967. Laser physics, computer sciences.
Dror Sarid, Professor; Ph.D., Hebrew (Jerusalem), 1972. Laser light scattering, electrooptics, acoustooptics, integrated optics.
Robert Schowengerdt, Associate Professor; Ph.D., Arizona, 1975; joint appointment in Office of Arid Lands Studies and Electrical and Computer Engineering. Thematic Mapper MTF evaluation, image restoration.
Bernhard O. Seraphin, Professor Emeritus; Ph.D., Humboldt State, 1951. Modulation spectroscopy, optical properties of solids, solar energy.
Roland V. Shack, Professor; Ph.D., London, 1965. Image formation and evaluation, optical design and testing, interference and diffraction, light scattering.
Richard L. Shoemaker, Professor; Ph.D., Illinois at Urbana-Champaign, 1971; joint appointment in Chemistry. Nonlinear optical effects, ultrahigh-resolution spectroscopy, optical transient effects, coherent molecular interactions.
Philip N. Slater, Professor; Ph.D., London, 1958. Photographic remote sensing, systems and techniques.
Orestes N. Stavroudis, Professor Emeritus; Ph.D., London, 1959. Computer design and analysis of optical systems, structure of the geometric image.
Robin N. Strickland, Associate Professor; Ph.D., Sheffield, 1979; joint appointment in Electrical and Computer Engineering. Signal processing, digital image processing, image coding for bandwidth compression.
A. Francis Turner, Professor Emeritus; Ph.D., Berlin, 1935. Thin films, infrared signatures, air pollution, integrated optics.
Donald R. Uhlmann, Professor; Ph.D., Harvard; joint appointment in Materials Science. Molecular composites, organic modified oxides, optical materials, solgel synthesis of ceramics.
William H. Wing, Professor; Ph.D., Michigan, 1968; joint appointment in Physics. Atomic and molecular physics, laser spectroscopy.
William L. Wolfe, Professor; M.S., 1956, M.S.E., 1966, Michigan; joint appointment in Radiology. Infrared systems and devices, radiometry, optical materials.
Ewan M. Wright, Assistant Professor; Ph.D., Heriot-Watt (Scotland), 1983. Nonlinear and integrated optics, quantum optics.
James C. Wyant, Professor; Ph.D., Rochester, 1968. Optical testing, interferometry, holography.

SECTION 7: PHYSICS

UNIVERSITY OF ARIZONA

Department of Physics

Programs of Study	The department prepares students for employment in universities, industry, and government. Graduate students are encouraged to participate in the teaching program of the department. Students are expected to take a qualifying examination (primarily on undergraduate physics) during their first week of residence. The M.S. degree requires passing this qualifying examination and completing at least 30 units (semester hours) of graduate work, of which at least 15 must be in physics. In addition, the M.S. degree candidates are required to choose one of three options: writing a thesis (up to 6 hours of credit) and passing an oral examination, completing 6 additional units of graduate physics and passing an oral examination, or passing the written and oral parts of the preliminary examination for the Ph.D. The Ph.D. degree requires the completion of at least 36 units of graduate work, exclusive of dissertation credits; a passing grade on the preliminary examination, which must be taken by the end of the fifth semester of residence; the demonstration of a reading knowledge of one foreign language; a dissertation based on independent research; and a defense of the dissertation in a final examination. Interdisciplinary work with other departments is possible, especially in the areas of chemical physics, materials and surface physics, astrophysics, space physics, and optical sciences.
Research Facilities	The department's general facilities include a shop for students and staff and a physics common room for preprints and current physics journals. A more extensive collection of physics journals and more than 5,000 physics books are located in the science-library building. The department's research and teaching are greatly facilitated by access to the University's Instrument Shop, Central Electronics Shop, and Liquid Cryogenics Facility. The University is part of an NSF consortium for supercomputers with a satellite link to the Princeton CYBER 205. The University also operates on campus an IBM 3090 Model 300E system, a Convex C240 minisupercomputer, and several VAX's, one of which is in the Department of Physics. IBM PC and Apple microcomputers and Sun Workstations are available in the department for general use. Facilities and equipment for particular research areas in physics include two Van de Graaff accelerators (2 MV and 6 MV), a 3-MV tandem accelerator, a mountain-altitude cosmic-ray laboratory, spectrographs, spectrometers, superconducting magnets for solid-state physics, electron microscopes, electron probes (including Auger, RHEED, and LEED), a materials-processing laboratory, thin-film molecular beam epitaxy (MBE) and sputter epitaxy equipment, thin-film X-ray diffraction facilities, high-pressure systems, cryogenic systems covering the entire temperature range of 300 K down to 0.02 K, a computerized Mössbauer spectroscopy system, ultrahigh-resolution infrared and visible lasers, atomic-beam machines, and an observatory for work in experimental relativity and solar seismology. There is an active program in theoretical physics covering atomic, nuclear, and condensed-matter physics, astrophysics, and high-energy theory. The Departments of Physics and Geosciences jointly operate a facility, sponsored by the National Science Foundation, that uses accelerators for radioisotope dating.
Financial Aid	The stipend for a teaching assistantship for the 1990–91 academic year ranged from $8468 to $9821 for 20 hours of work per week (preparation and grading included). Research assistantships are often available, especially after the first year of graduate work. The stipend is the same as that for teaching assistants with comparable qualifications. The out-of-state tuition fee of $2703 per semester is waived for teaching and research assistants. Tuition scholarships are available through the Scholarships and Financial Aids Office. Entering students are considered for Faculty of Science Fellowships of $10,000 each and U.S. Department of Education Fellowships with stipends up to $12,500.
Cost of Study	The registration fee for 7 or more units for both state residents and nonresidents is $795 per semester in 1991–92. Nonresidents also pay a nonresident tuition fee, which ranges from $1258 per semester for 7 units to $2703 per semester for 12 or more units.
Cost of Living	For single students in fall 1990, graduate dormitory rates were $1362 per academic year for an air-conditioned double-room unit. The cost of food at the University cafeterias was $224 per month. University housing facilities for married students cost $240–$455 per month.
Student Group	The University has a student body of about 35,735, including 8,213 graduate students. Over 2,000 students come from foreign countries. Approximately two thirds of the 145 graduate students in physics are supported by assistantships. Most of the graduate students live off campus and are in their late twenties. The atmosphere of the department is informal.
Location	Metropolitan Tucson is in a rapidly growing area with a population of about 680,000. It is surrounded by mountains and is about 65 miles north of Nogales, Mexico. Midday winter temperatures are in the 60s and 70s; low humidity and sunshine prevail. Tucson is a winter resort with many outdoor sports, including skiing an hour away in the 9,200-foot Catalina Mountains. The city has its own symphony orchestra with a resident director. Excellent museums, concerts, theater, entertainment areas (including those in the Community Center complex), and public lectures provide many interesting activities. Founded about A.D. 800 by an Indian tribe and involved in early Spanish exploration, Tucson is in an area of considerable historic interest.
The University and The Department	The University is a state-supported school founded in 1885. The spacious campus with its air-conditioned buildings is designed for comfortable year-round study and research. The University, which has its own 90-inch telescope, is in a center for astronomical research, being adjacent to the Kitt Peak National Observatory and 50 miles from the Kitt Peak telescopes. The department is large enough to have a strong and reasonably diversified research program, yet it maintains close and informal relationships between faculty members and graduate students and with people in different research areas through colloquia, teas, small informal seminars, scattered bull sessions, and occasional picnics and parties.
Applying	Application forms for admission and financial assistance may be obtained from the department at the address given below. The department requires that all applicants take the GRE General Test and the Subject Test in physics and have the scores sent to the department. Applications for assistantships should be made before February 1; notification of awards is made after March 15.
Correspondence and Information	Admissions and Assistantships Committee Department of Physics University of Arizona 1118 East 4th Street Tucson, Arizona 85721 Telephone: 602-621-2290

University of Arizona

THE FACULTY AND THEIR RESEARCH

Atomic, Molecular, and Optical Physics
Stanley Bashkin, Professor; Ph.D., Wisconsin. Spectroscopy of ions (beam-foil spectroscopy).
William S. Bickel, Professor; Ph.D., Penn State. Experimental optics and spectroscopy, polarized light scattering, acoustics.
Douglas J. Donahue, Professor; Ph.D., Wisconsin. Atomic and nuclear physics, accelerator mass spectrometry, radiocarbon dating.
Peter Franken, Professor; Ph.D., Columbia. Experimental atomic physics, quantum optics.
John D. McCullen, Professor; Ph.D., Colorado. Experimental atomic physics, quantum optics.
Laurence C. McIntyre Jr., Professor; Ph.D., Wisconsin. Experimental atomic physics, ion-beam analysis.
John O. Stoner Jr., Professor; Ph.D., Princeton. Experimental atomic physics, spectroscopy, thin-film optics.
William H. Wing, Professor; Ph.D., Michigan. Modern spectroscopy, high-precision metrology, astrophysics, chemical physics.

Biophysics
John O. Kessler, Professor; Ph.D., Columbia. Individual and collective phenomena of swimming cells, theory and experiment.
Rein Kilkson, Professor; Ph.D., Yale. Molecular and theoretical biophysics.

Condensed-Matter Physics
Robert H. Chambers, Professor; Ph.D., Carnegie Tech. Experimental solid-state physics, dislocation dynamics, internal friction.
Roy M. Emrick, Professor; Ph.D., Illinois at Urbana-Champaign. Experimental physics, thin films, Mössbauer spectroscopy.
Charles M. Falco, Professor; Ph.D., California, Irvine. Artificially structured materials, thin-film physics, superconductivity, magnetism.
Donald R. Huffman, Professor; Ph.D., California, Riverside. Experimental physics, spectroscopy of solids, astrophysics.
John A. Leavitt, Professor; Ph.D., Harvard. Experimental atomic physics, ion-beam analysis.
Royal W. Stark, Professor; Ph.D., Case Tech. Electronic properties of metals, low-temperature physics.
Wing Yim Tam, Assistant Professor; Ph.D., California, Santa Barbara. Experimental fluid mechanics.
Carl T. Tomizuka, Professor; Ph.D., Illinois at Urbana-Champaign. Study of defects at high pressures, use of computers in instruction.
Joseph J. Vuillemin, Professor; Ph.D., Chicago. Experimental Fermi surfaces.

Cosmic-Ray and High-Energy Physics
Theodore Bowen, Professor; Ph.D., Chicago. Experimental cosmic rays, elementary particle physics, medical physics.
Geoffrey E. Forden, Assistant Professor; Ph.D., Indiana. High-energy experimentalist.
Ke-Chiang Hsieh, Associate Professor; Ph.D., Chicago. Experimental space physics.
Edgar W. Jenkins, Professor; Ph.D., Columbia. Experimental elementary particle physics.
Kenneth A. Johns, Assistant Professor; Ph.D., Rice. High-energy experimentalist.
John P. Rutherfoord, Professor; Ph.D., Cornell. Experimental high-energy physics.
Michael A. Shupe, Associate Professor; Ph.D., Tufts. Experimental high-energy physics.

Experimental Relativity and Solar Seismology
Henry A. Hill, Professor; Ph.D., Minnesota. Experimental relativity, solar seismology.

Theoretical Physics
W. David Arnett, Professor; Ph.D., Yale. Theoretical astrophysics.
Bruce R. Barrett, Professor; Ph.D., Stanford. Nuclear many-body theory, nuclear structure theory, interacting-boson model.
Adam S. Burrows, Associate Professor; Ph.D., MIT. Astrophysics, neutron stars, supernovae.
Peter Carruthers, Professor and Head; Ph.D., Cornell. Elementary particle and field theory.
Jose D. Garcia, Professor; Ph.D., Wisconsin. Atomic collision theory.
Anna Hasenfratz, Associate Professor; Ph.D., Eötvös Loránd (Budapest). High-energy theory.
Kurt W. Just, Professor; Dr.rer.nat., Berlin. Theory of particles, fields, and gravity.
Stephan W. Koch, Associate Professor; Ph.D., Frankfurt. Condensed-matter theory.
Sigurd Köhler, Professor; Ph.D., Uppsala (Sweden). Nuclear many-body theory, heavy-ion collision theory.
Willis E. Lamb Jr., Professor; Ph.D., Berkeley. Theoretical atomic physics, quantum optics.
Sumit Mazumdar, Associate Professor; Ph.D., Princeton. Condensed-matter theory.
Richard L. Morse, Professor; Ph.D., California, San Diego. Theoretical plasma physics.
Robert H. Parmenter, Professor; Ph.D., MIT. Many-body theory, superconductivity, ferromagnetism.
Adrian Patrascioiu, Professor; Ph.D., MIT. Theoretical elementary particle physics.
Johann Rafelski, Professor; Ph.D., Frankfurt. Relativistic nuclear theory, muon-catalyzed fusion.
Ina Sarcevic, Assistant Professor; Ph.D., Minnesota. High-energy theory.
Michael D. Scadron, Professor; Ph.D., Berkeley. Theoretical elementary particle physics.
Alwyn C. Scott, Professor; D.Sc., MIT. Complex system.
Daniel Stein, Associate Professor; Ph.D., Princeton. Condensed-matter theory.
Robert L. Thews, Professor and Associate Head; Ph.D., MIT. Elementary particle theory.
Douglas Toussaint, Associate Professor; Ph.D., Minnesota. High-energy theory.

AFFILIATED STAFF
George Burr, Research Scientist; Ph.D., Chicago. Geophysical science.
Charles C. Curtis, Research Assistant Professor; Ph.D., Arizona. Space physics.
E. David Davis, Research Associate; Ph.D., Witwatersrand (Johannesburg). Nuclear theory.
John Dutcher, Research Associate; Ph.D., Simon Fraser. Brillouin light scattering.
Brad Engel, Research Associate; Ph.D., Florida. Surface magnetism and magnetooptics.
Lisa Ensman, Research Intern; Ph.D., California, Santa Cruz. Astronomy and astronomical physics.
Bruce A. Fryxell, Senior Research Associate; Ph.D., Illinois. Theoretical astrophysics.
Rajesh Gandhi, Research Associate; Ph.D., Wisconsin. High-energy theory.
Martin Greiner, Adjunct Assistant Professor; Ph.D., Giessen (Germany). Theoretical physics.
Anthony J. T. Jull, Research Scientist; Ph.D., Bristol (England). Geochemistry.
Aleksandar Kocic, Research Associate; Ph.D., Illinois. High-energy theory.
Nai Hang Kwong, Research Associate; Ph.D., Caltech. Many-body theory, atomic collision theory.
Kwan-Wu Lai, Research Professor; Ph.D., Michigan. Experimental high-energy physics.
Peter Lipa, Research Associate; Ph.D., International High Energy Physics, (Vienna). High-energy theory.
Fulvio Melia, Associate Professor; Ph.D., MIT. Theoretical astronomical physics.
Ludwig Neise, Adjunct Assistant Professor; Ph.D., Frankfurt. Nuclear theory.
Paul Oglesby, Research Associate; Ph.D., Arizona. Solar seismology.
Alburt Pifer, Adjunct Research Associate; Ph.D., Brown. Partical physics.
Jon Slaughter, Assistant Research Professor; Ph.D., Michigan State. Condensed-matter physics.
Eric Swartz, Adjunct Assistant Professor; Ph.D., Cornell. Low-temperature physics.
Alex Wilson, Adjunct Professor; Ph.D., Berkeley. Radiocarbon dating.

UNIVERSITY OF CALIFORNIA, LOS ANGELES
Department of Physics

Programs of Study

The department offers programs of study leading to the degrees of Master of Arts in Teaching, Master of Science, and Doctor of Philosophy in physics.

Required for the M.S. degree are the passing of a written examination and 9 quarter courses, of which 6 must be graduate courses in physics and 3 may be acceptable undergraduate or graduate courses. This degree is awarded only to students in the Ph.D. Program who have satisfied these requirements.

The requirements for the Ph.D. are (1) a written comprehensive examination on the level of a first-year core graduate physics curriculum, (2) an oral presentation based on a preliminary research project, (3) an oral qualifying examination given by the student's own doctoral committee, and (4) a dissertation based on original research.

Research Facilities

There are facilities for experimental research in accelerator physics, astrophysics, elementary particle physics, intermediate-energy nuclear physics, low-temperature physics, nuclear structure, physical acoustics, plasma physics, solid-state physics, and spectroscopy. The campus computing network is available with a variety of high-speed computers. The department has several on-line computer consoles oriented to problems of mathematical physics.

Financial Aid

Teaching assistantships are available and began at $12,015 for nine months in 1990–91. Fellowships and research assistantships are also available.

Cost of Study

In 1990–91, registration and other fees totaled $734 per quarter (subject to increase). Students who had not been bona fide residents of California for more than one year were charged an additional $2138 per quarter. Waivers, full and partial, of this out-of-state tuition are granted on the basis of academic excellence.

Cost of Living

The estimated cost of living for California residents is $13,000 for three quarters, including all fees, room and board, and books. Married student housing is available at costs that range from about $6000 to $12,000 for three quarters, including utilities.

Student Group

The total number of UCLA students is more than 32,000; about 11,500 are graduate students, and 180 of these are in physics. There are approximately 230 undergraduate majors in physics. About 95 percent of the graduate physics students support themselves on stipends from various fellowships and assistantships.

Location

UCLA is located in Westwood, a pleasant suburban area to the west of Los Angeles' Civic Center and just a few miles from the Pacific Ocean. Nearby beaches, mountains, and deserts offer diverse leisure-time activities. A campus recreation center, including an Olympic-size swimming pool, twenty-one tennis courts, and several sports fields, and a recreation club program offer opportunities for students to participate in some thirty different activities on and off campus. Public lectures, concerts, art exhibits, and the theater are an integral part of the University community.

The University

Established in 1919, UCLA moved in 1929 to its present 383-acre campus. The Department of Physics occupies two adjacent buildings, Knudsen Hall and Kinsey Hall, and has shared in the rapid growth of the University in the last ten years.

Applying

Applications for admission and financial aid (assistantships/fellowships) are obtainable from the Graduate Student Office, Department of Physics. They must be filed with a complete set of duplicate transcripts of records by January 15 for the fall quarter.

Correspondence and Information

Graduate Affairs Officer
Department of Physics
University of California
Los Angeles, California 90024
Telephone: 213-825-2307

SECTION 7: PHYSICS

University of California, Los Angeles

THE FACULTY AND THEIR RESEARCH

Professors: Maha Abdalla, Ph.D., Imperial College (London), 1971: space plasma physics. Ernest Abers, Ph.D., Berkeley, 1963: elementary particle theory. Schlomo Alexander, Ph.D., Hebrew (Jerusalem), Weizmann (Israel), 1958: solid-state theory. Eric Becklin, Ph.D., Caltech, 1968: astrophysics. Claude Bernard, Ph.D., Harvard, 1976: elementary particle theory. Rubin Braunstein, Ph.D., Syracuse, 1954: solid-state experimental. Robjin Bruinsma, Ph.D., USC, 1979: solid-state theory. Charles Buchanan, Ph.D., Stanford, 1966: elementary particle experimental. Nina Byers, Ph.D., Chicago, 1956: elementary particle theory. Sudip Chakravarty, Ph.D., Northwestern, 1976: solid-state theory. Marvin Chester, Ph.D., Caltech, 1961: solid-state experimental. W. Gilbert Clark, Ph.D., Cornell, 1961: solid-state experimental. David Cline, Ph.D., Wisconsin, 1965: accelerator, astroparticle, and elementary particle experimental. John Cornwall, Ph.D., Berkeley, 1963: elementary particle theory. Ferdinand Coroniti, Ph.D., Berkeley, 1969: space physics theory. John Dawson, Ph.D., Maryland, 1956: plasma theory. Eric D'Hoker, Ph.D., Princeton, 1981: elementary particle theory. Sergio Ferrara, Ph.D., Rome, 1968: elementary particle theory. Burton Fried, Ph.D., Chicago, 1952: plasma theory. Christian Fronsdal, Ph.D., UCLA, 1957: field theory. George Gruner, Ph.D., Hungarian Academy of Science, 1977: solid-state experimental. George Igo, Ph.D., Berkeley, 1953: intermediate energy; nuclear experimental. Charles Kennel, Chairman of the Department; Ph.D., Princeton, 1964: plasma and astrophysics theory. Steven Kivelson, Ph.D., Harvard, 1979: solid-state theory. Leon Knopoff, Ph.D., Caltech, 1949: geophysics theory. Ian McLean, Ph.D., Glasgow, 1974: astrophysics. George Morales, Ph.D., California, San Diego, 1973: plasma theory. Steven Moszkowski, Ph.D., Chicago, 1952: nuclear theory. Bernard Nefkens, Ph.D., Utrecht, 1959: particle physics experimental. Richard Norton, Ph.D., Pennsylvania, 1960: elementary particle theory. Raymond Orbach, Ph.D., Berkeley, 1960: solid-state theory and experimental. Roberto Peccei, Ph.D., MIT, 1969: elementary particle theory. Rene Pellat, D.Sc., Paris, 1967: plasma theory. Claudio Pelligrini, Ph.D., Rome, 1965: accelerator experimental. Seth Putterman, Ph.D., Rockefeller, 1970: low-temperature and acoustical theory. Joseph Rudnick, Ph.D., California, San Diego, 1970: solid-state theory. Peter Schlein, Ph.D., Northwestern, 1959: high-energy experimental. William Slater, Ph.D., Chicago, 1958: high-energy experimental. Reiner Stenzel, Ph.D., Caltech, 1970: experimental plasma physics. Terry Tomboulis, Ph.D., MIT, 1975: elementary particle theory. Charles Whitten, Ph.D., Princeton, 1966: nuclear experimental. Gary Williams, Ph.D., Berkeley, 1974: low-temperature experimental. Alfred Y. Wong, Ph.D., Princeton, 1963: plasma experimental. Chun W. Wong, Ph.D., Harvard, 1965: nuclear theory.
Professor in Residence: Walter Gekelman, Ph.D., Stevens, 1972: experimental plasma physics.
Associate Professors: Robjin Bruinsma, Ph.D., USC, 1979: solid-state theory. Robert Cousins, Ph.D., Stanford, 1981: high-energy experimental. Eric D'Hoker, Ph.D., Princeton, 1981: elementary particle theory. Graciela Gelmini, Ph.D., La Plata, 1981: elementary particles.
Assistant Professors: Katsushi Arisaka, Ph.D., Tokyo, 1985: high-energy experimental. Shechao Feng, Ph.D., Harvard, 1986: solid-state theory. Jay Hauser, Ph.D., Caltech, 1985: elementary particle experimental. Hong-Wen Jiang, Ph.D., Case Western, 1989: condensed-matter experimental. Thomas Mueller, Ph.D., Bonn, 1983: high-energy experimental. James Rosenzweig, Ph.D., Wisconsin, 1988: accelerator experimental. Hidenori Sonoda, Ph.D., Caltech, 1985: elementary particle theory.

RESEARCH ACTIVITIES

High-Energy Physics. Several active groups are carrying out experiments designed to study the strong, electromagnetic, and weak interactions of quarks and gluons using counter, chamber, and calorimeter techniques. The experiments are carried out with high-energy accelerators and storage rings at the Stanford Linear Accelerator Center, Fermilab, the Brookhaven National Laboratory, and CERN in Switzerland. Theoretical investigations lie in the general areas of continuum gauge field theories, lattice gauge theories, supersymmetry, strings and superstrings, source theory, and gravitation and deal with unified theories of strong, electroweak, and gravitational interactions. Work in these general areas includes analytical, phenomenological, and computational (i.e., computer-based) studies. The high-energy groups maintain a VAX 6320 with high-speed network links to major research laboratories worldwide.
Particle Physics and Nuclear Structure at Intermediate Energies. The interactions of mesons, photons, and protons with polarized protons, deuterons, and light nuclei are being studied to determine the pp, pd, πN, and KN scattering parameters and to test the validity of charge symmetry, isospin invariance, and time reversal invariance. Polarization and spin transfer measurements for polarized proton scattering on nuclei are being used to investigate intermediate-energy reaction mechanisms and nuclear structure. The experiments are performed at the Los Alamos Meson Physics Facility, the TRIUMF Meson Physics Facility at Vancouver, the National Laboratory for High Energy Physics at Tsukuba, Japan, the Saturne II Accelerator at Saclay, France, and the Bates-MIT Electron Accelerator. Theoretical studies of the structure and properties of nuclei and particles are being pursued, including the study of nuclear forces, effective interactions in nuclei, and the nuclear equation of state.
Relativistic Heavy Ions. At a bombarding energy of 2.5 Gev/A, the subthreshold production of strange particles and antiprotons is currently being studied at the Bevalac. Dilepton production experiments at 15 Gev/A are being planned for the Alternating Gradient Synchrotron facility.
Low-Temperature and Acoustical Physics. The principal interests of this group include the superfluid isotopes of liquid helium, ultrasonics, nonlinear acoustics, and nonlinear fluctuating hydrodynamics.
Plasma Physics. Research on a broad class of problems in basic plasma physics and space plasma physics is carried out by several theoretical, experimental, and computer simulation groups. Topics investigated include magnetic confinement of plasmas, RF heating, properties of nonneutral plasmas, ionospheric modification by radio waves, plasma techniques for particle acceleration, advanced radiation sources, properties of planetary magnetospheres, and plasmas near pulsars. Experimental facilities on campus include tokamaks and other magnetic confinement configurations; a large-scale reconnection experiment with advanced computer diagnostics; a 10-meter-long, high-magnetic field device for studying plasma waves and turbulence; and several smaller discharge plasma devices devoted to nonlinear wave studies. An ionospheric heating facility located in Alaska is operated by UCLA. Versatile computer terminal facilities and Sun Workstations provide access to VAX computers, an SCS40 mini-Cray, an Ardent graphics computer, and an IBM 3090 computer on campus and also to a number of supercomputers nationwide, including the one in San Diego and the National Magnetic Fusion Energy Computer Center in Livermore.
Condensed-Matter Physics. A wide range of topics are under active investigation. Theoretical work is being pursued in the areas of high-temperature superconductivity, interplay of magnetism and superconductivity, granular superconductivity, fractional quantum Hall effect, macroscopic quantum tunneling, Molt phenomena and strongly correlated systems, transport phenomena in solids, energy transfer, nonequilibrium superconductivity, random systems, phase transitions in polymers and gels, polymer glasses, colloids and colloidal crystals, macromolecular mesophases, and polymer solutions. Experimental work includes magnetic resonance in spin glasses; transient phenomena in thin superconducting strips; excitation transfer in glasses; single particle and Josephson tunneling in superconductors; proximity effect studies of normal metals; electron localization in one- and two-dimensional metals; thermal, frequency-dependent electrical, and nonlinear transport in quasi one-dimensional conductors; impurity and disorder effects in charge density wave systems; competition between magnetism and superconductivity; competition between charge density wave formation and superconductivity; structural phase transitions in colloidal crystals; hydrodynamic instabilities in strongly interacting colloids; phase transitions in polymeric glasses; magnetic, superconducting, and disorder properties of metallic glasses; very low temperature magnetic properties of random exchange Heisenberg antiferromagnetic chains; linear and nonlinear optics; semiconductor transport; photochromic and electrochromic glasses; phase transitions in disordered media; glasses; low-frequency modes; far-infrared cyclotron resonance; Brillouin scattering; band structure of metals and semiconductors; light scattering and absorption in biologically interesting molecules; and surface studies.

SECTION 7: PHYSICS

UNIVERSITY OF CALIFORNIA, RIVERSIDE

Department of Physics

Programs of Study

The department offers programs leading to the M.S. and Ph.D. degrees in physics.

All candidates for the M.S. degree must complete 36 quarter units of approved courses. Of these, at least 24 quarter units must be in the 200 series offered for a letter grade, with an average of B or better. The remaining requirement is completion of either of the following two plans: Plan I—satisfactory completion of a thesis; or Plan II—satisfactory performance on a comprehensive examination.

The Ph.D. requirements are a written comprehensive examination, satisfactory completion of seven basic graduate courses, an oral examination in the general area of the candidate's proposed research, a dissertation describing the results of the candidate's original research, and a final oral examination conducted by the candidate's doctoral committee.

Research Facilities

The department occupies a modern, specially designed building and auditorium complex. General facilities include machine and electronics shops, staging areas for construction of large equipment, a magnet laboratory, high-temperature and high-vacuum apparatus, and cryogenics facilities. There are facilities for doing experimental work in atomic physics, elementary particle physics, plasma physics, condensed-matter physics, and astrophysics and space physics. Details are given in the research activities section on the reverse of this page.

The Department of Physics operates a cluster of VAX computers and VAXstations as a general analysis facility. Special graphics devices include Evans & Sutherland and Megatek 3-D color systems and numerous laser printers. There are microcomputers available for undergraduate lab courses, research, and general student use. On campus there are a VAXcluster for academic computing, a campuswide computer network, and many microcomputers at central PC labs. UCR is a member of both the San Diego Supercomputer Consortium and the Illinois National Center for Supercomputer Applications offering CRAY Y-MPs.

Financial Aid

Teaching assistantships are available at a basic rate of $12,015 for the 1991–92 nine-month academic year. Research assistantships and fellowships are also available. Approximately 80 percent of the graduate students in physics receive some form of financial support.

Cost of Study

Registration and other fees are approximately $807 per quarter in 1991–92. An additional out-of-state tuition fee of approximately $3400 per quarter is charged to non-California residents. U.S. citizens and permanent residents of the United States may become California residents after living in the state for one year. The nonresident tuition fee may be waived for domestic teaching assistants and fellowship holders.

Cost of Living

Two- and three-bedroom houses, renting for approximately $260 and $280, respectively (1990–91 rates), are available for families at the Canyon Crest Student Village on campus. There are a wide variety of accommodations in the neighborhood of the campus, and many modern apartment buildings are within walking distance.

Student Group

The campus enrollment for 1990–91 was about 8,700 students; roughly one fifth of these students were enrolled in graduate programs. Graduate enrollment in physics numbers about 70, with students coming from many parts of the United States and abroad.

Location

Riverside is a city of approximately 236,000 people located 50 miles east of Los Angeles and 100 miles north of San Diego. The area has many orange groves, a superb winter climate, and abundant recreational facilities. The San Bernardino and San Jacinto mountains close by offer hiking and, in the winter, downhill and cross-country skiing. Palm Springs and the high desert are 60 miles to the east, and the Pacific Ocean is 50 miles to the west.

The University

The University occupies a spacious 1,200-acre modern campus at the foot of the Box Spring Mountains. The campus was established in 1907 with the founding of the Citrus Experimental Station to conduct research in the agriculture of southern California. In 1954, the College of Letters and Science was established, and in 1959 Riverside became a general campus of the University of California. Traditionally, the sciences have been strongly emphasized on this campus.

Applying

Requests for an application for admission should be directed to the graduate adviser, Department of Physics. Applications must be on file by June 1 for the fall quarter, September 1 for the winter quarter, and December 1 for the spring quarter for domestic applicants and two months earlier for foreign applicants. Applications for financial aid should be submitted by February 1 for consideration for the following academic year.

Correspondence and Information

Graduate Adviser
Department of Physics
University of California, Riverside
Riverside, California 92521
Telephone: 714-787-5332

Peterson's Guide to Graduate Programs in the Physical Sciences and Mathematics 1992

SECTION 7: PHYSICS

University of California, Riverside

THE FACULTY AND THEIR RESEARCH

In addition to the faculty members listed below, there are about 16 postdoctoral scholars and visiting professors.

Leon J. Bruner, Professor; Ph.D., Chicago, 1959. Electrical phenomena in biological systems, pattern recognition.
Byung-Ho Choi, Adjunct Professor; Ph.D., North Carolina at Chapel Hill, 1969. Theoretical atomic and molecular physics.
Shu-Yuan Chu, Visiting Professor; Ph.D., Berkeley, 1961. Theoretical high-energy physics.
Frederick W. Cummings, Professor; Ph.D., Stanford, 1960. Theoretical physics and quantum theory, cavity quantum electrodynamics, many-body theory, neural networks.
Bipin R. Desai, Professor; Ph.D., Berkeley, 1961. Theoretical high-energy physics.
Glen E. Everett, Professor; Ph.D., Chicago, 1962. Experimental solid-state physics and properties of magnetic materials.
Sun-Yiu Fung, Professor; Ph.D., Berkeley, 1963. Experimental high-energy physics and nuclear physics.
Peter E. Kaus, Professor; Ph.D., UCLA, 1955. Theoretical high-energy physics.
Anne Kernan, Professor; Ph.D., Ireland (Dublin), 1958. Experimental high-energy physics.
John Layter, Adjunct Professor; Ph.D., Columbia, 1972. Experimental high-energy physics.
Nai-Li Huang Lui, Professor; Ph.D., UCLA, 1966. Theoretical solid-state physics, surface physics.
Ernest Ma, Professor; Ph.D., California, Irvine, 1970. Theoretical high-energy physics.
Douglas E. MacLaughlin, Professor; Ph.D., Berkeley, 1966. Experimental solid-state physics, pulsed nuclear magnetic resonance.
Donald C. McCollum, Professor; Ph.D., Berkeley, 1960. Experimental atomic and molecular physics: electron-atom scattering.
John C. Nickel, Professor; Ph.D., Caltech, 1964. Experimental atomic and molecular physics: electron-atom scattering.
Michael Pollak, Professor; Ph.D., Pittsburgh, 1958. Theoretical and experimental physics of disordered systems.
Richard Seto, Assistant Professor; Ph.D., Columbia, 1983. Experimental relativistic heavy ion/particle physics.
Benjamin C. Shen, Professor and Chairman; Ph.D., Berkeley, 1965. Experimental high-energy physics, energy science.
Eugen Simanek, Professor; Ph.D., Czechoslovak Academy of Sciences, 1963. Theoretical solid-state physics.
Sandor Trajmar, Adjunct Professor; Ph.D., Berkeley, 1961. Atomic and molecular physics.
Gordon Van Dalen, Professor; Ph.D., California, Riverside, 1978. Experimental high-energy physics, computer analysis.
R. Stephen White, Professor; Ph.D., Berkeley, 1951. Space physics and astrophysics: radiation belts, celestial gamma-ray sources.
Robert L. Wild, Professor Emeritus; Ph.D., Missouri, 1950. Experimental solid-state physics.
Stephen J. Wimpenny, Associate Professor; Ph.D., Sheffield (England), 1980. Experimental high-energy physics.
Jose Wudka, Assistant Professor; Ph.D., MIT, 1986. Theoretical particle physics and astrophysics.
Jory A. Yarmoff, Associate Professor; Ph.D., UCLA, 1985. Experimental surface sciences.
Allen D. Zych, Professor; Ph.D., Case Western Reserve, 1968. Gamma-ray astrophysics, solar gamma rays and neutrons.

RESEARCH ACTIVITIES

Astrophysics and Space Physics. Facilities and equipment include a 100-kilojoule capacitor bank, a magnetosphere simulator chamber, Compton gamma-ray telescopes for balloon flights, and associated computers and ground support equipment. Current programs include astrophysical observations with balloon-borne detectors to measure gamma rays from pulsars, galactic sources, diffuse cosmic flux and cosmic bursts, and gamma rays and neutrons from the sun during solar flares and from the earth's atmosphere and air shower arrays; laboratory development of new gamma-ray astronomy detection techniques; very-high-energy celestial gamma-ray detection using atmospheric Cerenkov radiation with various large mirror assemblies; and laboratory magnetospheric, atmospheric, planet and comet simulation experiments carried out with the UCR-built simulation facility. Theoretical studies include gamma-ray bursts, magnetospheric and interplanetary magnetic field reconnection, and radiation belt particles. The Institute of Geophysics and Planetary Physics offers interdisciplinary research with other departments, such as anthropology and earth sciences.

Atomic Physics. An experimental program in electron collision physics. Topics of current interest include total electron impact cross-section measurements, differential elastic and inelastic cross-section measurements for electron impact on atoms and molecules, and metastable atomic beam development. Opportunities exist for studies of electron impact on laser-excited atoms and coincidence measurements through collaboration with the Jet Propulsion Laboratory in Pasadena.

Condensed-Matter Physics. Facilities and equipment include extensive electronic, microwave, and cryogenic apparatus; on-line data acquisition systems; NMR spectrometers; X-ray diffractometers; high-pressure facilities; apparatus for fabrication of metal and alloy samples; and high-temperature crystal-growing furnaces. A specially designed magnet laboratory houses six electromagnets, including an 87-kilogauss superconducting magnet. Experimental activities include studies of ferromagnetic, antiferromagnetic, and standing spin wave resonance; nuclear magnetic resonance and muon spin rotation measurements in high-temperature superconductors, heavy-fermion materials, and other highly correlated electron systems; high-temperature ferromagnetic and antiferromagnetic systems, with emphasis on phase transitions; superconductors and other magnetic systems; and optical properties of solids. Theoretical investigations include properties of disordered solids, properties of two-dimensional electron gas in a strong magnetic field, theory of superconductivity, and studies of tunneling junctions and their arrays.

Elementary Particle Physics. Theoretical studies involve gauge field theories, quantum chromodynamics, neutrino physics, quark confinement and phenomenology of heavy quark production, and vector bosons and inclusive hadron scattering at low- and high-P_T. Various extensions of the standard model of strong, weak, and electromagnetic interactions, including supersymmetry and grand unification, are pursued with emphasis on verifiable experimental implications. There are also investigations into the phenomenology of nuclear structure functions and the determination of quark and gluon distribution functions. The experimental high-energy physics group has four major activities: (1) ultra-high-energy proton-antiproton collisions in the DZERO detector at the Fermilab Tevatron Collider—including the search for the top quark and supersymmetric particles; (2) electron-positron collisions at SLAC and CERN—at SLAC, the group is in the TPC/Two-Gamma experiment studying the properties of quarks and leptons at the 30-GeV PEP; at CERN, the group is part of the OPAL collaboration taking data for precision tests of the standard model at the world's highest energy e^+e^- storage-ring accelerator, LEP; (3) neutrino physics at Los Alamos—the UCR group in the LSND experiment study neutrino interactions, neutrino oscillation, and neutrinos from extraterrestial sources with a large liquid scintillator imaging detector at the LAMPF meson facility; and (4) developing silicon microstrip tracking detectors for the SSC.

Relativistic Heavy-Ion Physics. The experimental group studies nucleus-nucleus collisions at high energies in order to understand the behavior of extended nuclear matter under extreme conditions of density and temperature. A main goal is to reach the phase transition from ordinary nuclear matter to the "quark-gluon plasma." Experiments are carried out with the E-802 spectrometer system at the Brookhaven National Laboratory. Future projects include the construction of a large detector at the Realistic Heavy Ion Collider where energies of 200 GeV/nucleon will be available.

Surface Physics. Experimental studies involve the electronic, geometric, and chemical structure and chemical reactions that occur at surfaces and the interaction of radiation with surfaces. UCR has ultrahigh-vacuum chambers equipped for low-energy electron diffraction, Auger electron spectroscopy, thermal desorption spectroscopy, electron-stimulated desorption, impact collision ion-scattering spectroscopy, and other techniques. Facilities at the National Synchrotron Light Source at Brookhaven National Laboratory are used for soft X-ray photoemission, photon-stimulated desorption, and near-edge X-ray absorption fine structure. Theoretical investigations include atom-molecule non-rigid-surface scattering and interaction of photons and atomic clusters with surfaces.

Theoretical Studies. Collective phenomena in atoms and quantized radiation, nonlinear dynamics, many-body theory, and bioelectric phenomena.

UNIVERSITY OF CINCINNATI
Department of Physics

Programs of Study

The Department of Physics offers programs of study leading to the Ph.D. and M.S. degrees. The M.S. degree is not a prerequisite for the Ph.D. degree, and incoming students are assumed to be preparing for the Ph.D. Major areas of research are experimental condensed-matter physics, theoretical condensed-matter physics, experimental elementary particle physics, and theoretical elementary particle physics. There is also representation in space physics, atomic physics, astronomy, and medical physics.

A student typically takes core courses during the first year, in addition to a seminar that serves as an introduction to the professors in the department and to their research. The student begins research, in addition to taking additional specialized course work, in the second year. A Ph.D. qualifying examination is given at the beginning of the second year and can be retaken in the spring of that year.

There are thesis and nonthesis options for the master's degree. The nonthesis-option requirements are the satisfactory completion of 56 credits of graduate courses and the passing of a final examination that has been designed to test basic knowledge of course work and related physics. The thesis-option requirements are the satisfactory completion of 36 credits of graduate courses and the writing of a master's thesis based on at least 9 credits of work in research for the master's degree.

The requirements for the Ph.D. degree include the successful completion of the 48-credit core course sequence, the passing of the Ph.D. qualifying examination, successful completion of several specialized courses in the student's area of research, and the writing of a dissertation based on at least 15 credits of research for the doctorate. The student must also present a final oral defense of the thesis.

The department has hired 10 professors in the last five years and will hire more in the coming years. These young and enthusiastic researchers are a particular stimulation to the graduate students.

Research Facilities

The Department of Physics has recently moved into a completely new, modern research building. The research laboratories are housed on the first two floors of the building and have state-of-the-art support systems. A technical-operations laboratory, housed in the building and supervised by a full-time technician, is available for sample preparation and characterization. A complete machine shop, also housed in the building and manned by four machinists, is capable of fabricating the most intricate materials required for the successful running of experiments. A student shop is also available. A helium liquefier and recovery system was built into the new building. PCs are available for student use, and all PCs and terminals are connected to the University's VAX system and, through it, to the Ohio Supercomputer. During their first year in residence, students are encouraged to begin their research as soon as possible and to become acquainted with the various possibilities available in the department.

Financial Aid

Most full-time graduate students receive financial support in the form of a tuition waiver and a graduate assistantship or a research assistantship. The average stipend level for the 1991–92 academic year is $9300 for ten months. A teaching assistant typically teaches two undergraduate laboratories or grades for an undergraduate course. These duties require approximately 10 to 15 hours per week. All students are supported during the two summer months at the same average level, either by performing teaching duties or by helping in the research laboratories.

Cost of Study

In 1991–92 the cost per credit hour for in-state students is $121 and for out-of-state students, $240. There are also fees of $432 per year. These costs are not borne by the student who receives a tuition waiver or an assistantship. Medical insurance is available to graduate students.

Cost of Living

The cost of living in the Cincinnati area is remarkably low, and graduate students are able to live quite comfortably in the vicinity of the campus. Apartments are plentiful and shopping is quite convenient.

Student Group

The Department of Physics has between 60 and 70 full-time students at any one time. The first-year class averages about 15, and typically 3 to 4 students are women.

Location

The University of Cincinnati is located about 3 miles from the downtown section of the city of Cincinnati and is surrounded by a pleasant residential area. Transportation to the downtown area is easy and efficient. The city is quite hilly and has a large number of parks and green spaces within its boundaries. It is bounded on the south by the majestic Ohio River. The city offers a wide variety of activities. The Cincinnati Symphony, the Art Museum, the Playhouse in the Park, the Zoo, the Cincinnati Bengals, the Cincinnati Reds, and the riverfront are all major attractions.

The University

The University of Cincinnati was founded in 1819 and is a completely state-funded institution. It is one of the two major comprehensive institutions in the state of Ohio. There are approximately 38,000 students enrolled. The University provides educational opportunities in a wide variety of disciplines.

Applying

Students with a bachelor's degree in physics are invited to apply for admission to the graduate program. Students with a background in other, related areas will also be considered. The GRE General Test and Subject Test in physics are recommended for domestic students and are required for foreign students. Foreign students are also required to take the Test of English as a Foreign Language (TOEFL) and earn a score of at least 540 to be considered for admission. Applications should be received by March 1, as decisions are made soon after that date, but an application submitted later will gladly be considered if there are positions available. The application fee is $20.

Correspondence and Information

Director of Graduate Admissions
Department of Physics
Mail Location 11
University of Cincinnati
Cincinnati, Ohio 45221
Telephone: 513-556-0511

University of Cincinnati

THE FACULTY AND THEIR RESEARCH

William Chow, Professor; Ph.D., Case Institute of Technology, 1964. Theoretical condensed-matter physics.
Robert Endorf, Professor; Ph.D., Carnegie Mellon, 1971. Medical physics.
F. Paul Esposito, Professor; Ph.D., Chicago, 1971. Theoretical condensed-matter physics.
Henry Fenichel, Professor; Ph.D., Rutgers, 1964. Experimental condensed-matter physics.
Bernard Goodman, Professor; Ph.D., Pennsylvania, 1955. Theoretical condensed-matter physics.
Howard Jackson, Professor; Ph.D., Northwestern, 1971. Experimental condensed-matter physics.
Mark Jarrell, Assistant Professor; Ph.D., California, Santa Barbara, 1987. Theoretical condensed-matter physics.
Randy Johnson, Associate Professor; Ph.D., Berkeley, 1975. Experimental elementary particle physics.
William Joiner, Professor; Ph.D., Rutgers, 1962. Experimental condensed-matter physics.
Dale Jones, Professor; Ph.D., Washington (St. Louis), 1953. Experimental atomic physics.
Young Kim, Assistant Professor; Ph.D., Florida, 1986. Experimental condensed-matter physics.
Michael Ma, Associate Professor; Ph.D., Illinois, 1983. Theoretical condensed-matter physics.
Freydoon Mansouri, Professor; Ph.D., Johns Hopkins, 1969. Theoretical elementary particle physics.
David Mast, Assistant Professor; Ph.D., Northwestern, 1982. Experimental condensed-matter physics.
Brian Meadows, Professor; D.Phil, Oxford, 1966. Experimental elementary particle physics.
Richard Newrock, Professor and Head of the Department; Ph.D., Rutgers, 1970. Experimental condensed-matter physics.
Mirko Nussbaum, Professor; Ph.D., Johns Hopkins, 1962. Experimental elementary particle physics.
Frank Pinski, Associate Professor; Ph.D., Minnesota, 1977. Theoretical condensed-matter physics.
James Russell, Professor; Ph.D., Yale, 1958. Theoretical atomic physics.
Joseph Scanio, Professor; Ph.D., Berkeley, 1967. Theoretical elementary particle physics.
Slava Serota, Assistant Professor; Ph.D., MIT, 1987. Theoretical condensed-matter physics.
Mike Sitko, Assistant Professor; Ph.D., Wisconsin, 1980. Astronomy and astrophysics.
Leigh Smith, Assistant Professor; Ph.D., Illinois, 1988. Experimental condensed-matter physics.
Mike Sokoloff, Assistant Professor; Ph.D., Berkeley, 1983. Experimental elementary particle physics.
Peter Suranyi, Professor; Ph.D., Joint Institute for Nuclear Research (USSR), 1964. Theoretical elementary particle physics.
Tai-Fu Tuan, Professor; Ph.D., Pittsburgh, 1959. Space physics.
Rohana Wijewardhana, Associate Professor; Ph.D., MIT, 1984. Theoretical elementary particle physics.
Fu Chun Zhang, Associate Professor; Ph.D., Virginia Tech, 1983. Theoretical condensed-matter physics.

SECTION 7: PHYSICS

UNIVERSITY OF CONNECTICUT

Department of Physics

Programs of Study

The Department of Physics offers programs of study and research leading to both the M.S. and Ph.D. degrees in the following areas: atomic and molecular physics (experimental and theoretical), condensed-matter physics (experimental and theoretical), elementary particle and field theory, theory of general relativity and cosmology, nuclear theory, and quantum optics and lasers (experimental and theoretical). The master's degree may be earned under either of two plans. One requires at least 15 credits and a thesis. The second plan, appropriate for those intending to pursue a Ph.D., requires 24 credits of course work but no thesis. Courses leading to the M.S. with a specialization in optics are available as evening classes in the Norwalk area. Ordinarily, 24 credits of course work beyond the master's degree are included in the doctoral plan of study. Students continuing toward the Ph.D. must pass a general examination, a portion of which may, at the student's option, be taken after one year of graduate study. For the Ph.D., the student must also have a reading knowledge of at least one foreign language appropriate to his or her area of study or at least 6 credits of advanced work in a related area. The doctoral dissertation, written under the immediate and continuous supervision of an advisory committee, is expected to represent a significant contribution to the field of physics.

Admission to the graduate programs is limited and selective. Ordinarily, students who do not qualify for assistantships are not admitted. The master's degree generally requires one to two years to complete. The Ph.D. degree represents the equivalent of at least an additional two years of full-time study and research.

Research Facilities

The physics department is located in the Edward Gant Science Complex, which also houses the Institute of Materials Science, the Departments of Mathematics and Statistics, and the University's Computer Center. A wide variety of research facilities are located in the physics department. A 2-MeV positive ion and a 2-MeV electron Van de Graaff accelerator and several smaller accelerators and ion sources are available for research in atomic collisions and in applications of ion beams to materials modification and surface analysis. Some of the more extensively used instruments available include argon-ion and krypton-ion lasers, YAG-pumped pulsed dye laser systems, an excimer laser, CW ring dye lasers, standing-wave dye lasers, a grazing-incidence XUV spectrometer, a closed-cycle helium refrigerator and helium cryostats for low-temperature research, spectrometers for electron spin resonance and both CW and pulsed nuclear magnetic resonance studies, a superconducting SQUID magnetometer, a SPEX double monochromator, a diamond-anvil high-pressure cell, and a fast transient digitizer. The Institute of Materials Science, where some students do their research, contains extensive facilities for materials research studies, including X-ray physics, surface physics, and investigations of biomaterials. The University Computer Center has a supercomputer, an IBM ES/9000-580 with triadic processor, available for research. In addition, the physics department has its own research computers and Macintosh and IBM PC/AT computers on an AppleTalk network linked to an Ethernet network for a Sun-4/300 SPARCstation and other Sun Workstations. A substantial main library and the department's own research library are conveniently available. Joint research programs have been established at the National Synchrotron Light Source at Brookhaven National Laboratory and the Los Alamos Scientific Laboratory.

Financial Aid

For the 1991–92 academic year, the nine-month graduate assistantship stipend ranges from $9565 for entering students to $11,770 for those who have passed the general examinations. In addition, health insurance and tuition waivers (but not a waiver of fees) are provided to those holding assistantships. Assistantships are awarded competitively, and fellowship support is available for exceptionally well qualified candidates. Summer support is also available.

Cost of Study

In 1991–92, the cost per semester for carrying six credits is $1140 (tuition) plus $156 (fees) for Connecticut residents and $2970 (tuition) plus $156 (fees) for out-of-state students. The tuition portion is waived for those holding assistantships.

Cost of Living

A limited amount of space is available in a graduate dormitory. The cost is $1210 per semester for 1991–92 (not including meals). In addition, rental units are available off campus. An Apartment Complex List (updated annually) and an Off Campus Housing Computer List (updated weekly) are available free of charge from the Department of Residential Life, U-22 Room 107; telephone 203-486-3032.

Student Group

There are 47 full-time graduate students, almost all of whom are enrolled in the doctoral program. Nearly all the students have financial support, teaching or research assistantships, or fellowships. The atmosphere in the department is friendly and informal; students conducting thesis research can expect to work individually and closely with their faculty advisers.

Location

The University of Connecticut is located in the scenic New England community of Storrs, 25 miles northeast of Hartford, near Interstate 84. Boston (90 miles), New York (130 miles), and the Rhode Island beaches are all within easy driving distance. Hiking, cross-country skiing, canoeing, and fishing may be enjoyed within a 10-minute drive from campus. Many students ski in Vermont and New Hampshire during the winter. The University hosts a number of cultural and social events during the academic year. These range from major concert series to a multitude of other activities popular with undergraduates.

The University and The Department

The University of Connecticut is a state-supported institution with an enrollment of approximately 13,000 undergraduate and 4,000 graduate students on its Storrs campus. It is the state's land-grant institution and has strong research programs in many areas. The physics faculty currently numbers 36. In addition to the core curriculum, the department provides a variety of courses and seminar series in several specialized areas. There is also a weekly colloquium series.

Applying

Potential students are encouraged to apply in January for the following fall semester. While applications are considered at any time, early application optimizes the probability of receiving financial aid. International applications must be complete by April 1 and must include a TOEFL score for those whose native language is not English. All applicants must submit GRE General Test scores, and the Subject Test in physics is recommended, though optional. There is a $30 application fee.

Correspondence and Information

Professor Ralph H. Bartram, Head
Department of Physics, U-46
University of Connecticut
2152 Hillside Road
Storrs, Connecticut 06269-3046
Telephone: 203-486-4924

Peterson's Guide to Graduate Programs in the Physical Sciences and Mathematics 1992

SECTION 7: PHYSICS

University of Connecticut

THE FACULTY AND THEIR RESEARCH

Ralph H. Bartram, Professor and Department Head; Ph.D., NYU. Theoretical solid-state physics: optical and magnetic properties of point defects and impurities.

Leonid V. Azaroff, Professor and Director of the Institute of Materials Science; Ph.D., MIT. Metal physics: X-ray crystallography; polymer physics.

Gary D. Bent, Assistant Department Head; Ph.D., Connecticut. Theoretical atomic and molecular physics.

Philip E. Best, Professor; Ph.D., Western Australia. Experimental solid-state physics: X-ray and electron spectrometry, surface physics.

Joseph I. Budnick, Professor; Ph.D., Rutgers. Experimental solid-state physics: nuclear magnetic resonance, superconductivity, X-ray studies using synchrotron radiation.

Dwight H. Damon, Professor; Ph.D., Purdue. Experimental polymer physics: transport properties.

Gayanath W. Fernando, Assistant Professor; Ph.D., Cornell. Condensed-matter theory: electronic structure calculations.

Otis R. Gilliam, Professor; Ph.D., Duke. Experimental solid-state physics: paramagnetic resonance, radiation damage in solids.

Phillip L. Gould, Assistant Professor; Ph.D., MIT. Experimental quantum optics: laser cooling and trapping of atoms.

Yukap Hahn, Professor; Ph.D., Yale. Theoretical atomic physics: scattering theory, hyperfine structure theory, electron-ion collisions in plasmas.

Kurt Haller, Professor; Ph.D., Columbia. Theoretical high-energy physics: particle theory, quantum field theory.

Douglas S. Hamilton, Professor; Ph.D., Wisconsin–Madison. Experimental solid-state physics: laser spectroscopy.

Howard C. Hayden, Professor; Ph.D., Denver. Experimental atomic physics, ion implantation.

William A. Hines, Professor; Ph.D., Berkeley. Experimental solid-state physics: nuclear magnetic resonance, magnetic susceptibility of metals and alloys.

Muhammad M. Islam, Professor; Ph.D., Imperial College (London). Theoretical high-energy physics: scattering, nucleon substructure.

Juha Javanainen, Professor; Ph.D., Helsinki Tech. Theoretical quantum optics, interaction of light with atoms.

Lawrence A. Kappers, Associate Professor; Ph.D., Missouri. Experimental solid-state physics: optical properties, color centers, radiation damage.

Quentin C. Kessel, Professor; Ph.D., Connecticut. Experimental atomic and molecular physics: ionization, X rays, and Auger electrons.

Paul G. Klemens, Professor Emeritus; D.Phil., Oxford. Theoretical solid-state physics: transport properties.

Frederick P. Lipschultz, Associate Professor; Ph.D., Cornell. Experimental low-temperature physics, physics instrumentation.

David P. Madacsi, Associate Professor; Ph.D., Connecticut. Condensed-matter physics: paramagnetic resonance, computer modeling of lattice defects.

Ronald L. Mallett, Professor; Ph.D., Penn State. Theory of general relativity and cosmology.

Philip D. Mannheim, Professor; Ph.D., Weizmann (Israel). Elementary particle theory, field theory, general relativity, cosmology and astrophysics.

David Markowitz, Associate Professor; Ph.D., Illinois at Urbana-Champaign. Theoretical solid-state physics: many-body theory.

Thomas I. Moran, Professor; Ph.D., Yale. Experimental atomic physics: atomic beams, lifetimes, thermal collisions.

Fred A. Otter, Professor; Ph.D., Illinois. Experimental solid-state physics: surface physics.

Douglas M. Pease, Associate Professor; Ph.D., Connecticut. Experimental solid-state physics: X-ray studies of alloys.

Cynthia W. Peterson, Professor; Ph.D., Cornell. Experimental solid-state physics.

Edward Pollack, Professor; Ph.D., NYU. Experimental atomic physics: atomic and molecular beams, atom-molecule collisions.

George H. Rawitscher, Professor; Ph.D., Stanford. Theoretical nuclear physics: nuclear reactions, electron-nucleus scattering.

Arnold Russek, Professor; Ph.D., NYU. Theoretical physics: atomic structure, atomic collision theory.

Robert Schor, Professor Emeritus; Ph.D., Michigan. Theoretical chemical physics: biophysics, macromolecules.

Winthrop W. Smith, Professor; Ph.D., MIT. Experimental atomic and molecular physics: ion-atom collisions, XUV and laser spectroscopy.

Ralph B. Snyder, Associate Professor; Ph.D., Harvard. Theoretical physics: atomic structure, atomic collision theory.

Frederick E. Steigert, Associate Professor; Ph.D., Indiana. Experimental nuclear physics.

Mark S. Swanson, Associate Professor; Ph.D., Missouri. Theoretical high-energy physics: quantum field theory.

A teaching assistant lends a helping hand in an introductory laboratory.

The physics building.

UNIVERSITY OF FLORIDA

Department of Physics

Programs of Study

The Department of Physics offers programs of study leading to the M.S. and Ph.D. degrees. The master's degree is not a prerequisite for the Ph.D., and a nonthesis M.S. program is available. Usually, a Ph.D. student devotes the first year to core courses, after which the comprehensive and qualifying exams are taken. Specialized courses and dissertation research are begun the following year. Some teaching is required of all degree candidates, and reading proficiency in a foreign language is required of Ph.D. candidates. Students generally need five years to complete the Ph.D. program.

Experimental research areas in the department include ultralow-temperature physics, condensed-matter physics, high-energy particle physics, development of advanced radiation detectors, synchrotron radiation, surface physics, X-ray and neutron scattering, NMR spectroscopy and imaging, nuclear physics, and atomic physics. Theoretical research is pursued in statistical mechanics, scattering, quantum fields, astrophysics and general relativity, and atomic, molecular, chemical, condensed-matter, and high-energy particle physics.

The student-faculty ratio in the department is low, allowing close interaction between students and their research supervisors. Arrangements with other departments provide opportunities for study in the interdisciplinary areas of astrophysics, chemical physics, material science, and medical physics. Through cooperative programs, research may be conducted at Oak Ridge National Laboratory, Los Alamos National Laboratory, Fermilab, Cornell University, and other DOE facilities.

Research Facilities

The department is housed in four modern air-conditioned buildings. Facilities include machine and electronics shops. Specialized equipment includes a 4-MV Van de Graaff accelerator, a helium liquefier and recovery system, cascade nuclear-demagnetization refrigerators, helium dilution refrigerators, high-field superconducting solenoids, magnetic resonance spectrometers, high-pressure equipment, an electron energy loss spectrometer, a rotating anode X-ray source, a small-sample calorimeter for low temperatures, and ultrahigh-vacuum surface analytical systems. Computing services are provided by the University's IBM 3090-600J/6VF, Sun and VAX networks, and the high-energy group's system of nineteen VAXstation 3100s and seventeen DECstation 5000s in a distributed computing environment, as well as by minicomputers in several research laboratories for on-line control and analysis. The central science library holds more than 800,000 books and periodicals and offers extensive computer-search facilities.

Financial Aid

Each year, based on undergraduate and graduate records, GRE scores, and letters of recommendation, the department provides approximately forty graduate teaching assistantships. In 1991–92, one-third-time stipends provide about $12,000 for twelve months. Duties, which require approximately 13 hours per week, include teaching undergraduate laboratory or discussion sections or grading in advanced courses. Advanced students are supported by research grants. Several fellowships, with no required teaching duties, are available.

Cost of Study

For Florida residents, fees are $77 per credit in 1991–92. The out-of-state surcharge of $239 per credit is normally waived for assistants or fellows. A health fee of $40 is required for each semester.

Cost of Living

On- and off-campus housing accommodations are readily available for both single and married students. University-operated apartment complexes provide varying environments and living quarters at costs ranging from $203 per month for a one-bedroom apartment to $260 per month for a two-bedroom town house. Privately owned housing, in either suburban or rural settings, is available at reasonable rates. The cost of living is generally lower than in the urban areas of the northern and western United States.

Student Group

The University enrollment is approximately 35,000, including about 5,000 graduate and professional students. The Department of Physics has about 80 graduate students, who come from all parts of the United States and from several foreign countries. Recent graduates are successfully pursuing careers both in the academic world and in industrial and government laboratories.

Location

Gainesville, a city of 95,000 people, is situated in north-central Florida, 65 miles from the Atlantic Ocean and 55 miles from the Gulf of Mexico; both offer excellent beaches and recreational opportunities. In addition, Gainesville is surrounded by numerous freshwater lakes, springs, and rivers. A moderate climate, with mean temperatures of 81°F in July and 58°F in January, permits year-round outdoor activities. Cultural offerings include two local theater groups, a civic ballet, a chorus, an orchestra, and a number of festivals. Tampa, the Kennedy Space Center, and Disney World are all less than 3 hours away by car.

The University

The University is the focus of a lively cultural, intellectual, and political life. Eminent scholars, artists, and public officials bring enlightening and stimulating ideas to the campus on a regular basis. Prominent musicians give concerts of classical and contemporary music, many of them free. Frontiers of Science, a campuswide lecture series sponsored by the Department of Physics, brings several distinguished researchers to the University each year. The Sanibel Symposia program, conducted jointly by physicists and chemists of the Quantum Theory Project, is an annual event that brings internationally known scientists to the area.

Applying

Graduate study may begin in any semester, and the application deadline is two months prior to the start of that semester. Assistantships are usually awarded for the fall semester, with the financial aid decision made the previous spring. Up-to-date transcripts and current GRE General Test scores should be on file with the University registrar as early as possible, preferably by February 1. The GRE Subject Test in physics is also required. Application forms can be obtained by writing to the address below.

Correspondence and Information

Graduate Coordinator
Department of Physics
University of Florida
Gainesville, Florida 32611
Telephone: 904-392-0521

SECTION 7: PHYSICS

University of Florida

THE FACULTY AND THEIR RESEARCH
E. Dwight Adams, Professor; Ph.D., Duke, 1960. Ultralow-temperature physics: liquid and solid ^{3}He.
E. Raymond Andrew, Graduate Research Professor; Ph.D., Cambridge, 1948. Nuclear magnetic resonance.
Paul R. Avery, Associate Professor; Ph.D., Illinois, 1980. Experimental high-energy physics.
Thomas L. Bailey, Professor Emeritus; Ph.D., Chicago, 1953. Atomic and molecular physics.
Stanley S. Ballard, Distinguished Service Professor (Emeritus); Ph.D., Berkeley, 1934. Optics, physics education.
Stuart E. Brown, Assistant Professor; Ph.D., UCLA, 1988. Condensed-matter experiment.
Arthur A. Broyles, Professor Emeritus; Ph.D., Yale, 1949. Quantum electrodynamics, nuclear explosions, foundations of physics.
J.-Robert Buchler, Professor; Ph.D., California, San Diego, 1969. Astrophysics.
Michael J. Burns, Assistant Professor; Ph.D., UCLA, 1984. Condensed-matter experiment.
Andrew W. Cumming, Assistant Professor; Ph.D., MIT, 1988. Condensed-matter experiment.
Steven L. Detweiler, Professor; Ph.D., Chicago, 1974. Relativistic astrophysics.
James W. Dufty, Professor; Ph.D., Lehigh, 1967. Nonequilibrium statistical mechanics.
F. Eugene Dunnam, Professor; Ph.D., LSU, 1958. Nuclear physics.
Richard D. Field Jr., Professor; Ph.D., Berkeley, 1971. Elementary particle theory.
James N. Fry, Associate Professor; Ph.D., Princeton, 1979. Theoretical astrophysics, cosmology.
Richard E. Garrett, Professor Emeritus; Ph.D., Virginia, 1953. Physics education.
Peter J. Hirschfeld, Assistant Professor; Ph.D., Princeton, 1984. Condensed-matter theory.
Charles F. Hooper Jr., Professor; Ph.D., Johns Hopkins, 1963. Statistical mechanics, dense plasma physics.
Gary G. Ihas, Professor; Ph.D., Michigan, 1971. Ultralow-temperature physics: superfluid helium.
James R. Ipser, Professor; Ph.D., Caltech, 1969. Relativistic astrophysics.
John R. Klauder, Professor; Ph.D., Princeton, 1959. Mathematical physics.
Pradeep Kumar, Associate Professor; Ph.D., California, San Diego, 1973. Condensed-matter theory.
Dov I. Levine, Assistant Professor; Ph.D., Pennsylvania, 1986. Condensed-matter theory.
Per-Olov Löwdin, Graduate Research Professor; Ph.D., Uppsala (Sweden), 1948. Quantum theory of matter.
Mark W. Meisel, Assistant Professor; Ph.D., Northwestern, 1983. Condensed-matter experiment, ultralow-temperature physics.
David A. Micha, Professor; Ph.D., Uppsala (Sweden), 1966. Theoretical chemical physics, quantum theory of matter.
Hendrik J. Monkhorst, Professor; Ph.D., Groningen (Netherlands), 1968. Theoretical chemical physics, quantum theory of matter.
Khandker A. Muttalib, Assistant Professor; Ph.D., Princeton, 1982. Condensed-matter theory.
Stephen E. Nagler, Associate Professor; Ph.D., Toronto, 1982. Condensed matter; synchrotron, X-ray, and neutron scattering.
Yngve N. Ohrn, Professor; Ph.D., Uppsala (Sweden), 1963. Quantum theory of matter.
Raymond Pepinsky, Professor; Ph.D., Chicago, 1940. Crystallography, biophysics.
Lennart R. Peterson, Professor; Ph.D., MIT, 1966. Atmospheric and atomic physics.
Zongan Qiu, Assistant Professor; Ph.D., Chicago, 1986. High-energy theory.
Pierre Ramond, Professor; Ph.D., Syracuse, 1969. Elementary particle theory.
John R. Sabin, Professor; Ph.D., New Hampshire, 1966. Quantum theory of matter.
L. Elizabeth Seiberling, Associate Professor; Ph.D., Caltech, 1980. Condensed-matter experiment.
Pierre Sikivie, Professor; Ph.D., Yale, 1975. Elementary particle theory.
Chris J. Stanton, Assistant Professor; Ph.D., Cornell, 1986. Condensed-matter theory.
Gregory R. Stewart, Professor; Ph.D., Stanford, 1975. Solid-state physics, novel materials.
Neil S. Sullivan, Professor and Chairman; Ph.D., Harvard, 1972. Condensed-matter physics, quantum crystals, axion search.
Yasumasa Takano, Associate Professor; Ph.D., Helsinki, 1978. Ultralow-temperature physics.
David B. Tanner, Professor; Ph.D., Cornell, 1972. Condensed-matter experiment, axion search.
Billy S. Thomas, Professor Emeritus; Ph.D., Vanderbilt, 1959. Supersymmetry in quantum mechanics.
Charles B. Thorn, Professor; Ph.D., Berkeley, 1971. Elementary particle theory.
Samuel B. Trickey, Professor; Ph.D., Texas A&M, 1968. Theory and computation of solid and ultrathin film properties.
Henri A. Van Rinsvelt, Professor; Ph.D., Utrecht (Netherlands), 1965. Nuclear physics, X-ray analysis, ion-solid interactions.
James K. Walker, Professor; Ph.D., Glasgow, 1960. Experimental high-energy particle physics.
Bernard F. Whiting, Assistant Professor; Ph.D., Melbourne, 1979. Theoretical astrophysics.
Richard P. Woodard, Assistant Professor; Ph.D., Harvard, 1984. Quantum gravity and quantum field theory.
John M. Yelton, Assistant Professor; Ph.D., Oxford, 1981. Experimental high-energy particle physics.

RESEARCH ACTIVITIES
Applied Physics. Investigations are carried out with particle-induced X-ray emission and on proton microprobe development.
Astrophysics. Research includes stellar evolution, stellar variability, nonlinear fluid dynamics and chaos, and effects of general relativity in problems of astrophysics and cosmology and large-scale structure.
Atomic and Molecular Physics. Studies are being made of collisions of low-energy electrons with isolated molecules.
Chemical Physics. This program is concerned with theoretical, experimental, and computational aspects of problems at the interface between chemistry and physics.
Condensed-Matter Experiment. Studies include: NMR of molecular dynamics in solids, liquids, and polymers. Optical spectroscopy of high-T_c superconductors, polymers, and low-dimensional conductors; X-ray scattering; ion scattering; calorimetry; UHV surface spectroscopy; and electronic properties of novel materials.
Condensed-Matter Theory. Research areas include properties of highly correlated quantum fluids and solids, semiconductors and superconductors, disordered metals and insulators, and quasicrystals.
Experimental Particle Physics. Colliding beam studies are conducted at Fermi National Accelerator Laboratory and also at the Cornell University Laboratory for Nuclear Studies. Nonaccelerator studies also include axion searches.
Low-Temperature Physics. Work focuses on quantum fluids and solids and nuclear magnets at temperatures down to 10 microkelvin. New experiments are planned for the new Microkelvin Laboratory, one of only two such facilities in the country.
Medical Physics. Research includes nuclear magnetic resonance imaging and in vivo NMR spectroscopy.
Nuclear Physics. Heavy ions in-beam spectroscopy is carried out at Oak Ridge National Laboratory and Florida State University.
Particle Physics. Elementary particles and their high-energy interactions are studied. Superstring theories are explored as a basis for the synthesis of all fundamental interactions. Implications of particle physics for cosmology are explored.
Quantum Theory Project. Formal theory and computational methodology for electronic structure, optical excitations, atomic and molecular scattering, and surface phenomena are studied.
Statistical Mechanics. Research on fluctuations and transport includes plasma line broadening, nonlinear dynamics of systems far from equilibrium, nonequilibrium phase transitions, the physics of amorphous- and lattice-spin systems, and heavy fermions.

UNIVERSITY OF GEORGIA
Department of Physics and Astronomy

Program of Study

Graduate programs are available that lead to a Ph.D. or an M.S. degree in physics. These programs are intended to prepare the student for a career in scientific research, teaching, or both.

The requirements for the Ph.D. in physics include completing the necessary course work, passing a preliminary examination, writing an approved dissertation, and passing a final oral examination. The Ph.D. program must be approved by an advisory committee.

Course work includes basic sequences in mechanics, electrodynamics, mathematical physics, and quantum mechanics and advanced courses in areas of interest. A student should expect to complete within two years a core of basic courses in preparation for the comprehensive preliminary examination.

The requirements for the M.S. in physics include 40 quarter hours of course work, an approved thesis, and a final oral examination. The program of courses depends on the background and interest of the student and is planned under the supervision of a major professor. The M.S. program must be approved by an advisory committee.

The research and instruction program is enriched by visiting professors who provide short courses on specialized topics. In addition, weekly colloquia and seminars are sponsored by the department to provide for the interaction of students and faculty with those working at the frontiers of physics.

Research Facilities

The department has several laser spectroscopy laboratories for the study of interactions in condensed matter. An atomic and molecular physics laboratory contains a 5-MeV Van de Graaff accelerator and a VAX 3100 computer for the study of atomic and molecular collision processes. Both laboratories contain instrumentation of the highest quality.

The University of Georgia provides one of the finest environments for computation in the world. Supercomputers available through the University's Advanced Computational Methods Center include a CDC CYBER 205 vector processing system, front-ended by a CDC CYBER 960, an Intel hypercube, and a four-processor IBM 3090 with two vector facilities.

Research is also done in association with facilities at Oak Ridge National Laboratory, Indiana University Cyclotron Facility, Los Alamos Meson Physics Facility, the TRIUMF facility, Kitt Peak National Observatory, Cerro Tololo Inter-American Observatory, the Very Large Array Radio Telescope, and various NASA orbiting telescopes.

Financial Aid

Graduate assistantships and teaching assistantships are available for nine- or twelve-month periods; these assistantships include remission of out-of-state fees. Research assistantships are also available. For highly qualified candidates, the department may award special teaching assistantships or graduate fellowships through the Graduate School.

The nine-month stipend for an assistantship is $9430 for an M.S. student and $10,060 for a Ph.D. candidate. The expected teaching load at this level of support averages 16 hours per week.

Cost of Study

Tuition and fees for students with an assistantship during the 1991–92 academic year are $143 per quarter. Students without an assistantship pay tuition and fees of $692 per quarter (in-state amount) or $1840 (out-of-state amount). The academic year is divided into three quarters. Books and supplies, depending on the student's course load, cost about $100 per quarter.

Cost of Living

Accommodations for single students in University residence halls cost $460–$565 per quarter in 1991–92. Both furnished and unfurnished apartments are available to married students at rates of $175–$230 per month. Affordable off-campus housing is plentiful.

Meal plans are available at a cost of $454 per quarter for the five-day meal plan (Monday–Friday) and $536 for the seven-day meal plan (Monday–Sunday).

Student Group

There are currently 34 students enrolled in the graduate physics and astronomy program, and almost all receive some financial aid. Total graduate enrollment at the University is 5,026. There are 28,395 students at the University of Georgia.

Location

Athens is a small, pleasant town that combines a southern heritage with the sophistication of a major university. The town's culture is strongly influenced by the University and features a lively arts scene with activities in music, the visual arts, and literature. It is also a convenient location for those who enjoy big cities, sports, and the outdoors. Atlanta is only 60 miles away, the mountains 80 miles away, and the ocean 190 miles distant.

The University and The Department

The University, the oldest chartered state university in the United States, is a major research institution with yearly research expenditures of over $145-million. Its library facilities are ranked among the top twenty in the country. The Science Library alone has more than 750,000 volumes and subscribes to nearly 300 physics and astronomy periodicals.

The Center for Simulational Physics offers students the opportunity to participate in a distinctive, vigorous program of research and training in simulational physics with an emphasis on the use of supercomputers. Because of their participation, students have access to lectures, seminars, and international collaborations.

Various weekly seminars expose students to work being done by the department and researchers at other institutions as well as give students the opportunity to polish their own lecturing skills.

Applying

Applications for admission are accepted throughout the year, but most students are admitted to begin in the fall quarter. GRE General Test scores are required of all applicants, and the Subject Test in physics is recommended. TOEFL scores are required of international students. International students with assistantships take the TSE or SPEAK test before being assigned teaching duties. Undergraduate preparation is assumed to be Symon-level mechanics, Reitz- and Milford-level electromagnetism, and Eisberg-level modern physics. Students requesting information also receive the *Graduate Bulletin*.

Correspondence and Information

Dr. Alan K. Edwards, Graduate Coordinator
Department of Physics and Astronomy
University of Georgia
Athens, Georgia 30602
Telephone: 404-542-2818

SECTION 7: PHYSICS

University of Georgia

THE FACULTY AND THEIR RESEARCH

Robert L. Anderson, Professor; Ph.D., Wayne State, 1963. Mathematical physics (theoretical).
F. Todd Baker, Professor and Department Head; Ph.D., Michigan, 1970. Intermediate energy nuclear physics (experimental).
Kurt Binder, Adjunct Professor; Ph.D., Vienna, 1969. Condensed-matter physics (theoretical and simulational).
Jean-Pierre Caillault, Assistant Professor; Ph.D., Columbia, 1985. Observational astronomy.
Tsu-Teh Chou, Professor; Ph.D., Iowa, 1965. High-energy physics (theoretical).
William M. Dennis, Assistant Professor; Ph.D., Dublin, 1985. Nonlinear optical and far-infrared properties of condensed matter (experimental).
M. M. Duncan, Professor; Ph.D., Duke, 1956. Atomic and ionic collisions (experimental).
Alan K. Edwards, Professor and Graduate Coordinator; Ph.D., Nebraska, 1968. Atomic and molecular collisions (experimental).
Moshé Flato, Adjunct Professor; Ph.D. (physics), Hebrew (Jerusalem), 1963; Ph.D. (mathematics), Paris IV (Sorbonne), 1965. Application of group theory to strong interactions (theoretical).
Rainer W. Gerling, Adjunct Assistant Professor; Ph.D., Erlangen-Nurenberg, 1981. Condensed-matter physics (theoretical and simulational).
Wolfgang Gerhard Grill, Adjunct Professor; Ph.D., Heidelberg, 1975. Condensed-matter physics, solid-state spectroscopy (experimental).
John Hegarty, Adjunct Professor; Ph.D., Ireland (Dublin), 1976. Optical properties of solids, laser spectroscopy, electrooptics (experimental).
Timothy G. Heil, Associate Professor; Ph.D., Illinois, 1977. Atomic and molecular physics, astrophysics (theoretical).
George F. Imbusch, Adjunct Lecturer; Ph.D., Stanford, 1964. Optical properties of solids, laser spectroscopy (experimental).
Weiyi Jia, Adjunct Lecturer; Ph.D., Chinese Academy of Sciences, 1963. Optical properties of solids, laser spectroscopy (experimental).
Coates R. Johnson, Associate Professor; Ph.D., Berkeley, 1963. General relativity (theoretical).
David P. Landau, Research Professor and Director, Center for Simulational Physics; Ph.D., Yale, 1967. Solid-state physics, critical phenomena, condensed-matter physics (theoretical and simulational).
M. Howard Lee, Professor; Ph.D., Pennsylvania, 1969. Statistical mechanics (theoretical).
W. Gary Love, Regents' Professor; Ph.D., Tennessee, 1968. Nuclear physics (theoretical).
Loris A. Magnani, Assistant Professor; Ph.D., Maryland, 1987. Radio astronomy.
Richard S. Meltzer, Professor; Ph.D., Chicago, 1968. Optical properties of solids (experimental).
Manuel G. Menendez, Professor; Ph.D., Florida, 1963. Atomic and electron collisions (experimental).
Kin-Keung Mon, Assistant Professor; Ph.D., Cornell, 1980. Condensed-matter physics (theoretical and experimental).
Kanzo Nakayama, Assistant Professor; Ph.D., Bonn, 1984. Nuclear physics (theoretical).
Dennis C. Rapaport, Adjunct Lecturer; Ph.D., King's College (London). Condensed-matter physics (theoretical and simulational).
John E. Rives, Professor; Ph.D., Duke, 1962. Low-temperature magnetism, optical properties of solids (experimental).
Heinz-Bernd Schüttler, Assistant Professor; Ph.D., UCLA, 1984. Condensed-matter physics (theoretical and simulational).
Alan Scott, Professor; Ph.D., London, 1960. Intermediate-energy nuclear physics (experimental).
J. Scott Shaw, Associate Professor; Ph.D., Pennsylvania, 1970. Astronomy (experimental and theoretical).
Malcolm F. Steuer, Professor; Ph.D., Virginia, 1957. Nuclear, atomic, and molecular physics (experimental).
George L. Strobel, Associate Professor; Ph.D., USC, 1965. High-energy and nuclear physics (theoretical).
Robert H. Swendsen, Adjunct Professor; Ph.D., Pennsylvania, 1971. Condensed-matter physics (theoretical and simulational).
Charles A. Uzes, Associate Professor; Ph.D., California, Riverside, 1967. High-energy and solid-state physics, physical oceanography (theoretical).
Robert M. Wood, Professor; Ph.D., Wisconsin, 1964. Nuclear, atomic, and molecular physics (experimental).
William M. Yen, Graham Perdue Professor of Physics; Ph.D., Washington (St. Louis), 1962. Optical properties of solids, laser spectroscopy (experimental).
Ming-Bao Yu, Adjunct Lecturer; Ph.D., Jilin (People's Republic of China), 1961. Statistical mechanics (theoretical).

RESEARCH SPECIALTIES

Theoretical

Astrophysics. Modeling of interstellar clouds and low-temperature ionized regions.
Atomic and Molecular Physics. Atomic and molecular processes within diffuse matter; charge transfer and ionic excitation collisions, using molecular electronic potential energy surfaces.
Condensed-Matter Physics. Equilibrium and nonequilibrium statistical mechanics, many-body theory, band calculations of ferroelectric crystals.
Elementary Particles and Fields. High-energy collisions, gauge theory, particle models.
Mathematical Physics. Group, phase-space, F-P-S Poincaré linearization and perturbation analyses of relativistic field equations.
Nuclear Physics. Scattering, reactions, and structure theory; applications to electron-, meson-, nucleon-, heavy-ion–, and antinucleon-induced reactions; nucleon-nucleon forces.
Relativity. Einstein's gravitational theory, Einstein's unified field theory.

Experimental

Astronomy. Binary stars, variable stars, multiwavelength observations of late-type stars and star formation regions.
Atomic and Molecular Physics. Electron and ion spectra from fast ion-atom and ion-molecule collisions, molecular electron affinities, electron attachment to molecules, multiphoton ionization, molecular ions in solids.
Nuclear Physics–Intermediate Energy. Studies of spin excitations, giant resonances, the structure of the continuum using 300–800 MeV polarized protons and 400 MeV polarized deuterons; studies of transitional nuclei; neutron structure of excited states, extracted from inelastic scattering experiments with polarized protons at 100–500 MeV and with density-dependent forces and from inelastic pion-nucleus scattering at 130–160 MeV.
Solid-State Physics. Laser spectroscopy of solids; dynamics of excited states of crystalline, amorphous, and magnetic materials; magnon and phonon dynamics and nonlinear phonon physics.
Center for Simulational Physics. Monte Carlo, Monte Carlo Renormalization Group, and molecular dynamics studies of phase transitions and critical behavior; quantum Monte Carlo studies of many-body systems; spin dynamics studies of elementary excitations in magnetic models.

UNIVERSITY OF HOUSTON
Department of Physics

Program of Study

The Department of Physics offers a program of study leading to the Master of Science and Doctor of Philosophy degrees.

A candidate for the M.S. degree must complete a minimum of 30 graduate credit hours, including courses in classical mechanics, quantum mechanics, statistical physics, electromagnetic theory, and mathematical physics; present an acceptable thesis; and defend the thesis before a thesis committee. A nonthesis M.S. degree can be awarded after the successful completion of 36 graduate credit hours and an examination.

A candidate for the Ph.D. degree must successfully complete a minimum of 48 semester hours and pass a comprehensive examination covering classical mechanics, quantum mechanics, statistical physics, and electromagnetic theory at the graduate level. As part of the comprehensive examination, a student is required to make a research proposal. The final oral examination is primarily, but not exclusively, a defense of the dissertation.

The comprehensive examination may be attempted a maximum of two times. There is no language requirement for the M.S. or Ph.D. degree.

Current research programs involve investigations in condensed matter, theoretical physics, high-energy and particle physics, intermediate-energy physics, solid-state physics, space physics, accelerator physics, and plasma physics. These programs are supported by University and federal research grants and involve students as research assistants usually after they have had a year or more of graduate study.

Research Facilities

The research facilities of the department are located on the fourth through seventh floors of the Science and Research One Building. The department has twenty-one large, well-equipped laboratories for carrying out experimental research in atmospheric physics, condensed-matter physics, high-pressure low-temperature physics, high-energy physics, intermediate-energy physics, experimental astrophysics, nonlinear dynamics, particle physics, plasma physics, space physics, statistical physics, and surface physics. The research in accelerator physics is carried out at the Texas Accelerator Center (The Woodlands, Texas), as well as at the Fermilab (Batavia, Illinois) and the proposed SSC Laboratory (Waxahachie, Texas). To support these activities, the department maintains a large, comprehensive machine shop and a particle detector development and production facility. There are several VAX and MicroVAX computers in the department, all connected by Ethernet to the rest of the University's computers as well as most national networks. Many faculty members and graduate students collaborate closely with scientists at other universities and national laboratories. The department is also closely aligned in its research endeavors with two major research centers at the University: the Space Vacuum Epitaxy Center and the Texas Center for Superconductivity. The two centers, focused on material research, have total research and support staff of over 200 personnel and laboratory space of over 100,000 square feet.

Financial Aid

In 1990–91, the department supported 20 students on teaching assistantships; awards averaged $1037 per month for the nine-month academic year. Approximately half of these were awarded to new graduate students. In addition, 24 students were supported by research assistantships, which also carried awards of $1037 per month for the nine-month academic year. The department usually supports about 90 percent of its full-time graduate students. Supported students pay Texas-resident tuition and fees.

Cost of Study

In 1991–92, the tuition and fees for graduate students who are Texas residents are approximately $510 per semester for the normal course load of 15 semester hours. Nonresident students who are not supported pay approximately $2010 per semester for tuition and fees.

Cost of Living

The University has limited space in residence halls, which cost $2415 for a single room and $2255 for a double room for the 1991–92 academic year. In addition, upper-level housing is available in Cougar Place for $260 per month or $280 per month with kitchenette; all utilities are included. The University offers optional meal plans. The cost is $1335 for fifteen meals per week and $1515 for nineteen meals per week for the academic year.

Student Group

In spring 1991, the University had 31,770 students, including 6,243 graduate students. The department had 63 graduate students. All of the full-time students in the department received financial support in the form of teaching or research assistantships.

Location

The University is located in the center of a growing science complex that includes the Johnson Space Center, chemical companies, petrochemical laboratories, the Texas Accelerator Center, and a large medical center. Houston has approximately 2,400 manufacturing companies, thus providing excellent employment opportunities. Many recreational opportunities are available; Galveston beach is only 45 miles away, and the central Texas hill country is only a 3-hour drive from Houston. There are also numerous sports and cultural events.

The University

The University of Houston was founded in 1927 and became a four-year institution in 1934. It has made rapid progress toward national recognition of its academic programs since becoming state supported in 1963.

Applying

Application forms for admission and financial aid may be obtained by writing to the Department of Physics. To be considered for a teaching assistantship for the fall semester, applicants should submit by March 15 an application, three letters of recommendation, official transcripts of previous college work, and GRE General Test scores (plus TOEFL scores, for international students). Texas residents must pay a $25 application fee; nonresidents pay $75. The application deadline for teaching assistantships for the spring semester is October 15.

Correspondence and Information

Chairman, Graduate Studies Committee
Department of Physics, 408 S&R 1
University of Houston
Houston, Texas 77204-5504

University of Houston

THE FACULTY AND THEIR RESEARCH

James R. Benbrook, Professor and Department Chairman; Ph.D., Washington (Seattle), 1969. Space physics, cosmic-ray physics.
Edgar A. Bering III, Professor; Ph.D., Berkeley, 1974. Space physics.
C. W. (Paul) Chu, Professor and T. L. L. Temple Chair of Science; Ph.D., California, San Diego, 1968. Solid-state physics, superconductivity, magnetism.
W. K. Chu, Distinguished University Professor; Ph.D., Baylor, 1969. Superconductivity, ion beam processing.
Terry Golding, Assistant Professor; Ph.D., Cambridge, 1989. Semiconductor physics.
Michael Gorman, Associate Professor; Ph.D., Chicago, 1977. Nonlinear phenomena, fluid dynamics, optics.
Gemunu Gunaratne, Assistant Professor; Ph.D., Cornell, 1986. Nonlinear phenomena, hydrodynamics, pattern formation.
Robert Helleman, Professor; Ph.D., Yeshiva, 1971. Theoretical physics, nonlinear dynamics.
Alvin F. Hildebrandt, Professor; Ph.D., Texas A&M, 1956. Solar energy.
Peiherng Hor, Assistant Professor; Ph.D., Houston, 1990. Condensed-matter physics, magnetism.
Bambi Hu, Professor; Ph.D., Cornell, 1974. Nonlinear dynamics, critical phenomena, quantum field theory.
Hugh T. Hudson, Professor; Ph.D., Virginia, 1962. Solid-state physics, physics education.
Ed V. Hungerford III, Professor; Ph.D., Georgia Tech, 1967. Intermediate-energy physics.
Alex Ignatiev, Distinguished University Professor; Ph.D., Cornell, 1972. Surface physics, solid-state physics, chemical physics.
Robert M. Kiehn, Professor; Ph.D., MIT, 1952. Theoretical physics, mathematical physics.
Donald J. Kouri, Professor; Ph.D., Wisconsin–Madison, 1965. Chemical physics.
Kwong Lau, Assistant Professor; Ph.D., Maryland, 1981. High-energy particle physics.
Billy W. Mayes II, Professor; Ph.D., MIT, 1969. Intermediate-energy physics.
Joseph L. McCauley, Professor; Ph.D., Yale, 1972. Theoretical physics.
John Miller, Assistant Professor; Ph.D., Illinois at Urbana-Champaign, 1985. Condensed-matter physics, superconductivity.
Simon C. Moss, Professor and M. D. Anderson Chair in Physics; Ph.D., MIT, 1962. Solid-state physics, materials science.
Shoroku Ohnuma, Professor; Ph.D., Rochester, 1956. Accelerator physics.
Lawrence S. Pinsky, Associate Professor; Ph.D., Rochester, 1973. Intermediate-energy physics.
George F. Reiter, Professor; Ph.D., Stanford, 1967. Nonlinear phenomena, condensed-matter theory.
William R. Sheldon, Professor; Ph.D., Missouri–Columbia, 1960. Cosmic-ray physics, space physics.
Wu-Pei Su, Associate Professor; Ph.D., Pennsylvania, 1981. Theoretical solid-state physics.
Chin-Sen Ting, Professor; Ph.D., California, San Diego, 1970. Theoretical solid-state physics.
Lorin L. Vant-Hull, Professor; Ph.D., Caltech, 1966. Solar energy.
Roy Weinstein, Professor; Ph.D., MIT, 1954. Particle physics.
Lowell T. Wood, Associate Professor; Ph.D., Texas at Austin, 1968. Physics of optical fibers, electrodynamics of composite media, optical properties of high-T_c superconductors.

SECTION 7: PHYSICS

UNIVERSITY OF KANSAS

Department of Physics and Astronomy

Programs of Study

The Department of Physics and Astronomy offers a program of study leading to the Ph.D. in physics and also M.S. programs in physics, computational physics/astronomy, geophysics, and atmospheric science.

The master's degree in physics requires 30 hours of advanced courses (up to 6 of which may be transferred from another accredited university) and at least 2 hours of master's research with satisfactory progress. A minimum average of B is required, as is a general examination in physics. The various master's programs differ in their detailed requirements.

The Ph.D. program begins with formal course work (which typically extends over two years for a well-prepared student) and, after admission to candidacy, is followed by Ph.D. research. The required courses include those needed for the M.S. in physics, so it is possible to obtain the M.S. on the way to the Ph.D. degree. Course work should average better than a B. There is no language requirement, but a demonstrated skill in computer programming related to the student's field of study is required. A comprehensive oral examination is required for admission to candidacy. Following the oral exam, the student may choose a research project from the broad spectrum of experimental and theoretical research areas represented within the department. These include high-energy particle physics, astrophysics and cosmology, space physics, plasma physics, solid-state/condensed-matter physics, and nuclear physics. After carrying out the research project under the guidance of a faculty member, the student must submit a dissertation showing the results of original research and must defend it in a final oral examination. A minimum of three full academic years of residency is required; the actual time taken to complete the Ph.D. varies considerably.

Research Facilities

Extensive computing facilities exist both in the department and at the University. The computer center, located in an adjacent building, houses a VAX 9000. The department houses a VAX 3800 computer to which a number of MicroVAX II computers are linked; there are also an IIS astronomical-image-processing computer and IRIS Workstations. A large collection of books and journals is contained in the adjacent Science Library, and the department maintains a preprint collection in its own reading room. Other facilities include the Tombaugh Astronomical Observatory, a professionally staffed machine shop plus "student" shop, and a well-equipped high-temperature superconductivity lab capable of fabricating and measuring properties of high-T_c superconductors. The high-energy physics and nuclear physics groups utilize experimental facilities at various universities and national laboratories as part of collaborative experiments. There is an extensive library of magnetic tapes of *Voyager* and *Explorer* data for space physics research. For use in atmospheric science research there is a satellite dish and NAFAX weather receiver, plus wind tunnel and vortex-generation laboratories.

Financial Aid

The principal form of financial aid is the graduate teaching assistantship; most first-year graduate students in the department have this type of support. A half-time teaching assistantship (the usual appointment) carries a 9-month stipend ranging from $7800 to $8500 plus a 75 percent tuition fee waiver. Summer support is also available. Beginning graduate students may also be considered for graduate school fellowships in a University-wide competition. A few research assistantships are available for qualified first-year students, although the tendency is to award such assistantships to more advanced students.

Cost of Study

All 1990–91 teaching assistants appointed half-time or more qualified for the 75 percent tuition fee waiver and paid tuition fees of $243.75 per semester, regardless of residence. Full-time students with private support or with fellowships from sources outside the University paid tuition fees of $975 each semester if they were Kansas residents and $2754 per semester if they were nonresidents. (University fees are set by the Board of Regents and are subject to change at any time; the fees listed here are not expected to change greatly for 1991–92.)

Cost of Living

Room and board are available in University dormitories; the cost of the 1990–91 academic year was $3350. There are a limited number of furnished one- and two-bedroom University apartments for married students and their families; the rent for 1990–91 was $187–$250 per month plus utilities. Many rooms and apartments, both furnished and unfurnished, are available off campus.

Student Group

The University of Kansas has an enrollment of approximately 26,000 students, including about 7,000 graduate students. The department contains approximately 50 graduate students drawn from throughout the United States and abroad. Most of these students are supported as either teaching assistants or research assistants.

Location

The University's main campus occupies 1,000 acres on and around Mount Oread in the city of Lawrence, a growing community of 60,000 located among the forested, rolling hills of eastern Kansas. Near Lawrence are four lake resort areas for boating, fishing, and swimming. Metropolitan Kansas City lies about 40 miles east of Lawrence via interstate highway and offers a variety of cultural and recreational activites.

The University

The University of Kansas is a state-supported school founded in 1866. Long known for its commitment to academic excellence, the University considers research an important part of the educational process. In addition to the College of Liberal Arts and Sciences and the Graduate School, the University houses a number of professional schools and programs, which include Engineering, Medicine, Law, Business, Journalism, and many others.

Applying

Completed applications should be received by March 1, for those requesting graduate teaching assistantships for the fall semester; applications are accepted until all the positions are filled, but preference is given to those received by the priority date. Applications for admissions that do not require assistantships should be completed by July 1. Application forms and additional information may be obtained by writing to the address given below or by calling the number indicated.

Correspondence and Information

Chairman, Admissions and Assistantships Committee
Department of Physics and Astronomy
University of Kansas
Lawrence, Kansas 66045
Telephone: 913-864-4626

SECTION 7: PHYSICS

University of Kansas

THE FACULTY AND THEIR RESEARCH

Raymond G. Ammar, Professor and Department Chairman; Ph.D., Chicago, 1959. Experimental high-energy physics.
Barbara J. Anthony-Twarog, Associate Professor; Ph.D., Yale, 1981. Observational astronomy, stellar evolution in open star clusters, CCD and photoelectric photometry, globular clusters.
Thomas P. Armstrong, Professor; Ph.D., Iowa, 1966. Space physics, plasma physics.
Raymond W. Arritt, Assistant Professor; Ph.D., Colorado State, 1985. Atmospheric science.
Scott R. Baird, Adjunct Professor; Ph.D., Washington (Seattle), 1979. Stellar spectroscopy, variable stars.
Philip S. Baringer, Assistant Professor; Ph.D., Indiana, 1985. Experimental high-energy physics.
David B. Beard, Professor Emeritus; Ph.D., Cornell, 1950. Theoretical physics, astrophysics, space physics, plasma physics.
Robert C. Bearse, Professor; Ph.D., Rice, 1964. Experimental nuclear physics, nuclear safeguards, materials control and accounting, computer database applications.
David A. Braaten, Assistant Professor; Ph.D., California, Davis, 1988. Atmospheric science.
Wai-Yim Ching, Adjunct Professor; Ph.D., LSU, 1974. Solid-state physics, electronic structures.
Thomas E. Cravens, Professor; Ph.D., Harvard, 1975. Space physics, plasma physics.
Jack W. Culvahouse, Professor; Ph.D., Harvard, 1957. Experimental condensed-matter physics, magnetic properties of solids, computer simulations of transport in solids.
John P. Davidson, Professor; Ph.D., Washington (St. Louis), 1952. Theoretical nuclear structure physics, atomic physics, astrophysics.
Robin E. P. Davis, Professor; D.Phil., Oxford, 1962. Experimental high-energy physics.
Gisela Dreschhoff, Courtesy Assistant Professor; Dr.Sc., Braunschweig Technical (Germany), 1972. Geophysics, energy storage in solids.
Joe R. Eagleman, Professor; Ph.D., Missouri, 1963. Atmospheric science.
Jacob Enoch, Associate Professor; Ph.D., Wisconsin, 1956. Theoretical physics.
Robert J. Friauf, Professor; Ph.D., Chicago, 1953. Experimental condensed-matter physics, diffusion and color centers, molecular dynamics and Monte Carlo simulations.
Paul Goldhammer, Professor Emeritus; Ph.D., Washington (St. Louis), 1956. Theoretical physics, nuclear structure physics, atomic physics.
Curtis Hall, Instructor; M.S., Wisconsin, 1977. Atmospheric science.
Ralph W. Knapp, Courtesy Assistant Professor; Ph.D., Indiana, 1977. Geophysics.
Ralph W. Krone, Professor Emeritus; Ph.D., Johns Hopkins, 1949. Experimental nuclear physics.
Nowhan Kwak, Professor; Ph.D., Tufts, 1962. Experimental high-energy physics.
Claude Laird, Visiting Assistant Professor; Ph.D., Kansas, 1986. Solar-terrestrial relations.
Dan Ling, Associate Professor; Ph.D., Michigan, 1955. Theoretical physics.
Carl D. McElwee, Courtesy Assistant Professor; Ph.D., Kansas, 1970. Geophysics, magnetic properties of solids.
Douglas W. McKay, Professor; Ph.D., Northwestern, 1968. Theoretical elementary-particle physics and particle astrophysics.
Adrian L. Melott, Associate Professor; Ph.D., Texas, 1981. Astrophysics and cosmology, computational physics.
Herman J. Munczek, Professor; Ph.D., Buenos Aires, 1958. Theoretical elementary-particle physics.
Francis W. Prosser, Professor and Associate Chairman; Ph.D., Kansas, 1955. Experimental nuclear physics.
John P. Ralston, Associate Professor; Ph.D., Oregon, 1980. Theoretical elementary-particle physics and particle astrophysics.
Stephen J. Sanders, Associate Professor; Ph.D., Yale, 1977. Experimental nuclear physics.
Richard C. Sapp, Professor; Ph.D., Ohio State, 1955. Experimental solid-state physics.
Sergei F. Shandarin, Professor; Ph.D., Moscow Physical Technical Institute, 1971. Astrophysics and cosmology, large-scale structure, nonlinear dynamics, computational physics.
Stephen J. Shawl, Professor; Ph.D., Texas, 1972. Observational astronomy, stellar astronomy, polarization, globular clusters.
Don W. Steeples, Courtesy Assistant Professor; Ph.D., Stanford, 1975. Geophysics.
Robert Stump, Professor Emeritus; Ph.D., Illinois, 1950. Experimental high-energy physics, atmospheric science.
Bruce A. Twarog, Associate Professor; Ph.D., Yale, 1980. Observational astronomy, stellar nucleosynthesis, chemical evolution of galaxies, stellar photometry.
Gordon G. Wiseman; Professor Emeritus; Ph.D., Kansas, 1950. Experimental solid-state physics.
Kai-Wai Wong, Professor; Ph.D., Northwestern, 1962. Many-body theory, superconductivity, liquid helium.
Edward J. Zeller, Professor; Ph.D., Wisconsin, 1951. Experimental physics, geophysics.

UNIVERSITY OF MASSACHUSETTS LOWELL

Department of Physics and Applied Physics

Programs of Study

The Department of Physics and Applied Physics offers programs leading to the degrees of Master of Science and Doctor of Philosophy.

The M.S. degree may be taken in physics or radiological science and protection (health physics) or in the applied physics option in optical sciences. Course requirements for the M.S. program consist of a total of 30 credits, including work on a thesis or project. The M.S. may serve as a basis for further study toward a Ph.D. degree. Students are expected to complete the M.S. program in two years.

The Ph.D. program requires 60 credits, including thesis research. Candidates for the degree must pass a doctoral research admission examination (taken after successfully completing two semesters of an advanced research project) and demonstrate a reading knowledge of a foreign language and proficiency in computer programming. Areas of research include experimental and theoretical nuclear physics, experimental and theoretical solid-state physics, laser physics and far-infrared spectroscopy, scattering of electromagnetic radiation from rough surfaces, scattering theory, quantum optics, field theory, astrophysics, relativity, particle physics, atmospheric and environmental physics, energy applications, applied mechanics, radiological sciences, and computational physics.

Research Facilities

The department is housed in the Olney Science Center and in the Pinanski Energy Center. Major pieces of equipment include a 5.5-MeV DC or pulsed beam accelerator with Mobley bunching system, a 2-MeV Van de Graaff accelerator, a 400-kilocurie Co-60 source, a 1-MW swimming pool reactor, a 100-kilogauss superconducting magnet with radial and vertical access, CO_2 and far-infrared lasers, an FT-IR spectrometer, a 5-W Argon-ion laser with a tunable dye laser, an He-gas closed-cycle temperature controller, a high-speed photon counting system, a rahman and fluorescence system, a nitrogen-pumped dye laser, CW and pulsed lasers, a UV-VIS-IR spectrophotometer and a wavelength scannable ellipsometer, a thin-film evaporator, Rutherford backscattering and PIXE spectrometers, and several minicomputers and personal computers. A University-wide VAX computer is also available for research use.

Financial Aid

In 1991–92, financial aid is available in the form of teaching assistantships at $7250–$8750 per academic year and research assistantships at $9750 for the academic year. Summer research stipends are available for qualified students at $1000 per month. In addition, INPO and Department of Energy fellowships are available for students in radiological sciences.

Cost of Study

In 1991–92, tuition for Massachusetts residents is approximately $2000 for full-time students. For nonresidents, tuition is approximately $6300 for full-time students. Fees and health insurance amount to approximately $1200 per year.

Cost of Living

The Lowell area offers a great variety of living accommodations. There is a limited amount of on-campus housing available for graduate students, including single-student accommodations with cooking facilities and unfurnished apartments for married students. Early application for these is essential. Cost per month for single students is $290. Married student apartments cost $440–$490 per month.

Student Group

In September of 1990, the total University enrollment was approximately 11,000, including about 3,000 graduate students. There are about 100 full-time graduate students in physics. Fifty-five of these students receive financial support. Approximately 20 percent of the physics graduate students are women, and about 30 percent are international students.

Location

The University is located in the city of Lowell, a community of about 100,000 residents, 30 miles northwest of Boston. The campus is located on both sides of the Merrimack River. Within an hour's drive of Lowell are the cultural, educational, and recreational activities of the Boston area, as well as many nationally known sites. The locale is also ideally situated for the pursuit of outdoor activities. The lake and mountain regions of New Hampshire and the Atlantic beaches are easily accessible. Because of Lowell's proximity to the ocean, the climate is not as severe as that of other areas of New England.

The University and The Department

The University of Massachusetts Lowell was created by an act of the 1973 Massachusetts legislature, calling for the merger of Lowell Technological Institute with Lowell State College. The Department of Physics and Applied Physics, originally part of Lowell Technological Institute, has offered the Ph.D. since 1964; the total number of students who have received graduate degrees from the department is about 200.

Applying

Applications must be submitted no later than June 1 preceding the fall term in which the applicant wishes to enroll. The General Test of the GRE, the TOEFL for international students whose native language is not English, transcripts in duplicate, and three letters of reference are required. There is an application fee of $10 for Massachusetts residents and $25 for nonresidents.

Correspondence and Information

Dr. James J. Egan
Physics Graduate Coordinator
Department of Physics and Applied Physics
University of Massachusetts Lowell
Lowell, Massachusetts 01854
Telephone: 508-934-3774

Peterson's Guide to Graduate Programs in the Physical Sciences and Mathematics 1992

University of Massachusetts Lowell

THE FACULTY AND THEIR RESEARCH

Professors
A. Altman, Ph.D., Maryland. Theoretical atomic physics.
A. Baker (Emeritus), Ph.D., Brandeis. Atomic theory.
L. E. Beghian (Emeritus), D.Phil., Oxford. Experimental nuclear physics, food irradiation.
G. Carr, Ph.D., Cornell. Physics education.
G. P. Couchell, Ph.D., Columbia. Experimental nuclear physics.
J. J. Egan, Ph.D., Kentucky. Experimental nuclear physics.
Z. Fried, Ph.D., Brandeis. Radiation theory, optics.
J. Y. Harris, Ph.D., Rutgers. Radiological science and protection, radiation biology.
L. C. Kannenberg, Ph.D., Northeastern. Relativity, radiation theory.
A. S. Karakashian, Chairman; Ph.D., Maryland. Theoretical solid-state physics, optics.
G. H. R. Kegel, Ph.D., MIT. Experimental nuclear physics, radiation effects in materials.
A. Liuzzi, Ph.D., NYU. Physics education.
A. Mittler, Ph.D., Kentucky. Experimental nuclear physics.
D. J. Pullen, D.Phil., Oxford. Experimental nuclear physics.
W. A. Schier, Ph.D., Notre Dame. Experimental nuclear physics.
K. J. Sebastian, Ph.D., Maryland. Particle physics theory.
E. Sheldon, Ph.D., D.Sc., London. Theoretical nuclear physics, astrophysics.
K. W. Skrable, Ph.D., Rutgers. Radiological science and protection, internal dosimetry.
R. W. Stimets, Ph.D., MIT. Experimental laser physics, astronomy, image processing.
Y. Y. Teng, Ph.D., Maryland. Experimental solid-state physics, thin films, optics, materials.
J. Waldman, Ph.D., MIT. Experimental laser physics, infrared spectroscopy.
M. Wilner, Ph.D., MIT. Theoretical solid-state physics, optics.

Associate Professors
G. E. Chabot, Ph.D., Massachusetts Lowell. Radiation science and protection, dosimetry, radiation shielding.
P. Harihar, Ph.D., Columbia. Experimental nuclear physics.
J. Kumar, Ph.D., Rutgers. Optics.
D. M. Larsen, Ph.D., MIT. Theoretical solid-state physics.
T. V. Marcella, Ph.D., Boston College. Physics education.
R. D. McLeod, M.S., Lowell Technological Institute. Theory of vision.
P. J. Ring, Ph.D., Brown. Physics education.
A. Sachs, Ph.D., New Hampshire. Theoretical mechanics.
C. Wong, Ph.D., Case Tech. Experimental solid-state physics, optics.

Assistant Professors
S. Broude, Ph.D., Academy of Sciences (Moscow). Experimental laser physics, semiconductor physics.
C. S. French, Ph.D., Massachusetts Lowell. Radiation science and protection, medical physics.

Adjunct Faculty
G. Crowley, Ph.D., Leicester (England). Atmospheric physics.
R. H. Giles, Ph.D., Massachusetts Lowell. Solid-state physics, optics.
W. Goodhue, Ph.D., Massachusetts Lowell. Submicron devices.
H. Haskal, Ph.D., Case Tech. Solid-state physics, optics.
G. B. Inglis, Ph.D., Delaware. Radiological science and protection, medical physics.
K. R. Kase, Ph.D., Stanford. Radiological science and protection, medical physics.
H. Luther, Ph.D., Penn State. Optics.
D. E. McCurdy, Ph.D., Colorado State. Radiological science and protection.
T. G. Tsuei, Ph.D., Clarkson. Optics.

RESEARCH ACTIVITIES AND PARTICIPATING FACULTY

Theoretical Physics
Atomic and molecular physics: electron emission and radiative transitions, inelastic electron-atom scattering, mesic atoms. (Altman, Fried)
Elementary particles and fields: current algebra, quark models, renormalization groups, quantum field theory, quantum electrodynamics, gauge theory, supersymmetry, general relativity. (Kannenberg, Sebastian)
Nuclear physics: nuclear cross-sections and reaction mechanisms, nuclear energy-level classifications, magnetic substate population studies. (Sheldon)
Optics and solid-state physics: quantum optics, dielectric waveguides, surface plasmons, ultraviolet and far-infrared spectra, electronic and vibrational cluster calculations, optical and electronic properties of semiconductors and multiple-quantum-well structures, theory of vision. (Fried, Kannenberg, Karakashian, Larsen, McLeod, Sachs, Stimets, Wilner)

Experimental Physics
Applied physics: radiation effects, Rutherford back-scattering, food irradiation, PIXE, nuclear instrumentation. (Beghian, Egan, Kegel, Mittler)
Nuclear physics: neutron cross-sections, fission reaction studies, inelastic neutron scattering, delayed-neutron spectral studies. (Beghian, Egan, Couchell, Kegel, Mittler, Pullen, Schier)
Optics and solid-state physics: holography, tunable visible-infrared and far-infrared lasers, optoelectronic materials and devices, image processing, surface plasmons, radiation damage in optical and electronic materials and devices. (Broude, Karakashian, Kumar, Stimets, Teng, Waldman, Wong)

Radiological Sciences
Internal and external radiation dosimetry, biological effects of radiation, radon and radioactive aerosol collection and measurement systems and techniques, environmental sampling and analysis. (Chabot, French, Harris, Ring, Skrable)

SECTION 7: PHYSICS

UNIVERSITY OF MICHIGAN

Department of Physics

Programs of Study

The department offers programs leading to the Master of Science and Doctor of Philosophy in physics. Students in the Ph.D. program are encouraged to participate in research at the earliest opportunity, usually no later than the summer following their first year. The department offers numerous opportunities for research in theoretical and experimental fields, including musical acoustics, atomic physics, astrophysics, biophysics, optical physics, condensed-matter physics, elementary particle physics, and nuclear physics. The requirements for the Ph.D. are as follows. Students must pass, with a grade of B or better, eight prescribed graduate physics courses (500 level) or show equivalent competence, two cognate courses, two advanced graduate physics courses (600 level) on special topics, and a 4-credit course of supervised nonthesis research. Students must also pass a written qualifying examination on advanced undergraduate material no later than the beginning of their third year, and they must demonstrate proficiency in a foreign language. Ph.D. students are expected to attain candidacy by the end of their fifth term. Completion of the degree involves writing a thesis based on independent research done under the supervision of a faculty adviser and passing a final oral exam.

Students in the M.S. degree program must have passing grades (C or above) in 24 credit hours of courses at the 400 level and above, including two cognate courses, and they must maintain a grade average of B or better. At least 12 hours of courses must be in physics, and the students must pass with a grade of B– or better two of a selected group of 500-level courses.

Research Facilities

Physics faculty and students collaborate in the new NSF Science and Technology Center for Ultrafast Optical Science, which has extensive facilities on the University's North Campus. Special research equipment in the physics department includes a positron microscope; a molecular-beam epitaxy system for the fabrication of superlattices and study of ultrathin films of metals and semiconductors; several laser systems, including frequency-stabilized dye lasers, Q-switched YAG pumped dye lasers, and synchronously pumped mode-locked dye lasers; a surface plasmon resonance probe; scanning tunneling microscopes; numerous spectrometers, including a Mössbauer spectrometer specially modified for the study of active sites in proteins; superconducting quantum interference devices (SQUIDs); and numerous devices for particle trapping and precision measurements. Computing facilities include a six-processor DEC VAX 6200 and eight LAVC VAXstations, a twelve-workstation Sun system running TeX, and five Apollo workstations with Mentor Graphics. Numerous PCs are connected to the departmental network. Access to BITNET, Internet, HEPNET, LSNET, SLAC, and NSFNET is available. Equipment for experiments in high-energy physics—involving work on proton decay, cosmic ray detection, polarized proton scattering, and p^+p^-, e^+e^-, and p^+p^+ collisions—is designed and constructed by Michigan physicists; recent collaborations have taken place at Argonne and Brookhaven national laboratories, CERN, Fermilab, and SLAC.

Financial Aid

The department's Regents'/H. R. Crane Fellowships pay a stipend of $18,000 for twelve months in 1991–92, plus full tuition and fees; students are released from teaching or research during their first year, to concentrate on course work and degree requirements, and are required to become seriously involved in research beginning in the summer following their first year. The University's Regents' Fellowships pay a stipend of $13,200, plus full tuition and fees. Several other merit-based scholarships are available through the department, to both incoming and advanced students.

Graduate teaching assistantships cover a period of eight months and paid a stipend of $8800 plus tuition in 1990–91; teaching loads usually consist of four 2-hour elementary lab sections per week. Graduate TAs are represented by a union. Graduate research assistantships cover eight months for precandidates and twelve months for candidates; stipends were $8800 and $13,200, respectively, in 1990–91. Summer RA appointments are available for most students; stipends ran up to $1100 per month for four months in summer 1990. A very small number of summer TA appointments are also available. Students who have at least a one-quarter-time appointment (half a normal appointment) as a research or teaching assistant are eligible to participate in the University's group health insurance plan.

Cost of Study

Tuition in 1990–91 for precandidates was $3037 per term for full-time in-state residents and $6352 per term for full-time out-of-state residents; candidates paid $1899 per term. Tuition is waived for students with one-quarter-time or more teaching or research assistantships. Most fees are included in the tuition; fees not included total about $70 each semester. Books and supplies cost approximately $265 per term.

Cost of Living

Living costs for the 1990–91 academic year of eight months, including room and board, transportation, and personal needs, were estimated at $6100 for a single student with no dependents.

Student Group

The University of Michigan has approximately 35,000 students, of whom 13,500 are graduate students. The Department of Physics enrolled approximately 180 graduate students in 1990–91.

Location

The University is in Ann Arbor, 40 miles west of Detroit in the Huron River Valley. The Huron River flows through the campus. Ann Arbor is a beautiful, tree-lined town that combines the charm of a small city with the sophistication of cities many times its size. Rightly regarded as a cultural center of the Midwest, it offers numerous opportunities for recreation and enjoyment.

The University and The Department

The University of Michigan, founded in 1817, is one of the nation's oldest public institutions of higher education. Consistently ranked among the best universities in the world, Michigan has a strong tradition of leadership in the development of the modern American research university—a tradition sustained by the wide-ranging interests and activities of its faculty and students. Exceptional facilities and programs, both academic and nonacademic, are available. The Department of Physics has played a leading role in the development of modern physics, with accomplishments ranging from the discovery of proton spin and the invention of the racetrack synchrotron and the bubble chamber to the recent landmark detection of neutrinos from supernova 1987A; in the 1930s, it hosted world-renowned summer symposia in theoretical physics.

Applying

Applications for admission in the fall term are due by February 1 of the preceding spring. The GRE General Test is required, and the GRE Subject Test in physics is recommended. For further information on admission requirements, students should write to the address given below.

Correspondence and Information

Dr. James W. Allen, Associate Chair for Graduate Studies
Department of Physics
1049 Randall Laboratory of Physics
University of Michigan
Ann Arbor, Michigan 48109-1120
Telephone: 313-764-4438

SECTION 7: PHYSICS

University of Michigan

THE FACULTY AND THEIR RESEARCH

Fred C. Adams, Assistant Professor; Ph.D., Berkeley, 1988. Theoretical astrophysics, star formation.
Carl W. Akerlof, Professor; Ph.D., Cornell, 1966. Experimental high-energy physics, astrophysics, cosmic rays.
Ratindranath Akhoury, Assistant Professor; Ph.D., SUNY at Stony Brook, 1980. Theoretical high-energy physics.
James W. Allen, Professor and Associate Chair for Graduate Studies; Ph.D., Stanford, 1968. Experimental condensed-matter physics.
Dante E. Amidei, Assistant Professor; Ph.D., Berkeley, 1984. Experimental high-energy physics.
Meigan C. Aronson, Assistant Professor; Ph.D., Illinois at Urbana-Champaign, 1987. Experimental condensed-matter physics.
Daniel Axelrod, Professor; Ph.D., Berkeley, 1974. Experimental biophysics.
Tofigh Azemoon, Assistant Research Scientist; Ph.D., University College (London), 1973. Experimental high-energy physics.
Robert Ball, Associate Research Scientist; Ph.D., Michigan State, 1979. Experimental high-energy physics, elementary particles.
Frederick D. Becchetti Jr., Professor; Ph.D., Minnesota, 1969. Experimental nuclear physics.
Michael Bretz, Professor; Ph.D., Washington (Seattle), 1971. Experimental physics, low-temperature physics, condensed matter.
Philip H. Bucksbaum, Professor and Science Director, Center for Ultrafast Optical Science; Ph.D., Berkeley, 1980. Experimental atomic and optical physics.
Myron Campbell, Assistant Professor; Ph.D., Yale, 1982. Experimental high-energy physics, elementary particles.
J. Wehrley Chapman, Professor; Ph.D., Duke, 1966. Experimental high-energy physics.
Timothy E. Chupp, Associate Professor; Ph.D., Washington (Seattle), 1983. Experimental atomic physics, precision measurements.
Roy Clarke, Professor and Director of the Applied Physics Program; Ph.D., London, 1973. Experimental solid state.
C. Tristram Coffin, Professor; Ph.D., Washington (Seattle), 1956. Experimental high-energy physics, biophysics.
Ralph S. Conti, Assistant Research Scientist; Ph.D., Berkeley, 1979. Experimental atomic physics.
Steven Dierker, Associate Professor; Ph.D., Illinois at Urbana-Champaign, 1983. Condensed-matter experiment, applied physics.
Thomas M. Donahue, Professor; Ph.D., Johns Hopkins, 1947. Astrophysics, atomic physics, space physics.
Martin B. Einhorn, Professor; Ph.D., Princeton, 1968. Theoretical high-energy physics.
August Evrard, Assistant Professor; Ph.D., SUNY at Stony Brook, 1987. Theoretical astrophysics, large-scale cosmic structures.
Stephen B. Fahy, Assistant Professor; Ph.D., Berkeley, 1987. Applied physics, experimental condensed-matter physics.
Stuart Field, Assistant Professor; Ph.D., Chicago, 1986. Applied physics, experimental condensed-matter physics.
G. W. Ford, Professor; Ph.D., Michigan, 1954. Theoretical physics.
Katherine Freese, Associate Professor; Ph.D., Chicago, 1984. Theoretical astrophysics, cosmology.
William E. Frieze, Research Investigator; Ph.D., Yale, 1979. Experimental solid-state physics.
David W. Gidley, Associate Professor; Ph.D., Michigan, 1979. Experimental physics, atomic physics, surface physics.
Walter S. Gray, Associate Professor; Ph.D., Colorado, 1964. Experimental nuclear physics.
Herold R. Gustafson, Associate Research Scientist; Ph.D., Washington (Seattle), 1968. Experimental high-energy physics.
Karl T. Hecht, Professor; Ph.D., Michigan, 1955. Theoretical nuclear physics, nuclear structure.
Dennis J. Hegyi, Professor; Ph.D., Princeton, 1968. Experimental physics, astrophysics.
Rombout Hoogerbeets, Assistant Professor; Ph.D., Amsterdam, 1984. Experimental condensed-matter physics.
Joachim W. Janecke, Professor; Dr.rer.nat., Heidelberg, 1955. Experimental nuclear physics.
Lawrence W. Jones, Professor; Ph.D., Berkeley, 1952. Experimental high-energy physics.
Gordon L. Kane, Professor; Ph.D., Illinois, 1963. Theoretical physics, elementary particles.
David A. Kessler, Associate Research Scientist; Ph.D., Princeton, 1981. Theoretical physics, condensed matter.
Samuel Krimm, Professor and Director of the Protein Structure and Design Program; Ph.D., Princeton, 1950. Experimental biophysics.
Alan D. Krisch, Professor; Ph.D., Cornell, 1964. Experimental high-energy physics.
Jean P. Krisch, Associate Professor and Associate Chair for Undergraduate Studies; Ph.D., Cornell, 1965. Theoretical physics, elementary particles, physics teaching.
Robert R. Lewis, Professor; Ph.D., Michigan, 1954. Theoretical low-energy physics, neutrinos.
Michael J. Longo, Professor; Ph.D., Berkeley, 1961. Experimental high-energy physics, instrumentation.
James M. Matthews, Associate Research Scientist; Ph.D., Wisconsin, 1984. Experimental high-energy physics, elementary particles.
Roberto D. Merlin, Professor; Dr.rer.nat., Stuttgart, 1978. Experimental solid state.
Donald I. Meyer, Professor; Ph.D., Washington (Seattle), 1953. Experimental high-energy physics.
Homer A. Neal, Professor and Chair of the Department; Ph.D., Michigan, 1966. Experimental high-energy physics.
David F. Nitz, Associate Research Scientist; Ph.D., Rochester, 1978. Experimental high-energy physics.
Franco Nori, Assistant Professor; Ph.D., Illinois at Urbana-Champaign, 1987. Condensed-matter theory.
Bradford G. Orr, Assistant Professor; Ph.D., Minnesota, 1985. Experimental condensed-matter physics, applied physics.
Oliver E. Overseth, Professor; Ph.D., Brown, 1958. Experimental high-energy physics, elementary particles.
Steven Rand, Associate Professor; Ph.D., Toronto, 1978. Optical physics, laser physics, applied physics.
Jaiwad Rasul, Assistant Professor; Ph.D., London, 1980. Theoretical condensed-matter physics.
Byron P. Roe, Professor; Ph.D., Cornell, 1959. Experimental high-energy physics.
Thomas Roser, Assistant Research Scientist; Ph.D., Swiss Federal Institute of Technology, 1984. Experimental high-energy physics.
Marc H. Ross, Professor; Ph.D., Wisconsin, 1952. Environmental physics.
Leonard M. Sander, Professor; Ph.D., Berkeley, 1968. Theoretical physics, condensed matter.
T. Michael Sanders, Professor; Ph.D., Columbia, 1954. Experimental physics, low temperatures, solid state.
Richard H. Sands, Professor; Ph.D., Washington (St. Louis), 1954. Biophysics.
Robert S. Savit, Associate Professor; Ph.D., Stanford, 1973. Theoretical solid state, elementary particles.
Daniel Sinclair, Professor; Ph.D., Glasgow, 1957. Experimental high-energy physics, particle astrophysics.
Mark A. Skalsey, Assistant Research Scientist; Ph.D., Michigan, 1982. Atomic physics, nuclear physics, weak interactions.
Gregory R. Snow, Assistant Professor; Ph.D., Princeton, 1983. Experimental high-energy physics.
Duncan G. Steel, Associate Professor; Ph.D., Michigan, 1976. Experimental physics, laser physics, atomic physics.
Gregory Tarlé, Associate Professor; Ph.D., Berkeley, 1978. Experimental astrophysics, particle physics, nuclear physics.
Rudolf P. Thun, Professor; Ph.D., SUNY at Stony Brook, 1972. Experimental high-energy physics.
Robert S. Tickle, Professor; Ph.D., Virginia, 1960. Experimental nuclear physics.
Yukio Tomozawa, Professor; D.Sc., Tokyo, 1961. Theoretical high-energy physics, elementary particles.
Ctirad Uher, Professor and Associate Chair for Research and Facilities; Ph.D., New South Wales, 1975. Experimental solid-state physics.
John C. van der Velde, Professor; Ph.D., Michigan, 1958. Experimental high-energy physics.
Martinus J. G. Veltman, MacArthur Professor of Theoretical Physics; Dr.rer.nat., Utrecht (Netherlands), 1963. Elementary particles.
John F. Ward, Professor; D.Phil., Oxford, 1961. Experimental physics, quantum electronics.
Gabriel Weinreich, Professor; Ph.D., Columbia, 1954. Experimental physics, atomic physics, musical acoustics.
David N. Williams, Professor; Ph.D., Berkeley, 1964. Theoretical physics, elementary particles.
Alfred C.-T. Wu, Professor; Ph.D., Maryland, 1960. Theoretical physics.
Y. P. Edward Yao, Professor; Ph.D., Harvard, 1964. Theoretical high-energy physics, elementary particles.
Jens C. Zorn, Professor; Ph.D., Yale, 1961. Experimental physics, atomic physics.

SECTION 7: PHYSICS

UNIVERSITY OF NOTRE DAME

Department of Physics

Programs of Study

The Department of Physics offers programs of study leading to the Ph.D. degree and the M.S. degree. Major areas of research include atomic physics, microelectronics, nuclear physics, solid-state physics, elementary particle physics, astrophysics, statistical physics, general relativity, and the history and philosophy of science. Interdisciplinary programs are available in radiation physics and chemical physics.

The requirements for the Ph.D. include a minimum of 72 credit hours of courses, seminars, and research. Each graduate student is expected to become actively involved in research during the first year. A candidacy examination consisting of both oral and written parts is to be completed, normally early in the third year. The candidate must demonstrate the ability to perform research and must show a broad understanding of physics. A thesis is required and must be approved by the student's doctoral committee and defended orally by the student before this committee.

Research Facilities

The Department of Physics is housed in Nieuwland Science Hall, which contains a well-equipped research library. Research facilities include a heavy-ion accelerator laboratory with 4-MV and 9-MV Van de Graaff accelerators. Associated nuclear facilities include a universal negative-ion source, a polarized-ion source, a 1-meter magnetic spectrometer, a multidetector array for gamma-ray spectroscopy, and a superconducting solenoid for radioactive beam studies. Active users programs in nuclear physics are also under way at the Argonne National Laboratory and the National Superconducting Cyclotron Laboratory. Facilities for accelerator-based atomic physics research include vacuum ultraviolet monochromators and high-resolution positron-sensitive photon detectors at Notre Dame and at Argonne National Laboratory. Laser and microwave spectroscopy of fast beams of atoms and molecules is carried out with CO_2, CO, Nd-yag, and color center laser systems and microwave sources from DC to 18 GHz. Neutral atom cooling and trapping is carried out with tuneable diode lasers. Elementary particle physics research is carried out at the Fermi National Accelerator Laboratory, Brookhaven National Laboratory, and the Superconducting Supercollider Laboratory. On-campus facilities are used for the development of new particle detection systems, including scintillating fiber and CsI detectors, counters, and chambers for use in experiments. A large cosmic-ray shower array, for investigating astronomical point sources of ultrahigh-energy gamma rays, such as Cygnus X-3, is under construction near campus. In solid-state physics, facilities are available for molecular-beam epitaxy (MBE) of semiconductor films, superlattices, and microstructures and for bulk crystal growth, low-temperature electron tunneling, microwave studies of high-temperature superconductors, resonance studies in ferromagnetic materials, surface physics, X-ray and fluorescence characterization of solids, and optical and far-infrared studies of semiconductors, including photoluminescence and magnetooptical measurements. EXAFS and X-ray-scattering experiments are also carried out at the Brookhaven National Laboratory, and neutron diffraction studies at the National Institute of Standards and Technology. Computing facilities include the University's Convex C-240 40 MFlop minisupercomputer, several Sun and MAC IIci workstation clusters, and an IBM 3081 K. Departmental facilities include a VAXstation cluster, several IBM RS/6000 UNIX workstations, two PE3220s, a number of desktop PCs and Macintoshes, a teaching cluster of fifteen Sun SPARCstation IIs, and a VAX 3600.

Financial Aid

Graduate teaching assistantships are normally available to all Ph.D. students and for 1991–92 include a minimum nine-month stipend of $9500, plus payment of tuition and fees. In 1991–92, most incoming graduate assistants are supported at a level of $10,000 for nine months. Higher stipends are available for exceptionally well qualified applicants. Summer support is normally provided from federal research funding. Research fellowships are available on a competitive basis. Advanced students often receive support as research assistants.

Cost of Study

Graduate tuition for 1991–92 is $13,385; summer tuition and fees are $137. Payment of tuition and fees is provided in addition to the student stipends.

Cost of Living

Accommodations for single students are available on or adjacent to the campus at a cost of $700–$1000 per semester. Accommodations for married students are available near the campus for $195–$268 per month. Privately owned rooms and apartments are also for rent near the campus.

Student Group

There are 82 graduate students in the Department of Physics. In 1990–91, the University had an enrollment of 10,026 students, of whom 2,481 were graduate students.

Location

The University of Notre Dame is located adjacent to the city of South Bend in northern Indiana. South Bend enjoys an active social and cultural life and has numerous parks and recreational areas. Chicago is readily accessible by plane, bus, or train and is less than 2 hours away by car. Nearby Lake Michigan provides excellent facilities for water sports. Skiing and other winter sports are also available in the area.

The University

The University of Notre Dame, founded in 1842, is a private, independent, fully coeducational school. The campus of 1,250 acres offers an uncrowded setting that includes two attractive lakes and numerous wooded areas. The intellectual, cultural, and athletic traditions at Notre Dame, coupled with the beauty of the campus, contribute to the University's fine reputation. The University offers a variety of cultural and recreational activities, including plays, concerts, and lecture series in many areas, bringing to Notre Dame world-famous personalities. On the campus are facilities for indoor and outdoor sports; tennis and ice-skating can be enjoyed the year round. The students and faculty represent a rich diversity of religious, racial, and ethnic backgrounds.

Applying

Applications are invited from qualified students without regard to sex, race, religion, or national or ethnic origin. Both the General Test and the Subject Test in physics of the Graduate Record Examinations are required. Complete applications should be submitted as early as possible in the academic year. A detailed departmental brochure and application information are available by writing to the address below.

Correspondence and Information

Chairman, Admissions Committee
Department of Physics
University of Notre Dame
Notre Dame, Indiana 46556
Telephone: 219-239-6386

SECTION 7: PHYSICS

University of Notre Dame

THE FACULTY AND THEIR RESEARCH

Ani Aprahamian, Ph.D., Clark, 1985. Experimental nuclear physics: gamma-ray spectroscopy.
Gerald B. Arnold, Ph.D., UCLA, 1977. Theoretical solid-state physics: magnetism, high-temperature superconductivity.
Ikaros I. Bigi, Ph.D., Munich, 1976. Theoretical high-energy physics.
Nripendra N. Biswas, Ph.D., Calcutta, 1954. Experimental physics: high-energy elementary particle physics, boson resonances and inclusive reactions.
Howard A. Blackstead, Ph.D., Rice, 1967. Experimental physics: solid-state physics, magnetism and acoustics.
Samir K. Bose, Ph.D., Rochester, 1962. Theoretical physics: black holes and other astrophysical objects in general relativity, applications of group theory to elementary particles.
Cornelius P. Browne, Ph.D., Wisconsin, 1951. Experimental physics: nuclear structure, reaction energies.
Bruce A. Bunker, Ph.D., Washington (Seattle), 1980. Experimental physics: X-ray, UV, and electron spectroscopy of solids and surfaces.
Neal M. Cason, Ph.D., Wisconsin, 1964. Experimental physics: high-energy elementary particle physics, particle spectroscopy and colliding beam experiments.
Paul R. Chagnon, Ph.D., Johns Hopkins, 1955. Experimental physics: nuclear structure, electromagnetic transitions.
James T. Cushing, Ph.D., Iowa, 1963. Theoretical physics: foundational problems in quantum theory, history and philosophy of twentieth-century physics.
Sperry E. Darden, Ph.D., Wisconsin, 1955. Experimental physics: nuclear structure, reactions and scattering of polarized particles.
Malgorzata Dobrowolska-Furdyna, Ph.D., Polish Academy of Sciences, 1979. Experimental solid-state physics.
John D. Dow, Ph.D., Rochester, 1967; Freimann Professor of Physics. Theoretical physics: solid-state physics and microelectronics, scanning tunneling microscopy.
Emerson G. Funk, Ph.D., Michigan, 1958. Experimental physics: gamma-ray spectroscopy, environmental radioactivity.
Jacek K. Furdyna, Ph.D., Northwestern, 1960; Marquez University Professor. Experimental solid-state physics: man-made materials.
Umesh Garg, Ph.D., SUNY at Stony Brook, 1978. Experimental nuclear physics: nuclear structure, giant resonances, gamma-ray spectroscopy, high-spin states.
Walter R. Johnson, Ph.D., Michigan, 1957. Theoretical physics: quantum electrodynamics, atomic physics.
Gerald L. Jones, Ph.D., Kansas, 1961; Chairman of the Department. Theoretical physics: statistical mechanics, phase transitions, transport properties.
V. Paul Kenney, Ph.D., Fordham, 1956. Experimental physics: high-energy elementary particle physics.
James J. Kolata, Ph.D., Michigan State, 1969. Experimental physics: nuclear structure, heavy-ion reactions.
A. Eugene Livingston, Ph.D., Alberta, 1974. Experimental physics: atomic physics, spectroscopy of highly ionized atoms.
John M. LoSecco, Ph.D., Harvard, 1976. Experimental and theoretical physics: high-energy elementary particle physics.
Stephen R. Lundeen, Ph.D., Harvard, 1975. Experimental physics: atomic physics.
Eugene R. Marshalek, Ph.D., Berkeley, 1962. Theoretical nuclear physics.
William D. McGlinn, Ph.D., Kansas, 1959. Theoretical physics: elementary particle physics.
Kathie E. Newman, Ph.D., Washington (Seattle), 1981. Theoretical physics: statistical mechanics, studies of ordering in semiconductors.
John A. Poirier, Ph.D., Stanford, 1959. Experimental physics: high-energy elementary particle physics.
Terrence W. Rettig, Ph.D., Indiana, 1976. Observational astronomy: globular clusters, comets.
Randal C. Ruchti, Ph.D., Michigan State, 1973. Experimental physics: high-energy elementary particle physics.
Steven T. Ruggiero, Ph.D., Stanford, 1981. Experimental physics: solid-state and low-temperature physics, superconductivity.
Jonathan R. Sapirstein, Ph.D., Stanford, 1979. Theoretical physics: quantum electrodynamics.
Paul E. Shanley, Ph.D., Northeastern, 1966. Theoretical physics: nuclear reactions, few-body problems, quantum chaos.
William D. Shephard, Ph.D., Wisconsin, 1962. Experimental physics: high-energy elementary particle physics, hadron and photon interactions and multiparticle systems.
Carol E. Tanner, Ph.D., Berkeley, 1985. Experimental physics: atomic physics.
Walter J. Tomasch, Ph.D., Case Institute of Technology, 1958. Experimental physics: solid-state physics, low-temperature physics, high-temperature superconductivity.
Mitchell R. Wayne, Ph.D., UCLA, 1985. Experimental high-energy elementary particle physics.
Michael Wiescher, Ph.D., Münster, 1980. Experimental nuclear physics: nuclear astrophysics.

RESEARCH ACTIVITIES

Theoretical

Atomic Physics: quantum electrodynamics, weak interactions, atomic many-body theory, photoionization and photoexcitation. (Johnson, Sapirstein, 1 postdoctoral fellow)
Elementary Particle Physics: formal properties of quantum field theories, structure of spontaneous breakdown of symmetry, phenomenology of strong and weak processes, rare decays, CP violation, supergravity and new particles. (Bigi, Bose, McGlinn, 1 postdoctoral fellow)
General Relativity: black holes in a magnetic field, charged black holes. (Bose)
History and Philosophy of Science: interpretative problems in quantum mechanics. (Cushing)
Nuclear Physics: many-body problem, nuclear reactions, few-body problem, boson expansions. (Marshalek, Shanley)
Solid State: many-body problem, superconductivity, tunneling phenomena, metal-metal interfaces, inhomogeneous and layered superconductors, hopping transport, studies of ordering in semiconductors. (Arnold, Newman, 1 postdoctoral fellow)
Statistical Mechanics: phase transitions, critical phenomena in fluids. (Jones, Newman, 1 postdoctoral fellow, 1 visiting scholar)

Experimental

Astrophysics: Neutrino and proton decay studies with IMB detector, air shower array measurements of ultrahigh-energy cosmic rays, fast data acquisition of spectra and images of comets, nuclear reaction rates relevant to nucleosynthesis. (LoSecco, Poirier, Rettig, Wiescher, 2 postdoctoral fellows, 1 faculty fellow and specialist)
Atomic Physics: atomic structure, parity violation, tests of fundamental symmetries, excitation mechanisms, and radiative decays in neutral and ionized atoms. (Livingston, Lundeen, Tanner, 1 postdoctoral fellow)
High-Energy Elementary Particle Physics: fixed target production of heavy and light particle states, including charm, beauty particles, and gluonic matter; $p\bar{p}$ colliding beam multiparticle production, including jets and quark-gluon plasma, scintillating fiber detector, and target development for SSC. (Biswas, Cason, Kenney, LoSecco, Poirier, Ruchti, Shephard, 4 faculty fellows and specialists, 2 postdoctoral fellows, 1 visiting research associate)
Nuclear Physics: nuclear structure, reaction energies, electromagnetic transitions, gamma-ray spectroscopy, high-spin states, polarized particles, giant resonances, heavy-ion reactions, nuclear astrophysics. (Aprahamian, Browne, Darden, Garg, Kolata, Wiescher, 3 faculty fellows and specialists, 1 postdoctoral fellow)
Solid-State Physics: low-temperature physics, superconducting microwave absorption, metal and semiconductor superlattices, magnetism, magnetic resonance, magnetoelastic effects, high-temperature superconductivity, optical and far-infrared spectroscopy of semiconductors, crystal growth and MBE of semiconductors, magnetostatic effects, layered superconductors, single-electron tunneling, X-ray absorption spectroscopy (EXAFS and XANES), condensed-matter systems. (Blackstead, Bunker, Dobrowolska-Furdyna, Furdyna, Ruggiero, Tomasch, 6 faculty fellows and specialists, 2 postdoctoral fellows, 1 visiting scholar)

SECTION 7: PHYSICS

UNIVERSITY OF PENNSYLVANIA

Department of Physics

Program of Study

The Department of Physics offers an outstanding program of graduate study leading to the Ph.D. degree in physics. The four primary areas of research in the department are particle physics, nuclear physics, condensed-matter physics, and astrophysics. Both experimental and theoretical research are performed in all four concentrations.

The program attempts to provide students with a comprehensive overview of the discipline while challenging them to explore in detail an area of particular interest. Students must pass seven required courses, including one-semester courses in elementary particle, nuclear, and solid-state physics. They must also complete a course in modern laboratory techniques. In addition, credit for thesis research partially fulfills the University requirement of twenty courses for the Ph.D. degree. Most students in the program participate in the teaching activities of the department at some time during their graduate career, usually serving as teaching assistants during their first year.

In order to gain admission to Ph.D. candidacy, students must pass a written examination on first-year courses and an oral examination in a subfield of their choice. Doctoral candidates must complete a dissertation based on independent scientific research and defend it orally before a committee of faculty members. There is no foreign language requirement. For students with normal undergraduate preparation, the Ph.D. degree requires four to six years to complete. Many of the department's graduates have contributed significantly to the field in the areas of research, industry, and academia.

Research Facilities

Research facilities are mainly located in three buildings: the Laboratory for Research in the Structure of Matter, David Rittenhouse Laboratory, and the Tandem Accelerator Laboratory. Research in all areas is done both on-site and elsewhere. The particle physics group has ongoing programs at Fermilab. Nuclear physics groups do experiments at the on-site tandem accelerator and at a number of national facilities, including Los Alamos Meson Physics Facility and heavy-ion facilities at Oak Ridge and Argonne. Condensed-matter experiments are done on campus and at national light sources.

Financial Aid

Generally, all students in physics are fully supported by teaching assistantships, research assistantships, or fellowships; 1991–92 stipends are $8530 for nine months. Summer support is included, usually in the form of research assistantships; 1992 summer stipends are $4470.

Cost of Study

Tuition and fees are provided for all graduate students.

Cost of Living

Students have a number of housing options, including on-campus and off-campus apartments. On-campus housing—generally single apartments—averages about $650 per month. Married students should expect to pay approximately $760 per month for on-campus housing. Off-campus rentals are slightly less expensive, ranging from $300 per month for an efficiency apartment to $700 per month for a three-bedroom apartment. The University offers a variety of meal plans for graduate students, and these are priced from $860 to $2200 per academic year. In addition, many on-campus and most off-campus apartments feature small kitchens and are within walking distance of several stores.

Student Group

The University has about 9,000 graduate and professional students, of whom about 2,250 are in Arts and Sciences. Currently, 120 graduate students are enrolled in physics. In the past twelve months, nineteen Ph.D.'s in physics were awarded. Approximately 25–30 new students arrive each year.

Location

The University is located in historic Philadelphia, a city comprised of many interesting and distinctive neighborhoods. This colorful array includes such sections as Colonial Society Hill, Chinatown, and University City, the area surrounding Penn. Center City, with its museums, theaters, concert halls, markets, and sports arenas, lies just east of Penn across the Schuylkill River, which challenges rowers from colleges and universities all over the world. Those who wish to explore still further can easily drive to the beaches of New Jersey, the mountains of Pennsylvania, or even to New York or Washington, D.C., which are equidistant from Philadelphia.

The University

A member of the Ivy League, the University was founded in 1740 by Benjamin Franklin, the noted statesman, author, and inventor. Since its inception and throughout its 250-year history, Penn has been at the forefront of educational and research developments. With its wide range of rigorous graduate programs, Penn has set new standards of excellence in the arts and sciences, engineering, medicine (including dental and veterinary), and business.

Applying

The GRE General Test is required for admission, and the Subject Test in physics is required for financial aid consideration. Applicants whose native language is not English must take the Test of English as a Foreign Language (TOEFL) and the Test of Spoken English (TSE). Applications should be sent to the address indicated below; those sent directly to the department will be delayed in processing. Completed applications and all supporting materials must be submitted by February 1 for financial aid consideration. Applicants who are admitted to the program will be notified during the months of March and April.

Correspondence and Information

For additional information:
Graduate Group Chairman
Department of Physics
University of Pennsylvania
209 South 33rd Street
Philadelphia, Pennsylvania 19104
Telephone: 215-898-3125

To submit completed applications:
Graduate Admissions Office
University of Pennsylvania
16 College Hall
Philadelphia, Pennsylvania 19104

Peterson's Guide to Graduate Programs in the Physical Sciences and Mathematics 1992

SECTION 7: PHYSICS

University of Pennsylvania

THE FACULTY AND THEIR RESEARCH

Fay Ajzenberg-Selove, Ph.D., Professor. Nuclear experiments.
Ralph D. Amado, D.Phil., Professor. Nuclear theory.
Kenneth R. Atkins, Ph.D., Professor. Gravitation and relativity.
David P. Balamuth, Ph.D., Professor. Nuclear experiments.
Eugene W. Beier, Ph.D., Professor. Elementary particle experiments.
Sidney Bludman, Ph.D., Professor. Astrophysics theory.
Howard Brody, Ph.D., Professor. Physics of sports.
Elias Burstein, A.M., Mary Amanda Wood Emeritus Professor. Condensed-matter experiments.
Herbert B. Callen, Ph.D., Professor Emeritus. Condensed-matter theory.
Max E. Caspari, Ph.D., Professor Emeritus. Condensed-matter experiments.
Jeffrey M. Cohen, Ph.D., Associate Professor. Astrophysics theory.
Michael Cohen, Ph.D., Professor. Condensed-matter theory.
Mirjam Cvetic, Ph.D., Assistant Professor. Elementary particle theory.
Gerald Dolan, Ph.D., Professor. Condensed-matter experiments.
H. Terry Fortune, Ph.D., Professor and Graduate Group Chairman. Nuclear experiments.
Sherman Frankel, Ph.D., Professor. High-energy and nuclear experiments.
William Frati, Ph.D., Research Associate Professor. Experimental particles.
Anthony F. Garito, Ph.D., Professor. Condensed-matter experiments.
Larry Gladney, Ph.D., Assistant Professor. Elementary particle experiments.
Keith A. Griffioen, Ph.D., Assistant Professor. Nuclear experiments.
A. Brooks Harris, Ph.D., Professor. Condensed-matter theory.
Paul Heiney, Ph.D., Associate Professor. Condensed-matter experiments.
Robert Hollebeek, Ph.D., Professor. Elementary particle experiments.
Abraham Klein, Ph.D., Professor. Nuclear theory.
Kenneth Lande, Ph.D., Professor of Physics and Astronomy. Astrophysics experiments.
Paul Langacker, Ph.D., Professor. Elementary particle theory.
Nigel Lockyer, Ph.D., Assistant Professor. Elementary particle experiments.
Tom C. Lubensky, Ph.D., Professor. Condensed-matter theory.
Alfred K. Mann, Ph.D., Professor. Elementary particle experiments.
Eugene Mele, Ph.D., Professor. Condensed-matter theory.
Roy Middleton, Ph.D., Professor of Physics and Mellon Term Professor in the Natural Sciences. Nuclear experiments.
Philip Nelson, Ph.D., Assistant Professor. Elementary particle theory.
Makoto Oka, Ph.D., Assistant Professor. Nuclear theory.
Burt Ovrut, Ph.D., Associate Professor. Elementary particle theory.
E. Ward Plummer, Ph.D., Professor. Condensed-matter experiments.
Gino Segre, Ph.D., Professor and Chairman. Elementary particle theory.
Walter Selove, Ph.D., Professor. Elementary particle experiments.
Pekke Sinervo, Ph.D., Assistant Professor. Elementary particle experiments.
Paul Soven, Ph.D., Professor. Condensed-matter theory.
Paul J. Steinhardt, Ph.D., Professor. Elementary particle and condensed-matter theory.
C. W. Ufford, Ph.D., Professor Emeritus.
Walter D. Wales, Ph.D., Professor. Elementary particle experiments.
Roger H. Walmsley, Ph.D., Associate Professor. Statistical physics.
Hugh H. Williams, Ph.D., Professor. Elementary particle experiments.
Thomas H. Wood, Ph.D., Professor Emeritus. Biophysics.
Arjun Yodh, Ph.D., Assistant Professor. Condensed-matter experiments.
Robert W. Zurmühle, Ph.D., Professor. Nuclear experiments.

Benjamin Franklin, founder of the University of Pennsylvania.

Philadelphia is an exciting array of skyscrapers, museums, markets, and theaters, surrounded by a number of colorful neighborhoods.

Begun in 1895, the Quadrangle, the University's oldest dormitory, houses almost 1,700 students.

SECTION 7: PHYSICS

UNIVERSITY OF PITTSBURGH

Department of Physics and Astronomy

Programs of Study The graduate programs in physics and astronomy are designed primarily for students who wish to obtain the Ph.D. degree, although the M.S. degree is also offered. The Ph.D. program attempts to minimize formal requirements and hasten their completion while at the same time providing high-quality training for the student. There is a basic sequence of courses to be taken by all graduate students unless the equivalent material has been demonstrably mastered in other ways. For physics, the basic sequence includes one term of classical mechanics and two terms each of quantum mechanics, electromagnetic theory, statistical mechanics, and modern physical laboratory methods. More specialized courses in advanced nuclear physics, quantum mechanics, condensed-matter and statistical physics, upper-atmosphere physics, particle physics, relativity, and other subjects are offered. In astronomy, the basic sequence includes one term each of astronomical instrumentation, spherical astronomy, celestial mechanics, astronomical spectroscopy, and statistical astronomy and two terms of astrophysics. Students who wish to conduct research in astrophysics may choose to pursue the Ph.D. in physics, astronomy, or both.

Graduate students normally take a written preliminary examination, based on undergraduate physics or astronomy, at the end of the first year and a written comprehensive examination, based on graduate physics or astronomy, at the end of the second year. They then commence research and are advanced to candidacy. Preparation and defense of a satisfactory dissertation complete the requirements. Most students with a baccalaureate degree should find it possible to attain the Ph.D. within five years.

In astronomy there are active research programs in both astrometry and astrophysics. Astrometric research is directed toward the detection of unseen companions in astrometric and spectroscopic-astrometric binaries and the determination of distances to various stars. Astrophysics research is directed toward studies of quasars, the formation of galaxies, and studies of the early universe. In physics, active areas of research are atomic physics (experiment and theory), elementary particle physics (experiment and theory), general relativity (theory), nuclear physics (experiment and theory), statistical mechanics, nonlinear dynamics, quantum chaos, solid-state physics (experiment and theory), and space physics (experiment and theory). Research in a variety of topics in applied physics is being pursued, and research in biophysics, geophysics, radiation physics, and X-ray crystallography is conducted in other departments.

Research Facilities The department's facilities include the physics library, a professionally staffed machine shop, a student shop, an electronics shop, and a glassblowing shop. Large computing programs are carried out either at the University central computing center (using a network of VAX 88 computers), or at the Pittsburgh Supercomputing Center (using a CRAY Y-MP), which is operated jointly by the University of Pittsburgh, Carnegie Mellon University, and Westinghouse Electric Corporation. (The two directors of the center, Ralph Roskies and Michael Levine, are members of the physics departments at the University of Pittsburgh and Carnegie Mellon University, respectively.) Other facilities include the Allegheny Observatory (for positional astronomy) and the Laurel Mountain Airglow Observatory, located east of Pittsburgh in the Allegheny Mountains. Particle and nuclear physics research is carried out at such national and international facilities as Brookhaven National Laboratory; Fermi National Laboratory in Chicago; CERN in Geneva, Switzerland; Bates Linear Accelerator Center at MIT; Oak Ridge National Laboratory; and Los Alamos National Laboratory. Similarly, the astrophysics observation programs are conducted at national and international observatories in Arecibo, Puerto Rico; in Green Bank, West Virginia and Socorro, New Mexico (NRAO); at Kitt Peak and Mt. Hopkins, Arizona; at Cerro Tololo and Las Campanis, Chile; and in Palomar, California. Atmospheric physics research is conducted from rocket launching sites at Wallops Island, Virginia, and Fort Churchill on Hudson's Bay, Canada; an additional airglow observatory is maintained in Arequipa, Peru.

Financial Aid Financial aid is normally provided through teaching assistantships during the first year and through research assistantships thereafter. University fellowships are also available. The department endeavors to continue to support each student admitted with aid throughout his or her entire graduate career, provided good academic standing is maintained. Teaching assistantship appointments carry a stipend of $8490 for two terms in 1991–92, and all tuition charges are exempted. Research assistantship appointments may be held in connection with most of the department's research programs. The stipend is $12,735 for twelve months, and all tuition charges are exempted.

Cost of Study For full-time students who are not Pennsylvania residents, tuition per term in 1991–92 is $5840. Part-time students pay $482 per credit. Pennsylvania residents pay $2920 per term and $241 per credit.

Cost of Living Most students live in rooms or apartments in the Oakland area of Pittsburgh and pay, typically, $335–$500 per month for housing and $300–$400 per month for meals.

Student Group The department's graduate student body in 1990–91 consisted of 88 students, of whom 74 were men; 80 students received financial aid. These figures are typical of the department's graduate enrollment.

Location Pittsburgh is situated in a hilly and wooded region of western Pennsylvania at the junction of two rivers. The region has a natural beauty, which has once again become enjoyable as a result of a strong program of air-pollution control. The University is located about 3 miles east of downtown Pittsburgh in the city's cultural center. Adjacent to the campus are Carnegie Mellon University, the Carnegie Institute, the Museums of Art and Natural History, the public library, and the Music Hall. Schenley Park adjoins the campus; it has picnic areas, playing fields, trails, and an excellent botanical conservatory.

The Department The department has long been active in research and in the training of Ph.D.'s. Close cooperation exists between this department and the physics department of Carnegie Mellon University. The graduate students of both institutions benefit from belonging to one of the largest communities of active physicists in the country.

Applying Students who wish to apply for admission or financial aid should take the Graduate Record Examinations, including the Subject Test in physics. Applicants should request that the registrars of their undergraduate and graduate schools send transcripts of their records to the department. Three letters of recommendation are required for admission with aid. All applications must be received by January 31.

Correspondence and Information
Professor Raymond S. Willey
Admissions Committee
Department of Physics and Astronomy
University of Pittsburgh
Pittsburgh, Pennsylvania 15260
Telephone: 412-624-9066

Peterson's Guide to Graduate Programs in the Physical Sciences and Mathematics 1992

SECTION 7: PHYSICS

University of Pittsburgh

THE FACULTY AND THEIR RESEARCH

John H. Anderson, Professor Emeritus; Ph.D., Chicago. Experimental solid-state physics.
Norman Austern, Professor; Ph.D., Wisconsin. Theoretical nuclear physics.
Elizabeth U. Baranger, Professor and Associate Provost of Graduate Studies; Ph.D., Cornell. Theoretical nuclear physics.
James E. Bayfield, Professor; Ph.D., Yale. Experimental atomic physics and quantum optics.
Iwo Bialynicki-Birula, Adjunct Professor; Ph.D., Warsaw. Theoretical high-energy physics.
Manfred A. Biondi, Professor Emeritus; Ph.D., MIT. Experimental atomic physics and aeronomy.
Daniel Boyanovsky, Assistant Professor; Ph.D., California, Santa Barbara. Condensed-matter physics.
Frank H. Briggs IV, Associate Professor; Ph.D., Cornell. Astrophysics, radio astronomy.
Robert D. Carlitz, Professor; Ph.D., Caltech. Theoretical high-energy physics.
Wolfgang J. Choyke, Research Professor; Ph.D., Ohio State. Solid-state physics, radiation damage, optics, semiconductor physics.
Wilfred W. Cleland, Professor; Ph.D., Yale. Experimental high-energy physics.
Bernard L. Cohen, Professor; Ph.D., Carnegie Mellon. Energy and environment.
John David Crawford, Assistant Professor; Ph.D., Berkeley. Nonlinear dynamics.
Wilfried A. W. Daehnick, Professor and Associate Provost of Research; Ph.D., Washington (St. Louis). Experimental nuclear physics.
Robert P. Devaty, Associate Professor; Ph.D., Cornell. Experimental solid-state physics.
Timothy R. Donoghue, Professor; Ph.D., Notre Dame. Experimental nuclear physics.
Richard M. Drisko, Professor; Ph.D., Carnegie Mellon. Theoretical nuclear physics.
H. E. Anthony Duncan, Professor; Ph.D., MIT. Theoretical high-energy physics.
Steven A. Dytman, Associate Professor; Ph.D., Carnegie Mellon. Experimental nuclear physics.
Eugene Engels Jr., Professor; Ph.D., Princeton. Experimental high-energy physics.
Peter W. Erdman, Research Associate Professor; Ph.D., Pittsburgh. Experimental atomic and atmospheric physics.
Wade L. Fite, Professor Emeritus; Ph.D., Harvard. Experimental atomic physics, applied physics.
Myron P. Garfunkel, Professor; Ph.D., Rutgers. Experimental low-temperature physics, superconductivity.
George D. Gatewood, Professor; Ph.D., Pittsburgh. Astronomy, astrometry, search for planetary systems orbiting neighboring stars.
Edward Gerjuoy, Professor Emeritus; Ph.D., Berkeley. Theoretical atomic physics.
Walter I. Goldburg, Professor; Ph.D., Duke. Experimental solid-state physics, phase transitions, light scattering, turbulence.
Yadin Y. Goldschmidt, Associate Professor; Ph.D., Hebrew (Jerusalem). Condensed-matter theory, statistical mechanics.
Dincer H. Gunduz, Instructor; Ph.D., Pittsburgh. Physics education.
David Halliday, Professor Emeritus; Ph.D., Pittsburgh. General physics.
Cyril Hazard, R. K. Mellon Professor; Ph.D., Manchester. Astrophysics.
Thomas J. Humanic, Assistant Professor; Ph.D., Pittsburgh. Experimental intermediate high-energy physics.
Allen I. Janis, Professor; Ph.D., Syracuse. General relativity, philosophy of science.
David M. Jasnow, Professor; Ph.D., Illinois. Theory of phase transitions, statistical mechanics.
Ranier Johnsen, Associate Professor; Ph.D., Kiel (Germany). Experimental atomic and plasma physics.
Frederic Keffer, Professor Emeritus; Ph.D., Berkeley. Theoretical solid-state physics.
Joost H. Kiewiet de Jonge, Associate Professor Emeritus; Ph.D., Harvard. Astronomy.
Peter F. M. Koehler, Professor; Ph.D., Rochester. Experimental high-energy physics.
Irving J. Lowe, Professor; Ph.D., Washington (St. Louis). Experimental solid-state physics, nuclear magnetic resonance, nuclear magnetic resonance imaging.
James V. Maher, Professor; Ph.D., Yale. Experimental solid-state physics, critical phenomena, physics of fluids.
Ezra T. Newman, Professor; Ph.D., Syracuse. General relativity, twistor theory.
Lorne A. Page, Professor Emeritus; Ph.D., Cornell. Experimental nuclear and particle physics.
Richard H. Pratt, Professor; Ph.D., Chicago. Theoretical atomic and low-energy particle physics, bremsstrahlung, hot plasma processes, photon scattering.
Ralph Z. Roskies, Professor; Ph.D., Princeton. Theoretical high-energy physics, use of computer in theoretical physics.
Carlo Rovelli, Assistant Professor; Dottorato di Ricerca, Padua. General relativity.
Jurg X. Saladin, Professor; Ph.D., Swiss Federal Institute of Technology. Experimental nuclear physics.
Regina E. Schulte-Ladbeck, Assistant Professor, Ph.D., Heidelberg. Astrophysics.
Paul F. Shepard, Professor; Ph.D., Princeton. Experimental high-energy physics.
Philip Stehle, Professor Emeritus; Ph.D., Princeton. Theoretical physics, quantum optics.
G. Alec Stewart, Associate Professor and Dean, University Honors College; Ph.D., Washington (Seattle). Experimental solid-state physics.
Frank Tabakin, Professor; Ph.D., MIT. Theoretical nuclear physics.
Julia A. Thompson, Professor; Ph.D., Yale. Experimental high-energy physics, optical instrumentation.
John R. Townsend, Professor Emeritus; Ph.D., Cornell. Experimental solid-state physics, radiation damage.
David A. Turnshek, Assistant Professor; Ph.D., Arizona. Observational extragalactic astronomy.
C. Martin Vincent, Professor; Ph.D., Witwatersrand (South Africa). Theoretical nuclear physics.
Raymond S. Willey, Professor; Ph.D., Stanford. Theoretical high-energy physics.
Jeffrey Winicour, Professor; Ph.D., Syracuse. General relativity.
Arthur M. Wolfe, Professor; Ph.D., Texas at Austin. Astrophysics, quasars, radio astronomy.
Edward C. Zipf, Professor; Ph.D., Johns Hopkins. Experimental atomic and atmospheric physics.

The department is centered in Allen Hall.

The 30-inch Thaw refractor of the Allegheny Observatory.

SECTION 7: PHYSICS

UNIVERSITY OF PUERTO RICO, RÍO PIEDRAS

Department of Physics

Programs of Study

The Department of Physics at the Río Piedras campus of the University of Puerto Rico offers a program leading to the M.S. degree in physics. It also offers, jointly with the Department of Chemistry, a program leading to the Ph.D. degree in chemical physics. Opportunities for postdoctoral research are available as well.

Candidates for the M.S. degree in physics must complete at least 24 credit hours, pass a written comprehensive examination, complete research for a thesis, and orally defend the thesis.

Candidates for the Ph.D. degree in chemical physics are required to complete a minimum of 45 credit hours in graduate courses in physics or chemistry with an overall grade index of no less than 3 and to pass three written qualifying examinations. One of the qualifying examinations must be in physical chemistry and the other two in areas of physics to be selected by the student with the approval of his or her adviser. Completion of original thesis research (24 credit hours) and an oral defense of the thesis are also required.

Research Facilities

The department has laboratories for experimental work in various research areas of solid-state physics, including thin films, crystallography, and spectroscopy. Several members of the faculty use the facilities of the National Astronomy and Ionosphere Center at Arecibo for their research, and some also use the SLAC and Brookhaven synchrotron radiation sources. Theoretical research is supported by extensive computer facilities on campus as well as access to supercomputer centers on the mainland.

Financial Aid

In 1991–92, teaching and laboratory assistantships pay $5000 for ten months for students in the M.S. program and $5500 for Ph.D. candidates and include exemption from tuition. Special fellowships are offered for qualified U.S. citizens. A limited support fund is available for students from Spanish-speaking countries; these students may also obtain financial aid from the OAS through its local offices. Most of the students receive some type of financial aid.

Cost of Study

Tuition is $55 per credit for residents of Puerto Rico and $3500 per year for nonresidents in 1991–92.

Cost of Living

Costs of housing and living expenses vary considerably but are usually similar to those in the main metropolitan areas of the United States.

Student Group

The department has more than 45 graduate students, who come from Puerto Rico and the continental United States, as well as from Central and South America and China and several other foreign countries.

Location

San Juan is the capital and cultural center of Puerto Rico. Old San Juan is a fascinating city, with numerous historic sites, excellent beaches, and other tourist attractions. The cultural life in the city, and especially around the University, is fairly active and includes frequent concerts, plays, lectures, and other events. The internationally known Casals Festival of Music is held each year in June.

The University

The University was founded in 1903. The Río Piedras campus is the oldest and largest and includes, among other divisions, the College of Natural Sciences and the Schools of Medicine and Architecture. Graduate programs in many disciplines are well established, including Ph.D. programs in chemistry and biology and a very active M.S. program in mathematics.

Applying

Applications should be filed no later than May 15 for the fall semester and November 15 for the spring semester. Students should apply to the Department of Physics by May 1 for fellowships and teaching and laboratory assistantships.

Correspondence and Information

Dr. Alfredo J. Torruella, Chairman
Department of Physics
P.O. Box 23343
University of Puerto Rico
Río Piedras, Puerto Rico 00931
Telephone: 809-764-0000 Ext. 2311 or 4746
             809-764-0620
Fax: 809-764-2890

Peterson's Guide to Graduate Programs in the Physical Sciences and Mathematics 1992

SECTION 7: PHYSICS

University of Puerto Rico, Río Piedras

THE FACULTY AND THEIR RESEARCH

D. R. Altschuler, Professor; Ph.D., Brandeis, 1974. Astronomy.
L. B. Bhuiyan, Associate Professor; Ph.D., London, 1977. Statistical mechanics.
L. Blum, Professor; Ph.D., Buenos Aires, 1956. Statistical mechanics and thermodynamics.
M. Gómez, Professor; Ph.D., Cornell, 1968. Theoretical solid-state physics.
C. A. Huber, Associate Professor; Ph.D., Brown, 1983. Experimental solid-state physics.
T. E. Huber, Associate Professor; Ph.D., Brown, 1983. Experimental solid-state physics.
R. S. Katiyar, Professor; Ph.D., Indian Institute of Science, 1967. Spectroscopy.
A. Martínez, Assistant Professor; Ph.D., American, 1990. Experimental solid-state physics.
J. A. Muir, Professor; Ph.D., Northwestern, 1966. Experimental solid-state physics, crystallography.
J. F. Nieves, Professor; Ph.D., Pennsylvania, 1980. Theoretical particle physics.
J. Ponce de Leon, Assistant Professor; Ph.D., Venezuela (Caracas), 1985. General relativity.
R. G. Selsby, Professor; Ph.D., Ohio State, 1969. Quantum chemistry.
J. M. Tharrats, Professor; Ph.D., Madrid, 1952. Theoretical physics.
A. J. Torruella, Professor and Chairman of the Department; Ph.D., Yale, 1965. Theoretical physics.
M. R. Ubriaco, Assistant Professor; Ph.D., North Carolina, 1989. Theoretical particle physics.
Z. S. Weisz, Professor; Ph.D., Hebrew (Jerusalem), 1962. Experimental solid-state physics.

SECTION 7: PHYSICS

UNIVERSITY OF ROCHESTER
Institute of Optics

Programs of Study

The Institute of Optics offers a full-time Ph.D. program, an M.S. in optics available through either full-time or part-time study, an M.S. pursued through a co-op program, and a part-time M.S. program in general studies with a concentration in optics. The doctoral program emphasizes fundamental principles of optical physics and engineering and their application in research. The master's programs emphasize specialized training in optical engineering. The M.S. is not a prerequisite for the Ph.D.

There are also interdisciplinary programs with the Center for Visual Science, the Medical Center, and the Laboratory for Laser Energetics (LLE). LLE conducts basic studies in laser fusion, X-ray optics, and mode-locked picosecond-pulse lasers.

M.S. degrees require 30 hours of course work, including 18–23 hours of required core courses, which form the basis for the comprehensive examination. This examination is taken by all M.S. candidates except students with better than a B average who are writing theses. The thesis is not a degree requirement. There is no language requirement. Master's degrees are normally completed in nine to twelve months. In a co-op program, students complete the first semester of course work, work full-time for twelve months, and then return to campus for the final semester of course work.

Most entering Ph.D. students receive financial aid, permitting them to take a full program of four courses per semester in the first year. The first-year curriculum normally consists of one-year courses in physical optics, geometrical and instrumental optics, and mathematical methods; a semester of quantum mechanics for optics; and one elective. A written preliminary examination is taken at the start of the second year, covering optics, mathematics, physics, and applied physics. Students continue with course work and begin independent research during the second year; they present an intended thesis problem at an oral qualifying examination within fifteen months after passing the written examination. General requirements for the Ph.D. are one year of full-time residence, 90 hours of graduate work (or 60 hours beyond the M.S.), a year of work as a teaching assistant, and completion and defense of a doctoral dissertation. There is no foreign language requirement. The Ph.D. usually takes four or five years to complete.

The academic year is September through May.

Research Facilities

Facilities for thesis research include laboratories for virtually all phases of optics and laser systems research, including Fourier optics, ultrahigh-resolution spectroscopy, quantum interactions, nonlinear optics, picosecond-pulse shaping, gradient index optics, fibers, interferometers, pattern recognition, speckle, optical coherence, electromagnetic theory, diffraction, X-ray production from laser-induced plasmas, Raman spectroscopy, optical materials, thin films, and infrared lasers. Support facilities include an optical fabrication and testing laboratory and a machine shop with full-time technicians, a student machine shop, darkrooms, and the services of design draftsmen. Computing systems include a departmental μVAX and terminal server network, a group of networked SPARCstations, a departmental Appleshare system, an Alliant FX/8 superminicomputer, and a variety of work group–based workstations. These systems are maintained by full-time college employees.

Financial Aid

Nearly all Ph.D. students receive full financial support. In 1991–92, first-year students may receive tuition remission plus a departmental assistantship or fellowship of $11,500 for a twelve-month period.

At the M.S. level, financial aid is available as a partial tuition scholarship plus a long-term loan for the balance of the tuition fee, with a partial stipend in addition. Awards are made on the basis of scholastic merit.

Cost of Study

Graduate tuition for 1991–92 is $15,136.

Cost of Living

The University owns furnished and unfurnished apartments. Many students rent privately owned housing. A register of apartments and houses for rent is maintained by the University's Apartment Office.

Student Group

In 1990–91, the Institute of Optics had 88 Ph.D. students, 47 full-time M.S. candidates, and 15 part-time M.S. students. In the fall of 1990, 47 new full-time students entered. All of the Institute's 1990 M.S. and Ph.D. recipients obtained jobs in optics.

The University has about 5,600 undergraduates and about 3,600 graduate students.

Location

Rochester, with a metropolitan area of 600,000 people, specializes in precision industry, much of which is related to optics (Eastman Kodak, Bausch & Lomb, and Xerox being the best known).

The Institute

The Institute of Optics has been a leader in the training of optical scientists and engineers for over fifty years. The faculty includes fourteen Fellows of the Optical Society of America, two recipients of the OSA's Lomb Medal, four winners of the Ives Medal, four past presidents of the OSA, a past editor of the *Journal of the Optical Society of America*, a past president of the Society of Photo-Optical Instrumentation Engineers, and editors or associate editors of *Optica Acta, Optics Communications, Optics Letters,* and *Physical Review A*.

Applying

Requests for application forms should be sent to the Optics Graduate Admissions Committee. Applicants ordinarily have an undergraduate degree in physics or engineering. GRE scores are strongly recommended. Those applying for financial aid must submit completed applications by February 15 for fall entrance. Midyear entrance is strongly discouraged. Students with undergraduate degrees from universities in which English is not the language of instruction must submit scores on the Test of English as a Foreign Language.

Correspondence and Information

Administrator
Optics Graduate Admissions Committee
Institute of Optics
University of Rochester
Rochester, New York 14627
Telephone: 716-275-7764

SECTION 7: PHYSICS

University of Rochester

THE FACULTY AND THEIR RESEARCH

Professors
Brian J. Thompson, Provost; Ph.D., Manchester. Holography, image processing, coherence, phase microscopy.
Duncan T. Moore, Director of the Institute of Optics; Ph.D., Rochester. Lens design, automatic phase measurements.

Robert Boyd, Ph.D., Berkeley. Nonlinear optics, infrared detection and generation, astronomical instrumentation.
L. Nevil Davy, M.A., Tennessee. Electronic imaging and image processing. (part-time)
Joseph H. Eberly, Ph.D., Stanford. Resonant interaction of light with atoms and molecules, multiphoton processes, quantum electrodynamics.
Nicholas George, Ph.D., Caltech. Electromagnetic theory, physical optics, optical systems, speckle, X-ray optics, pattern recognition.
M. Parker Givens, Ph.D., Cornell. Holography, optical data processing. (Emeritus)
Dennis G. Hall, Ph.D., Tennessee. Integrated optics, fiber optics, optical properties of solids, quantum electronics.
Robert E. Hopkins, Ph.D., Rochester. Design, assembly, and testing of optics for high-power laser fusion systems. (Emeritus)
Rudolf Kingslake, D.Sc., Imperial College (London). Design of lenses and optical systems, measurement of aberrations and their effects upon optical images. (Emeritus)
Erwin Loewen, Sc.D., MIT. Diffraction gratings, precision engineering and metrology. (part-time)
David L. MacAdam, Ph.D., MIT. Color measurement, color vision, color reproduction. (part-time; also Adjunct Professor of Visual Science)
Erich W. Marchand, Ph.D., Rochester. Geometrical optics, gradient index optics. (part-time)
Carlos R. Stroud Jr., Ph.D., Washington (St. Louis). Quantum and semiclassical radiation theory, high-resolution spectroscopy.
Kenneth J. Teegarden, Ph.D., Illinois at Urbana-Champaign. Optical and electronic properties of crystalline and noncrystalline semiconductors and insulators, development of new optical materials.
David Williams, Ph.D., California, San Diego. Sensitivity and resolution of the human visual system to patterns that are modulated in wavelength, space, and time.
Emil Wolf, Ph.D., Bristol; D.Sc., Edinburgh. Electromagnetic theory and physical optics, diffraction and theory of partial coherence.

Associate Professors
Govind Agrawal, Ph.D., Indian Institute of Technology (Delhi). Semiconductor lasers, solid-state physics, optical communications, laser physics, phase conjugation, optical bistability, nonlinear phenomena.
Jay M. Eastman, Ph.D., Rochester. Thin films, optical instrumentation. (part-time)
Stephen Jacobs, Ph.D., Rochester. Optical materials.
G. Michael Morris, Ph.D., Caltech. Holography, white light optical processing, coherence, electromagnetic wave propagation.
Wolf Seka, Ph.D., Texas. Lasers.
Gary Wicks, Ph.D., Cornell. III-V semiconductors: epitaxial growth, optical properties, optical devices.

Assistant Professors
Thomas G. Brown, Ph.D., Rochester. Guided wave optics, optical properties of semiconductors, optical communications.
Gregory W. Forbes, Ph.D., Australian National. Optical aberration theory, nonionizing optics, statistical estimation of parameters.
Susan Houde-Walter, Ph.D., Rochester. Microoptics, integrated optics, interferometry, holography.
R. J. Dwayne Miller, Ph.D., Manitoba. Ultrafast spectroscopy, laser development, physics of ultrafast phenomena.
Warren E. Smith, Ph.D., Arizona. Image processing, medical imaging, inverse problems involving shift-variant operators.
Ian Walmsley, Ph.D., Rochester. Nonlinear optics, optical probing of fundamental processes in solid-state materials.

SECTION 7: PHYSICS

UNIVERSITY OF ROCHESTER

Department of Physics and Astronomy

Programs of Study

The department offers programs of study leading to the Ph.D. degree in physics, physics and astronomy, or astronomy with an emphasis on observational astronomy. Students normally earn the M.A. or M.S. degree in physics en route to the Ph.D. The M.A. can be awarded after the completion of 30 semester hours of course work and a comprehensive examination; the M.S. degree in physics requires a thesis in addition to the course work. Students are not usually admitted to work toward a master's degree unless they are also working toward a Ph.D.

Most candidates for the Ph.D. degree take two years of course work, and a written and oral preliminary examination is given during the second year. Requirements for the Ph.D. include demonstrating competence in classical physics, quantum mechanics, mathematical methods, electromagnetic theory, and statistical physics, as well as in an advanced area of specialization. A doctoral thesis, based on a significant piece of original research, and a final oral thesis defense are required of all Ph.D. candidates. A typical program of study requires about five years to complete. A minor is not required, although students are encouraged to broaden their understanding of other subfields of physics or astronomy beyond the area of their thesis research. There is no foreign language requirement.

The department provides opportunities for Ph.D. thesis research in observational astronomy, theoretical astrophysics, biological physics, experimental and theoretical condensed-matter physics, experimental and theoretical elementary particle physics, experimental and theoretical nuclear physics, theoretical plasma physics, and experimental and theoretical quantum optics.

Research Facilities

The Nuclear Structure Research Laboratory houses an 18-MV tandem Van de Graaff accelerator. Other equipment available includes a high-precision magnetic spectrograph, a recoil mass spectrometer of unique design, and an in-beam laser spectroscopy line. The C. E. Kenneth Mees Observatory houses a 61-cm reflector equipped for spectroscopy, photometry, speckle interferometry, or photography. An infrared CCD array is in active use. Additional laboratories include a superconducting magnet system for studies of heavily doped semiconductors, a facility for investigations of subpicosecond energy transport in biomolecules and solids, facilities for the design and construction of large and sensitive high-energy-physics detectors, and a sophisticated photon-counting laboratory for experiments in quantum optics. Computing facilities include a CYBER 175, an IBM 3032, a DEC 2060, a VAX 8800, and a number of minicomputers. The Physics-Optics-Astronomy Library, within the physics building, provides ready access to more than 350 journals. The facilities of the University's Laboratory for Laser Energetics also are available.

Financial Aid

A few exceptional fellowships provide stipends of up to $17,200 for the calendar year. In 1991–92, graduate teaching assistantships, which require 15 hours of work per week during the academic year, carry stipends of $9095 for 8½ months. Additional support is available for participation in summer research. A teaching assistant who also takes part in full-time summer research receives a total of $13,190 for the calendar year. Graduate research assistantships, which require 15 hours of work per week during the academic year and full-time work during the summer, pay $13,900 for the calendar year. A number of Rush Rhees Fellowships for academic excellence are also available to supplement teaching or research assistantships.

Cost of Study

For students with fewer than 90 credit hours of accumulated graduate course work, tuition in 1991–92 is $15,136 for the academic year. Tuition for more advanced students is $660 per academic year. (Currently, all graduate students in the department receive special awards to cover tuition.) All full-time graduate students are charged a health service fee of approximately $735 per year. The cost of books and supplies is about $300 per year.

Cost of Living

The cost of living in Rochester is among the lowest for metropolitan areas. Supermarkets with moderate prices are located near the University, or meals may be obtained on campus. University-operated housing for either single or married graduate students is available within easy walking distance of the campus, with costs starting at $256 per person per month in 1991–92. Free shuttle-bus service is available within the University complex. Additional privately owned rooms and apartments are available in the residential areas near the University.

Student Group

There are approximately 115 graduate students in physics and astronomy; about 15 are women, and about 10 percent of the students are married. All full-time students receive some form of financial aid. Admission to graduate study is highly competitive, with only about 20 new students admitted each year; about 40 percent have undergraduate degrees from institutions outside the United States. The department has more than 1,000 alumni around the world, many of whom have achieved international eminence.

Location

With approximately 750,000 inhabitants, the Rochester metropolitan area is the third largest in the state. A city with high-technology industries, it is located on the southern shore of Lake Ontario. Niagara Falls, the scenic Finger Lakes district, and the rugged Adirondack Mountains are all within a few hours' drive. The Rochester Philharmonic Orchestra and the Rochester Americans ice-hockey team provide two examples of the range of experiences available. Rochester is readily accessible both by air and by car.

The University and The Department

The University of Rochester is a private institution with approximately 4,500 undergraduates, 3,000 graduate students, and 1,000 faculty members. The Department of Physics and Astronomy, one of the largest and strongest departments within the University, has a reputation for excellence in graduate education and research spanning more than fifty years. Within the past few years, 4 faculty members of the department have been awarded major national prizes in recognition of their research accomplishments.

Applying

Catalogs and application forms may be obtained upon request from the address below. Students are admitted in September only, and completed applications should be received by February 15 in order for applicants to be considered for financial aid. Applicants should take the GRE General and physics Subject tests in time for scores to arrive by February 15. TOEFL scores are required of foreign students whose native language is not English.

Correspondence and Information

Student Counselor
Department of Physics and Astronomy
University of Rochester
Rochester, New York 14627
Telephone: 716-275-4356

or

Chairman
Department of Physics and Astronomy
University of Rochester
Rochester, New York 14627
Telephone: 716-275-4344

SECTION 7: PHYSICS

University of Rochester

THE FACULTY AND THEIR RESEARCH

A. Bodek, Professor; Ph.D., MIT, 1972. Experimental elementary particle physics, neutrino physics.
T. G. Castner, Professor Emeritus; Ph.D., Illinois, 1958. Experimental condensed-matter physics, metal-insulator transition.
D. Cline, Professor and Director, Nuclear Structure Research Laboratory; Ph.D., Manchester, 1963. Experimental nuclear physics, shape and structure transitions in nuclei.
R. Colby, Adjunct Assistant Professor; Ph.D., Northwestern, 1985. Experimental polymer physics.
A. Das, Associate Professor; Ph.D., SUNY at Stony Brook, 1977. Theoretical elementary particle physics, GUTs, supergravity.
D. H. Douglass, Professor; Ph.D., MIT, 1959. Condensed-matter physics.
J. H. Eberly, Professor; Ph.D., Stanford, 1962. Theoretical quantum optics, resonant and multiphoton interactions with matter.
T. Ferbel, Professor; Ph.D., Yale, 1963. Experimental elementary particle physics, phenomenology of strong interactions.
W. J. Forrest, Associate Professor; Ph.D., California, San Diego, 1974. Observational infrared astronomy, infrared CCD detectors.
J. B. French, Andrew Carnegie Professor; Ph.D., MIT, 1948. Statistical nuclear physics, quantum chaos.
H. W. Fulbright, Professor Emeritus; Ph.D., Washington (St. Louis), 1944. Experimental nuclear physics, structures and reaction mechanisms, radio astronomy.
Y. Gao, Assistant Professor; Ph.D., Purdue, 1986. Experimental condensed-matter physics.
S. B. Gazes, Assistant Professor; Ph.D., MIT, 1983. Reaction mechanisms in experimental nuclear physics.
H. E. Gove, Professor; Ph.D., MIT, 1950. Experimental nuclear physics, heavy ions, TAMS.
C. R. Hagen, Professor; Ph.D., MIT, 1962. Theoretical elementary particle physics, quantum field theory.
H. L. Helfer, Professor; Ph.D., Chicago, 1953. Theoretical astrophysics, dense plasma equations of state, abundance determinations.
E. H. Jacobsen, Professor Emeritus; Ph.D., MIT, 1954. Experimental condensed-matter physics, high-resolution electron microscopy.
R. S. Knox, Professor and Director of Undergraduate Studies; Ph.D., Rochester, 1958. Theoretical condensed-matter physics, biological physics.
D. S. Koltun, Professor; Ph.D., Princeton, 1961. Theoretical nuclear physics, meson interactions with nuclei and gravitational interactions of fast particles and photons.
F. Lobkowicz, Professor; D.Sc., Zurich, 1960. Experimental elementary particle physics and physics instrumentation, investigation of QCD via direct photon production.
L. Mandel, Professor; Ph.D., London, 1951. Experimental and theoretical quantum optics, ring lasers and phase transitions in lasers.
V. S. Mathur, Senior Research Associate and Professor (part-time); Ph.D., Delhi, 1958. Theoretical elementary particle physics, phenomenology, weak interaction theory.
A. C. Melissinos, Professor; Ph.D., MIT, 1958. Experimental elementary particle physics, e^+e^- annihilations and gravitational physics at relativistic velocities.
S. Okubo, Professor; Ph.D., Rochester, 1958. Theoretical elementary particle physics, symmetries of weak and strong interactions.
S. L. Olsen, Professor; Ph.D., Wisconsin, 1970. Experimental elementary particle physics and sophisticated electronic detection techniques, production of charmed particles and search for free quarks.
J. L. Pipher, Professor and Director, C. E. Kenneth Mees Observatory; Ph.D., Cornell, 1971. Observational infrared CCD detectors.
S. J. Rajeev, Assistant Professor; Ph.D., Syracuse, 1984. Theoretical particle physics.
M. Rubinstein, Adjunct Assistant Professor; Ph.D., Harvard, 1983. Theoretical condensed-matter physics, polymer physics.
M. P. Savedoff, Professor Emeritus; Ph.D., Princeton, 1957. Theoretical astrophysics, stellar interiors, interstellar matter, hydrodynamics.
Y. Shapir, Associate Professor; Ph.D., Tel-Aviv, 1981. Theoretical condensed-matter physics.
S. L. Sharpless, Professor Emeritus; Ph.D., Chicago, 1952. Observational astronomy, galactic structure.
A. Simon, Professor; Ph.D., Rochester, 1950. Theoretical plasma physics, controlled thermonuclear fusion.
P. F. Slattery, Professor and Chairman; Ph.D., Yale, 1967. Experimental elementary particle physics, investigation of QCD via direct γ production.
R. L. Sproull, Professor Emeritus; Ph.D., Cornell, 1943.
C. R. Stroud, Professor; Ph.D., Washington (St. Louis), 1969. Quantum and semiclassical radiation theory, theoretical and experimental quantum optics.
S. Teitel, Assistant Professor; Ph.D., Cornell, 1981. Statistical and condensed-matter physics.
J. H. Thomas, Professor; Ph.D., Purdue, 1966. Solar physics.
E. H. Thorndike, Professor; Ph.D., Harvard, 1960. Experimental elementary particle physics, weak decays of b quarks.
P. Tipton, Assistant Professor; Ph.D., Rochester, 1987. Experimental elementary particle physics, proton-antiproton collisions at 1.8 TeV.
H. M. Van Horn, Professor; Ph.D., Cornell, 1965. Theoretical astrophysics, degenerate stars.
D. M. Watson, Assistant Professor; Ph.D., Berkeley, 1983. Experimental astrophysics, star formation, galactic structure.
E. Wolf, Wilson Professor; Ph.D., Bristol, 1948. Theoretical quantum optics, diffraction and theory of partial coherence.
F. Wolfs, Assistant Professor; Ph.D., Chicago, 1987. Experimental nuclear physics, solar neutrinos, nuclear astrophysics.

SOME RECENT PUBLICATIONS

Astronomy and Astrophysics
Van Horn, H. M., and T. Koupelis. A quasi-one-dimensional model for narrow astrophysical jets. *Ap. J.* 342:146, 1989.
Watson, D. M. Far-infrared and submillimeter spectroscopy of interstellar clouds: The physics and chemistry of star-forming regions. *Physica Scripta* T11:33 (invited review).

Condensed-Matter Physics
Gao, Y. L. Growth of Al oxide layers on GaAs (100) by reaction with condensed molecular oxygen. *J. Appl. Phys.*, in press.
Teitel, S., and Ying-Hong Li. Flux flow resistance in frustrated Josephson junction arrays. *Phys. Rev. Lett.* 65:2595, 1990.

High-Energy/Elementary Particle Physics
Okubo, S. Integrability condition and finite-periodic Toda lattice. *J. Math. Phys.* 31:1919, 1990.
Thorndike, E. H., S. Eno, et al. Search for a fourth-generation charge–1/3 quark. *Phys. Rev. Lett.* 63:1910, 1989.

Nuclear Physics
Kolton, D. S., and J. E. Villate. Nuclear medium modification of nucleons via a Bethe ansatz model. *Phys. Rev.* C43:1967, 1991.
Wolfs, F. L. H., et al. Measurements of 3He(n,gamma) 4He cross section at thermal neutron energies. *Phys. Rev. Lett.* 63:2721, 1989.

Plasma Physics and Fusion
Simon, A., and R. W. Short. Energy and nonlinearity considerations for the enhanced plasma wave model of Raman scattering. *Phys. Fluids* B1(5), 1989.

Quantum Optics
Eberly, J. H., and Q. Su. Suppression of ionization and atomic electron localization by short intense laser pulses. *Phys. Rev.* A43:2472, 1991.
Mandel, L., Z. Y. Ou, L. J. Wang, and X. Y. Zou. Evidence for phase memory in two-photon down-conversion through entanglement with the vacuum. *Phys. Rev.* A41:566, 1990.

SECTION 7: PHYSICS

UNIVERSITY OF SOUTH CAROLINA

Department of Physics and Astronomy

Programs of Study

The department offers comprehensive experimental research programs in intermediate-energy nuclear physics, elementary particle physics, superconductivity and surface physics, chemical physics, and atomic physics. There is a broad effort in theoretical research, with programs in the foundations of quantum theory, nuclear theory, quantum optics, atomic physics, quantum gravity, quantum electrodynamics, general relativity, cosmology, astrophysics, and critical phenomena.

A significant part of the research program is international in both scope and location. The intermediate-energy nuclear physics group performs experiments at accelerator laboratories in the United States and in France using pions, protons, and polarized gamma rays to study the properties of the hadronic force and nuclear structure. A major effort in double beta decay, in collaboration with Pacific Northwest Laboratory, maintains an ultralow background detector in the Homestake gold mine in Lead, South Dakota. The high-energy physics group participates in the AMY collaboration at the TRISTAN storage ring at KEK in Japan and in experiments at Fermi National Accelerator Laboratory. The theoretical effort in the foundations of quantum theory is a cooperative venture with Tel-Aviv University, and one of its faculty members holds a joint appointment. The large effort in electron spin resonance is carried out primarily on campus, in collaboration with several other universities in the United States and South America. Neutron diffraction studies of magnetic and ferroelectric samples are carried out at Oak Ridge and Argonne national laboratories.

The remainder of the experimental programs are carried out on or near the campus. They include the investigation of magnetic properties of materials using electron spin resonance, Mössbauer spectroscopy, and SQUID magnetometry; the development of instrumentation and applications in environmental studies; the interaction of low-energy ions with solids, using the University's low-energy Van de Graaff accelerator; and investigations of inner-shell atomic structure.

The program in the foundations of quantum theory is coupled with the theoretical efforts in quantum optics and atomic physics. The main thrust uses very simple models to investigate problems of renormalization, dissipation, and relaxation of a physical system coupled to a larger reservoir; potential effects in quantum phenomena; quantum optics; and the development of better techniques in perturbative calculations.

Research Facilities

An eight-story Physical Sciences Center houses research laboratories, a machine shop, classrooms, and office space. A research laboratory, adjacent to the center, houses a Van de Graaff accelerator used in solid-state studies. The high-energy and intermediate-energy nuclear physics programs are carried out at national and international laboratories, as described above.

Financial Aid

Graduate teaching assistantships are available that require 16 hours of work per week and carry a stipend of $9000 per year in 1991–92. Almost all full-time graduate students receive full graduate assistantships.

Cost of Study

Tuition for graduate students is $122 per semester hour in 1991–92. This rate is reduced to $425 per semester with teaching assistantships for the fall and spring semesters and $105 per semester with teaching assistantships in the summer.

Cost of Living

Costs for rooms on campus range from $460 to $1277 per term for single students in 1991–92. Apartments are available for married students at rents ranging from $259 to $461 per month.

Student Group

There are approximately 25,000 students, including about 8,000 graduate students, at the University. The graduate student body in the physics department numbers between 30 and 40. These students come from many states and foreign countries.

Location

The Columbia metropolitan area, with a population of 472,800, is situated in the center of South Carolina. The climate is moderate, with an average annual temperature of 64 degrees. Recreational opportunities include camping, swimming, golf, and tennis. Sesquicentennial State Park and ten other state parks are within a 75-mile radius of the city, and Lake Murray, with more than 500 miles of shoreline, is only a few miles northwest of the city.

The University and The Department

The University of South Carolina was founded in 1801 and is located in the center of Columbia, the state capital. The University offers graduate degrees in most of the departments of the Colleges of Humanities and Social Sciences and of Science and Mathematics and in the Colleges of Business Administration, Education, Engineering, Journalism, Pharmacy, and Public Health. The University's School of Medicine offers the degree of Doctor of Medicine. In addition, graduate professional degrees are offered in accountancy, audiology, business administration, criminal justice, engineering, fine arts, international business studies, librarianship, media arts, music, music education, social work, and speech pathology. The course offerings represent strong traditional curricula at both the undergraduate and graduate levels, with additional courses in the above research areas. The Department of Physics and Astronomy offers a variety of service courses (lower-level courses open to students not majoring in physics), which have total enrollments of more than 2,000 students from other departments and colleges on campus.

Applying

Application forms should be obtained from the Graduate School. Completed applications should be submitted directly to the dean of the Graduate School.

Correspondence and Information

Professor Horacio A. Farach
Department of Physics and Astronomy
University of South Carolina
Columbia, South Carolina 29208
Telephone: 803-777-6407

SECTION 7: PHYSICS

University of South Carolina

THE FACULTY AND THEIR RESEARCH

Yakir Aharonov, Professor (joint appointment with Tel-Aviv University); Ph.D., Bristol. Theoretical physics.
Jeeva S. Anandan, Professor; Ph.D., Pittsburgh. Theoretical physics.
Chi-Kwan Au, Professor; Ph.D., Columbia. Theoretical physics.
Frank T. Avignone III, Professor and Department Chairman; Ph.D., Georgia Tech. Nuclear and elementary particle physics.
Gary S. Blanpied, Professor; Ph.D., Texas at Austin. Nuclear physics.
Richard L. Childers, Professor; Ph.D., Tennessee. High-energy physics.
Richard J. Creswick, Associate Professor; Ph.D., Berkeley. Theoretical condensed-matter physics.
Colgate W. Darden III, Professor; Ph.D., MIT. High-energy physics.
Timir Datta, Associate Professor; Ph.D., Tulane. Condensed-matter physics.
Chaden Djalali, Assistant Professor; Ph.D., Paris. Nuclear physics.
Ronald D. Edge, Professor; Ph.D., Cambridge. Nuclear and solid-state physics.
Horacio A. Farach, Professor; Ph.D., Buenos Aires. Solid-state physics, magnetic resonance.
Joseph E. Johnson, Associate Professor; Ph.D., SUNY at Stony Brook. Theoretical physics.
Edwin R. Jones Jr., Professor; Ph.D., Wisconsin. Solid-state physics.
James M. Knight, Professor; Ph.D., Maryland. Theoretical physics.
Kuniharu Kubodera, Professor; Ph.D., Tokyo. Theoretical nuclear physics.
Fred Myhrer, Professor; Ph.D., Rochester. Theoretical nuclear physics.
Charles P. Poole Jr., Professor; Ph.D., Maryland. Solid-state physics, magnetic resonance.
Barry M. Preedom, Professor; Ph.D., Tennessee. Nuclear physics.
Carl Rosenfeld, Associate Professor; Ph.D., Caltech. High-energy physics.
John L. Safko, Professor; Ph.D., North Carolina. Theoretical physics.
O. F. Schuette, Professor; Ph.D., Yale. Isotope separation, history of physics.
Jeffrey R. Wilson, Assistant Professor; Ph.D., Purdue. High-energy physics.

A lecture on the foundations of quantum theory.

Detectors used in nuclear physics experiments at the LEGS facility at Brookhaven National Laboratory.

Surface scattering of ions experiment.

SECTION 7: PHYSICS

UNIVERSITY OF TENNESSEE, KNOXVILLE

Department of Physics and Astronomy

Programs of Study

The Department of Physics and Astronomy offers work leading to the degrees of Master of Science in physics and Doctor of Philosophy in physics on its Knoxville campus and in its branch graduate schools at Oak Ridge and Tullahoma. Ph.D. programs also exist in radiological health physics and in chemical physics.

Basic graduate courses are given in modern physics, classical mechanics, electromagnetic theory, thermodynamics, statistical mechanics, and quantum mechanics. More specialized courses are given in elementary particle physics, molecular physics, nuclear physics, solid-state physics, plasma physics, and relativity, and there are advanced courses and seminars on topics of interest to members of the faculty and students. Theoretical research is carried on in nuclear theory, elementary particles, atomic and molecular structure, statistical mechanics, kinetic theory, molecular vibrations and rotations, plasma physics, solid state, and quantum theory of fields. Experimental research is mainly in the fields of elementary particle physics, nuclear physics, ultrasonics, solid state, plasmas, low temperatures, atomic and molecular spectroscopy, and biophysics.

Research Facilities

The building in which the department is housed contains 66,000 square feet and is completely air conditioned. It provides the department with excellent teaching, research, and shop facilities. In addition to working in the department's own well-equipped research laboratories, faculty and students often use the accelerators, reactors, and other large devices at the nearby Oak Ridge National Laboratory. The facilities of the University computing center are available for work on theoretical problems and analysis of data.

Financial Aid

Financial aid is available in the form of graduate teaching assistantships and research assistantships. In 1991–92, teaching assistantships pay a stipend of $8468 for nine months plus a waiver of tuition and fees. Research assistantships offer a similar stipend with a waiver of tuition and fees. Additional supplements are available under the state-funded Science Alliance. These supplements can be as high as $9000 for exceptionally promising students.

There are Oak Ridge graduate fellowships available to advanced students working at the Oak Ridge National Laboratory. Other fellowships and scholarships are available. Information may be obtained from the Graduate School.

Cost of Study

In 1991–92, tuition and fees for full-time students are $1039 per semester for Tennessee residents and $2641 for out-of-state students.

Cost of Living

Estimated costs in 1991–92, including tuition, range from $11,300 to $16,400 for a single student and $17,300 to $19,500 for a married couple. University housing for married students rents for prices that range from $210 per month for a one-bedroom unfurnished apartment to $285 per month for a three-bedroom furnished apartment. A two-bedroom furnished apartment with all utilities rents for $300 per month.

Student Group

The present enrollment of the University of Tennessee, Knoxville, is 22,996. In 1990–91, there were 80 graduate students in physics. Of these, 12 were from Tennessee, and 39 were international students. Thirteen were women.

Location

The University is located in the city of Knoxville, with a metropolitan population of over 300,000. The Great Smoky Mountains and many TVA lakes are within easy driving distance.

The University

The University of Tennessee is the official state university and federal land-grant institution of Tennessee. The University offers programs of study leading to bachelor's degrees in 184 fields of study, master's degrees in 152 specialties, and doctorates in 63 specialties.

Applying

All students intending to do graduate work in physics at the University of Tennessee must submit Graduate Record Examinations General Test scores or obtain a waiver from the department. In order to ensure their consideration, all applications for teaching and research assistantships should be submitted by February 1.

Correspondence and Information

Department of Physics and Astronomy
University of Tennessee
Knoxville, Tennessee 37996-1200
Telephone: 615-974-3342

Peterson's Guide to Graduate Programs in the Physical Sciences and Mathematics 1992

SECTION 7: PHYSICS

University of Tennessee, Knoxville

THE FACULTY AND THEIR RESEARCH

Professor and Head
W. M. Bugg, Ph.D., Tennessee. Physics of elementary particles.

Professors
C. R. Bingham, Ph.D., Tennessee. Nuclear physics.
W. E. Blass, Ph.D., Michigan State. Molecular spectroscopy.
C. Bottcher, Ph.D., Queen's (Belfast). Computational physics.
M. A. Breazeale, Ph.D., Michigan State. Ultrasonics.
J. Burgdorfer, Ph.D., Berlin. Atomic physics.
T. A. Callcott, Ph.D., Purdue. Solid-state physics.
R. W. Childers, Ph.D., Vanderbilt. Theoretical physics, elementary particles.
L. G. Christophorou, Ph.D., Manchester. Experimental radiation physics.
F. E. Close, D.Phil., Oxford. Nuclear and elementary particle physics.
E. W. Colglazier, Ph.D., Caltech. Theoretical elementary particles.
T. C. Collins, Ph.D., Florida. Theoretical condensed matter.
G. T. Condo, Ph.D., Illinois at Urbana-Champaign. Physics of elementary particles.
W. E. Deeds, Ph.D., Ohio State. Theoretical physics, molecular vibrations.
K. E. Duckett, Ph.D., Tennessee. Textile physics.
K. Fox, Ph.D., Michigan. Molecular spectroscopy.
S. Georghiou, Ph.D., Manchester (England). Biophysics.
M. W. Guidry, Ph.D., Tennessee. Nuclear physics.
E. G. Harris, Ph.D., Tennessee. Theoretical physics, dynamics of ionized plasmas.
E. L. Hart, Ph.D., Cornell. Physics of elementary particles.
H. C. Jacobson, Ph.D., Yale. Molecular spectroscopy.
D. T. King, Ph.D., Bristol (England). Physics of elementary particles.
R. J. Lovell, Ph.D., Vanderbilt. Infrared and Raman spectroscopy.
J. H. Macek, Ph.D., RPI. Theoretical atomic physics.
G. D. Mahan, Ph.D., Berkeley. Solid-state physics.
A. H. Nielsen, Emeritus; Ph.D., Michigan. Infrared and molecular spectroscopy.
F. E. Obenshain, Ph.D., Pittsburgh. Solid-state physics.
L. R. Painter, Ph.D., Tennessee. Physics, optical properties.
D. J. Pegg, Ph.D., New Hampshire. Atomic physics.
J. J. Quinn, Ph.D., Maryland. Condensed matter.
L. L. Riedinger, Ph.D., Vanderbilt. Experimental nuclear physics.
R. H. Ritchie, Ph.D., Tennessee. Theoretical physics.
I. A. Sellin, Ph.D., Chicago. Atomic physics.
C. C. Shih, Ph.D., Cornell. Theoretical elementary particles.
P. H. Stelson, Ph.D., MIT. Nuclear physics.
M. R. Strayer, Ph.D., MIT. Computational physics.
J. R. Thompson, Ph.D., Duke. Experimental solid-state physics.
J. O. Thomson, Ph.D., Illinois at Urbana-Champaign. Low-temperature and solid-state physics.
B. F. L. Ward, Ph.D., Princeton. Theoretical elementary particles.
G. W. Wheeler, Ph.D., Yale. Accelerator physics.

Associate Professors
F. E. Barnes, Ph.D., Caltech. Physics of elementary particles.
M. Breinig, Ph.D., Oregon. Atomic physics.
S. B. Elston, Ph.D., Massachusetts. Atomic physics.
T. L. Ferrell, Ph.D., Clemson. Theoretical physics.
T. Handler, Ph.D., Rutgers. Physics of elementary particles.
R. W. Lide, Ph.D., Michigan. Low-energy nuclear physics.
S. Y. Shieh, Ph.D., Maryland. Theoretical solid-state physics.
S. P. Sorensen, Ph.D., Copenhagen (Denmark). Experimental nuclear physics.

Assistant Professor
S. J. Daunt, Ph.D., Queen's at Kingston. Molecular spectroscopy. (Visiting)

Research Professor
H. O. Cohn, Ph.D., Indiana. Elementary particles.

Research Associate Professors
Y. C. Du, Ph.D., Beijing. Physics of elementary particles.
D. L. McCorkle, Ph.D., Tennessee. Atomic and molecular physics.

Research Assistant Professors
H. Faidas, Ph.D., Tennessee. Atomic and molecular physics.
R. J. Warmack, Ph.D., Tennessee. Liquid-state physics.

SECTION 7: PHYSICS

UNIVERSITY OF TEXAS AT AUSTIN

Department of Physics

Programs of Study

The Department of Physics offers programs leading to the Master of Arts and Doctor of Philosophy degrees. The time required for the M.A. degree averages one calendar year plus one semester for students with a good undergraduate background. Many qualified students work directly toward a Ph.D. and can expect, with no outside duties, to spend a minimum of three years beyond the Bachelor of Science. Most students, however, assume additional duties that increase this period. To be admitted to candidacy for the Ph.D., a student must first score at least 700 on the GRE Subject Test in physics and complete a sequence of introductory graduate courses with a minimum grade average above a B. Also, with guidance from a faculty sponsor, each student must present a public seminar on a current field of research in physics and successfully answer questions on an accompanying oral examination. These requirements must be satisfied within 2½ years of full-time graduate registration. Further course requirements depend on the student's field of specialization and are determined by the student's research supervisor. In addition, the Graduate School requires at least three courses in a supporting area (two of which must be outside physics). For general information and additional requirements, students should consult the Graduate School catalog.

Major areas of research include atomic and molecular physics, nuclear physics, plasma physics, and solid-state physics, with both theoretical and experimental programs. Theoretical research programs exist in cosmology and relativity, statistical mechanics and thermodynamics, and elementary particle physics.

Research Facilities

The Department of Physics occupies an area of over 190,000 square feet in a modern physics-mathematics-astronomy complex, which includes the Physics Library and five well-equipped technical shops staffed with 26 competent technicians supporting graduate research. A CDC Dual CYBER 170/750 computer system is available on campus, and there is a network of data-acquisition equipment interfaced to the department's DEC VAX-11/780 computer. The research facilities are described more fully in the graduate brochure available from the department.

Financial Aid

Graduate students in the Department of Physics have several possible sources of financial support. Some are employed directly by the department as teaching assistants or graduate assistants, some as research assistants for individual faculty members, and others as physicists for local research organizations. In addition, a limited number of University and federal fellowships are available for students with superior records. The stipends for these positions are comparable to those of other major universities and, where applicable, also include a tuition scholarship for the out-of-state portion of the tuition. Application should be made in the early fall, if at all possible.

Cost of Study

For 1991–92, tuition and required fees for 15 semester hours are $518 for Texas residents and $2078 for nonresidents and international students. Tuition and required fees for 9 semester hours (a typical course load for graduate students) are $522 for Texas residents and $1566 for nonresidents and international students.

Cost of Living

University-owned housing is available for both unmarried and married students. During the 1991–92 Long Session, room and meal costs for single graduate students in Jester Center range from $3085 to $3261 for double rooms and from $4116 to $4292 for a single room. There are a limited number of single rooms. The University Apartments offer housing accommodations for student families and single graduate students that range in cost from $251 to $406 per month. Private housing, including co-ops, apartments, duplexes, and houses, is also available.

Student Group

The University of Texas had an enrollment in 1990–91 of 26,983 men and 22,634 women. The majority of students are Texas residents, but students from all other states and more than 100 foreign countries and U.S. possessions are also in attendance. The University is one of the nation's leading sources of doctors, dentists, scientists, and teachers.

Location

Austin is located in central Texas. It is the state capital and has a population of about 397,000. The principal employers are the University, federal and state government agencies, and industrial research laboratories. Austin is a city of many parks and recreational facilities, located in an area of rolling hills and lakes. The climate is moderate with mild winters and warm-to-hot summers.

The University and The Department

The University of Texas is a state-supported institution. In 1990–91, its enrollment was 49,617 at its Austin campus, with 10,867 in its Graduate School and 1,700 faculty members above the rank of instructor. The Department of Physics currently consists of 65 faculty members, 60 nonteaching research personnel, 285 graduate students, and 227 undergraduates majoring in physics. In addition to offering the basic graduate courses, the department has courses and seminars in each area of specialization, as well as frequent colloquia featuring distinguished scientists from other institutions.

Applying

Correspondence concerning admission to the Graduate School should be directed to the Office of Admissions and Records. The Graduate School requires that applicants have a bachelor's degree or the equivalent, a minimum grade average of B in junior- and senior-level work, and a minimum combined score of 1000 on the verbal and quantitative sections of the General Test of the Graduate Record Examinations. However, the average combined score of successful applicants to the program in physics is approximately 1350. Applications should be returned at least sixty days prior to the beginning of the semester, earlier for international applicants.

Correspondence and Information

Graduate Coordinator
Department of Physics
University of Texas at Austin
Austin, Texas 78712-1081
Telephone: 512-471-1664

Peterson's Guide to Graduate Programs in the Physical Sciences and Mathematics 1992

SECTION 7: PHYSICS

University of Texas at Austin

THE FACULTY AND THEIR RESEARCH

Atomic and Molecular Physics
R. D. Bengtson, Professor; Ph.D., Maryland, 1968. Atomic transition probabilities, Stark broadening.
M. C. Downer, Assistant Professor; Ph.D., Harvard, 1983. Condensed matter, atomic physics, femtosecond spectroscopy.
M. Fink, Professor; Ph.D., Karlsruhe, 1966. Electron diffraction.
L. W. Frommhold, Professor; Dr.Habil., Hamburg, 1964. Atomic and molecular physics, gas discharge.
D. Heinzen, Assistant Professor; Ph.D., MIT, 1988. Ion trapping, photon ion interactions.
J. W. Keto, Professor; Ph.D., Michigan, 1968. Reactions and radiative processes of excited atoms and molecules.
F. A. Matsen, Professor; Ph.D., Princeton, 1940. Quantum mechanics, groups, linear algebra.
C. F. Moore, Professor; Ph.D., Florida State, 1964. High-energy electron collision.
M. Raizen, Assistant Professor; Ph.D., Texas at Austin, 1989. Quantum optics, photon ion interactions.
W. W. Robertson, Professor; Ph.D., Texas, 1955. Experimental atomic and molecular physics.
C. W. Scherr, Professor; Ph.D., Chicago, 1954. Theoretical quantum-mechanical studies of high accuracy on simple systems.
G. O. Sitz, Assistant Professor; Ph.D., Stanford, 1987. Scattering of molecules from surfaces.
R. E. Wyatt, Professor; Ph.D., Johns Hopkins, 1965. Classical and quantum chaos.

Classical Physics
A. M. Gleeson, Professor and Chairman; Ph.D., Pennsylvania, 1966. Field theory, underwater acoustics.
T. A. Griffy, Professor; Ph.D., Rice, 1961. Wave propagation, underwater acoustics.
H. L. Swinney, Professor; Ph.D., Johns Hopkins, 1968. Light-scattering studies of hydrodynamic and thermal instabilities.

Condensed-Matter Physics
P. R. Antoniewicz, Professor; Ph.D., Purdue, 1965. Theoretical solid state.
A. L. de Lozanne, Associate Professor; Ph.D., Stanford, 1982. Low-temperature vacuum tunneling microscopy.
F. W. de Wette, Professor; Ph.D., Utrecht (Netherlands), 1959. Theoretical solid state, surface dynamics.
M. C. Downer, Assistant Professor; Ph.D., Harvard, 1983. Condensed matter, atomic physics, femtosecond spectroscopy.
J. L. Erskine, Professor; Ph.D., Washington (Seattle), 1973. Experimental studies of surface and surface adsorbate phenomena.
J. D. Gavenda, Professor; Ph.D., Brown, 1959. Properties of conduction electrons in metals.
L. Kleinman, Professor; Ph.D., Berkeley, 1960. Theoretical studies of thin-metal films.
M. P. Marder, Assistant Professor; Ph.D., California, Santa Barbara, 1986. Pattern formation, material science.
J. T. Markert, Assistant Professor; Ph.D., Cornell, 1987. Study of physical properties of bulk material, particularly high T_c ceramics.
W. D. McCormick, Professor; Ph.D., Duke, 1959. Experimental low-temperature and solid-state physics, phase transitions.
Q. Niu, Assistant Professor; Ph.D., Washington (Seattle), 1985. Field theory of condensed matter physics, theory of superconductivity.
A. W. Nolle, Professor; Ph.D., MIT, 1947. Magnetic resonance relaxation, spectroscopy, luminescence.
C. K. Shih, Assistant Professor; Ph.D., Stanford, 1988. Study of surface properties of microelectronic materials.
J. B. Swift, Professor; Ph.D., Illinois at Urbana-Champaign, 1968. Many-body theory, phase transitions.
H. L. Swinney, Professor; Ph.D., Johns Hopkins, 1968. Light-scattering studies of hydrodynamic and thermal instabilities.
J. C. Thompson, Professor; Ph.D., Rice, 1956. Transport in liquid metals, amorphous semiconductors, metal-to-nonmetal transition.

Elementary Particle Physics
A. Böhm, Professor; Ph.D., Marburg, 1966. Particle phenomena—algebraic and group-theoretic methods.
C. B. Chiu, Professor; Ph.D., Berkeley, 1965. Strong interaction physics.
D. Dicus, Professor; Ph.D., UCLA, 1968. Field theory of weak interactions.
W. Fischler, Professor; Ph.D., Brussels, 1976. Invisible axion, supersymmetry.
V. Kaplunovsky, Assistant Professor; Ph.D., Tel-Aviv, 1983. Phenomenology of string theory.
Y. Ne'eman, Professor; Ph.D., Imperial College (London), 1961. Symmetries in elementary particle physics.
J. G. Polchinski, Professor; Ph.D., Berkeley, 1980. Supersymmetry.
E. C. G. Sudarshan, Professor; Ph.D., Rochester, 1958. Theoretical particle physics.
S. Weinberg, Professor; Ph.D., Princeton, 1957. Theory of strong and weak particle interaction.

High-Energy Physics
K. Lang, Assistant Professor; Ph.D., Rochester, 1985. Experimental study of rare decay of the K-meson.
J. L. Ritchie, Assistant Professor; Ph.D., Rochester, 1983. Experimental study of rare decay of the K-meson.

Nuclear Physics
W. R. Coker, Professor; Ph.D., Georgia, 1966. Mechanisms of nuclear reactions, three-body final-state problem.
G. W. Hoffmann, Professor; Ph.D., UCLA, 1971. Experimental nuclear physics.
E. V. Ivash, Professor; Ph.D., Michigan, 1952. Theoretical nuclear physics, particularly direct reactions, quantum mechanics.
C. F. Moore, Professor; Ph.D., Florida State, 1964. Experimental nuclear physics.
P. J. Riley, Professor; Ph.D., Alberta, 1962. Nuclear reactions and nuclear structure physics.
T. Udagawa, Professor; Ph.D., Tokyo University of Education, 1962. Theoretical nuclear structure.
S. A. A. Zaidi, Associate Professor; Ph.D., Heidelberg, 1964. Experimental and theoretical nuclear physics.

Plasma Physics
David Baldwin, Professor; Ph.D., MIT, 1962. Theoretical aspects of plasma physics and magnetic fusion energy.
R. D. Bengtson, Professor; Ph.D., Maryland, 1968. Plasma spectroscopy, experimental plasma physics.
H. Berk, Professor; Ph.D., Princeton, 1964. Theoretical plasma physics, computer simulation of plasmas.
W. E. Drummond, Professor; Ph.D., Stanford, 1958. Theoretical plasma physics.
K. W. Gentle, Professor; Ph.D., MIT, 1966. Nonlinear plasma processes.
R. D. Hazeltine, Professor; Ph.D., Michigan, 1968. Theoretical plasma physics.
C. W. Horton Jr., Professor; Ph.D., California, San Diego, 1967. Theoretical plasma physics.
P. J. Morrison, Associate Professor; Ph.D., California, San Diego, 1979. Plasma physics.
M. E. Oakes, Professor; Ph.D., Florida State, 1964. Wave propagation in plasmas with emphasis on resonances.
T. Tajima, Professor; Ph.D., California, Irvine, 1975. Theoretical plasma physics.
J. B. Taylor, Professor; Ph.D., Birmingham (England), 1955. Theory of plasma stability.

Relativity, Cosmology, and Quantum Field Theory
P. Candelas, Professor; D.Phil., Oxford, 1977. General relativity, techniques of quantization in curved space-time.
B. S. DeWitt, Professor; Ph.D., Harvard, 1950. Quantum field theory.
C. DeWitt-Morette, Professor; Ph.D., Paris, 1947. Mathematical physics, relativity theory.
W. Fischler, Professor; Ph.D., Brussels, 1976. Cosmology, gravity.
R. A. Matzner, Professor; Ph.D., Maryland, 1967. General cosmology, gravitational radiation.
L. C. Shepley, Associate Professor; Ph.D., Princeton, 1965. Cosmology, interaction of matter with gravitation.
S. Weinberg, Professor; Ph.D., Princeton, 1957. Cosmology, astrophysics.

Statistical Mechanics and Thermodynamics
I. Prigogine, Professor; Ph.D., Brussels, 1941. Statistical mechanics and thermodynamics.
L. E. Reichl, Professor; Ph.D., Denver, 1969. Strong-coupling, nonequilibrium, and quantum statistical mechanics.
W. C. Schieve, Professor; Ph.D., Lehigh, 1959. Nonequilibrium statistical mechanics.
J. S. Turner, Associate Professor; Ph.D., Indiana, 1969. Self-organization in physics, chemistry, and biology.

SECTION 7: PHYSICS

UNIVERSITY OF VIRGINIA
Department of Physics

Programs of Study

The Department of Physics offers programs leading to the M.A., M.S., M.A.T., and Ph.D. degrees. The primary emphasis is on the Ph.D. program. For a master's degree, a student must complete, with satisfactory progress, at least 24 graduate credits in an approved program, write a thesis, and take an examination on the thesis. For the Ph.D. degree, satisfactory performance in an approved course program is required. In addition, students must take a comprehensive examination, write a dissertation, and take the dissertation exam. Doctoral students must spend two semesters in residence at the University.

A Ph.D. in biophysics is available through an interdisciplinary program associated with the physics department and other science departments of the University.

Research Facilities

The physics department has major facilities for atomic physics, biophysics, and solid-state physics research, including high magnetic fields; lasers; a helium liquefier; low-temperature cryostats, including a dilution refrigerator; electron microscopes; photoelectron spectrometers; and electron-scattering devices. Major machine shop and electronics shop facilities are also available. The Institute for Nuclear and Particle Physics supports research programs as well as design work for CEBAF, a 4-GeV electron accelerator. Both the nuclear physics and condensed-matter physics programs have access to such national facilities as the Los Alamos Meson Physics Facility, the Stanford Linear Accelerator Center (SLAC), the Oak Ridge National Laboratory, and the Francis Bitter National Magnet Laboratory. A recently formed high-energy-particle physics group is currently conducting active research at Fermi National Accelerator Laboratory.

Financial Aid

A number of well-qualified entering students are awarded departmental or University fellowships. For 1991–92, these fellowships carry stipends of $10,000 for nine months plus remission of tuition. Many other tuition-free teaching and research assistantships are available and carry stipends of $8600–$10,000 for nine months. Students making satisfactory progress are normally supported until the Ph.D. degree program has been completed. Possibilities also exist for summer research support.

Cost of Study

In 1991–92, tuition and fees are $1677 per semester for Virginia residents and $4782 per semester for out-of-state residents. Most students receive fellowships, which cover tuition and fees.

Cost of Living

In 1991–92, estimated expenses for single students are $1938 for a room in a residence hall, $1700 for personal expenses, and $500 for books and supplies. Housing for married students is available at rents ranging from $376 to $466 per month. Off-campus housing is also available for married and single graduate students. For further information, students should write to the director of housing at Station 1, Page House.

Student Group

Enrollment at the University is approximately 18,130, including 6,350 graduate students. The Department of Physics has approximately 80 graduate students.

Location

The city of Charlottesville and the county of Albemarle have a combined population exceeding 100,000 and are located beside the foothills of the Blue Ridge Mountains in central Virginia. Monticello, Thomas Jefferson's home, and Ash Lawn, James Monroe's residence, are located in the county, and the courthouse designed by Jefferson is in Charlottesville's Court Square. Outdoor recreational opportunities abound in the nearby mountains and Shenandoah National Park.

Charlottesville is 110 miles from Washington, D.C., and 70 miles from Richmond. Bus and railway service and direct airline service to Washington, Atlanta, Pittsburgh, and New York are available.

The University

The University of Virginia, founded in 1819 by Thomas Jefferson, is a coeducational state institution that recognizes the importance of having a student body drawn from many parts of the country.

Applying

Applications for admission and financial assistance may be obtained by writing to the Department of Physics. To be eligible for all forms of financial assistance, students should submit completed applications by April 15.

Correspondence and Information

Graduate Advisor
Department of Physics
University of Virginia
McCormick Road
Charlottesville, Virginia 22903
Telephone: 804-924-3782

SECTION 7: PHYSICS

University of Virginia

THE FACULTY AND THEIR RESEARCH

Louis A. Bloomfield, Ph.D., Stanford, 1983: experimental atomic and solid-state physics. Arthur S. Brill, Ph.D., Pennsylvania, 1956: experimental biophysics, proteins and transition-metal ions. Vittorio Celli, Ph.D., Pavia (Italy), 1958: theoretical solid-state physics, surface studies. Robert V. Coleman, Ph.D., Virginia, 1956: experimental solid-state physics, superconducting electronics. Sergio Conetti, Dottore in Fisica, Università degli Studi di Trieste, 1967: experimental high-energy-particle physics. Michael Coopersmith, Ph.D., Cornell, 1962: theoretical physics, statistical mechanics, phase transitions. Bradley B. Cox, Ph.D., Duke, 1967: experimental high-energy-particle physics. Donal B. Day, Ph.D., Virginia, 1979: experimental nuclear and particle physics. Bascom S. Deaver Jr., Ph.D., Stanford, 1962: experimental solid-state physics, superconducting devices. Alan T. Dorsey, Ph.D., Illinois at Urbana-Champaign, 1987: theoretical solid-state physics and statistical mechanics. Edmond C. Dukes, Ph.D., Michigan, 1984: experimental high-energy-particle physics. Paul M. Fishbane, Ph.D., Princeton, 1967: theoretical physics, elementary particles. Michael Fowler, Ph.D., Cambridge, 1962: theoretical physics, field theory and solid-state theory. Thomas F. Gallagher, Ph.D., Harvard, 1971: collisions and spectroscopy of atoms and molecules. Piet C. Gugelot, Ph.D., Swiss Federal Institute of Technology, 1945: experimental nuclear physics. Frank L. Hereford, Ph.D., Virginia, 1947: color vision. George B. Hess, Ph.D., Stanford, 1967: experimental solid-state physics, liquid helium, physisorption. Pham Q. Hung, Ph.D., UCLA, 1978: theoretical particle physics, cosmology. Prabahan K. Kabir, Ph.D., Cornell, 1957: theoretical physics, elementary particles, nuclear and atomic physics. Hugh P. Kelly, Ph.D., Berkeley, 1963: many-body theory, atomic and molecular physics. Doris Kuhlmann-Wilsdorf, Ph.D., Göttingen, 1947; D.Sc., Witwatersrand (Johannesburg), 1954: theoretical materials science. Daniel J. Larson, Ph.D., Harvard, 1971: experimental atomic and molecular physics. Richard A. Lindgren, Ph.D., Yale, 1969: experimental nuclear and particle physics. Robert W. Lourie, Ph.D., MIT, 1986: experimental nuclear and particle physics. James S. McCarthy, Ph.D., Stanford, 1968: experimental nuclear and particle physics. Ralph C. Minehart, Ph.D., Harvard, 1962: experimental nuclear and particle physics. John W. Mitchell, D.Phil., Oxford, 1938: experimental solid-state physics, crystal studies, photographic processes. Kenneth Nelson, Ph.D., Wisconsin, 1986: experimental high-energy-particle physics. Julian V. Noble, Ph.D., Princeton, 1966: theoretical physics, nuclear physics, intermediate-energy physics. Blaine E. Norum, Ph.D., MIT, 1979: experimental nuclear and particle physics. Dinko Počanić, Ph.D., Zagreb (Yugoslavia), 1981: experimental intermediate-energy nuclear physics. S. Joseph Poon, Ph.D., Caltech, 1978: experimental solid-state physics, disordered systems, quasicrystals, superconducting materials. Rogers C. Ritter, Ph.D., Tennessee, 1961: gravitation, precision measurements, biophysics. John Ruvalds, Ph.D., Oregon, 1967: theoretical solid-state physics. Stephen Schnatterly, Ph.D., Illinois, 1965: experimental solid-state physics; soft X-ray and inelastic electron-scattering spectroscopy of solids, atoms, and molecules. Bellave S. Shivaram, Ph.D., Northwestern, 1984: experimental solid-state physics; novel superconductors, ultrasonic measurements. Stanley E. Sobottka, Ph.D., Stanford, 1960: X-ray crystallography, X-ray detector development, experimental nuclear physics. John W. Stewart, Ph.D., Harvard, 1954: experimental solid-state physics. Harry B. Thacker Jr., Ph.D., UCLA, 1973: elementary particle physics and quantum field theory. Stephen T. Thornton, Ph.D., Tennessee, 1967: experimental nuclear physics. Hans J. Weber, Ph.D., Frankfurt, 1965: theoretical nuclear and particle physics. W. D. Whitehead, Ph.D., Virginia, 1949: experimental nuclear and particle physics. Klaus Ziock, Ph.D., Bonn, 1956: experimental nuclear and particle physics.

RESEARCH ACTIVITIES

Theoretical

Atomic and Molecular Physics. (Kelly, 1 research associate) Photoionization cross-section calculations for atoms and molecules, correlation effects on atomic and molecular properties calculated by many-body theory.

Condensed Matter. (Celli, Coopersmith, Dorsey, Fowler, Ruvalds, 3 research associates) Novel mechanisms for superconductivity, phase transitions, and coupled magnetic impurity effects in spin-glass alloys; role of coupled magnetic impurities in superconducting critical fields and the influence of disorder on superconductor dynamics; structure, dynamics, and optical properties of solid surfaces; gas surface interactions; wave propagation in random media; spin chains; Bethe Ansatz systems; field theoretic models for solid-state systems; theory of macroscopic quantum phenomena; pattern formation; transport in high-temperature superconductors.

High-Energy Physics. (Fishbane, Hung, Kabir, Thacker) Theoretical studies of high-energy physics, including properties of quantum chromodynamics, lattice gauge theory, solvable models and conformal field theory, electroweak interactions, grand unified theories, supersymmetry, and the Very Early Universe.

Nuclear Physics. (Noble, Weber) Structure of light nuclei, electron scattering, and photonuclear reactions; interaction of pions with complex nuclei; quark models of hadrons.

Experimental

Atomic and Molecular Physics. (Bloomfield, Gallagher, Larson, 6 research associates) Laser spectroscopy of atoms, ions, small molecules, and clusters, including photodetachment and photoionization; microwave-optical double resonance; studies of collisions, using spectroscopic techniques; measurement of the effects of strong optical and microwave fields; and development of new techniques in laser spectroscopy.

Biological Physics. (Brill, Ritter, Sobottka, 3 research associates) Magnetic and optical properties of transition-metal ions in proteins, precision measurement of biophysical changes in viscosity and density of biomacromolecular solutions, X-ray diffraction from crystallized proteins, development of X-ray detectors.

Condensed Matter. (Coleman, Deaver, Hess, Poon, Schnatterly, Shivaram, 3 research associates) Electronic properties of metals and alloys in magnetic fields up to 230 kG, studied via magnetoresistance, Hall effect, and transport in metals; molecular spectroscopy through inelastic tunneling techniques; surface studies by scanning tunneling microscopy; superconducting electronics and devices; inelastic electron-scattering spectroscopy as a probe of plasmons, excitons, interband transitions, and core excitations in solids; soft X-ray emission spectroscopy of solids and surfaces; superconducting and transport properties of disordered and quasi-periodic alloys; high-temperature superconducting materials; properties of physisorbed films; materials with novel symmetry properties; superconductivity in heavy electrons and other novel systems at very low temperatures; ultrasonic measurements.

Gravitational Physics. (Ritter) Use of precision measurements in laboratory experiments testing gravitation.

Nuclear Physics. (Day, Gugelot, Lindgren, Lourie, McCarthy, Minehart, Norum, Počanić, Thornton, Ziock, 5 research associates, 4 research scientists) Active research programs at various accelerator laboratories in the United States and abroad, with the largest effort in electronuclear physics at the Bates electron accelerator (MIT), LEGS facility at Brookhaven, Amsterdam, Saskatchewan, Saclay (France), and SLAC; nuclear structure, nuclear momentum distributions, quark structure, and properties of excited nuclei; medium-energy physics, using the intense beams of mesons from the Los Alamos Meson Physics Facility, the SIN ring accelerator in Switzerland, and the LEAR antiproton ring at CERN; search for free quarks and heavy neutrinos as well as study of various pion-decay modes.

High-Energy-Particle Physics. (Cox, Conetti, Dukes, Nelson, 1 senior scientist, 2 research associates) Presently involved in active research programs at Fermi National Accelerator Laboratory to study various aspects of quantum chromodynamics and to determine quark and gluon structure functions. Experiments include measurements of the production of heavy quark charmonium states as detected by J/Ψ-photon decays and measurements of the production of high p_t direct photons in high-energy interaction. In addition, experiments to measure the production and the weak decays of B mesons and baryons will soon begin at Fermilab. An active participation in the planning for facilities and experiments at the Superconducting Super Collider (SSC) is ongoing, and this group will be involved in experiments to measure charge-parity violation in beauty quark decays at the SSC.

Appendixes

This section contains two appendixes. The first, Institutional Changes Since the 1991 Edition, lists institutions that have closed, moved, merged, or changed their name or status since the last edition of the guides. The second, Abbreviations Used in the Guides, gives abbreviations of degree names, tests, and organizations and agencies along with what those abbreviations stand for. These appendixes are identical in all six volumes of the Graduate Guides.

Institutional Changes Since the 1991 Edition

Following is an alphabetical listing of institutions that have recently closed, moved, merged with other institutions, or changed their names or status. In the case of a name change, the former name appears first, followed by the new name.

American Conservatory Theatre (San Francisco, California); name changed to American Conservatory Theater.

American Institute of Psychotherapy (Huntsville, Alabama); closed.

Antioch University Los Angeles (Marina del Rey, California); name changed to Antioch Southern California/Santa Barbara.

Baptist College at Charleston (Charleston, South Carolina); name changed to Charleston Southern University.

Catholic University of Puerto Rico (Ponce, Puerto Rico); name changed to Pontifical Catholic University of Puerto Rico.

Faith Baptist Bible College and Seminary (Ankeny, Iowa); name changed to Faith Baptist Bible College and Theological Seminary.

Graduate School of Political Management (New York, New York); name changed to The Graduate School of Political Management.

Kearney State College (Kearney, Nebraska); name changed to University of Nebraska at Kearney.

Madonna College (Livonia, Michigan); name changed to Madonna University.

Mary Immaculate Seminary (Northampton, Pennsylvania); no longer grants degrees.

Marycrest College (Davenport, Iowa); name changed to Teikyo Marycrest University.

Maryville College-Saint Louis (St. Louis, Missouri); name changed to Maryville University of St. Louis.

Mercer University Atlanta (Atlanta, Georgia); name changed to Mercer University, Cecil B. Day Campus.

Mercy College of Detroit (Detroit, Michigan); merged with University of Detroit to become University of Detroit Mercy.

Nazareth College in Kalamazoo (Kalamazoo, Michigan); closed.

Polytechnic University, Westchester Campus (Hawthorne, New York); name changed to Polytechnic University, Westchester Graduate Center.

Saint Paul Seminary School of Divinity of the College of St. Thomas (St. Paul, Minnesota); information now reported within University of St. Thomas.

Saint Peter's College, Englewood Cliffs Campus (Englewood Cliffs, New Jersey); information now reported within Saint Peter's College.

Salve Regina College (Newport, Rhode Island); name changed to Salve Regina University.

Shenandoah College and Conservatory (Winchester, Virginia); name changed to Shenandoah University.

Southeastern Massachusetts University (North Dartmouth, Massachusetts); name changed to University of Massachusetts Dartmouth.

University of Detroit (Detroit, Michigan); name changed to University of Detroit Mercy.

University of Lowell (Lowell, Massachusetts); name changed to University of Massachusetts Lowell.

Abbreviations Used in the Guides

DEGREES

The following list includes abbreviations used in the profiles in the 1992 edition of the guides. Because some degrees (e.g., Doctor of Education) can be abbreviated in more than one way (e.g., D.Ed. or Ed.D.), and because the abbreviations used in the guides reflect the preferences of the individual colleges and universities, the list may include two or more abbreviations for a single degree.

AC	Advanced Certificate
AD	Artist's Diploma
	Doctor of Arts
ADP	Artist's Diploma
Adv C	Advanced Certificate
Adv M	Advanced Master
AE	Aerospace Engineer
	Agricultural Engineer
AEMBA	Advanced Executive Master of Business Administration
AGC	Advanced Graduate Certificate
AGSC	Advanced Graduate Specialist Certificate
ALM	Master of Liberal Arts
AM	Master of Arts
AMRS	Master of Arts in Religious Studies
A Mus D	Doctor of Musical Arts
APC	Advanced Professional Certificate
App ME	Applied Mechanics Engineer
Ap Sci	Applied Scientist
ATM	Master of Art Therapy
B Th	Bachelor of Theology
CAES	Certificate of Advanced Educational Specialization
CAGS	Certificate of Advanced Graduate Studies
CAL	Certificate of Advanced Librarianship
	Certificate in Applied Linguistics
CAMS	Certificate of Advanced Management Studies
CAPS	Certificate of Advanced Professional Studies
CAS	Certificate of Advanced Studies
CASPA	Certificate of Advanced Study in Public Administration
CASR	Certificate in Advanced Social Research
CCJA	Certificate in Criminal Justice Administration
CE	Civil Engineer
CG	Certificate in Gerontology
CGS	Certificate of Graduate Studies
Ch E	Chemical Engineer
Chem E	Chemical Engineer
CHSS	Counseling and Human Services Specialist
CIS	Certificate in Information Science
CITS	Certificate of Individual Theological Studies
CLIS	Certificate of Library and Information Science
CMS	Certificate in Museum Studies
CPC	Certificate in Professional Counseling
CPH	Certificate in Public Health
C Phil	Candidate in Philosophy
CPI	Certificate in Planning Information
CPM	Certificate in Public Management
CSD	Certificate in Spiritual Direction
CSS	Certificate of Special Studies
CTS	Certificate of Theological Studies
CURP	Certificate in Urban and Regional Planning
DA	Doctor of Arts
DA Ed	Doctor of Arts in Education
DAIS	Doctor of Arts in Information Science
D Arch	Doctor of Architecture
DAST	Diploma of Advanced Studies in Teaching
DBA	Doctor of Business Administration
DC	Doctor of Chiropractic
D Chem	Doctor of Chemistry
DCL	Doctor of Canon Law
	Doctor of Civil Law
DCM	Doctor of Church Music
DDN	Diplôme du Droit Notarial
DDS	Doctor of Dental Surgery
DE	Doctor of Engineering
D Ed	Doctor of Education
D Eng	Doctor of Engineering
D Env	Doctor of Environment
D Env Des	Doctor of Environmental Design
DES	Doctor of Engineering Science
DESP	Diplôme des Etudes Spécialisées
DF	Doctor of Forestry
DFA	Doctor of Fine Arts
DHL	Doctor of Hebrew Letters
	Doctor of Hebrew Literature
DHS	Doctor of Human Services
DHSA	Doctor of Health Sciences Administration
DH Sc	Doctor of Health Science
DIBA	Doctor of International Business Administration
Dip A	Diploma in Aquaculture
Dip Cs	Diploma in Christian Studies
Dip Ma	Diploma in Marine Affairs
DIT	Doctor of Industrial Technology
D Jur	Doctor of Jurisprudence
DLIS	Doctor of Library and Information Science
DM	Doctor of Music
DMA	Doctor of Musical Arts
DMD	Doctor of Dental Medicine
DME	Doctor of Music Education
DM Ed	Doctor of Music Education
D Med Sc	Doctor of Medical Science
DMFT	Doctor of Marriage and Family Therapy
D Min	Doctor of Ministry
D Miss	Doctor of Missiology
DML	Doctor of Modern Languages
DMM	Doctor of Music Ministry
DMS	Doctor of Medical Science
DM Sc	Doctor of Medical Science
D Mus	Doctor of Music
D Mus A	Doctor of Musical Arts
D Mus Ed	Doctor of Music Education
DNS	Doctor of Nursing Science
DN Sc	Doctor of Nursing Science
DO	Doctor of Osteopathy
DPA	Diploma in Public Administration
	Doctor of Public Administration
DPC	Doctor of Pastoral Counseling
DPE	Doctor of Physical Education
DPH	Doctor of Public Health

Peterson's Guide to Graduate Programs in the Physical Sciences and Mathematics 1992

ABBREVIATIONS USED IN THE GUIDES

DPM	Doctor of Podiatric Medicine	MAABS	Master of Arts in Applied Behavioral Sciences
DPS	Doctor of Professional Studies	MAACCD	Master of Arts in Adult Christian Community Development
D Ps	Diploma of Psychology Doctor of Psychology	MA ADAM	Master of Arts in Alcoholism and Drug Abuse Ministry
Dr DES	Doctor of Design	MAAE	Master of Arts in Aeronautical and Astronautical Engineering Master of Arts in Applied Economics Master of Arts in Art Education
DRE	Doctor of Religious Education		
Dr PH	Doctor of Public Health		
DSA	Diplôme en Sciences Administratives		
D Sc	Doctor of Science		
D Sc A	Doctor of Applied Science	MAAOM	Master of Arts in Applied Organizational Management
DSM	Doctor of Sacred Music		
DSN	Doctor of Science in Nursing	MA Art Ed	Master of Arts in Art Education
DSS	Diploma of Specialized Studies	MAAS	Master of Arts in Alcoholism Studies Master of Arts in Arab Studies
DS Sc	Doctor of Social Science		
DSW	Doctor of Social Work	MAAT	Master of Arts in Applied Theology Master of Arts in Art Therapy
D Th	Doctor of Theology		
DVM	Doctor of Veterinary Medicine	MABM	Master of Agribusiness Management
DV Sc	Doctor of Veterinary Science	MABS	Master of Arts in Biblical Studies Master of Arts in Behavioral Science
EAA	Engineer in Aeronautics and Astronautics		
EAS	Education Administration Specialist	MAC	Master of Arts in Communication
Ed D	Doctor of Education	M Ac	Master of Accounting
Ed M	Master of Education	MACA	Master of Arts in Computer Applications
Ed S	Specialist in Education	M Acc	Master of Accountancy
EE	Electrical Engineer	M Acct	Master of Accountancy Master of Accounting
EM	Mining Engineer		
EMBA	Executive Master of Business Administration	M Accy	Master of Accountancy
EMHA	Executive Master of Health Administration	MACE	Master of Arts in Christian Education
EMPA	Executive Master of Public Affairs	MAC Ed	Master of Arts in Continuing Education
EMST	Executive Master of Science in Taxation	MACL	Master of Arts in Classroom Psychology
Eng	Engineer	MACM	Master of Arts in Christian Ministries
Engr	Engineer	MA Comm	Master of Arts in Communication
Eng Sc D	Doctor of Engineering Science	MACP	Master of Arts in Community Psychology
En S	Specialist in English	MACS	Master of Arts in Christian Studies Master of Arts in Church Service Master of Arts in Communication Studies Master of Arts in Computer Science
Env E	Environmental Engineer		
Exec MBA	Executive Master of Business Administration		
Exec MPA	Executive Master of Public Administration	MACSS	Master of Arts in Church Social Services
Exec MPH	Executive Master of Public Health	MACST	Master of Arts in Christian School Teaching
Exec MS	Executive Master of Science	MACT	Master of Arts in College Teaching
Exec MSE	Executive Master of Science in Engineering	MACTM	Master of Applied Communication Theory and Methodology
GDT	Graduate Diploma in Theology	M Ad	Master of Administration
Geol E	Geological Engineer	M Ad Ed	Master of Adult Education
HSD	Doctor of Health and Safety	M Adm	Master of Administration
HS Dir	Director of Health and Safety	M Admin	Master of Administration
IMA	Interdisciplinary Master of Arts	M Adm J	Master in Administration of Justice
IMBA	Integrative Master of Business Administration International Master of Business Administration	M Adm Mgt	Master of Administration Management
		MAE	Master of Aerospace Engineering Master of Agricultural Economics Master of Agricultural Education Master of Agricultural Engineering Master of Art Education Master of Arts in Education
IOE	Industrial and Operations Engineer		
JCD	Doctor of Canon Law		
JCL	Licentiate in Canon Law		
JD	Doctor of Jurisprudence (Juris Doctor)	MA Ed	Master of Arts in Education
JSD	Doctor of Juridical Science	MAEE	Master of Arts in Elementary Education
JSM	Master of Science of Law	MAES	Master of Arts in Environmental Sciences
LCL	Licentiate in Canon Law	MAET	Master of Arts in English Teaching
LL B	Bachelor of Laws	MAETS	Master of Arts in Education and Theological Studies
LL CM	Master of Comparative Law		
LL D	Doctor of Laws	MAFIS	Master of Accountancy and Financial Information Systems
LL M	Master of Laws		
L Th	Licenciate in Theology	MAFLL	Master of Arts in Foreign Language and Literature
MA	Master of Arts		
MAA	Master of Administrative Arts Master of Aeronautics and Astronautics Master of Applied Arts	M Ag	Master of Agriculture
		MAG	Master of Applied Geography
		M Ag E	Master of Agricultural Engineering

ABBREVIATIONS USED IN THE GUIDES

M Ag Ed	Master of Agricultural Education	MAPM	Master of Arts in Pastoral Ministry
M Agr	Master of Agriculture	M Ap Ma	Master of Applied Mathematics
MAH	Master of Arts in Humanities	MAPP	Master of Arts in Public Policy
MAHE	Master of Arts in Hebrew Education	M Appl Stat	Master of Applied Statistics
MAHEFE	Master of Arts in Home Economics and Family Ecology	MAPRS	Master of Arts in Pacific Rim Studies
		MA Ps	Master of Arts in Psychology
MAHL	Master of Arts in Hebrew Letters	MAPS	Master of Arts in Pastoral Studies
	Master of Arts in Hebrew Literature	M Ap Stat	Master of Applied Statistics
MAHRM	Master of Arts in Human Resources Management	MA Psych	Master of Arts in Psychology
		MAPW	Master of Arts in Professional Writing
MAHS	Master of Arts in Human Services	M Aq	Master of Aquaculture
MAHT	Master of Arts in History Teaching	MAR	Master of Arts in Religion
MAI	Master of Agricultural Industries		Master of Arts in Research
MAIA	Master of Arts in Industrial Arts	MA(R)	Master of Arts (Research)
	Master of Arts in International Affairs	MARC	Master of Arts in Religious Communication
MAICS	Master of Arts in Intercultural Studies	M Arch	Master of Architecture
MAID	Master of Arts in Interior Design	M Arch E	Master of Architectural Engineering
	Master of Arts in International Diplomacy	M Arch H	Master of Architectural History
MAIPS	Master of Arts in International Policy Studies	M Arch Studies	Master of Architectural Studies
MAIR	Master of Arts in Industrial Relations	M Arch UD	Master of Architecture in Urban Design
MAIS	Master of Arts in Interdisciplinary Studies	MARE	Master of Arts in Religious Education
	Master of Arts in International Studies	Mar Eng	Marine Engineer
MAJ	Master of Arts in Journalism	MARS	Master of Arts in Religious Studies
MAJC	Master of Arts in Journalism and Communication	MAS	Master of Accounting Science
			Master of Administrative Science
MAJCS	Master of Arts in Jewish Communal Service		Master of Aeronautical Science
MAJE	Master of Arts in Jewish Education		Master of Applied Science
MAJ Ed	Master of Arts in Jewish Education		Master of Applied Spirituality
MAJS	Master of Arts in Jewish Studies		Master of Applied Statistics
			Master of Archival Studies
MALA	Master of Arts in Liberal Arts	MASA	Master of Advanced Studies in Architecture
	Master of Arts in Liturgical Arts	MASAC	Master of Arts in Subtance Abuse Counseling
MALAS	Master of Arts in Latin American Studies	MA Sc	Master of Applied Science
MALD	Master of Arts in Law and Diplomacy	MASM	Master of Arts in Special Ministries
MALIS	Master of Arts in Library and Information Science	MASS	Master of Arts in Social Science
			Master of Arts in Special Studies
MALL	Master of Arts in Liberal Learning	MAT	Master of Arts in Teaching
MALS	Master of Arts in Liberal Studies	MATC	Master of Arts in Textiles and Clothing
	Master of Arts in Library Science	MATE	Master of Arts for the Teaching of English
MAM	Master of Agriculture and Management	Mat E	Materials Engineer
	Master of Animal Medicine	MATESL	Master of Arts in Teaching English as a Second Language
	Master of Applied Mechanics		
	Master of Arts Management	MA Th	Master of Arts in Theology
	Master of Arts—Ministry	MATM	Master of Arts in Teaching of Mathematics
	Master of Avian Medicine	MATP	Master of Arts in Teaching Psychology
	Master of Aviation Management	MATS	Master of Arts in Teaching of Science
MAMB	Master of Applied Molecular Biology		Master of Arts in Theological Studies
MAMC	Master of Arts in Mass Communication	MAUA	Master of Arts in Urban Affairs
MAMFC	Master of Arts in Marriage and Family Counseling	MAUD	Master of Arts in Urban Design
		M Aud	Master of Audiology
MA Mgt	Master of Arts in Management	MAURP	Master of Arts in Urban and Regional Planning
MA Missions	Master of Arts in Missions		
MAMRD	Master of Agricultural Management and Resource Development	MAV Ed	Master of Administration in Vocational Education
MAMS	Master of Applied Mathematical Sciences	MAW	Master of Arts in Worship
	Master of Associated Medical Sciences		Master of Arts in Writing
M Am S	Master of American Studies	MAYM	Master of Arts in Youth Ministry
MAM Sc	Master of Applied Mathematical Science	MBA	Master of Business Administration
MAM Th	Master of Arts in Music Therapy	MBAA	Master of Business Administration in Aviation
M Anesth Ed	Master of Anesthesiology Education		
MAP	Master of Applied Psychology	MBAE	Master of Biological and Agricultural Engineering
MAPA	Master of Arts in Public Administration		
	Master of Arts in Public Affairs	MBA-Ex	Master of Business Administration-Executive
MAPC	Master of Arts in Pastoral Counseling	MBAIB	Master of Business Administration in International Business
MAPCC	Master of Arts in Pastoral Care and Counseling		
MAPE	Master of Arts in Physical Education		

Peterson's Guide to Graduate Programs in the Physical Sciences and Mathematics 1992

ABBREVIATIONS USED IN THE GUIDES

MBAIBK	Master of Business Administration in International Banking	**M Div**	Master of Divinity
MBAIT	Master of Business Administration in International Trade	**MDS**	Master of Decision Sciences Master of Dental Surgery
MBC	Master of Building Construction	**ME**	Master of Education Master of Engineering
MBE	Master of Business Economics Master of Business Education	**MEA**	Master of Engineering Administration Master of Engineering Architecture
M Biomath	Master of Biomathematics	**Mech E**	Mechanical Engineer
MBIS	Master of Business Information Science	**M Econ**	Master of Economics
MBS	Master of Basic Science Master of Behavioral Science Master of Building Science	**MED**	Master of Education of the Deaf
		M Ed	Master of Education
		M Educ	Master of Education
MBT	Master of Business Taxation	**MEE**	Master of Electrical Engineering
M Bus Ed	Master of Business Education	**MEM**	Master of Engineering Management
MC	Master of Communication Master of Counseling	**MEMS**	Master of Emergency Medical Service Master of Engineering in Manufacturing Systems
MCA	Master of Communication Arts		
MCAT	Master of Creative Arts in Therapy	**M Eng**	Master of Engineering
MCC	Master of Clinical Chemistry	**M Eng Mgt**	Master of Engineering Management
MCD	Master of Communications Disorders	**M Engr**	Master of Engineering
MCE	Master of Christian Education Master of Civil Engineering	**M En S**	Master of Environmental Sciences
		M Env	Master of Environment
MC Ed	Master of Continuing Education	**M Env Des**	Master of Environmental Design
MCG	Master of Clinical Gerontology	**M Env Sc**	Master of Environmental Science
MCH	Master of Community Health	**MEP**	Master of Engineering Physics Master of Environmental Planning
M Ch E	Master of Chemical Engineering		
MCIM	Master of Clinical Immunology and Microbiology	**MEPC**	Master of Environmental Pollution Control
		MEPD	Master of Education–Professional Development
MCIS	Master of Computer and Information Science		
MCJ	Master of Comparative Jurisprudence Master of Criminal Justice	**MER**	Master of Energy Resources
		MERM	Master of Earth Resources Management
MCJA	Master of Criminal Justice Administration	**MES**	Master of Engineering Science Master of Environmental Studies
MCL	Master of Canon Law Master of Civil Law Master of Comparative Law	**ME Sc**	Master of Engineering Science
		MET	Master of Education in Teaching
M Cl D	Master of Clinical Dentistry	**Met E**	Metallurgical Engineer
M Cl Sc	Master of Clinical Science	**M Ext Ed**	Master of Extension Education
MCLT	Master of Clinical Laboratory Technology	**MF**	Master of Finance Master of Forestry
MCM	Master of Christian Ministry Master of Church Management Master of Church Music Master of Clinical Microbiology Master of Construction Management	**MFA**	Master of Fine Arts
		MFAW	Master of Fine Arts in Writing
		MFCC	Marriage and Family Counseling Certificate
		MFE	Master of Financial Economics Master of Forest Engineering
MCOEN	Master of Computer Engineering		
M Comm	Master of Communication	**M Fin**	Master of Finance
M Comp Sci	Master of Computer Science	**M For**	Master of Forestry
M Coun	Master of Counseling	**MFR**	Master of Forest Resources
MCP	Master of City Planning Master of Community Planning Master of Counseling Psychology	**M Fr**	Master of French
		MFRC	Master of Forest Resources and Conservation
MCRP	Master of City and Regional Planning	**MFS**	Master of Family Studies Master of Forensic Studies Master of Forest Studies Master of French Studies
MCRPUD	Master of City and Regional Planning and Urban Development		
MCS	Master of Clinical Science Master of Computer Science	**MFT**	Master of Family Therapy
		MGA	Master of Government Administration
MC Sc	Master of Computer Science	**M Geo E**	Master of Geological Engineering
MCSD	Master of Communication Sciences and Disorders	**MGPGP**	Master of Group Process and Group Psychotherapy
MCSM	Master of Construction Science/Management	**MGS**	Master of General Studies Master of Gerontological Studies
MC Sp	Master of Christian Spirituality		
MD	Doctor of Medicine	**MH**	Master of Health Master of Humanities
MDA	Master of Development Administration		
MDE	Master of Developmental Economics	**MHA**	Master of Health Administration Master of Hospital Administration
M Dent Sc	Master of Dental Sciences		
M Des	Master of Design	**MHAMS**	Master of Historical Administration and Museum Studies
M Des S	Master of Design Studies		

ABBREVIATIONS USED IN THE GUIDES

MHD	Master of Human Development	MJS	Master of Juridical Science
MHDL	Master of Human Development and Learning	MLA	Master of Landscape Architecture Master of Liberal Arts
MHE	Master of Health Education Master of Higher Education Master of Home Economics	M Land Arch	Master of Landscape Architecture
		ML Arch	Master of Landscape Architecture
MH Ec	Master of Home Economics	ML Arch UD	Master of Landscape Architecture and Urban Development
MH Ed	Master of Health Education		
MHE Ed	Master of Home Economics Education	MLAS	Master of Laboratory Animal Science
MHK	Master of Human Kinetics	MLAUD	Master of Landscape Architecture in Urban Development
MHL	Master of Hebrew Literature		
MHM	Master of Health Management	MLHR	Master of Labor and Human Resources
MHOS	Master of Human Organizational Science	M Lib	Master of Librarianship
MHP	Master of Heritage Preservation Master of Historic Preservation Master of Humanities in Philosophy	MLIR	Master of Labor and Industrial Relations
		MLIS	Master of Library and Information Science
		M Lit M	Master of Liturgical Music
MHPE	Master of Health Professions Education	M Litt	Master of Letters
MHR	Master of Human Resources	MLM	Master of Library Media
MHRIM	Master of Hotel, Restaurant, and Institutional Management	MLRHR	Master of Labor Relations and Human Resources
MHRM	Master of Human Resources Management	MLS	Master of Legal Studies Master of Liberal Studies Master of Library Science Master of Library Services Master of Life Sciences
MHROD	Master of Human Resources and Organization Development		
MHRTA	Master in Hotel, Restaurant, and Tourism Administration		
		MLSIT	Master of Library Science and Instructional Technology
MHS	Master of Health Sciences Master of Hispanic Studies Master of Human Services Master of Humane Studies		
		MLSP	Master of Law and Social Policy
		MLT	Master of Law in Taxation
MHSA	Master of Health Services Administration	MM	Master of Management Master of Ministry Master of Modern Studies Master of Music
MH Sc	Master of Health Sciences		
MHSE	Master of Health Science Education Therapy		
M Hum	Master of Humanities	MMA	Master of Manpower Administration Master of Marine Affairs Master of Media Arts Master of Musical Arts
M Hum Svcs	Master of Human Services		
MI	Master of Instruction Master of Insurance		
		MMAE	Master of Mechanical and Aerospace Engineering
MIA	Master of Intercultural Administration Master of International Administration		
		M Math	Master of Mathematics
MI Arch	Master of Interior Architecture	M Mat SE	Master of Material Science and Engineering
MIB	Master of International Business	MMB	Master of Medical Biochemistry
MIBA	Master of International Business Administration	MMC	Master of Mass Communications
		MME	Master of Manufacturing Engineering Master of Mechanical Engineering Master of Music Education
MIBS	Master of International Business Studies		
MID	Master of Industrial Design Master of Interior Design		
		MM Ed	Master of Music Education
MIE	Master of Industrial Engineering	M Med Sc	Master of Medical Science
MIHM	Master of International Health Management	M Met Mat E	Master of Metallurgical and Materials Engineering
MIJ	Master of International Journalism		
MILR	Master of Industrial and Labor Relations	M Met Mat S	Master of Metallurgy and Material Science
MILS	Master of Information and Library Science	MMFCC	Master of Marriage, Family, and Child Counseling
MIM	Master of Industrial Management Master of International Management		
		MMFT	Master of Marriage and Family Therapy
M In Ed	Master of Industrial Education	M Mgmt	Master of Management
MinI E	Mineral Engineer	M Mgt	Master of Management
MIP	Master of Intellectual Property	M Mgt S	Master of Management Science
MIPP	Master of International Public Policy	MMHE	Master of Material Engineering
MIR	Master of Industrial Relations	MMHS	Master of Management in Human Services
MIS	Master of Individualized Studies Master of Information Science Master of Information Systems Master of Interdisciplinary Studies Master of International Studies		
		M Min	Master of Ministries
		MMIS	Master of Management Information Systems
		M Miss	Master of Missiology
		MML	Master of Modern Languages
MIT	Master of Industrial Technology Master in Teaching	MMP	Master of Marine Policy Master of Music Performance
		MMR	Master of Marketing Research
MJ	Master of Journalism	MMS	Master of Management Science Master of Management Studies
MJEA	Master of Jewish Educational Administration		
MJ Ed	Master of Jewish Education		

ABBREVIATIONS USED IN THE GUIDES

MMS	Master of Marketing Science Master of Materials Science Master of Modern Studies	MPP	Master of Public Policy
		MPPA	Master of Public Policy Administration
		MPPM	Master of Public and Private Management
MM Sc	Master of Medical Science	MPPPM	Master of Plant Protection and Pest Management
MMSE	Master of Manufacturing Systems Engineering		
MM St	Master of Museum Studies	M Pr A	Master of Professional Accountancy
MMT	Master of Movement Therapy Master of Music Teaching	M Pr Gph	Master of Professional Geophysics
		M Pr Met	Master of Professional Meteorology
M Mtl E	Master of Metal Engineering	MPRTM	Master of Parks, Recreation, and Tourism Management
MM Ty	Master of Music Therapy		
M Mu	Master of Music	MPS	Master of Pastoral Studies Master of Political Science Master of Professional Studies Master of Public Service
M Mus	Master of Music		
M Mus Ed	Master of Music Education		
MN	Master of Nursing		
MNA	Master of Nonprofit Administration Master of Nursing Administration	M Ps	Master of Psychology
		M Psych	Master of Psychology
M Nat Sci	Master of Natural Science	MPT	Master of Pastoral Theology Master of Physical Therapy
MNE	Master of Nuclear Engineering		
MNO	Master of Nonprofit Organization	M Pub L	Master of Public Law
MNRM	Master of Natural Resource Management	MPVM	Master of Preventive Veterinary Medicine
MNS	Master of Natural Science Master of Nursing Sciences Master of Nutritional Science	MPW	Master of Public Works
		MRA	Master of Recreational Administration Master of Rehabilitation Administration Master of Resource Administration
MN Sc	Master of Nursing Science		
M Nsg	Master of Nursing	MRC	Master of Rehabilitation Counseling
M Nuc Sci	Master of Nuclear Science	MRCM	Master of Real Estate and Construction Management
M Nurs	Master of Nursing		
MOA	Maître d'Orthophonie et d'Audiologie	MRCP	Master of Regional and Community Planning
MOB	Master of Organizational Behavior	M Rc Pk	Master of Recreation and Parks
M Oc E	Master of Oceanographic Engineering	MRE	Master of Religious Education
MOD	Master of Organizational Development	M Rec	Master of Recreation
MOE	Master of Occupational Education Master of Ocean Engineering	MRED	Master of Real Estate Development
		M Rehab A	Master of Rehabilitation Administration
MOH	Master of Occupational Health	M Rel	Master of Religion
MOT	Master of Occupational Therapy	M Rel Ed	Master of Religious Education
MP	Master of Planning	MRM	Master of Resources Management
MPA	Master of Professional Accountancy Master of Public Administration Master of Public Affairs	MRP	Master of Regional Planning
		MRRA	Master of Recreation Resources Administration
MPA–URP	Master of Public Affairs and Urban and Regional Planning	MRTP	Master of Rural and Town Planning
		MS	Master of Science
MP Acc	Master of Professional Accountancy Master of Professional Accounting	MSA	Master of Science in Accounting Master of Science in Administration Master of Science in Anthropology
MP Acct	Master of Professional Accounting		
MP Aff	Master of Public Affairs	MSAA	Master of Science in Astronautics and Aeronautics
MPC	Master of Pastoral Counseling Master of Public Communication		
		MSAAE	Master of Science in Aeronautical and Astronautical Engineering
MPD	Master of Product Design		
MPE	Master of Physical Education	MS Acct	Master of Science in Accounting
MPER	Master of Personnel and Employee Relations	MS Adm	Master of Science in Administration
M Perf A	Master of Performing Arts	MSAE	Master of Science in Aerospace Engineering Master of Science in Architectural Engineering Master of Science in Art Education
M Pet E	Master of Petroleum Engineering		
MPFM	Master of (Applied) Public Financial Management		
MPH	Master of Public Health	MS Ae E	Master of Science in Aerospace Engineering
M Pharm	Master of Pharmacy	MS Ag	Master of Science in Agriculture
M Phil	Master of Philosophy	MS Ag E	Master of Science in Agricultural Engineering
M Phil F	Master of Philosophical Foundations		
MPHTM	Master of Public Health and Tropical Medicine	MSAHA	Master of Science in Allied Health Administration
		MSAI	Master of Science in Artificial Intelligence
MPIA	Master of Public and International Affairs	MSAM	Master of Science in Applied Mathematics
M Pl	Master of Planning	MSAOR	Master of Science in Applied Operations Research
MPM	Master of Personnel Management Master of Pest Management Master of Professional Management Master of Public Management		
		MSAP	Master of Science in Applied Physics
		MS Arch	Master of Science in Architecture

ABBREVIATIONS USED IN THE GUIDES

MS Arch St	Master of Architectural Studies
MSAS	Master of Science in Architectural Studies
MSAT	Master of Science in Advanced Technology
MSB	Master of Science in Business
MSBA	Master of Science in Business Administration
MSBAE	Master of Science in Biological and Agricultural Engineering
MSBE	Master of Science in Biomedical Engineering
MSBM	Master of Science in Business Management
MSBME	Master of Science in Biomedical Engineering
MS Bus Ed	Master of Science in Business Education
MSC	Master of Science in Commerce Master of Science in Counseling Master of Speech and Communication
M Sc	Master of Science
M Sc A	Master of Science (Applied)
M Sc CS	Master of Science in Computer Science
MSCDIS	Master of Science in Communication Disorders
MSCE	Master of Science in Civil Engineering Master of Science in Clinical Engineering Master of Science in Computer Engineering
M Sc E	Master of Science in Engineering
MSCEM	Master of Science in Civil Engineering Management
M Sc Eng	Master of Science in Engineering
MS Cer E	Master of Science in Ceramic Engineering
M Sc F	Master of Science in Forestry
M Sc FE	Master of Science in Forest Engineering
MS Ch E	Master of Science in Chemical Engineering
MSCIS	Master of Science in Computer and Information Science Master of Science in Computer Information Systems
MSCJ	Master of Science in Criminal Justice
MSCJA	Master of Science in Criminal Justice Administration
MSCLS	Master of Science in Clinical Laboratory Studies
MSCN	Master of Science in Clinical Nutrition
M Sc N	Master of Science in Nursing
MS Coun	Master of Science in Counseling
MSCP	Master of Science in Counseling Psychology
MS Cp E	Master of Science in Computer Engineering
M Sc Pharm	Master of Science in Pharmacy
M Sc Pl	Master of Science in Planning
MSCRP	Master of Science in Community and Regional Planning
MSCS	Master of Science in Computer Science
MSCSE	Master of Science in Computer Science and Engineering
M Sc T	Master of Science in Teaching
MSD	Master of Science in Dentistry Master of Science in Design Master of Science in Dietetics
MSDD	Master of Science in Design and Development
MSE	Master of Science Education Master of Science in Education Master of Science in Engineering Master of Software Engineering Master of Special Education
MSEC	Master of Science in Economic Aspects of Chemistry
MS Econ	Master of Science in Economics
MS Ed	Master of Science in Education
MSEE	Master of Science in Electrical Engineering
	Master of Science in Environmental Engineering
MSEH	Master of Science in Environmental Health
MSEM	Master of Science in Engineering Management Master of Science in Engineering of Mines
MSE Mech	Master of Science in Engineering Mechanics
MSE Mgt	Master of Science in Engineering Management
MS En E	Master of Science in Environmental Engineering
MS Eng	Master of Science in Engineering
MS Engr	Master of Science in Engineering
MS Env E	Master of Science in Environmental Engineering
MSER	Master of Science in Energy Resources
MSES	Master of Science in Engineering Science Master of Science in Environmental Studies
MSESM	Master of Science in Environmental Systems Management
MSESS	Master of Science in Exercise and Sport Studies
MSET	Master of Special Education Technology
MSF	Master of Science in Finance Master of Science in Forestry
MSFM	Master of Financial Management
MSFS	Master of Science in Family Studies Master of Science in Financial Services Master of Science in Foreign Service Master of Science in Forensic Science
MSG	Master of Science in Gerontology
MS Geo E	Master of Science in Geological Engineering
MSHA	Master of Science in Health Administration
MSHE	Master of Science in Home Economics
MSH Ed	Master of Science in Health Education
MSHP	Master of Science in Health Professions
MSHR	Master of Science in Human Resources
MSHRM	Master of Science in Human Resource Management
MSHRMD	Master of Science in Human Resources Management/Development
MSHS	Master of Science in Health and Safety Master of Science in Health Science Master of Science in Health Systems
MSHSA	Master of Science in Human Service Administration
MSHSE	Master of Science in Health Science Education
MS Hyg	Master of Science in Hygiene
MSI	Master of Science in Insurance
MSIA	Master of Science in Industrial Administration Master of Science in International Administration Master of Science in International Affairs
MSIB	Master of Science in International Business
MSIE	Master of Science in Industrial Engineering
MSILR	Master of Science in Industrial and Labor Relations
MSIM	Master of Science in Information Management
MSIR	Master of Science in Industrial Relations
MSIS	Master of Science in Information Science Master of Science in Information Systems Master of Science in Interdisciplinary Studies
MSIST	Master of Science in Information Systems Technology
MSJ	Master of Science in Journalism

Peterson's Guide to Graduate Programs in the Physical Sciences and Mathematics 1992

ABBREVIATIONS USED IN THE GUIDES

MSJBS	Master of Science in Japanese Business Studies	MS Nuc E	Master of Science in Nuclear Engineering
MSJPS	Master of Science in Justice and Public Safety	MSO	Master of Science in Orthodontics
		MSOB	Master of Science in Organizational Behavior
MSJS	Master of Science in Jewish Studies	M Soc	Master of Sociology
MSL	Master of Studies in Law	M Soc Sc	Master of Social Science
	Master of Science in Librarianship	M Soc Wk	Master of Social Work
MSLA	Master of Science in Legal Administration	MSOD	Master of Science in Organizational Development
MSLP	Master of Speech-Language Pathology		
MSLS	Master of Science in Library Science	MSOE	Master of Science in Ocean Engineering
	Master of Science in Logistics Systems	MSOM	Master of Science in Organization and Management
MSM	Master of Sacred Ministry		
	Master of Sacred Music	MS Op R	Master of Science in Operations Research
	Master of Science in Management	MSOR	Master of Science in Operations Research
	Master of Service Management	MSOT	Master of Science in Occupational Therapy
MS Mat E	Master of Science in Materials Engineering	MSP	Master of School Psychology
MS Mat SE	Master of Science in Material Science and Engineering		Master of Science in Pharmacy
			Master of Science in Planning
MSMC	Master of Science in Marketing Communications		Master of Social Psychology
			Master of Speech Pathology
	Master of Science in Mass Communications	MSPA	Master of Science in Professional Accountancy
MSMCS	Master of Science in Management and Computer Science		Master of Science in Public Administration
			Master of Speech Pathology and Audiology
MSME	Master of Science in Mechanical Engineering	MSPD	Master of Science in Pediatric Dentistry
MSMER	Master of Science in Management/Employee Relations	MSPE	Master of Science in Petroleum Engineering
			Master of Science in Physical Education
MS Met E	Master of Science in Metallurgical Engineering	M Sp Ed	Master of Special Education
MS Met Mat S	Master of Science in Metallurgy and Material Science	MS Pet E	Master of Science in Petroleum Engineering
		MSP Ex	Master of Science in Exercise Physiology
MS Metr	Master of Science in Meteorology	MSPH	Master of Science in Public Health
MS Met S	Master of Science in Metallurgical Sciences	MS Pharm	Master of Science in Pharmacy
MS Mfg E	Master of Science in Manufacturing Engineering	MSPHR	Master of Science in Pharmacy
		MS Phys Ed	Master of Science in Physical Education
MSMFSE	Master of Science in Manufacturing Systems Engineering	MS Phys Op	Master of Science in Physiological Optics
		MSPM	Master of Science in Project Management
MSMFT	Master of Science in Marriage and Family Therapy	MS Poly	Master of Science in Polymers
		MSPS	Master of Science in Planning Studies
MS Mgt	Master of Science in Management		Master of Science in Psychological Services
MSMHCA	Master of Science in Management/Health Care Administration	MSPT	Master of Science in Physical Therapy
		MS(R)	Master of Science (Research)
MSMI	Master of Science in Medical Illustration	MSRA	Master of Science in Recreation Administration
MS Min E	Master of Science in Mining Engineering		
MS Minrl RE	Master of Science in Mineral Engineering	MS Rad Sci	Master of Science in Radiation Science
MSMIS	Master of Science in Management Information Systems		Master of Science in Radiological Sciences
		MSRC	Master of Science in Resource Conservation
MSMM	Master of Science in Manufacturing Management	MSRE	Master of Science in Religious Education
		MSRMP	Master of Science in Radiological Medical Physics
MSMOB	Master of Science in Management and Organizational Behavior		
		MS(R)PT	Master of Science (Research) in Physical Therapy
MSMPA	Master of Science in Management/Public Administration		
		MSRS	Master of Science in Recreational Studies
MSMS	Master of Science in Management Science	MSS	Master of Science in Safety
MSMSA	Master of Science in Management Systems Analysis		Master of Selected Studies
			Master of Social Science
MSMSE	Master of Science in Manufacturing Systems Engineering		Master of Social Services
			Master of Sports Science
	Master of Science in Material Science Engineering	MSSA	Master of Science in Social Administration
		MSSE	Master of Science in Systems Engineering
MSMT	Master of Science in Medical Technology	MSSH	Master of Science in Speech and Hearing
MS Mt E	Master of Science in Materials Engineering	MSSHS	Master of Science in Speech and Hearing Sciences
MSN	Master of Science in Nursing		
MSNA	Master of Science in Nurse Anesthesia	MSSL	Master of Science in Speech and Language
MSNE	Master of Science in Nuclear Engineering	MSSM	Master of Science in Science Management
MSN(R)	Master of Science in Nursing (Research)		Master of Science in Systems Management
MSNS	Master of Science in Natural Science	MSSPA	Master of Science in Speech Pathology and Audiology
MS Nsg	Master of Science in Nursing		

684

Peterson's Guide to Graduate Programs in the Physical Sciences and Mathematics 1992

ABBREVIATIONS USED IN THE GUIDES

MS Sp Ed	Master of Science in Special Education	M Vet Sc	Master of Veterinary Science
MS Stat	Master of Science in Statistics	MVS	Master of Valuation Sciences
MSSW	Master of Science in Social Work	MVTE	Master of Vocational-Technical Education
MS Sy Sc	Master of Science in Systems Science	MVT Ed	Master of Vocational and Technical Education
MS Sys E	Master of Science in Systems Engineering		
MST	Master of Science in Taxation	MWPS	Master of Wood and Paper Science
	Master of Science in Teaching	MZS	Master of Zoological Science
	Master of Science Teaching	NA	Naval Architecture
	Master of Science Technology	ND	Doctor of Naturopathic Medicine
	Master of Secondary Teaching		Doctor of Nursing
	Master of Speech Therapy	NE	Naval Engineer
M Stat	Master of Statistics		Nuclear Engineer
MSTE	Master of Science in Transportation Engineering	NS	Nursing Specialist
		Ocean E	Ocean Engineer
MST Ed	Master of Science in Technical Education	OD	Doctor of Optometry
MS Text	Master of Science in Textiles	PD	Doctor of Pharmacy
MS Text Chem	Master of Science in Textile Chemistry		Professional Diploma
MSTM	Master of Science in Teaching Mathematics	PED	Doctor of Physical Education
	Master of Science in Technology Management	PE Dir	Director of Physical Education
	Master of Science in Tropical Medicine	PGC	Post-Graduate Certificate
MSTrPl	Master of Science in Transportation Planning	Pharm D	Doctor of Pharmacy
		PhD	Doctor of Philosophy
MSUD	Master of Science in Urban Design	Phil M	Master of Philosophy
MSUESM	Master of Science in Urban Environmental Systems Management	Ph L	Licentiate of Philosophy
		Ph M	Master of Philosophy
MSUS	Master of Science in Urban Studies	PMC	Post Master's Certificate
MSVC	Master of Science in Vocational Counseling	Postgraduate D	Postgraduate Diploma
MSW	Master of Social Work	Psy D	Doctor of Psychology
MSWREE	Master of Science in Water Resources and Environmental Engineering	Psy S	Specialist in Psychology
		Re D	Doctor of Recreation
MT	Master of Taxation	Re Dir	Director of Recreation
	Master of Teaching	Rh D	Doctor of Rehabilitation
	Master of Textiles	SAS	School Administrator and Supervisor
MTA	Master of Tax Accounting	SCCT	Specialist in Community College Teaching
	Master of Teaching Art	Sc D	Doctor of Science
	Master of Theater Arts	Sc D Hyg	Doctor of Science in Hygiene
M Tax	Master of Taxation	Sc M	Master of Science
MTE	Master of Teacher Education	Sc S	Specialist in Science
M Tech	Master of Technology	SD	Doctor of Science
M Th	Master of Theology		Specialist Degree
MTM	Master in the Teaching of Mathematics	S Ed	Specialist in Education
	Master of Theology and Ministry	SJD	Doctor of Juridical Science
MTMH	Master of Tropical Medicine and Hygiene	SLS	Specialist in Library Science
M Tox	Master of Toxicology	SM	Master of Science
MTPW	Master of Technical and Professional Writing	SM Arch S	Master of Science in Architectural Science
		SMBT	Master of Science in Building Technology
MTS	Master of Teaching Science	SM Vis S	Master of Science in Visual Science
	Master of Theological Studies	SP	Specialist Degree
MTSC	Master of Technical and Scientific Communication	SPA	Specialist in Public Administration
		SPC	School Psychology Certificate
MTSW	Master of Teaching Social Work	Sp C	Specialist in Counseling
MUA	Master of Urban Affairs	Sp CG	Specialist Certificate in Gerontology
	Master of Urban Architecture	Sp Ed	Specialist in Education
MUP	Master of Urban Planning	Sp Ed S	Special Education Specialist
MUPDD	Master of Urban Planning, Design, and Development	Sp M	Specialist in Microbiology
		SPS	School Psychology Specialist
MUPP	Master of Urban Planning and Policy		Special Education Specialist
MURP	Master of Urban and Regional Planning	S Psy S	Specialist in Psychological Services
	Master of Urban and Rural Planning	SSP	Specialist in School Psychology
MUS	Master of Urban Studies	STB	Bachelor of Sacred Theology
Mus AD	Doctor of Musical Arts	STD	Doctor of Sacred Theology
Mus D	Doctor of Music	STL	Licentiate of Sacred Theology
Mus M	Master of Music	STM	Master of Sacred Theology
MU Sys E	Master of Urban Systems Engineering		
MVA	Master of Visual Arts		
MVE	Master of Vocational Education		

Peterson's Guide to Graduate Programs in the Physical Sciences and Mathematics 1992

ABBREVIATIONS USED IN THE GUIDES

Th D	Doctor of Theology
Th M	Master of Theology
V Ed S	Vocational Education Specialist
VMD	Doctor of Veterinary Medicine
XMBA	Executive Master of Business Administration

TESTS

DAT	Dental Admission Testing test
GMAT	Graduate Management Admission Test
GRE	Graduate Record Examinations
LSAT	Law School Admission Test
MAT	Miller Analogies Test
MCAT	Medical College Admission Test
NTE	NTE Programs tests
OAT	Optometry Admission Test
PAEG	Prueba de Admisiones para Estudios Graduados
PCAT	Pharmacy College Admission Test
TOEFL	Test of English as a Foreign Language
TSE	Test of Spoken English
VCAT	Veterinary College Admission Test

ORGANIZATIONS AND AGENCIES

AACSB	American Assembly of Collegiate Schools of Business
AALS	Association of American Law Schools
AAMC	Association of American Medical Colleges
AAMFT	American Association for Marriage and Family Therapy
AANA	American Association of Nurse Anesthetists
AARTS	Association of Advanced Rabbinical and Talmudic Schools
ABA	American Bar Association
ABET	Accreditation Board for Engineering and Technology
ACEHSA	Accrediting Commission on Education for Health Services Administration
ACEJMC	Accrediting Council on Education in Journalism and Mass Communications
ACNM	American College of Nurse-Midwives
ACPE	American Council on Pharmaceutical Education
ADA	American Dental Association
ALA	American Library Association
AMA	American Medical Association
AOA	American Optometric Association
	American Osteopathic Association
APA	American Psychological Association
APMA	American Podiatric Medical Association
APTA	American Physical Therapy Association
ASHA	American Speech-Language-Hearing Association
ASLA	American Society of Landscape Architects
ATS	Association of Theological Schools in the United States and Canada
AVMA	American Veterinary Medical Association
CACMS	Committee for the Accreditation of Canadian Medical Schools
CAHEA	Committee on Allied Health Education and Accreditation
CANAEP	Council on Accreditation of Nurse Anesthesia Educational Programs
CCE	Council on Chiropractic Education
CDA	Canadian Dental Association
CEPH	Council on Education for Public Health
CGS	Council of Graduate Schools
COPA	Council on Postsecondary Accreditation
CORE	Council on Rehabilitation Education
CSWE	Council on Social Work Education
FIDER	Foundation for Interior Design Education Research
LCME	Liaison Committee on Medical Education
NAAB	National Architectural Accrediting Board
NASAD	National Association of Schools of Art and Design
NASM	National Association of Schools of Music
NCATE	National Council for Accreditation of Teacher Education
NLN	National League for Nursing
SAF	Society of American Foresters

Indexes

There are three indexes in this section. The first, Index of Full Descriptions and Announcements, gives page references for all programs that have chosen to place full two-page descriptions and announcements in this volume. It is arranged alphabetically by institution; within institutions, the arrangement is alphabetical by subject area. This is not an index to all programs in the book's directories of profiles; readers must refer to the directories themselves for profile information on programs that have not submitted the additional, more individualized statements. The second index, Index of Directories and Subject Areas in Books 2–6, gives book references for the directories in Books 2–6, for example, "Vision Sciences—Book 3," and also includes cross-references for subject area names not used in the directory structure, for example, "Physiological Optics (*see* Physiology; Vision Sciences)." The third index, Index of Directories and Subject Areas in This Book, gives page references for the directories in this volume and cross-references for subject area names not used in this volume's directory structure, for example, "Limnology (*see* Marine Sciences/Oceanography)."

Index of Full Descriptions and Announcements

American University
   Mathematics and Statistics 387
Arizona State University
   Chemistry and Biochemistry 85
   Mathematics 389
Auburn University
   Geology (Announcement) 194
   Physics 569
Ball State University
   Geology (Announcement) 194
Baylor University
   Chemistry 87
Boston College
   Chemistry 89
Boston University
   Chemistry 91
   Chemistry (Announcement) 45
   Physics 571
Bowling Green State University
   Geology 217
Brandeis University
   Chemistry 93
   Mathematics (Announcement) 351
   Physics 573
Brigham Young University
   Chemistry and Biochemistry 95
Brooklyn College of the City University of New York
   Physics (Announcement) 546
Brown University
   Applied Mathematics 391
   Geological Sciences (Announcement) 215
   Mathematics 393
   Mathematics (Announcement) 351
   Physics 575
California Institute of Technology
   Applied Mathematics 395
   Geological and Planetary Sciences 219
   Physics 577
California State University, Fresno
   Geology (Announcement) 195
   Mathematics (Announcement) 351
   Physics (Announcement) 546
California State University, Fullerton
   Chemistry and Biochemistry (Announcement) 46
   Mathematics (Announcement) 352
California State University, Los Angeles
   Chemistry and Biochemistry (Announcement) 47
   Mathematics and Computer Science (Announcement) 352
   Physics and Astronomy 579
Carleton University
   Geoscience (Announcement) 187
Carnegie Mellon University
   Chemistry 97
   Mathematics 397
   Statistics 399
Case Western Reserve University
   Chemistry 99
   Epidemiology and Biostatistics (Announcement) 347
   Operations Research (Announcement) 386
Catholic University of America
   Chemistry (Announcement) 47
City College of the City University of New York
   Physics (Announcement) 547
City University of New York (see individual schools)
Claremont Graduate School
   Mathematics (Announcement) 353
Clarkson University
   Physics (Announcement) 547
Clark University
   Physics (Announcement) 547

Clemson University
   Mathematical Sciences (Announcement) 353
   Physics and Astronomy 581
   Statistics (Announcement) 376
College of William and Mary
   Marine Science 307
   Mathematics (Announcement) 353
Colorado School of Mines
   Physics (Announcement) 547
Colorado State University
   Atmospheric Science 525
   Mathematics 401
Columbia University
   Atmospheric and Planetary Science 527
   Biochemistry and Molecular Biophysics (Announcements) 83, 567
   Chemistry 101
   Mathematics 403
Cornell University
   Applied Mathematics (Announcement) 338
   Astronomy and Space Sciences (Announcement) 31
   Biometry (Announcement) 345
   Geological Sciences (Announcement) 196
   Mathematics 405
Courant Institute of Mathematical Sciences (see New York University)
Dalhousie University
   Oceanography 309
Dartmouth College
   Earth Sciences (Announcement) 196
   Mathematics and Computer Science 407
DePaul University
   Chemistry (Announcement) 49
Drexel University
   Biomedical Engineering and Science (Announcement) 347
   Chemistry, Mathematics and Computer Science, and Physics and Atmospheric Sciences 409
Duke University
   Geology (Announcement) 196
   Pharmacology (Announcement) 83
East Carolina University
   Physics (Announcement) 548
Eastern Michigan University
   Chemistry 103
East Texas State University
   Physics (Announcement) 548
Emory University
   Biostatistics 411
   Mathematics and Computer Science 413
   Mathematics and Computer Science (Announcement) 354
   Physics 583
Florida Institute of Technology
   Biological Sciences (Announcement) 305
   Oceanography and Ocean Engineering (Announcement) 302
Florida International University
   Mathematical Sciences (Announcement) 355
   Physics (Announcement) 549
Florida State University
   Chemistry 105
   Oceanography 311
   Physics 585
   Physics (Announcement) 549
Georgetown University
   Radiation Science (Announcement) 567
George Washington University
   Statistics/Computer and Information Systems 415
Georgia Institute of Technology
   Mathematics (Announcement) 355
   Physics (Announcement) 549

INDEX OF FULL DESCRIPTIONS AND ANNOUNCEMENTS

Georgia State University
   Geology (Announcement) — 197
Harvard University
   Biophysics (Announcements) — 83, 567
   Earth and Planetary Sciences — 221
   Physics — 587
   Statistics (Announcement) — 377
Howard University
   Mathematics — 417
Indiana University Bloomington
   Geological Sciences — 223
   Physics — 589
   Physics (Announcement) — 550
Indiana University-Purdue University at Indianapolis
   Chemistry — 107
Institute of Paper Science and Technology
   Chemistry — 109
   Chemistry (Announcement) — 51
Iowa State University
   Geological and Atmospheric Sciences — 225
Johns Hopkins University
   Mathematical Sciences — 419
   Mathematics — 421
   Molecular Biophysics (Announcements) — 83, 567
   Physics and Astronomy — 591
Kent State University
   Chemistry — 111
   Mathematics (Announcement) — 357
   Physics — 593
Lehigh University
   Chemistry — 113
   Mathematics — 423
Louisiana State University
   Chemical Engineering (Announcement) — 83
   Electrical and Computer Engineering (Announcement) — 567
   Geoscience — 227
   Oceanography and Coastal Sciences — 313
   Physics and Astronomy — 595
Louisiana State University Medical Center
   Biometry and Genetics (Announcement) — 345
Loyola University Chicago
   Mathematical Sciences (Announcement) — 357
Massachusetts Institute of Technology
   Chemistry — 115
   Oceanography/Oceanographic Engineering — 315
   Toxicology (Announcement) — 83
McGill University
   Geology and Geophysics — 229
   Meteorology (Announcements) — 305, 520
McMaster University
   Geography (Announcement) — 188
Medical University of South Carolina
   Biostatistics, Epidemiology, and Systems Science — 425
Miami University
   Geology (Announcement) — 198
   Mathematics and Statistics (Announcement) — 358
Michigan State University
   Geological Sciences — 231
   Geological Sciences (Announcement) — 198
   Mathematics — 427
   Physics and Astronomy — 597
Michigan Technological University
   Mathematical Sciences — 429
Montana College of Mineral Science and Technology
   Chemistry and Geochemistry (Announcement) — 191
Mount Sinai School of Medicine of the City University of New York
   Biomathematical Sciences — 431
New Jersey Institute of Technology
   Applied Physics (Announcement) — 542
   Mathematics (Announcement) — 340
New Mexico Institute of Mining and Technology
   Geochemistry (Announcement) — 192
   Geology (Announcement) — 199
   Geophysics (Announcement) — 211

New Mexico State University
   Astronomy (Announcement) — 32
   Physics — 599
New York University
   Chemistry — 117
   Mathematics — 433
North Carolina Central University
   Chemistry (Announcement) — 54
North Carolina State University
   Biomathematics (Announcement) — 346
   Marine, Earth, and Atmospheric Sciences — 317
   Statistics (Announcement) — 378
Northeastern University
   Chemistry — 119
   Physics — 601
Northern Illinois University
   Geology — 233
   Physics (Announcement) — 552
Northwestern University
   Applied Mathematics — 435
   Chemical Engineering (Announcements) — 83, 567
   Mathematics — 437
   Physics and Astronomy — 603
Nova University
   Marine Biology and Coastal Zone Management (Announcement) — 303
Ohio State University
   Chemistry — 121
   Geodetic Science and Surveying (Announcement) — 386
   Mathematics — 439
   Medicinal Chemistry and Pharmacognosy (Announcement) — 83
Ohio University
   Physics — 605
Old Dominion University
   Geological Sciences (Announcement) — 200
   Mathematics and Statistics — 441
   Oceanography (Announcement) — 303
   Physics (Announcement) — 553
Oregon State University
   Chemistry — 123
   Geosciences — 235
   Mathematics — 443
   Physics — 607
   Physics (Announcement) — 553
   Statistics — 445
Pennsylvania State University
   Geosciences — 237
   Geosciences (Announcements) — 192, 200, 211
   Meteorology — 529
   Physics — 609
   Physics (Announcement) — 553
   Statistics (Announcement) — 379
Portland State University
   Environmental Sciences and Resources/Chemistry (Announcement) — 56
   Environmental Sciences and Resources/Geology (Announcement) — 200
   Environmental Sciences and Resources/Physics (Announcement) — 553
Princeton University
   Chemistry — 125
   Statistics and Operations Research (Announcement) — 379
Purdue University
   Biochemistry (Announcement) — 83
   Mathematics — 447
   Statistics — 449
   Statistics (Announcement) — 379
Rensselaer Polytechnic Institute
   Chemistry — 127
   Decision Sciences and Engineering Systems — 451
   Earth and Environmental Sciences — 239
   Mathematical Sciences — 453
   Mathematical Sciences (Announcement) — 341
   Physics (Announcement) — 554
Rice University
   Physics — 611

INDEX OF FULL DESCRIPTIONS AND ANNOUNCEMENTS

Rush University
   Biochemistry (Announcement)   83
Rutgers, The State University of New Jersey, Newark
   Chemistry   129
   Chemistry (Announcement)   57
Rutgers, The State University of New Jersey, New Brunswick
   Chemistry   131
   Mathematics (Announcement)   361
   Physics and Astronomy   613
San Francisco State University
   Physics and Astronomy (Announcement)   555
Seton Hall University
   Chemistry (Announcement)   57
Simon Fraser University
   Mathematics and Statistics   455
South Dakota School of Mines and Technology
   Meteorology (Announcement)   521
South Dakota State University
   Mathematics (Announcement)   362
Southern Illinois University at Carbondale
   Geology (Announcement)   201
   Molecular Science (Announcement)   386
Southern Methodist University
   Geological Sciences   241
Stanford University
   Chemistry   133
   Geology   243
State University of New York at Albany
   Atmospheric Science   531
State University of New York at Buffalo
   Physics   615
State University of New York at Stony Brook
   Chemistry   135
   Marine Environmental Sciences and Coastal Oceanography   319
   Marine Sciences (Announcement)   303
   Mathematics   457
   Physics and Theoretical Physics   617
   Terrestrial and Planetary Atmospheres   533
State University of New York College at Brockport
   Mathematics (Announcement)   363
State University of New York College of Environmental Science and Forestry
   Chemistry   137
Stevens Institute of Technology
   Chemistry and Chemical Biology   139
   Physics and Engineering Physics   619
Sul Ross State University
   Geology (Announcement)   202
Syracuse University
   Geology (Announcement)   202
   Physics   621
Temple University
   Mathematics (Announcement)   364
   Physics   623
   Statistics and Biostatistics   459
Texas A&M University
   Chemistry   141
   Geology (Announcement)   202
   Geophysics   245
   Geophysics (Announcement)   212
   Physics   625
Tufts University
   Physics and Astronomy   627
Tulane University
   Geology   247
University of Akron
   Physics   629
University of Alabama
   Physics and Astronomy (Announcement)   557
University of Alabama at Birmingham
   Biomathematics (Announcement)   346
   Biostatistics (Announcement)   348
   Physics (Announcement)   557
University of Alaska Fairbanks
   Physics (Announcement)   557
University of Arizona
   Chemistry   143
   Ecology and Evolutionary Biology (Announcement)   386
   Geosciences   249
   Mathematics and Applied Mathematics   461
   Optics   631
   Physics   633
   Planetary Sciences   251
University of British Columbia
   Geological Sciences (Announcement)   203
University of Calgary
   Geology and Geophysics   253
University of California at Berkeley
   Business Administration (Announcement)   386
   Statistics (Announcement)   381
University of California, Davis
   Atmospheric Science (Announcement)   522
   Statistics (Announcement)   381
University of California, Los Angeles
   Biomathematics   463
   Biostatistics   465
   Biostatistics (Announcement)   386
   Earth and Space Sciences   255
   Physics   635
University of California, Riverside
   Geological Sciences (Announcement)   203
   Mathematics   467
   Physics   637
University of California, San Diego
   Oceanography   321
University of California, San Francisco
   Pharmaceutical Chemistry (Announcement)   83
University of California, Santa Barbara
   Geological Sciences   257
   Statistics and Applied Probability (Announcement)   381
University of California, Santa Cruz
   Mathematics (Announcement)   366
University of Chicago
   Geophysical Sciences   259
   Geophysical Sciences (Announcement)   523
University of Cincinnati
   Geology   261
   Mathematics (Announcement)   366
   Physics   639
University of Colorado at Boulder
   Physics (Announcement)   558
University of Connecticut
   Marine Sciences   323
   Physics   641
University of Delaware
   Geology (Announcement)   204
   Mathematical Sciences   469
University of Denver
   Mathematics and Computer Science (Announcement)   367
University of Florida
   Chemistry   145
   Geology (Announcement)   204
   Mathematics   471
   Mathematics (Announcement)   367
   Physics   643
University of Georgia
   Chemistry   147
   Geology   263
   Mathematics   473
   Physics and Astronomy   645
   Statistics   475
University of Hawaii at Manoa
   Geology and Geophysics   265
   Geology and Geophysics (Announcement)   204
   Oceanography   325
University of Houston
   Chemistry   149
   Geosciences   267
   Mathematics   477
   Physics   647

INDEX OF FULL DESCRIPTIONS AND ANNOUNCEMENTS

University of Idaho	
Geology and Geological Engineering (Announcement)	204
University of Illinois at Chicago	
Chemistry	151
Geological Sciences	269
Mathematics, Statistics, and Computer Science	479
Mathematics, Statistics, and Computer Science (Announcements)	343, 367, 382
University of Illinois at Urbana-Champaign	
Food Science (Announcement)	83
Statistics	481
University of Iowa	
Mathematics	483
University of Kansas	
Medicinal Chemistry (Announcements)	77, 83
Physics and Astronomy	649
University of Kentucky	
Biochemistry (Announcement)	84
Radiation Sciences (Announcement)	567
Statistics	485
University of Louisville	
Chemistry (Announcement)	64
University of Maine	
Chemistry (Announcement)	64
University of Maryland College Park	
Chemical Physics	153
Chemical Physics (Announcement)	567
Chemistry and Biochemistry	155
Geology	271
Meteorology	535
Physics and Astronomy (Announcement)	560
University of Maryland Graduate School, Baltimore	
Applied Mathematics and Statistics	487
Chemistry and Biochemistry	157
Physics (Announcement)	560
University of Massachusetts at Amherst	
Molecular and Cellular Biology (Announcement)	84
University of Massachusetts, Lowell	
Physics and Applied Physics	651
University of Miami	
Chemistry	159
Physics (Announcement)	560
University of Michigan	
Chemistry	161
Materials Science and Engineering (Announcements)	84, 567
Physics	653
University of Minnesota, Duluth	
Geology (Announcement)	205
Mathematics and Statistics (Announcement)	343
Physics (Announcement)	560
University of Minnesota, Twin Cities Campus	
Astronomy (Announcement)	34
Biostatistics (Announcement)	349
Chemistry	163
Geology and Geophysics	273
University of Missouri-Columbia	
Physics (Announcement)	560
University of Missouri-Rolla	
Geology and Geophysics (Announcement)	205
University of Missouri-St. Louis	
Chemistry (Announcement)	65
University of Montana	
Physics and Astronomy (Announcement)	561
University of Nebraska-Lincoln	
Chemistry	165
Physics and Astronomy (Announcement)	561
University of Nevada, Las Vegas	
Physics (Announcement)	561
University of Nevada, Reno	
Geological Sciences	275
University of New Mexico	
Geology	277
Physics and Astronomy (Announcements)	34, 561
University of North Carolina at Chapel Hill	
Biostatistics (Announcement)	349
Mathematics (Announcement)	369
Physics and Astronomy (Announcement)	562
University of North Florida	
Mathematics and Statistics (Announcement)	369
University of North Texas	
Physics (Announcement)	562
University of Notre Dame	
Chemical Engineering (Announcement)	84
Mathematics	489
Physics	655
University of Oklahoma	
Chemistry and Biochemistry (Announcement)	66
Geology and Geophysics	279
University of Oregon	
Geological Sciences (Announcement)	206
University of Pennsylvania	
Geology (Announcement)	207
Physics	657
University of Pittsburgh	
Chemistry	167
Geology and Planetary Science	281
Physics and Astronomy	659
University of Puerto Rico, Mayagüez Campus	
Marine Sciences	327
University of Puerto Rico, Río Piedras	
Chemistry	169
Physics	661
University of Rhode Island	
Computer Science and Statistics (Announcement)	384
Oceanography	329
University of Rochester	
Electrical Engineering (Announcement)	567
Geological Sciences (Announcement)	207
Optics	663
Physics and Astronomy	665
Statistics	491
University of San Francisco	
Chemistry	171
University of Saskatchewan	
Geology, Geophysics, and Geological Engineering	283
University of South Carolina	
Physics and Astronomy	667
University of Southern California	
Biometry (Announcement)	346
Geological Sciences (Announcement)	207
Mathematics	493
University of South Florida	
Chemistry	173
Marine Science	331
Mathematics	495
University of Southwestern Louisiana	
Mathematics and Statistics	497
University of Tennessee, Knoxville	
Physics and Astronomy	669
University of Texas at Arlington	
Physics (Announcement)	563
University of Texas at Austin	
Chemistry and Biochemistry	175
Geological Sciences	285
Marine Science	333
Physics	671
University of Texas at Dallas	
Geosciences	287
Mathematical Sciences (Announcement)	371
University of Texas at San Antonio	
Earth and Physical Sciences (Announcement)	208
University of Toledo	
Geology	289
University of Toronto	
Earth Sciences	291
University of Utah	
Mathematics	499
University of Virginia	
Astronomy (Announcement)	34
Biochemistry (Announcement)	84
Biophysics (Announcements)	84, 568
Chemistry (Announcement)	69

INDEX OF FULL DESCRIPTIONS AND ANNOUNCEMENTS

University of Virginia (continued)
Environmental Sciences (Announcement)	216
Mathematics (Announcement)	372
Physics	673

University of Washington
Applied Mathematics	501
Bioengineering (Announcements)	84, 568
Biostatistics	503
Biostatistics (Announcement)	386
Geophysics	293
Materials Science and Engineering (Announcements)	84, 568
Physics (Announcement)	564

University of Waterloo
Earth Sciences (Announcement)	190

University of Wisconsin-Madison
Biophysics (Announcements)	84, 568
Biotechnology (Announcement)	84
Pharmacy (Announcement)	84

University of Wyoming
Mathematics	505

Utah State University
Chemistry and Biochemistry	177
Mathematics and Statistics	507

Vanderbilt University
Mathematics	509

Virginia Commonwealth University
Biochemistry and Molecular Biophysics (Announcement)	84
Biomedical Engineering (Announcements)	84, 568
Human Genetics (Announcement)	386

Virginia Polytechnic Institute and State University
Chemistry (Announcement)	70
Mathematics	511
Statistics	513

Washington University
Earth and Planetary Sciences	295

Wesleyan University
Chemistry	179
Mathematics	515

Western Kentucky University
Coal Chemistry (Announcement)	70

West Virginia University
Geology and Geography (Announcement)	209

Wright State University
Geological Sciences	297
Physics (Announcement)	566

Yale University
Applied Physics (Announcement)	543
Chemistry	181
Geology and Geophysics	299
Geology and Geophysics (Announcement)	210
Mathematics	517

Index of Directories and Subject Areas in Books 2–6

Accounting—Book 6
Acoustics—Book 4
Actuarial Science (*see* Insurance and Actuarial Science)
Administration (*see* Arts Administration; Business Administration and Management; Educational Administration; Health Services Management and Hospital Administration; Public Policy and Administration; Sports Administration)
Adult Education—Book 6
Advertising and Public Relations—Book 6
Aeronautical Engineering (*see* Aerospace/Aeronautical Engineering)
Aerospace/Aeronautical Engineering—Book 5
Aerospace Studies (*see* Aerospace/Aeronautical Engineering)
African Languages and Literatures (*see* African Studies)
African Studies—Book 2
Afro-American Studies—Book 2
Agribusiness (*see* Agricultural Economics and Agribusiness)
Agricultural Economics and Agribusiness—Book 2
Agricultural Education—Book 6
Agricultural Engineering—Book 5
Agricultural Sciences—Book 3
Agronomy and Soil Sciences—Book 3
Alcohol Abuse Counseling (*see* Drug and Alcohol/Substance Abuse Counseling; Counselor Education)
Allied Health Professions (*see* Dental Hygiene; Medical Physics; Medical Technology; Occupational Therapy; Physical Therapy; Speech-Language Pathology and Audiology)
Allopathic Medicine—Book 6
American Indian Studies (*see* North American Studies)
American Studies (*see* North American Studies)
Analytical Chemistry—Book 4
Anatomy—Book 3
Animal Behavior (*see* Biopsychology; Neurobiology; Zoology)
Animal Sciences—Book 3
Anthropology—Book 2
Applied Arts and Design—Book 2
Applied History (*see* Public History)
Applied Mathematics—Book 4
Applied Mechanics (*see* Mechanics)
Applied Physics—Book 4
Applied Sciences (*see* Engineering and Applied Sciences)
Applied Statistics (*see* Statistics)
Arabic (*see* Near and Middle Eastern Languages)
Arab Studies (*see* Near and Middle Eastern Studies)
Archaeology—Book 2
Architectural Engineering—Book 5
Architecture—Book 2
Area and Cultural Studies (*see* African Studies; Afro-American Studies; Asian Studies; East European and Soviet Studies; Jewish Studies; Latin American Studies; Near and Middle Eastern Studies; North American Studies)
Art Education—Book 6
Art/Fine Arts—Book 2
Art History—Book 2
Arts Administration—Book 2
Art Therapy—Book 2
Asian Languages—Book 2
Asian Studies—Book 2
Astronautical Engineering (*see* Aerospace/Aeronautical Engineering)
Astronomy—Book 4
Astrophysical Sciences (*see* Astrophysics; Meteorology and Atmospheric Sciences; Planetary Sciences)
Astrophysics—Book 4
Athletics Administration (*see* Physical Education and Human Movement Studies)
Atmospheric Sciences (*see* Meteorology and Atmospheric Sciences)
Audiology (*see* Speech-Language Pathology and Audiology)
Bacteriology—Book 3

Banking (*see* Finance and Banking)
Behavioral Sciences (*see* Biopsychology; Neurobiology; Psychology; Zoology)
Bible Studies (*see* Religion; Theology)
Bilingual and Bicultural Education—Book 6
Biochemistry—Book 3
Bioengineering (*see* Biomedical Engineering)
Biological Chemistry (*see* Biochemistry)
Biological Engineering (*see* Biomedical Engineering)
Biological Oceanography (*see* Marine Biology; Marine Sciences/Oceanography)
Biology and Biomedical Sciences—Book 3
Biomathematics (*see* Biometrics)
Biomedical Engineering—Book 5
Biometrics—Book 4
Biophysics—Book 3
Biopsychology—Book 3
Biostatistics—Book 4
Biotechnology—Book 3
Black Studies (*see* Afro-American Studies)
Botany and Plant Sciences—Book 3
Breeding (*see* Animal Sciences; Botany and Plant Sciences; Genetics; Horticulture)
Broadcasting (*see* Communication; Radio, Television, and Film)
Business Administration and Management—Book 6
Business Education—Book 6
Canadian Studies (*see* North American Studies)
Cancer Biology (*see* Cell Biology; Immunology; Molecular Biology; Pathology)
Cell Biology—Book 3
Cellular Physiology (*see* Cell Biology; Physiology)
Celtic Languages—Book 2
Ceramic Engineering (*see* Ceramic Sciences and Engineering)
Ceramics (*see* Art/Fine Arts; Ceramic Sciences and Engineering)
Ceramic Sciences and Engineering—Book 5
Cereal Chemistry (*see* Food Science and Technology)
Chemical Engineering—Book 5
Chemistry—Book 4
Child and Family Studies—Book 2
Child-Care Nursing—Book 6
Child-Health Nursing (*see* Child-Care Nursing)
Chinese Studies (*see* Asian Languages; Asian Studies)
Chiropractic—Book 6
Christian Studies (*see* Missions and Missiology; Religion; Religious Education; Theology)
Cinema (*see* Art/Fine Arts; Radio, Television, and Film)
City and Regional Planning—Book 2
Civil Engineering—Book 5
Classical Languages and Literatures (*see* Classics)
Classics—Book 2
Clinical Psychology—Book 2
Clothing and Textiles—Book 2
Cognitive Psychology (*see* Psychology)
Communication—Book 2
Communication Disorders (*see* Speech-Language Pathology and Audiology)
Community Affairs (*see* City and Regional Planning; Urban Studies)
Community College Education—Book 6
Community Health (*see* Environmental and Occupational Health; Epidemiology; Public and Community Health)
Community Planning (*see* Architecture; City and Regional Planning; Environmental Design; Urban Design; Urban Studies)
Community Psychology (*see* Social Psychology)
Comparative Literature—Book 2
Composition (*see* Music)
Computer Education—Book 6
Computer Engineering—Book 5
Computer Science—Book 5

Peterson's Guide to Graduate Programs in the Physical Sciences and Mathematics 1992

Computing Technology (see Computer Science)
Conservation (see Environmental Biology; Environmental Sciences; Natural Resources)
Construction Engineering—Book 5
Consumer Economics—Book 2
Continuing Education (see Adult Education)
Corrections (see Criminal Justice/Criminology)
Counseling (see Counseling Psychology; Counselor Education; Pastoral Ministry and Counseling; Rehabilitation Counseling)
Counseling Psychology—Book 2
Counselor Education—Book 6
Crafts (see Art/Fine Arts)
Creative Arts Therapies (see Art Therapy; Dance, Drama, and Music Therapy)
Criminal Justice/Criminology—Book 2
Crop Sciences (see Agricultural Sciences; Agronomy and Soil Sciences; Botany and Plant Sciences)
Curriculum and Instruction—Book 6
Cytology (see Cell Biology)
Dairy Science (see Animal Sciences)
Dance—Book 2
Dance, Drama, and Music Therapy—Book 2
Demography and Population Studies—Book 2
Dental Hygiene—Book 6
Dentistry—Book 6
Design (see Applied Arts and Design; Architecture; Art/Fine Arts; Environmental Design; Graphic Design; Industrial Design; Interior Design; Textile Design; Urban Design)
Developmental Biology—Book 3
Developmental Psychology—Book 2
Dietetics (see Nutrition)
Diplomacy (see International Affairs)
Drama/Theater Arts—Book 2
Drama Therapy (see Dance, Drama, and Music Therapy)
Dramatic Arts (see Drama/Theater Arts)
Drawing (see Art/Fine Arts)
Drug Abuse Counseling (see Drug and Alcohol/Substance Abuse Counseling; Counselor Education)
Drug and Alcohol/Substance Abuse Counseling—Book 2
Early Childhood Education—Book 6
Earth Sciences—Book 4
East Asian Studies (see Asian Studies)
East European and Soviet Studies—Book 2
Ecology—Book 3
Economics—Book 2
Education—Book 6
Educational Administration—Book 6
Educational Measurement and Evaluation—Book 6
Educational Media/Instructional Technology—Book 6
Educational Psychology—Book 6
Educational Theater (see Dance, Drama, and Music Therapy; Drama/Theater Arts; Education)
Education of the Blind (see Special Education)
Education of the Deaf (see Special Education)
Education of the Gifted—Book 6
Education of the Hearing Impaired (see Special Education)
Education of the Learning Disabled (see Special Education)
Education of the Mentally Retarded (see Special Education)
Education of the Multiply Handicapped—Book 6
Education of the Physically Handicapped (see Special Education)
Education of the Visually Handicapped (see Special Education)
Electrical Engineering—Book 5
Electronics Engineering (see Electrical Engineering)
Elementary Education—Book 6
Embryology (see Developmental Biology)
Endocrinology (see Physiology)
Energy Engineering (see Electrical Engineering; Energy Management and Policy; Mechanical Engineering; Mineral/Mining Engineering; Nuclear Engineering; Petroleum Engineering; Power Engineering; Solar Engineering)
Energy Management and Policy—Book 5
Engineering and Applied Sciences—Book 5
Engineering and Public Affairs (see Technology and Public Policy)
Engineering and Public Policy (see Energy Management and Policy; Technology and Public Policy)
Engineering Design—Book 5
Engineering Management—Book 5
Engineering Metallurgy (see Metallurgy)
Engineering Physics—Book 5
English—Book 2
English Education—Book 6
Entomology—Book 3
Environmental and Occupational Health—Book 6
Environmental Biology—Book 3
Environmental Design—Book 2
Environmental Engineering—Book 5
Environmental Policy and Resource Management—Book 2
Environmental Sciences—Book 3
Epidemiology—Book 6
Ethnic Studies (see specific area and cultural studies fields)
Ethnomusicology (see Music)
Evolutionary Biology—Book 3
Exercise Physiology (see Physical Education and Human Movement Studies)
Experimental Psychology—Book 2
Experimental Statistics (see Statistics)
Family Studies (see Child and Family Studies)
Family Therapy (see Marriage and Family Therapy)
Filmmaking (see Art/Fine Arts; Radio, Television, and Film)
Finance and Banking—Book 6
Fine Arts (see Art/Fine Arts)
Fire Protection Engineering—Book 5
Fish, Game, and Wildlife Management—Book 3
Folklore—Book 2
Food Engineering—Book 5
Foods (see Food Science and Technology; Nutrition)
Food Science and Technology—Book 3
Food Services Management (see Hospitality Administration)
Foreign Languages (see appropriate languages)
Foreign Languages Education—Book 6
Foreign Service (see International Affairs)
Forensics (see Speech and Interpersonal Communication)
Forensic Science (see Criminal Justice/Criminology)
Forestry—Book 3
Foundations and Philosophy of Education—Book 6
French—Book 2
Game and Wildlife Management (see Fish, Game, and Wildlife Management)
Gas Engineering (see Petroleum Engineering)
General Studies (see Liberal Studies)
Genetics—Book 3
Geochemistry—Book 4
Geodetic Sciences—Book 4
Geography—Book 2
Geological Engineering—Book 5
Geological Sciences (see Geology)
Geology—Book 4
Geophysical Fluid Dynamics (see Geophysics)
Geophysics—Book 4
Geosciences (see Earth Sciences; Geochemistry; Geodetic Sciences; Geology; Geophysics)
Geotechnical Engineering—Book 5
German—Book 2
Gerontological Nursing—Book 6
Gerontology—Book 2
Government/Political Science—Book 2
Graphic Design—Book 2
Greek (see Classics)
Guidance and Counseling (see Counselor Education)
Health Education—Book 6
Health Physics/Radiological Health—Book 6
Health-related Professions (see individual allied health professions)
Health Sciences (see Public and Community Health)
Health Services Management and Hospital Administration—Book 6
Health Systems (see Safety Engineering; Systems Engineering)
Hearing Sciences (see Speech-Language Pathology and Audiology)
Hebrew (see Near and Middle Eastern Languages)
Hebrew Studies (see Jewish Studies)

INDEX OF DIRECTORIES AND SUBJECT AREAS IN BOOKS 2–6

Higher Education—Book 6
Highway Engineering (see Transportation and Highway Engineering)
Hispanic Studies (see Latin American Studies; Spanish)
Histology (see Anatomy; Cell Biology)
Historic Preservation—Book 2
History—Book 2
History of Art (see Art History)
History of Science—Book 2
Home Economics—Book 2
Home Economics Education—Book 6
Horticulture—Book 3
Hospital Administration (see Health Services Management and Hospital Administration)
Hospitality Administration—Book 6
Household Economics, Sciences, and Management (see Home Economics)
Human Development—Book 2
Human Ecology (see Home Economics; Human Development)
Human Genetics (see Genetics)
Humanities (see Interdisciplinary Programs in the Humanities and Social Sciences; Liberal Studies; specific areas)
Human Movement Studies (see Dance; Physical Education and Human Movement Studies)
Human Resources Management and Personnel—Book 6
Human Services (see Social Work)
Hydrology (see Water Resources)
Illustration—Book 2
Immunology—Book 3
Industrial Administration—Book 6
Industrial and Labor Relations—Book 2
Industrial and Organizational Psychology—Book 2
Industrial Design—Book 2
Industrial Education (see Vocational and Technical Education)
Industrial Engineering (see Industrial/Management Engineering)
Industrial/Management Engineering—Book 5
Information Science—Book 5
Inorganic Chemistry—Book 4
Instructional Technology (see Educational Media/Instructional Technology)
Insurance and Actuarial Science—Book 6
Interdisciplinary Programs in the Humanities and Social Sciences—Book 2
Interior Design—Book 2
International Affairs—Book 2
International Business—Book 6
International Commerce (see International Business)
International Economics (see Economics; International Affairs; International Business)
International Service (see International Affairs)
Interpersonal Communication (see Speech and Interpersonal Communication)
Interpretation (see Translation and Interpretation)
Investment and Securities (see Business Administration and Management; Finance and Banking)
Islamic Studies (see Near and Middle Eastern Studies; Religion)
Italian—Book 2
Japanese Studies (see Asian Languages; Asian Studies)
Jewelry (see Art/Fine Arts)
Jewish Studies—Book 2
Journalism—Book 2
Judaic Studies (see Jewish Studies; Religion; Religious Education)
Junior College Education (see Community College Education)
Kinesiology (see Physical Education and Human Movement Studies)
Laboratory Medicine (see Immunology; Medical Technology; Microbiology; Pathology)
Labor Relations (see Industrial and Labor Relations)
Landscape Architecture—Book 2
Latin (see Classics)
Latin American Studies—Book 2
Law—Book 6
Law Enforcement (see Criminal Justice/Criminology)
Legal Studies (see Law)

Leisure Services (see Recreation)
Liberal Studies—Book 2
Librarianship (see Library Science)
Library Science—Book 2
Life Sciences (see Biology and Biomedical Sciences)
Limnology (see Marine Sciences/Oceanography)
Linguistics—Book 2
Literature (see Classics; Comparative Literature; appropriate language)
Macromolecular Science (see Polymer Science/Plastics Engineering)
Management (see Business Administration and Management)
Management Engineering (see Engineering Management; Industrial/Management Engineering)
Management Information Systems—Book 6
Manufacturing Engineering—Book 5
Marine Affairs (see Marine Sciences/Oceanography)
Marine Biology—Book 3
Marine Engineering (see Ocean Engineering)
Marine Sciences/Oceanography—Book 4
Marine Studies (see Marine Sciences/Oceanography)
Marketing—Book 6
Marriage and Family Therapy—Book 2
Mass and Organizational Communication—Book 2
Materials Engineering—Book 5
Materials Sciences—Book 5
Maternity Nursing—Book 6
Mathematical Physics—Book 4
Mathematical Statistics (see Statistics)
Mathematics—Book 4
Mathematics Education—Book 6
Mechanical Engineering—Book 5
Mechanics—Book 5
Medical Illustration (see Illustration)
Medical Microbiology—Book 3
Medical Nursing (see Medical-Surgical Nursing)
Medical Physics—Book 6
Medical Sciences (see Biology and Biomedical Sciences)
Medical-Surgical Nursing—Book 6
Medical Technology—Book 6
Medicinal Chemistry (see Pharmaceutical Sciences)
Medicine (see Allopathic Medicine; Naturopathic Medicine; Osteopathic Medicine; Podiatric Medicine)
Medieval and Renaissance Studies—Book 2
Metallurgical Engineering—Book 5
Metallurgy—Book 5
Metalsmithing (see Art/Fine Arts)
Meteorology and Atmospheric Sciences—Book 4
Microbiology—Book 3
Middle Eastern Studies (see Near and Middle Eastern Studies)
Middle School Education—Book 6
Midwifery (see Nurse Midwifery)
Mineral Economics—Book 2
Mineral/Mining Engineering—Book 5
Ministry (see Pastoral Ministry and Counseling; Theology)
Missions and Missiology—Book 2
Molecular Biology—Book 3
Molecular Biophysics (see Biophysics)
Motion Pictures (see Art/Fine Arts; Radio, Television, and Film)
Movement Studies (see Dance; Physical Education and Human Movement Studies)
Multilingual and Multicultural Education (see Bilingual and Bicultural Education; Teaching English as a Second Language)
Museum Studies—Book 2
Music—Book 2
Music Education—Book 6
Musicology (see Music)
Music Therapy (see Dance, Drama, and Music Therapy)
Native American Studies (see North American Studies)
Natural Resources—Book 3
Natural Resources Management (see Environmental Policy and Resource Management; Natural Resources)
Naturopathic Medicine—Book 6
Near and Middle Eastern Languages—Book 2
Near and Middle Eastern Studies—Book 2

Peterson's Guide to Graduate Programs in the Physical Sciences and Mathematics 1992

INDEX OF DIRECTORIES AND SUBJECT AREAS IN BOOKS 2–6

Near Environment (see Home Economics; Human Development)
Neural Sciences (see Biopsychology; Neurobiology)
Neurobiology—Book 3
Neuroendocrinology (see Biopsychology; Neurobiology; Physiology)
Neuropharmacology (see Biopsychology; Neurobiology; Pharmacology)
Neurophysiology (see Biopsychology; Neurobiology; Physiology)
Neurosciences (see Biopsychology; Neurobiology)
Nonprofit Management (see Business Administration and Management; Public Policy and Administration)
North American Studies—Book 2
Nuclear Engineering—Book 5
Nuclear Medical Technology (see Medical Technology)
Nuclear Physics (see Physics)
Nurse Anesthesia—Book 6
Nurse Midwifery—Book 6
Nurse Practitioner Studies—Book 6
Nursery School Education (see Early Childhood Education)
Nursing—Book 6
Nursing Administration—Book 6
Nursing Education—Book 6
Nutrition—Book 3
Occupational Education (see Vocational and Technical Education)
Occupational Health (see Environmental and Occupational Health)
Occupational Therapy—Book 6
Ocean Engineering—Book 5
Oceanography (see Marine Sciences/Oceanography)
Oncology (see Cell Biology; Immunology; Molecular Biology; Pathology)
Oncology Nursing—Book 6
Operations Research—Book 5
Optical Sciences—Book 4
Optics (see Applied Physics; Optical Sciences; Physics)
Optometry—Book 6
Oral and Dental Sciences—Book 6
Organic Chemistry—Book 4
Organismal Biology (see Biology and Biomedical Sciences; Zoology)
Organizational Behavior—Book 6
Organizational Psychology (see Industrial and Organizational Psychology)
Oriental Languages (see Asian Languages)
Oriental Studies (see Asian Studies)
Osteopathic Medicine—Book 6
Painting (see Art/Fine Arts)
Paper and Pulp Engineering—Book 5
Paper Chemistry (see Chemistry)
Parasitology—Book 3
Parks Administration (see Recreation)
Pastoral Ministry and Counseling—Book 2
Pathobiology (see Pathology)
Pathology—Book 3
Performing Arts (see Dance; Drama/Theater Arts; Music)
Personnel (see Human Resources Management and Personnel; Organizational Behavior)
Petroleum Engineering—Book 5
Pharmaceutical Chemistry (see Pharmaceutical Sciences)
Pharmaceutical Sciences—Book 6
Pharmacology—Book 3
Pharmacy—Book 6
Philosophy—Book 2
Philosophy of Education (see Foundations and Philosophy of Education)
Photobiology of Cells and Organelles (see Botany and Plant Sciences; Cell Biology)
Photography—Book 2
Physical Chemistry—Book 4
Physical Education and Human Movement Studies—Book 6
Physical Therapy—Book 6
Physics—Book 4
Physiological Optics (see Physiology; Vision Sciences)
Physiology—Book 3
Planetary Sciences—Book 4
Plant Pathology—Book 3
Plant Physiology—Book 3
Plant Sciences (see Botany and Plant Sciences)
Plasma Physics—Book 4
Plastics Engineering (see Polymer Science/Plastics Engineering)
Playwriting (see Drama/Theater Arts; Writing)
Podiatric Medicine—Book 6
Policy Studies (see Public Policy and Administration; Technology and Public Policy)
Political Science (see Government/Political Science)
Polymer Science/Plastics Engineering—Book 5
Pomology (see Agricultural Sciences; Botany and Plant Sciences; Horticulture)
Population Studies (see Demography and Population Studies)
Portuguese—Book 2
Poultry Science (see Animal Sciences)
Power Engineering—Book 5
Preventive Medicine (see Public and Community Health)
Printmaking (see Art/Fine Arts)
Product Design (see Environmental Design; Industrial Design)
Psychiatric Nursing—Book 6
Psychobiology (see Biopsychology)
Psychology—Book 2
Psychopharmacology (see Biopsychology; Neurobiology; Pharmacology)
Public Address (see Speech and Interpersonal Communication)
Public Administration (see Public Policy and Administration)
Public Affairs (see Public Policy and Administration; Technology and Public Policy)
Public and Community Health—Book 6
Public Health (see Public and Community Health)
Public Health Nursing—Book 6
Public History—Book 2
Public Policy and Administration—Book 2
Public Relations (see Advertising and Public Relations)
Quantitative Analysis—Book 2
Radiation Biology—Book 3
Radiological Health (see Health Physics/Radiological Health)
Radiological Physics (see Physics)
Radio, Television, and Film—Book 2
Range Science—Book 3
Reading—Book 6
Real Estate—Book 6
Recreation—Book 6
Regional Planning (see Architecture; City and Regional Planning; Environmental Design; Urban Design; Urban Studies)
Rehabilitation Counseling—Book 6
Rehabilitation Nursing—Book 6
Rehabilitation Therapy (see Physical Therapy)
Religion—Book 2
Religious Education—Book 6
Religious Studies (see Religion; Theology)
Remedial Education (see Special Education)
Renaissance Studies (see Medieval and Renaissance Studies)
Resource Management (see Environmental Policy and Resource Management)
Restaurant Administration (see Hospitality Administration)
Rhetoric (see Speech and Interpersonal Communication)
Romance Languages—Book 2
Romance Literatures (see Romance Languages)
Russian—Book 2
Safety Engineering—Book 5
Scandinavian Languages—Book 2
School Psychology—Book 2
Science Education—Book 6
Sculpture (see Art/Fine Arts)
Secondary Education—Book 6
Security Administration (see Criminal Justice/Criminology)
Slavic Languages—Book 2
Slavic Studies (see East European and Soviet Studies; Slavic Languages)
Social Psychology—Book 2
Social Sciences (see Interdisciplinary Programs in the Humanities and Social Sciences; specific areas)

Social Sciences Education—Book 6
Social Studies Education (see Social Sciences Education)
Social Welfare (see Social Work)
Social Work—Book 2
Sociobiology (see Evolutionary Biology)
Sociology—Book 2
Software Engineering—Book 5
Soil Sciences and Management (see Agronomy and Soil Sciences)
Solar Engineering—Book 5
Solid-Earth Sciences (see Earth Sciences)
Solid-State Sciences (see Materials Sciences)
Southeast Asian Studies (see Asian Studies)
Soviet Studies (see East European and Soviet Studies; Russian)
Space Sciences (see Astronomy; Astrophysics; Planetary Sciences)
Spanish—Book 2
Special Education—Book 6
Speech and Interpersonal Communication—Book 2
Speech-Language Pathology and Audiology—Book 6
Sport Management (see Sports Administration)
Sport Psychology and Sociology (see Physical Education and Human Movement Studies)
Sports Administration—Book 6
Statistics—Book 4
Structural Engineering—Book 5
Studio Art (see Art/Fine Arts)
Substance Abuse Counseling (see Drug and Alcohol/Substance Abuse Counseling)
Surgical Nursing (see Medical-Surgical Nursing)
Systems Analysis (see Systems Engineering)
Systems Engineering—Book 5
Systems Management (see Business Administration and Management)
Taxation—Book 6
Teacher Education (see Education)
Teaching English as a Second Language—Book 6
Technical Education (see Vocational and Technical Education)
Technology and Public Policy—Book 5
Telecommunications—Book 5

Television (see Radio, Television, and Film)
Teratology (see Developmental Biology; Environmental and Occupational Health; Pathology)
Textile Design—Book 2
Textiles (see Clothing and Textiles; Textile Design; Textile Sciences and Engineering)
Textile Sciences and Engineering—Book 5
Theater Arts (see Drama/Theater Arts)
Theology—Book 2
Theoretical Biology (see Biology and Biomedical Sciences)
Theoretical Physics (see Physics)
Therapeutics (see Pharmaceutical Sciences; Pharmacology; Pharmacy)
Toxicology—Book 3
Translation and Interpretation—Book 2
Transportation and Highway Engineering—Book 5
Tropical Medicine (see Parasitology)
Urban Design—Book 2
Urban Education—Book 6
Urban Planning (see Architecture; City and Regional Planning; Environmental Design; Urban Design; Urban Studies)
Urban Studies—Book 2
Urban Systems Engineering (see Systems Engineering)
Veterinary Medicine—Book 6
Veterinary Sciences—Book 6
Video (see Radio, Television, and Film)
Virology—Book 3
Vision Sciences—Book 6
Visual Arts (see Applied Arts and Design; Art/Fine Arts; Graphic Design; Illustration; Photography; Radio, Television, and Film)
Vocational and Technical Education—Book 6
Vocational Counseling (see Counselor Education)
Water Resources—Book 5
Wildlife Biology (see Zoology)
Wildlife Management (see Fish, Game, and Wildlife Management)
Women's Studies—Book 2
Writing—Book 2
Zoology—Book 3

Index of Directories and Subject Areas in This Book

Acoustics	541
Analytical Chemistry	42
Applied Mathematics	338
Applied Physics	541
Applied Statistics (*see* Statistics)	
Astronomy	30
Astrophysical Sciences (*see* Astrophysics; Meteorology and Atmospheric Sciences; Planetary Sciences)	
Astrophysics	35
Atmospheric Sciences (*see* Meteorology and Atmospheric Sciences)	
Biological Oceanography (*see* Marine Sciences/Oceanography)	
Biomathematics (*see* Biometrics)	
Biometrics	345
Biostatistics	347
Chemistry	45
Earth Sciences	187
Experimental Statistics (*see* Statistics)	
General Science (*see* specific topics)	
Geochemistry	191
Geodetic Sciences	193
Geological Engineering (*see* Geology)	
Geological Sciences (*see* Geology)	
Geology	194
Geophysical Fluid Dynamics (*see* Geophysics)	
Geophysics	210
Geosciences (*see* Earth Sciences; Geochemistry; Geodetic Sciences; Geology; Geophysics)	
Inorganic Chemistry	71
Limnology (*see* Marine Sciences/Oceanography)	
Marine Affairs (*see* Marine Sciences/Oceanography)	
Marine Sciences/Oceanography	302
Marine Studies (*see* Marine Sciences/Oceanography)	
Mathematical Physics	544
Mathematical Statistics (*see* Statistics)	
Mathematics	350
Meteorology and Atmospheric Sciences	520
Nuclear Physics (*see* Physics)	
Ocean Engineering (*see* Marine Sciences/Oceanography)	
Oceanography (*see* Marine Sciences/Oceanography)	
Optical Sciences	544
Optics (*see* Applied Physics; Optical Sciences; Physics)	
Organic Chemistry	75
Paper Chemistry (*see* Chemistry)	
Physical Chemistry	79
Physics	545
Planetary Sciences	215
Plasma Physics	566
Radiological Physics (*see* Physics)	
Solid-Earth Sciences (*see* Earth Sciences)	
Space Sciences (*see* Astronomy; Astrophysics; Planetary Sciences)	
Statistics	375
Theoretical Physics (*see* Physics)	

MORE OUTSTANDING TITLES FROM PETERSON'S

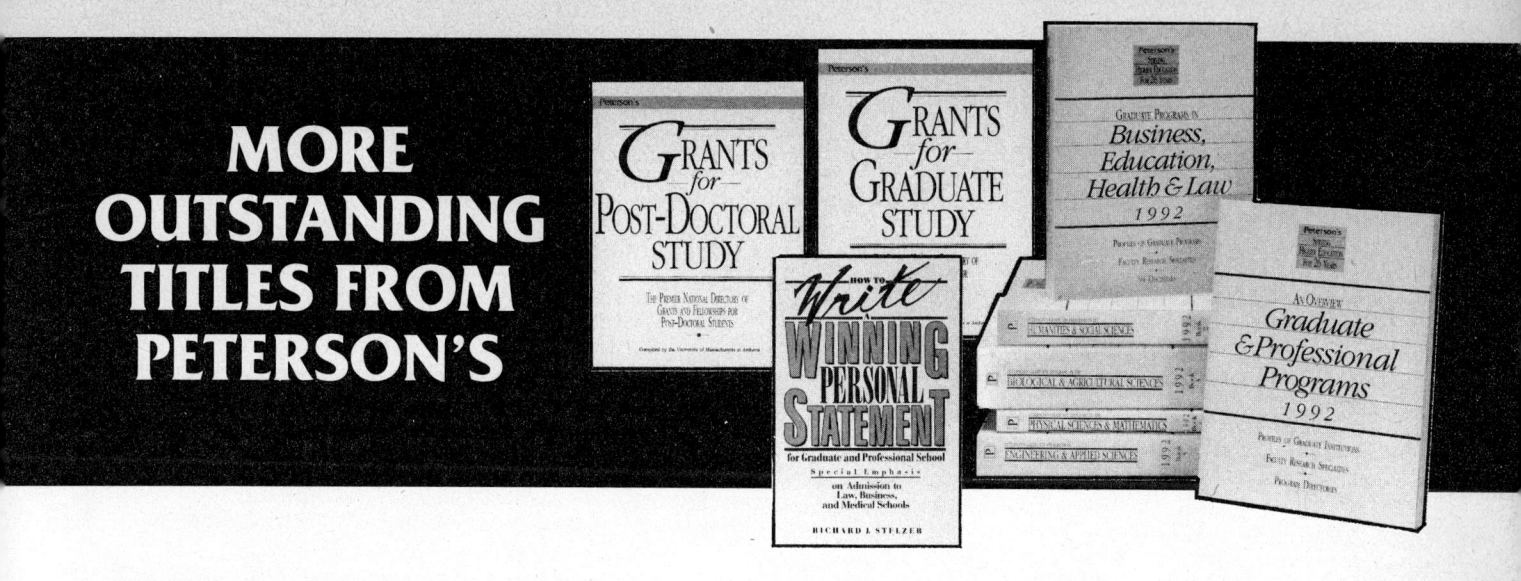

Peterson's Grants for Graduate Study and Grants for Post-Doctoral Study

Compiled by the University of Massachusetts at Amherst, these directories describe hundreds of research, teaching, and writing programs in the United States and abroad. The books' individual program profiles cover:

- Description and purpose of individual awards
- Eligibility requirements
- Application procedures and deadlines
- The stipend amount and other benefits
- Contact information

Both of these books are crucial for graduate students seeking sources of financial support for continued research and study.

PETERSON'S GRANTS FOR GRADUATE STUDY
$59.95 paperback

PETERSON'S GRANTS FOR POST-DOCTORAL STUDY
$54.95 paperback

HOW TO WRITE A WINNING PERSONAL STATEMENT FOR GRADUATE AND PROFESSIONAL SCHOOL

Richard J. Stelzer

For most students, writing a personal statement is the most daunting aspect of the graduate or professional school application. This book helps students write one that will put the admissions odds in their favor.

Samples of real-life, successful personal statements are augmented by advice and strategies that help students focus on relevant facts and overcome "writer's block."

A revealing feature of the book is its exclusive interviews with top admissions officials at leading graduate and professional schools. These interviews provide extraordinary insights on personal statements in a wide range of fields—including law, business, and medicine.

$9.95 paperback

Peterson's Annual Guides to Graduate Study
Complete Coverage of More Than 31,000 Graduate and Professional Programs in the U.S. and Canada

PETERSON'S GUIDE TO GRADUATE AND PROFESSIONAL PROGRAMS: AN OVERVIEW 1992
$21.95 paperback

PETERSON'S GUIDE TO GRADUATE PROGRAMS IN THE HUMANITIES AND SOCIAL SCIENCES 1992
$33.95 paperback

PETERSON'S GUIDE TO GRADUATE PROGRAMS IN THE BIOLOGICAL AND AGRICULTURAL SCIENCES 1992
$39.95 paperback

PETERSON'S GUIDE TO GRADUATE PROGRAMS IN BUSINESS, EDUCATION, HEALTH, AND LAW 1992
$21.95 paperback

PETERSON'S GUIDE TO GRADUATE PROGRAMS IN THE PHYSICAL SCIENCES AND MATHEMATICS 1992
$29.95 paperback

PETERSON'S GUIDE TO GRADUATE PROGRAMS IN ENGINEERING AND APPLIED SCIENCES 1992
$33.95 paperback

Look for these and other Peterson's titles in your local bookstore